Handbook of Tumor Syndromes

Edited by
Dongyou Liu

CRC Press
Taylor & Francis Group
Boca Raton London New York

CRC Press is an imprint of the
Taylor & Francis Group, an **informa** business

CRC Press
Taylor & Francis Group
6000 Broken Sound Parkway NW, Suite 300
Boca Raton, FL 33487-2742

First issued in paperback 2022

© 2020 by Taylor & Francis Group, LLC
CRC Press is an imprint of Taylor & Francis Group, an Informa business

No claim to original U.S. Government works

ISBN 13: 978-1-03-240004-4 (pbk)
ISBN 13: 978-0-8153-9380-1 (hbk)
ISBN 13: 978-1-351-18743-5 (ebk)

DOI: 10.1201/9781351187435

**Visit the Taylor & Francis Web site at
http://www.taylorandfrancis.com**

**and the CRC Press Web site at
http://www.crcpress.com**

Handbook of Tumor Syndromes

Contents

Section III Tumor Syndromes Affecting the Urogenitory System

Section IV Tumor Syndromes Affecting the Skin, Soft Tissue, and Bone

Section V Tumor Syndromes Affecting the Endocrine System

Section VI Tumor Syndromes Affecting the Hematopoietic and Lymphoreticular Systems

Section VII Overgrowth Syndromes, PTENopathies, and RASopathies

Preface

Tumors/cancers are characterized by uncontrolled growth of abnormal cells that extend beyond their usual boundaries and disrupt the normal functions of affected organs. It is estimated that tumors/cancers were responsible for 9.6 million deaths worldwide in 2018, mainly involving the lungs (1.59 million), liver (745,000), stomach (723,000), colorectal (694,000), breast (521,000), and esophagus (400,000).

Current evidence indicates that about 5%–10% of all tumors/cancers develop in individuals who acquire certain pathogenic variants that predispose to tumor syndromes, which are characterized by the presence of both tumor-related and syndromic symptoms. To date, over 300 tumor syndromes have been recognized. However, due to their rare occurrence, diverse clinical presentation, and complex pathogenic mechanisms, tumor syndromes are often overlooked, misdiagnosed, and improperly treated. To increase the awareness, improve the diagnostic accuracy, and enhance the treatment outcome for tumor syndromes, it is crucial to have a contemporary, comprehensive, and authoritative volume that outlines the fundamental principles and presents state-of-the-art details on these disorders in relation to their biology, pathogenesis, epidemiology, clinical features, diagnosis, treatment, and prognosis.

With chapters written by oncologists/clinicians in the forefront of tumor/cancer research and management, *Handbook of Tumor Syndromes* provides an invaluable reference for undergraduates and postgraduates majoring in medicine, dentistry, pharmacology, nursing, public health, and other biomedical disciplines; offers an indispensable guide for medical practitioners in the diagnosis and treatment of tumor syndromes; and supplies a reliable source of information for patients and their relatives in coping with tumor syndromes.

Undoubtedly, a volume that incorporates such a broad spectrum of disease identities requires a team's efforts. I am honored to have an international panel of experts as chapter contributors, whose willingness to share their in-depth knowledge and technical insights on tumor syndromes has greatly enriched this volume. I also thank Dr. Chuck Crumly, senior editor at CRC Press/Taylor & Francis Group, for his professionalism and dedication over the formulation of this volume. Finally, the understanding and support from my family—Liling Ma, Brenda, and Cathy—have been crucial to help maintain my focus and sanity during this all engrossing undertaking.

Editor

Dongyou Liu, PhD, studied veterinary science at Hunan Agricultural University, China and conducted postgraduate research on the generation and application of monoclonal antibodies for improved immunodiagnosis of human hydatidosis at the University of Melbourne, Australia. During the past three decades, he has worked at several research and clinical laboratories in Australia and the United States, with focus on molecular characterization and virulence determination of microbial pathogens such as ovine footrot bacterium (*Dichelobacter nodosus*), dermatophyte fungi (*Trichophyton, Microsporum,* and *Epidermophyton*) and listeriae (*Listeria* species), as well as the development of nucleic acid−based quality assurance models for securing sensitive and emerging viral pathogens. In addition, during the period of 1995–2001, he established and performed molecular tests at clinical laboratories for a range of human genetic disorders and cancers, including BRCA1, BRCA2, c-kit, B-/T-cell receptor gene rearrangements, t(11,14) chromosomal translocation, k-ras, fragile X syndrome, factor V Leiden, hemochromatosis, and prothrombin mutations. He is the primary author of more than 50 original research and review articles in peer-reviewed international journals, the contributor of 197 book chapters, and the editor of *Handbook of* Listeria monocytogenes (2008), *Handbook of Nucleic Acid Purification* (2009), *Molecular Detection of Foodborne Pathogens* (2009), *Molecular Detection of Human Viral Pathogens* (2010), *Molecular Detection of Human Bacterial Pathogens* (2011), *Molecular Detection of Human Fungal Pathogens* (2011), *Molecular Detection of Human Parasitic Pathogens* (2012), *Manual of Security Sensitive Microbes and Toxins* (2014), *Molecular Detection of Animal Viral Pathogens* (2016), *Laboratory Models for Foodborne Infections* (2017), *Handbook of Foodborne Diseases* (2018), all of which were released by CRC Press. He is also a co-editor for *Molecular Medical Microbiology,* Second Edition (Elsevier, 2014). Further, he is the author of the CRC *Pocket Guides to Biomedical Sciences* series books: *Tumors and Cancers: Central and Peripheral Nervous Systems* (2017), *Tumors and Cancers: Head−Neck−Heart−Lung−Gut* (2017), *Tumors and Cancers: Skin−Soft Tissue−Bone−Urogenitals* (2017), and *Tumors and Cancers: Endocrine Glands−Blood−Marrow−Lymph* (2017).

Contributors

N. Anitha, MDS
Department of Oral Pathology and Microbiology
Sree Balaji Dental College and Hospital
Bharath Institute of Higher Education and
    Research (BIHER)
Chennai, India

Adonye Banigo, MBChB, DO-HNS
Department of Otolaryngology
Head and Neck Surgery
University of Edinburgh
Edinburgh, United Kingdom

Raúl Barrera-Rodríguez, MD
Depto. de Bioquímica y Medicina Ambiental
Instituto Nacional de Enfermedades Respiratorias–SSA
México City, México

Jennifer Beebe-Dimmer, PhD
Population Studies and Disparities Research Program
Karmanos Cancer Institute
Detroit, Michigan

Elisabeth Bendstrup, MD, PhD
Department of Respiratory Diseases and Allergy
Aarhus University Hospital
Aarhus, Denmark

Fábio Guilherme Campos, MD, PhD
Division of Colorectal Surgery
Department of Gastroenterology
University of São Paulo
São Paulo, Brazil

Sirisak Chanprasert, MD
Division of Medical Genetics
Department of Medicine
University of Washington
Seattle, Washington

Jirat Chenbhanich, MD
Division of Medical Genetics
Department of Pediatrics
University of California, San Francisco
San Francisco, California

Wisit Cheungpasitporn, MD
Division of Nephrology
Department of Medicine
University of Mississippi Medical Center
Jackson, Mississippi

Priyanka Chhadva, MD
Department of Ophthalmology and Visual Sciences
Illinois Eye and Ear Infirmary
University of Illinois at Chicago
Chicago, Illinois

Luigia Cinque, PhD
Division of Medical Genetics
IRCCS Casa Sollievo della Sofferenza Hospital
San Giovanni Rotondo (FG), Italy

Kathleen A. Cooney, MD
Department of Medicine
Duke University School of Medicine
Durham, North Carolina

Valerie A. Fitzhugh, MD
Department of Pathology and Laboratory Medicine
Rutgers
and
State University of New Jersey
New Jersey Medical School
and
University Hospital
Newark, New Jersey

Edgar Garavito, MD
Histology Department
Fundación Universitaria Sanitas
Bogotá, Colombia

Marianne Geilswijk, MD
Department of Clinical Genetics
Aarhus University Hospital
Aarhus, Denmark

Morie A. Gertz, MD
Division of Hematology
Mayo Clinic
Rochester, Minnesota

Veda N. Giri, MD
Cancer Risk Assessment and Clinical Cancer Genetics
Departments of Medical Oncology and Cancer Biology
Sidney Kimmel Cancer Center
Thomas Jefferson University
Philadelphia, Pennsylvania

Gabriela Maria Abreu Gontijo, MD
Private Practice
Belo Horizonte (MG), Brazil

Sydney T. Grob, BS
Department of Pediatrics
University of Colorado Denver
and
The Morgan Adams Foundation Pediatric Brain Tumor
    Research Program
Children's Hospital Colorado
Aurora, Colorado

Vito Guarnieri, PhD
Division of Medical Genetics
IRCCS Casa Sollievo della Sofferenza Hospital
San Giovanni Rotondo (FG), Italy

Samantha Guhan, BA
Harvard Medical School
Boston, Massachusetts

**Andreas V. Hadjinicolaou, BChir (Cantab), MA (Oxon),
AFHEA**
Gastroenterology, Hepatology and
Internal Medicine
Cambridge University Hospitals NHS Trust
and
MRC Cancer Unit
Hutchison/MRC Research Centre
University of Cambridge
Cambridge, United Kingdom
and
MRC Human Immunology Unit, Radcliffe
Department of Medicine
University of Oxford
Oxford, United Kingdom

Tayfun Hakan, MD
Okan University
Vocational School of Health Sciences
İstanbul, Turkey

Claire Hartnett, MB, BCh, BAO, FRCSI, FEBO
Retinoblastoma Department
The Royal London Hospital
London, United Kingdom

Rina Kansal, MD
Blood Center of Wisconsin
Milwaukee, Wisconsin

Dongyou Liu, PhD
Royal College of Pathologists of Australasia
    Quality Assurance Programs
New South Wales, Australia

Mia Gebauer Madsen, MD, PhD
Department of Urology
Aarhus University Hospital
Aarhus, Denmark

Vikram K. Mahajan, MD
Department of Dermatology, Venereology and Leprosy
Dr. R.P. Govt. Medical College Kangra (Tanda)
Himachal Pradesh, India

Elizabeth E. Montgomery, MD
Department of Pathology
Johns Hopkins University School of Medicine
Baltimore, Maryland

Jean M. Mulcahy Levy, MD
Department of Pediatrics
University of Colorado Denver
and
The Morgan Adams Foundation Pediatric Brain Tumor
    Research Program
Children's Hospital Colorado
Aurora, Colorado

Iain Nixon, MBChB, FRCS (ORL-HND), PhD
Department of Otolaryngology
Head and Neck Surgery
University of Edinburgh
Edinburgh, United Kingdom

Somanath Padhi, MD
Department of Pathology and Laboratory Medicine
All India Institute of Medical Sciences
Bhubaneswar, India

Filipe Palavra, MD
Center for Child Development—Neuropediatrics Unit
Pediatric Hospital
Coimbra Hospital and University Center
and
Institute of Pharmacology and Experimental Therapeutics
    and Coimbra Institute for Clinical and Biomedical
    Research (iCBR)
University of Coimbra
and
Center for Innovative Biomedicine and
    Biotechnology (CIBB)
University of Coimbra
Coimbra, Portugal

Jonas Paludo, MD
Division of Hematology
Mayo Clinic
Rochester, Minnesota

Rafael Parra-Medina, MD, PhD
Research Institute
Department of Pathology
Fundación Universitaria de Ciencias de la Salud
Bogotá, Colombia

Susama Patra, MD
Department of Pathology and Laboratory Medicine
All India Institute of Medical Sciences
Bhubaneswar, India

Anneliese Velez Perez, MD
Department of Pathology and Genomic Medicine
Houston Methodist Hospital
Weill Medical College of Cornell University
Houston, Texas

Carlos Pérez-Malagón, MD
Hospital General Tercer Milenio
Instituto de Salud del Estado de Aguascalientes
Aguascalientes, México

Clóvis Antônio Lopes Pinto, MD
Hospital AC Camargo
São Paulo, Brazil

Paula Quintero-Ronderos, MD, PhD
School of Medicine and Health Sciences
Universidad del Rosario
Bogotá, Colombia

M. Ashwin Reddy, MA MBBChir, MD(Res), FRCOphth
The Judith Kingston Retinoblastoma Unit
Royal London Hospital
Barts Health NHS Trust
and
National Institute for Health Research
Biomedical Research Centre at Moorfields
Eye Hospital NHS Foundation Trust and UCL Institute
 of Ophthalmology
London, United Kingdom

Flávio Reis, PhD
Institute of Pharmacology and Experimental Therapeutics and
 Coimbra Institute for Clinical and Biomedical Research (iCBR)
and
Center for Innovative Biomedicine and Biotechnology (CIBB)
University of Coimbra
Coimbra, Portugal

Joana Jesus Ribeiro, MD
Neurology Department
Coimbra Hospital and University Centre
Coimbra, Portugal

Jae Ro, MD, PhD
Department of Pathology and Genomic Medicine
Houston Methodist Hospital
Weill Medical College of Cornell University
Houston, Texas

Mandeep S. Sagoo, MB, PhD, FRCS (Ed), FRCOphth
The Judith Kingston Retinoblastoma Unit
Royal London Hospital
Barts Health NHS Trust
and
National Institute for Health Research
Biomedical Research Centre at Moorfields
Eye Hospital NHS Foundation Trust and UCL Institute
 of Ophthalmology
London, United Kingdom

Alfredo Scillitani, MD
Unit of Endocrinology
IRCCS Casa Sollievo della Sofferenza Hospital
San Giovanni Rotondo (FG), Italy

Pete Setabutr, MD
Department of Ophthalmology and Visual Sciences
Illinois Eye and Ear Infirmary
University of Illinois at Chicago
Chicago, Illinois

Mashiko Setshedi, FCP (CMSA), PhD (UCT)
Division of Medical Gastroenterology
Department of Medicine
University of Cape Town
Cape Town, South Africa

and

MRC Human Immunology Unit
Radcliffe Department of Medicine
University of Oxford
Oxford, United Kingdom

Anne-Bine Skytte, MD, PhD
Department of Clinical Genetics
Aarhus University Hospital
Aarhus, Denmark

Mette Sommerlund, MD, PhD
Department of Dermatology
Aarhus University Hospital
Aarhus, Denmark

Alexandra Suttman, MS, CGC
Department of Pediatrics
University of Colorado Denver
Aurora, Colorado

Bente A. Talseth-Palmer, PhD
School of Biomedical Science and Pharmacy
Faculty of Health and Medicine
University of Newcastle and Hunter Medical Research
 Institute
Newcastle, Australia

and

Research and Development Unit
Møre and Romsdal Hospital Trust
Molde, Norway

Hensin Tsao, MD, PhD
Department of Dermatology
Wellman Center for Photomedicine
MGH Cancer Center
Massachusetts General Hospital
Boston, Massachusetts

Mariann Unhjem Wiik, Msc
Research Unit
Clinic for Medicine
Møre and Romsdal Hospital Trust
Ålesund, Norway

Dhaarna Wadhwa, MD
Department of Dermatology, Venereology and Leprosy
Dr. R.P. Govt. Medical College Kangra (Tanda)
Himachal Pradesh, India

1

Introductory Remarks

Dongyou Liu

CONTENTS

1.1 Preamble

Tumor (a term used interchangeably with cancer, malignancy, neoplasm, and lesion) is a much feared and possibly misconstrued disease due to its capacity to cause unusually high mortality among affected individuals (with global estimates of 18.1 million new cases and 9.6 million casualties in 2018) and also to its complex pathogenesis that remains inadequately elucidated so far.

Typically forming in tissues that acquire genetic mutations leading to sequential histological changes (from hyperplasia, metaplasia, dysplasia, neoplasia, to anaplasia), tumors demonstrate the hallmarks of (i) sustaining proliferative signaling due to abnormal activation of growth promoting signals, (ii) activating invasion and metastasis through Ecaderin receptor and possibly other mechanisms, (iii) resisting apoptosis by avoiding apoptosis (e.g., TP53 mutation) and autophagy, (iv) enabling replicative immortality through manipulation of telomerase and telomere, (v) avoiding immune destruction by recruitment of inflammatory T cells to prevent immune recognition, (vi) evading growth suppressors by genetic and epigenetic mechanisms (e.g., DNA methylation), (vii) deregulating cellular energetics via pseudohypoxia, (viii) inducing angiogenesis through gradation of angiogenetic switch, (ix) promoting inflammation leading to tumor necrosis, and (x) causing genome instability via telomere damage, centrosome amplification, epigenetic modification, and DNA damage [1].

While 75%–80% of tumors arise sporadically without any family connection (i.e., sporadic tumors), 20%–25% of tumors appear to be familial (including 10%–15% nonhereditary familial tumors [or familial tumors] and 5%–10% hereditary familial tumors [or hereditary tumors]) (Table 1.1). Compared to sporadic

and familial tumors, hereditary tumors tend to (i) occur at a younger age, (ii) produce multiple lesions (often in a distinct spectrum), (iii) produce bilateral lesions, (iv) involve multiple organs and systems, (v) associate with multiple hamartomatous lesions, (vi) locate in a specific site, (vii) display unique morphological features, (viii) induce precursor lesions, (ix) induce multiple benign lesions, and (x) affect multiple members (generations) of a family. In other words, hereditary tumors often present with tumor-related (nonsyndromic) symptoms as well as other (syndromic) symptoms. Considering the existence of syndromic symptoms, hereditary tumors are commonly referred to as hereditary tumor syndromes (also called hereditary cancer syndromes, inherited tumor/cancer syndromes, or tumor/cancer susceptibility syndromes) [2,3].

Nonetheless, as there are rules, so there are always exceptions. In fact, some sporadic tumors (e.g., Cronkhite–Canada syndrome [see Chapter 15], McCune–Albright syndrome [see Chapter 44], and CLOVES syndrome [see Chapter 88]) may display both tumor-related (nonsyndromic) and other (syndromic) symptoms, and are considered as tumor syndromes (and reflected by their names), although tumors associated with these nonhereditary tumor syndromes tend to appear later than those associated with hereditary counterparts. Therefore, apart from hereditary tumor syndromes, this volume also covers some sporadic and nonhereditary familial tumor syndromes for completeness as well as differential diagnostic purpose.

Given their distinct pathogenic mechanisms, complex inheritance patterns, and variable clinical expressivity, tumor syndromes (>300 described to date) pose tremendous intellectual, diagnostic, and therapeutic challenges for the biomedical community and others at large [4]. In the sections that follow, a concise

TABLE 1.1

Clinical and Molecular Characteristics of Sporadic, Familial, and Hereditary Tumors

Tumor	Sporadic	Familial	Hereditary
Proportion	75%–80%	10%–15%	5%–10%
Onset	Late (old age)	Earlier than sporadic	Earliest (young age)
Location	Unilateral location/single organ	Unilateral location/single organ	Bilateral locations/multiple organs
Clinical symptoms[a]	Tumor-related (non-syndromic)	Tumor-related (non-syndromic)	Tumor-related (nonsyndromic) and syndromic
First mutation[b]	De novo/somatic (acquired after birth)	De novo/somatic (acquired after birth)	Germline (acquired at birth)
Second mutation[c]	Somatic (acquired late in life)	Somatic (acquired late in life)	Somatic (acquired after birth)
Presence of mutations	Both first and second mutations in tumor tissue, but not germ cells (egg/sperm) nor blood (leukocytes)	Both first and second mutations in tumor tissue, but not germ cells (egg/sperm) nor blood (leukocytes)	First mutation in tumor tissue, germ cells (egg/sperm) and blood (leukocytes); second mutation in tumor tissue only
Inheritance	No	No	Yes
Familial occurrence	No	Yes	Yes
Risk factors	Advanced age, environmental exposure, hormone therapy, lifestyle	Same environmental and lifestyle factors within a family	Same genetic, environmental, and lifestyle factors within a family

[a] Some sporadic and nonhereditary familial tumors may show tumor-related (nonsyndromic) symptoms as well as other (syndromic) symptoms at times.

[b] Although inheritance of a germline mutation is implicated in a vast majority of hereditary tumors, de novo mutation that occurs occasionally in the germ cells may be also passed onto offspring.

[c] Based on Knudson's "two-hit" hypothesis, tumorigenesis in autosomal dominant disorders usually requires a second hit. However, recent evidence suggests that in some autosomal dominant disorders, tumors may emerge without a second hit (e.g., tuberous sclerosis complex) or require more than two hits.

overview on the molecular mechanisms, clinical features, diagnosis, treatment, and prognosis of tumor syndromes is presented in an attempt to assist in the comprehension of the remaining chapters in this volume.

1.2 Molecular Mechanisms of Tumor Syndromes

Tumor syndromes are largely attributed to inheritance of germline mutations, although a small number may evolve from de novo/sporadic mutations that occur in germ cells (egg or sperm) during fertilization (as in the case of hereditary tumor syndromes) or in non-germ cells sometime afterward (as in the case of sporadic/nonhereditary tumors syndromes). Besides mutations involving specific genes or related regions, chromosomal abnormalities may be also implicated in the development of tumor syndromes.

1.2.1 Genetic Mutations

Genetic mutation refers to a permanent change in the nucleotide sequence of DNA as a result of error in DNA replication and recombination, chemical damage, radiation, or failure in DNA repair (e.g., mismatch repair, nucleotide excision repair, direct repair, and recombination repair).

According its effect on structure, genetic mutation is divided into small-scale mutation (base substitution, insertion, deletion, and copy number variation) and large-scale mutation (chromosomal amplification/duplication, deletion, translocation, and inversion).

Representing the most common form of small-scale mutation (with small-scale mutation involving a single nucleotide being referred to as point mutation), *base substitution* consists of transition and transversion. Transition involves exchange of a purine for another purine (A ↔ G) or of a pyrimidine for another pyrimidine (T ↔ C), while transversion involves exchange of a purine for a pyrimidine or of a pyrimidine for a purine (C/T ↔ A/G). Base substitution leads to synonymous codon with no amino acid change is referred to as silent mutation, base substitution that generates a codon that encodes a different amino acid is referred to as missense mutation, and base substitution results in a stop codon with a truncating translation and nonfunctional protein is referred to as nonsense mutation. Small scale mutation due to *insertion* of additional base pairs in the coding region of a gene may alter splicing of the mRNA (spice-site mutation) or cause a shift in the reading frame (frameshift), resulting in an altered gene product. Small scale mutation due to *deletion* of one or more base pairs from the coding region of a gene may also cause frameshift, and possibly a nonfunctional product. Small-scale mutation that involves gene amplification and expansion of trinucleotide repeats is known as *copy number variation* (e.g., congenital central hypoventilation syndrome [CCHS], see Chapter 4) [4].

Large scale mutation includes chromosomal amplification/duplication, deletion, translocation, and inversion. Chromosomes are thread-like molecules that carry hereditary information from parents to offspring. Human cells (except germ cells) are diploid and contain 22 pairs of nonsex chromosomes (also called autosomes) and 1 pair of sex chromosomes (two X chromosomes in females, but an X chromosome and a Y chromosome in males). As human egg and sperm are produced by meiosis cell division, they have half the number of chromosomes (with a single set of 23 chromosomes) and thus haploid. During fertilization, a sperm combines with an egg to form a zygote, which again contains 2 sets of 23 chromosomes [5].

Structurally, each chromosome consists of two sister chromatids (or two identical chromosome copies) that are connected (aligned) in the centromeric region (or centromere), giving the appearance of an X (or H), and separating each chromatid (chromosome copy) into a short (p) arm and a long (q) arm, which together form about 400 total bands in a karyotype using Giemsa stain. At the ends of linear chromatids (chromosome copies) are telomeres (repetitive stretches of DNA) that lose a bit of the DNA with each cell division. A cell dies when all of the telomere DNA is lost [5]. However, some malignant cells manage to acquire the ability to keep their telomeres intact during division, and thus facilitate their continued/uncontrolled growth. Further, heterozygous germline mutation refers to change in a gene on one chromatid in germ cells (egg or sperm), which is inherited from a parent; homozygous germline mutations refer to similar changes in a gene on both chromatids in germ cells, which are inherited from two carrier parents; and compound heterozygous germline mutations refer to distinct changes in a gene involving both chromatids in germ cells, which are inherited from two carrier parents.

Any failure to produce the standard number of chromosomes (44 autosomes and 2 sex chromosomes) is known as aneuploidy (see Chapter 46 for details), which typically appears as monosomy (lacking one chromosome), trisomy (having one extra chromosome; e.g., 47 XXX, 47 XXY, 47 XYY), tetrasomy (having two extra chromosomes; e.g., XXXX, XXYY), and pentasomy (having three extra chromosomes; e.g., XXXXX, XXXXY, XYYYY)] (see Chapter 2, Figure 2.4 for an exemplary karyotype associated with Klinefelter syndrome) [6].

Chromosomal amplification/repetition refers to the presence of extra piece from another chromosome that leads to multiple copies of all chromosomal regions and increases the dosage of the genes located within. Chromosomal deletion refers to loss of large chromosomal regions including certain genes (with interstitial deletion being defined as an intra-chromosomal deletion that removes a segment of DNA from a single chromosome and thus apposes previously distant genes, and loss of heterozygosity being loss of one allele, either by a deletion or a genetic recombination event, in an organism with two different alleles). Chromosomal translocation refers to interchange of genetic parts from nonhomologous chromosomes, and chromosomal inversion refers to reversion of the orientation of a chromosomal segment [6].

According to its effect on function, genetic mutation is differentiated into loss-of-function (inactivating) mutation (leading to partial or complete loss of function in a gene product; an allele with a complete loss of function is known as null allele; phenotypes associated with such mutations are most often recessive; and when the reduced dosage of a normal gene product is not enough for a normal phenotype, it is called haploinsufficiency), gain of function (activating) mutation (leading to enhanced effect of a gene product; phenotypes associated with such mutations are most often dominant), dominant negative mutation (or antimorphic mutation, with an altered allele acting antagonistically to the wild-type allele), hypomorph (or hypomorphic mutation, which causes partial loss of function in relation to the normal function), neomorph (or neomorphic mutation, which causes a dominant gain of function in relation to the normal function), and lethal mutation (which causes death of an organism that carries the mutation).

Based on whether germ cells (egg/sperm) or non-germ cells are affected, genetic mutation is distinguished into germline, de novo, somatic, and sporadic mutations, along with germline (gonadal), and somatic (classic) mosaicisms.

Germline mutation (also known as germinal mutation or inherited mutation) originates from the germ cells (egg or sperm) of parents, and is transmitted (as constitutional mutation) via mitosis or germline to progeny. Given that germline mutation is present in every cell (including egg/sperm and non-germ cells) from birth, its associated tumors often emerge at a younger age than sporadic or nonhereditary familial tumors. Suitable samples for germline mutation detection include tumor tissue, leukocytes, and germ cells (egg or sperm).

De novo mutation refers to an alteration in a gene that occurs either in a germ cell (egg or sperm) of one parent, or in the fertilized egg itself during early embryogenesis. Therefore, some de novo mutations may be transmitted to progeny. De novo mutation is detectable in tumor tissue, leukocytes, and germ cells of affected individuals, but not in leukocytes (blood) of either parent.

Somatic mutation (also known as acquired mutation) is not inherited from a parent but acquired after conception. As it starts in a body cell, somatic mutation is passed on to any new cells created from that cell only. Due to its absence in germ cells (egg or sperm), somatic mutation is not transmitted to progeny. Somatic mutation is much more common than germline mutation, and accounts for 80%–85% of tumor cases. Suitable samples for somatic mutation detection include tumor tissue, but not leukocytes or germ cells (egg or sperm).

Sporadic mutation (also known as spontaneous mutation) results from natural processes in cells, usually after egg fertilization, and is thus considered as a form of somatic mutation. It represents the second hit for individuals with heterozygous germline mutation before tumorigenesis [7]. A tumor occurring in individuals without a family history is known as sporadic tumor.

Germline (gonadal) mosaicism initiates from a mutation that occurs after conception in an early stem cell that gives rise to all or part of germ cells. Therefore, some germ cells (egg or sperm) may carry a mutation, but others are normal.

Somatic (classic) mosaicism evolves from a mutation that takes place post-zygote or during a nondisjunction event in early mitosis, leading to the presence of more than one genotype in somatic cells. In contrast to germline mutation, somatic mosaic mutation affects only a portion of body cells and is not transmitted to progeny.

1.2.2 Inheritance Patterns

A genetic mutation (pathogenic variant) may be transmitted from parents onto offspring in several patterns, including autosomal dominant, autosomal recessive, X-linked dominant, X-linked recessive, or mitochondrial inheritance (Table 1.2) [8]. In autosomal dominant inheritance, an affected parent with a mutant gene (dominant gene) has a 50% chance of passing the mutant gene and the disorder onto offspring. In autosomal recessive inheritance, two unaffected parents, each with a mutant gene (recessive gene), have 25% chance of passing two copies of mutant gene (one from each parent), 50% chance of passing one mutant gene and one normal gene, and 25% chance of passing two copies of normal gene (one from each parent) onto their offspring, who become affected, unaffected carrier, and normal, respectively [9]. Inheritance patterns for genes located on sex chromosomes (chromosomes X and Y) differ notably from those located on autosomes (non-sex chromosomes) (chromosomes 1–22) in that females possess two copies of each X-linked gene but no copy of Y-linked genes whereas males possess only one copy each of X-linked and Y-linked genes [8]. Overall, most tumor syndromes undergo autosomal dominant and autosomal recessive inheritance, while some involve X-linked dominant, X-linked

TABLE 1.2

Inheritance Patterns Involved in Tumor Syndromes

Inheritance Pattern	Mutation	Characteristics	Exemplary Tumor Syndromes
Autosomal dominant	Heterozygous germline mutation on autosomal (non-sex) chromosome	An affected parent with one copy of mutant gene (dominant gene) and one copy of normal gene on a pair of autosomal chromosomes has a 50% chance of passing the mutant gene and thus the disorder onto each offspring. Patient's first-degree relatives (parents, children, and siblings) have a 50% risk of carrying the mutant gene themselves. Autosomal dominant disorder affects both males and females equally in every generation, and has a more common occurrence than autosomal recessive disorder.	Familial adenomatous polyposis; Beckwith–Wiedemann syndrome; Birt–Hogg–Dubé syndrome; Costello syndrome; Cowden syndrome
Autosomal recessive	Homozygous or compound heterozygous germline mutations on autosomal (non-sex) chromosome	Carrier parents with a mutant gene (recessive gene) and a normal gene (dominant gene) are usually unaffected themselves, but their offspring has 25% chance of acquiring two copies of mutant gene (one from each parent) and becoming affected, 50% chance of acquiring one mutant gene and one normal gene and becoming an unaffected carrier, and 25% chance of acquiring two copies of normal gene (one from each parent) and becoming unaffected. Autosomal recessive disorder typically affects both females and males equally, and tends to skip generations (i.e., the affected is usually offspring of unaffected carriers). Consanguinity may be apparent in parents of affected offspring.	Ataxia telangiectasia; Bloom syndrome; Fanconi anemia; Nijmegen breakage syndrome; Rothmund–Thomson syndrome; Werner syndrome
X-linked dominant	Heterozygous germline mutation on X chromosome	One mutant gene (dominant gene) on X chromosome is sufficient to cause the disorder when inherited from an affected parent. All daughters of affected fathers are affected. Offspring of an affected mother is affected when a dominant X chromosome is passed on but is unaffected when a recessive X chromosome is passed on. If the son is affected, the mother is always affected. No father-to-son transmission occurs. Overall, X-linked dominant disorder affects more females than males. Some X-linked dominant disorders are embryonically lethal in males, and are thus observed in females only.	Aicardi syndrome
X-linked recessive	Homozygous germline mutations on X chromosomes in female; hemizygous germline mutation on X chromosome in male	A hemizygous male with a mutant gene on X chromosome is affected. A homozygous female with two copies of mutant gene (one on each of two X chromosomes) is affected. A heterozygous female with a mutant gene on one X chromosome and a normal gene on another X chromosome is an unaffected carrier. Only son of heterozygous mother (carrier) is affected when mutant gene on X chromosome is passed on. All daughters of affected fathers are carriers. No father-to-son transmission occurs. On the whole, X-linked recessive disorder affects males in every generation, has a higher frequency in males than females, and is more common than X-linked dominant disorder.	Simpson Golabi Behmel syndrome; Wiskott–Aldrich syndrome
Mitochondrial	Heterozygous germline mutation in mitochondria	Mitochondrial transmission is strictly maternal and can occur in every generation. All offspring (both males and females) of affected mothers are affected.	Breast cancer linked to mitochondrial mutation

recessive, or mitochondrial inheritance. However, a few tumor syndromes (e.g., dyskeratosis congenital; see Chapter 69) show multiple patterns of inheritance (including autosomal dominant, autosomal recessive, and X-linked). Further, inheritance patterns for a number of tumor syndromes remain undefined.

1.2.3 Small- and Large-Scale Mutations Linked to Tumor Syndromes

Both small- and large-scale mutations are implicated in the initiation and development of tumor syndromes. A survey of about 100 tumor syndromes in this book indicates that genetic mutations involving various protein-coding genes and related regions are responsible for ~90% of tumor syndromes. These include alterations in genes that encode tumor suppressors (24%), oncogenic proteins (12%), transcription factors (7%), components of signaling pathways (5%), DNA repair proteins (9%), ubiquitination proteins (2%), enzymes (14%), and other proteins (17%) (Table 1.3) [10]. Further, chromosomal abnormalities (e.g., duplication/trisomy, deletion/monosomy, and translocation) are observed in ~10% of tumor syndromes (Table 1.3) [6]. The fact that tumor syndromes are associated with such a diversity of genetic and chromosomal alterations highlights their heterogeneous etiologies, and complicates their diagnosis and treatment [1,10].

TABLE 1.3

Molecular Mechanisms Underlying the Development of Tumor Syndromes

Gene Function/ Chromosomal Abnormality	Specific Gene/Chromosomal Abnormality	Molecular Mechanisms	Tumor Syndromes
Tumor suppressor	Adenomatous polyposis coli (*APC*, 5q22.2)	APC is a suppressor of the Wnt signaling pathway; mutations in *APC* (80% autosomal dominant inheritance, 20% de novo) compromise its suppressor function, leading to accumulation of β-catenin in the cells and alteration in gene expression related to cell migration, proliferation, differentiation, and apoptosis.	Familial adenomatous polyposis (FAP) (see Chapter 16); familial non-medullary thyroid carcinoma (FNMTC) (see Chapter 55)
	Additional sex combs-like 1 (*Asxl1*, 20q11.21)	Asxl1 is the obligate regulatory subunit of the deubiquitinase complex with BAP1 as catalytic subunit; loss-of-function mutations in *Asxl1* (autosomal dominant inheritance) induce abnormal sister chromatid separation and interrupt differentiation of bone marrow stromal cells (BMSC) and hematopoietic stem/progenitor cells (HSC/HPC) in the bone marrow.	Bohring–Opitz syndrome (BOS) (see Chapter 27)
	Breast cancer 1-associated protein (*BAP1*, 3p21.1)	*BAP1* encodes a protein that helps stall cells in S phase and promotes repair at sites of DNA double strand breaks through homologous recombination in the nucleus, and functions as a deubiquitinase in the cytoplasm; loss-of-function mutations in *BAP1* (autosomal dominant inheritance) result in its haploinsufficiency and prevent cells containing damaged DNA from apoptosis.	BAP1 tumor predisposition syndrome (BAP1-TPDS) (see Chapter 3)
	Breast cancer type 1 susceptibility protein (*BRCA1*, 17q21.31); breast cancer type 2 susceptibility protein (*BRCA2*, 13q12.3)	*BRCA1* and *BRCA2* encode proteins with tumor suppressor and DNA repair functions; loss-of-function mutations in *BRCA1* and *BRCA2* (autosomal dominant inheritance) compromise their capacity to regulate gene transcription and cell-cycle progression, and perform DNA repair and ubiquitination, leading to hereditary breast and ovarian cancer (>60% of cases).	Hereditary breast and ovarian cancer (HBOC) (see Chapter 30)
	Cell division cycle 73 (*CDC73* or HRPT2,1q31.2)	*CDC73* encodes a tumor suppressor, which regulates cyclin D1/PRAD1 expression during cell cycle progression; inactivating mutations in *CDC73* (autosomal dominant inheritance) are linked to hyperparathyroidism–jaw tumor syndrome, familial isolated hyperparathyroidism, and parathyroid carcinoma.	Hyperparathyroidism with jaw tumors (HPT-JT) (see Chapter 53)
	E-cadherin (*CDH1*, 16q22.1)	Loss of function mutations in *CDH1* (autosomal dominant inheritance) reduce E-cadherin expression and cell adhesion, and promote atypical cellular architecture and irregular cell growth; subsequent inactivation of the other *CDH1* allele through methylation, somatic mutations, and loss of heterozygosity abolishes E-cadherin expression and renders E-cadherin-deficient cells malignant and metastatic.	Hereditary diffuse gastric cancer (HDGC) (see Chapter 18)

(Continued)

TABLE 1.3 (*Continued*)

Molecular Mechanisms Underlying the Development of Tumor Syndromes

Gene Function/ Chromosomal Abnormality	Specific Gene/Chromosomal Abnormality	Molecular Mechanisms	Tumor Syndromes
	Cyclin-dependent kinase inhibitor 2A (*CDKN2A*, 9p21.3); cyclin-dependent kinase 4 (*CDK4*, 12q14.1); microphthalmia-inducing transcription factor (*MITF*, 3p13)	*CDKN2A* encodes two proteins (P16, which inhibits the complex of cyclin D1 with cyclin-dependent kinase 4 (CDK4) or 6 (CDK6), which promotes cellular proliferation and prevents Rb phosphorylation, facilitating the G1 to S transition; P14ARF, which inhibits MDM2, an E3 ubiquitin ligase involved in the degradation of p53, and thus enhances p53-dependent transactivation and apoptosis); CDK4 participates in the phosphorylation of Rb and contributes to cell-cycle G1 phase progression; MITF regulates melanocyte cell development, pigmentation, and neoplasia (all via autosomal dominant inheritance).	Cutaneous malignant melanoma (CMM) (see Chapter 38); familial atypical multiple mole melanoma syndrome (FAMMM) (see Chapter 39); familial pancreatic cancer (see Chapter 56)
	cAMP-response element binding protein BP (*CREBBP*, 16p13.3); *E1A-binding protein 300* (*EP300*, 22q13.2)	Loss of function mutations in *CREBBP* and *EP300* (autosomal dominant inheritance) disrupt p53 activation, stability, and transactivation of target genes, and are associated with up to 60% and 8% of Rubinstein–Taybi syndrome cases, respectively.	Rubinstein–Taybi syndrome (RTS) (see Chapter 9)
	Folliculin (*FLCN*, 17p11.2)	Mutations in *FLCN* (autosomal dominant inheritance) cause its haploinsufficiency related to tumor suppression, ciliogenesis, cell adhesion, and autophagy, resulting in familial genodermatosis and renal tumor.	Birt–Hogg–Dubé (BHD) syndrome (see Chapter 26)
	Homeobox B13 (*HOXB13*, 17q21.32)	*HOXB13* encodes a homeobox transcription factor which acts as a tumor suppressor and plays a role in skin development and maintenance; recurrent *HOXB13* G84E mutation (autosomal dominant inheritance) is strongly linked to hereditary prostate cancer, particularly in individuals of European descent.	Hereditary prostate cancer (see Chapter 33)
	Multiple endocrine neoplasia type 1 (*MEN1*, 11q13.1)	*MEN1* encodes a protein (menin) that participates in the specific methylation of Lys-4 of histone H3 (H3K4), and the regulation of cell proliferation, apoptosis, and genome integrity; loss-of-function mutations in *MEN1* (autosomal dominant inheritance) disrupt its interaction with other proteins, leading to decreased MEN1 expression and abnormal cell-cycle regulation and proliferation.	Multiple endocrine neoplasia 1 (MEN1) (see Chapter 60)
	Neurofibromatosis type 1 (*NF1*, 17q11.2); neurofibromatosis type 2 (*NF2*, 22q12.2)	*NF1* encodes a cell membrane protein (neurofibromin) belonging to the GTPase-activating family of tumor suppressor and acting as a negative regulator of RAS proteins; *NF2* encodes a membrane-cytoskeleton scaffolding protein (merlin) with unclear tumor suppressor function; mutations in *NF1* (autosomal dominant inheritance) reduce the capacity of neurofibromin to convert active RAS to inactive RAS, leading to increased activation of the RAS/MAPK signaling pathway (thus, NF1 is also considered as a RASopathy); mutations in *NF2* (autosomal dominant inheritance) either abrogate merlin synthesis or generate a defective protein that lacks the normal tumor-suppression function.	Neurofibromatosis types 1 and 2 (NF1 and NF2) (see Chapter 92)
	Paired-like homeobox 2B (*PHOX2B*, 4p13)	Loss of function polyalanine and nonpolyalanine repeat mutations in *PHOX2B* (90% de novo; 10% autosomal dominant inheritance) alter the expression of several downstream genes essential for the development of neural crest derivatives and hindbrain motor neuron.	Congenital central hypoventilation syndrome (CCHS) (see Chapter 4)

(Continued)

TABLE 1.3 (*Continued*)

Molecular Mechanisms Underlying the Development of Tumor Syndromes

Gene Function/ Chromosomal Abnormality	Specific Gene/Chromosomal Abnormality	Molecular Mechanisms	Tumor Syndromes
	Patched homolog 1 (*PTCH1*, 9q22.32)	*PTCH1* encodes a receptor for the secreted factor sonic hedgehog (SHH). Upon binding with SHH, PTCH1 represses transcription of genes encoding proteins belonging to the transforming growth factor (TGF)-β, thus controlling growth and development of normal tissue; loss-of-function mutations in *PTCH1* (autosomal dominant inheritance) abrogate its ability to bind with SHH and regulate cell growth.	Basal cell nevus syndrome (BCNS, or Gorlin syndrome) (see Chapter 85)
	Phosphatase and tensin homologue (*PTEN*, 10q23)	*PTEN* encodes a phosphatase, which, as a PI3K antagonist, negatively regulates the mitogen-activated protein kinase (MAPK) pathway to induce cell cycle arrest (when in the nucleus) and to elicit apoptosis (when in the cytoplasm); loss-of-function mutations in *PTEN* (autosomal dominant inheritance) result in uninhibited phosphorylation of AKT1 (leading to the inability to activate cell-cycle arrest and/or to undergo apoptosis) and dysregulated MAPK pathway (leading to abnormal cell survival).	Bannayan–Riley–Ruvalcaba syndrome (BRRS) (see Chapter 84); Cowden syndrome (see Chapter 90); Proteus syndrome (see Chapter 95)
	Retinoblastoma (*RB1*, 13q14)	Biallelic mutations in *RB1* (autosomal dominant inheritance) incapacitate its function in the regulation of cell cycle, differentiation, and control of genomic stability.	Retinoblastoma (see Chapter 7)
	SWI/SNF-related matrix-associated actin-dependent regulator of chromatin subfamily B member 1 (*SMARCB1*, 22q11.2); *SMARCB4* (19p13.1)	Loss-of-function mutations in *SMARCB1* or *SMARCB4* (autosomal dominant inheritance) reduce its ability to regulate the p16-CDK4/ cyclin D-Rb-E2F pathway and thus cell differentiation and proliferation.	Rhabdoid tumor predisposition syndrome (RTPS) (see Chapter 8); schwannomatosis (see Chapter 10)
	Serine/threonine kinase 11 (*STK11* or *LKB1*, 19p13.3)	Loss-of-function mutations in *STK11* (autosomal dominant inheritance) abrogate its ability to regulate cell differentiation and proliferation through interference with G1 cell-cycle arrest in a p53-independent manner, induction of epithelial cell apoptosis in a p53-dependent manner, and maintenance of cellular polarity, metabolism, and energy homeostasis.	Peutz–Jeghers syndrome (PJS) (see Chapter 22)
	Tumor protein 53 (*TP53*, 17p13.1)	TP53 is a multifunctional tumor suppressor controlling cell cycle and survival; mutations in *TP53* (autosomal dominant inheritance) cause loss of function in TP53 relating to induction of growth arrest, apoptosis, senescence, DNA repair, leaving DNA damages unrepaired and proliferation of cancerous cells unchecked.	Li–Fraumeni syndrome (LFS) (see Chapter 43)
	von Hippel–Lindau syndrome (*VHL*, 3p25.3)	VHL is a multifunctional protein involved in hypoxia response; loss-of-function mutations in *VHL* (autosomal dominant inheritance 80%; de novo 20%) increase expression, stabilization, and accumulation of hypoxia-inducible factor (HIF) proteins, leading to the formation of cysts and hypervascular tumors.	von Hippel–Lindau (VHL) syndrome (see Chapter 12)
	Wilms tumor 1 (*WT1*, 11p13)	Mutations in *WT1* (autosomal dominant inheritance) disrupt the DNA-binding domain of WT1 and reduce the ability of WT1 to regulate cell growth and apoptosis in related tissues (e.g., the kidneys and gonads), leading to a spectrum of clinical syndromes.	Familial Wilms tumor, WAGR (Wilms tumor, aniridia, genitourinary anomalies, mental retardation) syndrome, Denys–Drash syndrome (DDS), Frasier syndrome (see Chapter 29)

(Continued)

TABLE 1.3 (*Continued*)

Molecular Mechanisms Underlying the Development of Tumor Syndromes

Gene Function/ Chromosomal Abnormality	Specific Gene/Chromosomal Abnormality	Molecular Mechanisms	Tumor Syndromes
Oncogene	Anaplastic lymphoma kinase (*ALK, 2p23.2-1*); paired-like homeobox 2B (*PHOX2B, 4p13*)	Gain-of-function mutations in *ALK* (autosomal dominant inheritance) activate the tyrosine kinase domain that induces constitutive phosphorylation of ALK and increases expression of MYCN, accounting for 59% of familial neuroblastoma cases; loss-of-function mutations in *PHOX2B* are involved in 10% of familial neuroblastoma cases.	Familial neuroblastoma (see Chapter 5)
	B-Raf proto-oncogene serine/ threonine-protein kinase (*BRAF, 7q34*)	*BRAF* encodes a mitogen-activated protein kinase involved in the RAS/MAPK pathway, which regulates cell growth, proliferation, migration, and apoptosis; mutations in *BRAF* contribute to tumorigenesis and are implicated in several tumor syndromes (including serrated polyposis syndrome, for which inheritance pattern remains undefined).	Serrated polyposis syndrome (see Chapter 23)
	Casitas B-cell lymphoma (*CBL, 11q23.3*)	*CBL* encodes an E3 ubiquitin ligase and a multi-adaptor protein, which forms part of the RAS/mitogen-activated protein kinase (MAPK) pathway; like other RASopathies, mutations in *CBL* (autosomal dominant inheritance) lead to developmental disorders with neuro-cardio-facio-cutaneous manifestations.	CBL syndrome (also known as Noonan syndrome-like disorder with or without juvenile myelomonocytic leukemia) (see Chapter 87)
	Harvey rat sarcoma viral proto-oncogene homolog (*HRAS*, 11p15.5), Kirsten ras oncogene homolog (*KRAS*, 12p12.1), neuroblastoma RAS viral oncogene homolog (*NRAS*, 1p13.2), RAS-like protein in tissues (*RIT1*, 1q22)	*RAS* genes encode GTPases that bind to GDP/ GTP, and function as on–off binary switches for several downstream effectors such as RAF, phosphatidylinositol 3-kinases (PI3K), and ral guanine nucleotide exchange factor (RalGEF), which make up part of the MAPK/RAF/MEK/ERK pathway, the PI3K/AKT/mTOR pathway and the RalGEF/TBK1/IRF3/3-NF-κB pathway, respectively. Given their roles in the control of cell survival, proliferation, differentiation, senescence, and death, mutations in *RAS* as well as genes encoding components or regulators of the MAPK/RAF/MEK/ERK pathway are responsible for a group of genetic disorders collectively known as RASopathies. Specifically, Costello syndrome is due to germline *HRAS* mutations; Noonan syndrome is linked to germline *KRAS, NRAS* or *RIT1* mutations, Schimmelpenning–Feuerstein–Mims syndrome results from postzygotic somatic mutations in *HRAS, KRAS*, or *NRAS*; pancreatic cancer cases are caused by somatic *KRAS* mutations (all via autosomal dominant inheritance).	Costello syndrome (see Chapter 89); Noonan syndrome (see Chapter 93); Schimmelpenning–Feuerstein–Mims (SFM) syndrome (see Chapter 96); familial pancreatic cancer (see Chapter 56);
	KIT proto-oncogene (*KIT, 4q12*); *platelet-derived growth factor receptor* α (*PDGFRA*, 4q12)	Gain-of-function mutations in *KIT* and *PDGFRA* (autosomal dominant inheritance), which encode tyrosine kinase receptors, alter their regulation of the RAS/MAPK, PI3K/AKT/mTOR, and JAK/STAT3 signaling pathways, leading to tumorigenesis.	Familial gastrointestinal stromal tumor (GIST) (see Chapter 17)
	Mesenchymal-epithelial transition factor (*MET, 7q31.2*)	Mutations in *MET* (autosomal dominant inheritance) increase MET amplification and overexpression, attenuate MET receptor ubiquitination and degradation, prolong MET signaling, and promote oncogenesis, angiogenesis, and metastasis.	Hereditary papillary renal cell cancer (HPRCC) (see Chapter 32)

(Continued)

TABLE 1.3 (*Continued*)

Molecular Mechanisms Underlying the Development of Tumor Syndromes

Gene Function/ Chromosomal Abnormality	Specific Gene/Chromosomal Abnormality	Molecular Mechanisms	Tumor Syndromes
	Myeloproliferative leukemia virus oncogene (*MPL*, 1p34.2)	*MPL* encodes a thrombopoietin receptor; loss-of-function biallelic mutations in *MPL* compromise its ability to interact with its ligand thrombopoietin (THPO, a hematopoietic growth factor), leading to the development of thrombocytopenia and megakaryocytopenia without physical anomalies (autosomal recessive inheritance).	Congenital amegakaryocytic thrombocytopenia (CAMT) (see Chapter 66)
	Rearranged during transfection (*RET*, 10q11.21)	*RET* encodes a transmembrane tyrosine kinase, which interacts with ligands such as glial cell line–derived neurotrophic factor, neurturin, artemin, or persephin, leading to activation of downstream signaling pathways involved in cell differentiation, growth, migration, and survival; gain-of-function mutations in *RET* trigger cell growth, division, and development of MEN2 (autosomal dominant inheritance)	Multiple endocrine neoplasia 2 (MEN2) (see Chapter 60)
	SET-binding protein 1 (*SETBP1*, 18q12.3)	Gain- or loss-of-function mutations in *SETBP1* (autosomal dominant inheritance) may disturb its interaction with and regulation of proto-oncogene *SET*, which acts by inhibiting tumor suppressors such as PP2A and NM23-H1.	Schinzel–Giedion syndrome (see Chapter 10)
Transcription factor	Guanine-adenine-thymine-adenine 2 (*GATA2, 3q21.3*)	*GATA2* encodes a transcription factor (GATA binding protein 2) involved in the regulation of several downstream genes fundamental to the development of the lymphatics and hematopoietic systems; loss-of-function mutations in *GATA2* (autosomal dominant inheritance) limit the self-renewal capacity of hematopoietic stem cells, leading to the depletion of dendritic, monocyte, B, and natural killer cells, and subsequent development of cytopenia, myelodysplasia, myeloid leukemia, pulmonary alveolar proteinosis, lymphedema, and aggressive infections.	Guanine-adenine-thymine-adenine 2 (*GATA2*) deficiency (see Chapter 74)
	Regulators (e.g., *HLA-DQB1/ DQA1/DRB1, IRF5, STAT4, CXCR5, TNIP1, IL12A*, and *BLK*)	Mutations in genes encoding regulators of both adaptive and innate immunity alter glandular homeostasis, leading to glandular dysfunction (possibly non-inherited).	Sjögren syndrome (see Chapter 79)
	Microphthalmia-associated transcription (MIT) factors (*MITF, TFE3, TFEB*, and *TFEC*)	Mutations (e.g., missense or translocation variants) in *MITF, TFE3, TFEB*, and *TFEC* (autosomal dominant inheritance) alter the expression of various downstream genes, leading to renal cell cancer (RCC), melanoma, and hypopigmentation disorders.	Microphthalmia-associated transcription (MIT) family translocation renal cell cancer (tRCC) (see Chapter 34)
	Motor neuron and pancreas homeobox 1 gene (*MNX1*, or *HLXB9*, 7q36.3)	Mutations in *MNX1* (autosomal dominant inheritance) cause disruption in neural stem cell differentiation pathways as well as in pancreas development and function.	Currarino syndrome (see Chapter 37)
	Signal recognition particle 72 kDa (*SRP72*, 4q12)	*SRP72* encodes a transcription factor involved in signal recognition, translational arrest, and endoplasmic reticulum (ER) membrane targeting; mutation in *SRP72* (autosomal dominant inheritance) increase apoptosis after irradiation, leading to familial aplastic anemia/ myelodysplastic syndrome.	SRP72-associated bone marrow failure syndrome (see Chapter 80)

(*Continued*)

TABLE 1.3 (*Continued*)

Molecular Mechanisms Underlying the Development of Tumor Syndromes

Gene Function/ Chromosomal Abnormality	Specific Gene/Chromosomal Abnormality	Molecular Mechanisms	Tumor Syndromes
	T-box transcription factor T (*TBXT*)	*TBXT* is essential for proper development and maintenance of the notochord; gene duplication, single, and compound single nucleotide polymorphisms in *TBXT* confer significant susceptibility to chordoma (autosomal dominant inheritance).	Familial chordoma (see Chapter 40)
Signaling pathway	Fas cell surface death receptor (*FAS* or *TNFRSF6*, 10q23.31)	*FAS* encodes a protein involved in apoptosis; loss-of-function mutations in *FAS* (autosomal recessive inheritance) cause defective FAS-mediated apoptosis and account for 60%–70% of ALPS cases.	Autoimmune lymphoproliferative syndrome (ALPS) (see Chapter 63)
	FMS-like tyrosine kinase 4 (*FLT4*, 5q35.3)	Mutations in *FLT4* (or *VEGFR3*) (autosomal dominant inheritance) disrupt its interaction with VEGFC (a key regulator of blood vessel development in embryos and angiogenesis in adult tissues), and its activation of the MAPK1/ERK2, MAPK3/ERK1 signaling pathway, the MAPK8 and the JUN signaling pathway, and the AKT1 signaling pathway.	Milroy disease (congenital lymphedema) (see Chapter 45)
	Jagged 1 (*JAG1*, 20p12.2)	JAG1 and its NOTCH2 receptor are components of the Notch signaling pathway involved in determining cell differentiation and fate; loss-of-function mutations in *JAG1* (autosomal dominant inheritance) produce a severely truncated protein, which lacks the transmembrane region necessary for embedding in the cell membrane, and for trafficking through the cell, leading to tumor angiogenesis, neoplastic cell growth, and metastasis.	Alagille syndrome (ALGS) (see Chapter 13)
	Protein kinase cAMP-dependent type I regulatory subunit alpha (*PKAR1A*, 17q24.2)	Mutations in *PRKAR1A* (autosomal dominant inheritance) remove the regulatory subunit α and allow protein kinase A to turn on more often than usual, contributing to aberrant cell proliferation and heightened risk for endocrine and non-endocrine tumors and lesions.	Carney complex (see Chapter 51)
	SMAD family member 4 (*SMAD4*, 18q21.2); bone morphogenetic protein receptor type 1A (*BMPR1A*, 10q23.2)	Mutations in *SMAD4* and *BMPR1A* (autosomal dominant inheritance), which are components of the TGF-β signaling pathway, disrupt biological processes related to cell proliferation, apoptosis, angiogenesis, and epithelial–mesenchymal transition.	Juvenile polyposis syndrome (JPS) (see Chapter 19); familial pancreatic cancer (see Chapter 56);
DNA repair	Ataxia telangiectasia mutated (*ATM*, 11q22.3)	*ATM* encodes a serine/threonine kinase (ATM) involved in DNA repair; loss-of-function biallelic mutations in *ATM* (autosomal recessive inheritance) significantly compromise the ability to repair DNA damage caused by reactive oxygen species and other stressors.	Ataxia telangiectasia (see Chapter 62)
	Fanconi anemia (*FANCA, FANCC, FANCD1/BRCA2, FANCD2, FANCE, FANCF, FANCG/ XRCC9, FANCI, FANCJ/BACH1/ BRIP1, FANCL, FANCM, FANCN/PALB2, FANCO/ RAD51C, FANCP/SLX4, FANCQ(ERCC4), FANCB*)	FA pathway includes at least 22 genes that participate in cross-linked DNA repair, maintenance of genome stability, modulation of responses to oxidative stress, viral infection, and inflammation, and prevention of early-onset bone marrow failure and cancer (autosomal recessive inheritance).	Fanconi anemia (see Chapter 73)

(Continued)

TABLE 1.3 (*Continued*)

Molecular Mechanisms Underlying the Development of Tumor Syndromes

Gene Function/ Chromosomal Abnormality	Specific Gene/Chromosomal Abnormality	Molecular Mechanisms	Tumor Syndromes
	Human mutL homolog 1 (*MLH1*, 3p22.2); human mutS homolog 2 (*MSH2*, 2p21-p16.3); human mutS homolog 6 (*MSH6*, 2p16.3); human post-meiotic segregation increased 2 (*PMS2*, 7p22.1)	Biallelic (either homozygous or compound heterozygous) germline mutations in *MLH1*, *MSH2*, *MSH6*, and/or *PMS2* (autosomal recessive inheritance) disable DNA repair mechanisms and result in cutaneous features (e.g., café-au-lait spots) and increased risk for hematological, brain, and intestinal tumors during childhood.	Constitutional mismatch repair deficiency (CMMRD) syndrome (see Chapter 14)
	Human mutL homolog 1 (*MLH1*, 3p22.2); human mutS homolog 2 (*MSH2*, 2p21-p16.3); human mutS homolog 6 (*MSH6*, 2p16.3); human post-meiotic segregation increased 2 (*PMS2*, 7p22.1)	Monoallelic/heterozygous germline mutations in *MLH1*, *MSH2*, *MSH6*, and/or *PMS2* (autosomal dominant inheritance) alter DNA repair and cause gastrointestinal and genitourinary cancers during adulthood.	Lynch syndrome (see Chapter 20)
	MutY DNA glycosylase (*MUTYH*, 1p34.1)	Biallelic (homozygous or compound heterozygous) mutations in *MUTYH* (autosomal recessive inheritance) produce a truncated protein that fails to rectify somatic G>T transversions in multiple tumor suppressor genes (e.g., *APC* and *KRAS*), leading to cell overgrowth, serrated adenomas, hyperplastic polyps, and sporadic colorectal cancer.	MUTYH-associated polyposis (MAP) (see Chapter 21)
	RecQ helicase L2 (*RECQL2* or *WRN*, 8p12)	Homozygous or compound heterozygous germline mutations in *WRN* (autosomal recessive inheritance) produce loss-of-function WRN, leading to genomic instability and tumor susceptibility.	Werner syndrome (see Chapter 49)
	RecQ helicase L3 (*RECQL3* or *BLM*, 15q26.1)	BLM participates in the replication and repair of DNA through formation of a complex with DNA topoisomerase IIIα and RMI; biallelic mutations in *BLM* (autosomal recessive inheritance) either cause premature translation termination and mRNA destabilization and complete absence of detectable BLM levels, or produce a mutant BLM that is incapable of unwinding DNA, leading to increased cellular hyper-recombinability and hypermutability.	Bloom syndrome (see Chapter 64)
	RecQ helicase L4 (*RECQL4*, 8q 24.3)	Homozygous or compound heterozygous (frameshift/missense) mutations in *RECQL4* (autosomal recessive inheritance) disrupt its functions relating to DNA replication, UV-induced DNA damage repair, telomere and mitochondrial DNA maintenance, interaction with p53 tumor suppressor gene, and oxidative stress relief.	Rothmund–Thomson syndrome (RTS) (see Chapter 48)
	Complementation groups *XPA-XPG*; translesion synthesis variant type *XPV*	*XPA-XPG* as well *XPV* encode proteins in the nucleotide excision repair (NER) pathway that rectifies helix-distorting lesions due to UV radiation; mutations in these genes (autosomal recessive inheritance) cause defective DNA repair and sensitivity to sunlight (solar/UV radiation), leading to cutaneous symptoms, eye damage, neurodegenerative processes, and skin and central nervous system tumors at early age.	Xeroderma pigmentosum (see Chapter 50)

(Continued)

TABLE 1.3 (*Continued*)

Molecular Mechanisms Underlying the Development of Tumor Syndromes

Gene Function/ Chromosomal Abnormality	Specific Gene/Chromosomal Abnormality	Molecular Mechanisms	Tumor Syndromes
Ubiquitination	Cylindromatosis (*CYLD*, 16q12.1)	*CYLD* encodes a protein with deubiquitinase activity; mutations in *CYLD* (autosomal dominant inheritance) enhance constitutive activation of transcription factor NF-κ-B, increase resistance to apoptosis, and contribute to hyperproliferation and tumorigenesis.	Brooke–Spiegler syndrome (BSS) (see Chapter 36)
	Tripartite motif containing protein 37 (*TRIM37*, 17q22)	Homozygous or compound heterozygous mutations in *TRIM37* (autosomal recessive inheritance) alter the function of TRIM37 (which has E3 ubiquitin ligase activity) and disrupt several biological processes (e.g., oncogenesis, apoptosis, organelle transport, cell-cycle control, peroxisomal biogenesis, transcriptional repression, and ubiquitination.	Mulibrey nanism (see Chapter 35)
Enzyme	21-hydroxylase (P450c21) (*CYP21A2*, 6p21.3)	*CYP21A2* encodes a hydroxylase for conversion of cholesterol to aldosterone and cortisol in the zona fasciculata of the adrenal glands; mutations in *CYP21A2* (autosomal recessive inheritance) cause 21-hydroxylase deficiency and insufficient production of aldosterone and cortisol, leading to increased secretion of adrenocorticotropic hormone (ACTH) from hypophysis, enhanced biosynthesis of adrenocortical androgens, subsequent accumulation of progesterone, 17-hydroxyprogesterone, androstenedione, and testosterone, and ultimate changes in primary or secondary sex characteristics in some affected infants, children, or adults.	Congenital adrenal hyperplasia (CAH) (see Chapter 52)
	Dicer 1 ribonuclease III (*DICER1*, 14q32.13)	*DICER1* encodes a ribonuclease; loss-of-function mutations in *DICER1* (autosomal dominant inheritance) reduce or abolish its ability to cleave precursor microRNA (pre-miRNA) into mature effector miRNA and small interfering RNA (siRNA), which negatively regulate stem cell proliferation, organogenesis, cell cycle progression, and oncogenesis.	Pleuropulmonary blastoma (see Chapter 6)
	Yeast Dis3 homolog (*DIS3L2*, 2q37.1)	*DIS3L2* encodes a protein with 3-prime/5-prime exoribonucleolytic activity, which is critical for degradation of both mRNA and noncoding RNA; loss-of-function homozygous or compound heterozygous mutations in *DIS3L2* (autosomal recessive inheritance) cause insulin-like growth factor 2 (*Igf2*) upregulation, leading to overgrowth and/or Wilms tumor.	Perlman syndrome (see Chapter 94)
	Elastase, neutrophil expressed (*ELANE*, 19p13.3); HCLS1 associated protein X-1 (*HAX1*, 1q21.3); growth factor independent protein 1 (*GFI1*, 1p22.1)	*ELANE* encodes a neutrophil elastase belonging to a subfamily of serine proteases that hydrolyze elastin and other proteins within neutrophil lysosomes (azurophil granules) as well as of the extracellular matrix; mutations in *ELANE* (autosomal dominant inheritance) produce a nonfunctional neutrophil elastase that accumulates in the cytoplasm instead of the azurophil granules, leading to endoplasmic reticulum stress, accelerated apoptosis of neutrophil precursors, activation of the unfolded protein response, inability to phagosize invading pathogens and cell debris, and development of cyclic neutropenia and severe congenital neutropenia (45% of cases).	Severe congenital neutropenia (SCN) (see Chapter 77)

(*Continued*)

TABLE 1.3 (*Continued*)

Molecular Mechanisms Underlying the Development of Tumor Syndromes

Gene Function/ Chromosomal Abnormality	Specific Gene/Chromosomal Abnormality	Molecular Mechanisms	Tumor Syndromes
	Establishment of cohesion 1 homolog 2 (*ESCO2*, 8p21)	*ESCO2* encodes a protein with acetyltransferase activity; homozygous or compound heterozygous mutations in *ESCO2* (autosomal recessive inheritance) eliminate acetyltransferase activity and interfere with subsequent cohesion between sister chromatids.	Roberts syndrome (RBS) (see Chapter 47)
	Glycosyltransferases (*EXT1*, 8q24.11; *EXT2*, 11p11.2)	*EXT1* (exostosin-1) and *EXT2* (exostosin-2) encode glycosyltransferases; loss-of-function mutations in *EXT1* (exostosin-1) and *EXT2* (exostosin-2) (autosomal dominant inheritance) disrupt the synthesis and assembly of heparan sulfate chains for subsequent production of heparan sulfate proteoglycans, which influence bone and cartilage formation through interaction with bone morphogenetic proteins.	Hereditary multiple osteochondromas (HMO) (see Chapter 41)
	Zeste homolog 2 (*EZH2*, 7q36.1)	*EZH2* encodes a histone methyltransferase involved in chromatin remodeling and transcriptional regulation; germline mutations in *EZH2* (autosomal dominant inheritance) alter/block its methyltransferase activity and facilitate trimethylation of H3K27, leading to distinct craniofacial skeleton formation.	Weaver syndrome (see Chapter 100)
	Fumarate hydratase (*FH*, 1q42.1)	Loss-of-function mutations in *FH* (autosomal dominant inheritance) inactivate fumarate hydratase, leading to increased intracellular fumarate accumulation, decreased hypoxia-inducible factor (HIF) degradation, upregulation in transcriptional pathways mediated by HIF-1α, and ultimately epigenetic abnormalities (e.g., leiomyomas and renal cell cancer).	Hereditary leiomyomatosis and renal cell cancer (HLRCC) (see Chapter 31)
	Nuclear receptor-binding set domain protein 1 (*NSD1*, 5q35.3)	*NSD1* encodes a protein with histone methyltransferase activity, which preferentially catalyzes the transfer of methyl residues to lysine residue 36 of histone 3 (H3K36) and lysine residue 20 of histone 4 (H4K20); mutations in *NSD1* (autosomal dominant inheritance 5%; de novo 95%) contribute to *NSD1* haploinsufficiency, which prevents one copy of the gene from producing any functional protein, and subsequently disrupts the normal activity of genes involved in growth and development.	Sotos syndrome (see Chapter 98)
	Phosphoinositide-3-kinase (*PIK3CA*, 3q26.32)	*PIK3CA* encodes the p110α catalytic subunit of phosphoinositide-3-kinase heterodimer involved in the PI3K/AKT/mTOR signaling pathway; like other PIK3CA-associated overgrowth, CLOVES (*c*ongenital *l*ipomatous *o*vergrowth, *v*ascular malformations [typically truncal], *e*pidermal nevi, *s*coliosis/skeletal/spinal abnormalities, and seizures/central nervous system malformations) syndrome is attributed to activating somatic mutations in *PIK3CA* (non-inherited).	CLOVES syndrome (see Chapter 88); Klippel–Trenaunay syndrome (KTS) (see Chapter 91)
	Cationic trypsinogen serine protease 1 (*PRSS1*, 7q34)	Gain-of-function mutations in *PRSS1* (autosomal dominant inheritance) produce a misfolding cationic trypsinogen that is converted prematurely to active trypsin in the pancreas before excretion to small intestine, leading to heightened trypsin activity, stability, and autoactivation; autodigestion of the pancreatic parenchyma; strong inflammatory responses (pancreatitis).	Hereditary pancreatitis (see Chapter 57)

(*Continued*)

TABLE 1.3 (*Continued*)

Molecular Mechanisms Underlying the Development of Tumor Syndromes

Gene Function/ Chromosomal Abnormality	Specific Gene/Chromosomal Abnormality	Molecular Mechanisms	Tumor Syndromes
	Rhomboid 5 homolog 2 (*RHBDF2*, 17q25.1)	*RHBDF2* encodes a protease; gain-of-function mutations in *RHBDF2* (autosomal dominant inheritance) increase the protease stability, augment the maturation and activity of ADAM17 in epidermal keratinocytes, enhance EGFR activity, and accelerate wound healing and epithelial tumorigenesis, leading to tylosis (hyperproliferation of skin in the palms and soles, loss of hair), oral leukoplakia, esophageal cancer, and other abnormalities.	Tylosis with esophageal cancer (TOC) (see Chapter 24)
	Succinate dehydrogenase A, B, C, D (*SDHx*)	SDH is part of both the citric acid cycle (in which SDH oxidizes succinate to fumarate) and the aerobic electron transfer chain; mutations in *SDHx* (autosomal dominant inheritance) disrupt several biological processes (e.g., cell apoptosis, angiogenesis, proliferation, glycolysis), leading to tumorigenesis.	Hereditary pheochromocytoma and paraganglioma (SDHx) syndrome (see Chapter 58)
Other protein	Aryl hydrocarbon receptor interacting protein (*AIP*, 11q13.2)	*AIP* encodes a protein, which functions as a receptor for aryl hydrocarbons and a ligand-activated transcription factor that regulates the expression of many xenobiotic metabolizing enzymes, and prevents cells from uncontrolled growth and division; loss-of-function mutations in *AIP* (autosomal dominant inheritance) compromise the ability of AIP to control the growth and division of pituitary cells, leading to the development to pituitary adenomas.	Familial isolated pituitary adenoma (FIPA) (see Chapter 54)
	Bardet–Biedl syndrome-related (*BBS1-21*)	*BBS1-21* encode cilia proteins; mutations in *BBS1-21* (autosomal recessive inheritance) produce altered proteins that compromise the structures and functions of cilia and basal bodies, leading to the development of ciliopathies affecting multiple organs/systems.	Bardet–Biedl syndrome (BBS) (see Chapter 25)
	Budding uninhibited by benzimidazole-related-1 (*BUB1B*, 15q15.1)	*BUB1B* encodes a checkpoint protein involved in spindle assembly checkpoint (SAC), which provides a surveillance mechanism to ensure faithful chromosome segregation during mitosis; homozygous or compound heterozygous mutations in *BUB1B* (autosomal recessive inheritance) cause BubR1 insufficiency, leading to the development of MVA syndrome.	Mosaic variegated aneuploidy (MVA) syndrome (see Chapter 46)
	Cyclin dependent kinase inhibitor 1B (*CDKN1B*, 12p13.1)	*CDKN1B* encodes a cyclin dependent kinase inhibitor 1B, which binds to and prevents the activation of cyclin E-CDK2 or cyclin D-CDK4 complexes, and thus controls cell cycle progression at G1; loss-of-function mutations in *CDKN1B* (autosomal dominant inheritance) lead to unchecked cell growth and division, typically in MEN1 patients without *MEN1* mutations (20% of cases).	Multiple endocrine neoplasia 4 (MEN4) (see Chapter 60)
	CCAAT/enhancer binding protein alpha (*CEBPA*, 19q13.1); runt-related transcription factor 1 (*RUNX1*, 21q22.12); ankyrin repeat domain 26 (*ANKRD26*, 10p12.1); ETS translocation variant gene 6″ (*ETV6*, 12p13.2); DEAD-box helicase 41 (*DDX41*, 5q35.3)	*CEBPA* (encoding a transcription factor) and *DDX41* (encoding a DEAD-box protein showing RNA helicase activity) are implicated in nonsyndromic familial AML with no pre-existing disorders; while *RUNX1* (encoding a transcription factor), *ANKRD26* (encoding a protein) and *ETV6* (encoding a nuclear protein functioning as an ETS family transcription factor) are involved in nonsyndromic familial AML with pre-existing platelet disorders (all via autosomal dominant inheritance).	Familial acute myeloid leukemia (AML) (see Chapter 70)

(*Continued*)

TABLE 1.3 (*Continued*)

Molecular Mechanisms Underlying the Development of Tumor Syndromes

Gene Function/ Chromosomal Abnormality	Specific Gene/Chromosomal Abnormality	Molecular Mechanisms	Tumor Syndromes
	Chediak–Higashi syndrome (*CHS1*, formerly *LYST*, 1q42.3)	*CHS1* encodes a protein involved in the regulation of natural killer (NK) cell lytic activity; loss-of-function biallelic mutations in *CHS1* (autosomal recessive inheritance) incapacitate its role as a scaffold protein in membrane fusion and fission events, leading to the formation of giant lysosomes or lysosome-related organelles in host cells, and subsequent skin hypomelanosis/oculocutaneous albinism.	Chediak–Higashi (CHS) syndrome (see Chapter 65)
	G-protein coupled glucagon receptor (*GCGR*, 17q25.3)	*GCGR* encodes a glucagon receptor that plays a central role in the regulation of hepatic glucose production by promoting glycogen hydrolysis and gluconeogenesis; homozygous-inactivating mutations in *GCGR* (autosomal recessive inheritance) decrease its glucagon binding potential, cause abnormal receptor internalization and calcium mobilization, and increase apoptosis, leading to multiple pancreatic neuroendocrine tumors (PNET), pancreatic alpha cell hyperplasia (PACH), and hyperglucagonemia without glucagonoma syndrome.	Mahvash disease (see Chapter 59)
	Gap junction protein β2 (*GJB2*, 13q12.11)	*GJB2* encodes a transmembrane protein; missense mutations in *GJB2* (autosomal dominant inheritance) cause GJC permeability changes (hyperactive hemichannels), reduce translocation, and increase membrane flow and cell death.	Keratitis–ichthyosis–deafness (KID) syndrome (see Chapter 42)
	Guanine nucleotide binding protein, alpha stimulating (*GNAS*, 20q13.32)	*GNAS* encodes α-subunit of the stimulatory G protein (Gsα) associated with the cAMP-dependent pathway; early embryonic somatic mosaic activating or gain-of-function mutations in *GNAS* result in activation of cAMP, inhibition of GTPase activity, constitutive activation of adenylate cyclase, overproduction of several hormones, and disruption of the signaling turnoff mechanism (not inherited).	McCune–Albright syndrome (MAS) (see Chapter 44)
	Glypican (*GPC3*, Xq26.2)	*GPC3* encodes a glycosylphosphatidylinositol-linked cell surface heparan sulfate proteoglycan, which acts as an inhibitor of hedgehog signaling pathway; binding of hedgehog to GPC3 triggers endocytosis and degradation of the GPC3/hedgehog complex, inducing apoptosis of certain unwanted cells; loss-of-function mutations in *GPC3* (X-linked recessive inheritance) result in hyperactivation of hedgehog signaling pathway and increase cell proliferation, leading to overgrowth and embryonal tumor.	Simpson–Golabi–Behmel syndrome (SGBS) (see Chapter 97)
	TCR signaling (*RasGRP1*, *ZAP70*, *PI3K*, *ITK*), actin cytoskeleton arrangements (*CORO1A*, *WASP*), co-stimulation (*LRBA*, *MAGT1*), leukocyte development (*GATA2*, *MCM4*), lymphocyte cell death (*XIAP*, *STK4*, *CTPS1*), cytotoxic effector (*CD16*, *NKG2D*, *SLAM receptors like 2B4*, *CD27*),	Individuals with mutated genes related to B-cell development and/or function are susceptible to EBV infection, which takes over the control of B cells and causes a spectrum of clinical manifestations (e.g., severe infectious mononucleosis with fever, sore throat, lymphadenopathy, splenomegaly, Hodgkin and non-Hodgkin lymphomas, and hypo/dysgammaglobulinemia), collectively referred to LPD associated with EBV infection or EBV-driven LPD (autosomal recessive inheritance).	Lymphoproliferative disorders (LPD) associated with Epstein–Barr virus (EBV) infection (see Chapter 75)

(*Continued*)

TABLE 1.3 (*Continued*)

Molecular Mechanisms Underlying the Development of Tumor Syndromes

Gene Function/ Chromosomal Abnormality	Specific Gene/Chromosomal Abnormality	Molecular Mechanisms	Tumor Syndromes
	RNA binding motif protein 8 (*RBM8A*, 1q21.1)	*RBM8A* encodes a protein with a conserved RNA-binding motif that influences various downstream processes (e.g., nuclear mRNA export, subcellular mRNA localization, translation efficiency and nonsense-mediated mRNA decay; inheritance of a microdeletion at chromosome 1q21.1 from one parent and a noncoding SNP in *RBM8A* from the other results in loss of reduction *RBMBA* in platelets, leading to a combined defect of platelet production and function in thrombocytopenia with absent radii (autosomal recessive inheritance, 75%).	Thrombocytopenia-absent radius (TAR) (see Chapter 81)
	Ribosomal proteins (*RPS7, RPS10, RPS17, RPS19, RPS24, RPS26, RPL5, RPL11, RPL35A*)	Both small (e.g., RPS7, RPS10, RPS17, RPS19, RPS24, RPS26) and large (e.g., RPL5, RPL11, RPL35A) subunits of ribosomal proteins are required for ribosome biogenesis; mutations in related RP genes reduce overall numbers of ribosomes, interrupt erythroid progenitor cell division, self-destroy blood-forming cells in the bone marrow (anemia), and activate apoptosis-related signaling pathway (often autosomal dominant inheritance).	Diamond–Blackfan anemia (DBA) (see Chapter 67)
	Shwachman–Bodian–Diamond syndrome (*SBDS*, 7q11.21)	*SBDS* encodes a protein involved in ribosome biogenesis, microtubule stabilization, actin polymerization, stromal microenvironment and genome stability; mutations in *SBD* (autosomal recessive inheritance) compromise hematopoiesis, hinder but not arrest the regeneration of neutrophils, red cells, and platelets.	Shwachman–Bodian–Diamond syndrome (SBDS) (see Chapter 78)
	Solute carrier family 26 member 4 (*SLC26A4*, 7q22.3)	*SLC26A4* encodes an iodide-chloride transporter (pendrin) that contributes to homeostasis of ion concentration in the endolymphatic sac and hair cells in the inner ear, and formation of hormone in the thyroid; loss-of-function biallelic mutations in *SLC26A4* (autosomal recessive inheritance) cause endolymph acidification, auditory sensory transduction defects, impaired iodide organification, abnormal intracellular thyroid hormone synthesis, enlargement of the vestibular aqueduct (EVA), syndromic and nonsyndromic hearing loss, and thyroid goiter.	Pendred syndrome (see Chapter 61)
	Tuberous sclerosis complex 1 (*TSC1*, 9q34); tuberous sclerosis complex (*TSC2*, 16p13.3)	*TSC1* and *TSC2* encode hamartin (which regulates mTOR–S6K and cell adhesion through interaction with ezrin and Rho) and tuberin (which regulates mTOR–S6K and GTPase-activating proteins as well as cell cycle), respectively; mutations in *TSC1* and *TSC2* (autosomal dominant inheritance 33%; de novo 66%) disrupt the mTOR-inhibiting complex formed by TSC1 and TSC2, and compromise their ability to regulate cell proliferation and differentiation, leading to epilepsy, mental retardation, cortical tubers, renal angiomyolipomas, retinal hamartomas, and facial angiofibromas.	Tuberous sclerosis complex 1 (TSC1); tuberous sclerosis complex 2 (TSC2) (see Chapter 99)

(Continued)

TABLE 1.3 (*Continued*)

Molecular Mechanisms Underlying the Development of Tumor Syndromes

Gene Function/ Chromosomal Abnormality	Specific Gene/Chromosomal Abnormality	Molecular Mechanisms	Tumor Syndromes
	Wiskott–Aldrich syndrome (*WAS*, Xp11.23)	*WAS* encodes a protein (WASP) involved in actin polymerization, receptor engagement, signaling transduction, and cytoskeletal rearrangement; mutations in *WAS* (X-linked recessive inheritance) cause defective migration, anchoring, and localization of T and B lymphocytes, neutrophils, macrophages, and dendritic cells.	Wiskott–Aldrich syndrome (WAS) (see Chapter 83)
Chromosome abnormality	Chromosome 6q deletion	Chromosome 6q deletion occurs in 55% of Waldenström macroglobulinemia patients (mostly non-inherited).	Waldenström macroglobulinemia (see Chapter 82)
	Chromosomes 7 and 14 inversions and translocations; Nijmegen breakage syndrome (*NBN*, 8q21.3)	Chromosomes 7 and 14 inversions/translocations create chromosomal instability in peripheral T lymphocytes; biallelic mutations in *NBN* (autosomal recessive inheritance), encoding nibrin involved in DNA double strand break repair, underlie the development of a chromosomal instability disorder known as Nijmegen breakage syndrome.	Nijmegen breakage syndrome (NBS) (see Chapter 76)
	Chromosome 11p15.5 defects	Epigenetic or cytogenetic defects in chromosome 11p15.5 affect either the whole 1 Mb region or specifically one of the two imprinting control regions (ICRs): the telomeric *IGF2/H19* domain controlled by ICR1 and the centromeric *KCNQ1/CDKN1C* domain controlled by ICR2; these aberrations result in downregulation of maternally expressed growth-restraining genes (e.g., *CDKN1C*) and/or upregulation of paternally expressed growth-promoting genes (e.g., *IGF2*), leading to overgrowth and tumor predisposition (autosomal dominant inheritance).	Beckwith–Wiedemann syndrome (BWS) (see Chapter 86)
	Defective telomere (*DKC1, TERT, TERC, TINF2, RTEL1, NOP10, NHP2, WRAP53, CTC1, PARN*)	Mutations in genes encoding telomerase and related components (X-linked recessive, autosomal dominant, or autosomal recessive inheritance) alter the normal function of telomerase and cause premature telomere shortening and subsequent replicative senescence, leading to premature stem cell failure, aging, and cancer predisposition.	Dyskeratosis congenita (see Chapter 69)
	Monosomy 7	Monosomy 7 (loss of chromosome 7 or -7) or interstitial (partial) deletion of the long arm of chromosome 7 (del[7q] or 7q-) results in haploinsufficiency of several important genes (e.g., *SAMD9* [sterile α motif domain 9; 7q21.3], *SAMD9L* [SAMD9-like; 7q21.3], *EZH2* [enhancer of zeste homolog 2; 7q36.1], *MLL3* [mixed lineage leukemia 3; 7q36.1], and *CUX1* [homeobox transcription factor; 7q22.1]), leading to bone marrow failure/MDS/AML.	Familial monosomy 7 syndrome (see Chapter 71)
	Supernumerary X-chromosome	Random error in cell division (so-called nondisjunction) causes abnormal number of chromosomes in reproductive cell (egg or sperm), leading to gene dosage compensation, skewed X-inactivation, and genetic polymorphisms (not inherited).	Klinefelter syndrome (see Chapter 2)

(*Continued*)

TABLE 1.3 (*Continued*)

Molecular Mechanisms Underlying the Development of Tumor Syndromes

Gene Function/ Chromosomal Abnormality	Specific Gene/Chromosomal Abnormality	Molecular Mechanisms	Tumor Syndromes
	Trisomies, IgH translocations, combined IgH translocation and trisomies, isolated monosomy 14	Trisomies (involving odd-numbered chromosomes except for chromosomes 1, 13, and 21; 42% of cases), IgH translocations (immunoglobulin heavy chain (IgH) translocations including t(11;14)(q13;q32)—*CCND1* (*cyclin D1*); t(4;14)(p16;q32)—*FGFR-3* and *MMSET*; t(14;16) (q32;q23)—*C-MAF*; t(14;20)(q32;q11)—*MAFB*; other IgH translocations—*CCND3* (cyclin D3) in t(6;14)(p21;q32) translocation or other IgH translocations involving uncommon partner chromosomes; 30% of cases), combined IgH translocation and trisomies (trisomies plus one recurrent IgH translocation; 15% of cases), isolated monosomy 14 (14q32 translocations involving unknown partner chromosomes; 4.5% of cases) (generally not inherited).	Familial multiple myeloma (see Chapter 72)
	Trisomy 21	Trisomy 21 (with an extra copy deriving mostly from maternal side) results from failure of chromosome 21 pair to separate during formation of an egg (or sperm), and causes dosage imbalance of genes, leading to increased expression and regulation of many genes throughout the genome (not inherited).	Down syndrome (see Chapter 68)
	Xp22 abnormality	Xp22 between p22.2 and p22.3 contains 5 loci behaving as recessive traits, and 2 loci linked to death in nullisomic males, and mosaic pattern in monosomic females; genes encoding steroid sulfatase and serologically defined, male-specific antigen are found in this region (X-linked dominant inheritance).	Aicardi syndrome (see Chapter 2)
	Y chromosome *gr/gr* deletion	Y chromosome *gr/gr* deletion and mutations in other genes (e.g., *KITLG*, *PDE11A*, *SPRY4*, and *BAK1*) (autosomal recessive inheritance) disrupt biological processes relating to cell proliferation, migration, and anti-apoptosis.	Familial testicular germ cell tumor (TGCT) (see Chapter 28)

1.3 Clinical Features of Tumor Syndromes

Tumor syndromes typically display both tumor-related symptoms as well as syndromic symptoms. In hereditary tumor syndromes, syndromic symptoms often appear at birth and become apparent shortly after birth, while tumor or tumor-related symptoms may emerge somewhat later (after second mutation/hit). Nevertheless, hereditary tumors still have a much earlier occurrence than sporadic tumors and may harbor distinct biomarkers and gene mutations in comparison with sporadic tumors. For example, hereditary breast cancer often appears earlier, harbors BRCA1 and BRCA2 mutations, and shows strong expression of Ki67 and EGFR, whereas sporadic breast cancer appears later, possesses *ERBB2* and *myc* mutations, and has weak expression of Ki67 and EGFR [2].

It is noteworthy that some tumor syndromes have heterogeneous etiologies and contain genetic mutations that are linked to other syndromes. For instance, familial non-medullary thyroid carcinoma (FNMTC) may be syndromic (showing similarities to familial adenomatous polyposis [FAP], Cowden syndrome, Carney complex, Werner syndrome, DICER syndrome/pleuropulmonary blastoma) or non-syndromic (having distinct genetic mutations such as *SRGAP1* [12q14], *TITF-1/NKX2.1* [14q13], *FOXE1* [9q22.33], *HABP2* [10q25.3]) (see Chapter 55). Similarly, familial pancreatic cancer harbors multiple driver mutations (e.g., oncogene *KRAS*, and tumor suppressor genes *CDKN2A*, *TP53*, and *SMAD4*) (see Chapter 56).

Sporadic mutations may also be involved in some non-inherited disorders that display both tumor-related (nonsyndromic) and other (syndromic) symptoms. For instance, Cronkhite–Canada syndrome (CCS) is a rare sporadic/non-inherited disorder related to autoimmune dysfunction, infection, and allergic response, and typically displays epithelial disturbances in both the gastrointestinal tract (e.g., gastrointestinal hamartomas, juvenile type gastrointestinal polyps often associated with diarrhea, colorectal serrated adenomas, ectodermal dysplasia) and epidermis (e.g., skin hyperpigmentation, alopecia, onychodystrophy) in addition to peripheral edema, malabsorption, and increased malignant transformation (e.g., gastrointestinal cancer and multiple myeloma at adult age) (see Chapter 15). Because some of these sporadic, nonhereditary disorders manifest clinical features that may confuse the diagnosis of hereditary tumor syndromes, they are also discussed in this book.

1.4 Diagnosis of Tumor Syndromes

Diagnosis of tumor syndromes involves physical exam, family history review, biochemical assessment, imaging, histopathology, cytogenetics, and molecular tests (if available). When an unknown/uncharacterized tumor syndrome is suspected, identification of its genetic cause using karyotyping, linkage analysis, genome-wide association study (GWAS), and positional cloning should be a priority. On the other hand, when a known/characterized tumor syndrome is encountered, application of molecular genetic tests targeting relevant genes helps its confirmation.

1.4.1 Uncharacterized Tumor Syndromes

For an unknown/uncharacterized tumor syndrome, several molecular techniques (e.g., karyotyping, linkage analysis, GWAS, and positional cloning) can be utilized to pinpoint the genetic cause and characterize the culprit gene.

1.4.1.1 Karyotyping

Karyotyping is the process of taking micrograph images of mitotic cells from tumor biopsy, bone marrow, or leukocyte culture (after pretreatment with a hypotonic solution to swell the cells and spread the chromosomes, arresting mitosis in metaphase [when chromosomes are in their most condensed conformations] with colchicine, staining with a suitable stain, and then squashing the preparation on the slide forcing the chromosomes into a single plane, and visualizing the diploid set of chromosomes). By cutting up photomicrographs and grouping chromosomes in pairs to form a karyotype, it helps determine their number and aberrations (e.g., breakage, loss, duplication, translocation, or inversion involving several megabases or more of DNA). Depending on the stain used, chromosomes may show different band patterns (alternating light and dark stripes along the lengths of chromosomes), including G-bands (Giemsa stain following digestion of chromosomes with trypsin, yielding 300–400 light/TA-rich and dark/GC-rich bands), R-bands (reverse of G-bands, yielding light/GC-rich and dark/AT-rich bands), C-bands (undergoing alkaline denaturation prior to Giemsa staining, revealing centromeric or constitutive heterochromatin as dark bands), Q-bands (quinacrine stain, yielding yellow fluorescent bands of differing intensity similar to G-bands), T-bands (visualizing telomeres), and NOR-bands (silver stain, yielding a dark region denoting the activity of rRNA genes within the NOR) [11].

1.4.1.2 Linkage Analysis

Linkage is the tendency for genes and other genetic markers to be inherited together in a nearby location on the same chromosome. Linkage analysis searches for chromosomal segments that cosegregate with the disease phenotype through families. In parametric (or model-based) linkage analysis (if the relationship between phenotypic and genetic similarity is known), the probability that a gene important for a disease is linked to a genetic marker is estimated through the LOD (logarithm [base 10] of odds) score, with a LOD score of 3 or more being an indicator that the two loci are linked and are close to one another. In non-parametric (model-free) linkage analysis, the probability of an allele being identical is evaluated without particular model assumptions. Both parametric and nonparametric linkage analyses are useful for detecting regions of the genome harboring high penetrance risk variants [12,13].

1.4.1.3 Genome-Wide Association Studies

GWAS typically examine a genome-wide set of genetic variants in two large groups of individuals to see if any variant is associated with a trait. As single-nucleotide polymorphisms (SNP) tend to occur more frequently in people with a particular disease than in people without the disease, they constitute a useful target for GWAS. A genome-wide significance level of 0.05 has been shown to correspond to a LOD score of 3.3 or higher. GWAS is powerful for detecting common risk variants which have small individual effects on risk, whereas linkage analysis is valuable for detecting genes with rare, high penetrance risk variants [14,15].

1.4.1.4 Positional Cloning

Positional cloning is a technique used to locate the position of a disease-associated gene along the chromosome or the candidate region after linkage analysis is done, without prior knowledge about the biochemical basis of the disease or function of the gene. Positional cloning typically involves patient recruitment, DNA collection, and examination of genetic recombination frequencies generated by meiotic cross-overs and on genome-wide molecular studies. Following the definition of a critical genomic region, the causative gene mutation is identified upon analysis of available databases from genome browsers (e.g., Ensembl or Santa Cruz Genome Browser), and its functional role in the pathogenesis of the disorder is delineated with the use of cell-based or animal-based experiments [16].

1.4.2 Characterized Tumor Syndromes

For a known/characterized tumor syndrome, especially in an individual who has relatives with a confirmed tumor syndrome, molecular genetic tests (e.g., PCR, sequencing analysis) targeting relevant genes can be used for diagnosis and verification, along with other procedures.

1.5 Treatment and Prognosis of Tumor Syndromes

Current treatment options for tumors consist of surgery, radiotherapy, chemotherapy, and complementary therapies. As tumor syndromes develop both tumor-related (non-syndromic) and syndromic symptoms, a more comprehensive plan is required for their treatment. This should include not only specific therapies to target tumor-related symptoms but also appropriate measures to control and relieve syndromic symptoms.

Prognosis for tumor syndromes is dependent on tumor type, location, and grade; patient age and health status; and severity of syndromic symptoms. In general, patients with lower grade tumors have a more favorable prognosis than those with higher grade tumors. Further, patients with mild syndromic symptoms have a better outcome than those with severe syndromic symptoms.

1.6 Future Perspectives

Given their molecular heterogeneity and clinical complexity, tumor syndromes have proven difficult to diagnose and challenging to treat. Several things may help explain these conundra. First, tumor syndromes are relatively rare (making up only 5%–10% of clinical tumor cases) and few physicians are familiar with the full clinical spectrums of tumor syndromes. Second, tumor syndromes result from diverse genetic mutations or chromosomal abnormalities that demand individualized molecular tests for identification and diagnosis, which are sadly lacking at the moment. Third, tumor syndromes often affect multiple organs and sites, which add complexity to surgical removal and radiotherapy. Fourth, tumor syndromes show both tumor-related symptoms and syndromic symptoms, which increase the difficulty of their treatment. While the publication of this book helps to enhance the profile and awareness of tumor syndromes, development of individualized diagnostic tests for these disorders is a necessity. In addition, improved understanding of molecular mechanisms of tumor syndromes is essential for designing innovative intervention measures to mitigate these disorders.

Acknowledgment

The author is indebted to Dr. John Pedersen, a specialist in anatomical pathology, for his introduction into and guidance on molecular diagnosis of tumors/cancers and genetic disorders between 1995 and 2001, which have inspired this work.

REFERENCES

1. Birner P, Prager G, Streubel B. Molecular pathology of cancer: How to communicate with disease. *ESMO Open.* 2016;1(5):e000085.

2. van der Groep P, Bouter A, van der Zanden R et al. Distinction between hereditary and sporadic breast cancer on the basis of clinicopathological data. *J Clin Pathol.* 2006;59(6):611–7.

3. Rahner N, Steinke V. Hereditary cancer syndromes. *Dtsch Arztebl Int.* 2008;105(41):706–14.

4. Ngeow J, Eng C. Precision medicine in heritable cancer: When somatic tumour testing and germline mutations meet. *NPJ Genom Med.* 2016;1:15006.

5. Trask BJ. Human cytogenetics: 46 chromosomes, 46 years and counting. *Nat Rev Genet* 2002;3:769–78.

6. Wu ZH. Phenotypes and genotypes of the chromosomal instability syndromes. *Transl Pediatr.* 2016;5(2):79–83.

7. Hino O, Kobayashi T, Mourning AGK. The two-hit hypothesis, tumor suppressor genes, and the tuberous sclerosis complex. *Cancer Sci.* 2017;108(1):5–11.

8. Genetic Alliance; The New York-Mid-Atlantic Consortium for Genetic and Newborn Screening Services. *Understanding Genetics: A New York, Mid-Atlantic Guide for Patients and Health Professionals.* Washington (DC): Genetic Alliance; 2009 Jul 8. APPENDIX E, INHERITANCE PATTERNS.

9. Gulani A, Weiler T. *Genetics, Autosomal Recessive. StatPearls* [Internet]. Treasure Island (FL): StatPearls Publishing; 2020. 2019 Aug 24.

10. Saletta F, Dalla Pozza L, Byrne JA. Genetic causes of cancer predisposition in children and adolescents. *Transl Pediatr.* 2015;4(2):67–75.

11. Speicher MR, Ballard SG, Ward DC. Karyotyping human chromosomes by combinatorial multi-fluor FISH. *Nat Genet.* 1996; 12, 368–75.

12. Pulst SM. Genetic linkage analysis. *Arch Neurol.* 1999;56(6): 667–72.

13. Ott J, Wang J, Leal SM. Genetic linkage analysis in the age of whole-genome sequencing. *Nat Rev Genet.* 2015;16(5):275–84.

14. Ku CS, Cooper DN, Wu M et al. Gene discovery in familial cancer syndromes by exome sequencing: Prospects for the elucidation of familial colorectal cancer type X. *Mod Pathol.* 2012;25(8):1055–68.

15. Machiela MJ, Grünewald TGP, Surdez D et al. Genome-wide association study identifies multiple new loci associated with Ewing sarcoma susceptibility. *Nat Commun.* 2018;9(1):3184.

16. Puliti A, Caridi G, Ravazzolo R, Ghiggeri GM. Teaching molecular genetics: Chapter 4-positional cloning of genetic disorders. *Pediatr Nephrol.* 2007;22(12):2023–9.

Section I

Tumor Syndromes Affecting the Brain, Head, and Lungs

2

Aicardi Syndrome and Klinefelter Syndrome

Dongyou Liu

CONTENTS

2.1 Introduction

First described by Aicardi et al. in 1965, Aicardi syndrome is a rare, X-linked dominant, neurodevelopmental disorder affecting almost exclusively females, and occasionally 47,XXY males. Clinically, Aicardi syndrome is characterized by a triad of features: agenesis or dysgenesis of the corpus callosum (absent or underdeveloped tissue connecting the left and right halves of the brain), infantile spasms (seizures/epilepsy), and chorioretinal lacunae (defects in the light-sensitive tissue at the retina) (Table 2.1). However, not all affected girls show the classic triad, and some may instead develop other neuronal and extraneuronal defects, including asymmetry between the two sides of the brain, reduced brain folds and grooves, enlarged ventricles (fluid-filled cavities), microcephaly (small head), microphthalmia (small or poorly developed eyes), coloboma (cat's eye, due to gap or hole in the optic nerve), philtrum (a short area between the upper lip and the nose), flat nose with an upturned tip, large ears, sparse eyebrows, small hands, scoliosis (abnormal curvature of the spine), developmental delay and intellectual disability, gastrointestinal problems (constipation/diarrhea, gastroesophageal reflux, and difficulty feeding), and increased cancer risk [1].

Aicardi syndrome is unrelated to Aicardi–Goutières syndrome (AGS), which is an autosomal recessive disorder characterized by early-onset encephalopathy, intellectual and physical handicap, calcification of the basal ganglia (particularly the putamen, globus pallidus, thalamus), leukodystrophy, cerebral atrophy, chronic cerebrospinal fluid (CSF) leukocytosis, and increased concentration of interferon-alpha in the CSF. At the genetic level, AGS is linked to variations in *TREX1*, *RNASEH2A*, *RNASEH2B*, *RNASEH2C*, *SAMHD1*, and *ADAR* genes [2].

Klinefelter syndrome, first reported by Klinefelter et al. in 1942, is a common genetic disorder that occurs in males with an extra X chromosome (XXY) instead of the usual male sex karyotype (XY). Clinically, Klinefelter syndrome is characterized by small testes (that do not produce as much testosterone as usual), delayed or incomplete puberty, gynecomastia (breast enlargement), sparse facial and body hair, azoospermia (inability to produce sperm), infertility, cryptorchidism (undescended testes), hypospadias (the opening of the urethra on the underside of the penis), micropenis (unusually small penis), tall stature, and increased risk for extragonadal germ cell tumor (GCT), breast cancer, and systemic lupus erythematosus (Table 2.1). Cytogenetically, 90% of Klinefelter syndrome patients possess non-mosaic 47,XXY karyotype (due to the aneuploidy of the sex chromosomes), 7% are mosaic (e.g., 47,XXY/46,XY), and 3% have variant (e.g., 48,XXXY or 48,XXYY) and structurally abnormal X chromosome (e.g., 47,iXq,Y) [3].

2.2 Biology

Of the 46 chromosomes found in human cells, two (i.e., X and Y chromosomes) are involved in the determination of male or female sex characteristics. Whereas females have two X chromosomes (46,XX), males possess one X chromosome and one Y chromosome (46,XY). Under normal circumstances, one of the two X chromosomes in females is permanently inactivated in somatic cells (i.e., non-egg/sperm cells) early in embryonic development to maintain only one active copy of X chromosome in each body cell, just like males.

Aicardi syndrome (agenesis of corpus callosum with chorioretinal abnormality, agenesis of corpus callosum with infantile spasms and ocular abnormalities, callosal agenesis and ocular abnormalities, or chorioretinal anomalies with agenesis of corpus callosum) is an X-linked dominant disorder that possibly arises from de novo mutation on one of the two X chromosomes in

TABLE 2.1

Key Features of Aicardi Syndrome and Klinefelter Syndrome

Features	Aicardi Syndrome	Klinefelter Syndrome
Non-tumor features	Chorioretinal lacunae, cleft lip, cleft palate, corpus callosum agenesis, developmental delay/mental deficiency/mental retardation, asymmetric facies, gross motor delay, plagiocephaly, rib anomalies, infantile spasms/seizures, vertebral anomalies	47,XXY and variants (constitutional), gynecomastia, hypergonadotropic hypogonadism, relatively long legs, obesity, small penis, small testicles, infertility, low IQ
Tumor features (possible)	Choroid plexus papilloma (angiosarcoma, colorectal polyps, gastric polyps, hepatoblastoma, scalp lipoma, medulloblastoma, benign teratoma of soft palate, parapharyngeal embryonal cell cancer)	Extragonadal germ cell tumor (astrocytoma, breast cancer, chordoma, Hodgkin lymphoma, acute myeloid leukemia (AML, incl. ANLL), non-Hodgkin lymphoma, retinoblastoma, sarcoma, papillary thyroid cancer)
Related gene (chromosome)	AIC (Xp22)	(extra X chromosome)
Mode of inheritance	X-linked dominant (XLD)	Not applicable
Estimated prevalence	~1 per 100,000 newborns	150 per 100,000 newborn males; 355 per 100,000 Asian males

46,XX females and occasionally 47,XXY males (who are associated with Klinefelter syndrome), leading to Xp22 abnormalities. Aicardi syndrome is not seen in 46,XY males since a mutation in the only copy of the X chromosome is nearly always lethal very early in development. Employing methylation-sensitive restriction analysis and segregation of the active X chromosome in somatic cell hybrids, skewed X inactivation (which often occurs in incontinentia pigmenti) was observed in lymphocytes from cytogenetically normal girls with Aicardi syndrome, while random X inactivation was found in children with higher neurologic severity and vertebral anomalies [4,5].

Klinefelter syndrome (Klinefelter's syndrome, XXY syndrome, or XXY trisomy) is due to the presence of one extra copy of the X chromosome in each cell (47,XXY), which escapes X-inactivation or displays polymorphisms in specific genes (e.g., the trinucleotide repeat length of androgen receptor gene), leading to subdued gonadal development, reduced levels of testosterone, and altered brain development and growth. Based on microarray gene expression analysis of lymphocytes and testis transcriptome analysis of Sertoli, Leydig, and germ cells from Klinefelter syndrome males, 480 autosomal genes are upregulated and >200 genes are downregulated. Epigenetic modulation of autosomal genes (e.g., sperm-associated antigen 1 [SPAG1] on the long arm of chromosome 8 that codes for a protein involved in signal transduction pathways in spermatogenesis) caused by aneuploidy may appear in a tissue-specific manner, manifesting as gonadal failure, insulin resistance, dyslipidemia, and coagulability and other symptoms. Due to the effects of extra genetic material, Klinefelter syndrome males with 48,XXXY or 49,XXXXY karyotype often show more severe signs and symptoms than those with classic 47,XXY karyotype. Apart from altered male sexual development, these variants are associated with congenital anomalies (e.g., inguinal hernia, congenital heart disease, cleft palate and velopharyngeal insufficiency, and kidney malformation), intellectual disability, distinctive facial features, skeletal abnormalities, poor coordination, severe speech problems, and dental conditions (e.g., taurodontism and caries) [3].

2.3 Pathogenesis

Although most patients with Aicardi syndrome possess apparently normal chromosomal and genomic profiles, some demonstrate a presumably balanced X;3 translocation, suggesting the involvement of Xp22 between p22.2 and p22.3 in Aicardi syndrome pathogenesis. The region between p22.2 and p22.3 contains loci that have been implicated in short stature, X-linked recessive chondrodysplasia punctata, mental retardation, X-linked ichthyosis, Kallmann syndrome, Aicardi syndrome, and focal dermal hypoplasia, including genes encoding steroid sulfatase and serologically defined, male-specific antigen. While the first five loci escape lyonization and behave as recessive traits; the last two lyonize, with nullisomic males being at a presumably lethal state, and monosomic females showing a mosaic pattern [5].

Aicardi syndrome appears to be genetically heterogeneous and its associated genetic changes are not restricted to the X chromosome. A recent genome-wide array comparative hybridization study revealed the presence in an Aicardi syndrome-affected girl of a ~550-kb deletion in the 6q27 region and a ~4.2-Mb duplication in the 12q24.32q24.33 region, derived from a maternal 6q;12q translocation [4]. In addition, de novo nonsense mutation in autosomal gene TEAD1 (encoding TEA domain transcription factor 1, which has previously been linked to chorioretinal atrophy) and missense mutation in OCEL1 may contribute to the retinal phenotypes in Aicardi syndrome [6]. Interestingly, TEA domain transcription factor 1 is a DNA-binding protein that acts as a ubiquitous transcriptional enhancer. Missense mutation in TEAD1 gene has been identified previously to cause Sveinsson's chorioretinal atrophy (also known as helicoid peripapillary chorioretinal dystrophy) through disrupted binding to YAP62 (a cotranscriptional factor) and altered binding to downstream genes. Sveinsson's chorioretinal atrophy is an autosomal dominant disease characterized by atrophy of the retina, retinal pigment epithelium (RPE), and radial and serpiginous retinal and peripapillary atrophy. Aicardi syndrome differs from Sveinsson's chorioretinal atrophy by its broad, multicentric, punched-out lacunae. Nonsense mutation in TEAD1 (rs926971) observed in Aicardi syndrome may affect transcription of downstream targets and influence the size of the retinal microvasculature.

Sex chromosomes usually evolve from an identical pair of autosomes, with the X chromosome retaining most of the original genes (649 genes), while the Y chromosome loses most, apart from 17 genes (which are shared with the X chromosome). These mutual genes are involved in the regulation of other genes throughout the entire genome.

Klinefelter syndrome results largely from the secondary effect of extra genetic material on the X chromosome arising from a random error in cell division called nondisjunction, in which chromosome fails to separate at anaphase during meiosis I, meiosis II, or mitosis, leading to a reproductive cell (egg or sperm) with an abnormal number of chromosomes. Aberrant partitioning of chromosomes or chromatid during maternal or paternal meiosis mainly occurs during oogenesis or spermatogenesis or during early division of the fertilized egg (3% of cases). In 50% of cases, the supernumerary X-chromosome is inherited from the mother, and in the other 50%, it comes from the father. In Klinefelter syndrome patients with an additional maternal X chromosome, nondisjunction takes places in either the first or second meiotic division. On the other hand, in Klinefelter syndrome patients with paternal supernumerary X chromosome, nondisjunction likely occurs in the first meiotic division, given that meiosis II error often results in either XX or YY gametes and XXX or XYY zygotes.

The impact of supernumerary X-chromosome(s) on Klinefelter syndrome phenotypic features and variability involves several mechanisms, including (i) gene dosage compensation, (ii) skewed X-inactivation, (iii) genetic polymorphisms, and (iv) parental origin of supernumerary X chromosome.

Gene dosage compensation involves genetic equalization through X-inactivation, which randomly silences one of the X-chromosomes in female cell and keeps the other X-chromosome transcriptionally active. In cases of X-chromosome aneuploidy, extra X-chromosome is similarly inactivated. However, two pseudoautosomal regions (PAR1 and PAR2) along with up to 15% of additional genes on the short arm of the X chromosome (Xp) escape inactivation and are expressed from both X-chromosomes. In males with Klinefelter syndrome, extra copies of these escapee genes are transcriptionally active, and their overexpressions modulate the related cellular and developmental pathways. Located at the terminal region of the short arms and consisting of 24 genes (including short-stature homeobox-containing gene on chromosome X or SHOX), PAR1 is required during male meiosis for X–Y chromosome pairing, a process which is known to have a critical function in spermatogenesis. PAR2 is situated at the tips of the long arms and contains four genes (Eif2s3x, Ddx3x, Kdm5c, Kdm6a).

Skewed X-inactivation refers to the preferential, non-random inactivation of one of the two X chromosomes in females. While random X-inactivation keeps each X chromosome active in about half of cells, skewed X-inactivation keeps one X chromosome active in more than half of cells. Significant correlation is observed between skewed X inactivation and smaller gray matter volume in the left insula of the brain, which is involved in social, emotional, and mental processing.

Gene polymorphisms often occur in the androgen receptor gene mapped to Xq11.2–12, which contains a highly polymorphic CAG trinucleotide repeat (range 9–37). The CAG repeat length is inversely related to receptor activity. Long CAG repeat length (low receptor activity) often correlates with tall stature, arm span, gynecomastia, small testes, HDL cholesterol, hematocrit, and later reactivation of pituitary–testicular axis; whereas short CAG repeat length (high receptor activity) correlates with longer penile length, higher bone density, and higher likelihood of having a stable partnership or professional employment [7].

Parental origin of supernumerary X chromosome, especially paternal supernumerary X-chromosome, is a contributing factor for developmental problems, altered steroidogenesis, increased hematocrit, later puberty, insulin resistance, and cardiac findings of a shorter QTc time, due to the presence of non-inactivated extra genes from the supranumerous X chromosome.

2.4 Epidemiology

Aicardi syndrome is a rare, X-linked dominant disorder, affecting ~1 in 100,000 newborns. There are estimated 800 and 4000 individuals with the disease in the United States and worldwide, respectively. The mean age at diagnosis is about 10 years (range 1–31 years); the mean and median ages of death are 8.3 years and 18.5 years, respectively.

Klinefelter syndrome is an non-inherited chromosomal disorder affecting 1 in 500–700 newborn males (or 355 cases per 100,000 Asian males), and accounting for up to 11% of nonobstructive azoospermic males. It is estimated that only 25%–35% of affected males are diagnosed (including 19% adults, 10% infants, and 6% young boys or adolescent males, with the mean age of mid-30s at diagnosis), while the remaining 65%–75% are unnoticed (due to mild phenotypes and significant overlapping with other conditions).

In the United States, about 3075 infants with Klinefelter syndrome are born annually. The prevalence of Klinefelter syndrome ranges from 0.1%–0.2% in newborn male infants, 3%–4% among infertile males, to 10%–12% in azoospermic patients. There is a notable increase in the prevalence of Klinefelter syndrome in recent decades, reflecting the increasing maternal age (fourfold increase in mothers aged >40 years, compared to mothers aged <24), poor sperm quality associated with increased hyperploidy of the sperm, environmentally derived increase of errors in paternal meiosis I, and decreasing rate of elective termination for prenatally diagnosed Klinefelter syndrome.

2.5 Clinical Features

As a rare neurodevelopmental disorder that affects almost exclusively females and occasionally 47,XXY males (associated with Klinefelter syndrome), Aicardi syndrome is initially characterized by the classic triad (agenesis of the corpus callosum, infantile spasms, and chorioretinal lacunae). Subsequent case reports have extended the clinical profiles of Aicardi syndrome, including pathological effects on the *head and neck* (microcephaly, facial asymmetry, microphthalmia, coloboma, bilateral chorioretinopathy, chorioretinal lacunae, retinal detachment, cataract, nystagmus, optic atrophy, eyelid twitching, sparse lateral eyebrows, upturned nasal tip, decreased angle of nasal bridge, prominent premaxilla, cleft lip, cleft palate) (Figures 2.1 and 2.2) [8,9], *lungs and chest* (recurrent pneumonia, absent ribs, extra ribs, fused ribs, bifid ribs), *abdomen* (hiatal hernia), *bones* (scoliosis, spina bifida, butterfly or block vertebrae, hemivertebrae, abnormal costovertebral articulation, proximally placed thumbs, small hands), *skin, nails, and hair* (scalp lipoma, multiple nevi, hypopigmented macules, skin tags, hemangiomas, sparse lateral eyebrows), *central nervous system* (profound mental retardation,

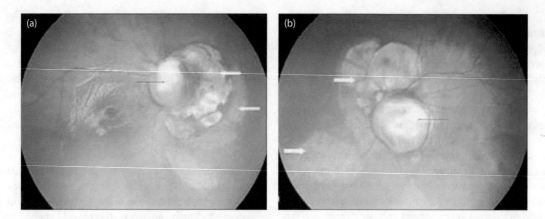

FIGURE 2.1 Aicardi syndrome in a 3-year-old girl showing (a) optic disc coloboma (black arrow) and dome-shaped loci of pale areas with sharp borders nasal to the optic disc suggestive of chorioretinal lacunae (white arrows) in the right eye; (b) optic disc coloboma (black arrow) and chorioretinal lacunae nasal to optic disc (white arrow) in the left eye. (Photo credit: Shah PK et al. *Indian J Ophthalmol.* 2009;57(3):234–6.)

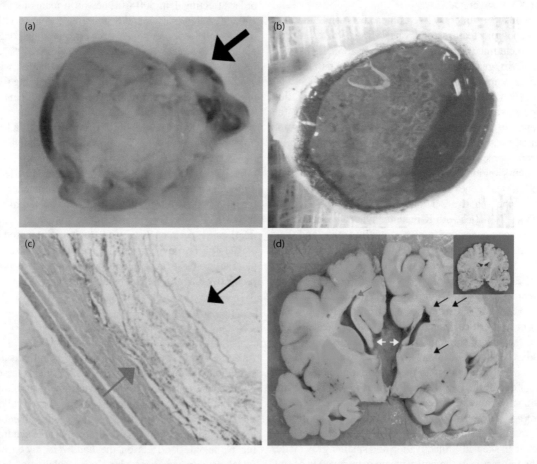

FIGURE 2.2 Aicardi syndrome in a 20-year-old female showing (a) coloboma posteriorly in left eye; (b) a calotte in right eye with myriad of lacunae, unusual papillary proliferations of the retinal pigment epithelium (RPE) with atrophic centers; (c) a regional lacuna (black arrow) at the equator, along with atrophic retina, loss of RPE and fibrosis at the base (red arrow); (d) multiple gray matter heterotopias (black arrows), absence of the corpus callosum with Probst bundle formation (white arrows) and abnormal shape of the lateral ventricles (red arrow) in cross-section of gross brain (upper right inset is a normal brain for comparison). (Photo credit: Mavrommatis MA et al. *Am J Ophthalmol Case Rep.* 2018;12:61–4.)

infantile spasms, seizures, axial hypotonia, Dandy–Walker malformation, Arnola–Chiari malformation, cavum septum pellucidum, choroid plexus cyst, intracranial cysts, delayed myelination, partial/total agenesis of corpus callosum, enlarged lateral and third ventricles, cortical heterotopias, subependymal heterotopias, pachygyria, hypoplastic cerebellar vermis,

dysplasia of the cerebellar hemispheres, frontal/perisylvian polymicrogyria, tectal enlargement, widening of the operculum) (Figure 2.2) [9], *growth and endocrine functions* (postnatal growth retardation, lower growth rate after 7–9 years of age, precocious or delayed puberty), and *neoplasia* (choroid plexus papilloma, choroid plexus carcinoma, lipoma, angiosarcoma, hepatoblastoma, intestinal polyposis, benign teratoma, embryonal carcinoma, retinoblastoma) [10,11].

Recent surveys indicate that the most commonly observed clinical symptoms of Aicardi syndrome are agenesis or dysgenesis of the corpus callosum (84%), chorioretinal lacunae (74%), seizures (71%), ophthalmological phenotypes (coloboma, optic nerve dysplasia, agenesis, atrophy, glial proliferation, pseudoadenomatous proliferation of retinal pigment epithelium, or posterior staphyloma) and microphthalmia (63%), classic triad, and other signs (chorioretinal lacunae, seizures, agenesis/dysgenesis of the corpus callosum, gray matter heterotopias, and polymicrogyria, 32%) [8–10,12].

Klinefelter syndrome is a relatively common genetic disorder affecting men with non-mosaic 47,XXY (90% of cases), mosaic (e.g., 47,XXY/46,XY, 7%), and other rare karyotypes (e.g., 48,XXXY, 48,XXYY, or 49,XXXXY, 3%). Depending on the supernumerary X chromosome and the effects of hypogonadism, Klinefelter syndrome may manifest as small testes, gynecomastia in late puberty, gynoid aspect of hips (broad hips), sparse body hair, tall stature, signs of androgen deficiency, low serum testosterone, elevated gonadotropins, azoospermia, oligospermia with hyalinization, and fibrosis of the seminiferous tubules. Further, genital anomalies (micropenis, undescended testis, bifid scrotum, and hypospadias) may appear at birth, while longer legs and speech disabilities may occur during infancy (Figure 2.3) [7,13,14].

Specifically, Klinefelter syndrome features before puberty are attributable to supernumerary X chromosome (Figure 2.4) [15] and include congenital malformations (cleft palate, hernia),

longer legs, and small testes, as well as speech and language disabilities and azoospermia.

Klinefelter syndrome features at puberty or during adulthood are caused by testosterone deficiency, and include sparse body and facial hair, female pubic escutcheon, reduced muscle mass, bilateral gynecomastia, eunuchoid skeleton, impaired estradiol/testosterone ratio, longer legs, impaired sexual desire, impaired erectile function, weakness and loss of vigor, and impaired well-being.

Klinefelter syndrome features before puberty with progressive worsening after puberty result from both supernumerary X chromosome and testosterone deficiency, and include eunuchoid skeletal proportions, gynoid hips, tall stature, genital abnormalities at birth, elevated gonadotropins, BMI in the range of overweight or obesity, metabolic abnormalities, reduced bone mineral density, and mood disturbances.

In general, patients with non-mosaic 47,XXY karyotype show more severe clinical symptoms and endocrine abnormalities than those with mosaic (e.g., 47,XXY/46,XY) karyotype. At puberty, the penis and secondary sexual characteristics progress in a normal fashion, but testes volume remains small (<4 mL in volume). Many Klinefelter syndrome features (e.g., hypogonadism) and comorbidities (e.g., diabetes, obesity, gynoid fat distribution, metabolic syndrome, reduced muscle strength, osteoporosis, cardiovascular disease, depression, paraphilia, autism, and obsessive–compulsive trait) appear at adulthood and increase with advancing age. Indeed, about 80% of men aged 25 or above suffer from decreased libido and erectile dysfunction. Further, Klinefelter males have a higher incidence of breast cancer (in

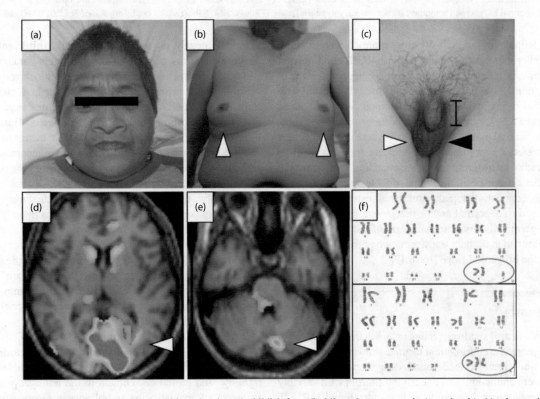

FIGURE 2.3 Klinefelter syndrome in a 10-year-old boy showing (a) childish face; (b) bilateral gynecomastia (arrowheads); (c) micropenis (bar), right normal testis (white arrowhead), and empty left scrotum (black arrowhead); (d, e) hypoperfusion in the bilateral occipital lobe (d, arrowhead) and cerebellum (e, arrowhead) in 99mTc-ethylcysteinate dimer single-photon emission computed tomography; (f) mosaic form of 47,XXY/48,XXXY in chromosome analysis. (Photo credit: Sasaki R et al. *Intern Med.* 2019;58(3):437–40.)

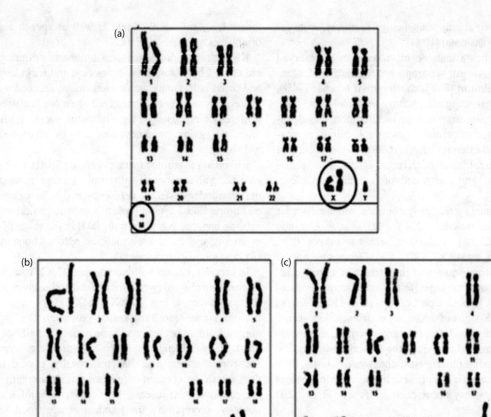

FIGURE 2.4 Klinefelter syndrome presenting with (a) an extra chromosome X and small supernumerary marker chromosomes (sSMCs) in fetal cell karyotype; (b) normal karyotype for mother; (c) normal karyotype for father. (Photo credit: Saberzadeh J et al. *Iran J Med Sci*. 2019;44(1):65–9.)

3%–7% of Klinefelter men, 20–50 times more common than in non-Klinefelter men), extragonadal GCT (in 0.1% of Klinefelter males; manifesting as precocious puberty, cough, dyspnea, or chest pain), non-Hodgkin lymphoma, and calcifying nested stromal-epithelial tumor (CNSET, an uncommon primary hepatic tumor) (Figure 2.5) [17], but a lower incidence of prostate cancer compared to normal males. The notable increase in breast cancer among Klinefelter syndrome men is thought to result from an altered estradiol:testosterone ratio, longstanding gynecomastia, obesity, genetic predisposition, and possibly testosterone administration. In contrast, the pathogenesis of extragonadal GCT may be related to abnormal and/or incomplete migration of the primordial germ cells from the endoderm of the yolk sac to the gonads, leading to malignant transformation of the midline germ cells along the urogenital ridge [16,18,19].

2.6 Diagnosis

Diagnosis of Aicardi syndrome relies on ophthalmological exam (e.g., pathognomonic chorioretinal lacunae showing white or yellow-white, well-circumscribed, round, depigmented areas of the retinal pigment epithelium and underlying choroid with variably dense pigmentation at their borders), brain ultrasound/MRI/CT (e.g., dysgenesis of the corpus callosum, gross cerebral asymmetry with polymicrogyria or pachygyria, periventricular and intracortical gray matter heterotopia, choroid plexus papillomas,

ventriculomegaly, intracerebral cysts at the third ventricle and in the choroid plexus, posterior fossa, and cerebellar abnormalities), EEG, and skeletal findings (hemivertebrae, block vertebrae, fused vertebrae, and missing ribs) [20–22]. Further, prenatal ultrasound or fetal MRI may reveal some features (e.g., agenesis of the corpus callosum) suggestive of the disease. Transfontanellar ultrasound provides a cheap and affordable means of confirming Aicardi syndrome in neonates, with similar accuracy to MRI.

While the presence of the classic triad (agenesis of the corpus callosum, distinctive chorioretinal lacunae, and infantile spasms) is diagnostic, the existence of two classic features plus at least two other major (e.g., cortical malformations [mostly polymicrogyria], periventricular and subcortical heterotopia, cysts around third cerebral ventricle and/or choroid plexus, optic disc/nerve coloboma, or hypoplasia) or supporting features (e.g., vertebral and rib abnormalities, microphthalmia, "split-brain" on electroencephalography, gross cerebral hemispheric asymmetry, vascular malformations, or vascular malignancy) is strongly indicative of Aicardi syndrome [10].

In a recent report by Akinfenwa et al. [11], a 16-year-old Caucasian girl, who was diagnosed with Aicardi syndrome 3 years earlier showing characteristic chorioretinal lacunae in the right eye and retinal detachment in the left eye, presented with recurrent left eye irritation and blindness. Gross examination of the left enucleated eye revealed an intact cornea with visible pupil and irregular margins; however, the posterior cavity was filled with a red-brown clot (evidence of a vitreous hemorrhage), and

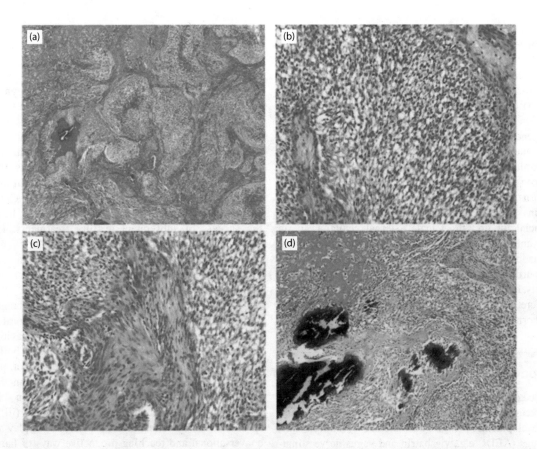

FIGURE 2.5 Klinefelter syndrome in a 20-year-old male with calcifying nested stromal-epithelial tumor (CNSET, an uncommon primary hepatic tumor) showing (a) nests of epithelial cells surrounded by a desmoplastic stroma with focal calcifications (×40); (b) epithelioid nests (×200); (c) desmoplastic stroma of spindle cells (×200); (d) necrosis and regions of calcification (×100). (Photo credit: Tsuruta S et al. *World J Surg Oncol.* 2018;16(1):227.)

the lens showed abnormal positioning. Histopathologic examination uncovered a partially degenerated, cataractous lens that was displaced anteriorly, a mass projecting into the vitreous, filling nearly the entire posterior cavity, and a detached retina. The mass was confirmed as a focally differentiated retinoblastoma with Homer Wright rosettes and positive staining for synaptophysin and negative staining for the wild-type Rb protein (suggestive of an *RB1* gene mutation). In addition, the left enucleated eye displayed background changes of Aicardi syndrome, including multiple areas of chorioretinal lacunae (which are composed of lakes of blood and proteinaceous material surrounded by proliferation of the retinal pigment epithelium, and nests of tumor cells), optic nerve hypoplasia, ectopic retinal pigment epithelium in the meninges, and a peripapillary coloboma. This study highlights the importance of conducting meticulous ophthalmic examination of Aicardi syndrome patients through the teenage years and early adulthood.

Differential diagnoses for Aicardi syndrome include (i) microcephaly with or without chorioretinopathy, lymphedema, or mental retardation (MCLMR, in which chorioretinal changes do not involve peripheral and optic nerves, and neuronal migration defects are uncommon, whereas chorioretinal lacunae in Aicardi syndrome are central and involve the optic nerves, and neuronal migration defects are almost universal); (ii) oculocerebrocutaneous syndrome (OCCS, which may display orbital cysts and anophthalmia or microphthalmia, focal skin defects, polymicrogyria, periventricular nodular heterotopias, enlarged lateral ventricles,

and agenesis of the corpus callosum, but predominant in males); (iii) tuberous sclerosis complex and Rett syndrome (which also shows infantile spasms); (iv) orofaciodigital syndrome type IX (OFD 9, which may show chorioretinal lacunae); (v) Goltz syndrome and microphthalmia with linear skin defects syndrome (which also have microphthalmia and other developmental eye defects, but differ from Aicardi syndrome by their characteristic skin defects and other features); (vi) Dandy–Walker syndrome, agenesis of the corpus callosum, neuronal migration disorders, Lennox–Gastaut syndrome, lissencephaly, West syndrome, and cyclin-dependent kinase-like 5 disorder, all of which may display seizure; and (vii) ocular toxoplasmosis (showing small or peripheral lacunae) [10].

Diagnosis of Klinefelter syndrome involves clinical observation (cryptorchidism, behavioral problems, learning disability, and tall stature before puberty; gynecomastia and micro-orchidism, poor muscular bulk with excessively long legs, small shoulders, and broad hips, a nonfamilial obesity in late adolescence; symptomatic hypogonadism, infertility, and/or sexual dysfunction in adulthood), karyotyping and florescence in situ hybridization of lymphocytes, skin fibroblasts, or testicular biopsy samples (47,XXY, 47,XXY/46,XY mosaicism, or multiple X chromosome aneuploidy, often with additional Y chromosomes, particularly in adolescents and young men with small testes; hypergonadotropic hypogonadism). Noninvasive prenatal testing (NIPT) of sex chromosome aneuploidies and autosomal trisomies from maternal blood (containing cell-free fetal DNA, typically in the first trimester of older pregnant women) is also valuable. Further confirmation is made by prenatal chorionic-villous sampling, amniocentesis, and/or postnatal blood testing [23].

(CAG)n repeat length of the androgen receptor gene appears positively associated with height, height-to-arm span ratio, gynecomastia, small testes, arm length, arm span, and leg length in Klinefelter syndrome patients. The growth-related short stature homeobox gene (SHOX) situated on the X and Y chromosomes is also informative for Klinefelter syndrome, Turner syndrome (growth retardation), and syndromes with sex hormone aneuploidy via the so-called gene-dose effect (increased growth).

In newborns, a postnatal endocrine surge with elevation of gonadotropins and sex hormones (testosterone) shortly after birth is crucial for growth of male external genitalia, sperm production, and sexual motivation. However, the development of hypergonadotropic hypogonadism (as indicated by increased second-to-fourth finger ratio) in Klinefelter karyotype is evident in midpuberty, leading to decreased levels of androgens, increased pituitary secretion of follicle-stimulating hormone (FSH) and luteinizing hormone, and elevated estrogen to androgen ratio.

2.7 Treatment

In the absence of a specific cure, treatment of Aicardi syndrome centers on measures that control seizures (infantile spasms) and deal with developmental delays. These include (i) multiple antiepileptic drugs (AEDs, e.g., vigabatrin and vagus nerve stimulators) to control seizures, (ii) resection of large choroid plexus papilloma to alleviate hydrocephalus, (iii) supporting therapies (e.g., physical, occupational, speech, and vision therapies), (iv) musculoskeletal treatment to prevent scoliosis-related complications, and (v) regular monitoring for vascular malformations and pigmentary lesions, gastrointestinal complications (e.g., constipation), and scoliosis [24].

Treatment of Klinefelter syndrome should center on key aspects of the disease: (i) low testosterone, (ii) enlarged breasts, (iii) infertility, (iv) language, learning, social, and behavioral symptoms, and (v) comorbidities [25–28].

Patients with Klinefelter syndrome typically show hypogonadism (androgen deficiency), as about half of XXY males have low testosterone levels (below 12 nmol/L between 20–30 years of age). Therefore, provision of supplemental testosterone to XXY males with low testosterone helps improve muscle mass, deepen the voice, promote growth of facial and body hair, help the reproductive organs to mature, build and maintain bone strength and thus prevent osteoporosis in later years, produce a more masculine appearance and relieve anxiety and depression, and increase focus and attention. However, testosterone supplementation will not increase testicular size, decrease breast growth, or correct infertility. Supplemental testosterone may be given through injections (every 2–3 weeks), pills, or transdermal application (e.g., wearing a testosterone patch or rubbing testosterone gel on the skin); restoration of testosterone level takes >12 months. Possible side effects of testosterone supplementation include acne, skin rashes from patches or gels, erythrocytosis, sleep apnea, aggressive behavior, and higher risk of an enlarged prostate gland or prostate cancer in older age [29–31].

XXY males with overdeveloped breast tissue (termed *gynecomastia*) are at higher risk for breast cancer than other men and are recommended for mastectomy to remove or reduce the breasts, lower cancer risk, and reduce psychological discomfort. Further, antiestrogens (tamoxifen) may be used to treat gynecomastia if the breast gland expansion is painful or occurred very rapidly.

Although XXY males make some healthy sperm during puberty, they do not seem to produce enough sperm to fertilize an egg naturally in adulthood. Assistive reproductive technology (ART, such as testicular sperm extraction with intracytoplasmic sperm injection [TESE-ICSI]) is helpful in improving the infertility of Klinefelter syndrome patients. Collecting and storing sperm from adolescent XXY males also helps increase the success of subsequent fertility treatments. In addition, early sperm cryopreservation may be undertaken for young adults with spermaturia and/or oligospermia in semen before testosterone administration which may suppress any remaining spermatogenesis [29–31].

For children with Klinefelter syndrome with language development and learning delays, social and behavioral symptoms, consultations with physical, occupational, behavioral, mental health, and family therapists are helpful in: (i) building motor skills and strength and improving muscle control, posture, and balance, (ii) building skills needed for daily functioning (e.g., social and play skills, interaction and conversation skills, and job or career skills that match interests and abilities), (iii) developing specific social skills (e.g., asking other kids to play and starting conversations) and teaching productive ways of handling frustration, shyness, anger, and other emotions that can arise from feeling different, (iv) finding ways to cope with feelings of sadness, depression, self-doubt, and low self-esteem as well as substance abuse problems, and (v) identifying relationship problems and developing communication skills and understanding other people's needs [29–31].

Klinefelter syndrome—related comorbidities include testicular fibrosis, congenital malformations, insulin resistance, obesity, loss of muscle and bone mass, metabolic syndrome, type 2 diabetes, heightened blood pressure, cardiovascular disease, psychiatric disorders, and increased cancer risk [3].

2.8 Prognosis and Prevention

Depending on the severity of seizures and involvement of other organ systems, the prognosis for girls with Aicardi syndrome is generally poor, with the estimated age of death at 8.8 years (range, 1 month—33 years), and the chance of surviving until age 27 years at 62%. The common causes of death range from respiratory infections, systemic infections, and malignancy to sudden unexpected death in epilepsy (SUDEP). Despite extensive retinal involvement with chorioretinal lacunae, most children with Aicardi syndrome have some useful visual behavior.

Given that nearly all known cases of Aicardi syndrome are sporadic, the condition affects people with no history of the disorder in their family. Its risk to sibs (except for identical twins) is less than 1%. For a female with Aicardi syndrome, the expected ratio of offspring at delivery is 33% unaffected female, 33% affected female, and 33% unaffected male.

Klinefelter syndrome reduces life expectancy by 2–5 years and increases risk for breast cancer, extragonadal GCT, and other diseases. While non-mosaic XXY infants younger than 2 years have normal external genitalia, facial features, height, and weight, they often show a delay in walking and/or speech. Early detection of Klinefelter syndrome is therefore important to monitor potential developmental problems and to ensure smooth transition from adolescence to adulthood with appropriate intervention measures [32].

2.9 Conclusion

Aicardi syndrome is a rare genetic disease affecting 1 in 100,000 newborn females and occasionally 47,XXY males. Initially characterized by a classical trio of agenesis of the corpus callosum, infantile spasms (seizures/epilepsy), and chorioretinal lacunae, the clinical profile of Aicardi syndrome has been extended to include other neuronal and extraneuronal defects, such as characteristic facial features (prominent premaxilla, upturned nasal tip, decreased angle of the nasal bridge, sparse lateral eyebrows, coloboma of the optic nerve), microcephaly, periventricular heterotopias, microgyria, enlarged ventricles or porencephalic cysts, moderate-to-severe developmental delay and intellectual disability, as well as increased cancer risk (e.g., choroid plexus papilloma, choroid plexus carcinoma, lipoma, angiosarcoma, hepatoblastoma, intestinal polyposis, benign teratoma, embryonal carcinoma, retinoblastoma) [11].

Klinefelter syndrome is the most common sex chromosomal disorder, affecting 1 in 500–700 newborn males who possess one or more extra X chromosomes. Clinically, the disease is characterized by primary infertility, atrophic testes, hypergonadotropic hypogonadism, gynecomastia, eunuchoidism, decreased facial and body hair, tall stature with eunuchoid body proportions, neurocognitive impairment, learning difficulties and behavior problems, and increased risk of autoimmune diseases and tumors (e.g., breast cancer and extragonadal GCT).

Owing to current inadequate understanding of the underlying pathophysiology of Aicardi syndrome and Klinefelter syndrome, the options for treating these genetic disorders and reducing the associated cancer risks are severely limited at this stage. There is an obvious need to conduct further research on the molecular basis of these syndromes in order to develop innovative and effective intervention measures.

REFERENCES

1. Sutton VR, Van den Veyver IB. Aicardi syndrome. In: Adam MP, Ardinger HH, Pagon RA et al. editors. *GeneReviews®* [Internet]. Seattle (WA): University of Washington, Seattle; 1993–2017. [updated 2014 Nov 6].
2. Crow YJ. Aicardi-Goutières syndrome. In: Adam MP, Ardinger HH, Pagon RA et al. editors. *GeneReviews®* [Internet]. Seattle (WA): University of Washington, Seattle; 1993–2017. [updated 2016 Nov 22].
3. Bonomi M, Rochira V, Pasquali D et al. Klinefelter syndrome (KS): Genetics, clinical phenotype and hypogonadism. *J Endocrinol Invest.* 2017;40(2):123–34.
4. Prontera P, Bartocci A, Ottaviani V et al. Aicardi syndrome associated with autosomal genomic imbalance: Coincidence or evidence for autosomal inheritance with sex-limited expression? *Mol Syndromol.* 2013;4(4):197–202.
5. Lund C, Striano P, Sorte HS et al. Exome sequencing fails to identify the genetic cause of Aicardi syndrome. *Mol Syndromol.* 2016;7(4):234–8.
6. Wong BK, Sutton VR, Lewis RA, Van den Veyver IB. Independent variant analysis of TEAD1 and OCEL1 in 38 Aicardi syndrome patients. *Mol Genet Genomic Med.* 2017;5(2):117–21.
7. Akcan N, Poyrazoğlu Ş, Baş F, Bundak R, Darendeliler F. Klinefelter syndrome in childhood: Variability in clinical and molecular findings. *J Clin Res Pediatr Endocrinol.* 2018;10(2):100–7.
8. Shah PK, Narendran V, Kalpana N. Aicardi syndrome: The importance of an ophthalmologist in its diagnosis. *Indian J Ophthalmol.* 2009;57(3):234–6.
9. Mavrommatis MA, Friedman AH, Fowkes ME, Hefti MM. Aicardi syndrome in a 20-year-old female. *Am J Ophthalmol Case Rep.* 2018;12:61–4.
10. Beres S. Aicardi syndrome. American Academy of Ophthalmology Website. www.aao.org/pediatric-center-detail/ neuro-ophthalmology-aicardi-syndrome
11. Akinfenwa PY, Chévez-Barrios P, Harper CA, Gombos DS. Late presentation of retinoblastoma in a teen with Aicardi Syndrome. *Ocul Oncol Pathol.* 2016;2(3):181–4.
12. Shirley K, O'Keefe M, McKee S, McLoone E. A clinical study of Aicardi syndrome in Northern Ireland: The spectrum of ophthalmic findings. *Eye (Lond).* 2016;30(7):1011–6.
13. Sasaki R, Ohta Y, Takahashi Y et al. A rare case of Klinefelter syndrome accompanied by spastic paraplegia and peripheral neuropathy. *Intern Med.* 2019;58(3):437–40.
14. Davis SM, Rogol AD, Ross JL. Testis development and fertility potential in boys with Klinefelter syndrome. *Endocrinol Metab Clin North Am.* 2015;44(4):843–65.
15. Saberzadeh J, Miri MR, Dianatpour M et al. The first case of a small supernumerary marker chromosome 18 in a Klinefelter fetus: A case report. *Iran J Med Sci.* 2019;44(1):65–9.
16. Barazani Y, Sabanegh E Jr. Rare case of monozygotic twins diagnosed with Klinefelter syndrome during evaluation for infertility. *Rev Urol.* 2015;17(1):42–5.
17. Tsuruta S, Kimura N, Ishido K et al. Calcifying nested stromal epithelial tumor of the liver in a patient with Klinefelter syndrome: A case report and review of the literature. *World J Surg Oncol.* 2018;16(1):227.
18. Pradhan D, Kaman L, Dhillon J, Mohanty SK. Mediastinal mixed germ cell tumor in an infertile male with Klinefelter syndrome: A case report and literature review. *J Cancer Res Ther.* 2015;11(4):1034.
19. Salzano A, Arcopinto M, Marra AM et al. Klinefelter syndrome, cardiovascular system, and thromboembolic disease: Review of literature and clinical perspectives. *Eur J Endocrinol.* 2016;175(1):R27–40.
20. Hergan B, Atar OD, Poretti A, Huisman TA. Serial fetal MRI for the diagnosis of Aicardi syndrome. *Neuroradiol J.* 2013;26(4):380–4.
21. Pires CR, Araujo Júnior E, Czapkowski A, Zanforlin Filho SM. Aicardi syndrome: Neonatal diagnosis by means of transfontanellar ultrasound. *World J Radiol.* 2014;6(7):511–4.

22. Gacio S, Lescano S. Foetal magnetic resonance images of two cases of Aicardi syndrome. *J Clin Diagn Res.* 2017;11(7):SD07–9.

23. Hodhod A, Umurangwa F, El-Sherbiny M. Prepubertal diagnosis of Klinefelter syndrome due to penoscrotal malformations: Case report and review of literature. *Can Urol Assoc J.* 2015;9(5–6):E333–6.

24. Podkorytova I, Gupta A, Wyllie E et al. Aicardi syndrome: Epilepsy surgery as a palliative treatment option for selected patients and pathological findings. *Epileptic Disord.* 2016;18(4):431–9.

25. Gies I, Unuane D, Velkeniers B, De Schepper J. Management of Klinefelter syndrome during transition. *Eur J Endocrinol.* 2014;171(2):R67–77.

26. Chang S, Skakkebæk A, Gravholt CH. Klinefelter syndrome and medical treatment: Hypogonadism and beyond. *Hormones (Athens).* 2015;14(4):531–48.

27. Davis S, Howell S, Wilson R et al. Advances in the interdisciplinary care of children with Klinefelter syndrome. *Adv Pediatr.* 2016;63(1):15–46.

28. Oates R. Adolescent Klinefelter syndrome: Is there an advantage to testis tissue harvesting or not? *F1000Res.* 2016;5. pii: F1000 Faculty Rev-1595.

29. Shiraishi K, Matsuyama H. Klinefelter syndrome: From pediatrics to geriatrics. *Reprod Med Biol.* 2018;18(2):140–50.

30. Bearelly P, Oates R. Recent advances in managing and understanding Klinefelter syndrome. *F1000Res.* 2019;8. pii: F1000 Faculty Rev-112.

31. De Sanctis V, Fiscina B, Soliman A, Giovannini M, Yassin M. Klinefelter syndrome and cancer: From childhood to adulthood. *Pediatr Endocrinol Rev.* 2013;11(1):44–50.

32. Los E, Ford GA. *Klinefelter Syndrome. StatPearls* [Internet]. Treasure Island (FL): StatPearls Publishing; 2019 Jan–2019 Mar 22.

3

BAP1 Tumor Predisposition Syndrome

Dongyou Liu

CONTENTS

3.1 Introduction

BAP1 tumor predisposition syndrome (BAP1-TPDS, also known as BAP1 hereditary cancer predisposition syndrome or BAP1 cancer syndrome) is a relatively new disorder that was first described in 2011. Showing an increased risk for a specific skin lesion (i.e., melanocytic *BAP1*-mutated atypical intradermal tumor [MBAIT]) and several malignancies (e.g., uveal melanoma, malignant mesothelioma, cutaneous melanoma, clear cell renal cell carcinoma (ccRCC), and basal cell carcinoma), BAP1-TPDS appears to be associated with germline mutations in the breast cancer 1 (BRCA1)-associated tumor protein (BAP1) gene on chromosome 3p21.1 [1].

3.2 Biology

Initially identified in 1998, BAP1 (BRCA1-associated protein) is a deubiquitinating hydrolase that binds to the RING finger domain of the BRCA1 protein and suppresses the growth of human breast cancer cells in soft agar. As a nuclear protein encoded by the *BAP1* gene located on chromosome 3p21.1, BAP1 partially interacts through its deubiquitinase activity with other

proteins (e.g., HCFC1, YY1, OGT, ASXL1/2, and FOXK1/2), and participates in several cellular processes, including chromatin remodeling, cell cycle progression, cell differentiation, and DNA damage responses [2,3]. Germline mutations in the *BAP1* gene result in haploinsufficiency of the BAP1 protein that either lacks the nuclear localization sequence or displays reduced deubiquitinase activity [4,5], thus confirming the role of BAP1 as a bona fide tumor suppressor [6,7].

The involvement of BAP1 in hereditary tumor predisposition syndrome (*BAP1*-TPDS, MIM 614327) was established in 2011, following detection of germline *BAP1* mutations from two unrelated families with malignant mesothelioma and uveal melanoma [8,9]. Subsequently, mutations in the *BAP1* gene have been found in families showing predisposition to basal cell carcinoma and renal cell carcinoma as well as melanocytic *BAP1*-mutated atypical intradermal tumor (MBAIT, also known as atypical Spitz tumor/nevus) [10–14].

It is clear now that BAP1-TPDS is attributable to germline inactivation of one *BAP1* copy followed by somatic mutation in the other *BAP1* copy (often through loss of the entire chromosome) in line with the two-hit hypothesis of cancer development. Indeed, most MBAIT (atypical Spitz tumor) seen in BAP1-TPDS contains mutation in the second *BAP1* allele as a consequence of loss of heterozygosity (LOH) or loss of expression of the wild-type allele.

Some sporadic tumors (including cholangiocarcinoma, mesothelioma, renal cell carcinoma, and uveal melanoma) may occur as single tumors without other findings of BAP1-TPDS. However, tumors harboring somatic *BAP1* variants that are not detected in the germline (blood) are non-inheritable [15].

3.3 Pathogenesis

Mapped to the short arm of chromosome 3 (3p21.1), the *BAP1* (BRCA1-associated protein) gene comprises 17 exons, which generate a transcript of 3717 bp, encoding a 729 aa, 90 kDa protein (BAP1). Structurally, BAP1 contains several domains organized in the order of: N terminus—ubiquitin carboxyl hydrolase (UCH) domain—BRCA1-associated RING domain protein 1 (BARD1) binding region-host cell factor 1 (HCF1) binding domain (HBM)—Ying Yang 1 (YY1) binding region—BRCA1 binding region—C-terminal domain (CTD)—nuclear localization signal (NLS)—C terminus. Among these, nuclear UCH is a deubiquitinating enzyme involved in the regulation of target genes in cell cycle control, cellular differentiation, and DNA damage repair; BRCA1 and BARD1 binding domains form a tumor suppressor heterodimeric complex, which acts as an essential DNA damage repair enzyme and promotes E3 ubiquitin ligase activity to regulate DNA damage response; HCF1 interacts with transcription factor YY1 for cell proliferation and cell cycle control; and binding between CTD and additional sex combs like (ASXL1/2) protein results in the formation of the polycomb group repressive deubiquitinase complex (PR-DUB), which contributes to stem cell pluripotency and other developmental processes [16–18].

In the nucleus, BAP1 activates transcription that helps stall cells in S phase and promotes repair at sites of DNA double strand breaks through homologous recombination [19]. In the cytoplasm, BAP1 deubiquitinases and stabilizes type 3 inositol-1,4,5-trisphosphate receptor (IP3R3), through which calcium (Ca^{2+}) release from the endoplasmic reticulum (where Ca^{2+} is normally stored) into the cytosol and mitochondria is increased. In the mitochondria, Ca^{2+} regulates aerobic respiration and programmed cell death, or apoptosis. Furthermore, through interaction with the tumor suppressor gene *BRCA1*, BAP1 contributes to genome stability and epigenetic modification [20–23].

Mutations in the *BAP1* gene may produce prematurely terminated BAP1 protein, reduce cellular BAP1 levels, and alter the ubiquitin carboxyl-terminal hydrolase domain, thus affecting the deubiquitinase activity of BAP1. Reduced nuclear and cytoplasmic activities of BAP1 decrease IP3R3 levels and Ca^{2+} flux, and prevent cells containing damaged DNA from apoptosis. There is evidence that carriers of germline BAP1 mutations with only one normal BAP1 allele have 50% less of cellular BAP1 than normal persons. Furthermore, cells from carriers of germline BAP1 mutations display reduced ability to repair DNA by homologous recombination (due to the reduced nuclear BAP1 levels) and impaired apoptosis upon exposure to asbestos, ultraviolet light, and irradiation (due to the reduced mitochondrial Ca^{2+} levels), leading to increased vulnerability to malignant transformation [24].

Alterations in the *BAP1* gene range from missense and nonsense mutations and splice site and frameshift mutations to truncating mutations (due to base substitutions, insertions, or deletions). Interestingly, about 70% of reported germline BAP1 mutations result from chromosomal deletion, leading to truncated BAP1 protein. As a consequence, the NLS and/or the C-terminal protein-binding domain in BAP1 may be disrupted. On the other hand, missense mutations (20% of cases) may affect the ubiquitin hydrolase function of BAP1.

The most notable mutations found in the affected families include c.2050C>T, p.Q684*; c.1882_1885delTCAC, p.S628Pfs*8; c.1717delC, p.L573Wfs*3; c.588G>A, p.Trp196*; and c.178C>T, p.R60*. Other mutations occasionally observed are c.2057-4G>T, c.932-58_59delTG, c.*45C>G, and c.1182C >G [25].

3.4 Epidemiology

3.4.1 Prevalence

As a rare genetic disorder, BAP1-TPDS was initially described in 2011 from two unrelated families in Louisiana and Wisconsin, showing high incidence of mesothelioma and uveal melanoma. BAP1 cancer syndrome has been subsequently identified in 57 families and 174 individuals (including 67 [39%] males, 95 [55%] females, and 12 with no gender detail) across the world by 2015. Among these is a very large multigeneration family that originates from a German couple who immigrated to the United States in the early 1700s. As approximately 3000 patients with mesothelioma and 2500 patients with uveal melanoma are diagnosed annually in the United States, the incidence for simultaneous occurrence of these malignancies in more than one individual in the same family is estimated at 36 per trillion population.

3.4.2 Inheritance

BAP1-TPDS follows an autosomal dominant inheritance pattern, and the progeny of affected individuals have a 50% chance of inheriting the BAP1 pathogenic variant. Based on two-hit hypothesis, an individual who inherits a nonfunctional BAP1 allele requires inactivation in the remaining allele later in life before malignancy emerges.

3.4.3 Penetrance

BAP1 mutations are highly penetrant, and most carriers of BAP1 mutations are expected to develop one or more malignancies during their lifetime. Of the 174 individuals harboring a heterozygous germline *BAP1* pathogenic variant, 148 (85%) are diagnosed with at least one type of cancer. In fact, most affected patients (90%) have two or more types of tumors in their parents or second-degree relatives.

3.5 Clinical Features

Clinically, individuals with BAP1-TPDS show an increased risk for developing a specific skin lesion (atypical Spitz tumor) and several malignancies (Table 3.1). Frequently, affected individuals may have more than one type of primary cancer [26].

TABLE 3.1

Percentage of Germline and Sporadic *BAP1* Mutations Observed in Tumors

Tumor	Germline Mutation	Somatic Mutation	Comments
Melanocytic *BAP1*-mutated atypical intradermal tumor (MBAIT)	18%	11%	Frequent biallelic *BAP1* inactivation and B-Raf proto-oncogene (*BRAF*^V600E^ mutation
Uveal melanoma	31%	50%	Somatic *BAP1* mutations show high correlation with monosomy of chromosome 3
Malignant mesothelioma	22%	~60%	Germline mutations tend to predominate peritoneal disease while sporadic mutations are often involved in pleural disease
Cutaneous melanoma	13%	5%	
Clear cell renal cell carcinoma (ccRCC)	10%	14%	Sporadic mutations are associated with a more aggressive disease (average survival of 31.2 months) than germline mutations (average survival of 78.2 months)
Basal cell carcinoma	6%		
Other malignancies	1%–9%		

3.5.1 Melanocytic *BAP1*-Mutated Atypical Intradermal Tumor

Melanocytic *BAP1*-mutated atypical intradermal tumor (MBAIT) is a benign atypical skin melanocytic lesion on the head, neck, trunk, and limbs (Figure 3.1). Clinically, MBAIT is a well-circumscribed, dome-shaped, skin-colored or reddish-brown papule or nodule of approximately 0.2–1.0 cm (average 5 mm) in diameter. Histologically, MBAIT sits between benign Spitz nevus and malignant melanoma, and is sometimes referred to as atypical Spitz tumor/nevus. However, MBAIT is mostly intradermal with occasional involvement of the junctional epidermis, and lacks epidermal hyperplasia (mitotic figures), clefting between melanocytes, and Kamino bodies, which are common in traditional Spitz nevus. Molecularly, MBAIT often harbors biallelic inactivation of *BAP1* and frequent B-Raf proto-oncogene (*BRAF*^V600E^) mutation, the latter of which is absent in Spitz nevus [27]. MBAIT is found in 18% of individuals harboring germline *BAP1* mutations. With an initial age of onset at 25 years and a median age of onset at 42 years (which are several years before development of characteristic malignancies), MBAIT may serve as a potential marker for *BAP1* cancer syndrome [28,29].

3.5.2 Uveal Melanoma

Affecting 31% of individuals with germline *BAP1* mutations, uveal melanoma is the most common malignancy identified in BAP1-TPDS [30]. With an initial age of diagnosis at 16 years and a median age of onset at 51 years, uveal melanoma appears much earlier in people with BAP1 mutations than those without BAP1 mutations (62 years). Interestingly, people with BAP1-expressing uveal melanoma have a mean survival of 9.97 years compared to 4.74 years in those with BAP1-non-expressing uveal melanoma [31–33]. Clinically, uveal melanoma may involve the choroid (90%), ciliary body (6%), and iris (4%), with the mean basal dimension of 11.1 mm and the mean thickness of 5.5 mm (Figure 3.2). While approximately 50% of the tumors is pigmented, 15% is nonpigmented, and about one-third has a mixed

(pigmented and nonpigmented) appearance. Further, 70% of choroidal uveal melanoma is dome shaped, 20% is mushroom-like (after breaking through the Bruch's membrane and growing into the subretinal space), and 5% remains flat [34–36].

3.5.3 Malignant Mesothelioma

Occurring in 22% of affected individuals, *BAP1*-related malignant mesothelioma has an initial age of diagnosis at 34 years and a median age of onset at 55 years (often without a history of asbestos exposure), compared to 72 years in sporadic malignant mesothelioma (usually with a history of asbestos exposure) (Figure 3.3) [37–39]. Interestingly, *BAP1*-related malignant mesothelioma tends to occur in the peritoneum of women with *BAP1* mutations whereas sporadic malignant mesothelioma often occurs in the peritoneum of men in the general population [40–42]. Further, while individuals with sporadic BAP1 mutations tend to be free of mesothelioma, those with germline BAP1 mutations often develop mesothelioma and always show loss of heterozygosity (i.e., biallelic BAP1 inactivation) [43,44].

3.5.4 Cutaneous Melanoma

Seen in 13% of affected individuals, *BAP1*-related cutaneous melanoma has an initial age of diagnosis at 25 years and a median age of onset at 46 years, compared to 58 years in the general population. Multiple primary cutaneous melanomas are not uncommon in people with *BAP1* mutations [45].

3.5.5 Clear Cell Renal Cell Carcinoma

About 10% of individuals with heterozygous *BAP1* germline pathogenic variants develop clear cell renal cell carcinoma (ccRCC), with an initial age of diagnosis at 36 years and a median age of onset at 46 years, compared to 64 years in patients without *BAP1* mutations. Further, patients with *BAP1*-related RCC have a shorter survival time that those without *BAP1* mutations (31.2 vs 78.2 months) [46].

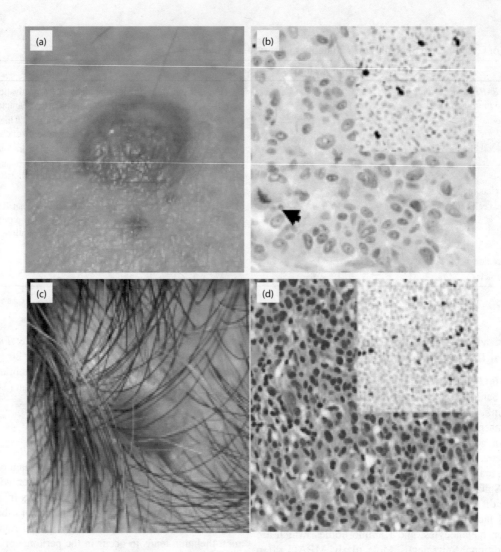

FIGURE 3.1 (a) Image of a nevoid melanoma-like melanocytic proliferation (NEMMP) from a patient with BAP1 mutation. (b) Atypical cytological features of lesion (a) include nuclear pleomorphism, prominent nucleoli, and a dermal mitotic figure (arrow) (H&E) as well as focal increases in Ki67 staining (inset). (c) Image of a distinct NEMMP from another patient with BAP1 mutation in the same family. (d) A proliferative area of lesion (c) shows marked nuclear atypia and hyperchromasia as well as elevated Ki 67 staining (inset). (Photo credit: Njauw CN et al. *PLOS ONE*. 2012;7(4):e35295.)

3.5.6 Basal Cell Carcinoma

Up to 6.3% of individuals with *BAP1* mutations may develop basal cell carcinoma, often at multiple sites, and having initial and median ages of diagnosis at 25 and 50 years, respectively.

3.5.7 Other Malignancies

Individuals with *BAP1* mutations may also develop other cancers, often at earlier ages than those in the general population. These include breast cancer (9.5% of cases), non-small cell lung adenocarcinoma (3.5%), ovarian cancer (3%), prostate cancer (3%), cholangiocarcinoma (2.3%), sarcoma (2.3%), meningioma (2%), neuroendocrine tumor (1.2%), colorectal cancer (1.2%), and thyroid cancer (1.2%) [47–55].

3.6 Diagnosis

Diagnosis of *BAP1* cancer syndrome involves physical examination (for skin abnormalities, tenderness, unusual mass, etc.), family history review, imaging (CT for omental stranding, laparoscopy for studding and multiple white plaques on the diaphragm and peritoneum with adhesions), histopathology (for characteristic tumor cells and BAP1 expression) and molecular testing (for germline *BAP1* mutations in blood sample) (see Figures 3.1–3.3) [56–59].

An individual who has two or more confirmed *BAP1*-related tumors (i.e., atypical Spitz tumor, uveal melanoma, malignant mesothelioma, cutaneous melanoma, ccRCC, and basal cell carcinoma), or one *BAP1*-relatd tumor, and a first- or second-degree relative with a confirmed *BAP1*-related tumor, should be suspected of *BAP1* cancer syndrome.

Identification of a heterozygous germline pathogenic variant in *BAP1* helps establish the diagnosis of *BAP1* cancer syndrome. Individuals with two or more *BAP1*-related tumors in themselves and/or first or second-degree relatives, with the exclusion of families with only multiple cutaneous melanomas, are advised to undertake *BAP1* testing, which can be conducted with a single-gene testing (e.g., sequence analysis and gene-targeted deletion/duplication analysis) or a multigene panel.

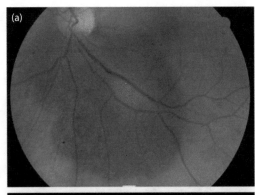

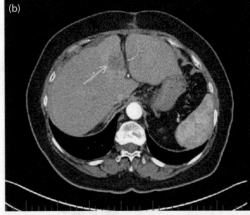

FIGURE 3.2 (a) Retinal imaging of a 72-year-old female with germline BAP1 mutation revealed a melanocytic tumor with irregular margins affecting a large area of the retina (shadow). (b) Preoperative CT of the abdomen identified a 32-mm lobulated lesion in segment 4B of the liver (arrow). (Photo credit: Klebe S et al. *Biomark Res.* 2015;3:14.)

Sequence analysis allows detection of *BAP1* pathogenic variants consisting of small intragenic deletions/insertions; missense, nonsense, and splice site variants; but generally not exon or whole-gene deletions/duplications.

Gene-targeted deletion/duplication analysis relies on quantitative PCR, long-range PCR, multiplex ligation-dependent probe amplification (MLPA), and gene-targeted microarray to detect intragenic or single-exon deletions or duplications.

Somatic *BAP1* variants may occur in *BAP1*-related tumors as well as in non-*BAP1*-related tumors. As these variants are not present in the germline (i.e., undetected in leukocyte DNA), they are not heritable [60].

Differential diagnoses include tumors that also occur in non-*BAP1*-related syndromes such as uveal melanoma (containing mutated *BRCA2*), cutaneous melanoma (containing mutated *CDKN2A, CDK4, MC1R, MITF*), and familial ccRCC (containing mutated *SDHA, SDHB, SDHC, SDHD, SDHAF2, MAX* in hereditary paraganglioma–pheochromocytoma syndrome; *VHL* in Von Hippel–Lindau syndrome).

3.7 Treatment

Treatment options for various BAP1-related malignancies include surgery, radiotherapy, and chemotherapy.

Uveal melanoma is an aggressive tumor that should be managed with transpupillary thermotherapy (TTT, which should be limited to flat tumor of 2.5 mm in thickness and can achieve tumor regression in 90% of cases, and tumor recurrence in 10% of cases), plaque radiotherapy (which provides excellent local tumor control when used in combination with TTT, and has a 3% recurrence rate after 5 years), charged particle irradiation (which is indicated for tumors of up to 14 mm in thickness and basal diameter of up to 28 mm; with recurrence rates of 4% and 10% and metastasis rates of 18.5% and 26.6% after 5 and 10 years, respectively), stereotactic radiotherapy (SRT, which does not require preoperative surgical marking to determine the tumor location, and can achieve a local tumor control in 98% of cases and tumor height reduction in 97% of cases), local resection (which represents an alternative, eye-sparing treatment for patient with choroidal melanoma), enucleation (which is indicated for large melanoma that occupies most of the intraocular space or that has invaded the optic nerve, as well as recurrent tumor), or orbital exenteration.

Malignant mesothelioma is refractory to surgical intervention and multimodality strategies, and a complete cure is not realistic.

As BAP1 protein deficiency often increases ubiquitinated histone 2A levels in tumor cells, histone deacetylase (HDAC) inhibitors such as valproic acid, trichostatin A, and suberoylanilide hydroxamic acid may be utilized to reverse this condition in tumor cells showing increased melanocytic differentiation [61].

3.8 Prognosis and Prevention

Patients with *BAP1*-related uveal melanoma, cutaneous melanoma, and renal cell carcinoma have more aggressive disease than those without *BAP1*-related tumors [62,63]. In fact, patients with uveal melanoma that lack BAP1 expression have a mean survival of 4.74 years as compared to 9.97 years in patients with BAP1 expression. Patients with BAP1-related uveal melanomas have mortality rates of 10%, 18%, and 21% at 5, 10, and 12 years after brachytherapy, respectively. Metastasis may occur hematogenously to the liver, lung, bone, and other organs. Patients with liver metastases have a survival time of 4–6 months, while patients with other metastases may survive for 19–28 months [64].

On the other hand, patients with *BAP1*-related malignant mesothelioma have less aggressive disease than those without *BAP1*-related tumors. This may be attributable to the possibility that malignant mesothelioma with germline *BAP1* mutation show better response to treatment, or that malignant mesothelioma with germline *BAP1* mutations have well-defined (thus low-grade) morphology compared to those without.

To prevent primary manifestations, individuals with BAP1 mutations should avoid arc-welding (uveal melanoma), asbestos exposure and smoking (malignant mesothelioma), sun exposure (basal cell carcinoma), and use sunscreen and protective clothing (basal cell carcinoma).

Families with a germline *BAP1* mutation should receive counseling about cancer risk management options (e.g., yearly examination for eye and skin tumors) and risk to family members (e.g., early detection of BAP1-associated cancers in at-risk family members, prenatal testing prior to pregnancy decision, and minimizing exposure to ultraviolet light and asbestos).

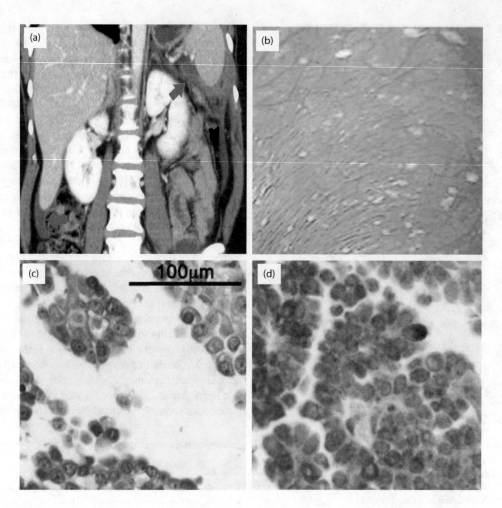

FIGURE 3.3 (a) CT imaging of the abdomen from a 45-year-old female harboring inactivating truncating germline BAP1 mutation, demonstrating omental thickening (arrows). (b) Laparoscopic view of white nodules/plaques on the surface of the diaphragm. (c) H&E staining of biopsy revealing epithelioid malignant mesothelioma. (d) BAP1 immunostaining of biopsy showing absence of nuclear staining indicative of biallelic BAP1 inactivation and accumulation of truncated inactive BAP1 protein in the cytoplasm, in addition to nuclear staining in a few inflammatory cells (which retain one wild-type BAP1 allele) infiltrating among tumor cells. (Photo credit: Kittaneh M et al. *J Transl Med.* 2018;16(1):194.)

3.9 Conclusion

BAP1 tumor predisposition syndrome (BAP1-TPDS) is an autosomal dominant disorder that confers increasing risk for characteristic benign melanocytic skin lesion known as "melanocytic *BAP1*-mutated atypical intradermal tumor" (MBAIT, formerly atypical Spitz tumor/nevus) and other malignancies (typically uveal melanoma, malignant mesothelioma, cutaneous melanoma, and ccRCC) [65]. Germline mutations in the tumor suppressor gene *BAP1* appears to the key mechanism for the development of BAP1-TPDS [66]. Clinically, patients with *BAP1*-related uveal melanoma, cutaneous melanoma, and renal cell carcinoma often have more aggressive disease than those without *BAP1*-related tumors [67]. However, patients with *BAP1*-related malignant mesothelioma tend to have less aggressive disease than those without *BAP1*-related tumors. Compared to the same types of tumors occurring in a sporadic setting, malignancies occurring in a setting of germline *BAP1* mutations are somewhat less aggressive [65,68]. Whereas BAP1-related uveal melanoma, cutaneous melanoma, ccRCC, and other malignancies are treatable by a combination of surgery, radiotherapy, and chemotherapy,

malignant mesothelioma containing *BAP1* mutations is currently refractory to surgical intervention and multimodality strategies.

REFERENCES

1. Abdel-Rahman MH, Pilarski R, Cebulla CM et al. Germline BAP1 mutation predisposes to uveal melanoma, lung adenocarcinoma, meningioma, and other cancers. *J Med Genet.* 2011;48(12):856–9.
2. Harbour JW, Onken MD, Roberson ED et al. Frequent mutation of BAP1 in metastasizing uveal melanomas. *Science.* 2010;330(6009):1410–3.
3. Harbour JW, Chao DL. A molecular revolution in uveal melanoma: Implications for patient care and targeted therapy. *Ophthalmology.* 2014;121:1281–8.
4. Eletr ZM, Wilkinson KD. An emerging model for BAP1's role in regulating cell cycle progression. *Cell Biochem Biophys.* 2011;60:3–11.
5. Pilarski R, Cebulla CM, Massengill JB et al. Expanding the clinical phenotype of hereditary BAP1 cancer predisposition syndrome, reporting three new cases. *Genes Chromosomes Cancer.* 2014;53(2):177–82.

6. Ventii KH, Devi NS, Friedrich KL et al. BRCA1-associated protein-1 is a tumor suppressor that requires deubiquitinating activity and nuclear localization. *Cancer Res.* 2008;68(17):6953–62.

7. Kadariya Y, Cheung M, Xu J et al. Bap1 is a bona fide tumor suppressor: Genetic evidence from mouse models carrying heterozygous germline Bap1 mutations. *Cancer Res.* 2016;76(9):2836–44.

8. Bott M, Brevet M, Taylor BS et al. The nuclear deubiquitinase BAP1 is commonly inactivated by somatic mutations and 3p21.1 losses in malignant pleural mesothelioma. *Nat Genet.* 2011;43(7):668–72.

9. Testa JR, Cheung M, Pei J et al. Germline BAP1 mutations predispose to malignant mesothelioma. *Nat Genet.* 2011;43(10):1022–5.

10. Carbone M, Ferris LK, Baumann F et al. BAP1 cancer syndrome: Malignant mesothelioma, uveal and cutaneous melanoma, and MBAITs. *J Transl Med.* 2012;10:179.

11. Carbone M, Yang H, Pass HI, Krausz T, Testa JR, Gaudino G. BAP1 and cancer. *Nat Rev Cancer.* 2013;13(3):153–9.

12. Popova T, Hebert L, Jacquemin V et al. Germline BAP1 mutations predispose to renal cell carcinomas. *Am J Hum Genet.* 2013;92(6):974–80.

13. Rai K, Pilarski R, Cebulla CM, Abdel-Rahman MH. Comprehensive review of BAP1 tumor predisposition syndrome with report of two new cases. *Clin Genet.* 2016;89(3):285–94.

14. Rai K, Pilarski R, Boru G et al. Germline BAP1 alterations in familial uveal melanoma. *Genes Chromosomes Cancer.* 2017;56(2):168–74.

15. Loeser H, Waldschmidt D, Kuetting F et al. Somatic BRCA1-associated protein 1 (BAP1) loss is an early and rare event in esophageal adenocarcinoma. *Mol Clin Oncol.* 2017;7(2):225–8.

16. Yu H, Mashtalir N, Daou S et al. The ubiquitin carboxyl hydrolase BAP1 forms a ternary complex with YY1 and HCF-1 and is a critical regulator of gene expression. *Mol Cell Biol.* 2010;30:5071–85.

17. Yu H, Pak H, Hammond-Martel I et al. Tumor suppressor and deubiquitinase BAP1 promotes DNA double-strand break repair. *Proc Natl Acad Sci USA.* 2014;111:285–90.

18. Hanpude P, Bhattacharya S, Kumar Singh A, Kanti Maiti T. Ubiquitin recognition of BAP1: Understanding its enzymatic function. *Biosci Rep.* 2017;37(5). pii: BSR20171099.

19. Eletr ZM, Yin L, Wilkinson KD. BAP1 is phosphorylated at serine 592 in S-phase following DNA damage. *FEBS Lett.* 2013;587:3906–11.

20. Amelio I. Genes versus Environment: Cytoplasmic BAP1 determines the toxic response to environmental stressors in mesothelioma. *Cell Death Dis.* 2017;8(6):e2907.

21. Bononi A, Giorgi C, Patergnani S et al. BAP1 regulates IP3R3-mediated Ca_{2+} flux to mitochondria suppressing cell transformation. *Nature.* 2017;546(7659):549–53.

22. Dai F, Lee H, Zhang Y et al. BAP1 inhibits the ER stress gene regulatory network and modulates metabolic stress response. *Proc Natl Acad Sci U S A.* 2017;114(12):3192–7.

23. Hebert L, Bellanger D, Guillas C et al. Modulating BAP1 expression affects ROS homeostasis, cell motility and mitochondrial function. *Oncotarget.* 2017;8(42):72513–27.

24. Farquhar N, Thornton S, Coupland SE et al. Patterns of BAP1 protein expression provide insights into prognostic significance and the biology of uveal melanoma. *J Pathol Clin Res.* 2017;4(1):26–38.

25. Lin M, Zhang L, Hildebrandt MAT, Huang M, Wu X, Ye Y. Common, germline genetic variations in the novel tumor suppressor *BAP1* and risk of developing different types of cancer. *Oncotarget.* 2017;8(43):74936–46.

26. Klebe S, Driml J, Nasu M et al. BAP1 hereditary cancer predisposition syndrome: A case report and review of literature. *Biomark Res.* 2015;3:14.

27. Wiesner T, Murali R, Fried I et al. A distinct subset of atypical Spitz tumors is characterized by BRAF mutation and loss of BAP1 expression. *Am J Surg Pathol.* 2012;36(6):818–30.

28. Ghosh K, Modi B, James WD, Capell BC. BAP1: Case report and insight into a novel tumor suppressor. *BMC Dermatol.* 2017;17(1):13.

29. Griewank KG, Müller H, Jackett LA et al. SF3B1 and BAP1 mutations in blue nevus-like melanoma. *Mod Pathol.* 2017;30(7):928–39.

30. Njauw CN, Kim I, Piris A et al. Germline BAP1 inactivation is preferentially associated with metastatic ocular melanoma and cutaneous-ocular melanoma families. *PLOS ONE.* 2012;7(4):e35295.

31. Aoude LG, Wadt K, Bojesen A et al. A BAP1 mutation in a Danish family predisposes to uveal melanoma and other cancers. *PLOS ONE.* 2013;8:e72144.

32. Baumann F, Flores E, Napolitano A et al. Mesothelioma patients with germline BAP1 mutations have 7-fold improved long-term survival. *Carcinogenesis.* 2015;36(1):76–81.

33. Grisanti S, Tura A. Uveal Melanoma. In: Scott JF, Gerstenblith MR, editors. *Noncutaneous Melanoma* [Internet]. Brisbane (AU): Codon Publications; 2018 Mar. Chapter 1.

34. Doherty RE, Alfawaz M, Francis J, Lijka-Jones B, Sisley K. Genetics of uveal melanoma. In: Scott JF, Gerstenblith MR, editors. *Noncutaneous Melanoma* [Internet]. Brisbane (AU): Codon Publications; 2018 Mar. Chapter 2.

35. Cebulla CM, Binkley EM, Pilarski R et al. Analysis of BAP1 germline gene mutation in young uveal melanoma patients. *Ophthalmic Genet.* 2015;36 (2):126–31.

36. Masoomian B, Shields JA, Shields CL. Overview of BAP1 cancer predisposition syndrome and the relationship to uveal melanoma. *J Curr Ophthalmol.* 2018;30(2):102–9.

37. Leblay N, Leprêtre F, Le Stang N et al. BAP1 is altered by copy number loss, mutation, and/or loss of protein expression in more than 70% of malignant peritoneal mesotheliomas. *J Thorac Oncol.* 2017;12(4):724–33.

38. Melaiu O, Gemignani F, Landi S. The genetic susceptibility in the development of malignant pleural mesothelioma. *J Thorac Dis.* 2018;10(Suppl 2):S246–52.

39. Vivero M, Bueno R, Chirieac LR. Clinicopathologic and genetic characteristics of young patients with pleural diffuse malignant mesothelioma. *Mod Pathol.* 2018;31(1):122–31.

40. Alakus H, Yost SE, Woo B et al. BAP1 mutation is a frequent somatic event in peritoneal malignant mesothelioma. *J Transl Med.* 2015;13:122.

41. Nasu M, Emi M, Pastorino S et al. High incidence of somatic BAP1 alterations in sporadic malignant mesothelioma. *J Thorac Oncol.* 2015;10(4):565–76.

42. Cheung M, Testa JR. *BAP1*, a tumor suppressor gene driving malignant mesothelioma. *Transl Lung Cancer Res.* 2017;6(3):270–8.

43. Alì G, Bruno R, Fontanini G. The pathological and molecular diagnosis of malignant pleural mesothelioma: A literature review. *J Thorac Dis.* 2018;10(Suppl 2):S276–84.

44. Bruno R, Alì G, Fontanini G. Molecular markers and new diagnostic methods to differentiate malignant from benign mesothelial pleural proliferations: A literature review. *J Thorac Dis.* 2018;10(Suppl 2):S342–52.

45. O'Shea SJ, Robles-Espinoza CD, McLellan L et al. A population-based analysis of germline BAP1 mutations in melanoma. *Hum Mol Genet.* 2017;26(4):717–28.

46. Hakimi AA, Ostrovnaya I, Reva B et al. Adverse outcomes in clear cell renal cell carcinoma with mutations of 3p21 epigenetic regulators BAP1 and SETD2: A report by MSKCC and the KIRC TCGA research network. *Clin Cancer Res.* 2013;19:3259–67.

47. Dey A, Seshasayee D, Noubade R et al. Loss of the tumor suppressor BAP1 causes myeloid transformation. *Science.* 2012;337:1541–6.

48. Pena-Llopis S, Vega-Rubin-de-Celis S, Liao A et al. BAP1 loss defines a new class of renal cell carcinoma. *Nat Genet.* 2012;44(7):751–9.

49. Farley MN, Schmidt LS, Mester JL et al. A novel germline mutation in BAP1 predisposes to familial clear-cell renal cell carcinoma. *Mol Cancer Res.* 2013;11:1061–71.

50. Jiao Y, Pawlik TM, Anders RA et al. Exome sequencing identifies frequent inactivating mutations in BAP1, *ARID*1A and PBRM1 in intrahepatic cholangiocarcinomas. *Nat Genet.* 2013;45(12):1470–3.

51. Ho TH, Kapur P, Joseph RW et al. Loss of PBRM1 and BAP1 expression is less common in non-clear cell renal cell carcinoma than in clear cell renal cell carcinoma. *Urol Oncol.* 2015;33:23e29–14.

52. Al-Shamsi HO, Anand D, Shroff RT et al. BRCA-associated protein 1 mutant cholangiocarcinoma: An aggressive disease subtype. *J Gastrointest Oncol.* 2016;7(4):556–61.

53. Eckel-Passow JE, Serie DJ, Cheville JC et al. BAP1 and PBRM1 in metastatic clear cell renal cell carcinoma: Tumor heterogeneity and concordance with paired primary tumor. *BMC Urol.* 2017;17(1):19.

54. Ge YZ, Xu LW, Zhou CC et al. A *BAP1* Mutation-specific microRNA signature predicts clinical outcomes in clear cell renal cell carcinoma patients with wild-type *BAP1*. *J Cancer.* 2017;8(13):2643–52.

55. Shankar GM, Abedalthagafi M, Vaubel RA et al. Germline and somatic BAP1 mutations in high-grade rhabdoid meningiomas. *Neuro Oncol.* 2017;19(4):535–45.

56. Righi L, Duregon E, Vatrano S et al. BRCA1-associated protein 1 (BAP1) immunohistochemical expression as a diagnostic tool in malignant pleural mesothelioma classification: A large retrospective study. *J Thorac Oncol.* 2016;11(11):2006–17.

57. Wang LM, Shi ZW, Wang JL et al. Diagnostic accuracy of BRCA1-associated protein 1 in malignant mesothelioma: A meta-analysis. *Oncotarget.* 2017;8(40):68863–72.

58. Kittaneh M, Berkelhammer C. Detecting germline BAP1 mutations in patients with peritoneal mesothelioma: Benefits to patient and family members. *J Transl Med.* 2018;16(1):194.

59. Kushitani K, Amatya VJ, Mawas AS et al. Utility of Survivin, BAP1, and Ki-67 immunohistochemistry in distinguishing epithelioid mesothelioma from reactive mesothelial hyperplasia. *Oncol Lett.* 2018;15(3):3540–7.

60. Chen Z, Gaudino G, Pass HI, Carbone M, Yang H. Diagnostic and prognostic biomarkers for malignant mesothelioma: An update. *Transl Lung Cancer Res.* 2017;6(3):259–69.

61. Parasramka M, Yan IK, Wang X et al. BAP1 dependent expression of long non-coding RNA NEAT-1 contributes to sensitivity to gemcitabine in cholangiocarcinoma. *Mol Cancer.* 2017;16(1):22.

62. Joseph RW, Kapur P, Serie DJ et al. Loss of BAP1 protein expression is an independent marker of poor prognosis in patients with low-risk clear cell renal cell carcinoma. *Cancer.* 2014;120:1059–67.

63. Song H, Wang L, Lyu J, Wu Y, Guo W, Ren G. Loss of nuclear BAP1 expression is associated with poor prognosis in oral mucosal melanoma. *Oncotarget.* 2017;8(17):29080–90.

64. Pulford E, Huilgol K, Moffat D, Henderson DW, Klebe S. Malignant mesothelioma, BAP1 immunohistochemistry, and VEGFA: Does BAP1 have potential for early diagnosis and assessment of prognosis? *Dis Markers.* 2017;2017: 1310478.

65. Pilarski R, Rai K, Cebulla C, Abdel-Rahman M. *BAP1* tumor predisposition syndrome. In: Adam MP, Ardinger HH, Pagon RA et al. editors. *GeneReviews®* [Internet]. Seattle (WA): University of Washington, Seattle; 1993–2018. 2016 Oct 13.

66. Wiesner T, Obenauf AC, Murali R et al. Germline mutations in BAP1 predispose to melanocytic tumors. *Nat Genet.* 2011;43(10):1018–21.

67. Soares de Sá BC, de Macedo MP, Torrezan GT et al. BAP1 tumor predisposition syndrome case report: pathological and clinical aspects of BAP1-inactivated melanocytic tumors (BIMTs), including dermoscopy and confocal microscopy. *BMC Cancer.* 2019;19(1):1077.

68. Chau C, van Doorn R, van Poppelen NM et al. Families with BAP1-tumor predisposition syndrome in The Netherlands: Path to identification and a proposal for genetic screening guidelines. *Cancers (Basel).* 2019;11(8):1114.

4

Congenital Central Hypoventilation Syndrome

Dongyou Liu

CONTENTS

4.1 Introduction

Congenital central hypoventilation syndrome (CCHS) is a rare autosomal dominant disorder of respiratory and autonomic regulation that typically affects newborns (classic CCHS) and occasionally toddlers, children, and adults (a milder later-onset CCHS [LO-CCHS]).

Clinically, classic CCHS displays (i) severe alveolar hypoventilation (i.e., monotonous respiratory rates and shallow breathing) asleep (during non-rapid eye movement sleep) or awake; (ii) autonomic nervous system dysregulation (ANSD); (iii) altered neural crest-derived structures (i.e., Hirschsprung disease [HSCR]) and neural crest tumors (neuroblastoma, ganglioneuroma, and ganglioneuroblastoma); and (iv) other signs and symptoms (e.g., reduced or absent central chemosensitivity and dyspnea sensations, heart rate and blood pressure dysregulation, esophageal dysmotility, ocular disorders, and sudden death). On the other hand, LO-CCHS presents with nocturnal alveolar hypoventilation and mild ANSD.

At the molecular level, isolated and syndromic CCHS is mostly attributable to the polyalanine repeat expansion mutations (PARM) in the paired-like homeobox 2B (*PHOX2B*) gene on chromosome 4p13, with more severe disease being typically linked to nonpolyalanine repeat expansion mutations (NPARM) in *PHOX2B*. In rare cases, CCHS may be associated with mutations in several other genes, including GDNF (chromosome 5p13.2), RET (chromosome 10q11.21), ASCL1 (chromosome 12q23.2), and EDN3 (chromosome 20q13.32) [1].

4.2 Biology

CCHS (also referred to as congenital failure of autonomic control, Ondine's curse, and Ondine-Hirschsprung disease) was first recognized in 1970 as an autonomic nervous system (ANS) dysfunction leading to central apnea crises as exemplified by hypoventilation during sleep (in less severe forms) and both during sleep and while awake (in more severe cases). In 1978, CCHS was linked to HSCR (a developmental disorder of the enteric system characterized by aganglionosis in the distal colon leading to the inability of affected neonates to pass first stool or meconium) in 15%–20% of cases (the association of CCHS and HSCR is referred to as Haddad syndrome, which is caused by heterozygous mutation in the *ASCL1* gene) and tumors of neural crest derivatives (e.g., neuroblastoma, ganglioneuroma, and ganglioneuroblastoma) in 5%–10% of cases, particularly in the first 2 years of life. Cohort studies from 1992 onward further expanded the scope of CCHS to clinical symptoms in other organs (e.g., cardiac arrhythmias, orthostatic hypotension, abnormal pupillary reflex, esophageal dysmotility, diaphoresis, decreased heart rate variability, chronic constipation, and excessive drowsiness after use of sedatives and antihistamines). In 2002, a mother–daughter transmission of CCHS suggesting a dominant mode of inheritance was observed. In 2003, the underlying gene responsible for CCHS was identified as *PHOX2B*, which is located on chromosome 4p13, and encodes a transcription factor that regulates genes involved in the development of the ANS and the neuronal structures controlling breathing. Heterozygous germline mutations in *PHOX2B* induce defective migration and/or differentiation of neural crest derivatives, leading to a lack of adequate autonomic control of respiration with decreased sensitivity to hypercapnia and hypoxia [1–3].

4.3 Pathogenesis

Mapped to chromosome 4p13, the *PHOX2B* gene (also known as neuroblastoma paired-type homeobox gene [*NBPHOX*] or paired mesoderm homeobox gene [*PMX2B*]) measures 4888 bp in length, and is organized in the order of 5′-UTR (361 bp), exon 1 (241 bp), exon 2 (188 bp), exon 3 (516 bp), and 3′-UTR (1725 bp). *PHOX2B* produces an mRNA transcript of 3218 nucleotides, and its coding region (945 nucleotides) encodes a highly conserved protein (PHOX2B) of 314 amino acids. Within the PHOX2B protein, residues 99–148 form a homeodomain region implicated in the binding to target DNA elements, while residues 159–186 and residues 241–300 comprise two stretches of 9 and 20 alanine repeats, encoded by GCN triplets (GCA, GCT, GCC, or GCG), whose functional role remains largely unknown [4].

Belonging to the Ensembl protein family, PHOX2B functions as a homeobox domain transcription factor of several genes essential for the development of neural crest derivatives and hindbrain motor neuron, including TH, DBH (dopamine β-hydroxylase), PHOX2A (paired-like homeobox 2A), PHOX2B itself, RET (rearranged during transfection), TLX2 (T-cell leukemia homeobox 2), ALK (anaplastic lymphoma kinase), SOX10, Hand1, SCG2, and MSX1. In addition, through interaction with PHOX2B, CREB-binding protein (CREBBP/CBP) and TRIM11 may co-mediate synergistic trans-activation.

Expressed during neural development in the nuclei of brainstem areas that contain pathways controlling breathing and auditory functions [5], PHOX2B is regulated transcriptionally by itself, as well as E2a and Hand2 at specific sympathetic and enteric nervous system developmental stages. Other proteins that may play a possible role in the regulation of PHOX2B are SOX10, PHOX2A, and HASH1. Further, by forming trimer with Pbx1 and Meis1, Hoxb1 and Hoxb2 also contribute to the regulation of PHOX2B transcription.

Heterozygous germline (constitutive) *PHOX2B* mutations affect the binding of the paired-homeodomain protein to the target sites on DNA, and predispose to diseases of the sympathetic nervous system such as CCHS (which causes decreased sensitivity to hypercarbia [elevated carbon dioxide {CO_2}] levels in the blood and hypoxia [tissues depleted of oxygen] in the brainstem leading to an impaired respiratory response) and neuroblastoma [6–8]. These mutations often come in the form of triplet repeat expansions including PARM and NPARM.

PARM are in-frame triplet duplications involving the second polyalanine repeat in *PHOX2B* exon 3 that codes for 20 alanines, resulting in expansion of 4–13 additional alanine repeats. PARM occur in 90% of CCHS cases, produce a variable clinical profile, and their sizes correlate to disease severity. While individuals who have 20 alanines or fewer are unaffected, those individuals heterozygous for 24–26 GCN repeats (genotypes 20/24–26) may have a mild phenotype only (e.g., gastroesophageal reflux, constipation, cardiac arrhythmias, ocular, and endocrinological disorders) manifesting during illness or exposure to respiratory depressants, and those heterozygous for 26 to 33 GCN

repeats (genotypes 20/26–33) are fully affected and often require continuous ventilatory support.

NPARM are missense, nonsense, frameshift, or truncation mutations that occur in *PHOX2B* exons 1, 2, or 3 [9]. PHOX2B frameshift and truncation mutants display reduced ability to bind to and induce nuclear translocation of neuronal calcium sensor protein HPCAL1 (VILIP-3), thus impeding the differentiation of immature sympathetic neurons and predisposing to neuroblastoma [10]. It appears that *DBH, TLX2, RET, PHOX2B*, and *GFAP* are target genes susceptible to the reduced activity of NPARM proteins. PARM occur in about 10% of CCHS cases that present with a more severe clinical profiles including CCHS and other neurocristopathies (i.e., HSCR and an increased risk for neuroblastoma) [11,12].

In addition, in frame deletion within the polyAla, stretch may occur in healthy individuals, people with schizophrenia, and individuals with some milder forms of CCHS (who may harbor smaller PARM such as 7, 13, 14, 15, or 20 repeats). However, somatic *PHOX2B* mutations are only detected in neuroblastoma cell lines and neuroblastoma samples, but generally not in CCHS patients.

CCHS patients possessing *PHOX2B* polyalanine expansion mutation may sometimes contain other genetic variants (e.g., *ASCL [HASH1], BDNF, BMP2, ECE1, EDN3, GFRA1, GDNF, PHOX2A, RET*).

4.4 Epidemiology

4.4.1 Prevalence

CCHS is estimated to occur at 1 per 200,0000 live births. With the introduction of molecular genetic test for the paired-like homeobox (PHOX) 2B gene (*PHOX2B*), at least 1000 individuals worldwide are considered to have the disease to date. However, this number may be an underestimate due to lack of recognition, clinical variation, and decreased penetrance [13].

4.4.2 Inheritance

Although 90% of CCHS patients appear to have de novo *PHOX2B* pathogenic variants, about 10% inherit from an affected parent or an asymptomatic parent (who is mosaic for a *PHOX2B* variant) in an autosomal dominant manner. Polyalanine repeat expansion in *PHOX2B* is meiotically stable since a stable number of repeats have been consistently documented during parent-to-child transmission, including instances of parental mosaicism for the expansion.

4.4.3 Penetrance

PHOX2B pathogenic variants show incomplete penetrance, as CCHS-affected individuals have variable clinical presentations.

4.5 Clinical Features

CCHS typically shows a reduced response to hypercapnia and hypoxia (leading to alveolar hypoventilation) accompanied by ANSD [14,15]. However, clinical severity of CCHS appears to correlate to the type and length of triplet expansions in the

PHOX2B gene, including PARM in exon 3, and NPARM in exons 1, 2, or 3 [16–19].

PARM involving in-frame duplications in exon 3 are responsible for 90% of CCHS cases, with newborns harboring larger polyAla expansions (compound heterozygosity for the normal 20 CGN and the abnormal 27–33 CGN alleles, i.e., 20/27-20/33) generally developing classic CCHS while others (toddlers, children, and adults) possessing smaller polyAla expansions (compound heterozygosity for the normal 20 CGN and the abnormal 24–25 CGN alleles, i.e., 20/24-20/25) showing a milder LO-CCHS. On the other hand, NPARM involving frameshift and nonsense mutations in exons 1, 2, or 3 account for about 10% of CCHS cases [20].

Classic CCHS typically occurs in neonates, who may have adequate ventilation while awake but develop apparent hypoventilation with monotonous respiratory rates and shallow breathing (diminished tidal volume) during sleep. Some severely affected individuals may hypoventilate both awake and asleep, and often suffer from apnea or hypercapnia requiring continuous ventilatory support. Some affected neonates do not breathe at birth (due to perinatal asphyxia). During sleep, the infant's oxygen levels fall, and CO_2 levels rise with no compensatory rise in respiratory rate or signs of agitation. If untreated, infants may develop fixed pulmonary hypertension and right-sided heart failure. Besides severe alveolar hypoventilation (i.e., monotonous respiratory rates and shallow breathing), classic CCHS may also present with ANSD, altered neural crest-derived structures (i.e., HSCR in 13%–20% cases, particularly in patients with 20/27 and above), neural crest tumors (neuroblastoma, ganglioneuroma, and ganglioneuroblastoma in 5%–6% of cases, particularly in patients with 20/29-20/33), cardiac arrhythmia (20/25–27), and other symptoms (e.g., reduced or absent central chemosensitivity and dyspnea sensations, heart rate and blood pressure dysregulation, esophageal dysmotility, ocular disorders, and sudden death secondary to central hypoventilation) [21].

LO-CCHS is associated with nocturnal alveolar hypoventilation and mild ANSD in toddlers, children, and adults, and may be precipitated by severe respiratory infection or exposure to sedatives or general anesthesia. Some patients require mechanical ventilation during sleep.

Patients harboring frameshift and nonsense mutations (NPARM) are commonly associated with a more severe clinical phenotype (e.g., HSCR in 87%–100% cases, tumors of neural crest origin such as neuroblastoma in 50% of cases) that requires continuous ventilatory support [22]. Specifically, of <100 patients with NPARM reported to date, most have frame shifts in *PHOX2B* exon 3, with extensive gut involvement, increased risk of peripheral neuroblastic tumors, cardiac arrhythmia, and need for continuous ventilator support (CVS). On the other hand, patients having frameshift variants in the 3′ region of exon 2 and the 5′ region of exon 3 are associated with a milder form of CCHS. In addition, missense pathogenic variants primarily affecting the homeodomain of the protein may induce a variable clinical presentation, whereas nonsense variants leading to a complete absence of the gene product may cause a severe clinical disease.

4.6 Diagnosis

Classic CCHS is suspected when a newborn displays alveolar hypoventilation (with absent or negligible ventilatory sensitivity to

hypercapnia and/or hypoxemia) and ANSD, ranging from severe breath-holding spells, lack of physiologic responsiveness to the challenges of exercise and environmental stressors, diminished pupillary light response, esophageal dysmotility, severe constipation even in the absence of HSCR, profuse sweating, and reduced basal body temperature, to altered perception of anxiety, in the absence of primary neuromuscular, lung, or cardiac disease or brainstem lesion that could account for ANSD and other signs [1].

LO-CCHS is suspect when an older individual (>1 month of age, late childhood, or adulthood) meets the criteria above for the newborn or shows apparent life-threatening events and cyanosis during sleep, recurrent severe pulmonic infections with related hypoventilation, unexplained seizures, respiratory depression after antiseizure medication, sedation, or anesthesia, unexplained neurocognitive delay with history of prior cyanosis, unexplained nocturnal hypercarbia and hypoxemia, unresolved central alveolar hypoventilation after treatment for obstructive sleep apnea, unresponsiveness to hypercarbia or hypoxemia (prolonged underwater swimming), pneumonia, sudden and unexpected death (sudden infant death syndrome [SIDS] or sudden unexplained death of childhood [SUDC], especially if there is a family history of CCHS).

Evaluation of patients with hypoventilation may include: (i) chest radiograph, (ii) diaphragm fluoroscopy, (iii) electrocardiogram and echocardiogram, (iv) magnetic resonance imaging and/or computed tomography scan of brain and brainstem for gross anatomic lesions, (v) tests for metabolic disorders, (vi) comprehensive neurological evaluation, and (vii) polysomnogram/polysomnography to establish the presence of hypoventilation and sleep-related breathing disorder (as polysomnography helps reveals worsening hypoxia and hypercapnia in CCHS patients during sleep than during wakefulness). Further, barium enema (an x-ray exam that detects abnormalities in the colon) and rectal biopsy may be considered for patients with abdominal distension or constipation-related HSCR (Figure 4.1) [23,24]. Histopathology is also useful for revealing neuronal loses associated with CCHS (Figure 4.2) [2,25].

However, identification of a PARM or NPARM in the *PHOX2B* gene is necessary to confirm the diagnosis of CCHS [26]. This is typically carried out by (i) targeted analysis for pathogenic variants (fragment length analysis or screening test), which amplifies the region encoding the polyalanine repeat and determines the polyalanine repeat length expansion (PARM), deletion, duplication (NPARM), or somatic mosaicism (upon sequence analysis) in 95% of individuals with CCHS; (ii) sequence analysis, which is performed on individuals with negative fragment analysis result and detects *PHOX2B* polyalanine repeats (PARM) and frameshifts in the polyalanine region (NPARM) in 92% of individuals with CCHS, as well as missense, nonsense, frameshift, or stop codon variants outside of the polyalanine region (NPARM) in 8% of individuals with CCHS; (iii) deletion/duplication (MLPA) analysis, which is performed on individuals with negative sequence analysis result and detects deletions of the entire *PHOX2B* gene, or a single or multiple exons in <1% of individuals with CCHS, who can be missed with sequencing and targeted analysis for pathogenic variants [1].

For PARM testing interpretation, individuals with 20 alanines (GCN repeats) on both *PHOXB* alleles in the second repeat region of exon 3 are considered normal or unaffected; individuals heterozygous for 24 or 25 alanine repeats (genotypes 20/24 or 20/25) are considered mild phenotypes, who may show symptoms when

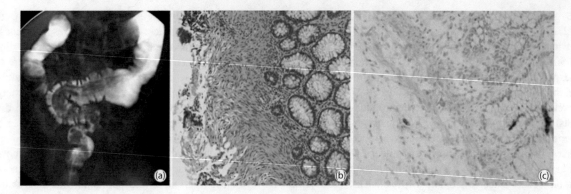

FIGURE 4.1   Congenital central hypoventilation syndrome (CCHS) with Hirschsprung disease in a 41-year-old male showing (a) transitional zone (arrow) in the middle of the sigmoid colon by barium enema; (b) absence of ganglion cells in frozen section of the biopsy; (c) aberrant acetylcholine esterase by enzyme histochemistry. (Photo credit: Lee CW et al. *J Korean Med Sci.* 2011;26(2):312–5.)

exposed to respiratory depressants or severe intercurrent pulmonary illness; individuals heterozygous for 25–33 alanine repeats (genotypes 20/25-20/33) are considered affected; individuals with 9, 13, 14, and 15 GCN repeats are considered benign.

For NPARM testing interpretation, individuals with out-of-frame deletions or duplications of 1–38 nucleotides located outside of the polyalanine repeat and frameshift variants affecting the region encoding the polyalanine repeats are considered affected, and typically have more severe phenotypes than individuals with PARM; on rare occasions, individuals with a small frameshift variant may have reduced but variable penetrance; individuals harboring deletions of 6216 bp (involving only *PHOX2B* exon 3) to 2.6 Mb (involving all of *PHOX2B* and 12 other genes) have been found with alveolar hypoventilation or HSCR.

Differential diagnoses for CCHS include primary neuromuscular, lung, or cardiac disease; identifiable brainstem lesion that produces symptoms characteristic of CCHS (e.g., ANSD, which is associated with altered thermoregulation, diaphoresis, pupillary

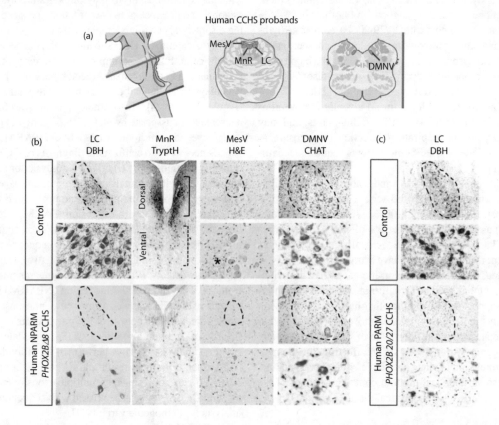

FIGURE 4.2   Pathological changes in human brainstem caused by congenital central hypoventilation syndrome (CCHS). (a) Schematic representation of human hindbrain at levels of pons (*pink*) and medulla (*blue*). (b) Dramatic losses in NPARM *PHOX2BΔ8* proband locus coeruleus (LC), dorsal median raphe (MnR), mesencephalic trigeminal nucleus (MesV), and dorsal motor nucleus of vagus (DMNV). Note lack of DBH expression in LC and tryptophan hydroxylase (TrypH) in dorsal MnR, indicating defects in synthesis of noradrenaline and serotonin production; DMNV with diminished cholinergic neurons, indicated by choline acetyltransferase (CHAT) expression; *asterisk* indicating diminished MesV fibers originating from the nucleus in the proband. (c) Dramatic loss in PARM *PHOX2B* 20/27 proband LC. Note a significant reduction of DBH expression. (Photo credit: Nobuta H et al. *Acta Neuropathol.* 2015;130(2):171–83.)

response, vasomotor function, and bradycardia); unexplained childhood and adult alveolar hypoventilation or adverse event (cyanosis or seizures) secondary to sedation; asphyxia; severe pulmonary infection; trauma; tumor; infarction; severe prematurity; ROHHAD (rapid-onset obesity with hypothalamic dysfunction, hypoventilation, and autonomic dysregulation); hypothalamic dysfunction (altered water balance, hyperprolactinemia, hypothyroidism, altered onset of puberty, growth hormone deficiency, and ACTH insufficiency) [27]. Since these conditions are not caused by mutations in the *PHOX2B* gene, they can be distinguished from CCHS by their negative results in molecular testing for *PHOX2B* pathogenic variants.

It should be noted that allelic *PHOX2B* variants in intron 2 and in exon 3 may occur in SIDS and HSCR, and polyalanine repeat contraction variants in the *PHOX2B* polyalanine repeat tract are sometimes found in patients with schizophrenia and strabismus.

4.7 Treatment

CCHS is a rare disorder in young children in which the main manifestations are alveolar hypoventilation and autonomic dysregulation. As CCHS patients do not outgrow the disorder, their continuing survival with a good quality of life depends largely on the adequacy of ventilatory support [28,29]. Therefore, treatment options for CCHS include mechanical ventilation (e.g., positive pressure ventilation [PPV] via tracheostomy, noninvasive positive pressure ventilation [NIPPV], and diaphragm pacing [DP] to compensate for the altered/absent ventilatory responses to hypoxemia and hypercarbia), surgery and chemotherapy (for neuroblastomas), and cardiac pacemaker (for cardiac problems such as asystoles) [24,30–36].

4.8 Prognosis and Prevention

CCHS is a genetic disorder that does not resolve spontaneously, respond to pharmacologic stimulants, or improve with advancing age. For affected newborns to survive into toddlers and young adults, constant ventilatory support is necessary. Suboptimal ventilatory support and development of asystoles are potential causes of sudden death in CCHS patients.

CCHS patients are advised to avoid swimming and breath-holding contests (asphyxia; death), alcohol (respiratory depression), recreational drugs (possible death), and medications/sedatives/anesthetics that could induce respiratory depression. In addition, CCHS patients should undergo yearly (every 6 months until age 3 years) comprehensive, multiple-day in-hospital physiologic evaluation (to optimize ventilatory support awake and asleep), yearly 72-hour Holter recording (to identify any prolonged sinus pauses), yearly echocardiogram (to identify right ventricular hypertrophy or cor pulmonale), yearly hemoglobin, hematocrit, and reticulocyte counts (to identify polycythemia), yearly neurocognitive testing (to evaluate the success of artificial ventilation), and evaluation of HSCR and tumors of neural crest origin in children with specific *PHOX2B* variants [25,37].

While about 90% of CCHS patients are heterozygous for a de novo *PHOX2B* pathogenic variant, 10% have an affected parent or an asymptomatic parent (who is mosaic for a *PHOX2B* variant). Given its autosomal dominant inheritance (with 50% chance of passing from parents to offspring, parents of children with a known *PHOX2B* pathogenic variant should go for *PHOX2B* screening test to determine their risk for LO-CCHS or mosaicism. Prenatal diagnosis and preimplantation genetic diagnosis may be considered if a *PHOX2B* pathogenic variant is known in an affected family member.

4.9 Conclusion

Congenital central hypoventilation syndrome (CCHS) is a rare genetic disorder that mainly affects newborns (classic CCHS) and occasionally toddlers and young adults (LO-CCHS), with alveolar hypoventilation and autonomic nervous system dysregulation (ANSD) being classic symptoms. Resulting from a decreased sensitivity to hypercapnia and hypoxia in the brainstem, CCHS-related hypoventilation is evident during non-rapid eye sleep in most patients (in contrast to other respiratory disorders), and also during wakeful states in more severely affected individuals, leading to monotonous respiratory rates and shallow breathing, cyanosis, pulmonary hypertension, cor pulmonale, and central nervous system hypoxic damage, as well as sudden death if left untreated. ANSD may induce cardiac arrhythmias, orthostatic hypotension, episodes of profuse diaphoresis, pupillary abnormalities, severe constipation, Hirschsprung disease (HSCR), and neural crest tumors (particularly neuroblastoma).

The molecular mechanism underscoring CCHS pathogenesis relates to germline mutations in the *PHOX2B* gene on chromosome 4.13, including PARM in exon 3 and NPARM (such as frameshift and nonsense mutations) exons 1, 2, or 3. Alterations in this neurogenesis regulator gene results in abnormal neural crest cell development, leading to CCHS, HSCR, and neuroblastoma. While individuals with large expansions (27–33 repeats) in exon 3 have classic CCHS and often require continuous ventilatory support, those with small expansions (24–26 repeats) have LO-CCHS and rarely require ventilation. About 10% of patients with NPARM (e.g., frameshift and nonsense mutations) also develop HSCR (due to aganglionosis in the distal colon causing bowel obstruction) and tumors of neural crest and need continuous ventilatory support.

Due to current lack of effective cures, patients with CCHS require at least nocturnal ventilation or continuous ventilator support (CVS) and regular surveillance in order to avoid the long-term complications of chronic hypoxemia and maintain quality of life. Therefore, if a *PHOX2B* pathogenic variant is identified in an affected family member, prenatal diagnosis and preimplantation genetic diagnosis may help reduce potential suffering in the family concerned.

REFERENCES

1. Weese-Mayer DE, Marazita ML, Rand CM, Berry-Kravis EM. Congenital central hypoventilation syndrome. In: Adam MP, Ardinger HH, Pagon RA et al. editors. *GeneReviews®* [Internet]. Seattle (WA): University of Washington, Seattle; 1993–2018. 2004 Jan 28 [updated 2014 Jan 30].

2. Nobuta H, Cilio MR, Danhaive O et al. Dysregulation of locus coeruleus development in congenital central hypoventilation syndrome. *Acta Neuropathol.* 2015;130(2):171–83.

3. Moreira TS, Takakura AC, Czeisler C, Otero JJ. Respiratory and autonomic dysfunction in congenital central hypoventilation syndrome. *J Neurophysiol.* 2016;116(2):742–52.

4. Di Lascio S, Belperio D, Benfante R, Fornasari D. Alanine expansions associated with congenital central hypoventilation syndrome impair PHOX2B homeodomain-mediated dimerization and nuclear import. *J Biol Chem.* 2016;291(25):13375–93.

5. Fu C, Xue J, Wang R et al. Chemosensitive Phox2b-expressing neurons are crucial for hypercapnic ventilatory response in the nucleus tractus solitarius. *J Physiol.* 2017;595(14):4973–89.

6. Sharman M, Gallea C, Lehongre K et al. The cerebral cost of breathing: An FMRI case-study in congenital central hypoventilation syndrome. *PLOS ONE.* 2014;9(9):e107850.

7. Harper RM, Kumar R, Macey PM, Harper RK, Ogren JA. Impaired neural structure and function contributing to autonomic symptoms in congenital central hypoventilation syndrome. *Front Neurosci.* 2015;9:415.

8. Trang H, Masri Zada T, Heraut F. Abnormal auditory pathways in PHOX2B mutation positive congenital central hypoventilation syndrome. *BMC Neurol.* 2015;15:41.

9. Unger SA, Guillot M, Urquhart DS. A case of "abnormally abnormal" hypoxic ventilatory responses: A novel NPARM *PHOX 2B* gene mutation. *J Clin Sleep Med.* 2017;13(8):1013–5.

10. Di Lascio S, Benfante R, Di Zanni E et al. Structural and functional differences in PHOX2B frameshift mutations underlie isolated or syndromic congenital central hypoventilation syndrome. *Hum Mutat.* 2018;39(2):219–36.

11. Sandoval RL, Zaconeta CM, Margotto PR et al. Congenital central hypoventilation syndrome associated with Hirschsprung's disease: Case report and literature review. *Rev Paul Pediatr.* 2016;34(3):374–8.

12. Kasi AS, Jurgensen TJ, Yen S, Kun SS, Keens TG, Perez IA. Three-generation family with congenital central hypoventilation syndrome and novel *PHOX2B* gene non-polyalanine repeat mutation. *J Clin Sleep Med.* 2017;13(7):925–7.

13. Almutairi A. Congenital central hypoventilation syndrome with PHOX2B mutation in Saudi Arabia: A- single center experience. *Int J Health Sci (Qassim).* 2014;8(3):311–4.

14. Patwari PP, Carroll MS, Rand CM, Kumar R, Harper R, Weese-Mayer DE. Congenital central hypoventilation syndrome and the PHOX2B gene: A model of respiratory and autonomic dysregulation. *Respir Physiol Neurobiol.* 2010;173(3):322–35.

15. Reverdin AK, Mosquera R, Colasurdo GN, Jon CK, Clements RM. Airway obstruction in congenital central hypoventilation syndrome. *BMJ Case Rep.* 2014;2014. pii: bcr2013200911.

16. Wang TC, Su YN, Lai MC. PHOX2B mutation in a Taiwanese newborn with congenital central hypoventilation syndrome. *Pediatr Neonatol.* 2014;55(1):68–70.

17. Attali V, Straus C, Pottier M et al. Normal sleep on mechanical ventilation in adult patients with congenital central alveolar hypoventilation (Ondine's curse syndrome). *Orphanet J Rare Dis.* 2017;12(1):18.

18. Bygarski E, Paterson M, Lemire EG. Extreme intra-familial variability of congenital central hypoventilation syndrome: A case series. *J Med Case Rep.* 2013;7:117.

19. Cain JT, Kim DI, Quast M et al. Nonsense pathogenic variants in exon 1 of PHOX2B lead to translational reinitiation in congenital central hypoventilation syndrome. *Am J Med Genet A.* 2017;173(5):1200–7.

20. Amimoto Y, Okada K, Nakano H, Sasaki A, Hayasaka K, Odajima H. A case of congenital central hypoventilation syndrome with a novel mutation of the PHOX2B gene presenting as central sleep apnea. *J Clin Sleep Med.* 2014;10(3):327–9.

21. Lee JP, Hung YP, O'Dorisio TM, Howe JR, Hornick JL, Bellizzi AM. Examination of PHOX2B in adult neuroendocrine neoplasms reveals relatively frequent expression in phaeochromocytomas and paragangliomas. *Histopathology.* 2017;71(4):503–10.

22. Fernández RM, Mathieu Y, Luzón-Toro B et al. Contributions of PHOX2B in the pathogenesis of Hirschsprung disease. *PLOS ONE.* 2013;8(1):e54043.

23. Lee CW, Lee JH, Jung EY et al. Haddad syndrome with PHOX2B gene mutation in a Korean infant. *J Korean Med Sci.* 2011;26(2):312–5.

24. Kasi AS, Perez IA, Kun SS, Keens TG. Congenital central hypoventilation syndrome: Diagnostic and management challenges. *Pediatr Health Med Ther.* 2016;7:99–107.

25. Di Zanni E, Adamo A, Belligni E et al. Common PHOX2B poly-alanine contractions impair RET gene transcription, predisposing to Hirschsprung disease. *Biochim Biophys Acta.* 2017;1863(7):1770–7.

26. Szymońska I, Borgenvik TL, Karlsvik TM et al. Novel mutation-deletion in the *PHOX2B* gene of the patient diagnosed with neuroblastoma, Hirschsprung's disease, and congenital central hypoventilation syndrome (NB-HSCR-CCHS) cluster. *J Genet Syndr Gene Ther.* 2015;6(3). pii: 269.

27. Rojnueangnit K, Descartes M. Congenital central hypoventilation syndrome mimicking mitochondrial disease. *Clin Case Rep.* 2018;6(3):465–8.

28. Carroll MS, Patwari PP, Kenny AS et al. Residual chemosensitivity to ventilatory challenges in genotyped congenital central hypoventilation syndrome. *J Appl Physiol.* 2014;116(4):439–50.

29. Verkaeren E, Brion A, Hurbault A et al. Health-related quality of life in young adults with congenital central hypoventilation syndrome due to PHOX2B mutations: A cross-sectional study. *Respir Res.* 2015;16:80

30. Amin R, Moraes TJ, Skitch A, Irwin MS, Meyn S, Witmans M. Diagnostic practices and disease surveillance in Canadian children with congenital central hypoventilation syndrome. *Can Respir J.* 2013;20(3):165–70.

31. Williams P, Wegner E, Ziegler DS. Outcomes in multifocal neuroblastoma as part of the neurocristopathy syndrome. *Pediatrics.* 2014;134(2):e611–6.

32. Diep B, Wang A, Kun S et al. Diaphragm pacing without Ttracheostomy in congenital central hypoventilation syndrome patients. *Respiration.* 2015;89(6):534–8.

33. Preutthipan A, Kuptanon T, Kamalaporn H, Leejakpai A, Nugboon M, Wattanasirichaigoon D. Using non-invasive bi-level positive airway pressure ventilator via tracheostomy in children with congenital central hypoventilation syndrome: Two case reports. *J Med Case Rep.* 2015;9:149.

34. Hirooka K, Kamata K, Horisawa S, Nomura M, Taira T, Ozaki M. Conscious sedation with dexmedetomidine for implantation of a phrenic nerve stimulator in a pediatric case of late-onset congenital central hypoventilation syndrome. *JA Clin Rep.* 2017;3(1):46.

35. Schirwani S, Pysden K, Chetcuti P, Blyth M. Carbamazepine improves apneic episodes in congenital central hypoventilation syndrome (CCHS) with a novel *PHOX2B* exon 1 missense mutation. *J Clin Sleep Med.* 2017;13(11):1359–62.

36. Wang A, Kun S, Diep B, Davidson Ward SL, Keens TG, Perez IA. Obstructive sleep apnea in Ppatients with congenital central hypoventilation syndrome ventilated by diaphragm pacing without tracheostomy. *J Clin Sleep Med.* 2018;14(2):261–4.

37. Hong SY, Hsin YL, Lee IC. An Infant with congenital central hypoventilation syndrome: Transient burst suppression electroencephalogram. *Pediatr Neonatol.* 2016;57(4):357–8.

5

Familial Neuroblastoma

Dongyou Liu

CONTENTS

5.1 Introduction

Deriving from embryonic cells that form the primitive neural crest and give rise to the adrenal medulla and the sympathetic nervous system, neuroblastoma affects various sites along the sympathoadrenal axis, most commonly the adrenal medulla and also the paraspinal sympathetic ganglia between the neck and pelvis (including the spinal cord, neck, chest, abdomen, and pelvis).

With a median age of 18 months at diagnosis, and 90% of cases recognized by 5 years of age, neuroblastoma accounts for 7%–10% of pediatric cancers and 15% of cancer-related deaths in children. Given its heterogeneous biology and diverse location, neuroblastoma exhibits varied clinical behavior and outcome, ranging from spontaneous regression, treatment-resistant progression, and metastasis to death.

While a vast majority of neuroblastoma cases do not have family connection (so-called sporadic neuroblastoma), about 2% of cases demonstrate a family history, form bilateral/multifocal primary tumors, and have an earlier median age of diagnosis, suggesting a hereditary predisposition (so-called familial neuroblastoma or hereditary neuroblastoma).

At the molecular level, sporadic neuroblastoma is attributable to de novo germline or somatic single-nucleotide polymorphisms (SNP) that induce gain or loss of function in a number of genes, each with low relative risk, but acting together to increase the chances of disease occurrence. On the other hand, 59% of familial neuroblastoma cases are associated with autosomal-dominant inheritance of germline, gain-of-function mutations in the anaplastic lymphoma kinase (*ALK*) gene located on chromosome 2p23.2-1. A subset (~10%) of familial cases are due to inheritance of germline, loss-of-function mutations in the paired-like homeobox 2B (*PHOX2B*) gene located on chromosome 4p13. In addition, neuroblastoma may constitute part of cancer predisposition syndromes. Nonetheless, genetic causes for about 15% of familial neuroblastoma cases remain unknown to date [1].

5.2 Biology

During embryonal development, the neural crest containing multipotent neural crest cells is formed in the dorsal part of the neural tube that will later become the brain and the spinal cord. With the ability to undergo epithelial-to-mesenchymal transition and migrate to various body locations, neural crest cells are committed to progressively restricted cell lineages, and eventually differentiate into a diversity of cell types, including the peripheral nervous system (neurons, glial cells, and Schwann cells), endocrine and paraendocrine cells, melanocytes in the epidermis, craniofacial cartilage, bone, and connective tissue [2].

In the trunk region of the neural tube, neural crest cells form the sympathoadrenal lineage, which subsequently differentiates into neurons and glial cells in the sympathetic ganglia, and chromaffin and supportive cells in the adrenal medulla and paraganglia. A well-maintained balance between cell proliferation, migration, differentiation, and death (apoptosis) is crucial for the embryonic development of the sympathoadrenal system, which is composed of the sympathetic nervous system, the adrenal medulla, and functionally related paraganglia [3].

First described by Wright in 1910, neuroblastoma (a name reflecting the association of cells with fibrils that are arranged similarly to neuroblasts) is thought to have resulted from abnormal genetic changes in the sympathoadrenal lineage or its early neural crest–derived precursors, leading to imperfect terminal differentiation, given its common appearance in the adrenal gland medulla and paraspinal sympathetic ganglia [4].

Specifically, SNP in the *BARD1* (chromosome 2q35), *CASC15*, *DDX4* (chromosome 5q11.2), *DUSP12* (chromosome 1q23.3), *KIF15*, *HACE1* (chromosome 6q16), *HSD17B12* (chromosome 11p11.2), *IL31RA* (chromosome 5q11.2), *LIN28B* (chromosome 6q16), *LMO1* (chromosome 11p15.4), *NEFL*, *TP53*, and possible other genes are implicated in the tumorigenesis of sporadic neuroblastoma. Furthermore, a heritable copy number variation (CNV) at chromosome 1q21.1 encompassing *NBPF53* also underlies the pathogenesis of sporadic neuroblastoma [5–9].

The *PHOX2B* gene was shown in 2003 to predispose neuroblastoma in a small proportion of families. Interestingly, loss-of-function *PHOX2B* mutations also occur in congenital central hypoventilation syndrome (CCHS) and Hirschsprung disease (HSCR, or aganglionosis of the colon) (see Chapter 4 in this volume for further details). The ALK gene was found in 2008 to cause 59% of familial neuroblastoma cases [10].

In addition, neuroblastoma may sometimes constitute part of cancer predisposition syndromes, including ROHHAD syndrome (rapid-onset obesity, hypothalamic dysfunction, hypoventilation, and autonomic dysfunction, which overlaps clinically with CCHS but lacks *PHOX2B* mutation), RASopathies (e.g., Costello syndrome [*HRAS* mutation], Noonan syndrome [*PTPN11*, *SOS1*, *KRAS*, *NRAS*, *RAF1*, *BRAF*, *MEK1*, and *RIT1* mutations], neurofibromatosis type 1 [*NF1* mutation], Beckwith–Wiedemann syndrome [BWS, *CDKN1C* mutation], Li–Fraumeni syndrome [LFS, *TP53* mutation], and hereditary pheochromocytoma/paraganglioma syndrome, etc.) [11–13]. However, genetic causes for about 15% of familial neuroblastoma cases have yet to be identified.

5.3 Pathogenesis

Familial neuroblastoma is mainly attributed to highly penetrant mutations in either the *ALK* gene or the *PHOX2B* gene.

Mapped to chromosome 2p23.2-1, the *ALK* gene is composed of 728 kb organized in 29 exons. It generates a transcript of 6222 nucleotides, including coding sequence of 4.9 kb, and encodes a 1620 aa, 177 kDa protein (ALK), which upon N-glycosylation produces a single-chain glycoprotein of 200 kDa, and after extracellular cleavage leads to a truncated protein of 140 kDa.

Being a membrane-associated tyrosine kinase receptor belonging to the insulin receptor family, ALK includes an extracellular region (19-1038, consisting of two MAM [meprin, A-5 protein and receptor protein-tyrosine phosphatase mu; 264–427 and 478–636] and one low-density lipoprotein class A [LDLa, 437–473] domains, and one glycine-rich region), a transmembrane region (1039–1059), an intracytoplasmic domain (1060–1620), and a tyrosine kinase domain (1116–1392).

Expressed mostly in the central and peripheral nervous systems during embryonic development, much less in adults, and the testis, ALK is located on the cell membrane. Fusion between the C-terminal region of ALK and the N-terminal region of nucleophosmin 1 (NPM1) initiates the processes of cell cycle progression, migration and evasion of apoptosis. Depending on the absence or presence of a ligand (ALKAL1 and ALKAL2), ALK may exert both proapoptotic and antiapoptotic functions [14–17].

Germline, gain-of-function, single-base missense mutations in key regulatory regions of the kinase domain of ALK promote ligand-independent signaling through disruption of the auto-inhibited conformation of the kinase, and result in activation of the tyrosine kinase domain that induces constitutive phosphorylation of ALK and increases expression of MYCN [18–30]. Germline ALK mutations are responsible for 59% of familial cases. Among them, R1275Q is present in 45% of familial cases and 33% of sporadic cases, whereas F1174L and F1245V mutations occur exclusively in sporadic cases at frequencies of around 30% and 12%, respectively. Another infrequent mechanism of ligand-independent ALK signaling relates to translocations and large deletions that cause truncation of the extracellular region of ALK. Further, ALK is also activated by somatic mutation or amplification in 10% of sporadic neuroblastoma in the absence of germline mutation.

The *PHOX2B* gene located on chromosome 4p13 encodes a master regulator of autonomic nervous system development. Germline, loss-of-function mutations in the *PHOX2B* gene are responsible for a smaller subset (~10%) of familial neuroblastoma cases. Interestingly, polyalanine repeat expansion mutations (PARM, typically duplication mutations) in the second polyalanine stretch of PHOX2B exon 3 are associated with CCHS, in which neuroblastoma is occasionally present (with 1%–2% neuroblastoma risk); whereas nonpolyalanine repeat expansion mutations (NPARM, typically missense, frameshift, or truncating mutations) in *PHOX2B* exons 1, 2 or 3 are found in patients with both CCHS and HSCR, in which neuroblastoma occurs at increased frequency (with 45% neuroblastoma risk) (see Chapter 4 for details).

Some additional familial cases are associated with germline mutations in RAS pathway and other known cancer predisposition genes (e.g., *TP53* and *CDKN1C*). The genetic basis for 15% of family-related neuroblastoma remains to be elucidated.

5.4 Epidemiology

5.4.1 Prevalence

As the most common solid extracranial malignancy of childhood, neuroblastoma has an incidence of 1 case per 7000 births, and 1 case per 100,000 children under 15 years of age. It accounts for 7%–10% of pediatric cancers and 15% of cancer-related deaths in children. The median age of diagnosis is around 18 months, with 40% of cases recognized before 1 year of age and 90% of cases before 5 years of age. There is a slight predominance of males among patients (male-to-female ratio of 1.3:1.1). Although a vast majority of clinical cases are attributed to sporadic neuroblastoma, which evolves from de novo gene mutations, about 2% are family related.

5.4.2 Inheritance

Making up of about 2% of clinical cases, familial neuroblastoma is inherited in an autosomal dominant fashion. While 59% of familial cases are linked to germline, gain-of-function mutations in *ALK*, a small proportion (10%) of familial cases are due to loss-of-function mutations in *PHOX2B*, which is also involved in the pathogenesis of CCHS and HSCR (refer to Chapter 4 in this book).

5.4.3 Penetrance

Familial neuroblastoma displays variable penetrance and expressivity. For example, R1275Q mutation in the *ALK* gene has a near complete penetrance in affected families, while G1128A mutation is more weakly activating and correlates to a 25% likelihood of developing neuroblastoma. The families harboring *PHOX2B* mutations also have variable penetrance. Like many other cancer predisposition syndromes, familial neuroblastoma demonstrates an earlier age of onset and multiple primary tumor sites.

5.5 Clinical Features

Clinical presentations of neuroblastoma are highly variable and mostly nonspecific, including fever (25% of cases), abdominal pain (22%), abdominal mass (21%), bone pain (19%), abdominal distension (18%), weight loss (15%), neurologic change (11%), bruising (6%), nodule on the skull (4%), adenomegaly (3%), systemic arterial hypertension (2%), and others (40%) [31]. In addition, hypertension may appear as a result of renal artery compression that stimulates the renin–angiotensin system and hyperaldosteronism, or increased catecholamine secretion. Seizures (in the context of hypertensive encephalopathy) and heart failure may be also observed. Some patients (13%) may be asymptomatic, from whom a tumor mass is found in the abdominal, mediastinum, and paravertebral regions during examination for other complaints (e.g., diffuse bone pain, anemia, pancytopenia, irritability, ocular proptosis, periorbitary bruising, and nodules in the skull due to distant metastases). The intense bone pain can sometimes prevent the child from moving freely [32,33].

About 5% of patients with neuroblastoma may display clinical symptoms suggestive of paraneoplastic syndrome, including ataxia (lack of coordination with regard to muscle movements), myoclonus (sudden muscle contractions), opsoclonus (involuntary, multidirectional, uncoordinated, and hyperkinetic ocular movements), irritability, intractable diarrhea (when neoplastic cells produce vasoactive intestinal peptide), Horner syndrome (unilateral ptosis, miosis, and anhidrosis, usually associated with tumor in the upper thoracic or cervical region), and Pepper syndrome (respiratory failure caused by liver metastasis).

The most common sites of primary neuroblastoma are the abdomen (48% adrenal gland, 25% retroperitoneum), thorax (16%), neck, chest, and pelvis. The most common sites of metastatic neuroblastoma are the bone marrow (37%), bones (33%), lymph nodes (13%), liver (10%), brain (4%), skin (0.4%), and other locations (3%). In 33% of cases, metastases occur in more than one site.

5.6 Diagnosis

Diagnosis of neuroblastoma involves medical history review (family history, earlier median age of diagnosis) physical examination (e.g., blood pressure; unusual mass in the abdominal, mediastinum, and paravertebral regions; bilateral/multifocal primary tumors), laboratory testing (e.g., gas chromatography and mass spectroscopy [GC-MS] detection of urinary catecholamine metabolites including vanillylmandelic acid [VMA] and homovanillic acid [HVA] in 90% of cases; metaiodobenzylguanidine [MIBG] mapping), imaging study (e.g., chest radiograph, abdominal ultrasound), histopathology (of tissue/bone marrow biopsy, bone marrow infiltration by tumor), and molecular confirmation (including cytogenetic evaluation of neoplastic cells; identification of *ALK* and *PHOX2B* mutations) [34].

Macroscopically, neuroblastoma in the paravertebral region (22%) may extend through the neural foramen, compress the spinal cord, and trigger neurological symptoms (e.g., radicular pain, paraplegia, and fecal/urinary incontinence).

Microscopically, neuroblastoma shows small, round, blue tumor cells with sparse cytoplasm and hyperchromatic nuclei (Figure 5.1) [35]. While neuroblastoma in the most aggressive form consists entirely of immature neural precursor cells, ganglioneuroma is composed entirely of mature neural tissue.

Molecular genetic identification of *ALK* and *PHOX2B* mutations provides a valuable means of confirming familial neuroblastoma. In addition, structural CNVs may be used to determine the aggressiveness of sporadic neuroblastoma, as localized neuroblastoma harboring a low number of structural CNV appears less aggressive than metastatic neuroblastoma containing a high number of structural CNV.

Based on localization, lymph node involvement, and metastasis formation, neuroblastoma is staged by the International Neuroblastoma Staging System (INSS) into five groups (1–4, and 4S). Of these, stage 1 specifies localized tumor that is completely resected; stage 2 is localized tumor that may or may not be completely resected; stage 3 refers to unilateral tumor that extends to the midline; stage 4 indicates primary tumor that metastasizes to distant lymph nodes, bones, bone marrow, liver, skin, and other organs (except as defined in stage 4S); and stage 4S (4 special) highlights primary localized tumor (stages 1–2) showing limited dissemination to the skin, liver, and bone marrow, which often undergoes natural involution and regression after minimal or no medical intervention (usually occurring in children under 1 year of age). Interestingly, the relative percentages of patients with stages 1, 2, 3, 4, and 4S disease are 15%, 5%, 29%, 46%, and 5%, respectively.

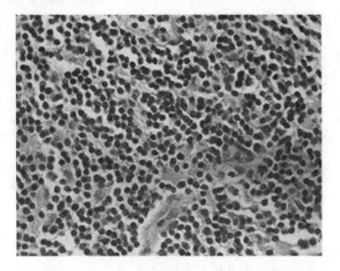

FIGURE 5.1 Photomicrograph of primary neuroblastoma showing the presence of small, round, blue tumor cells with sparse cytoplasm and hyperchromatic nuclei (HE staining). (Photo credit: Snapkov I et al. *BMC Cancer.* 2016;16:490.)

Another widely used grading system is Shimada grading, in which the level of tumor differentiation (i.e., the number of neuroblast cells in tumor tissue) indicates the level of aggressiveness of the tumor. Neuroblast cells are small to medium in size and have indiscernible to thin cytoplasm and vaguely defined cytoplasmic borders.

To further assist disease management, neuroblastoma patients may be stratified into low-, intermediate-, and high-risk groups, taking account of stage, age, *MYCN* status, histology, and chromosomal ploidy. In general, patients over 1 year of age with a metastatic disease at the time diagnosis and an unfavorable histopathological classification tend to have a poor prognosis.

5.7 Treatment

Current treatment options for patients with neuroblastoma comprise surgical resection, myeloablation, radiotherapy, intensive biologic/immunotherapy, and autologous stem cell rescue. Factors influencing treatment choices include patient age, histology, genetic abnormalities, and stage of disease, based on which low-, intermediate-, and high-risk groups are assigned.

Localized tumors in young patients have a tendency for spontaneous differentiation and regression, as the sympathoadrenal tissue undergoes a natural regression from 18 months of age. Surgical resection may be considered for patients with localized disease (stages 1–3), who do not show spontaneous regression. Preoperative chemotherapy enhances complete resection possibility for neuroblastomas that involve renal vessels, celiac trunk, or superior mesenteric artery. A combination of chemotherapy (e.g., cyclophosphamide and topotecan or doxorubicin, dexrazoxane, and vincristine), support care, and bone marrow transplantation help minimize surgical complication (e.g., hemorrhage) and improve overall survival.

However, surgical resection appears to have little benefit in high-risk patients (which are defined by older age, unfavorable histopathology, loss of heterozygosity for chromosome 1p or 11q, and amplification of *MYCN*). While high-risk patients are responsive to chemotherapy initially, many relapse and succumb to therapy-resistant disease.

A number of ALK inhibitors are being evaluated for neuroblastoma harboring ALK mutations. Crizotinib is a first-generation ALK inhibitor with the ability to induce complete and sustained tumor regression in R1275Q mutants, but it provides only partial growth inhibition to F1174L mutants. Second-generation ALK inhibitors (e.g., ceritinib, alectinib, and brigatinib) show superior efficacy against both crizotinib-naïve and resistant mutants, including ALK F1174L mutants, and help overcome de novo resistance to crizotinib. Third-generation ALK inhibitors (e.g., lorlatinib) are also effective against ALK inhibitor-resistant mutants including F1174L, F1245C, and R1275Q mutants, and they exhibit improved potency over crizotinib- and ceritinib-resistant ALK F1174L mutants [10,36,37].

5.8 Prognosis and Prevention

Neuroblastoma is a highly aggressive tumor, as metastasis to bone marrow, bone, lymph nodes, liver, and brain occurs in ~50% of neuroblastoma patients at diagnosis. It represents the second most deadly childhood cancer, with death caused by disease progression (72% of cases), toxicity from chemotherapy (23%), and surgical complications (4%) [38].

Prognosis for neuroblastoma patients is influenced by age of the patient, location of the tumor, surgical staging, and microscopic grading. Patients with low- and intermediate-risk disease are highly curable and have a 5-year survival rate of 95%. However, those with high-risk disease experience frequent relapse (40%–50% of cases) and have a 5-year survival rate of 40%–50%. Indeed, patients harboring highly penetrant, heritable *ALK* or *PHOX2B* (NPARM) mutations have a significant risk (45%–50%) to develop one or more tumors [39,40].

Surveillance for familial neuroblastoma may cover individuals with LFS (germline TP53-R337H mutations), BWS (germline CDKN1C mutations), Costello syndrome (HRAS mutations), and those with a strong family history of neuroblastoma or clearly bilateral/multifocal neuroblastoma [41]. Surveillance, consisting of ultrasonography and urinary catecholamines, may be conducted more frequently up to age 6 years, less frequently from ages 6 to 10, and stop afterward.

5.9 Conclusion

Neuroblastoma is an embryonal tumor of the sympathoadrenal system that more commonly affects young children than other population groups. Evolving from neural crest cells that experience imperfect terminal differentiation, neuroblastoma is distributed in various sites along the sympathoadrenal axis, including the adrenal medulla and the paraspinal ganglia (e.g., the spinal cord, neck, chest, abdomen, and pelvis). Neuroblastoma is noted for its diverse clinical behaviors (from spontaneous regression to fatal outcome) and high aggression, with 50% of patients showing metastases at diagnosis, and a mortality rate of more than 60% despite multimodal therapy. Making up a majority of clinical cases, sporadic neuroblastoma shows no family connection and appears to result from SNP in an expanding set of genes. By contrast, familial (hereditary) neuroblastoma accounts for about 2% of clinical cases and is mainly associated with gain-of-function mutations in the *ALK* gene and loss-of-function mutations in the *PHOX2B* gene [42]. Considering the current lack of specific treatments for neuroblastoma sufferers, it is important to conduct further research on the molecular tumorigenesis in order to develop innovative and highly effective therapies for this pediatric malignancy.

REFERENCES

1. Tolbert VP, Coggins GE, Maris JM. Genetic susceptibility to neuroblastoma. *Curr Opin Genet Dev.* 2017;42:81–90.
2. Tafavogh S, Catchpoole DR, Kennedy PJ. Cellular quantitative analysis of neuroblastoma tumor and splitting overlapping cells. *BMC Bioinformatics.* 2014;15:272.
3. Becker J, Wilting J. WNT signaling, the development of the sympathoadrenal-paraganglionic system and neuroblastoma. *Cell Mol Life Sci.* 2018;75(6):1057–70.
4. Schleiermacher G, Janoueix-Lerosey I, Delattre O. Recent insights into the biology of neuroblastoma. *Int J Cancer.* 2014;135(10):2249–61.

5. Bettinsoli P, Ferrari-Toninelli G, Bonini SA, Prandelli C, Memo M. Notch ligand delta-like 1 as a novel molecular target in childhood neuroblastoma. *BMC Cancer.* 2017;17(1):352.

6. Chikaraishi K, Takenobu H, Sugino RP et al. CFC1 is a cancer stemness-regulating factor in neuroblastoma. *Oncotarget.* 2017;8(28):45046–59.

7. Tonini GP. Growth, progression and chromosome instability of neuroblastoma: A new scenario of tumorigenesis? *BMC Cancer.* 2017;17(1):20.

8. Yuan LQ, Wang JH, Zhu K et al. A highly malignant case of neuroblastoma with substantial increase of single-nucleotide variants and normal mismatch repair system: A case report. *Medicine (Baltimore).* 2017;96(50):e8845.

9. Topcagic J, Feldman R, Ghazalpour A, Swensen J, Gatalica Z, Vranic S. Comprehensive molecular profiling of advanced/metastatic olfactory neuroblastomas. *PLOS ONE.* 2018;13(1):e0191244.

10. Trigg RM, Turner SD. ALK in Neuroblastoma: Biological and therapeutic implications. *Cancers (Basel).* 2018;10(4). pii: E113.

11. Capasso M, McDaniel LD, Cimmino F et al. The functional variant rs34330 of CDKN1B is associated with risk of neuroblastoma. *J Cell Mol Med.* 2017;21(12):3224–30.

12. Kamihara J, Bourdeaut F, Foulkes WD et al. Retinoblastoma and neuroblastoma predisposition and surveillance. *Clin Cancer Res.* 2017;23(13):e98–106.

13. McDaniel LD, Conkrite KL, Chang X et al. Common variants upstream of MLF1 at 3q25 and within CPZ at 4p16 associated with neuroblastoma. *PLOS Genet.* 2017;13(5):e1006787.

14. Liu M, Inoue K, Leng T, Zhou A, Guo S, Xiong ZG. ASIC1 promotes differentiation of neuroblastoma by negatively regulating Notch signaling pathway. *Oncotarget.* 2017;8(5):8283–93.

15. Wilkaniec A, Gąssowska M, Czapski GA, Cieślik M, Sulkowski G, Adamczyk A. P2X7 receptor-pannexin 1 interaction mediates extracellular alpha-synuclein-induced ATP release in neuroblastoma SH-SY5Y cells. *Purinergic Signal.* 2017;13(3):347–61.

16. Fadeev A, Mendoza-Garcia P, Irion U et al. ALKALs are *in vivo* ligands for ALK family receptor tyrosine kinases in the neural crest and derived cells. *Proc Natl Acad Sci U S A.* 2018;115(4):E630–8.

17. Tian X, Zhou D, Chen L et al. Polo-like kinase 4 mediates epithelial-mesenchymal transition in neuroblastoma via PI3K/Akt signaling pathway. *Cell Death Dis.* 2018;9(2):54.

18. Huang M, Weiss WA. Neuroblastoma and MYCN. *Cold Spring Harb Perspect Med.* 2013;3(10):a014415.

19. Gnanaprakasam JN, Wang R. MYC in regulating immunity: Metabolism and beyond. *Genes (Basel).* 2017;8(3). pii: E88.

20. Hadjidaniel MD, Muthugounder S, Hung LT et al. Tumor-associated macrophages promote neuroblastoma via STAT3 phosphorylation and up-regulation of c-MYC. *Oncotarget.* 2017;8(53):91516–29.

21. Kushner BH, LaQuaglia MP, Modak S et al. *MYCN*-amplified stage 2/3 neuroblastoma: Excellent survival in the era of anti-GD2 immunotherapy. *Oncotarget.* 2017;8(56):95293–302.

22. Niemas-Teshiba R, Matsuno R, Wang LL et al. MYC-family protein overexpression and prominent nucleolar formation represent prognostic indicators and potential therapeutic targets for aggressive high-MKI neuroblastomas: A report from the children's oncology group. *Oncotarget.* 2017;9(5):6416–32.

23. Ruiz-Pérez MV, Henley AB, Arsenian-Henriksson M. The MYCN protein in health and disease. *Genes (Basel).* 2017;8(4). pii: E113.

24. Yang XH, Tang F, Shin J, Cunningham JM. Incorporating genomic, transcriptomic and clinical data: A prognostic and stem cell-like MYC and PRC imbalance in high-risk neuroblastoma. *BMC Syst Biol.* 2017;11(Suppl 5):92.

25. Agarwal S, Milazzo G, Rajapakshe K et al. MYCN acts as a direct co-regulator of p53 in MYCN amplified neuroblastoma. *Oncotarget.* 2018;9(29):20323–38.

26. Chen L, Alexe G, Dharia NV et al. CRISPR-Cas9 screen reveals a MYCN-amplified neuroblastoma dependency on EZH2. *J Clin Invest.* 2018;128(1):446–62.

27. Chen Q, Deng R, Zhao X et al. Sumoylation of EphB1 suppresses neuroblastoma tumorigenesis via inhibiting PKCγ activation. *Cell Physiol Biochem.* 2018;45(6):2283–92.

28. Wang T, Liu L, Chen X et al. MYCN drives glutaminolysis in neuroblastoma and confers sensitivity to an ROS augmenting agent. *Cell Death Dis.* 2018;9(2):220.

29. Zammit V, Baron B, Ayers D. MiRNA influences in neuroblast modulation: An introspective analysis. *Genes (Basel).* 2018;9(1). pii: E26.

30. Zhong ZY, Shi BJ, Zhou H, Wang WB. CD133 expression and MYCN amplification induce chemoresistance and reduce average survival time in pediatric neuroblastoma. *J Int Med Res.* 2018;46(3):1209–20.

31. Kume A, Morikawa T, Ogawa M, Yamashita A, Yamaguchi S, Fukayama M. Congenital neuroblastoma with placental involvement. *Int J Clin Exp Pathol.* 2014;7(11):8198–204

32. Djougarian A, Kodsi S. Hypertensive retinopathy as the initial presentation of neuroblastoma. *Am J Ophthalmol Case Rep.* 2017;7:123–5.

33. Huang MD, Hsu LS, Chuang HC, Lin WY, Lin WH, Yen CW, Chen ML. Adult renal neuroblastoma: A case report and literature review. *Medicine (Baltimore).* 2018;97(14):e0345.

34. Davis J, Novotny N, Macknis J, Alpay-Savasan Z, Goncalves LF. Diagnosis of neonatal neuroblastoma with postmortem magnetic resonance imaging. *Radiol Case Rep.* 2016;12(1):191–5.

35. Snapkov I, Öqvist CO, Figenschau Y, Kogner P, Johnsen JI, Sveinbjørnsson B. The role of formyl peptide receptor 1 (FPR1) in neuroblastoma tumorigenesis. *BMC Cancer.* 2016;16:490.

36. Matthay KK, George RE, Yu AL. Promising therapeutic targets in neuroblastoma. *Clin Cancer Res.* 2012;18(10):2740–53.

37. Valter K, Zhivotovsky B, Gogvadze V. Cell death-based treatment of neuroblastoma. *Cell Death Dis.* 2018;9(2):113.

38. Lucena JN, Alves MTS, Abib SCV, Souza GO, Neves RPC, Caran EMM. Clinical and epidemiological characteristics and survival outcomes of children with neuroblastoma: 21 years of experience at the Instituto de Ooncologia Pediatrica, in Sao Paulo, Brazil. *Rev Paul Pediatr.* 2018;36(3):254–60.

39. Applebaum MA, Vaksman Z, Lee SM et al. Neuroblastoma survivors are at increased risk for second malignancies: A report from the International Neuroblastoma Risk Group Project. *Eur J Cancer.* 2017;72:177–85.

40. Alshareef A, Irwin MS, Gupta N et al. The absence of a novel intron 19-retaining *ALK* transcript (*ALK*-I19) and *MYCN* amplification correlates with an excellent clinical outcome in neuroblastoma patients. *Oncotarget.* 2018;9(12):10698–713.

41. Buhagiar A, Ayers D. Chemoresistance, cancer stem cells, and miRNA influences: The case for neuroblastoma. *Anal Cell Pathol (Amst).* 2015;2015:150634.

42. Barr EK, Applebaum MA. Genetic predisposition to neuroblastoma. *Children (Basel).* 2018;5(9):119.

6

Pleuropulmonary Blastoma

Raúl Barrera-Rodríguez and Carlos Pérez-Malagón

CONTENTS

6.1 Introduction

Pleuropulmonary blastoma (PPB) is an extremely rare and aggressive mesenchymal neoplasm of the lung which affects children from the newborn period to 12 years of age (mean age of 2–5 years). Most cases are diagnosed before 4 years of age with no predilection for gender [1–3]. It is also known as pulmonary blastoma of childhood or pneumoblastoma.

The term *blastoma* refers to a malignant neoplasm that occurs in immature or developing cells, suggesting a "primitive," almost embryo-like, appearance. Therefore, it is considered as a dysontogenic neoplasm [4] and can be an analog neoplasm to Wilms tumor, neuroblastoma, pancreatoblastoma, retinoblastoma, or hepatoblastoma [3,5]. PPB has also been known with different names, including embryonal rhabdomyosarcoma (arising within a congenital bronchogenic cyst), rhabdomyosarcoma (arising in congenital cystic adenomatoid malformation [CCAM]), pulmonary sarcoma (arising in mesenchymal cystic hamartoma), and pulmonary blastoma (associated with cystic lung disease of childhood) [3].

As one of the three subtypes of pulmonary blastoma (i.e., monophasic pulmonary blastoma [monophasic PB], classic biphasic pulmonary blastoma [CBPB], and PPB), PPB is purely mesenchymal in phenotype without malignant epithelial elements. In contrast, monophasic PB comprises epithelial malignant component only and is often referred to as well-differentiated fetal adenocarcinoma (WDFA), and CBPB has a biphasic (epithelial and mesenchymal) histologic pattern. While PPB usually affects children younger than 5 years of age, monophasic PB and CBPB affect adults (mean age of 43 years). In addition, PPB is linked to heterozygous mutation in the DICER1 gene on chromosome 14q32, while monophasic PB and CBPB do not contain such mutation.

6.2 Biology

Representing the most common primary malignancy of the lungs during childhood, PPB arises from either the tissue covering the lungs ("pleura") or in the lung tissue itself ("pulmonary") [6,7]. Commonly located in the lung periphery, involving the visceral and parietal pleura, PPB is also found in mediastinum, thoracic great vessels, regional lymph nodes, and diaphragm. PPB can occasionally produce distant metastases mainly in the brain, bone, liver, pancreas, kidney, and adrenal glands [5,8].

First described in 1988 by Manivel et al. [7] from a series of 11 intrathoracic pulmonary neoplasms in young children, PPB is now considered as part of an inherited cancer syndrome (PPB familial tumor and dysplasia syndrome [PPBFTDS], which also includes multinodular goiter-1 with or without Sertoli—Leydig cell tumors, among others). Following the suggestion of John Priest, the International Pleuropulmonary Blastoma Registry (IPPBR: http://www.ppbregistry.org) was established to facilitate research on the genetic basis for this familial cancer syndrome.

One of the early observations for the distinction of this syndrome was that in children, the PPB was commonly multifocal and could occur in young relatives. Many cases of heritable PPB reported to date indicate a clear inherited predisposition to various tumors and cancer-like conditions in the PPB patients or in their relatives. Current data collected by the IPPBR demonstrates that in approximately 35%–40% of PPB cases, the patient or young family members (e.g., siblings, parents, grandparents, uncles, aunts, cousins, etc.) are prone to develop both cancerous and noncancerous (benign) tumors.

6.3 Pathogenesis

Studies done by Harris et al. in 2006 [9] showed that the conditional loss of a gene called DICER1 in the developing mouse lung results in cystic airway dilation, disruption of branching morphogenesis, and mesenchymal expansion that resembles the early stage of PPB. In April 2009, Hill et al. [10] reported the presence of genetic mutations in DICER1 gene of 11 families with PPB, lung cysts, cystic nephroma, and/or rhabdomyosarcoma and concluded that germline DICER1 mutations may contribute to the process of PPB development. Then, the existence of such PPB patients was attributed to mosaic DICER1 mutations or to biallelic DICER1 mutations in tumor tissue only [11–13]. In 2011, these and other observations led by Slate et al. [14] suggested that the phenotypes associated with the constitutional (germline) DICER1 mutations should be collectively referred to as "DICER1 syndrome" in view of their potentially common pathogenesis.

Using a family-based linkage study in four families with inherited predisposition to PPB, Hill et al. in 2009 [10] mapped the PPB locus to chromosome 14q31.1-q32 in a region of 7-Mb flanked by rs12886750 and rs8008246. Then, by gene sequence analysis, it was demonstrated that germline mutations in DICER1 gene (OMIM, 601200; NCBI Gene 23405) are associated with susceptibility to familial PPB. The gene encodes a ~218 kDa cytoplasmic endonuclease that belongs to the ribonuclease III (RNase IIIb; PDB Structure, 2EB1) protein family, which is required to cleave the precursor microRNAs (pre-miRNAs) into ~22 nucleotide "mature" effector miRNAs and small interfering RNAs (siRNAs), both are the crucial modulators of gene expression at the post-transcriptional level [15–17]. The miRNAs negatively regulate gene expression and appear to play critical roles in stem cell proliferation, organogenesis, cell cycle progression, regulatory control of some lymphocyte subsets, and oncogenesis (Figure 6.1) [14,18].

The genetic mutation in DICER1 affects all cells of the body and is related with different conditions found in PPB patients and relatives. The majority of DICER1 mutations reported so far are biallelic germline mutations which appear to confer a loss of function. Messinger et al. in 2015 [18] sequenced 97 patients with PPB for DICER1 mutations and found that germline DICER1 mutation frequency was nearly 66%, with all mutations being heterozygous nonsense or frameshift mutations. Seki et al. [12] reported that five of eight Japanese patients with PPB carried germline DICER1 mutations, commonly biallelic. Slade et al. [14] reported that constitutional DICER1 mutations were identified in 11 of 14 patients with PPB.

Somatic mutations are also heavily concentrated in "hotspots" inserted between exon 24 and exon 25 [19], all of which are the base pairs that encode the metal-ion-binding residues or adjacent to such sites of the RNase IIIb domain which is the key component for the production of mature miRNAs (Figure 6.2). Variation in five hotspot codons are observed in virtually all DICER1-associated tumors; mosaicism for missense variants in these hotspot codons confers a more severe phenotype.

The importance of mutated DICER1 in the development of diseases has been also shown in several genetic models. For example, DICER1-*null* mouse embryos survive until embryonic day 7.5 but then die before axis formation, whereas murine DICER1-*null* embryonic stem cells proliferate slowly and are defective in differentiation, suggesting some aspects of differentiation (and stem cell proliferation) depend on this protein. DICER1-*null zebrafish* showed a slow growth and developmental delay before dying at 14–21 days post fertilization, whereas mutant DICER1-*null* embryos have abnormal morphogenesis of heart, brain, and other organs. In a mouse model in which a nonfunctional allele of DICER1 is lost in lung epithelium at day 10.5 in embryogenesis, the resultant phenotype shows a marked cystic dilation of the lung airspaces and mesenchymal expansion which resembles the early stage of PPB. In addition, the DICER1-*null* lung tissues showed increased expression of FGF-10 protein with an expanded distribution in lung mesenchyme.

There is a connection between "cysts" (air-filled lucencies or low-attenuating areas with a thin wall [usually <2 mm]) in the lung and the later development of PPB in these spaces. The reason such spaces are slightly prone to development of PPB is unknown. The early stage of disease is characterized by cystic expansion of the lung airspaces lined by benign-appearing epithelium. Mesenchymal cells susceptible to malignant transformation reside within the cyst walls and form a dense layer beneath the lining epithelium known as the cambium layer. Sarcomatous overgrowth by the mesenchymal cells growing as solid masses dictates a clinical outcome for children with this tumor; however, not all cystic PPBs are destined to progress to sarcomas. The natural history of the tumor suggests a multistep genetic pathogenesis. Thus, many children have lung cysts without ever developing PPB. This problem raises questions for physicians and families about how to treat children with lung cysts, and there is no consensus on this issue.

6.4 Epidemiology

PPB is an exceptional intrathoracic tumor that affects 1 in 250,000 live births and accounts for approximately 15% of all primary pediatric pulmonary tumors [18]. It is estimated that 30 to 40 children worldwide and 15 to 25 children in the United States are diagnosed with PPB annually.

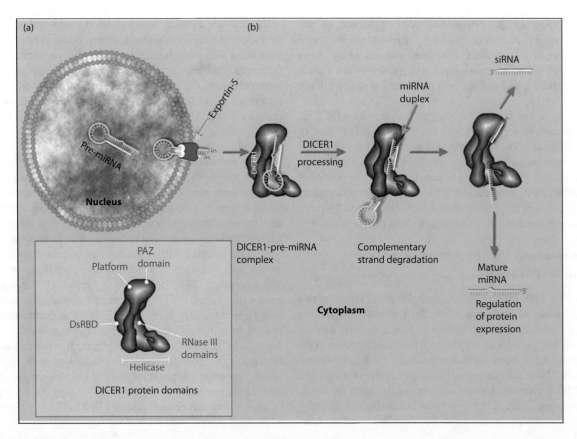

FIGURE 6.1 Schematic diagram of DICER1 in microRNA (miRNA) processing. (a) Premature double-stranded miRNAs (pre-miRNA) hairpins are exported into the cytoplasm through nuclear pores (exportin). (b) Cytoplasmic DICER1 protein recognizes short hairpin pre-miRNAs precursors to produce mature fragments of 21–23 nucleotides producing, respectively, small interfering RNAs (siRNA) and mature miRNAs.

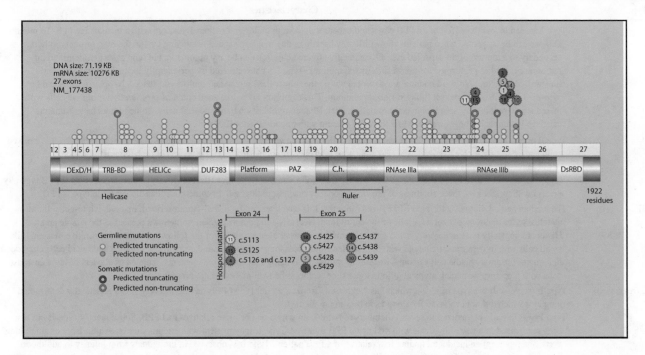

FIGURE 6.2 Schematic diagram of the DICER1 gene structure and domains in the unfolded protein. Yellow-filled circles represent nonsense and frameshift (truncating) substitutions, orange-filled circles represent missense (non-truncating) substitutions and color-filled polygons represent somatic "hotspot" substitutions affecting the RNase IIIb domain. The spectrum of germline mutations is dominated by truncating, loss-of-function (LOF) mutations. These are mainly single-nucleotide substitutions that produce new stop codons, and small insertions or deletions within exons that shift reading frame. Somatic DICER1 "hotspot" mutations occur in association with germline mutations, particularly LOF mutations. Mutant DICER1 proteins resulting from somatic mutations generally have reduced RNase IIIb activity which results in defective miRNA processing.

Until 2015, over 350 cases had been registered by the Pleuropulmonary Blastoma Registry (http://www.ppbregistry.org). The tumor affects mainly children with age ranges from 1 month to 12 years, but most cases are diagnosed before 4 years of age. It can be found prenatally or be present in older children and young adults.

Interestingly, the majority of obligate carriers with DICER1 mutations are phenotypically normal, indicating that loss of one DICER1 allele is compatible with normal development and insufficient for tumor formation. In addition, there is no evidence to show that DICER1 mutation is correlated with the gender of patients with PPB, even though it has been shown that female carriers seem to be more likely to be affected than males by other phenotypes of the DICER1 syndrome, such as thyroid disease.

6.5 Clinical Features

Patients with PPB are stratified in four different "types" which further describe different diseases, indicate disease progression, and predict clinical outcome (Table 6.1). This is because there is no strict dividing line between types, and there can be points along a biological continuum or spectrum. Although the stratification of PPB in a child is usually suggested by the appearance on x-rays, the type of PPB is properly determined after surgery, when the tumor can be examined through biopsy samples by histological studies.

Clinical presentation depends on the age and the type of PPB. At the time of diagnosis, almost one-fourth of cases are associated with other tumors such as cystic nephromas, ovarian teratomas, multiple intestinal polyps, thyroid malignancies, or medulloblastomas [3]. In the first year of life and even prenatally, the main clinical symptom is shortness of breath with or without pneumothorax. Although large lung cysts in the fetus can theoretically impair lung development, such occurrence appears to be uncommon.

According to the IPPBR, prenatal ultrasound examination has detected lung cysts in seven cases at 31–35 weeks' gestational age. Because large lung cysts can cause respiratory distress in newborns, it is recommended that prenatal identification of lung cysts lead to consultation with specialists in high-risk obstetrics and fetal medicine to monitor the pregnancy and manage the delivery.

In older children with advanced disease, the first symptoms are dyspnea which can be accompanied by cough, fever, malaise, and sometimes abdominal or chest pain. Also, they may have nonspecific symptoms, such as headache, vomiting, and weight loss. Physical findings may include absent or diminished breath sounds over the affected hemithorax. Occasionally, patients might develop pneumothorax. This clinical presentation can be commonly mistaken for respiratory infections such as pneumonia or a spontaneous pneumothorax of unknown cause. In some cases, it can be accidentally diagnosed during the investigation of other nonrelated clinical diseases [8,22,23].

TABLE 6.1

Pleuropulmonary Blastoma Types

Type	Characteristics
I	Type I is seen in less than 15% of all PPB cases and becomes evident in children less than 1 year old with a median age of 8–10 months. Disease can be suspected when patients show difficulty breathing due to a large space-occupying cyst in the lung or pneumothorax secondary to a rupture of the air-filled cyst. Occasionally affected children can be asymptomatic, but lung cysts are identified on radiographic studies performed for non-respiratory symptoms. Type I is characterized by peripherally located, multicystic, and thin-walled structure. No solid nodules are observed within the thin-walled septae and multicystic spaces. Neoplastic cells can be identified beneath a benign cyst lining surface epithelium. This cambium layer-like zone contains proliferating primitive cells with mixed small, rounded, or spindled hyperchromatic nuclei. Rhabdomyoblasts are seen occasionally. Some tumors have immature cartilage within their fibrous septae. More than half present with pneumothorax [20].
II	Type II occurs in 40%–50% of all PPB cases, affecting older children with a median age of 35 months at diagnosis. Typically, it is a mixed solid and cystic tumor characterized by varied amount of thickened or nodule-like areas composed of rhabdomyosarcomatous or blastematous components. When progression occurs, the mesenchymal cells in a type I PPB expand and overgrow the cyst septa, replacing the cyst with a cystic and solid sarcoma or a purely solid sarcoma (type II or III, respectively). Patients with type II PPB typically present with weight loss, fever, shortness of breath, and opacity on chest radiograph. Pneumothorax can also occur in type II PPB.
III	The median ages at diagnosis for type III PPB is 41 months. It is characterized by a well-circumscribed, mucoid, white-tan solid mass attached to the pleura and involves a lobe or the entire lung. Necrosis and hemorrhage are sometimes present in the friable areas. Histopathologically is an apparently heterogeneous tumor composed of one or more of the following elements: primitive blastema-like small cells with hyperchromatic nuclei with high nuclear-to-cytoplasmic ratio and abundant mitoses; spindled and ovoid cells embedded in a myxoid stroma; spindle cell sarcoma; nodules of immature or malignant chondroid elements. Isolated or clusters of large anaplastic cells with pleomorphic nuclei, atypical mitotic figures, or eosinophilic hyaline bodies are present.
Ir (regressed)	Type-Ir (regressed) is a type of lesion which has "regressed" from an earlier type I PPB or, alternatively, it is a genetically determined lung cyst which did not evolve so far along as to become malignant. Type-Ir was initially recognized in several members of families in which one or more relatives had PPB. Subsequently, type-Ir has also been recognized in a few individuals with no known PPB relatives. Type-Ir may present with pneumothorax, exhibit large or small lung cysts, and represent an incidental finding in a relative of a PPB patient. Type-Ir is observed in individuals from infancy to adulthood. The cysts of type-Ir are different from other lung cyst pathologies included in congenital pulmonary airway malformation or congenital cystic adenomatoid malformation categories [4].

Source: Yeh A, Edelman MC. *Atlas Genet Cytogenet Oncol Haematol.* 2011;15:374–7.

Note: Evidence based on patient age at diagnosis as well as recurrence with more advanced type suggests that type I PPB may evolve into the more malignant type II or type III. In general, this occurs when a PPB patient is not cured, and the tumor becomes more aggressive over time, leading to a transition from type I to II or III. Types II and III may be associated with metastasis, with the brain being the most common metastatic site [21].

6.6 Diagnosis

6.6.1 Imaging

Initial diagnostic workup includes a clinical history of the child and kindred as well as a chest x-ray. It is particularly important that when no clinical or chest x-ray improvement is observed in a treated patient, a computed tomography (CT) scan must be performed [21]. CT is the preferred imaging technique to classify PPB types because it can detect air-filled cysts/tumors and determine the site(s) of disease. In contrast, chest x-ray images are insensitive to small cystic changes and cannot reliably distinguish between masses and pneumonia consolidations. If tumors arise primarily in the chest wall or mediastinum, other diagnoses should be considered (see Section 6.6.3).

The most common PPB radiological image is a unilateral, or rarely bilateral, pulmonary cysts; there may also be a pneumothorax and, in some cases, an intrathoracic mass. Right hemithorax has been reported as the most affected side [4,21]. Septal thickening or intracystic mass is a clue.

IPPBR reports show that 49% of the cases of PPB have multiple lesions in the lung, 33% being bilateral cysts so that the complete removal of cysts is not possible. The most common affected lobe is the left upper followed by left lower lobe, right upper lobe, right middle lobe, and right lower lobe, in equal proportion. Some cases can develop a grossly bloody pleural effusion, or it can also be serosanguineous, cloudy, or purulent. Through cytology studies of pleural fluid and core needle biopsy of pleura, malignant cells can be detected [5]. Representative examples of CT-scan of PPB in patients with the DICER1 syndrome are shown in Figures 6.3 through 6.5.

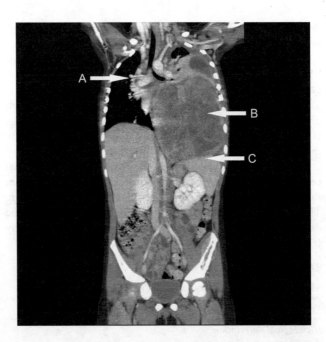

FIGURE 6.4 Pleuropulmonary blastoma type III in a patient with DICER1 syndrome. This coronal CT scan image shows mediastinal shift to the right side (arrow A) due to the large pleuropulmonary blastoma in the left hemithorax with hyperdense areas in its interior (arrow B). This tumor also lowers greatly the left hemidiaphragm (arrow C). (Photography courtesy of Gretchen Williams, BS, CCRP from the International PPB Registry.)

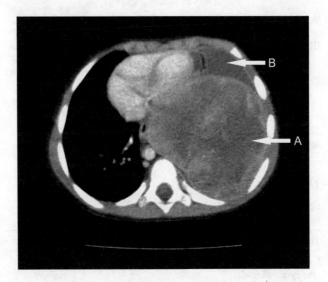

FIGURE 6.5 Pleuropulmonary blastoma type III in a patient with DICER1 syndrome. This transverse axial CT scan image shows a large well circumscribed pleuropulmonary blastoma in the left hemithorax characterized by different hyperdensities in its interior (arrow A). This size produces mediastinal shift to the right side. A small anterior pleural effusion is observed (arrow B). (Photography courtesy of Gretchen Williams, BS, CCRP from the International PPB Registry.)

6.6.2 Histopathology

Histopathologic examination of a biopsy or surgically excised cystic and/or solid lung tissue plays a critical role in the diagnosis of PPB. Usually located in the periphery of the lung, PPB may be also found in extrapulmonary sites, such as the mediastinum,

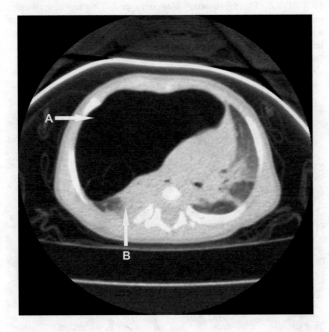

FIGURE 6.3 Pleuropulmonary blastoma type I in a patient with DICER1 syndrome. This transverse axial CT scan image shows a large pneumothorax (arrow A) in the right hemithorax with passive atelectasis (arrow B) and an important mediastinal shift to the left side. (Photography courtesy of Gretchen Williams, BS, CCRP from the International PPB Registry.)

diaphragm, and/or pleura, thoracic great vessels, regional lymph nodes, and diaphragm.

Type I PPB often shows the congenital lung malformations like CCAM. While type 1 PPB has no metastatic potential, type II or III PPB can lead to metastasis to the brain, bone, local thoracic lymph nodes, and liver. Pancreas, kidneys, and adrenal glands are rarely involved. In addition, children with type II or III PPB who failed therapy may have tumor recurrence locally in the thorax and/or distant metastatic disease.

Microscopically, PPB is composed of malignant mesenchymal cells which may be primitive blastema cells or large malignant sarcoma type cells. The epithelial component is typically benign. The histopathological description and immunoreactivity for the different of PPB are described in Tables 6.2 and 6.3, respectively (Figures 6.6 and 6.7).

TABLE 6.2

Microscopic Description of Different PPB Types

Type	Description
Histological description	
I	• Sections show a multicystic architecture. • Cysts are lined by respiratory-type epithelium above small primitive malignant cells. • Some cells may show rhabdomyoblastic differentiation, with a variable cambium layer. • Small nodules of primitive cartilage and hyalinized septal stroma may be present (Figure 6.6).
II	• Primitive small cells forming plaques or nodules. • May show mixed blastematous and sarcomatous features.
III	• Solid; may be composed of blastematous or sarcomatous appearing cells (Figure 6.7). • Foci of necrosis, hemorrhage, and fibrosis may be present.
Cytology description	
	• Moderately cellular smears with admixture of ovoid blastemal cells and spindle cells [24].

Source: http://www.pathologyoutlines.com/topic/pleurapleuropulmonary-blastoma.html

TABLE 6.3

Selected Cell-Typic Markers Commonly Used in Immunohistochemistry

Type of Cells	Antigen	Staining
Pneumocytes lining the cysts and small airspaces	Cytokeratin	Highlighted positive
Rhabdomyoblasts and primitive cells in subepithelial regions of cystic lesions	Muscle-specific actin and desmin	Positive
Tumor cells	Vimentin	Positive
	CD117 (*c-kit*), alpha-1-antitrypsin	Focally positive
	CD99	Weakly positive
	EMA, myogenin, S100, GFAP, neuron-specific enolase (NSE), TTF-1, alpha-fetoprotein, chromogranin, and synaptophysin	Negative

Source: http://www.pathologyoutlines.com/topic/lungtumorPPB.html

6.6.3 Differential Diagnosis

6.6.3.1 Lung Cysts or Pneumothorax

Children presenting with a family history of lung cysts should raise the possibility of PPB; nevertheless, multifocality or pneumothorax can be present in combination with multiple inherited or non-inherited disorders. Consequently, a differential diagnosis of PPB must be done on the basis of medical history and physical examination. Disorders to be considered in the differential diagnosis of PPB include the following:

1. *Birt–Hogg–Dubé (BHD) syndrome.* BHD is associated with multiple noncancerous (benign) skin tumors that affect the skin and lungs, especially if the patient

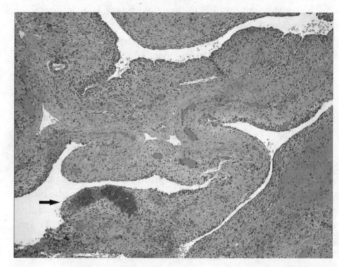

FIGURE 6.6 Pleuropulmonary blastoma type I. This low magnification image shows cystic airspaces lined by a modified respiratory epithelium with an underlying primitive mesenchymal stroma. Note in the bottom left, focal nodules of cartilage (arrow). (Photography courtesy of Megan K. Dishop MD, from the International PPB Registry.)

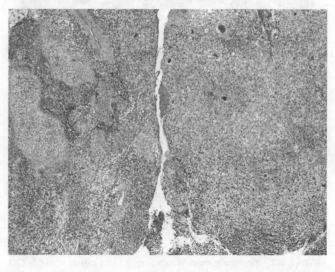

FIGURE 6.7 Pleuropulmonary blastoma type III. This low magnification image shows tumor composed of pleomorphic sarcoma and sarcomatoid elements of different types of differentiation. (Photography courtesy of Megan K. Dishop MD, from the International PPB Registry.)

or their relatives have a history of lung cysts, spontaneous pneumothorax, or kidney cancer. Consider that in patients with BHD, the characteristic skin lesions (fibrofolliculomas, trichodiscomas, and acrochordons) appear in the third and fourth decade of life.

2. *Marfan syndrome.* The pulmonary histology includes widespread or patchy cystic changes, emphysema, and spontaneous pneumothorax, focal pneumonia or bronchiectasis, bullae, congenital pulmonary malformations, and apical fibrosis. The inheritance of Marfan syndrome does not preclude the presence or development of a second pulmonary disease process. The differential diagnosis is based on clinical history and imaging studies.

3. *Cystic fibrosis (CF).* In patients with CF, mucus stasis eventually leads to airway plugging, chronic bacterial infection, inflammation, and bronchiectasis. As CF gets worse, pneumothorax is a common complication. Diagnosis may be difficult in CF patients when relying on plain radiography alone due to the presence of cysts and bullae. CT scan of the chest facilitates not only the diagnosis but also selection of appropriate management. CF is caused by mutation in the CFTR gene, which regulates anion transport and mucociliary clearance in the airways. Therefore, the diagnosis of cystic fibrosis is based on a positive chloride sweat test and the identification of mutations in the CFTR gene.

4. *Vascular Ehlers–Danlos syndrome (vEDS).* Spontaneous and/or recurrent pneumothoraces may be the first significant presenting feature of vEDS. Hemothorax and hemopneumothorax have been reported, often in association with pulmonary blebs, cystic lesions, and hemorrhagic or fibrous nodules. Hemoptysis can be severe and recurrent, even life-threatening. However, clinical diagnosis of vEDS is based on four criteria: a characteristic facial aspect (acrogeria) in most patients, thin and translucent skin with highly visible subcutaneous vessels, ecchymoses and hamartomas, and arterial, digestive, and obstetrical complications. The diagnosis of vEDS is based on clinical signs, noninvasive imaging, and the identification of a mutation of the COL3A1 gene.

5. *Tuberous sclerosis complex (TSC).* TSC is often detected during infancy or childhood but occasionally people with mild signs are not diagnosed until adulthood, or it goes undiagnosed. Affected patients have an overgrowth of abnormal smooth muscle-like cells in the lungs, resulting in the formation of lung cysts and the destruction of normal lung tissue. Lymphangioleiomyomatosis (LAM) is a frequent feature in this syndrome but it occurs almost exclusively in young women, typically presenting between 30 to 35 years of age. Symptoms have been reported to begin or worsen during pregnancy, suggesting that LAM may be hormonally influenced.

6. *Alpha-1 antitrypsin deficiency (AATD).* The initial symptoms of AATD include cough, sputum production, and wheezing. Symptoms are initially intermittent, and, if wheezing is the predominant symptom, patients often are told they have asthma. If recurrent episodes of cough are most prominent, patients may be treated with multiple courses of antibiotics and evaluated for sinusitis, postnasal drip, or gastroesophageal reflux. Although extremely rare, emphysema in children with AATD has been reported. The diagnosis of AATD can be established with the results of a simple blood test.

6.6.3.2 Congenital Pulmonary Airway Malformations

Various congenital lung lesions have been reported in young children, including congenital cystic adenomatoid malformation/congenital pulmonary airway malformation (CCAM/CPAM), the congenital peribronchial myofibroblastic tumor, fetal lung interstitial tumor (FLIT), pulmonary sequestration, bronchial atresia, and bronchogenic cyst. The differential diagnosis of CCAMs is very difficult using imaging studies alone and ultimately requires histological examination. In addition, pneumothoraces and the presence of multifocal or bilateral cysts are more common in PPB than in CCAMs, helping to make the diagnosis.

Pulmonary sequestrations and peripheral bronchogenic cysts are more complex lesions that are commonly diagnosed prenatally. Although their radiographic and histological features allow their differentiation from PPB, there is one report of a pulmonary sequestration in a patient with a germline DICER1 pathogenic variant.

6.6.3.3 Lung Tumors

Most tumors in newborns are solid lung tumors, whereas solid PPB has not been observed in a newborn. Differential diagnoses include fetal lung interstitial tumor, congenital peribronchial myofibroblastic tumor, and solid type 3 CCAM [25].

Thoracic tumors are rare under age 6 years. PPB is one of the most common lung tumors among children and is histologically characterized by blastemal and rhabdomyoblastic cells and anaplastic features. However, rhabdomyosarcoma is a malignant tumor that tends to originate in the chest wall or soft tissue of the diaphragm rather than the lung parenchyma.

Inflammatory myofibroblastic tumor originates in the lung, typically in children age >3–4 years, as a well-circumscribed, lobar-based mass. The tumor comprises myofibroblasts that can be demonstrated by immunostaining for smooth muscle actin (SMA). About 40%–50% of inflammatory myofibroblastic tumors have translocations involving ALK (the gene encoding anaplastic lymphoma kinase) and show immunostaining for the ALK protein.

Synovial sarcoma is the main differential diagnosis for PPB in young adults. The most common sites of origin are the thigh, knee, ankle, foot, and upper extremities, but in unusual cases, synovial sarcoma may arise within the chest wall, mediastinum, heart, lung, or pleura. Rarely are PPBs seen in teenagers and typically are more heterogeneous than synovial sarcomas, but the spindle cell components of PPB and synovial sarcoma can be remarkably similar. As synovial sarcoma is linked to a specific chromosomal translocation t(X;18)(p11;q11), which fuses the *SYT* gene from chromosome 18 to a homologous gene at Xp11 (*SSX1, SSX2,* or *SSX4*), immunohistochemical demonstration of airway epithelial cell markers or identification of fusion protein SYT–SSX is helpful for making a diagnosis of synovial sarcoma.

Classic biphasic pulmonary blastoma (CBPB) is a biphasic tumor with malignant epithelial elements and mesenchyme occurring in adults (mean age of 43 years).

6.7 Treatment

Three therapeutic modalities are currently used for the management of PPB: surgery, chemotherapy, and radiation therapy.

6.7.1 Surgery

Generally, a complete surgical resection is recommended as the initial approach for the treatment of PPB [22,26]. Surgery allows specific diagnosis of the tumor as well as improvement of symptoms caused by the presence of a large amount of abnormal tissue. When PPB presents as lung cysts and pneumothorax, removal of the cyst is recommended, but if PPB presents as a large mass in the chest, an attempt may be made to remove as much of the tumor as possible. Large PPBs often invade neighboring structures in the chest, such as the chest wall and mediastinum, compressing lung tissue on one side. Resection of these tumors should be performed with care, given that solid components of PPB are very friable, and spill is often inevitable. If a tumor mass is too large to be removed safely, occasional attempts are made to shrink it with neo-adjuvant chemotherapy and then try surgery later. In some circumstances, sites of unresectable residual disease may be titanium-clipped for radiographic localization and possible radiotherapy.

Involvement of the diaphragm may require excision of a portion of the diaphragm and use of a GORE-TEX patch. Drainage of pleural effusions should be approached with caution since the proper placement of needle and catheter can be difficult without radiographic guidance.

Brain parenchyma is the most common metastatic site for PPB. Resection is strongly suggested for intracranial mass lesions. High non-relapse mortality has been observed in several patients in whom cerebral PPB metastases were resected. Therapeutic resection of bone and liver metastases is rarely indicated.

6.7.2 Chemotherapy

To date, there are no proven regimens of chemotherapy for PPB. For type I PPB, chemotherapy may not be necessary, but less aggressive regimens can be used based on extent of resection and disease. Types II and III PPB are treated with aggressive surgical resection and intensive chemotherapy utilizing aggressive sarcoma-based regimens. The drugs used in adjuvant chemotherapy include vincristine, actinomycin-D, cyclophosphamide or ifosfamide, and doxorubicin [25–27]. Paclitaxel, Gemzar, and docetaxel have been also used in second- or third-line chemotherapy. Other agents that have been used with limited success include etoposide and cisplatin or carboplatin.

Occasionally, intracavitary cisplatin and intracavitary ^{32}P have been used in patients with PPB disease [5]. However, there are not enough data to document its effectiveness.

Delayed resection after chemotherapy is performed when the tumors are considered unresectable at the time of diagnosis. Individuals receiving neo-adjuvant chemotherapy may have marked tumor reduction; however, this response may be transient, and tumor can recur rapidly. The IPPBR recommends that surgical resection should occur no later than after the third cycle of chemotherapy. As with primary surgery, removal of all sites of gross disease should be attempted if possible. If gross total resection is not achieved, additional surgery may be required for local control. The 5-year overall survival rate in type II PPB is 71%, whereas type III PPB is 53%. To date, there are not sufficient data to evaluate the role of chemotherapy to prevent tumor progression in children up to 6.5 years of age with Type-Ir PPB, but it is not recommended for older children and adults with Type-Ir PPB.

6.7.3 Radiation Therapy

Radiation therapy is used primarily to treat PPB recurrence or metastasis, or in the setting of local control of residual unresectable tumor. Follow-up should include monitoring of brain and bone in addition to the primary site, as metastatic relapse frequently occurs at these sites. If radiation therapy is chosen, dosage appropriate for high-grade sarcoma (44 Gy or above) should be used [5,18,22,26].

Other treatments such as immunotherapy, stem cell replacement, or CRISPR/Cas9-mediated gene editing have not been reported for PPB.

6.8 Prognosis, Tumor Risk, and Testing Recommendation

6.8.1 Prognosis

Despite aggressive treatment protocols, the prognosis for patients with PPB is in relation to primary tumor size >5 cm, local or distant spread, especially to mediastinum, pleura, or central nervous system affection and histological subtypes II or III, and the age. Type I PPB has the most favorable prognosis with a 5-year survival rate of 89% (80–99), whereas it decreases for the other types: 71% (62%–81%) for type II and 53% (43%–65%) for type III. Analysis of 65 patients (13, type I; 24, type II; and 28, type III) who were studied for the European Cooperative Study Group for Paediatric Rare Tumors (EXPeRT) showed similar results (http://www.raretumors-children.eu/) [26].

6.8.2 Tumor Risk

Approximately 20% of children with PPB-DICER1 syndrome have a personal or family history of other childhood cancers. Cancers and other conditions described in these families include the following.

1. *Renal neoplasia.* Cystic nephroma (CN) is one of the most common tumors associated with PPB. It is observed in 9%–10% of PPB patients or relatives. In addition, 79% of cases occurred in children with PPB or in sibling of a PPB-affected child [10,28,29]. Typically, CN presents in the first 2 years of life as a painless enlarging abdominal or flank mass. The finding of bilateral tumors is a rare event that is highly suggestive of DICER1 mutation and of the associated

risk of PPB in the proband, even if the familial history is negative [29]. In addition, Wilms tumor also occurs in PPB patients, but much less frequently than CN. In Wilms tumor patients, genetic testing for DICER1 mutation must be offered only in children with other manifestations or with a familial history of the DICER1 syndrome.

2. *Gonadal tumors.* Ovarian sex-cord stromal tumors (OSCST) are a manifestation of the PPB familial cancer syndrome and may be an initial presentation of DICER1 mutations within a family. In particular, ovarian Sertoli—Leydig cell tumors and other sex-cord stromal tumors as well as ovarian dysgerminoma, testicular seminoma, and perhaps some germ cell tumors [30,31]. In the case of OSCST in PPB kindreds, the age at diagnosis is younger (median 11 years) than in "sporadic" cases [14,30–32]. These tumors are typically unilateral but can arise as bilaterally, often ≥10 cm, and predominantly solid.

3. *Soft tissue dysplasias.* These include hepatic cystic mesenchymal hamartoma, intestinal hamartomatous polyps (from esophagus to rectum, but often ileal and leading to intussusceptions [33], and nasal chondromesenchymal hamartoma (very rare, benign tumor of the sinonasal tract that has been observed in five PPB-DICER1 muted patients) [34].

4. *Ciliary-body medulloepithelioma (CBME).* This is a primitive neuroepithelial neoplasm arising in the anterior chamber of the eye that has been observed in four PPB-DICER1 muted kindreds [34,35].

5. *Botryoid-type embryonal rhabdomyosarcoma (ERMS).* This is a malignant tumor which arises from embryonic muscle cells. This very rare neoplasm of teenage and young adult women has occurred with PPB-DICER1 mutated patients and with PPB-related conditions [36,37].

6. *Nodular thyroid hyperplasia and differentiated thyroid carcinomas.* Thyroid goiter and nodules are common findings in PPB-DICER1 mutate family members and disproportionately affect female carriers. Thyroid abnormalities, especially multinodular goiter (MNG), might be the most frequent manifestation of DICER1 mutation [19,30]. However, only developing a thyroid abnormality is not the characteristic of DICER1 mutation.

7. *Other childhood cancers.* Pituitary blastoma and pineoblastoma have been reported in PPB-DICER1 mutated patients. In addition, some unusual leukemias have been reported in the setting of the familial PPB-DICER1 syndrome [1,38–40].

6.8.3 Testing Recommendation

To confirm PPB-DICER1 syndrome in a family member, a strong suspicion is necessary, as there is a 50% chance of inheriting the mutation. The identification of a heterozygous DICER1 mutation is necessary to establish the diagnosis in a proband. Sequence analysis should be performed first, but if no pathogenic variant is identified, deletion/duplication analysis may be considered. Of the 65% of individuals with PPB who have a detectable germline DICER1 mutation, 80% have inherited the germline pathogenic variant from a parent [19,41]. De novo germline variants account for approximately 20% of cases.

When germline DICER1 mutations have been identified in a PPB-affected family member, it is reasonable to offer molecular genetic testing to at-risk relatives of all ages to clarify their genetic status and to provide recommendations for age-appropriate surveillance and early intervention. If first-degree relatives are not able to undergo molecular genetic testing, surveillance should be based on clinical changes that would warrant further investigation. If the parents of the proband have a germline DICER1 mutation, the risk to siblings of inheriting the germline DICER mutation is 50%. If the parents are negative (20% of are de novo DICER1 mutation), it is not usually necessary to offer testing to the siblings, even if they have a risk slightly higher than the general population because of the possibility of germline mosaicism.

Mutation testing in PPB probands and kindred may eventually prove useful in the clinical management of the disease. The identification of which children have inherited a predisposition to the disease would help direct intensified screening CT scans to only those children who are at risk.

6.9 Conclusion and Future Perspectives

During the past decade, significant achievements have been made in the identification, treatment, and research about PPB-DICER1 syndrome, but much remains to be done [42,43].

The detection of DICER1 germline mutations in patients with PPB and their family history represents a unique opportunity to learn how DICER1 mutations preferentially affect the lung mesenchymal stem cells, which causes changes in the growth and differentiation of the lung. Likewise, these studies might be of benefit to learn how tissue-specific loss of DICER1 and altered miRNA processing could manifest in human disease as well as to discover new genes and that could be relevant for key cellular processes in lung development and oncogenesis. Thus, the results of these investigations could form the basis for the development of novel targetable molecules which can be used as a more rational (pathway-specific) therapeutic option.

To increase the awareness and improve the accuracy of the differential diagnosis of PPB, patients and relatives require multidisciplinary care, and should be encouraged to contact international cooperative groups (e.g., IPPBR, EXPeRT, or International OTST Registry) to facilitate their study and to develop registries, biospecimen banks, and clinical trials for this disease. The active participation in these organizations could potentially offer invaluable information with regard to the incidence and numbers of new patients and their families and could therefore facilitate the planning of epidemiologic, biologic, and therapeutic trials for this patient population.

Ultimately, a thorough understanding on the pathogenesis of the PPB-DICER1 syndrome will come from improved knowledge of the molecular mechanisms that govern the proliferation and cellular differentiation of mesenchymal tissue.

REFERENCES

1. Doros L, Schultz KA, Stewart DR et al. DICER1-Related Disorders. 2014 Apr 24. In: Adam MP, Ardinger HH, Pagon RA et al. editors. *GeneReviews®* [Internet]. Seattle (WA): University of Washington, Seattle; 1993–2018.

2. Schultz KA, Yang J, Doros L et al. DICER1-pleuropulmonary blastoma familial tumor predisposition syndrome: A unique constellation of neoplastic conditions. *Pathol Case Rev.* 2014;19:90–100.

3. Yeh A, Edelman MC. Pleuropulmonary blastoma. *Atlas Genet Cytogenet Oncol Haematol.* 2011;15:374–7.

4. Haider F, Al Saad K, Al-Hashimi F, Al-Hashimi H. It's rare so be aware: Pleuropulmonary blastoma mimicking congenital pulmonary airway malformation. *Thorac Cardiovasc Surg Rep.* 2017;6:e10–4.

5. Priest JR, McDermott MB, Bhatia S, Watterson J, Manivel JC, Dehner LP. Pleuropulmonary blastoma: A clinicopathologic study of 50 cases. *Cancer.* 1997;80:147–61.

6. Dehner LP. Pleuropulmonary blastoma is the pulmonary blastoma of childhood. *Semin Diagn Pathol.* 1994;11:144–51.

7. Manivel JC, Priest JR, Watterson J et al. Pleuropulmonary blastoma. The so-called pulmonary blastoma of childhood. *Cancer.* 1988;62:1516–26.

8. Priest JR, Andic D, Arbuckle S, Gonzalez-Gomez I, Hill DA, Williams G. Great vessel/cardiac extension and tumor embolism in pleuropulmonary blastoma: A report from the International Pleuropulmonary Blastoma Registry. *Pediatr Blood Cancer.* 2011;56:604–9.

9. Harris KS, Zhang Z, McManus MT, Harfe BD, Sun X. Dicer function is essential for lung epithelium morphogenesis. *Proc Natl Acad Sci USA.* 2006;103:2208–13.

10. Hill DA, Ivanovich J, Priest JR et al. DICER1 mutations in familial pleuropulmonary blastoma. *Science.* 2009;325:965.

11. Pugh TJ, Yu W, Yang J et al. Exome sequencing of pleuropulmonary blastoma reveals frequent biallelic loss of TP53 and two hits in DICER1 resulting in retention of 5p-derived miRNA hairpin loop sequences. *Oncogene.* 2014;33:5295–302.

12. Seki M, Yoshida K, Shiraishi Y et al. Biallelic DICER1 mutations in sporadic pleuropulmonary blastoma. *Cancer Res.* 2014;74:2742–9.

13. de Kock L, Wang YC, Revil T et al. *High-sensitivity sequencing reveals multi-organ somatic mosaicism causing DICER1 syndrome. J Med Genet.* 2016;53:43–52.

14. Slade I, Bacchelli C, Davies H et al. DICER1 syndrome: Clarifying the diagnosis, clinical features and management implications of a pleiotropic tumour predisposition syndrome. *J Med Genet.* 2011;48:273–8.

15. Carthew RW. Gene regulation by microRNAs. *Curr Opin Genet Dev.* 2006;16:203–8.

16. Ryan BM, Robles AI, Harris CC. Genetic variation in microRNA networks: The implications for cancer research. *Nat Rev Cancer.* 2010;10:389–402.

17. Bartel DP. MicroRNAs: Genomics, biogenesis, mechanism, and function. *Cell.* 2004;116:281–97.

18. Messinger YH, Stewart DR, Priest JR et al. Pleuropulmonary blastoma: A report on 350 central pathology-confirmed pleuropulmonary blastoma cases by the International Pleuropulmonary Blastoma Registry. *Cancer.* 2015;121:276–85.

19. Foulkes WD, Priest JR, Duchaine TF. DICER1: Mutations, microRNAs and mechanisms. *Nat Rev Cancer.* 2014;14:662–72.

20. Hill DA. USCAP Specialty Conference: Case 1-type I pleuropulmonary blastoma. *Pediatr Dev Pathol.* 2005;8:77–84.

21. Fosdal MB. Pleuropulmonary blastoma. *J Pediatr Oncol Nurs.* 2008;25:295–302.

22. Hill DA, Jarzembowski JA, Priest JR, Williams G, Schoettler P, Dehner LP. Type I pleuropulmonary blastoma: Pathology and biology study of 51 cases from the international pleuropulmonary blastoma registry. *Am J Surg Pathol.* 2008;32:282–95.

23. Priest JR, Williams GM, Hill DA, Dehner LP, Jaffé A. Pulmonary cysts in early childhood and the risk of malignancy. *Pediatr Pulmonol.* 2009;44:14–30.

24. Gelven PL, Hopkins MA, Green CA, Harley RA, Wilson MM. Fine-needle aspiration cytology of pleuropulmonary blastoma: Case report and review of the literature. *Diagn Cytopathol.* 1997;16:336–40.

25. Dishop MK, Kuruvilla S. Primary and metastatic lung tumors in the pediatric population: A review and 25-year experience at a large children's hospital. *Arch Pathol Lab Med.* 2008;132:1079–103.

26. Bisogno G, Brennan B, Orbach D et al. Treatment and prognostic factors in pleuropulmonary blastoma: An EXPeRT report. *Eur J Cancer.* 2014;50:178–84.

27. Zareifar S, Karimi M, Tasbihi M, Abdolkarimi B, Geramizadeh B. Pleuropulmonary blastoma is a rare malignancy in young adult and childhood. *Korean J Clin Oncol.* 2015;11:28–32.

28. Doros LA, Rossi CT, Yang J et al. DICER1 mutations in childhood cystic nephroma and its relationship to DICER1-renal sarcoma. *Mod Pathol.* 2014;27:1267–80.

29. Bahubeshi A, Bal N, Rio Frio T et al. Germline DICER1 mutations and familial cystic nephroma. *J Med Genet.* 2010;47:863–6.

30. Rio Frio T, Bahubeshi A, Kanellopoulou C et al. DICER1 mutations in familial multinodular goiter with and without ovarian Sertoli-Leydig cell tumors. *JAMA.* 2011;305:68–77.

31. Schultz KA, Pacheco MC, Yang J et al. Ovarian sex cord-stromal tumors, pleuropulmonary blastoma and DICER1 mutations: A report from the International Pleuropulmonary Blastoma Registry. *Gynecol Oncol.* 2011;122:246–50.

32. Foulkes WD, Bahubeshi A, Hamel N et al. Extending the phenotypes associated with DICER1 mutations. *Hum Mutat.* 2011;32:1381–4.

33. Sen D, Gulati YS, Majumder A, Bhattacharjee S, Chakrabarti R. Hepatic cystic mesenchymal hamartoma. *Med J Armed Forces India.* 2015;71:S574–7.

34. Priest JR, Williams GM, Manera R et al. Ciliary body medulloepithelioma: Four cases associated with pleuropulmonary blastoma--a report from the International Pleuropulmonary Blastoma Registry. *Br J Ophthalmol.* 2011;95:1001–5.

35. Mason KA, Navaratnam A, Theodorakopoulou E, Chokkalingam PG. Nasal chondromesenchymal hamartoma (NCMH): A systematic review of the literature with a new case report. *J Otolaryngol Head Neck Surg.* 2015;44:28.

36. Doros L, Yang J, Dehner L et al. DICER1 mutations in embryonal rhabdomyosarcomas from children with and without familial PPB-tumor predisposition syndrome. *Pediatr Blood Cancer.* 2012;59:558–60.

37. Dehner LP, Jarzembowski JA, Hill DA. Embryonal rhabdomyosarcoma of the uterine cervix: A report of 14 cases and a discussion of its unusual clinicopathological associations. *Mod Pathol.* 2012;25:602–14.

38. de Kock L, Plourde F, Carter MT et al. Germ-line and somatic DICER1 mutations in a pleuropulmonary blastoma. *Pediatr Blood Cancer.* 2013;60:2091–2.

39. Sabbaghian N, Hamel N, Srivastava A, Albrecht S, Priest JR, Foulkes WD. Germline DICER1 mutation and associated loss of heterozygosity in a pineoblastoma. *J Med Genet.* 2012;49:417–9.

40. Faure A, Atkinson J, Bouty A et al. DICER1 pleuropulmonary blastoma familial tumour predisposition syndrome: What the paediatric urologist needs to know. *J Pediatr Urol.* 2016;12:5–10.

41. Priest JR, Watterson J, Strong L et al. Pleuropulmonary blastoma: A marker for familial disease. *J Pediatr.* 1996;128: 220–4.

42. Parsons SK, Fishman SJ, Hoorntje LE et al. Aggressive multimodal treatment of pleuropulmonary blastoma. *Ann Thorac Surg.* 2001;72:939–42.

43. Romeo C, Impellizzeri P, Grosso M, Vitarelli E, Gentile C. Pleuropulmonary blastoma: Long-term survival and literature review. *Med Pediatr Oncol.* 1999;33:372–6.

7

Retinoblastoma

Claire Hartnett, Mandeep S. Sagoo, and M. Ashwin Reddy

CONTENTS

7.1 Introduction

As the most common primary intraocular tumor of childhood, retinoblastoma makes up 3% of pediatric malignancies worldwide. The global incidence of retinoblastoma is estimated at 1 in 16,000–20,000 live births per year [1,2]. Retinoblastoma is a cancer of early childhood and two-thirds are diagnosed before 2 years of age and 95% before 5 years. The principles discovered by studying retinoblastomas have led to the recognition that inactivation of tumor-suppressor genes broadly contributes to human tumorigenesis.

Retinoblastoma is initiated by mutation of the *RB1* gene that affects photoreceptor precursor cell in the developing retina. The cancer may be heritable or non-heritable and may involve one or both eyes. Untreated, retinoblastoma is fatal by spread to the central nervous system via the optic nerve and hematogenous route. A "white pupillary reflex," or leukocoria, and strabismus are the most common presenting symptoms. Early diagnosis, therefore, while the disease is still intraocular, is imperative. Intraocular disease is highly curable, with 5-year survival rates of 98% in high-income countries [3]. In countries without established specialized centers, primary care infrastructure, and educational strategies, the survival rate may be as low as 20%.

The treatment of retinoblastoma is multidisciplinary and aims, in order of priorities, at saving life, saving the eye, and preserving vision. Just over 50 years ago the treatment for retinoblastoma involved removal of the eye (enucleation); the prognosis was poor and survival rates were low. Today, there is a wide range of treatments available and retinoblastoma has one of the best cure rates of all childhood cancers [4]. However, the evidence base for clinical management is comprised of case series and cohort studies. Rigorous controlled clinical trials in retinoblastoma are lacking for a number of reasons: complex presentations (two eyes with different disease severity), too few patients in countries where resources are available to conduct clinical trials, and a high societal value on eyes and vision that imposes considerations beyond curing the cancer. The goal of using multicenter, randomized, controlled clinical trials to optimize and standardize treatments is laudable, but in reality is difficult to attain.

7.2 Biology

Retinoblastoma is a malignancy of the developing retina. It is a rare cancer usually initiated when both retinoblastoma gene (*RB1*) alleles are mutated in a single susceptible developing retinal cell. Inheritance of one mutant *RB1* allele strongly predisposes individuals to developing tumors that form after the second *RB1* allele is mutated [5,6]. The observation that children with bilateral retinoblastoma tended to be diagnosed at a younger age than those with non-heritable retinoblastoma led to Knudson's prediction in 1971 (the "two-hit" hypothesis) that two mutational events in a tumor suppressor gene are required to initiate a retinoblastoma tumor [6]. He proposed that heritable retinoblastoma results from a germline mutation (the "first hit," mutation 1, M1) and an acquired somatic mutation (the "second hit," M2), whereas non-heritable retinoblastoma arises when two somatic mutations are present in the same transformation suppressor gene in a susceptible cell [7].

RB1 is a large gene with 27 exons, encoding a 4.7 kb mRNA that translates into a 110 kDa phosphoprotein called the retinoblastoma protein (pRB) [8]. This protein is regarded as a cell cycle regulator that interacts with many proteins in the regulation of the cell cycle, differentiation, and control of genomic stability [9]. Loss of the tumor-suppressor functions of the pRB leads to genomic changes and instability and further mutations in oncogenes and other tumor suppressor genes that result in a retinal tumor [10]. In the 1980s, discovery of chromosomal deletions of some retinoblastoma patients pointed to a 13q14 locus [11,12].

Biallelic *RB1* inactivation is necessary to initiate most retinoblastomas, but alone it is not sufficient to form the full-blown cancerous growth. A benign retinal tumor, a retinoma, or a retinocytoma may also form following loss of both *RB1* alleles [13,14]. Further genetic or epigenetic changes are probably needed for malignant transformation. A retinoma may present as an elevated gray retinal mass with calcification and surrounding retinal pigment epithelium (RPE) proliferation and pigmentation, and is usually non-progressive. Finding of a retinoma on retinal examination of relative(s) of a patient with retinoblastoma indicates that they carry the mutant *RB1* allele.

Although both *RB1* alleles are mutated in nearly all retinoblastomas, a study analyzing the mutations of more than 1000 retinoblastomas identified a subset of unilateral tumors (1.4%) which showed no evidence of *RB1* mutation but had high-level amplification of the oncogene *MYCN* (amplified *MYCN*) [15]. This suggested that retinoblastoma can be initiated either by loss of tumor-suppressor function or by oncogene overexpression. This study also showed that another 1.5% of unilateral non-familial retinoblastomas are unexplained and have apparently normal *RB1* and *MYCN* genes. Children with this genetic form of retinoblastoma tend to have unilateral advanced intraocular disease at an earlier age.

The cellular origin of retinoblastoma has been elusive for many years, with the early hypothesis centering on a precursor photoreceptor cell in the developing retina. It is now believed that retinoblastoma evolves from a cell in the cone photoreceptor lineage [16–18]. It has been shown that retinoblastomas consistently express cone photoreceptor markers but not other retinal cell-type-specific proteins. In addition, depletion of *RB1* induces human cone precursor cell proliferation in vitro and pRB-depleted human cone photoreceptors form, typical of differentiated retinoblastomas in orthotopic xenografts [17].

7.3 Pathogenesis

The retinoblastoma tumor(s) arises from precursor retinal photoreceptor cells and so the majority of cases are diagnosed under the age of 5 years. In two-thirds of cases, only one eye is affected by retinoblastoma (unilateral) and in one-third of cases both eyes of the child are affected (bilateral).

All children with retinoblastoma tumors in both eyes have a germline *RB1* gene mutation on chromosome 13, either inherited or de novo, that predisposes them to develop retinal tumors in infancy but also potentially other cancers throughout life. These children are also at greater risk of developing multiple tumors in both eyes. Approximately 90% of these children have no family history of retinoblastoma and are the first affected in their family with a new germ line mutation. The risk to offspring is 50% by inheriting the mutant *RB1* gene, but 45% will develop retinoblastoma due to incomplete penetrance.

Most children with retinoblastoma in one eye (unilateral) and without a family history of the disease have normal constitutional *RB1* alleles, but somatic mutations occur on both *RB1* alleles in an immature retinal precursor cell. These children tend to develop one unilateral tumor. It is important to note, however, that 15% of children with unilateral retinoblastoma have constitutional *RB1* mutations that can be transmitted to their offspring. As a result, genetic testing is a key part in the management of all families affected by retinoblastoma.

Individuals with a germ line *RB1* mutation are at increased risk of developing secondary malignancies such as leiomyosarcoma, osteosarcoma, melanoma, lung cancer, and bladder cancer. This risk is further increased if they had received external beam radiotherapy (EBRT) for their retinoblastoma [19,20], and for this reason EBRT has largely been abandoned by most centers.

7.3.1 Trilateral Retinoblastoma

Approximately 5% of children with retinoblastoma associated with germline mutation of the *RB1* allele may develop trilateral retinoblastoma: i.e., retinoblastoma associated with a primary intracranial neuroectodermal tumor in the pineal or suprasellar region [21]. Most occur in the pineal region and are histologically pinealoblastomas but 20%–25% occur in the suprasellar region [21]. The prognosis is historically poor with 5-year survival rate of only 6% reported prior to 1995 [22,23]. Improvements in chemotherapy regimes and earlier detection rates however have resulted in 5-year survival rate of 44% reported from 1995 onward [22,24].

7.3.2 Retinoblastoma Associated with Chromosome 13q Deletion

Patients with chromosome 13q deletion syndrome are at increased risk of developing retinoblastoma [25] and require immediate referral after birth to an ophthalmologist for regular screening examination of the retinae. Such patients can have variable dysmorphic features including microcephaly, hypertelorism, downward slanting palpebral fissures, broad nasal bridge, and variable clinical features of hypotonia, developmental delay, and mental retardation [26]. They may also have gastrointestinal, cardiovascular, and genitourinary abnormalities. The severity of symptoms depends on the size and location of the deletion. Approximately 3%–5% of retinoblastoma patients carry a deletion of the q arm of chromosome 13 [27]. Patients with mosaic and non-mosaic deletions of chromosome 13q14 are both at risk of developing retinoblastoma.

7.4 Epidemiology

Retinoblastoma is an uncommon cancer that is diagnosed in approximately 1:16,000–1:20,000 global live births per year. This translates as approximately 8000 new cases each year worldwide [28,29]. It represents approximately 3% of all childhood cancers up to the age of 15 years [30]. There is no evidence of geographical or racial variations of retinoblastoma incidence [31,32].

Approximately 11% of all affected children reside in high-income countries, 69% in middle-income countries, and 20% in low-income countries [33]. Although prevalence is higher in middle- and low-income countries, most retinoblastoma treatment centers are in middle- and high-income countries. Retinoblastoma in low-income countries is associated with low patient survival at 30% [30,34,35]. In contrast, survival in high-income countries is 95%–98% [30,36]. This disparity in survival rates is a challenge that could potentially be reduced by a collaborative international effort [37]. Factors such as poor primary care facilities, late diagnosis, difficulty accessing retinoblastoma-specific health care, and socioeconomic issues may all lead to poor outcomes. In addition, poor understanding, education, and compliance, including a familial refusal to remove the affected eye (enucleation) and abandonment of therapy are difficulties clinicians face in low-income countries. With early diagnosis [38], many eyes can be safely treated and a lifetime of good vision in the unaffected eye is possible. Unfortunately, without timely diagnosis and appropriate treatment, fatal metastatic disease may develop.

7.5 Clinical Features

The majority of children with retinoblastoma without a family history are first noted because of a white pupil reflex or leukocoria. Parents may report a strange reflection or appearance in their child's eye, and this should alert a primary care physician to consider a diagnosis of retinoblastoma as their first differential. Unfortunately, in many instances the importance of this sign is not recognized, and referral and diagnosis may be delayed. Awareness campaigns to educate the lay public in the importance of a "white pupil" can be successful and beneficial. Digital images of a baby with retinoblastoma may show a white pupil "photoleukocoria" in contrast to the normal red reflex of the flash from a normal eye [39]. Retinoblastoma is the most important and dangerous condition to produce a leukocoria or a photoleukocoria. However other conditions such as congenital cataract,

TABLE 7.1

Frequency of Presenting Signs and Symptoms of Retinoblastoma

Sign/Symptom	Frequency
White reflex	56%
Strabismus	20%
Glaucoma	7%
Poor vision	5%
Orbital cellulitis	3%
Unilateral mydriasis	2%
Heterochromic iris	1%
Hyphema	1%
Other	2%

optic nerve coloboma, myelinated optic nerve fibers, high myopia, astigmatism, toxocariasis, and even the normal optic nerve head can also cause a similar reflex [40]. If such a sign is detected immediate referral to an ophthalmologist is advised [40].

The second most common presenting sign of retinoblastoma is strabismus, which may be either an esotropia or an exotropia (inward- or out-turning eye) [41]. This tends to occur when central vision is lost in an eye from the retinoblastoma. If a strabismus is suspected, the red reflex test should be applied to the child, and again, if abnormal prompt, urgent referral must be made to an ophthalmologist.

Other presentations of the condition include glaucoma or increased intraocular pressure, poor vision, change in iris color, or non-infective orbital inflammation (Table 7.1). In very late-stage disease, the eye may bulge from the orbit [41]. This can be a common presentation, unfortunately, in settings where awareness and resources are inadequate.

7.6 Diagnosis

7.6.1 Retinoblastoma

The referral of a child with possible retinoblastoma from an ophthalmologist to an ophthalmologist specializing in retinoblastoma is urgent, generally requiring an examination within a week [36]. The diagnosis of retinoblastoma is usually made following indirect ophthalmoscopy with the pupils pharmacologically dilated. Diagnosis does not rely on histopathological examination because biopsy incurs the risk of metastasis [42].

A detailed retinal examination under general anesthesia (EUA) is required for an accurate diagnosis to first confirm the condition by clinical phenotype from potential differential diagnoses such as Coats disease, persistent fetal vasculature, toxocara, medulloepithelioma, and vitreous hemorrhage. The anterior segment of the eye must also be examined carefully for signs of disease involvement including intraocular pressure and horizontal corneal diameter measurements. A thorough fundus examination with scleral depression must be performed with visualization of the retina to the ora serrata.

Retinoblastoma appears as a creamy white mass projecting into the vitreous or under the retina, with moderately dilated blood vessels running on the surface and penetrating the tumor. Hemorrhage may be present on the surface of the tumor or in the vitreous cavity. Clumped tumor cells in the vitreous are referred to as "seeds" and are highly indicative of retinoblastoma. Calcification, a common characteristic of retinoblastoma, is detected by ocular ultrasonography using brightness mode (b-scan). This demonstrates calcific flecks within the tumor. If the diagnosis is uncertain, the discovery of tumor calcification can be helpful in establishing the diagnosis of retinoblastoma.

A detailed fundus drawing is made by the clinician and is important to map tumor burden and location. Ideally, a wide-angle, handheld fundus camera, the RetCam (Clarity Medical Systems, Pleasanton, California, USA), is used to view and record the whole retina. It also allows for imaging of the anterior segment of the eye. Comparison of sequential images is useful to monitor and determine if the tumors are responding or relapsing. Good imaging and documentation of the entire retina supports classification and staging, documents treatment response, assists consultation with colleagues via telemedicine, and helps parents understand the disease and treatment options.

Careful examination of the retina and anterior structure of the eye is mandatory in order to classify the disease group of each eye. High-frequency ultrasound biomicroscopy (UBM) may be used to detect anterior disease and lens touch by the retinoblastoma tumor, which changes the classification of the disease to a more advanced group and hence the recommended treatment [43].

Optical coherence tomography can also be helpful to examine suspicious retinal lesions, discovering "invisible" early tumors in children with familial disease and in monitoring response to local treatments [44].

MRI is used to assess invasion of optic nerve and the presence of trilateral retinoblastoma [45] (pinealoblastoma and primitive neuroectodermal intracranial tumors associated with *RB1* mutations as well as metastatic retinoblastoma) (Figure 7.1) [22,46]. It is also helpful in determining invasion of the orbit. CT scans are avoided because radiation exposure in a child with a cancer syndrome (germline *RB1* mutation) risks second primary cancers [47].

Histologically, retinoblastomas are poorly differentiated malignant neuroblastic tumors, composed of cells with large hyperchromatic nuclei and scanty cytoplasm (Figure 7.2). Mitotic figures are common. In some tumors, differentiated cells form typical Flexner–Wintersteiner rosettes. Homer–Wright rosettes can also occur. Retinoblastomas often outgrow their blood supply, leading to cell necrosis. Apoptosis can also be evident in retinoblastomas. Calcification is also a common finding in retinoblastoma, but its etiology is not known.

The most important feature to ascertain from histopathological examination of the enucleated eye is the presence of high-risk histopathological features, such as extension of tumor into the choroid, sclera, optic nerve, and/or anterior ocular structures. These features are associated with risk of metastatic spread. The most common routes of dissemination are through the optic nerve to brain and through the choroid to bone marrow.

7.6.2 Heritable Retinoblastoma

Knowledge of the patient's *RB1* mutation profile enables precise screening of siblings and offspring of the proband. Without genetic testing, all children at risk (positive family history) should undergo regular EUA in the first 3 years of life to enable detection of early, small, easily treatable tumors. When genetic testing reveals the *RB1* mutation in a family, children at risk can be tested for that mutation. Of the unilateral retinoblastoma

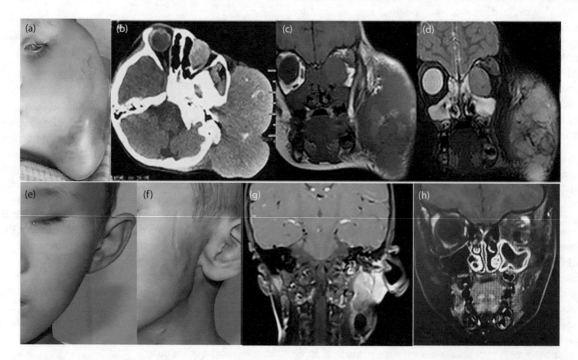

FIGURE 7.1 A child with metastatic retinoblastoma showing facial swelling on the left side (a). Axial CT shows attenuation to muscle, an ill-defined solitary mass in the left orbit, and a very large and moderately intense mass on the left side of the neck with extension to the parotid and submandibular glands (b). MRI shows iso-intensity on T1-weighted images (c), and moderate to marked enhancement post-contrast with fat suppression, hyperintensity on T2-weighted images (d) of the left orbit, and large mass. Postoperative photos of the child (e and f). MRI reveals a small residual tumor in the left side of the neck, which is hyperintense on T1 images (22 and 28 months post-operation, respectively) (g and h). (Photo credit: Wang P et al. *BMC Ophthalmol.* 2017;17:229.)

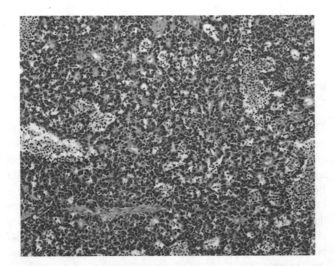

FIGURE 7.2 Histological examination of retinoblastoma reveals small undifferentiated cells that form rosettes (H&E, ×100). (Photo credit: Zafar SN et al. *BMC Res Notes.* 2013;6:304.)

patients, 85% will test negative, with <1% residual risk for undetectable low-level mosaicism, which reduces the intensity of surveillance for tumors in the normal eye and improves estimates of cancer risk of family members [48].

In familial retinoblastoma, prenatal screening can be performed on DNA obtained from amniotic fluid. In some cases, infants proven to carry the family's *RB1* mutation can be delivered a few weeks early, to optimize the chance of earliest possible detection of new tumors and allow for early treatment, but requires coordination with the family and obstetricians. Preimplantation genetic diagnosis is available in the UK, selecting babies without the mutation. Alternatively, where prenatal testing is unavailable, infants who are born with a risk of inheriting the *RB1* mutation should have genetic testing of the cord blood samples at birth. In at-risk infants, examination of the retina starts at birth, and continues at frequent intervals depending on the child's risk.

7.6.3 Classification and Staging

Classification of retinoblastoma severity/extent, as with other cancers, is the standard way to categorize eyes/patients best suited to particular therapies most likely to succeed, based on current evidence [48]. The optimal treatment and outcome for intraocular retinoblastoma balances the morbidity of treatment with the likelihood to cure the cancer.

The Reese−Ellsworth classification for intraocular retinoblastoma was used to predict outcomes following treatment with EBRT in 1963 [49]. In the 1990s, intravenous chemotherapy became available as a primary treatment, with radiotherapy used only as salvage. Kingston and colleagues in the London Retinoblastoma Service recognized that intravenous chemotherapy could be used to treat advanced intraocular retinoblastoma with relative success depending on the tumor stage [50]. Other centers worldwide reported similar results [51–53]. The most common chemotherapeutic agents used are vincristine, etoposide, and carboplatin. Across specialized centers, there is variability of therapeutic regimens of systemic chemotherapy,

including variations in the drugs utilized, the number of cycles administered, the addition of cyclosporine, and the use of focal therapies such as laser and cryotherapy to consolidate the response of chemotherapy.

An international collaboration led by Murphree developed the International Intraocular Retinoblastoma Classification (IIRC) [54] (Table 7.2) which better predicted responses to intravenous chemotherapy and focal therapies than the Reece−Ellsworth classification [55].

7.6.4 IIRC

Group A retinoblastomas: Small tumors, not threatening vision, are primarily treated with focal therapy such as cryotherapy or laser treatment. This may be the only treatment modality required but repeat treatments are often necessary.

Group B retinoblastomas: Medium-sized tumors or tumors near the macula and the optic nerve. These may be first treated with a small number of chemotherapy cycles to optimize the visual potential and then treated definitively with focal therapy.

Group C retinoblastomas: Medium to large tumors with limited vitreous and/or subretinal seeding—are primarily treated with chemotherapy (either intravenous or intra-arterial chemotherapy) followed by focal therapy. Intravitreal chemotherapy may be required to treat vitreous seeds of tumor.

Group D retinoblastomas: Large tumors with extensive vitreous and/or subretinal seeding are primarily treated with chemotherapy (either intravenous or intra-arterial chemotherapy depending on the center). Focal therapy may follow, including intravitreal chemotherapy to target vitreous seeds.

TABLE 7.2

International Intraocular Retinoblastoma Classification

Group	Definition
A	Very low risk. Eyes with small discrete retinal tumors (3 mm or smaller) at least 3 mm from the foveola and 1.5 mm from the optic nerve, without vitreous or subretinal seeding.
B	Low risk. Eyes without vitreous or subretinal seeding and discrete retinal tumor of any size or location. A small cuff of subretinal fluid extending no more than 5 mm from the base of the tumor is allowed.
C	Moderate risk. Eyes with discrete retinal tumors of any size and location and only focal vitreous or subretinal seeding (theoretically treatable with a radioactive plaque). Up to one quadrant of subretinal fluid may present.
D	High risk. Eyes with non-discrete endophytic or exophytic tumors accompanied by diffuse vitreous or subretinal seeding (more extensive seeding than Group C). Includes more than one quadrant of retinal detachment.
E	Very high-risk eyes that have been destroyed anatomically or functionally by the tumor with any one or more of the following: irreversible neovascular glaucoma, massive intraocular hemorrhage, aseptic orbital cellulitis, tumor anterior to anterior vitreous face, tumor touching the lens, diffuse infiltrating retinoblastoma, phthisis or pre-phthisis.

Group E retinoblastomas: Large tumors with high-risk features such as tumor touching the lens, neovascular glaucoma, orbital cellulitis, anterior segment involvement, total hyphema, suspected choroidal, scleral, optic nerve, or orbital involvement on ultrasonography or MRI. These eyes are often immediately enucleated because of the risk of metastatic spread.

The IIRC system has been and still is widely employed today. However, some centers have interpreted the classification differently, especially with regard to group D and E eyes, affecting assignment in over 25% of group E eyes [56], and has made interpretation of some study results difficult.

The American Joint Committee on Cancer (AJCC) and the Union for International Cancer Control (UICC) staging system developed the TNM classification for retinoblastoma. The 8th edition was recently published (Table 7.3) and for the first time included a hereditary component for the cancer. This was defined as bilateral Rb, Rb with an intracranial CNS midline embryonic

tumor, family history of Rb, or germline disease [57]. It remains to be seen whether this classification adds any practical value to outcomes reporting.

7.7 Treatment

The management and successful treatment of retinoblastoma depends on a collaborative integrated team approach between clinical specialists (ophthalmologists, pediatric oncologists and radiologists, nurses, geneticists), imaging specialists, child life (play) specialists, parents, and others involved in the care of the child. Specialized centers with interprofessional teams have developed expertise, resources, and equipment and specific treatment protocols which are evidence-based and are best-placed to deliver optimized outcomes for affected children and their parents.

TABLE 7.3

The 8th Edition AJCC cTNMH Retinoblastoma Clinical Staging Groups

Definition of primary tumor (cT)	
cTX	Unknown evidence of intraocular tumor
cT0	No evidence of intraocular tumor
cT1	Intraocular tumor(s) with subretinal fluid ≤5 mm from the base of any tumor
cT1a	Tumors ≤3 mm and further than 1.5 mm from the disc and fovea
cT1b	Tumors >3 mm or closer than 1.5 mm to the disc and fovea
cT2	Intraocular tumor(s) with retinal detachment, vitreous seeding or subretinal seeding
cT2a	Sub-retinal fluid >5 mm from the base of any tumor
cT2b	Tumors with vitreous seeding and/or subretinal seeding
cT3	Advanced intraocular tumor(s)
cT3a	Phthisis or pre-phthisis bulbi
cT3b	Tumor invasion of the pars plana, ciliary body, lens, zonules, iris or anterior chamber
cT3c	Raised intraocular pressure with neovascularization and/or buphthalmos
cT3d	Hyphema and/or massive vitreous hemorrhage
cT3e	Aseptic orbital cellulitis
cT4	Extraocular tumor(s) involving the orbit, including the optic nerve
cT4a	Radiological evidence of retrobulbar optic nerve involvement or thickening of the optic nerve or involvement of the orbital tissues
cT4b	Extraocular tumor clinically evident with proptosis and orbital mass
Definition of regional lymph nodes (cN)	
cNX	Regional lymph nodes cannot be assessed
cN0	No regional lymph nodes involvement
cN1	Evidence of preauricular, submandibular, and cervical lymph node involvement
Definition of distant metastasis (M)	
cM0	No signs or symptoms of intracranial or distant metastasis
cM1	Distant metastasis without microscopic confirmation
cM1a	Tumor(s) involving any distant site (e.g., bone marrow, liver) on clinical or radiological tests
cM1b	Tumor involving the central nervous system on radiological imaging (not including trilateral retinoblastoma)
pM1	Distant metastasis with microscopic confirmation
pM1a	Histopathological confirmation of tumor at any distant site (e.g., bone marrow, liver, or other)
pM1b	Histopathological confirmation of tumor in the cerebrospinal fluid or CNS parenchyma
Definition of heritable trait (H)	
HX	Unknown or insufficient evidence of a constitutional *RB1* gene mutation
H0	Normal *RB1* alleles in blood tested with demonstrated high sensitivity assays
H1	Bilateral retinoblastoma, retinoblastoma with an intracranial CNS midline embryonic tumor (i.e., trilateral retinoblastoma), patient with family history of retinoblastoma, or molecular definition of constitutional *RB1* gene mutation

Over the past 30 years, IIRC Group A−C eyes were successfully managed with >90% tumor control rates. Group E eyes were immediately enucleated for the most part. However, Group D eyes were the most challenging for clinicians, with many electing to enucleate immediately and others preferring to treat with intravenous chemotherapy and adjuvant focal therapies. In London, the success rate for salvage of Group D eyes is 63% using systemic chemotherapy initially and targeted chemotherapy treatments [58]. Many parents would like to save the eye if it is safe to do so. Parents need to be aware of the risk of enucleation after attempt to salvage, visual potential, and the treatment burden in terms of the number of EUA.

7.7.1 Enucleation

Globally, enucleation is a first-line therapy for the majority of eyes with retinoblastoma. It is the fastest and least costly treatment and results in fewer EUAs for children with Group D eyes versus those who have undergone salvage treatment [59,60]. Group E eyes are usually managed with enucleation. For intraocular retinoblastoma this operation is curative. Enucleation is also indicated for recurrent tumor that has failed all other treatment modalities. If both eyes are Group E, bilateral enucleation may be indicated, since attempts to save severely involved eyes may put the child's life in jeopardy from systemic metastasis [61].

Enucleation must be performed with great care to not penetrate the eye and spill tumor. A long optic nerve is important to ensure that the surgical margin is tumor-free, particularly in countries where deep invasion of the optic nerve is common. For genetic diagnosis, fresh tumor tissue harvest immediately after enucleation is very important for *RB1* mutation studies, since this is the best way to determine the exact mutations and thus know if the child has heritable or non-heritable retinoblastoma.

An orbital implant is placed in the socket and the four rectus muscles are sutured either in front of the implant or to the conjunctival fornices (the myoconjunctival technique) and both techniques allow for good movement of the implant which will mirror movement of the other eye. A conformer shell is placed on the conjunctiva directly following the surgery. This can be beneficial as the eye can look normal, and this is psychologically of importance to the parents and the child [62].

7.7.2 Post-Enucleation Adjuvant Chemotherapy

When high-risk histopathological features are present [63–67], post-enucleation adjuvant chemotherapy is indicated. Despite this prompt treatment, there is still a risk of metastases of up to 4% [68]. Only 24% of Group E eyes exhibit such high-risk features following primary enucleation [59], which is why there have been recent attempts to salvage Group E eyes. High-risk features have been shown to occur in 13% of Group D eyes that have undergone primary enucleation, and these can be associated with a lack of vitreous seeding [63]. Tumor extending into the optic nerve beyond the cribriform plate carries a significant risk of direct extension into the brain via the optic chiasm and seeding into the cerebrospinal fluid, subarachnoid space with leptomeningeal involvement. Tumor invasion of choroid and/or sclera carries a high risk of hematogenous spread via the choroidal vessels,

usually to the bone marrow. Direct spread along the ciliary vessels and nerves into the orbit may occur in advanced cases.

If retinoblastoma metastasizes, it generally becomes evident within 18 months of the last active tumor in the eye, and it is rare beyond 3 years. Bone marrow is the preferred site for retinoblastoma metastases. Only terminally are bone, lymph node, and liver involved. Lung metastases are rare and late.

7.7.3 Intravenous Chemotherapy

In 1996 intravenous chemotherapy become the standard primary treatment for smaller tumors that were not amenable to focal therapy. Following an initial response to the first few cycles of chemotherapy, focal therapy with cryotherapy or laser was performed to destroy residual or recurrent tumor [53]. Chemotherapy (carboplatin, etoposide, and vincristine) is given every 3 weeks via a central venous line. Increased intraocular concentrations of chemotherapy drugs may be induced in eyes with vitreous seeding by the application of a single-freeze cryotherapy at the peripheral retina in the vicinity of vitreous seeds [69]. EUAs with appropriate focal therapy if necessary every 4 to 6 weeks are important for at least 1 year after tumor activity, as local recurrences can occur up to 2 years following completion of intravenous chemotherapy treatment. The role of laser in addition to chemotherapy is discussed below.

Unlike radiation, chemotherapy is not associated with any long-term cosmetic deformity of the orbit and upper face, cataracts, or ocular complications. Acute toxicities of intravenous chemotherapy for retinoblastoma are as for other pediatric cancers, including short-term transient pancytopenia, hair loss, vincristine-induced neurotoxicity, and infections. Long-term toxicities include carboplatin-induced ototoxicity in some patients [70], second non-ocular cancer risk with alkylating agents [71], and secondary acute myeloid leukemia following intense chemotherapy with topoisomerase inhibitors [72].

7.7.4 Intra-Arterial Chemotherapy

Retinoblastoma treatment continues to evolve, primarily to reduce the morbidity of systemic side effects of intravenous chemotherapy and to increase salvage rates of eyes, especially Group D eyes. Targeted delivery of chemotherapy was thus introduced. Intra-ophthalmic artery chemotherapy was first introduced by Kaneko and colleagues in Tokyo [73], using a balloon to divert blood into the ophthalmic artery. It was then developed further into a super-selective treatment in the 2000s by Abramson in New York [74].

The procedure, performed under general anesthesia, involves the passing of a microcatheter through the femoral artery up to the orifice of the ophthalmic artery of the affected eye and chemotherapy (single drug or combination; melphalan, topotecan, and/or carboplatin) is infused in a pulsatile manner over 30 minutes. This technique results in a significantly higher drug concentration delivery to the ocular structures, up to 250-fold more than after intravenous chemotherapy [75]. Cure rates were impressive, with 85% of Group D treatment naïve eyes retained [76]. Recent retrospective nonrandomized studies investigated the success rate for primary intra-arterial chemotherapy and found it to be significantly higher in reaching control (>85%) than intravenous chemotherapy [77,78]. Performing meta-analysis on the reported outcomes of intra-arterial chemotherapy has been restricted by lack of comparative groups in many studies. One report suggested that 2%–3% of patients treated by intra-arterial chemotherapy developed metastatic disease [79]. This is a worrisome statistic which will have to be analyzed further in the years to come.

In many centers, intra-arterial chemotherapy has replaced intravenous chemotherapy as first-line treatment [80], including unilateral Group D eyes [81]. Understanding which phenotypes of Group D are at increased risk of adverse histopathology may result in an evidence-based approach minimizing the risk of metastases (see above).

Many retinoblastoma units were initially tentative in their use of intra-arterial chemotherapy due to complications including choroidal ischemia, visual loss, and third nerve palsy [82]. Vision in previously seeing eyes was initially lost in 42% of patients [83] and this was thought to be due to the learning curve for interventional neuroradiologists. However, Reddy et al. showed that patients with a similar catheterization method but with reduced dose of melphalan did not lose vision [84]. In addition, an appropriate dose was associated with fewer cranial nerve palsies.

7.7.5 Focal Therapy

Laser therapy and cryotherapy is the local application of therapy to the eye under direct visualization through the pharmacologically dilated pupil at the time of EUA. Focal therapy is the primary treatment for IIRC Group A eyes and can be used to consolidate responses of Group B, C, and D eyes following intravenous and possibly intra-arterial chemotherapy with the aim to physically destroy residual or recurrent tumor. The evidence for laser in addition to chemotherapy is not strong, and some units do not routinely treat tumors with laser after chemotherapy [85]. In addition, some phenotypes such as cavitary retinoblastoma have a low risk of relapse and are unlikely to flatten with multiple attempts at laser [86].

Two types of laser treatments are applied until the tumor is flat, atrophic, or completely calcified. Transpupillary thermotherapy involves directing an 810-nm diode laser directly to heat the tumor. Photocoagulation therapies with 532 nm, 810 nm, or continuous-wave 1064 nm laser beams are directly applied to small, active, or suspicious tumors until the tumor is destroyed by heat, appearing gray/white.

Cryotherapy involves freezing the tumor through the sclera with a nitrous oxide probe. Tumor cells are destroyed by the thawing of the ice ball; hence repeat applications are interspersed with 1-minute intervals to allow the tumor to thaw. It is common to apply three freeze-thaw cycles.

Retinoblastoma is a very radiosensitive tumor. Although ERBT carries the risk of further tumors in the radiation penumbra and facial disfigurement, radiation can be given focally by brachytherapy using plaque applicators, avoiding these complications. Initial attempts in London by Foster Moore and Stallard used a radon seed sutured into the tumor and allowed to decay naturally [87,88]. This was replaced by Cobalt-60 applicators [89], and now the major isotopes used are Ruthenium-106 and Iodine-131. Plaque radiotherapy allows localized radiation by suturing a radioactive plaque to the eye to deliver trans-scleral radiation dose of 40–50 Gy to the apex of the tumor over several days.

At the end of this period the plaque is removed. This method is effective in treating recurrences or as a single primary treatment in a location that is amenable for plaque placement.

7.7.6 Intravitreal Chemotherapy

Traditionally, one of the main reasons for treatment failure in retinoblastoma was the presence of vitreous seeds [90]. Intravitreal chemotherapy involves the direct injection of chemotherapy agents into the vitreous cavity of the eye. In retinoblastoma there was a concern that this may result in seeding outside the eye; however, advances by Munier and colleagues in Switzerland in 2010 in developing a safety-enhanced injection technique have reduced this risk [90]. Under anesthesia, the intraocular pressure of the eye is reduced by removing a small volume of aqueous fluid from the anterior chamber. Melphalan (with or without toptecan) is injected into the vitreous cavity behind the lens (in a tumor-free site confirmed by ultrasound biomicroscopy) through the sclera with a small gauge needle. Upon needle withdrawal the injection site is sealed and sterilized with cryotherapy and the eye is shaken gently to distribute the drug throughout the vitreous. The resulting high concentration of drug within the vitreous cavity can be successful in treating vitreous seeds [91]. This treatment, however, is an adjunctive and salvage treatment and it is only initiated after the source of the seeds is controlled.

7.7.7 Extraocular Retinoblastoma

Retinoblastoma may present with signs of extraocular disease, particularly in low-income countries. Orbital extension of an intraocular retinoblastoma may be detected clinically in children with proptosis or a fungating mass, or by imaging studies. Treatment includes neoadjuvant chemotherapy using carboplatin, etoposide, and vincristine. Other agents that are useful are cisplatin, cyclophosphamide, and anthracyclines [92]. This is followed by enucleation and limited excision of affected tissues, orbital radiation, and adjunctive chemotherapy [93]. Intrathecal chemotherapy may also be used, and high-dose chemotherapy with stem cell rescue may be added in metastatic disease. Children with orbital or metastatic spread without central nervous system invasion may be cured with intensive therapy [93].

7.8 Prognosis

With evolving methods of treatment, the prognosis for retinoblastoma is excellent with, at present, 98% survival cure rates in high-income countries [3]. Eye salvage rates have also improved with evolving treatments, particularly for Group D eyes, over the last 15 years.

The prognosis for vision is excellent in unilateral retinoblastoma for the normal eye but is very dependent on the location and size of tumor in the affected eye. Tumors located at the macula or fovea have a poor visual outcome but extrafoveal tumors have less impact on central visual acuity, and visual outcomes can be very good. Prognosis for vision in bilaterally affected eyes is also highly dependent on the size and location of the tumors but is better in those children with familial retinoblastoma who are screened from birth and treated early. Even in Group D eyes, 50% have better than 6/60 (20/200) vision, with laser being a risk factor for poor vision [94]. Bilateral enucleations are now rare due to improved awareness, prompt diagnosis, and new evolving treatments.

The most important factor in improving visual outcome for retinoblastoma children lies in the early recognition of the presenting signs of retinoblastoma by the parents and the primary care physician.

Following the initial management and treatment of active retinoblastoma, frequent monitoring of the tumor(s) by EUAs will be necessary, especially in the first 2 years following diagnosis, as recurrences may occur during this time. Regular EUAs are carried out until the child is old enough to cooperate with a fully dilated retinal examination in the clinic. Follow-up can then be carried out in the outpatient clinic.

Retinoblastoma patients who carry a *RB1* mutation are at risk of other cancers throughout their lives, including bone and soft tissue sarcomas during adolescence and early adulthood, malignant melanoma, and epithelial, bladder, and esophageal cancers. These risks are increased significantly by radiation exposure. Emphasizing the dangers of known carcinogenic factors such as radiation, smoking, excess UV light, and obesity is important for these patients. Routine x-rays and CT scans are also not advised. Many surviving patients who were previously treated for retinoblastoma as a child may be unaware of these risks.

All patients with a past history of retinoblastoma should be offered genetic testing for the *RB1* mutation and be offered counseling with regard to the risk of retinoblastoma occurring in their offspring. This is also of importance to those who had unilateral retinoblastoma, as many of these patients automatically make the assumption that their offspring cannot inherit the condition [95]. Accurate genetic counseling should be offered to the parents and the child when s/he reaches mature age.

It is important for all retinoblastoma patients with the heritable form to maintain contact with an oncologist throughout their lives for early detection and management of potential secondary malignancies.

7.9 Conclusion

Retinoblastoma is a unique cancer that has served an important role in the field of oncology with regard to understanding genetics, screening, treatment, and classification. Just over half a century ago, the main treatment for retinoblastoma was removal of the eye. Today, with such a large armamentarium of treatments available, eye salvage rates and patient survival have improved greatly. Genetic testing and screening have become indispensable for patients and have reduced the burden of disease. Unfortunately, survival rates for those affected in low-income countries remain low. International collaborations and twinning programs can build frameworks for knowledge and expertise exchange, filling gaps in specialized training and allowing local caregivers to deliver optimal health care. Adoption of first-line enucleation with implants and immediate prosthetic eyes, parent-to-parent discussion to allay fears of eye removal, education and information to parents can reduce the rate of noncompliance with treatment in those settings.

Most studies to date are primarily retrospective with no comparative groups, making research weak. Controlled clinical trials on the many aspects of retinoblastoma care are desirable to base best practice recommendations for treatments in the future, and global collaborations will make such research stronger.

Retinoblastoma is curable if diagnosed early. We are close to achieving 100% survival rate in the developed world; however, much work is required to achieve high global cure rates. Strategies aimed at early recognition of the signs of retinoblastoma have the potential to make a great impact, improving the survival rate and minimizing the burden of invasive treatments.

REFERENCES

1. Ortiz MV, Dunkel IJ. Retinoblastoma. *J Child Neurol.* 2016;31(2):227–36.
2. Moll AC, Kuik DJ, Bouter LM et al. Incidence and survival of retinoblastoma in the Netherlands: A register based study 1862–1995. *Br J Ophthalmol.* 1997;81(7):559–62.
3. MacCarthy A, Birch JM, Draper GJ et al. Retinoblastoma: treatment and survival in Great Britain 1963 to 2002. *Br J Ophthalmol.* 2009;93(1):38–9.
4. Fabian ID, Onadim Z, Karaa E et al. The management of retinoblastoma. *Oncogene.* 2018. doi: 10.1038/s41388–017–0050-x.
5. Knudson AG. Two genetic hits (more or less) to cancer. *Nat Rev Cancer.* 2001;1(2):157–62.
6. Hanahan D, Weinberg RA. The hallmarks of cancer. *Cell.* 2000;100(1):57–70.
7. Knudson AG. Mutation and cancer: Statistical study of retinoblastoma. *Proc Natl Acad Sci USA.* 1971;68(4):820–3.
8. Comings DE. A general theory of carcinogenesis. *Proc Natl Acad Sci USA.* 1973;70(12):3324–8.
9. Lohmann DR. RB1 Gene mutations in retinoblastoma. *Hum Mutat.* 1999;14:283–8.
10. Dick FA, Rubin SM. Molecular mechanisms underlying RB protein function. *Nat Rev Mol Cell Biol.* 2013;14(4):297–306.
11. Corson TW, Gallie BL. One hit, two hits, three hits, more? Genetic changes in the development of retinoblastoma. *Genes Chromosomes Cancer.* 2007;46(7):617–34.
12. Dryja, TP, Rapaport JM, Joyce JM, Petersen RA. Molecular detection of deletions involving band q14 of chromosome 13 in retinoblastomas. *Proc Natl Acad Sci USA.* 1986;83(19):7391–4.
13. Friend SH, Bernards R, Rogelj S et al. A human DNA segment with properties of the gene that predisposes to retinoblastoma and osteosarcoma. *Nature.* 1986;323(6089):643–6.
14. Dimaras H, Khetan V, Halliday W et al. Loss of *RB1* induces non-proliferative retinoma: Increasing genomic instability correlates with progression to retinoblastoma. *Hum Mol Genet.* 2008;17(10):1363–72.
15. Rushlow DE, Mol BM, Kennett JY et al. Characterisation of retinoblastomas without *RB1* mutations: Genomic, gene expression and clinical studies. *Lancet Oncol.* 2013;14(4):327–34.
16. Munier, FL, Balmer A, Van Melle et al. Radial asymmetry in the topography of retinoblastoma: Clues to the cell of origin. *Ophthalmic Genet.* 1994;15(3–4):101–6.
17. Xu, XL, Singh HP, Wang L et al. Rb suppresses human cone-precursor-derived retinoblastoma tumours. *Nature.* 2014;514(7522):385–88.
18. Bremner R, Sage J. The origin of human retinoblastoma. *Nature.* 2014;514(7522):312–3.
19. Kleinerman R, Schonfeld S, Tucker M. Sarcomas in hereditary retinoblastoma. *Clin Sarcoma Res.* 2012;2(1):15.

20. Maree T, Moll AC, Imhof SM et al. Risk of second malignancies in survivors of retinoblastoma: More than 40 years of follow up. *J Natl Cancer Inst*. 2008;100(24):1771–9.

21. Kivela T. Trilateral retinoblastoma: A meta-analysis of hereditary retinoblastoma associated with primary ectopic intracranial retinoblastoma. *J Clin Oncol*. 1999;17(6):1829–37.

22. de Jong MC, Kors WA, de Graaf P et al. Trilateral retinoblastoma: A systematic review and meta-analysis. *Lancet Oncol*. 2014;15(10):1157–67.

23. Blach LE, McCormick B, Aramson DH et al. Trilateral retinoblastoma-incidence and outcome: A decade of experience. *Int J Radiat Oncol Biol Phys*. 1994;29(4):729–33.

24. Wright KD, Qaddoumi I, Patay Z et al. Successful treatment of early detected trilateral retinoblastoma using standard infant brain tumor therapy. *Pediatr Blood Cancer*. 2010;55(3):570–2.

25. Kivela T, Tuppurainen K, Rilkonen P et al. Retinoblastoma associated with chromosomal 13q14 deletion mosaicism. *Ophthalmology*. 2003;110(10):1983–8.

26. Golde KT, Westerfeld CB, Mukai S. Ocular findings in patients with retinoblastoma and the 13q deletion syndrome. *Invest Ophthalmol Vis Sci*. 2009;50(13):4107.

27. Bunin GR, Emanuel BS, Meadows AT et al. Frequency of 13q abnormalities among 203 patients with retinoblastoma. *J Natl Cancer Inst*. 1989;81(5):370–4.

28. Kivela T. The epidemiological challenge of the most frequent eye cancer: Retinoblastoma, an issue of birth and death. *Br J Ophthalmol*. 2009;93(9):1129–31.

29. Seregard S, Lundall G, Svedberg H et al. Incidence of retinoblastoma from 1958 to 1998 in Northern Europe: Advantages of birth cohort analysis. *Ophthalmology*. 2004;111(6):1228–32.

30. Yun J, Li Y, Xu C et al. Epidemiology and *RB1* gene of retinoblastoma. *Int J Ophthalmol*. 2011;4(1):103–9.

31. Krishna SM, Yu GP, Finger PT. The effect of race on the incidence of retinoblastoma. *J Pediatr Ophthalmol Strabismus*. 2009;46(5):288–93.

32. Moreno F, Sinaki B, Fandino A et al. A population-based study of retinoblastoma incidence and survival in Argentine children. *Pediatr Blood Cancer*. 2014;61(9):1610–5.

33. One Retinoblastoma World. One Retinoblastoma World map. 1rbw (online). http://1rbw.org/.

34. Nyawira G, Kahaki K, Kariuki-Wanyoike M. Survival among the retinoblastoma patients at the Kenyatta National Hospital, Kenya. *J Ophthalmol. East Cent South Afr*. 2013;17:15–9.

35. Dean M, Bendfeldt G, Lou H et al. Increased incidence and disparity of diagnosis of retinoblastoma patients in Guatemala. *Cancer Lett*. 2014;351(1):59–63.

36. Canadian Retinoblastoma Society. National Retinoblastoma Strategy Canadian Guidelines for Care. *Can J Ophthalmol*. 2009;44:s1–88.

37. Dimaras H, Dimba EA, Gallie BL. Challenging the global retinoblastoma survival through a collaborative research effort. *Br J Ophthalmol*. 2010;94(11):1415–6.

38. Ascencio-LopezL,Torres-OjedaAA,Issac-OteroGetal.Treating retinoblastoma in the first year of life in a national tertiary pediatric hospital in Mexico. *Acta Pediatr*. 2015;104(9):e384–7.

39. Pesin N, Noble J, Gallie BL. Question: Can you identify this condition? Answer: Leukocoria. *Can Fam Physician*. 2010;56(2):155–6.

40. Muen W, Hindocha M, Reddy M. The role of education in the promotion of red reflex assessments. *JRSM Short Rep*. 2010;1(5):46.

41. Ellsworth RM. The practical management of retinoblastoma. *Trans Am Ophthalmol Soc*. 1969;67:462–534.

42. Karcioglu ZA. Fine needle aspiration biopsy (FNAB) for retinoblastoma. *Retina*. 2002;22(6):707–10.

43. Moulin AP, Gaillard MC, Balmer A et al. Ultrasound biomicroscopy evaluation of anterior extension in retinoblastoma: A clinicopathological study. *Br J Ophthalmol*. 2012;96(3):337–40.

44. Rootman DB, Gonzalez E, Mallipatna A et al. Hand-held high resolution spectral domain optical coherence tomography in retinoblastoma: Clinical and morphological considerations. *Br J Ophthalmol*. 2013;97(1):59–65.

45. de Jong MC, de Graaf P, Noij DP et al. Diagnostic performance of magnetic resonance imaging and computed tomography for advanced retinoblastoma: A systematic review and meta-analysis. *Ophthalmology*. 2014;121(5):1109–18.

46. Lee WH, Bookstein R, Hong F et al. Human retinoblastoma susceptibility gene: Cloning, identification and sequence. *Science*. 1987;235(4794):1394–99.

47. MacCarthy A, Bayne AM, Brownbill PA et al. Second and subsequent tumours among 1927 retinoblastoma patients diagnosed in Britain, 1951–2004. *Br J Cancer*. 2013;108(12):2455–63.

48. Finger PT, Harbour JW, Murphree et al. Retinoblastoma. In: Edge SB, Byrd DR, Carducci MA, Compton CC, editors. *AJCC Cancer Staging Manual*. New York, NY: Springer; 2010:561–8.

49. Reece AB, Ellsworth RM. The evaluation and current concept of retinoblastoma therapy. *Trans Am Acad Ophthal Otholaryngal*. 1963;67:164–72.

50. Kingston JE, Hungerford JL, Madreperla SA et al. Results of combined chemotherapy and radiotherapy for advanced intraocular retinoblastoma. *Arch Ophthalmol*. 1996;114(11):1339–43.

51. Murphree AL, Villablanca JG, Deegan WF 3rd et al. Chemotherapy plus local treatment in the management of intraocular retinoblastoma. *Arch Ophthalmol*. 1996;114(11):1348–56.

52. Shields CL, De Potter P, Himelstein BP et al. Chemoreduction in the initial management of intraocular retinoblastoma. *Arch Opththalomol*. 1996;114(11):1330–8.

53. Gallie BL, Budnig A, DeBoer G et al. Chemotherapy with focal therapy can cure intraocular retinoblastoma without radiotherapy. *Arch Ophthalmol*. 1996;114(11):1321–8.

54. Murphree AL. Intraocular retinoblastoma: The case for a new group classification. *Ophthalmol Clin North Am*. 2005;18(1):41–53.

55. Shields CL, Mashayekhi A, Au AK et al. The International Classification of Retinoblastoma predicts chemoreduction success. *Ophthalmology*. 2006;113(12):2276–80.

56. Novetsky DE, Abramson DH, Kim JW et al. Published international classification of retinoblastoma (ICRB) definitions contain inconsistencies- an analysis of impact. *Ophthalmic Genet*. 2009;30(1):40–4.

57. Mallipatna A, Gallie BL, Chevez-Barrios P et al. Retinoblastoma. In: Amin MB, Edge SB, Greene FL et al. editors. *AJJCC Cancer Staging Manual*. 8th ed. New York: Springer; 2017.

58. Fabain ID, Stacey AW, Johnson KP et al. Primary intravenous chemotherapy for group D retinoblastoma: A 13 year retrospective analysis. *Br J Ophthalmol*. 2017;101(1):82–8.

59. Fabian ID, Stacey AW, Johnson KC et al. Primary enucleation for group D retinoblastoma in the era of systemic and targated chemotherapy: The price of retaining an eye. *Br J Ophthalmol*. 2018;102(2):265–71.

60. Mallipatna AC, Sutherland JE, Gallie BL et al. Management and outcome of unilateral retinoblastoma. *J AAPOS.* 2009;13(6):546–50.

61. Zhao J, Dimaras H, Masssey C et al. Pre-enucleation chemotherapy for eyes severely affected by retinoblastoma masks risk of tumor extension and increases death from metastasis. *J Clin Oncol.* 2011;29(7):845–51.

62. Vincent AL, Webb MC, Gallie BL et al. Prosthetic conformers: A step towards improved rehabilitation of enucleated children. *Clin Exp Ophthalmol.* 2002;30:58–9.

63. Fabian ID, Stacey AW, Chowdhury T et al. High-risk histopathological features in primary and secondary enucleated international intraocular retinoblastoma classification group D eyes. *Ophthalmology.* 2017;124(6):854–58.

64. Kaliki S, Shields CL, Shaha SU et al. Postenucleation adjuvant chemotherapy with vincristine, etoposide and carboplatin for the treatment of high-risk retinoblastoma. *Arch Ophthalmol.* 2011;129(11):1422–7.

65. Brennan RC, Qaddoumi I, Billups CA et al. Comparison of high risk histopathological features in eyes with primary or secondary enucleation for retinoblastoma. *Br J Ophthalmol.* 2015;99(10):1366–71.

66. Kaliki S, Srinivasan V, Gupta A et al. Clinical features predictive of high-risk retinoblastoma in 403 Asia Indian patients: A case control study. *Ophthalmology.* 2015;122(6):1165–72.

67. Dimaras H, Hoen E, Budning A et al. Retinoblastoma CSF metastasis cured by multimodality chemotherapy without radiation. *Ophthalmic Genet.* 2009;30(3):121–6.

68. Honavar SG, Singh AD, Shields CL et al. Postenucleation adjuvant therapy in high-risk retinoblastoma. *Arch Ophthalmol.* 2002;120(7):923–21.

69. Wilson TW, Chan HS, Moselhy GM et al. Penetration of chemotherapy into the vitreous is increased by chemotherapy and cyclosporine in rabbits. *Arch Ophthalmol.* 1996;114(11):1390–5.

70. Jehanne M, Lumbroso-Le Rouic L, Savignoni A et al. Analysis of ototoxicity in young children receiving carboplatin in the context of conservative management of unilateral or bilateral retinoblastoma. *Pediatr Blood Cancer.* 2009;52(5):637–43.

71. Draper GJ, Sanders BM, Kingston JE. Second primary neoplasms in patients with retinoblastoma. *Br J Cancer.* 1986;53(5):661–71.

72. Gombos DS, Hungerford J, Abramson DH et al. Secondary acute myelogenous leukemia in patients with retinoblastoma: Is chemotherapy a factor? *Ophthalmology.* 2007;114(7):1378–83.

73. Yomane T, Kaneko A, Mohri M. The technique of ophthalmic arterial infusion therapy for patients with intraocular retinoblastoma. *Int J Clin Oncol.* 2004;9(2):69–73.

74. Abramson DH, Dunkel IJ, Brodie SE et al. A phase I/II study of direct intraarterial (ophthalmic artery) chemotherapy with melphalan for intraocualar retinoblastoma initial results. *Ophthalmology.* 2008;115(8):1398–404.

75. Taich P, Requejo F, Asprea M et al. Topotecan delivery to the optic nerve after ophthalmic artery chemosurgery. *PLOS ONE.* 2016;11(3):e0151343.

76. Abramson DH, Daniels AB, Marr BP et al. Intra-arterial chemotherapy (ophthalmic artery chemosurgery) for Group D retinoblastoma. *PLOS ONE.* 2016;11(3):e0146582.

77. Munier FL, Mosimann P, Houghton S et al. First-line intra-arterial versus intravenous chemotherapy in unilateral sporadic group D retinoblastoma: Evidence of better visual outcomes, ocular survival and shorter time to success with intra-arterial delivery from retrospective review of 20 years of treatment. *Br J Ophthalmol.* 2017;101(8):1086–93.

78. Shields CL, Jorge R, Say EA et al. Unilateral retinoblastoma managed with intravenous chemotherapy versus intra-arterial chemotherapy. Outcomes based on the Interantional Classification of Retinoblastoma. *Asia-Pac J Ophthalmol.* 2016;5(2):97–103.

79. Yousef YA, Soliman SE, Astudillo PP et al. Intra-arterial chemotherapy for retinoblastoma: A systematic review. *JAMA Ophthalmol.* 2016. doi: 10.1001/jamaophthalmol.2016.0244.

80. Grigorovski N, Lucena E, Mattosinho C et al. Use of intra-arterial chemotherapy for retinoblastoma: Results of a survey. *Int J Ophthalmol.* 2014;7(4):726–30.

81. Abramson DH, Shields CL, Munier FL et al. Treatment of retinoblastoma in 2015: Agreement and disagreement. *JAMA Ophthalmol.* 2015; 133(11):1341–7.

82. Muen WJ, Kingston, JE, Robertson F et al. Efficacy and complications of super-selective intra-ophthalmic artery melphalan for the treatment of refractory retinoblastoma. *Ophthalmology.* 2012;119(3):611–16.

83. Tsimpida M, Thompson DA, Liasis A et al. Visual outcomes following intraophthalmic melphalan for patients with refractory retinoblastoma and age appropriate vision. *Br J Ophthalmol.* 2013;97(11):1464–70.

84. Reddy MA, Naeem Z, Duncan C et al. Reduction of severe visual loss and complications following intra-arterial chemotherapy (IAC) for refractory retinoblastoma. *Br J Ophthalmol.* 2017;101(12):1704–08.

85. Fabian ID, Johnson KP, Stacey AW et al. Focal laser treatment in addition to chemotherapy for retinoblastoma. *Cochrane Database of Syst Rev.* 2017;6:CD012366.

86. Chaudhry S, Onadim Z, Sagoo MS et al. The recognition of cavitary retinoblastoma tumors: Implications for management and genetic analysis. *Retina.* 2017. doi: 10.1097/IAE.0000000000001597.

87. Stallard HB. Radiotherapy of malignant intraocular neoplasms. *Br J Ophthalmol.* 1948;32(9):618–39.

88. Moore RF, Stallard HB, Milner JG. Retinal gliomata treated by radon seeds. *Br J Ophthalmol.* 1931;15(12):673–96.

89. Stallard HB. The treatment of retinoblastoma. *Ophthalmologica.* 1996;151:214–30.

90. Munier FL, Gaillard MC, Balmer A et al. Intravitreal chemotherapy for vitreous disease in retinoblastoma revisited: From prohibition to conditional indications. *Br J Ophthalmol.* 2012;96(8):1078–83.

91. Shields CL, Douglas AM, Beggache M et al. Intravitreous chemotherapy for active vitreous seeding from retinoblastoma: Outcomes after 192 consecutive injections. The 2015 Howard Naquin Lecture. *Retina.* 2015;36(6):1184–90.

92. Chantada G, Luna-Fineman S, Sitorus RS et al. SIOP-PODC recommendations for graduated-intensity treatment of retinoblastoma in developing countries. *Pediatr Blood Cancer.* 2013;60(5):719–27.

93. Hanavar SG, Singh AD. Management of advanced retinoblastoma. *Ophthalmol Clin North Am.* 2005;18(1):65–73.

94. Fabian ID, Naeem Z, Stacey AW et al. Long-term visual acuity, strabismus and nystagmus outcomes following multimodality treatment in group D retinoblastoma eyes. *Am J Ophthalmol.* 2017;179:137–44.

95. Foster A, Boyes L, Burgess L et al. Patient understanding of genetic information influences reproductive decision making in retinoblastoma. *Clin Genet.* 2017;92(6):587–93.

8

Rhabdoid Tumor Predisposition Syndrome

Valerie A. Fitzhugh

CONTENTS

8.1 Introduction

Hereditary cancer syndromes can present as congenital syndromes affecting people at the earliest stages of life and extending into the adult years. Many examples, such as Gardner syndrome and neurofibromatoses, have been extensively studied with a wealth of literature documenting their association with various clinical diseases. Subsequent uncovering of new genes and their roles in specific diseases has facilitated the establishment of new syndromes.

One of these newly established syndromes is rhabdoid tumor predisposition syndrome. First described in 1999 by Sévenet et al. [1], the rhabdoid tumor predisposition syndrome includes malignancies that share common mutation in the *SMARCB1* gene. The most common tumor within the syndrome is malignant rhabdoid tumor, which was also the first tumor shown to contain a *SMARCB1* mutation. Malignant rhabdoid tumors (particularly of the kidney) were originally described in 1978 by Beckwith and Palmer [2], who felt these tumors were a variation on Wilms tumor. However, as pathologists became more and more familiar with the histologic features of malignant rhabdoid tumor, the diagnoses were being seen in sites outside the kidneys, most commonly the brain [1,3]. Other involved anatomic sites include the soft tissue, ovary, liver, and lung [3].

There are naming conventions that separate the renal and brain neoplasms from the other members of the malignant rhabdoid tumor group; in the kidney these neoplasms are referred to as malignant rhabdoid tumor of the kidney, and in the brain they are referred to as atypical teratoid/rhabdoid tumors [3–5]. Rorke et al. described the atypical teratoid/rhabdoid tumors in 1996 [5]. These tumors, prior to 1996, were being misdiagnosed as primitive neuroectodermal tumors or medulloblastomas; it was Rorke's group that established criteria for the atypical teratoid/rhabdoid tumors.

Regardless of how the tumor is named, rhabdoid tumors arising at any site within the body are recognized as the same entity, linked by morphology, biology, and clinical features [3].

This chapter focuses on the rhabdoid tumor predisposition syndrome in relation to its biology, pathogenesis, affected populations, clinical features, diagnosis, treatment, and prognosis. Future directions will also be considered.

8.2 Biology and Pathogenesis

SMARCB1 (also known as INI1, hSNF5, and BAF47) is a member of the switch/sucrose nonfermenting (SWI/SNF) complex, which is an ATPase-dependent, multi-subunit complex involved in transcriptional regulation and chromatin remodeling [3,6]. These complex subunits contribute to the regulation of differentiation and cell proliferation. SWI/SNF subunits play a role in tumor suppression. Inactivating mutations have been described in the rhabdoid tumor predisposition syndrome (and will be described in detail in this section) as well as several other tumor types [6]. The *SMARCB1* gene is composed of 9 exons spanning over 50 kb and encodes a 385 aa, 44 kDa subunit within the SWI/SNF complex that is important in transcription regulation by inhibition of entry into S-phase of mitosis [6].

Despite their rarity, there has been some emphasis on understanding the biology and pathogenesis of malignant rhabdoid tumors, which are intimately linked. Cytogenetic and molecular analyses have demonstrated that deletion of chromosome 22q11.2 is a recurrent genetic abnormality in these neoplasms [7–9]. Researchers began to turn their focus to the gene that was responsible for the development of the rhabdoid family of tumors. In 1998, Versteege et al. identified 6 homozygous deletions from a panel of 13 malignant rhabdoid tumor cell lines [10]. Within those six deletions, the authors identified a small focus of overlap involving the *SMARCB1* (*hSNF5/INI1*) region. These truncating mutations were identified in one allele and resulted in the loss of the other allele [10]. The authors went on to suggest that loss of function mutations of *SMARCB1* contribute to the oncogenesis of malignant rhabdoid tumors.

Several months later, Biegel et al. examined 29 malignant rhabdoid tumors (18 atypical teratoid/rhabdoid tumors, 7 malignant rhabdoid tumors of the kidney, and 4 extrarenal malignant rhabdoid tumors) for mutations in INI1 [11]. They utilized karyotype, microsatellite, and reverse transcript-polymerase chain reaction (RT-PCR) in their analysis. In addition, they sequenced the amplified PCR products. They found that 15 tumors had homozygous deletions of 1 or more exons of the *SMARCB1* gene; the other 14 tumors had mutations. Germline mutations were identified in four of the children in their study, three of whom had malignant rhabdoid tumors of the kidney and one of whom had an atypical teratoid/rhabdoid tumor. As a result, Biegel et al. also suggested that *SMARCB1* is a tumor suppressor gene implicated in the pathogenesis of malignant rhabdoid tumors of the kidney, brain, and extrarenal sites [11].

In light of these discoveries, Sévenet et al. published their landmark manuscript in 1999 in which they found that the loss of function mutations in SMARCB1 were not only seen in the tumors of the affected patients, but within their constitutional DNA as well; furthermore, apparently healthy family members within families affected with these tumors did not demonstrate this constitutional change in their DNA [1]. This led the authors to conclude that the constitutional mutation of *SMARCB1* defined a syndrome, which they termed the *rhabdoid predisposition syndrome* [1]. This syndrome, which predisposes to malignant rhabdoid tumors of the kidney, brain, and other extrarenal sites, also predisposes to malignant central nervous system tumors including medulloblastoma, central primitive neuroectodermal tumor, and choroid plexus carcinoma [1].

The rhabdoid tumor family is clearly distinct from other tumors, particularly within the central nervous system [12,13]. They do not contain the isochromosome 17q commonly identified in medulloblastoma. Instead, as discussed earlier, they are more often characterized by monosomy 22 or a deletion or translocation involving chromosome 22q11.2. Malignant rhabdoid tumor of the kidney also demonstrates deletions of chromosome 22. Fluorescence in situ hybridization (FISH) is thought to be the best method for detecting these deletions [12,13]. It is thought that *SMARCB1* acts as a tumor suppressor gene; it suppresses tumor formation by regulation of the p16-CDK4/cyclin D-Rb-E2F pathway. While the mechanism by which SMARCB1 activates p16 is not entirely understood, but what is known is that p16 expression is essential for SMARCB1-associated tumor suppression [12,13]. It also appears that expression of cyclin D1 is necessary for tumor development, confirmed by several murine knockout models [12,13]. Many exons of the SMARCB1 gene have been examined, and it appears that exons 5 and 9 are hot spots for the CNS rhabdoid tumors, whereas exons 2 and 6 of the gene appear to be the hot spots for malignant rhabdoid tumors of the kidney [12,13]. Genetic alterations identified throughout the literature include missense mutations, in-frame deletions, and splice site mutations, resulting in altered SMARCB1 proteins, whereas nonsense and frameshift mutations commonly result in

absent SMARCB1 protein [6]. Interestingly, while it is clear that patients who develop rhabdoid tumors have a poor prognosis, it does not appear that the type of genetic alteration affects overall survival. Also important is the age of the patient at presentation. Patients who develop disease at an early age often have germline *SMARCB1* mutations; these patients are at high risk for progression of their disease [14].

SMARCB1 is also closely linked to the development of schwannomatosis. Schwannomas are benign tumors of the nerve sheath that are most commonly solitary, but can be multifocal, particularly in several syndromes, most prominently neurofibromatosis type 2 [3]. Schwannomatosis is also a major type of neurofibromatosis, characterized by the development of cranial and peripheral nerve and spinal schwannomas. It is not related to NF2, as formerly thought, but instead related to SMARCB1. Unlike the rhabdoid tumor predisposition syndrome variant of the gene which expresses cyclin D1, the schwannomatosis variant represses cyclin D1, which may result in less lethal tumors [3].

Furthermore, mutations in the *SMARCB1* gene are implicated in a severe form of Coffin—Siris syndrome (known alternatively as fifth digit syndrome), which is a congenital multiple malformation disorder inherited in an autosomal dominant pattern and characterized by intellectual disability, coarse facial features, hypertrichosis, and hypoplastic or absent fifth digit nails or phalanges, in addition to malformations of the cardiac, gastrointestinal, genitourinary, and/or central nervous systems; sucking/feeding difficulties, poor growth, ophthalmologic abnormalities, hearing impairment, and spinal anomalies.

For a number of years, it was thought that *SMARCB1* was the only gene involved in the development of malignant rhabdoid tumors; however, studies began to show that there were other genetic pathways within the pathogenesis of these tumors. In 2006, Frühwald et al. published one of the first reports of rhabdoid tumor in a family that did not show a mutation in *SMARCB1* [15]. The patients were two children in a family of three children total. One of the children was diagnosed with an atypical teratoid/rhabdoid tumor, and a second was diagnosed with a malignant rhabdoid tumor of the kidney. The clinical and morphologic features were compatible in both cases, but no evidence of a *SMARCB1* mutation by comparative genomic hybridization (CGH), array CGH, FISH, gene dosage analysis, or DNA sequencing was found. Immunohistochemistry demonstrated normal expression of SMARCB1 as well. The authors thus concluded that a second locus must be implicated in the pathogenesis of malignant rhabdoid tumors and the rhabdoid tumor predisposition syndrome [15].

In 2010, a second candidate gene (i.e., *SMARCA4* located on chromosome 19p13.2) was uncovered. The *SMARCA4* gene (also known as Brahma related gene-1) encodes a 1647 aa, 185 kDa catalytic subunit (SMARCA4 or BRG1) of SWI/SNF complexes, which regulate gene expression by remodeling chromatin to alter nucleosome conformation, making it more accessible to transcriptional activation. Schneppenheim et al. sequenced four candidate genes from the SWI/SNF complex and noted that both children in the family demonstrated inactivation of *SMARCA4* due to a germline mutation and loss of heterozygosity by uniparental disomy [16]. This was the first report of a gene other than *SMARCB1* to be implicated in the pathogenesis of the rhabdoid tumor predisposition syndrome. A second report, also involving

SMARCA4, followed in 2011. This case involved a 9-month-old boy who developed atypical teratoid/rhabdoid tumor and succumbed to his disease 6 months after his diagnosis [17]. While the clinical significance of the *SMARCA4* mutation in malignant rhabdoid tumors is still unknown, there is some evidence that atypical teratoid/rhabdoid tumors that demonstrate this mutation may be associated with a poorer prognosis and higher germline mutation frequency when compared to *SMARCB1* [3]. In contrast to *SMARCB1*, *SMARCA4* appears to be involved in the pathogenesis of a mild to moderate form of Coffin−Siris syndrome.

Malignant rhabdoid tumors are well known to be biologically aggressive, particularly in very young children; the prognosis is uniformly poor [4]. In most patients, even with therapy (discussed later), death occurs within months of the initial diagnosis.

8.3 Epidemiology

The rhabdoid tumor family encompasses highly aggressive malignancies with a predilection for the pediatric population. These tumors generally present in infancy or early childhood [3,4,18]. Pediatric cancer is rare in and of itself, with the literature citing about 150 diagnosed cases per 1,000,000 under 20 years of age per annum in the United States and Europe [18]. Even rarer are childhood malignancies arising as part of a genetic syndrome, constituting only 1%–10% of all pediatric cancer cases [18]. Hence, the rhabdoid tumor predisposition syndrome is among the rarest of these already rare tumors, limiting the data available in terms of epidemiology.

It is well known, as described earlier in this chapter, that the tumors are most commonly identified in the kidney, where they are known as malignant rhabdoid tumor of the kidney, and in the central nervous system, where they are known as atypical teratoid/rhabdoid tumor [1,3,4,18]. The current literature suggests that up to one-third of patients who develop rhabdoid tumors have *SMARCB1* germline inactivating mutations [18]. Despite most of these mutations occurring de novo, some of these tumors are familial and are thus assigned to the rhabdoid tumor predisposition syndrome type 1 (RTPS1) [3]. Patients who have germline inactivating mutations in SMARCA4 and have family history of rhabdoid tumors are assigned to the rhabdoid tumor predisposition syndrome type 2 (RTPS2) [3].

Families who exhibit rhabdoid tumor predisposition syndrome have a pedigree that has a minimum of two individuals who have a germline mutation, most commonly in *SMARCB1*. These families typically have more than one individual affected by a rhabdoid tumor in any anatomic location [3]. Unfortunately, children with germline mutations of the *SMARCB1* gene have a higher incidence of multiple rhabdoid tumors; these children also develop tumors at a younger age when compared to cohorts with sporadic tumors [3]. In addition, 10%–15% of children who develop malignant rhabdoid tumors of the kidney, if they survive, go on to develop atypical teratoid/rhabdoid tumors [19].

Despite a thorough review of the literature, it is difficult to determine if there is a gender predilection for the rhabdoid tumor predisposition syndrome, likely due to the fact that this syndrome is very rare; the syndrome affects between 5 and 8.1 children per 1,000,000 [20]. However, studies have been undertaken of the two most common tumors identified within the rhabdoid tumor predisposition syndrome, malignant rhabdoid tumor of the kidney and atypical teratoid/rhabdoid tumor. The National Wilms Tumor Study looked at 142 patients with malignant rhabdoid tumor of the kidney and examined many variables, including age and gender of the affected children. In their study, 82 patients were male and 60 were female, leading to a male to female ratio of 1.37:1, a slight male predilection [19]. The age of the children in their study ranged from newborns to one patient who was older than 8 years (the child's exact age was not given in the study). As is commonly seen in these tumors, age at diagnosis was skewed toward the younger age groups [19]. Fifteen patients were diagnosed at stage I, 25 patients were diagnosed at stage II, 58 patients were diagnosed at stage III, 41 patients were diagnosed at stage IV, and 3 patients were diagnosed at stage V, classified as bilateral disease. This indicates that the disease tends to be diagnosed at higher stages [19]. In children with high stage disease, the lung appears to be a favored site. The lung metastases tend to be numerous and unresectable [19].

Atypical teratoid/rhabdoid tumor has also been studied, usually in national studies. In a study of the Pediatric Oncology Group [21], 55 patients were identified as eligible participants. Of these, 34 patients were boys and 21 were girls, again indicative of a slight female predilection. The ages of the patients ranged from 2 to 60 months with an average age of presentation of 17 months [21]. Thirty-six of the study patients had tumor located in the posterior fossa. Thirteen patients with tumors in the posterior fossa demonstrated extension into the supratentorial space. Seventeen cases were primary supratentorial tumors (4 suprasellar, 3 involving the pineal gland, 3 parietal lobe, 3 temporal lobe, 2 frontal lobe, and 2 CNS site unspecified). Two cases were multifocal at the time of presentation. Cerebrospinal fluid analysis was performed in 29 patients as part of their presentation; 9 demonstrated positivity for malignant cells [21]. Twenty-five patients went on to develop craniospinal metastases [21]. The metastases are most commonly unresectable. There is now some indication that the atypical teratoid/rhabdoid tumors should be classified into subgroups that may affect treatment. Atypical teratoid/rhabdoid tumor TYR, which is characterized by location in the posterior fossa, very young age at diagnosis, typically less than 1 year, demonstrates overexpression of the genes TYR and MITF [20]. The subgroup atypical teratoid/rhabdoid tumor MYC consists of predominately supratentorial tumors, and the patients are older than average at diagnosis (4−5 years old). The genes MYC, HOX, and HOTAIR are overexpressed [20]. The third subgroup, atypical teratoid rhabdoid tumor SHH, occurs in children between 2 and 5 years old, is either supratentorial or infratentorial, and is characterized by overexpression of the sonic hedgehog genes [20].

Another study examined the epidemiology of rhabdoid tumors of early childhood in relation to perinatal characteristics. They found that low birthweight, defined as <2.5 kg, was an independent risk factor for the development of both renal and extrarenal rhabdoid tumors as well as atypical teratoid/rhabdoid tumor [22]. Preterm delivery (defined as less than 37 weeks' gestation), postterm delivery (defined as >42 weeks' gestation), and multiple gestations, in particular twin gestations, were also found to be risk factors for the development of rhabdoid tumor [22]. The affected children were more likely to be white than any other racial group, and the parents were more likely to be older and well educated irrespective of race [22]. The mean age for all patients in their study was 12.3 months. The study group included 58 boys and 47 girls, again pointing to a slight male predominance.

As reported earlier, the rhabdoid tumor predisposition syndrome is most commonly the result of the inheritance of defective copies of either the SMARCB1 or SMARCA4 genes. There have been reports of reduced penetrance of the SMARCB1 gene in some cohorts with rhabdoid tumor predisposition syndrome. In rare familial cohorts, a pathogenic SMARCB1 variant is inherited from an unaffected parent or a parent with undiagnosed, late-onset rhabdoid tumor predisposition syndrome [20,23]. When examining newly diagnosed patients with rhabdoid tumors, 25%–35% will demonstrate a germline mutation in SMARCB1 [20]. In contrast, most patients whose rhabdoid tumor predisposition syndrome is due to inheritance of SMARCA4 inherited a defective copy of the gene from an unaffected healthy parent. While in most families the penetrance of rhabdoid tumors in the preceding generation was zero, in one family, two siblings both inherited defective copies of SMARCA4 and both developed rhabdoid tumor predisposition syndrome [16,20]. In the United States, the annual incidence among children <15 years old for rhabdoid tumor predisposition–related tumors is 0.19 per million for rhabdoid tumor of the kidney, 0.89 per million for atypical teratoid/rhabdoid tumor, and 0.32 for malignant rhabdoid tumors of other sites [20].

8.4 Clinical Features

Rhabdoid tumor predisposition syndrome, as discussed earlier, is defined as a syndrome with a marked propensity to develop rhabdoid tumors of the kidney, brain, and other soft tissue sites. Soft tissue sites that have been implicated as part of the syndrome including rhabdoid tumors of the head and neck, paravertebral muscles, mediastinum, heart, retroperitoneum, liver, and pelvis; small cell carcinoma of the ovary, hypercalcemic type, is also now being considered as possibly being part of the syndrome [20]. The clinical presentation of the rhabdoid tumor predisposition syndrome is most commonly determined and pursued based on the development of a rhabdoid tumor at an extremely young age. These will therefore be examined in detail for malignant rhabdoid tumor of the kidney and atypical teratoid/rhabdoid tumor; clinical features of the syndrome in context of the diagnosis of these tumors will also be discussed.

Malignant rhabdoid tumor of the kidney is a tumor that most commonly arises within the renal pelvis; invasion at this location is likely to present as hematuria. A comparison of clinical presentation of children with Wilms tumor (which more commonly arises peripherally) versus malignant rhabdoid tumor of the kidney demonstrated a higher incidence in hematuria in the rhabdoid tumor patients than the Wilms tumor patients [24]. In addition, anemia and fever were also seen in many of the patients diagnosed with rhabdoid tumor of the kidney. Hematuria, fever, and anemia are thus considered important considerations in a young child when entertaining the diagnosis of malignant rhabdoid tumor of the kidney [24].

Computed tomography (CT) imaging is considered the gold standard for radiologic examination of suspected malignant rhabdoid tumor of the kidney. CT findings that were examined in a series of eight tumors included location of tumor within the kidney, subcapsular hematoma, multiple tumor lobules, presence of calcification, enlarged vessels, vascular invasion, central tumor necrosis or hemorrhage, visibility of tumor margin, distant metastasis, and primary tumor size [25]. Of these, subcapsular hematoma and lobular appearance in a large centrally located heterogenous kidney mass containing calcifications was thought to be most consistent with rhabdoid tumor [25]. Since this study, other studies have proposed that there are no specific imaging findings for malignant rhabdoid tumor and that a series of findings must be taken into consideration [26].

Atypical teratoid rhabdoid tumors present as would be expected with a space-occupying lesion in the brain. Magnetic resonance imaging (MRI) is the most commonly used modality for imaging of these lesions. The tumors are large, most commonly have solid and cystic foci, and demonstrate focal calcifications [26]. Unfortunately, these imaging characteristics are nonspecific. It has been reported that approximately two-thirds of atypical teratoid/rhabdoid tumors demonstrate cellular elements similar to those seen in primitive neuroectodermal tumors; therefore, it can be difficult to differentiate these lesions by radiology alone [26]. Consequently, biopsy is important to differentiate these lesions and initiate appropriate treatment.

Any of the following diagnoses should prompt efforts directed at determining whether or not a patient might have rhabdoid tumor predisposition syndrome, particularly in a young child: (i) atypical teratoid/rhabdoid tumor, of which >50% are identified within the posterior fossa/cerebellum; (ii) rhabdoid tumor of the kidney; (iii) rhabdoid tumors of the head and neck, paravertebral muscles, mediastinum, heart, liver, retroperitoneum, pelvis, and bladder; (iv) small cell carcinoma, hypercalcemic type, of the ovary.

Most patients with rhabdoid tumor predisposition syndrome present at very young ages, often less than a year. Rhabdoid tumor predisposition syndrome should be considered in any of the following presentations: (i) prenatal detection of synchronous rhabdoid tumors; (ii) infantile/congenital diagnosis of rhabdoid tumor, where the average age of presentation is 4–7 months; (iii) synchronous rhabdoid tumors; (iv) family history of rhabdoid tumor or other SMARB1- or SMARCA4-associated tumors; (v) family history of small cell carcinoma, hypercalcemic type of the ovary, which suggests a SMARCA4 mutation since SMARCB1 has not been implicated in the ovarian tumors [20].

8.5 Diagnosis

The acquisition of tissue is paramount to proper diagnosis of rhabdoid tumors and can be used for genetic studies as well.

Histologically, atypical teratoid/rhabdoid tumor (Figure 8.1), malignant rhabdoid tumor of the kidney (Figure 8.2), and extrarenal/extracranial rhabdoid tumors share similar morphology. Rhabdoid cytomorphology is common in each of these tumor types. The rhabdoid morphology is characterized by large polygonal cells with vesicular, eccentrically placed nuclei. These nuclei often contain prominent nucleoli. The cytoplasm often contains a single eosinophilic cytoplasmic inclusion that corresponds to the presence of intermediate filaments [27,28]. The tumor classically grows as a discohesive mass of polygonal cells [27,28]. In addition to the classic pattern, some tumors also demonstrate spindled, epithelioid, ovoid, or polygonal shapes [27].

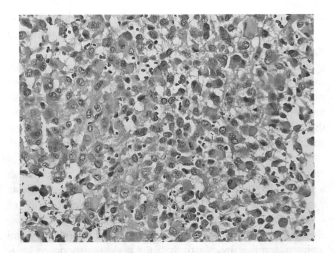

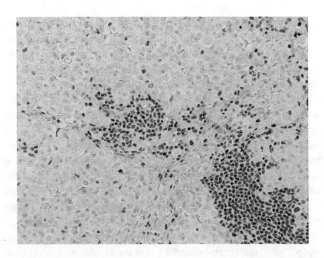

FIGURE 8.1 Atypical teratoid/rhabdoid tumor of the brain demonstrating classic rhabdoid morphology. Nuclei are vesicular with prominent nucleoli. The cytoplasm is eosinophilic and contains a single inclusion thought to contain intermediate filaments. The tumor cells are largely discohesive. Hematoxylin and eosin, 400×. (Photo credit: Jennifer Kasten, MD. Department of Pathology. Boston Children's Hospital, Boston, Massachusetts.)

FIGURE 8.3 Malignant rhabdoid tumor metastatic to a lymph node. This lymph node has been stained for INI-1. Notice that INI-1 is limited only to the normal mature lymphocytes (positive brown staining) and is lost in the tumor cells, demonstrated by negative (blue) staining. INI-1, 400×. (Photo credit: Jennifer Kasten, MD. Department of Pathology. Boston Children's Hospital, Boston, Massachusetts.)

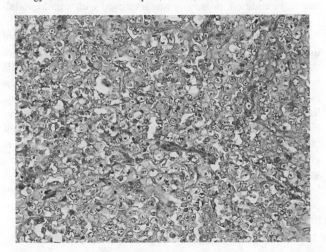

FIGURE 8.2 Malignant rhabdoid tumor of the kidney. This tumor almost demonstrates discohesion between the cells. Again seen are rhabdoid cells containing vesicular nuclei with prominent nucleoli. The cytoplasm again contains a single intracytoplasmic inclusion in many cells. Hematoxylin and eosin, 400×. (Photo credit: Jennifer Kasten, MD. Department of Pathology. Boston Children's Hospital, Boston, Massachusetts.)

The epithelioid differentiation can be seen as cords and ribbons of cells, to foci of squamoid and glandular epithelium, to poorly differentiated glandular structures [28]. These tumors are infiltrative with lymphovascular invasion, a high proliferative index, and tumor cell necrosis [27,28]. Atypical teratoid/rhabdoid tumors, as briefly mentioned earlier, can demonstrate features of primitive neuroectodermal tumor, which becomes an important differential diagnostic consideration [26,27].

Immunohistochemistry is an important ancillary tool in the diagnosis of rhabdoid tumors, particularly to differentiate it from many of its differential diagnostic counterparts, which include primitive neuroectodermal tumor, desmoplastic small round cell tumor, epithelioid sarcoma, and synovial sarcoma [27]. Because malignant rhabdoid tumors tend to exhibit epithelial, mesenchymal, and neuroectodermal differentiation, one would expect

markers encompassing those lineages to be immunoreactive. As one would expect, the tumors demonstrate immunoreactivity to epithelial membrane antigen and cytokeratin (epithelial markers), vimentin, smooth muscle actin, desmin, and CD99 (mesenchymal markers), and S-100, neuron-specific enolase, synaptophysin, and glial fibrillary acidic protein (neural markers) [27,29]. Most of the entities within the differential will react to only a subset of these markers rather than the entire panel, allowing for accurate diagnosis.

Immunohistochemistry for INI-1 has also been developed and is an important clue when considering the possibility of affliction with the rhabdoid tumor predisposition syndrome. In rhabdoid tumors, INI-1 is commonly lost (Figure 8.3), particularly in lesions that demonstrate a germline mutation for *SMARCB1* [27,28,30,31]. Since INI-1 is retained in normal tissues, any normal tissue that is submitted adjacent to or within the tumor (inflammatory cells for example) will retain INI-1 immunoreactivity. There have been reports of primitive neuroectodermal tumors that also demonstrate loss of INI-1; these tumors did not have a rhabdoid phenotype [31]. It is important in these cases to submit the remainder of the tissue (minus that submitted for cytogenetic or molecular studies) biopsied or resected in order to ensure that there is no rhabdoid component present, particularly in atypical teratoid/rhabdoid tumors which are known to have a primitive neuroectodermal component. In cases where the tumor has a rhabdoid morphology but INI-1 is retained, there should be concern for a germline mutation in *SMARCA4*, which has also been previously reported [17]. Immunohistochemistry for SMARCA4 is now available [20].

INI-1 is retained in both rhabdomyosarcoma and desmoplastic small round cell tumor, which aids in the differential diagnostic consideration [27]. INI-1 can be lost in synovial sarcoma, which is commonly immunoreactive to CD99, cytokeratin, and epithelial membrane antigen [27]. TLE1, positive in synovial sarcoma, has yet to be examined in malignant rhabdoid tumors [27]. Cytogenetic studies can be helpful in difficult cases. Epithelioid sarcoma proves to be the most difficult to separate from the rhabdoid tumors. It often has rhabdoid morphology, is positive for epithelial and mesenchymal markers, and also demonstrates loss of INI-1 [27]. There are differences in the clinical presentation, as malignant rhabdoid tumors tend to occur in very young children and epithelioid sarcoma tends to occur over a wide age range. The sites of presentation also differ [27].

Cytogenetic and molecular studies are also helpful in the diagnosis of rhabdoid tumors. Despite their high-grade cytologic features, they have remarkably few cytogenetic changes [7]. This is important because tumors in the differential diagnosis, including rhabdomyosarcoma, desmoplastic small round cell tumor, and synovial sarcoma, have specific genetic changes that allow for their distinction from the malignant rhabdoid tumors [6,27]. This can be accomplished by analysis of a karyotype from the tumor tissue of a patient diagnosed with a malignant rhabdoid tumor.

Molecular genetic testing is also important in patients diagnosed with malignant rhabdoid tumors at any site. Common molecular testing approaches include serial single gene testing as part of a multigene panel [20]. Fresh frozen tissue is best for this process and should be collected at the time of tumor submission to the pathologist. Serial single gene testing is best used in patients who demonstrate loss of INI-1 or SMARCA4 by immunohistochemistry. The order in which the tests are performed is determined by which gene is lost. If INI-1 is lost, SMARCB1 analysis should performed first; the opposite is true if SMARCA4 is lost [20]. Tests to perform include sequence analysis and gene-targeted deletion/duplication analysis [20].

In order to establish the diagnosis of rhabdoid tumor predisposition syndrome in a patient diagnosed with a malignant rhabdoid tumor at any site, he or she must demonstrate both of the following: (i) A rhabdoid tumor and/or a family history of either rhabdoid tumor and/or multiple SMARCB1 or SMARCA4 deficient tumors. These tumors may be metachronous or synchronous. (ii) Identification of a germline mutation in either *SMARCB1* (which is attributed to 85%–95% of RTPS cases) or *SMARCA4* (which is attributed to 5%–15% of RTPS cases) by molecular testing.

Genetic counseling is an important part of the diagnostic process. Most patients who develop malignant rhabdoid tumors and go on to be diagnosed with the rhabdoid tumor predisposition syndrome have a de novo germline mutation in *SMARCB1*. Most patients with the *SMARCA4* germline mutation inherited their defective gene from an unaffected parent. Each child of a person carrying a *SMARCA4* or *SMARCB1* mutation has a 50% chance of inheriting a mutated gene. Penetrance appears to be incomplete [20]. It is therefore important to introduce early genetic counseling and support to these patients and their families.

8.6 Treatment

As mentioned previously, the rhabdoid tumor predisposition tumor is rare. Therefore, treatment standards continue to evolve. The common denominator in these cases is that patients have been treated with a multimodality approach, which commonly consists of surgery, chemotherapy, and radiotherapy [3,20] depending on the age of the patient.

The Children's Oncology Group presented a study where they used a combination of surgery, two cycles of induction chemotherapy which consisted of cyclophosphamide, cisplatinum, etoposide, methotrexate, and vincristine, followed by consolidation therapy

with high-dose chemotherapy (carboplatinum and thiotepa) with stem cell rescue. Radiotherapy was added according to the age of the patient and the stage of the tumor at presentation [32].

The Dana Farber Consortium has studied combination chemotherapy, surgery, and radiotherapy divided into five phases: pre-irradiation, chemoradiation, consolidation, maintenance, and continuation therapy. Chemotherapy was administered intrathecally, and radiotherapy was administered based on age of the patient and stage of the disease [33].

The European Registry for Rhabdoid Tumors (EU-RHAB) was created to establish a common European database and establish a standardized treatment regimen for rhabdoid tumors. Patients were recruited into the Rhabdoid 2007 study over a period of 4 years. Patients were treated with systemic and intraventricular chemotherapy. Some treatment arms included radiotherapy and maintenance therapy. Nearly half the patients experience long-term remission [34].

A group in China employed gamma knife surgery in combination with radiation therapy, chemotherapy, both, or surgery alone. Gamma knife surgery was found to be a safe, effective method in the surgical treatment of atypical teratoid/rhabdoid tumor. In addition, prognosis was improved with a multimodality approach to therapy [35].

While much of the available literature of treatment of rhabdoid tumor centers on atypical teratoid/rhabdoid tumor, reports of therapy for rhabdoid tumor of the kidney also exist. The GPOH nephroblastoma group studied reviewed neoadjuvant chemotherapy in patients with malignant rhabdoid tumor of the kidney. Roughly half were assigned to a treatment arm without doxorubicin, while the other half was assigned to a treatment arm containing doxorubicin. Both cohorts received actinomycin D and vincristine. They found that patients who received doxorubicin saw a significantly improved response [36].

In most treatment studies of the malignant rhabdoid tumor family, it appears that survival is positively influenced by multimodality therapy.

8.7 Prognosis, Prevention, and Surveillance

Survival rates are dismal for malignant rhabdoid tumors regardless of site and seem to be the most poor in patients who are the youngest. The National Wilms Tumor Study Group has previously reported that mortality is as high as 80%, and most of the affected are dead within a year of diagnosis [27].

While multimodality chemotherapy has improved survival, it appears there is quite a long way to go. For those who survive this disease, other disease manifestations may result. One group described a case of a boy who developed a rhabdoid tumor at 8 years of age. He was found to have a SMARCB1 germline mutation. Five years later, at 13 years of age, he developed a conventional chondrosarcoma of the mandible [37]. Pediatric chondrosarcomas are exceedingly rare; what made this case even more relevant was that his chondrosarcoma demonstrated loss of INI-1 as well. This was one of the earliest cases described where a survivor of a rhabdoid tumor went on to develop a secondary tumor [37]. A second group described a patient, also a boy, who was originally diagnosed with an atypical teratoid/rhabdoid tumor at age 3. He survived, but then went on to develop a mammary analog secretory carcinoma at age 14 [38]. In this case, however, INI-1 or SMARCA4 immunohistochemistry was not performed, so it is uncertain whether this is related to germline mutation of SMARCB1 or SMARCA4 [6,38].

Surveillance is key in patients and families who are known to have rhabdoid tumor predisposition syndrome. To date, no formal recommendations for surveillance have been made in regard to the syndrome because the penetrance is incomplete and the tumors can occur in every tissue of the body [39]. This syndrome is difficult to screen for due to the fact that the age of onset in these patients is often quite young and sometimes is in utero. The tumors are clinically aggressive, and onset is rapid. Second malignancies, as discussed earlier in this section, are also a risk for patients who survive their illness [39]. Familial cases are rare.

Despite these challenges, Foulkes et al. have given recommendations for surveillance of patients who have been diagnosed with rhabdoid tumors as follows [39]:

1. Patients with SMARCB1 germline truncating mutations should undergo MRI examination of the brain every 3 months until 5 years of age [39].

2. Patients with *SMARCB1* germline truncating mutations who are at risk for abdominal malignant rhabdoid tumors should be considered for whole-body MRI until 5 years of age; the frequency is undetermined. In addition, ultrasound should be performed every 3 months [39].

3. Patients with *SMARCA4* germline truncating mutations are considered to be at low to very low risk to develop malignant rhabdoid tumors at any site; there are no data available to suggest a surveillance regimen for these patients [39].

4. Patients with *SMARCA4* germline truncating mutations who are at risk for developing small cell carcinoma, hypercalcemic type, of the ovary may be justified to undergo abdominal ultrasound every 6 months. The role of MRI is unknown. Preventative oophorectomy may be justified outside the pediatric age range [39].

Teplick et al. have also discussed screening recommendations for rhabdoid tumor predisposition syndromes. Their recommendations are based upon criteria crafted at the 2009 meeting entitled "Pediatric Cancer Genetics: From Gene Discovery to Cancer Screening" [40]. Based on this meeting, the following recommendations were developed (but to date not formally studied) [40]:

1. From birth to 1 year old, patients should have thorough neurologic and physical examinations as well as monthly ultrasounds of the head to assess for the development of a tumor of the central nervous system. In addition, these same patients should undergo abdominal ultrasound primarily focused on the kidneys every 2–3 months to assess for renal lesions [40].

2. From 1–4 years of age, MRI of the brain and spine and ultrasound of the abdomen should be performed every 6 months [40].

Screening and surveillance criteria have not been studied in rhabdoid tumor predisposition patients, but in *SMARCB1* germline mutation patients, screening may prove beneficial. Early detection has the potential to improve overall survival [40].

8.8 Conclusion

Rhabdoid tumor predisposition syndrome is a tumor syndrome in which rhabdoid tumors can present in any organ of the body. Rhabdoid tumor of the kidney and atypical teratoid/rhabdoid tumor of the brain are named based upon where they arise; all others are considered extrarenal/extracranial malignant rhabdoid tumors. The diagnosis of one of these lesions, particularly in a very young patient, should lead to suspicion of this diagnosis. In many families, *SMARCB1*, responsible for the rhabdoid tumor predisposition syndrome type 1, is mutated de novo in the germline, whereas *SMARCA4*, responsible for the rhabdoid tumor predisposition syndrome type 2, is passed down to an affected individual from a healthy unaffected carrier.

Rhabdoid tumors are deadly, with a mortality rate as high as 80% at 5 years; most of these patients succumb to their disease within a year. Tissue is essential not only for the diagnosis, but to carry out cytogenetic and molecular genetic analysis. Immunohistochemistry also plays an important role in the diagnosis of these tumors, particularly when loss of INI-1 or SMARCA4 is demonstrated.

Multimodality therapy, consisting of surgery, chemotherapy, and radiotherapy depending on the age, patient, and stage of the tumor, has been shown in various studies to be more effective than single-modality therapy in treatment and remission of these tumors. Screening and surveillance guidelines have been introduced; however, these guidelines are not targeted at prevention of tumor development, but rather diagnosis at an early stage. Oophorectomy is indicated in patients who have completed childbearing as a preventative measure against the SMARCA4-associated small cell tumor, hypercalcemic type, of the ovary. Genetic counseling is also indicated in these families.

Much has been learned by studying the rhabdoid tumor predisposition, and more stands to be gleaned to enable adequate treatment and eradication of these tumors. Future testing on the surveillance guidelines will help promote an increased survival benefit for affected patients.

REFERENCES

1. Sévenet N, Sheridan E, Amram D, Schneider P, Handgretinger R, Delattre O. Constitutional mutations of the h5N5/INI1 gene predispose to a variety of cancers. *Am J Hum Genet*. 1999;65:1342–8.

2. Beckwith JB, Palmer NS. Histopathology and prognosis of Wilms tumor. *Cancer*. 1978;41:1937–48.

3. Sredni ST, Tomita T. Rhabdoid tumor predisposition syndrome. *Ped Devel Pathol*. 2015;18:49–58.

4. Fitzhugh VA. Rhabdoid tumor predisposition syndrome and pleuropulmonary blastoma syndrome. *J Pediatr Genet*. 2016;5(2):124–8.

5. Rorke LB, Packer RJ, Biegel JA. Central nervous system atypical teratoid/rhabdoid tumors of infancy and childhood: Definition of an entity. *J Neurosurg*. 1996;85:56–65.

6. Agaimy A. The expanding family of SMARCB1 (INI1) deficient neoplasia: Implications of phenotypic, biological, and molecular heterogeneity. *Adv Anat Pathol.* 2014;21:394–410.

7. Douglass EC, Valentine M, Rowe ST et al. Malignant rhabdoid tumor: A highly malignant childhood tumor with minimal karyotypic changes. *Genes Chromosomes Cancer.* 1990;2:210–6.

8. Schofield DE, Beckwith JB, Sklar J. Loss of heterozygosity at chromosome regions 2q11–12 and 11p15.5 in renal rhabdoid tumors. *Genes Chromosomes Cancer.* 1996;15:10–7.

9. Biegel JA, Allen CS, Kawasaki K, Shimizu N, Budarf ML, Bell CJ. Narrowing the critical region for a rhabdoid tumor locus in 22q11. *Genes Chromosomes Cancer.* 1996;16:94–105.

10. Versteege I, Sévenet N, Lange J, Rousseau-Merck MF, Ambros P, Handgretinger R, Aurias A, Delattre O. Truncating mutations of hSNF5/INI1 in aggressive pediatric cancer. *Nature.* 1998;394:203–6.

11. Biegel JA, Zhou JY, Rorke LB, Stenstrom C, Wainwright LM, Fogelgren B. Germ-line and acquired mutations of INI1 in atypical teratoid and rhabdoid tumors. *Cancer Res.* 1999;59:74–9.

12. Biegel JA. Molecular genetics of atypical teratoid/rhabdoid tumors. *Neurosurg Focus.* 2006;20:1–7.

13. Roberts CWM, Biegel JA. The role of SMARCB1/INI1 in development of rhabdoid tumor. *Cancer Biol Ther.* 2009;8:412–6.

14. Kordes U, Gesk S, Frühwald MC et al. Clinical and molecular features in patients with atypical teratoid rhabdoid tumor or malignant rhabdoid tumor. *Genes Chromosomes Cancer.* 2010;49:176–81.

15. Frühwald MC, Hasselblatt M, Wirth S et al. Non-linkage of familial rhabdoid tumors to SMARCB1 implies a second locus for the rhabdoid tumor predisposition syndrome. *Pediatr Blood Cancer.* 2006;47:273–8.

16. Schneppenheim R, Frühwald MC, Gesk S et al. Germline nonsense mutation and somatic inactivation of SMARCA4/BRG1 in a family with rhabdoid tumor predisposition syndrome. *Am J Hum Genetics.* 2010;86:279–84.

17. Hasselblatt M, Gesk S, Oyen F et al. Nonsense mutation and inactivation of SMARCA4 (BRG1) in an atypical teratoid/rhabdoid tumor showing retained SMARCB1 expression. *Am J Surg Pathol.* 2011;35:933–5.

18. Teplick A, Kowalski M, Biegel JA, Nichols KE. Educational paper: Screening in cancer predisposition syndromes: Guidelines for the general pediatrician. *Eur J Pediatr.* 2011;170:285–94.

19. Tomlinson GE, Breslow NE, Dome J et al. Rhabdoid tumor of the kidney in the National Wilms' Tumor Study: Age at diagnosis as a prognostic factor. *J Clin Oncol.* 2005;23:7641–45.

20. Nemes K, Bens S, Bourdeaut F et al. Rhabdoid tumor predisposition syndrome. In: Adam MP, Ardinger HH, Pagon RA et al. editors. *GeneReviews®* [Internet]. Seattle (WA): University of Washington, Seattle. 2017:1993–2018.

21. Burger PC, Yu IT, Tihan T et al. Atypical teratoid/rhabdoid tumor of the central nervous system: A highly malignant tumor of infancy and childhood frequently mistaken for medulloblastoma. *Am J Surg Pathol.* 1998;22:1083–92.

22. Heck JE, Lombardi CA, Cockburn M, Meyers TJ, Wilhelm M, Ritz B. Epidemiology of rhabdoid tumors of early childhood. *Pediatr Blood Cancer.* 2013;60:77–81.

23. Ammerlaan ACJ, Ararou A, Houben MPWA et al. Long term survival and transmission of INI1-mutation via nonpenetrant males in a family with rhabdoid tumour predisposition syndrome. *Br J Cancer.* 2008;98:474–9.

24. Amar AM, Tomlinson G, Green DM, Breslow NE, de Alarcon PA. Clinical presentation of rhabdoid tumors of the kidney. *J Pediatr Hematol Oncol.* 2001;23:105–8.

25. Chung CJ, Lorenzo R, Rayder S et al. Rhabdoid tumors of the kidney in children: CT findings. *Am J Roent.* 1995;164:697–700.

26. Harris TJ, Donahue JE, Shur N, Tung GA. Case 168: Rhabdoid predisposition syndrome-familial cancer syndromes in children. *Radiology.* 2011;259:298–302.

27. Hollmann TJ, Hornick JL. INI-1 deficient tumors: Diagnostic features and molecular genetics. *Am J Surg Pathol.* 2011;35:e47–63.

28. Margol AS, Judkins AR. Pathology and diagnosis of SMARCB1-deficient tumors. *Cancer Genet.* 2014;207:358–64.

29. Fanburg-Smith JC, Hengge M, Hengge UR, Smith JSC, Miettinen M. Extrarenal rhabdoid tumors of soft tissue: A clinicopathologic and immunohistochemical study of 18 cases. *Ann Diagn Pathol.* 1998;2:351–62.

30. Judkins AR, Mauger J, Rorke LB, Biegel JA. Immunohistochemical analysis of hSNF5/INI1 in pediatric CNS neoplasms. *Am J Surg Pathol.* 2004;28:644–50.

31. Haberler C, Laggner U, Slavc I et al. Immunohistochemical analysis of INI1 protein in malignant pediatric CNS tumors: Lack of INI1 in atypical teratoid/rhabdoid tumors and in a fraction of primitive rhabdoid tumors without rhabdoid phenotype. *Am J Surg Pathol.* 2006;30:1462–68.

32. Reddy A, Strother D, Judkins AR et al. Treatment atypical teratoid rhabdoid tumors (ATRT) of the central nervous system with surgery, intensive chemotherapy, and 3-D conformal radiation (ACNS0333). A report from the Children Oncology Group. *Neuro Oncol.* 2016;18:iii2.

33. Chi SN, Zimmerman M, Yao X et al. Intensive multimodality treatment for children with newly diagnosed CNS atypical teratoid rhabdoid tumor. *J Clin Oncol.* 2009;27:385–9.

34. Bartelheim K, Nemes K, Seeringer A et al. Improved 6-year overall survival in AT/RT-results of the registry study Rhabdoid 2007. *Cancer Med.* 2016;5:1765–75.

35. Ren YM, Wu X, You C, Zhang YK, Li Q, Ju Y. Multimodal treatments combined with gamma knife surgery for primary atypical teratoid/rhabdoid tumor of the central nervous system: A single institute experience of 18 patients. *Child's Nervous System.* 2018;34:627–38.

36. Furtwängler R, Nourkami-Tutdibi N, Leuchsner I et al. Malignant rhabdoid tumor of the kidney: Significantly improved response to pre-operative treatment intensified with doxorubicin. *Cancer Gen.* 2014;207:434–6.

37. Forest F, David A, Arrufat S et al. Conventional chondrosarcoma in a survivor of rhabdoid tumor: Enlarging the spectrum of tumors associated with SMARCB1 germline mutations. *Am J Surg Pathol.* 2012;36:1892–6.

38. Woo J, Seethala RR, Sirintrapun SJ. Mammary analogue secretory carcinoma of the parotid gland as a secondary malignancy in a childhood survivor of atypical teratoid rhabdoid tumor. *Head and Neck Pathol.* 2014;8:194–7.

39. Foulkes WD, Kamihara J, Evans DGR et al. Cancer surveillance in Gorlin syndrome and rhabdoid tumor predisposition syndrome. *Clin Cancer Res.* 2017;23:e62–7.

40. Teplick A, Kowalski M, Biegel JA, Nichols KE. Educational paper. Screening in cancer predisposition syndromes: Guidelines for the general pediatrician. *Eur J Pediatr.* 2011;170:285–94.

9

Rubinstein–Taybi Syndrome

Dongyou Liu

CONTENTS

9.1 Introduction

Rubinstein–Taybi syndrome (RTS, including RTS1 [OMIM #180849] and RTS 2 [OMIM #613684]) is a rare autosomal dominant condition that typically causes short stature, dysmorphic facial features (e.g., prominent forehead, arched eyebrows, downslanting palpebral fissures, strabismus, pouting lower lip, short upper lip, high-arched palate, broad nasal bridge, beaked nose with nasal septum below alae/low hanging columella, talon cusps, micrognathia, and grimacing smile), microcephaly, broad thumbs and halluces (great toes), persistent fetal finger pads, frequent fractures of bone, cryptorchism, shawl scrotum, hypospadias, congenital cardiovascular defect, keloid, moderate to severe intellectual disability, and an increased risk of brain tumors, neural crest cell-derived tumors, and tumors arising in other developmental defects (including astrocytoma, glioma of the brain, medulloblastoma, meningioma, adrenal neuroblastoma, extra-adrenal neuroblastoma, oligodendroglioma, pilomatrixoma, and pineal gland tumor, and possibly adrenal adenoma, leiomyosarcoma, acute lymphoblastic leukemia, acute myeloid leukemia, nasopharyngeal rhabdomyosarcoma, non-Hodgkin lymphoma, maxillar odontoma, pheochromocytoma, schwannoma/neurilemmoma, and peripheral nerve seminoma) [1].

The molecular mechanisms underlying RTS relate to point mutations, whole gene deletions, or chromosome rearrangements in either the cAMP-response element binding protein-BP (*CREBBP*) gene on chromosome 16p13.3 (RTS1, in 50%–60% of cases) or the homologous gene E1A-binding protein 300 (*EP300*) on chromosome

22q13.2 (RTS2, in 5%–8% of cases). While a vast majority of RTS cases evolve from de novo mutations, some cases are attributed to inheritance of a faulty gene from one of the affected parents [1].

9.2 Biology

Despite the fact that cases related to RTS were first observed in 1957, a distinct combination of broad thumbs and toes, craniofacial abnormalities, and mental retardation justified its recognition as a clinical disease syndrome by Rubinstein and Taybi in 1963 [2].

Subsequent studies have redefined RTS as a multiple congenital anomaly disorder that can be separated into classic RTS (RTS1) and atypical RTS (RTS2) on the basis of varied clinical features and differing gene involvement.

RTS1 (also known as Rubinstein syndrome, broad thumb–hallux syndrome, or broad thumbs and great toes, characteristic facies, and mental retardation) presents with broad thumbs and halluces, dysmorphic facial features (e.g., broad nasal bridge, beaked nose with the nasal septum, downward slanting palpebral fissures, highly arched eyebrows, long eyelashes, highly arched palate, mild micrognathia, and grimacing smile), mental retardation, microcephaly, postnatal growth deficiency, and increased risk of childhood malignancies. RTS1 is linked to heterozygous point mutations or deletions in the transcriptional coactivator CREB-binding protein gene (*CREBBP*) on chromosome 16p13, which are detected in 50%–60% of RTS patients. There is a severe form of RTS known as proximal chromosome 16p13.3

deletion syndrome (or RTS deletion syndrome), which results from a contiguous gene deletion involving the *CREBBP* gene as well as other neighboring genes (e.g., *DNASE1* and *TRAP1*) [1].

RTS2 is a mild form of disease, which is associated with less severe facial dysmorphism, better cognitive function, milder skeletal phenotype (broad thumbs and big toes with no radial deviation, and sometimes normal thumbs and toes), but may have more severe microcephaly and malformation of facial bone structures than RTS1. RTS2 is due primarily to de novo heterozygous mutations in the *EP300* gene on chromosome 22q13.2, which are detected in 5%–8% of RTS patients [3,4].

It appears that RTS mostly results from de novo mutations in the *CREBBP* and *EP300* genes. Familial RTS is extremely rare; of 1000–2000 RTS cases reported to date, only 11 familial cases are reported and most of these are likely linked to somatic mosaicism. In cases of inherited RTS, affected children tend to have a more severe phenotype [5]. Available evidence supports an autosomal dominant mode of inheritance for RTS [2].

9.3 Pathogenesis

The *CREBBP* gene on chromosome 16p13.3 comprises 31 transcribed exons spanning 154 kb. The 3-prime distal flanking region of the *CREBBP* gene contains the *DNASE1* and *TRAP1* (or *HSP75*) genes. The *CREBBP* gene encodes a 2442 aa nuclear transcriptional coactivator protein (CREBBP) of 265 kDa, which includes a central region (consisting of a bromodomain and a histone acetyltransferase [HAT] domain) flanked by two transactivation domains. While the bromodomain facilitates protein–protein interactions, the HAT domain demonstrates intrinsic histone acetyltransferase activity, which plays a vital role in the regulation of gene expression through acetylation of histones H3 and H4, de-condensing chromatin and allowing for transcription. In addition, the transactivation domains recruit and interact with transcriptional machinery such as RNA polymerase II (Pol II) complex, co-activators, and repressors. Through these functionally distinct domains, CREBBP acts as a mediator of different signaling pathways, a negative regulator of the cell cycle by repressing the transition from G1 to S phase, and also a scaffold to stabilize additional protein interactions with the transcription complex via chromatin remodeling [1].

Presence of an adequate level of CREBBP acetyltransferase activity is essential for normal development. Loss of one functional copy (haploinsufficiency) of the CREBBP gene through heterozygous mutations and subsequent aberrant chromatin regulation underlies the developmental abnormalities in RTS1, contributing to the unusual incidence of neoplasms and the propensity to form keloids. Specially, germline pathogenic variants in *CREBBP* may result in a truncated protein or a protein with an amino acid substitution that interferes with the acetylation of histones, an important step in transcription activation. Somatic pathogenic variants in *CREBBP* may weaken its tumor suppressor function, leading to malignant transformation. To date, >150 pathogenic mutations (e.g., frameshift, nonsense substitution, splice site substitution, missense, insertion, microdeletion, or complex rearrangement) in the *CREBBP* gene have been identified in 50%–60% of patients meeting clinical diagnostic criteria

for RTS [6,7]. Interestingly, RTS patients with larger deletions do not always have a more severe phenotype than those with smaller deletions or point mutations [8–11].

As CREBBP is activated by various signaling pathways (PKA, MAPK, and CaMKIV) in physiological processes such as learning and long-term memory formation, it is not surprising that mutations in *CREBBP* contribute to cognition changes. Further, monoallelic deletion of *CREBBP* induces ESET/SETDB1 (ERG-associated protein with SET domain), a histone-specific trimethyltransferase, leading to condensation of pericentromeric heterochromatin structure in neurons and striatal neuron atrophy and dysfunction. In addition, changes in the state of chromatin can affect the expression of specific genes involved in seizures [12].

The *EP300* gene on chromosome 22q13.2 consists of 31 exons in a stretch of 87.75 kb and encodes p300 that contains 3 cysteine- and histidine-rich regions, including the carboxyterminal region interacting specifically with E1A, and a centrally located bromodomain representing a hallmark of certain transcriptional coactivators. As a homolog to CREBBP, p300 shares >70% and 63% homology with CREBBP at nucleotide and amino acid levels, respectively, and demonstrates HAT activity. Strongly expressed in human cerebellum as well as in other brain regions, p300 functions as transcriptional coactivator in the regulation of gene expression via chromatin remodeling and plays an important role in the processes of cell proliferation and differentiation. CREBBP/p300 may also modulate with p53 pathway through P53 acetylation, and loss of CREBBP/p300 could disrupt p53 activation, stability, and transactivation of target genes. It has been shown that p300 knockdown by hammerhead ribozymes inhibits apoptosis, probably by disrupting the p53-mediated response to DNA damage [1].

Pathogenic variants (34 deletions, duplications, and single-nucleotide variants identified to date) may induce truncated p300 protein or abolish allele expression, leading to loss of HAT activity and RTS2, suggesting an aberrant chromatin regulation. As *EP300* mutations are found in 5%–8% of RTS patients who do not harbor *CREBBP* mutations, *EP300* likely has only a minor part in the pathogenesis of RTS [3,4]. The rarity of *EP300* mutations in RTS2 might reflect the lower mutation rate of *EP300*, and the milder phenotype of RTS2, leading to underdiagnosis or misdiagnosis with other congenital malformations such as Cornelia de Lange syndrome (CdLS) [13–15].

It is clear that heterogeneous mutations in CBP/CREBBP domains and subsequent epigenetic alterations contribute to characteristic phenotypic features and cancer risk associated with RTS. In view of the fact that about 30% of RTS cases do not contain detectable mutations in either *CREBBP* or *EP300,* further studies are required to uncover other epigenetic mechanisms affecting histone acetylation and subsequently gene transcription that may potentially contribute to the development of RTS belonging to neither RTS1 nor RTS2.

9.4 Epidemiology

9.4.1 Prevalence

RTS is a rare genetic disorder, with an estimated prevalence of 1 per 100,000–125,000 live births. While patients of European origin dominate clinical cases, some individuals of black and Asian heritage may be also affected. To date, between 1000 and 2000 cases of RTS have been reported, including 11 cases of familial RTS. Among patients with 16p13.11 microdeletion, the average age at diagnosis is 6.6 years (range 0.5–30 years). The mean age at presentation (e.g., developmental delay, autism spectrum disorder, dysmorphic features, multiple congenital anomalies, epilepsy, and seizure disorder) is about 1 year in patients with a deletion as opposed to 11 years in those without.

9.4.2 Inheritance

Most RTS cases appear to arise from sporadic mutations since many patients do not have affected relatives. A small number (11 to date) cluster in families and, demonstrate an autosomal dominant inheritance pattern. Affected individuals have a 50% risk of passing the disease to their offspring.

9.4.3 Penetrance

RTS may have a low penetrance, as some patients with pathogenic variants are non-symptomatic and certain pathogenic variants are also found in healthy parents.

9.5 Clinical Features

RTS is associated with a wide spectrum of clinical manifestations, ranging from distinctive facial features, growth and other anomalies, intellectual and behavioral problems, to tumor risk [16,17].

Distinctive facial features include low frontal hairline, arched/thick eyebrows, downslanting palpebral fissures, protruding beaked nose with columella below alae nasi, dysplastic and low-set ears, an arched palate, mild micrognathia, talon cusps (an accessory cusp-like structure on the lingual side of the tooth, usually occurring on the maxillary incisors of the permanent dentition), small oral opening, thin upper lip, pouting lower lip, grimacing smile, and nearly completely closed eyes, which are mostly recognizable at birth or in infancy (Figure 9.1).

Growth anomalies are exemplified by characteristic broad and often angulated thumbs and great toes, clinodactyly of the fifth finger, and short stature (with average height for adult males of 153 cm and for adult females of 146 cm) (Figures 9.1 and 9.2). Despite apparent normal prenatal growth, parameters for height, weight, and head circumference may fall below the fifth percentile during infancy (due primarily to hypo-feeding exacerbated by gastroesophageal reflux). Males and females often become overweight/obese during childhood and adolescence, respectively.

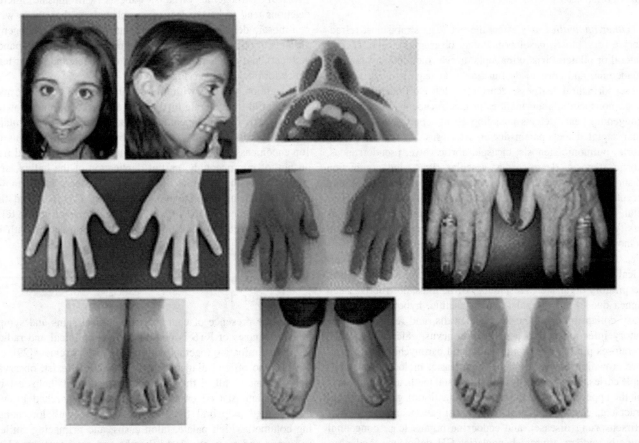

FIGURE 9.1 Rubinstein—Taybi syndrome in a 9-year-old female patient and her relatives (with heterozygous EP300 exon 31 deletion mutation c.7222_7223del, p.Gln2408Glufs*39), showing thickened and low-hanging columella of the patient (top left and center), talon cusp at an upper incisor of her mother (top right), normal thumbs and halluces of the patient (middle left, bottom left), and short and broad but not angulated thumbs of her mother (middle center, bottom center) and her grandmother (middle right, bottom right). (Photo credit: López M et al. *BMC Med Genet.* 2018;19(1):36.)

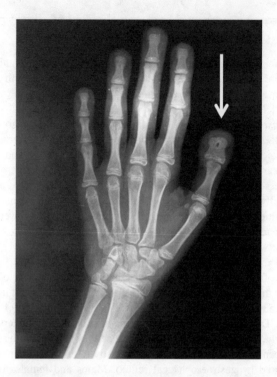

FIGURE 9.2 Rubinstein–Taybi syndrome in a 14-year-old female (with CREBBP insertion mutation c.3546insCC) showing a duplication of terminal phalanx of the thumb (arrow). (Photo credit: Marzuillo P et al. *BMC Med Genet.* 2013;14:28.)

Other anomalies may affect the eye (e.g., strabismus, refractory errors, ptosis, nasolacrimal duct obstruction, cataract, unilateral or bilateral iris/retinal/optic nerve coloboma, nystagmus, glaucoma, and corneal dysfunction), ear (e.g., conductive and/or sensorineural deafness, recurrent middle ear infections), heart (e.g., nonspecific abnormalities of electroencephalography, and congenital heart defects including atrial septal defect, ventricular septal defect, patent ductus arteriosus, coarctation of the aorta, pulmonic stenosis, bicuspid aortic valve, pseudotruncus, aortic stenosis, dextrocardia, vascular rings, and conduction disorders), genitourinary organs (e.g., renal abnormalities, and cryptorchidism/undescended testes in boys), orthopedic, skeletal, and spinal tissues (e.g., dislocated patellae, ligamentous laxity, spine curvatures, Legg–Perthes disease, slipped capital femoral epiphysis, craniospinal and posterior fossa abnormalities, Chiari malformation, syringomyelia, os odontoideum, cervical cord compression, and cervical vertebral abnormalities), respiratory organs (e.g., respiratory difficulties, obstructive sleep apnea due to narrow palate, micrognathia, hypotonia, obesity, easy collapsibility of the laryngeal walls, and recurrent respiratory infections), skin (e.g., keloids nevus, café-au-lait spots, keratoses pilaris, ingrown toenails, and paronychia), teeth (e.g., overcrowding of teeth, enamel hypoplasia, malocclusion, multiple caries, hypodontia, hyperdontia, natal teeth, and talon cusps on the upper incisors of the secondary dentition), gastrointestinal tract (e.g., gastroesophageal reflux, constipation, and megacolon/Hirschsprung disease), and endocrine organs (e.g., congenital hypothyroidism, thyroid hypoplasia, GH deficiency, and pituitary hypoplasia) [18–23].

Intellectual and behavioral problems include reduced IQ scores (between 25–79), short attention span, motor stereotypies, poor coordination, limited speech, decreased social interaction, worsening stamina and mobility, decreased tolerance for noise and crowds, hyperactivity, impulsivity, mood swings, seizures, and self-injury [24–26].

Tumor risk is evidenced by increased occurrence of meningioma, neuroblastoma, medulloblastoma, oligodendroglioma, pheochromocytoma, rhabdomyosarcoma, leiomyosarcoma, pilomatrixoma, seminoma, odontoma, choristoma, leukemia (e.g., acute myeloid leukemia and acute lymphoblastic leukemia) and lymphoma (e.g., follicular lymphoma and diffuse large B-cell lymphoma) in RTS patients [27,28].

Based on the frequency of occurrence, typical clinical features of RTS comprise distinct facial features (100%), intellectual disability (~100%), cryptorchidism (78%–100%), microcephaly (78%–100%), broad thumbs/halluces (96%), speech delay (90%), recurrent respiratory infections (75%), delayed bone age (74%), constipation (40%–74%), talon cusps (73%), gastroesophageal reflux (68%), EEG abnormalities (57%–66%), renal anomalies (52%), refractive defects, glaucoma, retinopathy (>50%), congenital heart defects (24%–38%), seizures (25%), keloids (24%), deafness (24%), growth retardation (21%), malignant tumors (3%–10%), and spinal cord tethering (<5%) [2].

Interestingly, congenital anomalies, tumors, dysmorphic features, or level of intellectual impairment show similar frequency between RTS patients with and without *CREBBP* pathogenic variants, seizures more commonly occur in patients with *CREBBP* pathogenic variants. Patients with mosaic microdeletions tend to have a less severe phenotype than those with non-mosaic deletions. Further, patients with *EP300* pathogenic variants often show milder skeletal findings (e.g., normal hands and feet), less affected intellectual development, and higher tendency for preeclampsia (Figure 9.1).

In addition, patients with microduplications of *CREBBP* may display mild to moderate intellectual disability, normal growth, characteristic facial appearance, minor extremity abnormalities, and variable other features. Individuals with deletions of 16p encompassing *CREBBP* may demonstrate an increased frequency of microcephaly, partial duplication of the hallux, and angulation of the first rays of the hands and feet. Furthermore, patients with large deletions encompassing *CREBBP* and the contiguous genes *DNASE1* and *TRAP1* may develop severe RTS (i.e., failure to thrive, life-threatening infections, and death in infancy).

9.6 Diagnosis

Due to the presence of characteristic clinical signs and symptoms, diagnosis of RTS is possible through clinical and radiological examination together with family history review [29].

While no official diagnostic criteria for RTS exist, observation of some or all of the following clinical signs helps establish its diagnosis: (i) craniofacial appearance (arched brows, downslanted palpebral fissures, beaked nose with low hanging columella, high palate, talon cusps, and grimacing smile); (ii) broad and often angulated thumbs and great toes, broad or abnormally shaped terminal phalanges; (iii) undescended testes in males; (iv) congenital heart defects; (v) obesity in childhood or adolescence; (vi) short stature in adulthood; (vii) reduced IQ scores (range 25–79) [30,31]. Radiography of the hands and feet facilitates detection of duplications of the first rays (Figure 9.2).

After establishing clinical diagnosis, molecular approaches (e.g., karyotyping to identify possible translocations, inversions, or deletions at 16p13.3 and 22q13.2; FISH to identify microdeletions; sequencing to detect *CREBBP* and *EP300* pathogenic variants; deletion/duplication analysis testing to identify exonic or whole-gene deletions/duplications of the coding and flanking intronic regions of genomic DNA that are not detected by sequencing) provide additional means of RTS confirmation.

Molecular testing is especially useful for assessing patients with indeterminate clinical diagnosis or atypical or very severe features. This usually begins with sequence analysis of *CREBBP*, then duplication/deletion analysis (if sequence analysis is negative), and finally sequence analysis and/or deletion/duplication analysis of *EP300* (if *CREBBP* pathogenic variants are not detected).

It should be noted that while *CREBBP* pathogenic variants are associated exclusively with RTS and occur in 50%–60% of RTS, *EP300* pathogenic variants are found in 5%–8% of RTS patients and occasionally in somatic colorectal cancer. Further, about 30% of RTS patients do not harbor *CREBBP* and *EP300* pathogenic variants. Therefore, the absence of *CREBBP* or *EP300* pathogenic variants does not absolutely rule out RTS.

Differential diagnoses for RTS included several syndromes that also show distinctive facial features and/or hand and foot abnormalities, such as *FGFR*-related craniosynostosis syndromes (i.e., Pfeiffer syndrome, Apert syndrome, Crouzon syndrome, Beare–Stevenson syndrome, *FGFR2*-related isolated coronal synostosis, Jackson–Weiss syndrome, Crouzon syndrome with acanthosis nigricans, and Muenke syndrome), Saethre–Chotzen syndrome, Greig cephalopolysyndactyly syndrome (GCPS), Floating–Harbor syndrome, Keipert syndrome, and CdLS [1,2].

Except for Muenke syndrome and *FGFR2*-related isolated coronal synostosis, all *FGFR*-related craniosynostosis syndromes show bicoronal craniosynostosis or cloverleaf skull, distinctive facial features, and variable hand and foot findings. On the other hand, Muenke syndrome and *FGFR2*-related isolated coronal synostosis have uni- or bicoronal craniosynostosis only. Muenke syndrome (or *FGFR3*-related coronal synostosis) is further distinguished by the presence of a pathogenic variant in *FGFR3*, and likewise, *FGFR2*-related isolated coronal synostosis harbors a pathogenic variant in *FGFR2*.

Saethre–Chotzen syndrome often displays coronal synostosis (unilateral or bilateral), facial asymmetry (particularly in patients with unicoronal synostosis), ptosis, and characteristic appearance of the ear (small pinna with a prominent crus). At the molecular level, 46%–80% of patients with Saethre–Chotzen syndrome harbor chromosome translocation involving 7p21 or ring chromosome 7. *TWIST1*.

GCPS shows preaxial polydactyly or mixed pre- and postaxial polydactyly, widely spaced eyes, and macrocephaly. Patients with severe GCPS may have seizures, hydrocephalus, and intellectual disability. About 75% of GCPS cases contain alterations in the *GLI3* gene.

Floating–Harbor syndrome often demonstrates typical craniofacial features, low birth weight, normal head circumference, short stature, bone age delay (until normalization between 6 and 12 years of age), skeletal anomalies (brachydactyly, clubbing, clinodactyly, short thumbs, prominent joints, clavicular abnormalities), severe receptive and expressive language impairment, hypernasality, high-pitched voice, and mild to moderate intellectual disability. Floating–Harbor syndrome is linked to pathogenic variant in *SRCAP,* that encodes an SNF2-related chromatin-remodeling factor with a coactivator function for CREBBP.

Keipert syndrome (nasodigitoacoustic syndrome) causes broad thumbs and halluces, brachydactyly, hypertelorism, hearing loss, facial dysmorphic features, and sensorineural deafness. It is X-linked and maps to Xq22.2-Xq28.

CdLS typically demonstrates slowed growth before and after birth, intellectual disability, developmental delay, autistic and/or self-destructive behaviors, skeletal abnormalities of the arms and hands, gastrointestinal problems, hirsutism (excess hair growth), hearing loss, myopia, congenital heart defects, cryptorchidism, and seizures. Molecularly, it is associated with heterozygous mutations in the *NIPBL* and *SMC3* genes and heterozygous (in females) or hemizygous (in males) mutations in *SMC1A* gene.

9.7 Treatment

As genetic mutations linked to RTS are essentially irreversible, treatment options for RTS are currently limited to symptomatic relief such as: (i) standard treatment for eye abnormalities, hearing loss, cardiac defects, cryptorchidism, and sleep apnea; (ii) surgical repair of significantly angulated thumbs or duplicated halluces; (iii) aggressive management of gastroesophageal reflux and constipation; and (v) use of sedation (e.g., 0.1 mg/kg midazolam, 2 mg/kg propofol) or general anesthesia for dental procedures [2,32–35].

Considering that CREBBP loss of function leads to a decrease in histone acetylation levels as well as CREBBP-dependent transcription, use of histone deacetylase (HDAC) inhibitors may potentially reverse CREBBP-dependent decrease in acetylation, and also increase acetylation levels at nonspecific promoters across a wide variety of genes, thus providing a potential treatment for RTS. HDAC inhibitors have been tested in experimental mice with genetic alterations of CREBBP leading to improvement in long-term potentiation and memory. In addition, use of HDAC inhibitors such as suberoylanilide hydroxamic acid (SAHA) or trichostatin A (TSA) improves deficits in synaptic plasticity and cognition in CREBBP mutant mice. Nonetheless, the effects of HDAC inhibitors on human RTS remain to be evaluated.

9.8 Prognosis, Surveillance, and Risk Assessment

9.8.1 Prognosis

Over 90% RTS patients with disabilities survive to adulthood, but their health care is complex and costly. The fact that RTS-affected individuals have an increased risk (up to 10%) of developing various tumors, including meningioma, neuroblastoma, medulloblastoma, rhabdomyosarcoma, leukemia, and lymphoma, further exacerbates the situation.

9.8.2 Surveillance

Regular surveillance (involving brain and medullary NMR; neuropsychiatric, audiologic, ophthalmologic, orthopedic, odonthoiatric, endocrinological, and dermatologic evaluation; pressure measurement; renal ultrasound scan; genetic counseling) plays a key role in maintaining the health and well-being of RTS patients. It is therefore it is important to monitor growth and feeding in the first year of life, conduct annual eye and hearing evaluation, and undertake routine monitoring for cardiac, dental, and renal anomalies in RTS patients [1].

9.8.3 Risk Assessment

RTS typically results from a de novo pathogenic variant, and most affected individuals represent simplex cases or the only affected member in their family. Therefore, if their parents are clinically unaffected and/or harbor no *CREBBP* or *EP300* pathogenic variants, their siblings have <1% chance of acquiring RTS. Inherited through an autosomal dominant pattern, RTS patients have a 50% risk of passing the disorder to their offspring. Prenatal testing is possible for pregnancies at increased risk if the pathogenic variant or deletion is known in the family.

9.9 Conclusion

Rubinstein–Taybi syndrome (RTS) is a rare congenital disorder characterized by dysmorphic facial features (e.g., arched eyebrows, slanted palpebral fissures, "beaked-shaped" nose with a broad fleshy bridge, long and deviated septum protruding below the level of the nasal alae with an associated short columella, labial commissures facing upward, teeth anomalies, and a grimacing smile with nearly completely closed eyes), microcephaly, broad thumbs and big toes as well as clinodactyly of the fifth finger, intellectual disability, postnatal growth retardation, and increased cancer risk [36,37]. In accordance with clinical variations and involvement of genes encoding *CREBBP* and *EP300*, RTS is distinguished into classic RTS (RTS1, 50%–60% of cases) and atypical RTS (RTS2, 5%–8% of cases). However, about 30% of RTS cases demonstrate typical clinical signs and symptoms but harbor no apparent mutations in *CREBBP* or *EP300*. Further investigations are necessary in order to unravel other potential epigenetic mechanisms underlying the development of RTS belonging to neither RTS1 nor RTS2. While diagnosis of RTS is feasible on the basis of clinical findings and family history review, use of molecular techniques adds a new dimension to the clarification of specific genotype–phenotype correlation, evaluation of somatic mosaicisms, discovery of novel candidate genes, and improved management for this rare disorder.

REFERENCES

1. Stevens CA. Rubinstein-Taybi syndrome. In: Adam MP, Ardinger HH, Pagon RA, Wallace SE, Bean LJH, Stephens K, Amemiya A, editors. *GeneReviews® [Internet]*. Seattle (WA): University of Washington, Seattle; 1993–2018. 2002 Aug 30 [updated 2014 Aug 7].

2. Milani D, Manzoni FM, Pezzani L et al. Rubinstein-Taybi syndrome: Clinical features, genetic basis, diagnosis, and management. *Ital J Pediatr.* 2015;41:4.

3. López M, Seidel V, Santibáñez P et al. First case report of inherited Rubinstein-Taybi syndrome associated with a novel EP300 variant. *BMC Med Genet.* 2016;17(1):97.

4. López M, García-Oguiza A, Armstrong J et al. Rubinstein-Taybi 2 associated to novel EP300 mutations: Deepening the clinical and genetic spectrum. *BMC Med Genet.* 2018;19(1):36.

5. Spena S, Gervasini C, Milani D. Ultra-rare syndromes: The example of Rubinstein-Taybi syndrome. *J Pediatr Genet.* 2015;4(3):177–86.

6. Park E, Kim Y, Ryu H, Kowall NW, Lee J, Ryu H. Epigenetic mechanisms of Rubinstein-Taybi syndrome. *Neuromol Med.* 2014;16(1):16–24.

7. Yoo HJ, Kim K, Kim IH et al. Whole exome sequencing for a patient with Rubinstein-Taybi syndrome reveals de novo variants besides an overt CREBBP mutation. *Int J Mol Sci.* 2015;16(3):5697–713.

8. Bentivegna A, Milani D, Gervasini C et al. Rubinstein-Taybi Syndrome: Spectrum of CREBBP mutations in Italian patients. *BMC Med Genet.* 2006;7:77.

9. Marzuillo P, Grandone A, Coppola R et al. Novel cAMP binding protein-BP (CREBBP) mutation in a girl with Rubinstein-Taybi syndrome, GH deficiency, Arnold Chiari malformation and pituitary hypoplasia. *BMC Med Genet.* 2013;14:28.

10. Dauwerse JG, van Belzen M, van Haeringen A et al. Analysis of mutations within the intron20 splice donor site of CREBBP in patients with and without classical RSTS. *Eur J Hum Genet.* 2016;24(11):1639–43.

11. de Vries TI, Monroe GR, van Belzen MJ et al. Mosaic CREBBP mutation causes overlapping clinical features of Rubinstein-Taybi and Filippi syndromes. *Eur J Hum Genet.* 2016;24(9):1363–6.

12. Wincent J, Luthman A, van Belzen M et al. CREBBP and EP300 mutational spectrum and clinical presentations in a cohort of Swedish patients with Rubinstein-Taybi syndrome. *Mol Genet Genomic Med.* 2015;4(1):39–45.

13. Crawford H, Moss J, McCleery JP, Anderson GM, Oliver C. Face scanning and spontaneous emotion preference in Cornelia de Lange syndrome and Rubinstein-Taybi syndrome. *J Neurodev Disord.* 2015;7(1):22.

14. Crawford H, Waite J, Oliver C. Diverse profiles of anxiety related disorders in fragile X, Cornelia de Lange and Rubinstein-Taybi syndromes. *J Autism Dev Disord.* 2017;47(12):3728–40.

15. Masuda K, Akiyama K, Arakawa M et al. Exome sequencing identification of EP300 mutation in a proband with coloboma and imperforate anus: Possible expansion of the phenotypic spectrum of Rubinstein-Taybi syndrome. *Mol Syndromol.* 2015;6(2):99–103.

16. Kumar S, Suthar R, Panigrahi I, Marwaha RK. Rubinstein-Taybi syndrome: Clinical profile of 11 patients and review of literature. *Indian J Hum Genet.* 2012;18(2):161–6.

17. Silva CC, Pedroso JL, Souza PV, Pinto WB, Barsottini OG. Broad thumbs and broad hallux: The hallmarks for the Rubinstein-Taybi syndrome. *Arq Neuropsiquiatr.* 2014;72(1):81–2.

18. Bienias W, Pastuszka M, Gutfreund K, Kaszuba A. Multiple keloids in a 16-year-old boy with Rubinstein-Taybi syndrome. *Arch Med Sci.* 2015;11(1):232–4.

19. Loomba RS, Geddes G. Tricuspid atresia and pulmonary atresia in a child with Rubinstein-Taybi syndrome. *Ann Pediatr Cardiol*. 2015;8(2):157–60.

20. Mishra S, Agarwalla SK, Potpalle DR, Dash NN. Rubinstein-Taybi syndrome with agenesis of corpus callosum. *J Pediatr Neurosci*. 2015;10(2):175–7.

21. Pasic S. Rubinstein-Taybi syndrome associated with humoral immunodeficiency. *J Investig Allergol Clin Immunol*. 2015;25(2):137–8.

22. Shilpashree P, Jaiswal AK, Kharge PM. Keloids: An unwanted spontaneity in Rubinstein-Taybi syndrome. *Indian J Dermatol*. 2015;60(2):214.

23. Zavras N, Mennonna R, Maris S, Vaos G. Circumscribed storiform collagenoma associated with Rubinstein-Taybi syndrome in a young adolescent. *Case Rep Dermatol*. 2016;8(1):59–63.

24. Philip J, Patil NM. Rubinstein Taybi Syndrome with psychosis. *Indian Pediatr*. 2016;53(8):750.

25. Waite J, Beck SR, Heald M, Powis L, Oliver C. Dissociation of cross-sectional trajectories for verbal and visuo-spatial working memory development in Rubinstein-Taybi syndrome. *J Autism Dev Disord*. 2016;46(6):2064–71.

26. Zheng F, Kasper LH, Bedford DC, Lerach S, Teubner BJ, Brindle PK. Mutation of the CH1 domain in the histone acetyltransferase CREBBP results in autism-relevant behaviors in mice. *PLOS ONE*. 2016;11(1):e0146366.

27. Olyaei Y, Sarmiento JM, Bannykh SI, Drazin D, Naruse RT, King W. Rubinstein-Taybi syndrome associated with pituitary macroadenoma: A case report. *Cureus*. 2017;9(4):e1151.

28. Romaniouk I, Romero A, Runza P, Nieto C, Mouzo R, Simal F. Management of neuroendocrine tumor in a patient with Rubinstein-Taybi syndrome in chronic hemodialysis. *Nefrologia*. 2017;pii: S0211–6995(17):30186–8.

29. Deepthi DA, Shaheen VS, Kumar MH, Ashraf S, Deepak JH. Broad thumb-hallux syndrome: A diagnosis made on clinical findings. *J Clin Diagn Res*. 2017;11(5):ZJ05–06.

30. Tirali RE, Sar C, Cehreli SB. Oro-facio-dental findings of Rubinstein-Taybi syndrome as a useful diagnostic feature. *J Clin Diagn Res*. 2014;8(1):276–8.

31. Iba K, Wada T, Yamashita T. Correction of thumb angulations after physiolysis of delta phalanges in a child with Rubinstein-Taybi syndrome: A case report. *Case Reports Plast Surg Hand Surg*. 2015;2(1):12–4.

32. Darlong V, Pandey R, Garg R, Pahwa D. Perioperative management of a patient of Rubinstein-Taybi syndrome with ovarian cyst for laparotomy. *J Anaesthesiol Clin Pharmacol*. 2014;30(3):422–4.

33. Bounakis N, Karampalis C, Sharp H, Tsirikos AI. Surgical treatment of scoliosis in Rubinstein-Taybi syndrome type 2: A case report. *J Med Case Rep*. 2015;9:10.

34. Karahan MA, Sert H, Ayhan Z, Ayhan B. Anaesthetic management of children with Rubinstein-Taybi syndrome. *Turk J Anaesthesiol Reanim*. 2016;44(3):152–4.

35. Lee KH, Park EY, Jung SW, Song SW, Lim HK. Hysterectomy due to abnormal uterine bleeding in a 15-year old girl with Rubinstein-Taybi syndrome. *J Lifestyle Med*. 2016; 6(2):76–8.

36. Pérez-Grijalba V, García-Oguiza A, López M et al. New insights into genetic variant spectrum and genotype-phenotype correlations of Rubinstein-Taybi syndrome in 39 CREBBP-positive patients. *Mol Genet Genomic Med*. 2019;7(11):e972.

37. Yu S, Wu B, Qian Y et al. Clinical exome sequencing identifies novel CREBBP variants in 18 Chinese Rubinstein-Taybi syndrome kids with high frequency of polydactyly. *Mol Genet Genomic Med*. 2019;7(12):e1009.

10

Schinzel–Giedion Syndrome

Dongyou Liu

CONTENTS

10.1 Introduction

Schinzel–Giedion syndrome (SGS, also referred to as Schinzel–Giedion midface retraction syndrome) is a very rare, autosomal dominant disorder that typically shows severe psychomotor retardation, seizures, distinctive facial features (e.g., high, prominent forehead; low nasal root; ocular hypertelorism; midface hypoplasia; and low-set ears), and multiple congenital anomalies (e.g., cardiac, genitourinary, renal, skeletal, and central nervous system malformations), along with a higher risk for embryonal tumors (e.g., malignant sacrococcygeal teratoma, neuroepithelial neoplasia, in 19% of cases), leading to a fatal outcome in 50% of affected children before the age of 2 years [1].

At the molecular level, SGS is attributed to heterozygous germline mutations clustering in a 12 bp hotspot in exon 4 of the SET-binding protein 1 (*SETBP1*) gene located on chromosome 18q12.3, which inhibits tumor suppressors (e.g., PP2A and NM23-H1) and enhances cell proliferation, differentiation, and transformation.

10.2 Biology

In 1978, Schinzel and Giedion described a new syndrome (so-called Schinzel–Giedion syndrome [SGS]) in two siblings, who presented with severe midface retraction (midfacial hypoplasia with coarse appearance and deep grooves under the eyes), multiple skull anomalies, clubfeet, cardiac and renal malformations, severe intellectual disability and low physical growth after birth, hydronephrosis, and genital anomalies [2,3]. Further studies also revealed its increased risk for embryonal tumors. In fact, of about 50 SGS patients reported to date, 9 have been found to have malignant sacrococcygeal teratoma, primitive neuroectodermal tumor, and hepatoblastoma.

As SGS occurs sporadically in most cases, the involvement of an autosomal dominant mutation in a single gene has been suspected. Indeed, a study published in 2010 pinpointed heterozygous de novo germline mutations in a 12 bp hotspot of *SETBP1* exon 4, encoding residues 868–871 of the SETBP1 protein, as the genetic cause of SGS.

10.3 Pathogenesis

The human *SETBP1* gene (also referred to as SEB, KIAA0437, NUP98 fusion gene), located on chromosome 18q12.3, consists of two isoforms. The isoform a is a 387 kb fragment organized in six exons, which generate a 9899 nt transcript, and a predicted protein of 1596 aa; the isoform b is a 197 kb fragment organized in four exons (with the first three exons being shared between the two isoforms), which encode a 1804 nt transcript, and a predicted protein of 242 aa.

SETBP1 isoform a is a nuclear protein with an estimated molecular mass of 170 kDa and includes a SET-binding region (amino acids 1292–1488), an oncoprotein SKI homologous region (amino acids 706–917), three bipartite NLS (nuclear localization signal) motifs (amino acids 462–477, 1370–1384, 1383–1399), three AT hook domains (amino acids 584–596, 1016–1028, 1451–1463), six PEST sequences (amino acids 1–13, 269–280, 548–561, 678–689, 806–830, 1502–1526), and four repeat domains (amino acids 1466–1473, 1474–1481, 1482–1489, 1520–1543).

Functionally, the SET-binding region interacts with and regulates the proto-oncogene *SET*, which acts by inhibiting tumor suppressors such as PP2A and NM23-H1. PP2A is a major protein phosphatase that inhibits cell proliferation, differentiation, and transformation via disruption of a G2/M cell-cycle checkpoint. As a major counterpart of SET, SETBP1 preferentially replaces PP2A in the SET/PP2A complex.

The oncoprotein SKI-homologous region shares homology with the proto-oncogene *SKI*, which inhibits the transcription of target genes downstream of the transforming growth factor-β (TGF-β) superfamily. Through regulation of the SKI/SKI homodimer and the SKI/SNON heterodimer, SETBP1 may play a part in cellular transformation.

Three putative bipartite NLS motifs are thought to assist the signal-dependent nuclear transport of SETBP1 across the nuclear pore. The AT-hook motifs increase the DNA-binding capacity of SETBP1 and thus enhance its potential in transcriptional regulation.

Given the role of SETBP1 in transcriptional activation and elongation as well as transcriptional repression (on tumor suppressor genes such as *RUNX1*), *SETBP1* mutations that generate a gain or loss of function may disturb the normal functions of SETBP1 and demonstrate a significant oncogenic potential. Gain-of-function mutations produce a new protein that often interferes with different downstream pathways (e.g., apoptosis, differentiation, and self-renewal) through altered functions of the PP2A, *HOXA* genes, and *TGF-β* signaling pathway. For example, translocation or microdeletion in the *SETBP1* gene locus is found to cause a gain-of-function effect, leading to overexpression and/or increased stability of SETBP1, and subsequent PP2A inhibition and uncontrolled cell proliferation. Germline de novo *SETBP1* mutations are often associated with a severe phenotype characterized by mental retardation associated with distorted neuronal layering, multi-organ development abnormalities, and higher than normal risk of tumors. Indeed, most SGS patients who develop embryonal tumors harbor activating germline *SETBP1* mutations. Loss-of-function *SETBP1* mutations producing haploinsufficiency may result in a milder phenotype, which is characterized by a complete lack of expressive speech with intact receptive language abilities, decreased fine motor skills, subtle dysmorphisms, and hyperactivity and autistic traits.

The most frequent de novo germline mutations (i.e., D868N, S869N, G870S, and I871T) associated with classic SGS occur in the 12 bp hotspot (also known as the canonical degron) of *SETBP1* exon 4, which contains a consensus-binding region critical for protein degradation through ubiquitin binding. Mutations in this hotspot disrupt the signal for protein degradation and result in

the accumulation of SETBP1 protein. It appears that mutations in residue D868 show a larger impact on degron/βTrCP1 interaction than those in residue G870. In fact, G870S has been shown to cause poor protein degradation and can be thus considered functionally equivalent to SETBP1 overexpression. Some atypical SGS may have mutations in other locations (e.g., E862, S867, and T873). Compared to variants D868N and G870S, which dramatically increase protein levels, variants E862K, S869N, and I871T cause a more modest increase in protein levels.

In contrast to germline variants that are responsible for the onset of SGS and increased risk for embryonal tumors, somatic *SETBP1* mutations tend to occur in patients with hematologic neoplasms, including acute myeloid leukemia (AML, <1% of cases), myeloproliferative neoplasm (MPN, 4%), myelodysplastic syndrome/myeloproliferative neoplasm overlap syndromes (MDS/MPN, 9%), myelodysplastic syndrome (MDS, 2–3%), juvenile myelomonocytic leukemia (JMML, 8–10%), chronic myelomonocytic leukemia (CMML, 15–19%), chronic neutrophilic leukemia (CNL, 10–38%), and atypical chronic myeloid leukemia (aCML, 30%) [4–16]. Most cases are associated with heterozygous missense mutations, and only a few cases show a homozygous mutation. SETBP1 involvement in leukemia transformation is mainly through activation of the *HOXA9* and *HOXA10* genes, and subsequent increase in leukemic cell proliferation [17]. Somatic *SETBP1* mutations observed in hematologic neoplasms appear to have a gain-of-function effect on the SETBP1 protein, leading to decreased binding of the βTrCP1 and increased protein levels. The somatic variants in hematologic neoplasms appear more disruptive than the germline variants in SGS, and patients with myeloid malignancies harboring *SETBP1*-mutations often have a significantly inferior overall survival and increased risk of disease progression [18–20].

10.4 Epidemiology

10.4.1 Prevalence

SGS is a very rare disorder characterized by severe midface retraction, multiple skull anomalies, clubfeet, and cardiac and renal malformations. Since its first description in 1978, about 50 cases have been documented in literature.

10.4.2 Inheritance

SGS is a single-gene disorder that demonstrates an autosomal dominant inheritance pattern. Siblings carrying the same disease-causing mutation in residue I871 have been reported.

10.4.3 Penetrance

Germline de novo *SETBP1* mutations demonstrate incomplete penetrance and variable expressivity. Classic SGS is associated with D868N, S869N, G870S, and I871T mutants, while atypical SGS may be linked to E862, S867, and T873. It appears that variants D868N and G870S often contribute to dramatically increased protein levels in comparison with variants E862K, S869N, and I871T. SGS patients carrying a mutation affecting SETBP1 residue S867 may show characteristic facial features,

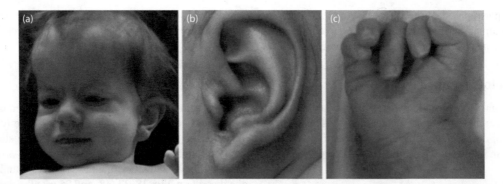

FIGURE 10.1   Classic features of Schinzel–Giedion syndrome (SGS) include (a) distinctive facies (large fontanels, prominent forehead, bitemporal narrowing, shallow orbits or prominent eyes); (b) typical question mark–shaped ear (low-set and posteriorly rotated ear with anteriorly angulated lobules); and (c) characteristic hand posture with clenched fingers. (Photo credit: Acuna-Hidalgo R et al. *PLOS Genet.* 2017;13(3):e1006683.)

genital anomalies, and seizures but no hydronephrosis. As hydronephrosis is considered one of the hallmark features, its absence in some individuals with mutations within the hotspot suggests that it should not be recognized as an obligatory feature for the diagnosis of SGS [21].

10.5 Clinical Features

Patients with Schinzel–Giedion syndrome typically display SGS facial gestalt, skeletal, brain, congenital, and neurologic anomalies, along with increased tumor risk and fatality [22,23].

10.5.1 SGS Facial Gestalt

Large fontanels, prominent/protruding forehead, midface hypoplasia (retracted and shortened midface and full cheeks, producing a facial frontal silhouette in the shape of "8"), bitemporal narrowing, shallow orbits or prominent eyes, deep groove under the eyes, upslanting palpebral fissures, hypertelorism, saddle nose with retracted root, upturned nose, low set ears (posteriorly rotating with anteriorly angulated lobules producing a question mark shape), large mouth with everted lower lip and protruding

tongue, micrognathia, philtrum groove, hypertrichosis, facial hemangioma, and short neck (Figure 10.1) [13].

10.5.2 Skeletal Anomalies

Sclerotic base of the skull with wide occipital synchondrosis, broad ribs, short pubic rami, wide pubic symphysis, hypoplastic distal phalanges in hands and feet, post-axial polydactyly, clenched fingers, and scoliosis (Figure 10.2) [23,24].

10.5.3 Structural Brain Anomalies

Underdeveloped corpus callosum, cortical atrophy or dysplasia, ventriculomegaly, and choroid plexus cysts.

10.5.4 Congenital Anomalies

Urogenital defects (hydronephrosis, double ureters, renal cysts and stones, hypospadias, small penis, retentio testis, hydrocele testis, displaced anus, and female extragenital hypoplasia), cardiac defects (faulty atrial septum, patent foramen ovale, patent ductus arteriosus, and cardiac hypertrophy), oral cavity/respiratory defects (tracheo/laryngomalacia, lung hypoplasia, micrognathia,

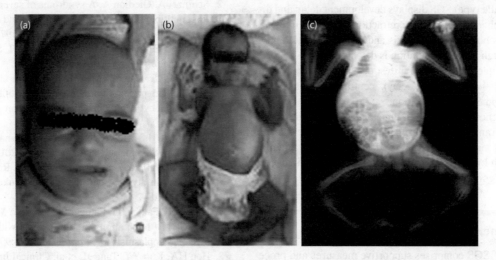

FIGURE 10.2   A newborn with Schinzel–Giedion syndrome (SGS) displaying (a) coarse facial features (midface retraction, frontal bossing, bitemporal narrowing, wide anterior fontanel, low nasal bridge); (b) abdominal distention and bilateral talipes equinovarus; and (c) long bones and broad ribs as revealed by babygram. (Photo credit: Bulut O et al. *Case Rep Genet.* 2017;2017:3740524.)

and gingiva hypertrophy, leading to swallowing and breathing difficulties), inguinal hernia, alacrima, talipe(s) equinovarus, hypoplasia of the pancreatic tail, or hepatosplenomegaly.

10.5.5 Neurologic Anomalies

Microcephaly, intractable epilepsy, seizures, spasticity and/or hypertonia, cerebral blindness or deafness (vision or hearing impairment), developmental delay or progressive failure to thrive, and severe intellectual disability (mental retardation).

10.5.6 Tumors

Embryonal tumors such as malignant sacrococcygeal teratoma, primitive neuroectodermal tumor, and hepatoblastoma [25].

10.5.7 Death

Most patients do not survive past childhood, and the longest documented survival is 15 years of age. The common causes of death during infancy are pneumonia, congenital cardiac defects, tumors, lung hypoplasia, intractable seizures, and sudden cardiac arrest.

10.6 Diagnosis

On the basis of reminiscent clinical features as well as *SETBP1* status, SGS is often recognized and diagnosed as three types [1].

Type I (complex and classic type) SGS is characterized by development delay and typical facial features (prominent forehead, midface retraction, and short and upturned nose) along with hydronephrosis or two of the characteristic skeletal anomalies (sclerotic skull base, wide occipital synchondrosis, increased cortical density or thickness, and broad ribs). Some patients may also show visual impairment, hearing loss, brain anomalies, neurological degeneration, and an increased incidence of embryonal tumors. The co-occurrence of progressive developmental retardation and multiple malformations indicates extreme disease severity and very poor prognosis [1].

Type II (middle type) SGS displays development delay and distinctive facial phenotype (midface retraction, short and upturned nose), but lacks both hydronephrosis and typical skeletal abnormalities, although *SETBP1* mutation is detected.

Type III (simple type) SGS manifest development delay (particularly expressive language delay), central nervous system involvement, and *SETBP1* alteration.

Prenatal diagnosis may rely on antenatal ultrasonography (to detect polyhydramnios or hydronephrosis and severe skeletal anomalies) and molecular testing (to identify *SETBP1* mutation in families with history of an affected child, and to rule out the empiric risk of gonadal mosaicism) [1,26].

10.7 Treatment

Management of SGS comprises supportive measures and procedures that target specific issues. As many SGS patients experience upper urinary tract obstruction due to hydronephritis, hydroureter, pyeloureteral junction (PUJ) stenosis, ureterovesical junction stenosis, or vesicoureteral reflux, they may be relieved with percutaneous nephrostomy, pyeloplasty, or ureteroneocystostomy.

10.8 Prognosis

Patients with SGS have a poor prognosis and most affected individuals do not survive beyond 2 years of age (with the longest documented survival being 15 years). In one study involving a limited number of patients, those harboring *SETBP1* substitutions in D868, S869, G870, and I871 were found to succumb to the disorder within 18, 32, 48, and 25 months, respectively.

10.9 Conclusion

Schinzel–Giedion syndrome (SGS) is a rare but clinically recognizable developmental disorder that typically presents with facial features, neurological alterations (e.g., severe intellectual disability, intractable epilepsy, and cerebral blindness and deafness), various congenital anomalies (e.g., heart defects, urogenital malformations, and bone abnormalities) and increased risk for embryonic tumors. Germline de novo mutations in a hotspot of *SETBP1* exon 4 underscore the development of SGS. In addition, activating germline *SETBP1* mutations are strongly associated with the occurrence of embryonal tumors in SGS patients. SETBP1 appears to act both as a negative regulator of PP2A activity and as transcriptional regulator (of the *HOXA* gene cluster, *RUNX1*, and *MYB*, as well as many TGF-β responsive genes). Further studies are clearly necessary to identify crucial pathways affected by alterations of *SETBP1* normal function, and to design innovative intervention measures for this rare genetic disease [27].

REFERENCES

1. Liu WL, He ZX, Li F, Ai R, Ma HW. Schinzel-Giedion syndrome: A novel case, review and revised diagnostic criteria. *J Genet.* 2018;97(1):35–46.
2. Schinzel A, Giedion A. A syndrome of severe midface retraction, multiple skull anomalies, clubfeet, and cardiac and renal malformations in sibs. *Am J Med Genet.* 1978;1:361–75.
3. Touge H, Fujinaga T, Okuda M, Aoshi H. Schinzel-Giedion syndrome. *Int J Urol.* 2001;8(5):237–41.
4. Laborde RR, Patnaik MM, Lasho TL et al. SETBP1 mutations in 415 patients with primary myelofibrosis or chronic myelomonocytic leukemia: Independent prognostic impact in CMML. *Leukemia.* 2013;27(10):2100–2.
5. Makishima H, Yoshida K, Nguyen N et al. Somatic SETBP1 mutations in myeloid malignancies. *Nat Genet.* 2013;45(8):942–6.
6. Piazza R, Valletta S, Winkelmann N et al. Recurrent SETBP1 mutations in atypical chronic myeloid leukemia. *Nat Genet.* 2013;45(1):18–24.
7. Fabiani E, Falconi G, Fianchi L, Criscuolo M, Leone G, Voso MT. SETBP1 mutations in 106 patients with therapy-related myeloid neoplasms. *Haematologica.* 2014;99(9):e152–3.
8. Hou HA, Kuo YY, Tang JL et al. Clinical implications of the SETBP1 mutation in patients with primary myelodysplastic syndrome and its stability during disease progression. *Am J Hematol.* 2014;89(2):181–6.

9. Choi HW, Kim HR, Baek HJ et al. Alteration of the SETBP1 gene and splicing pathway genes SF3B1, U2AF1, and SRSF2 in childhood acute myeloid leukemia. *Ann Lab Med.* 2015;35(1):118–22.

10. Bresolin S, De Filippi P, Vendemini F et al. Mutations of SETBP1 and JAK3 in juvenile myelomonocytic leukemia: A report from the Italian AIEOP study group. *Oncotarget.* 2016;7(20):28914–9.

11. Kanagal-Shamanna R, Luthra R, Yin CC et al. Myeloid neoplasms with isolated isochromosome 17q demonstrate a high frequency of mutations in SETBP1, SRSF2, ASXL1 and NRAS. *Oncotarget.* 2016;7(12):14251–8.

12. Nguyen N, Vishwakarma BA, Oakley K et al. Myb expression is critical for myeloid leukemia development induced by Setbp1 activation. *Oncotarget.* 2016;7(52):86300–12.

13. Acuna-Hidalgo R, Deriziotis P, Steehouwer M et al. Overlapping SETBP1 gain-of-function mutations in Schinzel-Giedion syndrome and hematologic malignancies. *PLOS Genet.* 2017;13(3):e1006683.

14. Coccaro N, Tota G, Zagaria A, Anelli L, Specchia G, Albano F. SETBP1 dysregulation in congenital disorders and myeloid neoplasms. *Oncotarget.* 2017;8(31):51920–35.

15. Linder K, Iragavarapu C, Liu D. SETBP1 mutations as a biomarker for myelodysplasia/myeloproliferative neoplasm overlap syndrome. *Biomark Res.* 2017;5:33.

16. Ouyang Y, Qiao C, Chen Y, Zhang SJ. Clinical significance of CSF3R, SRSF2 and SETBP1 mutations in chronic neutrophilic leukemia and chronic myelomonocytic leukemia. *Oncotarget.* 2017;8(13):20834–41.

17. Oakley K, Han Y, Vishwakarma BA et al. Setbp1 promotes the self-renewal of murine myeloid progenitors via activation of Hoxa9 and Hoxa10. *Blood.* 2012;119(25):6099–108.

18. Hu W, Wang X, Yang R et al. A novel mutation of SETBP1 in atypical chronic myeloid leukemia transformed from acute myelomonocytic leukemia. *Clin Case Rep.* 2015;3(6):448–52.

19. Stieglitz E, Troup CB, Gelston LC et al. Subclonal mutations in SETBP1 confer a poor prognosis in juvenile myelomonocytic leukemia. *Blood.* 2015;125(3):516–24.

20. Shou LH, Cao D, Dong XH et al. Prognostic significance of SETBP1 mutations in myelodysplastic syndromes, chronic myelomonocytic leukemia, and chronic neutrophilic leukemia: A meta-analysis. *PLOS ONE.* 2017;12(2):e0171608.

21. Albano LM, Sakae PP, Mataloun MM, Leone CR, Bertola DR, Kim CA. Hydronephrosis in Schinzel-Giedion syndrome: An important clue for the diagnosis. *Rev Hosp Clin Fac Med Sao Paulo.* 2004;59(2):89–92.

22. Hishimura N, Watari M, Ohata H et al. Genetic and prenatal findings in two Japanese patients with Schinzel-Giedion syndrome. *Clin Case Rep.* 2016;5(1):5–8.

23. Bulut O, Ince Z, Altunoglu U, Yildirim S, Coban A. Schinzel-Giedion syndrome with congenital megacalycosis in a Turkish patient: Report of SETBP1 mutation and literature review of the clinical features. *Case Rep Genet.* 2017;2017:3740524.

24. Sharma AK, Gonzales JA. Scoliosis in a case of Schinzel-Giedion syndrome. *HSS J.* 2009;5(2):120–2.

25. Matsumoto F, Tohda A, Shimada K, Okamoto N. Malignant retroperitoneal tumor arising in a multicystic dysplastic kidney of a girl with Schinzel-Giedion syndrome. *Int J Urol.* 2005;12(12):1061–2.

26. Volk A, Conboy E, Wical B, Patterson M, Kirmani S. Whole-exome sequencing in the clinic: Lessons from six consecutive cases from the clinician's perspective. *Mol Syndromol.* 2015;6(1):23–31.

27. Piazza R, Magistroni V, Redaelli S et al. SETBP1 induces transcription of a network of development genes by acting as an epigenetic hub. *Nat Commun.* 2018;9(1):2192.

11

Schwannomatosis

Tayfun Hakan

CONTENTS

11.1 Introduction

Schwannomatosis is a neurogenetic disorder characterized by the formation of multiple schwannomas and occasionally meningiomas, with schwannomas frequently occurring in the spine (74%), peripheral nerves (89%), and cranial nerve (mostly trigeminal, 8%) (Table 11.1). When the tumors are localized in an anatomically limited distribution on a single extremity or several contiguous segments of the spine, the disease is referred to as segmental (or mosaic) schwannomatosis, which is implicated in 33% of cases [1].

Schwannomatosis can be either sporadic or familial. Sporadic schwannomatosis affects individuals but not any other members of their families, while familial schwannomatosis affects individuals as well as their first-degree relative(s) who develop one or more schwannomas [2,3]. The available data indicate that sporadic schwannomatosis makes up a majority of clinical cases, and familial schwannomatosis accounts for 13%–25% of cases. In the 2016 World Health Organization classification, schwannomatosis was placed among familial tumor syndromes [4].

Schwannomatosis represents a third form of neurofibromatosis (the others being neurofibromatosis type 1 [NF1] and neurofibromatosis type 2 [NF2]) and shows a predisposition for developing multiple schwannomas and occasionally meningiomas. In particular, schwannomatosis demonstrates considerable overlapping with NF2 and was once considered a variant of NF2 (Table 11.1). However, schwannomatosis differs from NF2 by its absence of bilateral vestibular schwannomas (the hallmark of NF2) and other manifestations (Table 11.1), while NF2 stands out by its occurrence in a first-degree relative with unilateral vestibular schwannomas or any two of meningioma, glioma, schwannoma, or juvenile posterior lenticular opacities. At the molecular level, schwannomatosis is linked to germline mutations in the *SMARCB1* or leucine zipper-like transcriptional regulator 1

(*LZTR1*) genes [5], whereas NF2 contains germline mutation in the *NF2* gene (Table 11.1). Nonetheless, independent somatic mutations (instead of constitutional/germline mutations) affecting both *NF2* alleles have been detected in schwannomas from schwannomatosis patients. In view of their distinct phenotypic (nonvestibular schwannomas vs bilateral vestibular schwannomas) and genetic (*SMARCB1/LZTR1* vs *NF2*) features, schwannomatosis and NF2 are recognized as separate clinical entities (see Chapter 5 in this book) [6,7].

11.2 Biology

As the principle cells of the peripheral nervous system (PNS), Schwann cells are derived from neural crest, from which neural crest cells migrate during embryologic development to their target sites in the PNS and differentiate into Schwann cells. Schwann cells are involved in the myelination of peripheral axons, with each Schwann cell myelinating a single axon. This covering provides a protective barrier for the axon and increases salutatory conduction of the neuron.

Schwannomas are benign, encapsulated, slow-growing nerve sheath tumors composed of Schwann cells that surround the axons (Figure 11.1). Usually occurring singly in either peripheral or spinal nerves of otherwise normal individuals and having the capacity to enlarge and grow in previously healthy nerves, schwannomas are the most common benign peripheral nerve tumors in adults. However, the formation of multiple schwannomas in an individual suggests a genetic predisposition and possible existence of rare genetic diseases, such as schwannomatosis (which does not usually affect the vestibular or hearing nerves) or NF2 (which affects the vestibular or hearing nerves) (Table 11.1) [8].

First noted in 1973, schwannomatosis was thought to represent a mild form of neurofibromatosis since both schwannomatosis and

TABLE 11.1

Phenotypic and Genetic Characteristics of Schwannomatosis and Neurofibromatosis Type 2

Characteristics	Schwannomatosis	Neurofibromatosis Type 2
Peripheral nerve schwannoma	89%	68%
Spinal tumor	74%	63%–90%
Subcutaneous tumor	23%	43%–48%
Intracranial nonvestibular schwannoma	9%–10%	24%–51%
Intracranial meningioma	5%	45%–58%
Unilateral vestibular schwannoma	Rare	18%
Bilateral vestibular schwannoma	Absent	90%–95%
Ependymoma	Absent	18%–58%
Skin plaques	Absent	41%–48%
Intradermal tumor	Absent	27%
Retinal hamartoma	Absent	6%–22%
Epiretinal membrane	Absent	12%–40%
Subcapsular cataract	Absent	60%–81%
Germline mutation	*SMARCB1* (10%–50%), *LZTR1* (25%–40%)	*NF2* (>90%)

Source: Modified from Kehrer-Sawatzki H et al. *Hum Genet.* 2017;136(2):129–48.

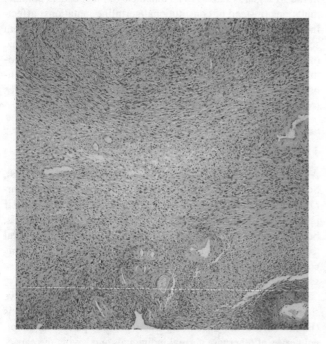

FIGURE 11.1 Schwannoma showing compact fascicles of elongated tumor cells with nuclear palisading and hyalinized vessels (HE ×100). (Adapted from Hakan T et al. *Turk Neurosurg.* 2008;18(3):320–3.)

NF2 induce multiple schwannomas. Following the description of additional cases and the expansion of its clinical spectrum (e.g., formation of multiple peripheral nerve schwannomas without concomitant involvement of the vestibular nerve), a consensus has emerged that schwannomatosis might represent a distinct identity from NF2. The isolation of the *NF2* locus in 2003 provided genetic evidence that NF2 is a different disease from schwannomatosis. Subsequent

identification of two tumor suppressor genes (*SMARCB1* and *LZTR1*) in 2007 and 2014 has put it beyond doubt that schwannomatosis is indeed a distinct clinical identity [5,7,9].

Although a vast majority of schwannomatosis cases are attributable to sporadic or de novo mutations in the *SMARCB1* and *LZTR1* genes, about 13%–25% are family-involved, and show an autosomal dominant inheritance pattern. While germline *SMARCB1* mutations are found in 10% of sporadic schwannomatosis patients, they are present in 50% of familial schwannomatosis patients. Similarly, germline *LZTR1* mutations occur in 10% and 40% of sporadic and familial schwannomatosis patients, respectively (Table 11.1). Overall, germline mutations of either *SMARCB1* or *LZTR1* have been identified in 86% of familial and 40% of sporadic schwannomatosis patients. However, it appears that *SMARCB1* mutations predispose to both schwannoma and meningioma, while *LZTR1* mutations predispose to schwannoma but not to meningioma [7,9,10].

Based on genetic analysis of blood and tumors from schwannomatosis patients, tumorigenesis of *SMARCB1*- and *LZTR1*-related schwannomatosis involves a 3-step, 4-hit mechanism, beginning with a germline *SMARCB1* or *LZTR1* mutation on allele one (hit 1), then somatic loss of heterozygosity on chromosome 22q containing second *SMARCB1* or *LZTR1* allele and one *NF2* allele (hits 2 and 3), and finally somatic mutation of the remaining wild-type *NF2* allele located on chromosome harboring the germline *SMARCB1* or *LZTR1* mutation (hit 4). This 3-step/4-hit model raises the possibility that germline *SMARCB1* or *LZTR1* mutation is associated with local or diffuse pain observed in schwannomatosis, while biallelic loss of *NF2* is responsible for tumor formation [11–13].

11.3 Pathogenesis

The *SMARCB1* (SWI/SNF-related, matrix-associated, actin-dependent regulator of chromatin, subfamily b, member 1; also termed *hSNF5 BAF47*, or integrase interactor 1 [*INI1*]) gene is located on 22q11.2, centromeric to the *NF2* gene, and consists of nine exons that encode a subunit of the SWI/SNF (SWItch/Sucrose Non-Fermentable) ATP-dependent chromatin-remodeling complex. Ubiquitously expressed in the nuclei of all normal cells, SMARCB1 targets the SWI/SNF complex to gene promoters and interacts with components in several pathways related to tumor proliferation and progression (e.g., p16-RB pathway, WNT-signaling pathway, sonic hedgehog-signaling pathway, and polycomb pathway), thus contributing to the regulation of cell cycle, lineage-specific gene expression and embryonic stem cell programming, and to the induction of senescence [14]. Mutations in the *SMARCB1* gene may result in complete loss, mosaic expression and reduced expression of the related proteins [7]. In the mouse model, SMARCB1-null embryos die between 3.5 and 5.5 days post-coitum, while SMARCB1 heterozygous-deficient mice develop aggressive cancer (e.g., rhabdoid-like tumors, T-cell lymphomas) within 11 weeks.

Interestingly, schwannomatosis-associated *SMARCB1* mutations are mainly found either at the 5′ or 3′ end of the gene, the most common of which is the c.*82C > T mutation located in the 3′ UTR. In addition, *SMARCB1* mutations in familial schwannomatosis are predominantly missense and splice site mutations as well as in-frame deletions, leading to the production of stable non-truncating transcripts with residual function, while *SMARCB1* mutations in sporadic schwannomatosis are likely frameshift or nonsense, leading to truncating or nonfunctional transcripts. On the other hand, *SMARCB1* mutations associated with rhabdoid tumor (a highly aggressive childhood cancer) of rhabdoid predisposition syndrome mostly occur in the central part of the gene and are either protein-truncating or alternatively whole-gene or multi-exon deletions, resulting in a complete absence of SMARCB1 expression in rhabdoid tumors. There is evidence that *Smarcb1* loss in early neural crest is necessary to initiate tumorigenesis in the cranial nerves and meninges with typical histological features and molecular profiles of human rhabdoid tumors, while *Smarcb1* loss at a later developmental stage in Schwann cells together with biallelic *Nf2* gene inactivation promotes schwannoma formation [15].

The *LZTR1* gene is situated approximately 3 Mb and 9 Mb centromeric to *SMARCB1* and *NF2*, respectively, on chromosome 22q11.21, and encodes a member of the BTB/POZ superfamily of proteins, which is present exclusively in the Golgi network and participates in multiple cellular processes (e.g., regulation of chromatin conformation and cell cycle). As a likely tumor suppressor gene, *LZTR1* may share a functional link with *SMARCB1* through nuclear receptor corepressor (N-CoR) interactions [9]. In up to 80% of *SMARCB1* mutation–negative schwannomatosis cases, *LZTR1* loss of function mutations (missense, splice site, nonsense, or frameshift) are identified across nearly all exons. Similar to *SMARCB1*-related schwannomatosis, schwannoma patients with *LZTR1* mutation also harbor somatic *NF2* mutations. Besides schwannomatosis, biallelic *LZTR1* mutations have been identified in glioblastoma [7,16].

In addition, a germline missense mutation has also been detected in the COQ6 gene on chromosome arm 14q in members of a family affected by schwannomatosis [17].

11.4 Epidemiology

Schwannomatosis is a rare tumor predisposition syndrome with an estimated annual incidence rate of 1 case per 40,000–80,000,

and familial schwannomatosis consists of 13%–25% of all clinical cases. Schwannomatosis affects both females and males equally and has an average age of 40 years (range 30–60 years) at diagnosis.

11.5 Clinical Features

Despite the fact that the onset of schwannomatosis often takes places in the second or third decade of the life, most cases remain unrecognized approximately 10 years after the initial symptoms emerge.

The clinical spectrum of schwannomatosis ranges from neuropathic pain, neurologic disability, to regional organ compression. Typically, patients complain of local, multifocal, or diffuse chronic pain (46%), masses (27%), or both (11%), as well as numbness or weakness of the muscles. Other common manifestations include headache, cranial nerve deficit (due to intracranial mass), neurological deficits (due to spinal cord compression), depression and anxiety (due to psychosocial stress), and bowel dysfunction or difficulty urinating [18–23].

Intracranial nonvestibular schwannomas and meningiomas, multiple non-intradermal peripheral, and spinal schwannomas are often found. Unilateral vestibular schwannoma has been described rarely. Spine and peripheral nerves are the most affected locations; cranial nerve schwannomas are rare. Schwannomas may be identified concurrently or emerge during follow-up in different anatomical locations (Figure 11.2) [24].

11.6 Diagnosis

Diagnostic workup for schwannomatosis consists of cranial and spinal magnetic resonance imaging (MRI), abdominal sonography, and detailed ophthalmology and otolaryngology examination, with final diagnosis based on lesions, histology, and genetics [12,25].

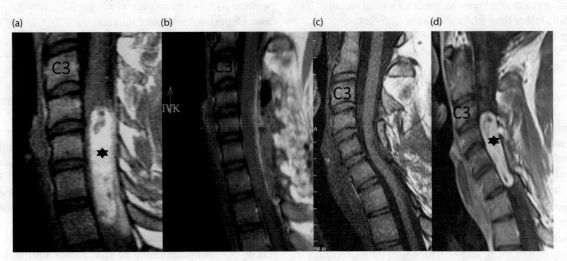

FIGURE 11.2 (a) A well-circumscribed, oval-shaped large mass (asterisk) with contrast enhancement at C4-7 levels on sagittal T1-weighted MRI scans. (b) Postoperative early follow-up MRI confirming total removal of the tumor on sagittal T1-weighted MRI with gadolinium. (c) Follow-up MRI 2 years after the tumor excision with no tumor residue and no newly raised tumor. (d) A new, well-circumscribed, oval-shaped large mass (asterisk) with contrast enhancement at C3–6 levels on sagittal T1-weighted MRI scans, 6 years after the previous tumor excision. (Adapted from Hakan T et al. *Turk Neurosurg.* 2017;27(1):163–4.)

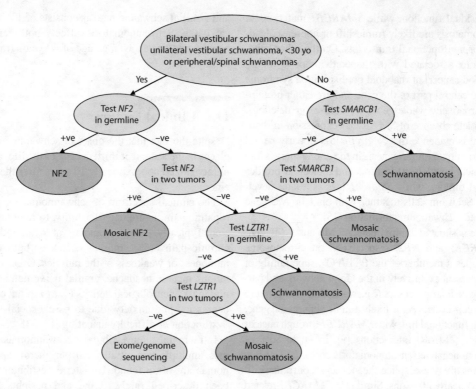

FIGURE 11.3 Optimal mutation screening strategy for schwannomatosis.

In patients with schwannomatosis, schwannomas are commonly located in the spine, peripheral nerves, and occasionally the cranial nerve (Table 11.1). Histologically, schwannomas display an intraneural growth pattern, peritumoral edema, myxoid change, and a mosaic INI1 staining by immunohistochemistry. Although vestibular schwannoma is pathognomonic for NF2 and represents a previous diagnostic criterion for exclusion, unilateral vestibular schwannoma may be detected in some schwannomatosis patients with germline mutations in *SMARCB1* and *LZTR1*. It should be noted that patients showing multiple nonvestibular schwannomas without vestibular schwannoma may have mosaic NF2 rather than schwannomatosis [26–28].

Occurring in ~5% of schwannomatosis patients, meningiomas often show a predilection for the cerebral falx.

Some schwannomatosis patients may develop additional tumors such as lipomas and angiolipomas due to genetic mosaicism.

Clinical criteria for definitive diagnosis of schwannomatosis are:

1. Two or more biopsy-proven non-intradermal schwannomas with no radiographic evidence of bilateral vestibular schwannomas on high-quality MRI (detailed study of internal auditory canal with slices no more than 3 mm thick), or
2. One biopsy-proven non-intradermal schwannoma or intracranial meningioma with a first-degree relative affected by schwannomatosis.

Molecular criteria for definitive diagnosis of schwannomatosis are:

1. Biopsy-proven schwannoma or meningioma and a germline *SMARCB1/LZTR1* mutation, or

2. At least two biopsy-proven schwannomas or meningiomas harboring a shared SMARB1/*LZTR1* mutation and differing N2 mutations.

Patients showing the following characteristics do not fulfill diagnosis for schwannomatosis: germline pathogenic NF2 mutation, meeting diagnostic criteria for NF2, first-degree relative with NF2, and schwannomas in previous field of radiation therapy only.

An optimal mutation screening strategy for schwannomatosis is presented in Figure 11.3 [27,29]. For confirmation of mosaic (segmental) schwannomatosis or mosaic NF2, positive molecular results from two tumors are required.

11.7 Treatment

Treatment options for schwannomatosis consist of symptomatic management and surgery [30–32].

Neuropathic medications (e.g., gabapentin and pregabalin), antidepressants (e.g., amitriptyline, nortriptyline, duloxetine), mood stabilizations (e.g., lamotrigine, valproate, opiate), anti-inflammatory medications, muscle relaxants, and lidocaine patches are used for chronic pain in patients. Spinal block, radiofrequency, and transcutaneous electrical nerve stimulation may be considered for pain control.

Surgery is the treatment of choice for symptomatic schwannomas or tumors demonstrating enlargement during the follow-up [2]. Surgical excision of the schwannomas may provide local pain relief by decompressing the neighboring tissues. However, the major risk of surgery is iatrogenic nerve injury.

There is no approved chemotherapy for schwannomas. Drugs that target cyclin D1, PLK1, Aurora A, p16, and interferons pathways are studied. Flavopiridol, which is a pan-cdk inhibitor and repressor of cyclin D1, and bevacizumab, which is a VEGF inhibitor, are some examples of medications that are in the spotlight.

Although stereotactic radiosurgery is a well-known treatment modality for sporadic vestibular and spinal schwannomas, the role of radiation in the management of schwannomatosis-related schwannomas is still unclear. Radiotherapy may be considered for patients who have life-threatening, enlarging schwannomas that cannot be surgically removed.

11.8 Prognosis

Although schwannomas are generally benign, their malignant transformation may occur occasionally. In schwannomatosis patients, newly developed malignant schwannomas may be found during follow-up. Schwannomatosis caused by *SMARCB1* germline mutations may have an increased risk of malignancy.

Patients with comorbid conditions tend to have a higher risk for malignancy, especially for malignant peripheral nerve sheath tumors and atypical teratoid/rhabdoid tumors that can be life threatening.

Emerging new schwannomas from the previously healthy nerves either in same or different anatomical locations during one's lifetime is common. Therefore, routine examination of the spine by MRI is recommended in patients with schwannomatosis.

Patients who undergo many operations during their adult life may develop new neurological deficits and depression.

11.9 Conclusion

Schwannomatosis is a rare form of neurofibromatosis characterized by the formation of multiple, often painful, schwannomas in the peripheral nervous system. Despite its shared phenotypic features with NF2, schwannomatosis is a distinct clinical entity. While vestibular schwannoma is absent in familial schwannomatosis, unilateral vestibular schwannoma may sometimes be observed. This makes the use of molecular testing techniques extremely valuable in order to achieve a definitive diagnosis of schwannomatosis. Given our limited understanding of schwannomatosis and NF2 in relation to their genetics, pathogenesis, clinical presentation, and outcome, further studies are necessary to improve the detection of additional predisposing, or modifying, genes, the determination of cellular pathways, genotype/phenotype correlations, and the design of novel therapies based on affected pathways [33]. As schwannomatosis tumors survive only a few passages before senescence, the recent establishment of immortalized Schwann cell lines from human tumors provides an impetus for advancing schwannomatosis research [34,35].

REFERENCES

1. MacCollin M, Woodfn W, Kronn D, Short MP. Schwannomatosis: A clinical and pathologic study. *Neurology*. 1996;46:1072–9.
2. Gonzalvo A, Fowler A, Cook RJ et al. Schwannomatosis, sporadic schwannomatosis, and familial schwannomatosis: A surgical series with long-term follow-up. Clinical article. *J Neurosurg*. 2011;114(3):756–62.
3. Mansukhani SA, Butala RP, Shetty SH, Khedekar RG. Familial schwannomatosis: A diagnostic challenge. *J Clin Diagn Res*. 2017;11(2):RD01–03.
4. Stemmer-rachmaninov AO, Hulsebos TJM, Wesseling P. Schwannomatosis, in: Louis DN, Ohgaki H, Wiestler OD, Cavenee WK (Eds), *WHO Classification of Tumours of the Central Nervous System*. 4th ed., Lyon, IARC Press, 2016; pp. 302–3.
5. Hulsebos TJ, Plomp AS, Wolterman RA, Robanus-Maandag EC, Baas F, Wesseling P. Germline mutation of INI1/SMARCB1 in familial schwannomatosis. *Am J Hum Genet*. 2007;80(4):805–10.
6. Kresak JL, Walsh M. Neurofibromatosis: A review of NF1, NF2, and schwannomatosis. *J Pediatr Genet*. 2016;5(2):98–104.
7. Kehrer-Sawatzki H, Farschtschi S, Mautner VF, Cooper DN. The molecular pathogenesis of schwannomatosis, a paradigm for the co-involvement of multiple tumour suppressor genes in tumorigenesis. *Hum Genet*. 2017;136(2):129–48.
8. Grapperon AM, Franques J, Roche PH, Battaglia F. Does hereditary neuropathy with liability to pressure palsy predispose to schwannomatosis? *J Clin Neurol*. 2014;10(4):371–2.
9. Piotrowski A, Xie J, Liu YF et al. Germline loss-of-function mutations in LZTR1 predispose to an inherited disorder of multiple schwannomas. *Nat Genet*. 2014;46(2):182–7.
10. Mehta GU, Feldman MJ, Wang H, Ding D, Chittiboina P. Unilateral vestibular schwannoma in a patient with schwannomatosis in the absence of LZTR1 mutation. *J Neurosurg*. 2016;125(6):1469–71.
11. Vitte J, Gao F, Coppola G, Judkins AR, Giovannini M. Timing of Smarcb1 and Nf2 inactivation determines schwannoma versus rhabdoid tumor development. *Nat Commun*. 2017;8(1):300.
12. Plotkin SR, Blakeley JO, Evans DG et al. Update from the 2011 International Schwannomatosis Workshop: From genetics to diagnostic criteria. *Am J Med Genet A*. 2013;161A(3):405–16.
13. Jordan JT, Smith MJ, Walker JA et al. Pain correlates with germline mutation in schwannomatosis. *Medicine (Baltimore)*. 2018;97(5):e9717.
14. Allen MD, Freund SM, Zinzalla G, Bycroft M. The SWI/SNF subunit INI1 contains an N-terminal winged helix DNA binding domain that is a target for mutations in schwannomatosis. *Structure*. 2015;23(7):1344–9.
15. Kohashi K, Oda Y. Oncogenic roles of SMARCB1/INI1 and its deficient tumors. *Cancer Sci*. 2017;108(4):547–52.
16. Paganini I, Chang VY, Capone GL et al. Expanding the mutational spectrum of LZTR1 in schwannomatosis. *Eur J Hum Genet*. 2015;23(7):963–8.
17. Zhang K, Lin JW, Wang J et al. A germline missense mutation in COQ6 is associated with susceptibility to familial schwannomatosis. *Genet Med*. 2014;16(10):787–92.
18. Hakan T, Celikoğlu E, Aker F, Barişik N. Spinal schwannomatosis: Case report of a rare condition. *Turk Neurosurg*. 2008;18(3):320–3.
19. Hakan T, Celikoglu E, Aker F, Barisik N. Addendum to spinal schwannomatosis: Case report of a rare condition. *Turk Neurosurg*. 2017;27(1):163–4.
20. Gosk J, Gutkowska O, Kuliński S, Urban M, Hałoń A. Multiple schwannomas of the digital nerves and superficial radial nerve: Two unusual cases of segmental schwannomatosis. *Folia Neuropathol*. 2015;53(2):158–67.
21. Kwon NY, Oh HM, Ko YJ. Multiple lower extremity mononeuropathies by segmental schwannomatosis: A case report. *Ann Rehabil Med*. 2015;39(5):833–7.

22. Lee SH, Kim SH, Kim BJ, Lim DJ. Multiple schwannomas of the spine: Review of the schwannomatosis or congenital neurilemmomatosis: A case report. *Korean J Spine*. 2015;12(2):91–4.

23. Abdulla FA, Sasi MP. Schwannomatosis of cervical vagus nerve. *Case Rep Surg*. 2016;2016:8020919.

24. Merker VL, Esparza S, Smith MJ, Stemmer-Rachamimov A, Plotkin SR. Clinical features of schwannomatosis: A retrospective analysis of 87 patients. *Oncologist*. 2012;17(10):1317–22.

25. Ahlawat S, Fayad LM, Khan MS et al. Current whole-body MRI applications in the neurofibromatoses: NF1, NF2, and schwannomatosis. *Neurology*. 2016;87(7 Suppl 1):S31–9.

26. Sulhyan KR, Deshmukh BD, Gosavi AV, Ramteerthakar NA. Neuroblastoma-like schwannoma in a case of schwannomatosis: Report of a rare case. *Int J Health Sci (Qassim)*. 2015;9(4):478–81.

27. Smith MJ, Isidor B, Beetz C et al. Mutations in LZTR1 add to the complex heterogeneity of schwannomatosis. *Neurology*. 2015;84(2):141–7.

28. Smith MJ, Bowers NL, Bulman M et al. Revisiting neurofibromatosis type 2 diagnostic criteria to exclude LZTR1-related schwannomatosis. *Neurology*. 2017;88(1):87–92. Erratum in: Neurology. 2017;89(2):215.

29. Hanemann CO, Blakeley JO, Nunes FP et al. Current status and recommendations for biomarkers and biobanking in neurofibromatosis. *Neurology*. 2016;87(7 Suppl 1):S40–8.

30. Reddy RG, Banda VR, Gunadal S, Banda NR. A rare occurrence and management of familial schwannomatosis. *BMJ Case Rep*. 2013;2013.

31. Blakeley J, Schreck KC, Evans DG et al. Clinical response to bevacizumab in schwannomatosis. *Neurology*. 2014;83(21):1986–7.

32. Blakeley JO, Plotkin SR. Therapeutic advances for the tumors associated with neurofibromatosis type 1, type 2, and schwannomatosis. *Neuro Oncol*. 2016;18(5):624–38.

33. Widemann BC, Acosta MT, Ammoun S et al. CTF meeting 2012: Translation of the basic understanding of the biology and genetics of NF1, NF2, and schwannomatosis toward the development of effective therapies. *Am J Med Genet A*. 2014;164A(3):563–78.

34. Ostrow KL, Donaldson K, Blakeley J, Belzberg A, Hoke A. Immortalized human schwann cell lines derived from tumors of schwannomatosis patients. *PLOS ONE*. 2015;10(12): e0144620.

35. Ostrow KL, Bergner AL, Blakeley J et al. Creation of an international registry to support discovery in schwannomatosis. *Am J Med Genet A*. 2017;173(2):407–13.

12

Von Hippel–Lindau Syndrome

Dongyou Liu

CONTENTS

12.1 Introduction

Von Hippel–Lindau (VHL) syndrome is a hereditary autosomal dominant disorder characterized by the formation of cysts and/or tumors in several organ systems, including hemangioblastoma (a benign tumor composed of newly formed blood vessels, and typically affecting the central nervous system [cerebellar hemangioblastoma] and retina [retinal hemangioblastoma or angioma]), clear cell renal cell carcinoma (ccRCC), pheochromocytoma, paraganglioma, pancreatic neuroendocrine tumor (PNET), endolymphatic sac tumor (ELST), papillary cystadenoma of the epididymis and broad ligament, and so on.

Depending on the types of tumors formed, VHL can be classified as type 1 (without pheochromocytoma), and type 2 (with pheochromocytoma). Type 2 is further divided into type 2A (with pheochromocytoma and cerebellar hemangioblastoma, but not renal cell carcinoma [RCC]), type 2B (with pheochromocytoma, cerebellar hemangioblastoma and RCC), and type 2C (with pheochromocytoma only) [1].

The underlying cause of VHL syndrome is attributable to heterozygous germline (80%) or de novo (20%) mutations in the namesake tumor suppressor gene (*VHL*) on chromosome 3p25.3. There is evidence that variation in the cyclin D1 gene (CCND1) on chromosome 11q13.3 may have a modifying effect on the phenotype. Interestingly, homozygous or compound heterozygous mutations in the VHL gene may be associated with familial erythrocytosis-2 (ECYT2), which is an autosomal recessive disorder characterized by increased red blood cell mass, increased serum levels of erythropoietin (EPO), and normal oxygen affinity, and high risk for peripheral thrombosis and cerebrovascular events [2].

12.2 Biology

Initial documentation of retinal hemangioblastoma (then called "angiomatosis retinae") from two siblings was made by Collins in 1894, and the clinical appearance and progression of similar lesions were described in detail by von Hippel in 1904. Subsequently, renal cysts and tumors as well as epididymal cysts were found in association with retinal hemangioblastoma by Brandt in 1921. With the addition of cerebellar hemangioblastoma, the term *central nervous system angiomatosis* was proposed for this group of disease by Lindau in 1927. Finally, the eponymous term *von Hippel–Lindau disease* (VHL disease; synonyms: von Hippel–Lindau syndrome, VHL syndrome) was adopted in honor of von Hippel (for his description of retinal hemangioblastomas or von Hippel tumors) and Lindau (for his description of cerebellar hemangioblastomas or Lindau tumors) [3].

Following its localization on the short arm of chromosome 3 (3p25–26) by Seizinger et al. in 1988, the *VHL* gene responsible for VHL syndrome was isolated and sequenced by Latif et al. in 1993. As a tumor suppressor gene, *VHL* encodes the VHL protein (pVHL) that keeps cells from growing and dividing too rapidly or in an uncontrolled way. Mutations in *VHL* gene result in the production of aberrant or no pVHL that loses its capacity to effectively regulate cell survival and division. Similar to other tumor suppressor gene disorders, cells grow and divide uncontrollably to form the cysts and tumors characteristic of VHL syndrome [4].

VHL syndrome conforms to the "two-hit" model of hereditary cancer, with a germline *VHL* mutation from an affected parent causing a defective allele in all cell types of the body (first hit) followed by a somatic event (second hit, frequently allelic loss/ loss of heterozygosity [49%], hypermethylation [35%] or point mutations) on a wild-type allele from an unaffected parent, leading to the tumorigenesis. While 80% of VHL cases are familial, resulting from constitutional/germline mutation in one allele (first hit) and somatic mutation in another allele (second hit), about 20% of VHL cases are sporadic, due to somatic mutation in one allele (first hit) and second somatic mutation in another allele (second hit). Thus, biallelic *VHL* inactivation underscores the tumorigenesis of VHL syndrome [5,6].

12.3 Pathogenesis

Mapped to chromosome 3p25.3, the *VHL* gene consists of over 10 kb organized in three exons, and encodes a 213 amino acid protein (pVHL30) with a molecular weight of 30 kDa (which is a glycan-anchored membrane protein responsible for signal transduction and present in both nuclear and cytoplasmic compartments), and a second, smaller 160 amino acid isoform (pVHL19, lacking the first 53 residues) with a molecular weight of 19 kDa as a result of alternate translation initiation from codon 54. Both isoforms demonstrate tumor suppressor function [7].

pVHL is found in both nuclear and cytoplasmic compartments, between which pVHL travels freely. pVHL interacts with elongins B, C, and Cullin-2 to form the VBC complex, an E3 ubiquitin ligase, which is primarily involved in the proteasomal degradation of hypoxia-inducible factor (HIF, a heterodimeric transcription factor responsible for oxygen regulation in the cells), regulation of apoptosis (p53 inactivation and increased

NF-κB activity), stabilization of microtubules, and regulation of extracellular matrix.

Abnormal or absent pVHL increases expression, stabilization, and accumulation of HIF proteins, which induce differential upregulation of >50 genes related to the uptake and metabolism of glucose (GLUT-1), angiogenesis (vascular endothelial growth factor [VEGF], platelet-derived growth factor [PDGF]), mitogenesis (transforming growth factor alpha [TGFα]), erythropoiesis (erythropoietin), control of extracellular pH (CA9), and tumor microenvironment (LOX, MMP1), leading to the formation of cysts and hypervascular tumors characteristic of VHL [8]. HIF-independent pathways implicated in the VHL tumorigenesis relate to the deposition of fibronectin and collagen IV within the extracellular matrix, stabilization of microtubules, maintenance of primary cilium, and regulation of apoptosis (p53 and Jun-B) [7].

To date, >500 germline pathogenic variants responsible for the development of VHL syndrome have been identified. These occur mainly between codon 54 and the carboxy terminal (with hot spot codon 167), affecting both isoforms of protein, and include missense mutations (52%), frameshift (13%), nonsense mutations (11%), large/complete deletions (11%), splice site variants (7%), and in-frame deletions/insertions (6%) [9–11].

In addition, de novo/somatic mutations in *VHL* are responsible for sporadic ccRCC and cerebellar hemangioblastoma in patients with no family connection [12].

Based on clinical phenotypic manifestations, VHL has been classified into types 1, 2A, 2B, or 2C (Table 12.1). At the genetic level, VHL type 1 is linked to deletions, nonsense mutations, microdeletions/insertions, and missense mutations that severely disrupt pVHL structure and activity, and has high risk for retinal hemangioblastoma (angioma), cerebellar hemangioblastoma, RCC, pancreatic cyst, and neuroendocrine tumor, but a low risk for pheochromocytoma; VHL type 2 is nearly always (78%–96%) associated with missense mutations resulting in a substitution of an amino acid on the surface of the protein, and conferring a high lifetime risk for pheochromocytoma. More specifically, VHL type 2A contains a distinct missense mutation (p.Tyr169His) and shows a high risk for pheochromocytoma, retinal angioma, and cerebellar hemangioblastoma, but a low risk for RCC; VHL type 2B harbors missense mutations of codon 167 (e.g., p.Arg167Gln or p.Arg167Trp) and shows a high risk for pheochromocytoma (82% at 50 years), RCC (60% at 60 years),

TABLE 12.1

Characteristics of VHL Types

Type	Associated Risk	Clinical Presentation	Genetic Mutation
Type 1	Decreased risk for pheochromocytoma	Retinal and cerebellar hemangioblastomas, renal cell carcinoma, pancreatic cyst, and neuroendocrine tumor	Deletions, nonsense mutations, microdeletions/insertions, and missense mutations
Type 2A	Increased risk for pheochromocytoma, low risk of renal cell carcinoma	Pheochromocytoma, retinal and cerebellar hemangioblastomas	Missense mutation (p.Tyr169His)
Type 2B	Increased risk for pheochromocytoma, high risk of renal cell carcinoma	Pheochromocytoma, renal cell carcinoma, retinal and cerebellar hemangioblastomas, pancreatic cyst and neuroendocrine tumor	Missense mutation (p.Arg167Gln or p.Arg167Trp)
Type 2C	Increased risk for pheochromocytoma	Pheochromocytoma only	Missense mutations at codons 238 and 259

retinal hemangioblastoma (angioma), cerebellar hemangioblastoma, pancreatic cysts, and neuroendocrine tumor; VHL type 2C has specific missense mutations at codons 238 and 259, and develops pheochromocytoma only (Table 12.1) [1].

12.4 Epidemiology

12.4.1 Prevalence

VHL syndrome has an estimated prevalence of 1 in 36,000–45,000 live births. The average age of onset is 26 years (most commonly at 18–30 years) [13].

12.4.2 Inheritance

VHL syndrome demonstrates an autosomal dominant inheritance pattern, and individuals with VHL disease have a 50% chance of passing the mutation to their offspring. About 80% of patients acquire an altered copy of the gene from an affected parent (including a mosaic parent who shows clinical VHL disease but no identifiable *VHL* gene mutation in peripheral blood lymphocytes), and 20% of patients are due to sporadic or de novo mutation that occurs during the formation of reproductive cells (eggs or sperm) or very early in development [14].

Mosaicism may be present in both somatic cells and gonadal cells, or be found only in gonadal, or germline, cells. Individuals with germline mosaicism are unaffected, while those with somatic mosaicism may be unaffected or mildly affected. Nevertheless, an individual who is mosaic for a mutation in the *VHL* gene still has a 50% chance of passing on the mutation to an offspring.

Unlike most other autosomal dominant disorders, in which one altered copy of a gene in each cell is sufficient to cause disease, two copies of the *VHL* gene have to be altered to trigger the formation of specific cysts and tumors in VHL syndrome, with the mutation in the second copy taking place during a person's lifetime in certain cells within the brain, retina, or kidneys.

12.4.3 Penetrance

VHL syndrome has a high penetrance (>90%), and almost everyone with an inherited *VHL* mutation will acquire a mutation in the second copy of the gene in some cells and become symptomatic by age 70 years. Due likely to shortening of telomere length, VHL syndrome demonstrates progressively earlier age of onset and more severe presentation in successive generations (so-called genetic anticipation).

12.5 Clinical Features

Depending on the organs involved, VHL syndrome may present a highly variable clinical picture [15–17].

12.5.1 Central Nerve System

Affecting up to 80% of patients with a mean onset age of 33 years, cerebellar hemangioblastoma exhibits linear growth, exponential growth, or periods of fluctuation and appears as a distinct, red, vascular mass (of >2 mm) with a thin layer of capsule. Partial germline mutation and male sex may contribute to high tumor burden. Commonly found in the cerebellum (69%), brainstem (22%), spinal cord (53%), cauda equina (11%), or supratentorial location (7%), this tumor compresses the neighboring neural structures and causes cerebrospinal fluid obstruction, leading to headache, vomiting, pain, sensory or motor deficits (e.g., weakness, paresthesias), gait disturbances (ataxia), dysmetria, nystagmus, slurred speech, and polyglobulia (due to paraneoplastic effect) [18,19].

12.5.2 Eyes

Affecting up to 62% of patients, with a mean onset age of 25 years, retinal hemangioblastoma (angioma) represents the first sign of VHL detected. Being a benign tumor located in the periphery (50%) or at the optic nerve (50%), its high expression of VEGF causes vascularization, vascular leakage, exudation, hemorrhage, and retinal detachment, leading to visual field defect or blindness (8% of cases) (Figure 12.1) [20].

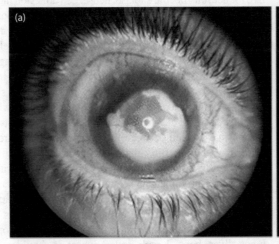 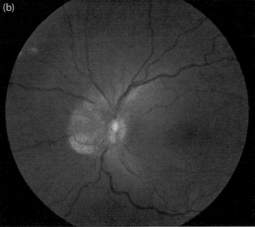

FIGURE 12.1 Von Hippel–Lindau disease in a 41-year-old male showing (a) neovascularization of the cornea and iris (rubeosis iridis) and matured cataract in the right eye; (b) retinal hemangioblastoma involving the optic nerve in the left eye. (Photo credit: Chen S et al. *J Med Case Rep.* 2015;9:66.)

12.5.3 Kidneys

Found in 60% of patients, with a mean onset age of 39 years, renal cysts may be asymptomatic but have the capacity to progress to solid ccRCC masses, which are seen in 30% of patients, leading to flank pain or hematuria.

12.5.4 Adrenal Glands

Affecting up to 20% of patients, with a mean onset age of 30 years, pheochromocytoma (67% sporadic, 33% familial) forms bilateral and occasionally multifocal masses in the adrenal medulla, producing excess catecholamine/norepinephrine (in 67% of cases) that is responsible for paroxysmal or sustained hypertension, palpitations, tachycardia, headaches, panic attacks, excessive sweating, pallor, and nausea. About 5% of pheochromocytoma is of malignant type. As a pheochromocytoma in extra-adrenal locations, mostly along the sympathetic axis in the abdomen or thorax, paraganglioma occurs in 15% of cases and is mostly nonfunctional. However, paraganglioma demonstrates a higher frequency of malignant transformation than pheochromocytoma [21,22].

12.5.5 Pancreas

Found in up to 70% of patients, with a mean onset age of 36 years, pancreatic cyst and neuroendocrine tumor are slow-growing and hormonally inactive, and their mass effect on the intestine or bile duct may be problematic at times [23]. In addition, through replacement of pancreatic parenchyma, these lesions may contribute to exocrine or endocrine deficiency. In 8% of patients, PNET may become malignant and metastatic.

12.5.6 Other Organs

Affecting up to 15% of patients, with a mean onset age of 31 years, ELST arising from the endolymphatic epithelium within the vestibular aqueduct is a benign cystadenoma that may extend into the extraosseous portion of the endolymphatic sac into the petrous bone, the inner ear and even the brain. As this tumor is locally invasive, it may cause intralabyrinthine hemorrhage, endolymphatic hydrops, or otic capsule invasion, leading to sudden uni- or bilateral hearing loss (100%), tinnitus (77%), disequilibrium (62%), and facial paresis (8%, due to lesions of >3 cm) [24].

Epididymal or papillary cystadenoma in males affects up to 50% of patients and is benign and largely asymptomatic. However, bilateral tumors may cause complete obstruction of efferent ductules and spermatic cords, leading to infertility. Papillary cystadenoma of the broad ligament in women is rare and mostly asymptomatic.

12.6 Diagnosis

Diagnosis of VHL syndrome involves medical history review, physical examination, biochemical tests, imaging study, histopathology, and molecular genetic testing [14].

As about 80% of cases result from inheritance of germline *VHL* pathogenic variants, medical history review helps identify susceptible individuals for imaging detection of VHL-associated tumors and subsequent laboratory confirmation. For 20% of cases with no known family history, imaging detection of multiple tumors assists in diagnosis [25].

Physical examination may be useful for identifying muscle weakness, sensory deficits, and ataxia related to cerebellar hemangioblastomas.

Contrast-enhanced magnetic resonance imaging (MRI) of the head and spine facilitates identification of cerebellar hemangioblastoma as an enhancing mural nodule in association with a pseudocyst or syrinx in the brain (80%) and spinal cord (20%). Ophthalmoscope in conjunction with pharmacological dilation of the iris helps identify most retinal lesions, while fundoscopy reveals retinal detachment, macular edema, or cataracts. Ultrasound, MRI, or computed tomography (CT) helps detect bilateral and multiple renal cysts of different sizes, whereas CT with contrast is preferred for identification of RCC as solid or cystic masses with pushing borders. Further, enhanced MRI of the abdomen is the preferred modality compared to CT for detection of pheochromocytoma, whereas meta-iodobenzylguanidine (MIBG) scintigraphy is useful for identification of extra-adrenal pheochromocytoma. On postcontrast CT, PNET appears as an enhancing mass. Endoscopic ultrasound and somatostatin receptor scintigraphy are also valuable for detecting pancreatic cyst and tumor. On CT, ELST is a small lesion found within the endolymphatic sac and may have focal areas of low and high attenuation, and large tumors may expand from the endolymphatic duct into the temporal bone. Audiogram is also of value for verifying hearing loss caused by ELST. Ultrasonography is the modality of choice in locating epididymal or papillary cystadenoma [26].

Laboratory test for serum and urinary catecholamines aids in the diagnosis of pheochromocytoma, with plasma free metanephrines being more sensitive (97%) than 24-h urinary measurement of catecholamines for pheochromocytoma detection.

Histopathologically, hemangioblastoma comprises polygonal-shaped stromal cells (with hyperchromatic nuclei and clear to eosinophilic foamy cytoplasm) flanked by a dense capillary network. While the stromal cells represent developmentally arrested VHL-deficient hemangioblast progenitor cells with hematopoietic differentiation potential, the capillary network is the outcome of reactive VEGF-driven angiogenesis. The stromal cells stain positive for neuron-specific enolase, neural cell adhesion molecule, and vimentin (Figure 12.2). A combination of inhibin-positivity and CD10-negativity in hemangioblastoma is helpful in its different ion from metastatic RCC [18,20]. Resembling cerebellar hemangioblastoma grossly and histologically, retinal hemangioblastoma may show different levels of gliosis and hemorrhage reflecting the severity of the lesion. RCC associated with VHL syndrome is invariably of clear-cell type and minimally invasive, with neoplastic cells showing clear cytoplasm and prominent vascular component interspersed between the clear cells. Pheochromocytoma arises from chromaffin cells in the medulla of the adrenal gland and typically displays polygonal neoplastic cells (with round to oval nuclei, a prominent nucleolus and finely granular amphophilic cytoplasm) arranged in small nests or "zellballen" pattern, which are surrounded by S-100-positive sustentacular cells. Malignant pheochromocytoma shows chromaffin cells at locations (e.g., the lungs, liver, bone, lymph nodes) where it should not normally occur [22]. Paraganglioma originates from extra-adrenal chromaffin cells, which are present in the sympathetic nervous system

(a)
(b)
(c)

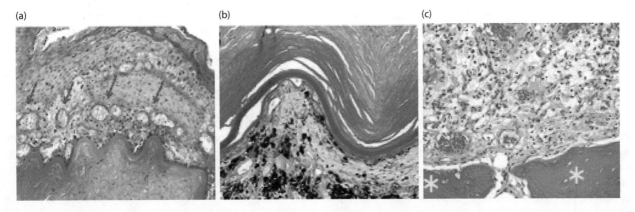

FIGURE 12.2 Von Hippel–Lindau disease in a 41-year-old male showing (a) neovascularization of the cornea and iris, including many small vascular lumina (arrows) in the corneal subepithelial region (a) and also in the surface of the iris (b) in the left eye, and retinal hemangioblastoma composed of classical foamy (tumor) cells admixed with small capillaries and osseous tissues (asterisks) adjacent to retinal hemangioblastoma in the right eye (c). (Photo credit: Chen S et al. *J Med Case Rep.* 2015;9:66.)

and parasympathetic ganglia, facilitating its development in the thorax, abdomen, pelvis, or head and neck regions, contains moderately sized uniform cells with pale to eosinophilic cytoplasm and round to oval nuclei showing characteristic neuroendocrine "salt-and-pepper" granular chromatin pattern. The neoplastic cells in PNET stain positive for pancreatic and gastrointestinal hormones. Pancreatic cysts and cystadenomas have clear glycogen-rich, uniform cuboidal epithelial cells with a round nucleus and fibrous stroma. ELST is a papillary cystic glandular neoplasm with variable patterns. Due to its adenomatous appearance and association with local bone erosion, ELST was once known as low-grade adenocarcinoma. Epididymal cystadenoma typically arises in the caput epididymides and contains cystic and adenomatous areas. Similar to epididymal cystadenoma with respect to the prominent papillary architecture, papillary cystadenoma shows a fibrovascular and hyaline stroma.

Molecular genetic tests include Southern blotting (for detecting whole gene deletions and gene rearrangements), fluorescence in situ hybridization (FISH, for confirming deletions), multiplex ligation-dependent probe amplification (MLPA, for detecting deletions and insertions) and sequencing (for determining the coding exons and intron–exon boundaries, and detecting most other mutations). Identification of a heterozygous pathogenic *VHL* variant confirms the diagnosis of VHL even if clinical and imaging findings are inconclusive. It should be noted that patients without a family history may be mosaic for a *VHL* mutation, which can be absent in peripheral blood leukocytes, but present in gonadal, or germline, cells. Careful clinical testing and mutation analysis in gonadal tissues will help reveal the pathologic variant. In addition, peripheral blood from the affected parent should be tested to verify the possible mosaicism [11].

Current clinical diagnostic criteria focus on finding the following VHL-related tumors: (i) cerebellar hemangioblastoma (and retinal hemangioblastoma), (ii) ELST, (iii) RCC, (iv) pheochromocytoma, paraganglioma, and/or glomus tumor, and (v) PNET and/or multiple pancreatic cysts.

A combination of the following clinical, genetic/family factors is diagnostic for VHL syndrome: (i) at least two cerebellar hemangioblastomas, (ii) at least one cerebellar hemangioblastoma and one other VHL-related tumor described above, (iii) at least one VHL-related tumor described above, and a pathogenic

mutation in *VHL* gene or a first-degree relative with VHL syndrome [27].

Apart from its association with VHL syndrome, pheochromocytoma may also occur in other hereditary syndromes, including MEN type 2A, 2B (adrenal lesion due to *RET* mutation), neurofibromatosis type 1 (adrenal lesion due to *NF1* mutation), familial paraganglioma syndrome type 1, 3, 4 (head and neck, adrenal and extra-adrenal lesions due to *SDHA, SDHB,* or *SDHC* mutation), and familial pheochromocytoma (adrenal and extra-adrenal lesions due to chromosomes 2 and 16 mutations).

12.7 Treatment

Treatment options for VHL syndrome consist of surgery, radiotherapy, and chemotherapy [28].

Patients with symptomatic cerebellar hemangioblastoma should undergo surgery, while those with asymptomatic lesions are monitored with annual imaging. Small solid tumors (<3 cm diameter) or those in inoperable sites are effectively removed by stereotactic radio surgery (SRS), and tumors with high vascular may need preoperative arterial remobilization to reduce bleeding risk [18,29–31].

Retinal hemangioblastoma may be treated by laser photocoagulation (for posterior lesions of <1.5 mm in diameter), cryotherapy (for anterior lesions of <3 mm in diameter), vitreoretinal surgery (in cases where retinal detachment occurs and exudation is present), and external beam radiotherapy (for refractory cases). In addition, intravitreal anti-VEGF therapy may help control growth and ameliorate edema and exudate formation.

RCC require early nephrectomy, while smaller lesions may be treated with cryoablation and radiofrequency ablation.

Pheochromocytoma and paraganglioma are treated by laparoscopic surgery or partial adrenalectomy (particularly for pediatric patients). Preoperative treatment with alpha-adrenergic blockade, and optional additional beta-adrenergic blockade may be utilized to prevent catecholamine-mediated life-threatening complications (e.g., cardiac arrhythmias, hypertensive crisis, and myocardial infarction) [22].

RCC (>3 cm) should be removed early by nephron-sparing or partial nephrectomy or radiofrequency ablation, while lesions

<3 cm require no intervention. Renal transplantation may be necessary for patients undergoing bilateral nephrectomy.

PNET (measuring ≥3 cm in size, doubling <500 days, or having a pathologic variant in exon 3) poses a high risk of metastasis and should be resected with enucleation by Whipple procedure or partial pancreatectomy. Pancreatic cysts that are asymptomatic can be left alone, although those causing obstructive symptoms require laparoscopic surgical decompression.

ELST detectable on radiographs or causing intralabyrinthine hemorrhage should be resected to relieve audiovestibular symptoms and to preserve hearing and vestibular function. ELST may recur in 3% of cases.

Epididymal or broad ligament papillary cyst adenomas generally require no surgery, but symptomatic or fertility-threatening lesions should be removed.

New drugs under development include intravitreal VEGF receptor inhibitor (ranibizumab and bevacizumab) for retinal hemangioblastoma, thalidomide for multifocal spinal hemangioblastoma, and tyrosine kinase inhibitor (sunitinib) for pheochromocytoma and ccRCC [32].

12.8 Prognosis and Prevention

Life expectancy for VHL patients has been hovering between 40 and 52 years prior to the availability of early detection and advanced treatment measures. Currently, patients with VHL syndrome have a life expectancy of 59.4 years in males, and 48.4 years in females.

Tobacco, chemicals, and industrial toxins as well as contact sports should be avoided to minimize damages to affected organs.

Implementation of a screening program for patients with VHL syndrome helps identify manifestations early before deficit or metastasis appears. This may include annual ophthalmoscopy (direct and indirect) from age 1 for retinal hemangioblastoma, MRI scan of the head every 12–36 months from age 12 for cerebellar hemangioblastoma, annual MRI (or ultrasound) scan of the abdomen every 12 months from age 12 for RCC and pancreatic tumor, annual blood pressure monitoring and 24-h urine evaluation of catecholamines from age 6 and annual MRI from age 12 for pheochromocytoma, ultrasonography of the testes from age 18 for epididymal cystadenoma [1].

While 80% of VHL patients have an affected parent, about 20% may develop a de novo pathogenic variant or acquire from a parent who is mosaic for *VHL* mutation. Prenatal testing for pregnancy may be considered if a family member has a known pathogenic *VHL* variant, which has a 50% chance of passing to the offspring.

12.9 Conclusion

Von Hippel–Lindau syndrome (VHL) is a hereditary autosomal dominant disorder characterized by the development of a variety of cysts and tumors, including cerebellar and retinal hemangioblastomas, RCC, pheochromocytoma, pancreatic tumors, and ELST [33–35]. Based on the types of tumors formed, VHL is separated into type 1 (without pheochromocytoma), type 2A (with pheochromocytoma and cerebellar hemangioblastomas but without RCC), type 2B (with pheochromocytoma, cerebellar hemangioblastomas, and RCC) and type 2C (with pheochromocytoma only). About 80% of patients have a positive family history and contain germline mutations in the tumor suppressor gene *VHL*. However, up to 20% of patients have no family history and are attributed to de novo *VHL* mutations. Diagnosis of VHL relies on a combination of clinical and molecular criteria, and treatment options include surgery, radiotherapy, and chemotherapy. Regular surveillance for common VHL manifestations helps monitor disease progression and control potential metastasis [36].

REFERENCES

1. Frantzen C, Klasson TD, Links TP, Giles RH. Von Hippel-Lindau syndrome. In: Pagon RA, Adam MP, Ardinger HH et al., eds. *GeneReviews® [Internet]*. Seattle (WA): University of Washington, Seattle; 1993–2017. 2000 May 17 [updated 2015 Aug 6].
2. Ben-Skowronek I, Kozaczuk S. Von Hippel-Lindau syndrome. *Horm Res Paediatr*. 2015;84(3):145–52.
3. Chittiboina P, Lonser RR. Von Hippel-Lindau disease. *Handb Clin Neurol*. 2015;132:139–56.
4. Findeis-Hosey JJ, McMahon KQ, Findeis SK. Von Hippel-Lindau disease. *J Pediatr Genet*. 2016;5(2):116–23.
5. Gläsker S, Neumann HPH, Koch CA, Vortmeyer AO. Von Hippel-Lindau disease. In: De Groot LJ, Chrousos G, Dungan K et al., eds. *Endotext [Internet]*. South Dartmouth (MA): MDText.com, Inc.; 2000. 2015 Jul 11.
6. Lou LH, Shen H, Lin J et al. T-cell lymphoma with von Hippel-Lindau disease: A rare case report and review of literature. *Int J Clin Exp Pathol*. 2015;8(5):5837–43.
7. Bader HL, Hsu T. Systemic VHL gene functions and the VHL disease. *FEBS Lett*. 2012;586(11):1562–9.
8. Lee JS, Lee JH, Lee KE et al. Genotype-phenotype analysis of von Hippel-Lindau syndrome in Korean families: HIF-α binding site missense mutations elevate age-specific risk for CNS hemangioblastoma. *BMC Med Genet*. 2016;17(1):48.
9. Dandanell M, Friis-Hansen L, Sunde L, Nielsen FC, Hansen TV. Identification of 3 novel VHL germ-line mutations in Danish VHL patients. *BMC Med Genet*. 2012;13:54.
10. Losonczy G, Fazakas F, Pfliegler G et al. Three novel germline VHL mutations in Hungarian von Hippel-Lindau patients, including a nonsense mutation in a fifteen-year-old boy with renal cell carcinoma. *BMC Med Genet*. 2013;14:3.
11. Alosi D, Bisgaard ML, Hemmingsen SN, Krogh LN, Mikkelsen HB, Binderup MLM. Management of gene variants of unknown significance: Analysis method and risk assessment of the *VHL* mutation p.P81S (c.241C>T). *Curr Genomics*. 2017;18(1):93–103.
12. Ding X, Zhang C, Frerich JM et al. De novo VHL germline mutation detected in a patient with mild clinical phenotype of von Hippel-Lindau disease. *J Neurosurg*. 2014;121(2):384–6.
13. Takayanagi S, Mukasa A, Nakatomi H et al. Development of database and genomic medicine for von Hippel-Lindau disease in Japan. *Neurol Med Chir (Tokyo)*. 2017;57(2):59–65.
14. Varshney N, Kebede AA, Owusu-Dapaah H, Lather J, Kaushik M, Bhullar JS. A review of Von Hippel-Lindau syndrome. *J Kidney Cancer VHL*. 2017;4(3):20–9.

15. Zhang C, Yang AI, Vasconcelos L et al. Von Hippel-Lindau disease associated pulmonary carcinoid with cranial metastasis. *J Clin Endocrinol Metab.* 2014;99(8):2633–6.

16. Valero E, Rumiz E, Pellicer M. Cardiac involvement in Von Hippel-Lindau disease. *Med Princ Pract.* 2016;25(2):196–8.

17. Yaghobi Joybari A, Azadeh P. Von Hippel-Lindau disease with multi-organ involvement: A case report and 8-year clinical course with follow-up. *Am J Case Rep.* 2017;18:1220–4.

18. Mills SA, Oh MC, Rutkowski MJ, Sughrue ME, Barani IJ, Parsa AT. Supratentorial hemangioblastoma: Clinical features, prognosis, and predictive value of location for von Hippel-Lindau disease. *Neuro Oncol.* 2012;14(8):1097–104.

19. Tucer B, Ekici MA, Kazanci B, Guclu B. Hemangioblastoma of the filum terminale associated with von Hippel-Lindau disease: A case report. *Turk Neurosurg.* 2013;23(5):672–5.

20. Chen S, Chew EY, Chan CC. Pathology characteristics of ocular von Hippel-Lindau disease with neovascularization of the iris and cornea: A case report. *J Med Case Rep.* 2015;9:66.

21. Bachurska S, Staykov D, Belovezhdov V et al. Bilateral pheochromocytoma/intra-adrenal paraganglioma in von Hippel-Lindau patient causing acute myocardial infarction. *Pol J Pathol.* 2014;65(1):78–82.

22. Rednam SP, Erez A, Druker H et al. Von Hippel-Lindau and hereditary pheochromocytoma/paraganglioma syndromes: Clinical features, genetics, and surveillance recommendations in childhood. *Clin Cancer Res.* 2017;23(12):e68–75.

23. Zhi XT, Bo QY, Zhao F, Sun D, Li T. Von Hippel-Lindau disease involving pancreas and biliary system: A rare case report. *Medicine (Baltimore).* 2017;96(1):e5808.

24. Yang X, Liu XS, Fang Y, Zhang XH, Zhang YK. Endolymphatic sac tumor with von Hippel-Lindau disease: Report of a case with atypical pathology of endolymphatic sac tumor. *Int J Clin Exp Pathol.* 2014;7(5):2609–14.

25. Mikhail MI, Singh AK. *Von Hippel Lindau Syndrome. StatPearls [Internet].* Treasure Island (FL): StatPearls Publishing; 2018. 2017 Oct 20.

26. Ilias I, Meristoudis G. Functional imaging of paragangliomas with an emphasis on Von Hippel-Lindau-associated disease: A mini review. *J Kidney Cancer VHL.* 2017;4(3):30–6.

27. Schmid S, Gillessen S, Binet I et al. Management of von Hippel-Lindau disease: An interdisciplinary review. *Oncol Res Treat.* 2014;37(12):761–71.

28. Agarwal R, Liebe S, Turski ML et al. Targeted therapy for genetic cancer syndromes: Von Hippel-Lindau disease, Cowden syndrome, and Proteus syndrome. *Discov Med.* 2015;19(103):109–16.

29. Kim BY, Jonasch E, McCutcheon IE. Pazopanib therapy for cerebellar hemangioblastomas in von Hippel-Lindau disease: Case report. *Target Oncol.* 2012;7(2):145–9.

30. Schunemann V, Huntoon K, Lonser RR. Personalized medicine for nervous system manifestations of von Hippel-Lindau disease. *Front Surg.* 2016;3:39.

31. Ordookhanian C, Kaloostian PE, Ghostine SS, Spiess PE, Etame AB. Management strategies and outcomes for VHL-related craniospinal hemangioblastomas. *J Kidney Cancer VHL.* 2017;4(3):37–44.

32. Kim E, Zschiedrich S. Renal cell carcinoma in von Hippel-Lindau disease-from tumor genetics to novel therapeutic strategies. *Front Pediatr.* 2018;6:16.

33. Pradhan R, George N, Mandal K, Agarwal A, Gupta SK. Endocrine manifestations of Von Hippel-Landau disease. *Indian J Endocrinol Metab.* 2019;23(1):159–64.

34. Ruppert MD, Gavin M, Mitchell KT, Peiris AN. Ocular manifestations of von Hippel-Lindau disease. *Cureus.* 2019;11(8):e5319.

35. Kim HS, Kim JH, Jang HJ, Han B, Zhang DY. Clinicopathologic significance of VHL gene alteration in clear-cell renal cell carcinoma: An updated meta-analysis and review. *Int J Mol Sci.* 2018;19(9):2529.

36. van Leeuwaarde RS, Ahmad S, Links TP et al. Von Hippel-Lindau Syndrome. 2000 May 17 [Updated 2018 Sep 6]. In: Adam MP, Ardinger HH, Pagon RA et al., editors. *GeneReviews®* [Internet]. Seattle (WA): University of Washington, Seattle; 1993–2020.

Section II

Tumor Syndromes Affecting the Digestive System

13

Alagille Syndrome

Dongyou Liu

CONTENTS

13.1 Introduction

Alagille syndrome (ALGS) is an autosomal dominant disorder affecting multiple systems, with presentations ranging from subclinical disease to life-threatening condition, including hepatic (cholestasis, leading to intrahepatic bile duct paucity), cardiac (particularly peripheral pulmonary artery stenosis), renal (dysplastic kidneys), skeletal (butterfly vertebrae), ophthalmologic (pigmentary retinopathy), and facial (frontal bossing/prominent forehead, flat/hypoplastic midface, deeply set eyes, upslanting/narrow palpebral fissures, hypertelorism, anterior chamber defects with posterior embryotoxon, bulbous or long straight nose, short philtrum, large ears, and triangular chin) abnormalities, in addition to increased risk for hepatocellular cancer (hepatoma), and possibly papillary thyroid cancer and Wilms tumor (nephroblastoma). The mortality rate of ALGS can reach 10% [1].

Genetically, ALGS is attributed to defects in the Notch signaling pathway, with mutation in the Jagged 1 (*JAG1*) gene on chromosome 20p12.2 accounting for a majority (~96%) of cases (ALGS type 1), and mutation in the neurogenic locus notch homolog protein 2 (*NOTCH2*) gene on chromosome 1p12 being responsible for a small proportion (~2%) (ALGS type 2, with prominent renal malformations). Both the *JAG1* and *NOTCH2* genes are components of the Notch signaling pathway [2].

13.2 Biology

Cases relating to ALGS (also known as arteriohepatic dysplasia, syndromic bile duct paucity, syndromic hepatic ductular hypoplasia, syndromic intrahepatic biliary hypoplasia, cholestasis with peripheral pulmonary stenosis, intrahepatic biliary atresia/dysgenesis, Alagille–Watson syndrome, or Watson–Miller syndrome) were initially reported by Alagille et al. in 1969 and then by Watson and Miller in 1973, with classic criteria for diagnosing this disorder established by Alagille et al. in 1975 (see Section 13.6) [1].

Once considered as a hepatic disease, the clinical spectrum of ALGS has subsequently expanded to include abnormalities in other organs, such as cholestasis with bile duct paucity on liver biopsy, congenital cardiac defects (pulmonary artery stenosis), posterior embryotoxon in the eye, characteristic facial features, and butterfly vertebrae, in addition to renal and vascular abnormalities. Subsequent identification of *JAG1* and *NOTCH2* mutations, which occur in 96% and 2% of ALGS patients, respectively, has highlighted haploinsufficiency (loss of function) as the main molecular mechanism for ALGS [3].

13.3 Pathogenesis

The *JAG1* gene on chromosome 20p12.2 spans over 36 kb consisting of 26 exons, which range from 28 bp to 2284 bp in size, while 25 introns in between vary from 89 bp to nearly 9 kb in size. It produces three different transcripts by alternative splicing, with the most important one being 5.901 kb in size, which encodes a 1218 aa cell surface protein (JAG1). Structurally, JAG1 comprises a relatively small intracellular domain, a transmembrane domain, and a larger extracellular component. In turn, the extracellular component consists of a 21 aa signal peptide, an N-terminal region, a 40 aa highly conserved DSL domain (named for the *Drosophila melanogaster* and *Caenorhabditis elegans* ligands, delta, serrate, and lag-2), 16 epidermal growth factor(EGF)-like repeats, and a cysteine-rich region [4].

In the developing embryo, JAG1 is expressed in the mesocardium, pulmonary artery, major arteries, distal cardiac outflow tract, metanephros, pancreas, branchial arches, around the major bronchial branches, portal vein, the neural tube optic vesicle, and the otocyst. In adults, JAG1 is found in the heart, placenta, pancreas, and prostate, and at lower levels in the lung, liver, kidney, thymus, testis, and leucocytes.

Functionally, JAG1 is a ligand for NOTCH2 transmembrane receptor (a key signaling molecule on the surface of a variety of cells), with the signal peptide orienting JAG1 to the cell surface, the DSL domain binding JAG1 to NOTCH2, and the EGF repeats increasing the affinity of JAG1 to NOTCH2.

As components of the highly conserved Notch signaling pathway, interaction between NOTCH2 receptor and its JAG1 ligand triggers a cascade of intracellular downstream effects that determine cell fate and differentiation. Alterations in the *JAG1* and *NOTCH2* genes have been implicated in pathological processes including tumor angiogenesis, neoplastic cell growth, cancer stem cell maintenance, and metastasis.

Haploinsufficiency of JAG1 appears to be the pathogenic mechanism behind the vast majority (96%) of ALGS cases. This is supported by the findings that most *JAG1* pathogenic variants produce a severely truncated protein, which lacks the transmembrane region necessary for embedding in the cell membrane and for trafficking through the cell. As a consequence, the altered JAG1 fails to appear on the cell surface, thus resulting in functional haploinsufficiency.

To date, >500 *JAG1* mutations have been identified in individuals with ALGS, of which 69% are truncating variants (frameshift and nonsense), 16% are splice site variants, 11% are missense variants, and 4% are whole-gene deletions (including chromosome 20p12.2 microdeletion). Interestingly, 67% of these pathogenic variants occur in exons 1–6, 9, 12, 17, 20, 23, and 24, while 33% are present in other exons [5].

Interestingly, mutations in the *JAG1* gene have been detected in 2% of tetralogy of Fallot (TOF, a complex congenital heart disease with highly heterogeneous etiology) cases and 4% of pulmonic stenosis/peripheral pulmonary stenosis (PT/PPT)

cases, which do not meet the clinical criteria for ALGS. This highlights the potential role of *JAG1* mutations in the pathogenesis of heart conditions, since JAG1 expression occurs in the critical vascular structures of heart development and cardiac disease.

The *NOTCH2* neurogenic locus notch homolog protein 2 (*NOTCH2*) gene on chromosome 1p12 consists of 34 exons and encodes a transmembrane protein (NOTCH2), which represents one of the Notch family of transmembrane receptors (i.e., NOTCH1, NOTCH2, NOTCH3, and NOTCH4) in humans. Structurally, Notch transmembrane receptors include an extracellular domain with multiple EGF-like repeats and an intracellular domain with multiple different domain types. The intracellular domain consists of seven ankyrin (ANK) repeats that participate in protein–protein interaction [6,7].

Through interaction with their ligands, Notch transmembrane receptors exert intracellular downstream effects that control cell fate decisions. There is evidence that alterations in *NOTCH1* contribute to isolated cardiac defects and T lymphoblastic leukemias, mutations in *NOTCH2* are associated with ALGS and Hajdu–Cheney syndrome (which is due to gain of function mutations in *NOTCH2* exon 34), and changes in *NOTCH3* cause cerebral autosomal dominant arteriopathy with subcortical infarcts and leukoencephalopathy (CADASIL) [8–11].

The most notable pathogenic variants identified in the NOTCH2 protein include seven missense (p.Cys444Tyr, p.Cys373Arg, p.Pro394Ser, p.Pro383Ser, p.Arg1953Cys, p.Arg1953His, p.Cys480Arg), one nonsense (p.Arg2003Ter), one frameshift (p.Ser856LeufsTer17), and one splice site (c.5930-1G>A) mutation, with six in the EGF-like repeats (extracellular domain) and four in the ANK repeats [5,12,13].

13.4 Epidemiology

13.4.1 Prevalence

ALGS has an estimated incidence of 1 per 30,000–50,000 live births. About 500 ALGS cases have been described in the literature so far, of which a vast majority possess pathogenic *JAG1* variants and only 2% (10) harbor pathogenic *NOTCH2* variants. Considering the past practice of ascertaining the disorder based solely on the presence of liver disease with conjugated hyperbilirubinemia, which occurs in 80%–90% of patients, most commonly in neonates, but at a reduced rate in older children and adults, ALGS is undoubtedly an underdiagnosed or underreported disease entity.

13.4.2 Inheritance

About 40% of ALGS patients have a clear family history of the disorder and likely acquire germline *JAG1* or *NOTCH2* mutations from one of their parents, while ~60% of individuals show no family connection and appear to develop the disease from de novo mutations or from asymptomatic patents who have germline mosaicism (which occurs at a frequency of up to 8%). Transmitted through an autosomal dominant pattern, the offspring of a patient with a pathogenic *JAG1* or *NOTCH2* variant has a 50% chance of inheriting the mutation.

13.4.3 Penetrance

ALGS displays high penetrance (97%) and variable expressivity, with clinical features ranging from subclinical to severe. In one study, 21% of patients met diagnostic criteria independent of family history, 32% were asymptomatic but met diagnostic criteria after additional testing, 43% displayed one or two features of ALGS, and 4% had no features of ALGS. Inter- and intrafamilial variabilities in clinical manifestations of ALGS are also common. Given the lack of strong correlation between the type and location of *JAG1* or *NOTCH2* mutation and severity of the disease, other genomic modifiers beyond the known *JAG1* or *NOTCH2* mutations may be responsible for variable expressivity of ALGS.

13.5 Clinical Features

ALGS is an autosomal dominant disorder involving multiple systems including the liver, heart, eyes, faces, skeleton, kidneys, blood vessels, and other organs, along with increased risk for malignancies. Mortality (~10%) may result from cardiac, hepatic, and vascular involvements [14].

13.5.1 Hepatic Features

Affecting ~95% of ALGS patients, mostly neonates or infants under 3 months of age, chronic cholestasis (conjugated hyperbilirubinemia, elevated direct bilirubin; elevated serum alkaline phosphatase, gamma-glutamyl transpeptidase, and aminotransferases; increased serum bile acids, cholesterol, and triglycerides) is responsible for causing jaundice/cirrhosis, intense pruritus, xanthomas (due to fatty deposits on the extensor surfaces), illthrift (due to fat malabsorption), and progressive liver failure. Liver biopsy often reveals paucity of the intrahepatic bile ducts (in 60% of infants of <6 months, and 95% after 6 months) as well as ductal proliferation (which may sometimes be confused with biliary atresia) (Figure 13.1) [15–17].

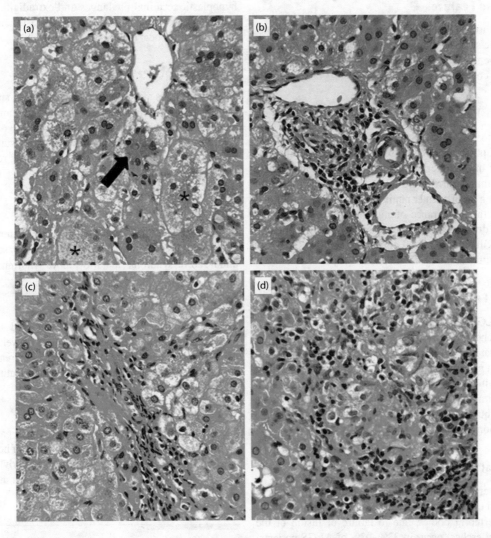

FIGURE 13.1   Alagille syndrome in a 2-year-old boy showing (a) canalicular cholestasis (arrow) with scattered hepatic giant cells containing eight to ten nuclei (asterisks); (b) two small dilated veins, a small artery, and no recognizable bile ducts consistent with Alagille syndrome at 8 weeks; (c) small portal tract with inconspicuous blood vessels and no recognizable bile ducts, no cholestasis, and no giant cell transformation in the hepatic parenchyma; (d) a granuloma with a central cluster of epithelioid histiocytes partially surrounded by a band of small mature lymphocytes, consistent with both Alagille syndrome and sarcoidosis at 26 months. (Photo credit: Mannion M et al. *Pediatr Rheumatol Online J.* 2012;10(1):32.)

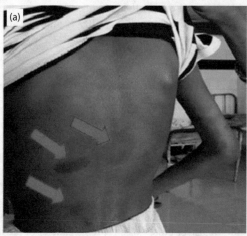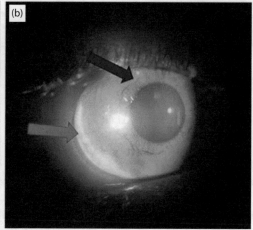

FIGURE 13.2   Alagille syndrome in a 10-year-old boy showing: (a) hyperpigmented patches (red arrow) in the skin; (b) posterior embryotoxon (red arrow) and an iris strand (blue arrow). (Photo credit: Pati GK et al. *J Med Case Rep.* 2016;10(1):342.)

13.5.2 Cardiac Features

Reported in up to 97% of ALGS patients, cardiac anomalies include pulmonic stenosis (peripheral and branch, 67% of cases), heart murmur, ventricular septal defect, atrial septal defect, aortic stenosis, and coarctation of the aorta [18].

13.5.3 Ocular Features

Occurring in up to 95% of ALGS patients but only 15% in the general population, posterior embryotoxon (a prominent Schwalbe ring) is a defect in the anterior chamber of the eye identifiable by slit-lamp examination (Figure 13.2) [16]. Axenfeld–Rieger anomaly (showing an off-center pupil [corectopia] or extra holes in the iris with the appearance of multiple pupils [polycoria], 45%), optic disk drusen (90%), diffuse fundus hypopigmentation (57%), and speckling of the retinal pigment epithelium (33%) are other ocular features associated with ALGS [19].

13.5.4 Facial Features

Patients with ALGS typically have characteristic facies consisting of a high forehead with frontal bossing or flattening, deep-set eyes with moderate hypertelorism, upslanting palpebral fissures, saddle or straight nose with a bulbous tip, depressed nasal bridge, large ears, prominent mandible, and pointed chin. These facial features are commonly seen in ALGS patients harboring a *JAG1* mutation, but less frequently in ALGS patients carrying a *NOTCH2* mutation [20].

13.5.5 Skeletal Features

Visible on anteroposterior radiograph, butterfly vertebrae (butterfly-shaped thoracic vertebrae secondary to clefting abnormality of the vertebral bodies, due to failure of fusion of the anterior vertebral arches) occur in 33%–93% of ALGS patients. Other axial skeletal features range from narrowing of the interpedicular distance in the lumbar spine, pointed anterior process of C1, spina bifida occulta and fusion of adjacent vertebrae, and hemivertebrae to absence of the 12th rib. Fusiform digits with

hypoplastic terminal phalanges and extradigital/supernumerary flexion creases (due possibly to cholestasis and/or an intrinsic defect of the bones) may be also observed.

13.5.6 Renal Features

Renal anomalies affecting the structures (e.g., small hyperechoic kidney, ureteropelvic obstruction, renal cysts) and functions (e.g., renal tubular acidosis) are found in up to 39% of ALGS patients (Figure 13.3) [21]. Hypertension, renal artery stenosis, vesicoureteral reflux, urinary obstruction, and chronic renal failure may also occur in some ALGS patients [22].

13.5.7 Vascular Features

Intracranial bleeding (15% of cases) after rupture of intracranial aneurysms is potentially devastating. Renovascular anomalies, middle aortic syndrome, and Moyamoya syndrome may also occur.

13.5.8 Additional Features

Short stature (due to poor growth associated with cholestasis), immunodeficiency, recurrent infection, developmental delay, delayed puberty, high-pitched voice, and pancreatic insufficiency (up to 40% of cases) are additional clinical features of ALGS.

13.5.9 Malignancies

ALGS has been shown to predispose affected patients to hepatocellular cancer (hepatoma, due likely to loss of heterozygosity for a cell cycle–regulating gene rather than underlying chronic liver disease), and possibly papillary thyroid cancer and Wilms tumor (nephroblastoma).

13.6 Diagnosis

Classic criteria for diagnosing ALGS are based on observing anomalies in three out of five organs/systems: (i) cholestasis (jaundice with conjugated hyperbilirubinemia, paucity of bile

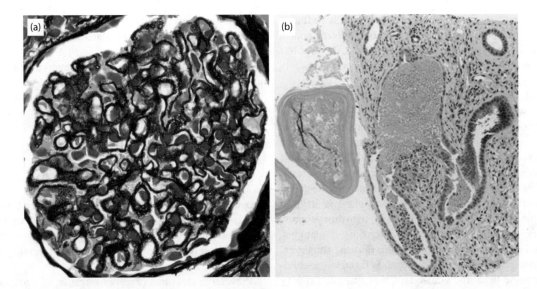

FIGURE 13.3 Alagille syndrome in a 19-year-old male showing: (a) diffuse vacuolization of the glomerular basement membranes (including mesangial cells and podocytes) and subepithelial "spike" formation, indicative of membranous nephropathy (Jones methenamine silver, ×600); (b) immature medullary tubules focally surrounded by spindled cells, consistent with renal dysplasia; and a large yellow–brown bile cast on the left (H&E, ×40). (Photo credit: Bissonnette MLZ et al. *Kidney Int Rep*. 2016;2(3):493–7.)

duct on liver biopsy), (ii) congenital heart disease (typically peripheral pulmonary artery stenosis, and also pulmonary atresia, atrial septal defect, ventricular septal defect), (iii) face—dysmorphic facies (broad forehead, deep-set eyes, upslanting palpebral fissures, prominent ears, straight nose with bulbous tip, and pointed chin), (iv) skeletal anomalies (butterfly vertebrae, and occasionally hemivertebrae, fusion of adjacent vertebrae, and spina bifida occulta), and (v) eye anomalies (posterior embryotoxon, with prominent Schwalbe ring at the junction of the iris and cornea) [1,2].

Recognizing the importance of renal and vascular anomalies in this complex disorder, the current criteria for ALGS have expanded such that finding anomalies in three out of seven organs/systems (which include five organ/system anomalies outlined in classic criteria plus anomalies in the kidneys and vasculature [usually around the head and neck]) (Table 13.1) are sufficient for a clinical diagnosis. Furthermore, fulfilling one or more classic criteria together with an affected first-degree relative is also sufficient for a diagnosis. In addition, with the finding of cholestasis, a liver biopsy (especially for infants under 6 months of age who may not present with a marked paucity of the bile ducts) is no longer mandatory for making a diagnosis of ALGS [1,2].

Thus, apart from medical history review and physical examination (for ophthalmologic, developmental, and vascular anomalies), clinical diagnosis of ALGS requires input from imaging (e.g., hepatic ultrasound, technetium 99 m scan; x-rays for butterfly vertebrae; gonioscopy for embryotoxon) and laboratory studies (e.g., liver function tests for direct bilirubin, serum aminotransferases, serum bile acids, cholesterol, triglycerides and gamma-glutamyl transpeptidase; urine analysis for renal tubular acidosis; stool exam for pancreatic insufficiency; echocardiogram for cardiac anomalies), and histopathology (for diagnosis of hepatocellular cancer from biopsy), and so on [1,2].

Use of molecular procedures to identify a pathogenic *JAG1* or *NOTCH2* variant helps confirm clinical diagnosis of ALGS

and also facilitates prenatal diagnosis or preimplantation genetic diagnosis. Specifically, sequence analysis detects pathogenic *JAG1* variants in 89% of individuals who fulfill clinical diagnostic criteria; deletion/duplication analysis detects *JAG1* exon and whole-gene deletions (including microdeletion of 20p12) in 7% of affected individuals. Pathogenic *NOTCH2* variants are identifiable by sequence analysis in 2% of individuals with ALGS [1,2].

For practical reasons, sequence analysis of the *JAG1* gene is carried out first. If no pathogenic *JAG1* variant is detected by sequence analysis, deletion/duplication analysis (including FISH, MLPA) is then conducted to identify deletions or duplications of *JAG1* exon(s) or of the whole gene. If a deletion involving the entire *JAG1* gene is identified in a patient with developmental

TABLE 13.1

Updated Clinical Criteria for Diagnosis of ALGS

No.	Organ/System Involvement	Clinical Characteristics
1	Hepatic manifestations	Cholestasis (due to paucity of biliary ducts), conjugated hyperbilirubinemia, pruritus, xanthomas, and cirrhosis
2	Cardiac defects	Peripheral pulmonic stenosis (67%), tetralogy of Fallot (16%), ventricular septal defect, atrial septal defect, aortic stenosis, and coarctation of the aorta
3	Dysmorphic facies	Prominent, broad forehead, deep-set eyes with moderate hypertelorism, prominent ears, triangular face with a pointed chin, and broad nasal bridge
4	Skeletal anomalies	Butterfly vertebrae, hemivertebrae, and pathological fractures of long bones
5	Ophthalmologic anomalies	Posterior embryotoxon with a prominent Schwalbe line
6	Renal anomalies	Renal dysplasia, glomerular mesangiolipidosis, and renal tubular acidosis
7	Vasculature anomalies	Aneurysms, abnormalities in cerebral arteries, reno-vascular abnormalities, Moyamoya syndrome, and middle aortic syndrome

delay and/or hearing loss, a full cytogenetic study may be undertaken to determine whether a rare chromosome rearrangement (translocation or inversion) or deletion is present. If no *JAG1* variant/deletion/duplication is identified, sequence analysis of the *NOTCH2* gene should be performed [23].

Differential diagnoses for ALGS include other conditions that present with *cholestasis* (neonates with biliary atresia, sepsis, galactosemia, tyrosinemia, choledochal cyst; patients with progressive familial intrahepatic cholestasis types 1 and 2, arthrogryposis–renal dysfunction–cholestasis syndrome, benign recurrent intrahepatic cholestasis, Norwegian cholestasis [Aagenaes syndrome], benign recurrent intrahepatic cholestasis), *interlobular bile duct paucity* (patients with alpha-1 antitrypsin deficiency, hypopituitarism, cystic fibrosis, trihydroxycoprostanic acid excess, childhood primary sclerosing cholangitis, mitochondrial disorders, congenital hepatic fibrosis, congenital syphilis, cytomegalovirus, rubella or hepatitis B infection, childhood autoimmune hepatitis, graft-versus-host disease, primary sclerosing cholangitis, Down syndrome, Zellweger syndrome, Ivemark syndrome, and Smith–Lemli–Opitz syndrome), *cardiac defects* (ventricular septal defect, tetralogy of Fallot), *pulmonary stenosis* (RAS-MAPK pathway disorders, deletion 22q11 syndrome, Williams syndrome), *pulmonic vascular abnormalities* (Noonan syndrome, Watson syndrome, William syndrome, Down syndrome, and LEOPARD syndrome), *posterior embryotoxon* (Rieger syndrome, Bannayan–Riley–Ruvalcaba syndrome, Axenfeld–Rieger syndrome, and 15% of the general population), *germline pathogenic variants* (Hajdu–Cheney syndrome with pathogenic gain-of-function variant in *NOTCH2* exon 34), and *somatic pathogenic variants* (splenic marginal zone lymphoma with recurrent somatic pathogenic gain-of-function variants in *NOTCH2*) [1,2].

13.7 Treatment

Due to its diverse clinical presentations, management of ALGS relies on a multidisciplinary approach that caters to the specific needs of individual patients. Given that liver disease and congenital heart defects exert an overwhelming influence on the outcome of ALGS, treatment should generally focus on the consequences of hepatic and cardiac involvements [3,24,25].

For patients with cholestasis, supportive treatment with choleretic agents (e.g., ursodeoxycholic acid, naltrexone, rifampin, colesevelam, and cholestyramine) helps ameliorate debilitating pruritus and disfiguring xanthomas. Surgical partial internal biliary diversion and ileal exclusion are also helpful in controlling itch and xanthomas. Kasai procedure (hepatic portoenterostomy), which is often used in patients with biliary atresia, seems to worsen the outcome of ALGS. Liver transplantation may be considered for end-stage liver disease (about 15% of ALGS patients, who tend to have a high serum total bilirubin between 12–24 months of age together with fibrosis on liver biopsy and xanthomata on physical examination), as it improves liver function and catch-up growth in 90% of cases and has a 5-year survival rate of 80% [26].

For patients with cardiac, renal, and vascular anomalies, treatment aims at symptomatic relief, with the application of standard surgical and medical measures.

For patients with neurovascular anomalies such as symptomatic Moyamoya disease, revascularization helps prevent ischemic events and neurologic disability.

For patients with ophthalmological and vertebral anomalies, intervention is rarely necessary.

For patients with malnutrition, provision of fat-soluble vitamin supplements is important to maintain optimal growth and development.

13.8 Prognosis and Prevention

Prognosis for individuals with ALGS reflects the type and severity of organ involvements. Currently, the mortality rate for ALGS stands at ~10%, with early mortality attributed mainly to cardiac disease or severe liver disease, and later mortality to vascular accidents.

Preventive measures (e.g., optimization of nutrition and vitamin replacement, use of a spleen guard during activities, avoidance of contact sports and alcohol, especially for those with splenomegaly, chronic liver disease, and vascular involvement) and regular surveillance (e.g., monitoring of plasma concentration of fat-soluble vitamins, growth, and heart) help improve the prognosis for ALGS patients [27].

Since ALGS is an autosomal dominant disorder that results from germline (40%) or de novo (60%) mutations in the *JAG1* (96%) or *NOTCH2* (2%) gene, offspring of an individual with ALGS have a 50% chance of inheriting the causative pathogenic variant. On the other hand, parents of a child with de novo pathogenic variant have low recurrence risk (but greater than general population due to the possibility of germline mosaicism) to subsequent offspring. Therefore, if an affected family member is known to harbor the pathogenic variant, prenatal for pregnancies at increased risk and preimplantation genetic diagnosis may be considered [28].

13.9 Conclusion

Alagille syndrome (ALGS) is an autosomal dominant disorder affecting multiple organs with hallmark presentations of paucity of intrahepatic bile ducts leading to cholestasis, peripheral pulmonary artery stenosis, vertebral arch defects, poor linear growth, characteristic facies, and embryotoxon posterior in the eye [29,30]. The identification of hepatocellular cancer, papillary thyroid cancer. and Wilms tumor in a number of ALGS patients suggests its link to increased cancer risk.

At the molecular level, ALGS is attributed to germline or de novo mutations in the *JAG1* and *NOTCH2* genes [31,32]. As components of the highly conserved Notch signaling pathway, interaction between NOTCH2 transmembrane receptor and its ligand JAG1 triggers a cascade of downstream genes that determine cell fate and differentiation. While 89% of patients are found to have mutations in the *JAG1* gene, an additional 7% harbor deletions in genomic region that incorporates *JAG1*, and 2% possess mutations in the *NOTCH2* gene. Further, 40% of patients show family history of ALGS and appear to have inherited pathogenic variant from affected parents, and 60% of patients are due to de novo mutations in the *JAG1* or *NOTCH2* gene.

Given its diverse manifestations and its variable expressivity, even within families, clinical diagnosis of ALGS focusing on bile duct paucity, chronic cholestasis, and congenital heart disease can be challenging and error prone. Molecular detection of pathogenic *JAG1* or *NOTCH2* variants in suspected patients and related family members removes this uncertainty and provides much-needed improvement in the management of ALGS. It is likely that ongoing research on the molecular mechanisms of ALGS will lead to development of novel, highly effective therapies for this underrecognized cancer predisposition syndrome.

REFERENCES

1. Spinner NB, Leonard LD, Krantz ID. Alagille syndrome. In: Adam MP, Ardinger HH, Pagon RA et al. eds. *GeneReviews®* *[Internet].* Seattle (WA): University of Washington, Seattle; 1993–2018. 2000 May 19 [updated 2013 Feb 28].

2. Diaz-Frias J, Kondamudi NP. *Alagille Syndrome. StatPearls [Internet].* Treasure Island (FL): StatPearls Publishing; 2018. 2018 Jun 18.

3. Turnpenny PD, Ellard S. Alagille syndrome: Pathogenesis, diagnosis and management. *Eur J Hum Genet.* 2012;20(3):251–7.

4. Grochowski CM, Loomes KM, Spinner NB. Jagged1 (JAG1): Structure, expression, and disease associations. *Gene.* 2016;576(1):381–4.

5. Brennan A, Kesavan A. Novel heterozygous mutations in *JAG1* and *NOTCH2* genes in a neonatal patient with Alagille syndrome. *Case Rep Pediatr.* 2017;2017:1368189.

6. Kamath BM, Baur RC, Loomes KM et al. NOTCH2 mutations in Alagille syndrome. *J Med Genet.* 2012;49(2):138–44.

7. Braune EB, Lendahl U. Notch—a goldilocks signaling pathway in disease and cancer therapy. *Discov Med.* 2016;21(115): 189–96.

8. Louvi A, Artavanis-Tsakonas S. Notch and disease: A growing field. *Semin Cell Dev Biol.* 2012;23(4):473–80.

9. Penton AL, Leonard LD, Spinner NB. Notch signaling in human development and disease. *Semin Cell Dev Biol.* 2012;23(4):450–7.

10. Kopan R, Chen S, Liu Z. Alagille, Notch, and robustness: Why duplicating systems does not ensure redundancy. *Pediatr Nephrol.* 2014;29(4):651–7.

11. Mašek J, Andersson ER. The developmental biology of genetic Notch disorders. *Development.* 2017;144(10):1743–63.

12. Zanotti S, Canalis E. Notch signaling in skeletal health and disease. *Eur J Endocrinol.* 2013;168(6):R95–103.

13. Reyes-de la Rosa ADP, Varela-Fascinetto G, García-Delgado C et al. A novel c.91dupG *JAG1* gene mutation is associated with early onset and severe Alagille syndrome. *Case Rep Genet.* 2018;2018:1369413.

14. Saleh M, Kamath BM, Chitayat D. Alagille syndrome: Clinical perspectives. *Appl Clin Genet.* 2016;9:75–82.

15. Mannion M, Zolak M, Kelly DR, Beukelman T, Cron RQ. Sarcoidosis in a young child with Alagille syndrome: A case report. *Pediatr Rheumatol Online J.* 2012;10(1):32.

16. Pati GK, Singh A, Nath P et al. A 10-year-old child presenting with syndromic paucity of bile ducts (Alagille syndrome): A case report. *J Med Case Rep.* 2016;10(1):342.

17. Kim J, Yang B, Paik N, Choe YH, Paik YH. A case of Alagille syndrome presenting with chronic cholestasis in an adult. *Clin Mol Hepatol.* 2017;23(3):260–4.

18. Gorospe Sarasúa L, Ayala-Carbonero AM, Fernández-Méndez MÁ. Calcified atherosclerosis of the pulmonary trunk, stenosis of the main pulmonary arteries, and post-stenotic dilation of segmental pulmonary arteries in a patient with Alagille syndrome. *Arch Bronconeumol.* 2017;53(2):73–4.

19. Ho DK, Levin AV, Anninger WV, Piccoli DA, Eagle RC Jr. Anterior chamber pathology in Alagille syndrome. *Ocul Oncol Pathol.* 2016;2(4):270–5.

20. Bresnahan JJ, Winthrop ZA, Salman R, Majeed S. Alagille syndrome: A case report highlighting dysmorphic facies, chronic illness, and depression. *Case Rep Psychiatry.* 2016;2016:1657691.

21. Bissonnette MLZ, Lane JC, Chang A. Extreme renal pathology in Alagille syndrome. *Kidney Int Rep.* 2016;2(3):493–7.

22. Di Pinto D, Adragna M. Renal manifestations in children with Alagille syndrome. *Arch Argent Pediatr.* 2018; 116(2):149–53.

23. Vozzi D, Licastro D, Martelossi S, Athanasakis E, Gasparini P, Fabretto A. Alagille syndrome: A new missense mutation detected by whole-exome sequencing in a case previously found to be negative by DHPLC and MLPA. *Mol Syndromol.* 2013;4(4):207–10.

24. Berniczei-Royko A, Chałas R, Mitura I, Nagy K, Prussak E. Medical and dental management of Alagille syndrome: A review. *Med Sci Monit.* 2014;20:476–80.

25. Xie X, Lu Y, Wang X, Wu B, Yu H. JAGGED1 gene variations in Chinese twin sisters with Alagille syndrome. *Int J Clin Exp Pathol.* 2015;8(7):8506–11.

26. Pavanello M, Severino M, D'Antiga L et al. Pretransplant management of basilar artery aneurysm and moyamoya disease in a child with Alagille syndrome. *Liver Transpl.* 2015;21(9):1227–30.

27. Kamath BM, Chen Z, Romero R et al. Quality of life and its determinants in a multicenter cohort of children with Alagille syndrome. *J Pediatr.* 2015;167(2):390–6.e3.

28. Chen CP, Yin CS, Wang LK et al. Molecular genetic characterization of a prenatally detected de novo interstitial deletion of chromosome 20p (20p12-p13) encompassing JAG1 and a literature review of prenatal diagnosis of Alagille syndrome. *Taiwan J Obstet Gynecol.* 2017;56(3):390–3.

29. Fukumoto M, Ikeda T, Sugiyama T, Ueki M, Sato T, Ishizaki E. A case of Alagille syndrome complicated by intraocular lens subluxation and rhegmatogenous retinal detachment. *Clin Ophthalmol.* 2013;7:1463–5.

30. Ennaifer R, Ben Farhat L, Cheikh M, Romdhane H, Marzouk I, Belhadj N. Focal liver hyperplasia in a patient with Alagille syndrome: Diagnostic difficulties. A case report. *Int J Surg Case Rep.* 2016;25:55–61.

31. Micaglio E, Andronache AA, Carrera P et al. Novel JAG1 deletion variant in patient with atypical Alagille syndrome. *Int J Mol Sci.* 2019;20(24):6247.

32. Gilbert MA, Bauer RC, Rajagopalan R et al. Alagille syndrome mutation update: Comprehensive overview of JAG1 and NOTCH2 mutation frequencies and insight into missense variant classification. *Hum Mutat.* 2019;40(12):2197–220.

14

Constitutional Mismatch Repair Deficiency Syndrome

Dongyou Liu

CONTENTS

14.1 Introduction

Constitutional mismatch repair deficiency (CMMRD) syndrome (also referred to as biallelic mismatch repair deficiency) is an autosomal recessive disorder presenting with cutaneous features (especially café-au-lait spots) and a strong predisposition to hematological, brain, and intestinal tumors during childhood.

Once regarded as a variant of Lynch syndrome, and referred to under Turcot syndrome for a number of years, CMMRD is now recognized as a distinct childhood cancer predisposition syndrome that results from biallelic (either homozygous or compound heterozygous) germline mutations in the key DNA mismatch repair genes: human mutL homolog 1 (*MLH1*), human mutS homolog 2 (*MSH2*), human mutS homolog 6 (*MSH6*), and human post-meiotic segregation increased 2 (*PMS2*) (Table 14.1) [1].

In contrast, Lynch syndrome is an autosomal dominant disorder caused by heterozygous/monoallelic germline mutation in one of the alleles of genes (*MLH1*, *MSH2*, *MSH6*, and *PMS2*) encoding components of the mismatch pathway, leading to the development of gastrointestinal and genitourinary cancers during adulthood (see Chapter 20 in this volume) [2].

Turcot syndrome may be separated into two types. Accounting for 33% of cases, Turcot syndrome type 1 is an autosomal recessive disorder attributed to homozygous or compound heterozygous germline-inactivating mutations in DNA mismatch repair genes *MLH1* and *PMS2*, with increased risk for brain tumor (e.g., glioma), colorectal adenomas, and relatively few colonic polyps. Representing 67% of cases, Turcot syndrome type 2 is an autosomal dominant disorder linked to heterozygous germline inactivating mutations in familial adenomatous polyposis (*FAP*) gene (see Chapter 16 in this volume), with increased risk for multiple colonic polyps, colorectal cancer, and brain tumor (e.g., medulloblastoma) [2].

14.2 Biology

Cases related to CMMRD were reported in 1999 from offspring of parents who both had colon cancer at an early age and carried *MLH1* mutations (indicative of Lynch syndrome). Possessing biallelic germline *MLH1* mutations, the offspring developed café-au-lait spots, which resembled those of neurofibromatosis type 1 (NF1), and hematologic malignancy during early childhood. Since then, about 200 CMMRD cases involving children and young adults have been described. It is notable that these patients all harbor homozygous or compound heterozygous mutations in one of the four DNA mismatch repair genes (i.e., *MLH1*, *MSH2*, *MSH6*, and *PMS2*) [3].

DNA mismatch repair genes encode proteins that have the ability to correct single-base nucleotide mismatches (insertions or deletions, especially in repetitive elements called microsatellites) erroneously introduced by DNA polymerases during meiosis and mitosis. Therefore, they play a vital role in repairing mutable mistakes within DNA, particularly after DNA replication, and maintaining the genomic stability of mammalian cells. Not surprisingly, mutations in DNA mismatch repair genes (e.g., *MLH1*, *MSH2*, *MSH6*, and *PMS2*) cause replication repair deficiency, leading to point mutations and microsatellite instability. In turn, alterations of microsatellites in gene-encoding regions may shift the translational reading frame and introduce new stop

TABLE 14.1

Characteristics of Mismatch Repair Genes Involved in CMMRD

Gene	Chromosome Location	Gene Structure	Protein	Function	Clinical Features
MLH1 (mutL homolog 1)	3p22.2	57 kb, 19 exons	756 aa, 84 kDa	MutL α (MLH1+PMS2) interacts with MutS β (MSH2+MSH3), facilitating PMS2 within MutL α to introduce a single-strand break near DNA mismatch, which provides an entry point for exonuclease to degrade the strand containing the mismatch. MutL α (MHL1 and PMS2) also interacts with MutS α (MSH2 +MSH6) to recruit DNA helicase that unwinds the mismatch containing DNA strand for DNA excision and repair.	Observed in 14% of cases, mutation in MLH1 (an obligate partner of PMS2) results in concurrent loss of MLH1/PMS2; patients with biallelic MLH1 mutations are less likely to survive their first hematologic tumor.
MSH2 (mutS homolog 2)	2p21-p16.3	80 kb, 16 exons	934 aa, 104 kDa	MutS α (MSH2+MSH6) recognizes single base mismatches, altered nucleotides, and dinucleotide insertion-deletion loops of up to 2 nucleotides in the DNA; and targets mono- and dinucleotide frameshift slippages for repair. MutS β (MSH2+MSH3) recognizes insertion-deletion loops of ≥2–13 nucleotides; and targets di-, tri-, and tetranucleotide frameshift slippages for repair.	Observed in 9% of cases, mutation in MSH2 (an obligate partner of MSH6) results in concurrent loss of MSH2/MSH6; patients with biallelic MSH2 mutations are less likely to survive their first hematologic tumor.
MSH6 (mutS homolog 6)	2p16.3	23 kb, 10 exons	1360 aa, 152 kDa	See MSH2	Observed in 19% of cases, mutation in MSH6 is associated with isolated loss; patients with biallelic MSH6 mutations are more likely to survive their first brain tumor.
PMS2 (post-meiotic segregation increased 2)	7p22.1	38 kb, 15 exons	862 aa, 95 kDa	See MLH1	Observed in 58% of cases, mutation in PMS2 is associated with isolated loss; patients with biallelic PMS2 mutations are more likely to survive their first brain tumor.

codon, leading to the production of truncated or functionally inactive proteins [4,5].

While heterozygous/monoallelic mutations in *MLH1*, *MSH2*, *MSH6*, and *PMS2* underscore the development of gastrointestinal and genitourinary cancers associated with Lynch syndrome, homozygous/biallelic mutations contribute to the pathogenesis of CMMRD, which display café-au-lait spots as well as increased risk for hematological, brain, and intestinal tumors. Further, compared to Lynch syndrome in which inheritance of a heterozygous germline mutation is followed by a somatic mutation in the second allele, CMMRD involves inheritance of a germline mutation from each parent to initiate the disease process, resulting in either homozygous mutation (indicated by the presence of identical mutation on both alleles of a specific gene) or compound heterozygous mutation (or genetic compound, indicated by the presence of two different mutations on both alleles of a specific gene) [1].

14.3 Pathogenesis

Human cells are exposed to both extrinsic (e.g., ultraviolet radiation, carcinogens, and oxidative stress, etc.) and intrinsic (e.g., replication error) factors that can damage DNA, impair gene expression, and induce genomic instability and mutagenesis. To maintain stability and prevent transmission of mutations to progeny, human cells have evolved a number of mechanisms to repair damaged DNA. Of these, DNA mismatch repair genes play a vital role in repairing mistakes that remain after DNA polymerases replicate DNA.

The DNA mismatch repair genes implicated in the pathogenesis of CMMRD are *MLH1*, *MSH2*, *MSH6*, and *PMS2*, which range from 23 kb to 80 kb in size, consist of 10–19 exons, and encode proteins (MLH1, MSH2, MSH6, and PMS2) of 84–152 kDa in molecular weight with tumor suppressor function (Table 14.1).

The MLH1 protein comprises an ATPase domain and two interaction domains (one for Mut homologs MSH2, MSH3, and MSH6, and the other for PMS2, MHL3, or PMS1. Typically, MLH1 heterodimerizes with PMS2 to form MutL α, which binds to MutS β (composed of MSH2 and MSH3) and some accessory proteins, facilitating PMS2 within MutL α to introduce a single-strand break near DNA mismatches, which provides a new entry point for the exonuclease EXO1 to degrade the strand containing the mismatch. MLH1 can also heterodimerize with MLH3 to form MutL γ, which is involved in meiosis.

The MSH2 protein possesses a DNA binding domain and two interaction domains (one for MSH3 or MSH6, and the other for MLH1 and PMS2). MSH2 heterodimerizes with MSH6 to form MutS α (which recognizes single-base mismatches [e.g., G/T, A/C], altered nucleotides [e.g., O^6-methyguanine, 5-fluorode-oxyuracil], and dinucleotide insertion-deletion loops of up to 2 nucleotides in the DNA), and with MSH3 to form MutS β (which recognizes insertion-deletion loops of ≥2–13 nucleotides, but not insertion-deletion loops of 1 or single-base mismatches) [6]. After binding to respective mismatch, MutS α or β forms a ternary complex with MutL α (composed of MHL1 and PMS2) to recruit proteins (e.g., DNA helicase MCM9, which unwinds the mismatch containing DNA strand) for DNA excision and repair or to trigger cell demise. While MutS α targets mono- and dinucleotide frameshift slippages for repair, MutS β targets di-, tri-, and tetranucleotide frameshift slippages for repair (e.g., microsatellites with insertion-deletion loops of 2, 3, or 4).

The MSH6 protein has a highly conserved helix-turn-helix domain associated with an adenine nucleotide and magnesium binding motif (or Walker-A motif) with ATPase activity. MSH6 forms a heterodimer (i.e., MutS α) with MSH2, thus participating in the rectification/repair of single base mismatches and dinucleotide insertion-deletion loops (IDL) in the DNA.

The PMS2 protein heterodimerizes with MLH1 to form MutL α, which is involved in DNA repair initiated by MutS α (MSH2-MSH6) or MutS β (MSH2-MSH3). It appears that germline *PMS2* mutations dominate clinical cases of CMMRD. In one study, 58% of CMMRD patients harbored mutations in *PMS2*, 19% in *MSH6*, 14% in *MLH1*, and 9% in *MSH2*. This may possibly be due to the fact that a mutation in *PMS2* or *MSH6* often produces isolated loss, while a mutation in *MLH1* or *MSH2* results in concurrent loss of MLH1/PMS2 or MSH2/MSH6, respectively, as *MLH1* and *MSH2* are the obligatory partners in the formation of MLH1/PMS2 and MSH2/MSH6 heterodimers. Consequently, patients with biallelic mutations in *PMS2* or *MSH6* are more likely to survive their first tumors than those with biallelic mutations in *MLH1* or *MSH2* (Table 14.2) [7–9].

14.4 Epidemiology

14.4.1 Prevalence

CMMRD is a rare disease, with an estimated incidence of approximately 1 per million patients. To date, about 200 CMMRD cases have been reported in the literature. The median age at diagnosis of CMMRD-related hematological and brain tumors in is 6.6 (1.2–30.8) and 10.3 (3.3–40) years, respectively, whereas the median age at diagnosis of Lynch syndrome−associated colorectal adenocarcinomas is 21.4 years (11.4–36.6). Because CMMRD patients tend to develop a variety of malignant tumors during early life, most fail to reach adulthood.

TABLE 14.2

Details of C4CMMRD Scoring System [23,24]

Indication for CMMRD Testing in Cancer Patients	≥3 Points
Malignancies/premalignancies: One is mandatory; if more than one is present in the patient, add the points	
Lynch syndrome−associated tumors (colorectal, endometrial, small bowel, ureter, renal pelvis, biliary tract, stomach, bladder carcinomas) at age <25 years	3 points
Multiple bowel adenomas at age <25 years and absence of APC/MUTYH mutation(s) or a single high-grade dysplasia adenoma at age <25 years	3 points
WHO grade III or IV glioma at age <25 years	2 points
NHL of T-cell lineage or sPNET at age <18 years	2 points
Any malignancy at age <18 years	1 point
Additional features: Optional; if more than one of the following is present, add the points	
Clinical sign of NF1 and/or ≥2 hyperpigmented and/or hypopigmented skin alterations diameter of >1 cm in the patient	2 points
Diagnosis of Lynch syndrome in a 1st- or 2nd-degree relative	2 points
Lynch syndrome−associated tumors (colorectal, endometrial, small bowel, ureter, renal pelvis, biliary tract, stomach, bladder carcinomas) before the age of 60 in 1st-, 2nd-, or 3rd-degree relative	1 point
A sibling with Lynch syndrome−associated tumors (colorectal, endometrial, small bowel, ureter, renal pelvis, biliary tract, stomach, bladder carcinomas), high-grade glioma, sPNET, or NHL	2 points
A sibling with any type of childhood malignancy	1 point
Multiple pilomatricomas in the patient	2 points
One pilomatricoma in the patient	1 point
Agenesis of the corpus callosum or non-therapy−induced cavernoma in the patient	1 point
Consanguineous parents	1 point
Deficiency/reduced levels of IgG2/4 and/or IgA	1 point

14.4.2 Inheritance

CMMRD shows an autosomal recessive pattern of inheritance. If a child inherits a normal gene from one parent and a mutated gene from the other parent, s/he will be a carrier for the disease without showing symptoms. A child from two carrier parents has a 25% chance of acquiring two defective genes and becomes affected, a 50% chance of acquiring a normal gene and a mutated gene and becomes an asymptomatic carrier, and a 25% chance of acquiring two normal genes and becomes genetically normal for that particular trait. Thus, if both parents have Lynch syndrome (heterozygous mutation), the offspring have a 25% chance of developing CMMRD (homozygous mutation), a 25% chance of acquiring no mutation, and a 50% chance of developing Lynch syndrome (heterozygous mutation). The carrier frequencies for mutations in the mismatch repair genes MLH1, MSH2, MSH6, and PMS2 are estimated to be 1 in 1946, 1 in 2841, 1 in 758, and 1 in 714, respectively.

14.4.3 Penetrance

CMMRD is a highly penetrating cancer predisposition syndrome, with most patients harboring biallelic mutations developing cancer by their 20s. Mutations in different mismatch repair genes (i.e., *MLH1*, *MSH2*, *MLH6*, and *PSM2*) may impact on clinical presentation. For example, patients with biallelic *MLH1* or *MSH2* mutations develop hematological malignancies in infancy or early childhood more frequently than those with biallelic *MSH6* or *PSM2* mutations [10]. On the other hand, patients with biallelic *MSH6* or *PSM2* mutations have a higher prevalence of brain tumors later during childhood than those with biallelic *MLH1* or *MSH2* mutations, and are more likely to survive their first brain tumors and develop colorectal cancer as a second or third primary malignancy in adolescence or young adulthood than those with biallelic mutations in *MSH1* and *MSH2*, whose first hematological tumors prove to be highly malignant and deadly (Table 14.2).

14.5 Clinical Features

CMMRD usually presents with two categories of clinical symptoms, i.e., cutaneous manifestations (e.g., café-au-lait macules and/or depigmented spots), and early-onset tumors (which encompass hematologic malignancies, brain tumors, Lynch syndrome−associated tumors, and other malignancies) [11].

Cutaneous manifestations, exemplified by café-au-lait macules and/or depigmented spots, are nonneoplastic and present in >80% of CMMRD patients. Typically, the number of café-au-lait macules over 1 cm in diameter range between 2 and >10 (Figure 14.1) [9]. Although café-au-lait macules observed in CMMRD resemble those in neurofibromatosis type I (NF1), they have irregular borders, display a segmental distribution, and vary in the degree of pigmentation. Other features of NF1 found in CMMRD patients consist of skinfold freckling, Lisch nodules, neurofibromas, tibial pseudarthrosis, and agenesis of the corpus callosum [12]. Nonetheless, café-au-lait macules are largely absent in Lynch syndrome, or only occasionally present in Turcot syndrome [13].

Hematologic malignancies occur in 30% of CMMRD patients during childhood and include non-Hodgkin lymphoma (NHL) and other lymphomas (median age of 6 years at diagnosis),

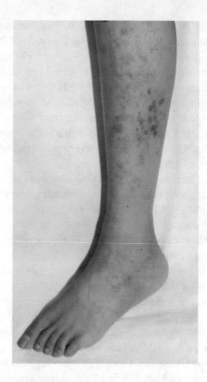

FIGURE 14.1 Constitutional mismatch repair deficiency syndrome in a 30-year-old female with homozygous germline mutation of the *PMS2* mismatch repair gene [c.1500del (p.Val501TrpfsTer94) in exon 11] showing café-au-lait lesions on a background of lichen planus in the left leg. (Photo credit: Ramchander NC et al. *BMC Med Genet.* 2017;18(1):40.)

acute lymphoblastic leukemia (ALL, median age of 5 years), acute myeloid leukemia (AML, median age of 9 years), atypical chronic myeloid leukemia (aCML, age of 1 year), and acute leukemia (age of 2 years) [14].

Brain tumors affect 50% of pediatric patients with CMMRD and may range from glioblastoma and other astrocytomas (median age of 9 years at diagnosis), supratentorial primitive neuroectodermal tumor (sPNET, median age of 8 years), to medulloblastoma (median age of 7 years).

Lynch syndrome−associated tumors consist of colorectal cancer (median age of 16 years at diagnosis), small bowel cancer (median age of 26 years), endometrial cancer (median age of 24 years), ureter/renal pelvis cancer (age of 15 years), multiple synchronous adenomas, and multiple intestinal polyps. Digestive tract tumors are detected in 30% of children suffering from CMMRD [15–17].

Other malignancies comprise neuroblastoma (age of 13 years), Wilms tumor (age of 4 years), rhabdomyosarcoma (age of 4 years), ovarian neuroectodermal tumor (age of 21 years), infant myofibromatosis (age of 1 year), breast cancer (age of 35 years), and sarcoma (age of 65 years) [18,19].

14.6 Diagnosis

Individuals with early-onset cancer (e.g., high-grade gliomas, T-lymphoblastic lymphoma, or colorectal carcinomas), cutaneous features (café-au-lait macules and/or depigmented spots) resembling NF1, and a family history of Lynch syndrome−associated malignancies, consanguinity, sibling suffering from childhood

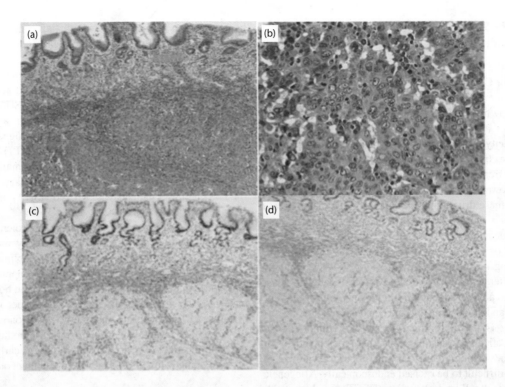

FIGURE 14.2 Constitutional mismatch repair deficiency syndrome in a 67-year-old male with a solitary gastric mass showing a well-demarcated, poorly differentiated carcinoma with an organoid growth pattern and pushing borders (a, H&E, 100×); abundant peritumoral lymphocytic response (b, H&E, 400×); loss of MLH1 expression (c, 100×), and loss of PMS2 expression (d, 100×). In c and d, the gastric medullary carcinoma is located at the bottom part of each field, while the overlying unremarkable gastric mucosa is located at the top part of each field. (Photo credit: Lowenthal BM et al. *Case Rep Pathol.* 2017;2017:3427343.)

cancer, and development of second malignancy, should be suspected of CMMRD [20].

Microscopic examination of tumor tissue allows identification of characteristic tumors associated with CMMRD. For example, gliomas contain large, multinucleated giant cells with clumped nuclei and cells with many smaller, eccentrically placed nuclei mimicking pleomorphic xanthroastrocytomas. Gastric medullary carcinoma displays an organoid growth pattern and pushing borders, abundant peritumoral lymphocytic response, and loss of MLH1 and PMS2 expression (Figure 14.2) [21].

Immunohistochemistry (IHC) revealing loss of mismatch repair protein expression in both normal and tumor cells (within the tissue or skin biopsy) is highly indicative of CMMRD, as normal cells within Lynch syndrome tumors are usually stained positive. Furthermore, while biallelic truncating mutations in *PMS2* or *MSH6* lead to isolated loss of these proteins, mutations in *MLH1* or *MSH2* tend to cause concurrent loss of MLH1/PMS2 or MSH2/MSH6, respectively. Since some missense mutations may result in retained staining of the protein, a positive staining outcome does not necessarily rule out CMMRD.

Molecular identification of biallelic mutations in one of the four mismatch genes (*MLH1, MSH2, MSH6, PMS2*) offers an invaluable means of confirming the diagnosis of CMMRD. As multiple pseudogenes exist for *PMS2* in the human genome, care should be taken to avoid false positive results when identifying *PMS2* mutations [22].

Realizing the clinical complexity of CMMRD, the European "Care for CMMRD (C4CMMRD)" consortium designed a highly sensitive scoring system to help streamline the decision making, with individuals who reach a score of ≥3 being likely affected, and for whom further testing for biallelic mutations in

mismatch genes (*MLH1, MSH2, MSH6, PMS2*) and/or or immunohistochemical staining of four mismatch repair proteins being recommended (Table 14.2) [23,24].

Differential diagnoses for CMMRD include Lynch syndrome (due to heterozygous germline mutations in *MSH2, MLH1, MSH6, PMS2, EPCAM* followed by somatic mutation of second allele within tissue, causing colorectal cancer and gastrointestinal tract cancer, female reproductive tract cancer, urinary tract cancer, and glioblastoma in patients of ~40 years of age), Turcot syndrome (due to homozygous germline mutations in *MLH1* and *PMS2*, with increased risk for glioma, colorectal adenomas, and relatively few colonic polyps, and also due to heterozygous germline mutations in *FAP*, with increased risk for multiple colonic polyps, colorectal cancer, and medulloblastoma, usually in adults), Lynch–like syndrome (due to biallelic somatic mutations in *MSH2, MLH1, MSH6, PMS2*), sporadic MSI-H colorectal cancer (due to biallelic somatic methylation of the *MLH1* promoter; affecting older age >70 years, and especially females), elevated microsatellite alterations at selected tetranucleotide (EMAST) repeats (due to IL-6 – induced, reversible, nuclear-to-cytosolic shift of MSH3; causing advanced staged colorectal cancer in African Americans), and NF1 (due to heterozygous germline mutations in *NF1* followed by somatic mutation of second allele within tissue; causing flat, light brown spots on the skin [café-au-lait spots], freckling in the armpits or groin area, tiny bumps on the iris of the eye [Lisch nodules], soft, pea-sized bumps on or under the skin [neurofibromas], bone deformities, tumor on the optic nerve [optic glioma], learning disabilities, larger than average head size, and short stature) [25].

14.7 Treatment

Depending on the type and stage of malignancies present, treatment options for CMMRD-affected patients range from chemotherapy, radiotherapy, and surgery to stem cell transplantation.

Chemotherapy represents a frontline treatment for CMMRD, and selection of appropriate drugs requires some understanding on the toxicity profile of and tumor resistance to a particular agent. For example, hematopoietic cancer and glioblastoma are commonly treated with mercaptopurine and temozolomide, respectively. However, mismatch repair–deficient cells appear to be resistant to these drugs. Indeed, temozolomide may increase the accumulation of somatic mutations, promote further mismatch repair deficiency, and strengthen temozolomide resistance in MSH2 or MSH6 negative glioblastoma. For this reason, nivolumab (3 mg/kg doses every 2 weeks for 36 weeks) may be used instead of temozolomide to treat CMMRD-related glioblastoma, leading to a 60% reduction in tumor size, improved clinical symptoms, and an ongoing durable response.

Radiotherapy is useful for treating some of tumors found in CMMRD patients. Colectomy may be recommended for CMMRD patients with high-grade dysplasia or large number of polyps that are difficult to be excised endoscopically. Allogeneic hematopoietic stem cell transplantation (HSCT) may be considered for selective CMMRD patients with AML.

Several types of agents have shown promise for treating CMMRD-related tumors. These include anti-inflammatory agents (e.g., aspirin and ibuprofen), tumor-maturing agents (e.g., retinoids), and checkpoint inhibitors.

Aspirin and ibuprofen are nonsteroidal anti-inflammatory drugs with the ability to inhibit cyclooxygenase (COX) enzymes and decrease prostaglandin synthesis. Through inhibition of prostaglandins that regulate apoptosis, angiogenesis, and tumor-cell invasiveness, aspirin may reduce the numbers of tumor-infiltrating lymphocytes, overcome suppression of T cell mediated antitumor immunity, and halt cancer development CMMRD.

Because CMMRD tumors are often of hypermutation phenotypes, they are sensitive to immune checkpoint inhibitors, which block the actions of proteins (e.g., PD1 and PD-L1) that impede the immune response to cancer, and thus enhance host anticancer T cell response and antitumor activity [26–29].

In addition, CMMRD tumors often contain mutated microsatellites in gene-encoding regions, leading to frameshifts and subsequent production of truncated and functionally inactive proteins that may be further processed into mutanome-derived epitopes (so called neoantigens). Increased amounts of neoantigens in CMMRD tumors render them readily recognized by cytotoxic T lymphocytes. Development of neoantigen-loaded cell vaccinations therefore offers another valuable approach for countering CMMRD tumors [30–32].

14.8 Prognosis and Prevention

CMMRD due to biallelic mutations in mismatch repair genes is associated with early onset tumors and has a much worse prognosis than Lynch syndrome that is caused by monoallelic mutations. In one study, patients with CMMRD were shown to have a median survival of 27 months after diagnosing the first malignancy (which was the cause of death in 45% of cases) and a 10-year overall survival rate of 39.5%.

The nature of mismatch repair gene mutations also plays a role in the development of tumor spectrum and the outcome of CMMRD. For instance, patients with MSH6 and/or PMS2 mutations tend to have relatively lenient brain tumors within 10 years of life, with >40% of patients developing second primary tumors. On the other hand, patients with MLH1/MSH2 mutations often develop aggressive hematological malignancies, with only 22% of patients developing second primary tumors (due to the fact that fewer patients with MLH1/MSH2 mutations survive their first malignancy).

Apart from the use of aspirin and ibuprofen as preventive agents, screening family members for heterozygous (59% chance in siblings) and homozygous (25% chance in siblings) mutations in peripheral blood DNA, and implementation of appropriate surveillance program (consisting of MRI brain, whole body MRI, complete blood count, abdominal ultrasound, upper gastrointestinal endoscopy, visual capsule endoscopy, ileocolonoscopy, gynecologic exam, transvaginal ultrasound, pipelle curettage, urine cytology, and dipstick) provide the opportunity to earlier recognition of asymptomatic tumors and improved treatment outcome [23,24].

14.9 Conclusion

Constitutional mismatch repair deficiency syndrome (CMMRD) is a rare autosomal recessive cancer predisposition syndrome characterized by (i) cutaneous features (particularly café-au-lait spots) reminiscent of NF1, and (ii) early-onset tumors (including leukemia/lymphoma, brain tumors, Lynch syndrome–associated tumors, and other malignancies). Since its first recognition in 1999, biallelic germline mutations in the mismatch repair genes *MLH1*, *MSH2*, *MSH6*, and *PMS2* have been shown to underscore the molecular pathogenesis of CMMRD. Compared to Lynch syndrome (caused by monoallelic mutations in the mismatch repair genes) showing increased tumor risk during adulthood, CMMRD is noted for the development of pediatric tumors that demonstrate extreme lethality. Given its clinical overlapping with other cancer predisposition syndromes caused by altered mismatch repair genes as well as other genes [33], use of molecular genetic techniques is essential to achieve accurate and early diagnosis of CMMRD. As current management of CMMRD focusing on symptomatic relief and regular surveillance is complex, costly, and often ineffective, development of innovative, one-for-all therapies at the molecular level is desirable and also keenly sought.

REFERENCES

1. Wimmer K, Kratz CP. Constitutional mismatch repair-deficiency syndrome. *Haematologica*. 2010;95(5):699–701.
2. Carethers JM. Hereditary, sporadic, and metastatic colorectal cancer are commonly driven by specific spectrums of defective DNA mismatch repair components. *Trans Am Clin Climatol Assoc*. 2016;127:81–97.
3. Ramachandra C, Challa VR, Shetty R. Constitutional mismatch repair deficiency syndrome: Do we know it? *Indian J Hum Genet*. 2014;20(2):192–4.

4. Kasi PM. Mutational burden on circulating cell-free tumor-DNA testing as a surrogate marker of mismatch repair deficiency or microsatellite instability in patients with colorectal cancers. *J Gastrointest Oncol.* 2017;8(4):747–8.

5. Suzuki S, Iwaizumi M, Yamada H et al. MBD4 frameshift mutation caused by DNA mismatch repair deficiency enhances cytotoxicity by trifluridine, an active antitumor agent of TAS-102, in colorectal cancer cells. *Oncotarget.* 2017;9(14):11477–88.

6. Adam R, Spier I, Zhao B et al. Exome sequencing identifies biallelic MSH3 germline mutations as a recessive subtype of colorectal adenomatous polyposis. *Am J Hum Genet.* 2016;99(2):337–51.

7. Lindsay H, Jubran RF, Wang L, Kipp BR, May WA. Simultaneous colonic adenocarcinoma and medulloblastoma in a 12-year-old with biallelic deletions in PMS2. *J Pediatr.* 2013;163(2):601–3.

8. Mork ME, Borras E, Taggart MW et al. Identification of a novel PMS2 alteration c.505C>G (R169G) in trans with a PMS2 pathogenic mutation in a patient with constitutional mismatch repair deficiency. *Fam Cancer.* 2016;15(4):587–91.

9. Ramchander NC, Ryan NA, Crosbie EJ, Evans DG. Homozygous germ-line mutation of the PMS2 mismatch repair gene: A unique case report of constitutional mismatch repair deficiency (CMMRD). *BMC Med Genet.* 2017;18(1):40.

10. Ripperger T, Beger C, Rahner N et al. Constitutional mismatch repair deficiency and childhood leukemia/lymphoma—report on a novel biallelic MSH6 mutation. *Haematologica.* 2010;95(5):841–4.

11. Rengifo-Cam W, Jasperson K, Garrido-Laguna I et al. A 30-year-old man with three primary malignancies: A case of constitutional mismatch repair deficiency. *ACG Case Rep J.* 2017;4:e34.

12. Baas AF, Gabbett M, Rimac M et al. Agenesis of the corpus callosum and gray matter heterotopia in three patients with constitutional mismatch repair deficiency syndrome. *Eur J Hum Genet.* 2013;21(1):55–61.

13. Polubothu S, Scott RH, Vabres P, Kinsler VA. Atypical dermal melanocytosis: A diagnostic clue in constitutional mismatch repair deficiency syndrome. *Br J Dermatol.* 2017;177(5):e185–6.

14. Alexander TB, McGee RB, Kaye EC et al. Metachronous T-lymphoblastic lymphoma and Burkitt lymphoma in a child with constitutional mismatch repair deficiency syndrome. *Pediatr Blood Cancer.* 2016;63(8):1454–6.

15. Hashmi AA, Ali R, Hussain ZF et al. Mismatch repair deficiency screening in colorectal carcinoma by a four-antibody immunohistochemical panel in Pakistani population and its correlation with histopathological parameters. *World J Surg Oncol.* 2017;15(1):116.

16. Ricker CN, Hanna DL, Peng C et al. DNA mismatch repair deficiency and hereditary syndromes in Latino patients with colorectal cancer. *Cancer.* 2017;123(19):3732–43.

17. Zhang Y, Sun Z, Mao X et al. Impact of mismatch-repair deficiency on the colorectal cancer immune microenvironment. *Oncotarget.* 2017;8(49):85526–36.

18. Modi MB, Patel PN, Modi VM et al. First reported case of alveolar soft part sarcoma in constitutional mismatch repair deficiency syndrome tumor spectrum—diagnosed in one of the siblings with constitutional mismatch repair deficiency. *South Asian J Cancer.* 2017;6(1):41–3.

19. Silva VW, Askan G, Daniel TD et al. Biliary carcinomas: Pathology and the role of DNA mismatch repair deficiency. *Chin Clin Oncol.* 2016;5(5):62.

20. Lavoine N, Colas C, Muleris M et al. Constitutional mismatch repair deficiency syndrome: Clinical description in a French cohort. *J Med Genet.* 2015;52(11):770–8.

21. Lowenthal BM, Chan TW, Thorson JA, Kelly KJ, Savides TJ, Valasek MA. Gastric medullary carcinoma with sporadic mismatch repair deficiency and a TP53 R273C mutation: An unusual case with wild-type BRAF. *Case Rep Pathol.* 2017;2017:3427343.

22. Takehara Y, Nagasaka T, Nyuya A et al. Accuracy of four mononucleotide-repeat markers for the identification of DNA mismatch-repair deficiency in solid tumors. *J Transl Med.* 2018;16(1):5.

23. Wimmer K, Kratz CP, Vasen HF et al. Diagnostic criteria for constitutional mismatch repair deficiency syndrome: Suggestions of the European consortium "care for CMMRD" (C4CMMRD). *J Med Genet.* 2014;51(6):355–65.

24. Tabori U, Hansford JR, Achatz MI et al. Clinical management and tumor surveillance recommendations of inherited mismatch repair deficiency in childhood. *Clin Cancer Res.* 2017;23(11):e32–7.

25. Duthie K, Bond K, Kaunelis D. *DNA Mismatch Repair Deficiency Tumour Testing for Patients with Colorectal Cancer: Ethical Issues [Internet].* Ottawa (ON): Canadian Agency for Drugs and Technologies in Health; 2016.

26. Lee V, Murphy A, Le DT, Diaz LA Jr. Mismatch repair deficiency and response to immune checkpoint blockade. *Oncologist.* 2016;21(10):1200–11.

27. Kim ST, Klempner SJ, Park SH et al. Correlating programmed death ligand 1 (PD-L1) expression, mismatch repair deficiency, and outcomes across tumor types: Implications for immunotherapy. *Oncotarget.* 2017;8(44):77415–23.

28. Le DT, Durham JN, Smith KN et al. Mismatch repair deficiency predicts response of solid tumors to PD-1 blockade. *Science.* 2017;357(6349):409–13.

29. Viale G, Trapani D, Curigliano G. Mismatch repair deficiency as a predictive biomarker for immunotherapy efficacy. *Biomed Res Int.* 2017;2017:4719194.

30. Bupathi M, Wu C. Biomarkers for immune therapy in colorectal cancer: Mismatch-repair deficiency and others. *J Gastrointest Oncol.* 2016;7(5):713–20.

31. Westdorp H, Kolders S, Hoogerbrugge N, de Vries IJM, Jongmans MCJ, Schreibelt G. Immunotherapy holds the key to cancer treatment and prevention in constitutional mismatch repair deficiency (CMMRD) syndrome. *Cancer Lett.* 2017;403:159–64.

32. Abedalthagafi M. Constitutional mismatch repair-deficiency: Current problems and emerging therapeutic strategies. *Oncotarget.* 2018;9(83):35458–69.

33. Michaeli O, Tabori U. Pediatric high grade gliomas in the context of cancer predisposition syndromes. *J Korean Neurosurg Soc.* 2018;61(3):319–32.

15

Cronkhite–Canada Syndrome

Dongyou Liu

CONTENTS

15.1 Introduction

Cronkhite–Canada syndrome (CCS) is a rare sporadic/non-inherited disorder with an adult-age onset. Clinically, CCS is characterized by epithelial disturbances in both the gastrointestinal tract and epidermis, ranging from gastrointestinal hamartomas, juvenile type gastrointestinal polyps (often associated with diarrhea), colorectal serrated adenomas, ectodermal dysplasia, skin hyperpigmentation (darkening skin on the hands, arms, palms, soles, neck, and face), alopecia (hair loss), onychodystrophy (nail atrophy), peripheral edema (excess fluid accumulation in arms and legs), to malabsorption. This is accompanied by increased malignant transformation (with gastrointestinal cancer and multiple myeloma or Kahler's disease occurring in up to 15% of cases) and unprovoked thromboembolism, leading to a high mortality of 55% [1,2].

15.2 Biology

Without an obvious familial predisposition, CCS is thought to have an autoimmune origin. This is underscored by the infiltration of inflammatory cells and expression of autoimmune-related IgG4 antibody in polyps and sera from CCS-affected individuals, along with a positive response in 90% of CCS patients to immunosuppressive treatments (e.g., corticosteroids).

In their initial report in 1955, Leonard Wolsey Cronkhite and Wilma Jeanne Canada described an unusual fatal syndrome affecting two females (42 and 75 years old). The patients experienced loss of hair, eyebrows, and axillar hair; diffuse brown discoloration of the face, neck, and hands; atrophic tongue with brown discoloration; and onychodystrophy. This was followed a few weeks later by diarrhea, nausea, vomiting, and abdominal pain. Laboratory examination revealed the presence of anemia while gastric and colonic histology showed benign adenomatous polyposis [3]. Subsequent studies by others extended clinical profiles of CCS, including protein-losing enteropathy with electrolyte disturbances (hypocalcemia, hypomagnesemia, and hypokalemia) and formation of nonadenomatous cystic polyps [4,5].

Despite consistent appearance of gastrointestinal polyposis and hyperpigmentation, CCS is notably variable in its initial symptoms, which have been exploited for its differentiation into five types: type 1 has diarrhea (35.4%); type 2 shows taste/gustatory abnormalities (dysgeusia, 40.9%); type 3 is dominated by abnormal sensation in the mouth with thirst (e.g., abnormal salt and sweet taste, 6.4%); type 4 includes hair loss (alopecia) and nail atrophy (9.1%); and type 5 displays abdominal symptoms (e.g., loss of appetite in the absence of diarrhea), malaise, dysgeusia, nail atrophy, and hair loss (8.2%) [6].

15.3 Pathogenesis

Being a sporadic, non-congenital disease, CCS appears to be linked to autoimmune dysfunction, infection, and allergic response [1].

First, CCS polyps often contain elevated antinuclear antibody (ANA) and IgG4 levels. Some sporadic juvenile CCS polyps are infiltrated by IgG4 plasma cells, which are typically observed in autoimmune disorders such as autoimmune pancreatitis, sclerosing cholangitis, and retroperitoneal fibrosis. Therefore, a proportion of CCS can be attributed to innate immune response enhanced by B cell activating factor (BAFF) and IL-13 that leads to the infiltration of IgG4 producing B cells. In addition, there is a clear association between CCS and hypothyroidism and various other autoimmune diseases (e.g., membranous glomerulonephritis, systemic lupus erythematosus, rheumatoid arthritis, and scleroderma). Moreover, skin and nail changes in CCS resemble those in autoimmune polyendocrinopathy–candidiasis–ectodermal dystrophy (APECED) syndrome. The fact that CCS manifestations are notably alleviated or lessened by immunosuppression agents (e.g., corticosteroids or long-term azathioprine) provides further evidence of an inflammatory cause of CCS.

Second, CCS patients tend to have inflammatory cell infiltration with mononuclear cells and eosinophils. Some CCS individuals may suffer combined infection with two or more pathogens (e.g., *Helicobacter pylori*, *Ascaris*).

Third, allergies to hair dye and topical medications tend to increase IgE and eosinophil levels and exacerbate clinical symptoms in CCS patients. Stresses (e.g., excessive physical exertion and mental strain) may also cause local inflammation in the gastrointestinal mucosa and trigger CCS.

15.4 Epidemiology

CCS is a rare, noninherited disorder affecting approximately 1 per million, with about 500 cases reported to date. It shows a mean age of onset at 60 years (range 31–85 years) and typically takes 3 months to 1 year from onset to diagnosis. There is a slight male predominance (male to female ratio of 3:2) among patients. Although all ethnic groups are susceptible, people of Asian (particularly Japanese, who account for 75% of documented cases) or European descent are more often affected [7].

15.5 Clinical Features

Clinically, CCS is a rare non-congenital disease with epithelial disturbances in both the gastrointestinal tract and epidermis, including diffuse multiple polyposis (in the stomach, small and large intestines, but not esophagus), diarrhea (several times per day), gustatory loss (dysgeusia/hypogeusia due to mucositis, oral infections, other abnormalities of mucosal surface, zinc and copper deficiencies), abdominal pain, weight loss (often >10 kg), cutaneous hyperpigmentation (e.g., brownish patches with a clear boundary and colored spots on the limbs, face, body, palms, and soles of the feet), nail atrophy (or onychodystrophy, such as thinning, splitting, and onycholysis), hair loss (or alopecia, on the scalp, eyebrows, eyelashes, axilla, pubic areas, and limbs, due potentially to shrinkage or atrophy of hair follicles), protein-losing enteropathy, edema, anemia, inflammation of tongue (glossitis), difficulty in swallowing (dysphagia), and dry mouth (xerostomia) [8,9]. Other comorbidities range from vitiligo, systemic lupus erythematosus, scleroderma, hypothyroidism, to membranous glomerulonephritis. As a consequence, gastrointestinal malignancies (primarily affecting the colon and less frequently the stomach) can develop in up to 15% of CCS patients [10–12].

Recently, Zong et al. [12] reported a case of CCS involving a 55-year-old Chinese male who presented with a 3-month history of frequent watery diarrhea (10–15 times per day), loss of taste, and weight loss (10 kg) in addition to a left heel bone fracture that had occurred about 2 weeks before the onset of diarrhea. Physical examination revealed marked alopecia, brownish macular pigmentation over the palms and soles, and onychodystrophy of the fingernails and toenails. Laboratory testing of blood specimen uncovered suboptimal levels of serum total protein, albumin,

hemoglobin, total cholesterol, serum calcium. Colonoscopy identified numerous polyps in the colonic and rectal mucosa, while gastroscopy detected multiple polyps in the stomach and duodenum. Histological examination of biopsy specimens from the colon and the stomach revealed adenomatous and inflammatory polyps. These findings supported a diagnosis of CCS. After receiving oral prednisone (20 mg/day), oral calcium carbonate (1.5 g/day), and vitamin D3 (125 U/day) for 3 months, the patient appeared to be free of clinical symptoms, and the number of small polyps in the colon was also reduced. In the subsequent 4 months, the patient was given prednisone at 10 mg/day. This led to the reemergence of clinical symptoms (e.g., atrophic fingernails, mild abdominal discomfort without diarrhea, and recurrence of the polyps). Following readministration with prednisone at 20 mg/day, his symptoms soon subsided, although the polyp in the sigmoid colon was found to be cancerated, which required surgery. Since then, the patient was treated with prednisone (5 mg/day) for 5 years and appeared to be fully recovered.

15.6 Diagnosis

CCS is a rare, non-hereditary gastrointestinal polyposis syndrome showing the cutaneous triad of alopecia, nail changes, and hyperpigmentation. Patients (especially of Japanese descent) with unexplained chronic diarrhea and ectodermal abnormalities should undergo medical history review, physical examination, endoscopy, and histopathology to confirm the diagnosis of CCS.

Physical examination reveals ectodermal abnormalities (e.g., cutaneous hyperpigmentation, dystrophic changes of fingernails as well as toenails, and alopecia), diarrhea, weight loss, and abdominal pain. Cutaneous hyperpigmentation may occur secondary to profound malabsorption, usually after (and occasionally before) the onset of gastrointestinal symptoms (e.g., diarrhea, protein-losing enteropathy). An improvement in diarrhea often helps lessen hyperpigmentation.

Endoscopy (colonoscopy) uncovers multiple polyps (measuring 2–40 mm in size, appearing diffuse sessile or pedunculated with either smooth or rough surface) throughout the digestive tract (commonly in the stomach and colon, less commonly in the small intestine and rectum, and rarely in the esophagus) (Figure 15.1) [13]. Typical endoscopic findings consist of: (i) mucosal edema and enlarged villi; (ii) reddened mucosa; (iii) white villus or scattered white spots; (iv) flat protuberances or small polyps (herpes-like lesions at the jejunum and strawberry-like lesions at the ileum); (v) elongated villi; and (vi) atrophied villi. Gastroscopy shows red and edematous granular polyps (strawberry-like) with giant mucosal folds or thickenings (carpet-like polyposis of the stomach), which may mimic Menetrier disease in some cases (Figures 15.1 and 15.2) [1,14]. Based on the magnified observation (using single-balloon enteroscopy [SIF-Y0007], which has the outer diameter of 9.9 mm and provides a magnification of up to ×80), white villi (previously detected by video capsule enteroscopy) reveal additional morphological details: (i) irregular villus structure; (ii) scattered white spots within the tips of villi (reflecting the pathological dilatation of the lymphatic ducts); (iii) small granular structure on the tips of villi; (iv) irregular caliber of the loop-like capillaries; and (v) spotted reddening within villi (reflecting interstitial bleeding) [15].

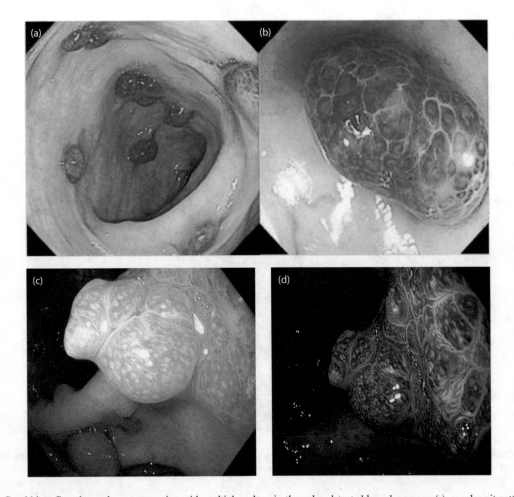

FIGURE 15.1 Cronkhite−Canada syndrome presenting with multiple polyps in the colon detected by colonoscopy (a); regular pit pattern of glandular tubes and type III−IV of microvessels in the mucous of the polyp detected by magnifying endoscopy with narrow band imaging (ME-NBI) (b); strawberry-like polyps in the stomach detected by high-resolution white light endoscopy (c) and narrow band imaging (NBI) (d). (Photo credits: Zhao R et al. *BMC Gastroenterol.* 2016;16:23; Kopáčová M et al. *Gastroenterol Res Pract.* 2013;2013:856873.)

Histopathological examination of gastric and intestinal specimens shows four types of polyps: hamartomatous (juvenile), hyperplastic, adenomatous (tubular or serrated), and inflammatory (with marked infiltration of eosinophils, lymphocytes, IgG-4 plasma cells, and scattered neutrophils) (Figure 15.3) [14]. Most of the polyps in CCS are hamartomatous and represent non-neoplastic and non-atypical ductal proliferation, with cystic dilatation, swelling of the lamina propria mucosa, and mononuclear infiltration (predominantly eosinophils). While gastric, duodenal, jejunal, and ileal polyps and some colonic polyps appear similar to juvenile/hamartomatous polyps without dysplastic changes, the remaining colonic polyps may form tubular or serrated adenomas, the latter of which resemble hyperplastic polyps in being of sessile type and possessing saw-toothed, elongated, and dilated crypts. However, hyperplastic polyps differ from serrated adenomas by displaying insignificant nuclear atypia. The gastric polyps display millet fundic glands, pyloric glandular stomach mucosa, chronic inflammation, and surface epithelial hyperplasia. Immunohistochemical staining identifies CD138-, IgG-, and IgG4-positive plasma cells in CCS polyps, and a regimen containing H2-receptor antagonists, cromolyn sodium, and loratadine permits detection of mast cells in intestinal biopsies.

Furthermore, laboratory tests may be used to detect anemia and hypoalbuminemia as well as elevated inflammatory markers ESR and CRP in CCS patients.

Differential diagnoses for CCS include other conditions that produce morphologically indistinguishable multiple hamartomatous polyps in the digestive tract (e.g., Peutz−Jeghers syndrome, juvenile polyposis syndrome, familial adenomatous polyposis, hyperplastic polyposis, Cowden syndrome, Gardner syndrome, Turcot syndrome, and Menetrier disease).

Peutz−Jeghers syndrome is an inherited disorder that develops hamartomatous polyps in the gastrointestinal tract and pigmented macules on the lips, buccal mucosa, and skin in people under 30 years of age.

Juvenile polyposis syndrome generates hamartomatous polyps containing an inflammatory component mainly in the colon before 10 years of age. Compared to juvenile polyps seen in CCS which show severe architectural changes in the surrounding mucosa, those in juvenile polyposis syndrome have normal-histological appearance in the surrounding mucosa. Further, juvenile polyposis syndrome does not have ectodermal features such as onychodystrophy and skin hyperpigmentation. In addition, juvenile polyposis syndrome is linked to mutation in BMPR1A or SMAD4 gene.

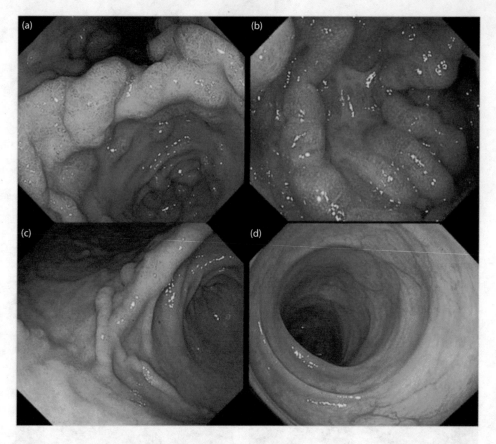

FIGURE 15.2   Cronkhite–Canada syndrome presenting with numerous polyps in the stomach (a) and colon (b), and markedly reduced numbers of gastric (c) and colon (d) polyps 8 months after cyclosporine A treatment. (Photo credit: Yamakawa K et al. *BMC Gastroenterol.* 2016;16(1):123.)

As an inherited syndrome linked to an abnormal autosomal dominant gene, adenomatous polyposis forms multiple adenomatous polyps in the colon that progress to colonic cancer in 100% of cases by the age of 50 years.

Hyperplastic polyposis produces abundant polyps in the colon, but not in the stomach or small bowel. Its diagnostic criteria consist of: (i) five or more hyperplastic polyps (two of which are over 1 cm) proximal to the sigmoid colon; (ii) any number of hyperplastic polyps proximal to the sigmoid colon in patients whose first-degree relatives have hyperplastic polyposis; or (iii) >30 hyperplastic polyps in the colon.

Cowden syndrome is an autosomal dominant disorder that displays hamartomatous polyposis and extraintestinal manifestations (facial trichilemmomas, macrocephaly, mucocutaneous lesions, acral keratoses, and thyroidal and breast diseases) (see Chapter 90 in this book). CCS differs from Cowden syndrome in having no facial mucocutaneous lesions, no association with breast hamartomas or carcinomas, no hyoid carcinomas, no PTEN mutations, and no obvious familial linkage.

15.7 Treatment

The mainstay treatment for CCS consists of corticosteroid therapy (particularly for ANA-positive patients), electrolyte rebalance, and nutritional support (including vitamins).

Approximately 90% of CCS patients respond well to corticosteroids (e.g., prednisolone, dexamethasone), while steroid-resistant CCS patients respond to cyclosporine (CyA, a calcineurin inhibitor) (Figures 15.2 and 15.3) [14]. Due to the side effects associated with the long-term use of corticosteroids, azathioprine (an immunosuppressant) instead of corticosteroids may be utilized at a later stage. Other useful therapies include infliximab (anti-TNFα therapy), and mesalazine (or 5-amino salicylic acid, an anti-inflammatory drug).

For symptomatic relief and management of complications, antibiotics (e.g., clarithromycin, amoxicillin, and albendazole for *H. pylori* and other pathogens), antihistamine receptor agonist agents (e.g., ranitidine), mast cell stabilizer (e.g., cromolyn sodium, when degranulating eosinophils and mast cells are found in biopsies), lansoprazole and sulfasalazine (for intestinal ulcer and rheumatic arthritis, respectively), and proton pump inhibitors (H2 receptor antagonists) may be considered [7].

Surgical intervention (e.g., gastrectomy, endoscopic polypectomy, polyp electrocision) is reserved for treating bowel obstruction, prolapse, severe protein-losing enteropathy, persistent hematochezia, and malignant transformation.

A combination of prednisone, proton pump inhibitors, antibiotics, 5-aminosalicylic acid, *Lactobacillus*, antacid drugs, nutritional therapy, and polypectomy over a period of 6–12 months has been shown to aid in the remission of diarrhea, skin hyperpigmentation, and atrophic nails as well as the reduction of gastrointestinal polyps.

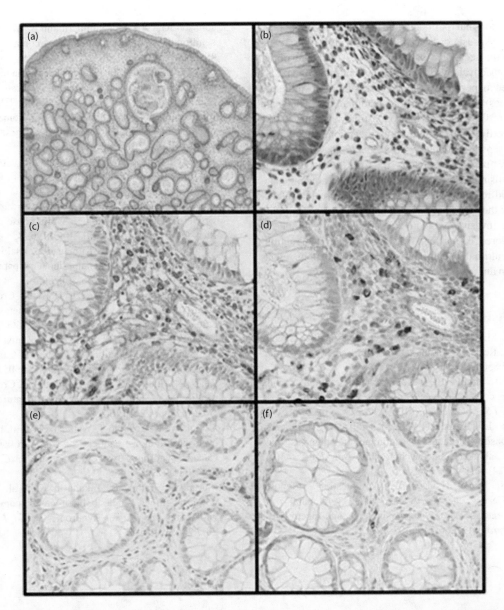

FIGURE 15.3 Cronkhite–Canada syndrome presenting with prominent cystic dilation of the crypts (a) (H&E, ×40); a mixed inflammatory infiltrate with prominent eosinophils (b) (H&E, ×400); IgG-positive plasma cells numbered 60/HPF (c) (IgG-immunostaining, ×400); a high number of IgG4-positive plasma cells in the same field (50 labeled cells per HPF) (83% IgG4/IgG plasma cells) (d) (IgG4-immunostaining, ×400); few IgG- and IgG4-positive plasma cells after treatment with CyA (e) (IgG-immunostaining, ×400), (f) (IgG4-immunostaining, ×400). (Photo credit: Yamakawa K et al. *BMC Gastroenterol.* 2016;16(1):123.)

15.8 Prognosis and Prevention

Despite its low incidence, CCS is a progressive disease showing a variable course with spontaneous regressions in 5%–10% of cases and a 5-year mortality rate of 55% due to potentially fatal complications (e.g., anemia, malnutrition, gastrointestinal bleeding, congestive heart failure, septicemia, and gastric and colonic cancers) and other comorbidities (e.g., recurrent severe acute pancreatitis, myelodysplastic syndrome, giant cell bone tumor, multiple rib fractures, cecal intussusception, schizophrenia, portal thrombosis, and membranous glomerulonephritis). Early diagnosis, appropriate treatment, and follow-up (e.g., annual endoscopic examination of the stomach, colon, and rectum) are therefore important to improve the poor prognosis of CCS.

As hair dye and topical pharmaceuticals as well as physical/mental stresses may expedite the onset or exacerbate the clinical outcome of CCS, individuals suspected of the condition should avoid use of these products, for preventive purposes.

15.9 Conclusion

CCS is a rare nonfamilial disorder characterized by epithelial disturbances in both the gastrointestinal tract and epidermis. Its clinical presentations range from diffuse polyposis of the gastrointestinal tract (with diffuse gastrointestinal inflammatory polyposis sparing the esophagus on endoscopy being a hallmark of the disease), nausea, vomiting, chronic diarrhea, protein-losing enteropathy, abdominal pain, malnutrition, weight loss, taste abnormalities (dysgeusia or ageusia), cutaneous hyperpigmentation (usually in the neck, face, palms, and soles), nail deformities (onychodystrophy/onycholysis, including loss of finger/toe nails), hair loss (alopecia), peripheral edema (excess fluid accumulation in arms and legs), to gastrointestinal malignancies. The main complications of CCS include gastrointestinal bleeding, severe cachexia, congestive heart failure, intussusception, osteoporotic fractures, and sepsis. Although CCS seems to have an autoimmune etiology, further research is required to elucidate the disease mechanism and develop more effective treatment, especially for steroid-resistant CCS. Future identification of molecular markers for the early detection of malignant transformation will permit less invasive and more cost-effective surveillance of gastric and colorectal (particularly sigmoid colon and rectum) cancer in CCS, ultimately contributing to the control of its relentless progression and high mortality.

REFERENCES

1. Kopáčová M, Urban O, Cyrany J et al. Cronkhite-Canada syndrome: Review of the literature. *Gastroenterol Res Pract.* 2013;2013:856873.
2. Nemade NL, Shukla UB, Wagholikar GD. Cronkhite Canada syndrome complicated by pulmonary embolism-A case report. *Int J Surg Case Rep.* 2017;30:17–22.
3. Cronkhite LW, Canada WJ. Generalised gastrointestinal polyposis—An unusual syndrome of polyposis, pigmentation, alopecia and onychotrophia. *New Engl J Med.* 1955;252(24):1011–5.
4. Jarnum S, Jensen H. Diffuse gastrointestinal polyposis with ectodermal changes. A case with severe malabsorption and enteric loss of plasma proteins and electrolytes. *Gastroenterology.* 1966;50(1):107–18.
5. Johnson JG, Gilbert E, Zimmermann B, Watne AL. Gardner's syndrome, colon cancer, and sarcoma. *J Surg Oncol.* 1972;4(4):354–62.
6. Goto A. Cronkhite-Canada syndrome: Epidemiological study of 110 cases reported in Japan. *Nihon Geka Hokan.* 1995;64(1):3–14.
7. She Q, Jiang JX, Si XM, Tian XY, Shi RH, Zhang GX. A severe course of Cronkhite-Canada syndrome and the review of clinical features and therapy in 49 Chinese patients. *Turk J Gastroenterol.* 2013;24(3):277–85.
8. Wen XH, Wang L, Wang YX, Qian JM. Cronkhite-Canada syndrome: Report of six cases and review of literature. *World J Gastroenterol.* 2014;20(23):7518–22.
9. Iqbal U, Chaudhary A, Karim MA, Anwar H, Merrell N. Cronkhite-Canada syndrome: A rare cause of chronic diarrhea. *Gastroenterol Res.* 2017;10(3):196–8.
10. Fuyuno Y, Moriyama T, Kumagae Y, Esaki M. Cronkhite-Canada syndrome with a major duodenal papillary adenocarcinoma. *Intern Med.* 2017;56(20):2805–7.
11. Wang J, Zhao L, Ma N, Che J, Li H, Cao B. Cronkhite-Canada syndrome associated with colon cancer metastatic to liver: A case report. *Medicine (Baltimore).* 2017;96(38):e7466.
12. Zong Y, Zhao H, Yu L, Ji M, Wu Y, Zhang S. Case report-malignant transformation in Cronkhite-Canada syndrome polyp. *Medicine (Baltimore).* 2017;96(6):e6051.
13. Zhao R, Huang M, Banafea O et al. Cronkhite-Canada syndrome: A rare case report and literature review. *BMC Gastroenterol.* 2016;16:23.
14. Yamakawa K, Yoshino T, Watanabe K et al. Effectiveness of cyclosporine as a treatment for steroid-resistant Cronkhite-Canada syndrome: two case reports. *BMC Gastroenterol.* 2016;16(1):123.
15. Murata M, Bamba S, Takahashi K et al. Application of novel magnified single balloon enteroscopy for a patient with Cronkhite-Canada syndrome. *World J Gastroenterol.* 2017;23(22):4121–6.

16

Familial Adenomatous Polyposis

Mariann Unhjem Wiik and Bente A. Talseth-Palmer

CONTENTS

16.1 Introduction

Familial adenomatous polyposis (FAP) is one of the known genetic predispositions for colorectal cancer (CRC) [1], that is characterized by the development of numerous polyps in the colorectum at an unusually young age [2]. Being responsible for <1% of all CRC, which causes almost 700.000 deaths worldwide each year, the annual incidents of FAP stand at approximately 13.000 [1,3].

The adenomatous tumor syndrome can be divided into two subtypes; classical FAP and attenuated FAP (AFAP), which are distinguished by varied clinical manifestations. In the earlier reports, Gardner syndrome (polyps in the colon and extracolonic tumors like osteomas of the skull, thyroid cancer, epidermoid cysts, fibromas, and desmoid tumors—invasive tumors of the soft tissue) and Turcot syndrome (increased risk of CRC and brain cancer) were thought to be subtypes of FAP. But these two terms should not be used in contexts other than historical ones as they are now both known to be part of the classic FAP spectrum [4–7].

FAP shows an autosomal dominant inheritance pattern, and mainly produces numerous adenomatous polyps in the colon and rectum, which will transform into CRC if left untreated [1,4]. The syndrome is caused by germline mutations in the adenomatous polyposis coli *(APC)* gene, which is a known tumor suppressor gene [1,2,8–10]. Approximately 70%–80% of FAP patients have a family history, while the remaining 20%–30% contain "de novo" mutations [1,2,11,12]. If FAP affected individuals are not detected and treated at an early stage, they will eventually develop CRC (almost in 100% of cases) (Table 16.1) [13].

TABLE 16.1

A Short Overview of FAP

Inheritance	Autosomal dominant [4,14,15]
Gene	*APC*, chromosome 5q21–22 [16]
Location of polyps and cancers	Primarily in the colorectal area, but the upper gastrointestinal tract can also be involved [7,9]
Most common extra-intestinal manifestations (classic FAP)	Thyroid cancer (2%, 5%–12%) [17–19]
	Hepatoblastoma (1%–2%) [2,4,20]
	Osteoma (20%) [4]
	Desmoid tumors (10%–30%) [21]
	Dental abnormalities (17%) [4]
	CHRPE (56%–92%) [4,22]

TABLE 16.2

Genotype–Phenotype Correlations in Classic FAP

Mutation Codon	Phenotype
At 1309	Earlier development of polyps and CRC, age of onset 20 yr [13,25,29]
From 976 to 1067	3- to 4-fold increased risk of duodenal adenomas [24,25]
From 543 to 1309	High risk of CHRPE [24]
Beyond 1309 (mainly between 1445–1580)	6-fold increased risk of developing desmoid tumors [24,25,29,54]
From 140 to 1309	Increased risk of thyroid cancer [55]
Beyond 1444	2-fold increased risk of osteoma [24,56]

16.2 Biological Background

16.2.1 FAP Subtypes

FAP consists of two subtypes, classic FAP and AFAP, both of which result from pathogenic variants of the *APC* gene, but have differing clinical features, diagnostic criteria, and surveillance programs.

16.2.2 Genetics

FAP and AFAP are usually attributable to germline mutations in the *APC* gene on chromosome 5q21–22, and over 800 pathogenic mutations have been identified between codons 156 and 2011 [7,13,23,24]. Comprising 18 coding exons, the *APC* gene has a reading frame of 8532 bp [16], which translates into a protein of 2843 aa, involved in cell migration and adhesion [25]. The most common mutation in *APC* leads to a premature stop codon, and is caused by frameshift mutations (68%), nonsense mutations (20%–30%), missense mutations (2%), or large deletions (2%) [25,26]. These mutations all lead to truncation of the protein in the C-terminal region [25,27,28]. The two mutation hotspots in the *APC* gene are codons 1061 and 1309 [13,25,29,30]. Studies have found that 17% of FAP patients have mutation in codon 1309 [31], and 11% in codon 1061 [31], and mutations in these codons account for approximately one-third of germline mutations found in FAP [28,30,32].

In ~30% of individuals with a clinical diagnosis of FAP, a mutation in *APC* cannot be detected [1,4]. Large genomic deletions have been identified in 12%–15% of these individuals, all of them with classical FAP [33,34]. The absence of an *APC* mutation in clinical diagnosed patients can also be due to differences in diagnostic criteria or use of different *APC* screening strategies [16]. However, the development and application of new genetic screening methods are helping reveal mutations in patients thought to be *APC* mutation-negative [16], which range from heterozygous mutations [35], promoter mutations [36], deep intronic mutations [37], and complex rearrangements of the gene [38] to somatic mutations and mosaicism [39–42].

16.2.3 Correlation between Genotype and Phenotype

Mutations in *APC* between codons 169 and 1393 will most likely result in classic FAP, and correlations between the position of mutation and the development of certain manifestations also exist (Table 16.2) [13,32]. There is also observed noticeable variability in disease expression within and between families with the same *APC* mutation [13,46].

FAP patients with mutations in hotspot codon 1309 have shown to express a particularly severe phenotype with a 10-year earlier presentation of bowel symptoms, and the mean age for development of symptoms being 19.8 years [25,29,46]. However, large variations have been found, with the vast range from 19 to 41 years at diagnosis of CRC [29]. FAP patients harboring mutation in codon 1309 also have, at average, 4000 colorectal polyps at time of colectomy compared with 600 polyps in other FAP patients [47]. Mutations in codon 1309 are also associated with earlier development of CRC, at a mean age of 28–35 years, but variations are also found here [24,29].

Although there are no clear lines regarding mutation site and whether a patient will develop FAP or AFAP, AFAP is associated with mutations 5′ to codon 168, 3′ to codon 1580, and in the spliced part of exon 9 [29,48,49]. Usually there are <100 colonic adenomas present, a much milder phenotype than classic FAP [50]. As with classic FAP, AFAP patients show variations in phenotype, particularly patients with 5′ mutations [49–51]. Severe cases of AFAP have been quite similar to classical FAP, and it is important to realize that although AFAP-affected patients usually have a milder phenotype with a higher age of onset, CRC are frequently found also in these families, even in patients with few polyps [51–53].

16.2.4 Modifier Alleles

There are several studies suggesting that low-penetrant susceptibility genes can affect development of sporadic CRC [57–61]. There is evidence showing that the action of modifier genes can result in differences in tumor initiation (number of adenomas) in FAP [62]. These differences are not correlated with *APC* mutations; the variations in intrafamilial pattern of colonic FAP severity seems to be consistent with the actions of the modifier genes [62–65]. Differences in tumor progression, however, do not seem to be affected by modifier alleles [62]. Several of these modifier alleles are found in the multiple intestinal neoplasia (*Min*) mouse model, with *Mom1* (modifier of Min 1) being one of the best known [16]. Another modifying allele, *pla2g2a*, has also shown to affect the growth rate of tumors but there is no evidence linking the PLA2G2A allele to tumor development in human FAP patients [64,66–68]. However, it has been shown that PLA2G2A might be important when it comes to the prevalence of fundic gland polyps in FAP patients [69]. Seven genes that are possible modifiers of *Mom5* have been identified, and recent reports suggest that a new *Xenopus tropicalis* (an aquatic tetrapod vertebrate) tumor model can be especially suitable for the identification or characterization of modifiers associated with tumor formation mediated by *APC* [70].

Approximately 40 CRC susceptibility loci have been identified through genome-wide association studies (GWAS) [71]. Each locus induces a small increase in the risk of developing CRC, but only the cumulative risk is expected to have any clinical significance [71]. Two of these single-nucleotide polymorphisms (SNPs) have been associated with adenoma number in FAP patients with a more severe phenotype [72]. Interesting results are also found in studies of the *Min* mouse model of FAP, among which strain variations in number of polyps exist [65,73]. It is also found that *atp5a1* on 18q21 can act as a modifier gene in mice (*APC^Min* mice) [74]. The mutant version of Atp5a1 in mice has shown to cause approximately 90% reduction in small intestinal and colonic polyps by suppression of polyp formation [75]. Mice carrying this mutation will also develop more carcinomas from adenomas [74].

16.3 Pathogenesis

Individuals with FAP are born with a disadvantage in the form of a germline mutation. According to Knudson's two-hit hypothesis, two mutational events are crucial for cancer development [76]. People with an inherited germline mutation already have one of these two mutations at birth and cancer can develop when a second, somatic mutation occurs [76]. In sporadic cancer, both of the mutations occur somatically, which explains the difference seen in age of onset [76].

APC plays a significant role as a tumor suppressor gene in the *Wnt* signaling pathway, which in turn regulates phosphorylation and degradation of β-catenin [16,25,77]. The protein β-catenin binds to the E-cadherin (a cell adhesion molecule) and links it to the actin cytoskeleton [25]. Not only FAP, but also the majority of sporadic cases of CRC (~80%) are associated with somatic mutations in *APC* [14,32,78].

Mutations in *APC* that are associated with cancer often cause expression of N-terminal fragments [77]. This domain of *APC* contains repeats of seven amino acids (heptad repeats) that are predicted to form coiled structures and regulate the expression of export signals involved in the transportation of *APC* between the nucleus and cytoplasm [77,79,80]. In the middle of the gene are domains that are important to the *Wnt* signaling pathway. These are three 15-residue repeats that bind to β-catenin, and seven 20-residue repeats with the same function, but these are regulated by phosphorylation [77].

16.3.1 Functions of *APC*

The APC protein has several functions, the most important of which is to work as a scaffolding protein in a complex of proteins modulated by the *Wnt* signaling pathway [81]. The protein complex regulates phosphorylation of β-catenin, which in turn affects how much β-catenin is available for transcriptional activation [77]. When an individual has a mutation in *APC*, the protein

loses its normal function causing β-catenin to accumulate in the body cells and bind to transcription factors [81]. This in turn leads to an alteration in gene expression in a number of genes affecting mechanisms such as migration, proliferation, differentiation, and apoptosis in the cells [25,82,83]. However, other pathways that lead to regulation of β-catenin do exist, and these are independent of *APC* and the *Wnt* signaling pathway [77,84,85]. This makes it even more challenging to understand exactly how a mutation in *APC* results in deregulation of β-catenin which in turn initiates the development of colon cancer.

The APC protein also has a role in controlling the cell cycle and stabilizing microtubules [25]. Mutations in the APC gene can cause defects in the structure of mitotic spindles, which have a crucial role in correct segregation of DNA in cell division, and furthermore chromosomal miss-segregation, which in turn can lead to cancer by inducing aneuploidy [86–88]. APC protein can also be found inside the nucleus and holds the information for a number of important nuclear import and export signals [80]. Potential functions for nuclear APC protein are nuclear shuttling of β-catenin and contribution in regulating transcription via its ability to directly bind DNA [89].

Thus, APC is a protein that contributes to several important processes controlling cell growth in a body cell. A normal cell found in the gut lumen has a life span of 3–5 days, while a mutation in the *APC* gene can lead to disruption of these processes involving differentiation and apoptosis, among others [25,77,82,83]. Knowing this can help us understand why a mutation in a single gene can initiate cancer development.

16.3.2 Gastric Adenocarcinoma and Proximal Polyposis of the Stomach

In 2012, another *APC*-associated polyposis condition was described: gastric adenocarcinoma and proximal polyposis of the stomach (GAPPS) [43]. Individuals with GAPPS have an increased risk of developing gastric cancer but a normal gastric antrum, pylorus, small intestine, and colon [43]. GAPPS patients have pathogenic variants in promoter 1B of *APC* [44,45]. Initially, three families were found to have pathogenic variants in the Ying Yang 1 (YY1) binding motif of promoter 1B of *APC* [44]. The reason that these variants result in gastric polyposis instead of colonic polyposis is thought to be that the *APC* 1B promoter drives expression exclusively in the stomach, while the 1A promoter will drive *APC* expression in the colon [4,31]. The penetrance is currently unknown for GAPPS [4].

There were several diagnostic criteria for GAPPS, but these were suggested prior to the identification of pathogenic mutations in promoter 1B of *APC* [4,34]. Currently, no consensus exists regarding management and surveillance of GAPPS [4,45]. GAPPS is an *APC*-associated polyposis condition but is not a part of the FAP spectrum and will therefore not be further discussed in this chapter.

16.4 Epidemiology

Worldwide, about 85% of CRC cases are sporadic, while the remaining 15% are familial. FAP accounts for <1% of the familial cases (Figure 16.1) [1].

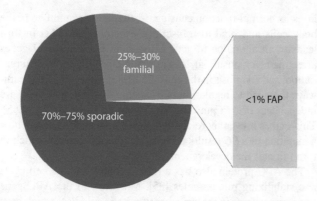

FIGURE 16.1 Approximate overview of sporadic cancers and familial contributions, including FAP.

It was estimated that approximately 14,000 new cases of FAP occurred globally in 2012 [3]. FAP is one of the most well-known and understood genetic diseases, and in many countries there are local FAP registries [3]. FAP affects both men and women equally, but considerable differences in incidence exist among various population groups, with Western and industrialized countries having the highest incidence [1,13]. In the UK the incidence is 1 per 19,000, while the incidence in the United States and other Western countries is reported to be 1 per 8000–14,000 [90,91]. Although 70%–80% of cases show a family history, the remaining 20%–30% are mutation carriers that contain de novo mutations with no clear family history [11,12,92–94].

16.5 Clinical Features

Classical FAP is characterized by the development of up to thousands of adenomatous polyps in the colon, which predisposes to adenocarcinoma (CRC) [1]. AFAP differs from classical FAP in clinical manifestation, and has different guidelines for management [4].

Classic FAP is recognized by an innumerable quantity of adenomas in the colon [1]. By definition >100 adenomatous polyps, but usually several thousands of polyps are present [7]. The age of onset is variable, but in general, the polyps start emerging in the second and third decade of life, and if left untreated, most of the affected individuals (93%) will have developed CRC by the fourth decade of life [4,16]. The number of polyps rapidly increases once they start appearing, and almost all of the affected individuals will develop colonic adenocarcinoma at some point. That is to say the penetrance is almost 100%, but variances in phenotype, both intra- and interfamilial, exist [4,13,16,95]. Development of upper gastrointestinal adenomas and adenocarcinomas in the small intestine and in the stomach may also occur [7]. In particular, the periampullary region is commonly involved in these upper gastrointestinal adenomas [7].

The attenuated form of FAP is, as the name indicates, characterized by a diminished number of adenomas [7]. On average there are about 30 polyps in the colon, but variations have been shown within families [7]. The risk for CRC is still significant, at 70% by 80 years of age [4]. It is not only the number of polyps that differs from classic FAP, but also the age of onset, which on average is 12 years later for patients with AFAP, at age 50–55 years [4,7].

16.5.1 Extracolonic Features of FAP

16.5.1.1 Small Bowel

Adenomatous polyps of the duodenum are seen in 50%–90% of individuals affected with classic FAP (Figure 16.2 and Table 16.3) [4,96]. The polyps are usually found in the second and third parts of the duodenum but can also appear in the distal small bowel with less frequency [15]. Only 5%–12% of individuals with duodenal polyps will develop duodenal cancer, but after prophylactic colectomy, duodenal desmoids and cancer are the leading causes of death in FAP patients [4,15,97]. There is no clear correlation between number of colonic polyps and number of polyps in the upper gastrointestinal tract [15].

At least 50% of the classic FAP patients develop adenomatous polyps of the periampullary region [4]. These polyps can obstruct the pancreatic duct, which in turn can lead to pancreatitis or biliary obstruction [2]. Both of these conditions are found in increased frequency in individuals with classic FAP [2].

Classic FAP patients have a lifetime risk between 5% and 12% of developing cancer in the duodenum [4]. The mean age of diagnosis is reported to be 45–52 years [15,98]. Together with thyroid cancer (lifetime risk of 2.5%–11.8%) [17,18], small bowel cancer is the most common extracolonic type of cancer seen in classic FAP patients [4].

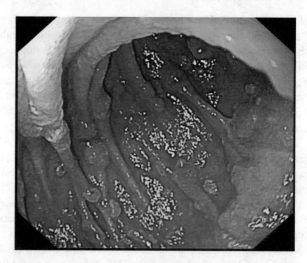

FIGURE 16.2 Duodenal adenomas in a patient with FAP [99].

TABLE 16.3

Extracolonic Manifestations of Classic FAP

Associated Findings	Benign	Malign
Desmoid tumors [21,114,115]	+	
Adenomas/carcinomas of small intestine [4,15,96]	+	+
CHRPE [4,22]	+	
Carcinoma of thyroid, pancreas, liver [2,17,18,20,100,103]		+
Adrenal masses [118,119]	+	+
Epidermoid cysts [4,25,118]	+	
CNS medulloblastoma [6,7]		+
Osteomas, fibromas, lipomas, nasopharyngeal angiofibromas, dental abnormalities [4,118]	+	
Fundic gland polyps [4,104,105]	+	+

16.5.1.2 Thyroid

Thyroid cancer is found at various frequencies in classic FAP patients, ranging from 2.5%–11.8% [17,18,100]. Female patients have a significantly higher risk of developing thyroid cancer than male patients, with a prevalence ratio of 80:1 [101]. Classic FAP thyroid cancers are usually diagnosed between 18–35 years of age, where the dominating form of carcinoma is papillary, which may have a cribriform (perforated) pattern [18]. Benign thyroid disease (e.g., thyroid nodules) is also found in classic FAP patients, but the data are limited [4,18,100].

16.5.1.3 Liver

Individuals affected with classic FAP have 750–7500 times higher risk of developing hepatoblastoma than the general population [4]. However, the absolute risk is estimated to be 1%–2%, and most of the hepatoblastomas occur during the first 5 years of life [2,4,20].

16.5.1.4 Pancreas

Data on pancreatic cancer as an extracolonic manifestation of FAP are very limited but exist [102]. A study of 197 classic FAP families did show an increased lifetime risk, which was estimated to be 1% [103].

16.5.1.5 Central Nervous System

Some classic FAP patients develop brain tumors in addition to colonic polyps [6,7]. These tumors are typically medulloblastomas, and although the risk is only 1% it is increased in FAP affected relative to the rest of the population [4].

16.5.1.6 Stomach

Hundreds of fundic gland polyps are found in the fundus and body of the stomach in 12.5%–84.0% of individuals with FAP (Figure 16.3) [4,104,105]. These are benign neoplasms which have been considered to be hamartomatous, but up to 25% of

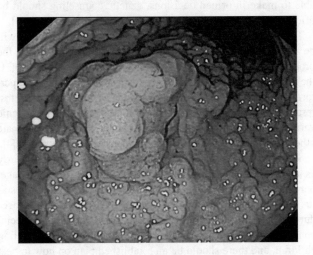

FIGURE 16.3 Fundic gland polyps found in the ventricle of an FAP patient [99].

the fundic glands found in FAP patients can show foveolar epithelial dysplasia [4,106,107]. In addition, the risk for adenomatous polyps in the gastric antrum is increased in FAP patients [98,108,109].

Gastric cancer has been reported in patients from Western countries, where fundic glands or adenomatous polyps have been detected, although the risk is low [109,110]. The lifetime risk is found to be about 0.6% in Western countries, whereas it may be higher in people of Korean or Japanese origin [4,104]. It is believed that the carcinoma arises more often from adenomatous polyps than fundic glands, but both cases have been reported [107,111,112].

16.5.1.7 Benign Extra-Intestinal Manifestations

About 20% of classic FAP patients will get osteomas [4]. These are bony growths that do not become malignant and normally do not cause clinical problems [4,113]. The osteomas are most commonly found in the skull and lower jaw but can occur everywhere in the body where there is bone tissue [4]. In children, this can be an early sign of FAP, even before adenomatous polyps develop in the colon [4].

Desmoid tumors are relatively common in classic FAP patients (10%–30%), who have 800 times greater risk than the general population [21,114,115]. The tumors are benign in themselves, but even so, they contribute to an increased mortality in patients with classic FAP [4,114]. As they grow, they can cause pressure on or obstruct vital organs like blood vessels or the bowel [21]. Possible complications are ischemia, perforation and hemorrhage, and obstruction of the ureters [116]. The tumors are usually located in the abdomen or abdominal wall (65%) and occur most frequently in the second and third decade of life (80% occurring by 40 years) [114]. Factors that can trigger the growth of desmoid tumors are surgical trauma and family history of desmoid tumors [116]. The desmoids are best evaluated by MRI or CT, the latter with a scoring system developed especially for desmoid tumors in FAP [117].

Dental abnormalities such as supernumerary teeth, absence of teeth, un-erupted teeth, odontomas, and dentigerous cysts have been found in about 17% of classic FAP patients, which is high compared to the 1%–2% observed in the general population [4].

Congenital hypertrophy of the retinal pigment epithelium (CHRPE) refers to a pigmented fundus lesion of the retina and has been seen in 56%–92% of classic FAP patients [4,22]. If multiple bilateral CHRPE is found, it can indicate that a person has classic FAP [22]. The prevalence in the general population ranges from 0.3% to 40%, but the appearance is usually one single and well-demarcated hyperpigmented lesion [22]. CHRPE is a benign abnormality but can be used as a diagnostic guide for classical FAP [22].

Benign epidermoid fibromas and cysts can be found all over the body of some patients [4]. They can cause cosmetic problems and can get infected and need surgical removal [4,118]. They usually occur at an early age [25]. In some FAP patients, there have been reported causes of nasopharyngeal angiofibromas, and they occur with a frequency 25 times higher than the general population [118].

Adrenal masses (abnormal growth in the adrenal gland) are found in 7%–13% of FAP patients, which is up to 4 times more frequent than in the general population [118,119]. Most of the adrenal masses are benign adenomas, but functional lesions and adrenal cortical carcinomas do occur [4,118]. Masses larger than 6 cm indicate removal by surgery [118].

16.5.2 Extracolonic and Extra-Intestinal Features of AFAP

Individuals with AFAP can also show some extracolonic features, but there are few reports on these [4,23]. Therefore, the extracolonic features in individuals with AFAP might be underestimated [23]. Polyps and cancers in the upper gastrointestinal tract as seen in classic FAP can also be found in patients with AFAP [4,23]. Gastric and duodenal adenomas are seen most frequently, and fundic gland polyps are described as premalignant in a few casuistic reports [23,107]. Other than that, cancers in the upper gastrointestinal tract do not seem to be prominent in AFAP [23]. CHRPE is not considered an extra-intestinal feature in AFAP individuals [23]. Also, desmoid tumors are rarely found in AFAP patients [23,51].

16.6 Diagnosis

The diagnosis of FAP is considered if an individual has clinical findings which characterize the tumor syndrome [4]. The National Comprehensive Cancer Network (NCCN) has published guidelines with criteria for consideration of FAP and AFAP [4]. According to these, both FAP and AFAP should be suspected if an individual has:

1. At least 10–20 colorectal adenomatous polyps.
2. A family history of colorectal adenomatous polyps. There are some factors that can influence the necessity of genetic counseling and/or testing, such as age of onset, number of affected family members, presence of extracolonic manifestations, and number of polyps.
3. A family history of extracolonic findings as shown in Table 16.3. This may be in addition to the other clinical findings listed above.
4. Early-onset CRC with few or no polyps can also influence the decision of offering genetic testing of the APC gene.

If an individual is shown to have a heterozygous germline pathogenic variant in the APC gene along with some of the clinical characteristic findings of *APC*-associated polyposis syndromes, the diagnosis is established.

16.6.1 Diagnosis of Classic FAP

An individual with the diagnosis of classic FAP has a heterozygous germline pathogenic variant in *APC* in addition to one of the following clinical symptoms [4]:

1. 100 or more colorectal adenomatous polyps (two exceptions are individuals who have had a colectomy and younger individuals)
2. <100 colorectal adenomatous polyps AND a relative with confirmed FAP

16.6.2 Diagnosis of AFAP

An individual with the diagnosis of AFAP has a heterozygous germline pathogenic variant in *APC* and one of the following [4]:

1. One or more relatives with confirmed AFAP
2. <100 colorectal adenomatous polyps
3. >100 colorectal adenomatous polyps after the fourth decade of life

16.6.3 Genetic Testing

Genetic testing is performed if a patient has a family history of FAP or personal clinical symptoms as described above [4]. Testing is a useful tool to differentiate between FAP, AFAP, other hereditary CRC (e.g., Lynch syndrome [HNPCC]), and early-onset sporadic CRC [23]. A molecular genetic test can confirm the diagnosis of FAP if a heterozygous germline pathogenic variant in *APC* is detected [4]. Genetic testing can be performed by testing a single gene or by using multigene panels [4]. Multigene panels include the *APC* gene *and* other genes that might be of interest to investigate due to differential diagnoses (i.e., Lynch syndrome, MUTYH-associated polyposis [MAP], Peutz–Jeghers Syndrome [PJS], juvenile polyposis syndrome, NTHL1-associated polyposis [NAP], and polymerase proofreading-associated polyposis [PPAP], to mention a few), and it is up to the clinician to determine which multigene panel to use. There are two ways of performing single-gene testing. One is sequencing of the APC gene ($\geq$90% of probands with pathogenic variant are detectable with this method), and another is deletion/duplication tests of *APC* (approximately 8%–12% of probands with pathogenic variant are detectable with this method) [4].

16.6.4 Genetic Counseling

The aim of genetic counseling is to provide individuals and families at risk of a genetic disorder with information on inheritance and implications such as risk assessment and surveillance programs [120]. This is done to help them make an informed choice of whether to accept an offer for genetic testing or not. To be able to make informed decisions, genetic counseling should be offered to the patients both pre- and post-genetic testing [121].

The proband, someone with a clinical diagnosis of FAP, will be tested first [122]. If a pathogenic variant in *APC* is found, the proband's relatives at risk can be offered predictive testing [4]. If the proband is not available for genetic testing, testing can be performed on a relative with or without clinical symptoms, but the test result must be interpreted with caution [4]. If an *APC* pathogenic variant is not detected, the test result cannot be used to eliminate other family members who might have a pathogenic mutation [4].

Surveillance programs for patients with FAP can start as early as age 10 years (age 18 years for AFAP). Due to this young age, parents tend to want to know their children's *APC* mutation status by this age to possibly avoid unnecessary screening procedures if the child does not have a pathogenic variant [4]. Special considerations need to be taken into account when counseling children, and there should be an established plan on how to best take care of these patients. However, it has been shown that most children harboring the mutation do not have more psychological distress after genetic testing than prior to it [123]. Nevertheless, long-term psychological support should be available for those who show symptoms of distress or anxiety (often found in children with mutation-positive siblings) [123].

When genetic testing shows a de novo mutation (neither of the parents have the identified variant), this is most likely due to parental mosaicism, although it could be caused by other mechanisms [4]. Mosaicism is the presence of more than one cell line with different genetic composition, and it may involve both somatic cells and germline cells [124]. However, possibilities like alternate paternity, maternity, or adoption may also be an explanation [4].

16.7 Treatment

Currently, there is no such thing as a cure for FAP, but manifestations can be treated. For individuals with an established FAP diagnosis, colectomy (removal of the colon) is recommended treatment for risk reduction of CRC [25]. A colectomy prevents the morbidity and mortality associated with FAP and CRC [2]. By screening patients with FAP, timely treatment (e.g., colectomy) can be offered, which in turn leads to improvement in survival for all individuals with FAP. Screening of FAP patients has shown to reduce development of CRC by 55% and contribute to an improvement in mortality associated with CRC [2,25]. Treatment and surveillance differ between FAP and AFAP; see Table 16.4 for a short overview of surveillance options.

16.7.1 Surveillance and Prophylaxis in FAP

16.7.1.1 Colorectum

By recommendation from the American Gastroenterological Association, sigmoidoscopy should be performed yearly from 10–12 years of age [25]. The recommendation is current for both

TABLE 16.4

Surveillance in Classic FAP and AFAP

	Classic FAP	**AFAP**
Colorectum	Yearly sigmoidoscopy from 10–12 years to 40 years. Every 3–5 years from age 40–50 if no adenomas are detected [2,4,25]	Colonoscopy every 2–3 years from age 18–20. Every 1–2 years if adenomas are detected [2,4,23]
Duodenum	Esophagogastroduodenoscopy from 25–30 years, frequency depending on polyp severity [2,134]	N/A
Desmoid tumors	Annual abdominal palpation [4]	N/A
Thyroid	Annual palpation in addition to ultrasound [17,18,25]	Annual screening, ultrasound of thyroid should also be considered even without clinical findings [4]
CNS	Annual evaluation for neurologic defects [4]	Annual evaluation for neurologic defects [4]
Liver	Palpation, ultrasound, and measurement of AFP every 3–6 months from age 0–8 years is suggested [4,133]	N/A

individuals with a confirmed molecular diagnosis of FAP and at-risk family members who have not undergone genetic testing. If adenomas are identified by sigmoidoscopy, full colonoscopy is indicated [2]. Sigmoidoscopy should be performed with 2-year intervals if the results are normal, and annually until colectomy if adenomas are detected. For high-risk members in families without a detected mutation, the recommendation is sigmoidoscopy every other year until 40 years of age [2]. If results are still normal, the intervals can be prolonged to every 3–5 years and discontinued at 50 years of age if no adenomas are detected by then.

16.7.1.1.1 Management of Colorectal Polyps

The recommended treatment for reducing the risk of CRC in FAP patients is colectomy. Two main types of colectomy exist: ileorectal anastomosis (IRA), and proctocolectomy with ileal pouch–anal anastomosis (IPAA) [2,25]. IRA is a simpler procedure compared with IPAA and is associated with fewer perioperative complications [25]. When IPAA is performed, pelvic dissection is necessary, which in turn presents the risk for hemorrhage, female reduction in fertility, and the possibility for nerve damage [2,125,126]. A study that has compared the two procedures has reported less bowel frequency and use of pads with IRA, although higher frequency in fecal urgency is reported [127]. For IPAA, reoperation within the first 30 days is more common, although rectal reoperation is more frequent after IRA (28%) than IPAA (3%) [127]. In the IRA group, rectal cancer has been observed (5%) [127].

If the patient has large numbers of polyps, IPAA is the treatment of choice [2,25]. This is because IRA has been associated with a higher risk for CRC, which again is associated with residual rectal mucosa [25,128]. The risk is estimated to 29% at 60 years of age but it is argued that, with a 15-year follow-up, the risk will be only 2%. This still makes IRA an acceptable option for some FAP patients. In patients with few or no polyps, both methods can be considered, and individual decisions can be made [2]. Also, studies have shown that mutation site at the *APC* gene is correlated with various clinical expressions, and it is recommended that genotype should impact the choice of treatment method [2,50]. However, a consensus regarding this has as of date not been reached by the European Hereditary Tumor Group (EHTG; formerly the Mallorca Group) [2].

Several other factors should be considered when deciding between IRA and IPAA, including fertility and desmoid development [2]. The risk of reduction in fertility in women with FAP is significant after IPAA compared to treatment by IRA. Therefore, if possible, IPAA should be either postponed or avoided in young women who want to have children [2]. For patients at risk of developing desmoid tumors, conversion from IRA to IPAA has been shown to be challenging. A primary IPAA may therefore be the best option for patients with a family history of desmoid tumors, or mutation beyond codon 1309 [2,25]. However, it is also argued that FAP patients with mutations distal to codon 1309 have been shown to have a milder form of polyposis, and IPAA may therefore be overtreatment [2].

As described above, the choice of surgery type depends on several factors, but the final choice should always be made by the patient after sufficient information about the diagnosis and pros and cons regarding the options [2]. There are no guidelines considering timing of surgery, but most classic FAP patients undergo surgery between 15–25 years of age [2]. Generally, there is indication for (procto)colectomy if there are a large number of adenomas [2].

16.7.1.2 Duodenum

There are five stages (O–IV) to describe the severity of duodenal polyps, also known as Spigelman classification [129]. Stage O indicates absence of polyps, stage I mild disease, while stages II–IV are associated with more severe duodenal polyposis [2]. The system is based on points, which are given relative to number of polyps, the size of polyps, histology, and the adenomas degree of dysplasia. Screening of the duodenum may contribute to detection of patients with advanced duodenal polyps (Spigelman stage III and IV) and reduction of mortality caused by duodenal cancer [2]. The most important factors contributing to duodenal polyps seem to be age and mutation site [2,4]. The risk of developing duodenal cancer among all FAP patients is relatively low at 3%–5% [2,4]. However, patients with stage III–IV duodenal polyps have a 7%–36% risk of duodenal cancer and it is therefore very important to identify these patients to give them necessary treatment [2,96]. Periampullary carcinoma is in fact the leading cause of death in patients with FAP who have had a colectomy [13].

Surveillance is done by esophagogastroduodenoscopy, and the frequency is determined by the severity of the polyps [13]. There is no consensus regarding the initiation of surveillance, but the EHTG recommends starting between 25–30 years of age, as duodenal cancer before 30 years of age has shown to be extremely rare [2]. The surveillance can also be adjusted according to changes in the Spigelman classification within the individual patient and necessary treatment initiated. A side-viewing endoscope should be used to get proper visualization of the papilla and a tissue biopsy of the papilla if it is enlarged, even if no polyps are detected [2]. In addition, endoscopic retrograde cholangiopancreatography may be necessary for the surveillance of adenomas of the common bile duct [13]. Routine small bowel screening distal to the duodenum is not recommended, as cancer rarely develops here [121].

16.7.1.2.1 Management of Duodenal Polyps

There are two options of treatment: endoscopic and surgical [2]. Several endoscopic treatments exist, including snare excision, thermal removal of tissue, argon plasma coagulation, and photodynamic therapy. The few studies that have been performed on endoscopic treatment have shown that occurrence of adenomas (50%) and complications (17%; hemorrhage, pancreatitis, and perforation) after treatment are high [2]. Of surgical options, duodenotomy (local surgery) or pancreas-sparing duodenectomy are preferable options when the pancreas is not involved [2,4]. If cancer or duodenal papilla is suspected or confirmed, pancreaticoduodenectomy (Whipple procedure) should be considered, although the procedure is associated with a higher rate of morbidity [4]. An important advantage of local surgical treatment is the postponement of major surgery in young FAP patients [2].

16.7.1.3 Desmoid Tumors

There are no established screening programs for desmoid tumors due to limited data and limitations in treatment, but an annual

abdominal palpation may be done [4]. As many as 10%–30% of classic FAP patients will develop desmoid tumors, and those who have undergone abdominal surgery, have mutation beyond codon 1309, or have a family history of desmoid tumors are especially at risk [2,21,25,97,130]. The majority of desmoid tumors in classic FAP patients are found in the abdominal or intra-abdominal wall [2]. The tumors can be detected and diagnosed by CT or MRI, and the latter can also be used to provide information about tumor activity. Surgical removal of small, well-defined tumors, which is associated with high risk of reoccurrence, can be done [131]. Intra-abdominal tumors should be treated relative to their presentation; tamoxifen and other estrogens can be used for slow-growing tumors with few symptoms, while more aggressive desmoids should be considered for treatment with chemotherapy. In addition, chemoprevention with nonsteroidal anti-inflammatory drugs (NSAIDs) may be used in patients who have had their first prophylactic surgery as a supplement to endoscopic surveillance [2]. NSAIDs might also be a temporary option in patients with mild polyposis who want to postpone surgery, as it has shown a trend to reduce number and size of polyps [132]. However, the use of NSAIDS is controversial, as cardiovascular side effects have been reported, and it should not be considered in patients with cardiovascular risk factors [2].

16.7.1.4 Thyroid and CNS

It is reasonable to do a thyroid palpation as part of an annual routine physical examination [25]. This should start in the late teenage years [4]. In addition to this, annual thyroid ultrasound, with fine-needle aspiration if nodules are found, may also be considered, as palpation has shown to not always be enough to detect cancer [17,18]. Evaluation for neurologic defects to screen for CNS neoplasm should also be part of the annual physical examination [4].

16.7.1.5 Liver

No clear guidelines exist, but the NCCN has suggested screening for hepatoblastoma using a combination of palpation, ultrasound, and measurement of alpha-fetoprotein (AFP) every 3–6 months from age 0–8 years [4]. This surveillance program has shown to detect hepatoblastomas at an early stage in patients with Beckwith–Wiedemann syndrome (increased risk of hepatoblastoma) [133].

16.7.1.6 Osteomas

Osteomas can be removed for cosmetic reasons, but they do not represent any danger in themselves [4].

16.7.2 Surveillance and Prophylaxis in AFAP

For AFAP-affected families, the surveillance and prophylactic recommendations differ slightly from those offered to classic FAP patients. With the mean age of diagnosis being 10–15 years later than in classic FAP, there is no purpose in starting surveillance programs at the same age as classic FAP patients [2].

16.7.2.1 Colorectum

Colonoscopy is recommended every 2–3 years from 18–20 years of age for individuals affected with AFAP [2,4]. Because AFAP patients have shown to develop fewer colonic adenomas, the recommended endoscopy is colonoscopy instead of sigmoidoscopy [23]. When adenomas are detected, the frequency of the colonoscopy should increase to every 1–2 years depending on the severity of the polyps [4]. The indications for colectomy are the same as for FAP.

16.7.2.2 Extracolonic Features

AFAP patients have a lower risk of developing both desmoid tumors and hepatoblastoma than FAP patients, and therefore screening is currently not recommended [4]. As with classic FAP, ultrasound may be reasonable even without clinical findings for thyroid and CNS, as it has shown that thyroid cancer can be difficult to detect in AFAP individuals [4].

16.8 Prevention

16.8.1 Primary Manifestations

To reduce the risk of developing CRC, colectomy is recommended in FAP patients [4]. Colectomy may also be necessary in patients with AFAP, but in most cases (approximately 2/3) periodic colonoscopic polypectomy is sufficient to prevent cancer development [4]. Colectomy removes the risk of CRC, but not the risk of extracolonic cancer. If duodenal and/or ampullary polyps have severe dysplasia, are larger than one centimeter in diameter, or cause any symptoms, they should be considered for endoscopic or surgical removal to reduce the risk of adenocarcinoma in the duodenum or the periampullary region [4].

16.8.2 Agents and Circumstances That Should Be Avoided

The risk of developing desmoid tumors does increase by abdominal surgery [4]. Individuals at risk of desmoids (e.g., family history and mutations beyond codon 1309) should delay surgery if possible in order to minimize the necessity for a second surgery.

For women who have undergone total colectomy with IPAA, there is a higher risk for decreased fertility than those who have had an IRA, which should be considered when surgery is needed [4].

16.8.3 Genetic and Prenatal Testing

Genetic testing can be seen as a preventive action, as it contributes to identifying *APC* mutation−positive families so that predictive testing can be offered to relatives at risk. Predictive testing makes it possible to determine who should be offered surveillance programs, as *APC* mutation−negative individuals are not at greater risk than the general population [122].

When a pathogenic mutation in *APC* is identified, the possibility of prenatal testing and preimplantation genetic diagnosis (PGD) arises [135]. The aim of PGD is to perform gene testing on fertilized embryos, followed by implantment of an embryo that does not harbor the pathogenic mutation [136]. PGD is usually performed when there is a known mutation in one of the parents, and it is important to keep in mind that the detection of an *APC* mutation in a fetus does not predict the severity of the disease or the age of onset [4].

16.9 Conclusion and Future Perspectives

Every year approximately 14,000 humans are born with a germline mutation in *APC*, which gives rise to the diagnosis of FAP [3,10]. Classic FAP is recognized by up to thousands of polyps in the colon, with age of onset usually in the second or third decade of life but can be seen as early as in childhood [4,7,16]. For AFAP, the average number of adenomas in the colon is 30, and age of onset is 50–55 years [6]. Both patients with classic FAP (93%) and AFAP (70%) will almost always develop cancer from their colonic adenomas if they are left untreated [4,16]. Frequently seen extracolonic manifestations in patients with classic FAP are polyps in the small bowel (50%–90%), fundic gland polyps (12.5%–84.0%), CHRPE (56%–92%), and desmoid tumors (10%–30%) [4,15,21,22,96,104,105,114,115]. The same extracolonic features are seen in patients with AFAP, except for cancer in the small bowel, desmoid tumors, and CHRPE, which are rarely found in these patients [23,51]. Surveillance programs and prophylaxis differ between classic FAP and AFAP.

In ∼30% of patients with a clinical diagnosis of FAP no germline mutation is detected [1,4]. This can be due to large genomic deletions, differences in diagnostic criteria, or differences in FAP screening strategies [33,34]. The continuous development of new genetic screening methods makes it possible to detect mutations in patients thought to be *APC* mutation negative, and this work will also be important in the future [16]. Due to many *APC* mutation-negative patients, several genes have recently been discovered to cause different polyposis syndromes [16,137].

It is also suggested that epigenetic factors may contribute in the explanation of clinical FAP patients without a molecular diagnosis [16]. A recent study found a reduction in the expression of *APC* in such FAP patients, but more research is required to understand how epigenetic factors affect phenotype in polyposis syndromes [138].

The action of modifier genes has shown to be of importance to the clinical phenotype in FAP [64]. However, more research is needed to be able to identify the specific genes. It is suggested that testing candidate loci will be a good start, these including COX2, MTHFR, and DNA repair genes [64]. Identification of modifier genes in FAP will be important to the understanding of the mechanisms of colorectal tumorigenesis [64].

Acknowledgments

We would like to thank Wenche Sjursen, Lars-Fredrik Engebretsen, Trond Are Johnsen, and Tone Seim Fuglset for valuable comments on drafts of the chapter, and Vemund Paulsen for letting us republish his pictures.

REFERENCES

1. Half E, Bercovich D, Rozen P. Familial adenomatous polyposis. *Orphanet J Rare Dis.* 2009;4:22.
2. Vasen HFA, Möslein G, Alonso A et al. Guidelines for the clinical management of familial adenomatous polyposis (FAP). *Gut.* 2008;57:704–13.
3. Torre LA, Bry F, Siegel RL, Ferlay J, Lortet-Tieulent, Jemal A. Global cancer statistics, 2012. *CA Cancer J Clin.* 2015;65:87–108.
4. APC-associated Polyposis Conditions. University of Washington, 1998 (updated 2017). (Accessed October 24, 2017, at https://www.ncbi.nlm.nih.gov/books/NBK1345/.)
5. Mercier KA, Al-Jazrawe M, Poon R, Acuff Z, Alman B. A metabolomics pilot study on desmoid tumors and novel drug candidates. *Sci Rep.* 2018;8(1):584.
6. Juhn E, Khachemoune A. Gardner syndrome. *Am J Clin Dermatol.* 2010;11:117–22.
7. Bronner MP. Gastrointestinal inherited polyposis syndromes. *Mod Pathol.* 2003;16:359–65.
8. Groden J, Thliveris A, Samowitz W et al. Identification and characterization of the familial adenomatous polyposis coli gene. *Cell.* 1991;66:589–600.
9. Russo A, Catania VE, Cavallaro A et al. Molecular analysis of the APC gene in Sicilian patients with familial adenomatous polyposis (F.A.P.). *Int J Surg* 2014;12:S125–9.
10. Kinzler KW, Nilbert MC, Su LK et al. Identification of FAP locus genes from chromosome 5q21. *Science.* 1991;253:661–5.
11. Campos FG. Surgical treatment of familial adenomatous polyposis: Dilemmas and current recommendations. *World J Gastroenterol.* 2014;20:16620–9.
12. Bisgaard ML, Fenger K, Bülow S, Niebuhr E, Mohr J. Familial adenomatous polyposis (FAP): Frequency, penetrance, and mutation rate. *Hum Mutat.* 1994;3:121–5.
13. Aihara H, Kumar N, Thompson CC. Diagnosis, surveillance, and treatment strategies for familial adenomatous polyposis: Rationale and update. *Eur J Gastroenterol Hepatol.* 2014;26:255–62.
14. Macrae F, du Sart D, Nasioulas S. Familial adenomatous polyposis. *Best Pract Res Clin Gastroenterol.* 2009;23:197–207.
15. Kadmon M, Tandara A, Herfarth C. Duodenal adenomatosis in familial adenomatous polyposis coli A review of the literature and results from the Heidelberg Polyposis Register. *Int J Colorectal Dis.* 2001;16:63–75.
16. Talseth-Palmer BA. The genetic basis of colonic adenomatous polyposis syndromes. *Heredit Cancer Clin Pract* 2017;15:5.
17. Herraiz M, Barbesine G, Faquin W et al. Prevalence of thyroid cancer in familial adenomatous polyposis syndrome and the role of screening ultrasound examinations. *Clin Gastroenterol Hepatol.* 2007;5:367–73.
18. Jarrar AM, Milas M, Mitchell J et al. Screening for thyroid cancer in patients with familial adenomatous polyposis. *Ann Surg.* 2011;253:515–21.
19. Truta B, Allen BA, Conrad PG et al. Genotype and phenotype of patients with both familial adenomatous polyposis and thyroid carcinoma. *Fam Cancer.* 2003;2:95–9.
20. Maire F, Hammel P, Terris B et al. Intraductal papillary and mucinous pancreatic tumour: A new extracolonic tumour in familial adenomatous polyposis. *Gut.* 2002;51:446–9.
21. Nieuwenhuis MH, Mathus-Vliegen EM, Baeten CG et al. Evaluation of management of desmoid tumours associated with familial adenomatous polyposis in Dutch patients. *Brit J Cancer* 2011;104:37–42.
22. Chen CS, Phillips KD, Grist S et al. Congenital hypertrophy of the retinal pigment epithelium (CHRPE) in familial colorectal cancer. *Fam Cancer.* 2006;5:397–404.
23. Knudsen AL, Bisgaard ML, Bülow S. Attenuated familial adenomatous polyposis (AFAP). A review of the literature. *Fam Cancer.* 2002;2:43–55.
24. Bertario L, Russo A, Sala P et al. Multiple approach to the exploration of genotype-phenotype correlations in familial adenomatous polyposis. *J Clin Oncol.* 2003;1:1698–707.
25. Galiatsatos P, Foulkes WD. Familial adenomatous polyposis. *Am J Gastroenterol.* 2006;101:385–98.
26. Zhang S, Qin H, Lv H et al. Novel and reported APC germline mutations in Chinese patients with familial adenomatous polyposis. *Gene.* 2016;577:187–92.
27. Friedl W, Aretz S. Familial adenomatous polyposis: Experience from a study of 1164 unrelated German polyposis patients. *Heredit Cancer Clin Pract* 2005;3:95–114.
28. Béroud C, Soussi T. APC gene: Database of germline and somatic mutations in human tumors and cell Llines. *Nucleic Acids Res.* 1996;24:121–4.
29. Friedl W, Caspari R, Sengteller M et al. Can APC mutation analysis contribute to therapeutic decisions in familial adenomatous polyposis? Experience from 680 FAP families. *Gut.* 2001;48:515–21.
30. Miyoshi Y, Ando H, Nagase H et al. Germ-line mutations of the APC gene in 53 familial adenomatous polyposis patients. *Proc Natl Acad Sci USA.* 1992;89:4452–6.
31. Jass JR. Colorectal polyposes: From phenotype to diagnosis. *Pathol Res Pract.* 2008;204:431–47.
32. Fearnhead NS, Britton MP, Bodmer WF. The ABC of APC. *Hum Mol Genet.* 2001;10:721–33.
33. Sieber OM, Lamlum H, Crabtree MD et al. Whole-gene APC deletions cause classical familial adenomatous polyposis, but not attenuated polyposis or "multiple" colorectal adenomas. *Proc Natl Acad Sci USA.* 2002;99:2954–8.
34. Michils G, Tejpar S, Thoelen R et al. Large deletions of the APC gene in 15% of mutation-negative patients with classical polyposis (FAP): A Belgian study. *Hum Mutat.* 2005;25:125–34.
35. Out AA, van Minderhout IJ, van der Stoep N et al. High-resolution melting (HRM) re-analysis of a polyposis patients cohort reveals previously undetected heterozygous and mosaic APC gene mutations. *Fam Cancer.* 2015;14:247–57.
36. Rohlin A, Engwall Y, Fritzell K et al. Inactivation of promoter 1B of APC causes partial gene silencing: Evidence for a significant role of the promoter in regulation and causative of familial adenomatous polyposis. *Oncogene.* 2011;15:4977–89.
37. Spier I, Horpaopan S, Vogt S et al. Deep intronic APC mutations explain a substantial proportion of patients with familial or early-onset adenomatous polyposis. *Hum Mutat.* 2012;33:1045–50.
38. Charames GS, Ramyar L, Mitri A et al. A large novel deletion in the APC promoter region causes gene silencing and leads to classical familial adenomatous polyposis in a Manitoba Mennonite kindred. *Hum Genet.* 2008;124:535–41.
39. Spier I, Drichel D, Kerick M et al. Low-level APC mutational mosaicism is the underlying cause in a substantial fraction of unexplained colorectal adenomatous polyposis cases. *J Med Genet.* 2016;53:172–9.
40. Hes FJ, Nielsen M, Bik EC et al. Somatic APC mosaicism: an underestimated cause of polyposis coli. *Gut.* 2008;57:71–6.

41. Yamaguchi K, Komura M, Yamaguchi R et al. Detection of APC mosaicism by next-generation sequencing in an FAP patient. *J Hum Genet.* 2015;60:227–31.

42. Aretz S, Stienen D, Friedrichs N et al. Somatic APC mosaicism: A frequent cause of familial adenomatous polyposis (FAP). *Hum Mutat.* 2007;28:985–92.

43. Worthley DL, Phillips KD, Wayte N et al. Gastric adenocarcinoma and proximal polyposis of the stomach (GAPPS): A new autosomal dominant syndrome. *Gut.* 2012;61:774–9.

44. Li J, Woods SL, Healey S et al. Point mutations in exon 1B of APC reveal gastric adenocarcinoma and proximal polyposis of the stomach as a familial adenomatous polyposis variant. *Am J Hum Genet.* 2016;98:830–42.

45. Repak R, Kohoutova D, Podhola M et al. The first European family with gastric adenocarcinoma and proximal polyposis of the stomach: Case report and review of the literature. *Gastrointest Endosc.* 2016;84:718–25.

46. Wachsmannova-Matelova L, Stevurkova V, Adamcikova Z, Holec V, Zajac V. Different phenotype manifestation of familial adenomatous polyposis in families with APC mutation at codon 1309. *Neoplasma.* 2009;56:486–9.

47. Nugent KP, Philips RK, Hodgson SV et al. Phenotypic expression in familial adenomatous polyposis: partial prediction by mutation analysis. *Gut.* 1994;35:1622–3.

48. Spirio L, Olschwang S, Groden J et al. Alleles of the APC gene: An attenuated form of familial polyposis. *Cell.* 1993;75:951–7.

49. Soravia C, Berk T, Madlensky L et al. Genotype–phenotype correlations in attenuated adenomatous polyposis coli. *Am J Hum Genet.* 1998;62:1290–301.

50. Nieuwenhuis MH, Vasen HFA. Correlations between mutation site in APC and phenotype of familial adenomatous polyposis (FAP): A review of the literature. *Critl Rev Oncol Hematol* 2007;61:153–61.

51. Burt RW, Leppert MF, Slattery ML et al. Genetic testing and phenotype in a large kindred with attenuated familial adenomatous polyposis. *Gastroenterology.* 2004;127:444–51.

52. van der Luijt R, Khan PM, Vasen HFA et al. Germline mutationsin the 3 part of APC exon 15 do not result in truncated proteins andare associated with attenuated adenomatous polyposis coli. *Hum Genet.* 1996;98:727–34.

53. Young J, Simms LA, Tarish J et al. A family with attenuated familialadenomatous polyposis due to a mutation in the alternatively splicedregion of APC exon 9. *Hum Mutat.* 1998;11:450–5.

54. Gebert JF, Dupon C, Kadmon M et al. Combined molecular and clinical approaches for the identification of families with familial adenomatous polyposis coli. *Ann Surg.* 1999;229:350–61.

55. Cetta F, Montalto G, Gori M, Curia MC, Cama A, Olschwang S. Germline mutations of the APC gene in patients with familial adenomatous polyposis-associated thyroid carcinoma: Results from a European cooperative study. *J Clin Endocrinol Metab.* 2000;85:286–92.

56. Davies DR, Armstrong JG, Thakker N et al. Severe Gardner syndrome in families with mutations restricted to a specific region of the APC gene. *Am J Hum Genet.* 1995;57:1151–8.

57. Jaeger E, Webb E, Howarth K et al. Common genetic variants at the CRAC1 (HMPS) locus on chromosome 15q13.3 influence colorectal cancer risk. *Nature Genet* 2008;40:26–8.

58. Tenesa A, Farrington SM, Prendergast JGD et al. Genome-wide association scan identifies a colorectal cancer susceptibility locus on 11q23 and replicates risk loci at 8q24 and 18q21. *Nature Genet* 2008;40:631–7.

59. Houlston RS, Cheadle J, Dobbins SE et al. Meta-analysis of three genome-wide association studies identifies susceptibility loci for colorectal cancer at 1q41, 3q26.2, 12q13.13 and 20q13.33. *Nature Genet* 2010;42:973–7.

60. COGENTstudy, Houlston RS, Webb E et al. Meta-analysis of genome-wide association data identifies four new susceptibility loci for colorectal cancer. *Nature Genet* 2008;40:1426–35.

61. Tomlinson IPM, Webb E, Carvajal-Carmona L et al. A genome-wide association study identifies colorectal cancer susceptibility loci on chromosomes 10p14 and 8q23.3. *Nature Genet* 2008;40:623–30.

62. Crabtree M, Tomlinson I, Talbot I, Phillips R. Variability in the severity of colonic disease in familial adenomatous polyposis results from differences in tumour initiation rather than progression and depends relatively little on patient age. *Gut.* 2001;49:540–3.

63. Crabtree MD, Fletcher C, Churchman M et al. Analysis of candidate modifier loci for the severity of colonic familial adenomatous polyposis, with evidence for the importance of the N-acetyl transferases. *Gut.* 2004;53:271–6.

64. Crabtree MD, Tomlinson IPM, Hodgson SV, Neale K, Phillips RKS, Houlston RS. Explaining variation in familial adenomatous polyposis: Relationship between genotype and phenotype and evidence for modifier genes. *Gut.* 2002;51(3):420–3.

65. Dietrich WF, Lander ES, Smith JS et al. Genetic identification of Mom-1, a major modifier locus affecting Min-induced intestinal neoplasia in the mouse. *Cell.* 1993;75:631–9.

66. Nimmrich I, Friedl W, Kruse R et al. Loss of the PLA2G2A gene in a sporadic colorectal tumor of a patient with a PLA2G2A germline mutation and abscence of PLA2G2A germline alterations in patients with FAP. *Hum Genet.* 1997;100:345–9.

67. Tomlinson IP, Beck NE, Neale K, Bodmer WF. Variants at the secretory phospholipase A2 (PLA2G2A) locus: Analysis of associations with familial adenomatous polyposis and sporadic colorectal tumors. *Ann Hum Genet.* 1996;60:369–76.

68. Cormier RT, Dove WF. Dnmt1N/+ reduces the net growth rate and multiplicity of intestinal adenomas in C57BL/6-multiple intestinal neoplasia (Min)/+ mice independently of p53 but demonstrates strong synergy with the modifier of Min 1(AKR) resistance allele. *Cancer Res.* 2000;60:3965–70.

69. Yanaru-Fujisawa R, Matsumoto T, Kukita Y et al. Impact of Phospholipase A2 group IIa gene polymorphism on phenotypic features of patients with familial adenomatous polyposis. *Dis Colon Rectum.* 2007;50:223–31.

70. Van Nieuwenhuysen T, Naert T, Tran HT et al. TALEN-mediated apc mutation in Xenopus tropicalis phenocopies familial adenomatous polyposis. *Oncoscience* 2015;2:555–66.

71. Short E, Thomas LE, Hurley J, Jose S, Sampson JR. Inherited predisposition to colorectal cancer: towards a more complete picture. *J Medl Genet* 2015;52:791–6.

72. Ghorbanoghli Z, Nieuwenhuis MH, Houwing-Duistermaat JJ et al. Colorectal cancer risk variants at 8q23.3 and 11q23.1 are associated with disease phenotype in APC mutation carriers. *Fam Cancer.* 2016;15:563–70.

73. Su LK, Kinzler KW, Vogelstein B et al. Multiple intestinal neoplasia caused by a mutation in the murine homolog of the APC gene. *Science.* 1992;256:668–70.

74. Baran AA, Silverman KA, Zeskand J et al. The modifier of Min 2 (Mom2) locus: embryonic lethality of a mutation in the Atp5a1 gene suggests a novel mechanism of polyp suppression. *Genome Res.* 2007;17:566–76.

75. Silverman KA, Koratkar R, Siracusa LD, Buchberg AM. Identification of the modifier of Min 2 (Mom2) locus, a new mutation that influences Apc-induced intestinal neoplasia. *Genome Res.* 2002;12:88–97.

76. Knudson AGJ. Mutation and cancer: Statistical study of retinoblastoma. *Proc Natl Acad Sci USA.* 1971;68:820–3.

77. Näthke I. APC at a glance. *J Cell Sci.* 2004;117:4873–5.

78. Rowan AJ, Lamlum H, Ilyas M et al. APC mutations in sporadic colorectal tumors: A mutational "hotspot" and interdependence of the "two hits". *Proc Natl Acad Sci USA.* 2000;97:3352–7.

79. Joslyn G, Richardson DS, White R, Alber T. Dimer formation by an N-terminal coiled coil in the APC protein. *PProc Natl Acad Sci USA* 1993;90:11109–13.

80. Henderson BR, Fagotto F. The ins and outs of APC and β-catenin nuclear transport. *EMBO Rep.* 2002;3:834–9.

81. Fodde R. The APC gene in colorectal cancer. *Eur J Cancer.* 2002;38:867–71.

82. van de Wetering M, Sancho E, Verweij C et al. The β-catenin/TCF-4 complex imposes a crypt progenitor phenotype on colorectal cancer cells. *Cell.* 2002;111:241–50.

83. Chen T, Yang I, Irby R et al. Regulation of caspase expression and apoptosis by adenomatous polyposis coli. *Cancer Res.* 2003;63:4368–74.

84. Xiao J-H, Ghosn C, Hinchman C et al. Adenomatous polyposis coli (APC)-independent regulation of β-catenin degradation via a retinoid X receptor-mediated pathway. *J Biol Chem.* 2003;278:29954–62.

85. Liu J, Stevens J, Rote CA et al. Siah-1 mediates a novel β-catenin degradation pathway linking p53 to the adenomatous polyposis coli protein. *Mol Cell.* 2001;7:927–36.

86. Goshima G, Wollman R, Goodwin SS et al. Genes required for mitotic spindle assembly in Drosophila S2 cells. *Science.* 2007;316:417–21.

87. Dikovskaya D, Newton IP, Näthke IS. The adenomatous polyposis coli protein is required for the formation of robust spindles formed in CSF xenopus extracts. *Mol Biol Cell.* 2004;15:2978–91.

88. Fodde R, Kuipers J, Rosenberg C et al. Mutations in the APC tumour suppressor gene cause chromosomal instability. *Nature Cell Biol* 2001;3:433–8.

89. Deka J, Herter P, Sprenger-Haußels M et al. The APC protein binds to A/T rich DNA sequences. *Oncogene.* 1999;18:5654–61.

90. Ginsberg GG, Kochman ML, Norton ID, Gostout CJ. *Clinical Gastrointestinal Endoscopy.* Philadelphia: Elsevier Saunders; 2006.

91. Andresen PA, Gedde-Dahl jr T, Fausa O, Eide TJ, Heiberg A. Genetiske analyser ved familiær adenomatøs polypose. *Tidsskr Nor Legeforen* 2001;121:64–8.

92. Kelley SR. Surgery for the polyposis syndromes. *Current Surg Therapy: Elsevier Health Sci.* 2016;12:216–22.

93. Gayther SA, Wells D, SenGupta SB et al. Regionally clustered APC mutations are associated with a severe phenotype and occur at a high frequency in new mutation cases of adenomatous polyposis coli. *Hum Molr Genet* 1994;3:53–6.

94. Zhang Y, Lu G, Hu Q et al. A de novo germline mutation of APC for inheritable colon cancer in a Chinese family using multigene next generation sequencing. *Biochem Biophysl Res Commun* 2014;447:503–7.

95. Heiskanen I, Kellokumpu I, Järvinen H. Management of duodenal adenomas in 98 patients with familial adenomatous polyposis. *Endoscopy.* 1999;31:412–6.

96. Bülow S, Björk J, Christensen IJ et al. Duodenal adenomatosis in familial adenomatous polyposis. *Gut.* 2004;53:381–6.

97. Sturt NJH, Gallagher MC, Basset P et al. Evidence for genetic predisposition to desmoid tumours in familial adenomatous polyposis independent of the germline APC mutation. *Gut.* 2004;53.

98. Wallace MH, Phillips RK. Upper gastrointestinal disease in patients with familial adenomatous polyposis. *Brit J Surg* 1998;85:742–50.

99. Bildequiz 1/2014 2014. (Accessed February 23, 2018, at https://gastroenterologen.no/2014/05/bildequiz-12014/.)

100. Steinhagen E, Guillem JG, Chang G et al. The prevalence of thyroid cancer and benign thyroid disease in patients with familial adenomatous polyposis may be higher than previously recognized. *Clin Colorectal Cancer.* 2012;11:304–8.

101. Cetta F. FAP associated papillary thyroid carcinoma: A peculiar subtype of familial nonmedullary thyroid cancer. *Pathol Res Int* 2015;2015:309348.

102. Seket B, Saurin J-C, Scoazec J-Y, Partensky C. Carcinome pancréatique à cellules acineuses et polypose adénomateuse familiale. *Gastroentérol Clin Biol* 2003;27:818–20.

103. Giardiello FM, Offerhaus GJA, Lee DH et al. Increased risk of thyroid and pancreatic carcinoma in familial adenomatous polyposis. *Gut.* 1993;34:1394–6.

104. Burt RW. Gastric fundical gland polyps. *Gastroenterology.* 2003;125:1462–9.

105. Abraham SC, Nobukawa B, Giardiello FM, Hamilton SR, Wu T-T. Sporadic fundic gland polyps. *Am J Pathol.* 2001;158:1005–10.

106. Wu T-T, Kornacki S, Rashid A, Yardley JH, Hamilton SR. Dysplasia and dysregulation of proliferation in foveolar and surface epithelia of fundic gland polyps from patients with familial adenomatous polyposis. *Am J Surg Pathol.* 1998;22:293–8.

107. Zwick A, Munir M, Ryan CK et al. Gastric adenocarcinoma and dysplasia in fundic gland polyps of a patient with attenuated adenomatous polyposis coli. *Gastroenterology.* 1997;113:659–63.

108. Bülow S, Alm T, Fausa O, Hultcrantz R, Järvinen H, Vasen H. Duodenal adenomatosis in familial adenomatous polyposis. DAF Project Group. *Int J Colorectal Dis.* 1995;10:43–6.

109. Offerhaus GJ, Entius MM, Giardiello FM. Upper gastrointestinal polyps in familial adenomatous polyposis. *Hepatogastroenterology.* 1999;46:667–9.

110. Garrean S, Hering J, Saied A, Jani J, Espat NJ. Gastric adenocarcinoma arising from fundic gland polyps in a patient with familial adenomatous polyposis syndrome. *Am Surg.* 2008;74:79–83.

111. Hofgärtner WT, Thorp M, Ramus MW et al. Gastric adenocarcinoma associated with fundic gland polyps in a patient with attenuated familial adenomatous polyposis. *Am J Gastroenterol.* 1999;94:2275–81.

112. Attard TM, Giardiello FM, Argani P, Cuffari C. Fundic gland polyposis with high-grade dysplasia in a child with attenuated familial adenomatous polyposis and familial gastric cancer. *J Pediatr Gastroenterol Nutr.* 2001;32:215–8.

113. Bilkay U, Erdem O, Ozek C et al. Benign osteoma with Gardner syndrome: review of the literature and report of a case. *J Craniofac Surg.* 2004;15:506–9.

114. Sinha A, Tekkis PP, Gibbons DC, Phillips RK, Clark SK. Risk factors predicting desmoid occurrence in patients with familial adenomatous polyposis: a meta-analysis. *Colorectal Dis.* 2011;13:1222–9.

115. Nieuwenhuis MH, Casparie M, Mathus-Vliegen LMH, Dekkers OM, Hogendoorn PCW, Vasen HFA. A nation-wide study comparing sporadic and familial adenomatous polyposis-related desmoid-type fibromatoses. *Int J Cancer.* 2011;129:256–61.

116. Santos M, Rocha A, Martins V, Santos M. Desmoid tumours in familial adenomatous polyposis: Review of 17 patients from a Portuguese tertiary center. *J Clin Diagn Res* 2016;10.

117. Middleton SB, Clark SK, Matravers P, Katz D, Reznek R, Philips RK. Stepwise progression of familial adenomatous polyposis-associated desmoid precursor lesions demonstrated by a novel CT scoring system. *Dis Colon Rectum.* 2003;46:481–5.

118. Groen E, Roos A, Muntinghe FL et al. Extra-intestinal manifestations of familial adenomatous polyposis. *Ann Surg Oncol.* 2008;15:2439–50.

119. Smith TG, Clark SK, Katz DE, Reznek RH, Philips RK. Adrenal masses are associated with familial adenomatous polyposis. *Dis Colon Rectum.* 2000;43:1739–42.

120. Berkenstadt M, Shiloh S, Barkai G, Katznelson MB-M-, Goldman B. Perceived personal control (PPC): A new concept in measuring outcome of genetic counseling. *Am J Med Genet.* 1999;82:53–9.

121. Syngal S, Brand RE, Church JM, Giardiello FM, Hampel HL, Burt RW. ACG clinical guideline: Genetic testing and management of hereditary gastrointestinal cancer syndromes. *Am J Gastroenterol.* 2015;110:223–63.

122. Rex DK, Johnson DA, Lieberman DA, Burt RW, Sonnenberg A. Colorectal cancer prevention 2000: Screening recommendations of the American College of Gastroenterology. *Am J Gastroenterol.* 2000;95:868–77.

123. Codori AM, Zawacki KL, Petersen GM et al. Genetic testing for hereditary colorectal cancer in children: Long-term psychological effects. *Am J Med Genet.* 2003;116(A):117–28.

124. Campbell IM, Yuan B, Robberecht C et al. Parental somatic mosaicism is underrecognized and influences recurrence risk of genomic disorders. *Am J Hum Genet.* 2014;95:173–82.

125. Kartheuser A, Stangherlin P, Brandt D, Remue C, Sempoux C. Restorative proctocolectomy and ileal pouch-anal anastomosis for familial adenomatous polyposis revisited. *Fam Cancer.* 2006;5:241–60.

126. Olsen KO, Joelsson M, Laurberg S, Oresland T. Fertility after ileal pouch-anal anastomosis in women with ulcerative colitis. *Brit J Surg* 1999;86:493–5.

127. Aziz O, Athanasiou T, Fazio VW et al. Meta-analysis of observational studies of ileorectal versus ileal pouch–anal anastomosis for familial adenomatous polyposis. *Brit J Surg* 2006;93:407–17.

128. Nugent KP, Phillips RKS. Rectal cancer risk in older patients with familial adenomatous polyposis and an ileorectal anastomosis: A cause for concern. *Brit J Surg* 1992;79:1204–6.

129. Spigelman AD, Talbot IC, Williams CB, Domizio P, Phillips RKS. Upper gastrointestinal cancer in patients with familial adenomatous polyposis. *Lancet.* 1989;334:783–5.

130. Vasen HF, Bülow S, Myrhøi T et al. Decision analysis in the management of duodenal adenomatosis in familial adenomatous polyposis. *Gut.* 1997;40:716–9.

131. Guillem JG, Wood WC, Moley JF et al. ASCO/SSO review of current role of risk-reducing surgery in common hereditary cancer syndromes. *J Clin Oncol.* 2006;24:4642–60.

132. Burn J, Bishop DT, Chapman PD et al. A randomized placebo-controlled prevention trial of aspirin and/or resistant starch in young people with familial adenomatous polyposis. *Cancer Prevn Res* 2011;4:655–65.

133. Tan TY, Amor DJ. Tumour surveillance in Beckwith-Wiedemann syndrome and hemihyperplasia: A critical review of the evidence and suggested guidelines for local practice. *J Pediatr Child Health* 2006;42:486–90.

134. Brosens LAA, Keller JJ, Offerhaus GJA, Goggins M, Giardiello FM. Prevention and management of duodenal polyps in familial adenomatous polyposis. *Gut.* 2005;54: 1034–43.

135. Moutou C, Gardes N, Nicod J-C, Viville S. Strategies and outcomes of PGD of familial adenomatous polyposis. *Mol Hum Reprod.* 2006;13:95–101.

136. Ruangvutilert P, Phophong P. Preimplantation genetic diagnosis (PGD). *Thai J Obstetr Gynaecol* 2001;13:47–53.

137. Spier I, Kerick M, Drichel D et al. Exome sequencing identifies potential novel candidate genes in patients with unexplained colorectal adenomatous polyposis. *Fam Cancer.* 2016;15:281–8.

138. Aceto GM, Fantini F, De Iure S et al. Correlation between mutations and mRNA expression of APC and MUTYH genes: New insight into hereditary colorectal polyposis predisposition. *J Exp Clin Cancer Res.* 2015;34:131.

17

Familial Gastrointestinal Stromal Tumor Syndrome

Dongyou Liu

CONTENTS

17.1 Introduction

Making up ~0.5% of all gastrointestinal malignancies, gastrointestinal stromal tumor (GIST) is the most common mesenchymal neoplasm of the gastrointestinal tract that originates from the interstitial cells of Cajal (ICC). Occurring in the stomach (60%), small intestine (30%), colon–rectum (5%), greater omentum and mesentery (4%), and esophagus (1%), GIST demonstrates notable pathogenic and genetic heterogeneity that presents diagnostic, therapeutic, and prognostic challenges even for experienced clinicians.

Approximately 95% of GIST cases show no family connection, and likely evolve from sporadic events, involving the KIT proto-oncogene (*KIT*, 75%), platelet-derived growth factor receptor α (*PDGFRA*) gene (~10%), succinate dehydrogenase (*SDHA, SDHB, SDHC, SDHD*) complex (5%), *BRAF* (2%), neurofibromatosis type 1 (*NF1*) gene (1.5%), epidermal growth factor receptor (*EGFR*) gene (0.9%), and other genes. While sporadic GIST often produces solitary lesions, multiple sporadic lesions in the form of synchronous or metachronous tumors are sometimes associated with familial GIST, NF1, Carney–Stratakis syndrome, and Carney triad [1–3].

On the other hand, 5% of GIST cases demonstrate a family history of disease and result mainly from inheritance of germline *KIT, PDGFRA, NF1*, or *SDH* mutations (Table 17.1). Clinically, familial (syndromic) GIST is characterized by multiple lesions (which are histologically similar to sporadic GIST, except for expansion of the myenteric plexus Cajal cell population), hyperpigmentation, mast cell tumor, and ICC hyperplasia–associated dysphagia (whose non-neoplastic polyclonal proliferation differs from monoclonal proliferation seen in GIST) [4,5].

17.2 Biology

GIST is a stromal tumor that evolves from the ICC in the myenteric plexus of the gastrointestinal tract, which serves as a pacemaker for gastrointestinal motility. Prior to 1998, GIST was regarded as smooth muscle or neural neoplasm due to its spindle cell morphology, similar to leiomyoma, leiomyosarcoma, or schwannoma. Following the detection of *KIT* mutation and related c-KIT (CD117) positivity in mast cell and germ cell tumors from a family with multiple GIST, and subsequent identification of platelet-derived growth factor receptor α (PDGFRA) gene mutation, GIST was recognized as distinct entity from leiomyoma, leiomyosarcoma, or schwannoma [6].

Besides mutations in *KIT* and *PDGFRA* that drive downstream intracellular pathways leading to tumorigenesis, several other genes (e.g., *SDH, NF1, BRAF, EGFR, RAS, TP53*) have been implicated in a small proportion (10%–15%) of GIST cases (which lack *KIT* and *PDGFRA* mutations and are referred to as *KIT/PDGFRA*-wild type) [7,8]. However,

TABLE 17.1

Characteristics of Major Familial (Syndromic) GIST-Related Genes

Gene	Chromosome Location	Inheritance	Gene Structure	Protein	Function	Clinical Features
KIT (c-KIT proto-oncogene)	4q12	Autosomal dominant, high penetrance	54.7 kb, 21 exons	971 aa, 145 kDa	Type III tyrosine kinase receptor involved in RAS/MAPK, PI3 K/AKT/mTOR and JAK/STAT3 signaling	GIST, skin hyperpigmentation, mast cell disease, ICCH, dysphagia, acute myelogenous lukemia, piebaldism Average diagnostic age of 48 years
PDGFRA (platelet-derived growth factor receptor α)	4q12	Autosomal dominant, high penetrance	54.3 kb, 28 exons	1,089 aa, 170 kDa	Type III tyrosine kinase receptor involved in RAS/MAPK, PI3 K/AKT/mTOR and JAK/STAT3 signaling	GIST, inflammatory fibroid polyps (gastrointestinal fibrous tumors), gastrointestinal lipoma, large hands Average diagnostic age of 41–48 years
SDH (succinate dehydrogenase)	*SDHA*: 5p15.33 *SDHB*: 1p36.13 *SDHC*: 1q23.3 *SDHD*: 11q23.1	Autosomal dominant, incomplete penetrance	*SDHA*: 5.2 kb, 16 exons *SDHB*:1.7 kb, 8 exons *SDHC*: 1.8 kb, 6 exons *SDHD*: 8.9 kb, 4 exons	*SDHA*: 664 aa, 72.7 kDa *SDHB*: 280 aa, 31.6 kDa *SDHC*: 169 aa, 18.6 kDa *SDHD*: 159 aa, 17 kDa	*SDHA* converts succinate to fumarate *SDHB* tunnels electrons through the complex *SDHC/SDHD* dimer provides binding sites for ubiquinone and water	GIST, pheocromocytoma/paraganglioma, renal cell carcinoma and pituitary adenoma *SDHC* and *SDHD* show a higher risk for head and neck paraganglioma; *SDHB* has a higher risk for phaeochromocytoma/paraganglioma, renal carcinoma Average diagnostic age of 22–24 years
NF1 (neurofibromatosis type 1)	17q11.2	Autosomal dominant, complete penetrance, variable expression	280 kb, 62 exons (58 constitutive, 4 alternative)	2,818 aa, 320 kDa	Neurofibromin increases the rate of GTP hydrolysis of Ras, and acts as a tumor suppressor by reducing Ras activity	GIST, neurofibroma, ICCH, dysphagia, juvenile myelomonocytic leukemia, Watson syndrome Average diagnostic age of 49 years

familial (syndromic) GIST is mostly attributable to germline mutations in one of the following genes: *KIT*, *PDGFRA*, *SDH*, or *NF1* (Table 17.1) [9].

For successful development of GIST, initial mutation (sporadic or germline) involving one of susceptible genes in an allele requires another genetic event that alters the second wild-type allele. This is often realized with loss of heterozygosity (LOH) in chromosome 14q (as well as monosomy 14; up to 70% of GIST cases), chromosome 22q (nearly 50% of GIST cases), and less frequently other chromosomes (e.g., 1p, 9p, 10q, 11p, 13q, 15q, and 17p). Interestingly, LOH in chromosome 14q predisposes gastric location, predominantly stable karyotypes, and favorable clinical outcomes; LOH in chromosome 22q confers increased cytogenetic complexity and poor disease-free survival; LOH in chromosome 1p specifies for intestinal location, increased cytogenetic complexity, and poor clinical outcome; and LOH in chromosomes 9p, 11p, and 17p potentiate GIST malignancy.

17.3 Pathogenesis

17.3.1 KIT

Mapped to chromosome 4q12, the *KIT* gene encodes the 145 kDa receptor tyrosine kinase c-KIT, which belongs to the type III receptor tyrosine kinase family (which also includes PDGFRA, PDGFRB, macrophage colony stimulating factor receptor [CSF1R], and FL cytokine receptor [FLT3]).

Structurally, c-KIT consists of an extracellular domain (encoded by exons 8 and 9), juxtamembrane domain (encoded by exon 11), tyrosine kinase domain I, and tyrosine kinase domain II (encoded by exons 13 and 17).

Maintained in an inactive form through autoinhibition of the kinase domain, c-KIT is activated upon binding to stem cell factor (SCF, a c-KIT ligand) and undergoing homodimerization. The activated c-KIT then participates in RAS/MAPK, PI3 K/AKT/mTOR, and JAK/STAT3 signaling [10].

Gain-of-function mutations in the *KIT* gene result in the production of altered c-KIT protein, leading to SCF-binding induced activation (of the MAP kinase cascade and PI3 K/AKT pathway, leading to upregulation of transcriptional factors MYC, ELK, CREB, and FOS; or downregulation of cell cycle inhibitors and promotion of anti-apoptotic effects; exons 8 and 9; 5%–10% of cases), constitutive activation (exon 11; 65%–70%), or an active conformation to TK domains (exons 13 and 17; 1%–2%).

In sporadic GIST, *KIT* mutations cluster in exon 11 (65% of cases), exon 9 (8%), exon 13 (1%), exon 17 (1%), and exon 8 (<1%), whereas in familial GIST, *KIT* mutations occur in exon 11 (70% of cases, e.g., p.Y553 K, p.W557R, p.V559A, p.V560del, p.Q575_577delinsH, p.L576P, c.D579del, c.1756_1758delGAT, p.L756_P577InsQL), exon 17 (11%, e.g., p.D816F, p.D816Y, p.D820Y, p.D820 V, p.N822H, p.N822 K, p.N822Y, and p.Y823D, of which D820Y substitution is homologous to *PDGFRA* D846), exon 13 (e.g., p.K642E), exon 9 (3%, e.g., p.A502_Y503dup), and exon 8 (e.g., p.D419del, p.TYD417–419Y), as well as region between the transmembrane and tyrosine kinase domains (e.g., deletion of one of two consecutive valine residues) [11]. Interestingly, exon 11 mutations in familial GIST are dominated by substitutions (61%, e.g., p.W557R, p.V559A, and p.L576P) and deletions and insertion/deletions (29%), compared to the 31%/60% ratio in sporadic GIST. It is of note that germline *KIT*-mutants K559I and D816 V may cause familial mastocytosis without detectable GIST [12].

During the progression of familial GIST containing germline *KIT*-mutations, a precursor, non-neoplastic polyclonal lesion (interstitial cells of Cajal hyperplasia [ICCH]), emerges first. Similar to sporadic GIST, deletions at chromosomes 14 and 22 and loss of heterozygosity (LOH) involving the *KIT* wild-type allele then turn ICCH into overt tumor. As patients harboring germline *KIT*-mutations suffer frequent/severe gut occlusion/hemorrhage, they often develop multiple lesions.

17.3.2 PDGFRA

Also mapped to chromosome 4q12, the *PDGFRA* gene encodes a type III tyrosine kinase receptor, which upon activation through binding to all PDGF, except PDGF-DD, triggers the same pathways (e.g., MAPK, AKT, STAT1, and STAT3) as c-KIT. Thus, activating mutations in *KIT* and *PDGFRA* are responsible for similar tumors, despite their mutual exclusivity [13].

PDFRA mutations commonly affect exon 12 (juxtamembrane domain; <1% of all GIST; e.g., p.Y555C, p.V561D), exon 14 (TK domain–ATP biding region; 1% of all GIST; e.g., p.P653L, p.N659 K, and p. N659Y; the latter two predisposing a gastric location, favorable clinical outcome, and epithelioid morphology), and exon 18 (TK domain-activation loop, with germline and sporadic mutations in this exon accounting for 25% and 82% of cases, respectively; p.D842 V alone representing 65% of *PDGFRA*-related GIST or 5% of all GIST) [14,15].

Familial GIST involving *PDGFRA* germline mutations features a spectrum of tumors that include GIST, inflammatory fibroid polyps (IFP, including gastrointestinal fibrous tumors), and gastrointestinal lipoma, in addition to large hands. This highlights the role of *PDGFRA* in GIST and IFP pathogenesis despite the fact that IFP often precedes GIST (mean diagnostic ages of 40.6 and 48.1 years, respectively). However, diffuse ICCH, which commonly occurs in KIT-related familial GIST, is not found in germline *PDGFRA* mutants, supporting the classification of *PDGFRA* mutants and *KIT* mutants as distinct entities.

17.3.3 SDH

The succinate dehydrogenase gene complex (*SDHx*) comprises four subunits (*SDHA*, *SDHB*, *SDHC*, and *SDHD*, mapped to chromosomes 5p15.33, 1p36.13, 1q23.3, and 11q23.1, respectively), which are components of the citric acid cycle and respiratory electron transfer chain and play a crucial role in succinate-to-fumarate conversion in the Krebs cycle [16–18].

Mutations in any one of the *SDH* subunits affect the entire complex, and the resultant SDH deficiency prevents succinate conversion to fumarate, leading to succinate accumulation that inhibits prolyl-hydroxylase (hampering hypoxia-inducible factor [HIF]-1α degradation, and increasing oncogenic transcription) and TET DNA hydroxylases (impairing 5-methylcytosine conversion to 5-hydroxymethylcytosine required for DNA demethylation, and thus enhancing aberrant DNA hypermethylation implied in oncogenesis) [19,20].

SDH deficiency underlies the development of GIST, paraganglioma, renal cell carcinoma, and pituitary adenoma, which may occur in familial GIST, SDH-deficient paraganglioma syndrome, Carney–Stratakis syndrome (an autosomal dominant disorder characterized by gastric GIST and paraganglioma containing germline *SDHB/C/D* mutations), and Carney triad (a nonheritable disease characterized by gastric stromal sarcoma, paraganglioma, pulmonary chondroma, and female predilection) [21]. SDH-deficient GISTs are notably restricted to stomach (especially antrum) and show a multinodular pattern or plexiform [22–25].

17.3.4 NF1

The *NF1* gene on 17q11.2 encodes neurofibromin that contains a GAP-related domain (GRD) responsible for converting active Ras-GTP to inactive Ras-GDP, thus negatively regulating RAS signaling. Functioning as a tumor suppressor gene, *NF1* inactivation accelerates the conversion from active GTP-bound to inactive GDP-bound RAS and increases RAS activity in the MAPK cascade downstream to KIT and PDGFRA, without the involvement of the PI3K-AKT and JAK-STAT signal pathways.

Individuals with *NF1* mutations are at high risk of developing GIST. Similar to germline *KIT* mutations, *NF1*-related GIST begins with preneoplastic ICCH (diffuse and focal hyperplastic foci of CD117-positive ICC, often found around nerve plexuses) followed by LOH in chromosomes 14q and 22q. Interestingly, these frequently deleted regions encompass a number of functionally important genes, including *PARP2* (regulating DNA repair and apoptosis and thus suppressing genomic instability), *APEX1* (encoding a DNA repair enzyme implicated in the base excision pathway), and *NDRG2* (inhibiting tumor proliferation and promotes apoptosis) at chromosome 14q11.2; *SIVA* (encoding a pro-apoptotic protein that binds to the tumor necrosis factor receptor CD27) at chromosome 14q32.33; *MAX* (encoding a basic helix-loop-helix leucine zipper transcription factor that interacts with MYC, its heterozygous or homozygous inactivating mutations accounting for 17% of sporadic GIST and 50% of sporadic and NF-1-associated GIST) at chromosome 14q23.3; and *NF2* (encoding merlin that inhibits the activities of RAS and RAC, and suppresses tumor cell growth) at chromosome 22q12.2.

NF1-associated GIST has a younger onset age, occurs in the duodenum and small intestine, and shows small size, tumor multiplicity, spindle cell morphology, low mitotic rates, and CD117-positivity [26].

17.4 Epidemiology

17.4.1 Prevalence

GIST represents the most common mesenchymal neoplasm of the gastrointestinal tract and accounts for 0.5% of all gastrointestinal malignancies. The estimated annual incidence for GIST is 14–20 cases per million, with a mean diagnostic age of 58 years (range 18–95 years). However, patients with familial GIST have a mean diagnostic age of 48 years [27].

17.4.2 Inheritance

Familial (syndromic) GIST linked to germline *KIT/PDGFRA/ SDH/NF1* mutations is inherited in an autosomal dominant pattern (Table 17.1).

17.4.3 Penetrance

Germline *KIT/PDGFRA/SDHB/SDHD* mutations are highly penetrant, while *SDHC* and especially *SDHA* variants show low penetrance. Further, pathogenic *NF1* variants demonstrate complete penetrance and variable expression.

17.5 Clinical Features

Patients with GIST often present with nonspecific symptoms such as abdominal pain, gastrointestinal bleeding, and abdominal fullness or mass [28]. In familial (syndromic) GIST, other clinical manifestations may be observed, including skin pigmentation alterations, lentigines, vitiligo, peristalsis disturbances (due to the presence of diffuse ICCH) and mast-cell disorders in germline KIT mutants, and large hands in germline *PDGFRA* mutants.

17.6 Diagnosis

Diagnosis of GIST relies on medical history review, physical examination, imaging study, histopathology/immunohistochemistry, and molecular genetic testing [29].

Compared to sporadic GIST, familial (syndromic) GIST demonstrates a clear family history of disease.

Imaging techniques (eg, video capsule endoscopy [VCE], double-balloon–assisted enteroscopy [DBE], spiral enteroscopy, labeled red blood cell nuclear scan, esophagogastroduodenoscopy, colonoscopy, computerized tomography enterography, and angiography) are useful for determining tumor location and size and assessing the overall tumor spectrum (Figures 17.1 and 17.2) [30–34].

Histopathological examination of formalin-fixed, paraffin-embedded tissue sections stained with hematoxylin/eosin, CD117, S100, vimentin, DOG1 (delay of germination 1), CD34, and PDGFRA helps verify the tumor types (Figures 17.1 and 17.2) [35,36]. For example, SDH-deficient GIST expresses DOG1, CD34 (≥75% of cases), and CD117 (strongly/diffusely), which can be exploited for diagnostic purpose.

Sequence analysis of *KIT* (exons 9, 11, 13, and 17), *PDGFRA* (exons 12, 14, and 18), *SDH*, and *NF1* genes conducted on paraffin-embedded tissues and/or buccal swab samples provides further genetic details of tumors. Indeed, assessment of *KIT* and *PDGFRA* mutational status is vital to distinguish metastases from synchronous multicentric GIST [37–39].

Although germline *KIT/PDGFRA/SDH/NF1* mutants often demonstrate variations in distribution and pattern of GIST and other tumors (Table 17.1), which are valuable for their initial identification, further confirmation on the basis of histopathologic, immunohistochemical, and/or molecular genetic evidence is essential for definitive diagnosis of familial GIST [40].

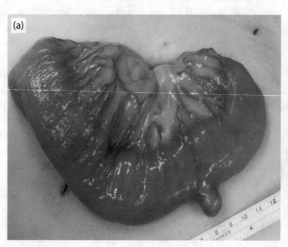
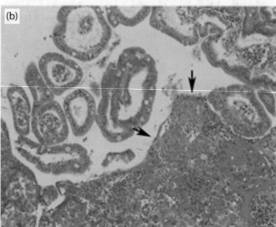

FIGURE 17.1 Familial gastrointestinal stromal tumor syndrome in a 64-year-old female presenting with a 2.5 cm tumor arising from the antimesenteric border of the proximal jejunum as revealed by laparoscopic exploration (a); an intramural gastrointestinal stromal tumor composed of fascicles of spindle cells with bland nuclei, causing ischemic erosion of the overlying mucosa (arrows) (H&E staining) (b). (Photo credit: Yuval JB et al. *BMC Res Notes*. 2014;7:695.)

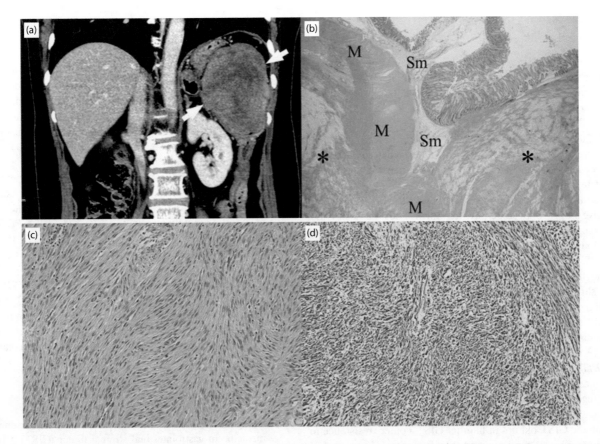

FIGURE 17.2 Familial gastrointestinal stromal tumor syndrome in a 57-year-old female presenting with (a) a large heterogenous mass (arrows) in the descending colon as revealed by coronal CT image; (b) spindle cell type tumor (*) invading submucosa (Sm) and proper muscle (M) by histopathology (H&E, ×10); (c) a highly cellular gastrointestinal stromal tumor composed of broad bundles of elongated cells, with mitotic count of 15/50 HPFs by high-power field microscopy (H&E, ×100); (d) diffusely c-kit positive tumor cells by immunohistochemistry. (Photo credit: Kang JH et al. *PLOS One*. 2016;11(12):e0166377.)

For example, familial *KIT*-mutant GIST occurs throughout the gastrointestinal tract (i.e., stomach 64.5%, small intestine 25.1%, and colorectal region 5.1%), shows a spindle, epithelioid, or spindle-and-epithelioid cytology, and expresses type III receptor tyrosine kinase (CD117, 96.4% of cases) and DOG1 (96.9% of cases) (Figure 17.3) [41,42].

Clinical criteria for diagnosis of NF1 center on finding of ≥6 café-au-lait macules >5 mm or >15 mm in prepubertal or post-pubertal subjects, respectively; ≥2 neurofibromas (one if plexiform); axillary/inguinal freckles; optic glioma; ≥2 iris

hamartomas; bony dysplasia; and a NF1-affected first-degree relative. Molecular identification of *NF1* mutation provides further diagnostic confirmation [26].

Based on tumor size and mitotic activity, the NIH consensus criteria define gastric GIST as very low risk, low risk, intermediate risk, and high risk. Furthermore, incorporating data on tumor size, mitotic rate, and presence of either nodal or distant metastasis, the 7th UICC/AJCC TNM staging system classifies gastric GIST into stages IA, IB, II, IIIA, IIIB, IIIC, and IV [41].

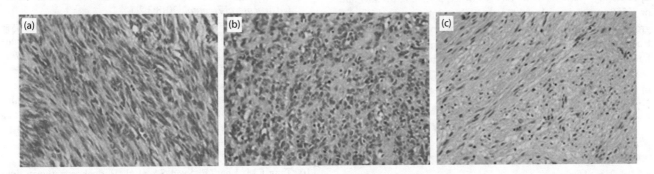

FIGURE 17.3 Familial gastrointestinal stromal tumor syndrome presenting with (a) spindle cell type tumor composed of fusiform spindle cells with a fascicular growth pattern; (b) epithelioid type tumor composed of large round or polygonal cells with abundant, often eosinophilic or clear cytoplasm; (c) mixed-type tumor composed of mixture of both types (H&E, ×400). (Photo credit: Park CH et al. *BMC Gastroenterol*. 2017;17(1):141.)

17.7 Treatment

Current treatment strategies for GIST consist mainly of surgery and/or chemotherapy (e.g., imatinib mesylate), as GIST shows a poor response to radiotherapy [43].

Surgery (e.g., wedge resection, segmental resection, and pancreaticoduodenectomy) may be considered for patients with GIST. Patients with very low- and low-risk GIST can be treated with surgery alone, while those with intermediate-high risk or advanced GIST benefit from combined surgery and imatinib mesylate therapy [44].

Imatinib mesylate is a tyrosine kinase inhibitor that proves effective for pre- or postoperative treatment of 80%–85% of GIST harboring *KIT* or *PDGFRA* (except PDGFRA exon 18 D842 V) mutations, and also for treatment of unresectable/metastatic disease [45,46]. With a tenfold more active favorable response than any agent examined for treatment of GIST, imatinib mesylate helps shrink tumor, converts patients from inoperable to operable, and dramatically reduces GIST metastases. Patients harboring *KIT* exons 8, 9, and 11 mutations are particularly susceptible to imatinib mesylate treatment, while those possessing *KIT* exon-13 and 17 mutations are somewhat resistant. For patients with *KIT* or *PDGFRA* mutations who have poor response to imatinib mesylate, alternative tyrosine kinase inhibitors (e.g., sunitinib, regorafenib, dasatinib, and crenolanib) may be considered. In fact, *PDGFRA* exon 18 mutation p.D842 V, commonly found in sporadic GIST but not in familial GIST, confers resistance to both imatinib and sunitinib, and therefore use of dasatinib and crenolanib is suggested [47–49].

For the remaining 10%–15% of GIST due to mutated *SDH*, *NF1*, and *BRAF* genes (so-called *KIT/PDGFRA* wildtype), PI3 K/AKT/mTOR inhibitors and other agents need to be explored [50].

17.8 Prognosis

Prognosis for GIST is determined by a number of factors, including tumor size, mitotic rate, risk grade, pathological type, CD34 expression, adjacent involvement, patient age and gender, etc. [51,52].

Large tumor size, high mitotic rate, and high-risk grade are unfavorable prognostic factors for GIST. GIST of >5 cm in size, >5 mitotic counts per 50 high-power fields, and epithelioid and mixed pathological type show a higher recurrence rate. Specifically, patients with KIT exon 11 deletions (W557 and/or K558) often have large tumor, high mitotic count, high risk grade, frequent metastasis, and poor postoperative recurrence-free survival. On the other hand, GIST patients of <50 years and female gender show a more favorable outcome [53].

Without imatinib mesylate treatment, GIST patients classified in the very low-, low-, intermediate- and high-risk groups have 5-year overall survival rates of 100%, 100%, 89.6%, and 65.9%, respectively.

17.9 Conclusion

Gastrointestinal stromal tumor (GIST) is a relatively uncommon malignancy of the gastrointestinal tract that affects the stomach (60%), small intestine (30%), colon–rectum (5%), greater omentum and mesentery (4%), and esophagus (1%). While sporadic GIST makes up 95% of cases that likely evolve from sporadic events without apparent family connection, familial (syndromic) GIST represent about 5% of cases due to autosomal dominant inheritance of germline mutations. Clinically, sporadic GIST typically produces solitary lesions, although it may also cause multiple sporadic lesions (so-called synchronous or metachronous tumors) in association with familial syndromes. By contrast, familial (syndromic) GIST often presents with multiple lesions, hyperpigmentation, mast cell tumor, and ICC hyperplasia-associated dysphagia in patients younger than their counterparts with sporadic GIST. Molecularly, GIST is attributed to de novo or germline mutations involving the *KIT* (75%), *PDGFRA* (~10%), *SDH* (5%), *NF1* (1.5%), and other genes. The fact that *KIT* and *PDGFRA* all encode receptor tyrosine kinases enables the use of receptor tyrosine kinase inhibitors for treating 85% of GIST cases. However, for the remaining 15% of *KIT/PDGFRA* intact cases (wild-type GIST), specific treatment is still lacking. This clearly calls for further experimental investigation into the molecular mechanisms of wild-type GIST, with the goal of identifying novel, specific therapeutic agents for improved management of GIST.

REFERENCES

1. Ravegnini G, Nannini M, Sammarini G et al. Personalized medicine in gastrointestinal stromal tumor (GIST): Clinical implications of the somatic and germline DNA analysis. *Int J Mol Sci.* 2015;16(7):15592–608.
2. Torous VF, Su A, Lu DY, Dry SM. Adult patient with synchronous gastrointestinal stromal tumor and Xp11 translocation-associated renal cell carcinoma: A unique case presentation with discussion and review of literature. *Case Rep Urol.* 2015;2015:814809.
3. Comandini D, Damiani A, Pastorino A. Synchronous GISTs associated with multiple sporadic tumors: A case report. *Drugs Context.* 2017;6:212307.
4. Ricci R. Syndromic gastrointestinal stromal tumors. *Hered Cancer Clin Pract.* 2016;14:15.
5. Mei L, Du W, Idowu M, von Mehren M, Boikos SA. Advances and challenges on management of gastrointestinal stromal tumors. *Front Oncol.* 2018;8:135.
6. Niinuma T, Suzuki H, Sugai T. Molecular characterization and pathogenesis of gastrointestinal stromal tumor. *Transl Gastroenterol Hepatol.* 2018;3:2.
7. Kelly L, Bryan K, Kim SY et al. Post-transcriptional dysregulation by miRNAs is implicated in the pathogenesis of gastrointestinal stromal tumor [GIST]. *PLOS One.* 2013;8(5):e64102.
8. Lasota J, Xi L, Coates T, Dennis R et al. No KRAS mutations found in gastrointestinal stromal tumors (GISTs): Molecular genetic study of 514 cases. *Mod Pathol.* 2013;26(11):1488–91.
9. Lynch HT, Lynch JF, Shaw TG. Hereditary gastrointestinal cancer syndromes. *Gastrointest Cancer Res.* 2011;4(4 Suppl 1):S9–17.
10. Tang CM, Lee TE, Syed SA et al. Hedgehog pathway dysregulation contributes to the pathogenesis of human gastrointestinal stromal tumors via GLI-mediated activation of KIT expression. *Oncotarget.* 2016;7(48):78226–41.

11. Huss S, Kunstlinger H, Wardelmann E et al. A subset of gastrointestinal stromal tumors previously regarded as wild-type tumors carries somatic activating mutations in KIT exon 8 (p.D419del). *Mod Pathol.* 2013;26:1004–12.

12. Sanchez-Hidalgo JM, Duran-Martinez M, Molero-Payan R et al. Gastrointestinal stromal tumors: A multidisciplinary challenge. *World J Gastroenterol.* 2018;24(18):1925–41.

13. Geramizadeh B, Jowkar Z, Mousavi SJ. Molecular and immunohistochemical study of platelet derived growth factor receptor alpha in KIT negative gastrointestinal stromal tumors; the first report from Iran. *Middle East J Dig Dis.* 2016;8(3):226–31.

14. Pantaleo MA, Astolfi A, Urbini M et al. Dystrophin deregulation is associated with tumor progression in KIT/PDGFRA mutant gastrointestinal stromal tumors. *Clin Sarcoma Res.* 2014;4:9.

15. Ricci R, Martini M, Cenci T et al. PDGFRA-mutant syndrome. *Mod Pathol.* 2015;28(7):954–64.

16. Miettinen M, Lasota J. Succinate dehydrogenase deficient gastrointestinal stromal tumors (GISTs) - a review. *Int J Biochem Cell Biol.* 2014;53:514–9.

17. Pantaleo MA, Astolfi A, Urbini M et al. Analysis of all subunits, SDHA, SDHB, SDHC, SDHD, of the succinate dehydrogenase complex in KIT/PDGFRA wild-type GIST. *Eur J Hum Genet.* 2014;22:32–9.

18. Belinsky MG, Cai KQ, Zhou Y et al. Succinate dehydrogenase deficiency in a PDGFRA mutated GIST. *BMC Cancer.* 2017;17(1):512.

19. Killian JK, Kim SY, Miettinen M et al. Succinate dehydrogenase mutation underlies global epigenomic divergence in gastrointestinal stromal tumor. *Cancer Discov.* 2013;3:648–57.

20. Mason EF, Hornick JL. Succinate dehydrogenase deficiency is associated with decreased 5-hydroxymethylcytosine production in gastrointestinal stromal tumors: Implications for mechanisms of tumorigenesis. *Mod Pathol.* 2013;26(11):1492–7.

21. Wang JH, Lasota J, Miettinen M. Succinate dehydrogenase subunit B (SDHB) is expressed in neurofibromatosis 1-associated gastrointestinal stromal tumors (Gists): Implications for the SDHB expression based classification of GISTs. *J Cancer.* 2011;2:90–3.

22. Miettinen M, Wang ZF, Sarlomo-Rikala M, Osuch C, Rutkowski P, Lasota J. Succinate dehydrogenase-deficient GISTs: A clinicopathologic, immunohistochemical, and molecular genetic study of 66 gastric GISTs with predilection to young age. *Am J Surg Pathol.* 2011;35:1712–21.

23. Italiano A, Chen CL, Sung YS et al. SDHA loss of function mutations in a subset of young adult wild-type gastrointestinal stromal tumors. *BMC Cancer.* 2012;12:408.

24. Jiang Q, Zhang Y, Zhou YH et al. A novel germline mutation in SDHA identified in a rare case of gastrointestinal stromal tumor complicated with renal cell carcinoma. *Int J Clin Exp Pathol.* 2015;8:12188–97.

25. Urbini M, Astolfi A, Indio V et al. SDHC methylation in gastrointestinal stromal tumors (GIST): A case report. *BMC Med Genet.* 2015;16:87.

26. Agaimy A, Vassos N, Croner RS. Gastrointestinal manifestations of neurofibromatosis type 1 (Recklinghausen's disease): Clinicopathological spectrum with pathogenetic considerations. *Int J Clin Exp Pathol.* 2012;5:852–62.

27. Güller U, Tarantino I, Cerny T, Schmied BM, Warschkow R. Population-based SEER trend analysis of overall and cancer-specific survival in 5138 patients with gastrointestinal stromal tumor. *BMC Cancer.* 2015;15:557.

28. Al-Maghrabi H, Meliti A. Gastric gastrointestinal stromal tumor with osseous differentiation and stromal calcification: A case report and review of literature. *SAGE Open Med Case Rep.* 2017;5:2050313X17746310.

29. Shen C, Chen H, Yin Y et al. Duodenal gastrointestinal stromal tumors: Clinicopathological characteristics, surgery, and long-term outcome. *BMC Surg.* 2015;15:98.

30. Boguszewski CL, Fighera TM, Bornschein A et al. Genetic studies in a coexistence of acromegaly, pheochromocytoma, gastrointestinal stromal tumor (GIST) and thyroid follicular adenoma. *Arq Bras Endocrinol Metabol.* 2012;56(8):507–12.

31. Stemate A, Filimon AM, Tomescu M, Negreanu L. Colon capsule endoscopy leading to gastrointestinal stromal tumor (GIST) diagnosis after colonoscopy failure. *BMC Res Notes.* 2015;8:558.

32. Kang JH, Kim SH, Kim YH, Rha SE, Hur BY, Han JK. CT features of colorectal schwannomas: Differentiation from gastrointestinal stromal tumors. *PLOS One.* 2016;11(12): e0166377.

33. Yuval JB, Almogy G, Doviner V, Bala M. Diagnostic and therapeutic approach to obscure gastrointestinal bleeding in a patient with a jejunal gastrointestinal stromal tumor: A case report. *BMC Res Notes.* 2014;7:695.

34. Fujimoto S, Muguruma N, Okamoto K et al. A novel theranostic combination of near-infrared fluorescence imaging and laser irradiation targeting c-KIT for gastrointestinal stromal tumors. *Theranostics.* 2018;8(9):2313–28.

35. Miettinen M, Killian JK, Wang ZF et al. Immunohistochemical loss of succinate dehydrogenase subunit A (SDHA) in gastrointestinal stromal tumors (GISTs) signals SDHA germline mutation. *Am J Surg Pathol.* 2013;37:234–40.

36. Hirota S. Differential diagnosis of gastrointestinal stromal tumor by histopathology and immunohistochemistry. *Transl Gastroenterol Hepatol.* 2018;3:27.

37. Maier J, Lange T, Kerle I et al. Detection of mutant free circulating tumor DNA in the plasma of patients with gastrointestinal stromal tumor harboring activating mutations of CKIT or PDGFRA. *Clin Cancer Res.* 2013;19(17):4854–67.

38. Astolfi A, Urbini M, Indio V, Nannini M et al. Whole exome sequencing (WES) on formalin-fixed, paraffin-embedded (FFPE) tumor tissue in gastrointestinal stromal tumors (GIST). *BMC Genomics.* 2015;16:892.

39. Chen Q, Li R, Zhang ZG et al. Oncogene mutational analysis in Chinese gastrointestinal stromal tumor patients. *Onco Targets Ther.* 2018;11:2279–86.

40. Wang M, Xu J, Zhang Y et al. Gastrointestinal stromal tumor: 15-years' experience in a single center. *BMC Surg.* 2014;14:93.

41. Park CH, Kim GH, Lee BE et al. Two staging systems for gastrointestinal stromal tumors in the stomach: Which is better? *BMC Gastroenterol.* 2017;17(1):141.

42. Boonstra PA, Ter Elst A, Tibbesma M et al. A single digital droplet PCR assay to detect multiple *KIT* exon 11 mutations in tumor and plasma from patients with gastrointestinal stromal tumors. *Oncotarget.* 2018;9(17):13870–83.

43. Kameyama H, Kanda T, Tajima Y et al. Management of rectal gastrointestinal stromal tumor. *Transl Gastroenterol Hepatol.* 2018;3:8.

44. Maki RG, Blay JY, Demetri GD et al. Key issues in the clinical management of gastrointestinal stromal tumors: An expert discussion. *Oncologist.* 2015;20:823–30.

45. Reichardt P, Demetri GD, Gelderblom H et al. Correlation of KIT and PDGFRA mutational status with clinical benefit in patients with gastrointestinal stromal tumor treated with sunitinib in a worldwide treatment-use trial. *BMC Cancer.* 2016;16:22.

46. Peixoto A, Costa-Moreira P, Silva M et al. Gastrointestinal stromal tumors in the imatinib era: 15 years' experience of a tertiary center. *J Gastrointest Oncol.* 2018;9(2):358–62.

47. Heinrich MC, Griffith D, McKinley A et al. Crenolanib inhibits the drug-resistant PDGFRA D842 V mutation associated with imatinib-resistant gastrointestinal stromal tumors. *Clin Cancer Res.* 2012;18(16):4375–84.

48. Lee JH, Kim Y, Choi JW, Kim YS. Correlation of imatinib resistance with the mutational status of KIT and PDGFRA genes in gastrointestinal stromal tumors: A meta-analysis. *J Gastrointestin Liver Dis.* 2013;22(4):413–8.

49. Brohl AS, Demicco EG, Mourtzikos K, Maki RG. Response to sunitinib of a gastrointestinal stromal tumor with a rare exon 12 PDGFRA mutation. *Clin Sarcoma Res.* 2015;5:21.

50. Akahoshi K, Oya M, Koga T, Shiratsuchi Y. Current clinical management of gastrointestinal stromal tumor. *World J Gastroenterol.* 2018;24(26):2806–17.

51. Lv A, Li Z, Tian X et al. SKP2 high expression, KIT exon 11 deletions, and gastrointestinal bleeding as predictors of poor prognosis in primary gastrointestinal stromal tumors. *PLOS One.* 2013;8(5):e62951.

52. Liu X, Qiu H, Zhang P et al. Prognostic factors of primary gastrointestinal stromal tumors: A cohort study based on high-volume centers. *Chin J Cancer Res.* 2018;30(1):61–71.

53. Kramer K, Knippschild U, Mayer B et al. Impact of age and gender on tumor related prognosis in gastrointestinal stromal tumors (GIST). *BMC Cancer.* 2015;15:57.

18

Hereditary Diffuse Gastric Cancer

Dongyou Liu

CONTENTS

18.1 Introduction

The stomach is affected by both benign and malignant tumors, including *primary epithelial tumors* (intraepithelial neoplasia [adenoma], carcinoma [adenocarcinoma—intestinal/diffuse, papillary adenocarcinoma, tubular adenocarcinoma, mucinous adenocarcinoma, signet-ring cell carcinoma, adenosquamous carcinoma, squamous cell carcinoma, small cell carcinoma, undifferentiated carcinoma], carcinoid [well-differentiated endocrine neoplasm]), *primary mesenchymal tumors* (leiomyoma, schwannoma, granular cell tumor, glomus tumor, leiomyosarcoma, gastrointestinal stromal tumor—benign/uncertain malignant potential/malignant, Kaposi sarcoma), *primary malignant lymphomas* (marginal zone B-cell lymphoma of MALT-type, mantle cell lymphoma, diffuse large B-cell lymphoma), and *secondary tumors*.

The most important tumors of the stomach (i.e., stomach cancer or gastric cancer) belong to the histological type of adenocarcinoma, which contains several subtypes and accounts for about 90% of all malignant gastric cancer identified. Along with other malignant tumors (e.g., leiomyosarcoma, MALT lymphoma), gastric adenocarcinoma (cancer) represents the second leading cause of malignancy-related deaths and the fifth most frequently diagnosed neoplasm in adults.

While 90% of gastric cancers are sporadic, about 10% show familial aggregation, which is promoted by shared exposure to carcinogens (i.e., nitrogen, cigarette smoke, and alcohol consumption), similar hygiene/dietary habits (salty and spicy food, smoked or cured food, high intake of meats, low vitamin A and C diet), and inherited genetic susceptibility (e.g., hereditary cancer syndromes).

The main hereditary cancer syndromes related to gastric cancer are hereditary diffuse gastric cancer (HDGC), gastric adenocarcinoma and proximal polyposis of the stomach (GAPPS), and familial intestinal gastric cancer (FIGC), which are all autosomal-dominant disorders with a Mendelian inheritance pattern, and which are implicated in up to 3% of all gastric cancers [1].

HDGC is the principal familial gastric cancer syndrome frequently linked to heterozygous germline mutations in the E-cadherin gene (*CDH1*) mapped to 16q22.1. Individuals harboring *CDH1* mutations demonstrate a heightened risk for diffuse gastric cancer (67%–70% in males, and 56%–83% in females), which is characterized by invasion and multiplication of cancerous (malignant) cells underneath the stomach lining (i.e., the submucosa), making the lining thick and rigid (rubber-like) instead of forming solid tumor. This cancer has a tendency to spread (metastasize) from the stomach to other tissues (e.g., the liver or nearby bones). In addition, 42% of females harboring *CDH1* mutations may develop lobular carcinoma of the breast, and some *CDH1* mutation carriers may form colorectal cancer. Further, some families with nonmalignant HDGC phenotype may show cleft lip/palate [2]. Another HDGC-related gene is *CTNNA1* located at 5q31.2, which encodes catenin alpha 1 (alpha-E-cadherin or CTNNA1). Through its association with both E- and N-cadherins, CTNNA1 mediates the linkage of cadherins on the plasma membrane to the actin filaments inside the cell, and thus plays a crucial role in cell adhesion and differentiation.

GAPPS is a gastric cancer predisposition syndrome associated with mutations in the *APC* promoter 1B region. While GAPPS nonmalignant phenotype produces fundic gland polyps of the proximal stomach, GAPPS malignant phenotype confers heightened risk for gastric cancer. Due to the fact that familial adenomatous polyposis (FAP) often contains germline mutations in the *APC* gene, GAPPS with mutations in the *APC* promoter 1B region is considered a phenotypic variant of FAP. Nonetheless, FAP poses limited gastric cancer risk (<1%) and has a tendency to develop colorectal duodenal/ampullary, thyroid, and desmoid tumors, and hepatoblastoma and medulloblastoma instead (see Chapter 16 in this book).

FIGC is another hereditary cancer syndrome for which no germline mutation has been identified to date. It typically involves two or more cases of intestinal-type gastric cancer in first- or second-degree relatives with at least one younger than age 50, or in families with three or more cases of intestinal-type gastric cancer.

18.2 Biology

The stomach is a "J" shaped, muscular sac located in the superior aspect of the abdomen, whose main functions are to store and digest ingested food for the nutrient supply to the body and to eliminate most microbial pathogens with acidic gastric juice prior their infiltration into the intestines and other abdominal organs.

Anatomically, the stomach is divided into four regions: cardia (surrounding the superior opening of the stomach that connects to the esophagus), fundus (the rounded portion superior to and left of the cardia), corpus (the large central portion inferior to the fundus), and pylorus (connecting the stomach to the duodenum).

The stomach wall comprises four layers (from superficial to deep): Serosa (peritoneum), muscularis externa, submucosa, and mucosa. The serosa (peritoneum) consists of an epithelial layer and connective tissue, which cover most of the stomach and connect to the abdominal wall. The muscularis externa contains three layers of smooth muscles: inner oblique (unique to stomach), middle circular (forming the pylorus), and outer longitudinal. The submucosa comprises loose connective tissue, blood vessels, lymphatic vessels, and Meissner's nerve plexus, and supports the mucosa during its contraction and relaxation (i.e., peristalsis). The mucosa lines the stomach cavity and consists of columnar epithelium (1 mm in height), lamina propria, and muscularis mucosa, covered by a 100- to 200-μm thick mucus. Gastric foveolae extend to the muscularis mucosa (at which the tubular glands formed by different exocrine cells are located).

The columnar cells in the mucosa secrete mucin; the chief (zymogenic) cells in the fundus generate protein-digesting pre-enzyme pepsinogen; the parietal (oxyntic) cells in the fundus, cardia, and pylorus secrete acid, hydrochloric acid (H+ ions), and intrinsic factor, and the G cells in the fundus, pylorus, and corpus secrete gastrin (which stimulates parietal cells for hydrochloric acid production).

Based on Lauren classification, gastric cancer (adenocarcinoma) is separated into three main histological subtypes: intestinal, diffuse, and mixed, which are thought to involve different pathways of tumorigenesis [3].

Intestinal gastric cancer (adenocarcinoma), mainly occurring in a distal location to cardia, is a well-differentiated tumor that likely evolves from intestinal cells and shows morphologic similarity to adenocarcinomas arising in the intestinal tract, with the tumor cells being cohesive and forming tubular or glandular structures on a background of intestinal metaplasia. Intestinal gastric cancer of Lauren classification encompasses the "tubular" (with prominent, dilated, or slit-like branching tubules and acini), "papillary" (with well-differentiated tumor cells and exophytic growth/fibrovascular stalks), and "mucinous" (with extracellular mucinous pools) subtypes of the WHO classification. Constituting just over 50% of gastric cancer, intestinal gastric cancer appears to undergo a multistep progression from chronic gastritis through gastric atrophy, metaplasia, dysplasia and ultimately carcinoma, shows a clear link to *Helicobacter pylori* infection (which is found in nearly 90% of the noncancerous gastric mucosa in intestinal gastric cancer) as well as diet and environmental factors, and has the ability to metastasize to the lymph nodes and liver. About 10% of intestinal gastric cancer are attributed to FIGC, which is noted for the absence of germline mutations [4].

Diffuse gastric cancer (adenocarcinoma), often present in a proximal location to cardia, is an undifferentiated or poorly differentiated tumor arising from the gastric mucosa. It typically shows a lack of intercellular adhesion, which disrupts the formation of glandular structures, with round small cells diffusely infiltrating the stomach wall (stroma), and little or no gland. It correlates to the poor cohesive subtype (signet-ring cell carcinoma, which displays predominantly or exclusively signet ring cells formed after intracellular mucus pushing the nucleus to the side) of the WHO classification. Making up 33% of gastric cancer, diffuse gastric cancer results directly from chronic active inflammation (active gastritis) without a clear manifestation of intermediate premalignant steps (i.e., gastric atrophy, metaplasia, and dysplasia), and has the ability to metastasize to the lymph nodes, ovaries, and serosa. Among diffuse gastric cancers, over two-thirds involve germline mutations in the E-cadherin gene (*CDH1*) on chromosome 16q22.1 (so-called hereditary diffuse gastric cancer [HDGC], also known as familial diffuse gastric cancer [FDGC], hereditary diffuse gastric adenocarcinoma, E-cadherin-associated hereditary gastric cancer), while fewer than one-third are related to *H. pylori* infection (which may cause mutations in the *IL1B* and *IL1RN* genes on chromosome 2q14.1) as well as somatic mutations in the *KRAS* gene on chromosome 12p12.1 [2].

Mixed-type gastric cancer displays non-homogenous mixtures of both intestinal and diffuse type architectures, including a deeper infiltration of the gastric wall, and a higher metastatic rate to regional lymph nodes. Accounting for about 15% of gastric cancer, mixed-type gastric cancer correlates to the mix subtype of the WHO classification, and may be subdivided into four histologically distinct groups, with the first group showing a combination of intestinal and diffuse features, the second showing more diffuse structures with a nodular growth pattern, the third group showing both a glandular pattern and signet ring cells, and the fourth group revealing an excessive production of mucin. Overall, mixed-type gastric cancer appears to be more aggressive than intestinal and diffuse types and requires gastrectomy (including lymph node dissection) instead of endoscopic resection [5].

18.3 Pathogenesis

Although both dietary and environmental risk factors are implicated in the pathogenesis of intestinal and diffuse gastric cancer, environmental factors and *H. pylori* infection appear to be more closely associated with intestinal gastric cancer (which makes up >50% of gastric cancer), and genetic alteration is more often involved in diffuse gastric cancer (which represents 33% of gastric cancer) [6–10].

Currently, three molecular mechanisms (chromosome instability, microsatellite instability, and epigenetic alterations) are thought to drive gastric carcinogenesis, involving activation of oncogenes, inactivation of tumor suppressor genes, and deregulation of signaling pathways, leading to aberrant cell cycle regulation and changes in the expression of growth factors and cytokines that regulate differentiation and survival of tumor cells [11,12].

In the case of diffuse gastric cancer, the cell adhesion protein E-cadherin (*CDH1*) gene located on chromosome 16q22.1 is critically important, and its germline mutations are responsible for over two-thirds of diffuse gastric cancer (so-called HDGC).

The *CDH1* gene is a 100 kb fragment organized in 16 exons, and encodes a 120 kDa calcium-dependent cellular adhesion protein called E-cadherin (or Cadherin-1), which has three domains (extracellular domain on the N-terminal, transmembrane domain in the center, and highly conserved cytoplasmic domain on the C-terminal) and demonstrates tumor suppressor functions. As a transmembrane glycoprotein predominantly expressed at the basolateral membrane of epithelial cells, E-cadherin mediates calcium-dependent cell–cell adhesion and invasion–suppression through interaction of its C-terminal with the actin cytoskeleton via undercoat proteins called catenins (α-, β-, and γ-), and contributes to the establishment and maintenance of polarized and differentiated epithelia during development. In addition, E-cadherin participates in signal transduction, differentiation, gene expression, cell motility, and inflammation [13].

Germline mutations in the *CDH1* gene decrease E-cadherin expression, reduce cell adhesion, and promote atypical cellular architecture and irregular cell growth (as exemplified by so-called nonmalignant HDGC phenotype that may show cleft lip/palate). Subsequent inactivation of the other *CDH1* allele through methylation, somatic mutations, and loss of heterozygosity abolishes E-cadherin expression and renders E-cadherin-deficient cells malignant and metastatic [14].

To date, at least 155 germline pathogenic variants (e.g., small intragenic deletions/insertions, missense, nonsense, and splice site variants) leading to truncating (80%) or altered (20%) E-cadherin have been identified from patients with HDGC. Interestingly, in HDGC patients who harbor germline *CDH1* alterations, the second hit mainly involves somatic alterations in *CDH1* (75%), promoter hypermethylation of *CDH1* (32%), loss of heterozygosity (LOH, 25%), both promoter hypermethylation and LOH (18%), and neither promoter hypermethylation nor LOH (25%) [15].

Germline CTNNA1 truncating mutations have been detected in a small number of patients with HDGC and colorectal cancer and may contribute to disease susceptibility. Since CTNNA1 functions in the same complex as E-cadherin, it draws attention to the role of a broader signaling network in the pathogenesis of HDGC. Other potential HDGC susceptibility genes include *MAP3K6* (encoding a serine/threonine protein kinase), *INSR* (encoding a receptor tyrosine kinase), *CD44* (encoding a cell-surface glycoprotein), *PALB2*, *BRCA1*, and *RAD51C* [16].

Fewer than one-third of diffuse gastric cancers are attributed to *H. pylori* infection (which induces mutations in the *IL1B* and *IL1RN* genes on chromosome 2q14.1) and somatic mutations in the *KRAS* gene on chromosome 12p12.1. The *IL1B* (interleukin 1-beta) gene on chromosome 2q14.1 encodes a subunit of interleukin-1 (i.e., IL1B), a cytokine involved in physiologic and pathophysiologic immune and inflammatory responses. The *IL1RN* (interleukin-1 receptor antagonist) gene on chromosome 2q14.1 encodes a protein that binds to IL1 receptors (IL1R1) and inhibits the biologic activity of IL1-alpha (IL1A) and IL1-beta (IL1B) [17].

Infection with *H. pylori* may alter the *IL1B* and *IL1RN* genes, which in turn interfere with the normal activity of IL1-alpha (IL1A) and IL1-beta (IL1B) cytokines, leading to the development of diffuse and intestinal gastric cancer, especially the latter (which is mostly distal or non-cardia gastric cancer). Chronic *H. pylori* infection causes epithelial damage, leading to active gastritis, atrophic gastritis (e.g., loss of parietal cells and chief cells and glandular atrophy), intestinal metaplasia (which replaces the gastric epithelium), low-grade dysplasia, high-grade dysplasia, and finally adenocarcinoma. The pathogenesis of *H pylori*–mediated gastric cancer appears to involve several mechanisms, including cytotoxin-associated gene A, vacuolating cytotoxin A–induced chronic inflammation, oxidative damage, genomic instability, and epigenetic changes in gastric epithelial cells. Diffuse gastric cancer–related to *H. pylori* infection can be also inherited in an autosomal dominant manner [18].

18.4 Epidemiology

18.4.1 Prevalence

Gastric cancer (adenocarcinoma) constitutes the second leading cause (behind lung cancer) of malignancy-related deaths and is the fifth most frequently diagnosed cancer (behind lung, breast, and colorectal cancers) in the world, with an estimated 841,000 deaths and 984,000 new cases in 2013. The incidence of gastric cancer ranges from 2.0–62.2 cases per 100,000, including 6.9 cases per 100.000 males in Norway; 41.3, 46.8, 48.2, and 62.2 cases per 100,000 males in China, Japan, Mongolia, and Korea, respectively. There is a notable male dominance among gastric cancer patients (with a male:female ratio of 2.3:1.0) [19].

Making up about 3% of gastric cancer, HDGC usually occurs between the late 30s and early 40s of life, with the average age of onset at 38 years (range 14–69 years). Males and females are equally susceptible to HDGC. In contrast, intestinal gastric cancer is diagnosed in older patients and tends to affect males more than females.

18.4.2 Inheritance

HDGC is an autosomal dominant disorder, and inheritance of an abnormal copy of the tumor suppressor gene *CDH1* from an affected person is sufficient to increase lifetime risk for diffuse gastric cancer by 70%–80%, and lobular breast cancer by 40%–60%.

18.4.3 Penetrance

HDGC shows a high but incomplete penetrance, and up to 70% and 83% of males and females harboring *CDH1* mutations are likely to develop advanced gastric cancer by age 80.

18.5 Clinical Features

Clinically, patients with gastric cancer often display loss of appetite, difficulty swallowing (dysphagia), weight loss, persistent abdominal (belly) pain, vague discomfort in the abdomen (usually above the navel), postprandial fullness (a sense of fullness in the upper abdomen after eating a small meal), heartburn, indigestion, nausea, vomiting (with or without blood), swelling or fluid buildup in the abdomen, and low red blood cell count (anemia).

Besides possible non-tumor features (e.g., cleft lip and cleft palate), patients with HDGC tend to develop diffuse type gastric cancer, mixed intestinal and diffuse type gastric cancer, lobular breast cancer, colorectal cancer, cutaneous melanoma, and prostate cancer.

If cancer metastasizes to other sites, symptoms may include hepatomegaly (enlarged liver), yellowing of the eyes and skin (jaundice), Virchow node (left supraclavicular adenopathy), Sister Mary Joseph node (periumbilical nodule), Irish node (left axillary node), Krukenberg tumor (ovary mass), Blumer shelf (cul-de-sac mass), ascites (peritoneal carcinomatosis, an abnormal buildup of fluid in the peritoneum), skin nodules (firm lumps under the skin), and fractures (broken bones).

18.6 Diagnosis

Diagnosis of gastric cancer generally involves family history review, endoscopy (e.g., narrow-band imaging with or without magnification [NBI-ME], chromoendoscopy [CE], flexible spectral imaging color enhancement [FICE] endoscopy with or without magnification [FIME], and confocal laser endomicroscopy [CLE]), histology, and molecular genetic testing [20–23].

Histologically, diffuse gastric cancer/adenocarcinoma (also known as poorly cohesive, linitis plastica, or signet ring adenocarcinoma) typically shows extensive infiltration of poorly differentiated discohesive malignant cells under the normal-looking mucosa, causing widespread thickening and rigidity of the gastric wall (so-called linitis plastica) without forming a tumor mass (in contrast to intestinal gastric cancer) and leading to pyloric obstruction. Microscopically, diffuse gastric cancer contains gastric-type mucus cells, infiltrating (sometimes transmurally) as individual cells or small clusters, numerous signet ring cells (due to accumulation of abundant intracellular mucin that pushes the nucleus toward the periphery), submucosal fibrosis, mucosal ulceration, hypertrophic muscularis propria, and marked desmoplastic and inflammatory reaction (Figure 18.1) [24]. Immunohistochemically, diffuse gastric cancer stains positive for mucicarmine, alcian blue-PAS, CEA, EMA, keratin, and villin, but negative for TTF1 and p53.

In the diffuse gastric cancer linked to *H. pylori* infection (which induces gastric atrophy and intestinal metaplasia), the

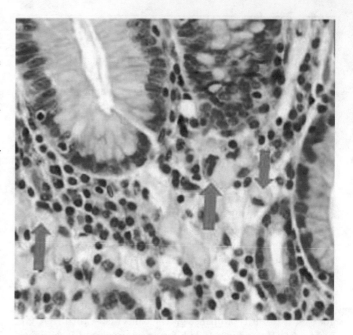

FIGURE 18.1 Hereditary diffuse gastric cancer in a 50-year-old male with *CDH1* missense variant c.1679C>G (p.T560R) showing infiltrating adenocarcinoma poorly differentiated with mucinous and signet ring cell features (large amount of mucin pushing the nucleus to the cell periphery) (arrows). (Photo credit: Yelskaya Z et al. *PLOS ONE*. 2016;11(11):e0165654.)

tumor cells are arranged in a gland-like formation and begin in the mucosa (Figure 18.2) [25].

A clinical diagnosis of HDGC is established in a patient with histologically confirmed diffuse gastric cancer and one of the following: (i) a family history of one or more first- or second-degree relatives with gastric cancer; (ii) a personal and/or family history of one individual with diffuse gastric cancer diagnosed before age 40 years; or (iii) a personal and/or family history of diffuse gastric cancer and lobular breast cancer, one diagnosed before age 50 years.

Families with aggregation of gastric cancer and an index case with diffuse gastric cancer, but not fulfilling the above criteria for HDGC, are classified as FDGC.

If clinical features and family history are inconclusive, molecular genetic testing for a heterozygous pathogenic variant in *CDH1* is recommended. In addition, molecular genetic testing is advised for individuals with a clinical diagnosis of HDGC in order to facilitate family studies [26].

Molecular genetic testing is currently performed by using single-gene testing (sequence analysis of *CDH1* followed by gene-targeted deletion/duplication analysis if no pathogenic variant is found in sequence analysis) and multigene panel (including *CDH1* and other genes of interest). The proportion of patients with HDGC harboring *CDH1* pathogenic variants ranges from 50% by sequence analysis to 4% by gene-targeted deletion/duplication analysis (e.g., quantitative PCR, long-range PCR, multiplex ligation-dependent probe amplification [MLPA], and a gene-targeted microarray).

As about one-third of families with HDGC do not contain identifiable *CDH1* germline pathogenic variants, molecular genetic testing for mutations in the *IL1B, IL1RN,* and *KRAS* genes may be conducted if necessary.

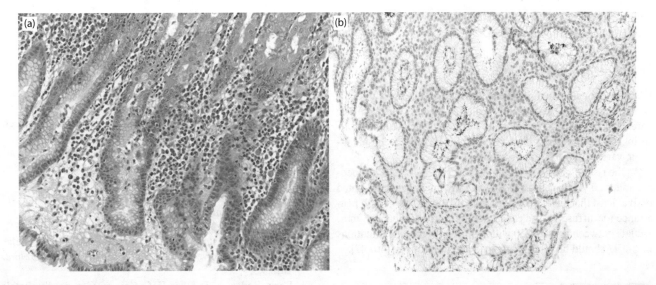

FIGURE 18.2 Hereditary diffuse gastric cancer presenting with: (a) *H. pylori*–induced non-atrophic chronic gastritis, with areas of active inflammation containing characteristic granulocytic infiltration in the crypt epithelium (arrow); (b) immunohistological staining of *H. pylori* (brown color) with anti-Helicobacter antibody in diffuse gastric cancer. (Photo credit: Eftang LL et al. *BMC Cancer.* 2013;13:586.)

Differential diagnoses for HDGC include intestinal gastric cancer (which forms a large protruded, ulcerated, or infiltrative lesion in the stomach, and shows tubular or glandular formations of variable differentiation resembling adenocarcinoma of the intestinal tract), and other hereditary cancer syndromes that also cause gastric cancer (e.g., hereditary nonpolyposis colon cancer [HNPCC or Lynch syndrome, predominantly intestinal type], 13% risk, containing germline pathogenic variants in *MLH1*, *MSH2*, *MSH6*, *PMS2*, *EPCAM*), familial adenomatous syndrome (FAP, <1% risk, containing germline pathogenic variants in *APC*), GAPPS (intestinal type, >1% risk, containing heterozygous germline pathogenic variants in the promoter 1B of *APC*), Peutz–Jeghers syndrome (PJS, 29% risk, containing pathogenic variants in *STK11*), juvenile polyposis syndrome (JPS, 21% risk, containing pathogenic variants in *BMPR1A*, *SMAD4*), Li–Fraumeni syndrome (5% risk, containing pathogenic variants in *TP53*, *CHEK2*, and *CDKN2A*), hereditary breast and ovarian cancer syndrome (containing pathogenic variants in *BRCA1* and *BRCA2*), Carney complex (containing pathogenic variants in *PRKAR1A*), Carney–Stratakis syndrome (containing germline pathogenic variants in *SDHB*, *SDHC*, and *SDHD*), and phosphatase and tensin homolog (PTEN, or Cowden syndrome)] [27].

18.7 Treatment

Treatment options for gastric cancer are dependent on preoperative staging, and may include endoscopic resection for early local disease (superficial, limited mucosa disease <T1b, N0); surgical resection for resectable disease (lymphadenectomy <T3, any N); neoadjuvant (>T2, particularly bulky T3/T4, perigastric nodes, linitis plastica, or positive peritoneal cytology) and adjuvant (>T1N1 or >T3N0) chemotherapy or chemoradiation for locally advanced resectable disease, and palliative therapy for locally advanced unresectable and advanced metastatic disease (T4, any N, or M1) [28].

Endoscopic resection for patients with early local disease can be either endoscopic mucosal resection (EMR) or endoscopic submucosal dissection (ESD). ESD outperforms EMR for en bloc, complete, and curative resection with lower recurrence rate. Alternatively, surgical resection with total or subtotal gastrectomy may be considered.

Targeted therapy for gastric cancer includes inhibitors of human epidermal growth factor receptor 2 (HER2) (e.g., trastuzumab) and vascular endothelial growth factor receptor 2 (VEGF2) (e.g., ramucirumab). Trastuzumab is a monoclonal antibody that inhibits HER2-mediated signaling and prevents proliferation of HER2-dependent tumors. Ramucirumab is a human monoclonal antibody against VEGF2, and demonstrates survival benefit in patients with advanced gastric cancer [29–37].

18.8 Prognosis and Risk Management

18.8.1 Prognosis

Early gastric cancer (with malignancy limited to the mucosa or submucosa) has a 5-year survival (5-y overall survival [OS]) rate of between 85%–100%, while advanced gastric cancer (showing polypoid lesions, ulcerated with well-defined border, ulcerated with ill-defined borders, or infiltrating diffuse without evidence of mass or ulcers) has a 5-y OS rate of only 5%–20%. Based on pathological stage and intervention, the 5-y OS includes surgery only (IA 93.6%, IIA 81.8%, and IIIA 54.2%) or with neoadjuvant (I 76.5%, II 46.3%, III 18.3%, and IV 5.7%). Furthermore, the median OS for patients with advanced gastric cancer after first-line chemotherapy is less than 12 months, and the OS for patients after second-line therapy is only ~6 months. Interestingly, patients with affected relatives have a better prognosis (including disease-free survival rate, recurrence-free survival rate, and overall survival rate) than those without a family history. Overall, diffuse type gastric cancer (adenocarcinoma) is more aggressive and has a worse prognosis than intestinal type gastric cancer (adenocarcinoma) [38,39].

18.8.2 Risk Management

HDGC is an autosomal dominant disorder, and most individuals containing a *CDH1* pathogenic variant have inherited from one parent and have a 50% chance of passing the mutation to offspring. Prenatal testing for pregnancies is possible if the pathogenic variant in the family is known.

As the risk of gastric cancer related to HDGC is <1% for people under 20 years, genetic testing and treatment can be initiated until at least that age. Further, given that HDGC shows high penetrance for diffuse gastric cancer and lobular breast cancer, individuals between ages 20–30 with confirmed germline mutations in *CDH1* should be offered prophylactic gastrectomy [40–42].

18.9 Conclusion

Being an autosomal-dominant familial gastric cancer syndrome, hereditary diffuse gastric cancer (HDGC) is responsible for causing over two-thirds of diffuse gastric cancers and shows a clear link to heterozygous germline mutations in the E-cadherin gene (*CDH1*). Patients harboring *CDH1* mutations have an increased risk for diffuse gastric cancer (up to 70% in males, and 83% in females) as well as lobular breast cancer (42% in females). Compared to intestinal gastric cancer, which tends to form a solid mass in a distal location to cardia, with the tumor cells arranged in tubular or glandular formation on a background of intestinal metaplasia, and strong link to *H. pylori* infection; diffuse gastric cancer induces rubber-like thickening on the stomach wall in a proximal location to cardia, with numerous signet ring cells (in which intracellular mucin pushes the nucleus to the side), occasional link to *H. pylori* infection, and propensity to metastasize to other tissues. Although diagnosis of HDGC is possible on the basis of clinical features and family history review, use of molecular genetic testing to identify heterozygous pathogenic variants in *CDH1* is extremely helpful if clinical features and family history are inconclusive or further family studies are required [43]. Treatment of HDGC involves endoscopic resection, lymphadenectomy, neoadjuvant/adjuvant chemotherapy, and chemoradiation as well as palliative therapy [44].

REFERENCES

1. Marqués-Lespier JM, González-Pons M, Cruz-Correa M. Current perspectives on gastric cancer. *Gastroenterol Clin North Am.* 2016;45(3):413–28.
2. Kaurah P, Huntsman DG. Hereditary diffuse gastric cancer. 2002 Nov 4 [updated 2018 Mar 22]. In: Adam MP, Ardinger HH, Pagon RA et al. editors. *GeneReviews® [Internet].* Seattle (WA): University of Washington, Seattle; 1993–2018.
3. Boland CR, Yurgelun MB. Historical perspective on familial gastric cancer. *Cell Mol Gastroenterol Hepatol.* 2017;3(2):192–200.
4. Choi JM, Kim SG, Kim J et al. Clinical implications of preexisting adenoma in endoscopically resected early gastric cancers. *PLOS ONE.* 2017;12(5):e0178419.
5. Komatsu S, Ichikawa D, Miyamae M et al. Histological mixed-type as an independent prognostic factor in stage I gastric carcinoma. *World J Gastroenterol.* 2015;21(2):549–55.
6. Carrasco-Avino G, Riquelme I, Padilla O, Villaseca M, Aguayo FR, Corvalan AH. The conundrum of the Epstein-Barr virus-associated gastric carcinoma in the Americas. *Oncotarget.* 2017;8(43):75687–98.
7. Sokolova O, Naumann M. NF-κB signaling in gastric cancer. *Toxins (Basel).* 2017;9(4). pii: E119.
8. Song Y, Wang Y, Tong C et al. A unified model of the hierarchical and stochastic theories of gastric cancer. *Br J Cancer.* 2017;116(8):973–89.
9. Bizzaro N, Antico A, Villalta D. Autoimmunity and gastric cancer. *Int J Mol Sci.* 2018;19(2). pii: E377.
10. Díaz P, Valenzuela Valderrama M, Bravo J, Quest AFG. *Helicobacter pylori* and gastric cancer: Adaptive cellular mechanisms involved in disease progression. *Front Microbiol.* 2018;9:5.
11. Recio-Boiles A, Babiker HM. *Cancer, Gastric. StatPearls [Internet].* Treasure Island (FL): StatPearls Publishing; 2017 Jun–2017 Oct 22.
12. Todisco A. Regulation of gastric metaplasia, dysplasia, and neoplasia by bone morphogenetic protein signaling. *Cell Mol Gastroenterol Hepatol.* 2017;3(3):339–47.
13. Melo S, Figueiredo J, Fernandes MS et al. Predicting the functional impact of CDH1 missense mutations in hereditary diffuse gastric cancer. *Int J Mol Sci.* 2017;18(12). pii: E2687.
14. Cho SY, Park JW, Liu Y et al. Sporadic early-onset diffuse gastric cancers have high frequency of somatic CDH1 alterations, but low frequency of somatic RHOA mutations compared with late-onset cancers. *Gastroenterology.* 2017;153(2):536–49.
15. Lee YS, Cho YS, Lee GK et al. Genomic profile analysis of diffuse-type gastric cancers. *Genome Biol.* 2014;15(4):R55.
16. Zhang H, Feng M, Feng Y et al. Germline mutations in hereditary diffuse gastric cancer. *Chin J Cancer Res.* 2018;30(1):122–30.
17. Zhang XY, Zhang PY, Aboul-Soud MA. From inflammation to gastric cancer: Role of *Helicobacter pylori*. *Oncol Lett.* 2017;13(2):543–8.
18. Lim B, Kim JH, Kim M, Kim SY. Genomic and epigenomic heterogeneity in molecular subtypes of gastric cancer. *World J Gastroenterol.* 2016;22(3):1190–201.
19. Ang TL, Fock KM. Clinical epidemiology of gastric cancer. *Singapore Med J.* 2014;55(12):621–8.
20. Choi YJ, Kim N. Gastric cancer and family history. *Korean J Intern Med.* 2016;31(6):1042–53.
21. Sung JK. Diagnosis and management of gastric dysplasia. *Korean J Intern Med.* 2016;31(2):201–9.
22. Feroce I, Serrano D, Biffi R et al. Hereditary diffuse gastric cancer in two families: A case report. *Oncol Lett.* 2017 14(2):1671–4.
23. Zylberberg HM, Sultan K, Rubin S. Hereditary diffuse gastric cancer: One family's story. *World J Clin Cases.* 2018;6(1):1–5.
24. Yelskaya Z, Bacares R, Salo-Mullen E et al. CDH1 missense variant c.1679C>G (p.T560R) completely disrupts normal splicing through creation of a novel 5' splice site. *PLOS ONE.* 2016;11(11):e0165654.
25. Eftang LL, Esbensen Y, Tannæs TM, Blom GP, Bukholm IR, Bukholm G. Up-regulation of CLDN1 in gastric cancer is correlated with reduced survival. *BMC Cancer.* 2013;13:586.

26. Hamilton LE, Jones K, Church N, Medlicott S. Synchronous appendiceal and intramucosal gastric signet ring cell carcinomas in an individual with CDH1-associated hereditary diffuse gastric carcinoma: A case report of a novel association and review of the literature. *BMC Gastroenterol.* 2013;13:114.

27. Dai M, Yuan F, Fu C, Shen G, Hu S, Shen G. Relationship between epithelial cell adhesion molecule (EpCAM) overexpression and gastric cancer patients: A systematic review and meta-analysis. *PLOS ONE.* 2017;12(4):e0175357.

28. PDQ Adult Treatment Editorial Board. *Gastric Cancer Treatment (PDQ®): Health Professional Version. PDQ Cancer Information Summaries [Internet].* Bethesda (MD): National Cancer Institute (US); 2002-. 2016 Jun 30.

29. Chen T, Xu XY, Zhou PH. Emerging molecular classifications and therapeutic implications for gastric cancer. *Chin J Cancer.* 2016;35:49.

30. Choi YY, Noh SH, Cheong JH. Molecular dimensions of gastric cancer: Translational and clinical perspectives. *J Pathol Transl Med.* 2016;50(1):1–9.

31. Ma J, Shen H, Kapesa L, Zeng S. Lauren classification and individualized chemotherapy in gastric cancer. *Oncol Lett.* 2016;11(5):2959–64.

32. Lei X, Wang F, Ke Y et al. The role of antiangiogenic agents in the treatment of gastric cancer: A systematic review and meta-analysis. *Medicine (Baltimore).* 2017;96(10):e6301.

33. Liu X, Meltzer SJ. Gastric cancer in the era of precision medicine. *Cell Mol Gastroenterol Hepatol.* 2017;3(3):348–58.

34. Macedo F, Ladeira K, Longatto-Filho A, Martins SF. Gastric cancer and angiogenesis: Is VEGF a useful biomarker to assess progression and remission? *J Gastric Cancer.* 2017;17(1):1–10.

35. Wang HB, Liao XF, Zhang J. Clinicopathological factors associated with HER2-positive gastric cancer: A meta-analysis. *Medicine (Baltimore).* 2017;96(44):e8437.

36. Zheng Y, Zhu XQ, Ren XG. Third-line chemotherapy in advanced gastric cancer: A systematic review and meta-analysis. *Medicine (Baltimore).* 2017;96(24):e6884.

37. Li D, Lo W, Rudloff U. Merging perspectives: Genotype-directed molecular therapy for hereditary diffuse gastric cancer (HDGC) and E-cadherin-EGFR crosstalk. *Clin Transl Med.* 2018;7(1):7.

38. Gu L, Chen M, Guo D et al. PD-L1 and gastric cancer prognosis: A systematic review and meta-analysis. *PLOS ONE.* 2017;12(8):e0182692.

39. Luu C, Thapa R, Woo K et al. Does histology really influence gastric cancer prognosis? *J Gastrointest Oncol.* 2017;8(6): 1026–36.

40. PDQ Screening and Prevention Editorial Board. *Stomach (Gastric) Cancer Prevention (PDQ®): Patient Version. PDQ Cancer Information Summaries [Internet].* Bethesda (MD): National Cancer Institute (US); 2002-. 2015 Dec 1.

41. van der Post RS, Vogelaar IP, Carneiro F, Hereditary diffuse gastric cancer: Updated clinical guidelines with an emphasis on germline CDH1 mutation carriers. *J Med Genet.* 2015;52(6):361–74.

42. Mi EZ, Mi EZ, di Pietro M et al. Comparative study of endoscopic surveillance in hereditary diffuse gastric cancer according to CDH1 mutation status. *Gastrointest Endosc.* 2018;87(2):408–18.

43. Kumar S, Long JM, Ginsberg GG, Katona BW. The role of endoscopy in the management of hereditary diffuse gastric cancer syndrome. *World J Gastroenterol.* 2019;25(23):2878–86.

44. Shenoy S. CDH1 (E-Cadherin) mutation and gastric cancer: Genetics, molecular mechanisms and guidelines for management. *Cancer Manag Res.* 2019;11:10477–86.

19

Juvenile Polyposis Syndrome

Rafael Parra-Medina, Elizabeth E. Montgomery, Paula Quintero-Ronderos, and Edgar Garavito

CONTENTS

19.1 Introduction

Juvenile polyposis syndrome (JPS), considered one of the most common hamartomatous polyposis syndromes, is a rare autosomal dominant hereditary disease with notable genetic and phenotypic heterogeneity [1]. First described by McColl in 1964 [2], JPS is characterized by the formation of 1–100 hamartomatous polyps throughout the gastrointestinal tract, mostly in the colon (98%), stomach (60%–85%), duodenum (14%–33%), and less frequently in the small bowel (7%) [1,3,4]. The term "juvenile" refers to the type of polyps rather than to the age of disease onset [5], although symptoms usually begin in the first and second decades. Some patients may have only four or five polyps over their lifetime, whereas others in the same family may have more than 100 [5,6,7,8]. Many cases of JPS are caused by germline mutations in one of the two genes associated with transformation growth factor β (TGF β): SMAD4 located on chromosome 18q21.1 or BMPR1A on chromosome 10q23.2 [9]. JPS patients have an increased risk of cancer (colon, stomach, pancreas, and small bowel) and congenital disorders [6,10]. Juvenile polyposis of infancy, a variant of JPS showing phenotypic overlap with PTEN hamartoma syndromes, is caused by contiguous deletion of BMPR1A and PTEN, both located on chromosome 10q23 [4].

19.2 Biology and Pathogenesis

The key genes involved in the pathogenesis of JPS comprise BMP1A (coding for the bone morphogenetic protein receptor type IA) and SMAD4 (or DPC4, coding for an intracellular signaling protein), which participate as components of the TGF-β signaling pathway in a variety of biological processes such as cell proliferation, apoptosis, angiogenesis, cell differentiation, and epithelial–mesenchymal transition (EMT) [11–14]. Alterations

identified in these genes include nonsense and missense mutations, deletions and insertions, as well as mutations within the regulatory regions (e.g., promoters). In about 40% of JPS patients no mutations are detected, suggesting a polygenic inheritance (other genes not yet studied) or epigenetic factors underlying the abnormal clinical phenotype. Some patients lack a causative mutation, although tumor cells from these patients are void of BMP1A and SMAD4 expression as assessed by immunohistochemistry [12,15,16]. Moreover, studies have shown that somatic mutations may act together with germline mutations, particularly in SMAD4. This suggests that loss of heterozygosity (LOH) as a second hit is important for polyp development and/or the transition to carcinoma [16].

The TGF-β receptor has been shown to maintain homeostasis between diverse biological processes such as cell proliferation, differentiation, and apoptosis [17,18]. This signaling pathway is activated by the interaction of a variety of ligands such as BMPs and other TGF-β family members with their receptors. The ligands interact with a dimeric receptor, homo- or heterodimer, composed of one of the type I and type II receptors. The binding of ligand to type II receptors produces structural changes that permit their auto-activation and the trans-activation of type I receptors, which results in phosphorylation of various signaling molecules thus affecting gene expression. The main pathway activated by these receptors (canonic pathway) involves many SMAD family members, particularly the R-SMADs [17,18]. SMAD2 and 3 relay the signals from the TGFR1A receptor while SMAD1, 5, and 8 relay from BMPR1A. These R-SMADs bind to SMAD4 in the cytoplasm and translocate into the nucleus where they act as transcriptional co-activators or co-repressors helped by proteins that mediate binding to DNA. The molecular complex formed by R-SMADs, SMAD4, and other co-factors, binds to GC-rich DNA sequences and to the SMADs response element, known as SMAD binding element (SME). The SME consensus binding sequencing is CAGAC (Figure 19.1) [18–20]. Many genes have SME within their promoter regions, thus the TGF-β pathway regulates a variety of biological processes involved in cell/tissue growth and differentiation. In addition to its role during embryogenesis and development, the TGF-β pathway participates in hormone regulation, immune response, bone formation, cell proliferation, apoptosis, tissue remodeling and repair, angiogenesis, and erythropoiesis [18–20].

The canonic pathway through the SMADs is usually associated with tumor development control mechanisms such as cell cycle inhibition, apoptosis induction, and inhibition of cell immortalization (Figure 19.1) [17,18,21,22], and this applies to juvenile polyps and polyposis as well. There is an inhibition of SMAD4 downstream molecules related to cell cycle progression such as cdc25, c-Myc, ID1, ID2, CDK2/Cyclin A, E complexes, and CDK4,6/Cyclin D. Also, there is overexpression of proteins associated with cell cycle arrest such as p15, p16, p21, p27, and menin [22]. All these alterations cause a tumor suppression effect. Moreover, this pathway enhances the activity of pro-apoptotic molecules such as DAXX, DAPK, TIEG1, SAPK, Apaf1, caspases 3 and 7, Smac/DIABLO, and p73. There is also an inhibition of anti-apoptotic proteins such as survivin and molecules from the AKT-signaling pathway [21,22]. The canonic pathway suppresses hTERT expression, which is the polymerase subunit of the telomerase. It is related to telomere elongation, thus contributing to cell immortalization [22]. Considering all these pathways, there is a potentially causative effect between gene mutations and the loss of function for TGF-β and SMAD4 receptors in the pathogenesis of polyps. Furthermore, studies have shown a dual role of TGF-β

pathway during tumorigenesis. On one hand, the loss of tumor suppressive activities allows the development of polyps and well-differentiated neoplasms. On the other hand, the loss of proliferative control produces malignant transition of dysplasia/adenomatous change or tumor progression of some carcinomas (Figure 19.1) [17,18,21,22]. Bearing in mind the tumor suppressive characteristics from the canonic pathways, it is important to note how the loss of function from these molecules may be related to the development of dysplastic/adenoma-like lesions and polyps. However, it does not explain dual behavior. Recently, studies have described that the canonic pathways not only suppressed tumor development, but also promoted tumor growth through EMT and cell migration, invasion, and tumor metastasis [22]. These processes depend on the autocrine and paracrine activities, respectively. The autocrine activity of TGF-β results in phenotype change within epithelial cells. It is characterized by cell organization and adhesion mediated by cadherin E, claudins, occludins, and cytokeratins. The disorganized cells have a mesenchymal and migratory phenotype, which increases their interaction with the extracellular matrix due to the expression of cadherin N, vimentin, fibronectin, and vitronectin [21]. This EMT results in the loss of cell−cell interaction, the acquisition of a fibroblastic morphology in the epithelial cells, and the gain of migration and invasion capacity of these cells. The paracrine activity of TGF-β results in the suppression of local immune response, increased angiogenesis, and myofibroblast generation [17,18,21,22]. The overexpression of TGF-β causes an increase in apoptosis, inhibition of proliferation, and suppression of lymphocytic cytotoxic activity as well as an increase in the number of regulatory T cells.

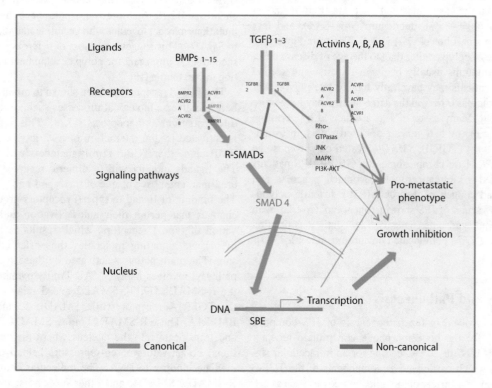

FIGURE 19.1 Genes at play in juvenile polyposis syndrome. In green, the canonical pathway is shown downstream of the receiver for the BMPs and in blue the non-canonical pathways. In red, the molecules whose genes are mutated in patients with JPS.

This permits tumor cell evasion of immune surveillance, thus facilitating tumor progression. TGF-β increases expression of vascular growth factors such as VEGF and CTGF, promoting angiogenesis and expression of genes related to vascular integrity (*ANGPT1*) to enhance oxygen and nutrient acquisition by tumor cells and capacity to penetrate the endothelium to reach the blood vessels [21,22]. Also, TGF-β favors the development of myofibroblasts in charge of cytokine, growth factors, and extracellular matrix component production, which in turn promotes tumor cell development. Thus, the SMADs-dependent canonic pathway has a dual role of promoting both tumor progression and suppression (Figure 19.1) [22]. TGF-β receptors also take part in other intracellular signaling pathways such as pI3K-AKT, RhoGTPases, p38/JNK, and MAP kinase to suppress cell growth and to enhance tumor progression due to the EMT [17,18,21,22]. Hence, non-canonic pathways also have a dual role. However, tumor progression activity predominates.

Mutations in the *SMAD4* gene in patients with JPS are associated with an early appearance of polyps and a higher risk of gastrointestinal tumor and malignancy [23]. This may be due to the fact that *BMP1A* mutations affect both pathways (canonic and non-canonic), which are related to the tumor progression, while *SMAD4* mutations only affect the canonic pathway [17,18,21,22]. The unanswered question concerns the molecular mechanisms behind the alteration of this pathway. Of note, normal or early tumor cells show alterations typically associated with suppression of tumor growth, while advanced lesions are associated with alterations that promote cell growth and tumor dissemination. A possible hypothesis could be the second hit theory, which states that cancer results at least from two events related to the genome. The first one is when there is a de novo or inherited mutation in a gene controlling the cell cycle. The second one is when a somatic event complements a possible alteration in the control of the cell cycle, thus producing tumor progression, dissemination, and metastasis. The second hit could result from LOH of the affected gene, the functional inactivation of this gene, or mutations and/ or inactivation in genes that are related to the affected gene or genes controlling the cell cycle (e.g., oncogenes, tumor suppressor, genome repair, and apoptosis genes). However, no positive results were obtained after analysis of thousands of genes that may be associated with the second hit by next-generation sequencing (NGS). This could be explained by the requirement of studying an extensive group of genes or because the second hit may not be associated with a mutation—instead it may be associated with a suppression of gene expression and/or gene functional inactivation (a post-translational alteration such as promoter methylation) [16].

The fact that some tumor samples do not show expression of certain proteins, even though there is evidence of gene structural alteration, suggests epigenetic factors acting as a second hit. They may contribute to the inactivation of genes controlling cell cycle or promoting the activity of tumor progression genes [16]. The relationship of hypoxia with modification of cell behavior is also well described. When a tumor cell divides, the oxygen supply diminishes, thus activating the cell response to hypoxia which depends on the stability of hypoxia-inducible factor 1-alpha (HIF1a). HIF1a is a transcription factor that trans-activates the genes related to erythrocyte production, angiogenesis, and reprogramming of cell metabolism by the increase in aerobic glycolysis [19]. HIF1a also induces expression of genes associated with EMT such as Twist, Snail, Slug, and ZEB1/2, which act during tumor progression, invasion, and metastasis. Therefore, hypoxia modulates the cell behavior, increasing the survival and tumor progression [19]. Finally, the EMT is also related to epigenetic factors that alter the cadherin E expression due to its promoter methylation and the expression of Snail and Slug due to the DNA methylation in their first intronic region. Moreover, miRNAs such as miRNA-200, which is associated with epithelial differentiation, are inhibited by ZEB1 (an EMT activator). Contrarily, p53 increases the expression of miRNA-200. Therefore, a loss of function of this gene is associated with a greater EMT [22]. Any of these factors could be at play in the progression of juvenile polyps to carcinoma.

19.3 Epidemiology

The incidence of JPS ranges from approximately 1:100.000 to 1:160.000 live births [24]. About 50%–60% of JPS patients harbor a germline mutation in the *SMAD4* or *BMPR1A* gene [25] and up to 86% of affected patients with a germline *SMAD4* mutation have extracolonic involvement [26,27]. The most common mutation in *SMAD4* is a four-base deletion at the exon 9 [28]. A gastric phenotype (juvenile gastric polyposis) is more frequently found in patients carrying a *SMAD4* mutation (73%) than in those carrying a *BMPR1A* mutation (8%) [26,29].

Patients with JPS should be distinguished from those with sporadic juvenile polyps, which are detected in approximately 2% of the children who have endoscopy. Polyps are most commonly found in children 4–5 years of age [8]. Sporadic juvenile polyps are not associated with an increased risk of gastrointestinal cancer [31]. A family history of juvenile polyps is found in 20%–50% of patients with JPS, showing an autosomal dominant inheritance pattern with variable penetrance [8].

Patients with JPS are at increased risk of cancer, especially colorectal and stomach cancer. The reported cumulative lifetime risk of gastrointestinal cancer in patients with JPS ranges from 9% to 86% [32–40]. Höfting et al. were the first to report the presence of colorectal and gastric cancer in 98 patients, 13.6% and 8.8%, respectively [33]. Howe et al. reported a 38% lifetime risk of developing colorectal cancer and 21% risk of upper gastrointestinal cancer [37]. Brosens et al. reported a 34-fold increase in risk of developing colorectal cancer among JPS patients in comparison with the general population, with the estimated cumulative lifetime risk of 38.7% [38]. A review in a Japanese population revealed the lifetime risk (at 70 years) of 86.2% for developing any malignant tumor without a gender predilection [40]. Pancreatic and small intestinal carcinomas have been reported in a small number of patients with JPS [34,35,40]. Other tumors that have been documented in JPS patients include breast cancer, bladder cancer, thyroid cancer, sarcomatoid carcinoma, and choriocarcinoma [40–42].

Congenital defects occur in approximately 15% of JPS patients, who typically lack germline mutations in characteristic genes. A subset of patients with JPS also has hereditary hemorrhagic telangiectasias, gastrointestinal vascular malformations, mucocutaneous telangiectasias, and pulmonary arteriovenous malformations [6].

19.4 Clinical Features

While many JPS patients are asymptomatic, some may manifest as anemia, rectal bleeding, abdominal pain, diarrhea, passage of tissue per rectum, bowel obstruction, intussusception, and polyp prolapse [5–8,43]. The polyps are most commonly in the colon (98%) but also in the stomach (60%–85%), duodenum (14%–33%), and small bowel (7%) [1,3,4]. Polyps and symptoms begin to appear in the first decade of life. The average age at diagnosis is 18.5 years, but may be later.

JPS has been phenotypically classified into three categories [44]: (i) juvenile polyposis coli, in which juvenile polyps are restricted to the colorectum, rarely small bowel; (ii) generalized juvenile polyposis, in which polyps arise in the colon, stomach, and small bowel (these first two forms appear to be variable expressions of the same disease and may be sporadic or inherited, and usually present later in childhood or in adult life, and patients have an increased risk of gastrointestinal cancer [25]); and (iii) juvenile polyposis of infancy, characterized by gastric, small bowel, and colorectal polyps. The polyps vary in size from 1 to 30 mm and may be sessile or pedunculated. These patients suffer from diarrhea, protein-losing enteropathy, bleeding, and rectal prolapse. Many of these patients have congenital abnormalities, including macrocephaly and generalized hypotonia. Death usually occurs at an early age [25].

Extraintestinal manifestations of JPS are common (30%) and sometimes clinically significant with associated with mobility and mortality [45]. These manifestations include midgut malrotation, heart defects, vascular malformations, cranial abnormalities, cleft palate, and polydactyly [45,46]. Some patients with *SMAD4* mutations have a variant of the syndrome termed JPS/HHT (JPS-hereditary hemorrhagic telangiectasia [HHT] syndrome (21%–81%) [16,26,47,48]. These patients had visceral arteriovenous malformations (AVM) (86%), pulmonary AVMs (58%–81%), epistaxis (61%–71%), intrapulmonary shunting on echocardiogram (61%), telangiectasias (57%), mucocutaneous telangiectasias (48%), hepatic AVMs (38%), aortopathy (38%), and intracranial AVMs (4%) [5,48–50].

Macroscopically, the polyps vary in size from 5 to 50 mm (mean 1.06 cm), and typically have spherical, lobulated, and pedunculated appearance with surface erosion (Figure 19.2a). Most large polyps are pedunculated, but small polyps, mainly in the stomach, are sessile (Figure 19.2b). On cut section, there are cystic spaces filled with mucin [25,51]. Microscopically, juvenile polyps are characterized by strikingly edematous and inflamed lamina propria, often with surface erosion and marked chronic and active inflammation with granulation tissue (Figures 19.3a–k). Regardless of the site, juvenile polyps almost always show glandular distortion with dilated cystic glands filled with mucin, which is frequently thick. Juvenile polyps show the least degree of smooth muscle proliferation in the lamina propria [52], especially compared to Peutz–Jeghers polyps (Figures 19.3l and m); almost half of juvenile polyps show no smooth muscle proliferation. In general, sporadic juvenile polyps consist of one rounded structure, whereas large syndromic juvenile polyps often have multiple lobes. Importantly, the flat mucosa between the polyps is normal, which helps distinguish them from inflammatory pseudopolyps that are seen in patients with inflammatory bowel disease, Cronkhite–Canada syndrome, or in the stomach, Ménétrier disease. A subset has dysplasia (e.g., 15% of gastric polyps [4]). Lymphoid follicles can be present [52]. Polyps from individuals with *SMAD4* mutations can show loss of protein on immunohistochemistry [53] and often have a more proliferative phenotype and less stroma compared to those from patients with *BMPR1A* mutations [54].

19.5 Diagnosis

The diagnostic criteria for JPS were established in 1975 and revised by Jass et al. in 1988 [36]. According to these, one of the following must be present: (i) multiple juvenile polyps throughout the gastrointestinal tract and/or (ii) more than five juvenile polyps in the colorectum and/or (iii) any number of juvenile polyps with a family history of juvenile polyposis. The 2010 WHO classification changed the number of polyps to "more than three to five" for the first criterion [55].

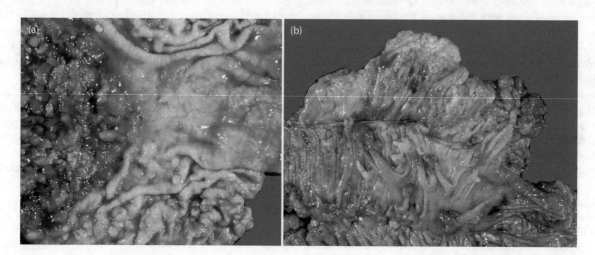

FIGURE 19.2 (a) Juvenile polyposis syndrome, resection of colon. Many of the polyps are multilobulated on long stalks, but close inspection reveals smaller dome-like polyps. (b) Juvenile polyposis, gastric resection. Numerous polyps are present, ranging in size from tiny nodules to large pedunculated mucoid lesions.

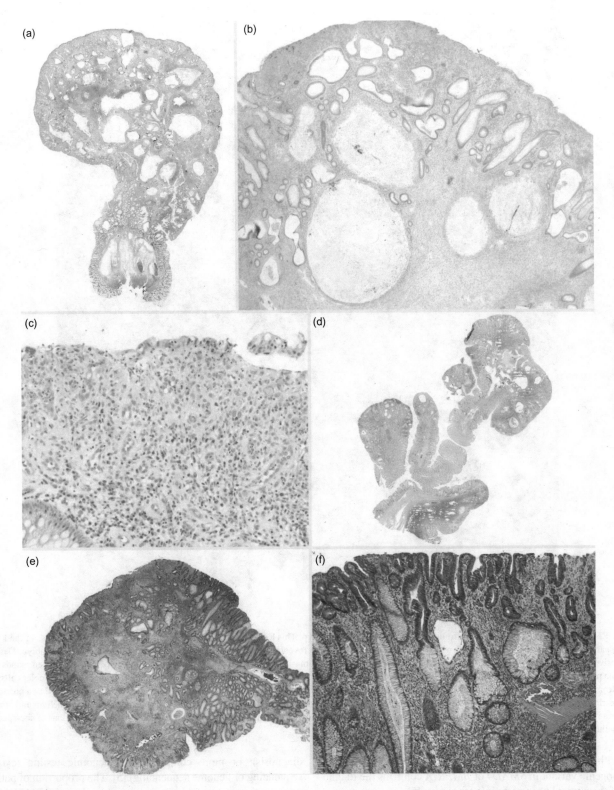

FIGURE 19.3 (a) Sporadic juvenile polyp. Note that the surface is smooth and only one lobule is present. The glands in the polyp are dilated and separated by abundant stroma. This type of polyp differs from the inflammatory pseudopolyps in inflammatory bowel disease because the adjoining mucosa in the stalk is wholly normal. (b) Sporadic juvenile polyp. Note the dilated glands that have burst and extruded their mucin into the lamina propria. (c) Sporadic colorectal juvenile polyp. The surface of this polyp is eroded with underlying granulation tissue. Note the reactive epithelial changes at the upper part of the image. (d) Syndromic colorectal juvenile polyp. This polyp has complex multilobulated architecture, although the adjoining flat mucosa at the base is normal appearing. (e) Syndromic colorectal juvenile polyp. This example has low-grade dysplasia (adenoma-like change) on the surface at the upper part of the image. (f) Syndromic colorectal juvenile polyp. This is a higher magnification of the lesion seen in Figure 19.2. The dysplastic nuclei are hyperchromatic.

(Continued)

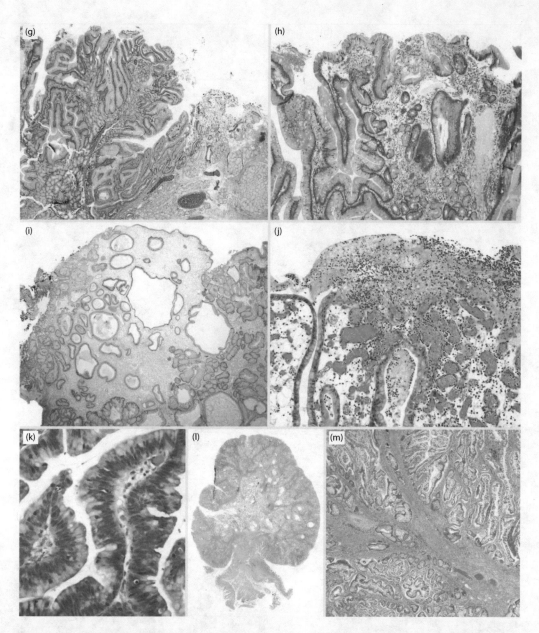

FIGURE 19.3 (CONTINUED)   (g) Syndromic small intestinal juvenile polyp. This lesion arose in the duodenum. There are Brunner glands at the lower right part of the image. Note that the polyp at the left has extensive gastric foveolar metaplasia. (h) Syndromic small intestinal juvenile polyp. Despite extensive gastric foveolar metaplasia, the lesion shows dilated glands and stromal edema. (i) Syndromic gastric juvenile polyp. Note the dilated glands and lamina propria edema. (j) Syndromic gastric juvenile polyp. The surface is eroded and there are striking reactive epithelial changes in the foveolar cells. (k) Syndromic gastric juvenile polyp. This lesion has low-grade dysplasia with gastric foveolar differentiation. This dysplasia can be considered as a precursor to gastric carcinoma in these patients. (l) Small intestinal Peutz–Jeghers polyp. Note that cords of smooth muscle separate zones of epithelium and lamina propria into nests. (m) Small intestinal Peutz–Jeghers polyp. Note the thick cords of eosinophilic smooth muscle that partitions the sections of site-specific epithelium and lamina propria.

With newer techniques, the identification of a heterozygous pathogenic variant in *SMAD4* or *BMPR1A* confirms the diagnosis if the clinical phenotype is inconclusive [5].

Molecular genetic testing approaches can include *SMAD4* or *BMPR1A* concurrent testing (deletion/duplication analysis); serial single-gene testing can be considered in patients with manifestations suggestive of JPS/HHT (*SMAD4* is performed first; *BMPR1A* should be considered next if *SMAD4* pathogenic variant is absent; and finally, molecular genetic testing of HHT-related genes may be considered); multigene panel (including *SMAD4*, *BMPR1A*, *PTEN*, and other genes in the differential diagnosis); or more comprehensive genomic testing (exome sequencing or genome sequencing) [5]. The proportion of pathogenic variants detectable by sequence analysis in *BMPR1A* is 69%–85% and in *SMAD4* 83%, while by the gene-targeted deletion/duplication analysis in *BMPR1A* is 15% and in *SMAD4* 17% [5,13,14,26,45,56].

There is an incomplete correlation between genotype and phenotype. Some members of the same family can have more than 100 polyps and others just a few. Of note, 1%–2% of individuals in the general population develop a solitary juvenile polyp, but they do not meet any diagnostic criteria for JPS [5].

Other genetically related disorders may also contain *BMPR1A* and *SMAD4* mutations. Patients with hereditary mixed polyposis syndrome have *BMPR1A* germline mutations. These patients have adenomatous, hyperplastic, and atypical juvenile polyps. Myhre syndrome is caused by a heterozygous gain of function pathogenic variant located in the amino acids 496 or 500 of SMAD4 protein [5,57,58].

Further differential diagnoses for JPS include the PTEN hamartoma tumor syndrome (*PTEN*), nevoid basal cell carcinoma syndrome (*PTCH1, SUFU*), Peutz–Jeghers syndrome (*STK11*), hereditary mixed polyposis syndrome (15q13q14, duplications lead to overexpression of *GREM*1), familial adenomatous polyposis (*APC*), MUTYH-associated polyposis (*MUTYH*), and Lynch syndrome (*MLH1, MSH2, MSH6, PMS2, EPCAM*) [5]. Differential diagnosis for JPS/HHT is hereditary hemorrhagic telangiectasia (not associated with *SMAD4*), which is characterized by gastrointestinal bleeding and anemia without polyposis [5].

19.6 Treatment

The management of patients with JPS has the goals of assessment, surveillance, and/or treatment of the nonmalignant conditions and monitoring and addressing the risk of development of malignant tumors [43]. The most effective management is routine colonoscopy with endoscopic polypectomy. This may reduce morbidity, reducing the risk of cancer, bleeding, or intestinal obstruction [5].

Current guidelines [51] suggest (i) surveillance of the gastrointestinal tract in affected or at-risk JPS patients including screening for colon, stomach, and small bowel cancers (strong recommendation, very low quality of evidence); (ii) colectomy and ileorectal anastomosis or proctocolectomy and ileal pouch anal anastomosis is indicated for polyp-related symptoms, or when the polyps cannot be managed endoscopically (strong recommendation, low quality of evidence); and (iii) cardiovascular examination for and evaluation for hereditary hemorrhagic telangiectasia should be considered for *SMAD4* mutation carriers (conditional recommendation, very low quality of evidence).

Follow-up recommendations include surveillance with colonoscopy every year, beginning at age 12 years or earlier if symptoms occur, especially rectal bleeding. Colonoscopy should be repeated every 1–3 years depending on polyp burden, and polyps ≥5 mm should be removed [45,51]. Upper endoscopy is recommended every 1–3 years beginning at age 12 years, or earlier for symptoms, and should be repeated every 1–3 years, depending on severity, with removal of polyps ≥5 mm [51]. Patients with mild polyposis can be managed by frequent endoscopic examinations and polypectomy [25,44,59,60]. Intraoperative enteroscopy to evaluate small intestinal polyps can be considered at the time of colorectal surgery [61].

Prophylactic total or subtotal colectomy or gastrectomy should be considered in patients with multiple polyps (>50–100 polyps), severe symptoms, juvenile polyps with dysplasia, or a family history of colorectal cancer. The decision between subtotal versus total proctocolectomy is made on the basis of the rectal polyp burden [60,62,63]. Patients who have undergone proctocolectomy and subtotal colectomy with ileorectal anastomosis continue to require endoscopic follow-up [62].

Other screening should include annual complete blood count, cardiovascular examination, and assessment for hereditary hemorrhagic telangiectasia if the patient has a *SMAD4* mutation [45,51].

High expression of COX-2 has been demonstrated in JPS polyps and has been correlated with polyp size and dysplasia [64,65]. JPS patients with *BMPR1A* germline defects have higher COX-2 expression than do JPS patients in whom no germline mutation is detected [64]. However, to date there are no available clinical trials to support the use of nonsteroidal anti-inflammatory drugs as chemoprevention.

19.7 Prevention

The key issue for JPS patients is to avoid progression to cancer and to diagnose related diseases early. Prenatal testing for pregnant women at risk is possible if the pathogenic variants in the family are known. Approximately 20%–50% of individuals with JPS have an affected parent. However, 25%–50% of probands with JPS have no previous history of polyps in the family and may have a de novo alteration. Each sibling of an affected parent has a 50% chance of inheriting the pathogenic variant and developing JPS. It is necessary to evaluate the relatives of an affected individual to prompt the initiation of treatment and preventive measures. The evaluation can include molecular genetic approaches [5,66]. Furthermore, all JPS patients with an identified *SMAD4* mutation or family history suggestive of HHT should actively screen for HHT. It is also necessary to assess for pulmonary AV-malformation and thoracic aortic dilation [67,68].

19.8 Conclusion

JPS is a rare autosomal dominant hereditary disease that is genetically and phenotypically heterogeneous. Many examples of JPS are caused by germline mutations in *SMAD4* or *BMPR1A*. These alterations increase the risk of cancer and congenital disorders.

Early diagnosis of JPS has important consequences for the patients and their relatives [69]. Therefore, the gastroenterologist and pathologist confronted with the diagnosis of a juvenile polyp should consider the possibility of JPS and recommended genetic counseling if it is likely [70]. Currently, multigene panels are available for the diagnosis of polyposis syndromes and colorectal cancers [71,72]. However, more data are needed to clarify the importance of genes with uncertain significance that are sometimes detected on such panels [30,73].

Future studies in JPS are likely to focus on molecular pathways to allow better understanding of gastrointestinal carcinogenesis with the goal of offering improved detection and treatment to JPS patients and their relatives.

REFERENCES

1. Schreibman IR, Baker M, Amos C, McGarrity TJ. The hamartomatous polyposis syndromes: A clinical and molecular review. *Am J Gastroenterol.* 2005;100(2):476–90.
2. Mccoll I, Busxey HJ, Veale AM, Morson BC. Juvenile polyposis coli. *Proc R Soc Med.* 1964;57:896–7.

3. Chow E, Macrae F. A review of juvenile polyposis syndrome. *J Gastroenterol Hepatol.* 2005;20(11):1634–40.

4. Brosens LAA, Wood LD, Offerhaus GJ et al. Pathology and genetics of syndromic gastric polyps. *Int J Surg Pathol.* 2015;24(3):185–99.

5. Larsen Haidle J, Howe JR. Juvenile polyposis syndrome. In: Adam MP, Ardinger HH, Pagon RA et al., editors. *GeneReviews® [Internet].* Seattle (WA): University of Washington, Seattle; 1993–2018. 2003 May 13.

6. Jasperson KW, Tuohy TM, Neklason DW, Burt RW. Hereditary and familial colon cancer. *Gastroenterology.* 2010;138(6):2044–58.

7. Jung I, Gurzu S, Turdean GS. Current status of familial gastrointestinal polyposis syndromes. *World J Gastrointest Oncol.* 2015;7(11):347–55.

8. Manfredi M. Hereditary hamartomatous polyposis syndromes: Understanding the disease risks as children reach adulthood. *Gastroenterol Hepatol (N Y).* 2010;6(3):185–96.

9. Howe JR, Sayed MG, Ahmed AF et al. The prevalence of MADH4 and BMPR1A mutations in juvenile polyposis and absence of BMPR2, BMPR1B, and ACVR1 mutations. *J Med Genet.* 2004;41(7):484–91.

10. Ma H, Brosens LAA, Offerhaus GJA, Giardiello FM, de Leng WWJ, Montgomery EA. Pathology and genetics of hereditary colorectal cancer. *Pathology.* 2018;50(1):49–59.

11. Dahdaleh FS, Carr JC, Calva D, Howe JR. Juvenile polyposis and other intestinal polyposis syndromes with microdeletions of chromosome 10q22-23. *Clin Genet.* 2012;81(2):110–6.

12. Ahmed A, Alsaleem B. Nonfamilial juvenile polyposis syndrome with exon 5 novel mutation in SMAD 4 gene. *Case Rep Pediatr.* 2017;2017:5321860.

13. Calva-Cerqueira D, Chinnathambi S, Pechman B, Bair J, Larsen-Haidle J, Howe JR. The rate of germline mutations and large deletions of SMAD4 and BMPR1A in juvenile polyposis. *Clin Genet.* 2009;75(1):79–85.

14. Calva-Cerqueira D, Dahdaleh FS, Woodfield G et al. Discovery of the BMPR1A promoter and germline mutations that cause juvenile polyposis. *Hum Mol Genet.* 2010;19(23):4654–62.

15. Wosiak A, Wodziński D, Kolasa M, Sałagacka-Kubiak A, Balcerczak E. SMAD-4 gene expression in human colorectal cancer: Comparison with some clinical and pathological parameters. *Pathol Res Pract.* 2017;213(1):45–9.

16. Jelsig A, Qvist N, Brusgaard K, Nielsen C, Hansen T, Ousager L. Hamartomatous polyposis syndromes: A review. *Orphanet J Rare Dis.* 2014;9(1):101.

17. Wrana JL. Signaling by the TGFβ superfamily. *Cold Spring Harb Perspect Biol.* 2013;5(10):a011197.

18. Jung B, Staudacher JJ, Beauchamp D. Transforming growth factor β superfamily signaling in development of colorectal cancer. *Gastroenterology.* 2017;152(1):36–52.

19. Yokobori T, Nishiyama M. TGF-β signaling in gastrointestinal cancers: Progress in basic and clinical research. *J Clin Med.* 2017;6(1).

20. Weiss A, Attisano L. The TGFbeta superfamily signaling pathway. *Wiley Interdiscip Rev Dev Biol.* 2013;2(1):47–63.

21. Bachman KE, Park BH. Duel nature of TGF-beta signaling: Tumor suppressor vs. tumor promoter. *Curr Opin Oncol.* 2005;17(1):49–54.

22. Lebrun J-J. The dual role of TGFβ in human cancer: From tumor suppression to cancer metastasis. *ISRN Mol Biol.* 2012;2012:381428.

23. Aytac E, Sulu B, Heald B et al. Genotype-defined cancer risk in juvenile polyposis syndrome. *Br J Surg.* 2015;102(1):114–8.

24. Burt RW, Bishop DT, Lynch HT, Rozen P, Winawer SJ. Risk and surveillance of individuals with heritable factors for colorectal cancer. WHO Collaborating Centre for the Prevention of Colorectal Cancer. *Bull World Health Organ.* 1990;68(5):655–65.

25. Brosens LAA, Langeveld D, van Hattem WA, Giardiello FM, Offerhaus GJA. Juvenile polyposis syndrome. *World J Gastroenterol.* 2011;17(44):4839–44.

26. Aretz S, Stienen D, Uhlhaas S et al. High proportion of large genomic deletions and a genotype phenotype update in 80 unrelated families with juvenile polyposis syndrome. *J Med Genet.* 2007;44(11):702–9.

27. Sayed MG, Ahmed AF, Ringold JR et al. Germline SMAD4 or BMPR1A mutations and phenotype of juvenile polyposis. *Ann Surg Oncol.* 2002;9(9):901–6.

28. Wu TT, Rezai B, Rashid A et al. Genetic alterations and epithelial dysplasia in juvenile polyposis syndrome and sporadic juvenile polyps. *Am J Pathol.* 1997;150(3):939–47.

29. Friedl W, Uhlhaas S, Schulmann K et al. Juvenile polyposis: Massive gastric polyposis is more common in MADH4 mutation carriers than in BMPR1A mutation carriers. *Hum Genet.* 2002;111(1):108–11.

30. LaDuca H, Stuenkel AJ, Dolinsky JS et al. Utilization of multigene panels in hereditary cancer predisposition testing: Analysis of more than 2,000 patients. *Genet Med.* 2014;16(11):830–7.

31. Nugent KP, Talbot IC, Hodgson SV, Phillips RK. Solitary juvenile polyps: Not a marker for subsequent malignancy. *Gastroenterology.* 1993;105(3):698–700.

32. Järvinen H, Franssila KO. Familial juvenile polyposis coli; increased risk of colorectal cancer. *Gut.* 1984;25(7):792–800.

33. Höfting I, Pott G, Stolte M. [The syndrome of juvenile polyposis]. *Leber Magen Darm.* 1993;23(3):107–8, 111–2.

34. Coburn MC, Pricolo VE, DeLuca FG, Bland KI. Malignant potential in intestinal juvenile polyposis syndromes. *Ann Surg Oncol.* 1995;2(5):386–91.

35. Agnifili A, Verzaro R, Gola P et al. Juvenile polyposis: Case report and assessment of the neoplastic risk in 271 patients reported in the literature. *Dig Surg.* 1999;16(2):161–6.

36. Jass JR, Williams CB, Bussey HJ, Morson BC. Juvenile polyposis--a precancerous condition. *Histopathology.* 1988;13(6):619–30.

37. Howe JR, Mitros FA, Summers RW. The risk of gastrointestinal carcinoma in familial juvenile polyposis. *Ann Surg Oncol.* 1998;5(8):751–6.

38. Brosens LAA, van Hattem A, Hylind LM et al. Risk of colorectal cancer in juvenile polyposis. *Gut.* 2007;56(7):965–7.

39. Ma C, Giardiello FM, Montgomery EA. Upper tract juvenile polyps in juvenile polyposis patients: Dysplasia and malignancy are associated with foveolar, intestinal, and pyloric differentiation. *Am J Surg Pathol.* 2014;38(12):1618–26.

40. Ishida H, Ishibashi K, Iwama T. Malignant tumors associated with juvenile polyposis syndrome in Japan. *Surg Today.* 2018;48(3):253–63.

41. Silva-Smith R, Sussman DA. Co-occurrence of Lynch syndrome and juvenile polyposis syndrome confirmed by multigene panel testing. *Fam Cancer.* 2018;17(1):87–90.

42. Parra-Medina R, Correa PL, Moreno JJ, Lucero PM, Yaspe E, Polo F. Carcinosarcoma with choriocarcinomatous and

osteosarcomatous differentiation in a patient with juvenile polyposis syndrome. *Rare Tumors.* 2015;7(3):5778.

43. Rustgi AK. Hereditary gastrointestinal polyposis and nonpolyposis syndromes. *N Engl J Med.* 1994;331(25):1694–702.

44. Sachatello CR, Griffen WO. Hereditary polyploid diseases of the gastrointestinal tract: A working classification. *Am J Surg.* 1975;129(2):198–203.

45. Latchford AR, Neale K, Phillips RKS, Clark SK. Juvenile polyposis syndrome. *Dis Colon Rectum.* 2012;55(10):1038–43.

46. Desai DC, Murday V, Phillips RK, Neale KF, Milla P, Hodgson SV. A survey of phenotypic features in juvenile polyposis. *J Med Genet.* 1998;35(6):476–81.

47. Gallione CJ, Repetto GM, Legius E et al. A combined syndrome of juvenile polyposis and hereditary haemorrhagic telangiectasia associated with mutations in MADH4 (SMAD4). *Lancet* (London, England). 2004;363(9412):852–9.

48. O'Malley M, LaGuardia L, Kalady MF et al. The prevalence of hereditary hemorrhagic telangiectasia in juvenile polyposis syndrome. *Dis Colon Rectum.* 2012;55(8):886–92.

49. Nishida T, Faughnan ME, Krings T et al. Brain arteriovenous malformations associated with hereditary hemorrhagic telangiectasia: Gene-phenotype correlations. *Am J Med Genet A.* 2012;158A(11):2829–34.

50. Wain KE, Ellingson MS, McDonald J et al. Appreciating the broad clinical features of SMAD4 mutation carriers: A multicenter chart review. *Genet Med.* 2014;16(8):588–93.

51. Syngal S, Brand RE, Church JM et al. ACG clinical guideline: Genetic testing and management of hereditary gastrointestinal cancer syndromes. *Am J Gastroenterol.* 2015;110(2):223–62; quiz 263.

52. Shaco-Levy R, Jasperson KW, Martin K et al. Morphologic characterization of hamartomatous gastrointestinal polyps in Cowden syndrome, Peutz-Jeghers syndrome, and juvenile polyposis syndrome. *Hum Pathol.* 2016;49:39–48.

53. Langeveld D, van Hattem WA, de Leng WWJ et al. SMAD4 immunohistochemistry reflects genetic status in juvenile polyposis syndrome. *Clin Cancer Res.* 2010;16(16):4126–34.

54. van Hattem WA, Langeveld D, de Leng WWJ et al. Histologic variations in juvenile polyp phenotype correlate with genetic defect underlying juvenile polyposis. *Am J Surg Pathol.* 2011;35(4):530–6.

55. Bosman FT, Carneiro F, Hruban RH, Theise ND. *WHO Classification of Tumours of the Digestive System.* 4th ed. France: IARC; 2010.

56. van Hattem WA, Brosens LAA, de Leng WWJ et al. Large genomic deletions of SMAD4, BMPR1A and PTEN in juvenile polyposis. *Gut.* 2008;57(5):623–7.

57. O'Riordan JM, O'Donoghue D, Green A et al. Hereditary mixed polyposis syndrome due to a BMPR1A mutation. *Colorectal Dis.* 2010;12(6):570–3.

58. Caputo V, Bocchinfuso G, Castori M et al. Novel SMAD4 mutation causing Myhre syndrome. *Am J Med Genet A.* 2014;164A(7):1835–40.

59. Scott-Conner CE, Hausmann M, Hall TJ, Skelton DS, Anglin BL, Subramony C. Familial juvenile polyposis: Patterns of recurrence and implications for surgical management. *J Am Coll Surg.* 1995;181(5):407–13.

60. Howe JR, Ringold JC, Hughes JH, Summers RW. Direct genetic testing for Smad4 mutations in patients at risk for juvenile polyposis. *Surgery.* 1999;126(2):162–70.

61. Rodriguez-Bigas MA, Penetrante RB, Herrera L, Petrelli NJ. Intraoperative small bowel enteroscopy in familial adenomatous and familial juvenile polyposis. *Gastrointest Endosc.* 1995;42(6):560–4.

62. Oncel M, Church JM, Remzi FH, Fazio VW. Colonic surgery in patients with juvenile polyposis syndrome: A case series. *Dis Colon Rectum.* 2005;48(1):49–55–6.

63. Cairns SR, Scholefield JH, Steele RJ et al. Guidelines for colorectal cancer screening and surveillance in moderate and high risk groups (update from 2002). *Gut.* 2010;59(5):666–89.

64. van Hattem WA, Brosens LAA, Marks SY et al. Increased cyclooxygenase-2 expression in juvenile polyposis syndrome. *Clin Gastroenterol Hepatol.* 2009;7(1):93–7.

65. Gupta S Das, Das RN, Ghosh R et al. Expression of COX-2 and p53 in juvenile polyposis coli and its correlation with adenomatous changes. *J Cancer Res Ther.* 2016;12(1):359–63.

66. Bronner MP. Gastrointestinal inherited polyposis syndromes. *Mod Pathol.* 2003;16(4):359–65.

67. McDonald J, Bayrak-Toydemir P, Pyeritz RE. Hereditary hemorrhagic telangiectasia: An overview of diagnosis, management, and pathogenesis. *Genet Med.* 2011;13(7):607–16.

68. Heald B, Rigelsky C, Moran R et al. Prevalence of thoracic aortopathy in patients with juvenile polyposis syndrome-hereditary hemorrhagic telangiectasia due to SMAD4. *Am J Med Genet A.* 2015;167A(8):1758–62.

69. Hussain T, Church JM. Juvenile polyposis syndrome. *Clin Case Rep.* 2019;8(1):92–5.

70. Rosty C. The role of the surgical pathologist in the diagnosis of Ggastrointestinal polyposis syndromes. *Adv Anat Pathol.* 2018;25(1):1–13.

71. Gallego CJ, Shirts BH, Bennette CS et al. Next-generation sequencing panels for the diagnosis of colorectal cancer and polyposis syndromes: A cost-effectiveness analysis. *J Clin Oncol.* 2015;33(18):2084–91.

72. Domchek SM, Bradbury A, Garber JE, Offit K, Robson ME. Multiplex genetic testing for cancer susceptibility: Out on the high wire without a net? *J Clin Oncol.* 2013;31(10):1267–70.

73. Susswein LR, Marshall ML, Nusbaum R et al. Pathogenic and likely pathogenic variant prevalence among the first 10,000 patients referred for next-generation cancer panel testing. *Genet Med.* 2016;18(8):823–32.

20

Lynch Syndrome

Andreas V. Hadjinicolaou and Mashiko Setshedi

CONTENTS

20.1 Introduction

Lynch syndrome (LS, also referred to as hereditary nonpolyposis colorectal cancer [HNPCC]) is a rare genetic condition with a typically autosomal dominant inheritance pattern, and as the most prevalent form of hereditary colorectal cancer (CRC) it accounts for 2%–7% of all CRC. The prototype abnormality in LS is a defect in one of the mismatch repair (MMR) genes [1]; consequently, >90% of all CRC in LS patients have high microsatellite instability (MSI-high).

Characteristically, carriers of mutations in one of the MMR pathway components have a very high risk of developing early-onset (by age 50) CRC (25%–70%), endometrial cancer (EC) (30%–70%) and other tumors (e.g., ovarian, renal, ureteral, gastric, and biliary cancers) in comparison to the general population (0.2% for CRC). They also have a significantly higher lifetime risk for cancer development compared to the general population (>80% vs 2% for CRC, 71% vs 1.5% for EC, 10%–15% vs <1% for ovarian cancer). In the case of CRC, this high risk relates to precancerous polyps appearing earlier in life and progressing to cancer faster in LS patients (2–3 years) compared to the general population (>10 years). Screening of patients and affected family members by colonoscopy and early removal of polyps is currently the only proven approach by which CRC in LS can be prevented. Tumor and genetic testing are also offered to patients for diagnosis, surveillance, and family planning. Aspirin appears to be the most effective chemotherapy agent that offers some protection.

20.2 History

LS was one of the first hereditary cancer syndromes to be described. It was originally reported in 1913 by Aldred S. Warthin, a pathologist who mapped the family history and pathology of three cancer-prone families, also noting the autosomal mendelian pattern of inheritance. The syndrome, however, is named after Henry T. Lynch, who in 1962 encountered a patient with a similar family history to that described by Warthin. Over the ensuing 39 years, by careful review of patient and pathology records, Lynch and his colleagues identified more families with this inheritance pattern in association with increased cancer risk [2–4]. Despite strong evidence suggesting a genetic basis for LS, due to the prevailing dogma (that diseases were caused by environmental factors), a genetic link for cancer and thus the notion of cancer heritability had hitherto not been accepted. It was only in the late twentieth century, with the explosion and availability of molecular genetic tools, that the long-described clinical syndrome of LS was recognized as an inherited condition, supported by scientific evidence. To contrast it to the already known condition of familial adenomatous polyposis (FAP), characterized by at least 100 polyps at colonoscopy, LS was termed HNPCC; this, however, was later dropped as a misnomer because of the noted occurrence of both extracolonic cancers and colonic polyps in this condition (albeit fewer and less prominent than in FAP). By consensus, therefore, it is now known as LS [4–6].

20.3 Pathophysiology

20.3.1 Molecular Mechanism of Disease

The primary abnormality in LS is an autosomal dominant heterozygous germline mutation in one of the alleles of genes encoding for components of the MMR pathway. As such, carriers have a high cancer risk without necessarily developing cancer. Those predisposed individuals that eventually develop a cancer suffer an additional somatic mutation in the remaining wild-type allele, resulting in complete loss of function of the MMR gene in question. When fully functional, the MMR pathway is able to repair single base–pair mismatches and small insertion or deletion loops (IDLs) that often form during DNA replication when the polymerase enzyme attempts to replicate small repeat sequences [7,8]. IDLs lead to length alterations in nucleotide sequences causing frameshift mutations and DNA mismatches if not repaired by the MMR machinery. As IDLs commonly occur within regions of repetitive nucleotide sequences called microsatellites, LS-associated cancers with a defective MMR pathway typically manifest with what is known as microsatellite instability (MSI). The presence of MSI is itself independently associated with a 100-fold increase in the mutation rate. This means that tumors with MSI have a mutator phenotype (>12 mutations per 10^6 bases) [8] and carriers of MMR gene mutations have a rapid adenoma-carcinoma progression sequence. In the absence of any MMR gene mutations, tumors are said to be microsatellite stable, whereas tumors are MSI-low in the presence of one mutation, and MSI-high in the presence of two or more mutations [9]. Accordingly, MSI tumors are mutation-prone and are linked to a more aggressive phenotype and consequently a worse prognosis [10]. Mutations often reported within the aforementioned tandem repeats include those of genes related to carcinogenesis, e.g. transforming growth factor-β type II receptor (TGFBR2), adenomatous polyposis coli (APC), and BCL2-associated X protein (BAX) [11,12].

LS has been associated with functional defects in four distinct MMR proteins, namely hMLH1, hMSH2, hMSH6, and hPMS2. Based on profiling of germline mutations found in LS-associated tumors, hMSH2 and hMLH1 are the most frequently mutated MMR genes (40%–50% and 30%–37% of tumors, respectively) [13], whereas hMSH6 defects are found in approximately 7%–13% of individuals with LS. hPMS2 mutations are considered rare, but in some studies up to 9% of LS-associated tumors were reported to carry germline mutations in the hPMS2 gene. In addition, a constitutional 3′ end deletion of EPCAM, which is immediately upstream of the MSH2 gene, may cause LS through

epigenetic silencing of MSH2 [14,15]. With this, MSH2 may be implicated in up to 60% of LS-associated tumors. Genes responsible for the remaining cases of LS remain to be discovered but there have been case reports of other MMR components that might be implicated, such as hPMS1. Notably, carriers of a MSH2 gene mutation have the highest general cancer risk across the spectrum, especially for the development of urological cancers. On the other hand, MSH6 mutation carriers have the highest risk of developing EC [16,17]. Surprisingly, EC with MSH6 mutations show low MSI [18,19], although instability, specifically at mononucleotide repeat sequences, is consistently observed [20]. The general risk of developing cancer in MLH1 gene mutation carriers lies between that reported for MSH6 and MSH2 carriers [21–25]. Patients with PMS2 mutation develop CRC with MSI, but atypically there may be absence of a family history or a late age of onset [26–29]. The risk of CRC in EPCAM deletion is similar to that seen with mutations in MLH1, MSH2, MSH6, or a combined EPCAM–MSH2 deletion [30]. In contrast to CRC, the risk of developing EC is lower in women with EPCAM deletion compared to that of women with mutated MHL1, MSH2, or MSH6 genes or a combined EPCAM–MSH2 deletion. Notably, MSI is not only observed in LS-associated CRC, as 15% of sporadic CRC demonstrate this feature. Sporadic MSI-high CRC typically develops through somatic promoter methylation of MLH1, leading to silencing of MLH1. Key differences between LS CRC and sporadic CRC are depicted in Table 20.1.

20.3.2 Heterogeneity of Genotype and Phenotype

Variations in LS phenotype, even in cases with the same germline mutation, are common. Although heterozygous mutations cause typical LS, the presence of biallelic MMR mutations is associated with constitutional mismatch repair deficiency (CMMRD) syndrome, a more severe form of disease. CMMRD is characterized by a high incidence of hematological malignancies, glioblastoma, urinary tract tumors, and neurofibromatosis typically in the pediatric age group [31,32]. Other phenotypic variations include age of onset, tumor morphology, MMR protein expression profile, and MSI stability not only between family members, but also between metachronous or even synchronous lesions in the same individual [33,34]. The variability in presentation has been ascribed to the effect of environmental factors and possibly the nature of the genes acquiring somatic mutations due to a mutator phenotype; however, this is largely speculative. This variability has led to genome-wide association studies (GWAS) that have

TABLE 20.1

Differences between LS CRC and Sporadic CRC

Variable	LS CRC	Sporadic CRC
Main defective pathway	MSI pathway	Chromosomal instability pathway
MMR	Deficient	Normal in 85%, deficient in 15%
BRAF mutation	Absent	Present
KRAS mutation	Absent	Present
Epigenetic inactivation of MLH1	Absent	Present
Typical age of onset	44–60 y	69 y
Stage-adjusted survival	Better	Poorer

in fact identified 20 variants. However, data from patients with mutations gave inconsistent results, failing to verify this variability and suggesting a need for further investigation [35,36]. Some variants do have functional significance, whereas the pathogenic potential of missense variants, collectively termed variants of uncertain significance (VUS) is unknown.

20.4 Epidemiology

20.4.1 Prevalence

A fifth of cases of LS CRC have a family history of CRC in at least one first-degree relative. The prevalence of LS in the population has been difficult to estimate mainly because of the variation in the criteria used to define the syndrome in each study. However, based on recent studies looking at CRC patients and their first-degree relatives, 1 in 279 (0.036%) of the population carry mutations in one of the main DNA mismatch repair genes [37]. This suggests that the true prevalence of LS might be higher than previously thought, and a significant portion of the population have LS go undiagnosed as a result of low penetrance. Race does not seem to influence the prevalence of LS, but certain ethnicities or groups, such as individuals of an Icelandic, Scandinavian, French Canadian, or Ashkenazi Jewish origin are more often affected by specific founder mutations in one or more of the four principal MMR genes.

The incidence of LS in the United States is estimated to be 2%–5%, with around 7500 new cases each year. Gender does not seem to be a factor in developing LS, as the syndrome affects both males and females to a similar extent.

20.4.2 Tumor Risk

All in all, LS is associated with an increased lifetime risk of developing CRC with estimates ranging from 22% to 66% [38]. It accounts for 2%–4% of all CRC cases. Patients with LS develop CRC at an early age, around 44–60 years, compared with an average of 69 years in sporadic CRC cases [39]. As a result of a more rapid adenoma–carcinoma sequence progression, the evolution from polyp to malignancy is at least 3 times faster than that observed in sporadic cancer, i.e., 35 months compared to 10–15 years [39–41]. Approximately 70% of CRC in LS arise in the proximal colon [42], although in women these tend to occur in the left colon [40]. Histologically, the CRC observed in LS are poorly differentiated, with a predominant lymphocytic and mucinous phenotype and a near-diploid DNA content (Figure 20.1) [38,43,44]. Patients with LS CRC are known to have better survival than their stage-matched sporadic counterparts, a feature that may be attributed to the presence of tumor-infiltrating lymphocytes, and thus strong tumor-specific immune responses in the former [45,46].

In the case of EC, LS-associated cases account for approximately 2.5% of the total number and are diagnosed at a mean age of between 46 and 49 years compared to 60 years for sporadic EC cases.

LS classically presents with CRC or other cancers at an earlier age than the general population. Tumors of the proximal colon are most common; in fact, one-third are located in the cecum, necessitating full colonoscopy with cecal intubation to identify the lesion [47]. The colorectal tumors in LS are generally non-polypoidal [48]; however, when present, they are significantly reduced in number when compared to FAP. In addition to the well-described occurrence of colorectal and endometrial tumors, an increased incidence of cancers of the stomach, small bowel, hepatobiliary system, upper urinary tract, ovary, and later in life, glioblastoma, pancreas, breast, prostate, and adrenocortical has been noted [49–53]. Notably, as a result of the variable phenotypic expression, individuals with a heterozygous MMR gene mutation may only develop a solitary cancer later in life or even not develop a cancer at all. When skin tumors (including sebaceous gland tumors) occur, this is regarded as a variant of

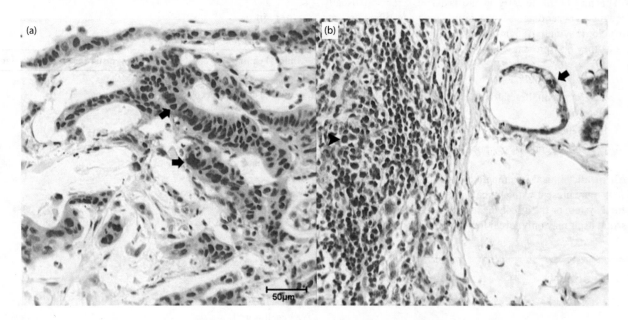

FIGURE 20.1 Mucinous adenocarcinoma of the colon from a 42-year-old patient with Lynch syndrome, showing a strong nuclear staining (red color) of the tumor cells (arrows) for MLH1 (a) and a negative staining of the tumor cells (arrow) for MSH2 (b). As an internal positive control, lymphocytes (arrowhead) in an adjacent lymph follicle shows a positive nuclear staining for MSH2. (Photo credit: Kunstmann E et al. *BMC Med Genet.* 2004;5:16.)

LS, called the Muir–Torre syndrome, initially described in 1981 [54]. These typically occur in patients carrying a mutation in the MSH2 gene [55].

The risk of EC is 30%–40% by age 70 years, with an average age at diagnosis of 46–49 years. In 50% of patients with both CRC and EC, the latter often precedes the diagnosis of the former. In contrast to the high lifetime risks seen with CRC and EC, the risk for developing ovarian, gastric, and glioblastoma by the age of 70 years is only 9%–12%, 13%, and 1%–4%, respectively. The average age at diagnosis is 42.5 years with approximately 30% of these tumors presenting before the age of 40. The mean age at diagnosis of gastric cancer specifically is 56 years, with the intestinal-type adenocarcinoma being the most commonly reported type, especially in Asian countries such as Japan, Korea, and China. Interestingly, Asian patients with LS have a higher lifetime risk of gastric cancer and a lower lifetime risk of EC compared to the global values.

For transitional cell carcinoma of the upper urinary tract (ureters and renal pelvis), general lifetime risk is 4%–10%. Certain groups of LS patients (e.g., those with MSH2 mutations) are at an increased risk not only for upper urinary tract tumors but also bladder cancer. Lifetime risk for development of small bowel adenocarcinoma in LS patients is 1%–3%, significantly higher than the estimated 0.3% for the general population. Small intestinal tumors in LS mainly occur in the duodenum and jejunum.

In the case of glioblastoma, the lifetime risk is 1%–4%. Its presence along with CRC gives rise to Turcot syndrome, which is another variant of LS. Finally, malignancies of the larynx, breast, prostate, liver, biliary tree, pancreas, and the hematopoietic system are all more common in patients with LS compared to the general population.

20.5 Clinical Features

Patients with LS are generally asymptomatic at time of diagnosis. This may be due to early diagnosis during screening, usually following a cancer diagnosis in a family member. When patients have colonic symptoms, however, typically in the second or third decade, they may present with:

1. Change in bowel habits
2. Gastrointestinal/rectal bleeding or fecal occult blood on testing
3. Iron deficiency anemia
4. Loss of weight, loss of appetite

All in all, clinical examination is relatively nonspecific, as patients may present with clinical signs relevant to the gastrointestinal system or the affected extracolonic site. Screening colonoscopy itself may only reveal few or even no polyps.

20.6 Diagnosis

20.6.1 General Considerations

The diagnosis of LS is complex and is based on comprehensive history taking and clinical feature profiling followed by extensive molecular tests including immunohistochemistry (IHC), genetic testing (for MSI), and germline testing in order to characterize the disease phenotype and tumor pathology.

Accurate and detailed history taking is crucial to a diagnosis of LS. Specifically, a family history of CRC and the finding of polyps at previous colonoscopy are essential to make the diagnosis. Further probing about a history of extracolonic cancers or related symptoms is also crucial. A family tree should be drafted, going back to as many generations as possible with details of the age of the patient or their family members at which cancer was diagnosed, the lineage (maternal/paternal), and the occurrence of synchronous or metachronous lesions. Thorough history taking will differentiate between a cancer syndrome and a diagnosis of sporadic cancer. This will prompt further genetic testing and extracolonic screening where appropriate to confirm LS, investigations that would otherwise be neglected leading to underdiagnosis [56] and poor patient care. To improve detection of LS, tumor testing is recommended for all patients under 70 years of age with CRC and/or EC [57].

20.6.2 Diagnostic Criteria

The diagnosis is made on the basis of the Amsterdam 1 or the revised Amsterdam 2 criteria as shown in Table 20.2. Diagnosis according to the Amsterdam 1 criteria requires all of the following [58]:

1. Three or more family members with a confirmed diagnosis of CRC, one of whom must be a first-degree relative of the other two.
2. Two successive affected generations (i.e., one of the patients must be a parent or offspring of another patient).
3. One or more CRCs diagnosed in a relative younger than 50 years.
4. FAP must have been excluded.

TABLE 20.2

Amsterdam 1 and 2 Criteria

Clinical Feature	Amsterdam 1	Amsterdam 2
Three or more family members with a confirmed diagnosis of colorectal cancer, one of whom is a first-degree relative of the other two	Yes	Yes
Two successive affected generations (one of the patients is a first-degree family member of the other patients)	Yes	Yes
One or more colon cancers diagnosed in a relative younger than 50 years	Yes	Yes
FAP has been excluded	Yes	Yes
LS-associated cancers, i.e., colorectal, endometrial, small bowel, transitional cell carcinoma of the upper urinary tract, stomach, ovarian, brain (Turcot syndrome), and sebaceous gland adenomas or keratoacanthomas (Muir–Torre syndrome)	No	Yes
Year published	1991	1999
Sensitivity	–	22%
Specificity	–	98%

In contrast to the Amsterdam 1 criteria, the Amsterdam 2 criteria were formulated to include both colorectal and extracolonic cancers, thus increasing the sensitivity [39] for identifying potential LS patients that require further genetic investigation. They require all of the following [59]:

1. Three or more family members with a confirmed diagnosis of LS-related cancer (i.e., colorectal, endometrial, small bowel, transitional cell carcinoma of the upper urinary tract, stomach, ovarian, brain (Turcot syndrome), and sebaceous gland adenomas, or keratoacanthomas (Muir–Torre syndrome [MTS]). Similar to Amsterdam 1, one of these relatives must be a first-degree relative of the other two.

2. Two successive affected generations (i.e., one of the patients must be a parent or offspring of another patient).

3. One or more CRCs must be diagnosed in a relative younger than 50 years.

4. FAP must have been excluded.

5. Tumors should be verified by pathology whenever possible.

Due to the fact that high MSI can be found in a small proportion of sporadic CRC, guidelines have been designed to determine whether an individual is at high risk of LS-associated tumors and therefore needs genetic testing for MSI. These less stringent guidelines are the Bethesda guidelines, devised in 1997 (and revised in 2004) by the National Cancer Institute (NCI) [60]:

1. CRC diagnosed in a patient who is younger than 50 years.

2. Presence of synchronous or metachronous CRC or another LS-associated tumor, regardless of age.

3. CRC with evidence of high MSI on histology (i.e., tumors with changes in two or more of the five NCI-recommended panels of MSI markers (see investigations below), diagnosed in a patient who is younger than 60 years.

4. CRC diagnosed in one or more first-degree relatives with an LS-related tumor with one of the cancers diagnosed in a patient younger than 50 years.

5. CRC diagnosed in two or more first- or second-degree relatives with LS-related tumors, regardless of age.

Each of the individual Bethesda guidelines has a sensitivity of 87.8% and a specificity of 97.5%. This sensitivity is superior to that of the Amsterdam 2 criteria (22%), and therefore the Bethesda guidelines are the best clinical guidelines tool to use for suspecting a likely diagnosis of LS. In patients where LS is suspected however, a confirmatory germline testing should be performed (see investigations below).

20.6.3 Differential Diagnosis

Despite thorough screening procedures, germline mutations of the *MMR* genes are undetected in up to 30% of families with a clinical suspicion of LS. In patients with a compelling family history and MMR negative screening, other hereditary cancers need to be excluded. These are:

1. LS mimickers (e.g., MLH1 constitutional epimutation [61]; constitutional mismatch repair deficiency [CMMD], with both alleles being affected [usually PMS2 and MSH6] [62]; BRAF mutation [found in 50% of sporadic MMR deficient CRC]; polymerase proof-reading-associated polyposis syndrome [PPAP] [63]; familial CRC type X syndrome [FCCTX] [intact DNA MMR system and absence of MSI] [64]; extra-dermal phenotype of MTS harboring mutated MLH1, MSH2, and MSH6 genes; other mutations in the MMR genes)

2. Other considerations (e.g., sporadic colon cancer, attenuated familial adenomatous polyposis, hyperplastic polyps, juvenile polyposis syndrome, nodular lymphoid hyperplasia, Turcot syndrome, Cronkite–Canada syndrome, familial clustering of late onset of colorectal neoplasm, lymphomatous polyposis, MYH-associated polyposis).

20.6.4 Diagnostic Investigations

20.6.4.1 MSI Testing

The use of the Bethesda guidelines can miss up to 28% of LS patients [39]; therefore tumor testing is indicated. Tumor testing identifies MSI as well the loss of expression of MMR proteins by IHC. MSI testing can be performed on fresh, frozen, or paraffin-embedded tumor tissue using PCR to amplify a standard panel of DNA sequences containing nucleotide repeats [65,66]. Five microsatellite markers are necessary to determine MSI; these include two mononucleotide (BAT25, BAT26) and three dinucleotide markers (D2S123, D5S346, D17S250) [67]. Expansion or contraction of 30% or more of repetitive sequences in the tumor compared to healthy control mucosa from the same patient is indicative of high MSI levels (MSI-high).

20.6.4.2 Immunohistochemistry

IHC is used an as alternative to MSI testing to demonstrate MMR gene defects, by loss of staining/expression of the affected protein [68,69]. IHC testing has the added advantage that loss of a specific mismatch gene product (MLH1, MSH2, MSH6, and PMS2) can direct subsequent germline testing to that specific gene. IHC has excellent concordance with DNA-based MSI testing, with a sensitivity of >90% and specificity of 100% [66]. Identification of a mutation should prompt germline testing for patients and their families. BRAF V600E mutation testing and evaluation of the MLH1 promoter methylation status are recommended to differentiate a true germline MLH1 mutation from a somatic MLH1 inactivation secondary to epigenetic silencing via promoter hypermethylation. This is because this specific BRAF mutation has been shown to rarely coexist with germline MLH1 mutations, whereas it is found in around 70% of CRC with MLH1 hypermethylation, which are most likely sporadic in nature. Therefore when a CRC patient is MLH1 deficient with promoter hypermethylation (suggesting a sporadic CRC), BRAF mutation analysis should also be done before germline testing, as

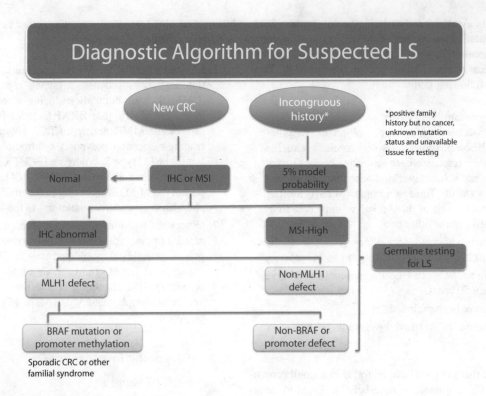

FIGURE 20.2   Investigational algorithm for diagnosis of suspected Lynch syndrome.

the presence of this combination almost certainly excludes LS. An investigational algorithm based on these tests is outlined in Figure 20.2.

20.6.4.3 Universal and Selective Screening

Although the Revised Bethesda Criteria have excellent specificity for picking up CRC patients with LS, their sensitivity is quite low. Universal genetic testing of all patients with new diagnosis of CRC (without constraints as to the age or presence of family history of LS) has gained popularity as a method to evaluate high-risk individuals for LS [22,70]; it has a sensitivity of 100% and a specificity of 93% and is associated with a reduction in morbidity and mortality of biological relatives of the affected patients. Concerns about this screening approach include mainly the cost and the possibility of false-positive tests, which may have economic and psychological impact [13]. On the other hand, selective testing detects the majority but not all of the LS cases and is indicated for patients with CRC diagnosed before the age of 70 (age-targeted selective screening) or older patients meeting the Revised Bethesda Criteria (Bethesda clinical-based selective screening). The former selective strategy has a sensitivity of 95.1% and a specificity of 95.5%. As a result of relative equivalence and cost-effectiveness, current guidelines recommend the use of either the universal or age-targeted selective strategy for screening tests (Table 20.3).

20.6.4.4 Prediction Models

Three models are currently in use: the MMRpredict, MMRpro, and PREMM models, each incorporating different variables. These models are used to quantify the risk of germline mutations

and therefore the need for patients to undergo germline testing. Their main function is in patients without a personal history of colorectal or other cancer but with a family history suggestive of LS, unknown mutation status, or unavailable tissue for testing (Figure 20.2). The American Gastroenterology Association (AGA) recommends offering these, which are superior to no action, or directly proceeding to germline testing, albeit the level of evidence for doing so is low [71].

20.6.4.5 Germline Mutation Testing

To confirm a diagnosis of LS, germline testing of mutations in the *MLH1, MSH2, MSH6, PMS2,* or *EPCAM* genes is required

TABLE 20.3

Universal versus Selective Screening

Variable	Universal	Selective Age Targeted	Selective Bethesda Based
Population	Patients with new diagnosis of CRC and their biological relatives (regardless of family history, clinical features, and age)	Patients with new diagnosis of CRC and <70 y of age (age cutoffs vary in studies; 50, 60, and 70 y have been proposed) and their biological relatives with same age restrictions	Patients with new diagnosis of CRC fulfilling the Revised Bethesda clinical criteria for LS and their biological relatives fulfilling the Revised Bethesda clinical criteria for LS
Sensitivity	100%	95.1%	87.8%
Specificity	93%	95.5%	97.5%
Diagnostic yield	2.2%	2.1%	2%

(Figure 20.2). This should be performed in all patients with MSI unstable tumors, and those meeting the Bethesda or Amsterdam criteria. Appropriate germline testing in family members of the patient will be dictated by a known mutation in the proband. Testing should be offered 10 years before the earliest age of cancer onset in the family or by age 20–25 years. Germline testing is not without complications and therefore a skilled molecular geneticist, where available, is key to the management of these patients.

20.7 Management

20.7.1 Screening

20.7.1.1 Colorectal Carcinoma

In LS patients and family members screening is best done by colonoscopy, which has been shown to decrease cancer mortality by 72%, decrease incidence of CRC by 62% (lifetime risk reduction from 60%–80% without colonoscopy screening to 10% with colonoscopy screening) [4,72–74] and increase life expectancy by 62%. LS patients or first-degree relatives should have colonoscopic screening every 1–2 years starting at ages 20–25 or 2–5 years before the youngest relative to be diagnosed with CRC if the diagnosis was before age 25 [39]. After the age of 35–40 years, colonoscopy should be performed annually.

20.7.1.2 Endometrial Cancer

Despite this being the most common extracolonic cancer, data on efficiency and effectiveness of screening in terms of cancer reduction is limited, hence the guidelines are mainly based on consensus expert advice rather than strong evidence [75]. The chief aim of screening for EC is to detect premalignant lesions that can be treated. Screening can be started in young women (who are mutation carriers) between the ages of 30–35 with pelvic examination. More invasive tests such as endometrial sampling and transvaginal ultrasound can be done on a case-by-case basis based on clinical indication [39], and ideally in the clinical trial setting.

20.7.1.3 Gastric Cancer

Screening for gastric cancer in mutation carriers is achieved by esophagogastroduodenoscopy (OGD) including antral biopsies to exclude *Helicobacter pylori* infection. This should start at age 30–35 years and may be repeated every 2–3 years in patients with risk factors for gastric cancer (gastric atrophy, intestinal metaplasia, family history of gastric cancer, and immigrants from highly endemic areas). Otherwise, due to the relatively low risk of gastric cancer in LS patients, surveillance may be undertaken in the research setting.

20.7.1.4 Small Bowel Cancer

Screening for small bowel cancer is not routinely recommended, although some suggest capsule endoscopy.

20.7.1.5 Other Cancers

Similar to small bowel cancer, screening for pancreatic, prostate, and ovarian cancer is not recommended as there is no identifiable benefit. Breast cancer screening in female carriers with an MMR gene mutation should be done under the current population screening programs, i.e., by annual mammography from the age of 45 or 50 years. In the case upper urinary tract cancer screening, there are no consensus guidelines, as no benefit has clearly been established [57]. As a result, urinalysis is sometimes suggested to be performed yearly starting at age 30–35. The presence of significant red blood cell numbers on urine dipstick can elicit further investigation with urine cytology and urinary tract imaging.

20.7.2 Surgical Treatment

The only effective therapy for CRC in patients with LS is colectomy. The choice of procedure is largely determined by tumor stage and location, patient choice, age, and anticipated quality of life. The most important consideration, however, is the estimated risk of metachronous lesions, which is significant at 30% at 10 years after initial diagnosis [76] and 8% at 5 years post-resection in patients with LS [47]. The cumulative risk of CRC at 10, 20, and 30 years following partial colectomy and frequent colonoscopy surveillance is 19%, 41%, and 62%, respectively [39,77]. These are far less favorable numbers when compared to a subtotal or total colectomy, which lead to a substantially reduced post-surgical risk of 0%–3.4%. Three operative procedures can be performed:

1. Subtotal colectomy with ileorectal anastomosis
2. Total colectomy with ileostomy
3. Total colectomy with ileoanal pouch

Subtotal colectomy with ileorectal anastomosis is generally the procedure of choice, on the condition that postoperative rectal surveillance is carried out annually. In this case post-resection surveillance sigmoidoscopy is recommended every 1–2 years. Total (procto)colectomy with ileoanal pouch or ileostomy eliminates the need for endoscopic surveillance of the rectal stump, but these procedures are generally reserved for LS patients with rectal cancers, due to concerns relating to increased postoperative morbidity and poor quality of life. Quality of life is similar in patients who undergo partial colectomy compared to subtotal colectomy; the latter, however, is associated with poorer functional outcomes (i.e., stool frequency and social impact) [78]. At older ages where life expectancy is reduced, partial colectomy may often be an option [39].

20.7.2.1 Prophylactic Colectomy

Prophylactic total or subtotal colectomy is controversial but may be an alternative to surveillance colonoscopy for individuals with confirmed mutations. Because of incomplete penetrance, it is likely that up to 20% of these colectomies may be unnecessary, a value often used to argue against the idea of prophylactic colectomies. In addition, prophylactic colectomy does not eliminate the risk of extracolonic malignancies and if subtotal, does not eliminate the risk of developing metachronous rectal cancers. Based on a decision analysis tool, total or subtotal colectomy is superior to surveillance without surgery but only offers a marginal survival benefit. The expected gain in life expectancy is 15.6 years after total colectomy, 15.3 years after subtotal colectomy, and 13.5 years for surveillance only [79]. However, it is worth noting that this slight benefit in life expectancy reduces with increasing age. In terms of superiority between total versus subtotal colectomy, a study with 6 years of postoperative follow-up showed that 0% of patients undergoing total colectomy had metachronous lesions compared to 21% of patients undergoing subtotal colectomy. This difference was statistically significant, suggesting that total colectomy should be the first-line surgical prophylaxis option despite the fact that there were no significant differences between the two groups in terms of overall survival [80].

20.7.2.2 Prophylactic Hysterectomy and Bilateral Salpingo-Oophorectomy

Prophylactic hysterectomy with bilateral salpingo-oophorectomy is the only intervention that has been proved to be effective against the prevention of endometrial and ovarian cancer and is recommended to women with LS who have completed their families or are ≥40 years of age [39].

20.7.3 Medical Treatment

20.7.3.1 Aspirin

Similar to NSAIDs in FAP and sporadic CRC, there is ample interventional data to show that aspirin might cause a reduction in the high risk of colonic adenomas seen in patients with LS, and observational data confirm that the observed lowered risk of CRC is due to aspirin use [81–87]. In one study, patients with LS, receiving aspirin at a dose of 600 mg per day, were followed up for a mean of 55.7 months after which 4% developed CRC compared to 7% for patients in the placebo group [88]. In patients with a personal history of adenomas, aspirin at doses of 75–1200 mg daily reduces the incidence of new lesions, including in patients with previous CRC. The optimal dose of aspirin is still under study; however, patients should be screened for *H. pylori* infection and, if positive, should be treated before commencement of aspirin. The recommendation therefore is that patients with mutations should consider aspirin in addition to surveillance.

20.7.3.2 NSAIDs, Estrogens, Folic Acid, and Calcium

Although several uncontrolled studies are available for the use of these agents in predisposed individuals at high risk for LS, data are not convincing and as such, the use of these agents is not validated or recommended.

20.8 Prognosis

Compared to sporadic CRC (with a 5-year cumulative survival rate of 44%), patients with LS have an improved prognosis with 65% 5-year survival [89]. Other factors associated with better prognosis are age younger than 65 years and patients with CRC from families with a history of LS compared to those with sporadic CRC [90–93]. Factors associated with worse prognosis are tumors with elevated microsatellite alterations at selected tetranucleotide (EMAST) repeats. These tumors tend to be poorly differentiated, with mucinous features and located in the right colon. These EMAST repeats have become a biomarker of prognosis and tumor type. In addition, BRAF-positive mutations are associated with lower 5-year survival rates [94].

Patients with LS who are part of a screening program have a better outcome than those not screened. In a study following up patients with LS for 15 years, the risk of developing CRC was 62% less for those undergoing screening compared to those who were not screened. This was broadly ascribed to the removal of polyps by polypectomy during screening colonoscopy. Furthermore, there were no cancer-related deaths in the screened cohort [21,95]. The best evidence that colonoscopic screening is beneficial for preventing colon cancer in patients with LS has come from observational studies of 22 LS families that were followed for 15 years [8,9]. During the study period, 133 family members were voluntarily screened every 3 years, whereas 119 declined colonoscopic surveillance. A significant reduction in the incidence of CRC was observed in those screened compared those who were not, an effect attributed to the impact of polypectomies performed in the intervention group. In addition to this, the same study reported no CRC-related deaths occurring in the group that underwent regular colonoscopic screening compared with a 36% CRC-related mortality rate in the unscreened group [96].

Prognosis for LS patients diagnosed with endometrial, ovarian, and prostate cancer is relatively good, with a 10-year survival of approximately 80%, 81%, and 71%, respectively. The prognosis of bladder cancer is largely dependent on the stage and grade of the tumor. Noninvasive low-grade tumors have a 5-year survival of greater than 90% compared to high-grade cancers (60%–70%) [57]. As with sporadic pancreatic cancer, the prognosis in patients with LS-associated pancreatic cancer is dismal, with an average life expectancy of 6 months after diagnosis.

20.9 Conclusion and Future Perspectives

Although the combination of clinical guidelines and genetic testing has resulted in an increase in the number of Lynch syndrome (LS) diagnoses made, this condition is still underdiagnosed. Given this underdiagnosis as a result of the complexity of detecting the disease and the various differentials that mimic the disease, clinicians are required to be extremely vigilant in order to suspect LS. It is crucial to raise awareness in both health care professionals and the public in order for LS and LS-associated cancers to be prevented or at least diagnosed early. Future methods of diagnosis will include more targeted molecular genetic analysis to exclude mimickers of LS and gain a better understanding of the multiple phenotype-genotype profiles of LS, which are punctuated by their marked heterogeneity, making this condition an ever-evolving entity. The management of a patient with LS should be undertaken by a multidisciplinary team consisting of gastroenterologists, oncologists, surgeons, radiologists, geneticists, and if appropriate gynecologists, urologists, and neurologists. Several issues regarding the diagnosis (e.g., ideal investigation algorithm and verified criteria for investigation) and management (e.g., optimal dose of aspirin) of LS still remain to be elucidated and as such, ongoing research in these areas remains important.

REFERENCES

1. Warthin AS. Heredity with reference to carcinoma as shown by the study of the cases examined in the pathological laboratory of the University of Michigan, 1895–1913. *Arch Intern Med.* 1913;12:546–55. Reprinted as Classics in oncology. *CA Cancer J.* 1985;35:348–59.

2. Lynch HT, Shaw MW, Magnuson CW, Larsen AL, Krush AJ. Hereditary factors in cancer. Study of two large midwestern kindreds. *Arch Intern Med.* 1966;117:206–12.

3. Lynch HT, Krush AJ. Cancer family "G" revisited: 1895–1970. *Cancer.* 1971;27:1505–11.

4. Douglas JA, Gruber SB, Meister KA et al. History and molecular genetics of Lynch syndrome in family G: A century later. *JAMA.* 2005;294:2195–202.

5. Boland CR. Evolution of the nomenclature for the hereditary colorectal cancer syndromes. *Fam Cancer.* 2005;4:211–8.

6. Jass JR. Hereditary non-polyposis colorectal cancer: The rise and fall of a confusing term. *World J Gastroenterol.* 2006;12:4943–50.

7. Hsieh P, Yamane K. DNA mismatch repair: Molecular mechanism, cancer, and ageing. *Mech Ageing Dev.* 2008;129:391–407.

8. Kunkel TA, Erie DA. DNA mismatch repair. *Annu Rev Biochem.* 2005;74:681–710.

9. Boland CR, Thibodeau SN, Hamilton SR et al. A National Cancer Institute Workshop on Microsatellite Instability for cancer detection and familial predisposition: Development of international criteria for the determination of microsatellite instability in colorectal cancer. *Cancer Res.* 1998;58:5248–57.

10. Stoffel EM, Boland CR. Genetics and genetic testing in hereditary colorectal cancer. *Gastroenterology.* 2015;149:1191–203.e2.

11. Markowitz SD, Roberts AB. Tumor suppressor activity of the TGF-beta pathway in human cancers. *Cytokine Growth Factor Rev.* 1996;7:93–102.

12. Rampino N, Yamamoto H, Ionov Y et al. Somatic frameshift mutations in the BAX gene in colon cancers of the microsatellite mutator phenotype. *Science.* 1997;275:967–9.

13. Moreira L, Balaguer F, Lindor N et al. Identification of Lynch syndrome among patients with colorectal cancer. *JAMA.* 2012;308:1555–65.

14. Ligtenberg MJ, Kuiper RP, Chan TL et al. Heritable somatic methylation and inactivation of MSH2 in families with Lynch syndrome due to deletion of the 3′ exons of TACSTD1. *Nature Genet.* 2009;41:112–7.

15. Kovacs ME, Papp J, Szentirmay Z, Otto S, Olah E. Deletions removing the last exon of TACSTD1 constitute a distinct class of mutations predisposing to Lynch syndrome. *Hum Mutat.* 2009;30:197–203.

16. Wagner A, Hendriks Y, Meijers-Heijboer EJ et al. Atypical HNPCC owing to MSH6 germline mutations: Analysis of a large Dutch pedigree. *J Med Genet.* 2001;38:318–22.

17. Hendriks YM, Wagner A, Morreau H et al. Cancer risk in hereditary nonpolyposis colorectal cancer due to MSH6 mutations: Impact on counseling and surveillance. *Gastroenterology.* 2004;127:17–25.

18. Wu Y, Berends MJ, Mensink RG et al. Association of hereditary nonpolyposis colorectal cancer-related tumors displaying low microsatellite instability with MSH6 germline mutations. *Am J Hum Genet.* 1999;65:1291–8.

19. Berends MJ, Wu Y, Sijmons RH et al. Molecular and clinical characteristics of MSH6 variants: An analysis of 25 index carriers of a germline variant. *Am J Hum Genet.* 2002;70:26–37.

20. Verma L, Kane MF, Brassett C et al. Mononucleotide microsatellite instability and germline MSH6 mutation analysis in early onset colorectal cancer. *J Med Genet.* 1999;36:678–82.

21. Aarnio M, Mecklin JP, Aaltonen LA, Nystrom-Lahti M, Jarvinen HJ. Life-time risk of different cancers in hereditary non-polyposis colorectal cancer (HNPCC) syndrome. *Int J Cancer.* 1995;64:430–3.

22. Hampel H, Frankel WL, Martin E et al. Screening for the Lynch syndrome (hereditary nonpolyposis colorectal cancer). *New Engl J Med.* 2005;352:1851–60.

23. Barrow E, Robinson L, Alduaij W et al. Cumulative lifetime incidence of extracolonic cancers in Lynch syndrome: A report of 121 families with proven mutations. *Clin Genet.* 2009;75:141–9.

24. Bonadona V, Bonaiti B, Olschwang S et al. Cancer risks associated with germline mutations in MLH1, MSH2, and MSH6 genes in Lynch syndrome. *JAMA.* 2011;305:2304–10.

25. Baglietto L, Lindor NM, Dowty JG et al. Risks of Lynch syndrome cancers for MSH6 mutation carriers. *J Natl Cancer Inst.* 2010;102:193–201.

26. Nakagawa H, Lockman JC, Frankel WL et al. Mismatch repair gene PMS2: Disease-causing germline mutations are frequent in patients whose tumors stain negative for PMS2 protein, but paralogous genes obscure mutation detection and interpretation. *Cancer Res.* 2004;64:4721–7.

27. Worthley DL, Walsh MD, Barker M et al. Familial mutations in PMS2 can cause autosomal dominant hereditary nonpolyposis colorectal cancer. *Gastroenterology.* 2005;128:1431–6.

28. Hendriks YM, Jagmohan-Changur S, van der Klift HM et al. Heterozygous mutations in PMS2 cause hereditary nonpolyposis colorectal carcinoma (Lynch syndrome). *Gastroenterology.* 2006;130:312–22.

29. Senter L, Clendenning M, Sotamaa K et al. The clinical phenotype of Lynch syndrome due to germ-line PMS2 mutations. *Gastroenterology.* 2008;135:419–28.

30. Kempers MJ, Kuiper RP, Ockeloen CW et al. Risk of colorectal and endometrial cancers in EPCAM deletion-positive Lynch syndrome: A cohort study. *Lancet Oncol.* 2011;12:49–55.

31. Ricciardone MD, Ozcelik T, Cevher B et al. Human MLH1 deficiency predisposes to hematological malignancy and neurofibromatosis type 1. *Cancer Res.* 1999;59:290–3.

32. Wang Q, Lasset C, Desseigne F et al. Neurofibromatosis and early onset of cancers in hMLH1-deficient children. *Cancer Res.* 1999;59:294–7.

33. Palomaki GE, McClain MR, Melillo S, Hampel HL, Thibodeau SN. EGAPP supplementary evidence review: DNA testing strategies aimed at reducing morbidity and mortality from Lynch syndrome. *Genet Med.* 2009;11:42–65.

34. Halvarsson B, Muller W, Planck M et al. Phenotypic heterogeneity in hereditary non-polyposis colorectal cancer: Identical germline mutations associated with variable tumour morphology and immunohistochemical expression. *J Clin Pathol.* 2007;60:781–6.

35. Dunlop MG, Tenesa A, Farrington SM et al. Cumulative impact of common genetic variants and other risk factors on colorectal cancer risk in 42,103 individuals. *Gut.* 2013;62:871–81.

36. Wijnen JT, Brohet RM, van Eijk R et al. Chromosome 8q23.3 and 11q23.1 variants modify colorectal cancer risk in Lynch syndrome. *Gastroenterology.* 2009;136:131–7.

37. Win AK, Jenkins MA, Dowty JG et al. Prevalence and penetrance of major genes and polygenes for colorectal cancer. *Cancer Epidemiol Biomarkers Prev.* 2017;26:404–12.

38. Kastrinos F, Stoffel EM. History, genetics, and strategies for cancer prevention in Lynch syndrome. *Clin Gastroenterol Hepatol.* 2014;12:715–27; quiz e41–3.

39. Giardiello FM, Allen JI, Axilbund JE et al. Guidelines on genetic evaluation and management of Lynch syndrome: A consensus statement by the US Multi-society Task Force on colorectal cancer. *Am J Gastroenterol.* 2014;109:1159–79.

40. Edelstein DL, Axilbund J, Baxter M et al. Rapid development of colorectal neoplasia in patients with Lynch syndrome. *Clin Gastroenterol Hepatol.* 2011;9:340–3.

41. Toribara NW, Sleisenger MH. Screening for colorectal cancer. *New Engl J Med.* 1995;332:861–7.

42. Lynch HT, de la Chapelle A. Hereditary colorectal cancer. *New Engl J Med.* 2003;348:919–32.

43. Jass JR, Smyrk TC, Stewart SM, Lane MR, Lanspa SJ, Lynch HT. Pathology of hereditary non-polyposis colorectal cancer. *Anticancer Res.* 1994;14:1631–4.

44. Lengauer C, Kinzler KW, Vogelstein B. Genetic instabilities in human cancers. *Nature.* 1998;396:643–9.

45. Watson P, Lin KM, Rodriguez-Bigas MA et al. Colorectal carcinoma survival among hereditary nonpolyposis colorectal carcinoma family members. *Cancer.* 1998;83:259–66.

46. Deschoolmeester V, Baay M, Van Marck E et al. Tumor infiltrating lymphocytes: An intriguing player in the survival of colorectal cancer patients. *BMC Immunol.* 2010;11:19.

47. Lanspa SJ, Jenkins JX, Cavalieri RJ et al. Surveillance in Lynch syndrome: How aggressive? *Am J Gastroenterol.* 1994;89:1978–80.

48. Rondagh EJ, Gulikers S, Gomez-Garcia EB et al. Nonpolypoid colorectal neoplasms: A challenge in endoscopic surveillance of patients with Lynch syndrome. *Endoscopy.* 2013;45:257–64.

49. Watson P, Vasen HFA, Mecklin JP et al. The risk of extracolonic, extra-endometrial cancer in the Lynch syndrome. *Int J Cancer.* 2008;123:444–9.

50. Watson P, Lynch HT. The tumor spectrum in HNPCC. *Anticancer Res.* 1994;14:1635–9.

51. Kastrinos F, Mukherjee B, Tayob N et al. Risk of pancreatic cancer in families with Lynch syndrome. *JAMA.* 2009;302:1790–5.

52. Win AK, Young JP, Lindor NM et al. Colorectal and other cancer risks for carriers and noncarriers from families with a DNA mismatch repair gene mutation: A prospective cohort study. *J Clin Oncol.* 2012;30:958–64.

53. Bauer CM, Ray AM, Halstead-Nussloch BA et al. Hereditary prostate cancer as a feature of Lynch syndrome. *Fam Cancer.* 2011;10:37–42.

54. Lynch HT, Lynch PM, Pester J, Fusaro RM. The cancer family syndrome. Rare cutaneous phenotypic linkage of Torre's syndrome. *Arch Intern Med.* 1981;141:607–11.

55. Mangold E, Pagenstecher C, Leister M et al. A genotype-phenotype correlation in HNPCC: Strong predominance of msh2 mutations in 41 patients with Muir-Torre syndrome. *J Med Genet.* 2004;41:567–72.

56. Trano G, Wasmuth HH, Sjursen W, Hofsli E, Vatten LJ. Awareness of heredity in colorectal cancer patients is insufficient among clinicians: A Norwegian population-based study. *Colorectal Dis.* 2009;11:456–61.

57. Vasen HF, Blanco I, Aktan-Collan K et al. Revised guidelines for the clinical management of Lynch syndrome (HNPCC): Recommendations by a group of European experts. *Gut.* 2013;62:812–23.

58. Vasen HF, Mecklin JP, Khan PM, Lynch HT. The international collaborative group on hereditary non-polyposis colorectal cancer (ICG-HNPCC). *Dis Colon Rectum.* 1991;34:424–5.

59. Vasen HF, Watson P, Mecklin JP, Lynch HT. New clinical criteria for hereditary nonpolyposis colorectal cancer (HNPCC, Lynch syndrome) proposed by the International Collaborative group on HNPCC. *Gastroenterology.* 1999;116:1453–6.

60. Umar A, Boland CR, Terdiman JP et al. Revised Bethesda Guidelines for hereditary nonpolyposis colorectal cancer (Lynch syndrome) and microsatellite instability. *J Natl Cancer Inst.* 2004;96:261–8.

61. Gazzoli I, Loda M, Garber J, Syngal S, Kolodner RD. A hereditary nonpolyposis colorectal carcinoma case associated with hypermethylation of the MLH1 gene in normal tissue and loss of heterozygosity of the unmethylated allele in the resulting microsatellite instability-high tumor. *Cancer Res.* 2002;62:3925–8.

62. Bakry D, Aronson M, Durno C et al. Genetic and clinical determinants of constitutional mismatch repair deficiency syndrome: Report from the constitutional mismatch repair deficiency consortium. *Eur J Cancer.* 2014;50:987–96.

63. Carethers JM, Stoffel EM. Lynch syndrome and Lynch syndrome mimics: The growing complex landscape of hereditary colon cancer. *World J Gastroenterol.* 2015;21:9253–61.

64. Carethers JM, Koi M, Tseng-Rogenski SS. EMAST is a form of microsatellite instability that is initiated by inflammation and modulates colorectal cancer progression. *Genes.* 2015;6:185–205.

65. Thibodeau SN, Bren G, Schaid D. Microsatellite instability in cancer of the proximal colon. *Science.* 1993;260:816–9.

66. Lindor NM, Burgart LJ, Leontovich O et al. Immunohistochemistry versus microsatellite instability testing in phenotyping colorectal tumors. *J Clin Oncol.* 2002;20:1043–8.

67. Buhard O, Cattaneo F, Wong YF et al. Multipopulation analysis of polymorphisms in five mononucleotide repeats used to determine the microsatellite instability status of human tumors. *J Clin Oncol.* 2006;24:241–51.

68. Shia J. Immunohistochemistry versus microsatellite instability testing for screening colorectal cancer patients at risk for hereditary nonpolyposis colorectal cancer syndrome. Part I. The utility of immunohistochemistry. *J Mol Diagn.* 2008;10:293–300.

69. Weissman SM, Bellcross C, Bittner CC et al. Genetic counseling considerations in the evaluation of families for Lynch syndrome—A review. *J Genet Couns.* 2011;20:5–19.

70. Hampel H, Frankel WL, Martin E et al. Feasibility of screening for Lynch syndrome among patients with colorectal cancer. *J Clin Oncol.* 2008;26:5783–8.

71. Rubenstein JH, Enns R, Heidelbaugh J, Barkun A. American gastroenterological association institute guideline on the diagnosis and management of Lynch syndrome. *Gastroenterology.* 2015;149:777–82; quiz e16–7.

72. Dove-Edwin I, Sasieni P, Adams J, Thomas HJ. Prevention of colorectal cancer by colonoscopic surveillance in individuals with a family history of colorectal cancer: 16 year, prospective, follow-up study. *BMJ.* 2005;331:1047.

73. Jarvinen HJ, Mecklin JP, Sistonen P. Screening reduces colorectal cancer rate in families with hereditary nonpolyposis colorectal cancer. *Gastroenterology.* 1995;108:1405–11.

74. Vasen HF, van Ballegooijen M, Buskens E et al. A cost-effectiveness analysis of colorectal screening of hereditary

nonpolyposis colorectal carcinoma gene carriers. *Cancer.* 1998;82:1632–7.

75. Schmeler KM, Lynch HT, Chen LM et al. Prophylactic surgery to reduce the risk of gynecologic cancers in the Lynch syndrome. *New Engl J Med.* 2006;354:261–9.

76. Hampel H. Point: Justification for Lynch syndrome screening among all patients with newly diagnosed colorectal cancer. *J Natl Compr Cancer Netw.* 2010;8:597–601.

77. Parry S, Win AK, Parry B et al. Metachronous colorectal cancer risk for mismatch repair gene mutation carriers: The advantage of more extensive colon surgery. *Gut.* 2011;60:950–7.

78. Haanstra JF, de Vos Tot Nederveen Cappel WH, Gopie JP et al. Quality of life after surgery for colon cancer in patients with Lynch syndrome: Partial versus subtotal colectomy. *Dis Colon Rectum.* 2012;55:653–9.

79. Syngal S, Weeks JC, Schrag D, Garber JE, Kuntz KM. Benefits of colonoscopic surveillance and prophylactic colectomy in patients with hereditary nonpolyposis colorectal cancer mutations. *Ann Intern Med.* 1998;129:787–96.

80. Stupart DA, Goldberg PA, Baigrie RJ, Algar U, Ramesar R. Surgery for colonic cancer in HNPCC: Total vs segmental colectomy. *Colorectal Dis.* 2011;13:1395–9.

81. Baron JA, Cole BF, Sandler RS et al. A randomized trial of aspirin to prevent colorectal adenomas. *New Engl J Med.* 2003;348:891–9.

82. Sandler RS, Halabi S, Baron JA et al. A randomized trial of aspirin to prevent colorectal adenomas in patients with previous colorectal cancer. *New Engl J Med.* 2003;348:883–90.

83. Bertagnolli MM, Eagle CJ, Zauber AG et al. Celecoxib for the prevention of sporadic colorectal adenomas. *New Engl J Med.* 2006;355:873–84.

84. Giovannucci E, Rimm EB, Stampfer MJ, Colditz GA, Ascherio A, Willett WC. Aspirin use and the risk for colorectal cancer and adenoma in male health professionals. *Ann Intern Med.* 1994;121:241–6.

85. Markowitz SD. Aspirin and colon cancer--targeting prevention? *New Engl J Med.* 2007;356:2195–8.

86. Thun MJ, Namboodiri MM, Heath CW, Jr. Aspirin use and reduced risk of fatal colon cancer. *New Engl J Med.* 1991;325:1593–6.

87. Ait Ouakrim D, Dashti SG, Chau R et al. Aspirin, ibuprofen, and the risk of colorectal cancer in Lynch syndrome. *J Natl Cancer Inst.* 2015;107.

88. Burn J, Gerdes AM, Macrae F et al. Long-term effect of aspirin on cancer risk in carriers of hereditary colorectal cancer: An analysis from the CAPP2 randomised controlled trial. *Lancet.* 2011;378:2081–7.

89. Sankila R, Aaltonen LA, Jarvinen HJ, Mecklin JP. Better survival rates in patients with MLH1-associated hereditary colorectal cancer. *Gastroenterology.* 1996;110:682–7.

90. Boland CR, Goel A. Microsatellite instability in colorectal cancer. *Gastroenterology.* 2010;138:2073–87.e3.

91. Vilar E, Gruber SB. Microsatellite instability in colorectal cancer-the stable evidence. *Nature Rev Clin Oncol.* 2010;7:153–62.

92. Ismael NE, El Sheikh SA, Talaat SM, Salem EM. Mismatch repair proteins and microsatellite instability in colorectal carcinoma (MLH1, MSH2, MSH6 and PMS2): Histopathological and immunohistochemical study. *Open Access Maced J Med Sci.* 2017;5:9–13.

93. Chen W, Swanson BJ, Frankel WL. Molecular genetics of microsatellite-unstable colorectal cancer for pathologists. *Diagn Pathol.* 2017;12:24.

94. Toon CW, Chou A, DeSilva K et al. BRAFV600E immunohistochemistry in conjunction with mismatch repair status predicts survival in patients with colorectal cancer. *Mod Pathol.* 2014;27:644–50.

95. Houlston RS, Murday V, Harocopos C, Williams CB, Slack J. Screening and genetic counselling for relatives of patients with colorectal cancer in a family cancer clinic. *BMJ.* 1990;301:366–8.

96. Carethers JM. Microsatellite instability pathway and EMAST in colorectal cancer. *Curr Colorectal Cancer Rep.* 2017;13:73–80.

21

MUTYH-Associated Polyposis

Dongyou Liu

CONTENTS

21.1 Introduction

First described in 2002, MUTYH-associated polyposis (MAP) is an autosomal recessive disorder characterized by adult onset of multiple colorectal adenomas and adenomatous polyps (generally numbering between 10 and 100), and a predisposition for colorectal cancer. In addition, patients with MAP may show increased risk for ovarian cancer, small intestinal cancer, urinary bladder cancer, and cutaneous melanoma, along with breast cancer, endometrial cancer, gastroduodenal polyps, duodenal cancer, papillary thyroid cancer, carcinoid, chondrosarcoma, astrocytoma, pilomatrixoma, and sebaceous adenoma. Molecularly, MAP is linked to homozygous (biallelic) or compound heterozygous mutation in the *MUTYH* gene mapped to 1p32.1–p34.3 [1].

21.2 Biology

Inherited colorectal cancer predisposition syndromes consist of two major phenotypic groups: polyposis and nonpolyposis. In turn, the polyposis group is divided into four subgroups: adenomatous, hamartomatous, mixed, and serrated, while the nonpolyposis group is separated into two subgroups: MMR-deficient and MMR-proficient (Table 21.1) [2,3].

Constituting one of the five inherited colorectal cancer predisposition syndromes (i.e., familial adenomatous polyposis [FAP],

polymerase proofreading–associated polyposis [PPAP], MAP, NTHL1-associated polyposis [NAP], and biallelic MMR) within the adenomatous subgroup, MAP (also known as familial adenomatous polyposis 2 or FAP2) is an autosomal recessive disorder noted for its phenotype resemblance to attenuated familial adenomatous polyposis (which is a variant of FAP), including the presence of <100 colorectal polyps of adenomatous morphology (and occasional sessile-serrated adenomas, hyperplastic polyps, and mixed polyps), early-onset colorectal cancer (average age of mid-50s), and clear linkage to germline biallelic mutations in *MUTYH* on chromosome 1p34.1, leading to defective base excision repair [4–11].

In contrast, FAP is an autosomal dominant disorder that presents with 100–1000 s synchronous colorectal polyps of adenomatous morphology (in addition to polyps of the upper gastrointestinal tract, desmoid tumors, and osteomas) from early childhood to mid-30s (typically 16), early-onset colorectal cancer (at average age of mid-50s), and linkage to germline mutations in the *APC* (adenomatous polyposis coli) tumor suppressor gene on chromosome 5q22.2. A milder form of FAP, i.e., atypical familial adenomatous polyposis (AFAP or oligopolyposis), produces <100 colorectal polyps (typically 30) at between 40 and 70 years (average 55), and has a 70% risk of colorectal cancer by age 80 (see Chapter 16 in this volume) [12–15].

Biallelic MUTYH mutations are found in up to 40% of patients with FAP-like and AFAP-like phenotypes in which *APC* mutation is absent or when evidence of vertical transmission is unclear. Given its autosomal recessive inheritance, MAP is often

TABLE 21.1

Classification of Inherited Colorectal Cancer Predisposition Syndromes

	Phenotype	Syndrome	Gene (Chromosome)	Inheritance
Polyposis	Adenomatous	Familial adenomatous polyposis (FAP)	*APC* (5q22.2)	AD
		Polymerase proofreading-associated polyposis (PPAP)	*POLE* (12q24.33) or *POLD1* (19q13.33)	AD
		MUTYH-associated polyposis (MAP)	*MUTYH* (1p34.1)	AR
		NTHL1-associated polyposis (NAP)	*NTHL1* (16p13.3)	AR
		Biallelic MMR	*MSH2, MSH6, MLH1, PMS2, MSH3*	AR
	Hamartomatous	Peutz–Jeghers syndrome (PJS)	*STK11* (19q13.3)	AD
		Juvenile polyposis syndrome (JPS)	*SMAD4* (18q21.2), *BMPR1A* (10q23.2)	AD
		PTEN hamartoma tumor syndrome (PHTS)	*PTEN* (10q23.31)	AD
	Mixed	Hereditary mixed polyposis syndrome (HMPS)	*GREM1* (15q13.3)	AD
	Serrated	Sessile serrated polyposis cancer syndrome (SSPCS)	*RNF43* (17q22)	AD
Nonpolyposis	MMR-deficient	Lynch syndrome (hereditary nonpolyposis colorectal cancer or HNPCC)	*MSH2, MSH6, MLH1, PMS2, EPCAM*	AD
		Lynch-like syndrome	*hMLH1, hMSH2, hMSH6, hPMS2*	AD
	MMR-proficient	Familial colorectal cancer type X (FCCTX)	*RPS20* (8q12.1), *BMPR1A* (10q23.2), *SEMA4A* (1q22)	AD

Abbreviations: AD, autosomal dominant; AR, autosomal recessive.

diagnosed after colorectal cancer has already developed, in comparison with FAP [16,17]. Indeed, without treatment, patients with biallelic MUTYH mutation have a colorectal cancer of approximately 80% by the age of 70 years [18].

21.3 Pathogenesis

The *MUTYH* gene located on the short (p) arm of chromosome 1 at position 34.1 (i.e., chromosome 1p34.1) consists of 16 exons spanning 11.2 kb. Its transcribed mRNA is 1854 bp long and encodes a 546 aa, 52 kDa adenine DNA glycosylase (i.e., MUTYH, also known as mutY homolog, MYH, and hMYH), which includes several different isoforms (e.g., α, β, and γ) due to alternative splicing of pre-mRNA, and also use of alternate transcriptional start and polyadenylation sites. The MUTYH protein includes an N-terminal domain on the 5′ side (the catalytic region), a C-terminal domain on the 3′ side (with base excision repair function), binding sites for a DNA binding domain, an adenine binding motif, and several interaction domains. MUTYH is involved in base excision repair (BER) of damaged DNA caused by ionizing radiation, chemical oxidants, and reactive oxygen species are generated during aerobic metabolism [19–21]. Composed of paired nucleotides (i.e., adenine thymine [A:T], and guanine:cytosine [G:C]), human DNA is vulnerable to alteration by oxidation of a guanine to 8-oxo-7,8-dihydro-2′-deoxyguanosine (8-oxoG), leading to mispair with adenine (instead of cytosine) in the normal guanine-cytosine pair. MUTYH utilizes a base-flipping mechanism to recognize and excise the misincorporated adenine from the oxoG:A mismatch [22,23]. This facilitates DNA polymerases and OGG1 (another BER glycosylase) to restore the oxoG:C pair and replace the oxidized guanine with a guanine, respectively, preventing the G:C > T:A transversion from occurring [24–27]. Thus, MUTYH works in synergy with OOG1 to maintain genome stability and eliminate potential DNA mutations

during cell division. In addition, MUTYH is a potential mediator of p53 tumor suppression [28–30].

Mutations in the *MUTYH* gene produce a dysfunctional truncated MUTYH protein that fails to rectify somatic G > T transversions in multiple tumor suppressor genes (including *APC* and *KRAS*), leading to cell overgrowth and characteristic MUTYH-associated polyposis (e.g., serrated adenomas, hyperplastic polyps, and sporadic colorectal cancer). Compared to the general population, carriers with biallelic (compound heterozygous or homozygous) MUTYH mutations inherited from both parents are 19 and 17 times more likely to develop urinary bladder cancer and ovarian cancer, respectively, whereas carriers with monoallelic MUTYH mutations inherited from only one parent are 9.3, 4.5, 2.1, and 1.4 times more likely to have gastric cancer, hepatobiliary cancer, endometrial cancer, and breast cancer, respectively [31,32].

To date, >300 unique *MUTYH* pathogenic variants have been identified from various parts of the world. Of these, missense mutations p.Y179C (c.536A > G; p.Tyr179Cys) in exon 7 and p.G396D (c.1187g > A; p.Gly396Asp; formerly c.494A4G/p.Tyr165Cys and c.1145G4A/p.Gly382Asp) in exon 13 are found in Eastern, Southern, and Central Europe, North America, European inhabitants from Canada, and Sephardic Jews, but are absent in Finland, India, Pakistan, Tunisia, Singapore, and Ashkenazi Jewish. Further, deletion mutation c.1147delC (p.Ala385ProfsTer23) is specific for Northern Europe; c.1214C > T (p.Pro405Leu), c.1437_1439del (p.Glu480del) and c.1227_1228dup (p.Glu410GlyfsTer43) are limited to the Netherlands, Italy, and Spain/Portugal/Tunisia, respectively; deletion of exons 4–16 is commonly found in Spain, Brazil, and France. On the other hand, heterozygous mutations p.Arg19, p.Arg109Trp, and c.857G > A (p.Gly286Glu) are specific for Asia (Japan, Taiwan, South Korea), while heterozygous mutations p.Glu480del, p.Tyr104, and p.Glu480 are specific for Southern Europe, Pakistan, and India, respectively. Interestingly, homozygosity for the c.536A > G/p.Tyr179Cys pathogenic variant appears to confer risk for a more severe phenotype and earlier onset of colorectal cancer than homozygosity for the c.1187G > A/p.Gly396Asp pathogenic variant [33–38].

21.4 Epidemiology

21.4.1 Prevalence

MAP represents 0.7% of all colorectal cancer and up to 2% of familial or early-onset colorectal cancer (which typically contains up to 20 adenomas in affected individuals) [39]. Biallelic germline *MUTYH* pathogenic variants have an incidence of 1 per 10,000 and 40,000 births, affecting 0.005% of population, but 0.01%–0.04% of the Caucasian population. On the other hand, monoallelic (heterozygous) *MUTYH* pathogenic variants are found in 1%–2% of the global population.

21.4.2 Inheritance

Being an autosomal recessive disorder, biallelic (compound heterozygous or homozygous) *MUTYH* mutations (causing MAP) are inherited from both parents. However, monoallelic (heterozygous) *MUTYH* mutations are inherited from only one parent.

21.4.3 Penetrance

MAP demonstrates close to 100% penetrance for colon polyps (with 10–100 adenomatous polyps/adenomas of the colon and rectum) and increases risk of colorectal cancer by 60%–70%. Up to one-third of individuals harboring biallelic germline *MUTYH* pathogenic variants develop colorectal cancer in the absence of polyposis. Heterozygous *MUTYH* mutation carriers (which are more common than biallelic *MUTYH* mutation carriers) may have a slightly increased risk for colorectal cancer. In addition, MAP confers a lifetime risk of 38% for developing extraintestinal malignancies (e.g., ovarian, breast, bladder, and skin cancers, which have a median age of onset between 51 and 61 years).

21.5 Clinical Features

MAP patients who harbor biallelic (homozygous or compound heterozygous) *MUTYH* pathogenic variants typically develop ten to hundreds of polyps at a mean presentation age of 50 years and show increased lifetime risk of colorectal cancer (43%–100%, sometimes in the absence of polyposis), duodenal polyps/adenoma (17%–25%), duodenal cancer (4%), serrated adenoma, hyperplastic/sessile serrated polyps, mixed (hyperplastic and adenomatous) polyps, malignancies of the ovary (14%, mean age 51 years), bladder (25% for males and 8% for females, mean age 61 years), skin, breast (mean age 51 years), and endometrium (mean age 51 years), sebaceous glands, thyroid (multinodular goiter, single nodules, and papillary thyroid cancer), and retina (congenital hypertrophy of retinal pigment epithelium [CHRPE], 5.5%). In fact, the incidence of extraintestinal malignancies in MAP patients is almost twice that of the general population [40–43].

On the other hand, monoallelic *MUTYH* mutation carriers (who harbor a heterozygous germline *MUTYH* pathogenic variant) display approximately a one- to twofold risk above the general population for colorectal cancer (7.2% for males and 5.6% for females), gastric cancer (5% for males and 2.3% for females), hepatobiliary cancer (3% for males and 1.4% for females), endometrial cancer (3%), and breast cancer (11%).

Due to the fact that p.Y179C mutation significantly reduces MUTYH glycosylase activity in relation to pG396D mutation, patients with homozygous p.Y179C mutation often display more severe disease (including earlier presentation and higher colorectal cancer risk) than those with biallelic p.G396D or those with compound heterozygous p.G396D/p.Y179C [1].

21.6 Diagnosis

Diagnosis of MAP requires observation of characteristic clinical and molecular findings [44].

Clinical findings (as obtained through medical history review, physical examination, colonoscopy, upper endoscopy, baseline thyroid ultrasound, histopathology) indicative of MAP include: (i) colonic adenomas and/or hyperplastic/serrated sessile polyps amounting to 1–10 for people <40 years, <19 for people between 40–60 years, >20 for people >60 years, 20 to a few hundred for people of any age; (ii) >100 colonic polyps in the absence of a heterozygous germline *APC* pathogenic variant; (iii) colorectal cancer diagnosed in individuals <40 years; (iv) family history of colon cancer (± polyps) showing an autosomal recessive inheritance pattern.

Histologically, MAP-associated colorectal cancer often display poor (26%) or moderate (71%) differentiation (Figure 21.1a), are predominately mucinous (Figure 21.1b), have Crohn's-like infiltrate (Figure 21.1c), contain tumor-infiltrating lymphocytes (Figure 21.1d), show p53 dysfunction as assessed by nuclear staining indicative (Figure 21.1e), and their tumor infiltrating lymphocytes may stain positive for $CD3^+$, $CD8^+$, and $CD57^+$ (Figure 21.1f) [45].

Molecular findings suggestive of MAP include: (i) identification of a specific somatic *KRAS* pathogenic variant (c.34G > T in codon 12), which occurs in 64% of MAP patients; (ii) microsatellite stable colorectal cancer; (iii) identification of biallelic *MUTYH* pathogenic variants (definitive diagnosis). It should be noted that 1%–2% of the general population may carry monoallelic MUTYH mutations, and combined MSH6 and monoallelic MUTYH mutations may possibly contribute to colorectal cancer risk.

Molecular testing for MAP often begins with sequence analysis of *MUTYH* (which uncovers small intragenic deletions/insertions and missense, nonsense, and splice site variants, with detection rate of ~99%) followed by gene-targeted deletion/duplication analysis (which detects intragenic deletions or duplications through the use of quantitative PCR, long-range PCR, or multiplex ligation-dependent probe amplification [MLPA]) if only one or no pathogenic variant is found by sequence analysis [46–50].

Differential diagnoses for MAP include other inherited polyposis and colon cancer conditions such as *APC*-associated polyposis (heterozygous germline *APC* pathogenic variant, autosomal dominant inheritance), *NTHL1*-associated polyposis (homozygous germline *NTHL1* pathogenic variants, autosomal

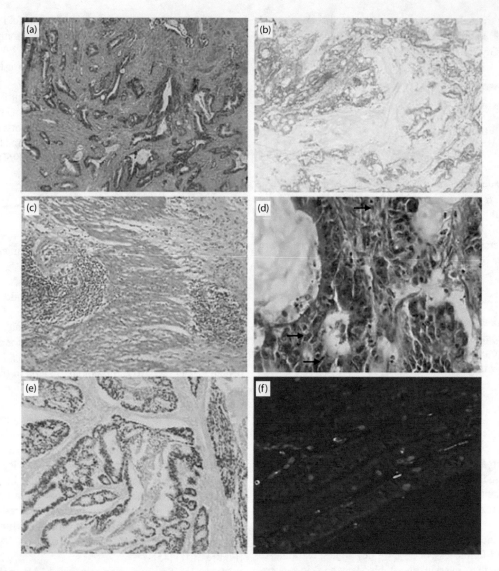

FIGURE 21.1 Histology of MAP-associated colorectal cancer. (a) Moderate differentiation, 5×; (b) >50% mucinous, 5×; (c) Crohn's-like infiltrate, 5×; (d) tumor infiltrating lymphocytes, 40×; (e) p53 staining, >75% nuclear staining, 5×; (f) CD3/CD8/CD57 immunofluorescent staining, red cells: CD3+, purple cells: CD3+ CD8+, white cells: CD3+ CD8+ CD57+, 40×. (Photo credit: Nielsen et al., *BMC Cancer.* 2009;9:184.)

recessive inheritance), Lynch syndrome (hereditary nonpolyposis colon cancer [HNPCC], colon cancer exhibiting microsatellite instability, heterozygous germline pathogenic variant in *MLH1*, *MSH2*, *MSH6*, or *PMS2*, germline deletion in *EPCAM*, autosomal dominant inheritance), Peutz–Jeghers syndrome (PJS, gastrointestinal hamartomatous polyps, and mucocutaneous pigmentation, heterozygous pathogenic variant in *STK11* in 94% of cases, autosomal dominant inheritance), juvenile polyposis syndrome (JPS, mutation in *BMPR1A* and *SMAD4*, autosomal dominant inheritance), *PTEN* hamartoma tumor syndromes (PHTS, presence of *PTEN* pathogenic variant, autosomal dominant inheritance), hereditary mixed polyposis syndrome (HMPS, heterozygous duplication on chromosome 15q13-q14 or heterozygous pathogenic variant in *BMPR1A*, autosomal dominant inheritance), and serrated polyposis syndrome (SPS, heterozygous germline pathogenic variants in *PTEN*, occasional biallelic germline pathogenic variants in *MUTYH*) [51–56].

21.7 Treatment

Suspicious colon polyps and colon cancer identified by colonoscopy should be removed by polypectomy, subtotal colectomy with ileorectal anastomosis (IRA), total proctocolectomy with ileostomy, and proctocolectomy with or without mucosectomy and ileal pouch anal anastomosis (IPAA). Duodenal polyps showing dysplasia or villous changes should be excised during endoscopy [57,58].

21.8 Prognosis, Surveillance, and Genetic Counseling

21.8.1 Prognosis

The risk for MAP patients to develop colorectal cancer is 19%, 43%, and 80% by age 50, 60, and 70 years, respectively, with an average age of onset of 48 years. Individuals with MAP-associated colorectal cancer have a 5-year survival rate of 78% compared to 63% for individuals with sporadic colorectal cancer.

21.8.2 Surveillance

Patients harboring biallelic germline *MUTYH* pathogenic variants should undergo pan colonoscopy (from 18–25 years of age), upper endoscopy, and side-viewing duodenoscopy (from 30–35 years of age). Individuals with a heterozygous germline *MUTYH* pathogenic variant may be offered average moderate-risk colorectal screening based on family history. Molecular genetic testing for familial pathogenic variants is offered to all sibs of an individual with genetically confirmed MAP; this helps reduce morbidity and mortality through early diagnosis and treatment.

21.8.3 Genetic Counseling

As MAP shows autosomal recessive inheritance, each sib of an affected individual has a 25% chance of being affected, a 50% chance of being a carrier, and a 25% chance of being unaffected and not a carrier. Offspring of individuals with one or two *MUTYH* pathogenic variants have a 0.5%–1.0% chance of inheriting two *MUTYH* pathogenic variants [1].

21.9 Conclusion

MUTYH-associated polyposis (MAP) represents an autosomal recessive form of intestinal polyposis that predisposes patients to colorectal cancer as well as extracolonic neoplasms. Clinically, MAP resembles AFAP (a milder form of FAP) in terms of polyp number and onset age. However, MAP differs from AFAP genetically by having biallelic (homozygous or compound heterozygous) mutations in the base excision repair gene *MUTYH*, instead of germline mutations in the *APC* gene (Table 21.1) [54,59]. Compared to patients harboring biallelic *MUTYH* mutations, those carrying monoallelic *MUTYH* mutations have a somewhat lower risk of developing colorectal cancer and other malignancies. Diagnosis of MAP is dependent on observation of typical clinical symptoms and application of molecular tests for mutations in *MUTYH* gene provides additional confirmation of this rare disorder. Given the lack of a specific cure, treatment options for MAP consist of symptomatic management (e.g., surgical resection of gastrointestinal polyps) in combination with regular surveillance.

REFERENCES

1. Nielsen M, Lynch H, Infante E, Brand R. MUTYH-associated polyposis. In: Adam MP, Ardinger HH, Pagon RA et al. editors. *GeneReviews*® [Internet]. Seattle (WA): University of Washington, Seattle; 1993–2017. 2012 Oct 4 [updated 2015 Sep 24].
2. Jung I, Gurzu S, Turdean GS. Current status of familial gastrointestinal polyposis syndromes. *World J Gastrointest Oncol.* 2015;7(11):347–55.
3. Leoz ML, Carballal S, Moreira L, Ocaña T, Balaguer F. The genetic basis of familial adenomatous polyposis and its implications for clinical practice and risk management. *Appl Clin Genet.* 2015;8:95–107.
4. Half E, Bercovich D, Rozen P. Familial adenomatous polyposis. *Orphanet J Rare Dis.* 2009;4:22.
5. Hes FJ, Ruano D, Nieuwenhuis M et al. Colorectal cancer risk variants on 11q23 and 15q13 are associated with unexplained adenomatous polyposis. *J Med Genet.* 2014;51(1):55–60.
6. Hurley JJ, Ewing I, Sampson JR, Dolwani S. Gastrointestinal polyposis syndromes for the general gastroenterologist. *Frontline Gastroenterol.* 2014;5(1):68–76.
7. Schlussel AT, Gagliano RA Jr, Seto-Donlon S et al. The evolution of colorectal cancer genetics-Part 2: Clinical implications and applications. *J Gastrointest Oncol.* 2014;5(5):336–44.
8. Cheng TH, Gorman M, Martin L et al. Common colorectal cancer risk alleles contribute to the multiple colorectal adenoma phenotype, but do not influence colonic polyposis in FAP. *Eur J Hum Genet.* 2015;23(2):260–3.
9. Khan N, Lipsa A, Arunachal G, Ramadwar M, Sarin R. Novel mutations and phenotypic associations identified through APC, MUTYH, NTHL1, POLD1, POLE gene analysis in Indian familial adenomatous polyposis cohort. *Sci Rep.* 2017;7(1):2214.
10. Talseth-Palmer BA. The genetic basis of colonic adenomatous polyposis syndromes. *Hered Cancer Clin Pract.* 2017;15:5.
11. Valle L. Recent discoveries in the genetics of familial colorectal cancer and polyposis. *Clin Gastroenterol Hepatol.* 2017;15(6):809–19.
12. Lynch HT, Shaw TG. Practical genetics of colorectal cancer. *Chin Clin Oncol.* 2013;2(2):12.
13. Torrezan GT, da Silva FC, Santos EM et al. Mutational spectrum of the APC and MUTYH genes and genotype-phenotype correlations in Brazilian FAP, AFAP, and MAP patients. *Orphanet J Rare Dis.* 2013;8:54.
14. Stigliano V, Sanchez-Mete L, Martayan A, Anti M. Early-onset colorectal cancer: A sporadic or inherited disease? *World J Gastroenterol.* 2014;20(35):12420–30.
15. Roncucci L, Pedroni M, Mariani F. Attenuated adenomatous polyposis of the large bowel: Present and future. *World J Gastroenterol.* 2017;23(23):4135–9.
16. Poulsen ML, Bisgaard ML. MUTYH associated polyposis (MAP). *Curr Genomics.* 2008;9(6):420–35.
17. Aceto GM, Fantini F, De Iure S et al. Correlation between mutations and mRNA expression of APC and MUTYH genes: New insight into hereditary colorectal polyposis predisposition. *J Exp Clin Cancer Res.* 2015;34:131.
18. Kantor M, Sobrado J, Patel S, Eiseler S, Ochner C. Hereditary colorectal tumors: A literature review on MUTYH-associated polyposis. *Gastroenterol Res Pract.* 2017;2017:8693182.
19. Venesio T, Balsamo A, D'Agostino VG, Ranzani GN. MUTYH-associated polyposis (MAP), the syndrome implicating base excision repair in inherited predisposition to colorectal tumors. *Front Oncol.* 2012;2:83.
20. Venesio T, Balsamo A, Errichiello E, Ranzani GN, Risio M. Oxidative DNA damage drives carcinogenesis in MUTYH-associated-polyposis by specific mutations of mitochondrial and MAPK genes. *Mod Pathol.* 2013;26(10):1371–81.
21. Luncsford PJ, Manvilla BA, Patterson DN et al. Coordination of MYH DNA glycosylase and APE1 endonuclease activities via physical interactions. *DNA Repair (Amst).* 2013;12(12):1043–52.
22. Markkanen E, Dorn J, Hübscher U. MUTYH DNA glycosylase: The rationale for removing undamaged bases from the DNA. *Front Genet.* 2013;4:18.

23. Viel A, Bruselles A, Meccia E et al. A specific mutational signature associated with DNA 8-oxoguanine persistence in MUTYH-defective colorectal cancer. *EBioMedicine*. 2017;20:39–49.

24. Raetz AG, Xie Y, Kundu S, Brinkmeyer MK, Chang C, David SS. Cancer-associated variants and a common polymorphism of MUTYH exhibit reduced repair of oxidative DNA damage using a GFP-based assay in mammalian cells. *Carcinogenesis*. 2012;33(11):2301–9.

25. Shinmura K, Goto M, Tao H, Matsuura S, Matsuda T, Sugimura H. Impaired suppressive activities of human MUTYH variant proteins against oxidative mutagenesis. *World J Gastroenterol*. 2012;18(47):6935–42.

26. Shinmura K, Goto M, Tao H et al. Impaired 8-hydroxygua-nine repair activity of MUTYH variant p.Arg109Trp found in a Japanese patient with early-onset colorectal cancer. *Oxid Med Cell Longev*. 2014;2014:617351.

27. Manlove AH, McKibbin PL, Doyle EL, Majumdar C, Hamm ML, David SS. Structure-activity relationships reveal key features of 8-oxoguanine: A mismatch detection by the MutY glycosylase. *ACS Chem Biol*. 2017;12(9):2335–44.

28. Hwang BJ, Shi G, Lu AL. Mammalian MutY homolog (MYH or MUTYH) protects cells from oxidative DNA damage. *DNA Repair (Amst)*. 2014;13:10–21.

29. Isoda T, Nakatsu Y, Yamauchi K et al. Abnormality in Wnt signaling is causatively associated with oxidative stress-induced intestinal tumorigenesis in MUTYH-null mice. *Int J Biol Sci*. 2014;10(8):940–7.

30. Grasso F, Di Meo S, De Luca G et al. The MUTYH base excision repair gene protects against inflammation-associated colorectal carcinogenesis. *Oncotarget*. 2015;6(23):19671–84.

31. Brinkmeyer MK, David SS. Distinct functional consequences of MUTYH variants associated with colorectal cancer: Damaged DNA affinity, glycosylase activity and interaction with PCNA and Hus1. *DNA Repair (Amst)*. 2015;34:39–51.

32. Komine K, Shimodaira H, Takao M et al. Functional complementation assay for 47 MUTYH variants in a MutY-disrupted *Escherichia coli* strain. *Hum Mutat*. 2015;36(7):704–11.

33. Aretz S, Genuardi M, Hes FJ. Clinical utility gene card for: MUTYH-associated polyposis (MAP), autosomal recessive colorectal adenomatous polyposis, multiple colorectal adenomas, multiple adenomatous polyps (MAP)—Update 2012. *Eur J Hum Genet*. 2013;21(1).

34. Aretz S, Tricarico R, Papi L et al. MUTYH-associated polyposis (MAP): Evidence for the origin of the common European mutations p.Tyr179Cys and p.Gly396Asp by founder events. *Eur J Hum Genet*. 2014;22(7):923–9.

35. Pin E, Pastrello C, Tricarico R et al. MUTYH c.933+3A > C, associated with a severely impaired gene expression, is the first Italian founder mutation in MUTYH-associated polyposis. *Int J Cancer*. 2013;132(5):1060–9.

36. Turco E, Ventura I, Minoprio A et al. Understanding the role of the Q338H MUTYH variant in oxidative damage repair. *Nucleic Acids Res*. 2013;41(7):4093–103.

37. Boesaard EP, Vogelaar IP, Bult P et al. Germline MUTYH gene mutations are not frequently found in unselected patients with papillary breast carcinoma. *Hered Cancer Clin Pract*. 2014;12(1):21.

38. Guarinos C, Juárez M, Egoavil C et al. Prevalence and characteristics of MUTYH-associated polyposis in patients with multiple adenomatous and serrated polyps. *Clin Cancer Res*. 2014;20(5):1158–68.

39. Brosens LA, Offerhaus GJ, Giardiello FM. Hereditary colorectal cancer: Genetics and screening. *Surg Clin North Am*. 2015;95(5):1067–80.

40. Morak M, Heidenreich B, Keller G et al. Biallelic MUTYH mutations can mimic Lynch syndrome. *Eur J Hum Genet*. 2014;22(11):1334–7.

41. Rashid M, Fischer A, Wilson CH et al. Adenoma development in familial adenomatous polyposis and MUTYH-associated polyposis: Somatic landscape and driver genes. *J Pathol*. 2016;238(1):98–108.

42. Tieu AH, Edelstein D, Axilbund J et al. Clinical characteristics of multiple colorectal adenoma patients without germline APC or MYH mutations. *J Clin Gastroenterol*. 2016;50(7):584–8.

43. Win AK, Reece JC, Dowty JG et al. Risk of extracolonic cancers for people with biallelic and monoallelic mutations in MUTYH. *Int J Cancer*. 2016;139(7):1557–63.

44. Borras E, Taggart MW, Lynch PM, Vilar E. Establishing a diagnostic road map for MUTYH-associated polyposis. *Clin Cancer Res*. 2014;20(5):1061–3.

45. Nielsen M, de Miranda NF, van Puijenbroek M et al. Colorectal carcinomas in MUTYH-associated polyposis display histopathological similarities to microsatellite unstable carcinomas. *BMC Cancer*. 2009;9:184.

46. Mancini-DiNardo D, Judkins T, Woolstenhulme N et al. Design and validation of an oligonucleotide microarray for the detection of genomic rearrangements associated with common hereditary cancer syndromes. *J Exp Clin Cancer Res*. 2014;33:74.

47. Simbolo M, Mafficini A, Agostini M et al. Next-generation sequencing for genetic testing of familial colorectal cancer syndromes. *Hered Cancer Clin Pract*. 2015;13(1):18.

48. Cohen SA, Tan CA, Bisson R. An individual with both MUTYH-associated polyposis and Lynch syndrome identified by multi-gene hereditary cancer panel testing: A case report. *Front Genet*. 2016;7:36.

49. Tezcan G, Tunca B, Ak S, Cecener G, Egeli U. Molecular approach to genetic and epigenetic pathogenesis of early-onset colorectal cancer. *World J Gastrointest Oncol*. 2016;8(1):83–98.

50. Lorans M, Dow E, Macrae FA, Winship IM, Buchanan DD. Update on hereditary colorectal cancer: Improving the clinical utility of multigene panel testing. *Clin Colorectal Cancer*. 2018;17(2):e293–e305.

51. Clendenning M, Young JP, Walsh MD et al. Germline mutations in the polyposis-associated genes BMPR1A, SMAD4, PTEN, MUTYH and GREM1 are not common in individuals with serrated polyposis syndrome. *PLOS ONE*. 2013;8(6):e66705.

52. Oka S, Leon J, Tsuchimoto D, Sakumi K, Nakabeppu Y. MUTYH, an adenine DNA glycosylase, mediates p53 tumor suppression via PARP-dependent cell death. *Oncogenesis*. 2014;3:e121.

53. Spier I, Holzapfel S, Altmüller J et al. Frequency and phenotypic spectrum of germline mutations in POLE and seven other polymerase genes in 266 patients with colorectal adenomas and carcinomas. *Int J Cancer*. 2015;137(2):320–31.

54. Papp J, Kovacs ME, Matrai Z et al. Contribution of APC and MUTYH mutations to familial adenomatous polyposis susceptibility in Hungary. *Fam Cancer*. 2016;15(1):85–97.

55. Petronio M, Pinson S, Walter T et al. Type 1 serrated polyposis represents a predominantly female disease with a high

prevalence of dysplastic serrated adenomas, without germ-line mutation in MUTYH, APC, and PTEN genes. *United European Gastroenterol J.* 2016;4(2):305–13.

56. Lv XP. Gastrointestinal tract cancers: Genetics, heritability and germ line mutations. *Oncol Lett.* 2017;13(3):1499–508.

57. Yurgelun MB, Hornick JL, Curry VK et al. Therapy-associated polyposis as a late sequela of cancer treatment. *Clin Gastroenterol Hepatol.* 2014;12(6):1046–50.

58. Syngal S, Brand RE, Church JM et al. ACG clinical guideline: Genetic testing and management of hereditary gastrointestinal cancer syndromes. *Am J Gastroenterol.* 2015;110(2):223–62; quiz 263.

59. Kashfi SM, Golmohammadi M, Behboudi F, Nazemalhosseini-Mojarad E, Zali MR. MUTYH the base excision repair gene family member associated with colorectal cancer polyposis. *Gastroenterol Hepatol Bed Bench.* 2013;6(Suppl 1):S1–10.

22

Peutz–Jeghers Syndrome

Fábio Guilherme Campos

CONTENTS

22.1 Introduction

Peutz–Jeghers syndrome (PJS) is a rare autosomal dominant disorder characterized by the formation of melanocytic macules (dark-colored spots) on the lips, around and inside the mouth, the eyelid, near the nostrils, around the anus, dorsal aspect of the fingers and sole of the foot, multiple noncancerous growths (hamartomatous polyps) in the gastrointestinal tract, and an increased risk for various tumors of the gastrointestinal tract, cervix, ovary, pancreas, and breast (e.g., gastrointestinal cancer, adenoma malignum of the cervix, breast cancer, biliary tract/gallbladder cancer, endometrial cancer, ovarian sex cord tumor, ovarian and testicular Sertoli–Leydig cell tumor, esophageal cancer, adrenal cancer, fallopian tube cancer, anal cancer, intra-abdominal desmoplastic small cell tumor, hepatocellular cancer, bone tumor, glioma of the brain, liposarcoma, lung/bronchial cancer, cutaneous melanoma, multiple myeloma/Kahler disease, non-Hodgkin lymphoma, pancreatic adenocarcinoma, paraganglioma, and papillary thyroid cancer). At the molecular level, PJS is linked to mutations in the serine/threonine kinase 11 gene (*STK11*) located on chromosome 19p13.3 [1].

22.2 Biology

Although clinical findings (i.e., melanocytic macules, intussusception, and breast cancer) related to PJS were first observed in a pair of twins by Connor in 1895 and Hutchinson in 1896, it was Peutz and Jeghers who recognized and highlighted the familial association of gastrointestinal polyposis and mucocutaneous pigmentation in 1921 and 1949, respectively, and in whose honor the disease is named. Peutz–Jeghers syndrome (PJS, also known as Peutz–Jeghers polyposis, inherited hamartomatous intestinal polyps, polyps and spots syndrome, Hutchinson Weber–Peutz syndrome, and perioral lentiginosis) produces a unique type of hamartomatous polyps (with a characteristic smooth muscle core arising from the muscularis mucosae, which extends to the polyps) in the gastrointestinal tract (particularly the jejunum, ileum, duodenum, colon, and stomach, but not the esophagus) along with mucocutaneous pigmentation. The resulting polyps (ranging from 1 mm to >5 cm in size, and up to 20 per segment of the intestinal tract) often induce recurrent abdominal pain and intestinal obstruction/intussusception, bleeding, or chronic anemia (due to polyp ulceration of polyps). Occasionally, PJS polyps may occur in the nostrils, lungs, renal pelvis, or urinary bladder [2].

While PJS' association with tumors (e.g., breast cancer and jejunal adenocarcinoma) was regarded coincidental in early days, subsequent demonstration of heterozygous mutations in the tumor suppressor *STK11* gene on chromosome 19p13.3 has unquestionably pointed to its tumor predisposition capacity. It is estimated that PJS increases the risk of gastrointestinal malignancies (frequently in the stomach, duodenum, and colon) by up to 39%, and extraintestinal malignancies (particularly in the pancreas, breast, ovary, and testis) by up to 54%. Based on classic two-hit model (Knudsen hypothesis), tumorigenesis in most familial cancer syndromes involves inheritance of a germline mutation followed by a subsequent chromosomal deletion (loss of heterozygosity), chromosomal rearrangement, hypermethylation, or somatic mutation. Indeed, somatic inactivation of the unaffected allele of *STK11* is often observed in polyps and cancers from PJS patients. Nonetheless, data generated from *STK11*$^{+/-}$ mouse polyps and human PJS cases support both one-hit (haploinsufficiency, as seen in PJS associated nonsmall cell lung cancer) and two-hit (as seen in biallelic *STK11* inactivation in PJS related cervical cancer) models in the transition from hamartoma → low grade dysplasia → high grade dysplasia → carcinoma in the tumorigenesis of PJS. On the whole, about 55% of PJS cases have family history, while others may evolve from de novo *STK11* mutations without obvious family connection [1].

22.3 Pathogenesis

The *STK11* gene (also known as *LKB1*) located on chromosome 19p13.3 is composed of 10 exons spanning 22.6 kb and encodes a 433 aa serine/threonine-protein kinase (STK11) with a central kinase catalytic domain (residues 49–309), regulatory N- and C-terminal domains, and a nuclear localization signal near the N-terminal. Present in the cytoplasm and translocated to mitochondria during apoptosis, STK11 regulates cell differentiation and proliferation by interfering G1 cell cycle arrest in a p53-independent manner, induces epithelial cell apoptosis in a p53-dependent manner, and maintains cellular polarity (via tubulin stabilization, tight junction formation, and E-cadherin localization), metabolism, and energy homeostasis through phosphorylation/activation of adenosine monophosphate–activated protein kinase (AMPK) [3,4]. Furthermore, use of *STK11*$^{+/-}$ mouse model confirms upregulation of cyclooxygenase-2 (COX-2) in polyp tissue, in line of overexpression of COX-2 in 60%–80% of PJS hamartomas and PJS-associated tumors.

Mutations in the *STK11* gene result in the production of a malfunctioning STK11 protein that abrogates its kinase activity, causes aberrant signal transduction, and eliminates its ability to control cell growth and division, leading to the formation of noncancerous polyps and cancerous tumors. To date, about 300 heterozygous mutations (including deletions [34%], missense [21%], insertions [14.5%], splice site [14%], nonsense [12%], and deletions/insertions, inversions, genomic rearrangements [4.5%]) in the *STK11* gene have been identified in 95% of PJS patients using sequencing and multiplex ligation–dependent probe amplification (MLPA). In particular, a 1-bp deletion or 1-bp insertion in a 6-cysteine repeat mutation hotspot

(c.837–c.842) is observed in 7% of PJS families. JPS patients harboring pathogenic variants that predict premature truncation in STK11 protein have similar ages of onset for polyps to patients who are negative for pathogenic variants. In addition, these patients often have more gastrointestinal surgeries, a higher polyp count, an earlier age of first polypectomy, and a greater risk of melanoma than patients harboring missense mutation and other pathogenic variants [5].

Besides PJS polyps, somatic mutations in *STK11* have been identified in pancreatic cancer, cervical cancer, testicular tumor, malignant melanoma, and non-small cell lung cancers. In fact, at least 20% of cervical cancers and up to one-third of non-small cell lung cancers have sporadic *STK11* pathogenic variants in addition to the usual *p53* and *K-RAS* gene mutations.

22.4 Epidemiology

22.4.1 Prevalence

PJS is a rare disease with an estimated incidence of 1 per 50,000–200,000 live births, including 1 per 50,000–100,000 in Finland, and 1 per 60,000–300,000 in North America. PJS affects all racial or ethnic groups and occurs equally in males and females, although several mutations of the *STK11/LKB1* gene may be more common in certain ethnic groups. Over 90% of PJS patients develop polyps during their lifetime, and about 50% of affected individuals show gastrointestinal symptoms (e.g., transient intussusception, small-bowel obstruction, and bleeding) by the age of 20 (median onset age of 13 years). As a consequence, approximately 30% of PJS patients require laparotomy before the age of 10% and 68% before the age of 18 [1,2].

22.4.2 Inheritance

PJS is inherited in an autosomal dominant pattern, in which one copy of the altered *STK11* gene in each cell is sufficient to increase the risk by up to 54% of developing noncancerous polyps and cancerous tumors. While about 55% of PJS patients inherit the mutation from one affected parent, the rest (45%) do not have a family history and appear to acquire de novo mutation in the *STK11* gene. The latter may possibly reflect the low reproductive fitness of PJS patients prior to the introduction of effective treatment for intussusception.

22.4.3 Penetrance

PJS demonstrates high penetrance with variable expression. Virtually all individuals harboring pathogenic variants in *STK11* manifest clinically (i.e., melanocytic macules and PJS-type intestinal polyps, along with a PJS-related cancer), with only one case of nonpenetrance of a *STK11* mutation being reported to date.

22.5 Clinical Features

Clinically, PJS is associated with mucocutaneous macules, gastrointestinal polyposis, and malignancy [6,7].

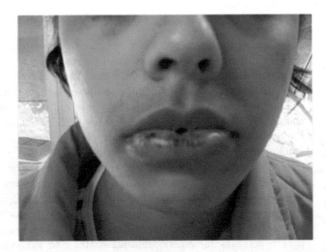

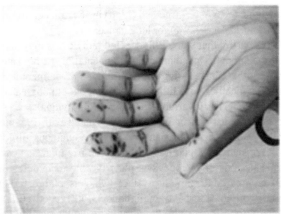

FIGURE 22.1 Peutz–Jeghers syndrome in a young female exhibiting mucocutaneous pigmentation on the lip (left) and hand (right).

22.5.1 Mucocutaneous Macules

Mucocutaneous macules are small, flat, brown or dark-blue spots of 1–4 mm in size that occur in >95% of PJS patients. Melanocytic macules are rarely present at birth and usually develop on the lips by the end of the first year, and other sites by 5 years of age. Melanocytic macules are commonly found around the mouth crossing the vermilion border (or lips, a cardinal feature of PJS, 94% of cases), perianal area (94%), fingertips (73%), toe tips (62%), the dorsal and volar aspects of the hands and feet, buccal mucosa (65%), surrounding the eyes and ears, on the eyelids, and nostrils (15%) (Figure 22.1). Some macules on the extremities may fade during the onset of puberty, although buccal mucosa lesions tend to persist. Therefore, the absence of melanocytic macules in an adult does not rule out the diagnosis of PJS. Melanocytic macules are considered as freckles (ephilides) when a diagnosis of PJS is not made [8].

22.5.2 Gastrointestinal Polyposis

PJS polyps typically emerge within the first few years of life in the small intestine (typically jejunum followed by ileum and duodenum, 64% of cases), colon (63%), stomach (48%), and rectum

(32%), but not esophagus (Figure 22.2) [9]. Sometimes, polyps may occur in the renal pelvis, urinary bladder, ureters, gallbladder, lungs, and nostrils. Showing erratic growth, PJS polyps may remain the same size for many years, and some polyps may regress or autoamputate spontaneously. While their potential for malignant transformation remains to be determined, PJS polyps are responsible for small intestinal obstruction and intussusception (42.8% of cases, due mainly to polyps of ≥ 1 .5 cm in diameter, usually between the ages of 6 and 18 years), abdominal pain due to infarction (23%), hematochezia (rectal bleeding) due to ulceration (13.5%), prolapse of colonic polyp (7%), nausea, vomiting, and secondary iron deficiency/anemia (typically occurring in the second and third decades of life) [10–12].

22.5.3 Malignancy

PJS poses an increased risk for gastrointestinal and extra-intestinal cancer. The cumulative risk in PJS patients by the age of 60 is 39% (vs 5% in general population) for colorectal cancer, 29% (vs <1%) for stomach cancer, 13% (vs <1%) for small bowel cancer, 54% (vs 12.4%) for breast cancer, 21% (vs 1.6%) for ovarian tumors (e.g., ovarian sex cord tumors with annular tubules [SCTAT]), 10% (vs <1%) for cervical cancer, 9% (vs 2.71%) for

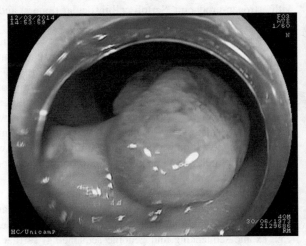

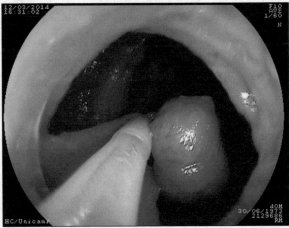

FIGURE 22.2 Peutz–Jeghers syndrome in a 17-year-old patient presenting with PJS polyps in the jejunum.

uterine cancer, 36% (vs 1.5%) for pancreatic cancer, 9% (vs <1%) for testicular cancer (e.g., large-cell calcifying Sertoli cell tumors [LCST]), and 17% (vs 6.9%) for lung cancer.

It is noteworthy that some PJS-affected individuals may show isolated melanocytic mucocutaneous pigmentation without polyps, while others have polyps in the absence of pigmentation. Melanocytic macules pose little malignancy risk, although a certain type of cancer is invariably found in PJS individuals.

Clinically, breast cancer in women associated with PJS may appear at an early age. Ovarian SCTAT and mucinous tumors of the ovaries and fallopian tubes may cause irregular or heavy menstrual periods and precocious puberty (due to hyperestrogenism). Compared to sporadic SCTAT (which is a large, unilateral tumor with a 20% risk of malignancy), PJS-related SCTAT is a bilateral multifocal small tumor with focal calcification and a typically benign course. Symptoms of adenoma malignum (a rare well-differentiated adenocarcinoma of the uterine cervix) comprise bleeding or a mucoid, watery vaginal discharge [13]. LCST of the testes may secrete estrogen, leading to gynecomastia, advanced skeletal age, and short stature.

22.6 Diagnosis

Patients displaying some or the following clinical characteristics (e.g., family history of PJS, repeated bouts of abdominal pain in patients under 25 years, unexplained intestinal bleeding in a young patient, prolapse of tissue from the rectum, menstrual irregularities in females, gynecomastia in males, precocious puberty, gastrointestinal intussusception with bowel obstruction, mucocutaneous pigmentation, melena or rectal bleeding, hematemesis, and anemia) should be suspected of PJS and undergo further investigation [14,15].

Diagnostic workup for PJS includes family history review, physical examination (for mucocutaneous pigmentation, etc.), imaging studies (e.g., upper gastrointestinal endoscopy, capsule endoscopy, magnetic resonance imaging [MRI] enteroclysis, computed tomography [CT] scanning with oral contrast medium, CT enterography, colonoscopy, push enteroscopy, intraoperative enteroscopy [IOE], double-balloon enteroscopy [DBE], endoscopic ultrasonography [EUS], endoscopic retrograde cholangiopancreatography [ERCP], chest radiography or CT scanning in smokers), laboratory studies (e.g., complete blood cell [CBC] count for potential anemia, iron studies, hemoccult for occult blood in the stool, carcinoembryonic antigen [CEA] for screening and monitoring of cancer degeneration), histopathological investigation, and genetic testing (for heterozygous mutations in the *STK11* gene) [16,17].

22.6.1 Melanocytic Macules

Melanocytic macules from PJS patients often contain increased melanocytes with long pigment-filled dendrites at the epidermal–dermal junction and show increased melanin in the basal cells. Macroscopically, pigmentation occurs mainly in vertical bands interrupted by unpigmented areas. Some dark-colored spots may have a stippled appearance under magnification. Electron microscopy reveals a blockage in pigment transfer from melanocytes to keratinocytes in melanocytic macules of the fingers and toes.

22.6.2 Gastrointestinal Polyps

Rectal polyp (rectal mass) may sometimes be felt during rectal examination, and some large rectal polyps may even prolapse outside the anus. Imaging techniques help determine the presence and the location of small intestinal polyps. For distal small-bowel polyps that are beyond the reach of conventional endoscopy, use of video capsule endoscopy (VCE) or magnetic resonance enterography (MRE) is invaluable [18,19].

Radiography shows intussusception as a soft tissue mass located on the right upper quadrant containing concentric circular lucency (target sign) due to mesenteric fat, and showing meniscus sign due to a crescent of air from the intussuscipiens at the apex of intussusception. Ultrasound reveals a "pseudokidney" or "doughnut" appearance of intussusception, consisting of inner hyperechoic mesenteric fat inside and the outer hypoechoic rim. CT shows a "bowel-within-bowel" appearance of intussusception, with the inner loop of the bowel separated from the outer loop by a crescent of fat-attenuation mesentery [20–23].

Macroscopically, PJS polyps are pedunculated with a coarse lobulated surface and resemble hamartomatous polyps (Figure 22.2). Histologically, PJS polyps show an elongated, frond-like epithelium with cystic dilatation of glands (which contain deeply eosinophilic mucin) overlying an arborizing network (branching tree-like core) of smooth muscle derived from the muscularis mucosae throughout the lamina propria (typically occurring in small bowel polyps) or lobular organization (as in colonic crypts), along with prominent hypermucinous goblet cells, and pseudoinvasion by histopathologically benign epithelium without transforming into cancer (as in PJS small intestine polyps, but not in PJS colon or stomach polyps, possibly reflecting the role of *STK11* in cell polarity) (Figure 22.3). Indeed, the formation of a unique smooth muscle core that arborizes throughout PJS polyps helps differentiate from sporadic hamartomatous polyps (with a mean age of 55 years at diagnosis) and hamartomatous polyps associated with other syndromes, and the absence of cytologic atypia in PJS polyps separates it from typical adenomas (which display cytologic atypia and lack of differentiation). The prominent smooth muscle component in PJS polyps suggests the role of smooth muscle cells in polyposis development.

Individuals with two or more PJS-type intestinal polyps, mucocutaneous macules, gynecomastia in males (associated with estrogen-producing Sertoli cell testicular tumors) and history of intussusception should be suspected of PJS.

Well-established clinical diagnostic criteria for PJS comprise: (i) three histopathologically proven PJS polyps, (ii) classic mucocutaneous pigmentation, and (iii) a positive family history.

World Health Organization (WHO) diagnostic criteria for PJS includes any of the following: (i) three or more histologically confirmed PJS polyps, or (ii) any number of PJS polyps with a positive family history, or (iii) characteristic, prominent, mucocutaneous pigmentation with a positive family history, or (iv) any number of PJS polyps and characteristic, prominent, mucocutaneous pigmentation [1].

Molecular identification of a *STK11* pathogenic variant using sequence analysis, gene-targeted deletion/duplication analysis, multigene panel, and more comprehensive genomic testing (e.g., exome sequencing and genome sequencing) further supports the diagnosis [24–30]. While it is not absolutely required

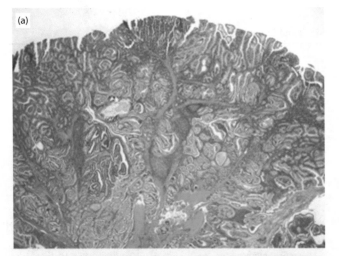

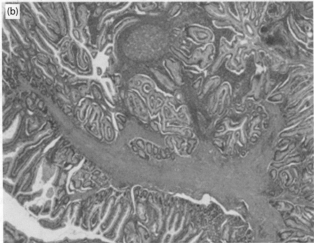

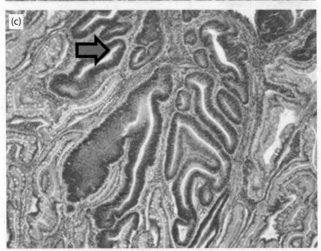

FIGURE 22.3 Histology of Peutz–Jeghers polyp. (a) Multilobulated polyps with a papilliferous arrangement; (b) muscular fibers forming bridges with mucosal glandular hyperplasia; (c) areas of high-grade dysplasia (arrow).

for establishing a definitive diagnosis of PJS, genetic testing for *STK11* pathogenic variants helps differentiate tumors (e.g., breast cancer) caused by other mutated genes (e.g., *BRCA1* and *BRCA2*), facilitates implementation of appropriate treatment strategies, and enables accurate identification of at-risk relatives

who will benefit from early treatment and preventive measures as well as pregnancy planning. About 95% of PJS patients have been shown to harbor *STK11* pathogenic variants, of which 80% are small intragenic deletions/insertions, and missense, nonsense, and splice site variants (detected by sequencing analysis), and 15% are intragenic deletions or duplications (detected by gene-targeted deletion/duplication analysis or MLPA) [31–34]. For a small proportion (4%) of PJS patients who do not have germline mutations in the *STK11* gene, involvement of other unidentified genes is possible [35].

Cancers showing a special association with PJS are adenoma malignum (ADM, a rare tumor of the cervix, 5% of cases), sex cord tumor with annular tubules (SCTAT, a rare tumor of the ovary), and Sertoli cell testicular tumor in the testes [36].

As a very rare, highly differentiated adenocarcinoma of the endocervical glands, ADM causes watery vaginal discharge or vaginal bleeding. Macroscopically, ADM has a firm or nodular appearance, resembling a polypoid mass, and presents as multiple cervical cysts in imaging studies. Due to its histological resemblance to normal endocervical glands, ADM is sometimes referred to as minimal deviation adenocarcinoma. Nevertheless, ADM is distinguishable by its associated desmoplastic response, nuclear atypia, deep invasion of the cervical wall, and identification of a focus of undifferentiated adenocarcinoma. In addition, ADM stains positive for Alcian blue periodic acid Schiff, while normal endocervical glands are positive for HIK1803 monoclonal antibody to gastric gland mucous cell-mucin endocervical glands only.

SCTAT is an asymptomatic adnexal cyst or mass occurring in about 10% of female PJS patients. As it sometimes produces estrogen, SCTAT may contribute to precocious puberty. Macroscopically, PJS-associated SCTAT is a bilateral, multifocal lesion with focal calcifications, in contrast to sporadic SCTAT, which is large and unilateral. PJS Histologically, SCTAT shows simple or complex tubules lined by cells that possess peripherally placed nuclei surrounded a hyaline-filled lumen. SCTAT has a low malignant potential and a good prognosis [37].

Sertoli cell testicular tumors typically present as asymptomatic, bilateral multifocal lesions with calcifications or prepubertal gynecomasty (due to aromatase production, which converts testosterone to an estrogen precursor, causing prepubertal gynecomasty) in PJS children (of 1 to 14 years). They rarely progress to invasive large calcifying Sertoli cell tumors (ILCST) [38].

Differential diagnoses for PJS include several hereditary tumor syndromes that show overlapping signs and symptoms with PJS, i.e. juvenile polyposis syndrome (JPS), Cowden syndrome (CS), Carney complex (CC), familial adenomatous polyposis (FAP), and hereditary nonpolyposis colorectal cancer (HNPCC) (Table 22.1) [39–42]. Other notable differential diagnosis consists of Laugier–Hunziker syndrome (LHS, presence of perioral, digit, and nailbed hyperpigmentation/lentiginosis of dark-brown to blue-black color in adults; absence of polyposis and cancer risk; lack of family history).

22.7 Treatment

Treatment options for PJS include surgery, chemotherapy, and symptomatic management.

TABLE 22.1

Characteristics of Hereditary Tumor Syndromes Showing Overlapping Signs and Symptoms with PJS

Syndrome	Pigmentation	Gastrointestinal Tumor	Sertoli Cell Tumor	Other Neoplasm	Other Notable Sign	Altered Gene (Frequency)
Peutz–Jeghers syndrome (PJS)	Facial, mucosal	Adenoma, hamartoma	+/–	Colon, gastric, cervical, ovarian, breast, pancreatic, lung cancer	Hyperestrogenism	*STK11* (95%)
Juvenile polyposis syndrome (JPS)	–	Adenoma, hamartoma	–	Colon cancer	Heart defects?	*SMAD4* (20%), *BMPR1A* (20%)
Cowden syndrome (CS)	Axillary, inguinal, facial	Adenoma, hamartoma	–	Breast, thyroid, and endometrial cancer	Macrocephaly, breast fibrosis	*PTEN* (70%)
Carney complex (CC)	Facial, mucosal	–	+	Thyroid cancer; skin and heart myxomas		*PRKAR1A* (60%)
Familial adenomatous polyposis (FAP)	–	Adenoma	–	Colon and brain cancer, desmoid tumors, osteomas, congenital hypertrophy of the retinal pigment epithelium (CHRPE)		*APC* (80%)
Hereditary nonpolyposis colorectal cancer (HNPCC)		Adenoma	–	Endometrial, gastric, renal pelvis, and ureter, ovarian tumors; sebaceous adenoma		*MLH1, MS H2, MSH3, MSH6, PMS1, PMS2* (50%)

Source: Modified from Kohlmann W, Gruber SB. In: Adam MP et al. editors. *GeneReviews®* [Internet]. Seattle (WA): University of Washington, Seattle; 1993–2018. 2004 Feb 5 [updated 2018 Feb 1].

22.7.1 Surgery

Although PJS-associated melanocytic macules do not develop into melanoma or other malignancy, they may be removed with laser treatment for cosmetic reasons. Given their potential to cause intussusception/obstruction/intestinal bleeding, undergo malignant transformation, or become too large for endoscopic excision, PJS polyps of >1.0 cm in size should be removed by standard endoscopy. Laparoscopic-assisted enteroscopy offers a less invasive option for polypectomy. Double-balloon endoscopy (DBE) or balloon-assisted enteroscopy with or without laparotomy is an effective and safe procedure for removing large (3.0–5.0 cm in size) and distal small-bowel polyps, obviating the need for intraoperative enteroscopy or enterotomy. Instead of endoscopic approach, surgical technique (reduction, enterotomy and polyp resection) is recommended for treatment of acute intussusception with intestinal obstruction caused by large PJS polyps. Gonadal tumors in males and females are managed through conservative measures. Prophylactic hysterectomy and bilateral salpingo-oophorectomy may be considered for women with gynecologic malignancy [1].

22.7.2 Chemotherapy

Cyclooxygenase 2 (COX-2) is highly upregulated in a murine model of PJS, and COX inhibitors (e.g., celecoxib) are being investigated for treating gastric polyposis. Indeed, use of celecoxib leads to a decrease in the formation of new polyps and the size of preexisting polyps in murine model, and a significant reduction in gastric polyp size in two of the six PJS patients under investigation. However, further studies are required on the side effects (e.g., risk of myocardial infarction and stroke) of celecoxib or other COX-2 inhibitors prior to their clinical application for PJS treatment. Being a proximal member of the mammalian target of rapamycin (mTOR) pathways, STK11 is upregulated in

PJS neoplasia. As a suppressor of the mTOR pathway, rapamycin has been shown to reduce polyp burden in a murine model of PJS and may have the potential for JPS treatment. Further, everolimus appears to be effective in inducing a partial remission in PJS patients with advanced pancreatic cancer. Aromatase inhibitor anastrozole has been also used for successful treatment of ILCST [43].

22.8 Prognosis, Surveillance, and Risk Evaluation

PJS is known to increase the lifetime risk of both gastrointestinal (e.g., stomach, duodenum, and colon) and extraintestinal malignancies, including cancer of the breast (by 54%, mean age of 37), colon (by 39%, mean age of 46), pancreas (by 36%, mean age of 41), stomach (by 29%, mean age of 30), ovary (by 21%, mean age of 28), lung (by 15%), small intestine (by 13%, mean age of 42), cervix (by 10%, mean age of 34), uterus (by 9%), and testis (by 9%, mean age of 9). The overall relative risk for cancer is greater in females than in males. A survey of 72 PJS patients revealed that 48% had died from cancer by the age of 57 years.

Surveillance for gastric and small-bowel polyposis and removal of large polyps play a vital part in reducing the likelihood of complications in PJS patients, prolong life expectancy, and improve outcomes through the early detection of carcinomas. Surveillance based on imaging and other techniques should start from birth to teenage (for testis polyposis/tumor at 1–2 year interval), age 8 (for gastric and intestinal polyposis/tumor, at 3-year intervals), age 18 (for breast, ovary, uterus polyposis/tumor, at 1-year intervals), and age 30 (for pancreas polyposis/tumor at 1- to 2-year intervals). Clean-sweep enteroscopy (double-balloon or intraoperative assisted and push enteroscopy) may be used to reduce the risk of obstruction, surgical resection, and short bowel syndrome [44].

Being an autosomal dominant disorder, inheritance of a mutated copy of the *STK11* gene from one PJS parent is sufficient to increase the cancer risk by up to 54%. As 55% of PJS patients have a family connection, apparently asymptomatic older and younger at-risk relatives of an affected individual (proband) should be assessed for hyperpigmented macules in the buccal mucosa and skin of the digits and genital area, gastrointestinal polyps, and *STK11* pathogenic variant with the goal to identify those who would benefit from early treatment and preventive measures. In addition, discussion of potential risks to offspring and reproductive options should be held with affected or at-risk young adults who are planning for pregnancy.

22.9 Conclusion

Peutz–Jeghers syndrome (PJS) is an autosomal dominant disorder that demonstrates hallmark features of hamartous gastrointestinal polyps (which display differentiation defects both in the epithelial and stromal components with an increase in myofibroblasts, and differ from juvenile type polyps by their smooth muscle components and adenomatous changes), mucocutaneous pigmentation (dark-colored spots) on the lips, buccal region, vulva, toes, and fingers (which appear in the first decade and usually disappear from the third decade), and a heightened risk of intestinal and extraintestinal neoplasms (which is estimated at 6%, 18%, 31%, 41%, and 67% for people aged 30, 40, 50, 60, and 70, respectively). Molecular studies have pinpointed mutations in the serine/threonine kinase 11 gene (*STK11*) on chromosome 19p13.3 as the underlying cause of PJS. While about 55% of PJS cases result from inheritance of mutated *STK11*, others likely arise from sporadic *STK11* mutations. In contrast to many other autosomal dominant disorders that fit into the classic Knudsen hypothesis, PJS appears to involve both one-hit (haploinsufficiency) and two-hit models in its tumorigenesis. Although PJS can be adequately diagnosed on the basis of well-established clinical criteria without involvement of genetic testing, molecular identification of a pathogenic variant in the *STK11* gene helps distinguishes PJS related tumors from those harboring other mutated genes, which may have different treatment options, and permits accurate identification of at-risk relatives for early treatment and preventive measures as well as pregnancy planning. Current strategies for management of PJS consist of regular screening for polyps and surgical interventions. There is a notable absence of effective chemotherapies for PJS. Further research is certainly warranted in this area.

REFERENCES

1. McGarrity TJ, Amos CI, Baker MJ. Peutz-Jeghers Syndrome. In: Adam MP, Ardinger HH, Pagon RA et al. editors. *GeneReviews*® [Internet]. Seattle (WA): University of Washington, Seattle; 1993–2018.2001 Feb 23 [updated 2016 Jul 14].
2. Riegert-Johnson D, Gleeson FC, Westra W et al. Peutz-Jeghers syndrome. In: Riegert-Johnson DL, Boardman LA, Hefferon T, Roberts M, editors. *Cancer Syndromes* [Internet]. Bethesda (MD): National Center for Biotechnology Information (US); 2009. No abstract available. 2008 Jul 18 [updated 2008 Aug 9].
3. Vaahtomeri K, Mäkelä TP. Molecular mechanisms of tumor suppression by LKB1. *FEBS Lett.* 2011;585(7):944–5.
4. Wang YS, Chen J, Cui F et al. LKB1 is a DNA damage response protein that regulates cellular sensitivity to PARP inhibitors. *Oncotarget.* 2016;7(45):73389–401.
5. Jang MS, Lee YM, Ko BM, Kang G, Kim JW, Hong YH. Complete STK11 deletion and atypical symptoms in Peutz-Jeghers syndrome. *Ann Lab Med.* 2017;37(5):462–4.
6. Wang R, Qi X, Liu X, Guo X. Peutz-Jeghers syndrome: Four cases in one family. *Intractable Rare Dis Res.* 2016;5(1):42–3.
7. Wang R, Qi X, Shao X, Guo X. A Large intracolonic mass in a patient with Peutz-Jeghers syndrome. *Middle East J Dig Dis.* 2017;9(3):173–5.
8. Medina-Murillo GR, Rodríguez-Medina U, Rodríguez-Wong U. Disseminated plantar lentigines associated with Peutz-Jeghers syndrome. *Rev Gastroenterol Mex.* 2016;81(3):168–9.
9. Zou BC, Wang FF, Zhao G et al. A giant and extensive solitary Peutz-Jeghers-type polyp in the antrum of stomach: Case report. *Medicine (Baltim).* 2017;96(49):e8466.
10. De Silva WS, Pathirana AA, Gamage BD, Manawasighe DS, Jayasundara B, Kiriwandeniya U. Extra-ampullary Peutz-Jeghers polyp causing duodenal intussusception leading to biliary obstruction: A case report. *J Med Case Rep.* 2016;10:196.
11. Rathi CD, Solanke DB, Kabra NL, Ingle MA, Sawant PD. A rare case of solitary Peutz Jeghers type hamartomatous duodenal polyp with dysplasia. *J Clin Diagn Res.* 2016;10(7):OD03–4.
12. Duan SX, Wang GH, Zhong J et al. Peutz-Jeghers syndrome with intermittent upper intestinal obstruction: A case report and review of the literature. *Medicine (Baltim).* 2017;96(17):e6538
13. Kim EN, Kim GH, Kim J et al. A pyloric gland-phenotype ovarian mucinous tumor resembling lobular endocervical glandular hyperplasia in a patient with Peutz-Jeghers syndrome. *J Pathol Transl Med.* 2017;51(2):159–64.
14. Matini E, Houshangi H, Jangholi E, Farjad Azad P, Najibpour R, Farshad A. Peutz-Jeghers syndrome with diffuse gastrointestinal polyposis: Three cases in a family with different manifestations and no evidence of malignancy during 14 years follow up. *Iran Red Crescent Med J.* 2015;17(12):e19271.
15. Zhang LJ, Su Z, Liu X, Wang L, Zhang Q. Peutz-Jeghers syndrome with early onset of pre-adolescent gynecomastia: A predigree case report and clinical and molecular genetic analysis. *Am J Transl Res.* 2017;9(5):2639–44.
16. Suzuki K, Higuchi H, Shimizu S, Nakano M, Serizawa H, Morinaga S. Endoscopic snare papillectomy for a solitary Peutz-Jeghers-type polyp in the duodenum with ingrowth into the common bile duct: Case report. *World J Gastroenterol.* 2015;21(26):8215–20.
17. Adán-Merino L, Aldeguer-Martínez M, Lozano-Maya M, Hernández-García-Gallardo D, Casado-Fariñas I. Lights and shadows in the diagnosis and surveillance of a young asymptomatic patient with Peutz-Jeghers syndrome. *Rev Gastroenterol Mex.* 2016;81(1):59–61.
18. Huang ZH, Song Z, Zhang P, Wu J, Huang Y. Clinical features, endoscopic polypectomy and STK11 gene mutation in a nine-month-old Peutz-Jeghers syndrome Chinese infant. *World J Gastroenterol.* 2016;22(11):3261–7.
19. Cheng W, Liu H, Gu Z, Hu Z, Wang L, Wang X. Narrow-band imaging endoscopy is advantageous over conventional white light endoscopy for the diagnosis and treatment of children with Peutz-Jeghers syndrome. *Medicine (Baltim).* 2017;96(19):e6671.

20. Tomas C, Soyer P, Dohan A, Dray X, Boudiaf M, Hoeffel C. Update on imaging of Peutz-Jeghers syndrome. *World J Gastroenterol*. 2014;20(31):10864–75.

21. Krishnan V, Chawla A, Wee E, Peh WC. Clinics in diagnostic imaging. 159. Jejunal intussusception due to Peutz-Jeghers syndrome. *Singapore Med J*. 2015;56(2):81–5; quiz 86.

22. Kılıç S, Atıcı A, Soyköse-Açıkalın Ö. Peutz-Jeghers syndrome: an unusual cause of recurrent intussusception in a 7-year-old boy. *Turk J Pediatr*. 2016;58(5):535–7.

23. Kalavant AB, Menon P, Mitra S, Thapa BR, Narasimha Rao KL. Solitary Peutz-Jeghers polyp of jejunum: A rare cause of childhood intussusception. *J Indian Assoc Pediatr Surg*. 2017;22(4):245–7.

24. JH, Yoo JH, Choi YJ et al. A novel de novo mutation in the serine-threonine kinase STK11 gene in a Korean patient with Peutz-Jeghers syndrome. *BMC Med Genet*. 2008;9:44.

25. Brito S, Póvoas M, Dupont J, Lopes AI. Peutz-Jeghers syndrome: Early clinical expression of a new STK11 gene variant. *BMJ Case Rep*. 2015;2015. pii: bcr2015211345.

26. Huang Z, Miao S, Wang L, Zhang P, Wu B, Wu J, Huang Y. Clinical characteristics and STK11 gene mutations in Chinese children with Peutz-Jeghers syndrome. *BMC Gastroenterol*. 2015;15:166.

27. Chen C, Zhang X, Wang D et al. Genetic screening and analysis of LKB1 gene in Chinese patients with Peutz-Jeghers syndrome. *Med Sci Monit*. 2016;22:3628–40.

28. Masuda K, Kobayashi Y, Kimura T et al. Characterization of the STK11 splicing variant as a normal splicing isomer in a patient with Peutz-Jeghers syndrome harboring genomic deletion of the STK11 gene. *Hum Genome Var*. 2016;3:16002.

29. Tan H, Mei L, Huang Y et al. Three novel mutations of STK11 gene in Chinese patients with Peutz-Jeghers syndrome. *BMC Med Genet*. 2016;17(1):77.

30. Chen JH, Zheng JJ, Guo Q et al. A novel mutation in the STK11 gene causes heritable Peutz-Jeghers syndrome—a case report. *BMC Med Genet*. 2017;18(1):19.

31. Chiang JM, Chen TC. Clinical manifestations and STK11 germline mutations in Taiwanese patients with Peutz-Jeghers syndrome. *Asian J Surg*. 2017 Aug 28. pii: S1015-9584(17)30254-3. [Epub ahead of print].

32. Linhart H, Bormann F, Hutter B, Brors B, Lyko F. Genetic and epigenetic profiling of a solitary Peutz-Jeghers colon polyp. *Cold Spring Harb Mol Case Stud*. 2017;3(3):a001610.

33. Zhao ZY, Jiang YL, Li BR et al. A novel germline mutation (c.A527G) in STK11 gene causes Peutz-Jeghers syndrome in a Chinese girl: A case report. *Medicine (Baltim)*. 2017;96(49):e8591.

34. Zhao ZY, Jiang YL, Li BR et al. Sanger sequencing in exonic regions of STK11 gene uncovers a novel de-novo germline mutation (c.962_963delCC) associated with Peutz-Jeghers syndrome and elevated cancer risk: case report of a Chinese patient. *BMC Med Genet*. 2017;18(1):130.

35. Zhang Y, Ke Y, Zheng X, Liu Q, Duan X. Correlation between genotype and phenotype in three families with Peutz-Jeghers Syndrome. *Exp Ther Med*. 2017;13(2):507–14.

36. Neyaz A, Husain N, Deodhar M, Khurana R, Shukla S, Arora A. Synchronous cervical minimal deviation adenocarcinoma, gastric type adenocarcinoma and lobular endocervical glandular hyperplasia along with STIL in Peutz-Jeghers syndrome: Eliciting oncogenesis pathways. *Turk Patoloji Derg*. 2017;1(1):1–7.

37. Zhou F, Lv B, Dong L, Wan F, Qin J, Huang L. Multiple genital tract tumors and mucinous adenocarcinoma of colon in a woman with Peutz-Jeghers syndrome: A case report and review of literatures. *Int J Clin Exp Pathol*. 2014;7(7):4448–53.

38. Bellfield EJ, Alemzadeh R. Recurrent ovarian Sertoli-Leydig cell tumor in a child with Peutz-Jeghers syndrome. *Oxf Med Case Reports*. 2016;2016(8):omw048.

39. Chew MH, Tan WS, Liu Y, Cheah PY, Loi CT, Tang CL. Genomics of hereditary colorectal cancer: Lessons learnt from 25 years of the Singapore Polyposis Registry. *Ann Acad Med Singapore*. 2015;44(8):290–6.

40. Jung I, Gurzu S, Turdean GS. Current status of familial gastrointestinal polyposis syndromes. *World J Gastrointest Oncol*. 2015;7(11):347–55.

41. Tomlinson I. The Mendelian colorectal cancer syndromes. *Ann Clin Biochem*. 2015;52(6):690–2.

42. Kohlmann W, Gruber SB. Lynch syndrome. In: Adam MP, Ardinger HH, Pagon RA et al. editors. *GeneReviews®* [Internet]. Seattle (WA): University of Washington, Seattle; 1993–2018. 2004 Feb 5 [updated 2018 Feb 1].

43. Koç Yekedüz M, Şıklar Z et al. Response to anastrozole treatment in a case with Peutz-Jeghers syndrome and a large cell calcifying Sertoli cell tumor. *J Clin Res Pediatr Endocrinol*. 2017;9(2):168–71.

44. Mozaffari HR, Rezaei F, Sharifi R, Mirbahari SG. Sevenyear follow-up of Peutz-Jeghers syndrome. *Case Rep Dent*. 2016;2016:6052181.

23

Serrated Polyposis Syndrome

Dongyou Liu

CONTENTS

23.1 Introduction

First described in 1970s to avoid misclassification as familial adenomatous polyposis (FAP), serrated polyposis syndrome (SPS) is a colorectal cancer predisposition disorder characterized by the formation of numerous serrated polyps throughout the colon, which may transform into colorectal adenomas, serrated adenomas, and cancer. The World Health Organization (WHO) defines SPS as having: (i) at least five serrated polyps (i.e., sessile serrated adenomas) proximal to the sigmoid colon, with two or more being >10 mm in diameter; (ii) any number of serrated polyps proximal to the sigmoid colon, with a first-degree relative diagnosed with this disease; or (iii) >20 serrated polyps of any size distributed throughout the colon. Recent studies have indicated that the molecular mechanism underlying SPS pathogenesis is likely attributed to specific *BRAF, KRAS,* and *RNF43* mutations followed by epigenetic silencing of tumor suppressor genes through promoter hypermethylation (termed CpG island methylator phenotype [CIMP]) [1–3].

23.2 Biology

Extending from the cecum (which connects the ileum of the small intestine) to the anal canal, the colon (large intestine) is a tube of 150 cm in length that can be separated into ascending, transverse, descending, and sigmoid sections. The ascending colon (sometimes referred to as the proximal or right colon) starts from the cecum, ascends to the right lobe of the liver, and turns 90° at the right colic flexure (or hepatic flexure) to become the transverse colon, which crosses the abdomen horizontally to the spleen. The transverse colon then turns another 90° at the left colic flexure (or splenic flexure) to become the descending colon, which moves inferiorly and connects the sigmoid colon. The sigmoid colon (sometimes referred to as the distal or left colon) is a 40-cm long tube forming a characteristic "S" shape that connects the rectum, then the anus.

Cancers affecting the colon and rectum (i.e., colorectal cancer) represent the third most common malignancy and the third leading cause of cancer-related death worldwide. Although many colorectal cancer cases arise sporadically, about 30% show a family history, and only 5% are clearly linked to hereditary disorders, including Lynch syndrome (mutated *MLH1, MSH2, MSH6, PMS2,* or *EpCAM*), familial adenomatous polyposis (FAP, mutated *APC*), MUTYH-associated polyposis (MAP, mutated *MUTYH*), juvenile polyposis syndrome (JPS, mutated *STK11*), Peutz–Jeghers syndrome (PJS, mutated *SMAD4* or *BMPR1A*), hereditary mixed polyposis syndrome (HMPS, mutated *GREM1*), and serrated polyposis syndrome (SPS, mutated *RNF43* and other genes) [4–8].

SPS (also known as sessile serrated polyposis cancer syndrome [SSPCS] or hyperplastic polyposis syndrome [HPS]) typically induces serrated polyps (which are noted for their serrated or "saw-tooth-like" appearance of the crypt epithelium) in the large intestine, which may be distinguished histologically into hyperplastic polyps (HP), sessile serrated adenomas/polyps (SSA/P, with "sessile serrated adenoma" and "sessile serrated polyp" being considered synonymous), and traditional serrated adenomas (TSA) (Figure 23.1 and Table 23.1) [9].

Accounting for over three-quarters of all serrated lesions, HP are small (usually <0.5 cm), flat polyps with more pronounced

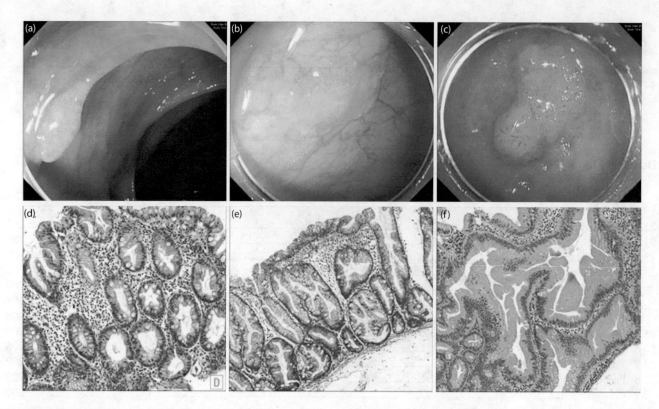

FIGURE 23.1 (a) Endoscopy of hyperplastic polyp (HP): a 6-mm transverse colon polyp with a smooth and pale appearance. (b) Endoscopy of sessile serrated adenoma (SSA): a 12-mm ascending colon polyp with a flat or sessile appearance and indistinct borders as well as a characteristic rim of debris. (c) Endoscopy of traditional serrated adenoma (TSA): a 15-mm descending colon polyp with a granulonodular and lobular appearance. (d) Histopathology of HP showing serrated crypts confined to the upper crypt. (e) Histopathology of SSA showing enlarged serrated surfaces and epithelial serration extending to the crypt bases. (f) Histopathology of TSA showing characteristic serrated crypt with pseudostratified, elongated nuclei and abundant eosinophilic cytoplasm (d–f, H&E). (Photo credit: Kim ER et al. *Intest Res.* 2017;15(3):402–10.)

serration in the upper half and surface of the polyps than at the base. The straight crypts extend symmetrically from the surface of the polyp to the muscularis mucosae without significant distortion. Typically occurring in the distal colon (left side), HP may be further divided on the basis of the characteristics of lining epithelium into microvesicular hyperplastic polyps (MVHP or type 2 hyperplastic polyp), goblet cell hyperplastic polyps (GCHP or type 1 hyperplastic polyp), and mucin poor polyps (MPHP) (Table 23.1). HP are previously believed to have no malignant potential; however, MVHP of >0.5 cm in size and showing atypical morphology may represent a precursor to SSA/P, with the capacity to develop into adenoma. Etiologically, HP may possibly evolve from crypt fusion and defects in apoptosis in combination with hypermaturation of the upper crypt and surface epithelium. Further, MPHP may represent injured MVHP with inflammation and reactive epithelial changes [1].

Accounting for almost one-quarter of all serrated lesions, SSA/P are flat polyps with a yellow, green, or rust-colored mucus cap, and show disorganized and distorted crypt growth pattern, with dilated and/or branched crypts at the basal portion of the polyp appearing "boot-," "L-," or "anchor"-shaped. Other features include nuclear atypia, dystrophic goblet cells, and an absence of neuroendocrine cells. Displaying a right colonic predominance (proximal [66%] and distal [33%] to the splenic flexure), SSA/P have potential for dysplasia and malignant transformation and are responsible for 20%–30% of colon cancer [1,10].

Representing <1% of all serrated polyps, TSA are often pedunculated and located distally (left side). TSA are similar to adenomas in appearance and have potential for dysplasia, although their protruding morphology and low numbers render them easy to detect and of insignificant clinical concern compared to SSA/P [1].

Evolution from normal mucosa to SSA/P likely involves an activating mutation in *BRAF* or *KRAS* oncogenes or other genes, permitting apoptosis evasion and leading to small aberrant crypt foci with serrated glands. Further transformation of SSA/P into malignant lesions through an epigenetic hypermethylation mechanism of CpG islands within promoter regions of tumor suppressor genes (the so-called serrated pathway) adds to the proliferative effect of a *BRAF* mutation to produce SSA/P, promotes microsatellite instability (MSI) in SSA/P, and leads to the formation of CIMP tumors (or serrated adenocarcinomas), which account for a third of all colorectal cancer [11].

CpG islands are 0.5-2-kb regions rich in cytosine guanine dinucleotides and are found in the 5′ region of nearly 50% of human genes. Addition of a methyl group (CH_3) to the cytosine nucleotide in a CpG dinucleotide context of gene promoters provides an epigenetic mechanism for regulating gene expression without altering the DNA sequence. Aberrant DNA methylation of CpG islands within the promoter regions of multiple genes (so-called CIMP) can result in transcriptional silencing of important tumor suppressor genes, leading to neoplastic growth [12].

TABLE 23.1

Phenotypic and Genotypic Features of Serrated Polyps

	Hyperplastic Polyps (HP)			Sessile Serrated Adenoma/ Polyp (SSA/P)	Traditional Serrated Adenoma (TSA)
	Microvesicular Hyperplastic Polyp (MVHP)	Goblet Cell Hyperplastic Polyp (GCHP)	Mucin Poor Hyperplastic Polyp (MPHP)		
Endoscopic feature	Flat, sessile, small (2–5 mm), pale, glistening polyps with indistinct borders; weak vascular network	Flat, sessile, small (2–5 mm), pale, glistening polyps with indistinct borders; weak vascular network	Flat, sessile, small (2–5 mm), pale, glistening polyps with indistinct borders; weak vascular network	Flat, sessile, pale polyps with a yellow, green or rust-colored mucus cap and a rim of bubbles or debris at the borders; larger than HP (often >10 mm)	Sessile, pedunculated polyps; larger and bulkier than HP and SSA/P
Histological feature	Straight crypts; stellate-shaped crypt lumens; narrow and uniform crypt bases; vesicular or "frothy" eosinophilic cytoplasm containing small droplet (microvesicular) mucin; scattered goblet cells	Straight crypts, with few or no luminal serrations; predominantly goblet cells	Straight crypts; micropapillary architecture; prominent nuclear atypia; marked decrease or complete absence of mucin and goblet cells; luminal serration pattern similar to MVHP	Serrated crypts; dilated crypts and serration extending into the lower third of the crypt; inverted T- or L-shaped crypt bases; focal nuclear stratification, mild nuclear atypia, or dystrophic goblet cells in the crypt bases; increased mucin production, absence of enteroendocrine cells, and absence of a thickened basement membrane under the surface; small foci of pseudostratification and eosinophilic change of the surface epithelium; small prominent nucleoli, open chromatin, and irregular nuclear contours; mitoses	Filiform crypts; ectopic crypt formation (crypts with bases not seated adjacent to the muscularis mucosae); abundant eosinophilic cytoplasm; centrally placed elongated nuclei that are hyperchromatic and display pseudostratification; unequivocal traditional adenomatous dysplasia
Clinical feature	Most common HP, together with GCHP and MPHP, making up 75% of all serrated polyps; distal/ left colon (10%–15% in proximal and transverse colon); possible dysplastic potential as a precursor to SSA/P if of large size (>5 mm), with atypical morphology, and in proximal colon	Less common HP, together with MVHP and MPHP, making up 75% of all serrated polyps; distal/ left colon; no dysplastic potential	Least common HP (together with MVHP and GCHP, making up 75% of all serrated polyps); distal/left colon; no dysplastic potential	Common, making up 25% of all serrated polyps; proximal/right colon; dysplastic potential	Rare, making up <1% of all serrated polyps; distal/left colon; dysplastic potential
Molecular feature	*BRAF* mutation (76%); CIMP (68%)	*KRAS* mutation (54%)	*BRAF* mutation; CIMP	*BRAF* mutation (75%–82%); *RNF43* mutation (19%); CIMP (92%); MSI-high	*KRAS*, *BRAF* or *RNF43* mutation; MSS; CIMP-low

23.3 Pathogenesis

In the classical adenoma-adenocarcinoma sequence, the stepwise progression of premalignant lesion to carcinoma is characterized by chromosomal instability (CIN) and *APC, KRAS,* and *TP53* mutations. In the serrated pathway, the transformation from normal mucosa to SSA/P and serrated adenocarcinomas likely involves specific *BRAF* and *KRAS* mutations, MSI, and CIMP, which may cause *MLH1* or *MGMT* (O6-methylguanine-DNA methyltransferase) silencing [13–17].

The *BRAF* and *KRAS* oncogenes are two members of the mitogen-activated protein kinase (MAPK) cascade. Both sporadic and SPS-associated serrated lesions have shown to contain activating somatic mutations in *BRAF* and *KRAS,* in addition to CIMP, MSI, and unique expression signature. Indeed, the *BRAF* missense mutation V600E is observed in 67%–88% of HP (notably right-sided MVHP), 61%–83% of SSP/A, 0% of conventional adenomas, and 49% of colorectal cancer, while the *KRAS* missense mutations (e.g., G12D, G12 V, G13D) are present in 6%–17% of HP (notably left-sided GCHP), 7%–25% of SSP/A, 3% of conventional adenomas, and 6% of colorectal cancer. Further, CIMP is detected in 41%–73% of HP and 44%–77% of SSP/A, and MSI is found in 36% of SSP/A and 40% of colorectal cancer arising in the setting of serrated polyposis. Moreover, the *BRAF* mutation, CIMP, and MSI occur in >80% of serrated adenocarcinomas. These findings highlight the direct correlation of the frequency of *BRAF* and *KRAS* mutations and the number/location of polyps in SPS specimens [18,19].

The *RNF43* (*R132X*) tumor suppressor gene on chromosome 17q22 consists of 11 exons spanning 60 kb and encodes a 783 aa, 90 kDa transmembrane E3 ubiquitin protein ligase (RNF43), which acts as a Wnt signaling inhibitor by targeting the Frizzled receptor for degradation. Consisting of a transmembrane domain 5-prime of the RING finger motif and two C-terminal nuclear localization signals, RNF43 is a HAP95 (AKAP8L) binding ubiquitin ligase occurring in the endoplasmic reticulum that promotes cell growth and is upregulated in colon cancer. Truncating mutations in *RNF43* has been identified in sporadic SSP/A and TSA, and heterozygosity for a germline nonsense mutation in *RNF43* is associated with a high risk of developing SSA/P. In fact, RNF43 mutations (e.g., frameshift mutations, Gly659fs and Arg117fs) are found in 18.9% of patients with colorectal and endometrial cancer and appear significantly enriched in microsatellite instability-high (MSI-H) tumors in both tumor types. Interestingly, truncating *RNF43* mutations and inactivating *APC* mutations are mutually exclusive in colorectal cancer [2,20].

Several other factors (e.g., cigarette smoking, alcohol intake, fiber intake, calcium intake, NSAID use, obesity, and diabetes mellitus) may also contribute to the development of both syndromic and sporadic SSA/P. In fact, smoking has been linked to CIMP high cancers, *BRAF* mutations, and tumors with MSI. Given its association with smaller lesions known as aberrant crypt foci, smoking may play a part in the initiation and subsequent growth of serrated lesions. In addition, there is a stronger correlation between smoking and distally located serrated polyps in comparison with proximal polyps [21,22].

23.4 Epidemiology

23.4.1 Prevalence

SPS has an estimated prevalence of 1 per 3000 and is detected 2% of patients undergoing their first screening colonoscopy. However, the true prevalence of serrated lesions is likely higher than that reported in the literature, as serrated lesion detection is highly variable and operator-dependent [23,24]. In one study that focused on the prevalence and detection rates of serrated polyps of all subtypes located proximal to the splenic flexure, the proportion of colonoscopies with at least one proximal serrated polyp ranged from 1% to 18%, and the detection rates per colonoscopy ranged from 0.01% to 0.26%. This suggests that substantial numbers of endoscopists miss more than half of the serrated lesions in the proximal colon [25]. Further, SPS displays no sex preference among patients, and appears to be restricted to individuals of Northern European ancestry. The mean age at diagnosis is 55 years (range 41–60 years).

23.4.2 Inheritance

SPS is possibly inherited, but its inheritance pattern remains undefined. Up to 50% of SPS patients show a family history of colorectal cancer, and about 32% of first-degree relatives (i.e., parents, siblings and children) of SPS patients are found to have polyps on screening colonoscopy.

23.4.3 Penetrance

SPS demonstrates a high penetrance, with 25%–70% of patients having colorectal cancer at time of diagnosis or during follow-up. First-degree relatives have a 32% risk of developing multiple serrated polyps and a fivefold increased risk of colon cancer. In addition, the presence of serrated lesions is also associated with an increased occurrence of synchronous adenomas [26].

23.5 Clinical Features

Clinically, patients with SPS may display nonspecific symptoms such as acute large bowel obstruction, diffuse abdominal pain or discomfort, weight loss, bowel habit changes, rectal bleeding episodes (particularly after emergence of colorectal cancer), and anemia. However, some patients may be asymptomatic [13].

23.6 Diagnosis

Given its nonspecific clinical presentation, diagnosis of SPS relies on endoscopic detection and histological examination of serrated polyps, accompanied by family history review and molecular testing [27].

While conventional endoscopy is useful for evaluating serrated polyps (Figure 23.1 and Table 23.1), it has an inferior sensitivity compared to chromoendoscopy. The latter combines indigo carmine as a contrast (which accumulates in pits and innominate grooves of the colonic mucosa, outlining the limits of flat lesions and drawing the described Kudo patterns) with a magnification endoscope, permitting identification of hyperplastic and serrated polyps as Kudo type I (normal) and type II (stellate or papillary) [28].

Serrated polyps are epithelial lesions that give a serrated appearance due to infolding of crypt epithelium. Histological examination allows further differentiation of serrated polyps into HP, SSA/P, and TSA (Figure 23.1 and Table 23.1) [13,29,30].

According to the WHO, clinical diagnosis of SPS requires fulfillment of at least one of the following criteria: (i) at least five serrated polyps (i.e., sessile serrated adenomas [SSA]) in the area proximal to the sigmoid colon, two of which measure >10 mm in diameter; (ii) any number of serrated polyps in the area proximal to the sigmoid colon, with a first-degree family history of SPS; and (iii) >20 serrated polyps of any size distributed throughout the colon.

Given the frequent occurrence of mutations in *BRAF* and *KRAS* oncogenes as well as *RNF43* tumor suppressor gene, MSI, and CIMP in SPS (Table 23.1), molecular detection of these genetic alterations provides a valuable approach for diagnosis of SPS and related colorectal cancer. In general, SSA/P are associated with proximal colorectal cancer and show high levels of CIMP, *BRAF* mutations, and MSI-high. In contrast, TSA occur in distal location, and relate to MSS, CIMP-low colorectal cancer with KRAS mutations. Polyps from SPS patients often display a significantly higher frequency of *BRAF, KRAS,* or *RNF43* mutations than sporadic HP. SPS may be excluded if both *BRAF* and *KRAS* mutations are present in <10% of HP from one patient, or if <5% of HP are MSI [31].

Differential diagnoses for SPS include other rare colonic polyposis syndromes, ranging from Cowden syndrome (mutations in *PTEN*; macrocephaly, tumors of the endometrium, thyroid, and breast; hamartomatous polyposis of the gastrointestinal tract), JPS (mutations in *SMAD4* or *BMPR1A*; serrated polyps), hereditary mixed polyposis syndrome (mutation in *GREM1*; conventional adenoma, atypical hamartomatous polyps, serrated polyps, and frequent colorectal carcinoma), and *MUTYH*-associated polyposis (biallelic mutations in *MUTYH*; serrated polyps) [6,32–35].

23.7 Treatment

Considering the potential of serrated polyps (SSA/P) for malignant transformation, SPS patients with high polyp burden or large serrated polyps (5 mm) should undertake colonoscopy, polypectomy, or surgery (e.g., colectomy and ileorectal anastomosis, especially when colorectal cancer is diagnosed). Patients with smaller serrated polyps (<5 mm) should be monitored by yearly colonoscopies and removed by cold snaring or electrocautery if necessary. Traditional serrated adenomas (pedunculated polyps) can be removed by conventional electrocautery snare polypectomy. Large, flat HP can be removed by endoscopic mucosal resection, together with the use of argon plasma coagulation in the lesion borders to destroy residual tissue and reduce the risk of recurrence [14,36,37].

23.8 Prognosis and Surveillance

Being a genetic disease, SPS is associated with multiple serrated lesions, younger age of lesion onset, family history of colorectal cancer, restricted ethnicity, and increased cancer risk. Without intervention, up to 70% of SPS patients will develop colorectal cancer, while first-degree relatives (including mother, father, sister, brother, children) have a 32% risk of developing multiple serrated polyps and a fivefold increased risk of colon cancer. The prognosis of early-manifesting lesions in SPS can be improved through timely diagnosis, appropriate treatment, and implementation of specific surveillance programs.

A typical surveillance program for SPS includes: (i) colonoscopy with pancolonic chromoendoscopy every 1–2 years with removal of all polyps ≥5 mm; (ii) colectomy for large or multiple tumors, or high grade dysplasia, when colonoscopy is ineffective; (iii) screening colonoscopy every 1–2 years for first-degree relatives of 10 years younger than the index case [38].

Specifically, MVHP and GCHP in proximal location and >5 mm should be monitored by colonoscopy at 5-year intervals. SSA/P of <10 mm should be monitored by colonoscopy at 5-year intervals; those of ≥10 mm or any size and ≥3 in number should be monitored by colonoscopy at 3-year intervals; those of ≥10 mm and two or more or dysplasias present should be monitored by colonoscopy at 1–3 year intervals. Traditional serrated adenomas of <10 mm should be monitored by colonoscopy at 5-year intervals and those of >10 mm or any size and >2 in number should be monitored at 3-year intervals [1,14,39,40].

23.9 Conclusion

Serrated polyposis syndrome (SPS) is a genetic disorder characterized by the formation of a heterogeneous group of serrated lesions including HP (75%), SSP/A (up to 25%), and TSA (<1%) in the colon. Despite their common occurrence, HP show limited malignant potential. In contrast, SSP/A and occasionally TSA tend to undergo malignant transformation through the serrated pathway, which involves an activating mutation in *BRAF* or *KRAS* oncogenes and *RNF43* tumor suppressor gene followed by epigenetic hypermethylation of CpG islands within promoter regions of tumor suppressor genes, leading to serrated adenocarcinomas (colorectal cancer). Compared to other colorectal cancer predisposition syndromes, SPS is noted for its capacity to pose a much higher cancer risk, with up to 70% of patients developing colorectal cancer during their lifetime, and 32% of first-degree relatives developing multiple serrated polyps as well as colon cancer. Early diagnosis, appropriate treatment, and regular surveillance represent the current measures for the management of SPS and help improve the quality of life for SPS patients.

REFERENCES

1. Rex DK, Ahnen DJ, Baron JA et al. Serrated lesions of the colorectum: Review and recommendations from an expert panel. *Am J Gastroenterol.* 2012;107(9):1315–30.
2. Giannakis M, Hodis E, Mu XJ et al. RNF43 is frequently mutated in colorectal and endometrial cancers. *Nature Genet.* 2014;46:1264–6.
3. González N, Caballero M, Cannesa C. Serrated polyposis syndrome. *Rev Gastroenterol Mex.* 2018;83(1):62–3.

4. Cancer Genome Atlas Network. Comprehensive molecular characterization of human colon and rectal cancer. *Nature*. 2012;487:330–7.

5. Patel SG, Ahnen DJ. Familial colon cancer syndromes: An update of a rapidly evolving field. *Curr Gastroenterol Rep*. 2012;14(5):428–38.

6. Brosens LA, Offerhaus GJ, Giardiello FM. Hereditary colorectal cancer: Genetics and screening. *Surg Clin North Am*. 2015;95(5):1067–80.

7. Castro J, Cuatrecasas M, Balaguer F, Ricart E, Pellisé M. Polyposis syndrome associated with long-standing inflammatory bowel disease. *Rev Esp Enferm Dig*. 2017;109(11):796–8.

8. Lv XP. Gastrointestinal tract cancers: Genetics, heritability and germ line mutations. *Oncol Lett*. 2017;13(3):1499–508.

9. Herreros de Tejada A, González-Lois C, Santiago J. Serrated lesions and serrated polyposis syndrome. *Rev Esp Enferm Dig*. 2017;109(7):516–26.

10. Sugumar A, Sinicrope FA. Serrated polyps of the colon. *F1000 Med Rep*. 2010;2:89.

11. Clendenning M, Young JP, Walsh MD et al. Germline mutations in the polyposis-associated genes BMPR1A, SMAD4, PTEN, MUTYH and GREM1 are not common in individuals with serrated polyposis syndrome. *PLOS ONE*. 2013;8(6):e66705.

12. Drini M, Wong NC, Scott HS et al. Investigating the potential role of genetic and epigenetic variation of DNA methyltransferase genes in hyperplastic polyposis syndrome. *PLOS ONE*. 2011;6(2):e16831.

13. Guarinos C, Sánchez-Fortún C, Rodríguez-Soler M, Alenda C, Payá A, Jover R. Serrated polyposis syndrome: Molecular, pathological and clinical aspects. *World J Gastroenterol*. 2012;18(20):2452–61.

14. Anderson JC. Pathogenesis and management of serrated polyps: Current status and future directions. *Gut Liver*. 2014;8(6):582–9.

15. Gala MK, Mizukami Y, Le LP et al. Germline mutations in oncogene-induced senescence pathways are associated with multiple sessile serrated adenomas. *Gastroenterology*. 2014;146:520–9.

16. He EY, Wyld L, Sloane MA, Canfell K, Ward RL. The molecular characteristics of colonic neoplasms in serrated polyposis: A systematic review and meta-analysis. *J Pathol Clin Res*. 2016;2(3):127–37.

17. Horpaopan S, Kirfel J, Peters S et al. Exome sequencing characterizes the somatic mutation spectrum of early serrated lesions in a patient with serrated polyposis syndrome (SPS). *Hered Cancer Clin Pract*. 2017;15:22.

18. Delker DA, McGettigan BM, Kanth P et al. RNA sequencing of sessile serrated colon polyps identifies differentially expressed genes and immunohistochemical markers. *PLOS ONE*. 2014;9(2):e88367.

19. Kanth P, Bronner MP, Boucher KM et al. Gene signature in sessile serrated polyps identifies colon cancer subtype. *Cancer Prev Res (Phila)*. 2016;9(6):456–65.

20. Taupin D, Lam W, Rangiah D et al. A deleterious RNF43 germline mutation in a severely affected serrated polyposis kindred. *Hum Genome Var*. 2015;2:15013.

21. Toyoshima N, Sakamoto T, Makazu M et al. Prevalence of serrated polyposis syndrome and its association with synchronous advanced adenoma and lifestyle. *Mol Clin Oncol*. 2015;3(1):69–72.

22. Wu Y, Mullin A, Stoita A. Clinical predictors for sessile serrated polyposis syndrome: A case control study. *World J Gastrointest Endosc*. 2017;9(9):464–70.

23. Kim HK, Seo KJ, Choi HH et al. Clinicopathological characteristics of serrated polyposis syndrome in Korea: Single center experience. *Gastroenterol Res Pract*. 2015;2015:842876.

24. Kim ER, Jeon J, Lee JH et al. Clinical characteristics of patients with serrated polyposis syndrome in Korea: Comparison with Western patients. *Intest Res*. 2017;15(3):402–10.

25. Crowder CD, Sweet K, Lehman A, Frankel WL. Serrated polyposis is an underdiagnosed and unclear syndrome: The surgical pathologist has a role in improving detection. *Am J Surg Pathol*. 2012;36(8):1178–85.

26. Boparai KS, Reitsma JB, Lemmens V et al. Increased colorectal cancer risk in first-degree relatives of patients with hyperplastic polyposis syndrome. *Gut*. 2010; 59: 1222–5.

27. Pyleris E, Koutsounas IS, Karantanos P. Three colon adenocarcinomas arising in a patient with serrated polyposis syndrome: Case report and review of the literature. *Viszeralmedizin*. 2014;30(2):136–9.

28. Kamiński MF, Hassan C, Bisschops R et al. Advanced imaging for detection and differentiation of colorectal neoplasia: European Society of Gastrointestinal Endoscopy (ESGE) Guideline. *Endoscopy*. 2014;46(5):435–49.

29. Rosty C, Buchanan DD, Walsh MD et al. Phenotype and polyp landscape in serrated polyposis syndrome: A series of 100 patients from genetics clinics. *Am J Surg Pathol*. 2012;36(6):876–82.

30. Jasperson KW, Kanth P, Kirchhoff AC et al. Serrated polyposis: Colonic phenotype, extracolonic features, and familial risk in a large cohort. *Dis Colon Rectum*. 2013;56(11):1211–6.

31. Rosty C, Walsh MD, Walters RJ et al. Multiplicity and molecular heterogeneity of colorectal carcinomas in individuals with serrated polyposis. *Am J Surg Pathol*. 2013;37(3):434–42.

32. Ngeow J, Heald B, Rybicki LA et al. Prevalence of germline PTEN, BMPR1A, SMAD4, STK11, and ENG mutations in patients with moderate-load colorectal polyps. *Gastroenterology*. 2013;144(7):1402–9, 1409.e1–5.

33. Hurley JJ, Ewing I, Sampson JR, Dolwani S. Gastrointestinal polyposis syndromes for the general gastroenterologist. *Frontline Gastroenterol*. 2014;5(1):68–76.

34. Schlussel AT, Gagliano RA Jr, Seto-Donlon S et al. The evolution of colorectal cancer genetics-Part 2: Clinical implications and applications. *J Gastrointest Oncol*. 2014;5(5):336–44.

35. Davis H, Irshad S, Bansal M et al. Aberrant epithelial GREM1 expression initiates colonic tumorigenesis from cells outside the stem cell niche. *Nat Med*. 2015;21(1):62–70.

36. Rosty C, Hewett DG, Brown IS, Leggett BA, Whitehall VL. Serrated polyps of the large intestine: Current understanding of diagnosis, pathogenesis, and clinical management. *J Gastroenterol*. 2013;48(3):287–302.

37. Suzuki D, Matsumoto S, Mashima H. A case with serrated polyposis syndrome controlled by multiple applications of endoscopic mucosal resection and endoscopic submucosal dissection. *Am J Case Rep*. 2017;18:304–7.

38. Hazewinkel Y, Koornstra J-J, Boparai KS et al. Yield of screening colonoscopy in first-degree relatives of patients with serrated polyposis syndrome. *J Clin Gastroenterol*. 2015;49:407–12.

39. Syngal S, Brand RE, Church JM et al. ACG clinical guideline: Genetic testing and management of hereditary gastrointestinal cancer syndromes. *Am J Gastroenterol*. 2015;110(2):223–62.

40. East JE, Atkin WS, Bateman AC et al. British Society of Gastroenterology position statement on serrated polyps in the colon and rectum. *Gut*. 2017;66(7):1181–96.

24

Tylosis with Esophageal Cancer

Dongyou Liu

CONTENTS

24.1 Introduction

Tylosis with esophageal cancer (TOC) is a rare, autosomal dominant disorder characterized by tylotic changes (focal thickening of the skin, callus formation, or hyperkeratosis) of the palms and soles (so-called palmoplantar keratoderma [PPK]), oral leukokeratosis, and a high lifetime risk (95% at the age of 65) of esophageal squamous cell carcinoma. While PPK usually begins around age 10, esophageal cancer occurs after age 20. At the molecular level, TOC is linked to missense mutations in the rhomboid 5 homolog 2 (*RHBDF2*) gene located on chromosome 17q25.1 [1].

24.2 Biology

First described by Howell-Evans et al. in 1958, TOC (also referred to as Howell-Evans syndrome, Clarke–Howell-Evans–McConnell syndrome, Bennion–Patterson syndrome, tylosis-esophageal carcinoma, palmoplantar hyperkeratosis-esophageal carcinoma syndrome, PPK with esophageal cancer, familial keratoderma with carcinoma of the esophagus, focal non-epidermolytic PPK with carcinoma of the esophagus, keratosis palmaris et plantaris with esophageal cancer, palmoplantar ectodermal dysplasia type III, and keratosis palmoplantaris–esophageal carcinoma syndrome) is a genetic disorder that typically shows non-epidermolytic palmoplantar keratoderma, oral leukoplakia, and an increased risk of developing squamous cell carcinoma of the esophagus [1,2].

PPK is a thickening of the stratum corneum of the palms and soles that can be divided into four subtypes: diffuse, punctate, focal, and striate. In general, diffuse PPK emerges at birth or shortly thereafter, affects the entire palm and sole with a sharp cutoff at an erythematous border, and shows no follicular or oral lesions [3]. In contrast, focal PPK tends to have a late onset, often in response to mechanical trauma, and is accompanied by the development of oral and follicular hyperkeratosis.

Depending on the presence or absence of cytolysis in the upper spinous and granular layers of the epidermis, familial tylosis palmoplantaris (PPK) is separated histologically into two forms: epidermolytic and non-epidermolytic. The epidermolytic PPK is linked to heterozygous mutation in the keratin-9 gene (*KRT9*) on chromosome 17q12.2, and its mild form may be associated with mutation in the keratin-1 gene (*KRT1*) on chromosome 12q13.13, both of which do not seem to confer heightened risk to esophageal cancer [4–7]. In contrast, the non-epidermolytic PPK (i.e., TOC) is caused by genetic defect in the *RHBDF2* (homolog of Drosophila rhomboid 5) gene on chromosome 17q25.1 and shows a marked risk of developing esophageal cancer [1].

Based on the time when PPK first appears, TOC may be differentiated into two forms: A and B. In type A TOC, PPK first arises between ages 5 and 15, and its lesions are not well demarcated, often contain fissuring, and are commonly associated with ectodermal dysplasia. For this reason, TOC is occasionally called palmoplantar ectodermal dysplasia type III. In type B TOC, PPK develops at birth or in the first years of life, and its lesions are well demarcated, rarely complicated by painful fissuring, and infrequently associated with ectodermal dysplasia [8].

Linkage mapping and targeted next-generation sequencing studies have identified gain-of-function (GOF) mutations (p.I186T, p.P189L, and p.D188N) in the human rhomboid family protein RHBDF2 (or inactive rhomboid protease iRhom), which is encoded by the *RHBDF2* gene on 17q25.1, and which is involved in epidermal growth factor receptor (EGFR) shedding [9].

24.3 Pathogenesis

Located on the long arm of chromosome 17 (i.e., 17q25.1), the *RHBDF2* gene encodes a catalytically inactive rhomboid (iRhom) protease, iRhom2 (also referred to as RHBDF2 or RHBDL6), one of a family of enzymes possessing a core of six transmembrane helices with the active site residues lying in a hydrophilic cavity [9,10]. Being highly conserved and predominantly expressed in the skin, RHBDF2 (iRhom2) regulates the maturation of the multi-substrate ectodomain sheddase enzyme ADAM17 (whose substrates include TNF-alpha, amphiregulin [AREG], and HBEGF), inhibits the rhomboid like protease 2 (RHBDL2), which cleaves thrombomodulin at the transmembrane domain, releasing soluble thrombomodulin, and also acts on epidermal growth factor (EGF) [11]. In addition, RHBDF2 modulates the epithelial response to physical stress and injury in the esophagus and skin through its interaction with K16 and its binding partner K6 [12,13].

GOF mutations (c.557T > C [p.Ile186Thr], c.566C > T [p.Pro189Leu] and c.562 G > A [p.Asp188Asn]) in RHBDF2 (iRhom2), which is a short-lived protein, increase its stability and augment the maturation and activity of ADAM17 in epidermal keratinocytes, significantly upregulating the shedding of ADAM17 substrates (e.g., EGF-family growth factors, pro-inflammatory cytokines, EGF family ligand AREG, which is a functional driver of tylosis). This increases EGFR activity, desmosome turnover, and immature epidermal desmosomes, upregulates epidermal transglutaminase activity, and induces accelerated wound healing as well as epithelial tumorigenesis, resulting in tylosis (hyperproliferation of skin in the palms and soles, loss of hair), oral leukoplakia, esophageal cancer, and other abnormalities (e.g., ovarian epithelial cancer, gastric cancer, lung cancer, corneal defects, congenital pulmonary stenosis, total anomalous pulmonary venous connection, deafness, and optic atrophy) [14–17].

Interestingly, several genes related to *RHBDF2* are also implicated in the development of cancer and other pathological conditions, including rhomboid domain containing 2 (RHBDD2) in the development of breast cancer, rhomboid family 1 (RHBDF1) in head and neck cancer, and RHBDD1 in B cell lymphoma, as RHBDD1 cleaves Bcl-2-interacting killer (BIK), which is a pro-apoptotic member of the B cell lymphoma 2 (Bcl-2) family [18].

24.4 Epidemiology

24.4.1 Prevalence

TOC is a rare disorder, with an estimated prevalence of <1 per 1,000,000 in the general population. The disease tends to affect individuals of Western European descent and American families of similar ancestry. Of the 345 family members identified in the UK, 89 have been diagnosed with tylosis [1].

24.4.2 Inheritance

TOC is an autosomal dominant disorder, for which inheritance of one copy of the altered *RHBDF2* gene from one parent is sufficient to cause the disease.

24.4.3 Penetrance

TOC shows complete penetrance of the cutaneous features (e.g., well-limited palmoplantar hyperkeratosis, accentuated in pressure areas) from 7–8 years of age to puberty, with esophageal cancer (predominantly squamous esophageal cancer) appearing from mid-50s onwards, affecting 95% of patients by the age of 65 [1].

24.5 Clinical Features

Clinically, TOC is typically associated with palmoplantar keratoderma, oral leukokeratosis, and esophageal cancer [19].

Palmoplantar keratoderma (hyperkeratosis palmaris et plantaris) is a focal thickening of the skin that often appears as areas of yellowish thickened plaques restricted to areas of weight bearing and/or friction on the palms and soles (e.g., the heels and forefeet). Occasionally, it may be confluent over the palms and soles. It usually emerges by 7–8 years of age but may occur as late as puberty. Other complications include discomfort, pruritus, deep and painful fissures, infection, follicular papules, and cutaneous horns [1].

Oral leukokeratosis (which is largely a benign, precursor lesion with the potential to transform into squamous cancer of the oropharynx) and follicular hyperkeratosis are also features of TOC.

Esophageal lesions appear as small (2–5 mm), white, polyploid dots throughout the esophagus, and carcinoma usually develops in the lower two-thirds of the esophagus at an average age of 45 years. TOC patients with *esophageal cancer* may show dysphagia (difficulty swallowing), odynophagia, anorexia, vomiting, loss of appetite, sitophobia (fear of eating), choking, coughing, and weight loss. Occasionally, the esophageal cancer may spread to the lungs (breathlessness, cough), liver (jaundice, abdominal swelling due to fluid, ascites), bones (pain or unexpected fractures) or local lymph nodes (glandular swellings in the neck) [1].

24.6 Diagnosis

Diagnosis of TOC is based on a positive family history, characteristic clinical features (e.g., focal palmar and plantar hyperkeratosis and esophageal lesions) (Figure 24.1), and mutations in *RHBDF2*.

Esophagogastroscopy (fiberoptic examination of the gullet and stomach) of squamous cell esophageal cancer often reveals a flat, plaque-like or proliferative swelling extending around the esophageal wall and down the length of the esophagus, reducing the lumen of the esophagus (Figure 24.2). CT scan of the chest and abdomen may help determine the presence of any local or distal spread.

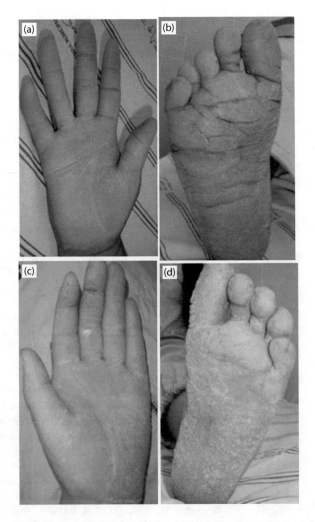

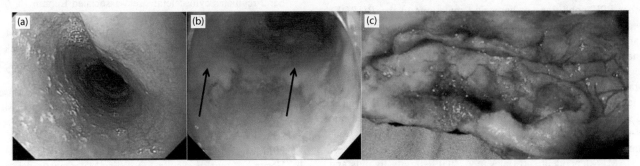

FIGURE 24.1 Palmoplantar hyperkeratosis in a familial case of tylosis with esophageal cancer involving a 37-year-old woman (a and b) and her 8-year-old daughter (c and d). (Photo credit: de Souza CA et al. *An Bras Dermatol.* 2009;84(5):527–9.)

Histological examination of esophageal biopsies (four equally spaced apart at any particular level from the upper, middle, and lower esophagus) helps uncover prominent keratohyalin granules, inflammatory cell infiltrate, and parakeratosis, suggestive of dysplasia. In addition, histological features of the affected skin include acanthosis, hyperkeratosis, and hypergranulosis but no parakeratosis or spongiosis.

Molecular testing has uncovered three disease-related missense mutations in *RHBDF2*: c.557 T → C (p.Ile186Thr), c.566C → T (p.Pro189Leu), and c.562 G → A (p.Asp188Asn) [1].

Differential diagnoses for TOC include other PPKs (e.g., diffuse epidermolytic PPK [due to mutations in the gene encoding the palmoplantar specific keratin, *KRT9*, and in some cases *KRT1*], diffuse, non-epidermolytic PPK [due to mutations in the gene encoding the water channel protein, Aquaporin 5] (Figure 24.3), and non-TOC focal PPK [showing abnormalities in other organs including hearing loss and cardiomyopathy due to defective gap junctions or desmosomes, respectively]), Buschke–Fischer–Brauer disease, Curth–Macklin ichthyosis, Gamborg Nielsen syndrome, Greither disease, Haber syndrome, hereditary punctate palmoplantar keratoderma, Jadassohn–Lewandowsky syndrome (pachyonychia congenita type I), keratosis follicularis spinulosa decalvans, keratosis linearis with ichthyosis, congenital and sclerosing keratoderma syndrome, Meleda disease, mucosa hyperkeratosis syndrome, Naegeli–Franceschetti–Jadassohn syndrome, Naxos disease, Olmsted syndrome (TRPV3 mutation), PPK and leukokeratosis anogenitalis, pandysautonomia, papillomatosis of Gougerot and Carteaud, Papillon–Lefèvre syndrome, punctate porokeratotic keratoderma, Richner–Hanhart syndrome, Schöpf–Schulz–Passarge syndrome, Unna Thost disease, Vohwinkel syndrome, Wong dermatomyositis, acrokeratosis neoplastica (Bazex syndrome); and acanthosis nigricans maligna [20–33].

24.7 Treatment

For PPK, treatment options consist of oral retinoids (e.g., etretinate, isotretinoin, and acitretin), topical therapies (e.g., soaking in salt water and gentle removal of dead tissue [debridement]; 50% propylene glycol in water under plastic dressing overnight weekly), and symptomatic management of the tylosis (e.g., regular application of emollients, specialist footwear, and early treatment of fissures and super-added infection, particularly tinea pedis). Side effects of oral retinoids are nasal excoriation and bleeding, hyperlipidemia, and abnormal liver function tests [34].

For resectable esophageal cancer, surgery is advised; for non-resectable esophageal cancer, radiotherapy with or without chemotherapy is used. Insertion of a mesh stent may enable the patient to swallow while awaiting treatment or if only palliative treatment is required [35,36].

FIGURE 24.2 (a) Conventional imaging of esophageal lesions in a 32-year-old female with TOC; (b) dysplastic esophageal lesions (arrows) in a 57-year-old male with TOC; (c) part of a resected esophagus from a 37-year-old female with TOC. (Photo credits: (a and b) Ellis A et al. *Orphanet J Rare Dis.* 2015;10:126. (c) de Souza CA et al. *An Bras Dermatol.* 2009;84(5):527–9.)

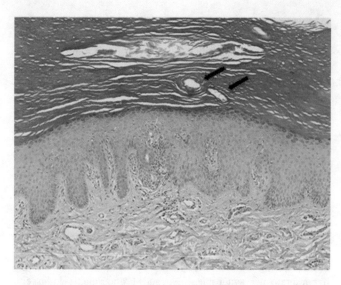

FIGURE 24.3 Histology of a punch biopsy from the foot of a patient presenting with diffuse non-epidermolytic palmoplantar keratoderma linked to an AQP5 mutation. Note a markedly thickened stratum corneum with a prominent stratum granulosum and a moderate acanthosis, dilated acrosyringial intracorneal ducts (arrows), and focally miliaria rubra–like changes with spongiosis and exocytosis of lymphocytes around the intraepidermal acrosyringial ducts (H&E). (Photo credit: Krøigård AB et al. *BMC Dermatol.* 2016;16(1):7.)

24.8 Prognosis and Prevention

The prognosis of TOC depends on the stage of cancer. Patients with cancer restricted to the mucosa (stage 1) have about a 5-year survival rate of 80%, those with submucosal involvement of <50%, those with extension to the muscularis propria of 20%, those with extension to adjacent structures (stage 3) of 7%, and those with distant metastases (stage 4) of <3%.

Lifestyle and diet modifications (e.g., smoking cessation, alcohol restriction, fresh fruit and vegetable diet) are helpful to improve the prognosis of TOC.

24.9 Conclusion

As a rare, autosomal dominant disorder showing focal hyperkeratotic skin on the palms and soles, oral leukoplakia, and esophageal cancer, tylosis with esophageal cancer (TOC) differs from other PPK conditions in harboring mutations in the *RHBDF2* gene and having the tendency to develop and squamous cell carcinoma of the esophagus. The RHBDF2 (iRhom2) protein encoded by the *RHBDF2* gene appears to play a key role in the regulation of the trafficking and activation of ADAM17, a membrane bound sheddase, which in turn participates in the cleavage and release of membrane-bound TNFα and EGFR ligands. Gain-of-function mutations in the highly conserved cytoplasmic N-terminus of RHBDF2 (iRhom2) increase its stability, leading to increased ADAM17 activity and upregulated shedding of ADAM17-dependent substrates (e.g., EGFR-family ligands and TNFα). Given the involvement of EGFR pathway dysregulation in epithelial malignancies (e.g., sporadic squamous cell carcinoma of the esophagus), the aberrant EGFR signaling and TNFα

shedding may underlie the tumorigenesis of TOC. In the absence of effective therapy for TOC, further investigation on the EGFR signaling pathway may offer a potential approach for mitigating this genetic condition.

REFERENCES

1. Ellis A, Risk JM, Maruthappu T, Kelsell DP. Tylosis with oesophageal cancer: Diagnosis, management and molecular mechanisms. *Orphanet J Rare Dis.* 2015;10:126.
2. Howel-Evans W, McConnell RB, Clarke CA, Sheppard PM. Carcinoma of the oesophagus with keratosis palmaris et plantaris (tylosis): A study of two families. *Quart J Med.* 1958; 27:413–29.
3. Das A, Kumar D, Das NK. Diffuse non-epidermolytic palmoplantar keratoderma. *Indian Pediatr.* 2013;50(10):979.
4. Shimomura Y, Wajid M, Weiser J, Kraemer L, Christiano AM. Mutations in the keratin 9 gene in Pakistani families with epidermolytic palmoplantar keratoderma. *Clin Exp Dermatol.* 2010;35(7):759–64.
5. Liu WT, Ke HP, Zhao Y et al. The most common mutation of KRT9, c.C487T (p.R163W), in epidermolytic palmoplantar keratoderma in two large Chinese pedigrees. *Anat Rec (Hoboken).* 2012;295(4):604–9.
6. Fu DJ, Thomson C, Lunny DP et al. Keratin 9 is required for the structural integrity and terminal differentiation of the palmoplantar epidermis. *J Invest Dermatol.* 2014;134(3):754–63.
7. Guo Y, Shi M, Tan ZP, Shi XL. Possible anticipation in familial epidermolytic palmoplantar keratoderma with the p.R163W mutation of Keratin 9. *Genet Mol Res.* 2014;13(4):8089–93.
8. Hinterberger L, Pföhler C, Vogt T, Müller CS. Diffuse epidermolytic palmoplantar keratoderma (Unna-Thost-). *BMJ Case Rep.* 2012;2012. pii: bcr2012006443.
9. Blaydon DC, Etheridge SL, Risk JM et al. RHBDF2 mutations are associated with tylosis, a familial esophageal cancer syndrome. *Am J Hum Genet.* 2012;90(2):340–6.
10. Risk JM, Evans KE, Jones J et al. Characterization of a 500 kb region on 17q25 and the exclusion of candidate genes as the familial Tylosis Oesophageal Cancer (TOC) locus. *Oncogene.* 2002;21(41):6395–402.
11. Brooke MA, Etheridge SL, Kaplan N et al. iRHOM2-dependent regulation of ADAM17 in cutaneous disease and epidermal barrier function. *Hum Mol Genet.* 2014;23(15):4064–76.
12. Lee MY, Nam KH, Choi KC. iRhoms; Its functions and essential roles. *Biomol Ther (Seoul).* 2016;24(2):109–14.
13. Maruthappu T, Chikh A, Fell B et al. Rhomboid family member 2 regulates cytoskeletal stress-associated Keratin 16. *Nat Commun.* 2017;8:14174.
14. Saarinen S, Vahteristo P, Lehtonen R et al. Analysis of a Finnish family confirms RHBDF2 mutations as the underlying factor in tylosis with esophageal cancer. *Fam Cancer.* 2012;11(3):525–8.
15. Hosur V, Johnson KR, Burzenski LM, Stearns TM, Maser RS, Shultz LD. Rhbdf2 mutations increase its protein stability and drive EGFR hyperactivation through enhanced secretion of amphiregulin. *Proc Natl Acad Sci U S A.* 2014;111(21):E2200–9.
16. Hosur V, Low BE, Shultz LD, Wiles MV. Genetic deletion of amphiregulin restores the normal skin phenotype in a mouse model of the human skin disease tylosis. *Biol Open.* 2017;6(8):1174–9.

17. Hosur V, Lyons BL, Burzenski LM, Shultz LD. Tissue-specific role of RHBDF2 in cutaneous wound healing and hyperproliferative skin disease. *BMC Res Notes*. 2017;10(1):573.

18. Al-Musalhi B, Shehata N, Billick R. Small cell variant of T-cell prolymphocytic leukemia with acquired palmoplantar keratoderma and cutaneous infiltration. *Oman Med J*. 2016;31(1):73–6.

19. de Souza CA, Santos Ada C, Santos Lda C, Carneiro AL. Hereditary tylosis syndrome and esophagus cancer. *An Bras Dermatol*. 2009;84(5):527–9.

20. Lee JY, Sung-II In, Kim HJ, Jeong SY, Choung YH, Kim YC. Hereditary palmoplantar keratoderma and deafness resulting from genetic mutation of Connexin 26. *J Korean Med Sci*. 2010;25(10):1539–42.

21. Lestre S, Lozano E, Meireles C, Barata Feio A. Autoimmune thyroiditis presenting as palmoplantar keratoderma. *Case Rep Med*. 2010;2010:604890.

22. Attia AM, Bakry OA. Olmsted syndrome. *J Dermatol Case Rep*. 2013;7(2):42–5.

23. Blaydon DC, Lind LK, Plagnol V et al. Mutations in AQP5, encoding a water-channel protein, cause autosomal-dominant diffuse nonepidermolytic palmoplantar keratoderma. *Am J Hum Genet*. 2013;93(2):330–5.

24. Kubo A, Shiohama A, Sasaki T et al. Mutations in SERPINB7, encoding a member of the serine protease inhibitor superfamily, cause Nagashima-type palmoplantar keratosis. *Am J Hum Genet*. 2013;93(5):945–56.

25. Kubo A. Nagashima-type palmoplantar keratosis: a common Asian type caused by SERPINB7 protease inhibitor deficiency. *J Invest Dermatol*. 2014;134(8):2076–9.

26. Cao X, Yin J, Wang H et al. Mutation in AQP5, encoding aquaporin 5, causes palmoplantar keratoderma Bothnia type. *J Invest Dermatol*. 2014;134(1):284–7.

27. He Y, Zeng K, Zhang X et al. A gain-of-function mutation in TRPV3 causes focal palmoplantar keratoderma in a Chinese family. *J Invest Dermatol*. 2015;135(3):907–9.

28. Iqtadar S, Mumtaz SU, Abaidullah S. Papillon-Lèfevre syndrome with palmoplantar keratoderma and periodontitis, a rare cause of pyrexia of unknown origin: a case report. *J Med Case Rep*. 2015;9:288.

29. Plassais J, Guaguère E, Lagoutte L et al. A spontaneous KRT16 mutation in a dog breed: A model for human focal non-epidermolytic palmoplantar keratoderma (FNEPPK). *J Invest Dermatol*. 2015;135(4):1187–90.

30. Sreeramulu B, Shyam ND, Ajay P, Suman P. Papillon-Lefèvre syndrome: Clinical presentation and management options. *Clin Cosmet Investig Dent*. 2015;7:75–81.

31. Krøigård AB, Hetland LE, Clemmensen O, Blaydon DC, Hertz JM, Bygum A. The first Danish family reported with an AQP5 mutation presenting diffuse non-epidermolytic palmoplantar keratoderma of Bothnian type, hyperhidrosis and frequent Corynebacterium infections: A case report. *BMC Dermatol*. 2016;16(1):7.

32. Lovgren ML, McAleer MA, Irvine AD et al. Mutations in desmoglein 1 cause diverse inherited palmoplantar keratoderma phenotypes: Implications for genetic screening. *Br J Dermatol*. 2017;176(5):1345–50.

33. Maruthappu T, McGinty LA, Blaydon DC, Fell B et al. Recessive mutation in FAM83G associated with palmoplantar keratoderma and exuberant scalp hair. *J Invest Dermatol*. 2018;138(4):984–7.

34. Lernia VD, Ficarelli E, Zanelli M. Ineffectiveness of tumor necrosis factor-α blockers and ustekinumab in a case of type IV pityriasis rubra pilaris. *Indian Dermatol Online J*. 2015;6(3):207–9.

35. Barret M, Prat F. Diagnosis and treatment of superficial esophageal cancer. *Ann Gastroenterol*. 2018;31(3):256–65.

36. PDQ Adult Treatment Editorial Board. *Esophageal Cancer Treatment (PDQ®): Health Professional Version*. PDQ Cancer Information Summaries [Internet]. Bethesda (MD): National Cancer Institute (US); 2002. 2018 Feb 6.

Section III

Tumor Syndromes Affecting the Urogenitory System

25

Bardet–Biedl Syndrome

Dongyou Liu

CONTENTS

25.1 Introduction

Bardet–Biedl syndrome (BBS) is a rare autosomal recessive disorder belonging to the group of ciliopathies in which defects in cilia structure and/or function underscore the pathogenesis. Clinically, BBS is characterized by retinal dystrophy, obesity (fat deposition along the abdomen), polydactyly (extra digit on the hands), cognitive impairment, and renal and urogenital anomalies as well as increased tumor risk [1].

At least 21 genes relating to primary cilia function have been identified in patients (~80%) with BBS, of which *BBS1* and *BBS10* account for the majority of genotypes in Northern Europe (51%) and North America (20%) (Table 25.1) [2]. Since up to 50% of affected families possess pathogenic variants related to *BBS6/MKKS*, *BBS10*, and *BBS12* (which all encode chaperonin-like proteins), and often develop more severe phenotypes than families with other *BBS* gene mutations, BBS is sometimes referred to as a type of chaperonopathy. Nonetheless, in about 20% of BBS-affected individuals, a culprit gene mutation has yet to be found [3].

25.2 Biology

Initial cases related to BBS were first reported by Laurence and Moon in 1866 from a family who presented a combined clinical picture of retinitis pigmentosa, obesity, and cognitive impairment with later development of paraparesis (so-called Laurence–Moon syndrome). In 1920, Bardet described the triad of obesity, polydactyly, and retinitis pigmentosa from another family, and

in 1922, Biedl observed retinitis pigmentosa, polydactyly, obesity, hypogenitalism, and intellectual impairment in two siblings. These findings thus expanded the clinical spectrum of Laurence–Moon syndrome to polydactyly and hypogenitalism (i.e., Bardet–Biedl syndrome). Despite their obvious differences (i.e., progressive spastic paraparesis and distal muscle weakness in Laurence–Moon syndrome and polydactyly in BBS), these two overlapping phenotypes share common genetic background (both containing mutations in genes involved ciliary biogenesis and trafficking). The term *Bardet–Biedl syndrome* (BBS) is often used to describe both syndromes [4].

Located on the surface of different types of cells, cilia are the tiny hair-like structures (either motile or immotile) that utilize an architectural element known as basal body to anchor to a cell. Possessing central microtubule pair necessary for ciliary mobility, motile cilia propel fluid (e.g., mucus) through the local environment and assist in cell motility. Immotile (primary) cilia have unique $9 + 0$ structure with nine microtubule triplets arranged in a circle with an outer membrane, and participate in cell signaling, left-right asymmetry, tissue formation, and homeostasis. In the eyes, rod and cone photoreceptor cells of the retina are known to utilize their immotile cilia for light-perceiving function [5].

Most of the genes implicated in BBS encode proteins that are building blocks for cilia and basal bodies. Apart from their involvement in ciliogenesis, BBS proteins also help transport G-protein coupled receptor, somatostatin receptor, and other molecules across cilia [6]. Mutations in *BBS* genes result in altered proteins that disturb the normal functions of immotile cilia, leading to cone-rod dystrophy, renal abnormalities (e.g., hyposthenuria), anosmia (inability to smell), hearing loss, and situs inversus associated with BBS [5,7].

TABLE 25.1

Characteristics of Genes Involved in Bardet–Biedl Syndrome (BBS)

Gene	Chromosome Location (Gene Structure)	Protein Function (Size)	Recurrent Variants	Clinical Features	Comments
BBS1	11q13.2 (23 kb/ 17 exons)	BBSome (593 aa/ 60 kDa)	c.1169T>G (p.M390R), Northern European descent; c.1091+3G>C, Faroe Islands	Macrocephaly, obesity, high palate, nystagmus, diabetes mellitus, hypertension, intellectual disability, ataxia, hearing/speech impairment, cataract, global developmental delay, nephrogenic diabetes insipidus, strabismus, short/broad foot, high/ narrow palate, myopia, hepatic fibrosis, hypogonadism, aganglionic megacolon, left ventricular hypertrophy, abnormality of the kidney, glaucoma, biliary tract abnormality, dental crowding, brachydactyly, polydactyly, asthma, hypodontia, rod-cone dystrophy, astigmatism, hirsutism, syndactyly, vaginal atresia, abnormal ovary, small testicles, micropenis, retinal degeneration	Representing 23% of pathologic variants; associated with milder phenotype than BBS2 and BBS10, better visual acuity and larger ERG amplitudes than other BBS types; p.M390R homozygotes showing a more severe ocular phenotype than compound heterozygotes; missense mutations causing a lower level of biochemical cardiovascular disease markers than BBS 10 and other BBS1 mutations; also implicated in nonsyndromic retinitis pigmentosa
BBS2	16q21 (54 kb/ 17 exons)	BBSome (721 aa/ 79 kDa)	c.472−2A>G, Hutterites; c.565C>T (p.R189*), Tunisia; c.311A>C (p.D104A) and c.1895G>C, Ashkenazi Jews	Obesity, intellectual disability, global developmental delay, atrial septal defect, bicuspid aortic valve, dilated cardiomyopathy, hypogonadism, rod-cone dystrophy, postaxial hand polydactyly, retinal degeneration, external genital hypoplasia; absence of myopia	Representing 8.1% of pathologic variants; having a higher frequency in Iran (29%); biallelic *BBS2* mutations in some antenatal cases displaying cystic kidneys and polydactyly and/or hepatic fibrosis but no encephalocele; also implicated in Meckel or Meckel-like syndrome, nonsyndromic retinitis pigmentosa
BBS3 (ARL6)	3q11.2 (36 kb/ 9 exons)	ADP-ribosylation factor-like GTPase 6 (186 aa/21 kDa); involved in BBSome assembly	c.272T>C (p.I91T), India	Obesity, intellectual disability, global developmental delay, brachydactyly, renal hypoplasia, rod-cone dystrophy, tricuspid regurgitation, external genital hypoplasia, polydactyly; presence of myopia	Having a higher frequency in India (18%); also implicated in nonsyndromic retinitis pigmentosa
BBS4	15q22.3 (52 kb/16 exons)	BBSome (519 aa/ 58 kDa)	c.77_220del144, Iran	Obesity, intellectual disability, abnormality of the dentition, cryptorchidism, nyctalopia, hypogonadism, brachydactyly, rod-cone dystrophy, renal cyst, syndactyly, retinal degeneration, external genital hypoplasia, polydactyly; presence of myopia	Representing 2.3% of pathogenic variants; alternate splicing resulting in multiple transcript variants; biallelic *BBS4* mutations in some antenatal cases displaying cystic kidneys and polydactyly and/or hepatic fibrosis but no encephalocele; also implicated in Meckel or Meckel-like syndrome
BBS5	2q31.1 (24 kb/ 12 exons)	BBSome (341 aa/ 38 kDa)		Obesity, cognitive impairment, hypogonadism, brachydactyly, rod-cone dystrophy, macular dystrophy, syndactyly, external genital hypoplasia, polydactyly	Alternate transcriptional splice variants observed
BBS6 (MKKS)	20p12.2 (33 kb/ 6 exons)	Chaperonin complex (570 aa/62 kDa)		Obesity, diabetes mellitus, intellectual disability, hypospadias, rod-cone dystrophy, renal cyst, syndactyly, external genital hypoplasia, polydactyly; more severe renal disease	Representing 5.8% pathogenic variants; alternative splicing resulting in multiple transcript variants; biallelic *BBS6* mutations in some antenatal cases displaying cystic kidneys and polydactyly and/or hepatic fibrosis but no encephalocele; also implicated in McKusick–Kaufman syndrome, Meckel or Meckel-like syndrome

(Continued)

TABLE 25.1 (Continued)

Characteristics of Genes Involved in Bardet–Biedl Syndrome (BBS)

Gene	Chromosome Location (Gene Structure)	Protein Function (size)	Recurrent Variants	Clinical Features	Comments
BBS7	4q27 (46 kb/ 19 exons)	BBSome (715 aa/ 80 kDa)	c.1967_1968delTAinsC (p.L656Pfs*18), Russia	Obesity, intellectual disability, rod-cone dystrophy, external genital hypoplasia, polydactyly	Representing 1.5% pathogenic variants; two transcript variants encoding distinct isoforms
BBS8 (TTC8)	14q32.11 (56 kb/ 15 exons)	Tetratricopeptide repeat protein 8/BBSome (541 aa/61 kDa)	c.459+1G>A, Tunisia	Obesity, intellectual disability, global developmental delay, cognitive impairment, brachycephaly, hypogonadism, hypospadias, rod-cone dystrophy, situs inversus totalis, renal dysplasia, polydactyly	Representing 1.2% of pathogenic variants; alternative splicing resulting in multiple transcript variants; also implicated in nonsyndromic retinitis pigmentosa
BBS9	7p14.3 (506 kb/ 25 exons)	BBSome (887 aa/ 99 kDa)		Obesity, intellectual disability, rod-cone dystrophy, polydactyly	Representing 6% pathogenic variants; alternatively spliced transcript variants encoding different isoform
BBS10	12q21.2 (3.9 kb/ 2 exons)	Chaperonin complex (723 aa/80 kDa)	c.271_272insT (p.C91Lfs*5), European descent; K243IfsX15), South Africa	Obesity, cognitive impairment, renal insufficiency, hypogonadism, rod-cone dystrophy, renal cyst, polydactyly	Representing 20% pathogenic variants; causing more severe renal disease, significantly higher BMI-Z, greater visceral adiposity, greater insulin resistance, and higher frequency of urogenital anomalies than BBS1
BBS11 (TRIM32)	9q33.1 (13 kb/ 2 exons)	E3 ubiquitin-protein ligase TRIM32 (653 aa/71 kDa/E3) involved in membrane trafficking		Obesity, retinopathy, hypogonadism, abnormality of the kidney, polydactyly	Also implicated in Limb-girdle muscular dystrophy type 2H, sarcotubular myopathy
BBS12	4q27 (44 kb/ 2 exons)	Chaperonin complex (710 aa/79 kDa)	C.1156-1157CG>TA (p.Arg386*), Iran	Obesity, cognitive impairment, hypogonadism, abnormality of the kidney, rod-cone dystrophy, polydactyly	Representing 5% pathogenic variants; two transcript variants encoding the same protein; showing more severe renal disease, higher frequency of cognitive impairment than BBS1
BBS13 (MKS1)	17q23 (14 kb/ 17 exons)	Meckel syndrome type 1 protein (559 aa/64 kDa); basal body/ centriole migration involved in organization of the transition zone		Obesity, intellectual disability, global developmental delay, rod-cone dystrophy, polydactyly	Representing 4.5% pathogenic variants; multiple transcript variants encoding different isoforms; also implicated in Meckel syndrome-1
BBS14 (CEP290)	12q21.32 (93 kb/ 54 exons)	Centrosomal protein 290 (2479 aa/290 kDa); basal body involved in organization of the transition zone and ciliary entry of BBSome		Obesity, intellectual disability, global developmental delay, rod-cone dystrophy	Also implicated in Joubert syndrome, nephronophthisis, Senior-Loken syndrome, Meckel syndrome-4, Leber congenital amaurosis, several forms of cancer
BBS15 (C2ORF86/ WDPCP)	2p15 (720 kb/ 12 exons)	WD repeat-containing and planar cell polarity effector protein (746 aa/85 kDa); basal body involved in regulation of septins localization and ciliogenesis		Similar to BBS2 and BBS4	Compound heterozygous mutation associated with poly syndactyly, coarctation of the aorta, and tongue hamartoma; alternative splicing in BBS15 resulting in multiple transcript variants; also implicated in Mecke syndrome

(Continued)

TABLE 25.1 (*Continued*)

Characteristics of Genes Involved in Bardet–Biedl Syndrome (BBS)

Gene	Chromosome Location (Gene Structure)	Protein Function (size)	Recurrent Variants	Clinical Features	Comments
BBS16 (*SDCCAG8*)	1q43 (244 kb/ 18 exons)	Serologically defined colon cancer antigen 8 (713 aa/82 kDa); basal body involved in interaction with OFD1, and regulation of pericentriolar material recruitment to the centrosomal region		Obesity, intellectual disability, hearing impairment, global developmental delay, recurrent respiratory infections, cognitive impairment, renal insufficiency, respiratory distress, recurrent otitis media, hypogonadism, rod–cone dystrophy, renal agenesis, renal cyst, renal dysplasia, retinal degeneration, external genital hypoplasia, bronchiolitis; notable absence of polydactyly	Mutations in *BBS16* associated with retinal–renal ciliopathy; also implicated in Senior-Loken syndrome-7
BBS17 (*LZTFL1*)	3p21.31 (92 kb/ 10 exons)	Leucine zipper transcription factor-like protein 1/ BBSome (299 aa/34 kDa); involved in regulation of BBSome ciliary trafficking and Shh signalling		Obesity, global developmental delay, cognitive impairment, hypogonadism, brachydactyly, situs inversus totalis, renal cyst, stage 5 chronic kidney disease, cone/cone–rod dystrophy, retinal degeneration, external genital hypoplasia, mesoaxial polydactyly	Nonsense mutations in *BBS17* associated with polydactyly, obesity, cognitive impairment, hypogonadism, and kidney failure; also functioning as a tumor suppressor through interaction with E-cadherin and the actin cytoskeleton; alternative splicing of *BBS17* resulting in multiple transcript variants
BBS18 (*BBIP1*)	10q25.2 (20 kb/ 4 exons)	BBSome (92 aa/10 kDa)		Obesity, cataract, cognitive impairment, renal insufficiency, brachydactyly, rod–cone dystrophy	Alternative splicing of *BBS18* resulting in multiple transcript variants
BBS19 (*IFT27*)	22q12.3 (18 kb/ 7 exons)	Intraflagellar transport protein 27/G-protein (186 aa/20 kDa); involved in intraflagellar transport		Obesity, intellectual disability, renal insufficiency, hypogonadism, rod–cone dystrophy, hyposmia, external genital hypoplasia, polydactyly	Alternative splicing of *BBS19* resulting in multiple transcript variants
BBS20 (*IFT172*)	2p23.3 (45 kb)	Intraflagellar transport protein 172 (1749 aa/197 kDa); involved in intraflagellar transport		Obesity, intellectual disability, microcephaly, hypogonadism, rod–cone dystrophy, polydactyly	Mutations in *BBS20* associated with skeletal ciliopathies, with or without polydactyly (e.g., short-rib thoracic dysplasias 1, 9 or 10); also implicated in Jeune syndrome, Mainzer-Saldino syndrome; nonsyndromic retinitis pigmentosa
BBS21 (*C8ORF37*)	8q22.1 (25 kb/ 6 exons)	Chromosome 8 open reading frame 37 protein (207 aa/23 kDa); involved in mediation of the basal body anchoring to the plasma membrane and assembly of cilium	290-kb deletion, Northern European descent	Obesity, delayed speech and language development, blindness, horseshoe kidney, myopia, retinal thinning, hypodontia, rod–cone dystrophy, constriction of peripheral visual field, postaxial hand polydactyly, hypoplasia of the fovea, cone/ cone–rod dystrophy, retinal atrophy, elevated hepatic transaminase, reduced amplitude of dark-adapted bright flash electroretinogram a-wave, hyperautofluorescent macular lesion	Mutations in *BBS21* associated with autosomal recessive cone-rod dystrophy and retinitis pigmentosa; also implicated in nephronophthisis, Senior-Loken syndrome, Joubert syndrome

25.3 Pathogenesis

The identification of 21 genes involved in BBS (i.e., *BBS1*, *BBS2*, *BBS3 [ARL6]*, *BBS4*, *BBS5*, *BBS6 [MKKS]*, *BBS7*, *BBS8 [TTC8]*, *BBS9*, *BBS10*, *BBS11 [TRIM32]*, *BBS12*, *BBS13 [MKS1]*, *BBS14 [CEP290]*, *BBS15 [WDPCP]*, *BBS16 [SDCCAG8]*, *BBS17 [LZTFL1]*, *BBS18 [BBIP1]*, *BBS19 [IFT27]*, *BBS20 [IFT72]*, and *BBS21 [C8ORF37/NPHP1]*) through linkage analysis and other approaches have revealed important insights on the pathogenesis of this clinically and genetically heterogeneous ciliopathy [8–10].

Functionally, the protein products of the 21 BBS-causing genes can be separated into BBSome (which is a multiprotein complex composed of protein subunits encoded by *BBS1*, *BBS2*, *BBS4*, *BBS5*, *BBS7*, *BBS8*, *BBS9*, *BBS17*, and *BBS18*), chaperonin complex (*BBS6*, *BBS10*, and *BBS12*), basal body (*BBS13*, *BBS14*, *BBS15*, and *BBS16*), and other related biological function (*BBS3*, *BBS11*, *BBS19*, *BBS20*, and *BBS21*) [11].

BBSome (comprising BBS1, BBS2, BBS4, BBS5, BBS7, BBS8, BBS9, and BBS18 subunits) is involved in the trafficking of molecules (e.g., multiple G protein-coupled receptors, melanin-concentrating hormone receptor 1, and somatostatin receptor 3) to the cilium and also in the assembly of intraflagellar transport particles, which mediate bidirectional movement of nonmembrane molecules along the axoneme and between the axoneme and the membrane. The chaperonin-like BBS proteins encoded by the *BBS6/MKKS*, *BBS10*, and *BBS12* genes share structural homology with the CCT family of group II chaperonins and are indispensable for the formation of BBSome [12]. BBS17 and BBS20 negatively regulate BBSome trafficking. BBS11 encodes an E3 ubiquitin ligase that helps recruit of BBSome. BBS3 mediates the transition between vesicular and intraciliary trafficking, restricts the entry of ciliary vesicle into the cilium, and modulates Wnt signaling. BBS19 encodes a component of the IFT-B complex that links the BBS cargo to IFT machinery [13].

Mutations in the BBS genes produce altered proteins that compromise the structures and functions of cilia and basal bodies, leading to the development of ciliopathies affecting multiple organs/systems. Although the disruption of any of the 21 genes can result in cilia impairment, biallelic mutations in *BBS1* and *BBS10* are responsible for a majority of clinical cases, including 51% of Northern European cases and 20% of North American cases. Furthermore, mutations in the *BBS6/MKKS*, *BBS10*, and *BBS12* genes (which all encode chaperonin-like proteins) contribute to 50% of BBS-related morbidity.

Homozygous or compound heterozygous *BBS* mutations (e.g., frameshifts, premature stop codons, or alterations at splice sites, leading to amino acid substitutions) are found in ~80% of cases. While patients with *BBS1* mutations show milder ophthalmologic involvement, those with *BBS2*, *BBS3*, and *BBS4* mutations often experience classic deterioration of their vision. On the other hand, patients with *BBS10* mutations have a tendency to develop obesity and insulin resistance.

Given the role of consanguinity/inbreeding in the development and maintenance of BBS, a number of founder mutations (which originate from a single identifiable ancestor and proliferate in a kinship or community) have been identified in various parts of world. These include *BBS1* M390R in Europe, *BBS1* c.1091+3G>C in the Faroe Islands, *BBS2* c.472-2A>G in Hutterites, *BBS2* p.R189* and *BBS8* c.459+1G>A in Tunisia, *BBS2* c.311A>C (p.D104A) and c.1895G>C in Ashkenazi Jews, *BBS3* c.272T>C (p.I91T) in India, *BBS4* c.77_220del144 and c.1156-1157 CG>TA (p.Arg386*) in Iran, and *BBS7* c.1967_1968delTAinsC in Russia, and *BBS10* p.C91Lfs*5 truncation in several ethnic groups [14,15]. Founder mutations that share a common ancestor and repeatedly occur in a single haplotype in a population may be referred to as recurrent mutations. However, recurrent mutations also include non-founder mutations that do not share a common ancestor and repeatedly occur in more than one haplotype in a population.

Besides BBS, gene mutations that lead to ciliary dysfunction are also implicated in other ciliopathies. For example, genetic changes involving BBS14 (CEP290) can also cause Joubert syndrome, Leber congenital amaurosis, Meckel syndrome, and Senior−Loken syndrome (Table 25.1). In addition, BBS may overlap clinically to other syndromes with distinct gene defects (e.g., Alström syndrome containing *ALMS1* mutations, and Joubert syndrome containing *NPHP1* gene mutations) [13].

25.4 Epidemiology

25.4.1 Prevalence

BBS shows a varied prevalence among different population groups in the world. Relatively high incidence is observed in consanguineous/inbred populations (e.g., 1:3,700 among inhabitants in the Faroe Islands, 1: 13,500 among the Bedouin peoples of Kuwait, 1:17,500 among the island population of Newfoundland, 1:65,000 in the Middle East), while comparatively low incidence is noted in non-consanguineous/outbred populations (e.g., 1:100,000 in North America, and 1 per 160,000 in Switzerland). Interestingly, *BBS1* (e.g., p.M390R) and *BBS10* (e.g., p.C91Lfs*5) mutations commonly occur in Europe and North America (accounting for 33% of BBS cases); *BBS1*, *BBS2*, and *BBS8* mutations in Tunisia (17%); *BBS1*, *BBS3*, and *BBS4* in Saudi Arabia (17%). The median age of BBS patients at diagnosis is 9 years [16,17].

25.4.2 Inheritance

BBS displays an autosomal recessive inheritance pattern, and involves inheritance of two abnormal variants/alleles (one from each parent) to develop overt disease. An individual who receives one normal allele and one mutated allele is a carrier for the disease and is usually asymptomatic. Two carrier parents have a 25% chance of having an affected child (containing two abnormal/mutated alleles), a 50% chance of having a carrier child (containing one normal allele and one abnormal allele), and a 25% chance of having a normal child (containing two normal alleles). As BBS mutations exert dominant-negative effects on the function of the remaining (wild-type) gene allele, some carriers with heterozygous mutations may sometimes become symptomatic, and others acquire additional mutations in another gene. The inheritance of multiple mutations involving multiple genes in autosomal recessive disorder is known as oligogenic inheritance [18]. A BBS-affected individual who carries two mutations (i.e., homozygous mutations) in one gene and a third mutation (i.e., heterozygous mutation) in a separate gene is defined as showing triallelic inheritance pattern [19].

25.4.3 Penetrance

BBS shows incomplete penetrance and variable expressivity, as some biallelic BBS gene mutation carriers appear to be asymptomatic [20]. The fact that BBS patients sometimes show triallelic or oligogenic inheritance further increases the clinical complexity of the disorder, with mutation at a second gene modulating the penetrance and/or expressivity of recessive mutations at a primary locus. For example, a BBS patient with kidney impairment was found to harbor mutations in both *BBS6* (*MKKS*, compound heterozygous) and *NPHP4* genes, whereas his sibling with no kidney impairment contained mutation only in *BBS6* (*MKKS*).

25.5 Clinical Features

Being a clinically heterogeneous disorder, BBS is responsible for causing both primary (retinal degeneration [typically rod-cone dystrophy, 90%–100% of cases], obesity [72%–92%], polydactyly [63%–81%], hypogenitalism [59%–98%], intellectual impairment [50%–61%], and renal anomalies [20%–53%]) and secondary (brachydactyly/syndactyly [46%–100%], developmental delay [50%–91%], ataxia/poor coordination [40%–86%], speech deficit [54%–81%], anosmia/hyposmia [60%], orodental anomalies [51%], diabetes mellitus [6%–48%], deafness [11%–12%], congenital heart disease/cardiopathy [7%–10%], hepatic fibrosis, olfactory deficit) manifestations. While BBS-related symptoms are not always apparent at birth, they nonetheless become progressively worse and complete by the first and second decades of life. In addition, clinical symptoms may vary between individuals, even within the same family, and also among patients harboring distinct gene mutations (Table 25.1).

Among the primary (cardinal) features of BBS, *retinal degeneration* includes rod-cone dystrophy, choroidal dystrophy, and global severe retinal dystrophy, of which rod-cone dystrophy (also referred to as retinitis pigmentosa due to defects in the transport of phototransduction proteins from the inner to the outer segments of photoreceptors causing rod and cone cell death) is most common, affecting 90%–100% of patients. Rod-cone dystrophy is a progressive retinal degeneration that usually manifests as night blindness by age 7 or 8, loss of color discrimination, and progressive tunnel vision (lose of peripheral vision) by the first decade of life, then loss of central vision and legal blindness by the second or third decade of life (Figure 25.1) [21–23]. Additional ophthalmologic features consist of nystagmus (rapid, involuntary eye movements), strabismus (lazy eye), high myopia, cataract (clouding of the lens), and glaucoma (damage to the optic nerve conducting signals to the brain).

Obesity is characterized by the disproportionate distribution of adipose tissue on the abdomen and chest (so-called "truncal obesity') rather than the arms and legs, producing an apple-shape body type (with mean body mass index of 31.5 mg/m^2 in females and 36.6 mg/m^2 in males). Emerging in 90% of patients from the first year of life, obesity becomes most prominent in the trunk and proximal limbs in adulthood. Factors contributing to the development of obesity may include (i) deregulation of appetite, (ii) altered leptin resistance, (iii) altered neuroendocrine signaling from ciliated neurons to fat storage tissues, (iv) impaired leptin receptor signaling, (v) reduced number of cilia due to BBS gene mutations and alteration in the sonic hedgehog (Shh) and Wnt signaling pathway in the differentiating preadipocytes [24]. At the molecular level, *BBS4* (rs7178130) and *BBS6* (rs6108572 and rs221667) are linked to early-onset childhood obesity and

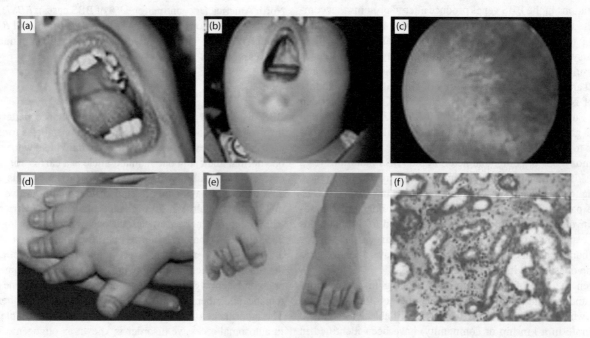

FIGURE 25.1 Clinical manifestations of Bardet–Biedl syndrome include: (a) dental crowding, (b) high palate, (c) rod-cone dystrophy, (d) brachydactyly, (e) postaxial polydactyly, and (f) renal degeneration (dilated tubules, disruption of the tubular basement membrane, and extensive interstitial fibrosis). (Photo credits: [a–c] Forsythe E et al. *Front Pediatr.* 2018;6:23; [d–f] Halbritter J et al. *Am J Hum Genet.* 2013;93(5):915–25.)

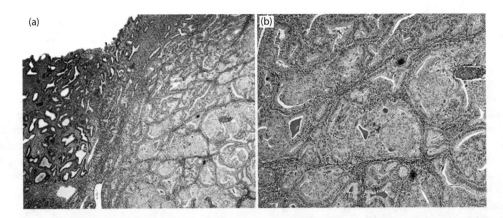

FIGURE 25.2 Bardet–Biedl syndrome in a 26-year-old female causing endometrioid adenocarcinoma with squamous metaplasia: (a) sharp demarcation between adenocarcinoma and adjacent endometrium containing atypical complex hyperplasia; (b) endometrioid adenocarcinoma with prominent areas of squamous metaplasia. (Photo credit: Grechukhina O et al. *Case Rep Obstet Gynecol.* 2018;2018:1952351.)

common adult morbidity, while *BBS2* (rs4784675) is associated with common adult obesity.

Polydactyly occurs in 68%–81% of BBS patients, including postaxial polydactyly (an extra toe near the fifth "little" toe) as a result of aberrant Shh signaling. Other related features comprise brachydactyly (abnormally short fingers and toes), partial syndactyly (webbing between the second and third toes), fifth-finger clinodactyly (inwardly curved little finger), prominent "sandal gap" (between the first and second toes), and broad feet with a flat arch (Figure 25.1) [25,26].

Hypogenitalis manifests as hypogonadism in males (e.g., small penis buried in the adipose tissue, undescended testes or cryptorchidism in 9% of cases), and genital anomalies (e.g., hypoplastic fallopian tubes, uterus, and ovaries; partial and complete vaginal atresia; septate vagina; duplex uterus; hematocolpos; persistent urogenital sinus; vesicovaginal fistula; absent vaginal orifice; and absent urethral orifice) in females, leading to delayed puberty and invariable infertility in males, and irregular cycles, polycystic ovaries, and reduced fertility in females.

Intellectual impairment is predominated by learning difficulties (e.g., emotional immaturity, attention difficulties, and slow thought processes) along with other psychiatric problems (anxiety, mood disorders, depression, bipolar disorder, obsessive compulsive behavior, and psychosomatic manifestations).

Renal anomalies include calyceal clubbing or calyceal cysts, parenchymal cysts, fetal lobulation and diffuse cortical scarring, unilateral agenesis, and renal dysplasia, leading to hydronephrosis, chronic glomerulonephritis, defective tubular concentrating ability, renal tubular acidosis, hypertension, and end-stage renal disease requiring dialysis or transplantation (Figure 25.1) [26,27]. Occurring in 53%–82% of cases, renal anomalies represent a major cause of BBS-related morbidity and mortality [28,29].

The most notable secondary features of BBS include craniofacial dysmorphism (brachycephaly, macrocephaly, male frontal balding, large ears, deep-set eyes, hypertelorism, downslanting palpebral fissure, long smooth philtrum, depressed nasal bridge, anteverted nares, thin upper lip, high-arched palate, dental crowding, midface retrusion, and mild retrognathia), ataxia/impaired coordination, mild hypertonia affecting all four limbs, speech defects (delayed speech until four ages of age, prolonged

syllable repetition times, high-pitched, nasal voice), anosmia/hyposmia, noninsulin-dependent diabetes mellitus, hearing loss (due to acute and cardiac anomalies (e.g., premature stenosis of heart valves, arrhythmias, chronic otitis media), hepatic anomalies (perilobular fibrosis, periportal fibrosis, biliary cirrhosis, portal hypertension, hepatic tumor), olfactory deficit (decreased ability to sense smells due to a change in the size to a brain center called the olfactory bulb), and other genotype-related features (e.g., extra digits in BBS4; characteristic ocular phenotypes in BBS2, BBS3, and BBS4; less severe ophthalmologic in BBS1; higher insulin resistance and visceral adiposity in BBS10 than BBS1) (Table 25.1) [23,30].

In addition, BBS appears to predispose to cancer development as demonstrated by the identification of clear cell renal cell carcinoma (ccRCC) in three BBS patients [31], ovarian teratoma in a 22-year-old woman with confirmed BBS since the age of 11 [32], and endometrial adenocarcinoma in a 26-year-old woman harboring BBS mutation and having abnormal uterine bleeding (Figure 25.2) [33].

25.6 Diagnosis

Diagnosis of BBS involves medical history review, physical examination (e.g., blood pressure, speech assessment), laboratory evaluations (serum urea and creatinine levels, blood sugar/lipid profile, liver function tests, urine analysis), imaging study (e.g., full-field rod and cone electroretinogram [ERG] to measure the electrical response of the retina to light stimulation, renal ultrasound, echocardiography, IVP), histopathology and molecular genetic testing.

BBS typically shows six primary/cardinal features (i.e., retinal degeneration [typically rod-cone dystrophy], obesity, polydactyly, hypogenitalism, intellectual impairment, and renal anomalies) and a number of secondary features (e.g., speech defects, developmental delay, diabetes mellitus, hypercholesterolemia, short stature, hepatic fibrosis, etc.). While association of polydactyly and renal anomalies is suggestive of BBS, observation of four out of six primary features is necessary for a diagnosis. Alternatively, co-presence of three primary features plus two secondary features is also adequate for confirming the diagnosis.

Molecular genetic testing of BBS gene mutations (e.g., *BBS1* p.M390R, *BBS2* p.Y24X, *BBS2* p.R275X, *BBS10* c.91fsX5) provides a useful means of confirming the diagnosis in about 80% of cases, especially for young patients who do not show/develop sufficient symptoms to meet the diagnostic criteria [34]. For instance, although retinal degeneration (rod-cone dystrophy) occurs in 93% of cases, it does not usually appear until 8 years of age. Common genetic screening strategies for BBS include (i) SNP arrays for homozygosity mapping and gene-targeted deletion/duplication analysis for intragenic or single exon deletions or duplications (using quantitative PCR, long-range PCR, multiplex ligation-dependent probe amplification, and gene-targeted microarray), (ii) direct sequencing for BBS gene mutations (small intragenic deletions/insertions; missense, nonsense, and splice site variants), and (iii) next-generation sequencing (NGS) for screening all ciliopathy genes (e.g., *BBS*, nephronophthisis genes, *ALMS1* gene, *CCDC28B* gene) and identifying the modifiers/epistatic effect of other genes [35,36]. The available data indicates that pathogenic variants responsible for BBS are mostly commonly detected in *BBS1* (23.2%), *BBS10* (20%), *BBS2* (8.1%), *BBS9* (6.0%), *BBS6* (5.8%), *BBS12* (5%), *BBS13* (4.5%), *BBS4* (2.3%), *BBS7* (1.5%), and BBS8 (1.2%).

Differential diagnoses for BBS include conditions which demonstrate clinical resemblances, including *McKusick-Kaufman syndrome* (MKKS, autosomal recessive disorder; causing the triad of hydrometrocolpos, postaxial polydactyly, and congenital heart disease, as well as genitourinary abnormalities, underdeveloped lungs, gastrointestinal abnormalities, and kidney defects; due to mutations in the *BBS6/MKKS* gene), *Alstrom syndrome* (autosomal recessive disorder; causing cone-rod dystrophy, obesity, progressive sensorineural hearing impairment, dilated cardiomyopathy, insulin-resistant diabetes mellitus syndrome, and developmental delay; due to *ALMS1* mutations), *Joubert syndrome* (autosomal recessive disorder; causing episodic hyperpnea, developmental delay, intellectual disability, hypotonia, oculomotor apraxia, ataxia, vermis hyoplasia or agenesis, characteristic molar tooth sign on cranial magnetic resonance imaging, retinal dystrophy, cystic dysplasia and nephronophthisis, ocular colobomas, occipital encephalocele, hepatic fibrosis, polydactyly, oral hamartomas, and endocrine abnormalities; due to mutations in *NPHP1, AHI1, CEP290/NPHP6, TMEM67/MKS3, RPGRIP1L, CC2D2A, ARL13B, INPP5E, OFD1, TMEM216, KIF7, TCTN1, TCTN2, TMEM237, CEP41, TMEM138, CPLANE1,* and *TTC21B*), *Senior–Løken syndrome* (autosomal recessive disorder; causing retinitis pigmentosa, cystic renal dysplasia, nephronophthisis, medullary cystic kidneys, polycystic kidneys., cerebellar vermis hypoplasia, ataxia, developmental delay, intellectual disability, occipital encephalocele, and oculomotor apraxia; due to mutations in *CEP290, NPHP1, NPHP3, NPHP4, IQCB1,* and *SDCCAG8*), *Meckel syndrome* (autosomal recessive disorder causing the triad of occipital encephalocele, large polycystic kidneys, and postaxial polydactyly as well as orofacial clefting, genital anomalies, CNS malformations, fibrosis of the liver, pulmonary hypoplasia; due to distinct mutations in the *BBS2, BBS4,* and *BBS6* genes), *Leber congenital amaurosis* (causing severe dystrophy of the retina, nystagmus, sluggish or near-absent pupillary responses, photophobia, high hyperopia, and keratoconus, eye poking, pressing,

and rubbing; due to mutations n *GUCY2D, RPE65, SPATA7, AIPL1, LCA5, RPGRIP1, CRX, CRB1, IMPDH1, RD3, RDH12,* and *CEP290*; mutations in *LRAT* and *TULP1* may be associated with an LCA-like phenotype), *Biemon syndrome type II* (autosomal recessive disorder; causing iris coloboma, intellectual disability, obesity, polydactyly, hypogonadism, hydrocephalus, and facial dysostosis), *Prader–Willi syndrome* (genetic disorder; causing hypotonia, feeding difficulties, failure to thrive, short stature, genital abnormalities, excessive appetite, progressive obesity, cognitive impairment, temper tantrums, obsessive/compulsive behavior, skin picking; due to nonfunctional genes in a region of chromosome 15) [26,37].

25.7 Treatment

Treatment for BBS centers largely on symptomatic relief, with the goal of ensuring that people with BBS reach their greatest potential, as current therapeutic measures are incapable of completely eliminating already established organ anomalies [23,38,39].

Surgery is useful for removing extra digits and correcting hydrocolpos, vaginal atresia, hypospadias, and congenital heart defects. Antibiotic prophylaxis is indicated after surgical and dental procedures. Kidney transplantation may be considered for 10% of BBS patients who develop end-stage renal disease (ESRD).

BBS patients with diabetes, hypertension, and metabolic disturbances should be given appropriate medications to minimize their impact on vulnerable organ systems (particularly the eyes and kidneys). Diet and exercise programs along with active lifestyle are recommended for BBS patients with obesity. Hormone supplements may be prescribed to BBS patients with low gonadotropin and sex hormone to reduce puberty-related stress.

Age-appropriate vitamin and mineral supplements help support best function of the eyes but do not cure the retinal dystrophy or halt the development of blindness.

Ongoing research and development on novel therapeutic modalities (e.g., gene therapy, exon skipping therapy, nonsense suppression therapy, gene editing, gene repurposing, targeted therapies, and nonpharmacological interventions) for BBS will greatly enhance our capability in the management of this rare genetic disorder [23,40].

25.8 Prognosis and Prevention

BBS is a genetic disorder in which homozygous or compound heterozygous mutations in the genes relating to primary (immotile) cilia function are responsible for inducing multiorgan anomalies. Given the current lack of specific treatments that help rectify or halt established or progressing organ anomalies, the prognosis for BBS patients is moderate to poor (particularly with regard to vision). In fact, children with BBS will likely develop night blindness apparent by age 7–8 years, and legal blindness by age 15 years.

Regular assessment of body weight, blood pressure and glucose level, endocrine and lipid profile, liver and renal function, and eyesight is important to help BBS-affected individuals maintain good health and improve the quality of life. If a pathogenic

BBS variant is detected in an affected family member, prenatal testing and preimplantation genetic diagnosis may form part of prevention strategies for BBS.

25.9 Conclusion

Bardet–Biedl syndrome (BBS) is a rare multisystem disorder that results from homozygous or compound heterozygous mutations in one of the 21 genes relating to cilia structure and/or function. Apart from six primary/cardinal features (retinal degeneration, obesity, polydactyly, hypogenitalism, intellectual impairment, and renal anomalies), BBS also induces a number of secondary features (e.g., developmental delay, speech deficit, brachydactyly or syndactyly, dental defects, ataxia or poor coordination, olfactory deficit, diabetes mellitus, and congenital heart disease; some authors also mention hypertension, liver abnormalities, bronchial asthma, otitis, rhinitis, craniofacial dysmorphism), which often vary among patients and also depend upon which BBS gene is involved. Furthermore, BBS-affected patients show an increased risk for tumor development involving the kidneys, ovary, endometrium, and possibly other organs. Considering its clinical and genetic heterogeneity, diagnosis of BBS is greatly facilitated by the application of molecular genetic techniques. In the absence of specific treatment, further study on the molecular mechanisms of BBS is justified in order to design new therapeutic approaches for the effective management of this disease.

REFERENCES

1. Forsythe E, Beales PL. Bardet-Biedl syndrome. In: Adam MP, Ardinger HH, Pagon RA et al. editors. *GeneReviews®* *[Internet]*. Seattle (WA): University of Washington, Seattle; 1993–2018 2003 Jul 14 [updated 2015 Apr 23].
2. Guo DF, Rahmouni K. Molecular basis of the obesity associated with Bardet-Biedl syndrome. *Trends Endocrinol Metab.* 2011;22(7):286–93.
3. Castro-Sánchez S, Álvarez-Satta M, Valverde D. Bardet-Biedl syndrome: A rare genetic disease. *J Pediatr Genet* 2013;2(2):77–83.
4. Forsythe E, Beales PL. Bardet-Biedl syndrome. *Eur J Hum Genet.* 2013;21:8–13.
5. Hildebrandt F, Benzing T, Katsanis N. Ciliopathies. *N Engl J Med.* 2011;364:1533–43.
6. Zaghloul NA, Katsanis N. Mechanistic insights into Bardet-Biedl syndrome, a model ciliopathy. *J Clin Invest.* 2009;119(3):428–37.
7. Novas R, Cardenas-Rodriguez M, Irigoín F, Badano JL. Bardet-Biedl syndrome: Is it only cilia dysfunction? *FEBS Lett.* 2015;589(22):3479–91.
8. Badano JL, Ansley SJ, Leitch CC, Lewis RA, Lupski JR, Katsanis N. Identification of a novel Bardet-Biedl syndrome protein, BBS7, that shares structural features with BBS1 and BBS2. *Am J Hum Genet.* 2003;72:650–8.
9. Aldahmesh MA, Li Y, Alhashem A et al. *IFT27*, encoding a small GTPase component of IFT particles, is mutated in a consanguineous family with Bardet-Biedl syndrome. *Hum Mol Genet.* 2014;23:3307–15.
10. Bujakowska KM, Zhang Q, Siemiatkowska AM et al. Mutations in *IFT172* cause isolated retinal degeneration and Bardet-Biedl syndrome. *Hum Mol Genet.* 2015;24:230–42.
11. Suspitsin EN, Imyanitov EN. Bardet-Biedl syndrome. *Mol Syndromol* 2016;7(2):62–71.
12. Álvarez-Satta M, Castro-Sánchez S, Valverde D. Bardet-Biedl syndrome as a chaperonopathy: dissecting the major role of chaperonin-like BBS proteins (BBS6-BBS10-BBS12). *Front Mol Biosci* 2017;4:55.
13. M'hamdi O, Ouertani I, Chaabouni-Bouhamed H. Update on the genetics of Bardet-Biedl syndrome. *Mol Syndromol* 2014;5(2):51–6.
14. Muller J, Stoetzel C, Vincent MC et al. Identification of 28 novel mutations in the Bardet-Biedl syndrome genes: the burden of private mutations in an extensively heterogeneous disease. *Hum Genet.* 2010;127:583–93.
15. Lindstrand A, Davis EE, Carvalho CM et al. Recurrent CNVs and SNVs at the *NPHP1* locus contribute pathogenic alleles to Bardet-Biedl syndrome. *Am J Hum Genet.* 2014;94:745–54.
16. M'hamdi O, Ouertani I, Maazoul F, Chaabouni-Bouhamed H. Prevalence of Bardet-Biedl syndrome in Tunisia. *J Community Genet* 2011;2:97–9.
17. Hirano M, Satake W, Ihara K et al. The first nationwide survey and genetic analyses of Bardet-Biedl syndrome in Japan. *PLOS ONE* 2015;10:e0136317.
18. Zaghloul NA, Liu Y, Gerdes JM, Gascue C, Oh EC. Functional analyses of variants reveal a significant role for dominant negative and common alleles in oligogenic Bardet-Biedl syndrome. *Proc Natl Acad Sci USA.* 2010;107:10602–7.
19. Abu-Safieh L, Al-Anazi S, Al-Abdi L et al. In search of triallelism in Bardet-Biedl syndrome. *Eur J Hum Genet.* 2012;20:420–7.
20. Schaefer E, Zaloszyc A, Lauer J et al. Mutations in *SDCCAG8/NPHP10* cause Bardet-Biedl syndrome and are associated with penetrant renal disease and absent polydactyly. *Mol Syndromol* 2011;1:273–81.
21. Aldahmesh MA, Safieh LA, Alkuraya H et al. Molecular characterization of retinitis pigmentosa in Saudi Arabia. *Mol Vis.* 2009;15:2464–9.
22. Andrade LJ, Andrade R, França CS, Bittencourt AV. Pigmentary retinopathy due to Bardet-Biedl syndrome: case report and literature review. *Arq Bras Oftalmol.* 2009;72(5):694–6.
23. Forsythe E, Kenny J, Bacchelli C, Beales PL. Managing Bardet-Biedl syndrome-now and in the future. *Front Pediatr* 2018;6:23.
24. Feuillan PP, Ng D, Han JC et al. Patients with Bardet-Biedl syndrome have hyperleptinemia suggestive of leptin resistance. *J Clin Endocrinol Metab.* 2011;96:E528–35.
25. Baker TM, Sturm EL, Turner CE, Petersen SM. Diagnosis of Bardet-Biedl syndrome in consecutive pregnancies affected with echogenic kidneys and polydactyly in a consanguineous couple. *Case Rep Genet* 2013;2013:159143.
26. Halbritter J, Bizet AA, Schmidts M et al. Defects in the IFT-B component IFT172 cause Jeune and Mainzer-Saldino syndromes in humans. *Am J Hum Genet.* 2013;93(5):915–25.
27. Zacchia M, Di Iorio V, Trepiccione F, Caterino M, Capasso G. The kidney in Bardet-Biedl syndrome: Possible pathogenesis of urine concentrating defect. *Kidney Dis (Basel)* 2017;3(2):57–65.
28. Imhoff O, Marion V, Stoetzel C et al. Bardet-Biedl syndrome: a study of the renal and cardiovascular phenotypes in a French cohort. *Clin J Am Soc Nephrol.* 2011;6:22–9.
29. Forsythe E, Sparks K, Best S et al. Risk factors for severe renal disease in Bardet-Biedl syndrome. *J Am Soc Nephrol.* 2017; 28(3):963–70.

30. Forsythe E, Sparks K, Hoskins BE et al. Genetic predictors of cardiovascular morbidity in Bardet-Biedl syndrome. *Clin Genet.* 2015;87:343–9.

31. Beales PL, Reid HA, Griffiths MH, Maher ER, Flinter FA, Woolf AS. Renal cancer and malformations in relatives of patients with Bardet-Biedl syndrome. *Nephrol Dial Transplant.* 2000;15(12):1977–85.

32. Tica I, Tica OS, Nicoară AD, Tica VI, Tica AA. Ovarian teratomas in a patient with Bardet-Biedl syndrome, a rare association. *Rom J Morphol Embryol.* 2016;57(4):1403–8.

33. Grechukhina O, Gressel GM, Munday W, Wong S, Santin A, Vash-Margita A. Endometrial carcinoma in a 26-year-old patient with Bardet-Biedl syndrome. *Case Rep Obstet Gynecol* 2018;2018:1952351.

34. Redin C, Le Gras S, Mhamdi O et al. Targeted high-throughput sequencing for diagnosis of genetically heterogeneous diseases: efficient mutation detection in Bardet-Biedl and Alström syndromes. *J Med Genet.* 2012;49:502–12.

35. Ajmal M, Khan MI, Neveling K et al. Exome sequencing identifies a novel and a recurrent *BBS1* mutation in Pakistani families with Bardet-Biedl syndrome. *Mol Vis.* 2013;19:644–53.

36. Scheidecker S, Etard C, Pierce NW et al. Exome sequencing of Bardet-Biedl syndrome patient identifies a null mutation in the BBSome subunit *BBIP1 (BBS18). J Med Genet.* 2014;51:132–6.

37. Farmer A, Aymé S, de Heredia ML et al. EURO-WABB: an EU rare diseases registry for Wolfram syndrome, Alström syndrome and Bardet-Biedl syndrome. *BMC Pediatr.* 2013;13:130.

38. Priya S, Nampoothiri S, Sen P, Sripriya S. Bardet-Biedl syndrome: Genetics, molecular pathophysiology, and disease management. *Indian J Ophthalmol.* 2016;64(9):620–7.

39. Panny A, Glurich I, Haws RM, Acharya A. Oral and craniofacial anomalies of Bardet-Biedl syndrome: Dental management in the context of a rare disease. *J Dent Res.* 2017;96(12):1361–9.

40. Seo S, Mullins RF, Dumitrescu AV et al. Subretinal gene therapy of mice with Bardet-Biedl syndrome type 1. *Invest Ophthalmol Vis Sci.* 2013;54:6118–32.

26

Birt–Hogg–Dubé Syndrome

**Marianne Geilswijk, Mette Sommerlund, Mia Gebauer Madsen,
Anne-Bine Skytte, and Elisabeth Bendstrup**

CONTENTS

26.1 Introduction

Birt–Hogg–Dubé syndrome (BHD, OMIM#135150) is an autosomal dominantly inherited cancer predisposition syndrome caused by germline pathogenic variants in the folliculin gene, *FLCN* [1]. The first report of BHD is generally attributed to Canadian physicians Birt, Hogg, and Dubé, who in 1977 described a familial genodermatosis characterized by fibrofolliculomas, trichodiscomas, and acrochordons [2]. Named in honor of Birt, Hogg, and Dubé, this condition was later linked to an increased risk of developing benign and malignant renal neoplasms, cystic lung disease, and recurrent spontaneous pneumothoraces, besides specific cutaneous findings [3–5].

26.2 Biological Background

BHD is an autosomal dominant, monogenic condition, caused by germline pathogenic *FLCN* variants. Germline homozygosity or compound heterozygosity for pathogenic *FLCN* variants is considered incompatible with life. *FLCN* maps to chromosome 17p11.2 and contains 14 exons. The encoded protein, folliculin, spans 579 amino acids and is highly conserved among species. So far, >150 different disease-causing *FLCN* variants have been reported, with over 50% of these occurring in a mutational hot spot in exon 11 [6,7]. The majority of pathogenic variants are nonsense substitutions, deletions, and insertions but larger genomic rearrangements have also been described. Most cases of BHD are caused by the introduction of a premature stop codon or by frameshift, both leading to loss of function of the *FLCN* gene product [8,9]. However, missense variants have also been described in families with phenotypical BHD [10,11]. In these cases, the detected variants have been shown to contribute to instability of the folliculin protein. Hence, the BHD phenotype is caused by *FLCN* haploinsufficiency and, as described below, in BHD-associated renal tumorigenesis by further biallelic *FLCN* loss of function.

26.3 Pathogenesis

The functions of the folliculin protein (FLCN) are not fully understood. In silico, in vitro, and in vivo studies of both human and animal materials have shown that FLCN is involved in several fundamental cellular mechanisms. In fact, FLCN acts not only as a tumor suppressor but also a protein required for ciliogenesis, cell adhesion, and autophagy, through its interaction with folliculin-interaction-protein-1 and -2 (FNIP1 and FNIP2) separately or in conjunction [12,13].

26.3.1 Renal Tumorigenesis

In support of the tumor suppressor role of FLCN, somatic second hits in *FLCN* have been identified in BHD-associated renal tumor tissue samples, and in murine studies tumor growth of a *FLCN*/null cell line was reverted as wild-type *FLCN* was reintroduced [14,15]. By contrast, increased frequencies of somatic pathogenic *FLCN* variants have not been detected in sporadic renal tumors [16]. As a tumor suppressor, FLCN interacts with 5′-AMP-activated protein kinase (AMPK) through FNIP1 and FNIP2 and regulates the AKT-mTOR-pathway, which is a major determinant of cell growth and protein synthesis [12]. However, FLCN's actions in the AKT-mTOR-pathway are incompletely understood, as various studies have given conflicting results regarding m-TOR up- or downregulation as a consequence of FLCN deficiency. It is generally agreed that FLCN-associated tumor suppression through the AKT-m-TOR-pathway is complex and possibly dependent on circumstances and other genetic factors [17,18]. Thus, the tumorigenic impact of FLCN-deficiency in BHD through the AKT-mTOR-pathway requires further clarification. Besides the AKT-mTOR-pathway, other cellular signaling pathways involved in tumor suppression are also regulated by FLCN. Studies have shown that loss of *FLCN* affects both TGF-β- and EGFR-signaling pathways, which play critical roles in tumorigenesis [15,19,20]. Further, FLCN is involved in the regulation of mitochondrial metabolism and of lysosome function [21,22].

26.3.2 Cutaneous Findings

Examination of skin biopsies from BHD-affected individuals has not revealed somatic second hits in *FLCN* [23]. Thus, the cutaneous phenotype is presumably caused by *FLCN* haploinsufficiency. It is not understood how *FLCN* haploinsufficiency contributes to the formation of fibrofolliculomas.

26.3.3 Cystic Lung Disease

Similar to BHD-related kidney cancer, the mechanisms by which FLCN deficiency leads to cystic lung disease are not fully elucidated and it is still debated whether the two-hit-model is applicable or if cyst formation is the result of *FLCN* haploinsufficiency only. It has been suggested that FLCN is involved in cell-to-cell adhesion in part by interaction with FNIP1/FNIP2 and AMPK, which is also the target of the E-cadherin-LKB1-pathway involved in the formation of cell junctions in bronchial epithelial cells [24]. Other investigators have reported increased cell-to-cell adhesion through the regulation of PKP4-signaling in both kidney and lung FLCN-deficient cells, although the effect of FLCN deficiency in the latter was cell-type dependent [25,26]. Hence, Khabibulli et al. observed that cell-to-cell adhesion in human bronchial epithelial cells increased in connection with FLCN deficiency. Based on this, it has been proposed that lung cyst formation in BHD is a result of stretch-induced mechanical stress of the alveolar wall. The "stretch hypothesis" has later been supported by findings by other investigators and is found to be in accordance with the well-established observation that BHD-associated lung cysts are localized in the basilar rather than the apical compartments of the lungs [27,28]. With the stretch hypothesis it is assumed that pneumothorax in BHD is the result of cysts rupturing because of increased cell-to-cell adhesion and subsequent decreased tolerance to mechanical stress.

26.4 Epidemiology

The exact prevalence of BHD is not known. So far, more than 600 BHD families have been described worldwide without evidence of ethnical or geographical predominance [29]. Any combination of clinical features can be seen among members of the same family as well as affected non-symptomatic individuals. Thus, because of incomplete penetrance and marked inter- and intrafamilial variability in expression, it is generally assumed that BHD is underdiagnosed [30]. Further, unawareness of the syndrome among clinicians in relevant medical specialties may also cause the syndrome to be overlooked. In large cohort studies, clinical manifestations are described with equal frequencies in males and females and rarely before the age of 20 years, although some reports of BHD-associated spontaneous pneumothoraces and/or renal cell carcinoma (RCC) in childhood do exist [31–33]. General differences in prevalence of specific clinical manifestations have been reported across cohorts originating from Western and Asian countries, respectively. Thus, cutaneous findings are reported with higher frequency in American BHD populations, while pneumothorax-only families are mainly reported in Asia. It is not settled whether these discrepancies are actual phenotypical variations between ethnic groups or the result of selection bias. To date, no clear evidence of genotype−phenotype correlations exists, although trends toward phenotypes being dominated by lung findings in correlation to pathogenic variants in the 3′ end of *FLCN* have been reported [34,35]. Furthermore, it is continuously debated if BHD is associated with an increased risk of malignant diseases other than RCC.

26.5 Clinical Features

26.5.1 Cutaneous Findings

The skin manifestations of BHD include a variety of benign tumors. The most frequent tumors are fibrofolliculomas, trichodiscomas, and acrochordons. Fibrofolliculomas and trichodiscomas are clinically indistinguishable, small 2- to 4-mm dome-shaped pale flesh-colored/whitish tumors primarily localized on the face, neck, retroauricular area, and upper body (Figure 26.1) [30]. The number of tumors varies substantially from few to numerous, also in familial cases of BHD harboring the same pathogenic variant in *FLCN*.

Fibrofolliculomas and trichodiscomas normally develop from the second decade, and the number of tumors increases in the third to fourth decade [30]. Many patients request medical treatment for the facial tumors at general practitioners and

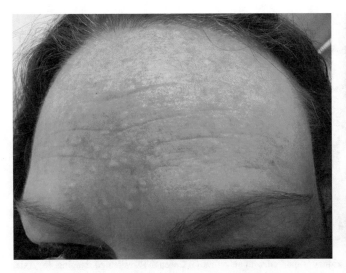

FIGURE 26.1 Fibrofolliculomas.

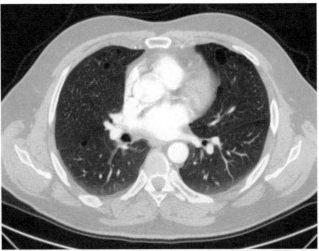

FIGURE 26.2 Chest CT scan of BHD-affected individual.

dermatologists. Patients are often misdiagnosed with acne vulgaris or acne rosacea and receive local and systemic antibiotics or retinoids without any effect [36]. Diagnosis is made by a 3- to 4-mm punch biopsy from skin lesions, from which histology will show either fibrofolliculomas or trichodiscomas. Some pathologists argue that fibrofolliculoma and trichodiscoma represent a single pathological process at different stages of development [37]. Recently, a variant of fibrofolliculomas named comedonal or cystic fibrofolliculoma has been described in association with BHD [38]. Other skin tumors described in BHD patients include lipomas, collagenomas, angiofibromas, and melanomas [37,39]. Currently, there is not sufficient available data to evaluate whether there is a true association between BHD and malignant melanoma. However, in a recent research letter, regular dermatological examination for cutaneous malignant melanoma in BHD patients is recommended [40].

Facial tumors with similarities to fibrofolliculoma/trichodiscoma may occur in association with other genodermatoses. Important differential diagnoses are angiofibromas in tuberous sclerosis complex and MEN1 (multiple endocrine neoplasia type 1), and trichoepitheliomas in Brooke–Spiegler syndrome (cylindroma syndrome) and familial trichoepitheliomas [39,41].

Many patients only present with fibrofolliculomas/trichodiscomas without a history of renal tumors or pneumothorax, and therefore skin biopsy, family history, and further investigations of the lungs and kidneys as well as genetic testing should be considered.

26.5.2 Pulmonary Findings

The majority of BHD patients have cystic lung disease (77%–89%) [42]. The relationship between cystic lung disease and pneumothorax has been recognized for more than a decade, and it was originally reported that 24%–38% of patients with BHD experience a pneumothorax [34,43]. However, in a recent questionnaire survey, more than 75% of BHD patients reported at least one spontaneous pneumothorax, with an 80% risk of recurrence [44].

BHD patients normally have no respiratory symptoms. When experiencing pneumothorax, BHD patients suffer similar symptoms as other pneumothorax patients. Pulmonary function

parameters are normal or only slightly impaired unless recurrent pneumothorax has caused scarring [34,45]. BHD patients have an up to 50-fold increased risk of experiencing a pneumothorax compared to an age-matched general population [5,46]. Pneumothorax in BHD has no gender predilection. The earliest reported age of an initial pneumothorax is 7 years, the median age of pneumothorax is 38 years (range, 22–71 years), and the median age of the last pneumothorax is 42 years (range, 22–75 years) [34,47]. A spontaneous pneumothorax may be the first presentation of BHD and also the only manifestation [46]. Spontaneous pneumothoraces have been reported in BHD patients without radiological apparent cysts [48].

Several risk factors for BHD-associated spontaneous pneumothorax have been identified. A family history of spontaneous pneumothorax results in a significantly increased risk of a spontaneous pneumothorax compared to patients without a family history, as did the number of lung cysts, total lung cyst volume, largest cyst diameter, and largest cyst volume [34,42].

Chest CT studies have found the cysts to be bilateral, varying in size and shape, and to have a lower lung–predominant localization, in accordance with the stress hypothesis. The cysts can be round, oval, irregular, or multiseptated. The cyst size varies from 0.2 to 8 cm. Patients can have a few to more than 100 cysts (Figure 26.2) [43,49,50].

Only few studies have described the histopathological changes in BHD-associated lung cysts. Cyst walls are very thin and translucent, and surrounded by normal lung parenchyma without inflammation. Intracystic septa, seen in less than 20% of patients, are composed of interlobular septa and venules protruding into the cysts. Cysts often abut on interlobular septa, and there is an even distribution of the cysts between the subpleural and the intrapulmonary area [51]. There are no other pathologic findings than cysts in the lung parenchyma in BHD patients.

26.5.3 Renal Findings

Patients with BHD are at risk of developing renal cysts and renal tumors [3,4].

Simple renal cysts are a benign condition with a high prevalence in the older population. Usually, renal cysts do not cause

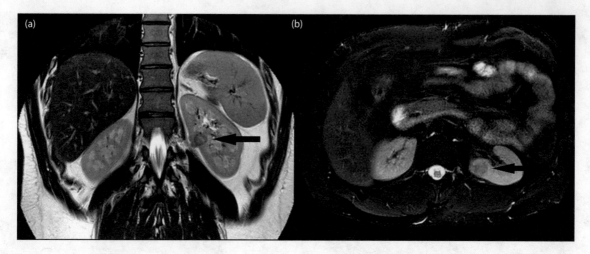

FIGURE 26.3 MRI scan of BHD-affected individual. A tumor in the left kidney is marked with an arrow. (a) Coronal T2-weighted imaging. (b) Axial T2-weighted fat-saturated imaging.

symptoms unless they grow so big that they cause pain by compressing on other organs, or they get infected, causing pain and fever. Renal cysts can have multiple septa, thickening of the wall, calcifications, or contrast enhancement (so-called complex cysts). Complex cysts have a potential to become malignant, and a follow-up imaging or biopsy is therefore mandatory [52].

According to the American Cancer Society, the overall lifetime risk for developing kidney cancer is about 1.6% [53]. In the BHD population, the risk is increased considerably. In comparison to unaffected siblings, the risk of developing renal tumors is sevenfold greater in a BHD patient [5]. Between 16% and 34% of BHD patients develop renal tumors, and in more than half of these patients, bilateral and multifocal tumors are described (Figure 26.3) [9,42,54–56].

The average age of renal tumor onset is just below the age of 50 but has been described in patients between 15 and 83 years [57,58].

In sporadic renal tumors, aggressive clear cell renal cell carcinoma (ccRCC) is the most common subtype (75%), followed by papillary RCC (pRCC, 10%), low-aggressive chromophobe RCC (chRCC, 5%), benign oncycotoma (5%), and other rare subtypes (<5%) [56]. This is in contrast to the BHD-associated renal tumors, where the majority are oncocytoma, chRCC, or hybrid chromophobe-oncocytoma type. Pavlovich et al. were the first to present a systematic pathologic analysis of BHD-associated renal tumors. They reviewed 130 renal tumors resected from 30 BHD patients, including 65 hybrid chromophobe-oncocytoma type (50%), 44 chRCC (34%), 12 ccRCC (9%), 7 oncocytoma (5%), and 2 pRCC (2%) [56]. In a cohort of French BHD patients, Benusiglio et al. reported a total of 124 patients in which 33 renal tumor (27%), 23 oncocytoma or hybrid chromophobe-oncocytoma type (70%), 3 ccRCC (9%), and 1 pRCC (3%) were identified histopathologically, while the remaining patients had no histopathological data [57]. This pattern is remarkably different from the histologic spectrum in sporadic renal tumors confirmed by several studies [35,42,58]. It is recommended by The International Society of Urological Pathology Tumor Panel that a diagnosis of BHD should be considered when a hybrid chromophobe-oncocytoma tumor is found [59]. Oncocytosis is observed histologically in about half of the BHD-associated tumors [60]. Renal oncocytosis is characterized by a spectrum of oncocytic

changes that diffusely involve the renal parenchyma, and it is possibly a precancerous lesion [54,60,61].

After examining 124 BHD patients, Pavlovich et al. noted that tumor histology was not clearly concordant within affected families, indicating that there is no obvious genotype–phenotype correlation with regard to tumor histology [62].

Based on previous studies, it has been suggested that BHD-associated tumors due to the diverse histology tend to be low-aggressive and not likely to metastasize, compared to sporadic RCCs. Thus Pavlovich et al. reported 2 cases of metastatic disease in a cohort of 14 BHD patients with RCCs, and Benusiglio et al. reported 4 cases of metastatic disease in a cohort of 33 BHD patients with malignant renal tumors [57,62]. However, in a recent cohort study Houweling et al. reported clinical data from a cohort of 115 BHD patients, of which 14 had malignant renal tumors, and 5 of these had metastatic disease [9]. In all studies, the majority of the metastatic cases were found to have the more aggressive ccRCC, and in most cases the tumors were 4 cm and locally advanced at presentation [9,55,62]. Data on the risk of metastatic disease in a cohort of BHD patients followed in a screening program are still missing.

26.5.4 Other Clinical Manifestations

Studies on the possible correlation between BHD and the development of colorectal adenomas and/or adenocarcinomas have showed conflicting results. In 2002, Zbar et al. examined 111 BHD cases and their unaffected siblings, and found no evidence for an increased risk of colorectal cancer in BHD. In contrast, Nahorski et al. reported in 2010 an increased lifetime risk of colorectal cancer in BHD in comparison to the background population [63]. Other studies have also suggested an increased risk of colorectal cancer in BHD, but clear evidence for causality is still lacking [14].

A wide range of other conditions have also been reported in association with BHD, including parotid oncocytomas [34,58,64,65], thyroid nodules and cancer [8,66–68], cerebral and pulmonary vascular malformations [69,70], breast cancer [71], cutaneous and choroidal melanomas [40,72–74], multiple lipomas, and parathyroid adenomas [75]. Whether these clinical features are actual manifestations of BHD or not is still unsettled [76].

26.6 Diagnosis

In 2009, the European BHD Consortium proposed a set of major and minor criteria for the diagnosis of BHD [30]. In these criteria, clinical findings, family history, and the results of molecular genetic testing of *FLCN* are included. Histologically verified multiple fibrofolliculomas are considered diagnostic for the syndrome as is the detection of a pathogenic *FLCN* variant by molecular genetic testing. The presence of one minor criterion (multiple basilar lung cyst of no other origin, early onset/multifocal/bilateral/hybrid chromophobe-oncocytoma renal cancer, or one first-degree relative with BHD) suggests the diagnosis, and two minor criteria are diagnostic. These diagnostic criteria were subsequently modified by Schmidt et al., who consider the detection of a pathogenic *FLCN* variant confirmative of BHD, while two or more histologically proven fibrofolliculomas are highly suggestive of the diagnosis, and other clinical findings alone or in combination should raise suspicion of the syndrome (Table 26.1) [7].

As for other inherited conditions, molecular genetic screening for BHD can be performed using DNA extracted from leucocytes,

TABLE 26.1

Schmidt's Revised BHD Diagnostic Criteria

Major criteria (high likelihood of BHD)

- Two or more cutaneous papules clinically compatible with fibrofolliculoma/trichodiscoma with at least one biopsy proven/histologically confirmed fibrofolliculoma

Minor criteria (suspicious of BHD)

- Multiple bilateral basilar lung cysts with or without spontaneous pneumothorax (<40 years of age), especially with family history of these lung manifestations
- Bilateral multifocal chromophobe or hybrid oncocytic renal tumors, especially with family history of this histological subtype
- Any combination of cutaneous, lung, or renal manifestations known to be associated with BHD in an individual or his or her family

Definitive diagnosis

- Positive germline FLCN mutation test

Source: Schmidt LS, Linehan WM. *Expert Opinion on Orphan Drugs.* 2015;3(1):15–29. Reprinted with permission from Taylor & Francis Group.

or other easily obtainable tissue. In general, tumor tissue should not be used for germline genetic screening. Screening *FLCN*, using sequencing and quantitative analysis in combination, in individuals with clinically diagnosed BHD, has a diagnostic yield >90% [9].

To date, only one de novo pathogenic *FLCN* variant has been reported [77]. This implies that in most cases, BHD is inherited from an affected parent. As BHD is an autosomal dominantly inherited condition, offspring of an affected individual is at a 50% risk of inheriting the causative pathogenic *FLCN* variant. Thus, diagnosis of BHD impacts not only the examined individual but also potentially affected symptomatic and non-symptomatic family members. Genetic counseling should be provided and diagnostic or predictive genetic testing offered for relevant relatives. For ethical reasons, it is not common to offer predictive testing to children, as the cancer risk in BHD is generally believed to be low before age 20 years.

Relevant differential diagnoses to consider are other inherited conditions which predispose to the development of either RCC (e.g., von Hippel Lindau disease, tuberous sclerosis complex, hereditary papillary renal cancer), lung cysts, and/or spontaneous pneumothorax (lymphangioleiomyomatosis [LAM] and tuberous sclerosis complex, Marfan syndrome, Ehlers–Danlos syndrome, Langerhans cell histiocytosis, lymphocytic interstitial pneumonia [LIP], amyloidosis, *Pneumocystis jiroveci* pneumonia, and emphysema) or cutaneous hamartomas (tuberous sclerosis complex, Cowden syndrome) [78]. Often BHD can be distinguished from the above-mentioned conditions by family history as well as clinical and radiological findings. Table 26.2 presents an overview of pulmonary differential diagnoses for BHD.

26.7 Treatment

26.7.1 Cutaneous Findings

The benign facial skin tumors can cause cosmetic nuisances. Especially female patients request dermatological assistance. The principal treatment of fibrofolliculomas/trichodiscomas is CO_2 laser treatment, but ErbiumYAG laser is also used [79,80]. Others use hyfrecation (electrodessication) [81]. The treatment can be performed in local analgesia. Treatment is not curative, and

TABLE 26.2

Radiologic Differential Diagnoses in Birt–Hogg–Dubé Syndrome

Disease	Distribution of Cysts on Chest CT	Associated Radiologic Findings	Extrapulmonary Involvement
Birt–Hogg–Dubé	Lower lung predominance	None	Kidneys
Pulmonary Langerhans cell histiocytosis	Upper lung predominance	Nodules with upper lobe predominance	
Lymphangioleiomyomatosis	Diffuse	Effusion	Kidneys
Lymphocytic interstitial pneumonia	Lower lung predominance	Ground glass opacities	
Amyloidosis	Lower lung predominance	Nodules with lower lobe predominance	

Note: Normally, the clinical context distinguishes BHD from *Pneumocystis jiroveci* pneumonia. Paraseptal emphysema can manifest as bullae of varying size and shape, most often has an upper lung predominant localization, and the pulmonary function shows an obstructive pattern. In LIP, a pattern of ground glass opacities, nodules, and septal thickening are present together with cysts. Also, LIP is usually associated with autoimmune diseases. LAM is almost only seen in women and is characterized by diffuse, uniform, and round cysts, and most patients have elevated levels of serum vascular endothelial growth factor-D contrary to BHD [43,101]. Langerhans cell histiocytosis is characterized by irregular cysts and nodules and has an upper lobe predominance.

repetitive treatments are often necessary. The treatment creates minor depigmented scars, which can easily be covered by makeup. The m-TOR inhibitor rapamycin is used for topical treatment for facial angiofibromas in patients with tuberous sclerosis complex. Recently, topical rapamycin 0.1% was tested in a double-blind placebo-controlled randomized split-face study on facial fibrofolliculomas in 19 patients with BHD for 6 months. However, no effect of treatment was found [82]. Future elucidation of the molecular pathways involved in fibrofolliculoma/trichodiscoma tumor formation is vital for developing topical pathogenesis-directed therapy [83].

26.7.2 Pulmonary Findings

BHD patients do not normally develop progressive or significant lung function decline or respiratory impairment [45]. Although there is no association between the use of tobacco and pneumothorax in BHD, patients should be advised against smoking [84].

Prevention and treatment of pneumothorax is the main principle of pulmonary management. Patients should be informed of the increased risk of pneumothorax in general, and that diving and air travel can increase the risk further. Patients should be educated on the symptoms and signs of pneumothorax and how to react in case they experience such symptoms. As the number and size of cysts are related to the risk of pneumothorax, a baseline chest CT can be helpful to characterize the extent of pulmonary involvement, and a demonstration of the cysts may also facilitate patient education.

The treatment of pneumothorax in BHD patients does not differ from the treatment of pneumothorax in any other disease. Depending on the severity of the pneumothorax, treatment ranges from observation with oxygen treatment to the insertion of a tube, thoracostomy, or pleurodesis [34]. The recurrence rate after a sentinel of pneumothorax is up to 75%, and therefore surgical intervention with resection and pleurodesis should be considered even in patients with a first episode of pneumothorax to avoid morbidity with repeated pneumothoraces [34,46,84,85].

26.7.3 Renal Findings

BHD patients are likely to experience tumor recurrence and require multiple renal interventions throughout their lifetime. The challenge is to preserve kidney function and to secure cancer control. Small renal tumors (<4 cm) have a low growth rate (0–0.08 cm/year), and active surveillance is a reasonable option since metastases are rarely seen in renal tumors <4 cm [86–88]. In BHD-associated RCCs, the general recommendation is to observe the tumor until it reaches 3 cm. As mentioned above, the majority of BHD-associated RCCs are the low-aggressive chRCC and hybrid chromophobe-oncocytoma types, which are unlikely to metastasize.

The approach for management depends on the tumor size as well as location and number of tumors, and the choice of surgical approach is dependent on the experience and preference of the surgeon [55]. It is also mandatory to look at the general condition of the patient, since surgery entails surgical stress and may deteriorate the kidney function. Nephron-sparing surgery is recommended whenever possible, and a partial nephrectomy (PN) can be performed open, laparoscopic, or robot-assisted laparoscopic. If the tumor is large or central, a radical nephrectomy is necessary.

Thermal ablation techniques are minimally invasive techniques which include cryotherapy and radiofrequency ablation. These techniques are equal in terms of the different outcomes defined by complications, metastatic progression, and cancer-specific survival [89]. Cryotherapy ablation is an established minimally invasive technique for the treatment of small renal tumors (<4 cm), but there is a lack of long-term efficacy data from large prospective or randomized studies. Most of the data available on cryotherapy ablation are limited to treatment of renal tumors in elderly patients who are poor surgical candidates [90].

Because of the lack of robust evidence in the literature, the European Association of Urology guidelines on RCC specifically conclude no recommendation can be made on cryotherapy ablation, whereas the American Urological Association (AUA) considers thermal ablation techniques as an alternate approach for the management of small renal tumors [91–93]. The AUA guidelines state that thermal ablation is associated with a higher local recurrence rate, but with the use of salvage therapy, the metastasis-free survival and cancer-specific survival rates are comparable to PN [93]. The renal functional outcome after a thermal ablation is also comparable to PN [93]. Many reports/reviews do not recommend the use of thermal ablation techniques in BHD patients [94]. Post-ablation imaging can be challenging to interpret, and previous thermal ablation may complicate subsequent surgery with higher rates of perioperative complications and conversion to radical nephrectomy [95,96]. Since BHD patients are at lifelong risk of developing new tumors, it is suggested that previous thermal ablation can complicate the long-term evaluation and additional surgery, if this is needed [55]. Conversely, Yang et al. reported a study in which they evaluated the efficacy and safety of thermal ablation as salvage treatment for renal tumor in von Hippel–Lindau disease patients after previous PN. Comparable to BHD, von Hippel–Lindau disease leads to a significant increased risk of developing multifocal and bilateral renal tumors. The study concluded that thermal ablation seems to represent a suitable treatment option for von Hippel–Lindau disease patients, with a high technical success rate and causing minor changes in renal function [97].

The use of thermal ablation techniques for renal tumors in BHD requires a setup in specialized institutions with dedicated radiologists to secure a correct interpretation of the post-ablation imaging. In an experienced thermal ablation setup, BHD patients with renal tumors should be offered thermal ablation as an alternative to a surgical approach.

If BHD-associated RCC progresses to metastatic disease, it requires a multidisciplinary management, and the treatment is comparable to the treatment of metastatic disease from sporadic RCC. Nevertheless, it is hypothesized that metastatic RCC in BHD patients is more sensitive to tyrosine kinase and mTOR-inhibitors due to the mTOR signaling pathway activation seen in BHD-associated renal tumors [57]. However, the evidence for the use of these agents in BHD-associated metastatic disease is lacking [94].

26.8 Prognosis and Prevention

26.8.1 Cutaneous Findings

Skin tumors are benign and do not constitute a risk for patients with BHD. As the skin tumors associated with BHD often present before pneumothoraces and renal tumors, increased awareness by general practitioners and dermatologists on skin tumors (fibrofolliculoma/trichodiscoma) may help identifying individuals with BHD and enable an early diagnosis and renal cancer surveillance [98].

26.8.2 Pulmonary Findings

Apart from the risk of pneumothorax, BHD patients have a good prognosis with respect to pulmonary involvement. Even though some patients have more than 100 cysts, lung function parameters are normal or only slightly impaired [45]. The number, size, and total cyst volume is positively correlated to an increased risk of pneumothorax [34].

There are limited data on the risk of pneumothorax during air travel or scuba diving. Patients have complained of chest pressure (52%), anxiety (9%–50%), headache (3%–31%), shortness of breath (4%–28%), chest pain (6%–28%), nausea (4%–20%), fatigue (3%–7%), oxygen desaturation by handheld pulse oximetry (4%), palpitations (2.8%), peripheral cyanosis (2%), abnormal chills (1.4%), and dizziness (0.7%) during air travel. Symptoms were similar between patients with and without a previous spontaneous pneumothorax. The flight-related pneumothorax risk has been calculated to range between 0.12%–0.63% per flight during or within 1 month after air travel. Patients with a prior pleurodesis are less likely to develop a flight-related pneumothorax [44,99]. Patients may develop a pneumothorax up to 1 month after air travel, and accordingly, they should be informed to consult a physician if they experience any symptoms such as dyspnea or chest pain during or shortly after a flight. If experiencing such symptoms before air travel, they should be advised to seek medical consultancy and to have a clinical checkup before flying [44,99]. There are no firm guidelines on the interval from a spontaneous pneumothorax to air travel. Recommendations vary from no time to 3 weeks after radiographic remission.

In a questionnaire survey among 54 BHD patients who reported diving, 10 experienced symptoms such as shortness of breath (11.1%), anxiety (3.7%), dizziness (1.9%), abnormal fatigue (1.9%), abnormal chills (1.9%), and hemoptysis (1.9%). The diving-related pneumothorax risk was calculated to 0.33% per episode of diving session [99].

The British Thoracic Society (BTS) guideline recommends against diving after an episode of spontaneous pneumothorax unless the patient has undergone bilateral surgical pleurodesis and the lung function and postoperative thoracic CT are normal [100].

26.8.3 Renal Findings

The increased risk of developing RCC is the most serious and life-threatening complication associated with BHD. Screening and follow-up programs are therefore mandatory to prevent metastatic RCC and chronic renal insufficiency.

As BHD-associated RCC has been described in young age, at-risk individuals are recommended genetic testing in the beginning of their 20s [32,57]. All BHD-affected individuals should have an abdominal imaging at least every 36 months [55,94]. Generally, MRI is preferred as the screening tool for long-term surveillance. Renal ultrasonography is investigator-dependent

and might not detect small tumors. CT with intravenous contrast is a sensitive and informative imaging but should be saved for special situations (e.g., CT-guided cryotherapy) to reduce the exposure of radiation to the patient [55]. An individual surveillance program should be followed after the identification of a renal tumor, depending on the size and histology of the tumor as well as the age and comorbidities of the patient.

BHD-associated RCCs usually have a favorable clinical course due to the low-aggressive behavior of the chromophobe and hybrid chromophobe-oncocytoma tumors. Still, a minority of BHD patients develop ccRCC, which has a higher potential to metastasize [9,62].

Data on the prognosis of BHD-associated RCC are missing. A long-term follow-up of a BHD population is much needed to elucidate the differences in the clinical course and prognosis between sporadic and BHD-associated RCC.

26.9 Conclusion

Birt–Hogg–Dubé syndrome is an autosomal dominant inherited condition caused by pathogenic variants in the *FLCN* gene, predisposing affected individuals to the development of benign skin tumors, lung cysts, and recurrent spontaneous pneumothoraces as well as benign and malignant renal tumors. BHD is characterized by marked intra- and interfamilial variability in expression and is likely underdiagnosed. Awareness of the diagnosis is warranted, as affected individuals are recommended regular renal imaging surveillance to prevent RCC and for specialized management of spontaneous pneumothoraces. The suspicion of BHD should be prompted in the case of early-onset or familial renal cancer, chRCC or hybrid chromophobe-oncocytoma RCC, early-onset or familial spontaneous pneumothorax, chest CT showing characteristic lung cysts, occurrence of fibrofolliculomas, or any combination of these findings. Confirmatory genetic testing should be offered, and affected individuals and relatives referred for genetic counseling.

REFERENCES

1. Nickerson ML, Warren MB, Toro JR et al. Mutations in a novel gene lead to kidney tumors, lung wall defects, and benign tumors of the hair follicle in patients with the Birt-Hogg-Dube syndrome. *Cancer Cell*. 2002;2:157–64.
2. Birt AR, Hogg GR, Dube WJ. Hereditary multiple fibrofolliculomas with trichodiscomas and acrochordons. *Arch Dermatol*. 1977;113:1674–7.
3. Roth JS, Rabinowitz AD, Benson M, Grossman ME. Bilateral renal cell carcinoma in the Birt-Hogg-Dube syndrome. *J Am Acad Dermatol*. 1993;29:1055–6.
4. Toro JR, Glenn G, Duray P et al. Birt-Hogg-Dube syndrome: A novel marker of kidney neoplasia. *Arch Dermatol*. 1999;135:1195–202.
5. Zbar B, Alvord WG, Glenn G et al. Risk of renal and colonic neoplasms and spontaneous pneumothorax in the Birt-Hogg-Dubé syndrome. *Cancer Epidemiol Biomarkers Prev*. 2002;11:393–400.
6. Leiden Open Variation Database 3.0. (https://grenada.lumc.nl/LOVD2/shared1/home.php?select_db=FLCN). Accessed July 06, 2017.

7. Schmidt LS, Linehan WM. Clinical features, genetics and potential therapeutic approaches for Birt-Hogg-Dube syndrome. *Expert Opin Orphan Drugs* 2015;3:15–29.

8. Kluger N, Giraud S, Coupier I et al. Birt-Hogg-Dube syndrome: Clinical and genetic studies of 10 French families. *Br J Dermatol.* 2010;162:527–37.

9. Houweling AC, Gijezen LM, Jonker MA et al. Renal cancer and pneumothorax risk in Birt-Hogg-Dube syndrome; an analysis of 115 FLCN mutation carriers from 35 BHD families. *Br J Cancer.* 2011;105:1912–9.

10. Nahorski MS, Reiman A, Lim DH et al. Birt Hogg-Dubé syndrome-associated FLCN mutations disrupt protein stability. *Hum Mutat.* 2011;32:921–9.

11. Rossing M, Albrechtsen A, Skytte AB et al. Genetic screening of the FLCN gene identify six novel variants and a Danish founder mutation. *J Hum Genet.* 2017;62:151–7.

12. Baba M, Hong S-, Sharma N et al. Folliculin encoded by the BHD gene interacts with a binding protein, FNIP1, and AMPK, and is involved in AMPK and mTOR signaling. *Proc Natl Acad Sci U S A.* 2006;103:15552–7.

13. Hasumi H, Baba M, Hong SB et al. Identification and characterization of a novel folliculin-interacting protein FNIP2. *Gene.* 2008;415:60–7.

14. Khoo SK, Giraud S, Kahnoski K et al. Clinical and genetic studies of Birt-Hogg-Dube syndrome. *J Med Genet.* 2002;39:906–12.

15. Hong S-, Oh H, Valera VA, Baba M, Schmidt LS, Linehan WM. Inactivation of the FLCN tumor suppressor gene induces TFE3 transcriptional activity by increasing its nuclear localization. *PLOS ONE* 2010;5.

16. Davis CF, Ricketts CJ, Wang M et al. The somatic genomic landscape of chromophobe renal cell carcinoma. *Cancer Cell.* 2014;26:319–30.

17. Hartman TR, Nicolas E, Klein-Szanto A et al. The role of the Birt-Hogg-Dube protein in mTOR activation and renal tumorigenesis. *Oncogene.* 2009;28:1594–604.

18. Hudon V, Sabourin S, Dydensborg AB et al. Renal tumour suppressor function of the Birt-Hogg-Dube syndrome gene product folliculin. *J Med Genet.* 2010;47:182–9.

19. Cash TP, Gruber JJ, Hartman TR, Henske EP, Simon MC. Loss of the Birt-Hogg-Dube tumor suppressor results in apoptotic resistance due to aberrant TGFbeta-mediated transcription. *Oncogene.* 2011;30:2534–46.

20. Laviolette LA, Mermoud J, Calvo IA et al. Negative regulation of EGFR signalling by the human folliculin tumour suppressor protein. *Nat Commun.* 2017;8:15866.

21. Klomp JA, Petillo D, Niemi NM et al. Birt-Hogg-Dube renal tumors are genetically distinct from other renal neoplasias and are associated with up-regulation of mitochondrial gene expression. *BMC Med Genomics* 2010;3:59,8794–3-59.

22. Petit CS, Roczniak-Ferguson A, Ferguson SM. Recruitment of folliculin to lysosomes supports the amino acid-dependent activation of Rag GTPases. *J Cell Biol.* 2013;202:1107–22.

23. Van Steensel MAM, Verstraeten VLRM, Frank J et al. Novel mutations in the BHD gene and absence of loss of heterozygosity in fibrofolliculomas of Birt–Hogg–Dubé patients. *J Invest Dermatol.* 2007;127:588–93.

24. Goncharova EA, Goncharov DA, James ML et al. Folliculin controls lung alveolar enlargement and epithelial cell survival through E-cadherin, LKB1, and AMPK. *Cell Rep* 2014;7:412–23.

25. Medvetz DA, Khabibullin D, Hariharan V et al. Folliculin, the product of the Birt-Hogg-Dube tumor suppressor gene, interacts with the adherens junction protein p0071 to regulate cell-cell adhesion. *PLOS ONE* 2012;7:e47842.

26. Khabibullin D, Medvetz DA, Pinilla M et al. Folliculin regulates cell-cell adhesion, AMPK, and mTORC1 in a cell-type-specific manner in lung-derived cells. *Physiol Rep* 2014;2. doi:10.14814/phy2.12107.

27. Johannesma PC, Houweling AC, van Waesberghe JH et al. The pathogenesis of pneumothorax in Birt-Hogg-Dube syndrome: A hypothesis. *Respirology.* 2014;19:1248–50.

28. Kennedy JC, Khabibullin D, Henske EP. Mechanisms of pulmonary cyst pathogenesis in Birt-Hogg-Dube syndrome: The stretch hypothesis. *Semin Cell Dev Biol.* 2016;52:47–52.

29. BHD Foundation. (https://www.bhdsyndrome.org/for-researchers/what-is-bhd/introduction/published-bhd-families/). Accessed 02 October, 2017.

30. Menko FH, van Steensel MA, Giraud S et al. Birt-Hogg-Dube syndrome: Diagnosis and management. *Lancet Oncol.* 2009;10:1199–206.

31. Johannesma PC, Van Den Borne B, Nagelkerke A et al. Clinical cases; spontaneous pneumothorax at the age of 14. Radiological evidence of birt-hogg-dube syndrome. *Am J Respir Crit Care Med.* 2014;189.

32. Schneider M, Dinkelborg K, Xiao X et al. Early onset renal cell carcinoma in an adolescent girl with germline FLCN exon 5 deletion. *Fam Cancer.* 2018; 17:135–9.

33. Geilswijk M, Bendstrup E, Madsen MG, Sommerlund M, Skytte AB. Childhood pneumothorax in Birt-Hogg-Dube syndrome: A cohort study and review of the literature. *Mol Genet Genomic Med* 2018;6(3):332–8.

34. Toro JR, Pautler SE, Stewart L et al. Lung cysts, spontaneous pneumothorax, and genetic associations in 89 families with Birt-Hogg-Dube syndrome. *Am J Respir Crit Care Med.* 2007;175:1044–53.

35. Kunogi M, Kurihara M, Ikegami TS et al. Clinical and genetic spectrum of Birt-Hogg-Dube syndrome patients in whom pneumothorax and/or multiple lung cysts are the presenting feature. *J Med Genet.* 2010;47:281–7.

36. Del Rosso JQ, Silverberg N, Zeichner JA. When acne is not acne. *Dermatol Clin.* 2016;34:225–8.

37. Adley BP, Smith ND, Nayar R, Yang XJ. Birt–Hogg–Dubé syndrome: Clinicopathologic findings and genetic alterations. *Arch Pathol Lab Med.* 2006;130:1865–70.

38. Aivaz O, Berkman S, Middelton L, Linehan WM, DiGiovanna JJ, Cowen EW. Comedonal and cystic fibrofolliculomas in Birt-Hogg-Dube syndrome. *JAMA Dermatol* 2015;151:770–4.

39. Schaffer JV, Gohara MA, McNiff JM, Aasi SZ, Dvoretzky I. Multiple facial angiofibromas: A cutaneous manifestation of Birt-Hogg-Dube syndrome. *J Am Acad Dermatol.* 2005;53:S108–11.

40. Sattler EC, Ertl-Wagner B, Pellegrini C et al. Cutaneous melanoma in Birt-Hogg-Dube syndrome: Part of the clinical spectrum? *Br J Dermatol.* 2018;178:e132–3.

41. Kazakov DV. Brooke-Spiegler syndrome and phenotypic variants: An update. *Head Neck Pathol* 2016;10:125–30.

42. Toro JR, Wei MH, Glenn GM et al. BHD mutations, clinical and molecular genetic investigations of Birt-Hogg-Dube syndrome: A new series of 50 families and a review of published reports. *J Med Genet.* 2008;45:321–31.

43. Agarwal PP, Gross BH, Holloway BJ, Seely J, Stark P, Kazerooni EA. Thoracic CT findings in Birt–Hogg–Dubé syndrome. *Am J Roentgenol* 2011;196:349–52.

44. Gupta N, Kopras EJ, Henske EP et al. Spontaneous pneumothoraces in patients with Birt-Hogg-Dube syndrome. *Ann Am Thorac Soc* 2017;14:706–13.

45. Tomassetti S, Carloni A, Chilosi M et al. Pulmonary features of Birt-Hogg-Dube syndrome: Cystic lesions and pulmonary histiocytoma. *Respir Med.* 2011;105:768–74.

46. Rehman HU. Birt–Hogg–Dubé syndrome: Report of a new mutation. *Can Respir J.* 2012;19:193–5.

47. Bessis D, Giraud S, Richard S. A novel familial germline mutation in the initiator codon of the BHD gene in a patient with Birt–Hogg–Dubé syndrome. *Br J Dermatol.* 2006;155:1067–9.

48. Onuki T, Goto Y, Kuramochi M et al. Radiologically indeterminate pulmonary cysts in Birt–Hogg–Dubé syndrome. *Ann Thorac Surg.* 2014;97:682–5.

49. Tobino K, Hirai T, Johkoh T et al. Differentiation between Birt–Hogg–Dubé syndrome and lymphangioleiomyomatosis: Quantitative analysis of pulmonary cysts on computed tomography of the chest in 66 females. *Eur J Radiol.* 2012;81:1340–6.

50. Ayo DS, Aughenbaugh GL, Yi ES, Hand JL, Ryu JH. Cystic lung disease in Birt–Hogg–Dubé syndrome. *Chest.* 2007;132:679–84.

51. Kumasaka T, Hayashi T, Mitani K et al. Characterization of pulmonary cysts in Birt–Hogg–Dubé syndrome: Histopathological and morphometric analysis of 229 pulmonary cysts from 50 unrelated patients. *Histopathology.* 2014;65:100–10.

52. Israel GM, Bosniak MA. An update of the Bosniak renal cyst classification system. *Urology.* 2005;66:484–8.

53. American Cancer Society. (http://www.cancer.org). Accessed 02 February, 2018.

54. Hasumi H, Baba M, Hasumi Y, Furuya M, Yao M. Birt–Hogg–Dubé syndrome: Clinical and molecular aspects of recently identified kidney cancer syndrome. *Int J Urol.* 2016;23:204–10.

55. Stamatakis L, Metwalli AR, Middelton LA, Marston Linehan W. Diagnosis and management of BHD-associated kidney cancer. *Fam Cancer.* 2013;12:397–402.

56. Pavlovich CP, Walther MM, Eyler RA et al. Renal tumors in the Birt-Hogg-Dube syndrome. *Am J Surg Pathol.* 2002;26:1542–52.

57. Benusiglio PR, Giraud S, Deveaux S et al. Renal cell tumour characteristics in patients with the Birt-Hogg-Dube cancer susceptibility syndrome: A retrospective, multicentre study. *Orphanet J Rare Dis.* 2014;9:163.

58. Schmidt LS, Nickerson ML, Warren MB et al. Germline BHD-mutation spectrum and phenotype analysis of a large cohort of families with Birt-Hogg-Dube syndrome. *Am J Hum Genet.* 2005;76:1023–33.

59. Srigley JR, Delahunt B, Eble JN et al. The International Society of Urological Pathology (ISUP) Vancouver classification of renal neoplasia. *Am J Surg Pathol.* 2013;37:1469–89.

60. Kuroda N, Furuya M, Nagashima Y et al. Review of renal tumors associated with Birt-Hogg-Dube syndrome with focus on clinical and pathobiological aspects. *Pol J Pathol.* 2014;65:93–9.

61. Przybycin CG, Magi-Galluzzi C, McKenney JK. Hereditary syndromes with associated renal neoplasia: A practical guide to histologic recognition in renal tumor resection specimens. *Adv Anat Pathol.* 2013;20:245–63.

62. Pavlovich CP, Grubb III RL, Hurley K et al. Evaluation and management of renal tumors in the Birt–Hogg–Dubé syndrome. *J Urol.* 2005;173:1482–6.

63. Nahorski MS, Lim DH, Martin L et al. Investigation of the Birt-Hogg-Dube tumour suppressor gene (FLCN) in familial and sporadic colorectal cancer. *J Med Genet.* 2010;47:385–90.

64. Maffe A, Toschi B, Circo G et al. Constitutional FLCN mutations in patients with suspected Birt-Hogg-Dube syndrome ascertained for non-cutaneous manifestations. *Clin Genet.* 2011;79:345–54.

65. Pradella LM, Lang M, Kurelac I et al. Where Birt–Hogg–Dubé meets Cowden syndrome: Mirrored genetic defects in two cases of syndromic oncocytic tumours. *Eur J Hum Genet.* 2013;21:1169–72.

66. Benusiglio PR, Gad S, Massard C et al. Case report: Expanding the tumour spectrum associated with the Birt-Hogg-Dubé cancer susceptibility syndrome. *F1000 Res* 2014;3.

67. Hao S, Long F, Sun F, Liu T, Li D, Jiang S. Birt–Hogg–Dubé syndrome: A literature review and case study of a Chinese woman presenting a novel FLCN mutation. *BMC Pulm Med.* 2017;17.

68. Dong L, Gao M, Hao WJ et al. Case report of Birt-Hogg-Dube syndrome: Germline mutations of FLCN detected in patients with renal cancer and thyroid cancer. *Medicine (Baltim).* 2016;95:e3695.

69. Kapoor R, Evins AI, Steitieh D, Bernardo A, Stieg PE. Birt-Hogg-Dube syndrome and intracranial vascular pathologies. *Fam Cancer.* 2015;14:595–7.

70. Matsutani N, Dejima H, Takahashi Y et al. Birt-Hogg-Dube syndrome accompanied by pulmonary arteriovenous malformation. *J Thorac Dis* 2016;8:E1187–9.

71. Palmirotta R, Savonarola A, Ludovici G et al. Association between Birt Hogg Dubé syndrome and cancer predisposition. *Anticancer Res.* 2010;30:751–8.

72. Mota-Burgos A, Acosta EH, Marquez FV, Mendiola M, Herrera-Ceballos E. Birt-Hogg-Dube syndrome in a patient with melanoma and a novel mutation in the FCLN gene. *Int J Dermatol.* 2013;52:323–6.

73. Fontcuberta IC, Salomão DR, Quiram PA, Pulido JS. Choroidal melanoma and lid fibrofoliculomas in Birt–Hogg–Dubé syndrome. *Ophthalmic Genet.* 2011;32:143–6.

74. Marous CL, Marous MR, Welch RJ, Shields JA, Shields CL. Choroidal melanoma, zector melanocytosis, and retinal pigment epithelial microdetachments in Birt-Hogg-Dube syndrome. *Retin Cases Brief Rep* 2017;13(3):202–6.

75. Chung JY, Ramos-Caro FA, Beers B, Ford MJ, Flowers F. Multiple lipomas, angiolipomas, and parathyroid adenomas in a patient with Birt-Hogg-Dube syndrome. *Int J Dermatol.* 1996;35:365–7.

76. Whitworth J, Stausbøl-Grøn B, Skytte A. Genetically diagnosed Birt–Hogg–Dubé syndrome and familial cerebral cavernous malformations in the same individual: A case report. *Fam Cancer.* 2017;16:139–42.

77. Menko FH, Johannesma PC, van Moorselaar RJ et al. A de novo FLCN mutation in a patient with spontaneous pneumothorax and renal cancer; a clinical and molecular evaluation. *Fam Cancer.* 2013;12:373–9.

78. Toro JR. Birt-Hogg-Dube syndrome. In: Pagon RA, Adam MP, Ardinger HH et al. editors. *GeneReviews(R).* Seattle (WA): University of Washington, Seattle. GeneReviews is a registered trademark of the University of Washington, Seattle. All rights reserved; 1993.

79. Jacob CI, Dover JS. Birt-Hogg-Dube syndrome: Treatment of cutaneous manifestations with laser skin resurfacing. *Arch Dermatol.* 2001;137:98–9.

80. Gambichler T, Wolter M, Altmeyer P, Hoffman K. Treatment of Birt-Hogg-Dube syndrome with erbium:YAG laser. *J Am Acad Dermatol.* 2000;43:856–8.

81. Pritchard SE, Mahmoudizad R, Parekh PK. Successful treatment of facial papules with electrodessication in a patient with Birt-Hogg-Dube syndrome. *Dermatol Online J.* 2014;20.

82. Gijezen LMC, Vernooij M, Martens H et al. Topical rapamycin as a treatment for fibrofolliculomas in Birt–Hogg–Dubé syndrome: A double-blind placebo-controlled randomized split-face trial. *PLOS ONE* 2014;9.

83. Schmidt LS. Birt–Hogg–Dubé syndrome: From gene discovery to molecularly targeted therapies. *Fam Cancer.* 2013;12:357–64.

84. Gupta N, Sunwoo BY, Kotloff RM. Birt-Hogg-Dube Syndrome. *Clin Chest Med.* 2016;37:475–86.

85. Gupta N, Seyama K, McCormack FX. Pulmonary manifestations of Birt-Hogg-Dube syndrome. *Fam Cancer.* 2013;12:387–96.

86. Wehle MJ, Thiel DD, Petrou SP, Young PR, Frank I, Karsteadt N. Conservative management of incidental contrast-enhancing renal masses as safe alternative to invasive therapy. *Urology.* 2004;64:49–52.

87. Pierorazio PM, Johnson MH, Ball MW et al. Five-year analysis of a multi-institutional prospective clinical trial of delayed intervention and surveillance for small renal masses: The DISSRM registry. *Eur Urol.* 2015;68:408–15.

88. Mason RJ, Abdolell M, Trottier G et al. Growth kinetics of renal masses: Analysis of a prospective cohort of patients undergoing active surveillance. *Eur Urol.* 2011;59:863–7.

89. Pierorazio PM, Johnson MH, Patel HD et al. Management of renal masses and localized renal cancer: Systematic review and meta-analysis. *J Urol.* 2016;196:989–99.

90. Zargar H, Atwell TD, Cadeddu JA et al. Cryoablation for small renal masses: Selection criteria, complications, and functional and oncologic results. *Eur Urol.* 2016;69:116–28.

91. Ljungberg B, Bensalah K, Canfield S et al. EAU guidelines on renal cell carcinoma: 2014 update. *Eur Urol.* 2015;67:913–24.

92. Rodriguez Faba O, Akdogan B, Marszalek M et al. Current status of focal cryoablation for small renal masses. *Urology.* 2016;90:9–15.

93. Campbell S, Uzzo RG, Allaf ME et al. Renal mass and localized renal cancer: AUA guideline. *J Urol.* 2017;198:520–9.

94. Schmidt LS, Linehan WM. Molecular genetics and clinical features of Birt-Hogg-Dube syndrome. *Nat Rev Urol* 2015;12:558–69.

95. Wile GE, Leyendecker JR, Krehbiel KA, Dyer RB, Zagoria RJ. CT and MR imaging after imaging-guided thermal ablation of renal neoplasms. *Radiographics.* 2007;27:325,39; discussion 339–40.

96. Kowalczyk KJ, Hooper HB, Linehan WM, Pinto PA, Wood BJ, Bratslavsky G. Partial nephrectomy after previous radio frequency ablation: The National Cancer Institute experience. *J Urol.* 2009;182:2158–63.

97. Yang B, Autorino R, Remer EM et al. Probe ablation as salvage therapy for renal tumors in von Hippel-Lindau patients: The Cleveland Clinic experience with 3 years follow-up. *Urol Oncol.* 2013;31:686–92.

98. Wofford J, Fenves AZ, Jackson JM, Kimball AB, Menter A. The spectrum of nephrocutaneous diseases and associations: Genetic causes of nephrocutaneous disease. *J Am Acad Dermatol.* 2016;74:231,44; quiz 245–6.

99. Johannesma PC, van de Beek I, van der Wel JW et al. Risk of spontaneous pneumothorax due to air travel and diving in patients with Birt-Hogg-Dube syndrome. *Springerplus* 2016;5:1506.

100. MacDuff A, Arnold A, Harvey J, BTS Pleural Disease Guideline Group. Management of spontaneous pneumothorax: British Thoracic Society pleural disease guideline 2010. *Thorax.* 2010;65(Suppl 2):ii18–31.

101. Young LR, Vandyke R, Gulleman PM et al. Serum vascular endothelial growth factor-D prospectively distinguishes lymphangioleiomyomatosis from other diseases. *Chest.* 2010;138:674–81.

27

Bohring–Opitz Syndrome

Dongyou Liu

CONTENTS

27.1 Introduction

Bohring–Opitz syndrome (BOS) is a rare, autosomal dominant disorder that displays typical craniofacial features (microcephaly, trigonocephaly, prominent metopic ridge, retrognathia, prominent eyes with hypoplastic supraorbital ridges, upslanting palpebral fissures, depressed nasal bridge, anteverted nares, low-set and posteriorly rotated ears, palatal abnormalities and broad alveolar ridges, flammeus nevus, low anterior hairline), distinct posture, joint abnormalities, intrauterine growth restriction (IUGR), short stature, abnormal tone, severe intellectual disability, feeding problems, small size at birth, susceptibility to infections, failure to thrive, and high infant mortality as well as tumor risk. Molecularly, approximately 75% of BOS cases are linked to germline mutations in the additional sex combs–like 1 (*Asxl1*) gene located on chromosome 20q11.21 [1].

27.2 Biology

BOS (also referred to as Oberklaid–Danks syndrome, Opitz trigonocephaly-like syndrome, Opitz C syndrome, C-like syndrome, and Bohring syndrome) is a malformation disorder, which was first described by Bohring et al. in 1999 from four patients who presented with a prominent metopic suture, hypertelorism, exophthalmos, cleft lip and palate, and limb anomalies. The clinical spectrum of BOS has subsequently expanded to include nevus flammeus of the face, upslanting palpebral fissures, hirsutism, trigonocephaly, prominent metopic suture,

severe intrauterine growth retardation, poor feeding, profound mental retardation, exophthalmos, flexion of the elbows and wrists with deviation of the wrists, and metacarpophalangeal joints, in addition to tumor risk [2,3]. The identification of de novo frameshift and nonsense mutations in the *Asxl1* gene on chromosome 20q11.21 in 2011 suggests that loss of function in this gene underscores the molecular mechanism of BOS [4].

BOS appears to be distinct from C syndrome (a phenotypically similar disorder caused by heterozygous mutation in the CD96 gene on chromosome 3q13), Shashi–Pena syndrome (due to *Asxl2* gene mutation on chromosome 2p23.3; with the presence of macrocephaly and episodic hypoglycemia), and Bainbridge–Ropers syndrome (due to de novo heterozygous nonsense and frameshift mutations mutation in the *ASXL3* gene on chromosome 18q12.1) [5]. While Bainbridge–Ropers syndrome shows severe psychomotor retardation, feeding problems, severe postnatal growth retardation, arched eyebrows, anteverted nares, and ulnar deviation of the hands as well as anteverted nares), it differs from BOS by having no "BOS posture" of elbow and wrist flexion, myopia or trigonocephaly (Figure 27.1 and Table 27.1) [6].

27.3 Pathogenesis

Initially noted by its role in the activation and silencing of homeobox (*Hox*) genes via binding to chromatin and regulating ubiquitination of specific histones (e.g., histone H2A) in *Drosophila*, the additional sex combs (*Asx*) gene encodes the Asx protein, which belongs to the enhancer of Trithorax and Polycomb group, with the ability to modify chromatin structures and thus control the

FIGURE 27.1 Clinical presentation of a 41.5-month-old girl with Bainbridge—Ropers syndrome (*ASXL3* mutation) includes prominent tall forehead, arched eyebrows with subtle synophrys and periorbital fullness, prominent columella with hypoplastic alae nasi, thin upper lip, and borderline low-set ears. However, trigonocephaly and prominent metopic ridge, which usually occur in Boehring—Opitz syndrome patients (*ASXL1* mutation), are absent. (Photo credit: Bainbridge et al. *Genome Med.* 2013;5(2):11.)

active/repressive transcriptional states and significantly impact on various biological processes.

In humans, the *Asx* gene consists of three orthologs, i.e., the additional sex combs-like 1, 2, and 3 genes (*Asxl1, Asxl2, Asxl3*), which contain 13, 13, and 12 exons, spanning 81, 144, and 172 kb on chromosomes 20q11.21, 2p23.3, and 18q12.1, respectively, and encode putative polycomb proteins Asxl1 (1,541 aa, 165 kDa), Asxl2 (1,435 aa, 153 kDa), and Asxl3 (2,248 aa, 242 kDa) that share conserved domains, including N-terminal ASXN, ASXH domains, and a C-terminal plant homeodomain (Table 27.1). The size differences among Asxl1 (1,541 aa), Asxl2 (1,435 aa), and Asxl3 (2,248 aa) reflect the different lengths of the protein-coding parts of the two last exons of the genes, i.e., exons 11 (1,957 bp) and 12 (3,708 bp) in *Asxl3* versus exons 12 (634 and 718 bp) and 13 (2,907 and 2,448 bp) of *Asxl1* and *Asxl2*, respectively. Through formation of complexes with other proteins and recruitment of histone H3 in specific cell types and deubiquitination of specific histone H2A, *Asxl1, Asxl2*, and *Asxl3* act as histone methyltransferases and partake in epigenetic regulation by activating or repressing the transcription of genes involved in either differentiation or proliferation [3].

Specifically, as the obligate regulatory subunit of a deubiquitinase complex whose catalytic subunit is BAP1 (BRCA1-associated protein 1), Asxl1 demonstrates multifaceted functions, both as an essential cofactor for the histone H2A deubiquitinase BAP1 and as a critical mediator of the function of polycomb repressive complex 2 (PRC2), and provides a novel way to maintain normal sister chromatid separation and to regulate gene expression by assembling epigenetic regulators and transcription factors to specific gene loci. There is evidence that Asxl1 plays a regulating role in the self-renewal and differentiation of bone marrow stromal cells (BMSC) and hematopoietic stem/progenitor cells (HSC/HPC) in the bone marrow (BM) [3].

The involvement of the *Asxl* gene in BOS was confirmed in 2011, after identification of de novo heterozygous frameshift or nonsense mutations in the *Asxl1* gene from patients with BOS, some of whom have Wilms tumor [4]. Asxl1 truncating mutations appear to confer loss-of-function on the Asxl-BAP1 complex [7,8]. While germline *Asxl1* mutations lead to BOS-like phenotypes (e.g., severe developmental defects, early childhood mortality), somatic *Asxl1* mutations are associated with non-heritable myelodysplastic syndrome (MDS)-like disease [9,10].

TABLE 27.1

Characteristics of Additional Sex Combs–Like (*Asxl*) Gene and Related Diseases

Gene	Chromosome Location	Gene Length (kb)	No. of Exons	Encoded Protein	Related Disease	Key Clinical Features
Asxl1	20q11.21	81	13	Asxl1 (1,541 aa, 165 kDa)	Bohring–Opitz syndrome or myeloid malignancy	Germline *Asxl1* mutation: presence of microcephaly, distinctive facial features, BOS posture, myopia, nevus flammeus, seizures, apnea, and recurrent infections; risk of Wilms tumor and medulloblastoma Somatic *Asxl1* mutation: risk of myeloid malignancies
Asxl2	2p23.3	144	13	Asxl2 (1,435 aa, 153 kDa)	Shashi–Pena syndrome	*Asxl2* gene mutation: presence of macrocephaly and episodic hypoglycemia; no obvious tumor risk
Asxl3	18q12.1	172	12	Asxl3 (2,248 aa, 242 kDa)	Bainbridge–Ropers syndrome	*Asxl3* gene mutation: absence of distinctive facial features, BOS posture, myopia, nevus flammeus, seizures, apnea, and recurrent infections; no obvious tumor risk

Indeed, several myeloid malignancies (e.g., MDS, chronic myelo-monocytic leukemia, and acute myeloid leukemia) have been detected in elderly patients who harbor somatic *Asxl1* alterations. In addition, BMSC derived from chronic myelomonocytic leukemia patients show lower expression of *Asxl1*, leading to depletion of total H2AK119Ub by ~90% and reduction of H3K27me3 levels by ~50% in myeloid progenitor cells, impairs HSC/HPC pool, and skews cell differentiation with a bias to granulocytic/monocytic lineage and thus myeloid differentiation [11–13]. In experimental mouse model, *Asxl1* global loss as well as conditional deletion in osteoblasts and their progenitors contributes to significant bone loss and a markedly decreased number of BMSC compared with wild-type mice. The *Asxl1*(−/−) BMSC also show impaired self-renewal and skewed differentiation, away from osteoblasts to adipocytes, highlighting that Asxl1 negatively regulates adipogenesis. Reintroduction of *Asxl1* into *Asxl1*(−/−) BMSC appears to help restore their self-renewal capacity and lineage commitment [14,15].

27.4 Epidemiology

27.4.1 Prevalence

BOS is a rare disease with unknown prevalence. To date, about 60 clinically diagnosed cases have been described in the literature, of which approximately 75% are shown to harbor mutations in the *Asxl1* gene. Females appear to be more often affected that males, and up to 50% of BOS patients often die before the age of 2 years.

27.4.2 Inheritance

BOS phenotype appears to be inherited in an autosomal dominant pattern.

27.4.3 Penetrance

BOS shows a complete penetrance, as children harboring *Asxl1* pathogenic variants are usually clinically evident (showing typical craniofacial features and posture, growth failure, intellectual disability, and other anomalies). Furthermore, many (up to 40%) elderly who have somatic *Asxl1* mutations develop myeloid malignancies in the absence of typical BOS symptoms.

27.5 Clinical Features

Clinically, BOS is associated with distinctive craniofacial features and posture, growth failure, intellectual disability, and other anomalies [16–19].

Craniofacial features include microcephaly (100%), wide prominent forehead (89%), trigonocephaly (due to prominent metopic ridge), glabellar and eyelid nevus flammeus (simplex, 89%), synophrys, hypotonic face with full cheeks, prominent or proptotic globes (100%) with high myopia (87%), micro/retrognathia (89%), hypertrichosis with rapidly growing hair and nails (89%), cleft lip with or without cleft palate, broad alveolar ridges or high narrow palate (87%), upslanting palpebral fissures

(67%), low-set ears with increased posterior angulation (67%), low posterior hairline (67%), widely set eyes (hypertelorism, 55%), depressed and wide nasal bridge (50%), and anteverted nares (50%).

Skeletal features are exemplified by typical BOS posture (i.e., flexion at the elbows with ulnar deviation, and flexion of the wrists and metacarpophalangeal joints), which is most striking in early childhood but may become less apparent with age. Other skeletal changes include truncal hypotonia with hypertonia of the extremities, congenital contractures (60%), hip and radial head dislocations (33%), and pectus excavatum.

Feeding difficulties (due to cyclic vomiting with possible poor gastric motility) may result in failure to thrive (100%), deep palmar creases (57%), aspiration, and dehydration. Affected children often require G-tubes, and feeding may improve with age.

Developmental delays (100%) include IUGR (83%), poor growth in the first year of life (due to chronic emesis and feeding intolerance), speech and walking difficulty, happy and pleasant demeanor.

Neurologic features consist of seizures (which are common and typically responsive to standard epileptic medications), and primary brain anomalies (e.g., slender corpus callosum and brainstem, Dandy–Walker malformation, delayed myelination, and enlarged ventricles).

Other anomalies include cardiovascular (e.g., arrhythmia, idiopathic and transient bradycardia, apnea, atrial septal defects, patent ductus arteriosus, and cardiac hypertrophy), renal (e.g., renal stone), genital, respiratory (e.g., tongue-based airway obstruction) and ophthalmologic (e.g., myopia, strabismus, colobomas, retinal and optic nerve atrophy, and abnormal coloration of the retinas) abnormalities. Recurrent infections (62%) may cause infant mortality (40%) and usually decline with age.

Malignancy: Children with BOS appear to have a higher risk for Wilms tumor than the general population. Medulloblastoma may also occur in BOS patients. In addition, up to 40% of the elderly who harbor somatic *Asxl1* mutations without showing BOS phenotype are found to have myeloid malignancies (e.g., MDS, chronic myelomonocytic leukemia, and acute myeloid leukemia) [3,11].

27.6 Diagnosis

Patients showing the following features are suggestive of BOS: (i) craniofacial (microcephaly or trigonocephaly/prominent metopic ridge; glabellar and eyelid nevus flammeus; prominent globes; cleft lip; palatal anomalies: cleft palate, high-arched palate, or prominent palatine ridges; micrognathia and/or retrognathia); (ii) growth and feeding (BOS posture; IUGR; severe feeding difficulties with chronic emesis; poor postnatal weight gain and linear growth); (iii) neurologic (global developmental delay or intellectual disability; seizures); (iv) respiratory (recurrent respiratory infections); (v) sleep (sleep disturbance; obstructive sleep apnea); and (vi) ophthalmologic (high myopia; variable optic nerve and retinal anomalies). Further confirmation of BOS is through molecular identification of a constitutional heterozygous pathogenic variant (e.g., small intragenic deletion/insertion, missense, nonsense, and splice site mutation) in *Asxl1*, which is present in about 75% of BOS patients using sequence analysis.

Interestingly, BOS patients with an identifiable *Asxl1* mutation have a higher incidence of myopia (87% vs 40%) and hypertrichosis (89% vs 17%) compared to those without identified mutations [1].

It should be noted that somatic mosaicism for *Asxl1* variants may be found in the elderly BOS patients or in other non-BOS cohorts (e.g., elderly who have myeloid malignancies) [3].

Differential diagnoses for BOS consist of several disorders that display overlapping features with BOS, such as C syndrome (Opitz trigonocephaly syndrome due to mutation in the *CD96* or *TACTILE* gene on chromosome 3q13 or the *IFT140* gene; presence of trigonocephaly, severe mental retardation, hypotonia, variable cardiac defects, redundant skin, and dysmorphic facial features, including upslanted palpebral fissures, epicanthal folds, depressed nasal bridge, and low-set, posteriorly rotated ears, but absence of nevus flammeus over glabella, BOS posture, high myopia), Shashi–Pena syndrome (*Asxl2* syndrome due to *Asxl2* mutation; presence of macrocephaly and episodic hypoglycemia), Bainbridge–Ropers syndrome (*Asxl3* syndrome due to *Asxl3* mutation; absence of BOS posture, myopia, nevus flammeus, distinctive constellation of facial features, seizures, apnea, and recurrent infections), Cornelia de Lange syndrome (CdLS, due to mutations in the *HDAC8*, *NIPBL*, *RAD21*, *SMC1A*, or *SMC3* genes; presence of small hands and feet), *KLHL7*-associated syndrome (Crisponi syndrome/cold-induced sweating syndrome type 1-like syndrome, an autosomal recessive disorder due to *KLHL7* mutation; absence of high myopia and retinal/optic nerve atrophy; occurrence of retinitis pigmentosa in childhood, hyperthermia, and oropharyngeal muscle contraction), Smith–Lemli–Opitz syndrome (due to mutations in the *DHCR7* gene; presence of distinctive facial features, microcephaly, intellectual disability/behavioral problems, hypotonia [weak muscle tone], feeding difficulties, syndactyly [fused second and third toes], and polydactyly [extra fingers or toes]), desmosterolosis (due to mutations in the *DHCR24* gene; brain abnormalities/developmental delay, delayed speech and motor skills, spasticity [muscle stiffness], arthrogryposis [stiff, rigid joints], short stature, abnormal head size, micrognathia [small lower jaw], cleft palate [opening in the roof of the mouth], nystagmus [involuntary eye movements], strabismus [eyes that do not look in the same direction], heart defects, and seizures), and lathosterolosis (due to mutation in SC5D gene; presence of multiple congenital anomalies, mental retardation, and liver disease) [20–22].

27.7 Treatment

Treatment options for BOS are largely symptom based, ranging from identification and avoidance of triggers (e.g., vaccine, infection, and anesthesia), daily maintenance medication (e.g., cyproheptadine), and early abortive treatment (e.g., lorazepam, ondansetron, and acetaminophen or some combination of an antiemetic, pain reliever, and sedative) for cyclic vomiting; nasogastric tube or percutaneous endoscopic gastrostomy tube (G-tube or GJ-tube) with advanced formula containing vitamin supplementation for aspiration, chronic severe emesis and malnutrition; polysomnography for obstructive sleep apnea; tracheostomy for patients with recurrent aspiration who develop secondary lung disease; noninvasive pressure support (e.g., CPAP, BiPAP) or surgical intervention (e.g., mandibular distraction) for patients with cleft lip or palate, micrognathia, or obstructive sleep apnea; sodium valproate and other antiepileptic medications for control of seizures; intravenous antibiotic therapy and assisted ventilation for recurrent lung infections; other measures for congenital heart defects, intellectual disability, myopia, urinary tract infections, urinary retention, and renal stones [23].

Management of global developmental disability, intellectual disability and/or motor dysfunction in BOS affected children is complex and age-dependent, involving occupational, physical, speech, and feeding therapies.

27.8 Prognosis and Prevention

Although BOS is rare, many (40%) affected children die early (from ages 23 hours to 6 years) due to feeding difficulty and respiratory infections [1].

Prevention of primary manifestations of BOS requires adequate treatment of severe emesis, regular surveillance (e.g., renal ultrasound every 3 months from birth to age 8 to screen for the development of Wilms tumor; frequent monitoring of growth and development; close monitoring of feeding intolerance; regular follow-up for vision optimization), and agent/circumstance avoidance (e.g., triggers for vomiting) [1].

As the vast majority of BOS occurs as the result of a de novo variant in *Asxl1*, the absence of an *Asxl1* pathogenic variant in the leukocyte DNA of either parent (suggestive of parental germline mosaicism) portend a low recurrence risk to sibs (~1%) [1].

Nonetheless, molecular genetic testing is useful to evaluate a pregnancy at theoretically increased risk due to constitutional and/or germline mosaicism for an *Asxl1* pathogenic variant in a clinically unaffected parent.

27.9 Conclusion

Bohring–Opitz syndrome (BOS) is a rare genetic disorder characterized by distinctive craniofacial appearance, abnormal posture, feeding difficulties, severe developmental delay and intellectual disability, failure to thrive, and early mortality (as a consequence of unexplained bradycardia, obstructive apnea, or pulmonary infections). Of about 60 BOS cases described to date, 75% are shown to contain de novo mutations in the *Asxl1* gene. While germline *Asxl1* mutations are linked to BOS phenotype, somatic *Asxl1* mutations disrupt hematopoiesis, impair self-renewal capacity, and skew differentiation away from osteoblasts to adipocytes, leading to myeloid malignancies (particularly in the elderly) without BOS symptoms. In the absence of effective therapy, there is an obvious need to further uncover the molecular mechanisms of BOS, with the goal of developing novel countermeasures against this genetic condition.

REFERENCES

1. Russell B, Tan WH, Graham JM Jr. Bohring-Opitz Syndrome. In: Adam MP, Ardinger HH, Pagon RA et al., editors. *GeneReviews®* [Internet]. Seattle (WA): University of Washington, Seattle; 1993–2018. 2018 Feb 15.

2. Russell B, Graham JM Jr. Expanding our knowledge of conditions associated with the ASXL gene family. *Genome Med.* 2013;5(2):16.

3. Dangiolo SB, Wilson A, Jobanputra V, Anyane-Yeboa K. Bohring-Opitz syndrome (BOS) with a new ASXL1 pathogenic variant: Review of the most prevalent molecular and phenotypic features of the syndrome. *Am J Med Genet A.* 2015;167A(12):3161–6.

4. Hoischen A, van Bon BW, Rodríguez-Santiago B et al. De novo nonsense mutations in ASXL1 cause Bohring-Opitz syndrome. *Nat Genet.* 2011;43(8):729–31.

5. Bainbridge MN, Hu H, Muzny DM et al. De novo truncating mutations in ASXL3 are associated with a novel clinical phenotype with similarities to Bohring-Opitz syndrome. *Genome Med.* 2013;5(2):11.

6. Kuechler A, Czeschik JC, Graf E et al. Bainbridge–Ropers syndrome caused by loss-of-function variants in ASXL3: A recognizable condition. *Eur J Hum Genet.* 2017;25(2):183–91.

7. Avila M, Kirchhoff M, Marle N et al. Delineation of a new chromosome 20q11.2 duplication syndrome including the ASXL1 gene. *Am J Med Genet A.* 2013;161A(7):1594–8.

8. Dinan AM, Atkins JF, Firth AE. ASXL gain-of-function truncation mutants: Defective and dysregulated forms of a natural ribosomal frameshifting product? *Biol Direct.* 2017;12(1):24.

9. Arunachal G, Danda S, Omprakash S, Kumar S. A novel de-novo frameshift mutation of the ASXL1 gene in a classic case of Bohring-Opitz syndrome. *Clin Dysmorphol.* 2016;25(3):101–5.

10. Carlston CM, O'Donnell-Luria AH, Underhill HR et al. Pathogenic ASXL1 somatic variants in reference databases complicate germline variant interpretation for Bohring-Opitz Syndrome. *Hum Mutat.* 2017;38(5):517–23.

11. Wang J, Li Z, He Y et al. Loss of Asxl1 leads to myelodysplastic syndrome-like disease in mice. *Blood.* 2014;123(4):541–53.

12. Balasubramani A, Larjo A, Bassein JA et al. Cancer-associated ASXL1 mutations may act as gain-of-function mutations of the ASXL1-BAP1 complex. *Nat Commun.* 2015;6:7307.

13. Zhang P, Chen Z, Li R et al. Loss of ASXL1 in the bone marrow niche dysregulates hematopoietic stem and progenitor cell fates. *Cell Discov.* 2018;4:4.

14. Abdel-Wahab O, Gao J, Adli M et al. Deletion of Asxl1 results in myelodysplasia and severe developmental defects *in vivo. J Exp Med.* 2013;210(12):2641–59.

15. Zhang P, Xing C, Rhodes SD et al. Loss of Asxl1 alters self-renewal and cell fate of bone marrow stromal cell, leading to Bohring-Opitz-like syndrome in mice. *Stem Cell Reports.* 2016;6(6):914–25.

16. Hastings RW, Newbury-Ecob R, Lunt PW. A case of probable Bohring-Opitz syndrome with medulloblastoma. *Clin Dysmorphol.* 2010;19(4):202–5.

17. Hastings R, Cobben JM, Gillessen-Kaesbach G et al. Bohring-Opitz (Oberklaid-Danks) syndrome: Clinical study, review of the literature, and discussion of possible pathogenesis. *Eur J Hum Genet.* 2011;19(5):513–9.

18. Magini P, Della Monica M, Uzielli ML et al. Two novel patients with Bohring-Opitz syndrome caused by de novo ASXL1 mutations. *Am J Med Genet A.* 2012;158A(4):917–21.

19. Visayaragawan N, Selvarajah N, Apparau H, Kamaru Ambu V. Bohring-opitz syndrome - A case of a rare genetic disorder. *Med J Malaysia.* 2017;72(4):248–9.

20. Hori I, Miya F, Ohashi K et al. Novel splicing mutation in the ASXL3 gene causing Bainbridge–Ropers syndrome. *Am J Med Genet A.* 2016;170(7):1863–7.

21. Urreizti R, Roca-Ayats N, Trepat J et al. Screening of CD96 and ASXL1 in 11 patients with Opitz C or Bohring-Opitz syndromes. *Am J Med Genet A.* 2016;170A(1):24–31.

22. Bruel AL, Bigoni S, Kennedy J et al. Expanding the clinical spectrum of recessive truncating mutations of *KLHL7* to a Bohring-Opitz-like phenotype. *J Med Genet.* 2017;54(12):830–5.

23. Russell B, Johnston JJ, Biesecker LG et al. Clinical management of patients with ASXL1 mutations and Bohring-Opitz syndrome, emphasizing the need for Wilms tumor surveillance. *Am J Med Genet A.* 2015;167A(9):2122–31.

28

Familial Testicular Germ Cell Tumor

Dongyou Liu

CONTENTS

28.1 Introduction

Primary tumors affecting the testis consist of six histological categories: (i) *germ cell tumors* (intratubular germ cell neoplasia unclassified [IGCNU], seminoma—seminoma with syncytiotrophoblastic cells, spermatocytic seminoma—spermatocytic seminoma with sarcoma, embryonal carcinoma, yolk sac tumor, trophoblastic tumors [choriocarcinoma, trophoblastic neoplasms other than choriocarcinoma—monophasic choriocarcinoma/placental site trophoblastic tumor, teratoma, dermoid cyst, monodermal teratoma, teratoma with somatic type malignancies], mixed embryonal carcinoma and teratoma, mixed teratoma and seminoma, choriocarcinoma and teratoma/embryonal carcinoma); (ii) *sex cord/gonadal stromal tumors—pure forms* (Leydig cell tumor, malignant Leydig cell tumor, Sertoli cell tumor (Sertoli cell tumor lipid rich variant, sclerosing Sertoli cell tumor, large cell calcifying Sertoli cell tumor), malignant Sertoli cell tumor, granulosa cell tumor (adult type granulosa cell tumor, juvenile type granulosa cell tumor), tumors of the thecoma/fibroma group (thecoma, fibroma), sex cord/gonadal stromal tumor, incompletely differentiated, sex cord/gonadal stromal tumors—mixed forms, malignant sex cord/gonadal stromal tumors, tumors containing both germ cell and sex cord/gonadal stromal elements (gonadoblastoma, germ cell-sex cord/gonadal stromal tumor—unclassified); (iii) *miscellaneous tumors of the testis* (carcinoid tumor, tumors of ovarian epithelial types, serous tumor of borderline malignancy, serous carcinoma, well-differentiated endometrioid carcinoma, mucinous cystadenoma, mucinous cystadenocarcinoma, Brenner tumor, nephroblastoma, paraganglioma); (iv) *hematopoietic tumors*; (v) *tumors of collecting ducts and rete* (adenoma, carcinoma); (vi) *tumors of paratesticular structures* (adenomatoid tumor, malignant mesothelioma, benign mesothelioma [well-differentiated papillary mesothelioma, cystic mesothelioma], adenocarcinoma of the epididymis, papillary cystadenoma of the epididymis, melanotic neuroectodermal tumor, desmoplastic small round cell tumor); and (vii) *mesenchymal tumors of the spermatic cord and testicular adnexae*. In addition, *secondary tumors* (mainly lymphomas that metastasize from other parts of the body) may also be encountered in the testis [1].

Clinically, germ cell tumors (GCT) account for 95% and 77% of testicular neoplasms in adults and children, respectively, with seminoma (60%) being the most dominant adult subtype and teratoma (62%) being the most common pediatric subtype, while gonadal stromal tumor (<5%), other testicular tumors and secondary tumors make up the remainder.

In the United States, testicular germ cell tumors (TGCT) are often grouped into seminoma (including classic, anaplastic, and spermatocytic variants) and nonseminoma (including choriocarcinoma, embryonal carcinoma, teratoma, and yolk sac tumor) histotypes (Figure 28.1). Out of these, seminoma, yolk sac tumor, choriocarcinoma, and embryonal carcinoma are malignant, whereas teratoma is benign in prepubertal children but may be malignant in adolescents and adults. In contrast, most gonadal stromal tumors (e.g., Leydig cell tumor, Sertoli cell tumor, juvenile granulosa cell tumor, and gonadoblastoma) are benign, although they can occasionally be malignant, especially in older children.

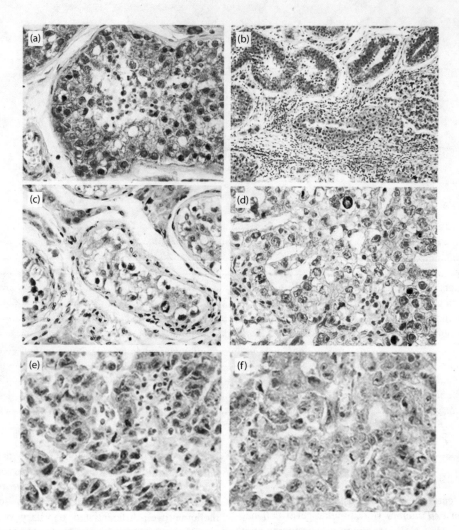

FIGURE 28.1 Immunohistochemical staining for CCDC6 in human normal testis (a), normal testis adjacent to an area of seminoma (b), intratubular germ cell neoplasia unclassified (IGCNU) (c), seminoma (d), yolk sac tumor (e), and embryonal carcinoma (f). Note that nuclear reactivity for CCDC6 is mainly detected in primordial germinal cells and Sertoli cells of normal testis, but almost absent in IGCNU, seminoma, yolk sac tumor, and embryonal carcinoma (200×, 106×, 250×, 250×, 250×, 250× magnifications, respectively). (Photo credit: Staibano S et al. *BMC Cancer.* 2013;13:433.)

Although a majority of TGCT cases appear to occur sporadically and have no family history, approximately 1.4% of newly diagnosed male patients show obvious familial segregations, with their sons and siblings demonstrating four- to sixfold and eight- to tenfold increases in TGCT risk, respectively. Interestingly, familial TGCT often involve two members, who are on average 2.5 years younger than sporadic patients and tend to have bilateral disease as opposed to unilateral disease in sporadic patients. Further analyses suggest that familial TGCT is predominantly an autosomal recessive disorder that is attributable to the combined effects of multiple common alleles (e.g., Y chromosome *gr/gr* deletion, and mutations in *KITLG, PDE11A, SPRY4,* and *BAK1* genes), each conferring modest risk [2].

28.2 Biology

The testes (singular: testis, commonly called testicles) are a pair of ovoid glandular organs of 5 × 3 × 2.5 cm in dimension and 10–15 g in weight that are suspended outside the body in a fleshy sac called the scrotum, which attaches to the body between the base of the penis and anus. Besides production of male sex hormone testosterone, the testes generate as many as 12 trillion sperm in a male's lifetime, about 400 million of which are released in a single ejaculation.

Structurally, the testes are covered by the tunica vaginalis (an extension of the peritoneum of the abdomen), the tunica albuginea (a tough, protective sheath of dense irregular connective tissue), and the tunica vasculosa (consisting of blood vessels and connective tissue). The seminiferous tubules (150–250 μm in diameter, totaling 800 in each testis) open into a series of uncoiled, interconnected channels called the rete testis, which is connected via ducts (or tubes) to a tightly coiled tube called the epididymis. The epididymis joins to a long, large duct called the vas deferens (which exits via the spermatic cord).

At the histological level, the testes are composed of germ cells and stromal cells. Germ cells line the seminiferous tubules and are capable of multiplication and differentiation into spermatocytes and spermatids, which move from the lining to the epididymis, where they mature into spermatozoa or sperm cells. Mature sperm cells travel through the vas deferens and combine with fluids made by the seminal vesicles and the prostate gland to create semen, which is pushed out of the body through the urethra during ejaculation.

Stromal cells are supportive cells that are separated into Sertoli cells (or sustentacular cells, which are located in the seminiferous tubules and help make and transport sperm) and Leydig cells (or interstitial cells, which are located between the seminiferous tubules and secrete male sex hormones, mostly testosterone, contributing to sperm production and maintenance of sex drive, or libido, and other male features).

The testicular collecting system, testicular tunics, and spermatic cord as well as rete testis (principally of intratesticular location) are sometimes referred to as the paratesticular region.

Human germ cells are derived from precursor cells known as primordial germ cells (PGC), which are reproductive cells with the capacity to differentiate into sperm in males and eggs in females. Arising from a pluripotent cell population during weeks 5–6 of embryologic development, PGC migrate along the midline of the body to the genital ridge, actively proliferate, and express several factors (e.g., OCT3/4). Some PGC may fail to reach the genital ridge and become trapped along the migratory path, where the surviving extragonadal germ cells may remain at risk of tumorigenesis. PGC reaching the genital ridge stop proliferating and begin to differentiate inside the developing gonads into either sperm or eggs.

During its transition into germ cell tumor, human germ cell has to overcome the normal cellular regulatory processes and move along the differentiation pathways that mimic embryogenesis. As atypical cells with enlarged hyperchromatic nuclei, clumped chromatin, and prominent nucleoli, aligned along the basement membrane of seminiferous tubules in the spermatogonial niche, IGCNU, or germ cell neoplasia in situ (GCNIS) then undergo (i) abortive spermatogenesis leading to the formation of seminoma, and (ii) a caricature of embryonic development resulting in tumors of totipotent cells (or embryonal carcinoma cells). This tumorigenetic process seems to coincide with the onset of puberty in both sexes. The relatively high number of germ cells in the testis compared with the ovaries at this time is a key factor for increased incidence of GCT in the males [3,4].

GCT appear to reflect their histologic stages of development, with testicular seminomas and ovarian dysgerminomas resembling primitive undifferentiated PGC and displaying an invasive phenotype, while nonseminomas represent more differentiated forms such as somatic tissue (teratoma) or extraembryonic structures (yolk sac tumor) or placenta (choriocarcinoma) and show a relatively benign phenotype. Furthermore, >90% of GCT occur in the gonads, and the remainder are extragonadal (located along the midline of the body such as the mediastinum, retroperitoneum, and central nervous system), and mainly affect children.

28.3 Pathogenesis

Risk factors for testicular tumors include age (highest incidence of GCT is among in men between 15 and 35 years old), cryptorchidism (undescended testis), testicular microlithiasis, family history of testicular cancer among first-degree relatives (father/ brothers, isochromosome of the short arm of chromosome 12), Klinefelter syndrome (testicular dysgenesis), Down syndrome, Li–Fraumeni syndrome, persistent Müllerian duct syndrome,

contralateral testicular tumor or IGCNU (also known as carcinoma in situ), subfertility/infertility, hypovirilization, testicular atrophy, increased adult height and body mass index, diet rich in cheese, previous marijuana exposure, vasectomy, trauma, mumps, HIV infection, in utero exposure to diethylstilbestrol, gynecomastia, maternal smoking during pregnancy, and occupational exposure to polyvinyl chloride, pesticides, and nonsteroidal estrogens (e.g., diethylstilbestrol or DES) [5–10].

Specific gene analyses revealed the Y chromosome *gr/gr* deletion and *PDE11A* gene mutations as genetic modifiers of familial TGCT risk. Genome-wide association studies (GWAS) further identified the *KITLG* (KIT ligand on chromosome 12), *SPRY4* (chromosome 5), and *BAK1* (BCL2-antagonist/killer 1 on chromosome 6) genes as potential TGCT risk modifiers [2,11–14].

Commonly observed in males suffering from infertility, the Y chromosome *gr/gr* deletion (i.e., a 1.6 Mb deletion of the AZFc region on the Y chromosome) is found in 3.0% of familial TGCT cases, 2% sporadic TGCT cases, and 1.3% of unaffected males, corresponding to threefold and twofold increases in TGCT risk, respectively. In addition, mosaic loss of chromosome Y (mLOY) in blood-derived DNA is common in older men [15].

Phosphodiesterase (PDE) 11A (encoded by the *PDE11A* gene) is a critical regulator of cyclic AMP signaling in adrenal and other steroidogenic tissues and is highly expressed in the normal human testis. Germline-inactivating mutations (including 20 missense, 4 splice site, 2 nonsense, 7 synonymous, and 22 intronic) of the *PDE11A* gene decreases *PDE11A* protein expression, reduces PDE activity, and increases cAMP levels in testicular tumor samples, with 5 nonsynonymous substitutions (p.F258Y, p.G291R, p.V820M, p.R545X, and p.K568R) being unique to familial TGCT subjects [16].

Belonging to the family of short-chain helical cytokines, KITLG has the ability to activate its receptor tyrosine kinase KIT and initiates multiple cellular responses related to the normal proliferation and migration of primordial germ cells and the development of hematopoietic cells, melanocytes, and germ cells. While germline *KITLG* mutations are implicated in familial progressive hyperpigmentation, germline *KIT* mutations lead to familial gastrointestinal stromal tumors and familial diffuse cutaneous mastocytosis. In addition, somatic mutations (or overexpression) of the *KIT* oncogene are involved in both familial and sporadic testicular cancers. Not surprisingly, use of tyrosine kinase inhibitor imatinib (which includes KIT among its ligands) results in complete response in pretreated disseminated testicular seminoma with *KIT* overexpression [17].

The *SPRY4* gene on chromosome 5 encodes an inhibitor of the mitogen-activated protein kinase (MAPK) pathway, which is activated by the KITLG–KIT pathway. *SPRY4* has been shown to increase methylation of the maternal allele promoter in TGCT patients, and a reduced expression of *SPRY4* may lead to enhanced survival of abnormal PGC. On the other hand, the *BAK1* (BCL2-antagonist/killer 1) gene on chromosome 6 encodes a pro-apoptotic protein that controls the death of mislocalized PGC during migration, and its activity is repressed by the KITLG–KIT pathway.

Moreover, RAS association domain family protein 1A (*RASSF1A*), a tumor suppressor gene located on the chromosome region 3p21, is also implicated in TGCT tumorigenesis. This gene is involved in cell-cycle control, microtubule stabilization, cellular adhesion, motility, and apoptosis. Altered methylation of the RASSF1A promoter is detected in the blood of patients with TGCT (including both seminomas and nonseminomas) [18–20].

Other chromosomal loci (e.g., 2q14.2, 3q26.2, 4q35.2, 7q36.3, 9p24.3, 10q26.13, 15q21.3, 15q22.31, 16q22.3, 19p12, and Xq28) may also play a possible role in TGCT tumorigenesis [21]. In addition, tripartite motif (TRIM) family proteins contain the RING finger domain, which mainly functions as E3 ubiquitin ligases, and modulates ubiquitination leading to degradation, activation, or functional modification of target proteins. Of the 70 known genes within the TRIM family, approximately 20 (including TRIM44) are associated with malignancies through their involvement in cell proliferation, migration, and anti-apoptosis [22–33].

Thus, although a rare, highly penetrant familial TGCT susceptibility locus has not been identified by linkage analyses, it appears that multiple low-penetrance genes may play a part in the pathogenesis of familial multiple-case phenotype.

28.4 Epidemiology

28.4.1 Prevalence

With an annual incidence of 11.5 cases per 100,000 males in Nordic countries, 3–8 cases per 100,000 males in western countries, and 0.5–3 cases per 100,000 males in Africa, Asia, and South America, testicular cancer is responsible for approximately 1.5% of male neoplasms and 5% of urologic malignancies globally [34,35]. Typically, testicular cancer demonstrates three incidence peaks in white males: the first at 2–4 years of age (mainly non-germ cell histologies), the second (and largest) between ages 25 and 35 years (seminoma and nonseminoma histologies), and the third manifesting in the mid-60s–70s (primarily spermatocytic seminoma) [36]. The lifetime risk of TGCT in Caucasian men is estimated to be 1 in 230 [37].

While GCT in adults represent about 95% of all testicular malignancies, those in children only make up 60%–75%. Further, most adult testicular neoplasms are malignant, and 25%–33% of pediatric testicular neoplasms are benign.

28.4.2 Inheritance

In 18 international centers, 985 familial TGCT cases from 461 families have been described. A vast majority (88%) of these families have only two cases, with 49% cases involving sibling pairs (suggesting an autosomal recessive inheritance), and 19% cases involving father/son, grandfather/father/son (suggesting autosomal dominant inheritance, as well as female relatives (suggesting X-linked inheritance) [38,39]. The overall heritability ranges from 37% to 49%, with a four- to sixfold increase in risk in sons of men with TGCT, and an eight- to tenfold increase in risk among men with an affected brother [40–42].

28.4.3 Penetrance

Familial TGCT is associated with a low penetrance, as the condition generally affects two family members only and reflects the combined effects of multiple common alleles. This differs from other susceptibility disorders (e.g., *BRCA*-related hereditary breast/ovarian cancer), in which multiple-case, multigeneration families are commonly observed.

28.5 Clinical Features

Patients with testicular cancer may show painless unilateral lump or swelling, pain or discomfort in a testicle or the scrotum, dull ache in the lower abdomen or groin, sudden buildup of fluid in the scrotum (hydrocele), breast tenderness or growth (gynecomastia), lower back pain, shortness of breath, chest pain, and bloody sputum or phlegm (later-stage testicular cancer), swelling of one or both legs, and blood clot.

About 1.5% of patients develop bilateral testicular tumors, of which synchronous (simultaneous) tumors account for 17%, and metachronous (developed at different time points) tumors 83%. In familial TGCT, bilateral disease is found in 9.8% cases compared to 2.8% cases in sporadic TGCT [43–48].

28.6 Diagnosis

Diagnosis of testicular cancer involves physical examination (for palpable scrotal mass, which may be painless and hardened, and which needs to be differentiated from epididymitis, hydrocele, hernia, and spermatic cord torsion in pediatric patients), ultrasonography (benign tumors are often well circumscribed with sharp borders and decreased blood flows on Doppler studies), biopsy in high-risk patients, immunohistochemical detection of tumor markers, and genetic testing [49].

Seminoma (or dysgerminoma in the ovary) usually presents as a bulky, painless, gray-white, lobulated mass of the testis in men of 40 years (compared to nonseminomatous GCT in men of 25 years). Seminoma may sometimes occur in the mediastinum, pineal gland (germinoma), and retroperitoneum. Histologically, seminoma displays sheets of relatively uniform, large, round-polyhedral cells divided into poorly demarcated lobules by delicate fibrous septa with T lymphocytes and plasma cells, abundant clear/watery cytoplasm (glycogen), large central nuclei, 1–2 prominent, elongated, and irregular nucleoli, minimal mitotic figures, occasional infarction and edema, rare fibrosis, and osseous metaplasia. Immunohistochemically, seminoma is positive for placental alkaline phosphatase (PLAP), ferritin, PAS (with and without diastase), vimentin, and angiotensin-1-converting enzyme, but negative for cytokeratin (syncytiotrophoblastic giant cells are positive), alpha-fetoprotein (AFP), human chorionic gonadotropin (hCG; syncytiotrophoblastic giant cells are positive), CD30, and EMA.

As a common prepubertal testicular tumor of germ cell origin, yolk sac tumor occurs primarily in children of <2 years of age, and often presents as a solid, stage I testicular mass associated with increased levels of AFP.

Teratoma is uniformly benign in children but often malignant in postpubertal/adult cases. Macroscopically, teratoma is a large (5–10 cm), multinodular, heterogenous (solid, cartilaginous, cystic) lesion. Histologically, mature teratoma shows mixture of elements of ectoderm, endoderm, and mesoderm; immature teratoma shows foci resembling embryonic or fetal structures, primitive neuroectoderm, poorly formed cartilage, neuroblasts, loose mesenchyme, and primitive glandular structures; teratoma with malignant transformation has focal malignancy of somatic type (e.g., squamous cell carcinoma, adenocarcinoma, sarcoma).

Useful serum tumor markers for TGCT include hCG (which does not increase in yolk sac tumors), AFP (which increases by up to 92% in yolk sac tumors), and lactate dehydrogenase.

Immunohistochemical markers for TGCT consist of PLAP, OCT3/4 (POU5F1), NANOG, SOX2, REX1, AP-2γ (TFAP2C), LIN28, etc., which are expressed in primordial germ cells/gonocytes and embryonic pluripotency-related cells but not in normal adult germ cells. Other diagnostic markers include HMGA1, HMGA2, kinase Aurora-B (which is present in IGCNU, seminomas, and embryonal carcinomas but not in teratomas and YST) [50,51].

Pediatric patients with testicular tumors are often designated according to the Children's Oncology Group staging system as Stages 1–4, whereas adolescents with GCT are staged as adults using the TNM system of the American Joint Committee on Cancer and the International Union Against Cancer.

28.7 Treatment

Treatment options for TGCT consist of surgery (e.g., partial or bilateral orchiectomy, testis-sparing surgery for children with benign GCT or infertile patients with bilateral or solitary testicular tumor), radiotherapy (at a dose of 18–20 Gy), and/or chemotherapy for advanced disease (e.g., several cycles of antineoplastic drugs cisplatin, bleomycin, and etoposide [BEP] chemotherapy). In addition, testosterone replacement therapy (e.g., controlled-release transdermal patches) should be considered for patients with potential androgen insufficiency after bilateral or partial orchiectomy [52,53].

If diagnosed early, >90% of patients with TGCT are curable, and only a small percentage of patients diagnosed with metastatic tumors are not permanently cured. For malignant stages I and II testes tumors, surgery alone yields excellent survival rates, with recurrence rates of 15.5% and 75%, respectively. Platinum-based multiagent chemotherapy provides an effective treatment for recurrent disease. Retroperitoneal lymph node dissection (RPLND) offers an important staging and therapeutic approach for mixed GCT in adolescents [34,54,55].

28.8 Prognosis

Seminoma and nonseminomatous GCT have 5-year survival rates of 96% and 90%, respectively; whereas synchronous and metachronous tumors have 5-year survival rates of 88% and 95%, respectively [56]. Patients with stage I seminoma have a 15%–20% of risk of developing a tumor relapse in their lifetime [57,58]. In particular, tumor of >4 cm in diameter and with infiltration of the rete testis has a tendency to relapse. A combination of surveillance, adjuvant chemotherapy, and adjuvant radiation of the posterior abdominal wall may help lower the relapse risk in patients with clinical stage I seminoma to <5%. Long-term morbidity associated with the use of radiation and chemotherapies in treatment for TGCT includes cardiovascular disease,

metabolic syndrome, infertility, nephrotoxicity, neuropathy, and ototoxicity (due to cisplatin), as well as pulmonary toxicity (due to bleomycin) [59,60].

Prognostic factors for TGCT include age (e.g., young children vs adolescents), stage of disease, primary site of disease, tumor marker decline (AFP and beta-hCG) in response to therapy, histology (e.g., seminomatous vs nonseminomatous), TRIM44 expression, and MGMT and CALCA promoter methylation [61–64].

Patients with TGCT who want to be fathers should undergo preoperative sperm freezing procedure and should be informed about assisted reproductive technologies.

28.9 Conclusion

Representing approximately 1.5% of male malignancies, testicular cancer often affects children of 2–4 years in age, adults between ages 25 and 35 years, and elderly of the mid-60s–70s. Being the most common testicular cancer, GCT is responsible for 95% and 77% of testicular neoplasms in adults and children, respectively. While a majority TGCT arises sporadically without family history, about 1.4% of newly diagnosed male patients demonstrate familial segregation, with their sons and siblings showing four- to sixfold and eight- to tenfold increases in TGCT risk, respectively. Compared with sporadic TGCT, familial TGCT often involve two members, who are on average 2.5 years younger and have a higher percentage of bilateral disease. In the absence of a highly penetrant susceptibility locus, familial TGCT appears to be attributable to the combined effects of multiple low-penetrance alleles (e.g., Y chromosome *gr/gr* deletion, and mutations in *KITLG*, *PDE11A*, *SPRY4*, and *BAK1* genes) that underlie the familial phenotype [65]. If diagnosed early, >90% of TGCT cases are curable using a combination of surgery, radiotherapy, and chemotherapy. Testosterone replacement therapy is helpful for patients with potential androgen insufficiency after bilateral or partial orchiectomy.

REFERENCES

1. Moch H, Cubilla AL, Humphrey PA, Reuter VE, Ulbright TM. The 2016 WHO classification of tumours of the urinary system and male genital organs-part a: Renal, penile, and testicular tumours. *Eur Urol.* 2016;70:93–105.
2. Pyle LC, Nathanson KL. Genetic changes associated with testicular cancer susceptibility. *Semin Oncol.* 2016;43(5):575–81.
3. Giambartolomei C, Mueller CM, Greene MH, Korde LA. A mini-review of familial ovarian germ cell tumors: An additional manifestation of the familial testicular germ cell tumor syndrome. *Cancer Epidemiol.* 2009;33(1):31–6.
4. von Eyben FE, Jensen MB, Høyer S. Frequency and markers of precursor lesions and implications for the pathogenesis of testicular germ cell tumors. *Clin Genitourin Cancer.* 2017. pii: S1558-7673(17)30265-3.
5. Greene MH, Kratz CP, Mai PL et al. Familial testicular germ cell tumors in adults: 2010 summary of genetic risk factors and clinical phenotype. *Endocr Relat Cancer.* 2010;17(2):R109–21.
6. Biggs ML, Doody DR, Trabert B, Starr JR, Chen C, Schwartz SM. Consumption of alcoholic beverages in adolescence and adulthood and risk of testicular germ cell tumor. *Int J Cancer.* 2016;139(11):2405–14.

7. Togawa K, Le Cornet C, Feychting M et al. Parental occupational exposure to heavy metals and welding fumes and risk of testicular germ cell tumors in offspring: A registry-based case-control study. *Cancer Epidemiol Biomarkers Prev.* 2016;25(10):1426–34.

8. Cools M, Looijenga L. Update on the pathophysiology and risk factors for the development of malignant testicular germ cell tumors in complete androgen insensitivity syndrome. *Sex Dev.* 2017;11(4):175–81.

9. Le Cornet C, Fervers B, Pukkala E et al. Parental occupational exposure to organic solvents and testicular germ cell tumors in their offspring: NORD-TEST study. *Environ Health Perspect.* 2017;125(6):067023.

10. Liang W, Song L, Peng Z, Zou Y, Dai S. Possible association between androgenic alopecia and risk of prostate cancer and testicular germ cell tumor: A systematic review and meta-analysis. *BMC Cancer.* 2018;18(1):279.

11. Kratz CP, Edelman DC, Wang Y, Meltzer PS, Greene MH. Genetic and epigenetic analysis of monozygotic twins discordant for testicular cancer. *Int J Mol Epidemiol Genet.* 2014;5(3):135–9.

12. Taylor-Weiner A, Zack T, O'Donnell E et al. Genomic evolution and chemoresistance in germ-cell tumours. *Nature.* 2016;540(7631):114–8.

13. Boccellino M, Vanacore D, Zappavigna S et al. Testicular cancer from diagnosis to epigenetic factors. *Oncotarget.* 2017;8(61):104654–63.

14. Wang Z, McGlynn KA, Rajpert-De Meyts E et al. Meta-analysis of five genome-wide association studies identifies multiple new loci associated with testicular germ cell tumor. *Nat Genet.* 2017;49(7):1141–7.

15. Machiela MJ, Dagnall CL, Pathak A et al. Mosaic chromosome Y loss and testicular germ cell tumor risk. *J Hum Genet.* 2017;62(6):637–40.

16. Pathak A, Stewart DR, Faucz FR et al. Rare inactivating PDE11A variants associated with testicular germ cell tumors. *Endocr Relat Cancer.* 2015;22(6):909–17.

17. Azevedo MF, Horvath A, Bornstein ER et al. Cyclic AMP and c-KIT signaling in familial testicular germ cell tumor predisposition. *J Clin Endocrinol Metab.* 2013;98(8):E1393–400.

18. Killian JK, Dorssers LC, Trabert B et al. Imprints and DPPA3 are bypassed during pluripotency- and differentiation-coupled methylation reprogramming in testicular germ cell tumors. *Genome Res.* 2016;26(11):1490–504.

19. Hacioglu BM, Kodaz H, Erdogan B et al. K-RAS and N-RAS mutations in testicular germ cell tumors. *Bosn J Basic Med Sci.* 2017;17(2):159–63.

20. Markulin D, Vojta A, Samaržija I et al. Association between RASSF1A promoter methylation and testicular germ cell tumor: A meta-analysis and a cohort study. *Cancer Genomics Proteomics.* 2017;14(5):363–72.

21. Ruark E, Seal S, McDonald H et al. Identification of nine new susceptibility loci for testicular cancer, including variants near DAZL and PRDM14. *Nat Genet.* 2013;45(6):686–9.

22. Davis-Dao CA, Siegmund KD, Vandenberg DJ et al. Heterogenous effect of androgen receptor CAG tract length on testicular germ cell tumor risk: Shorter repeats associated with seminoma but not other histologic types. *Carcinogenesis.* 2011;32(8):1238–43.

23. Heidenberg DJ, Barton JH, Young D, Grinkemeyer M, Sesterhenn IA. N-cadherin expression in testicular germ cell and gonadal stromal tumors. *J Cancer.* 2012;3:381–9.

24. Stadler ZK, Esposito D, Shah S et al. Rare *de novo* germline copy-number variation in testicular cancer. *Am J Hum Genet.* 2012;91(2):379–83.

25. Juliachs M, Castillo-Ávila W, Vidal A et al. ErbBs inhibition by lapatinib blocks tumor growth in an orthotopic model of human testicular germ cell tumor. *Int J Cancer.* 2013;133(1):235–46.

26. Schumacher FR, Wang Z, Skotheim RI et al. Testicular germ cell tumor susceptibility associated with the UCK2 locus on chromosome 1q23. *Hum Mol Genet.* 2013;22(13):2748–53.

27. Zechel JL, Doerner SK, Lager A, Tesar PJ, Heaney JD, Nadeau JH. Contrasting effects of Deadend1 (Dnd1) gain and loss of function mutations on allelic inheritance, testicular cancer, and intestinal polyposis. *BMC Genet.* 2013;14:54.

28. Ferreira HJ, Heyn H, Garcia del Muro X et al. Epigenetic loss of the PIWI/piRNA machinery in human testicular tumorigenesis. *Epigenetics.* 2014;9(1):113–8.

29. Gainetdinov IV, Skvortsova YV, Stukacheva EA et al. Expression profiles of PIWIL2 short isoforms differ in testicular germ cell tumors of various differentiation subtypes. *PLOS ONE.* 2014;9(11):e112528.

30. Rounge TB, Furu K, Skotheim RI, Haugen TB, Grotmol T, Enerly E. Profiling of the small RNA populations in human testicular germ cell tumors shows global loss of piRNAs. *Mol Cancer.* 2015;14:153.

31. Romano FJ, Rossetti S, Conteduca V et al. Role of DNA repair machinery and p53 in the testicular germ cell cancer: A review. *Oncotarget.* 2016;7(51):85641–9.

32. Loveday C, Litchfield K, Levy M et al. Validation of loci at 2q14.2 and 15q21.3 as risk factors for testicular cancer. *Oncotarget* 2017;9(16):12630–8.

33. Yamada Y, Takayama KI, Fujimura T et al. A novel prognostic factor TRIM44 promotes cell proliferation and migration, and inhibits apoptosis in testicular germ cell tumor. *Cancer Sci.* 2017;108(1):32–41.

34. Vasdev N, Moon A, Thorpe AC. Classification, epidemiology and therapies for testicular germ cell tumours. *Int J Dev Biol.* 2013;57(2–4):133–9.

35. Ghazarian AA, Trabert B, Graubard BI, Schwartz SM, Altekruse SF, McGlynn KA. Incidence of testicular germ cell tumors among US men by census region. *Cancer.* 2015;121(23):4181–9.

36. Lin X, Wu D, Zheng N, Xia Q, Han Y. Gonadal germ cell tumors in children: A retrospective review of a 10-year single-center experience. *Medicine (Baltimore).* 2017;96(26):e7386.

37. Damjanov I, Wewer-Albrechtsen N. Testicular germ cell tumors and related research from a historical point of view. *Int J Dev Biol.* 2013;57(2–4):197–200.

38. Mai PL, Friedlander M, Tucker K et al. The international testicular cancer linkage consortium: A clinicopathologic descriptive analysis of 461 familial malignant testicular germ cell tumor kindred. *Urol Oncol.* 2010;28(5):492–9.

39. Mueller CM, Korde LA, McMaster ML et al. Familial testicular germ cell tumor: No associated syndromic pattern identified. *Hered Cancer Clin Pract.* 2014;12(1):3.

40. Chia VM, Li Y, Goldin LR et al. Risk of cancer in first- and second-degree relatives of testicular germ cell tumor cases and controls. *Int J Cancer.* 2009;124(4):952–7.

41. Litchfield K, Thomsen H, Mitchell JS et al. Quantifying the heritability of testicular germ cell tumour using both population-based and genomic approaches. *Sci Rep.* 2015;5:13889.

42. McMaster ML, Heimdal KR, Loud JT, Bracci JS, Rosenberg PS, Greene MH. Nontesticular cancers in relatives of testicular germ cell tumor (TGCT) patients from multiple-case TGCT families. *Cancer Med.* 2015;4(7):1069–78.

43. Sarıcı H, Telli O, Eroğlu M. Bilateral testicular germ cell tumors. *Turk J Urol.* 2013;39(4):249–52.

44. James FV, Mathews A. Familial testicular germ cell tumor with bilateral disease. *J Cancer Res Ther.* 2016;12(1):422–3.

45. Arda E, Cakiroglu B, Cetin G, Yuksel I. Metachronous testicular cancer after orchiectomy: A rare case. *Cureus.* 2017;9(11):e1833.

46. Boudaoud N, Loron G, Pons M et al. Bilateral methachronous testicular germ cell tumor and testicular microlithiasis in a child: Genetic analysis and insights. A case report. *Int J Surg Case Rep.* 2017;41:76–9.

47. Buck DA, Smith TD, Montana WN. An uncommon presentation of a metachronous testicular primary nonseminoma and seminoma separated by two decades and a testicular cancer literature review. *Case Rep Oncol.* 2017;10(3):832–9.

48. Kopp RP, Chevinsky M, Bernstein M et al. Bilateral testicular germ cell tumors in the era of multimodal therapy. *Urology.* 2017;103:154–60.

49. Silveira SM, da Cunha IW, Marchi FA, Busso AF, Lopes A, Rogatto SR. Genomic screening of testicular germ cell tumors from monozygotic twins. *Orphanet J Rare Dis.* 2014;9:181.

50. Pathak A, Adams CD, Loud JT et al. Prospectively identified incident testicular cancer risk in a familial testicular cancer cohort. *Cancer Epidemiol Biomarkers Prev.* 2015;24(10):1614–21.

51. Chieffi P. An up-date on epigenetic and molecular markers in testicular germ cell tumors. *Intractable Rare Dis Res.* 2017;6(4):319–21.

52. PDQ Pediatric Treatment Editorial Board. *Childhood Extracranial Germ Cell Tumors Treatment (PDQ®): Health Professional Version. PDQ Cancer Information Summaries [Internet].* Bethesda (MD): National Cancer Institute (US); 2002-. 2018 Apr 5.

53. Aparicio J, Terrasa J, Durán I et al. SEOM clinical guidelines for the management of germ cell testicular cancer (2016). *Clin Transl Oncol.* 2016;18(12):1187–96.

54. Joshi A, Zanwar S, Shetty N et al. Epidemiology of male seminomatous and nonseminomatous germ cell tumors and response to first-line chemotherapy from a tertiary cancer center in India. *Indian J Cancer.* 2016;53(2):313–6.

55. Loh KP, Fung C. Novel therapies in platinum-refractory metastatic germ cell tumor: A case report with a focus on a PD-1 inhibitor. *Rare Tumors.* 2017;9(2):6867.

56. Kvammen Ø, Myklebust TÅ, Solberg A et al. Long-term relative survival after diagnosis of testicular germ cell tumor. *Cancer Epidemiol Biomarkers Prev.* 2016;25(5):773–9.

57. Alatassi H, O'Bryan BE, Messer JC, Wang Z. Nephroblastoma arising from primary testicular germ cell tumor: A case report and literature review. *Case Rep Pathol.* 2016;2016:7318672.

58. Johnson K, Brunet B. Brain metastases as presenting feature in 'burned out' testicular germ cell tumor. *Cureus.* 2016;8(4):e551.

59. Gil T, Sideris S, Aoun F et al. Testicular germ cell tumor: Short and long-term side effects of treatment among survivors. *Mol Clin Oncol.* 2016;5(3):258–64.

60. Ghezzi M, De Toni L, Palego P et al. Increased risk of testis failure in testicular germ cell tumor survivors undergoing radiotherapy. *Oncotarget.* 2017;9(3):3060–8.

61. Kojima T, Kawai K, Tsuchiya K et al. Identification of a subgroup with worse prognosis among patients with poor-risk testicular germ cell tumor. *Int J Urol.* 2015;22(10):923–7.

62. Kalavska K, Chovanec M, Zatovicova M et al. Prognostic value of serum carbonic anhydrase IX in testicular germ cell tumor patients. *Oncol Lett.* 2016;12(4):2590–8.

63. Martinelli CMDS, Lengert AVH, Cárcano FM et al. MGMT and CALCA promoter methylation are associated with poor prognosis in testicular germ cell tumor patients. *Oncotarget.* 2016;8(31):50608–17.

64. Bogefors K, Giwercman YL, Eberhard J et al. Androgen receptor gene CAG and GGN repeat lengths as predictors of recovery of spermatogenesis following testicular germ cell cancer treatment. *Asian J Androl.* 2017;19(5):538–42.

65. Basiri A, Movahhed S, Parvin M, Salimi M, Rezaeetalab GH. The histologic features of intratubular germ cell neoplasia and its correlation with tumor behavior. *Investig Clin Urol.* 2016;57(3):191–5.

29

Familial Wilms Tumor and Related Syndromes

Dongyou Liu

CONTENTS

29.1 Introduction

Wilms tumor (WT, also known as nephroblastoma) is a renal malignancy that often presents with single-nodule, multifocal unilateral lesions or bilateral disease in the kidneys. Rarely occurring in adults, WT is responsible for nearly 90% of renal cancer cases in children (median age 3.5 years). At the genetic level, WT is mostly attributed to aberrations in the *WT1* gene located on chromosome 11p13 (Table 29.1). Although a majority of WT cases arise sporadically, about 5%–10% of cases occur as a part of genetic predisposition syndromes such as familial WT, WAGR (WT, aniridia, genitourinary anomalies, and mental retardation), Denys–Drash syndrome (DDS), Beckwith–Wiedemann syndrome (BWS), and asymmetric overgrowth [1].

29.2 Biology

The kidneys are a pair of bean-shaped organs (of 10–12 cm in length, 5–7 cm in width, 2–3 cm in thickness, and 135–150 g in weight) that are located along the posterior muscular wall of the abdominal cavity. Covered by a thin layer of fibrous connective tissue (the renal capsule), the kidneys comprise an outlayer of soft, dense, vascular renal cortex, and an inner renal medulla, the latter of which is composed of seven cone-shaped renal pyramids separated by the cortical tissue (called renal columns of Bertin). Each kidney contains around 1 million individual nephrons (the functional units), which are made of renal corpuscle and renal tubule. The renal corpuscle comprises the capillaries of the glomerulus that is surrounded by the glomerular capsule

(or Bowman's capsule, a cup-shaped double layer of simple squamous epithelium with a hollow space between the layers). The glomerulus consists of podocytes and a basement membrane allowing water and certain solutes to be filtered across. Podocytes form a thin filter with the endothelium of the capillaries to separate urine from blood passing through the glomerulus. The outer layer of the glomerular capsule keeps the urine separated from the blood within the capsule. At the far end of the glomerular capsule is the mouth of the renal tubule, which carries urine from the glomerular capsule to the renal pelvis.

In embryological terms, the kidneys are originated from two distinct lineages, the nephrogenic (metanephric mesenchyme [MM]) and the ductogenic (ureteric bud [UB]). The MM develops into the epithelial nephric or Wolffian duct, Six2-positive mesenchymal cells (that form the nephrons), and Foxd1-positive cells (that give rise to the stromal cells); whereas the UB turns into the collecting ducts. WT is thought to arise from residual nephrogenic blastemal cells (also called nephrogenic rests [NR]), which are abnormally retained embryonic kidney precursor cells arranged in clusters and located either perilobarly (at the periphery of the renal lobules) or intralobarly (in the central part of the lobe), and which fail to mature into the normal renal parenchyma [2]. NR are found in about 1% of unselected pediatric autopsies, 35% of the kidneys with unilateral WT, and nearly 100% of the kidneys with bilateral WT. A higher proportion of perilobar rests (52%) are observed in patients with bilateral WT compared to patients with unilateral WT (18%). Intralobar nephrogenic rests (ILNR) appear to arise earlier in the development when compared with perilobarly nephrogenic rests (PLNR), which are associated with stromal-type WT and diagnosed at younger age. The presence of diffuse or multifocal NR throughout the kidneys is referred to as

TABLE 29.1

Confirmed and Putative Genes Involved in the Pathogenesis of Wilms Tumor

Phenotype	Gene	Mutation	Chromosome Location	Inheritance	Related Syndromes
Wilms tumor type 1 (or nephroblastoma)	*WT1*	Heterozygous mutation in the *WT1* gene	11p13	Autosomal dominant	Wilms tumor, familial Wilms tumor, WAGR syndrome, Denys–Drash syndrome, Frasier syndrome
Wilms tumor type 2	*WT2 (H19)*	Mutation of the *H19/IGF2*-imprinting control region (ICR1)	11p15.5	Autosomal dominant	Beckwith–Wiedemann syndrome
Wilms tumor type 3	*WT3*		16q	Autosomal dominant	Wilms tumor
Wilms tumor type 4 (familial Wilms tumor 1)	*WT4 (FWT1)*		17q12-q21	Autosomal dominant	Familial Wilms tumor
Wilms tumor susceptibility 5	*POU6F2*	Mutation in the *POU6F2* gene	7p14.1	Autosomal dominant	Wilms tumor
Wilms tumor susceptibility 6	*REST*	Mutation in the *REST* gene	4q12		Familial Wilms tumor
Familial Wilms tumor 2	*FWT2*		19q13.4	Autosomal dominant	Familial Wilms tumor

TABLE 29.2

Comparison between Wilms Tumor and Nephrogenic Rest

Feature	Wilms Tumor	Nephrogenic Rest
Shape	Spherical	Oval
Fibrous capsule	Present	Absent
Skeletal muscle differentiation	Common	Uncommon
Quantity	Usually solitary	Often multifocal

nephroblastomatosis. As a unique category of nephroblastomatosis, diffuse hyperplastic perilobar nephroblastomatosis forms a thick rind around one or both kidneys and is considered a preneoplastic condition. Careful examination of the juncture between the lesion and the surrounding renal parenchyma is crucial to identify diffuse hyperplastic perilobar NR from WT (Table 29.2) [3].

Mimicking the differentiation of the developing kidney, WT classically consists of triphasic components: blastemal, epithelial, and stromal (Figure 29.1). First noted by Thomas F. Rance

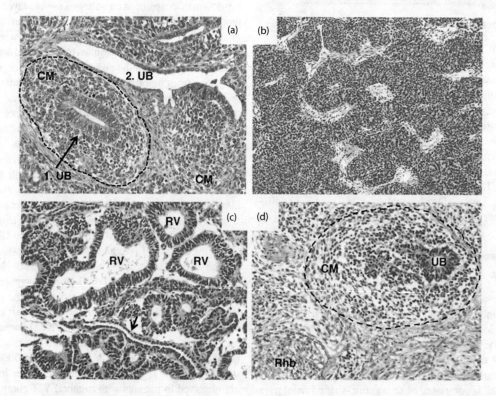

FIGURE 29.1 Histology of Wilms tumor. (a) Wilms tumor with blastemal-predominant histology showing a typical serpentine growth pattern with few or absent ureteric bud (UB)-like structures (1.UB, cross-section, showing a central epithelial blastema pattern indicated by arrow, and a UB-like structure surrounded by condensing mesenchyme [CM, demarcated by broken lines]; 2.UB, longitudinal section). (b and c) Wilms tumor with epithelial-predominant histology showing renal vesicle (RV)-like epithelial structures and thin collecting duct-like epithelial structures. (d) Wilms tumor with a stromal-predominant histology harboring a *WT1* mutation showing a central epithelial blastema pattern (demarcated by broken lines) with rhabdomyogenesis (Rhb, demarcated by red broken lines) (H&E, ×400). (Photo credit: Fukuzawa R et al. *PLOS ONE.* 2017;12(10):e0186333.)

in 1814, the histological description of WT was detailed by Carl Max Wilhelm Wilms in 1899. Although WT is primarily a sporadic disease, 1%–2% of cases show clear family history and are apparently linked to WAGR (due to large deletion in the *WT1* gene on chromosome 11p13), DDS (progressive renal disease, male pseudohermaphroditism, and WT), and BWS (macrosomia, macroglossia, omphalocele, and growth retardation, and hemihypertrophy; due to a mutation in the *WT2* gene on chromosome 11p15) [4]. WT appears to follow the two-hit model for tumor suppressor genes, with the initial loss of *WT1* exacerbated by mutation in another allele or other genes, leading to the transformation of NRs into WT [5].

29.3 Pathogenesis

WT (or nephroblastoma as the currently preferred term) has been shown to form part of clinical phenotypes of nearly 50 conditions. Conditions associated with a high risk (>20%) for WT include *WT1* deletions (e.g., WAGR syndrome), truncating and pathogenic missense *WT1* mutations (e.g., DDS), familial WT, Perlman syndrome, mosaic variegated aneuploidy, and Fanconi anemia D1/biallelic *BRCA2* mutations; conditions associated with a moderate risk (5%–20%) for WT comprise *WT1* intron 9 splice mutations (Frasier syndrome [FS]), BWS (due to 11p15 uniparental disomy, isolated *H19* hypermethylation), Simpson–Golabi–Behmel syndrome (due to *GPC3* mutations/deletions); conditions associated with a low risk (<5%) for WT are isolated hemihypertrophy (individuals with hemihypertrophy due to 11p15 uniparental disomy or isolated *H19* hypermethylation are at moderate risk), Bloom syndrome, Li–Fraumeni syndrome/Li–Fraumeni-like syndrome, hereditary hyperparathyroidism–jaw tumor syndrome, Mulibrey nanism, trisomy 18, trisomy 13, and 2q37 deletions [1,3].

The *WT1* gene located on the short (p) arm of chromosome 11 at position 13 (11p13) encodes a 449 aa, 49 kDa zinc finger transcription factor (WT1) that comprises a C-terminal zinc finger DNA binding domain and an N-terminal transactivational domain, and has multiple alternatively spliced isoforms [6]. Expressed mainly in the kidney, gonad, uterus, and mesothelium, WT1 may function as a transcriptional activator or repressor, contributing significantly to the development of the kidneys and gonads (ovaries in females and testes in males). Constitutional and somatic mutations in the *WT1* gene disrupt the DNA-binding domain of WT1 and reduce the ability of WT1 to regulate cell growth and apoptosis in related tissues, leading to fewer functional nephrons at birth, smaller glomerulus size, and various combinations of three cardinal features (i.e., WT, genitourinary abnormalities, and renal dysfunction) [7–9]. Mutations in the *WT1* gene have been identified in patients with WT, WAGR syndrome, DDS, and FS [10]. In addition, mutations in *WT1* exon 7 and 9 have been recurrently identified in acute myeloid leukemia [11].

Patients who have WT without syndromic features (which encompass both sporadic and familial WTs) may harbor germline *WT1* pathogenic variants (2%–4% of cases) that tend to affect the sex determination and genital tract development in males, and cause bilateral or multicentric tumors at an early age. Individuals with nonsyndromic WT may be also caused by alterations of chromosome 11p15.5 (including gain of methylation at IC1, paternal uniparental disomy of 11p15.5, microdeletion,

and microinsertion) and the X chromosome (*WTX* or *FAM123B* at Xq11.1). Recently, two autosomal dominant WT predisposition genes, *WT4* (*FWT1*) and *FWT2*, have been mapped to chromosomes 17q12-q21 and 19q13, respectively, although these genes have yet to be identified (Table 29.1). Interestingly, WT in *FWT1*-related families is diagnosed at a later age (mean age of 6 years) and more advanced stage than sporadic WT, and *FWT1* mutations have a 30% penetrance rate.

WAGR syndrome is one of the main syndromic conditions associated with germline *WT1* mutations (the others being DDS and FS). Having an incidence of 7–8 cases per 1000 individuals with WT, WAGR syndrome displays WT, complete or partial aniridia (an eye problem), ambiguous external genitalia and cryptorchidism in males, and mental retardation (intellectual disability). Molecularly, the condition is attributed to heterozygous contiguous gene deletion of chromosome 11p13 that includes both *WT1* (whose deletion causes genitourinary features and WT predisposition) and *PAX6* (which lies within 0.6 Mb of *WT1* and whose deletion is responsible for aniridia) (first hit) followed by the mutation in the second *WT1* allele (second hit). With a 45%–60% lifetime risk of WT, WAGR syndrome has an early age at diagnosis (90% patients developing WT by age 4 and 98% by age 7), and a frequent bilateral disease [1,3].

DDS (including nephrotic syndrome type 4, and Meacham syndrome) results from heterozygous germline missense mutation in the exon 8 or 9 (zinc finger region) of the *WT1* gene that affects its DNA-binding properties, and subsequent biallelic inactivation of *WT1*. Characterized by diffuse mesangial sclerosis that leads to hypertension, proteinuria, and early-onset renal failure, as well as gonadal dysgenesis (from ambiguous to normal-appearing female in both XY and XX individuals), DDS poses a 74%–90% lifetime risk of WT.

FS and DDS possibly represent two ends of a phenotypic spectrum, with the former characterized by undermasculinized external genitalia (ranging from ambiguous to normal-appearing female in an individual with a 46,XY karyotype), focal segmental glomerulosclerosis (which progresses to renal failure by the second or third decade), and gonadoblastoma. Linked to heterozygous single-nucleotide variants in the *WT1* intron 9 donor splice site that alter *WT1* splicing and prevent the formation of WT1 isoforms, FS poses a relatively low risk for WT.

BWS is an overgrowth disorder caused by gain-of-function mutations (e.g., methylation at imprinting center [IC1]) in the *WT2* gene on chromosome 11p15.5, resulting in alteration in beta-catenin in the WNT (wingless) signaling pathway. Manifesting as macrosomia, macroglossia, visceromegaly, hemihyperplasia, omphalocele, neonatal hypoglycemia, ear creases/pits, adrenocortical cytomegaly, renal abnormalities, and embryonal tumors e.g.(e.g., WT, hepatoblastoma, neuroblastoma, and rhabdomyosarcoma), BWS increases the risk for WT by 7%–25%, with 81% of affected individuals developing WT by age 5 and 93% by age 8.

29.4 Epidemiology

29.4.1 Prevalence

WT (or nephroblastoma) is the most common type of pediatric renal cancer, with an annual incidence of 7.1 cases per 1 million

TABLE 29.3

Age-Related Renal Tumors

Year	Common	Sometimes	Rare
Birth	Mesoblastic nephroma	Wilms tumor	Rhabdoid tumor
<1	Wilms tumor, mesoblastic nephroma	Rhabdoid tumor, clear cell sarcoma	
1–5	Wilms tumor	Clear cell sarcoma	Mesoblastic nephroma (<3 years), rhabdoid tumor
5–10	Wilms tumor	Clear cell sarcoma, renal cell carcinoma	
11–15	Wilms tumor, renal cell carcinoma	Primitive neuroectodermal tumor	

children before the age of 15 years (Table 29.3). With a median age of 3.5 years at diagnosis, WT demonstrates a male to female ratio of 0.92:1.00 for unilateral disease (mean age of 44 months at diagnosis), but 0.60:1.00 for bilateral disease (mean age of 31 months at diagnosis). Compared to the Caucasian population, Asians have a substantially lower incidence of WT.

29.4.2 Inheritance

Although WT is mostly sporadic (with an onset age of 42–47 months), about 10% of cases occur in patients who have an affected relative (i.e., familial WT, with an onset age of 30–33 months). Germline *WT1* pathologic variants demonstrate an autosomal dominant inheritance pattern.

29.4.3 Penetrance

Familial WT shows variable expressivity and reduced penetrance estimated at 30%.

29.5 Clinical Features

Clinical symptoms of WT range from abdominal mass (firm, nontender swelling; 90% of cases), abdominal pain (due to a rupture or intratumoral hemorrhage; 40%), macroscopic hematuria (blood in the urine as tumor extends to the collecting system; 18%), microscopic hematuria (24%) urinary disturbances, anemia, hypertension (due to activation of the renin-angiotensin system; 25%), hypercalcemia, hemihyperplasia, cryptorchidism, hypospadias, varicocele (due to inferior vena cava [IVC] obstruction), to constitutional symptoms (fever, anorexia, and weight loss; 10%).

WT is usually unilateral, but 5%–10% of affected individuals (especially those with a WT predisposition) may have bilateral or multicentric tumors, which suggest a genetic predisposition [12].

Patients with WAGR syndrome may show aniridia, genitourinary anomalies (including ambiguous genitalia), renal insufficiency or failure, and intellectual disability; those with BWS have macrosomia, ear creases/pits, macroglossia, omphalocele, visceromegaly, hemihyperplasia, adrenocortical cytomegaly, renal abnormalities, and other embryonal tumors (e.g., hepatoblastoma, neuroblastoma, and rhabdomyosarcoma).

29.6 Diagnosis

Procedures for diagnosing WT and related syndromes include physical exam (for abdominal mass and signs of associated syndromes such as aniridia, developmental delay, hypospadias, cryptorchidism, pseudohermaphrodism, overgrowth, and hemihyperplasia), medical history review, biochemical tests (complete blood count [CBC], liver function test, renal function test, urinalysis), abdominal imaging (abdominal x-ray, ultrasonography [US], computed tomography [CT] scan, and/or magnetic resonance imaging [MRI]), biopsy (for histologic diagnosis of WT and its subtypes), and molecular testing [1,3].

US facilitates initial investigation of the tumor size and origin, the relationship between vena cava and aorta, and possible IVC or renal vein (as 4% of WT patients have IVC or atrial involvement and 11% have renal vein involvement). An abdominal Doppler US helps check for possible thrombus in the renal vein and the IVC.

CT scan of the abdomen is a sensitive technique for confirming the renal origin of the mass, determining bilateral tumors, and identifying cavoatrial thrombus. CT scan of the chest helps identify metastatic lung nodules, which occur in 15% of WT patients. Furthermore, fluorine F 18-fludeoxyglucose (18F-FDG) positron emission tomography (PET)-CT is also useful for detecting bilateral disease.

MRI with contrast of the abdomen enables detection of the vascular involvement, accurate imaging of the kidney, and discrimination between WT and nephroblastomatosis.

Laboratory tests (for total blood count, renal and liver function, calcium level, urinalysis, 24-h urine catecholamine, neuron-specific enolase, lactate dehydrogenase, alpha-fetoprotein, β-human chorionic gonadotropin, and ferritin) provide a valuable approach for preoperative assessment of WT as well as identification of von Willebrand disease, which affects 1%–8% of WT patients.

Macroscopically, WT presents as well circumscribed, large, solid, pale gray to slightly pink or yellow-gray masses with pseudocapsule, soft consistency, and heterogeneous cut surface (including areas of viable tumor, hemorrhage, and necrosis). In 5% of cases, multicentric tumors are present.

Histologically, WT typically shows a classic triphasic pattern of epithelial (tubular), stromal, and blastemal components, with a central epithelial blastemal pattern containing aggregations of blastemal cells around the UB-like epithelial structure, forming a nodular condensed structure (characteristic of *WT1*-mutant WT) (Figure 29.1a) [2]. The *blastemal component* is the least differentiated and contains small round blue cells with overlapping nuclei and brisk mitotic activity in diffuse, serpentine, nodular, or basaloid patterns. While the serpentine variant contains broad bands of undifferentiated cells surrounded by fibromyxoid stroma, the basaloid variant forms nests or cords of blastema showing a distinctive peripheral palisading of elongated cells with epithelial differentiation. It is of note that the diffuse variant may show marked infiltrative growth with no pseudocapsule between the tumor and adjacent tissues. The *epithelial component* displays a spectrum of differentiation, including tubular formation with primitive epithelial rosette- and glomerulus-like structures, squamous epithelial islands, and mucinous epithelium. The *stromal component* comprises densely packed undifferentiated mesenchymal cells or loose cellular myxoid areas. In some cases, neoplastic stroma may contain well-differentiated smooth or skeletal muscle cells, adipose tissue, cartilage, bone, mature ganglion cells, and glial tissue [13]. The representative features of blastemal-predominant tumors, epithelial-predominant tumors, and stromal-predominant tumors are shown in Figure 29.1b through d [2].

Observed in 10% of WT cases (particularly older patients aged 10–16 years), anaplastic WT often shows large, atypical multipolar mitotic figures and markedly enlarged hyperchromatic nuclei (hyperchromasia), and contains genetic signatures of WT (e.g., *P53* mutation and *MYCN* dysregulation). Anaplastic histology occurs in 12%–14% of patients with bilateral disease.

Patients showing physical, radiologic, or histologic features of WT should be assessed by molecular genetic testing (e.g., single-gene testing, gene-targeted deletion/duplication analysis, methylation studies, multigene panel [consisting of *WT1*, *REST*, and *DICER1*, etc.], and chromosomal microarray). The existence of a relative with WT is suggestive of a genetic predisposition. Individuals with WT but in the absence of syndromic features should be tested for *WT1* by sequence analysis; this is followed by gene-targeted deletion/duplication analysis of *WT1* for intragenic deletions if a pathogenic variant is absent. Individuals with WT and physical features of BWS should undergo methylation studies of 11p15.5 imprinting center 1 (IC1) if a germline *WT1* pathogenic variant is absent. Individuals suspected of WAGR may be tested for 11p13 deletions encompassing *WT1* and *PAX6* by chromosomal microarray (CMA). Patients harboring uniparental disomy at 11p15.5 may be assessed by CMA-SNP array.

The stage of WT is usually determined on the basis of the imaging studies and the surgical and pathologic findings at nephrectomy. According to the staging systems of Children's Oncology Group (COG) and Société Internationale d'Oncologie Pédiatrique (SIOP), WT is classified into five stages, including stage I (tumor limited to the kidney, renal capsule intact; 43% of cases), stage II (regional extension of the tumor; lymph node negative; 20% of cases), stage III (tumor penetrating through the peritoneal surface, lymph nodes in the abdomen or pelvis involved; 21% of cases), stage IV (hematogenous metastases, lymph node metastases outside the abdominopelvic region; 11% of cases), and stage IV (bilateral involvement; 5% of cases).

Differential diagnoses for WT showing a monophasic histologic pattern include non-WT renal tumors (e.g., renal cell carcinoma, metanephric adenoma, and hyperplastic NR for epithelial elements; clear cell sarcoma of the kidney, mesoblastic nephroma, and synovial sarcoma for stromal elements; other embryonal "small round blue cell tumors" such as congenital mesoblastic nephroma, neuroblastoma, primitive neuroectodermal tumor/Ewing sarcoma, desmoplastic small round cell tumor, and lymphoma for blastemal elements). Immunohistochemical stains may aid in the differential diagnosis, as the blastemal elements show focal positivity for vimentin, the epithelial elements react for keratin and epithelial membrane antigen (EMA), and the mesenchymal elements show a heterogeneous reactivity. In 90% of WT, immunoreactivity for WT1 antigen is detected [14].

29.7 Treatment

Treatment options for WT consist of surgery, chemotherapy, and radiotherapy. In North America, surgery is usually performed prior to chemotherapy, whereas in Europe, preoperative chemotherapy is preferred in order to decrease the risk of intraoperative rupture, downstage the tumor, and reduce the need for irradiation. In particular, WT with a renal vein and caval extension should undergo chemotherapy before surgery is conducted [5,15–17].

For patients with unilateral WT, the standard operation involves a transperitoneal radical nephrectomy, while for patients with bilateral WT (which is located in one pole and infiltrates <1/3 of the kidney but does not invade the renal vein), nephron-sparing surgery (NSS) such as partial nephrectomy or enucleation may be performed (after 9–12 weeks of preoperative chemotherapy) [18].

Preoperative chemotherapy (consisting of doxorubicin, vincristine, and dactinomycin) is indicated for WT in a solitary kidney, synchronous bilateral WT, extension of tumor thrombus in the IVC above the level of the hepatic veins, tumor involving contiguous structures, inoperable WT, or pulmonary compromise due to extensive pulmonary metastases. A biopsy through a flank approach is necessary before preoperative chemotherapy.

Postoperative chemotherapy for WT may be administered according to regimen EE-4A (vincristine, dactinomycin × 18 weeks postnephrectomy), regimen DD-4A (vincristine, dactinomycin, doxorubicin × 24 weeks; baseline nephrectomy or biopsy with subsequent nephrectomy), and regimen I (vincristine, doxorubicin, cyclophosphamide, etoposide × 24 weeks).

Postoperative radiotherapy is not necessary for children with stage I or stage II tumors, although it is required in the setting of local tumor stage III (e.g., 10.8 Gy of radiation to the flank together with dactinomycin, vincristine, and doxorubicin or 20 Gy of radiation to the flank together with dactinomycin and vincristine).

29.8 Prognosis

WT in most affected children is curable, with a cure rate in most WT showing triphasic patterns (i.e., blastemal, epithelial, and stromal) being close to 90%. Long-term survival rates for WT patients currently stand at 90% in high-income countries, 80% in middle-income countries, and 20%–50% in low-income countries [19]. Pediatric patients have a better overall survival rate of 88% than adults (69%).

Prognostic factors for WT comprise the stage of disease, histological subtype (e.g., anaplastic and blastemal) of tumor, 1q gain or 1p/16q loss of heterozygosity (LOH) in chemotherapy-naive WT, and age at diagnosis (children <2 or >4 years; adults >18 years). Anaplastic type represents 11.5% of all WT cases but is responsible for 52% of WT mortalities. Gain of 1q confers inferior survival in unilateral WT and is a more powerful predictor of outcome than 1p or 16q loss. Older patients show a higher risk of recurrence due to delay in diagnosis, advanced tumor stage at presentation, and a higher frequency of anaplasia [20].

Patients with clinical features suggestive of a genetic predisposition (bilateral or multifocal WT, syndromic features, congenital anomalies, or familial WT) should undergo molecular testing and genetic counseling. If the *WT1* pathogenic variant identified in the patient cannot be detected in the DNA of either parent, the risk to the sibs is likely to be low. However, offspring of a patient are at a 50% risk of inheriting the *WT1* germline pathogenic variant. It is noteworthy that besides *WT1*, several genes (e.g., *REST*, *CTR9*, and *BRCA2*) may be also involved in familial WT.

29.9 Conclusion

Predominantly affecting children, Wilms tumor (WT, or nephroblastoma) is a renal malignancy that typically produces an asymptomatic abdominal mass, gross hematuria, abdominal pain, or hypertension. While a majority of WT cases appear to evolve sporadically, about 5%–10% form a part of genetic predisposition syndromes (e.g., familial WT, WAGR, DDS, FS, and BWS, among others). At the molecular level, WT is attributed to aberrations in the *WT1* gene located on chromosome 11p13 in addition to several other genes (Table 29.1). Initial diagnosis of WT involves abdominal US, CT, or MRI; confirmation of the condition requires histopathological examination of biopsy specimens. Management of WT relies on a combination of surgery, chemotherapy, and radiotherapy, which currently has a success rate of >90%.

REFERENCES

1. Leslie SW, Bhimji SS. *Cancer, Wilms (Nephroblastoma). StatPearls [Internet].* Treasure Island (FL): StatPearls Publishing; 2017.
2. Fukuzawa R, Anaka MR, Morison IM, Reeve AE. The developmental programme for genesis of the entire kidney is recapitulated in Wilms tumour. *PLOS ONE.* 2017;12(10):e0186333.
3. van den Heuvel-Eibrink MM, editor. *Wilms Tumor [Internet].* Brisbane (AU): Codon Publications; 2016.
4. Scott RH, Stiller CA, Walker L, Rahman N. Syndromes and constitutional chromosomal abnormalities associated with Wilms tumour. *J Med Genet.* 2006;43(9):705–15.
5. Brok J, Treger TD, Gooskens SL, van den Heuvel-Eibrink MM, Pritchard-Jones K. Biology and treatment of renal tumours in childhood. *Eur J Cancer.* 2016;68:179–95.
6. Bielińska E, Matiakowska K, Haus O. Heterogeneity of human WT1 gene. *Postepy Hig Med Dosw (Online).* 2017;71(0):595–601.
7. Carraro DM, Ramalho RF, Maschietto M. Gene expression in Wilms tumor: Disturbance of the Wnt signaling pathway and microRNA biogenesis. In: van den Heuvel-Eibrink MM, editor. *Wilms Tumor [Internet].* Brisbane (AU): Codon Publications; 2016. Chapter 10.
8. Lee KY. A unified pathogenesis for kidney diseases, including genetic diseases and cancers, by the protein-homeostasis-system hypothesis. *Kidney Res Clin Pract.* 2017; 36(2):132–44.
9. Maturu P. The Inflammatory microenvironment in Wilms tumors. In: van den Heuvel-Eibrink MM, editor. *Wilms Tumor [Internet].* Brisbane (AU): Codon Publications; 2016. Chapter 12.
10. Dome JS, Huff V. Wilms tumor predisposition. In: Adam MP, Ardinger HH, Pagon RA et al. editors. *GeneReviews® [Internet].* Seattle (WA): University of Washington, Seattle; 1993–2018. 2003 Dec 19 [updated 2016 Oct 20].
11. Hosen N, Maeda T, Hashii Y et al. Wilms tumor 1 peptide vaccination after hematopoietic stem cell transplant in leukemia patients. *Stem Cell Investig.* 2016;3:90.
12. Kaneko Y, Okita H, Haruta M et al. A high incidence of WT1 abnormality in bilateral Wilms tumours in Japan, and the penetrance rates in children with WT1 germline mutation. *Br J Cancer.* 2015;112(6):1121–33.
13. Diniz G. Histopathological and molecular characteristics of Wilms tumor. In: van den Heuvel-Eibrink MM, editor. *Wilms Tumor [Internet].* Brisbane (AU): Codon Publications; 2016. Chapter 3.
14. Popov SD, Sebire NJ, Vujanic GM. Wilms' tumour – Histology and differential diagnosis. In: van den Heuvel-Eibrink MM, editor. *Wilms Tumor [Internet].* Brisbane (AU): Codon Publications; 2016. Chapter 1.
15. Millar AJW, Cox S, Davidson A. Management of bilateral Wilms tumours. In: van den Heuvel-Eibrink MM, editor. *Wilms Tumor [Internet].* Brisbane (AU): Codon Publications; 2016. Chapter 5.
16. Lopes RI, Lorenzo A. Recent advances in the management of Wilms' tumor. *F1000Res.* 2017;6:670.
17. PDQ Pediatric Treatment Editorial Board. *Wilms Tumor and Other Childhood Kidney Tumors Treatment (PDQ®): Health Professional Version. PDQ Cancer Information Summaries [Internet].* Bethesda (MD): National Cancer Institute (US); 2002-. 2017 Dec 6.
18. Eriksen KO, Johal NS, Mushtaq I. Minimally invasive surgery in management of renal tumours in children. *Transl Pediatr.* 2016;5(4):305–14.
19. Njuguna F, Martijn HA, Kuremu RT et al. Wilms tumor treatment outcomes: Perspectives from a low-income setting. *J Glob Oncol.* 2016;3(5):555–62.
20. Stefanowicz J, Kosiak M, Kosiak W, Lipska-Ziętkiewicz BS. Chronic kidney disease in Wilms tumour survivors – What do we know today? In: van den Heuvel-Eibrink MM, editor. *Wilms Tumor [Internet].* Brisbane (AU): Codon Publications; 2016. Chapter 9.

30

Hereditary Breast and Ovarian Cancer

Dongyou Liu

CONTENTS

30.1 Introduction

Hereditary breast and ovarian cancer (HBOC) is an autosomal dominant syndrome characterized by the early onset of multiple/bilateral tumors in the breast and ovary on a family setting. Apart from causing 5%–7% of all breast cancer cases and 10% of all ovarian cancer cases, HBOC is also implicated in the cancer development involving other organs (e.g., the fallopian tube, endometrium, pancreas, prostate, colorectum, and skin).

At the molecular level, heterozygous germline mutations in the *BRCA1* and *BRCA2* genes significantly increase the risk for breast and ovarian cancer, contributing to >60% of hereditary breast and ovarian cancer cases [1]. In addition, germline mutations in *PTEN, TP53, CDH1, STK11, NBS1, NF1, ATM, CHK2, FANCJ, FANCM, PALB2,* and *RAD51C* may be implicated in the tumorigenesis of familial cases of breast and ovarian cancer at times (Table 30.1).

30.2 Biology

The breast is a specialized organ involved in milk production (lactation). Covered by the skin studded with sweat glands and hair follicles and encircled by a thin layer of connective tissue (called fascia), the breast is supplied with the internal mammary artery, and protected immunologically with lymph nodes located under the arms, above the collarbone, and deep inside the breast.

Structurally, the breast can be separated into 15–20 lobes, each of which is subdivided into multiple lobules, which end in dozens of tiny bulbs. The lobes, lobules, and bulbs are connected by thin tubes (or ducts) that merge at the nipple. The breast in children and males is nearly identical to that in adult females, although it lacks well-developed lobules and bulbs involved in milk production.

Histologically, the breast (mammary gland) is lined up with the epithelium composed of luminal and basal/myoepithelial cells. Located on the surface of the ductal lumen, luminal cells undergo terminal differentiation into lobuloalveolar cells for milk secretion (which is regulated by cyclical production of various steroids and hormones). Basal/myoepithelial cells sit below luminal cells and help maintain ductal contractility for milk release. Occupying the spaces between the lobes, lobules, bulbs, and ducts, and consisting of adipocytes, fibroblasts, and endothelial cells, the stroma (a connective tissue) supports the extensive system of ducts and alveoli.

Based on histological criteria, breast cancer is classified into two groups: carcinoma (or adenocarcinoma) and sarcoma (Figure 30.1). Accounting for >95% of primary breast cancer cases, breast carcinoma originates from the epithelial cells that line the lobules and terminal ducts (sometimes referred to as the terminal duct lobular unit [TDLU]) of the breast. In contrast, making up <1% of primary breast cancer, breast sarcoma (e.g.,

TABLE 30.1

Characteristics of Key Genes Implicated in the Development of Breast and Ovarian Cancer

Gene	Chromosome Location	Function	Inheritance/ Penetrance	Syndrome with Germline Mutations	Cancer Type
BRCA1	17q21.31	Double strand break repair through homologous recombination.	AD/high	HBOC	Breast and ovarian cancer
BRCA2	13q12.3	Double strand break repair through homologous recombination.	AD/high	HBOC	Breast and ovarian cancer
PTEN	10q23.31	Phosphatidylinositol 3-phosphate, suppresses AKT signaling	AD/high	Cowden syndrome, PTEN hamartoma	Breast cancer
TP53	17p13.1	Transcription factor, regulates cell cycle, apoptosis, senescence	AD/high	Li–Fraumeni syndrome	Breast and ovarian cancer
CDH1	16q22.1	E-cadherin gene, maintains cell adherence	AD/high	Hereditary diffuse gastric cancer syndrome	Breast and ovarian cancer
STK11	9p13.3	Serine/threonine kinase, regulates cell polarity	AD/high	Peutz–Jeghers syndrome	Breast and ovarian cancer
NBS1	8q21.3	Cell cycle checkpoint after DNA damage, member of the MRN complex	AD/high	Nijmegen breakage syndrome	Breast cancer
NF1	17q11.2	Negative regulator of Ras signaling	AD/high	Neurofibromatosis type I	Breast cancer
ATM	11q22.3	PI3 kinase-related kinase, cell cycle checkpoint and DSB repair	AD/moderate	Ataxia telangiectasia	Breast cancer
CHK2	22q12.1	Activation of cell cycle checkpoint after DNA damage	AD/moderate	Li–Fraumeni syndrome	Breast and ovarian cancer
FANCJ	17q23.2	Interstrand crosslink repair	AD/moderate	Fanconi anemia	Breast and ovarian cancer
FANCM	14q21.2	Interstrand crosslink repair	AD/moderate	Fanconi anemia	Breast cancer
PALB2	16p12.2	Interstrand crosslink repair, homologous recombination	AD/moderate	Fanconi anemia	Breast and ovarian cancer
RAD51C	17q22	Interstrand crosslink repair, homologous recombination	AD/moderate	FA-like syndrome	Breast and ovarian cancer

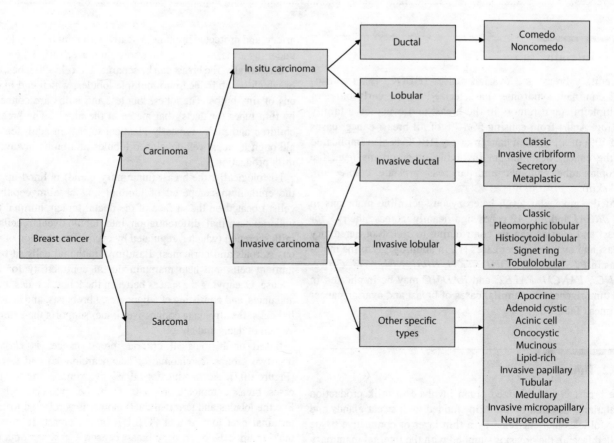

FIGURE 30.1   Histological classification of breast cancer.

angiosarcoma and phyllodes tumor) evolves from the stromal cells (including myofibroblasts and blood vessel cells) of the breast [2].

According to the location involved, breast carcinoma is further distinguished into ductal or lobular carcinoma. While ductal carcinoma begins in the cells of the ducts that carry milk to the nipple, lobular carcinoma arises in the lobes or lobules that produce milk. Furthermore, carcinoma that remains in situ is known as in situ carcinoma, and carcinoma that invades/metastasizes nearby tissue/organ, lymph nodes, and/or another part of the body is known as invasive carcinoma. Invasive carcinoma consists of invasive ductal carcinoma (IDC), invasive lobular carcinoma (ILC) and invasive carcinoma of other special types. Representing 75%–80% of breast carcinoma, IDC shows malignant ductal proliferation along with stromal invasion, and comprises classic IDC NST (no specific type), invasive cribriform carcinoma, secretory carcinoma (juvenile carcinoma), and metaplastic carcinoma. Constituting 10%–15% of breast carcinoma, ILC usually grows as single cells arranged individually, in single file, or in sheets, and includes classic type, pleomorphic lobular carcinoma, histiocytoid lobular carcinoma, signet ring ILC, and tubulo-lobular carcinoma. Making up 10% of breast carcinoma, invasive carcinoma of other special types displays unique histopathologic features relating to cell type (apocrine, adenoid cystic, actinic cell, and oncocytic carcinoma), amount, type and location of secretion (mucinous/colloid, and lipid-rich carcinoma), architectural features (invasive papillary, tubular, medullary, and invasive micropapillary carcinoma), and immunohistochemical profile (neuroendocrine carcinoma) (Figure 30.1) [2–4].

The ovaries are a pair of oval-shaped, unevenly surfaced, grayish organs of 4×2 cm $\times 8$ mm in dimension and 3.5 g in weight (during childbearing) that stay within the broad ligament below the uterine (fallopian) tubes on either side of the uterus. Covered by a single-cell mesothelial layer (known as the ovarian surface epithelium or the germinal epithelium of Waldeyer), the ovary consists of outer cortex (which contains spindle-shaped fibroblasts and houses the follicles and oocytes, with the ability to grow and mature into eggs/ova) and inner medulla (which comprises loose stromal tissue including ovarian vasculature).

Deriving from different cell types in the ovaries, ovarian cancer is differentiated into epithelial ovary carcinoma (65%), germ cell tumors (15%), and sex cord-stromal tumors (10%). Arising from the coelomic epithelium, not the Müllerian ducts, in the ovarian surface epithelium (OSE), which also covers the serosa of the fallopian tubes, uterus, and peritoneal cavity, epithelial ovary carcinoma (EOC) is separated into two types, on the basis of clinical and molecular characteristics. Type 1 EOC includes low-grade serous, mucinous (both probably evolving from benign adenomas and borderline tumors), clear cell, endometrioid (both arising from endometriosis), and transitional cell carcinomas, which rarely contain TP53 mutations, often present at earlier stages and have an indolent clinical course. Type 2 EOC comprises high-grade serous, endometrioid carcinosarcoma and undifferentiated carcinomas, which develop from the surface epithelium of the ovary or from inclusion cysts, accounting for 75% of EOC and a vast majority of ovarian cancer-related mortalities, contain TP53 mutations, and behave aggressively [5].

Affecting the germ cells, germ cell tumors include dysgerminoma (30%–50%), teratoma (20%), embryonal carcinoma (rare), and choriocarcinoma (very rare). Involving stromal cells (theca cells, fibroblasts, and Leydig cells) and gonadal primitive sex cords (granulosa cells and Sertoli cells), which are engaged in steroid hormone production (androgens, estrogens, and corticoids), sex cord–stromal tumors consist of pure stromal, pure sex cord, and mixed sex cord–stromal tumors. Of these, granulosa cell tumor (GCT) within the pure sex cord tumor category is most common, accounting for 5% of all malignant ovarian tumors [5].

Breast and ovarian cancer rank first and fifth (after breast, colorectum, lung, and corpus uteri) in incidence among all women's malignancies. Although approximately 90%–95% of breast and ovarian cancer initiate with sporadic/de novo mutations, 5%–10% of breast and ovarian cancer are linked to germline mutations in the *BRCA1* and *BRCA2* as well as other genes to a lesser extent (Table 30.1). Subsequent loss of heterozygosity (LOH) or haploinsufficiency-inducing mutation at the wild-type allele ultimately leads to malignancy transformation [6,7].

30.3 Pathogenesis

Identified in 1994, the breast cancer type 1 susceptibility protein (*BRCA1*) gene on chromosome 17q21.31 spans about 110 kb (including 30 kb of tandem duplication) with 24 exons, which encode a 1863 aa, 220-kda nuclear phosphoprotein (BRCA1). Structurally, BRCA1 consists of a RING finger domain (residues 1–109, encoded by exons 2–7, partly overlapping with the BARD1-binding region) near the N-terminus, two nuclear localization signals (residues 501–507 and 607–614, in exon 11), a serine cluster domain (located in exons 11–13), and a BRCT domain (residues 1650–1863, encoded by exons 16–24) at the C-terminus.

Functionally, the RING finger motif (residues 24–64) contained within the RING domain (residues 1–109) interacts with BARD1 (BRCA1-associated RING domain protein 1) and dramatically increases the ubiquitin ligase activity of BRCA1, which is fundamental to cell-cycle progression, gene transcription regulation, DNA damage response, and ubiquitination. The two nuclear localization signals (NLS, encoded by exon 11) interact with importin-alpha (involved in BRCA1 transport from the cytosol to the nucleus) and other proteins (e.g., retinoblastoma protein, c-Myc, RAD50, and RAD51) relating to transcription regulation, DNA repair, and cell cycle progression. The C-terminal (BRCT) domain interacts with substrates of DNA damage-activated kinases (e.g., ATM, ATR, Abraxas, CtIP, and BACH1), transcription regulators (e.g., p53 and BACH1), and DNA damage repair proteins (e.g., CtIP and CCDC98). The serine cluster domain (in exons 11–13) contains serine residues (988, 1189, 1387, 1423, 1457, 1524, and 1542) that can be phosphorylated *by ATM, ATR, CHK1,* or *CHK2* [8].

Germline mutations (typically frameshift variants) in *BRCA1* result in a missing or nonfunctional protein. These loss-of-function mutations compromise DNA damage repair, induce transcriptional activation, alter cell-cycle arrest and apoptosis, create genome instability, and ultimately predispose to cancer development (especially breast and ovarian cancer, pancreatic, and prostate cancer) [9].

Among the >1800 pathogenic variants [e.g., c.68_69delAG (185delAG or 187delAG) yielding p.Glu23ValfsTer17; c.5096G>A yielding p.Arg1699Gln; c.5266dupC (5385insC or 5382insC) yielding p.Gln1756ProfsTer74] have been identified in *BRCA1*, most of which affect the N-terminal RING domain, coding regions of exons 11–13, and BRCT domain (BRCA1 C-terminus). Together, these mutations are responsible for 35% of familial breast cancer cases [10–12].

Identified in 1995, the breast cancer type 2 susceptibility protein (*BRCA2*) gene on chromosome 13q12.3 consists of a 85 kb DNA stretch organized in 27 exons, which encode a 3418 aa, 380-kda phosphoprotein (BRCA2). Structurally, BRCA2 includes a PALB2 binding domain in the N-terminus (residues 1–990), a RAD51 binding domain (residues 990–2100, consisting of eight tandem BRC repeats of about 30 aa each) and a DNA binding domain in the C-terminus (residues 2100–3100, for interacting with single-stranded DNA as well as DSS1).

The PALB2 binding domain in the N-terminus binds to PALB2, which facilitates BRCA2 localization and RAD51 chromatin loading at the damage site [13]. The RAD51 binding domain consists of 8 repeats of 30 residues long (called the BRC motif, in exon 11), which mediate the binding of BRCA2 to RAD51 recombinase and stimulate RAD51 assembly onto ssDNA for subsequent recombinational DNA repair. The DNA binding domain in the C-terminus contains single-stranded DNA and double-stranded DNA binding motifs, and interacts with DSS1, a 70 aa protein identified from the genomic region on chromosome 7q21.3 that is deleted in split-hand/split-foot syndrome. DSS1 appears to play a role in the replacement of RPA with RAD51 on ssDNA [14].

Germline mutations in *BRCA2* [including frameshift deletions, insertions, or nonsense variants such as c.771_775delTCAAA (999del5) yielding p.Asn257LysfsTer17; c.5946delT (6174delT) yielding p.Ser1982ArgfsTer22; c.9976A>T yielding p.Lys3326Ter] result in prematurely a truncated, loss-of-function protein, which introduces errors in DNA phosphorylation and repair and contributes to 25% of familial breast cancer cases. In fact, missense mutations (e.g., D2723H) in the C-terminal region (residues 2500–2850) interfering with the binding of BRCA2 to DSS1 or disrupting RAD51 loading onto damaged chromatin have been detected in >25% of cancer cases. Phosphorylation at Ser3291 by cyclin-dependent kinase (CDK) also interrupts the C-terminal BRCA2–RAD51 interaction [15].

30.4 Epidemiology

30.4.1 Prevalence

Being the most common malignancy in women, who have a 12.5% chance of developing breast cancer throughout their lifetime, breast cancer has an estimated incidence of up to 90 cases per 100,000 women. Occasionally, breast cancer is found in men and children/adolescents. Ovarian cancer ranks fifth in incidence (after breast, colorectum, lung, and corpus uteri) among all malignancies in women, who have 1%–2% chance of developing ovarian cancer at some stage of their life [16].

Although sporadic/de novo mutations contribute to 95% of breast cancer and 90% of ovarian cancer, germline mutations in *BRCA1* and *BRCA2* are implicated in 5% of breast cancer and 10% ovarian cancer. Indeed, HBOC associated with germline *BRCA1* and *BRCA2* mutations is responsible for 35% of familial breast cancer and 25% of familial ovarian cancer, respectively [17].

BRCA1/2 pathogenic variants have an estimated prevalence of 1:400–500 in the general population (excluding Ashkenazim), but 1:40 in the Ashkenazi Jewish population. In fact, three founder *BRCA* mutations [185delAG, 5382insC (*BRCA1*); and 6174delT (*BRCA2*)] account for 90% of hereditary breast and/or ovarian cancer among Ashkenazi Jews. Further, the c.5266dupC (BIC: 5382insC) pathogenic variant in the 3′ end of *BRCA1* confers a higher risk for breast cancer, whereas the c.68_69delAG (BIC: 185delAG) pathogenic variant in the 5′ end of *BRCA1* has a higher risk for ovarian cancer in the Ashkenazi Jewish population. Similarly, 9 *BRCA1/2* mutations (including 185delAG) are responsible for 53% of cancers in the US Hispanic population; a founder mutation [999del5 (*BRCA2*)] occurs in 8% of patients with female breast cancer, 40% of patients with male breast cancer, and 6% of patients with ovarian cancer in one particular group of Iceland; and six *BRCA* founder mutations account for 75%–85% of hereditary breast and/or ovarian cancers among a group of French Canadians.

Overall, *BRCA1* mutations increase the risk for breast cancer (46%–87%) and ovarian cancer (39%–63%) in women by age 70. Similarly, *BRCA2* mutations show 38%–84% risk for breast cancer and 16.5%–27% risk for ovarian cancer in women by age 70. In addition, *BRCA1* or *BRCA2* mutations demonstrate increased risk for breast cancer (1.2% or up to 8.9%, respectively) in males by age 70 [18].

30.4.2 Inheritance

HBOC associated with *BRCA1* and *BRCA2* is inherited in an autosomal dominant pattern. A parent harboring a *BRCA1* or *BRCA2* germline variant has a 50% chance of passing the variant to offspring.

30.4.3 Penetrance

Germline *BRCA1* and *BRCA2* mutations display high penetrance and variable expressivity. A high percentage of women harboring *BRCA1* or *BRCA2* pathogenic variants are likely to develop breast and/or ovarian cancer by their 70s, compared to general population.

30.5 Clinical Features

Patients with breast cancer typically display (i) a lump or thickening in or near the breast or in the underarm area; (ii) a change in the size or shape of the breast; (iii) a dimple or puckering in the skin of the breast; (iv) a nipple turned inward into the breast; (v) fluid, other than breast milk, from the nipple, especially if bloody; (vi) scaly, red, or swollen skin on the breast, nipple, or areola (the dark area of skin around the nipple); and (vii) dimples in the breast that look like the skin of an orange (called peau d'orange). Some patients may be asymptomatic and are only diagnosed after a mass in the breast or the axilla (underarm), or changes to the breast skin or nipple become noticed accidentally.

Clinical symptoms of ovarian cancer range from abnormal uterine bleeding, abdominal pain (due to torsion, rupture, or hemorrhage), and fever to isosexual precocity in children. Sex cord-stromal tumors of the ovary are associated with hyperestrogenic and occasionally hyperandrogenic signs.

30.6 Diagnosis

Diagnosis of breast and ovarian cancer involves medical history review (for health habits, past illnesses and treatments), physical exam (for signs of lumps or anything else unusual in the breasts, under the arms and other regions; unusual bleeding), laboratory evaluation (for abnormal amounts of substances in blood), imaging study (x-rays, ultrasound, MRI), histopathology (for characteristic histologic features in biopsy or other samples), and molecular testing (for a heterozygous germline pathogenic variant in *BRCA1* or *BRCA2*) [1].

30.6.1 Clinical Diagnosis of Breast Cancer

Histologically, ductal carcinoma in situ (DCIS) is a neoplastic proliferation of epithelial cells limited to the ducts or lobules, and characterized by cellular and nuclear atypia and potential malignant capacity. DCIS may spread from the ducts into the lobular acini, producing extensive lesions (so-called lobular cancerization). Immunohistochemically, DCIS is consistently positive for E-cadherin and β-catenin, but weakly positive or negative for high molecular weight (HMW) keratin, which is normally expressed in the ductal basal cell layer, with typical peripheral CK8-18 expression.

Lobular carcinoma in situ (LCIS) often shows a relatively uniform population of round, small-to-medium-sized cells that have normal chromatic nuclei filling the distended lobules in a noncohesive pattern. Pleomorphism, mitosis, and necrosis are absent or rarely present. Intracellular mucin droplets, sometimes with signet ring nuclei, are observed. Immunohistochemically, LCIS lacks E-cadherin (due to genetic mutation) and β-catenin expressions but is positive for HMW keratin.

IDC has no specific histologic characteristics other than invasion through the basement membrane of a breast duct, and shows a tendency to metastasize via lymphatics. DCIS is a frequently associated finding on pathologic examination.

ILC typically displays round, small, relatively uniform, and noncohesive cells and has characteristic growth pattern with single-file infiltration of the stroma (so called the "Indian file" arrangement of small tumor cells). Molecular changes associated with ILC (particularly the pleomorphic subtype) include inactivation of E-cadherin by mutation, loss of heterozygosity, or methylation.

After histologic and imaging assessments, breast cancer is graded. The Nottingham histologic score system (which is the Elston–Ellis modification of Scarff–Bloom–Richardson grading system) is mostly commonly used. This system takes three factors into consideration: (i) glandular (acinar)/tubular differentiation (how well the tumor cells try to recreate normal glands), (ii) nuclear pleomorphism (how "ugly" the tumor cells look), and (iii) mitotic count (how much the tumor cells are dividing). With each of these features scored from 1–3, the final added scores range from 3–9, with Grade 1 (low grade or well differentiated) tumors having a score of 3, 4, or 5, Grade 2 (intermediate grade or moderately differentiated) tumors having a score of 6 or 7, and Grade 3 (high grade or poorly differentiated) tumors having a score of 8 or 9.

Further, the staging of breast cancer is determined as stage 0 (noninvasive tumor such as DCIS), stage I (tumor measures <2 cm or invades normal surrounding breast tissue), stage II (tumor measures between 2 cm and 5 cm or invades lymph nodes in the armpit, or both), stage III (tumor measures between 2 and 5 cm and attaches to the skin or surrounding tissues, and invades lymph nodes in the armpit and near the breastbone), and stage IV (tumor of any size has spread beyond the breast and nearby lymph nodes to other organs of the body).

A more elaborate staging system (the TNM system) designed by the American Joint Committee on Cancer may be utilized to determine the stages of breast cancer. This system incorporates the **T**umor size, lymph **N**ode involvement, and **M**etastasis (or spread of the cancer) into the staging, namely, pT (primary tumor) 0 (no evidence of primary tumor), pTis (DCIS), pT1 (tumor ≤20 mm in greatest dimension), pT1mi (tumor ≤1 mm), pT1a (tumor >1 mm but ≤5 mm), pT1b (tumor >5 mm but ≤10 mm), pT1c (tumor >10 mm but ≤20 mm), pT2 (tumor >20 mm but ≤50 mm), pT3 (tumor >50 mm), pT4 (tumor of any size with direct extension to the chest wall and/or to the skin), pT4a (extension to chest wall, not including only pectoralis muscle adherence/invasion), pT4b (ulceration and/or ipsilateral satellite nodules and/or edema of the skin), pT4c (both T4a and T4b), and pT4d (inflammatory carcinoma); pN (regional lymph nodes) X (regional lymph nodes cannot be assessed), pN0 (no regional lymph node metastasis histologically), pN0 (i−) (no regional lymph node metastases histologically, negative IHC), pN0 (i+) [malignant cells in regional lymph node(s) <0.2 mm and <200 cells (detected by H&E or IHC including ITC)], pN1mi [micrometastases (>0.2 mm and/or >200 cells, but none >2.0 mm)], pN1a (metastases in 1–3 axillary lymph nodes, pN2a: metastases in 4–9 axillary lymph nodes), and pN3a (metastases in 10 or more axillary lymph nodes); pM (distant metastasis) X (metastatic sites cannot be assessed), pM0 (no metastases), and pM1 (distant detectable metastasis by classic clinical and radiographic means and/or histologically proven >0.2 mm).

Once a clinical diagnosis of breast cancer is made, further tests (e.g., hormone receptor assessment and gene expression profiling) are conducted with the cancer tissue to determine the growth potential, likelihood to spread through the body, and best treatment option and chance of recurrence.

Based on the presence of hormone receptors (ER and PR) and the quantity of HER2 detected by immunohistochemical (IHC) and other related procedures, invasive breast cancer can be classified as: (i) hormone receptor-positive, (ii) hormone receptor-negative, (iii) HER2-positive, (iv) HER2-negative, (v) triple-negative, (vi) triple-positive. The availability of this information helps design tailor-made treatments for invasive breast cancer types/subtypes, and provides valuable insights on the prognosis of invasive breast cancer [19,20].

Gene expression profiling investigates on the activity of multiple genes (21 in Oncotype DX test, 70 in MammaPrint test, and 58 in Prosigna breast cancer prognostic gene signature assay or PAM50 test), which consist of not only *BRCA1/2*, but also *ATM*, *CHEK2*, and *PALB2*, etc. This helps predict likelihood of cancer spread to other parts of the body or recurrence as well as potential response to chemotherapy.

Moreover, tests (not usually recommended as part of a routine breast cancer workup) may be conducted for ploidy and cell proliferation rate. The cells containing normal amount of DNA are described as *diploid*, and the cells containing abnormal amount of DNA are described as *aneuploid*. The rate of cancer cell division can be estimated by the S-phase fraction or a Ki-67 test. Ki-67 is a nuclear protein of 359 kDa encoded by Ki-67 gene located on the long arm of chromosome 10 (10q$^{26.2}$). Ki-67 expression increases with cell growth and is required for maintaining cell proliferation, DNA metabolic process, cellular response to heat, meiosis, and organ regeneration. The cancer cells with a high S-phase fraction or Ki-67 labeling index divide more rapidly, suggesting a more aggressive cancer [21]. Further, S100A14 and S100A16 are recently identified cancer biomarkers that play a potential role in tumor progression. Higher expression levels of S100A14 and S100A16 proteins in breast cancer appear to be significantly associated with a younger age (<60 years), ER-negative status, HER2-positive status, and a poorer prognosis (Figure 30.2) [22].

30.6.2 Clinical Diagnosis of Ovarian Cancer

Histologically, ovarian serous tumor contains tall, columnar, ciliated epithelial cells filled with clear serous fluid, and frequent psammoma bodies (concentric calcifications) (Figure 30.3) [23]. It stains positive for CK7, CK8/18, CK19, EMA, B72.3, S100, amylase (25%), ER (50%), PR (50%), androgen receptor (50%), N-cadherin, but negative for CEA, CDX2, and CK20. Cancer antigen (CA)-125 shows high sensitivity and high specificity for postmenopausal women, but high sensitivity and low specificity for premenopausal women [24,25].

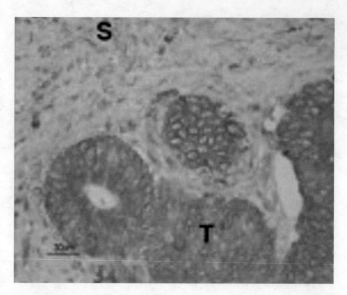

FIGURE 30.3 Light microscope image of mucinous ovarian tumor tissue (T) and stroma (S) showing secreted frizzled-related protein 4 (sFRP4) expression (brown color) (DAB stain, H&E counterstain, scale bar = 30 μm). (Photo credit: Saran U et al. *BMC Cell Biol.* 2012;13:25.)

Dysgerminoma displays nests of round and ovoid cells separated by fibrous stroma with T lymphocytes. It stains positive for OCT4 (strong nuclear staining in 90% cells), c-kit (87%), CAM5.2 (20%), and AE1-AE3 (8%), but negative for CK7, CK20, EMA, HMW keratin, CD30, and vimentin.

Granulosa cell tumor (GCT) includes small, bland, cuboidal to polygonal cells with a coffee bean–like longitudinal nuclear groove and a microfollicular structure (the Call-Exner body) and demonstrates microfollicular, macrofollicular, trabecular, insular, diffuse, and watered-silk (gyriform) growth patterns.

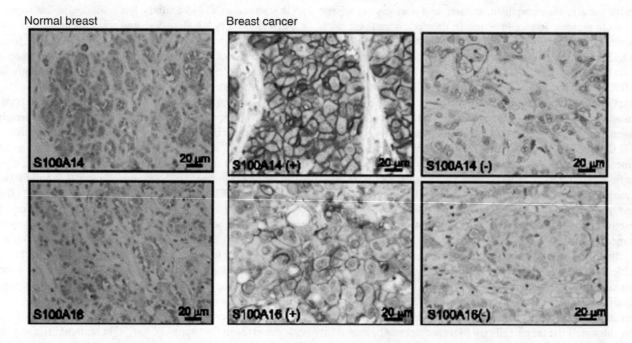

FIGURE 30.2 Immunohistochemical examination of S100A14 and S100A16 expression in normal and cancerous breast tissues. No or faint signals from these proteins were detected in normal epithelial cells of the breast, whereas both proteins were strongly expressed along the cell membrane in about half of breast cancer tissues. (Photo credit: Tanaka M et al. *BMC Cancer.* 2015;15:53.)

Microfollicular GCT contains pathognomonic Call-Exner bodies (small rings of granulosa cells surrounding eosinophilic fluid and basement membrane material). Macrofollicular GCT contains one or more large cysts lined with granulosa cells. Trabecular and insular GCT have granulosa cells organized into nests and bands, with an intervening fibrothecomatous stroma present in trabecular GCT. Diffuse GCT contains sheets of cells arranged in no pattern. Watered-silk GCT contains cells arranged in single file in line. Adult GCT displays large, pale, ovoid or angular nuclei with nuclear grooves, mild nuclear atypia, few mitotic figures, little cytoplasm, and occasional luteinization. In contrast, juvenile GCT shows round hyperchromatic nuclei without nuclear grooves, severe nuclear atypia, more mitotic figures, and more cytoplasm (which is dense). GCT stains positive for inhibin alpha, vimentin, calretinin, CD99, smooth muscle actin, desmoplakin, S100 (50%), keratin (30%–50%), anti-Mullerian hormone (focally), desmin (35%), and silver stains (reticulin surrounding cluster of cells), but negative for EMA. Molecularly, GCT shows monosomy 22 (~40%), trisomy 12 (~30%), +14 (~30%), monosomy X (~10%), monosomy 17 (5%), although most tumors (80%) are diploid or near-diploid.

Staging of ovarian cancer is based on the tumor, node, metastasis (TNM) or the International Federation of Gynecology and Obstetrics (FIGO) systems. In the FIGO system, stage I describes tumors confined to ovaries, stage II has pelvic extension or primary peritoneal cancer, stage III indicates spread to the peritoneum outside the pelvis and/or metastasis to the retroperitoneal lymph nodes, and stage IV contains distant metastasis [26].

30.6.3 Molecular Confirmation of HBOC

Molecular identification of heterozygous germline *BRCA1* or *BRCA2* pathogenic variants through sequence analysis (for small intragenic deletions/insertions, missense, nonsense, and splice site variants in ≥80% of cases), and deletion/duplication analysis (for intragenic deletions or duplications in ~10% of cases) helps establish or confirm the diagnosis of HBOC in ≥90% of patients. The expected proportions of *BRCA1* and *BRCA2* pathogenic variants implicated in HBOC stand at 66% and 34%, respectively [27–30].

Molecular genetic testing should be carried out on individuals suspected of *BRCA1* or *BRCA2*- associated HBOC, including those with a personal or family history (first-, second-, or third-degree relative in either lineage) and any of the following characteristics: (i) breast cancer diagnosed at or before age 50 years; (ii) ovarian cancer; (iii) multiple primary breast cancers either in one or both breasts; (iv) male breast cancer; (v) triple-negative (estrogen receptor-negative, progesterone receptor-negative, and HER2/neu [human epidermal growth factor receptor 2]-negative) breast cancer, particularly when diagnosed before age 60 years; (vi) the combination of pancreatic cancer and/or prostate cancer (Gleason score ≥7) with breast cancer, and/or ovarian cancer; (vii) breast cancer diagnosed at any age in an individual of Ashkenazi Jewish ancestry; (viii) two or more relatives with breast cancer, one under age 50; (ix) three or more relatives with breast cancer at any age; (x) a previously identified *BRCA1* or *BRCA2* pathogenic variant in the family [31].

Individuals of Ashkenazi Jewish ancestry should be targeted for three *BRCA1* and *BRCA2* pathogenic founder variants:

BRCA1 c.68_69delAG (BIC: 185delAG) *BRCA1* c.5266dupC (BIC: 5382insC), and *BRCA2* c.5946delT (BIC: 6174delT), which are responsible for up to 99% of pathogenic variants identified in this group. If targeted analysis yields a negative result, sequence and deletion/duplication analyses of *BRCA1* and *BRCA2* or a multigene panel should be utilized [32,33].

30.6.4 Differential Diagnoses

Differential diagnoses for HBOC consist of autosomal dominant breast cancer syndromes (e.g., Li—Fraumeni syndrome [due to *TP53* mutation; increased risk for breast cancer ≤79%, soft tissue sarcoma, osteosarcoma, brain tumor, adrenocortical carcinoma, leukemia], Cowden syndrome/*PTEN* hamartoma tumor syndrome [due to *PTEN* mutation; increased risk for breast cancer 25–≤85%, thyroid cancer, renal cell carcinoma, endometrial carcinoma, colorectal cancer], hereditary diffuse gastric cancer [due to *CDH1* mutation, increased risk for breast cancer 39%–52%, diffuse gastric cancer], CHEK2 [due to *CHEK2* mutation; increased risk for breast cancer 25%–39%, prostate cancer, stomach cancer, sarcoma, kidney cancer], ATM heterozygotes/ataxia-telangiectasia [due to *ATM* mutation; increased risk for breast cancer 17%–52%, other cancer], PALB2 [due to *PALB2* mutation; increased risk for breast cancer ≤58%, male breast cancer, pancreatic cancer], Peutz—Jeghers syndrome [due to *STK11* mutation; increased risk for breast cancer 32%–54%, gastrointestinal malignancies, ovarian cancer/sex cord tumor with annular tubules, cervical cancer/adenoma malignum, uterine cancer, pancreatic cancer, Sertoli cell testicular cancer, lung cancer], RAD51C [due to *RAD51C* mutation; increased risk for ovarian cancer], and Lynch syndrome [due to *MLH1, MSH2, MSH6, PMS2, EPCAM* mutations, increased risk for ovarian cancer, nonpolyposis colorectal cancer, endometrial cancer, other cancer]), as well as autosomal recessive breast cancer syndromes (e.g., Bloom's syndrome [due to *BLM* mutation; increased risk for breast cancer, epithelial carcinoma, lymphoma, leukemia, other cancer], Werner syndrome [due to mutation; increased risk for breast cancer risk, sarcoma, melanoma, thyroid cancer, hematologic malignancies]). In addition, germline BRCA2 mutations may be found in familial pancreatic cancer and Fanconi anemia complementation group FANCD1 [1].

30.7 Treatment

Treatment options for patients with breast and/or ovarian cancer consist of surgery, chemotherapy, radiotherapy, hormone therapy, and targeted therapy [34–36].

For breast cancer patients, surgery includes (i) breast-conserving surgery (or lumpectomy, partial mastectomy, segmental mastectomy, quadrantectomy, or breast-sparing surgery) for removing the cancer and some normal tissue around it, but not the breast itself; (ii) total mastectomy (or simple mastectomy) for removing the whole breast that has cancer; breast reconstruction with the patient's own (nonbreast) tissue or by using implants filled with saline or silicone gel may be undertaken at the time of the mastectomy or at some time after; (iii) modified radical mastectomy for removing the whole breast that has cancer, many of the lymph nodes under the arm, the lining over the chest muscles, and sometimes part of the chest wall muscles.

Chemotherapy consists of (i) preoperative therapy (or neoadjuvant therapy), given before surgery to shrink the tumor and reduce the amount of tissue that needs to be removed during surgery; (ii) postoperative therapy (or adjuvant therapy), given after surgery to kill any cancer cells that are left and to lower the risk that the cancer will come back.

Radiotherapy includes external radiation therapy (using a machine outside the body to send radiation toward the cancer) and internal radiation therapy (using a radioactive substance such as strontium-89 sealed in needles, seeds, wires, or catheters that are placed directly into or near the cancer) to kill cancer cells or keep them from growing.

Hormone therapy removes hormones (made by glands in the body and circulated in the bloodstream) or blocks their action and stops cancer cells from growing. It is most often used as an adjuvant therapy to help reduce the post-surgery relapse risk as well as metastases.

Targeted therapy uses drugs or other substances to identify and attack specific cancer cells without harming normal cells.

For ovarian cancer patients, surgery alone is adequate for stage I ovarian epithelial cancer (OEC), while surgery and chemotherapy (e.g., polyadenosine diphosphate-ribose polymerase inhibitor [PARPi:olaparib]) are required for later stage, *BRCA*-mutated OEC. Surgery and chemotherapy are appropriate for dysgerminoma, but not radiotherapy. Use of BEP (bleomycin, etoposide, and cisplatin) has been shown to result in a 5-year survival rate of up to 100% for dysgerminomas and 85% for non-dysgerminomatous tumors. A combination of surgery (transabdominal hysterectomy or bilateral salpingo-oophorectomy for women beyond childbearing age, and unilateral oophorectomy for younger women) and chemotherapy is useful for treating GCT.

For individuals with *BRCA1/BRCA2* pathogenic variants, bilateral prophylactic mastectomy reduces the risk of developing breast cancer by 90%–95%, while chemoprevention with tamoxifen or raloxifene also contributes to lower breast cancer risk. Prophylactic oophorectomy, salpingectomy with ovarian retention, and tubal ligation also help decrease risk for ovarian cancer/ fallopian tube cancer [37].

30.8 Prognosis and Prevention

Prognosis for breast cancer depends upon the type, location, grade, and stage of the cancer, levels of ER, PG, HER2/neu in the tumor tissue, presence of tumor markers, general condition (including menopausal status), age of the patient, etc. Patients with stages 0, I, I, III, and IV breast cancer have a 5-year overall survival rates of 93%, 88%, 74%–81%, 41%–67%, and 15%, respectively [38].

As ovarian cancer is often diagnosed at an advanced stage (III or IV), it often has a poor prognosis, with a 5-year survival rate of only 45% (stage III: 40%–60%; stage IV: 17%). Patients with stage I or II GCT have 5-year a survival rate of 95%, and those with stage III or IV GCT have 5-year a survival rate of 59%. Juvenile GCT often recurs within 3 years and is rapidly fatal.

Prophylactic bilateral mastectomy and chemoprevention (e.g., tamoxifen) may be used for breast cancer prevention, while prophylactic oophorectomy is beneficial for ovarian cancer prevention [39].

Since germline pathogenic variants in *BRCA1* and *BRCA2* follow an autosomal dominant inheritance pattern and show incomplete penetrance, the relatives of patients harboring such pathogenic variants should undergo molecular genetic testing, which will help implement early intervention and regular surveillance measures to reduce potential disease burden. Prenatal testing and preimplantation diagnosis may be considered for reducing cancer risk in offspring of individuals harboring *BRCA1/BRCA2* pathogenic variants [40].

30.9 Conclusion

Hereditary breast and ovarian cancer (HBOC) is an autosomal dominant syndrome that increases the risk for female and male breast cancer, ovarian cancer (including fallopian tube and primary peritoneal cancers), and other malignancies (e.g., prostate cancer, pancreatic cancer, and melanoma). Apart from its early appearance and obvious family connection, hereditary breast and ovarian cancer often produces multiple/bilateral lesions that harm the human body in more than one way (e.g., higher nutrient demand, more frequent physical compression, and more opportunity for spreading/metastasis). The molecular basis of HBOC lies in heterozygous germline mutations in the *BRCA1* and *BRCA2* genes, which are responsible for >60% of familial breast and ovarian cancer cases. Identification of pathogenic *BRCA1* and *BRCA2* variants provides a valuable means of establishing and confirming the diagnosis of HBOC. Given the limited options for treating breast and ovarian cancer at the moment, further research on the pathogenesis of these neoplasms is crucial for designing novel, effective measures against this important cancer predisposition syndrome.

REFERENCES

1. Casaubon JT, Regan JP. *BRCA 1 and 2. StatPearls [Internet]*. Treasure Island (FL): StatPearls Publishing; 2018 Jan-. 2018 Mar 25.
2. Lakhani S, Ellis I, Schnitt S et al. *4th. WHO Classification of Tumours of the Breast*. Lyon: IARC Press; 2012.
3. Sinn HP, Kreipe H. A brief overview of the WHO classification of breast tumors, 4th Edition, focusing on issues and updates from the 3rd Edition. *Breast Care (Basel)*. 2013;8(2):149–54.
4. Mego M, Cierna Z, Janega P et al. Relationship between circulating tumor cells and epithelial to mesenchymal transition in early breast cancer. *BMC Cancer*. 2015;15:533.
5. Kurman RJ, editor. *WHO classification of tumours of female reproductive organs. International Agency for Research on Cancer; World Health Organization*. Lyon: International Agency for Research on Cancer, 2014.
6. Burgess M, Puhalla S. BRCA 1/2-mutation related and sporadic breast and ovarian cancers: More alike than different. *Front Oncol*. 2014;4:19.
7. Phelan CM, Kuchenbaecker KB, Tyrer JP et al. Identification of 12 new susceptibility loci for different histotypes of epithelial ovarian cancer. *Nat Genet*. 2017;49(5):680–91.
8. Jiang Q, Greenberg RA. Deciphering the BRCA1 tumor suppressor network. *J Biol Chem*. 2015;290(29):17724–32.

9. Katsuki Y, Takata M. Defects in homologous recombination repair behind the human diseases: FA and HBOC. *Endocr Relat Cancer.* 2016;23(10):T19-37.

10. Gambino G, Tancredi M, Falaschi E, Aretini P, Caligo MA. Characterization of three alternative transcripts of the BRCA1 gene in patients with breast cancer and a family history of breast and/or ovarian cancer who tested negative for pathogenic mutations. *Int J Mol Med.* 2015;35(4):950–6.

11. Cao WM, Gao Y, Yang HJ et al. Novel germline mutations and unclassified variants of BRCA1 and BRCA2 genes in Chinese women with familial breast/ovarian cancer. *BMC Cancer.* 2016;16:64.

12. Jara L, Morales S, de Mayo T, Gonzalez-Hormazabal P, Carrasco V, Godoy R. Mutations in BRCA1, BRCA2 and other breast and ovarian cancer susceptibility genes in Central and South American populations. *Biol Res.* 2017;50(1):35.

13. Yang C, Arnold AG, Trottier M et al. Characterization of a novel germline PALB2 duplication in a hereditary breast and ovarian cancer family. *Breast Cancer Res Treat.* 2016;160(3):447–56.

14. Petrucelli N, Daly MB, Pal T. *BRCA1-* and *BRCA2*-associated hereditary breast and ovarian cancer. In: Adam MP, Ardinger HH, Pagon RA et al. editors. *GeneReviews® [Internet].* Seattle (WA): University of Washington, Seattle; 1993–2018. 1998 Sep 4 [updated 2016 Dec 15].

15. Silver DP, Livingston DM. Mechanisms of BRCA1 tumor suppression. *Cancer Discov.* 2012;2(8):679–84.

16. Wittersheim M, Büttner R, Markiefka B. Genotype/Phenotype correlations in patients with hereditary breast cancer. *Breast Care (Basel).* 2015;10(1):22–6.

17. Alemar B, Gregório C, Herzog J et al. BRCA1 and BRCA2 mutational profile and prevalence in hereditary breast and ovarian cancer (HBOC) probands from Southern Brazil: Are international testing criteria appropriate for this specific population? *PLOS ONE.* 2017;12(11):e0187630.

18. Paul A, Paul S. The breast cancer susceptibility genes (BRCA) in breast and ovarian cancers. *Front Biosci (Landmark Ed).* 2014;19:605–18.

19. Hahnen E, Hauke J, Engel C, Neidhardt G, Rhiem K, Schmutzler RK. Germline mutations in triple-negative breast cancer. *Breast Care (Basel).* 2017;12(1):15–9.

20. Mundhofir FE, Wulandari CE, Prajoko YW, Winarni TI. BRCA1 gene mutation screening for the hereditary breast and/or ovarian cancer syndrome in breast cancer cases: A first high resolution DNA melting analysis in Indonesia. *Asian Pac J Cancer Prev.* 2016;17(3):1539–46.

21. Khabaz MN, Abdelrahman A, Butt N et al. Immunohistochemical staining of leptin is associated with grade, stage, lymph node involvement, recurrence, and hormone receptor phenotypes in breast cancer. *BMC Womens Health.* 2017;17(1):105.

22. Wang J, Ma S, Ma R et al. KIF2A silencing inhibits the proliferation and migration of breast cancer cells and correlates with unfavorable prognosis in breast cancer. *BMC Cancer.* 2014;14:461.

23. Tanaka M, Ichikawa-Tomikawa N, Shishito N et al. Co-expression of S100A14 and S100A16 correlates with a poor prognosis in human breast cancer and promotes cancer cell invasion. *BMC Cancer.* 2015;15:53.

24. Saran U, Arfuso F, Zeps N, Dharmarajan A. Secreted frizzled-related protein 4 expression is positively associated with responsiveness to cisplatin of ovarian cancer cell lines *in vitro* and with lower tumour grade in mucinous ovarian cancers. *BMC Cell Biol.* 2012;13:25.

25. Tuhkanen H, Soini Y, Kosma VM et al. Nuclear expression of Snail1 in borderline and malignant epithelial ovarian tumours is associated with tumour progression. *BMC Cancer.* 2009;9:289.

26. Wang S, Wei H, Zhang S. Dickkopf-4 is frequently overexpressed in epithelial ovarian carcinoma and promotes tumor invasion. *BMC Cancer.* 2017;17(1):455.

27. Walker JL, Powell CB, Chen LM et al. Society of Gynecologic Oncology recommendations for the prevention of ovarian cancer. *Cancer.* 2015;121(13):2108–20.

28. Eccles DM, Mitchell G, Monteiro AN et al. BRCA1 and BRCA2 genetic testing-pitfalls and recommendations for managing variants of uncertain clinical significance. *Ann Oncol.* 2015;26(10):2057–65.

29. Li J, Meeks H, Feng BJ et al. Targeted massively parallel sequencing of a panel of putative breast cancer susceptibility genes in a large cohort of multiple-case breast and ovarian cancer families. *J Med Genet.* 2016;53(1):34–42.

30. Grindedal EM, Heramb C, Karsrud I et al. Current guidelines for BRCA testing of breast cancer patients are insufficient to detect all mutation carriers. *BMC Cancer.* 2017;17(1):438.

31. Capoluongo E, Ellison G, López-Guerrero JA et al. Guidance statement on BRCA1/2 tumor testing in ovarian cancer patients. *Semin Oncol.* 2017;44(3):187–97.

32. Eliade M, Skrzypski J, Baurand A et al. The transfer of multigene panel testing for hereditary breast and ovarian cancer to healthcare: What are the implications for the management of patients and families? *Oncotarget.* 2017;8(2):1957–71.

33. Tedaldi G, Tebaldi M, Zampiga V et al. Multiple-gene panel analysis in a case series of 255 women with hereditary breast and ovarian cancer. *Oncotarget.* 2017;8(29):47064–75.

34. Agarwal R, Liebe S, Turski ML et al. Targeted therapy for hereditary cancer syndromes: Hereditary breast and ovarian cancer syndrome, Lynch syndrome, familial adenomatous polyposis, and Li-Fraumeni syndrome. *Discov Med.* 2014;18(101):331–9.

35. Iyevleva AG, Imyanitov EN. Cytotoxic and targeted therapy for hereditary cancers. *Hered Cancer Clin Pract.* 2016;14(1):17.

36. Parkes A, Arun BK, Litton JK. Systemic treatment strategies for patients with hereditary breast cancer syndromes. *Oncologist.* 2017;22(6):655–66.

37. Rhiem K, Schmutzler R. Impact of prophylactic mastectomy in BRCA1/2 mutation carriers. *Breast Care (Basel).* 2014;9(6):385–9.

38. Engel C, Fischer C. Breast cancer risks and risk prediction models. *Breast Care (Basel).* 2015;10(1):7–12.

39. Kast K, Rhiem K. Familial breast cancer - targeted therapy in secondary and tertiary prevention. *Breast Care (Basel).* 2015;10(1):27–31.

40. Zeichner SB, Stanislaw C, Meisel JL. Prevention and screening in hereditary breast and ovarian cancer. *Oncology (Williston Park).* 2016;30(10):896–904.

31

Hereditary Leiomyomatosis and Renal Cell Cancer

Dongyou Liu

CONTENTS

31.1 Introduction

Hereditary leiomyomatosis and renal cell cancer (HLRCC) is an autosomal dominant cancer predisposition syndrome that typically presents with cutaneous leiomyoma, uterine leiomyoma (fibroids), and type 2 papillary renal cell carcinoma. Occasionally, other unusual forms of tumors (e.g., basal cell carcinoma and melanoma) may be observed.

Heterogeneous germline mutations in the fumarate hydratase (*FH*) gene located on chromosome 1q42.1, which converts fumarate to malate in the Kreb cycle, form the molecular basis of HLRCC. In contrast, homozygous mutations in *FH* resulting in fumarate hydratase deficiency (FMRD) are associated with an autosomal recessive hereditary phenotype characterized by severe neonatal encephalopathy, poor feeding, failure to thrive, hypotonia, lethargy, seizures, and early death [1].

31.2 Biology

A hereditary form of multiple cutaneous leiomyoma (or leiomyomata) was first described from an Italian family by Kloepfer et al. in 1958 [2]. Subsequent study by Reed et al. in 1973 demonstrated the autosomal dominant inheritance of multiple cutaneous leiomyoma (MCL) in association with uterine leiomyoma and/or leiomyosarcoma, justifying the nomination of the disease as Reed syndrome (i.e., multiple cutaneous and uterine leiomyomatosis [MCUL]) [3]. Following the finding of renal cell carcinoma with papillary architecture in Reed syndrome patients by Launonen et al. and Kiuru et al. in 2001, the name "hereditary leiomyomatosis renal cell cancer/carcinoma (HLRCC)" was adopted [4,5]. The mapping of the culprit gene to chromosome 1q42.1 in 2001 and identification of the *FH* gene in 2002 have helped uncover the underlying cause of HLRCC [6].

With the description of >300 families to date, the clinical spectrum of HLRCC has expanded from the triad of cutaneous leiomyoma, uterine leiomyoma, and type 2 papillary renal cell carcinoma to include basal cell carcinoma, melanoma, leiomyosarcoma, and other renal cell carcinoma types as atypical presentation of this order.

Leiomyoma (synonym: leiomyomata) is a benign neoplasm that evolves from the erector pili muscle of the pilosebaceous unit (cutaneous leiomyoma/piloleiomyoma), the cutaneous vascular smooth muscle fibers (angioleiomyoma), or the dartos muscle (genital leiomyoma/uterine leiomyoma). Histologically, leiomyoma displays whorled (fascicular) pattern of smooth muscle bundles separated by well vascularized connective tissue, elongated smooth muscle cells with eosinophilic or occasional fibrillar cytoplasm and distinct cell membranes, low (<5) mitotic figures per 10 high power, absence of significant atypia, and lack of hemorrhage and necrosis (Figures 31.1b and 31.2b). However,

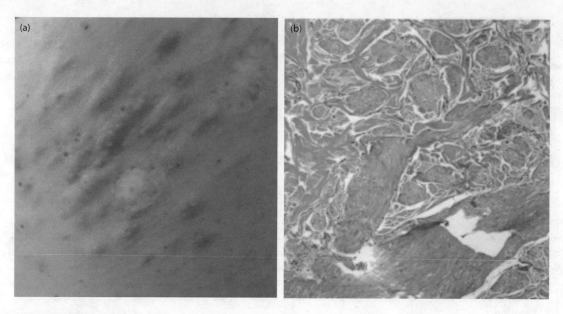

FIGURE 31.1 Pink-to-brown papules and nodules on the left subscapular region of a 37-year-old female suffering from multiple cutaneous and uterine leiomyomatosis syndrome (a). Hematoxylin-eosin stained section reveals a poorly circumscribed lesion from interlacing bundles of benign smooth muscle cells (b). (Photo credit: Diluvio L et al. *BMC Dermatol.* 2014;14:7.)

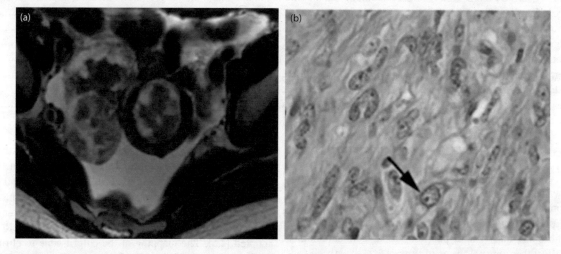

FIGURE 31.2 MRI axial T2 image of abdomen and pelvis of a 27-year-old female with HLRCC shows both 4 cm intramural and 8 cm right subserosal fibroids (a). Histology of resected uterine leiomyoma demonstrates diffuse mild-moderate nuclear atypia and large cherry-red nucleoli with distinct clearing of the coarsely granular chromatin around the nucleoli (arrow) (b). (Photo credit: Bortoletto P et al. *Case Rep Womens Health.* 2017;15:31–34.)

leiomyoma has the potential to undergo malignant transformation and turn into leiomyosarcoma, which demonstrates obvious nuclear atypia, high mitotic activity, large size, frequent hemorrhage, and necrosis [7].

Renal cell cancer (RCC) arises from the cells in the parenchyma of the kidneys, and consists of several histological types (e.g., clear cell RCC, papillary RCC, chromophobe RCC). Of these, papillary RCC is a peripheral lesion with necrosis, hemorrhage, and calcification, and shows tumor cells organized in a spindle-shaped pattern and possible areas of internal hemorrhage and cystic alterations. Papillary RCC can be further distinguished into two subtypes, with subtype 1 (basophilic, 73% of papillary RCC cases) being low grade and consisting of papillae lined by a single layer of small cells with scanty clear-to-basophilic cytoplasm and hyperchromatic nuclei, and subtype 2 (eosinophilic, 27% of papillary RCC cases) being high grade, and consisting

of papillae lined by cells with abundant granular eosinophilic cytoplasm and prominent nucleoli, which is commonly found in HLRCC (Figure 31.3b) [8].

Although germline mutation in *FH* predisposes to HLRCC (which is detected in 76%–100% of families with suggestive clinical features), another genetic event (e.g., loss of heterozygosity [LOH] at the *FH* gene locus [which is found in 80% of HLRCC renal cancer] or somatic mutation of the *FH* gene) in the other allele is necessary for the initiation of this order.

31.3 Pathogenesis

Mapped to on chromosome 1q42.1, the *FH* gene comprises a 22.15 kb stretch of DNA with 10 exons, and encodes 510 aa, 54 kDa FH that catalyzes the conversion of fumarate to

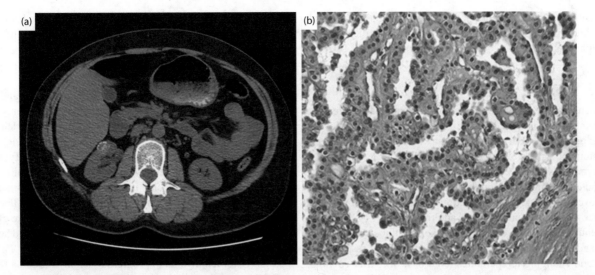

FIGURE 31.3 Noncontrast CT reveals a cortically based, partially exophytic mass with peripheral calcifications, located in the lower pole of the right kidney of a 51-year-old HLRCC-affected female (a). Histology of partial nephrectomy specimen displays renal cell cancer with papillary type 2 morphology, noting predominantly eosinophilic cells forming several cell strata on the axi with large cytoplasms, pseudostratified nuclei, prominent eosinophilic nucleoli, and perinucleolar halo (H&E, 400×) (b). (Photo credit: Fondriest SA et al. *Radiol Case Rep.* 2015;10(1):962.)

L-malate in the tricarboxylic acid (Krebs) cycle. Depending on the translation start site used, FH consists of both a cytosolic form and an N-terminal extended form (which targets the mitochondrion, where removal of the extension yields the same form as in the cytoplasm).

Heterozygous germline *FH* mutation (e.g., missense, nonsense, frameshift, and splice site variants) followed by second hit (de novo mutation or loss of heterozygosity) on the wildtype allele results in *FH* inactivation and loss of function of FH. The reduced FH activity increases intracellular fumarate accumulation, which competitively inhibits the 2-OG-dependent hypoxia-inducible factor (HIF) hydroxylases, decreases HIF degradation, and upregulates transcriptional pathways mediated by HIF-1α, ultimately leading to epigenetic abnormalities (e.g., leiomyomas and RCC) [9].

The pathogenic variant 905–1G>A appears to be a founder mutation affecting four families of Jewish Iranian origin, the pathogenic variant c.302G>C occurs in a German and English family, and the pathogenic variant p.Glu404Ter is detected in three families in the Netherlands [10].

It is of interest to note that heterogeneous germline mutations observed in HLRCC chiefly occur in the 5′ end of the *FH* gene, in comparison with homozygous or compound heterozygous germline mutations implicated in autosomal recessive hereditary FH-deficiency causing rapidly progressive neurological impairment (e.g., hypotonia, seizures, and cerebral atrophy) and early mortality, which involve the 3′ end of the *FH* gene [11].

Furthermore, a small number of patients with paraganglioma/pheochromocytoma have been shown to harbor germline *FH* pathogenic variants. Occasionally, sporadic uterine leiomyosarcoma and cutaneous leiomyoma without other associated tumors characteristic of the heritable disease may be caused by biallelic inactivation (somatic mutation plus LOH) of the *FH* gene.

Chromosomal changes observed in HLRCC renal tumors with papillary type 2 morphology include losses in chromosomes 1q, 13q12.3-q21.1, 14, and gains in chromosomes 2, 7, 17, and X.

31.4 Epidemiology

31.4.1 Prevalence

HLRCC is rare cancer predisposition syndrome, with over 300 families from various populations described to date. Though rare, over 200 families worldwide have been reported to carry germline mutations in *FH*.

31.4.2 Inheritance

HLRCC demonstrates an autosomal dominant pattern of inheritance and afflicts multiple generations of the same family. Individual with HLRCC has a 50% chance of passing pathogenic FH variant to offspring, with one copy of the altered gene in each cell being sufficient to cause the disorder. A recent study indicates that the median age at diagnosis for RCC between the first and second generation, and between the first and third generation stand at 18.6 and 36.2 years, respectively. The risk for HLRCC patients to develop RCC is 6.5 times higher than that for the general population [12].

31.4.3 Penetrance

HLRCC shows high but incomplete penetrance and wide phenotypic variation ranging from minor skin involvement to fatal metastatic RCC. Some individuals harboring an *FH* pathogenic variant may be asymptomatic.

31.5 Clinical Features

HLRCC is associated with the development of cutaneous leiomyoma, uterine leiomyoma (fibroids), and type 2 papillary renal cell carcinoma, although other forms of neoplasms may occur occasionally. Depending on the location, these tumors may be responsible for causing abdominal/flank mass, abdominal/flank pain, abdominal discomfort, hematuria, male infertility, fatigue, or weight loss. Some patients may be asymptomatic [13,14].

31.5.1 Cutaneous Leiomyoma

Occurring in 76% of cases, cutaneous leiomyoma is often the first manifestation of HLRCC, with a mean onset age of 25 years (range: age 10–47 years). Appearing as firm, skin-colored, erythematous or light-brown soft dermal papules or nodules (multiple or single) of 0.2–2.0 cm in diameter and 1–150 in number distributed in single, clustered, segmental, or disseminated patterns over the trunk/shoulders, extremities, and occasionally on the face or neck, more often in adulthood than in childhood, cutaneous leiomyoma tends to increase in number and dimension, and cause pain, paresthesia, and itching in response to touch and changing temperatures (cold or heat). After rubbing, the lesion may develop transient piloerection or elevation, giving a pseudo Darier sign. Sometimes, it may undergo malignant transformation and develop into leiomyosarcoma [15,16].

31.5.2 Uterine Leiomyoma (Fibroids)

Affecting approximately 80% of HLRCC females (with a mean age of 28 years, range age 18–52 years, which is average 10 years earlier than sporadic cases), uterine leiomyoma tends to produce numerous and large lesions, and causes abnormal uterine bleeding (menorrhagia and metrorrhagia), lower abdomen pressure, and pelvic pain (30% of cases), starting in women from the mid-teens to early 20s [17].

31.5.3 Papillary Renal Cell Cancer/Carcinoma

Found in up to 20% of patients (median age of 41 years, range age 11–90 years) with HLRCC, RCC often forms as unilateral, solitary, palpable mass that causes hematuria and lower back pain. In about 7% of patients, RCC appears before 20 years of age [18].

31.5.4 Other Tumors

Besides cutaneous and uterine leiomyomas and RCC, patients with HLRCC may sometimes develop cutaneous basal cell carcinoma, melanoma, adrenocortical hyperplasia/tumor, thyroid follicular carcinoma, bladder cancer, liver hemangioma, Leydig cell tumor, ovarian cystadenoma, gastrointestinal stromal tumor, breast cancer, leukemia, cutis verticis gyrata, eruptive collagenomas, Charcot–Marie–Tooth disease, renal cysts, etc. [19–24].

31.6 Diagnosis

Clinical diagnosis of HLRCC centers on identification of multiple cutaneous leiomyomas, severely symptomatic early-onset uterine leiomyomas, and type 2 papillary renal carcinoma before age 40, along with a first-degree relative who shows one of these features.

Cutaneous leiomyomas typically form skin-colored to light brown papules or nodules on the trunk, extremities, and face, although they may appear single, grouped/clustered, segmental, and disseminated at times, with 40% of HLRCC patients having five or fewer lesions. Confirmation of cutaneous leiomyoma requires histological examination, which reveals proliferation of interlacing bundles of smooth muscle fibers with centrally located, long blunt-edged nuclei, but general absence of cytologic atypia (Figure 31.3a) [25].

Uterine leiomyomas (fibroids) often show numerous and large lesions and correlate to the presence of cutaneous leiomyomas. Imaging techniques are useful for locating uterine leiomyomas. However, histological examination of biopsy or surgically resected tumor is important for identification. Typically, HLRCC-related uterine leiomyoma displays prominent eosinophilic nucleoli, perinucleolar haloes, cytoplasmic eosinophilic globules, frequent atypia, multinucleated giant cells, fibrillary cytoplasm, epithelioid growth pattern, schwannoma-like growth pattern, "Orphan Annie nuclei" with optical clearing, and hemangiopericytomatous blood vessels (Figure 31.3b) [26]. There is possibility for leiomyoma to progress to leiomyosarcoma [27].

RCC appears as a solitary and unilateral lesion of 2.3–20 cm in size with partial cystic area, and occasionally as multifocal or bilateral lesions. It frequently invades capsular and perinephric adipose tissue as well as renal vein and vena cava. Abdominal CT scan helps reveal hypodensity lesion in the kidney, while contrast enhanced CT shows homogenous or nonhomogenous and less enhanced mass. Histologically, RCC is composed of neoplastic cells organized in papillary, tubulopapillary, solid, cystic tubulocystic, vacuolated/cribriform, or mixed pattern. The most common RCC type linked to HLRCC is papillary predominated type 2, which demonstrates large tumor cells with eosinophilic cytoplasm, pseudostratified nuclei containing prominent inclusion-like eosinophilic nucleoli, and perinucleolar clearing/haloes (Figure 31.3c) [28]. Immunohistochemically, tumor cells stain positive for S-(2-succino)-cysteine (2SC), PAX8, CD10, vimentin, cytokeratin 8/18, HIF-1a, and GLUT1 but show reduced expression of FH. Other RCC types (e.g., tubulopapillary RCC, collecting-duct RCC, clear cell RCC, unclassified RCC, oncocytic tumor, cystic tumor, angiomyolipoma, and Wilms tumor) may also be observed. Besides RCC, solitary or multifocal renal cysts or tubular cells with hobnail patterns in the renal parenchyma adjacent to the main tumor may be present [29].

Molecular identification of a heterozygous pathogenic variant in *FH* establishes the diagnosis of HLRCC, along with the following clinical findings: (i) multiple cutaneous leiomyomas (with ≥1 histologically confirmed leiomyoma) without family history of HLRCC, (ii) a single leiomyoma with family history of HLRCC, and/or (iii) one or more papillary type 2, tubulopapillary, collecting-duct renal tumors with/out family history of HLRCC [30,31].

As FH enzyme activity is reduced by ≤60% in HLRCC patients, it may be examined alternatively in patients with atypical presentation and without detectable FH pathogenic variant [32]. In addition, due to the high prevalence of uterine leiomyomas in the general population, a solitary uterine leiomyoma even in the presence of a heterozygous *FH* pathogenic variant is insufficient for the diagnosis of HLRCC.

Found in 76%–100% of families with suggestive clinical features of HLRCC, germline mutations in the *FH* gene encompass missense (57%), nonsense, and frameshift (27%), large deletion (4%, including whole gene deletion), small deletion (4%), duplication (2%), and splice site (6%) alterations.

Molecular testing for pathogenic *FH* variants usually begins with sequence analysis of the *FH* gene (which identifies small intragenic deletions/insertions and missense, nonsense, and splice site variants in 70%–90% of cases), followed by gene-targeted deletion/duplication analysis using multigene panel (which detects intragenic deletions or duplications in 5%–14% of cases) if no pathogenic variant is found.

Differential diagnoses for cutaneous leiomyomas related to HLRCC consist of dermatofibroma, eccrine spiradenoma, neurofibroma (due to neurofibromatosis type 1, which causes café-au-lait macules, axillary freckling, and soft, rarely painful neurofibromas in the first decade of life, instead of HLRCC's firm, often painful leiomyomas between the ages of 10 and 30 years), angiolipoma, neurilemmoma, glomus tumor, keloid, hamartomas, and blue rubber bleb nevus syndrome when painful, or dermal nevus, trichoepithelioma, lipoma, cylindroma, or poroma among the asymptomatic tumors.

HLRCC-related uterine leiomyoma differs from sporadic and nonsyndromic uterine leiomyoma that commonly occurs in women by its co-presence of cutaneous leiomyoma, and germline FH mutation.

Further, HLRCC-related papillary type 2 RCC (which demonstrates the hallmarks of prominent eosinophilic nucleoli and perinucleolar halo) requires discrimination form other RCC types (e.g., collecting duct carcinoma [CDC], clear cell RCC, tubulocystic carcinoma, mucinous tubular and spindle- cell carcinoma [MTSCC, abundant mucin deposition on the background stroma and presence of elongated/anastomosing tubules], Xp11.2 RCC [TFE3 mutation], ALK renal cancer [ALK mutation] and renal oncocytoma).

Further, several familial cancer syndromes may also present with RCC as part of clinical features. These include von Hippel–Lindau syndrome (VHL, predisposition to clear cell RCC, CNS hemangioblastoma, retinal angioma, pheochromocytoma, and endolymphatic sac tumors), Birt–Hogg–Dubé syndrome (BHDS, predisposition to renal oncocytoma, chromophobe RCC, oncocytic hybrid tumor, cutaneous fibrofolliculomas and/or with multiple lung cysts and spontaneous pneumothorax), and hereditary papillary renal cancer (HPRC, predisposition to papillary type RCC). It should be noted that HLRCC-related RCC (even <1 cm in diameter) is prone to early invasion and metastasis, compared to most sporadic and other inherited RCC, which tends to be well circumscribed with a very low probability of metastasis if <3 cm in diameter [1].

31.7 Treatment

Treatment options for HLRCC patients with cutaneous leiomyomas include (i) surgical excision, cryoablation, and/or laser excision to remove painful cutaneous leiomyomas, and (ii) pain medication (e.g., calcium channel blockers, alpha blockers, nitroglycerin, antidepressants, or antiepileptic drugs) to reduce pain [33].

Treatment options for HLRCC patients with uterine leiomyomas (fibroids) consist of gonadotropin-releasing hormone agonists (GnRHa), antihormonal medications, pain relievers, myomectomy, and hysterectomy (when necessary) [34].

Treatment options for HLRCC patients with RCC comprise open partial nephrectomy/nephron-sparing therapy (with wide surgical margins and consideration of retroperitoneal lymphadenectomy, particularly for small bilateral and multifocal renal tumors), and total nephrectomy (if there is doubt that a partial nephrectomy would be curative) [35]. Patients with advanced HLRCC-associated RCC may be offered bevacizumab plus erlotinib therapy, which resulted in objective response rate (ORR) of 50%, median progression-free survival (PFS) of 13.3 months, and median overall survival (OS) of 14.1 months as reported recently [36].

31.8 Prognosis and Prevention

About 15%–20% HLRCC patients develop papillary type 2 RCC and CDC, which may metastasize to regional lymph node, distant lymph node, lung, bone, liver, peritoneum, pleura, and meninges, and are responsible for death within 5 years after initial diagnosis.

Annual surveillance for individuals with heterozygous *FH* mutations and at-risk family members should include full skin examination (for changes suggestive of leiomyosarcoma), annual gynecologic consultation (for uterine fibroids), and abdominal MRI (for renal lesion starting at age 8–10 years) [37,38]. Prenatal testing and preimplantation genetic diagnosis may be considered for individuals harboring pathogenic *FH* variant.

31.9 Conclusion

Hereditary leiomyomatosis and renal cell cancer (HLRCC) is an autosomal dominant syndrome predisposing to multiple cutaneous and uterine leiomyomas as well as type 2 papillary RCC in adulthood and occasionally in childhood (with the risk of developing RCC before age 20 estimated at 1%–2%, and lifetime risk of RCC around 15%). As heterozygous germline mutations in the *FH* gene underpin the development of HLRCC, molecular identification of pathogenic *FH* variants provides a valuable approach for establishing/confirming its diagnosis, even in individuals who do not show all syndromic features or clear family history [39,40]. The fact that pathogenic *FH* variants often compromise FH enzyme expression makes the use of immunohistochemical detection for low or absent FH protein levels and increased 2SC levels in renal tumor possible [41]. Treatment of HLRCC involves surgery, pain medication, hormone therapy, and chemotherapy, etc. [40]. Predictive germline testing of at-risk family members from the age of 8 years, and annual renal MRI for carriers of pathogenic *FH* germline variants from the age of 8 years will facilitate early diagnosis and treatment of type 2 papillary RCC, which is an aggressive tumor with metastatic potential, leading to death in 70% of affected individuals within five years of diagnosis [42].

REFERENCES

1. Pithukpakorn M, Toro JR. Hereditary leiomyomatosis and renal cell cancer. In: Adam MP, Ardinger HH, Pagon RA et al. editors. *GeneReviews®* [Internet]. Seattle (WA): University of Washington, Seattle; 1993–2018. 2006 Jul 31 [updated 2015 Aug 6].

2. Kloepfer HW, Krafchuk J, Derbres V et al. Hereditary multiple leiomyoma of the skin. *Am J Hum Genet.* 1958;10:48–52.

3. Reed WB, Walker R, Horowitz R. Cutaneous leiomyomata with uterine leiomyomata. *Acta Derm Venereol.* 1973;53:409–16.

4. Launonen V, Vierimaa O, Kiuru M et al. Inherited susceptibility to uterine leiomyomas and renal cell cancer. *Proc Natl Acad Sci USA.* 2001;98:3387–92.

5. Kiuru M, Lehtonen R, Arola J et al. Few FH mutations in sporadic counterparts of tumor types observed in hereditary leiomyomatosis and renal cell cancer families. *Cancer Res.* 2002;62(16):4554–7.

6. Almeida FT, Santos RP, Carvalho SD, Brito MC. Reed's syndrome. *Indian J Dermatol.* 2018;63(3):261–3.

7. Tulandi T, Foulkes WD. Hereditary leiomyomatosis and renal cell cancer syndrome. *CMAJ.* 2016;188(2):140.

8. Ristau BT, Kamat SN, Tarin TV. Abnormal cystic tumor in a patient with hereditary leiomyomatosis and renal cell cancer syndrome: Evidence of a precursor lesion? *Case Rep Urol.* 2015;2015:303872.

9. Kerins MJ, Milligan J, Wohlschlegel JA, Ooi A. Fumarate hydratase inactivation in hereditary leiomyomatosis and renal cell cancer is synthetic lethal with ferroptosis induction. *Cancer Sci.* 2018;109(9):2757–6.

10. Gardie B, Remenieras A, Kattygnarath D et al. Novel FH mutations in families with hereditary leiomyomatosis and renal cell cancer (HLRCC) and patients with isolated type 2 papillary renal cell carcinoma. *J Med Genet.* 2011;48(4):226–34.

11. Harrison WJ, Andrici J, Maclean F et al. Fumarate hydratase-deficient uterine leiomyomas occur in both the syndromic and sporadic settings. *Am J Surg Pathol.* 2016;40(5):599–607.

12. Mandal RK, Koley S, Banerjee S, Kabiraj SP, Ghosh SK, Kumar P. Familial leiomyomatosis cutis affecting nine family members in two successive generations including four cases of Reed's syndrome. *Indian J Dermatol Venereol Leprol.* 2013;79(1):83–7.

13. Adamane S, Desai S, Menon S. Hereditary leiomyomatosis and renal cell cancer syndrome associated renal cell carcinoma. *Indian J Pathol Microbiol.* 2017;60(1):108–10.

14. Çaliskan E, Bodur S, Ulubay M et al. Hereditary leiomyomatosis and renal cell carcinoma syndrome: A case report and implications of early onset. *An Bras Dermatol.* 2017;92(5 Suppl 1):88–91.

15. Lencastre A, Cabete J, Gonçalves R, João A, Fidalgo A. Cutaneous leiomyomatosis in a mother and daughter. *An Bras Dermatol.* 2013;88(6 Suppl 1):124–7.

16. Bell RC, Austin ET, Arnold SJ, Lin FC, Walker JR, Larsen BT. Rare leiomyoma of the tunica dartos: A case report with clinical relevance for malignant transformation and HLRCC. *Case Rep Pathol.* 2016;2016:6471520.

17. Natália F, Tiago O, Pedro O, Sandro G. Hereditary leiomyomatosis and renal cell carcinoma: Case report and review of the literature. *Urol Ann.* 2018;10(1):108–10.

18. Kuroda N, Ohe C, Kato I et al. Review of hereditary leiomyomatosis renal cell carcinoma with focus on clinical and pathobiological aspects of renal tumors. *Pol J Pathol.* 2017;68(4):284–90.

19. Behnes CL, Schlegel C, Shoukier M et al. Hereditary papillary renal cell carcinoma primarily diagnosed in a cervical lymph node: A case report of a 30-year-old woman with multiple metastases. *BMC Urol.* 2013;13:3.

20. Shuch B, Ricketts CJ, Vocke CD, Valera VA et al. Adrenal nodular hyperplasia in hereditary leiomyomatosis and renal cell cancer. *J Urol.* 2013;189(2):430–5.

21. Udager AM, Alva A, Chen YB et al. Hereditary leiomyomatosis and renal cell carcinoma (HLRCC): A rapid autopsy report of metastatic renal cell carcinoma. *Am J Surg Pathol.* 2014;38(4):567–77.

22. Sommer LL, Schnur RE, Heymann WR. Melanoma and basal cell carcinoma in the hereditary leiomyomatosis and renal cell cancer syndrome. An expansion of the oncologic spectrum. *J Dermatol Case Rep.* 2016;10(3):53–5.

23. Bevans SL, Mayo TT, Pavlidakey PG, Cannon AD, Korf BR, Mercado PJ. Unusual presentation of hereditary leiomyomatosis mimicking neurofibromatosis. *JAAD Case Rep.* 2018;4(5):440–1.

24. Chaucer B, Stone A, Demanes A, Seibert SM. Nivolumab-induced encephalitis in hereditary leiomyomatosis and renal cell cancer syndrome. *Case Rep Oncol Med.* 2018;2018 :4273231.

25. Diluvio L, Torti C, Terrinoni A et al. Dermoscopy as an adjuvant tool for detecting skin leiomyomas in patient with uterine fibroids and cerebral cavernomas. *BMC Dermatol.* 2014;14:7.

26. Bortoletto P, Lindsey JL, Yuan L et al. Hereditary leiomyomatosis and renal cell cancer: Cutaneous lesions & atypical fibroids. *Case Rep Womens Health.* 2017;15:31–4.

27. Sanz-Ortega J, Vocke C, Stratton P, Linehan WM, Merino MJ. Morphologic and molecular characteristics of uterine leiomyomas in hereditary leiomyomatosis and renal cancer (HLRCC) syndrome. *Am J Surg Pathol.* 2013;37(1):74–80.

28. Fondriest SA, Gowdy JM, Goyal M, Sheridan KC, Wasdahl DA. Concurrent renal-cell carcinoma and cutaneous leiomyomas: A case of HLRCC. *Radiol Case Rep.* 2015; 10(1):962.

29. Miettinen M, Felisiak-Golabek A, Wasag B et al. Fumarase-deficient uterine leiomyomas: An immunohistochemical, molecular genetic, and clinicopathologic study of 86 cases. *Am J Surg Pathol.* 2016;40(12):1661–9.

30. Kuwada M, Chihara Y, Lou Y et al. Novel missense mutation in the FH gene in familial renal cell cancer patients lacking cutaneous leiomyomas. *BMC Res Notes.* 2014;7:203.

31. Chan MMY, Barnicoat A, Mumtaz F et al. Cascade fumarate hydratase mutation screening allows early detection of kidney tumour: A case report. *BMC Med Genet.* 2017;18(1):79.

32. Carter CS, Skala SL, Chinnaiyan AM et al. Immunohistochemical characterization of fumarate hydratase (FH) and succinate dehydrogenase (SDH) in cutaneous leiomyomas for detection of familial cancer syndromes. *Am J Surg Pathol.* 2017;41(6):801–9.

33. Hsu T, Cornelius LA, Rosman IS, Nemer KM. Treatment of cutaneous leiomyomas with 5% lidocaine patches in a patient with hereditary leiomyomatosis and renal cell cancer (Reed syndrome). *JAAD Case Rep.* 2017;3(5):407–9.

34. Menko FH, Maher ER, Schmidt LS et al. Hereditary leiomyomatosis and renal cell cancer (HLRCC): Renal cancer risk, surveillance and treatment. *Fam Cancer.* 2014;13(4):637–44.

35. Kamai T, Abe H, Arai K et al. Radical nephrectomy and regional lymph node dissection for locally advanced type 2 papillary renal cell carcinoma in an at-risk individual from a family with hereditary leiomyomatosis and renal cell cancer: A case report. *BMC Cancer.* 2016;16:232.

36. Choi Y, Keam B, Kim M et al. Bevacizumab plus Erlotinib combination therapy for advanced hereditary leiomyomatosis and renal cell carcinoma-associated renal cell carcinoma: A multicenter retrospective analysis in Korean patients. *Cancer Res Treat.* 2019;51(4):1549–56.

37. van Spaendonck-Zwarts KY, Badeloe S, Oosting SF et al. Hereditary leiomyomatosis and renal cell cancer presenting as metastatic kidney cancer at 18 years of age: Implications for surveillance. *Fam Cancer.* 2012;11(1):123–9.

38. Kopp RP, Stratton KL, Glogowski E et al. Utility of prospective pathologic evaluation to inform clinical genetic testing for hereditary leiomyomatosis and renal cell carcinoma. *Cancer.* 2017;123(13):2452–8.

39. Gunnala V, Pereira N, Irani M et al. Novel fumarate hydratase mutation in siblings with early onset uterine leiomyomas and hereditary leiomyomatosis and renal cell cancer syndrome. *Int J Gynecol Pathol.* 2018;37(3):256–61.

40. Forde C, Lim DHK, Alwan Y et al. Hereditary leiomyomatosis and renal cell cancer: Clinical, molecular, and screening features in a cohort of 185 affected individuals. *Eur Urol Oncol.* 2019 Dec 9. pii: S2588-9311(19)30161-0.

41. Kamai T, Tomosugi N, Abe H, Kaji Y, Oyama T, Yoshida K. Protein profiling of blood samples from patients with hereditary leiomyomatosis and renal cell cancer by surface-enhanced laser desorption/ionization time-of-flight mass spectrometry. *Int J Mol Sci.* 2012;13(11):14518–32.

42. Schultz KAP, Rednam SP, Kamihara J et al. *PTEN, DICER1, FH,* and their associated tumor susceptibility syndromes: Clinical features, genetics, and surveillance recommendations in childhood. *Clin Cancer Res.* 2017;23(12):e76–82.

32

Hereditary Papillary Renal Cell Cancer

Dongyou Liu

CONTENTS

32.1 Introduction

The kidneys are affected by a range of tumors, among which renal cell cancer/carcinoma (RCC) stands out and accounts for 85% of all kidney malignancies alone. Derived from cells of the proximal renal tubular epithelium and comprising multiple histological types, RCC is predominated by clear cell renal cell cancer (ccRCC), which makes up 80% of all RCC cases, followed by papillary renal cell cancer (pRCC), which represents 10%–15% of all RCC cases.

Histologically, pRCC consists of two morphologically different types. While type 1 pRCC contains predominantly basophilic cells with scanty cytoplasms and arranged in a single layer along the papillary axi, type 2 pRCC shows mostly eosinophilic cells forming several strata on the axis, that contain large cytoplasms, pseudostratified nuclei, prominent eosinophilic nucleoli, and perinucleolar halo (Figure 32.1). Furthermore, type 1 pRCC occurs frequently (constituting 73% of pRCC) and often produces multiple, bilateral, indolent, low-grade lesions, in comparison with type 2 pRCC, which tends to cause high grade lesions, with tumor cells displaying large-sized nuclei and prominent nucleoli.

Hereditary papillary renal cell cancer (HPRCC) is an autosomal dominant genetic disorder characterized by the formation of multifocal or bilateral renal lesions with papillary type 1 morphology in patients of relatively young age (around 46 years). Molecularly, heterozygous germline mutations in the mesenchymal-epithelial transition factor gene (*MET*) on chromosome 7q31.2 underscore the tumorigenesis of HPRCC [1].

32.2 Biology

As one of the essential organs in the body involved in the discharge of excess water, minerals, and unwanted substances, the kidney is susceptible to a diversity of tumors, ranging from ccRCC (accounting for 80% of RCC cases), pRCC (10%–15% of RCC cases), chromophobe RCC (chRCC, 4%–5% of RCC cases), collecting duct carcinoma (<1% of RCC cases), and translocation RCC (tRCC, including Xp11 tRCC and t(6;11) RCC; <1% of RCC cases) to other rare histological types (e.g., mucinous tubular and spindle cell carcinoma, multilocular cystic ccRCC, carcinoma associated with neuroblastoma, tubulocystic carcinoma, thyroid-like follicular carcinoma of kidney, acquired cystic kidney disease-associated RCC, and clear cell papillary RCC) [2].

Although sporadic/de novo mutations are responsible for over 90% of RCC cases, germline mutations involving renal cancer predisposing genes (e.g., *VHL, MET, FLCN, FH, SDHB/C/D, TSC1, TSC2, BAP1,* and *MITF*) are implicated in 5%–8% of RCC cases [3,4]. Among these, HPRCC due to germline *MET* mutations predisposes to pRCC type 1; hereditary leiomyomatosis and renal cell cancer (HLRCC) caused by germline *FH* mutations increases risk for pRCC type 2; von Hippel–Lindau syndrome (VHL), BAP1 tumor predisposition syndrome (BAP1), and PTEN hamartomatous syndrome (PTHS or Cowden syndrome) include ccRCC as part of their clinical spectrums; SDH-associated renal cancer (SDH-RCC) and tuberous sclerosis complex (TSC) are associated with ccRCC and other tumors; Birt–Hogg–Dubé syndrome (BHD) is linked to chRCC and oncocytic RCC; and microphthalmia-associated transcription

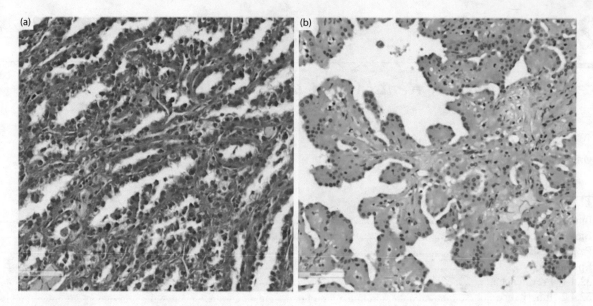

FIGURE 32.1 Morphological characteristics of papillary renal cell carcinoma (pRCC) types 1 and 2. While type 1 pRCC shows predominantly basophilic cells with scanty cytoplasms and arranged in a single layer along the papillary axi (a), type 2 pRCC contains mostly eosinophilic cells forming several cell strata on the axi, with large cytoplasms, pseudostratified nuclei, prominent eosinophilic nucleoli, and perinucleolar halo (b). (Photo credit: Yin X et al. *PLOS ONE*. 2015;10(12):e0143468.)

factor syndrome (MITF) is involved in tRCC, including Xp11 tRCC and t(6;11) RCC (Table 32.1) [5,6].

Heterogeneous germline mutations in the *MET* gene lay the foundation for HPRCC, however, the presence of a second copy of the mutant *MET* allele is necessary for tumor initiation in the kidneys. Over 3000 microscopic papillary tumors (incipient lesions) may occur in apparently normal renal parenchyma within a single kidney of HPRCC patient, suggesting multiple independent early events.

32.3 Pathogenesis

Mapped to chromosome 7q31.2, the *MET* gene spans 125 kb with 21 exons, and encodes a 1390 aa, 155 kDa precursor (MET) that is proteolytically cleaved at a furin site to yield a highly glycosylated extracellular α-subunit (50 kDa) and a transmembrane β-subunit (145 kDa) linked via a disulfide bond that function as the mature receptor for hepatocyte growth factor (HGF, also known as scatter factor, which is produced by mesenchymal cells).

Expressed by epithelial cells at the plasma membrane, the extracellular portion of MET consists of the Sema domain (homology to semaphorins, including the full α-subunit and the N-terminal part of the β-subunit), cysteine-rich MET-related sequence (MRS domain), glycine-proline-rich repeats (G-P repeats), and four immunoglobulin-plexin-transcription (IPT) repeats (Ig domains); while the intracellular region of MET comprises the juxtamembrane (JM) domain (including a serine residue Ser 985 for inhibiting the receptor kinase activity upon phosphorylation, a tyrosine Tyr 1003 for MET polyubiquitination, endocytosis, and degradation upon interaction with the ubiquitin ligase CBL), tyrosine kinase domain (consisting of Tyr 1234 and Tyr 1235, which are transphosphorylated after MET activation), and carboxyterminal docking sites (containing two crucial tyrosines,

Tyr 1349 and Tyr 1356, which are inserted into the multisubstrate docking site, capable of recruiting downstream adapter proteins with Src homology-2 [SH2] domains) [7].

Upon binding (via residues 25–307 of the α-subunit and residues 308–519 of the β-subunit in the N-terminal Sema domain) to its HGF ligand at the cell surface, MET undergoes dimerization and transphosphorylation of tyrosines Tyr 1234 and Tyr 1235 on the intracellular docking sites. The phosphorylated docking sites allow MET to interact with downstream effectors such as STAT3, PIK3R1, SRC, PCLG1, GRB2 (growth factor receptor–bound protein 2) and GAB2 (GRB2-associated binding protein), leading to the activation of the PI3 K/AKT/mTOR, STAT3, and CDC42 as well as VEGF/RAS/RAF/MEK, signaling pathways, and thus contributing to cytoskeletal changes, cell proliferation, migration/invasion (PI3 K/AKT/mTOR), angiogenesis/branching morphogenesis (VEGF/RAS/RAF/MEK, STAT3), and survival/apoptosis (PI3 K/AKT/mTOR) [8].

During embryonic stage, MET expression in stem cells and progenitor cells only facilitates gastrulation, growth and migration of muscles and neuronal precursors, angiogenesis, and kidney formation. In adults, MET contributes to wound healing, organ regeneration, and tissue remodeling. In addition, MET participates in the differentiation and proliferation of hematopoietic cells, and regulation of cortical bone osteogenesis [9].

Germline mutations in the *MET* gene destabilize the inactive (or stabilize the active) MET kinase conformation, leading to ligand-independent constitutive kinase activation. For example, mutations in the Sema domain likely affect the structure of the ligand-binding domain, resulting in the receptor activity of MET independent of HGF and increased activation of downstream genes. Mutations in the JM domain attenuate MET receptor ubiquitination and degradation and prolong MET signaling, leading to subsequent overactivation. Missense variants (e.g., M1149 T, V1206L, V1238I, D1246N, and Y1248C) in the regions that flank the critical tyrosine residues Y1234

TABLE 32.1

Characteristic of Inherited Tumor Syndromes Predisposing to Renal Cancer

Syndrome	Chromosome Location	Gene	Protein	Renal Cancer Histology	Other Tumors	Non-Neoplastic Feature	Treatment Options
Hereditary papillary renal cell cancer (HPRCC)	7q31.2	*MET*	MET	Papillary type 1			Active surveillance <3 cm; surgical excision ≥3cm; Met kinase agents
Hereditary leiomyomatosis and renal cell cancer (HLRCC)	1q42.1	*FH*	Fumarate hydratase	Papillary type 2		Cutaneous leiomyomas, uterine leiomyomas	Wide margin surgical excision; HIF-VEGF, antioxidant response, and reductive carboxylation pathway agents
BAP1 tumor predisposition syndrome (BAP1)	3p21.2	*BAP1*	BRCA associated protein	Clear cell	Melanoma, uveal melanoma, mesothelioma	Epithelioid atypical Spitz tumors	Surgical excision
PTEN hamartoma syndrome (PTHS) or Cowden syndrome	10q23.31	PTEN	PTEN	Clear cell	Breast, endometrial and thyroid cancer	Mucocutaneous papules (acral keratosis, facial trichilemmomas), hamartomas, lipomas, macrocephaly	AKT signaling pathway agents
von Hippel–Lindau syndrome (VHL)	3p25.3	*VHL*	pVHL	Clear cell	Hemangioblastoma (brain, spine, retina), pheochromocytoma (adrenal glands), endolymphatic sac tumors (inner ear), neuroendocrine tumors (pancreas)	Pancreatic, renal cysts	Active surveillance <3 cm; surgical excision ≥3cm; HIF-VEGF pathway agents
SDH associated renal cancer (SDH-RCC)	1p36.13 1q23.3 11q23.1	*SDHB* *SDHC* *SDHD*	Succinate dehydrogenase subunits B, C, D	Clear cell, chromophobe, oncocytoma	Paraganglioma, pheochromocytoma		Surgical excision; HIF-VEGF, and reductive carboxylation pathway agents
Tuberous sclerosis complex (TSC)	9q34 16p13.3	*TSC1* *TSC2*	Hamartin Tuberin	Clear cell, angiomyolipoma, epithelioid angiomyolipoma	Angiomyolipomas, subependymal giant cell astrocytomas	Facial angiofibroma, hypomelanotic macule, connective tissue nevus, forehead plaque, ungual and peri-ungual fibromas	RCC, surgical excision; AML, embolization; mTOR pathway agents
Birt–Hogg–Dubé syndrome (BHD)	17p11.2	*FLCN*	Folliculin	Oncocytic, chromophobe		Fibrofolliculomas, lung cysts, pneumothorax	Active surveillance <3 cm; surgical excision ≥3cm; mTOR pathway agents
Microphthalmia-associated transcription factor (MITF)	3p13	*MITF*	MITF	Translocation renal cell [Xp11 tRCC and t(6;11) RCC]	Melanoma		MAPK pathway agents

and Y1235 within the kinase domain constitutively activate the MET protein. Further, point mutations in the multifunctional docking site affect signal transduction. Overall, MET amplification and overexpression promote oncogenesis, angiogenesis, and metastasis [10–12].

Apart from germline mutations in the *MET* gene, trisomy of chromosome 7 and non-random duplication of the chromosome harboring the mutated *MET* allele occur in hereditary pRCC type 1. Chromosome 7 gain drives *MET* gene copy number increase in pRCC, and subsequently increases in MET protein overexpression [13]. In addition, somatic missense *MET* mutations (D1246H, Y1228C, and M1268 T) in the kinase domain causing MET amplification are observed in <10% of sporadic pRCC [12].

Besides their involvement in pRCC, mutations (e.g., R970C, R988C, T992I, T1010I, S1058P in the JM domain; E168D, L299F, S323G, and N375S in the Sema domain) have been implicated in small cell lung cancers (SCLC), non-small cell lung cancers (NSCLC, 5.6%), lung adenocarcinomas (8%), squamous cell lung cancer (3%), advanced breast cancer (9%), advanced ovarian cancer (7.4%), and colorectal cancer. There is evidence that MET kinase domain mutations may be a mechanism of therapeutic resistance in refractory lung cancer [14].

Furthermore, deletion of *MET* exon 14 results in the exclusion of an ubiquitination target site within the cytoplasmic domain and persistent presence of MET at the cell surface, hampering osteoblastic differentiation as shown in four families with osteofibrous dysplasia (OSFD, manifesting as painless swelling or anterior bowing of the long bones, commonly the tibia and fibula). Change in the promoter region of the MET leading to decreased MET promoter activity and altered binding of specific transcription factor complexes is linked to susceptibility to autism in some families.

32.4 Epidemiology

32.4.1 Prevalence

Tumors of the kidneys represent 2%–3% of all adult malignancies, with RCC alone accounting for 85% of kidney neoplasms. There is a male predominance among RCC patients (male:female ratio of 2:1). Among various RCC histologic types, ccRCC (75%) predominates. This is followed by pRCC (10%–15%), chRCC (4%–5%), collecting duct carcinoma (<1% of RCC cases), tRCC (<1% of RCC cases), and other histologic types. Interestingly, over 90% of RCC cases do not show family connection and appear to emerge from sporadic/de novo mutations, whereas 5%–8% of RCC cases are attributed to germline mutations that occur in families, leading to bilateral or multifocal tumors in patients of relatively young age (up to 46 years) [1]. HPRCC has an estimated incidence of <1:1500.00, and its rarity is highlighted by the fact that only about 35 affected families have been reported worldwide.

32.4.2 Inheritance

Hereditary HPRCC is a familial tumor predisposition syndrome showing an autosomal dominant pattern of inheritance.

32.4.3 Penetrance

HPRCC demonstrates a high penetrance. While patients with early disease may be asymptomatic, those with later disease typically present with various symptoms. However, HPRCC shows complete penetrance in patients harboring pathogenic *MET* variants by age 80.

32.5 Clinical Features

Early RCC is often small and does not produce any signs and symptoms. Later RCC is associated with hematuria (blood in the urine), low back pain (on one side or both sides) that does not go away, a lump on the side, belly, or lower back, anemia (low red blood cell counts), fatigue (feeling tired, sluggish, and run-down due to anemia), fever (that does not get better), loss of appetite and weight, night sweats, high blood pressure, and high level blood calcium.

32.6 Diagnosis

Diagnosis of HPRCC involves medical history review, physical exam, laboratory assays, imaging study (eg, x-rays, computed tomography [CT], magnetic resonance imaging [MRI], ultrasound), histopathology, and molecular genetic testing.

Due to its hypovascularity, hereditary pRCC may be confused with cysts, and use of MRI or CT rather than ultrasound is recommended for its visualization. Macroscopically, pRCC is a relatively slow-growing, peripheral lesion with necrosis, hemorrhage, and calcification. Histological examination of pRCC reveals tumor cells arranged in a spindle pattern and possible areas of internal hemorrhage and cystic alterations. More specifically, pRCC type 1 is a low nuclear grade lesion with papillae lined by a single layer of predominantly basophilic cells that contain scanty clear-to-basophilic cytoplasm and hyperchromatic nuclei; pRCC type 2 is a high nuclear grade lesion with papillae lined by several strata of mostly eosinophilic cells that have abundant granular eosinophilic cytoplasm, pseudostratified nuclei, prominent eosinophilic nucleoli, and perinucleolar halo [13].

On the other hand, ccRCC shows solid, alveolar, acinar, cystic growth patterns of cells with clear or eosinophilic cytoplasm, forming characteristic network of small, thin-walled, "chicken-wire" vasculature; chRCC contains tumor cells with prominent cell borders, finely reticular pale cytoplasm, perinuclear halos, and wrinkled raisinoid nuclei; collecting duct carcinoma has irregular, infiltrating tubules with high grade cells often containing mucin, marked stromal desmoplasia, and inflammatory infiltrate; RCC with Xp11 comprises epithelioid cells with clear to eosinophilic, abundant voluminous cytoplasm, and frequent psammoma bodies (in stromal hyaline nodules); and RCC with t(6;11) translocation demonstrates large epithelioid cells with abundant clear cytoplasm, mimicking a typical ccRCC, and clusters of small cells surrounding hyaline material produce distinctive "rosette-forming" pattern.

Immunohistochemical staining reveals higher MET protein in pRCC type 1 than in pRCC type 2, and absence of FH protein in pRCC type 2.

Molecular identification of germline mutations in the MET gene helps confirm the diagnosis of HPRCC.

Further determination of pRCC stages yields additional information about whether cancer has spread from one area of the body to another. In one study, the percentages of pRCC tumor grades were 8.2% (Grade 1), 48.0% (Grade 2), 32.6% (Grade 3), and 11.2% (Grade 4), whereas the percentages of pRCC clinical stages were 67.3% (Stage 1), 9.2% (Stage 2), 11.2% (Stage 3), 9.2% (Stage 4), and 3.1% (unknown). Type 1 pRCC correlated with lower tumor grades and clinical stages than type 2 pRCC. In addition, patients with papillary (17.6%), chromophobe (16.9%), collecting duct (55.7%), and sarcomatoid (82.8%) variants were less likely to have T3 or greater disease than those with clear cell histology [1].

32.7 Treatment

Treatment options for renal cancer typically include a combination of surgery, cryoablation, chemotherapy, hormone therapy, or immunotherapy.

For localized RCC, primary treatment is nephrectomy (e.g., partial nephrectomy, multiplex partial nephrectomy, repeat renal surgery, or salvage renal surgery) when the largest tumor reaches 3 cm in size. Partial nephrectomy (or nephron-sparing surgery) spares the kidney but removes the tumor and some of the surrounding tissue. It is particularly suited for patients with solitary RCC lesions of 4 cm or less. Multiplex partial nephrectomy is useful for patients with multifocal RCC, while repeat renal surgery and salvage renal surgery are considered for patients with hereditary RCC if necessary. In addition, cryoablation (freezing cancer cells) or radiofrequency ablation (heating the tumor with high-energy radio waves) may be an alternative for unresectable tumor that is solid and in a contained area [15].

For advanced or disseminated RCC, radical nephrectomy (removal of an entire affected kidney) may be necessary. Locoregional therapy (e.g., antivascular endothelial growth factor receptor [VEGFR] tyrosine kinase inhibitors [TKI] sunitinib, pazopanib, and foretinib, and mammalian target of rapamycin [mTOR] inhibitor temsirolimus) may help reduce tumor size and palliate symptoms, while systemic therapy shows only limited efficacy [16]. Kidney transplantation may be considered for patients with severe kidney dysfunction. Currently, the 5-year living donor graft survival rate stands at nearly 90%.

A number of novel drugs targeting MET are under development for improved treatment of HPRCC [17]. These include kinase inhibitors (low molecular weight molecules are used to inhibit MET transphosphorylation and downstream effector recruitment), HGF inhibitors (truncated HGF, anti-HGF neutralizing antibodies, and uncleavable HGF are used to bind or displace MET without inducing receptor activation), decoy MET (soluble truncated MET receptor is used to prevent both ligand binding and MET receptor homodimerization and thus inhibit MET activation in HGF-dependent and independent manners), immunotherapy agents (either enhancing the immunologic response to MET-expressing tumor cells or stimulating immune cells and altering differentiation/growth of tumor cells; the former involves use of cytokines to trigger nonspecific stimulation of numerous immune cells, and the latter relies on monoclonal antibody to facilitate destruction of tumor cells by complement-dependent cytotoxicity and cell-mediated cytotoxicity) [18,19].

32.8 Prognosis

Prognosis for patients diagnosed with RCC depends on the tumor size, nodal status, and pathologic stage. Five-year disease-specific survival for patients with TNM stage I RCC is 81%, TNM stage II RCC is 74%, TNM stage III RCC is 53%, and TNM stage IV RCC is 8% [20]. Patients with localized RCC may metastasize (25%–30%) and recur (30%). Patients undergoing partial resection have 5-year survival of 100% [21].

32.9 Conclusion

Hereditary papillary renal cell cancer (HPRCC) is an autosomal dominant syndrome that typically causes multifocal and bilateral pRCC type 1 in patients aged about 40 years and with family history of renal cancer. Heterogeneous germline mutations in the *MET* gene on chromosome 7q31.2 that induce aberrant activity of the intracellular tyrosine kinase domain of the membrane-bound MET and subsequent activation of downstream signal pathways underlie the molecular basis of HPRCC pathogenesis. Apart from physical exam, medical history review, imaging, and histopathology, molecular genetic testing for MET pathogenic variant helps confirm HPRCC diagnosis [22]. Current management of HPRCC involves a combination of surgery, chemotherapy, and kidney transplantation. Further development of novel drugs is crucial to improve the outcome of patients suffering from HPRCC.

REFERENCES

1. Haas NB, Nathanson KL. Hereditary kidney cancer syndromes. *Adv Chronic Kidney Dis.* 2014;21(1):81–90.
2. Crumley SM, Divatia M, Truong L, Shen S, Ayala AG, Ro JY. Renal cell carcinoma: Evolving and emerging subtypes. *World J Clin Cases.* 2013;1(9):262–75.
3. Durinck S, Stawiski EW, Pavía-Jiménez A et al. Spectrum of diverse genomic alterations define non-clear cell renal carcinoma subtypes. *Nat Genet.* 2015;47(1):13–21.
4. Schmidt LS, Linehan WM. Genetic predisposition to kidney cancer. *Semin Oncol.* 2016;43(5):566–74.
5. Rao Q, Xia QY, Cheng L, Zhou XJ. Molecular genetics and immunohistochemistry characterization of uncommon and recently described renal cell carcinomas. *Chin J Cancer Res.* 2016;28(1):29–49.
6. Costa WH, Netto JG, Cunha IW. Urological cancer related to familial syndromes. *Int Braz J Urol.* 2017;43(2):192–201.
7. Fay AP, Signoretti S, Choueiri TK. MET as a target in papillary renal cell carcinoma. *Clin Cancer Res.* 2014;20(13):3361–3.
8. Cojocaru E, Lozneanu L, Giușcă SE, Căruntu ID, Danciu M. Renal carcinogenesis--insights into signaling pathways. *Rom J Morphol Embryol.* 2015;56(1):15–9.
9. Wala SJ, Karamchandani JR, Saleeb R et al. An integrated genomic analysis of papillary renal cell carcinoma type 1 uncovers the role of focal adhesion and extracellular matrix pathways. *Mol Oncol.* 2015;9(8):1667–77.
10. Yap NY, Rajandram R, Ng KL, Pailoor J, Fadzli A, Gobe GC. Genetic and chromosomal aberrations and their clinical significance in renal neoplasms. *Biomed Res Int.* 2015;2015:476508.

11. Petejova N, Martinek A. Renal cell carcinoma: Review of etiology, pathophysiology and risk factors. *Biomed Pap Med Fac Univ Palacky Olomouc Czech Repub*. 2016;160(2):183–94.

12. Tovar EA, Graveel CR. MET in human cancer: Germline and somatic mutations. *Ann Transl Med*. 2017;5(10):205.

13. Yin X, Zhang T, Su X et al. Relationships between chromosome 7 gain, MET gene copy number increase and MET protein overexpression in Chinese papillary renal cell carcinoma patients. *PLOS ONE*. 2015;10(12):e0143468.

14. Diamond JR, Salgia R, Varella-Garcia M et al. Initial clinical sensitivity and acquired resistance to MET inhibition in MET-mutated papillary renal cell carcinoma. *J Clin Oncol*. 2013;31(16):e254–8.

15. Baiocco JA, Metwalli AR. Multiplex partial nephrectomy, repeat partial nephrectomy, and salvage partial nephrectomy remain the primary treatment in multifocal and hereditary kidney cancer. *Front Oncol*. 2017;7:244.

16. Stein MN, Hirshfield KM, Zhong H, Singer EA, Ali SM, Ganesan S. Response to crizotinib in a patient with MET-mutant papillary renal cell cancer after progression on tivantinib. *Eur Urol*. 2015;67(2):353–4.

17. Albiges L, Guegan J, Le Formal A et al. MET is a potential target across all papillary renal cell carcinomas: Result from a large molecular study of pRCC with CGH array and matching gene expression array. *Clin Cancer Res*. 2014;20(13):3411–21.

18. Yang OC, Maxwell PH, Pollard PJ. Renal cell carcinoma: Translational aspects of metabolism and therapeutic consequences. *Kidney Int*. 2013;84(4):667–81.

19. Courthod G, Tucci M, Di Maio M, Scagliotti GV. Papillary renal cell carcinoma: A review of the current therapeutic landscape. *Crit Rev Oncol Hematol*. 2015;96(1):100–12.

20. Keegan KA, Schupp CW, Chamie K, Hellenthal NJ, Evans CP, Koppie TM. Histopathology of surgically treated renal cell carcinoma: Survival differences by subtype and stage. *J Urol*. 2012;188(2):391–7.

21. Sidana A, Kadakia M, Friend JC et al. Determinants and prognostic implications of malignant ascites in metastatic papillary renal cancer. *Urol Oncol*. 2017;35(3):114.e9–14.

22. Mikhaylenko DS, Klimov AV, Matveev VB et al. Case of hereditary papillary renal cell carcinoma type I in a patient with a germline MET mutation in Russia. *Front Oncol*. 2020;9:1566.

33

Hereditary Prostate Cancer

Veda N. Giri, Jennifer Beebe-Dimmer, and Kathleen A. Cooney

CONTENTS

33.1 Introduction

Hereditary prostate cancer is characterized by familial clustering of prostate cancer, particularly father, brothers, or in a vertical generation in families, as well as early onset (EO) of disease [1,2]. Men in families with other hereditary cancer syndromes have also been observed to be at increased risk of prostate cancer, including hereditary breast and ovarian cancer (HBOC) syndrome and Lynch syndrome [2–5]. Furthermore, insights into inherited prostate cancer predisposition are arising from men with metastatic prostate cancer through tumor sequencing [5–7]. Thus, the spectrum of genes contributing to inherited prostate cancer has expanded. Here we discuss genetic predisposition to inherited prostate cancer and emerging insights into the biology, epidemiology, clinical features, diagnosis, treatment, and management.

33.2 Genetic Predisposition, Biology, and Pathogenesis

In the latter part of the twentieth century, investigators began to observe that prostate cancer clusters within families [8]. Initial segregation studies suggested that the best model for the disease was autosomal dominant inheritance of one or more rare alleles which contributes preferentially to EO disease [9–12.] This observation inspired clinicians and researchers around the world to collect pedigree information and DNA samples from families with multiple cases of prostate cancer. Genome-wide linkage scans were completed on a large number of families which led to the identification of a number of potential prostate cancer risk loci; however, many of these loci could not be consistently confirmed in follow-up studies (see reviews [13,14]). There are a number of reasons why prostate cancer linkage studies have

been challenging. First, prostate cancer is very clinically heterogeneous, and no specific phenotype has been associated with hereditary disease, though recently up to 12% of men with metastatic disease have been reported to have inherited genetic mutations [6,7]. Second, the majority of men with prostate cancer are diagnosed at an older age, making it often challenging to obtain clinical histories and DNA samples from men in vertical generations. Finally, use of prostate-specific antigen (PSA) to identify asymptomatic prostate cancer cases was introduced around 1990, resulting in a dramatic increase in the number of cases detected in countries where testing was available. Unfortunately, many of these cancers were, in hindsight, low-volume and/or low-grade cancers which are extremely prevalent in the general population and may not need treatment [15]. Since family history is a well-recognized risk factor for prostate cancer, PSA testing has been preferentially used in men with a positive family history, leading to the overdiagnosis of sporadic cases in families [16] and subsequent loss of statistical power for prostate cancer linkage studies.

The only prostate cancer susceptibility gene identified through linkage studies is *HOXB13* on the long arm of chromosome 17 (i.e., 17q21.32). In 2003, a genome-wide linkage scan conducted on 175 prostate cancer pedigrees from the University of Michigan (UM) identified a novel linkage region on chromosome 17q near *BRCA1* [17]. Through mutation screening, *BRCA1* mutations were excluded as the cause of this linkage signal [18]. Our investigative team subsequently narrowed the candidate region to a 10 cM interval using a large number of pedigrees from UM and Johns Hopkins University (JHU) by focusing on the 147 families with ≥4 cases of prostate cancer AND early age of prostate cancer diagnosis (LOD = 5.49) [19]. Using emerging next-generation sequencing technologies, targeted sequencing of all 202 genes in the candidate region was performed and only the youngest family member from 94 UM and JHU multiplex prostate cancer families with evidence for linkage to the candidate region was tested [20]. A recurrent *HOXB13* missense mutation (G84E) was identified that was predicted to be damaging by bioinformatic prediction algorithms in 4 families, and perfect segregation of the mutation in 14 additional men with prostate cancer in their respective families was observed. The G84E mutation was only identified in individuals of European descent, although other mutations have been identified in African American [20] and Chinese men with prostate cancer [21]; notably all observed mutations are in functional domains of the molecule. Importantly, the carrier frequency observed in men with a positive family history (2.2%) was identical to men with an early age of prostate cancer diagnosis (2.2%), and the highest carrier frequency was observed in the subset of men with BOTH a positive family history and EO disease (3.1%). In contrast, the carrier frequency in men diagnosed with prostate cancer above the age of 65 was only 0.65%. These results emphasize the enrichment of contributing genetic factors to EO familial prostate cancer and highlight the challenges of using unselected prostate cancer cases in gene discovery studies. Over a dozen studies have confirmed our *HOXB13* findings, including a study by Karlsson et al., who found that almost 10% of men in Sweden with EO familial prostate cancer are carriers of the G84E mutation [22]. There is evidence that G84E mutation occurs on a common haplotype consistent with a founder effect [23]. Finally, extensive pathological studies revealed evidence of pseudohyperplastic-type features and a markedly low prevalence of ERG+ tumors in men carrying the *HOXB13* mutation [24]. This suggests that cancers arising in *HOXB13* G84E carriers may have unique molecular pathways implicated in the development of prostate cancer (Figure 33.1).

Most of the other genes that increase prostate cancer risk are associated with known cancer syndromes marked by a preponderance of cancers other than prostate cancer in the family. Deleterious mutations in both *BRCA1* and *BRCA2*, associated with hereditary breast and/or ovarian cancer, have been shown to increase the risk of prostate cancer (see review [25]). There are data demonstrating that men with deleterious *BRCA1/2* mutations experience more aggressive forms of prostate cancer [26]. Studies of families with ONLY prostate cancer, however, have failed to uncover a significant number of *BRCA1/2* mutations, leading to the conclusion that the mutations in *BRCA1/2* contribute to a small number of classic hereditary prostate cancer families. More recently, several lines of evidence have suggested that male carriers of mismatch repair (MMR) gene mutations (associated with Lynch syndrome) have an increased risk of prostate cancer [27,28]. Prostate cancers arising in men with Lynch syndrome demonstrate classic microsatellite instability, which is not commonly observed in classic hereditary or sporadic prostate cancer cases [29]. Notably, in families with a cancer syndrome (e.g., HBOC) including prostate cancer, the age of diagnosis of prostate cancer has been shown to be younger than expected [27,30].

The importance of germline DNA repair gene mutations in prostate cancer was significantly heightened based on results from several recent publications. Leongamornlert et al. [31] sequenced 22 tumor suppressor genes in germline DNA from 191 men with hereditary prostate cancer. Fourteen men (7.3%) were found to have a loss of function mutation in one of eight genes, all in the DNA repair pathway, and the prostate cancers in these men were statistically more likely to be clinically advanced disease. Robinson et al. [32] reported integrative tumor sequencing data from 150 men with castrate-resistant prostate cancer (CRPCa) and identified DNA repair/recombination gene

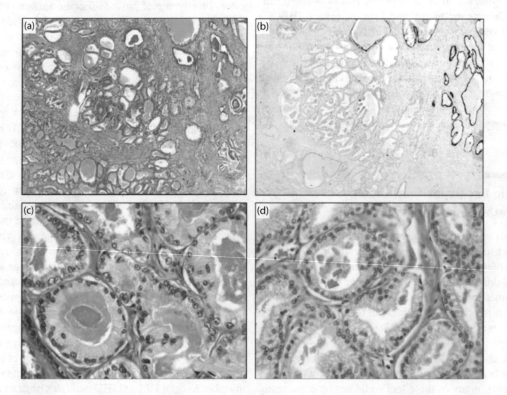

FIGURE 33.1 Histopathologic view of pseudohyperplastic features of HOXB13 G84E-related prostate adenocarcinoma, including salient features of irregular/tufted luminal borders and cystically dilated glands with papillary infoldings, and less specific features of tall/columnar epithelium, pale to granular cytoplasm, and basally oriented nucleoli. (Smith et al. *Am J Surg Pathol.* 2014;38:615–26.)

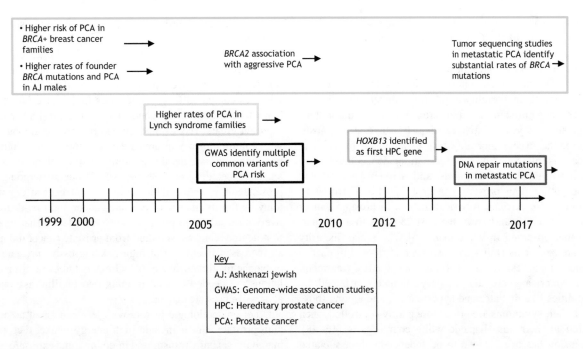

FIGURE 33.2 Timeline of discovery of genetic predisposition to inherited prostate cancer.

alterations in 23% of cases with the majority of samples having biallelic alterations. Germline mutations in *BRCA1/2* or *ATM* occurred in 13/150 (8.7%) of cases. A larger multisite study of 692 men with metastatic prostate cancer found an even higher rate of germline mutations (11.8%) [6]. These studies emphasize the genetic enrichment resulting from selecting for study men with clinically aggressive, late-stage disease, as opposed to most previous studies which included either unselected prostate cancer cases, some of which were clinically insignificant, or men solely selected on the basis of positive family history, regardless of clinical aggressiveness.

Genome-wide association studies (GWAS) have been performed using thousands of prostate cancer cases and controls, leading to the identification of over 100 single-nucleotide polymorphisms (SNPS) associated with prostate cancer. As expected, most of the variants are common and result in minor risk elevation. Furthermore, the majority of these variants are in noncoding regions and the functional implications of many of these variants are unknown. Taken together, the GWAS SNPs account for only about one-third of the familial clustering of prostate cancer [33]. Use of SNPs for prediction of prostate cancer shows promise, especially when combined with other factors [34–36].

Figure 33.2 highlights discovery and insights into genetic predisposition to inherited prostate cancer over time.

33.3 Epidemiology

Prostate cancer continues to be the most commonly diagnosed invasive cancer among men in the United States (US) with an estimated 164,690 new cases expected in 2018 [37]. Incidence has declined significantly in the most recent 5 years of complete case ascertainment from the Surveillance, Epidemiology and End Results (SEER) Cancer Registry program, presumably in response to the US Preventive Services Task Force (USPSTF)

statement in 2012 recommending against PSA screening for men with the general population regardless of age [38,39]. In 2018, the USPSTF revised this statement based upon an updated review of the scientific evidence, and has recommended that for men in the general population aged 55–69 years, screening should be the patient's decision after a discussion with their physician about the benefits and harms of testing [40]. It is premature to predict how much this revision will influence future incidence of disease, however over the same 5-year period (2011–2015), mortality in men with prostate cancer declined only slightly [39]. Prostate cancer remains the second leading cause of cancer mortality in the US with an estimated 29,430 deaths in 2018 [37]. Nearly 95% of patients are diagnosed with either organ-confined or regionally advanced disease, and for these men the 5-year relative survival rate is virtually 100%. However, for the 5% of men first diagnosed with metastatic prostate cancer, 5-year survival drops to just 30% [37].

Prostate cancer is a heterogeneous disease, both clinically and genetically, which complicates the epidemiologic investigation of determinants of disease. Older age, African American race, and a family history of prostate cancer, particularly among first-degree relatives, remain as the only consistent established risk factors for the disease useful to clinicians in risk prediction [41]. The lifetime probability of being diagnosed with prostate cancer is approximately 11%, or 1 in 9 men. Prostate cancer is primarily a disease of elderly men, with most diagnoses occurring after age 65, however the age-specific incidence for prostate cancer increases sharply after age 50 [37]. It has been estimated that about 10% of prostate cancer cases diagnosed in the US occur in men ≤55 years [42]. The likelihood of a significant genetic component is greater among men diagnosed with earlier-onset disease, supported by investigations in known prostate cancer genes (*HOXB13*, *BRCA2*) [20,43]. There is evidence to suggest that there are inherent biologic differences in men diagnosed with EO disease. Among men with similar Gleason grade or stage at diagnosis, younger men (<45 years at diagnosis) have been shown to be more likely to succumb to their disease compared with older men (65 and older) [44].

The incidence of prostate cancer is approximately 70% higher in African American men compared with non-Hispanic whites in the US, and their mortality is 2.4 times higher [45]. There is no longer any significant racial difference in either tumor stage at diagnosis or 5-year relative survival by stage at diagnosis. Therefore, the excess mortality risk experienced by African Americans is influenced by the excess in incidence, must occur farther out from time of diagnosis, and is likely be attributed to causes other than prostate cancer [37,39,45]. Fortunately, the mortality for African American men with prostate cancer has declined significantly over the past 25 years to a rate parallel to that observed in white men [45]. The racial disparity in prostate cancer mortality has also been attributed in part to differences in access to medical care and quality treatment. African Americans are also disproportionately affected by comorbidities like diabetes and hypertension, and among those affected, their symptoms are often more poorly controlled when compared to their non-Hispanic white counterparts [46–48]. Hypertension has been shown to be independently associated with prostate cancer risk [49], and more aggressive prostate cancer clinical features in African American men [49] as well as an increase in risk for biochemical recurrence after radical prostatectomy in both African Americans and whites [50,51]. Interestingly, African migrants to the US have a 50% higher risk of being diagnosed with prostate cancer compared with African American men born in the US, with incidence rates similar to those in sub-Saharan West Africa suggesting the contribution of genetic factors to disparities in risk [52]. To date, no unique gene has been identified to explain familial prostate cancer in African Americans and this continues to be an area of active investigation but is hindered by low minority recruitment into familial prostate cancer genetic studies. Novel mutations in a number of DNA-repair genes (*BRCA1/2*, *BRIP1*, *ATM*) have been observed in African American men with EO prostate cancer [53]. Likewise, common germline genetic variants discovered to be important in prostate cancer among men of European descent to date are similarly distributed in African American men with prostate cancer [54]. Understanding the distinctive genetic landscape of prostate cancer in African American men will hinge upon increasing our knowledge of the contribution of gene−environment interaction and epigenetic changes as well as sporadic mutations in tumor tissue in this high-risk population.

Men with a positive family history in a father, brother, or son have a two- to threefold higher risk of prostate cancer compared with men in the general population. Risk is increased further in men with multiple affected relatives or relatives diagnosed earlier than age 60 [55,56]. In fact, a family history that includes three or more affected relatives diagnosed with EO prostate cancer has been shown to be associated with an approximate tenfold increase in risk [57]. A positive family history is often considered a strong marker of genetic susceptibility; however, the contribution of shared environment, particularly environmental exposures that occur early in life, as well as gene−environment interactions, must be considered in understanding the cause of familial disease. A study of more than 80,000 adoptees and their adoptive parents suggests that about 5% of prostate cancers

diagnosed among family members are attributed to shared environment and not genetics [58].

The association of family history of prostate cancer with aggressiveness of disease has been inconsistent. Some studies reported higher cancer-specific and overall mortality in familial cases compared to sporadic cases [59]. These findings were consistent with a study conducted in the pre-PSA era which reported greater risk of biochemical recurrence among patients with a family history of prostate cancer [60]. Other studies have not found an association of family history of prostate cancer and advanced disease. One study of 471 men undergoing radical prostatectomy evaluated biochemical recurrence after prostatectomy by family history of prostate cancer and first-degree relative with death from prostate cancer and found that those with first-degree relatives who died from prostate cancer did not have an increased likelihood of high-risk/aggressive prostate cancer or biochemical recurrence [61]. Greater studies in diverse populations are needed to lend insights into familial association to aggressive prostate cancer.

Other epidemiologic factors with less consistent associations with prostate cancer include high energy intake, diet, physical inactivity, sexually transmitted infections, and exposure to heavy metals such as cadmium or pesticides [62,63]. More recently, obesity and obesity-related comorbidities such as hypertension and diabetes appear to be associated with aggressive clinical characteristics at time of diagnosis and/or an increase in risk of disease progression after treatment and disease-specific mortality [49,64–66.]

33.4 Identification of Men with Potential Inherited Prostate Cancer

A spectrum of features may indicate that a man with prostate cancer may have inherited the disease. Typical pedigree features of inherited prostate cancer include: (1) prostate cancer in close family members such as father, brother, or son; (2) generational prostate cancer: three generations in a row (grandfather, father, son) with prostate cancer on either the maternal or paternal side; (3) young age at diagnosis (age <60) in a man with prostate cancer or his relatives; (4) multiple potentially linked cancers in families: relatives with cancers of the breast, ovary, pancreas, uterus, or colon, particularly if diagnosed with these cancers at younger ages; (5) advanced or metastatic prostate cancer unselected for family history [2,67]. Figure 33.3 highlights pedigrees with classic hereditary cancer syndromes in which prostate cancer has been implicated.

Consideration of men with prostate cancer for genetic testing has undergone significant change over the past year. Prior to 2017, genetic testing for hereditary prostate cancer primarily centered around identification of HBOC in a family to proceed with *BRCA1/2* testing [4]. These features included men with prostate cancer (Gleason > = 7) with > = 1 close blood relative with ovarian cancer at any age or breast cancer< = 50 or > = 2 close blood relatives with breast, pancreatic, prostate (Gleason> = 7) cancers at any age, metastatic prostate cancer, or tumor sequencing with *BRCA1/2* mutations [4]. A recent consensus conference in 2017 expanded consideration of men

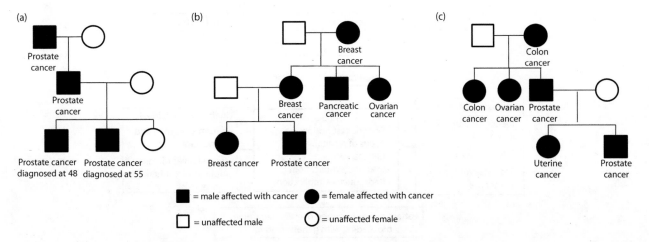

FIGURE 33.3 Pedigrees highlighting (a) hereditary prostate cancer, (b) hereditary breast and ovarian cancer syndrome, and (c) Lynch syndrome with prostate cancer cases included.

for genetic testing to include (i) family history of HBOC based upon Clinical Practice Guidelines in Oncology (NCCN) criteria [4], classic hereditary prostate cancer with generational prostate cancer, EO prostate cancer in two brothers or prostate cancer in a nuclear family [1], or Lynch syndrome [68,69]; (ii) family history of two or more close relatives with cancers in these syndromes; (iii) tumor sequencing with mutations in inherited cancer genes; and (iv) metastatic castration-resistant prostate cancer [3]. The NCCN 2018 prostate cancer guidelines subsequently expanded consideration of genetic testing for men with very low to unfavorable intermediate risk disease if they have a strong family history (brother, father, or multiple family members with prostate cancer diagnosed <60; known DNA repair/*BRCA* mutation in a family; known Lynch syndrome mutation in a family; >1 relative with breast, ovarian, or pancreatic cancer [suggestive of *BRCA2* mutation] or colorectal, endometrial, gastric, ovarian, pancreatic, small bowel, urothelial, kidney, or bile duct cancer [suggestive of Lynch syndrome]) or men with high-risk metastatic disease [5].

Furthermore, the NCCN 2018 prostate guideline update states to consider tumor testing for mutations in homologous recombination repair genes and MSI/immunohistochemistry (IHC) for potential Lynch syndrome in men with regionally advanced (local lymph nodes [N1]) or metastatic prostate cancer [5].

33.5 Diagnosis

Histopathology represents the gold standard for diagnosing prostate cancer (Figure 33.4). Like other hereditary syndromes, the definitive diagnosis of inherited prostate cancer is based upon genetic testing that reveals a mutation in a cancer risk gene. For men with a family history suggestive of HBOC, the definitive diagnosis of inherited prostate cancer is based upon identification of a *BRCA1* or *BRCA2* mutation. In men with classic hereditary prostate cancer, the diagnosis can be confirmed by a positive *HOXB13* mutation. If the family history suggests Lynch syndrome, the confirmatory diagnosis is based upon identification of a mutation in any of the DNA MMR genes (*MLH1*, *MSH2*, *MSH6*, *PMS2*, and *EPCAM*) [3,4,69]. Germline testing should be pursued if tumor testing shows microsatellite instability or MMR deficiency on IHC to confirm Lynch syndrome.

Tumor sequencing for targeted therapy or clinical trial determination may uncover somatic mutations in cancer risk genes that could be inherited. The current NCCN Hereditary Breast and Ovarian Cancer guideline states to perform *BRCA1/2* testing for tumor showing a *BRCA* mutation [4]. A recent consensus conference advocated germline testing for men with tumor sequencing revealing mutations in *BRCA1/2*, DNA MMR genes, *ATM*, and *HOXB13* [3].

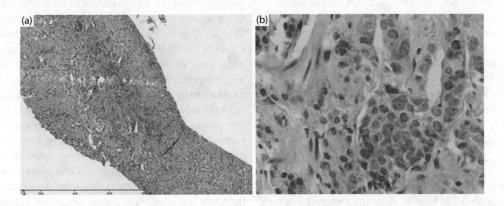

FIGURE 33.4 Prostate adenocarcinoma, predominant Gleason 4 pattern and secondary Gleason 3 pattern (Gleason 4 + 3 = 7) (a: H&E, 10× magnification; b: H&E, 400× magnification). (Photo credit: Somwaru AS et al. *BMC Med Imaging.* 2016;16:16.)

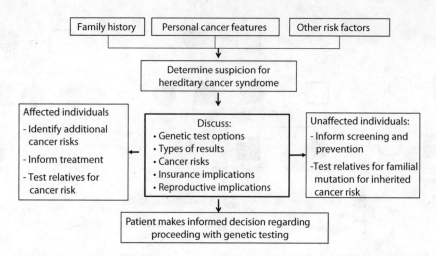

FIGURE 33.5 Model for genetic evaluation for inherited cancer risk.

Inherited contribution to advanced or metastatic prostate cancer in a man should also be confirmed by germline genetic testing. A recent consensus conference advocated testing *BRCA1/2* and *ATM* in men with metastatic castration-resistant prostate cancer [3]. The NCCN 2018 Prostate guideline advocates testing *BRCA1*, *BRCA2*, *ATM*, *PALB2*, and *FANCA* for men with regionally advanced or metastatic prostate cancer [5].

Several professional organizations advocate genetic testing be performed with appropriate genetic counseling. The purpose of genetic counseling is to educate patients about the basic principles of cancer inheritance, genetic test options, types of test results, potential cancer risks, genetic protection laws, insurance and reproductive considerations, screening and management based on test results, and cascade testing in a family (when a mutation is identified) [70]. Genetic counseling involves intake of a patient's family history, personal medical and cancer history, and risk factors to consider appropriate genetic testing based upon the patient making an informed decision for testing. Figure 33.5 displays a schema for genetic counseling for inherited cancer assessment. Multiple genetic test options exist for diagnosis of inherited prostate cancer, including single gene testing, focused prostate cancer panels, or comprehensive cancer panels. Each of these test options have benefits and considerations which patients need to understand prior to proceeding with genetic testing [71]. The most appropriate delivery of genetic counseling to men with or at risk for prostate cancer is under current debate. A classic model of referral to a genetic counselor may be appropriate for men considering genetic testing for risk assessment. However, when genetic testing is being considered for management or decision-making of prostate cancer treatment, alternate models of delivery of genetic education to patients is being considered, such as by phone, video, or telehealth. These modalities of counseling delivery need further study for optimized patient experience.

33.6 Treatment

Prostate cancer is a clinically heterogenous disease and a number of different approaches exist to determine prognosis which influences choice of therapy. Risk stratification uses a variety of clinical and laboratory data, including PSA at time of diagnosis,

Gleason grade, and American Joint Commission on Cancer staging information [72]; risk categories extend from very low to very high risk localized disease to regional and metastatic disease [5]. The choice of treatment for localized disease has changed dramatically over the past decade due to the recognition that low-grade cancers may not cause morbidity and/or mortality in an individual's lifetime. For men with Gleason grade 6 and low-volume disease with life expectancies <20 years, active surveillance including every 6-month assessments and protocol-driven biopsies can be considered. For patients with intermediate risk disease, the patient and provider have a number of treatment choices to consider including radical prostatectomy (robotic or open) and external beam radiation therapy (often including androgen deprivation therapy). Many factors enter into the choice of therapy including life expectancy, tumor characteristics, and patient/provider preferences. Regardless of treatment choice for both localized and regional prostate cancer, 5-year survivals are ≥9%.

Prostate cancer metastasizes to regional lymph nodes and bone and visceral sites. Although treatment for metastatic disease has improved significantly over the past 30 years, this stage of disease is still considered uniformly fatal. For those men who are destined to develop metastatic disease, serum PSA values can detect recurrence years before radiographic and/or clinical evidence of metastatic disease, providing opportunities for early intervention. The mainstay of treatment for advanced and/or metastatic prostate cancer involves suppression of androgens. Primary approaches include use of luteinizing hormone-releasing hormone (LHRH) agonist ± first generation androgen antagonists such as bicalutamide, LHRH agonist, or orchiectomy. Second-generation androgen antagonists such as apalutamide or enzalutamide and androgen metabolism inhibitors such as abiraterone initially showed promise in treatment of castrate-resistant prostate cancer and have now been moved forward in the treatment of the disease. Unfortunately, these new medications are quite expensive and contribute to financial toxicity for men with advanced prostate cancer. When modulation of the androgen axis no longer is successful, there are several other lines of treatment for men with castrate-resistant disease, all showing survival benefit, including chemotherapy (docetaxel and cabazitaxel) and radiopharmaceuticals (radium-223). Immunotherapy has also been studied, and sipuleucel-T was demonstrated to improve survival for men with metastatic prostate cancer which is slowly progressing on androgen deprivation therapy. Overall, the introduction of many new agents for the treatment of advanced prostate cancer has resulted in significant improvement in both morbidity and mortality. Challenges remain in identifying the optimal treatment approach for individual patients [73]. There is some optimism that precision medicine, in which molecular analysis of the patient's tumor is performed using CLIA-certified procedures, may help guide therapy selection.

Genetic testing is increasingly informing treatment, particularly for men with advanced or metastatic prostate cancer [5]. Men with metastatic prostate cancer and inherited mutations in DNA repair genes, such as *BRCA2* and *ATM*, may have improved clinical response to PARP inhibitors such as olaparib [74]. Immunotherapy may also increasingly play a role in treatment of men with metastatic prostate cancer who have evidence of aberrant DNA mismatch repair [75]. Multiple therapeutic clinical trials have now emerged that incorporate germline genetic testing, which will be expected to expand indications for genetic testing in oncology to identify men with prostate cancer for optimized treatments.

33.7 Prognosis and Prevention

Studies attempting to address whether men with familial or hereditary prostate cancer have a poorer prognosis have produced mixed results. Some suggest that men with familial or hereditary prostate cancer possess more aggressive clinical features and have higher rates of mortality compared to men diagnosed with sporadic disease [76,77]. However, others show no difference in tumor characteristics or survival [78–80.] While there are data to support the idea that men with prostate cancer who harbor a mutation in *BRCA1* have a more aggressive disease course and poorer prognosis compared to non-carriers, the evidence is much stronger and more consistent in *BRCA2* mutation carriers [81,82]. *BRCA2* mutation carriers in particular tend to present at a younger age, have higher rates of both lymph node and distant metastases at time of diagnosis, and have higher mortality rates compared with non-carriers [26,83–86.] A systematic review of studies of *BRCA2* in prostate cancer patients found that the median age at diagnosis among mutation carriers was 62 years, with a median PSA at time of diagnosis of 15 ng/dL. Forty-one percent of carriers presented with stage T3 and higher disease compared to 29% of non-carriers. Twenty-six percent of carriers initially presented with metastatic disease compared to just 8% of non-carriers [87]. This suggests that early and more aggressive PSA screening of *BRCA1/2* mutation carriers would have a significant impact on their survival. In fact, early results from the IMPACT (**I**dentification of **M**en with a genetic predisposition to **P**rost**A**te **C**ancer: **T**argeted screening in *BRCA1/2* mutation carriers and controls) study supported this notion. Trial investigators observed a significantly greater proportion of *BRCA2* mutation carriers diagnosed with prostate cancer after PSA testing had intermediate- to high-risk disease. Of the prostate cancers diagnosed among *BRCA1* mutation carriers (n = 10), none were considered high-risk (defined as having clinical or pathologic stage T3 and higher and/or Gleason pattern 4+3 and higher) [88].

In addition to *BRCA1/2*, a number of other DNA repair genes have been examined as prognostic biomarkers in prostate cancer. Pritchard and colleagues sequenced the exomes of 20 DNA repair genes associated with cancer in men with metastatic prostate cancer and compared mutation frequency in these genes with expected frequency in the Exome Aggregation Consortium and men with localized prostate cancer from the Cancer Genome Atlas cohort [6]. At least one mutation was observed in approximately 12% of 692 men with metastatic disease compared with 4.6% of men with localized disease and just 2.7% of individuals without a known cancer diagnosis. With the exception of *BRCA1/2*, only mutations in *ATM*, *CHEK2*, and *RAD51D* were more frequent in metastatic prostate cancer compared with the other two groups. Germline mutations in *ATM* have also been linked to lethal prostate cancer [81].

33.8 Conclusion

New knowledge gained from precision oncology studies of men with metastatic prostate cancer has added significantly to our understanding of germline variation that contributes to prostate cancer susceptibility. The clinical application of genetic testing is evolving with expansion of genetic testing guidelines and testing capability. Research to identify additional genetic determinants of inherited susceptibility to prostate cancer is needed to address "missing heritability" in families suspected of inherited prostate cancer but with negative clinical testing. Greater understanding of modifiers of prostate cancer risk is also needed. Men are particularly interested in understanding modifiable factors that can decrease their future risk of disease. Knowledge of germline genetics may play an increasing role in the management and treatment of this potentially fatal cancer.

REFERENCES

1. Carter BS, Bova GS, Beaty TH et al. Hereditary prostate cancer: Epidemiologic and clinical features. *J Urol.* 1993;150:797–802.
2. Genetics of Prostate Cancer (PDQ®). National Cancer Institute. (Accessed April 15, 2018, at https://www.cancer.gov/types/prostate/hp/prostate-genetics-pdq.)
3. Giri VN, Knudsen KE, Kelly WK et al. Role of genetic testing for inherited prostate cancer risk: Philadelphia Prostate Cancer Consensus Conference 2017. *J Clin Oncol.* 2018;36:414–24.
4. NCCN Clinical Practice Guidelines in Oncology (NCCN Guidelines®). Genetic/Familial High-risk Assessment: Breast and Ovarian (Version 1.2018). (Accessed April 23, 2018, at https://www.nccn.org/professionals/physician_gls/pdf/genetics_screening.pdf.)
5. NCCN Clinical Practice Guidelines in Oncology (NCCN Guidelines®). Prostate (Version 1.2018). (Accessed April 23, 2018, at https://www.nccn.org/professionals/physician_gls/pdf/prostate.pdf.)
6. Pritchard CC, Mateo J, Walsh MF et al. Inherited DNA-repair gene mutations in men with metastatic prostate cancer. *N Engl J Med.* 2016;375:443–53.
7. Mandelker D, Zhang L, Kemel Y et al. Mutation detection in patients with advanced cancer by universal sequencing of cancer-related genes in tumor and normal dna vs guideline-based germline testing. *JAMA.* 2017;318:825–35.

8. Steinberg GD, Carter BS, Beaty TH, Childs B, Walsh PC. Family history and the risk of prostate cancer. *Prostate.* 1990;17:337–47.

9. Carter BS, Beaty TH, Steinberg GD, Childs B, Walsh PC. Mendelian inheritance of familial prostate cancer. *Proc Natl Acad Sci USA.* 1992;89:3367–71.

10. Schaid DJ, McDonnell SK, Blute ML, Thibodeau SN. Evidence for autosomal dominant inheritance of prostate cancer. *Am J Hum Genet.* 1998;62:1425–38.

11. Valeri A, Briollais L, Azzouzi R et al. Segregation analysis of prostate cancer in France: Evidence for autosomal dominant inheritance and residual brother-brother dependence. *Ann Hum Genet.* 2003;67:125–37.

12. Verhage BA, Baffoe-Bonnie AB, Baglietto L et al. Autosomal dominant inheritance of prostate cancer: A confirmatory study. *Urology.* 2001;57:97–101.

13. Schaid DJ. The complex genetic epidemiology of prostate cancer. *Hum Mol Genet.* 2004;13 Spec No 1:R103–R21.

14. Ostrander EA, Stanford JL. Genetics of prostate cancer: Too many loci, too few genes. *Am J Hum Genet.* 2000; 67:1367–75.

15. Shieh Y, Eklund M, Sawaya GF, Black WC, Kramer BS, Esserman LJ. Population-based screening for cancer: Hope and hype. *Nat Rev Clin Oncol* 2016;13(9):550–65.

16. Bratt O, Drevin L, Akre O, Garmo H, Stattin P. Family history and probability of prostate cancer, differentiated by risk category: A nationwide population-based study. *J Natl Cancer Inst.* 2016;108.

17. Lange EM, Gillanders EM, Davis CC et al. Genome-wide scan for prostate cancer susceptibility genes using families from the University of Michigan Prostate Cancer Genetics Project finds evidence for linkage on chromosome 17 near BRCA1. *Prostate.* 2003;57:326–34.

18. Zuhlke KA, Madeoy JJ, Beebe-Dimmer J et al. Truncating BRCA1 mutations are uncommon in a cohort of hereditary prostate cancer families with evidence of linkage to 17q markers. *Clin Cancer Res.* 2004;10:5975–80.

19. Lange EM, Robbins CM, Gillanders EM et al. Fine-mapping the putative chromosome 17q21–22 prostate cancer susceptibility gene to a 10 cM region based on linkage analysis. *Hum Genet.* 2007;121:49–55.

20. Ewing CM, Ray AM, Lange EM et al. Germline mutations in HOXB13 and prostate-cancer risk. *N Engl J Med.* 2012;366:141–9.

21. Lin X, Qu L, Chen Z et al. A novel germline mutation in HOXB13 is associated with prostate cancer risk in Chinese men. *Prostate.* 2013;73:169–75.

22. Karlsson R, Aly M, Clements M et al. A population-based assessment of germline HOXB13 G84E mutation and prostate cancer risk. *Eur Urol.* 2014;65(1):169–76.

23. Xu J, Lange EM, Lu L et al. HOXB13 is a susceptibility gene for prostate cancer: Results from the International Consortium for Prostate Cancer Genetics (ICPCG). *Hum Genet.* 2013;132:5–14.

24. Smith SC, Palanisamy N, Zuhlke KA et al. HOXB13 G84E-related familial prostate cancers: A clinical, histologic, and molecular survey. *Am J Surg Pathol.* 2014; 38:615–26.

25. Li D, Kumaraswamy E, Harlan-Williams LM, Jensen RA. The role of BRCA1 and BRCA2 in prostate cancer. *Front Biosci.* 2013;18:1445–59.

26. Castro E, Goh C, Olmos D et al. Germline BRCA mutations are associated with higher risk of nodal involvement, distant metastasis, and poor survival outcomes in prostate cancer. *J Clin Oncol.* 2013;31:1748–57.

27. Grindedal EM, Moller P, Eeles R et al. Germ-line mutations in mismatch repair genes associated with prostate cancer. *Cancer Epidemiol Biomarkers Prev.* 2009;18:2460–7.

28. Raymond VM, Mukherjee B, Wang F et al. Elevated risk of prostate cancer among men with Lynch syndrome. *J Clin Oncol.* 2013;31:1713–8.

29. Bauer CM, Ray AM, Halstead-Nussloch BA et al. Hereditary prostate cancer as a feature of Lynch syndrome. *Fam Cancer.* 2011;10:37–42.

30. Breast Cancer Linkage C. Cancer risks in BRCA2 mutation carriers. *J Natl Cancer Inst.* 1999;91:1310–6.

31. Leongamornlert D, Saunders E, Dadaev T et al. Frequent germline deleterious mutations in DNA repair genes in familial prostate cancer cases are associated with advanced disease. *Brit J Cancer* 2014;110:1663–72.

32. Robinson D, Van Allen EM, Wu YM et al. Integrative clinical genomics of advanced prostate cancer. *Cell.* 2015;161:1215–28.

33. Eeles RA, Olama AA, Benlloch S et al. Identification of 23 new prostate cancer susceptibility loci using the iCOGS custom genotyping array. *Nat Genet.* 2013;45:385–91, 91e1–2.

34. Agalliu I, Wang Z, Wang T et al. Characterization of SNPs associated with prostate cancer in men of Ashkenazic descent from the set of GWAS identified SNPs: Impact of cancer family history and cumulative SNP risk prediction. *PLOS ONE* 2013;8:e60083.

35. MacInnis RJ, Antoniou AC, Eeles RA et al. A risk prediction algorithm based on family history and common genetic variants: Application to prostate cancer with potential clinical impact. *Genet Epidemiol.* 2011;35:549–56.

36. Lecarpentier J, Silvestri V, Kuchenbaecker KB et al. Prediction of breast and prostate cancer risks in male BRCA1 and BRCA2 mutation carriers using polygenic risk scores. *J Clin Oncol.* 2017;35:2240–50.

37. Siegel RL, Miller KD, Jemal A. Cancer statistics, 2018. *CA Cancer J Clin.* 2018;68:7–30.

38. Moyer VA. Screening for prostate cancer: U.S. Preventive Services Task Force recommendation statement. *Ann Intern Med.* 2012;157:120–34.

39. Noone AM, Howlader N, Krapcho M et al. *SEER Cancer Statistics Review, 1975–2015.* Bethesda, MD: National Cancer Institute; 2018.

40. Bibbins-Domingo K, Grossman DC, Curry SJ. The US preventive services task force 2017 draft recommendation statement on screening for prostate cancer: An invitation to review and comment. *JAMA.* 2017;317:1949–50.

41. Attard G, Parker C, Eeles RA et al. Prostate cancer. *Lancet.* 2016;387:70–82.

42. Salinas CA, Tsodikov A, Ishak-Howard M, Cooney KA. Prostate cancer in young men: An important clinical entity. *Nat Rev Urol* 2014;11:317–23.

43. Kote-Jarai Z, Leongamornlert D, Saunders E et al. BRCA2 is a moderate penetrance gene contributing to young-onset prostate cancer: Implications for genetic testing in prostate cancer patients. *Brit J Cancer* 2011;105:1230–4.

44. Lin DW, Porter M, Montgomery B. Treatment and survival outcomes in young men diagnosed with prostate cancer: A population-based cohort study. *Cancer.* 2009;115:2863–71.

45. DeSantis CE, Siegel RL, Sauer AG et al. Cancer statistics for African Americans, 2016: Progress and opportunities in reducing racial disparities. *CA Cancer J Clin*. 2016; 66:290–308.

46. Ford ES, Giles WH, Dietz WH. Prevalence of the metabolic syndrome among US adults: Findings from the third National Health and Nutrition Examination Survey. *JAMA*. 2002;287:356–9.

47. Smith SC, Jr., Clark LT, Cooper RS et al. Discovering the full spectrum of cardiovascular disease: Minority Health Summit 2003: Report of the Obesity, Metabolic Syndrome, and Hypertension Writing Group. *Circulation*. 2005;111:e134–9.

48. Clark LT, El-Atat F. Metabolic syndrome in African Americans: Implications for preventing coronary heart disease. *Clin Cardiol*. 2007;30:161–4.

49. Beebe-Dimmer JL, Nock NL, Neslund-Dudas C et al. Racial differences in risk of prostate cancer associated with metabolic syndrome. *Urology*. 2009;74:185–90.

50. Post JM, Beebe-Dimmer JL, Morgenstern H et al. The metabolic syndrome and biochemical recurrence following radical prostatectomy. *Prostate Cancer* 2011;2011:245642.

51. Asmar R, Beebe-Dimmer JL, Korgavkar K, Keele GR, Cooney KA. Hypertension, obesity and prostate cancer biochemical recurrence after radical prostatectomy. *Prostate Cancer Prostatic Dis*. 2013;16:62–6.

52. Medhanie GA, Fedewa SA, Adissu H, DeSantis CE, Siegel RL, Jemal A. Cancer incidence profile in sub-Saharan African-born blacks in the United States: Similarities and differences with US-born non-Hispanic blacks. *Cancer*. 2017;123:3116–24.

53. Beebe-Dimmer JL, Zuhlke KA, Johnson AM, Liesman D, Cooney KA. Rare germline mutations in African American men diagnosed with early-onset prostate cancer. *Prostate*. 2018;78:321–6.

54. Han Y, Signorello LB, Strom SS et al. Generalizability of established prostate cancer risk variants in men of African ancestry. *Int J Cancer*. 2015;136:1210–7.

55. Stanford JL, Ostrander EA. Familial prostate cancer. *Epidemiol Rev*. 2001;23:19–23.

56. Johns LE, Houlston RS. A systematic review and meta-analysis of familial prostate cancer risk. *BJU Int*. 2003; 91:789–94.

57. Brandt A, Bermejo JL, Sundquist J, Hemminki K. Age-specific risk of incident prostate cancer and risk of death from prostate cancer defined by the number of affected family members. *Eur Urol*. 2010;58:275–80.

58. Sundquist K, Sundquist J, Ji J. Contribution of shared environmental factors to familial aggregation of common cancers: An adoption study in Sweden. *Eur J Cancer Prev*. 2015;24:162–4.

59. Westerman ME, Gershman B, Karnes RJ, Thompson RH, Rangel L, Boorjian SA. Impact of a family history of prostate cancer on clinicopathologic outcomes and survival following radical prostatectomy. *World J Urol*. 2016;34:1115–22.

60. Kupelian PA, Kupelian VA, Witte JS, Macklis R, Klein EA. Family history of prostate cancer in patients with localized prostate cancer: An independent predictor of treatment outcome. *J Clin Oncol*. 1997;15:1478–80.

61. Raheem OA, Cohen SA, Parsons JK, Palazzi KL, Kane CJ. A family history of lethal prostate cancer and risk of aggressive prostate cancer in patients undergoing radical prostatectomy. *Sci Rep*. 2015;5:10544.

62. Giri VN, Cassidy AE, Beebe-Dimmer J, Smith DC, Bock CH, Cooney KA. Association between Agent Orange and prostate cancer: A pilot case-control study. *Urology*. 2004;63:757–60.

63. Tangen CM, Neuhouser ML, Stanford JL. Prostate cancer. In: Thun MJ, Linet MS, Cerhan JR, Haiman CA, Schottenfeld D, editors. *Cancer Epidemiology and Prevention*, 4th ed. New York: Oxford University Press; 2017: 997–1018.

64. Cao Y, Giovannucci E. Obesity and prostate cancer. *Recent Results Cancer Res*. 2016;208:137–53.

65. Cai H, Xu Z, Xu T, Yu B, Zou Q. Diabetes mellitus is associated with elevated risk of mortality amongst patients with prostate cancer: A meta-analysis of 11 cohort studies. *Diabetes Metab Res Rev*. 2015;31:336–43.

66. Allott EH, Masko EM, Freedland SJ. Obesity and prostate cancer: Weighing the evidence. *Eur Urol*. 2013;63:800–9.

67. Giri VN, Beebe-Dimmer JL, Riley BD et al. Familial prostate cancer. *Semin Oncol*. 2016;43:560–5.

68. Umar A, Boland CR, Terdiman JP et al. Revised Bethesda Guidelines for Hereditary Nonpolyposis Colorectal Cancer (Lynch Syndrome) and Microsatellite Instability. *J Natl Cancer Inst*. 2004;96:261–8.

69. NCCN Clinical Practice Guidelines in Oncology (NCCN Guidelines®): Genetic/Familial High-Risk Assessment: Colorectal (Version 3.2017). (Accessed April 23, 2018, at https://www.nccn.org/professionals/physician_gls/pdf/genetics_colon.pdf.)

70. Riley BD, Culver JO, Skrzynia C et al. Essential elements of genetic cancer risk assessment, counseling, and testing: Updated recommendations of the National Society of Genetic Counselors. *J Genet Couns*. 2012;21:151–61.

71. Giri VN, Obeid E, Gross L et al. Inherited mutations in men undergoing multigene panel testing for prostate cancer: Emerging implications for personalized prostate cancer genetic evaluation. *JCO Precision Oncology* 2017:1–17. doi: 10.1200/PO.16.00039.

72. Amin MB, Edge SB, Greene FL et al. (Eds). *AJCC Cancer Staging Manual*, 8th ed. New York: Springer International Publishing; 2017.

73. Nuhn P, De Bono JS, Fizazi K et al. Update on systemic prostate cancer therapies: Management of metastatic castration-resistant prostate cancer in the era of precision oncology. *Eur Urol*. 2019;75(1):88–99.

74. Mateo J, Carreira S, Sandhu S et al. DNA-repair defects and olaparib in metastatic prostate cancer. *N Engl J Med*. 2015;373:1697–708.

75. Le DT, Durham JN, Smith KN et al. Mismatch-repair deficiency predicts response of solid tumors to PD-1 blockade. *Science*. 2017;357(6349):409–13.

76. Brandt A, Bermejo JL, Sundquist J, Hemminki K. Age at diagnosis and age at death in familial prostate cancer. *Oncologist*. 2009;14:1209–17.

77. Pakkanen S, Kujala PM, Ha N, Matikainen MP, Schleutker J, Tammela TL. Clinical and histopathological characteristics of familial prostate cancer in Finland. *BJU Int*. 2012;109:557–63.

78. Gronberg H, Damber L, Tavelin B, Damber JE. No difference in survival between sporadic, familial and hereditary prostate cancer. *Br J Urol* 1998;82:564–7.

79. Bratt O, Damber JE, Emanuelsson M, Gronberg H. Hereditary prostate cancer: Clinical characteristics and survival. *J Urol*. 2002;167:2423–6.

80. Cremers RG, Aben KK, van Oort IM et al. The clinical phenotype of hereditary versus sporadic prostate cancer: HPC definition revisited. *Prostate*. 2016;76:897–904.

81. Na R, Zheng SL, Han M et al. Germline mutations in ATM and BRCA1/2 distinguish risk for lethal and indolent prostate cancer and are associated with early age at death. *Eur Urol.* 2017;71:740–7.

82. Mateo J, Cheng HH, Beltran H et al. Clinical outcome of prostate cancer patients with germline DNA repair mutations: Retrospective analysis from an international study. *Eur Urol.* 2018;73:687–93.

83. Gallagher DJ, Gaudet MM, Pal P et al. Germline BRCA mutations denote a clinicopathologic subset of prostate cancer. *Clin Cancer Res.* 2010;16:2115–21.

84. Tryggvadottir L, Vidarsdottir L, Thorgeirsson T et al. Prostate cancer progression and survival in BRCA2 mutation carriers. *J Natl Cancer Inst.* 2007;99:929–35.

85. Thorne H, Willems AJ, Niedermayr E et al. Decreased prostate cancer-specific survival of men with BRCA2 mutations from multiple breast cancer families. *Cancer Prev Res (Phila)* 2011;4:1002–10.

86. Mitra A, Fisher C, Foster CS et al. Prostate cancer in male BRCA1 and BRCA2 mutation carriers has a more aggressive phenotype. *Brit J Cancer* 2008;98:502.

87. Gleicher S, Kauffman EC, Kotula L, Bratslavsky G, Vourganti S. Implications of high rates of metastatic prostate cancer in BRCA2 mutation carriers. *Prostate.* 2016;76:1135–45.

88. Bancroft EK, Page EC, Castro E et al. Targeted prostate cancer screening in BRCA1 and BRCA2 mutation carriers: Results from the initial screening round of the IMPACT study. *Eur Urol.* 2014;66:489–99.

34

Microphthalmia-Associated Transcription Family Translocation Renal Cell Cancer

Dongyou Liu

CONTENTS

34.1 Introduction

The microphthalmia-associated transcription (MIT) family comprises four basic helix-loop-helix (bHLH) zipper transcription factors (i.e., MITF, TFE3, TFEB, and TFEC) that regulate the expression from promoters containing a DNA response element that includes specific flanking nucleotides in addition to a core E-box element usually bound by bHLH zipper transcription factors.

Mutations (e.g., missense or translocation variants) in the underlying genes are responsible for causing renal cell cancer (RCC), melanoma, and hypopigmentation disorders (Table 34.1). For example, fusion of *TFE3* or *TFEB* with respective translocation partners results in chromosome translocations (X;1)(p11.2;q21) or t(6;11)(p21.1;q12), and subsequent overexpression of TFE3 or TFEB, leading to Xp11 translocation RCC (Xp11 tRCC) or translocation (6;11) RCC [t(6,11) RCC], which are collectively referred to as MIT family tRCC [1,2]. On the other hand, mutation in MITF E318 (i.e., c.952G→A; p.E318 K) increases the risk for RCC and/or melanoma, whereas mutations in other regions of MITF are associated with Waardenburg syndrome type 2 (WS2A; deafness, minor defects, and pigmentation abnormalities) and Tietz albinism-deafness syndrome (deafness and leucism). Furthermore, germline mutation in TFEC is implicated in the pathogenesis of WS2A (Table 34.1).

Relative to clear cell RCC (80% of all RCC cases), papillary RCC (10%), and chromophobe RCC (4%–5%), MIT family tRCC is rare, and represents about 1% of all RCC cases. However, considering the fact that Xp11 tRCC and t(6,11) RCC may occasionally display similar microscopic features to RCC and papillary RCC, correct identification of uncommon RCC types is fundamental to the selection and implementation of appropriate management strategies for patients with RCC [3].

34.2 Biology

RCC is a heterogeneous disease whose clinical behavior and prognosis are influenced by distinct histologic types that demonstrate characteristic morphological, immunohistochemical, ultrastructural, and cytogenetic features [e.g., clear cell RCC, papillary RCC, chromophobe RCC, Xp11 tRCC and t(6;11) RCC] (Table 34.2) [4].

While clear cell RCC and papillary RCC account for over 90% of clinical cases and occur in both adults and children, Xp11 tRCC and t(6;11) RCC are rare renal malignancies that preferentially affect young patients [5–7].

Xp11 tRCC (also known as *TFE3*-rearranged RCC) is a brownish-yellow solid mass with an occasional gray-white cut surface (resembling papillary RCC), frequent necrosis, and hemorrhage. It is responsible for 20%–40% of pediatric RCC and only 1% of adult RCC (with an average onset age of 50 years). Histologically, Xp11 tRCC has large epithelioid cells in papillary and nested growth pattern and containing clear to eosinophilic cytoplasm, prominent nucleoli, and psammoma bodies (Figure 34.1). Cytogenetically, Xp11 tRCC contains a fusion of *TFE3* in Xp11 with *ASPSCR1* (also known as *ASPL*), *PRCC*, *SFPQ* (also known as *PSF*), *CLTC*, *NONO*, *RBM10*, *PARP14*, *LUC7L3*, *KHSRP*, *DVL2*, *MED15*, or *GRIPAP1* in

TABLE 34.1

Characteristics of the Microphthalmia-Associated Transcription (MIT) Family of Transcription Factors

Factor	Chromosome Location	Gene	Protein	Clinical Disease
MITF	3p13	228 kb	526 aa/58 kDa	Germline missense mutation (p.E318 K) causes RCC and/or melanoma; other germline mutations are involved in Waardenburg syndrome type 2 (deafness, minor defects, and pigmentation abnormalities) and Tietz albinism-deafness syndrome (deafness and leucism)
TFE3	Xp11.23	14 kb	575 aa/61 kDa	Fusion of *TFE3* in Xp11 with *ASPSCR1* (also known as *ASPL*), *PRCC*, *SFPQ* (also known as *PSF*), *CLTC*, *NONO*, *RBM10*, *PARP14*, *LUC7L3*, *KHSRP*, *DVL2*, *MED15*, or *GRIPAP1* in 1q22 results in chromosome translocation t (X;1)(p11.2;q21) and Xp11 tRCC
TFEB	6p21.1	52 kb	476 aa/52 kDa	Fusion of *TFEB* in 6p21 with *MALAT1* (also known as *Alpha*) in 11q12 results in chromosome translocation t(6;11)(p21.1;q12) and t(6;11) RCC; additional translocation partners for *TFEB* include *COL21A1,- CADM2*, and *KHDRBS2*, which may be unrelated to RCC
TFEC	7q31.2	224 kb	347 aa/38 kDa	Germline mutations lead to Waardenburg syndrome type 2a (WS2A), presenting as wide nasal bridge, underdeveloped nasal alae, heterochromia iridis, premature graying of hair, white forelock, synophrys, white eyebrow and eyelashes, congenital sensorineural hearing impairment, partial albinism, and hypoplastic iris stroma

1q22, resulting in chromosome translocation t(X;1)(p11.2;q21). Interestingly, Xp11 tRCC with a *PRCC-TFE3* fusion shows a smaller structure of nested or papillary cells, less abundant cytoplasm, and less conspicuous nuclei than Xp11 tRCC with an *ASPSCR1-TFE3* fusion, but the latter (which also occurs in alveolar soft part sarcoma) has a worse prognosis than the former. Due to its frequent lymph node metastasis, Xp11 tRCC confers a similar prognosis to clear cell RCC and a worse prognosis than papillary RCC [2].

Similarly, t(6;11) RCC (also known as *TFEB*-rearranged RCC) is cystic or solid mass with occasionally mahogany-brown cut surface and has an average onset age of 30 years. Histologically, t(6;11) RCC displays nests of large epithelioid cells with clear to eosinophilic cytoplasm and clusters of small cells with chromatin condensed nuclei surrounding the hyaline material (of basement membrane origin, appearing microscopically as pink hyaline globule), a morphological feature that is also occasionally observed in Xp11 tRCC (Figure 34.2). Cytogenetically, t(6;11) RCC has a fusion of *TFEB* in 6p21 with *MALAT1* (also known as *Alpha*) in 11q12, resulting in chromosome translocation t(6;11)(p21.1;q12) [2].

TFE3 and TFEB belong to the MIT family consisting of MITF, TFE3, TFEB, and TFEC. These transcription factors share a bHLH DNA-binding domain and similar target genes. Fusion of *TFE3* or *TFEB* with *ASPSCR1* or *MALAT1* (or related genes) leads to their overexpression, which underlies the development of Xp11 tRCC and t(6;11) RCC (collectively known as MIT family tRCC) as well as other tumors (e.g., alveolar soft part sarcoma, melanoma, clear cell sarcoma, angiomyolipoma, and perivascular epithelioid cell tumor [PEComa]) [8]. Indeed, *TFE3* gene fusions have been identified in PEComas of the kidney and soft tissue [9].

34.3 Pathogenesis

The MIT family comprises four transcription factors (i.e., MITF, TFE3, TFEB, and TFEC) that all possess an N-terminal domain,

a central transcriptional activation domain, a bHLH zipper domain, and a C-terminal transactivation region.

The microphthalmia-associated transcription factor (*MITF*) gene located on chromosome 3p13 spans 228 kb and contains 12 exons, with the first 4 exons differentially spliced to encode the unique N termini of MITF isoforms, and all MIT variants (MITF-A [526 aa, 58 kDa], MITF-B, MITF-C, MITF-D, MITF-E, MITF-H, MITF-J, MITF-M, MITF-CX, MITF-MC) consisting of 8 common downstream exons. The N- and C-terminal domains of MITF facilitate its transactivation activity as well as its oncogenic role [10]. Upon binding to M-boxes (5'-TCATGTG-3') and symmetrical DNA sequences (E-boxes) (5'-CACGTG-3') in the promoters of target genes (e.g., BCL2, tyrosinase [TYR], tyrosinase-related protein 1 [TYRP1]), MITF regulates the differentiation of neural crest−derived melanocytes, mast cells, osteoclasts, and optic cup−derived retinal pigment epithelium [11–14]. MITF modulates pigment cell-specific transcription of melanogenesis via the MAPK-Erk signaling pathway (which is deregulated in >90% of melanomas) mediated by p38-alpha and p38-beta, and regulates retinal pigment epithelium migration with the help of pigment epithelium-derived factor (PEDF, an antiangiogenic protein) [15–20]. In addition, MITF upregulates the transcription of hypoxia-inducible factor (HIF1-α) (which drives renal tumorigenesis) [21,22]. The transcriptional activity of MITF is suppressed by SUMOylation, a post-translational modification, at a SUMO consensus site involving the E318 codon [23]. However, the MITF p.E318 K variant severely impairs the negative regulation of SUMOylation on MITF, upregulates MITF and HIF-1α, and promotes invasive and tumorigenic behaviors in melanocytes and renal cells, leading to RCC and/or melanoma (including familial melanoma) [24–30]. Interestingly, MITF-A, MITF-B, MITF-C, MITF-H, and MITF-M are expressed in many cell lines and the kidneys. MITF-M is found only in melanoma cell lines, whereas MITF-B does not seem to occur in melanoma cell lines [31–33]. Heterozygous *MITF* mutations (occurring at S298, S397, S401, and S405 sites) are implicated in the pathogenesis of auditory-pigmentary syndromes such as Waardenburg syndrome type 2

TABLE 34.2

Comparative Molecular/Cytogenetic, Morphological, and Immunohistochemical Properties of RCC Types

RCC Type	Molecular/Cytogenetic	Morphological	Immunohistochemical							
			TFE3	TFEB	Cathepsin K	HMB45	Melan A	CAIX	CK7	AMACR
Clear cell RCC	Inactivation of *VHL*; upregulation of hypoxia inducible factor (HIF); 5q21+ (70%); 14q−(41%); *BAP1* mutation in high grade lesion	Cells with clear or eosinophilic cytoplasm, displaying solid, alveolar, acinar, cystic growth patterns, and forming characteristic network of small, thin-walled, "chickenwire" vasculature	−	−	−	−	−	+	−	−
Papillary RCC	Heterozygous germline mutation in *MET* (type 1) or *FH* (type2)	Type 1: basophilic cells with scanty cytoplasms, and arranged in a single layer along the papillary axis Type 2 eosinophilic cells with large cytoplasms, pseudostratified nuclei, prominent eosinophilic nucleoli and perinucleolar halo, and arranged in several strata along the papillary axis	−	−	−	−	−	−	+	+/−
Chromophobe RCC	Multiple losses of whole chromosomes (e.g., 1, 2, 6, 10, 13, 17, 21 or Y); mutation in *TP53* (32%) and *PTEN* (9%)	Cells with prominent borders, finely reticular pale cytoplasm, perinuclear halo, nuclei with preserved chromatin and irregular, wrinkled nuclear membrane (raisinoid nuclei)	−	−	−	−	−	−	+	−
Xp11 tRCC	Chromosome translocation t(X;17) (p11.2;q25)	Nested and alveolar growth (papillary architecture) of polygonal, clear and eosinophilic cells with clear to pale pink fluffy cytoplasm and abundant psammoma bodies	+	−	+/−	−/+ focally	+focally/−	−/+ focally	−	+
t(6;11) RCC	Chromosome translocation t(6;11) (p21.1;q12)	Nests of polygonal, clear to eosinophilic cells occasionally forming papillae; clusters of small eosinophilic cells with small hyperchromatic nuclei surrounding hyaline material	−	+	+	+/−	+	−/+ focally	−	+

(causing deafness, minor defects, and abnormalities in pigmentation) and Tietz albinism-deafness syndrome (displaying deafness and leucism).

The transcription factor binding to IGHM enhancer 3 (or transcription factor E3) gene (*TFE3*) located on chromosome Xp11.23 spans 14 kb and encodes a 575 aa, 61 kDa protein (TFE3). Acting as bHLH domain-containing transcription factor, TFE3 binds MUE3-type E-box sequences (5'-CANNTG-3') in the promoter of genes and enhances the expression of genes downstream of transforming growth factor beta (TGF-beta) signaling. Fusion of *TFE3* with translocation partner (e.g., alveolar soft part sarcoma chromosome region, candidate 1 [*ASPSCR1*], papillary renal cell carcinoma [*PRCC*], non-POU domain containing, octamer-binding [*NONO*]) produces chromosome translocation t(X;1)(p11.2;q21), leading to Xp11 tRCC, alveolar soft part sarcoma and other tumors [34,35]. In association with TFEB, TFE3 activates the expression of CD40L in T cells and contributes to T-cell-dependent antibody responses in activated CD4(+) T cells and thymus-dependent humoral immunity.

The transcription factor EB gene (*TFEB*) located on chromosome 6p21.1 spans 52 kb and encodes a 476 aa, 52 kDa protein (TFEB) with specificity for E-box sequences (5'-CANNTG-3') and involvement in PI3K-AKT-mTOR signaling pathway. Fusion between *TFEB* and translocation partner *MALAT1* produces chromosome translocation t(6;11)(p21.1;q12) and t(6;11) RCC [36]. Also, in association with TFE3, TFEB activates the expression of CD40L in T cells, and thus participates in T-cell-dependent antibody responses in activated CD4(+) T cells and thymus-dependent humoral immunity. Further, TFEB recognizes and binds the CLEAR-box sequence (5'-GTCACGTGAC-3') present in the regulatory region of many lysosomal genes, thus activating the expression of lysosomal genes. TFEB also promotes expression of genes involved in autophagy, and specifically recognizes the gamma-E3 box present in the heavy-chain immunoglobulin enhancer. Finally, TFEB takes part in the signal transduction processes for normal vascularization of the placenta.

The transcription factor EC gene (*TFEC*) located on chromosome 7q31.2 spans 224 kb and encodes a 347 aa, 38 kDa protein (TFEC). By binding to E-box recognition sequences as homo- or heterodimers, TFEC regulate via C-MYB transcription factor network the expression of non-muscle myosin II heavy chain-A gene, and also coregulates target genes in osteoclasts and other cellular processes (e.g., differentiation, growth, and survival). Mutation in TFEC is apparently linked to the development of WS2A.

34.4 Epidemiology

34.4.1 Prevalence

Xp11 tRCC accounts for 20%–40% of childhood RCC and 1% of adult RCC, with average onset age of 50 years. t(6;11) RCC is very rare, and only about 60 cases have been described in the literature. The average onset age for t(6;11) RCC is about 30 years old.

The prevalence of p.E318 K mutation is estimated at 0.3% among the general population, 1.4% among patients with melanoma only, 1.5% among patients with RCC only, and 4% among patients with melanoma and RCC.

34.4.2 Inheritance

MIT family tRCC demonstrates autosomal dominant inheritance.

34.4.3 Penetrance

MIT family tRCC and missense MITF mutation p.E318 K show incomplete penetrance.

34.5 Clinical Features

Early RCC is often asymptomatic, while late RCC may present with hematuria (blood in the urine), low back pain (on one side or both sides, that does not go away), a lump on the side, belly, or lower back, anemia (low red blood cell counts), fatigue (feeling tired, sluggish, and run-down due to anemia), fever (that does not get better), loss of appetite and weight, night sweats, high blood pressure, and high level blood calcium.

34.6 Diagnosis

A combination of medical history review, physical exam, laboratory assays, imaging studies (e.g., x-rays, CT, MRI, ultrasound), histopathology, and molecular genetic testing are utilized for diagnosis of RCC.

Specific diagnosis of MIT family tRCC relies on the use of (i) immunohistochemistry, (ii) break-apart fluorescence in situ hybridization (FISH), and (iii) reverse transcriptase-polymerase chain reaction (RT-PCR)/5'-rapid amplification of cDNA ends (5'-RACE)/karyotyping. Whereas FFPE specimens are applicable to the first two approaches, fresh specimens are necessary for the third approach [37].

Histologically, Xp11 tRCC shows clear cells with papillary architecture and abundant psammomatous bodies (Figure 34.1) [38,39], whereas t(6;11) RCC tends to give a biphasic appearance with both large and small epithelioid cells and nodules of hyaline material of basement membrane origin (Figure 34.2) [40,41].

Due to the fact that Xp11 tRCC shares morphological features with clear cell RCC and papillary RCC at times, and that the characteristic morphology of small cells clustering around hyaline material observed in t(6;11) RCC is occasionally found in Xp11 tRCC, use of immunohistochemical stains (e.g., TFE3, TFEB, HMB45, melan A, and cathepsin K) and preferably break-apart FISH for detecting fusions of *TFE3* or *TFEB* in FFPE specimens (or RT-PCR, 5'-RACE, and karyotyping in fresh specimens) is beneficial to confirmation of Xp11 tRCC and t(6;11) RCC [42–44].

Specifically, Xp11 tRCC with a *PRCC-TFE3* fusion displays a high rate of cathepsin K positivity (86%), while Xp11 tRCC with an *ASPSCR1-TFE3* fusion is negative for cathepsin K. By contrast, t(6;11) RCC demonstrates strong and diffuse cytoplasmic immunostaining for cathepsin K (100%) [45,46].

Differential diagnoses for MIT family tRCC [i.e., Xp11 tRCC and t(6;11) RCC] should include other renal tumors with eosinophilic cytoplasm (e.g., oncocytoma, chromophobe RCC, hybrid tumor, tubulocystic carcinoma, papillary RCC, clear cell RCC with predominant eosinophilic cell morphology, follicular thyroid-like carcinoma, hereditary leiomyomatosis–associated

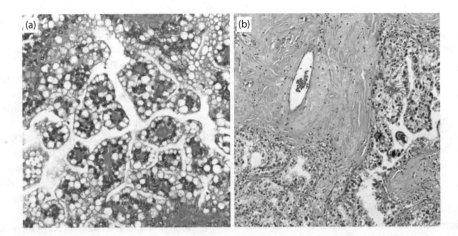

FIGURE 34.1 Histology of Xp11 tRCC. (a) Clear cells with voluminous cytoplasm and distinct cell borders showing typical papillary architecture and hyalinized fibrovascular cords; (b) clear cells arranged in tubule-alveolar pattern laying in hyalinized stroma (bottom left corner), and psammoma bodies (right side). (Photo credit: Kmetec A, Jeruc J. *Radiol Oncol.* 2014;48:197–202.)

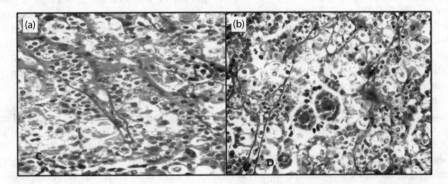

FIGURE 34.2 Histology of t(6;11) RCC. (a) Biphasic population of large epithelioid cells with voluminous eosinophilic cytoplasm and small cells with scant cytoplasm and small rounded nuclei containing condensed chromatin; (b) clusters of small cells around hyaline pink material forming "pseudorosettes" (H&E). (Photo credit: Arneja SK, Gujar N. *Int J Surg Case Rep.* 2015;7C:16–9.)

RCC, acquired cystic kidney disease–associated RCC, rhabdoid RCC, epithelioid angiomyolipoma, epithelioid angiomyolipoma, and unclassified RCC).

34.7 Treatment

Treatment options for RCC include surgery, cryoablation, chemotherapy, hormone therapy, or immunotherapy.

Patients with localized RCC may undergo partial nephrectomy when the largest tumor reaches 3 cm in size. Patients with unresectable tumor that is solid and in a contained area may have cryoablation or radiofrequency ablation.

Patients with advanced or disseminated RCC may require radical nephrectomy. Locoregional therapy (e.g., mTOR inhibitors) may help reduce MIT family tRCC tumor size and palliate symptoms, while kidney transplantation may be considered for patients with severe kidney dysfunction.

34.8 Prognosis

In general, children with MIT family tRCC have a more favorable prognosis than adults. Xp11 tRCC has a tendency to metastasize to lymph nodes and other organs as late as 20–30 years

after diagnosis and confers a worse prognosis than papillary RCC and similar prognosis to clear cell RCC. In addition, Xp11 tRCC with an *ASPSCR1-TFE3* fusion may have a worse prognosis than Xp11 tRCC with other fusion partners. Most t(6;11) RCC cases behave indolently and have good prognosis [47].

34.9 Conclusion

Transcription factors (i.e., MITF, TFE3, TFEB, and TFEC) in the microphthalmia-associated transcription (MIT) family possess a bHLH zipper domain that specifically recognizes the DNA response element contained in the promoters of a number of downstream genes involved in the differentiation, proliferation, migration, and survival of melanocytes and renal cells. Mutations (e.g., missense and translocation variants) in the underlying genes contribute to the overexpression of MITF, TFE3, TFEB, and TFEC, and subsequent development of distinct MIT family translation RCC [i.e., Xp11 tRCC and t(6;11) RCC] and malignant melanoma (including familial melanoma) as well as other syndromic diseases [48]. Due to their histologic resemblances to clear cell RCC and papillary RCC, Xp11 tRCC and t(6;11) RCC provide a notable challenge for even experienced pathologists to accurately diagnose. Recent elucidation of the molecular basis of MIT family tRCC permits the design of highly specific approaches for confirmative diagnosis. Further studies on the molecular pathogenesis of these uncommon RCC types will undoubtedly yield new insights from which innovative therapeutics can be derived.

REFERENCES

1. Moch H, Humphrey PA, Ulbright TM, Reuter VE. *WHO Classification of Tumours of the Urinary System and Male Genital Organs.* 4th ed. Lyon, France: IARC Press; 2016.
2. Inamura K. Translocation renal cell carcinoma: An update on clinicopathological and molecular features. *Cancers (Basel).* 2017;9(9). pii: E111.
3. Durinck S, Stawiski EW, Pavia-Jimenez A et al. Spectrum of diverse genomic alterations define non-clear cell renal carcinoma subtypes. *Nat Genet.* 2015;47:13–21.
4. Linehan WM, Spellman PT, Ricketts CJ et al. Comprehensive molecular characterization of papillary renal-cell carcinoma. *N Engl J Med.* 2016;374:135–15.
5. Argani P, Antonescu CR, Illei PB et al. Primary renal neoplasms with the ASPL-TFE3 gene fusion of alveolar soft part sarcoma: A distinctive tumor entity previously included among renal cell carcinomas of children and adolescents. *Am J Pathol.* 2001;159:179–92.
6. Argani P, Hawkins A, Griffin CA et al. A distinctive pediatric renal neoplasm characterized by epithelioid morphology, basement membrane production, focal HMB45 immunoreactivity, and t(6;11)(p21.1;q12) chromosome translocation. *Am J Pathol.* 2001;158:2089–96.
7. Argani P, Antonescu CR, Couturier J et al. PRCC-TFE3 renal carcinomas: Morphologic, immunohistochemical, ultrastructural, and molecular analysis of an entity associated with the t(X;1)(p11.2;q21). *Am J Surg Pathol.* 2002;26:1553–66.
8. Chen YB, Xu J, Skanderup AJ et al. Molecular analysis of aggressive renal cell carcinoma with unclassified histology reveals distinct subsets. *Nat. Commun.* 2016;7:13131.
9. Davis IJ, Kim JJ, Ozsolak F et al. Oncogenic MITF dysregulation in clear cell sarcoma: Defining the MiT family of human cancers. *Cancer Cell.* 2006;9(6):473–84.
10. Pogenberg V, Ogmundsdóttir MH, Bergsteinsdóttir K et al. Restricted leucine zipper dimerization and specificity of DNA recognition of the melanocyte master regulator MITF. *Genes Dev.* 2012;26(23):2647–58.
11. Hsiao JJ, Fisher DE. The roles of microphthalmia-associated transcription factor and pigmentation in melanoma. *Arch Biochem Biophys.* 2014;563:28–34.
12. Lauss M, Haq R, Cirenajwis H et al. Genome-wide DNA methylation analysis in melanoma reveals the importance of CpG methylation in MITF regulation. *J Invest Dermatol.* 2015;135(7):1820–8.
13. Dar AA, Majid S, Bezrookove V et al. BPTF transduces MITF-driven prosurvival signals in melanoma cells. *Proc Natl Acad Sci U S A.* 2016;113(22):6254–8.
14. Wang C, Zhao L, Su Q et al. Phosphorylation of MITF by AKT affects its downstream targets and causes TP53-dependent cell senescence. *Int J Biochem Cell Biol.* 2016;80:132–42.
15. Tachibana M. Cochlear melanocytes and MITF signaling. *J Investig Dermatol Symp Proc.* 2001;6(1):95–8.
16. Dadras SS, Lin RJ, Razavi G et al. A novel role for microphthalmia-associated transcription factor-regulated pigment epithelium-derived factor during melanoma progression. *Am J Pathol.* 2015;185(1):252–65.
17. Wellbrock C, Arozarena I. Microphthalmia-associated transcription factor in melanoma development and MAP-kinase pathway targeted therapy. *Pigment Cell Melanoma Res.* 2015;28(4):390–406.
18. Alver TN, Lavelle TJ, Longva AS, Øy GF, Hovig E, Bøe SL. MITF depletion elevates expression levels of ERBB3 receptor and its cognate ligand NRG1-beta in melanoma. *Oncotarget.* 2016;7(34):55128–40.
19. Fane ME, Chhabra Y, Hollingsworth DEJ et al. NFIB mediates BRN2 driven melanoma cell migration and invasion through regulation of EZH2 and MITF. *EBioMedicine.* 2017;16:63–75.
20. Simmons JL, Pierce CJ, Al-Ejeh F, Boyle GM. MITF and BRN2 contribute to metastatic growth after dissemination of melanoma. *Sci Rep.* 2017;7(1):10909.
21. Nooron N, Ohba K, Takeda K, Shibahara S, Chiabchalard A. Dysregulated expression of MITF in subsets of hepatocellular carcinoma and cholangiocarcinoma. *Tohoku J Exp Med.* 2017;242(4):291–302.
22. Oh TI, Lee YM, Lim BO, Lim JH. Inhibition of NAT10 suppresses melanogenesis and melanoma growth by attenuating microphthalmia-associated transcription factor (MITF) expression. *Int J Mol Sci.* 2017;18(9). pii: E1924.
23. Bertolotto C, Lesueur F, Giuliano S et al. A SUMOylation-defective MITF germline mutation predisposes to melanoma and renal carcinoma. *Nature.* 2011;480:94–8. Erratum: Nature.2011; 531: 126.
24. Yokoyama S, Woods S L, Boyle GM et al. A novel recurrent mutation in MITF predisposes to familial and sporadic melanoma. *Nature.* 2011;480:99–103.
25. Hartman ML, Talar B, Noman MZ, Gajos-Michniewicz A, Chouaib S, Czyz M. Gene expression profiling identifies microphthalmia-associated transcription factor (MITF) and Dickkopf-1 (DKK1) as regulators of microenvironment-driven alterations in melanoma phenotype. *PLOS ONE.* 2014;9(4):e95157.
26. Potrony M, Badenas C, Aguilera P et al. Update in genetic susceptibility in melanoma. *Ann Transl Med.* 2015;3(15):210.
27. Potrony M, Puig-Butille JA, Aguilera P et al. Prevalence of MITF p.E318 K in patients with melanoma independent of the presence of CDKN2A causative mutations. *JAMA Dermatol.* 2016;152(4):405–12.
28. Wadt KA, Aoude LG, Krogh L et al. Molecular characterization of melanoma cases in Denmark suspected of genetic predisposition. *PLOS ONE.* 2015;10(3):e0122662.
29. Stoehr CG, Walter B, Denzinger S et al. The microphthalmia-associated transcription factor p.E318 K mutation does not play a major role in sporadic renal cell tumors from Caucasian patients. *Pathobiology.* 2016;83(4):165–9.
30. Bianchi-Smiraglia A, Bagati A, Fink EE et al. Microphthalmia-associated transcription factor suppresses invasion by reducing intracellular GTP pools. *Oncogene.* 2017;36(1):84–96.
31. Tsao H, Chin L, Garraway LA, Fisher DE. Melanoma: From mutations to medicine. *Genes Dev.* 2012;26(11):1131–55.
32. Grill C, Bergsteinsdóttir K, Ogmundsdóttir MH et al. MITF mutations associated with pigment deficiency syndromes and melanoma have different effects on protein function. *Hum Mol Genet.* 2013;22(21):4357–67.
33. Bourseguin J, Bonet C, Renaud E et al. FANCD2 functions as a critical factor downstream of MiTF to maintain the proliferation and survival of melanoma cells. *Sci Rep.* 2016;6:36539.

34. Argani P, Zhong M, Reuter VE et al. TFE3-Fusion variant analysis defines specific clinicopathologic associations among Xp11 translocation cancers. *Am J Surg Pathol.* 2016;40:723–37.

35. Argani P, Zhang L, Reuter VE, Tickoo SK, Antonescu CR. RBM10-TFE3 renal cell carcinoma: A potential diagnostic pitfall due to cryptic intrachromosomal Xp11.2 inversion resulting in false-negative TFE3 FISH. *Am J Surg Pathol.* 2017;41:655–62.

36. Argani P, Reuter VE, Zhang L et al. TFEB-amplified renal cell carcinomas: An aggressive molecular subset demonstrating variable melanocytic marker expression and morphologic heterogeneity. *Am J Surg Pathol.* 2016;40:1484–95.

37. Kryvenko ON, Jorda M, Argani P, Epstein JI. Diagnostic approach to eosinophilic renal neoplasms. *Arch Pathol Lab Med.* 2014;138(11):1531–41.

38. Chaste D, Vian E, Verhoest G, Blanchet P. Translocation renal cell carcinoma t(6;11)(p21;q12) and sickle cell anemia: First report and review of the literature. *Korean J Urol.* 2014;55(2):145–7.

39. Kmetec A, Jeruc J. Xp 11.2 translocation renal carcinoma in young adults; recently classified distinct subtype. *Radiol Oncol.* 2014;48:197–202.

40. Rao Q, Zhang XM, Tu P et al. Renal cell carcinomas with t(6;11)(p21;q12) presenting with tubulocystic renal carcinoma-like features. *Int J Clin Exp Pathol.* 2013;6(7):1452–7.

41. Arneja SK, Gujar N. Renal cell carcinoma with t(6:11) (p21;q12). A case report highlighting distinctive immunohistologic features of this rare tumor. *Int J Surg Case Rep.* 2015;7C:16–9.

42. Argani P, Yonescu R, Morsberger L et al. Molecular confirmation of t(6;11)(p21;q12) renal cell carcinoma in archival paraffin-embedded material using a break-apart TFEB FISH assay expands its clinicopathologic spectrum. *Am J Surg Pathol.* 2012;36:1516–26.

43. Gotoh M, Ichikawa H, Arai E et al. Comprehensive exploration of novel chimeric transcripts in clear cell renal cell carcinomas using whole transcriptome analysis. *Genes Chromosomes Cancer.* 2014;53:1018–32.

44. Perrino CM, Wang JF, Collins BT. Microphthalmia transcription factor immunohistochemistry for FNA biopsy of ocular malignant melanoma. *Cancer Cytopathol.* 2015;123(7):394–400.

45. Argani P, Hicks J, De Marzo AM et al. Xp11 translocation renal cell carcinoma (RCC): Extended immunohistochemical profile emphasizing novel RCC markers. *Am J Surg Pathol.* 2010;34:1295–303.

46. Smith NE, Illei PB, Allaf M et al. t(6;11) renal cell carcinoma (RCC): Expanded immunohistochemical profile emphasizing novel RCC markers and report of 10 new genetically confirmed cases. *Am J Surg Pathol.* 2014;38:604–14.

47. De Velasco G, Culhane AC, Fay AP et al. Molecular subtypes improve prognostic value of international metastatic renal cell carcinoma database consortium prognostic model. *Oncologist.* 2017;22:286–92.

48. Schmidt LS, Linehan WM. Genetic predisposition to kidney cancer. *Semin Oncol.* 2016;43(5):566–74.

35

Mulibrey Nanism

Dongyou Liu

CONTENTS

35.1 Introduction

Mulibrey nanism is a rare autosomal recessive disorder characterized by prenatal onset growth failure, abnormalities of the face (dysmorphic features), eye (yellowish dots in the ocular fundi), heart (pericardial constriction), muscle (hypotonia), liver (hepatomegaly), and brain (J-shaped sella turcica, enlargement of cerebral ventricles), along with increased tumor risk (including cutaneous nevi flammei and Wilms tumor).

At the molecular level, mulibrey nanism is attributed to homozygous or compound heterozygous mutations in the tripartite motif containing protein 37 (*TRIM37*) gene on chromosome 17q22–q23 encoding an E3 ubiquitin ligase (TRIM37), which is a member of the tripartite motif protein family, and which performs several cellular functions such as developmental patterning and oncogenesis [1].

35.2 Biology

Cases relating to mulibrey (muscle-liver-brain-eye) nanism (dwarfism) (also known as Perheentupa syndrome or pericardial constriction with growth failure) were initially described by Perheentupa et al. in 1973 involving 23 Finnish patients (including 3 pairs of affected sibs born of consanguineous parents, suggestive of autosomal recessive inheritance), with clinical manifestations ranging from congenital growth failure, hydrocephaloid skull, triangular face, yellowish dots and pigment dispersion in the ocular fundi, peculiar voice, muscular hypotonia, enlarged liver (hepatomegaly), and raised venous pressure to cutaneous nevi flammei

and cystic dysplasia of the tibia. In subsequent years, sporadic cases of mulibrey nanism reported in Egypt, Canada, the United States, and other parts of the world extended the clinical spectrum of this disorder to fibrous dysplasia of the tibia, hypoplasia of the choroid, J-shaped (long shallow) sella turcica, low birth weight and length, infertility, constrictive pericarditis (overgrowth of the fibrous sac surrounding the heart that restricts normal filling of the heart and raises venous pressure), myocardial hypertrophy, fibrosis, insulin resistance with type 2 diabetes, and increased tumor risk (e.g., Wilms tumor) [2,3].

The cloning and identification of the *TRIM37* gene on chromosome 17q22-q23 in 2001 helped unravel the molecular pathogenesis of mulibrey nanism [4–6]. It appears that homozygous or compound heterozygous mutations in *TRIM37* cause loss of function in TRIM, which is an E3 ubiquitin ligase (TRIM) involved in diverse cellular functions including developmental patterning and oncogenesis. Given its frequent peroxisomal localization in the cytoplasm, TRIM is also referred to as a peroxisomal protein, and mulibrey nanism as a peroxisomal disorder [7].

Being single-membrane–bound subcellular organelles, peroxisomes participate in numerous metabolic processes (e.g., β-oxidation of long- and very-long-chain fatty acids; biosynthesis of plasmalogens, cholesterol, and bile acid; and the degradation of amino acids and purine). Mutations in genes encoding peroxisomal proteins may result in single peroxisomal enzyme defect as well as defective biogenesis of the peroxisome (i.e., peroxisome biogenesis disorders). Not surprisingly, many clinical features of mulibrey nanism (e.g., prenatal-onset growth failure, facial dysmorphism, hepatomegaly, pigmentary changes in retina, and muscular weakness) are also found in other peroxisome biogenesis disorders, including Zellweger syndrome, neonatal adrenoleukodystrophy,

TABLE 35.1

Wilms Tumor Predisposition Syndromes and Conditions

Syndrome/Condition	Gene	Estimated Wilms Tumor Risk	Comments
9q22.3 microdeletion syndrome	9q22.3	Rare	Overgrowth phenotype
Alagille syndrome	*JAG1,* *NOTCH2*	Rare	
Beckwith–Wiedemann syndrome	*WT2*	Up to 22%	Overgrowth phenotype
Bloom syndrome	*BLM*	3%	
Bohring–Opitz syndrome	*ASXL1*	7%	
CLOVES syndrome	*PIK3CA*	1%–2%	Congenital lipomatous overgrowth, vascular malformations, epidermal nevi, and skeletal/spinal abnormalities
Denys–Drash syndrome	*WT1*	8%–10%	
DICER1-related disorders	*DICER1*	Low risk with most *DICER1* variants; higher risk (18%) in individuals with Gly803Arg variant	
Familial Wilms tumor	*FWT1, FWT2,* *CTR9*	2%	
Fanconi anemia with biallelic mutations in *BRCA2* (FANCD1) or *PALB2* (FANCN)	*BRCA2,* *PALB2*	20 ~40%	
Frasier syndrome	*WT1*	8%–10%	
Genitourinary anomalies	*WT1*	Rare	
Hyperparathyroid–jaw tumor syndrome	*CDC73*	3%	
Isolated hemihyperplasia		3%–4%	Overgrowth phenotype
Li–Fraumeni syndrome	*TP53, CHEK2*	Rare	
Mulibrey nanism	*TRIM37*	6%	
Mosaic variegated aneuploidy (MVA)	*BUB1B*	25% overall in MVA; >85% in individuals with *BUB1B* pathogenic variants	
Perlman syndrome	*DIS3L2*	30%	Overgrowth phenotype
Simpson–Golabi–Behmel syndrome	*GPC3, GPC4*	4%–9%	Overgrowth phenotype
Sotos syndrome	*NSD1*	Rare	Overgrowth phenotype
Sporadic aniridia	*WT1, PAX6*	Rare	
WAGR syndrome	*WT1*	15%	Wilms tumor, aniridia, genitourinary anomaly, and mental retardation
Trisomy 13		Rare	
Trisomy 18		Rare	

infantile Refsum disease, and rhizomelic chondrodysplasia punctata type 1. However, compared to other peroxisomal disorders, mulibrey nanism tends to have a mild course and does not induce neurological symptoms. The life span of patients with mulibrey nanism is largely influenced by accompanying cardiopathy.

In addition, given a clear association between mulibrey nanism and Wilms tumor (with a 6% risk), mulibrey nanism is considered as one of most important differential diagnoses for syndromes and conditions that predispose to Wilms tumor (Table 35.1) [8,9].

35.3 Pathogenesis

Mapped to chromosome 17q22–q23, *TRIM37* gene consists of an open reading frame of 2892 bp and encodes a RING-B-box-coiled-coil (RBCC) protein (TRIM37) of 964 amino acids in length and 130-kDa in mass weight [10].

As a zinc finger protein in the tripartite motif (TRIM) protein family, TRIM37 comprises an N-terminal RING finger domain (a cysteine-rich, zinc-binding domain), a B-box motif (another cysteine-rich, zinc-binding domain), two coiled-coil regions, a central TNF-receptor-associated factor (TRAF)-like domain, a third

coiled-coil domain, an acidic domain, two nuclear localization signals, and a second acidic domain. The RING finger and B-box domains chelate zinc and participate in protein–protein and/or protein–nucleic acid interactions The TRAF-like domain interacts with TRAF and other proteins, and acts as scaffold molecules for receptors, kinases, and various regulators in signaling pathways [1].

Expressed in the dorsal root and trigeminal ganglia, liver, and epithelia of multiple tissues during early embryogenesis and located in peroxisomes, TRIM37 functions both as molecular building block and as molecular modifier, and mediates diverse cellular processes (e.g., oncogenesis, apoptosis, organelle transport, cell-cycle control, peroxisomal biogenesis, transcriptional repression, and ubiquitination). TRIM37 possesses E3 ubiquitin-protein ligase activity that prevents centriole reduplication events. In addition, TRIM37 monoubiquitinates 'Lys-119' of histone H2A (H2AK119Ub), a specific tag for epigenetic transcriptional repression [11,12].

Homozygous or compound heterozygous mutations in *TRIM37* lead to loss-of-function in TRIM37. To date, 23 different disease-associated mutations in *TRIM37* have been identified, including most notable frameshift mutations resulting in truncated TRIM37 proteins from Finland [i.e., the Finn major (c.493–2A→G) and the Finn minor mutation (c.2212delG)],

Czech (c.838–842delACTTT), America (c.1346–1347insA), and Turkey (c.855_862delTGAATTAG) [13–15].

Specifically, by retaining 164 residues of the wild-type proteins, and introducing 10 nonsense residues, the Finn major mutant TRIM37 loses its peroxisomal localization and shows diffuse cellular staining. On the other hand, by retaining 737 residues of the wild-type protein despite the loss of 227 residues at C-terminus, the Finn minor mutant TRIM37 maintains its peroxisomal localization. In addition, compound heterozygosity for two mutations in the TRIM37 gene is found in an Australian girl with mulibrey nanism.

Inherited biallelic inactivation of TRIM37 predisposes to both mesenchymal and epithelial ovarian tumors, while dysregulation of TRIM37 (e.g., alternatively spliced variants that lack exon 23 or exon 2, leading to reduced or absent expression of TRIM37, allelic loss at the TRIM37 locus 17q22–23, and CpG promoter methylation) is implicated in the pathogenesis of sporadic fibrothecomas. Apart from syndromic signs, female patients with mulibrey nanism often develop benign mesenchymal tumors of ovarian sex cord–stromal origin, fibrothecoma, epithelial neoplasias (e.g., ovarian adenofibromas, ovarian poorly differentiated adenocarcinoma, and endometrial adenocarcinoma) [16–18].

35.4 Epidemiology

35.4.1 Prevalence

Mulibrey nanism is an extremely rare disorder, with approximately 140 cases reported in the literature to date, including 110 from Finland (where the incidence of mulibrey nanism is estimated at 1:40,000), and 30 from Spain, France, North, Central, and South America, Iran, Syria, and Egypt [19,20]. Affecting males and females in equal numbers, mulibrey nanism has a median diagnostic age of 2.1 years (range 0.02–52 years).

35.4.2 Inheritance

Mulibrey nanism is an autosomal recessive genetic disorder, with two copies of the defective *TRIM37* gene (one from each parent) required for disease initiation and development. The parents of an affected individual are often closely related (consanguineous), and carry one copy of the defective gene each without displaying any symptoms of the disorder. An individual who inherits one normal gene and one mutated gene becomes a carrier for the disease and is usually asymptomatic. Thus, two carrier parents have a 25% chance of having an affected child, 50% chance of having a carrier child, and 25% chance of having a normal child.

35.4.3 Penetrance

Mulibrey nanism shows variable penetrance, and individuals harboring homozygous or compound heterozygous mutations in *TRIM37* often develop a diverse spectrum of clinical symptoms. Even siblings who suffer from mulibrey nanism sometimes do not have identical symptoms.

35.5 Clinical Features

Mulibrey nanism is an congenital multi-organ disorder that typically manifests as *prenatal onset growth failure* (thin extremities

99% of cases, accentuated lumbar lordosis 96%, general gracility 95%, narrow shoulders 94%, small bell-shaped thoracic cage 94%, barrel-like trunk 92%, hypoplastic buttocks 52%), and abnormalities of the *face* (facial triangularity, high and broad forehead, low nasal bridge, telecanthus 90%, scaphocephaly 81%, hypoplastic tongue 80%, low-set located posteriorly rotated ears 54%, dental crowding 50%, high hairline 45%, wide fontanels and sutures in infancy 37%, abnormally wide metopic suture 31%, cleft palate 2%), eye (yellowish dots in the ocular fundi 79%), *heart* (cardiac anomaly 15%, cardiomegaly 14%, congenital heart failure 12%, dyspnea 12%, prominent veins on upper body 11%, pericardial constriction 6%, cyanosis on the lips and fingertips), *muscle* (muscular hypotonia 68%), liver (hepatomegaly 45%, hepatosplenomegaly 11%, ascites 11%), *brain* (J-shaped sella turcica—a depression in the sphenoid bone at the base of the skull 89%, large cerebral ventricles and basal cisternas 43%), *urogenitals* (renal anomaly 18%, hydronephrosis 6%, abnormally located kidney 5%, hypoplastic kidneys 5%, horseshoe kidney 2%, cryptorchidism 6%, prominent clitoris 2%, hyperplastic labiae 1%, spontaneous pubarche with menarche at median age 14.7 years), *high-pitched voice* (96%), *skeletal asymmetry* (15%), in addition to high risk for *tumors* (e.g., cutaneous nevi flammei 65%, Wilms tumor 6%, benign renal neoplasia 3%, hepatoblastoma 2%, fibrothecoma, ovarian mesenchymal tumor, ovarian adenofibroma, ovarian poorly differentiated adenocarcinoma, endometrial adenocarcinoma, gastrointestinal carcinoid tumor, neuropituitary Langerhans cell histiocytosis, acute lymphoblastic leukemia, thyroid tumor) (Figure 35.1) [21,22]. *Other symptoms* include feeding difficulties, respiratory tract infections, congestion in the lungs, buildup of fibrous tissue in the walls of the lungs (pulmonary fibrosis), swelling of the arms and/or legs (peripheral edema), unusually thin shinbone (fibrous tibia dysplasia), and early onset type 2 diabetes [23].

Further, mulibrey nanism patients who develop Wilms tumor may show swelling in the abdomen, blood in the urine and abnormal urine color, fever, poor appetite, high blood pressure, abdominal or chest pain, nausea, constipation, large and distended veins across the abdomen, malaise or feeling unwell, vomiting, unexplained weight loss.

35.6 Diagnosis

Mulibrey nanism is noted for its early-onset growth failure and its cardiac involvement, which contributes significantly to morbidity and mortality. Therefore, early diagnosis is crucial for prompt implementation of intervention measures to control disease progression, limit potential complications, and improve life expectancy of affected individuals.

Diagnosis of mulibrey nanism requires thorough clinical evaluation (for characteristic physical findings), imaging study (e.g., x-ray for abnormal calcium deposits in the sac surrounding the heart or pericardium; abdominal ultrasonography), electrocardiogram (for electrical activity of the heart), ocular examination (for ocular abnormalities), laboratory assays (e.g., 3-h oral glucose tolerance test), histopathology, and molecular genetic testing (for *TRIM37* pathogenic variants).

In accordance with recently updated clinical criteria, observation of three major and one minor or two major and three minor

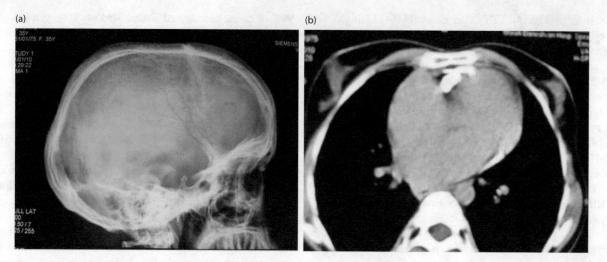

FIGURE 35.1 Mulibrey nanism in a 35-year-old female showing characteristic J-shaped sella turcica (a, skull x-ray) and extensive pericardial calcification in anterior and inferior surfaces of the heart (b, chest CT scan). (Photo credit: Behzadnia N et al. *Tanaffos*. 2011;10(1):48–51.)

signs below is sufficient for making the diagnosis of mulibrey nanism [24]:

Major signs

1. Growth failure (small for gestational age [SGA] and lacking catch-up growth 95%, or height in children 2.5 SDS below the population's mean for age 94%, or 8 in adults 3.0 SDS below the population's mean 90%)
2. Characteristic radiological findings (slender long bones with thick cortex and narrow medullar channels 93%, or low and shallow [J-shaped] sella turcica, 89%)
3. Characteristic craniofacial features (scaphocephaly, triangular face, high and broad forehead, low nasal bridge, and telecanthus 90%)
4. Characteristic ocular findings (yellowish dots in the retinal midperipheral region 79%)
5. Mulibrey nanism in a sibling (17%)

Minor signs

1. Peculiar high-pitched voice (96%)
2. Hepatomegaly (70%)
3. Cutaneous nevi flammei (65%)
4. Fibrous dysplasia of the long bone (25%)

Molecular identification of homozygous or compound heterozygous mutations in *TRIM37* provides a clear confirmation of mulibrey nanism. While PCR targeting specific mutation is adequate for Finnish patients, sequencing of the whole *TRIM37* gene is necessary for non-Finnish patients.

Differential diagnoses for mulibrey nanism include Silver–Russell syndrome (growth failure/dwarfism and facial dysmorphism in the absence of cardiomyopathy, hepatomegaly, and ophthalmologic involvements, chromosomes 11p15 and 7 abnormalities), 3M-syndrome (short stature/dwarfism, triangle-shaped face with a broad, prominent forehead, pointed chin, hypoplastic midface, *CUL7* and *OBSL1* mutations), Robinow syndrome (mild to moderate short stature due to postnatal growth retardation, distinctive craniofacial abnormalities, additional skeletal malformations, and/or genital abnormalities, *ROR2*, *FZD2*, *WNT5A*, *DVL1*, and *DVL3* mutations), and primary empty sella syndrome (an empty space filled with cerebrospinal fluid in the sella turcica area of the brain due to a defect in the sella diaphragm, unusual facial features, a high-arched palate, moderate short stature, increased bone density/osteosclerosis, and normal pituitary function).

35.7 Treatment

Treatment options for mulibrey nanism include symptomatic relief (for feeding or respiratory difficulty during infancy), surgery (pericardiectomy for constrictive pericarditis), hormone replacement therapy (for children with growth hormone deficiency, delayed puberty, infrequent or very light menstrual periods, hypothyroidism, hypoadrenocorticism, and abnormal ovaries or testes), and chemotherapy (diuretics and digoxin for progressive heart failure) [25].

Mulibrey nanism patients presenting with stages 1 or 2 Wilms tumor may undergo surgery to remove the kidney, surrounding tissues, and some nearby lymph nodes, followed by chemotherapy and/or radiotherapy. Those with stages 3, 4, or 5 Wilms tumor may require a combination of surgery, radiotherapy, and chemotherapy (to shrink the tumor prior to surgery). Kidney transplant may be considered for patients following removal of both kidneys [8].

35.8 Prognosis

Constrictive pericarditis with restrictive cardiomyopathy plays a large part in the prognosis of mulibrey nanism, as its often leads to congestive heart failure (51% during a 25-year follow-up) and early death (median age of 6.8 years, range 1.4–48 years). In contrast, noncardiac death (10% during a 25-year follow-up) has a median age of 33.7 years (range 29.1–43.6 years). With early diagnosis and appropriate treatment, it is possible for individuals with mulibrey nanism to live until 70 years of age.

35.9 Conclusion

Mulibrey nanism is an autosomal recessive multi-organ disorder that manifests as prenatal-onset growth failure (infants born small for gestational age with progressive growth failure and poor weight gain), dimorphic facial features (scaphocephaly, triangular face, high and broad forehead, low nasal bridge, and yellowish dots in retinal midperipheral region), constrictive pericardium/progressive cardiomyopathy, muscle hypotonia, J-shaped sella turcica, hypoplasia of various endocrine glands, infertility, insulin resistance with type 2 diabetes, peculiar high-pitched voice, hepatomegaly, cutaneous nevi flammei, and fibrous dysplasia of long bones, along with increased risk for Wilms tumor and ovarian fibrothecomas. Although mulibrey nanism occurs at a relatively high frequency in the Finnish population (estimated at 1:40,000), it is comparatively rare elsewhere. Following the identification of homozygous or compound heterozygous mutations in the *TRIM37* gene on chromosome 17q22–q23 as the underlying cause of mulibrey nanism, it becomes feasible to achieve a rapid and accurate diagnosis for this disorder. This permits remarkable improvement in the management of patients with mulibrey nanism, including early attention to cardiac disturbances and application of pericardiectomy for symptom relief and enhanced survival. There is no doubt that further research on the molecular pathogenesis of mulibrey nanism will likely lead to innovative countermeasures against this rare, but no longer ignorable hereditary disease.

REFERENCES

1. Brigant B, Metzinger-Le Meuth V, Rochette J, Metzinger L. TRIMming down to TRIM37: Relevance to inflammation, cardiovascular disorders, and cancer in Mulibrey nanism. *Int J Mol Sci.* 2018;20(1). pii: E67.
2. Hämäläinen RH, Mowat D, Gabbett MT, O'brien TA, Kallijärvi J, Lehesjoki AE. Wilms' tumor and novel TRIM37 mutations in an Australian patient with mulibrey nanism. *Clin Genet.* 2006;70(6):473–9.
3. Sivunen J, Karlberg S, Lohi J, Karlberg N, Lipsanen-Nyman M, Jalanko H. Renal findings in patients with Mulibrey nanism. *Pediatr Nephrol.* 2017;32(9):1531–6.
4. Avela K, Lipsanen-Nyman M, Perheentupa J et al. Assignment of the mulibrey nanism gene to 17q by linkage and linkage-disequilibrium analysis. *Am J Hum Genet.* 1997;60(4):896–902.
5. Paavola P, Avela K, Horelli-Kuitunen N et al. High-resolution physical and genetic mapping of the critical region for Meckel syndrome and Mulibrey Nanism on chromosome 17q22-q23. *Genome Res.* 1999;9(3):267–76.
6. Kallijärvi J, Avela K, Lipsanen-Nyman M, Ulmanen I, Lehesjoki AE. The TRIM37 gene encodes a peroxisomal RING-B-box-coiled-coil protein: Classification of mulibrey nanism as a new peroxisomal disorder. *Am J Hum Genet.* 2002;70(5):1215–28.
7. Wang W, Xia ZJ, Farré JC, Subramani S. TRIM37, a novel E3 ligase for PEX5-mediated peroxisomal matrix protein import. *J Cell Biol.* 2017;216(9):2843–58.
8. PDQ Pediatric Treatment Editorial Board. *Wilms Tumor and Other Childhood Kidney Tumors Treatment (PDQ®): Health Professional Version. PDQ Cancer Information Summaries* [Internet]. Bethesda (MD): National Cancer Institute (US); 2002–2018 Oct 29.
9. Dome JS, Huff V. Wilms Tumor Predisposition. In: Adam MP, Ardinger HH, Pagon RA et al. editors. *GeneReviews®* [Internet]. Seattle (WA): University of Washington, Seattle; 1993–2019. 2003 Dec 19 [updated 2016 Oct 20].
10. Karlberg N, Jalanko H, Kallijärvi J, Lehesjoki AE, Lipsanen-Nyman M. Insulin resistance syndrome in subjects with mutated RING finger protein TRIM37. *Diabetes.* 2005;54(12):3577–81.
11. Lehesjoki AE, Reed VA, Mark Gardiner R, Greene ND. Expression of MUL, a gene encoding a novel RBCC family ring-finger protein, in human and mouse embryogenesis. *Mech Dev.* 2001;108(1–2):221–5.
12. Kettunen KM, Karikoski R, Hämäläinen RH et al. Trim37-deficient mice recapitulate several features of the multi-organ disorder Mulibrey nanism. *Biol Open.* 2016;5(5):584–95.
13. Kumpf M, Hämäläinen RH, Hofbeck M, Baden W. Refractory congestive heart failure following delayed pericardectomy in a 12-year-old child with Mulibrey nanism due to a novel mutation in TRIM37. *Eur J Pediatr.* 2013;172(10):1415–8.
14. Mozzillo E, Cozzolino C, Genesio R et al. Mulibrey nanism: Two novel mutations in a child identified by Array CGH and DNA sequencing. *Am J Med Genet A.* 2016;170(8):2196–9.
15. Jobic F, Morin G, Vincent-Delorme C et al. New intragenic rearrangements in non-Finnish mulibrey nanism. *Am J Med Genet A.* 2017;173(10):2782–8.
16. Karlberg N, Karlberg S, Karikoski R et al. High frequency of tumours in Mulibrey nanism. *J Pathol.* 2009a;218(2):163–71.
17. Karlberg S, Lipsanen-Nyman M, Lassus H, Kallijärvi J, Lehesjoki AE, Butzow R. Gynecological tumors in Mulibrey nanism and role for RING finger protein TRIM37 in the pathogenesis of ovarian fibrothecomas. *Mod Pathol.* 2009b;22(4):570–8.
18. Karlberg S, Toppari J, Karlberg N et al. Testicular failure and male infertility in the monogenic Mulibrey nanism disorder. *J Clin Endocrinol Metab.* 2011;96(11):3399–407.
19. Al Saadi T, Alkhatib M, Turk T, Turkmani K, Abbas F, Khouri L. Report of two Syrian siblings with Mulibrey nanism. *Oxf Med Case Reports.* 2015;2015(12):367–70.
20. Davarpasand T, Sotoudeh Anvari M, Naderan M, Boroumand MA, Ahmadi H. Constrictive pericarditis and primary amenorrhea with syndactyly in an Iranian female: Mulibrey nanism syndrome. *J Tehran Heart Cent.* 2016;11(4):187–91.
21. Karlberg S, Tiitinen A, Lipsanen-Nyman M. Failure of sexual maturation in Mulibrey nanism. *N Engl J Med.* 2004;351(24):2559–60.
22. Behzadnia N, Sharif-Kashani B, Ahmadi ZH, Mirhosseini SM. Mulibrey nanism in a 35 year-old Iranian female with constrictive pericarditis. *Tanaffos.* 2011;10(1):48–51.
23. Tikanoja T, Taipale P. Ascites and nuchal fold as the first signs of progressive cardiac diastolic dysfunction in a fetus with fetal growth restriction due to mulibrey nanism. *Ultrasound Obstet Gynecol.* 2004;23(4):414–5.
24. Karlberg N, Jalanko H, Perheentupa J, Lipsanen-Nyman M. Mulibrey nanism: Clinical features and diagnostic criteria. *J Med Genet.* 2004b;41(2):92–8.
25. Christov G, Burch M, Andrews R et al. Thoracoscopic pericardiectomy for constrictive pericarditis in a pediatric patient with mulibrey nanism. *World J Pediatr Congenit Heart Surg.* 2013;4(4):442–3.

Section IV

Tumor Syndromes Affecting the Skin, Soft Tissue, and Bone

36

Brooke–Spiegler Syndrome

Dongyou Liu

CONTENTS

36.1 Introduction

Brooke–Spiegler syndrome (BSS) is an autosomal dominant disorder that predisposes to multiple appendageal tumors (i.e., cylindroma, spiradenoma, spiradenocylindroma, and trichoepithelioma) in the skin, occasionally the salivary glands, and rarely the breast (mammary cylindroma). In 5%–10% of cases, malignant tumors may arise from preexisting benign cutaneous lesions. Phenotypic variants of BSS include familial cylindromatosis (FC) and multiple familial trichoepithelioma (MFT), which only cause cylindroma and trichoepithelioma (cribriform trichoblastoma), respectively.

Heterozygous germline mutations in the cylindromatosis (*CYLD*) gene located on chromosome 16q12.1, which encodes a protein with deubiquitinase activity, underpin the molecular pathogenesis of BBS, FC, and MFT, as 80%–85% of patients with classical BSS phenotype, 100% of patients with FC phenotype, and 40%–50% of patients with MFT phenotype harbor mutations in the *CYLD* gene [1].

36.2 Biology

Although Ancell was credited with introducing the term *cylindroma* for cases related to BSS in 1842, it was Henry G. Brooke and Eduard Spiegler, whose precise clinical and histopathological descriptions of trichoepitheliomas under the designation "epithelioma adenoides cysticum" in 1892 and "cylindromas" in 1899, respectively, helped establish this genodermatosis

displaying multiple cylindromas, spiradenomas, spiradenocylindromas, and trichoepitheliomas as a distinct autosomal dominant disorder that bears their names [2].

Subsequent studies revealed remarkable phenotypic variations among individuals with BSS, with some patients developing multiple cylindromas only (so-called familial cylindromatosis [FC]), while others having multiple trichoepitheliomas only (so-called multiple familial trichoepithelioma [MFT]). In addition, an individual may display varied tumor spectrums, and individuals within the same family may also show different tumor spectrums. By 2003, it became apparent that mutations in the *CYLD* gene on 16q12.1 form the molecular basis of BSS (85%), FC (100%), and MFT (45%), putting it beyond doubt that FC and MFT represent BSS despite their varied tumor spectrums. For this reason, the term "*CYLD* cutaneous syndrome" has been proposed by some authors to cover the genodermatosis consisting of BSS, FC, and MFT [3].

Cylindroma is a benign skin adnexal tumor of apocrine differentiation and appears to originate from the pluripotent stem cells in the folliculo-sebaceous-apocrine unit. Syndromic cylindroma, as in the case of BSS, has an early onset age and forms multiple smooth/pedunculated or ulcerated/disfiguring lesions, which can cover the entire scalp like a turban (so-called turban tumor) [4].

Trichoepithelioma is a benign neoplasm with follicular differentiation and may exist in a solitary or familial form. The solitary form occurs in the head and neck region as well as any portion of hair bearing skin, while the familial/multiple form emerges in adolescence or adulthood and shows a predilection for central facial distribution [5].

36.3 Pathogenesis

Mapped to chromosome 16q12.1, the *CYLD* gene spans 56 kb and contains 20 exons, of which the first three are untranslated, exon 3 (in the 5-prime untranslated region) and the 9-bp exon 7 (which is coding) show alternative splicing, and exon 4 contains the ATG start codon. The encoded protein CYLD is a 956 aa, 120 kDa deubiquitinating enzyme while a splice variant lacking exon 7 produces a 953 aa protein.

Structurally, the CYLD protein consists of three cytoskeletal-associated protein-glycine-conserved (CAP-GLY) domains, which show sequence homology to the catalytic domain of ubiquitin carboxy-terminal hydrolases and appear to coordinate the attachment of organelles to microtubules.

By removing lys63-linked ubiquitin chains from several specific substrates, CYLD deubiquitinates several NF-kappa-B regulators (e.g., TRAF2, TRAF6, NEMO [IKBKG], and BCL3), and negatively regulates the nuclear factor-kappa B and c-Jun N-terminal kinase pathways [6]. Therefore, inhibition or loss of CYLD deubiquitinating activity enhances constitutive activation of transcription factor NF-kappa-B, increases resistance to apoptosis, and contributes to hyperproliferation and tumorigenesis. Indeed, reduced CYLD level induces B-cell lymphoma-3/p50/p52-dependent nuclear factor-κB activation and triggers the expression of genes encoding cyclin D1 and N-cadherin. In turn, elevated levels of cyclin D1 and N-cadherin promote melanoma proliferation and invasion [7,8].

Germline *CYLD* mutations (including frameshift due to small insertions or deletions 41%; nonsense 35%; missense 14%; and splice site 10%) often result in truncated and catalytically inactive proteins that show reduced deubiquitinase activity and antineoplastic effects, leading to tumor formation. To date, 93 mutations [e.g., c.1628_1629delCT, c.1783C>T (pGln 595*), ic.1821_1826+1delinsCT/L607Ffs*9, c.2146C>A (p.Gln716 K), c.2272C/T (p.R758X), c.2350+5G>A, c.2662_2664delTTT (p.Phe888del), c.2666A>T/p.D889 V, c.2686+60_*3340del5362, c.2712delT/p.905Kfs*8, c. 2806C>T (p.R936X)] have been described in the *CYLD* gene, most of which occur in *CYLD* exons 9–20 that encode the NEMO binding site and the catalytic domain, while spared large deletions are seldom found in *CYLD* exons 4–8 [9–13].

Overall, 85% of individuals with classical BSS phenotype, 100% of individuals with FC phenotype, and 45% of individuals with the MFT phenotype are shown to harbor germline *CYLD* pathogenic variants. A recurrent heterozygous nonsense *CYLD* mutation (c.2272C/T, p.R758X) has been shown in the patients with BSS, FC (the Netherlands), and MFT (Spain) worldwide.

Further, loss of heterozygosity (LOH) in chromosome 16q involving the wild-type allele is found in familial cylindromas (70% of cases) and trichoepitheliomas, indicating that *CYLD* acts as a tumor suppressor gene (i.e., a recessive oncogene), and many sporadic cylindromas also contain LOH in chromosome 16q. In addition, somatic mutations (both LOH and sequence alterations) in the *CYLD* gene are occasionally identified in patients with sporadic and familial diseases. In some MFT cases, a second locus at chromosome 9p21 has been implicated, and sporadic trichoepithelioma may be also caused by mutations in the *PTCH* gene (which is involved in nevoid basal-cell carcinoma syndrome) [14].

36.4 Epidemiology

36.4.1 Prevalence

BSS is a rare genodermatosis, with about 200 cases reported in the literature to date. The prevalence of *CYLD* mutations in the UK population is estimated at 1:100,000. BSS appears to be underdiagnosed, as its lesions are phenotypically variable and not easily distinguished from cutaneous lesions in the head and neck caused by other diseases [15,16]. While BSS tumors usually emerge in adolescence, they can occur from after adrenarche in childhood up to the fourth decade. There is a notable female predominance among BSS patients (female:male ratio of 6–9:1).

36.4.2 Inheritance

BSS and its phenotypic variants FC and MFT show autosomal dominant transmission, with one copy of the mutated *CYLD* gene inherited from affected or carrier parents and a second copy caused by another genetic event during a person's lifetime. In addition, sporadic cases due to loss of both alleles of the *CYLD* gene occur occasionally. However, unlike inherited cases that are associated with multiple lesions, sporadic cases usually cause solitary lesion.

36.4.3 Penetrance

BSS demonstrates variable penetrance (between 60%–100%), as individuals harboring *CYLD* gene mutations may develop cylindroma, spiradenoma, spiradenocylindroma, and trichoepithelioma (BSS); cylindroma only (FC); or trichoepithelioma (cribriform trichoblastoma) only (MFT). Furthermore, these tumors can occur in the same person or in different persons within the same family.

36.5 Clinical Features

BSS typically manifests as multiple flesh-colored, pink and bluish 0.5- to 3.0-cm papules and nodules (varying from 10–30 to several hundred) on the face, ears, and scalp. The so-called "turban tumor" refers to multiple confluent lesions on the scalp. These tumors often emerge around puberty, grow slowly throughout life, and increase in number with age. In about 5%–10% of cases, tumors may show rapid enlargement, ulceration, and bleeding, which are suggestive of malignant transformation [17].

Besides the scalp lesions (which correspond histologically to cylindroma, spiradenoma, or spiradenocylindroma), classical BSS phenotype may cause bilateral small (0.2–1 cm) discrete and/or confluent skin colored papules on the nasolabial folds (so-called trichoepithelioma). In <2% of patients with BSS, salivary gland tumor (affecting the parotid gland, but rarely the submandibular glands and intranasal minor salivary glands) may develop after the age of 40 years along with skin lesions.

As phenotypic variants of BSS, FC and MFT produce only cylindroma and trichoepithelioma (cribriform trichoblastoma), respectively. Trichoepithelioma associated with MFT may extend from the nasolabial folds to the inner aspects of the eyebrow, eyelids, and external auditory canal, causing visual impairment and hearing loss. Further, malignant transformation of trichoepithelioma into basal cell carcinoma may take place [18].

36.6 Diagnosis

Current diagnosis for BSS and its phenotypic variants FC and MFT relies on clinical-pathological correlation, positive familial association, and detection of *CYLD* mutations [19].

Histological identification of distinct tumor spectrum (i.e., cylindroma, spiradenoma, spiradenocylindroma, and trichoepithelioma in BSS, cylindroma in FC, and trichoepithelioma or cribriform trichoblastoma in MFT) plays a crucial part in the diagnosis of BSS.

Cylindroma is a benign adnexal tumor with no attachment to the epidermis and shows a folliculosebaceous distribution (specifically the areas that are rich in hair follicles and sebaceous glands). It appears as solitary or multiple skin-colored or erythematous, smooth/pedunculated or ulcerated/disfiguring lesions of a few mm to several cm in size on the scalp, neck, face, and sometimes trunk and extremities (<10%), usually in adults. Solitary cylindroma often occurs sporadically in middle-aged and elderly persons who have no family history of similar lesions; whereas multiple cylindromas begin in early adulthood and grow in size and number throughout the life, coalescing in the scalp to form the so-called turban tumor that is associated with alopecia. In FC, multiple, erythematous nodules typically arise on the scalp but also on the forehead or pubic area. Histologically, cylindroma is a circumscribed, nonencapsulated dermal nodule composed of islands and cords of monomorphic basaloid cells surrounded by a thick, eosinophilic, PAS-positive hyaline sheath closely resembling a basement membrane and arranged in an interlocking "jigsaw puzzle"-like architecture. The cells at the periphery are small and dark with a tendency for palisading, while those in the center are large and pale, with occasional presence of lumina or pseudolumina (Figure 36.1) [20–22].

Spiradenoma forms single or several variably sized, well-demarcated nodules of blue/black appearance, commonly in the extremities, trunk, and less frequently in the scalp. Due to its common co-occurrence with cylindroma, giving rise to spiradenocylindroma, spiradenoma appears to have a similar apocrine origin to cylindroma. Histologically, spiradenoma shows small epithelial basaloid cells with hyperchromatic nuclei at the periphery in a trabecular, reticular, solid or diffuse growth pattern, along with large cells containing pale nuclei and sparse lymphocytic infiltrate in the center, in addition to droplets of PAS-positive hyaline basal membrane material (Figure 36.2) [23].

Spiradenocylindroma commonly occurs in the scalp and face and occasionally in the head and neck, and contains areas of both spiradenomatous and cylindromatous elements that may closely intermingle or sharply segregate. In spiradenomatous areas, cells with small, dark, hyperchromatic nuclei in the periphery are surrounded by a thin basement membrane, and cells with large, pale nuclei in the center are arranged in discrete aggregates, with lymphocyte infiltration around the blood vessels and within the neoplastic nodules. In cylindromatous areas, small cell aggregates of distinctively rectangular, triangular, and polyhedral shape form a jigsaw puzzle architecture and are surrounded by thick, homogeneous eosinophilic basal membrane. Furthermore, the cells at the periphery are aligned in a palisade with small, dark nuclei and a small amount of cytoplasm, while the remaining cells have large, plump nuclei and copious, pale cytoplasm. Scanty lymphoid cells are found within the cylindromatous area (Figure 36.3) [24].

Trichoepithelioma (cribriform trichoblastoma or follicular tumor) is a benign neoplasm that presents as firm, elevated, flesh-colored nodules of <2 cm in diameter in the head and neck region as well as midface (mainly on the nose and nasolabial folds) in adolescence or adulthood. The defining lesions of MFT are bilaterally and symmetrically distributed, flesh-colored papules or small nodules on the face (especially the nose), neck, or chest. Histologically, trichoepithelioma is a symmetric lesion showing dual differentiation toward follicular germinative

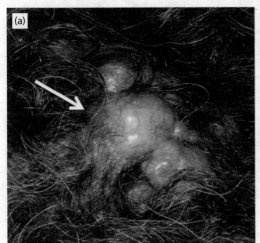
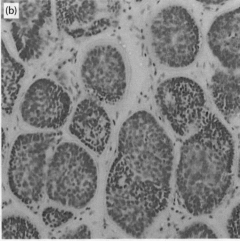

FIGURE 36.1 Cylindroma in the setting of Brooke–Spiegler syndrome. (a) Cylindromas on the scalp appear well-circumscribed, pink, nodular with arborizing blood vessels. (b) Histology of cylindroma reveals well-defined nests (lobules) of basaloid cells (the outer layer of cells having small hyperchromatic nuclei and the inner section of cells containing oval vesicular nuclei) separated by an eosinophilic basement membrane. (Photo credit: Dubois A et al. *PLOS Curr.* 2015;7. pii: ecurrents.eogt.45c4e63dd43d62e12228cc5264d6a0db.)

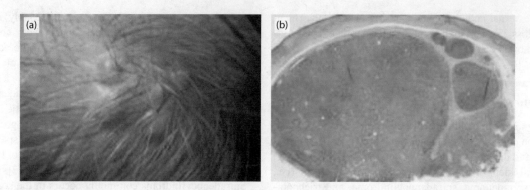

FIGURE 36.2   Spiradenoma from a 52-year-old male with Brooke−Spiegler syndrome. (a) Spiradenoma forms firm, pink-to-skin-colored, smooth, exophytic papulonodules of 5−10 mm in size on the vertex of the scalp. (b) Histology of spiradenoma shows circumscribed aggregates of basaloid cells (including small, basaloid cells with hyperchromatic nuclei located toward the periphery and large cells with pale nuclei located closer to the center) within the dermis and subcutis. (Photo credit: Tran K et al. *Dermatol Online J*. 2012;18(12):15.)

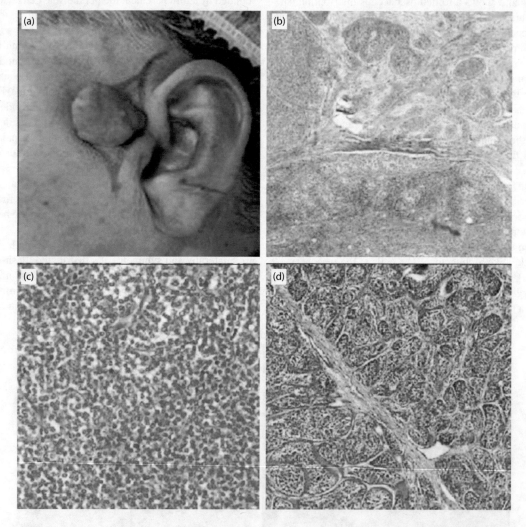

FIGURE 36.3   Spiradenocylindroma in a 70-year-old female. (a) A nodular subcutaneous mass of 3 cm × 2.5 cm appears in the left preauricular region. (b) Spiradenocylindroma consists of both spiradenoma and cylindroma components (H&E, ×100). (c) Spiradenomatous part of tumor includes lymphoid infiltrate within the neoplastic nodules (H&E, ×400). (d) Cylindromatous part of tumor has small cell aggregates of distinctively rectangular, triangular, and polyhedral shape that form a jigsaw puzzle architecture (H&E, ×100). (Photo credit: Baliyan A et al. *Indian J Dermatopathol Diagn Dermatol*. 2018;5:66−8.)

epithelium and specific follicular stroma. While the epithelial component comprises bland basaloid cells arranged in a cribriform pattern, the stroma includes delicate fibrillary collagen bundles and numerous spindled fibroblasts, resembling follicular papillae (papillary mesenchymal bodies) and perifollicular sheath. Trichoepithelioma related to MFT may consist of areas with ductal differentiation and intratumoral lymphocytes, resembling a spiradenomatous moiety (Figure 36.4) [25,26].

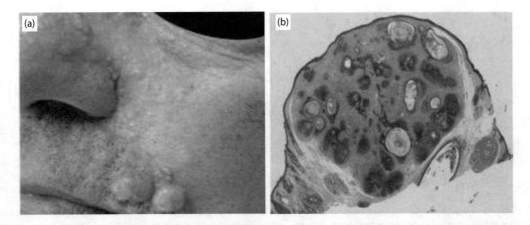

FIGURE 36.4 Multiple familial trichoepitheliomas in a 62-year-old male. (a) Skin-colored papules occur in the periorbital region, nose, nasolabial folds, and upper lip. (b) Histopathology reveals islands of basaloid cells with peripheral palisading and small horny cysts. (Photo credit: Farkas K et al. *BMC Genet.* 2016;17:36.)

Although tumors in the setting of BSS/FC/MFT are generally indistinguishable from their sporadic counterparts, they tend to be multifocal, have several different neoplasm types in a single biopsy, and show hybrid histopathological features.

Malignant transformation of preexisting cylindroma, spiradenoma, and spiradenocylindroma may assume one or more of four morphological patterns: (i) salivary gland type basal cell adenocarcinoma-like pattern, low-grade (BCAC-LG, which displays small to medium-sized basaloid cells arranged in nodules or sheets, sometimes with focal peripheral palisading, replacing the dual epithelial populations of benign tumor); (ii) salivary gland type basal cell adenocarcinoma-like pattern, high-grade (BCAC-HG, which shows medium to large pleomorphic basaloid cells arranged in confluent sheets and nodules, with scant cytoplasm and vesicular nuclei containing conspicuous nucleoli, and atypical mitoses, replacing the dual cell population of benign tumor); (iii) invasive adenocarcinoma, not otherwise

specified (IAC-NOS); (iv) sarcomatoid (metaplastic) carcinoma [27,28].

Malignant cylindroma (i.e., cylindrocarcinoma) displays rapid growth, pink-blue color, ulceration and hemorrhage, and contains irregular aggregates of mostly enlarged cells with marked anaplasia and mitotic figures, and without hyaline sheath (Figure 36.5). Similarly, malignant spiradenoma (i.e., spiradenocarcinoma) is a flesh-colored nodule that lacks a dual cell population but has mitotic figures within a malignant component (Figure 36.6) [29,30].

Basal cell adenoma of membranous type (which is histologically similar to dermal cylindroma) is a common benign tumor occurring in the salivary glands, and appears to have evolved from precursor lesions (either hyperplasia of reserve cells in intercalated ducts or proliferative lobular buds with prominent hyaline material, and microadenomatous foci adjacent to intercalated ducts or acini). Malignant transformation of a preexisting

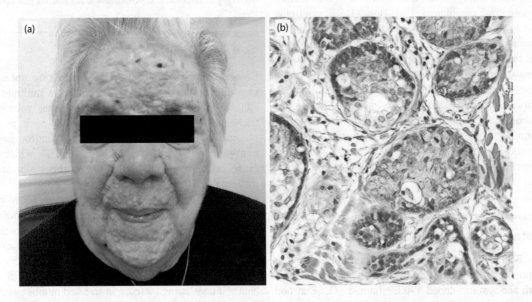

FIGURE 36.5 Malignant transformation of cylindroma to cylindrocarcinoma in an 83-year-old female with Brooke–Spiegler syndrome. (a) Cylindrocarcinoma forms multiple nodules on the face. (b) Irregular aggregates of cylindrocarcinoma display mostly enlarged cells with mitotic figures and absence of hyaline sheath (H&E, ×400). (Photo credit: Borik L et al. *Dermatol Pract Concept.* 2015;5(2):61–5.)

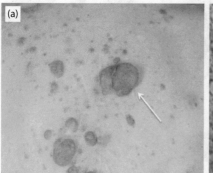
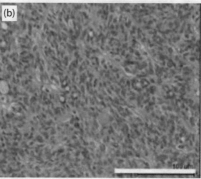
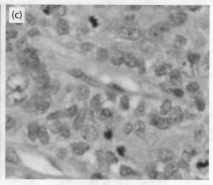

FIGURE 36.6   Malignant spiradenoma (spiradenocarcinoma) in a 50-year-old male with CYLD cutaneous syndrome. (a) Spiradenocarcinoma appears as flesh-colored nodules (white arrow). (b) Low-grade spiradenocarcinoma lacks a dual cell population (H&E, ×200). (c) Mitotic figures occur within malignant component (H&E, ×400). (Photo credit: Hoyle A et al. *Cutan Pathol.* 2018;45(10):760–3.)

membranous basal cell adenoma leads to either low-grade or high-grade basal cell adenocarcinoma [31].

Molecular detection of *CYLD* pathogenic variants from blood, buccal swabs, mouthwash samples, or solid tissue by PCR (targeting exons 4–20) and sequencing analysis (focusing on exons 4–20, exonic sequence plus at least 10 bp at the 5′ and 3′ of each intron) helps confirm the diagnosis of BSS, FC, and MFT. This approach is particularly useful for patients [32].

Differential diagnoses for BSS include several autosomal dominant inherited tumor syndromes that produce lesions in the head and neck, i.e., Cowden syndrome (tricholemmoma in the skin, papillomatous papule in the oral mucosa; plus breast, thyroid, and endometrial carcinoma), tuberous sclerosis complex (angiofibroma and fibrous plaque in the skin, enamel pitting in the teeth; plus angiomyolipoma in the kidneys, tumor, and cortical dysplasia in CNS, learning difficulties, and seizure), nevoid basal cell carcinoma syndrome (odontogenic keratocyst in the jaws, basal cell carcinoma in the skin; plus ovarian fibroma, medulloblastoma), Muir–Torre syndrome (sebaceous adenoma and carcinoma in the skin; plus colorectal, urothelial, and endometrial carcinoma), Birt–Hogg–Dubé syndrome (fibrofolliculoma in the skin; plus cyst and pneumothorax in the lungs, carcinoma in the kidneys), Peutz–Jeghers syndrome (pigmentation in the skin and mucosa; plus gastrointestinal polyps and carcinoma), Gardner syndrome (osteoma and odontoma in the jaws, supernumerary teeth; plus colorectal adenocarcinoma), multiple endocrine neoplasia 2B (mucosal neuroma in the lips and oral mucosa; plus pheochromocytoma and medullary thyroid carcinoma), and hyperparathyroidism–jaw tumor syndrome (fibro-osseous lesions of the jaws; plus adenoma and carcinoma in the parathyroid glands; Wilms tumor, carcinoma and cyst in the kidneys) [33].

36.7 Treatment

Treatment of BSS centers on removal of adnexal tumors by surgical excision, dermabrasion, electrodessication, cryotherapy, ablation (with neodymium-doped YAG, erbium-YAG, or carbon dioxide lasers), photodynamic therapy, and radiotherapy [34–38]. Wide local excision or laser ablation represents a preferred treatment that helps reduce risk of recurrence and malignant

transformation. Erbium-YAG laser appears to yield few scars and recurrences. Sodium salicylate and prostaglandin A1 inhibit NF-kB activity and may be considered for controlling tumor growth. Aspirin and its derivatives may contribute to rapid formation of new lesions.

36.8 Prognosis

Prognosis for cylindroma and spiradenoma is reasonably good. BSS lesions (particularly longstanding ones) have a 5%–10% risk of malignant transformation (which is indicated by ulceration, bleeding, fast growth, blue nodules, or pain). Spiradenoma shows a higher propensity to malignancy than cylindroma. Malignant spiradenoma shows a 50% risk of metastasis and a 39% risk of mortality if left untreated. Therefore, follow-up care of patients with BSS tumors is recommended to monitor the potential development of new lesions and malignant degeneration. Genetic testing allows individuals with affected relatives to ascertain their own risk and to use the information for family planning purposes.

36.9 Conclusion

Brooke–Spiegler syndrome (BSS) is a rare autosomal dominant disorder associated with formation of multiple benign skin adnexal tumors/nodules (e.g., cylindroma, spiradenoma, and trichoepithelioma) on the scalp, face, and neck, as well as neoplasms in the salivary glands. Familial cylindromatosis (FC) and multiple familial trichoepitheliomas (MFT), which respectively cause cylindroma (predominately on the scalp) and trichoepithelioma (particularly around the nose) only, are considered the phenotypic variants of BSS. Heterogeneous germline mutations in the *CYLD* gene on chromosome 16q12.1, encoding a deubiquitinase and tumor suppressor that downregulates the transcription factor NF-κB, are implicated in the tumorigenesis of BSS (85% of cases), FC (100% of cases), and MFT (45% of cases). In general, BSS tumors are slow-growing and not life-threatening, but their continuing increase in size and number with age creates a cosmetic dilemma for patients and also carries a 5%–10% risk of malignant transformation that can be deadly. Contemporary diagnosis of BSS relies on histopathological differentiation of cylindroma, spiradenoma, and trichoepithelioma. Given their notable variations in clinical presentations even among members of the same family, molecular identification of *CYLD* pathogenic variants is indispensable for confirmation of BSS, FC, and MFT. At the present, treatment of BSS is limited to lesion removal, and further research is required for developing innovative intervention measures that resolve the underlying issues associated with this disorder.

REFERENCES

1. Trufant J, Robinson M, Patel R. Brooke-Spiegler syndrome. *Dermatol Online J.* 2012;18(12):16.
2. Mohiuddin W, Laun J, Cruse W. Brooke-Spiegler syndrome. *Eplasty.* 2018;18:ic14.
3. Kazakov DV. Brooke-Spiegler syndrome and phenotypic variants: An update. *Head Neck Pathol.* 2016;10(2):125–30.
4. Manicketh I, Singh R, Ghosh PK. Eccrine cylindroma of the face and scalp. *Indian Dermatol Online J.* 2016;7(3):203–5.
5. Stoica LE, Dascălu RC, Pătrașcu V et al. Solitary trichoepithelioma: Clinical, dermatoscopic and histopathological findings. *Rom J Morphol Embryol.* 2015;56(2 Suppl):827–32.
6. Alameda JP, Fernández-Aceñero MJ, Moreno-Maldonado R et al. CYLD regulates keratinocyte differentiation and skin cancer progression in humans. *Cell Death Dis.* 2011;2:e208.
7. Brinkhuizen T, Weijzen CA, Eben J et al. Immunohistochemical analysis of the mechanistic target of rapamycin and hypoxia signalling pathways in basal cell carcinoma and trichoepithelioma. *PLOS ONE.* 2014;9(9):e106427.
8. Rajan N, Andersson MK, Sinclair N et al. Overexpression of MYB drives proliferation of CYLD-defective cylindroma cells. *J Pathol.* 2016;239(2):197–205.
9. Nagy N, Rajan N, Farkas K, Kinyó A, Kemény L, Széll M. A mutational hotspot in CYLD causing cylindromas: A comparison of phenotypes arising in different genetic backgrounds. *Acta Derm Venereol.* 2013;93(6):743–5.
10. Pinho AC, Gouveia MJ, Gameiro AR, Cardoso JC, Gonçalo MM. Brooke-Spiegler syndrome - an underrecognized cause of multiple familial scalp tumors: Report of a new germline mutation. *J Dermatol Case Rep.* 2015;9(3):67–70.
11. Zhang QG, Liang YH. A recurrent R936X mutation of CYLD gene in a Chinese family with multiple familial trichoepithelioma. *Indian J Dermatol Venereol Leprol.* 2015;81(2):192–4.
12. Aguilera CA, De la Varga Martínez R, García LO, Jiménez-Gallo D, Planelles CA, Barrios ML. Heterozygous cylindromatosis gene mutation c.1628_1629delCT in a family with Brook-Spiegler syndrome. *Indian J Dermatol.* 2016; 61(5):580.
13. Tantcheva-Poór I, Vanecek T, Lurati MC et al. Report of three novel germline CYLD mutations in unrelated patients with Brooke-Spiegler syndrome, including classic phenotype, multiple familial trichoepitheliomas and malignant transformation. *Dermatology.* 2016;232(1):30–7.
14. van den Ouweland AM, Elfferich P, Lamping R et al. Identification of a large rearrangement in CYLD as a cause of familial cylindromatosis. *Fam Cancer.* 2011;10(1):127–32.
15. Manchanda K, Bansal M, Bhayana AA, Pandey S. Brooke-Spiegler syndrome: A rare entity. *Int J Trichology.* 2012;4(1):29–31.
16. Rathi M, Awasthi S, Budania SK, Ahmad F, Dutta S, Kumar A. Brooke-Spiegler syndrome: A rare entity. *Case Rep Pathol.* 2014;2014:231895.
17. Lavorato FG, Miller MD, Obadia DL, Nery NS, Silva RS. Syndrome in question. Brooke-Spiegler syndrome. *An Bras Dermatol.* 2014;89(1):175–6.
18. Sicinska J, Rakowska A, Czuwara-Ladykowska J et al. Cylindroma transforming into basal cell carcinoma in a patient with Brooke-Spiegler syndrome. *J Dermatol Case Rep.* 2007;1(1):4–9.
19. Rato M, Filipe Monteiro A, Aranha J, Tavares E. Multiple papulonodular lesions on central area of the face: What is your diagnosis? *Dermatol Online J.* 2017;23(6). pii: 13030/qt67m6h7tb.
20. Dhir G, Makkar M, Suri V, Dubey V. Familial dermal eccrine cylindromatosis with emphasis on histology and genetic mapping. *Ann Med Health Sci Res.* 2013;3(Suppl 1):S3–6.
21. Cohen YK, Elpern DJ. Dermatoscopic pattern of a cylindroma. *Dermatol Pract Concept.* 2014;4(1):67–8.
22. Dubois A, Wilson V, Bourn D, Rajan N. CYLD genetic testing for Brooke-Spiegler syndrome, familial cylindromatosis and multiple familial trichoepitheliomas. *PLOS Curr.* 2015;7. pii: ecurrents.eogt.45c4e63dd43d62e12228cc5264d6a0db.
23. Tran K, DeFelice T, Robinson M, Patel R, Sanchez M. Spiradenomas. *Dermatol Online J.* 2012;18(12):15.
24. Baliyan A, Dhingra H, Kumar M. Spiradenocylindroma of skin: A hybrid tumor. *Indian J Dermatopathol Diagn Dermatol.* 2018;5:66–8.
25. Farkas K, Deák BK, Sánchez LC et al. The CYLD p.R758X worldwide recurrent nonsense mutation detected in patients with multiple familial trichoepithelioma type 1, Brooke-Spiegler syndrome and familial cylindromatosis represents a mutational hotspot in the gene. *BMC Genet.* 2016;17:36.
26. Dubois A, Alonso-Sanchez A, Bajaj V, Husain A, Rajan N. Multiple facial trichoepitheliomas and vulval cysts: Extending the phenotypic spectrum in CYLD cutaneous syndrome. *JAMA Dermatol.* 2017;153(8):826–8.
27. Akgul GG, Yenidogan E, Dinc S, Pak I, Colakoglu MK, Gulcelik MA. Malign cylindroma of the scalp with multiple cervical lymph node metastasis-A case report. *Int J Surg Case Rep.* 2013;4(7):589–92.
28. Mapar MA, Ranjbari N, Afshar N, Karimzadeh I, Karimzadeh A. Severely disfiguring multiple familial trichoepitheliomas with basal cell carcinoma. *Indian J Dermatol Venereol Leprol.* 2014;80(4):349–52.
29. Borik L, Heller P, Shrivastava M, Kazlouskaya V. Malignant cylindroma in a patient with Brooke-Spiegler syndrome. *Dermatol Pract Concept.* 2015;5(2):61–5.
30. Hoyle A, Davies K, Rajan N, Melly L. p63 and smooth muscle actin expression in low-grade spiradenocarcinomas in a case of CYLD cutaneous syndrome. *J Cutan Pathol.* 2018;45(10):760–3.
31. Kalina P, El-Azhary R. Brooke-Spiegler syndrome with multiple scalp cylindromas and bilateral parotid gland adenomas. *Case Rep Radiol.* 2012;2012:249583.
32. Li ZL, Guan HH, Xiao XM et al. Germline mutation analysis in the CYLD gene in Chinese patients with multiple trichoepitheliomas. *Genet Mol Res.* 2014;13(4):9650–5.
33. Kennedy RA, Thavaraj S, Diaz-Cano S. An overview of autosomal dominant tumour syndromes with prominent features in the oral and maxillofacial region. *Head Neck Pathol.* 2017;11(3):364–76.

34. Chaudhary S, Dayal S. Radiofrequency ablation: A safe and economical modality in treatment of Brooke-Spiegler syndrome. *Dermatol Online J.* 2012;18(8):7.

35. Karalija A, Andersson MN. The surgical treatment of familial cylindromatosis through subgaleal scalp excision. *Case Reports Plast Surg Hand Surg.* 2015;2(3–4):57–9.

36. Dhua S, Sekhar DR. A rare case of eccrine spiradenoma-treatment and management. *Eur J Plast Surg.* 2016;39:143–6.

37. Kuphal S, Schneider N, Massoumi R, Hellerbrand C, Bosserhoff AK. UVB radiation represses CYLD expression in melanocytes. *Oncol Lett.* 2017;14(6):7262–8.

38. Portincasa A, Cecchino L, Trecca EMC et al. A rare case of Brooke-Spiegler syndrome: Integrated surgical treatment of multiple giant eccrine spiradenomas of the head and neck in a young girl. *Int J Surg Case Rep.* 2018;51:277–81.

37

Currarino Syndrome

Dongyou Liu

CONTENTS

37.1 Introduction

Currarino syndrome (or Currarino triad) is a rare autosomal dominant disorder defined by the classical triad of anorectal malformation (imperforate anus 52%, anorectal stenosis 48%), sacral anomaly (hemisacrum 68%, partial sacral agenesis 16%), and presacral mass (anterior sacral meningocele 60%, presacral teratoma 25%, presacral lipoma and other tumors/cysts 15%), leading to severe and intractable constipation (in 95% of cases) and urinary incontinence.

The molecular basis of Currarino syndrome relates to germline mutations in the motor neuron and pancreas homeobox 1 gene (*MNX1*, or *HLXB9*) on chromosome 7q36.3, which are found in nearly all familial cases and also about 30% of sporadic cases [1].

37.2 Biology

Currarino syndrome (synonyms: Currarino triad, sacral agenesis syndrome, hereditary presacral teratoma, presacral teratoma with sacral dysgenesis) was first described by Currarino in 1981 from infants showing anorectal malformation (hindgut anomaly), sacral anomaly (sacral bone defect), and presacral mass [2].

As the most frequent type of sacral anomaly, hemisacrum (also known as "scimitar sacrum" because of peculiar radiological findings such as sickle-shaped or crescent-shaped sacral deformities) represents a pathognomonic sign of Currarino syndrome. Another common sacral anomaly is sacral agenesis, which is defined as the congenital absence of the whole or part of the sacrum.

Although anterior sacral meningocele (a unilocular or multilocular extension or herniation of the dura mater and arachnoid out of the sacral spinal canal through the defect in the sacrum into the retroperitoneum and intraperitoneal space, and containing cerebrospinal fluid) is a presacral mass commonly associated with Currarino syndrome, teratoma (a neoplasm originating from primordial germ cells and all three embryonal germ layers) and lipoma are sometimes observed.

Histogenetically, Currarino syndrome is thought to evolve from errors in progenitor cells at the region of the caudal eminence, leading to malformation of the caudal notochord and aberration of secondary neurulation with incomplete separation of the ectodermal and endodermal layers during embryonic development. Subsequent studies indicated that the underlying mechanisms for Currarino syndrome relate to mutations in the *MNX1* (or *HLXB9*) gene on chromosome 7q36.3, which cause haploinsufficiency (loss of function) of transcription factor motor neuron and pancreas homeobox 1, and which show an autosomal dominant inheritance in the majority of patients [1,3].

37.3 Pathogenesis

Located on chromosome 7q36.3, the *MNX1* gene (also called homeobox gene *HLXB9*) is a 16 kb stretch of DNA that encodes a nuclear protein (MNX1) of 401 aa in length and 40 kDa in mass weight. Through alternative splicing, *MNX1* may produce several transcript variants and isoforms [4,5].

Structurally, MNX1 comprises a polyalanine repeat region and a homeobox domain, which underscore its DNA-binding transcription factor activity and contribute to pancreas development and function through interaction with neural stem cell differentiation pathways and lineage-specific markers.

Mutations (including frameshift [45%], missense [25%], nonsense [12%], and splice site variants) in the *MNX1* gene causing haploinsufficiency have been found in almost all familial cases and 30% of sporadic cases. In addition, large deletion or complex rearrangement involving the 7q36 region may also occur in Currarino syndrome. To date, at least 82 *MNX1* heterozygous pathogenic variants have been reported. Most missense variants are located in the homeobox domain, and only a few (p.Met1Ile, p.Pro-27Leu, and p.Arg243Trp) are outside of this region. Interestingly, patients with missense variant may show milder phenotypes (anorectal stenosis, cystic formation, and anterior angulation of coccyx) than those with other null variants [6–8].

There is evidence that factors other than *MNX1* may play accessory roles in the pathogenesis of Currarino syndrome. These include proneuronal basic-helix-loop-helix proteins and LIM (named after Lin-11, Isl-2, and Mec-3 protein) domain proteins, *ETV3L*, *ARID3A*, and *NCAPD3*.

37.4 Epidemiology

37.4.1 Prevalence

Currarino syndrome is a rare disorder, with an estimated incidence of 1–9 in 100,000 people. To date, >350 cases of Currarino syndrome have been reported in the literature, of which >80% involve infancy or childhood. Currarino syndrome shows a gender bias, with a male:female ratio of 2:1 in pediatric cases, and a female:male ratio of 6:1 in adult cases

37.4.2 Inheritance

Currarino syndrome follows an autosomal dominant inheritance pattern involving *MNX1*. About 50%–60% of Currarino syndrome cases show a family history of triad-associated anomalies, although 30% of cases are linked to sporadic mutations.

37.4.3 Penetrance

Currarino syndrome demonstrates incomplete penetrance as siblings harboring pathogenic *MNX1* variants develop variable clinical and imaging features. While some Currarino syndrome patients display the complete triad of anorectal malformation, sacral anomaly, and presacral mass, others may present with a partial phenotype (mild Currarino syndrome showing hemisacrum and one of the two anomalies, either anorectal malformation or presacral mass; minimal Currarino syndrome having only hemisacrum), and about one-third may have few or atypical symptoms.

37.5 Clinical Features

Currarino syndrome typically manifests as the classic triad of anorectal malformation, sacral anomaly, and presacral mass in infants or children (80%), and rarely in adults, leading to intractable/chronic constipation, urinary retention, incontinence,

bowel obstruction, and other clinical symptoms. However, some patients do not develop the full spectrum of clinical diseases (with the mild form showing hemisacrum and one of the two anomalies, either anorectal malformation or presacral mass; the minimal form having only hemisacrum), and about one-third may be asymptomatic [9–12].

Anorectal malformation affects all patients with Currarino syndrome and mainly consists of imperforate anus/rectal fistulae (52%) and anorectal stenosis/anal atresia (48%). While imperforate anus dominantly occurs in males (82%), anorectal stenosis is commonly observed in females (71%) [13].

Sacral anomaly varies from slightly dysplastic sacrum to a complete sacral agenesis (or caudal regression), with hemisacrum/sickle-shaped sacrum (68%) and partial sacral agenesis/sacral agenesis below S2 (16%) predominating.

Presacral mass is found in 92%–100% of patients with Currarino syndrome, and ranges from anterior sacral meningocele (60%) and benign or malignant teratoma (25%) to other tumors/cysts (e.g., lipoma, leiomyosarcoma, dermoid cyst, and epidermoid cyst; 15%) (Figure 37.1) [14–16]. Presacral teratoma is a benign tumor with about 1% risk of malignant transformation (into neuroendocrine carcinoma), which is somewhat lower than that of sacrococcygeal teratoma (in association with congenital anomalies of the lower vertebrae, genitourinary system, and anorectum) (Figure 37.2) [16–18]. Presacral mass creates local pressure and results in constipation, urinary incontinence, sacral anesthesia, paresthesia of the lower extremities, disturbance of anal sphincter control, etc. A long-lasting anal fistula is suggestive of neural tube anomaly (e.g., anterior sacral teratoma).

Other complications include meningitis, congenital single kidney, hydronephrosis, developmental delays, neural tube defects, sensorineural deafness, depressed nasal bridge, joint hyperextensibility, cataracts, ptosis, microcephaly, facial dysmorphism, Hirschsprung disease, vertebral anomalies, vesicoureteral reflux, voiding dysfunction, duplex ureters, tethered cord, and leiomyomatosis peritonealis disseminata [19–26].

37.6 Diagnosis

Clinical diagnosis of Currarino syndrome is based on observation of anorectal malformation (imperforate anus 52%, anorectal stenosis 48%), sacral anomaly (hemisacrum, partial sacral agenesis), and presacral mass (anterior sacral meningocele, presacral teratoma, presacral lipoma, and other tumors/cysts), especially in infants or children, leading to chronic constipation (due to ventral meningocele and/or presacral mass), dysganglionosis (due to malformation of the enteric nervous system as seen in Hirschsprung disease), and tethered cords (which is associated with meningitis; 14%–57%).

Imaging techniques (e.g., x-ray, ultrasound, CT, and MRI) may be used for identifying sacral anomaly, presacral mass (e.g., anterior sacral meningocele and presacral teratoma), and other defects. High-pressure distal colostogram may reveal a rectourethral fistula; MRI may show anterior sacral meningoceles as a fluid signal collection arising from the anterior aspect of the spine at the level of a vertebral defect, while presacral teratomas

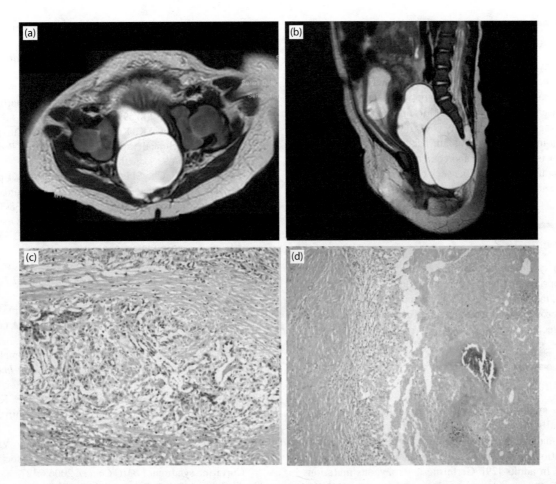

FIGURE 37.1 Currarino syndrome in a 3-month-old girl showing (a) a septated cystic lesion located within the pelvis; (b) the lesion in continuity with the spinal canal through a hiatus located in the anterior and right aspect of the sacrum, along with the significantly compressed urinary bladder containing a Foley catheter; (c) viable yolk sac tumor with solid growth pattern and glandular formation; (d) necrosis, dystrophic calcification, and histiocytes inside tumor. (Photo credit: Hage P et al. *Int J Surg Case Rep.* 2019;57:102–5.)

as multiseptated cystic masses containing soft tissue and fat that give heterogeneous appearance [27].

As *MNX1* mutations are present in >90% of familial cases and 30% of sporadic cases, identification of frequently occurring pathogenic variants such as frameshift [c.293delG (p.Gly98Alafs*124), c.434delG (p.Gly145Alafs*77), c.450_472d elGGGGCCTCCCGGCGCAGGCGGCGC (p.Gly151Leufs* 67),

c.645_648delCGCG (p.Ala216Profs*5)], nonsense [c.558_ 575delinsGGAGTAGCGGGCCA (p.Tyr186*), c.634C>T* (p.Gln212*)], and missense [c.883A>C (p.Lys295Gln)] variants by sequence analysis provides molecular evidence for Currarino syndrome. Further, large deletions in the 7q terminal region (e.g., a 5.1-Mb deletion in 7q36 deletion) may be examined by MLPA analysis [6].

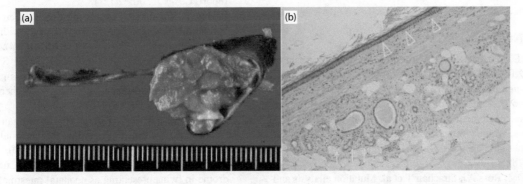

FIGURE 37.2 Mature teratoma in a 44-year-old man with Currarino syndrome. (a) Macroscopic appearance of resected teratoma. (b) Microscopic examination of resected teratoma reveals the keratinizing stratified squamous epithelium (△) in the cyst wall; fat, sweat glands (arrows), and peripheral nerves (▲) around the cyst wall. Scale bar = 200 μm. (Photo credit: Emoto S et al. *Surg Case Rep.* 2018;4(1):9.)

37.7 Treatment

Currarino syndrome is noted for the presence of presacral mass (teratoma, hamartoma, neuroenteric cyst, anterior meningocele), anorectal malformation, and fistula between the colon and spinal canal; its treatment requires a multidisciplinary approach. While posterior sagittal anorectoplasty is useful for removing teratoma of <10 cm in diameter, sacral laminectomy or anterior abdominal surgery is applicable to anorectal malformation and presacral mass, and dural ligation of the neck of the meningocele is appropriate for anterior sacral meningocele [28].

37.8 Prognosis

Currarino syndrome has a favorable prognosis. Early diagnosis and appropriate treatment can reduce life-threatening complications (e.g., meningitis, rectal fistulas, and malignant transformation of sacrococcygeal teratoma) and further improve quality of life for affected patients.

37.9 Conclusion

Currarino syndrome is a rare autosomal dominant disorder that typically presents with sacral agenesis (or sacral anomaly, the only mandatory clinical feature for diagnosis), anorectal malformations, and presacral mass in infants and children, but rarely in adults [29]. Germline heterozygous mutations in the motor neuron and pancreas homeobox 1 gene (*MNX1*) on chromosome 7q36.3 appear to underline the molecular pathogenesis of Currarino syndrome. Clinical diagnosis of Currarino syndrome relies on finding anorectal malformation (imperforate anus 52%, anorectal stenosis 48%), sacral anomaly (hemisacrum, partial sacral agenesis), and presacral mass (anterior sacral meningocele, presacral teratoma, presacral lipoma, and other tumors/cysts) [30]. Molecular identification of *MNX1* pathogenic variants, which occur in >90% of familial cases and 30% of sporadic cases, provides further confirmation of Currarino syndrome. Surgical removal of presacral teratoma of <10 cm in diameter and anorectal malformation reduces potential risk of malignant transformation and is curative for some affected patients.

REFERENCES

1. Belloni E, Martucciello G, Verderio D et al. Involvement of the HLXB9 homeobox gene in Currarino syndrome. *Am J Hum Genet*. 2000;66(1):312–9.
2. Currarino G, Coln D, Votteler T. Triad of anorectal, sacral, and presacral anomalies. *Am J Roentgenol*. 1981;137:395–8.
3. Lynch SA, Wang Y, Strachan T, Burn J, Lindsay S. Autosomal dominant sacral agenesis: Currarino syndrome. *J Med Genet*. 2000;37(8):561–6.
4. Hagan DM, Ross AJ, Strachan T et al. Mutation analysis and embryonic expression of the HLXB9 Currarino syndrome gene. *Am J Hum Genet*. 2000;66(5):1504–15.

5. Coutton C, Poreau B, Devillard F et al. Currarino syndrome and HPE microform associated with a 2.7-Mb deletion in 7q36.3 excluding SHH gene. *Mol Syndromol*. 2014; 5(1):25–31.
6. Köchling J, Karbasiyan M, Reis A. Spectrum of mutations and genotype-phenotype analysis in Currarino syndrome. *Eur J Hum Genet*. 2001;9(8):599–605.
7. Kumar B, Sinha AK, Kumar P, Kumar A. Currarino syndrome: Rare clinical variants. *J Indian Assoc Pediatr Surg*. 2016;21(4):187–9.
8. Lee S, Kim EJ, Cho SI et al. Spectrum of MNX1 pathogenic variants and associated clinical features in Korean patients with Currarino syndrome. *Ann Lab Med*. 2018;38(3):242–8.
9. Monclair T, Lundar T, Smevik B, Holm I, Ørstavik KH. Currarino syndrome at Rikshospitalet 1961-2012. *Tidsskr Nor Laegeforen*. 2013;133(22):2364–8.
10. Kassir R, Kaczmarek D. A late-recognized Currarino syndrome in an adult revealed by an anal fistula. *Int J Surg Case Rep*. 2014;5(5):240–2.
11. Shoji M, Nojima N, Yoshikawa A et al. Currarino syndrome in an adult presenting with a presacral abscess: A case report. *J Med Case Rep*. 2014;8:77.
12. Akay S, Battal B, Karaman B, Bozkurt Y. Complete currarino syndrome recognized in adulthood. *J Clin Imaging Sci*. 2015;5:10.
13. Buyukbese Sarsu S, Parmaksiz ME, Cabalar E, Karapur A, Kaya C. A very rare cause of anal atresia: Currarino syndrome. *J Clin Med Res*. 2016;8(5):420–3.
14. Nappi C, Di Spiezio Sardo A, Mandato VD et al. Leiomyomatosis peritonealis disseminata in association with Currarino syndrome? *BMC Cancer*. 2006;6:127.
15. Al Qahtani HM, Suliman Aljoqiman K, Arabi H, Al Shaalan H, Singh S. Fatal meningitis in a 14-month-old with Currarino triad. *Case Rep Radiol*. 2016;2016:1346895.
16. Hage P, Kseib C, Adem C, Chouairy CJ, Matta R. Atypical presentation of currarino syndrome: A case report. *Int J Surg Case Rep*. 2019;57:102–5.
17. Lin YH, Huang RL, Lai HC. Presacral teratoma in a Currarrino syndrome woman with an unreported insertion in MNX1 gene. *Taiwan J Obstet Gynecol*. 2011;50(4):512–4.
18. Emoto S, Kaneko M, Murono K et al. Surgical management for a huge presacral teratoma and a meningocele in an adult with Currarino triad: A case report. *Surg Case Rep*. 2018;4(1):9.
19. Saberi H, Habibi Z, Adhami A. Currarino's syndrome misinterpreted as Hirschsprung's disease for 17 years: A case report. *Cases J*. 2009;2(1):118.
20. Pavone P, Ruggieri M, Lombardo I et al. Microcephaly, sensorineural deafness and Currarino triad with duplication-deletion of distal 7q. *Eur J Pediatr*. 2010;169(4):475–81.
21. Kansal R, Mahore A, Dange N, Kukreja S. Epidermoid cyst inside anterior sacral meningocele in an adult patient of Currarino syndrome manifesting with meningitis. *Turk Neurosurg*. 2012;22(5):659–61.
22. Patel RV, Shepherd G, Kumar H, Patwardhan N. Neonatal Currarino's syndrome presenting as intestinal obstruction. *BMJ Case Rep*. 2013;2013. pii: bcr2013200310.
23. Patel RV, De Coppi P, Kiely E, Pierro A. Currarino's syndrome in twins presenting as neonatal intestinal obstruction-identical presentation in non-identical twins. *BMJ Case Rep*. 2014;2014. pii: bcr2014204276.

24. Versteegh HP, Feitz WF, van Lindert EJ, Marcelis C, de Blaauw I. "This bicycle gives me a headache," a congenital anomaly. *BMC Res Notes*. 2013;6:412.

25. Kim SH, Paek SH, Kim HY, Jung SE, Park KW. Currarino triad with Müllerian duct anomaly in mother and daughter without MNX1 gene mutation. *Ann Surg Treat Res*. 2016;90(1):49–52.

26. Cococcioni L, Paccagnini S, Pozzi E et al. Currarino syndrome and microcephaly due to a rare 7q36.2 microdeletion: A case report. *Ital J Pediatr*. 2018;44(1):59.

27. Lee JK, Towbin AJ. Currarino syndrome and the effect of a large anterior sacral meningocele on distal colostogram in an anorectal malformation. *J Radiol Case Rep*. 2016;10(6):16–21.

28. Chakhalian D, Gunasekaran A, Gandhi G, Bradley L, Mizell J, Kazemi N. Multidisciplinary surgical treatment of presacral meningocele and teratoma in an adult with Currarino triad. *Surg Neurol Int*. 2017;8:77.

29. Vinod MS, Shaw SC, Devgan A, Mukherjee S. The Currarino triad. *Med J Armed Forces India*. 2018;74(4):374–6.

30. Chatani S, Onaya H, Kato S, Inaba Y. Adenocarcinoma and neuroendocrine tumor arising within presacral teratoma associated with Currarino syndrome: A case report. *Indian J Radiol Imaging*. 2019;29(3):327–31.

38

Cutaneous Malignant Melanoma

Dongyou Liu

CONTENTS

38.1 Introduction

The skin is affected by several categories of tumors, including keratinocytic, melanocytic, appendageal, hematolymphoid, soft tissue, and neural tumors as well as hereditary tumor syndromes. Of these, melanocytic tumor is further separated into *malignant melanoma* (superficial spreading melanoma, nodular melanoma, lentigo maligna, acral-lentiginous melanoma, desmoplastic melanoma, melanoma arising from blue nevus, melanoma arising in a giant congenital nevus, melanoma of childhood, nevoid melanoma, and persistent melanoma); and *benign melanocytic tumor* (congenital melanocytic nevi [superficial type, proliferative nodules in congenital melanocytic nevi], dermal melanocytic lesions [Mongolian spot, nevus of Ito and Ota], blue nevus [cellular blue nevus], combined nevus, melanotic macules, simple lentigo and lentiginous nevus, dysplastic nevus, site-specific nevi [acral, genital, Meyerson nevus], persistent [recurrent] melanocytic nevus, Spitz nevus, pigmented spindle cell nevus [Reed], and halo nevus).

Malignant melanoma is a neoplasm of melanocytes, which are melanin (pigment)-producing cells commonly present in the skin, as well as the eyes, ears, gastrointestinal tract, leptomeninges, oral and genital mucous membranes. Melanin (pigment) produced by melanocytes gives the skin its natural color. Exposure to the sun increases melanin production by melanocytes, rendering the skin tan or dark.

Affecting melanocytes in the skin, cutaneous malignant melanoma (CMM) may arise de novo or evolve from preexisting lesions such as congenital, acquired, or dysplastic (atypical) nevus. Risk factors for CMM range from history of melanoma or nonmelanoma skin cancer, family history of CMM, presence of atypical nevi or numerous nevi, history of severe sunburns (blisters) or intense intermittent sun exposures, and light skin/blond hair to giant melanocytic nevus. In addition, mutations in certain genes may predispose to malignant melanoma, including some associated with high risk (e.g., *CDKN2A*, *CDK4*, *MITF*, *BRCA2*, *NBS1*, *CHK2*, *MLH1*, *MSH2*, *XP*) and a few with relatively low risk (e.g., *MC1R*) [1,2].

38.2 Biology

CMM represents malignant transformation of melanocytes in the epidermis as a consequence of sporadic mutation or inheritance of specific germline pathogenic variants [3].

Familial cases of CMM were described by Norris as early as 1820, involving three generations in a family who had moles on various parts of their bodies, which later turned into deadly skin tumors [4].

Subsequent studies have identified a number of risk factors (e.g., phenotype [fair complexion, blue eyes, blond or red hair], cutaneous sensitivity to sun exposure [freckling, inability to tan, sunburn tendency], history of severe sunburn [blistering] or intense intermittent sun exposure, number and subtypes of existing nevi [atypical nevi or giant melanocytic nevi], history of melanoma, and immunosuppression) that may trigger the development of CMM.

More recent molecular investigations have highlighted the roles of certain genes in the pathogenesis of CMM, such as high susceptibility gene (e.g., *CDKN2A*, *CDK4*, *MITF*, *BRCA2*, *NBS1*, *CHK2*, *MLH1*, *MSH2*, *XP*) as well as low susceptibility gene (e.g., *MC1R*) [5,6].

38.3 Pathogenesis

38.3.1 CDKN2A

The *CDKN2A* (cyclin-dependent kinase inhibitor 2A) gene on chromosome 9p21.3 spans 27.5 kb with four exons (1β, 1α, 2, and 3) and encodes two proteins, p16 (by exon 1α, exon 2, and exon 3) and p14ARF (by exon 1 β, exon 2, and exon 3, an *a*lternate *r*eading *f*rame protein product of the *CDKN2A* locus) [7,8].

P16 (also known as p16INK4a or CDKN2A) is a 156 aa, 16.5 kDa protein belonging to the CDKN2 cyclin-dependent kinase inhibitor family. P16 inhibits the complex of cyclin D1 with cyclin-dependent kinase 4 (CDK4) or 6 (CDK6), which promotes cellular proliferation and prevents Rb (retinoblastoma gene product) phosphorylation, which facilitates the G1-to-S transition.

P14ARF functions as an inhibitor of MDM2, an E3 ubiquitin ligase responsible for degradation of p53, and thus enhances p53-dependent transactivation and apoptosis. Further, p14ARF induces cell cycle arrest in G2 phase and apoptosis in a p53-independent manner by preventing the activation of cyclin B1/CDC2 complexes. Thus, loss of P14ARF activity has a similar effect as loss of P53.

While homozygous deletions contribute to *CDKN2A* inactivation in a majority of cases, hypermethylation of its promoter region may also cause occasional *CDKN2A* inactivation [9]. Germline *CDKN2A* mutations have been identified in 46% of familial melanoma cases in France, 18% in the United States, 8% in Sweden, 6% in Poland [10–14].

38.3.2 CDK4

The *CDK4* (cyclin dependent kinase 4) gene on chromosome 12q14.1 spans 8.2 kb and encodes a 303 aa, 33.7 kDa protein (CDK4) belonging to the Ser/Thr protein kinase family. As the catalytic subunit of the protein kinase complex, CDK4 along with its partner CDK6 is involved in the phosphorylation of Rb, and thus plays an important role in cell cycle G1 phase progression. CDK4 activity is controlled by the regulatory subunits D-type cyclins and CDK inhibitor p16. Alterations in *CDK4* are implicated in a small proportion of familial malignant melanoma.

38.3.3 MITF

The *MITF* (microphthalmia-inducing transcription factor) gene on chromosome 3p13 spans 228 kb and encodes a 526 aa, 58.7 kDa transcription factor (MITF) containing both basic helix-loop-helix and leucine zipper structural features. MITF binds to the promoter site of multiple target genes and regulates melanocyte cell development, pigmentation, and neoplasia. Heterozygous mutations in *MITF* cause auditory-pigmentary syndromes (e.g., Waardenburg syndrome type 2 and Tietz albinism deafness syndrome). *MITF* amplification is present in about 10% of primary melanomas, and an even higher incidence among metastatic melanomas.

38.3.4 BRCA2

The *BRCA2* (breast cancer type 2 susceptibility protein or BRCA2 DNA repair-associated) gene on chromosome 13q13.1 measures 85 kb in length and encodes a 3418 aa, 384 kDa protein (BRCA2) involved in double-strand break repair and/or homologous recombination. BRCA2 contains several 70 aa motifs (BRC motifs) that mediate binding to the RAD51 recombinase involved in DNA repair through homologous recombination. Germline mutations in *BRCA2* predispose to a range of cancer types including melanoma. While BRCA2 pathogenic variants are found in 3% of patients with familial ocular melanoma, they do not seem to be involved in familial CMM.

38.3.5 NBS1

The *NBS1* (Nijmegen breakage syndrome 1) gene on chromosome 8q21.3 measures 69.8 kb in length and encodes a 754 aa, 84.9 kDa protein (NBN or nibrin, also referred to as p95). Nibrin interacts with MRE11 and RAD50 to form the MRN complex, which possesses single-strand endonuclease activity and double-strand-specific 3'-5' exonuclease activity and participates in double-strand break (DSB) repair. In addition, nibrin plays a role in the maintenance of telomere length and thus chromosome integrity, and also in the control of intra-S-phase checkpoint. Besides its implication in Nijmegen breakage syndrome, an autosomal recessive disorder characterized by spontaneous chromosomal instability, immunodeficiency, and predisposition to cancer, the *NBS1* gene may be also involved in malignant melanoma.

38.3.6 CHK2

The *CHK2* (checkpoint kinase 2) gene on chromosome 22q12.1 is 54.6 kb in length and encodes a 543 aa, 60.9 kDa protein (CHK2), which is a serine/threonine-protein kinase required for checkpoint-mediated cell cycle arrest, activation of DNA repair, and apoptosis in response to DNA double-strand breaks. Specifically, CHK2 regulates cell cycle checkpoint arrest through inhibition of CDC25 phosphatase activity and subsequent blockage of cell cycle progression; regulates DNA repair by homologous recombination through phosphorylation of BRCA2, and increased association of RAD51, and regulates apoptosis through the phosphorylation of p53/TP53, MDM4, and PML, leading to increased accumulation and reduced degradation of p53/TP53. Germline mutation in *CHK2* exon 10 (1100delC) is observed in patients with malignant melanoma.

38.3.7 MLH1

The *MLH1* (MutL homolog 1) gene on chromosome 3p22.2 is 57 kb in length and encodes a 756 aa, 84 kDa protein (MLH1), which heterodimerizes with PMS2 to form MutL alpha, a component of the post-replicative DNA mismatch repair system (MMR). MutL alpha (MLH1-PMS2) is involved in cell cycle arrest and apoptosis in case of major DNA damages.

38.3.8 MSH2

The *MSH2* (MutS homolog 2) gene on chromosome 3p21 spans 260 kb and encodes a 934 aa, 104 kDa protein (MSH2), which heterodimerizes with MSH6 to form MutS alpha or with MSH3 to form MutS beta, both involved in post-replicative DNA mismatch repair (MMR). MutS alpha or beta forms a ternary complex with the MutL alpha heterodimer, which directs the downstream MMR events. In melanocytes, MSH2 may modulate both UV-B-induced cell cycle regulation and apoptosis. Mutations in *MSH2* often occur in hereditary nonpolyposis colon cancer (HNPCC).

38.3.9 XP

The *XP* (xeroderma pigmentosum) genes (*XPA*, *XPC*, and *XPD*) encode proteins involved in nucleotide excision repair (NER), a specialized type of DNA repair for UV radiation-induced photoproducts and DNA adducts. Mutations in XP genes are implicated in xeroderma pigmentosum syndrome, which presents with malignant melanoma, basal cell carcinoma, and squamous cell cancer of the skin.

38.3.10 MC1R

The *MC1R* (melanocortin 1 receptor) gene on chromosome 16q24.3 spans 8.8 kb and encodes a 317 aa, 34.7 kDa protein (MC1R), which is a receptor for melanocyte-stimulating hormone (MSH). Mutations in *MC1R* produce loss of function of MC1R, leading to increased pheomelanin production, lighter skin and hair color, and increased risk for melanoma and nonmelanoma skin cancer.

38.4 Epidemiology

38.4.1 Prevalence

CMM makes up 3%–5% of all skin cancer but causes 75% of all skin cancer–related deaths. The incidence of CMM is estimated at 20 per 100,000 worldwide, 26.7 and 16.7 per 100,000 males and females, respectively, in the United States, but 32 and 35 per 100,000 males and females, respectively, in Denmark. Furthermore, data from the United States indicates that white (30.9 per 100,000 males, 19.7 per 100,000 females) are more susceptible to CMM than Hispanic (4.0 males, 3.9 females), American Indian/Alaska Native (3.9 males, 3.7 females), Asian/Pacific Islander (1.6 males, 1.3 females), and black (1.2 males, 0.9 females).

38.4.2 Inheritance

Familial melanoma constitutes 5%–10% of cutaneous melanoma cases and is attributable to germline mutations in several autosomal dominant genes including *CDKN2A*. In fact, *CDKN2A* pathogenic mutations are detected in 20%–40% of families with three or more cases of cutaneous melanoma, and in 10%–15% of patients with multiple primary melanomas [15].

38.4.3 Penetrance

CMM associated with *CDKN2A* germline mutations shows 30% penetrance by 50 years of age and 67% by 80 years of age (ranging from 58% in Europe, 76% in North America, and 91% in Australia, indicating that UV increases the penetrance of *CDKN2A* mutations) [16].

38.5 Clinical Features

Clinically, CMM may appear as *melanoma in situ* (malignant melanocytes confined to the epidermis; observed in 40% of cases; curative excision with 5-mm margins), *superficial spreading melanoma* (beginning as an asymptomatic brown or black spot/macule with asymmetry, irregular borders, and variations in color; becoming invasive after an initial slow radial growth phase; most common melanoma in people with fair skin, often on the trunk of men and legs of women aged 30–50 years; 70% of cases), *nodular melanoma* (progressing rapidly from superficial spreading melanoma to a vertically growing lesion over months, and appearing blue to black, sometimes pink or red in color, with ulceration and bleeding; second most common melanoma in fair-skinned people, typically on the trunk, head, or neck of men; 5% of cases), *lentigo maligna* (beginning as an irregularly shaped tan spot/macule that slowly grows to form a larger spot/patch over 5–15 years, with 3%–5% becoming invasive lentigo maligna melanoma, risk of progression proportional to lesion size; 4%–15% of cases), *lentigo maligna melanoma* (formation of bumps/papules in lentigo maligna; often in persons of 60–80 years with sun-damaged skin; 3% of cases), *amelanotic melanoma* (rare, nonpigmented melanoma resembling eczema, fungal infection, basal cell carcinoma, and squamous cell carcinoma; often diagnosed at advanced stage; <5% of cases), *acral-lentiginous melanoma* (lentiginous melanoma on acral surfaces on body protrusions [e.g., the fingertips, knuckles, elbows, knees, buttocks, toes, heels, and ears] and specialized acral skin on the palms and soles that lacks hair follicles; accounting for 75% of cases involving black and Asian persons, and 5% of cases involving white persons), *desmoplastic melanoma* (showing scant spindle cells, lymphoid aggregates, and minimal cellular atypia; on sun-exposed skin of the head and neck of older adults; <4%

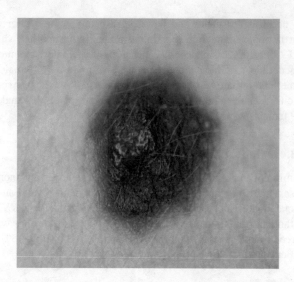

FIGURE 38.1 Invasive melanoma in a 5-year-old boy manifesting as a pigmented lesion with a peripheral halo of hypopigmentation in the upper back. (Photo credit: Yuen J et al. *SAGE Open Med Case Rep.* 2019;7:2050313X19829630.)

of cases) and *subungual melanoma* (forming a single longitudinal line of pigmentation on a nail or longitudinal melanonychia; 0.7%–3.5% of cases) (Figure 38.1) [17,18].

38.6 Diagnosis

Diagnosis of CMM involves family history review, physical exam (ABCDE mnemonic, dermoscopy), histopathology (biopsy of suspicious pigmented lesion) and molecular testing (*CDKN2A* pathogenic variants) [19–25].

ABCDE mnemonic takes account of *a*symmetry (half of lesion is different from other half), *b*order irregularities (irregular or poorly defined border), *c*olor variation (varied from one area to another; different shades of tan, brown, black; sometimes red, white, or blue within the same lesion), *d*iameter (>6 mm, bigger than the size of a pencil eraser), and *e*volution ("ugly duckling" sign—a mole looks different from surrounding moles, or changes in size, shape, or color), and provides a very useful tool for melanoma evaluation [26].

Histopathological examination of biopsy samples reveals cytologic atypia, amplified cellularity, and the number of dermal mitotic figures that help distinguish benign disease from malignant melanoma, and determines histologic subtype of the lesion on the basis of cellular features such as microsatellitosis, tumor-infiltrating lymphocytes, cellular regression, angiolymphatic invasion, vertical growth phase, neurotropism, and pure desmoplasia (Figures 38.2 and 38.3) [18,27].

Melanoma may be staged as 0, I (A/B), II (A/B/C), III (A/B/C/D), and IV according to the TNM classification system, which involves a combination of histologic and clinical features, such as the thickness of the tumor (Breslow depth), degree of nodal involvement, and the extent of metastatic spread to other areas of the body, of which Breslow depth, mitotic rate (mitoses/mm²), and ulceration are particularly useful. At the time of diagnosis, 85% of patients have localized disease (stages I and II), 15% have regional nodal disease, and 2% have distant metastases [26,28].

Differential diagnoses for cutaneous melanoma include nevus or solar lentigo, atypical nevus, squamous cell carcinoma, pigmented basal cell carcinoma, pyogenic granuloma, and seborrheic keratosis.

38.7 Treatment

Surgical removal represents the primary treatment for CMM, with Breslow depth/thickness of the lesion determining definitive surgical margins (i.e., melanoma in situ: margin 0.5 cm; Breslow depth ≤2 mm: margin 1 cm; Breslow depth ≥2 mm: margin 2 cm). A patient showing Breslow depth of ≥1.0 mm is often referred for a sentinel node biopsy (which drains the site of the primary melanoma in the presence of nodal metastases) and removal (if nodal micrometastatic disease detected by positive biopsy study or ultrasonography).

Systemic chemotherapy (dacarbazine) may be used for melanoma at advanced stage III (unresectable regional metastases) or stage IV (distant metastases), with response rate in the range of 10%–20%. Isolated limb perfusion with melphalan (alternatively dacarbazine, cisplatin, carboplatin, thiotepa, tumor necrosis factor alpha) may be considered as adjuvant treatment or as treatment of locally recurrent melanoma of an extremity. Talimogene laherparepvec (T-VEC) that produces granulocyte-monocyte colony stimulating factor (GM-CSF) in situ and incites the host immune response against tumor may be used as intralesional injection into nonresectable cutaneous, subcutaneous, or nodal lesions in patients with metastatic melanoma with limited systemic disease. Targeted therapies that have some effects on advanced/metastatic melanoma include BRAF inhibitors, MEK inhibitors, NRAS inhibitors, and cKIT inhibitors. Stereotactic radiosurgery may be also employed to ablate limited metastatic disease, used locally after excision of a primary melanoma with high-risk features (e.g., desmoplastic tumor) to reduce the rate of local recurrence, or delivered to the nodal basin with metastatic risk to reduce the rate of lymphatic disease [29–36].

38.8 Prognosis and Prevention

Patients with stage I, II, III, and IV melanoma have 5-year overall survival rates of >98%, 53%–81%, 40%–78%, and 15%–20%, respectively [26].

Younger, female patients with extremity stage I and II melanoma have a more favorable prognosis. Patients with melanoma of increasing Breslow depth (thickness), along with ulceration, diminished lymphoid response, evidence of tumor regression, microscopic satellites, lymphovascular invasion, and non–spindle-cell type tumors, have a poorer prognosis.

Patients with regional lymph node metastases have melanoma-specific survival rate of 77% for 5 years and 69% for 10 years. The number of positive lymph nodes is a key prognostic factor for stage III melanoma, whereas the increasing number of metastatic sites, visceral location of metastases (lung, liver, brain, bone), absence of resectable metastases, male sex, and shorter duration of remission portend poor prognosis for stage IV melanoma [27,36].

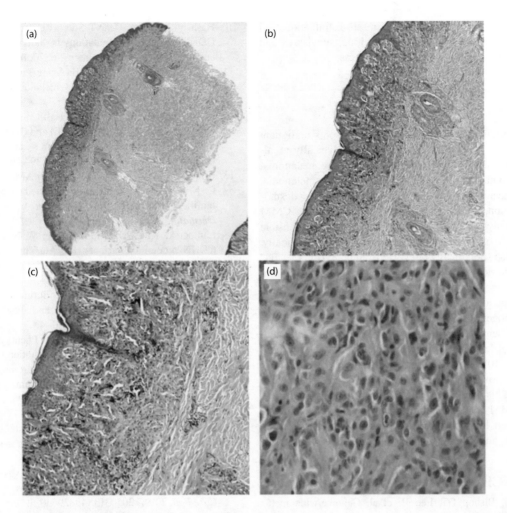

FIGURE 38.2 Histopathology of melanoma biopsy from a 5-year-old boy revealing (a) symmetrical, compound melanocytic proliferation involving the base of the epidermis and superficial dermis without an inflammatory response (20×); (b) confinement of melanocytes to the epidermal base, with only focal intra-epidermal ascent of melanocytes (40×); (c) maturation of dermal melanocytes in the base (100×); (d) some multinucleate melanocytes of variable nuclear size (>4× variation), along with occasional nucleoli and mitoses (400×). (Photo credit: Yuen J et al. *SAGE Open Med Case Rep.* 2019;7:2050313X19829630.)

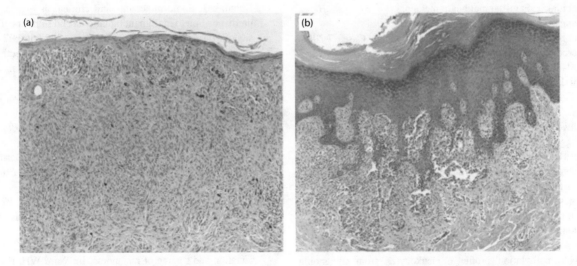

FIGURE 38.3 (a) Histopathology of superficial spreading melanoma showing an *in situ* component in the epidermis with underlying dermal invasion. (b) Histopathology of acral melanoma showing a lentiginous (linear) *in situ* growth pattern along the epidermal ridges with underlying invasion into the dermis. (Photo credit: Rabbie R et al. *J Pathol.* 2019;247(5):539–51.)

Reduction of ultraviolet exposure, skin self-examination, and use of sunscreen products (protective clothing, sunglasses and sun creams) are helpful in preventing melanoma.

38.9 Conclusion

Cutaneous malignant melanoma (CMM) represents malignant transformation of melanocytes in the skin and typically manifests as melanoma in situ, superficial spreading melanoma, nodular melanoma, lentigo maligna, lentigo maligna melanoma, amelanotic melanoma, acral-lentiginous melanoma, desmoplastic melanoma, and subungual melanoma. Although CMM accounts for only 3%–5% of all skin cancer, it is responsible for causing 75% of all skin cancer-related deaths. Etiologically, CMM may arise sporadically or evolve from preexisting lesions (e.g., congenital, acquired, or dysplastic/atypical nevus), with a number of risk factors contributing to the pathogenic processes. Recent investigations uncovered the role of certain genes in the molecular pathogenesis of CMM, among which *CDKN2A* encoding a cyclin-dependent kinase inhibitor appears to be implicated in 20%–40% of families with three or more cases of cutaneous melanoma, and 10%–15% of patients with multiple primary melanomas. Diagnosis of CMM is based on family history review, physical exam, histopathology, and molecular testing. Treatment options for CMM consist of surgical removal, systemic chemotherapy, isolated limb perfusion with melphalan, T-VEC, targeted therapies, and stereotactic radiosurgery.

REFERENCES

1. Law MH, Bishop DT, Lee JE et al. Genome-wide meta-analysis identifies five new susceptibility loci for cutaneous malignant melanoma. *Nat Genet.* 2015;47(9):987–95.
2. Wadt KA, Aoude LG, Krogh L et al. Molecular characterization of melanoma cases in Denmark suspected of genetic predisposition. *PLOS ONE.* 2015;10(3):e0122662.
3. Mort RL, Jackson IJ, Patton EE. The melanocyte lineage in development and disease. *Development.* 2015;142(4):620–32.
4. Debniak T. Familial malignant melanoma — overview. *Hered Cancer Clin Pract.* 2004;2(3):123–9.
5. Koopmann J, Goggins M, Hruban RH. Case 7-2004: Hereditary melanoma and pancreatic cancer. *N Engl J Med.* 2004;350(25):2623–4; author reply 2623–4.
6. Shi J, Yang XR, Ballew B et al. Rare missense variants in POT1 predispose to familial cutaneous malignant melanoma. *Nat Genet.* 2014;46(5):482–6.
7. Dilworth D, Liu L, Stewart AK, Berenson JR, Lassam N, Hogg D. Germline CDKN2A mutation implicated in predisposition to multiple myeloma. *Blood.* 2000;95(5):1869–71.
8. Debniak T, Górski B, Scott RJ et al. Germline mutation and large deletion analysis of the CDKN2A and ARF genes in families with multiple melanoma or an aggregation of malignant melanoma and breast cancer. *Int J Cancer.* 2004;110(4):558–62.
9. de Araújo É, Marchi FA, Rodrigues TC et al. Genome-wide DNA methylation profile of leukocytes from melanoma patients with and without CDKN2A mutations. *Exp Mol Pathol.* 2014;97(3):425–32.
10. Hashemi J, Platz A, Ueno T, Stierner U, Ringborg U, Hansson J. CDKN2A germ-line mutations in individuals with multiple cutaneous melanomas. *Cancer Res.* 2000;60(24):6864–7.
11. Yang G, Niendorf KB, Tsao H. A novel methionine-53-valine mutation of p16 in a hereditary melanoma kindred. *J Invest Dermatol.* 2004;123(3):574–5.
12. Laud K, Marian C, Avril MF et al. Comprehensive analysis of CDKN2A (p16INK4A/p14ARF) and CDKN2B genes in 53 melanoma index cases considered to be at heightened risk of melanoma. *J Med Genet.* 2006;43(1):39–47.
13. Erlandson A, Appelqvist F, Wennberg AM, Holm J, Enerbäck C. Novel CDKN2A mutations detected in western Swedish families with hereditary malignant melanoma. *J Invest Dermatol.* 2007;127(6):1465–7.
14. Lamperska KM, Przybyła A, Kycler W, Mackiewicz A. The CDKN2a common variants: 148 Ala/Thr and 500 C/G in 3' UTR, and their association with clinical course of melanoma. *Acta Biochim Pol.* 2007;54(1):119–24.
15. Gruis NA, van der Velden PA, Bergman W, Frants RR. Familial melanoma; CDKN2A and beyond. *J Investig Dermatol Symp Proc.* 1999;4(1):50–4.
16. Lynch HT, Brand RE, Hogg D et al. Phenotypic variation in eight extended CDKN2A germline mutation familial atypical multiple mole melanoma-pancreatic carcinoma-prone families: The familial atypical mole melanoma-pancreatic carcinoma syndrome. *Cancer.* 2002;94(1):84–96.
17. Ashton-Prolla P, Bakos L, Junqueira G Jr, Giugliani R, Azevedo SJ, Hogg D. Clinical and molecular characterization of patients at risk for hereditary melanoma in southern Brazil. *J Invest Dermatol.* 2008;128(2):421–5.
18. Yuen J, AlZahrani F, Horne G et al. Invasive melanoma in a 5-year-old Canadian patient: A case report. *SAGE Open Med Case Rep.* 2019;7:2050313X19829630.
19. Debniak T, Górski B, Huzarski T et al. A common variant of CDKN2A (p16) predisposes to breast cancer. *J Med Genet.* 2005;42(10):763–5.
20. Debniak T, Scott RJ, Huzarski T et al. CDKN2A common variant and multi-organ cancer risk--a population-based study. *Int J Cancer.* 2006;118(12):3180–2. Erratum in: Int J Cancer. 2006;119(10):2502.
21. Debniak T. Some molecular and clinical aspects of genetic predisposition to malignant melanoma and tumours of various site of origin. *Hered Cancer Clin Pract.* 2007;5(2):97–116.
22. Leachman SA, Carucci J, Kohlmann W et al. Selection criteria for genetic assessment of patients with familial melanoma. *J Am Acad Dermatol.* 2009;61(4):677.e1–14.
23. van der Rhee JI, Boonk SE, Putter H et al. Surveillance of second-degree relatives from melanoma families with a CDKN2A germline mutation. *Cancer Epidemiol Biomarkers Prev.* 2013;22(10):1771–7.
24. Kato J, Horimoto K, Sato S, Minowa T, Uhara H. Dermoscopy of melanoma and non-melanoma skin cancers. *Front Med (Lausanne).* 2019;6:180.
25. Puckett Y, Wilson AM, Thevenin C. *Cancer, Melanoma Pathology. StatPearls* [Internet]. Treasure Island (FL): StatPearls Publishing; 2019 Jan–2019 Jul 26.
26. Ward WH, Lambreton F, Goel N, Yu JQ, Farma JM. Clinical presentation and staging of melanoma. In: Ward WH, Farma JM, editors. *Cutaneous Melanoma: Etiology and Therapy* [Internet]. Brisbane (AU): Codon Publications; 2017 Dec 21. Chapter 6.

27. Rabbie R, Ferguson P, Molina-Aguilar C, Adams DJ, Robles-Espinoza CD. Melanoma subtypes: Genomic profiles, prognostic molecular markers and therapeutic possibilities. *J Pathol.* 2019;247(5):539–51.

28. Wang W, Niendorf KB, Patel D et al. Estimating CDKN2A carrier probability and personalizing cancer risk assessments in hereditary melanoma using MelaPRO. *Cancer Res.* 2010;70(2):552–9.

29. Soura E, Eliades PJ, Shannon K, Stratigos AJ, Tsao H. Hereditary melanoma: Update on syndromes and management: Emerging melanoma cancer complexes and genetic counseling. *J Am Acad Dermatol.* 2016;74(3):411–20; quiz 421–2.

30. Soura E, Eliades PJ, Shannon K, Stratigos AJ, Tsao H. Hereditary melanoma: Update on syndromes and management: Genetics of familial atypical multiple mole melanoma syndrome. *J Am Acad Dermatol.* 2016;74(3):395–407; quiz 408–10.

31. Leachman SA, Lucero OM, Sampson JE et al. Identification, genetic testing, and management of hereditary melanoma. *Cancer Metastasis Rev.* 2017;36(1):77–90.

32. Gonzalez A. Sentinel lymph node biopsy: Past and present implications for the management of cutaneous melanoma with nodal metastasis. *Am J Clin Dermatol.* 2018;19(Suppl 1):24–30.

33. Dimitriou F, Braun RP, Mangana J. Update on adjuvant melanoma therapy. *Curr Opin Oncol.* 2018;30(2):118–24.

34. Dummer R, Mangana J, Frauchiger AL, Lang C, Micaletto S, Barysch MJ. How I treat metastatic melanoma. *ESMO Open.* 2019;4(Suppl 2):e000509.

35. Fujimura T, Fujisawa Y, Kambayashi Y, Aiba S. Significance of BRAF kinase inhibitors for melanoma treatment: From bench to bedside. *Cancers (Basel).* 2019;11(9). pii: E1342.

36. Levin T, Mæhle L. Uptake of genetic counseling, genetic testing and surveillance in hereditary malignant melanoma (CDKN2A) in Norway. *Fam Cancer.* 2017;16(2):257–65.

39

Familial Atypical Multiple Mole Melanoma Syndrome

Samantha Guhan and Hensin Tsao

CONTENTS

39.1 Introduction

Malignant melanoma has high morbidity and mortality rates if left untreated. It is the sixth most common cancer and accounts for greater than 47,000 deaths annually worldwide [1]. Melanoma can be sporadic or hereditary, with about 7%–15% patients diagnosed with melanoma having a positive family history [1]. While many family members share melanoma history due to similar level of sun exposure, hereditary mutations and familial syndromes also play an important role. One such syndrome is familial atypical multiple mole melanoma syndrome (FAMMM), which is characterized by multiple atypical nevi (usually greater than 50) with characteristic histological features, and a history of cutaneous melanoma in at least one first- or second-degree family member (Figure 39.1a). Approximately 40% of patients with FAMMM syndrome have a mutation in the gene *CDKN2A*, which is passed in an autosomal dominant fashion with variable penetrance and expressivity. Patients with FAMMM also have an increased risk for other systemic cancers, especially pancreatic cancer (PC) (Figure 39.1d).

39.2 Biological Background and History

Norris published the first reported case in 1820, when he described a 59-year-old man with multiple nevi who passed away due to metastatic melanoma. The patient had a strong family history of melanoma, and several of his family members had high total body nevus counts. In 1952, Cawley et al. suggested a hereditary basis to melanoma after publishing a case on a father and his three children suffering from the disease [2]. Several other published reports documented a possible hereditary basis to melanoma [3–10], including Anderson et al. in 1967 who reported a family in which 15 individuals developed the cancer [11]. In 1968, Lynch et al. suggested an autosomal inheritance pattern to hereditary melanoma [12]. Ten years later, in 1978, Lynch et al. and Clark concurrently described the clinical features of FAMMM (although Clark officially named the disorder B-K mole syndrome) [13,14]. In the 1980s, segregation analysis confirmed the inheritance pattern to be autosomal dominant with varying penetrance and expressivity [15,16]. In 1991, Lynch and Fusaro noticed an association of FAMMM with PC [17].

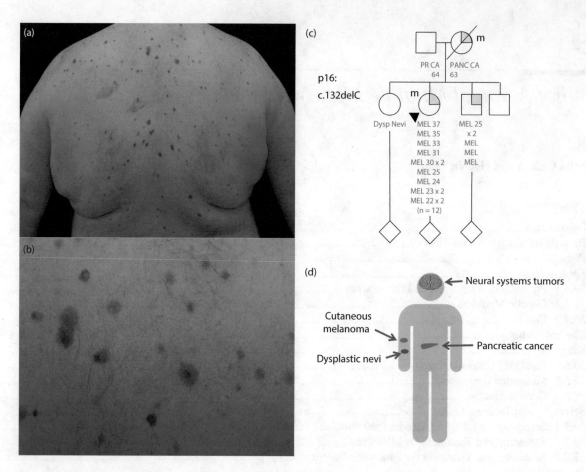

FIGURE 39.1 The familial atypical multiple mole-melanoma phenotype. (a) Skin phenotype of a FAMMM patient. (b) Clinically atypical moles frequently associated with FAMMM syndrome. (c) Pedigree of a FAMMM kindred showing multiple early onset cutaneous melanomas (proband and brother) and pancreatic cancer (PANC CA; mother). The patient and mother are carriers of a p16 mutation (m). (d) FAMMM patients who carry the *CDKN2A* mutation are at a higher risk for multiple dysplastic nevi, melanoma, neural tumors (melanoma-astrocytoma syndrome), and pancreatic cancer.

In the mid-1990s, several researchers published a correlation of FAMMM with a mutation in the gene *CDKN2A* [18–20]. Caldas reported in 1994 that PC was also associated with a *CDKN2A* mutation [21], and in 2000 Vasen made the connection between FAMMM and PC because of their similar *CDKN2A* mutation profile [22].

39.3 Pathogenesis

FAMMM follows an autosomal dominant inheritance pattern (Figure 39.1c) with variable penetrance and expressivity [1,15,23,24]. Penetrance is affected by a number of genetic and environmental factors (see Section 39.4 for details). While hereditary melanoma has been associated with SNPs and mutations in a variety of genes, FAMMM is most closely associated with a mutation in the gene *CDKN2A* and occasionally *CDK4*. In fact, hereditary melanoma accompanied by dysplastic nevi indicates the likelihood of a *CDKN2A* mutation [1].

CDKN2A is a tumor suppressor gene located on chromosome 9p21.3 It is composed of four exons (1α, 1β, 2, 3), which are differentially spliced to encode two proteins with distinct reading frames. The first reading frame (exons 1α, 2, 3) encodes the protein p16INK4a, while the second reading frame (1β, 2, 3) encodes the protein p14ARF [1,20,25]. P16INKa inhibits the

proteins cyclnD1/CDK4, which decreases phosphorylation of the protein RB1. The hypophosphorylated RB1 inhibits transcription factor E2F, ultimately leading to an arrest in the G1 phase of the cell cycle and limiting the G-S transition [26]. On the other hand, p14ARF binds to the N terminus of human double minute-2 (HDM2) protein, targeting it for immediate degradation. HDM2 not only inhibits RB1, but also is an E3 ubiquitin ligase which, through ubiquitination of tumor suppressor p53, marks it for proteasome destruction. The degradation of HDM2 saves p53, allowing it to activate its downstream targets such as CDKN1A, which, similarly to *CDKN2A*, inhibits *CDK4* (Figure 39.2) [27].

Thus, mutations in *CDK2A* can inactivate two distinct tumor suppressor pathways. While a majority (70%) of *CDKN2A* mutations are nonsense or missense, insertions/deletions (23%), splicing mutations (5%) and regulatory mutations (2%) have been documented as well [28]. Most *CDKN2A* mutations affect p16INK4a, with a majority of the deletions occurring in exon 1α (present only in p16INK4a) and 2 (present in both P16INK4a and p14ARF) [23,29]. Arap et al. demonstrated that mutations in exon 2 preferentially affect p16INK4a more than the p14ARF protein product [27,30]. Goldstein et al. found that specific mutations in the *CDKN2A* gene vary from region to region. The vast majority of family mutations in Sweden and the Netherlands are p.R112_L113*ins*R, while France, Spain, and Italy share p.G101W.

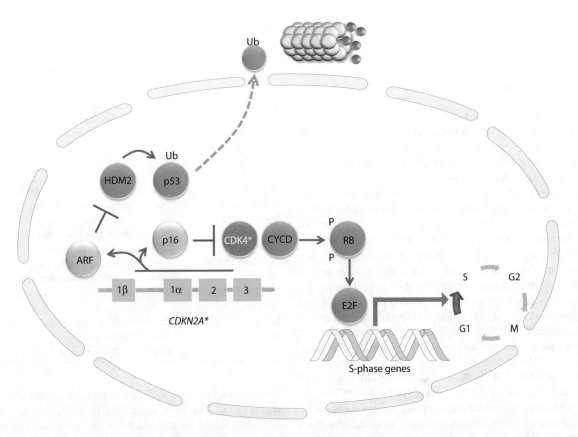

FIGURE 39.2 Pathways linked to FAMMM predisposition. *CDKN2A* is composed of four exons (1α, 1β, 2, 3), which are differentially spliced to encode two proteins with distinct reading frames. The first reading frame (exons 1α, 2, 3) encodes the protein p16, while the second reading frame (1β, 2, 3) encodes the protein ARF. P16 inhibits the proteins CYCD/CDK4, which decreases phosphorylation of the protein RB1. The hypophosphorylated RB1 inhibits transcription factor E2F, ultimately leading to an arrest in the G1 phase of the cell cycle. ARF binds to the N terminus of human double minute-2 (HDM2) protein, targeting it for immediate degradation. HDM2 is a E3 ubiquitin ligase which marks p53 for proteasome destruction. The degradation of HDM2 saves p53, allowing it to activate its downstream targets. Mutations in *CDKN2A* lead to inactivation of these two tumor suppressor pathways.

p.M53I, c.IVS2-105A>G, p.R24P, and p.L32P are most frequent in the United States and Australia [28].

While p14ARF mutations are rare, researchers have reported deletions and insertions in exon 1β only in cases of hereditary melanoma [27,29,31]. Randerson-Moor et al. demonstrated that families suffering from these mutations are at a higher risk of developing a subtype of FAMMM called melanoma astrocytoma [1,32]. Melanoma astrocytoma was first described by Kaufman et al. in 1993, and later by Azizi et al., when they noticed that a subset of families with hereditary melanoma also suffered from numerous types of nervous system tumors [33,34]. Patients are generally less than 30 years old and develop the nervous system tumors either before or after their cutaneous malignant melanoma (CMM) [1]. Multiple research studies have shown that this syndrome is caused by larger scale, contiguous deletions [32,35,36].

As described by Tsao et al., several animal models have supported the notion that *CDKN2A* plays a causative role in carcinogenesis and melanoma formation [29]. Serrano et al. demonstrated that mice with *CDKN2A* mutations have a higher frequency of fibrosarcomas and lymphomas [29,37]. Chin et al. showed that mice with p16INK4a inactivation and activating HRAS mutations quickly developed CMM with high penetrance [38]. VanBrocklin showed similar results when *CDKN2A* null mice also receive an activating NRAS mutation [39]. Yang et al. showed that mice null for *CDKN2A* and DNA excision repair

genes and exposed to UV radiation developed multiple cutaneous melanomas [40].

Other than *CDKN2A*, another high-risk locus for the development of FAMMM is *CDK4*, which is located on chromosome 12q14.1, and is also transmitted in an autosomal dominant fashion [1]. While germline *CDK4* mutations are very rare, several familial cases have been reported [41–44]. Currently, all reported mutations have occurred at the p16 binding site, substituting arginine 24 with either histidine or cysteine [44,45]. When altered, *CDK4* cannot be inhibited by p16INK4a, and thus promotes a G-S cell cycle transition. Since mutations affect interaction with p16INK4a, *CDK4* mutations have a very similar phenotype to p16INK4a mutations [1].

39.4 Epidemiology

39.4.1 Prevalence, Penetrance, and Expressivity

Approximately 39% of familial melanoma patients harbor a *CDKN2A* mutation, but the prevalence varies from region to region [46]. In 2007, Goldstein et al. found the prevalence to be 20% in Australia, 45% in North America, and 57% in Europe. The mutation also had a higher prevalence in areas where other risk factors like tanning and fair skin were absent [46]. In another study, Goldstein et al. reported that 92% of familial melanoma cases harbored a p.R112_L113insR or c.331_332insGTC *CDKN2A* mutation. The same two mutations accounted for 15% of French mutations and 60% of Mediterranean mutations. Australia and UK shared similar mutation profiles (most common mutations being p.M53I, c.IVS2-105a>G, p.R24P, and p.L32P), with these mutations comprising 42% and 50%, respectively, of familial cases [28]. Nikolaou et al. found that the Greek familial melanoma cases had the highest number of *CDKN2A* mutations and hypothesized that the genetic profile has a larger impact in regions with fewer melanoma cases [47].

CDKN2A mutations in FAMMM demonstrate variable penetrance and expressivity. Bishop et al. found the mutation to have approximately 30% penetrance by age 50 and approximately 67% penetrance by age 80 [48]. Penetrance also varied significantly by geographic location. In Europe, the mutation has 13% penetrance in 50 year olds and 58% penetrance in 80 year olds. In the United States, penetrance for 50 and 80 year olds was 50% and 76%, respectively. Australia has penetrance values of 32% by 50 years and 91% by 80 years. Bishop did not find gender to be a significant modifier. In 2005, Begg et al. concluded that the penetrance was lower than previously thought [49]. He found that the *CDKN2A* melanoma risk was 14% by 50 years (95% confidence interval [CI] = 8%–22%), 24% by 70 years (95% CI = 15%–34%), and 28% by 80 years (95% CI = 18%–40%). In 2011, Cust et al. concluded that the real penetrance was somewhere in between the two values generated from the aforementioned studies. They found that the relative risk, accounting for other modifiers, of developing melanoma after a *CDKN2A* mutation was 4.3. The authors furthermore found that the mutation had very low prevalence in primary melanoma cases [50].

39.4.2 Genetic Modifiers

Several modifiers, both genetic and environmental, affect the penetrance of the *CDKN2A* mutation in FAMMM. Major genetic modifiers include mutations in MC1R (melanocortin-1 receptor), MITF (microphthalmia-associated transcription factor), glutathione S-transferase, and various SNPs (Table 39.1).

39.4.2.1 MC1R Variants

MC1R is a critical player in the pigmentation of human skin. Specifically, it increases the proportion of eumelanin to pheomelanin. Eumelanin is the major workhorse in protecting darker skin from UV damage. Healy et al. found that MC1R contributes to tanning ability in non-red-haired individuals [51]. α-MSH binds to the 7-pass transmembrane protein MC1R, activating adenylate cyclase, and thus increasing cAMP. Increased levels of cAMP trigger expression of microphthalmia transcription factor (MITF), which plays a key role in melanin production. Increased cAMP also leads to the protein kinase A (PKA) phosphorylation of cAMP response-element binding protein (CREB).

Caucasian population harbors many variations of the MC1R locus. Some variants, deemed R, correlate significantly with freckles, sun sensitivity, and red hair, while others, deemed r, do not have significant correlations. R alleles were hypothesized to reduce the cell surface expression of MC1R [52,53]. As expected, mutations in these gene are more prevalent in

TABLE 39.1

List of Genetic Variants Associated with FAMMM Syndrome

Associated Variant	Normal Function
MC1R	Increases proportion of eumelanin to pheomelanin
MITF (E318K)	Regulate differentiation, development, and cell cycle of melanocytes
Glutathione S-transferase theta 1	Detoxify and inactivate ROS produced during UV exposure
20q11.22 (MYH7B/ PIGU/ASIP)	PIGU involved in cell cycle, ASIP involved in creation of pheomelanin
11q14.3 (TYR)	Role in rate-limiting step of melanin creation
5p13.2 (SLC45A2)	Involved in eumelanin synthesis
15q131(HERC2/ OCA2)	HER2 associated with DNA repair of damaged chromosomes after UV radiation
1q21.3 (ARNT/ ANXA9/LASS2)	ARNT associated with response to P450;
	ANXA9 associated with TH2 response; LASS associated with apoptosis after ionizing radiation
15q13.1 (rs1129038 and rs129138)	Multiple roles
MTAP	Tumor suppressor, adjacent to CDKN2A
BCL7A	Tumor suppressor
IL9	Involved in TH1:TH2 ratio

fair-skinned individuals. Eighty percent of individuals with fair skin and red hair have a mutation in MC1R, while only 20% of darker-haired individuals have the mutation. The mutation is even less common (4%) in individuals who tan easily [23]. Kennedy et al. found that the Asp84Glu mutation has the highest risk of leading to melanoma, independent of hair and skin color [54]. In 2008, Raimondi et al. conducted a meta-analysis on the relationship between prevalent MC1R variants (p.V60L, p.D84E, p.V92M, p.R142H, p.R151C, p.I155T, p.R160W, p.R163Q, and p.D294H) and melanoma and fair skin/red hair [55]. They found that some variants were associated with melanoma only (p.I155T, p. R163Q), and some were associated with red hair and melanoma (p.R151C, p.R160W and p.D294H, p.D84E, p.R142H). Raimondi et al. concluded that the variants associated with melanoma and not red hair/fair skin must contribute to risk via non-pigmentary pathways, such as "immunomodulary and anti-inflammatory" effects [55]. In 2011, Mitra et al. provided further evidence that MC1R variants predisposed melanoma risks through pathways independent of UV exposure, and hypothesized that this was due to the increased levels of pheomelanin in MC1R mutants [23,56]. Newton et al. hypothesized that non-pigmentary pathways include reduced c-Fos transcription factor (involved in DNA repair) and p38 MAPK signaling [53]. p38 is responsible for downstream pathways involving cell differentiation and cell apoptosis.

Numerous studies have specifically studied the effect of MC1R mutations on *CDKN2A* penetrance [57]. Van der Velden found that the R151C variant was significantly associated with the disease in Dutch populations. Twenty-one percent of FAMMM family members had the mutation, compared to 5% of the general population studied [58]. Box et al. found that the R160W, D294H MC1R alleles additionally affected penetrance in Australian populations [59]. Debniak saw similar results in Polish populations [60]. Fargnoli et al. found that *CDKN2A* mutation carriers with additional multiple MC1R variants had a four times higher risk of developing melanoma across multiple populations [61]. In 2010, Demenais et al. studied the association of MC1R variants in *CDKN2A* mutations carriers in a large scale GenoMEL study. The study included 815 carriers from North America, Europe, and Australia, and showed that both red hair phenotype variants and non-red hair phenotype variants were associated with increased risk. They have two variants rather than one, which increased the risk of developing melanoma by 2.6-fold [62]. Multiple nevi and hair color increased the effects on penetrance. Furthermore, the MC1R risk seemed to have a greater effect when p14 was not mutated [52,62].

The effect of MC1R mutation on the age of onset of melanoma has yielded inconsistent results. Box et al. found that the MC1R mutation increased the *CDKN2A* penetrance to 84% in 15 melanoma pedigrees and decreased the mean age of onset of melanoma from 58.1 years to 37.8 years [59]. However, Chaudru et al. found that the presence of MC1R variants did not affect age of onset [63]. Palmer et al. found that the CMM and MC1R relationship was strongest in individuals with olive skin [64].

39.4.2.2 MITF (E318K) Variant

As MC1R affects pigmentation through the MITF gene, mutations in MITF can also increase the chances of an individual developing melanoma. MITF is an important player in the regulation of melanocytes, as its downstream targets affect differentiation, development, and cell-cycle progression of melanocytes. While not a lot of research has been done on the effect of MITF mutations specifically on *CDKN2A* penetrance in FAMMM, increased MITF expression has been found in 15% of metastatic melanoma cases [65]. Yokoyama et al. used whole-genome sequencing to identify the p.E381K MITF mutation as a melanoma-specific oncogene, with a LOD score of 2.7 [66]. The mutation is located in a small-ubiquitin-like modifier (SUMO) consensus site, negatively affecting SUMOlyation of MITF, increasing transcriptional activity of MITF targets like HIF1α (associated with renal cell carcinoma) [67]. Large case-control samples from the UK and Australia showed that this mutation was not only significantly associated with melanoma, but also with a higher nevus count and non-blue eye color [66]. Potrony et al. noted a MITF mutation in 2/69 (2.9%) of probands with familial *CDKN2A* mutations [68]. Much more work has been done on the role of MITF on multiple primary melanomas and non-FAMMM familial melanomas.

39.4.2.3 Glutathione S-Transferase Theta 1

Glutathione S-transferase protein products detoxify and inactivate reactive oxygen species (ROS) produced during UV exposure. Chaudru et al. examined the effect of GST variants on families with *CDKN2A* mutations, correcting for the presence of MITF mutations, phenotype variations, and sun exposure [69]. They studied the effect of SNPs in the coding region and gene deletions (i.e., GSTT1) and surprisingly found that the GSSTT1 gene deletion was protective against melanoma (OR adjusted for age, sex, *CDKN2A*, MC1R = 0.24). Chaudru et al. addressed this paradoxical result. GST catalyzes GSH conjugation in order to detoxify ROS and metabolites from UV exposure. GSH itself is protective against toxic metabolites in the cell (destruction of

hydroperoxides electrophilic centers) and its depletion could thus lead to cell injury [69,70]. Contradictory data has also been published. For example, Mossner et al. found that a GSST1 deletion had no significant effect on melanoma cases [71].

39.4.2.4 Additional Loci

While not a lot of *genome-wide association studies* (GWAS) have specifically linked loci to FAMMM, multiple studies have identified low risk loci for cutaneous melanoma, which may affect the penetrance of a *CDKN2A* mutation. These SNPs are usually associated with DNA repair, cell death, and immune responses. Significantly associated loci included 20q11.22 (MYH7B/PIGU/ASIP), 11q14.3 (TYR), and 5p13.2 (SLC45A2) [72,73]. PIGU is involved in regulation of the cell cycle, while ASIP is involved in the creation of pheomelanin. TYR is a tyrosinase which plays a critical role in the rate-limiting step of melanin creation. The TYR SNP reported by Chatzinasiou (R402Q) has also been associated with skin sensitivity and eye color [72]. SCL45A2 is involved in eumelanin synthesis, and mutations in this gene have not only been associated with variations in skin color but also with ocular albinism in mice [72]. Additional loci include 15q131(HERC2/OCA2), 1q21.3 (ARNT/LASS2/ANXA9), and 15q13.1 (rs1129038 and rs129138) [74,75]. HERC2 has played a role in DNA repair of damaged chromosomes in light of UV radiation and in post-transcriptional modification of proteins involved in excision repair [76,77]. ANXA9 has been associated with immune (TH2) responses during pemphigus. LASS has been associated with apoptosis following ionizing radiation, and ARNT plays a role in response to P450 [74]. Falchi et al. identified a gene adjacent to *CDKN2A* that may separately influence nevi and melanoma formation (MTAP), which is independently a tumor suppressor [78,79]. Finally, BCL7A (a tumor suppressor) and IL9 (involved in TH1:TH2) were found to have a stronger association with melanoma in families with a positive *CDKN2A* mutation [80].

39.4.3 Cancer Association

39.4.3.1 FAMMM and Pancreatic Cancer

FAMMM is significantly associated with PC. In 1994, Bergman et al. observed increased incidence of GI neoplasms in Dutch FAMMM families [81]. In 1995, Whelan et al. presented a case report in which they noted increased incidence of PC and melanoma in families with Gly93Trp mutations in *CDKN2A* [82]. Schutte et al. found that the p16/*CDKN2A* pathway was inactivated 98% of studied sporadic pancreatic carcinomas [83]. Goldstein et al. found that families with a *CDKN2A* mutation have a 22-fold increase of PC risk, and Vasen et al. concluded that families have a 17% chance of developing PC by age 75 [22,84]. Vasen et al. found 86 cases of melanoma in 27 families with *CDKN2A* mutations and 15 cases in 7 families with the mutation. Bartsch et al. reported 2/5 families with *CDKN2A* mutations to have both melanoma and PC (with mutations Q50X and E119X), but observed that the mutations were not observed in families with PC but not melanoma [85]. Rulyak et al. found that offspring from parents with multiple nevi, melanoma, or PC had a 48.9% chance (compared to 16.7 without any risk factors) of developing PC [86].

Various reports show conflicting data as to whether the age of onset of PC changes in the presence of FAMMM. Lynch et al. reported that while some families developed PC at a younger age (35, 45, 46, and 49 years), other FAMMM families reported PC at a later age [87]. Goldstein et al. also found that the median age of PC onset did not differ in families with FAMMM (71 years). Interestingly, that study also noted that FAMMM cases without PC reported a higher nevi count [88]. In contrast, Bartsch et al. found that PC had an earlier onset (median age 45 compared to median age of 55 in the general population) [85].

Familial atypical multiple melanoma-PC syndrome is slightly different from FAMMM. Patients with FAMMM-PC have a chance of developing either PC or melanoma, and the dominant disease varies from case to case [87]. Moskaluk et al. identified a case in which a family with a CDNK2A mutation developed PC but not the FAMMM phenotype [89]. Harinck et al. found a *CDKN2A* mutation in 6/28 (21%) of familial PC cases and found that three of those families had no evidence of cutaneous melanoma [90].

The development of PC seems to be caused by a sequence of mutations that advance "normal ductal epithelium to ductal hyperplasia to invasive ductal carcinoma" [91]. According to Parker et al., "the progression from level 1 to level 3 disease is often associated with a sequence of genetic alterations, first in K-ras and HER-2/meu, followed by p16 altercations, and finally abnormalities in p53, DPC4, BRCA2, and other tumor suppressor genes" [91,92].

A patient's susceptibility for PC is also greatly influenced by additional personal and environmental factors. For example, there is a significant association between obesity and physical exercise with PC [93]. Furthermore, smoking is also an avoidable risk factor. A study by Lowenfels et al. showed that smoking increased the risk of PC 154-fold in patients with hereditary pancreatitis, and PC diagnosis was on average 20 years earlier [94]. Rulyak et al. also found that smokers in PC-prone families had the high risk of developing the disease (OR 7.6) [95].

39.4.3.2 FAMMM and Other Cancers

There has been conflicting research regarding the association of FAMMM with other types of tumors. Greene and Bergman found that no other cancers associated with FAMMM [96,97]. However, in 1981, Lynch et al. observed increased carcinomas of the lung, skin, larynx, and breast in FAMMM patients [98]. In 1983, Lynch and Fusaro claimed to observe a fivefold increase in all types of carcinomas in FAMMM families [15]. Another study by Lynch et al. in 2002 noted breast cancer and esophageal carcinoma in two FAMMM patients [87]. Additional literature analysis suggests that three other cancer types may be associated with FAMMM: head/neck squamous carcinoma (HNSCC), breast cancer, and lung cancer.

Since HNSCC is also significantly associated with a *CDKN2A* mutation, it is not surprising that multiple FAMMM case reports have noted cases of HNSCC [99]. Vinarsky et al. noted a FAMMM patient who suffered from multiple dysplastic nevi, multiple cases of melanoma by age 40, PC, and a case of squamous cell carcinoma (SCC) of the tongue at age 22 [100]. Tissue from the SCC did reveal a *CDKN2A* mutation in the tumor cells. Yu et al. studied a family with a high incidence of melanoma and

SCC and found a *CDKN2A* R87P mutation in family members [101]. In a study of 8 FAMMM families with a *CDKN2A* mutation, Oldenburg also reported a family with two cases of oral SCC [102]. Schneider-Stock et al. described a 53-year-old male nonsmoker who developed three pharynx and oral cavity tumors, all positive for the *CDKN2A* mutation [103]. Whelan and Yarborough described similar results in other FAMMM families [82,104].

Borg et al. and Ghiorizo et al. correlated FAMMM with breast cancer. Borg et al. found that six families with a 113ins-Arg *CDKN2A* mutation not only had a high incidence of PC and melanoma but also breast cancer (8 observed cased of breast cancer compared to the expected 2.1; $P = 0.0014$) [105]. Ghiorzio et al. studied 14 melanoma-prone families from Liguria, Italy and observed three cases of breast cancer (SRR 2.04%, 95% CI = 0.42–6.00) in 3 out of 7 Gly101Trp families [106]. Of note, however, the relative risk of developing breast cancer in this study was higher in melanoma-prone families without the *CDKN2A* mutation.

Along with numerous published case reports of lung cancer in FAMMM patients, Vasen et al. showed a significant correlation of FAMMM with lung cancer. Vasen et al. studied 19 families with a *CDKN2A* mutation and noted 86 cases of melanoma, 15 cases of PC, and 11 cases of lung cancer. They calculated a 14.3% (95% CI = 1.4–27.2) risk of developing lung cancer in *CDKN2A*-positive families [22].

39.5 Clinical Features

A typical patient with the FAMMM syndrome has a family or personal history of melanoma and a large number of typical and atypical nevi on their body [1]. The characteristics of the nevi are quite diverse. The total number of nevi often ranges from 50–100, with lesions of varying shapes and sizes [107]. Patients are documented to have numerous atypical nevi with sizes greater than 5 mm, with colors varying from black to brown, tan, or red. The border can be irregular with color leakage or sharp round/oval with no color leakage. The contour can be macular or centrally raised. Lynch et al. observed nevi develop mostly on the trunk and proximal limbs, but also observed cases on the scalp. Children with FAMMM did develop higher numbers of nevi at an earlier age than controls, but the nevi were phenotypically indistinguishable from benign controls. Dysplastic nevi were often first recognized during puberty, with additional dysplastic nevi appearing until the fifth decade of life. During the seventh and eighth decades the nevi started to regress [107].

Histologically, Lynch and Fusaro defined an atypical nevus by the following characteristics: architectural atypia, cellular atypia, stromal fibroplasia with angiogenesis, and a chronic lymphocytic infiltrate [107]. However, the histologic appearance of the atypical nevi is not important, for multiple reasons. First, there is much debate over the definition of an "atypical" nevi. Piepkorn et al. graded atypical nevi with more scrutiny. By their definition, an early stage of an atypical nevi is characterized by "perivascular lymphohistiocytic infiltrates, condensation of hyalinized collagen, and pigmentary incontinence." Later-stage atypical nevi had hyperpigmented, elongated or bulbous rete ridges with melanocytic atypia and hyperplasia [108]. Steijlen et al. found the "presence of dust like melanin, irregular nevoid nests, markedly increased junctional activity and melanocytic nuclei equal or larger in size than overlying keratinocyte nuclei" to be the defining features [109]. Stejlen et al. and Ackerman both find the definition of "atypical" to be inadequate, with low sensitivity, specificity, and predictive value [109,110]. Second, a great deal of research has demonstrated that histologic features are not unique to the FAMMM syndrome. In a study of 521 individuals from Utah, Piepkorn et al. found that 53% of Caucasians had histological dysplasia in at least one of two nevi examined by biopsy [108].

Patients with FAMMM get melanomas at a much younger age than the general population. In a study of 182 FAMMM patients compared to 7512 control patients, van der Rhee et al. found that FAMMM patients had their first melanoma on average 15.3 years before the normal population [111]. Furthermore, patients with FAMMM had a 21.1% higher 5-year cumulative incidence for a second melanoma. The risk of a second melanoma was doubled in patients who had their first melanoma when they were less than 40 years old. Van der Rhee also found that the majority of younger patients developed melanomas on the head and neck rather than on the trunk and upper extremities [111]. In a study of 2608 cutaneous melanomas, Gillgren et al. confirmed that a majority of hereditary melanomas occurred on the head and neck [112]. Interestingly, despite the fact that atypical nevi have a higher chance of developing into a melanoma, a majority of FAMMM melanomas developed from normal skin, in between dysplastic nevi [1].

In 2002, Masback et al. examined the histological traits of CMMs in 26 FAMMM individuals compared to 78 matched controls [113]. Specifically, they studied inflammation, regression, reaction, ulceration, Clark invasion level, Breslow thickness, and the presence of benign dermal nevus cells. They found the tumors to have more ulceration but less inflammation and regression. While they did find that the FAMMM kindreds had a lower average Clark invasion level (Level II vs III/IV), they did not find difference in Breslow thickness to be statistically significant [113]. Kopf et al., Grange et al., and Aitken et al. also found similar results [114–116]. However, Barnhill et al., Duke et al., and Greene et al., found that familial melanoma cases had a smaller Breslow thickness than the controls [117–119].

Current work shows a higher prevalence of superficial spreading melanoma in FAMMM patients. Ford et al.'s combined analysis of 2952 melanoma patients and 3618 controls indicates more superficial spreading and lentigo maligna melanoma than acral lentiginous or nodular melanoma in FAMMM kindreds [120]. Sargen et al. compared patients with sporadic melanomas (N = 81), and melanoma families with and without the *CDKN2A* mutation (N = 123/N = 120), and also found a higher prevalence of superficial spreading melanoma in the *CDKN2A*-positive individuals [121]. Sargen et al. defined the superficial spreading melanoma with increased pagetoid scatter, higher pigmentation, and non-spindle-like morphology in vertical growth phase. They also found that these patients often had a lower VGP cytologic grade and non-mitogenic VGP. Van der Rhee also observed more superficial spreading melanoma in his study [111]. Overall, however, no histopathologic differences are significant enough to establish a diagnosis [24].

Furthermore, as explained in Section 39.4, FAMMM kindreds also often have a personal or family history of other systemic cancers, especially PC and neural tumors.

39.6 Diagnosis

39.6.1 FAMMM Clinical Diagnosis

FAMMM is a clinical diagnosis and is based upon clinical evaluation and family history. Although there is no single consensus definition, one accepted criterion includes [24]:

1. Cutaneous melanoma in ≥ 1 first- or second-degree relatives
2. High total body nevi count (>50), with multiple atypical nevi (variations in ABCDE morphology)
3. Characteristic histologic features: asymmetry, subepidermal fibroplasia, and lentiginous melanocytic hyperplasia with spindle or epithelioid melanocytes gathering in nests of variable size and fusing with adjacent rete ridges to form bridges; variable dermal lymphocyte infiltration, and the "shouldering phenomenon wherein intraepidermal melanocytes extend alone or in groups beyond the main dermal component"

Czajkowski et al. emphasized that the necessitated histologic features are used to "exclude melanoma and not to confirm the diagnosis of FAMMM syndrome" [93]. The current gold standard clinicians use to evaluate pigmented moles is the "ABCD(E)" rule (Asymmetric shape, Border irregularity, Color variability, Diameter greater than 6 mm, and Elevation or Evolution) [93].

39.6.2 Melanoma Diagnosis

As described in Section 39.5, pathologic/histologic features of melanomas in FAMMM kindreds are not specific enough to warrant a diagnosis. Melanomas are evaluated using the same techniques used to diagnose sporadic melanomas, using the ABCD rule (asymmetry, border irregularity, color variations, and different structural components on dermoscopy such as branch streaks, dots, pigment network, globules, and homogenous areas) [93,122]. Technologies significantly aid diagnosis. As outlined in an education review article by Mayer et al., techniques like dermatoscopy, total body photography, sequential digital dermatoscopy, and confocal microscopy improve detection time, sensitivity, and specificity of diagnosis [123].

Dermatoscopy has greatly increased the sensitivity of screening and has significantly decreased the amount of biopsied benign lesions [123]. Total body photography (TBP) is increasing in popularity as it tracks nevus morphology over time, but more research needs to be done to ensure its clinical benefit. Goodson et al. reported that TBP was associated with lower biopsy rates than dermatoscopy, and Rademaker et al. found that TBP allowed the detection of thinner lesions than would otherwise be found [124,125]. Salerni et al. found that TBP combined with digital dermoscopy identified 40% of melanomas not under initial surveillance over a 10-year period [126]. Further studies also showed earlier diagnosis through TBP, but Risser et al. reported that TBP did not reduce the total number of biopsies done on patients [127,128]. Confocal microscopy creates 3-D images on the skin using a low power laser. While it improves estimates of sensitivity and specificity to 90% and 86% respectively, it is extremely expensive, and more studies need to be done to evaluate its efficacy [123].

According to official guidelines from the National Comprehensive Cancer Network and the American Academy of Dermatology, lesions which are suspect for melanoma should fully excised for complete histological evaluation [129]. In selected cases where an excision may be difficult due to the size of the lesion, an incisional biopsy may be appropriate. Histopathology is then needed to make the final definitive diagnosis of melanoma.

39.6.3 Genetic Testing

Leachman et al. outline specific selection criteria for genetically assessing individuals for FAMMM. In a low melanoma-incidence population, one should only be screened if they have had ≥2 primary melanomas or come from a family with at least invasive melanoma and 1+ diagnosis of melanoma/PC in a first- or second-degree relative from the same side of the family. Individuals from moderate/high melanoma incidence areas (like Australia and the United States) should follow slightly different criteria. They should only be genetically screened if they have had ≥3 primary melanomas or come from a family with at least invasive melanoma and 2+ diagnosis of melanoma/PC in same-side

first- or second-degree relatives ("rule of 3's") [130]. Soura et al. outlines caveats with this diagnosis, claiming that many individuals might develop more than three melanomas simply because they have accumulated a lot of photo-damage over the years. Thus, they recommend the criteria be amended to only being screened if one had greater than three melanomas by age 40 [1]. Soura et al. also describes the benefit of a 3+ generation pedigree analysis, which analyzes the incidence of melanoma compared to family size. The presence of PC also is a strong determinant of genetic risk, since the main similarity between melanoma and PC is *CDKN2A* mutation [1]. Genetic testing would be based mostly on the mutation status of *CDKN2A*, but since *CDK4* is implicated in <1% of patients, that locus would be tested as well. One important benefit for *CDKN2A* screening is the identification of high-risk candidates who might develop FAMMM in the future. Furthermore, since FAMMM is passed in an autosomal dominant manner, there is a 50% chance that family members will share the same mutation, and thus genetic counseling also sheds light on family member risk [1]. The genetic counseling algorithm suggested by Soura et al. is shown in Figure 39.3 [1].

However, there is controversy regarding the utility of the genetic result. As described earlier, only 39% of FAMMM patients harbor the *CDKN2A* mutation. Furthermore, *CDKN2A* mutations have variable penetrance and expressivity, further

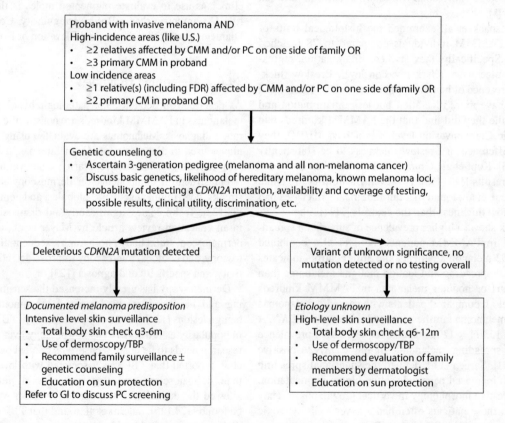

FIGURE 39.3 Genetic counseling algorithm for FAMMM patients. Patients with a personal history of melanoma may be considered for genetic counseling if certain criteria from high- and low-incidence area are met. The United States and Australia would be considered high-incidence areas, while England and Greece would be considered low-incidence areas. A genetic risk counselor would ascertain a three-generation pedigree and discuss the likelihood of hereditary melanoma, the molecular genetics related to familial melanoma risk, testing options, costs, risks of discrimination, and possible test results. If patient undergoes *CDKN2A* testing and a deleterious mutation is detected, intensive skin surveillance is recommended along with a referral to GI for discussion of pancreatic cancer screening. If testing is not pursued or if a normal or variant of unknown significance result is returned, the etiology of the familial pattern remains unknown. Given the family history, the patient is considered high risk and should undergo high-level skin surveillance. *Abbreviations:* PC, pancreatic cancer; TBP, total body photography, CM, cutaneous melanoma.

reducing the clinical utility of mutation status. A positive result would not necessarily mean one would develop FAMMM, while a negative result also does guarantee that one will not develop the syndrome. Multiple other genetic loci and environmental characteristics are risk factors for developing FAMMM, and many FAMMM patients will have a negative *CDKN2A* mutation status. Thus, a particular mutational status cannot be used diagnostically, and should not significantly alter the management plan of patients.

39.7 Screening and Treatment

39.7.1 Screening for *CDKN2A* Carrier Probability

Current estimates of hereditary melanoma risk rely on tools that estimate the probability of carrying a *CDKN2A* mutation. Currently two models exist: MELPREDICT and MelaPRO. MELPREDICT was created by Niendorf et al. in 2005 and is based on data from a cohort of 116 familial melanoma patients from Massachusetts General Hospital [131]. From this cohort, Niendorf et al. created a logistic regression model in which the independent variables included the number of proband primary melanomas, the amount of family primary melanomas, and the age of the proband [131]. MelaPRO was created in 2010 by Wang et al. and combines Mendelian inheritance and Bayesian probability theories [132]. MelaPRO is superior to MELPREDICT in multiple ways. Unlike MELPREDICT, it can be applied to patients who have not yet had a primary melanoma. It also takes into account geographic location, mutation penetrance, and full pedigree information [132]. When using MelaPRO, an individual inputs their detailed personal and family history, and based on its algorithm, MelaPRO discloses the carrier probability and the chance the individual will develop melanoma if they have not yet had any previous cancers [23]. Wang et al. found that MelaPRO surpassed MELPREDICT (C,0.82; 95% CI = 0.61–0.93), with a difference of 0.05 (95% CI = 0.007–0.17) [132]. Two pitfalls of MelaPRO include not including PC as a risk factor, and not taking into account other hereditary syndromes with a melanoma component [23].

39.7.2 Screening and Treatment for Melanoma

Due to their increased chance of developing melanoma, FAMMM patients should receive regular dermatologic surveillance. Since evidence shows that patients with FAMMM develop melanomas when they are much younger, skin surveillance should begin during late adolescence [1]. A health care provider should examine patients every 3–6 months. Most clinicians recommend that a 6-year screening interval is adequate, but Haenssle et al. recommends a 3-month interval, since patients can develop one new mole every 3 months [1,133]. Heightened screening should take place during puberty and pregnancy because moles become more unstable during those times [1]. Surveillance of the entire body is necessary, including the nails, genital area, oral mucosa, and scalp [1]. Screening and surveillance should use the tools outlined in Section 39.6 of this chapter. Once the diagnosis of melanoma has been made, surgical removal is determined by the disease stage and the depth of neoplastic infiltration. The majority of

FAMMM patients have macular melanocytic lesions which can be removed using deep razor blade excision [93]. As described by Gambichler et al., this method is not only efficacious, easy, and inexpensive, but also leads to a better cosmetic outcome, which is critical in patients who will probably require multiple excisions [134].

Patients also need to be educated about regular self-surveillance. Patients or their loved one should perform a monthly self-skin examination [24,93]. Patients and their family members should be counseled on the characteristics of possible melanocytic lesions, such as the ABCD rules and the ugly duck sign. Patients also need to be educated about the importance of skin and sun care. This includes avoiding as much solar radiation as possible (between 10 am−4 pm), including wearing sun protective clothing and glasses [93]. Furthermore, patients should always be applying UVA and UVB SPF15+ sunscreen. Adequate application of sunscreen is critical, as inadequate application leads to a false sense of security that might increase an individual's risk of sun damage [135]. Brian Diffey advises that sunscreen should be applied 15–30 minutes before going into the sun, and should be reapplied 15–30 minutes after initial exposure. Additional application is needed if patients participate in activities that could remove the sunscreen such as "swimming, toweling, or excessive sweating and rubbing" [136]. Jeffrey Schneider advocates for the using 2 mg/cm of body surface and the "teaspoon rule": more than half a teaspoon on the right arm, left arm, and head/neck, and a full teaspoon on the right leg, left leg, anterior torso, and posterior torso [93,137].

39.7.3 Screening and Treatment for Pancreatic Cancer

Due to the overall low incidence in the general population, it is unreasonable to screen everyone for PC [91]. There are multiple biological limitations to successful detection of a tumor at an early stage, including metastasis before a majority of symptoms develop, such as epigastric pain, obstructive jaundice, and weight loss, and difficulty imagining the pancreas due to its deep location in the abdominal cavity [91]. There is currently no gold standard method for detecting early signs of PC, and it is unclear whether current methods of screening in high-risk individuals actually increase survival.

In 2011, a 49-expert multidisciplinary panel met to establish guidelines for the screening of PC [138]. The panel decided that only high-risk individuals who are candidates for pancreatectomy should be screened, including *CDKN2A* mutation carriers with one or more first-degree relatives with PC. The consortium was unable to come to an agreement regarding minimum age of screening. Fifty-one percent of the consortium recommended screening starting at age 50, while others recommend screening 10 years earlier than the youngest family member who developed PC. Earlier screening for smokers with a family history of PC was considered, since smoking is a significant risk factor [138].

Screening targets include early invasive tumors and pre-invasive lesions, including intraductal papillary mucinous neoplasms (IPMNS) and pancreatic intraepithelial neoplasia (PanIN). IPMNS are usually greater than 1 cm and are derived from columnar cells. PanINs are derived from pancreatic ducts and are usually less than 5 mm in size [139–142].

The consortium decided that initial screening should include endoscopic ultrasound (EUS), MRI/magnetic resonance cholangiopancreatography (MRCP), but not CT or endoscopic retrograde cholangiopancreatogram (ERCP) [138]. EUS is one of the most sensitive and specific methods for identifying pancreatic lesions, and MRI/MRCP has the advantage of detecting extra-pancreatic lesions as well. A 2016 study by Harinck et al. showed that EUS and MRI screening is complementary rather than interchangeable [90]. The researchers concluded that EUS was sensitive for detecting small solid lesions, while MRI was beneficial to detect small cystic lesions [90]. A study in 2012 by Canto et al. demonstrated that EUS and MRI were far superior to CT in PC screening [143]. While CT was only able to identify abnormalities in 11% of patients, MRI and EUS were able to detect abnormalities in 33% and 42.6%, respectively [143]. Canto et al. found the when ERCP was performed as follow-up for EUS results, it did not provide significant benefit and was associated with numerous complications including a 7% pancreatitis rate [144]. No consensus was reached regarding the use of EUS with fine needle aspiration.

Follow-up surveillance for patients with detected lesions depends on characteristics of the lesions, such as type (solid vs cystic) and size. Individual counseling is recommended for further management and treatment. A pancreatectomy is not recommended for asymptomatic individuals due to its significant risks of mortality and morbidity, including lifelong diabetes and exocrine insufficiency. The consortium did not reach a consensus over the particular characteristics of pancreatic lesions that necessitated a pancreatectomy [91,138].

Currently a great deal of research is being done on the identification of potential markers to aid the screening process. One hot topic of interest is sialyated Lewis[a] antigen 19-9 (CA 19-9) [91]. In 2011 Zubarik et al. studied CA 19-9 in 546 high risk individuals with PC [145]. While CA 19-9 was elevated in 27 patients, only 5 patients were found to have neoplastic signs on EUS follow-up. Furthermore, CA 19-9 has low specificity because it is also associated with other abdominal malignancies, pancreatitis, and hepatitis [91]. Kim et al. also argued against the use of CA 19-9 as a screening marker since they found the positive predictive value to be less than 1% [146]. Liu et al. found that the combination of microRNA and CA 19-9 expression is more efficacious for the detection of PC, but more follow-up studies need to be completed [147]. Pancreatic fluid from ERCP and cyst fluid with EUS fine needle aspiration can also be analyzed for molecular markers. Some evidence suggests that mutant TP53, GNAS, and KRAS are associated with PC [141,148,149]. Additional work has also been done on the use CpG island methylation and mutation load distribution analysis for future screening methods [150,151].

39.8 Prognosis

Unfortunately, there are few studies on the prognosis of melanoma patients in the FAMMM context. To our knowledge, the latest study published on FAMMM prognosis was in 1998 by Hille et al. [152]. In that study, researchers examined morality in six Dutch pedigrees with proven *CDKN2A* mutations and FAMMM syndrome from 1830 to 1994. They found that Standard Mortality Ratio for FAMMM patients to be 1.6 (95%

CI = 1-2-2.9). The excess mortality was attributed specifically to cancer, with a majority of deaths from malignant melanoma and PC. Of the 44 cancer deaths, 10 were due to melanoma and 12 were due to PC [152].

Given that a FAMMM patient generally develops primary cutaneous melanoma, outcome is still dependent on the usual parameters defined for sporadic melanomas, e.g. tumor thickness, ulceration, mitotic rate (mitoses/mm^2) and lymphovascular invasion [153]. The T stage classification is defined by the presence of ulceration and mitoses $<1/mm^2$. The N category depends on the number of metastatic lymph nodes, and the M category depends on the number of distant metastases. Stage groupings based on T, N, and M criteria is outlined in the article by Balch et al. [153]. Overall, Stage I/II melanoma is localized, Stage III has region metastasis, and Stage IV has distant metastasis. The 10-year survival rate from Stage I/II melanomas ranges from 93% (Stage IA) to 39% (Stage IIC). The 5-year survival rate for Stage III melanomas ranges from 70% to 39%. The 1-year survival rate for Stage IV melanoma ranges from 62% to 33% [153], although the survival rates for metastatic melanomas are rapidly improving given all the new molecular and immune checkpoint inhibitors.

Development of PC has a very poor prognosis and is responsible for more than 200,000 deaths a year worldwide [154,155]. The overall 5-year survival rate is less than 5%, and mortality of this disease almost equates its incidence (approximately 97%). The poor prognosis is primarily due to the late diagnosis of the disease, beyond a point at which the tumor can be successfully resected [154,155]. In fact, approximately 50% of patients have metastatic disease by the time tumor is identified [154,155]. The lack of successful screening programs and late onset of symptoms are key contributors to the late diagnosis.

39.9 Conclusion

Familial atypical multiple mole melanoma (FAMMM) syndrome is an autosomal dominant phenotype that typically shows multiple atypical nevi (usually >50 and having specific histologic features such as asymmetry, subepidermal fibroplasia, lentiginous melanocytic hyperplasia with spindle or epithelioid melanocytes, variable dermal lymphocyte infiltration, and presence of "shouldering" phenomenon) and a dramatically increased risk of melanoma and PC. At the genetic level, FAMMM syndrome is frequently linked to germline mutations (frequently missense or nonsense mutations) in the *CDKN2A* gene located on chromosome 9p21, which impair the inhibitory functions of p16 and/or p14ARF, leading to enhanced cellular proliferation and reduced apoptosis. Indeed, patients with a germline *CDKN2A* mutation tend to develop cutaneous malignant melanoma earlier (33–45 years) than those without a *CDKN2A* mutation (53–61 years). Another gene implicated in FAMMM syndrome is *CDK4*, which is the target for p16 inhibition and plays an important role in normal cell cycle progression. Germline *CDK4* mutations are responsible for a phenotype that overlaps significantly with FAMMM syndrome, i.e., high burden of atypical nevi, early age of disease onset, and predilection for multiple primary melanoma. Therefore, apart from reviewing personal or family history of melanoma or other internal organ cancer, molecular identification of *CDKN2A* and *CDK4* mutations helps achieve an unequivocal diagnosis of FAMMM syndrome in patients with multiple atypical nevi. Furthermore, genetic counseling with the possibility of testing can be beneficial to familial melanoma patients in their cancer risk analysis and prevention.

Acknowledgment

This work was made possible in part by a grant from the National Institutes of Health (to H.T., K24-CA149202).

REFERENCES

1. Soura E, Eliades PJ, Shannon K, Stratigos AJ, Tsao H. Hereditary melanoma: Update on syndromes and management: Genetics of familial atypical multiple mole melanoma syndrome. *J Am Acad Dermatol.* 2016;74:395–407; quiz 8–10.

2. Cawley EP, Kruse WT, Pinkus HK. Genetic aspects of malignant melanoma. *AMA Arch Derm Syphilol* 1952;65:440–50.

3. Katzenellenbogen I, Sandbank M. Malignant melanomas in twins. *Arch Dermatol.* 1966;94:331–2.

4. Miller TR, Pack GT. The familial aspect of malignant melanoma. *Arch Dermatol.* 1962;86:35–9.

5. Moschella SL. A report of malignant melanoma of the skin in sisters. *Arch Dermatol.* 1961;84:1024–5.

6. Salamon T, Schnyder UW, Storck H. A contribution to the question of heredity of malignant melanomas. *Dermatologica* 1963;126:65–75.

7. Turkington RW. Familial factor in malignant melanoma. *JAMA.* 1965;192:77–82.

8. Smith FE, Henly WS, Knox JM, Lane M. Familial melanoma. *Arch Intern Med.* 1966;117:820–3.

9. Schoch EP Jr., Familial malignant melanoma. A pedigree and cytogenetic study. *Arch Dermatol.* 1963;88:445–56.

10. Munro DD. Multiple active junctional naevi with family history of malignant melanoma. *Proc R Soc Med* 1974;67:594–5.

11. Anderson DE, Smith JL, Jr., McBride CM. Hereditary aspects of malignant melanoma. *JAMA.* 1967;200:741–6.

12. Lynch HT, Anderson DE, Krush AJ. Heredity and intraocular malignant melanoma. Study of two families and review of forty-five cases. *Cancer.* 1968;21:119–25.

13. Lynch HT, Frichot BC, 3rd, Lynch JF. Familial atypical multiple mole-melanoma syndrome. *J Med Genet.* 1978;15:352–6.

14. Clark WH, Jr., Reimer RR, Greene M, Ainsworth AM, Mastrangelo MJ. Origin of familial malignant melanomas from heritable melanocytic lesions. 'The B-K mole syndrome'. *Arch Dermatol.* 1978;114:732–8.

15. Lynch HT, Fusaro RM, Kimberling WJ, Lynch JF, Danes BS. Familial atypical multiple mole-melanoma (FAMMM) syndrome: Segregation analysis. *J Med Genet.* 1983;20:342–4.

16. Lynch HT, Fusaro RM, Albano WA, Pester J, Kimberling WJ, Lynch JF. Phenotypic variation in the familial atypical multiple mole-melanoma syndrome (FAMMM). *J Med Genet.* 1983;20:25–9.

17. Lynch HT, Fusaro RM. Pancreatic cancer and the familial atypical multiple mole melanoma (FAMMM) syndrome. *Pancreas.* 1991;6:127–31.

18. Hussussian CJ, Struewing JP, Goldstein AM et al. Germline p16 mutations in familial melanoma. *Nat Genet.* 1994;8:15–21.

19. Gruis NA, Sandkuijl LA, van der Velden PA, Bergman W, Frants RR. CDKN2 explains part of the clinical phenotype in Dutch familial atypical multiple-mole melanoma (FAMMM) syndrome families. *Melanoma Res.* 1995;5:169–77.

20. Kamb A, Shattuck-Eidens D, Eeles R et al. Analysis of the p16 gene (CDKN2) as a candidate for the chromosome 9p melanoma susceptibility locus. *Nat Genet.* 1994;8:23–6.

21. Caldas C, Hahn SA, da Costa LT et al. Frequent somatic mutations and homozygous deletions of the p16 (MTS1) gene in pancreatic adenocarcinoma. *Nat Genet.* 1994;8:27–32.

22. Vasen HF, Gruis NA, Frants RR, van Der Velden PA, Hille ET, Bergman W. Risk of developing pancreatic cancer in families with familial atypical multiple mole melanoma associated with a specific 19 deletion of p16 (p16-Leiden). *Int J Cancer.* 2000;87:809–11.

23. Marzuka-Alcala A, Gabree MJ, Tsao H. Melanoma susceptibility genes and risk assessment. *Methods Mol Biol.* 2014;1102:381–93.

24. Eckerle Mize D, Bishop M, Resse E, Sluzevich J. Familial atypical multiple mole melanoma syndrome. In: Riegert-Johnson DL, Boardman LA, Hefferon T, Roberts M, eds. *Cancer Syndromes.* Bethesda (MD): 2009.

25. Kamb A, Gruis NA, Weaver-Feldhaus J et al. A cell cycle regulator potentially involved in genesis of many tumor types. *Science.* 1994;264:436–40.

26. Koh J, Enders GH, Dynlacht BD, Harlow E. Tumour-derived p16 alleles encoding proteins defective in cell-cycle inhibition. *Nature.* 1995;375:506–10.

27. Hewitt C, Lee Wu C, Evans G et al. Germline mutation of ARF in a melanoma kindred. *Hum Mol Genet.* 2002;11:1273–9.

28. Goldstein AM, Chan M, Harland M et al. High-risk melanoma susceptibility genes and pancreatic cancer, neural system tumors, and uveal melanoma across GenoMEL. *Cancer Res.* 2006;66:9818–28.

29. Tsao H, Chin L, Garraway LA, Fisher DE. Melanoma: From mutations to medicine. *Genes Dev.* 2012;26:1131–55.

30. Arap W, Knudsen E, Sewell DA et al. Functional analysis of wild-type and malignant glioma derived CDKN2Abeta alleles: Evidence for an RB-independent growth suppressive pathway. *Oncogene.* 1997;15:2013–20.

31. Harland M, Taylor CF, Chambers PA et al. A mutation hotspot at the p14ARF splice site. *Oncogene.* 2005;24:4604–8.

32. Randerson-Moor JA, Harland M, Williams S et al. A germline deletion of p14(ARF) but not CDKN2A in a melanoma-neural system tumour syndrome family. *Hum Mol Genet.* 2001;10:55–62.

33. Kaufman DK, Kimmel DW, Parisi JE, Michels VV. A familial syndrome with cutaneous malignant melanoma and cerebral astrocytoma. *Neurology.* 1993;43:1728–31.

34. Azizi E, Friedman J, Pavlotsky F et al. Familial cutaneous malignant melanoma and tumors of the nervous system. A hereditary cancer syndrome. *Cancer.* 1995;76:1571–8.

35. Pasmant E, Laurendeau I, Heron D, Vidaud M, Vidaud D, Bieche I. Characterization of a germ-line deletion, including the entire INK4/ARF locus, in a melanoma-neural system tumor family: Identification of ANRIL, an antisense noncoding RNA whose expression coclusters with ARF. *Cancer Res.* 2007;67:3963–9.

36. Frigerio S, Disciglio V, Manoukian S et al. A large *de novo* 9p21.3 deletion in a girl affected by astrocytoma and multiple melanoma. *BMC Med Genet.* 2014;15:59.

37. Serrano M, Hannon GJ, Beach D. A new regulatory motif in cell-cycle control causing specific inhibition of cyclin D/CDK4. *Nature.* 1993;366:704–7.

38. Chin L, Pomerantz J, Polsky D et al. Cooperative effects of INK4a and ras in melanoma susceptibility *in vivo. Genes Dev.* 1997;11:2822–34.

39. VanBrocklin MW, Robinson JP, Lastwika KJ, Khoury JD, Holmen SL. Targeted delivery of NRASQ61R and Cre-recombinase to post-natal melanocytes induces melanoma in Ink4a/Arflox/lox mice. *Pigment Cell Melanoma Res* 2010; 23:531–41.

40. Yang G, Curley D, Bosenberg MW, Tsao H. Loss of xeroderma pigmentosum C (Xpc) enhances melanoma photocarcinogenesis in Ink4a-Arf-deficient mice. *Cancer Res.* 2007; 67:5649–57.

41. Zuo L, Weger J, Yang Q et al. Germline mutations in the p16INK4a binding domain of CDK4 in familial melanoma. *Nat Genet.* 1996;12:97–9.

42. Soufir N, Avril MF, Chompret A et al. Prevalence of p16 and CDK4 germline mutations in 48 melanoma-prone families in France. The French Familial Melanoma Study Group. *Hum Mol Genet.* 1998;7:209–16.

43. Tsao H, Benoit E, Sober AJ, Thiele C, Haluska FG. Novel mutations in the p16/CDKN2A binding region of the cyclin-dependent kinase-4 gene. *Cancer Res.* 1998;58:109–13.

44. Molven A, Grimstvedt MB, Steine SJ et al. A large Norwegian family with inherited malignant melanoma, multiple atypical nevi, and CDK4 mutation. *Genes Chromosomes Cancer.* 2005;44:10–8.

45. Puntervoll HE, Yang XR, Vetti HH et al. Melanoma prone families with CDK4 germline mutation: Phenotypic profile and associations with MC1R variants. *J Med Genet.* 2013;50:264–70.

46. Goldstein AM, Chan M, Harland M et al. Features associated with germline CDKN2A mutations: A GenoMEL study of melanoma-prone families from three continents. *J Med Genet.* 2007;44:99–106.

47. Nikolaou V, Kang X, Stratigos A et al. Comprehensive mutational analysis of CDKN2A and CDK4 in Greek patients with cutaneous melanoma. *Br J Dermatol.* 2011;165:1219–22.

48. Bishop DT, Demenais F, Goldstein AM et al. Geographical variation in the penetrance of CDKN2A mutations for melanoma. *J Natl Cancer Inst.* 2002;94:894–903.

49. Begg CB, Orlow I, Hummer AJ et al. Lifetime risk of melanoma in CDKN2A mutation carriers in a population-based sample. *J Natl Cancer Inst.* 2005;97:1507–15.

50. Cust AE, Harland M, Makalic E et al. Melanoma risk for CDKN2A mutation carriers who are relatives of population-based case carriers in Australia and the UK. *J Med Genet.* 2011;48:266–72.

51. Healy E, Jordan SA, Budd PS, Suffolk R, Rees JL, Jackson IJ. Functional variation of MC1R alleles from red-haired individuals. *Hum Mol Genet.* 2001;10:2397–402.

52. Funari G, Menin C, Elefanti L, D'Andrea E, Scaini MC. Familial melanoma in Italy: A review. In: Armstrong A, ed. *Advances in Malignant Melanoma - Clinical and Research Perspectives.* InTech; 2011.

53. Newton RA, Roberts DW, Leonard JH, Sturm RA. Human melanocytes expressing MC1R variant alleles show impaired activation of multiple signaling pathways. *Peptides.* 2007;28:2387–96.

54. Kennedy C, ter Huurne J, Berkhout M et al. Melanocortin 1 receptor (MC1R) gene variants are associated with an increased risk for cutaneous melanoma which is largely independent of skin type and hair color. *J Invest Dermatol.* 2001;117:294–300.

55. Raimondi S, Sera F, Gandini S et al. MC1R variants, melanoma and red hair color phenotype: A meta-analysis. *Int J Cancer.* 2008;122:2753–60.

56. Mitra D, Xi L, Haber DA, Fisher DE. Why redheads are at increased risk of melanoma: A novel BRAF mutant mouse model. *Cancer Res.* 2011;71:PR6-PR.

57. Debniak T, Scott R, Masojc B et al. MC1R common variants, CDKN2A and their association with melanoma and breast cancer risk. *Int J Cancer.* 2006;119:2597–602.

58. van der Velden PA, Sandkuijl LA, Bergman W et al. Melanocortin-1 receptor variant R151C modifies melanoma risk in Dutch families with melanoma. *Am J Hum Genet.* 2001;69:774–9.

59. Box NF, Duffy DL, Chen W et al. MC1R genotype modifies risk of melanoma in families segregating CDKN2A mutations. *Am J Hum Genet.* 2001;69:765–73.

60. Debniak T. Some molecular and clinical aspects of genetic predisposition to malignant melanoma and tumours of various site of origin. *Hered Cancer Clin Pract* 2007;5:97–116.

61. Fargnoli MC, Gandini S, Peris K, Maisonneuve P, Raimondi S. MC1R variants increase melanoma risk in families with CDKN2A mutations: A meta-analysis. *Eur J Cancer.* 2010;46:1413–20.

62. Demenais F, Mohamdi H, Chaudru V et al. Association of MC1R variants and host phenotypes with melanoma risk in CDKN2A mutation carriers: A GenoMEL study. *J Natl Cancer Inst.* 2010;102:1568–83.

63. Chaudru V, Laud K, Avril MF et al. Melanocortin-1 receptor (MC1R) gene variants and dysplastic nevi modify penetrance of CDKN2A mutations in French melanoma-prone pedigrees. *Cancer Epidemiol Biomarkers Prev.* 2005;14:2384–90.

64. Palmer JS, Duffy DL, Box NF et al. Melanocortin-1 receptor polymorphisms and risk of melanoma: Is the association explained solely by pigmentation phenotype? *Am J Hum Genet.* 2000;66:176–86.

65. Garraway LA, Widlund HR, Rubin MA et al. Integrative genomic analyses identify MITF as a lineage survival oncogene amplified in malignant melanoma. *Nature.* 2005;436:117–22.

66. Yokoyama S, Woods SL, Boyle GM et al. A novel recurrent mutation in MITF predisposes to familial and sporadic melanoma. *Nature.* 2011;480:99–103.

67. Bertolotto C, Lesueur F, Giuliano S et al. A SUMOylation-defective MITF germline mutation predisposes to melanoma and renal carcinoma. *Nature.* 2011;480:94–8.

68. Potrony M, Puig-Butille JA, Aguilera P et al. Prevalence of MITF p.E318K in Patients with melanoma independent of the presence of CDKN2A causative mutations. *JAMA Dermatol* 2016;152:405–12.

69. Chaudru V, Lo MT, Lesueur F et al. Protective effect of copy number polymorphism of glutathione S-transferase T1 gene on melanoma risk in presence of CDKN2A mutations, MC1R variants and host-related phenotypes. *Fam Cancer.* 2009;8:371–7.

70. Reed DJ. Glutathione: Toxicological implications. *Annu Rev Pharmacol Toxicol.* 1990;30:603–31.

71. Mossner R, Anders N, Konig IR et al. Variations of the melanocortin-1 receptor and the glutathione-S transferase T1 and M1 genes in cutaneous malignant melanoma. *Arch Dermatol Res.* 2007;298:371–9.

72. Chatzinasiou F, Lill CM, Kypreou K et al. Comprehensive field synopsis and systematic meta-analyses of genetic association studies in cutaneous melanoma. *J Natl Cancer Inst.* 2011;103:1227–35.

73. Bishop DT, Demenais F, Iles MM et al. Genome-wide association study identifies three loci associated with melanoma risk. *Nat Genet.* 2009;41:920–5.

74. Amos CI, Wang LE, Lee JE et al. Genome-wide association study identifies novel loci predisposing to cutaneous melanoma. *Hum Mol Genet.* 2011;20:5012–23.

75. Macgregor S, Montgomery GW, Liu JZ et al. Genome-wide association study identifies a new melanoma susceptibility locus at 1q21.3. *Nat Genet.* 2011;43:1114–8.

76. Bekker-Jensen S, Rendtlew Danielsen J, Fugger K et al. HERC2 coordinates ubiquitin-dependent assembly of DNA repair factors on damaged chromosomes. *Nat Cell Biol.* 2010;12:80–6; sup pp 1–12.

77. Kang TH, Lindsey-Boltz LA, Reardon JT, Sancar A. Circadian control of XPA and excision repair of cisplatin-DNA damage by cryptochrome and HERC2 ubiquitin ligase. *Proc Natl Acad Sci U S A.* 2010;107:4890–5.

78. Falchi M, Bataille V, Hayward NK et al. Genome-wide association study identifies variants at 9p21 and 22q13 associated with development of cutaneous nevi. *Nat Genet.* 2009;41:915–9.

79. Behrmann I, Wallner S, Komyod W et al. Characterization of methylthioadenosin phosphorylase (MTAP) expression in malignant melanoma. *Am J Pathol.* 2003;163:683–90.

80. Yang XR, Pfeiffer RM, Wheeler W et al. Identification of modifier genes for cutaneous malignant melanoma in melanoma-prone families with and without CDKN2A mutations. *Int J Cancer.* 2009;125:2912–7.

81. Bergman W, Gruis NA, Sandkuijl LA, Frants RR. Genetics of seven Dutch familial atypical multiple mole-melanoma syndrome families: A review of linkage results including chromosomes 1 and 9. *J Invest Dermatol.* 1994;103:122S–5S.

82. Whelan AJ, Bartsch D, Goodfellow PJ. Brief report: A familial syndrome of pancreatic cancer and melanoma with a mutation in the CDKN2 tumor-suppressor gene. *N Engl J Med.* 1995;333:975–7.

83. Schutte M, Hruban RH, Geradts J et al. Abrogation of the Rb/p16 tumor-suppressive pathway in virtually all pancreatic carcinomas. *Cancer Res.* 1997;57:3126–30.

84. Goldstein AM, Fraser MC, Struewing JP et al. Increased risk of pancreatic cancer in melanoma-prone kindreds with p16INK4 mutations. *N Engl J Med.* 1995;333:970–4.

85. Bartsch DK, Sina-Frey M, Lang S et al. CDKN2A germline mutations in familial pancreatic cancer. *Ann Surg.* 2002;236:730–7.

86. Rulyak SJ, Brentnall TA, Lynch HT, Austin MA. Characterization of the neoplastic phenotype in the familial atypical multiple-mole melanoma-pancreatic carcinoma syndrome. *Cancer.* 2003;98:798–804.

87. Lynch HT, Brand RE, Hogg D et al. Phenotypic variation in eight extended CDKN2A germline mutation familial atypical multiple mole melanoma-pancreatic carcinoma-prone families: The familial atypical mole melanoma-pancreatic carcinoma syndrome. *Cancer.* 2002;94:84–96.

88. Goldstein AM, Struewing JP, Chidambaram A, Fraser MC, Tucker MA. Genotype-phenotype relationships in U.S. melanoma-prone families with CDKN2A and CDK4 mutations. *J Natl Cancer Inst.* 2000;92:1006–10.

89. Moskaluk CA, Hruban H, Lietman A et al. Novel germline p16(INK4) allele (Asp145Cys) in a family with multiple pancreatic carcinomas. Mutations in brief no. 148. Online. *Hum Mutat.* 1998;12:70.

90. Harinck F, Konings IC, Kluijt I et al. A multicentre comparative prospective blinded analysis of EUS and MRI for screening of pancreatic cancer in high-risk individuals. *Gut.* 2016;65:1505–13.

91. Lynch HT, Fusaro RM, Lynch JF, Brand R. Pancreatic cancer and the FAMMM syndrome. *Fam Cancer.* 2008;7:103–12.

92. Parker JF, Florell SR, Alexander A, DiSario JA, Shami PJ, Leachman SA. Pancreatic carcinoma surveillance in patients with familial melanoma. *Arch Dermatol.* 2003;139:1019–25.

93. Czajkowski R, Placek W, Drewa G, Czajkowska A, Uchanska G. FAMMM syndrome: Pathogenesis and management. *Dermatol Surg.* 2004;30:291–6.

94. Lowenfels AB, Maisonneuve P, Whitcomb DC, Lerch MM, DiMagno EP. Cigarette smoking as a risk factor for pancreatic cancer in patients with hereditary pancreatitis. *JAMA.* 2001;286:169–70.

95. Rulyak SJ, Lowenfels AB, Maisonneuve P, Brentnall TA. Risk factors for the development of pancreatic cancer in familial pancreatic cancer kindreds. *Gastroenterology.* 2003; 124:1292–9.

96. Greene MH, Tucker MA, Clark WH, Jr., Kraemer KH, Elder DE, Fraser MC. Hereditary melanoma and the dysplastic nevus syndrome: The risk of cancers other than melanoma. *J Am Acad Dermatol.* 1987;16:792–7.

97. Bergman W, Watson P, de Jong J, Lynch HT, Fusaro RM. Systemic cancer and the FAMMM syndrome. *Br J Cancer.* 1990;61:932–6.

98. Lynch HT, Fusaro RM, Pester J et al. Tumour spectrum in the FAMMM syndrome. *Br J Cancer.* 1981;44:553–60.

99. Ai L, Stephenson KK, Ling W et al. The p16 (CDKN2a/INK4a) tumor-suppressor gene in head and neck squamous cell carcinoma: A promoter methylation and protein expression study in 100 cases. *Mod Pathol.* 2003;16:944–50.

100. Vinarsky V, Fine RL, Assaad A et al. Head and neck squamous cell carcinoma in FAMMM syndrome. *Head Neck.* 2009;31:1524–7.

101. Yu KK, Zanation AM, Moss JR, Yarbrough WG. Familial head and neck cancer: Molecular analysis of a new clinical entity. *Laryngoscope.* 2002;112:1587–93.

102. Oldenburg RA, de Vos tot Nederveen Cappel WH, van Puijenbroek M et al. Extending the p16–Leiden tumour spectrum by respiratory tract tumours. *J Med Genet.* 2004;41:e31.

103. Schneider-Stock R, Giers A, Motsch C et al. Hereditary p16-Leiden mutation in a patient with multiple head and neck tumors. *Am J Hum Genet.* 2003;72:216–8.

104. Yarbrough WG, Aprelikova O, Pei H, Olshan AF, Liu ET. Familial tumor syndrome associated with a germline nonfunctional p16INK4a allele. *J Natl Cancer Inst.* 1996; 88:1489–91.

105. Borg A, Sandberg T, Nilsson K et al. High frequency of multiple melanomas and breast and pancreas carcinomas in CDKN2A mutation-positive melanoma families. *J Natl Cancer Inst.* 2000;92:1260–6.

106. Ghiorzo P, Ciotti P, Mantelli M et al. Characterization of ligurian melanoma families and risk of occurrence of other neoplasia. *Int J Cancer.* 1999;83:441–8.

107. Lynch HT, Fusaro RM. Genetic epidemiology and the familial atypical multiple mole melanoma syndrome. In: Lynch HT, Tautu P, eds. *Recent Progress in the Genetic Epidemiology of Cancer.* Berlin, Heidelberg: Springer; 1991.

108. Piepkorn M, Meyer LJ, Goldgar D et al. The dysplastic melanocytic nevus: A prevalent lesion that correlates poorly with clinical phenotype. *J Am Acad Dermatol.* 1989;20:407–15.

109. Steijlen PM, Bergman W, Hermans J, Scheffer E, Van Vloten WA, Ruiter DJ. The efficacy of histopathological criteria required for diagnosing dysplastic naevi. *Histopathology.* 1988;12:289–300.

110. Ackerman AB. What naevus is dysplastic, a syndrome and the commonest precursor of malignant melanoma? A riddle and an answer. *Histopathology.* 1988;13:241–56.

111. van der Rhee JI, Krijnen P, Gruis NA et al. Clinical and histologic characteristics of malignant melanoma in families with a germline mutation in CDKN2A. *J Am Acad Dermatol.* 2011;65:281–8.

112. Gillgren P, Brattstrom G, Frisell J, Palmgren J, Ringborg U, Hansson J. Body site of cutaneous malignant melanoma—A study on patients with hereditary and multiple sporadic tumours. *Melanoma Res.* 2003;13:279–86.

113. Masback A, Olsson H, Westerdahl J et al. Clinical and histopathological features of malignant melanoma in germline CDKN2A mutation families. *Melanoma Res.* 2002;12:549–57.

114. Kopf AW, Hellman LJ, Rogers GS et al. Familial malignant melanoma. *JAMA.* 1986;256:1915–9.

115. Grange F, Chompret A, Guilloud-Bataille M et al. Comparison between familial and nonfamilial melanoma in France. *Arch Dermatol.* 1995;131:1154–9.

116. Aitken JF, Duffy DL, Green A, Youl P, MacLennan R, Martin NG. Heterogeneity of melanoma risk in families of melanoma patients. *Am J Epidemiol.* 1994;140:961–73.

117. Barnhill RL, Roush GC, Titus-Ernstoff L, Ernstoff MS, Duray PH, Kirkwood JM. Comparison of nonfamilial and familial melanoma. *Dermatology.* 1992;184:2–7.

118. Duke D, Castresana J, Lucchina L et al. Familial cutaneous melanoma and two-mutational-event modeling. *Cancer.* 1993;72:3239–43.

119. Greene MH, Clark WH, Jr., Tucker MA, Kraemer KH, Elder DE, Fraser MC. High risk of malignant melanoma in melanoma-prone families with dysplastic nevi. *Ann Intern Med.* 1985;102:458–65.

120. Ford D, Bliss JM, Swerdlow AJ et al. Risk of cutaneous melanoma associated with a family history of the disease. The International Melanoma Analysis Group (IMAGE). *Int J Cancer.* 1995;62:377–81.

121. Sargen MR, Kanetsky PA, Newton-Bishop J et al. Histologic features of melanoma associated with CDKN2A genotype. *J Am Acad Dermatol.* 2015;72:496–507 e7.

122. Guibert P, Mollat F, Ligen M, Dreno B. Melanoma screening: Report of a survey in occupational medicine. *Arch Dermatol.* 2000;136:199–202.

123. Mayer JE, Swetter SM, Fu T, Geller AC. Screening, early detection, education, and trends for melanoma: Current status (2007–2013) and future directions: Part I. Epidemiology, high-risk groups, clinical strategies, and diagnostic technology. *J Am Acad Dermatol.* 2014;71:599 e1–12; quiz 610, 599 e12.

124. Goodson AG, Florell SR, Hyde M, Bowen GM, Grossman D. Comparative analysis of total body and dermatoscopic photographic monitoring of nevi in similar patient populations at risk for cutaneous melanoma. *Dermatol Surg.* 2010;36:1087–98.

125. Rademaker M, Oakley A. Digital monitoring by whole body photography and sequential digital dermoscopy detects thinner melanomas. *J Prim Health Care* 2010;2:268–72.

126. Salerni G, Carrera C, Lovatto L et al. Characterization of 1152 lesions excised over 10 years using total-body photography and digital dermatoscopy in the surveillance of patients at high risk for melanoma. *J Am Acad Dermatol.* 2012;67:836–45.

127. Oliveria SA, Chau D, Christos PJ, Charles CA, Mushlin AI, Halpern AC. Diagnostic accuracy of patients in performing skin self-examination and the impact of photography. *Arch Dermatol.* 2004;140:57–62.

128. Risser J, Pressley Z, Veledar E, Washington C, Chen SC. The impact of total body photography on biopsy rate in patients from a pigmented lesion clinic. *J Am Acad Dermatol.* 2007;57:428–34.

129. Bichakjian CK, Halpern AC, Johnson TM et al. Guidelines of care for the management of primary cutaneous melanoma. American Academy of Dermatology. *J Am Acad Dermatol.* 2011;65:1032–47.

130. Leachman SA, Lucero OM, Sampson JE et al. Identification, genetic testing, and management of hereditary melanoma. *Cancer Metastasis Rev.* 2017;36:77–90.

131. Niendorf KB, Goggins W, Yang G et al. MELPREDICT: A logistic regression model to estimate CDKN2A carrier probability. *J Med Genet.* 2006;43:501–6.

132. Wang W, Niendorf KB, Patel D et al. Estimating CDKN2A carrier probability and personalizing cancer risk assessments in hereditary melanoma using MelaPRO. *Cancer Res.* 2010;70:552–9.

133. Haenssle HA, Korpas B, Hansen-Hagge C et al. Selection of patients for long-term surveillance with digital dermoscopy by assessment of melanoma risk factors. *Arch Dermatol.* 2010;146:257–64.

134. Gambichler T, Senger E, Rapp S, Alamouti D, Altmeyer P, Hoffmann K. Deep shave excision of macular melanocytic nevi with the razor blade biopsy technique. *Dermatol Surg.* 2000;26:662–6.

135. Diffey B. Sunscreen isn't enough. *J Photochem Photobiol B.* 2001;64:105–8.

136. Diffey BL. When should sunscreen be reapplied? *J Am Acad Dermatol.* 2001;45:882–5.

137. Schneider J. The teaspoon rule of applying sunscreen. *Arch Dermatol.* 2002;138:838–9.

138. Canto MI, Harinck F, Hruban RH et al. International Cancer of the Pancreas Screening (CAPS) Consortium summit on the management of patients with increased risk for familial pancreatic cancer. *Gut.* 2013;62:339–47.

139. Meckler KA, Brentnall TA, Haggitt RC et al. Familial fibrocystic pancreatic atrophy with endocrine cell hyperplasia and pancreatic carcinoma. *Am J Surg Pathol.* 2001;25:1047–53.

140. Hruban RH, Canto MI, Yeo CJ. Prevention of pancreatic cancer and strategies for management of familial pancreatic cancer. *Dig Dis.* 2001;19:76–84.

141. Wu J, Matthaei H, Maitra A et al. Recurrent GNAS mutations define an unexpected pathway for pancreatic cyst development. *Sci Transl Med* 2011;3:92ra66.

142. Matthaei H, Schulick RD, Hruban RH, Maitra A. Cystic precursors to invasive pancreatic cancer. *Nat Rev Gastroenterol Hepatol* 2011;8:141–50.

143. Canto MI, Hruban RH, Fishman EK et al. Frequent detection of pancreatic lesions in asymptomatic high-risk individuals. *Gastroenterology.* 2012;142:796–804; quiz e14–5.

144. Canto MI, Goggins M, Hruban RH et al. Screening for early pancreatic neoplasia in high-risk individuals: A prospective controlled study. *Clin Gastroenterol Hepatol.* 2006;4:766–81; quiz 665.

145. Zubarik R, Gordon SR, Lidofsky SD et al. Screening for pancreatic cancer in a high-risk population with serum CA 19–9 and targeted EUS: A feasibility study. *Gastrointest Endosc.* 2011;74:87–95.

146. Kim JE, Lee KT, Lee JK, Paik SW, Rhee JC, Choi KW. Clinical usefulness of carbohydrate antigen 19–9 as a screening test for pancreatic cancer in an asymptomatic population. *J Gastroenterol Hepatol.* 2004;19:182–6.

147. Liu J, Gao J, Du Y et al. Combination of plasma microRNAs with serum CA19-9 for early detection of pancreatic cancer. *Int J Cancer.* 2012;131:683–91.

148. Kanda M, Knight S, Topazian M et al. Mutant GNAS detected in duodenal collections of secretin-stimulated pancreatic juice indicates the presence or emergence of pancreatic cysts. *Gut.* 2013;62:1024–33.

149. Kanda M, Sadakari Y, Borges M et al. Mutant TP53 in duodenal samples of pancreatic juice from patients with pancreatic cancer or high-grade dysplasia. *Clin Gastroenterol Hepatol.* 2013;11:719–30 e5.

150. Sato N, Fukushima N, Hruban RH, Goggins M. CpG island methylation profile of pancreatic intraepithelial neoplasia. *Mod Pathol.* 2008;21:238–44.

151. Tarafa G, Tuck D, Ladner D et al. Mutational load distribution analysis yields metrics reflecting genetic instability during pancreatic carcinogenesis. *Proc Natl Acad Sci U S A.* 2008;105:4306–11.

152. Hille ET, van Duijn E, Gruis NA, Rosendaal FR, Bergman W, Vandenbroucke JP. Excess cancer mortality in six Dutch pedigrees with the familial atypical multiple mole-melanoma syndrome from 1830 to 1994. *J Invest Dermatol.* 1998;110:788–92.

153. Balch CM, Gershenwald JE, Soong SJ et al. Final version of 2009 AJCC melanoma staging and classification. *J Clin Oncol.* 2009;27:6199–206.

154. Jemal A, Siegel R, Ward E et al. Cancer statistics, 2006. *CA Cancer J Clin.* 2006;56:106–30.

155. Riker AI, Hagmaier R. The familial atypical multiple mole melanoma (FAMMM)-pancreatic carcinoma (PC) syndrome. In: Ellis CN. ed. *Inherited Cancer Syndromes.* New York, NY: Springer; 2011, pp. 135–44.

40

Familial Chordoma

Alexandra Suttman, Sydney T. Grob, and Jean M. Mulcahy Levy

CONTENTS

40.1 Introduction

Chordomas are rare bony tumors that develop from persistent notochord remnants in less than 1/100,000 individuals [1]. Median age at diagnosis is in the sixth decade of life [2], with only about 5% of chordomas presenting in childhood or adolescence [3], and overall chordoma in pediatric patients has been reported as fewer than 1 in 10,000,000 [4].

These slow-growing tumors primarily arise along the axial spine [5]. Pediatric chordomas are more likely to present in the skull base (Figure 40.1) and have underlying hereditary predispositions [2].

Drs. Foote, Albin, and Hall reported the first "familial" case of chordoma occurring in siblings in 1958 [6], and at least 10 families with multiple members affected by chordoma have been reported since [7]. Familial chordoma, defined as having >1 individual in a family diagnosed with chordoma, is exceedingly rare, with an approximated incidence of less than 1% of all chordoma diagnoses [7].

The susceptibility to chordoma is multivariate and complex, and it involves multiple mechanisms of common and rare genetic variants [8]. Hereditary predisposition to chordoma can occur due to T-box transcription factor T (*TBXT*) gene duplication, single and compound single-nucleotide polymorphisms (SNPs), as well as in the context of tuberous sclerosis complex (TSC). Hereditary predisposition to chordoma should be considered in individuals with early-onset chordoma, features of TSC, and/or in families with >1 diagnosis of chordoma.

40.2 Biological Background and Pathogenesis

TBXT (previously identified as T brachyury transcription factor) is the 2018 updated nomenclature endorsed by the HUGO Gene Nomenclature Committee for the gene that encodes a tissue-specific transcription factor that is expressed in the nucleus of notochord cells [9]. It is essential for proper development and maintenance of the notochord [10]. Its expression in chordomas mimics expression in the embryonic notochord [2]. A recent study of 104 cases of sporadic chordoma found somatic duplication of *TBXT* in up to 27% of cases [11].

In contrast, heterozygous germline duplications in *TBXT* have been identified in approximately 40% families with multiple chordoma diagnoses. These duplications are thought to confer a significant susceptibility to the development of chordoma [2], although it is currently still unclear how this gene contributes to the pathogenesis of chordoma. In vitro, silencing brachyury halts chordoma tumor growth [8]. As noted previously, at least partial allelic gain of *TBXT* is a common somatic finding in sporadic chordomas, further implicating this gene in the pathogenesis of chordoma [12].

In addition to germline *TBXT* duplications, particular SNPs have been implicated in increasing the susceptibility to chordoma in a non-mendelian manner. One SNP, rs2305089, commonly found in those with European ancestry, is thought to alter the expression of *TBXT* and its downstream targets. This SNP is expected to increase an individual's risk of chordoma about 2.5–5× per allele [8,13]. While rs2305089 may increase the risk of chordoma in European populations, it has not been found to be implicated in the risk of skull-based chordoma in Chinese populations [14], and

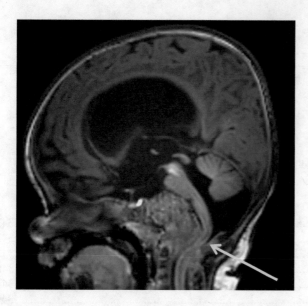

FIGURE 40.1 Magnetic resonance imaging (MRI), sagittal T1-weighted image. Large, extra-axial, heterogenous skull base mass (chordoma) with distortion of the brainstem and resultant severe obstructive hydrocephalus in a pediatric patient.

as such, the pathogenesis of this SNP across different ethnic populations has yet to be characterized. Various other SNPs have also been reported to increase the risk of chordoma [8]. SNP rs1056048 may only be associated with families who have a concurrent *TBXT* duplication [8] and is thought to impact splicing. rs3816300, which is found to have stronger association in patients diagnosed under age 20 with skull-based chordoma, increases risk for chordoma tenfold when found in conjunction with rs2305089 [8].

As evidenced, non-syndromic predisposition to chordoma is multivariant, complex, and incompletely characterized. In addition, penetrance and expressivity have yet to be defined, and as such, predictive testing for predisposition to chordoma does not currently impart poignant clinical utility with regard to medical management at this time.

40.2.1 Chordoma as a Feature of Tuberous Sclerosis Complex

Chordomas may also appear as a rare feature of TSC [15]. TSC is an autosomal dominant neuro-cutaneous syndrome with an incidence of approximately 1/5800 live births [16,17]. Features of TSC can be highly variable, may affect every organ system, and are noted in Table 40.1 (modified from the TSC Consensus Group) [16]. Chordomas associated with TSC typically occur before the age of 15, with a median age at diagnosis around 6 months. TSC-related chordomas are more likely to be diagnosed in the skull base or sacrum, with sacral diagnoses associated with a particularly early age of onset in the fetal or neonatal period. While sacral TSC-related chordomas exhibit early age of onset, they may have a better clinical outcome than sacral chordomas in patients without TSC [3].

A diagnosis of TSC is made molecularly when an individual has a pathogenic mutation in the gene *TSC1* or *TSC2*. Alternatively, a diagnosis can be made when an individual has two major features or one major feature with at least two minor features as described in Table 40.1.

TABLE 40.1

Clinical Features of Tuberous Sclerosis Complex [18]

Major Features

Angiofibromas (≥3) or fibrous cephalic plaque

Cardiac rhabdomyoma

Cortical dysplasias, including tubers and cerebral white matter migration lines

Hypomelanotic macules (≥3; ≥5 mm in diameter)

Lymphangioleiomyomatosis (LAM)

Multiple retinal nodular hamartomas

Renal angiomyolipoma

Shagreen patch

Subependymal giant cell astrocytoma (SEGA)

Subependymal nodules (SENs)

Ungual fibromas (≥2)

Minor Features

"Confetti" skin lesions (numerous 1–3 mm hypopigmented macules scattered over regions of the body such as the arms and legs)

Dental enamel pits (>3)

Multiple renal cysts

Nonrenal hamartomas

Retinal achromic patch

Note: The combination of LAM and angiomyolipomas without other features does not meet the clinical diagnostic criteria for a definite diagnosis of TSC.

TSC1 and *TSC2* encode for the proteins hamartin and tuberin, respectively. Together they form a complex that is thought to function as a "tumor suppressor" with roles in cellular signaling and cell growth and proliferation. Alteration and/or inactivation of the hamartin-tuberin protein complex results in unregulated activation of mTOR pathway, which is crucial for cell-cycle regulation [19]. Increased mTOR activity results in phosphorylation of S6 ribosomal protein and eukaryotic translation initiation factor 4E (eIF-4E). This results in the abnormal growth of multiple tissues including facial angiofibroma, cardiac rhabdomyoma, subependymal nodules, and subependymal giant cell astrocytomas, which are all clinical characteristics of patients with TSC [20].

While it is currently unclear exactly how TSC predisposes individuals to chordoma, Lee-Jones et al. demonstrated that sacrococcygeal chordomas in patients with TSC show somatic inactivation of the wild type *TSC1* or *TSC2*, further suggesting the pathogenesis of these genes in chordoma and following Knudsen's two-hit hypothesis [21].

40.2.2 Rhabdoid Tumor Predisposition Syndrome

Loss of SMARCB1/INI protein expression has been identified somatically in chordomas, particularly those diagnosed in the pediatric period. Germline *SMARCB1* mutations result in rhabdoid tumor predisposition syndrome (RTPS). It is currently unclear whether individuals with germline *SMARCB1* mutations are at an increased risk to develop chordoma, as chordoma has not been reported as a feature of RTPS thus far, and the phenotypic spectrum of RTPS is incompletely defined [22].

40.3 Epidemiology

40.3.1 Prevalence

Possibly arising from the remnants of the embryonic notochord, chordoma is a rare, slow growing, and locally invasive bone tumor that has an incidence of 0.08 per 100,000, shows a male predilection (male to female ratio of 1.7:1), and affects people of 40–60 years of age (with a median diagnostic age of 59 years, and a marginally younger mean age for cases at the skull base). While a majority of chordoma cases are sporadic, familial chordoma involving approximately 40 individuals from 10 families has been reported in the literature. Interestingly, familial chordoma tends to occur in the area of the skull base, exhibits an early onset of symptoms (with the mean diagnostic age of 29 years), and demonstrates a female predominance (male to female ratio of 1:1.8).

40.3.2 Inheritance

Familial chordoma is inherited in an autosomal dominant pattern, in which one copy of the altered gene is sufficient to increase the risk of the disorder.

40.3.3 Penetrance

Familial chordoma displays a high penetrance, and most patients harboring related pathogenic variants develop chordoma at certain stages of their life.

40.4 Clinical Features

Given that chordomas are slow-growing and can be locally destructive, the clinical presentation of chordomas often depends on where chordoma lesions are found. Previous studies have shown that 69.6% of pediatric chordomas involve the clivus/skull base and cervical spine compared to 29.3% in adults. On the other hand, only 1.8% of pediatric chordomas are found in the sacrococcygeal region compared to 36.0% in adults.

Patients with cranial chordomas often present with headaches and diplopia, followed by cranial nerve palsies, visual loss, seizures, extremity weakness or numbness, neck pain, nasal congestion, or dysphagia. Chordomas of the spine and sacrococcygeal region are usually associated with local pain or signs of neural compression (e.g., radiculitis, bowel or urinary incontinence, or sexual dysfunction) as well as palpable mass (in the case of advanced sacrococcygeal chordomas).

40.5 Diagnosis

There are approximately 300 new cases of chordoma in the United States each year [23]. In pediatric patients, there have been less than 300 cases reported in the literature [24]. The diagnosis of this tumor is made based on clinical, topographical, radiological, and histological findings [24]. Young children with an intracranial chordoma will often experience headaches as a clinical symptom before diagnosis. Infants and very young children, those below 5, diagnosed with this tumor often too will present with long tract

signs, lower cranial nerve palsy, and torticollis. Sacrococcygeal chordoma is another form of chordoma in which patients will present clinically with perineal pain as well as potential bladder or bowel problems [24]. Lastly, vertebral column chordoma is often characterized by pain and deformity in the spine. More general symptoms like hydrocephalus occur in approximately one-third of patients [25]. Adult patients similarly present with symptoms related to location of their primary lesion [26].

These diagnoses are often made according to a combined imaging approach incorporating computed tomography (CT) as well as magnetic resonance imaging (MRI). Both of these techniques must be used in tandem due to the intricate location of this tumor type [27]. Pre-imaging is used to determine the location and makeup of the tumor. These tumors occur midline [28] and often present as destructive bone lesions with a surrounding soft tissue mass [26]. Post-surgical imaging techniques are used to help inform what adjuvant therapies are needed next [27]. Creating a diagnosis strictly based on imaging is often difficult and can lead to inconsistencies, which is why taking a multimodal approach to diagnosing a lesion is necessary [28].

As revealed by contrasted CT, chordomas often appear as a contrast-enhancing, lytic, or mixed bony lesion with a myxoid component. On MRI, chordomas are hypointense or isointense on T1-weighted sequences and hyperintense on T2-weighted sequences, and show heterogeneous enhancement with gadolinium, generating a lobulated "honeycomb" appearance (due to the presence of gelatinous mucoid substance, old hemorrhage, or areas of necrosis within the tumor).

Histologically, three chordoma variants are recognized. Classic chordoma consists of groups of cells with fibrous septa (Figure 40.2a); chondroid chordoma is present when focal or extensive areas of the matrix mimic hyaline cartilaginous tumors, and dedifferentiated chordoma is a biphasic tumor with features of chordoma juxtaposed to a high-grade undifferentiated spindle cell tumor or osteosarcoma. Typical immunostaining for chordoma includes epithelial membrane antigen, S100 protein, AE1/AE3, retention of nuclear INI-1 expression, nuclear brachyury, and epidermal growth factor receptor (EGFR) expressions (Figure 40.2b) [26]. Loss of SMARCB1/INI1 protein expression has been described in poorly differentiated chordomas [22,29]. In comparison to adult chordoma, pediatric cases show in higher frequency in the cranial region and are often characterized by increased cellularity, solid growth, cytologic atypia, and increased mitotic activity. Pediatric chordomas are often classified as being "poorly differentiated chordoma," and as described previously, they are often associated with the loss of SMARCB1/INI1 [30].

40.6 Treatment

Standard treatment for chordoma at present includes surgical resection followed by adjuvant therapies including radiotherapy and rarely followed by chemotherapy [31]. Maximal safe resection is the first-line therapy for chordoma. If gross total resection is not feasible, maximal tumor debulking should be achieved. Conventional radiation therapy is considered to be problematic, as this set of tumors is often radio-resistant even though high-dose radiation could be used to obtain control of the lesion.

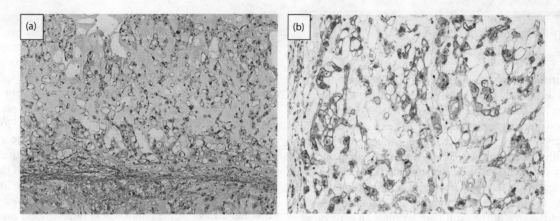

FIGURE 40.2 Microscopic examination of chordoma revealing round cells with vacuolated cytoplasm arranged in cord-like fashion in a myxoid stroma (a; H&E, magnification ×100), and strong membranous and cytoplasmic staining of tumor cells (brown colored) by anti-EGFR antibody (b; magnification ×200). (Photo credit: Launay SG et al. *BMC Cancer.* 2011;11:423.)

Unfortunately, the dose of radiation that could be feasibly used on chordoma is limited by the tissues that surround the brain tumor including the brainstem, optic chiasma, and temporal lobe. These intricate and delicate structures limit viable radiation dosing to 50 Gy in 25 fractions as to not damage these organs. Adult studies have shown that both high-dose (70–78 Gy) image-guided intensity-modulated radiation [32] and hyperfractionated proton beam radiation [33] are able to provide local control with a low rate of adverse events.

Proton beam therapy with doses between 72–76–80 GyRBE have been shown to be successful in controlling tumor growth in pediatric patients as well [34]. A study of 30 children treated with photon-proton or proton-alone high-dose fractionated radiotherapy all had positive responses to this treatment [35]. Surgery followed by proton beam therapy has resulted in an overall control rate between 10%–40% in pediatric patients [36].

Chordoma lesions often display resistance to chemotherapy agents, making this treatment option of low efficacy. Chemotherapy may be used as a last resort option when surgical resection and radiation have both failed [37]. Improved technology has become available which has made understanding the molecular underpinnings and genetic aberrations of tumors much more apparent [38]. This has resulted in the more recent development of targeted therapies. For instance, the activation of phosphorylated forms of platelet-derived growth factor hormone (PDGFR)-b and of KIT receptors have been targeted with drugs such as imatinib, which has shown success in clinical trial when used by themselves or with sirolimus. Another example of using targeted therapies in pediatric chordoma include using inhibitory drugs linked to EGFR like cetuximab, gefitinib, and erlotinib [31]. The further development of targeted therapy and its use in patients with chordoma remains an area of evolving research.

40.7 Prognosis

Morbidity still remains of great concern when discussing surgical resection and radiation therapy, particularly in children whose brains are still developing. Aggressive surgical resection often comes at the expense of significant morbidity. Radiation therapy still remains of concern, as high-dose radiation therapy

is known to lead to physical and functional impairment, particularly in the pediatric population.

Adult studies have demonstrated a 5-year survival rate of 65% for skull base chordoma [39]. The median overall survival of both intracranial and extracranial chordoma is reported as 7.7 years for men, and 7.8 years for women [40].

Overall, pediatric patients who present with chordoma often have lesions that are very aggressive and associated with a shorter overall survival and a higher percentage of recurrence [29]. In pediatric patients that do show evidence of more aggressive tumors, there are reports of these lesions being more metastatic and atypical in their histology [41]. Many researchers have noted a positive correlation between chordomas with atypical features and the likelihood of metastasis [29]. Most patients who undergo less than a gross total resection (nearly all patients fall into this category) will experience at least one recurrence. Therefore, the overall rate of tumor control in pediatric patients hovers right below 30% 10 years out from diagnosis [36].

Given the low incidence of this tumor type in children, there is little known about the overall survival of patients with this diagnosis. Benk et al. conducted a study of 18 children with a skull base or cervical spine chordoma who had received surgery followed by radiation, and reported that a 5-year disease-free survival was at 63%, with a 5-year overall survival increasing to 68% [42]. Location of the tumor is another prognostic factor in which patients with cervical spine chordomas have a worse overall prognosis than patients who have tumors arising at the skull base [42]. Due to the paucity of reported familial chordoma patients, it is not possible to determine how this would contribute to their overall prognosis outside of what is seen in differences between adult and pediatric patients.

40.8 Conclusion

Familial chordoma has a heterogeneous genetic background with the potential for compound heterozygous processes. While the identification of familial chordoma provides important risk information for probands and family members, consensus guidelines regarding the management of these individuals and families in regard to the risk for chordoma do not currently exist. As such, management should be based on personal and family history with consideration of the benefits and limitations of imaging modalities. Management recommendations for individuals with TSC do exist for extra-chordoma related risks [16]. The possibility of TSC should be considered for any individual diagnosed with chordoma in the neonatal/postnatal period, even in the context of a negative family history. TSC can frequently arise de novo, and the phenotype is variable, particularly in the pediatric period [3,15,43]. Integration of genetics specialty providers may be considered for anyone diagnosed with chordoma given the rarity of this tumor, the ability to provide risk assessment and genetic test options, and the possibility of a larger syndromic disorder such as TSC.

REFERENCES

1. Stiller CA, Trama A, Serraino D et al. Descriptive epidemiology of sarcomas in Europe: Report from the RARECARE project. *Eur J Cancer.* 2013;49:684–95.

2. Yang XR, Ng D, Alcorta DA et al. T (brachyury) gene duplication confers major susceptibility to familial chordoma. *Nat Genet.* 2009;41:1176–8.

3. McMaster ML, Goldstein AM, Parry DM. Clinical features distinguish childhood chordoma associated with tuberous sclerosis complex (TSC) from chordoma in the general paediatric population. *J Med Genet.* 2011;48:444–9.

4. McMaster ML, Goldstein AM, Bromley CM, Ishibe N, Parry DM. Chordoma: Incidence and survival patterns in the United States, 1973–1995. *Cancer Causes Control.* 2001;12:1–11.

5. *World Health Organization Classification of Tumours: Pathology and Genetics of Tumours of Soft Tissue and Bone.* Lyon: IARC Press; 2013.

6. Foote RF, Ablin G, Hall WW. Chordoma in siblings. *Calif Med.* 1958;88:383–6.

7. Wang KE, Wu Z, Tian K et al. Familial chordoma: A case report and review of the literature. *Oncol Lett.* 2015;10:2937–40.

8. Kelley MJ, Shi J, Ballew B et al. Characterization of T gene sequence variants and germline duplications in familial and sporadic chordoma. *Hum Genet.* 2014;133:1289–97.

9. Kispert A, Koschorz B, Herrmann BG. The T protein encoded by Brachyury is a tissue-specific transcription factor. *EMBO J.* 1995;14:4763–72.

10. Kispert A, Herrmann BG. Immunohistochemical analysis of the Brachyury protein in wild-type and mutant mouse embryos. *Dev Biol.* 1994;161:179–93.

11. Tarpey PS, Behjati S, Young MD et al. The driver landscape of sporadic chordoma. *Nat Commun.* 2017;8:890.

12. Presneau N, Shalaby A, Ye H et al. Role of the transcription factor T (brachyury) in the pathogenesis of sporadic chordoma: A genetic and functional-based study. *J Pathol.* 2011;223:327–35.

13. Pillay N, Plagnol V, Tarpey PS et al. A common single-nucleotide variant in T is strongly associated with chordoma. *Nat Genet.* 2012;44:1185–7.

14. Wu Z, Wang K, Wang L et al. The brachyury Gly177Asp SNP is not associated with a risk of skull base chordoma in the Chinese population. *Int J Mol Sci.* 2013;14:21258–65.

15. Dahl NA, Luebbert T, Loi M et al. Chordoma occurs in young children with tuberous sclerosis. *J Neuropathol Exp Neurol.* 2017;76:418–23.

16. Krueger DA, Northrup H. Tuberous sclerosis complex surveillance and management: Recommendations of the 2012 International Tuberous Sclerosis Complex Consensus Conference. *Pediatr Neurol.* 2013;49:255–65.

17. O'Callaghan FJ, Shiell AW, Osborne JP, Martyn CN. Prevalence of tuberous sclerosis estimated by capture-recapture analysis. *Lancet.* 1998;351:1490.

18. Northrup H, Krueger DA. Tuberous sclerosis complex diagnostic criteria update: Recommendations of the 2012 IInternational Tuberous Sclerosis Complex Consensus Conference. *Pediatr Neurol.* 2013;49:243–54.

19. Narayanan V. Tuberous sclerosis complex: Genetics to pathogenesis. *Pediatr Neurol.* 2003;29:404–9.

20. Presneau N, Shalaby A, Idowu B et al. Potential therapeutic targets for chordoma: PI3 K/AKT/TSC1/TSC2/mTOR pathway. *Br J Cancer.* 2009;100:1406–14.

21. Lee-Jones L, Aligianis I, Davies PA et al. Sacrococcygeal chordomas in patients with tuberous sclerosis complex show somatic loss of TSC1 or TSC2. *Genes, Chromosomes Cancer.* 2004;41:80–5.

22. Antonelli M, Raso A, Mascelli S et al. SMARCB1/INI1 Involvement in Pediatric Chordoma: A Mutational and Immunohistochemical Analysis. *Am J Surg Pathol.* 2017;41:56–61.

23. Sebro R, DeLaney T, Hornicek F et al. Differences in sex distribution, anatomic location and MR imaging appearance of pediatric compared to adult chordomas. *BMC Med Imaging.* 2016;16:53.

24. Beccaria K, Sainte-Rose C, Zerah M, Puget S. Paediatric chordomas. *Orphanet J Rare Dis.* 2015;10:116.

25. Habrand JL, Datchary J, Bolle S et al. Chordoma in children: Case-report and review of literature. *Rep Pract Oncol Radiother.* 2016;21:1–7.

26. Walcott BP, Nahed BV, Mohyeldin A, Coumans JV, Kahle KT, Ferreira MJ. Chordoma: Current concepts, management, and future directions. *Lancet Oncol.* 2012;13:e69–76.

27. Dincer A. Imaging findings of the pediatric clivus chordomas. In: Özek M, Cinalli G, Maixner W, Sainte-Rose C. (eds) *Posterior Fossa Tumors in Children.* p. 683–92. Springer, 2015.

28. Kunimatsu A, Kunimatsu N. Skull base tumors and tumor-like lesions: A pictorial review. *Pol J Radiol.* 2017;82:398–409.

29. Mobley BC, McKenney JK, Bangs CD et al. Loss of SMARCB1/INI1 expression in poorly differentiated chordomas. *Acta Neuropathol.* 2010;120:745–53.

30. Owosho AA, Zhang L, Rosenblum MK, Antonescu CR. High sensitivity of FISH analysis in detecting homozygous SMARCB1 deletions in poorly differentiated chordoma: A clinicopathologic and molecular study of nine cases. *Genes, Chromosomes Cancer.* 2018;57:89–95.

31. Nibu Y, Jose-Edwards DS, Di Gregorio A. From notochord formation to hereditary chordoma: The many roles of Brachyury. *BioMed Res Int.* 2013;2013:826435.

32. Sahgal A, Chan MW, Atenafu EG et al. Image-guided, intensity-modulated radiation therapy (IG-IMRT) for skull base chordoma and chondrosarcoma: Preliminary outcomes. *Neuro Oncol.* 2015;17:889–94.

33. Hayashi Y, Mizumoto M, Akutsu H et al. Hyperfractionated high-dose proton beam radiotherapy for clival chordomas after surgical removal. *Brit J Radiol.* 2016;89:20151051.

34. Takagi M, Demizu Y, Nagano F et al. Treatment outcomes of proton or carbon ion therapy for skull base chordoma: A retrospective study. *Radiat Oncol.* 2018;13(1):232.

35. Habrand JL, Schneider R, Alapetite C et al. Proton therapy in pediatric skull base and cervical canal low-grade bone malignancies. *Int J Radiat Oncol Biol Phys.* 2008;71:672–5.

36. Dhall G, Traverso M, Finlay JL, Shane L, Gonzalez-Gomez I, Jubran R. The role of chemotherapy in pediatric clival chordomas. *J Neurooncol.* 2011;103:657–62.

37. Tsai EC, Santoreneos S, Rutka JT. Tumors of the skull base in children: Review of tumor types and management strategies. *Neurosurg Focus.* 2002;12:e1.

38. Grob ST, Levy JMM. Improving diagnostic and therapeutic outcomes in pediatric brain. *Mol Diagn Ther.* 2018;22:25–39.

39. Bohman LE, Koch M, Bailey RL, Alonso-Basanta M, Lee JY. Skull base chordoma and chondrosarcoma: Influence of clinical and demographic factors on prognosis: A SEER analysis. *World Neurosurg.* 2014;82:806–14.

40. Smoll NR, Gautschi OP, Radovanovic I, Schaller K, Weber DC. Incidence and relative survival of chordomas: The standardized mortality ratio and the impact of chordomas on a population. *Cancer.* 2013;119:2029–37.

41. Ridenour RV, 3rd, Ahrens WA, Folpe AL, Miller DV. Clinical and histopathologic features of chordomas in children and young adults. *Pediatr Dev Pathol.* 2010;13:9–17.

42. Benk V, Liebsch NJ, Munzenrider JE, Efird J, McManus P, Suit H. Base of skull and cervical spine chordomas in children treated by high-dose irradiation. *Int J Radiat Oncol Biol Phys.* 1995;31:577–81.

43. Lountzis NI, Hogarty MD, Kim HJ, Junkins-Hopkins JM. Cutaneous metastatic chordoma with concomitant tuberous sclerosis. *J Am Acad Dermatol.* 2006;55:S6–10.

41

Hereditary Multiple Osteochondromas

Dongyou Liu

CONTENTS

41.1 Introduction

Hereditary multiple osteochondromas (HMO, also referred to as hereditary multiple exostoses [HME]) represent a rare autosomal dominant disorder characterized by the growth of benign cartilage-capped bone tumors (osteochondromas or exostoses) within the perichondrium flanking the growth plates of long bones, ribs, hip, and vertebrae in very young and adolescent patients, leading to pain, functional problems, and deformities, especially of the forearm.

Molecularly, HMO is linked to heterozygous loss-of-function mutations in *EXT1* (exostosin-1; 8q24.11; 56%–78% of cases) and *EXT2* (exostosin-2; 11p11.2; 21%–44% of cases), which encode Golgi-resident glycosyltransferases involved in the synthesis and assembly of heparan sulfate (HS) chains for subsequent production of heparan sulfate proteoglycans (HSPG), which through interaction with bone morphogenetic proteins (BMP) influence bone and cartilage formation [1,2].

41.2 Biology

Osteochondroma (previously referred to as osteocartilaginous exostosis, meaning outgrowth of bone, whereas osteochondroma specifies cartilaginous growth in the child that ossifies at skeletal maturity) is a cartilage-capped, benign surface bone neoplasm that characteristically involves both cortical and medullary bone with the underlying bone. Arising and growing in size in the first decade of life until skeletal maturation or shortly thereafter, osteochondroma forms solitary (sporadic) or multiple (hereditary) lesions typically in the metaphyseal region of the long bones of the limbs (e.g., the distal femur, upper humerus, upper tibia, and fibula), and occasionally of the flat bones (e.g., the ilium and scapula). Common complications of osteochondroma range from osseous deformities, fracture, bursa formation with or without bursitis, vascular compromise, and neurologic symptoms to malignant transformation [3].

Multiple osteochondromas are a multiple (hereditary) form of disease that demonstrates autosomal dominant inheritance and is hence called hereditary multiple osteochondroma (HMO, also known as hereditary multiple exostoses [HME], diaphyseal aclasis, or familial osteochondromatosis). Besides multiple osteochondroma lesions, patients with HMO also develop other orthopedic deformities (e.g., shortening of the ulna with secondary bowing of the radius 39%–60%, inequality of the limbs 10%–50%, varus or valgus angulation of the knee 8%–33%, deformity of the ankle 2%–54%, disproportionately short stature 40%). Further, multiple osteochondromas associated with HMO have a higher risk (0.5%–3%) of malignant transformation than solitary osteochondromas (<1%) [4].

Although cases related to HMO were reported as early as in 1814, the culprit genes *EXT1* and *EXT2* underlying its molecular pathogenesis were identified only in 1997. These genes encode Golgi-associated glycosyltransferases that take part in the chain elongation step of HS biosynthesis. Heparan sulfate is a key component of HSPG, which through interaction with BMP influence bone and cartilage formation. Loss-of-function mutations in *EXT1* and *EXT2* hinder HS and HSPG synthesis, and subsequent

HSPG deficiency compromises the functionality of BMP, leading to aberrant bone and cartilage formation and development of multiple osteochondromas [5].

41.3 Pathogenesis

The *EXT1* (exostosin-1) gene on chromosome 8q24.11 comprises 350 kb of genomic DNA organized into 11 exons and encodes a 746 aa glycosyltransferase (EXT1), which plays an important role in the elongation of the HS backbone [6].

The *EXT2* (exostosin-2) gene on chromosome 11p11.2 spans 108 kb with 14 exons (plus two alternative exons) and encodes a 718 aa glycosyltransferase (EXT2), which assists in the folding and transport of *EXT1* to the Golgi apparatus [6].

Located in the endoplasmic reticulum, EXT1 forms a heterooligomeric complex with EXT2 in the Golgi apparatus, which demonstrates a much higher glycosyltransferase activity than EXT1 or EXT2 alone and functions as HP polymerase in the chain elongation step of HP biosynthesis by the addition of N-acetylglucosamine and glucuronic acid. After deacetylation and sulfation of most N-acetylglucosamines, epimerization of the glucuronic into iduronic acid and further sulfation, HP chains emerge. Heparan sulfate is an essential component of HSPG, which interact with BMP and indirectly affect skeletogenesis and skeletal growth and morphogenesis [7–9].

Germline loss-of-function mutations (e.g., nonsense, frameshift, splice site, missense variants) in the *EXT1* and *EXT2* genes leading to premature terminations of the EXT proteins cause HS deficiency and subsequent cytoskeletal abnormalities (e.g., actin accumulation, excessive bundling by alpha-actinin, and abnormal presence of muscle-specific alpha-actin) [1].

To date, >400 *EXT1* pathogenic variants are observed in 56%–78% of HMO cases. Interestingly, polypyrimidine tracts in exons 1 and 6 are targeted by frameshift variants, while codons 339 and 340 are impacted by missense variants. Further, promoter hypermethylation is involved in the epigenetic loss of *EXT1* activity, resulting in leukemia and other cancer [2].

On the other hand, >200 *EXT2* pathogenic variants are found in 21%–44% of HMO cases. These mutations are mostly loss-of-function variants (frameshift, in-frame deletion, nonsense, splice site) located in the first eight exons [2,10].

Given that *EXT1* mutations have a more common occurrence in HMO than *EXT2* mutations, *EXT1* appears to pose a somewhat higher risk of malignant transformation than *EXT2* [11].

41.4 Epidemiology

41.4.1 Prevalence

As the most common bone neoplasm, osteochondroma makes up as much as 50% of all benign osseous tumors and 10%–15% of all benign and malignant osseous tumors. Osteochondroma may appear solitary (sporadic; 85%) or multiple (hereditary; 15%). Solitary osteochondroma is usually asymptomatic and represents incidental findings in 1%–2% of patients undergoing radiographic evaluation. HMO have an estimated incidence of 1:50,000, a predilection for males (male-to-female ratio 1.5:1), and a median diagnostic age of 3 years (with nearly all affected individuals diagnosed by age 12 years) [12].

41.4.2 Inheritance

HMO display autosomal dominant inheritance, with germline mutations accounting for 95% of cases, and de novo pathogenic variants causing 10% of cases. About 62% of HMO patients show a positive family history [13].

41.4.3 Penetrance

The penetrance of HMO is 100% in males and 96% in females.

41.5 Clinical Features

HMO typically manifest as multiple benign cartilage-capped bone tumors that grow outward from the metaphyses of long bones (e.g., the femur 30%, radius/ulna 13%, tibia 20%, fibula 13%) in the first decade of life until puberty, causing chronic site-specific pain in a majority of patients and generalized pain in 50% of patients (due to compression of the nerves, blood vessels, and tendons). Complications of HMO range from reduced skeletal growth, bony deformity (scoliosis, erosion of the adjacent bone, spontaneous hemothorax, rib exostoses), restricted joint motion, shortened stature (40%), premature osteoarthrosis, compression of peripheral nerves, obstetric problem, psychosocial problems, and 5% risk for fracture to 3%–5% risk for malignant degeneration (secondary to peripheral chondrosarcoma) [5,14–18].

41.6 Diagnosis

Diagnostic workup for HMO consists of family history review, physical examination, imaging study, histopathology, and molecular testing.

X-rays reveal dense structures of bones, and bone growth associated with osteochondromas. Computed tomography (CT) scan displays detailed bony lesion and presence of calcification. Magnetic resonance imaging (MRI) delineates bone masses and osteochondromas far more precisely than other techniques. Ultrasound helps clarify aneurysms/pseudoaneurysms and thrombosis, and clearly depicts the cartilaginous cap of osteochondroma. Angiography detects vascular lesions caused by osteochondroma and characterizes neovascularity associated with malignant transformation of osteochondroma (Figures 41.1 and 41.2) [4,19–21].

Observation of two or more radiologically confirms osteochondromas in the juxta-metaphyseal region of the long bones, with or without a positive family history, is indicative of HMO, as about 10% of affected individuals do not have clear family connection.

Macroscopically, osteochondroma is a lobulated sessile (broad based) or pedunculated (stemmed) tumor projecting from the bone surface, with a shiny glistening bluish to grey cartilage cap, which measures 1–3 cm in thickness in children, but only a few mm in the fully grown skeleton. On cross-section, the perichondrium (thin fibrous capsule), cortex, and the medulla of the osteochondroma show continuity with the periosteum of the underlying bone. On T2-weighted MRI, a cartilage cap >1.5 cm after puberty may indicate possible malignant transformation [19].

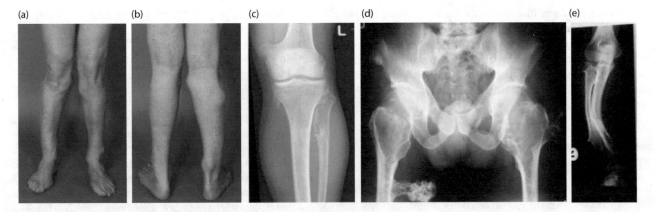

FIGURE 41.1 Hereditary multiple osteochondroma (HMO) in a 26-year-old male showing multiple lumps leading to deformity in the legs (a,b); multiple osteochondromas around the knee (c); the pelvis and proximal femur (d); the deformed forearm (shortening of the ulna with secondary bowing of radius) (e). (Photo credit: Bovée JV. *Orphanet J Rare Dis*. 2008;3:3.)

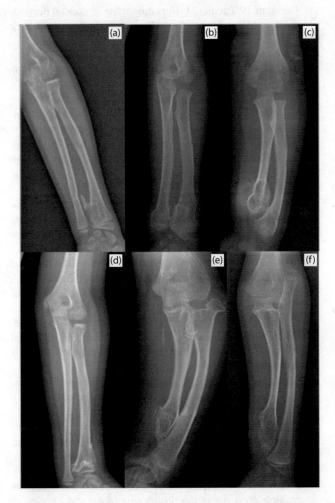

FIGURE 41.2 Forearm deformities associated with hereditary multiple osteochondroma (HMO). (a,b) Grade 2 osteochondromas of the distal radius or ulna without obvious shortening of either bone or proportionate shortening; (c) Grade 3, Masada type I, osteochondromas of the distal ulna or radius resulting in a relatively shortened ulna and a bowing radius; (d) Grade 3, Masada type III, osteochondromas affecting distal radius and resulting in a relatively shortened radius; (e) Grade 4, Masada type IIa, osteochondromas at the distal ulna and the proximal metaphysis of the radius, leading to a relatively shortened ulna and dislocated radial head; (f) Grade 4, Masada type IIb, osteochondromas at the radial head dislocated without a proximal radial osteochondroma. (Photo credit: Li Y et al. *BMC Med Genet*. 2017;18(1):126.)

Microscopically, osteochondroma contains endochondral ossification at the junction of the cartilaginous cap and the underlying bone, and yellow marrow in the medullary part.

Molecular identification of *EXT1* (56%–78%) and/or *EXT2* (21%–44%) pathogenic variants on peripheral blood by PCR and sequencing analysis provides further confirmation of HMO in 70%–95% of affected individuals. This is particularly useful for family members presenting a mild or no phenotype [22–24].

Differential diagnoses for HMO include *solitary osteochrondroma* (juxtacortical osteosarcoma, soft tissue osteosarcoma, and heterotopic ossification; absence of the continuity of cancellous and cortical bone from the host bone to the lesion; affecting 1%–2% of the general population), *metachondromatosis* (osteochondromas in the digits pointing toward the adjacent growth plate and not shortening or bowing of the long bone, joint deformity, or subluxation; intraosseous enchondromas; autosomal dominant disorder due to *PTPN11* mutation), *Langer–Giedion syndrome* (contiguous gene deletion syndrome involving *EXT1*; intellectual disability, characteristic craniofacial and digital anomalies; skeletal abnormalities; due to *TRPS1* haploinsufficiency), *11p11 deletion syndrome* (formerly DEFECT 11 or Potocki–Shaffer syndrome; contiguous gene deletion syndrome involving *EXT2* and *ALX4*; craniofacial abnormalities, syndactyly, intellectual disability; parietal foramina and ossification defects of the skull in case of *ALX4* deletion), *dysplasia epiphysealis hemimelica* (DEH, Trevor disease, tarso-epiphysial aclasis; developmental disorder with cartilaginous overgrowth of a portion of one or more epiphyses; the lower extremity on one side of the body; restricted to either the medial or lateral side of the limb [hemimelic]; usually diagnosed prior to the age of 15 years, more often in boys than in girls; lesion growth until puberty), and *enchondromatosis* (Ollier disease and Maffucci syndrome; multiple cartilage tumors in the medulla of bone, predilection for the short tubular bones and a unilateral predominance; nonhereditary syndromes) [2].

41.7 Treatment

Surgical resection is the mainstay treatment for HMO, especially for large, symptomatic lesions or lesions with suspicious imaging features (e.g., growth in a skeletally mature patient, irregular or indistinct margins, focal areas of radiolucency, osseous erosions or destruction), which induce pain (due to compression of tendons and muscles), cause severe osseous deformities, and increase risk for malignant transformation. Postoperative recurrence is 2%, and affected patients undergo an average of 2.7 surgical operations [1,25].

41.8 Prognosis and Prevention

Solitary osteochondroma is a benign lesion and does not affect life expectancy. About 1% of osteochondroma patients may experience malignant transformation [25].

HMO may cause severe osseous deformities, affecting patients' daily living activities, and have a higher risk (3%–5%) of malignant transformation.

Resulting from malignant transformation in the cartilage cap of osteochondroma, secondary peripheral chondrosarcoma confers 10-year survival rates of 83% and 29% for grade I and III chondrosarcomas, respectively. Further, osteosarcoma and spindle cell sarcoma may develop in the stalk of the osteochondroma occasionally.

Prenatal testing and preimplantation genetic diagnosis may be considered for at-risk pregnancies if *EXT1* and/or *EXT2* pathogenic variants are known in the family.

41.9 Conclusion

Hereditary multiple osteochondromas (HMO) represent a rare autosomal dominant condition that induces two or more (mean 15–18) benign cartilage-capped bony outgrowths (tumors) projecting from the external surface of bone containing a marrow cavity in continuity with the underlying bone, primarily at the juxta-epiphyseal region of long bones, in very young and adolescent patients, leading to chronic pain, nerve impingement, skeletal deformities, maligning degeneration, and other complications. The molecular mechanisms underlining the development of HMO are attributed to germline loss-of-function mutations in *EXT1* and *EXT2*, located at 8q24.11 and 11p11.2, respectively, causing HP deficiency, affecting HSPG biosynthesis, and compromising the efficiency of BMP during bone and cartilage formation. Diagnosis of HMO relies on clinical and radiological findings together with histopathological evaluation. Management of HMO centers on surgical removal of osteochondromas that cause pain and/or deformities.

REFERENCES

1. Pacifici M. Hereditary multiple exostoses: New insights into pathogenesis, clinical complications, and potential treatments. *Curr Osteoporos Rep.* 2017;15(3):142–52.
2. Wuyts W, Schmale GA, Chansky HA, Raskind WH. Hereditary multiple osteochondromas. In: Adam MP, Ardinger HH, Pagon RA et al. editors. *GeneReviews®* [Internet]. Seattle (WA): University of Washington, Seattle; 1993–2018. 2000 Aug 3 [updated 2013 Nov 21].
3. Alabdullrahman LW, Byerly DW. Osteochondroma. (eds) In: *StatPearls* [Internet]. Treasure Island (FL): StatPearls Publishing; 2020 Jan-. 2019 Nov 24.
4. Bovée JV. Multiple osteochondromas. *Orphanet J Rare Dis.* 2008;3:3.
5. Jones KB. Glycobiology and the growth plate: Current concepts in multiple hereditary exostoses. *J Pediatr Orthop.* 2011;31(5):577–86.
6. Hameetman L, Bovée JV, Taminiau AH, Kroon HM, Hogendoorn PC. Multiple osteochondromas: Clinicopathological and genetic spectrum and suggestions for clinical management. *Hered Cancer Clin Pract.* 2004;2(4):161–73.
7. Cuellar A, Reddi AH. Cell biology of osteochondromas: Bone morphogenic protein signalling and heparan sulphates. *Int Orthop.* 2013;37(8):1591–6.
8. Huegel J, Sgariglia F, Enomoto-Iwamoto M, Koyama E, Dormans JP, Pacifici M. Heparan sulfate in skeletal development, growth, and pathology: The case of hereditary multiple exostoses. *Dev Dyn.* 2013;242(9):1021–32.
9. Busse-Wicher M, Wicher KB, Kusche-Gullberg M. The exostosin family: Proteins with many functions. *Matrix Biol.* 2014;35:25–33.
10. Ruan W, Cao L, Chen Z, Kong M, Bi Q. Novel exostosin-2 mutation identified in a Chinese family with hereditary multiple osteochondroma. *Oncol Lett.* 2018;15(4):4383–9.
11. Akbaroghli S, Balali M, Kamalidehghan B et al. Identification of a new mutation in an Iranian family with hereditary multiple osteochondromas. *Ther Clin Risk Manag.* 2016;13:15–9.
12. Li Y, Han B, Tang J, Chen M, Wang Z. Identification of risk factors affecting bone formation in gradual ulnar lengthening in children with hereditary multiple exostoses: A retrospective study. *Medicine (Baltim).* 2019;98(5):e14280.
13. Mărginean CO, Meliţ LE, Mărginean MO. Daughter and mother diagnosed with hereditary multiple exostoses: A case report and a review of the literature. *Medicine (Baltim).* 2017;96(1):e5824.
14. Makhdom AM, Jiang F, Hamdy RC, Benaroch TE, Lavigne M, Saran N. Hip joint osteochondroma: Systematic review of the literature and report of three further cases. *Adv Orthop.* 2014;2014:180254.
15. Woodside JC, Ganey T, Gaston RG. Multiple osteochondroma of the hand: Initial and long-term follow-up study. *Hand (N Y).* 2015;10(4):616–20.
16. Beltrami G, Ristori G, Scoccianti G, Tamburini A, Capanna R. Hereditary multiple exostoses: A review of clinical appearance and metabolic pattern. *Clin Cases Miner Bone Metab.* 2016;13(2):110–8.
17. Mazza D, Fabbri M, Calderaro C et al. Chest pain caused by multiple exostoses of the ribs: A case report and a review of literature. *World J Orthop.* 2017;8(5):436–40.
18. Zoboski RJ. Osteochondroma and spinal cord compression in a patient with hereditary multiple exostoses: A case report. *J Chiropr Med.* 2017;16(1):72–7.
19. Kitsoulis P, Galani V, Stefanaki K et al. Osteochondromas: Review of the clinical, radiological and pathological features. *In Vivo.* 2008;22(5):633–46.

20. Gavanier M, Blum A. Imaging of benign complications of exostoses of the shoulder, pelvic girdles and appendicular skeleton. *Diagn Interv Imaging.* 2017;98(1):21–8.

21. Li Y, Wang J, Wang Z, Tang J, Yu T. A genotype-phenotype study of hereditary multiple exostoses in forty-six Chinese patients. *BMC Med Genet.* 2017;18(1):126.

22. Hong G, Guo X, Yan W et al. Identification of a novel mutation in the EXT1 gene from a patient with multiple osteochondromas by exome sequencing. *Mol Med Rep.* 2017;15(2):657–64.

23. Medek K, Zeman J, Honzík T et al. Hereditary multiple exostoses: Clinical, molecular and radiologic survey in 9 families. *Prague Med Rep.* 2017;118(2–3):87–94.

24. Long X, Li Z, Huang Y et al. Identification of pathogenic mutations in 6 Chinese families with multiple exostoses by whole-exome sequencing and multiplex ligation-dependent probe amplification: Case series. *Medicine (Baltim).* 2019;98(20):e15692.

25. D'Arienzo A, Andreani L, Sacchetti F, Colangeli S, Capanna R. Hereditary multiple exostoses: Current insights. *Orthop Res Rev.* 2019;11:199–211.

42

Keratitis–Ichthyosis–Deafness Syndrome

Dongyou Liu

CONTENTS

42.1 Introduction

Keratitis–ichthyosis–deafness (KID) syndrome is a rare autosomal dominant genodermatosis that typically manifests as keratitis (inflammation of the cornea causing pain, photophobia, neovascularization, blindness), bilateral sensorineural hearing loss, and ichthyosis/erythrokeratoderma (patches of red, thickened, scaly, and dry skin) along with partial hair loss and squamous cell carcinoma (SCC) of the skin and oral mucosa in 15% of patients.

Heterozygous missense mutations in the *GJB2* (gap junction protein β2) gene on chromosome 13q12.11 encoding connexin26 (gap junction protein) underlie the molecular pathogenesis of KID syndrome, although 64% of cases appear to evolve from sporadic mutations in *GJB2* [1].

42.2 Biology

Connexins (Cx) are a family of 21 structurally related transmembrane proteins (subunits) that contain four transmembrane domains (TM1-TM4) linked by two extracellular loops (ECL) and one intracellular loop (ICL), with the amino terminus (NT) and the carboxyl terminus (CT) located inside the cytoplasm. Posttranslational modification through phosphorylation on serines enables connexins to travel to membranes, assemble, degrade, and gate functional gap junction channels (GJC). Based on their molecular masses (ranging from 25 to 62 kDa), connexins are designated as connecxin26 (Cx26), connecxin30 (Cx30), connecxin32 (Cx32), connecxin43 (Cx43), connecxin46 (Cx46), connecxin50 (Cx50), etc. Further, according to sequence similarities at the nucleotide and amino acid levels, connexins are separated into three categories α, β, or γ. While most cell types express >1 connexin (e.g., keratinocytes make nine connexins, including Cx26, Cx30, Cx31, and Cx43), one connexin protein type may be produced by the cells of different tissues (e.g., Cx26, Cx30, Cx31, and Cx43 are secreted by the epithelia of the inner ear, cornea, and the epidermis and its appendages) [2,3].

With a short half-life of 2–4 h, connexins participate in intracellular connexin−protein interaction, cell−extracellular space exchange, and cell−cell communication through formation of hemichannels and GJC Specifically, six connexin subunits gather together as hemichannel (or gap junction hemichannel, also known as connexon) in the endoplasmic reticulum or Golgi body and then move to cellular membranes, where two hemichannels join through hydrophobic interactions to form GJC, which is an aqueous pore between the cytoplasm of two adjacent cells, facilitating the exchange of ions (K+, Ca2+), signaling molecules (IP3, cAMP, cGMP, ATP) and metabolites (e.g., glucose, sugar, amino acid, glutathione) (Figure 42.1). Via these activities, connexins activate signaling pathways and affect cellular phenotypes. Not surprisingly, total or partial connexin dysfunctions may lead to a variety of genetic disorders such as skin abnormalities, cardiopathies, neurodegenerative and developmental diseases, cataracts, hereditary deafness, and cancer (collectively known as connexinopathies) (Table 42.1) [4–6].

42.3 Pathogenesis

The *GJB2* (gap junction protein β2) gene on chromosome 13q12.11 spans 5.5 kb with 2 exons. While exon 1 is untranslated, the exon 2 encodes a 226 aa, 26 kDa gap junction protein (connexin 26 [Cx26]).

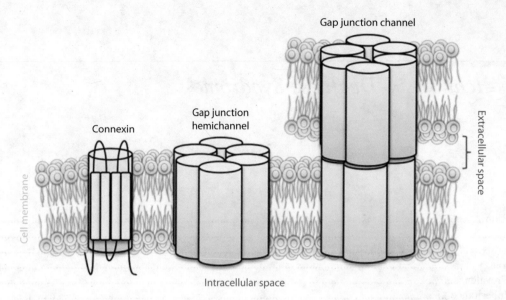

FIGURE 42.1 Schematic representation of connexin (each including four domains linked by two extracellular loops and one intracellular loop, with the N-terminus and the C-terminus located inside the cytoplasm), hemichannel (or gap junction hemichannel, each comprising six connexin subunits) and gap junction channel (each consisting of two hemichannels forming an aqueous pore between the cytoplasm of two neighboring cells). (Photo credit: Gudmundsson S et al. *Hum Mol Genet*. 2017;26(6):1070–7.)

Cx26 is expressed strongly in the stria vascularis of the cochlea, weakly in keratinocytes of the stratum basale and stratum granulosum as well as other organ systems, and in epithelial cells of the mammary gland. Cx26 coexpression with Cx30, Cx31, and Cx43 is also noted in human epidermal keratinocytes and cochlea [7].

Mutations in *GJB2* alter the structure and function of Cx26, leading to GJC permeability changes, reduced translocation, increased cell death, increased membrane flow, lower levels and slower diffusion of molecules, and hyperactive hemichannels. Apart from KID syndrome, Cx mutations are also implicated in the development of Vohwinkel syndrome (syndromic hearing loss), Bart–Pumphrey syndrome (syndromic hearing loss), syndromic sensorineural hearing loss with keratoderma (PPK; syndromic hearing loss), hystrix-like ichthyosis-deafness syndrome (HID; syndromic hearing loss), nonsyndromic deafness (autosomal dominant *DFNA3*, autosomal recessive *DFNB1*, *which is responsible for* as much as 50% of pre-lingual recessive deafness), and porokeratotic eccrine ostial and dermal duct nevus (PEODDN; skin manifestation only) [8–10].

To date, >100 *GJB2* mutations have been identified, with 18 implicated in KID syndromic phenotype (e.g., G11E, G12R,

N14 K, N14Y, S17F in the N-terminus, A40 V, G45E, D50A, D50N, D50Y in the TM1, A88 V in the TM2). Interestingly, G11E appears to enhance Ca^{2+} uptake; G12R, N14Y, and S17F increase intracellular Ca^{2+} concentration and ATP release; N14 K induces higher outward currents; A40 V impairs extracellular Ca^{2+}, pH, and Zn^{2+} regulation, increases whole cell currents and causes leaky hemichannels; G45E induces aberrant hemichannel activity, increases voltage sensitive gating, and enhances Ca^{2+} permeability; D50A enhances hemichannel activity; D50N reduces extracellular Ca^{2+} inhibition, decreases unitary conductance, increases open hemichannel current rectification and voltage-shifted activation; D50Y alters extracellular Ca^{2+} gating; A88 V increases hemichannel activity and CO_2 insensitivity. Furthermore, heterozygous missense mutations D50Y and D50N appear to cause more severe skin anomalies, and carcinoma of the tongue in Japanese patients [11–13].

In addition, revertant mosaicism (RM), in which the pathogenic effect of a germline mutation is corrected by a second somatic event, may be involved in the development of healthy-looking skin in a patient with KID syndrome (with a second-site de novo mutation independently inhibiting Cx26-Asp50Asn

TABLE 42.1

Examples of Genetic Disorders Linked to Connexin Dysfunctions

Connexin	Gene	Disease	Tumor (Mechanism)
Cx26	*GJB2*	Keratitis–ichthyosis–deafness syndrome	Breast (protein-connexin interaction, gap junction channel), cervical (hemichannel activity, gap junction channel)
Cx30	*GJB6*	Non-syndromic hearing loss	Glioma (gap junction channel), gastric (gap junction channel)
Cx32	*GJB1*	Charcot–Marie–Tooth neuropathy, X-linked 1	Breast (gap junction channel), renal cell carcinoma (protein–connexin interaction), ovarian (protein–connexin interaction)
Cx43	*GJA1*	Oculodentodigital dysplasia, keratoderma–hypotrichosis–leukonychia totalis syndrome	Brain (gap junction channel), breast (hemichannel activity), ovarian (protein–connexin interaction)
Cx46	*GJA3*	Congenital cataracts	Brain (gap junction channel)
Cx50	*GJA8*	Autosomal dominant congenital cataract	Cervical (gap junction channel)

expression in gap junction channels and reverting the dominant negative effect of the p.Asp50Asn mutation) [5].

42.4 Epidemiology

42.4.1 Prevalence

KID syndrome is a rare genodermatosis with a reported prevalence of <1 per 1,000,000. To date, approximately 100 KID syndrome cases (including 11 families) are described [14,15].

42.4.2 Inheritance

Although KID syndrome is considered an autosomal dominant disorder, which indicates that one copy of the altered *GJB2* gene in each cell is sufficient to cause the disorder, 64% of cases appear result from sporadic mutations in *GJB2*. There is also possible involvement of germinal mosaicism in the development of KID syndrome.

42.4.3 Penetrance

KID syndrome demonstrates a varied penetrance as patients harboring *GJB2* pathogenic variants may suffer from deafness without other syndromic features. Further, patients with certain pathogenic variants (e.g., the p.Ser17Phe mutation) may have a more severe phenotype and a higher risk for tongue carcinoma.

42.5 Clinical Features

KID syndrome is a congenital disorder that typically causes keratitis, bilateral deafness, and ichthyosis/erythrokeratoderma as well as occasional inflammatory nodules, skin infections, squamous cell carcinoma and other neoplasms, cyst formation, peripheral neuropathy, scoliosis, short stature, scarring alopecia, and scars of the face, axillae, and groin (Figures 42.2 and 42.3) [16–19]. Ocular involvements in 95% of patients include photophobia and blepharitis in early childhood, vascularizing keratitis, neovascularization, scarring, progressive decline in visual acuity, and finally blindness [20,21].

42.6 Diagnosis

Diagnosis of KID syndrome involves physical exam, histopathology, and molecular testing.

Clinical diagnosis of CID syndrome requires fulfilling five major criteria (erythrokeratoderma 100%; neurosensorial deafness 100%; vascularizing keratitis; reticulated palmoplantar hyperkeratosis; alopecia) and four minor criteria (susceptibility to infections; dental dysplasia; hypohidrosis; growth delay). Please note that minor criteria may not be always present.

Molecular identification of *GJB2* mutations provides further confirmation of KID syndrome [22].

42.7 Treatment

KID syndrome is associated with hearing loss and visual and keratinization anomalies, for which no specific remedies are available. Therefore, management of KID syndrome should aim for the rehabilitation of hearing loss (by removing blockage of the external ear canal with epithelial debris, and by use of cochlear implantation), and symptomatic relief of skin problems, visual disturbances, and other complications [23–26].

42.8 Prognosis and Prevention

In the absence of effective treatment, patients with KID syndrome rely on cochlear implantation for improved hearing and other measures for skin and visual disturbances.

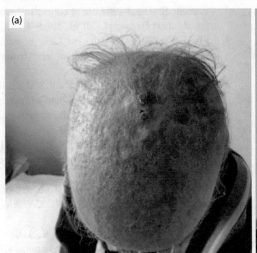 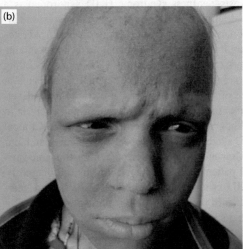

FIGURE 42.2 Keratitis–ichthyosis–deafness syndrome in a 13-year-old female showing (a) diffuse alopecia of scalp with cicatricial lesions at places, multiple hyperkeratotic hyperpigmented papulo-plaque lesions, and few lesions of folliculitis in the healing phase; (b) photophobia and erythrokeratoderma of both cheeks with hypotrichosis of the eyelashes and eyebrows. (Photo credit: Shanker V et al. *Indian Dermatol Online J*. 2012;3(1):48–50.)

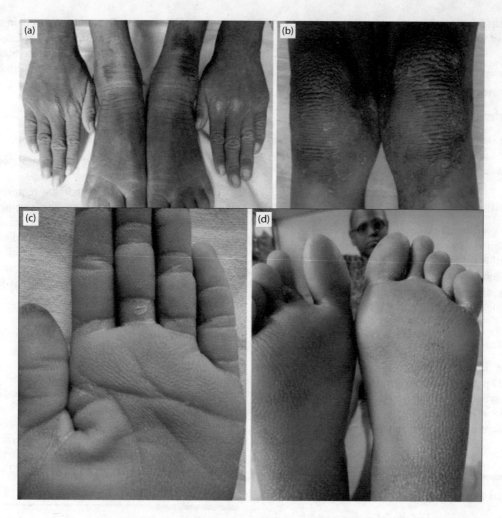

FIGURE 42.3 Keratitis–ichthyosis–deafness syndrome in a 13-year-old female showing diffuse hyperpigmentation and hyperkeratosis of the dorsa of hands and feet (a) and knees (b); keratoderma on the palm (c) and soles (d). (Photo credit: Shanker V et al. *Indian Dermatol Online J*. 2012;3(1):48–50.)

42.9 Conclusion

Keratitis–ichthyosis–deafness syndrome (KID syndrome) is a rare autosomal dominant disorder characterized by vascularizing keratitis, bilateral deafness, and ichthyosis/erythrokeratoderma as well as increased risk for squamous cell carcinoma. Heterozygous missense mutations in the *GJB2* gene encoding gap junction protein connexin26 (Cx26), which participates in the maintenance of tissue homeostasis, growth control, development, and synchronized response of cell stimuli, appear to underlie the pathogenesis of this congenital disease. Clinical diagnosis of KID syndrome relies on meeting five major and four minor criteria. Molecular identification of *GJB2* pathogenic variants in affected patients (36%) further confirms the diagnosis. Treatment of KID syndrome centers on rehabilitation of hearing loss, and standard procedures for symptomatic relief of skin problems, visual disturbances, and other complications.

REFERENCES

1. Lee JR, White TW. Connexin-26 mutations in deafness and skin disease. *Expert Rev Mol Med*. 2009;11:e35.
2. Xu J, Nicholson BJ. The role of connexins in ear and skin physiology - functional insights from disease-associated mutations. *Biochim Biophys Acta*. 2013;1828(1):167–78.
3. Bruzzone R. The double life of connexin channels: Single is a treat. *J Invest Dermatol*. 2015;135(4):940–3.
4. Lilly E, Sellitto C, Milstone LM, White TW. Connexin channels in congenital skin disorders. *Semin Cell Dev Biol*. 2016;50:4–12.
5. Gudmundsson S, Wilbe M, Ekvall S et al. Revertant mosaicism repairs skin lesions in a patient with keratitis-ichthyosis-deafness syndrome by second-site mutations in connexin 26. *Hum Mol Genet*. 2017;26(6):1070–7.
6. Sinyuk M, Mulkearns-Hubert EE, Reizes O, Lathia J. Cancer connectors: Connexins, gap Junctions, and communication. *Front Oncol*. 2018;8:646.
7. García IE, Prado P, Pupo A et al. Connexinopathies: A structural and functional glimpse. *BMC Cell Biol*. 2016;17(Suppl 1):17.
8. Meigh L, Hussain N, Mulkey DK, Dale N. Connexin26 hemichannels with a mutation that causes KID syndrome in humans lack sensitivity to CO2. *Elife*. 2014;3:e04249.
9. Sanchez HA, Verselis VK. Aberrant Cx26 hemichannels and keratitis-ichthyosis-deafness syndrome: Insights into syndromic hearing loss. *Front Cell Neurosci*. 2014;8:354.

10. García IE, Bosen F, Mujica P et al. From hyperactive connexin26 hemichannels to impairments in epidermal calcium gradient and permeability barrier in the keratitis-ichthyosis-deafness syndrome. *J Invest Dermatol*. 2016;136(3):574–83.

11. Mazereeuw-Hautier J, Chiaverini C, Jonca N et al. Lethal form of keratitis-ichthyosis-deafness syndrome caused by the GJB2 mutation p.Ser17Phe. *Acta Derm Venereol*. 2014;94(5):591–2.

12. Kutkowska-Kaźmierczak A, Niepokój K, Wertheim-Tysarowska K et al. Phenotypic variability in gap junction syndromic skin disorders: Experience from KID and Clouston syndromes' clinical diagnostics. *J Appl Genet*. 2015;56(3):329–37.

13. Hamadah I, Haider M, Chisti M. A novel homozygous mutation of GJB2-A new variant of keratitis-ichthyosis-deafness syndrome? *JAAD Case Rep*. 2019;5(3):283–7.

14. Al Fahaad H. Keratitis-ichthyosis-deafness syndrome: First affected family reported in the Middle East. *Int Med Case Rep J*. 2014;7:63–6.

15. Dalamón VK, Buonfiglio P, Larralde M et al. Connexin 26 (GJB2) mutation in an Argentinean patient with keratitis-ichthyosis-deafness (KID) syndrome: A case report. *BMC Med Genet*. 2016;17(1):37.

16. Fozza C, Poddie F, Contini S et al. Keratitis-ichthyosis-deafness syndrome, atypical connexin GJB2 gene mutation, and peripheral T-cell lymphoma: More than a random association? *Case Rep Hematol*. 2011;2011:848461.

17. Sakabe J, Yoshiki R, Sugita K et al. Connexin 26 (GJB2) mutations in keratitis-ichthyosis-deafness syndrome presenting with squamous cell carcinoma. *J Dermatol*. 2012;39(9):814–5.

18. Shanker V, Gupta M, Prashar A. Keratitis-Ichthyosis-Deafness syndrome: A rare congenital disorder. *Indian Dermatol Online J*. 2012;3(1):48–50.

19. Homeida L, Wiley RT, Fatahzadeh M. Oral squamous cell carcinoma in a patient with keratitis-ichthyosis-deafness syndrome: A rare case. *Oral Surg Oral Med Oral Pathol Oral Radiol*. 2015;119(4):e226–32.

20. Esmer C, Salas-Alanis JC, Fajardo-Ramirez OR, Ramírez B, Hua R, Choate K. Lethal keratitis, ichthyosis, and deafness syndrome due to the A88 V connexin 26 mutation. *Rev Invest Clin*. 2016;68(3):143–6.

21. Godillot C, Severino-Freire M, Michaud V et al. Keratitis-ichthyosis-deafness syndrome: Early death caused by the GJB2 mutation p.Gly12Arg. *Acta Derm Venereol*. 2019;99(10):921–2.

22. Levit NA, Sellitto C, Wang HZ et al. Aberrant connexin26 hemichannels underlying keratitis-ichthyosis-deafness syndrome are potently inhibited by mefloquine. *J Invest Dermatol*. 2015;135(4):1033–42.

23. Werchau S, Toberer F, Enk A, Helmbold P. Keratitis-ichthyosis-deafness syndrome: Response to alitretinoin and review of literature. *Arch Dermatol*. 2011;147(8):993–5.

24. Prasad SC, Bygum A. Successful treatment with alitretinoin of dissecting cellulitis of the scalp in keratitis-ichthyosis-deafness syndrome. *Acta Derm Venereol*. 2013;93(4):473–4.

25. Gumus B, Incesulu A, Pinarbasli MO. Cochlear implantation in patients with keratitis-ichthyosis-deafness syndrome: A report of two cases. *Case Rep Otolaryngol*. 2017;2017:3913187.

26. Wolfe CM, Davis A, Shaath TS, Cohen GF. Visual impairment reversal with oral acitretin therapy in keratitis-ichthyosis-deafness (KID) syndrome. *JAAD Case Rep*. 2017;3(6):556–8.

43

Li–Fraumeni Syndrome

Dongyou Liu

CONTENTS

43.1 Introduction

Li–Fraumeni syndrome (LFS) is an autosomal dominant disorder that predisposes to the development of multiple tumors (e.g., soft tissue sarcoma, osteosarcoma, premenopausal breast cancer, brain tumor, and adrenal cortical carcinoma), often at an early-onset age.

Molecularly, germline heterozygous mutations in the tumor suppressor gene *TP53* on chromosome 17p13.1 cause loss of function in its encoded protein p53, leaving DNA damages unrepaired and proliferation of cancerous cells unchecked [1].

43.2 Biology

Early evidence of a new familial cancer syndrome was obtained by Frederick Li and Joseph F. Fraumeni, Jr. in 1969 from four families in which sibs or cousins developed rhabdomyosarcoma or other soft tissue sarcomas in infancy while parents and relatives had early-onset breast cancer and other malignancies. Further investigations expanded its clinical profile to include early-onset sarcoma and premenopausal breast cancer along with osteosarcoma, adrenocortical carcinoma, brain tumor, leukemia, lymphoma, lung cancer, and other neoplasms, and linked this sarcoma, breast, leukemia, and adrenal gland (SBLA) cancer syndrome to mutations in the tumor suppressor gene *TP53* [2].

While a majority (>80%) of LFS cases develop in the background of germline *TP53* mutations, a small proportion (7%–20%) appear to evolve from sporadic mutation in one of the parent's germ cells, or de novo mutation during embryogenesis. *TP53* pathogenic variants may cause either loss-of-function or gain-of-function/dominant negative effect. Dominant negative mutations in the DNA-binding domain may block the function of the wild-type allele and often initiate cancer debut at an earlier age than loss-of-function mutations. Based on the well-known two-hit model, tumorigenesis frequently requires additional genetic damage causing loss of heterozygosity (LOH) and/or gain of function involving the *TP53* gene in the second allele [3].

Given that around 20%–30% of affected families do not carry *TP53* mutations, the involvement of other genes or mechanisms in the development of LFS is suspected. This led to the identification of checkpoint kinase 2 (*CHEK2*) gene on chromosome 22q12.1, encoding a key negative regulator of TP53, in Li–Fraumeni-like syndrome (LFL) cases, which show increased risk for multiple cancers in childhood but have a different pattern of specific cancers in comparison with classic LFS. In addition, alteration in *TP53* gene promoter, posttranslational modification of p53 protein, and genetic modifiers (e.g., MDM2-SNP309, 16 bp duplication in *TP53* intron 3 or PIN3, miR-605, short telomere, copy number variation) may also play accessory roles in the development of LFS [4,5].

Relative to other hereditary cancer syndromes that often develop few specific cancer types, LFS is noted for inducing five core cancers (i.e., premenopausal breast cancer 28%; soft tissue sarcoma 14%; brain tumor 13%; adrenocortical carcinoma 11%; osteosarcoma 8%) in children and adolescents, along with a large variety of other tumors (e.g., leukemia/lymphoma 4%; colorectal cancer 3%; lung adenocarcinoma, melanoma, pancreatic, renal cell, and thyroid tumors, as well as gonadal germ cell tumors) [6].

43.3 Pathogenesis

Located on the short arm of chromosome 17 (i.e., 17p13.1), the *TP53* (tumor protein 53) gene spans 25.7 kb with 11 exons and encodes a 393 aa, 53 kDa protein (TP53) with a multiplicity of cellular functions (e.g., induction of growth arrest, apoptosis, senescence, DNA repair; negative regulation of cell division).

Because of its critical role in rectifying DNA damages, *TP53* is dabbed "guardian of the genome." Under physiologic conditions, p53 protein levels remain low. During cellular stress and DNA damage, p53 is stabilized, activated, and subsequently translocated to the nucleus. Inside the nucleus, p53 acts as a transcription factor regulating the genes involved in cell-cycle arrest, differentiation, senescence, apoptosis, DNA repair, genomic stability, autophagy, angiogenesis, metabolism, and drug sensitivity [7].

Mutation in and deregulation of the *TP53* gene result in total or partial loss-of-function p53 protein. Of >700 germline mutations in the *TP53* gene identified to date, some are gain-of-function mutations (missense variants; 74%) that lead to accumulation of stable but inactive p53 protein in the nucleus of tumor cells, have a dominant-negative effect or promote an oncogenic effect, and are implicated in early-onset cancer; others are loss-of-function mutations (nonsense 9% or frameshift variants) that do not lead to accumulation of p53 protein and thus fail to carry out DNA repair and/or control the proliferation of tumor cells. Interestingly, many missense mutations are found in a hot-spot region of exons 5–8 (i.e., the DNA binding region of the gene), including hotspots at codons 125, 158, 175, 196, 213, 220, 245, 248, 273, 282, and 337. Further, missense mutations at codons 164–194 and codons 237–250 often occur in brain cancer, while those at codons 115–135, S2-S2-H2 motif: codons 273–286 are found in adrenocortical carcinoma. In addition, missense mutation p.Val31Ile is observed in patients with late-onset cancer, and frameshift mutation p.Pro98Leufs*25 is detected in patients with early-onset cancer. Partial loss of alleles is associated with decreased numbers of tumors and late-onset disease [8–11].

The *CHEK2* (checkpoint kinase 2) gene on chromosome 22q12.1 is 54 kb in length with 14 exons and encodes a 543 aa, 60.9 kDa serine/threonine-protein kinase involved in checkpoint-mediated cell cycle arrest, activation of DNA repair (through phosphorylation of BRCA2, enhancing the association of RAD51 with chromatin) and apoptosis (through phosphorylation of p53/TP53, MDM4, and PML) in response to DNA double-strand breaks, as well as negative regulation of cell cycle progression. Mutations in *CHEK2* are implicated in LFL, and also confer susceptibility to sarcoma, breast cancer, and brain tumor. However, carriers of *CHEK2* mutations have lower incidence of cancer in their families than those of *TP53* mutations [12–14].

43.4 Epidemiology

43.4.1 Prevalence

LFS is a rare hereditary cancer syndrome, with the frequency of germline *TP53* mutation estimated at 1:5000 to 1:20,000. Over 500 LFS families and 1827 confirmed carriers are recorded as of April 2016 in the database of the International Association for Research on Cancer (IARC). This clearly indicates possible underreporting of *TP53* germline mutations from diagnostic genetic laboratories or underdiagnosis of LFS globally [15].

The occurrence of specific malignancies in LFS-affected families differs during childhood (predominated by soft tissue sarcoma, brain tumor, and adrenocortical carcinoma between 0–10 years of age), adolescence (osteosarcoma between 11–20 years of age), and adulthood (breast cancer and brain tumor after 20 years of age). A bimodal distribution of LFS-related cancer is noted, including one peak before the age of 10 (e.g., osteosarcoma, adrenocortical carcinoma, glioma, and soft tissue sarcoma) and a second peak between the ages of 30 and 50 years (e.g., breast cancer and brain tumor). Males who have no childhood cancer tend to develop multiple primary tumors in their 50s [6].

43.4.2 Inheritance

LFS is an autosomal dominant disorder, for which one copy of the altered *TP53* gene in each cell is sufficient to pass the disease to 50% of offspring. Approximately 80% of patients harboring germline mutations in *TP53* have parents and other family members with similar cancer profile, while 7%–20% of patients appear to result from de novo mutations without family connection [3].

43.4.3 Penetrance

LFS demonstrates high penetrance and age-dependent expressivity. The risk for *TP53* mutation carriers to develop cancer by ages 20, 30, 40, and 50 years is calculated at 12%, 35%, 52%, and 80%, respectively. Female carriers of *TP53* pathogenic variants have a higher overall risk (93%) of developing breast cancer by age 50 years than male carriers (68%–73%). Further, affected women often present at an earlier age (29 years) than affected men (40 years). Genetic anticipation may also influence clinical presentation of LFS, where the inheritance of a germline *TP53* mutation is associated with earlier-onset tumors in successive generations [1,6].

43.5 Clinical Features

LFS typically causes sarcomas (e.g., osteosarcoma, rhabdomyosarcoma, leiomyosarcoma, orbital liposarcoma, atypical fibroxanthoma, spindle cell sarcoma, histiosarcoma, undifferentiated pleomorphic sarcoma; 25%–30% of cases), brain tumors (e.g., astrocytoma, glioblastoma, medulloblastoma, ependymoma, choroid plexus carcinoma, malignant triton tumor; 9%–16%), adrenocortical carcinoma (ACC; 10%–14%), breast cancer (25%–30%), and other tumors (e.g., colorectal, esophageal, pancreatic, stomach cancers; renal cell carcinoma; prostate, endometrial, ovarian, gonadal germ cell tumors; leukemia, Hodgkin and non-Hodgkin lymphomas; lung cancer; melanoma and non-melanoma skin cancer; non-medullary thyroid cancer). Of these, pre-menopausal breast cancer 28%, soft tissue sarcoma 14%, brain tumor 13%, adrenocortical carcinoma 11%, and osteosarcoma 8% are core cancers most closely associated with LFS. Interestingly, brain tumor, soft tissue sarcoma, and adrenocortical carcinoma commonly appear within the first decade of life, osteosarcoma predominates during the second decade, and breast cancer is often premenopausal. Further, male patients tend to develop brain tumor and hematopoietic and stomach cancers, whereas female patients are often diagnosed with adrenocortical carcinoma and skin cancer. Both males and females are equally affected by soft tissue and bone sarcomas [16–18].

Affected patients may show a range of cancer-related symptoms, including headaches, fever, fatigue, pain, new moles, lumps, or swellings, bleeding gums, changes in vision or nerve function, loss of appetite, and unexplained weight loss.

43.6 Diagnosis

Diagnostic workup for LFS ranges from medical history review, physical examination, imaging studies, laboratory assessment, histopathology, to molecular testing (Figures 43.1 through 43.3) [19–21].

Establishment of LFS diagnosis requires fulfillment of clinical criteria or identification of a germline pathogenic variant in *TP53* with or without family cancer history.

Clinical criteria for LFS include: (i) a sarcoma diagnosed before age 45 years; (ii) a first-degree relative (parent, sibling or child) with any cancer before age 45 years; (iii) first- or second-degree relative (grandparent, aunt/uncle, niece/nephew, or grandchild) with any cancer before age 45 years or a sarcoma at any age. Individuals meeting all three criteria are considered as having LFS. For those who only meet some of these criteria, a positive *TP53* test is sufficient for establishing the diagnosis of LFS [22].

Alternatively, Chompret criteria for diagnosing LFS and implementing germline *TP53* mutation screening program require meeting at least one of the following criteria: (i) familial presentation (patient with tumor belonging to LFS tumor spectrum [e.g., soft tissue sarcoma, osteosarcoma, CNS tumor, breast cancer, adrenocortical carcinoma, leukemia, bronchoalveolar lung cancer] before age 46 years AND at least one first- or second-degree relative with an above LFS tumor [except breast cancer if patient has breast cancer] before age 56 years or with multiple tumors at any age), (ii) multiple primary tumors (patient with adrenocortical carcinoma, choroid plexus carcinoma, or rhabdomyosarcoma of embryonal anaplastic subtype, irrespective of family history), (iii) rare tumors (patient with adrenocortical carcinoma, choroid plexus carcinoma, or rhabdomyosarcoma of embryonal anaplastic subtype, irrespective of family history), and (iv) early-onset breast cancer (breast cancer before age 31 years) [22].

Since 80% of families with features of LFS have an identifiable *TP53* pathogenic variant, molecular detection of *TP53* mutations provides a valuable tool for confirmation of LFS cases. By sequencing the entire *TP53* coding region (exons 2–11), about 95% of *TP53* pathogenic variants (including many missense variants) are identified. In general, a negative *TP53* result does not rule out of LFS; however, a negative test in individual with a family history of known *TP53* mutation does rule out LFS [1,22].

Individuals or families who do not meet clinical criteria or Chompret criteria for LFS (particularly individuals who are older than those with LFS and who are *TP53* negative) may be suspected of LFL.

Clinical criteria for LFL consist of two suggested definitions. LFL definition 1 (Birch definition) stipulates (i) a person diagnosed with any childhood cancer, sarcoma, brain tumor, or adrenal cortical tumor before age 45; (ii) a first- or second-degree relative diagnosed with a typical LFS cancer (e.g., sarcoma, breast cancer, brain cancer, adrenal cortical tumor, or leukemia) at any age; (iii) a first- or second-degree relative diagnosed with any cancer before age 60. LFL definition 2 (ELES criteria) stipulates two first- or second-degree relatives diagnosed with a

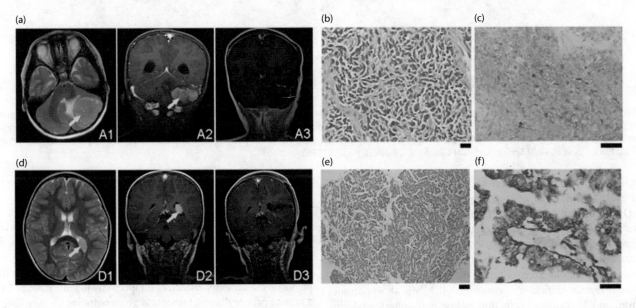

FIGURE 43.1 Li—Fraumeni syndrome in a 5-year-old girl harboring *TP53* mutation (c.730G > A; p.G244S) showing (a) tumor mass (indicated with arrows) located in the left cerebellar hemisphere (A1-A2), no recurrence at 18 months post-surgery (A3) on MRI; (b) medulloblastoma stained with H&E; (c) medulloblastoma stained with neuron-specific enolase (NSE) antibody. Li—Fraumeni syndrome in a 3-year-old boy (brother of above) harboring the same *TP53* mutation showing (d) tumor mass (indicated with arrows) in the posterior of left lateral ventricle (D1-D2), no recurrence at 18 months post-surgery (D3) on MRI; (e) choroid plexus papilloma stained with H&E; (f) choroid plexus papilloma stained with pan cytokeratin antibody. (Photo credit: Hu H et al. *Sci Rep.* 2016;6:20221.)

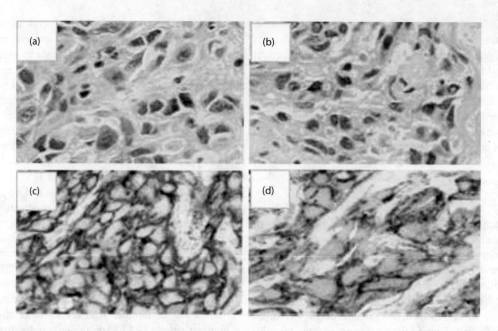

FIGURE 43.2 Li–Fraumeni syndrome in a 32-year-old female with TP53 mutation (IVS6 + 1G > T) showing epithelioid angiosarcoma composed of spindle cells with pleomorphic epithelioid eosinophilic cytoplasm and large hyperchromatic nuclei (a and b); diffuse expression of CD 31 (c) and CD 34 (d). (Photo credit: Barbosa OV et al. *Sao Paulo Med J.* 2015;133(2):151–3.)

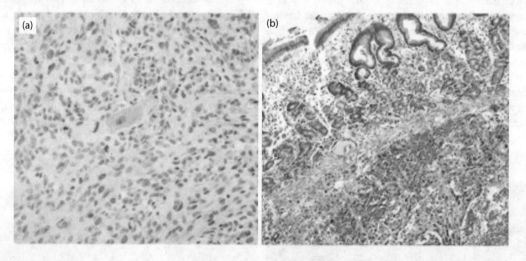

FIGURE 43.3 Li–Fraumeni syndrome in a 4-year-old-boy with *TP53* missense mutation (c.844C > T, p.Arg282Trp) showing diffusely infiltrative growing astrocytoma in brain sample (a). Li–Fraumeni syndrome a 27-year-old-female (mother of above) with the same *TP53* missense mutation displaying metastatic choriocarcinoma (marked nuclear pleomorphism, degenerative changes, syncytiotrophoblastic giant cells) in submucosal gastric mass (b). (Photo credit: Cotter JA et al. *Cold Spring Harb Mol Case Stud.* 2018;4(2). pii: a002576.)

typical LFS cancer (e.g., sarcoma, breast cancer, brain cancer, adrenal cortical tumor, or leukemia) at any age [14,15,23].

Differential diagnoses for LFS include *hereditary breast-ovarian cancer syndrome* (cancers of the breast, ovary, pancreas, prostate, and melanoma; due to *BRCA1* or *BRCA2* instead of *TP53*; inheritance of two mutated *BRCA2* alleles relates to Fanconi anemia type D1), and *constitutional mismatch repair deficiency syndrome* (CMMR-D syndrome; childhood leukemia, brain tumors, or early-onset gastrointestinal cancer; due to inheritance of two mutated alleles of a mismatch repair gene, such as *MLH1*, *MSH2*, *MSH6*, *PMS1*, and *PMS2*; inheritance of one mutated allele of an MMR gene relates to hereditary nonpolyposis colorectal cancer [HNPCC] or Lynch syndrome) [6,22].

43.7 Treatment

Treatment of LFS relies on standard procedures that provide symptomatic relief and prevent malignant transformation or spread of existing tumors [6].

Women with *TP53* positive breast cancer may be offered prophylactic mastectomy instead of lumpectomy to reduce the risk of a second primary breast tumor and avoid radiation therapy [24].

Combined therapy incorporating surgical resection and chemotherapy may be utilized for treatment of LFS-associated tumors. For example, two preoperative courses of neoadjuvant chemotherapy (a combination of cisplatin [CDDP] and tetrahydropyranyl-adriamycin [THP-ADR]) for hepatoblastoma (transitional liver

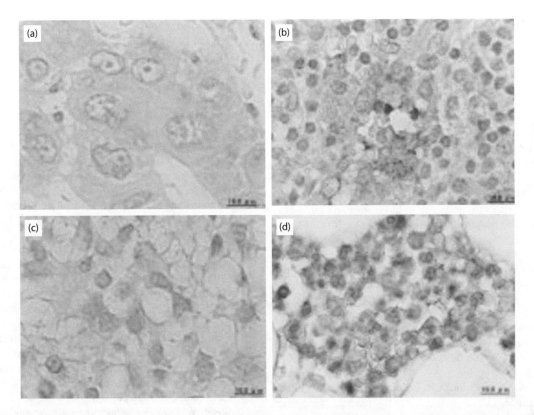

FIGURE 43.4 Li–Fraumeni syndrome in an 8-year-old-girl harboring *TP53* missense mutation (c.722 C > T, p.Ser241Phe) showing (a) transitional liver cell tumor (TLCT) sample prior to chemotherapy; (b) post-chemotherapy TLCT sample; (c) osteosarcoma sample prior to chemotherapy; (d) post-chemotherapy osteosarcoma sample (stained with antibody against CD44v8-10). (Photo credit: Yoshida GJ et al. *BMC Cancer.* 2012;12:444.)

cell tumor [TLCT], an intermediate between blastomatous tumor and adult-type tumor) followed by complete removal of residual liver tumor by surgical extended lobectomy was successful in treating a 8-year-old LFS patient harboring *TP53* missense mutation (c.722 C > T, p.Ser241Phe) (Figure 43.4a). Similarly, preoperative neoadjuvant chemotherapy (a combination of CDDP, ifosfamide [IFO], and methotrexate [MTX]) for osteosarcoma followed by total resection of the diminished osteosarcoma and replacement of an artificial joint yielded successful outcome in the same patient as assessed by specific antibody against CD44v8–10, with no sign of relapse or latent metastasis about 2 years after last surgery (Figure 43.4b) [25].

43.8 Prognosis and Prevention

LFS patients show a 50% risk of developing cancer before age 30, and 90% lifetime risk of developing any cancer type. Patients may develop second primary malignancies within 6–12 years after the first cancer, especially upon radiation exposure [26].

LFS patients with *TP53* genetic mutation are particularly sensitive to UV radiation and should avoid or minimize exposure to diagnostic and therapeutic radiation. They should also limit exposure to tobacco smoking and alcohol intake.

Psychological monitoring and support are helpful for LFS patients, who are prone to anxiety, depression, and clinically relevant distress.

Surveillance measures for children or adults with a germline *TP53* pathogenic variant include: (i) comprehensive annual physical examination; (ii) prompt evaluation of lingering symptoms and illnesses (e.g., headaches, bone pain, or abdominal discomfort) by physician; (iii) annual breast MRI and twice-annual clinical breast examination from age 20–25 years; (iv) routine screening for colorectal cancer with colonoscopy every 2–3 years from age 25 years [15].

Prenatal testing and preimplantation genetic diagnosis may be considered for families with known *TP53* pathogenic variant.

43.9 Conclusion

Li–Fraumeni syndrome (LFS) is an autosomal dominant disorder that predisposes to early development of five core cancers (premenopausal breast cancer, soft tissue sarcoma, brain tumor, adrenocortical carcinoma, and osteosarcoma) along with a variety of other tumors. The key mechanism underlying the pathogenesis of LFS relates to germline heterozygous mutations in the *TP53* gene, which are found in 80% of affected patients. Diagnosis of LFS requires fulfillment of clinical criteria and/or Chompret criteria, with molecular identification of *TP53* pathogenic variant providing further confirmation [5]. Treatment of LFS is based on standard procedures that provide symptomatic relief and prevent malignant transformation or spread of existing tumors [27]. As LFS confers 70% and nearly 100% lifetime risk of cancer in men and women, respectively, research and development of innovative approaches to halt the cancer progression would greatly improve the quality of life for affected patients.

REFERENCES

1. Correa H. Li-Fraumeni syndrome. *J Pediatr Genet*. 2016; 5(2):84–8.
2. Agarwal R, Liebe S, Turski ML et al. Targeted therapy for hereditary cancer syndromes: Hereditary breast and ovarian cancer syndrome, Lynch syndrome, familial adenomatous polyposis, and Li-Fraumeni syndrome. *Discov Med*. 2014;18(101):331–9.
3. Guha T, Malkin D. Inherited TP53 mutations and the Li-Fraumeni syndrome. *Cold Spring Harb Perspect Med*. 2017;7(4). pii: a026187.
4. Silva AG, Krepischi AC, Pearson PL, Hainaut P, Rosenberg C, Achatz MI. The profile and contribution of rare germline copy number variants to cancer risk in Li-Fraumeni patients negative for TP53 mutations. *Orphanet J Rare Dis*. 2014;9:63.
5. Batalini F, Peacock EG, Stobie L et al. Li-Fraumeni syndrome: Not a straightforward diagnosis anymore-the interpretation of pathogenic variants of low allele frequency and the differences between germline PVs, mosaicism, and clonal hematopoiesis. *Breast Cancer Res*. 2019;21(1):107.
6. Schneider K, Zelley K, Nichols KE, Garber J. Li-Fraumeni syndrome. In: Pagon RA, Adam MP, Ardinger HH et al. editors. *GeneReviews®* [Internet]. Seattle (WA): University of Washington, Seattle; 1993–2017. 1999 Jan 19 [updated 2013 Apr 11].
7. Sorrell AD, Espenschied CR, Culver JO, Weitzel JN. Tumor protein p53 (TP53) testing and Li-Fraumeni syndrome: Current status of clinical applications and future directions. *Mol Diagn Ther*. 2013;17(1):31–47.
8. Xu J, Qian J, Hu Y, Wang J, Zhou X, Chen H, Fang JY. Heterogeneity of Li-Fraumeni syndrome links to unequal gain-of-function effects of p53 mutations. *Sci Rep*. 2014;4:4223.
9. Borges LM, Ayres FM. R337H mutation of the TP53 gene as a clinical marker in cancer patients: A systematic review of literature. *Genet Mol Res*. 2015;14(4):17034–43.
10. Pantziarka P. Primed for cancer: Li Fraumeni syndrome and the pre-cancerous niche. *Ecancermedicalscience*. 2015;9:541.
11. AlHarbi M, Mubarak N, AlMubarak L et al. Rare TP53 variant associated with Li-Fraumeni syndrome exhibits variable penetrance in a Saudi family. *NPJ Genom Med*. 2018;3:35.
12. Cho Y, Kim J, Kim Y, Jeong J, Lee KA. A case of late-onset Li-Fraumeni-like syndrome with unilateral breast cancer. *Ann Lab Med*. 2013;33(3):212–6.
13. Chao A, Lai CH, Lee YS, Ueng SH, Lin CY, Wang TH. Molecular characteristics of endometrial cancer coexisting with peritoneal malignant mesothelioma in Li-Fraumeni-like syndrome. *BMC Cancer*. 2015;15:8.
14. Giacomazzi CR, Giacomazzi J, Netto CB et al. Pediatric cancer and Li-Fraumeni/Li-Fraumeni-like syndromes: A review for the pediatrician. *Rev Assoc Med Bras*. 2015;61(3):282–9.
15. Aedma SK, Kasi A. *Li Fraumeni Syndrome. StatPearls [Internet]*. Treasure Island (FL): StatPearls Publishing; 2019 Jan–2019 Apr 5.
16. Park KJ, Choi HJ, Suh SP, Ki CS, Kim JW. Germline TP53 mutation and clinical characteristics of Korean patients with Li-Fraumeni syndrome. *Ann Lab Med*. 2016;36(5):463–8.
17. Nandikolla AG, Venugopal S, Anampa J. Breast cancer in patients with Li-Fraumeni syndrome - a case-series study and review of literature. *Breast Cancer (Dove Med Press)*. 2017;9:207–15.
18. Swaminathan M, Bannon SA, Routbort M et al. Hematologic malignancies and Li-Fraumeni syndrome. *Cold Spring Harb Mol Case Stud*. 2019;5(1). pii: a003210.
19. Barbosa OV, Reiriz AB, Boff RA, Oliveira WP, Rossi L. Angiosarcoma in previously irradiated breast in patient with Li-Fraumeni syndrome. A case report. *Sao Paulo Med J*. 2015;133(2):151–3.
20. Hu H, Liu J, Liao X et al. Genetic and functional analysis of a Li Fraumeni syndrome family in China. *Sci Rep*. 2016;6:20221.
21. Cotter JA, Szymanski L, Karimov C et al. Transmission of a TP53 germline mutation from unaffected male carrier associated with pediatric glioblastoma in his child and gestational choriocarcinoma in his female partner. *Cold Spring Harb Mol Case Stud*. 2018;4(2). pii: a002576.
22. Vogel WH. Li-Fraumeni syndrome. *J Adv Pract Oncol*. 2017;8(7):742–6.
23. Zhuang X, Li Y, Cao H, Wang T et al. Case report of a Li-Fraumeni syndrome-like phenotype with a de novo mutation in CHEK2. *Medicine (Baltim)*. 2016;95(29):e4251.
24. Langan RC, Lagisetty KH, Atay S, Pandalai P, Stojadinovic A, Rudloff U, Avital I. Surgery for Li Fraumeni syndrome: Pushing the limits of surgical oncology. *Am J Clin Oncol*. 2015;38(1):98–102.
25. Yoshida GJ, Fuchimoto Y, Osumi T et al. Li-Fraumeni syndrome with simultaneous osteosarcoma and liver cancer: Increased expression of a CD44 variant isoform after chemotherapy. *BMC Cancer*. 2012;12:444.
26. Mai PL, Best AF, Peters JA et al. Risks of first and subsequent cancers among TP53 mutation carriers in the National Cancer Institute Li-Fraumeni syndrome cohort. *Cancer*. 2016;122(23):3673–81.
27. Aedma SK, Kasi A Li Fraumeni Syndrome. [Updated January 13, 2020]. In: StatPearls [Internet]. Treasure Island (FL): StatPearls Publishing; 2020 Jan-.

44

McCune–Albright Syndrome

N. Anitha

CONTENTS

44.1 Introduction

McCune–Albright syndrome (MAS) is a rare, non-inherited disorder linked to early embryonic somatic mosaic activating or gain-of-function mutations in the *GNAS* gene that encodes guanine nucleotide binding protein, alpha stimulating. Originally defined by the classic triad of polyostotic fibrous dysplasia, café au-lait skin pigmentation, and precocious puberty, the clinical spectrum of MAS has been subsequently expanded to include hyperthyroidism, Cushing syndrome, pituitary gigantism/acromegaly (if mutated cells are present in thyroid, adrenal, and/or pituitary tissues), renal phosphate wasting with or without rickets/osteomalacia, and hepatic and cardiac involvement [1].

44.2 Biological Background

Initially described in 1937 by McCune and Bruch as well as Albright and colleagues, MAS (also known as Albright–McCune–Sternberg syndrome, Albright-Sternberg syndrome, Albright syndrome, Albright's disease, Albright's disease of bone, Albright's syndrome with precocious puberty, fibrous dysplasia with pigmentary skin changes and precocious puberty, osteitis fibrosa disseminata, polyostotic fibrous dysplasia) is a condition affecting skeletal

(fibrous dysplasia), skin (café-au-lait spots), and certain endocrine organs (hormone-producing tissues) [2]. However, MAS is clinically heterogeneous, and its presentation may include various other endocrinologic anomalies such as thyrotoxicosis, pituitary gigantism, and Cushing syndrome [3].

Representing one type of fibro-osseous lesions that are characterized by the excessive proliferation of cellular fibrous connective tissue containing foci of mineralization and irregular bony trabeculae that replace normal bone, fibrous dysplasia may affect a single bone (monostotic, particularly the cranium) or multiple bones (polyostotic, especially the femur, tibia, ribs, and facial bones) [4]. Polyostotic fibrous dysplasia has abnormal areas (lesions) in multiple bones confined to one side of the body. Replacement of bone with fibrous tissue often leads to fractures, uneven growth, and deformity (e.g., asymmetric growth of the face or leg, and abnormal curvature of the spine or scoliosis) as well as bone cancer (<1% of MAS cases) [5–7]. Apart from MAS, fibrous dysplasia also occurs in Mazabraud syndrome and Jaffe–Lichtenstein syndrome (Table 44.1).

Fibro-osseous lesions are a diverse group of processes that may be developmental (hamartomatous), reactive, dysplastic, or neoplastic [8,9]. Based on Waldron's classification system, fibro-osseous lesions of jaws consist of: (i) fibrous dysplasia, (ii) cemento-osseous dysplasia (fibro-osseous or cemental lesions presumably arising in the periodontal ligament), and (iii) ossifying

TABLE 44.1

Syndromes That Include Fibrous Dysplasia as Part of Clinical Presentation

McCune–Albright Syndrome	Mazabraud Syndrome	Jaffe–Lichenstein Syndrome
• Fibrous dysplasia • Cafe-au-lait spots • Endocrine dysfunction (sexual precocity, pituitary adenoma or hyperthyroidism)	• Fibrous dysplasia • Myxomas	• Fibrous dysplasia • Cafe-au-lait spots • No endocrine dysfunction

TABLE 44.2

Waldron Classification of Fibro-Osseous Lesions of Jaws [10]

Lesion	Characteristic
Fibrous dysplasia	Monostotic
	Polyostotic
Cemento-osseous dysplasia	Periapical cemental dysplasia
	Localized fibro-osseous cemental lesions (probably reactive in nature)
	Florid cemento-osseous dysplasia (gigantiform cementoma)
	Ossifying and cementifying fibroma
Ossifying fibroma	Cementoblastoma, osteoblastoma, and osteoid osteoma
	Juvenile active ossifying fibroma and other so-called aggressive, active ossifying/cementifying fibromas

fibroma (fibro-osseous neoplasms of uncertain or detectable relationship to those arising in the periodontal ligament) (Table 44.2) [10]. Ossifying fibroma differs from fibrous dysplasia by having a well-circumscribed and sharply defined margin [11].

At the molecular level, MAS is attributed to random, postzygotic, somatic, activating or gain-of-function mutations in the *GNAS* gene, leading to the formation of a monoclonal population of mutated cells harboring an altered copy of the *GNAS* gene within variously affected tissues, alongside normal cells with an unchanged *GNAS* gene. Furthermore, the demonstration of the mutation in peripheral blood leukocytes but not in DNA from biopsies of clinically normal skin highlights that MAS is a disorder of mosaicism resulting from postzygotic somatic cell mutation. The relative number and location of cells harboring the mutated *GNAS* gene ultimately determines the features and severity of MAS, while the non-mosaic state for most activating mutations is presumably lethal to the embryo [1].

Reflecting a state of coexistence of genetically distinct cell populations in a single individual arising from a single fertilized egg as a consequence of post-zygotic genetic events, mosaicism may occur in somatic or germline cells (and referred to as somatic mosaicism or germline mosaicism). Somatic mosaicism may evolve from errors in replication of segments of or whole chromosome (chromosomal aneuploidy), copy-neutral reciprocal gains and losses (acquired uniparental disomy or loss of heterozygosity), changes in nuclear or mitochondrial DNA, point mutation, or spontaneous reversion of an existing DNA mutation. Whereas somatic mosaicism taking place in early development is associated with widespread disease, that in late development may

be silent or induce limited disease. However, somatic mosaic mutations do not pass onto offspring. By contrast, germline mosaicism has no phenotypic consequences on an individual, but will pass onto offspring [1,12].

44.3 Pathogenesis

Located on the long arm of chromosome 20 (i.e., 20q13.32), the *GNAS* gene consists of 13 exons spanning over 20-kb, and encodes a 394 aa, 46 kDa α-subunit of the stimulatory G protein (Gsα) [13,14]. The Gsα protein comprises two domains: a guanosine triphosphate hydrolase (GTPase) domain involved in the binding and hydrolysis of GTP, and a helical domain that buries GTP within core of protein and determines the intrinsic GTPase activity of the α subunit [15]. Being ubiquitously expressed, Gsα couples hormone receptors to adenyl cyclase for the generation of intracellular cyclic adenosine $3',5'$-monophosphate (cAMP) that mediates the activity of peptide hormones (e.g., PTH, TSH, gonadotropins, ACTH, GHRH, ADH, glucagon, calcitonin) [16,17].

MAS represents an outcome of mosaic somatic activating (gain-of-function) mutations in *GNAS* that induce amino acid substitution of either residue arg201 (e.g., R201C, R201H, R201S, and R201G; >95% of cases) or gln227 (e.g., Q227R and Q227 K; <5% of cases) in the α subunit of Gsα associated with the cAMP-dependent pathway. Specifically, arg201 mutations result in activation of cAMP, while mutations at arg201 or gln 227 inhibit the GTPase activity and maintain Gsα active form. Together, these missense mutations generate a constitutively activated form of Gsα with reduced GTPase catalytic ability leading to failure in controlling the Gsα activation and cAMP production, constitutive activation of adenylate cyclase, overproduction of several hormones, and disruption of the signaling turnoff mechanism. Enhanced Gsα signaling and cAMP production accelerate the commitment of bone marrow stromal cells into the osteoblastic lineage and their further differentiation into osteoblasts, resulting in the formation of fibrous dysplastic lesions consisting of fibrous cells that express early osteoblastic markers (e.g., alkaline phosphatase). Furthermore, increased Gsα signaling expedites the action of alpha-MSH for melanin production and skin lesion formation. Persistently high levels of cAMP also activate PKA and mitogen-activated protein kinase (MAPK, as seen in cardiac hypertrophy), induce overexpression of cellular oncogene *fos* (*c-fos* gene, as seen in fibrous dysplasia), and alter the processing of the FGF23 protein by GALNT3 and furin, leading to elevated levels of the inactive C-terminal fragment of FGF23. Given the ubiquitous expression of Gsα, MAS can affect tissues derived from ectoderm, mesoderm and endoderm, including the skin, skeleton, and endocrine organs [18].

Besides somatic activating (gain-of-function) *GNAS* mutations seen in MAS, germline inactivating (loss-of-function) *GNAS* mutations contribute to multiple phenotypes, ranging from pseudohypoparathyroidism (PPHP; due to an inactivating mutation of the paternal *GNAS* allele resulting in expression of the protein product Gsα only from the maternal allele), pseudohypoparathyroidism 1A (due to an inactivating mutation of the maternal *GNAS* allele resulting in expression of the protein product Gsα only from the paternal allele), pseudohypoparathyroidism IB (due to deletion of the regulatory differentially methylated region of the *GNAS* locus), to progressive osseous heteroplasia (POH; due possibly to an inactivating *GNAS* mutation of the paternal allele) [19,20].

44.4 Epidemiology

44.4.1 Prevalence

MAS is a rare disease and has an estimated prevalence of between 1/100,000 and 1/1,000,000 worldwide. Fibrous dysplasia has been reported to account for about 7% of all benign bone tumors. In a retrospective study of 80 cases with benign fibro-osseous lesions over 20 years, fibrous dysplasia was found in 24.4%, with a majority occurring in the second decade (65%), and a clear male predilection (male to female ration of 1.8:1) [1].

44.4.2 Inheritance

MAS is a non-inherited disorder resulting from a random, somatic, activating mutation in the *GNAS* gene that occurs in the very early stage of development. This creates a monoclonal population of mutated cells containing an altered *GNAS* gene alongside normal cells containing an intact *GNAS* gene. While affected individuals may also have reproductive cells (eggs or sperm) with the mutation, the resulting embryo is likely to harbor the mutation in every cell. As this non-mosaic state involving most activating mutations is lethal to the embryo, MAS is not passed to the next generation [1].

44.5 Clinical Features

MAS shows involvement of skeleton (fibrous dysplasia of bone), skin (skin pigmentation or café-au-lait spots), and certain endocrine organs (dysfunction of certain endocrine glands) as well as additional tissues [21,22].

44.5.1 Fibrous Dysplasia

Fibrous dysplasia is a fibro-osseous lesion that may affect a single bone (monostotic), multiple bones (polyostotic) or multiple bones, with skin pigmentation and endocrine abnormalities (Table 44.3) [23].

TABLE 44.3

Clinical Severity Is Dependent upon the Time of *GNAS* Mutation

Mutation	Manifestation
At early embryologic life (mutation in one undifferentiated stem cells)	Multiple bone lesions, skin pigmentation (café-au-lait spots), endocrine disturbance
At later stage of embryologic life (mutation in skeletal progenitor cells)	Multiple bone lesions (fibrous dysplasia)
At postnatal life (progeny of mutated cell confined to one site)	Fibrous dysplasia affecting a single bone

44.5.1.1 Monostotic Fibrous Dysplasia

The monostotic fibrous dysplasia is limited to single bone, and jaws (particularly the maxilla) are the more common site. It is mostly diagnosed in the second decades of life and shows an equal gender predilection. The condition is manifested as painless swelling. Maxillary lesions often involve the adjacent zygoma, sphenoid, and occiput and is designated as craniofacial fibrous dysplasia, and mandibular lesions are usually monostotic [24].

44.5.1.2 Polyostotic Fibrous Dysplasia/ Jaffe–Lichtenstein Syndrome/ McCune–Albright Syndrome

Polyostotic fibrous dysplasia may involve minimal to 75% of entire skeleton (Table 44.1).

In fibrous dysplasia, the bone is replaced by abnormal connective tissue. This weakens the bone and makes it fragile and prone to fracture. Fibrous dysplasia often presents unilaterally. Depending on the bone involved, the symptoms may vary (Table 44.4). Usually affected parts of skeleton are the long bones of arms and legs, bones of the face and skull (craniofacial areas), and the ribs (Figure 44.1).

44.5.2 Skin Manifestations

Skin manifestation in MAS is an abnormal skin pigmentation known as café-au-lait spots, which appear as light brown macules with irregular outline. They are present at birth or neonatal period and become more pronounced with age (Figure 44.2).

44.5.3 Endocrine Abnormalities

The endocrine system comprises the system of glands that regulates the body's rate of growth, sexual development, and certain metabolic functions. MAS is associated with early onset of puberty (gonadotropin independent precocious puberty) and also development of secondary sexual characteristics at an early

TABLE 44.4

Clinical Features of Fibrous Dysplasia

Site of Involvement	Symptoms
Long bones	• Frequent fractures and bowing (weight bearing bones) • Leg length discrepancy (in children) • Eventually affect the ability to walk (abnormal gait, e.g., walking with a limp)
Craniofacial region	• Pain • Nasal congestion • Malaligned teeth • Facial asymmetry • Frontal bossing • Proptosis • Vertical dystopia
Lesions compressing nearby nerves (rare)	• Compression of optic nerve (vision loss) • Compression of auditory nerve (hearing impairment)

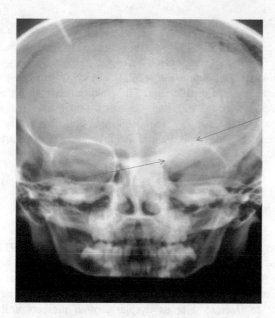

FIGURE 44.1 Ground glass appearance involving roof, lateral and medial walls of left orbit. (Photo credit: Gutch M et al. *J ASEAN Fed Endocr Soc.* 2015;30(1):40–3.)

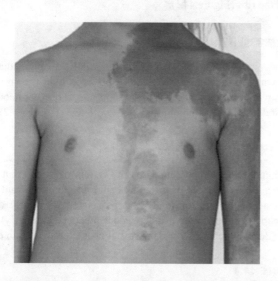

FIGURE 44.2 Café-au-lait skin pigmentation on the chest and arm of a 5-year-old girl with McCune–Albright syndrome. (Photo credit: Dumitrescu CE, Collins MT. *Orphanet J Rare Dis.* 2008;3:12.)

age. This is due to early activation of glands that secrete sex hormones.

In females, there may be early vaginal bleeding or development of breasts, ranging from the first few months of life to 6 or 7 years of age. The early onset of menstruation is possibly caused by excess estrogen produced by cysts that develop in one of the ovaries. Less commonly, boys may also experience precocious puberty, with enlargement of penis and one or both testicles, in addition to thickened scrotum and growth of pubic hair and hair under the armpits at an early age.

Endocrine disorders also affect the thyroid, pituitary, and adrenal glands. Enlargement of the thyroid gland (goiter), development of thyroid nodules, and increased secretion of thyroid hormone (hyperthyroidism) may occur. The symptoms of hyperthyroidism are fast heart rate, high blood pressure, weight loss, tremors, sweating, anxiety, and heat intolerance. Growth hormone levels are increased in some individuals, affecting growth and muscle mass. Increased growth hormone may cause large head (macrocephaly, due to excess expansion of fibrous dysplasia in the skull), vision problems, and acromegaly (a condition of growth excess after closure of epiphyseal plates, leading to large hands and feet, arthritis, and coarse facial features) [25].

Cortisol (adrenal gland hormones) regulates metabolism of glucose and modulates stress. In MAS, cortisol levels are elevated and lead to Cushing syndrome (showing weight gain in the face and upper body, slowed growth in children, fragile skin, fatigue, and other health problems) before age 2. Certain individuals show elevated phosphate levels in blood, as renal absorption of phosphate is impaired. Fibrous dysplasia tissue produces a protein (fibroblast growth factor 23 [FGF 23]) that causes inability of the kidney to metabolize phosphate. Hypophosphatemia is seen likely in fibrous dysplasia individuals, leading to rickets (childhood) or osteomalacia (adults). Children with rickets have bowing deformity of the legs and short stature, while adults with osteomalacia suffer from softening of bone, fracture at younger age, and bone pain [1].

44.5.4 Other Tissues

Less common symptoms of MAS include gastroesophageal reflux, gastrointestinal polyps, pancreatitis, tachycardia, aortic root dilatation, and high-output heart failure [1].

44.6 Diagnosis

44.6.1 Diagnostic Workup

Diagnosis of MAS relies on the finding of two or more typical clinical features (i.e., café-au-lait spots, bone and endocrine abnormalities) along with patient medical history, radiography, blood tests, histopathology, and molecular testing [26,27].

X-ray studies evaluate the presence and extent of fibrous dysplasia.

Blood tests examine elevated hormone levels (estrogen, testosterone, cortisol, thyroid hormone, growth hormone, prolactin, somatomedin C) and increased body activity (elevated alkaline phosphatase).

Bone scan (bone scintigraphy) is useful for determining the extent of bone disease. A harmless radioactive dye is injected into the affected bone and a special camera is used to track the dye as it travels through bone.

Bone biopsy reveals the characteristic features of bone affected by fibrous dysplasia. Also, biopsy helps distinguish fibrous dysplasia from growth or tumors [28].

Radiography. Plain radiographic study helps reveal the characteristic "ground glass appearance" (resembling the small fragments of a shattered windshield) in MAS (Figure 44.1). Long bone has "lytic appearance." Long and short bones show radiolucent lesion in diaphysis or metaphysis with scalloping of endosteum with or without bone expansion and absence of periosteal reaction. The radiolucent lesion displays thick sclerotic border called as the "rind sign." The frontal bone is commonly involved

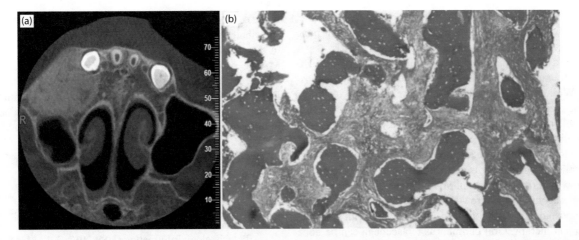

FIGURE 44.3 Cone beam computed tomography scan reveals typical granular trabecular pattern and bicortical expansion with more than half of maxillary sinus involvement in relation to the maxillary right back tooth region (a). Histological examination shows immature woven bone with fibrous stroma (b). (Photo credit: Nair SN et al. *Case Rep Dent.* 2016;2016:6439026.).

among the facial and skull bones with obliteration of frontal and sphenoid sinuses. The maxillary and mandibular bone involvements produce mixed radiolucency and radiopacity with displacement of the teeth and distortion of the nasal cavities [29].

Histopathology. Biopsy of lesional bone and its histopathologic study confirm the radiographic diagnosis. Lesional tissue shows fibrillar connective tissue containing numerous trabeculae of coarse woven immature bone, irregular in shape and evenly spaced (Figure 44.3). Osteocytes are large and bone formation by stellate osteoblasts. Bone formation can be seen but cuboidal osteoblastic rimming is not seen. Osteoclastic activity may also be seen in areas of calcification of osteoid, which extends to the surface of the trabeculae [30].

Molecular testing. Polymerase chain reaction (PCR) can be used to detect somatic mutation in the *GNAS* gene. Due to the mosaic nature of MAS, a negative genetic test result (e.g., in blood) does not exclude the presence of the mutation in other tissues. Individuals with only monostotic fibrous dysplasia require identification of a *GNAS* somatic activating mutation for confirmation, based on sequencing analysis of exons 8 and 9, with mutations p.Arg201His and p.Arg201Cys detected in 8%–90% and 75%–100% of cases, respectively [31,32].

44.6.2 Diagnostic Consideration

Café-au-lait spots associated with MAS tend to have irregular borders, while those observed in other disorders have smooth borders. Like the bone lesions, café-au-lait spots in MAS may appear on only one side of the body.

Patients with recurrent fracture with minimal trauma or without trauma can be considered for diagnosis of MAS. If multiple fractures of bone with deformity is seen, a milder form of osteogenesis imperfecta should be considered.

Fibrous dysplasia of the skeleton involving the skull base is common. Nuclear medicine bone scan helps in detecting the extent of disease and prediction of functional outcome. Plain radiographs show classical ground glass appearance, but small microfractures causing pain can be detected by CT scan or MRI. Fibrous dysplasia of proximal femur demonstrates shepherd crook deformity (Figure 44.4) [33].

Precocious puberty commonly occurs in girls. Testicular masses of Leydig/Sertoli cell hyperplasia are seen in boys with MAS, and ultrasound is used to detect the same. These testicular masses cannot be detected by physical examination, and malignant transformation is rare.

Hyperthyroidism is common. Thyroid involvement without frank hyperthyroidism is more common. Thyroid-stimulating hormone (TSH) will be suppressed and triiodothyronine will be elevated. Ultrasound reveals abnormal thyroid gland. Regular follow-up of thyroid hormones and thyroid scan should be done.

44.7 Treatment

In the absence of a cure, MAS is usually managed with symptomatic relief. Precocious puberty can be treated with aromatase inhibitors, gonadotropin-releasing hormone, ketoconazole,

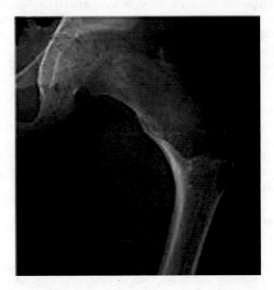

FIGURE 44.4 Radiography of fibrous dysplasia in a 10-year-old child showing a proximal femur with typical ground glass appearance and shepherd's crook deformity. (Photo credit: Dumitrescu CE, Collins MT. *Orphanet J Rare Dis.* 2008;3:12.)

estrogen receptor agonists, spironolactone, cyproterone acetate, and medroxyprogesterone acetate [34]. This helps prevent bone age advancement and compromise of adult height in girls.

Fibrous dysplasia has no proven medical therapy. The incidence of fractures and pain intensity can be reduced by bisphosphates. While surgical resection is effective, it is difficult in the skull and facial regions. Surgery can be performed when there is functional cranial nerve deficit to prevent loss of function.

Hyperthyroidism in MAS can be treated with thioamides (propylthiouracil), methimazole, and radioiodine (for tissue ablation), while thyroidectomy is useful for persistent hyperthyroidism. Growth hormone excess in MAS is treated with octreotide, dopamine agonists, and growth hormone receptor antagonists (e.g., pegvisomant). Hypogonadism can be treated with hormone replacement therapy. Oral phosphorus replacement can be done for hypophosphatemia, and for hypophosphatemia rickets, calcitriol therapy can be given.

Certain symptoms of MAS require surgical intervention, such as ovarian cystectomy or oophorectomy (precocious puberty). Traction or fixation for fracture caused by polyostotic fibrous dysplasia, thyroidectomy for hyperthyroidism, bilateral adrenalectomy for infantile cushing syndrome can be the treatment options. Pituitary adenoma (gigantism or acromegaly) causing vision problems should be surgically removed with skilled neurosurgeons [1].

44.8 Prognosis and Surveillance

Prognosis for individuals with MAS depends on disease location and severity. While patients with skin lesions require minimal medical attention, those with extensive bone disease needed to be treated promptly to reduce potential sequelae such as loss of mobility, progressive scoliosis, facial deformity, and loss of vision and/or hearing. With effective symptomatic management, very few patients may develop Cushing syndrome, which could be fatal [1]. Patients with MAS should avoid contact sports and other high-risk activities with significant skeletal involvement.

Surveillance for individual with MAS includes monitoring (i) infants for clinical signs of hypercortisolism; (ii) children for growth acceleration (related to precocious puberty and/or growth hormone excess); (iii) children age <5 years for thyroid function; (iv) males for testicular lesions with physical examination and testicular ultrasound; (v) existing fibrous dysplasia and development of new lesions; (vi) phosphorus levels for the development of hypophosphatemia; (vii) vision and hearing in craniofacial fibrous dysplasia; (viii) progressive scoliosis with periodic skull CT in spine fibrous dysplasia; (ix) children with thyroid abnormalities by ultrasound examination and thyroid function test [1].

44.9 Conclusion

McCune–Albright syndrome (MAS) is a rare genetic disorder characterized by fibrous dysplasia, skin pigmentation (café-au-lait macules), precocious puberty, and other clinical symptoms (e.g., gastrointestinal polyps). The molecular pathogenesis of MAS is underlined by post-zygotic somatic activating mutations in the *GNAS* gene, which alter either residue arg201 or residue gln227

in the α-subunit of the stimulatory G protein (Gsα), which in turn contributes to the activation of the Gsα-cAMP signaling pathway in the mutation-containing cells [35,36]. Although diagnosis of MAS is possible on the basis of characteristic clinical findings, use of molecular tests provides additional means of confirmation. Due to a lack of specific therapy for MAS, current treatment centers on symptomatic management. Further research is necessary in order to develop novel therapeutic strategies against MAS.

REFERENCES

1. Dean L. McCune-Albright Syndrome. In: Pratt V, McLeod H, Dean L, Malheiro A, Rubinstein W. editors. *Medical Genetics Summaries* [Internet]. Bethesda (MD): National Center for Biotechnology Information (US); 2012. 2012 Mar 8 [updated 2017 Mar 6].
2. Albright F, Butler AM, Hampton AO. Syndrome characterized by osteitisfibro-sadisseminate, areas of pigmentation and endocrine dysfunction with precocious puberty in females. *N Eng J Med*. 1937;216:727–47.
3. Dumitrescu CE, Collins MT. McCune Albright syndrome. *Orphanet Journal of Rare Diseases*. 2008;3:12.
4. Basaran R, Kaksi M, Gur E, Efendioglu M, Balkuv E, Sav A. Monostotic fibrous dysplasia involving occipital bone: A case report and review of literature. *Pan Afr Med J*. 2014;19:124.
5. Boyce AM. Fibrous dysplasia. In: De Groot LJ, Chrousos G, Dungan K et al. editors. *Endotext* [Internet]. South Dartmouth (MA): MDText.com, Inc.; 2000.
6. Boyce AM, Collins MT. Fibrous dysplasia/McCune Albright syndrome. *Gene Reviews*. February 26, 2015. http://ncbi.nlm.nih.gov
7. Anitha N, Sankari SL, Malathi L, Karthick R. Fibrous dysplasia-recent concepts. *J Pharm Bioallied Sci*. 2015;7(Suppl 1):S171–2.
8. Prabhu S, Sharanya S, Naik PM et al. Fibroosseous lesions of oral and maxillofacial region:Reetrospective analysis for 20 years. *J Oral Maxillofac Pathol*. 2013;17(1);36–40.
9. Robinson C, Collins MT, Boyce AM. Fibrous dysplasia/McCune-Albright syndrome: Clinical and translational perspectives. *Curr Osteoporos Rep*. 2016;14(5):178–86.
10. Waldron CA. Fibro-osseous lesions of the jaws. *J Oral Maxillofac Surg*. 1993;51(8):828–35.
11. El-Mofty SK. Fibro-osseous lesions of the craniofacial skeleton: An update. *Head Neck Pathol*. 2014;8(4):432–44.
12. National organization for rare disorders. Available at http://rarediseases.org/
13. Kozasa T, Itoh H, Tsukamoto T, Kaziro Y. Isolation and characterization of human Gs alpha gene. *Proc Natl Acad Sci U S A*. 1988;85(7):2081–5.
14. Gejman PV, Weinstein LS, Martinez M et al. Genetic mapping of the Gs alpha subunit gene (GNAS1) to the distal long arm of chromosome 20 using a polymorphism detected by denaturing gradient gel electrophoresis. *Genomics*. 1991;9(4):782–3.
15. Landis CA, Masters SB, Spada A et al. GTPase inhibiting mutations activate the alpha chain of Gs and stimulate adenyl cyclase in human pituitary tumours. *Nature*. 1989;340(6236):692–6.
16. De Sanctis L, Romagholo D, Greggi ON et al. Searching for Arg 201 mutation in GNAS gene in Italian patients with McCune Albright syndrome. *J Pediatr Endocrinol Metab*. 2002;15(Suppl 3):883–9.

17. Cabrera Vera TM, Vanhauwe J, Thomas TO et al. Insight into G protein structure, function and regulation. *Endocr Rev.* 2003;24(6):765–81.

18. Lumbroso S, Paris F, Sultan C. Activating Gs alpha mutations: Analysis of 113 patients with signs of McCune-Albright syndrome-a European collaborative study. *J Clin Endocrinol Metab.* 2004;89:2107–13.

19. Spiegel AM, Weinstein LS. Inherited diseases involving G-protein and G-protein coupled receptors. *Annu Rev Med.* 2004;55:27–39.

20. Turan S, Bastepe M. GNAS spectrum of disorders. *Curr Osteoporos Rep.* 2015;13(3):146–58.

21. Rajpal K, Agarwal R, Chhabra R, Bhattacharya M. Updated classification schemes for fibro-osseous lesions of the oral and maxillofacial region: A review. *IOSR J Dent Med Sci.* 2014; 13 (2): 99–103.

22. Rarediseases.org. Available at www.rarediseases.org/ rare-diseases/mccune-albright-syndrome/

23. Collins MT, Singer FR, Eugster E. McCune-Albright syndrome and the extraskeletal manifestations of fibrous dysplasia. *Orphanet J Rare Dis.* 2012;7(Suppl 1):S4.

24. Muthusamy S, Subhawong T, Conway SA, Temple HT. Locally aggressive fibrous dysplasia mimicking malignancy: A report of four cases and review of the literature. *Clin Orthop Relat Res.* 2015;473(2):742–50.

25. Salenave S, Boyce AM, Collins MT, Chanson P. Acromegaly and McCune-Albright syndrome. *J Clin Endocrinol Metab.* 2014;99(6):1955–69.

26. Emedicine.medscape.com. Available at https://emedicine. medscape.com/article/127233-overview

27. Emedicine.medscape.com. Available at http://emedicine.med-scape.com/-McCune Albright syndrome Differential diagnosis

28. Neville BW, Damm DD, Allen CM, Chi A. *Textbook of Oral and Maxillofacial Pathology First South Asia edition,* 2016.

29. Bulakbaşi N, Bozlar U, Karademir I, Kocaoğlu M, Somuncu I. CT and MRI in the evaluation of craniospinal involvement with polyostotic fibrous dysplasia in McCune-Albright syndrome. *Diagn Interv Radiol.* 2008;14(4):177–81.

30. Nair SN, Kini R, Rao PK et al. Fibrous dysplasia versus juvenile ossifying fibroma: A dilemma. *Case Rep Dent.* 2016;2016:6439026.

31. Idowu BD, AL-Adnani M, O Donnell P et al. A sensitive mutation-specific screening technique for GNAS1 mutations in cases of fibrous dysplasia:The first report of a codon 227 mutation in bone. *Histopathology.* 2007;50:691–704.

32. Zhou J, Sun LH, Cui B et al. Genetic diagnosis of multiple affected tissues in a patient with McCune-Albright syndrome. *Endocrine.* 2007;31(2):212–7.

33. Patil S, Prajapathi P, Prajapati S et al. A case of McCune Albright syndrome with fibrous dysplasia and endocrinopathies. *RJPBCS.* 2017;8(2);352–7.

34. Mieszczak J, Eugster EA. Treatment of precocious puberty in McCune-Albright syndrome. *Pediatr Endocrinol Rev.* 2007;4(Suppl 4):419–22.

35. Weinstein LS, Shenker A, Gejman PV et al. Activating mutations of the stimulatory G protein in the McCune Albrightsyndrome. *N Engl J Med.* 1991;325(24):1688–95.

36. Weinstein LS. G(s)alpha mutations in fibrous dysplasia and McCune-Albright syndrome. *J Bone Miner Res.* 2006; 21(Suppl 2):120–4.

45

Milroy Disease

Dongyou Liu

CONTENTS

45.1 Introduction

Milroy disease is an autosomal dominant disorder of the lymphatic system characterized by congenital lymphedema (accumulation of protein-rich fluid in tissues, particularly involving the lower part of the body), hydrocele (37% of males), prominent veins (23%), cellulitis (20%, often in males), upslanting toenails (14%), papillomatosis (10%), and urethral abnormalities in males (4%), along with increased risk for angiosarcoma, lymphangiosarcoma, and kaposiform hemangioendothelioma.

Being a chronic and progressive disease, lymphedema may be primary or secondary. Accounting for about 10% of lymphedema cases, primary lymphedema represents a failure in the development of the lymphatic system (lymphangiogenesis), leading to structural or functional abnormalities in the lymphatic system and impaired maintenance of interstitial fluid balance. Being an important cause of primary lymphedema, Milroy disease (i.e., congenital lymphedema) is attributed to heterozygous mutations in the *FLT4* (5q35.3) gene. Other potential causes of primary lymphedema include syndromic disorders (e.g., Turner or Noonan syndrome), systemic/visceral lymphatic abnormalities (localized or generalized lymphedema), disturbed growth and/or cutaneous/vascular anomalies (e.g., Proteus syndrome), and late-onset primary lymphedema [1].

Making up 90% of lymphedema cases, secondary lymphedema is an acquired condition that compromises the function of the existing lymphatic system (e.g., trauma, surgery [e.g., removal of lymph nodes for cancer treatment in developed countries], infection [e.g., filariasis in developing countries], and tumor [e.g., breast cancer]).

45.2 Biology

As a drainage network that begins in the interstitial spaces and ends in the great veins of the neck or thorax, the lymphatic system is composed of lymphatic vessels (including initial lymphatics, pre-collectors, and collecting lymphatics, with unidirectional valves dividing the collecting vessels into segments called *lymphangions, and* collecting lymphatics pumping the lymph through the regional lymph nodes and reaching the thoracic duct or the right lymphatic trunk), lymph nodes (which clean and filter lymph, a clear fluid containing infection-fighting white blood cells originated from plasma), tonsils (which act as the first line of immune defense against invading pathogens), adenoids, spleen (which as a blood filter controls red blood cells and blood storage in the body, and helps fight infection) and thymus (which stores immature lymphocytes/specialized white blood cells before maturing into active T cells to destroy infected or cancerous cells). Functionally, the lymphatic system is involved in maintenance of interstitial fluid balance (an increase in filtration or a reduction in lymphatic removal or both contributing to edema), immune surveillance (and maturation of immune cells) and absorption of fat (which is taken up by the intestinal lymphatics and transported to the venous circulation). Besides removing circulating fluid and large molecules from the extracellular spaces of almost all body tissues and transporting them to the lymph nodes, lymphatic vessels carry antigens and immune cells from the tissues as part of host defense [2–4].

The most common diseases affecting the lymphatic system are lymphadenopathy (enlargement of the lymph nodes caused by infection, inflammation, or lymphoma), lymphedema (swelling on both sides of the body due to lymph node blockage, and accumulation of lymph fluid containing plasma proteins, extravascular blood cells, excess water, and parenchymal products in the interstitial tissue, leading to enlargement and fibrosis of the subcutaneous compartment and hyperkeratosis of the skin or elephantiasis), and cancers [5,6].

Hereditary lymphedema of the legs was initially observed by Nonne in 1891, and a detailed description was made by Milroy in 1892 involving a 6-generation, 97-member family with leg edema. Subsequent reports expanded the clinical profile of Milroy disease to congenital chylous ascites, swelling of the scrotum (hydrocele), marked loss of albumin into the intestinal tract (leading to hypoproteinemia), persistent bilateral pleural effusion, congenital edema of the hands, bilateral swelling of the legs and feet, wart-like growths (papillomas), and prominent leg veins, upslanting toenails, deep creases in the toes, and cellulitis (non-contagious skin infection) [1].

The identification of heterozygous mutations in the *FLT4* (5q35.3) gene yielded critical insights into the molecular pathogenesis of Milroy disease.

45.3 Pathogenesis

The *FLT4* (FMS-like tyrosine kinase 4) gene on chromosome 5q35.3 is 48.7 kb in length with 31 exons and encodes a 1,363 aa, 152 kDa tyrosine-protein kinase (FLT4, also known as vascular endothelial growth factor receptor 3 [VEGFR3]), which is a lymphatic endothelial cell-specific receptor for vascular endothelial growth factors C (VEGFC) and D (VEGFD) [7].

Expressed mainly in lymphatic endothelia, FLT4 interacts with VEGFC (which is a key regulator of blood vessel development in embryos and angiogenesis in adult tissues), enhances VEGFC production, promotes growth, survival, and migration of endothelial cells, and contributes to adult lymphangiogenesis and the buildup of the vascular network and the cardiovascular system during embryogenesis. In addition, FLT4 mediates activation of the MAPK1/ERK2, MAPK3/ERK1 signaling pathway, the MAPK8 and the JUN signaling pathway, and the AKT1 signaling pathway, and promotes phosphorylation of PIK3R1 (regulatory subunit of phosphatidylinositol 3-kinase) and MAPK8 at "Thr-183" and "Tyr-185", and of AKT1 at "Ser-473" [8,9].

Mutations in *FLT4* cause defective blood vessel development in early embryos and formation of abnormally organized large vessels with faulty lumens, leading to fluid accumulation and development of Milroy disease (hereditary lymphedema type IA). While *FLT4* benign variants (e.g., p.Pro641Ser, p.Asn199Asp, p.Thr494Ala, p.Gln890His, p.Pro954Ser, p.Pro1008Leu, p.Arg1146His, p.Arg1342Leu) do not alter the tyrosine kinase activity, *FLT4* pathogenic variants (e.g., p.Gly857Arg, p.His1035Arg, p.Arg1041Pro, p.Leu1044Pro, p.Pro1114Leu, p.Ser1235Cys) reduce or eliminate the tyrosine kinase activity and impede blood vessel development [10,11].

The *VEGFC* (vascular endothelial growth factor C) gene on chromosome 4q34.3 spans 109 kb, and encodes a 419 aa, 46.8 kDa protein (VEGFC) that acts as a ligand for vascular endothelial growth factor-3 (VEGFR3 or FLT4).

Expressed in the heart, placenta, ovary, and small intestine, VEGFC is a member of the platelet-derived growth factor/vascular endothelial growth factor (PDGF/VEGF) family that promotes angiogenesis and endothelial cell growth during embryogenesis,

maintains differentiated lymphatic endothelium in adults, and affects the permeability of blood vessels. *VEGFC* exon 4 mutation c.571_572insTT mutation (VEGFCinsTT) induces a frameshift from codon 191 and a stop codon 10 aa further downstream and is responsible for causing a Milroy-like primary lymphedema. Therefore, for Milroy disease patients with no identifiable *FLT4/VEGFR3* mutation, further test for *VEGFC* mutation may be considered, especially if the lymphoscintigram demonstrates poor uptake with tortuous lymphatics and rerouting [12–14].

Other genes possibly involved in the pathogenesis of primary lymphedema include *CCBE1*, *FOXC2*, *GATA2*, *GJC2*, *KIF11*, and *SOX18* [15].

45.4 Epidemiology

45.4.1 Prevalence

Lymphedema is an external (or internal) manifestation of lymphatic system insufficiency and deranged lymph transport, with an incidence of 1 per 30 individuals worldwide. Primary lymphedema has a prevalence of 1.15 cases per 100,000 (for which Milroy disease is a major cause), while secondary lymphedema affects 10 cases per 100,000.

45.4.2 Inheritance

Milroy disease demonstrates autosomal dominant inheritance, with one copy of the altered gene in each cell being sufficient to cause the disorder. While many cases involve inheritance of the mutation from one affected parent, some cases may evolve from new mutations in the *FLT4* gene.

45.4.3 Penetrance

Approximately 85%–90% of patients with a *FLT4* pathogenic variant develop lower-limb lymphedema by age 3 years, while 10%–15% of patients with a *FLT4* pathogenic variant are clinically unaffected.

45.5 Clinical Features

Clinically, Milroy disease typically manifests as bilateral, occasionally asymmetric lower-limb lymphedema that appears on the dorsum of the feet at (or before) birth, and that may improve or progress to affect the whole lower leg with age, along with sensations of heaviness, achiness, decreased range of motion, and skin changes.

Other features may include hydrocele (37% of males), prominent veins (23%), upslanting toenails (14%), papillomatosis (10%), urethral abnormalities in males (4%), recurrent cellulitis (20% of cases), prenatal pleural effusion, and hydrops fetalis as well as tumors [16–18].

45.6 Diagnosis

Diagnostic workup for Milroy disease consists of medical history review, physical examination (circumferential [>2 cm] and/or volume [>200 mL or 5%] differences and skin changes between the affected and non-affected extremity), imaging study (e.g., duplex ultrasound, lymphoscintigraphy, magnetic resonance lymphangiography), laboratory assessment, histopathology, and molecular analysis [19,20].

Individuals showing the following signs should be suspected of Milroy disease: (i) lower-limb swelling (usually but not always bilateral, present at birth or develops soon after; the swelling predominantly affecting the dorsum of the feet of neonates, but improving or progressing to affect the whole lower leg with age); (ii) large-caliber veins; (iii) upslanting, "ski-jump" toenails (Figure 45.1) [21].

Based on severity, lymphedema is classified into three stages: stage I (spontaneously reversible and typically pitting edema), stage II (spongy consistency of the tissue without signs of pitting edema), and stage III (or lymphostatic elephantiasis). Another approach (the Common Terminology Criteria for Adverse Events v3.0 [CTCAE]) differentiates lymphedema into four grades: grade 1 (5%–10% interlimb discrepancy in volume or circumference at point of greatest visible difference; swelling or obscuration of anatomic architecture on close inspection; pitting edema), grade 2 (>10%–30% interlimb discrepancy in volume or circumference at point of greatest visible difference; readily apparent obscuration of anatomic architecture; obliteration of skin folds; readily apparent deviation from normal anatomic contour), grade 3 (>30% interlimb discrepancy in volume; lymphorrhea; gross deviation from normal anatomic contour; interfering with activities of daily living), and grade 4 (progression to malignancy such as lymphangiosarcoma; amputation indicated; disabling lymphedema). Furthermore, the International Society of Lymphology classifies lymphedema into four stages: stage 0 (latent; some damage to lymphatics, but no visible edema), stage I (spontaneously reversible, acute phase; pitting edema; reversible with limb elevation; normal or almost normal limb size upon waking in the morning), stage II (spontaneously irreversible, chronic phase; spongy consistency, non-pitting, and fibrosis leading to the hardening of the limbs and increasing size), and stage III (lymphostatic elephantiasis; irreversible, end-stage;

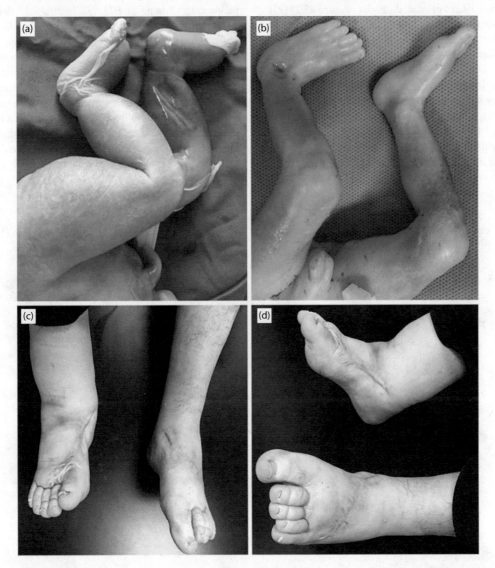

FIGURE 45.1 (a) Milroy disease in neonate harboring *FLT4* exon 22 mutation (c.3075G>A) showing: (a) edema in both lower limbs with slightly purple skin; (b) no obvious abnormality in matched child; (c and d) Milroy disease in father of affected neonate also possessing the same *FLT4* mutation showing edema in right lower limb, and normal left lower limb after plastic surgery. (Photo credit: Zhang S et al. *Front Genet.* 2019;10:206.)

irreversible, large, hard/fibrotic and unresponsive limbs with trophic skin changes; most commonly associated with the filarial cause of secondary lymphedema). Each of these stages may be subclassified as minimal (<20% volume excess), moderate (20%–40% volume excess), or severe (>40%) volume excess [1].

Duplex ultrasound is a simple, noninvasive technique for identification of lymphedema by specific tissue characteristics and therapy response. Lymphoscintigraphy (radionuclide imaging) may reveal a lack of uptake of radioactive colloid in the ilioinguinal lymph nodes due to a paucity or impairment of lymphatic vessels in the lower limbs. Compared to lymphangiography (which involves direct injection of dye into the lymphatic vessels in the foot before x-ray, and requires locating lymphatic vessels for cannulation), lymphoscintigraphy is less invasive and easier to perform. Lymphoscintigraphy may be omitted if molecular test for Milroy disease is available [1,22].

Histopathology helps uncover inflammatory cell infiltration and cellular changes in skin tissues from patients with Milroy disease in comparison with unaffected controls and reveals morphological characteristics of related tumors (Figure 45.2) [21].

Molecular identification of *FLT4 (VEGFR3)* pathogenic variants provides further confirmation of Milroy disease. Sequence analysis of *FLT4* exons 17–26 allows detection of pathogenic variants in ≤75% in well-phenotyped patients with Milroy disease. It is possible that affected individuals without *FLT4* mutations may have genetic alterations in other genes (e.g., *VEGFC*) [1,9].

Differential diagnoses for Milroy disease ("woody" swelling of the dorsum of the feet with few associated features) include *microcephaly with or without chorioretinopathy, lymphedema, or mental retardation* (MCLMR; autosomal dominant condition due to *KIF11* mutation), *VEGFC-related lymphedema* (*VEGFC* mutations), *Turner syndrome* (short stature, stature disproportion, primary amenorrhea, neck webbing, congenital lymphedema of the hands and feet, high-arched palate, short metacarpals, scoliosis, Madelung deformity, hearing difficulties, cardiac and renal anomalies, hypothyroidism, glucose intolerance, and extremity lymphedema that improves over time; females with one normal X chromosome and either absence of the second sex chromosome [X or Y] with or without mosaicism or partial deletion of the X chromosome), *Noonan syndrome* (short stature, congenital heart defect, and developmental delay of variable degree; broad or webbed neck, unusual chest shape with superior pectus carinatum and inferior pectus excavatum, cryptorchidism, and characteristic facies; varied coagulation defects and lymphatic dysplasias; pulmonary valve stenosis, often with dysplasia, in 20%–50% of individuals; hypertrophic cardiomyopathy in 20%–30% of individuals; congenital lymphedema of the feet and legs with lower-limb and genital edema, chylous reflux, intestinal lymphangiectasia, or chylothoraxes; autosomal dominant disorder due to pathogenic variants in *PTPN11* in 50% of affected individuals, *SOS1* in approximately 13%, *RAF1* in 3%–17%, *KRAS* in <5%, *NRAS*, *BRAF*, and *MAP2K1* in <1%), *hypotrichosis-lymphedema-telangiectasia syndrome* (childhood-onset lymphedema in the lower limbs, loss of hair, and telangiectasia, particularly on the palms; autosomal dominant or autosomal recessive disorder due to pathogenic variants in *SOX18*), *lymphedema-distichiasis syndrome* (lower-limb lymphedema typically in late childhood or puberty; edema and cellulitis often in males; distichiasis or aberrant eyelashes arising from the Meibomian glands ranging from a full set of extra eyelashes to a single hair at birth in 94% of affected

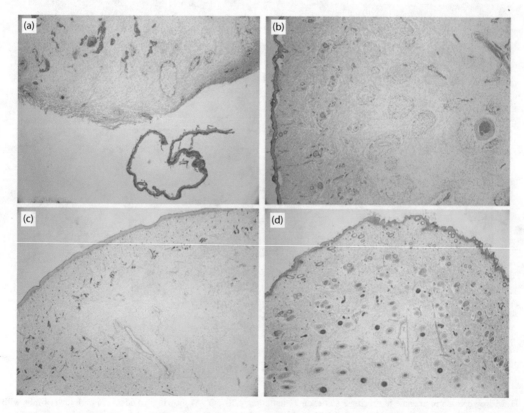

FIGURE 45.2 (a) Foot dorsum of Milroy disease neonate with *FLT4* exon 22 mutation (c.3075G>A); (b) lateral low-limb skin of Milroy disease neonate; (c) foot dorsum of matched child; (d) lateral low-limb skin of matched child (H&E, 40×). (Photo credit: Zhang S et al. *Front Genet.* 2019;10:206.)

individuals; corneal irritation, recurrent conjunctivitis, and photophobia in 75% of patients due to aberrant eyelashes; early-onset varicose veins 50%, congenital heart disease 7%, and ptosis 30%; asymptomatic in 25%; autosomal dominant disorder to *FOXC2* pathogenic variants), *Meige disease* (pubertal-onset lymphedema; autosomal dominant disorder with reduced penetrance), *lymphedema with yellow nails* (yellow nail syndrome; very slow growing, nails with transverse over-curvature and hardening of the nail plate; after age 50 years autosomal dominant disorder), *deep venous thrombosis, lipedema, obesity, injury, rheumatologic disease, vascular malformations, malignancy,* and *infection* [1,23].

45.7 Treatment

Treatment options for Milroy disease include a conservative approach (known as complete decongestive therapy [CDT]) and surgery [19,24,25].

CDT offers the first-line treatment for Milroy disease and comprises physical therapy, manual lymphatic drainage, and skin care that decrease lymphatic fluid accumulation in the tissues and prevent disease progression [26].

Surgery utilizing axillary reverse mapping to delineate arm lymphatics from axillary lymphatics helps prevent lymphedema [27].

Other measures to prevent secondary complications such as secondary cellulitis may be implemented (e.g., prevention of foot infections, particularly athlete's foot/infected eczema; prompt treatment for early cellulitis with appropriate antibiotics; prophylactic antibiotics in recurrent cases such as penicillin V 500 mg daily) [25].

45.8 Prognosis and Prevention

Milroy disease—associated lymphedema negatively impacts the patient's quality of life through psychological morbidity, but not necessarily life span.

Prevention measures include education about lymphedema, exercise, hygiene (skin and nail care), extremity positioning (keeping the arm or leg elevated above the level of the heart when possible; avoiding constrictive pressure on the affected arm or leg), avoiding the pooling of blood in the involved extremity [28].

45.9 Conclusion

Milroy disease is an autosomal dominant disorder that is responsible for causing congenital primary lymphedema (manifesting as bilateral, painless, and chronic edema at birth and predominantly on the dorsum of the feet, together with hydrocele, prominent veins, cellulitis, upslanting toenails, papillomatosis, urethral abnormalities, and risk for angiosarcoma, lymphangiosarcoma and kaposiform hemangioendothelioma). Molecularly, heterozygous germline mutations in the *FLT4* gene are involved in 70% of Milroy disease cases. Diagnosis of Milroy disease is based on clinical criteria, and molecular identification of *FLT4* pathogenic variants permits further confirmation. In the absence of specific remedy, management of Milroy disease involves conservative palliative approach supplemented by surgical removal.

REFERENCES

1. Brice GW, Mansour S, Ostergaard P et al. Milroy disease. In: Adam MP, Ardinger HH, Pagon RA et al. editors. *GeneReviews® [Internet].* Seattle (WA): University of Washington, Seattle; 1993–2018. 2006 Apr 27 [Updated 2014 Sep 25].
2. Zheng W, Aspelund A, Alitalo K. Lymphangiogenic factors, mechanisms, and applications. *J Clin Invest.* 2014;124(3): 878–87.
3. Adamczyk LA, Gordon K, Kholová I et al. Lymph vessels: The forgotten second circulation in health and disease. *Virchows Arch.* 2016;469(1):3–17.
4. Kazenwadel J, Harvey NL. Morphogenesis of the lymphatic vasculature: A focus on new progenitors and cellular mechanisms important for constructing lymphatic vessels. *Dev Dyn.* 2016;245(3):209–19.
5. Richter GT, Friedman AB. Hemangiomas and vascular malformations: Current theory and management. *Int J Pediatr.* 2012;2012:645678.
6. Padera TP, Meijer EF, Munn LL. The lymphatic system in disease processes and cancer progression. *Annu Rev Biomed Eng.* 2016;18:125–58.
7. Secker GA, Harvey NL. VEGFR signaling during lymphatic vascular development: From progenitor cells to functional vessels. *Dev Dyn.* 2015;244(3):323–31.
8. Mendola A, Schlögel MJ, Ghalamkarpour A et al. Mutations in the VEGFR3 signaling pathway explain 36% of familial lymphedema. *Mol Syndromol.* 2013;4:257–66.
9. Dixon JB, Weiler MJ. Bridging the divide between pathogenesis and detection in lymphedema. *Semin Cell Dev Biol.* 2015;38:75–82.
10. Irrthum A, Karkkainen MJ, Devriendt K, Alitalo K, Vikkula M. Congenital hereditary lymphedema caused by a mutation that inactivates VEGFR3 tyrosine kinase. *Am J Hum Genet.* 2000;67(2):295–301.
11. Verstraeten VL, Holnthoner W, van Steensel MA et al. Functional analysis of FLT4 mutations associated with Nonne-Milroy lymphedema. *J Invest Dermatol.* 2009;129(2):509–12.
12. Gordon K, Schulte D, Brice G et al. Mutation in vascular endothelial growth factor-C, a ligand for vascular endothelial growth factor receptor-3, is associated with autosomal dominant Milroy-like primary lymphedema. *Circ Res.* 2013;112:956–60.
13. Balboa-Beltran E, Fernandez-Seara MJ, Perez-Munuzuri A et al. A novel stop mutation in the vascular endothelial growth factor-C gene (VEGFC) results in Milroy-like disease. *J Med Genet.* 2014;51:475–8.
14. Nadarajah N, Schulte D, McConnell V et al. A novel splice-site mutation in VEGFC is associated with congenital primary lymphoedema of Gordon. *Int J Mol Sci.* 2018;19(8). pii: E2259.
15. Brouillard P, Boon L, Vikkula M. Genetics of lymphatic anomalies. *J Clin Invest.* 2014;124(3):898–904.
16. Tabareau-Delalande F, de Muret A, Miquelestorena-Standley E, Decouvelaere AV, de Pinieux G. Cutaneous epithelioid clear cells angiosarcoma in a young woman with congenital lymphedema. *Case Rep Pathol.* 2013;2013:931973.
17. Mortimer PS, Rockson SG. New developments in clinical aspects of lymphatic disease. *J Clin Invest.* 2014;124:915–21.
18. Ly CL, Kataru RP, Mehrara BJ. Inflammatory manifestations of lymphedema. *Int J Mol Sci.* 2017;18(1). pii: E171.

19. Kayıran O, De La Cruz C, Tane K, Soran A. Lymphedema: From diagnosis to treatment. *Turk J Surg*. 2017;33(2):51–7.

20. O'Donnell TF Jr, Rasmussen JC, Sevick-Muraca EM. New diagnostic modalities in the evaluation of lymphedema. *J Vasc Surg Venous Lymphat Disord*. 2017;5(2):261–73.

21. Zhang S, Chen X, Yuan L et al. Immunohistochemical evaluation of histological change in a Chinese Milroy disease family with venous and skin abnormities. *Front Genet*. 2019;10:206.

22. Modi S, Stanton AWB, Svensson WE, Peters AM, Mortimer PS, Levick JR. Human lymphatic pumping measured in healthy and lymphoedematous arms by lymphatic congestion lymphoscintigraphy. *J Physiol Lond*. 2007;583:271–85.

23. Mansour S, Brice GW, Jeffery S, Mortimer P. Lymphedema-distichiasis syndrome. In: Adam MP, Ardinger HH, Pagon RA et al. editors. *GeneReviews®* [Internet]. Seattle (WA): University of Washington, Seattle; 1993–2018. 2005 Mar 29 [updated 2012 May 24].

24. Saito Y, Nakagami H, Kaneda Y, Morishita R. Lymphedema and therapeutic lymphangiogenesis. *Biomed Res Int*. 2013; 2013:804675.

25. PDQ Supportive and Palliative Care Editorial Board. *Lymphedema (PDQ®): Health Professional Version. PDQ Cancer Information Summaries [Internet]*. Bethesda (MD): National Cancer Institute (US); 2002. 2019 Aug 28.

26. Oremus M, Dayes I, Walker K, Raina P. Systematic review: Conservative treatments for secondary lymphedema. *BMC Cancer*. 2012;12:6.

27. Garza R 3rd, Skoracki R, Hock K, Povoski SP. A comprehensive overview on the surgical management of secondary lymphedema of the upper and lower extremities related to prior oncologic therapies. *BMC Cancer*. 2017;17(1):468.

28. Douglass J, Graves P, Gordon S. Self-care for management of secondary lymphedema: A systematic review. *PLoS Negl Trop Dis*. 2016;10(6):e0004740.

46

Mosaic Variegated Aneuploidy Syndrome

Dongyou Liu

CONTENTS

46.1 Introduction

Mosaic variegated aneuploidy (MVA) syndrome is a rare, autosomal recessive disorder that manifests as mosaic aneuploidies (predominantly trisomies and monosomies, involving multiple chromosomes and tissues as a result of defective cell division), diverse phenotypic abnormalities (e.g., growth retardation, microcephaly, facial dysmorphism, congenital heart defects, developmental delay) and predisposition to cancer (e.g., early-onset Wilms tumor, rhabdomyosarcoma, leukemia).

Molecularly, homozygous or compound heterozygous mutations in the *BUB1B* (15q15.1), *CEP57* (11q21), or *TRIP13* (5p15.33) genes underlie the development of MVA syndrome [1,2].

46.2 Biology

Aneuploidy refers to abnormal gain or loss of chromosomes (i.e., deviation from the diploid/euploid state of having 22 pairs of autosomes and a pair of sex chromosomes [XY, male; XX, female]) during cell division in which chromosomes do not separate/duplicate properly as a consequence of failure(s) in mitotic checkpoint, chromatid attachment mechanism, centrosome, kinetochore, and microtubule functions. Aneuploidy occurring in a fraction of cells is known as chromosomal mosaicism. Representing an important chromosome aberration, aneuploidy is responsible for causing spontaneous abortion and other disorders (including cancer) in humans (Figure 46.1) [3–5].

Mitotic checkpoint such as spindle assembly checkpoint (SAC) is a surveillance mechanism that ensures faithful chromosome segregation during mitosis. A weakened SAC leads to nondisjunction, with the assembly of incorrect number of chromosomes in a daughter cell (e.g., one lacking a copy, another having an extra copy). Loss of SAC function contributes to premature chromatid separation (PCS) in >50% of metaphase cells and presence of mosaic aneuploidy as in the case of mosaic variegated aneuploidy syndrome (MVA syndrome, also known as premature chromatid separation [PCS] syndrome) [6–10].

Of the three genes [i.e., *BUB1B* (15q15.1), *CEP57* (11q21), or *TRIP13* (5p15.33)] involved in the pathogenesis of MVA syndrome, *BUB1B* (budding uninhibited by benzimidazole-related-1) encodes a key protein in the mitotic spindle checkpoint that inhibits anaphase-promoting complex/cyclosome (APC/C) activity until all chromosomes establish proper attachment to the mitotic spindle, and thus plays a critical role in SAC. Mutations in *BUB1B* as well as an intergenic mutation 44-kb upstream of *BUB1B*, which suppresses the transcription level of *BUB1B*, cause deficiency/insufficiency of BubR1, leading to the development of MVA syndrome (cataracts, uncontrollable chronic seizures, polycystic kidneys, obesity, and cancer such as Wilms tumor and rhabdomyosarcoma in 75% of cases) [11,12].

CEP57 (centrosomal protein 57 kDa) encodes a centrosomal protein (CEP57) involved in nucleating and stabilizing microtubules. Mutations in *CEP57* occur in MVA syndrome patients in which cancer is notably absent [13].

TRIP13 (thyroid hormone receptor interactor 13) encodes an ATPase that converts the mitotic checkpoint protein MAD2 from a closed (active) to an open (inactive) conformation in prometaphase (to sustain mitotic checkpoint signaling) and inactivates MAD2 in metaphase (to silence mitotic checkpoint and trigger

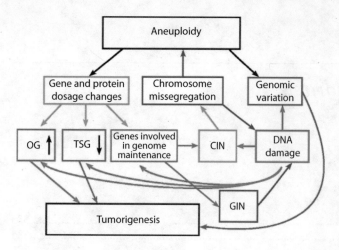

FIGURE 46.1 The role of aneuploidy in tumorigenesis. Aneuploidy induces gene and protein dosage changes within aneuploid chromosome, which increase oncogene (OG) activity, decrease tumor suppressor gene (TSG) activity, leading or tumorigenesis; and which also disrupt genes involved in genome maintenance, leading to chromosome instability (CIN) and genomic instability (GIN). Further, aneuploidy induces genomic variation, which contributes to tumorigenesis. GIN may introduce DNA damage, which causes CIN (and subsequent chromosome misaggregation) and genomic variation, further enhancing aneuploidy or tumorigenesis. Chromosome missegregation may also cause DNA damage, leading to increased OG activity, decreased TSG activity, and alterations in genes involved in genome maintenance. (Photo credit: Giam M, Rancati G. *Cell Div.* 2015;10:3.)

anaphase onset). Thus, *TRIP13* mutations causing TRIP13 deficiency impair the mitotic checkpoint and increase chromosome missegregation and aneuploidy [14].

46.3 Pathogenesis

MVA syndrome stems from defective cell division (specifically during mitosis) that results in >10% of cells with missing (monosomy) or extra (trisomy) genetic material in multiple chromosomes and tissues (mosaic aneuploidies) [15]. Among various genes implicated in the pathogenesis of MVA syndrome, *BUB1B* (15q15.1), *CEP57* (11q21), and *TRIP13* (5p15.33) are most notable standouts [1].

The *BUB1B* (budding uninhibited by benzimidazoles 1 homolog beta) gene on chromosome 15q15.1 spans 60 kb with 23 exons and encodes a 1050 aa, 120 kDa kinase (BubR1), which is homologous to yeast Mad3, with mitotic spindle checkpoint function.

Preferentially expressed in tissues with a high mitotic index, BubR1 blocks the binding of CDC20 to the APC/C and delays the onset of anaphase until proper attachment of metaphase spindles to kinetochores before sister chromatids segregate toward opposing spindles. BubR1 also inhibits premature activation of polo-like kinase-1 (PLK1) in interphase and suppresses centrosome amplification. Further, BubR1 binds the motor protein CENPE for regulation of kinetochore-microtubule interactions and checkpoint signaling. In addition, BubR1 may trigger apoptosis in polyploid cells that exit aberrantly from mitotic arrest [2,6].

Homozygous or compound heterozygous germline mutations in *BUB1B* cause premature protein truncation, absent transcript, or decreased protein stability, and ultimately BUBR1 deficiency, which in turn reduces elasticity in arteries, decreases neurogenesis in the brain, and induces tumorigenesis [2,6].

The *CEP57* gene on chromosome 11q21 is >42 kb in length with 11 exons, and encodes a 500 aa, 57 kDa protein (CEP57 or translokin) involved in microtubule stabilization.

Structurally, CEP57 consists of an N-terminal coiled-coil domain (for centrosome localization and for multimerization) and a C-terminal coiled-coil domain (for nucleating, bundling, and anchoring microtubules to the centrosomes).

Through specific interactions with fibroblast growth factor 2 (FGF2), sorting nexin 6, Ran-binding protein M, the kinesins KIF3A and KIF3B, CEP57 mediates the nuclear translocation and mitogenic activity of FGF2, which is a pleiotropic growth factor involved in embryonic development, wound healing, angiogenesis, and tumor progression. Besides its role in the control of nucleocytoplasmic distribution of the cyclin D1 in quiescent cells, CEP57 also helps maintain stable interactions between microtubules and the kinetochore to ensure correct chromosomal number during cell division [16,17].

Biallelic, loss-of-function mutations in *CEP57* result in loss of CEP57, leading to aneuploidy, as seen in patients with MVA syndrome. *CEP57* 520_521delGA deletion, c.915_925dup11 insertion and c.241C>T (p.R81X) are observed in MVA syndrome patients who are *BUB1B*-negative [13].

The *TRIP13* gene on chromosome 5p15.33 measures 26.5 kb and encodes a 432 aa, 48.5 kDa ATPase (TRIP13) that interacts with thyroid hormone receptors (hormone-dependent transcription factors).

TRIP13 helps sustain mitotic SAC by converting the mitotic checkpoint protein MAD2 from a closed (active) to an open (inactive) conformation in prometaphase, silences the mitotic checkpoint in metaphase through inactivation of MAD2 when all kinetochores are attached, and triggers anaphase onset. In addition, TRIP13 promotes early steps of DNA double-strand breaks (DSBs) repair process.

Homozygous truncating mutations (e.g., R354X) in *TRIP13* impair the mitotic checkpoint function of TRIP1 and cause chromosome missegregation, premature chromatid separation, and aneuploidy as well as early-onset Wilms tumor [14].

46.4 Epidemiology

46.4.1 Prevalence

MVA syndrome has an estimated prevalence of <1 per 1,000,000, with about 50 cases reported in the literature to date.

46.4.2 Inheritance

MVA syndrome displays autosomal recessive inheritance, and affected patients possess homozygous or compound heterozygous mutations in the *BUB1B*, *CEP57*, or *TRIP13* genes. Although parents of MVA syndrome patients each carry one copy of the mutated gene, they are typically asymptomatic.

46.4.3 Penetrance

MVA syndrome shows high penetrance and variable expressivity, and its clinical spectrum ranges from a severe and even lethal course (caused mainly by *BUB1B* mutations) to a mild phenotype without microcephaly or mental retardation (often containing *CEP57* or *TRIP13* pathogenic variants).

46.5 Clinical Features

Clinically, patients with MVA syndrome (especially those associated with *BUB1B* mutations) typically display pre- and/or postnatal growth retardation, microcephaly, developmental delay/mental retardation, neurological abnormalities (e.g., Dandy–Walker malformation, seizures), ophthalmological anomalies (e.g., cataract, strabismus, corneal opacity, microphthalmia, glaucoma), mild dysmorphic features (e.g., triangular facies, micrognathia, epicanthic folds, broad nasal bridge, low-set ears), skeletal/hand/foot abnormalities (e.g., clinodactyly), dermatological anomalies (e.g., café-au-lait patches), gastrointestinal defects, renal anomalies, cardiac defects, and predisposition to tumors (e.g., embryonal rhabdomyosarcoma, Wilms tumor, acute

lymphoid leukemia, granulosa cell malignant tumor of the ovary before the age of 5 years; 33% of cases) [18–21]. However, some MVA syndrome patients (particularly those harboring *CEP57* or *TRIP13* pathogenic variants) may show a mild phenotype without microcephaly or mental retardation [22].

46.6 Diagnosis

Diagnostic workup for MVA syndrome comprises medical history review, physical exam, laboratory assessment (>25% aneuploid cells), histopathology (tumor), cytogenetic analysis (premature chromatid separation [PCS]; mosaic/variegated aneuploidy), and molecular testing (*BUB1B*, *CEP57*, and *TRIP13* mutations).

Cytogenetic analysis reveals mosaic aneuploidies, predominantly trisomies and monosomies, in patients with MVA syndrome, with the proportion of aneuploid cells being >10% (Figure 46.2) [23].

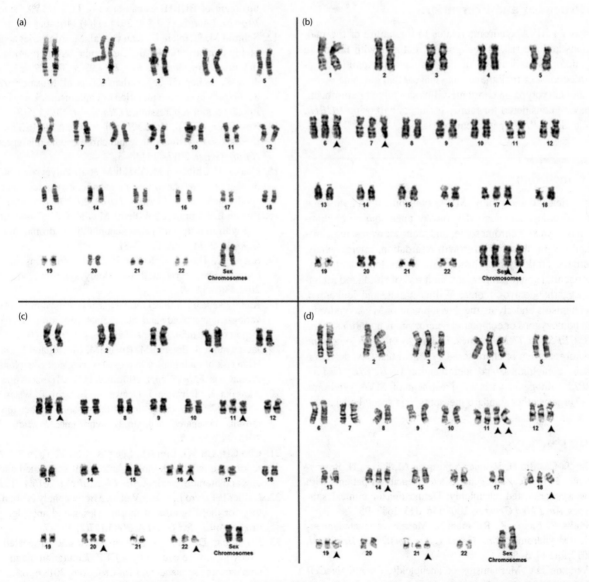

FIGURE 46.2 Mosaic variegated aneuploidy syndrome in a 11-year 6-month-old female with a frameshift mutation in *CEP57* (c.697delA, p.Lys235Argfs*31) showing 46,XX karyotype in 17 of 22 cells (a), and various aneuploidies in the remaining five cells, including 51,XXXX,+6,+7,+17 (b), 51,XX,+6,+11,+16,+20,+22 (c), 55,XX,+3,+4,+11,+11,+12,+14,+18,+19,+21 (d), with the aneuploidies indicated by arrowhead. (Photo credit: Brightman DS et al. *Clin Case Rep.* 2018;6(8):1531–4.)

Molecular identification of pathogenic variants in the *BUB1B*, *CEP57*, and *TRIP13* genes helps confirm the diagnosis of MVA syndrome.

Differential diagnoses for MVA syndrome include Roberts syndrome, ataxia telangiectasia, xeroderma pigmentosum, Bloom syndrome, Werner syndrome and Nijmegen breakage syndrome (aneuploidy), and Fanconi anemia (microcephaly with chromosome instability).

46.7 Treatment

Treatment options for MVA syndrome consist of surgery, other standard measures (e.g., growth hormone therapy for the treatment of growth failure), and special education (for developmental delay) [24].

46.8 Prognosis and Prevention

Prognosis for MVA syndrome relates to the nature of the malformations and the risk of malignancy. Patients with failure to thrive and/or complications of congenital abnormalities, epilepsy, infections, or malignancy may have lethal outcome.

Prenatal karyotype on chorionic villus sampling or amniocentesis may be considered for family members harboring *BUB1B*, *CEP57*, and/or *TRIP13* pathogenic variants.

46.9 Conclusion

Mosaic variegated aneuploidy (MVA) syndrome is an autosomal recessive disorder characterized by mosaic aneuploidies (predominantly trisomy/extra chromosome, and monosomy/missing chromosome), severe intrauterine growth retardation, microcephaly, eye anomalies, mild dysmorphism, variable developmental delay, other congenital abnormalities, and high risk of childhood malignancy (e.g., rhabdomyosarcoma, Wilms tumor, and leukemia). The mechanisms underlying the development of MVA syndrome relate to homozygous or compound heterozygous mutations in the *BUB1B*, *CEP57*, or *TRIP13* genes. Diagnosis of MVA syndrome relies on observation of characteristic clinical features, detection of mosaic aneuploidies, and identification of *BUB1B*, *CEP57*, or *TRIP13* pathogenic variants. Treatment of MVA syndrome involves surgery, other standard measures, and special education.

REFERENCES

1. García-Castillo H, Vásquez-Velásquez AI, Rivera H, Barros-Núñez P. Clinical and genetic heterogeneity in patients with mosaic variegated aneuploidy: Delineation of clinical subtypes. *Am J Med Genet A.* 2008;146A(13):1687–95.
2. Hanks S, Snape K, Rahman N. Mosaic variegated aneuploidy syndrome. *Atlas Genet Cytogenet Oncol Haematol.* 2012;16(5):376–80.
3. Compton DA. Mechanisms of aneuploidy. *Curr Opin Cell Biol.* 2011;23(1):109–13.
4. Giam M, Rancati G. Aneuploidy and chromosomal instability in cancer: A jackpot to chaos. *Cell Div.* 2015;10:3.
5. Potapova T, Gorbsky GJ. The consequences of chromosome segregation errors in mitosis and meiosis. *Biology (Basel).* 2017;6(1):12.
6. Hanks S, Coleman K, Reid S et al. Constitutional aneuploidy and cancer predisposition caused by biallelic mutations in BUB1B. *Nature Genet.* 2004;36:1159–61.
7. Hanks S, Coleman K, Summersgill B et al. Comparative genomic hybridization and BUB1B mutation analyses in childhood cancers associated with mosaic variegated aneuploidy syndrome. *Cancer Lett.* 2006;239(2):234–8.
8. Kato M, Kato T, Hosoba E et al. PCS/MVA syndrome caused by an *Alu* insertion in the *BUB1B* gene. *Hum Genome Var.* 2017;4:17021.
9. Miyamoto T, Matsuura S. Ciliopathy in PCS (MVA) syndrome. *Oncotarget.* 2015;6(28):24582–3.
10. Lane S, Kauppi L. Meiotic spindle assembly checkpoint and aneuploidy in males versus. Females. *Cell Mol Life Sci.* 2019;76(6):1135–50.
11. Ochiai H, Miyamoto T, Kanai A et al. TALEN-mediated single-base-pair editing identification of an intergenic mutation upstream of BUB1B as causative of PCS (MVA) syndrome. *Proc Natl Acad Sci U S A.* 2014;111(4):1461–6.
12. Schmid M, Steinlein C, Tian Q et al. Mosaic variegated aneuploidy in mouse BubR1 deficient embryos and pregnancy loss in human. *Chromosome Res.* 2014;22(3):375–92.
13. Aziz K, Sieben CJ, Jeganathan KB et al. Mosaic-variegated aneuploidy syndrome mutation or haploinsufficiency in Cep57 impairs tumor suppression. *J Clin Invest.* 2018;128(8):3517–34.
14. Yost S, de Wolf B, Hanks S et al. Biallelic TRIP13 mutations predispose to Wilms tumor and chromosome missegregation. *Nature Genet.* 2017;49:1148–51.
15. Chaker F, Chihaoui M, Yazidi M et al. Polycystic ovary syndrome: A new phenotype in mosaic variegated aneuploidy syndrome? *Ann Endocrinol (Paris).* 2017;78(1):58–61.
16. Pinson L, Mannini L, Willems M et al. CEP57 mutation in a girl with mosaic variegated aneuploidy syndrome. *Am J Med Genet A.* 2014;164A(1):177–81.
17. Snape K, Hanks S, Ruark E et al. Mutations in CEP57 cause mosaic variegated aneuploidy syndrome. *Nat Genet.* 2011;43(6):527–9.
18. Kajii T, Ikeuchi T, Yang ZQ et al. Cancer-prone syndrome of mosaic variegated aneuploidy and total premature chromatid separation: Report of five infants. *Am J Med Genet.* 2001;104(1):57–64.
19. Jacquemont S, Bocéno M, Rival JM, Méchinaud F, David A. High risk of malignancy in mosaic variegated aneuploidy syndrome. *Am J Med Genet.* 2002;109(1):17–21; discussion 16.
20. Akasaka N, Tohyama J, Ogawa A, Takachi T, Watanabe A, Asami K. Refractory infantile spasms associated with mosaic variegated aneuploidy syndrome. *Pediatr Neurol.* 2013;49(5):364–7.
21. Cho CH, Oh MJ, Lim CS, Lee CK, Cho Y, Yoon SY. A case report of a fetus with mosaic autosomal variegated aneuploidies and literature review. *Ann Clin Lab Sci.* 2015;45(1):106–9.
22. Callier P, Faivre L, Cusin V et al. Microcephaly is not mandatory for the diagnosis of mosaic variegated aneuploidy syndrome. *Am J Med Genet A.* 2005;137(2):204–7.
23. Brightman DS, Ejaz S, Dauber A. Mosaic variegated aneuploidy syndrome caused by a CEP57 mutation diagnosed by whole exome sequencing. *Clin Case Rep.* 2018;6(8):1531–4.
24. Laberko A, Balashov D, Deripapa E et al. Hematopoietic stem cell transplantation in a patient with type 1 mosaic variegated aneuploidy syndrome. *Orphanet J Rare Dis.* 2019;14(1):97.

47

Roberts Syndrome

Dongyou Liu

CONTENTS

47.1 Introduction

Roberts syndrome (RBS) is a rare autosomal recessive disorder that typically manifests as pre/postnatal growth retardation, limb defects (bilateral symmetric tetraphocomelia or hypomelia due to mesomelic shortening, oligodactyly with thumb aplasia or hypoplasia, syndactyly, clinodactyly, elbow and knee flexion contractures), craniofacial anomalies (e.g., microcephaly, cleft lip/palate, premaxillary prominence, micrognathia, microbrachycephaly, malar flattening, downslanted palpebral fissures, widely spaced eyes, exophthalmos resulting from shallow orbits, corneal clouding, underdeveloped ala nasi, beaked nose, ear malformations), mental retardation, cardiac and renal abnormalities, along with characteristic cytogenetic findings ("railroad track" chromatids and premature centromere separation in metaphase spreads) and increased risk for malignancy (cutaneous hemangioma, possibly cavernous hemangioma, cutaneous melanoma, rhabdomyosarcoma) [1].

Molecularly, RBS and its milder form SC phocomelia syndrome are attributed to homozygous or compound heterozygous mutations in the *ESCO2* (establishment of cohesion 1 homolog 2) gene on chromosome 8p21.1, which eliminate the acetyltransferase activity of the encoded protein (ESCO2) and interfere subsequent cohesion between sister chromatids.

47.2 Biology

Although cases of tetra phocomelia and facial anomalies (bilateral cleft lip) were documented by Deboze in the early 1670s (translated by Bouchard in 1672) and by Virchow in 1898, a detailed description of RBS (formerly pseudothalidomide syndrome, based on its similar limb malformations to those observed in thalidomide syndrome) was only made by Roberts in 1919, who noted phocomelia, bilateral cleft lip and cleft palate, and protrusion of the intermaxillary region in three affected sibs of an Italian consanguine couple [1].

Subsequent studies revealed the diverse clinical spectrum of RBS, ranging from growth retardation, limb defects, and craniofacial anomalies to multi-organ dysfunction, which led to the creation of SC phocomelia syndrome (for the initials of the surnames of the two families described) by Herrmann and coworkers in 1969 to cover part of this complex disease. Compared to RBS, SC phocomelia syndrome is a milder phenotype with a lesser degree of limb reduction and a higher likelihood of survival to adulthood. However, it has become clear that RBS and SC phocomelia syndrome are different names for the same disease, with the former being a severe, often fatal form of disease (as a result of respiratory and cardiac failure) and the latter a milder form [2].

The mapping of RBS-associated gene region to chromosome 8p21.1 by Strausberg et al. in 2002 and the identification of the *ESCO2* gene from affected patients by Vega et al. in 2005 helped elucidate the key mechanisms underlining the development of Roberts syndrome. It appears that elimination of the acetyltransferase activity of ESCO2 compromises cohesion between sister chromatids (sister chromatid cohesion [SCC], which is required for high fidelity chromosome segregation, efficient gene transcription, and DNA replication as well as DNA damage repair), leading to increased mitotic failure and apoptosis, limited progenitor cell proliferation, and eventual development of RBS. The observation of characteristic heterochromatin repulsion (HR) and the detection of *ESCO2* mutations in patients with Roberts syndrome and SC phocomelia syndrome have put beyond doubt that these two identities represent different phenotypic expressions of the same disease [3].

47.3 Pathogenesis

The *ESCO2* gene on chromosome 8p21.1 measures 30 kb organized in 11 exons (with the coding region located between exons 2–11), which transcribes into a 3376 nt mRNA and a 1806 nt open reading frame, and translates into a 601 aa, 68.3 kDa acetyltransferase (ESCO2), which represents one of two establishments of cohesion factors (the other being ESCO1) necessary for proper SCC and high fidelity chromosome segregation [4,5].

Structurally, ESCO2 comprises a C-terminal domain with acetyltransferase activity (which is involved in the establishment of SCC during S phase of mitosis after DNA replication and postreplicative SCC induced by double-strand breaks) and an N-terminal domain for chromatin binding. Besides its role in SCC formation, ESCO2 also participates in gene expression, chromosome condensation, and DNA damage repair [6].

Inside the cell, four proteins (i.e., SMC1 [structural maintenance of chromosomes], SMC3, RAD21, and STAG1) interact to form a ring-like structure (cohesin ring or complex). With the help of NIPBL and Mau-2 proteins, this cohesin ring is loaded onto DNA during the G1 phase of the cell cycle. In S-phase, ESCO1/ESCO2 acetylate SMC3 (which may be deacetylated by HDAC8) and lock it in the cohesion ring that surrounds sister chromatids (forming so-called SCC); this ensures that only sister chromatids are paired together and that chromosomes are properly segregated later. During transition from prophase to prometaphase, cohesion between chromatid arms is cleaved through the antiestablishment pathway involving WAPL. During metaphase-to-anaphase transition, centromeric cohesin rings are removed by enzyme separase after proper bipolar attachment of all chromosomes to allow sister chromatid segregation to opposing spindle poles. The cofactors, PDS5, WAPL, and Sororin, are involved in association and/or dissociation with chromatin (Figure 47.1) [7–12].

Homozygous mutations in *ESCO2* cause loss of function, truncation, or single amino acid changes in ESCO2. As the acetyltransferase activity of ESCO2 is required for buildup and maintenance of the cohesin complex, elimination of this activity impairs gene expression, ribosomal RNA production, nucleolar form and function, and phosphorylation of S6K1, S6, and 4EBP1 (which results in mTOR signaling inhibition and p53 pathway

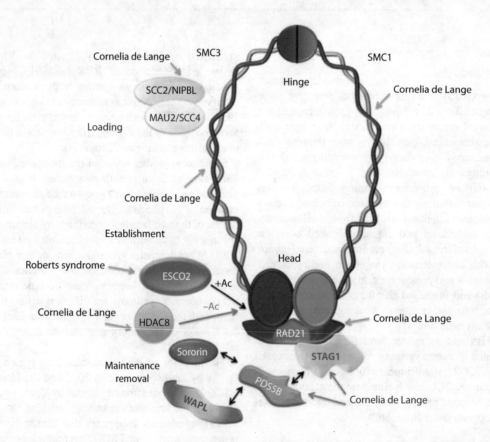

FIGURE 47.1 Cohesin and cohesin regulators in human cohesinopathies. Human Cornelia de Lange syndrome (CdLS) and Roberts syndrome (RBS) are attributed to mutations in specific genes (orange arrows) as well as genes encoding cohesin and cohesin regulators (green arrows). While the most severe phenotype of CdLS results from mutations in the *SCC2/NIPBL* gene, mild variants of CdLS are associated with mutations in genes encoding the cohesin subunits SMC1α, SMC3, and RAD21, and the histone deacetylase, HDAC8. The precise role of *PDS5A/PDS5B* and *STAG1* mutations in human CdLS requires further determination. RBS appears to be caused by mutations in *ESCO2*, and it has a phenotype closely resembling CdLS. (Photo credit: Barbero JL. *Appl Clin Genet.* 2013;6:15–23.)

activation), leading to lack of cohesion at heterochromatic regions, activation of the mitotic spindle checkpoint, mitotic delay, reduced cell proliferation, and ultimately development of RBS.

Most *ESCO2* pathogenic variants detected in RBS patients are frameshift or nonsense mutations that induce protein truncation or nonsense-mediated decay, whereas missense mutations (e.g., c.1615T>G [p.Trp539Gly]) generate amino acid substitutions in the acetyltransferase domain [13].

Apart from *ESCO2*-related RBS, mutations in SMC1, SMC3, RAD21, NIPBL, HDAC8, and cohesin-associated PDS5/APRIN causing defective SCC (the process in which sister chromatids are paired during the cell cycle) are implicated in the development of Cornelia de Lange syndrome (CdLS). In contrast to RBS, CdLS is a transcription-based disorder that does not exhibit elevated levels of apoptosis or mitotic failure, and its chromatin-bound cohesin is not only involved in sister chromatid tethering, but also transcription regulation (Figure 47.1) [14].

47.4 Epidemiology

47.4.1 Prevalence

RBS is a rare disease, with around 150 cases reported in literature to date.

47.4.2 Inheritance

RBS shows an autosomal recessive disorder exclusively involving the *ESCO2* gene. Parental consanguinity is common.

47.4.3 Penetrance

RBS shows variable clinical severity, from prenatal lethal to viable beyond 30 years of age. While many patients develop hallmark manifestations of microcephaly, craniofacial defects, mental retardation, limb deformities and growth retardation, some affected individuals also display cardiac defects and corneal opacity.

47.5 Clinical Features

RBS is known to cause growth retardation, limb deformities, craniofacial dysmorphism, and other anomalies [15,16].

Growth retardation consistently appears at birth, while postnatal growth delay also occurs in some patients.

Limb deformities range from absent arms and legs, arms and legs with rudimentary digits (e.g., brachydactyly, oligodactyly, clinodactyly), mesomelic shortening, anterior-posterior axis involvement, and bone fusions to flexion contractures. The upper limbs (radii, ulnae, and humeri) are often more severely affected than the lower limbs (fibulae, tibiae, and femur) (Figure 47.2) [17].

Craniofacial dysmorphism consists of microcephaly, micrognathia, cleft lip/palate, encephaloceles, prominent maxilla, exophthalmos, hypertelorism, corneal clouding, downslanted palpebral fissures, wide nasal bridge, telecanthus, hypoplastic nasal alae, beaked nose, malar flattening, ear malformations, and midfacial capillary hemangioma.

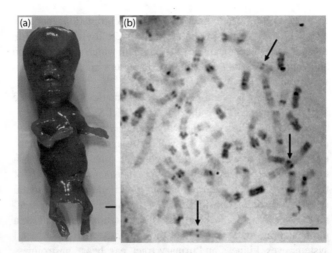

FIGURE 47.2   (a) Fetus of 18 weeks' gestation affected by Roberts syndrome showing multiple congenital anomalies: hypertelorism, micrognathia, tetraphocomelia, and oligodactyly. (b) C-banded metaphase chromosomes from the affected fetus showing the pathognomonic cytogenetic anomaly in Roberts syndrome: chromosomes with premature centromere separation (black arrows) and heterochromatin puffing; scale bar = 5 μm. (Photo credit: Dupont C et al. *Mol Cytogenet.* 2014;7(1):59.)

Other abnormalities may include atrial septal defect, ventricular septal defect, patent ductus arteriosus; polycystic kidney, horseshoe kidney; enlarged phalluses and clitorises, cryptorchidism; sparse hair, silvery blonde scalp hair; intellectual disability; premature separation of centromeres (PCS, heterochromatin push); and tumors (cutaneous hemangioma, possibly cavernous hemangioma, cutaneous melanoma, rhabdomyosarcoma).

47.6 Diagnosis

Diagnostic workup for RBS includes medical history review, physical exam, laboratory assessment, imaging study, cytogenetic analysis (for PCS and HR), histopathology (for highly fragmented nucleoli), and molecular testing (for *ESCO2* pathogenic variants) [18].

Individuals showing the following clinical symptoms should be suspected of RBS: (i) prenatal growth retardation (mild to severe), (ii) limb malformations (bilateral symmetric tetraphocomelia or hypomelia caused by mesomelic shortening; upper limbs are more severely affected than lower limbs; other limb malformations include oligodactyly with thumb aplasia or hypoplasia, syndactyly, clinodactyly, and elbow and knee flexion contractures), (iii) craniofacial abnormalities (bilateral cleft lip and/or palate, micrognathia, widely spaced eyes, exophthalmos, downslanted palpebral fissures, malar flattening, underdeveloped ala nasi, and ear malformation).

Cytogenetic analysis based on Giemsa or C-banding techniques reveals characteristic chromosomal abnormality of PCS and separation of the HR in most chromosomes in all metaphases. Many chromosomes may give a "railroad track" appearance due to the absence of primary constriction, and presence of "puffing" or "repulsion" at the heterochromatic regions around the centromeres and nucleolar organizers (Figure 47.2). In RBS cell cultures, aneuploidy, micronucleation, and multilobulated nuclei are commonly observed. It should be noted that PCS/HR in RBS differs from premature sister chromatid separation (PSCS) in CdLS and premature centromere division (PCD) in mosaic variegated aneuploidy syndrome in which separation and splaying involve both the centromeric regions and the entire sister chromatids [1].

Molecular identification of *ESCO2* mutations by sequence analysis further confirms the diagnosis of RBS.

Differential diagnoses for RBS include *Baller—Gerold syndrome* (coronal craniosynostosis, brachycephaly, ocular proptosis, prominent forehead; radial ray defect, oligodactyly, aplasia or hypoplasia of the thumb, and/or aplasia or hypoplasia of the radius; growth retardation and poikiloderma; autosomal recessive disorder due to *RECQL4* mutations), *Fanconi anemia* (FA; physical abnormalities, short stature; abnormal skin pigmentation; malformations of the thumbs, forearms, skeletal system, eyes, kidneys and urinary tract, ear, heart, gastrointestinal system, oral cavity, and central nervous system; hearing loss; hypogonadism; developmental delay; bone marrow failure, increased risk of malignancy, due to mutations in FA complementation group genes), *thrombocytopenia-absent radius (TAR) syndrome* (bilateral absence of the radii with the presence of both thumbs and transient thrombocytopenia [<50 platelets/nL]; anomalies of the skeleton, heart, and genitourinary system [renal anomalies and agenesis of uterus, cervix, and upper part of the vagina]; autosomal recessive disorder due to compound heterozygosity of *RBM8A*), *X-linked tetra-amelia* (Zimmer tetraphocomelia; tetra-amelia; facial clefts, absence of ears and nose, and anal atresia; absence of frontal bones; pulmonary hypoplasia with adenomatoid malformation; absence of thyroid; dysplastic kidneys, gallbladder, spleen, uterus, and ovaries; imperforate vagina), *tetra-amelia syndrome* (absence of all four limbs and anomalies involving the cranium and the face [cleft lip/cleft palate, micrognathia, microtia, single naris, choanal atresia, absence of nose]; eyes [microphthalmia, microcornea, cataract, coloboma, palpebral fusion]; urogenital system [renal agenesis, persistence of cloaca, absence of external genitalia, atresia of vagina]; anus [atresia]; heart; lungs [hypoplasia/aplasia]; skeleton [hypoplasia/absence of pelvic bones, absence of ribs, absence of vertebrae], and central nervous system [agenesis of olfactory nerves, agenesis of optic nerves, agenesis of corpus callosum, hydrocephalus]; due to *WNT3* mutations), *splenogonadal fusion with limb defects and micrognathia* (abnormal fusion between the spleen and the gonad or the remnants of the mesonephros); tetramelia, mild mandibular and oral abnormalities [micrognathia; multiple unerupted teeth; crowding of the upper incisors; and deep, narrow, V-shaped palate without cleft]; autosomal dominant disorder), *DK phocomelia syndrome* (phocomelia, thrombocytopenia, encephalocele, and urogenital abnormalities; cleft palate, absence of radius and digits, anal atresia, abnormal lobation of the lungs, and diaphragmatic agenesis; autosomal recessive disorder), *Holt—Oram syndrome* (HOS; upper-extremity malformations involving radial, thenar, or carpal bones; history of congenital heart malformation, ostium secundum atrial septal defect [ASD] and ventricular septal defect [VSD], especially those occurring in the muscular trabeculated septum; cardiac conduction disease; occasional phocomelia; autosomal dominant disorder due to *TBX5*.mutation in <70% of cases), *thalidomide embryopathy* (abnormalities of the long bones of the extremities such as the thumb, radius, humerus, ulna, fingers on the ulnar side of the hand; absence of the radius, ulna, and humerus in extreme cases; the hand bud arising from the shoulders; defects involving the ears [anotia, microtia, accessory auricles], eyes [coloboma of the iris, anophthalmia, microphthalmia], heart, kidneys, and urinary, alimentary, and genital tracts), *Cornelia de Lange syndrome* (CdLS; distinctive facial features [synophrys, arched eyebrows, long eyelashes, small nose with anteverted nares, small widely spaced teeth, and microcephaly], growth retardation, hirsutism, and upper limb reduction defects; cardiac septal defects, gastrointestinal dysfunction, hearing loss, myopia, and cryptorchidism or hypoplastic genitalia; premature sister chromatid separation [PSCS]; autosomal dominant disorder due to mutations in *NIPBL*, *SMC1A* and *SMC3*; X-linked CdLS due to mutations in *SMC1A*), *mosaic variegated aneuploidy syndrome* (severe microcephaly, growth deficiency, intellectual disability, childhood cancer predisposition, and constitutional mosaicism for chromosomal gains and losses; PCD, in which mitotic cells show split centromeres and splayed chromatids; autosomal recessive disorder due to pathogenic variants in *BUB1B*) [1,19].

47.7 Treatment

Treatment options for RBS include surgery (for cleft lip and/or palate, for correction of limb abnormalities, and to improve proper development of the prehensile hand grasp) and other standard procedures (prostheses, speech assessment/therapy, aggressive treatment of otitis media, special education for developmental delays, and standard treatment for ophthalmologic, cardiac, and renal abnormalities) [1,20].

As RBS is linked to mTOR signaling inhibition and p53 pathway activation, stimulation of the TOR pathway with L-leucine or other reagents may help rescue developmental defects caused by ESCO2 deficiency [1].

47.8 Prognosis and Prevention

Prognosis for RBS patients is dependent on malformations present, with high mortality among severely affected pregnancies and newborns. Without treatment, severely affected patients may die at young age, while mildly affected individuals (with SC phocomelia syndrome) often survive to adulthood.

Prenatal testing (ultrasound detection of skeletal and renal abnormalities; cytogenetic analysis of amniocentesis [at 15–18 weeks' gestation] or chorionic villus sample [at 10–12 weeks' gestation]) and preimplantation genetic diagnosis may be considered for pregnancies at risk.

47.9 Conclusion

Roberts syndrome (RBS) is an autosomal recessive disorder characterized by pre/postnatal growth retardation (mild to severe), limb malformations (e.g., bilateral symmetric tetraphocomelia or hypomelia due to mesomelic shortening), and craniofacial findings (e.g., microcephaly and cleft lip and/or palate), along with predisposition to cutaneous hemangioma, possibly cavernous hemangioma, cutaneous melanoma, and rhabdomyosarcoma. The pathogenesis of RBS relates to homozygous or compound heterozygous germline mutations in the *ESCO2* gene, which eliminate the acetyltransferase activity of ESCO2 required for buildup and maintenance of the cohesin complex, leading to impaired gene expression, ribosomal RNA production, nucleolar form and function, phosphorylation of S6K1, S6, and 4EBP1 (which causes mTOR signaling inhibition and p53 pathway activation). Diagnosis of RBS is based on observation of a combination of clinical features (including growth retardation, symmetric mesomelic shortening of the limbs, and characteristic facies with microcephaly), cytogenetic analysis (for characteristic chromosomal abnormality of PCS and separation of the HR), and molecular identification of *ESCO2* pathogenic variants. Treatment of RBS involves surgery and other standard procedures.

REFERENCES

1. Gordillo M, Vega H, Jabs EW. Roberts syndrome. In: Adam MP, Ardinger HH, Pagon RA, Wallace SE, Bean LJH, Stephens K, Amemiya A, editors. *GeneReviews®* [Internet]. Seattle (WA): University of Washington, Seattle; 1993–2018. 2006 Apr 18 [updated 2013 Nov 14].
2. Keypour F, Naghi I, Behnam B. Roberts-SC Phocomelia syndrome (pseudothalidomide syndrome): A case report. *J Family Reprod Health*. 2013;7(1):45–7.
3. Schüle B, Oviedo A, Johnston K, Pai S, Francke U. Inactivating mutations in ESCO2 cause SC phocomelia and Roberts syndrome: No phenotype-genotype correlation. *Am J Hum Genet*. 2005;77:1117–28.
4. van der Lelij P, Godthelp BC, van Zon W et al. The cellular phenotype of Roberts syndrome fibroblasts as revealed by ectopic expression of ESCO2. *PLOS ONE*. 2009;4(9):e6936.
5. Brooker AS, Berkowitz KM. The roles of cohesins in mitosis, meiosis, and human health and disease. *Methods Mol Biol*. 2014;1170:229–66.
6. Banerji R, Skibbens RV, Iovine MK. Cohesin mediates Esco2-dependent transcriptional regulation in a zebrafish regenerating fin model of Roberts Syndrome. *Biol Open*. 2017;6(12):1802–13.
7. Xu B, Lu S, Gerton JL. Roberts syndrome: A deficit in acetylated cohesin leads to nucleolar dysfunction. *Rare Dis*. 2014;2:e27743.
8. Barbero JL. Genetic basis of cohesinopathies. *Appl Clin Genet*. 2013;6:15–23.
9. Trainor PA, Merrill AE. Ribosome biogenesis in skeletal development and the pathogenesis of skeletal disorders. *Biochim Biophys Acta*. 2014;1842(6):769–78.
10. Percival SM, Thomas HR, Amsterdam A et al. Variations in dysfunction of sister chromatid cohesion in esco2 mutant zebrafish reflect the phenotypic diversity of Roberts syndrome. *Dis Model Mech*. 2015;8(8):941–55.
11. Banerji R, Skibbens RV, Iovine MK. How many roads lead to cohesinopathies? *Dev Dyn*. 2017;246(11):881–8.
12. Rivera-Colón Y, Maguire A, Liszczak GP, Olia AS, Marmorstein R. Molecular basis for cohesin acetylation by establishment of sister chromatid cohesion N-acetyltransferase ESCO1. *J Biol Chem*. 2016;291(51):26468–77.
13. Xu B, Gogol M, Gaudenz K, Gerton JL. Improved transcription and translation with L-leucine stimulation of mTORC1 in Roberts syndrome. *BMC Genomics*. 2016;17:25.
14. Skibbens RV, Colquhoun JM, Green MJ et al. Cohesinopathies of a feather flock together. *PLoS Genet*. 2013;9(12):e1004036
15. Al Kaissi A, Csepan R, Klaushofer K, Grill F. Femoral-tibial-synostosis in a child with Roberts syndrome (Pseudothalidomide): A case report. *Cases J*. 2008;1(1):109.
16. Dogan M, Firinci F, Balci YI et al. The Roberts syndrome: A case report of an infant with valvular aortic stenosis and mutation in ESCO2. *J Pak Med Assoc*. 2014;64(4):457–60.
17. Dupont C, Bucourt M, Guimiot F et al. 3D-FISH analysis reveals chromatid cohesion defect during interphase in Roberts syndrome. *Mol Cytogenet*. 2014;7(1):59.
18. Socolov RV, Andreescu NI, Haliciu AM et al. Intrapartum diagnostic of Roberts syndrome–case presentation. *Rom J Morphol Embryol*. 2015;56(2):585–8.
19. van der Lelij P, Oostra AB, Rooimans MA, Joenje H, de Winter JP. Diagnostic overlap between Fanconi anemia and the cohesinopathies: Roberts syndrome and Warsaw breakage syndrome. *Anemia*. 2010;2010:565268.
20. Zhou J, Yang X, Jin X, Jia Z, Lu H, Qi Z. Long-term survival after corrective surgeries in two patients with severe deformities due to Roberts syndrome: A Case report and review of the literature. *Exp Ther Med*. 2018;15(2):1702–11.

48

Rothmund–Thomson Syndrome

Vikram K. Mahajan and Dhaarna Wadhwa

CONTENTS

48.1 Introduction

Rothmund–Thomson syndrome (RTS, syn. poikiloderma congenital) is a rare heterozygous genophotodermatosis characterized by early-onset poikiloderma and heterogeneous clinical features including skeletal abnormalities, short stature, premature aging, sparse scalp hair, sparse or absent eyelashes and/or eyebrows, juvenile cataracts, and an increased susceptibility to develop neoplasia.

48.2 Biological Background

Auguste Rothmund, a German ophthalmologist, as early as 1868 described 10 Bavarian children having poikiloderma, growth retardation, and juvenile cataract that was rapidly progressive. Later, in 1936, British dermatologist Sydney Thomson applied the term "poikiloderma congenitale" for cases involving three children who displayed skeletal defects (e.g., bilateral thumb aplasia and hypoplastic radii and ulnae) in addition to poikiloderma and growth retardation and no cataract. William Taylor, after reviewing their original reports in 1957, coined the eponym "Rothmund–Thomson syndrome" and suggested that the two disorders were actually the same except that cases described by Thomson had no cataract, consanguinity, and hypogonadism. Based on clinical features and subsequent molecular studies revealing homozygous or compound heterozygous mutations in the human helicase gene *RECQL4* in a subset of RTS patients, two clinical forms of RTS have emerged. The RTS-I (Thomson type) is characterized by absence of consanguinity or genetic mutations, hypogonadism, and juvenile cataract [1,2]. The RTS-II (Rothmund type) marked

by congenital bone defects and an increased risk for osteosarcoma, is associated with homozygous or compound heterozygous (frameshift/missense) mutations in the *RECQL4* DNA helicase gene [2,3]. However, this gene mosaicism is detectable in only 65% RTS cases, reflecting a genetic heterogeneity [2]. Thus, the exact role of *RECQL4* mutations in the pathogenesis of RTS or whether the two RTS forms are distinct or not requires further delineation at the molecular level.

48.3 Pathogenesis

Resulting from defective DNA damage repair, RTS is inherited as an autosomal recessive disorder. That means only individuals who inherit two copies of an abnormal gene for the same trait, one from each parent, will develop the disease, whereas those who inherit one normal gene and one abnormal gene will be carriers for the disease without showing symptoms themselves. However, two carrier parents have a 25% risk of getting an affected child, a 50% risk of getting a carrier child, and a 25% chance of getting a normal child with each pregnancy. Most cases occur in isolation and the affected children are usually born after uneventful gestation, and with some exceptions, consanguinity is not a consistent feature. The syndrome is attributed to homozygous or compound heterozygous (frameshift/missense) mutation in the *RECQL4* gene mapped to chromosome 8q 24.3 spanning 21 exons that is important in maintaining genomic integrity. It encodes a 133-kDa protein of 1208 amino acids, the ATP-dependent DNA helicase Q4 (also known as RecQ helicase-like 4, RecQ protein-like 4, RecQ protein 4, RECQ4) [1,4]. Helicases are enzymes that bind to DNA and temporarily unwind the double helix of DNA molecule. Although the precise function of ATP-dependent DNA helicase Q4 remains to be clarified, its involvement in DNA replication, UV-induced DNA damage repair, telomere and mitochondrial DNA maintenance, interaction with p53 tumor suppressor gene, and oxidative stress by effectively clearing reactive oxygen species especially in proliferating tissues (e.g., developing skin and bone effecting skeletal development and hematopoiesis) has been demonstrated [1–12]. Mutations (about 40 identified so far) in the *RECQL4* gene increase the risk of defective DNA replication, unrepaired DNA and sustained genomic instability, enhanced oxidant sensitivity leading to premature aging, and predisposition to skeletal, cutaneous, and/or hematological abnormalities/malignancies in RTS patients.

48.4 Epidemiology

The prevalence of this rare genomic instability syndrome remains undefined due to its variable clinical spectrum and relative scarcity (with fewer than 400 cases being reported to date). Given its high phenotypic variation, cases showing atypical or borderline clinical presentations may be overlooked or undiagnosed. While no population is spared from RTS, which equally affects both genders of all ethnicities and many nationalities, specific mutations may exist within certain ethnic groups. As an autosomal recessive disorder, most affected patients appear as isolated cases, except those from consanguineous families.

48.5 Clinical Features

The photosensitivity and development of poikiloderma in early childhood and variable clinical features of growth retardation, bilateral juvenile cataract by the age of 7 years in approximately 40% children, normal life expectancy and mental development, and high incidence of malignancy are the hallmarks of RTS. A mild to intense erythema, edema, and blistering after brief sun exposure in infancy occurs due to photosensitivity. The photosensitivity tends to decrease with age or may persist until adult life. The more typical poikilodermatous skin changes appear between 3 and 6 months or as late as 2 years of life in 90% cases and mainly involve cheeks, malar area, helicis, and light-exposed skin of extremities (Figure 48.1). It tends to become generalized with advancing age, spreading to flexure surfaces of extremities and buttocks but sparing trunk and abdomen in most cases, and persist for life. Sparse, brittle, or thin scalp hair, sparse or absent eyelashes and/or eyebrows in 70%, beard, pubic and axillary hair in 50%, and occasional dystrophic nails are other heterogeneous features [13]. Café-au-lait spots may develop later, while few patients have premature skin aging and one-third of cases will manifest palmoplantar hyperkeratosis.

Ocular lesions observed in 10%–50% cases seem to be less frequent than thought previously and are now considered as a minor feature only [13,14]. Bilateral subcapsular cataract developing in the early years of life with rapid onset in 2–3 months is a frequent feature and will develop fully by 4–7 years of age in 10%–40% patients. Other ophthalmic abnormalities include exophthalmos, strabismus, photophobia, Meibomian gland dysfunction,

keratoconus, corneal atrophy/scleralization, prominent Schwalbe's lines, absence of mesodermal layer of iris, tilted optic discs, temporetinal degeneration/atrophy or coloboma, pigment deposits over cornea and conjunctiva, blue sclera, congenital bilateral glaucoma, and occasional total blindness in later life [14–17].

Physical growth is frequently deficient in at least two-thirds of patients and manifests as short stature sometimes to an extent of dwarfism, delicate limbs, small hands, and bird-like skull. Skeletal abnormalities occur in 68% of patients particularly with *RECQL4* mutations, with a majority of patients presenting with either obvious congenital skeletal defects or subtle abnormalities on radio-imaging [3,18]. The common skeletal anomalies include frontal bossing, saddle nose, short stubby fingers, syndactyly, absent or hypoplastic thumbs, hypoplasia or agenesis of patella, and congenital radial ray defects, absent or short dysmorphic one or both radii and ulnae, diffuse or localized osteoporosis, and osteogenesis imperfecta. Abnormal metaphyseal trabeculation, brachymesophalangy, thumb aplasia/hypoplasia, osteopenia, destructive bone lesions, radial aplasia, hypoplasia or its head dislocation, and patella ossification are common findings on radio-imaging.

Other isolated and less common abnormalities are calcinosis and porokeratosis [17], suppurative otitis media, lower respiratory tract infection and bronchiectasis [19,20], aminoaciduria, myelodysplasia, leukemia, progressive leucopenia, or aplastic anemia [21–23], gastrointestinal anomalies (pyloric stenosis, anal atresia, annular pancreas, rectovaginal fistula, chronic emesis, diarrhea) [24], defective dentition (rudimentary or hypoplastic teeth, microdontia, short roots, unusual crown formations, early caries) in 27%–59% cases [25–27], sensorineural deafness [28], mental retardation and delayed speech [1], growth hormone deficiency [29,30], anhidrosis and immune dysfunction [31], hypoparathyroidism, infertility, and hypogonadism (in 30% cases) [32].

RTS is a genomic instability syndrome due to defects in DNA damage repair, particularly *RECQL4* mutations, which predispose patients for increased risk of malignancies. In contrast, obligate heterozygous parents of RTS cases do not show increased cancer risk.

RTS patients are prone to develop at least one neoplasm, have a second malignancy, or develop multiple primary cancers [13,33]. However, the overall prevalence of cancers in adults with RTS is unknown. Skin cancers and both isolated and multicentric osteosarcoma are the most common malignancies in RTS, with an estimated prevalence of 5% and 30%, respectively [13]. RTS-associated osteosarcoma has an earlier onset at the mean age of 14 years compared to 17 years in sporadic cases in general population, but with no difference in the site of tumor development (mostly in femur, tibia) and the histological (mostly osteoblastic) subtype [1,13,33]. However, the proportion of 17.9%–25.6% for RTS-associated multicentric osteosarcoma is higher compared to 0.4%–10% for all sporadic cases, suggesting an increased risk for multicentric osteosarcoma [1,34]. RTS patients with pathogenic variants in *RECQL4* notably have increased risk for developing osteosarcoma [3]. Malignant fibrous histiocytoma is another mesenchymal tumor reported occasionally [35]. Skin cancers such as basal cell carcinoma, squamous cell carcinoma, Bowen's disease, malignant eccrine poroma, and melanoma occur with an estimated prevalence of 5% and early onset at the mean age of 34 years in

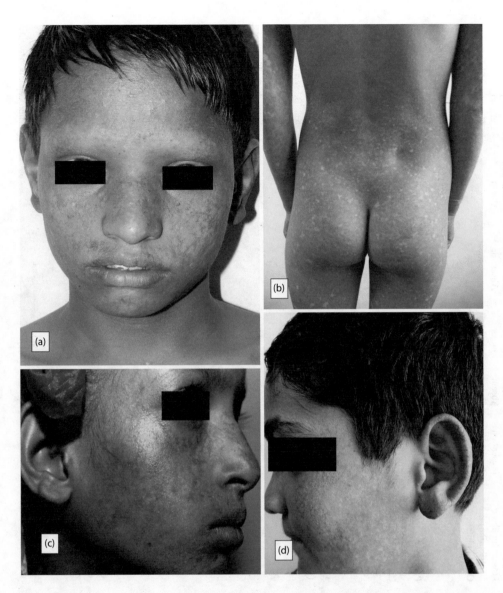

FIGURE 48.1 (a) Photo distribution of poikiloderma involving facial skin having reticulate hyperpigmentation, marked atrophy, and telangiectatic erythema in a 13-year-old boy. He also had poikiloderma of non-sun-exposed skin of trunk, back, and extremities. Also note his normal scalp hair and absent eyebrows and eyelashes. (b) Poikiloderma involving non-sun-exposed skin of back, buttocks, and extremities. (c) Characteristic photo-distributed poikiloderma involving face in an 8-year-old girl. Note: thin delicate facies, thin eye brows and normal eye lashes. (d)Poikilodermatous skin changes over cheek and ear pinna in a 14-year-old boy.

individuals with RTS compared to the general population but later than osteosarcoma [13,34,36,37]. These cancers involve non-sun-exposed surfaces, and squamous cell carcinoma is the most common epithelial tumor.

Myelodysplasia, acute myeloid leukemia, Hodgkin disease, anaplastic large-cell lymphoma, large-B cell lymphoma, and acute lymphoblastic leukemia may occur as either primary or secondary malignancies [22,23,38].

48.6 Diagnosis

No diagnostic criteria are currently available due to great morphologic and genetic heterogeneities. The diagnosis of RTS is largely clinical, based on photosensitivity with onset in early childhood, appearance and spreading of poikiloderma over time,

and presence of other more consistent clinical features. In the absence of more robust molecular diagnostic techniques, screening for *RECQL4* gene mutations can be offered to all patients with characteristic skin changes despite its presence in only two-thirds of RTS cases. This is important for early identification and managing RTS-associated manifestations, genetic counseling, and recommending cancer surveillance, as patients with pathogenic *RECQL4* variants show high risk of developing malignancies, particularly osteosarcoma. This will also differentiate RTS from two distinct *RECQL4* mutation disorders (i.e., RAPADILINO [Radial hypoplasia, Patella hypoplasia and cleft or Arched palate, Diarrhea and dislocated joints, Little size and limb malformation, slender Nose and normal intelligence] syndrome and Baller–Gerold syndrome) [39] and other syndromes that have overlapping clinical signs and that are usually considered in the differential diagnosis (Table 48.1).

TABLE 48.1

Characteristics of Major Hereditary Poikiloderma Syndromes Considered in Differential Diagnosis of Rothmund–Thomson Syndrome

S. No.	Syndrome	Inheritance	Bullae	Photo Sensitivity	Poikiloderma	Keratotic Lesions	Eczema	Nail Changes	Palmoplantar Keratoderma	Mucosal Lesions	Additional and Isolated Features
1	Rothmund–Thomson syndrome	AR	Occasional (over photo exposed skin)	Since infancy	Present (in photo distribution) Onset in infancy	Present on exposed skin in many patients	Absent	None or occasional dystrophy	Absent or occasional	Absent	Cataract, Dwarfism, Hypogonadism, Normal mental growth, Growth retardation, Bone abnormalities, Osteosarcoma
2	Bloom syndrome	AR	Occasional (over lower lip)	Present	Present (not a true poikiloderma) Café-au-lait macules are more characteristic	Absent	Absent	None	Absent	Absent	Pre- and post natal growth retardation, Recurrent infections (otitis media), chronic pulmonary disease, death is early from cancers
3	Werner syndrome (segmental progeria)	AR	Absent	Absent	Absent	Absent	Absent	None	Absent	Absent	Short stature, Premature aging and graying of hair, Bilateral cataract, Bird-like facies after 10 years of age, Scleroderma-like acral skin, Diabetes mellitus, hypogonadism, Subcutaneous calcification and ulceration, Increased risk for mesenchymal cancer and osteosarcoma, Causes of death include cancer or myocardial infarction
4	Xeroderma pigmentosa	AR	Occasional (over photo exposed skin)	Acute and present since infancy	Present in photo distribution and evolves in late childhood	Present (not limited to Acral areas)	Absent	Absent	Absent	Absent	Lentigen-like macules, Skin cancers, Neurologic changes, Photo phobia/Corneal opacities, No congenital cataract
5	Cockayne syndrome	AR	Absent	Present	Present in photo distribution	Absent	Absent	Absent	Absent	Absent	Dwarfism, Microcephaly, Mental retardation, Retinitis pigmentosa, Severe tooth decay, Bone abnormalities, Blindness, Conductive deafness

(Continued)

TABLE 48.1 (Continued)

Characteristics of Major Hereditary Poikiloderma Syndromes Considered in Differential Diagnosis of Rothmund–Thomson Syndrome

S. No.	Syndrome	Inheritance	Bullae	Photo Sensitivity	Poikiloderma	Keratotic Lesions	Eczema	Nail Changes	Palmoplantar Keratoderma	Mucosal Lesions	Additional and Isolated Features
6	Kindler syndrome	AR/AD	Acral with onset before one year of age Trauma induced	Since infancy, Improves with age leading to poikiloderma onset	Onset at 2–3 years of age, Widespread, Involves Sun-exposed and Non-sun-exposed body areas. Diffuse skin atrophy	Absent	Absent	Mild or none	Diffuse	Leukokeratosis	Defective gene mapped to chromosome 20p12.3. Stenosis of esophagus, anus and urethra, Phimosis, Webbing of digits, Anhidrosis Dental abnormalities, Mucosal/gingival inflammation
7	Weary's hereditary acrokeratotic poikiloderma	AD	Acral with onset after one year of age	Absent	Limited to neck and upper chest. Onset within 1st year of life	Present in Acral distribution	Present	None	Punctate or absent	Absent	Poor dentition, Enteropathy, cardiac abnormalities
8	Hereditary sclerosing poikiloderma	AD	Absent	Absent	Generalized poikiloderma with onset at childhood Accentuated in flexors with sclerotic bands	Absent	Absent	Clubbing is occasional	Sclerodermatous plaques	Absent	Poor dentition, Calcinosis cutis
9	Dyskeratosis congenita	XR, AD or AR	Absent (or occasional)	Absent	Poikiloderma with onset at childhood is more of dyschromatic variety. Involves face, neck, trunk, thighs	Absent	Absent	Always present and severe Precedes poikiloderma	Absent	Premalignant Leukokeratosis/ Leukoplakia	Blood dyscrasias, Growth and mental retardation, Poor dentition
10	Poikiloderma with neutropenia (clericuzio type)	AR	Absent	Absent	Poikiloderma since infancy over face and limbs without sparing flexural skin and trunk	Absent	Present	Hyperkeratotic nails	Absent	Absent	Neutropenia, Recurrent (pulmonary) infections
11	Mendes da costa syndrome	XR	Present (Acral bullae)	Absent	Present (spares the trunk)	Absent	Absent	Occasional	Absent	Absent	Acrocynosis, Scarring alopecia, Dwarfism, Micro cephaly, Mental retardation, Short conical digits, Bone abnormalities
12	Dermatopathia pigmentosa-reticularis	AD	Present (Acral bullae)	Absent	Dyschromatic (Trunkal)	Present (scattered Keratotic lesions)	Absent	Dystrophy or Hypoplasia	Hyperkeratosis	Absent	Corneal opacities, Alopecia, Axillary papillomatosis Pseudoainhum, Flexon contractures, Hypertrophic scars
13	Franceschetti–Jadassohn syndrome	AD	Absent	Absent	More of reticulated hyperpigmentation	Absent	Absent	Absent	Present	Absent	Hypo/Anhidrosis, Poor dentition
14	Degos–Touraine syndrome	?	Present (Acral or facial)	Absent	Photo distributed	Present in Acral distribution	Absent	Absent	Absent	Absent	Enteropathy

Note: Acrogeria (Gottron syndrome), Fanconi anemia, and ataxia telangiectasia are occasional mimickers.

Abbreviations: AD, autosomal dominant; AR, autosomal recessive; XR, X-linked recessive.

The carrier frequency for RTS is unknown, but being an autosomal recessive disorder there is a 50% risk of being a carrier of an *RECQL4* pathogenic variant for each sibling of the proband's parents. Prenatal testing for pregnancies at increased risk and pre-implantation genetic diagnosis for RTS is also possible provided the *RECQL4* pathogenic variants in the family members are known. Although skeletal abnormalities may be detected by ultrasonography at 16–18 weeks of gestation, their absence in a fetus at risk will not exclude the possibility of RTS by itself.

48.7 Treatment

The management in RTS is need based, and a lifelong multidisciplinary approach is recommended. All patients need complete clinical evaluation for cutaneous, ocular, or bone abnormalities by clinicians conversant with the natural history and for precise follow-up. Presence of focal epidermal flattening and hydropic degeneration of basal layer with proliferation of blood vessels, and melanin incontinence/melanophages in skin biopsy will mark various clinical changes of atrophy, telangiectasia, and hyperpigmentation in poikiloderma (Figure 48.2). Histopathological confirmation is required for suspected cancers. A skeletal survey of the long bones at infancy for underlying skeletal dysplasia and diagnosis of osteogenic sarcoma, pulmonary function tests, and computed tomography for bronchitis and bronchiectasis should be considered in the diagnostic workup. Patients with anemia or cytopenias need laboratory evaluation by complete blood counts, and bone marrow biopsy.

Visual impairment from cataracts needs surgical management. Individuals with cancers are treated by surgical excisions and according to standard chemotherapy and/or radiation regimens. Telangiectatic component of the rash has been treated with pulsed dye laser with good cosmetic outcome [40]. Though umbilical cord blood stem cell transplantation in a patient with RTS and combined immunodeficiency and allogenic bone marrow transplantation in RTS with myelodysplastic syndrome have been performed successfully, their long-term efficacy remains unevaluated [41,42].

48.8 Prevention, Genetic Counseling, and Prognosis

48.8.1 Prevention

Avoidance of excessive sun exposure and use of broad-spectrum sunscreens with both UVA and UVB protection are recommended to prevent skin cancer. Calcium and vitamin D supplements may also be warranted in individuals with osteopenia or a history of fractures. Surveillance includes annual physical examination for monitoring and detection of osteoarticular or skin malignancies by skeletal radio-imaging or dermatopathologic examination, ocular abnormalities, and other growth parameters.

48.8.2 Genetic Counseling

Genetic counseling is recommended for all RTS patients along with their parents and siblings for early detection and treatment of syndrome-associated manifestations and possibility of transmission to the offspring. Identification of parental consanguinity and affected family members or who are at risk of being carriers of RECQL4 mutation is imperative. Both parents of the proband are obligate heterozygotes for a disease-causing mutation and the siblings have a 25% chance of being affected or non-carriers, and a 50% chance of being asymptomatic carriers. Heterozygous carriers of a RECQL4 mutation are asymptomatic.

48.8.3 Prognosis

The life span in the absence of malignancy is normal. Death from metastatic osteosarcoma and other cancers remains the major cause in most cases. However, the osteosarcoma remains responsive to standard chemotherapy and clinical outcome is similar in RTS and non-RTS patients, with a 60%–70% 5-year survival rate [43].

48.9 Conclusion and Future Perspectives

Rothmund–Thomson syndrome (RTS) is a rare genetically and phenotypically heterozygous autosomal recessive genophotodermatosis

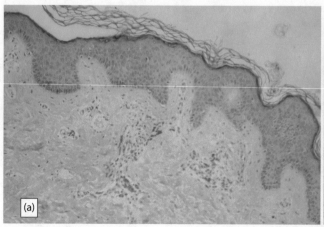
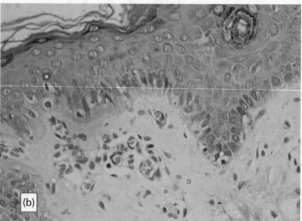

FIGURE 48.2 Histology of biopsy from gluteal skin. Poikiloderma features of focal epidermal flattening, hydropic degeneration of basal layer, proliferation of blood vessels, perivascular and periappendageal inflammatory infiltrate of lymphohistiocytes, melanin incontinence, and melanophages. (a) H&E, ×10, (b) H&E, ×40.

featuring a DNA damage repair defect. The onset is in early infancy with photosensitivity evolving into a photo-distributed poikiloderma in childhood in most cases and occasionally until adulthood. The other heterogeneous clinical features include short stature, hair abnormalities (sparse scalp hair, sparse or absent eyelashes and/or eyebrows and pubic and axillary hair), juvenile cataract, skeletal abnormalities (frontal bossing, saddle nose, congenital radial ray defects), and premature aging. In addition, gastrointestinal and hematological signs may occur in a few patients. There is a predisposition to extracutaneous (osteosarcoma, myelodysplastic syndrome) and cutaneous malignancies (squamous cell or basal cell carcinomas, malignant fibrous histiocytoma), particularly in patients associated with pathogenic mutations of *RECQL4* DNA helicase gene. The diagnosis is mainly from characteristic clinical features, and ultimate prognosis is dictated by development and type of malignancy.

Based on clinical and molecular studies, the two clinical forms are identified. The RTS-I has poikiloderma, hypogonadism, juvenile cataract, and no identified gene mutation, while RTS-II having poikiloderma, skeletal abnormalities, and no cataract is due to homozygous or compound heterozygous mutations in the *RECQL4* DNA helicase gene mapped to chromosome 8q24.3. The veracity of its pathogenetic role in RTS or whether its two forms are distinct or not remains debatable, as only two-thirds of the RTS patients have homozygous or compound heterozygous mutations for this gene. To date, only one of the 56 different RECQL4 mutations, 39 of which are identified in RTS, is linked mainly to skeletal defects and increased susceptibility for development of osteosarcoma and skin cancer. On the other hand, like poikiloderma, the presence of bilateral cataracts is not a diagnostic feature for RTS-II. Besides, the most common exon 9c.1573delT (p.Cys525AlafsX33) mutation in the RECQL4 gene is also identified in RAPADILINO syndrome and Baller–Gerold syndrome, characterized by radial hypoplasia and craniosynostosis [44–46]. These have several overlapping features (short stature, radial ray abnormalities) but cataract is specific to RTS, joint dislocation and patellar hypoplasia to RAPADILINO, and craniosynostosis to Baller–Gerold syndrome. The development of either osteosarcoma or lymphoma in RAPADILINO patients and NK/T-cell lymphoma in patients with Baller–Gerold syndrome [47,48] is also indicative of predisposition to malignancies in RECQL4-associated syndromes and reflects the spectrum of distinct syndromes with RECQL4 mutations. On the other hand, Clericuzio-type poikiloderma with neutropenia, a clinical sub-entity within the RTS spectrum, lacks *RECQL4* mutations and is genetically distinct from RTS [3,49,50]. Clearly, mutation(s) of other as yet unknown or minor gene(s) loci for locus heterogeneity in the other one-third clinically diagnosed RTS patients need to be identified. However, it is possible that the entire phenotypic variations of RTS still remain unexplained as less defined or atypical variants described might be as a result of rare combinations of RECQL4 mutations or novel genomic or epigenetic mechanisms that have remained undetected currently for technical limitations.

The genotype–phenotype correlation in RTS patients also remains unresolved, as all identified mutant proteins in patients are partially active. The complete deletion of RECQL4 function is lethal in humans and it remains unidentified. Only two

mutations, intron 1c.84+6del16 and intron 2c.118+27del25, detected upstream of exon5, are present in compound heterozygous patients and both lead to mis-splicing [1]. However, genomic identification of both mutations is insufficient to predict the consequences of these mutations. Studies on animal models together with genetic analysis in human RTS patients demonstrate that RECQL4 is a multifunctional protein and each specific function requires a different domain, indicating that different mutations in this gene can lead to different phenotypes. However, the observations need to be validated by elucidating the differences in response of RECQL4-deficient cells to different genotoxic agents.

The relevance of elaborating the defective cellular pathways and involvement of somatic RECQL4 mutations is also important for the putative cancer risk of unaffected obligate heterozygote carriers and in sporadic osteosarcoma cases. Development of standard diagnostic criteria and guidelines for surveillance and assessment for cancer risk in affected patients is imperative for early diagnosis and comprehensive health care.

REFERENCES

1. Larizza L, Roversi G, Volpi L. Rothmund-Thomson syndrome. *Orphanet J Rare Dis.* 2010;5:2.
2. Kitao S, Shimamoto A, Goto M et al. Mutations in RECQL4 cause a subset of cases of Rothmund-Thomson syndrome. *Nat Genet.* 1999;22:82–4.
3. Wang LL, Gannavarapu A, Kozinetz CA et al. Association between osteosarcoma and deleterious mutations in the RECQL4 gene in Rothmund-Thomson syndrome. *J Natl Cancer Inst.* 2003;95:669–74.
4. Kitao S, Lindor NM, Shiratori M, Furuichi Y, Shimamoto A. Rothmund-Thomson syndrome responsible gene, RECQL4: genomic structure and products. *Genomics.* 1999; 61:268–76.
5. Sengupta S, Shimamoto A, Koshiji M et al. Tumor suppressor p53 represses transcription of RECQ helicase. *Oncogene.* 2005; 24:1738–48.
6. Woo LL, Futami K, Shimamoto A, Furuichi Y, Frank KM. The Rothmund- Thomson gene product RECQL4 localizes to the nucleolus in response to oxidative stress. *Exp Cell Res.* 2006;312:3443–57.
7. Kumata Y, Tada S, Yamanada Y et al. Possible involvement of RecQL4 in the repair of double-strand DNA breaks in Xenopus egg extracts. *Biochim Biophys Acta.* 2007; 1773:556–64.
8. Schurman SH, Hedayati M, Wang Z et al. Direct and indirect roles of RECQL4 in modulating base excision repair capacity. *Hum Mol Genet.* 2009;18:3470–83.
9. Smeets MF, DeLuca E, Wall M et al. The Rothmund-Thomson syndrome helicase RECQL4 is essential for hematopoiesis. *J Clin Invest.* 2014;124:3551–65.
10. Lu L, Harutyunyan K, Jin W et al. RECQL4 regulates p53 function *in vivo* during skeletogenesis. *J Bone Miner Res.* 2015;30:1077–89.
11. Ng AJ, Walia MK, Smeets MF et al. The DNA helicase RECQL4 is required for normal osteoblast expansion and osteosarcoma formation. *PLoS Genet.* 2015;11:e1005160.

12. Werner SR, Prahalad AK, Yang J, Hock JM. RECQL4-deficient cells are hypersensitive to oxidative stress/damage: Insights for osteosarcoma prevalence and heterogeneity in Rothmund-Thomson syndrome. *Biochem Biophys Res Commun.* 2006;345:403–9.

13. Wang LL, Levy ML, Lewis RA et al. Clinical manifestations in a cohort of 41 Rothmund-Thomson syndrome patients. *Am J Med Genet.* 2001;102:11–7.

14. Dollfus H, Porto F, Caussade P et al. Ocular manifestations in the inherited DNA repair disorders. *Surv Ophthalmol.* 2003;48:107–22.

15. Chinmayee JT, Meghana GR, Prathiba RK, Ramesh TK. Ophthalmic manifestations in Rothmund-Thomson syndrome: Case report and review of literature. *Indian J Ophthalmol.* 2017;65:1025–7.

16. Nathanson M, Dandine M, Gaudelus J, Mousset S, Lasry D, Perelman R. Rothmund-Thomson syndrome with glaucoma. *Ann Pediatr (Paris).* 1983;30:520–5.

17. Mak RK, Griffiths WA, Mellerio JE. An unusual patient with Rothmund- Thomson syndrome, porokeratosis and bilateral iris dysgenesis. *Clin Exp Dermatol.* 2006;31:401–3.

18. Mehollin-Ray AR, Kozinetz CA, Schlesinger AE, Guillerman RP, Wang LL. Radiographic abnormalities in Rothmund-Thomson syndrome and genotype-phenotype correlation with RECQL4 mutation status. *AJR Am J Roentgenol.* 2008;191:W62–6.

19. Reix P, Derelle J, Levrey-Hadden H, Plauchu H, Bellon G. Bronchiectasis in two pediatric patients with Rothmund-Thomson syndrome. *Pediatr Int.* 2007;49:118–20.

20. Mahajan VK, Sharma V, Chauhan PS, Mehta KS, Raina R. Rothmund-Thomson syndrome with bronchiectasis: an uncommon phenotype? *Indian J Dermatol Venereol Leprol.* 2015;81:190–2.

21. Knoell KA, Sidhu-Malik NK, Malik RK. Aplastic anemia in a patient with Rothmund-Thomson syndrome. *J Pediatr Hematol Oncol.* 1999;21:444–6.

22. Narayan S, Fleming C, Trainer AH, Craig JA. Rothmund-Thomson syndrome with myelodysplasia. *Pediatr Dermatol.* 2001;18:210–2.

23. Pianigiani E, De Aloe G, Andreassi A, Rubegni P, Fimiani M. Rothmund-Thomson syndrome (Thomson type) and myelodysplasia. *Pediatr Dermatol.* 2001;18:422–5.

24. Blaustein HS, Stevens AW, Stevens PD, Grossman ME. Rothmund-Thomson syndrome associated with annular pancreas and duodenal stenosis: A case report. *Pediatr Dermatol.* 1993;10:159–63.

25. Pujol LA, Erickson RP, Heidenreich RA, Cunniff C. Variable presentation of Rothmund-Thomson syndrome. *Am J Med Genet.* 2000;95:204–7.

26. Haytaç MC, Oztunç H, Mete UO, Kaya M. Rothmund-Thomson syndrome: A case report. *Oral Surg Oral Med Oral Pathol Oral Radiol Endod.* 2002;94:479–84.

27. Roinioti TD, Stefanopoulos PK. Short root anomaly associated with Rothmund-Thomson syndrome. *Oral Surg Oral Med Oral Pathol Oral Radiol Endod.* 2007;103:e19–22.

28. Beghini A, Castorina P, Roversi G, Modiano P, Larizza L. RNA processing defects of the helicase gene RECQL4 in a compound heterozygous Rothmund-Thomson patient. *Am J Med Genet A.* 2003;120A:395–9.

29. Sznajer Y, Siitonen HA, Roversi G et al. Atypical Rothmund-Thomson syndrome in a patient with compound heterozygous mutations in RECQL4 gene and phenotypic features in RECQL4 syndromes. *Eur J Pediatr.* 2008;167:175–81.

30. Kaufmann S, Jones M, Culler FL, Jones KL. Growth hormone deficiency in the Rothmund-Thomson syndrome. *Am J Med Genet.* 1986;23:861–8.

31. Snels DG, Bavinck JN, Muller H, Vermeer BJ. A female patient with the Rothmund-Thomson syndrome associated with anhidrosis and severe infections of the respiratory tract. *Dermatology.* 1998;196:260–3.

32. Werder EA, Mürset G, Illig R, Prader A. Hypogonadism and parathyroid adenoma in congenital poikiloderma (Rothmund–Thomson syndrome). *Clin Endocrinol.* 1975;4:75–82.

33. Spurney C, Gorlick R, Meyers PA, Healey JH, Huvos AG. Multicentric osteosarcoma, Rothmund-Thomson syndrome, and secondary nasopharyngeal non-Hodgkin's lymphoma: a case report and review of the literature. *J Pediatr Hematol Oncol.* 1998;20:494–7.

34. Stinco G, Governatori G, Mattighello P, Patrone P. Multiple cutaneous neoplasms in a patient with Rothmund-Thomson syndrome: Case report and published work review. *J Dermatol.* 2008;35:154–61.

35. Ilhan I, Arikan U, Buyukpamukcu M. Rothmund–Thomson syndrome and malignant fibrous histiocytoma: A case report. *Pediatr Hematol Oncol.* 1995;12:103–5.

36. Van Hees CLM, Duinen CMV, Bruijin JA, Vermeer BJ. Malignant eccrine poroma in a patient with Rothmund-Thomson syndrome. *Br J Dermatol.* 1996;134:813–5.

37. Howell SM, Bray DW. Amelanotic melanoma in a patient with Rothmund-Thomson syndrome. *Arch Dermatol.* 2008;144:416–7.

38. Simon T, Kohlhase J, Wilhelm C, Kochanek M, De Carolis B, Berthold F. Multiple malignant diseases in a patient with Rothmund-Thomson syndrome with RECQL4 mutations: Case report and literature review. *Am J Med Genet A.* 2010;152A:1575–9.

39. Dietschy T, Shevelev I, Stagljar I. The molecular role of the Rothmund- Thomson-, RAPADILINO- and Baller-Gerold-gene product, *RECQ*L4: Recent progress. *Cell Mol Life Sci.* 2007;64:796–802.

40. Potozkin JR, Geronemus RG. Treatment of the poikilodermatous component of the Rothmund-Thomson syndrome with the flashlamp-pumped pulsed dye laser: A case report. *Pediatr Dermatol.* 1991;8:162–5.

41. Broom MA, Wang LL, Otta SK et al. Successful umbilical cord blood stem cell transplantation in a patient with Rothmund-Thomson syndrome and combined immunodeficiency. *Clin Genet.* 2006;69:337–43.

42. Rizzari C, Cacchiocchi D, Rovelli A, Biondi A, Cantù-Rainoldi A, Uderzo C, Masera G. Myelodysplastic syndrome in a child with Rothmund-Thomson syndrome. A case report. *J Pediatr Hematol Oncol.* 1996;18:96–7.

43. Hicks MJ, Roth JR, Kozinetz CA, Wang LL. Clinicopathologic features of osteosarcoma in patients with Rothmund-Thomson syndrome. *J Clin Oncol.* 2007; 25:370–5.

44. Siitonen HA, Kopra O, Kääriäinen H et al. Molecular defect of RAPADILINO syndrome expands the phenotype spectrum of RECQL diseases. *Hum Mol Genet.* 2003;12:2837–44.

45. Mergarbane, Mégarbané A, Melki I et al. Overlap between Baller-Gerold and Rothmund-Thomson syndrome. *J Clin Dysmorphol.* 2000;9:303–5.

46. Van Maldergem L, Siitonen HA, Jalkh N et al. Revisiting the craniosynostosis-radial ray hypoplasia association: Baller-Gerold syndrome caused by mutations in the RECQL4 gene. *J Med Genet.* 2005;43:142–52.

47. Siitonen HA, Sotkasiira J, Biervliet M et al. The mutation spectrum in RECQL4 diseases. *Eur J Hum Genet.* 2009; 17:151–8.

48. Debeljak M, Zver A, Jazbec J. A patient with Baller-Gerold syndrome and midline NK/T lymphoma. *Am J Med Genet.* 2009;15:755–9.

49. Wang LL, Gannavarapu A, Clericuzio CL, Erickson RP, Irvine AD, Plon SE. Absence of RECQL4 mutations in poikiloderma with neutropenia in Navajo and non-Navajo patients. *Am J Med Genet.* 2003;118A:299–301.

50. Van Hove JL, Jaeken J, Proesmans M et al. Clericuzio type poikiloderma with neutropenia is distinct from Rothmund-Thomson syndrome. *Am J Med Genet A.* 2005;132A:152–8.

49

Werner Syndrome

Dongyou Liu

CONTENTS

49.1 Introduction

Werner syndrome is a rare autosomal recessive disorder that typically displays normal development until the end of the first decade followed by accelerated growth and aging from the 20s–30s (e.g., loss and graying of hair, hoarseness, scleroderma-like skin, bilateral ocular cataracts, type 2 diabetes mellitus, atherosclerosis, osteoporosis, skin ulcers, hypogonadism, and cancer), with myocardial infarction and cancer being the most common causes of death.

The gene responsible for Werner syndrome was discovered in 1996 as *WRN* (*RECQL2*) on chromosome 8p12, which encodes a homolog of the *E. coli* RecQ DNA helicase (RECQL2) critically important for genomic integrity. Homozygous or compound heterozygous germline mutations in *WRN* occur in approximately 90% of patients with Werner syndrome [1].

49.2 Biology

Initially documented by Werner in 1904 from four siblings of 31–40 years in age who presented with "cataracts in connection with scleroderma," short stature, and premature graying of hair. Subsequent descriptions of additional cases by Oppenheimer and Kugel in 1934 and Thannhauser in 1945 led to the proposal of Werner syndrome (or Werner's syndrome, formerly progeria of adults, to distinguish from Hutchinson–Gilford progeria syndrome, or progeria of childhood) for this adult premature aging syndrome [2].

The localization by linkage study of the culprit gene region to chromosome 8 in 1992 and the identification by positional cloning of the *WRN* gene in 1996 revealed molecular insights on the

pathogenesis of Werner syndrome. Indeed, these new findings helped clarify the premature aging seen in Werner syndrome as distinct from normal aging on a cellular level [3–5]. Further, the characterization of the *WRN* encoded product as RecQ helicase (RECQL2), one of five members (i.e., RECQL1, BLM, WRN, RECQL4, and RECQL5) in the RecQ helicase family, helped link Werner syndrome to Bloom syndrome (BLM) and Rothmund–Thomson syndrome (RECQL4), each of which features genomic instability and susceptibility to cancer, and each of which demonstrates notable differences in the characteristics of genomic instability and the sites and types of cancers associated [6,7].

Genomic instability represents one of the key mechanisms (the others being telomere attrition, epigenetic alterations [e.g., DNA methylation], loss of proteostasis, deregulation of nutrient sensing, mitochondrial dysfunction, cellular senescence, stem cell exhaustion, altered intercellular communication) that underline the aging process as well as cancer susceptibility. DNA repair machinery and telomeres play a crucial role in ensuring genome integrity and stability. Functioning as DNA-dependent ATPase and ATP-dependent DNA unwinding enzyme, RecQ helicase protects genome stability by regulating DNA repair pathways and telomeres. Mutations in RecQ helicase-encoding genes have serious consequences [8–10].

49.3 Pathogenesis

The *WRN* (RECQL2) gene on chromosome 8p12 spans 142 kb with 35 exons (including 34 coding exons) and encodes a 1,432 aa, 162 kDa RecQ helicase L2 (WRN or RECQL2), which belongs to the RecQ subfamily of DNA helicase proteins (consisting of RECQL1, BLM, WRN/RECQL2, RECQL4, and RECQL5 in humans), and which is the only RecQ helicase with 3′→5′ exonuclease activity [11].

Structurally, WRN consists of an N-terminal 3′→5′ exonuclease domain (60–288 aa), an ATP-dependent 3′→5′ helicase domain (551–859 aa), a RQC (RecQ helicase conserved region) domain (956–1,064 aa), and a C-terminal HRDC (helicase RNase D C-terminal) domain (1,142–1,235 aa) and nuclear localization signal (NLS, 1,370–1,375 aa). The region between RQC and HRDC also shows single-strand DNA annealing activity and may influence oligomerization of the WRN protein. WRN helicase preferentially unwinds complex DNA structures such as tetraplex DNA, double-strand DNA with mismatch bubbles, and Holliday junctions, whereas WRN exonuclease preferentially digests single strands in complex DNA structures [12].

Through its magnesium and ATP-dependent DNA-helicase activity (which unwinds and separates double-stranded DNA) and 3′→5′ exonuclease activity (which trims the broken ends of damaged DNA by removing nucleotides) toward double-stranded DNA with a 5′-overhang, WRN contributes to DNA repair (homologous recombination for DSB repair using the intact sister chromatid during late S and G_2 phases, non-homologous end joining [NHEJ] for exogenous DSB repair during the G_0 and G_1 phases, and base excision for single strand DNA repair), replication, transcription, telomere maintenance, and genome stability. Further, WRN interacts with tumor suppressor p53 via its C-terminus [2,11].

Homozygous or compound heterozygous mutations in the *WRN* gene produce functional null alleles and loss of function WRN. Of 83 *WRN* pathogenic variants identified from patients with Werner syndrome to date, many are stop codons, small indels, or splicing mutations yielding truncated WRN with loss of the nuclear localization signal at the C-terminus (e.g., c.1105C>T, c.2089-3024A>G, c.2179dupT, c.3139-1G>C, c.3460-2A>C, c.3590delA). Amino acid substitutions (e.g., p.Lys125Asn and Lys135Glu) located in the exonuclease domain are functionally null variants that generate unstable WRN lacking helicase activity [1,13].

Ethnicity-specific *WRN* pathogenic variants consist of those affecting Japanese (c.3139-1G>C, r.3139 3233del95; c.1105C>T, p.R369*; c.3446delA, p.E1149fs), Sardinian (c.2089-3024A>G, r.2088 2089ins106), Indian/Pakistani (c.561A>G, r.557-654del98), Moroccan (c.2179dupT, p.C727fs), Turkish (c.3460-2A>G, r.3460 3572del113), and Dutch (c.3590delA, pN1197fs) populations [1,11,13].

49.4 Epidemiology

49.4.1 Prevalence

Werner syndrome is estimated to affect 1:380,000-1:1,000,000 globally, based on analysis of the most common pathogenic variant c.1105C>T, which accounts for 20%–25% of pathogenic variants in the European and Japanese populations. In Japan and Sardinia, Werner syndrome has an incidence of 1 in 20,000–40,000 and 1 in 50,000, respectively. Werner syndrome has a median age of diagnosis between late 30s–40s, and a mean age of death at 54 years due mainly to myocardial infarction and cancer [1].

49.4.2 Inheritance

Werner syndrome shows autosomal-recessive inheritance.

49.4.3 Penetrance

Werner syndrome demonstrates high penetrance. While some patients harboring WRN pathogenic variants may fail to fulfill clinical diagnostic criteria, they still display other symptoms for a confident diagnosis of Werner syndrome to be made.

49.5 Clinical Features

Werner syndrome is a segmental progeroid syndrome that resembles accelerated aging clinically, with normal development until the end of the first decade, lack of a growth spurt during the early teen years, loss and graying of hair, hoarseness, and scleroderma-like skin in the 20s; bilateral ocular cataracts (median onset age of 31 years), type 2 diabetes mellitus, hypogonadism, arteriosclerosis (e.g., coronary artery atherosclerosis, leading to myocardial infarction), skin ulcers (Achilles tendon, medial malleolus, lateral malleolus), and osteoporosis (the long bones) in the 30s, and characteristic "bird-like" appearance at the bridge of the nose during the third or fourth decade (Figures 49.1) [14] along with decline in fertility and increased risk for cancer (e.g., soft tissue sarcoma, osteosarcoma, melanoma, thyroid carcinoma, and acral lentiginous melanomas [on the feet and nasal mucosa]) (Figure 49.2) [15–17].

49.6 Diagnosis

Clinical diagnosis of Werner syndrome is based on observing the four cardinal signs: (i) bilateral ocular cataracts (99% of cases), (ii) premature graying and/or thinning of scalp hair (100%), (iii) characteristic dermatologic pathology (e.g., indolent deep ulcerations around Achilles tendons and at elbows are almost pathognomonic to Werner syndrome; 96%), and (iv) short stature (95%); and nine additional signs: (i) thin limbs (98%), (ii) pinched facial features (96%), (iii) osteoporosis (91%), (iv) voice change (89%), (v) hypogonadism (80%), (vi) type 2 diabetes mellitus (71%), (vii) soft tissue calcification (67%), (viii) tumor (44%), and (ix) atherosclerosis (30%) [1,18,19].

Patients showing all four cardinal signs and two additional signs (definite) or the first three cardinal signs and two additional signs (probable) are diagnosed as having Werner syndrome.

Molecular identification of biallelic *WRN* pathogenic variants by sequence analysis helps confirm the diagnosis if clinical features are inconclusive.

Differential diagnoses for Werner syndrome include atypical Werner syndrome (early age of onset at early 20s or earlier, faster rate of progression; normal WRN proteins, heterozygous pathogenic missense variants in *LMNA* in 15% of cases), mandibular hypoplasia, deafness, progeroid features, and lipodystrophy syndrome (MDPL; progeroid features, lipodystrophy, characteristic facial features, sensorineural hearing loss; absence of ocular cataracts), mandibulo-acral dysplasia (MAD; short stature, type A lipodystrophy, loss of fat in the

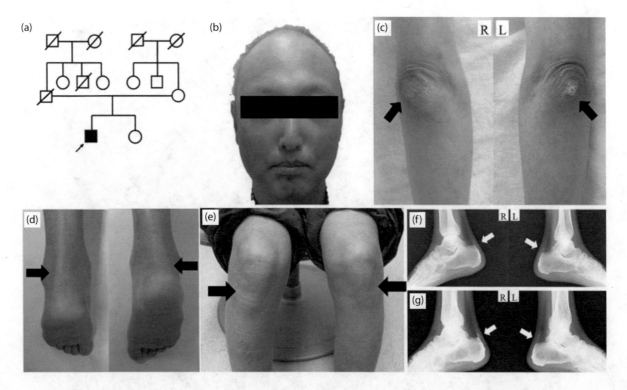

FIGURE 49.1 Werner syndrome in an 8-year-old boy with compound heterozygous *WRN* mutations c.1720+1G>A plus c.3139-1G>C showing: (a) patient's family tree; (b) characteristic "bird-like" face and senile appearance; (c) cutaneous hyperkeratosis in bilateral elbows (arrows); (d) knees (arrows); (e) ankles (arrows); (f,g) x-ray showing the segmented calcification of the Achilles tendons at the diagnosis and 5 years later. (Photo credit: Matsumoto N et al. *Intern Med.* 2019;58(7):1033–6.)

extremities but accumulation of fat in the neck and trunk, thin, hyperpigmented skin, partial alopecia, prominent eyes, convex nasal ridge, tooth loss, micrognathia, retrognathia, and short fingers; biallelic pathogenic variants in *LMNA*, and zinc metalloproteinase *ZMPSTE24*), Hutchinson–Gilford progeria syndrome (HGPS, progeria of childhood; accelerated aging, profound failure to thrive during the first year, characteristic facies, partial alopecia progressing to total alopecia, loss of subcutaneous fat, progressive joint contractures, bone changes, abnormal tightness and/or small soft outpouchings of the skin over the abdomen and upper thighs during the second to third year; severe atherosclerosis; death due to cardiac or cerebrovascular disease between age 6 and 20 years; average life span of approximately 14.6 years; autosomal dominant disorder due to *LMNA* pathogenic variant c.1824C>T), early-onset type 2 diabetes with secondary complications (mimicking some features of Werner syndrome), myotonic dystrophy type 1 or myotonic dystrophy type 2 (young adult-onset cataracts, muscle wasting in adults), scleroderma, mixed connective tissue disorders, and lipodystrophy (similar skin features), Charcot–Marie–tooth hereditary neuropathy or familial leg ulcers of juvenile onset (distal atrophy and skin ulcerations in the absence of other manifestations characteristic of Werner syndrome), Rothmund–Thomson syndrome (RTS; autosomal recessive disorder due to pathogenic variants in *RECQL4*), BLM (increased sister chromatid exchange; autosomal recessive disorder due to pathogenic variants in *BLM*), Li–Fraumeni syndrome (multiple cancers, absence of juvenile-onset cataracts, autosomal dominant disorder due to pathogenic variants in *TP53*), Flynn–Aird syndrome

(cataracts, skin atrophy and ulceration; neurologic abnormalities), brachiooculofacial syndrome (premature graying in adults; strabismus, coloboma, and microphthalmia; dysmorphic facial features; autosomal dominant disorder due to *TFAP2A* pathogenic variants), SHORT syndrome (*s*hort stature, *h*yperextensibility, he*r*nia, *o*cular depression, *R*ieger anomaly, and *t*eething delay; progeria-like facies and lipodystrophy, type 2 diabetes mellitus, cataracts and glaucoma; autosomal dominant disorder due to pathogenic variants in *PIK3R1* [1,18,19].

49.7 Treatment

Management of Werner syndrome involves aggressive treatment of skin ulcers (e.g., Bosentan), control of type 2 diabetes mellitus (e.g., pioglitazone, sitagliptin), cholesterol-lowering drugs for abnormal lipid profile, surgical treatment of ocular cataracts, and treatment of malignancies (Figure 49.3) [11,20,21].

49.8 Prognosis and Prevention

Prognosis for Werner syndrome is generally poor, with mean age of death at 54 years as a result of myocardial infarction and cancer.

Affected patients should avoid smoking and trauma to the skin and undertake regular exercise and weight control to reduce atherosclerosis risk.

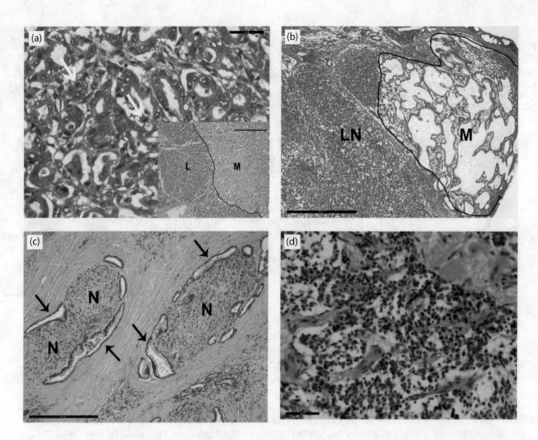

FIGURE 49.2 Werner syndrome in an adult male with *WRN* mutation c.1336C>T c.3139-1G>C showing (a) H&E staining of pancreatic adenocarcinoma liver metastasis in Patient 3. Arrows indicate mitotic figures. Scale bar, 50 μm. Inset shows liver (L) and adjacent metastasis (M) separated by a sharp boundary (dark line). Inset scale bar, 500 μm. (b) H&E staining of pancreatic adenocarcinoma lymph node metastasis in Patient 4 growing in and expanding the subcapsular sinus of an abdominal lymph node (LN). Scale bar, 500 μm. (c) H&E staining of pancreatic adenocarcinoma perineural invasion in Patient 4. Arrows mark invading carcinoma. N, nerve. Scale bar, 500 μm. (d) H&E staining of a pulmonary carcinoid in Patient 4. Scale bar, 100 μm. (Photo credit: Tokita M et al. *Sci Rep.* 2016;6:32038.)

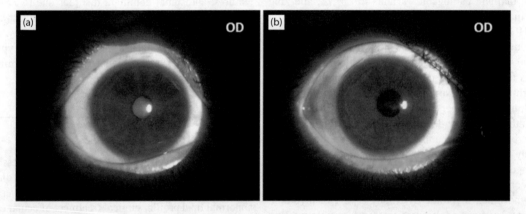

FIGURE 49.3 Werner syndrome in a 26-year-old male with *WRN* frameshift mutation (p. Y 1157 Cfs * 7) showing: (a) opaque and cloudy lens; (b) clear IOL in place after surgery in the anterior segment of the right eye. OD, right eye. (Photo credit: Chen CL et al. *BMC Ophthalmol.* 2018;18(1):199.)

Prenatal testing and preimplantation genetic diagnosis may be considered for pregnancy at increased risk and individuals with an affected family member.

49.9 Conclusion

Werner syndrome is a rare autosomal recessive disorder characterized by premature aging, chromosomal instability, and cancer predisposition. Molecular mechanisms of Werner syndrome relate to homozygous or compound heterozygous mutations of the *WRN* gene, which encodes a DNA helicase (RECQL2). Given its roles in the maintenance of chromosome integrity through DNA replication, repair, and recombination, the effect of *WRN* mutations leading to truncated, loss-of-function DNA helicase (RECQL2) is not surprisingly broad and severe, with myocardial infarction and cancer being the main causes of death in affected patients (mean age of death at 54 years). Clinical diagnosis of

Werner syndrome relies on observing four cardinal features and nine additional features [22,23]. As 90% of patients with Werner syndrome harbor *WRN* pathogenic variants, molecular testing provides a valuable means of confirmation if clinical features are inconclusive [24]. Management of Werner syndrome currently involves standard procedures that aim to reduce or alleviate the damages and harms associated with this adult premature aging disorder [25].

REFERENCES

1. Oshima J, Martin GM, Hisama FM. Werner syndrome. In: Adam MP, Ardinger HH, Pagon RA et al. editors. *GeneReviews®* [Internet]. Seattle (WA): University of Washington, Seattle; 1993–2018. 2002 Dec 2 [updated 2016 Sep 29].

2. Shamanna RA, Croteau DL, Lee JH, Bohr VA. Recent advances in understanding Werner syndrome. *F1000Res.* 2017;6:1779.

3. Goto M, Rubenstein M, Weber J, Woods K, Drayna D. Genetic linkage of Werner's syndrome to five markers on chromosome 8. *Nature.* 1992; 355(6362):735–8.

4. Yu CE, Oshima J, Fu YH et al. Positional cloning of the Werner's syndrome gene. *Science.* 1996;272(5259):258–62.

5. Maierhofer A, Flunkert J, Oshima J et al. Accelerated epigenetic aging in Werner syndrome. *Aging (Albany NY).* 2017;9(4):1143–52.

6. Ishikawa N, Nakamura K, Izumiyama-Shimomura N et al. Accelerated *in vivo* epidermal telomere loss in Werner syndrome. *Aging (Albany NY).* 2011;3(4):417–29.

7. de Renty C, Ellis NA. Bloom's syndrome: Why not premature aging?: A comparison of the BLM and WRN helicases. *Ageing Res Rev.* 2017;33:36–51.

8. Croteau DL, Popuri V, Opresko PL et al. Human RecQ helicases in DNA repair, recombination, and replication. *Annu Rev Biochem.* 2014;83:519–52.

9. Shimamoto A, Yokote K, Tahara H. Werner syndrome-specific induced pluripotent stem cells: recovery of telomere function by reprogramming. *Front Genet.* 2015;6:10.

10. Yamaga M, Takemoto M, Shoji M et al. Werner syndrome: A model for sarcopenia due to accelerated aging. *Aging (Albany NY).* 2017;9(7):1738–44.

11. Oshima J, Sidorova JM, Monnat RJ Jr. Werner syndrome: Clinical features, pathogenesis and potential therapeutic interventions. *Ageing Res Rev.* 2017;33:105–14.

12. Mukherjee S, Sinha D, Bhattacharya S, Srinivasan K, Abdisalaam S, Asaithamby A. Werner syndrome protein and DNA replication. *Int J Mol Sci.* 2018;19(11). pii: E3442.

13. Yokote K, Chanprasert S, Lee L et al. *WRN* mutation update: Mutation spectrum, patient registries, and translational prospects. *Hum Mutat.* 2017;38(1):7–15.

14. Matsumoto N, Ohta Y, Deguchi K et al. Characteristic clinical features of Werner syndrome with a novel compound heterozygous *WRN* mutation c.1720+1G>A plus c.3139-1G>C. *Intern Med.* 2019;58(7):1033–6.

15. Goto M, Ishikawa Y, Sugimoto M, Furuichi Y. Werner syndrome: A changing pattern of clinical manifestations in Japan (1917~2008). *Biosci Trends.* 2013;7(1):13–22.

16. Lauper JM, Krause A, Vaughan TL, Monnat RJ Jr. Spectrum and risk of neoplasia in Werner syndrome: A systematic review. *PLOS ONE.* 2013;8(4):e59709.

17. Tokita M, Kennedy SR, Risques RA et al. Werner syndrome through the lens of tissue and tumour genomics. *Sci Rep.* 2016;6:32038.

18. Sickles CK, Gross GP. *Progeria (Werner syndrome). StatPearls* [Internet]. Treasure Island (FL): StatPearls Publishing; 2019 Jan–2019 May 6.

19. Oshima J, Hisama FM. Search and insights into novel genetic alterations leading to classical and atypical Werner syndrome. *Gerontology.* 2014;60(3):239–46.

20. Chen CL, Yang JS, Zhang X et al. A case report of Werner's syndrome with bilateral juvenile cataracts. *BMC Ophthalmol.* 2018;18(1):199.

21. Rincón A, Mora L, Suarez-Obando F, Rojas JA. Werner's syndrome: Understanding the phenotype of premature aging-First case described in Colombia. *Case Rep Genet.* 2019;2019:8538325.

22. Mazzarello V, Ferrari M, Ena P. Werner syndrome: Quantitative assessment of skin aging. *Clin Cosmet Investig Dermatol.* 2018;11:397–402.

23. Kaur A, Grover P, Albawaliz A, Chauhan M, Barthel B. Growing old too fast: A rare case of Werner syndrome. *Cureus.* 2019;11(5):e4743.

24. Maezawa Y, Kato H, Takemoto M et al. Biallelic WRN mutations in newly identified Japanese Werner syndrome patients. *Mol Syndromol.* 2018;9(4):214–8.

25. Hayashi K, Tasaka T, Kondo T et al. Successful cord blood transplantation in a Werner syndrome patient with high-risk myelodysplastic syndrome. *Intern Med.* 2019;58(1):109–13.

50

Xeroderma Pigmentosum

Dongyou Liu

CONTENTS

50.1 Introduction

Xeroderma pigmentosum (XP) is a rare autosomal recessive disorder characterized by increased photosensitivity. Upon exposure to sunlight (or solar/ultraviolet [UV] radiation, specifically UVB and UVC), patients develop cutaneous symptoms (e.g., lentiginous pigmentation), eye damage, neurodegenerative processes, skin and central nervous system tumors (e.g., basal cell carcinoma, squamous cell carcinoma, keratoacanthoma) at an early age, and decreased life span due to skin cancer and neurodegenerative sequelae.

As a genetically heterogeneous disorder, XP is linked to defective DNA repair involving the nucleotide excision repair (NER) pathway (complementation groups XPA–XPG) and translesion synthesis (variant type XPV) (Table 50.1) [1,2].

50.2 Biology

XP (formerly xeroderma or parchment skin) was first described by Moriz Kaposi in 1874 from four patients with thin, dry skin (xeroderma) that had wrinkling, checkered pigmentation, small dilatations of the vessels, skin contraction, and skin-based tumors. Subsequent reports extended its clinical profile to neurologic abnormalities, dwarfism, gonadal hypoplasia, and mental deficiency, with patients of complementation groups XPA, XPB, XPD, and XPG exhibiting neurological abnormalities, while XPC, XPE, and XPF patients rarely developing neurological symptoms (Table 50.1) [3].

Although the role of UV radiation in the initiation of XP was recognized in 1926, the link between UV-induced DNA damage,

faulty DNA repair, and human cancer in XP was established 40 years later. Indeed, XP patients with neurological degeneration display an even lower rate of post-UV DNA repair than those with no neurological abnormalities. Further, XP patients often develop non-melanoma skin cancer at 8 years of age, which is >50 years earlier than the general population. In addition, XP patients of 29 years or under show >2000-fold increase in basal cell and squamous cell carcinoma of the skin, cutaneous melanoma, cancer of the anterior eye, and cancer of the anterior tongue compared to the general population [4].

Based on cell complementation studies (cell fusion technique followed by assessment of DNA repair), XP is distinguished into seven complementation groups (XPA–XPG) in addition to a milder variant form known as XP-variant (XPV). Specifically, XPA shows <5% of normal DNA repair rate; XPB 3%–40%, XPC 10%–30%, XPD 15%–50%, XPE 50%, XPF 15%–30%, XPG 5%–30%, and XPV 100% (normal). Sequencing analysis enables identification of distinct causative gene for each of these complementation groups (i.e., *XPA, ERCC3, XPC, ERCC2, DDB2, ERCC4,* and *ERCC5* for XPA–XPG, respectively, and *POLH* for XPV) (Table 50.1). Of these, complementation groups XPA–XPG are due to defective DNA repair involving the NER pathway, whereas XP variant type XPV is linked to defective DNA repair involving during translesion synthesis [2,5].

50.3 Pathogenesis

Mammalian cells are constantly exposed to UV/ionizing radiation, reactive oxygen species (ROS), replication errors, and chemotherapy that cause DNA damage, cell death, premature

411

TABLE 50.1

Molecular and Clinical Characteristics of Xeroderma Pigmentosum Groups

Phenotype	Gene (Chromosome/ Coding Sequence)	Protein Function	Frequency of Occurrence (USA/Europe/ Japan)	Clinical Features	Syndromes
Xeroderma pigmentosum, complementation group A (XPA or XP1)	*XPA* (9q22.33/1,377 bp)	DNA repair protein complementing XPA cells that assists with DNA unwinding	9%/20%/55%	Photosensitivity, poikiloderma, lentigines, skin cancer, neurodegeneration (mild to severe)	XPA
Xeroderma pigmentosum, complementation group B (XPB or XP2)	*ERCC3* (2q14.3/2,751 bp)	TFIIH complex helicase XPB involved in DNA unwinding	1%/2%/0%	Photosensitivity, poikiloderma, lentigines, skin cancer, neurodegeneration	XPB, XP/CS, TTD
Xeroderma pigmentosum, complementation group C (XPC or XP3)	*XPC* (3p25.1/3,558 bp)	DNA repair protein complementing XPC cells that recognizes global genome defects	43%/31%/3%	Photosensitivity, poikiloderma, lentigines, skin cancer	XPC
Xeroderma pigmentosum, complementation group D (XPD or XP4 or XPH or XP8)	*ERCC2* (19q13.32/2,400 bp)	TFIIH complex helicase XPD involved in DNA unwinding	28%/16%/5%	Photosensitivity, poikiloderma, lentigines, skin cancer, neurodegeneration, brain tumor	XPD, XP/CS, TTD, COFS
Xeroderma pigmentosum, complementation group E, DDB-negative subtype (XPE or XP5)	*DDB2* (11p11.2/4,193 bp)	DNA damage-binding protein 2 that recognizes global genome defects	3%/0%/3%	Photosensitivity, poikiloderma, lentigines, skin cancer, neurodegeneration	XPE
Xeroderma pigmentosum, complementation group F (XPF or XP6)	*ERCC4* (16p13.12/2,881 bp)	DNA repair endonuclease XPF that forms an endonuclease with ERCC1 to incise damaged DNA for repair	0%/3%/7%	Photosensitivity, poikiloderma, lentigines, skin cancer, neurodegeneration, brain tumor	XPF, Fanconi anemia
Xeroderma pigmentosum, complementation group G (XPG or XP7)	*ERCC5* (13q33.1/4,091 bp)	DNA repair protein complementing XPG cells (endonuclease) that incises damaged DNA	3%/9%/1%	Photosensitivity, poikiloderma, lentigines, skin cancer, neurodegeneration	XPG, XP/CS
Xeroderma pigmentosum, variant type (XPV or pigmentary xerodermoid)	*POLH* (6p21.1/2,140 bp)	DNA polymerase eta that performs translesion DNA synthesis after UV exposure	7%/13%/25%	Milder photosensitivity, poikiloderma	XPV

Abbreviations: COFS, cerebrooculofacioskeletal syndrome; ERCC, excision repair, complementing defective, in Chinese hamster; POLH, DNA polymerase eta; TFIIH complex, transcription factor II H complex; TTD, trichothiodystrophy; XP/CS, xeroderma pigmentosum-Cockayne syndrome complex.

aging, and tumorigenesis. Several DNA repair mechanisms are employed by mammalian cells to prevent the consequences of DNA injuries and to preserve genetic integrity. These include (i) base excision repair (BER) for oxidative lesions, (ii) NER for helix-distorting lesions caused by UV radiation, (iii) translesion synthesis for various lesions, (iv) mismatch repair (MMR) for replication errors, (v) single-strand break repair (SSBR) for single-strand breaks caused by ionizing radiation and ROS, (vi) homologous recombination (HR) for double-strand breaks caused by ionizing radiation and ROS, (vii) non-homologous end joining (NHEJ) for double-strand breaks caused by ionizing radiation and ROS, and (viii) DNA interstrand crosslink repair pathway for interstrand crosslinks due to chemotherapy [6].

XP is a genetically heterogeneous disorder that increases sensitivity to sunlight (solar/UV radiation). The culprit genes underlying the development of XP are shown to encode proteins involved in the NER for helix-distorting lesions (i.e., *XPA*, *ERCC3*, *XPC*, *ERCC2*, *DDB2*, *ERCC4*, and *ERCC5*) and translesion synthesis for various lesions (i.e., *POLH*) [6].

As the most important repair pathway in mammals for removal of UV light-induced lesions (including cyclobutane pyrimidine dimers [CPD], 6–4 photoproducts, and helix-distorting chemical adducts), the NER pathway consists of two subpathways, i.e., global genome repair (GGR) and transcription-coupled repair (TCR). The GGR subpathway is a slow process that utilizes XPC and DDB2/XPE to identify/mark DNA injuries/lesions anywhere in the genome. The TCR subpathway relies on CSA/*ERCC8* and CSB/*ERCC6* to detect DNA damages occurring at transcribed strands of active genes that block RNA polymerase II transcription/elongation and that are inefficiently recognized by the GGR subpathway, allowing rapid resumption of the vital process of RNA synthesis. Once detected, the DNA lesions are removed and repaired by the multi-subunit TFIIH complex (transcription factor II H complex, consisting of XPA, ERCC3/XPB, ERCC2/XPD, ERCC4/XPF, ERCC5/XPG, and other molecules) in the NER pathway. Specifically, XPB and XPD helicases in the TFIIH complex open the DNA double helix around the lesion, and XPA and replication protein A (RPA) help assemble and properly orientate XPF and XPG endonucleases, which excise the damaged strand around the lesion (5′ and 3′, respectively), leaving an excised stretch of ~30 nucleotides for DNA polymerase δ/ε and auxiliary factors to fill, and ligase 1 to seal (Figure 50.1) [1,3,7].

In variant type XPV (accounting for 20% of XP cases), the NER pathway is intact, while DNA polymerase eta (η) gene (*POLH*) required for the translesion synthesis is defective, leading to postreplication failure and ultimate disease development [3].

Homozygous or compound heterozygous mutations in the *XPA*, *ERCC3*, *XPC*, *ERCC2*, *DDB2*, *ERCC4*, and *ERCC5* as well as *POLH* genes produce loss-of-function proteins and compromise the efficiency of the NER pathway to remove and repair DNA lesions caused by UV radiation, leading to the development of hereditary disorders XP, Cockayne syndrome (CS), and trichothiodystrophy (TTD). While XP patients show an increased risk for skin cancer, CS (due to *CSA* and *CSB* mutations) and trichothiodystrophy (due to *TTDA* mutation) patients do not develop skin cancer. However, XP-CS complex patients (due to loss of function

ERCC3/XPB, ERCC2/XPD, ERCC4/XPF, ERCC5/XPG) may still have skin cancer similar to XP patients (Table 50.1) [8].

50.4 Epidemiology

50.4.1 Prevalence

XP has an estimated prevalence of 1:250,000 in the United States, 1:400,000 in Europe, 1:20,000–100,000 in Japan, and 1:10000–30000 in North Africa (Tunisia, Algeria, Morocco, Libya, and Egypt and the Middle East (Turkey, Israel, and Syria), especially in communities in which consanguinity is common [9].

50.4.2 Inheritance

XP shows autosomal recessive inheritance, with each sib of an affected individual having a 25% chance of being affected, a

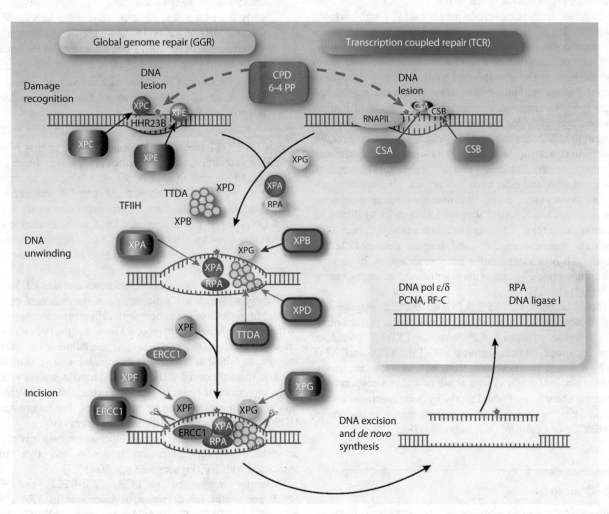

FIGURE 50.1 Schematic representation of the nucleotide excision repair (NER) pathway. While the transcription-coupled repair (TCR) subpathway targets DNA lesions from actively transcribing genes, the global genome repair (GGR) subpathway targets DNA lesions from the remainder of the genome. In GGR, DNA lesions such as UV-induced cyclobutane pyrimidine dimers (CPD) or 6-4 photoproducts (6-4 PP) are recognized by proteins XPE (DDB2) and XPC. In TCR, such lesions are marked by CSA and CSB. Once detected, XPB (ERCC3) and XPD (ERCC2) helicases in the TFIIH complex unwind the region surrounding the lesion along with XPA and XPG (ERCC5) and replication protein A (RPA). XPF and XPG (ERCC5) endonucleases perform incisions to remove the lesion in a fragment of about 30 nucleotides. The resulting gap is filled in by de novo DNA synthesis. Mutation affecting one part of the pathway compromises the functionality of the entire pathway. Mutations in the genes in rectangles are associated with clinical diseases. (Photo credit: DiGiovanna JJ, Kraemer KH. *J Invest Dermatol.* 2012;132(3):785–96.)

50% chance of being an asymptomatic carrier, and a 25% chance of being unaffected and non-carrier.

50.4.3 Penetrance

XP demonstrates high penetrance and variable expressivity. While patients of complementation groups XPA, XPB, XPD, and XPG tend to have neurological abnormalities, those of complementation groups XPC, XPE, and XPF rarely show neurological symptoms.

50.5 Clinical Features

Clinically, XP typically presents with severe sunburn after exposure to only small amounts of sunlight, blistering or freckling on minimum sun exposure, cutaneous irregular dark spots and freckle-like pigmentation of the face before 2 years of age (lentigines, 86%), scaly skin, xeroderma (dry skin), rough-surfaced growths (solar keratoses), skin cancer (basal cell carcinoma 71%, squamous cell carcinoma 42%, malignant melanoma, neurofibroma, trichilemmoma, keratoacanthoma); eyes painfully sensitive to the sun (photophobia) and easily becoming irritated, bloodshot, and clouded; corneal ulcerations; telangiectasia (poikiloderma/spider veins); limited growth of hair on chest and legs; neurologic abnormalities (acquired microcephaly, diminished/absent deep tendon stretch reflexes, progressive sensorineural hearing loss, progressive cognitive impairment, abnormal speech, areflexia, ataxia, peripheral neuropathy, inability to walk and talk, medulloblastoma, glioblastoma, spinal cord astrocytoma, and schwannoma; reflecting the underlying neuronal loss, cortical atrophy, and ventricular dilatation without inflammation; 25%), and death (due to skin cancer 34%, neurologic degeneration 31%, and internal cancer 17%; XP patients with neurodegeneration have mean age of 29 years at death compared to 37 years in those without neurodegeneration) (Table 50.1) [10–15].

It should be noted that XPB, XPD, and XPG displaying neurologic abnormalities are referred to as xeroderma pigmentosum (XP)-Cockayne syndrome (CS) complex (XP-CS complex), which combines clinical features of XPB, XPD, and XPG (photosensitivity, poikiloderma, lentigines, skin cancer, neurodegeneration, brain tumor) with those of CS (photosensitivity, neurodegeneration, intellectual disability, joint contractures/dislocations, hearing loss, and other defects, due to mutations in *CSA/ERCC8* or *CSB/ERCC6*) (Table 50.1) [8].

50.6 Diagnosis

Diagnosis of XP may be achieved by clinical findings (extreme sensitivity to UV, or appearance of lentiginosis on the face at an early age), DNA repair tests (e.g., measurement of post-UV unscheduled DNA synthesis [UDS], UV survival by colony formation, analysis of the recovery of post-UV DNA/RNA synthesis), laboratory assessment (photoallergy), histopathology (cutaneous or ocular tumors), and molecular tests (e.g., PCR, plasmid host cell reactivation [HCR]) [16].

Specifically, individuals presenting with the following skin, eye, nervous system, and family history findings should be suspected of XP:

i. *Skin*: Acute sun sensitivity, severe sunburn with blistering or persistent erythema on minimal sun exposure; marked freckle-like pigmentation or lentigines on the face before 2 years of age; skin cancer within the first decade of life (Figure 50.2) [16].

ii. *Eye*: Photophobia with prominent conjunctival injection; severe keratitis, corneal opacification and vascularization; increased pigmentation of the lids with loss of lashes; atrophy of the skin of the lids resulting in ectropion, entropion, or complete loss of the lids.

iii. *Nervous system*: Diminished or absent deep tendon stretch reflexes, EMG and nerve conduction velocities showing an axonal or mixed neuropathy; progressive sensorineural hearing loss and audiometry revealing early high-tone hearing loss; acquired microcephaly, CT and MRI of the brain showing enlarged ventricles with thinning of the cortex and thickening of the bones of the skull; progressive cognitive impairment.

iv. *Family history*: Consistent with autosomal recessive inheritance, although absence of family history does not preclude the diagnosis) [2].

Post-UV UDS is measured in cultured skin fibroblasts. Namely, skin fibroblast cultures are established from a 3- to 4-mm punch biopsy at an unexposed area of the skin (e.g., the upper inner arm or the buttocks). Fibroblasts are then UV-irradiated in a Petri dish. As nucleotides are incorporated into newly synthesized DNA to replace the damaged DNA in the irradiated cells, UDS is measured by autoradiography, liquid scintillation counting, or fluorescence assay. A reduced level of UDS confirms the diagnosis of XP [16].

Complementation assay for NER defects analyzes UDS in heterodikaryons, which are generated from the fusion of primary dermal fibroblasts of the patient with cells representative of each of the XP complementation groups. Given that the two cell strains are labeled with beads of different size, the failure of the heterodikaryons, identified as binuclear cells containing beads of different sizes, to recover normal UDS levels and remain at the low levels in the mononuclear cells are classified in the same complementation group. In contrast, the heterodikaryons that restore normal UDS levels possess genetically different defects [16].

Histopathology using H&E and other stains help reveal characteristic morphology of cutaneous, ocular, and other tumors associated with XP (Figures 50.3 and 50.4) [17].

Molecular tests such as PCR, PCR-RFLP, and PCR-SSCP are useful for detection of mutations in *XPA* (India: c.335_338delTTATinsCATAAGAAA; Japan: c.390-1G>C [carrier frequency of 1%]; Tunisia: p.Arg228Ter), *XPC* (North Africa: c.1643_1644delTG), *ERCC2* (Iraqi Jewish: p.Arg683Gln), *POLH* (Tunisia/North Africa: exon 10 deletion; Japan: c.490G>T [splice site variant]; p.Ser242Ter; p.Glu306Ter and c.1661delA; Basque/Northern Spain: c.764+1G>A), and other genes [2,18–20].

Further, plasmid HCR utilizes UV-treated plasmid pRSVcat that contains the sequence of reporter gene (i.e., chloramphenicol

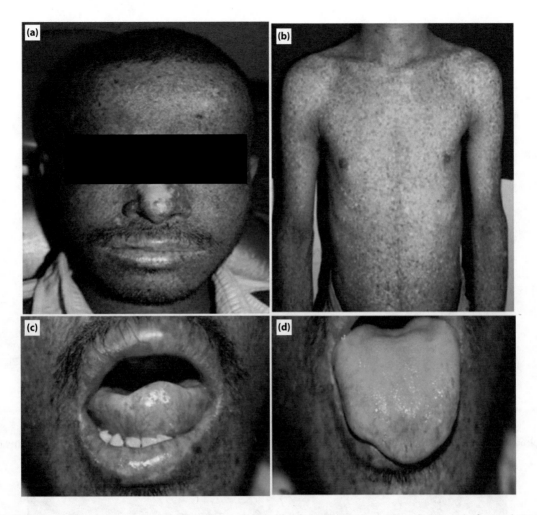

FIGURE 50.2 Xeroderma pigmentosum manifests as: (a) lentigines (freckle-like pigmentation) on sun-exposed areas on the face, and atrophic, hypopigmented skin on the nose; (b) mottled hyperpigmented and hypopigmented areas giving salt-and-pepper appearance to skin; (c) actinic cheilitis of upper and lower lips and decreased mouth opening affecting oral hygiene; (d) precancerous lesion affecting the tip of the tongue. (Photo credit: Mareddy S et al. *Sci World J.* 2013;2013:534752.)

acetyltransferase [CAT]). After treatment with UV and transfection into XP cell lines from patients with cloned XP genes (XPA-XPG, control plasmid), pRSVcat is allowed for repair and expression for 2 days, CAT activity is then measured and DNA repair capacity (DRC) is calculated as the percentage of the residual CAT gene expression after repair of damaged DNA in comparison with undamaged DNA (considered as 100%). Reduced CAT activity correlates to increased UV doses in all of the cell lines. HCR offers an easier, more rapid, and more sensitive assay for the diagnosis of XPB-XPG patients and non-Japanese XPA cases [16].

Differential diagnoses for XP include other syndromes associated with defective NER and cutaneous photosensitivity (e.g., CS; growth failure/short status, microcephaly, pigmentary retinal degeneration, photosensitivity, kyphoscoliosis, impaired neurodevelopment, gait defects, sensorineural deafness, premature aging, distinct facial features such as deep-set eyes, prominent ears, and a wizened appearance; no increased association with malignancy nor abnormal pigmentation; due to *CSA/ERCC8* and *CSB/ERCC6* mutations), trichothiodystrophy (TTD; short stature, brittle hair, intellectual impairment, photosensitivity,

ichthyosis, decreased fertility; no increased association with malignancy nor abnormal pigmentation; due to *GTF2H5/TTDA* mutation), cerebrooculofacioskeletal syndrome (COFS; growth failure, microcephaly, congenital cataracts and microphthalmia, arthrogryposis, severely impaired neurodevelopment, photosensitivity, premature aging, severe developmental delay, facial dysmorphism with a prominent nasal root and overhanging upper lip; due to mutations in *CSB*, *XPD*, *XPG*, and *ERCC1*)] [8], and other diseases exhibiting cutaneous photosensitivity (e.g., Rothmund–Thomson syndrome [red-brown pigmentation or pigment loss in the face, limbs, neck, and hips; abnormal hair, mental retardation, and early cataract], Baller–Gerold syndrome, Hartnup disorder [a disorder of amino acid absorption due to biallelic pathogenic variants in *SLC6A19*, a non-polar amino acid transporter; reduced levels of niacin; pellagra-like symptoms of photosensitivity with dermatitis, diarrhea, and dementia], Carney complex [lentigines without skin damage such as atrophy and telangiectasia/poikiloderma, cutaneous findings are not limited to sun-exposed sites], porphyrin disease [abnormality in porphyrin metabolism; presence of porphyrin in blood, stool, and urine], squamous cell carcinoma, freckles, sallow or taupe spots) [2].

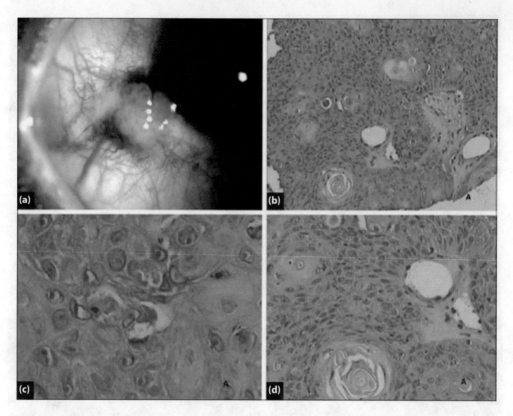

FIGURE 50.3 Xeroderma pigmentosum in an 18-year-old female showing: (a) right eye photo; (b) squamous cell carcinoma stage T2 with various degrees of atypia; (c) pleomorphism; (d) keratinization with the formation of squamous pearls. (Photo credit: Caso ER et al. *Indian J Ophthalmol*. 2019;67(7):1190–2.)

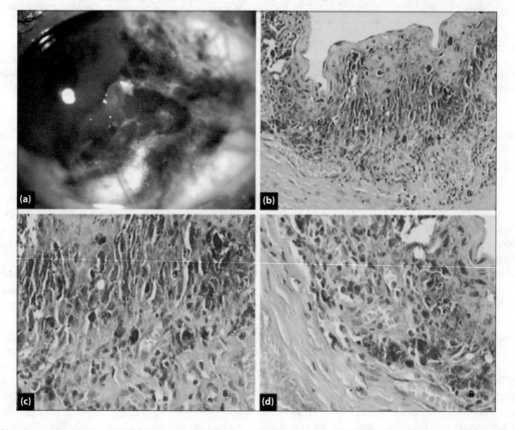

FIGURE 50.4 Xeroderma pigmentosum in an 18-year-old female showing: (a) left eye photo; (b) conjunctival melanoma stage T1b; (c,d) melanocytes with nuclear pleomorphism. (Photo credit: Caso ER et al. *Indian J Ophthalmol*. 2019;67(7):1190–2.)

50.7 Treatment

Management of XP involves protective measures (sun avoidance, sunscreen application, vitamin D supplement), medical care (oral retinoids such as isotretinoin and acitretin, topical 5-fluorouracil, DNA repair enzymes such as bacteriophage T4 endonuclease, photolysis), surgical resection (removal of skin cancer; dermatome shaving or dermabrasion), consultations (ophthalmic, neurologic, dental), and follow-up programs (3-monthly evaluation, genetic counseling) [16,21,22].

50.8 Prognosis and Prevention

Prognosis for XP is generally poor, with about two-thirds of affected patients dying before age 20 years (due mainly to metastatic melanoma or invasive squamous cell carcinoma), while one-third of patients with less severe XP managing to survive into their 40s [23,24].

Prevention of XP centers on avoidance of sunlight/UV radiation exposure and cigarette smoke, vigilance of screening, dietary supplementation with oral vitamin D (to compensate for strict sun avoidance), and extent of any neurological involvement.

Prenatal diagnosis (e.g., DNA repair tests or PCR-RFLP on chorionic villus-derived cells or amniocytes in affected families) and preimplantation genetic testing may be offered to individuals harboring related pathogenic variants [2].

50.9 Conclusion

Xeroderma pigmentosum (XP) is a rare autosomal recessive photosensitive disorder that typically presents with cutaneous lesions (e.g., lentigines, poikiloderma/telangiectasia), ocular abnormalities, neurodegeneration, early-onset skin and central nervous system tumors (e.g., basal cell carcinoma, squamous cell carcinoma, keratoacanthoma), and a reduced life span as a result of skin cancer and neurologic degeneration. Using cell fusion procedure, DNA repair assessment, and sequencing analysis, XP is separated into seven complementation groups (XPA-XPG) as well as an XP variant type (XPV). While the genes underlying complementation groups XPA-XPG (i.e., *XPA, ERCC3, XPC, ERCC2, DDB2, ERCC4*, and *ERCC5*, respectively) participate in the NER pathway, that implicated in the development of XPV (i.e., *POLH*) represents a key component in the translesion synthesis pathway [25,26]. Homozygous or compound heterozygous mutations in these genes result in loss of function proteins, which decrease/eliminate the capacity of the NER and translesion synthesis pathways to remove and repair DNA lesions due to UV radiation, contributing to the development of XP. Diagnosis of XP is possible on the basis of characteristic clinical findings, DNA repair assessment, and molecular analysis. Treatment of XP relies on a multidisciplinary approach consisting of protective measures, medical care, surgical resection, consultations, and follow-up programs. There is considerable scope for further elucidation of molecular mechanisms of XP to facilitate the design of innovative, effective, and specific countermeasures against this genetic disorder.

REFERENCES

1. Black JO. Xeroderma pigmentosum. *Head Neck Pathol.* 2016;10(2):139–44.
2. Kraemer KH, DiGiovanna JJ. Xeroderma pigmentosum. In: Adam MP, Ardinger HH, Pagon RA et al. editors. *GeneReviews®* [Internet]. Seattle (WA): University of Washington, Seattle; 1993–2018. 2003 Jun 20 [updated 2016 Sep 29].
3. DiGiovanna JJ, Kraemer KH. Shining a light on xeroderma pigmentosum. *J Invest Dermatol.* 2012;132(3):785–96.
4. Lehmann AR, McGibbon D, Stefanini M. Xeroderma pigmentosum. *Orphanet J Rare Dis.* 2011;6:70–145.
5. Bowden NA, Beveridge NJ, Ashton KA, Baines KJ, Scott RJ. Understanding xeroderma pigmentosum complementation groups using gene expression profiling after UV-light exposure. *Int J Mol Sci.* 2015;16(7):15985–96.
6. Torgovnick A, Schumacher B. DNA repair mechanisms in cancer development and therapy. *Front Genet.* 2015;6:157.
7. Fadda E. Role of the XPA protein in the NER pathway: A perspective on the function of structural disorder in macromolecular assembly. *Comput Struct Biotechnol J.* 2015;14:78–85.
8. Natale V, Raquer H. Xeroderma pigmentosum-Cockayne syndrome complex. *Orphanet J Rare Dis.* 2017;12(1):65.
9. Kraemer KH, DiGiovanna JJ. Forty years of research on xeroderma pigmentosum at the US National Institutes of Health. *Photochem Photobiol.* 2015;91(2):452–9.
10. Halkud R, Shenoy AM, Naik SM, Chavan P, Sidappa KT, Biswas S. Xeroderma pigmentosum: Clinicopathological review of the multiple oculocutaneous malignancies and complications. *Indian J Surg Oncol.* 2014;5(2):120–4.
11. Zheng JF, Mo HY, Wang ZZ. Clinicopathological characteristics of xeroderma pigmentosum associated with keratoacanthoma: A case report and literature review. *Int J Clin Exp Med.* 2014;7(10):3410–4.
12. Shalabi N, Galor A, Dubovy SR, Thompson J, Bermudez-Magner JA, Karp CL. Atypical fibroxanthoma of the conjunctiva in xeroderma pigmentosum. *Ocul Oncol Pathol.* 2015;1(4):254–8.
13. Alwatban L, Binamer Y. Xeroderma pigmentosum at a tertiary care center in Saudi Arabia. *Ann Saudi Med.* 2017;37(3):240–4.
14. Han C, Huang X, Hua R et al. The association between XPG polymorphisms and cancer susceptibility: Evidence from observational studies. *Medicine (Baltim).* 2017;96(32):e7467.
15. Ribeiro MG, Zunta GL, Santos JS, Moraes AM, Lima CSP, Ortega MM. Clinical features related to xeroderma pigmentosum in a Brazilian patient diagnosed at advanced age. *Appl Clin Genet.* 2018;11:89–92.
16. Mareddy S, Reddy J, Babu S, Balan P. Xeroderma pigmentosum: Man deprived of his right to light. *Sci World J.* 2013;2013:534752.
17. Caso ER, Marcos AA, Morales M, Belfort RN. Simultaneous squamous cell carcinoma and malignant melanoma of the conjunctiva in a teenager with xeroderma pigmentosum: Case report. *Indian J Ophthalmol.* 2019;67(7):1190–2.
18. Schubert S, Lehmann J, Kalfon L, Slor H, Falik-Zaccai TC, Emmert S. Clinical utility gene card for: Xeroderma pigmentosum. *Eur J Hum Genet.* 2014;22(7).

19. Bensenouci S, Louhibi L, De Verneuil H, Mahmoudi K, Saidi-Mehtar N. Diagnosis of xeroderma pigmentosum groups A and C by detection of two prevalent mutations in West Algerian population: A rapid genotyping tool for the frequent XPC mutation c.1643_1644delTG. *Biomed Res Int*. 2016;2016:2180946.

20. Fang X, Sun Y. Whole-exome sequencing enables the diagnosis of variant-type xeroderma pigmentosum. *Front Genet*. 2019;10:495.

21. Rouanet S, Warrick E, Gache Y et al. Genetic correction of stem cells in the treatment of inherited diseases and focus on xeroderma pigmentosum. *Int J Mol Sci*. 2013;14(10):20019–36.

22. Kraemer KH, Tamura D, Khan SG. Pembrolizumab treatment of a patient with xeroderma pigmentosum with disseminated melanoma and multiple nonmelanoma skin cancers. *Br J Dermatol*. 2018;178(5):1009.

23. Narang A, Reddy JC, Idrees Z, Injarie AM, Nischal KK. Long-term outcome of bilateral penetrating keratoplasty in a child with xeroderma pigmentosum: Case report and literature review. *Eye (Lond)*. 2013;27(6):775–6.

24. Lucero R, Horowitz D. Xeroderma Pigmentosum. [Updated 2019 Nov 27]. In: StatPearls [Internet]. Treasure Island (FL): StatPearls Publishing; 2020 Jan-.

25. Donahue BA, Yin S, Taylor JS, Reines D, Hanawalt PC. Transcript cleavage by RNA polymerase II arrested by a cyclobutane pyrimidine dimer in the DNA template. *Proc Natl Acad Sci USA*. 1994;91:8502–6.

26. Puumalainen MR, Rüthemann P, Min JH, Naegeli H. Xeroderma pigmentosum group C sensor: Unprecedented recognition strategy and tight spatiotemporal regulation. *Cell Mol Life Sci*. 2016;73(3):547–66.

Section V

Tumor Syndromes Affecting the Endocrine System

51

Carney Complex

Dongyou Liu

CONTENTS

51.1 Introduction

Carney complex (CNC) is a rare autosomal dominant multiple neoplasia disorder characterized by cardiac, endocrine, cutaneous, and neural myxomas, large-cell calcifying Sertoli cell tumor, psammomatous melanotic schwannoma, and pigmented mucosal and skin lesions.

At the molecular level, Carney complex is largely attributed to germline heterozygous mutations in the *PRKAR1A* gene on chromosome 17q24.2, which encodes protein kinase cAMP-dependent type I regulatory subunit alpha (PKAR1A). In about 20% of cases, sporadic mutations without affected relatives appear to be responsible [1,2].

51.2 Biology

Cases related to CNC (also known as Carney syndrome) were initially reported by Schweizer-Cagianut et al. in 1982 and Rhodes et al. in 1984 as a pleiotropic syndrome of cutaneous, cardiac, and endocrine involvement. Further definition by J. Aidan Carney in 1985 as the complex of myxomas, spotty skin pigmentation (lentigines), and endocrine overactivity led to the creation of acronyms LAMB (lentigines, atrial myxoma, mucocutaneous myxoma, and blue nevi) and NAME (nevi, atrial myxoma, myxoid neurofibroma, ephelides) for this autosomal dominant disorder. The disorder was redesignated as Carney complex in 1986 and Carney syndrome in 1994 to differentiate from Carney triad reported by Carney in 1983 and another disorder Carney–Stratakis syndrome described in 2002 [2,3].

Carney triad (also called Carney syndrome by some authors) is noted for the coexistence of three types of neoplasms (gastric stromal sarcoma/gastrointestinal stromal tumor, pulmonary chondroma, and extra-adrenal paraganglioma/pheochromocytoma) along with other tumors (e.g., adrenocortical adenoma), which mainly affect young women (mean onset age of 20 years) in a nonfamilial setting, and for the association with gain-of-function mutations of c-kit (*KIT*) and platelet-derived growth factor receptor A (*PDGFRA*) in gastric gastrointestinal stromal tumor, and occasional mitochondrial complex II succinate dehydrogenase (*SDH*) enzyme subunits *SDHA*, *SDHB*, or *SDHC* mutations in paraganglioma/pheochromocytoma (9.5%) [4].

On the other hand, Carney–Stratakis syndrome (or Carney–Stratakis dyad) often induces paraganglioma/pheochromocytoma and gastrointestinal stromal tumor, which affect males and females equally, often in a familial setting, and contain loss-of-function mutations in *SDHB, SDHC*, and *SDHD*, but not *KIT* and *PDGFRA* [4].

Additional studies indicate that primary pigmented nodular adrenocortical disease (PPNAD) is a most notable endocrine tumor associated with CNC (up to 60% of cases), and cardiac myxoma (occurring in 53% of cases) represents an important cause of CNC-related death. After localization of the CNC susceptibility gene region in chromosome 17q24.2 in 1998, and subsequent characterization of the *PRKAR1A* gene in 2000, molecular mechanisms of CNC have become apparent [5,6].

51.3 Pathogenesis

The *PRKAR1A* gene on chromosome 17q24.2 comprises 11 exons (1A, 1B, 2, 3, 4A, 4B, 5-10), and encodes 384 aa type 1 alpha subunit of protein kinase A (*PRKAR1A*) involved in the cyclic AMP-dependent protein kinase (PKA) signaling pathway.

Specifically, as the regulatory subunit of the protein kinase A, *PRKAR1A* interacts with two catalytic subunits to form the PKA heterotetramer. *PRKAR1A* defects cause *PRKAR1A* haploinsufficiency and compromise its regulatory function, leading to unrestrained catalytic subunit activity, increased cell proliferation in cAMP-responsive tissues, and tumor formation [7].

Germline heterozygous mutations (e.g., single base substitutions, small deletions/insertions, combined rearrangements, or large deletion) in the *PRKAR1A* gene produce frame shifts and/or premature nonsense stop codons (83%), and result in shorter or defective *PRKAR1A* mRNAs, which are readily degraded by the nonsense-mediated mRNA decay (NMD) surveillance mechanism. Without type 1 alpha, protein kinase A turns on more often than normal, and contributes to aberrant cell proliferation and a heightened risk of developing various endocrine and non-endocrine tumors and lesions [8–14].

Among >125 *PRKAR1A* pathogenic mutations (http://prkar1a. nichd.nih.gov) identified so far, single base substitutions (e.g., c.82C>T, c.491_492delTG, c.709-2_709-7 delATTTTT), small (≤15 bp) deletions/insertions, and combined rearrangements occur throughout the whole open reading frame, whereas large deletions cover most of the exons or the whole gene locus [15–18].

Defects of other PKA subunits (e.g., the catalytic subunits PRKACA on chromosome 19p13.1 [adrenal hyperplasia] and PRKACB on chromosome 1p31.1 [pigmented spots, myxoma, pituitary adenoma]) may be also implicated in CNC pathogenesis [19,20]. Some 20% CNC cases without *PRKAR1A* mutations may have somatic alterations at chromosome 2p16, which often lead to gene amplifications, suggesting the yet to be identified gene located at 2p16 as a potential oncogene [2].

51.4 Epidemiology

51.4.1 Prevalence

CNC is a rare disorder, with over 750 cases reported by 2013. Females predominate among CNC patients (female:male ratio of 60:40; median diagnostic age of 20 years). PPNAD frequently

occurs in infants and individuals of the second and third decade of life and has a median diagnostic age of 34 years. Unlike sporadic cardiac myxoma that is found exclusively in the left atrium of older females, CNC-related cardiac myxoma affects all chambers and has a median diagnostic age of 20 years [1,2].

51.4.2 Inheritance

Carney complex shows autosomal dominant inheritance, and affected females are almost fivefold more likely to transmit the disorder than males. The fact that males often develop large cell calcified Sertoli cell tumors (LCCSCT), which may cause infertility, helps partly explain for the female dominance among CNC patients. Approximately 80% of CNC cases have a familial history, and the remaining 20% are due to de novo germline mutation [1,2].

51.4.3 Penetrance

CNC demonstrates high penetrance of 70%–80% by the age of 40 years, and >95% by the age of 50 years, and its clinical manifestations vary significantly between patients, even those within the same family.

51.5 Clinical Features

CNC affects multiple organs and typically causes skin lesions, PPNAD, myxomas, and other diseases [1,21,22].

Skin lesions (e.g., lentigines, café-au-lait spots, blue nevi, freckling; 77% of cases) may emerge at birth and usually represent the first manifestation of CNC (Figure 51.1) [23]. Lentigines appear as multiple small (0.2–2 mm), flat, poorly circumcised, brown to black macules (or slightly raised macules, similar to nevi, in African Americans) typically around the vermilion border of the lips, on the eyelids, ears, and the genital area. Blue nevi are larger (up to 8 mm), blue to black, dome-shaped lesions that are less common than lentigines. A subtype of blue nevus is epithelioid blue nevus (EBN), which shows intensive pigmentation and poorly circumscribed proliferative regions with associated

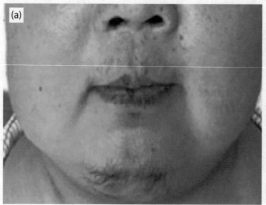
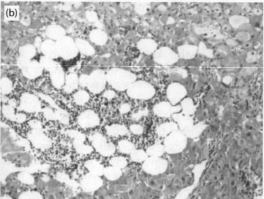

FIGURE 51.1 Carney complex in a 26-year-old male with *PRKAR1A* pathogenic variant showing pigmented spots on the lips (a), and primary pigmented nodular adrenocortical disease (PPNAD) with local myelolipoma like change in the right adrenal gland (b, H&E, 100×). (Photo credit: Li S et al. *World J Clin Cases.* 2018;6(14):800–6.)

dermal fibrosis. Although EBN is very rare in the general population, it is relatively commonly found in patients with CNC. Café-au-lait spots and depigmented lesions are occasionally observed in CNC. In general, cutaneous pigmentary lesions tend to increase in intensity, number, and distribution around puberty, and fade after the fourth decade [1,2].

PPNAD is an endocrine tumor that appears after skin lesions and shows a peak during infancy, and another peak between the second and third decade of life (median diagnostic age of 34 years, female:male ratio of 2.4:1). As the most common endocrine lesion in CNC (up to 60% of cases), PPNAD affects bilateral adrenal glands (with adrenal cortex peppered by small pigmented nodules of <1 cm surrounded by usually atrophic cortex) and represents a cause of adrenocorticotropic hormone (ACTH)-independent overproduction of cortisol. PPNAD may progress to an atypical form of Cushing syndrome (Figure 51.1) [23,27,28].

Myxomas of the heart (53%), skin (33%), breast (20%), and genitalia are benign tumors that may damage critical organs through mass or blocking effects (Figures 51.2 and 51.3) [24,25]. In particular, cardiac myxoma (of a few mm to 8 cm in diameter) may begin as early as 3 years of age in any cardiac chamber, is often multiple and recurrent, and causes a triad of symptoms (stroke, peripheral artery occlusions due to myxoma embolization; heart failure due to reduced cardiac output or complete occlusion of a valvular orifice; emaciation, recurrent fevers probably related to production of cytokines such as interleukin [IL-6] and tumor), accounting for >50% of CNC-related deaths. Skin myxoma is a white, flesh-colored, opalescent, or pink subcutaneous nodule (often <1 cm in diameter) with a smooth surface that commonly emerges on the eyelids, external ear canal, breast nipples, and genitalia, but not hands and feet. Breast myxoma (of

2 mm – 2 cm in diameter) may be multicentric or bilateral, pink or white masses with a mucoid appearance and usually occurs in females after puberty. It gives a diffuse nodularity without dominant masses in physical examination [1,2,26].

Other diseases include LCCSCT (33% of male patients), thyroid nodule (cystic or nodular disease, benign thyroid adenoma, thyroid cancer; 5%), acromegaly (10%), pituitary tumor (somatomammotroph hyperplasia, growth hormone [GH]−producing adenoma with acromegaly, prolactinoma), pancreatic tumor (acinar cell carcinoma, adenocarcinoma, intraductal pancreatic mucinous neoplasia), ovarian tumor (ovarian cyst, seruscysadenoma, cystic teratoma), breast tumor (breast and nipple myxoma, myxoid fibroadenoma, ductal adenoma; 3% of female patients), psammomatous melanotic schwannoma (PMS, 10%), and osteochondromyxoma (also referred to as Carney bone tumor; osteochondroma containing myxoid elements; originating from bony cortices and commonly found in the nasal region, tibia, and radius; 1%) [1,2,29].

51.6 Diagnosis

Diagnosis of CNC on the basis of 12 major clinical criteria and 2 supplemental criteria has a sensitivity of 98%, and molecular identification of *PRKAR1A* pathogenic variants provides confirmation in about 80% of cases [1,2].

The 12 major clinical criteria for CNC diagnosis are: (i) spotty skin pigmentation on the lips, conjunctiva and inner or outer canthi, vaginal and penile mucosa; (ii) blue nevus or multiple epithelioid blue nevi; (iii) cutaneous and mucosal myxomas; (iv) cardiac myxoma; (v) breast myxomatosis or fat-suppressed MRI findings; (vi) osteochondromyxoma; (vii) PPNAD or a

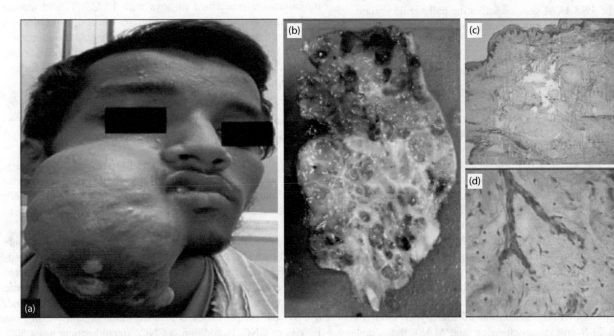

FIGURE 51.2 Carney complex in an 18-year-old male showing (a) large nodular and focally ulcerated swelling on the right cheek (15 cm × 8 cm × 6 cm); (b) circumscribed, nodular, solid-cystic mass with gelatinous and hemorrhagic areas in cut surface of cheek mass; (c) tumor in the dermis and subcutaneous plane of cheek mass composed of multiple hypocellular nodules with abundant myxoid stroma and spindle to stellate cells, alongside focally increased cellularity (H&E, ×4); (d) small proliferating blood vessels with stellate cells and myxoid stroma in the background (H&E, ×100). (Photo credit: Karegar M et al. *J Lab Physicians.* 2018;10(3):354−6.)

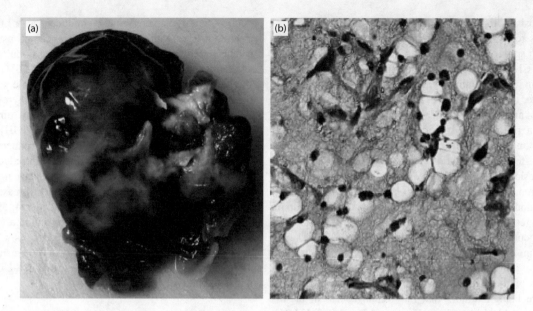

FIGURE 51.3   Carney complex in a 21-year-old female showing an excised right ventricle mass (a); and myxomatous cellular proliferations with sparse collagen fibers consistent with myxoma in heart mass (b). (Photo credit: Saleh Y et al. *Case Rep Cardiol.* 2018;2018:2959041.)

paradoxical positive response of urinary glucocorticosteroids to dexamethasone administration during Liddle's test; (viii) acromegaly due to GH-producing adenoma or evidence of excess GH production; (ix) LCCSCT or characteristic calcification on testicular ultrasonography; (x) thyroid carcinoma or multiple, hypoechoic nodules on thyroid ultrasonography, in a young patient; (xi) psammomatous melanotic schwannoma; and (xii) breast ductal adenoma [1,2].

The two supplemental criteria for CNC diagnosis are (i) affected first-degree relative, and (ii) inactivating mutation of the *PRKAR1A* gene or activating pathogenic variants of *PRKACA* (single base substitutions and copy number variation) and *PRKACB* [1,2].

A diagnosis of CNC is established when a patient meets either two of the major clinical criteria confirmed by histology, imaging, or biochemical testing, or one major clinical criterion and one supplemental criterion [1,2].

Differential diagnoses for CNC include disorders producing lentigines (benign familial lentiginosis, Peutz–Jeghers syndrome, LEOPARD syndrome, Noonan syndrome with lentiginosis, Bannayan–Riley–Ruvalcaba syndrome, and *PTEN* hamartoma tumor syndrome), those causing café-au-lait spots (McCune–Albright syndrome [MAS], neurofibromatosis type 1 [NF1], neurofibromatosis type 2 [NF2], and Watson syndrome), that causing cardiac myxoma (sporadic myxoma cardiomyopathy), those causing thyroid tumor (Cowden syndrome, *PTEN* hamartoma tumor syndrome, and sporadic thyroid tumors), that causing LCCSCT (Peutz–Jeghers syndrome), those causing adrenal cortical tumor (Beckwith–Wiedemann syndrome [BWS], Li–Fraumeni syndrome [LFS], multiple endocrine neoplasia type 1 [MEN1], congenital adrenal hyperplasia resulting from 21-hydroxylase deficiency, and MAS), those causing GH-secreting pituitary adenoma or somatotropinoma (MEN1, isolated familial somatotropinoma [IFS]), and those causing schwannoma (NF1, NF2, and isolated familial schwannomatosis) [1,2,30].

51.7 Treatment

Given its broad spectrum of clinical diseases, management of CNC requires a multidisciplined approach that is tailor-made for specific complications from individual patients [31,32].

Surgical resection may be necessary for cardiac, cutaneous, and mammary myxomas; bilateral adrenalectomy or inhibitors of steroidogenesis (e.g., ketoconazole or mitotane) may be used for selected cases of PPNAD; orchiectomy and/or aromatase inhibitors are prescribed to boys with LCCSCT (and associated gynecomastia, premature epiphyseal fusion, and induction of central precocious puberty); transsphenoidal/transcranial surgery or somatostatin analogs may be applied to SH and/or GH-producing pituitary adenomas; and surgery is useful for primary and/or metastatic lesions of PMS [1,2].

51.8 Prognosis and Prevention

Patients with CNC have an average life span of 50–55 years. Complications of heart myxoma (e.g., emboli/strokes, postoperative cardiomyopathy, and cardiac arrhythmias), metastatic PMS, PPNAD, pancreatic and other tumors represent common causes of CNC-related death [1,2,32].

Upon molecular identification of *PRKR1A* pathogenic variants in CNC patients, genetic screening is recommended for first-degree relatives (parents, siblings, and offspring). In the event that a parent of CNC patient contains a *PRKR1A* pathogenic variant, the sibling has a 50% risk of CNC. On the other hand, if CNC patient possesses a de novo mutation, the sibling has 1% risk of CNC. In addition, a CNC-affected patient has a 50% chance of passing the disorder on to offspring [1,2].

Prenatal testing for CNC may be conducted via chorionic villous sampling (CVS) at 10–12 weeks of gestation or amnioparacentesis at 15–18 weeks of gestation. Preimplantation genetic diagnosis allows the selection of disease-free embryos for implantation [1,2].

51.9 Conclusion

Carney complex (CNC) is an autosomal dominant multisystem disorder that results mainly from germline mutations in the *PRKAR1A* gene on chromosome on 17q24.2, encoding the regulatory subunit 1 alpha of the protein kinase A (PRKAR1A). Germline mutations in *PRKAR1A* produce a deficiency of the R1a subunit (i.e., PRKAR1A haploinsufficiency), and enhance intracellular signaling by PKA, leading to the upregulation of D-type cyclins or activation of the mTOR pathway, which favor cell proliferation. Besides cutaneous lesions (e.g., lentigines, café-au-lait spots, blue nevi, freckling) that appear at birth, CNC is associated with the development of myxomas in the heart, skin, and breast as early as 3 years of age, and PPNAD, in addition to various tumors in other organs. Diagnosis of CNC relies on meeting either 2 of the 12 major clinical criteria or 1 of the 12 major clinical criteria and 1 of the 2 supplemental criteria. Molecular identification of *PRKAR1A* pathogenic variants allows confirmation of CNC in 80% of cases and provides a valuable means of prenatal testing and preimplantation genetic diagnosis. Tailor-made approaches are utilized for the treatment of various complications associated with CNC [31,32].

REFERENCES

1. Kaltsas G, Kanakis G, Chrousos G. Carney's complex. In: De Groot LJ, Chrousos G, Dungan K et al. (editors). *Endotext* [Internet]. South Dartmouth (MA): MDText.com, Inc.; 2000–2013 Jul 31.
2. Stratakis CA. Carney complex: A familial lentiginosis predisposing to a variety of tumors. *Rev Endocr Metab Disord.* 2016;17(3):367–71
3. Correa R, Salpea P, Stratakis CA. Carney complex: An update. *Eur J Endocrinol.* 2015;173(4):M85–97.
4. Boikos SA, Xekouki P, Fumagalli E et al. Carney triad can be (rarely) associated with germline succinate dehydrogenase defects. *Eur J Hum Genet.* 2016;24(4):569–73.
5. Salpea P, Horvath A, London E et al. Deletions of the PRKAR1A locus at 17q24.2-q24.3 in Carney complex: Genotype-phenotype correlations and implications for genetic testing. *J Clin Endocrinol Metab.* 2014;99(1):E183–8.
6. Berthon AS, Szarek E, Stratakis CA. PRKACA: The catalytic subunit of protein kinase A and adrenocortical tumors. *Front Cell Dev Biol.* 2015;3:26.
7. Cazabat L, Ragazzon B, Varin A et al. Inactivation of the Carney complex gene 1 (PRKAR1A) alters spatiotemporal regulation of cAMP and cAMP-dependent protein kinase: A study using genetically encoded FRET-based reporters. *Hum Mol Genet.* 2014;23(5):1163–74.
8. Birla S, Aggarwal S, Sharma A, Tandon N. Rare association of acromegaly with left atrial myxoma in Carney's complex due to novel PRKAR1A mutation. *Endocrinol Diabetes Metab Case Rep.* 2014;2014:140023.
9. Iwata T, Tamanaha T, Koezuka R et al. Germline deletion and a somatic mutation of the PRKAR1A gene in a Carney complex-related pituitary adenoma. *Eur J Endocrinol.* 2015;172(1):K5–10.
10. Rhayem Y, Le Stunff C, Abdel Khalek W et al. Functional characterization of PRKAR1A mutations reveals a unique molecular mechanism causing acrodysostosis but multiple mechanisms causing Carney complex. *J Biol Chem.* 2015;290(46):27816–28.
11. Papanastasiou L, Fountoulakis S, Voulgaris N et al. Identification of a novel mutation of the PRKAR1A gene in a patient with Carney complex with significant osteoporosis and recurrent fractures. *Hormones (Athens).* 2016;15(1):129–35.
12. He J, Sun M, Li E et al. Recurrent somatic mutations of *PRKAR1A* in isolated cardiac myxoma. *Oncotarget.* 2017;8(61):103968–74.
13. Hernández-Ramírez LC, Tatsi C, Lodish MB et al. Corticotropinoma as a component of Carney complex. *Endocr Soc.* 2017;1(7):918–25.
14. Saloustros E, Salpea P, Starost M et al. Prkar1a gene knockout in the pancreas leads to neuroendocrine tumorigenesis. *Endocr Relat Cancer.* 2017;24(1):31–40.
15. Bataille MG, Rhayem Y, Sousa SB et al. Systematic screening for PRKAR1A gene rearrangement in Carney complex: Identification and functional characterization of a new in-frame deletion. *Eur J Endocrinol.* 2013;170(1):151–60.
16. Guo H, Xu J, Xiong H, Hu S. Case studies of two related Chinese patients with Carney complex presenting with extensive cardiac myxomas and PRKAR1A gene mutation of c.491_492delTG. *World J Surg Oncol.* 2015;13:83.
17. Jang YS, Moon SD, Kim JH, Lee IS, Lee JM, Kim HS. A novel PRKAR1A mutation resulting in a splicing variant in a case of Carney complex. *Korean J Intern Med.* 2015; 30(5):730–4.
18. Cai XL, Wu J, Luo YY, Chen L, Han XY, Ji LN. A novel mutation of PRKAR1A caused Carney complex in a Chinese patient. *Chin Med J (Engl).* 2017;130(24):3009–10.
19. Basso F, Rocchetti F, Rodriguez S et al. Comparison of the effects of PRKAR1A and PRKAR2B depletion on signaling pathways, cell growth, and cell cycle control of adrenocortical cells. *Horm Metab Res.* 2014;46(12):883–8.
20. Forlino A, Vetro A, Garavelli L, Ciccone R, London E, Stratakis CA, Zuffardi O. PRKACB and Carney complex. *N Engl J Med.* 2014;370(11):1065–7.
21. Kim H, Cho HY, Lee JN, Park KY. Carney complex with multiple cardiac myxomas, pigmented nodular adrenocortical hyperplasia, epithelioid blue nevus, and multiple calcified lesions of the testis: A case report. *J Pathol Transl Med.* 2016;50(4):312–4.
22. Kiriakopoulos A, Linos D. Carney syndrome presented as a pathological spine fracture in a 35-year-old male. *Am J Case Rep.* 2018;19:1366–9.
23. Li S, Duan L, Wang FD, Lu L, Jin ZY. Carney complex: Two case reports and review of literature. *World J Clin Cases.* 2018;6(14):800–6.
24. Karegar M, Sarwate M, Kothari K, Rojekar A, Naik L. Cytologic diagnosis of unusual, large multiple cutaneous myxomas in a case of Carney complex. *J Lab Physicians.* 2018;10(3):354–6.
25. Saleh Y, Hammad B, Almaghraby A et al. Carney complex: A rare case of multicentric cardiac myxoma associated with endocrinopathy. *Case Rep Cardiol.* 2018;2018:2959041.
26. Serio A, Favalli V, Giuliani L et al. Cardio-oncology: The Carney complex type I. *J Am Coll Cardiol.* 2016; 68(17):1921–3.

27. Korpaisarn S, Trachoo O, Panthan B, Aroonroch R, Suvikapa-kornkul R, Sriphrapradang C. A novel *PRKAR1A* mutation identified in a patient with isolated primary pigmented nodular adrenocortical disease. *Case Rep Oncol.* 2017;10(2):769–76.

28. Zhang CD, Pichurin PN, Bobr A, Lyden ML, Young WF, Bancos I. Cushing syndrome: Uncovering Carney complex due to novel PRKAR1A mutation. *Endocrinol Diabetes Metab Case Rep.* 2019;2019. pii: EDM180150.

29. Schernthaner-Reiter MH, Trivellin G, Stratakis CA. MEN1, MEN4, and Carney complex: Pathology and molecular genetics. *Neuroendocrinology.* 2016;103(1):18–31.

30. Salpea P, Stratakis CA. Carney complex and McCune Albright syndrome: An overview of clinical manifestations and human molecular genetics. *Mol Cell Endocrinol.* 2014;386(1–2):85–91.

31. Kamilaris CDC, Faucz FR, Voutetakis A, Stratakis CA. Carney complex. *Exp Clin Endocrinol Diabetes.* 2019;127(2–03):156–64.

32. Vindhyal MR, Elhomsy G. Carney Complex. [Updated 2019 Dec 6]. In: StatPearls [Internet]. Treasure Island (FL): StatPearls Publishing; 2020 Jan.

52

Congenital Adrenal Hyperplasia

Dongyou Liu

CONTENTS

52.1 Introduction

Congenital adrenal hyperplasia (CAH) is an autosomal recessive disorder resulting from inherited defects in one of the steroidogenic enzymes (especially 21-hydroxylase, which is implicated in 90% of CAH cases) that take part in the adrenal synthesis of cortisol, aldosterone, and testosterone from cholesterol.

Based on the severity of the disease, CAH is separated into classic and non-classic forms. Attributed to decreased production of cortisol and aldosterone and increased production of adrenal cortical androgens, classic CAH is a severe disease that can be further divided into salt-wasting (SW, 76%) and simple virilizing (SV, 23%) subtypes, with presentations ranging from prenatal virilization in girls, adrenocortical failure, and early puberty in both sexes to hyperpigmentation (only in salt-wasting subtype). On the other hand, non-classic CAH (NCCAH) is a mild, often asymptomatic disease that is linked to males and females with normal genitalia at birth, and postnatal hyperandrogenism in adult females [1–3].

Encoded by the *CYP21A2* gene on chromosome 6p21.3, 21-hydroxylase (P450c21) plays a key role in the conversion of cholesterol to aldosterone and cortisol in the zona fasciculata of the adrenal glands (Figure 52.1). Mutations in the *CYP21A2* gene cause 21-hydroxylase deficiency and insufficient production of aldosterone and cortisol, leading to increased secretion of adrenocorticotropic hormone (ACTH) from hypophysis, enhanced biosynthesis of adrenocortical androgens, subsequent accumulation of progesterone, 17-hydroxyprogesterone, androstenedione and testosterone, and ultimate changes in primary or secondary sex characteristics in some affected infants, children, or adults (90%). Further, mutations in *CYP11B1* gene on chromosome 8q24.3

encoding 11β-hydroxylase (P450c11β) contribute to the buildup of 11β-deoxycorticosterone and 11β-deoxycortisol, and development of CAH (females virilized with ambiguous genitalia, males unchanged in the classic form; males and females with normal genitalia at birth, hyperandrogenism postnatally in non-classic form; 5%). In addition, mutations in *CYP17A1* gene on chromosome 10q21-q22 encoding 17α-hydroxylase/17,20-lyase (P450c17) lead to accumulation of 17OH-pregnenolone and 17OH-progesterone and subsequent development of CAH (variable sexual development, infantile female genitalia; 5%) (Table 52.1) [4–6].

52.2 Biology

Biochemical pathways of adrenal steroidogenesis consist of three major routes that all start with cholesterol: mineralocorticoids (end product: aldosterone), glucocorticoids (end product: cortisol), and sex steroids (end product: testosterone) (Figure 52.1). Multiple proteins/enzymes are required for these biochemical pathways, including steroidogenic acute regulatory protein (StAR), cytochrome P450 cholesterol side-chain cleavage (CYP11A1), 3β-hydroxysteroid dehydrogenase type 2 (HSD3B2), 17α-hydroxylase/17,20-lyase (CYP17A1), cytochrome *b5* (CYB5A), 21-hydroxylase (CYP21A2), 11β-hydroxylase (CYP11B1), aldosterone synthase (CYP11B2), and 17β-hydroxysteroid dehydrogenase type 5 (AKR1C3) (Figure 52.1). Although mutations in the underlying genes encoding any of these proteins/enzymes may cause disruptions in steroidogenesis, potentially leading to CAH, those affecting 21 hydroxylase (P450c21), 11ß hydroxylase (P450c11β) and 17α-hydroxylase/17,20-lyase (P450c17), stand out most

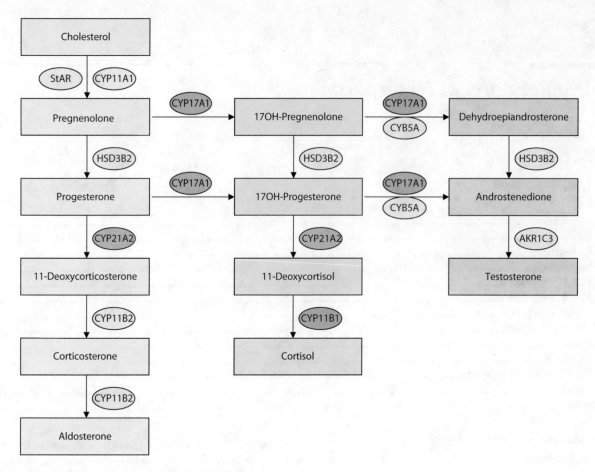

FIGURE 52.1 Biochemical pathways of adrenal steroid hormone synthesis. StAR, steroidogenic acute regulatory protein; CYP11A1, cytochrome P450 cholesterol side-chain cleavage; HSD3B2, 3β-hydroxysteroid dehydrogenase type 2; CYP17A1, 17α-hydroxylase/17,20-lyase; CYB5A, cytochrome b5; CYP21A2, 21-hydroxylase; CYP11B1, 11β-hydroxylase; CYP11B2, aldosterone synthase; AKR1C3, 17β-hydroxysteroid dehydrogenase type 5.

(Table 52.1). In fact, mutations in the *CYP21A2* (6p21.3), *CYP11B1* (8q24.3), and *CYP17A1* (10q21-q22) genes encoding P450c21, P450c11β, and P450c17 account for approximately 90%, 5%, and 5% of CAH cases, respectively (Table 52.1) [7–10].

CAH (also referred to as adrenogenital syndrome) was initially described by Luigi de Crecchio in 1865 from a prematurely deceased virilized female with enlarged adrenal glands, which was later diagnosed as non-salt-wasting congenital adrenal hyperplasia. Since then, the clinical profiles of CAH have been expanded to include phenotypes that belong to classic CAH (salt-wasting or simple virilizing) and NCCAH. Subsequent identification and characterization of the genes encoding various proteins/enzymes involved in the biochemical pathways of adrenal steroid hormone synthesis have helped uncover the molecular pathogenesis of CAH and enabled development of highly specific and sensitive procedures for improved definition and diagnosis of CAH. Specifically, as 90% of CAH are attributed to 21-hydroxylase (P450c21) deficiency, adrenogenital syndrome and congenital adrenocortical hyperplasia are collectively called the classic form of 21-OHD CAH (which include the salt-wasting form of 21-OHD CAH and the simple virilizing form of 21-OHD CAH), while the "attenuated" or "late-onset" congenital adrenocortical hyperplasia are known as the non-classic form of 21 OHD CAH [5,11].

52.3 Pathogenesis

Located in the human leukocyte antigen (HLA) gene cluster on chromosome 6p21.3, the *CYP21A2* (cytochrome P450 family 21 subfamily A member 2) gene spans 3.4 kb with 10 exons and encodes a 494 aa, 55.8 kDa protein (cytochrome P-450c21 or 21-hydroxylase), which constitutes a member of the cytochrome P450 superfamily of enzymes involved in the canalization of pathways for drug metabolism and synthesis of cholesterol, steroids, and other lipids [8–10].

In addition, a pseudogene *CYP21A2P*, which shares 98% sequence homology in exons and 96% sequence homology in introns with *CYP21A2*, is found about 30 kb downstream of *CYP21A2*. Recombination resulting from unequal crossing over during meiosis between these two genes may produce gross *CYP21A2* deletions or duplications, which account for 20%–30% of *CYP21A2* pathogenic variants implicated in 21-OHD CAH [8,9].

Present in the endoplasmic reticulum of the adrenal glands, 21-hydroxylase (cytochrome P-450c21) hydroxylates steroids at the 21 position, and is crucial for the synthesis of steroid hormones such as cortisol (which helps maintain blood sugar levels, protects the body from stress, and suppresses inflammation) and aldosterone (sometimes called the salt-retaining hormone, which regulates the retention of salt by the kidneys, and indirectly the fluid levels and blood pressure in the body).

TABLE 52.1

Genetic and Phenotypic Characteristics of Congenital Adrenal Hyperplasia (CAH) due to Defective Adrenal Steroidogenesis

Gene	Chromosome Location	Defective Enzyme	Disorder	Clinical Features	Incidence
CYP21A2	6p21.3	21-Hydroxylase (P450c21)	21-Hydroxylase deficiency (CAH)	*Classic CAH salt wasting subtype* is a very severe disease with no 21-hydroxylase activity and often presents with failure to thrive, cortisol deficiency, mineralocorticoid deficiency, hyponatremia, hyperkalemia, high plasma renin activity (PRA), hypovolemic shock, excess androgen production early in life, and virilization of external genitalia in females *Classic CAH simple virilizing subtype* is a disease of intermediate severity with 1%–2% of 21-hydroxylase activity, and typically manifests as virilization of external genitalia in females, progressive premature pubarche, progressive virilization with clitoromegaly (female) or increased penile size (male), elevated androgen levels leading to accelerated growth velocity and advanced bone age, or premature fusion of the epiphyses causing final short stature *Non-classic CAH* is a mild disease with 20%–50% of 21-hydroxylase activity, and may be asymptomatic or display signs of androgen excess later in life (acne, hirsutism, menstrual irregularities, anovulation, infertility)	1:15,000 newborns; accounting for 90% of CAH cases
CYP17A1	10q21-q22	17α-Hydroxylase/ 17,20-lyase (P450c17)	17-Hydroxylase deficiency (CAH)	Hypertension, hypokalemia, hypogonadism; 46,XX: primary amenorrhea; absence of secondary sexual characteristics; 46,XY: undervirilization; abdominal testes	1:50,000 newborns; accounting for 5% of CAH cases
CYP11B1	8q24.3	11β-Hydroxylase (P450c11β)	11β-Hydroxylase deficiency (CAH)	*Classic form:* females virilized with ambiguous genitalia, males unchanged; low renin hypertension with normal aldosterone levels; adrenal insufficiency; prenatal androgen excess (46,XX DSD); *Non-classic form:* males and females with normal genitalia at birth; hyperandrogenism postnatally; mild/absent hypertension; various degrees of partial adrenal insufficiency; various degrees of androgen excess after birth;	1:200,000 newborns, high prevalence in Moroccan Jewish; accounting for up to 5% of CAH cases
StAR	8p11.2	Lipoid CAH	Congenital lipoid adrenal hyperplasia (CLAH)	*Classic form:* hyperreninemic, hypokalaemic hypotension/neonatal salt-wasting crisis; neonatal adrenal insufficiency; female external genitalia in 46,XY; spontaneous ovarian sex steroid production in 46,XX *Non-classic form:* variable degrees/normal; late-onset adrenal insufficiency; variable degrees of 46,XY DSD	Rare; more frequent in Japanese, Palestinians, Koreans
HSD3B2	1p13.1	3β-Hydroxysteroid dehydrogenase type 2	3β-Hydroxysteroid dehydrogenase deficiency (CAH)	*Classic form:* hyperreninemic, hypokalaemic hypotension/neonatal salt-wasting crisis; neonatal adrenal insufficiency; DSD in both sexes *Non-classic form:* various/partial adrenal insufficiency; premature adrenarche, precocious pseudopuberty, irregular menstrual cycles	Rare
CYP11A1	15q23-q24	Cholesterol side-chain cleavage enzyme (P450scc)	P450 side chain cleavage syndrome	Adrenal insufficiency or absence; 46,XY DSD, gonadal insufficiency	Rare
POR	7q11.2	P450-oxidoreductase (POR)	P450 oxidoreductase deficiency (CAH)	Volume depletion: skeletal malformations (Antley–Bixler): maternal virilization; 46,XX: mild-to-moderate virilization; 46,XY: undervirilization from, hypospadias to female-appearing	Rare; more common in Japan and Korea
CYB5A	18q22.3	Cytochrome b5	Cytochrome b5 deficiency (CAH)	46,XY: DSD; 46,XX: absence of puberty	Rare
CYP11B2	8q24.3	Aldosterone synthase	Aldosterone synthase deficiency	Isolated mineralocorticoid deficiency	Rare

Mutations in the *CYP21A2* gene resulting in decease or absence of 21-hydroxylase exert negative impacts on the conversion of cholesterol to aldosterone and cortisol, ranging from blockage of the cortisol production pathway (and insufficient production of cortisol and aldosterone), accumulation of 17-hydroxyprogesterone and other metabolic intermediates, to biosynthesis of androgen in the adrenal cortex (which contributes to variable degrees of virilization in the external genitalia of affected female fetuses as well as rapid postnatal growth in male and female newborns). Further, insufficient mineralocorticoid causes salt-wasting CAH, while lack of cortisol secretion (a hallmark of CAH) induces elevated ACTH production, which stimulates excessive synthesis of adrenal products and buildup of precursor molecules, leading to hyperplasia of the steroid-producing cells of the adrenal cortex and testicular adrenal rest tumor [12,13].

The >140 *CYP21A2* pathogenic variants identified so far include point mutations, small deletions, small insertions, complex rearrangements, and gross deletion/duplication. Interestingly, nine pathogenic variants due to gene conversion from pseudogene *CYP21A2P* to *CYP21A2*, together with *CYP21A2* deletion, account for about 95% of all disease-causing alleles. Furthermore, deletion of the entire *CYP21A2* gene (representing about 15% of mutations), 8-bp deletion in exon 3, mutations in the exon 6 cluster and Arg356Trp are associated with salt-wasting CAH; intron 2 mutations may cause either salt-wasting or simple virilizing phenotype; Ile172Asn mutation is implicated in simple virilizing CAH, and Pro30Leu, Val281Leu, and Pro454Ser often occur in non-classic form of CAH [14].

52.4 Epidemiology

52.4.1 Prevalence

CAH caused by 21-hydroxylase (P450c21) deficiency accounts for 90% of cases, and has an estimated incidence of 1:16,000 live births (classic CAH, including salt-wasting phenotype 76%, simple virilizing phenotype 23%), and <1:1000 live births (NCCAH). Ashkenazi Jews (1:27; 1:3 carriers), Hispanics (1:53), Yugoslavs (1:62), New Yorkers (1:100), Italians (1:300), and Yupik Eskimos (1:300) appear to be more susceptible to NCCAH than people in Saudi Arabia (1:5,000), Europe and North

America (1:10,000-1:16,000), Japan (1:21,000), and New Zealand (1:23,000) [5].

CAH due to 11β-hydroxylase (P450c11β) deficiency makes up 5% of cases and affects 1:100,000 live births in Caucasians, and 1:7000 live births in Moroccan Jews. Similarly, CAH owing to 17-hydroxylase/17,20-lyase (P450c17) deficiency represents about 5% of cases, affecting 1:50,000 live births worldwide, and probably more in Brazil and Asia [5].

52.4.2 Inheritance

CAH shows an autosomal recessive inheritance. While classic 21-hydroxylase deficiency often results from inheritance of two severely affected alleles, non-classic 21-hydroxylase deficiency is due to inheritance of either two mild 21-hydroxylase deficiency alleles or one severe and one mild allele (compound heterozygote) [5].

52.4.3 Penetrance

CAH demonstrates high genotype–phenotype concordance (90.5% for salt-wasting CAH, 85.1% for simple-virilizing CAH, and 97.8% for NCCAH). Although both sexes are equally affected by CAH, the phenotypic expression may show some gender differences, due to varied accumulation of testosterone or precursor hormones [5].

52.5 Clinical Features

CAH due to 21-hydroxylase deficiency (21 OHD CAH) consists of two clinical forms: classic and non-classic.

Characterized clinically by prenatal virilization in females (resulting in genital ambiguity at birth), postnatal virilization in both males and females (Figure 52.2), classic CAH (or classic 21-OHD CAH) is further distinguished into salt-wasting and simple virilizing subtypes [3,15].

Classic CAH salt-wasting subtype is a very severe, potentially life-threatening disease with no 21-hydroxylase activity (leading to inadequate aldosterone production) and accounts for 76% of classic CAH cases. Its key clinical features include failure to thrive, cortisol deficiency, mineralocorticoid deficiency,

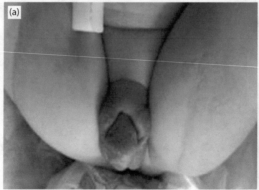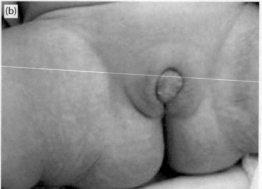

FIGURE 52.2 Congenital adrenal hyperplasia in a newborn showing (a) significant virilization of the external genitalia (Prader stage 3); (b) regression of the phallus at the seventh month of the hormone replacement treatment (prior to constructive operation). (Photo credit: Kırmızıbekmez H et al. *Case Rep Pediatr.* 2015;2015:196374.)

hyponatremia, hyperkalemia, high plasma renin activity (PRA), hypovolemic shock, excess androgen production early in life, and virilization of external genitalia in females (Table 52.1).

Classic CAH simple virilizing subtype is a disease of intermediate severity with 1%–2% of 21-hydroxylase activity and represents 23% of classic CAH cases. Its notable clinical features comprise virilization of external genitalia in females, progressive premature pubarche, progressive virilization with clitoromegaly (female) or increased penile size (male), elevated androgen levels leading to accelerated growth velocity and advanced bone age, or premature fusion of the epiphyses causing final short stature.

NCCAH is a mild form of disease with 20%–50% of 21-hydroxylase activity, leading to absent prenatal virilization, variable postnatal virilization (hyperandrogenism in adult females), absent salt wasting, and rare cortisol deficiency, along with signs of androgen excess later in life (acne, hirsutism, menstrual irregularities, anovulation, infertility) [1].

Furthermore, CAH is associated with adrenal hyperplasia and testicular adrenal rest tumor (TART), due to insufficient cortisol secretion that contributes to increased ACTH production and adrenal products. TART typically presents as bilateral masses in over 20% of boys and up to 94% of adults with classic 21 OHD CAH, and causes obstruction of seminiferous tubules, gonadal dysfunction, and infertility (Figure 52.3) [16,17].

52.6 Diagnosis

Diagnostic workup for CAH involves medical history review, physical examination (ambiguous genitalia because of exposure to high concentrations of androgens in utero; hyperpigmentation and penile enlargement), laboratory assessment (hypoglycemia due to hypocortisolism, hyponatremia due to hypoaldosteronism, hyperkalemia due to hypoaldosteronism, blood 17α-hydroxyprogesterone >242 nmol/L in salt-wasting patients, oligomenorrhea polycystic ovaries and hirsutism), ultrasound (anomalies in the urogenital tract), histopathology (adrenal hyperplasia, testicular adrenal rest tumor), and cytogenetic and molecular analysis (*CYP21A2* pathogenic variant) [18–21].

Differential diagnoses for CAH caused by 21-hydroxylase-deficiency include CAH due to 11-β-hydroxylase deficiency (hypertension in most patients, hypokalemia, virilization), CAH due to 17-hydroxylase/17,20-lyase deficiency (hypertension, hypokalemia, and hypogonadism; 46,XX: primary amenorrhea and absence of secondary sexual characteristics; 46,XY: under-virilization, abdominal testes), CAH due to 3β-hydroxysteroid dehydrogenase deficiency (salt wasting in congenital cases), congenital lipoid adrenal hyperplasia (CLAH) due to StAR deficiency (salt wasting), P450 side chain cleavage syndrome due to CYP11A1 deficiency (salt wasting), 5-α-reductase deficiency, adrenal hypoplasia, adrenal insufficiency, androgen insensitivity syndrome, bilateral adrenal hemorrhage, other causes of hyperkalemia/hyponatremia, hypertrophic pyloric stenosis, obstructive uropathy, polycystic ovarian syndrome, defects in testosterone synthesis, Denys–Drash syndrome, disorders of gender development, and familial glucocorticoid deficiency (Table 52.1) [22].

52.7 Treatment

Treatment options for CAH include surgery (for ambiguous genitalia, severe virilization) and steroid replacement therapy

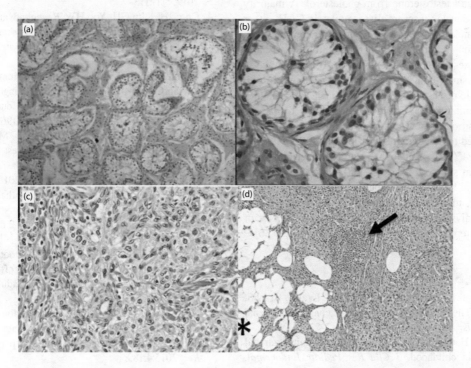

FIGURE 52.3 Congenital adrenal hyperplasia-associated testicular adrenal rest tumors showing seminiferous tubules with thickened basement membranes, and predominantly germ cell aplasia with an occasional tubule containing spermatogonia in testicular tissue (a, 4×; b, 40×); large nuclear-to-cytoplasmic ratio, round to elliptical nuclei with a loose chromatin, and eosinophilic and granulated cytoplasm (c); adipose tissue metaplasia (asterisk) and focal lymphocytic infiltrates (arrow) (d). (Photo credits: (a and b) Jayarajah U et al. *J Surg Case Rep.* 2018;2018(10):rjy255; (c and d) Lolis E et al. *J Endocr Soc.* 2018;2(6):513–7.)

(mineralocorticoids, glucocorticoids, cortisone, and hydrocortisone/cortisol) to suppress ACTH production and to prevent further virilization [9,20,23–30].

52.8 Prognosis and Prevention

Prognosis for most patients with CAH is good with early diagnosis and appropriate intervention. Issues such as ambiguous genitalia, infertility, short stature (due either to glucocorticoid-induced growth suppression caused by overtreatment with glucocorticoids or to advanced skeletal maturation caused by inadequate glucocorticoid treatment), female gender identity issues, and virilization in women may create emotional stress on CAH patients [26].

CAH patients should avoid physical stresses (e.g., febrile illness, gastroenteritis with dehydration, surgery accompanied by general anesthesia, and major trauma), which can be relieved by increased doses of glucocorticoids.

As CAH shows autosomal dominant inheritance, prenatal diagnosis (by ultrasound or chorionic villus sampling or amniocentesis for elevated levels of 17-ketosteroids, pregnanetriol, and 17-OHP) and preimplantation genetic testing (*CYP21A2* pathogenic variant) may be offered to prospective parents with affected relatives.

52.9 Conclusion

Congenital adrenal hyperplasia (CAH) is an autosomal recessive disorder associated with impaired steroidogenic enzyme activity in the adrenal cortex, leading to reduced synthesis of cortisol, aldosterone, and testosterone from cholesterol. A majority of patients (90%) develop a severe disease (classic CAH, either salt-wasting or simple virilizing), with symptoms ranging from prenatal virilization in girls, adrenocortical failure, and early puberty in both sexes to hyperpigmentation (only in salt-wasting subtype), while a small number of patients (5%) show a mild, often asymptomatic disease (NCCAH), with males and females having normal genitalia at birth, and adult females displaying postnatal hyperandrogenism. Molecular pathogenesis of CAH is mainly attributed to biallelic mutations in the *CYP21A2* gene 21-hydroxylase (P450c21), and to a lesser extent, other genes involved in adrenal steroidogenesis [31–35]. Diagnosis of CAH is based on medical history review, physical examination, laboratory assessment, ultrasound, histopathology, and cytogenetic and molecular analysis. Treatment options for CAH consist of surgery, hormone replacement, and symptomatic relief [29,36].

REFERENCES

1. Ayalon-Dangur I, Segev-Becker A, Ayalon I, Eyal O, Israel S, Weintrob N. The many faces of non-classic congenital adrenal hyperplasia. *Isr Med Assoc J.* 2017;19(5):317–22.
2. Kurtoğlu S, Hatipoğlu N. Non-classical congenital adrenal hyperplasia in childhood. *J Clin Res Pediatr Endocrinol.* 2017;9(1):1–7.
3. Nermoen I, Husebye ES, Myhre AG, Løvås K. Classic congenital adrenal hyperplasia. *Tidsskr Nor Laegeforen.* 2017;137(7):540–3.
4. Speiser PW. Congenital adrenal hyperplasia. *F1000Res.* 2015;4(F1000 Faculty Rev):601.
5. New M, Yau M, Lekarev O et al. Congenital adrenal hyperplasia. In: De Groot LJ, Chrousos G, Dungan K et al. editors. *Endotext* [Internet]. South Dartmouth (MA): MDText.com, Inc.; 2000-. 2017 Mar 15.
6. Bhimji SS, Sinha V. *Adrenal Congenital Hyperplasia. StatPearls* [Internet]. Treasure Island (FL): StatPearls Publishing; 2018 Jan-. 2017 Nov 29.
7. Kim CJ. Congenital lipoid adrenal hyperplasia. *Ann Pediatr Endocrinol Metab.* 2014;19(4):179–83.
8. Turcu AF, Auchus RJ. Adrenal steroidogenesis and congenital adrenal hyperplasia. *Endocrinol Metab Clin North Am.* 2015;44(2):275–96.
9. Turcu AF, Auchus RJ. Novel treatment strategies in congenital adrenal hyperplasia. *Curr Opin Endocrinol Diabetes Obes.* 2016;23(3):225–32.
10. Nimkarn S, Gangishetti PK, Yau M, New MI. 21-Hydroxylase-deficient congenital adrenal hyperplasia. In: Adam MP, Ardinger HH, Pagon RA et al. editors. *GeneReviews®* [Internet]. Seattle (WA): University of Washington, Seattle; 1993–2018. 2002 Feb 26 [updated 2016 Feb 4].
11. Sahakitrungruang T. Clinical and molecular review of atypical congenital adrenal hyperplasia. *Ann Pediatr Endocrinol Metab.* 2015;20(1):1–7.
12. Flück CE. Mechanisms in endocrinology: Update on pathogenesis of primary adrenal insufficiency: Beyond steroid enzyme deficiency and autoimmune adrenal destruction. *Eur J Endocrinol.* 2017;177(3):R99–R111.
13. Piskinpasa H, Ciftci Dogansen S, Kusku Cabuk F et al. Bilateral adrenal and testicular mass in a patient with congenital adrenal hyperplasia. *Acta Endocrinol (Buchar).* 2019;-5(1):113–7.
14. Choi JH, Kim GH, Yoo HW. Recent advances in biochemical and molecular analysis of congenital adrenal hyperplasia due to 21-hydroxylase deficiency. *Ann Pediatr Endocrinol Metab.* 2016;21(1):1–6.
15. Kırmızıbekmez H, Yesiltepe Mutlu RG, Moralıoğlu S, Tellioğlu A, Cerrah Celayir A. Concurrence of meningomyelocele and salt-wasting congenital adrenal hyperplasia due to 21-hydroxylase deficiency. *Case Rep Pediatr.* 2015;2015:196374.
16. Jayarajah U, Herath KB, Fernando MH, de Silva VC, Goonewardena S. Testicular adrenal rest tumour in an adult patient with congenital adrenal hyperplasia: A case report and review of literature. *J Surg Case Rep.* 2018;2018(10):rjy255.
17. Lolis E, Juhlin CC, Nordenström A, Falhammar H. Extensive bilateral adrenal rest testicular tumors in a patient with 3β-hydroxysteroid dehydrogenase type 2 deficiency. *J Endocr Soc.* 2018;2(6):513–7.
18. Kok HK, Sherlock M, Healy NA, Doody O, Govender P, Torreggiani WC. Imaging features of poorly controlled congenital adrenal hyperplasia in adults. *Br J Radiol.* 2015;88(1053):20150352.
19. Kolahdouz M, Mohammadi Z, Kolahdouz P, Tajamolian M, Khanahmad H. Pitfalls in molecular diagnosis of 21-hydroxylase deficiency in congenital adrenal hyperplasia. *Adv Biomed Res.* 2015;4:189.
20. Yau M, New M. Congenital adrenal hyperplasia: Diagnosis and emergency treatment. In: De Groot LJ, Chrousos G, Dungan K et al. editors. *Endotext* [Internet]. South Dartmouth (MA): MDText.com, Inc.; 2000-.2015 Apr 12.

21. Pignatelli D, Carvalho BL, Palmeiro A, Barros A, Guerreiro SG, Maçut D. The complexities in genotyping of congenital adrenal hyperplasia: 21-Hydroxylase deficiency. *Front Endocrinol (Lausanne)*. 2019;10:432.

22. Burdea L, Mendez MD. *21 Hydroxylase Deficiency. StatPearls* [Internet]. Treasure Island (FL): StatPearls Publishing; 2019 Jan-.2019 Jun 16.

23. Speiser P, Azziz R, Baskin L et al. Congenital adrenal hyperplasia due to 21-hydroxylase deficiency: An endocrine society clinical practice guideline. *J Clin Endocrinol Metab*. 2010;95(9):4133–60.

24. Turcu AF, Auchus RJ. The next 150 years of congenital adrenal hyperplasia. *J Steroid Biochem Mol Biol*. 2015; 153:63–71.

25. Bachelot A, Grouthier V, Courtillot C, Dulon J, Touraine P. Management of endocrine disease: Congenital adrenal hyperplasia due to 21-hydroxylase deficiency: Update on the management of adult patients and prenatal treatment. *Eur J Endocrinol*. 2017;176(4):R167–81.

26. Choi JH, Yoo HW. Management issues of congenital adrenal hyperplasia during the transition from pediatric to adult care. *Korean J Pediatr*. 2017;60(2):31–7.

27. Improda N, Barbieri F, Ciccarelli GP, Capalbo D, Salerno M. Cardiovascular health in children and adolescents with congenital adrenal hyperplasia due to 21-hydroxilase deficiency. *Front Endocrinol (Lausanne)*. 2019;10:212.

28. Livadas S, Bothou C. Management of the female with non-classical congenital adrenal hyperplasia (NCCAH): A patient-oriented approach. *Front Endocrinol (Lausanne)*. 2019;10:366.

29. Speiser PW. Emerging medical therapies for congenital adrenal hyperplasia. *F1000Res*. 2019;8:363.

30. Reisch N. Review of health problems in adult patients with classic congenital adrenal hyperplasia due to 21-hydroxylase deficiency. *Exp Clin Endocrinol Diabetes*. 2019;127(2–03):171–7.

31. Wang D, Wang J, Tong T, Yang Q. Non-classical 11β-hydroxylase deficiency caused by compound heterozygous mutations: A case study and literature review. *J Ovarian Res*. 2018;11(1):82.

32. Baranowski ES, Arlt W, Idkowiak J. Monogenic disorders of adrenal steroidogenesis. *Horm Res Paediatr*. 2018;89(5):292–310.

33. Miller WL, Merke DP. Tenascin-X, congenital adrenal hyperplasia, and the CAH-X syndrome. *Horm Res Paediatr*. 2018;89(5):352–61

34. Zhao X, Su Z, Liu X et al. Long-term follow-up in a Chinese child with congenital lipoid adrenal hyperplasia due to a StAR gene mutation. *BMC Endocr Disord*. 2018;18(1):78.

35. Al Alawi AM, Nordenström A, Falhammar H. Clinical perspectives in congenital adrenal hyperplasia due to 3β-hydroxysteroid dehydrogenase type 2 deficiency. *Endocrine*. 2019;63(3):407–21.

36. Witchel SF. Congenital adrenal hyperplasia. *J Pediatr Adolesc Gynecol*. 2017;30(5):520–34.

53

Familial Hyperparathyroidism

Luigia Cinque, Alfredo Scillitani, and Vito Guarnieri

CONTENTS

53.1 Introduction

Hyperparathyroidism (HPT) is an endocrine condition characterized by abnormal secretion of parathyroid hormone (PTH) from the chief cells of the parathyroid glands [1]. In its "primary" type (pHPT), the hypersecretion results from an autonomous enlargement of the glands in the form of a pathologic adenoma (85% of cases), hyperplasia (15%), or rarely a malignant carcinoma (<1%) [1]. PTH overproduction causes hypercalcemia that is not usually mitigated by the calcium-sensing receptor (CaSR, 3q13.33-q21.1), the master regulator of the narrow blood calcium levels, that in normal physiological conditions compensates the hypercalcemia by blocking the bone and kidney reabsorption of calcium, in turn shutting down PTH secretion [2]. As a consequence, patients with pHPT may suffer from hypercalcemia symptoms (fatigue, muscle weakness, bone/joint pain/osteopenia, nephrolithiasis, constipation) and need surgical treatment to remove the parathyroid lesion(s) [1].

Usually, in the case of sporadic pHPT, surgical treatment is conclusive with immediate relief of the symptomatology. However, in a small fraction of cases, HPT may be familial and display an incomplete manifestation of more complicated syndromic forms with a genetic origin (Table 53.1).

Less than the 10% of pHPT is familial and inherited as part of more specific calcium metabolism disorders such as familial hypercalcemia hypocalciuria (FHHI-II-III, MIM145980-145981-600740) caused by heterozygous inactivating mutations of the CaSR, GNA11 and AP2S1 [3] genes, respectively, or neonatal severe hyperparathyroidism (NSHPT, MIM239200) due to homozygous inactivating mutations of the CaSR [3] gene. Moreover, pHPT may be inherited in tumor syndromes such as multiple endocrine neoplasia (MEN) type I (MIM131100), II (MIM171400), or IV (MIM610755) or hyperparathyroidism with jaw tumor (HPT-JT, MIM145001), caused by inactivating mutations in MEN1 [4], RET [4], CDKN1b [5], and CDC73 [6] genes, respectively (Table 53.1). In these syndromes, pHPT accompanies others, not just parathyroid lesions, affecting diverse parenchymas: mostly pituitary gland or pancreas in MEN1 [4] and IV [5], thyroid in MEN2 [4], jaw, kidney, and uterus in HPT-JT [6]. Finally, familial pHPT featured just by parathyroid lesions is defined as familial isolated hyperparathyroidism (FHIP): inactivating mutations in the above-mentioned genes (MEN1, CaSR, and CDC73) were found involved in this syndrome in no more than the 10% of all cases [7,8]. More recently, activating mutations of the GCM2 gene were also identified in 18% of FIHP cases [9], previously resulted negative at the screening of the previously identified genes (Table 53.1).

53.2 Hyperparathyroidism with Jaw Tumor Syndrome

HPT-JT syndrome is characterized by pHPT (familial or sporadic, up to 95% of cases [10]), ossifying fibroma of the maxilla/jaw (up to 50% of cases [10]), renal lesions (cysts, hamartoma, or Wilms tumors; up to 20% of cases [10]), uterine leiomyomata (up

TABLE 53.1

Syndromic and Isolated Genetic Disorders Having Primary Hyperparathyroidism as the Main Clinical Symptom

	Acronym	Name of the Syndrome	OMIM	Gene	Chromosome	Inheritance	Inactivating/Activating
Syndromic	MEN1	Multiple endocrine neoplasia type I	131100	MEN1	11q13.1	AD	Inactivating
	MEN2A	Multiple endocrine neoplasia type II	171400	RET	10q11.2	AD	Inactivating
	MEN4	Multiple endocrine neoplasia type IV	610755	CDKN1B	12p13.1	AD	Inactivating
	NSHPT	Neonatal severe hyperparathyroidism	239200	CaSR	3q13.3-q21.1	AR	Inactivating
	FHH1	Familial hypercalciuria hypocalcemia type I	145980	CaSR	3q13.3-q21.1	AD	Inactivating
	FHH2	Familial hypercalciuria hypocalcemia type III	145981	GNA11	19p13.3	AD	Inactivating
	FHH3	Familial hypercalciuria hypocalcemia type III	600740	AP2S1	19q13.3	AD	Inactivating
	HPT-JT	Hyperparathyroidism with jaw tumors	145001	CDC73	1q32.2	AD	Inactivating
Isolated	FIHP	Familial isolated hyperparathyroidism	145000	CDC73	1q32.2	AD	Inactivating
				MEN1	11q13.1	AD	Inactivating
				CaSR	3q13.3-q21.1	AD	Inactivating
				GCM2	6p24.2	AD	Activating

to 35% of cases [11]) and parathyroid lesions (adenoma in 60%–70%, parathyroid carcinoma or PC, up to 15% of cases [10]). Onset of the disease may vary, with the earliest case at 7 years and the latest at 65 years [12,13]; in general, first manifestations appear in the fourth decade with HPT due to a solitary adenoma in the 70% of cases [14].

53.2.1 History

In 1958, Jackson reported seven cases of HPT-JT in association with recurrent pancreatitis [15]. Interestingly, the authors did not recognize the syndrome and they supposed it could be a *fruste* MEN1 form. Only in 1990, starting from the study of the third generation of the same family, the authors were able to define this condition as a clinically and genetically distinct entity, in the same way, ascribing the jaw tumors as a specific clinical feature [16]. Moreover, they proved that the genes causing this condition and the MEN1 syndrome were located on different chromosomal positions [16], thus offering definitive molecular evidence that these two diseases are distinct identities.

53.2.2 Genetics

HPT-JT syndrome is caused by inactivating mutations of the tumor suppressor *CDC73* gene, formerly known by the name of the locus where the gene is located, *HRPT2* (1q31.2 [17]). The gene was identified by positional candidate approach after several previous linkage studies narrowed the critical genomic region to 12 cM surrounding the locus 1q25.32 [18–21]. Mutations of this gene were further confirmed in up to 70% of HPT-JT cases [17]. The unexpected and apparent discordance between the frequency of parathyroid carcinoma in pHPT (less than 1% [1]) and the HPT-JT syndrome (up to 15%) suggested a possible bias of selection of patients; instead, it was later proved that the CDC73 is the master gene of the sporadic parathyroid carcinoma (being resulted mutated in up to 30% of these specific cases [22]). No definitive data are available about the penetrance of CDC73 mutations, although it could possibly reach 70% [17] in HPT-JT, while, as stated above, in sporadic parathyroid carcinoma is about 30% [22], and in FHIP is less than 10% [7,8].

53.2.3 Epidemiology

53.2.3.1 Prevalence

While primary hyperparathyroidism (pHPT) is relatively common and affects 0.3% of the general population, 1%–3% of postmenopausal women, and an overall incidence of 21.6 cases per 100 000 person-years, HPT-JT syndrome has unknown prevalence, with only about 200 cases being reported in the literature.

53.2.3.2 Inheritance

HPT-JT syndrome is inherited in an autosomal dominant pattern, in which one copy of the altered CDC73 gene in each cell is sufficient to cause the disorder.

53.2.3.3 Penetrance

HPT-JT syndrome shows an incomplete penetrance of 80%–90%, but about 70% in females. The age-related penetrance was estimated as 11% at age 25, 65% at age 50, and 83% at age 70 in the Dutch population.

53.2.4 Clinical Features

As hyperparathyroidism disrupts the normal balance of calcium in the blood, patients with HJT syndrome often present with nausea, vomiting, high blood pressure (hypertension), weakness, fatigue, kidney stones, and thinning of the bones (osteoporosis), accompanied by parathyroid adenoma and carcinoma, fibroma in the jaw, and tumors of the uterus and kidneys (including Wilms tumor).

53.2.5 Diagnosis

53.2.5.1 Clinical Diagnosis

HPT-JT should be deduced based on the following clinical and biochemical features:

1. HPT (familial or sporadic) due to a single (rarely multiple [23]) parathyroid lesion (usually a benign adenoma in 60%–70% of cases, carcinoma in 15%), in association with other clinical signs

2. Ossifying, also termed cementifying, fibroma of the maxilla or mandible

3. Renal tumors: Cysts, hamartoma, Wilms tumors

4. Uterine leiomyomata

Jaw tumors may be often confused with "brown" tumors (osteitis fibrosa cystica) typical of severe pHPT and characterized by active osteoclastic activity caused by HPT [24–26]. However, unlike brown tumors, ossifying fibromas of the maxilla/mandible in HPT-JT do not contain giant cells, being made up of fibrous tissues containing cementum, mature bone, and other *bony* materials in different amounts, without distinct trabeculae [27]. Moreover, they do not resolve after parathyroidectomy, and this helps to distinguish pHPT from HPT-JT [28].

Renal involvement was found in HPT-JT as renal hamartomas, Wilms tumors or polycystic kidney disease [18,29], renal cortical adenoma, and papillary renal cell carcinoma [30]. After identification of the CDC73 gene, mutations were also found in sporadic cases of clear cell, papillary, and chromophobe renal carcinomas and Wilms tumors [31], negative for VHL gene mutations, suggestive of a mutual exclusivity of VHL or CDC73 mutations in these tumors, that, however was never further confirmed.

Uterine tumors are likewise common in the general population, so without other stigmata, although in combination with HPT, are not evocative enough of the disease, which needs to be carefully ascertained by additional imaging studies (panoramic radiography, neck and renal ultrasound). However, they were first associated with the syndrome after a retrospective analysis of 33 kindreds in 13 HPT-JT families [11]. The study revealed a high percentage of women (up to 91%) suffering for menorrhagia and requiring hysterectomy (median age of 35 years). The authors properly concluded that the early detection of the genetic mutation could be of great help in preventing a possible low reproductive fitness of HPT-JT female patients [11].

Parathyroid neoplasia. Particular attention should be taken when imaging is used to detect a parathyroid lesion. An approach based on at least two different techniques, such as sestamibi scan and ultrasound, is warranted, not only in initial diagnosis but also (and mostly) in the case of recurrence, to determine loco-regional (neck, mediastinum, and chest) metastatic spread [32,33]. At the neck ultrasound, specific morphologic features seem to be ascribable to benign adenomas, that appear smooth, small, and homogeneous, while malignant neoplasias are more irregular in shape, with different lobes, larger and weightier [34,35]. Fine needle aspiration (FNA) of the nodule is not recommended in initial diagnosis, either because cytology specimens do not always give definitive results [36,37], or mostly because of the high risk of seeding tumor cells along the needle track [38,39]. Instead, FNA is usually utilized in the case of clinical confirmation of metastatic spread.

It is important to highlight that first clinical manifestations of HPT-JT may be, as stated above, just the pHPT due to a parathyroid tumor but in absence of jaw neoplasia. This may complicate the diagnosis due to the symptomatic overlapping with other pHPT associated disorders such as (multiple endocrine neoplasia type I [MEN1], described in a different chapter of this book) or FHH or FHIP. In this case the differential diagnosis may be based on:

1. The level of hypercalcemia, usually higher in the first decades in FHH, rather than in HPT-JT [40], and the presence of the hypercalciuria.

2. Careful imaging studies useful either to search for specific (jaw or parathyroid, or renal and uterine) HPT-JT lesions or other (pituitary, thyroidal, or pancreatic) neoplasias, in order to exclude a different syndrome.

Clinical diagnosis needs to be further confirmed by genetic testing to search for CDC73 gene mutation. Genetic screening should take into account not only the classic mutations of the coding regions but should also search for large genomic deletions at the CDC73 genomic locus (1q31.2). Indeed, a recent analysis of more than 180 subjects affected by different hypercalcemic disorders (PHPT, HPT-JT, FIHP, or sporadic PC) indicated that, irrespective of the first diagnosis, up to one-third of the cases carried a CDC73 gene mutation, 30% of which were large genomic deletions [41]. Familial genetic screening once the mutation is ascertained in the proband is beneficial in order to identify asymptomatic subjects (carriers) to address for closer clinical follow-up [42].

53.2.5.2 Challenging Diagnosis in HPT-JT

As mentioned, in HPT-JT, parathyroid lesions may be malignant up to 15% of cases. However, in the absence of metastasis or local recurrences, recognizing a malignant from a benign parathyroid neoplasia in the initial diagnosis can be challenging even for an expert endocrinologist or surgeon [43,44]. In general, parathyroid carcinoma presents with profound hypercalcemia accompanied by fatigue, depression, nausea, vomiting, abdominal and bone pain, and often appears as a cervical palpable mass [45]. During the operation, an experienced surgeon may recognize a carcinoma by some specific features such as:

1. Size with an average of 3 cm and weight ranging from 2 to 10 g, up to 250 times weightier than a normal parathyroid gland [46]

2. Color ranging from grayish to white [46]

3. Firm adherence and/or invasion of surrounding structures [47]

Conversely, benign lesions cause just mild hypercalcemia and are not usually detectable by neck palpation, due to their smaller size and weight. However, it is important to highlight that the definitive diagnosis of a carcinoma can only be made upon histopathological examination of resected tumor tissue (see below). Thus, if performed before the operation, genetic screening may greatly help the surgery choice, since the presence of a CDC73 mutation that associates to a parathyroid carcinoma in up to 15% of HPT-JT cases could point toward the better option: demolitive in case of suspicion of a malignant form, with the removal of the whole thyroid, annexes, and laryngeal nerve (if compromised), or conservative, consisting of removal of the local affected parathyroid tissue [48]. This reflects dramatically on the management of the patient, since the en bloc resection approach (in the case of carcinoma) leads to up the 90% of long-term overall survival with 10% local failure, compared to the conservative approach, with 50% local recurrence and 46% mortality [48,49].

53.2.5.3 Histology of Parathyroid Lesions and Parathyroid Carcinoma

Along with parathyroid carcinoma and "classic" benign adenoma, atypical parathyroid adenoma (APA) was recently established as a novel histology entity, defined as having the most frequent features of malignancy such as high proliferation index, presence of mitotic figures, peritumoral fibers, and necrosis, but lacking the distinctive and exclusive malignant feature of invasion of the capsule and the surrounding parenchymal tissues [50]. It is important to note that not enough epidemiologic data about the outcome of the APA are available in literature in order to suggest or refute that the APA could be prone to outcome as a next parathyroid carcinoma (PC) [44,52]. In line with the hypothesis that the APA could be considered a histology entity different from the PC, some authors reported a lower frequency of CDC73 gene mutations in APA compared to the PC, and a consistently lower negative fraction of immunohistochemistry (IHC) for the parafibromin protein [52].

However, parafibromin IHC reported controversial results in classifying/identifying/confirming a malignant PC. Some authors found it hard to deal with the antibody widely used by the scientific community and decided to define a scoring system based on a gradient of negative staining, from total to focal loss, in order to better interpret and be consistent with molecular results [53–56], while others rigorously decided to classify as "normal/positive" all specimens even with lower (or focal) expression, and "pathologic/negative" the ones with complete absence of expression [52,57–59].

More recently, a comprehensive analysis of a large collection of more than 789 different histology samples was finally able to define more stringent pathologic criteria for a definitive diagnosis of parathyroid malignancy [60]: benign parathyroid neoplasms present with a typical grouped structure, while parathyroid carcinomas-CDC73 mutated present:

1. A sheet-like proliferation of tumor cells characterized by an arborizing vasculature
2. A specific eosinophilic cytoplasm
3. Nuclear enlargement and perinuclear cytoplasmic empty spaces typical of aged atypia (Figure 53.1)

53.2.6 Pathogenesis: Role of the *CDC73* Gene

The *CDC73* gene encodes for a ubiquitously expressed protein, namely parafibromin, whose role is described below. Inactivating mutations are scattered throughout the coding sequence with particular high frequencies for the exon 1, 2, and 7 [61], that could be considered such mutational hot-spots. Mutations can be missense, nonsense, frameshift, or affecting the splicing sites [61]. In the vast majority, whether translated, mutated proteins are instable and thus degraded by through a proteasome mediated process [62,63], indicating that the loss of expression typical of tumor suppressor genes is the base of the tumorigenesis CDC73-induced. In a few cases a different mutational effect, affecting the subcellular localization, has been reported: instead the mutation alters the nucleolar localization signal (NoLS), making it impossible for the protein to reach the right subcellular compartment [62,63]. More recently, large genomic deletions at the CDC73 locus (1q31.2) were also identified [41,64,65,67–72]. The deletions may involve single or multiple exons or even encompass the whole genomic region along with flanking genes. In these latter cases, additional neurodevelopmental symptoms were recorded [71] that could be ascribed to the other genes deleted.

The gene consists of 17 exons encoding a ubiquitously expressed and evolutionary conserved 531 amino-acid protein known as parafibromin. To date, more than 120 different coding mutations were identified, and the majority of these reported CDC73 mutations are frameshift and nonsense [61], and as stated earlier, large or whole-gene deletions have also been described [41,64,65,67–72]. Germline coding mutations of CDC73 gene have been identified in 75% of the reported HPT-JT families, while 25% of these families, who do not harbor CDC73 mutations or deletions, may have defects involving the promoter or untranslated regions, uncharacterized alternative transcripts, or epigenetic modifications [17,73,74].

Parafibromin shares no homologies to known protein domains. However, about 200 amino acids of the parafibromin C-terminal domain share 27% sequence identity with the yeast Cdc73 protein, a component of the yeast polymerase-associated factor 1 (PAF1), a key transcriptional regulatory complex that directly interacts with RNA polymerase II subunit A [75]. The PAF1 complex interacts with RNA polymerase II at both the promoter and the coding regions of transcriptionally active genes regulating key transcriptional events of histone modification, chromatin remodeling, initiation, and elongation. The yeast Paf1 complex is composed of five subunits: Paf1, Ctr9, Leo1, Cdc73, and Rtf1. Deletion of Rtf1 or Cdc73 results in the loss of association of the remaining Paf1 complex members with chromatin and a significant reduction in binding of the complex to RNA pol II [75].

The human PAF1 complex consists of five subunits (hPaf1, hCtr9, hLeo1, hCdc73, shared with yeast Paf1 complex, and hSki8, an higher eukaryotic-specific subunit) [76]. Although a human homolog of Rtf1 does exist, it seems not to be part of the human PAF1 complex [77].

The human PAF1 complex was reported to hold several functions (reviewed in [6]):

1. It plays a role in both initiation and elongation of the transcription, associating with nonphosphorylated-Ser2 and phosphorylated-Ser5 on the C-terminal domain (CTD) of the major subunit of RNA polymerase II.
2. In the process of transcriptional initiation, the PAF1 complex is required for the histone H2B monoubiquitination mediated by Rad6/Bre1, a prerequisite for both H3K4 and H3K79 methylation mediated by Set1 and Dot1, respectively.
3. At initiation, the PAF1 complex associates with the Set1-like HMTase complex required for H3K4 methylation.
4. The PAF1 complex controls elongating RNA polymerase II CTD serine 2 phosphorylation.

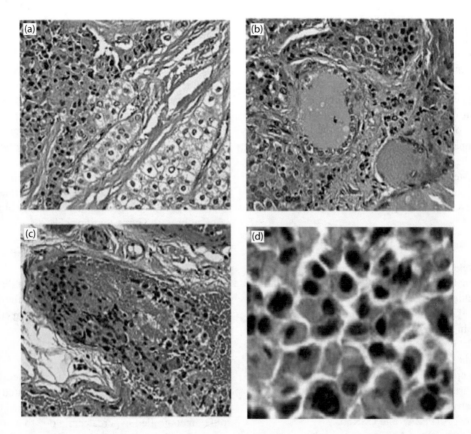

FIGURE 53.1 Histology of parathyroid carcinoma. (a) Tumor overview; (b) thyroid invasion; (c) vascular invasion; (d) nuclear atypia. (Photo credit: Guarnieri V et al. *BMC Med Genet.* 2017;18(1):83.)

5. During elongation, the PAF1 complex associates with elongation factors in conjunction with the histone chaperon FACT [79].

6. It directly interacts with 3′-end cleavage and polyadenylation factors at an early point in the transcription cycle [77].

As part of the human PAF1 complex that binds at the residues 223–415, parafibromin is required for expression of several cellular genes involved in cell cycle regulation, cell growth, protein synthesis, and lipid and nucleic acid metabolism. The protein has multiple functions, including (i) behind the regulation of genetic transcription, parafibromin interacts with the histone methyltransferase complex for histone modifications and chromatin remodeling [75,80]; (ii) it regulates cell growth via the (iii) downregulation of cyclin D1 expression and Wnt signaling [81,82] and (iv) inhibition of the c-myc proto-oncogene [83,84].

Parafibromin is also implicated in embryonic development directly regulating genes of cell growth and survival, such as the H19 gene, insulin-like growth factors 1 and 2 (IGF1 and IGF2), insulin-like growth factor binding protein 4 (IGFBP4), high mobility AT-hook 1 and 2 (HMGA1 and HMGA2), and 3-hydroxy-3-methylglutaryl-coenzyme A synthase 2 (HMGCS2) [85]. Parafibromin was shown to have a dual role as a tumor suppressor and oncoprotein depending on the cellular environment. As a tumor suppressor, parafibromin overexpression seems to result in (i) inhibition of NIH3T3 and HEK293 cell proliferation;

(ii) increase in G1 arrest and (iii) apoptosis in Hela cells, and (iv) downregulation of the cell cycle regulator cyclin D1.

As oncogene, overexpression of parafibromin enhances growth in cells expressing the SV40 large T-antigen, and vice versa: it inhibits growth in cells that do not express the SV40 large T-antigen [86].

The functional role of parafibromin in the nucleus is consistent with the identification of highly conserved nuclear localization signal (NLS) at residues 125–139. Parafibromin localizes into the nucleolar compartment; thereby three nucleolar localization signals (NoLSs) at residues 76–92, 192–194, and 393–409 have been reported [87,88]. Absence of parafibromin nucleolar expression was found in sporadic KP with CDC73 mutations even showing a nuclear expression. Thus, the disruption of nucleolar localization might cause parathyroid tumorigenesis independent of nuclear parafibromin expression [89].

Although parafibromin is a predominantly nuclear protein, some evidence suggests the cytoplasmatic interaction with the actin-binding proteins, actinin-2 and actinin-3, which are involved in organization of the cytoskeletal structure [90].

Furthermore, a recent work revealed parafibromin as a transcriptional scaffold whose distinct morphogen intracellular signals converge on, coordinating the expression of specific target genes and generating appropriate cellular responses [91]. Parafibromin competitively interacts with the Wnt and Hedgehog signal effectors, β-catenin and GLI1, respectively, and binds Notch intracellular domain (NICD), promoting downstream target genes activation. The platform function of parafibromin is regulated by its phosphorylation and dephosphorylation status on Y290/293/315, and mediated by PTK6 and SHP2, respectively [91].

The tyrosine-dephosphorylated form of parafibromin at Y290/293/315 is required to bind β-catenin, GLI1, and NICD, and seems to enhance the Wnt, Hh, and Notch signaling activation. Because β-catenin and GLI1 binding sites overlap in the N-terminal domain of parafibromin, these two effectors compete intracellularly for the same parafibromin region and interact with the PAF1 complex to increase the expression of their target genes [91], while the interaction between NICD and parafibromin requires the C-terminal region [91]. A reciprocal inhibition of the Wnt and Hh signals has been observed, and also reported in human gastric cancer cells [92].

53.2.6.1 Parafibromin and Wnt

Among the most relevant parafibromin functions is the activation of the Wnt signaling pathway [82,84,93,94], involved in the development of multicellular organisms, in cell proliferation, differentiation, survival, cell motility, and apoptosis via its central component β-catenin [95].

In absence of the Wnt ligand, most endogenous β-catenin is located at the cell membrane, where it associates with α-catenin and E-cadherin in epithelial cell adherent junctions. The cytoplasmatic counterparts is marked for proteasomal degradation by a multi-subunit destruction complex consisting of Axin, APC, protein phosphatase 2 A (PP2A), and GSK3b.

When the Wnt ligand binds the receptor complex, composed of Frizzled and LRP5/6, the inhibitory complex is dissociated and active unphosphorylated β catenin enters the nucleus and interacts with TCF/LEF DNA binding proteins to initiate transcription of Wnt target genes [95,96].

Studies in *Drosophila* show that Hyrax homolog of human parafibromin is a component of the Wnt/Wg signaling. Parafibromin/Hyrax has been proved to be a positive regulator in the Wnt pathway and is required for nuclear transduction of the WNT/Wg signaling in HEK293 cells. Parafibromin binds directly to the C-terminal region of nuclear β-catenin/Armadillo via its conserved N-terminal domain and recruits other components of the PAF1 complex in order to regulate the transcription of Wnt target genes encoding, for example, the c-Myc oncoprotein and the cell cycle protein cyclin D1 [82,97].

Dysregulation of the Wnt signaling pathway causes many different human tumor types, including colon, hepatocellular carcinoma, leukemia, and melanoma [98], and aberrant activation of Wnt/β catenin signaling pathway was observed in a subset of parathyroid carcinomas due to epigenetic APC loss of expression [99]. Furthermore, it has also been identified in pHPT and in parathyroid tumors associated with chronic renal failure [100].

53.2.6.2 Parafibromin and Hedgehog

Parafibromin appears to be involved in the activation of Hedgehog pathway (Hh) and is required for the expression of specific Hh target genes.

The detailed way that nuclear GLI cofactors are recruited and functionally cooperate to regulate transcription and mediate direct chromatin remodeling of the Hh target genes remains doubtful; the C-terminal activation domain of GLI3 and Ci, the

single *Drosophila* Gli, have been reported to interact with CREB binding protein (CBP) which has HAT activities involved in transcription control [101,102]. Moreover, the C-terminal GLI/Ci domain contains a motif comparable to a consensus binding element for the TATA-box recognition component TAFII31 [103]. Mosimann et al. provided evidence for the involvement of parafibromin/Hyrax as a novel GLI/Ci binding partner and as a positive component in Hh signaling. Impairment of Hyrax function decreases Hh signaling activity in *Drosophila* cell culture and leads to a decrease of Hh target gene expression in vivo, while RNAi-mediated knockdown of parafibromin decreased the transcriptional activity of GLI1 and GLI2 in human cell cultures. As happens for the β-catenin, GLIs/Ci directly binds the N-terminus of parafibromin/Hyrax (amino acids 200–343) via the domain of interaction with Su (fu) known as Region 1, a highly conserved recruitment site in the N-terminal portion of all the GLIs/Ci factors [104].

This evidence suggests the role of parafibromin as a GLI/Ci auxiliary factor and the involvement of the PAF1 complex to control Hh target gene expression in human cells.

53.2.7 Treatment

Hyperparathyroidism per se is a usually benign non-life-threatening disease; however, major risks associated with HPT-JT syndrome consist of the formation of jaw or renal tumors and mostly to the uncontrolled hypercalcemia due to a possible parathyroid carcinoma. Regarding ossifying fibroma of the mandible/maxilla or for uterine lesions, no specific recommendations are suggested other than close monitoring of the tumors' evolution with the possible surgical removal, and, in the case of bone tumors, bone grafting and reconstruction are also encouraged [27]. Conversely, the treatment of extreme hypercalcemia, especially in the case of malignant parathyroid neoplasia, is often a challenge to deal with, mostly because although parathyroid cancers show all the stigmata related to very high hypercalcemia (while benign lesions often present with mild symptoms or patients are asymptomatic), they usually manifest at advanced stages, thus with scarce possibilities to be recognized earlier, or even present as normal (or asymptomatic) hyperparathyroidism [105,106].

Surgical removal is the gold standard for the treatment of parathyroid disease, being the adequate surgical approach applied depending on the type of the lesion, for the reasons explained above about the long-term overall survival (in cases of parathyroid carcinoma) and also because recurrence and the need for a second operation usually accompanies all the comorbidities associated with relapse of the disease, with a higher risk for the patient [107]. Possible repeated surgery, defined as "debulking," is indicated in cases of metastatic disease, for removal of PTH-secreting lesions that represent the real cause of mortality, rather than the tumor mass, so that survival is possible only if the hypercalcemia is controlled [108].

Apart from common emergency treatments for severe hypercalcemia such as saline infusion and use of diuretics, in order to possibly chemically lower the blood calcium level, use of bisphosphonates (BP) was introduced in clinical practice more than four decades ago [109]. BP are compounds resembling the structure of inorganic pyrophosphate with the function to bind the active sites of bone remodeling, decreasing the osteoclastic activity and (to a different extent) inducing osteoclast apoptosis [110]. Several studies have been conducted with different drugs, administered orally or by intravenous dose, and a recent meta-analysis confirmed the efficacy of BPs in the acute phase of hypercalcemia and for a short time (<6 months) [111]; on the other hand, BPs seem to increase the BMD. The need for long-acting drugs pushed toward the use of allosteric modulators of the calcium sensing receptor. Initially developed and used for the treatment of patients affected by classic pHPT [112], cinacalcet (in its second-generation form) binds to the transmembrane domains of the CaSR protein and, by inducing a conformational change of the protein structure, increases the sensitivity of the receptor for the calcium [113]. This in turn decreases PTH secretion from the parathyroid chief cells. Unlike BPs, cinacalcet demonstrated to have a long-term effect (up to 44 months, for one of the longer studies reported so far [114]) in rectifying the serum calcium levels, while the effect on BMD is controversial.

53.2.7.1 Chemotherapy and Radiotherapy

With regard to chemotherapy or radiotherapy treatments, few anecdotal cases recorded variable and not conclusive efficacy, although it was reported that radiotherapy might help in reducing the metastatic mass [115]. All these data should be taken with caution since they refer to a very limited number of patients.

53.3 GCM2: From Isolated Hypoparathyroidism to Familial Hyperparathyroidism

The glial cells missing 2 gene (chromosome 6p24.2), homolog of the *dgmc* in *Drosophila* is one of the genes mutated in familial isolated hypoparathyroidism, along with the CaSR, GNA11, and PTH [116] genes.

53.3.1 History and Genetics

GCM2 belongs to a family of transcription factors that shares a common highly conserved DNA binding domain in the N-terminal region known as GCM motif. This transcriptional regulator also contains two transcriptional activation domains (TAD1, TAD2), one following the DNA binding domain and the other at the C-terminus, and one inhibitory domain [117].

Bowl et al., elucidated the molecular mechanism underlying DNA-binding and transactivation required for this transcription factor, characterizing four GCM2 homozygous germline mutations found in eight families with autosomal recessive hypoparathyroidism: 3D-modeling of the GCMB DNA-binding domain revealed that the R110 residue is important for the structural integrity of helix 2, which forms part of the GCMB/DNA binding interface. Mutations that disrupt the GCMB/DNA binding domain abolish the DNA binding capacity, while mutations that result in C-terminal deletion instead reduce the transactivating property of the protein [118].

In humans, GCM transcripts are present in two isoforms, namely GCMA/1 and GCMB/2 with/without a nuclear localization signal (NLS) sequence, respectively [119]. The GCMB/2 protein shows a significantly weaker transcriptional activity as compared with GCMA/1, depending on the presence of an additional inhibitory domain in GCMB/2 (1). The human GCM transcripts are expressed in both developmental and post-developmental stages, in fetal brain and in adult kidney but also in pathological conditions such as embryonic and pediatric brain tumors [119].

GCM2 is the unique gene involved in the parathyroid gland development during embryogenesis, and studies on murine models showed that its expression is restricted to parathyroid glands. It follows that it is considered the master regulator of parathyroid development, promoting survival and differentiation of committed precursors toward parathyroid cells, although it is not involved in the expression of differentiation markers, such as CaSR, in developing embryo [120].

Studies of patients with hypoparathyroidism and mouse models revealed that the development of parathyroid glands from endodermal cells of the third and fourth pharyngeal pouches and neural crest arising from embryonic mid- and hind-brain is coordinated by several genes encoding transcription factors including Gcm2 and other transcription factors such as Tbx1, Gata3, Sox3, Aire1, Hoxs, and Paxs. Their expression has been proved to be normal in the third pharyngeal pouch of Gcm2$^{-/-}$ mouse embryos and they act upstream of Gcm2 in a transcriptional regulatory cascade [121]. These studies were able to elucidate the molecular mechanism to promote the parathyroid development:

1. cDNA microarray analysis of mice lacking Tbx1 showed downregulation of Gcm2 in the pharyngeal region, indicating that Tbx1, regulated in turn by Sonic Hedgehog (Shh), is upstream of Gcm2 [122,123].

2. Grigorieva et al. demonstrated that GCMB is transcriptionally regulated by GATA3, a transcription factor responsible of HDR syndrome (MIM146255), that binds the promoter of the human GCMB and is implicated in the maintained differentiation and survival of parathyroid progenitor cells [124].

3. Hoxa3 seems to be required for the initiation of Gcm2 expression in the third pharyngeal pouches. Moreover, Hoxa3 and Pax1 are required for the maintenance of Gcm2 expression [125].

Gunther et al. demonstrated that Gcm2$^{-/-}$ deficient mice, despite the lack of parathyroid glands, showed a PTH serum level similar to the WT, indicating that an alternative source of PTH secretion to regulate the serum calcium concentration does exist. While the parathyroid glands secrete the majority of PTH, the thymus contains specific PTH-secreting cells whose differentiation is regulated by Gcm1 [126].

Studies in human parathyroid cells from hyperplastic glands of patients with chronic disease made it possible to identify genes regulated by the GCM2 gene. By a gene silencing approach, Mizobuchi et al. demonstrated that the downregulation of GCM2 expression in human parathyroid cells resulted in downregulation of the CaSR expression, and GCM2 transcription factor is responsible for the high level of the CaSR receptor in the parathyroid glands, thereby suggesting the CaSR gene is one of the GCM2 target genes [127]. To reinforce these findings, functional GCM-response elements were identified in both CaSR promoters

(P1 and P2), proving that the receptor is transactivated by the GCM2 gene [128].

Moreover, the regulation of GCM2 by GATA3 probably involved the CaSR. Instead, parathyroid chief cells express the CaSR on their surface membrane, which, in response to small changes in the ionized calcium level, activates the MAPK pathway to negatively regulate PTH secretion by inhibiting the secretion of preformed PTH, the proliferation of parathyroid cells, and the transcription of the pre-pro PTH gene [129]. Low serum calcium level reduces GCMB expression and enhances PTH secretion and proliferation of parathyroid glands [130], while an increase of serum calcium concentration leads to upregulation of GCMB, decreases PTH secretion, and parathyroid cell proliferation [124]. As several studies showed that the regulator of GCMB gene expression, GATA3, is phosphorylated by MAPKs cascade in human T cells, it is likely that in parathyroid cells the CaSR activates MAPK pathway, causing phosphorylation of GATA3 and its translocation to the nucleus to initiate GCMB expression [124].

Several studies in *Drosophila* have demonstrated that one of the downstream target genes regulated by Gcm encodes for a protein with regulator of G-protein signaling (RGS) domain that promotes the GTP hydrolysis in the heterotrimeric G-protein α subunits and the rapid attenuation of signaling pathway by GPCR [131]. This would suggest the possibility that, also in mammals, GCMB regulates expression of an RGS protein, having the role to shut down the CaSR signaling pathway, in turn facilitating GATA3-regulated expression of GCMB [124].

Canaff et al. investigated GCM2 mutations inherited in an autosomal dominant or recessive manner in two families with hypoparathyroidism. While recessive GCM2 mutations were unable to influence WT GCM2 activity, dominantly inherited GCM2 mutants seemed to exhibit a dominant-negative effect on WT GCM2 activity. Because GCM2 transcription factor does not function as a dimer, they proposed that the dominant-negative effect might be exerted by the binding with other, not yet identified proteins [128].

53.3.2 Pathogenesis: Role of GCM2 in pHPT and FIHP

The first inactivating homozygous GCM2 mutations in ADH were reported in 2001 [132] and for 10 years its role in hypocalcemic states was confirmed by the identification of several mutations inherited either by a recessive pattern [118,133,135,136] or in an autosomal dominant fashion, in the latter case due to a dominant negative effect of the mutated copy on the wild-type counterpart [128,137,138].

These studies contributed to the careful characterization of the structure of this transcription factor consisting of: (i) a N-ter DNA binding domain (residues 19–174) common to all other *gcm* proteins [139–142]; (ii) two transactivation domains, TAD1 and TAD2 (residues 174–263 and 426–504, respectively [117]); and (iii) an inhibitory domain (ID, residues 264–352 [117]).

In 2004, the role in hyperparathyroid states was hypothesized by Kebebew et al. By RT-qPCR they evaluated GCM2 mRNA expression in normal, hyperplastic, adenomatous, and carcinomatous parathyroid tissues from patients with primary and secondary hyperparathyroidism. Although no difference was found between parathyroid adenoma, hyperplasia, and carcinoma, the level of

GCM2 mRNA expression was higher in pathologic glands causing primary and secondary hyperparathyroidism as compared to normal parathyroid tissue [130]. This result was in contrast with Correa et al., who had reported under-expression of GCM2 mRNA in parathyroid adenoma of pHPT as compared to normal and hyperplastic parathyroid glands from patients with uremia [143]. However, it seemed that while hypoparathyroidism was characterized by the lack of GCMB gene expression, overexpression or constitutive activation of this gene was somehow associated with parathyroid tumorigenesis in hyperparathyroidism. Alternatively, the aberrant expression of GCMB gene in parathyroid tumors might be due to the deregulation of upstream signals or other transcription factors such as Hoxa3, Pax1, Pax9, and Eya1 (reviewed in [130]).

In 2011, on a preliminary analysis of 30 parathyroid adenomas, Mannstadt et al. identified several polymorphisms (among these, the p.Y282D) and one putative mutation (p.V382M), but the assays they performed on these variants failed to prove any functional role in the onset of sporadic (in this case) hyperparathyroidism [144]. In 2014, a large correlation analysis on three different Italian pHPT cohorts actually confirmed that the 282D was more prevalent in patients with respect to the controls (blood donors) and that the variant leads the protein to be more active compared to the wild-type form [145]. Thus, a dualistic function, similarly proved for the CaSR and GNA11 genes (inactive/overactive in the hyperpara/hypopara disorders, respectively) was delineating, although in the opposite direction, since the GCM2 gene is inactive in hypoparathyroidism and active in hyperparathyroidism (Table 53.1). However, only in 2016, the conclusive confirmation of the involvement of the GCM2 gene as one of the FIHP causing genes (along with the CaSR, CDC73, and MEN1) was achieved by Guan et al., who applied next-generation sequencing to a panel of eight FHIP probands belonging to eight different, apparently unrelated FIHP families, resulted negative for coding mutations of the previously known causing genes. They identified three GCM2 gene variants that they confirmed on a second validation cohort of 40 FHIP kindreds [9]. By an elegant functional approach, they proved that the protein encoded by the gene contained an additional domain renamed C-terminal conserved inhibitory domain (CCID, residues 379–395), whose mutations cause an overactivation of the protein, and in turn a possible derangement in the parathyroid cell hyperproliferation. In this way, the GCM2 gene in the FHIP can be considered a parathyroid proto-oncogene [9].

Further studies elucidated the prevalence of the Y282D-GCM2 variant in a subset of sporadic parathyroid carcinoma (n = 23) and atypical adenoma (n = 39). The authors reported a relatively high (due the small size of their survey) frequency of the 282D variant in both groups (17.4% PC vs 5.1% APA) with a positive correlation with higher calcium correlation. It is important to note that, for the first time, this paper found a possible correlation between the GCM2 gene variants/mutations and aggressive parathyroid neoplasia [146].

53.4 Prognosis

The prognosis of HPT-JT syndrome depends on the histology of the lesion(s) and, as stated earlier, on the surgery approach, especially in cases of parathyroid carcinoma. In these cases, the survival rate reported is 78% at 5 years and 49% at 10 years [107,134], while multiple lesions and metastasis are among the negative prognostic factors. However, despite the presence of a CDC73, gene mutation gives a risk of malignant potential only in about the 15% of cases [17]; however, some also reported that cancers carrying a CDC73 coding mutation and corresponding negative parafibromin staining may have a worse course, with respect to the tumors having just the loss of parafibromin expression but with no apparent coding mutation [53,78].

With regard to the FIHP-GCM2 induced, in a retrospective analysis on 18 FHIP-GCM2 mutated patients and more than 450 sporadic pHPT, the authors observed that patients carrying a GCM2 mutation were statistically more prone to develop a multi-gland parathyroid disease with substantial risk of carcinoma [66] compared to sporadic pHPT cases. Thus, these patients need to be recognized (hopefully in initial diagnosis) by specific clinical features (multiple parathyroid lesions in association with higher PTH level) or (better) by genetic testing, in order to drive and improve the clinical/surgical management. Thus, although clinical data on parathyroid carcinoma due to a GCM2 gene mutation are scarce and thus the prognosis is not known, in these aggressive at-risk cases, the same precautions and guidelines of classic CDC73-induced lesions should be applied.

53.5 Conclusion

Familial isolated hyperparathyroidism (FHIP) is a subtle syndrome that might hide a more complex genetic disorder. Thus, genetic screening becomes of basic importance for follow-up and patient management. Instead, the probability of positive genetic testing in FIHP, not just for the GCM2 gene but also for the other three genes (CaSR, CDC73, and MEN1), was defined on a retrospective analysis on more than 650 pHPT patients. The authors showed that in cases of family history of pHPT, male sex, age at onset <45 years, in association with multi-gland disease, the percentage of having a putative genetic mutation ranged from 88% to 92% [51].

Acknowledgments

The authors gratefully acknowledge Flavia Pugliese and Antonio Salcuni for insightful contribution and comments and Andreina Guarnieri for the helpful revision of the English language.

REFERENCES

1. Bilezikian JP, Bandeira L, Khan A et al. Hyperparathyroidism. *Lancet.* 2018;391:168–78.
2. Hendy GN, Guarnieri V, Canaff L. Calcium-sensing receptor and associated diseases. *Prog Mol Biol Transl Sci.* 2009;89:31–95.
3. Thakker RV. Genetics of parathyroid tumours. *J Intern Med.* 2016;280:574–83.
4. Hyde SM, Cote GJ, Grubbs EG. Genetics of multiple endocrine neoplasia type 1/multiple endocrine neoplasia type 2 syndromes. *Endocrinol Metab Clin North Am.* 2017;46:491–502.
5. Alrezk R, Hannah-Shmouni F, Stratakis CA. MEN4 and CDKN1B mutations: The latest of the MEN syndromes. *Endocr Relat Cancer.* 2017;24:T195–208.
6. Newey PJ, Bowl MR, Thakker RV. Parafibromin—functional insights. *J Intern Med.* 2009;266:84–98.
7. Hyde SM, Rich TA, Waguespack SG, Perrier ND, Hu MI. CDC73-related disorders. In: Adam MP, Ardinger HH, Pagon RA et al., editors. *GeneReviews®* [Internet]. Seattle (WA): University of Washington, Seattle; 1993–2018. 2008 Dec 31 [updated 2018 Apr 26].
8. Cetani F, Pardi E, Ambrogini E et al. Genetic analyses in familial isolated hyperparathyroidism: Implication for clinical assessment and surgical management. *Clin Endocrinol (Oxf).* 2006;64:146–52.
9. Guan B, Welch JM, Sapp JC et al. GCM2-activating mutations in familial isolated hyperparathyroidism. *Am J Hum Genet.* 2016;99:1034–44.
10. Chen JD, Morrison C, Zhang C et al. Hyperparathyroidism-jaw tumour syndrome. *J Intern Med.* 2003;253:634–42.
11. Bradley KJ, Hobbs MR, Buley ID et al. Uterine tumours are a phenotypic manifestation of the hyperparathyroidism-jaw tumour syndrome. *J Intern Med.* 2005;257:18–26.
12. Pichardo-Lowden AR, Manni A, Saunders BD et al. Familial hyperparathyroidism due to a germline pathogenic variant of the CDC73 gene: Implications for management and age-appropriate testing of relatives at risk. *Endocr Pract.* 2011;17:602–9.
13. Bradley KJ, Cavaco BM, Bowl MR et al. Parafibromin pathogenic variants in hereditary hyperparathyroidism syndromes and parathyroid tumours. *Clin Endocrinol (Oxf).* 2006;64:299–306.
14. Cavaco BM, Guerra L, Bradley KJ et al. Hyperparathyroidism-jaw tumor syndrome in Roma families from Portugalis due to a founder mutation of the HRPT2 gene. *J Clin Endocrinol Metab.* 2004;89:1747–52.
15. Jackson CE. Hereditary hyperparathyroidism associated with recurrent pancreatitis. *Ann Intern Med.* 1958;49:829–36.
16. Jackson CE, Norum RA, Boyd SB et al. Hereditary hyperparathyroidism and multiple ossifying jaw fibromas: A clinically and genetically distinct syndrome. *Surgery.* 1990;108:1006–12.
17. Carpten JD, Robbins CM, Villablanca A et al. HRPT2, encoding parafibromin, is mutated in hyperparathyroidism-jaw tumor syndrome. *Nat Genet.* 2002;32:676–80.
18. Szabó J, Heath B, Hill VM et al. Hereditary hyperparathyroidism-jaw tumor syndrome: The endocrine tumor gene HRPT2 maps to chromosome 1q21-q31. *Am J Hum Genet.* 1995;56:944–50.
19. Teh BT, Farnebo F, Kristoffersson U et al. Autosomal dominant primary hyperparathyroidism and jaw tumor syndrome associated with renal hamartomas and cystic kidney disease: Linkage to 1q21-q32 and loss of the wild type allele in renal hamartomas. *J Clin Endocrinol Metab.* 1996;81:4204–11.
20. Hobbs MR, Pole AR, Pidwirny GN et al. Hyperparathyroidism-jaw tumor syndrome: The HRPT2 locus is within a 0.7-cM region on chromosome 1q. *Am J Hum Genet.* 1999;64:518–25.
21. Hobbs MR, Rosen IB, Jackson CE. Revised 14.7-cM locus for the hyperparathyroidism-jaw tumor syndrome gene, HRPT2. *Am J Hum Genet.* 2002;70:1376–7.
22. Shattuck TM, Välimäki S, Obara T et al. Somatic and germline mutations of the HRPT2 gene in sporadic parathyroid carcinoma. *N Engl J Med.* 2003;349:1722–9.

23. DeLellis RA, Mangray S. Heritable forms of primary hyperparathyroidism: A current perspective. *Histopathology.* 2018;72:117–32.

24. Keyser JS, Postma GN. Brown tumor of the mandible. *Am J Otolaryngol.* 1996;17(6):407–10.

25. Bandeira F, Cusano NE, Silva BC, Cassibba S, Almeida CB, Machado VC, Bilezikian JP. Bone disease in primary hyperparathyroidism. *Arq Bras Endocrinol Metabol.* 2014;58:553–61.

26. Ennazk L, El Mghari G, El Ansari N. Jaw tumor in primary hyperparathyroidism is not always a brown tumor. *Clin Cases Miner Bone Metab.* 2016;13:64–6.

27. du Preez H, Adams A, Richards P et al. Hyperparathyroidism jaw tumour syndrome: A pictoral review. *Insights Imaging.* 2016;7:793–800.

28. Jackson MA, Rich TA, Hu MI et al. "CDC73"-related disorders. In: Adam MP, Ardinger HH, Pagon RA et al., editors. *GeneReviews®* [Internet]. Seattle (WA): University of Washington, Seattle; 1993–2001.

29. Kakinuma A, Morimoto I, Nakano Y et al. Familial primary hyperparathyroidism complicated with Wilms' tumor. *Intern Med.* 1994;33:123–6.

30. Haven CJ, Wong FK, van Dam EW et al. A genotypic and histopathological study of a large Dutch kindred with hyperparathyroidism-jaw tumor syndrome. *J Clin Endocrinol Metab.* 2000;85:1449–54.

31. Zhao J, Yart A, Frigerio S et al. Sporadic human renal tumors display frequent allelic imbalances and novel mutations of the HRPT2 gene. *Oncogene.* 2007;26:3440–9.

32. Kebebew E, Arici C, Duh QY et al. Localization and reoperation results for persistent and recurrent parathyroid carcinoma. *Arch Surg.* 2001;136:878–85.

33. Patel CN, Salahudeen HM, Lansdown M et al. Clinical utility of ultrasound and 99mTc sestamibi SPECT/CT for preoperative localization of parathyroid adenoma in patients with primary hyperparathyroidism. *Clin Radiol.* 2010; 65:278–87.

34. Hara H, Igarashi A, Yano Y et al. Ultrasonographic features of parathyroid carcinoma. *Endocr J.* 2001;48:213–7.

35. Araujo Castro M, López AA, Fragueiro LM et al. Giant parathyroid adenoma: Differential aspects compared to parathyroid carcinoma. *Endocrinol Diabetes Metab Case Rep.* 2017;2017:17–0041.

36. Kassahun WT, Jonas S. Focus on parathyroid carcinoma. *Int J Surg.* 2011;9:13–9.

37. Papanicolau-Sengos A, Brumund K, Lin G, Hasteh F. Cytologic findings of a clear cell parathyroid lesion. *Diagn Cytopathol.* 2013;41:725–8.

38. Shah KS, Ethunandan M. Tumour seeding after fine-needle aspiration and core biopsy of the head and neck—a systematic review. *Br J Oral Maxillofac Surg.* 2016;54:260–5.

39. Kim J, Horowitz G, Hong M et al. The dangers of parathyroid biopsy. *J Otolaryngol Head Neck Surg.* 2017;46:4.

40. Marx SJ, Lourenco DM. Familial hyperparathyroidism—disorders of growth and secretion in hormone-secretory tissue. *Horm Metab Res.* 2017;49:805–15.

41. Bricaire L, Odou MF, Cardot-Bauters C et al. Frequent large germline HRPT2 deletions in a French National cohort of patients with primary hyperparathyroidism. *J Clin Endocrinol Metab.* 2013;98:E403–8.

42. Guarnieri V, Scillitani A, Muscarella LA et al. Diagnosis of parathyroid tumors in familial isolated hyperparathyroidism with HRPT2 mutation: Implications for cancer surveillance. *J Clin Endocrinol Metab.* 2006;91:2827–32.

43. Delellis RA. Challenging lesions in the differential diagnosis of endocrine tumors: Parathyroid carcinoma. *Endocr Pathol.* 2008;19:221–5. Winter.

44. Christakis I, Bussaidy N, Clarke C et al. Differentiating atypical parathyroid neoplasm from parathyroid cancer. *Ann Surg Oncol.* 2016;23:2889–97.

45. Wei CH, Harari A. Parathyroid carcinoma: Update and guidelines for management. *Curr Treat Options Oncol.* 2012;13:11–23.

46. Clark O. Parathyroid carcinoma. In: Doherty GM, Way LW, editors. *Current Surgical Diagnosis and Treatment.* Michigan: McGraw-Hill Companies; 2006; 284–93.

47. Rodgers SE, Perrier ND. Parathyroid carcinoma. *Curr Opin Oncol.* 2006;18:16–22.

48. Koea JB, Shaw JH. Parathyroid cancer: Biology and management. *Surg Oncol.* 1999;8:155–65.

49. Wang P, Xue S, Wang S et al. Clinical characteristics and treatment outcomes of parathyroid carcinoma: A retrospective review of 234 cases. *Oncol Lett.* 2017;14:7276–82.

50. DeLellis RA. Parathyroid carcinoma: An overview. *Adv Anat Pathol.* 2005;12:53–61.

51. El Lakis M, Nockel P, Gaitanidis A et al. Probability of positive genetic testing results in patients with family history of primary hyperparathyroidism. *J Am Coll Surg.* 2018;226(5):933–8.

52. Guarnieri V, Battista C, Muscarella LA et al. CDC73 mutations and parafibromin immunohistochemistry in parathyroid tumors: Clinical correlations in a single-centre patient cohort. *Cell Oncol (Dordr).* 2012;35:411–22.

53. Witteveen JE, Hamdy NA, Dekkers OM et al. Downregulation of CaSR expression and global loss of parafibromin staining are strong negative determinants of prognosis in parathyroid carcinoma. *Mod Pathol.* 2011;24:688–97.

54. Juhlin CC, Villablanca A, Sandelin K et al. Parafibromin immunoreactivity: Its use as an additional diagnostic marker for parathyroid tumor classification. *Endocr Relat Cancer.* 2007;14:501–12.

55. Kim HK, Oh YL, Kim SH et al. Parafibromin immunohistochemical staining to differentiate parathyroid carcinoma from parathyroid adenoma. *Head Neck.* 2012;34:201–6.

56. Wang O, Wang C, Nie M et al. Novel HRPT2/CDC73 gene mutations and loss of expression of parafibromin in Chinese patients with clinically sporadic parathyroid carcinomas. *PLOS ONE.* 2012;7:e45567.

57. Fernandez-Ranvier GG, Khanafshar E, Tacha D et al. Defining a molecular phenotype for benign and malignant parathyroid tumors. *Cancer.* 2009;115:334–44.

58. Howell VM, Gill A, Clarkson A, Nelson AE et al. Accuracy of combined protein gene product 9.5 and parafibromin markers for immunohistochemical diagnosis of parathyroid carcinoma. *J Clin Endocrinol Metab.* 2009;94:434–41.

59. Ozolins A, Narbuts Z, Vanags A et al. Evaluation of malignant parathyroid tumours in two European cohorts of patients with sporadic primary hyperparathyroidism. *Langenbecks Arch Surg.* 2016;401:943–51.

60. Gill AJ, Lim G, Cheung VKY, Andrici J et al. Parafibromin-deficient (HPT-JT type, CDC73 mutated) parathyroid tumors demonstrate distinctive morphologic features. *Am J Surg Pathol.* 2019;43(1):35–46.

61. Newey PJ, Bowl MR, Cranston T et al. Cell division cycle protein 73 homolog (CDC73) mutations in the hyperparathyroidism-jaw tumor syndrome (HPT-JT) and parathyroid tumors. *Hum Mutat.* 2010;3:295–307.

62. Pazienza V, la Torre A, Baorda F et al. Identification and functional characterization of three NoLS (nucleolar localisation signals) mutations of the CDC73 gene. *PLOS ONE.* 2013;8:e82292.

63. Masi G, Iacobone M, Sinigaglia A et al. Characterization of a new CDC73 missense mutation that impairs Parafibromin expression and nucleolar localization. *PLOS ONE.* 2014;9:e97994.

64. Cascón A, Huarte-Mendicoa CV, Javier Leandro-García L et al. Detection of the first gross CDC73 germline deletion in an HPT-JT syndrome family. *Genes Chromosomes Cancer.* 2011;50:922–9.

65. Domingues R, Tomaz RA, Martins C, Nunes C, Bugalho MJ, Cavaco BM. Identification of the first germline HRPT2 whole-gene deletion in a patient with primary hyperparathyroidism. *Clinl Endocrinol (Oxf).* 2012;76:33–8.

66. El Lakis M, Nockel P, Guan B et al. Familial isolated primary hyperparathyroidism associated with germline GCM2 mutations is more aggressive and has a lesser rate of biochemical cure. *Surgery.* 2018;163:31–4.

67. Korpi-Hyövälti E, Cranston T, Ryhänen E et al. CDC73 intragenic deletion in familial primary hyperparathyroidism associated with parathyroid carcinoma. *J Clin Endocrinol Metab.* 2014;99:3044–8.

68. Kong J, Wang O, Nie M et al. Familial isolated primary hyperparathyroidism/hyperparathyroidism-jaw tumour syndrome caused by germline gross deletion or point mutations of CDC73 gene in Chinese. *Clin Endocrinol (Oxf).* 2014;81:222–30.

69. Davidson JT, Lam CG, McGee RB, Bahrami A, Diaz-Thomas A. Parathyroid cancer in the pediatric patient. *J Pediatr Hematol Oncol.* 2016;38:32–7.

70. Guarnieri V, Seaberg RM, Kelly C et al. Erratum to: Large intragenic deletion of CDC73 (exons 4-10) in a three-generation hyperparathyroidism-jaw tumor (HPT-JT) syndrome family. *BMC Med Genet.* 2017;18:99.

71. Rubinstein JC, Majumdar SK, Laskin W et al. Hyperparathyroidism-jaw tumor syndrome associated with large-scale 1q31 deletion. *J Endocr Soc.* 2017;1:926–30.

72. Mamedova E, Mokrysheva N, Vasilyev E et al. Primary hyperparathyroidism in young patients in Russia: High frequency of hyperparathyroidism-jaw tumor syndrome. *Endocr Connect.* 2017;6:557–65.

73. Cetani F, Pardi E, Borsari S et al. Genetic analyses of the HRPT2 gene in primary hyperparathyroidism: Germline and somatic mutations in familial and sporadic parathyroid tumors. *J Clin Endocrinol Metab.* 2004;11:5583–91.

74. Bradley KJ, Cavaco BM, Bowl MR et al. Parafibromin mutations in hereditary hyperparathyroidism syndromes and parathyroid tumors. *Clin Endocrinol (Oxf).* 2006;3:299–306.

75. Rozenblatt-Rosen O, Hughes CM, Nannepaga SJ et al. The parafibromin tumor suppressor protein is part of a human Paf1 complex. *Mol Cell Biol.* 2005;2:612–20.

76. Zhu B, Mandal SS, Pham AD et al. The human PAF complex coordinates transcription with events downstream of RNA synthesis. *Genes Dev.* 2005;19:1668–73.

77. Nordick K, Hoffman MG, Betz JL et al. Direct interactions between the Paf1 complex and a cleavage and polyadenylation factor are revealed by dissociation of Paf1 from RNA polymerase II. *Eukaryot Cell.* 2008;7:1158–67.

78. Cetani F, Banti C, Pardi E et al. CDC73 mutational status and loss of parafibromin in the outcome of parathyroid cancer. *Endocr Connect.* 2013;4:186–95.

79. Pavri R, Zhu B, Li G, Trojer P et al. Histone H2B monoubiquitination functions cooperatively with FACT to regulate elongation by RNA polymerase II. *Cell.* 2006;125:703–17.

80. Yart A, Gstaiger M, Wirbelauer C et al. The HRPT2 tumor suppressor gene product parafibromin associates with human PAF1 and RNA polymerase II. *Mol Cell Biol.* 2005;12:5052–60.

81. Woodard GE, Lin L, Zhang JH et al. Parafibromin, product of the hyperparathyroidism-jaw tumor syndrome gene HRPT2, regulates cyclin D1/PRAD1 expression. *Oncogene.* 2005;7:1272–6.

82. Mosimann C, Hausmann G, Basler K. Parafibromin/Hyrax activates Wnt/Wg target gene transcription by direct association with beta-catenin/Armadillo. *Cell.* 2006;2:327–41.

83. Cardoso L, Stevenson M, Thakker RV. Molecular genetics of syndromic and non-syndromic forms of parathyroid carcinoma. *Hum Mutat.* 2017;12:1621–48.

84. Lin L, Zhang JH, Panicker LM et al. The parafibromin tumor suppressor protein inhibits cell proliferation by repression of the c-mycprotooncogene. *Proc Natl Acad Sci USA.* 2008;105:17420–5.

85. Wang P, Bowl MR, Bender S et al. Parafibromin, a component of the human PAF complex, regulates growth factors and is required for embryonic development and survival in adult mice. *Mol Cell Biol.* 2008;9:2930–40.

86. Iwata T, Mizusawa N, Taketani Y et al. Parafibromin tumor suppressor enhances cell growth in the cells expressing SV40 large T antigen. *Oncogene.* 2007;42:6176–83.

87. Hahn MA, Marsh DJ. Nucleolar localization of parafibromin is mediated by three nucleolar localization signals. *FEBS Lett.* 2007;26:5070–4.

88. Lin L, Czapiga M, Nini L et al. Nuclear localization of the parafibromin tumor suppressor protein implicated in the hyperparathyroidism-jaw tumor syndrome enhances its pro-apoptotic function. *Mol Cancer Res.* 2007;2:183–93.

89. Juhlin CC, Haglund F, Obara T et al. Absence of nucleolar parafibromin immunoreactivity in subsets of parathyroid malignant tumors. *Virchows Arch.* 2011;1:47–53.

90. Agarwal SK, Simonds WF, Marx SJ. The parafibromin tumor suppressor protein interacts with actin-binding proteins actinin-2 and actinin-3. *Mol Cancer.* 2008;7:65.

91. Kikuchi I, Takahashi-Kanemitsu A, Sakiyama N et al. Dephosphorylated parafibromin is a transcriptional coactivator of the Wnt/Hedgehog/Notch pathways. *Nat Commun.* 2016;7:12887.

92. Yanai K, Nakamura M, Akiyoshi T et al. Crosstalk of hedgehog and Wnt pathways in gastric cancer. *Cancer Lett.* 2008;1:145–56.

93. Bradley KJ, Bowl MR, Williams SE et al. Parafibromin is a nuclear protein with a functional monopartite nuclear localization signal. *Oncogene.* 2007;8:1213–21.

94. Zhang C, Kong D, Tan MH et al. Parafibromin inhibits cancer cell growth and causes G1 phase arrest. *Biochem Biophys Res Commun.* 2006;350:17–24.

95. Willert K, Jones KA. Wnt signaling: Is the party in the nucleus? *Genes Dev.* 2006;11:1394–404.

96. Gordon MD, Nusse R. Wnt signaling: Multiple pathways, multiple receptors, and multiple transcription factors. *J Biol Chem.* 2006;32:22429–33.

97. Juhlin CC, Haglund F, Villablanca A et al. Loss of expression for the Wnt pathway components adenomatous polyposis coli and glycogen synthase kinase 3-beta in parathyroid carcinomas. *Int J Oncol.* 2009;2:481–92.

98. Polakis P. Wnt signaling in cancer. *Cold Spring Harb Perspect Biol.* 2012;5. pii: a008052.

99. Svedlund J, Aurén M, Sundström M et al. Aberrant WNT/β-catenin signaling in parathyroid carcinoma. *Mol Cancer.* 2010;9:294.

100. Björklund P, Akerström G, Westin G. Accumulation of non-phosphorylated beta-catenin and c-myc in primary and uremic secondary hyperparathyroid tumors. *J Clin Endocrinol Metab.* 2007;92:338–44.

101. Akimaru H, Chen Y, Dai P et al. Drosophila CBP is a co-activator of cubitus interruptus in hedgehog signalling. *Nature.* 1997;6626:735–8.

102. Dai P, Akimaru H, Tanaka Y et al. Sonic Hedgehog-induced activation of the Gli1 promoter is mediated by GLI3. *J Biol Chem.* 1999;12:8143–52.

103. Yoon JW, Liu CZ, Yang JT et al. GLI activates transcription through a herpes simplex viral protein 16-like activation domain. *J Biol Chem.* 1998;6:3496–501.

104. Mosimann C, Hausmann G, Basler K. The role of Parafibromin/Hyrax as a nuclear Gli/Ci-interacting protein in Hedgehog target gene control. *Mech Dev.* 2009;5–6:394–405.

105. Campennì A, Ruggeri RM, Sindoni A et al. Parathyroid carcinoma presenting as normocalcemic hyperparathyroidism. *J Bone Miner Metab.* 2012;30:367–72.

106. Campennì A, Ruggeri RM, Sindoni A et al. Parathyroid carcinoma as a challenging diagnosis: Report of three cases. *Hormones (Athens).* 2012;11:368–76.

107. Harari A, Waring A, Fernandez-Ranvier G et al. Parathyroid carcinoma: A 43-year outcome and survival analysis. *J Clin Endocrinol Metab.* 2011;96:3679–86.

108. Sharretts JM, Kebebew E, Simonds WF. Parathyroid cancer. *Semin Onco.* 2010;37:580–90.

109. Russell RG, Mühlbauer RC, Bisaz S et al. The influence of pyrophosphate, condensed phosphates, phosphonates and other phosphate compounds on the dissolution of hydroxyapatite *in vitro* and on bone resorption induced by parathyroid hormone in tissue culture and in thyroparathyroidectomised rats. *Calcif Tissue Res.* 1970;3:183–96.

110. Drake MT, Clarke BL, Khosla S. Bisphosphonates: Mechanism of action and role in clinical practice. *Mayo Clin Proc.* 2008;83:1032–45.

111. Leere JS, Karmisholt J, Robaczyk M et al. Contemporary medical management of primary hyperparathyroidism: A systematic review. *Front Endocrinol (Lausanne).* 2017;8:79.

112. Marcocci C, Cetani F. Update on the use of cinacalcet in the management of primary hyperparathyroidism. *J Endocrinol Invest.* 2012;35:90–5.

113. Nemeth EF, Heaton WH, Miller M et al. Pharmacodynamics of the type II calcimimetic compound cinacalcet HCl. *J Pharmacol Exp Ther.* 2004;308:627–35.

114. Marotta V, Di Somma C, Rubino M et al. Potential role of cinacalcet hydrochloride in sporadic primary hyperparathyroidism without surgery indication. *Endocrine.* 2015;49:274–8.

115. Munson ND, Foote RL, Northcutt RC et al. Parathyroid carcinoma: Is there a role for adjuvant radiation therapy? *Cancer.* 2003;98:2378–84.

116. Mannstadt M, Bilezikian JP, Thakker RV et al. Hypoparathyroidism. *Nat Rev Dis Primers.* 2017;3:17055.

117. Tuerk EE, Schreiber J, Wegner M. Protein stability and domain topology determine the transcriptional activity of the mammalian glial cells missing homolog, GCMb. *J Biol Chem.* 2000;275:4774–82.

118. Bowl MR, Mirczuk SM, Grigorieva IV et al. Identification and characterization of novel parathyroid-specific transcription factor Glial Cells Missing Homolog B (GCMB) mutations in eight families with autosomal recessive hypoparathyroidism. *Hum Mol Genet.* 2010;19:2028–38.

119. Kanemura Y, Hiraga S, Arita N et al. Isolation and expression analysis of a novel human homologue of the Drosophila glial cells missing (gcm) gene. *FEBS Lett.* 1999;442:151–6.

120. Liu Z, Yu S, Manley NR. Gcm2 is required for the differentiation and survival of parathyroid precursor cells in the parathyroid/thymus primordia. *Dev Biol.* 2007;305:333–46.

121. Grigorieva IV, Thakker RV. Transcription factors in parathyroid development: Lessons from hypoparathyroid disorders. *Ann N Y Acad Sci.* 2011;1237:24–38.

122. Ivins S, Lammerts van Beuren K, Roberts C et al. Microarray analysis detects differentially expressed genes in the pharyngeal region of mice lacking Tbx1. *Dev Biol.* 2005;285:554–69.

123. Garg V, Yamagishi C, Hu T et al. Tbx1, a Di George syndrome candidate gene, is regulated by sonic hedgehog during pharyngeal arch development. *Dev Biol.* 2001;235:62–73.

124. Grigorieva IV, Mirczuk S, Gaynor KU et al. Gata3-deficient mice develop parathyroid abnormalities due to dysregulation of the parathyroid-specific transcription factor Gcm2. *J Clin Invest.* 2010;120:2144–55.

125. Su D, Ellis S, Napier A et al. Hoxa3 and pax1 regulate epithelial cell death and proliferation during thymus and parathyroid organogenesis. *Dev Biol.* 2001;236:316–29.

126. Günther T, Chen ZF, Kim J et al. Genetic ablation of parathyroid glands reveals another source of parathyroid hormone. *Nature.* 2000;406:199–203.

127. Mizobuchi M, Ritter CS, Krits I et al. Calcium-sensing receptor expression is regulated by glial cells missing-2 in human parathyroid cells. *J Bone Miner Res.* 2009;24:1173–9.

128. Canaff L, Zhou X, Mosesova I et al. Glial cells missing-2 (GCM2) transactivates the calcium-sensing receptor gene: Effect of a dominant-negative GCM2 mutant associated with autosomal dominant hypoparathyroidism. *Hum Mutat.* 2009;30:85–92.

129. Brown EM, MacLeod RJ. Extracellular calcium sensing and extracellular calcium signaling. *Physiol Rev.* 2001; 81:239–97.

130. Kebebew E, Peng M, Wong MG et al. GCMB gene, a master regulator of parathyroid gland development, expression, and regulation in hyperparathyroidism. *Surgery.* 2004;136:1261–6.

131. Granderath S, Stollewerk A, Greig S et al. loco encodes an RGS protein required for Drosophila glial differentiation. *Development.* 1999;126:1781–91.

132. Ding C, Buckingham B, Levine MA. Familial isolated hypoparathyroidism caused by a mutation in the gene for the transcription factor GCMB. *J Clin Invest.* 2001;108:1215–20.

133. Baumber L, Tufarelli C, Patel S et al. Identification of a novel mutation disrupting the DNA binding activity of GCM2 in autosomal recessive familial isolated hypoparathyroidism. *J Med Genet.* 2005;42:443–8.

134. Hundahl SA, Fleming ID, Fremgen AM, Menck HR. Two hundred eighty-six cases of parathyroid carcinoma treated in the U.S. between 1985–1995: A National Cancer Data Base Report. The American College of Surgeons Commission on Cancer and the American Cancer Society. *Cancer.* 1999;3:538–44.

135. Mitsui T, Narumi S, Inokuchi M et al. Comprehensive next-generation sequencing analyses of hypoparathyroidism: Identification of novel GCM2 mutations. *J Clin Endocrinol Metab.* 2014;99:E2421–8.

136. Doyle D, Kirwin SM, Sol-Church K et al. A novel mutation in the GCM2 gene causing severe congenital isolated hypoparathyroidism. *J Pediatr Endocrinol Metab.* 2012;25:741–6.

137. Mirczuk SM, Bowl MR, Nesbit MA et al. A missense glial cells missing homolog B (GCMB) mutation, Asn502His, causes autosomal dominant hypoparathyroidism. *J Clin Endocrinol Metab.* 2010;95:3512–6.

138. Yi HS, Eom YS, Park IeB et al. Identification and characterization of C106R, a novel mutation in the DNA-binding domain of GCMB, in a family with autosomal-dominant hypoparathyroidism. *Clin Endocrinol (Oxf).* 2012;76:625–33.

139. Akiyama Y, Hosoya T, Poole AM et al. The gcm-motif: A novel DNA-binding motif conserved in Drosophila and mammals. *Proc Natl Acad Sci USA.* 1996;93:14912–6.

140. Schreiber J, Sock E, Wegner M. The regulator of early gliogenesis glial cells missing is a transcription factor with a novel type of DNA-binding domain. *Proc Natl Acad Sci USA.* 1997;94:4739–44.

141. Altshuller Y, Copeland NG, Gilbert DJ et al. Gcm1, a mammalian homolog of Drosophila glial cells missing. *FEBS Lett.* 1996;393:201–4.

142. Schreiber J, Enderich J, Wegner M. Structural requirements for DNA binding of GCM proteins. *Nucleic Acids Res.* 1998;26:2337–43.

143. Correa P, Akerström G, Westin G. Underexpression of Gcm2, a master regulatory gene of parathyroid gland development, in adenomas of primary hyperparathyroidism. *Clin Endocrinol (Oxf).* 2002;57:501–5.

144. Mannstadt M, Holick E, Zhao W et al. Mutational analysis of GCMB, a parathyroid-specific transcription factor, in parathyroid adenoma of primary hyperparathyroidism. *J Endocrinol.* 2011;210:165–71.

145. D'Agruma L, Coco M, Guarnieri V et al. Increased prevalence of the GCM2 polymorphism, Y282D, in primary hyperparathyroidism: Analysis of three Italian cohorts. *J Clin Endocrinol Metab.* 2014;99:E2794–8.

146. Marchiori E, Pelizzo MR, Herten M et al. Specifying the molecular pattern of sporadic parathyroid tumorigenesis-The Y282D variant of the GCM2 gene. *Biomed Pharmacother.* 2017;92:843–8.

54

Familial Isolated Pituitary Adenoma

Dongyou Liu

CONTENTS

54.1 Introduction

Familial isolated pituitary adenoma (FIPA) is a recently defined autosomal dominant clinical entity that predisposes to pituitary adenoma (a benign tumor in the pituitary gland), and which differs clinically and genetically from previously recognized syndromic diseases (e.g., multiple endocrine neoplasia type 1 and Carney complex).

Germline heterozygous mutations in the *AIP* (aryl-hydrocarbon interacting protein) gene on chromosome 11q13.2 are found in 25%–30% of FIPA patients, who develop predominantly growth hormone (GH)-secreting adenomas (or somatotropinomas, which cause acromegaly), along with adrenocorticotropic hormone (ACTH)-secreting adenomas (which cause Cushing disease), thyroid hormone (TSH)-secreting thyrotropinoma, and prolactin (PRL)-secreting prolactinomas. Further, *AIP* mutations are also detected in >10% of patients under <30 years of age with sporadic macroadenomas, and in >20% of children with macroadenomas [1–3].

54.2 Biology

Pituitary adenomas are mostly benign neoplasms (>99%) of the anterior pituitary gland that impact on the host through local compression and altered patterns of hormone secretion. However, <0.2% of pituitary adenomas are malignant with the potential to metastasize.

According to the types of hormones produced, pituitary adenomas are separated into *lactotrophinomas* (prolactin [PRL] secreting; including densely granulated, sparsely granulated lactotrophinomas, acidophil stem cell adenoma; ~50% of cases), *gonadotrophinomas* (follicle-stimulating hormone [FSH] or luteinizing hormone [LH] secreting but predominantly nonfunctioning gonadotrophinomas; ~30% of cases), *somatotrophinomas* (GH secreting; including densely granulated, sparsely granulated somatotrophinomas; 15%–20% of cases), *corticotrophinomas* (ACTH secreting; including densely granulated, sparsely granulated corticotrophinomas, Crooke's cell adenomas; 5%–10% of cases), and *thyrotrophinomas* (TSH secreting; <1% of cases). However, some pituitary adenomas may secrete more than one hormone (so-called mixed or plurihormonal adenomas), and others may be hormone-negative (null cell) adenomas (13% of FIPA cases) [2].

Although 95% of pituitary adenomas arise sporadically without family connection, about 5% may occur as part of hereditary syndromes (e.g., multiple endocrine neoplasia type 1 [MEN1] or type 4 [MEN4], Carney complex), or as an isolated (non-syndromic) disorder (e.g., FIPA).

FIPA was first noted by Beckers in 1999 from families with pituitary adenoma and acromegaly in the absence of clinical and genetic features linked to previously recognized syndromes (e.g., MEN1 and Carney complex). The mapping of the susceptible gene region to chromosome 11q13.2 by Soares et al. in 2005 and identification of heterozygous germline mutations in the aryl-hydrocarbon interacting protein (*AIP*) gene by several groups in 2007 provided valuable insights into the molecular mechanisms of FIPA. Given that germline heterozygous *AIP* mutations are found in only 25%–30% of FIPA cases, further research is essential to decipher the genetic basis of non-*AIP*-related FIPA [1,4–6].

54.3 Pathogenesis

The *AIP* gene on chromosome 11q13.2 is a 8 kb DNA with six exons and encodes a 330-aa, 37.6 kDa protein (AIP, previously known as hepatitis B virus [HBV] X-associated protein [XAP2] or aryl hydrocarbon receptor [AhR]-associated protein [ARA9]), which functions as a receptor for aryl hydrocarbons and a ligand-activated transcription factor.

The AIP protein shares a structural homology with immunophilin proteins by having three tetratricopeptide repeats (TPR), conserved anti-parallel pair of α helices, and a final seventh α helix at the C-terminus, which mediates the protein–protein interactions. Forming part of a multiprotein complex in the cytoplasm, AIP moves to the nucleus upon binding to its ligand (e.g., AhR, which is a ligand-inducible transcription factor that modulates cellular responses to various xenobiotic toxins such as dioxins, and some endogenous compounds such as cAMP), where it regulates the expression of many xenobiotic metabolizing enzymes and helps prevent cells from uncontrolled growth and division [1].

Mutations in the *AIP* gene alter the structure of AIP or reduce the amount of functional AIP, hamper AIP interactions with AhR, phosphodiesterases (PDE) 4A5 and PDE2A3 as well as other proteins (e.g., survivin—an inhibitor of apoptosis; cardiac troponin interacting kinase 3 [TNNI3 K]; transmembrane receptors RET and EGFR; nuclear receptors such as estrogen receptor α [ERα], glucocorticoid receptor [GR], peroxisome proliferator-activated receptor α [PPARα], thyroid hormone receptor β1 [TRβ1]; actin; heat shock protein 90 [Hsp90]; heat shock cognate 70 [Hsc70]; hepatitis B virus X protein [HBV X]; Epstein Barr virus nuclear antigen 3 [EBNA3]), increase intracellular cAMP concentrations, and compromise the ability of AIP to control the growth and division of pituitary cells, leading to the development to pituitary adenomas [7].

Of the >70 *AIP* pathogenic variants (including nonsense, missense, frameshift, and splice site variants, intragenic deletions and insertions, and deletions of one or more exons) identified in FIPA families, nonsense or frameshift mutations, small or large deletions generate premature stop codons with a resultant truncated protein (21% of cases), while missense mutations affect structurally conserved amino acids of the TPR domains or the C-terminal α-helix and induce change in protein folding or loss of partner protein binding sites, resulting in unstable proteins, rapid degradation, and thus loss of function [3].

Most notable *AIP* pathogenic variants include c.40C>T (p.Gln14Ter), c.241C>T (p.Arg81Ter), c.721A>G (p.Lys241Glu), c.805_825dup (p.Phe269_His275dup), c.807C>T (p.F269F), c.811C>T (p.Arg271Trp), c.910C>T (p.Arg304Ter), and c.911G>A (p.Arg304Gln). Of these, several (e.g., p c.910C>T, c.911G>A, c.811C>T, c.807C>T, and c.721A>G) are recurrent/hotspot pathogenic variants involved in familial as well as simplex cases [3,8,9].

To date, *AIP* gene mutations have been identified in 25%–30% of FIPA patients, who often develop somatotropinomas in childhood, which are larger than FIPA tumors without *AIP* gene mutations. Further, *AIP* gene mutations are also detected in >10% of patients under <30 years of age with sporadic macroadenomas, and >20% of children with macroadenomas. For example, de novo *AIP* mutations (c.721A>T/p.Lys241*, and p.R304*) have been described in a few patients with corticotroph adenoma, but not in their parents or siblings [3].

Apart from the involvement of *AIP* in FIPA and sporadic pituitary adenomas, several other genes [e.g., *CDH23* (10q22.1), *GPR101* (Xq26.3), *GNAS1* (20q13.32), and *USP8* (15q21.2)] are also implicated in the development of sporadic pituitary adenomas. Some pituitary adenomas may result from duplication on the X chromosome comprising G protein-coupled receptor gene *GPR101* (Xq26.3) (so-called X-linked acrogigantism [XLAG]).

Located in a region containing the human deafness loci DFNB12 and USH1D on chromosome 10q22.1, the *CDH23* (cadherin-related 23) gene spans 419 kb and encodes a 3354 aa, 369 kDa protein (CDH23), which is a calcium-dependent cell–cell adhesion glycoprotein. Required for the maintenance of the proper organization of the stereocilia bundle of hair cells in the cochlea and the vestibule during late embryonic/early postnatal development, CDH23 constitutes part of the functional network formed by USH1C, USH1G, CDH23, and MYO7A that mediates mechanotransduction in cochlear hair cells. Mutations in the *CDH23* gene are observed in 12% of patients with sporadic pituitary adenomas. Interestingly, heterozygous and homozygous *CDH23* mutations appear to play a role in the development of Usher syndrome 1D and nonsyndromic autosomal recessive deafness DFNB12 [3,5,10].

54.4 Epidemiology

54.4.1 Prevalence

Pituitary adenomas have an estimated incidence of 4 cases per 100,000 (as high as 78–94 cases per 100,000 in the province of Liège, Belgium, and Banbury, UK), and make up 10%–15% of all intracranial tumors. About 95% of pituitary adenomas are sporadic (non-familial) neoplasms, and 5% may form part of hereditary syndromes (e.g., MEN1 or MEN4, Carney complex), or appear as an isolated (non-syndromic) disorder (e.g., FIPA). FIPA is rare, accounting for approximately 2% of pituitary adenomas. Of 211 FIPA-associated families reported in the literature, about 50 families and 50 simplex cases (i.e., a single occurrence in a family) harbor *AIP* pathogenic variants, with an onset age of 20–24 years (range 6–78 years). The average number of patients (onset age 18–24 years; male:female ratio of 2:1) with pituitary adenoma in *AIP* mutation positive families is 3–4 in comparison with 2–3 (onset age 40 years; male:female ratio of 1:1) in *AIP* mutation negative families [11,12].

54.4.2 Inheritance

FIPA is an autosomal dominant disorder, in which one copy of the altered *AIP* gene in each cell is sufficient to cause the disease.

54.4.3 Penetrance

FIPA demonstrates incomplete penetrance (10%–90%), with almost half of the *AIP* mutation positive patients showing no family history. This suggests possible existence of genetic or environmental modifying factors for the full disease to occur.

54.5 Clinical Features

Clinically, patients with *AIP*-related FIPA often develop somatotropinoma (growth hormone-secreting; 35%–53%), prolactinoma (prolactin-secreting; 22%–37.5%), nonfunctioning pituitary adenomas (NFPA; 14%), and rarely, ACTH-secreting (Cushing disease 2.9%–4.8%), TSH-secreting (0.2%), and FSH-secreting (0.2%) adenomas. Further, somatomammotropinoma (growth hormone- and prolactin-secreting adenoma, 3.2%), ACTH/PRL-secreting (0.1%), and FSH/TSH-secreting (0.1%) adenomas may be observed. The presence of these tumors may cause headaches and visual field loss (due to mass effect), acromegaly, Cushing disease, and other symptoms (due to altered hormone secretion) [2,3].

Acromegaly occurs in 80% of FIPA cases and shows coarse facial features, protruding jaw, and enlarged extremities. The occurrence of at least two cases of acromegaly or gigantism in a family without features of other endocrine syndromes is defined as familial isolated somatotropinoma (FIS), which tends to have onset about 4 to 10 years earlier than sporadic disease [2,3].

Cushing disease results from ACTH hypersecretion and excess cortisol secretion, and displays central obesity, moon facies, diabetes, "buffalo hump," hypertension, fatigue, easy bruising, depression, and reproductive disorders, in addition to increased morbidity and mortality as a result of cardiovascular or cerebrovascular disease and infections [2,3].

Patients with increased GH and/or IGF-1 secretion often have acromegaly and high risk for cardiovascular, cerebrovascular, rheumatologic/orthopedic, and metabolic complications. Patients with prolactinoma show increased plasma prolactin levels, which may induce amenorrhea/oligomenorrhea, galactorrhea, infertility, impotence, headaches, and visual disturbance [2,3,13].

54.6 Diagnosis

Diagnosis of FIPA involves family history review, laboratory evaluation of hormone levels, pituitary MRI, histopathology, and molecular identification of heterozygous *AIP* pathogenic variants (Figure 54.1) [12,14].

Histologically, GH-producing somatotropinoma has two major subtypes: densely and sparsely granulated. Densely granulated somatotroph adenoma (DGSA) is acidophilic tumor displaying diffuse and intense GH immunopositivity for low molecular weight keratin (LMWK), Cam5.2, in a diffuse perinuclear pattern. Sparsely granulated somatotroph adenoma (SGSA) is a chromophobic or mildly acidophilic tumor, showing scarce and weak GH immunopositivity, and its fibrous bodies, which correspond to accumulation of paranuclear cytokeratins, are characteristic for this subtype [2,3].

ACTH-secreting corticotrophinoma has three morphological subtypes: densely granulated (containing basophilic or amphophilic, PAS positive cells with strong immunopositivity for ACTH and LMWK/Cam5.2 in a perinuclear pattern; associated with Cushing disease and Nelson syndrome), sparsely granulated (immunopositivity for LMWK/Cam5.2 and variable for ACTH), and Crooke's cell adenoma (ACTH secreting adenoma, non-neoplastic corticotroph often with intracytoplasmic, perinuclear hyaline material, and strong immunostaining for LMWK/Cam5.2 in ring-like pattern around the nucleus) [2,3].

Differential diagnoses for FIPA include other causes of pituitary tumors, such as *multiple endocrine neoplasia type 1* (MEN1; presence of two endocrine tumors such as parathyroid, pituitary, or gastro-entero-pancreatic tract tumors; hypercalcemia by age 50 years; autosomal dominant disorder due to *MEN1* mutations), *MEN1-like syndrome* (or MEN4; due to mutation in *CDKN1B* [encoding cyclin dependent kinase inhibitor 1B]), *Carney complex* (skin pigmentary abnormalities, myxomas of the heart and other organs, endocrine tumors or overactivity, and schwannomas; primary pigmented nodular adrenocortical disease [PPNAD], which causes Cushing syndrome; large-cell calcifying Sertoli cell tumors [LCCSCTs]; multiple thyroid nodules; somatotroph cell hyperplasia; acromegaly in 10% of adults; autosomal dominant disorder due to mutation in *PRKAR1A*

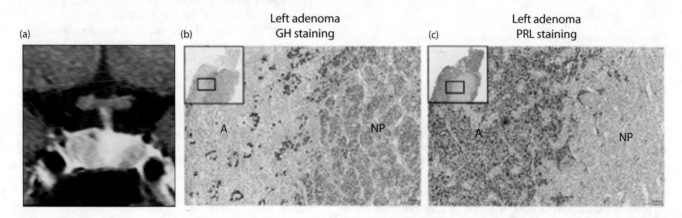

(a) (b) Left adenoma GH staining (c) Left adenoma PRL staining

A NP A NP

FIGURE 54.1 Familial isolated pituitary adenoma (FIPA) in a 27-year-old female showing (a) pituitary adenomas (left 4 mm, right 6 mm) on pituitary MRI; (b) scattered staining of adenoma tissue (A) compared to normal pituitary tissue (NP) from left adenoma by GH stain; (c) strong staining of adenoma tissue (A) compared to normal pituitary tissue (NP) from left adenoma by PRL stain. (Photo credit: Marques P et al. *Int J Endocrinol.* 2018;2018:8581626.)

[encoding protein kinase A1A regulatory subunit]), *McCune–Albright syndrome* (polyostotic fibrous dysplasia, café-au-lait patches, multiple endocrine disorders [e.g., multinodular goiters, multinodular adrenal hyperplasia, precocious puberty, and pituitary adenomas]; increased secretion of growth hormone and prolactin; due to somatic mosaicism resulting in gain-of-function pathogenic variants in *GNAS* [guanine nucleotide-binding protein G(s) subunit alpha isoforms short]), *sporadic pituitary tumors*, and *craniopharyngiomas* (mass effects on the normal pituitary and surrounding tissues leading to hormonal deficiencies) [2,3].

54.7 Treatment

Treatment options for pituitary adenomas consist of surgery, chemotherapy (e.g., somatostatin analogs, growth hormone receptor antagonists, and dopamine agonists), and/or radiotherapy, with the ultimate goals of achieving endocrinological remission and tumor size reduction [3,15].

54.7.1 Surgery

Large (>10 mm in diameter), aggressive/growing, or recurring pituitary adenomas associated with FIPA, especially *AIP*-related FIPA, require surgery and possibly radiotherapy. Microadenomas (<10 mm in diameter) with normal clinical and biochemistry findings are monitored closely.

54.7.2 Chemotherapy

Somatotropinomas containing germline *AIP* pathogenic variant often show poor response to somatostatin analogs and may be treated with growth hormone receptor antagonist. Prolactinomas respond well to dopamine agonist (e.g., cabergoline, bromocriptine; 70%–90% remission rate). NFPAs do not respond to traditional somatostatin analogs and are treated by surgery and radiotherapy if necessary.

54.7.3 Radiotherapy

Growing or recurring pituitary adenomas are treated with radiotherapy (conventional or radiosurgery).

54.8 Prognosis and Prevention

Prognosis of FIPA is influenced by tumor morphology, aggressiveness, and treatment responsiveness. FIPA containing AIP mutations are more aggressive and have less favorable prognosis than sporadic pituitary tumors. Further, *AIP* mutation-related pituitary adenomas (e.g., somatotropinomas/somatolactotropinomas or prolactinomas) have a poor response to chemotherapy and often require frequent reoperation and even radiotherapy [2,3].

Prenatal testing and preimplantation diagnosis may be considered if *AIP* pathogenic variant has been identified in the family.

54.9 Conclusion

Familial isolated pituitary adenoma (FIPA) is an autosomal dominant disorder characterized by the presence of pituitary tumors without other endocrine or associated abnormalities. Germline heterozygous mutations in the aryl hydrocarbon receptor interacting protein *(AIP)* gene occur in 25%–30% of FIPA patients, who may develop growth hormone-secreting somatotropinoma, prolactin-secreting prolactinoma, growth hormone and prolactin co-secreting somatomammotropinoma, nonfunctioning pituitary adenomas (NFPA), and rarely TSH- or ACTH-secreting thyrotropinoma and corticotropinoma [16,17]. The *AIP*-mutated FIPA is often larger (macroadenoma), shows more aggressive behavior, and has an earlier onset of disease (18–23 years) than non-*AIP*-related FIPA or sporadic pituitary adenomas. Clinical presentations of FIPA include headaches and visual field loss (due to mass effects) and other symptoms (due to excessive or insufficient hormone secretion). Diagnosis and treatment of FIPA rely on a multidisciplinary approach. As FIPAs are frequently resistant to somatostatin analog, they may require radiotherapy.

REFERENCES

1. Beckers A, Aaltonen LA, Daly AF, Karhu A. Familial isolated pituitary adenomas (FIPA) and the pituitary adenoma predisposition due to mutations in the aryl hydrocarbon receptor interacting protein (AIP) gene. *Endocr Rev.* 2013;34(2):239–77.
2. Stiles CE, Korbonits M. Familial isolated pituitary adenoma. In: De Groot LJ, Chrousos G, Dungan K et al. editors. *Endotext [Internet].* South Dartmouth (MA): MDText.com, Inc.; 2000. 2016 Nov 11.
3. Korbonits M, Kumar AV. AIP-related familial isolated pituitary adenomas. In: Adam MP, Ardinger HH, Pagon RA et al. editors. *GeneReviews® [Internet].* Seattle (WA): University of Washington, Seattle; 1993–2018. 2012 Jun 21.
4. Fukuoka H, Takahashi Y. The role of genetic and epigenetic changes in pituitary tumorigenesis. *Neurol Med Chir (Tokyo).* 2014;54(12):943–57.
5. Caimari F, Korbonits M. Novel genetic causes of pituitary adenomas. *Clin Cancer Res.* 2016;22(20):5030–42.
6. Wei ZQ, Li Y, Li WH, Lou JC, Zhang B. Research advances in pituitary adenoma and DNA methylation. *Zhongguo Yi Xue Ke Xue Yuan Xue Bao.* 2016;38(4):475–9.
7. Lines KE, Stevenson M, Thakker RV. Animal models of pituitary neoplasia. *Mol Cell Endocrinol.* 2016;421:68–81.
8. Demir H, Donner I, Kivipelto L et al. Mutation analysis of inhibitory guanine nucleotide binding protein alpha (GNAI) loci in young and familial pituitary adenomas. *PLOS ONE.* 2014;9(10):e109897.
9. Cansu GB, Taşkıran B, Trivellin G, Faucz FR, Stratakis CA. A novel truncating AIP mutation, p.W279*, in a familial isolated pituitary adenoma (FIPA) kindred. *Hormones (Athens).* 2016;15(3):441–4.
10. Zhou Y, Zhang X, Klibanski A. Genetic and epigenetic mutations of tumor suppressive genes in sporadic pituitary adenoma. *Mol Cell Endocrinol.* 2014;386(1-2):16–33.
11. Guaraldi F, Storr HL, Ghizzoni L, Ghigo E, Savage MO. Paediatric pituitary adenomas: A decade of change. *Horm Res Paediatr.* 2014;81(3):145–55.

12. Syro LV, Rotondo F, Ramirez A et al. Progress in the diagnosis and classification of pituitary adenomas. *Front Endocrinol (Lausanne).* 2015;6:97.

13. Vasilev V, Daly A, Naves L, Zacharieva S, Beckers A. Clinical and genetic aspects of familial isolated pituitary adenomas. *Clinics (Sao Paulo).* 2012;67(Suppl 1):37–41.

14. Marques P, Barry S, Ronaldson A et al. Emergence of pituitary adenoma in a child during surveillance: Clinical challenges and the family members' view in an *AIP* mutation-positive family. *Int J Endocrinol.* 2018;2018:8581626.

15. Oki Y. Medical management of functioning pituitary adenoma: An update. *Neurol Med Chir (Tokyo).* 2014;54(12): 958–65.

16. Tatsi C, Stratakis CA. The genetics of pituitary adenomas. *J Clin Med.* 2019;9:(1). pii: E30.

17. Daly AF, Cano DA, Venegas-Moreno E et al. AIP and MEN1 mutations and AIP immunohistochemistry in pituitary adenomas in a tertiary referral center. *Endocr Connect.* 2019;8(4):338–48.

55

Familial Non-Medullary Thyroid Carcinoma

Adonye Banigo and Iain Nixon

CONTENTS

55.1 Introduction

Carcinomas affecting the thyroid consist of three broad categories: non-medullary differentiated (main subtypes papillary and follicular), non-medullary undifferentiated (anaplastic), and medullary (Figure 55.1) [1,2].

Non-medullary differentiated thyroid carcinoma is more often simply referred to as differentiated thyroid carcinoma because it is much more common than anaplastic thyroid carcinoma [3] and it forms the bulk of the thyroid work for clinicians in endocrinology, oncology, and surgery, but for clarity in this chapter we will refer to it as non-medullary thyroid carcinoma (NMTC).

Although beyond the scope of this chapter, we will briefly discuss medullary thyroid carcinoma (MTC) as it is the type of the thyroid carcinoma classically recognized as having the potential for familial inheritance. Twenty-five percent of MTC are familial, associated with the multiple endocrine neoplasia-type 2 (MEN2) and multiple endocrine neoplasia type 3 (MEN3) syndromes (see Chapter 60 in this book) [2]. In the past, calcitonin and carcinogenic embryonic antigen (CEA) levels were used to screen family members, but now genetic screening is used, looking for recognized point mutations consistent with MTC, e.g. point mutations of the ret oncogene on chromosome 20 are found in patients with MEN2A.

55.1.1 Familial Non-Medullary Thyroid Carcinoma

In contrast to MTC, NMTC is increasingly common and can be divided into familial and non-familial. The terminology for defining and categorizing familial non-medullary thyroid carcinoma (FNMTC) in the literature is not strict and leads to confusion. Authors have defined FNMTC as non-medullary thyroid carcinoma occurring in one or two family members. The problem with this definition is that the probability of a sporadic event of thyroid carcinoma in two first-degree relatives is up to 70% [4]. This means that potentially 70% of patients diagnosed with FNMTC using these criteria are being subject to unnecessary further investigation and anxiety that they may be carrying a genetic disease, when this is not the case. However, the probability of a sporadic event in three first-degree relatives with NMTC is less than 5%. Therefore, the most appropriate clinical definition of FNMTC is NMTC occurring in three first-degree relatives [5]. Sporadic NMTC shows no recurring pattern in families. If two first-degree relatives were diagnosed with NMTC, then they are said to have a family history of NMTC as opposed to true FNMTC. FNMTC is further classified into syndromic and non-syndromic FNMTC

55.1.2 Syndromic FNMTC

This is defined as NMTC occurring in three first-degree relatives in the presence of another known syndrome. These are also known as familial tumor syndromes with a preponderance of non-thyroidal tumors, or familial thyroid carcinoma syndromes (Table 55.1). The criteria for three first-degree relatives are less important in syndromic FNMTC because of the existence of another known syndrome and the better characterization of its genetic landscape.

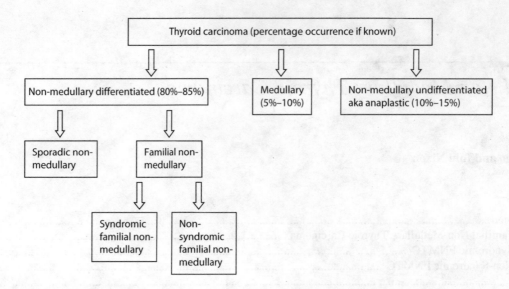

FIGURE 55.1 Thyroid carcinoma categories.

TABLE 55.1

Features of Familial Thyroid Carcinoma Syndromes

Familial Thyroid Carcinoma Syndrome	Features
FAP including Gardner syndrome (variant of FAP)	Female preponderance <30 years old [6]. 100-fold increased risk of papillary thyroid carcinoma (ref [1]), 160-fold in young women with FAP. Prevalence of PTC 2%–12% of patients. FAP's main feature is gastrointestinal tract lesions with high malignant transformation rates, whereas Gardner's is characterized by extra-intestinal malformations like osteomas, fibromas, and epithelial cysts [14].
PTEN hamartoma tumor syndrome (PHTS) aka Cowden	Multiple hamartomas and mucocutaneous lesions [14]. Thyroid disease affects over two-thirds of these patients and is the most common extracutaneous manifestation [6]. Thyroid disease could be benign (adenomas, thyroiditis) or malignant (carcinomas).
Carney syndrome	Skin and mucosal pigmentation. A range of non-endocrine and endocrine tumors including thyroid [6,14].
Werner syndrome	Premature aging (bilateral cataracts, gray hair, skin atrophy), benign and malignant thyroid tumors. [6,14]. Higher risk of anaplastic thyroid carcinomas.
DICER 1	Pleuropulmonary blastomas, ovarian tumors (Sertoli–Leydig cell tumors) and cystic nephromas. Although there is an association with multinodular goiter disease, the prevalence of thyroid cancer is low.
Peutz–Jeghers syndrome	Hamartomatous polyps, hyperpigmented macules on lips and oral mucosa (see Chapter 22).
McCune–Albright syndrome	Fibrous dysplasia, café-au-lait macules, hyperfunctioning endocrine disease (see Chapter 44).

TABLE 55.2

Genes with Potential FNMTC Association[a]

Gene Name	Gene Locus
SRGAP1	Chromosome 12q14
TITF-1/NKX2–1	Chromosome 14q13.3
FOXE1	Chromosome 9q22.33
HABP2 G534E variant	Chromosome 10q25.3
MNG1	Chromosome 14q31
TCO	Chromosome 19p13.2
fPTC/PRN	Chromosome 1q21
NMTC1	Chromosome 2q21
DICER1	Chromosome 14q32
FTEN	Chromosome 8p23.1-p22

[a] Summarized from Refs [5,6,14].

55.1.3 Non-Syndromic FNMTC

This is defined as NMTC occurring in three first-degree relatives in the absence of another known syndrome. Some literature refers to this category as familial tumor syndromes characterized by the preponderance of NMTC [6]. Non-syndromic FNMTC is a heterogenous entity with several genes identified with inconsistent associations with NMTC and poor validation in some cases (Table 55.2). Hence the clinical diagnostic criteria of three first-degree relatives should be strictly adhered to for correct diagnosis of syndromic FNMTC.

55.2 Biology

The thyroid gland lies in the lower anterior neck and consists of two lateral lobes joined by an isthmus which usually overlies the second and third tracheal rings. It contains many follicular and parafollicular cells (Figure 55.2). The former cells store the thyroid hormones thyroxine and triiodothyronine following their production from iodine. These hormones, often referred to as the major metabolic hormones, regulate processes in nearly every cell in the body. The latter cells secrete calcitonin and are thus

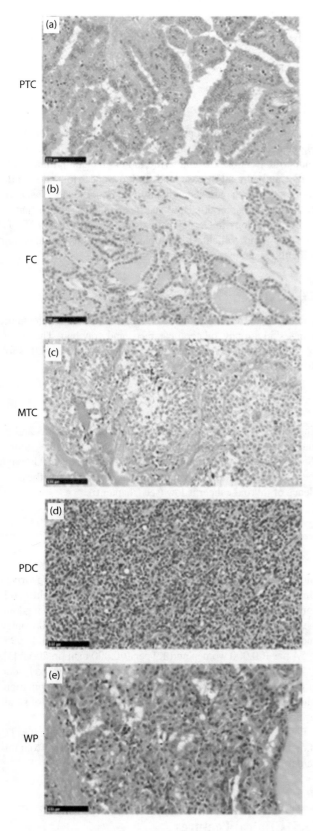

PTC

FC

MTC

PDC

WP

FIGURE 55.2 Histological features of thyroid tumors: (a) papillary thyroid carcinoma (PTC), (b) follicular carcinoma (FC), (c) medullary thyroid carcinoma (MTC), (d) poorly differentiated carcinoma (PDC), (e) well-differentiated tumor with uncertain malignant potential (WP). Scale bar: 100 μm. (Photo credit: Teshima M et al. *BMC Cancer.* 2019;19(1):245.)

also known as C cells. Thyroid carcinoma refers to cancers that originate from these cells, and it is the most common endocrine tumor [1]. There have been many terms used to describe and classify thyroid carcinomas, some of which are still in use today. For simplicity and clarity, thyroid carcinomas are often grouped into three broad categories (Figure 55.1) [2].

Non-medullary differentiated thyroid carcinoma arises from follicular cells; examples include papillary, follicular, and Hurthle cell carcinoma. Non-medullary undifferentiated thyroid carcinoma is synonymous with anaplastic thyroid carcinoma. MTC arises from the para-follicular cells (or C-cells), which produce the hormone calcitonin.

NMTC in a familial setting was first reported in identical 24-year-old twins in 1955 in Kansas City, USA [7]. The patients underwent total thyroidectomy and radical neck dissection, and their histological analysis showed foci of papillary thyroid carcinoma with lymph node metastases.

55.3 Pathogenesis

55.3.1 Non-Syndromic FNMTC

In contrast to syndromic FNMTC, the genetic background of non-syndromic FNMTC has not been definitively concluded as no distinct gene has yet been identified. Studies suggest an autosomal dominant behavior with incomplete penetrance and variable expressivity.

Some of the genetic anomalies have been associated with a specific histological type, e.g., papillary thyroid carcinoma with or without oxyphilia (chromosome 19p13.2), papillary thyroid carcinoma with papillary renal cell hyperplasia (chromosome 1q21), and multinodular goiter with papillary thyroid carcinoma (chromosome 14q32) [6]. Due to the heterogeneity of genetic alterations associated with non-syndromic FNMTC, no single genetic test is available to diagnose the condition. Below we discuss four susceptibility genes that have the most evidence for their role in non-syndromic FNMTC.

55.3.1.1 SRGAP1 Gene (12q14)

The SRGAP1 gene identified at the 12q14 locus regulates CDC42, which is a small G-protein in neurons that in turn affects cell mobility. CDC42 mediates multiple signaling pathways and acts as a signal transduction convergence point in signaling networks within cells. These functions enable it to play a role in tumorigenesis [8,9,10]. A genome-wide linkage analysis in 38 FNMTC families identified this gene with a 30% probability of linkage with FNMTC [10].

55.3.1.2 TITF-1/NKX2.1 Gene (14q13)

Thyroid transcription factor-1 (TITF-1) is the protein that activates the transcription of thyroglobulin, thyrotropin, and thyroperoxidase receptors. Targeted DNA sequencing of 653 participants (304 with papillary thyroid cancer and 349 controls) found the TITF-1/NKX2.1 mutation in 4 patients with papillary thyroid cancer [11]. The tumorigenesis capability of this gene mutation was confirmed when animal studies showed increased proliferation of thyroid cells in rats transfected with this mutation. Further validation studies are required to determine the clinical utility of testing for this gene.

55.3.1.3 FOXE1 Gene (9q22.33)

The FOXE1 gene (Forkhead box E1) is located at chromosome 9q22.33 and it encodes for the thyroid transcription factor-2, which regulates the gene expression of thyroglobulin and thyroperoxidase, and the migration of the thyroid from the pharynx to the neck [12]. Studies showed that mutant forms of FOXE1 promoted cell proliferation and migration, suggesting a role in tumor formation [13]. There are several inconsistent studies regarding FOXE1 gene, overall suggesting it might be a low-risk susceptibility gene for FNMTC [14].

55.3.1.4 HABP2 Gene (10q25.3)

The hyaluronan-binding protein 2 gene (HAPB2) is located on 10q25.3. Increased expression of this gene is consistent with FNMTC possibly because of a loss of tumor suppression [14]. Its evidence is conflicting. One study in 64 patients demonstrated a link between HABP2 gene and papillary thyroid carcinoma [15]. This is in contrast to another study in over 4000 patients which showed that cases and control patients had similar frequencies of HABP2 gene [16]. Its role in FNMTC remains to be validated.

55.3.2 Syndromic FNMTC

The conditions associated with syndromic FNMTC are discussed in more detail in other chapters of this book. An outline of the most common syndromic FNMTC conditions is provided here.

55.3.2.1 Familial Adenomatous Polyposis

This is an autosomal dominant disorder caused by mutations in the adenomatous polyposis coli (APC) gene on chromosome 5q21 leading to multitudes of adenomatous polyps in the colon that develop during early adulthood. There are numerous extra-colonic manifestations in familial adenomatous polyposis (FAP), papillary thyroid carcinoma being one of them. Historically, papillary thyroid carcinoma is said to occur in about 2% of FAP patients, but a recent study showed a prevalence of 12% [17]. For full manifestations of FAP, thyroid carcinomas are usually bilateral and multifocal with a cribriform-morular variant of papillary thyroid carcinoma on histology [6]. This histological appearance in a patient not known to have FAP should raise the possibility of this syndrome (see Chapter 16).

55.3.2.2 PTEN-Harmatoma Syndrome, aka Cowden Syndrome

This arises from mutations of the PTEN tumor suppressor gene on chromosome 10q23.3. Its features are manifested in affected patients by the age of 30 years including its pathognomic mucocutaneous lesions [18]. The International Cowden's Consortium CD Diagnosis Criteria/National Comprehensive Cancer Network set up major and minor diagnostic criteria. Follicular carcinoma is more common in these patients and a major diagnostic criterion, but they can be afflicted with a range of benign and malignant thyroid pathologies, including papillary carcinoma, adenomatous nodules, multinodular goiter, and thyroiditis, to name a few. Patients who have multicentric thyroid pathology, especially young patients with previous thyroiditis, should be tested for this syndrome (see Chapter 90) [6].

55.3.2.3 Carney Complex

This is an autosomal dominant disease with a diverse range of pigmented and skin and mucosal lesions, endocrine and non-endocrine tumors. The endocrine tumors include pituitary adenomas and adrenal disease, and so they appear to share similar components with patients with multiple endocrine neoplasia syndromes (see Chapter 60). However, the main difference is that the thyroid pathology in Carney patients is characterized by NMTC (papillary and follicular), multiple adenomatous nodules, and follicular adenomas (see Chapter 51) [6].

55.3.2.4 Werner Syndrome

This autosomal recessive connective tissue disease causes premature aging, skin atrophy, bilateral cataracts, and gray hair. This is one syndrome associated with a higher risk of anaplastic as well as follicular and papillary carcinomas (see Chapter 49).

55.3.2.5 DICER 1 Syndrome

The DICER1 gene is located on chromosome 14g32.13 and its mutations cause pleuropulmonary blastomas, ovarian tumors (Sertoli–Leydig cell tumors) and cystic nephromas. Although there is an association with multinodular goiter disease, the prevalence of thyroid cancer is low (see Chapter 6).

55.4 Epidemiology

Globally, the incidence of thyroid carcinoma is increasing, driven almost entirely by an increase in differentiated thyroid carcinoma, particularly the papillary subtype. In the United Kingdom, the annual incidence has increased from 2.3 per 100,000 women and 0.9 per 100,000 men (in the years between 1971 and 1995) to 5.1 per 100,000 women and 1.9 per 100,000 men (in 2008) [3]. This trend has been reported internationally and is likely due in large part to an increase in screening within the population, which is identifying a previously undiagnosed reservoir of small lesions of questionable clinical significance.

FNMTC accounts for around 5%–10% of patients with NMTC [4,6] and up to 7% of all thyroid carcinomas originating from follicular epithelial cells [19].

55.5 Clinical Features

FNMTC, like other thyroid tumors, tend to present as a solitary nodule either as a thyroid lump or increasingly as an incidental finding on scans performed for other reasons. For example, a hot spot in the thyroid on a PET scan is known to be associated with a 50% thyroid malignancy rate.

Although some groups have described a trend toward more aggressive features in FNMTC, not all data support this. Poor prognostic features like larger tumors, higher rates of extra-thyroidal extension, and higher rates of nodal metastases have been reported in some case series in sporadic versus FNMTC.

Some data suggest that FNMTC has an association with multicentric disease in younger patients with increased rates of both extra thyroidal extension and lymph node involvement with higher recurrence rates as a result [20,21]. However, this is not supported by other published data [22,23]. The difficulty with the data is the heterogeneity of the studies. The rarity of the diagnosis has led groups to use less strict criteria as a definition of FNMTC. In these studies, the definition was based on between one and three first-order relatives affected by disease.

As discussed earlier, several familial syndromes are recognized as having an increased frequency of thyroid carcinoma; FAP (aka Gardner syndrome), Cowden syndrome, Werner syndrome, and Carney complex type 1, to name a few. These represent a mixed group of diseases, but they do share some similar characteristics like young age and multicentric disease. Their non-thyroid disease may be their cause for initial presentation and so they may be referred from other medical and surgical specialties.

55.6 Diagnosis

Diagnosis of non-syndromic FNMTC is largely dependent on NMTC occurring in three first-degree relatives in the absence of other known associated syndromes, because there is no test for FNMTC. There is a role for preventive screening in patients, as discussed in Section 55.7. The diagnostic workup is the same as for patients with any thyroid mass: history, examination, ultrasound scanning (USS), and fine needle aspiration (FNA). Recent USS guidelines have helped to standardize USS grading of thyroid lumps from U1 to U5, with only U3 and above requiring FNA. However, for patients with suspected FNMTC, it is advisable to have a low threshold for FNA [24].

Clearly the first, and indeed second, case of non-syndromic FNMTC diagnosed within a single family group cannot be defined as familial in nature. Indeed, increased scrutiny of affected family members of patients diagnosed with PTC is likely to lead to higher than population levels of diagnosis of sporadic disease. For this reason, the definition of a further two affected first-degree relatives (rather than one) is favored to minimize the chance of falsely diagnosing sporadic as familial disease.

55.7 Treatment

Similar to sporadic NMTC, surgery is the mainstay for treatment of FNMTC. Patients undergo either a total thyroidectomy or a lobectomy. The topic of extent of treatment for differentiated thyroid cancer in general (familial and non-familial) is controversial. However, there are well-recognized principles in the management of non-familial NMTC. Small, low risk tumors can safely be managed with thyroid lobectomy alone. Higher risk disease, particularly with multinodular glands and metastatic nodal disease, are managed with total thyroidectomy. Elective neck dissection is generally not recommended unless the primary lesion is considered at high risk of being associated with occult nodal disease (T3/T4). How to translate these treatment recommendations to the familial setting is unclear. Whether the presence of a familial inheritance confers a high-risk nature to the entire gland is not clear.

As discussed earlier, FNMTC seem more likely to have multifocal disease, extra-thyroidal extension, and neck metastases. When these aggressive features are present, then the decision to proceed to a total thyroidectomy, appropriate neck dissection, and postoperative radio-iodine ablation is widely accepted, as would be the recommendation for a patient with sporadic NMTC with the same features. The decision is more challenging for patients with unilateral, otherwise low-risk disease. For sporadic patients, lobectomy would be advised, and it is unclear whether FNMTC should be considered high risk and therefore mandate total thyroidectomy. There is also a debate regarding elective neck surgery in patients with no evidence of nodal disease given that prophylactic central neck dissection may have a role in patients with adverse features. However, as the evidence stands, the consensus for FNMTC should probably follow that of sporadic NMTC which says that there is no role for prophylactic neck dissection in a node-negative neck. Decisions regarding adjuvant therapy (RAI and TSH suppression) should follow recommendations for sporadic disease [25].

55.8 Prognosis and Prevention

There is conflicting evidence as to whether FNMTC is a predictor of poor outcome in terms of recurrence rates and deaths. There is some evidence to suggest there is no difference in prognosis [22,23], but there appears to be a larger body of published evidence showing that FNMTC patients have higher rates of recurrence and metastasis, lower rates of cure, and higher rates of death [26,27]. However, in low-risk FNMTC which lacks aggressive features, survival and recurrence rates are excellent.

In terms of prevention, there is a potential role for screening. This would be directed at all remaining first-degree relatives once FNMTC has been diagnosed. Screening should take place with ultrasound examination. If the initial ultrasound is normal, then it should be repeated annually. It remains unclear whether screening does translate to improved clinical outcomes overall. Furthermore, as part of the screening process, false positives will lead to unnecessary surgery with its associated risks. It is also worth mentioning that, unlike in some variants of familial medullary thyroid cancer, there is no role for prophylactic thyroidectomy.

55.9 Conclusion

While syndromic FNMTC has been well researched with validated genes, non-syndromic FNMTC is currently poorly understood. Future research should focus on standardizing its definition to three first-degree relatives, on genetic identification with the development of newer genetic testing techniques, and refinement of treatment recommendations in these at-risk family groups.

REFERENCES

1. Ramsden J, Watkinson JC. Thyroid cancer. In: Gleeson M, Browning GC, Burton MJ et al. eds. *Scott Brown's Otorhinolaryngology, Head and Neck Surgery*. Vol. 2. 7th ed. Hodder Arnold; 2008. 2663–701.
2. British Thyroid Association guidelines for the management of thyroid cancer. *Clin Endocrinol*. 2014;81(S1).
3. Cancer Research UK. Thyroid cancer statistics. *Thyroid*. 2006;16:181–6. Accessed on 26/02/2018 at http://www.cancerresearchuk.org/health-professional/cancer-statistics/statistics-by-cancer-type/thyroid-cancer in multiply affected kindreds.
4. Charkes ND. On the prevalence of familial nonmedullary thyroid cancer.
5. Nixon IJ, Suarez C, Simo R et al. The impact of family history on non-medullary thyroid cancer. *Eur J Surg Oncol*. 2016;42(10):1455–63.
6. Nose V. Familial thyroid cancer: A review. *Mod Pathol*. 2011;24(S2):S19–33.
7. Robinson DW, Orr TG. Carcinoma of the thyroid and other diseases of the thyroid in identical twins. *AMA Arch Surg*. 1955;70:923–8.
8. Wong K, Ren XR, Huang YZ et al. Signal transduction in neuronal migration. *Cell*. 2001;107:209–11.
9. Etienne-Manneville S. CDC42- the centre of polarity. *J Cell Sci*. 2004;117:1291–300.
10. He H, Bronisz A, Liyanarachchi S et al. SRGAP1 is a candidate gene for papillary thyroid carcinoma suspectibility. *J Clin Endocrinol Metab*. 2013;98:E973–80.
11. Ngan ESW, Lang BHH, Liu T et al. A germline mutation (A339V) in thyroid transcription factor-1 (TITF-1/NKX2.1) in patients with multinodular goiter and papillary thyroid carcinoma. *J Natl Cancer Inst*. 2009;101:162–75.
12. De Felice M, Ovitt C, Biffali E et al. A mouse model for hereditary thyroid dysgenesis and cleft palate. *Nat Genet*. 1998;19:395–8.
13. Pereira J, da Silva J, Tomaz R et al. Identification of a novel germline FOXE1 variant in patients with familial non-medullary thyroid carcinoma (FNMTC). *Endocrine*. 2015;49:204–14.
14. Yang SP, Ngeow J. Familial non-medullary thyroid cancer: Unravelling the genetic maze. *Endocr Relat Cancer*. 2016;23:R577–95.
15. Zhang T, Xing M. HABP2 GS34E mutation in familial nonmedullary thyroid cancer. *J Natl Cancer Inst*. 2016;108;djv415.
16. Sahasrabudhe R, Stultz J, Williamson J et al. The HABP2 GS34E variant is an unlikely cause of familial non-medullary thyroid cancer. *J Clin Endocrinol Metab*. 2015;101:1098–103.
17. Herraiz M, Barbesino G, Faquin W et al. Prevalence of thyroid cancer in familial adenomatous polyposis syndrome and the role of screening ultrasound examinations. *Clin Gastroenterol Hepatol*. 2007;5(3):367–73.
18. Zbuk M, Eng C. Cancer phenomics: RET and PTEN as illustrative models. *Nat Rev Cancer*. 2007;7(1):35–45.
19. Mazeh H. Sippel RS. Familial nonmedullary thyroid carcinoma. *Thyroid*. 2013;23:1049–56.
20. Lei S, Wang D, Ge J et al. Single-center study of familial papillary thyroid cancer in China: Surgical considerations. *World J Surg Oncol*. 2015;13:115.
21. Wang X, Cheng W, Li J et al. Endocrine tumours: Familial nonmedullary thyroid carcinoma is a more aggressive disease: A systematic review and meta-analysis. *Eur J Endocrinol*. 2015;1172:R253–62.
22. Ito Y, Kakudo K, Hirokawa M et al. Biological behaviour and prognosis of familial papillary thyroid carcinoma. *Surgery*. 2009;145:100–5.
23. Rosario PW, Calsolari MR. Should a family history of papillary thyroid carcinoma indicate more aggressive therapy in patients with this tumour? *Arq Bras Endocrinol Metabol*. 2014;58:812–6.
24. Mitchell AL, Gandhi A, Coombes D, Perros P. Management of thyroid cancer: United Kingdom National Multidisciplinary Guidelines. *J Laryngol Otol*. 2016;130(S2):S150–60.
25. Haugen BR, Sawka AM, Alexander EK et al. American Thyroid Association guidelines on the management of thyroid nodules and differentiated thyroid cancer task force review and recommendation on the proposed renaming of encapsulated follicular variant papillary thyroid carcinoma without invasion to noninvasive follicular thyroid neoplasm with papillary-like nuclear features. *Thyroid*. 2017;27(4):481–3.
26. Mazeh H, Benavidez J, Poehls JL, Youngwirth L, Chen H, Sippel RS. In patients with thyroid cancer of follicular cell origin, a family history of nonmedullary thyroid cancer in one first-degree relative is associated with more aggressive disease. *Thyroid*. 2012;22:3–8.
27. Capezzone M, Marchisotta S, Cantara S et al. Familial non-medullary thyroid carcinoma displays the features of clinical anticipation suggestive of a distinct biological entity. *Endocr Relat Cancer*. 2008;15:1057–81.

56

Familial Pancreatic Cancer

Dongyou Liu

CONTENTS

56.1 Introduction

Familial pancreatic cancer (FPC) refers to a subset of pancreatic cancer (particularly pancreatic ductal adenocarcinoma [PDAC], which accounts for 90% of pancreatic cancer) affecting at least two first-degree relatives that does not meet the criteria of other hereditary cancer syndromes. Accounting for about 7% of pancreatic cancer cases, FPC differs from other inherited cancer syndromes (e.g., hereditary breast-ovarian cancer syndrome [HBOC], Peutz–Jeghers syndrome, familial atypical multiple mole melanoma [FAMMM], Lynch syndrome [hereditary nonpolyposis colorectal carcinoma or HNPCC], ataxia-telangiectasia, hereditary pancreatitis and familial adenomatous polyposis [FAP]; 3% of pancreatic cancer cases) by its absent syndromic features, limited genetic alterations (<20%), and modest familial aggregation (Table 56.1) [1,2].

56.2 Biology

The pancreas is a J-shaped, flattened, soft, lobulated organ of 12–15 cm in length located on the posterior abdominal wall behind the stomach. Comprising five macroscopic regions (head, uncinate process, neck, body, and tail), the pancreas is separated into two functional cellular compartments: exocrine and endocrine.

Making up >95% of the pancreatic parenchyma, the exocrine pancreas (serous gland) is composed of a million "berry-like" clusters of zymogenic cells (acini) connected by ductules, which join to form intralobular ducts, which then form interlobular ducts that drain into branches of the main pancreatic duct. The main pancreatic duct unites with the common bile duct, forming the hepatopancreatic ampulla of Vater, which opens into the duodenum and controls the flow of the bile and pancreatic fluid with a muscular valve (the sphincter of Oddi). Under the influence of secretin and cholecystokinin, the zymogenic cells (acini) secrete trypsin (digesting proteins), lipase (digesting fats), amylase (digesting carbohydrates), and other enzymes, while ductular cells generate bicarbonate to make the pancreatic fluid alkaline.

Representing about 2% of pancreatic parenchyma, the endocrine pancreas consists of pancreatic islets (clusters) (or islets of Langerhans) scattered throughout the pancreas. The islets include several cell types (α, β, δ, A, B, C, D, E, F), of which α cells (20% of islets) secrete glucagon; β cells (70% of islets) secrete insulin and amylin; δ cells (<10% of islets) secrete somatostatin; A cells (and also D cells) secrete adrenocorticotropin (ACTH), serotonin (5-HT), melanocyte-stimulating hormone (MSH), and vasoactive intestinal peptide (VIP); D cells secrete gastrin; E cells (or epsilon cells, <1% of islets) secrete ghrelin; and F cells (or PP/gamma cells, <5% of islets) secrete pancreatic polypeptide. The secreted hormones are transferred via the blood circulation to control energy metabolism and storage throughout the body.

Although both benign (e.g., serous cystadenoma [SCA], intraductal papillary-mucinous neoplasm [IPMN], pancreatic intraepithelial neoplasia [PanIN], mucinous cystic neoplasm [MCN]) and malignant tumors affect the pancreas, the former are overshadowed by the latter with their capacity to induce severe clinical diseases and to confer poor prognosis.

Among malignant tumors of the pancreas, 95% occur in the exocrine pancreas, and include pancreatic ductal adenocarcinoma

TABLE 56.1

Key Features of Inherited Cancer Syndromes Related to Pancreatic Cancer

Inherited Syndrome	Causative Gene	Chromosome Location	Lifetime Risk of PC	FPC Patients with Mutation	Other Associated Cancers	Other Syndromic Features
Hereditary pancreatitis	PRSS1	7q34	40% (70 years)		None	None
	SPINK1	5q32	53% (75 years)			
Peutz–Jeghers syndrome (PJS)	STK11/LKB1	19p13.3	36% (70 years)		Esophagus, stomach, small intestine, colon, breast, lung, ovary, uterus	Gastrointestinal hamartomatous; perioral and buccal hyperpigmented macules
Familial atypical multiple mole melanoma (FAMMM)	CDKN2A (MTS1)	9p21.3	17% (75 years)	2.5%	Melanoma	Multiple melanocytic nevi (often >50)
Hereditary breast-ovarian cancer syndrome (HBOC)	BRCA1	17q21.31	2%–7% (70 years)	1.2% 3.7%	Breast, ovary, prostate, melanoma	
	BRCA2	13q13.1				
Lynch syndrome (HNPCC)	MLH1	3p22.2	3.7% (70 years)	0.5%–1.9%	Colon, endometrium, ovary, stomach, small intestine, urinary tract, brain, cutaneous sebaceous glands	
	MSH2	2p21				
	MSH6	2p16.3				
	PMS2	7p22.1				
Familial adenomatous polyposis (FAP)	APC	5q22.2	1.7% (80 years)		Colon, medulloblastoma, papillary thyroid carcinoma, hepatoblastoma, desmoid tumor	Extra, missing, or unerupted teeth; congenital hypertrophy of the retinal pigment epithelium
Familial pancreatic cancer (FPC)	KRAS	12p12.1	8%–12% (2 first-degree relatives)	<20%		None
	CDKN2A	9p21.3				
	TP53	17p13.1	16%–38% (>3 first-degree relatives)			
	SMAD4	18q21.2				

(PDAC, arising from the ductal cells of the exocrine pancreas, particularly in the head region; 90%), acinar cell carcinoma (originating from acinar cells; 1%–2%), pancreatoblastoma (originating from acinar cells; <1%), and solid-pseudopapillary neoplasm (1%–2%). Due to its clinical dominance, exocrine pancreatic cancer, specifically PDAC, is commonly referred to as pancreatic cancer.

In contrast, malignant tumors affecting the endocrine pancreas constitute 5% of pancreatic malignancies, which are predominated by pancreatic neuroendocrine tumors (PanNET) or islet cell tumors (e.g., glucagonoma [arising from glucagon-producing α cells], insulinoma [arising from insulin-producing β cells], gastrinoma [arising from gastrin-producing D cells], somatostatinoma, VIPoma, and carcinoid [arising from serotonin-producing cells]).

While 90% of pancreatic cancer may result from sporadic events, about 10% are attributed to hereditary factors, including inherited cancer syndromes (3%) and FPC (7%).

It appears that infiltrating ductal carcinoma of the pancreas evolves from three different types of noninvasive precursor lesions (i.e., intraductal papillary mucinous neoplasms [IPMN], mucinous cystic neoplasms [MCN], and pancreatic intraepithelial neoplasia [PanIN]). IPMN is a macroscopic noninvasive mucin-producing epithelial lesion containing long finger-like papillae. MCN is a macroscopic mucin-producing cyst-forming epithelial lesion with a peculiar stromal tissue. PanIN is a microscopic non-invasive small epithelial neoplasm in the smaller pancreatic ducts,

with cytological and architectural atypia. Based on the degree of atypia in the ductal epithelium, PanIN is separated into stages PanIN1a (flat), PanIN1b (papillary without dysplasia), PanIN2 (papillary with dysplastic changes), and PanIN3 (carcinoma *in situ*, which is frequently associated with invasive cancer).

More specifically, histological subtype PDAC arises from the ductal cells and progresses from low grade PanIN 1a, 1b, and 2, to high grade PanIN3, to eventual invasive PDAC (Figures 56.1 and 56.2). Molecular mechanisms underlying the development of PDAC relate to the accumulation of *KRAS* mutation in PanIN1, *CDKN2A* mutation in PanIN2, *TP53* mutation in PanIN3, and *SMAD4* mutation in invasive PDAC. Alternatively, acquisition of *KRAS* mutation in PanIN1 is followed by mitotic errors that create genome instability/copy number alterations enabling rapid acquisition of *CDKN2A*, *TP53*, and *SMAD4* mutations in invasive PDAC. The whole process may take a decade or longer to complete. Available evidence suggests that about 40% of PDAC cases have undergone the first route, and about 60% of PDAC cases have taken the second route (Figure 56.1) [3–5].

56.3 Pathogenesis

Most (90%) of pancreatic cancer appear to arise from de novo mutations; about 10% are tracked to hereditary factors such as inherited cancer syndromes (3%) and FPC (7%) [6].

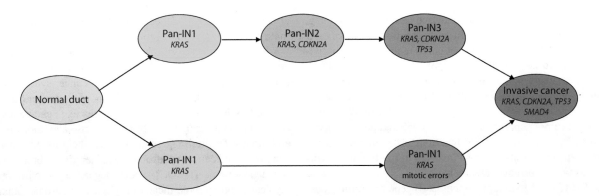

FIGURE 56.1 Development of pancreatic ductal adenocarcinoma (PDAC) involves either progressive accumulation of *KRAS*, *CDKN2A*, and *TP53* mutations at different stages of pancreatic intraepithelial neoplasms (PanIN) and *SMAD4* mutation in invasive cancer, or acquisition of *KRAS* mutation followed by mitotic errors that create genomic instability/copy number alterations enabling rapid acquisition of *CDKN2A*, *TP53*, and *SMAD4* mutations in invasive cancer. Available evidence indicates that the former accounts for about 40% of PDAC cases, and the latter is responsible for about 60% of PDAC cases.

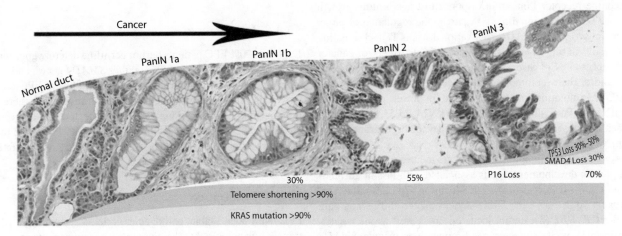

FIGURE 56.2 Morphological changes from normal duct, low grade pancreatic intraepithelial neoplasia (PanIN) 1a, 1b, and 2, to high grade PanIN3 before development of invasive pancreatic ductal adenocarcinoma (PDAC). (Photo credit: Hackeng WM et al. *Diagn Pathol*. 2016;11:47.)

Among inherited cancer syndromes that include pancreatic cancer as part of clinical features, hereditary pancreatitis (*PRSS1*), Peutz−Jeghers syndrome (*STK11*), familial atypical multiple mole melanoma (*CDKN2A*), hereditary breast-ovarian cancer syndrome (*BRCA1*, *BRCA2*), Lynch syndrome (*MLH1*, *MSH2*, *MLH6*, *PMS2*), and familial adenomatous polyposis (*APC*) are most notable, with lifetime risk of pancreatic cancer as high as 40% (Table 56.1) [4,7].

FPC, in particular familial PDAC, appears to share the same driver mutations in the oncogene *KRAS* and tumor suppressor genes *CDKN2A*, *TP53*, and *SMAD4*, which are detected in <20% of FPC samples, but in a much higher proportion of pancreatic cancer samples. A number of other genes (e.g., *MLL3*, *TGFBR2*, *ARID1A*, *SF3B1*, *EPC1*, *ARID2*, *ZIM2*, *MAP2K4*, *NALCN*, *SLC16A4*, *MAGE/A6*, *RNF43*, *GNAS*, *RREB1*, and *PBRM1*) may be also implicated in the development of pancreatic cancer [5,8−11].

The *KRAS* (Kirsten ras oncogene homolog) gene on chromosome 12p12.1 spans 46 kb and encodes a 189 aa, 21.6 kDa protein (KRAS) that is a member of the small GTPase superfamily. KRAS bind GDP/GTP and participates in the regulation of cell proliferation. Mutations in the *KRAS* gene alter the encoded protein (commonly at G12, G13, and Q61) and contribute to uncontrolled cell proliferation. *KRAS* pathogenic variants occur in >90% of pancreatic cancer. In 5% of pancreatic cancer without *KRAS* mutation, *BRAF* mutation (1.5%) is sometimes present. Besides pancreatic cancer, KRAS mutations are also involved in Noonan syndrome and cardiofaciocutaneous syndrome.

The *CDKN2A* (cyclin-dependent kinase inhibitor 2A) gene on chromosome 9p21.3 is a 27.5 kb DNA and encodes a 156 aa, 16 kDa protein (p16), that functions as a negative regulator of the G_1-to-S transition in the cell cycle through strong interaction with CDK4 and CDK6, and also as a tumor suppressor through interaction with, and sequestration of the E3 ubiquitin-protein ligase MDM2 in the nucleolus, which induces degradation of tumor suppressor protein p53, and enhances p53-dependent transactivation and apoptosis. Mutation/deletion (or hypermethylation of the promoter region) in the *CDKN2A* gene inactivates p16, and results in stimulation of the proliferative activity in >90% of pancreatic cancer. Some pancreatic cancer with intact *CDKN2A* may have somatic mutations in other cell-cycle regulators (e.g., *FBXW7* or *ANAPC2*) [12]. *CDKN2A* mutations are also observed in cutaneous malignant melanoma, and melanoma-pancreatic cancer syndrome.

The *TP53* (tumor protein 53) gene on chromosome 17p13.1 measures 25.7 kb in length and encodes 393 aa, 43.6 kDa protein (p53) that consists of transcriptional activation, DNA binding, and oligomerization domains. In response to diverse cellular stresses, p53 controls a set of target genes required for cell division, leading to cell cycle arrest, apoptosis, senescence, DNA repair, or changes in metabolism. Mutations (e.g., frameshift, nonsense, or premature stop codons affecting the DNA binding domain at residues 175, 245, 248, 273, and 282) in *TP53* gene deprive the tumor suppressor function of p53 and permit uncontrolled cell proliferation. *TP53* mutations occur in nearly 85% of pancreatic cancer and are also involved in Li−Fraumeni syndrome, as well as osteogenic sarcoma.

The *SMAD4* (SMAD family member 4) gene on chromosome 18q21.2 spans 56.6 kb and encodes a 552 aa, 60 kDa protein (SMAD4) that is a member of the Smad family of signal transduction proteins. SMAD4 forms homomeric complexes and heteromeric complexes with other activated Smad proteins, which, upon phosphorylation and activation by transmembrane serine-threonine receptor kinases in response to transforming growth factor (TGF)-beta signaling, take part in the regulation of target gene transcription. *SMAD4* inactivation disrupts TGF-β signaling and results in tumorigenesis and metastatic recurrence of pancreatic cancer (60% of cases) as well as juvenile polyposis syndrome.

Besides their common occurrence in PDAC, KRAS and *CDKN2A* mutations are often found in lower grade lesions PanIN1 and PanIN2, while *KRAS, CDKN2A, TP53,* and *SMAD4* genes are observed with high frequency in high grade lesion PanIN3 [13].

Acinar cell carcinoma (ACC) appears to have a slightly different genetic development, and is associated with loss of chromosome 11p (~50%), loss of the *TP53* locus on 17p (25%), the *APC* locus on 5q21 (50%), the *SMAD4* locus on 18q (60%), and gain of the *CTNNB1* (β-catenin) locus on 3p as well as microsatellite instability (10%). Gene-specific mutations include *SMAD4* (25%), *JAK1* (20%), *BRAF* (13%), *RB1* (13%), *TP53* (13%), *APC* (9%), *ARID1A* (9%), *GNAS* (9%), *MLL3* (9%), *PTEN* (9%), *ATM* (4%), *BAP1* (4%), *BRCA2* (4%), *PALB2* (4%), *MEN1* (4%), and *RNF43* (4%) [13,14].

PanNET mostly occur sporadically (90%), and some are associated with inherited familial syndromes (e.g., multiple endocrine neoplasia type 1 [MEN1], von Hippel−Lindau syndrome [VHL], neurofibromatosis type 1 [NF1], tuberous sclerosis complex [TSC], and glucagon cell adenomatosis GCA]). Sporadic PanNET often show loss of heterozygosity at the *MEN1* locus (20%–45%) or somatic mutations in the *MEN1* gene (45%), somatic inactivating mutations in *ATRX* or *DAXX* (45%), and mutations in mTOR pathway genes (*TSC1/2*) (15%) [13,15].

56.4 Epidemiology

56.4.1 Prevalence

Representing the fourth most common cause (after breast cancer, prostate cancer, and colorectal cancer) of cancer-related deaths (41,000 in 2015) in the United States and the twelfth most common cancer related mortality (227,000 in 2015) worldwide, pancreatic cancer has an estimated incidence of 8–12 per 100,000 (14.8 per 100,000 in black males; 68 per 100,000 in people of >55 years; 8.8 per 100,000 in the general population). Mainly

affecting people of >45 years, pancreatic cancer shows a slight male bias (male to female ratio of 1.3:1) [16,17].

Although a vast majority (90%) of pancreatic cancers are due to de novo mutations, about 10% are associated with inheritance of pathogenic variants, including inherited cancer syndromes (3%) and FPC (7%). Making up 7% of all pancreatic cancer cases, FPC is noted by its absent syndromic features (e.g., cutaneous, skeletal and growth anomalies), limited genetic alterations (<20%), and modest familial aggregation, which help its differentiation from inherited cancer syndromes (e.g., hereditary breast-ovarian cancer syndrome [HBOC], Peutz−Jeghers syndrome, familial atypical multiple mole melanoma, Lynch syndrome, ataxia-telangiectasia, hereditary pancreatitis, and FAP) that includes pancreatic cancer as part of syndromic features [1,2]. In comparison with sporadic pancreatic cancer, FPC has an early onset (58–68 years vs 61–74 years in sporadic cases), affects more Ashkenazi Jews than Caucasians, and has worse prognosis.

56.4.2 Inheritance

Less than 20% FPC patients harbor germline heterozygous mutations in *KRAS, CDKN2A, TP53,* and/or *SMAD4* genes, and show autosomal dominant inheritance. However, 80% of FPC do not have identifiable culprit genes and their transmission therefore remains unknown. Individuals with two affected first-degree relatives have a 6.4-fold greater lifetime risk (8%–12%), and those with three first-degree relatives have a 32-fold greater lifetime risk (40%) of developing pancreatic cancer [18].

56.4.3 Penetrance

FPC with identifiable pathogenic variants shows high penetrance. It is notable that the age of death from pancreatic cancer is 10 years younger with each generation (a phenomenon known as anticipation).

56.5 Clinical Features

FPC typically manifests as pancreatic cancer in the absence of chronic pancreatitis, other types of cancer, and clinical features (e.g., cutaneous, skeletal, and growth anomalies) that are associated with inherited cancer syndromes (e.g., hereditary breast-ovarian cancer syndrome, Peutz−Jeghers syndrome, familial atypical multiple mole melanoma, Lynch syndrome, ataxia-telangiectasia, hereditary pancreatitis, and FAP). However, in rare FPC cases, other tumors (e.g., melanoma, endometrial cancer, breast cancer, ovarian cancer, and bile duct cancer) may be occasionally observed [19].

Early-stage pancreatic cancer is usually asymptomatic, although late-stage pancreatic cancer may display jaundice, light-colored stools, dark urine, pain in the upper or middle abdomen and back, weight loss, loss of appetite, and fatigue.

56.6 Diagnosis

Diagnosis of FPC involves family/personal history review (for pancreatitis), physical examination, laboratory evaluation, imaging study, histopathology, and molecular analysis [20,21].

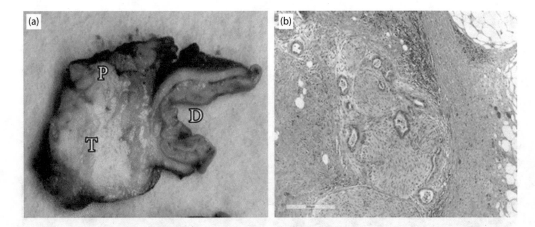

FIGURE 56.3 (a) Macroscopic appearance of pancreatic ductal adenocarcinoma (PDAC) showing a poorly demarcated, firm, white tumor in the pancreatic parenchyma (T, tumor; P, pancreatic parenchyma; D, duodenum). (b) Histopathology of PDAC revealing haphazardly arranged infiltrating glandular and ductal structures, as well as perineural invasion. (Photo credit: Hackeng WM et al. *Diagn Pathol.* 2016;11:47.)

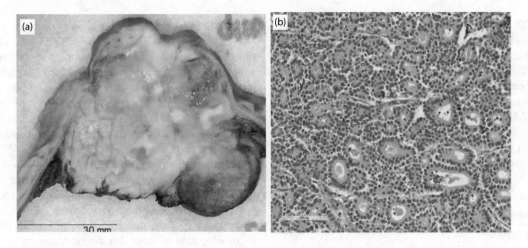

FIGURE 56.4 (a) Macroscopic appearance of acinar cell carcinoma (ACC). (b) Histopathology of ACC revealing tumor cells with granular cytoplasm and round to oval uniform nuclei forming small acinar structures. (Photo credit: Hackeng WM et al. *Diagn Pathol.* 2016;11:47.)

Pancreatic ductal adenocarcinoma (PDAC) is a firm, poorly demarcated, white-yellow mass with the adjacent non-neoplastic pancreatic parenchyma being atrophic and fibrotic, and the main pancreatic duct dilated. The tumor contains haphazardly arranged infiltrating glandular and ductal structures, abundant desmoplastic stroma, eosinophilic to clear cytoplasm, enlarged pleomorphic nuclei, along with perineural, lymphatic, and blood vessel invasion. Immunohistochemically, it is positive for mucin (MUC1, MUC3, MUC4, MUC5AC) and glycoprotein tumor antigens (e.g., CA19-9), with aberrant TP53 expression or SMAD4 loss (Figure 56.3) [3,22].

ACC is a relatively soft and well-circumscribed tumor and often displays seemingly normal exocrine pancreatic cells with enlarged uniform nuclei, prominent nucleoli, and finely granular eosinophilic cytoplasm. Immunohistochemically, it is positive for pancreatic exocrine enzymes (e.g., trypsin, chymotrypsin, and lipase), BCL10, and monoclonal antibody 2P-1-2-1 (Figure 56.4) [3].

PanNET is a soft, red or white, well-demarcated lesion. The neoplastic cells possess granular amphophilic to eosinophilic cytoplasm and typical coarsely clumped "salt-and-pepper" chromatin

(distinct neuroendocrine morphology). Immunohistochemically, it is positive for neuroendocrine markers (synaptophysin, chromogranin A), and peptide hormones (e.g., insulin, glucagon) (Figure 56.5) [3].

56.7 Treatment

Treatment options for pancreatic cancer consist of surgery, chemotherapy, and/or radiotherapy [23].

Surgical resection is curative for small, localized exocrine pancreatic cancer (in 10%–20% of patients), but not for unresectable, metastatic, or recurrent disease.

Use of systemic preoperative non-adjuvant chemotherapy alone or in combination with radiotherapy for 3–6 months before undergoing surgery helps reduce the size of the tumor, increase the likelihood of negative resection margins, and tests the effects of cytotoxic medications *in vivo*. Following surgery, adjuvant chemotherapy with gemcitabine (also known as dFdC: 2′,2′-difluorodeoxycytidine, which remains a cornerstone treatment for all stages of PDAC) or 5-fluorouracil is recommended.

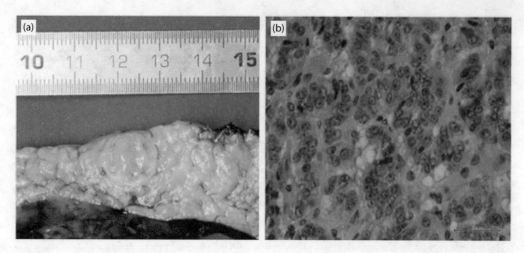

FIGURE 56.5 (a) Macroscopic appearance of pancreatic neuroendocrine tumor (PanNET) showing a well-demarcated pinkish tumor surrounded by normal pancreatic parenchyma. (b) Histopathology of PanNET revealing typical salt-and-pepper chromatin (b). (Photo credit: Hackeng WM et al. *Diagn Pathol*. 2016;11:47.)

FOLFIRINOX (fluorouracil, folic acid, irinotecan, and oxaliplatin; a platinum-based chemotherapy) and protein-bound paclitaxel (nab-paclitaxel) chemotherapy are more effective than gemcitabine but appear to cause significant side effects. Tumors harboring somatic or germline pathogenic variants in genes related to DNA double strand damage repair (e.g., *BRCA1*, *BRCA2*, *PALB2*, or *ATM*) show better responses to platinum-based chemotherapy [24–29].

Application of palliative measures relieves the symptoms of pancreatic cancer and improves quality of life but not overall survival.

56.8 Prognosis

Pancreatic cancer is a very aggressive neoplasm with poor prognosis. Patients with pancreatic ductal adenocarcinoma (PDAC) have a median survival time of 6 months after diagnosis, and 1- and 5-year survival rates of 29% and 7%, respectively. One reason for this is the tendency of PDAC to quickly spread and metastasize to distant organs, and tumors in most diagnosed patients are surgically unresectable. Patients with ACC also have distant metastasis (60%, similar to PDAC) and overall 5-year survival rate of 45%. Patients with PanNET have an overall 5-year survival rate of 42% [28].

56.9 Conclusion

Pancreatic cancer is a comparatively rare but most devastating malignancy in the world. Although 90% of pancreatic cancer arise from de novo mutations, about 10% are hereditary, involving inherited cancer syndromes (3%) and FPC (FPC, 7%). Affecting at least two first-degree relatives, FPC differs from other inherited cancer syndromes by its lack of syndromic features, limited genetic alterations (<20%), and modest familial aggregation. The most frequently mutated genes in FPC include oncogene *KRAS*, and tumor suppressor genes *CDKN2A*, *TP53*, and *SMAD4*, which are detected in <20% of FPC samples, but

60%–95% of pancreatic cancer samples. It appears that progressive acquisition of *KRAS*, *CDKN2A*, *TP53*, and *SMAD4* mutations underlines the evolution from precursor lesions to invasive pancreatic ductal adenocarcinoma (PDAC), which is the most common histological subtype and accounts for 90% of pancreatic cancer alone. As molecular testing provides confirmation for 1/5 of cases only, current diagnosis of FPC relies heavily on clinical criteria, which result in many patients with surgically unresectable lesions upon diagnosis [30]. Management of FPC involves a combination of surgery, chemotherapy, and radiotherapy. Given its relatively low prevalence, large scale screening for pancreatic cancer is economically infeasible [13]. Targeted testing of high-risk populations is a preferred approach for screening. Further understanding of the molecular mechanisms of FPC is crucial for development of novel therapeutic measures for reducing the mortality of this deadly neoplasm [31].

REFERENCES

1. Carrera S, Sancho A, Azkona E, Azkuna J, Lopez-Vivanco G. Hereditary pancreatic cancer: Related syndromes and clinical perspective. *Hered Cancer Clin Pract*. 2017;15:9.
2. Chen F, Roberts NJ, Klein AP. Inherited pancreatic cancer. *Chin Clin Oncol*. 2017;6(6):58.
3. Hackeng WM, Hruban RH, Offerhaus GJ, Brosens LA. Surgical and molecular pathology of pancreatic neoplasms. *Diagn Pathol*. 2016;11(1):47.
4. Pelosi E, Castelli G, Testa U. Pancreatic cancer: Molecular characterization, clonal evolution and cancer stem cells. *Biomedicines*. 2017;5(4). pii: E65.
5. Waters AM, Der CJ. KRAS: The critical driver and therapeutic target for pancreatic cancer. *Cold Spring Harb Perspect Med*. 2018;8(9). pii: a031435.
6. Ghiorzo P. Genetic predisposition to pancreatic cancer. *World J Gastroenterol*. 2014;20(31):10778–89.
7. Underhill ML, Germansky KA, Yurgelun MB. Advances in hereditary colorectal and pancreatic cancers. *Clin Ther*. 2016;38(7):1600–21.
8. Rustgi AK. Familial pancreatic cancer: Genetic advances. *Genes Dev*. 2014;28(1):1–7.

9. Amundadottir LT. Pancreatic cancer genetics. *Int J Biol Sci.* 2016;12(3):314–25.

10. Sikdar N, Saha G, Dutta A, Ghosh S, Shrikhande SV, Banerjee S. Genetic alterations of periampullary and pancreatic ductal adenocarcinoma: An overview. *Curr Genomics.* 2018;19(6):444–63.

11. Natale F, Vivo M, Falco G, Angrisano T. Deciphering DNA methylation signatures of pancreatic cancer and pancreatitis. *Clin Epigenetics.* 2019;11(1):132.

12. García-Reyes B, Kretz AL, Ruff JP et al. The emerging role of cyclin-dependent kinases (CDKs) in pancreatic ductal adenocarcinoma. *Int J Mol Sci.* 2018;19(10). pii: E3219.

13. Ohmoto A, Yachida S, Morizane C. Genomic features and clinical management of patients with hereditary pancreatic cancer syndromes and familial pancreatic cancer. *Int J Mol Sci.* 2019;20(3). pii: E561.

14. Roberts NJ, Klein AP. Genome-wide sequencing to identify the cause of hereditary cancer syndromes: With examples from familial pancreatic cancer. *Cancer Lett.* 2013;340(2):227–33.

15. Goral V. Pancreatic cancer: Pathogenesis and diagnosis. *Asian Pac J Cancer Prev.* 2015;16(14):5619–24.

16. Matsubayashi H, Takaori K, Morizane C et al. Familial pancreatic cancer: Concept, management and issues. *World J Gastroenterol.* 2017;23(6):935–48.

17. Rawla P, Sunkara T, Gaduputi V. Epidemiology of pancreatic cancer: Global trends, etiology and risk factors. *World J Oncol.* 2019;10(1):10–27.

18. Welinsky S, Lucas AL. Familial pancreatic cancer and the future of directed screening. *Gut Liver.* 2017;11(6):761–70.

19. Petersen GM. Familial pancreatic cancer. *Semin Oncol.* 2016;43(5):548–53.

20. Kanno A, Masamune A, Hanada K, Kikuyama M, Kitano M. Advances in early detection of pancreatic cancer. *Diagnostics (Basel).* 2019;9(1). pii: E18.

21. Singhi AD, Koay EJ, Chari ST, Maitra A. Early detection of pancreatic cancer: Opportunities and challenges. *Gastroenterology.* 2019;156(7):2024–40.

22. Haeberle L, Esposito I. Pathology of pancreatic cancer. *Transl Gastroenterol Hepatol.* 2019;4:50.

23. PDQ Adult Treatment Editorial Board. *Pancreatic cancer treatment (PDQ®): Health professional version.* PDQ Cancer Information Summaries [Internet]. Bethesda (MD): National Cancer Institute (US); 2002-.2019 Jul 15.

24. Amrutkar M, Gladhaug IP. Pancreatic cancer chemoresistance to gemcitabine. *Cancers (Basel).* 2017;9(11). pii: E157.

25. Rahman SH, Urquhart R, Molinari M. Neoadjuvant therapy for resectable pancreatic cancer. *World J Gastrointest Oncol.* 2017;9(12):457–65.

26. Garcia-Carbonero N, Li W, Cabeza-Morales M, Martinez-Useros J, Garcia-Foncillas J. New hope for pancreatic ductal adenocarcinoma treatment targeting endoplasmic reticulum stress response: A systematic review. *Int J Mol Sci.* 2018;19(9). pii: E2468.

27. Martinez-Bosch N, Vinaixa J, Navarro P. Immune evasion in pancreatic cancer: From mechanisms to therapy. *Cancers (Basel).* 2018;10(1). pii: E6.

28. Orth M, Metzger P, Gerum S et al. Pancreatic ductal adenocarcinoma: Biological hallmarks, current status, and future perspectives of combined modality treatment approaches. *Radiat Oncol.* 2019;14(1):141.

29. Klaiber U, Hackert T, Neoptolemos JP. Adjuvant treatment for pancreatic cancer. *Transl Gastroenterol Hepatol.* 2019;4:27.

30. Matsubayashi H, Kiyozumi Y, Ishiwatari H, Uesaka K, Kikuyama M, Ono H. Surveillance of individuals with a family history of pancreatic cancer and inherited cancer syndromes: A strategy for detecting early pancreatic cancers. *Diagnostics (Basel).* 2019;9(4):169.

31. González-Borja I, Viúdez A, Goñi S et al. Omics approaches in pancreatic adenocarcinoma. *Cancers (Basel).* 2019;11(8). pii: E1052.

57

Hereditary Pancreatitis

Dongyou Liu

CONTENTS

57.1 Introduction

Hereditary pancreatitis refers to acute recurrent pancreatitis or chronic pancreatitis that involves at least two members of a family (at least one brother or sister under 40 years) without a history of alcohol abuse. Clinical symptoms of hereditary pancreatitis often appear at an early age (before the second decade of life) and may include strictures, fluid collections, exocrine and endocrine insufficiency, as well as significantly increased risk for pancreatic cancer (estimated at 10%, 18.7%, and 53.5% for patients 50, 60, and 75 years of age, respectively) [1].

Molecularly, gain-of-function mutation in *PRSS1* (7q34), and loss-of-function mutations in *SPINK1* (5q32), *CFTR* (7q31.2), and *CTRC* (1p36.21) that impact on the structure, function, and level of digestive enzymes in the pancreas are implicated in the pathogenesis of hereditary pancreatitis [1,2].

57.2 Biology

Hereditary pancreatitis was first described by Comfort and Steinberg in 1952 from a six-member family spanning three generations, among which four developed chronic relapsing pancreatitis at or before the third decade of life, and two additional members had likely disease, suggesting a genetic linkage of this disorder. Subsequent documentation of >200 additional families with multiple generational members permitted a clear definition of hereditary pancreatitis as the occurrence of acute recurrent pancreatitis or chronic pancreatitis in at least two first-degree or three second-degree relatives, in two or more generations (autosomal dominant inheritance), without precipitating factors (e.g., alcohol abuse, tobacco smoking) and with a negative workup for known causes of chronic pancreatitis (e.g., gallstones, microlithiasis, hypertriglyceridemia, malignancy, autoimmune pancreatitis, pancreas divisum, pancreatic/biliary sphincter dysfunction, adverse medication reactions) [1,3]. As a subset of familial pancreatitis (which refers to pancreatitis from any cause (nongenetic or genetic) in a family with an incidence that is greater than would be expected by chance alone), hereditary pancreatitis is a gene disorder that has clear inheritance patterns (autosomal dominant or autosomal recessive) and displays no syndromic features (e.g., cutaneous, skeletal or growth anomalies).

Following the localization of several chromosomal markers on the long arm of chromosome 7, a genetic defect involving an arginine to histidine substitution in codon 122 (R122H) was identified in the cationic trypsinogen serine protease 1 (*PRSS1*) gene on chromosome 7q34 by Whitcomb and coworkers in 1996. Detection of further *PRSS1* mutations (e.g., A16 V, D22G, K23R, N29I, N29 T, R122C) lent support for the role of this trypsin-encoding gene in the pathogenesis of hereditary pancreatitis, and provided impetus for the search of additional genes involved in hereditary pancreatitis, including the *SPINK1* (serine protease inhibitor Kazal type 1) on chromosome 5q32 in 2009, the *CFTR* (cystic fibrosis transmembrane conductance regulator) gene on chromosome 7q31.2 in 2010, and the *CTRC* (chymotrypsin C) gene on chromosome 1p36.21 in 2012 [1,4,5].

It is apparent that gain-of-function *PRSS1* mutation facilitates premature conversion of trypsinogen to active trypsin and also premature activation of inactive pancreatic zymogens in the

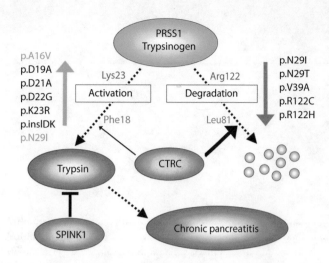

FIGURE 57.1 Schematic presentation of trypsin-dependent pathological pathway in chronic pancreatitis. Gain-of-function *PRSS1* mutation facilitates premature conversion of trypsinogen to active trypsin, loss-of-function *SPINK1* mutation decreases inhibition on trypsin, and loss-of-function *CTRC* mutation hinders degradation of trypsinogen and trypsin. Consequently, the availability of elevated levels of active trypsin causes autodigestion of pancreatic parenchyma and contributes to the development of recurrent acute pancreatitis and chronic pancreatitis. (Photo credit: Hegyi E, Sahin-Tóth M. *Dig Dis Sci*. 2017;62(7):1692–701.)

pancreas instead of the duodenum, leading to autodigestion of pancreatic parenchyma, formation of pancreatic fibrosis, and ultimately development of recurrent acute pancreatitis/chronic pancreatitis. Similarly, loss-of-function *SPINK1* mutation reduces inhibition of active trypsin in the pancreas while loss-of-function *CTRC* mutation hampers degradation of trypsinogen and trypsin. Further, loss-of-function *CFTR* mutation hinders degradation of trypsinogen and trypsin. Thus, continuing presence of elevated levels of active trypsin in the pancreas causes autodigestion of pancreatic parenchyma and provokes inflammatory responses, leading to the development of chronic pancreatitis and eventually pancreatic cancer (Figure 57.1) [6–8].

57.3 Pathogenesis

Past attempts to decipher the molecular mechanisms of hereditary pancreatitis have resulted in the identification of a number of genes that affect digestive enzymes in the pancreas, including the well-known *PRSS1* (7q34), *SPINK1* (5q32), *CFTR* (7q31.2), and *CTRC* (1p36.21) as well as several other less prominent genes (e.g., *CTSB*, *LDN2*, *CPA1*, *PRSS2*) [1].

The *PRSS1* gene on chromosome 7q34 spans 12.8 kb and encodes a 247 aa, 26.5 kDa cationic trypsinogen (or trypsinogen-1, an inactive form of trypsin), which is a member of the trypsin family of serine proteases. Secreted by the pancreas, trypsinogen is cleaved of an eight-amino-acid trypsinogen activation peptide (TAP) by enterokinase to form active trypsin in the small intestine. TAP may be also cleaved by trypsin in the presence of calcium and association with a binding site formed in the activation region. The resultant active (mature) trypsin is an endopeptidase that cleaves peptide chains following a lysine or arginine, and converts inactive zymogens produced by the pancreas into active digestive enzymes in the duodenum upon food

exposure. A second calcium-binding site in trypsinogen as well as trypsin may be occupied by calcium, which prevents trypsin degradation by trypsin on the autolysis site (Arg-122), and by chymotrypsin C (CTRC) at Leu-81 within the calcium binding site [9].

Germline heterozygous mutations in *PRSS1* gene produce a defective trypsinogen that is converted prematurely to active trypsin in the pancreas before excretion to the small intestine, leading to heightened trypsin activity, stability, and autoactivation; autodigestion of the pancreatic parenchyma; strong inflammatory responses (pancreatitis); and increased resistance to multiple physiologic defense mechanisms that prevent the premature conversion of trypsinogen to trypsin. *PRSS1* mutations that generate misfolding cationic trypsinogen may result in diminished secretion, intracellular retention, endoplasmic reticulum stress, and eventually chronic pancreatitis [10].

Over 30 gain-of-function *PRSS1* mutations are identified in 60%–100% of patients with hereditary pancreatitis. Of these, R122H (found in 78% of cases) alters the destruction regulatory site, thus preventing autolysis of trypsin, and increasing trypsin stability and quantity in the pancreas; N29I (present in 12% of cases) targets the other regulatory site, leading to increased and inappropriate autoactivation of trypsin and increased autodigestion of the pancreas, and A16 V induces an abnormal amino acid in the cleavage site, hindering conversion of trypsinogen to trypsin [8,10,11].

The *SPINK1* gene on chromosome 5q32 measures 14.6 kb and encodes a 79 aa, 8.5 kDa protein (SPINK1) which is a pancreatic secretory trypsin inhibitor. Secreted by pancreatic acinar cells, SPINK1 hampers trypsin-catalyzed premature activation of zymogens in the pancreas and pancreatic duct and prevents autodigestion of the pancreatic parenchyma. Biallic loss-of-function mutations (e.g., N34S, P55S) in *SPINK1* reduce available trypsin inhibitor and increase intrapancreatic trypsin activity, resulting in an imbalance of proteases, autodigestion of pancreatic parenchyma, and increased susceptibility (by up to 23%) to pancreatitis. *SPINK1* mutations are detected in 80% of hereditary pancreatitis cases and about 20% of small families with hereditary pancreatitis who do not possess a *PRSS1* germline pathogenic variant. Interestingly, *SPINK1* p.Asn34Ser (c.101A>G) is commonly identified in the United States, Europe, and India, whereas *SPINK1* splice variant (c.194+2T>C, also known as IVS3+2T>C) is often detected in China, Japan, and Korea [8,12].

The *CFTR* gene on chromosome 7q31.2 spans 250 kb and encodes a 1480 aa, 168 kDa protein (CFTR) belonging to the ATP-binding cassette (ABC) transporter superfamily. Functioning as a chloride channel, CFTR controls sodium, chloride, and bicarbonate transport across epithelial surfaces in the respiratory system, sweat glands, gallbladder, and pancreas. Biallelic *CFTR* mutations (>2000 identified so far) cause impaired folding and trafficking of CFTR, and abnormal sodium and chloride transport, leading to defective pancreatic secretion and hereditary pancreatitis (e.g., R75Q) and cystic fibrosis (e.g., F508-delta/F508-delta). *CFTR* pathogenic variants are found in >50% of hereditary pancreatitis cases and in about 25% of families who do not possess a *PRSS1* pathogenic variant [1,8].

The *CTRC* gene on chromosome 1p36.21 is 10.8 kb in length and encodes a 268 aa, 29 kDa protein (CTRC), which is a member of the peptidase S1 family. Secreted by pancreatic acinar cells, CTRC functions as a serum calcium-decreasing factor with chymotrypsin-like protease activity. By targeting specific cleavage sites within zymogen precursors, CTRC regulates activation and degradation of trypsinogen and procarboxypeptidase, and prevents trypsin activation of digestive enzymes and autodigestion of the pancreas. *CTRC* mutations (e.g., A73 T, V235I, R253W, and K247_R254del) reduce CTRC secretion, increase CTRC degradation by trypsin, and cause catalytic defect in CTRC. *CTRC* pathogenic variants occur at a frequency of 1%–3% in hereditary pancreatitis cases and are identified in a small number of patients with idiopathic chronic pancreatitis often in association with another trypsin-activating variant [1,8].

Overall, disease-causing gain-of-function mutation in *PRSS1* along with disease-modifying loss-of-function mutations in *SPINK1*, *CFTR*, and *CTRC* (although the presence of isolated pathogenic variants in these three genes is insufficient to cause pancreatitis) underscores the molecular pathogenesis of hereditary pancreatitis, which reflects the outcomes of pancreatic injury, inflammatory response, healing/regeneration, and complications. Interestingly, *PRSS1*-related hereditary pancreatitis tends to have an earlier onset of disease (by 10 years) than non-*PRSS1*-related hereditary pancreatitis [8,13].

57.4 Epidemiology

57.4.1 Prevalence

Chronic pancreatitis has an estimated incidence of 5–14 in 100,000 in the United States, with a mean diagnostic age of 64 years. Acute pancreatitis affects both sexes equally, while chronic pancreatitis shows a predilection for men. Hereditary pancreatitis has an estimated incidence of 0.3 per 100,000 people in France, 0.125 per 100,000 in Germany, and 0.57 per 100,000 in Denmark, with a median diagnostic age of 5–19 years [14].

57.4.2 Inheritance

Hereditary pancreatitis due to *PRSS1* pathogenic variants shows autosomal dominant inheritance. While most *SPINK1* mutations undergo autosomal recessive transmission, some (e.g., *SPINK1* c.27delC) may be inherited in an autosomal dominant manner. *CFTR* transmission is autosomal recessive, while *CTRC* transmission remains largely unknown [1].

57.4.3 Penetrance

Hereditary pancreatitis shows incomplete penetrance, ranging from 40% in Spain, 80% in the United States, and 93% in France to 80%–96% in England. Further, distinct *PRSS1* mutations may also vary in penetrance (e.g., A16 V, 43%; R122H, 80%; N29I, 93%) [8,10].

57.5 Clinical Features

Hereditary pancreatitis may manifest initially as acute pancreatitis (pancreatic inflammation, sudden onset of abdominal pain in the epigastric region with radiation to the back, nausea, vomiting), then recurrent pancreatitis (due to more parenchymal damage and duct distortion), and finally chronic pancreatitis (fibrosis, parenchymal calcification, ductal stricture, peripancreatic fluid collections; biliary obstruction, pancreatic duct stone formation, malabsorption, steatorrhea due to pancreatic exocrine insufficiency, diabetes mellitus due to pancreatic endocrine insufficiency, weight loss, chronic pain) [15,16].

The most frequent symptoms of hereditary pancreatitis are epigastric pain (83%), calcifications (61%), pancreatic exocrine insufficiency (35%–37%), diabetes mellitus (26%–32%, of which 60% are insulin dependent), and pseudocysts (23%).

Hereditary pancreatitis often has an earlier age of onset (9 years of age in maternal inheritance pattern, 14 years of age in paternal inheritance pattern) compared to other pancreatitis. In addition, hereditary pancreatitis shows a significantly increased risk of developing pancreatic cancer [17–19].

57.6 Diagnosis

Diagnostic workup for hereditary pancreatitis involves personal/familial medical history review, physical examination, laboratory assessment (elevated amylase and lipase in acute pancreatitis; normal to mildly elevated amylase and lipase in chronic pancreatitis due to loss of functional exocrine pancreatic tissue from pancreatic fibrosis; elevated serum bilirubin and alkaline phosphatase due to compression of the intrapancreatic portion of the bile duct by edema, fibrosis, or pancreatic cancer), imaging studies (e.g., transabdominal ultrasound, endoscopic ultrasound [EUS], computerized tomography [CT], magnetic resonance imaging [MRI], maximum intensity projection 3-dimensional magnetic resonance cholangiopancreatography [MRCP] and endoscopic retrograde cholangiopancreatography [ERCP]), histopathology, and molecular analysis (*PRSS1, SPINK1, CFTR,* and *CTRC* pathogenic variants) [20].

Imaging techniques such as CT, MRCP, and ERCP have proven useful in helping to assess pathological changes (e.g., parenchymal calcifications, dilatation or irregular stricture of the pancreatic duct, parenchymal atrophy, pseudocyst, and pancreas divisum) associated with hereditary pancreatitis (Figure 57.2) [21].

Hereditary pancreatitis often displays an acute phase and a chronic phase. Acute pancreatitis is indicated by the presence of two of the following three findings: (i) sudden onset of typical epigastric abdominal pain; (ii) elevation of serum amylase or lipase more than three times the upper limits of normal; (iii) characteristic findings of acute pancreatitis on abdominal imaging. Chronic pancreatitis is indicated by pancreatic inflammation lasting >6 months with irreversible pancreatic changes documented by one of the following: (i) histology (atrophy; fibrosis, sclerosis); (ii) abdominal imaging (inflammatory masses; pancreatic parenchyma and ductal calcifications; pseudocysts); (iii) functional studies (pancreatic exocrine insufficiency with maldigestion of food; pancreatic endocrine insufficiency with diabetes mellitus) [1,8].

Clinical diagnosis of hereditary pancreatitis is based on the presence of chronic pancreatitis in two first-degree or three second-degree relatives, in two or more generations (autosomal

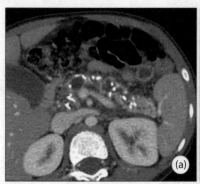
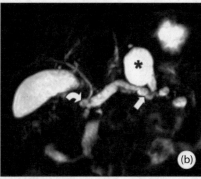
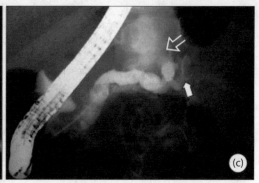

FIGURE 57.2   Hereditary pancreatitis in a 7-year-old boy with chronic pancreatitis and *SPINK1* mutation showing: (a) multiple parenchymal calcifications, irregular dilatation of the pancreatic duct, and parenchymal atrophy on computed tomography image; (b) diffuse dilatation of the pancreatic duct, connection between the pseudocyst (asterisk) and pancreatic duct (arrow), minor duct crossing over the common bile duct (curved arrow), and pancreas divisum on maximum intensity projection 3-dimensional magnetic resonance cholangiopancreatography image; (c) connection between the pseudocyst and pancreatic duct, contrasting agent–filled pseudocyst (open arrow), and irregular stricture of the pancreatic duct (arrow) on endoscopic retrograde cholangiopancreatography image. (Photo credit: Hwang JY et al. *Pediatr Gastroenterol Hepatol Nutr.* 2015;18(2):73–84.)

dominant inheritance), without precipitating factors and with a negative workup for known causes of chronic pancreatitis [1].

Molecular identification of *PRSS1 or SPINK1* pathogenic variants further confirms the diagnosis. Individuals meeting one or more of the following criteria should consider genetic testing for hereditary pancreatitis: (i) unexplained documented pancreatitis in a child; (ii) a family history of idiopathic chronic pancreatitis, recurrent acute pancreatitis, or pancreatitis in childhood without a known etiology; (iii) recurrent acute pancreatitis where no cause can be determined; (iv) relatives with known mutations associated with hereditary pancreatitis; (v) idiopathic chronic pancreatitis in patients who are younger than age 25; (vi) patients who fulfill criteria for participation in approved research projects [1].

Differential diagnoses for hereditary pancreatitis include alcohol-related chronic pancreatitis, tropical pancreatitis, idiopathic chronic pancreatitis (a single occurrence of pancreatitis in a family without identifiable etiology), familial pancreatitis, syndromes with pancreatitis as a finding (e.g., Pearson marrow pancreas syndrome, CEL maturity-onset diabetes of the young [CEL-MODY], Johanson–Blizzard syndrome), and syndromes associated with pancreatic exocrine insufficiency (e.g., Shwachman–Diamond syndrome) [8].

57.7 Treatment

Management of hereditary pancreatitis requires a multidisciplinary approach that incorporates surgery (for pain, necrosis, pseudocyst formation, resectable or worrisome mass lesions, pancreatectomy with islet cell transplant), therapeutic endoscopy (for biliary obstruction, pancreatic duct stricture, pancreatic duct stones, pseudocyst formation, pancreas divisum), chemotherapy, radiotherapy, and other standard medical procedures (e.g., nutritional/dietary modification and support, pancreatic enzyme supplementation for exocrine insufficiency, pain medication, diabetes mellitus control, antidepressants) [22].

57.8 Prognosis and Prevention

Prognosis for hereditary pancreatitis patients without pancreatic cancer (adenocarcinoma) is favorable with median survival age of 74 years if diagnosed early and treated appropriately. However, hereditary pancreatitis patients have a high risk of developing pancreatic cancer, which often occurs 20 years earlier than the general population, leading to earlier and increased mortality.

Patients with hereditary pancreatitis should avoid environmental (e.g., tobacco smoking, alcohol use, dietary fat), emotional (stress, depression) and medicinal (ACE inhibitors, HMG co-A reductase inhibitors, and selective serotonin reuptake inhibitors) triggers that exacerbate and worsen pancreatitis and increase risk for pancreatic cancer. Screening and surveillance for diabetes mellitus, and exocrine insufficiency pancreatic cancer in hereditary pancreatitis patients facilitates early and preventive intervention. Prenatal testing and preimplantation genetic diagnosis may be considered for individuals harboring *PRSS1, SPINK1, CFTR,* and *CTRC* pathogenic variants [8,23–25].

57.9 Conclusion

Hereditary pancreatitis is a rare disorder that is defined as the occurrence of acute recurrent pancreatitis or chronic pancreatitis in at least two first-degree or three second-degree relatives, in two or more generations, without precipitating factors and with a negative workup for known causes of chronic pancreatitis [26,27]. At the molecular level, gain-of-function mutation in *PRSS1* (7q34) and loss-of-function mutations in *SPINK1* (5q32), *CFTR* (7q31.2), and *CTRC* (1p36.21) affecting the structure, function, and level of digestive enzymes in the pancreas underlie the development of hereditary pancreatitis [28,29]. Specifically, gain-of-function *PRSS1* mutation induces premature conversion of trypsinogen to active trypsin and also premature activation of inactive pancreatic zymogens in the pancreas; loss-of-function *SPINK1* reduces inhibition of active trypsin in the pancreas, and loss-of-function

CTRC and *CFTR* mutations hinders degradation of trypsinogen and trypsin. Together, a continuing presence of elevated levels of active trypsin in the pancreas induces autodigestion of pancreatic parenchyma, formation of pancreatic fibrosis, and development of recurrent acute pancreatitis/chronic pancreatitis as well as pancreatic cancer at an earlier age. Diagnosis of hereditary pancreatitis is based on meeting certain clinical criteria and/or identification of disease-causing or -modifying *PRSS1*, *SPINK1*, *CFTR*, and *CTRC* pathogenic variants. Treatment of hereditary pancreatitis involves a multidisciplinary approach consisting of surgery, therapeutic endoscopy, chemotherapy, radiotherapy, and other standard medical procedures [30].

REFERENCES

1. LaRusch J, Solomon S, Whitcomb DC. Pancreatitis overview. In: Adam MP, Ardinger HH, Pagon RA, Wallace SE, Bean LJH, Stephens K, Amemiya A, editors. *GeneReviews®* [Internet]. Seattle (WA): University of Washington, Seattle; 1993–2018. 2014 Mar 13.
2. Masamune A. Genetics of pancreatitis: The 2014 update. *Tohoku J Exp Med*. 2014;232(2):69–77.
3. Raphael KL, Willingham FF. Hereditary pancreatitis: Current perspectives. *Clin Exp Gastroenterol*. 2016;9:197–207.
4. Szabó A, Sahin-Tóth M. Increased activation of hereditary pancreatitis-associated human cationic trypsinogen mutants in presence of chymotrypsin C. *J Biol Chem*. 2012;287(24):20701–10.
5. Jancsó Z, Sahin-Tóth M. Tighter control by chymotrypsin C (CTRC) explains lack of association between human anionic trypsinogen and hereditary pancreatitis. *J Biol Chem*. 2016;291(25):12897–905
6. Hegyi E, Sahin-Tóth M. Genetic risk in chronic pancreatitis: The typsin-dependent pathway. *Dig Dis Sci*. 2017;62(7):1692–701.
7. Sahin-Tóth M. Genetic risk in chronic pancreatitis: The misfolding-dependent pathway. *Curr Opin Gastroenterol*. 2017;33(5):390–5.
8. Zou WB, Tang XY, Zhou DZ et al. SPINK1, *PRSS*1, CTRC, and CFTR genotypes influence disease onset and clinical outcomes in chronic pancreatitis. *Clin Transl Gastroenterol*. 2018;9(11):204.
9. Rygiel AM, Beer S, Simon P et al. Gene conversion between cationic trypsinogen (PRSS1) and the pseudogene trypsinogen 6 (PRSS3P2) in patients with chronic pancreatitis. *Hum Mutat*. 2015;36(3):350–6.
10. Solomon S, Whitcomb DC, LaRusch J. PRSS1-related hereditary pancreatitis. In: Adam MP, Ardinger HH, Pagon RA et al. editors. *GeneReviews®* [Internet]. Seattle (WA): University of Washington, Seattle; 1993–2018. 2012 Mar 1.
11. Pelaez-Luna M, Robles-Diaz G, Canizales-Quinteros S, Tusié-Luna MT. PRSS1 and SPINK1 mutations in idiopathic chronic and recurrent acute pancreatitis. *World J Gastroenterol*. 2014;20(33):11788–92.
12. Patel MR, Eppolito AL, Willingham FF. Hereditary pancreatitis for the endoscopist. *Therap Adv Gastroenterol*. 2013;6(2):169–79.

13. Rosendahl J, Bödeker H, Mössner J, Teich N. Hereditary chronic pancreatitis. *Orphanet J Rare Dis*. 2007;2:1.
14. Solomon S, Whitcomb DC. Genetics of pancreatitis: An update for clinicians and genetic counselors. *Curr Gastroenterol Rep*. 2012;14(2):112–7.
15. Dai LN, Chen YW, Yan WH, Lu LN, Tao YJ, Cai W. Hereditary pancreatitis of 3 Chinese children: Case report and literature review. *Medicine (Baltim)*. 2016;95(36):e4604.
16. Dytz MG, Marcelino PA, de Castro Santos O et al. Clinical aspects of pancreatogenic diabetes secondary to hereditary pancreatitis. *Diabetol Metab Syndr*. 2017;9:4.
17. Becker AE, Hernandez YG, Frucht H, Lucas AL. Pancreatic ductal adenocarcinoma: Risk factors, screening, and early detection. *World J Gastroenterol*. 2014;20(32):11182–98.
18. Weiss FU. Pancreatic cancer risk in hereditary pancreatitis. *Front Physiol*. 2014;5:70.
19. Matsubayashi H, Takaori K, Morizane C et al. Familial pancreatic cancer: Concept, management and issues. *World J Gastroenterol*. 2017;23(6):935–48.
20. Singhi AD, Pai RK, Kant JA et al. The histopathology of PRSS1 hereditary pancreatitis. *Am J Surg Pathol*. 2014;38(3):346–53.
21. Hwang JY, Yoon HK, Kim KM. Characteristics of pediatric pancreatitis on magnetic resonance cholangiopancreatography. *Pediatr Gastroenterol Hepatol Nutr*. 2015;18(2):73–84. (correction: Pediatr Gastroenterol Hepatol Nutr. 2015; 18(3):216].
22. Lew D, Afghani E, Pandol S. Chronic pancreatitis: Current status and challenges for prevention and treatment. *Dig Dis Sci*. 2017;62(7):1702–12.
23. Bruenderman E, Martin RC 2nd. A cost analysis of a pancreatic cancer screening protocol in high-risk populations. *Am J Surg*. 2015;210(3):409–16.
24. Welinsky S, Lucas AL. Familial pancreatic cancer and the future of directed screening. *Gut Liver*. 2017;11(6):761–70.
25. Kikuyama M, Kamisawa T, Kuruma S et al. Early diagnosis to improve the poor prognosis of pancreatic cancer. *Cancers (Basel)*. 2018;10(2). pii: E48.
26. National Guideline Centre (UK). Pancreatitis. London: National Institute for Health and Care Excellence (UK); 2018 Sep. (NICE Guideline, No. 104.)
27. Barry K. Chronic pancreatitis: Diagnosis and treatment. *Am Fam Physician*. 2018;97(6):385–93.
28. Dytz MG, Mendes de Melo J, de Castro Santos O et al. Hereditary pancreatitis associated with the N29 T mutation of the PRSS1 fene in a Brazilian family: A case-control study. *Medicine (Baltim)*. 2015;94(37):e1508.
29. Patel J, Madan A, Gammon A, Sossenheimer M, Samadder NJ. Rare hereditary cause of chronic pancreatitis in a young male: SPINK1 mutation. *Pan Afr Med J*. 2017;28:110.
30. Uc A, Andersen DK, Bellin MD et al. Chronic pancreatitis in the 21st century–research challenges and opportunities: Summary of a National Institute of Diabetes and Digestive and Kidney Diseases Workshop. *Pancreas*. 2016;45(10):1365–75.

58

Hereditary Pheochromocytoma and Paraganglioma Syndrome

Dongyou Liu

CONTENTS

58.1 Introduction

Pheochromocytoma and paraganglioma (PPGL) are rare tumors of the paraganglia, which are formed by the aggregation of the cell nuclei of the autonomic nervous system (including sympathetic and parasympathetic nerves), symmetrically distributed along the paravertebral axis from the base of the skull and neck to the pelvis, with the largest found in the adrenal medulla. Specifically, pheochromocytoma (also known as adrenal chromaffin cell tumor) is catecholamine-secreting paraganglioma confined to the adrenal medulla. Sympathetic paraganglioma located along the paravertebral axis (and not in the adrenal gland) is known as extra-adrenal sympathetic paraganglioma. Since most PPGL (especially those affecting the adrenal glands) secrete excessive amounts of catecholamines that predispose to elevated blood pressure, palpitations, sweats, anxiety, and gastrointestinal disease, they are also known as neuroendocrine tumors [1].

Molecularly, a majority (60%) of PPGL evolve from sporadic events, while about 40% contain germline pathogenic variants and are commonly referred to as hereditary pheochromocytoma and paraganglioma (hereditary PPGL) due to their hereditary nature. Some hereditary PPGL form a standalone disease without other syndromic features (initially known as hereditary paraganglioma syndrome before its link to pheochromocytoma is established; now as hereditary PPGL syndrome [SDHx]), and others are associated with hereditary tumor syndromes (e.g., multiple endocrine neoplasia type 2A and 2B, neurofibromatosis type 1, and Von Hippel–Lindau syndrome) (Table 58.1) [2].

58.2 Biology

Pheochromocytoma (*pheo*, brown; *chromo*, chromo stain; *cytoma*, tumor) is a neuroendocrine tumor that arises from chromaffin cells (i.e., brown-black colored cells due to oxidization and polymerization of catecholamines by potassium dichromate) of the adrenal medulla (which contains the largest paraganglia made up of the aggregated cell nuclei of the autonomic nervous system) and accounts for 80% of catecholamine-secreting tumors. Paraganglioma (tumor of the paraganglia) is a neuroendocrine tumor that develops at the paraganglia located outside of the adrenal gland [3].

Based on their location/origin, PPGL are separated into sympathetic PPGL (tumors derived from sympathetic tissue in adrenal or extra-adrenal abdominal locations; frequent catecholamine production; 80% in the adrenal medulla, 20% in extra-adrenal sites such as the prevertebral and paravertebral sympathetic ganglia of the chest, abdomen, and pelvis), and parasympathetic PPGL (tumors derived from parasympathetic tissue in the thorax or head and neck along the glossopharyngeal and vagal nerves) [4,5].

Extra-adrenal PPGL in the abdomen originate mainly from chromaffin tissue around the inferior mesenteric artery (the organ of Zuckerkandl) or aortic bifurcation, and frequently produce catecholamines, whereas head and neck PGL (HNPGL; formerly glomus tumor or carotid body tumor) are mostly chromaffin-negative, with only 4% secreting catecholamines [6].

Approximately 60% of PPGL result from de novo mutations, and 40% are due to inheritance of one of the tumor susceptibility genes whose mutations either cause abnormal activation

TABLE 58.1

Molecular and Clinical Features of Hereditary Pheochromocytoma and Paraganglioma

Pathway	Hereditary Syndrome	Gene	Mutation Type (Transmission)	Clinical Features	Malignancy Risk
Kinase signaling	Multiple endocrine neoplasia type 2 (MEN2)	*RET*	Germline or somatic (autosomal dominant)	MEN2A: Medullary thyroid carcinoma, PCC (60% bilateral; epinephrine; first manifestation in 10%–30% of cases), hyperparathyroidism, cutaneous lichen amyloidosis; earliest age of diagnosis at 5–8 years MEN2B: Medullary thyroid carcinoma (~50% penetrance), PCC (60% bilateral; epinephrine), multiple neuromas, marfanoid habitus; earliest age of diagnosis at 12 years FMTC: Familial medullary thyroid carcinoma	Low (2.9%)
	Syndrome not defined	*TMEM127*	Germline (autosomal dominant)	PCC (40% bilateral), HNPGL, extra-adrenal PGL.; no pediatric cases reported	Low/moderate (4.3%–12%)
	Neurofibromatosis type 1 (NF1)	*NF1*	Germline or somatic (autosomal dominant)	Neurofibroma, café-au-lait spots, PCC (15% bilateral; present in 4% of cases; epinephrine), Lisch nodules, optic pathway/CNS gliomas, GIST; earliest age of diagnosis at 7 years	Low/moderate (9.3%–33%)
	Syndrome not defined	*MAX*	Germline or somatic (paternal)	PCC (60% bilateral); earliest age of diagnosis at 17 years	Low/moderate (9%–25%)
Pseudohypoxia	Von Hippel–Lindau syndrome type 2 (VHL2)	*VHL*	Germline or somatic (autosomal dominant)	VHL2a: Retinal and CNS hemangioblastomas, PCC (40% bilateral; norepinephrine; present in 10%–20% of adult cases, 6%–49% of pediatric cases), endolymphatic sac tumor, epididymal cystadenoma VHL2b: renal-cell cyst and carcinoma (norepinephrine), retinal and CNS hemangioblastomas, pancreatic neoplasm and cyst, PCC (40% bilateral, norepinephrine), endolymphatic sac tumor, epididymis cystadenoma VHL2c: PCC (40% bilateral, norepinephrine l) Earliest age of diagnosis at 5 years	Low (3%)
	Paraganglioma–pheochromocytoma syndrome (SDHx)	*SDHA*	Germline (autosomal dominant)	Single extra-adrenal PGL: autosomal dominant inheritance; earliest age of diagnosis at 8 years	Low (0%–14.3%)
		SDHB	Germline or rare somatic (autosomal dominant)	HNPGL (20% multiple), PCC, extra-adrenal PGL, GIST, RCC, earliest age of diagnosis at 6 years; *SDHB* mutations in 12.5%–20% of pediatric cases	High (34%–97%)
		SDHC	Germline (autosomal dominant)	HNPGL (20% multiple), GIST, earliest age of diagnosis at 12 years; norepinephrine/rarely dopamine	Low/moderate
		SDHD	Germline (paternal)	HNPGL (50% multiple), PCC, extra-adrenal PGL, GIST, papillary thyroid carcinoma (rarely); norepinephrine/ rarely dopamine; earliest age of diagnosis at 5 years; *SDHD* mutation in 10% pediatric PCC cases	Low (3.5%)
		SDHAF2	Germline (paternal)	HNPGL (90% multiple); norepinephrine/rarely dopamine; earliest age of diagnosis at 15 years	Low

Abbreviations: PCC, pheochromocytoma; PGL, paraganglioma; HNPGL, head and neck paraganglioma; RCC, renal cell carcinoma; GIST, gastrointestinal stromal tumor.

TABLE 58.2

Molecular Characteristics and Clinical Involvement of Succinate Dehydrogenase Genes

Gene	Chromosome Location	Gene Size, Exon Number	Protein Size	Missense/Nonsense Mutations Identified	Deletion/Insertion Mutations Identified	Clinical Involvement
SDHA	5p15.33	46.5 kb, 15 exons	664 aa, 72.6 kDa	29	0	PCC 0.5%–3%; TAPGL 0.5%–3%; HNPGL 0.5%–3%; GIST; PA
SDHB	1p36.13	35.4 kb, 8 exons	280 aa, 31.6 kDa	212	82	PCC 20%–25%; TAPGL 50%; HNPGL 20%–30%; RCC 14%; GIST; PA
SDHC	1q23.3	61 kb, 6 exons	169 aa, 18.6 kDa	26	5	TAPGL 4%–8%; RCC rare; GIST
SDHD	11q23.1	33 kb, 4 exons	159 aa, 17 kDa	164	71	PCC 10%–25%; TAPGL 20%–25%; HNPGL 85%; RCC 8%; GIST; PA
SDHAF2	11q12.2	17.4 kb, 2 exons	166 aa, 19.5 kDa	4	1	HNPGL 100%

Abbreviations: PCC, pheochromocytoma; TAPGL, thoracoabdominal paraganglioma; HNPGL, head and neck paraganglioma; RCC, renal cell carcinoma; PA, pituitary adenoma; GIST, gastrointestinal stromal tumor.

of kinase signaling pathways (e.g., PI3Kinase/AKT, RAS/RAF/ERK, and mTOR pathways) or induce overexpression of vascular endothelial growth factor (VEGF) (due to pseudohypoxia), leading to impaired DNA methylation/increased vascularization (Table 58.1) [7–9].

58.3 Pathogenesis

As hereditary tumor syndromes MEN II, NF1, and VHL that present with pheochromocytoma and/or paraganglioma as part of their clinical manifestations are covered elsewhere in this volume, the current chapter will focus on hereditary PPGL syndrome (hereditary PPGL syndrome or SDHx).

Succinate dehydrogenase (SDH) is part of both the citric acid cycle (in which SDH oxidizes succinate to fumarate) and aerobic electron transfer chain. In eukaryotes, succinate dehydrogenase complex (also known as mitochondrial complex II) consists of four subunits (SDHA, SDHB, SDHC, and SDHD) encoded by the *SDHx* genes (*SDHA, SDHB, SDHC, SDHD*) in the nuclear genome. While SDHA and SDHB are the enzymatic subunits, SDHC and SDHD anchor the heterotetrameric complex to the mitochondrial membrane. In addition, SDHAF2 encoded by *SDHAF2* is a factor involved in the assembly of the SDH complex (Table 58.2) [10]. SDH mutation as well as TRAP1 upregulation contributes to succinate accumulation in the cytoplasm, which in turn stabilizes HIF, activates GPR91, and initiates downstream signaling cascades. Further, SDH mutation enhances ROS production, leading to DNA damage and NRF2 activation, and also inhibits PHDs causing HIF catabolism. Thus, these activities work in concert toward cell apoptosis, angiogenesis, proliferation, glycolysis, and tumorigenesis (Figure 58.1) [11].

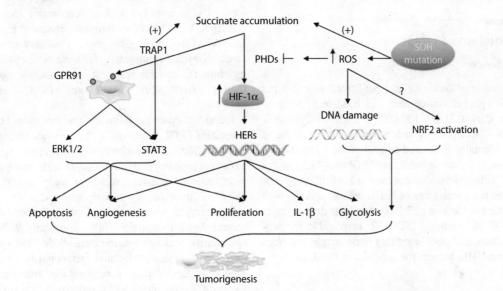

FIGURE 58.1 Role of succinate dehydrogenase (SDH) mutation in tumorigenesis. SDH mutation as well as TRAP1 upregulation induces succinate accumulation in the cytoplasm, which stabilizes HIF, activates GPR91, and initiates downstream signaling cascades. SDH mutation also stimulates ROS production, which causes DNA damage and NRF2 activation, and inhibits the activity of PHDs, leading to HIF catabolism. Together, these activities promote cell apoptosis, angiogenesis, proliferation, glycolysis, and ultimately tumorigenesis. (Photo credit: Zhao T et al. *Oncotarget.* 2017;8(32):53819–28.)

The *SDHA* (succinate dehydrogenase complex flavoprotein subunit A) gene encodes a major catalytic subunit of succinate-ubiquinone oxidoreductase, which participates in the transfer of electrons from succinate to ubiquinone in the mitochondrial electron transport chain. Heterozygous *SDHA* mutations are occasionally associated with pheochromocytoma (PCC), thoracoabdominal paraganglioma (TAPGL), head and neck paraganglioma (HNPGL), gastrointestinal stromal tumor (GIST) and pituitary adenoma (PA). In contrast, homozygous germline *SDHA* mutations are implicated in Leigh syndrome (an inherited mitochondrial respiratory chain deficiency). Homozygous germline mutations in *SDHA* cause neurological disorder (Leigh syndrome), as well as a syndrome of optic atrophy, ataxia, and myopathy [12].

The *SDHB* (succinate dehydrogenase complex iron sulfur subunit B) gene encodes a protein specifically involved in the oxidation of succinate, transporting electrons from FADH to ubiquinone. Sporadic and germline mutations in *SDHB* cause paraganglioma and pheochromocytoma [13,14].

The *SDHC* (succinate dehydrogenase complex subunit C) gene encodes one of two integral membrane proteins that anchor other subunits of the complex to the inner mitochondrial membrane. Heterozygous mutations in *SDHC* are linked to paragangliomas.

The *SDHD* (succinate dehydrogenase complex subunit D) gene encodes one of two integral membrane proteins anchoring the complex to the matrix side of the mitochondrial inner membrane. Mutations in *SDHD* lead to hereditary paraganglioma, which is transmitted almost exclusively through the paternal allele. It is notable that three *SDHD* pathogenic variants (p.Asp92Tyr, p.Leu95Pro, p.Leu139Pro) are implicated in nearly all hereditary paraganglioma cases in the Dutch population, whereas *SDHD* pathogenic variant p.Met1Ile is a founder variant in the Chinese population [2].

The *SDHAF2* (succinate dehydrogenase complex assembly factor 2) gene encodes a mitochondrial protein involved in the flavination of SDHA. Mutations in *SDHAF2* are implicated in the development of HNPGL (Table 58.2).

58.4 Epidemiology

58.4.1 Prevalence

PPGL have an estimated incidence of 1 per 300,000 (or 2–8 per 1 million), while pheochromocytoma and head/neck paraganglioma occur in 0.04–0.21 per 100,000. Pheochromocytoma is observed in 0.1%–1% of patients with hypertension and in 5% of patients with incidentally discovered adrenal masses. The average age at diagnosis is 24.9 years in hereditary PPGL and 43.9 years in sporadic PPGL, with peak incidence at the third to fifth decades of life. In pediatrics, the average age of PPGL presentation is 11–13 years, with a male predilection of 2:1 [2]. Available data suggests that hereditary PPGL syndrome (SDHx) is responsible for two-thirds of PPGL cases, and other hereditary tumor syndromes (e.g., MEN II, NF1, and VHL) account for one-third of PPGL cases.

58.4.2 Inheritance

PPGL due to *SDHA*, *SDHB*, and *SDHC* mutations show autosomal dominant transmission, whereas those due to *SDHD* and *SDHAF2* mutations have paternal transmission.

58.4.3 Penetrance

PPGL demonstrate variable penetrance, which appears to be age-gene-, and site-dependent. For example, penetrance of *SDHA*-related manifestation is 13% at 40 years and 10% at 70 years; penetrance of *SDHB*-related manifestation is 29%, 50%, 45%, and 77% at age 30, 35, 40, and 45 years, respectively; penetrance of *SDHD*-related manifestation is 48%, 50%, 73%, and 86% at age 30, 31, 40, and 50 years, respectively. Further, the penetrance of *SDHB* mutations is 15% for skull base and neck paraganglioma and 69% for extra-adrenal abdominal or thoracic paraganglioma, and the penetrance of *SDHD* mutations is 68% for skull base and neck paraganglioma and 35% for extra-adrenal abdominal or thoracic paraganglioma [2].

58.5 Clinical Features

Clinical manifestations of PPGL include sustained hypertension (60%–90% of pediatric cases), paroxysmal hypertension (50% of adult cases), headaches and hypertension (67% of pediatric cases), palpitations, sweating, pallor, nausea, and flushing (47%–57% of pediatric cases), anxiety, weight loss, visual disturbance polyuria, polydipsia, dysphagia, hoarseness, hearing disturbances, and pain (due to mass effect of nonfunctional HNPGL). Patients with epinephrine-secreting tumors may show hypoglycemia and hypotensive shock (due to excess catecholamine production and circulatory collapse), and those with dopamine-secreting tumors are often asymptomatic [15–17].

58.6 Diagnosis

Diagnostic workup for hereditary PPGL syndrome involves family history review, physical examination, biochemical evaluation (excess catecholamine secretion, or plasma/urinary metanephrine and 3-methoxytyramine, an indicator of dopamine secretion, for 30% of tumors that lack catecholamine secretion), imaging studies (CT and MRI for tumor localization; ^{18}F-FDOPA or ^{68}Ga DOTATATE scanning for primary solitary or metastatic disease; ^{123}I-MIBG scintigraphy for metastatic PPGL), histopathology (distinct cellular features), and molecular genetic analysis (*SDHA*, *SDHB*, *SDHC*, *SDHD*, and *SDHAF2* pathogenic variants) [18,19].

Individuals with the following findings should be suspected of hereditary PPGL syndrome: (i) tumors that are multiple (i.e., >1 paraganglioma or pheochromocytoma, including bilateral adrenal pheochromocytoma), multifocal with multiple synchronous or metachronous lesions, recurrent, early onset (i.e., <45 years), and (ii) a family history of such tumors [2].

Biochemical tests based on plasma normetanephrine and metanephrine (sensitivity 100%, specificity 94%), plasma norepinephrine and epinephrine (sensitivity 92%, specificity 91%), primary normetanephrine and metanephrine (sensitivity 100%, specificity 95%), urinary norepinephrine and epinephrine (sensitivity 100%, specificity 83%), and urinary vanillylmandelic acid (sensitivity 63%–75%, specificity 94%) are useful for diagnosis of pediatric pheochromocytoma [16].

Histological examination of PPGL reveals characteristic small nests of uniform polygonal chromaffin cells (Zellballen)

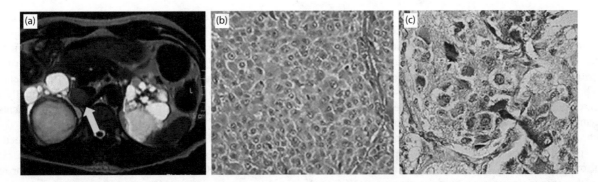

FIGURE 58.2 Composite pheochromocytoma in a 53-year-old male showing: (a) acquired cystic kidney disease and a right adrenal mass (arrow) on magnetic resonance image; (b) pheochromocytoma component in resected tumor (H&E, ×400); (c) ganglioneuroblastoma component in resected tumor (Nissl staining, ×400). (Photo credit: Shida Y et al. *BMC Res Notes*. 2015;8:257.)

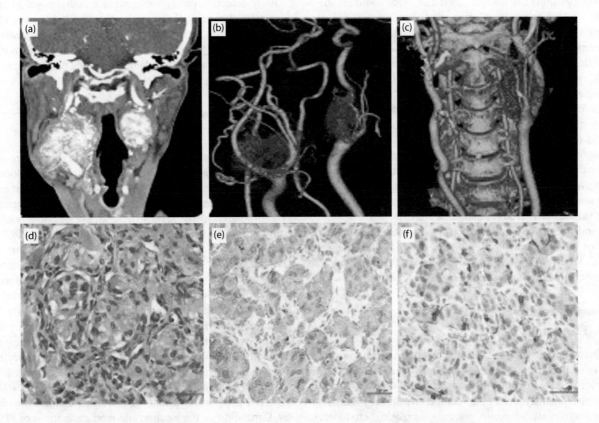

FIGURE 58.3 Paraganglioma in the neck of a 59-year-old male showing: (a) bilateral masses at the bifurcation of both sides of the carotid artery on coronal contrast enhanced multiple-planner reconstruction (MPR) image; (b) tumor vascularity on computed tomographic angiography (CTA) image; (c) the right mass excised and stent implantation in bilateral internal carotid arteries on postoperative CTA image; (d) the nested (Zellballen) pattern of neoplastic cells and their neuroendocrine appearance (H&E, ×400); (e) the chief and sustentacular cells in paraganglioma specimen by synaptophysin stain (×400); (f) the chief and sustentacular cells in paraganglioma by S-100 stain (×400). (Photo credit: Xiao Z et al. *BMC Med Imaging*. 2015;15:38.)

(Figures 58.2 and 58.3) [20,21]. Observation of tumor deposits in tissues that do not normally contain chromaffin cells (e.g., lymph nodes, liver, bone, lung, and other distant metastatic sites) offers an unequivocal evidence of paraganglioma [22].

Molecular identification of *SDHA*, *SDHB*, *SDHC*, *SDHD*, and *SDHAF2* pathogenic variants further confirms the diagnosis of PPGL in the adrenal as well as extra-adrenal tissues (Table 58.2). Genetic testing is recommended for (i) patients with a personal or family history of clinical features suggestive of hereditary PPGL syndrome, (ii) patients with bilateral or multifocal tumors,

(iii) patients with sympathetic or malignant extra-adrenal paragangliomas, (iv) patients diagnosed before age 40 years, and (v) patients between 40–50 years of age with a unilateral pheochromocytoma and no personal or family history suggestive of hereditary disease, but not for patients >50 years [2,23,24].

Differential diagnosis of hereditary PPGL syndrome include multiple endocrine neoplasia type 2 (MEN2A: medullary thyroid carcinoma, pheochromocytoma in 50% cases, hyperparathyroidism; 80% of MEN2 cases; MEN2B: mucocutaneous neuroma or diffuse ganglioneuromatosis of the gastroenteric mucosa, pheochromocytoma in 50% cases, slender body habitus, joint laxity, and skeletal malformations, 5% of MEN2 cases; familial medullary thyroid carcinoma (FMTC): medullary thyroid carcinoma as its only feature; autosomal dominant disorder due to *RET* pathogenic variants), VHL syndrome (hemangioblastomas of the brain, spinal cord, and retina; renal cysts, clear cell renal cell carcinoma; pancreatic cysts and neuroendocrine tumors; endolymphatic sac tumors; epididymal and broad ligament cysts; pheochromocytoma in 6%–9% of VHL type 1%, 40%–59% VHL type 2, as sole manifestation of VHL type 2C; mean onset age of onset of 30 years, autosomal dominant disorder due to *VHL* pathogenic variants), neurofibromatosis type 1 (NF1; multiple café-au-lait spots, axillary and inguinal freckling, multiple cutaneous neurofibromas, iris Lisch nodules; learning disabilities in 50% cases; GIST, carcinoid tumor; pheochromocytoma and hypertension in 20%–50% cases; autosomal dominant disorder due to *NF1* mutation), Carney triad (extra-adrenal paraganglioma, GIST, pulmonary chondroma); adrenal cortical adenoma and esophageal leiomyoma; affecting young women; due to *KIT* and *PDGFRA* mutations (GIST) and occasional *SDHA, SDHB, SDHC*, mutations (9.5% of paraganglioma), Carney–Stratakis dyad (paraganglioma and GIST; autosomal dominant disorder due to *SDHB, SDHC*, or *SDHD* pathogenic variants), renal cell carcinoma (occasional germline *SDHB* pathogenic variants), thyroid carcinoma (occasional *SDHB* or an *SDHD* pathogenic variants), Leigh syndrome (a neurodegenerative mitochondrial encephalomyopathy; homozygous and compound heterozygous pathogenic variants in *SDHA*), other undescribed syndromes presenting with paraganglioma and pheochromocytoma (pathogenic variants in *TMEM127* and *MAX*) [2,25].

Overall, germline pathogenic variants in *SDHB, SDHD, VHL, RET*, or *NF1* account for 90% of hereditary PPGL cases, and *SDHC, SDHA, SDGAF2, TMEM127*, or *MAX* (1.2%) mutations are responsible for only 10% of such cases [2].

58.7 Treatment

Patients with localized and regional PPGL (including localized disease recurrence) are treated with alpha- and beta-adrenergic blockade (e.g., non-competitive alpha-1 and -2 adrenoreceptor antagonist phenoxybenzamine; competitive alpha-1 adrenoreceptor antagonists doxazosin, prazosin, terazosin; beta blockers atenolol, metonolol, propranolol) followed by surgery [26,27].

Patients with unresectable or metastatic PPGL are treated with a combination of catecholamine blockade, surgery, chemotherapy (e.g., cyclophosphamide, vincristine, and dacarbazine), radiofrequency ablation, cryoablation, and radiotherapy (e.g., external beam radiotherapy) [28–33].

58.8 Prognosis and Prevention

Prognosis for 85% of patients with nonmalignant (benign) PPGL is reasonably favorable, with 15-year survival rate of 100%. Prognosis for 15% of patients with malignant/metastatic PPGL is generally poor (due to local recurrence or metastasis), with 5-, 10-, and 15-year survival rates of 78%, 62%, and 31%, respectively.

About 12%–38% of pediatric patients may experience recurrence in extra-adrenal (18%), contralateral (13%), or ipsilateral (16%) sites 25 years after diagnosis. Patients with germline mutations (e.g., *SDHD*) may have recurrence 10 years earlier than those with sporadic mutations. Further, 72% of patients with *SDHB* mutation develop malignant disease (especially extra-adrenal and thoracic PGL), compared to 9.4% of patients with *SDHD* mutation who have malignant disease [34–36].

Patients undergoing PCC resection are associated with improved cardiomyopathy/ hypertension (96% of cases), and those without PCC resection may experience death or require cardiac transplantation (44% of cases).

Patients should avoid of habitation at high altitudes and activities that promote long-term exposure to hypoxia (including cigarette smoking).

Early detection through surveillance and removal of tumors help prevent or minimize complications related to mass effects, catecholamine hypersecretion, and malignant transformation.

Prenatal testing and preimplantation genetic diagnosis may be considered for prospective parents harboring pathogenic variants in *SDHA, SDHB, SDHC, SDHD*, and *SDHAF2*.

58.9 Conclusion

Pheochromocytoma and paraganglioma (PPGL) are rare tumors that affect the adrenal medulla and extra-adrenal locations, respectively. Although a majority (60%) of PPGL appear to arise from sporadic mutations, 40% are attributed to germline mutations in genes involved mainly in the kinase signaling or hypoxia pathways. Of all hereditary PPGL cases, hereditary PPGL syndrome (SDHx) account for two-thirds and other hereditary tumor syndromes (e.g., MEN II, NF1, and VHL) represent one-third [37,38]. Clinical symptoms of PPGL range from sustained/paroxysmal hypertension, headaches, palpitations, sweating, pallor, nausea and flushing, anxiety, weight loss, visual disturbance polyuria, polydipsia, dysphagia, hoarseness, and hearing disturbances to pain. Diagnostic of hereditary PPGL syndrome is based on finding multiple, recurrent, early-onset pheochromocytomas/paragangliomas and a family history of such tumors [39,40]. Molecular identification of *SDHA, SDHB, SDHC, SDHD*, and *SDHAF2* pathogenic variants helps further confirm the diagnosis. Depending on the location and metastatic status of PPGL, a combination of catecholamine blockade, surgery, chemotherapy, radiofrequency ablation, cryoablation, and radiotherapy may be used for treatment [41–43].

REFERENCES

1. Conzo G, Pasquali D, Colantuoni V et al. Current concepts of pheochromocytoma. *Int J Surg.* 2014;12(5):469–74.
2. Kirmani S, Young WF. Hereditary paraganglioma-pheochromocytoma syndromes. In: Adam MP, Ardinger HH, Pagon RA et al. editors. *GeneReviews®* [Internet]. Seattle (WA): University of Washington, Seattle; 1993–2018. 2008 May 21 [updated 2014 Nov 6].
3. Else T. 15 Years of paraganglioma: Pheochromocytoma, paraganglioma and genetic syndromes: A historical perspective. *Endocr Relat Cancer.* 2015;22(4):T147–59.

4. Lam AK. Update on paragangliomas and pheochromocytomas. *Turk Patoloji Derg.* 2015;31(Suppl 1):105–12.

5. Lloyd RV, Osamura RY, Kloppel G, Rosai J. *WHO Classification of Tumours: Pathology and Genetics of Tumours of Endocrine Organs.* Lyon, France: IARC Press. 2017.

6. Williams MD, Tischler AS. Update from the 4th edition of the World Health Organization classification of head and neck tumours: Paragangliomas. *Head Neck Pathol.* 2017;11:88–95.

7. Amorim-Pires D, Peixoto J, Lima J. Hypoxia pathway mutations in pheochromocytomas and paragangliomas. *Cytogenet Genome Res.* 2016;150(3–4):227–41.

8. Jochmanová I, Zelinka T, Widimský J Jr, Pacak K. HIF signaling pathway in pheochromocytoma and other neuroendocrine tumors. *Physiol Res.* 2014;63(Suppl 2):S251–62.

9. Henegan JC Jr, Gomez CR. Heritable cancer syndromes related to the hypoxia pathway. *Front Oncol.* 2016;6:68.

10. Baysal BE, Maher ER. 15 Years of paraganglioma: Genetics and mechanism of pheochromocytoma-paraganglioma syndromes characterized by germline SDHB and SDHD mutations. *Endocr Relat Cancer.* 2015;22(4):T71–82.

11. Zhao T, Mu X, You Q. Succinate: An initiator in tumorigenesis and progression. *Oncotarget.* 2017;8(32):53819–28.

12. Neumann HP, Young WF Jr, Krauss T et al. 65 Years of the double helix: Genetics informs precision practice in the diagnosis and management of pheochromocytoma. *Endocr Relat Cancer.* 2018;25(8):T201–19.

13. Rijken JA, Niemeijer ND, Leemans CR et al. Nationwide study of patients with head and neck paragangliomas carrying SDHB germline mutations. *BJS Open.* 2018;2(2):62–9.

14. Yamanaka M, Shiga K, Fujiwara S et al. A novel SDHB IVS2-2A>C mutation is responsible for hereditary pheochromocytoma/paraganglioma syndrome. *Tohoku J Exp Med.* 2018;245(2):99–105.

15. Benn DE, Robinson BG, Clifton-Bligh RJ. 15 Years of paraganglioma: Clinical manifestations of paraganglioma syndromes types 1–5. *Endocr Relat Cancer.* 2015;22(4):T91–103.

16. Bholah R, Bunchman TE. Review of pediatric pheochromocytoma and paraganglioma. *Front Pediatr.* 2017;5:155.

17. Kavinga Gunawardane PT, Grossman A. The clinical genetics of phaeochromocytoma and paraganglioma. *Arch Endocrinol Metab.* 2017;61(5):490–500.

18. Williams MD. Paragangliomas of the head and neck: An overview from diagnosis to genetics. *Head Neck Pathol.* 2017;11(3):278–87.

19. Pacak K, Tella SH. Pheochromocytoma and paraganglioma. In: De Groot LJ, Chrousos G, Dungan K et al. editors. *Endotext* [Internet]. South Dartmouth (MA): MDText.com, Inc.; 2000-.2018 Jan 4.

20. Shida Y, Igawa T, Abe K et al. Composite pheochromocytoma of the adrenal gland: A case series. *BMC Res Notes.* 2015;8:257.

21. Xiao Z, She D, Cao D. Multiple paragangliomas of head and neck associated with hepatic paraganglioma: A case report. *BMC Med Imaging.* 2015;15:38.

22. Tischler AS, deKrijger RR. 15 Years of paraganglioma: Pathology of pheochromocytoma and paraganglioma. *Endocr Relat Cancer.* 2015;22(4):T123–33.

23. Eisenhofer G, Klink B, Richter S, Lenders JW, Robledo M. Metabologenomics of phaeochromocytoma and paraganglioma: An integrated approach for personalised biochemical and genetic testing. *Clin Biochem Rev.* 2017;38(2):69–100.

24. Zhikrivetskaya SO, Snezhkina AV, Zaretsky AR et al. Molecular markers of paragangliomas/pheochromocytomas. *Oncotarget.* 2017;8(15):25756–82.

25. Settas N, Faucz FR, Stratakis CA. Succinate dehydrogenase (SDH) deficiency, Carney triad and the epigenome. *Mol Cell Endocrinol.* 2018;469:107–11

26. Fishbein L, Orlowski R, Cohen D. Pheochromocytoma/paraganglioma: Review of perioperative management of blood pressure and update on genetic mutations associated with pheochromocytoma. *J Clin Hypertens (Greenwich).* 2013;15(6):428–34.

27. Smith JD, Harvey RN, Darr OA et al. Head and neck paragangliomas: A two-decade institutional experience and algorithm for management. *Laryngoscope Investig Otolaryngol.* 2017;2(6):380–9.

28. Lowery AJ, Walsh S, McDermott EW, Prichard RS. Molecular and therapeutic advances in the diagnosis and management of malignant pheochromocytomas and paragangliomas. *Oncologist.* 2013;18(4):391–407.

29. Castinetti F, Taieb D, Henry JF et al. Management of endocrine disease: Outcome of adrenal sparing surgery in heritable pheochromocytoma. *Eur J Endocrinol.* 2016;174(1):R9–18.

30. Chew WHW, Courtney E, Lim KH et al. Clinical management of pheochromocytoma and paraganglioma in Singapore: Missed opportunities for genetic testing. *Mol Genet Genomic Med.* 2017;5(5):602–7.

31. Jimenez C. Treatment for patients with malignant pheochromocytomas and paragangliomas: A perspective from the hallmarks of cancer. *Front Endocrinol (Lausanne).* 2018;9:277.

32. PDQ Adult Treatment Editorial Board. *Pheochromocytoma and Paraganglioma Treatment (PDQ®): Health Professional Version.* PDQ Cancer Information Summaries [Internet]. Bethesda (MD): National Cancer Institute (US); 2002-. 2018 Feb 8.

33. PDQ Pediatric Treatment Editorial Board. *Childhood Pheochromocytoma and Paraganglioma Treatment (PDQ®): Health Professional Version.* 2019 Dec 23. In: PDQ Cancer Information Summaries [Internet]. Bethesda (MD): National Cancer Institute (US); 2002.

34. Shuch B, Ricketts CJ, Metwalli AR, Pacak K, Linehan WM. The genetic basis of pheochromocytoma and paraganglioma: Implications for management. *Urology.* 2014;83(6):1225–32.

35. Nicolas M, Dahia P. Predictors of outcome in phaeochromocytomas and paragangliomas. *F1000Res.* 2017;6:2160.

36. Mamilla D, Araque KA, Brofferio A et al. Postoperative management in patients with pheochromocytoma and paraganglioma. *Cancers (Basel).* 2019;11(7):936.

37. Koopman K, Gaal J, de Krijger RR. Pheochromocytomas and Paragangliomas: New developments with regard to classification, genetics, and cell of origin. *Cancers (Basel).* 2019;11(8):1070.

38. Pereira BD, Luiz HV, Ferreira AG et al. Genetics of pheochromocytoma and paraganglioma. In: Mariani-Costantini R, editor. *Paraganglioma: A Multidisciplinary Approach* [Internet]. Brisbane (AU): Codon Publications; 2019 Jul 2. Chapter 1.

39. Shen Y, Cheng L. Biochemical diagnosis of pheochromocytoma and paraganglioma. In: Mariani-Costantini R, editor. *Paraganglioma: A Multidisciplinary Approach* [Internet]. Brisbane (AU): Codon Publications; 2019 Jul 2. Chapter 2.

40. Itani M, Mhlanga J. Imaging of pheochromocytoma and para-ganglioma. In: Mariani-Costantini R, editor. *Paraganglioma: A Multidisciplinary Approach* [Internet]. Brisbane (AU): Codon Publications; 2019 Jul 2. Chapter 3.

41. García MIDO, Palasí R, Gómez RC et al. Surgical and phar-macological management of functioning pheochromocy-toma and paraganglioma. In: Mariani-Costantini R, editor. *Paraganglioma: A Multidisciplinary Approach* [Internet]. Brisbane (AU): Codon Publications; 2019 Jul 2. Chapter 4.

42. Pang Y, Liu Y, Pacak K, Yang C. Pheochromocytomas and paragangliomas: From genetic diversity to targeted therapies. *Cancers (Basel)*. 2019;11(4):436.

43. Nölting S, Ullrich M, Pietzsch J et al. Current management of pheochromocytoma/paraganglioma: A guide for the practic-ing clinician in the era of precision medicine. *Cancers (Basel)*. 2019;11(10):1505.

59

Mahvash Disease

Anneliese Velez Perez and Jae Ro

CONTENTS

59.1 Introduction

The vast majority of pancreatic neuroendocrine tumors (PNET) are sporadic, but the recognition of an inherited PNET highlights the evolving clinical complexity in the era of personalized medicine. The initially well-defined inherited PNET syndromes comprise multiple endocrine neoplasia type 1 (MEN-1), von Hippel–Lindau disease (VHL), neurofibromatosis (NF) type 1, and tuberous sclerosis (TS). Over the past decade, the spectrum of inherited PNET has expanded with the inclusion of new presentations such as MEN type 4 (MEN-4), glucagon cell adenomatosis (GCA; Mahvash disease), along with germline mutations in DNA repair genes (MUTYH, BRCA2, and CHEK2) in a subset of seemingly sporadic PNET as well as isolated case reports of succinate dehydrogenase deficient (SDH)- and mismatch repair-deficient PNET occurring in the setting of SDH-related familial paraganglioma syndrome and Lynch syndrome, respectively. These findings provide evidence that the spectrum of inherited PNET is indeed larger than what most physicians generally anticipated, likely reaching a rate greater than 15% of the overall presentations. From a morphological perspective, the identification of multifocal/bilateral PNET or multiple organ involvement of PNET should prompt the attention of clinicians and pathologists to the possibility of an underlying genetic susceptibility. Since not all inherited PNET manifest with multifocal disease, careful morphological assessment of the tumor and the non-tumorous tissue, and application of immunohistochemical (IHC) biomarkers are essential in the determination of inherited PNET.

Mahvash disease is a newly described entity in the inherited PNET syndrome spectrum. It is characterized by inactivating glucagon receptor mutations resulting in multiple PNET, pancreatic α cell hyperplasia (PACH) and hyperglucagonemia without glucagonoma syndrome. These unique qualities phenotypically resemble those of G-protein-coupled glucagon receptor (GCGR) knockout mice, which triggered the discovery of the novel molecular findings and the development of a novel research model for Mahvash disease [1].

The WHO Classification of Tumors of Endocrine Organs (2017) includes Mahvash disease and GCA under the classification of glucagon cell hyperplasia and neoplasia (GCHN) [2]. These lesions have also been referred as PACH and nesidioblastosis. The terminology of GCA was first suggested by Henopp and colleagues in 2009 to describe PACH associated with multiple PNET unrelated to MEN1, VHL, or p27 MEN syndromes [3]. Later in the same year, Zhou and colleagues identified an inactivating GCGR gene mutation in a patient presenting with PACH, multiple PNET, and hyperglucagonemia without glucagonoma syndrome [4]. This specific clinical, pathologic, and molecular presentation was named Mahvash disease after the index patient in which the distinct etiology was first described [1,5].

However, subsequent testing of previously described patients with GCA showed that not all GCA cases harbored the characteristic GCGR gene mutation of Mahvash disease [6]. Therefore, two subtypes of GCA have been proposed according to their histologic morphology and GCGR mutational status: GCA positive for GCGR mutation and GCA negative for GCGR mutation [6].

59.2 Biology

The pancreas is a J-shaped, flattened, lobulated, soft organ (of 12–15 cm in length) located on the posterior abdominal wall behind the stomach. Structurally, the pancreas contains two functional compartments: exocrine and endocrine. The exocrine pancreas (serous gland) makes up >95% of the pancreatic parenchyma and includes a million "berry-like" clusters of zymogenic cells (acini) connected by ductules with associated connective tissue, vessels, and nerves. Regulated by secretin and cholecystokinin, the zymogenic cells are capable of producing trypsin (for digesting proteins), lipase (for digesting fats), amylase (for digesting carbohydrates), and other enzymes, whereas the ductular cells generate bicarbonate, which renders the pancreatic fluid alkaline.

The endocrine pancreas constitutes about 2% of pancreatic parenchyma and comprises clusters of pancreatic islets scattered throughout the pancreas. The islets include several cell types (α, β, δ, A, B, C, D, D1, E, F, enterochromaffin), collectively known as neuroendocrine cells. Of these, α cells (accounting for 20% of islets) secrete glucagon; β cells (68% of islets) secrete insulin and amylin; δ cells (10% of islets) secrete somatostatin, A cells secrete adrenocorticotropin (ACTH) and melanocyte-stimulating hormone (MSH), D cells secrete gastrin, D1 cells secrete vasoactive intestinal peptide (VIP), E cells (or epsilon cells, <1% of islets) secrete ghrelin, F cells (or PP/gamma cells, 2% of islets) secrete pancreatic polypeptide, and enterochromaffin cells secrete serotonin (5-HT). Upon secretion, these hormones are transferred via the blood to control energy metabolism and storage throughout the body.

While pancreatic neuroendocrine tumor (PNET) arising from glucagon-producing α cells is called glucagonoma, that arising from insulin-producing β cells is called insulinoma, that arising from somatostatin-producing δ cells is called somatostatinoma, that arising from gastrin-producing D cells is called gastrinoma, that arising from pancreatic polypeptide-producing F (PP/gamma) cells is associated with multiple hormonal syndromes, and that arising from serotonin-producing cells is called carcinoid tumor. Further, nonfunctioning PNET produces no or insufficient hormones but is associated with nonspecific clinical symptoms (e.g., vague abdominal pain).

Glucagon (abbreviated from *glucose agon*ist) is a 29-aa, 3.485 kDa polypeptide generated from the cleavage of proglucagon by proprotein convertase 2 in pancreatic islet α cells. Alternate source of proglucagon comes from intestinal L cells, including glicentin, GLP-1 (an incretin), IP-2, and GLP-2 (promotes intestinal growth). While hypoglycemic conditions induce α-cell secretion of glucagon, hyperglycemic conditions increase β-cell release of insulin.

Glucagon is a counterregulatory hormone for insulin, which lowers the extracellular glucose (usually stored in the liver as a polymer of glucose molecules or polysaccharide glycogen), and works to raise the concentration of glucose in the bloodstream through its binding to the glucagon receptor (GCGR, a G protein-coupled receptor of 485 aa) located in the plasma membranes of the liver (hepatocytes or liver cells) as well as the kidney, pancreas, heart, brain, and smooth muscle. This activates the stimulatory G protein, and then adenylate cyclase, triggering cAMP production and converting stored polysaccharide glycogen into glucose (i.e., glycogenolysis). After exhaustion of stored glycogen, glucagon promotes synthesis of additional glucose (i.e., gluconeogenesis) in the liver and kidneys. In addition, glucagon may shut off glycolysis in the liver, turning glycolytic intermediates into gluconeogenesis.

Pancreatic tumors, such as glucagonoma (an amino acid deficiency syndrome), are known to induce abnormally elevated levels of glucagon, with symptoms of necrolytic migratory erythema, reduced amino acids, and hyperglycemia.

The novel molecular etiology first described in Mahvash disease was a homozygous missense mutation on exon 4 of the GCGR gene producing an exchange of a proline to a serine at position 86 (P86S) [4]. The specific P86S mutations resulted in decreased glucagon binding potential, abnormal receptor internalization and calcium mobilization, and increased apoptosis [4,7]. Additional germline mutations producing missense mutations and stop codons were found in patients with similar clinical and pathological characteristics in exons 2, 8, 11, 12 [6,8]. No mutations in MEN1 or VHL genes were found in any of the described patients.

Patients with GCA but without GCGR mutations have also been described. The genetic background and pathophysiology of such cases have not been clearly elucidated.

59.3 Pathogenesis

Glucagon is an important hormone secreted by pancreatic α cells which regulates glucose metabolism by increasing glycolysis and gluconeogenesis in response to hypoglycemia. Glucagon uses a GCGR that results in the production of cAMP [7,9]. Its receptor, GCGR, is widely distributed in an array of tissues, mainly in the liver and kidney, but also in the pancreas, gastrointestinal tract, and other sites [8].

Murine models with GCGR knockout mice have provided valuable insight into the possible pathogenesis and natural disease progression of Mahvash disease. Mice with a null mutation of GCGR exhibited markedly elevated levels of glucagon and glucagon-like peptide 1 in association with PACH [10]. Sophisticated studies with GCGR deficient mice by Yu and colleagues demonstrated that defective glucagon signaling leads to GCA with the eventual development of well-differentiated PNET and micro PNET with rare metastasis [7].

The development of PACH after GCGR inhibition raises the possibility of a feedback mechanism between glucagon and pancreatic α cells. Murine studies showed that inhibition of liver-specific GCGR led to hyperglucagonemia and PACH despite normal function of pancreatic GCGR [11]. These findings, along with other published data, led to the realization that the liver likely produces or secretes a factor that increases á cell density in the pancreas [12–14].

Additional studies performed by Solloway and colleagues demonstrated that glucagon also regulates hepatic amino acid metabolism and serum amino acid (sAA) levels [15]. Moreover, they showed that inhibition of GCGR led to increased sAA levels and PACH by activation of the mammalian target of rapamycin (mTOR). These findings provide supporting evidence that amino acids are the missing link in the feedback mechanism between the liver and the pancreas. Therefore, a liver-to-pancreatic α cell axis has been proposed where increased sAA levels promote secretion of glucagon by α cells, which in return accelerates ureagenesis and increases amino acid clearance [16]. It is then hypothesized that disruption or inactivation of hepatic GCGR leads to elevated levels of sAA, causing PACH and hyperglucagonemia [16].

It is unclear how PACH progresses to PNET in patients with Mahvash disease. Some authors speculate that PNET arising in

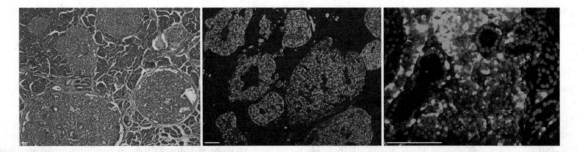

FIGURE 59.1   Pancreas in Mahvash disease. Left: pancreatic islet cell hyperplasia (H&E). Center: glucagon expression in islet cells (green). Right: inverted glucagon (green) to insulin (red) secreting cell ratio (counterstained with nuclear dye blue). (With permission from Lucas MB et al. *J Mol Genet Med.* 2013;7:84.)

a background of PACH acquire additional mutations that contribute to their tumorigenesis [1]. Further studies are needed to elucidate the progression from hyperplasia to neoplasia.

59.4 Epidemiology

Mahvash disease is a very rare disease which was termed by Ouyang and colleagues in 2011 after identifying the novel GCGR mutation in their index patient [5]. To date, at least 14 possible cases of GCA have been identified in the English literature, with 5 cases demonstrating GCGR mutations [3–6,8,17–22]. However, it is debatable if all of these cases truly represent GCA and/or Mahvash disease.

Both males and females appear to be affected with a probable slight male predominance (male to female ratio 1.6:1). Patient age ranges from 25 to 68 years with a mean age of 47. In some cases, a history of consanguinity is present [4,6]. Genetic and genealogical studies support an autosomal recessive mode of inheritance.

Since Mahvash disease is a newly recognized entity, the true incidence of the disease is not yet known. A few cases have been identified in patients of Middle Eastern descent [6]. Based on the estimated frequency of deleterious GCGR polymorphisms, Lucas and colleagues predicted that the prevalence may be approximately 4 per million, with 0.4% frequency of heterozygous carrier state and 0.2% frequency of inactivating GCGR mutations [1].

59.5 Clinical Features

As described earlier, patients with Mahvash disease present with multiple PNET in a background of GCA. There is mild to severe hyperglucagonemia that can reach levels up to 60,000 pg/mL (normal less than 150 pg/mL). However, due to deficient GCGR, patients with Mahvash syndrome, which is sAA deficiency syndrome caused by high glucagon levels, do show high glucagon level (hyperglucagonemia) but do not have signs of glucagonoma syndrome, such as skin rash, hyperglycemia, thromboembolism, or stomatitis. Glucagon levels may normalize after tumor resection in mild cases or remain markedly high in severely affected patients.

The most common presenting symptom is abdominal pain or discomfort. Some patients may be asymptomatic, with a pancreatic tumor incidentally found by imaging performed for other reasons. Other symptoms may include weight loss, mild hypoglycemia, constipation or diarrhea, fatigue, recurrent pancreatitis, polyuria, and polydipsia.

The PNET in GCA and Mahvash disease can occur in any location of the pancreas. They are well differentiated, demonstrate very low Ki67 proliferation index (generally <2%), and show glucagon expression (Figure 59.1). However, they behave as nonfunctional PNET due to GCGR defects. They exhibit very low metastatic potential with only very rare cases exhibiting metastasis to the lymph nodes.

59.6 Diagnosis

PNET in Mahvash disease can be visualized by different imaging modalities, including CT scan, magnetic resonance imaging, ultrasound, and radiolabeled octreotide scans. If the lesion is too small, it may not be captured by conventional imaging studies. Fine needle aspiration may be used for tissue diagnosis in patients with a well-visualized tumor nodule.

Blood glucagon levels should be measured before and after resection of PNET. However, a high level of suspicion is required, which may not be achieved due to the rarity of the condition and the lack of symptoms usually produced by glucagonoma syndrome. Additional laboratory tests that may be ordered in conjunction to blood glucagon levels include blood glucose, insulin levels, gastrin, vasoactive peptide, pancreatic polypeptide, etc.

In suspected cases of Mahvash disease, blood or tissue samples should be sent for GCGR sequencing to stablish a definite diagnosis. Additional testing to rule out other inherited PNET syndrome, such as MEN1, VHL, and p57 MEN1, should be performed. In cases where the patient presents with all the classic features of Mahvash disease, and where other neuroendocrine tumor syndromes have been ruled out by history and/or imaging, a diagnosis can be made without the need of GCGR sequencing [1]. In established cases, genetic counseling for siblings and relatives should be strongly considered.

Pathologically, Mahvash disease is a unique syndrome characterized by the presence of multiple PNET in a background of GCA without hyperglucagonemic symptoms. These findings are appreciated by the histopathologic examination of surgical resection specimens obtained by pylorus-sparing pancreaticoduodenectomy or distal pancreatectomy procedures. The resected pancreas is either of normal size to hypertrophic and may demonstrate one or more well-circumscribed, round to lobulated, solid tumors with tan and homogenous cut surfaces. The tumors may range from 0.5 to 10 cm in size (mean size 3.7 cm). Occasionally, cystic degeneration is present, and patients may present as a cystic pancreatic mass [19]. The tumors are composed of monomorphic round cells with eosinophilic cytoplasm, fine stippled chromatin, and a lobular or trabecular growth pattern. The mitotic index is very low (<2/10 high power fields) and Ki67 proliferation index (<3%) is also low. These tumors show expression of synaptophysin, chromogranin, and neuron-specific enolase by immunohistochemistry, confirming their neuroendocrine differentiation. They may also express glucagon or demonstrate negative expression of all other pancreatic hormones, including glucagon, insulin, somatostatin, or pancreatic polypeptide.

The histopathological correlation of GCA is the presence of diffuse PACH in 60%–80% of islet cells (Figure 59.1). The enlarged islets have a mean size of 324 μm (range 176–726 μm) compared to normal pancreatic tissue (mean 117 μm) [3]. The large nests can produce confluent and expanded areas which may imperceptibly merge into a microadenoma or micro-PNET. In fact, the pancreas of these patients can contain multiple micro-PNET of less than 5 mm, which may be very difficult to differentiate from a giant islet cell island. Immunohistochemistry of enlarged islet cells shows an inverted glucagon to insulin secreting cell ratio, demonstrating the increased number of glucagon producing α cells when compared to insulin-producing α cells (Figure 59.1).

The PACH to neoplastic progression sequence has not been well elucidated. However, we can appreciate histologically a seemingly indiscernible transition from CGA to microadenoma to PNET. These findings suggest a type of reactive PACH resulting in hyperglucagonemia produced by the inactivating GCGR mutation [5]. The genetic landscape of GCA harboring wild-type GCGR remains an area of interest; however, GCA with wild-type GCGR and mutant GCGR exhibits some phenotypical differences. GCA patients harboring germline GCGR mutations were found to display larger (>5 mm) and increased number of PNET (including micro PNET and PNET) when comparing with those lacking the GCGR mutation [6].

GCA-related micro PNET and PNET lack immunohistochemical plurihormonality that is frequently identified in MEN-1-related hyperplasia-neoplasia sequence, though some micro PNET can display scattered pancreatic polypeptide cells in GCA. The use of menin (human MEN1) immunohistochemistry may be useful in the distinction of some challenging cases from MEN1. In a given case, the identification of CGA, multifocal microadenomas, and PNET and the use of immunohistochemical biomarkers (menin, p27, tuberin, alpha-inhibin, CAIX, MMR proteins, and SDHB) allow the practicing pathologist to raise the possibility of a familial disorder.

59.7 Treatment

Since Mahvash disease is a newly described entity, the optimal clinical management has not been well established. It usually entails surgical resection of PNET and close clinical follow-up with radiographic imaging and measurement of blood glucose levels. Somatostatin analogs may also be used to suppress glucagon production if symptoms of hypoglycemia

or other complications related to glucagon levels persist [17,23]. Pharmacological chaperones have shown to help in increasing GCGR functionality in cell models; however, feasible clinical agents that can be used in human subjects are still in development [1,24].

59.8 Prognosis

There is limited information on the long-term implications of Mahvash disease. Follow-up data from confirmed and possible cases show no recurrence of PNET after excision up to 18 years. However, it is not unfathomable to think that tumor recurrence may occur in any remnant of pancreatic tissue of these patients. Of all the cases, only one patient developed lymph node micrometastasis. All case subjects were well after excision except for three patients who passed away due to complications which included necrotizing pancreatitis and sepsis [6]. Of the deceased patients, one patient had glucagonoma syndrome, which may argue against its inclusion in the case group.

Murine models have provided some insight into the natural disease progression of Mahvash disease. Yu and colleagues performed a sophisticated study observing pancreas from GCGR-deficient mice compared to wild-type and heterozygous mice. In their study, they found that at around 10–12 months, GCGR-deficient mice developed PNET and microtumors with occasional liver metastasis [7]. In a second experiment from Yu and colleagues, GCGR-deficient mice were observed for a period of 22 months, and relevant findings were compared to those observed in wild-type and heterozygous mice. It was noted that GCGR-deficient mice had markedly decreased survival when compared to their normal counterparts [24]. It remains to be seen if these findings can be extrapolated to human subjects.

59.9 Conclusion and Future Perspectives

Mahvash disease is a newly described entity with at least 14 cases published in the English literature. Patients with Mahvash disease present with multiple PNET and micro PNET arising in a background of GCA, and hyperglucagonemia without glucagonoma syndrome. The novel molecular finding in these patients is an inactivating mutation in the GCGR gene [25]. However, not all patients with similar clinical findings harbor GCGR mutations. These findings suggest that not all cases are caused by defects in GCGR. Further studies are needed to elucidate if GCGR wild-type GCA cases have a different mechanism of disease.

Additional research is needed to expand our understanding of this interesting entity. There is a need to establish objective diagnostic criteria to recognize patients with GCA and Mahvash disease. It is essential to rule out other PNET syndromes by a thorough clinical history, radiographic imaging, and laboratory testing. An optimal treatment plan must be established by a multidisciplinary approach involving treating physicians, radiologists, and pathologists. Current patients with Mahvash disease should be closely followed to detect any new early PNET before progressing to tumor metastasis. Also, siblings and relatives from established Mahvash disease patients may benefit from blood glucose screening and genetic counseling.

REFERENCES

1. Lucas MB, Yu VE, Yu R. Mahvash disease: Pancreatic neuroendocrine tumor syndrome caused by inactivating glucagon receptor mutation. *J Mol Genet Med.* 2013;7:84.

2. Lloyd RV, Osamura RY, Klöppel G, Rosai J. WHO Classification of Tumours of Endocrine Organs. *Lyon, Int Agency Res Cancer.* 2017;282–3.

3. Henopp T, Anlauf M, Schmitt A. et al. Glucagon cell adenomatosis: A newly recognized disease of the endocrine pancreas. *J Clin Endocrinol Metab.* 2009;94(1):213–7.

4. Zhou C, Dhall D, Nissen NN, Chen CR, Yu R. Homozygous P86S mutation of the human glucagon receptor is associated with hyperglucagonemia, á cell hyperplasia, and islet cell tumor. *Pancreas.* 2009;38(8):941–6.

5. Ouyang D, Dhall D, Yu R. Pathologic pancreatic endocrine cell hyperplasia. *World J Gastroenterol.* 2011;17(2):137–43.

6. Sipos B, Sperveslage J, Anlauf M et al. Glucagon cell hyperplasia and neoplasia with and without glucagon receptor mutations. *J Clin Endocrinol Metab.* 2015;100(5):E783–8.

7. Yu R, Dhall D, Nissen NN, Zhou C, Ren SG. Pancreatic neuroendocrine tumors in glucagon receptor-deficient mice. *PLOS ONE.* 2011;6(8):e23397.

8. Miller HC, Kidd M, Modlin IM et al. Glucagon receptor gene mutations with hyperglucagonemia but without the glucagonoma syndrome. *World J Gastrointest Surg.* 2015;7(4):60–6.

9. Mayo KE, Miller LJ, Bataille D et al. International Union of Pharmacology. XXXV. The Glucagon Receptor Family. *Pharmacol Rev.* 2003;55(1):167–94.

10. Gelling RW, Du XQ, Dichmann DS et al. Lower blood glucose, hyperglucagonemia, and pancreatic á cell hyperplasia in glucagon receptor knockout mice. *Proc Natl Acad Sci U S A.* 2003;100(3):1438–43.

11. Longuet C, Robledo AM, Dean ED et al. Liver-specific disruption of the murine glucagon receptor produces á-cell hyperplasia: Evidence for a circulating á-cell growth factor. *Diabetes.* 2013;62(4):1196–205.

12. Chen M, Gavrilova O, Zhao WQ et al. Increased glucose tolerance and reduced adiposity in the absence of fasting hypoglycemia in mice with liver-specific Gs alpha deficiency. *J Clin Invest.* 2005;115(11):3217–27.

13. Lee Y, Berglund ED, Wang MY et al. Metabolic manifestations of insulin deficiency do not occur without glucagon action. *Proc Natl Acad Sci U S A.* 2012;109(37):14972–6.

14. Yu R, Zheng Y, Lucas MB, Tong YG. Elusive liver factor that causes pancreatic á cell hyperplasia: A review of literature. *World J Gastrointest Pathophysiol.* 2015;6(4):131–9.

15. Solloway MJ, Madjidi A, Gu C et al. Glucagon couples hepatic amino acid catabolism to mTOR-dependent regulation of á-cell mass. *Cell Rep.* 2015;12(3):495–510.

16. Holst JJ, Wewer Albrechtsen NJ, Pedersen J, Knop FK. Glucagon and amino acids are linked in a mutual feedback cycle: The liver-á-cell axis. *Diabetes.* 2017;66(2):235–40.

17. Yu R, Nissen NN, Dhall D, Heaney AP. Nesidioblastosis and hyperplasia of á cells, microglucagonoma, and nonfunctioning islet cell tumor of the pancreas. Review of the literature. *Pancreas.* 2008;36(4):428–31.

18. Al-Sarireh B, Haidermota M, Verbeke C, Rees DA, Yu R, Griffiths AP. Glucagon cell adenomatosis without glucagon receptor mutation. *Pancreas.* 2013;42(2):360–2.

19. Brown K, Kristopaitis T, Yong S, Chejfec G, Pickleman J. Cystic glucagonoma: A rare variant of an uncommon neuroendocrine pancreas tumor. *J Gastrointest Surg.* 1998; 2(6):533–6.

20. Martignoni ME, Kated H, Stiegler M et al. Nesidioblastosis with glucagon-reactive islet cell hyperplasia: A case report. *Pancreas.* 2003;26(4):402–5.

21. Balas D, Senegas-Balas F, Delvaux M et al. Silent human pancreatic glucagonoma and "A" nesidioblastosis. *Pancreas.* 1988;3(6):734–9.

22. Azemoto N, Kumagi T, Yokota T et al. An unusual case of subclinical diffuse glucagonoma coexisting with two nodules in the pancreas: Characteristic features on computed tomography. *Clin Res Hepatol Gastroenterol.* 2012;36(3):e43–7.

23. Ro C, Chai W, Yu VE, Yu R. Pancreatic neuroendocrine tumors: Biology, diagnosis, and treatment. *Chin J Cancer.* 2013;32(6):312–24.

24. Yu R, Chen CR, Liu X, Kodra JT. Rescue of a pathogenic mutant human glucagon receptor by pharmacological chaperones. *J Mol Endocrinol.* 2012;49(2):69–78.

25. Yu R, Ren SG, Mirocha J. Glucagon receptor is required for long-term survival: A natural history study of the Mahvash disease in a murine model. *Endocrinol Nutr.* 2012;59(9):523–30.

60

Multiple Endocrine Neoplasia

Dongyou Liu

CONTENTS

60.1 Introduction

Multiple endocrine neoplasia (MEN) is an autosomal dominant disorder of the hormone-producing glands (i.e., the endocrine system) characterized by the formation of histologically similar tumors in two or more endocrine glands, and the presence of different gene mutations.

According to the specific genes involved, type of hormones produced, and glandular and extraglandular sites affected, MEN is distinguished into five subtypes: MEN type 1 (MEN1), MEN type 2A (MEN2A), MEN type 2B (MEN2B), familial medullary thyroid carcinoma (FMTC), and MEN type 4 (MEN4), for which germline heterozygous mutations involving the *MEN1* (MEN1), *RET* (MEN2A, MEN2B, and FMTC), and *CDKN1B* (MEN4) genes are fundamentally responsible (Table 60.1) [1–3].

60.2 Biology

The endocrine system consists of several hormone-producing glands (e.g., adrenal, pituitary, parathyroid, thyroid, and pancreatic) that secrete various hormones for the regulation of cellular and tissue functions throughout the body.

Evolving from neoplastic proliferations of neuroendocrine cells, MEN is a tumor predisposition disorder of the endocrine system that was initially documented by Erdheim in 1903 from a patient with acromegaly and four enlarged parathyroid glands. After the report of classic MEN1 tumor triad (i.e., nonmalignant neoplasms of the pituitary gland, the parathyroid gland, and the pancreas) by Cushing and Davidhoff in the 1920s, and further

description by Underdahl and coworkers in 1953, Wermer in 1954, and Zollinger and Ellison in 1955, the term multiple endocrine neoplasia type I (MEN1) was proposed by Lulu and coworkers in 1968 as the preferred designation for this autosomal dominant syndrome. The gene region underlying the development of MEN1 was mapped to chromosome 11q13.1 in 1988, and the *MEN1* gene was confirmed in 1997 by Chandrasekharappa et al. and Lemmens et al. as the cause of disease [1,2].

MEN2 was first noted in the nineteenth century, and its association with bilateral pheochromocytoma and medullary thyroid cancer (MTC, a carcinoma of the parafollicular calcitonin secreting C cells) and was observed by Sipple in 1961. Further extension of its clinical profile to hyperparathyroidism and Cushing's disease allowed a precise definition of MEN2A (or Sipple syndrome) as a subtype that encompasses both benign and malignant endocrine neoplasia with other nonendocrine diseases (i.e., MTC, pheochromocytoma, hyperparathyroidism, and Cushing's disease) in 1968. Differences in the combination of endocrine neoplasia with or without nonendocrine diseases gave rise additional MEN2 subtypes (i.e., MEN2B [also known as MEN3 or mucosal neuroma syndrome, formerly Wagenmann–Froboese syndrome] and FMTC [familial medullary thyroid carcinoma, a variant of MEN2A]). The observation of germline *RET* mutations in MEN2A (MTC, pheochromocytoma, hyperparathyroidism, and Cushing's disease), MEN2B (multicentric bilateral MTC, and bilateral pheochromocytoma in 50% of cases, in association with a marfanoid habitus, mucosal neuromas, medullated corneal fibers, and intestinal autonomic ganglion dysfunction leading to megacolon) and FMTC (MTC only) confirmed their genetic relatedness [3].

MEN4 (also referred to as MENX) was recognized as a subtype that covers some MEN1-like cases without mutations in the *MEN1* gene. Presenting with parathyroid and anterior pituitary tumors, in possible association with tumors of the reproductive organs, adrenals, and kidneys, MEN4 was linked to germline mutation in the cyclin-dependent kinase inhibitor (*CDKN1B*) gene which encodes p27 protein in 2006 [4,5].

60.3 Pathogenesis

MEN encompasses five clinically distinct subtypes that are attributable to germline mutations in the *MEN1* (MEN1), *RET* (MEN2A, MEN2B, and FMTC), and *CDKN1B* (MEN4) genes (Table 60.1) [6].

The *MEN1* (multiple endocrine neoplasia type 1) gene on chromosome 11q13.1 spans 7.7 kb with 10 exons (as the first

exon and part of exon 10 in the original 9 kb gene are not translated) and encodes a 615 aa, 68 kDa protein (menin). As a component of the histone methyltransferase complex (consisting of methyltransferase, RNA polymerase II, and menin) that specifically methylates "Lys-4" of histone H3 (H3K4), menin participates in the regulation of cell proliferation, apoptosis, and genome integrity through several pathways and processes (e.g., TGFB1-mediated inhibition of cell-proliferation, SMAD3 transcriptional activity, telomerase expression, JUND-mediated transcriptional activation on AP1 sites, DNA repair, and p27 expression) [7,8].

Mutations in *MEN1* produce a loss-of-function menin, disrupting its interaction with other proteins, incapacitating its function in cell cycle regulation and proliferation, and resulting in decreased p27 expression, MEN1 (*MEN1* pathogenic variants observed in 80% of cases), isolated hyperparathyroid tumor, gastrinoma, insulinoma, and bronchial carcinoid.

TABLE 60.1

Molecular and Clinical Characteristics of Multiple Endocrine Neoplasia (MEN) Subtypes

Subtype (Alternative Name)	Gene (Size)	Chromosome Location	Protein (Size)	Organ Involvement	Clinical Features
MEN1	*MEN1* (7.7 kb)	11q13.1	Menin (615 aa, 68 kDa)	*Glands*: parathyroid, pancreas, pituitary, thyroid, adrenal *Extraglands*: bronchus, thymus, stomach, lipomas, skin tumors	Parathyroid adenoma (90%); pituitary adenoma (30%–40%; including prolactinoma 20%, somatotrophinoma 10%, corticotrophinoma <5%, nonfunctioning <5%); enteropancreatic tumor (30%–70%; including gastrinoma 40%, insulinoma 10%, nonfunctioning and PPoma 20–55%, glucagonoma <1%, VIPoma <1%); other tumors (adrenal cortical tumor 40%, pheochromocytoma <1%, bronchopulmonary NET 2%, thymic NET 2%, gastric NET 10%, lipoma 30%, angiofibroma 85%, collagenoma 70%, meningioma 8%); death due to pancreatic NET, bronchial or thymic NET
MEN2A	*RET* (53 kb)	10q11.21	RET (1,114 aa, 124 kDa)	*Glands*: thyroid, adrenal, parathyroid *Extraglands*: none	Medullary thyroid carcinoma (MTC, 90%), pheochromocytoma (50%) and parathyroid adenoma (25%–30%); cutaneous lichen amyloidosis (10%); Hirschsprung disease (2%); 70%–80% of all MEN2 cases; death due to MTC
MEN2B (MEN3)	*RET* (53 kb)	10q11.21	RET (1,114 aa, 124 kDa)	*Glands*: thyroid, adrenal *Extraglands*: marfanoid habitus, mucosal neuromas, corneal nerve hypertrophy, intestinal ganglioneuromatosis	MTC (95%), pheochromocytoma (40%–50%); mucosal/intestinal neuroma (100%); marfanoid habitus (100%); mucosal/intestinal ganglioneuroma (60%); 5% of all MEN2 cases; death due to MTC
FMTC (MEN2A variant)	*RET* (53 kb)	10q11.21	RET (1,114 aa, 124 kDa)	*Glands*: thyroid *Extraglands*: none	MTC (100%), late onset, low penetrance; 10%–20% of all MEN2 cases
MEN4 (MENX, MEN1 variant)	*CDKN1B* (7.3 kb)	12p13.1	p27 (198 aa, 22 kDa)	*Glands*: parathyroid, gastroenteropancreatic (GEP), pituitary *Extraglands*: none	Parathyroid adenoma; pituitary adenoma; reproductive organ tumors (testicular cancer, neuroendocrine cervical carcinoma); adrenal, and renal tumors

To date, >1300 heterozygous germline *MEN1* germline mutations (including deletions 45%, nonsense mutations 25%, insertions 15%, missense mutations 10%, and splice site mutations <5%) have been identified in patients with MEN1. Of these, >70% of the mutations cause premature stop codon and truncated protein. Notable *MEN1* pathogenic variants include c.249_252delGTCT (deletion at codons 83–84), c.1546_1547insC (insertion at codon 516), c.1378C>T (Arg460Ter) and c.628_631delACAG (deletion at codons 210–211), which are found in 4.5%, 2.7%, 2.6% and 2.5% of MEN1-affected families [9].

The *RET* (rearranged during transfection) gene on chromosome 10q11.21 spans 53 kb with 21 exons and encodes a 1,114 aa, 124 kDa protein (transmembrane tyrosine kinase), which is composed of an extracellular portion (including signal peptide, cadherin-like domain, and cysteine-rich region), a single transmembrane region, and an intracellular portion (including two tyrosine kinase domains, TK1 and TK2). Interaction with ligands such as glial cell line-derived neurotrophic factor (GDNF), neurturin (TNT), artemin, or persephin stimulates RET dimerization, cross-autophosphorylation, and subsequent phosphorylation of intracellular substrates, leading to activation of downstream signaling pathways involved in cell differentiation, growth, migration, and survival [10].

Germline gain-of-function *RET* mutations (e.g., missense mutation at codons 609, 611, 618, 620, or 634; point mutation at codon 918) overactivate the protein's signaling function, triggering cell growth and division, and development of MEN2. While mutations in the cysteine-rich region of RET extracellular domain (coded by the genes in exon10 and 11) are associated with MEN2A, mutations in the intracellular TK2 domain cause MEN2B-associated tumors [11].

By contrast, germline loss-of-function *RET* mutations cause Hirschsprung disease (HSCR), with enlargement of the bowel and constipation or obstipation in neonates.

The *CDKN1B* (cyclin dependent kinase inhibitor 1B or P27Kip1) gene on chromosome 12p13.1 measures 7.3 kb and encodes a 198 aa, 22 kDa protein (p27), which binds to and prevents the activation of cyclin E-CDK2 or cyclin D-CDK4 complexes and thus controls the cell cycle progression at G1. Heterozygous loss-of-function mutations in the *CDKN1B* gene reduce the amount of functional p27, leading to unchecked cell growth and division. Typically occurring in MEN 1 patients without *MEN1* mutations (20% of cases), germline *CDNK1B* mutations may be occasionally associated with sporadic primary hyperparathyroidism [12,13].

60.4 Epidemiology

60.4.1 Prevalence

MEN1 has an estimated incidence of 1 per 30,000 and affects men and women equally. Thymic tumors are exclusively found in men, pancreatic tumors occur more often in men, and pituitary tumors are slightly higher in women. MEN2 has an incidence of 1 per 35,000, with MEN2A, FMTC, and MEN2B subtypes making up 70%–80%, 10%–20%, and 5% of all MEN2 cases, respectively. MEN4 is rare and accounts for 1/10 of MEN1 cases (which have no *MEN1* mutations) [14].

60.4.2 Inheritance

All MEN subtypes (i.e., MEN1, MEN2A, MEN2B, FMTC, and MEN4) demonstrate autosomal dominant inheritance. About 90% of MEN1 patients inherit a germline heterozygous pathogenic variant from one parent (familial cases) and 10% of MEN1 patients acquire de novo mutation at an early embryonic stage (sporadic cases), which is present in all cells at birth. For tumor development, an *MEN1* somatic mutation (second hit) leading loss of heterozygosity (LOH) in the predisposed endocrine cell is required.

60.4.3 Penetrance

MEN1 shows a high degree of penetrance (>95% on biochemical signs, and 80% on clinical signs by the fifth decade of life), which appears to be age-dependent, ranging from 7%, 52%, 87%, and 98% to 99% and 100% in patients of 10, 20, 30, 40, 50, and 60 years of age, respectively. MEN2 also displays very high penetrance and variable expressivity. Indeed, MTC shows 100% penetrance in MEN2A and MEN2B.

60.5 Clinical Features

MEN1 is associated with 20 mostly benign endocrine and non-endocrine tumors, especially neoplasms of the pituitary gland, the parathyroid gland, and the pancreas (the 3 "P"s). MEN1 parathyroid tumor may induce hypercalcemia, lethargy, depression, confusion, anorexia, constipation, nausea, vomiting, diuresis, dehydration, hypercalciuria, kidney stones, increased bone resorption/fracture risk, hypertension, and shortened QT interval. MEN1 pituitary tumor is adenoma of anterior cells, such as growth hormone-secreting prolactinoma, which may cause oligomenorrhea/amenorrhea and galactorrhea in females and sexual dysfunction in males. MEN1 pancreatic tumor involving the islet cells may appear as gastrinoma or insulinoma. MEN1 adrenocortical tumor may show primary hypercortisolism or hyperaldosteronism (Table 60.1) [15,16].

MEN2 causes both malignant and benign tumors, including MTC (a neuroendocrine tumor of the thyroid gland due to hyperplasia of calcitonin-producing parafollicular C-cells, the only cells in the thyroid gland derived from neural crest cells), and C cell hyperplasia (a preneoplastic lesion) (Figure 60.1). MTC may be very aggressive in MEN2B (often in early childhood, with affected patients rarely surviving after the adolescence), variably aggressive in MEN2A (often in early adulthood), and almost indolent in FTMC (often in middle age). In addition, MEN2A and MEN2B may develop pheochromocytoma (a benign bilateral/multicentric adrenal medullary tumor; 40%–50% of cases; presenting as paroxysmal attacks of headache, anxiety, diaphoresis, and palpitations; mean age of presentation at 25–32 years, and as early as 8–12 years) (Figure 60.2); MEN2A may have parathyroid adenoma or hyperplasia, mild or asymptomatic primary hyperparathyroidism (10%–25% of cases), cutaneous lichen amyloidosis (CLA; presenting as pruritic, pigmented, scaly papules usually in extensor surfaces of extremities and interscapular region due to amyloid deposition), Hirschsprung disease (HSCR or chronic aganglionic megacolon; with the absence of autonomic ganglion

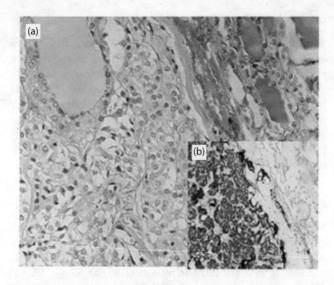

FIGURE 60.1 MEN2A in a 40-year-old woman with medullary thyroid carcinoma showing: (a) sheets of polygonal cells with amphophilic cytoplasm, oval nuclei, "salt-and-pepper" granular chromatin, and inconspicuous nucleoli in the resected thyroid gland (hematoxylin-eosin-safran, ×200); (b) tumor cells with strong positivity for chromogranin. (Photo credit: Efared B et al. *J Med Case Rep.* 2017;11:208.)

cells in the parasympathetic chain of the sigmoid colon, leading to peristalsis, chronic obstruction, and megacolon); and MEN2B shows mucosal neuromas of the lips and tongue, distinctive facies with enlarged lips, ganglioneuromatosis of the gastrointestinal tract, marfanoid habitus, kyphoscoliosis/lordosis, joint laxity, and lack of tears. FMTC is noted for the occurrence of MTC in more than one individual in families without pheochromocytoma or parathyroid adenoma/hyperplasia (Table 60.1) [3,17–20].

MEN4 consists of MEN1 cases that lack *MEN1* mutations, and thus resembles MEN1 clinically (Table 60.1) [12].

60.6 Diagnosis

Diagnostic workup for MEN involves family history review, clinical findings, biochemical evaluation (e.g., intact PTH, serum calcium, prolactin, somatomedin C, glucose, insulin, pro-insulin, gastrin, pancreatic polypeptide, glucagon, and ingestion of a test meal followed by measurement of pancreatic polypeptide and gastrin), imaging study, histopathology, and molecular analysis [21,22].

Establishment of MEN1 diagnosis requires meeting any one of three criteria: (i) an individual with a known *MEN1* gene

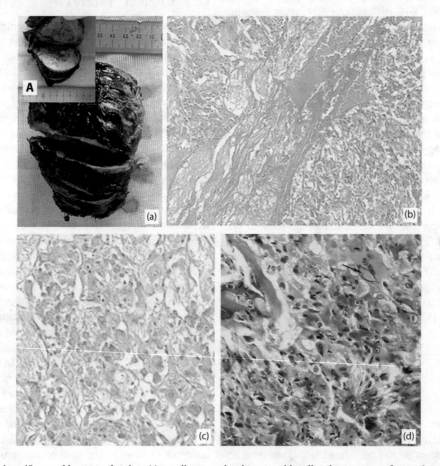

FIGURE 60.2 MEN2A in a 40-year-old woman showing: (a) a well-encapsulated tumor with yellow-brown cut surface and areas of hemorrhage in the resected right adrenal gland; (b) two components of the tumors in the right adrenal gland: ganglioneuroma (left portion) and pheochromocytoma (right portion) separated by dilated blood vessels (hematoxylin-eosin-safran [HES], ×100); (c) multiple mature ganglion cells with abundant eosinophilic cytoplasm, round nuclei, conspicuous nucleoli, within a fibrillary background in the ganglioneuroma component of the tumor (HES ×200); (d) polygonal, oval-shaped, and spindle cells with amphophilic cytoplasm, round to spindle-shaped nuclei, inconspicuous nucleoli (including a bizarre cell with pseudoinclusion at the top part of the image [black arrow]) in the pheochromocytoma component of the tumor (HES, ×400). (Photo credit: Efared B et al. *J Med Case Rep.* 2017;11:208.)

mutation but does not have clinical or biochemical evidence of disease, (ii) an individual with one MEN1-associated tumor and a first-degree relative diagnosed with MEN1, and (iii) an individual with at least two MEN1-associated tumors [1,23,24].

Establishment of MEN2 diagnosis depends on fulfilling relevant clinical criteria. MEN2A displays two or more specific endocrine tumors (MTC, pheochromocytoma, or parathyroid adenoma/hyperplasia) in a single individual or in close relatives. MEN2B shows early-onset MTC, mucosal neuromas of the lips and tongue, medullated corneal nerve fibers, distinctive facies with enlarged lips, and an asthenic "marfanoid" body habitus. FMTC has four or more cases of MTC in the absence of pheochromocytoma or parathyroid adenoma/hyperplasia in families [3].

Establishment of MEN4 diagnosis is based on identification of *CDKN1B* pathogenic variant from MEN1 patients who harbor no *MEN1* mutation [12].

Molecular identification of heterozygous germline *MEN1*, *RET*, and *CDKN1B* pathogenic variant is confirmative for MEN1, MEN2 (including MEN2A, MEN2B, and FMTC), and MEN4 even if clinical features are inconclusive.

Differential diagnoses for MEN include von Hippel–Lindau syndrome (pheochromocytoma, renal cell carcinoma, cerebellar and spinal hemangioblastoma, and retinal angioma; autosomal dominant disorder due to germline *VHL* pathogenic variant), tuberous sclerosis complex, hereditary paraganglioma-pheochromocytoma syndrome (due to *SDHA*, *SDHB*, *SDHC*, *SDHD*, and *SDHAF2* pathogenic variants), *TMEM127*-associated susceptibility to pheochromocytoma (due to germline *TMEM127* pathogenic variant), *MAX*-associated susceptibility to pheochromocytoma (due to *MAX* germline pathogenic variant), neurofibromatosis type 1 (NF1; pheochromocytoma, multiple café-au-lait macules, neurofibromas, Lisch nodules, axillary or inguinal freckling, and/or positive family history; due to *NF1* pathogenic variant), polycythemia and paraganglioma/pheochromocytoma (due to germline *DNMT3A*, *EGLN1*, *EGLN2*, *EPAS1*, *FH*, *HIF2A*, *IDH1*, *KIF1B*, *MDH2*, and *SLC25A11* pathogenic variants) [3,19,25].

60.7 Treatment

Treatment options for MEN range from surgery and radiotherapy to chemotherapy [26–30].

Surgical removal of the thyroid gland and lymph node dissection is recommended for MTC, while resection of one or more parathyroid glands is applicable to primary hyperparathyroidism. Adrenalectomy is a treatment of choice for pheochromocytoma. Prophylactic thyroidectomy is considered for individuals with an identified germline RET pathogenic variant [29].

Chemotherapy (e.g., kinase inhibitors) is valuable for treating metastatic MTC, and medications may be used to reduce parathyroid hormone secretion.

External beam radiation therapy (EBRT) or intensity-modulated radiation therapy (IMRT) are useful for advanced locoregional disease.

60.8 Prognosis and Prevention

MEN1 patients have a reasonably good prognosis, with a 15-year survival rate of 93% compared to 68% in sporadic patients. Presence of glucagonoma, VIPoma, somatostatinoma, and nonfunctioning pancreatic endocrine tumors in MEN1 is associated with a three- to fourfold increased risk of death. MEN2A and FMTC patients with MTC have 5- and 10-year survival rates of 90% and 75%, respectively. MEN2B patients with clinically apparent, often aggressive MTC have a poorer prognosis than MEN2A patients. MEN4 patients have a similar prognosis to MEN1 patients.

Prenatal testing and preimplantation diagnosis may be considered for prospective parents harboring *MEN1*, *RET*, and *CDKN1B* pathogenic variants [31].

60.9 Conclusion

Multiple endocrine neoplasia (MEN) is an autosomal dominant disorder of the endocrine glands that encompasses five clinically distinct subtypes (i.e., MEN1, MEN2A, MEN2B, FMTC, and MEN4) caused by germline heterozygous mutations in *MEN1* (MEN1), *RET* (MEN2A, MEN2B, and FMTC), and *CDKN1B* (MEN4). While MEN1 and MEN4 present with mainly nonmalignant tumors in the parathyroid, pituitary, and pancreatic glands, MEN2 is responsible for causing both benign and malignant tumors, including malignant MTC, pheochromocytoma, and parathyroid adenoma, along with nontumor associations (e.g., cutaneous lichen amyloidosis and Hirschsprung disease in MEN2A; ganglioneuroma and absence of tears in MEN2B). Diagnosis of MEN subtypes relies on fulfilling relevant clinical criteria as well as identification of *MEN1*, *RET*, and *CDKN1B* pathogenic variants. Treatment of MEN involves a combination of surgery, radiotherapy, and chemotherapy [30].

REFERENCES

1. Feliberti E, Perry RR, Vinik A. Multiple enddocrine neoplasia type I and MEN II. In: De Groot LJ, Chrousos G, Dungan K et al. editors. *Endotext [Internet]*. South Dartmouth (MA): MDText.com, Inc.; 2000-.2013 Jul 19.
2. Giusti F, Marini F, Brandi ML. Multiple endocrine neoplasia type 1. In: Pagon RA, Adam MP, Ardinger HH et al. editors. *GeneReviews® [Internet]*. Seattle (WA): University of Washington, Seattle; 1993–2017. 2005 Aug 31 [updated 2017 Dec 14].
3. Eng C. Multiple endocrine neoplasia type 2. In: Adam MP, Ardinger HH, Pagon RA, Wallace SE, Bean LJH, Stephens K, Amemiya A, editors. *GeneReviews® [Internet]*. Seattle (WA): University of Washington, Seattle; 1993–2019. 1999 Sep 27 [updated 2019 Aug 15].
4. Pellegata NS. MENX and MEN4. *Clinics (Sao Paulo)*. 2012;67(Suppl 1):13–8.
5. Thakker RV. Multiple endocrine neoplasia type 1 (MEN1) and type 4 (MEN4). *Mol Cell Endocrinol*. 2014;386(1–2):2–15.
6. Norton JA, Krampitz G, Jensen RT. Multiple endocrine neoplasia: Genetics and clinical management. *Surg Oncol Clin N Am*. 2015;24(4):795–832.
7. Dreijerink KMA, Timmers HTM, Brown M. Twenty years of menin: Emerging opportunities for restoration of transcriptional regulation in MEN1. *Endocr Relat Cancer*. 2017;24(10):T135–45.

8. Feng Z, Ma J, Hua X. Epigenetic regulation by the menin pathway. *Endocr Relat Cancer*. 2017;24(10):T147–59.

9. Falchetti A. Genetics of multiple endocrine neoplasia type 1 syndrome: What's new and what's old. *F1000Res*. 2017;6. pii: F1000 Faculty Rev-73.

10. Krampitz GW, Norton JA. RET gene mutations (genotype and phenotype) of multiple endocrine neoplasia type 2 and familial medullary thyroid carcinoma. *Cancer*. 2014;120(13):1920–31.

11. Lu F, Chen X, Bai Y, Feng Y, Wu J. A large Chinese pedigree of multiple endocrine neoplasia type 2A with a novel C634Y/D707E germline mutation in RET exon 11. *Oncol Lett*. 2017;14(3):3552–8.

12. Schernthaner-Reiter MH, Trivellin G, Stratakis CA. MEN1, MEN4, and Carney complex: Pathology and molecular Genetics. *Neuroendocrinology*. 2016;103(1):18–31.

13. Alrezk R, Hannah-Shmouni F, Stratakis CA. MEN4 and *CDKN1B* mutations: The latest of the MEN syndromes. *Endocr Relat Cancer*. 2017;24(10):T195–208.

14. Pacheco MC. Multiple endocrine neoplasia: A genetically diverse group of familial tumor syndromes. *J Pediatr Genet*. 2016;5(2):89–97.

15. Cinque L, Sparaneo A, Salcuni AS et al. MEN1 gene mutation with parathyroid carcinoma: First report of a familial case. *Endocr Connect*. 2017;6(8):886–91.

16. Kamilaris CDC, Stratakis CA. Multiple endocrine neoplasia type 1 (MEN1): An update and the significance of early genetic and clinical diagnosis. *Front Endocrinol (Lausanne)*. 2019;10:339.

17. Wells SA Jr, Pacini F, Robinson BG, Santoro M. Multiple endocrine neoplasia type 2 and familial medullary thyroid carcinoma: An update. *J Clin Endocrinol Metab*. 2013;98(8):3149–64.

18. Efared B, Atsame-Ebang G, Tahirou S et al. Bilateral pheochromocytoma with ganglioneuroma component associated with multiple neuroendocrine neoplasia type 2A: A case report. *J Med Case Rep*. 2017;11(1):208.

19. Marquard J, Eng C. Multiple endocrine neoplasia type 2. In: Pagon RA, Adam MP, Ardinger HH et al. editors. *GeneReviews® [Internet]*. Seattle (WA): University of Washington, Seattle; 1993–2017. 1999 Sep 27 [updated 2015 Jun 25].

20. Yasir M, Kasi A. Multiple endocrine neoplasias, type 2 (MEN II, pheochromocytoma and amyloid producing medullary rhyroid carcinoma, Sipple syndrome). In: *StatPearls [Internet]*. Treasure Island (FL): StatPearls Publishing; 2019 Jan. [Updated 2019 Apr 3].

21. Marini F, Giusti F, Brandi ML. Genetic test in multiple endocrine neoplasia type 1 syndrome: An evolving story. *World J Exp Med*. 2015;5(2):124–9.

22. Wang C, Yun T, Wang Z et al. Pathological characteristics and genetic features of melanin-producing medullary thyroid carcinoma. *Diagn Pathol*. 2018;13(1):86.

23. Manoharan J, Albers MB, Bartsch DK. The future: Diagnostic and imaging advances in MEN1 therapeutic approaches and management strategies. *Endocr Relat Cancer*. 2017;24(10):T209–25.

24. Singh G, Jialal I. Multiple endocrine neoplasia type 1 (MEN I, Wermer Syndrome). In: *StatPearls [Internet]*. Treasure Island (FL): StatPearls Publishing; 2019 Jan-. 2019 Apr 3.

25. Stratakis CA. Hereditary syndromes predisposing to endocrine tumors and their skin manifestations. *Rev Endocr Metab Disord*. 2016;17(3):381–8.

26. Agarwal SK. The future: Genetics advances in MEN1 therapeutic approaches and management strategies. *Endocr Relat Cancer*. 2017;24(10):T119–34.

27. Marini F, Giusti F, Tonelli F, Brandi ML. Management impact: Effects on quality of life and prognosis in MEN1. *Endocr Relat Cancer*. 2017;24(10):T227–42.

28. van Leeuwaarde RS, de Laat JM, Pieterman CRC, Dreijerink K, Vriens MR, Valk GD. The future: Medical advances in MEN1 therapeutic approaches and management strategies. *Endocr Relat Cancer*. 2017;24(10):T179–93.

29. Sadowski SM, Cadiot G, Dansin E, Goudet P, Triponez F. The future: Surgical advances in MEN1 therapeutic approaches and management strategies. *Endocr Relat Cancer*. 2017;24(10):T243–60.

30. PDQ Pediatric Treatment Editorial Board. *Childhood Multiple Endocrine Neoplasia (MEN) Syndromes Treatment (PDQ®): Health Professional Version*. 2019 Dec 23. In: PDQ Cancer Information Summaries [Internet]. Bethesda (MD): National Cancer Institute (US); 2002-.

31. Correa FA, Farias EC, Castroneves LA, Lourenço DM Jr, Hoff AO. Quality of life and coping in multiple endocrine neoplasia type 2. *J Endocr Soc*. 2019;3(6):1167–74.

61

Pendred Syndrome

Dongyou Liu

CONTENTS

61.1 Introduction

Pendred syndrome (also known as goiter-deafness syndrome) is a relatively rare autosomal recessive disorder characterized by congenital sensorineural deafness, enlargement of the vestibular aqueduct (EVA), childhood-onset goiter/hypothyroidism, and incomplete iodide organization, in addition to increased risk of thyroid adenoma, thyroid cancer, and follicular thyroid hyperplasia (due to prolonged hypothyroidism and TSH stimulation).

Molecularly, homozygous or compound heterozygous mutations in the *SLC26A4* (solute carrier family 26 member 4) gene on chromosome 7q22.3 encoding pendrin (an iodide-chloride transporter) appear to underline the development of Pendred syndrome (about 50% of cases) [1].

61.2 Biology

Hearing loss is an etiologically heterogeneous disorder that is responsible for causing sensory impairment in approximately 5% of the world's population. Mechanistically, both nongenetic and genetic factors are implicated in the pathogenesis of hearing loss. Nongenetic factors range from infection and head injury to environmental exposure (e.g., excessive noise or chemicals), while genetic causes consist of >400 syndromic hearing loss conditions (inducing hearing loss as well as other symptoms) and >150 nonsyndromic hearing loss loci (inducing hearing loss only). More often than not, nongenetic factors may expedite or exacerbate progressive hearing loss in individuals with relevant genetic defects [2].

Transmitted via autosomal dominant (e.g., Waardenburg syndrome, branchiootorenal syndrome, Stickler syndrome, CHARGE syndrome, Treacher–Collins syndrome, neurofibromatosis type II), autosomal recessive (Usher syndrome, Pendred syndrome, Jervell and Lange–Nielsen syndrome, Cockayne syndrome, biotinidase deficiency, Refsum disease), X-linked recessive (Alport syndrome, Norrie disease, X-linked congenital stapes fixation with perilymph gusher) and mitochondrial (mitochondrial encephalopathy, lactic acidosis and stroke-like episodes [MELAS], maternally inherited diabetes and deafness [MIDD], Kearns–Sayre syndrome, myoclonic epilepsy and ragged red fibers [MERRF]) manners, hearing loss syndromes are responsible for 30% hereditary hearing losses [3,4].

Similarly, nonsyndromic hearing loss loci also undergo autosomal dominant (about 20%), autosomal recessive (about 80%), X-linked (1%), Y-linked (<1%), and mitochondrial transmissions, and account for 70% of hereditary hearing losses. The most notable nonsyndromic hearing loss locus is autosomal recessive DFNB1, which contains the *GJB2* gene encoding the gap-junction protein connexin 26 (Cx26) with the ability to recycle potassium ions back to the endolymph of the cochlear duct. Being the most common deafness-causing gene, *GJB2* alone accounts for up to 50% of non-syndromic sensorineural hearing loss (SNHL) in certain population groups [4].

Implicated in approximately 10% of hereditary deafness cases, Pendred syndrome was first described by Vaughan Pendred in 1896, and its clinical triad of deafness, goiter, and thyroid dysfunction was defined by Fraser in 1965. The gene region associated with Pendred syndrome was mapped to the DFNB4 locus on chromosome 7q22.3 in 1996, and the culprit gene was identified in 1997 as *SLC26A4* (solute carrier family 26 member 4), which encodes an iodide-chloride transporter. Subsequent studies indicate that homozygous or compound heterozygous *SLC26A4* mutations occur in about 50% of Pendred syndrome cases, whereas heterozygous *SLC26A4* pathogenic variant together with heterozygous mutation in another gene (e.g., the FOXI1 gene on chromosome 5q35.1 encoding Forkhead box protein I1 or the *KCNJ10* gene on chromosome 1q23.2 encoding the ATP-sensitive inward rectifier potassium channel 10) may be present in other Pendred syndrome cases [5,6].

61.3 Pathogenesis

The *SLC26A4* (solute carrier family 26 member 4, or human pendrin polypeptide) gene on chromosome 7q22.3 spans 67 kb and encodes a 780 aa, 85 kDa anion transporter protein (pendrin) belonging to the SLC26 anion transporter family.

Pendrin is capable of mediating sodium-independent chloride-iodide exchange, and also accepting formate and bicarbonate as substrates. Specifically, pendrin takes part in apical iodide transport in the thyroid, transfers bicarbonate into endolymph in exchange for chloride in the inner ear, and engages in urinary bicarbonate excretion with tubular chloride reabsorption in renal collecting duct β-intercalated cells. Through these activities, pendrin contributes to the homeostasis of ion concentration in the endolymphatic sac and hair cells in the inner ear, and the formation of hormone in the thyroid [7].

Biallelic mutations in the *SLC26A4* gene reduce or eliminate pendrin, leading to endolymph acidification, auditory sensory transduction defects, impaired iodide organification, abnormal intracellular thyroid hormone synthesis, EVA, syndromic and non-syndromic hearing loss (due to loss of endocochlear potential), and thyroid goiter.

To date, >360 *SLC26A4* mutations (including splice site, frameshift, missense, and nonsense mutations, as well as occasional large deletions) have been identified. Among them, pathogenic variants p.Leu236Pro (26%), p.Thr416Pro (15%), and c.1001+1G>A (14%) are found in 50% of Pendred syndrome patients of Northern European descent; pathogenic variants c.919-2A>G, p.His723Arg, and p.Val239Asp are prevalent among Chinese, Japanese, Korean, and Pakistani populations; pathogenic variant p.Glu384Gly frequently occurs in Northern Europeans; and pathogenic variant p.Gln514Lys is common in Spanish patients [8–13].

Interestingly, in East Asian (Chinese, Korean, and Japanese) patients with EVA (also known as dilation of the vestibular aqueduct [DVA], the most penetrant manifestation of Pendred syndrome), 80% possess biallelic *SLC26A4* pathogenic variants, slightly >10% have one pathogenic variant, and <10% have no pathogenic variants [14–16]. In contrast, among European or North American Caucasian patients with EVA, 25% harbor two mutant alleles, 25% have one mutant allele, and 50% show no obvious mutations in *SLC26A4* gene, suggesting the potential involvement of other genes (e.g., *FOXI1, KCNJ10, GJB2, SCARB2, DUOX2*) in the pathogenesis of hereditary hearing loss [1].

There is tangible evidence that mutations in the *FOXI1* gene on chromosome 5q35.1 encoding forkhead box protein I1 and

the *KCNJ10* gene on chromosome 1q23.2 encoding the ATP-sensitive inward rectifier potassium channel 10 may be implicated in about 2% of non-classic Pendred syndrome (also known as nonsyndromic enlarged vestibular aqueduct [NSEVA]), but not classic Pendred syndrome. Further, biallelic *KCNJ10* pathogenic variants are involved in SeSAME syndrome (seizures, sensorineural deafness, ataxia, mental retardation, and electrolyte imbalance) and EAST syndrome (epilepsy, ataxia, sensorineural deafness, and tubulopathy) [17,18].

61.4 Epidemiology

61.4.1 Prevalence

Pendred syndrome due to *SLC26A4* mutation is responsible for approximately 10% of hereditary deafness, second only to connexin 26 (*GJB2*) gene, with an estimated incidence of 7.5–10 in 100,000.

61.4.2 Inheritance

Pendred syndrome shows autosomal dominant inheritance.

61.4.3 Penetrance

Pendred syndrome displays high penetrance in relation to syndromic deafness with EVA, but low penetrance in goiter (40% occurring in late childhood or early puberty and 60% in early adult life).

61.5 Clinical Features

Classic Pendred syndrome (due to homozygous *SLC26A4* mutations) typically manifests as bilateral, severe to profound SNHL (>60 dB) with congenital/prelingual onset, vestibular dysfunction (ranging from mild unilateral canal paresis to gross bilateral absence of function; often involving infants with normal motor development but episodical walking difficulty; 66% of cases), and temporal bone abnormalities (Mondini malformation or dysplasia, i.e., presence of both cochlear hypoplasia and bilateral EVA), as well as euthyroid goiter in late childhood to early adulthood (75% of cases) (Figure 61.1). Other symptoms include abdominal distension (bloating), cerebellar ataxia (lack of coordination between muscles, limbs, and joints; lack of ability to judge distances that can lead to under- or overshoot in grasping movements; inability to perform rapid movements requiring antagonizing muscle groups to be switched on and off repeatedly), coarse facial appearance, and increased risk of thyroid adenoma, thyroid cancer, and follicular thyroid hyperplasia (due to prolonged hypothyroidism and TSH stimulation) (Figure 61.2) [19–23].

Non-classic Pendred syndrome (due to heterozygous *SLC26A4* mutation along with either *FOXI1* or *KCNJ10* mutation), which is also referred to as nonsyndromic enlarged vestibular aqueduct (NSEVA), may display sensorineural hearing impairment without other obvious abnormalities (i.e., nonsyndromic hearing loss), bilateral or unilateral EVA, and absence of thyroid defects [1].

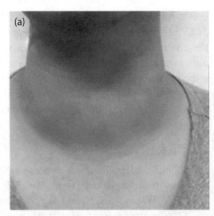
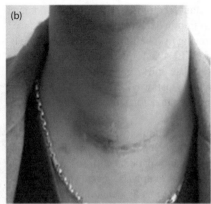

FIGURE 61.1 Pendred syndrome in a 26-year-old male showing: (a) an enlarged, firm goiter with no palpable nodules, and (b) post-thyroidectomy image. (Photo credit: Hu EW et al. *Oncol Lett.* 2014;8(5):2059–62.)

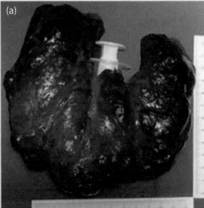
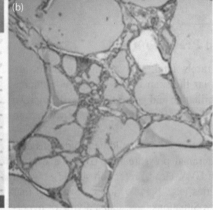

FIGURE 61.2 Pendred syndrome in a 26-year-old male showing: (a) abnormal thyroid specimen, and (b) short thyroid follicular epithelial cells and reduced hyperplasia in thyroid specimen. (Photo credit: Hu EW et al. *Oncol Lett.* 2014;8(5):2059–62.)

61.6 Diagnosis

Diagnosis of Pendred syndrome involves assessment of hearing loss (which may appear at birth or during childhood, and progress from a mixed conductive/SNHL to total hearing loss) and thyroid goiter (apparent from the first decade toward the end of the second decade), imaging study (MRI scanning of the inner ear for widened or large vestibular aqueducts with enlarged endolymphatic sacs [EVA] and abnormalities of the cochleae such as Mondini dysplasia), histopathology, genetic testing (for *SLC26A4* pathogenic variants, which occur in 50% of patients) and further functional tests (e.g., perchlorate discharge test, thyroid function test for mild cases of thyroid dysfunction) [24–26].

Individuals presenting with the following findings should be suspected of Pendred syndrome: (i) clinical findings (congenital or prelingual, non-progressive, severe to profound sensorineural hearing impairment based on auditory brain stem response testing or pure tone audiometry), (ii) temporal bone imaging findings (Mondini malformation or dysplasia [presence of both cochlear hypoplasia and bilateral EVA detected on thin-cut CT]), (iii) endocrine findings (euthyroid goiter due to an organification defect of iodide) [1].

Presence of SNHL, characteristic temporal bone abnormalities identified on thin-cut CT, and euthyroid goiter establishes the clinical diagnosis of Pendred syndrome.

Identification of biallelic pathogenic variants in *SLC26A4* establishes the molecular diagnosis of Pendred syndrome On the other hand, identification of double heterozygosity for one pathogenic variant in *SLC26A4* and one pathogenic variant in either *FOXI1* or *KCNJ10* confirms the diagnosis of NSEVA [25].

Differential diagnoses for Pendred syndrome include congenital inherited hearing impairment, congenital cytomegalovirus, branchiootorenal syndrome, congenital hypothyroidism with sensorineural hearing loss, resistance to thyroid hormone, autoimmune thyroid diseases (e.g., Graves disease, Hashimoto thyroiditis, and primary idiopathic myxedema) [1].

61.7 Treatment

Pendred syndrome is a genetic disorder for which no specific treatment is available. Therefore, management of Pendred syndrome should focus on hearing habilitation (hearing aids, cochlear implants, auditory brainstem implants) and goiter remedy (medical/surgical treatment of abnormal thyroid function/thyromegaly) [24].

Hearing aids are important for children with mild to moderate hearing loss as they help improve their academic performance. Due to its reduced swallowing risk, behind the ear (BTE) style is preferred over in-the-ear (ITE) style for pediatric patients.

Cochlear implants stimulate spiral ganglion cells (the first-order neurons of the auditory pathway) directly and provide a clinically and cost-effective option for children with severe to profound bilateral hearing loss.

Auditory brainstem implants stimulate the second-order auditory neurons (instead of the cochlear nerve) in the cochlear nucleus and may be considered for patients with severe cochlear or cochlear nerve malformations or aplasia, severe cochlear ossification, and temporal bone fractures associated with traumatic cochlear nerve avulsion.

Goiter remedy consists of medical/surgical treatment of abnormal thyroid function/thyromegaly. Thyroid hormone deficiency can be rectified with hormone supplement, and goiter may be removed by total thyroidectomy (Figure 61.1) [21,24].

61.8 Prognosis and Prevention

Pendred syndrome is rarely fatal, and affected patients are expected to have a similar life span to the general population. Attention to hearing habilitation and goiter remedy will improve the quality of life for individuals with hearing impairment and abnormal thyroid function.

Weightlifting and/or contact sports should be avoided, as dramatic increases in intracranial pressure may trigger a decline in hearing in those with EVA. Head protection should be worn when engaged in activities (e.g., bicycle riding or skiing) that might lead to head injury. Situations (e.g., scuba diving or hyperbaric oxygen treatment) that can lead to barotrauma (extreme, rapid changes in pressure) should also be avoided.

Low-dose rapamycin may be considered to decrease acute symptoms and prevent progression of hearing loss in Pendred syndrome patients [27].

Prenatal testing and preimplantation genetic diagnosis (targeting *SLC26A4* pathogenic variants 1494C > T and 1555A > G or other related mutations) may be considered for pregnant women.

61.9 Conclusion

Pendred syndrome is an autosomal recessive disorder that typically causes congenital SNHL and goiter (hypothyroidism) in early puberty (40%) or adulthood (60%) as well as thyroid adenoma, thyroid cancer, and follicular thyroid hyperplasia. Accounting for about 10% of hereditary hearing impairment, Pendred syndrome is due mainly to homozygous or compound heterozygous mutations in the *SLC26A4* gene that encodes a multifunctional anion exchanger (pendrin), resulting in endolymph acidification, auditory sensory transduction defects, impaired organification of iodide in the thyroid and goiter, abnormal intracellular thyroid hormone synthesis, EVA, and syndromic and nonsyndromic hearing loss. Clinical diagnosis of Pendred syndrome relies on observation of SNHL, characteristic temporal bone abnormalities (with bilateral EVA) on thin-cut CT, and euthyroid

goiter. Molecular identification of biallelic pathogenic variants in *SLC26A4* provides further confirmation of Pendred syndrome. In the absence of specific treatment, hearing habilitation (hearing aids, cochlear implants, auditory brainstem implants) and goiter remedy (medical/surgical treatment of abnormal thyroid function/thyromegaly) offers a practical and effective approach to improve the well-being of Pendred syndrome patients [28].

REFERENCES

1. Smith RJH. Pendred syndrome/Nonsyndromic enlarged vestibular aqueduct. In: Adam MP, Ardinger HH, Pagon RA, Wallace SE, Bean LJH, Stephens K, Amemiya A, editors. *GeneReviews® [Internet]*. Seattle (WA): University of Washington, Seattle; 1993–2018. 1998 Sep 28 [updated 2017 Oct 19].
2. Rehman AU, Friedman TB, Griffith AJ. Unresolved questions regarding human hereditary deafness. *Oral Dis.* 2017;23(5):551–8.
3. Koffler T, Ushakov K, Avraham KB. Genetics of hearing loss: Syndromic. *Otolaryngol Clin North Am.* 2015;48(6):1041–61.
4. Shearer AE, Hildebrand MS, Smith RJH. Hereditary hearing loss and deafness overview. In: Adam MP, Ardinger HH, Pagon RA et al. editors. *GeneReviews® [Internet]*. Seattle (WA): University of Washington, Seattle; 1993–2019. 1999 Feb 14 [Updated 2017 Jul 27].
5. Soh LM, Druce M, Grossman AB et al. Evaluation of genotype-phenotype relationships in patients referred for endocrine assessment in suspected Pendred syndrome. *Eur J Endocrinol.* 2015;172(2):217–26.
6. Zhao X, Cheng X, Huang L et al. Analysis of mutations in the FOXI1 and KCNJ10 genes in infants with a single-allele *SLC26A4* mutation. *Biosci Trends.* 2019;13(3):261–6.
7. Sharma AK, Krieger T, Rigby AC, Zelikovic I, Alper SL. Human SLC26A4/Pendrin STAS domain is a nucleotide-binding protein: Refolding and characterization for structural studies. *Biochem Biophys Rep.* 2016;8:184–91.
8. Shin JW, Lee SC, Lee HK, Park HJ. Genetic screening of GJB2 and SLC26A4 in Korean cochlear implantees: Experience of Soree Ear Clinic. *Clin Exp Otorhinolaryngol.* 2012;5(Suppl 1):S10–3.
9. Ito T, Muskett J, Chattaraj P et al. *SLC26A4* mutation testing for hearing loss associated with enlargement of the vestibular aqueduct. *World J Otorhinolaryngol.* 2013;3(2):26–34.
10. Yazdanpanahi N, Tabatabaiefar MA, Farrokhi E et al. Compound heterozygosity for two novel SLC26A4 mutations in a large Iranian pedigree with Pendred syndrome. *Clin Exp Otorhinolaryngol.* 2013;6(4):201–8.
11. Miyagawa M, Nishio SY, Usami S; Deafness Gene Study Consortium. Mutation spectrum and genotype-phenotype correlation of hearing loss patients caused by SLC26A4 mutations in the Japanese: A large cohort study. *J Hum Genet.* 2014;59(5):262–8.
12. Gonçalves AC, Santos R, O'Neill A, Escada P, Fialho G, Caria H. Further characterisation of the recently described SLC26A4 c.918+2T>C mutation and reporting of a novel variant predicted to be damaging. *Acta Otorhinolaryngol Ital.* 2016;36(3):233–8.
13. Nonose RW, Lezirovitz K, de Mello Auricchio MTB, Batissoco AC, Yamamoto GL, Mingroni-Netto RC. Mutation

analysis of SLC26A4 (Pendrin) gene in a Brazilian sample of hearing-impaired subjects. *BMC Med Genet.* 2018;19(1):73.

14. Yuan Y, Guo W, Tang J et al. Molecular epidemiology and functional assessment of novel allelic variants of SLC26A4 in non-syndromic hearing loss patients with enlarged vestibular aqueduct in China. *PLOS ONE.* 2012;7(11):e49984.

15. Fu C, Zheng H, Zhang S et al. Mutation screening of the SLC26A4 gene in a cohort of 192 Chinese patients with congenital hypothyroidism. *Arch Endocrinol Metab.* 2016;60(4):323–7.

16. Ideura M, Nishio SY, Moteki H et al. Comprehensive analysis of syndromic hearing loss patients in Japan. *Sci Rep.* 2019;9(1):11976.

17. Landa P, Differ AM, Rajput K, Jenkins L, Bitner-Glindzicz M. Lack of significant association between mutations of KCNJ10 or FOXI1 and SLC26A4 mutations in Pendred syndrome/enlarged vestibular aqueducts. *BMC Med Genet.* 2013;14:85.

18. Pique LM, Brennan ML, Davidson CJ, Schaefer F, Greinwald J Jr, Schrijver I. Mutation analysis of the SLC26A4, FOXI1 and KCNJ10 genes in individuals with congenital hearing loss. *PeerJ.* 2014;2:e384.

19. Son EJ, Nosé V. Familial follicular cell-derived thyroid carcinoma. *Front Endocrinol (Lausanne).* 2012;3:61.

20. Dror AA, Lenz DR, Shivatzki S et al. Atrophic thyroid follicles and inner ear defects reminiscent of cochlear hypothyroidism in Slc26a4-related deafness. *Mamm Genome.* 2014;25(7–8):304–16.

21. Hu EW, Liu LB, Jiang RY, He XH. Goiter and hearing impairment: A case of a male patient with Pendred syndrome. *Oncol Lett.* 2014;8(5):2059–62.

22. Cherian KE, Kapoor N, Mathews SS, Paul TV. Endocrine glands and hearing: Auditory manifestations of various endocrine and metabolic conditions. *Indian J Endocrinol Metab.* 2017;21(3):464–9.

23. Chao JR, Chattaraj P, Munjal T et al. SLC26A4-linked CEVA haplotype correlates with phenotype in patients with enlargement of the vestibular aqueduct. *BMC Med Genet.* 2019;20(1):118.

24. Chen MM, Oghalai JS. Diagnosis and management of congenital sensorineural hearing loss. *Curr Treat Options Pediatr.* 2016;2(3):256–65.

25. Smith N, U-King-Im JM, Karalliedde J. Delayed diagnosis of Pendred syndrome. *BMJ Case Rep.* 2016;2016. pii: bcr2016215271.

26. Chow YP, Abdul Murad NA, Mohd Rani Z et al. Exome sequencing identifies SLC26A4, GJB2, SCARB2 and DUOX2 mutations in 2 siblings with Pendred syndrome in a Malaysian family. *Orphanet J Rare Dis.* 2017;12(1):40.

27. Hosoya M, Saeki T, Saegusa C et al. Estimating the concentration of therapeutic range using disease-specific iPS cells: Low-dose rapamycin therapy for Pendred syndrome. *Regen Ther.* 2018;10:54–63.

28. Garabet Diramerian L, Ejaz S. Pendred Syndrome. [Updated 2019 Oct 18]. In: *StatPearls [Internet].* Treasure Island (FL): StatPearls Publishing, Florida. 2020 Jan.

Section VI

Tumor Syndromes Affecting the Hematopoietic and Lymphoreticular Systems

62

Ataxia Telangiectasia

Dongyou Liu

CONTENTS

62.1 Introduction

Ataxia telangiectasia (AT, also known as ATM syndrome) is an autosomal recessive disorder that typically presents with progressive cerebellar ataxia from ages 1–4, oculomotor apraxia, choreoathetosis, telangiectasia of the conjunctivae, immune defects, recurrent sinopulmonary infections, radiation sensitivity, premature aging, poor growth, gonadal atrophy, delayed pubertal development, insulin-resistant diabetes, and a predisposition to malignancy (particularly leukemia and lymphoma). Apart from classic form of early-onset AT, mild or non-classic form of adult-onset AT or AT with early-onset dystonia is also recognized [1].

Molecularly, homozygous or compound heterozygous mutation in the *ATM* (ataxia telangiectasia mutated) gene on chromosome 11q22.3 appears to be responsible for the pathogenesis of ataxia telangiectasia. Indeed, experimental evidence suggests that homozygous or heterozygous *ATM* frameshift or splice site mutations causing protein truncation are linked to classic AT, whereas *ATM* missense mutations that leave residual amount of functional ATM protein are implicated in mild or non-classic AT (which often presents with normal brain MRI and infrequent extraneurological features) (Table 62.1) [2].

62.2 Biology

Early cases of AT (also referred to ATM syndrome, Louis-Bar Syndrome, genome instability syndrome, chromosomal instability syndrome, DNA repair disorder, DNA damage response syndrome, and neurocutaneous syndrome) were described by

Syllaba and Henner in 1926 (from three adolescent Czech siblings with progressive chorea and dystonia in association with ocular telangiectasia) and by Louis-Bar in 1941 (involving a 9-year-old boy with progressive cerebellar ataxia and extensive cutaneous telangiectasia). Studies by Boder and Sedgwick in 1957 and others in subsequent years helped define AT as a familial syndrome of progressive cerebellar ataxia, oculocutaneous telangiectasia, and frequent pulmonary infection. Upon analysis of clinical data from 101 cases, Boder and Sedgwick in 1963 listed the most common manifestations of AT as cerebellar ataxia (100%), oculocutaneous telangiectasia (100%), characteristic facies (98%), choreoathetosis (91%), progeric changes of the skin and hair (88%), eye movement apraxia (84%), sinopulmonary infections (83%), familial occurrence (45%), and mental retardation (33%). Since then, the clinical profile of AT has been further expanded to include atrophy of the thymus, adrenals, spleen, and lymphoid tissues; bronchiectasis and the presence of bilateral ovarian dysgerminoma; elevation of serum alpha-fetoprotein; dystonia, chorea, dementia, peripheral neuropathy, IgE deficiency, chromosomal abnormalities; dystonia, myoclonus, pyramidal signs, seizures, and persistent lymphopenia. Considering the differences in disease onset, two major forms of ataxia telangiectasia (i.e., classic form of early-onset AT, and non-classic or mild form of adult-onset AT or AT with early-onset dystonia) are now recognized [3,4].

After mapping the AT-related gene region to chromosome 11q22.3 in 1988, the culprit gene (*ATM*, standing for ataxia telangiectasia mutated) was identified by Savitsky and coworkers in 1995. The *ATM* encoded product appears to be a kinase, as mutations (truncating, missense, or splice site) in the *ATM* gene contribute to the total or partial loss of ATM kinase activity. Additional findings

TABLE 62.1

Key Clinical and Molecular Characteristics of Classic and Mild Ataxia Telangiectasia

	Classic AT	**Mild AT**
Neurological deficits	Neurological deficits affecting toddlers create wheelchair dependency by 10 years of age	Neurological deficits in childhood are associated with myoclonus, dystonia, choreoathetosis, tremor, and later ataxia
Immune defects	Immunodeficiency and/or lymphopenia affect 2/3 of patients	Immunodeficiencies occur infrequently
Pulmonary disease	Common	Uncommon
Tumor	Lymphoid malignancies occur at a younger age; hematopoietic and non-hematopoietic malignancies may appear at an older age	Mostly non-hematopoietic malignancies emerge later in life
ATM protein	Truncating mutations yielding a nonfunctional ATM with no kinase activity	Missense or split-site mutations yielding a marginally functional ATM with residual kinase activity

indicate that the absence of ATM kinase activity correlates to classical phenotype of AT, and the presence of residual ATM kinase activity is associated with a mild and atypical phenotype (which is associated with normal endocrine and pulmonary function, normal immunoglobulins, absence of telangiectasia, reduced x-ray hypersensitivity in lymphocytes, and extended life span) [5].

62.3 Pathogenesis

Located on chromosome 11q22.3, the *ATM* gene spans 150 kb with 66 exons (including 62 coding and 4 non-coding), and encodes a 3056 aa, 315 kDa protein (ATM), which is a serine/threonine kinase belonging to the phosphatidylinositol 3-kinase family. Structurally, the ATM protein consists of four regions: N-terminal Huntingtin, elongation factor 3, protein phosphatase 2A, and yeast kinase TOR (HEAT) repeat, a FRAP, ATM, TRRAP (FAT) domain, a protein kinase domain and C-terminal FAT-C domain. Located primarily in the nucleus, along with small amounts in the cytoplasm (in association with mitochondria and peroxisomes), and activated by double-stranded DNA break, oxidative stress, hypoxia, hypotonic stress, hyperthermia, chloroquine, and agents affecting chromatin organization, the ATM protein phosphorylates/activates various downstream effectors (including p53, Chk2, MDM2, 53BP1, SMC1, BRCA1, FANCD2, H2AX, c-abl, nibrin, Mre11, KAP1) involved in DNA repair, cell cycle control, apoptosis, and other pathways [6–9].

As nearly one million DNA lesions are introduced daily through the activity of reactive oxygen species and other mechanisms into human cells, the capacity of human cells to respond to and repair DNA damage is vital for genomic stability. Homozygous or compound heterozygous (due to inheritance of different pathogenic variants from each parent) mutations in the *ATM* gene result in the production of a truncated or altered ATM protein, which shows no or only partial kinase activity and significantly compromises the ability of human cells to repair DNA damages caused by reactive oxygen species and other stressors [10].

To date, >800 mutations [e.g., c.1A>G (p.Met1Val), c.103C>T (p.Arg35Ter), c.3894dupT (p.Ala1299CysfsTer3), c.5763-1050A>G (formerly 5762ins137), c.6154G>A (p.Glu2052Lys), c.6200C>A (p.Ala2067Asp), c.6679C>T (p.Arg2227Cys), c.7271T>G (p.Val2424Gly), c.7886_7890delTATTA (p.Ile2629SerfsTer25), c.8147T>C (p.Val2716Ala), c.8494C>T (p.Arg2832Cys)] have been identified in the proximal, central, and distal regions of the human *ATM* gene, leading to a total absence of the ATM protein in 95% of individuals with AT as examined by immunoblotting. Although normal expression of ATM protein remains in 1% of patients harboring ATM pathogenic variants, the resultant protein appears to lack kinase activity (so-called "kinase-dead" ATM protein) [11–13].

It is clear that homozygous or heterozygous *ATM* frameshift or splice site mutations that produce truncated ATM protein without kinase activity are responsible for causing classic AT. By contrast, *ATM* missense mutations that produce altered ATM protein with residual kinase activity are involved in the development of mild or non-classic AT. Interestingly, the ATM pathogenic variant c.5762-1050A>G (formerly 5762ins137) often causes a slower rate of neurologic deterioration, later onset of manifestations, intermediate radiosensitivity, and little or no tumor risk. Furthermore, pathogenic variants c.1A>G, c.7271T>G, and c.8494C>T are implicated in a milder phenotype and longer life span, while pathogenic variants c.8147T>C and c.6679C>T confer increased tumor risk [14].

Besides ataxia telangiectasia, changes in chromosome 11q22.3 are also observed in somatic B-cell non-Hodgkin lymphoma, somatic mantle cell lymphoma, and somatic T-cell prolymphocytic leukemia.

62.4 Epidemiology

62.4.1 Prevalence

AT affected 1 per 40,000-300,000 live births and represents the most common cause of progressive cerebellar ataxia in childhood in countries with low coefficients of inbreeding. There is no racial or ethnic predilection among AT patients.

62.4.2 Inheritance

AT is an autosomal recessive disorder. Heterozygous parents harboring *ATM* pathogenic variant have a 25% chance of getting an affected offspring, who acquire biallelic (homozygous or compound heterozygous) mutations in the *ATM* gene; a 50% chance of getting a heterozygous carrier offspring, who is asymptomatic but at increased risk for breast cancer (especially in individuals with missense pathogenic variant) and coronary artery disease; and a 25% chance of getting an unaffected offspring. Affected individuals are often sterile with gonadal dysgenesis and do not reproduce.

62.4.3 Penetrance

AT shows a high penetrance and variable expressivity. Affected patients often develop a varying spectrum of clinical manifestations.

62.5 Clinical Features

Based on its clinical variations, AT is separated into classic (typical, early onset, childhood onset) and mild (non-classic, variant, atypical, late onset, adult onset) forms.

Classic AT is an early-onset disorder that becomes apparent between 1 and 4 years of age, with a diverse range of clinical signs and symptoms, including (i) neurologic anomalies (progressive cerebellar ataxia, leading to staggering walk, slurred speech, and oculomotor apraxia, reduced writing and drawing ability, difficulty in horizontal and vertical saccadic eye movements, choreoathetosis, myoclonic jerking, intention tremors, difficulties in dressing, eating, washing, and toileting, wheelchair confinement by 10 years of age, decreased or absent deep tendon reflexes in older patients, learning difficulties, dystonia, and adult-onset spinal muscular atrophy); (ii) immunodeficiency (poor antibody response to pneumococcal polysaccharide vaccines; reduced serum concentration of IgA, IgE, and IgG2; occasionally elevated NK lymphocyte levels; low ATM kinase levels; frequent sinopulmonary infections; 60%–80% of cases); (iii) infection (chronic bronchiectasis); (iv) pulmonary disease (pulmonary failure, lymphocytic infiltration); (v) other anomalies (elevated liver enzyme levels without apparent liver pathology); (vi) tumors (38% increase in tumor risk, particularly B-cell lymphoma, acute lymphocytic leukemia [ALL] of T-cell origin in younger children, aggressive T-cell leukemia in older children; ovarian cancer, breast cancer, gastric cancer, melanoma, leiomyomas, and sarcomas in older patients) (Figure 62.1) [15–17]; (vii) life expectancy (most patients live beyond age 25 years, some survived into their 50s) [1].

Mild AT is associated with less severe, later onset phenotype and longer survival (Table 62.1) [18].

62.6 Diagnosis

Individuals displaying a combination of neurologic anomalies (gait and truncal ataxia, head tilting, slurred speech, oculomotor apraxia, and abnormal ocular saccades) and one or more of the following features: telangiectasia, frequent sinopulmonary infections, IgA/IgG/IgE deficiency, lymphopenia affecting T lymphocytes, elevated alpha-fetoprotein (AFP) levels of above 10 ng/mL, chromosomal breaks at 14q11 (T-cell receptor-alpha locus) and at 14q32 (B-cell immunoglobulin heavy chain receptor locus) or t(7;14) translocations in peripheral blood or cultured lymphocytes and fibroblasts, and atrophy of the frontal and posterior vermis and both hemispheres detected by MRI should be suspected of AT (Figure 62.2) [19,20].

Confirmation of AT diagnosis is made possible by laboratory detection of absent or deficient ATM protein or its kinase activity in cultured cell lines, by identification of biallelic (homozygous or compound heterozygous) *ATM* pathological variant (e.g., c.1564_1565delAG Amish; c.103C>T (p.Arg35Ter) in North African Jewish; c.3894dupT in Sardinian, using sequence analysis (~90% of cases) and deletion/duplication analysis (1%–2% of cases), or by immunoblotting detection of absent (90% of cases)

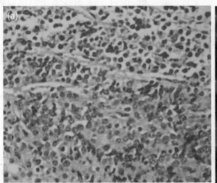
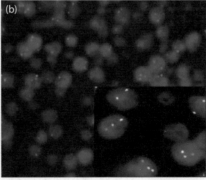

FIGURE 62.1 Ataxia telangiectasia in 7-year-old male with two ATM gene variants [c.742C>T (p.R248X, 2809) in exon 7 and c.6067-c.6068 ins GAGGGAAGAT (p.G2023Gfs*13) in exon 41], manifesting as Burkitt leukemia. (a) Bone marrow biopsy staining revealed bone trabeculae with diffuse lymphocytic infiltration (H&E); (b) Fluorescence in situ hybridization was positive for C-MYC gene rearrangement (insert: red, green, and yellow signals, with overlapping red and green signals). (Photo credit: Ye F et al. *Mol Clin Oncol.* 2018;9(5):493–8.)

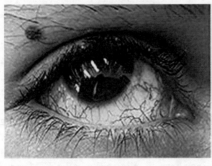
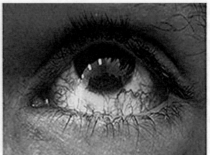

FIGURE 62.2 Ataxia telangiectasia in a 9-year-old boy who presented with gait difficulties since age 2, showing bilateral ocular telangiectasias. (Photo credit: Sauma L et al. *Arq Neuropsiquiatr.* 2015;73(7):638.)

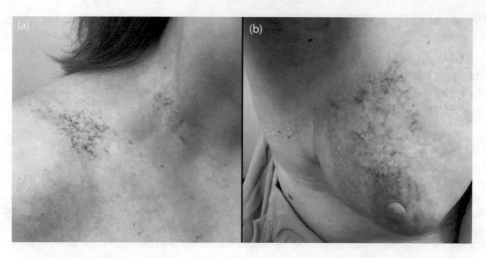

FIGURE 62.3 Ataxia telangiectasia in a 55-year-old female who harbored a heterozygous germline ATM pathogenic variant (c.8565_8566delTGinsAA) and developed severe telangiectasia with dense fibrosis after adjuvant breast and regional nodal radiation treatment following lumpectomy and axillary lymph node dissection for stage II invasive ductal carcinoma of the breast, as shown on the supraclavicular fossa (a), and the right breast and axilla (b). (Photo credit: Dosani M et al. *Cureus.* 2017;9(7):e1458.)

or reduced (10% of cases) ATM protein in a lymphoblastoid cell line (95% sensitivity and >98% specificity). The presence of >15% ATM protein is suggestive of another diagnosis, an *ATM* pathogenic missense variant, a leaky splicing *ATM* variant, an *ATM* pathogenic variant resulting in "kinase-dead" protein, or a pathogenic variant in a gene encoding an ancillary ATM-activating phosphatase like *PP2A*.

Moreover, patients and their cultured cells with AT and other XCIND (*x*-ray sensitivity, *c*ancer, *i*mmunodeficiency, *n*europathology, and *D*NA repair deficiency) are sensitive to x-rays and DNA-damaging agents. Radiation treatment for cancer in these patients will provoke development of telangiectasia (Figure 62.3) [21].

Differential diagnoses for AT include ataxia with oculomotor apraxia type 1 (AOA1 or aprataxin deficiency; childhood-onset ataxia), ataxia with oculomotor apraxia type 2 (AOA2 or senataxin deficiency; childhood-onset ataxia, elevated serum AFP concentration), autosomal recessive spinocerebellar ataxia 9 (SCAR9; childhood-onset ataxia, CoQ_{10} deficiency), infantile-onset spinocerebellar ataxia (IOSCA, childhood-onset ataxia, due to biallelic pathogenic variants in *TWNK* or *C10orf2*), sensory ataxic neuropathy with dysarthria and ophthalmoplegia (SANDO, childhood-onset ataxia; due to biallelic pathogenic variants in *POLG*), X-linked sideroblastic anemia and ataxia (childhood-onset ataxia; due to mutation in *ABCB7*), microcephaly, seizures, and developmental delay (MCSZ, early seizures and microcephaly, due to biallelic pathogenic variants in *PNKP* [polynucleotide kinase phosphatase]), Nijmegen breakage syndrome [intellectual impairment and microcephaly; autosomal recessive inheritance; biallelic pathogenic variants in *NBN* (nibrin); t(7;14) translocation], Mre11 deficiency [ataxia, normal serum AFP concentration; autosomal recessive inheritance; biallelic pathogenic variants in *MRE11*; t(7;14) translocation], or RAD50 deficiency [microcephaly, developmental delay, mild spasticity, non-progressive ataxia, T cell ALL (Ph+); autosomal recessive inheritance; biallelic pathogenic variants in *RAD50*; heterozygous *NBN* pathogenic variant; t(7;14) translocation], RNF168 deficiency (RIDDLE [*r*adiosensitivity, *i*mmunodeficiency, *d*ysmorphic features, and *le*arning difficulties] syndrome; ataxia and

telangiectasia, growth retardation, microcephaly, immunodeficiency, elevated serum AFP; autosomal recessive inheritance; due to biallelic pathogenic variants in *RNF168*), cerebral palsy (CP, non-progressive disorder of motor function due to malformation or early damage to the brain; distinctive regional or diffuse spasticity, absence of AT-related laboratory abnormalities), congenital ocular motor apraxia (COMA; delayed development of visual saccades), Friedreich ataxia (FA or FRDA, a common autosomal recessive cerebellar ataxia; ataxia typically appearing between 10 and 15 years of age, absence of telangiectasia and oculomotor apraxia, early absence of tendon reflexes, normal AFP, frequent scoliosis, abnormal EKG features) [1,22].

62.7 Treatment

Patients with AT are generally treated with standard procedures for symptomatic relief and quality of life improvement. These include supportive therapy and medications (e.g., trihexyphenidyl or Artane, amantadine, baclofen and BOTOX® injections; clonazepam, gabapentin and pregabalin or Lyrica; Riluzole) for neurologic anomalies (drooling, choreoathetosis, myoclonus/tremor, and ataxia), IVIG replacement therapy for immunodeficiency, high potency topical corticosteroids and/or cyclosporine A for small superficial chronic cutaneous granulomas, topical steroids plus IVIG therapy, systemic inhibitors of tumor necrosis factor (TNF-alpha) or direct injection of steroids for extensive granulomas, antibiotics for bronchiectasis and other infections (>7 days), monitoring of recurrent infection, pulmonary function, swallowing, nutrition, scoliosis, and immune function for pulmonary disease, careful monitoring of ionizing radiation and chemotherapeutic agents for tumor (due to a 30% increase in ionizing radiation sensitivity), eye muscle surgery for strabismus and 4-amino pyridine for nystagmus and vestibular deficits, gastrostomy tube (G-tube or feeding tube) for pulmonary and nutritional complications of dysphagia, glucocorticoids for inducing alternate splicing site in the *ATM* gene in order to partly restore its activity, and histone deacetylase inhibitors/EZH2 for minimizing effect of ATM absence [21,23–28].

62.8 Prognosis and Prevention

Patients with AT have a life expectancy of 25 years, which does not seem to correlate well with severity of neurological impairment, and which has been prolonged by antibiotic treatment. The most common causes of death are chronic lung disease (one-third of cases) and cancer (one-third of cases) [28,29]. Regular injection of immunoglobulins appears to be beneficial for preventing infection-related death.

Prenatal testing and preimplantation genetic diagnosis may be considered for prospective parents who harbor *ATM* pathogenic variants.

62.9 Conclusion

Ataxia-telangiectasia (AT) is an autosomal recessive neurodegenerative disorder that typically manifests as progressive neurologic impairment with cerebellar atrophy, ocular and cutaneous telangiectasia, immunodeficiency, increased sensitivity to ionizing radiation, and susceptibility to developing lymphoreticular malignancy, accompanied by elevated serum alpha-fetoprotein levels, chromosomal instability, absence of functional ATM protein, and homozygous or compound heterozygous "null" mutations in the *ATM* gene. As a large serine/threonine protein kinase, ATM functions as master controller of signal transduction for DNA damage response to double-strand break and other stresses. Presence of functional ATM is crucial for maintaining genome integrity and prolonging life span [30]. Current management of AT relies on standard procedures, and several promising reagents (e.g., glucocorticoids and histone deacetylase inhibitors/EZH2) are being evaluated for improving the outcome of patients with this disorder.

REFERENCES

1. Gatti R, Perlman S. Ataxia-Telangiectasia. In: Adam MP, Ardinger HH, Pagon RA et al. editors. *GeneReviews® [Internet]*. Seattle (WA): University of Washington, Seattle; 1993–2018. 1999 Mar 19 [updated 2016 Oct 27].
2. Rothblum-Oviatt C, Wright J, Lefton-Greif MA, McGrath-Morrow SA, Crawford TO, Lederman HM. Ataxia telangiectasia: A review. *Orphanet J Rare Dis.* 2016;11(1):159.
3. Chaudhary MW, Al-Baradie RS. Ataxia-telangiectasia: Future prospects. *Appl Clin Genet.* 2014;7:159–67.
4. Teive HA, Moro A, Moscovich M et al. Ataxia-telangiectasia - A historical review and a proposal for a new designation: ATM syndrome. *J Neurol Sci.* 2015;355(1–2):3–6.
5. Ambrose M, Gatti RA. Pathogenesis of ataxia-telangiectasia: The next generation of ATM functions. *Blood.* 2013;121(20):4036–45.
6. Farooqi AA, Attar R, Arslan BA, Romero MA, ul Haq MF, Qadir MI. Recently emerging signaling landscape of ataxia-telangiectasia mutated (ATM) kinase. *Asian Pac J Cancer Prev.* 2014;15(16):6485–8.
7. Awasthi P, Foiani M, Kumar A. ATM and ATR signaling at a glance. *J Cell Sci.* 2015;128(23):4255–62.
8. Ui A, Yasui A. Collaboration of MLLT1/ENL, Polycomb and ATM for transcription and genome integrity. *Nucleus.* 2016;7(2):138–45.
9. Clouaire T, Marnef A, Legube G. Taming tricky DSBs: ATM on duty. *DNA Repair (Amst).* 2017;56:84–91.
10. Herrup K. ATM and the epigenetics of the neuronal genome. *Mech Ageing Dev.* 2013;134(10):434–9.
11. Jeong H, Huh HJ, Youn J, Kim JS, Cho JW, Ki CS. Ataxia-telangiectasia with novel splicing mutations in the ATM gene. *Ann Lab Med.* 2014;34(1):80–4.
12. Zhang Y, Liu Z, Wang M et al. Single nucleotide polymorphism rs1801516 in ataxia telangiectasia-mutated gene predicts late fibrosis in cancer patients after radiotherapy: A PRISMA-compliant systematic review and meta-analysis. *Medicine (Baltim).* 2016;95(14):e3267.
13. Kuznetsova MV, Trofimov DY, Shubina ES et al. Two novel mutations associated with ataxia-telangiectasia identified using an ion AmpliSEquation inherited disease panel. *Front Neurol.* 2017;8:570.
14. Dahl ES, Aird KM. Ataxia-telangiectasia mutated modulation of carbon metabolism in cancer. *Front Oncol.* 2017;7:291.
15. Choi M, Kipps T, Kurzrock R. ATM mutations in cancer: Therapeutic implications. *Mol Cancer Ther.* 2016;15(8):1781–91.
16. Judge SJ, Plescia TA, Bateni CP, Darrow MA, Evans CP, Canter RJ. Retroperitoneal extramedullary hematopoietic pseudotumor in ataxia-telangiectasia. *Rare Tumors.* 2018;10:2036361318789724.
17. Ye F, Chai W, Yang M, Xie M, Yang L. Ataxia-telangiectasia with a novel ATM gene mutation and Burkitt leukemia: A case report. *Mol Clin Oncol.* 2018;9(5):493–8.
18. Newrick L, Sharrack N, Hadjivassiliou M. Late-onset ataxia telangiectasia. *Neurol Clin Pract.* 2014;4(4):365–7.
19. Kumar N, Aggarwal P, Dev N, Kumar G. Ataxia telangiectasia: Learning from previous mistakes. *BMJ Case Rep.* 2012;2012.
20. Sauma L, Teixeira KC, Montenegro MA. Ataxia telangiectasia. *Arq Neuropsiquiatr.* 2015;73(7):638.
21. Dosani M, Schrader KA, Nichol A et al. Severe late toxicity after adjuvant breast radiotherapy in a patient with a germline ataxia telangiectasia mutated gene: Future treatment decisions. *Cureus.* 2017;9(7):e1458.
22. Maciejczyk M, Mikoluc B, Pietrucha B et al. Oxidative stress, mitochondrial abnormalities and antioxidant defense in ataxia-telangiectasia, Bloom syndrome and Nijmegen breakage syndrome. *Redox Biol.* 2017;11:375–83.
23. Lefton-Greif MA, Crawford TO, McGrath-Morrow S, Carson KA, Lederman HM. Safety and caregiver satisfaction with gastrostomy in patients with ataxia telangiectasia. *Orphanet J Rare Dis.* 2011;6:23.
24. Lin DD, Barker PB, Lederman HM, Crawford TO. Cerebral abnormalities in adults with ataxia-telangiectasia. *AJNR Am J Neuroradiol.* 2014;35(1):119–23.
25. Bhatt JM, Bush A, van Gerven M et al. ERS statement on the multidisciplinary respiratory management of ataxia telangiectasia. *Eur Respir Rev.* 2015;24(138):565–81.
26. Karnitz LM, Zou L. Molecular pathways: Targeting ATR in cancer therapy. *Clin Cancer Res.* 2015;21(21):4780–5.
27. Weber AM, Ryan AJ. ATM and ATR as therapeutic targets in cancer. *Pharmacol Ther.* 2015;149:124–38.
28. Riboldi GM, Samanta D, Frucht S. Ataxia Telangiectasia (Louis-Bar Syndrome) [Updated 2019 Dec 31]. In: *StatPearls [Internet]*. Treasure Island (FL): StatPearls Publishing, Florida; 2020 Jan-.
29. Crawford TO, Skolasky RL, Fernandez R, Rosquist KJ, Lederman HM. Survival probability in ataxia telangiectasia. *Arch Dis Child.* 2006;91(7):610–1.
30. Khoronenkova SV. Mechanisms of non-canonical activation of ataxia telangiectasia mutated. *Biochemistry (Mosc).* 2016;81(13):1669–1675.

63

Autoimmune Lymphoproliferative Syndrome

Dongyou Liu

CONTENTS

63.1 Introduction

Autoimmune lymphoproliferative syndrome (ALPS) is a rare genetic disorder of apoptosis characterized by accumulation of autoreactive lymphocytes and childhood-onset chronic lymphadenopathy, splenomegaly, multilineage cytopenias, along with predisposition to Hodgkin and non-Hodgkin lymphomas.

Germline mutations in the *FAS* (*TNFRSF6 or* CD95, chromosome 10q23.31) gene causing defective FAS-mediated apoptosis are responsible for 60%–70% of ALPS cases, although somatic *FAS* mutations in circulating lymphocyte subsets occur in 0.5%–10% ALPS cases. Further, germline mutations in the *FASLG* (FAS ligand, *TNFSF6* or CD95L, chromosome 1q24.3) and *CASP10* (caspase 10, chromosome 2q33.1) genes account for 1% and <1% of ALPS cases, respectively, while genetic mechanisms for about 20% of ALPS cases remain undetermined. In addition, several relatively uncommon ALPS-related disorders (e.g., CEDS, RALD, DALD, and XLP1) have been described (Table 63.1). Once considered as ALPS variants, these disorders share similar clinical features with ALPS but appear to evolve from distinct genetic mechanisms [1].

63.2 Biology

ALPS (also known as Canale–Smith syndrome) was first recognized in the early 1990s from a cohort of patients who displayed chronic lymphoproliferation and elevated levels of double negative T cells (DNT) that express CD3 and alpha/beta T-cell receptor (TCRα/β) but not CD4 and CD8. Further studies identified additional signature abnormalities associated with ALPS (e.g., elevated levels of interleukin 10 [IL-10], IL-18, vitamin B_{12}, soluble FAS ligand [sFASL], IgG in plasma or sera, and defective FAS-mediated apoptosis), and helped expand its clinical spectrum that encompasses classic ALPS and ALPS-related disorders [2,3].

Subsequent characterization of several ALPS-associated genes (e.g., *FAS, FASLG, CASP10*) revealed valuable insights into the molecular pathogenesis of ALPS and ALPS-related disorders. Under normal circumstances, interaction between FAS and FAS ligand activates the FAS-associated death domain (FADD) protein and triggers the caspase cascade, leading to cellular apoptosis. However, germline or somatic mutations in *FAS, FASLG,* and *CASP10* (which participates in the formation of death-inducing signaling complex comprising FADD, caspase-8, and caspase-10) cause defects in RAS-mediated apoptosis and thus increase lymphocyte accumulation and autoimmune reactivity [4,5].

The availability of these molecular details enabled accurate classification of ALPS subtypes (i.e., ALPS-FAS due to homozygous germline mutations, ALPS-FAS due to heterozygous germline mutations, ALPS-sFAS, ALPS-FASLG, ALPS-CASP10, ALPS-U), and separation of ALPS-related disorders from ALPS (Table 63.1). As rare conditions that resemble ALPS clinically and add complexity to the diagnosis of ALPS, ALPS-related disorders include caspase-8 deficiency state (CEDS, due to germline *CASP8* mutation); Dianzani autoimmune lymphoproliferative disease (DALD, due possibly to germline *PRF1* or *SPP1* mutations); *RAS*-associated leukoproliferative disorder (RALD, due to germline or somatic *KRAS* or *NRAS* mutations, which is considered one of RASopathies), X-linked lymphoproliferative disease type 1 (XLP1, also known as classic XLP or Duncan syndrome, due to *SH2D1A* mutation); FADD deficiency (due to biallelic *FADD*

TABLE 63.1

Classification and Characteristics of Autoimmune Lymphoproliferative Syndrome (ALPS) and ALPS-Related Disorders

	Current Classification	Previous Classification	Gene (Chromosome)	Mutation (Frequency)	Diagnostic Criteria
ALPS	ALPS-FAS	ALPS type 0	*FAS* (10q23.31)	Germline homozygous mutations in *FAS gene*	*Required criteria* Chronic (>6 months), nonmalignant, noninfectious lymphadenopathy or splenomegaly or both
	ALPS-FAS	ALPS type Ia	*FAS* (10q23.31)	Germline heterozygous mutations in *FAS gene* (60%–70%)	Elevated CD3⁺TCRab⁺CD4⁻CD8⁻ "double negative T cells" (DNT) cells > 1.5% of total lymphocytes or 2.5% of CD3⁺ lymphocytes with normal or elevated lymphocyte counts
	ALPS-sFAS	ALPS type Im	*FAS* (10q23.31)	Somatic mutations in *FAS gene* (0.5%–10%)	
	ALPS-FASLG	ALPS type Ib	*FASLG* (1q24.3)	Germline homozygous or heterozygous mutations in FAS ligand gene (<1%)	*Primary accessory criteria* 1. Defective lymphocyte apoptosis (2 separate assays) 2. Somatic or germline mutation in *FAS*, *FASLG* or *CASP10*
	ALPS-CASP10	ALPS type IIa	*CASP10* (2q33.1)	Germline heterozygous mutations in caspase 10 gene (2%)	*Secondary accessory criteria* 1. Elevated plasma soluble FAS ligand (sFASL) (>200 pg/mL) or elevated plasma interleukin-10 (IL-10) levels or elevated serum or plasma vitamin B₁₂ levels or elevated plasma IL-18 levels
	ALPS-U	ALPS type III	Undefined	Undefined (but no mutations in FAS, FASLG, and CASP10 genes). (15%–20%)	2. Typical immunohistological findings as reviewed by an experienced hematopathologist 3. Autoimmune cytopenias and elevated IgG levels 4. Family history of a nonmalignant/noninfectious lymphoproliferation with or without autoimmunity
ALPS-related disorders	CEDS	ALPS type IIb	*CASP8* (2q33.1)	Loss-of-function germline mutations in caspase 8 gene (2%)	Lymphadenopathy and/or splenomegaly, marginal DNT elevation, recurrent infections, mutation in *CASP8*
	RALD	ALPS type IV	*KRAS* (12p12.1) *NRAS* (1p13.2)	Somatic mutations in KRAS and NRAS genes (6%)	Autoimmunity, lymphadenopathy and/or splenomegaly, persistent absolute or relative monocytosis, hypergammaglobulinemia, B lymphocytosis, normal DNTs and vitamin B₁₂, activating somatic mutations in *KRAS* or *NRAS*
	DALD	DALD	*PRF1* (10q22.1) *SPP1* (4q22.1)	Germline mutations in *PRF1* or *SPP1* (<1%)	Autoimmunity, lymphadenopathy and/or splenomegaly, normal DNTs, defective in vitro FAS-mediated apoptosis (FAS resistance) without *FAS* or *FASLG* mutations, germline *PRF1* or *SPP1* mutations as possible predisposition factors
	XLP1	XLP1	*SH2D1A* (Xq25)	Mutations in *SH2D1A* gene	Fulminant Epstein−Barr virus infection, hypogammaglobulinemia, lymphoma, mutation in *SH2D1A*

Abbreviations: CEDS, caspase 8 deficiency state; DALD, Dianzani autoimmune lymphoproliferative disease; αβ DNT, TCRαβ⁺ double-negative (CD4⁻CD8⁻) T (DNT) cells; LPD, lymphadenopathy; RALD, RAS-associated autoimmune leukoproliferative disorder; XLP1, X-linked lymphoproliferative disease type 1 (also known as classic XLP or Duncan's syndrome).

mutation), common variable immunodeficiency 9 (PRKCD deficiency, due to *PRKCD* mutation); *p110delta* activating mutation causing senescent T cells, lymphadenopathy, and immunodeficiency (PASLI or activated PI3 K delta syndrome); *CTLA-4* haploinsufficiency with autoimmune infiltration (CHAI); gain-of-function signal transducer and activator of transcription 3 mutations (GOF STAT3 mutations); and lipopolysaccharide-responsive vesicle trafficking, beach and anchor containing deficiency with autoantibodies, regulatory T-cell defects, autoimmune infiltration, and enteropathy (LRBA, due to *LRBA* mutations), etc [1].

63.3 Pathogenesis

Apoptosis (programmed cell death) is an important mechanism that helps maintain immune system homeostasis. Lymphocytes expand to counter against an invading pathogen, and undergo apoptosis after the danger of infection subsides, with only a small number of memory cells remaining. Not surprisingly, any defects in the key components of the apoptotic pathways may result in uncontrolled expansion and accumulation of lymphocytes including self-antigen-specific populations, and subsequent development of autoimmunity and other immune disorders.

The pathogenesis of ALPS is largely attributed to defective apoptosis of lymphocytes mediated via the FAS/FASLG (FAS ligand) signaling pathway. As CASP10 participates in the formation of a death-inducing signaling complex composed of FADD, caspase-8, and caspase-10, which triggers the downstream effector caspase cascade of the FAS/FASLG signaling pathway, it may also impact on lymphocyte apoptosis [6,7].

The *FAS* (*TNFRSF6* or CD95) gene on chromosome 10q23.31 is a 25 kb stretch of DNA with nine exons and encodes a protein (FAS) of the TNF-receptor superfamily, that takes part in the physiologic regulation of programmed cell death (apoptosis).

The FAS protein includes a 16-aa signal sequence (which is encoded by exons 1 and 2 and which is cleaved off after the FAS protein is trafficked to the cell surface) and a 319 aa, 36 kDa mature protein [which contains three extracellular cysteine-rich domains (CRD) encoded by exons 3, 4, and 5; a transmembrane domain encoded by exon 6; and intracellular domains encoded by exons 7–9].

The intracellular domain encoded by exon 9 is known as the death domain, which interacts with its ligand to form a death-inducing signaling complex consisting of FADD, caspase-8, and caspase-10, and triggers downstream effector caspase cascade, resulting in apoptosis. In addition, FAS is also capable of activating NF-kappaβ, MAPK3/ERK1, and MAPK8/JNK, and transducing proliferating signals in normal diploid fibroblast and T cells [8].

Mutations in the *FAS* gene alter the structure of its protein product, which has diminishing ability to carry out physiologic induction of apoptosis. Indeed, heterozygous pathogenic variants located in the CRD and the death domain exert dominant-negative interference on FAS-mediated apoptosis. On the other hand, homozygous or compound heterozygous pathogenic variants produce FAS loss of function or haploinsufficiency, with absent or reduced surface expression on lymphocytes, leading to a defective FAS-mediated apoptosis [9].

To date, >130 pathogenic variants (including missense, nonsense, splicing site variants, small deletions/insertions, gross deletions, and complex deletion/duplications) have been identified from patients with ALPS.

The *FASLG* gene on chromosome 1q24.3 spans 8 kb with four exons, and encodes a 281 aa, 31.4 kDa type II transmembrane protein (FASLG, FAS ligand, *TNFSF6*, or CD95L) belonging to the tumor necrosis factor family. FASLG binds to FAS, and activates the FAS/FASLG signaling pathway, causing activation-induced cell death (AICD) of T cells and cytotoxic T lymphocyte induced cell death [10].

A number of *FASLG* pathogenic variants [c.203dupT (p.Ala69fsTer138), c.263delT (p.Phe87fsTer95), c.466A>G (p.Arg156Gly), c.605G>C (p.Cys202Ser), c.740C>A (p.Ala247Glu), c.829G>A (p.Gly277Ser)] have been identified from patients with ALPS. These mutations reduce AICD of T cells and increase the levels of double-negative T cells. While pathogenic variant c.466A>G (p.Arg156Gly) involves autosomal dominant inheritance, other pathogenic variants show autosomal recessive heritance [1].

Interestingly, homozygous *FASLG* pathogenic missense variant c.740C>A (p.Ala247Glu) decreases FAS-mediated cell death

and FAS-dependent cytotoxicity. Heterozygous *FASLG* pathogenic variant c.466A>G (p.Arg156Gly) produces a dominant interfering FASLG protein that alters its extracellular FAS-binding region and that through its binding to wild-type prevents FAS-mediated apoptosis.

The *CASP10* (caspase 10) gene on chromosome 2q33.1 is of 48 kb in length with 11 exons, and encodes caspase 10, which participates in the FAS/FASLG signaling pathway through formation of a death-inducing signaling complex composed of FAS-associated death domain protein (FADD), caspase-8, and caspase-10, which triggers downstream caspase cascade, and apoptosis.

Three *CASP10* pathogenic variants [c.853C>T (p.Leu285Phe), c.1216A>T (p.Ile406Leu) and c.1337A>G (p.Tyr446Cys)] have been found from ALPS patients. These mutations reduce FAS expression and CASP10 activity and induce abnormal lymphocyte and dendritic cell homeostasis and immune regulatory defects.

63.4 Epidemiology

63.4.1 Prevalence

ALPS is a rare disorder, with approximately 1000 cases reported worldwide to date. While ALPS shows no racial and ethnic predilection, it does seem to have a slight male preponderance (male to female ratio of 1.6-2.2 to 1). Despite its common occurrence in early childhood, with a median age of onset of 3 years, ALPS is sometimes found in people between 18 and 35 years of age.

63.4.2 Inheritance

ALPS is a complex genetic disorder that may be transmitted through autosomal dominant (ALPS-FAS due to germline heterozygous *FAS* mutations, 60%–70% of cases; a few ALPS-FASLG due to germline heterozygous *FASLG* mutations, <1%; ALPS-CASP10 due to germline *CASP10* mutations, 2%) or autosomal recessive (ALPS-FAS due to germline homozygous *FAS* mutations; some ALPS-FASLG due to germline homozygous *FASLG* mutations, <1%) inheritance. ALPS-sFAS (due to somatic *FAS* mutation, 0.5%–10%) appears to evolve from somatic mosaicism and is clinically indistinguishable from ALPS-FAS (although ALPS-sFAS patients are typically older than ALPS-FAS patients). About 20% of ALPS cases remain genetically undefined (Table 63.1).

In autosomal dominant ALPS, an individual inherits a single copy of the mutated gene from one parent and a normal copy from the other parent and becomes a heterozygous carrier. Not all heterozygous carriers with a mutated copy of the gene develop ALPS, and about 40% are asymptomatic. In autosomal recessive ALPS, an individual acquires homozygous mutations after inheriting two copies of the mutated gene, one from each parent, and will develop clinical symptoms. In some ALPS cases, an individual with no family history of the disorder acquires a sporadic mutation that occurs at the time of conception (in the eggs or sperm) and can be passed on to offspring, or a somatic mutation that occurs later in the embryo development, affects blood cells. and cannot be passed on to offspring.

63.4.3 Penetrance

ALPS demonstrates incomplete penetrance and variable expressivity. Individuals heterozygous for an inherited/germline *FAS* pathogenic variant may have defective FAS-mediated apoptosis, but not necessarily clinical symptoms of ALPS. The fact that a subset of ALPS patients harbor both germline and somatic mutations in the *FAS* gene suggests the requirement of a "second hit" for clinical disease onset [11].

63.5 Clinical Features

ALPS is associated with a diverse spectrum of clinical manifestations.

  i. *Chronic nonmalignant lymphoproliferation:* Chronic and/or recurrent lymphadenopathy; splenomegaly with or without hypersplenism; hepatomegaly; lymphocytic interstitial pneumonia.
 ii. *Autoimmune disease:* Cytopenia, particularly combinations of autoimmune hemolytic anemia (AIHA), immune thrombocytopenia (ITP), and autoimmune neutropenia; autoimmune hepatitis, autoimmune glomerulonephritis, autoimmune thyroiditis, encephalomyelitis or Guillain–Barré syndrome, uveitis, iridocyclitis, aplastic anemia, vasculitis, pancreatitis, angioedema, and alopecia.
iii. *Malignancy:* Hodgkin and non-Hodgkin lymphomas (which are 50-fold and 14-fold more common in ALPS patients than the general population), carcinomas of the thyroid, breast, skin, tongue, liver; multiple neoplastic lesions such as thyroid/breast adenomas, and gliomas
 iv. *Other symptoms:* Urticaria and other skin rashes; vasculitis; panniculitis; arthritis and arthralgia; recurrent oral ulcers; humoral immunodeficiency; pulmonary infiltrates; premature ovarian insufficiency; hydrops fetalis; organic brain syndrome such as mental status changes, seizures, headaches. [12–18].

The relative frequencies of ALPS findings are lymphadenopathy (96%), splenomegaly (95%), hepatomegaly (72%), AIHA (29%), immune-mediated ITP (23%), neutropenia (19%), liver dysfunction (5%), infiltrative lung lesions (4%), glomerulonephritis (1%), and eye lesions (0.7%).

As lymphadenopathy and splenomegaly are largely asymptomatic, they are often identified during physical examination for other unrelated illnesses. On the other hand, cytopenias (particularly combined AIHA, ITP, and autoimmune neutropenia) may induce episodes of fatigue, pallor, and icterus (due to hemolytic anemia) initially, followed by easy bruising and mucocutaneous bleeding (due to thrombocytopenia) as well as bacterial infections (due to neutropenia).

63.6 Diagnosis

Diagnosis of ALPS relies on observation of relevant clinical findings (chronic nonmalignant lymphoproliferation, autoimmune disease, Hodgkin lymphoma and non-Hodgkin lymphoma, skin rashes, family history of ALPS or ALPS-related disorders), detection of laboratory abnormalities (abnormal soluble IL-10, FASLG, IL-18, and vitamin B_{12}; defective in vitro FAS-mediated apoptosis; elevated levels of α/β-double negative T cells (α/β-DNT cells) in peripheral blood or tissue that express alpha/beta T-cell receptor but lack both CD4 and CD8), and identification of germline pathogenic variants in *FAS*, *CASP10* and then *FASLG* (with somatic *FAS* pathogenic variants detected from sorted α/β-DNT cells by sequence analysis) [19–21].

According to the 2010 diagnostic criteria, a definitive diagnosis of ALPS is made when required criteria plus one primary accessory criterion are met, while a probable diagnosis is made when required criteria plus one secondary accessory criterion are met (Table 63.1) [22] (Figure 63.1).

Differential diagnoses of ALPS include *caspase-8 deficiency state* (CEDS, lymphadenopathy, splenomegaly, marginal elevation of DNT cells, defective FAS-mediated apoptosis, recurrent sinopulmonary infections, immunodeficiency instead of autoimmunity, increased risk for malignancy; autosomal recessive

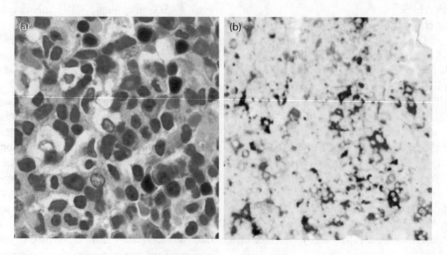

FIGURE 63.1 Immunohistochemistry showing lymphohistiocytic infiltrate with features of Rosai–Dorfman disease (a, H&E), and prominent S-100-positive histiocytes with evidence of emperipolesis scattered around lymphoid aggregates (b, S-100 stain) in a bone marrow of ALPS patient harboring heterozygous germline *FAS* mutation (case 23). (Photo credit: Xie Y et al. *Haematologica*. 2017;102(2):364–72.)

immunodeficiency syndrome due to loss of function *CASP8* mutation), *Dianzani autoimmune lymphoproliferative disorder* (DALD, autoimmunity, lymphoproliferation, splenomegaly, defective FAS function without expansion of DNT cells, overexpression of osteopontin and/or perforin) [23,24], *RAS-associated autoimmune leukoproliferative disorder* (RALD, mild peripheral lymphadenopathy, hepato/splenomegaly, autoimmunity, recurrent respiratory tract infections, defective non-FAS-mediated apoptosis; due to somatic gain-of-function variants in *NRAS* and *KRAS* detectable only in blood cells), *X-linked lymphoproliferative disease* (XLP, infectious mononucleosis, dysgammaglobulinemia, lymphoproliferative disorder, near-fatal or fatal Epstein−Barr virus (EBV) infection, hypogammaglobulinemia, non-Hodgkin lymphoma/high-grade B-cell lymphoma; due to hemizygous pathogenic variants in *SH2D1A*), *FAS-associated death domain deficiency* (FADD deficiency, severe infections, congenital heart defects, fever, liver dysfunction, seizures; autosomal recessive primary immunodeficiency syndrome due to biallelic pathogenic variants in *FADD*), *common variable immunodeficiency 9* (PRKCD deficiency, lymphadenopathy, hepato/splenomegaly, autoimmunity, NK cell dysfunction, recurrent infections; autosomal recessive primary immunodeficiency due to pathogenic variants in *PRKCD*), *common variable immune deficiency LRBA* (LRBA deficiency; antibody deficiency, autoimmunity, lymphoproliferation, lymphocyte infiltration, infection, enteropathy or inflammatory bowel disease; due to mutations in lipopolysaccharide-responsive and beige-like anchor protein gene *LRBA*), CTLA4 haploinsufficiency with autoimmune infiltration (CHAI, recurrent infections, autoimmune thrombocytopenias, CD4+ T-cell lymphopenia, B cell abnormalities, hypogammaglobulinemia, abnormal lymphocytic infiltration of nonlymphoid organs, diffuse lymphadenopathy, hepatosplenomegaly, EBV-associated Hodgkin lymphoma; autosomal dominant immune dysregulation syndrome due to heterozygous pathogenic loss-of-function variants in *CTLA4*), *common variable immune deficiency* (CVID, humoral immune deficiency, increased susceptibility to infections, diminished responses to protein and polysaccharide vaccines, autoimmune disease, lymphadenopathy, splenomegaly, lymphoma; due to heterozygous pathogenic variants in *CASP8* and *CTLA4*), *hyper IgM syndrome* (HIGM syndrome; neutropenia, thrombocytopenia, hemolytic anemia, lymphomas, gastrointestinal complications, elevated serum IgM concentration, defective specific antibody responses, recurrent infections; due to pathogenic variants in *CD40LG*), *HIGM2* (lymphoid hyperplasia, recurrent infections; autosomal recessive disorder due to pathogenic variants in *AICDA*), *WAS-related disorders* (including Wiskott−Aldrich syndrome, X-linked thrombocytopenia, and X-linked congenital neutropenia; thrombocytopenia, mucosal bleeding, bloody diarrhea; petechiae, eczema, immune thrombocytopenic purpura, hemolytic anemia, immune-mediated neutropenia, arthritis, vasculitis, immune-mediated damage to the kidneys and liver, recurrent infections, increased risk of lymphomas; due to *WAS* mutation), *gain-of-function mutations in signal transducer and activator of transcription 1 defect* (GOF STAT1 defect; chronic mucocutaneous candidiasis, recurrent *Staphylococcus aureus* infections, cerebral aneurysms, multiple autoimmune features; due to gain of function mutation in *STAT1*), *gain-of-function mutations in signal transducer and activator of transcription*

3 (GOF *STAT3* defect; lymphadenopathy, elevated α/β-DNT cells, autoimmune cytopenia, recurrent infections, short stature, multiorgan autoimmunity; due to gain of function mutation in *STAT3*), *loss-of-function STAT3 mutations* (immunodeficiency, hyper-IgE, recurrent infection), *activated PI3 K delta syndrome* (APDS, also known as PASLI; recurrent respiratory infections, both B- and T-cell defects; due to heterozygous gain-of-function mutations in *PI3KCD* or *PI3KR1*), *lymphoma* (both B-cell and T-cell lymphoma), *Castleman disease* (fever, unintended weight loss, fatigue, night sweats, nausea, hepatosplenomegaly), *Rosai−Dorfman disease* (massive painless cervical lymphadenopathy, leukocytosis, elevated erythrocyte sedimentation rate, hypergammaglobulinemia, sinus histiocytosis fever, pallor, unintended weight loss, malaise, rhinitis, hepatosplenomegaly), *Kikuchi−Fujimoto disease* (cervical lymphadenopathy, mild fever, night sweats, weight loss, nausea, vomiting, sore throat), *Evans syndrome* (a combination of AIHA and ITP), and *hemophagocytic lymphohistiocytosis* (HLH, fever, hepatosplenomegaly, cytopenia, neurological abnormalities) [1].

63.7 Treatment

Management of ALPS should center on (i) *treatment of lymphoproliferation, hypersplenism, and lymphomas* (corticosteroids and immunosuppressive drugs for treating airway obstruction, significant hypersplenism associated with splenomegaly, or autoimmune manifestations; sirolimus for treating lymphoproliferation; cyclophosphamide, anti-thymocyte globulin, and alemtuzumab for controlling lymphoproliferative manifestations), and (ii) *treatment of cytopenias and other autoimmune diseases* (corticosteroids and corticosteroid-sparing agents such as sirolimus and mycophenolate mofetil for autoimmune cytopenias and refractory cytopenias; combined IVIG and corticosteroids for severe autoimmune hemolytic anemia; rituximab for refractory cytopenias; low-dose G-CSF for isolated chronic neutropenia; thrombopoietin receptor agonist for isolated immune thrombocytopenia) [25−27].

Bone marrow (hematopoietic stem cell) transplantation (BMT/HSCT) represents the only curative treatment for ALPS and has been applied to ALPS-FAS with homozygous or compound heterozygous pathogenic variants, ALPS with severe or refractory autoimmune cytopenias, ALPS with lymphoma, and ALPS with complications from immunosuppressive therapy.

Splenectomy may be considered for life-threatening refractory cytopenias or severe hypersplenia. Pre- and post-splenectomy boost vaccinations (with consideration of vaccinations) and penicillin prophylaxis are recommended for individuals with splenectomy. Nonetheless, given that post-splenectomy sepsis represents the most lethal complication of ALPS, a decision to undertake splenectomy should be made with caution [28].

63.8 Prognosis and Prevention

In general, patients with ALPS have a favorable prognosis, and clinical symptoms often improve after puberty even without treatment. However, ALPS appears to confer a much higher risk of lymphoma in affected patients than in the general population.

Further, the location of specific gene mutation may also have a bearing on overall risk of complications including lymphoma. For example, patients harboring mutations that affect the intracellular domain of FAS tend to develop severe disease and show increased risk for Hodgkin and non-Hodgkin B-cell lymphomas. On the other hand, patients harboring mutations that affect the extracellular domain of FAS often develop lymphomas.

Regular monitoring the sizes of glands and the number of blood cells for patients harboring *FAS*, *FASLG*, or *CASP10* pathogenic variants helps determine whether or when it is necessary to control the overgrowth of lymphocytes or autoimmune complications.

Avoidance of aspirin and other nonsteroidal anti-inflammatory drugs by individuals with immune thrombocytopenia is recommended as these drugs may interfere with platelet function.

Prenatal testing and preimplantation genetic diagnosis may be offered to prospective parents with known *FAS*, *FASLG*, or *CASP10* pathogenic variants.

63.9 Conclusion

Autoimmune lymphoproliferative syndrome (ALPS) is a genetic disorder with the clinical hallmarks of chronic nonmalignant lymphoproliferation (lymphadenopathy and/or splenomegaly), autoimmune disease (multilineage cytopenias), and secondary malignancies (Hodgkin and non-Hodgkin lymphomas), and the defining laboratory finding of elevated $CD3^+TCR$-$\alpha\beta^+CD4^-CD8^-$ lymphocytes (so-called DNT cells, which may drive abnormal B cell activity and autoimmunity), along with some other common but less specific findings (e.g., polyclonal elevation of IgG and IgA, presence of autoantibodies to blood cell elements, elevated serum levels of vitamin B12, IL-10, soluble FAS ligand, and IL-18).

The underlying pathogenesis of ALPS relates to mutations in the *FAS*, *FASLG*, and *CASP10* genes, which encodes proteins involved in the formation of a death-inducing signaling complex composed of FADD, caspase-8, and caspase-10. This death-inducing signaling complex triggers the downstream effector caspase cascade of the FAS/FASLG signaling pathway, leading to lymphocyte apoptosis. Alterations in the *FAS*, FASLG, and *CASP10 proteins compromise their capacity to induce* FAS-mediated apoptosis *and result in the accumulation of unwanted lymphocytes and increased* autoimmune reactivity.

Diagnosis of ALPS is largely based on meeting the required criteria plus primary or secondary accessory criteria. Molecular identification of *FAS*, *FASLG*, or *CASP10* pathogenic variants helps further confirm and differentiate ALPS from ALPS-related disorders and other autoimmunity or immune deficiency syndromes. Treatment of ALPS involves symptomatic relief using standard procedures (e.g., surgery, chemotherapy). Further research is warranted in order to develop novel and effective therapeutics for the eventual control of this genetic disorder.

REFERENCES

1. Bleesing JJH, Nagaraj CB, Zhang K. Autoimmune lymphoproliferative syndrome. In: Adam MP, Ardinger HH, Pagon RA, Wallace SE, Bean LJH, Stephens K, Amemiya A, editors. *GeneReviews® [Internet].* Seattle (WA): University of Washington, Seattle; 1993–2018. 2006 Sep 14 [updated 2017 Aug 24].
2. Oliveira JB. The expanding spectrum of the autoimmune lymphoproliferative syndromes. *Curr Opin Pediatr.* 2013;25(6):722–9.
3. Justiz Vaillant AA, Stang CM. Lymphoproliferative disorders. In: *StatPearls [Internet].* Treasure Island (FL): StatPearls Publishing; 2019 Jan-.2019 Feb 22.
4. Vlachaki E, Diamantidis MD, Klonizakis P, Haralambidou-Vranitsa S, Ioannidou-Papagiannaki E, Klonizakis I. Pure red cell aplasia and lymphoproliferative disorders: An infrequent association. *Sci World J.* 2012;2012:475313.
5. Shah S, Wu E, Rao VK, Tarrant TK. Autoimmune lymphoproliferative syndrome: An update and review of the literature. *Curr Allergy Asthma Rep.* 2014;14(9):462.
6. Bride K, Teachey D. Autoimmune lymphoproliferative syndrome: More than a FAScinating disease. *F1000Res.* 2017;6:1928.
7. Lisco A, Wong CS, Price S et al. Paradoxical CD4 lymphopenia in autoimmune lymphoproliferative syndrome (ALPS). *Front Immunol.* 2019;10:1193.
8. Vavassori S, Galson JD, Trück J et al. Lymphadenopathy driven by TCR-Vγ8Vδ1 T-cell expansion in FAS-related autoimmune lymphoproliferative syndrome. *Blood Adv.* 2017;1(15):1101–6.
9. Marega LF, Teocchi MA, Dos Santos Vilela MM. Differential regulation of miR-146a/FAS and miR-21/FASLG axes in autoimmune lymphoproliferative syndrome due to FAS mutation (ALPS-FAS). *Clin Exp Immunol.* 2016;185(2):148–53.
10. Magerus-Chatinet A, Stolzenberg MC et al. Autoimmune lymphoproliferative syndrome caused by a homozygous null FAS ligand (FASLG) mutation. *J Allergy Clin Immunol.* 2013;131(2):486–90.
11. Mazerolles F, Stolzenberg MC, Pelle O et al. Autoimmune lymphoproliferative syndrome-FAS patients have an abnormal regulatory T cell (Treg) phenotype but display normal natural Treg-suppressive function on T cell proliferation. *Front Immunol.* 2018;9:718.
12. Hansford JR, Pal M, Poplawski N et al. In utero and early postnatal presentation of autoimmune lymphoproliferative syndrome in a family with a novel FAS mutation. *Haematologica.* 2013;98(4):e38–9.
13. Langan RC, Gill F, Raiji MT et al. Autoimmune pancreatitis in the autoimmune lymphoproliferative syndrome (ALPS): A sheep in wolves' clothing? *Pancreas.* 2013;42(2):363–6.
14. Rudman Spergel A, Walkovich K et al. Autoimmune lymphoproliferative syndrome misdiagnosed as hemophagocytic lymphohistiocytosis. *Pediatrics.* 2013;132(5):e1440–4.
15. Leal-Seabra F, Costa GS, Coelho HP, Oliveira A. Unexplained lymphadenopathies: Autoimmune lymphoproliferative syndrome in an adult patient. *BMJ Case Rep.* 2016;2016. pii: bcr2016216758.
16. Sriram S, Joshi AY, Rodriguez V, Kumar S. Autoimmune lymphoproliferative syndrome: A rare cause of disappearing HDL syndrome. *Case Reports Immunol.* 2016;2016:7945953.
17. Ucar D, Kim JS, Bishop RJ, Nussenblatt RB, Rao VK, Sen HN. Ocular inflammatory disorders in autoimmune lymphoproliferative syndrome (ALPS). *Ocul Immunol Inflamm.* 2017;25(5):703–9.
18. Pavlič A, Vrecl M, Jan J, Bizjak M, Nemec A. Case report of a molar-root incisor malformation in a patient with an autoimmune lymphoproliferative syndrome. *BMC Oral Health.* 2019;19(1):49.

19. Teachey DT, Lambert MP. Diagnosis and management of autoimmune cytopenias in childhood. *Pediatr Clin North Am.* 2013;60(6):1489–511.

20. Seidel MG. Autoimmune and other cytopenias in primary immunodeficiencies: Pathomechanisms, novel differential diagnoses, and treatment. *Blood.* 2014;124(15):2337–44.

21. Xie Y, Pittaluga S, Price S et al. Bone marrow findings in autoimmune lymphoproliferative syndrome with germline FAS mutation. *Haematologica.* 2017;102(2):364–72.

22. Oliveira JB, Bleesing JJ, Dianzani U et al. Revised diagnostic criteria and classification for the autoimmune lymphoproliferative syndrome (ALPS): Report from the 2009 NIH International Workshop. *Blood.* 2010;116(14):e35–40.

23. Chiocchetti A, Indelicato M, Bensi T et al. High levels of osteopontin associated with polymorphisms in its gene are a risk factor for development of autoimmunity/ lymphoproliferation. *Blood.* 2004;103(4):1376–82.

24. Clementi R, Chiocchetti A, Cappellano G et al. Variations of the perforin gene in patients with autoimmunity/lymphoproliferation and defective Fas function. *Blood.* 2006;108(9): 3079–84.

25. Arora S, Singh N, Chaudhary GK, John MJ. Autoimmune lymphoproliferative syndrome: Response to mycophenolate mofetil and pyrimethamine/sulfadoxine in a 5-year-old child. *Indian J Hematol Blood Transfus.* 2011;27(2):101–3.

26. Rao VK, Oliveira JB. How I treat autoimmune lymphoproliferative syndrome. *Blood.* 2011;118(22):5741–51.

27. Rao VK. Approaches to managing autoimmune cytopenias in novel immunological disorders with genetic underpinnings like autoimmune lymphoproliferative syndrome. *Front Pediatr.* 2015;3:65.

28. Teachey DT. New advances in the diagnosis and treatment of autoimmune lymphoproliferative syndrome. *Curr Opin Pediatr.* 2012;24(1):1–8.

64

Bloom Syndrome

Dongyou Liu

CONTENTS

64.1 Introduction

Bloom syndrome is a rare autosomal recessive genodermatosis that typically causes pre- and postnatal growth deficiency, sun-sensitive, telangiectatic, hypo- and hyperpigmented skin (facial lupus-like skin lesions), mild immunodeficiency, type 2 diabetes mellitus, hypogonadism, genomic instability, and predisposition to cancer.

Homozygous or compound heterozygous mutations in the *BLM* (or *RECQL3*) gene on chromosome 15q26.1 encoding RecQ protein-like-3 with DNA helicase activity (BLM or RECQL3) underlie the molecular pathogenesis of Bloom syndrome. By forming a complex with DNA topoisomerase IIIα and RMI, BLM participates in the replication and repair of DNA. Structural and functional changes in BLM introduced by *BLM* mutations often disrupt the DNA replication process, leading to genomic instability (e.g., chromosomal rearrangements and breakages) [1,2].

64.2 Biology

Bloom syndrome (also known as Bloom−Torre−Machacek syndrome or congenital telangiectatic erythema) was initially described by David Bloom in 1954 from three children presenting with telangiectatic erythema and short stature. Further observations by James German in 1965 revealed the association of chromosome breakage (isochromatic breaks, displaced acentric fragments, sister chromatid reunions, transverse breakage at the centromere) and cancer predisposition with Bloom syndrome. The high rate of sister chromatid exchanges (SCE) in Bloom syndrome−affected individuals has been subsequently used for its diagnosis [3].

Following mapping the genetic locus for Bloom syndrome to chromosome 15q26.1 in 1994, the causative gene for this disorder was identified as *BLM*, which encodes a 1417 aa protein (BLM) with homology to RecQ helicases. While mutated RecQ helicases have been associated with several genetic disorders (e.g., Rothmund−Thomson syndrome, RAPADILINO syndrome, Baller−Gerold syndrome, Werner syndrome, etc.), alteration in BLM appears to result in a different phenotype that is linked to chromosomal instability and predisposition to cancer (e.g., leukemia, lymphoma, colorectal cancer, Wilms tumor, osteogenic sarcoma, medulloblastoma) [4–6].

64.3 Pathogenesis

Located on chromosome 15q26.1, the *BLM* gene is a 4528-bp cDNA sequence with 22 exons and encodes a 1417-amino-acid protein (BLM or RECQL3) that constitutes one of the five RecQ helicases (BLM, WRN, RECQL1, RECQL4, and RECQL5) identified in humans. These DNA structure-specific enzymes are involved in DNA damage signaling, repair, replication, and telomere maintenance, and mutations within three of the five human *RECQ* genes (*BLM*, *WRN*, *RECQL4*) result in distinct heritable, cancer-predisposing diseases [4,7,8].

Structurally, BLM consists of the N-terminus (consisting of several short acid patches and also a region required for strand exchange), the topoisomerase IIIα-binding region, the ssDNA strand-annealing and strand exchange domain, the DNA helicase domain, the RecQ C-terminal domain (including the Zn²⁺-binding

(a)   (b)

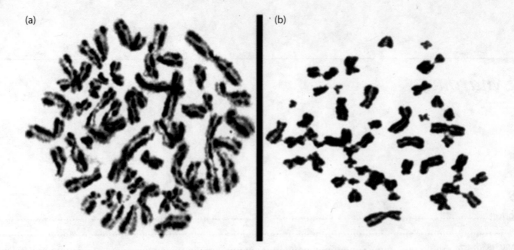

FIGURE 64.1 Bloom syndrome in a 22-year-old male showing increased sister chromatid exchange (SCE), with an average of 73 SCE/100 cells in patient bone marrow aspirate (a), as compared to 5 SCE/100 cell in normal control (b). (Photo credit: Aljarad S et al. *Oxf Med Case Reports*. 2018;2018(12):omy096.)

motif, winged helix), the helicase and ribonuclease D C-terminal domain, and the nuclear localization signal. Functionally, the N-terminus binds to interaction partners such as TOP3α, RMI1/2, and RAD51; the DNA helicase domain is responsible for ATP binding and hydrolysis; the RecQ C-terminal domain acts as a dsDNA-binding site; the -hairpin located in the winged helix domain is responsible for the duplex separation; the helicase and ribonuclease D C-terminal domain interacts with the helicase domain to facilitate efficient ATP hydrolysis and DNA unwinding [9].

As a DNA helicase distributed diffusely throughout the nucleus, BLM participated in the maintenance of the fork stability, separation of the DNA duplex into ssDNA substrate, regulation of nucleoprotein filaments, involvement of the synthesis-dependent strand annealing pathway to repair DSB, disentanglement of under-replicated DNA strands during metaphase, handling of unusual DNA structures, and keeping of genome stability. Specifically, by traversing along ssDNA in a 3' to 5' direction and breaking the hydrogen bonds that hold the two DNA strands together, BLM opens (unwinds) DNA and facilitates DNA replication and repair, RNA transcription, homologous recombination, and genomic stability [10–13].

Biallelic mutations in the *BLM* gene may either cause premature translation termination and mRNA destabilization and complete absence of detectable BLM protein levels or produce a mutant BLM protein that is incapable of unwinding DNA and is thus considered a catalytic null, leading to increased cellular hyper-recombinability and hypermutability. Among >60 *BLM* mutations identified, nucleotide insertions and deletions cause frameshifts and eliminate the C terminus of BLM containing the nuclear localization signals; nonsense variants (representing ∼1/3 of all pathogenic variants) change sense codons to nonsense or chain-terminating codons, yielding a truncated BLM; intron variants (representing ∼1/6 of all pathogenic variants) lead to splicing defects; missense variants (representing ∼1/6 of all pathogenic variants) contribute to the production of catalytically inactive BLM [2,14,15].

Notable BLM pathogenic variants include c.1933C>T (p.Gln645*) from Europeans and European Americans, a large deletion of exons 20–22 found mainly in patients from Portugal and Brazil, Gln700* among patients of Italians or American

Italians, Ser186* and Asn515fs from Japanese patients, Gln548* from Slavic patients, c.2207_2212delinsTAGATTC (2281del6/ins7 or blm^Ash, a 6-bp deletion/7-bp insertion in exon 10) leading to p.Tyr736LeufsTer5, and c.2407dupT (insT2407) causing a frameshift starting at Trp803 (p.Trp803LeufsTer4), which mainly affect Ashkenazi Jews [2].

SCE is a process employed by sexually reproducing organisms to promote genetic diversity in offspring and maintain genome stability at the same time. During meiosis that generates haploid gametes from a diploid progenitor, one round of DNA replication (which segregates homologous chromosomes inherited from different parents) is followed by two successive nuclear divisions (the second of which separates sister chromatids). Triggered by self-inflicted DNA DSB, recombination takes place between homologous chromosomes, and two sister chromatids rejoin with one another and exchange regions of the parental strands in the duplicated chromosomes. Alterations in the BLM protein hinder its ability to unwind DNA for subsequent DNA replication and repair, leading to an increased frequency of SCE (Figure 64.1) [16].

64.4 Epidemiology

64.4.1 Prevalence

Bloom syndrome is a rare disorder in most populations apart from Ashkenazi Jews, in which a carrier frequency of about 1% was estimated. This disorder is thought to affect both sexes equally, although males are often identified more frequently than females (with a male to female ratio of 4:1), due possibly to prenatal loss or excess early mortality that disproportionately affects females. So far, 281 cases are recorded in the Bloom syndrome registry, of which 67 are descendants of Ashkenazi Jews [3].

64.4.2 Inheritance

Bloom syndrome is an autosomal recessive disorder. While heterozygotes (carriers) are asymptomatic, they have a 25% chance of getting a normal child with no pathogenic variant, a 50% chance of getting an asymptomatic heterozygous carrier child with one mutated allele, and a 25% chance of getting an affected child harboring homozygous or compound heterozygous pathogenic variants.

64.4.3 Penetrance

Bloom syndrome demonstrated high penetrance and varied expressivity. Some affected patients (especially of Latin and Asian origins) may have clinical features suggestive of Bloom syndrome (e.g., micrognathia, dolichocephaly, multiple café-au-lait macules, and immunodeficiency at young age), but show no characteristic photosensitivity and facial telangiectasias/erythema [17,18].

64.5 Clinical Features

Bloom syndrome is associated with a spectrum of clinical symptoms: (i) growth deficiency (affecting height, weight, and head circumference); (ii) dysmorphic facies (narrow face with underdeveloped malar and mandibular prominences and retrognathia or micrognathia); (iii) feeding problems (slow feeding, decreased appetite); (iv) skin lesions (red, sun-sensitive rash/telangiectasia/poikiloderma on the nose, cheeks, the dorsa of the hands and forearms; cheilitis, blistering and fissuring of the lips, eyebrow and eyelash hair loss, alopecia areata, and vesicular and bullous lesions with excessive or intense sun exposure; café-au-lait macules and areas of hypopigmented skin); (v) immunodeficiency (low plasma IgM and IgA levels); (vi) infections; (vii) reduced fertility (azoospermia or severe oligospermia; premature menopause); (viii) other anomalies (tracheoesophageal fistula, cardiac malformation, absent thumbs, and absence of a toe and malformation of a thumb); (ix) medical complications (chronic bronchitis and bronchiectasis, pulmonary failure; myelodysplasia; diabetes mellitus; cancer) (Figures 64.2 and 64.3) [3,16,19–21].

Patients with Bloom syndrome show high susceptibility to a variety of tumors, which occur at younger age than the general population, and which represent the main cause of death). The most common tumors are leukemia (acute myelocytic leukemia [AML], acute lymphocytic leukemia [ALL]) and lymphoma (B-cell-including Burkitt lymphoma and T-cell; but not Hodgkin lymphoma) followed by solid tumors (e.g., adenocarcinoma of the upper and lower intestinal tract; squamous cell carcinomas of the head and neck, especially at the base of the tongue, epiglottis, and esophagus; Wilms tumor) (Figure 64.3) [3,20,22–24].

64.6 Diagnosis

Diagnosis of Bloom syndrome involves medical history review, physical examination, laboratory assessment, histopathology, and cytogenetic and molecular testing.

Individuals displaying following clinical or cytogenetic features should be suspected of Bloom syndrome.

Clinical features include (i) prenatal and postnatal growth deficiency impacting linear growth, weight gain, and head circumference; (ii) sun-sensitive, erythematous rash on the face

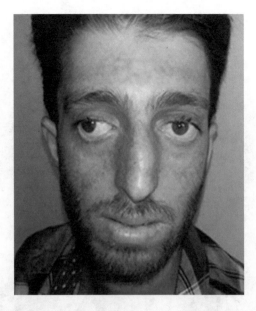

FIGURE 64.2 Bloom syndrome in a 22-year-old male showing dolichocephaly, ectropion of inferior eyelids, right lateral strabismus, right amblyopia, malar rash, and blistered fissured lower lip. (Photo credit: Aljarad S et al. *Oxf Med Case Reports.* 2018;2018(12):omy096.)

with a butterfly distribution; (iii) development of cancer at an earlier age than the general population.

Cytogenetic features include (i) increased numbers of sister-chromatid exchanges; (ii) increased quadriradial configurations in cultured blood lymphocytes (1%–2% in patient vs none in normal control); (iii) chromatid gaps, breaks, and rearrangements (Figure 64.1) [3,16].

Identification of biallelic *BLM* pathogenic variants by DNA sequencing analysis provides confirmation of Bloom syndrome. In cases where molecular analysis of *BLM* pathogenic variant is inconclusive, cytogenetic detection of increased SCE is helpful. Nonetheless, SCE analysis alone is insufficient to establish a diagnosis of Bloom syndrome given that increased SCE may also occur in patients harboring biallelic pathogenic variants in *RMI1*, *RMI2*, and *TOP3A* [3].

Differential diagnosis for Bloom syndrome consists of other chromosome breakage syndromes caused by pathogenic variants in genes encoding enzymes involved DNA replication and repair. These include RECQ-mediated genome instability 1 (SCE, small size; autosomal recessive disorder due to *RMI1* mutations), RECQ-mediated genome instability 2 (SCE, small size, multiple café-au-lait macules; autosomal recessive disorder due to *RMI2* mutations), microcephaly, growth restriction, and increased sister-chromatid exchange 2 (SCE, small size, multiple café-au-lait macules, no malar rash; autosomal recessive disorder due to *TOP3A* mutations), ataxia-telangiectasia (small stature, evidence of excessive genomic instability, telangiectasis, sinopulmonary infection, immunodeficiency, progressive cerebellar ataxia from early childhood, elevated alpha-fetoprotein levels; autosomal recessive disorder due to *ATM* pathogenic variants), ataxia telangiectasia-like disorder (small stature, evidence of excessive genomic instability, progressive cerebellar degeneration, no telangiectasias nor immunodeficiency; autosomal recessive disorder due to *MRE11* pathogenic variants), Fanconi anemia (small stature, evidence of

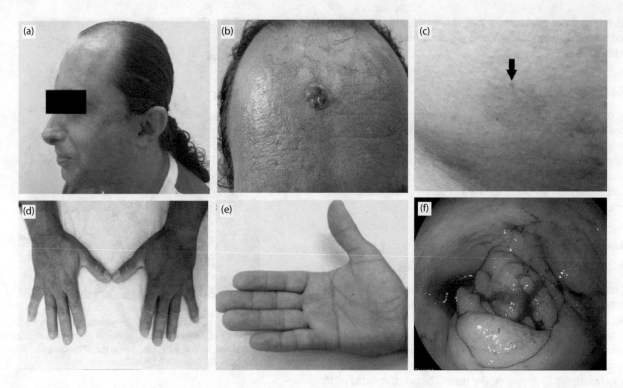

FIGURE 64.3 Bloom syndrome in a 34-year-old male showing: (a) dolichocephaly, beaked nose, and prominent ears; (b) basal cell carcinoma and telangiectatic erythema in the forehead; (c) café-au-lait macules in the anterior abdominal skin (arrow); (d) clinodactyly and dystrophic nails; (e) palmar transverse crease; and (f) polypoid-ulcerated mass in ascending colon (which was confirmed by biopsy as a well-differentiated adenocarcinoma). (Photo credit: Martinez CA et al. *Case Rep Surg.* 2016;2016:3176842.)

excessive genomic instability, cancer susceptibility, café-au-lait macules, hyper- or hypopigmentation, reduced fertility, endocrinopathy, skeletal malformations, bone marrow failure; autosomal recessive, autosomal dominant or X-linked disorder due mutations in *BRCA2, BRIP1, FANCA, FANCB, FANCC, FANCD2, FANCE, FANCF, FANCG,* or *FANCI*), Nijmegen breakage syndrome (small stature, evidence of excessive genomic instability, immunodeficiency, café-au-lait macules, predisposition to lymphoid malignancy, decline in intellectual performance, no telangiectasias; autosomal recessive disorder due to *NBN* mutations), Rothmund–Thomson syndrome (poikiloderma/erythema, telangiectasias, blistering, café-au-lait, hypopigmented spots, skeletal abnormalities, hypoplastic teeth, microdontia, missing and supernumerary teeth, juvenile cataracts, sparse scalp hair/alopecia, subfertility, anemia, neutropenia, aplastic anemia, osteosarcoma, melanoma, lymphoma, myelodysplasia, leukemia), and Werner syndrome (small stature, evidence of excessive genomic instability, increased incidence of diabetes, premature atherosclerosis, prematurely aged appearance, soft tissue sarcoma, osteosarcoma, melanoma, thyroid carcinoma; autosomal recessive disorder due to *WRN* mutations) [25–31].

64.7 Treatment

Management of Bloom syndrome requires a multidisciplinary approach that deals with skin anomalies, growth and nutrition problems, endocrinological abnormalities, malignancy, and other complications (e.g., infections, diabetes mellitus).

Non-disfiguring skin anomalies (e.g., café-au-lait spots, telangiectatic rash) do not usually require medical attention; growth problems may be treated with growth hormone (although the potential of growth hormone to increase cancer risk should be considered); feeding difficulty may be treated with the use of nasogastric tube; cancer may be treated with reduced doses of DNA-damaging chemicals (e.g., alkylating drugs) and ionizing radiation (due to increased UV photosensitivity in affected patients); infections may be controlled by antibiotic prophylaxis and therapy; insulin-resistant type 2 diabetes or hypothyroidism is managed with relevant medications [32,33].

64.8 Prognosis and Prevention

Bloom syndrome confers a shortened life expectancy and compromises the quality of life in affected individuals. Cancer is the main cause of death for Bloom syndrome patients. High-risk carriers will benefit from prenatal testing and preimplantation diagnosis [3].

As telangiectatic rash is usually exacerbated by sun exposure, barriers (e.g., clothing and hats) and UV-protecting sunscreens (30 or higher) should be applied. Exposure to radiation should be limited.

Appearance of constitutional signs and symptoms (e.g., malaise, fever, night sweats, adenopathy, unintentional weight loss, dark stools, rectal bleeding, abnormal growths or nodules, chronic pain, or fatigue) may be possible indicators of malignancies (lymphoma, colorectal cancer), and should be evaluated promptly.

64.9 Conclusion

Bloom syndrome is a rare autosomal recessive disorder characterized by prenatal and postnatal growth deficiency, photosensitive skin lesions, moderate immune deficiency, insulin-resistant diabetes, and predisposition to early-onset cancer. Homozygous or compound heterozygous mutations in the *BLM* gene, which encodes for a RecQ helicase, contribute to excessive homologous recombination, increased frequency of SCE, and high chromosomal instability. Diagnosis of Bloom syndrome relies on observation of characteristic clinical and cytogenetic features (e.g., increased frequency of SEC). Molecular identification of *BLM* pathogenic variants helps confirm its diagnosis. In the absence of specific treatment for the underlying genetic abnormality, a multidiscipline approach consisting of sun protection, infection control, insulin-resistance surveillance, and early cancer detection provides benefits for affected patients. Further research and development on novel therapeutics that restore *BLM* function and/or block the excessive homologous recombination will greatly improve outcomes for Bloom syndrome sufferers.

REFERENCES

1. Blackford AN, Nieminuszczy J, Schwab RA, Galanty Y, Jackson SP, Niedzwiedz W. TopBP1 interacts with BLM to maintain genome stability but is dispensable for preventing BLM degradation. *Mol Cell*. 2015;57(6):1133–41.
2. Cunniff C, Bassetti JA, Ellis NA. Bloom's syndrome: Clinical spectrum, molecular pathogenesis, and cancer predisposition. *Mol Syndromol*. 2017;8(1):4–23.
3. Flanagan M, Cunniff CM. Bloom syndrome. In: Adam MP, Ardinger HH, Pagon RA et al. editors. *GeneReviews®* [Internet]. Seattle (WA): University of Washington, Seattle; 1993–2019. 2006 Mar 22 [updated 2019 Feb 14].
4. Croteau DL, Popuri V, Opresko PL, Bohr VA. Human RecQ helicases in DNA repair, recombination, and replication. *Annu Rev Biochem*. 2014;83:519–52.
5. de Voer RM, Hahn MM, Mensenkamp AR et al. Deleterious germline BLM mutations and the risk for early-onset colorectal cancer. *Sci Rep*. 2015;5:14060.
6. Fu W, Ligabue A, Rogers KJ, Akey JM, Monnat RJ Jr. Human RECQ helicase pathogenic variants, population variation and "missing" diseases. *Hum Mutat*. 2017;38(2):193–203.
7. Hatkevich T, Kohl KP, McMahan S, Hartmann MA, Williams AM, Sekelsky J. Bloom syndrome helicase promotes meiotic crossover patterning and homolog disjunction. *Curr Biol*. 2017;27(1):96–102.
8. van Wietmarschen N, Merzouk S, Halsema N, Spierings DCJ, Guryev V, Lansdorp PM. BLM helicase suppresses recombination at G-quadruplex motifs in transcribed genes. *Nat Commun*. 2018;9(1):271.
9. Sun L, Huang Y, Edwards RA et al. Structural insight into BLM recognition by TopBP1. *Structure*. 2017;25(10):1582–8.e3.
10. Grierson PM, Lillard K, Behbehani GK et al. BLM helicase facilitates RNA polymerase I-mediated ribosomal RNA transcription. *Hum Mol Genet*. 2012;21(5):1172–83.
11. Böhm S, Bernstein KA. The role of post-translational modifications in fine-tuning BLM helicase function during DNA repair. *DNA Repair (Amst)*. 2014;22:123–32.
12. Drosopoulos WC, Kosiyatrakul ST, Schildkraut CL. BLM helicase facilitates telomere replication during leading strand synthesis of telomeres. *J Cell Biol*. 2015;210(2):191–208.
13. Qin W, Bazeille N, Henry E, Zhang B, Deprez E, Xi XG. Mechanistic insight into cadmium-induced inactivation of the Bloom protein. *Sci Rep*. 2016;6:26225.
14. Shastri VM, Schmidt KH. Cellular defects caused by hypomorphic variants of the Bloom syndrome helicase gene BLM. *Mol Genet Genomic Med*. 2015;4(1):106–19.
15. Priyadarshini R, Hussain M, Attri P et al. BLM potentiates c-jun degradation and alters its function as an oncogenic transcription factor. *Cell Rep*. 2018;24(4):947–61.e7.
16. Aljarad S, Alhamid A, Rahmeh AR et al. Bloom syndrome with myelodysplastic syndrome that was converted into acute myeloid leukaemia, with new ophthalmologic manifestations: The first report from Syria. *Oxf Med Case Reports*. 2018;2018(12):omy096.
17. Suspitsin EN, Sibgatullina FI, Lyazina LV, Imyanitov EN. First two cases of Bloom syndrome in Russia: Lack of skin manifestations in a *BLM* c.1642C>T (p.Q548X) homozygote as a likely cause of underdiagnosis. *Mol Syndromol*. 2017;8(2):103–6.
18. Chavez-Alvarez S, Villarreal-Martinez A, Velasco-Campos MDR et al. Bloom syndrome sans characteristic facial features in a Mestizo patient- a diagnostic challenge. *Indian J Dermatol Venereol Leprol*. 2019;85(1):130.
19. Relhan V, Sinha S, Bhatnagar T, Garg VK, Kochhar A. Bloom syndrome with extensive pulmonary involvement in a child. *Indian J Dermatol*. 2015;60(2):217.
20. Martinez CA, Pinheiro LV, Rossi DH et al. Adenocarcinoma of the right colon in a patient with Bloom syndrome. *Case Rep Surg*. 2016;2016:3176842.
21. Vekaria R, Bhatt R, Saravanan P, de Boer RC. Bloom's syndrome in an Indian man in the UK. *BMJ Case Rep*. 2016;2016. pii: bcr2015212297.
22. Sassi A, Popielarski M, Synowiec E, Morawiec Z, Wozniak K. BLM and RAD51 genes polymorphism and susceptibility to breast cancer. *Pathol Oncol Res*. 2013;19(3):451–9.
23. Arora A, Abdel-Fatah TM, Agarwal D et al. Transcriptomic and protein expression analysis reveals clinicopathological significance of Bloom syndrome helicase (BLM) in breast cancer. *Mol Cancer Ther*. 2015;14(4):1057–65.
24. Tzfoni I, Chayo J, Shaked M et al. Pancreatic cancer in bloom syndrome. *SAGE Open Med Case Rep*. 2019;7:2050313X 19855587.
25. Kamath-Loeb A, Loeb LA, Fry M. The Werner syndrome protein is distinguished from the Bloom syndrome protein by its capacity to tightly bind diverse DNA structures. *PLOS ONE*. 2012;7(1):e30189.
26. Suhasini AN, Brosh RM Jr. Fanconi anemia and Bloom's syndrome crosstalk through FANCJ-BLM helicase interaction. *Trends Genet*. 2012;28(1):7–13.
27. Kitano K. Structural mechanisms of human RecQ helicases WRN and BLM. *Front Genet*. 2014;5:366.
28. Panneerselvam J, Wang H, Zhang J, Che R, Yu H, Fei P. BLM promotes the activation of Fanconi Anemia signaling pathway. *Oncotarget*. 2016;7(22):32351–61.

29. de Renty C, Ellis NA. Bloom's syndrome: Why not prema-ture aging?: A comparison of the BLM and WRN helicases. *Ageing Res Rev.* 2017;33:36–51.

30. Maciejczyk M, Mikoluc B, Pietrucha B et al. Oxidative stress, mitochondrial abnormalities and antioxidant defense in Ataxia-telangiectasia, Bloom syndrome and Nijmegen break-age syndrome. *Redox Biol.* 2017;11:375–83.

31. Hafsi W, Badri T. *Bloom Syndrome (Congenital Telangiectatic Erythema). StatPearls [Internet].* Treasure Island (FL): StatPearls Publishing; 2018 Jan-. 2017 Oct 13.

32. Adams M, Jenney M, Lazarou L et al. Acute myeloid leukae-mia after treatment for acute lymphoblastic leukaemia in girl with Bloom syndrome. *J Genet Syndr Gene Ther.* 2013;4(8). pii: 1000177.

33. Campbell MB, Campbell WC, Rogers J et al. Bloom syn-drome: Research and data priorities for the development of precision medicine as identified by some affected families. *Cold Spring Harb Mol Case Stud.* 2018;4(2). pii: a002816.

65

Chediak–Higashi Syndrome

Dongyou Liu

CONTENTS

65.1 Introduction

Chediak–Higashi syndrome (CHS) is a rare, autosomal recessive disorder that typically presents with hematologic, infectious, pigmentary, and neurologic manifestations. While classic CHS occurring in early childhood is associated with severe infectious or hematologic complications, atypical CHS affecting adolescents and adults induces mild to moderate hematologic and infectious manifestations.

The molecular pathogenesis of CHS is attributed to homozygous or compound heterozygous mutations in the lysosomal trafficking regulator gene (*CHS1*, formerly *LYST*) on chromosome 1q42.3, leading to the formation of giant lysosomes or lysosome-related organelles in several cell types, and subsequent lysosomal dysfunction in several cell types, including melanocytes, lymphocytes, and platelets that perform lysosomal secretion [1].

65.2 Biology

CHS (also known as Chediak–Higashi disease or CHD) was initially described by Beguez Cesar in 1943, and further delineated by Chediak in 1952 and Higashi in 1954 as a genetic disorder with salient hematologic features (including leukocyte anomaly [Chediak] and congenital gigantism of peroxidase granules [Higashi]). This facilitated adoption of the term "Chediak and Higashi's disease" in 1955 and the current designation "Chediak–Higashi syndrome" in 1957 for this genetic disorder [2].

Besides the classic form of CHS, which is a fatal childhood disease manifesting as partial albinism (decreased pigmentation of hair and eyes), photophobia, nystagmus, large eosinophilic, peroxidase-positive inclusion bodies in the myeloblasts and promyelocytes of the bone marrow, neutropenia, recurrent/severe infections, and malignant lymphoma, a milder/atypical form of CHS is observed in young adults, who develop pigmentary abnormalities along with neurodegenerative disease [1].

The genetic region implicated in the development of CHS was mapped to chromosome 1q42.3 in 1996, and the gene (*CHS1*, originally called *LYST*) was shown in 1997 to encode a lysosomal trafficking regulator protein, which is implicated in the regulation of lysosome-related organelle size, fission, and secretion.

Examination of various cell types from CHS patients revealed giant lysosomes or lysosome-related organelles (LRO), which appear to have limited impact on the structure and function of some cell types but show huge impact on others, particularly those (e.g., melanocytes, lymphocytes, platelets, fibroblasts) involved in lysosomal secretion.

For example, epidermal melanocytes from CHS patients contain enlarged melanosomes in the perinuclear area that are somehow not transferred to surrounding keratinocytes, leading to skin hypomelanosis/oculocutaneous albinism. Further, CHS cytotoxic T-lymphocytes with oversized mature lytic granules lose the ability to mediate targeted cell killing and display unexpected delay in delivering peptide to MHC class II-expressing antigen presenting cells, underlining the development of hemophagocytic lymphohistiocytosis (HLH) and possibly lymphoma. Whereas platelets from patients with early onset CHS contain few or no dense bodies and display impaired ability to form platelet aggregation and hemostatic plugs, those from patients with late-onset CHS often have abundant granules and exhibit a reduced functional impairment. In addition, CHS fibroblasts with enlarged lysosomes produce inefficient fusion with the plasma membrane during the wound healing process [3,4].

65.3 Pathogenesis

The *CHS1* gene on chromosome 1q42.3 consists of 55 exons (including 2 alternative, mutually exclusive 5-prime untranslated regions), and generates a 13.5 kb transcript that encodes a 3801 aa, 430 kDa cytoplasmic protein (CHS1) involved in the regulation of lysosomal trafficking and the synthesis, fusion, and transport of cytoplasmic granules.

Structurally, the CHS protein includes the N-terminal domain (with several ARM/HEAT α-helix repeats for membrane interaction), the BEACH (beige and Chediak–Higashi) domain, and the C-terminal domain (with a WIDL-enriched sequence and seven WD40 repeats for protein–protein interaction).

The CHS protein is involved in the regulation of natural killer (NK) cell lytic activity, including the governance of lytic granule size, and control of polarization, exocytosis, and endo-lysosomal compartment [5,6].

Biallelic mutations in the *CHS1* gene produce loss-of-function protein that is incapable of acting as a scaffold protein in membrane fusion and fission events, and subsequent formation of giant lysosomes or lysosome-related organelles in host cells. The resultant lysosomal dysfunction disrupts the transfer of enlarged melanosomes from the perinuclear area to surrounding keratinocytes, leading to skin hypomelanosis/oculocutaneous albinism, and impairs downregulation of immune responses and promotes sustained activation and proliferation of cytotoxic T lymphocytes (CTL) and NK cells, leading to excessive macrophage activation characteristic of HLH. Interestingly, while a reduction of CTL cytotoxicity occurs in CHS patients with early-onset HLH, CTL cytotoxicity retains unchanged in patients who later or never develop HLH [7–9].

At least 64 mutations have been found in the *CHS1* coding sequence so far, including c.118_119insG [p.Ala40GlyfsTer24 (Ala40fsTer63)], c.148C>T [p.Arg50Ter], c.772T>C (961T>C) [p.Cys258Arg], c.925C>T [p.Arg309Ter], c.1467delG [p.Glu489 AspfsTer78 (Glu489fsTer566)], c.1540C>T [p.Arg514Ter], c.1902dupA (1897_1898insA) [p.Ala635SerfsTer4 (Lys633fs Ter638)], c.2413delG [p.Glu805AsnfsTer2 (Glu805fsTer806)], c.2454delA [p.Ala819Hisfs5 (Lys818fsTer823)], c.2623delT (2620delT) [p.Tyr875MetfsTer24 (Phe874fsTer898)], c.3073_3074delAA [p.Asn1025GlnfsTer6 (Asn1025fsTer1030)], c.3085C>T [p.Gln1029Ter], c.3310C>T [p.Arg1104Ter], c.3434dupA (3434_3435insA) [p.His1145GlnfsTer9 (His1145fs Ter1153)], c.3622C>T [p.Gln1208Ter], c.3944_3945insC [p.Thr1315fsTer1331], c.4052C>G [p.Ser1351Ter], c.4274delT [p.Leu1425TyrfsTer2 (Leu1425fsTer1426)], c.4361C>A [p.Ala1454Asp], c.4688G>A [p.Arg1563His], c.5061T>A [p.Tyr1687Ter], c.5317delA [p.Arg1773AspfsTer13 (Arg1773fs Ter1785)], c.5506C>T [p.Arg1836Ter], c.5541_5542delAA [p.Gln1847fsTer1850], c.5996T>A [p.Val1999Asp], c.6078C>A [p.Tyr2026Ter], c.7060_7066delCTATTAG [p.Leu2354Metfs Ter16 (Leu2354fsTer2369)], c.7555delT [p.Tyr2519IlefsTer10 (Tyr2519fsTer2528)], c.7982C>G [p.Ser2661Ter], c.8281A>T [p.Arg2761Ter], c.8428G>A [p.Glu2810Lys], c.8583G>A [p.Trp2861Ter], c.9107_9162del56 [p.Gly3036GlufsTer16

(Gly3036fsTer3051)], c.9228_9229insTTCTTTCAGT [p.Lys30 77PhefsTer4 (Lys3077fsTer3080)], c.9590delA [p.Tyr3197Leufs Ter62 (Tyr3197fsTer3258)], c.9827_9832ATACAA [p.Asn3276_ Thr3277del],c.9893delT[p.Phe3298Serfs7(Phe3298fsTer3304)], c.10127A>G[p.Asn3376Ser],c.10395delA[p.Gly3466AlafsTer2 (Lys3465fsTer3467)], c.10747G, c.11102G>T [p.Glu3668Ter], and c.11173G>A (11362G>A) [p.Gly3725Arg]. Of these pathogenic variants, 32 are substitutions (20 nonsense, 12 missense), 19 are deletions, 9 are insertions, and 4 are splice sites [1,10,11].

While frameshift, nonsense, and splice site mutations abrogate CHS1 protein and are responsible for causing severe childhood CHS, missense mutations may produce partially functioning, truncating protein and are associated with atypical, milder, later-onset CHS in adolescents or adults. Nonetheless, some individuals with two missense mutations have been shown to develop severe, childhood-onset CHS and HLH [1].

65.4 Epidemiology

65.4.1 Prevalence

CHS is a rare disorder, with >500 cases reported in the literature worldwide. Due to the fact that CHS demonstrates considerable phenotypic variability, it is probably unrecognized and underdiagnosed. CHS shows no racial or sexual predilection and typically emerges after birth and under the age of 5 years.

65.4.2 Inheritance

CHS is an autosomal recessive disorder. While heterozygotes (carriers) are asymptomatic and are not at risk of developing the disorder, they may have 25% chance of getting an affected child (homozygote), 50% chance of getting an asymptomatic carrier child (heterozygote), and 25% chance of getting a normal, non-carrier child.

65.4.3 Penetrance

CHS shows incomplete penetrance and variable expressivity. Individuals harboring *CHS1* pathogenic variants often develop clinical symptoms of differing severity. Some affected patients may display gait abnormalities (due to spastic paraplegia), cerebellar ataxia, peripheral neuropathy, and cerebellar atrophy, without pigmentary abnormalities of the skin or eyes, clinical features of immunodeficiency, and bleeding tendency, despite the fact that their peripheral blood demonstrates giant granules in granulocytes and reduced NK cell activity, and that they harbor homozygous *CHS1* missense mutation (c.4189T-G, F1397 V). While missense mutations are generally linked to atypical, milder, later-onset CHS, there are incidences involving individuals who possess two missense mutations, but present severe, childhood-onset CHS and HLH [12].

65.5 Clinical Features

CHS is responsible for causing two forms of clinical diseases: classic CHS (affecting children and accounting for up to 85% of cases), and atypical (mild) CHS (occurring in young adults and adults).

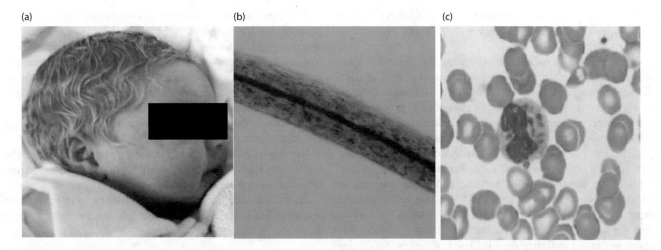

FIGURE 65.1 Chediak–Higashi syndrome in a 3-month-old male showing: (a) silvery-colored hair and hypopigmented skin; (b) pigment granules grouped together and scattered on the hair shaft; (c) peripheral blood with giant intracytoplasmic granules in leukocytes. (Photo credit: Fantinato GT et al. *An Bras Dermatol.* 2011;86(5):1029.)

Children with classic CHS tend to display: (i) partial oculocutaneous albinism (OCA, pigment dilution in hair, skin, and eyes, results in silvery or metallic hair [observable under the light microscope], reduced skin, iris, and retinal pigmentation, photophobia, and nystagmus); (ii) immunodeficiency (frequent/severe infections of the skin, upper respiratory tract, and periodontal region); (iii) immunodeficiency (recurrent infections); (iv) bleeding tendency (epistaxis, gum/mucosal bleeding, and easy bruising); (v) HLH (or accelerated phase; manifesting as fever, hepatosplenomegaly, lymphadenopathy, neutropenia, anemia, and sometimes thrombocytopenia, in addition to diffuse lymphohistiocytic infiltration of the liver, spleen, bone marrow, lymph nodes, and central nervous system; possibly triggered by Epstein–Barr virus infection and absence of NK cell function); (vi) neurologic features (low cognitive abilities, balance abnormalities, stroke, coma, ataxia, tremor, absent deep-tendon reflexes, and motor and sensory neuropathies) (Figure 65.1) [1,13–22].

Adolescents and adults with atypical or milder CHS may develop (i) subtle or absent oculocutaneous albinism; (ii)

infrequent infections; (iii) decreased platelet-dense bodies accompanied by subtle bleeding manifestations; (iv) progressive neurologic disease (intellectual disabilities, peripheral neuropathy, gait disturbances, tremor, parkinsonism, and spastic paraplegia) (Figure 65.2) [23,24].

65.6 Diagnosis

Diagnosis of CHS involves medical history review (recurrent or severe infections), physical exam (ophthalmologic findings), laboratory assessment (abnormal platelet aggregation), histopathology (abnormal WBC granules on blood smear), and molecular testing (*CHS1* pathogenic variants).

Individuals should be suspected of CHS with any of the following findings: (i) partial OCA; (ii) immunodeficiency; (iii) mild bleeding tendency; (iv) neurologic features; (v) pathognomic giant inclusions in polymorphonuclear neutrophils (PMN) and lymphocytes, or peroxidase-positive giant inclusions in leukocytes, megakaryocytes, and other bone marrow

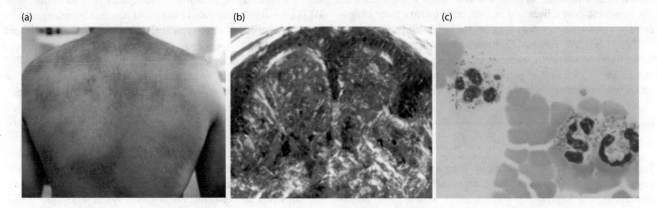

FIGURE 65.2 Atypical Chediak–Higashi syndrome in a 33-year-old male with homozygous 6 bp in-frame deletion c.9827_9832ATACAA (p.Asn3276_Thr3277del) in exon 43 and also homozygous intronic sequence variant (c.10701 + 8C>T) distal to the deletion, showing: (a) abnormal skin pigmentation on the back; (b) extensive deposition of amyloid in the papillary and superficial reticular dermis on skin biopsy, as indicated by positive Congo red staining, and apple green birefringence in the expanded papillary dermis with polarization; (c) polymorphonuclear leukocytes with enlarged, variably sized cytoplasmic granules in peripheral blood smear (Wright Giemsa, oil immersion). (Photo credit: Weisfeld-Adams JD et al. *Orphanet J Rare Dis.* 2013;8:46.)

precursors; (vi) pigment clumping on hair (polarized light microscopy of hair from patients with enlarged granules in blood smear) [1].

Evolving from uncontrolled proliferation of cytotoxic T lymphocytes and NK cells, HLH is observed in up to 85% of CHS cases. Diagnosis of HLH is based on either fulfillment of five of the eight criteria (fever, splenomegaly, cytopenias affecting at least two or three cell lineages, hypertriglyceridemia and/or hypofibrinogenemia, hemophagocytosis, low/absent NK-cell activity, hyperferritinemia, and high soluble interleukin-2-receptor [soluble CD25] levels) or detection of a *CHS1* pathogenic variant [7,25–29].

Molecular identification of biallelic pathogenic variants in *CHS1* helps confirm the diagnosis of CHS. Due to the variable expressivity of CHS, detection of a *CHS1* pathogenic variant does not necessarily prove the acuteness of current clinical diseases, but rather indicates the likelihood of certain conditions that may emerge eventually [1].

Differential diagnoses for CHS include *oculocutaneous albinism* (visual impairment, iris/retinal depigmentation, absent or reduced skin and hair pigment, lack of neutropenia, or neurologic abnormalities), *Hermansky Pudlak syndrome* (HPS; partial oculocutaneous albinism, bleeding diathesis secondary to absent platelet-dense bodies, congenital neutropenia, recurrent infections, developmental delay, balance abnormalities, tremor; general lack of giant intracellular granules in neutrophils; possible presence of giant granules in patients with acute and chronic myeloid leukemia, referred to as pseudo-Chediak–Higashi anomaly), *Griscelli syndrome* (mild skin hypopigmentation, silvery gray hair, normal platelet function; due to pathogenic variants in *MYO5A* [GSI, severe neurologic involvement], *RAB27A* [GSII, immunodeficiency and lymphohistiocytosis], *MLPH* (GSIII, hypopigmentation]), *Cross syndrome* (hypopigmentation, ocular defects, and severe developmental delay, lack of infectious component), *deficiency of endosomal adaptor p14* (partial albinism, short stature, congenital neutropenia, and lymphoid deficiency, altered azurophilic granule ultrastructure, absent giant inclusions in neutrophils), *familial hemophagocytic lymphohistiocytosis* (FLH; proliferation and infiltration of hyperactivated macrophages and T-lymphocytes, prolonged fever, cytopenias, hepatosplenomegaly, increased intracranial pressure, irritability, neck stiffness, hypotonia, hypertonia, convulsions, cranial nerve palsies, ataxia, hemiplegia, quadriplegia, blindness, coma, liver dysfunction, bone marrow hemophagocytosis; autosomal recessive disorder fur to biallelic mutations in *FHL1–FLH5*), *Vici syndrome* (cutaneous hypopigmentation, combined immunodeficiency, agenesis of the corpus callosum, bilateral cataracts, cleft lip and palate, cognitive impairment, seizures, and severe respiratory infections) [1,2].

65.7 Treatment

Treatment of CHS centers on ameliorating and/or rectifying the key clinical anomalies of the disorder; namely, hematologic and immunologic defects, ocular abnormalities, skin hypopigmentation, and neurologic abnormalities [30].

Hematologic and immunologic defects associated with CHS may be effectively treated with allogenic hematopoietic stem cell transplantation (HSCT, from an HLA-matched sibling or from an unrelated donor or cord blood), which has a remission rate of 75% within 8 weeks and which gives an optional outcome when performed prior to the emergence of HLH. Patients showing abnormal CTL cytotoxicity have high risk for HLH and should be considered for early HSCT. Chemotherapy (e.g., combination of etoposide, dexamethasone, and cyclosporine A, or intrathecal methotrexate and prednisolone) may be also beneficial [31,32].

Ocular abnormalities such as refractive errors may be rectified with corrective lenses to improve visual acuity. Sensitive eyes can be protected from UV light with sunglasses.

Skin hypopigmentation in CHS patients will benefit from the use of sunscreen that helps reduce further damage from sun and potential development of skin cancer.

Neurologic abnormalities have a tendency to progress with age, and CHS patients consider rehabilitation programs early in the course of the disease.

65.8 Prognosis and Prevention

Prognosis for classic CHS affecting children is poor. The most common cause of death is recurrent infections or HLH (or accelerated phase, with lymphoproliferation spreading to major organs), the latter of which accounts for nearly 90% of CHS-related deaths in the first decade of life. Patients who survive into adulthood develop progressive neurological symptoms. The overall 5-year survival rate for children who underwent HSCT is 62%.

Nonsteroidal anti-inflammatory drugs (NSAIDs, such as aspirin, ibuprofen) can exacerbate the bleeding tendency and should be avoided.

Prenatal testing and preimplantation diagnosis should be considered for prospective parents who harbor CHS1 pathogenic variants.

65.9 Conclusion

Chediak–Higashi syndrome (CHS) is a rare autosomal recessive disorder characterized by partial oculocutaneous albinism (OCA), increased susceptibility to infection, a mild bleeding tendency, and/or late-onset progressive neurological impairment. In its classic form (classic CHS), affected children develop severe infectious or hematologic complications, including fatal HLH (so-called accelerated phase) and possible lymphoma. In its atypical or milder form (atypical CHS), affected adolescents and adults have mild to moderate hematologic and infectious abnormalities, with later development of neurological anomalies. The pathogenesis of CHS is attributed to biallelic mutations in the *CHS1* gene that encode lysosomal trafficking regulator protein, leading to the formation of giant lysosomes or lysosome-related organelles in several cell types, and subsequent impairment in the function of cytotoxic lymphocytes and NK cells [33]. Diagnosis of CHS requires fulfilling a number of clinical criteria (including observation of pathognomic giant inclusions in neutrophils and lymphocytes). Molecular detection of *CHS1* pathogenic variants provides further confirmation of CHS. While allogenic HSCT offers a highly effective treatment for hematologic and immune defects in affected patients, efficient therapies for neurologic anomalies are currently lacking. Further research on the molecular aspects of CHS is essential for developing improved measures for the ultimate control of this disorder.

REFERENCES

1. Ajitkumar A, Ramphul K. *Chediak Higashi Syndrome. StatPearls [Internet].* Treasure Island (FL): StatPearls Publishing; 2019 Jan-. 2019 Apr 8.

2. Introne WJ, Westbroek W, Golas GA, Adams D. Chediak-Higashi syndrome. In: Adam MP, Ardinger HH, Pagon RA et al. editors. *GeneReviews® [Internet].* Seattle (WA): University of Washington, Seattle; 1993–2018. 2009 Mar 3 [updated 2015 Jan 15].

3. Wang L, Kantovitz KR, Cullinane AR et al. Skin fibroblasts from individuals with Chediak-Higashi syndrome (CHS) exhibit hyposensitive immunogenic response. *Orphanet J Rare Dis.* 2014;9:212.

4. Chiang SCC, Wood SM, Tesi B et al. Differences in granule morphology yet equally impaired exocytosis among cytotoxic T cells and NK cells from Chediak-Higashi syndrome patients. *Front Immunol.* 2017;8:426.

5. Holland P, Torgersen ML, Sandvig K, Simonsen A. LYST affects lysosome size and quantity, but not trafficking or degradation through autophagy or endocytosis. *Traffic.* 2014;15(12):1390–405.

6. Sepulveda FE, Burgess A, Heiligenstein X et al. LYST controls the biogenesis of the endosomal compartment required for secretory lysosome function. *Traffic.* 2015;16(2):191–203.

7. Bharti S, Bhatia P, Bansal D, Varma N. The accelerated phase of Chediak-Higashi syndrome: The importance of hematological evaluation. *Turk J Haematol.* 2013;30(1):85–7.

8. Beken B, Unal S, Gümrük F. Lysosomal vesicles, giant granules, and erythrophagocytosis in Chédiak-Higashi syndrome. *Turk J Haematol.* 2014;31(2):209–10.

9. Gil-Krzewska A, Saeed MB, Oszmiana A et al. An actin cytoskeletal barrier inhibits lytic granule release from natural killer cells in patients with Chediak-Higashi syndrome. *J Allergy Clin Immunol.* 2018;142(3):914–27.e6.

10. Helmi MM, Saleh M, Yacop B, ElSawy D. Chédiak-Higashi syndrome with novel gene mutation. *BMJ Case Rep.* 2017;2017. pii: bcr2016216628.

11. Jin Y, Zhang L, Wang S et al. Whole genome sequencing identifies novel compound heterozygous lysosomal trafficking regulator gene mutations associated with autosomal recessive Chediak-Higashi syndrome. *Sci Rep.* 2017;7:41308.

12. Elevli M, Hatipoğlu HU, Çivilibal M, Selçuk Duru N, Celkan T. Chediak-Higashi syndrome: A case report of a girl without silvery hair and oculocutaneous albinism presenting with hemophagocytic lymphohistiocytosis. *Turk J Haematol.* 2014;31(4):426–7.

13. Tanabe F, Kasai H, Morimoto M et al. Novel heterogenous CHS1 mutations identified in five Japanese patients with Chediak-Higashi syndrome. *Case Rep Med.* 2010;2010:464671.

14. Fantinato GT, Cestari Sda C, Afonso JP, Sousa LS, Enokihara MM. Do you know this syndrome? Chediak-Higashi syndrome. *An Bras Dermatol.* 2011;86(5):1029.

15. Patne SC, Kumar S, Bagri NK, Kumar A, Shukla J. Chédiak-higashi syndrome: A case report. *Indian J Hematol Blood Transfus.* 2013;29(2):80–3.

16. Rezende KM, Canela AH, Ortega AO, Tintel C, Bönecker M. Chediak-Higashi syndrome and premature exfoliation of primary teeth. *Braz Dent J.* 2013;24(6):667–70.

17. Karabel M, Kelekçi S, Sen V, Karabel D, Aliosmanoğlu C, Söker M. A rare cause of recurrent oral lesions: Chediak-Higashi syndrome. *Turk J Haematol.* 2014;31(3):313–4.

18. Karande S, Agarwal S, Gandhi B, Muranjan M. Chediak-Higashi syndrome in accelerated phase masquerading as severe acute malnutrition. *BMJ Case Rep.* 2014;2014.

19. Jaiswal P, Yadav YK, Bhasker N, Kushwaha R. Accelerated phase of Chediak-Higashi syndrome at initial presentation: A case report of an uncommon occurrence in a rare disorder. *J Clin Diagn Res.* 2015;9(12):ED13–4.

20. Raghuveer C, Murthy SC, Mithuna MN, Suresh T. Silvery hair with speckled dyspigmentation: Chediak-Higashi syndrome in three indian siblings. *Int J Trichology.* 2015;7(3):133–5.

21. Rudramurthy P, Lokanatha H. Chediak-Higashi syndrome: A case series from Karnataka, India. *Indian J Dermatol.* 2015; 60(5):524.

22. Lehky TJ, Groden C, Lear B, Toro C, Introne WJ. Peripheral nervous system manifestations of Chediak-Higashi disease. *Muscle Nerve.* 2017;55(3):359–65.

23. Weisfeld-Adams JD, Mehta L, Rucker JC et al. Atypical Chédiak-Higashi syndrome with attenuated phenotype: Three adult siblings homozygous for a novel LYST deletion and with neurodegenerative disease. *Orphanet J Rare Dis.* 2013;8:46.

24. Introne WJ, Westbroek W, Cullinane AR et al. Neurologic involvement in patients with atypical Chediak-Higashi disease. *Neurology.* 2016;86(14):1320–8.

25. Pullarkat ST. Accelerated phase of Chediak-Higashi syndrome. *Blood.* 2012;119(1):5.

26. Bouatay A, Hizem S, Tej A, Moatamri W, Boughamoura L, Kortas M. Chediak-Higashi syndrome presented as accelerated phase: Case report and review of the literature. *Indian J Hematol Blood Transfus.* 2014;30(Suppl 1):223–6.

27. Sood S, Biswas B, Kaushal V, Mandal T. Chediak-Higashi syndrome in accelerated phase: A rare case report with review of literature. *Indian J Hematol Blood Transfus.* 2014; 30(Suppl 1):195–8.

28. Jain M, Kumar A, Singh US, Kushwaha R. Chediak-Higashi syndrome in accelerated phase masquerading as acute leukemia. *Turk J Haematol.* 2016;33(4):349–50.

29. Maaloul I, Talmoudi J, Chabchoub I et al. Chediak-Higashi syndrome presenting in accelerated phase: A case report and literature review. *Hematol Oncol Stem Cell Ther.* 2016;9(2):71–5.

30. Lozano ML, Rivera J, Sánchez-Guiu I, Vicente V. Towards the targeted management of Chediak-Higashi syndrome. *Orphanet J Rare Dis.* 2014;9:132.

31. White JG, Hess RA, Gahl WA, Introne W. Rapid ultrastructural detection of success or failure after bone marrow transplantation in the Chediak-Higashi syndrome. *Platelets.* 2013;24(1):71–4.

32. Wu XL, Zhao XQ, Zhang BX, Xuan F, Guo HM, Ma FT. A novel frameshift mutation of Chediak-Higashi syndrome and treatment in the accelerated phase. *Braz J Med Biol Res.* 2017;50(4):e5727.

33. Gil-Krzewska A, Wood SM, Murakami Y et al. Chediak-Higashi syndrome: Lysosomal trafficking regulator domains regulate exocytosis of lytic granules but not cytokine secretion by natural killer cells. *J Allergy Clin Immunol.* 2016;137(4):1165–77.

66

Congenital Amegakaryocytic Thrombocytopenia

Dongyou Liu

CONTENTS

66.1 Introduction

Congenital amegakaryocytic thrombocytopenia (CAMT) is a rare autosomal recessive disorder that typically manifests as severe thrombocytopenia with reduced/absent megakaryocytes at birth, and pancytopenia due to trilineage bone marrow aplasia during childhood. CAMT follows a fatal clinical course in children that can only be reverted by allogeneic hematopoietic stem cell transplantation (HSCT).

Homozygous or compound heterozygous mutations in the myeloproliferative leukemia virus oncogene (*MPL*) on chromosome 1p34.2 encoding a thrombopoietin receptor (MPL) underlie the molecular pathogenesis of CAMP. An altered MPL loses its ability to interact with its ligand thrombopoietin (THPO, a hematopoietic growth factor), affecting the production of multipotent hematopoietic progenitor cells and platelets, and leading to thrombocytopenia and megakaryocytopenia without physical anomalies. While nonsense *MPL* mutations cause a complete loss of function in MPL (type I CAMT, with early onset of severe pancytopenia, decreased bone marrow activity, and very low platelet counts), missense *MPL* mutations affect the extracellular domain of MPL and partially impair its function (type II CAMT, with transient increases of platelet during the first year of life and an onset of bone marrow failure from 3 years of age) [1].

66.2 Biology

Platelets are small anucleate cells of 2.4–4.0 μm in size that move through the blood vessels and respond to vessel wall damages with the formation of aggregation, microparticles, and finally clot, which plugs the damaged vessel wall and prevents bleeding. Thus, platelets play a vital role in the preservation of vessel wall integrity and hemostatic processes.

During thrombopoiesis, a pluripotent hematopoietic stem cell divides and forms myeloid stem cell which makes early commitment to megakaryoblast. After undergoing further division, megakaryoblast becomes immature megakaryocyte which completes its terminal differentiation into mature megakaryocytes and finally platelet that possesses many polyploid nuclei, extensive intracellular organelles (for platelet function), and membrane structures (for platelet shedding) [2,3].

Many genes (e.g., transcription factors, cytokines, cell surface receptor molecules, molecular motors, and regulators of the cell cycle) participate in the thrombopoiesis process. Among these, THPO and its receptor (MPL) are the principal regulators of the THPO/MPL pathway involved in early and late thrombopoiesis and hematopoietic stem cell maintenance. Loss-of-function *MPL* mutations hinder the interaction between MPL and THPO and induce hypomegakaryocytic thrombocytopenias in childhood and bone marrow failure at a later stage of life, which are

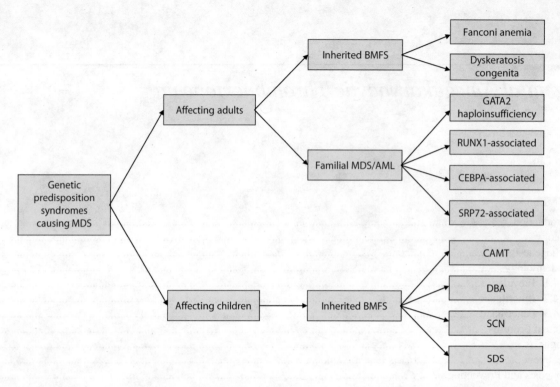

FIGURE 66.1 Classification of genetic predisposition syndromes causing myelodysplastic syndrome (MDS). *Abbreviations:* AML, acute myeloid leukemia; BMFS, bone marrow failure syndrome; CAMT, congenital amegakaryocytic thrombocytopenia; DBA, Diamond–Blackfan anemia; SCN, severe congenital neutropenia; SDS, Shwachman–Diamond syndrome.

characteristic of CAMT, an autosomal recessive disorder first recognized in 1999 [4,5].

Like many other genetic predisposition syndromes (e.g., acute myeloid leukemia [AML], bone marrow failure syndrome [BMFS], Diamond–Blackfan anemia [DBA], severe congenital neutropenia [SCN], and Shwachman–Diamond syndrome [SDS]) that cause myelodysplastic syndrome (MDS), CAMT is also associated with such hematologic malignancy (Figure 66.1) [6,7].

66.3 Pathogenesis

The *MPL* gene on chromosome 1p34.2 spans 17 kb with 12 exons and encodes a 635 aa protein (MPL) that function as the receptor for thrombopoietin (THPO).

Structurally, MPL consists of three functional domains: an extracellular portion (for cytokine binding), a transmembrane domain (TMD), and a cytoplasmic domain (for binding JAKs and other signaling molecules, such as STATs) [8].

MPL is a homodimeric type 1 receptor for hematopoietic growth factor THPO, and interaction between MPL and THPO activates tyrosine kinase (Tyk2) and Janus kinase (JAK2), which in turn phosphorylate Stat5 and Stat3, leading to the upregulation of JAK-STAT, MAPK-ERK1/2, and PI3K-AKT signaling pathways, and subsequent development/maturation of megakaryocytes, platelet production from birth, and maintenance of pluripotent hematopoietic stem cell compartment in postnatal hemopoiesis [9,10].

Of more than 50 pathogenic variants in *MPL* identified to date, homozygous or compound heterozygous loss-of-function

mutations occurring throughout the *MPL* gene (R102P, F104S, P106L, R119C, C268 T) result in absence or reduction of functional MPL at the cell surface, and subsequent decrease in THPO clearance by platelets, leading to thrombocytopenia and absent bone marrow megakaryocytes, and progressing into bone marrow failure/aplastic anemia prior to leukopenia during early childhood, typical of CAMT (Figure 66.2). It appears that frameshift or nonsense *MPL* mutations causing disruption or absence of MPL function are associated with more severe disease and rapider progression of bone marrow failure (type I CAMT), whereas missense *MPL* mutations affecting the extracellular domain of MPL and leaving residual MPL function are linked to a milder phenotype (type II CAMT). In addition, the presence of *MPL* double mutation G117 T (K39N or Baltimore mutation)/T814C (W272R) produces in cis loss-of-function MPL W272R and partially functional MPL K39N. Due to its trafficking defect, MPL W272R is not expressed on cell surface, but is retained in the ER (where it cannot respond to extracellular ligand). Effectively overriding the activating MPL K39N mutation, *MPL* K39N/W272R mutant causes highly elevated THPO levels in blood and deficit in two hematopoietic lineages [11–14].

On the other hand, heterozygous gain-of-function *MPL* mutations affecting the transmembrane domain of MPL (e.g., W515 K/L and S505N) are responsible for causing myeloproliferative neoplasms (essential thrombocythemia, primary myelofibrosis) and hereditary thrombocytosis. A patient harboring germline *MPL* R102P heterozygous mutation was shown to produce high levels of THPO in the serum but show incomplete clinical phenotype of hereditary thrombocytosis, in contrast to the presence of homozygous *MPL* R102P mutation in CAMT cases [15].

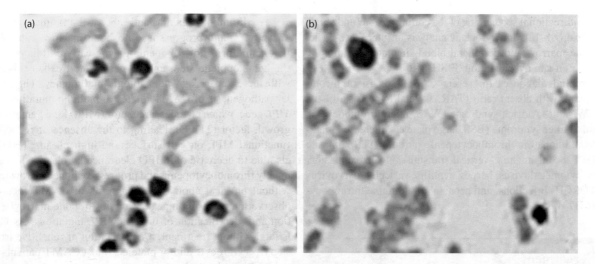

FIGURE 66.2 Congenital amegakaryocytic thrombocytopenia in a 2-year-old girl with a missense mutation in *MPL* exon 4 showing progression of bone marrow failure. (a) Bone marrow aspiration contained no megakaryocytes in a cellular particle with erythroid and myeloid precursors without dysplasia at 2 years of age. (b) Bone marrow aspirate displayed very hypocellular results with few lymphocytes at 2.5 years of age. (Photo credit: Ok Bozkaya İ et al. *Turk J Haematol.* 2015;32(2):172–4.)

66.4 Epidemiology

66.4.1 Prevalence

CAMT is a rare disorder, with about 100 cases reported in the literature to date. It is likely that the incidence of CAMT is underestimated/underdiagnosed due to difficulty and inconsistency in its diagnosis.

66.4.2 Inheritance

CAMT is an autosomal recessive disorder involving inheritance of biallelic (homozygous or compound heterozygous) *MPL* pathogenic variants from carrier (heterozygote) parents. A high rate of consanguinity is noted in parents of CAMT-affected children. Individuals with germline heterozygous *MPL* mutation do not develop CAMT but present with myeloproliferative neoplasms (which are defined by the presence of acquired chromosome change and persistent peripheral cytopenia and aplastic or hypoplastic bone marrow, without conclusive morphologic features) and hereditary thrombocytosis (high serum THPO levels). Nonetheless, a proportion of patients displaying the clinical picture of CAMT do not seem to carry *MPL* pathogenic variants, but possess THPO mutations (e.g., p.R38C, p.R99W, and p.R157*) instead, further highlighting the genetic heterogeneity of this order.

66.4.3 Penetrance

CAMT shows high penetrance but variable expressivity. Individuals carrying biallelic germline *MPL* mutations invariably develop CAMT, and the spectrum and severity of clinical manifestations may vary from one to another.

66.5 Clinical Features

CAMT typically presents with petechiae, purpura, gastrointestinal/pulmonary/intracranial hemorrhage (due to isolated thrombocytopenia), near absence of megakaryocytes, severe thrombocytopenia, and anemia prior to leukopenia, from the first day or within the first month of life. Congenital physical abnormalities are notably absent.

Type I-CAMT is a more severe form of the disease that shows persistently low platelet counts (21×10^9/L or below, leading to spontaneous bleeding) and early progression (by the age of 2 years) to bone marrow aplasia and pancytopenia.

Type II-CAMT is a milder form of the disease that causes transient increase of platelet counts $35–132 \times 10^9$/L from age 1 to 3–6 years or absence of pancytopenia. Type II CAMT shows a slightly delayed onset of bone marrow failure (mean age 48 months) in comparison with type I CAMT, which has a relatively early onset of bone marrow failure (mean age 22 months). Other possible symptoms associated with type II CAMT include cardiac defects (atrial and ventricular septal defects), abnormalities of the central nervous system (cerebral and cerebellar hypoplasia), and retardation of psychomotor development.

66.6 Diagnosis

Diagnosis of CAMT relies on physical exam (for relevant clinical signs), laboratory tests (e.g., CBC and MPV; review of PBS; coagulation profiles; bleeding time; platelet adhesion; platelet aggregation; platelet granule content and release reactions; platelet flow cytometry; electron microscopy; analysis of receptor expression, protein phosphorylation, and formation of second messengers; for evidence of hypomegakaryocytic thrombocytopenia [platelet count below 50×10^9/L] with a normal mean platelet volume and of highly elevated serum levels of THPO), peripheral blood smear (for morphologically normal platelets) and bone marrow aspiration smear (for absent or scant megakaryocytes and leukocytes) (Figure 66.2), and molecular genetic analysis (for lack of *MPL* RNA in bone marrow monocellular cells, for *MPL* mutation from whole blood in ethylenediaminetetraacetic acid [EDTA], buccal brushes or 10 mL of amniotic fluid by bidirectional sequencing of the coding regions and splice sites of exons 1–12), and cytogenic analysis of interphase nuclei from bone marrow and peripheral blood-simulated cultures by fluorescent in-situ hybridization (FISH)] [16].

Differential diagnoses for severe CAMT consist of thrombocytopenia with absent radii (TAR, skeletal hypoplasia of the arms), Wiskott–Aldrich syndrome (WAS, microthrombocytes), Bernard Soulier syndrome (BSS, large platelets, increased platelet levels), idiopathic thrombocytopenic purpura (ITP), aplastic anemia, Fanconi anemia, vertical transmission of toxoplasma, rubella, cytomegalovirus, herpes, syphilis, varicella, parvovirus B19 (TORCH) infections, and neonatal alloimmune thrombocytopenia [17–19].

66.7 Treatment

Treatment of CAMT includes supportive care (unrelated donor platelet transfusions, leukocyte-induced, and irradiated; fibrinolytic agents for minor bleeding), and HSCT (from human leukocyte antigen [HLA]-matched siblings and relatives), which is the only curative therapy [20,21].

Post-HSCT survival in CAMT patients is much improved with the use of conditioning regimen with busulfan and cyclophosphamide plus antithymocyte globulin and graft-versus-host prophylaxis with cyclosporine and methotrexate.

Thrombopoietin receptor agonists recognizing a site distinct from thrombopoietin may stimulate a receptor with an extracellular domain mutation and offer a potential treatment for a small subset of patients with CAM. These include Romiplostim (a peptibody interacting with the extracellular domain of the receptor), Eltrombopag (a small molecule binding to the transmembrane region of the receptor), and LGD 4665 (transmembrane domain binding agent), none of which is structurally related to THPO [22].

As CAMT patients often show high levels of circulating THPO, a strategy that helps partial recovery of MPL surface expression is clearly beneficial to hematopoietic normalization, or at least gives patients additional time to find match donors for bone marrow transplantation.

66.8 Prognosis and Prevention

CAMT has a poor prognosis, as nearly all affected children develop full marrow failure (tri-linear marrow aplasia) within 5–10 years, which is fatal without treatment, and 30% of CAMT patients die from bleeding complications and 20% from HSCT.

Nonsteroidal anti-inflammatory medications and aspirin should be avoided. While desmopressin acetate (DDAVP) is useful in older children and adults, it should be avoided in small infants due to the risk of hyponatremia.

66.9 Conclusion

Congenital amegakaryocytic thrombocytopenia (CAMT) is a rare, autosomal recessive disorder presenting at birth with severe thrombocytopenia and absent bone marrow megakaryocytes, and progressing into bone marrow failure/aplastic anemia prior to leukopenia during early childhood (due to the hematopoietic stem and progenitor cell exhaustion) along with increased risk of hematologic malignancy. The molecular pathogenesis of CAMT relates to biallelic mutations in the *MPL* gene, which encodes a receptor (MPL) for a hematopoietic growth factor (THPO), leading to the absence or reduction of functional MPL on cell surface and high plasma THPO levels (due to decrease of THPO clearance by platelets), and ultimately thrombocytopenia and progressive bone marrow failure, without physical anomalies. Diagnosis of CAMT focuses on observation of absent megakaryocytes in bone marrow aspiration and identification of biallelic *MPL* mutations. As CAMT is fatal without treatment, a combination of repetitive transfusion of erythrocyte and/or platelets (most CAMT patients show reduced platelet and erythrocyte counts prior to decrease in leukocyte counts) and HSCT (a curative measure) is necessary for effective management of this congenital disorder.

REFERENCES

1. Hirata S, Takayama N, Jono-Ohnishi R et al. Congenital amegakaryocytic thrombocytopenia iPS cells exhibit defective MPL-mediated signaling. *J Clin Invest.* 2013;123(9):3802–14.
2. Maserati E, Panarello C, Morerio C et al. Clonal chromosome anomalies and propensity to myeloid malignancies in congenital amegakaryocytic thrombocytopenia (OMIM 604498). *Haematologica.* 2008;93(8):1271–3.
3. Erlacher M, Strahm B. Missing cells: Pathophysiology, diagnosis, and management of (pan)cytopenia in childhood. *Front Pediatr.* 2015;3:64.
4. Al-Qahtani FS. Congenital amegakaryocytic thrombocytopenia: A brief review of the literature. *Clin Med Insights Pathol.* 2010;3:25–30.
5. Marletta C, Valli R, Pressato B et al. Chromosome anomalies in bone marrow as primary cause of aplastic or hypoplastic conditions and peripheral cytopenia: Disorders due to secondary impairment of RUNX1 and MPL genes. *Mol Cytogenet.* 2012;5(1):39.
6. Babushok DV, Bessler M. Genetic predisposition syndromes: When should they be considered in the work-up of MDS? *Best Pract Res Clin Haematol.* 2015;28(1):55–68.
7. Khincha PP, Savage SA. Neonatal manifestations of inherited bone marrow failure syndromes. *Semin Fetal Neonatal Med.* 2016;21(1):57–65.
8. Varghese LN, Defour JP, Pecquet C, Constantinescu SN. The thrombopoietin receptor: Structural basis of traffic and activation by ligand, mutations, agonists, and mutated calreticulin. *Front Endocrinol (Lausanne).* 2017;8:59.
9. Ballmaier M, Germeshausen M, Schulze H et al. c-mpl mutations are the cause of congenital amegakaryocytic thrombocytopenia. *Blood.* 2001;97:139–46.
10. Cleyrat C, Girard R, Choi EH et al. Gene editing rescue of a novel MPL mutant associated with congenital amegakaryocytic thrombocytopenia. *Blood Adv.* 2017;1(21):1815–26.
11. Passos-Coelho JL, Sebastião M, Gameiro P et al. Congenital amegakaryocytic thrombocytopenia--report of a new c-mpl gene missense mutation. *Am J Hematol.* 2007;82(3):240–1.

12. Savoia A, Dufour C, Locatelli F et al. Congenital amegakaryocytic thrombocytopenia: Clinical and biological consequences of five novel mutations. *Haematologica*. 2007;92(9): 1186–93.

13. Fox NE, Chen R, Hitchcock I, Keates-Baleeiro J, Frangoul H, Geddis AE. Compound heterozygous c-Mpl mutations in a child with congenital amegakaryocytic thrombocytopenia: Functional characterization and a review of the literature. *Exp Hematol*. 2009;37(4):495–503.

14. Ok Bozkaya İ, Yaralı N, Işık P, Ünsal Saç R, Tavil B, Tunç B. Severe clinical course in a patient with congenital amegakaryocytic thrombocytopenia due to a missense mutation of the c-MPL gene. *Turk J Haematol*. 2015;32(2):172–4.

15. Bellanné-Chantelot C, Mosca M, Marty C, Favier R, Vainchenker W, Plo I. Identification of MPL R102P mutation in hereditary thrombocytosis. *Front Endocrinol (Lausanne)*. 2017;8:235.

16. Ballmaier M, Holter W, Germeshausen M. Flow cytometric detection of MPL (CD110) as a diagnostic tool for differentiation of congenital thrombocytopenias. *Haematologica*. 2015;100(9): e341–4.

17. Geddis AE. Congenital amegakaryocytic thrombocytopenia and thrombocytopenia with absent radii. *Hematol Oncol Clin North Am*. 2009;23(2):321–31.

18. Arzanian MT. Inherited thrombocytopenia with a different type of gene mutation: A brief literature review and two case studies. *Iran J Pediatr*. 2016;26(5):e4105.

19. Germeshausen M, Ancliff P, Estrada J et al. MECOM-associated syndrome: A heterogeneous inherited bone marrow failure syndrome with amegakaryocytic thrombocytopenia. *Blood Adv*. 2018;2(6):586–96.

20. Wicke DC, Meyer J, Buesche G et al. Gene therapy of MPL deficiency: Challenging balance between leukemia and pancytopenia. *Mol Ther*. 2010;18(2):343–52.

21. Pecci A, Ragab I, Bozzi V et al. Thrombopoietin mutation in congenital amegakaryocytic thrombocytopenia treatable with romiplostim. *EMBO Mol Med*. 2018;10(1):63–75.

22. Fox NE, Lim J, Chen R, Geddis AE. F104S c-Mpl responds to a transmembrane domain-binding thrombopoietin receptor agonist: Proof of concept that selected receptor mutations in congenital amegakaryocytic thrombocytopenia can be stimulated with alternative thrombopoietic agents. *Exp Hematol*. 2010;38(5):384–91.

67

Diamond–Blackfan Anemia

Dongyou Liu

CONTENTS

67.1 Introduction

Diamond–Blackfan anemia (DBA) is a congenital disorder of the bone marrow characterized by macrocytic anemia, reticulocytopenia, severely reduced erythroid precursors, distinct craniofacial features (microcephaly [unusually small head size], low frontal hairline, hypertelorism [wide-set eyes], ptosis [droopy eyelids], broad/flat bridge of the nose, small/low-set ears, micrognathia [small lower jaw], cleft palate [an opening in the roof of the mouth] with or without cleft lip [a split in the upper lip]), growth abnormalities (short stature, short/webbed neck; smaller and higher shoulder blades, malformed hands, absent thumbs), cataracts (clouding of the lens of the eyes), glaucoma (increased pressure in the eyes), strabismus (eyes that do not look in the same direction), kidney and heart abnormalities, hypospadias (the opening of the urethra on the underside of the penis), along with predisposition to myelodysplastic syndrome (MDS; with immature blood cells failing to develop normally), acute myeloid leukemia (AML), and osteosarcoma. Some individuals may develop non-classical DBA, with less severe symptoms such as mild anemia beginning in adulthood [1].

As a clinically and genetically heterogeneous erythroblastopenia, 70% of DBA cases are attributed to heterozygous haploinsufficient mutations in the small (e.g., RPS7, RPS10, RPS17, RPS19, RPS24) and large (e.g., RPS26, RPL5, RPL11, and RPL35A) subunits of the ribosomal protein (RP) genes (Table 67.1), while about 30% of cases are due to alterations in non RP genes [2].

67.2 Biology

Although cases related to Diamond–Blackfan anemia (DBA, also known as Diamond–Blackfan disease, Blackfan–Diamond syndrome, chronic congenital agenerative anemia) were first published by Hugh Joseph in 1936, it was Louis Diamond and Kenneth Blackfan whose detailed description in 1938 established this congenital red cell hypoplasia (showing macrocytic anemia with reticulocytopenia and a normocellular bone marrow with a paucity of erythroid precursors) as a specific clinical entity. Subsequent studies further extended the clinical spectrum of DBA, which range from severe anemia, mild macrocytosis, reticulocytopenia, normocellular bone marrow with near absence of erythroid precursors, elevated erythrocyte adenosine deaminase activity, and fetal hemoglobin indicative of stress erythropoiesis to craniofacial, urogenital, upper limb, and cardiac malformations (50% of cases) in the first year of life [3].

Use of linkage analysis, candidate gene sequencing, whole-genome sequencing and microarray-based comparative genomic hybridization enabled identification of the RPS19 gene in 1999, and other related genes in the following years. It is apparent that nearly 70% of DBA cases are caused by mutations in the genes encoding RP of the small (e.g., RPS7, RPS10, RPS15A, RPS17, RPS19, RPS24, RPS26, RPS27, RPS28, RPS29) and large (e.g., RPL5, RPL11, RPL15, RPL18, RPL26, RPL27, RPL31, RPL35, RPL35A) subunits, leading to pre-ribosomal RNA (rRNA) maturation defects such as faulty ribosome biogenesis and/or inability of ribosomes to properly translate mRNA into protein. In particular, mutations in RPS19 (25%), RPL5 (9%), RPL11 (6.5%),

TABLE 67.1

Molecular and Clinical Characteristics of Diamond–Blackfan Anemia (DBA) Subtypes

DBA Subtype	Gene	Chromosome Locus	Protein	Notable Pathogenic Variants	Subtype Specific Features	% of Cases
DBA1	*RPS19*	19q13.2	40S ribosomal protein S19	p.Met1Val, p.Gln11X, p.Arg56Gln, pArg62Trp, pArg101His		25
DBA2	*8p23.3-p22*	8p23.3-p22				
DBA3	*RPS24*	10q22.3	40S ribosomal protein S24	p.Arg16Ter, p.Gln106Ter, p.Asn1_Met23del		2
DBA4	*RPS17*	15q25.2	40S ribosomal protein S17	p.Gly68TyrfsTer19		1
DBA5	*RPL35A*	3q29	60S ribosomal protein L35a	p.Leu28del, p.Val33Ile, p.Arg102Ter		3.5
DBA6	*RPL5*	1p22.1	60S ribosomal protein L5	p.Met1Arg, p. Arg58Lys, p.Arg58LysfsX55, p.Arg179X	Causing more severe craniofacial, congenital heart, and thumb defects than *RPL11* and *RPS19*; cleft lip/palate in 45% of cases	9
DBA7	*RPL11*	1p36.11	60S ribosomal protein L11	p.Arg18Pro, p.Gly30fs	Predominant thumb abnormalities	6.5
DBA8	*RPS7*	2p25.3	40S ribosomal protein S7			<1
DBA9	*RPS10*	6p21.31	40S ribosomal protein S10	p.Arg98Ser, p.Arg98Cys		6.4
DBA10	*RPS26*	12q13.2	40S ribosomal protein S26	p.Met1Val, p.Asp33Asn, p.Arg87Ter		2.6
DBA11	*RPL26*	17p13.1	60S ribosomal protein L26			1
DBA12	*RPL15*	3p24.2	60S ribosomal protein L15	p. Met70Val, p.Gly105Ser, p.Ser111Phe, p.Pro131Ser, p.Gly132Ala		<1
DBA13	*RPS29*	14q21.3	40S ribosomal protein S29			<1
DBA14	*TSR2*	Xp11.22	Pre-rRNA-processing protein TSR2 homolog		Mandibulofacial dysostosis	<1
DBA15	*RPS28*	19p13.2	40S ribosomal protein S28		Mandibulofacial dysostosis	<1
DBA16	*RPL27*	17q21.31	60S ribosomal protein L27			<1
DBA17	*RPS27*	1q21.3	40S ribosomal protein S27			<1
DBA	*GATA1*	Xp11.23	Erythroid transcription factor	Splice site variants IVS2+1delG and p.Leu74Va)	Profound anemia	<1

RPL10 (6.4%), *RPL35A* (3.5%), *RPS26* (2.6%), and *RPS24* (2%) represent the most common causes of DBA (Table 67.1) [4–7].

In approximately 30% of DBA cases, the underlying genetic defects remain undetermined, apart from five cases that are attributed to mutations in *TSR2* on chromosome Xp11.2 and *GATA1* on Xp11.23 [8].

67.3 Pathogenesis

Ribosome is a cellular structure involved in the translation of messenger RNA (mRNA) to an amino acid sequence (protein). In eukaryotes, ribosome (measuring 80S in size) is separated into the small (40S) and the large (60S) subunit, each of which consists of ribosomal RNA (rRNA) and RP. The large 60S subunit comprises a 5S rRNA, a 28S rRNA, a 5.8S subunit, and ~46 RP (or RPL, i.e., RP associated with large ribosomal subunit); whereas the small 40S subunit contains 18S rRNA and ~33 RP (or RPS, i.e., RP associated with small ribosomal subunit). During ribosome biogenesis, RP are synthesized by RP genes in pre-existing ribosomes in the cytoplasm and transferred into the nucleus to assemble with rRNA for new ribosomes. In addition, some RP take part in

signaling pathways within the cell that regulate cell division and control apoptosis [9–13].

Missense/nonsense/splice site mutations, microinsertion, microdeletion, or single-copy deletion affecting the RP genes (especially *RPS19, RPL5, RPL11, RPL10, RPL35A, RPS26, RPS24*) abrogate or reduce functional RP, decrease the overall numbers of ribosomes, block erythroid progenitor cell division, cause self-destruction of blood-forming cells in the bone marrow (anemia), and activate the apoptosis-related signaling pathway (e.g., TP53 tumor suppressor pathway), leading to the development of various congenital syndromes (or ribosomopathies, which are characterized by birth defects and anemia as well as increased risk of malignancies) [14–18].

67.4 Epidemiology

67.4.1 Prevalence

DBA has an estimated incidence of 1 in 150,000 (7 per 1,000,000) live births worldwide, with over 1000 cases reported in the literature. While DBA shows no racial or ethnic predilection, it does have a male predominance (male to female ratio of 54% to 46%) among patients.

67.4.2 Inheritance

Homozygous mutation in the ribosomal protein (*RP*) genes is likely lethal to embryos, as only heterozygous pathogenic variants are observed in DBA-affected individuals. Heterozygous germline *RP* mutations occur in about 70% of cases and demonstrate autosomal dominant inheritance, in which one copy of the altered gene in each cell is sufficient to cause the disorder. While 40%–45% of patients with autosomal dominant DBA inherit the pathogenic variant from a parent, 55%–60% of patients appear to evolve from a de novo mutation (with neither parent possessing pathogenic variant nor showing clinical evidence of the disorder). A few DBA cases are caused by mutations in *TSR2* on chromosome Xp11.2 and *GATA1* on Xp11.23, which indicate maternal inheritance. Further, about 30% of DBA cases appear to contain mutations in other unidentified genes [1].

67.4.3 Penetrance

DBA shows incomplete penetrance and variable expressivity. Even within the same family, some individuals harboring *RP* pathogenic variant may exhibit classic disease (i.e., profound isolated normochromic and usually macrocytic anemia with normal leukocytes and platelets in addition to other abnormalities), partial phenotype (presence of macrocytosis, but absence of anemia and/or an elevated eADA), or no clinical symptoms (so-called silent carriers). This indicates the potential influence of yet to be identified genetic modifiers on the clinical profile of DBA.

67.5 Clinical Features

Clinical presentations of DBA typically include anemia, congenital malformations, developmental delay, and malignancies.

Anemia is observed in 90% of patients during the first year of life (median age of onset at 2 months and median age of diagnosis at 3 months) and typically manifests as a profound isolated normochromic and usually macrocytic anemia with normal leukocytes and platelets (as indicated by isolated pallor without organomegaly and dyspnea), leading to nonimmune hydrops fetalis. In some patients, hematologic disease may be mild (mild anemia, no anemia with only subtle erythroid abnormalities, physical malformations without anemia) [19].

Congenital malformations occur in about 50% of patients and often affect: (i) head and face (50% of cases; microcephaly; hypertelorism, epicanthus, ptosis; microtia, low-set ears; broad, depressed nasal bridge; cleft lip/palate, high-arched palate; micrognathia; low anterior hairline), (ii) eye (congenital glaucoma, congenital cataract, strabismus), (iii) neck (webbing, short neck), (iv) upper limb and hand including thumb (38% of cases; absent radial artery, flat thenar eminence, triphalangeal/duplex/bifid/hypoplastic/absent thumb), (v) genitourinary (19% of cases; absent kidney, horseshoe kidney, hypospadias), (vi) heart (15% of cases; ventricular septal defect, atrial septal defect, coarctation of the aorta, other cardiac anomalies), and (vii) growth (25%–30% of cases; low birth weight, growth retardation/short stature) (Figures 67.1 and 67.2) [20–22].

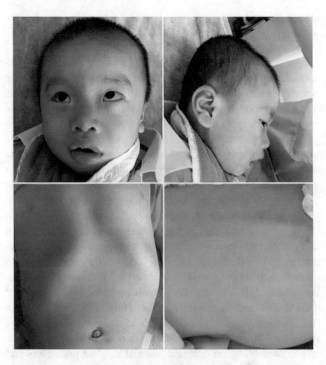

FIGURE 67.1 Diamond–Blackfan anemia in a 2-year-old boy showing cranial deformities, mild craniosynostosis, broad forehead, auricle dysplasia, arched and sparse eyebrows, hypertelorism, strabismus, broad nose with depressed nasal bridge, thick lips, micrognathia, open-mouthed expression, rib protrusion, and kyphosis. (Photo credit: Yuan H et al. *Mol Cytogenet.* 2016;9:58.)

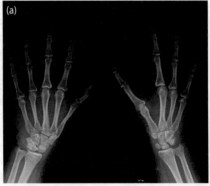 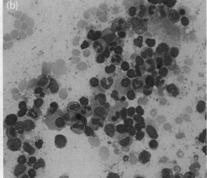

FIGURE 67.2 Diamond–Blackfan anemia in a 35-year-old Spanish male showing: (a) fusion of the phalanges of the finger with the thumb metacarpal in the left hand and absence of the first finger in the right hand on skeleton plain radiographs; (b) hypoplasia of the erythroid lineage with myeloid and megakaryocytic cells of a normal aspect and without any signs of dysplasia in bone marrow aspiration. (Photo credit: Flores Ballester E et al. *Clin Case Rep.* 2015;3(6):392–5.)

Development delay is sometimes observed in children with DBA.

Malignancies occur at an increasing frequency in DBA patients. Specifically, DBA patients have a 28- to 36-fold higher lifetime risk of developing AML, osteosarcoma, or colon cancer, and a 5-fold higher lifetime risk of developing MDS or female genital tumors than the general population.

Some DBA patients may develop non-classic symptoms, including (i) mild or absent anemia with macrocytosis, elevated erythrocyte adenosine deaminase activity (eADA), and/or elevated fetal hemoglobin or HbF concentration, (ii) later onset of disease, and (iii) congenital anomalies or short stature and minimal or no evidence of abnormal erythropoiesis [23,24].

67.6 Diagnosis

Diagnosis of DBA involves medical history review (for family history of anemia, and/or syndromic features), physical examination (for growth and craniofacial abnormalities), laboratory assessment (complete blood count with reticulocyte/ immature erythrocyte count, erythrocyte adenosine deaminase activity, fetal hemoglobin) histopathological examination (for content of erythroid precursors or hypocellularity with dysplasias and megaloblastic changes in bone marrow aspirate and biopsy), and molecular genetic testing (for *RP* and other pathogenic variants) [2,25].

Individuals displaying the following features should be suspected of DBA: (i) *clinical features* (pallor, weakness, failure to thrive, growth retardation, and congenital malformations such as craniofacial, upper-limb, heart, and genitourinary malformations); (ii) *laboratory features* [macrocytic anemia with no other significant cytopenias; increased red-cell mean corpuscular volume (MCV); reticulocytopenia ($<20 \times 10^9$/L); elevated eADA; elevated hemoglobin F or HbF concentration]; and (iii) *histopathological features* (normal marrow cellularity, erythroid hypoplasia, marked reduction in normoblasts, persistence of pronormoblasts on occasion, normal myeloid precursors and megakaryocytes) [1].

Diagnosis of DBA is established in an individual who fulfills all four diagnostic criteria (age younger than 1 year, macrocytic anemia with no other significant cytopenias, reticulocytopenia, and normal marrow cellularity with a paucity of erythroid precursors) [1].

Major supporting criteria for diagnosing DBA consist of *RPS19* mutation and positive family history.

Minor supporting criteria for diagnosing DNA include elevated eADA, elevated HbF concentration, one or more congenital anomalies, no evidence of another inherited disorder of bone marrow function [1].

Differential diagnoses for DBA include conditions that fail to meet all four of the diagnostic criteria and contain no *RP* or other related pathogenic variant such as:

Fanconi anemia: Short stature; abnormal skin pigmentation; malformations of the thumbs, forearms, skeletal system, eyes; abnormalities of the kidneys, urinary tract, ear, heart, gastrointestinal system, oral cavity, central nervous system; hearing loss; hypogonadism;

developmental delay; bone marrow failure; increased risk for non-hematologic malignancies; mutations in *FAN A, B, C, D1 (BRCA2), D2, E, F, G, I, J (BRIP1), L, M, N (PALB2), O (RAD51C), P (SLX4)*; with *FANCB* being X-linked and other types being autosomal recessive.

Pearson syndrome: Sideroblastic anemia of childhood, pancytopenia, exocrine pancreatic failure, renal tubular defects; progressive liver failure and intractable metabolic acidosis leading to infant death; neurologic symptoms in survivors; due to de novo deletions in mitochondrial DNA or large-scale partial deletions and duplications; maternal inheritance.

Dyskeratosis congenita: Classic triad of dysplastic nails, lacy reticular pigmentation of the upper chest and/or neck, and oral leukoplakia; increased risk for progressive l, MDS, or AML, solid tumors such as squamous cell carcinoma of the head/neck or anogenital cancer; pulmonary fibrosis; telomere biology disorder via maternal, autosomal recessive, or autosomal dominant inheritance.

Transient erythroblastopenia of childhood (TEC): Acquired anemia due to decreased production of red blood cell precursors; self-resolving within one and several months; affecting children >1 year compared to children <1 year in DBA, only 10% of children with elevated eADA compared to 85% in DBA, normocytic anemia compared to macrocytic anemia in DBA.

Shwachman–Diamond syndrome: Exocrine pancreatic dysfunction with malabsorption, malnutrition, and growth failure; hematologic abnormalities with single- or multi-lineage cytopenias and susceptibility to MDS and AML; bone abnormalities; persistent or intermittent neutropenia; short stature and recurrent infections; autosomal recessive disorder due to *SBDS* mutations.

GATA1-related X-linked cytopenia: Thrombocytopenia related easy bruising and mucosal bleeding such as epistaxis; anemia ranging from mild dyserythropoiesis to severe hydrops fetalis; platelet dysfunction, mild β-thalassemia, neutropenia, congenital erythropoietic porphyria or CEP in males, menorrhagia in females.

Cartilage-hair hypoplasia (CHH): Short tubular bones at birth; fine sparse blond hair; anemia; macrocytosis with or without anemia; defective T-cell-mediated responses resulting in severe immunodeficiency; autosomal recessive disorder due to *RMRP* mutation.

Human immunodeficiency virus (pure red cell aplasia), *parvovirus B19* (mild red cell aplasia), *viral hepatitis, human T-cell lymphotropic virus type 1, mononucleosis, drugs and toxins:* Antiepileptic drugs: diphenylhydantoin, sodium valproate, carbamazepine, sodium dipropylacetate; azathioprine; chloramphenicol and thiamphenicol; sulfonamides; isoniazid; procainamide.

Immune-mediated diseases: Thymoma, myasthenia gravis, systemic lupus erythematosus, and multiple endocrinopathies [26–32].

67.7 Treatment

Treatment options for DBA include glucocorticoid, transfusion, and hematopoietic stem cell transplantation (HSCT) [33,34].

Corticosteroid has an antiapoptotic effect among erythroid precursors and downregulates the expression of non-erythroid genes; it has long been used to improve the red blood cell count and increase hemoglobin levels in most DBA patients, particularly children >1 year of age. Side effects of corticosteroids range from osteoporosis, weight gain, cushingoid appearance, hypertension, diabetes mellitus, growth retardation, pathologic bone fractures, gastric ulcers, cataracts and glaucoma to increased susceptibility to infection.

Transfusion with packed red blood cells aims to reach a red cell hemoglobin concentration of 80–100 g/L in patients who are resistant to corticosteroids.

HSCT (or allogenic bone marrow transplantation) from an unaffected and HLA-identical sibling represents the only curative treatment for the hematologic manifestations of DBA and is recommended for patients with transfusion dependence or other cytopenias (Figure 67.3) [35].

Compounds that generate positive responses to DBA include cyclosporine A, deferasirox (iron chelator), interleukin 3, metoclopramide, and valproic acid [36].

In addition, a novel gene editing therapy known as CRISPR/Cas9, which harnesses the cell's own machinery to target specific DNA sequences and generates small deletions for the cell to repair, is being investigated for correcting point mutations and small indels in DBA patient cells, with the goal of finding an ultimate solution for the long-suffering patients with DBA [37].

67.8 Prognosis and Prevention

Prognosis for DBA is generally favorable, with 20% of patients going into remission, 40% becoming steroid dependent, and 40% being transfusion dependent (with transfusion-related iron overload). For patients who undergo bone marrow transplantation, the 3-year overall survival is 64% (range 50%–74%).

Early treatment of iron overload in DBA patients who fail to respond to steroid therapy and require RBC transfusions helps improve the prognosis. However, deferiprone may cause neutropenia, and should be avoided as a treatment of iron overload in persons with DBA.

For DBA with autosomal dominant inheritance, an individual carrying a pathogenic variant has a 50% chance of passing it to offspring. While males with *GATA1* or *TSR2*-related DBA pass the pathogenic variant to daughters but not sons, females heterozygous for a *GATA1* or *TSR2* pathogenic variant have a 50% chance of passing it to offspring (with male offspring becoming affected, and female offspring being carrier). Prenatal testing and preimplantation genetic diagnosis are valuable for prospective parents with familial pathogenic variant.

67.9 Conclusion

Diamond–Blackfan anemia (DBA) is a rare genetic disorder of erythroid hypoplasia, typically manifesting as macrocytic anemia (reticulocytopenia and erythroblastopenia with normal myeloid and megakaryocytic lineages; pale pallor, failure to thrive, feeding difficulties), congenital anomalies (triphalangeal, bifid, subluxed thumbs, subtle flattening of the thenar eminence with a normal radius, genitourinary and heart defects, webbed

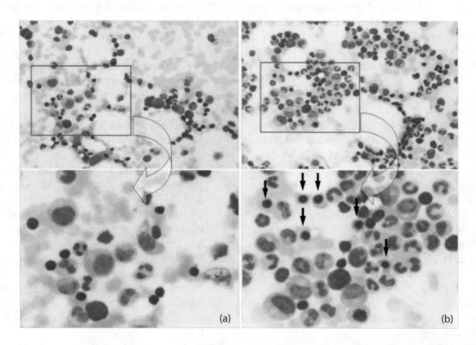

FIGURE 67.3 Diamond–Blackfan anemia in a 7-year-old girl showing: (a) normocellular marrow with markedly reduced erythropoiesis in bone marrow aspirates before bone marrow transplantation; (b) active trilineage hematopoiesis accompanied with presence of erythropoietic precursors (arrows) in bone marrow aspirates obtained 28 days after bone marrow transplantation. (Photo credit: Park JA et al. *J Korean Med Sci*. 2010;25(4):626–9.)

neck, fusion of cervical vertebrae, congenital asymmetric high scapula), and predisposition to malignancy (MDS, AML, colon carcinoma, female genital tumors, and osteosarcoma) [38]. Heterozygous haploinsufficient mutations in the small and large subunits of the *RP* genes appear to be the key mechanism that underlies the pathogenesis of most DBA cases. Diagnosis of DBA is based on meeting four clinical criteria (age younger than 1 year; macrocytic anemia with no other significant cytopenias; reticulocytopenia; normal marrow cellularity with a paucity of erythroid precursors), and identification of *RP* pathogenic variant provides further evidence of this disorder [39]. Treatment of DBA involves glucocorticoid, transfusion, and HSCT [40,41]. Given many unanswered questions surrounding DBA (e.g., the role of non-*RP* genes and modifiers in its pathogenesis), further research is necessary to improve our understanding of and develop innovative therapeutic measures against this congenital disorder.

REFERENCES

1. Clinton C, Gazda HT. Diamond–Blackfan anemia. In: Adam MP, Ardinger HH, Pagon RA et al. editors. *GeneReviews® [Internet]*. Seattle (WA): University of Washington, Seattle; 1993–2018. 2009 Jun 25 [updated 2016 Apr 7].
2. Da Costa L, O'Donohue MF, van Dooijeweert B et al. Molecular approaches to diagnose Diamond–Blackfan anemia: The EuroDBA experience. *Eur J Med Genet.* 2018;61(11):664–73.
3. Sakamoto KM, Narla A. Perspective on Diamond–Blackfan anemia: lessons from a rare congenital bone marrow failure syndrome. *Leukemia.* 2018;32(2):249–51.
4. Danilova N, Gazda HT. Ribosomopathies: How a common root can cause a tree of pathologies. *Dis Model Mech.* 2015;8(9):1013–26.
5. Ikeda F, Yoshida K, Toki T et al. Exome sequencing identified RPS15A as a novel causative gene for Diamond–Blackfan anemia. *Haematologica.* 2017;102(3):e93–6.
6. Ulirsch JC, Verboon JM, Kazerounian S et al. The genetic landscape of Diamond–Blackfan anemia. *Am J Hum Genet.* 2018;103(6):930–47.
7. Shi X, Huang X, Zhang Y, Cui X. Identification of a novel RPS26 nonsense mutation in a Chinese Diamond–Blackfan anemia patient. *BMC Med Genet.* 2019;20(1):120.
8. Ludwig LS, Gazda HT, Eng JC et al. Altered translation of GATA1 in Diamond–Blackfan anemia. *Nat Med.* 2014;20(7):748–53.
9. McGowan KA, Mason PJ. Animal models of Diamond Blackfan anemia. *Semin Hematol.* 2011;48(2):106–16.
10. Ellis SR. Nucleolar stress in Diamond Blackfan anemia pathophysiology. *Biochim Biophys Acta.* 2014;1842(6):765–8.
11. Nakhoul H, Ke J, Zhou X, Liao W, Zeng SX, Lu H. Ribosomopathies: mechanisms of disease. *Clin Med Insights Blood Disord.* 2014;7:7–16.
12. Ruggero D, Shimamura A. Marrow failure: A window into ribosome biology. *Blood.* 2014;124(18):2784–92.
13. Goudarzi KM, Lindström MS. Role of ribosomal protein mutations in tumor development (Review). *Int J Oncol.* 2016;48(4):1313–24.
14. Jiang H, Wu MY, Li DZ. A novel mutation of ribosomal protein S19 gene in a Chinese child with Diamond–Blackfan anemia. *Indian J Hematol Blood Transfus.* 2016;32(Suppl 1):233–4.

15. Yang Z, Keel SB, Shimamura A et al. Delayed globin synthesis leads to excess heme and the macrocytic anemia of Diamond Blackfan anemia and del(5q) myelodysplastic syndrome. *Sci Transl Med.* 2016;8(338):338ra67.
16. Gastou M, Rio S, Dussiot M et al. The severe phenotype of Diamond–Blackfan anemia is modulated by heat shock protein 70. *Blood Adv.* 2017;1(22):1959–76.
17. Venugopal P, Moore S, Lawrence DM et al. Self-reverting mutations partially correct the blood phenotype in a Diamond Blackfan anemia patient. *Haematologica.* 2017;102(12):e506–9.
18. Noel CB. Diamond–Blackfan anemia RPL35A: A case report. *J Med Case Rep.* 2019;13(1):185.
19. Muir C, Dodds A, Samaras K. Mid-life extra-haematopoetic manifestations of Diamond–Blackfan anaemia. *Endocrinol Diabetes Metab Case Rep.* 2017;2017. pii: 16–0141.
20. Flores Ballester E, Gil-Fernández JJ, Vázquez Blanco M et al. Adult-onset Diamond–Blackfan anemia with a novel mutation in the exon 5 of RPL11: Too late and too rare. *Clin Case Rep.* 2015;3(6):392–5.
21. Gomes RF, Munerato MC. The stomatological complications of Diamond–Blackfan anemia: A case report. *Clin Med Res.* 2016;14(2):97–102.
22. Yuan H, Meng Z, Liu L, Deng X, Hu X, Liang L. A de novo 1.6Mb microdeletion at 19q13.2 in a boy with Diamond–Blackfan anemia, global developmental delay and multiple congenital anomalies. *Mol Cytogenet.* 2016;9:58.
23. Farrar JE, Dahl N. Untangling the phenotypic heterogeneity of Diamond Blackfan anemia. *Semin Hematol.* 2011;48(2):124–35.
24. Farruggia P, Quarello P, Garelli E et al. The spectrum of non-classical Diamond–Blackfan anemia: A case of late beginning transfusion dependency associated to a new RPL5 mutation. *Pediatr Rep.* 2012;4(2):e25.
25. Steinberg-Shemer O, Keel S et al. Diamond Blackfan anemia: A nonclassical patient with diagnosis assisted by genomic analysis. *J Pediatr Hematol Oncol.* 2016;38(7):e260–2.
26. Chirnomas SD, Kupfer GM. The inherited bone marrow failure syndromes. *Pediatr Clin North Am.* 2013;60(6):1291–310.
27. Khincha PP, Savage SA. Genomic characterization of the inherited bone marrow failure syndromes. *Semin Hematol.* 2013;50(4):333–47.
28. Khincha PP, Savage SA. Neonatal manifestations of inherited bone marrow failure syndromes. *Semin Fetal Neonatal Med.* 2016;21(1):57–65.
29. Chung NG, Kim M. Current insights into inherited bone marrow failure syndromes. *Korean J Pediatr.* 2014;57(8):337–44.
30. Wilson DB, Link DC, Mason PJ, Bessler M. Inherited bone marrow failure syndromes in adolescents and young adults. *Ann Med.* 2014;46(6):353–63.
31. Babushok DV, Bessler M. Genetic predisposition syndromes: When should they be considered in the work-up of MDS? *Best Pract Res Clin Haematol.* 2015;28(1):55–68.
32. Bannon SA, DiNardo CD. Hereditary predispositions to myelodysplastic syndrome. *Int J Mol Sci.* 2016;17(6). pii: E838.
33. Narla A, Vlachos A, Nathan DG. Diamond Blackfan anemia treatment: Past, present, and future. *Semin Hematol.* 2011;48(2):117–23.

34. Chai KY, Quijano CJ, Chiruka S. Danazol: An effective and underutilised treatment option in Diamond–Blackfan anaemia. *Case Rep Hematol.* 2019;2019:4684156.

35. Park JA, Lim YJ, Park HJ, Kong SY, Park BK, Ghim TT. Normalization of red cell enolase level following allogeneic bone marrow transplantation in a child with Diamond–Blackfan anemia. *J Korean Med Sci.* 2010;25(4):626–9.

36. Sjögren SE, Flygare J. Progress towards mechanism-based treatment for Diamond–Blackfan anemia. *Scientific World Journal.* 2012;2012:184362.

37. Aspesi A, Monteleone V, Betti M et al. Lymphoblastoid cell lines from Diamond Blackfan anaemia patients exhibit a full ribosomal stress phenotype that is rescued by gene therapy. *Sci Rep.* 2017;7(1):12010.

38. Gadhiya K, Budh DP. Diamond blackfan anemia. In: *StatPearls [Internet].* Treasure Island (FL): StatPearls Publishing, Florida. 2020 Jan [updated 2019 Nov 23].

39. Da Costa L, Narla A, Mohandas N. An update on the pathogenesis and diagnosis of Diamond-Blackfan anemia. *F1000Res.* 2018;7(F1000 Faculty Rev):1350.

40. Aspesi A, Borsotti C, Follenzi A. Emerging Therapeutic Approaches for Diamond Blackfan Anemia. *Curr Gene Ther.* 2018;18(6):327–35.

41. Bartels M, Bierings M. How I manage children with Diamond-Blackfan anaemia. *Br J Haematol.* 2019;18(2):123–33.

68

Down Syndrome

Dongyou Liu

CONTENTS

68.1 Introduction

Down syndrome is a genetic disorder characterized by mental retardation/intellectual disability, growth anomalies (e.g., short stature, a single deep crease across the center of the palm), characteristic facial features (e.g., small nose, upward-slanting eyes), congenital heart diseases, hypotonia (weak muscle tone) in infancy, gastroesophageal reflux, celiac disease (gluten intolerance), hypothyroidism (underactive thyroid gland), ophthalmologic disorders, Alzheimer disease, and predisposition to myeloid leukemia (particularly acute megakaryoblastic leukemia [AMKL] and acute lymphoblastic leukemia [ALL]) (Figure 68.1).

Molecularly, Down syndrome is attributed to the presence of an extra, critical portion of chromosome 21 in all or some of the cells, leading to disruption of normal development. More specifically, trisomy 21 (constitutional trisomy 21 or trisomy Down syndrome) involving an extra chromosome 21 accounts for about 95% of Down syndrome cases; mosaic trisomy 21 (mosaic Down syndrome) due to the presence of an extra chromosome 21 in some cells but not in other cells causes 1%–2% of Down syndrome cases, and translocation trisomy 21 (translocation Down syndrome) involving translocation (attachment) of extra chromosome 21 material onto another chromosome (e.g., 14) is responsible for 3%–4% of Down syndrome cases (Figures 68.2 and 68.3) [1].

68.2 Biology

Down syndrome (also referred to as Down's syndrome or trisomy 21) is an ancient disease that is evidenced from pottery artifacts belonging to the Tolteca culture of Mexico and the Tumaco-La Tolita culture on the border of present-day Colombia and Ecuador over 2000 years ago. Distinct facial features (e.g., short palpebral fissures, oblique eyes, midface hypoplasia, and open mouth with macroglossia) depicted by these pottery artifacts reflect unquestionably the clinical phenotype of Down syndrome. Paintings from the fifteenth and sixteenth centuries provide additional clues on the historic prevalence of Down syndrome [1].

Following the association of Down syndrome with heart malformation in 1894, atrioventricular septal defects (AVSDs) in 1924, and leukemia in 1930, this genetic disorder has been increasingly recognized as a major cause of mental retardation, congenital heart diseases, dysmorphic facies, and myeloid leukemia. The identification of chromosomal rearrangement in Down syndrome–affected patients in 1959 revealed critical insights into its molecular pathogenesis. It is apparent that Down syndrome is due to the presence of an extra, critical portion of chromosome 21 in all or some of the cells, which induces dosage imbalance of genes, thus increasing the activation and expression of many genes throughout the genome [1].

68.3 Pathogenesis

Chromosome 21 (Hsa 21) represents the smallest of the 23 chromosomes pairs in humans. Spanning over 46 Mb DNA with 256 protein coding genes, 356 non-coding RNA genes and 207 pseudogenes, which together make up about 1.5% of total DNA in cells, chromosome 21 is divided into two portions: a short arm (p, which consists of 12 Mb DNA and is separated into 4 bands, i.e. 21p11.1, 21p11.2, 21p12, and 21p13), and a long arm (q, which contains 34 Mb DNA and is separated into 10 bands, i.e. 21q11.1,

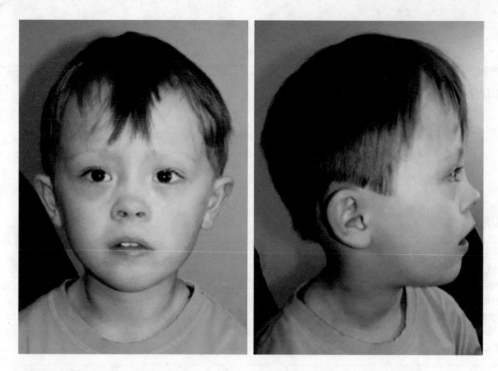

FIGURE 68.1   Down syndrome in a 5½-year-old boy with a mosaic microduplication of chromosome 21q22 displaying posterior plagiocephaly, flat face, epicanthus, upslanted palpebral fissures, and anteverted nares. (Photo credit: Schnabel F et al. *Mol Cytogenet*. 2018;11:62.)

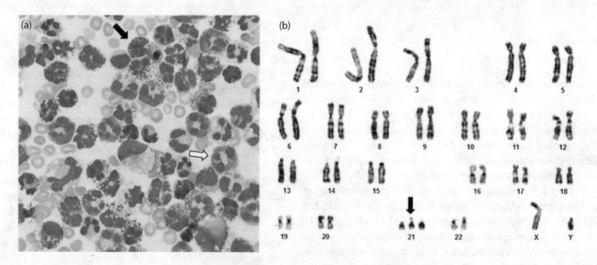

FIGURE 68.2   Down syndrome in a 14-year-old male with transient abnormal myelopoiesis showing large numbers of basophils (black arrow) and eosinophils (white arrow) in the pericardial fluid (Leishman staining) (a); and trisomy of chromosome 21 (black arrow) by karyotype (b). (Photo credit: Falasco BF et al. *Clin Case Rep*. 2019;7(7):1280–4.)

21q11.2, 21q21.1, 21q21.2, 21q21.3, 21q22.11, 21q22.12, 21q22.13, 21q22.2, and 21q22.3) [2].

While healthy individuals possess two copies of chromosome 21, a majority (95%) of Down syndrome patients have three copies of chromosome 21 (so called trisomy 21, constitutional trisomy 21, or trisomy Down syndrome), with the extra copy of chromosome 21 deriving mostly from the maternal side. Trisomy 21 is a chromosomal abnormality that results from the failure of the chromosome 21 pair to separate during the formation of an egg (or sperm), a process known as nondisjunction, which seems to occur more often as women age. When the egg with two copies of chromosome 21 is fertilized by a sperm with one copy of chromosome 21, the resulting embryo contains three copies

of chromosome 21 (trisomy 21), which are copied in every cell, leading to Down syndrome (Figure 68.2) [3,4].

Mosaic trisomy 21 (or mosaic Down syndrome) is responsible for 1%–2% of clinical cases. While the fertilized egg may have two copies of chromosome 21, a random error occurring during cell division early in the embryo development results in some cells with three copies of chromosome 21 (47 chromosomes in total), and other cells with two copies of chromosome 21 (46 chromosomes in total) [3].

Translocation trisomy 21 (translocation Down syndrome, Robertsonian translocation, or balanced translocation) causes 3%–4% of clinical cases. Occurring in 0.1% of the general population, balanced translocation (two chromosomes stuck together

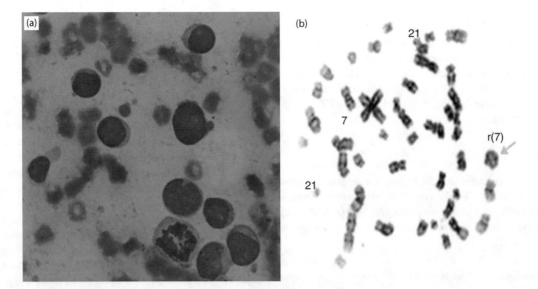

FIGURE 68.3 Down syndrome in a 21-month-old boy with minimally differentiated acute myeloid leukemia and chromosomal breaks at 7p21.2 and 7q34 showing undifferentiated blast cells in bone marrow aspirate (a); and bone marrow karyotype of 47,XY,−7, +r, +21c7/15, including ring chromosome 7 (arrow) (b). (Photo credit: Borges MLRDR et al. *Hematol Transfus Cell Ther.* 2019;41(1):84–7.)

during the production of an egg or sperm) induces no obvious phenotypic changes in carriers but increases risk (by as much as 15%) for trisomy in a pregnancy. Consequently, a seemingly unaffected carrier with 46 chromosomes may have extra chromosome 21 material attached (translocated) onto another chromosome (e.g., 14), without gaining or losing genetic material (i.e., balanced translocation). However, when this translocation is passed to offspring, it may become unbalanced and gain extra genetic material from chromosome 21, leading to Down syndrome (Figure 68.3) [3,5].

The presence of an extra copy of chromosome 21 in the cells causes dosage imbalance of genes (or an increased dosage or copy number of genes), leading to increased expression and regulation of many genes throughout the genome. Further, translocation that occurs at Down syndrome critical regions (DSCR), such as a 3.8–6.5 Mb region on band 21q21.22 and another critical region on band 21q22, which contain many genes responsible for Down syndrome phenotype (e.g., craniofacial abnormalities, congenital heart defects of the endocardial cushions, clinodactyly of the fifth finger, mental retardation) [1].

Besides other abnormalities, children born with constitutional trisomy 21 are highly susceptible to AMKL and ALL, which are usually preceded by a clonal neonatal preleukemic syndrome or transient abnormal myelopoiesis (TAM) unique to Down syndrome [6,7].

During the course of myeloid leukemogenesis in Down syndrome, trisomy 21 represents the first hit, which occurs prenatally in the fetal liver (the site of definitive fetal hematopoiesis), and sporadic N-terminal truncating mutation in the chromosome X-inked transcription factor gene *GATA1* (Xp11.23), which is essential for erythroid and megakaryocytic differentiation, represents the second

hit. The combination of trisomy 21 and *GATA1* mutation perturbs megakaryocyte progenitors and induces transient myeloproliferative disorder (TMD) with increased levels of circulating blast cells. In fact, 10%–15% of Down syndrome neonates develop TAM with >10% increase in circulating blast cells, while another 10%–15% of Down syndrome neonates harboring *GATA1* mutations develop silent TAM (clinically and hematologically silent disease) and produce <10% increase in circulating blast cells. Although 85%–90% of TMD cases resolve spontaneously by age of 6 months or after use of low-dose chemotherapy, some TMD clones persist and undergo further genetic/epigenetic changes and clonal expansion, leading to the development of myeloid leukemia such as AMKL and ALL by the age of 5 years. Indeed, frequent observation of somatic *GATA1* mutations in Down syndrome−related AMKL lends support on the natural progression of TAM to AMKL [8].

ALL is linked to mutations in *JAK2*, *NRAS*, and *KRAS*, mutation or overexpression of *CRLF2*, and trisomy 21. Acquired gain-of-function mutation in *JAK2* (Janus kinase 2) is observed in 30% of Down syndrome−related ALL. Several other genes (e.g., *HMGN1*, *DYRK1A*, *IKZF1*, *PAX5*, *ETV6-IGH*) implicated in ALL pathogenesis are also altered in patients with Down syndrome [9–11].

68.4 Epidemiology

68.4.1 Prevalence

As the most common cytogenetic abnormality affecting neonates, Down syndrome has an estimated incidence of 1 in 319–1000 live births, depending on maternal age and population groups (Table 68.1).

TABLE 68.1

Relationship between Maternal Age at the Expected Time of Delivery and Risk of Down Syndrome in Newborn

Maternal Age	20	25	30	35	36	37	38	39	40	41	42	43	44	45	46	47	48	49
Risk of Down syndrome	1/1667	1/1250	1/952	1/385	1/295	1/227	1/175	1/137	1/106	1/82	1/64	1/50	1/38	1/30	1/23	1/18	1/14	1/11

68.4.2 Inheritance

Trisomy 21 (trisomy Down syndrome), occurring in about 95% of clinical cases, is nonhereditary. Similarly, mosaic trisomy 21 (mosaic Down syndrome), causing 1%–2% of clinical cases, is also nonhereditary. On the other hand, translocation trisomy 21 (translocation Down syndrome), responsible for 3%–4% of clinical cases, is hereditary.

68.4.3 Penetrance

Down syndrome demonstrates high penetrance and extensive phenotypic variability. While cognitive impairment, muscle hypotonia at birth, and dysmorphic features are present in all affected individuals, other traits occur only in a portion of affected individuals.

68.5 Clinical Features

Clinically, Down syndrome is characterized by several conserved features (mental retardation/learning disability, craniofacial anomalies, and hypotonia/ poor muscle tone in early infancy) that occur in all affected individuals, along with other anomalies that may appear in some patients. In addition, Down syndrome significantly increases the risk for malignancies.

Craniofacial anomalies range from small chin, slanted eyes, flat nasal bridge, and protruding mouth (due to small mouth and large tongue) (Figure 68.1) [12].

Other anomalies include growth defects (a single crease of the palm, abnormal pattern of fingerprint, and short fingers and big toe), congenital cardiac defects (endocardial cushion defect, 40%; atrioventricular septal defect [AVSD, presence of a common atrioventricular junction as compared to the separate right and left atrioventricular junction in the normal heart], 35%), Alzheimer disease (70% risk after age 50), hypertension, gastrointestinal problems (duodenal stenosis and imperforate anus), celiac disease (gluten intolerance), hypothyroidism, and ophthalmologic disorders.

Malignancies associated with Down syndrome are predominated by myeloid leukemias (both AMKL and ALL), with a cumulative risk of 2% by age 5 and 2.7% by age 30, which represent a 10- to 20-fold increase relative to the general population (Figure 68.3) [5]. AMKL and ALL are usually preceded by TAM, which is a clonal neonatal preleukemic syndrome unique to Down syndrome, and which is defined by the presence of blasts in the peripheral blood and *GATA1* mutation. TAM may be asymptomatic in 10%–25% neonates (silent TAM, which is often noted as an incidental finding on review of a blood film) or cause disseminated leukemic infiltration in 10%–20% of neonates, with massive hepatosplenomegaly, effusions, coagulopathy, and multiorgan failure. Myeloid leukemias in Down syndrome patients often have an indolent presentation (evolving from myelodysplasia, progressive pancytopenia, especially thrombocytopenia and leucopenia and low percentage of circulating blasts to more serious disease). With the exception of germ cell tumor in children and testicular/gastric/liver tumors in adults, Down syndrome patients appear to have significantly lower-than-expected age-adjusted incidence rates of other solid tumors (e.g., neuroblastoma, PNET, and medulloblastoma in children; cancers of the breast, lung, prostate, colon, and oral cavity as well as malignant melanoma in adults) (Figures 68.2 and 68.4) [4,13–16].

68.6 Diagnosis

Diagnosis of Down syndrome involves medical history review (for relevant family medical history), physical examination (for growth defects, characteristic craniofacial features, etc.), laboratory assessment (for changing blast cell levels, etc.), cytogenetic (karyotyping) and molecular analyses (for chromosomal and genetic alterations such as *GATA1* mutation) [1].

Cytogenetic analysis involves Giemsa banding (G-banding) of fetal cells at metaphase stage on amniocytes (grown in vitro) or chorionic villus sample (CVS), and is considered the gold standard for identification of trisomy 21, balanced translocation, and all other aneuploidies. Use of Hsa 21–specific probes or whole-Hsa 21 in fluorescence in situ hybridization (FISH) of interphase nuclei greatly improves the accuracy of cytogenetic analysis. Recent application of quantitative fluorescent-PCR (QF-PCR) based on DNA polymorphic markers (microsatellites) from Hsa 21 permits

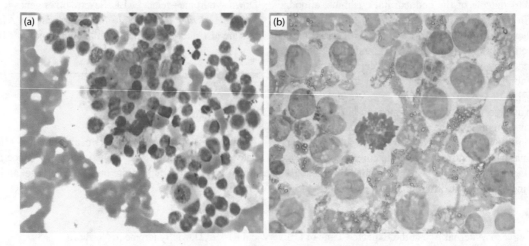

FIGURE 68.4   Down syndrome in a 2-year-old boy showing: (a) myeloid sarcoma with small, blue, round cells forming rosettes from fine needle aspirate of the mass lesion in left temporal region (Leishman); (b) myeloid blasts with abundant cytoplasm in peripheral blood film. (Photo credit: Marwah N et al. *Int J Surg Case Rep.* 2019;58:77–80.)

rapid, sensitive, and precise determination of three different alleles. PCR-based methods targeting polymorphic STR markers, molecular techniques that measure copy number of DNA sequences or Hsa 21 (e.g., multiplex probe ligation assay [MLPA], paralogous sequence quantification [PSQ]), and next-generation sequencing (NGS) further aid in the diagnosis of Down syndrome [1].

68.7 Treatment

As Down syndrome is a clinically and genetically heterogeneous disorder, its treatment requires a multidisciplined approach that focuses on relieving and rectifying the clinical manifestations and complications that compromise the patient's full potential and/or endanger the patient's life.

For example, cardiac defects can be surgically rectified to prevent serious complications; congenital cataracts are extracted after birth to allow light to reach the retina; feeding problems and failure to thrive are improved with a combination of cardiac surgery and balanced diet; and myeloid leukemias are treated with chemotherapy (e.g., cytotoxic drugs methotrexate, doxorubicin, vincristine, cytarabine, and etoposide as well as busulfan prior to a bone marrow transplant in certain leukemia patients) [17–20].

68.8 Prognosis and Prevention

Prognosis for Down syndrome is much improved with early diagnosis and appropriate therapeutic intervention. The average life span for Down syndrome patients currently stands at 55 years in developed countries.

Down syndrome patients with acute myeloid leukemia (AML) often have better outcomes than those with ALL, as Down syndrome–associated ALL is relatively resistant to cytotoxic drugs of trisomic lymphoblasts.

Prenatal diagnosis for high risk pregnancies at 14–24 weeks of gestation via amniocentesis and CVS and ultrasound (for increased fetal nuchal translucency) provides a reliable way to reduce the incidence of Down syndrome. Molecular techniques (e.g., FISH, QF-PCR, and MLPA) are increasingly applied for prenatal diagnosis of Down syndrome.

68.9 Conclusion

As the most common chromosomal abnormality among liveborn infants and one of the most common leukemia-predisposing disorders with unique clinical features, Down syndrome is associated with several congenital abnormalities (e.g., mental retardation/intellectual disability, growth defects, craniofacial dysmorphism, congenital heart diseases, hypotonia, gastroesophageal reflux, celiac disease, and hypothyroidism) and significantly increased risk (10- to 30-fold) of myeloid leukemia (e.g., AMKL, ALL), which is often preceded by clinically silent or overt transient abnormal myelopoiesis (TAM) [21–23]. The molecular basis of Down syndrome lies in the existence of an extra, critical portion of chromosome 21 in all or some of the cells, which, through its dosage imbalance effects, upregulates the expression of various downstream genes, perturbs fetal hematopoiesis, and facilitates the development of TAM and myeloid leukemia. While cytogenetic analysis of metaphase karyotype remains the method of choice for diagnosing Down syndrome, nucleic acid detection techniques (e.g., FISH and PCR) provide further improvement in terms of diagnostic speed, sensitivity, and accuracy. Management of Down syndrome relies on multidisciplined efforts that aim to provide symptomatic relief as well as rectify the underlying defects that improvise patient's quality of life. Apart from autism and epilepsy, most health comorbidities are not associated with poorer cognitive outcomes in Down syndrome. Thus, it is important for clinicians to consider these differences in provision of prognostic information [24].

REFERENCES

1. Asim A, Kumar A, Muthuswamy S, Jain S, Agarwal S. Down syndrome: An insight of the disease. *J Biomed Sci.* 2015;22:41.
2. Satgé D, Seidel MG. The pattern of malignancies in Down syndrome and its potential context with the immune system. *Front Immunol.* 2018;9:3058.
3. Ganguly BB, Kadam NN, Mandal PK. Complexity of chromosomal rearrangements in Down syndrome leukemia. *J Cancer Res Ther.* 2017;13(2):381–3.
4. Falasco BF, Durante B, Faria DK, Faria CS, Rosolen DCB, Antonangelo L. Transient abnormal myelopoiesis with pericardial effusion in Down syndrome: Case report. *Clin Case Rep.* 2019;7(7):1280–4.
5. Borges MLRDR, Soares-Ventura EM, Liehr T, Marques-Salles TJ. Minimally differentiated acute myeloid leukemia with ring/marker derived from chromosome 7 in a child with Down syndrome. *Hematol Transfus Cell Ther.* 2019;41(1):84–7.
6. Mateos MK, Barbaric D, Byatt SA, Sutton R, Marshall GM. Down syndrome and leukemia: Insights into leukemogenesis and translational targets. *Transl Pediatr.* 2015;4(2):76–92.
7. Talsma CE, Boyer DF. Myeloid leukemia associated with Down syndrome. *Blood.* 2017;130(2):230.
8. Bhatnagar N, Nizery L, Tunstall O, Vyas P, Roberts I. Transient abnormal myelopoiesis and AML in Down syndrome: An update. *Curr Hematol Malig Rep.* 2016;11(5):333–41.
9. Fernández-Martínez P, Zahonero C, Sánchez-Gómez P. DYRK1A: The double-edged kinase as a protagonist in cell growth and tumorigenesis. *Mol Cell Oncol.* 2015;2(1):e970048.
10. Arber DA, Orazi A, Hasserjian R et al. The 2016 revision to the World Health Organization classification of myeloid neoplasms and acute leukemia. *Blood.* 2016;127(20):2391–405.
11. Lee P, Bhansali R, Izraeli S, Hijiya N, Crispino JD. The biology, pathogenesis and clinical aspects of acute lymphoblastic leukemia in children with Down syndrome. *Leukemia.* 2016;30(9):1816–23.
12. Schnabel F, Smogavec M, Funke R, Pauli S, Burfeind P, Bartels I. Down syndrome phenotype in a boy with a mosaic microduplication of chromosome 21q22. *Mol Cytogenet.* 2018;11:62.
13. Almouhissen T, Badr H, AlMatrafi B, Alessa N, Nassir A. Testicular cancer in Down syndrome with spinal cord metastases. *Urol Ann.* 2016;8(4):503–5.
14. Dey N, Krie A, Klein J et al. Down's syndrome and triple negative breast cancer: A rare occurrence of distinctive clinical relationship. *Int J Mol Sci.* 2017;18(6). pii: E1218.
15. Marwah N, Bhutani N, Budhwar A, Sen R. Isolated myeloid sarcoma of the temporal bone: As the first clinical manifestation of acute myeloid leukemia in a patient of Down's syndrome. *Int J Surg Case Rep.* 2019;58:77–80.

16. Petruzzellis G, Valentini D, Del Bufalo F et al. Vemurafenib treatment of pleomorphic xanthoastrocytoma in a child with Down syndrome. *Front Oncol.* 2019;9:277.

17. Caldwell JT, Ge Y, Taub JW. Prognosis and management of acute myeloid leukemia in patients with Down syndrome. *Expert Rev Hematol.* 2014;7(6):831–40.

18. Izraeli S, Vora A, Zwaan CM, Whitlock J. How I treat ALL in Down's syndrome: pathobiology and management. *Blood.* 2014;123(1):35–40.

19. Hefti E, Blanco JG. Anthracycline-related cardiotoxicity in patients with acute myeloid leukemia and Down syndrome: A literature review. *Cardiovasc Toxicol.* 2016;16(1): 5–13.

20. Hefti E, Blanco JG. Pharmacokinetics of chemotherapeutic drugs in pediatric patients with Down syndrome and leukemia. *J Pediatr Hematol Oncol.* 2016;38(4):283–7.

21. Reiche L, Küry P, Göttle P. Aberrant oligodendrogenesis in Down Syndrome: Shift in gliogenesis?. *Cells.* 2019;8(12):1591.

22. Rodrigues M, Nunes J, Figueiredo S, Martins de Campos A, Geraldo AF. Neuroimaging assessment in Down syndrome: A pictorial review. *Insights Imaging.* 2019;10(1):52.

23. Zhang H, Liu L, Tian J. Molecular mechanisms of congenital heart disease in down syndrome. *Genes Dis.* 2019;6(4):372–7.

24. Startin CM, D'Souza H, Ball G et al. Health comorbidities and cognitive abilities across the lifespan in Down syndrome. *J Neurodev Disord.* 2020;12(1):4.

69

Dyskeratosis Congenita

Somanath Padhi and Susama Patra

CONTENTS

69.1 Introduction

Dyskeratosis congenita (DC) is an inherited, cancer-prone syndrome characterized by *diagnostic triad* of reticulated skin hyperpigmentation, nail dystrophy, oromucosal premalignant leukoplakia and subsequent increased risk of premature bone marrow failure and both solid and hematolymphoid malignancies [1]. The disease was first described by Zinsser in 1906 and was recognized as a clinical entity by Engman in 1926 and Cole in 1930; hence this is also known as Zinsser–Engman–Cole syndrome [2]. DC is a disease of defective telomere maintenance, and patients with DC have premature telomere shortening and subsequent replicative senescence, leading to premature stem cell failure, aging, and cancer predisposition [3]. Evidence exists for telomerase dysfunction, ribosome deficiency, and protein synthesis dysfunction in this disorder. Early mortality is often associated with bone marrow failure, infections, fatal pulmonary complications, or malignancy (Table 69.1) [4].

69.2 Epidemiology

DC is rare and has an estimated incidence of 1 case per 1 million, with over 200 affected individuals reported in the literature.

Approximately 2%–5% of the patients with bone marrow failure are identified to have DC. In 1995, a Dyskeratosis Congenita Registry was established at the Hammersmith Hospital, London, UK where 46 families including 83 patients (76 males) were recruited [5]. The DC registry includes patients from all over the world, with families from at least 40 different countries currently in the registry. DC has a male predilection (male to female ratio of nearly 3:1), reflecting an X-linked inheritance pattern. The disease manifests clinically between 5 and 12 years of age, with the skin hyperpigmentation and nail changes typically appearing first [6]. There has been no racial predilection reported.

69.3 Telomere Biology

69.3.1 Telomere Attrition

Telomeres are specialized structures at the end of chromosomes, consisting of nucleic acid and protein components that maintain the integrity of chromosome ends, protecting the natural ends of chromosomes from loss of DNA, abnormal fusion to other chromosomes, and from activation of DNA damage pathway response which may otherwise occur at free ends of DNA created by strand breaks. Telomeres shorten (attrition) with each cell division as a result of loss of 50–100 bp of the telomeric

TABLE 69.1

List of Abbreviations

ACD	Adrenocortical dysplasia homolog
AD	Autosomal dominant
AR	Autosomal recessive
AML	Acute myeloid leukemia
BMF	Bone marrow failure
BMT	Bone marrow transplantation
CTC1	CTS telomere maintenance complex component 1
CST complex	CTC1/STN1/TEN1 complex
DC	Dyskeratosis congenita
DCX	X linked dyskeratosis congenita
DKC1	Dyskerin1
DNA	Deoxyribonucleic acid
DDRA	DNA damage response activation
FISH	Fluorescence in situ hybridization
GAR1	Glycine-arginine-rich ribonucleoprotein
G-CSF	Granulocyte colony stimulating factor
GM-CSF	Granulocyte monocyte colony stimulating factor
HHS	Hoyeraal–Hreidarsson syndrome
HSC	Hematopoietic stem cells
HCT	Hematopoietic cell transplantation
HbF	Fetal hemoglobin
HLA	Human leukocyte antigen
IPF	Idiopathic pulmonary fibrosis
IUGR	Intrauterine growth retardation
MDS	Myelodysplastic syndrome
NAF1	Nuclear assembly factor 1 ribonucleoprotein
NHP2	Non-histone protein 2
NOP10	Nucleolar protein 10
NF-κB	Nuclear factor kappa beta
NCI BMFS	National Cancer Institute bone marrow failure syndrome
NGS	Next generation sequencing
POT1	Protection of telomere 1
PARN	Polyadenylate specific ribonuclease
ROS	Reactive oxygen species
RNA	Ribonucleic acid
RAP1	Repressor/activator site binding protein 1
RTEL1	Regulator of telomere elongation helicase
RS	Revesz syndrome
STN1	Oligonucleotide/oligosaccharide binding fold containing 1
SnRNP	Small nucleolar ribonucleoprotein
TEN1	Telomeric pathways with STN1
TINF2	TRF1 interacting nuclear factor 2
TRF1	Telomere repeat factor 1
TPP1	Telomere protection protein 1
TCAB1	Telomere Cajal body protein 1
TERT	Telomerase reverse transcriptase
TERC	Telomerase RNA component
USB1	U6 SnRNA biogenesis phosphodiesterase 1
WRAP53	WD repeat containing protein antisense to TP53
WES	Whole exome sequencing
XLR	X-linked recessive

DNA due to incomplete replication of 3′ ends by conventional DNA polymerases, which is called the end-replication problem. When telomeres reach a critically short threshold, the cell can no longer divide properly and undergo apoptosis or senescence. In somatic cells, telomere loss is a normal consequence of aging.

For example, average lymphocyte telomere length declines from approximately 11 kb at birth to only approximately 4 kb in elderly population. Hence, telomere is regarded as "biological clock" within a cell that determines the number of possible cell divisions [7,8].

In contrast to limited normal replicative capacity of most somatic cells, rapidly proliferating cell types such as hematopoietic stem cells in the bone marrow, epithelial cells of the gastrointestinal tract, skin, testes, etc., can restore and maintain longer telomere length to enable a greater number of cell divisions. Telomere attrition/loss is compensated by telomerase through de novo addition of TTAGGG repeats onto chromosome ends in those cells where it is normally expressed, as mentioned above [9,10].

In genetically susceptible individuals, germline loss of one or more of the several components of telomerase protein complex (described below) leads to rapid loss of telomere and abnormal loss of DNA material with each cell division. Hence, telomeropathies produce premature senescence (progeria) (Figure 69.1) [3,11].

69.3.2 Telomere Homeostasis

The nucleoprotein complex responsible for telomere homeostasis and the genes encoding these factors in which mutations lead to telomere biology disorders (so-called "telomeropathies") are broadly classified into five categories, listed here (Figure 69.2) [12,13].

69.3.2.1 Telomerase Complex

Telomerase is a ribonucleoprotein enzyme complex that compensates for telomere attrition through de novo addition of TTAGGG repeats onto chromosome in those cells where it is normally expressed. The complex includes the telomerase enzyme (a reverse transcriptase encoded by the gene telomerase reverse transcriptase [TERT]), an RNA template (encoded by TERC), and four associated proteins that affect assembly, trafficking, recruitment of telomerase to telomeres, and stability of telomerase, such as dyskerin, NOP10, NHP2, and GAR [14].

69.3.2.2 The Shelterin Complex

Telomeres are coated by shelterin complex (telomere end protection/capping), an assembly of six associated proteins: (i) telomeric repeat binding factor 1 (TRF1; also known as TERF1) and (ii) TRF2 (also known as TERF2), (iii) TRF1-interacting nuclear factor 2 (TIN2), (iv) protection of telomeres (POT1), (v) TIN2-interacting protein 1 (TPP1) (also known as TINT1, PToP, and PIP1), and (vi) repressor/activator protein 1 (RAP1). RAP1 in mammals is involved in subtelomeric gene silencing and transcriptional regulation, and it also acts as an essential modulator of the nuclear factor-κB (NF-κB)-mediated pathway. Shelterin components play multiple roles in maintaining telomere length homeostasis by forming the T loops, preventing DNA damage response activation (DDRA), and recruiting the telomerase complex and modulating its activity. Although the regulation of shelterin components in stem cells and during nuclear reprogramming is still unexplored, some recent evidence suggests shelterin components as key factors in "stemness." Some cases of premature aging in

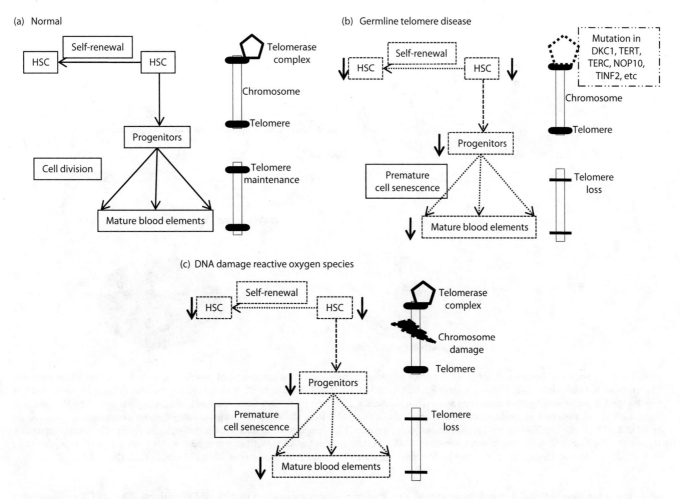

FIGURE 69.1 Mechanism of telomere attrition in health and disease. (a) Telomere attrition in physiologic state is the result of a balance between telomerase-mediated telomere elongation and telomere loss with each cell division. (b) Germline mutations involving several proteins of the telomerase complex and other telomere maintenance proteins lead to decreased self-renewal of the stem cell pool, enhanced telomere loss with each cell division, which finally leads to premature stem cell failure/premature aging (hence, all telomeropathies are considered as "progeric"). (c) Telomere loss can also be a result of reactive oxygen species (ROS) (generated as a part of inflammation, radiation, toxins, etc.)-mediated direct damage to the DNA. Regenerative/replicative stress, which is seen during bone marrow failure, stem cell regeneration following chemotherapy as well as following stem cell transplant, can also lead to telomere loss (not represented in the figure) (Modified from Townsley DM et al. *Blood.* 2014;124:2775–83. [3])

human syndromes have been linked to shelterin mutations, such as in TRF1, TRF2, and TIN2 (also known as TINF2).

69.3.2.3 Telomerase Trafficking

Once telomerase is assembled, it must be recruited to the telomere. Trafficking mediated by a protein TCAB1, which binds to the telomerase RNA template, is critical to this process.

69.3.2.4 Telomere Replication

The CTC1/STN1/TEN1 (CST) complex promotes extension of the C-rich strand of telomere DNA after telomerase has elongated the G-rich strand of the telomere DNA, creating extension of the duplex telomere DNA component.

69.3.2.5 Telomere Stability

RTEL1 (regulator of telomere length 1) is a DNA helicase that contributes both to integrity of duplex telomere DNA replication

as well as to the dismantling of structures called D-loops that could otherwise result in telomere DNA loss through excision repair mechanisms.

Compared to normal rate of telomere shortening in unaffected individuals of approximately 60 bp per year, individuals with telomere disorders lose telomeric DNA at approximately 120 bp per year. Furthermore, successive generations of affected individuals may be born with progressively shorter telomeres (a phenomenon known as disease anticipation) [15].

69.4 Genetics, Inheritance, and Pathophysiology

To date, there are nearly 14 genes that have been identified with DC (*ACD, DKC1, TERC, TERT, NOP10, NHP2, TINF2, USB1, TCAB1, CTC1, PARN, RTEL1, WRAP53,* and *C16orf57*). DC is genetically heterogeneous, with X-linked recessive, autosomal dominant, autosomal recessive, and rarely, sporadic subtypes. All mutations identified to date in DC patients affect telomerase components or telomere-stabilizing components

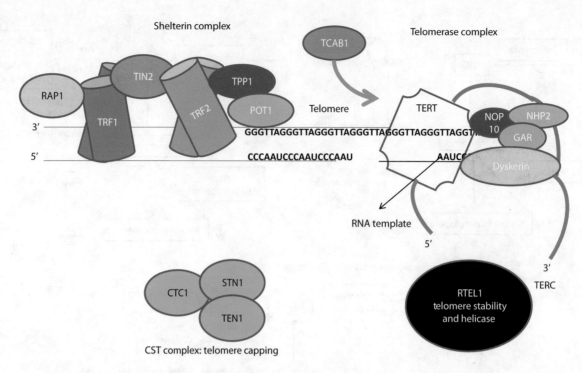

FIGURE 69.2 Shelterin complex and telomerase complex in telomere homeostasis and the associated proteins implicated in human telomeropathies. Shelterin complex comprises of six proteins; namely RAP1, TRF1, TRF2, TIN2, TPP1, and POT1, which play multiple roles in telomere homeostasis by forming the T loops, preventing DNA damage response (DDR) activation, recruiting the telomerase complex and modulating its activity. The 3′ strand overhangs the 5′ strand, folds back to invade the telomere double helix forming the T loop. Unfolding of T loop, necessary for telomere elongation, is carried out by RTEL1 (a DNA helicase). Telomerase, a reverse transcriptase enzyme (TERT) responsible for telomere elongation, is comprised of four integral proteins: Dyskerin1, NOP10, NHP2, GAR, and an RNA template (TERC). TCAB1 ensures trafficking of the telomerase complex to the telomeric ends. (Modified from Townsley DM et al. *Blood*. 2014;124:2775–83 [3]; Wegman-Ostrosky T, Savage A. *Br J Haematol*. 2017; 177(4):526–42. [14])

with the alteration of the normal function of the enzyme (Table 69.2) [16–22].

The first DC-associated gene, X-linked recessive DKC1, was discovered by linkage analysis in 1998. With the identification of mutations in DKC1, the first diagnostic tests became available. DKC1 (DKC gene 1 at Xq28) mutations are the most frequent,

appearing in up to 40% of DC patients. The DKC1 gene encodes the nucleolar protein dyskerin, which is ubiquitously expressed, and this protein is involved in ribosomal RNA processing telomere maintenance. Over 50 mutations have been found in *DKC1*. DKC1 mutation has been associated with DC phenotypes as well as and its variants such as Hoyeraal−Hreidarsson syndrome

TABLE 69.2

Genetic Abnormalities Commonly Associated with Dyskeratosis Congenita [2,12,14,16]

Gene	Protein	Function	Inheritance	% of Cases	Associated Diseases
DKC1	Dyskerin	Telomere maintenance and ribosomal biogenesis	XLR	∼ 40%	DC, HHS, IPF, aplastic anemia
TERT	TERT	Telomere maintenance	AD, AR	<5%	DC, HHS, IPF, aplastic anemia, liver disease
TERC	TERC	Telomere maintenance	AD	∼ 5%	DC, IPF, aplastic anemia, liver disease
TINF2	TIN2	Shelterin components	AD, S	<11%	DC, HHS
ACD	TPP1	Shelterin components	AD, AR		DC, HHS, aplastic anemia
NOP10	NOP10	Telomere maintenance and ribosomal biogenesis	AR	<1%	DC, IPF, aplastic anemia
NHP2	NHP2	Ribosomal biogenesis	AR	<1%	DC, IPF, aplastic anemia
TCAB1	TCAB1	Telomerase trafficking	AR	<1%	DC
RTEL1	Helicase	Telomeric DNA synthesis	AD, AR	–	DC, HHS, IPF
PARN	–	TERC RNA processing	AR	–	DC, IPF, HHS
C16orf57	C16orf57	–	AR	2%	–
NAF1	–	Telomere maintenance	AD	–	DC
STN		Telomere capping	AR	–	Coats plus syndrome
Unknown				40%	–

Note: For abbreviations, please refer to Table 69.1.

(HHS) and idiopathic pulmonary fibrosis (IPF). *DKC1* has been found to be a direct target of the *c-MYC* oncogene, strengthening the connection between DC and malignancy.

In the autosomal dominant (AD) form, mutations in the RNA component of telomerase (*TERC*) or *TERT* are responsible for diverse disease phenotypes which range from conventional DC to HHS, IPF, and liver disease. NOP10 is an autosomal recessive (AR) DKC1 gene which encodes small nucleolar ribonucleoprotein (snoRNP) associated with the telomerase complex. Patients with mutations in NOP10 have reduced telomere length and reduced TERC levels. Biallelic mutations in NHP2 have been identified in a group of AR DC patients, too. In persons with autosomal dominant DC and in terc-/- knockout mice, genetic anticipation (i.e., increasing severity and/or earlier disease presentation with each successive generation) has been reported.

Autosomal dominant mutations in TINF2, which encodes the shelterin protein TIN2, were suggested to cause the disease in 11% of patients, with high penetrance and an early age of onset. TIN2 maintains the composition and protein interaction of the different components of the shelterin complex. Patients with TIN2 mutations develop severe forms of the disease with very short telomere lengths.

TCAB1 is another AR DC gene. TCAB1 is a telomerase holoenzyme protein that facilitates trafficking of telomerase to Cajal bodies, the nuclear sites of nucleoprotein complex modification and assembly. Compound heterozygous mutations in TCAB1 cause alteration of the nuclear localization of telomerase, so it cannot elongate telomeres, thereby resulting in short telomeres. A heterozygous mutation was found on the conserved telomere maintenance component 1 gene (*CTC1*). This implication is also associated with a pleiotropic syndrome, Coats plus [18,22].

Homozygous autosomal recessive mutations in *RTEL1* lead to similar phenotypes that parallel with HHS. It is associated with short, heterogeneous telomeres. In the presence of functional DNA replication, *RTEL1* mutations produce a large amount of extra-chromosomal T-circles. Enzymes remove the T-circles and therefore shorten the telomere. *RTEL1* has a role in managing DNA damage by increasing sensitivity; therefore, mutations on this gene cause both telomeric and nontelomeric causes of DC [20].

Analysis of 270 families in the DC registry found that mutations in dyskerin (*DKC1*), *TERT*, and *TERC* only account for 64% of patients, with an additional 1% due to *NOP10*, suggesting that other genes associated with this syndrome are as yet unidentified. In addition to the mutations that directly affect telomere length, studies also indicate that a DC diagnosis should not be based solely on the length of the telomere, but also the fact that there are defects in telomere replication and protection. In addition, revertant mosaicism has been a new recurrent event in DC.

69.5 Clinical Features

Dyskeratosis congenita is a multisystemic disease with diverse system wise clinical presentations which is summarized in Table 69.3 [6,23,24–30].

69.5.1 Classic DKC

The classic and initial form is usually characterized by the mucocutaneous triad of abnormal skin pigmentation, nail dystrophy, and oromucosal leukoplakia which is described in 90%, 90%, and 80% of the cases, respectively (Figure 69.3) [3]. DC occurs mostly in males and clinically manifests between 5 and 12 years of age. Due to the heredity pattern, females may have less severe clinical features.

TABLE 69.3

System-Wise Presentations in Patients with Dyskeratosis Congenita[a]

System	Percentage	Findings
Skin	90%	Reticular hypo- and hyperpigmentation
Hairs	20%	Alopecia, premature graying of hairs, loss of eyebrows and eyelashes, hyperhidrosis
Nails	90%	Dystrophy, small thin nail plates with ridging
Oral cavity	80%	Leukoplakia (pathognomonic)
Teeth	15%	Dental caries, poor/abnormal denture, periodontitis
Hematological	80%	Cytopenia(s), bone marrow failure (aplastic anemia), myelodysplastic syndrome, acute myeloid leukemia
Ophthalmic	30%	Lacrimal duct stenosis (causing tearing), epiphora, blepharitis, retinopathy, cataract, corneal ulcers
Gastrointestinal	15%	Dysphagia (esophageal web/stenosis), esophagitis, esophageal intraepithelial lymphocytes, gastric parietal cell dropout, atrophic gastritis (proliferative senescence), lamina propria fibrosis, bloody diarrhea, failure to thrive, enterocolitis/pancolitis, villous blunting, atrophy, cell/gland drop-outs, increased apoptosis, duodenal neutrophil infiltrate, duodenal intraepithelial lymphocytes
Hepatic	0.7%	Telomerase deficiency impairs replicative potential of hepatocytes, and this reduces the ability of mature as well as progenitor cells to regenerate by the process of replicative senescence, so hepatocyte telomere shortening leads to progressive fibrosis and cirrhosis in chronic hepatic injury. However, it is not clear whether telomere deficiency only restricts the regenerative capacity of normal hepatocytes or has certain role in diseased cells also
Pulmonary	20%	Idiopathic pulmonary fibrosis
Endocrine/growth	10%	Short stature, hypogonadism/undescended testes, osteoporosis, avascular necrosis of hip
Genitourinary	10%	Meatal stenosis, hypospadias, penile leukoplakia, urethral stricture
Developmental	25%	Intrauterine growth retardation, microcephaly, cerebellar hypoplasia
Cancer risk		Head and neck squamous cell carcinoma, 40%; stomach/esophageal carcinoma, 17%; anorectum, 12%; skin, 12%; hematological: acute leukemia, myelodysplasia, 8%; liver, 5%

[a] Summary of findings from 118 patients in the London DC Registry [6,23]; and from other series [24–30].

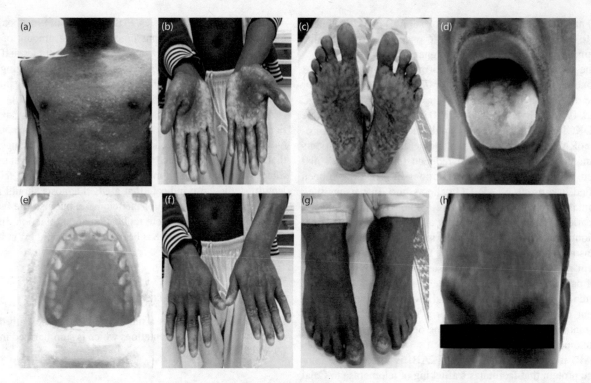

FIGURE 69.3 Phenotypic characteristics in dyskeratosis congenita. The characteristic reticulated hypo- and hyperpigmented macules all over the body, more marked on the trunk (a), palms (b), and soles (c); oromucosal lesions such as leukoplakia involving the dorsum of tongue (d), and hard palate (e), along with presence of dental caries (f), marked dystrophy of all the finger- and toenails (f and g) is quite evident in this case. There is frontal baldness, graying of scalp hairs (more on the temporal aspects), and loss of eyebrows and eyelashes (more so on the lateral aspect) (g). This patient also had coexistent severe aplastic anemia and gastric adenocarcinoma. (Reproduced with permission from Sahoo N et al. *Turk J Gastroenterol.* 2017; 28: 319–21. [24])

The primary cutaneous finding is abnormal skin pigmentation, with tan-to-gray hyperpigmented or hypopigmented macules and patches in a mottled or reticulated pattern. Reticulated pigmentation occurs in approximately 90% of patients. Poikilodermatous changes with atrophy and telangiectasia are common. The cutaneous presentation may clinically and histologically resemble graft-versus-host disease. The typical distribution involves the sun-exposed areas, including the upper trunk, neck, and face. Other cutaneous findings may include alopecia of the scalp, eyebrows, and eyelashes; premature graying of the hair; hyperhidrosis; hyperkeratosis of the palms and soles; and adermatoglyphia (loss of dermal ridges on fingers and toes).

Nail dystrophy is seen in approximately 90% of patients, with fingernail involvement often preceding toenail involvement. Progressive nail dystrophy begins with ridging and longitudinal splitting. Progressive atrophy, thinning, pterygium, and distortion eventuate in small, rudimentary, or absent nails.

Mucosal leukoplakia is a pathognomonic feature and occurs in approximately 80% of patients. It typically involves the buccal mucosa, tongue, and oropharynx. However, approximately 30% of the leukoplakic areas progress to malignant transformation with the development of a squamous cell carcinoma in 10–30 years, so they require frequent monitoring and biopsy of suspicious areas in order to detect a possible oral cancer.

In DC, severe periodontal destruction occurs due to anomalies in ectodermal-derived structures and a poor response in the patient caused by neutropenia. Patients may have gingival inflammation, bleeding, recession, and bone loss that simulates juvenile periodontitis. In addition, there may be defects in decreased root/

crown ratio and mild taurodontism. The evidence of multiple permanent teeth with decreased root/crown ratios may suggest a diagnosis of DC. These patients also may have more incidence of and more severe periodontal disease.

The BM findings in DC are variable and range from normal to different severity of aplasia depending on the stage of the disease. Aplastic anemia occurs in patients with a median age of presentation of 11 years. The anemia is macrocytic. Fetal hemoglobin levels are increased. Approximately 90% have peripheral cytopenia of one or more lineages. Initially, the patient usually presents thrombocytopenia, and during the evolution of the disease this becomes more global and they develop severe BMF. BMF (in up to 90% of patients) is one of the major causes of premature mortality in DC due to opportunistic infections because of the significant reduction in mature blood cells. The BM abnormalities can progress in different forms with the appearance of myelodysplasia in one or more lineages or leukemia. It has been reported that patients with DC have an observed expected ratio of 2663 to develop myelodysplasia (95% confidence interval 858–6215), with a mean age of onset of 35 years.

69.5.2 Variant Syndromes

69.5.2.1 Hoyeraal–Hreidarsson Syndrome

HHS is a clinically severe form of DC with disease manifestation beginning in early childhood. HHS was originally associated with X-linked mutations in DKC1, referred to as X-linked dyskeratosis congenita (DCX; also called as Zinsser-Engman-Cole syndrome); the syndrome was subsequently found to be associated with recessive mutations in ACD, RTEL1, TERT, and PARN, along with autosomal dominant mutations in TINF2 (Table 69.2).

In addition to the classic mucocutaneous and somatic features of DC, patients with HHS have the following clinical features [20]:

- IUGR
- Cerebellar hypoplasia (diagnostic feature)
- Microcephaly
- Developmental delay
- Severe immunodeficiency, much more severe than DC
- Early-onset progressive BMF

69.5.2.2 Revesz Syndrome

In addition to the associated features of DC, patients with Revesz syndrome (RS) are defined on the presence of bilateral exudative retinopathy, often in association with intracranial calcification and IUGR. Mutation in *TINF2* is the most common identified genetic cause of this syndrome.

69.5.2.3 Coats Plus Syndrome

Coats plus syndrome is an autosomal recessive disorder characterized by retinal telangiectasia with exudates, intracranial calcifications, cerebellar movement disorder, osteopenia, leukodystrophy, poor growth, and BMF. It is caused by autosomal recessive mutation in CTC1 or STN1 components of the CST telomere replication complex involved in duplex telomere DNA elongation [22].

69.5.3 Dyskeratosis Congenita and Cancer

In common with other inherited bone marrow failure syndromes, DC is associated with a 40%–50% increased susceptibility to cancer. Although many of the more severe cases of DC/HHS do not live long enough for cancer to develop, those that survive into their 30s and later are prone to develop malignant disease, most commonly myelodysplastic syndrome (MDS), acute myeloid leukemia (AML), and squamous cell carcinoma of the head and neck. Cancer is more common in DC caused by TERT and TERC mutation and least common in DC caused by TINF2, likely reflecting the fact that patients with TERT and TERC disease tend to live longer, and those with TINF2 much shorter. Disease caused by DKC1 is intermediate in age of presentation and cancer incidence. Data from the National Cancer Institute Bone Marrow Failure Syndrome (NCI BMFS) cohort suggests an overall increase in incidence of 11-fold [29].

Association between DC and malignancy is puzzling because increased telomerase activity is a signature of malignant cells, with most cancers exhibiting upregulated telomerase activities compared to their cells of origin. Short telomeres can lead to malignant transformation because when telomeres get critically short, they trigger a cell cycle arrest via a signaling pathway involving P53, most likely because the short telomeres are recognized as DNA damage by the cellular DNA repair machinery. Rarely, mutations in the signaling pathway or in genes mediating cell cycle arrest may occur in these cells, resulting in further replication and cell division and giving rise to chromosome instability as telomeres from different chromosomes are fused together by DNA repair proteins. Circular chromosomes formed by this process will break during mitosis and cells will enter a phase of chromosome fusion/breakage that will produce translocations, amplifications, and deletions, with the possibility of tumorigenic cells evolving. *DKC1* has been found to be a direct target of the c-*MYC* oncogene, strengthening the connection between DC and malignancy [30].

69.6 Diagnosis

The diagnosis of DC is most easily suspected in individuals with the classic mucocutaneous findings, bone marrow failure, and other somatic abnormalities, or in those from a family with known DC mutation. The combination of bone marrow failure and pulmonary fibrosis is highly suggestive of an underlying telomere disorder.

69.6.1 Clinical Criteria

Initially, registration in the DC registry required a family to have an index case with the classic triad of mucocutaneous findings. Subsequently, two different sets of diagnostic criteria have been adopted.

69.6.1.1 First Set of Criteria

Patients with any of the following combinations of features are defined as having DC [31]:

- All three classic mucocutaneous findings of abnormal skin pigmentation, nail dystrophy, and leukoplakia.
- One of three mucocutaneous features *plus* bone marrow failure and at least two other somatic features known to occur in DC (see Table 69.2).
- Four or more features of HHS (such as IUGR, developmental delay, severe immune deficiency, bone marrow failure, cerebellar hypoplasia).
- Aplastic anemia, MDS, or pulmonary fibrosis in the setting of a known pathogenic genetic variant affecting telomere function.
- Two or more features of DC *and* laboratory evidence of short telomeres (i.e., less than first percentile for age by multicolor flow cytometry with fluorescence in situ hybridization (flow-FISH) in several subsets of lymphocytes.

69.6.1.2 Second Set of Criteria

The second set of criteria differentiates findings suggestive of DC versus testing required to definitively establish the diagnosis [32].

- At least two of three classic mucocutaneous features
- One classic mucocutaneous feature plus two somatic features

- Progressive bone marrow failure, MDS, or AML
- Solid tumors associated with DC and that are otherwise atypical for age
- Pulmonary fibrosis
- Short lymphocyte telomeres (below the first percentile for age)

According to these criteria, a pathogenetic variant must be identified in a DC-associated gene to definitively establish the DC diagnosis. In addition to the above criteria, a family member of an index patient may be diagnosed with DC or at least with a telomere biology disorder if they share the same genetic variant with the proband and are identified to have short telomeres.

69.6.2 Laboratory Diagnosis

The diagnostic testing for DC consists of telomere length analysis and genetic testing for specific mutations (Figure 69.4).

69.6.2.1 Telomere Length Analysis

In all individuals with a de novo presentation and/or family history consistent with a telomere disorder, flow-FISH is performed on peripheral blood lymphocytes (not on granulocytes) using peptide nucleic acid probes for telomeric DNA

after ruling out Fanconi anemia by stress cytogenetics testing. Average telomere length below the first percentile for age is considered indicative of abnormally short telomeres and is consistent with DC or a related telomere biology disorder. Telomere length below first percentile for age in lymphocytes is 97% sensitive and 91% specific for DC in comparison with healthy relatives of people with DC or people with non-DC-inherited BMF syndromes [3,33]. In individuals with complex or atypical DC, it is recommended to analyze the six-cell panel more than the two-panel test of total lymphocytes and granulocytes. Even in cases with a known familial mutation, telomere length analysis is recommended to help predict the degree to which an individual family member may be affected by DC related complications.

69.6.2.2 Assessment of Bone Marrow Failure and Myelodysplasia

- Complete blood count, mean corpuscular volume, and absolute reticulocyte count. Macrocytosis and reticulocytopenia are characteristic features in DC.
- Elevated fetal hemoglobin (HbF); commonly seen in inherited BMF syndromes.
- Unilateral adequate trephine biopsy (>1 cm in length) for morphologic review and cellularity assessment.

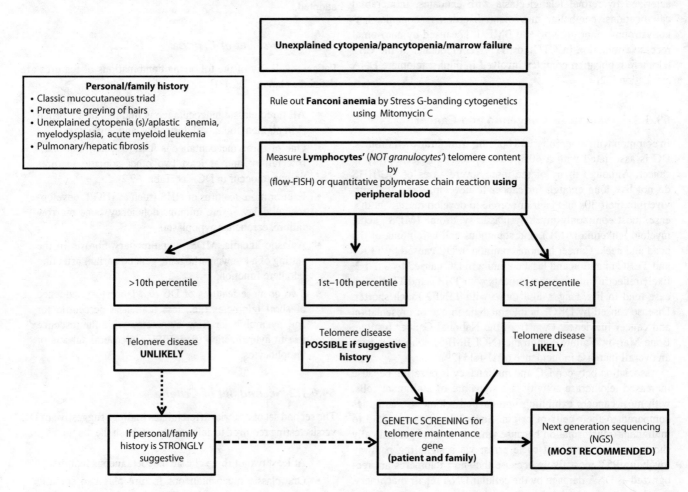

FIGURE 69.4   Algorithmic approach to laboratory diagnosis of germline telomere disease. (Adapted from Townsley DM et al. *Blood*. 2014;124:2775–83. [3])

- G-banding cytogenetics to assess acquired chromosomal aberrations.
- FISH analysis for specific aberrations like 5q-, monosomy 7, 7q-, trisomy 8, 20q-.

69.6.2.3 Molecular Sequencing

All patients who are suspected to have DC based on clinical criteria and telomere length analysis should have genetic testing performed to identify a causative mutation. This is critical for definitive establishment of a genetic diagnosis and ensuring the testing of first-degree relatives for potential carrier or disease status, and to determine family member eligibility to be a hematopoietic cell transplantation (HCT) donor.

Of all the techniques available nowadays, the next-generation sequencing (NGS) panel is the most recommended approach to identify gene mutations in patients with DC. Whole-exome sequencing (WES), which uses NGS methods to sequence the entire exome (coding regions of genes), is also readily commercially available. WES has the advantage of being able to identify for mutations in novel genes not previously associated with telomere disorders. However, not all mutations like TERC mutations can be identified by WES. It is noteworthy that although identification of pathogenic DC mutation is considered diagnostic in the appropriate clinical setting, negative results do not eliminate the possibility of DC, since a significant proportion of patients lack identifiable mutations.

69.7 Management

69.7.1 Treatment of Specific Complications

- Bone marrow examination at diagnosis; and subsequently on an annual basis.
- Supportive care
  - Judicious use of transfusions for severe anemia and thrombocytopenia, as extensive transfusions may be associated with worse outcomes from HCT due to transfusional iron overload and/or alloimmunization against red cells or HLA antigens.
  - Erythropoietin is usually not used due to the potential concerns it may increase the risk of MDS or AML in susceptible individuals.
  - Granulocytic colony stimulating factor (G-CSF) and/or granulocytic monocytic colony stimulating factor (GM-CSF) should be used with caution (due to the same potential concerns of development of MDS and AML) only when absolute neutrophil count below 200/cm or in those who are at risk of invasive bacterial or fungal infections.
  - Anabolic-androgenic steroids (e.g., oxymetholone, fluoxymesterone, danazole) may improve blood counts, though this may be ineffective with drug cessation. Danazole is useful in the setting of an anticipated delay in starting HCT or if a patient is ineligible to receive HCT.
- HCT: Allogenic HCT is the only curative option for bone marrow failure in patients with DC.

69.7.2 Therapies for Telomere Elongation

- Androgen therapy: A 2016 study has reported a beneficial effect of Danazol (400 mg twice daily administered for 2 years) in telomere length elongation. Telomere elongation was documented in 89% of patients at 12 months and 92% at 24 months. Hematological responses were seen in 79% of patients at 3 months and 83% at 24 months. Among eight patients for whom telomere measurements were performed after danazol was stopped, all reverted to telomere shortening, suggesting an indefinite period of therapy with this drug [15].

  In another 2014 study involving 16 younger patients with classic DC (median age; 11 years), the impact of androgen therapy with oxymetholone was less pronounced. While 69% showed a hematological response, none had changes in the expected, age related telomere decline [34,35].
- Therapies treating Wnt signaling (under experimentation) [36].
- Gene therapy (under experimentation) [37].

69.8 Prognosis

DC is a multisystem disorder that carries a poor prognosis (mean survival of 30 years), with most deaths related to infections, bleeding, and malignancy. In the DC registry, approximately 70% of affected individuals died of bone marrow failure or its complications, and these deaths occurred at a median age of 16 years. Eleven percent died from sudden pulmonary complications; a further 11% died of pulmonary disease in the bone marrow transplantation (BMT) setting. Seven percent died from malignancy (e.g., Hodgkin lymphoma, pancreatic carcinoma). Fatal opportunistic infections such as *Pneumocystis carinii* pneumonia and cytomegalovirus infection have been reported. Therapeutic interventions are mostly palliative, but HCT for aplastic anemia have been tried with variable success. The prognosis is worse for the X-linked and autosomal forms compared with the autosomal dominant form.

69.9 Conclusion and Future Perspectives

- In dyskeratosis congenita and related telomere biology disorders, mutations in genes that maintain telomere homeostasis in rapidly dividing cells lead to premature cell death, progeria, genomic instability, premature bone marrow failure, and increased risk of solid and hematolymphoid malignancies.
- The spectrum of phenotypic manifestations in DC can be extremely heterogenous. While bone marrow failure is the leading cause of mortality in childhood, pulmonary and hepatic fibrosis contribute to morbidity and mortality in adult life. HHS represents the most severe form of DC which manifests in early infancy even before the appearance of mucocutaneous changes, and this carries a very high mortality.

- The diagnosis of DC is based upon a set of clinical criteria supplemented with a laboratory evaluation of telomere length analysis by flow-FISH technique performed on peripheral blood lymphocytes. Telomere length analysis should be corroborated by appropriate genetic testing of both the patients and their family members, preferably by next-gene-sequencing technique.

- All patients with DC should have a comprehensive assessment, performed in a dedicated center with expertise in management of patients with telomere biology disorders. If a pathological mutation has been identified, first-degree relatives (especially siblings) should be tested to identify the occult disease and to determine their eligibility as a donor for hematopoietic cell transplantation.

- A variety of approaches to increase telomerase activity is under investigation, including androgen therapy, targeting Wnt signaling pathway, and possible gene therapy.

- The overall survival of individuals with DC has improved but continues to lag behind other bone marrow failure syndromes.

REFERENCES

1. Barbaro PM, Ziegler DS, Reddel RR. The wide-ranging clinical implications of the short telomere syndromes. *Intern Med J.* 2016;46:393.
2. Fernández García MS, Teruya-Feldstein J. The diagnosis and treatment of dyskeratosis congenita: A review. *J Blood Med.* 2014;5:157–7.
3. Townsley DM, Dumitriu B, Young NS. Bone marrow failure and telomeropathies. *Blood.* 2014;124:2775–83.
4. Martínez P, Blasco MA. Telomere-driven diseases and telomere-targeting therapies. *J Cell Biol.* 2017;216:875–87.
5. Knight S, Vulliamy T, Copplestone A, Gluckman E, Mason P, Dokal I. Dyskeratosis Congenita (DC) Registry: Identification of new features of DC. *Br J Haematol.* 1998;103:990–6.
6. Dokal I. Dyskeratosis congenita. *Hematology Am Soc Hematol Educ Program.* 2011;2011:480–6.
7. Liu L, Bailey SM, Okuka M et al. Telomere lengthening early in development. *Nat Cell Biol.* 2007;9:1436–41.
8. López-Otín C, Blasco MA, Partridge L, Serrano M, Kroemer G. The hallmarks of aging. *Cell.* 2013;153:1194–217.
9. Bertuch AA. The molecular genetics of the telomere biology disorders. *RNA Biol.* 2016;13:696.
10. Lansdrop PM. Telomere, stem cells, and hematology. *Blood.* 2008;111:1759.
11. Aubert G, Baerlocher GM, Vulto I, Poon SS, Lansdorp PM. Collapse of telomere homeostasis in hematopoietic cells caused by heterozygous mutations in telomerase genes. *PLoS Genet.* 2012;8:e1002696.
12. Mason PJ, Bessler M. The genetics of dyskeratosis congenita. *Cancer Genet* 2011;204:635–45.
13. Calado RT, Young NS. Telomere diseases. *N Engl J Med.* 2009;361:2353–65.
14. Wegman-Ostrosky T, Savage A. The genomics of inherited bone marrow failure syndrome: From mechanism to clinic. *Br J Haematol.* 2017;177(4):526–42.
15. Townsley DM, Dumitriu B, Liu D et al. Danazol treatment for telomere diseases. *N Engl J Med.* 2016;374:1922–31.
16. Kelmenson DA, Hanley M. Dyskeratosis Congenita. *N Engl J Med.* 2017;376:1460.
17. Bär C, Blasco MA. Telomeres and telomerase as therapeutic targets to prevent and treat age-related diseases. *F1000Res.* 2016;5. pii: F1000Faculty Rev-89.
18. Zhong F, Savage SA, Shkreli M et al. Disruption of telomerase trafficking by TCAB1 mutation causes dyskeratosis congenita. *Genes Dev.* 2011;25:11.
19. Jones M, Bisht K, Savage SA, Nandakumar J, Keegan CE, Maillard I. The shelterin complex and hematopoiesis. *J Clin Invest.* 2016;126:1621.
20. Glousker G, Touzot F, Revy P, Tzfati Y, Savage SA. Unraveling the pathogenesis of Hoyeraal-Hreidarsson syndrome, a complex telomere biology disorder. *Br J Haematol.* 2015;170:457.
21. Guo Y, Kartawinata M, Li J et al. Inherited bone marrow failure associated with germline mutation of ACD, the gene encoding telomere protein TPP1. *Blood.* 2014;124:2767.
22. Crow YJ, McMenamin J, Haenggeli CA et al. Coats' plus: A progressive familial syndrome of bilateral Coats' disease, characteristic cerebral calcification, leukoencephalopathy, slow pre- and post-natal linear growth and defects of bone marrow and integument. *Neuropediatrics.* 2004;35:10.
23. Kirwan M, Dokal I. Dyskeratosis congenita: A genetic disorder of many faces. *Clin Genet.* 2008;73:103.
24. Sahoo N, Padhi S, Patra S, Mishra P, Kumar R, Panigrahi M. Dyskeratosis congenita, bone marrow failure, and co-existent gastric adenocarcinoma: An insight into telomere biology. *Turk J Gastroenterol.* 2017;28:319–21.
25. Shimamura A, Alter BP. Pathophysiology and management of inherited bone marrow failure syndromes. *Blood Rev.* 2010;24:101.
26. Jonassaint NL, Guo N, Califano JA, Montgomery EA, Armanios M. The gastrointestinal manifestations of telomere-mediated disease. *Aging Cell.* 2013;12:319–23.
27. Rudolph KI, Chang S, Millard M et al. Inhibition of experimental liver cirrhosis in mice by telomerase gene delivery. *Science.* 2000;287:1253–8.
28. Wiemann SU, Satyanarayana A, Tsahuridu M et al. Hepatocyte telomere shortening and senescence are general markers of human liver cirrhosis. *FASEB J.* 2002;16:935–42.
29. Alter BP, Giri N, Savage SA, Rosenberg PS. Cancer in dyskeratosis congenita. *Blood.* 2009;113:6549.
30. Alter BP, Giri N, Savage SA et al. Malignancies and survival patterns in the National Cancer Institute inherited bone marrow failure syndromes cohort study. *Br J Haematol.* 2010;150:179.
31. Dokal I, Vulliamy T, Mason P, Bessler M. Clinical utility gene card for: Dyskeratosis congenita-update 2015. *Eur J Human Genet* 2015:23(4).
32. Savage SA. Dyskeratosis congenita. In: Pagon RA, Adam MP, Ardinger H et al. editors. *GeneReviews® [Internet].* Seattle (WA): University of Washington, Seattle; 1993–2019. Last revision: November 21, 2019.

33. Baerlocher GM, Lansdorp PM. Telomere length measurements in leukocyte subsets by automated multicolor flow-FISH. *Cytometry A.* 2003;55:1.

34. Dietz AC, Mehta PA, Vlachos A et al. Current knowledge and priorities for future research in late effects after hematopoietic cell transplantation for inherited bone marrow failure syndromes: Consensus statement from the second pediatric blood and Marrow Transplant Consortium International Conference on Late Effects after pediatric hematopoietic cell transplantation. *Biol Blood Marrow Transplant.* 2017;23:726.

35. Khincha PP, Wentzensen IM, Giri N, Alter BP, Savage SA. Response to androgen therapy in patients with dyskeratosis congenita. *Br J Haematol.* 2014;165:349–57.

36. Yang TB, Chen Q, Deng JT et al. Mutual reinforcement between telomere capping and canonical Wnt signalling in the intestinal stem cell niche. *Nat Commun.* 2017;8:14766.

37. Jäger K, Walter M. Therapeutic targeting of telomerase. *Genes (Basel).* 2016;7(7). pii: E39.

70

Familial Acute Myeloid Leukemia

Rina Kansal

CONTENTS

70.1 Introduction

Acute myeloid leukemia (AML) is a hematologic malignancy that results from a clonal proliferation of myeloid blasts, with extreme underlying genetic heterogeneity in the disease, involving sporadic or de novo genesis, and secondary to preexisting myelodysplastic syndromes. The latter are pre-leukemic disorders characterized by ineffective hematopoiesis, cytopenias, often with detected clonal cytogenetic abnormalities, and a propensity to progress to secondary AML [1]. A clinical history of AML affecting more than one family member is uncommon but may occur in both de novo and secondary AML, with the latter

including preexisting myelodysplastic syndromes in familial cases. While families with multiple individuals affected by acute leukemia were described in the 1960s and 1970s, a genetic basis for familial AML was not identified until 1990, when germline aberrations in the *TP53* gene were linked to Li–Fraumeni syndrome, wherein multiple hematopoietic and non-hematopoietic malignancies may occur, including frequently AML [2,3].

"Familial acute myeloid leukemia" in the absence of other non-hematopoietic malignancies was first recognized in rare families with thrombocytopenic individuals who had pre-leukemic myelodysplasia or developed overt AML. Such rare families with platelet disorders and propensity to develop leukemia were described at least four decades ago [4,5], with subsequent elucidation of haploinsufficiency of Runt-related transcription factor 1 gene (*RUNX1*) as the genetic cause of the hereditary germline defect [6]. In 2004, another genetic form of familial AML due to germline mutations in *CEBPA* was recognized in patients with no previous hematologic abnormality and clinical presentation similar to sporadic *CEBPA*-mutated AML [7].

The above non-syndromic forms of familial AML differ in clinical presentation from well-known heritable and genetic disorders that are typically diagnosed in pediatric patients but may also be diagnosed in adults, and all of these disorders have the common feature of increased lifetime risk for developing AML. Due to space constraints, for these disorders, the reader is referred to related chapters in this book as well as excellent reviews elsewhere [8–13], from which Table 70.1 summarizes a few clinical and genetic features. These disorders include classical inherited bone marrow failure syndromes, which are a heterogeneous group of disorders often with associated clinical features including one or more somatic abnormalities, and caused by genetic aberrations in different pathways, including DNA repair (Fanconi anemia), ribosome biogenesis (Schwachman–Diamond syndrome, Diamond–Blackfan syndrome), and telomere maintenance (dyskeratosis congenita). Two additional inherited bone marrow failure disorders, severe congenital neutropenia and congenital amegakaryocytic thrombocytopenia, present in infancy, usually without somatic abnormalities, and both may progress to myelodysplasia and AML. In addition, Down syndrome (trisomy 21) increases the risk for developing AML and is included in the World Health Organization classification of AML [14,15]. Further, constitutional mismatch repair deficiency, a rare autosomal recessive disease characterized by homozygous germline mutations in DNA mismatch repair genes, *MLH1*, *MSH2*, *MSH6*, *PMS2*, and *EPCAM* and clinical features reminiscent of neurofibromatosis type 1, may very rarely present in pediatric patients with AML that comprises de novo and therapy-related forms [16,17], although the typical presentation for this disease is with non-Hodgkin lymphoid (and not myeloid) high grade hematologic malignancies, and brain and colonic tumors [18,19].

In addition to the two non-syndromic familial forms of AML caused by aberrations in transcription factors *RUNX1* and *CEBPA*, germline mutations in the *GATA2* gene located in chromosome 3q21.3 band, and which encodes for a zinc-finger hematopoietic transcription factor, were reported in 2011 to predispose to myelodysplasia and AML [20]. Germline *GATA2* mutations are heritable and can be familial [20–22], with the myeloid neoplasm presentation primarily as myelodysplastic syndromes in pediatric age groups [22], or myelodysplasia-related AML in adults [23,24]. Diagnosis of *GATA2*-related myeloid neoplasms may be facilitated by the presence of associated clinical features, including immunodeficiencies, neutropenia, monocytopenia, prolonged or severe systemic infections, and features such as lymphedema and congenital deafness in previously described syndromes [21,25].

The revised 4th edition of the World Health Organization classification of hematologic malignancies includes "myeloid neoplasms with germline predisposition" as a separate category to encourage identification and diagnosis of these myeloid malignancies [14,26], classified as follows into three main groups: (i) Myeloid neoplasms with germline predisposition without a preexisting disorder or organ dysfunction, which include AML with germline *CEBPA* and myelodysplasia/AML with *DDX41* mutations (ii) Myeloid neoplasms with germline predisposition and preexisting platelet disorders, which include neoplasms with germline *RUNX1* or *ANKRD26* or *ETV6* mutations (iii) Myeloid neoplasms with germline predisposition and other organ dysfunction, which include the aforementioned inherited bone marrow failure syndromes (i.e., Fanconi anemia, Schwachman–Diamond syndrome, Diamond–Blackfan syndrome, dyskeratosis congenital, severe congenital neutropenia, and congenital amegakaryocytic thrombocytopenia), myeloid neoplasms with germline *GATA2* mutation, myeloid neoplasms associated with Down syndrome and those associated with neurofibromatosis, Noonan syndrome or Noonan syndrome–like disorders [14,26]. Further, in November 2017, a novel cancer predisposition syndrome was described due to germline loss of the *MBD4* gene that has a key function in recognizing T:G mispairing and initiating base excision repair, which contributed to early onset (age <35 years) AML in 3 patients, including 2 siblings, as a result of a biallelic inactivation of *MBD4* [27].

With our current state of knowledge, *non-syndromic familial acute myeloid leukemia* may be defined as multiple genetically and clinically heterogeneous disorders, each caused by genetic aberrations in a single gene, including *CEBPA* and *DDX41* (both with no preexisting disorders), or *RUNX1*, *ANKRD26*, and *ETV6* (all with preexisting platelet disorders), all of which co-segregate in family members leading to an increased and heritable predisposition to develop AML. These non-syndromic forms of familial AML, including those with previous platelet abnormalities, and including de novo and secondary AML, form the subject for this chapter.

70.2 Biology

70.2.1 Familial Platelet Disorders with Propensity to Acute Myeloid Leukemia

The molecular genetic aberration underlying *familial platelet disorders with propensity to myeloid malignancies* [online Mendelian inheritance in man (OMIM) #601399] *including acute myeloid leukemia (FPD/AML)* was identified in 1999 as germline mutations in the *RUNX1* gene [6], located in chromosome 21q22.12 band at the breakpoint in the t(8;21) translocation, the source from where *RUNX1* was originally cloned [28]. *RUNX1*, previously known as *AML1*, is critical for embryogenesis and hematopoiesis [29]. For the latter functions, *RUNX1* encodes for a protein that complexes with core binding factor β to form a heterodimeric core binding factor complex that regulates the expression of several genes critical for hematopoiesis [30]. In FPD/AML, mutations located in the highly conserved runt homology domain of the *RUNX1* gene cause a loss of RUNX1 function and are associated with a high incidence of hematologic myeloid malignancies including myelodysplasia and acute myeloid leukemia [31], and with a lower incidence of lymphoid malignancies including acute lymphoblastic leukemia and hairy cell leukemia [32,33]. At least 40 different mutations have been identified thus far in individuals with FPD/AML, with the majority being missense, and including nonsense, insertions, deletions, or splice site mutations [34].

TABLE 70.1

Clinical and Genetic Features of Major Inherited Bone Marrow Failure Syndromes with Significant Increased Risk of Developing Acute Myeloid Leukemia [8–13]

Characteristics of Syndromic Disorder	Fanconi Anemia	Diamond–Blackfan Anemia	Schwachman–Diamond Syndrome	Dyskeratosis Congenita	Severe Congenital Neutropenia	Amegakaryocytic Thrombocytopenia
Clinical features [8–13]						
Male:female [8]	1.2:1	1.1:1	1.5:1	4:1	1:2	0.8:1
Median (range) age, diagnosis [8]	6.6 (0–49) years	0.25 (0–64) years	1 (0–41) years	15 (0–75) years	3 (0–70) years	0.1 (0–11) years
% diagnosed ≥16 yrs age [8]	9%	1%	5%	46%	13%	0%
Main feature, hematologic	Pancytopenia	Anemia	Neutropenia	Pancytopenia	Neutropenia	Thrombocytopenia
Major non-hematologic features and physical anomalies [8,13]	Abnormal thumbs, radii, skin hyperpigmentation, short stature, deafness, bony deformities, congenital dislocation of hips, microcephaly, microphthalmia, gastrointestinal, renal and pituitary anomalies, cardiopulmonary rare, some developmental delay; **~25% normal**	Abnormal thumbs, flat thenars, short stature, webbed neck, fused cervical vertebrae, asymmetric high scapula, hypertelorism, epicanthal folds, cardiac defects, cleft lip, palate, skeletal abnormalities, rare developmental delay; **~70% normal**	Exocrine pancreatic insufficiency, neurodevelopmental and skeletal abnormalities	Pigmentation, dysplastic nails, oral leukoplakia, microcephaly, pulmonary fibrosis, esophageal stenosis, liver disease, sparse and early gray hair, osteoporosis; **~10% normal**	Severe infections; no physical anomalies; **all normal**	No physical anomalies; **all normal**
Solid tumors and lymphoid malignancies	Squamous cell cancer head and neck, anogenital; other solid malignancies in *FANCD2*	Osteosarcoma, colon cancer, female genital cancer; acute lymphoblastic leukemia	Acute lymphoblastic leukemia; no solid tumors	Squamous cell cancer head and neck	No solid tumors	No solid tumors
Development of AML or MDS	Yes; AML may present in undiagnosed Fanconi anemia	Yes	Yes	Yes	Yes	Yes
Incidence or risk of MDS or AML	Incidence MDS: 40% at age 50; AML: 15%–20% at age 40 [13]	Observed: expected ratio of 352 for MDS, 28 for AML [13]	Risk MDS or AML 19% at 20 years; 36% at 30 years [12,13]	MDS or AML in 7 of 50 patients [12]; observed:expected ratio for AML: 195 [9]	11% at 20 years; 22% after 15 yrs of G-CSF [13]	Risk 53% by age 17 [8]
Genetic features [8,11–13]						
Major mode of inheritance	Autosomal recessive; X-linked rare (*FANCB*)	Autosomal dominant; X-linked rare	Autosomal recessive	X-linked recessive	Autosomal and X-linked	Autosomal recessive
Aberrant genes,[a] chromosomal locus, % of patients [8,11–13]	*FANCA*, 16q24.3, ~70% [8] *FANCC*. 9q22.3, ~10% *FANCE*. 6p21.3, ~10% *FANCG*, 9p13, ~10% Rare: *FANCB*, Xp22.31; *FANCD1 (BRCA2)*,13q12.3; *FANCD2*, 3p25.3; *FANCF*, 11p15; *FANCI*, 15q25; *FANCJ (BRIP1)*, 17q22.3; *FANCL (PHF9)*, 2p16.1; *FANCM*, 14q21.3; *FANCN (PALB2)*, 16p12.1; *FANCO (RAD51C)*,19q22; *FANCP (SLX4)*,16p13.3; *FANCQ (ERCC4)*,16p13.12; *FANCR (RAD51)*,15q15.1; *FANCT (UBE2T)*, 1q32.1; *FANCU (XRCC2)*, 7q36.1; *FANCV (REV7)*, 1p36.22	Autosomal dominant and de novo: *RPS19*, 19q13.2%, 25% *RPS24*, 10q22.3%, 2%; *RPS17*, 15q25.2%, 1%; *RPL35A*, 3q29, 3.5%; *RPL5*, 1p22.1, 6.6%; *RPL11*, 1p36.11, 4.8%; *RPS7*, 2p25.3%, 1%; *RPS26*, 12q13.2, 2.6%; *RPS10*, 6p21.31, 6.4%; *RPL26*, 17p13.1%, 1%; *RPL27*, 17q21.31; *RPS27*, 1q21.3; *RPS28*, 19p13.2 X-linked recessive: *GATA1*, Xp11.23; *TSR2*, Xp11.22 Genetic causes in ~40% patients unknown	*SBDS*, 7q11.21, >90%	*DKC1*, Xq28%, 36%; [8,12] Autosomal dominant: *TERC*, 3q26.2%, 10%; *TERT*, 5p15.33%, 1%; *TINF2*, 14q12%, 15%; includes de novo; *RTEL1*, 20q13.33 Autosomal recessive: *NOP10 (NOLA3)*, 15q14, <1%; *NHP2 (NOLA2)*, 5q35.2, <1%; *WRAP53 (TCAB1)*, 17p13.1, rare; *TERT*; *RTEL1*; *CTC1*, 17p13.1; *TERT* all autosomal and de novo 5%–10%; genetic causes in ~30% patients unknown	Autosomal dominant: *ELANE*, 19p13.3%, 75%; *GFI1*, 1p22.1, <1% Autosomal recessive: *HAX1*, 1q21.3%, 1%; *G6PC3*, 17q21.31, <1%; *CSF3R*, 1p34.3, <1% X-linked recessive: *WAS*, Xp11.23, <1%	*MPL*, 1p34.2
Pathogenetic pathway	FA/BRCA DNA repair	Ribosome biogenesis	Ribosome biogenesis	Telomere maintenance	Myeloid maturation arrest	Megakaryocytes absent

[a] Additional information obtained from respective # OMIM webpages at www.omim.org, last accessed February 21, 2018.

70.2.2 Familial *CEBPA* Mutated Acute Myeloid Leukemia

The *CCAAT/enhancer binding protein (C/EBP), alpha* gene *(CEBPA)* is a single exon gene located on chromosome 19q13.1 band. The gene belongs to a family of basic leucine zipper proteins, and encodes for a transcription factor that is crucial for maturation of hematopoietic myeloid cells; no mature granulocytes are observed in *CEBPA*-mutant mice and the t(8;21) translocation downregulates *CEBPA* to lead to AML [35–37]. The *CEBPA* gene has 2 transactivation domains at the N-terminus, and a basic region and leucine zipper region at the C-terminus. Mutations in *CEBPA* may be present in sporadic AML in 5%–14% cases. Most commonly, mutations in the N-terminus cause a frameshift, leading to premature termination and absence of the normal, full-length 42 kDa protein, but still allowing downstream expression for formation of the smaller 30 kDa protein, a dominant-negative isoform. Mutations in the C-terminus disrupt the leucine zipper region, which prevents dimerization and loss of DNA activity [36].

70.2.3 Inherited Thrombocytopenias with Germline Mutations Including in *ANKRD26*

Among *inherited thrombocytopenias*, which are reviewed elsewhere [38], germline mutations in three genes, *RUNX1*, *ANKRD26* and *ETV6*, are known thus far to predispose to hematologic malignancies, including AML. All three disorders are inherited in an autosomal dominant mode. The *ANKRD26* (ankyrin repeat domain 26) gene (OMIM #610855), located in chromosome 10p12.1 band and composed of 34 exons, encodes for a 1709 amino acid protein containing four N-terminal ankyrin repeats, which function in protein-protein interactions, and a C-terminal coiled-coil region [39]. Mutations in *ANKRD26* cause familial thrombocytopenia 2 [THC2, (OMIM #88000)].

70.2.4 Inherited Thrombocytopenias with Mutations in *ETV6*

The *ETS translocation variant gene 6 (ETV6)* gene, located in chromosome 12p13.2 band and composed of 8 exons spanning 240 kb, encodes for a nuclear protein that functions as an ETS family transcription factor and plays a key role in hematopoiesis and development before birth [40,41]. The protein product contains an N-terminal pointed domain that is involved in protein-protein interactions with itself and other proteins, a C-terminal DNA-binding domain that is highly conserved among ETS family transcription factors, and an intervening linker region that indirectly affects the DNA binding domain. The *ETV6* gene (OMIM #600618) is well-known for its involvement in several somatic chromosomal translocations in myeloid and lymphoid leukemias, in synergy with several partner genes for chromosomal fusion [41], with the *ETV6-RUNX1* fusion being the most common translocation in pediatric B-lymphoblastic leukemia. In 2015, germline mutations in *ETV6* in families with inherited thrombocytopenia 5 [THC5 (OMIM # 616216)] were found to be associated with diverse hematologic malignancies, including acute lymphoblastic leukemia, and occasionally, AML and myelodysplasia, multiple myeloma and non-hematologic cancer [42,43].

70.2.5 *DEAD*-Box Helicase 41 (*DDX1*) Gene

The DEAD-box proteins are named after the strictly conserved sequence Asp-Glu-Ala-Asp. The *DEAD-box helicase 41 (DDX41)* gene, located on chromosome 5q35.3 band, encodes for a DEAD-box protein family member, which functions as a RNA helicase [44]. Little is known about the molecular mechanisms and functions of DDX41 in hematopoiesis. Wildtype DDX41 interacts with RNA spliceosomal proteins in a normal process that is disrupted by mutations in *DDX41*, which were first identified in 2015 to co-segregate in patients with familial AML and myelodysplasia [45,46].

70.3 Pathogenesis

70.3.1 Familial Platelet Disorder with Predisposition to Acute Myeloid Leukemia

Familial platelet disorder with predisposition to acute myeloid leukemia (FPD/AML) is inherited in an autosomal dominant mode. The penetrance is incomplete since the inherited heterozygous germline *RUNX1* mutations are not sufficient to cause myeloid malignancy. After a long latency period with 33 years as the average age at diagnosis, acquisition of additional somatic mutations or clonal chromosomal abnormalities leads to the development of AML in 20%–60% of individuals with the inherited germline *RUNX1* mutations [47–52].

Regarding mutations in *RUNX1*, firstly, those that are dominant negative are more likely to cause AML than those that lead to haploinsufficiency, since the mutant forms may also inhibit wildtype RUNXI [48,53]. Secondly, additional acquired somatic mutations in a subset of FPD patients lead to leukemia, including biallelic *RUNX1* mutations, which are often present in familial FPD at the time of diagnosis of AML [50,54,55], and appear to be the most frequent genetic event for leukemic progression in FPD/AML patients. Further, in de novo AML, somatic *RUNX1* mutations, which are Class II mutations [56], co-exist with mutations in *FLT3* (Class I mutations) in a majority of cases [57], leading to leukemia according to the two-hit model of leukemogenesis [56]. Similar to the de novo cases, in familial FPD/AML with germline *RUNX1* mutations, acquisition of additional somatic mutations such as observed in *FLT3*, *MLL*-PTD [57,58], and other genes including *ASXL1*, *IDH1*, *IDH2*, *TET2* [57–59], likely lead to the development of AML. Notably, in de novo AML cases, *RUNX1* mutations only rarely co-occur with mutations in *NPM1* or *CEBPA* [60], and *NPM1* mutations have not been reported to date in FPD/AML patients.

Interestingly, microdeletions in chromosome 21 with allelic loss of *RUNX1* may be associated with FPD/AML [6], and may also be observed in patients with constitutional syndromic defects and inherited thrombocytopenia harboring de novo allelic *RUNX1* loss in the absence of mutations in the residual wild-type allele [61], with progression to myelodysplasia or AML [62,63]. Further, in FPD patients with allelic loss of *RUNX1*, the genetic events causing transformation from germline *RUNX1* haploinsufficiency to myelodysplasia or AML are suggested to be different than those in FPD cases with *RUNX1* point mutations. In the former group with allelic loss, the transforming events include acquired trisomy 21 with haploinsufficiency of RUNX1 and additional acquired mutations [64]. In the latter group with point mutations in *RUNXI*, acquired mutations in *CDC25C*, which allow cell cycle progression despite DNA damage, followed by mutations in *GATA2* have been described as pre-leukemic events leading to AML in Japanese FPD/AML patients [65], but *CDC25C* mutations have not been confirmed in cohorts from France [50,54], Ireland [59] and the USA [55,66], suggesting possible differences in pathogenesis related to ethnic origin.

Notably, individuals with inherited germline *RUNX1* mutations often have a higher incidence of somatically acquired clonal abnormalities that may occur at an earlier age [66], as compared with age-matched normal individuals with age-related clonal hematopoiesis involving *DNMT3A*, *TET2*, and *ASXL1* genes [67–69]. Nonetheless, additional studies are necessary to establish whether detecting those clonal abnormalities at an earlier age in FPD carriers is associated with an increased risk for developing AML.

70.3.2 Familial *CEBPA*-Mutated Acute Myeloid Leukemia

Similar to mutations in *RUNXI* that may be acquired somatically or inherited as germline in AML, mutations in *CEBPA* may also be present in both sporadic and familial *CEBPA*-mutant AML, with the latter inherited in an autosomal dominant manner. In contrast with sporadic forms of AML that occur primarily in older adults, familial *CEBPA*-mutated AML occurs at younger ages, most often under 40 years, with the shortest reported latency period of 1.75 years [70].

Table 70.2 summarizes the clinical features, *CEBPA* mutations and cytogenetics at onset of AML in 42 patients, and numbers of tested carrier individuals, from 17 families with AML due to a germline *CEBPA* mutation reported from 2004 to 2017, including from European and Asian countries [70–82]. Pure familial *CEBPA*-mutated AML harbors heritable mono-allelic germline mutations that are present in the N-terminal part of the *CEBPA* gene in the families (12/17, 70%) reported thus far. In individuals with germline N-terminal mutations in *CEBPA*, acquisition of a somatic *CEBPA* mutation, often in the C-terminus, leads to the development of a biallelic *CEBPA* mutant AML [71,74]. As shown in Table 70.2, in 100% (n = 20) of patients from 12 families with N-terminal *CEBPA* germline mutations, the 2nd somatic mutation was in the C-terminal domain of *CEBPA*, including with dual C-terminal somatic mutations in 2/20 patients [70,71,74]. Intriguingly, an additional N-terminal somatic *CEBPA* mutation was reported in only one unique example of a donor-derived leukemia, wherein the donor (proband's healthy sister) also harbored the same N-terminal germline *CEBPA* mutation [79]. Familial *CEBPA*-mutated cases with N-terminal germline mutations have a near-complete penetrance for developing leukemia, with germline N-terminal mutations identified in 4 of 14 asymptomatic members tested from 6 families [7,71,74,76,77,79].

Germline C-terminal *CEBPA* mutations were first identified in 3 AML cases with no family history of leukemia [77]. In 2016, a germline C-terminal *CEBPA* mutation was identified to be the cause for developing AML in 10 familial members at ages between 2.8–62 years [81]. In that family, 7 of 12 examined members with no history of leukemia were carriers of the heterozygous germline C-terminal *CEBPA* mutation, with ages ranging 27–88 years [81]. A second family with an asymptomatic carrier of a germline C-terminal *CEBPA* mutation was reported subsequently [82]. Both reports suggested that germline C-terminal *CEBPA* mutations may be present in asymptomatic individuals even without a family history of leukemia. In addition, as shown in Table 70.2, a 2nd *CEBPA* somatic mutation was present in 3 of 4 examined patients with germline C-terminal *CEBPA*-mutated acute myeloid leukemia, including 2 N-terminal and 1 C-terminal *CEBPA* somatic mutations [77,82].

70.3.3 Familial *ANKRD26*-Related Thrombocytopenia

Inherited thrombocytopenia due to germline ANKRD26 mutations may predispose to AML. In a subset of familial inherited thrombocytopenia, point mutations clustering in a highly conserved 22-nucleotide region in the 5′ untranslated (UTR) region of the *ANKRD26* gene exhibit complete penetrance for thrombocytopenia since the mutations occur only in affected members and are absent in all healthy family members [83–85]. The germline mutations induce persistent overexpression of the *ANKRD26* gene in megakaryopoiesis, which causes a profound defect in proplatelet formation, leading to thrombocytopenia [83,86]. Furthermore, germline mutations in *ANKRD26* may alter the regulation of RUNX1 and ETS family transcription factor, FLI1, and cause heritable thrombocytopenia [86].

70.3.4 Familial *ETV6*-Related Thrombocytopenia

Inherited thrombocytopenia due to germline ETV6 mutations may predispose to diverse hematologic malignancies, including AML. In a small subset of families with inherited thrombocytopenia, germline mutations in *ETV6* co-segregate in the thrombocytopenic individuals, contributing to an increased risk of diverse hematologic malignancies, most commonly AML. In addition, among myeloid malignancies, myelodysplastic syndromes, AML, T-lymphocytic/myeloid mixed phenotype leukemia and polycythemia vera have been reported in 6 families [42,87–89]. The germline *ETV6* mutations have thus far been reported to be missense or frameshift, located primarily in the C-terminal DNA-binding domain, except for the recurring p.P214L mutation, which is located in the linker region between the N-terminus and C-terminus, and indirectly affects DNA binding [42,43,88,89]. All of the familial germline *ETV6* mutations cause decreased or absent nuclear localization of the ETV6 protein, with abnormal megakaryopoiesis and decreased proplatelet formation [42,43,87,89]. In the hematologic malignancies examined thus far in the individuals harboring germline *ETV6* mutations, the remaining *ETV6* allele is wildtype. Somatic mutations other than in *ETV6* were present in one patient, including in *BCOR* and *RUNX1* at time of progression from thrombocytopenia to myelodysplasia, and additionally in *KRAS* at further progression to higher grade myelodysplasia [42].

70.3.5 Inherited Mutations in *DDX41*

Inherited germline mutations in DDX41 predispose to myelodysplasia, and additional somatic mutations, including in the second

TABLE 70.2

Clinical and Genetic Features at Diagnosis of Acute Myeloid Leukemia in All Reported Patients in Pure Familial (and Non-Familial) Germline *CEBPA*-Mutated Acute Myeloid Leukemia during 2004 to 2017 [7,71–82]

Publication for Families with Germline *CEBPA* Mutations, Year Reported, Ethnic Origin	Germline *CEBPA* Mutations in Family Members with AML	Generation: Number of Family Members with AML	Gender, Age in Years at Onset of AML	Anticipation in AML Family, Paternal/ Maternal Inheritance	Cytogenetic Karyotype with Clonal Abnormality in AML at Diagnosis	Somatic *CEBPA* Mutations at AML Diagnosis	N Carriers of Germline Mutation/ N Healthy Members Tested	Age of Carrier in Years (Generation)
Familial acute myeloid leukemia with germline N-terminal CEBPA mutations								
Smith et al. (2004) [7]	c.212delC	II:1 father of III:1&2	Male, 10		Not done	NA	0/5	None
	c.212delC	III:1 proband	Male, 30	Present, III:2 and IV:1, maternal	Normal	c.1054–1089dup [71]		
	c.212delC	III:2, sister of III:1	Female, 18		Normal	c.1063_1089dup [71]		
	c.212delC	IV:1 son of III:2	Male, 2			c.1087_1089dup [71]		
Sellick et al. (2005) [72]. De Lord et al. (1997) [73], Caucasian	NA	III:1 father of IV:1&2	Male, 34	Present, III:1 and IV:1, III:1 and	NA	NA	NA/None tested	NA
	c.218_219insC [71]	IV:1 father of V:1	Male, 25	IV:2 paternal; IV:1 and V:1, paternal	del(6)(q21)	c.1075_1081delinsCTGGAGGCCA [71]		
	c.218_219insC [71]	IV:2	Female, 24		Normal	c.1075_1077dup [71]		
	c.217insC	V:1	Male, 4		Normal	NA		
Pabst et al. (2008) [74], pedigree A	c.291delC	I:1 mother of II:1	Female, 46	Present, I:1 and II:1, maternal	Monosomy 7	NA	1 male/3 children of II:1	19 [71] (III)
	c.291delC	II:1	Female, 40		Normal	c.1085_1087dup [71]		
Pabst et al. (2008) [74], pedigree B	c.464_465insT	I:1 father of II:1	Male, 42	Present, I:1 and II:1, paternal	NA	c.G1207C; c.A1210C [71]	NA	NA
	c.464_465insT	II:1 proband	Female, 27		Normal	c.1087_1089dup [71]		
Renneville et al. (2009) [75]	c.218insC	I:1 mother of II:1	Female, 23	Present, I:1 and II:1, maternal	Normal	c. c.991_992insGA [71]	None tested/2	NA
	c.218insC[a]	II:1	Male, 5		Normal	c.1067_1068insGCG [71]		
Nanri et al. (2010) [76], Japanese	c.351_352insCTAC [71]	I:1, father of II:1	Male, 39	Present, I:1 and II:1, paternal	NA	c.1067_1068insGGGCCCTCGCCCCC CCGCCG [71]	1 male/1	24 (II)
	c.351_352insCTAC [71]	II:1	Male, 26		NA	c.1087_1089dup [71]		
Taskesen et al. (2011) [77]	c.308delG [71]	I:1 proband	Female, 25	NA	NA	c.1126_1127ins1079_1227 [71]	NA; 1 obligate [71]	NA
	NA, 2 others	NA	Female and male, NA	NA	NA	NA		NA
Taskesen et al. (2011) [77], Stelljes et al. (2011) [78]	c.338delC	I:1, mother of II:1	Female, 28	Present, I:1 and II:1, maternal	Normal	c.1087_1089dup [71]	0/3	None
	c.338delC	II:1	Female, 2		Normal	c.1076_1087dup [71]		
Xiao et al. (2011) [79], Chinese study	c.584_589dup	I:1 proband	Male, 36	NA	del(9)(q11q34)	c.247dupC (N-terminal) c.914_916dup	1/1, sister, donor for transplant	33
Debeljak et al. (2013) [70]	c.297_315del [71]	I:1 twin A mono-zygotic	Female, 1.75 (21 months)	NA	Normal	c.1087_1089dup [71] c.1061_1210del [71]	NA	NA
	c.297_315del [71]	I:2 twin B	Female, 15	NA	Normal	c.1087_1089dup [71]		

(Continued)

TABLE 70.2 (Continued)

Clinical and Genetic Features at Diagnosis of Acute Myeloid Leukemia in All Reported Patients in Pure Familial (and Non-Familial) Germline *CEBPA*-Mutated Acute Myeloid Leukemia during 2004 to 2017 [7,71–82]

Publication for Families with Germline *CEBPA* Mutations, Year Reported, Ethnic Origin	Germline *CEBPA* Mutations in Family Members with AML	Generation: Number of Family Members with AML	Gender, Age in Years at Onset of AML	Anticipation in AML Family, Paternal/ Maternal Inheritance	Cytogenetic Karyotype with Clonal Abnormality in AML at Diagnosis	Somatic *CEBPA* Mutations at AML Diagnosis	N Carriers of Germline Mutation/N Healthy Members Tested	Age of Carrier in Years (Generation)
Tawana et al. (2015) [71]b	NA	I:1, mother of II:1	Female, 32	Present, I:1 and II:1, maternal	NA	NA	1/1, aunt of II:1	41
	NA	I:2, sister of I:1	Female, 3	I:1, maternal	NA	NA		
	c.218_219insC	II:1 proband	Female, 18	NA	NA; failed	NA		
Yan et al. (2016) [80], Vietnamese	c.134insC	I:1	Male, 33	NA	del(9)(q13q22)	insertion of 33 bases c.937_938	NA	NA

Familial and non-familial (no family history) acute myeloid leukemia patients with germline C-terminal CEBPA mutations

Publication for Families with Germline *CEBPA* Mutations, Year Reported, Ethnic Origin	Germline *CEBPA* Mutations in Family Members with AML	Generation: Number of Family Members with AML	Gender, Age in Years at Onset of AML	Anticipation in AML Family, Paternal/ Maternal Inheritance	Cytogenetic Karyotype with Clonal Abnormality in AML at Diagnosis	Somatic *CEBPA* Mutations at AML Diagnosis	N Carriers of Germline Mutation/N Healthy Members Tested	Age of Carrier in Years (Generation)
Taskesen et al. (2011) [77], patient 1	C-terminal c.T1096C	No family history	51, gender NA	NA	NA	N-terminal c.478_485del	NA/none tested	NA
Taskesen et al. (2011) [77], patient 2	C-terminal c.G1164A	No family history	33, gender NA	NA	NA	None in *CEBPA*c	NA/none tested	NA
Taskesen et al. (2011) [77], patient 3	C-terminal c.G1036T	No family history	69, gender NA	NA	NA	C-terminal c.1086insAAG	NA/none tested	NA
Pathak et al. (2016) [81]	NA	II:5	Male, 62	Present, II:8 and III:7, maternal;	NA	NA	7/12; III:2 mother of IV:1,2,3,4;	44 at deathd (III:2);
	NA	II:8 sister of II:5	Female, 53	III:2 and	NA	NA	III:5 mother of	88 (III:5);
	NA	III:7, son of II:8	Male, 36	IV:1,2,3,4, maternal;	NA	NA	IV:7; III:10	III:10 dead at 80;
	c.A1932C	III:8, son of II:8	Male, 58	NA	NA	NA	brother of III:8;	54(IV:5,6);
	c.A1932C	IV:1	Female, 20	III:5 and IV:7, maternal;	Normal	NA	IV:5,6, twin sons	60(IV:11);
	NA	IV:2	Male, 2.8 (34 mos)	IV:7 maternal; IV:7 and V:2, paternal	NA	NA	of III:2; IV:11, sister of IV:7; V:1	27 (V:1)
	NA	IV:3	Male, 6		NA	NA	sister of V:2	
	c.A1932C; p.Q311P	IV:4 proband	Female, 11		Normal	NA		
	NA	IV:7 father of V:2	Male, 41		NA	NA		
	c.A1932C	V:2	Female, 22		Normal	NA		
Ram et al. (2017) [82]	c.G442T	I:1	Female, 36, proband	Present, maternal	Normal	c.68dupC	2 (mother and sister)/4	66 and 37

Abbreviations: AML, acute myeloid leukemia; N, number; NA, information not available; mo, months.

a Not present 2 years after receiving allogeneic transplant [75].
b Mutations reported by Tawana et al. [71] as per *CEBPA* transcript: ENST00000498907 (hg19).
c Germline *CEBPA* mono-allelic mutated AML does not qualify for a diagnosis of biallelic *CEBPA* mutated AML [14].
d Death due to carcinoma with no history of any leukemia.

allele for *DDX41*, act as driver mutations to lead to the development of overt myeloid neoplasia [46], similar to the somatic acquisition of a second mutation in familial *CEBPA*-mutated AML and *RUNX1* mutated FPD/AML, as described above. However, in contrast with *CEBPA* and *RUNX1*, germline mutations in familial *DDX1* driven AML/myelodysplasia have a long latency, with onset of disease in 4th to 8th decade, similar to the onset of somatically driven myeloid neoplasia. Germline *DDX41* mutations include frameshift or missense [46,90], and in about 50% of these, additional missense somatic mutations occur in the other *DDX41* allele at onset of disease [46], which may include myeloid or lymphoid neoplasms. The most frequently reported germline mutation is a frameshift leading to a premature stop codon (p.D140Gfs*2), and the most frequent co-existing somatic *DDX41* mutation at presentation is a missense mutation, p.R525H [91]. A subset of families with myelodysplasia/AML with inherited *DDX41* mutations have been shown to have short telomere lengths, similar to patients with inherited *TERC* or *TERT* mutations in familial myelodysplasia/AML, suggesting that instead of inheritance by Mendelian genetics, germline *DDX41* mutations represent a strong inherited risk factor predisposing to myeloid neoplasia [92].

70.4 Epidemiology

Pure non-syndromic familial AML is considered rare, but the true incidence and prevalence may be currently underestimated due to variations in clinical features, latency, and small numbers of reported families for the currently known genetic causes. In addition to pediatric patients, familial causes of leukemia may also be diagnosed in adults at onset of AML. The entire spectrum of genetic causes in familial AML is yet to be defined, with uncharacterized causes for 62% (17/27) families in one previous study [93].

In contrast with sporadic AML, which is primarily a disease of older adults with median age of 68 years at diagnosis [94], familial AML develops at younger ages.

70.4.1 Familial Platelet Disorder with Propensity for Acute Myeloid Leukemia

For FPD/AML, the median age for presentation of germline *RUNX1* related hematologic malignancy (myelodysplasia, acute myeloid or lymphoblastic leukemia) is in the early-to-mid 30s (range 6–72 years), with at least 50 families with genetically confirmed FPD/AML reported thus far. The true incidence and prevalence of FPD/AML among newly diagnosed AML patients is currently unknown. In a Taiwanese cohort of 470 de novo AML, 63 distinct *RUNX1* mutations were observed in 13.2% (n = 62) patients. In that study, 5 (1.1% of all AML) patients with no prior hematologic abnormalities or familial history harbored missense *RUNX1* mutations that could have been germline, since those mutations were present at diagnosis and also at remission [57]. Further, in a European cohort of 945 patients aged 18–65 years at diagnosis of AML, 59 *RUNX1* mutations were observed in 5.6% (n = 53), with two distinct mutations in each of six (<1%) patients [6], but it is unclear whether those 6 patients represented FPD/AML.

70.4.2 Familial *CEBPA*-Mutated Acute Myeloid Leukemia

As discussed above and summarized in Table 70.2, there were 17 reported families for this entity as of January 2018. However, in a cohort of 187 consecutively diagnosed AML patients including 9.6% *CEBPA* mutated, 2 (1.1% of all, or 11% of *CEBPA* mutated) patients had familial *CEBPA*-mutated AML, suggesting that the true incidence and prevalence might be higher than is currently reported [74]. Given that 21 380 new cases of AML were expected to occur in 2017 in the United States [94], ~200 new cases could potentially represent germline *CEBPA*-mutated AML.

In another larger but selected cohort of 1182 cytogenetically normal AML patients, the *CEBPA*-mutated subset comprised 12.8% (n = 151), with 5 (3% of *CEBPA* mutated) harboring germline *CEBPA* mutations. Interestingly, a family history of AML was present in only 2 of those 5 patients, with both harboring germline N-terminus *CEBPA* mutations, while the remaining 3 patients with no family history of leukemia harbored germline C-terminus *CEBPA* mutations [77]. Collectively, in contrast with N-terminal germline *CEBPA* mutations, familial C-terminal germline *CEBPA* mutations appear to be present in greater numbers of asymptomatic individuals, with lower penetrance for developing AML as reported in 2 families [81,82], further suggesting that the prevalence and spectrum of heterogeneity in clinical presentation and genetic mutations in familial *CEBPA*-mutated AML is yet to be defined.

70.4.3 Familial Inherited Thrombocytopenia, Including Due to *ANKRD26* Mutations

While the genetic causes for almost 50% of *familial inherited thrombocytopenia* are yet unknown [95], the known germline mutations that may lead to familial AML include mutations of *RUNX1*, *ANKRD26* and *ETV6*, which comprise 3%, 18% and 5%, respectively, of all inherited thrombocytopenia [38]. Differences in prevalence may exist related to ethnic origin. A nationwide Japanese study of genetic causes of inherited thrombocytopenia in 43 families with available clinical samples for genetic analysis underscored the importance of molecular genetic analysis to confirm the diagnosis of FPD due to germline mutations in *RUNX1* in 16% (n = 7) families [95].

Inherited thrombocytopenia due to ANKRD26 mutations is relatively common, with 21 (10%) families diagnosed in a cohort of 210 families with inherited thrombocytopenia, including families from Italy, Spain, North America, Argentina and Senegal [84]. However, ethnic differences may exist since the same above mentioned Japanese nationwide survey showed only 1 (2%) of 43 families with *ANKRD26*-related inherited thrombocytopenia [95]. Important to note, there is a high incidence of progression to myeloid malignancies, as was evidenced in 10 (8.5%) of 118 patients with *ANKRD26* related thrombocytopenia, including 4.9% AML, 2.2% myelodysplastic syndromes and 1.3% chronic myeloid leukemia [85].

From a different perspective (other than inherited thrombocytopenias), in a cohort of 250 consecutive non-familial AML patients, mutations in *ANKRD26* were found in 4 (1.6%), and 1 of those 4 (25% of all *ANKRD26* mutated; <1% of all AML) was found to be a member of a family with typical but undiagnosed inherited thrombocytopenia due to *ANKRD26* mutation [96].

70.4.4 Inherited Thrombocytopenia Due to *ETV6* Mutations

Inherited thrombocytopenia due to ETV6 mutations may be less common than that due to germline *ANKRD26* mutations, with reports of 2 (8.6%) *ETV6*-related in 23 European [43], and 7 (2.6%) of 274 European [88], and none identified in 43 Japanese families with inherited thrombocytopenia [95]. In contrast with inherited thrombocytopenia due to germline *RUNX1* or *ANKRD26* mutations, germline *ETV6* mutations predispose most frequently to acute lymphoblastic leukemia [43,88]. In the 7 *ETV6*-related families within the unselected cohort of 274 consecutive families with inherited thrombocytopenia, progression to hematologic malignancies was observed in 25% (5 of 20 affected patients in 5 of 7 *ETV6*-related families), including acute lymphoblastic leukemia in 4 patients and polycythemia vera in one [88].

70.4.5 Germline *DDX41* Mutations

Germline DDX41 mutations predisposing to myeloid neoplasms were recognized by whole exome sequencing analysis in families affected by AML and myelodysplasia, and by targeted sequencing for somatic versus germline mutations in paired tumor and normal DNA samples from patients with myelodysplasia and secondary AML [46]. Individuals with germline *DDX41* mutations develop myeloid malignancies at older ages, similar to the development of sporadic hematologic malignancies [46,90,91]. The onset of disease was at 44–73 years in 11 patients from 7 families with genetically confirmed *DDX41* mutations [46]. In addition, in a cohort of 1034 patients with myelodysplastic syndromes and secondary AML, targeted sequencing revealed germline *DDX41* mutations in 8 (0.7%) patients. Familial germline *DDX41* mutations were also identified in healthy carriers, and are notably present very rarely in publicly available exome sequencing databases from healthy individuals [46]. At least 70 families with germline *DDX41* mutations have been reported, such that families with *DDX41* or *RUNX1* mutations are the most frequent germline causes identified thus far, for familial myeloid neoplasms, including acute myeloid leukemia. Ethnic differences have been observed in the types of germline *DDX41* mutations, with specific mutations exclusive to Caucasian (p.D140Gfs*2) and Asian (p.A500Cfs*9) populations, while the *DDX41* mutation, p.R525H, is the most common somatic mutation in both populations, suggesting that these germline mutations could represent inherited events through generations from a founder germline event instead of Mendelian inheritance.

70.5 Clinical Features

The clinical features of familial AML are varied, depending upon the underlying germline genetic etiology. In addition, there may be significant inter-familial and intra-familial differences in clinical features, particularly in families with autosomal dominant FPD/AML.

70.5.1 Familial Platelet Disorder with Propensity for Acute Myeloid Leukemia

In FPD/AML families, individual carriers with germline *RUNX1* mutations may be completely asymptomatic and may even lack a history of thrombocytopenia or bleeding. Patients with FPD/AML characteristically have mild or moderate thrombocytopenia, and secondarily, may have a qualitative reduction in dense granule secretion [97]. Eczema has been reported in a rare FPD/AML family [98]. Further, in a French cohort, germline *RUNX1* mutations have been identified by exome sequencing of germline tissues in rare families with inherited bone marrow failure and previously unidentified genetic causes [99]. The median age of onset for FPD/AML is 33 years, but the disease may manifest with a malignancy even at the age of 72 years, with 6 years as the earliest age for onset of AML [31].

70.5.2 Familial *CEBPA*-Mutated Acute Myeloid Leukemia

Familial *CEBPA*-mutated AML occurs at younger ages than sporadic cases of AML. As calculated for all patients described in Table 70.2, for patients with N-terminal germline mutations, the age at onset of AML is lower [median 25 (range 1.75–46) years] than for patients with C-terminal germline mutations [median 36 (range 2.8–69) years], with overall median age of 26.5 (range 1.75–69) for all germline *CEBPA* mutations at onset of AML. The male to female ratio is similar (20M:19F) overall and for both N-terminal (14:14) and C-terminal (6M:5F) germline *CEBPA*-mutated individuals showing penetrance to AML. With the cases reported thus far, AML may not manifest until at least 88 years of age with a C-terminal germline *CEBPA* mutation [81], and at least 41 years for N-terminal mutations [71].

The clinical presentation in individual patients is very similar to that of sporadic *CEBPA* biallelic mutated AML [100,101], including presentation as de novo disease, usually as FAB subtypes M1, M2, M4, often with Auer rods, aberrant expression of CD7, normal cytogenetic karyotype, and frequent co-occurrence of *GATA1* and *WT1* somatic mutations at diagnosis [71].

Interestingly, families with *CEBPA* mutant AML demonstrate the phenomenon of *anticipation* previously described in familial AML [73,102], wherein disease presents with increasing severity or at earlier ages in successive generations. As detailed in Table 70.2, anticipation was present in 100% (10 of 10) families with confirmed germline *CEBPA* mutations, and interestingly, showing both paternal and maternal inheritance, with one family harboring C-terminal germline mutations showing both inheritance modes [81]. Intriguingly, although the mutations include insertions with varying nucleotide lengths, the mechanism for anticipation is not yet elucidated in familial AML. Penetrance is near-complete in cases of N-terminal germline mutations, and incomplete with C-terminal mutations with several healthy carriers reported [81] or with no family history of leukemia [82]. Since healthy carriers without disease may not have been reported for C-terminal germline mutations, and not all family members in the reported families with N-terminal germline mutations were tested, the true prevalence of germline *CEBPA*-mutated carriers is yet to be defined.

In contrast with FPD/AML patients in whom there may be a myelodysplastic phase preceding leukemia, or even the development of non-myeloid neoplasms, familial *CEBPA*-mutated patients present with pure de novo AML in the absence of a preceding myelodysplastic or cytopenic phase, and in the absence of any reported progression to any non-myeloid malignancies. Further, the natural post-remission course of familial *CEBPA*-mutated AML is re-occurrence of leukemia due to a somatic mutation that is different from the initial pathogenic mutation, indicating a new leukemic clone rather than a relapse of the original leukemia [71].

70.5.3 Inherited Thrombocytopenia Due to *ANKRD26* Mutations

Individuals with *inherited thrombocytopenia due to ANKRD26* mutations usually have mild bleeding tendency, in the absence of platelet macrocytosis. In the Italian study of 78 individuals from 21 families, platelets were deficient in glycoprotein Ia and α-granules, with normal in vitro platelet aggregation in the majority of cases [84]. Since there is a high incidence of myeloid neoplasms in *ANKRD26*-related thrombocytopenia, diagnosis of this entity is essential as an underlying cause in AML, including screening of potential related donors for hematopoietic transplantation.

70.5.4 *ETV6*-Related Inherited Thrombocytopenia

Similarly, individuals with *ETV6-related inherited thrombocytopenia* may present with mild or minimal bleeding tendencies, usually mild or moderate thrombocytopenia, with normal or slightly reduced platelet size and reduced ability of platelets to spread with collagen [88]. Red cell macrocytosis may be present [43] but is not a constant feature [88]. Myeloid malignancies in the *ETV6*-related families may present at any age, with range 8–82 years in the 6 reported patients thus far in 20 families.

Importantly, the clinical features are nonspecific in families with inherited thrombocytopenia and any of the three genetic causes (*RUNX1, ANKRD26, ETV6*) predisposing to hematologic malignancies. Examination of the bone marrow may show dysmegakaryopoiesis in all three entities [55,84,88,89,103], and therefore, molecular genetic analysis is necessary for diagnosis.

70.5.5 Germline *DDX41* Mutations

Individuals with *germline DDX41 mutations* present with myelodysplastic syndromes or AML at older ages (median age 62 years), similar to the ages for sporadic cases, with no specific clinical feature to suggest a germline cause. Since the latency can be as long as the seventh and eighth decades, the presentation is very similar to sporadic forms of disease if a family history is not evident. In addition to myeloid neoplasms, there may be predisposition to lymphoid neoplasms, including early-onset follicular lymphoma and Hodgkin lymphoma, multiple myeloma [90], and non-hematologic malignancies including non-small cell lung cancer [104]. Clinical manifestations in healthy carrier individuals, if examined, in families with germline *DDX41* mutations may be very subtle, including cytopenia or mild monocytosis. Given the recent discovery of *DDX41* as a familial predisposing gene for myeloid and lymphoid malignancies and the currently limited knowledge of functions of *DDX41*, including in cancer, it is likely that much is yet to be learned for these genetic mutations.

70.6 Diagnosis

Diagnosis of familial disease in a patient presenting with AML requires awareness of the possibility of familial disease and of the heterogeneous clinical features and genetic etiologies. A careful personal and family history is critical to diagnose germline etiology in apparently sporadic cases, in any patient with acute myeloid leukemia or myelodysplastic syndrome, including for personal and family history of cancer (hematologic or non-hematologic), any previous hematologic disorders, bleeding tendencies, and clinical examination for any associated abnormalities that may suggest a known inherited disorder. In particular, a history of myeloid neoplasia in two or more individuals in a family, young age (<50 years) at onset, prior history of bleeding tendency or thrombocytopenia, congenital or constitutional abnormalities, and the presence of mutations that could be germline (*CEBPA, RUNX1, ANKRD26, DDX41, ETV6*) in a multigene panel performed for detecting somatic mutations should prompt further evaluation for differentiation between somatic and germline mutants, and genetic counseling and clinically indicated genetic testing of family members. Moreover, the diagnosis of any familial germline mutation that predisposes to AML or pre-leukemic myelodysplastic syndromes is critical when a patient is being considered for an allogeneic hematopoietic stem cell transplant from a related donor, who must be tested for the genetic mutation to avoid reintroducing the pathogenic variant from an undiagnosed carrier in the family.

Specifically, as noted above, familial and sporadic *CEBPA* biallelic mutated AML may be very similar in clinical features, except for a family history, if present, and therefore germline testing is necessary to diagnose cases with possible familial inheritance. In a similar concept, in individuals with inherited thrombocytopenia, accurate diagnosis of FPD/AML requires genetic testing, as was shown in the nationwide Japanese study wherein clinical history and examination alone were unreliable for diagnosis [95]. The clinical features for germline *ANKRD26*-related or *RUNX1*-related inherited thrombocytopenia may also be similar, with normal-sized platelets, with confirmation possible only by molecular genetic testing.

Pathologically, *CEBPA* biallelic mutant AML (sporadic or familial) represent a molecularly defined entity as recognized by the 2016 WHO classification of AML [14,105]. Importantly, both sporadic and familial cases present with similar features, with the driver being germline in familial, as compared with somatically acquired in sporadic cases. It is therefore critical to test for the possibility of germline mutations in all patients with biallelic *CEBPA*-mutated AML, and similarly, in all patients with biallelic *RUNX1* mutated AML.

In a patient with AML, the diagnosis of familial *CEBPA*-mutated leukemia is made by identifying the germline *CEBPA* mutation in non-neoplastic cells and neoplastic (leukemic) cells in the proband, and by observing segregation of the pathogenic germline *CEBPA* variant across family members. In familial members with no clinical disease, or for a patient in remission, the presence of a heterozygous germline mutation would diagnose a carrier for the mutation examined, e.g. for *RUNX1* in FPD/AML families, or for *CEBPA* in *CEBPA*-mutated AML families.

In the molecular genetic laboratory, although next-generation sequencing assays are commonly used in several institutions in the United States, the diagnosis of *CEBPA* mutations is often made by traditional bidirectional Sanger sequencing procedures, since *CEBPA* is a GC-rich, single exon gene that is not easily analyzable by all available next-generation sequencing assays. Similar assays for *RUNX1* are also available in clinical laboratories, which can be performed on diagnostic and post-therapy samples. For demonstration of a germline variant in a patient with a hematologic neoplasm involving peripheral blood, such as a leukemia, samples other than peripheral blood are necessary. If there is hematologic disease, i.e., if there are leukemic cells (myeloblasts) in peripheral blood, then a sample such as a buccal swab would likely be contaminated with blood; thus genetic testing of a buccal swab would also detect somatic mutations, and therefore genetic testing of fibroblasts obtained by culture of a skin biopsy sample is preferred for germline testing. However, such testing requires an additional biopsy if not obtained at the time of a bone marrow biopsy, and time (few weeks) for fibroblast culture before results can be obtained. Alternatively, in patients in remission and in healthy carriers, buccal swabs or buccal mucosal samples, peripheral blood, T-lymphocytes, and hair have been used to test for germline testing.

For germline *RUNX1* mutations, the spectrum of genetic aberrations is diverse, including point mutations, microdeletions, and duplications of aberrant alleles. Therefore, in addition to bidirectional sequencing, molecular genetic techniques such as chromosomal arrays or multiplex ligation-dependent probe amplification to detect chromosomal losses and gains may be necessary to elucidate the genetic defect in individuals or families suspected of harboring a germline *RUNX1* aberration.

Importantly, germline mutations may also be detected in multigene next-generation sequencing analyses that are now routinely performed in several institutions in the United States for the detection of somatic mutations in leukemic or neoplastic samples, with guidelines for reporting germline variants from neoplastic (tumor) tissue analyses [106,107]. In these tumor mutational analyses, variant allele frequencies of close to 50% (40%–60%) or 100% should prompt consideration of the possibility of the variant being germline, and additional testing of a germline tissue sample if clinically indicated.

70.7 Treatment

Familial AML may require different management strategies from sporadic AML wherein the driver mutations are somatic in origin, in contrast to being germline in familial cases. Familial *CEBPA*-mutated AML may involve hematopoietic stem cell transplantation to eliminate the germline variant and prevent recurrence, whereas sporadic *CEBPA*-mutated cases may not require a transplant since they have a good prognosis in the absence of other unfavorable prognostic factors. The reader is referred to expert reviews for clinical management of AML with evaluation of prognostic risk factors [108], and for specific guidelines for different disease entities described above [109–112].

70.8 Prognosis

The prognosis varies with the etiologic genetic aberration causing familial AML and prognostic risk factors applicable for any AML, as previously reviewed [1]. Accurate recognition of the familial leukemia predisposition gene allows for management specific for inherited myeloid neoplasms and for the specific entity, with possible prevention of undesirable clinical consequences. As a common feature for all inherited myeloid neoplasms, knowledge of the inherited genetic cause allows screening of potential related donors, in case of the need for a hematopoietic stem cell transplant, to avoid re-introduction of the pathogenic mutant gene, as has been observed with germline *CEBPA* and *DDX41* mutations [79,113].

Familial *CEBPA*-mutated AML have biallelic *CEBPA* mutations, which have a favorable prognosis in the absence of unfavorable factors such as *FLT3*-ITD. In a comparison of all sporadic and familial *CEBPA*-mutated cases, the overall survival was better in familial than in sporadic cases, although similar when the comparison was only for patients <45 years in age [71]. However, familial *CEBPA*-mutated leukemia re-occurs after achievement of remission due to a new causative somatic mutation, and therefore hematopoietic stem cell transplant may be considered to prevent re-emergence of disease in familial cases. Despite recurrences, familial *CEBPA*-mutated AML has a median survival of 8 years after relapse, in contrast with a worse outcome after relapse in sporadic *CEBPA*-mutated cases [71].

In contrast with germline-mutated *CEBPA*, AML arising in FPD/AML with mutated *RUNX1* have an unfavorable prognosis [108], with hematopoietic stem cell transplant as the possible treatment to eradicate the mutant gene. Further, since only up to 60% of carriers with the germline *RUNX1* mutation in FPD/AML families develop disease, and there is significant inter- and intrafamilial variation in the degree of disease penetrance, the risk of developing neoplasms (myeloid or lymphoid) cannot be predicted for carrier individuals in affected families. Since a subset of carriers have been shown to have earlyonset age-related clonal hematopoiesis, it has been suggested that monitoring of such clonal mutations may identify *RUNX1* carrier individuals at higher risk for developing neoplasia [66]. The risk of progression in healthy carriers of *DDX41* mutations is currently unknown. For healthy carriers in families with familial AML, uniform monitoring guidelines irrespective of genetic cause are recommended, including baseline bone marrow morphology examination and annual blood counts, with specific guidelines depending upon the clinical features of the germline genetic cause and in the familial presentation [110].

70.9 Conclusion

Although familial acute myeloid leukemia (AML) has been known to occur for decades, increasing investigation of the molecular genetic causes of AML has also led to increasing recognition of germline causes of genetic predisposition to familial leukemia. Familial cases may appear as sporadic, with no obvious history, and awareness of the clinical entities and the clinical heterogeneity that may be present is critical for diagnosis, which is essential for appropriate patient management. A recent United States single institution study of consecutive newly diagnosed AML patients discovered the occurrence of at least 4 (18%) familial AML cases among 22 AML cases definitively classified as AML with mutated *NPM1* [114]. The latter AML subtype comprises the largest proportion of all newly diagnosed AML by the 2016 World Health Organization classification. Intriguingly, that study also found 75% non-smokers among the 4 molecularly confirmed NPM1-mutated familial AML patients; further, at least one additional young AML patient with familial leukemia was noted to be a non-smoker. In addition, the same study also showed a 74-year-old AML patient with biallelic *CEBPA* mutations to have a family history of AML, indicating that familial biallelic *CEBPA* mutated AML may first be diagnosed even in the eighth decade of life [114]. At the present time, clearly, much is yet to be learned in elucidating the heterogeneity in genetic causes, the clinical spectrum of disease, and causes of progression to overt malignancy in individuals in families with predisposition to AML, including pertaining to appropriate counseling and management of healthy family members who harbor predisposing genetic mutations, to eventually and hopefully prevent progression of germline predisposition in healthy individuals to overt malignancy in the future.

REFERENCES

1. Kansal R. Acute myeloid leukemia in the era of precision medicine: Recent advances in diagnostic classification and risk stratification. *Cancer Biol Med.* 2016;13:41–54.
2. Li FP, Fraumeni JF Jr, Mulvihill JJ et al. A cancer family syndrome in twenty-four kindreds. *Cancer Res.* 1988;48: 5358–62.
3. Garber JE, Goldstein AM, Kantor AF, Dreyfus MG, Fraumeni JF Jr, Li FP. Follow-up study of twenty-four families with Li-Fraumeni syndrome. *Cancer Res.* 1991;51:6094–7.
4. Luddy RE, Champion LA, Schwartz AD. A fatal myeloproliferative syndrome in a family with thrombocytopenia and platelet dysfunction. *Cancer.* 1978;41(5):1959–63.
5. Dowton SB, Beardsley D, Jamison D, Blattner S, Li FP. Studies of a familial platelet disorder. *Blood.* 1985;65(3):557–63.
6. Song WJ, Sullivan MG, Legare RD et al. Haploinsufficiency of CBFA2 causes familial thrombocytopenia with propensity to develop acute myelogenous leukaemia. *Nat Genet.* 1999;23:166–75.
7. Smith ML, Cavenagh JD, Lister TA, Fitzgibbon J. Mutation of CEBPA in familial acute myeloid leukemia. *N Engl J Med.* 2004;351(23):2403–7.
8. Alter BP. Diagnosis, genetics, and management of inherited bone marrow failure syndromes. *Hematology Am Soc Hematol Educ Program.* 2007;2007:29–39.
9. Alter BP, Giri N, Savage SA, Rosenberg PS. Cancer in dyskeratosis congenita. *Blood.* 2009;113(26):6549–57.
10. Dokal I, Vulliamy T. Inherited bone marrow failure syndromes. *Haematologica.* 2010;95(8):1236–40.
11. Parikh S, Bessler M. Recent insights into familiar inheritable bone marrow failure syndromes. *Curr Opin Pediatr.* 2012;24(1):23–32.
12. Stieglitz E, Loh ML. Genetic predispositions to childhood leukemia. *Ther Adv Hematol.* 2013;4:270–90.
13. Savage SA, Dufour C. Classical inherited bone marrow failure syndromes with high risk for myelodysplastic syndrome and acute myelogenous leukemia. *Sem Hematol.* 2017;54:105–14.
14. Arber DA, Orazi A, Hasserjian R et al. The 2016 revision to the World Health Organization classification of myeloid neoplasms and acute leukemia. *Blood.* 2016;127(20):2391–405.
15. Arber DA, Baumann I, Niemeyer CM, Brunning RD, Porwit A. Myeloid proliferations associated with Down syndrome. In: Swerdlow SH, Campo E, Harris NL et al. *WHO Classification of Tumours of Hematopoietic and Lymphoid Tissues.* Lyon: IARC Press; 2017: 169–71.
16. Scott RH, Mansour S, Pritchard-Jones K, Kumar D, MacSweeney F, Rahman N. Medulloblastoma, acute myelocytic leukemia and colonic carcinomas in a child with biallelic MSH6 mutations. *Nat Clin Pract Oncol.* 2007;4(2):130–4.
17. Ripperger T, Beger C, Rahner N et al. Constitutional mismatch repair deficiency and childhood leukemia/lymphoma—report on a novel biallelic MSH6 mutation. *Haematologica.* 2010;95(5):841–4.
18. Wimmer K, Etzler J. Constitutional mismatch repair-deficiency syndrome: Have we so far seen only the tip of an iceberg? *Hum Genet.* 2008;124(2):105–22.
19. Wimmer K, Kratz CP, Vasen HF et al. Diagnostic criteria for constitutional mismatch repair deficiency syndrome: Suggestions of the European consortium "care for CMMRD" (C4CMMRD). *J Med Genet.* 2014;51(6):355–65.
20. Hahn CN, Chong CE, Carmichael CL et al. Heritable GATA2 mutations associated with familial myelodysplastic syndrome and acute myeloid leukemia. *Nat Genet.* 2011;43(10):1012–7.
21. Hsu AP, Sampaio EP, Khan J et al. Mutations in GATA2 are associated with the autosomal dominant and sporadic monocytopenia and mycobacterial infection (MonoMAC) syndrome. *Blood.* 2011;118(10):2653–5.
22. Wlodarski MW, Hirabayashi S, Pastor V et al. Prevalence, clinical characteristics, and prognosis of GATA2-related myelodysplastic syndromes in children and adolescents. *Blood.* 2016;127(11):1387–97.
23. Pasquet M, Bellanné-Chantelot C, Tavitian S et al. High frequency of GATA2 mutations in patients with mild chronic neutropenia evolving to MonoMac syndrome, myelodysplasia, and acute myeloid leukemia. *Blood.* 2013;121(5):822–9.
24. Gao J, Gentzler RD, Timms AE et al. Heritable *GATA2* mutations associated with familial AML-MDS: A case report and review of literature [published erratum appears in J Hematol Oncol 2015;8:131]. *J Hematol Oncol.* 2014;7:36.
25. Ostergaard P, Simpson MA, Connell FC et al. Mutations in GATA2 cause primary lymphedema associated with a predisposition to acute myeloid leukemia (Emberger syndrome). *Nat Genet.* 2011;43(10):929–31.
26. Peterson LC, Bloomfield CD, Niemeyer CM, Dohner H, Godley LA. Myeloid neoplasms with germline predisposition. In: Swerdlow SH, Campo E, Harris NL et al. *WHO Classification of Tumours of Hematopoietic and Lymphoid Tissues.* Lyon: IARC Press; 2017: 122–8.
27. Sanders MA, Chew E, Flensburg C et al. Germline loss of MBD4 predisposes to leukemia due to a mutagenic cascade driven by 5 mC. BioRxiv preprint first posted online November 1, 2017. http://dx.doi.org/10.1101/180588.
28. Miyoshi H, Shimizu K, Kozu T, Maseki N, Kaneko Y, Ohki M. t(8;21) breakpoints on chromosome 21 in acute myeloid

leukemia are clustered within a limited region of a single gene, AML1. *Proc Natl Acad Sci USA.* 1991;88(23):10431–4.

29. Okuda T, van Deursen J, Hiebert SW, Grosveld G, Downing JR. AML1, the target of multiple chromosomal translocations in human leukemia, is essential for normal fetal liver hemato-poiesis. *Cell.* 1996;84(2):321–30.

30. Michaud J, Simpson KM, Escher R et al. Integrative analy-sis of RUNX1 downstream pathways and target genes. *BMC Genomics.* 2008;31:363.

31. Owen CJ, Toze CL, Koochin A et al. Five new pedigrees with inherited RUNX1 mutations causing familial plate-let disorder with propensity to myeloid malignancy. *Blood.* 2008;112(12):4639–45.

32. Nishimoto N, Imai Y, Ueda K et al. T cell acute lympho-blastic leukemia arising from familial platelet disorder. *Int J Hematol.* 2010;92(1):194–7.

33. Toya T, Yoshimi A, Morioka T et al. Development of hairy cell leukemia in familial platelet disorder with predisposition to acute myeloid leukemia. *Platelets.* 2014;25(4):300–2.

34. Morgan NV, Daly ME. Gene of the issue: RUNX1 mutations and inherited bleeding, *Platelets.* 2017;28(2):208–10.

35. Radomska HS, Huettner CS, Zhang P, Cheng T, Scadden DT, Tenen DG. CCAAT/enhancer binding protein alpha is a regu-latory switch sufficient for induction of granulocytic devel-opment from bipotential myeloid progenitors. *Mol Cell Biol.* 1998;18(7):4301–14.

36. Pabst T, Mueller BU, Zhang P et al. Dominant-negative muta-tions of CEBPA, encoding CCAAT/enhancer binding protein-alpha (C/EBPalpha), in acute myeloid leukemia. *Nat Genet.* 2001;27:263–70.

37. Pabst T, Mueller BU, Harakawa N et al. AML1-ETO down-regulates the granulocytic differentiation factor C/EBPalpha in t(8;21) myeloid leukemia. *Nat Med.* 2001;7(4):444–51.

38. Noris P, Pecci A. Hereditary thrombocytopenias: A growing list of disorders. *Hematology.* 2017;2017:385–99.

39. https://www.ncbi.nlm.nih.gov/gene/22852, last accessed February 9, 2018.

40. Hock H, Meade E, Medeiros S et al. Tel/Etv6 is an essential and selective regulator of adult hematopoietic stem cell sur-vival. *Genes Dev.* 2004;18(19):2336–41.

41. https://www.ncbi.nlm.nih.gov/gene/2120, last accessed February 9, 2018.

42. Zhang, MY, Churpek JE, Keel SB et al. Germline ETV6 mutations in familial thrombocytopenia and hematologic malignancy. *Nat Genet.* 2015;47:180–5.

43. Noetzli L, Lo RW, Lee-Sherick AB et al. Germline mutations in ETV6 are associated with thrombocytopenia, red cell mac-rocytosis and predisposition to lymphoblastic leukemia. *Nat Genet.* 2015;47:535–8.

44. https://www.ncbi.nlm.nih.gov/gene/51428, last accessed February 10, 2018.

45. Antony-Debré I, Steidl U. Functionally relevant RNA helicase mutations in familial and sporadic myeloid malignancies. *Cancer Cell.* 2015;27(5):609–11.

46. Polprasert C, Schulze I, Sekeres MA et al. Inherited and somatic defects in DDX41 in myeloid neoplasms. *Cancer Cell.* 2015;27(5):658–70.

47. Ganly P, Walker LC, Morris CM. Familial mutations of the transcription factor RUNX1 (AML1, CBFA2) predispose to acute myeloid leukemia. *Leuk Lymphoma.* 2004;45(1):1–10.

48. Michaud J, Wu F, Osato M et al. In vitro analyses of known and novel RUNX1/AML1 mutations in dominant familial platelet disorder with predisposition to acute myelogenous leukemia: Implications for mechanisms of pathogenesis. *Blood.* 2002;99:1364–72.

49. Minelli A, Maserati E, Rossi G et al. Familial platelet disor-der with propensity to acute myelogenous leukemia: Genetic heterogeneity and progression to leukemia via acquisition of clonal chromosome anomalies. *Genes Chromosomes Cancer.* 2004;40(3):165–71.

50. Preudhomme C, Renneville A, Bourdon V et al. High fre-quency of RUNX1 biallelic alteration in acute myeloid leukemia secondary to familial platelet disorder. *Blood.* 2009;113(22):5583–7.

51. Ripperger T, Steinemann D, Göhring G et al. A novel pedi-gree with heterozygous germline RUNX1 mutation causing familial MDS-related AML: Can these families serve as a multistep model for leukemic transformation? *Leukemia.* 2009;23(7):1364–6.

52. Jongmans MC, Kuiper RP, Carmichael CL et al. Novel RUNX1 mutations in familial platelet disorder with enhanced risk for acute myeloid leukemia: Clues for improved identification of the FPD/AML syndrome. *Leukemia.* 2010;24(1):242–6.

53. Antony-Debré I, Manchev VT, Balayn N et al. Level of RUNX1 activity is critical for leukemic predisposition but not for thrombocytopenia. *Blood.* 2015;125(6):930–40.

54. Antony-Debré I, Duployez N, Bucci M et al. Somatic muta-tions associated with leukemic progression of familial plate-let disorder with predisposition to acute myeloid leukemia. *Leukemia.* 2016;30(4):999–1002.

55. Kanagal-Shamanna R, Loghavi S, DiNardo CD et al. Bone marrow pathologic abnormalities in familial platelet disor-der with propensity for myeloid malignancy and germline RUNX1 mutation. *Haematologica.* 2017;102:1661–70.

56. Kelly LM, Gilliland DG. Genetics of myeloid leukemias. *Annu Rev Genomics Hum Genet.* 2002;3:179–98.

57. Tang JL, Hou HA, Chen CY et al. *AML1/RUNX1* mutations in 470 adult patients with de novo acute myeloid leukemia: Prognostic implication and interaction with other gene altera-tions. *Blood.* 2009;114:5352–61.

58. Gaidzik VI, Bullinger L, Schlenk RF et al. RUNX1 muta-tions in acute myeloid leukemia: Results from a comprehen-sive genetic and clinical analysis from the AML study group. *J Clin Oncol.* 2011;29(10):1364–72.

59. Haslam K, Langabeer SE, Hayat A, Conneally E, Vandenberghe E. Targeted next-generation sequencing of familial platelet disorder with predisposition to acute myeloid leukaemia. *Br J Haematol.* 2016;175(1):161–3.

60. Gaidzik VI, Teleanu V, Papaemmanuil E et al. RUNX1 mutations in acute myeloid leukemia are associated with distinct clinico-pathologic and genetic features. *Leukemia.* 2016;30:2160–8.

61. Beri-Dexheimer M, Latger-Cannard V, Philippe C et al. Clinical phenotype of germline RUNX1 haploinsufficiency: From point mutations to large genomic deletions. *Eur J Hum Genet.* 2008;16(8):1014–8.

62. Shinawi M, Erez A, Shardy DL et al. Syndromic thrombo-cytopenia and predisposition to acute myelogenous leukemia caused by constitutional microdeletions on chromosome 21q. *Blood.* 2008;112(4):1042–7.

63. Latger-Cannard V, Philippe C, Bouquet A et al. Haematological spectrum and genotype-phenotype correlations in nine unrelated families with RUNX1 mutations from the French network on inherited platelet disorders. *Orphanet J Rare Dis*. 2016;11:49.

64. Sakurai M, Kasahara H, Yoshida K et al. Genetic basis of myeloid transformation in familial platelet disorder/acute myeloid leukemia patients with haploinsufficient RUNX1 allele. *Blood Cancer J*. 2016;6:e392.

65. Yoshimi A, Toya T, Kawazu M et al. Recurrent CDC25C mutations drive malignant transformation in FPD/AML. *Nat Commun*. 2014;5:4770.

66. Churpek JE, Pyrtel K, Kanchi KL et al. Genomic analysis of germ line and somatic variants in familial myelodysplasia/ acute myeloid leukemia. *Blood*. 2015;126:2484–90.

67. Busque L, Patel JP, Figueroa ME et al. Recurrent somatic TET2 mutations in normal elderly individuals with clonal hematopoiesis. *Nat Genet*. 2012;44:1179–81.

68. Xie M, Lu C, Wang J et al. Age-related mutations associated with clonal hematopoietic expansion and malignancies. *Nat Med*. 2014;20:1472–8.

69. McKerrell T, Park N, Moreno T et al. Leukemia-associated somatic mutations drive distinct patterns of age-related clonal hemopoiesis. *Cell Rep*. 2015;10:1239–45.

70. Debeljak M, Kitanovski L, Pajic T, Jazbec J. Concordant acute myeloblastic leukemia in monozygotic twins with germline and shared somatic mutations in the gene for CCAAT enhancer-binding protein a with 13 years difference at onset. *Haematologica*. 2013;98(7):e73–4.

71. Tawana K, Wang J, Renneville A et al. Disease evolution and outcomes in familial AML with germline CEBPA mutations. *Blood*. 2015;126(10):1214–23.

72. Sellick GS, Spendlove HE, Catovsky D, Pritchard-Jones K, Houlston RS. Further evidence that germline CEBPA mutations cause dominant inheritance of acute myeloid leukaemia. *Leukemia*. 2005;19(7):1276–8.

73. De Lord C, Powles R, Mehta J et al. Familial acute myeloid leukaemia: Four male members of a single family over three consecutive generations exhibiting anticipation. *Br J Haematol*. 1998;100(3):557–60.

74. Pabst T, Eyholzer M, Haefliger S, Schardt J, Mueller BU. Somatic CEBPA mutations are a frequent second event in families with germline CEBPA mutations and familial acute myeloid leukemia. *J Clin Oncol*. 2008;26(31):5088–93.

75. Renneville A, Mialou V, Philippe N et al. Another pedigree with familial acute myeloid leukemia and germline CEBPA mutation. *Leukemia*. 2009;23(4):804–6.

76. Nanri T, Uike N, Kawakita T, Iwanaga E, Mitsuya H, Asou N. A family harboring a germ-line N-terminal C/EBPalpha mutation and development of acute myeloid leukemia with an additional somatic C-terminal C/EBPalpha mutation. *Genes Chromosomes Cancer*. 2010;49(3):237–41.

77. Taskesen E, Bullinger L, Corbacioglu A et al. Prognostic impact, concurrent genetic mutations, and gene expression features of AML with CEBPA mutations in a cohort of 1182 cytogenetically normal AML patients: Further evidence for CEBPA double mutant AML as a distinctive disease entity. *Blood*. 2011;117(8):2469–75.

78. Stelljes M, Corbacioglu A, Schlenk RF et al. Allogeneic stem cell transplant to eliminate germline mutations in the gene for CCAAT-enhancer binding protein a from hematopoietic cells in a family with AML. *Leukemia*. 2011;25(7):1209–10.

79. Xiao H, Shi J, Luo Y et al. First report of multiple CEBPA mutations contributing to donor origin of leukemia relapse after allogeneic hematopoietic stem cell transplantation. *Blood*. 2011;117(19):5257–60.

80. Yan B, Ng C, Moshi G et al. Myelodysplastic features in a patient with germline CEBPA-mutant acute myeloid leukaemia. *J Clin Pathol*. 2016;69(7):652–4.

81. Pathak A, Seipel K, Pemov A et al. Whole exome sequencing reveals a C-terminal germline variant in CEBPA-associated acute myeloid leukemia: 45-year follow up of a large family. *Haematologica*. 2016;101(7):846–52.

82. Ram J, Flamm G, Balys M et al. Index case of acute myeloid leukemia in a family harboring a novel CEBPA germ line mutation. *Blood Adv*. 2017;1(8):500–3.

83. Pippucci T, Savoia A, Perrotta S et al. Mutations in the 5′ UTR of ANKRD26, the ankyrin repeat domain 26 gene, cause an autosomal-dominant form of inherited thrombocytopenia, THC2. *Am J Hum Genet*. 2011;88(1):115–20.

84. Noris P, Perrotta S, Seri M et al. Mutations in ANKRD26 are responsible for a frequent form of inherited thrombocytopenia: Analysis of 78 patients from 21 families. *Blood*. 2011;117(24):6673–80.

85. Noris P, Favier R, Alessi MC et al. ANKRD26-related thrombocytopenia and myeloid malignancies. *Blood*. 2013;122(11):1987–9.

86. Bluteau D, Balduini A, Balayn N et al. Thrombocytopenia-associated mutations in the ANKRD26 regulatory region induce MAPK hyperactivation. *J Clin Invest*. 2014;124(2):580–91.

87. Topka S, Vijai J, Walsh MF et al. Germline ETV6 mutations confer susceptibility to acute lymphoblastic leukemia and thrombocytopenia. *PLoS Genet*. 2015;11(6):e1005262.

88. Melazzini F, Palombo F, Balduini A et al. Clinical and pathogenic features of ETV6-related thrombocytopenia with predisposition to acute lymphoblastic leukemia. *Haematologica*. 2016;101(11):1333–42.

89. Poggi M, Canault M, Favier M et al. Germline variants in ETV6 underlie reduced platelet formation, platelet dysfunction and increased levels of circulating CD34+ progenitors. *Haematologica*. 2017;102(2):282–94.

90. Lewinsohn M, Brown AL, Weinel LM et al. Novel germ line DDX41 mutations define families with a lower age of MDS/AML onset and lymphoid malignancies. *Blood*. 2016;127(8):1017–23.

91. Cheah JJC, Hahn CN, Hiwase DK, Scott HS, Brown AL. Myeloid neoplasms with germ line DDX41 mutation. *Int J Hematol*. 2017;106(2):163–74.

92. Cardoso SR, Ryan G, Walne AJ et al. Germline heterozygous DDX41 variants in a subset of familial myelodysplasia and acute myeloid leukemia. *Leukemia*. 2016;30(10):2083–6.

93. Holme H, Hossain U, Kirwan M, Walne A, Vulliamy T, Dokal I. Marked genetic heterogeneity in familial myelodysplasia/acute myeloid leukaemia. *Br J Haematol*. 2012; 158:242–8.

94. Howlader N, Noone AM, Krapcho M et al., eds. *SEER Cancer Statistics Review, 1975–2014, National Cancer Institute*. Bethesda, MD, https://seer.cancer.gov/csr/1975_2014/, based on November 2016 SEER data submission, posted to the SEER web site, April 2017. Last accessed February 4, 2018.

95. Yoshimi A, Toya T, Nannya Y et al. Spectrum of clinical and genetic features of patients with inherited platelet disorder

with suspected predisposition to hematological malignancies: A nationwide survey in Japan. *Ann Oncol.* 2016;27(5):887–95.

96. Marconi C, Canobbio I, Bozzi V et al. 5′ UTR point substitutions and N-terminal truncating mutations of ANKRD26 in acute myeloid leukemia. *J Hematol Oncol.* 2017;10(1):18.

97. Stockley J, Morgan NV, Bem D et al. Enrichment of FLI1 and RUNX1 mutations in families with excessive bleeding and platelet dense granule secretion defects. *Blood.* 2013;122(25):4090–3.

98. Sorrell A, Espenschied C, Wang W et al. Hereditary leukemia due to rare RUNX1c splice variant (L472X) presents with eczematous phenotype. *Int J Clin Med.* 2012;3(7).

99. Bluteau O, Sebert M, Leblanc T et al. A landscape of germ line mutations in a cohort of inherited bone marrow failure patients. *Blood.* 2018;131(7):717–32.

100. Green CL, Koo KK, Hills RK, Burnett AK, Linch DC, Gale RE. Prognostic significance of CEBPA mutations in a large cohort of younger adult patients with acute myeloid leukemia: Impact of double CEBPA mutations and the interaction with FLT3 and NPM1 mutations. *J Clin Oncol.* 2010;28(16):2739–47.

101. Fasan A, Haferlach C, Alpermann T et al. The role of different genetic subtypes of CEBPA mutated AML. *Leukemia.* 2014;28(4):794–803.

102. Horwitz M, Goode EL, Jarvik GP. Anticipation in familial leukemia. *Am J Hum Genet.* 1996;59:990–8.

103. Latger-Cannard V, Philippe C, Jonveaux P, Lecompte T, Favier R. Dysmegakaryopoiesis, a clue for an early diagnosis of familial platelet disorder with propensity to acute myeloid leukemia in case of unexplained inherited thrombocytopenia associated with normal-sized platelets. *J Pediatr Hematol Oncol.* 2011;33(7):e264–6.

104. Maciejewski JP, Padgett RA, Brown AL, Müller-Tidow C. DDX41-related myeloid neoplasia. *Semin Hematol.* 2017;54(2):94–7.

105. Arber DA, Brunning RD, Le Beau MM et al. Acute myeloid leukemia with recurrent genetic abnormalities. In: Swerdlow SH, Campo E, Harris NL et al. *WHO Classification of Tumours of Hematopoietic and Lymphoid Tissues.* Lyon: IARC Press; 2017: 130–49.

106. Raymond VM, Gray SW, Roychowdhury S et al. Germline findings in tumor-only sequencing: Points to consider for clinicians and laboratories. *J Natl Cancer Inst.* 2015;108(4). pii: djv351.

107. Li MM, Datto M, Duncavage EJ et al. Standards and guidelines for the interpretation and reporting of sequence variants in cancer: A joint consensus recommendation of the Association for Molecular Pathology, American Society of Clinical Oncology, and College of American Pathologists. *J Mol Diagn.* 2017;19(1):4–23.

108. Döhner H, Estey E, Grimwade D et al. Diagnosis and management of AML in adults: 2017 ELN recommendations from an international expert panel. *Blood.* 2017;129(4):424–47.

109. Ripperger T, Tauscher M, Haase D, Griesinger F, Schlegelberger B, Steinemann D. Managing individuals with propensity to myeloid malignancies due to germline RUNX1 deficiency. *Haematologica.* 2011;96:1892–4.

110. University of Chicago Hematopoietic Malignancies Cancer Risk Team. How I diagnose and manage individuals at risk for inherited myeloid malignancies. *Blood.* 2016; 128(14):1800–13.

111. Furutani E, Shimamura A. Germline genetic predisposition to hematologic malignancy. *J Clin Oncol.* 2017;35:1018–28.

112. Kratz CP, Achatz MI, Brugières L et al. Cancer screening recommendations for individuals with Li-Fraumeni syndrome. *Clin Cancer Res.* 2017;23:e38–45.

113. Kobayashi S, Kobayashi A, Osawa Y et al. Donor cell leukemia arising from preleukemic clones with a novel germline DDX41 mutation after allogenic hematopoietic stem cell transplantation. *Leukemia.* 2017;31(4):1020–2.

114. Kansal R. Classification of acute myeloid leukemia by the revised fourth edition World Health Organization criteria: A retrospective single-institution study with appraisal of the new entities of acute myeloid leukemia with gene mutations in *NPM1* and biallelic *CEBPA. Hum Pathol.* 2019;90:80–96.

71

Familial Monosomy 7 Syndrome

Dongyou Liu

CONTENTS

71.1 Introduction

Familial monosomy 7 syndrome (also referred to as bone marrow monosomy 7) is an autosomal dominant disorder characterized by early-childhood onset of bone marrow insufficiency/failure and predisposition to hematologic malignancies (e.g., myelodysplastic syndrome [MDS] and acute myeloid leukemia [AML]).

At the molecular level, monosomy 7 (loss of chromosome 7 or −7) or interstitial (partial) deletion of the long arm of chromosome 7 [del(7q) or 7q-] results in haploinsufficiency of several important genes [e.g., *SAMD9* (sterile α motif domain 9; 7q21.3), *SAMD9L* (SAMD9-like; 7q21.3), *EZH2* (enhancer of zeste homolog 2; 7q36.1), *MLL3* (mixed lineage leukemia 3; 7q36.1), and *CUX* (homeobox transcription factor; 7q22.1)], and represents a frequent cytogenetic abnormality in patients with bone marrow failure/MDS/AML. Nevertheless, monosomy 7 appears to have a more common occurrence in myeloid malignancies than del(7q) [1].

Although monosomy 7 is not uncommon and has been associated with MDS, AML, and other disorders (e.g., cerebellar ataxia/atrophy-pancytopenia syndrome, familial platelet disorder with propensity to AML, aplastic anemia, juvenile myelomonocytic leukemia [JMML], neurofibromatosis type 1, Noonan syndrome, Bloom syndrome, paroxysmal nocturnal hemoglobinuria, dyskeratosis congenita, Fanconi anemia, and Shwachman–Diamond syndrome), familial monosomy 7 syndrome is quite rare, with patients from <20 families diagnosed so far [2].

71.2 Biology

Monosomy 7 was first recognized in 1964 as a new clinical syndrome in the form of "sporadic adult MDS carrying monosomy 7 as the sole chromosomal anomaly that frequently develops into overt leukemia." Subsequent studies have confirmed monosomy 7 as the sole cytogenetic anomaly in children with MDS and JMML, with nearly half of affected children having refractory anemia, in contrast to the occurrence of either monosomy 7 or interstitial deletion of the long arm of chromosome 7 [del(7q)] in adult leukemia (which is often at more advanced disease stages with increased blast counts) [3].

Familial monosomy 7 syndrome (or bone marrow monosomy 7) refers to the occurrence of monosomy 7 as the sole anomaly in >2 siblings from a single family. To date, <20 families are shown to be affected by familial monosomy 7 syndrome, with most patients being children or adolescents [4].

Further research efforts toward pinpointing culprit genes located in chromosome 7 revealed *SAMD9* (sterile α motif domain 9), *SAMD9L* (SAMD9-like), *EZH2* (enhancer of zeste homolog 2), *MLL3* (mixed lineage leukemia 3), and *Miki/HEPACAM2* as commonly deleted genes in patients with myeloid neoplasia. Recent identification of germline heterozygous gain-of-function *SAMD9L* mutations (p.H880Q, p.I891T, p.R986C, and p.C1196S) in patients with neurological symptoms (ataxia, balance impairment, nystagmus, hyperreflexia, dysmetria, dysarthria) and hematologic abnormalities (single to tri-lineage cytopenias, MDS/-7), and *de novo* gain-of-function mutations in

SAMD9 in patients with MIRAGE syndrome (myelodysplasia, infection, restriction of growth, adrenal hypoplasia, genital phenotypes, and enteropathy) as well as MDS/-7 highlighted the role of these genes in the pathogenesis of MDS/-7 and other genetic disorders [5,6].

71.3 Pathogenesis

Evolving from early events, monosomy 7 (loss of chromosome 7 or −7) and interstitial (partial) deletion of the long arm of chromosome 7 [del(7q)] represent frequent acquired aberrations in sporadic MDS, AML, JMML, and other myeloid tumors, that are not detected in the germline. Commonly associated with secondary MDS/AML following mutagenic exposures (e.g., benzene, solvents and radiation; chemotherapy with alkylating agents), these cytogenetic abnormalities may be also provoked by antecedent aplastic anemia, Fanconi anemia, neurofibromatosis type I, cyclic neutropenia, and Schwachman syndrome as well as long-term treatment with granulocyte colony-stimulating factor (G-CSF) [7–9].

Mechanistically, monosomy 7 (−7) results from chromosome dissegregation in mitosis and is often associated with other unfavorable chromosome translocations [e.g., *EVI1* (3q26.2)]. On the other hand, del(7q) is due to chromosome rearrangement, and is often linked to more favorable karyotypes in AML [e.g., t(8;21), inv(16), t(15;17), and t(9;11)].

Familial monosomy 7 syndrome possibly represents a bone marrow environment conducive to the expansion of a −7 clone, and cells with −7 demonstrate a relative survival advantage over the surrounding bone marrow cells. This acquired abnormality may recur within the family, with relevant lesions emerging in the course of hematologic disease [10].

Among the genes identified in chromosome 7, *SAMD9* (sterile α motif domain 9), *SAMD9L* (SAMD9-like), *EZH2* (enhancer of zeste homolog 2), *MLL3* (mixed lineage leukemia 3), and *CUX* (mammalian ortholog of *Drosophila melanogaster* cut) appear to play a critical role in bone marrow failure/MDS/AML [6,11,12].

SAMD9 (7q21.3) and *SAMD9L* (7q21.3) encode endosomal proteins, which function as tumor suppressors and participate in homotypic fusion of primary/early endosomes for endosomal trafficking and metabolism of cytokine receptors. Germline gain-of-function mutations in *SAMD9/SAMD9L* are associated with infantile MDS/monosomy 7, with the age of onset at <5 years. In addition, gain-of-function mutations in the *SAMD9* gene are implicated in MIRAGE syndrome, whereas *SAMD9L* mutations also occur in ataxia pancytopenia syndrome. Further, germline gain-of-function mutations in *SAMD9/Samd9L* are linked to inherited bone marrow failure in children and adolescents [13–18].

EZH2 (7q36.1)) encodes a methyltransferase participates in the formation of the polycomb repressive complex 2 (PRC2), binds to lysine 27 of histone H3 (H3K27) and catalyzes the trimethylation of H3K27. Mutations in the *EZH2* gene eliminate its methyltransferase activity and are observed in 10% of MDS cases. In synergy with *TET2* or *RUNX1* mutations, *EZH2* haploinsufficiency linked to monosomy 7/del(7q) contributes to MDS pathogenesis.

MLL3 (7q36.1) encodes a protein with histone methyltransferase activity that targets lysine 4 of histone 3 (H3K4)., and acts as

a haploinsufficient tumor suppressor. Loss-of-function mutations in *MLL3* often co-occur with N-RAS, K-RAS, and PTPN11 activation, *NF1* deletion, and TP53 inactivation in MDS and AML.

CUX1 (7q22.1) encodes a homeobox transcription factor, which acts as a haploinsufficient tumor suppressor. Loss of heterozygosity or inactivating point mutation in *CUX* is associated with myeloid tumor, uterine leiomyoma, breast cancer, and tumors of the endometrium, large intestine, and lung [19].

71.4 Epidemiology

71.4.1 Prevalence

Monosomy 7 is a relatively common chromosomal abnormality with strong association with MDS, AML, and other disorders, and predominantly affects males (male to female ratio of 60 to 40). In a recent analysis of 6565 patients with cytogenetic abnormalities from the Mayo Clinic, 3192 (49%) have sole abnormality, including monosomy 7 ($n = 98$), 7q- ($n = 51$; such as 7q22, and 7q31-35), der(1;7)(q10;p10) ($n = 44$), balanced translocations ($n = 15$), ring 7 ($n = 13$), and 7p- ($n = 9$). Of these, monosomy 7 represents the most common abnormality in MDS, AML, and secondary/therapy-related MDS/AML [20]. Interestingly, monosomy 7 as sole abnormality accounts for 36% of cases, monosomy 7 with one additional abnormality for 14% of cases, and monosomy 7 as part of complex abnormalities for 50% of cases in the German–Austrian dataset. On the other hand, familial monosomy 7 syndrome is rare, with individuals from <20 families diagnosed so far.

71.4.2 Inheritance

Monosomy 7 may arise de novo, secondarily, or constitutionally. While monosomy 7 is absent in the germline, it appears an acquired abnormality that recurs within the family. In familial monosomy 7 syndrome, an autosomal dominant inheritance is noted [21].

71.4.3 Penetrance

Familial monosomy 7 syndrome shows incomplete penetrance and variable expressivity.

71.5 Clinical Features

Clinically, familial monosomy 7 syndrome resembles monosomy 7 or interstitial deletion of the long arm of chromosome 7 [del(7q) or 7q-], with typical presentations of early-childhood onset bone marrow insufficiency/failure along with increased tumor risk and rapid progression of MDS or AML. However, familial monosomy 7 syndrome tends to involve two or more individuals in a family whereas monosomy 7 or interstitial deletion of the long arm of chromosome 7 affects only one in a family [22].

71.5.1 Early-Childhood Onset

Abnormal hematologic findings related to familial monosomy 7 syndrome are noted in children as young as age 9 months, with

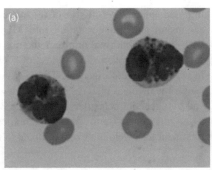
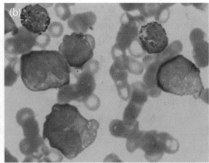

FIGURE 71.1 Characterization of basophils and myeloblasts from a patient suffering acute myeloid leukemia with basophilic differentiation transformed from myelodysplastic syndrome (e.g., acute basophilic leukemia or ABL) and having a loss of chromosome 7. (a) Peripheral blood smear showing medium-to large-sized abnormal cells with lobulated nucleus, and many basophilic granules (May-Giemsa staining). (b) Bone marrow smear showing large-sized myeloblasts with clear nucleoli, basophilic cytoplasm, and some vacuoles, along with some mature basophils (May-Giemsa staining). (Photo credit: Tanaka Y et al. *Case Rep Hematol.* 2017;2017:4695491.)

neurologic disorder (cerebellar ataxia or atrophy) sometimes appearing before the onset of hematologic anomalies.

71.5.2 Bone Marrow Insufficiency/Failure

Most patients with familial monosomy 7 syndrome show petechiae, easy bruising, and/or anemia, which are indicative of bone marrow insufficiency/failure.

71.5.3 Increased Tumor Risk and Rapid Progression

Familial monosomy 7 syndrome confers increased risk for bone marrow failure/MDS/AML, which often progresses rapidly after identification of monosomy 7 in peripheral blood. Most patients succumb to the disease within months to 3 years (Figure 71.1) [23].

Furthermore, monosomy 7 in association with one or more additional abnormalities may induce other symptoms and have poorer clinical outcomes than isolated monosomy 7.

71.6 Diagnosis

Diagnosis of familial monosomy 7 syndrome is based on medical history review, physical examination, cytogenetic analysis, and molecular testing.

Cytogenetic and interphase FISH analyses of bone marrow (without PHA or other mitogen stimulation) help reveal chromosomal abnormality, with bone marrow karyotype of 45,XX,-7 in females or 45,XY,-7 in males, often mosaic with normal cells (i.e., 46,XX in females and 46,XY in males) confirming the presence of monosomy 7 (with at least 3 out of 20 cells lacking a chromosome 7) (Figure 71.2) [24]. It is noteworthy that some individuals with a monosomy 7-positive family member may have a normal karyotype in peripheral blood and/or bone marrow but show mosaic monosomy 7 in peripheral blood and/or bone marrow later on [2].

Molecular testing also permits identification of monosomy 7. For example, chromosome microarray analysis (CMA) using oligonucleotide arrays or SNP genotyping arrays detect mosaic monosomy 7 readily. Genome sequencing or chromosome-targeted approach is also useful [25].

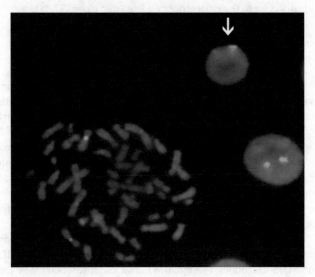

FIGURE 71.2 FISH using centromeric probe for chromosome 7 revealed monosomy 7 on bone marrow interphase cell (only one green signal) in a patient with therapy-related abnormality. (Photo credit: Tanizawa RS et al. *Rev Bras Hematol Hemoter.* 2011;33(6):425–31.)

Familial monosomy 7 syndrome should be suspected in an individual with (i) a relative having a confirmed hematologic disorder with monosomy 7, and (ii) evidence of bone marrow dysfunction (red cell macrocytosis and increased hemoglobin F concentration; bone marrow insufficiency such as thrombocytopenia, neutropenia, anemia, or bone marrow aplasia; MDS) and/or AML; monosomy 7 in peripheral blood and/or bone marrow cells). Parameters for severe aplastic anemia consists of granulocyte count <500/mL, platelet count <20,000/mL, reticulocyte count <1% after correction for hematocrit, and bone marrow biopsy with <25% of the normal cellularity for age [2].

Confirmation of familial monosomy 7 syndrome relies on identifying all of the following features: (i) monosomy 7 cells in peripheral blood or presence of MDS/AML; (ii) monosomy 7 cells in bone marrow; (iii) family member with characteristic hematologic findings and monosomy 7; (iv) exclusion of other hematologic disorders with monosomy 7 (e.g., normal chromosome breakage studies and telomere length assay) [2].

Differential diagnoses for familial monosomy 7 syndrome include disorders that also show monosomy 7, including cerebellar ataxia/atrophy–pancytopenia syndrome (cerebellar ataxia. hypoplastic anemia or AML with monosomy 7), familial platelet disorder with propensity to AML (due to *RUNX1* pathogenic variants), aplastic anemia, JMML, neurofibromatosis type 1, Noonan syndrome, Bloom syndrome, paroxysmal nocturnal hemoglobinuria, dyskeratosis congenita, Fanconi anemia, and Shwachman–Diamond syndrome [2].

71.7 Treatment

Treatment options for familial monosomy 7 syndrome consist of bone marrow transplantation (BMT, or allogeneic stem cell transplantation), and cytoreductive chemotherapy (e.g., demethylating agent 5-azacytidine, or imatinib for cytopenia) before transplant [26]. BMT is curative for monosomy 7 and should be performed ideally prior to the emergence of a leukemic clone [27,28].

71.8 Prognosis and Prevention

Monosomy 7 confers an unfavorable prognosis, with median survival of 14 months for sole abnormality, 11 months for cases with one additional abnormality, and 8 months for monosomy 7 as part of complex abnormalities. Furthermore, MDS with monosomy 7 appears to fare worse than MDS with del(7q) [29]. While patients <60 years of age with AML and sole monosomy 7 have median disease-free survival of 6 months and median overall survival of 9.6 months, patients >60 years of age with AML and sole monosomy 7 have median disease-free survival of 13.2 months and median overall survival of 8.4 months [20].

Identification of bone marrow abnormalities (cytopenias and bone marrow dysplasia) prior to the development of AML or MDS facilitates prompt BMT and improves outcomes for patients with monosomy 7. Due to the fact that monosomy 7 is absent in tissues sampled prenatally, and that specific gene mutation for familial monosomy 7 syndrome is unknown, prenatal diagnosis is not feasible.

71.9 Conclusion

Monosomy 7 (loss of chromosome 7) along with interstitial (partial) deletion of the long arm of chromosome 7 [del(7q)] is a genetic disorder that typically manifests as early-childhood onset bone marrow insufficiency/failure and predisposition to hematologic malignancies (particularly MDS and AML). While monosomy 7 is often identified as the sole cytogenetic anomaly in children with MDS and JMML, either monosomy 7 or del(7q) occurs in adult leukemia. Clinically, familial monosomy 7 syndrome resembles monosomy 7, mainly affecting children and adolescents. However, familial monosomy 7 syndrome tends to involve two or more individuals in a family, whereas monosomy 7 affects only one in a family. The pathogenesis of monosomy 7 or del(7q) is attributed to haploinsufficiency that results from the deletion of several important genes (e.g., *SAMD9, SAMD9L, EZH2,* and *MLL3*) within chromosome 7. Specifically, loss of chromosome 7 in hematopoietic stem and progenitor cells (HSPC) at an early stage followed by deletion of *SAMD9/SAMD9L* puts these cells

in an advantaged position over the surrounding cells for further expansion and transformation into MDS. Alternatively, HSPC with existing abnormal or even leukemic changes is disturbed by subsequent monosomy 7 and loss of *EZH2* and/or *MLL3*, leading to myeloid tumorigenesis. Diagnosis of familial monosomy 7 syndrome involves medical history review, physical examination, and cytogenetic/molecular assessment. Bone marrow transplantation (allogeneic stem cell transplantation) represents a curative treatment for myeloid leukemia linked to monosomy 7 or del(7q).

REFERENCES

1. Pezeshki A, Podder S, Kamel R, Corey SJ. Monosomy 7/del (7q) in inherited bone marrow failure syndromes: A systematic review. *Pediatr Blood Cancer.* 2017;64(12).
2. Morrissette JJD, Wertheim G, Olson T. Familial monosomy 7 syndrome. In: Adam MP, Ardinger HH, Pagon RA et al. editors. *GeneReviews®* [Internet]. Seattle (WA): University of Washington, Seattle; 1993–2018. 2010 Jul 8 [updated 2016 Jan 21].
3. Haase D. Cytogenetic features in myelodysplastic syndromes. *Ann Hematol.* 2008;87(7):515–26.
4. Eisfeld AK, Kohlschmidt J, Mrózek K et al. Mutational landscape and gene expression patterns in adult acute myeloid leukemias with monosomy 7 as a sole abnormality. *Cancer Res.* 2017;77(1):207–18.
5. Schwartz JR, Ma J, Lamprecht T et al. The genomic landscape of pediatric myelodysplastic syndromes. *Nat Commun.* 2017;8(1):1557.
6. Inaba T, Honda H, Matsui H. The enigma of monosomy 7. *Blood.* 2018;131(26):2891–8.
7. Olnes MJ, Poon A, Miranda SJ et al. Effects of granulocyte-colony-stimulating factor on Monosomy 7 aneuploidy in healthy hematopoietic stem cell and granulocyte donors. *Transfusion.* 2012;52(3):537–41.
8. Dimitriou M, Woll PS, Mortera-Blanco T et al. Perturbed hematopoietic stem and progenitor cell hierarchy in myelodysplastic syndromes patients with monosomy 7 as the sole cytogenetic abnormality. *Oncotarget.* 2016;7(45):72685–98.
9. Ullman D, Baumgartner E, Wnukowski N et al. Therapy-associated myelodysplastic syndrome with monosomy 7 arising in a Muir-Torre syndrome patient carrying SETBP1 mutation. *Mol Clin Oncol.* 2018;8(2):306–9.
10. Liew E, Owen C. Familial myelodysplastic syndromes: A review of the literature. *Haematologica.* 2011;96(10):1536–42.
11. Jerez A, Sugimoto Y, Makishima H et al. Loss of heterozygosity in 7q myeloid disorders: Clinical associations and genomic pathogenesis. *Blood.* 2012;119(25):6109–17.
12. Pastor VB, Sahoo SS, Boklan J et al. Constitutional SAMD9L mutations cause familial myelodysplastic syndrome and transient monosomy 7. *Haematologica.* 2018;103(3):427–37.
13. Nagamachi A, Matsui H, Asou H et al. Haploinsufficiency of SAMD9L, an endosome fusion facilitator, causes myeloid malignancies in mice mimicking human diseases with monosomy 7. *Cancer Cell.* 2013;24(3):305–17.
14. Buonocore F, Kühnen P, Suntharalingham JP et al. Somatic mutations and progressive monosomy modify SAMD9-related phenotypes in humans. *J Clin Invest.* 2017;127(5):1700–13.
15. Phowthongkum P, Chen DH, Raskind WH, Bird T. *SAMD9L*-related ataxia-pancytopenia syndrome. In: Adam

MP, Ardinger HH, Pagon RA et al. editors. *GeneReviews®* [Internet]. Seattle (WA): University of Washington, Seattle; 1993–2018. 2017 Jun 1.

16. Schwartz JR, Wang S, Ma J et al. Germline SAMD9 mutation in siblings with monosomy 7 and myelodysplastic syndrome. *Leukemia.* 2017;31(8):1827–30.

17. Davidsson J, Puschmann A, Tedgård U, Bryder D, Nilsson L, Cammenga J. SAMD9 and SAMD9L in inherited predisposition to ataxia, pancytopenia, and myeloid malignancies. *Leukemia.* 2018;32(5):1106–15.

18. Wong JC, Bryant V, Lamprecht T et al. Germline SAMD9 and SAMD9L mutations are associated with extensive genetic evolution and diverse hematologic outcomes. *JCI Insight.* 2018;3(14). pii: 121086.

19. Thoennissen NH, Lasho T, Thoennissen GB, Ogawa S, Tefferi A, Koeffler HP. Novel CUX1 missense mutation in association with 7q- at leukemic transformation of MPN. *Am J Hematol.* 2011;86(8):703–5.

20. Hussain FT, Nguyen EP, Raza S et al. Sole abnormalities of chromosome 7 in myeloid malignancies: Spectrum, histopathologic correlates, and prognostic implications. *Am J Hematol.* 2012;87(7):684–6.

21. Karimata K, Masuko M, Ushiki T et al. Myelodysplastic syndrome with Ph negative monosomy 7 chromosome following transient bone marrow dysplasia during imatinib treatment for chronic myeloid leukemia. *Intern Med.* 2011;50(5):481–5.

22. Rathi S, Kondekar S, Kadakia P, Sawardekar S, De T. Familial monosomy 7 syndrome associated with myelodysplasia. *Indian J Pediatr.* 2019;86(11):1059.

23. Tanaka Y, Tanaka A, Hashimoto A, Hayashi K, Shinzato I. Acute myeloid leukemia with basophilic differentiation transformed from myelodysplastic syndrome. *Case Rep Hematol.* 2017;2017:4695491.

24. Tanizawa RS, Kumeda CA, de Azevedo Neto RS, Leal Ade M, Ferreira Pde B, Velloso ED. Karyotypic and fluorescent *in situ* hybridization study of the centromere of chromosome 7 in secondary myeloid neoplasms. *Rev Bras Hematol Hemoter.* 2011;33(6):425–31.

25. Dwivedi AC, Lyons MJ, Kwiatkowski K et al. Clinical utility of chromosomal microarray analysis in the diagnosis and management of monosomy 7 mosaicism. *Mol Cytogenet.* 2014;7(1):93.

26. Jawad MD, Go RS, Ketterling RP, Begna KH, Reichard KK, Shi M. Transient monosomy 7 in a chronic myelogenous leukemia patient during nilotinib therapy: A case report. *Clin Case Rep.* 2016;4(3):282–6.

27. Otero L, de Souza DC, de Cássia Tavares R et al. Monosomy 7 in donor cell-derived leukemia after bone marrow transplantation for severe aplastic anemia: Report of a new case and review of the literature. *Genet Mol Biol.* 2012;35(4):734–6.

28. Ruiz-Gutierrez M, Bölükbaşı ÖV, Alexe G et al. Therapeutic discovery for marrow failure with MDS predisposition using pluripotent stem cells. *JCI Insight.* 2019;5. pii: 125157.

29. Cordoba I, González-Porras JR, Nomdedeu B et al. Better prognosis for patients with del(7q) than for patients with monosomy 7 in myelodysplastic syndrome. *Cancer.* 2012;118(1):127–33.

72

Familial Multiple Myeloma

Dongyou Liu

CONTENTS

72.1 Introduction

Multiple myeloma is a fatal B-cell malignancy characterized by clonal proliferation of malignant plasma cells in the bone marrow, bones, and other tissues and production of monoclonal protein in the blood or urine, leading to lytic bony lesions, fractures, anemia, renal failure, hypercalcemia, and immune dysfunction. Multiple myeloma is usually preceded by a precursor condition known as monoclonal gammopathy of undetermined significance (MGUS) followed by an intermediate stage called smoldering multiple myeloma (SMM). MGUS is asymptomatic and present in 3% of the general population over the age of 50 years, with potential to progress at a rate of 1%–2% per year. SMM is also asymptomatic and carries a higher risk of progression at 10% a year for the first 5 years, 3% per year for the next 5 years, and 1% per year thereafter [1].

Familial multiple myeloma accounts for a subset of multiple myeloma that shows familial clustering, with two or more members in each family affected, and presentation of various hematologic and solid tumors (e.g., MGUS, multiple myeloma, pancreatic cancer, acute myelogenous leukemia, and myelofibrosis).

Molecularly, progression from a premalignant condition MGUS to multiple myeloma is associated with the acquisition of chromosome abnormalities and genetic mutations such as *trisomies* (involving odd-numbered chromosomes except for chromosomes 1, 13, and 21; 42% of cases), *IgH translocations* [immunoglobulin heavy chain (IgH) translocations including t(11;14)(q13;q32)—*CCND1* (cyclin D1); t(4;14)(p16;q32)—*FGFR-3* and *MMSET*; t(14;16)(q32;q23)—*C-MAF*; t(14;20)

(q32;q11)—*MAFB*; other IgH translocations—*CCND3* (cyclin D3) in t(6;14)(p21;q32) translocation or other IgH translocations involving uncommon partner chromosomes; 30% of cases], *combined IgH translocation and trisomies* (trisomies plus one recurrent IgH translocation; 15% of cases), *isolated monosomy 14* (14q32 translocations involving unknown partner chromosomes; 4.5% of cases), *other cytogenetic abnormalities* (5.5% of cases), and *normal* (3% of cases) [1,2].

72.2 Biology

Although multiple myeloma has been known to affect humans since ancient times, it was Samuel Solly who provided a first description of this disease (including fatigue, bone pain from multiple fractures, replacement of the bone marrow by a red substance) in 1844, and Henry Bence Jones who associated copious protein excretion with this disorder in the middle of the nineteenth century. Subsequent studies revealed the involvement of monoclonal protein-producing plasma cells (terminally differentiated B-lymphocytes) in the causation of multiple myeloma [1].

Nowadays, multiple myeloma is recognized as a malignancy of bone marrow–residing plasma cells, which originate from the post-germinal lymphoid B-cell lineage and have the ability to produce immunoglobulin (Ig)/antibody and survive indefinitely in the bone marrow microenvironment, thus providing long-term support to host humoral immunity. Beginning from a precursor stage (MGUS) and progressing via an intermediate stage (SMM) to overt disease multiple myeloma, plasma cells undergo significant changes to become malignant cells that generate copious but completely nonfunctional Ig. In fact, many complications associated with multiple myeloma are attributed not only to invasive malignant cell growth in the bone and bone marrow, but also to the production of aberrant Ig. The pathologies of multiple myeloma are best summarized by the mnemonic CRAB (C, elevated calcium; R, renal failure; A, anemia; B, bone lesions), which was utilized previously as clinical diagnostic criteria to prevent patients with MGUS and SMM from receiving unnecessary and toxic chemotherapy. Molecularly, initiating somatic mutation accompanied by additional oncogenic mutations underlies the pathogenesis of multiple myeloma [3].

While early cases of multiple myeloma had no family connection, reports in the 1920s pointed to possible familial predisposition of this disorder. Following detailed documentation of two sisters with proven multiple myeloma in 1954 and further descriptions of familial cases in 1965, it becomes apparent that multiple myeloma has the ability to run within families [4,5].

72.3 Pathogenesis

Multiple myeloma represents an uncontrolled proliferation of plasma cell clones that evolves from a premalignant condition, MGUS, then an intermediate stage, SMM, and finally to an overt disease, multiple myeloma [6].

MGUS is an asymptomatic precursor pathogenic state that produces a plasma cell population of <10% in the bone marrow. MGUS is characterized by myeloma cell growth and alteration in the bone, which cause deterioration of both auxiliary and appendicular microarchitecture, contributing to skeletal fragility and increased risk of fracture and osteoporosis. Although isotype class switching and somatic hypermutation in the germinal center leading to malignant plasma cells initiate at MGUS, they are insufficient for evolution to multiple myeloma. Occurring in 3% of the general population over the age of 50 years, MGUS has potential of progression to multiple myeloma at a rate of 1%–2% per year. Additionally, some MGUS cases may turn into Waldenstrom macroglobulinemia, primary AL amyloidosis, or a lymphoproliferative disorder.

SMM is an asymptomatic intermediate state that meets the laboratory criteria of multiple myeloma (i.e., high serum or urinary monoclonal protein, clonal bone marrow plasma cells in the range of 10%–60%) without end-organ damage (i.e., hypercalcemia, renal insufficiency, anemia, or bone lesions). Compared to MGUS, SMM demonstrates a higher risk of progression at 10% a year for the first 5 years, 3% per year for the next 5 years, and 1% per year thereafter. However, all multiple myeloma cases evolve from SMM.

Multiple myeloma progressing from MGUS (at 1%–2% per year) and SMM (at 10%, 3% and 1% per year for 1, 5, and >6 years, respectively) often manifests as a plasmacytoma, a single tumor, and then multiple lesions after the acquisition of chromosome abnormalities and genetic mutations (from trisomies, IgH translocations, combined IgH translocation and trisomies, isolated monosomy 14, other cytogenetic abnormalities, to normal) [1,2].

Trisomies (trisomic multiple myeloma) refer to the presence of trisomies in the neoplastic plasma cells in the bone marrow and usually involve odd-numbered chromosomes except for chromosomes 1, 13, and 21. As chromosomes 1 and 13 are known to harbor important tumor suppressor genes [e.g., *RB1* (13q), *DIS3* (13q), *CDKN2C* (1q) and *FAF1* (1q)], loss of *CDKN2C* and *RB1* induces unregulated cell cycle within multiple myeloma, and loss of *FAF1* disrupts the process of multiple myeloma cell apoptosis. Trisomies are observed in 42% of multiple myeloma cases [1,2].

IgH translocations involving the immunoglobulin heavy chain (IgH) locus on chromosome 14q32 (IgH translocated multiple myeloma) are found in 30% of cases, and are due to errors of IgH switch recombination and somatic hypermutation resulting in the juxtaposition of IgH gene sequences located at chromosome 14q32 with non-immunoglobulin DNA loci of 11q13, 4p16.3, 16q23, 20q11, and 6p21. Specifically, t(11;14)(q13;q32) leads to the dysregulation of *CCND1* (*cyclin D1*, which is responsible for 15% of cases); t(4;14)(p16;q32) puts oncogenes *FGFR3* and *MMSET* under control of the IgH gene locus and increases the expression of *cyclin D2* (6% of cases); t(14;16)(q32;q23) and t(14;20)(q32;q11) also bring oncogenes *C-MAF* and *MAFB* within the IgH gene locus and enhance *cyclin D2* expression (representing 4% and <1% of cases, respectively), and t(6;14) (p21;q32) as well as other IgH translocations contributes to the increased expression of *CCND3* (*cyclin D3*, 5% of cases) [1,2].

Combined IgH translocation and trisomies (trisomies plus one of the recurrent IgH translocations) are primary cytogenetic abnormalities that appear at the time of establishment of MGUS and are found in 15% of multiple myeloma cases.

Isolated monosomy 14 such as 14q32 translocations involving unknown partner chromosomes is detected in 4.5% of multiple myeloma cases.

Other cytogenetic abnormalities in absence of IgH translocations or trisomy or monosomy 14 are fond in 5.5% of multiple myeloma cases.

Normal (no change) is found in 3% of multiple myeloma cases.

Chromosome abnormalities and genetic mutations alter chromosomal copy numbers, change gene expression signatures, and induce secondary mutations [e.g., gain(1q), del(1p), del(17p), del(13), *RAS* mutations, secondary translocations involving *MYC*], leading to gain of function in *K-Ras*, *N-Ras*, *FAM46C*, *MYC*, and *BRAF*, loss of function in *p53*, deregulation of the RAS/MAPK/NF-κB pathway, epigenetic changes with histone methylation/deacetylation, and genomic instability within multiple myeloma cells. In fact, overexpression of *MYC* associated with chromosomal rearrangement is a force driving the shift from MGUS to multiple myeloma [1,2].

72.4 Epidemiology

72.4.1 Prevalence

Multiple myeloma is a clonal plasma cell malignancy that makes up 0.8% of cancer cases and 13% (second after lymphoma) of hematologic malignancies worldwide. The estimated annual incidence of multiple myeloma is 0.4–5 cases per 100,000. Multiple myeloma tends to affect blacks more than whites (black to white ratio of 2 to 1) and has a much lower prevalence in Asians. Displaying a slight male predominance, multiple myeloma has a median diagnostic age of 66–70 years, with 0.02%–0.3% patients <30 years of age at diagnosis [1,2].

Relatives of patients with MGUS display a higher relative risk of developing MGUS (2.8-fold) and multiple myeloma (2.9-fold) than normal controls. The fact that multiple members from >100 families have been diagnosed with multiple myeloma or other plasma cell dyscrasias strongly supports the familiar clustering of multiple myeloma (i.e., familial multiple myeloma).

72.4.2 Inheritance

Multiple myeloma shows an autosomal dominant inheritance.

72.4.3 Penetrance

Multiple myeloma demonstrates incomplete penetrance and variable expressivity, which are influenced by the patient age, gender, racial and ethnic background, underlying immunodeficiency, exposure to radiation or dioxin-related compounds, family history of multiple myeloma and other hematolymphoid neoplasms, as well as obesity, cardiovascular disease, or type II diabetes mellitus.

72.5 Clinical Features

Multiple myeloma is an end stage of disease that progresses from MGUS and SMM.

MGUS is an asymptomatic precursor pathogenic state associated with myeloma cell growth, alterations in the bone, and increased risk of fracture and osteoporosis. It occurs in 3% of the general population over the age of 50 years and shows a plasma cell population of <10% in the bone marrow.

SMM is an asymptomatic intermediate state between MGUS and multiple myeloma and displays high serum or urinary monoclonal protein as well as clonal bone marrow plasma cells in the range of 10%–60%, without causing hypercalcemia, renal insufficiency, anemia, or bone lesions.

Multiple myeloma typically manifests as fatigue (due to anemia, 75% of cases), bone pain, osteolytic skeletal lesions (80%), hypercalcemia (15%), elevated serum creatinine ≥2 mg/dL (20%), and extramedullary disease (EMD) at the time of initial diagnosis (1%–2%), or later in the disease course (8%) [1,2].

72.6 Diagnosis

Diagnostic workup for MGUS, SMM and multiple myeloma involves complete blood count, serum calcium, serum creatinine, serum and urine protein electrophoresis with immunofixation, serum free light chain (SFLC) assay, and bone marrow examination. Further, low-dose whole-body computed tomography (CT), or fluorodeoxyglucose (FDG) positron emission tomography/CT (PET/CT), or plain radiographs of the entire skeleton help detect osteolytic bone lesions. Magnetic resonance imaging (MRI) of the whole body or spine/pelvis is necessary for patients with suspected SMM, and with attention directed to focal bone marrow lesions for patients with multiple myeloma. Moreover, FISH probes are useful for detecting trisomies, IgH translocations, *MYC* translocations, and abnormalities of chromosomes 1, 13, and 17 in bone marrow specimens [7].

Diagnosis of MGUS relies on quantification of immunoglobulin present in both the bloodstream and bone marrow, using serum protein electrophoresis (SPEP), urine protein electrophoresis (UPEP), serum and urine immunofixation electrophoresis (IFE), and SFLC assay. A plasma cell population of <10% in the bone marrow is indicative of MGUS, while a plasma cell content exceeding 10% in the bone marrow suggests either SMM or multiple myeloma (clinically evident) (Figures 72.1 and 72.2) [8,9].

Presence of monoclonal (M) protein (consisting of 50% IgG, 20% IgA, 20% immunoglobulin light chain, 2% IgD, and 0.5% IgM) in the serum or urine is suggestive of multiple myeloma, with detection sensitivity ranging from 82% by SPEP and 93% by serum IFE to 97% by either adding SFLC assay or a 24-h urine protein electrophoresis with immunofixation. About 2%–3% of multiple myeloma has no detectable M protein (so-called nonsecretory multiple myeloma) [10].

Previous diagnostic criteria for multiple myeloma required identification of CRAB features (hypercalcemia, renal failure, anemia, and osteolytic bone lesions) caused by the neoplastic clone of plasma cells. The availability of specific biomarkers and modern imaging tools permitted an improved definition of multiple myeloma and enabled the International Myeloma Working Group (IMWG) to update of the diagnostic criteria for multiple myeloma and related disorders in 2014, with the addition of three

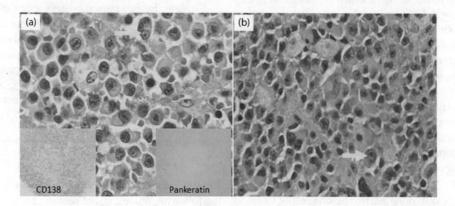

FIGURE 72.1 Microscopic examination of samples from a 39-year-old male with dural plasmacytoma and multiple myeloma revealed neoplastic proliferation of differentiated plasma cells of the middle cranial fossa mass (a) and epidural mass (b). Note the "clock face" pattern of nuclear chromatin and perinuclear halo (arrows). Immunochemistry (a, inset) was positive for CD138 (a plasma cell marker) and negative for pankeratin (marker for other tumors like glial tumors, malignant meningioma, or metastasis). (Photo credit: Gregorio LM, Soyemi TO. *Radiol Case Rep.* 2019;14(8):1007–13.)

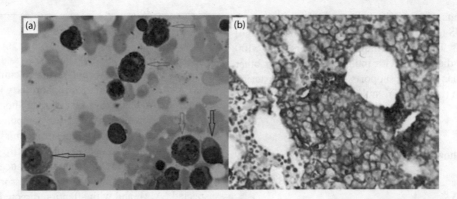

FIGURE 72.2 Multiple myeloma in a 46-year-old male showing plasma cells with eccentric nuclei and basophilic cytoplasm (black arrow), and abnormal plasma cells with numerous azurophilic granules (red arrow) in bone marrow aspirate (a, Giemsa stain); large sheets of myeloma cells in fixed paraffin-embedded bone marrow stained with CD 138 antibody using peroxidase-conjugated chromogen (b). (Photo credit: Sadeghi A et al. *Int J Hematol Oncol Stem Cell Res.* 2019;13(2):58–60.)

highly specific biomarkers (clonal bone marrow plasma cells ≥60%, SFLC ratio ≥100, >1 focal lesion on MRI) to existing markers of end-organ damage (hypercalcemia, renal insufficiency, anemia, or bone lesions). Therefore, a diagnosis of multiple myeloma can be made with 10% or more plasma cells in the bone marrow or a biopsy proven plasmacytoma plus one or more myeloma defining events (MDE), which consist of the presence of one or more CRAB features, or one or more biomarkers of malignancy [11].

Multiple myeloma can be stratified into standard risk [trisomies, t(11;14), t(6;14); accounting for 75% of new diagnoses], intermediate risk [t(4;14), gain(1q); accounting for 10% of new diagnoses] and high risk [t(14:16), t(14;20), del(17p); accounting for 15% of new diagnoses] on the basis of cytogenetic abnormalities in bone marrow specimens determined by FISH probes [12]. In addition, SMM is considered as high risk if bone marrow clonal plasma cells ≥10% plus any one or more of the following: serum M protein ≥30 g/L; IgA SMM; immunoparesis with reduction of two uninvolved immunoglobulin isotypes; serum involved/uninvolved free light chain ratio ≥8 (but <100); progressive increase in M protein level (evolving type of SMM); bone marrow clonal plasma cells 50%–60%; abnormal plasma cell immunophenotype (≥95% of bone marrow plasma cells are clonal) and reduction of one or more uninvolved immunoglobulin isotypes; t(4;14) or del 17p or 1q gain; increased circulating plasma cells; MRI with diffuse abnormalities or 1 focal lesion; PET-CT with focal lesion with increased uptake without underlying osteolytic bone destruction [1,2,12].

Differential diagnoses of multiple myeloma include:

IgM MGUS: Serum IgM monoclonal protein <3 g/dL; bone marrow lymphoplasmacytic infiltration <10%; no evidence of anemia, constitutional symptoms, hyperviscosity, lymphadenopathy, or hepatosplenomegaly that can be attributed to the underlying lymphoproliferative disorder.

Non-IgM MGUS: Serum monoclonal protein (non-IgM type) <3 g/dL; clonal bone marrow plasma cells <10%; absence of end-organ damage such as hypercalcemia, renal insufficiency, anemia, and bone lesions (CRAB) that can be attributed to the plasma cell proliferative disorder.

Light chain MGUS: Abnormal FLC ratio (<0.26 or >1.65); increased level of the appropriate involved light chain (increased kappa FLC in patients with ratio >1.65 and increased lambda FLC in patients with ratio <0.26); no immunoglobulin heavy chain expression on immunofixation; absence of end-organ damage that can be attributed to the plasma cell proliferative disorder; clonal bone marrow plasma cells <10%; urinary monoclonal protein <500 mg/24 h.

SMM: serum monoclonal protein (IgG or IgA) ≥3 gm/dL, or urinary monoclonal protein ≥500 mg per 24 h and/or clonal bone marrow plasma cells 10%–60%; absence of myeloma defining events or amyloidosis.

Solitary plasmacytoma: Biopsy-proven solitary lesion of bone or soft tissue with evidence of clonal plasma cells; normal bone marrow with no evidence of clonal plasma cells; normal skeletal survey and MRI (or CT) of spine and pelvis (except for the primary solitary lesion); absence of end-organ damage such as hypercalcemia, renal insufficiency, anemia, or bone lesions (CRAB) that can be attributed to a lympho-plasma cell proliferative disorder.

Solitary plasmacytoma with minimal marrow involvement: Biopsy-proven solitary lesion of bone or soft tissue with evidence of clonal plasma cells; clonal bone marrow plasma cells <10%; normal skeletal survey and MRI (or CT) of spine and pelvis (except for the primary solitary lesion); absence of end-organ damage such as hypercalcemia, renal insufficiency, anemia, or bone lesions (CRAB) that can be attributed to a lympho-plasma cell proliferative disorder [7,11,13].

72.7 Treatment

Current treatment for multiple myeloma may be organized in the following phases: initial therapy, autologous stem cell transplantation, consolidation/maintenance therapy, and treatment of relapse [14,15].

Initial therapy for standard and intermediate risk patients consists of four cycles of VRD (bortezomib, lenalidomide, dexamethasone). For high-risk patients, four cycles of KRD (carfilzomib, lenalidomide, dexamethasone) is offered in place of VRD. For patients who are ≥75 years of age, frail and unable to tolerate a triplet regimen, lenalidomide plus dexamethasone (Rd) is a preferred choice for initial therapy (especially for standard risk patients).

Autologous stem cell transplantation (ASCT) for eligible patients is performed after four cycles of VRD. On the other hand, transplant ineligible patients are kept on initial therapy for 12–18 months.

Consolidation/maintenance therapy (lenalidomide for standard risk patients; bortezomib for intermediate or high-risk patients) is prescribed after ASCT or initial therapy, with the duration of therapy determined by the presence or absence of high-risk cytogenetic features [16–18].

Treatment of relapse consists of Rd (lenalidomide, dexamethasone) or VRD/KRD/bortezomib, thalidomide, dexamethasone (VTD)/bortezomib, cyclophosphamide, dexamethasone (VCD) combinations (indolent relapse) or a triplet regimen or a combination of multiple active agents (more aggressive relapse) [19,20].

The above regimens may be also supplemented with new therapeutic approaches, such as specific proteasome inhibitors (PIs) and immunomodulatory drugs (IMiDs), monoclonal antibodies, and histone deacetylase inhibitors [21–26].

72.8 Prognosis

Patients with low, standard, and high-risk multiple myeloma have overall survival of >10, 7, and 2 years, respectively, and patients eligible for ASCT have 4-year survival rates of >80%. Host factors, tumor burden (stage), biology [presence or absence of secondary cytogenetic abnormalities such as del(17p), gain(1q), or del(1p); elevated lactate dehydrogenase level], and response to therapy may impact on patient survival [27].

Patients with stage I disease (serum albumin >3.5, serum beta-2-microglobulin <3.5; no high risk cytogenetics; normal LDH; making up 28% of cases) have 5-year survival rate of 82%; those with stage II disease (neither stage I or III; making up 62% of cases) have 5-year survival rate of 63%; and those with stage III disease [serum beta-2-microglobulin >5.5; high risk cytogenetics t(4;14), t(14;16), or del(17p)] or elevated LDH; making up 10% of cases] have 5-year survival rate of 40% [28].

72.9 Conclusion

Multiple myeloma results from a clonal expansion of malignant plasma cells that reside in the bone and bone marrow and produce copious nonfunctional immunoglobulins, leading to osteolytic lesions, bone pain, hypercalcemia, cytopenias, renal failure, and neuropathy. Evolving from a precursor stage, monoclonal gammopathy of unknown significance (MGUS) and intermediate stage, smoldering multiple myeloma (SMM), multiple myeloma is an aggressive disease (manifesting as a plasmacytoma, a single tumor, and then multiple lesions) with significant morbidity and mortality. The acquisition of various chromosome abnormalities and genetic mutations (e.g., trisomies, IgH translocations, combined IgH translocation and trisomies, isolated monosomy 14, and other cytogenetic abnormalities), which induce activation and upregulation of various pro-oncogenes and downregulation of tumor suppressor genes, appears to underlie the malignant transformation of plasma cells and pathogenesis of multiple myeloma [29,30]. Due to its nonspecific clinical symptoms, diagnosis of multiple myeloma relies on laboratory quantification of immunoglobulins in serum and urine specimens and observation of increased plasma cell population in bone marrow. Management of multiple myeloma involves initial therapy, autologous stem cell transplantation, consolidation/maintenance therapy, and treatment of relapse [30,31]. Recent development of new therapeutic agents (e.g., specific proteasome inhibitors and immunomodulatory drugs, monoclonal antibodies, and histone deacetylase inhibitors) provide additional arsenal for combating this important hematologic malignancy [32].

REFERENCES

1. Hari P. Recent advances in understanding multiple myeloma. *Hematol Oncol Stem Cell Ther.* 2017;10(4):267–71.
2. Dhakal B, Girnius S, Hari P. Recent advances in understanding multiple myeloma. *F1000Res.* 2016;5. pii: F1000 Faculty Rev-2053.
3. Halvarsson BM, Wihlborg AK, Ali M et al. Direct evidence for a polygenic etiology in familial multiple myeloma. *Blood Adv.* 2017;1(10):619–23.
4. Kazandjian D. Multiple myeloma epidemiology and survival: A unique malignancy. *Semin Oncol.* 2016;43(6):676–81.
5. Koura DT, Langston AA. Inherited predisposition to multiple myeloma. *Ther Adv Hematol.* 2013;4(4):291–7.
6. Fairfield H, Falank C, Avery L, Reagan MR. Multiple myeloma in the marrow: Pathogenesis and treatments. *Ann N Y Acad Sci.* 2016;1364:32–51.
7. Landgren O, Rajkumar SV. New developments in diagnosis, prognosis, and assessment of response in multiple myeloma. *Clin Cancer Res.* 2016;22(22):5428–33.
8. Gregorio LM, Soyemi TO. Multiple myeloma presenting as dural plasmacytoma. *Radiol Case Rep.* 2019;14(8):1007–13.
9. Sadeghi A, Nematollahi P, Moghaddas A, Darakhshandeh A. Multiple myeloma with intracytoplasmic azurophilic franules. *Int J Hematol Oncol Stem Cell Res.* 2019;13(2):58–60.
10. Dupuis MM, Tuchman SA. Non-secretory multiple myeloma: From biology to clinical management. *Onco Targets Ther.* 2016;9:7583–90.
11. Rajkumar SV, Dimopoulos MA, Palumbo A et al. International Myeloma Working Group updated criteria for the diagnosis of multiple myeloma. *Lancet Oncol.* 2014;15:e538–48.
12. Rajkumar SV. Updated diagnostic criteria and staging system for multiple myeloma. *Am Soc Clin Oncol Educ Book.* 2016;35:e418–23.
13. Musto P, Anderson KC, Attal M et al. Second primary malignancies in multiple myeloma: An overview and IMWG consensus. *Ann Oncol.* 2017;28(2):228–45.
14. Larocca A, Mina R, Gay F, Bringhen S, Boccadoro M. Emerging drugs and combinations to treat multiple myeloma. *Oncotarget.* 2017;8(36):60656–72.
15. PDQ Adult Treatment Editorial Board. *Plasma Cell Neoplasms (Including Multiple Myeloma) Treatment (PDQ®): Health Professional Version. PDQ Cancer Information Summaries [Internet].* Bethesda (MD): National Cancer Institute (US); 2002. 2017 Oct 20.

16. Lee HS, Min CK. Optimal maintenance and consolidation therapy for multiple myeloma in actual clinical practice. *Korean J Intern Med*. 2016;31(5):809–19.

17. Sengsayadeth S, Malard F, Savani BN, Garderet L, Mohty M. Posttransplant maintenance therapy in multiple myeloma: The changing landscape. *Blood Cancer J*. 2017;7(3):e545.

18. Ying L, YinHui T, Yunliang Z, Sun H. Lenalidomide and the risk of serious infection in patients with multiple myeloma: A systematic review and meta-analysis. *Oncotarget*. 2017;8(28):46593–600.

19. Dingli D, Ailawadhi S, Bergsagel PL et al. Therapy for relapsed multiple myeloma: Guidelines from the Mayo stratification for myeloma and risk-adapted therapy. *Mayo Clin Proc*. 2017;92(4):578–98.

20. Cook G, Zweegman S, Mateos MV, Suzan F, Moreau P. A question of class: Treatment options for patients with relapsed and/or refractory multiple myeloma. *Crit Rev Oncol Hematol*. 2018;121:74–89.

21. Nikesitch N, Ling SC. Molecular mechanisms in multiple myeloma drug resistance. *J Clin Pathol*. 2016;69(2):97–101.

22. Brayer J, Baz R. The potential of ixazomib, a second-generation proteasome inhibitor, in the treatment of multiple myeloma. *Ther Adv Hematol*. 2017;8(7):209–20.

23. Gonsalves WI, Milani P, Derudas D, Buadi FK. The next generation of novel therapies for the management of relapsed multiple myeloma. *Future Oncol*. 2017;13(1):63–75.

24. Issa ME, Takhsha FS, Chirumamilla CS, Perez-Novo C, Vanden Berghe W, Cuendet M. Epigenetic strategies to reverse drug resistance in heterogeneous multiple myeloma. *Clin Epigenetics*. 2017;9:17.

25. Raza S, Safyan RA, Rosenbaum E, Bowman AS, Lentzsch S. Optimizing current and emerging therapies in multiple myeloma: A guide for the hematologist. *Ther Adv Hematol*. 2017;8(2):55–70.

26. Zhang T, Wang S, Lin T et al. Systematic review and meta-analysis of the efficacy and safety of novel monoclonal antibodies for treatment of relapsed/refractory multiple myeloma. *Oncotarget*. 2017;8(20):34001–17.

27. Bustoros M, Mouhieddine TH, Detappe A, Ghobrial IM. Established and novel prognostic biomarkers in multiple myeloma. *Am Soc Clin Oncol Educ Book*. 2017;37: 548–60.

28. Hanbali A, Hassanein M, Rasheed W, Aljurf M, Alsharif F. The evolution of prognostic factors in multiple myeloma. *Adv Hematol*. 2017;2017:4812637.

29. Wang W, Zhang Y, Chen R et al. Chromosomal instability and acquired drug resistance in multiple myeloma. *Oncotarget*. 2017;8(44):78234–44.

30. Rajkumar SV. Multiple myeloma: 2018 update on diagnosis, risk-stratification, and management. *Am J Hematol*. 2018;93(8):981–1114.

31. Mateos MV, Ludwig H, Bazarbachi A et al. Insights on Multiple myeloma treatment strategies. *Hemasphere*. 2018;3(1):e163.

32. Nishihori T, Shain K. Insights on genomic and molecular alterations in multiple myeloma and their incorporation towards risk-adapted treatment strategy: Concise clinical review. *Int J Genomics*. 2017;2017:6934183.

73

Fanconi Anemia

Dongyou Liu

CONTENTS

73.1 Introduction

Fanconi anemia (FA) is a rare inherited disorder characterized by congenital abnormalities (short stature, radial ray defects, abnormal skin pigmentation, skeletal malformations of the upper and lower limbs, microcephaly, ophthalmic and genitourinary tract anomalies), progressive bone marrow failure (BMF) with pancytopenia (initially with thrombocytopenia or leukopenia), and predisposition to acute myeloid leukemia (AML) and epithelial malignancies (in the head and neck, skin, gastrointestinal tract, and genitourinary tract) in early childhood.

Molecularly, FA relates to mutations in FA genes encoding proteins that function together in the FA pathway to repair DNA interstrand crosslink damage, to maintain genome stability, to modulate responses to oxidative stress, viral infection, and inflammation, to facilitate mitophagic responses, to enhance signals that promote stem cell function and survival, and to prevent early-onset BMF and cancer (Table 73.1) [1,2].

73.2 Biology

Fanconi anemia was first reported by Guido Fanconi in 1927 involving three brothers who presented with congenital microcephaly, café-au-lait spots, cutaneous hemorrhage, and hypoplasia of the testes, and later died of a severe condition resembling pernicious anemia. However, the connection between FA and chromatid breaks in blood lymphocytes was not established until nearly four decades later [1,2].

Following detailed analyses of many additional cases, it has become clear that FA is a tumor-prone chromosomal instability disorder characterized by diverse clinical features, ranging from physical anomalies, progressive BMF (thrombocytopenia and later pancytopenia), and tumor susceptibility to occasional developmental delay/intellectual disability. Given the highly variable clinical presentations of FA, which overlaps phenotypically with other BMF syndromes to some extent, a chromosomal breakage assay was developed in 1976 to assesses lymphocyte or fibroblast hypersensitivity to chromosomal breakage induced by DNA crosslinking agents such as mitomycin C (MMC), diepoxybutane (DEB), cyclophosphamide, and cisplatinum, and this method is still considered the gold standard for diagnosing FA (Figure 73.1) [3,4].

Since the cloning of the first Fanconi anemia-related gene (*FANCC*) in 1992, a total of 22 genes (i.e., *FANCA, FANCB, FANCC, FANCD1/BRCA2, FANCD2, FANCE, FANCF, FANCG, FANCI, FANCJ, FANCL, FANCM, FANCN, FANCO, FANCP, FANCQ, FANCR/RAD51, FANCS, FANCT, FANCU, FANCV,* and *FANCW*) have been isolated and characterized by using complementation analysis of cell lines from affected patients, positional cloning, biochemical purification, and sequencing analysis (Table 73.1). These genes encode components of the FA (or FA/BRCA) pathway that work in coordination to repair DNA damage and ensure genome stability [5]. Homozygous or compound heterozygous mutations in 20 of these genes (except for *FANCB* and *FANCR/RAD51*), or heterozygous mutations in *FANCR/RAD51*, or hemizygous mutations in X-chromosome-linked FANCB result in nonfunctional proteins that cause disruption in the FA pathway, compromise DNA damage repair, and induce various chromosome breaks (Figure 73.1) [6–9].

TABLE 73.1

Characteristics of Fanconi Anemia Genes and Proteins

Gene	Synonym	Chromosomal Location	Protein	Functions	Effects
FANCA		16q24.3	Fanconi anemia group A protein	FA core complex	Pathogenic variants in 60%–70% of cases
FANCB	FAAP95	Xp22.31	Fanconi anemia group B protein	FA core complex	Pathogenic variants in ~2% of cases
FANCC		9p22.3	Fanconi anemia group C protein	FA core complex	Pathogenic variants in ~14% of cases
FANCD1	BRCA2	13q12.3	Breast cancer type 2 susceptibility protein	Homologous recombination repair, recruiting RAD51 onto DNA, interaction with FANCN, stalled replication fork protection	Pathogenic variants in ~3% of cases; not all patients have bone marrow failure
FANCD2		3p25.3	Fanconi anemia group D2 protein	ID complex (FANCI/FANCD2 heterodimer), ubiquitinated after DNA damage, stalled replication fork protection	Pathogenic variants in ~3% of cases
FANCE		6p21.3	Fanconi anemia group E protein	FA core complex	Pathogenic variants in ~3% of cases
FANCF		11p14.3	Fanconi anemia group F protein	FA core complex	Pathogenic variants in ~2% of cases
FANCG	XRCC9	9p13.3	Fanconi anemia group G protein	FA core complex	Pathogenic variants in ~10% of cases
FANCI		15q26.1	Fanconi anemia group I protein	ID complex (FANCI/FANCD2 heterodimer), ubiquitinated after DNA damage, required for FA core complex activation	Pathogenic variants in ~1% of cases
FANCJ	BRIP1, BACH1	17q23.2	Fanconi anemia group J protein	Homologous recombination repair, interstrand crosslink repair, 3' to 5' helicase, interaction with BRCA1	Pathogenic variants in ~2% of cases
FANCL	PHF9	2p16.1	Fanconi anemia group L protein	FA core complex, E3 ubiquitin ligase	No cancer
FANCM	Hef	14q21.3	Fanconi anemia group M protein	FA core complex, helicase, interstrand crosslink repair, DNA translocase, required for FANCI-D2 ubiquitination	The only known patient also has a FANCA mutation
FANCN	PALB2	16p12.2	Partner and localizer of BRCA2	Homologous recombination repair, interstrand crosslink repair, interaction with BRCA1 and BRCA2, facilitating BRCA2 function	
FANCO	RAD51C	17q25.1	DNA repair protein RAD51 homolog 3	Homologous recombination repair, interstrand crosslink repair, RAD51 paralog, RAD51 nucleoprotein filament stability	No bone marrow failure and cancer
FANCP	SLX4	16p13.3	Structure-specific endonuclease subunit SLX4	Interstrand crosslink unhooking, coordination with XPF-ERCC1, interaction with MUS81-EME1 and SLX1 nucleases	
FANCQ	ERCC4, XPF	16p13.12	DNA repair endonuclease XPF	Interstrand crosslink unhooking, structure-specific endonuclease, association with ERCC1	
FANCR	RAD51	15q15.1	DNA repair protein RAD51 homolog 1	Homologous recombination repair, stalled fork protection	No bone marrow failure and cancer
FANCS	BRCA1	17q21.31	Breast cancer type 1 susceptibility protein	Homologous recombination repair, promoting RAD51 recruitment, interaction with FANCN	No bone marrow failure
FANCT	UBE2T	1q32.1	Ubiquitin-conjugating enzyme E2 T	E2 ubiquitin-conjugating enzyme for FANCD2 complex, interaction with FANCL	
FANCU	XRCC2	7q36.1	DNA repair protein XRCC2	Homologous recombination repair, RAD51 paralog, RAD51 nucleoprotein filament stability	No bone marrow failure
FANCV	REV7, MAD2L2	1p36.22	Mitotic spindle assembly checkpoint protein MAD2B	Translesion synthesis	
FANCW	RFWD3	16q23.1	E3 ubiquitin-protein ligase RFWD3	Homologous recombination repair, ubiquitin ligase	

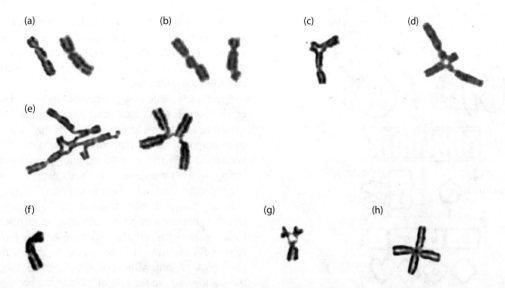

FIGURE 73.1 Chromosomal aberrations in Fanconi anemia as revealed by chromosomal breakage assay in the presence of mitomycin C (MMC). (a) Chromatid gap (broken piece in place); (b) chromatid break (broken piece dislocated); (c) chromatid interchange figure (triradial); (d) chromatid interchange figure (quadriradial); (e) other chromatid interchange figures. In the final analysis, (a) and (b) are counted as one, (c) and (d) as two break events. The left figure in (e) is counted as 8 break events (5 centromeres plus 3 open breaks); the right figure is equivalent to a quadriradial as in (d) (2 break events, in which two break points remain disconnected). (f) A gap is not 100% convincing and should be ignored. (g) An association of three acrocentric chromosomes showing satellite association, not to be confused with a triradial, as in (c). (h) Two overlapping chromosomes, not to be confused with a true quadriradial, as in (d). (Photo credit: Oostra AB et al. *Anemia.* 2012;2012:238731.)

73.3 Pathogenesis

The FA (or FA-BRCA) pathway is a nuclear multi-protein network that regulates cellular damage response to DNA interstrand crosslinks, and disruption of this pathway results in the cellular and clinical abnormalities associated with FA [10,11].

The identification of 22 genes encoding components of the FA pathway has uncovered critical insights into the mechanisms of crosslinked DNA repair (Table 73.1). The encoded FA proteins may be separated into three categories on the basis of their participation in interstrand crosslink (ICL) repair: (i) FA core complex (a homodimeric nuclear complex with ubiquitin ligase activity formed by FANCA, FANCB, FANCC, FANCE, FANCF, FANCG, FANCL, and FANCM, along with FA-associated proteins FAAP100, FAAP24, HES1, MHF1, and MHF2); (ii) ID complex (a FANCI/FANCD2 heterodimer formed after monoubiquitination mediated by FA core complex), and (iii) downstream repair factors (including DNA endonuclease FANCQ, nuclease scaffolding protein FANCP, translesion synthesis factor FANCV, and homologous recombination proteins FANCD1/BRCA2, FANCJ, FANCN, FANCO, FANCR, FANCS, and FANCU, and FANCW).

In response to crosslinking agents (e.g., MMC, diepoxybutane, and nitrogen mustards), FA core complex interacts with FANCT (a ubiquitin-conjugating enzyme). Together with ID complex, FA core complex/FANCT acts on ICLs, which covalently link the two DNA strands, preventing their separation, blocking critical cellular processes such as transcription and replication, and leading to gross-chromosomal aberrations like chromosome deletion, chromosome loss and DNA breaks. FANCP and FANCQ then cooperate to create DNA strand incisions on either side of ICL (a process called unhooking), and FANCV (an accessory subunit of DNA polymerase) engages in translesion synthesis over unhooked ICL. At the final step of ICL repair, a homologous recombination-mediated process involving FANCD1, FANCJ, FANCN, FANCO, FANCR, FANCS, FANCU, and FANCW restores the fidelity of the genome (Figure 73.2) [12–16].

Biallelic (homozygous or compound heterozygous) mutations in one of the 20 *FANC* genes, or heterozygous mutations in *FANCR* (*RAD51*), or hemizygous mutations in *FANCB* (X-linked) generate loss of function FA proteins, disrupt the FA pathway, result in hypersensitivity to agents that cause interstrand DNA crosslinks, leaving spontaneous and DNA damage-induced chromosomal breaks in the cells unrepaired, and increasing chromosomal fragility and cancer susceptibility [17,18].

It appears that pathogenic variants of differing types may have a bearing on the phenotypic presentations of FA. For example, patients harboring *FANCD1* (*BRCA2*) IVS7 pathogenic variant often develop AML and solid tumors by age 3 years, while those with other *BRCA2* pathogenic variants do so by age 6 years; patients with homozygous *FANCA* pathogenic variants tend to have early onset of anemia and high incidence of leukemia compared to those with other *FANCA* pathogenic variants; patients having *FANCC* c.456+4A>T, p.Arg548Ter, and p.Leu554Pro may have early onset hematologic abnormalities and severe congenital anomalies in comparison with those having other *FANCC* pathogenic variants (e.g., del22G, p.Asp23IlefsTer23, p.Gln13Ter, c.456+4A>T). Further, patients with FANCG pathogenic variants are associated with severe marrow failure and a high incidence of leukemia compared to those with *FANCC* pathogenic variants, and patients with *FANCN* (*PALB2*) pathogenic variants often develop solid tumors (e.g., medulloblastoma, Wilms tumor) [1,2,19].

FA gene mutations commonly detected in clinical cases include *FANCA* (60%–70% of cases), *FANCC* (14%), *FANCG*

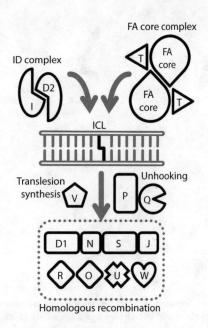

FIGURE 73.2 Schematic presentation of the FA pathway. FA core complex contains a homodimeric module for ubiquitination. FANCT acts as a ubiquitin-conjugating enzyme associated with FA core complex. FANCD2 and FANCI form a heterodimer (ID complex), which acts as target for mono-ubiquitination mediated by FA core complex. FANCP and FANCQ are involved in DNA strand incisions on either side of interstrand crosslink (ICL) (a process called "unhooking"). FANCV is an accessory subunit of DNA polymerase involved in translesion synthesis over unhooked ICL. At the final step of ICL repair, a homologous recombination-mediated process restores the fidelity of the genome. (Photo credit: Sakai W, Sugasawa K. *Genes Environ.* 2019;41:6.)

(10%), *FANCD1* (BRCA2, 3%), *FANCD2* (3%), *FANCE* (3%), *FANCB* (2%), *FANCF* (2%), *FANCJ* (2%), and *FANCI* (1%). In contrast, mutations in other FA genes (*FANCL, FANCM, FANCN, FANCO, FANCP, FANCQ, FANCR, FANCT, FANCU, FANCV,* and *FANCW*) are infrequent, and found in <1% of clinical cases [20,21].

73.4 Epidemiology

73.4.1 Prevalence

Fanconi anemia is an importance genetic cause of aplastic anemia and BMF, with an estimated incidence of 1 in 360,000 live births and a carrier frequency of approximately 1 in 181 in North Americans and <1 in 100 in certain population groups, such as Ashkenazi Jews (*FANCC, FANCD1/BRCA2,* 1 in 89), northern Europeans (*FANCC*), Afrikaners (*FANCA,* 1 in 83), sub-Saharan blacks (*FANCG*), and Spanish gypsies (*FANCA,* 1 in 64. There is a slight male predominance among patients (male to female ratio of 1.2:1).

73.4.2 Inheritance

Fanconi anemia caused by homozygous or compound heterozygous mutations involving one of 20 *FANC* genes (*FANCA, FANCC, FANCG, FANCD1/BRCA2, FANCD2, FANCE, FANCF, FANCJ, FANCI, FANCL, FANCM, FANCN, FANCO,*

FANCP, FANCQ, FANCT, FANCU, FANCV, FANCW) demonstrates autosomal recessive inheritance, FA due to heterozygous mutations in *FANCR* (*RAD51*) shows autosomal dominant inheritance, and FA resulting from hemizygous mutations in *FANCB* has X-linked inheritance.

Specifically, autosomal recessive transmission involves two carrier parents (both harboring heterozygous *FANC* pathogenic variants), who have a 25% risk of producing an affected offspring (with homozygous or compound heterozygous *FANC* pathogenic variant), a 50% risk of producing an asymptomatic carrier offspring (with heterozygous *FANC* pathogenic variant) and 25% risk of producing a normal offspring (no *FANC* pathogenic variant).

In autosomal dominant transmission, heterozygous *FANCR* (*RAD51*) mutation involving one allele is sufficient to cause disease. Interestingly, all affected patients with *FANCR* (*RAD51*)-related FA reported to date appear to evolve from a de novo pathogenic variant, without obvious family connection. However, these patients have the risk of passing the pathogenic variant to their offspring [22].

In X-linked transmission, a carrier female has a 50% of risk of passing the *FANCB* pathogenic variant to their offspring, with male offspring becoming affected, and female offspring becoming an unaffected carrier.

73.4.3 Penetrance

Fanconi anemia displays incomplete penetrance and variable expressivity. Patients harboring *FANC* pathogenic variants may not always show the whole spectrum of clinical diseases (i.e., physical anomalies, progressive BMF/pancytopenia, and tumor predisposition), and some individuals may have neither physical anomalies nor pancytopenia).

73.5 Clinical Features

Clinically, FA often presents with physical anomalies (75% of cases), progressive BMF (from thrombocytopenia to pancytopenia), and tumor susceptibility.

Physical anomalies range from growth deficiency (short stature/reduced growth, low birth weight); abnormal skin pigmentation (café-au-lait macules, hypopigmentation; 40% of cases); skeletal malformations (absent, hypoplastic, bifid, duplicated, triphalangeal, long, proximally placed thumbs, 35%; absent or hypoplastic in case with abnormal thumbs, absent or weak pulse radii, 7%; flat thenar eminence, absent first metacarpal, clinodactyly, polydactyly in hands, 5%; abnormal toes, club feet, syndactyly, 5%; dysplastic, short ulnae, 1%; congenital hip dislocation); microcephaly (20%); ophthalmic anomalies (microphthalmia, cataracts, astigmatism, strabismus, epicanthal folds, hypotelorism, hypertelorism, ptosis, 20%); genitourinary tract anomalies (horseshoe, ectopic, pelvic, hypoplastic, dysplastic, or absent kidney, hydronephrosis or hydroureter, 20%; hypospadias, micropenis, cryptorchidism, anorchia, hypo- or azoospermia, reduced fertility in males, 25%; bicornuate or uterus malposition, small ovaries in females, 2%); endocrine anomalies (hypothyroidism, diabetes mellitus, glucose abnormality); deafness.(due usually to middle ear bony anomalies) and abnormal ear shape (dysplastic, narrow ear canal, abnormal pinna, 10%); congenital

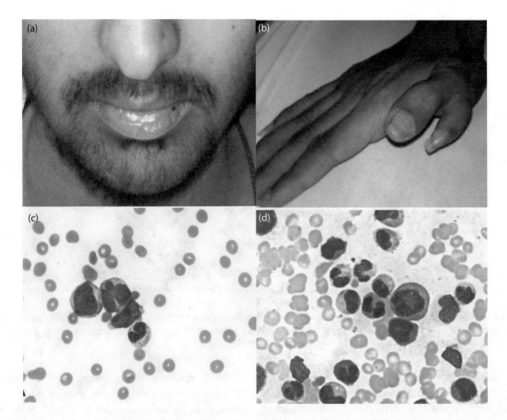

FIGURE 73.3 Fanconi anemia in a 17-year-old male showing Fanconi facies with tapering jaw (a); thumb polydactyly (b); blast cells in peripheral blood smear (c); and Auer rod-containing blast cell in bone marrow aspirate (d), indicating the transformation of Fanconi anemia to acute myeloid leukemia. (Photo credit: Hussain S, Adil SN. *BMC Res Notes.* 2013;6:316.)

heart defect (patent ductus arteriosus, atrial septal defect, ventricular septal defect, coarctation of the aorta, truncus arteriosus, situs inversus, 6%); gastrointestinal anomalies (esophageal, duodenal, or jejunal atresia, imperforate anus, tracheoesophageal fistula, annular pancreas, malrotation, 5%); central nervous system (small pituitary, pituitary stalk interruption syndrome, absent corpus callosum, cerebellar hypoplasia, hydrocephalus, dilated ventricles, 3%); to other anomalies (triangular, micrognathia, mid-face hypoplasia, 2%; spina bifida, scoliosis, hemivertebrae, rib anomalies, 2%; Sprengel deformity, Klippel–Feil anomaly, short or webbed neck, low hairline, 1%) (Figure 73.3) [1,2,23].

Bone marrow failure appears as thrombocytopenia or leukopenia (associated with macrocytosis and elevated fetal hemoglobin) at 5–10 years of age followed by pancytopenia (which generally worsens over time, with absolute neutrophil count changing from <1500/mm³ to <500/mm³; platelet count from 150,000–50,000/mm³ to <30,000/mm³; hemoglobin level from ≥8 g/dL to <8 g/dL) and sweet syndrome (neutrophilic skin infiltration, a natural progression of hematologic disease), leading to extreme fatigue, shortness of breath, pale skin, feeling dizzy, headaches, chest pain, frequent infections, easy bruising, and nosebleeds (Figure 73.3) [1,2,23].

Tumor susceptibility is reflected by a 500- to 800-fold increase of risk for AML at age 5–15 years typically after the onset of marrow failure, with the cumulative incidence of 13% by age 50 years, and a 500- to 700-fold increase of risk for head and neck squamous cell carcinomas (particularly the tongue) and tumors of the vulva and vagina (Figure 73.4) [24–26].

Developmental delay or intellectual disability may occur in 10% of patients with FA.

73.6 Diagnosis

Diagnosis of FA involves medical history review, physical examination, laboratory assessment, imaging, histopathology, and cytogenetic/molecular analysis.

Patients showing the following clinical and laboratory features should be suspected of FA: (i) physical anomalies (short stature; abnormal skin pigmentation such as café-au-lait macules, hypopigmentation; skeletal malformations such as hypoplastic thumb, hypoplastic radius; microcephaly; ophthalmic anomalies; genitourinary tract anomalies); (ii) laboratory findings (macrocytosis, increased fetal hemoglobin, cytopenia especially thrombocytopenia, leukopenia and neutropenia); (iii) pathology findings (progressive BMF, adult-onset aplastic anemia, myelodysplastic syndrome, AML, early-onset solid tumors such as squamous cell carcinomas and liver tumors, inordinate toxicities from chemotherapy or radiation).

Further evidence of FA is obtained by cytogenetic analysis of lymphocytes with DEB and MMC, which reveals increased chromosome breakage and radial forms (Figure 73.1) [3].

Confirmation of FA diagnosis is achieved with molecular identification of biallelic pathogenic variants in one of the 20 genes, or heterozygous pathogenic variant in *FANCR (RAD51)*, or hemizygous pathogenic variant in *FANCB* [27].

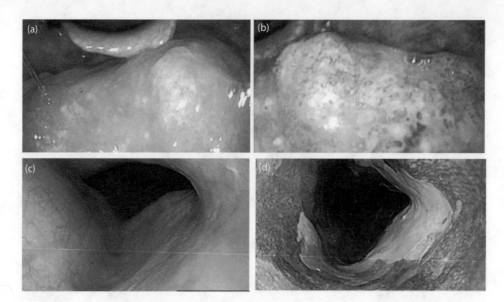

FIGURE 73.4 Fanconi anemia in a 30-ish male with squamous cell carcinoma displaying a white, elevated lesion in the left tongue root (a); corresponding brownish area on narrow-band imaging (NBI) (b); a poor vascular permeability pattern in the upper thoracic esophagus (c), and corresponding unstained area by Lugol staining (d). (Photo credit: Onishi S et al. *Intern Med.* 2019;58(4):529–33.)

Differential diagnoses for FA include hereditary breast and ovarian cancer (heterozygous pathogenic variants in *FANCD1/ BRCA2*, *FANCJ/BRIP1*, and *FANCN/PALB2*), pancreatic cancer (heterozygous pathogenic variants in *FANCN/PALB2*), xeroderma pigmentosum (*FANCN*), Cockayne syndrome (*FANCN*), XFE progeroid syndrome (*FANCN*), Bloom syndrome (spontaneous chromosome breakage independent of diepoxybutane), ataxia-telangiectasia (spontaneous chromosome breakage independent of diepoxybutane), Nijmegen breakage syndrome (NBS; short stature, progressive microcephaly with loss of cognitive skills, premature ovarian failure in females, recurrent sinopulmonary infections, and an increased risk for lymphoma; increased chromosome breakage with MMC; autosomal recessive disorder due to NBN pathogenic variants), Seckel syndrome (growth retardation, microcephaly with intellectual disability, characteristic "bird-headed" facial appearance, pancytopenia or AML, increased chromosome breakage with DNA crosslinking agents such as MMC and DEB, autosomal recessive disorder due to biallelic pathogenic variants in *ATR*, *NIN*, *ATRIP*, *RBBP8*, *CEP152*, *CENPJ*, and *CEP63*), neurofibromatosis type 1 (café-au-lait macules), TAR syndrome (thrombocytopenia with absent radii), dyskeratosis congenita, Diamond–Blackfan anemia, Shwachman–Diamond syndrome, severe congenital neutropenia, amegakaryocytic thrombocytopenia, Baller–Gerold syndrome, Rothmund–Thomson syndrome, Roberts syndrome, Warsaw breakage syndrome, DK-phocomelia, VACTERL hydrocephalus syndrome (radial ray defects), and Wiskott–Aldrich syndrome [1,2,28–30].

73.7 Treatment

Treatment options for FA include oral androgens (e.g., oxymetholone, to improve red cell and platelet counts in 50% of patients), G-CSF (to improve neutrophil count), hematopoietic stem cell transplantation (HSCT), and surgery/radiation and chemotherapy (for solid tumors) [31–33].

HSCT represents the only curative modality for the hematologic manifestations and use of reduced doses of alkylating agents (e.g., cyclophosphamide [CY] with or without antithymocyte globulin [ATG]) and radiation (e.g., thoracoabdominal irradiation [TAI]/or total body irradiation [TBI]) for conditioning increases its efficiency. Recent incorporation of fludarabine in the conditioning regimen further improves its outcome [31,33].

73.8 Prognosis and Prevention

Prognosis for FA is poor, with 1-, 5-, and 10-year overall survival rates of 92.5%, 89%, and 86%, respectively. The average life span for people with FA is in the range of 20–30 years, with about 80% of patients living to age 18 or older. The most common causes of death for FA patients are BMF, leukemia, and solid tumors. Children between 5 and 15 years of age have an increased tendency to develop AML and myelodysplasia, and those surviving to adulthood often develop solid tumors (e.g., tumors of the reproductive organs in affected females). Patients undergoing allogeneic HSCT have a 5-year survival rate of 83% [34].

Human papilloma virus (HPV) vaccination starting from 9 years of age may help reduce the risk of gynecologic cancer in females, and oral cancer in all individuals.

Avoidance of smoking, secondhand smoke, and alcohol, and safe sex practices are helpful in reducing the cancer risk in FA patients.

Early diagnosis (based on DEB/MMC or molecular testing) of FA in suspected individuals facilitates prompt treatment and improves patient survival.

Prenatal diagnosis for and preimplantation testing (including cytogenetic analysis of chromosome breakage in fetal cells obtained by chorionic villus sampling or amniocentesis in the presence of DEB/MMC or molecular test) may be offered to prospective parents who harbor pathogenic variants in one of 22 *FANC* genes [35].

73.9 Conclusion

Fanconi anemia a rare genetic disorder that typically manifests as congenital anomalies, progressive BMF (from thrombocytopenia or leukopenia to pancytopenia), and increased susceptibility to leukemia and solid tumors. The molecular mechanisms of FA lie in mutations in the 22 genes encoding components of the FA pathway, leading to nonfunctional FA proteins that increase sensitivity to crosslinking agents, hamper interstrand crosslink repair, and contribute to genomic instability. While observation of characteristic clinical features aids in FA diagnosis, examination of chromatid breaks induced by DNA crosslinking agents is invaluable for confirming its identity. Furthermore, use of molecular techniques allows identification of specific gene mutations, yielding genetic insights that can be exploited for improved treatment and prevention of this fatal syndrome.

REFERENCES

1. Mehta PA, Tolar J. Fanconi anemia. In: Adam MP, Ardinger HH, Pagon RA et al. editors. *GeneReviews® [Internet]*. Seattle (WA): University of Washington, Seattle; 1993–2018. 2002 Feb 14 [updated 2018 Mar 8].
2. Moore CA, Krishnan K. *Bone Marrow Failure. StatPearls* [Internet]. Treasure Island (FL): StatPearls Publishing; 2018. 2017 Oct 6.
3. Oostra AB, Nieuwint AW, Joenje H, de Winter JP. Diagnosis of Fanconi anemia: Chromosomal breakage analysis. *Anemia*. 2012;2012:238731.
4. Cantor SB, Brosh RM Jr. What is wrong with Fanconi anemia cells? *Cell Cycle*. 2014;13(24):3823–7.
5. Katsuki Y, Takata M. Defects in homologous recombination repair behind the human diseases: FA and HBOC. *Endocr Relat Cancer*. 2016;23(10):T19–37.
6. Duxin JP, Walter JC. What is the DNA repair defect underlying Fanconi anemia? *Curr Opin Cell Biol*. 2015;37:49–60.
7. Bluteau D, Masliah-Planchon J, Clairmont C et al. Biallelic inactivation of REV7 is associated with Fanconi anemia. *J Clin Invest*. 2016;126:3580–4.
8. Brosh RM Jr, Bellani M, Liu Y, Seidman MM. Fanconi anemia: A DNA repair disorder characterized by accelerated decline of the hematopoietic stem cell compartment and other features of aging. *Ageing Res Rev*. 2017;33:67–75.
9. Mamrak NE, Shimamura A, Howlett NG. Recent discoveries in the molecular pathogenesis of the inherited bone marrow failure syndrome Fanconi anemia. *Blood Rev*. 2017;31(3):93–9.
10. Michl J, Zimmer J, Tarsounas M. Interplay between Fanconi anemia and homologous recombination pathways in genome integrity. *EMBO J*. 2016;35(9):909–23.
11. Bhattacharjee S, Nandi S. DNA damage response and cancer therapeutics through the lens of the Fanconi anemia DNA repair pathway. *Cell Commun Signal*. 2017;15(1):41.
12. Huang Y, Leung JW, Lowery M et al. Modularized functions of the fanconi anemia core complex. *Cell Rep*. 2014;7:1849–57.
13. Lopez-Martinez D, Liang CC, Cohn MA. Cellular response to DNA interstrand crosslinks: The Fanconi anemia pathway. *Cell Mol Life Sci*. 2016;73(16):3097–114.
14. Palovcak A, Liu W, Yuan F, Zhang Y. Maintenance of genome stability by Fanconi anemia proteins. *Cell Biosci*. 2017;7:8.
15. Datta A, Brosh RM Jr. Holding all the cards-How Fanconi anemia proteins deal with replication stress and preserve genomic stability. *Genes (Basel)*. 2019;10(2). pii: E170.
16. Sakai W, Sugasawa K. Importance of finding the bona fide target of the Fanconi anemia pathway. *Genes Environ*. 2019;41:6.
17. Bagby G. Recent advances in understanding hematopoiesis in Fanconi Anemia. *F1000Res*. 2018;7:105.
18. Sumpter R Jr, Levine B. Emerging functions of the Fanconi anemia pathway at a glance. *J Cell Sci*. 2017;130(16):2657–62.
19. Chen H, Zhang S, Wu Z. Fanconi anemia pathway defects in inherited and sporadic cancers. *Transl Pediatr*. 2014;3(4):300–4.
20. Cheung RS, Taniguchi T. Recent insights into the molecular basis of Fanconi anemia: Genes, modifiers, and drivers. *Int J Hematol*. 2017;106(3):335–44.
21. Knies K, Inano S, Ramirez MJ et al. Biallelic mutations in the ubiquitin ligase RFWD3 cause Fanconi anemia. *J Clin Invest*. 2017; 127: 3013–27.
22. Ameziane N, May P, Haitjema A et al. A novel Fanconi anaemia subtype associated with a dominant-negative mutation in RAD51. *Nat Commun*. 2015;6:8829.
23. Hussain S, Adil SN. Rare cytogenetic abnormalities in acute myeloid leukemia transformed from Fanconi anemia—a case report. *BMC Res Notes*. 2013;6:316.
24. Alter BP. Fanconi anemia and the development of leukemia. *Best Pract Res Clin Haematol*. 2014;27(3-4):214–21.
25. Degrolard-Courcet E, Sokolowska J, Padeano MM et al. Development of primary early-onset colorectal cancers due to biallelic mutations of the FANCD1/BRCA2 gene. *Eur J Hum Genet*. 2014;22(8):979–87.
26. Onishi S, Tajika M, Tanaka T et al. Superficial esophageal cancer in a Fanconi anemia patient that was treated successfully by endoscopic submucosal resection. *Intern Med*. 2019;58(4):529–33.
27. Solomon PJ, Margaret P, Rajendran R et al. A case report and literature review of Fanconi Anemia (FA) diagnosed by genetic testing. *Ital J Pediatr*. 2015;41:38.
28. Zhu X. Current insights into the diagnosis and treatment of inherited bone marrow failure syndromes in China. *Stem Cell Investig*. 2015;2:15.
29. Wu ZH. Phenotypes and genotypes of the chromosomal instability syndromes. *Transl Pediatr*. 2016;5(2):79–83.
30. Iwafuchi H. The histopathology of bone marrow failure in children. *J Clin Exp Hematop*. 2018;58(2):68–86.
31. Chao MM, Ebell W, Bader P et al. Consensus of German transplant centers on hematopoietic stem cell transplantation in Fanconi anemia. *Klin Padiatr*. 2015;227(3):157–65.
32. Petryk A, Kanakatti Shankar R, Giri N et al. Endocrine disorders in Fanconi anemia: recommendations for screening and treatment. *J Clin Endocrinol Metab*. 2015;100(3):803–11.
33. Ayas M. Hematopoietic cell transplantation in Fanconi anemia and dyskeratosis congenita: A minireview. *Hematol Oncol Stem Cell Ther*. 2017;10(4):285–9.
34. Van Wassenhove LD, Mochly-Rosen D, Weinberg KI. Aldehyde dehydrogenase 2 in aplastic anemia, Fanconi anemia and hematopoietic stem cells. *Mol Genet Metab*. 2016;119(1-2):28–36.
35. Sorbi F, Mecacci F, Di Filippo A, Fambrini M. Pregnancy in Fanconi anaemia with bone marrow failure: a case report and review of the literature. *BMC Pregnancy Childbirth*. 2017;17(1):53.

74

GATA2 Deficiency

Dongyou Liu

CONTENTS

74.1 Introduction

Guanine-adenine-thymine-adenine 2 (*GATA2*) deficiency is a rare autosomal dominant disorder associated with aggressive infections, respiratory problems, hearing loss, leg swelling, cytopenia, myelodysplasia, myeloid leukemia, pulmonary alveolar proteinosis (PAP) and lymphedema. *GATA2* deficiency can be separated into several distinct phenotypes, including monoMAC syndrome (monocytopenia and *Mycobacterium avium* complex infections), dendritic cell, monocyte, and lymphocyte (DCML) deficiency, familial myelodysplastic syndrome/acute myeloid leukemia (MDS/AML) (myeloid neoplasms), deafness-lymphedema-leukemia syndrome (Emberger syndrome), and natural killer (NK) cell deficiency [1].

Heterozygous mutations in the *GATA2 gene* on chromosome 3q21.3 impair the function (either protein dysfunction or uniallelic reduced transcription/translation) of its encoded protein (GATA2), which participates in the regulation of several downstream genes fundamental to the development of the lymphatics and hematopoietic systems, leading to GATA2 deficiency [1].

74.2 Biology

GATA2 deficiency (also known as GATA2 deficiency syndrome or GATA2 haploinsufficiency) is a recently defined disorder that covers several seemingly unrelated diseases, each with a specific constellation of signs and symptoms (including monoMAC syndrome, Emberger syndrome, familial MDS/AML) that were found in 2011 to contain inactivating mutations in the *GATA2*

gene. In subsequent years, a number of newly or previously described diseases (i.e., DCML; formerly immunodeficiency 21), NK cell deficiency, chronic myelomonocytic leukemia (CMML), aplastic anemia, and chronic neutropenia were also grouped under GATA2 deficiency due to their apparent link to inactivating *GATA2* mutations [2–6].

Interestingly, peripheral blood examination reveals that DCML mainly affects people between 20–35 years of age, while MDS occurs in people of 30–35 years of age. In contrast, clinical assessment shows that HPV and upper respiratory tract infection (URTI) typically affect 20 years old, mycobacteria and pulmonary infection and autoimmunity affect 25 years old, refractory infection, solid tumor, and abnormal cytogenetics are found in 30–35 years old.

Since the *GATA2* gene encodes a transcription factor (GATA2) that plays a critical role in the development, proliferation, maintenance, and functionality of hematopoietic stem cells (HSC) that later differentiate into blood cells, lymphatic and other tissues, genetic mutations resulting in reduction (haploinsufficiency) of the GATA2 protein will undoubtedly have a wide-ranging impact on various organs and systems, inducing apparently benign abnormalities initially, which often progress to severe organ failure, MDS, leukemia, and aggressive infections [7–10].

74.3 Pathogenesis

Located on chromosome 3q21.3, the guanine-adenine-thymine-adenine 2 (*GATA2*) gene spans 13.7 kb with seven exons (including two untranslated 5-prime exons) and encodes a 480 aa, 50 kDa protein (GATA2 or GATA binding protein 2), which constitutes one of six GATA transcription factors in humans (i.e., GATA1–6), with GATA1–3 largely expressed in hematopoietic cells, while GATA4–6 is often expressed in nonhematopoietic tissues such as heart and gut. Specifically, GATA-2 predominantly occurs in adult and developing HSC, myeloid progenitors and mast cells, and represents a pivotal regulator of HSC and their progeny, hematopoietic progenitor cells (HPC). By contrast, GATA-1 often targets erythroid cells, megakaryocytes, and eosinophils, and GATA3 affects T cells [11].

Structurally, GATA2 consists of N-terminus, transactivation domain (TAD), negative regulatory domain (NRD), two zinc finger domains (i.e., N-finger at residues 294–344, and C-finger at residues 349–398, which serve as DNA-binding domains and sites for protein–protein interaction), nuclear localization signal (NLS), TAD, and C-terminus, within which several sites of post-translational modifications (i.e., serine phosphorylation, acetylation, SUMOlyation, and ubiquitination) exist [11].

GATA2 transcription is regulated by CEBPA, HOXA9, ETS1, BMP4, NOTCH1, SPI1, EVI1, and cytokines IL1 and TNFα as well as ATPase component of the SWI/SNF complex in cooperation with LDB1. Post-translational modifications (PTM) of GATA2 consist of phosphorylation, acetylation, SUMOlyation, and ubiquitination [11].

Expressed in mature megakaryocytes, mast cells, and monocytes, GATA2 binds directly to the consensus DNA sequence (A/T)GATA(A/G)GATA2 in its downstream effectors (e.g., SPI1, FLI1, TAL1, LMO2, RUNX1, GATA1, and CEBPA) via its two highly conserved zinc finger domains. Specifically, GATA2 forms a core heptad regulatory unit (consisting GATA2, TAL1, LYL1, LMO2, ERG, FLI1, and RUNXI) that is found over 1000 loci in primitive hematopoietic cells, and plays an essential role in the proliferation and differentiation of HSC, including the endothelial to hematopoietic transition (yielding the first adult HSC) in embryo and HSC survival and self-renewal in adult hematopoiesis [12].

Heterozygous mutations in *GATA2* inactivate one allele and reduce the available amounts of GATA2 (so-called haploinsufficiency). Insufficient GATA2 still allows HSC to enter cell cycle and differentiate, but limits their self-renewal capacity, leading to the depletion of dendritic, monocyte, B, and NK cells, and subsequent development of cytopenia, myelodysplasia, myeloid leukemia, PAP, lymphedema, and aggressive infections [13–15]. On the other hand, GATA2 overexpression blocks the differentiation of HSC [16].

To date, nearly 100 germline (inherited, 33%) or somatic (de novo, 66%) *GATA2* mutations have been described, ranging from frameshift, nonsense (stop codon), or missense (amino acid substitution) mutations, single nucleotide variants, and in-frame insertion or deletion to whole-gene deletions, with two-thirds of mutations located in the zinc finger domains. Interestingly, *GATA2* pathogenic variant gT354M is frequently found in AML (60% of cases) instead of MDS (38% of cases), whereas *GATA2* pathogenic variants gR396Q and gR398W commonly occur in MDS (78% and 50% of cases, respectively) compared to AML (22% and 17% of cases, respectively) [17–20].

It is not unusual to find germline heterozygous *GATA2* mutations co-occurring with other genetic alterations [e.g., *ASXL1, monosomy 7, trisomy 8, trisomy 21, der(1;7), +1q, −7q, EZH2, HECW2, GATA1; with increased* risk for MDS/AML], and

somatic heterozygous *GATA2* mutations with biallelic *CEBPA* mutation and t(9;11), leading to intermediate to high risk for AML and CML. Indeed, GATA2 deficiency patient with MDS may acquire additional pathogenic variant (e.g., *NRAS* Q61 K) during transition to AML [1,11].

74.4 Epidemiology

74.4.1 Prevalence

GATA2 deficiency is a rare disorder, with >200 patients molecularly confirmed so far. According to a recent survey, GATA2 deficiency shows a slight male predilection (male to female ratio of 56 to 44), with a median diagnostic age of 30 years, affects white (75%) more than Hispanic (16%), African American (7%), and Asian (2%) subjects, and typically manifests as severe viral infections (32%), disseminated NTM infections (28%), MDS/AML (21%), lymphedema (9%), invasive fungal infections (4%), or no symptoms (7%) [1].

74.4.2 Inheritance

GATA2 deficiency displays autosomal dominant inheritance.

74.4.3 Penetrance

GATA2 deficiency displays high (>90%) penetrance and variable expressivity. Despite their shared genetic background (all with inactivating *GATA2* mutations), diseases grouped under GATA2 deficiency (i.e., MonoMAC syndrome, Emberger syndrome, familial MDS/AML, DCML, NK cell deficiency, CMML, aplastic anemia, and chronic neutropenia) each shows a specific constellation of signs and symptoms. In addition, individuals with identical *GATA2* gene mutations may exhibit vastly different presentations.

74.5 Clinical Features

As GATA2 deficiency covers several previously unrelated disorders that all contain mutations in the transcription factor gene *GATA2*, its clinical presentations are not surprisingly diverse, including:

MDS/AML often associated with secondary mutations, median onset age of 30 years, occurring in 30%–50% of cases, lifetime risk of 90%.

Warts, severe viral infection (human papilloma virus, herpesviruses) present in 60%–70% of cases, 10%–20% cases due to disseminated CMV, EBV, or VZV.

Pulmonary alveolar proteinosis or decreased lung function (loss of volume or diffusion) accounting for 18% proven PAP, usually GM-CSF antibody negative; 10% pulmonary artery hypertension (PAH), 50% abnormal PFT, 14% pneumonia.

Mycobacterial or fungal infection representing 20%–50% non-tuberculous mycobacterial infection (NTM), 16% aspergillosis, 9% histoplasmosis.

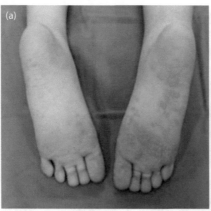
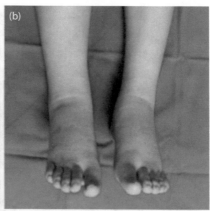

FIGURE 74.1 GATA2 deficiency in a 20-year-old female showing (a) multiple warts on both feet; (b) lymphedema in both lower legs. (Photo credit: Seo SK et al. *Korean J Intern Med.* 2016;31(1):188–90.)

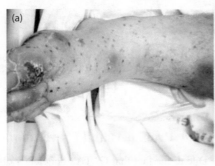
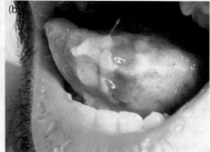

FIGURE 74.2 GATA2 deficiency in a 26-year-old male showing herpetic lesions in his left leg (a) and massive aphthous-like ulcerations in oral mucosa with pseudomembranes (b). (Photo credit: Ortueta-Olartecoechea AI. *Saudi J Ophthalmol.* 2018;32(2):164–6.)

Recurrent upper respiratory tract infection (otitis, sinusitis; 10%–20% of cases); autoimmune manifestation (panniculitis, arthritis, lupus-like hypothyroidism, hepatitis; accounting for 30%–50% panniculitis).

Solid tumors (HPV-related breast cancer, EBV+ mesenchymal skin cancer; representing 20%–35% intra-epithelial neoplasia, 22% of breast cancer in women >35 years, 10% skin cancer.

Lymphedema (11%–20% of childhood or adolescent case); thrombosis (DVT, PE, catheter-related; 25% risk).

Congenital deafness responsible for 20% abnormal audiograms.

Preterm labor (maternal trait, 33% of cases) (Figures 74.1 and 74.2) [21–23].

74.6 Diagnosis

Diagnostic workup for GATA deficiency involves the following:

Medical history review: Personal and family history of warts, mycobacterial infection, autoimmunity, chest disease, cytopenia, acute myeloid leukemia.

Physical examination: For warts and other abnormalities.

Laboratory tests: Blood count for monocytopenia; lymphocyte subsets for B cell and NK cell deficiency,

CD4:CD8 inversion (<1·0), absence of CD1c+, CD141+ and plasmacytoid blood DC; occasional IgA or IgG deficiency; bone marrow for megakaryocyte dysplasia, hypocellularity, fibrosis; blood for lupus anticoagulant, elevated FLT3 ligand (10- to 100-fold), absent transitional B cells and CD56^bright NK cells; bone marrow flow cytometry for loss of primitive multi-lymphoid progenitors (MLP), lymphoid-primed multipotent progenitors (LMPP), and reduction of granulocytic monocytic progenitors (GMP); normal CD56+ plasma cells present; lungs for diminished lung volumes and transfer factor; pulmonary infiltrates on CT; PAP on biopsy without GM-CSF antibodies.

Histopathology: Bone marrow for megakaryocyte dysplasia, hypocellularity, fibrosis; tissues biopsy with special stains for mycobacteria and fungi; neoplastic lesions for HPV and herpes virus nucleic acid or antigens.

Cytogenetic and molecular analyses: Monosomy 7; trisomy 8; *ASXL1* mutation; *GATA-2* gene deletion, mutation in codons 1–398 or intron 5 enhancer (Figure 74.3) [24,25].

It appears that GATA2 deficiency differs from sporadic myelodysplasia in a number of aspects, including median onset age (21–33 years vs 66 years), family connection (1 in 3 vs rare), Hb level (115–123 g/L vs 96 g/L), neutrophils (1·8–2·6 × 10⁹/L vs 0·55 × 10⁹/L), platelets (127–160 × 10⁹/L vs 49 × 10⁹/L), mDC (0 vs 0.008 × 10⁹/L), pDC (0 vs 0.009 × 10⁹/L),

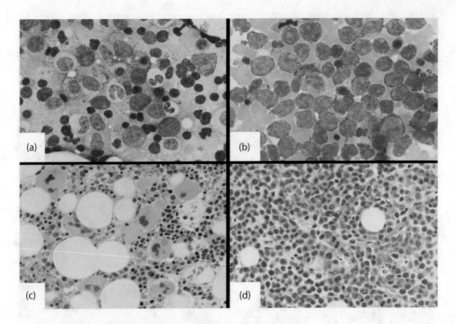

FIGURE 74.3 GATA2 deficiency in a 25-year-old female showing MDS in bone marrow at the time of presentation and identification of *GATA2* mutation (a and b); M5a immature monoblastic AML and identification of *NRAS* Q61 K mutation 4 months after initial MDS diagnosis (c and d). (Photo credit: McReynolds LJ et al. *Leuk Res Rep.* 2019;12:100176.)

monocytes (0.01×10^9/L vs 0.1×10^9/L), B cells (0.01×10^9/L vs 0.041×10^9/L), NK cells (0.002×10^9/L vs 0.111×10^9/L), CD4:CD8 ratio (<1 vs >1), bone marrow cellularity (hypocellular in 72% cases vs hypocellular in 95% cases), fibrosis (73% cases vs 10% cases), megakaryocytes (90% atypia vs 60%–70% atypia), FLT3 ligand (4752 pg/mL, median 294–8750 pg/mL vs 163 pg/mL, median 57–613 pg/mL), and infection (warts/mycobacteria/herpesviruses vs neutropenic fever). Of these, disparities in platelets (normal 150–400; GATA2 127–160 $\times 10^9$/L; sporadic myelodysplasia 49 $\times 10^9$/L) and is most striking [26].

Differential diagnoses for GATA2 deficiency (characteristic combination of warts with monocytopenia) include other conditions associated with papillomatosis or generalized verrucosis (e.g., epidermodysplasia verruciformis [EV] warts, hypogammaglobulinemia, immunodeficiency, myelokathexis [WHIM] syndrome warts, immunodeficiency, lymphoedema and anogenital dysplasia [WILD] syndrome, DOCK8 deficiency syndrome, and idiopathic CD4 lymphocytopenia).

74.7 Treatment

Treatment of GATA2 deficiency should aim to control infections, manage pulmonary disease, and rectify bone marrow dysfunction.

Control of infections involves administration of antibiotics (e.g., azithromycin to treat and/or prevent mycobacterial and routine bacterial infections) and vaccination (HPV).

Management of pulmonary diseases such as PAP and PAH provides symptomatic relief for patients with GATA2 deficiency.

Bone marrow dysfunction represents the underlying defect of GATA deficiency, for which non-myeloablative hematopoietic stem cell transplantation (HSCT) is a curative modality that helps reconstitute B-cell, NK, and monocyte populations, and improve immune competency.

74.8 Prognosis and Prevention

Due to the fact that GATA2 deficiency often manifests as MDS/AML, its prognosis is generally poor. After onset of symptoms, affected patients have a 5-year survival rate of 91%, 10-year survival rate of 84%, and 20-year survival rate of 67%. Patients with GATA2 deficiency who undergo allogeneic HSCT for MDS/AML have overall survival rate of 96% by age 20 years, 77% by age 40 years, and 45% by age 60 years; those without HSCT have overall survival rate of 89% by age 20 years, 76% by age 30 years, and 53% by age 40 years. Posttransplant survival rate for patients with GATA2 deficiency is 72% by 1 year, 65% by 2 years, and 54% by 4 years [27].

Members of related families are advised to undergo GATA2 deficiency screening for pediatric neutropenia, monocytopenia, B-cell or NK cell cytopenia, dendritic cell deficiency, MDS with hypocellular bone marrow, familial MDS/AML, PAP in absence of anti-GM-CSF autoantibodies, recurrent extragenital HPV warts or severe refractory genital HPV, severe viral infection (HSV, EBV), later onset lymphedema, sensorineural deafness with immunodeficiency, disseminated NTM, and disseminated or severe fungal infections. This will facilitate early intervention and reduce unnecessary morbidity and mortality [28].

74.9 Conclusion

GATA2 deficiency is a recently defined genetic disorder that unifies several previously considered unrelated diseases, including monoMAC and Emberger syndromes, familial MDS/AML, DCML, and NK cell deficiency, on the basis of their shared, haploinsufficient genetic defects in the *GATA2* gene [29,30]. Specifically, heterozygous germline or somatic *GATA2* mutations reduce GATA2 protein production and hinder the self-renewal capacity of hematopoietic stem cells, leading to subsequent depletion of dendritic, monocyte, B, and NK cells. Consequently, a broad spectrum of clinical symptoms emerge, from MDS/AML, warts, severe viral and bacterial infections, PAP or decreased lung function, autoimmune manifestation, solid tumors, lymphedema, thrombosis, and congenital deafness to preterm labor. Diagnosis of GATA2 deficiency requires a multifaceted approach, encompassing medical history review, physical examination, laboratory assessment, histopathology, and cytogenetic and molecular analyses. Treatment of GATA2 deficiency includes control of infections, management of pulmonary disease, and rectification of bone marrow dysfunction.

REFERENCES

1. Wlodarski MW, Collin M, Horwitz MS. GATA2 deficiency and related myeloid neoplasms. *Semin Hematol.* 2017;54(2):81–6.
2. Hahn CN, Chong C-E, Carmichael CL et al. Heritable GATA2 mutations associated with familial myelodysplastic syndrome and acute myeloid leukemia. *Nature Genet.* 2011;43:1012–7.
3. Hsu AP, Sampaio EP, Khan J et al. Mutations in GATA2 are associated with the autosomal dominant and sporadic monocytopenia and mycobacterial infection (MonoMAC) syndrome. *Blood.* 2011; 118: 2653–5.
4. Mir MA, Kochuparambil ST, Abraham RS et al. Spectrum of myeloid neoplasms and immune deficiency associated with germline GATA2 mutations. *Cancer Med.* 2015;4(4):490–9.
5. Babushok DV, Bessler M, Olson TS. Genetic predisposition to myelodysplastic syndrome and acute myeloid leukemia in children and young adults. *Leuk Lymphoma.* 2016;57(3):520–36.
6. Bannon SA, DiNardo CD. Hereditary predispositions to myelodysplastic syndrome. *Int J Mol Sci.* 2016;17(6). pii: E838.
7. Bresnick EH, Katsumura KR, Lee HY, Johnson KD, Perkins AS. Master regulatory GATA transcription factors: Mechanistic principles and emerging links to hematologic malignancies. *Nucleic Acids Res.* 2012;40(13):5819–31.
8. Ganapathi KA, Townsley DM, Hsu AP et al. GATA2 deficiency-associated bone marrow disorder differs from idiopathic aplastic anemia. *Blood.* 2015;125(1):56–70.
9. Baptista RLR, Dos Santos ACE, Gutiyama LM, Solza C, Zalcberg IR. Familial myelodysplastic/acute leukemia syndromes-myeloid neoplasms with germline predisposition. *Front Oncol.* 2017;7:206.
10. Koeffler HP, Leong G. Preleukemia: One name, many meanings. *Leukemia.* 2017;31(3):534–42.
11. Collin M, Dickinson R, Bigley V. Haematopoietic and immune defects associated with GATA2 mutation. *Br J Haematol.* 2015;169(2):173–87.
12. Dickinson RE, Milne P, Jardine L et al. The evolution of cellular deficiency in GATA2 mutation. *Blood.* 2014;123(6):863–74.
13. Kazenwadel J, Secker GA, Liu YJ et al. Loss-of-function germline GATA2 mutations in patients with MDS/AML or MonoMAC syndrome and primary lymphedema reveal a key role for GATA2 in the lymphatic vasculature. *Blood.* 2012;119:1283–91.
14. Hsu AP, McReynolds LJ, Holland SM. GATA2 deficiency. *Curr Opin Allergy Clin Immunol.* 2015;15(1):104–9.
15. Nováková M, Žaliová M, Suková M et al. Loss of B cells and their precursors is the most constant feature of GATA-2 deficiency in childhood myelodysplastic syndrome. *Haematologica.* 2016;101(6):707–16.
16. Zhang S-J, Ma L-Y, Huang Q-H et al. Gain-of-function mutation of GATA-2 in acute myeloid transformation of chronic myeloid leukemia. *Proc Natl Acad Sci USA.* 2008;105:2076–81.
17. Gao J, Gentzler RD, Timms AE et al. Heritable GATA2 mutations associated with familial AML-MDS: A case report and review of literature. *J Hematol Oncol.* 2014;7:36.
18. Cortés-Lavaud X, Landecho MF, Maicas M et al. GATA2 germline mutations impair GATA2 transcription, causing haploinsufficiency: Functional analysis of the p.Arg396Gln mutation. *J Immunol.* 2015;194(5):2190–8.
19. Ding LW, Ikezoe T, Tan KT et al. Mutational profiling of a MonoMAC syndrome family with GATA2 deficiency. *Leukemia.* 2017;31(1):244–5.
20. Chong CE, Venugopal P, Stokes PH et al. Differential effects on gene transcription and hematopoietic differentiation correlate with GATA2 mutant disease phenotypes. *Leukemia.* 2018;32(1):194–202.
21. Seo SK, Kim KY, Han SA et al. First Korean case of Emberger syndrome (primary lymphedema with myelodysplasia) with a novel GATA2 gene mutation. *Korean J Intern Med.* 2016;31(1):188–90.
22. Ortueta-Olartecoechea AI, Torres-Peña JL, Palacios-Hípola AI, Mencia-Gutierrez E. Herpetic ocular manifestations in a patient with GATA2 deficiency. *Saudi J Ophthalmol.* 2018;32(2):164–6.
23. Mendes-de-Almeida DP, Andrade FG, Borges G et al. GATA2 mutation in long stand *Mycobacterium kansasii* infection, myelodysplasia and MonoMAC syndrome: A case-report. *BMC Med Genet.* 2019;20(1):64.
24. Sanyi A, Jaye DL, Rosand CB, Box A, Shanmuganathan C, Waller EK. Diagnosis of GATA2 haplo-insufficiency in a young woman prompted by pancytopenia with deficiencies of B-cell and dendritic cell development. *Biomark Res.* 2018;6:13.
25. McReynolds LJ, Zhang Y, Yang Y et al. Rapid progression to AML in a patient with germline *GATA2* mutation and acquired *NRAS* Q61 K mutation. *Leuk Res Rep.* 2019;12:100176.
26. Fisher KE, Hsu AP, Williams CL et al. Somatic mutations in children with GATA2-associated myelodysplastic syndrome who lack other features of GATA2 deficiency. *Blood Adv.* 2017;1(7):443–8.
27. Wlodarski MW, Hirabayashi S, Pastor V et al. Prevalence, clinical characteristics, and prognosis of GATA2-related myelodysplastic syndromes in children and adolescents. *Blood.* 2016;127(11):1387–97; quiz 1518.
28. Spinner MA, Sanchez LA, Hsu AP et al. GATA2 deficiency: A protean disorder of hematopoiesis, lymphatics, and immunity. *Blood.* 2014;123(6):809–21.
29. McReynolds LJ, Yang Y, Yuen Wong H et al. MDS-associated mutations in germline GATA2 mutated patients with hematologic manifestations. *Leuk Res.* 2019;76:70–5.
30. Bao EL, Cheng AN, Sankaran VG. The genetics of human hematopoiesis and its disruption in disease. *EMBO Mol Med.* 2019;11(8):e10316.

75

LPD Associated with Epstein–Barr Virus Infection

Dongyou Liu

CONTENTS

75.1 Introduction

Individuals harboring mutations in certain genes (e.g., *SH2D1A, BIRC4, MAGT1, ITK, CTPS1, CD27, CORO1A*) that encode proteins involved in B cell development and/or function have intrinsic immune deficiency and show particular susceptibility to Epstein–Barr virus (EBV) infection, which occurs in 95% of adults without producing any symptoms, and which takes advantage of host immune deficiency and induces a group of severe diseases collectively known as EBV-driven lymphoproliferative disorders (LPD) or LPD associated with EBV infection (Table 75.1).

Clinically, EBV-driven LPD not only causes severe infectious mononucleosis with fever, sore throat, lymphadenopathy, and splenomegaly (due to EBV infection of B cells leading to cell lysis), but also predisposes to various B cell-related malignancies (e.g., Hodgkin and non-Hodgkin lymphomas) as well as hypo/dysgammaglobulinemia (due to worsening B cell malfunction) [1].

75.2 Biology

Primary immunodeficiencies are a group of heterogeneous disorders with immune system abnormalities that confer susceptibility to recurrent infections, autoimmunity, lymphoproliferation, granulomatous process, atopy, and malignancy [2]. Out of approximately 250 primary immunodeficiencies recognized so far, a small subgroup is associated with mutations in the genes that encode proteins involved in B cell development and/or function. These primary immunodeficiencies compromise the function of cytotoxic lymphocytes through alterations in TCR signaling (RasGRP1, ZAP70, PI3 K, ITK), actin cytoskeleton arrangements

(CORO1A, WASP), co-stimulation (LRBA, MAGT1), leukocyte development (GATA2, MCM4), lymphocyte cell death (XIAP, STK4, CTPS1), and cytotoxic effector (CD16, NKG2D, SLAM receptors like 2B4, CD27), allowing effective EBV control of B cells and subsequent development of a spectrum of clinical manifestations (including the development of lymphomas), that are collectively referred to as EBV-driven LPD or LPD associated with EBV infection (Figure 75.1) [3].

The first evidence on the connection between EBV infection and apparent immune deficiency was gained in 1974 by two independent groups. The affected patients were males from related families and showed vigorous proliferation of polyclonal T and B lymphocytes and histiocytes, cervical adenopathy, hepatosplenomegaly, hepatitis, elevated antibody titers to EBV, bone marrow failure with a hemophagocytic component leading to rapidly fatal infectious mononucleosis in 50% of cases, and development of dysgammaglobulinemia (from agammaglobulinemia to polyclonal hypergammaglobulinemia, in 30% of cases) and B cell lymphomas (often extranodal non-Hodgkin lymphomas of the Burkitt type, involving the ileocecal region of the intestine in 25% of surviving patients). Following the identification of the culprit gene *SH2D1A* on X chromosome in 1988, this hereditary immunodeficiency associated with EBV infection was named X-linked lymphoproliferative disorder or syndrome (XLP 1) [1].

Since then, several EBV-driven LPD linked to other genes (e.g., *BIRC4, MAGT1, ITK, CTPS1, CD27, CORO1A*) have been described. A common theme of these disorders is the involvement of a gene, mutations of which affect B cell development and/or function and contribute to immune deficiency and increased susceptibility to EBV infection, which in turn elicits an exaggerating response and leads to EBV-driven LPD or LPD associated with EBV infection [1].

TABLE 75.1

Characteristics of Lymphoproliferative Disorders Associated with EBV Infection

Disorder	Gene	Chromosome Location	Protein	Transmission	Defective Pathway/Function	EBV-Associated Symptoms	Noninfectious Diseases in the Absence of EBV Infection	No. of Cases Reported
XLP1	SH2D1A	Xq25	SH2 domain containing 1A/SLAM-associated protein	X-linked	SLAMR/SAP pathway (T and NK cytotoxicity and AICD)	Fulminant infectious mononucleosis, B cell lymphoma, HLH, lymphomatoid granulomatosis	EBV-negative B cell lymphoma, aplastic anemia, vasculitis	>100
XLP2	BIRC4	Xq25	Baculoviral IAP repeat-containing protein 4/ X-linked inhibitor of apoptosis/ E3 ubiquitin-protein ligase XIAP	X-linked	Excess of apoptosis (AICD, TRAIL-R, Fas) NOD1/2 signaling/function	Fulminant infectious mononucleosis, HLH, splenomegaly, cytopenias	Colitis, HLH, inflammatory bowel disease	>100
CTPS1 deficiency	CTPS1	1p34.2	CTP synthase 1	Autosomal recessive	*De novo* pyrimidine synthesis T- and B-cell proliferation	Severe infectious mononucleosis; lymphoproliferative disease, lymphoma	none	12
STK4 deficiency	STK4	20q13.12	Serine threonine kinase 4	Autosomal recessive	Hippo signaling pathway	B cell lymphoma, lymphoproliferative disease, autoimmune hemolytic anemia	Dermatitis, autoimmune cytopenias	
RASGRP1 deficiency	RASGRP1	15q14	RAS guanyl releasing protein 1	Autosomal recessive	MAPK pathway (ERK1/2, T-, B-cell proliferation); actin/cytoskeleton dynamics	LPD/B lymphoma, SMT, cytopenias, lung infections, disseminated tuberculosis	EBV-negative LPD	6
ZAP70 deficiency	ZAP70	2q11.2	70 KDa zeta-associated protein	Autosomal recessive	T-cell receptor (TCR) signaling	Absent CD8+ T cells in an individual with normal CD3+ and CD4+ T-cell counts		10
PASLI/APDS	PIK3CD	1p36.22	PI3K catalytic subunit 110δ: gain of function mutation	Autosomal dominant	PI3K/Akt signaling pathway	Lymphoma	Lymphoid nodules in upper airway and gastrointestinal tract, autoimmune cytopenias	
	PIK3CD	1p36.22	PI3K catalytic subunit 110δ: loss of function mutation	Autosomal recessive	PI3K/Akt signaling pathway	EBV viremia	Lymphadenopathy, hepatosplenomegaly, autoantibodies	
ITK deficiency	ITK	5q33.3	IL-2 inducible T cell kinase	Autosomal recessive	TCR induced calcium flux; T cell proliferation	Lymphoproliferation, Hodgkin lymphoma, HLH. hepatosplenomegaly, lung disease, lymphomatoid granulomatosis	Autoimmune kidney disease	13

(Continued)

TABLE 75.1 *(Continued)*

Characteristics of Lymphoproliferative Disorders Associated with EBV Infection

Disorder	Gene	Chromosome Location	Protein	Transmission	Defective Pathway/ Function	EBV-Associated Symptoms	Noninfectious Diseases in the Absence of EBV Infection	No. of Cases Reported
LRBA deficiency	*LRBA*	4q31.3	LPS-responsive beige-like anchor protein	Autosomal recessive	CTLA4 *pathway*	B cell lymphoproliferative disease, EBV viremia	Inflammatory bowel disease, chronic diarrhea, autoimmune cytopenias	
Coronin 1A deficiency	*CORO1A*	16p11.2	Coronin actin binding protein 1A	Autosomal recessive	Actin regulation T-cell survival NK cytotoxicity	B cell lymphoma, lymphoproliferative disease	Neurocognitive impairment	9
MAGT1 deficiency/XMEN disease	*MAGT1*	Xq21.1	Magnesium transporter 1 protein	X linked	NKG2D-dependent cytotoxicity	B cell lymphoma	Autoimmune cytopenias	11
CD27 deficiency	*CD27*	12p13.31	CD27	Autosomal recessive	CD27–CD70 pathway (T cell proliferation) NK cytotoxicity	Lymphoproliferative disease, HLH, lymphoma, aplastic anemia	none	18
FHL2 deficiency	*PRF1*	10q22.1	Perforin	Autosomal recessive	TGF-β *pathway*	HLH, CAEBV, splenomegaly	HLH	
FHL3 deficiency	*UNC13D*	17q25.3	Munc13–4	Autosomal recessive	CRM1 pathway	CAEBV, vasculitis, hepatitis, splenomegaly	HLH	
FHL5 deficiency	*STXBP2*	19p13.2	Munc18–2	Autosomal recessive	CREM-signaling pathway	CABEV, lymphoma, splenomegaly	HLH, colitis, bleeding	
FCGR3A deficiency	*FCGR3A*	1q23.3	Fcγ receptor 3A (CD16a)	Autosomal recessive	Fc-gamma receptor signaling pathway	EBV-positive Castleman disease	none	
MonoMac syndrome	*GATA2*	3q21.3	GATA binding protein 2	Autosomal dominant	STAT5-GATA2 pathway	Severe infectious mononucleosis, EBV-positive smooth muscle tumors, CAEBV, HLH	Myelodysplastic syndrome, autoimmune disease, pulmonary alveolar proteinosis, primary lymphedema	
Classical NK cell deficiency type 2	*MCM4*	8q11.21	Minichromosome maintenance complex component 4	Autosomal recessive	WAVE2-dependent pathway	EBV lymphproliferative disease, lymphoma	Adrenal insufficiency, growth retardation	
CARD11 deficiency	*CARD11*	7p22.2	Caspase recruitment domain-containing protein 11: gain of function mutations	Autosomal dominant	NF-κB activation pathway	EBV viremia	Autoimmune neutropenia	

Abbreviations: APDS, activated PI3Kδ syndrome; CAEBV, chronic active EBV disease; CTP, cytidine 5′ triphosphate; FHL, familial hemophagocytic lymphohistiocytosis; HLH, hemophagocytic lymphohistiocytosis; PASLI, p110δ-activating mutation causing senescent T cells, lymphadenopathy, and immune deficiency; PI3 K, phosphoinositide 3′ kinase; XLP: X linked lymphoproliferative disease; XMEN, X-linked immunodeficiency with magnesium defect, EBV infection, and neoplasia.

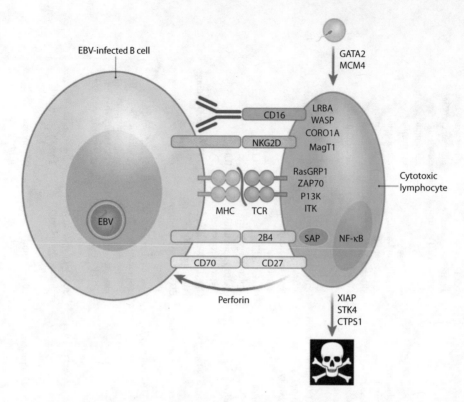

FIGURE 75.1 Schematic presentation of primary immunodeficiencies and susceptibility to EBV-associated diseases. Primary immunodeficiencies compromise the functions of cytotoxic lymphocytes through alterations in TCR signaling (RasGRP1, ZAP70, PI3 K, ITK), actin cytoskeleton arrangements (CORO1A, WASP), co-stimulation (LRBA, MAGT1), leucocyte development (GATA2, MCM4), lymphocyte cell death (XIAP, STK4, CTPS1) and cytotoxic effector (CD16, NKG2D, SLAM receptors like 2B4, CD27), allowing effective EBV control of B cells. (Photo credit: Damania B, Münz C. *FEMS Microbiol Rev.* 2019;43(2):181–92.)

75.3 Pathogenesis

Epstein–Barr virus (EBV) is a ubiquitous, oncogenic double-stranded DNA virus belonging to the Herpesviridae family, which comprises several well-known human-infecting viral pathogens including herpes simplex virus-1 (HSV-1, or human herpesvirus 1), herpes simplex virus-2 (HSV-2, or human herpesvirus 2), varicella zoster virus (VZV, or human herpesvirus 3), Epstein–Barr virus (EBV, or human herpesvirus 4), cytomegalovirus (CMV, or human herpesvirus 5), Roseolovirus (human herpesviruses 6 and 7), and Kaposi sarcoma–associated herpesvirus (KSHV, or human herpesvirus 8). Of these, EBV and KSHV are classified in the Gammaherpesvirinae subfamily, and primarily target B lymphocytes, epithelial cells, and other cells, with B cells being the site of latency.

Occurring in nearly 95% of the adult population, EBV typically enters human body via the oral route (saliva or droplets) and rarely, transfusion during infancy and adolescence, and establishes primary infection in the epithelial cells of the oropharynx and subsequently spreads to B lymphocytes in pharyngeal lymphoid tissues, where EBV drives the activation and transformation of B cells into latently infected B lymphoblasts, enabling EBV persistence. Depending on the viral gene expression pattern, EBV shows three latency patterns: latency type I expresses Epstein–Barr nuclear antigen I (EBNA-1) and two small noncoding Epstein–Barr- encoded RNAs (EBER) and is associated with Burkitt lymphoma; latency type II expresses EBNA-1, EBER

and the latent membrane proteins (LMP), LMP-1, LMP-2A, and LMP-2B, and is associated with Hodgkin lymphoma (HL), T cell non-Hodgkin lymphoma, and EBV-positive (EBV+) diffuse large B cell lymphoma (DLBCL); latency type III expresses the entire EBV repertoire (EBNA, EBER, and LMP) and is associated with post-transplant lymphoproliferative disorders (PTLD) and EBV-positive (EBV+) DLBCL. While most latently infected B lymphoblasts are eliminated by specific CD8$^+$ T cells, whose expansion reflects a robust immune response to EBV, some downregulate latent gene expression to escape T-cell immunosurveillance, and acquire a memory phenotype (absence of latent viral protein expression and EBV genome remaining in episomal form), which acts a reservoir for EBV without being invisible to the immune system. Peripheral EBV-infected memory B cells may return to the Waldeyer's ring and undergo reactivation to produce an infectious virus for shedding in the saliva. Although latent EBV infection is largely asymptomatic, it may sometimes cause infectious mononucleosis in adolescents, a self-limiting lymphoproliferative disease characterized by a polyclonal expansion of infected B cells that triggers a cytotoxic T cell response resulting in a transient, antigen-driven oligoclonal expansion of CD8+ T cells, with symptoms ranging from fever, sore throat, body aches, and swollen lymph nodes to general fatigue [4].

Existence of immune deficiency (e.g., primary immune deficiency, AIDS) or altered immune status (e.g., pregnancy, transplantation) may upset this delicate balance, facilitating the change of EBV-containing reservoir memory B cells from latency into the lytic cycle. This promotes EBV infection and expansion of new B cells, and leads to the development of hemophagocytic lymphohistiocytosis (HLH, or virus-associated hemophagocytic syndrome), nonmalignant B-cell LPD, and B-cell lymphomas (HL and non-Hodgkin lymphomas including Burkitt lymphoma and DLBCL), particularly in European and North American populations. Occasionally, EBV-infected new B cells may spread to T and NK cells, resulting in chronic viremia, EBV-positive lymphocyte infiltration of organs, and life-threatening LPD (e.g., hemophagocytic syndrome, EBV-positive T/NK-cell lymphoma) in Asian and South American populations [1].

Primary immunodeficiencies with increased susceptibility to EBV infection appear to result from mutations in genes that encode proteins involved in B-cell development and/or function, such as TCR signaling (RasGRP1, ZAP70, PI3 K, ITK), actin cytoskeleton arrangements (CORO1A, WASP), co-stimulation (LRBA, MAGT1), leukocyte development (GATA2, MCM4), lymphocyte cell death (XIAP, STK4, CTPS1) and cytotoxic effector (CD16, NKG2D, SLAM receptors like 2B4, CD27) (Figure 75.1). Homozygous or heterozygous mutations in these genes compromise the function of cytotoxic lymphocytes, allowing excessive B-cell growth and development of various noninfectious diseases (e.g., myelodysplastic syndrome, autoimmune

disease, pulmonary alveolar proteinosis, primary lymphedema, adrenal insufficiency, growth retardation) in the absence of EBV infection, and enabling EBV-infected B cells to dominate host immune responses leading to EBV-driven LPD of B, T, and NK cell derivation in the presence of EBV infection (Table 75.1) [1].

It appears that EBV-driven LPD typically presents as B cell lymphoproliferative diseases (e.g., infectious mononucleosis, chronic active EBV of B cell type, EBV–positive DLBCL lymphoma associated with chronic inflammation, lymphomatoid granulomatosis) in European and North American populations (Figure 75.2), and as T cell and NK cell lymphoproliferative diseases (e.g., EBV-positive T cell and NK cell lymphoproliferative diseases of childhood [chronic active EBV infection of T- and NK-cell type, systemic form or cutaneous form—chronic active EBV infection of T- and NK-cell type/severe mosquito bite allergy; systemic EBV-positive T cell lymphoma of childhood], aggressive NK-cell leukemia, nasal type extranodal NK/T cell lymphoma, primary EBV-positive nodal T- or NK-cell lymphoma) in Asian and Lain American populations [5–8].

In addition, individuals (95% of which have latent EBV infection) undergoing transplantation may develop PTLD, including nondestructive PTLD (plasmacytic hyperplasia, infectious

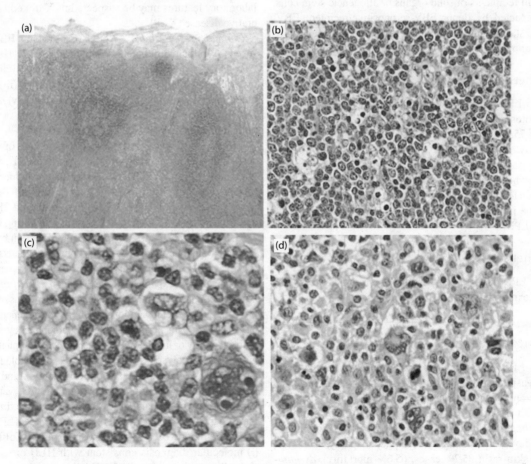

FIGURE 75.2 Lymphoproliferative disorders associated with EBV infection presenting as (a) infectious mononucleosis (retained architecture, hyperplastic lymphoid follicles and expanded paracortex in lymph node, H&E, 100×); (b) Burkitt lymphoma (regular, medium-sized cells with basophilic cytoplasm, coarse nuclear chromatin, and small peripheral nucleoli; tangible body macrophages scattered through the tumor generating a starry sky appearance; H&E, 100×); (c) classic Hodgkin lymphoma (infiltration of lymph node by Hodgkin and Reed–Sternberg cells or HRS cells in a mixed reactive background composed of eosinophils, small lymphocytes, histiocytes, and plasma cells; H&E, 400×); (d) EBV positive DLBCL, not otherwise specified (polymorphous infiltrate of HRS-like cells in a background of lymphocytes and histiocytes; H&E, 400×). (Photo credit: Dojcinov SD, Fend F, Quintanilla-Martinez L. *Pathogens*. 2018;7(1). pii: E28.)

mononucleosis-like PTLD, florid hyperplasia), polymorphic PTLD (polyclonal and monoclonal proliferations), monomorphic PTLD (monoclonal non-Hodgkin lymphomas such as DLBCL, Burkitt lymphoma, plasma cell myeloma, T-cell lymphoma), and HL (monoclonal) [9,10].

75.4 Epidemiology

75.4.1 Prevalence

EBV-driven LPD or LPD associated with EBV infection are rare diseases, with estimated incidences of 1 in 1–5 million people. To date, <1000 cases of EBV-driven LPD have been reported, including XLP1 (>100 cases), XLP2 (>100 cases), CD27 deficiency (18 cases), ITK deficiency (13 cases), CTPS1 deficiency (12 cases), *MAGT1 deficiency* (11 cases), ZAP70 deficiency (10 cases), and coronin 1A deficiency (9 cases) (Table 75.1). However, EBV-driven LPD may be underdiagnosed due to its severity and often rapidly fatal initial presentation, variable expression, and clinical overlap with other immunologic disorders, in addition to the absence of functional diagnostic assays [1].

On the other hand, the incidence of PTLD may vary from 2% to 20% in recipients of solid organs or allogeneic stem cells, which is impacted by donor EBV-seropositive/recipient EBV-seronegative status, ongoing immunosuppressive regimen, duration of immunosuppression, and status of EBV infection [11].

75.4.2 Inheritance

EBV-driven LPD may show X-linked (XLP1, XLP2, and MAGT1), autosomal recessive, or autosomal dominant inheritance (Table 75.1).

Specifically, XLP1 and XLP2 affect about 1/1 million and 1/5 million males (who possess one copy of X chromosome), respectively, who do not pass X-linked traits to their sons. However, in rare cases, a female who carries one altered copy of the *SH2D1A* or *XIAP* gene in each cell may develop signs and symptoms of EBV-driven LPD.

75.4.3 Penetrance

EBV-driven LPD has a varied penetrance (from 50% to 100%) and clinical expressivity.

75.5 Clinical Features

EBV-driven LPD is a heterogeneous disease that often presents with nonspecific, highly variable symptoms (e.g., malaise, fatigue, fever, night sweats, weight loss, and lymphadenopathy).

Males with XLP1 (*SH2D1A*-related XLP) typically display *hemophagocytic lymphohistiocytosis* (HLH; continuous high-grade fever, splenomegaly, lymphadenopathy, jaundice, edema, and skin rash; 35.2% cases, 65.6% mortality), *dysgammaglobulinemia* (often in EBV negative individuals, either prior to EBV infection or survivors of EBV infection; 50.5% cases, 13% mortality), *malignant lymphoma* (often in EBV negative individuals; high-grade B cell lymphomas of non-Hodgkin type and often extranodal/intestinal involvement, such as Burkitt lymphoma, immunoblastic lymphoma, small cleaved or mixed-cell lymphoma, and unclassifiable lymphoma; 24.2% cases, 9% mortality), *fulminant infectious mononucleosis* (a fatal or near-fatal EBV infection with widespread proliferation of cytotoxic T cells, EBV-infected B cells, and macrophages; lymphadenopathy, hepatosplenomegaly, fulminant hepatitis, hepatic necrosis, and profound bone marrow failure; 9.9% cases, 22.2% mortality generally secondary to liver failure), and *other manifestations* (aplastic anemia, vasculitis, and lymphoid granulomatosis; 15.4% cases, 28.6% mortality) [12,13].

Males with XLP2 (*BIRC4*-related XLP) may show *hemophagocytic lymphohistiocytosis* (HLH; often without EBV infection; 83% cases; 33% mortality), *recurrent HLH* (67% cases), *splenomegaly* (85% cases), *hypogammaglobulinemia* (30% cases), and *colitis ± liver disease* (13% cases, 60% mortality) [13].

75.6 Diagnosis

Diagnostic workup for EBV-driven LPD involves medical history review, physical examination, laboratory evaluation, histopathology, and molecular testing.

A male showing any of the following clinical presentations or laboratory features may be suspected of X-linked lymphoproliferative disease (XLP).

Clinical presentations: (i) fatal or near-fatal EBV infection/severe fulminant infectious mononucleosis; (ii) HLH resulting from EBV or other viral illness, especially in childhood or adolescence, or HLH without an identifiable trigger, (iii) dysgammaglobulinemia, (iv) malignant lymphoma (XLP1), (v) inflammatory bowel disease (XLP2), (vi) family history of one or more maternally related males with an XLP phenotype [14].

Laboratory features: (i) variable numbers of lymphocyte subsets including decreased or increased T cells, B cells, and NK cells, (ii) impaired T-cell re-stimulation-induced cell death (males with XLP1) or increased susceptibility to T-cell re-stimulation-induced cell death (XLP2), (iii) impaired 2B4-mediated cytotoxicity (XLP1), (iv) impaired NOD2 signaling (XLP2) [14].

Evidence of *acute EBV infection* (EBV detection by PCR; positive heterophile antibodies or monospot testing; detection of EBV-specific IgM antibodies; atypical lymphocytosis on peripheral blood smear with expansion of CD8 T cells), *HLH/fulminant infectious mononucleosis* (markedly elevated liver transaminases and/or liver dysfunction/coagulopathy, hypofibrinogenemia; inverted CD4:CD8 ratio in peripheral blood; hemophagocytosis on bone marrow biopsy or in other tissues such as CSF, lymph node; cytopenias; splenomegaly; elevated plasma levels of soluble IL-2 receptor alpha; hypertriglyceridemia; hyperferritinemia), and *dysgammaglobulinemia* (low serum concentration of IgG, variable serum concentrations of IgM and/or IgA, which may sometimes be abnormally increased) provide further confirmation of XLP [15].

Establishment of HLH diagnosis requires fulfilling either (i) molecular diagnosis consistent with HLH or (ii) five of the eight diagnostic criteria for HLH below:

- Fever ≥38.5°C
- Splenomegaly
- Cytopenia involving ≥2 cell lines (hemoglobin <90 g/L or <100 g/L in infants <4 weeks

- Platelets $<100 \times 10^9$/L
- Neutrophils $<1.0 \times 10^9$/L
- Hypertriglyceridemia (fasting, ≥ 265 mg/dL) or hypofibrinogenemia (≤ 1.5 g/L)
- Hemophagocytosis in bone marrow, spleen, or lymph nodes
- Low or absent natural killer cell activity; serum ferritin > 500 µg/L; elevated CD25 (soluble IL-2 receptor) levels (>2400 U/mL) [14]

Molecular testing is especially valuable for patients with low or absent SH2 domain protein 1A (SAP) expression by flow cytometry (XLP1) or low or absent XIAP expression by flow cytometry (XLP2), and normal perforin flow cytometric screening and CD107a flow cytometric screening.

Definitive diagnostic criteria for XLP include a male patient with severe/fatal EBV infection, HLH, immunodeficiency involving hypogammaglobulinemia of uncertain etiology, recurrence of a B-cell (typically non-Hodgkin) lymphoma, and a hemizygous *SH2D1A* (83%–97% of XLP1 cases) or *BIRC4* (12% of XLP2 cases) *pathogenic variant* [13,16].

Differential diagnoses for XLP include Bruton's agammaglobulinemia, common variable immunodeficiency (CVID; increased susceptibility to infections, diminished responses to protein and polysaccharide vaccines), Langerhans cell histiocytosis (LCH, histiocytosis X), non-LCH histiocytic disorder, malignant histiocytic disorder, familial HLH (fever, hepatosplenomegaly, skin rash, lymphadenopathy, cytopenias, hypertriglyceridemia, hypofibrinogenemia, and abnormal liver function; autosomal recessive disorder due to defects in perforin gene at 10q21–22), secondary hemophagocytic histiocytosis, infection-associated hemophagocytic syndrome, malignancy-associated hemophagocytic syndrome (acute lymphoblastic leukemia and other tumors before or during treatment), malignant histiocytosis (proliferation of large, atypical clear histiocyte-like cells with activated macrophages and lymphocytes; linked to Ki-1 or CD30[+] lymphoma with a translocation involving chromosomes 2 and 5), lymphomatoid granulomatosis, erythrophagocytic T cell lymphoma (hepatosplenomegaly, jaundice,

fever, and weight loss), familial hemophagocytic lymphohistiocytosis (FHL; excessive immune activation with uncontrolled T-lymphocyte and macrophage activation; autosomal recessive disorder due to mutations in *PRF1*, *UNC13D/MUNC13-4*, *STXBP2*, and *STX11*), Chediak–Higashi syndrome (partial oculocutaneous albinism, mild bleeding tendency, severe immunodeficiency, presence of huge secretory lysosomes in the neutrophils and lymphocytes and giant melanosomes on skin biopsy; autosomal recessive disorder due to *LYST* mutation), Griscelli syndrome type 2 (GS2; neurologic abnormalities, partial albinism with fair skin and silvery-grey hair; autosomal recessive disorder of cytotoxic T lymphocytes due to *RAB27A*), ITK deficiency (fatal HLH, hypogammaglobulinemia, HL, autoimmune-mediated renal disease often following EBV infection; autosomal recessive disorder due to *ITK* pathogenic variant), and CD27 deficiency (EBV viremia, hypogammaglobulinemia, and T cell dysfunction; autosomal recessive disorder due to pathogenic variants in *TNFRSF7*) [17].

75.7 Treatment

As EBV-driven LPD is associated with a diverse spectrum of clinical diseases, its management requires a multidisciplined approach, including the use of immunosuppressive agents (e.g., steroids and etoposide or anti-thymocyte globulin), rituximab (monoclonal antibody to CD20, when HLH is associated with EBV infection), regular intravenous immunoglobulin (IVIG) replacement (for hypogammaglobulinemia), chemotherapy (e.g., CHOP—cyclophosphamide, doxorubicin, vincristine, and prednisone for lymphoma), immunosuppression (for colitis associated with XLP2), and allogeneic hematopoietic cell transplantation (HSCT, the only curative therapy, often undertaken after lymphoma remission) (Figure 75.3) [17,18].

EBV-associated T- and/or NK-cell (EBV T/NK-cell) LPD are curable with allogeneic HSCT. Primary-EBV infection-associated HLH associated with EBV T/NK-cell lymphoproliferation are managed with steroids, cyclosporine A, and etoposide [19–21].

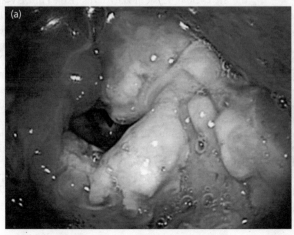 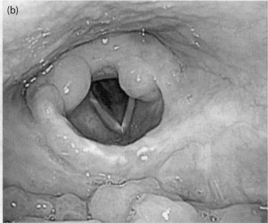

FIGURE 75.3 Epstein–Barr virus-positive T/NK-lymphoproliferative disease (LPD) with *SH2D1A/XIAP* hypomorphic gene variant c.1045_1047delGAG, p.Glu349del in a 21-year-old woman showing (a) the laryngeal CD4+T-cell LPD lesion prior to cancer chemotherapy; (b) improvement of the laryngeal LPD lesion after bone marrow transplantation. (Photo credit: Ishimura M et al. *Front Pediatr.* 2019;7:183.)

75.8 Prognosis and Prevention

EBV-driven LPD has a generally poor prognosis. Patients with XLP may have an improved overall survival (as high as 80%) after undertaking HSCT.

Molecular testing of at-risk relatives for the family-specific pathogenic variant facilitates makes it possible to implement early treatment. Prenatal testing for a pregnancy at increased risk and preimplantation genetic diagnosis for XLP may be considered once *SH2D1A* or *XIAP* pathogenic variant has been identified in an affected family member [17,22].

75.9 Conclusion

Lymphoproliferative disorders associated with Epstein–Barr virus (EBV) infection (also known as EBV-driven lymphoproliferative disorders) represent a heterogeneous group of diseases that result from interaction between primary immunodeficiency and EBV infection, leading to a severe combination of clinical features, ranging from HLH and hypogammaglobulinemia/dysgammaglobulinemia to LPD of B, T, and NK cell derivation, often with a fatal consequence. Specifically, primary immunodeficiencies linked to mutations in genes involved in B cell development and/or function compromise the function of cytotoxic lymphocytes through alterations in TCR signaling, actin cytoskeleton arrangements, co-stimulation, leukocyte development, lymphocyte cell death, and cytotoxic effector, allowing EBV-infected B cells to proliferate and spread to other B cells and occasionally T and NK cells [23,24]. Diagnosis of EBV-driven LPD is based on medical history review, physical examination, laboratory evaluation, histopathology, and molecular testing. Treatment of EBV-driven LPD relies on a combination of immunosuppressive agents, rituximab (monoclonal antibody to CD20), regular IVIG replacement, chemotherapy, immunosuppression, and allogeneic HSCT.

REFERENCES

1. Latour S, Winter S. Inherited immunodeficiencies with high predisposition to Epstein-Barr virus-driven lymphoproliferative diseases. *Front Immunol.* 2018;9:1103.
2. Raje N, Dinakar C. Overview of immunodeficiency disorders. *Immunol Allergy Clin North Am.* 2015;35(4):599–623.
3. Damania B, Münz C. Immunodeficiencies that predispose to pathologies by human oncogenic γ-herpesviruses. *FEMS Microbiol Rev.* 2019;43(2):181–92.
4. Justiz Vaillant AA, Stang CM. Lymphoproliferative disorders. In: *StatPearls [Internet]*. Treasure Island (FL): StatPearls Publishing; 2019 Jan-.2019 Jun 18.
5. Stepensky P, Weintraub M, Yanir A et al. IL-2-inducible T-cell kinase deficiency: Clinical presentation and therapeutic approach. *Haematologica.* 2011;96:472–6.
6. Kim HJ, Ko YH, Kim JE et al. Epstein-Barr virus-associated lymphoproliferative disorders: Review and update on 2016 WHO classification. *J Pathol Transl Med.* 2017;51(4):352–8.
7. Dojcinov SD, Fend F, Quintanilla-Martinez L. EBV-positive lymphoproliferations of B- T- and NK-cell derivation in non-immunocompromised hosts. *Pathogens.* 2018;7(1). pii: E28.
8. Kim WY, Montes-Mojarro IA, Fend F, Quintanilla-Martinez L. Epstein-Barr virus-associated T and NK-cell lymphoproliferative diseases. *Front Pediatr.* 2019;7:71.
9. Aida N, Ito T, Maruyama M et al. A case of Epstein-Barr virus-associated leiomyosarcoma concurrently with posttransplant lymphoproliferative disorders after renal transplantation. *Clin Med Insights Case Rep.* 2019;12:1179547619867330.
10. Crombie JL, LaCasce AS. Epstein Barr virus associated B-cell lymphomas and Iatrogenic lymphoproliferative disorders. *Front Oncol.* 2019;9:109.
11. Samant H, Kothadia JP. Transplantation posttransplantation lymphoproliferative disorders. In: *StatPearls [Internet]*. Treasure Island (FL): StatPearls Publishing; 2019 Jan-. 2019 Jul 1.
12. Yang X, Kanegane H, Nishida N et al. Clinical and genetic characteristics of XIAP deficiency in Japan. *J. Clin Immun.* 2012;32:411–20.
13. Latour S, Aguilar C. XIAP deficiency syndrome in humans. *Semin Cell Dev Biol.* 2015;39:115–23.
14. Pei Y, Lewis AE, Robertson ES. Current progress in EBV-associated B-cell lymphomas. *Adv Exp Med Biol.* 2017;1018: 57–74.
15. Shannon-Lowe C, Rickinson AB, Bell AI. Epstein-Barr virus-associated lymphomas. *Philos Trans R Soc Lond B Biol Sci.* 2017;372(1732). pii: 20160271.
16. Kelsen JR, Dawany N, Martinez A et al. A *de novo* whole gene deletion of XIAP detected by exome sequencing analysis in very early onset inflammatory bowel disease: A case report. *BMC Gastroenterol.* 2015;15:160.
17. Zhang K, Wakefield E, Marsh R. Lymphoproliferative disease, X-Linked. 2004 Feb 27 [Updated 2016 Jun 30]. In: Adam MP, Ardinger HH, Pagon RA et al. editors. *GeneReviews® [Internet]*. Seattle (WA): University of Washington, Seattle; 1993–2018.
18. Ishimura M, Eguchi K, Shiraishi A et al. Systemic Epstein-Barr virus-positive T/NK lymphoproliferative diseases with *SH2D1A/XIAP* hypomorphic gene variants. *Front Pediatr.* 2019;7:183.
19. Dugan JP, Coleman CB, Haverkos B. Opportunities to target the life cycle of Epstein-Barr virus (EBV) in EBV-associated lymphoproliferative disorders. *Front Oncol.* 2019;9:127
20. Tokuhira M, Tamaru JI, Kizaki M. Clinical management for other iatrogenic immunodeficiency-associated lymphoproliferative disorders. *J Clin Exp Hematop.* 2019;59(2):72–92.
21. Watanabe T, Sato Y, Masud HMAA et al. Antitumor activity of cyclin-dependent kinase inhibitor alsterpaullone in Epstein-Barr virus-associated lymphoproliferative disorders. *Cancer Sci.* 2020;111(1):279–87.
22. Dziadzio M, Ammann S, Canning C et al. Symptomatic male and female carriers in a large Caucasian kindred with XIAP deficiency. *J. Clin. Immun.* 2015;35:439–44.
23. Kimura H, Fujiwara S. Overview of EBV-associated T/NK-cell lymphoproliferative diseases. *Front Pediatr.* 2019;6:417.
24. Shannon-Lowe C, Rickinson A. The global landscape of EBV-associated tumors. *Front Oncol.* 2019;9:713.

76

Nijmegen Breakage Syndrome

Dongyou Liu

CONTENTS

76.1 Introduction

Nijmegen breakage syndrome (NBS) is a rare autosomal recessive disorder characterized by microcephaly, dysmorphic facial features, mild growth retardation, mild-to-moderate intellectual disability, premature ovarian insufficiency (hypergonadotropic hypogonadism), cellular and humoral immunodeficiency with recurrent sinopulmonary infections, radiosensitivity, and predisposition to lymphoid malignancies at an early age.

The molecular pathogenesis of NBS relates to inversions and translocations involving chromosomes 7 and 14 that create chromosomal instability in peripheral T lymphocytes, and biallelic (homozygous or compound heterozygous) mutations in the *NBN* gene on chromosome 8q21.3, which encodes nibrin involved in DNA double-strand break repair, underlie the development of this chromosomal instability disorder [1].

76.2 Biology

The first case related to NBS was described in 1979 from a Dutch boy who presented with microcephaly, growth and developmental retardation, IgA deficiency, and chromosomal rearrangements, which appeared similar to those found in ataxia telangiectasia (A-T) [2]. The observation of breakpoints in four chromosome sites (7p13, 7q35, 14q11, and 14q32) in the affected patient led the researchers at the University of Nijmegen in the Netherlands to name this disorder *Nijmegen breakage syndrome* in 1981. Although both NBS and A-T contain chromosomes 7 and 14 rearrangements, NBS causes characteristic microcephaly,

which is rare in A-T. Subsequent studies indicated that apart from microcephaly, developmental delay, and immunodeficiency, NBS confers a strong predisposition to lymphoreticular malignancies. In addition, cell cultures from NBS patients were noted by Jasper and coworkers in 1988 and others afterward to have hypersensitivity to ionizing radiation (radiosensitivity), chromosomal instability, and abnormal p53-mediated cell cycle regulation [3,4].

The causative gene for NBS was mapped to chromosome 8q21.3 by Saar and others in 1977 and identified by Varon et al. and Carney et al. in 1998 as NBS, which shares weak homology to *Saccharomyces cerevisiae* telomere and DNA repair protein Xrs2 in the N-terminus. The encoded protein nibrin appears to participate in the formation of a trimeric protein complex with MRE11 and RAD50, which is condensed as foci in the nucleus after irradiation, suggesting its potential role in DNA double strand break repair and signal transduction for cell-cycle checkpoints as a substrate of A-T mutated (ATM) kinase [5].

76.3 Pathogenesis

The *NBS* gene on chromosome 8q21.3 spans 51 kb with 16 exons and encodes a 754 aa, 95 kDa protein (nibrin or p95) that comprises three functional regions: the N-terminus (1–196 aa; damage recognition by binding to histone), the central region (278–343 aa; signal transduction by phosphorylation of serine residues), and the C-terminus (665–693 aa; complex formation by binding to MRE11), with one forkhead-associated (FHA) domain (24–108 aa) and two breast cancer carboxy-terminal (BRCT) domains (108–196 aa). The FHA domain is a phospho-specific protein–protein interaction motif that recognizes phosphorylation of the target protein, and the FHA/BRCT domains play a role in recognition of the sites of DSB. In response to ionizing radiation, the serine residues at 278 and 343 are phosphorylated by ATM kinase, facilitating intra-S phase checkpoint control.

Nibrin interacts with MRE11 and RAD50 to form a trimeric protein complex, which acts as a primary sensor of DNA double-strand breaks (DSB) after exposure to ionizing radiation. Acting as a regulatory subunit, nibrin plays a key role in the localization, signal transduction and catalytic activation of the complex. The combination of the highly conserved FHA and BRCT domains in nibrin is crucial for recognition of damaged sites, whereas the domain for MRE11-binding at the C-terminus is involved in both normal radiation sensitivity and nuclear localization of the complex (in a process known as nuclear foci formation). Following translocation to the vicinity of DSB, the trimeric protein complex collaborates with ATM kinase for homologous recombination DNA repair, checkpoint control, DNA replication, and telomere maintenance [6–9].

Homozygous or compound heterozygous mutations in *NBN* exons 6–10 produce truncation in the nibrin protein that lacks a C-terminal protein fragment (p70). The inability of this altered protein to undertake translational re-initiation appears to underlie the development to NBS. Specifically, absence or mutation of nibrin disrupts the ATM interaction site, impairs G2 checkpoint control, allows continued DNA synthesis in the presence of DSB (so-called radioresistant DNA synthesis [RDS]), and causes severe apoptotic defect. Besides frequent chromosome aberrations at the sites of T cell receptor (TCR), NBS is also associated with immunoglobulin heavy-chain (IgH) rearrangement, where V(D)J recombination spontaneously induces DSB, leading to B cell lymphomas [10–13].

The most notable NBS pathogenic variants include c.330T>G (p.Tyr110Ter), c.643C>T (p.Arg215Trp), c.657_661del5 (657del5) (p.Lys219AsnfsTer15), c.681delT (p.Phe227LeufsTer4), c.698_701del4 (p.Lys233SerfsTer4), c.741_742dup (742insGG) (p.Glu248GlyfsTer5), c.835_838del4 (p.Gln279ProfsTer1), c.842insT (p.Leu281PhefsTer3), c.900del25 (p.Gly301LysfsTer5), c.976C>T (p.Gln326Ter), c.1089C>A (p.Tyr363Ter), c.1125G>A (p.Trp375Ter), and c.1142delC (p.Pro381GlnfsTer22). Among these, the founder mutation c.657_661del5, of Slavic origin, creates a five base pair deletion in exon 6 and inserts a premature termination signal at codon 219, resulting in a 26 kDa amino-terminal fragment (p26-nibrin, containing the FHA domain and one BRCT domain) and a 70 kDa carboxy-terminal fragment (p70-nibrin, containing the second BRCT domain), which is responsible for a majority of clinical cases reported. Homozygous c.1089C>A pathogenic variant may show features of Fanconi anemia. Compound heterozygous mutations involving c.657_661del5 and c.643C>T are observed in monozygotic twins, who presented with severe neurologic features (microcephaly, mildly asymmetric lateral ventricles, enlarged subarachnoid areas and poor gyrification of the brain, and retarded psychomotor development) without chromosomal instability and radiation sensitivity [1].

76.4 Epidemiology

76.4.1 Prevalence

NBS is a rare disease with an estimated incidence of 1 per 100,000 live births in North and South America, Morocco, and New Zealand. The carrier frequency for *NBN* founder mutation c.657_661del5 (p.K219fsX19) may reach 1 in 34–155 in Eastern European/Slavic countries (e.g., Serbia, Czech Republic, Poland, Ukraine, and Russia). In addition to >150 published cases, many more cases are recorded in the European registry [1].

76.4.2 Inheritance

NBS shows autosomal recessive inheritance. Individuals harboring heterozygous pathogenic variants have 25% chance of producing an affected offspring (with homozygous or compound heterozygous pathogenic variants), 50% chance of producing an heterozygous carrier (who may be asymptomatic or show increased risk of breast cancer, prostate cancer, medulloblastoma, and melanoma), and 25% chance of producing an unaffected offspring (without pathogenic variant).

76.4.3 Penetrance

NBS demonstrates high penetrance and variable expressivity. While c.657_661del5 and many other loss-of-function pathogenic variants induce a classic phenotype, some biallelic truncating variants (e.g., c.741_742dupGG, c.330T>G, c.1125G>A) are known to cause mild or partial symptoms.

76.5 Clinical Features

Clinically, NBS often manifests as *progressive microcephaly* (a hallmark symptom of NBS, with head circumference >2 SD below the mean for age and gender), *craniofacial dysmorphism* (sloping forehead, receding mandible, upwardly slanted palpebral fissures, long beaked nose, upturned/anteverted nostrils, cleft lip/palate, choanal atresia, receding chin, relatively large ears), *mild growth delay and intellectual disability*, *premature ovarian failure*, *skin anomalies* (café-au-lait spots or vitiligo spots in 50%–70% of cases, progressive sarcoid-like granulomas), *skeletal anomalies* (clinodactyly of the fifth fingers and partial syndactyly of the second and third toes in 50% of cases), *congenital kidney anomalies* (renal hypoplasia/aplasia, the horseshoe or double kidney, ectopic/dystopic kidneys), *cellular and humoral immunodeficiency* (lymphopenia/leucopenia, agammaglobulinemia/ hypogammaglobulinemia), *recurrent infections* (sinusitis, pneumonia, bronchitis, chronic diarrhea, urinary tract infection, otitis media, mastoiditis), *hematological disorders* (aplastic anemia), and *malignancies* (non-Hodgkin lymphomas [NHL] of B and T cells such as diffuse large B cell lymphoma [DLBCL], T cell lymphoblastic lymphoma [TLBL], Burkitt and Burkitt-like lymphomas, Hodgkin lymphoma, prolymphocytic leukemia, acute lymphoblastic leukemia [ALL]; medulloblastoma, rhabdomyosarcoma, glioma, papillary thyroid carcinoma, gonadoblastoma, meningioma, neuroblastoma, and Ewing sarcoma; 40% of cases) at an early age (Figure 76.1) [1,14–18].

76.6 Diagnosis

Diagnosis of NBS is based on observation of characteristic clinical manifestations, chromosomal instability (spontaneous and

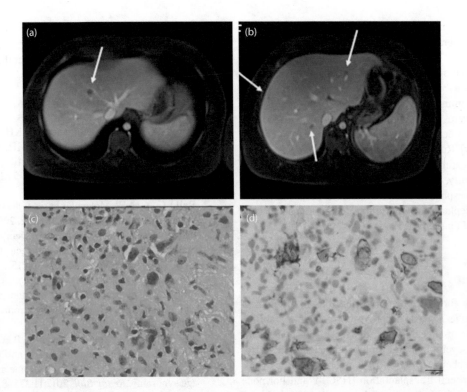

FIGURE 76.1 Nijmegen breakage syndrome in a 26-year-old male with Hodgkin lymphoma showing multiple hypointense liver lesions in the volumetric interpolated breath-hold sequence (VIBE) acquired in portal venous phase (a and b); histology of classical Hodgkin lymphoma with lacunar cells and mixed inflammatory background in liver biopsy (c, H&E staining; d, CD15 staining). (Photo credit: Engel K et al. *BMC Hematol.* 2014;14(1):2.)

induced), increased cellular sensitivity to ionizing radiation in vitro, combined cellular and humoral immunodeficiency, mutations in both alleles of the *NBN* gene, and complete absence of full length nibrin [1].

Individuals presenting with the following clinical features and laboratory findings should be suspected of NBS:

Clinical features: (i) Progressive microcephaly, (ii) recurrent infections (pneumonia, bronchitis, sinusitis, otitis media, mastoiditis), (iii) malignancies (NHL, ALL, solid tumors), (iv) craniofacial features (sloping forehead, prominent midface, receding mandible, upward-slanted palpebral fissures, prominent nose, relatively large ears, retrognathia), (v) growth retardation, (vi) decline in intellectual ability, (vii) premature ovarian insufficiency, (viii) family pedigree with microcephaly or hydrocephaly, malignancies, and early death due to severe/recurring infections or malignancy [1].

Laboratory findings: (i) Immunodeficiency (severe hypogammaglobulinemia in 20%–24% of cases and IgA deficiency in 50%–57% of cases; deficiencies of IgG2 and IgG4; reduced absolute numbers of total B cells, CD3+ T cells, and CD4+ cells, in 80%–89% of cases; increased frequency of T cells with a memory phenotype CD45RO+ and concomitant decrease in naïve T cells CD45RA+; greatly reduced in vitro proliferation of T and B lymphocytes to antigen and/or mitogenic stimuli); (ii) chromosome instability (inversions and translocations involving chromosomes 7 and 14 in PHA-stimulated lymphocytes in 10%–50% of

metaphases; breakpoints involving the loci for immunoglobulin and T cell-receptor genes such as 7p13, 7q35, 14q11, and 14q32); (iii) radiation sensitivity (cells from affected individuals showing decreased colony-forming ability upon exposure to ionizing radiation and radiomimetics in vitro) [1].

Molecular identification of biallelic pathogenic variants (c.657_661del5 and other) in *NBN* or absence of nibrin protein in a lymphoblastoid cell line by immunoblotting assay helps confirm the diagnosis of NBS. The pathogenic variant c.657_661del5 is present in ~100% of the Slavic population (Poland, Czech Republic, Ukraine), and ~70% of the North American population [1].

Differential diagnoses for NBD include several inherited disorders that also display microcephaly, growth delay, immunodeficiency, and bone marrow failure due to defective sensing, processing, and repair of DNA DSB (*LIG4* syndrome [microcephaly, dysmorphic face, growth retardation, combined cellular and humoral immunodeficiency], *NHEJ1* syndrome [microcephaly, dysmorphic facies, severe growth retardation, severe combined cellular and humoral immunodeficiency], Nijmegen breakage syndrome-like disorder [NBSLD or RAD50 deficiency; microcephaly, dysmorphic face, growth retardation, mild spasticity, non-progressive ataxia, normal puberty, chromosome instability at bands 7p13, 7q34, 14q11, and 14q32, absence of immunodeficiency and malignancy; due to mutation in RAD50 on chromosome 5q31.1], A-T [ovarian failure, neurodegeneration, telangiectasia, increased alpha fetal protein, immunodeficiency, cancer predisposition; due to biallelic mutations in the A-T mutated or *ATM* gene], A-T like disorder [ATLD; neurodegeneration, increased alpha fetal protein; due to *MRE11* mutation], Fanconi anemia [occasional microcephaly, growth retardation, skeletal abnormalities—radial defect, reduced fertility—hypergonadotropic hypogonadism in males, pancytopenia, no immunodeficiency], Seckel syndrome [severe microcephaly, severe prenatal and postnatal growth retardation, developmental delay, mental retardation, pancytopenia, no immunodeficiency], Rubinstein–Taybi syndrome [microcephaly, distinctive facial features, mild growth restriction, short stature, intellectual disability, recurrent infections, defect in polysaccharide antibody response, leukemia, brain tumor], and Bloom syndrome [microcephaly, growth failure, increased cancer incidence; due to *BLM* mutation]) [1].

76.7 Treatment

Treatment options for NBS comprise nutritional supplement (vitamin E and folic acid) to improve chromosome stability, intravenous (IVIG) or subcutaneous (SCIG) Ig replacement for individuals with severe humoral immunodeficiency and frequent infections, chemotherapy for lymphoid malignancies, hematopoietic stem cell transplantation (HSCT), and hormone replacement therapy for females with hypergonadotropic hypogonadism. Since x-irradiation may induce malignancies in NBS patients, radiation therapy should be avoided [19–21].

76.8 Prognosis and Prevention

Prognosis for NBS patients with malignancies is poor. NBS patients with B cell NHL fare better than those with T cell lymphoma. NBS patients treated for one cancer may have increased risk of developing different consecutive malignant disease.

Individuals harboring NBS pathogenic variants should use MR imaging and ultrasound if needed and avoid ionizing radiation to reduce the risk of developing malignancy.

Prenatal testing and preimplantation genetic diagnosis may be considered for prospective parents harboring *NBN* pathogenic variants.

76.9 Conclusion

Nijmegen breakage syndrome (NBS) is an autosomal recessive chromosome instability disorder that typically causes microcephaly, immunodeficiency, and predisposition for lymphoma at an early age, in addition to skin manifestations (e.g., café-au-lait spots, vitiligo, telangiectasia, flat hemangioma, and sun sensitivity of the eyelids). The underlying mechanism of NBS relates to homozygous or compound heterozygous mutations in the NBS gene encoding a protein that forms a trimeric complex with MRE11 and RAD50 for repairing DNA DSB after exposure to ionizing radiation [22]. Diagnosis of NBS requires observation of characteristic clinical manifestations, chromosomal instability (spontaneous and induced), increased cellular sensitivity to ionizing radiation in vitro, combined cellular and humoral immunodeficiency, mutations in both alleles of the *NBN* gene,

and complete absence of full-length nibrin. Treatment of NBS involves nutritional supplementation, Ig replacement, chemotherapy, HSCT, and hormone replacement therapy.

REFERENCES

1. Varon R, Demuth I, Chrzanowska KH. Nijmegen breakage syndrome. In: Adam MP, Ardinger HH, Pagon RA et al. editors. *GeneReviews®* [Internet]. Seattle (WA): University of Washington, Seattle; 1993–2018. 1999 May 17 [updated 2017 Feb 2].
2. Ball LG, Xiao W. Molecular basis of ataxia telangiectasia and related diseases. *Acta Pharmacol Sin.* 2005;26(8):897–907.
3. Chrzanowska KH, Gregorek H, Dembowska-Bagińska B, Kalina MA, Digweed M. Nijmegen breakage syndrome (NBS). *Orphanet J Rare Dis.* 2012;7:13.
4. Halevy T, Akov S, Bohndorf M et al. Chromosomal instability and molecular defects in induced pluripotent stem cells from Nijmegen breakage syndrome patients. *Cell Rep.* 2016;16(9):2499–511.
5. Maciejczyk M, Mikoluc B, Pietrucha B et al. Oxidative stress, mitochondrial abnormalities and antioxidant defense in Ataxia-telangiectasia, Bloom syndrome and Nijmegen breakage syndrome. *Redox Biol.* 2017;11:375–83.
6. Krenzlin H, Demuth I, Salewsky B et al. DNA damage in Nijmegen breakage syndrome cells leads to PARP hyperactivation and increased oxidative stress. *PLoS Genet.* 2012;8(3):e1002557.
7. Alster O, Bielak-Zmijewska A, Mosieniak G et al. The role of nibrin in doxorubicin-induced apoptosis and cell senescence in Nijmegen breakage syndrome patients lymphocytes. *PLOS ONE.* 2014;9(8):e104964.
8. Komatsu K. NBS1 and multiple regulations of DNA damage response. *J Radiat Res.* 2016;57(Suppl 1):i11–7.
9. Zhou H, Kawamura K, Yanagihara H, Kobayashi J, Zhang-Akiyama QM. NBS1 is regulated by two kind of mechanisms: ATM-dependent complex formation with MRE11 and RAD50, and cell cycle-dependent degradation of protein. *J Radiat Res.* 2017;58(4):487–94.
10. Wang Y, Hong Y, Li M et al. Mutation inactivation of Nijmegen breakage syndrome gene (NBS1) in hepatocellular carcinoma and intrahepatic cholangiocarcinoma. *PLOS ONE.* 2013;8(12):e82426.
11. Larsen DH, Hari F, Clapperton JA et al. The NBS1-treacle complex controls ribosomal RNA transcription in response to DNA damage. *Nat Cell Biol.* 2014;16(8):792–803.
12. Schröder-Heurich B, Bogdanova N, Wieland B et al. Functional deficiency of NBN, the Nijmegen breakage syndrome protein, in a p.R215W mutant breast cancer cell line. *BMC Cancer.* 2014;14:434.
13. Saito Y, Komatsu K. Functional role of NBS1 in radiation damage response and translesion DNA synthesis. *Biomolecules.* 2015;5(3):1990–2002.
14. Liana RA, Dan G, Nicolae M. Cutaneous sarcoid-like granulomas in a child known with Nijmegen breakage syndrome. *Iran J Pediatr.* 2013;23(1):100–4.
15. Pasic S, Kandolf-Sekulovic L, Djuricic S, Zolotarevski L, Simic R, Abinun M. Necrobiotic cutaneous granulomas in Nijmegen breakage syndrome. *J Investig Allergol Clin Immunol.* 2012;22(2):138–40.

16. Pasic S, Cupic M, Jovanovic T, Djukic S, Kavaric M, Lazarevic I. Nijmegen breakage syndrome and chronic polyarthritis. *Ital J Pediatr*. 2013;39:59.

17. Engel K, Rudelius M, Meinel FG, Peschel C, Keller U. An adult patient with Nijmegen breakage syndrome and Hodgkin's lymphoma. *BMC Hematol*. 2014;14(1):2.

18. Kocheva SA, Martinova K, Antevska-Trajkova Z, Coneska-Jovanova B, Eftimov A, Dimovski AJ. T-lymphoblastic leukemia/lymphoma in macedonian patients with Nijmegen breakage syndrome. *Balkan J Med Genet*. 2016;19(1):91–4.

19. Deripapa E, Balashov D, Rodina Y et al. Prospective study of a cohort of Russian Nijmegen breakage syndrome patients demonstrating predictive value of low kappa-deleting recombination excision circle (KREC) numbers and beneficial effect of hematopoietic stem cell transplantation (HSCT). *Front Immunol*. 2017;8:807.

20. Gałązka P, Czyżewski K, Szaflarska-Popławska A, Dębski R, Krenska A, Styczyński J. Complex profile of multiple hepatobiliary and gastrointestinal complications after hematopoietic stem cell transplantation in a child with Nijmegen breakage syndrome. *Cent Eur J Immunol*. 2019;44(3):327–31.

21. Wolska-Kuśnierz B, Gennery AR. Hematopoietic stem cell transplantation for DNA double strand breakage repair disorders. *Front Pediatr*. 2020;7:557.

22. Syed A, Tainer JA. The MRE11-RAD50-NBS1 complex conducts the orchestration of damage signaling and outcomes to stress in DNA replication and repair. *Annu Rev Biochem*. 2018;87:263–94.

77

Severe Congenital Neutropenia

Dongyou Liu

CONTENTS

77.1 Introduction

Severe congenital neutropenia (SCN) represents a heterogeneous group of rare hematological disorders characterized by neutropenia (shortage of neutrophils due to impaired maturation of neutrophil granulocytes), susceptibility to recurrent infections (in the sinuses, lungs, and liver) early in life, decreased bone density (osteopenia, 40% of cases) and increased risk for myelodysplastic syndromes (MDS) or acute myeloid leukemia (AML) during adolescence (20% of cases).

Molecularly, at least 29 different genes are implicated in the pathogenesis of SCN, including *ELANE* (encoding neutrophil elastase, 45% of cases), *SBDS* (encoding protein involved in ribosome function or assembly, 14%), *SLC37A4* (*G6PT*, 12%), *HAX1* (encoding component of the granulocyte-colony stimulating factor signaling pathway, 7%), and *G6PC3* (encoding glucose-6-phosphatase catalytic subunit, 2%) (Table 77.1) [1].

77.2 Biology

Neutrophils (or neutrocytes) are one of three types of white blood cells or leukocytes (the others being basophils and eosinophils) that possess highly lobulated nuclei with four to five separate lobes (so-called polymorphonuclear leukocytes [PMN]), and various cytoplasmic granules (lysosomes) (thus referred to as granulocytes). In hematoxylin and eosin staining, neutrophils appear neutral pink (so-called neutrophils), compared to basophils' dark blue and eosinophils' bright red appearance. With an average life span of 5–135 hours, neutrophils do not reside in normal, healthy tissues, but instead circulate inactive in the blood (making up 40%–60% of circulating leukocytes). As one of the first innate immune responders to infection, injury, or cancer, neutrophils can move within minutes to sites of inflammation to phagocytose cellular debris and any invading pathogens [2].

Originating from hematopoietic stem cells within the bone marrow, neutrophils undergo several developmental stages, including myeloblast (prominent nucleoli), promyelocyte (large cell, prominent granules), myelocyte (secondary granules), metamyelocyte (kidney bean–shaped nucleus), and band form (condensed, band-shaped nucleus), with each successive division resulting in more nuclear constriction and less cytoplasmic RNA or blue color in cytoplasm, before maturation (condensed, multilobed nucleus). SCN is attributed to impairment in neutrophil granulopoiesis that generate fewer mature neutrophils [2].

The initial description of SCN (or cyclic neutropenia) was made in 1910. This was followed by the reports of agranulocytosis (Schultz syndrome) in 1922, infantile genetic agranulocytosis (Kostman syndrome) in 1956, and preleukemic syndrome in 1970. The phenotypic heterogeneity of this disorder was highlighted in 1975 with the observations that bone marrow from some affected patients form loose promyelocyte colonies, while others produce normal neutrophil colonies in soft agar cultures. The identification of the *SLC37A4* (*G6PT*) and *ELANE* genes in 1999 and >20 other neutropenia-causing genes in subsequent years yielded invaluable insights on the genetic heterogeneity, molecular pathogenesis, and inheritance pattern of SCN [1].

It is now clear that mutations in the genes encoding proteins involved in receptor binding (G-CSFR, CXCR4), endoplasmic reticulum (G6PT, G6PC3, JAGN1, VPS13B), ribosomes (SBDS), nucleus (GF11, GATA2), mitochondria (HAX1, TAZ, AK2),

TABLE 77.1

Molecular and Clinical Characteristics of Congenital Neutropenias

Gene	Protein Function	Pathogenesis	Inheritance	Hematologic Abnormalities	Non-Hematologic Abnormalities	% of Cases
ELANE	Neutrophil elastase; secondary granule protease	Activation of the UPR and apoptosis of myeloid progenitor cells	Autosomal dominant	Severe congenital neutropenia type 1, monocytosis, eosinophilia, cyclic neutropenia, cyclic hematopoiesis, conversion to AML or MDS	Osteopenia	45
CXCR4	C-X-C chemokine receptor type 4; HSC homing and myeloid cell retention to the bone marrow	Reduced egress of HSC and mature neutrophils from the bone marrow	Autosomal dominant	Congenital neutropenia, warts, hypogammaglobulinemia, infections, myelokathexis (WHIM) syndrome, B cell defects, hypogammaglobulinemia	Warts	2
TCIRG1	V-type proton ATPase; pH regulation of intracellular compartments and organelles, including neutrophil phagocytic vacuoles; T-lymphocyte activation; immune response	Possible defective bone marrow niche regulation of granulopoiesis due to osteopetrosis or osteoclasts activation or direct effects on promyelocyte differentiation	Autosomal dominant	Severe congenital neutropenia	In some patients, prominent hemangiomas during G-CSF treatment	2
GFI1	Zinc finger protein Gfi-1; transcriptional repressor interacting with myeloid-specific transcription factors C/EBPα, C/EBPβ and PU.1	Diminished myeloid differentiation	Autosomal dominant	Severe congenital neutropenia type 2, lymphopenia, increased numbers of immature myeloid cells in the peripheral blood	None	
GATA2	Endothelial transcription factor GATA-2; embryonic and definitive hematopoiesis and lymphatic angiogenesis	Deregulated proliferation and differentiation of HSCs and reduced numbers of HSCs pool	Autosomal dominant	Congenital neutropenia, severe monocytopenia, dendritic cells and natural killer cells deficiencies, aplastic anemia, conversion to AML or MDS	Mycobacteria, fungal or human papilloma virus infections; pulmonary dysfunction including pulmonary alveolar proteinosis; warts and leg lymphedema	
SBDS	Ribosome maturation protein SBDS; regulation of later steps of ribosome biogenesis and mitotic spindle stabilization	Mitotic spindle destabilization, genomic instability and enhanced apoptosis of HSCs	Autosomal recessive	Shwachman–Diamond syndrome, thrombocytopenia, anemia, aplastic anemia, conversion to AML/MDS	Exocrine pancreatic insufficiency, cardiomyopathy, metaphyseal dysplasia, mental retardation and hepatic disease	14
SLC37A4	Glucose-6-phosphate, exchanger; transport of glucose-6-phosphate from the cytoplasm to the ER lumen, maintenance of glucose homeostasis and ATP-mediated calcium sequestration in the ER	Defective trans-ER transport, abnormal glycolysis and gluconeogenesis, elevated apoptosis of neutrophils and neutrophil dis-function	Autosomal recessive	Congenital neutropenia, glycogen storage disease type Ib (GSDIb)	Hypoglycemia, fasting hyper-lactacidemia, glycogen overload of the liver, colitis, pancreatitis and osteoporosis	12
HAX1	HCLS1-associated protein X-1; activation of HCLS1 adaptor protein in G-CSF signaling and anti-apoptotic functions	Reduced mitochondrial membrane potential, elevated apoptosis and abrogated G-CSFR signaling	Autosomal recessive	Severe congenital neutropenia type 3, conversion to AML or MDS	Neurological phenotype in patients with two specific mutations	7
G6PC3	Glucose-6-phosphatase 3; hydrolyzation of glucose-6-phosphate to glucose and phosphate in the ER	Impaired intracellular glucose homeostasis, UPR activation and elevated apoptosis of myeloid cells	Autosomal recessive	Severe congenital neutropenia type 4, thrombocytopenia, conversion to AML or MDS	Cardiac defects, increased superficial veins visibility, urogenital malformations, endocrine abnormalities and skin hyper-elasticity	2
VPS13B	Vacuolar protein sorting-associated protein 13B is a transmembrane protein involved in vesicle-mediated transport and sorting of proteins within the cell	Mutations alter the development and function of the eye, hematological system, and central nervous system	Autosomal recessive	Cohen syndrome	Psychomotor retardation, truncal obesity, microencephaly, skeletal dysplasia, hypotonia, myopia	2

(Continued)

TABLE 77.1 (Continued)

Molecular and Clinical Characteristics of Congenital Neutropenias

Gene	Protein Function	Pathogenesis	Inheritance	Hematologic Abnormalities	Non-Hematologic Abnormalities	% of Cases
JAGN1	Protein jagunal homolog 1; early secretory pathway, cell adhesion and cytotoxicity	Aberrant N-glycosylation of multiple proteins, elevated apoptosis and poor to no G-CSF response	Autosomal recessive	Severe congenital neutropenia type 6	Short stature and bone and teeth defects	
STK4	Serine/threonine protein kinase 4; upstream component of the mitogen-activated protein kinase pathway	Enhanced loss of mitochondrial membrane potential and increased susceptibility to apoptosis	Autosomal recessive	Congenital neutropenia, monocytopenia, T- and B-lymphopenia	Warts and atrial septal defects	
CLPB	Caseinolytic peptidase B protein homolog with ATPase activity, and involvement in DNA replication, protein degradation and reactivation of misfolded proteins	Allelic variant is associated with 3-methylglutaconic aciduria, leading to cataracts and neutropenia	Autosomal recessive	3-Methylglutaconic aciduria type VII, conversion to AML or MDS	Cataracts, 3-methylglutaconic aciduria, facial dysmorphism, cardiomyopathy or hypertrophy and hypothyroidism	
AP3B1	AP-3 complex subunit beta-1; cargo protein in endosomal trafficking	Defective endosome formation and processing and endosomal or lysosomal defects in immune cells	Autosomal recessive	Hermansky–Pudlak syndrome 2, impaired function of T and NK cells	Oculocutaneous albinism and hemorrhagic diathesis	
LAMTOR2	Regulator complex protein LAMTOR2; regulation of endosomal trafficking and sorting, growth factor signaling and cell proliferation	Defective MAPK and ERK signaling, diminished phagocytosis and disturbed endosomal trafficking	Autosomal recessive	p14 deficiency, accumulation of neutrophils in the bone marrow, defective cytotoxicity, lymphoid immunodeficiency	Oculocutaneous albinism and stunted growth	
USB1	U6 snRNA Phosphodiesterase; exoribonuclease in RNA processing from pre-RNA	Diminished biogenesis of U6 small nuclear RNA and elevated apoptosis	Autosomal recessive	Clericuzio type poikiloderma	Poikiloderma, generalized hyperkeratosis on palms and soles, short stature and recurrent pulmonary infections	
VPS45	Vacuolar protein sorting-associated protein 45; regulation of assembly of the SNARE (soluble N ethylma-leimide sensitive factor attachment protein receptor) complex, which plays an essential part in the trafficking and recycling of proteins through lysosomes, other endosomes and trans-Golgi complex	Degradation of key components of the SNARE complex, defective transport of proteins from the trans-Golgi network to endosomes, impaired cell motility, increased apoptosis, NADPH-oxidase dysfunction, diminished superoxide production by neutrophils, lack of surface expression of β1 integrin	Autosomal recessive	Severe congenital neutropenia type 5, anisocytosis and poikilocytosis, hypergammaglobulinemia, renal extramedullary hematopoiesis, bone marrow fibrosis, progressive anemia, thrombocytopenia	Nephromegaly, splenomegaly, osteosclerosis, neurologic abnormalities such as delayed development, cortical blindness, hearing loss and thin corpus callosum	
CXCR2	C-X-C chemokine receptor type 2; chemotaxis of neutrophils to the site of inflammation and migration from the bone marrow to peripheral blood	Abolished IL8-induced Erk1/2 phosphorylation and chemotaxis	Autosomal recessive	Congenital neutropenia, myelokathexis due to impaired neutrophil release from the bone marrow to the peripheral blood	NA	
EIF2AK3	Eukaryotic translation initiation factor 2-alpha kinase 3; phosphorylation and inactivation of eIF2A; type I ER membrane protein induced by ER stress owing to misfolded proteins	Failure in translational initiation and repression of global protein synthesis and mitochondrial functions	Autosomal recessive	Wolcott–Rallison syndrome	Early infancy-onset insulin-dependent diabetes mellitus, epiphyseal dysplasia, growth retardation, hepatic and renal dysfunction, developmental delay, exocrine pancreatic deficiency	

(Continued)

TABLE 77.1 (Continued)

Molecular and Clinical Characteristics of Congenital Neutropenias

Gene	Protein Function	Pathogenesis	Inheritance	Hematologic Abnormalities	Non-Hematologic Abnormalities	% of Cases
LYST	Lysosomal trafficking regulator; endosomal protein trafficking	Defective sorting of endosomal resident proteins	Autosomal recessive	Chédiak–Higashi syndrome, defective NK cell function; lysosomal inclusion bodies in myeloblasts, promyelocytes and granulocytes; macrophage activation, lymphoproliferative syndrome	Oculocutaneous albinism and neurodegeneration	
RAB27A	Ras-related protein Rab-27A; protein transport, small GTPase mediated signaling and lytic granule release	Defective vesicular trafficking, protein transport, endocytic and secretory pathway	Autosomal recessive	Griscelli syndrome, type 2, defective cytotoxicity, hypogammaglobulinemia, thrombocytopenia, anemia, hemophagocytosis	Oculocutaneous albinism	
AK2	Adenylate kinase 2; transfer of the terminal phosphate group between ATP and AMP	Aberrant mitochondrial metabolism and regulation of apoptosis	Autosomal recessive	Adenylate kinase 2 deficiency, severe lymphopenia	Inner ear hearing loss	
RMRP	RNA component of mitochondrial RNA processing endoribonuclease	Defective mitochondrial RNA cleavage and small interfering RNA production	Autosomal recessive	Cartilage-hair hypoplasia, immunodeficiency, anemia	Hypoplastic hair, skeletal dysplasia and cartilage hypoplasia	
TCN2	Transcobalamin-2; transport of cobalamin (vitamin B12) from the blood stream to the cells	Defective plasma transport of vitamin B12 resulting in B12 deficiency	Autosomal recessive	Transcobalamin II deficiency, megaloblastic anemia, pancytopenia	Methylmalonic aciduria, failure to thrive, recurrent infections, mental retardation and neurologic abnormalities	
CSF3R	G-CSF receptor	no G-CSF response owing to the absent expression of G-CSFR on the cell surface	Autosomal dominant and recessive	Severe congenital neutropenia type 7	None	
TAZ	Tafazzin; acyltransferase in lipid metabolism and regulation of phospholipid membrane homeostasis	Destabilization of mitochondrial respiratory chain complexes and elevated apoptosis	X-linked	Barth syndrome	Cardiomyopathy, skeletal myopathy, stunted growth, cardiolipin abnormalities and 3-methylglutaconic aciduria	4
WAS	Wiskott–Aldrich syndrome protein; regulation of actin rearrangement	Enhanced actin polymerization, altered cytoskeletal responses and genomic instability	X-linked	Severe congenital neutropenia type x, monocytopenia, lymphopenia, reduced numbers of natural killer cells, abrogated phagocyte activity	None	2
CD40LG	CD40 ligand; expressed on activated T cells, necessary for T cells to induce B cells to undergo Ig class-switching	Defective Ig class switch and defective *in vivo* clonal expansion of antigen-specific CD4+ T cells	X-linked	CD40 ligand deficiency, hyper-IgM syndrome type I (HIGM1), combined immunodeficiency; T, B, and dendritic cell deficiencies; defective B cell class switch; markedly reduced levels of IgG, IgA and IgE; reduced macrophage effector functions	Increased susceptibility to infections, increased risk for autoimmune disorders and malignancies	
mtDNA	Mitochondrial DNA converts chemical energy from food into ATP	mtDNA deletion disrupts cellular energy generation and negatively impacts on multiple organ systems	Exclusively maternal	Pearson syndrome, pancytopenia, refractory sideroblastic anemia, vacuolization of bone marrow precursors and macrophages	Exocrine pancreas and renal insufficiency or fibrosis, endocrine abnormalities, lactic acidosis, neuromuscular degeneration, mitochondrial myopathy	

endosomes and lysosomes (AP3B1, LYST, RAB27A), and cytoskeleton (WAS, HAX1), increase endoplasmic reticulum stress and apoptosis, deregulate transcription factors expression, alter granulocyte colony-stimulating factor receptor (G-CSFR) signal transduction, leading to arrest of granulopoiesis and acquisition of second leukemia-associated mutation (e.g., *RUNX1*) before conversion to MDS and AML (Table 77.1) [3–11].

77.3 Pathogenesis

The *ELANE* (elastase, neutrophil expressed) gene on chromosome 19p13.3 spans 5.2 kb and encodes a 267 aa, 28.5 kDa protein (neutrophil elastase) belonging to a subfamily of serine proteases that hydrolyze elastin and other proteins within specialized neutrophil lysosomes (called azurophil granules) as well as of the extracellular matrix (e.g., G-CSFR, VCAM, c-kit, CXCR4). Normally processed in the Golgi apparatus, stored in azurophil granules, and released after neutrophil activation, *ELANE* is capable of lysing elastin and collagen-IV, and degrading *Escherichia coli* outer membrane protein A (OmpA) and other bacterial virulence factors.

Heterozygous mutations in *ELANE* produce a nonfunctional neutrophil elastase that accumulates in the cytoplasm instead of the azurophil granules, leading to endoplasmic reticulum stress, accelerated apoptosis of neutrophil precursors, activation of the unfolded protein response, inability to phagosize invading pathogens and cell debris, and ultimate development of cyclic neutropenia and severe congenital neutropenia (45% of cases). Of >200 *ELANE* pathogenic variants identified to date, some (e.g., p.C151Y or p.G214R) are associated with a more severe phenotype (increased risk of leukemogenesis, poor G-CSF response, and risk of severe infections), while others may cause milder diseases [12,13].

The *SBDS* (Shwachman−Bodian−Diamond syndrome) gene on chromosome 7q11.21 is a 7.9 kb DNA fragment that encodes a 250 aa, 28.7 kDa protein (SBDS) involved in the assembly of mature ribosomes and ribosome biogenesis as well as cellular response to DNA damage and cell proliferation.

Specifically, through interaction with elongation factor-like GTPase 1, SBDS triggers release of eukaryotic initiation factor 6 (EIF6) from 60S pre-ribosomes in the cytoplasm and facilitates 80S ribosome assembly and EIF6 recycling to the nucleus for 60S rRNA processing and nuclear export. Homozygous mutations within *SBDS* cause Shwachman−Bodian−Diamond syndrome, which manifests as thrombocytopenia, anemia, and aplastic anemia (14% of SCN cases).

The *SLC37A4* (solute carrier family 37 member 4, or glucose-6-phosphate transporter *G6PT*) gene on chromosome 11q23.3 is a stretch of 6.8 kb DNA that encodes a 429 aa, 46 kDa protein (SLC37A4 or G6PT) with glucose-6-phosphate transmembrane transporter activity. Through regulation of glucose-6-phosphate transport from the cytoplasm to the lumen of the endoplasmic reticulum, this protein helps maintain glucose homeostasis and ATP-mediated calcium sequestration in the lumen of the endoplasmic reticulum. Homozygous mutations in *SLC37A4* disrupt carbohydrate digestion and absorption and glucose metabolism pathways, leading to severe congenital neutropenia (12% of cases), glycogen storage disease type Ib (GSDIb).

The *HAX1* (HCLS1 associated protein X-1) gene on chromosome 1q21.3 measures 3.3 kb in length, and encodes a 279 aa, 31 kDa protein (*HAX1*), which is an adaptor protein of the G-CSFR signaling pathway. Through interaction with hematopoietic cell-specific Lyn substrate 1 (a substrate of Src family tyrosine kinases), *HAX1* participates in the clathrin-mediated endocytosis pathway, promotes GNA13-mediated cell migration, and regulates the cortical actin cytoskeleton via its interaction with KCNC3 and Arp2/3 complex.

Expressed in mitochondria and cytoskeleton, *HAX1* functions as an anti-apoptotic protein. Homozygous mutations in *HAX1* abrogate the activation of HCLS1 and its downstream signaling, disturb mitochondrial function, and interrupt granulopoiesis, leading to severe congenital neutropenia (Kostmann disease, 7% of cases). While pathogenic variants (e.g., p.Q190X and p.R86X) affecting both *HAX1* isoforms induce SCN with neurological involvement, pathogenic variants (e.g., p.W44X) targeting one *HAX1* isoform cause SCN without neurological involvement. On the other hand, heterozygous mutations in *HAX1* are linked to polycystic kidney disease [14,15].

The *G6PC3* (glucose-6-phosphatase catalytic subunit 3) gene on chromosome 17q21.31 spans 5.6 kb with 6 exons and encodes the 346 aa, 38.7 kDa catalytic subunit of glucose-6-phosphatase (G6Pase). Located in the endoplasmic reticulum, G6Pase catalyzes the hydrolysis of glucose-6-phosphate to glucose and phosphate in the final step of the gluconeogenic and glycogenolytic pathways. Homozygous or compound heterozygous *G6PC3* mutations [e.g., c.130C>T (p.Phe44Ser, of Pakistani origin), c.210delC (p.Phe71SerfsTer46 or Ile70fsTer46, of Hispanic origin), c.758G>A (p.Arg253His. of Middle Eastern origin), c.778G>C (p.Gly260Arg, of European origin), c.829C>T (p.Gly277Ter, of European origin), c.935dupT (p.Asn313GlnfsTer74 or Asn313fs, of Iranian origin)] decrease cytoplasmic glucose and glucose-6-phosphate levels, which in turn induce GSK-3β and phosphorylation-mediated inactivation of the anti-apoptotic molecule Mcl-1, activate the endoplasmic reticulum stress mechanism, increase susceptibility to cellular apoptosis, and cause aberrant glycosylation of NADPH oxidase subunit, gp91phox, leading to deficits in neutrophil function, and a phenotypic continuum (nonsyndromic isolated SCN, classic G6PC3 deficiency [SCN plus cardiovascular and/or urogenital abnormalities], and severe G6PC3 deficiency [classic G6PC3 deficiency plus involvement of non-myeloid hematopoietic cell lines, additional extra-hematologic features, and pulmonary hypertension]) [16,17].

Besides severe neutropenia, mutations in neutropenia-related genes often cause non-hematopoietic organ failures, such as the heart (*G6PC3*, *TAZ*), urogenital system (*G6PC3*), bone (*SBDS*), exocrine pancreas (*SBDS*), skin (*LAMTOR2*, *RAB27A*), and liver (*SLC37A4*). Further, some affected patients may harbor combined gene mutations (e.g., *G6PC3* and *ELANE* or *HAX1* and *ELANE*), which may induce symptoms associated with both mutations [1].

77.4 Epidemiology

77.4.1 Prevalence

SCN has an estimated incidence of 1 in 200,000. Interestingly, *HAX1* mutations are responsible for 11% of SCN cases in Europe (due possibly to the large number of consanguineous families of Turkic or Arabic origin), but 7% in other parts of the world and none in the United States. Similarly, *G6PC3* mutations make up 25% of SCN cases in Israel, but 2% in other parts of the world. So far, nearly 100 individuals with molecularly confirmed G6PC3 deficiency have been documented.

77.4.2 Inheritance

SCN demonstrates autosomal dominant (e.g., ELANE), autosomal recessive (e.g., HAX1), X-linked and mtDNA (exclusively maternal) inheritance. Some SCN cases appear to arise from sporadic mutations without obvious family connection.

In autosomal dominant inheritance, one copy of the altered gene in each cell is sufficient to cause the disorder; in autosomal recessive inheritance, both copies of the gene in each cell have mutations; in X-linked inheritance, the mutated gene is located on the X chromosome, in which one altered copy of the gene in each cell is sufficient to cause the condition in males (who have only one X chromosome, and who cannot pass X-linked traits to their sons), and mutations in both copies of the gene is required for the disorder in females.

77.4.3 Penetrance

SCN shows high penetrance and variable expressivity. Although cyclic neutropenia has complete penetrance, its clinical severity varies among affected family members.

77.5 Clinical Features

Patients harboring neutropenia-causing genes may develop both hematological and non-hematologic abnormalities (Table 77.1) [1].

Specifically, *ELANE*-related neutropenia is associated with recurrent fever, skin, and oropharyngeal inflammation (mouth ulcer, gingivitis, sinusitis, and pharyngitis), and cervical adenopathy, and may appear as congenital neutropenia (omphalitis, diarrhea, pneumonia, and deep abscesses in the liver, lungs, and subcutaneous tissues in the first year of life; 15%–25% risk of MDS and AML by age 15 years) or cyclic neutropenia (3-week

intervals of fever and oral ulcerations; regular oscillations of blood cell counts; perianal cellulitis during neutropenic periods; generally healthy between neutropenic periods; no risk of malignancy or conversion to leukemia). Compared to persistent neutropenia (normocellular bone marrow and arrest of myeloid maturation) in congenital neutropenia, neutropenia tends to wax and wane in cyclic neutropenia [18].

G6PC3 deficiency typically displays a phenotypic spectrum, ranging from nonsyndromic isolated SCN (comprising only severe congenital neutropenia), classic G6PC3 deficiency (also known as severe congenital neutropenia type 4; consisting of SCN plus intermittent thrombocytopenia, cardiovascular defects [congenital heart defects, prominent superficial venous pattern], and urogenital abnormalities [cryptorchidism]) to severe G6PC3 deficiency (including classic G6PC3 deficiency plus primary pulmonary hypertension, non-myeloid cell involvement—severe lymphopenia, thymic hypoplasia) (Figures 77.1 and 77.2) [17,19–21].

77.6 Diagnosis

Diagnosis of SCN involves medical history review (consanguineous parents), physical examination (fever, recurrent infections, acute and severe umbilical infection in neonates), laboratory assessment (blood neutrophil counts, bone marrow examination for severe/persistent neutropenia with an absolute neutrophil count $<0.5 \times 10^9/L$), histopathology (bone marrow aspirate, biopsy to rule out or confirm leukemia, aplastic anemia or myelodysplasia), and molecular testing (for *ELANE* and other pathogenic variants) (Table 77.1) [1,22].

Differential diagnoses for severe congenital neutropenia include inherited conditions that present neutropenia as part of clinical phenotype, such as Barth syndrome, cartilage-hair hypoplasia, Charcot–Marie–Tooth disease (due to DNM2 mutation), Chediak–Higashi syndrome, Clericuzio poikiloderma with neutropenia, Cohen syndrome, glycogen storage disease type 1b, Griscelli syndrome type 2, Hermansky–Pudlak syndrome type 2, immunodeficiency due to defect in MAPBP-interacting protein (P14 deficiency), and WHIM syndrome [13,17,23,24].

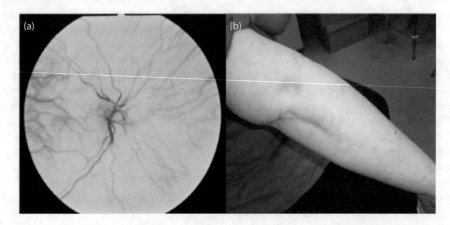

FIGURE 77.1   Severe congenital neutropenia in a 38-year-old female with homozygous G6PC3 mutation c.829C > T, p.Gln277X showing severe hypopigmentation and absence of macular differentiation in retina (a), and prominent superficial venous pattern in right arm (b). (Photo credit: Fernandez BA et al. *BMC Med Genet.* 2012;13:111.)

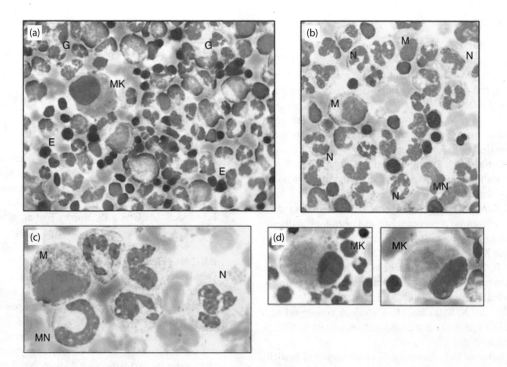

FIGURE 77.2 Congenital neutropenia in patients with G6PC3 mutations showing (a) rich cellularity with predominant granulopoiesis (G), some erythroblasts (E), and one micromegakaryocyte (MK) in bone marrow; (b) predominant granulopoiesis with some myelocytes (M), few metamyelocytes (MN), and many mature neutrophils (N); (c) neutrophils with hypersegmented appearance and thin opening between lobes and chromatin clumps; (d) examples of micromegakaryocytes. (Photo credit: Desplantes C et al. *Orphanet J Rare Dis*. 2014;9:183.)

77.7 Treatment

Management of SCN relies on antibiotics to control infections, G-CSF to keep absolute neutrophil counts above 0.5×10^9/L (ideally $>1 \times 10^9$/L), and hematopoietic stem cell transplantation (HSCT) for individuals who are refractory to high-dose G-CSF or who undergo malignant transformation (MDS/AML) [25–27].

As the treatment of choice for SCN, G-CSF increases the stimulation of G-CSFR signaling and provides compensatory mechanisms of granulopoiesis in patients with SCN.

77.8 Prognosis and Prevention

SCN used to confer a very poor prognosis prior to the availability of G-CSF and HSCT. Overall survival is much improved in 80% of patients who respond to G-CSF treatment. G-CSF non-responders who undergo HSCT have a 3-year event-free survival of 71%, and 5-year survival of 80%, although they are prone to HSCT-related morbidities (e.g., acute or chronic graft rejection, bacterial or viral infections, and increased risk for AML and secondary malignancies after chemotherapy or irradiation) that may have a negatively impact on disease outcome [26].

Along with early G-CSF treatment, efforts to maintain excellent oral hygiene, good nutritional status, and healthy living patterns are useful to reduce the risk of infections, which represent the most common cause of mortality in SCN.

Prenatal genetic diagnosis (through amniocentesis) is possible for autosomal dominant disorders, while DNA sequencing of the potential parents in addition to genetic counseling may be helpful for autosomal recessive disorders.

77.9 Conclusion

Severe congenital neutropenia (SCN) is a phenotypically and genetically heterogeneous disease characterized by impaired differentiation of neutrophilic granulocytes and maturation arrest of myelopoiesis at the level of promyelocytes, leading to reduced absolute neutrophil counts ($<0.5 \times 10^9$/L) in peripheral blood, and elevated numbers of atypical promyelocytes and myelocytes accompanied by a paucity of metamyelocytes, band cells, and mature neutrophils in bone marrow. As a consequence, affected patients show increased susceptibility to recurrent infections (e.g., otitis, gingivitis, skin infections, pneumonia, deep abscesses, and septicemia) and leukemia. The molecular basis of SCN relates to mutations in genes that encode proteins involved in receptor binding, endoplasmic reticulum, ribosomes, nucleus, mitochondria, endosomes/lysosomes, and cytoskeleton, resulting in endoplasmic reticulum stress, accelerated apoptosis of neutrophil precursors, arrest of granulopoiesis, inability to phagosize invading pathogens and cell debris, and acquisition of secondary mutation before conversion to MDS and AML [28]. Diagnosis of SCN relies on laboratory assessment of neutrophils and precursors in peripheral blood and bone marrow, and molecular identification of underlying gene mutations. Treatment options for SCN include antibiotics, G-CSF, and HSCT.

REFERENCES

1. Skokowa J, Dale DC, Touw IP, Zeidler C, Welte K. Severe congenital neutropenias. *Nat Rev Dis Primers*. 2017;3:17032.

2. Lawrence SM, Corriden R, Nizet V. The ontogeny of a neutrophil: Mechanisms of granulopoiesis and homeostasis. *Microbiol Mol Biol Rev*. 2018;82(1). pii: e00057-17.

3. Klein C, Grudzien M, Appaswamy G et al. HAX1 deficiency causes autosomal recessive severe congenital neutropenia (Kostmann disease). *Nature Genet* 2007;39:86–92.

4. Beel K, Cotter MM, Blatny J et al. A large kindred with X-linked neutropenia with an I294T mutation of the Wiskott-Aldrich syndrome gene. *Brit J Haemat* 2008;144:120–6.

5. Boztug K, Ding X-Q, Hartmann H et al. HAX1 mutations causing severe congenital neuropenia (sic) and neurological disease lead to cerebral microstructural abnormalities documented by quantitative MRI. *Am J Med Genet*. 2010;152A:3157–63.

6. Beekman R, Valkhof MG, Sanders MA et al. Sequential gain of mutations in severe congenital neutropenia progressing to acute myeloid leukemia. *Blood*. 2012;119(22):5071–7.

7. Vilboux T, Lev A, Malicdan MCV et al. A congenital neutrophil defect syndrome associated with mutations in VPS45. *New Eng J Med*. 2013;369:54–65.

8. Triot A, Jarvinen PM, Arostegui JI et al. Inherited biallelic CSF3R mutations in severe congenital neutropenia. *Blood*. 2014;123:3811–7.

9. Klimiankou M, Mellor-Heineke S, Zeidler C, Welte K, Skokowa J. Role of CSF3R mutations in the pathomechanism of congenital neutropenia and secondary acute myeloid leukemia. *Ann N Y Acad Sci*. 2016;1370:119–25.

10. Dwivedi P, Greis KD. Granulocyte colony-stimulating factor receptor signaling in severe congenital neutropenia, chronic neutrophilic leukemia, and related malignancies. *Exp Hematol*. 2017;46:9–20.

11. Veiga-da-Cunha M, Chevalier N, Stephenne X et al. Failure to eliminate a phosphorylated glucose analog leads to neutropenia in patients with G6PTand G6PC3 deficiency. *Proc Natl Acad Sci U S A*. 2019;116(4):1241–50.

12. Cho HK, Jeon IS. Different clinical phenotypes in familial severe congenital neutropenia cases with same mutation of the ELANE gene. *J Korean Med Sci*. 2014;29(3):452–5.

13. Dale DC. ELANE-related neutropenia. 2002 Jun 17 [Updated 2011 Jul 14]. In: Adam MP, Ardinger HH, Pagon RA et al. editors. *GeneReviews® [Internet]*. Seattle (WA): University of Washington, Seattle; 1993–2018.

14. Boztug K, Jarvinen PM, Salzer E et al. JAGN1 deficiency causes aberrant myeloid cell homeostasis and congenital neutropenia. *Nature Genet*. 2014;46:1021–7.

15. Tran TT, Vu QV, Wada T, Yachie A, Le Thi Minh H, Nguyen SN. Novel *HAX1* gene mutation in a Vietnamese

boy with severe congenital neutropenia. *Case Rep Pediatr*. 2018;2018:2798621.

16. Banka S, Newman WG. A clinical and molecular review of ubiquitous glucose-6-phosphatase deficiency caused by G6PC3 mutations. *Orphanet J Rare Dis*. 2013;8:84.

17. Banka S. G6PC3 deficiency. In: Adam MP, Ardinger HH, Pagon RA et al. editors. *GeneReviews® [Internet]*. Seattle (WA): University of Washington, Seattle; 1993–2018. 2015 Apr 16.

18. Vu QV, Wada T, Tran TT et al. Severe congenital neutropenia caused by the ELANE gene mutation in a Vietnamese boy with misdiagnosis of tuberculosis and autoimmune neutropenia: A case report. *BMC Hematol*. 2015;15:2.

19. Fernandez BA, Green JS, Bursey F et al. Adult siblings with homozygous G6PC3 mutations expand our understanding of the severe congenital neutropenia type 4 (SCN4) phenotype. *BMC Med Genet*. 2012;13:111.

20. Desplantes C, Fremond ML, Beaupain B et al. Clinical spectrum and long-term follow-up of 14 cases with G6PC3 mutations from the French Severe Congenital Neutropenia Registry. *Orphanet J Rare Dis*. 2014;9:183.

21. Notarangelo LD, Savoldi G, Cavagnini S et al. Severe congenital neutropenia due to G6PC3 deficiency: Early and delayed phenotype in two patients with two novel mutations. *Ital J Pediatr*. 2014;40:80.

22. van den Broek L, van der Werff-Ten Bosch J, Cortoos PJ, van Steijn S, van den Akker M. Severe neutropenia in a breastfed infant: A case report and discussion of the differential diagnosis. *Int Med Case Rep J*. 2018;11:333–7.

23. Wilson DB, Link DC, Mason PJ, Bessler M. Inherited bone marrow failure syndromes in adolescents and young adults. *Ann Med*. 2014;46(6):353–63.

24. Khincha PP, Savage SA. Neonatal manifestations of inherited bone marrow failure syndromes. *Semin Fetal Neonatal Med*. 2016;21(1):57–65.

25. Connelly JA, Choi SW, Levine JE. Hematopoietic stem cell transplantation for severe congenital neutropenia. *Curr Opin Hematol*. 2012;19(1):44–51.

26. Fioredda F, Iacobelli S, van Biezen A et al. Stem cell transplantation in severe congenital neutropenia: An analysis from the European Society for Blood and Marrow Transplantation. *Blood*. 2015;126(16):1885–92.

27. Fioredda F, Lanza T, Gallicola F et al. Long-term use of peg-filgrastim in children with severe congenital neutropenia: Clinical and pharmacokinetic data. *Blood*. 2016;128(17): 2178–81.

28. Liu Q, Sundqvist M, Li W et al. Functional characteristics of circulating granulocytes in severe congenital neutropenia caused by ELANE mutations. *BMC Pediatr*. 2019;19(1):189.

78

Shwachman–Diamond Syndrome

Dongyou Liu

CONTENTS

78.1 Introduction

Shwachman–Diamond syndrome (SDS) is a rare autosomal recessive multisystem disorder associated with poor growth, exocrine pancreatic insufficiency, skeletal abnormalities (e.g., metaphyseal dysplasia, narrow thorax, flared ribs, osteopenia), cognitive impairment, bone marrow failure (neutropenia, anemia, thrombocytopenia), and predisposition to myelodysplastic syndrome (MDS), aplastic anemia, and acute myeloid leukemia (AML).

Homozygous or compound heterozygous mutations in the Shwachman–Bodian–Diamond syndrome (*SBDS*) gene located on chromosome 7q11.21, which encodes a novel protein involved in ribosomal maturation, cell proliferation, and mitosis, appear to be the underlying defects of SDS, with a negative impact on hematopoiesis, hindering but not arresting the regeneration of neutrophils, red cells, and platelets [1].

78.2 Biology

SDS was initially described by Bartholomew and coworkers in 1959 as primary atrophy of the pancreas, and by Nezelof and Watchi in 1961 as congenital lipomatosis of the pancreas from two children who displayed exocrine pancreatic insufficiency (fatty involution, chronic diarrhea with fat stools, and low fecal elastase) and leucopenia (neutropenia). Subsequent observation of pancreatic insufficiency and pancytopenia in three kindred by Shwachman and coworkers in 1964 highlighted the syndromic feature of this disorder. Further studies extended the clinical profile of SDS to poor growth, skeletal abnormalities (metaphyseal dysplasia, narrow thorax, flared ribs, osteopenia), cutaneous involvement

(usually eczema, sometimes ichthyosis), cognitive impairment/psychomotor retardation, bone marrow failure (intermittent or persistent neutropenia, moderate anemia, mild to moderate thrombocytopenia), and susceptibility to MDS and AML [1].

After mapping the responsible gene region to chromosome 7q11.21 by Popovic and coworkers in 2002, the culprit gene for SDS was identified by Boocock and coworkers in 2003 as *SBDS* (for Shwachman–Bodian–Diamond syndrome). Occurring in 92% of SDS patients, *SBDS* pathogenic variants contribute to partial loss of function protein that compromises the regeneration of neutrophils, red cells, and platelets, leading to characteristic pathological changes associated with his autosomal recessive disorder [2].

As *SBDS* encodes a protein involved in ribosome biogenesis, SDS can be considered a ribosomopathy. Being highly evolutionarily conserved organelles, ribosomes are assembled from four types of ribosomal RNAs (rRNA) and about 80 ribosomal proteins (RP), and involved in protein synthesis. Disruption of ribosomal biogenesis by extracellular or intracellular stimuli contributes to ribosomal stress and accumulation of unincorporated/free RP, which bind to and inhibit MDM2 and impair its ubiquitin-mediated degradation of TP53 tumor suppressor, leading to the development of so-called ribosomopathies (characterized by early onset bone marrow failure, variable developmental abnormalities, and a lifelong cancer predisposition) [3–5].

78.3 Pathogenesis

The *SBDS* gene on chromosome 7q11.21 spans 7.9 kb with five exons and encodes a highly conserved 250 aa, 28.7 kDa ribosome maturation protein (SBDS), which is involved in ribosome biogenesis, microtubule stabilization, actin polymerization, stromal microenvironment, and genome stability.

SBDS has a predominantly intranucleolar localization, which is dependent on active ribosomal RNA transcription. Through interaction with elongation factor-like GTPase 1 (EFL1), SBDS triggers release of eukaryotic initiation factor 6 (EIF6) from 60S pre-ribosomal subunit in the cytoplasm, and facilitates 80S ribosomal subunit assembly, and EIF6 recycling to the nucleus for 60S rRNA processing and nuclear export. SBDS is also capable of associating with RPL3, RPL4, RPL6, RPL7, RPL7A, and RPL8; this helps remove free RP and reduce ribosomal stress [6–8].

Observed in about 90% of SDS cases, biallelic mutations (e.g., splice site or missense alterations) in *SBDS* produce prematurely truncated, partial loss-of-function SBDS protein that impairs association of the 40S and 60S subunits and subsequent rRNA biogenesis, and activates the mTOR/STAT3 pathway (which participates in neutrophil granulogenesis), hindering (but not arresting) the regeneration of neutrophils as well as erythrocytes and platelets [9]. Indeed, lymphoblast cell lines derived from SDS patients show hypersensitivity to RNA polymerase I-inhibiting actinomycin D, and patients harboring *SBDS* pathogenic variants often have low neutrophil (<1500 neutrophils/mm^3), hemoglobin (below normal range), and platelet (<150,000 platelets/mm^3) counts [10].

Since neutrophils along with monocytes are capable of phagocytosis and releasing proteases, antimicrobial peptides, and reactive oxygen species, they provide a powerful innate immune mechanism against bacterial and fungal infections. Not surprisingly, SDS-related reduction of neutrophils, erythrocytes, and platelets incapacitates host defense against infectious agents, leading to recurrent infections and other malaises. In addition, *SBDS* mutations cause severe depletion of pancreatic acinar cells and extensive fatty infiltration; the resultant exocrine deficiency (inadequate production of enzymes for breaking down and utilizing the nutrients from food) leads to frequent passage of fatty, foul-smelling stools (steatorrhea) by affected infants [1].

Among notable *SBDS* pathogenic variants [c.119delG (p.Ser41AlafsTer18), c.183_184delinsCT or 183TA>CT (p.Lys 62Ter), c.183_184delinsCT or 258+2T>C (p.Lys62Ter), c.258+ 1G>C, c.258+2T>C (p.Cys84TyrfsTer4), c.297_300delAAGA (p.Glu9AspfsTer20), c.377G>C (p.Arg126Thr), c.505C>T (p.Arg 169Cys), c.624+1G>C, and c.652C>T (p.Arg218Ter)], c.183_184 delinsCT, c.258+2T>C and c.(183_184delinsCT; 258+2T>C) are the most common and account for >76% of clinical cases, involving >200 families [1,11].

Recent studies have shown that affected patients harboring no *SBDS* pathogenic variant may carry biallelic mutations in *DNAJC21* (<1% of cases) or *EFL1* (<1% of cases), or monoallelic/heterozygous mutation in *SRP54* (<1% of cases), suggesting alternative molecular mechanisms for SDS pathogenesis [12–19].

78.4 Epidemiology

78.4.1 Prevalence

SDS affects 1 in 75,000 births. To date, a few hundred SDS cases have been reported, involving European, Indian, Chinese, Japanese, African, and American Indian population groups. There is a slight male bias (male to female ratio of 1.7:1) among SDS patients, with median age of diagnosis at 17 years (range 4–39 years) [20].

78.4.2 Inheritance

SDS shows autosomal recessive inheritance, with both copies of the *SBDS* gene in each cell being mutated for disease to occur. Parents of children with SDS are mostly asymptomatic carriers (of heterozygous *SBDS* mutations) and have a 25% chance of getting an affected offspring, a 50% chance of getting an unaffected carrier offspring, and a 25% chance of getting an unaffected, non-carrier offspring [1].

78.4.3 Penetrance

SDS demonstrates high penetrance and variable expressivity, with individuals harboring *SBDS* pathogenic variants displaying a broad spectrum of clinical symptoms.

78.5 Clinical Features

Attributed to the underlying defects in blood/bone marrow maturation (neutrophil, erythrocyte, and platelet regeneration), SDS is associated with exocrine pancreatic dysfunction, growth failure (short stature), bone abnormalities (metaphyseal dysplasia, narrow thorax, flared ribs/rib cage dysplasia, osteopenia), single- or multilineage cytopenias (intermittent or persistent neutropenia, moderate anemia, mild to moderate thrombocytopenia), fatigue/weakness, easy bruising/abnormal bleeding, recurrent infections in the lungs (pneumonia) and ear (otitis media), gastrointestinal abnormalities (gluten intolerance, malabsorption, malnutrition, steatorrhea, failure to thrive), abnormalities in the central nervous system (mental retardation), heart (cardiomyopathy), and skin (xerosis eczema), and predisposition to MDS and AML (with a cumulative transformation rate of 8%–20%) [21,22].

78.6 Diagnosis

Diagnostic workup for SDS consists of medical history review, physical examination, laboratory assessment (for exocrine pancreatic dysfunction, bone marrow failure), and molecular testing (for biallelic pathogenic variants in *SBDS*) [23].

Identification of biallelic *SBDS* mutation or fulfilling at least one criterion each from Category I and II helps establish the diagnosis of SDS:

Category I: (i) low levels of trypsinogen (age <3 years) or low pancreatic isoamylase levels (age >3 years); (ii) low levels of fecal elastase; (iii) supportive features (pancreatic lipomatosis; elevated 72-hour fecal fat excretion and absence of intestinal pathologic condition) [24].

Category II: (i) hypoproductive cytopenias; (ii) neutropenia (absolute neutrophil count <1,500 neutrophils/mm^3); (iii) anemia or idiopathic macrocytosis; (iv)

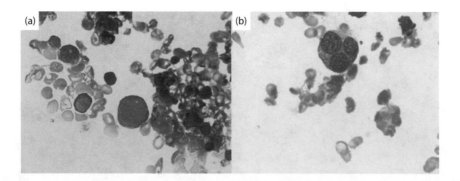

FIGURE 78.1 Shwachman−Diamond syndrome in a 19-year-old male showing markedly hypocellularity (10% cellularity) with features of dyserythropoiesis and dysmegakaryopoiesis and 3% blasts: (a) numerous blasts; (b) >50% dysplastic erythroid precursors in bone marrow aspirate. (Photo credit: Ong SY et al. *Leuk Res Rep.* 2018;9:54−7.)

thrombocytopenia (<150,000 platelets/mm^3); (v) bone marrow examination with any of the following: myelodysplasia, leukemia, hypocellularity for age, cytogenetic abnormalities (Figure 78.1) [24,25].

Supporting features: (i) first- or second-degree blood relative with SDS; (ii) personal history of congenital skeletal abnormalities consistent with chondrodysplasia or a congenital thoracic dystrophy, height 3% or less, of unclear cause, deficiency in two or more fat-soluble vitamins (A, 25-OHD, and E) [24].

Molecular identification of homozygous or compound heterozygous *SBDS* pathogenic variants (e.g., c.183_184delinsCT, c.258+2T>C), which occur in 92% of affected patients, is confirmative to SDS diagnosis [1,26]. In addition, patients who are negative for *SBDS* may sometimes harbor biallelic mutations in *DNAJC21* (<1%) and *EFL1* (<1%), or monoallelic/heterozygous mutation in *SRP54* (<1%) [19].

Differential diagnoses of SDS include cystic fibrosis (both upper respiratory infections and exocrine pancreatic dysfunction; sweat chloride testing and absence of primary bone marrow failure), Johanson−Blizzard syndrome (characteristic anomalies, severe developmental delays, and absence of hematologic abnormalities), Pearson bone marrow−pancreas syndrome (a rare mitochondrial disorder with both exocrine pancreatic dysfunction and bone marrow dysfunction), Diamond−Blackfan anemia (neutropenia appearing several years after disease onset), Fanconi anemia (anemia and thrombocytopenia, but rarely neutropenia), dyskeratosis congenita (anemia and thrombocytopenia, but rarely neutropenia), Kostmann syndrome (profound neutropenia during the first weeks of life, maturation arrest of granulopoiesis at the promyelocyte stage, and death due to bacterial infections; due to *HAX1* mutations), *ELANE*-related neutropenia (cyclic neutropenia and severe congenital neutropenia), cartilage-hair hypoplasia syndrome (dwarfism, metaphyseal chondrodysplasia, sparse hair, lymphopenia, hypogammaglobulinemia, and neutropenia; autosomal recessive disease due to mutations of the *RMRP* gene encoding a ribonuclease), Cohen syndrome (congenital neutropenia, mental retardation, microcephaly, facial abnormalities/moon face, myopia, pigmentary retinitis, trunk obesity, and ligament hyperlaxity, chronic gingivostomatitis; autosomal recessive disorder due to mutations of the *VPS13B* gene encoding an endoplasmic reticulum protein), neutropenia associated with poikiloderma Clericuzio type (skin atrophy, papular erythematous rash,

pachyonychia, recurrent pneumonia, severe neutropenia; due to *C16ORF57* mutations) [1,27–30].

78.7 Treatment

Treatment of SDS requires a multidisciplined approach encompassing (i) antibiotic control and prevention of infections (e.g., trimethoprim-sulfamethoxazole); (ii) oral pancreatic enzymes and fat-soluble vitamin A, D, E, and K supplementation to rectify exocrine pancreatic insufficiency; (iii) packed red blood cell (PRBC) and/or platelet transfusions for individuals with anemia and/or thrombocytopenia due to bi- or trilineage cytopenia; (vi) granulocyte-colony-stimulating factor (G-CSF) treatment for individuals with absolute neutrophil counts of 500/mm^3 or less; (v) hematopoietic stem cell transplantation (HSCT) for G-CSF-nonresponding individuals with severe pancytopenia, MDS, or AML; (vi) surgical intervention for rib and joint abnormalities, asymmetric growth, and joint deformities (Figure 78.2) [1,18,31].

78.8 Prognosis and Prevention

SDS has a median overall survival of 41 years of age. Most children with SDS are able to lead normal lives, with modern treatment options and ongoing management. Affected children and adults have 5% and 25% risk of developing leukemia, respectively.

Implementation of aggressive dental hygiene routine and prophylactic antibiotics helps promote oral health and reduce the incidence of mouth ulcers and gingivitis.

Prenatal testing and preimplantation genetic diagnosis may be considered for prospective parents with an affected family member.

78.9 Conclusion

Shwachman−Diamond syndrome (SDS) is a rare autosomal recessive disorder characterized by single- or multilineage cytopenias (neutropenia, anemia, and thrombocytopenia), exocrine pancreatic insufficiency (unique among congenital neutropenia diseases), skeletal malformations, hepatic and cognitive anomalies, recurrent infections, and predisposition to hematologic

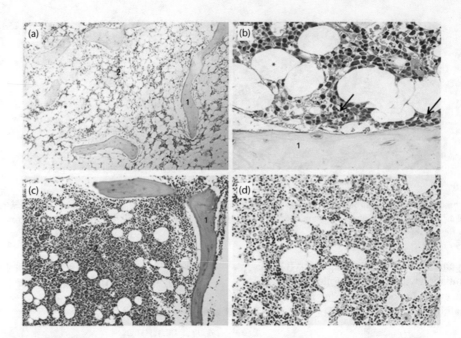

FIGURE 78.2 Shwachman–Diamond syndrome in a 6-year-old girl showing complete hemopoietic recovery after allogeneic transplantation of cord blood hematopoietic stem cells and bone marrow. Trephine bone marrow biopsy: (a,b) prior HSCT from HLA-identical donor; (c,d) 120 days after HSCT. *, adipocytes; arrow, atypical paratrabecular localization of erythrocaryocytes, with dysplastic and megaloblastoid features of individual ones; 1, bone trabeculae; 2, hematopoietic tissue. (a–c) Hematoxylin and eosin; (d) azure and eosin. A × 40, B × 400, C × 100, D × 100. (Photo credit: Isaev AA et al. *Bone Marrow Transplant.* 2017;52(9):1249–52.)

malignancies (particularly MDS and AML). The molecular pathogenesis of SDS relates mainly to biallelic mutations in the *SBDS* gene encoding a protein involved in rRNA biogenesis that disrupt ribosomal formation and hinder but not arrest blood cell regeneration. Diagnosis of SDS involves laboratory assessment of blood cell counts and identification of homozygous or compound heterozygous mutations in *SBDS* (92%), *DNAJC21* (<1%), and *EFL1* (<1%), or heterozygous mutation in *SRP54* (<1%). Treatment of SDS is based on a multidisciplined approach that incorporates antibiotic therapy, oral pancreatic enzymes and fat-soluble vitamin supplementation, blood and/or platelet transfusions, G-CSF treatment, and HSCT as well as surgical intervention.

REFERENCES

1. Myers K. Shwachman-diamond syndrome. In: Adam MP, Ardinger HH, Pagon RA et al. editors. *GeneReviews®* [Internet]. Seattle (WA): University of Washington, Seattle; 1993–2018. 2008 Jul 17 [Updated 2014 Sep 11].
2. Boocock GRB, Morrison JA, Popovic M, Richards N, Ellis L, Durie PR, Rommens JM. Mutations in SBDS are associated with Shwachman-Diamond syndrome. *Nature Genet.* 2003;33:97–101.
3. Burwick N, Coats SA, Nakamura T, Shimamura A. Impaired ribosomal subunit association in Shwachman-Diamond syndrome. *Blood.* 2012;120(26):5143–52.
4. Shenoy N, Kessel R, Bhagat TD et al. Alterations in the ribosomal machinery in cancer and hematologic disorders. *J Hematol Oncol.* 2012;5:32.
5. Ruggero D, Shimamura A. Marrow failure: A window into ribosome biology. *Blood.* 2014;124(18):2784–92.
6. In K, Zaini MA, Müller C, Warren AJ, von Lindern M, Calkhoven CF. Shwachman–Bodian–Diamond syndrome (SBDS) protein deficiency impairs translation re-initiation from C/EBPα and C/EBPβ mRNAs. *Nucleic Acids Res.* 2016;44(9):4134–46.
7. Liu Y, Liu F, Cao Y et al. Shwachman-Diamond syndrome protein SBDS maintains human telomeres by regulating telomerase recruitment. *Cell Rep.* 2018;22(7):1849–60.
8. Warren AJ. Molecular basis of the human ribosomopathy Shwachman-Diamond syndrome. *Adv Biol Regul.* 2018;67:109–27.
9. Carvalho CM, Zuccherato LW, Williams CL et al. Structural variation and missense mutation in SBDS associated with Shwachman-Diamond syndrome. *BMC Med Genet.* 2014;15:64.
10. Bezzerri V, Vella A, Calcaterra E et al. New insights into the Shwachman-Diamond syndrome-related haematological disorder: Hyper-activation of mTOR and STAT3 in leukocytes. *Sci Rep.* 2016;6:33165.
11. Alves C, Fernandes JC, Sampaio S, Paiva Rde M, Calado RT. Shwachman-Diamond syndrome: First molecular diagnosis in a Brazilian child. *Rev Bras Hematol Hemoter.* 2013;35(4):290–2.
12. Tummala H, Walne AJ, Williams M et al. DNAJC21 mutations link a cancer-prone bone marrow failure syndrome to corruption in 60S ribosome subunit maturation. *Am J Hum Genet.* 2016;99:115–24.
13. Dhanraj S, Matveev A, Li H et al. Biallelic mutations in DNAJC21 cause Shwachman-Diamond syndrome. *Blood.* 2017;129(11):1557–62.
14. Godley LA. DNAJC21: The new kid on the SDS block. *Blood.* 2017;129(11):1413–4.

15. Calamita P, Gatti G, Miluzio A, Scagliola A, Biffo S. Translating the game: Ribosomes as active players. *Front Genet.* 2018;9:533.

16. Tan QK, Cope H, Spillmann RC et al. Further evidence for the involvement of *EFL1* in a Shwachman-Diamond-like syndrome and expansion of the phenotypic features. *Cold Spring Harb Mol Case Stud.* 2018;4(5).

17. Bellanné-Chantelot C, Schmaltz-Panneau B, Marty C et al. Mutations in the *SRP54* gene cause severe congenital neutropenia as well as Shwachman-Diamond-like syndrome. *Blood.* 2018;132(12):1318–31.

18. Farooqui SM, Zulfiqar H, Aziz M. Shwachman-Diamond syndrome. In: *StatPearls* [Internet]. Treasure Island (FL): StatPearls Publishing; 2019 Jan-.2019 Jun 9.

19. Nelson A, Myers K. Shwachman-diamond syndrome. In: Adam MP, Ardinger HH, Pagon RA, Wallace SE, Bean LJH, Stephens K, Amemiya A, editors. *GeneReviews®* [Internet]. Seattle (WA): University of Washington, Seattle; 1993–2019. 2008 Jul 17 [updated 2018 Oct 18].

20. Alter BP, Giri N, Savage SA, Rosenberg PS. Cancer in the National Cancer Institute inherited bone marrow failure syndrome cohort after fifteen years of follow-up. *Haematologica.* 2018;103(1):30–9.

21. Myers KC, Bolyard AA, Otto B et al. Variable clinical presentation of Shwachman-Diamond syndrome: Update from the North American Shwachman-Diamond Syndrome Registry. *J Pediatr.* 2014;164(4):866–70.

22. Bucciol G, Cassiman D, Roskams T et al. Liver transplantation for very severe hepatopulmonary syndrome due to vitamin A-induced chronic liver disease in a patient with Shwachman-Diamond syndrome. *Orphanet J Rare Dis.* 2018;13(1):69.

23. Keogh SJ, McKee S, Smithson SF, Grier D, Steward CG. Shwachman-Diamond syndrome: A complex case demonstrating the potential for misdiagnosis as asphyxiating thoracic dystrophy (Jeune syndrome). *BMC Pediatr.* 2012;12:48.

24. Myers KC, Davies SM, Shimamura A. Clinical and molecular pathophysiology of Shwachman-Diamond syndrome: An update. *Hematol Oncol Clin North Am.* 2013;27(1):117–28.

25. Ong SY, Li ST, Wong GC, Ho AYL, Nagarajan C, Ngeow J. Delayed diagnosis of Shwachman diamond syndrome with short telomeres and a review of cases in Asia. *Leuk Res Rep.* 2018;9:54–7.

26. Cho WK, Jung IA, Kim J et al. Two cases of Shwachman-Diamond syndrome in adolescents confirmed by genetic analysis. *Ann Lab Med.* 2015;35(2):269–71.

27. Chung NG, Kim M. Current insights into inherited bone marrow failure syndromes. *Korean J Pediatr.* 2014;57(8): 337–44.

28. Wilson DB, Link DC, Mason PJ, Bessler M. Inherited bone marrow failure syndromes in adolescents and young adults. *Ann Med.* 2014;46(6):353–63.

29. Babushok DV, Bessler M. Genetic predisposition syndromes: When should they be considered in the work-up of MDS? *Best Pract Res Clin Haematol.* 2015;28(1):55–68.

30. Bannon SA, DiNardo CD. Hereditary predispositions to myelodysplastic syndrome. *Int J Mol Sci.* 2016;17(6). pii: E838.

31. Isaev AA, Deev RV, Kuliev A et al. First experience of hematopoietic stem cell transplantation treatment of Shwachman-Diamond syndrome using unaffected HLA-matched sibling donor produced through preimplantation HLA typing. *Bone Marrow Transplant.* 2017;52(9):1249–52.

79

Sjögren Syndrome

Dongyou Liu

CONTENTS

79.1 Introduction

Sjögren syndrome is an autoimmune disorder characterized by diffuse infiltration of autoreactive lymphocytes into exocrine glands (salivary and lacrimal glands) and other tissues, leading to glandular dysfunction. Based on the presence of particular clinical symptoms, Sjögren syndrome is distinguished into primary and secondary forms. Primary Sjögren syndrome (pSS) only displays xerophthalmia/keratoconjunctiva sicca (dry eyes) and xerostomia (dry mouth), which are collectively referred to as sicca symptoms. However, patients with pSS are prone to non-Hodgkin B-cell lymphoma and possibly other tumors (e.g., breast cancer, lung adenocarcinoma, leukemia) compared to the general population. Secondary Sjögren syndrome (sSS) shows sicca symptoms (i.e., keratoconjunctiva sicca [dry eyes] and xerostomia [dry mouth]) as well as extraglandular manifestations (i.e., autoimmune diseases such as rheumatoid arthritis, lupus erythematosus, and scleroderma) [1].

Although the precise molecular mechanisms for the development of Sjögren syndrome remain to be elucidated, several susceptibility genes have been implicated. These include *HLA-DQB1/HLA-DRB*, *IRF5*, *STAT4*, *CXCR5*, *TNIP1*, *IL12A*, and *BLK*, all of which play a role in the immune homeostasis, and some of which have been linked to other autoimmune diseases.

79.2 Biology

Early cases of Sjögren syndrome (or Sjögren's syndrome) were reported by Gougerot in 1926 and Sjögren in 1933 involving patients with oral and conjunctival dryness (so called sicca symptoms or sicca syndrome). The tendency for Sjögren syndrome patients to develop non-Hodgkin lymphoma, particularly involving mucosa-associated lymphoid tissue (MALT), was noted by Bunim and Talal in 1963. Additional studies allowed a clear definition of Sjögren syndrome as a chronic autoimmune disorder in which epithelial inflammation causes failure of lacrimal and salivary secretion and destruction of airway, biliary, pancreatic, and renal epithelia, with symptoms that may appear in isolation (pSS) or secondary to other autoimmune diseases such as rheumatoid arthritis, systemic lupus erythematosus, and scleroderma (sSS) [2].

From a pathobiological perspective, Sjögren syndrome is thought to arise through several events that may include (i) initiation by an exogenous factor (e.g., environmental insults, estrogen, Epstein–Barr virus [EBV]/hepatitis C virus [HCV]/human T cell leukemia virus [HTLV] infection), (ii) disruption of salivary gland epithelial cells, (iii) T lymphocyte migration and lymphocytic infiltration of exogenous glands, (iv) B lymphocyte hyperreactivity and production of rheumatoid factor (RF) and classic autoantibodies to Sjögren syndrome type A (Ro or SS-A) and Sjögren syndrome type B (La or SS-B) antigens, which are transcription modulator/ubiquitin E3 ligase/RNA-degrading protein and RNA-binding phosphoprotein, respectively [3,4].

With recent molecular findings, the potential role of genetic alterations in the pathogenesis of Sjögren syndrome has been increasingly recognized. It appears that mutations in *HLA-DQB1/DQA1/DRB1*, *IRF5*, *STAT4*, *CXCR5*, *TNIP1*, *IL12A*, and *BLK* as well as other yet to be identified genes, which participate in the control of immune homeostasis and gene expression, create structural/conformational/epitopic changes in the encoded proteins. As these altered proteins are no long considered as self-antigens, the immune network mounts both effector cytotoxicity responses mediated by T-lymphocytes and antibody responses mediated by B-lymphocytes in attempts to eliminate them. While this may or may not be effective in cleaning out the non-self-antigens, it often produces collateral damage in various tissues and organs. Loss of immunologic tolerance to self-antigens is a well-known factor in the development of human autoimmune diseases, including Sjögren syndrome. By taking advantage of host genetic defects, various exogenous factors (e.g., environmental insults, EBV/HCV/HTLV infection) exert a larger impact than they could otherwise on a healthy host and initiate a chain of events that lead ultimately to pathologies characteristic of Sjögren syndrome [5–7].

79.3 Pathogenesis

The pathological hallmark of Sjögren syndrome is a chronic inflammatory infiltrate consisting mainly of activated T and B cells into the exocrine glands, which induce apoptosis of glandular epithelial cells and alter glandular homeostasis, leading to glandular dysfunction, as evidenced by keratoconjunctiva sicca (dry eyes) and xerostomia (dry mouth). After migration to exogenous glands and other tissues, hyperreactive B lymphocytes generate RF and antibodies that contribute to various autoimmune disorders (e.g., rheumatoid arthritis, lupus erythematosus, and scleroderma). Not surprisingly, genome-wide association and candidate gene studies have pinpointed a number of genes (e.g., *HLA-DQB1/DQA1/DRB1*, *IRF5*, *STAT4*, *CXCR5*, *TNIP1*, *IL12A*, and *BLK*) involved in the regulation of both adaptive and innate immunity as the likely source of genetic susceptibility for Sjögren syndrome [6,8–10].

79.3.1 HLA-DQB1/DQA1/DRB1

The *HLA-DQB1*, *HLA-DQA1*, and *HLA-DRB1* genes from the human leukocyte antigen (HLA) class II region on chromosome 6p21.3 are strongly implicated in the pathogenesis of Sjögren syndrome. Upregulated expression of HLA-DR antigen and intracellular adhesion molecule-1 (ICAM1) is observed in the conjunctival epithelium of Sjögren syndrome patients with dry eyes, and this upregulation may be controlled by interferon-γ (IFN-γ) through the activation of transcription factor NFκB (nuclear factor κ-B). Further, there is a clear association of *HLA-DQB1/DQA1/DRB1* with anti-Ro and anti-La positivity in pSS Caucasians. Among various *HLA* pathogenic variants, haplotype

HLA-DRB1/DRB3/DQA1/DQB1 is identified in Caucasian patients with Sjögren syndrome, haplotype HLA-DRB1/DQA1/DQB1/DRB1 occurs in Japanese and Chinese patients, and haplotype HLA-DRB1/DRB4/DQA1/DQB1 in Japanese patients. In addition, several genes (e.g., *HLA-C*, *HLA-A*, *HLA-H*, and *HLA-G*) from the HLA Class I region may play an accessory role in the development of Sjögren syndrome [5].

79.3.2 IRF5

The *IRF5* (interferon regulatory factor 5) gene encodes a transcription factor that mediates type I interferon responses in monocytes, dendritic cells, and B cells upon viral infection, leading to the transcription of interferon-α genes and the production of proinflammatory cytokines (e.g., IL-12, p40, IL-6, and TNFα). *IRF5* (possibly in association with the neighboring gene *TNPO3* s) represents an established risk locus in systemic lupus erythematosus (SLE), rheumatoid arthritis, ulcerative colitis, primary biliary cirrhosis, and systemic sclerosis (SSc). Among known *IRF5* pathogenic variants, a CGGGG indel polymorphism in the promoter region is strongly linked to Sjögren syndrome [5].

79.3.3 STAT4

The *STAT4* (signal transducer and activator of transcription 4, particularly the third intron) gene encodes a transcription factor that is vital for type I interferon-initiated cellular responses. After induction by IL-12 in lymphocytes, *STAT4* contributes to the transcription of interferon-γ, and plays a partial role (along with IRF5 and TNIP1) in the development of SLE and rheumatoid arthritis [5].

79.3.4 IL12A

The *IL12A* (interleukin 12, particularly the promoter region) gene encodes the p35 subunit that interacts with the p40 subunit encoded by *IL12B* to form the IL-12 heterodimer, which functions as an immunomodulatory cytokine. Secreted by monocytes and dendritic cells, IL-12 is critical for T-helper 1 cell differentiation and for interferon-γ production by T cells and NK cells. Pathogenic variants in the *IL12A* 5′ end are involved in celiac disease, and those in the *IL12A* 3′ end are linked to primary biliary cirrhosis. As a component of the IL-12 signaling pathway, *IL12A* along with STAT4-dependent signaling demonstrates a clear association with non-Hodgkin lymphoma, SLE, EBV infection, dry eye syndrome, and Sjögren syndrome [5].

79.3.5 BLK

The *BLK* (B lymphoid kinase, in association with the neighboring gene *FAM167A*) gene encodes a non-receptor src family tyrosine kinase involved in B cell receptor signaling (which is essential for proper immune function, deletion of autoreactive B cells, and subsequent receptor editing) and B cell development. *BLK* pathogenic variants are associated with SLE and the shared promoter region and the first intron of BLK are likely involved in the development of Sjögren syndrome [5].

79.3.6 CXCR5

The *CXCR5* (chemokine [C-X-C motif] receptor 5, in association with the neighboring gene *DDX6*) gene encodes a membrane-bound protein present on some memory B cells and on follicular helper T cells. CXCR5 acts as a receptor for CXCL13, which directs B cells to lymphoid follicles, and which is expressed in 90% of pSS patients and only 10% of healthy individuals. *CXCR5* pathogenic variants are implicated in multiple sclerosis and primary biliary cirrhosis. CXCR5 dysregulation has been observed in B cells of the salivary gland and periphery tissues from Sjögren syndrome patients [5].

79.3.7 TNIP1

The *TNIP1* (TNFAIP3 interaction protein 1) gene encodes a protein that binds TNFAIP3 (or A20, an ubiquitin-editing enzyme encoded by *TNFAIP3* on chromosome 6), which acts as a negative feedback regulator of NFκB activation via its ovarian tumor and zinc finger domains and suppresses Toll-like receptor-mediated apoptosis. *TNIP1* pathogenic variants increase risk for SLE, rheumatoid arthritis, psoriasis, and SSc. About 77% of patients with primary Sjögren syndrome and mucosa-associated lymphoid tissue lymphoma have germline and somatic variations in A20 [5,11].

Interestingly, the involvement of the type II interferon pathway, acting through IFNγ downstream of IL-12 and STAT4, and supported by multiple genes (e.g., *TNIP1* and *TNFAIP3*) in the NFκB pathway, highlights that the pathogenesis of Sjögren syndrome is not limited to the type I interferon pathway as previously recognized. There is no doubt that many other yet to be identified genes are involved in the underlying genetic pathophysiology of Sjögren syndrome and require further characterization [12–14].

79.4 Epidemiology

79.4.1 Prevalence

Sjögren syndrome represents the second most common autoimmune disease after rheumatoid arthritis. The prevalence of primary Sjögren syndrome is estimated at 0.043% worldwide, including 0.004% (0.007% females and 0.0005% males) in the United States, 0.01% in France, 0.03% in Japan, 0.05% in Norway, 0.09%–0.23% in Greece, 0.21%–0.72% in Turkey, and 0.21%–0.77% in China. The prevalence of sSS and SLE varies from 6.5%–19%, that of sSS and rheumatoid arthritis is between 4%–31%, and that of sSS and systemic sclerosis is between 14%–20.5%, due to the application of different criteria. People 50 years or older show a relatively high incidence of 3%, and menopausal females (peak age 56 years) appear to be predominantly affected (female to male ratio of 9:1) [15].

79.4.2 Inheritance

While the inheritance pattern of Sjögren syndrome is presently unclear, familial clustering of autoimmune thyroid disease/SLE/rheumatoid arthritis (30%–35% of cases) and systemic/multiple sclerosis in Sjögren syndrome is frequently noted. Further investigations in Sjögren syndrome susceptibility genes (e.g., *HLA-DQB1/DQA1/DRB1*, *IRF5*, *STAT4*, *CXCR5*, *TNIP1*, *IL12A*, and *BLK*) will help yield new insights on the inheritance pattern of this autoimmune disorder.

79.4.3 Penetrance

Sjögren syndrome shows high penetrance and variable expressivity, with clinical presentations ranging from glandular manifestations to extraglandular features (including peri-epithelial lesions in pSS, and autoimmune diseases in sSS). Such a contrasting clinical expression pattern appears to reflect the amount and organization of mononuclear cells and the ratio of T/B lymphocytes in the infiltrates of target tissues, the levels of gamma-globulin and antibody in the serum, and cytokine expression in both peripheral blood and glands.

79.5 Clinical Features

Sjögren syndrome may display both exocrinopathy (sicca symptoms, i.e., dry eyes [xerophthalmia/keratoconjunctivitis sicca] and dry mouth [xerostomia]; due to inflammation and resultant pathology of the lacrimal and salivary glands) and extraglandular manifestations (malaise, fatigue, fibromyalgia, fever, arthralgia, synovitis, Raynaud phenomenon, peripheral neuropathy, autoimmune thyroiditis, renal tubular acidosis, myositis, chronic hepatitis, purpura, vasculitis, primary biliary cirrhosis, gastrointestinal symptoms, respiratory diseases, psychosis, lymphadenopathy, splenomegaly, lymphoma; occurring in about 60% of pSS cases) [16,17].

Specifically, patients with dry eyes (xerophthalmia/keratoconjunctivitis sicca) often have a "gritty" or "sandy" feeling in their eyes and may develop corneal ulceration and infection of the eyelids as complications (Figure 79.1) [18].

Patients with dry mouth (xerostomia) may experience difficulties in swallowing dry foods without fluid and require frequent small sips of water. In the absence of saliva, patients are prone

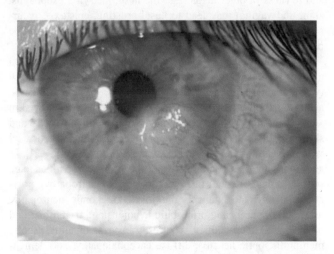

FIGURE 79.1 Sjögren syndrome in a patient with dry eyes showing severe ocular surface damage. (Photo credit: Tincani A et al. *BMC Med.* 2013;11:93.)

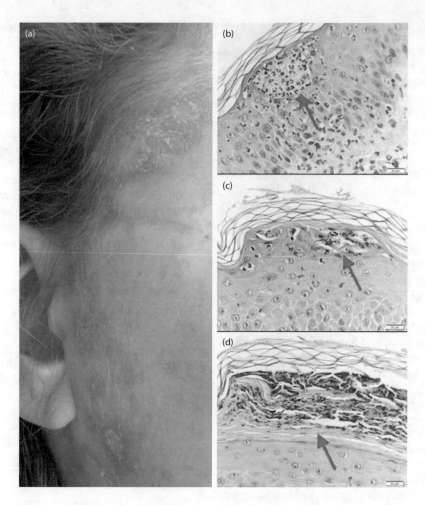

FIGURE 79.2 Sjögren syndrome in a 41-year-old Caucasian female with generalized exanthema and acute extensive transverse myelitis showing maculopustular and squamous exanthema in the cheek and forehead (a); neutrophils and lymphocytes forming intraepithelial pustules (b, red arrow); aggregated neutrophilic granulocytes in parakeratotic foci (c, red arrow); orthokeratosis associated with parakeratosis and mild superficial perivascular lymphocytic infiltrates around dilated vessels (d, red arrow) in punch biopsied erythematous macules and papulosquamous lesions from the left upper arm (H&E). (Photo credit: Kurz C et al. *BMC Res Notes*. 2014;7:580.)

to dental caries and oral candidiasis. Further, parotid swelling and dryness in other areas (e.g., the nose, pharynx, esophagus, skin, and vagina) may lead to nasal crusting, epistaxis, recurrent sinusitis, dry cough, dyspnea, nausea, dysphagia, epigastric pains, esophageal dysmotility, and gastritis [19].

In patients with pSS, interstitial nephritis (which may even precede the onset of sicca symptoms), distal renal acidosis (both type I and II), glomerulonephritis, mixed cryoglobulinemia, arthralgias, myalgias, and cutaneous manifestations (e.g., skin xerosis, angular cheilitis, erythema annulare, chilblain lupus, and skin vasculitis consisting of flat or palpable purpura and urticarial vasculitis) are often observed (Figures 79.2 and 79.3) [20,21]. In addition, pure or predominantly sensory polyneuropathies (e.g., sensory ataxic or small fiber sensory painful neuropathy), polyradiculopathy, mononeuritis multiplex, autonomic neuropathy (e.g., Adie's pupils and orthostatic hypotension), trigeminal and other cranial neuropathies, seizures, transverse myelitis, aseptic meningitis, optic neuritis, diffuse encephalopathy, and dementia may occur. With a 10- to 50-fold increase of risk relative to the general population, 2%–9% of Sjögren syndrome patients develop lymphoma (mainly non-Hodgkin B cell lymphoma in the form of marginal zone lymphoma of the mucosa-associated

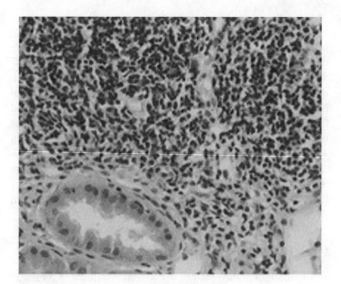

FIGURE 79.3 Primary Sjögren syndrome in a patient with tubulointerstitial nephritis showing typical mononuclear lymphocytic inflammatory infiltrate in renal biopsy (H&E). (Photo credit: Evans RD et al. *BMC Musculoskelet Disord*. 2016;17:2.)

lymphoid tissue [MALT] located in the parotid gland) and possibly other tumors (e.g., breast cancer, lung adenocarcinoma, leukemia) [22–30].

In patients with sSS (or Sjögren-overlap syndrome), extraglandular manifestations are represented by autoimmune diseases that target epithelial component (so-called autoimmune epithelitis) such as SLE (15%–36%), rheumatoid arthritis (20%–32%), scleroderma (11%–24%), autoimmune thyroiditis (Hashimoto thyroiditis and Graves disease), autoimmune hepatitis, and autoimmune cholangitis.

79.6 Diagnosis

Diagnosis of Sjögren syndrome involves medical history review, physical examination (for oral and ocular dryness and function, based on Schirmer test, slit-lamp exam with vital dye staining, salivary flow rate, and/or nuclear scintigraphic evaluation of the salivary glandular function), laboratory assessment (for autoantibodies ANA, RF, SS-A, and SS-B), histopathology (minor salivary gland/lip biopsy via an incision in the inner lip, for FLS in positive specimens) [31].

Sjögren syndrome is an autoimmune disease that induces strong activation of polyclonal B cells and production of circulating autoantibodies (frequently anti-nuclear antibodies [ANA], anti-Ro [SS-A], anti-La [SS-B], RF; occasionally anti-centromere, anti-Ki/SL, anti-Ku, anti-p80 coilin, anti-α fodrin, anti-carbonic anhydrase, anti-muscarinic receptor). Among these, anti-Ro (60% of cases) and anti-La (40% of cases) antibodies are indicators of consistent extraglandular manifestations and active immunological status (as exemplified by severe hypergammaglobulinemia, cryoglobulins, and increased risk for lymphoma), and are regarded the classical hallmarks of Sjögren syndrome [32].

Histopathology of Sjögren syndrome lip biopsy typically reveals chronic periductal sialoadenitis, with aggregates of >50 lymphocytes (including predominant CD4 positive cells, and CD5 positive B cells that produce IgG and IgM antibodies), and some macrophages and plasma cells, initially in the space around small interlobular-intralobular ducts, and subsequently spreading from the periductal position to the parenchyma, resulting in a diffuse infiltration of lymphocytes and loss of tissue architecture. Identification of at least one periductal lymphoid focus composed of at least 50 lymphocytes in 4 mm^2 of the periductal area is diagnostic for Sjögren syndrome (Figure 79.4) [18,33].

According to the 2016 classification criteria for pSS developed by the American College of Rheumatology/European League Against Rheumatism (ACR/EULAR), an individual achieving a total score of ≥4 points, derived from the weighted sum of the five items, and after application of inclusion and exclusion criteria, is classified as having pSS: (i) abnormal unstimulated salivary flow rate (≤0.1 mL/min) (1 point); (ii) abnormal Schirmer test (<5 mm in 5 min) (1 point); (iii) abnormal findings with lissamine green or fluorescein staining: ≥5 in ocular staining score or ≥4 in Van Bijsterveld score (1 point); (iv) presence of anti-SSA/Ro antibody positivity (3 points); (v) histological evidence of focal lymphocytic sialadenitis in lower lip biopsy, with a focus score ≥1 focus/4 mm^2 (1 focus = 50 lymphocytes/4 mm^2) (3 points); inclusion criteria: dryness of eyes or mouth for at least

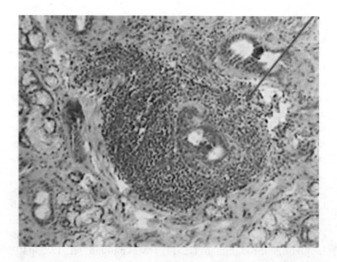

FIGURE 79.4 Sjögren syndrome in a patient showing periductal lymphoid focus (defined by having >50 CD4+ lymphocytes in the periductal area) in the minor salivary gland; finding at least one periductal lymphoid focus (arrow) in 4 mm^2 of tissue is diagnostic. (Photo credit: Tincani A et al. *BMC Med.* 2013;11:93.)

3 months, not explained otherwise (e.g., medications, infection); exclusion criteria: status post head/neck radiation, HIV/AID; sarcoidosis, active infection with hepatitis C virus (PCR replication rate), amyloidosis, graft versus host disease, IgG4-related disease [34,35].

Differential diagnoses for Sjögren syndrome include drug therapy (anticholinergic drugs), past radiation treatment (head and neck), systemic diseases (sarcoidosis, HCV, HIV/AIDS, graft-versus-host disease, preexisting lymphoma, rheumatoid arthritis, systemic lupus erythematosus, scleroderma, primary biliary cirrhosis, diabetes mellitus, cytomegalovirus and other herpes viruses, and ectodermal dysplasia).

79.7 Treatment

Current treatment for Sjögren syndrome centers on symptomatic relief for the exocrinopathy, and control of the extraglandular features [35].

Patients with dry eyes (xerophthalmia or keratoconjunctivitis sicca) may receive tear substitutes (artificial tears [preservative-free products, hypotonic solutions, and emulsions], autologous serum eye drops and platelet releasate); secretagogues (pilocarpine, cevimeline); cyclosporine A eye drops 0.1%; short-term topical corticosteroids; and/or punctal plugs.

Patients with dry mouth (xerostomia) may have topical fluorides for caries prevention; secretagogues (pilocarpine, cevimeline); saliva substitutes (mucin, caboxymethycellulose, hydroxymethilcellulose in the forms of lubricating gels, mouthwashes, lozenges, toothpastes, intraoral long-release inserts, and mucin spray), sugar-free chewing gum and electrostimulation of salivary glands, in addition to avoidance of drugs or substances (e.g., coffee, alcohol, nicotine) that promote xerostomia and patient education.

Complications associated with primary Sjögren syndrome may be treated with standard protocols: (i) parotid swelling (short-term oral corticosteroids; antibiotic treatment); (ii) arthritis (hydroxychloroquine; nonsteroidal anti-inflammatory drug [NSAIDs]; short-term oral/intraarticular corticosteroids; other disease-modifying antirheumatic drug [DMARDs] as with rheumatoid arthritis); (iii) interstitial lung disease (corticosteroids, oral or intravenous; cyclophosphamide for active alveolitis; pirfenidone, nintedanib); (iv) tubulointerstitial nephritis (potassium and bicarbonate replacement); (v) glomerulonephritis (corticosteroids, oral or intravenous; cyclophosphamide; mycofenolate mofetil); (vi) peripheral neuropathy (gabapentinoids; corticosteroids; intravenous immunoglobulins [IVIg]); (vi) cryoglobulinemic vasculitis (corticosteroids, plasmapheresis) [36–38].

Therapeutics targeting Th17 cytokines, cytokine receptors, and transcription factors may help correct the Th17/Treg imbalance associated with Sjögren syndrome. These include rituximab (B cells), abatacept (T cells), tocilizumab (IL-6R), secukinumab/ixekizumab/brodalumab (IL-17), ustekinumab (IL-12/23p40), tildarakizumab/guselkumab (IL-23p19), and digoxin target (transcription factor RORγT) [13,39].

Treatment options for Sjögren syndrome–related hematologic abnormities consist of glucocorticoids (first line), plasmapheresis (severe cases), cyclosporin A (cytopenia), and hematopoietic stem cell transplantation (HSCT) [40].

79.8 Prognosis and Prevention

Sjögren syndrome has a generally favorable prognosis, and pSS patients with no lymphoproliferative disorder have a life expectancy comparable to the general population. The main causes of death are cardiovascular disease, infections, solid tumors, and lymphoma [41].

Maintenance of oral hygiene (fluoride application, frequent dental examinations, prompt treatment of candida infections) reduces oral infection; use of sugar-free gums, lozenges, and maltose lozenges increases salivary flow, and avoidance of aggravating drugs (diuretics, beta blockers, tricyclic antidepressants, antihistamines) reduces irritants to dry eyes.

79.9 Conclusion

Sjögren syndrome is a chronic autoimmune disorder resulting from lymphocytic infiltration and subsequent destruction of the exocrine glands (salivary and lacrimal glands) and other organs/tissues (e.g., the nose, ears, skin, vagina, and respiratory and gastrointestinal systems). While glandular/sicca symptoms (i.e., dry eyes [xerophthalmia or keratoconjunctivitis sicca] and dry mouth [xerostomia]) predominate clinical presentations, they often associate with other extraglandular manifestations such as non-Hodgkin B cell lymphoma (in pSS) or autoimmune diseases (in sSS) [42–45]. Even though the molecular pathogenesis of Sjögren syndrome is far from clear, the recent identification of several susceptibility genes (e.g., *HLA-DQB1/HLA-DRB*, *IRF5*, *STAT4*, *CXCR5*, *TNIP1*, *IL12A*, and *BLK*) involved in the regulation of the immune homeostasis reaffirms the role of immune dysfunction in the development of this autoimmune disorder [46,47]. Diagnosis of Sjögren syndrome relies on detection of anti-SSA/Ro antibody in blood and histopathologic evidence of focal lymphocytic

sialadenitis in lower lip biopsy, along with abnormal unstimulated salivary flow rate, abnormal Schirmer test, and abnormal findings with lissamine green or fluorescein staining. Treatment options for Sjögren syndrome include symptomatic relief for the exocrinopathy (replacing moisture at affected glandular sites) and control of the extraglandular manifestations (suppressing the autoimmune response locally as well as systemically) [48]. Additional research is warranted on the molecular pathogenesis of Sjögren syndrome, the complete elucidation of which is crucial for designing countermeasures against this autoimmune disease.

REFERENCES

1. Carsons SE, Bhimji SS. Sjögren syndrome. In: *StatPearls* [Internet]. Treasure Island (FL): StatPearls Publishing; 2018 Jan-. 2017 Oct 1.
2. Both T, Dalm VASH, van Hagen PM et al. Reviewing primary Sjögren's syndrome: Beyond the dryness – from pathophysiology to diagnosis and treatment. *Int J Med Sci*. 2017;14:191–200.
3. Ferro F, Marcucci E, Orlandi M, Baldini C, Bartoloni-Bocci E. One year in review 2017: Primary Sjögren's syndrome. *Clin Exp Rheumatol*. 2017;35(2):179–91.
4. Tong L, Koh V, Thong BY. Review of autoantigens in Sjögren's syndrome: An update. *J Inflamm Res*. 2017;10:97–105.
5. Lessard CJ, Li H, Adrianto I et al. Variants at multiple loci implicated in both innate and adaptive immune responses are associated with Sjögren's syndrome. *Nat Genet*. 2013;45(11):1284–92.
6. Holdgate N, St Clair EW. Recent advances in primary Sjögren's syndrome. *F1000Res*. 2016;5. pii: F1000 Faculty Rev-1412.
7. Nair JJ, Singh TP. Sjögren's syndrome: Review of the aetiology, pathophysiology & potential therapeutic interventions. *J Clin Exp Dent*. 2017;9(4):e584–9.
8. Matsui K, Sano H. T Helper 17 cells in primary Sjögren's syndrome. *J Clin Med*. 2017;6(7). pii: E65.
9. Nakamura H, Horai Y, Shimizu T, Kawakami A. Modulation of apoptosis by cytotoxic mediators and cell-survival molecules in Sjögren's syndrome. *Int J Mol Sci*. 2018;19(8). pii: E2369.
10. Verstappen GM, Corneth OBJ, Bootsma H, Kroese FGM. Th17 cells in primary Sjögren's syndrome: Pathogenicity and plasticity. *J Autoimmun*. 2018;87:16–25.
11. Nocturne G, Boudaoud S, Miceli-Richard C et al. Germline and somatic genetic variations of TNFAIP3 in lymphoma complicating primary Sjögren's syndrome. *Blood*. 2013;122(25):4068–76.
12. Nezos A, Gravani F, Tassidou A et al. Type I and II interferon signatures in Sjögren's syndrome pathogenesis: Contributions in distinct clinical phenotypes and Sjögren's related lymphomagenesis. *J Autoimmun*. 2015;63:47–58.
13. Nezos A, Mavragani CP. Contribution of genetic factors to Sjögren's syndrome and Sjögren's syndrome related lymphomagenesis. *J Immunol Res*. 2015;2015:754825.
14. Sandhya P, Kurien BT, Danda D, Scofield RH. Update on pathogenesis of Sjögren's syndrome. *Curr Rheumatol Rev*. 2017;13(1):5–22.
15. Bolstad AI, Skarstein K. Epidemiology of Sjögren's syndrome-from an oral perspective. *Curr Oral Health Rep*. 2016;3(4):328–36.

16. Baer AN, Walitt B. Sjögren syndrome and other causes of sicca in older adults. *Clin Geriatr Med.* 2017;33(1):87–103.

17. Leone MC, Alunno A, Cafaro G et al. The clinical spectrum of primary Sjögren's syndrome: Beyond exocrine glands. *Reumatismo.* 2017;69(3):93–100.

18. Tincani A, Andreoli L, Cavazzana I et al. Novel aspects of Sjögren's syndrome in 2012. *BMC Med.* 2013;11:93.

19. Azuma N, Katada Y, Sano H. Deterioration in saliva quality in patients with Sjögren's syndrome: Impact of decrease in salivary epidermal growth factor on the severity of intraoral manifestations. *Inflamm Regen.* 2018;38:6.

20. Kurz C, Wunderlich S, Spieler D et al. Acute transverse myelitis and psoriasiform dermatitis associated with Sjoegren's syndrome: A case report. *BMC Res Notes.* 2014;7:580.

21. Evans RD, Laing CM, Ciurtin C, Walsh SB. Tubulointerstitial nephritis in primary Sjögren syndrome: Clinical manifestations and response to treatment. *BMC Musculoskelet Disord.* 2016;17:2.

22. Fragkioudaki S, Mavragani CP, Moutsopoulos HM. Predicting the risk for lymphoma development in Sjögren syndrome: An easy tool for clinical use. *Medicine (Baltim).* 2016;95(25):e3766.

23. Gupta S, Gupta N. Sjögren syndrome and pregnancy: A literature review. *Perm J.* 2017;21. pii: 16-047.

24. Katayama I. Dry skin manifestations in Sjögren syndrome and atopic dermatitis related to aberrant sudomotor function in inflammatory allergic skin diseases. *Allergol Int.* 2018;67(4):448–54.

25. Liu X, Li H, Yin Y, Ma D, Qu Y. Primary Sjögren's syndrome with diffuse cystic lung changes developed systemic lupus erythematosus: A case report and literature review. *Oncotarget.* 2017;8(21):35473–9.

26. Manzo C, Kechida M. Is primary Sjögren's syndrome a risk factor for malignancies different from lymphomas? What does the literature highlight about it? *Reumatologia.* 2017;55(3):136–9.

27. Yu W, Qu W, Wang Z et al. Sjögren's syndrome complicating pancytopenia, cerebral hemorrhage, and damage in nervous system: A case report and literature review. *Medicine (Baltim).* 2017;96(50):e8542.

28. Alunno A, Leone MC, Giacomelli R, Gerli R, Carubbi F. Lymphoma and lymphomagenesis in primary Sjögren's syndrome. *Front Med (Lausanne).* 2018;5:102.

29. Baldini C, Ferro F, Mosca M, Fallahi P, Antonelli A. The association of Sjögren syndrome and autoimmune thyroid disorders. *Front Endocrinol (Lausanne).* 2018;9:121.

30. Perzyńska-Mazan J, Maślińska M, Gasik R. Neurological manifestations of primary Sjögren's syndrome. *Reumatologia.* 2018;56(2):99–105.

31. Stefanski A, Tomiak C, Pleyer U, Dietrich T, Burmester GR, Dörner T. The diagnosis and treatment of Sjögren's syndrome. *Dtsch Arztebl Int.* 2017;114(20):354–61;

32. Katsiougiannis S, Wong DT. The proteomics of saliva in Sjögren's syndrome. *Rheum Dis Clin North Am.* 2016;42(3):449–56.

33. Fisher BA, Jonsson R, Daniels T et al. Standardisation of labial salivary gland histopathology in clinical trials in primary Sjögren's syndrome. *Ann Rheum Dis.* 2017;76(7):1161–8.

34. Shiboski CH, Shiboski SC, Seror R et al. International Sjögren's Syndrome Criteria Working Group. 2016 American College of Rheumatology/European League Against Rheumatism classification criteria for primary Sjögren's syndrome: A consensus and data-driven methodology involving three international patient cohorts. *Arthritis Rheumatol.* 2017;69:35–45.

35. Del Papa N, Vitali C. Management of primary Sjögren's syndrome: Recent developments and new classification criteria. *Ther Adv Musculoskelet Dis.* 2018;10(2):39–54.

36. Jin L, Li C, Li Y, Wu B. Clinical efficacy and safety of total glucosides of paeony for primary Sjögren's syndrome: A systematic review. *Evid Based Complement Alternat Med.* 2017;2017:3242301.

37. Wang SQ, Zhang LW, Wei P, Hua H. Is hydroxychloroquine effective in treating primary Sjögren's syndrome: A systematic review and meta-analysis. *BMC Musculoskelet Disord.* 2017;18(1):186.

38. van der Heijden EHM, Kruize AA, Radstake TRDJ, van Roon JAG. Optimizing conventional DMARD therapy for Sjögren's syndrome. *Autoimmun Rev.* 2018;17(5):480–92.

39. Barone F, Colafrancesco S. Sjögren's syndrome: From pathogenesis to novel therapeutic targets. *Clin Exp Rheumatol.* 2016;34(4 Suppl 98):58–62.

40. Carsons SE, Vivino FB, Parke A et al. Treatment guidelines for rheumatologic manifestations of Sjögren's syndrome: Use of biologic agents, management of fatigue, and inflammatory musculoskeletal pain. *Arthritis Care Res (Hoboken).* 2017;69(4):517–27.

41. Zhang Q, Wang X, Chen H, Shen B. Sjögren's syndrome is associated with negatively variable impacts on domains of health–related quality of life: Evidence from short form 36 questionnaire and a meta-analysis. *Patient Prefer Adherence.* 2017;11:905–11.

42. Ambrus JL, Suresh L, Peck A. Multiple roles for B-lymphocytes in Sjögren's syndrome. *J Clin Med.* 2016;5(10). pii: E87.

43. Ibrahem HM. B cell dysregulation in primary Sjögren's syndrome: A review. *Jpn Dent Sci Rev.* 2019;55(1):139–44. doi:10.1016/j.jdsr.2019.09.006

44. Sebastian A, Szachowicz A, Wiland P. Classification criteria for secondary Sjögren's syndrome. Current state of knowledge. *Reumatologia.* 2019;57(5):277–80.

45. Bjordal O, Norheim KB, Rødahl E, Jonsson R, Omdal R. Primary Sjögren's syndrome and the eye. *Surv Ophthalmol.* 2020;65(2):119–32.

46. Reale M, D'Angelo C, Costantini E, Laus M, Moretti A, Croce A. MicroRNA in Sjögren's syndrome: Their potential roles in pathogenesis and diagnosis. *J Immunol Res.* 2018;2018:7510174.

47. Ibáñez-Cabellos JS, Seco-Cervera M, Osca-Verdegal R, Pallardó FV, García-Giménez JL. Epigenetic regulation in the pathogenesis of Sjögren syndrome and rheumatoid arthritis. *Front Genet.* 2019;10:1104.

48. Brito-Zerón P, Retamozo S, Kostov B et al. Efficacy and safety of topical and systemic medications: A systematic literature review informing the EULAR recommendations for the management of Sjögren's syndrome. *RMD Open.* 2019;5(2):e001064.

80

SRP72-Associated Bone Marrow Failure Syndrome

Dongyou Liu

CONTENTS

80.1 Introduction

Bone marrow failure syndromes (BMFS) represent a group of inherited or acquired genetic disorders that typically cause aplastic anemia (due to failure of the bone marrow to produce adequate numbers of peripheral blood cells) and myelodysplasia (due to production of malformed and dysfunctional immature cells by clonal hematopoietic stem cells in the bone marrow) along with other syndromic phenotypic abnormalities.

Among inherited BMFS, amegakaryocytic thrombocytopenia, Fanconi anemia, dyskeratosis congenita, Shwachman−Diamond syndrome, congenital amegakaryocytic thrombocytopenia, Blackfan−Diamond anemia, reticular dysgenesis, Pearson syndrome, severe congenital neutropenia, and thrombocytopenia absent radii are relatively common and well known. However, there are a few BMFS that have caught our attention only in recent years after identification of the culprit genes (e.g., *SRP72* on chromosome 4q12, *ERCC6L2* on chromosome 9q22.32, and *DNAJC21* on chromosome 5p13.2) [1–4].

SRP72-associated BMFS (also referred to familial aplastic anemia/myelodysplastic syndrome with *SRP72* mutation, or BMFS1) is an autosomal dominant disorder caused by germline heterozygous mutations in the SRP72 (signal recognition particle 72 kDa) gene on chromosome 4q12, which encodes a transcription factor (SRP72). Besides early-onset aplastic anemia/pancytopenia and adult-onset myelodysplasia, this disorder may display other syndromic phenotypic abnormalities including congenital nerve deafness or labyrinthitis [5].

80.2 Biology

As the soft, flexible connective tissue within bone cavities, bone marrow is divided into a vascular section and nonvascular section. While the vascular section consists of blood vessels for nutrient supply and blood cell transport, the nonvascular section is involved in the production of various blood cells (e.g., erythrocytes [red blood cells], granulocytes [basophils, eosinophils, neutrophils, mast cells], monocytes, macrophages, dendritic cells, thrombocytes [platelets], B cells/plasma cells, T cells, and natural killer cells) in processes known as hematopoiesis and lymphopoiesis (Figure 80.1).

Bone marrow failure refers to a suboptimal functioning or malfunctioning in the bone marrow that causes decreased production of one or more major hematopoietic lineages, leading to diminished or absent hematopoietic precursors in the bone marrow, and mature blood cells in peripheral blood, which manifests clinically as aplastic anemia and myelodysplasia along with other extra-marrow features (commonly called bone marrow failure syndrome [BMFS]) [6].

BMFS may evolve from de novo mutation (so-called acquired BMFS) or develop through inheritance of germline mutation (so-called inherited BMFS). To date, >30 inherited BMFS involving >80 different genes have been described. Among the BMFS predisposition genes identified, many directly affect cell survival and function during hematopoiesis or modify biological processes in relevant hematopoietic lineage pathways, through ribosome biogenesis (e.g., Diamond−Blackfan anemia, Shwachman−Diamond syndrome), telomere biology (e.g.,

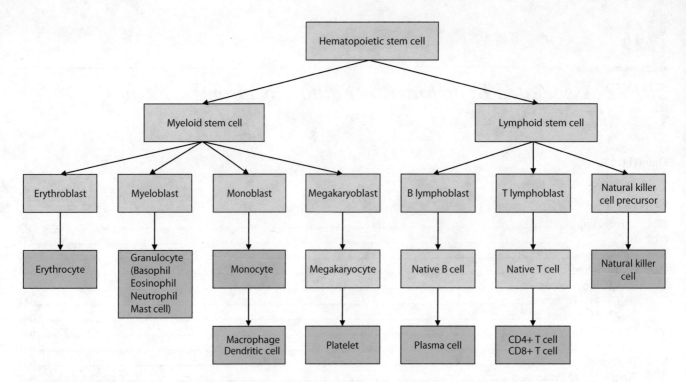

FIGURE 80.1 Schematic presentation of hematopoiesis and lymphopoiesis. Lodged within the bone marrow, hematopoietic stem cell (HSC, also referred to as hematopoietic stem and progenitor cell or HSPC) differentiate into myeloid and lymphoid stem cells. Myeloid stem cell then produces myeloid lineage cell (i.e., red blood cell [erythrocyte], white blood cell [basophil, eosinophil, neutrophil, mast cell, monocyte/macrophage/dendritic cell, but not lymphocyte] and thrombocyte [platelet]) in a process known as hematopoiesis (hemopoiesis). Lymphoid stem cell gives rise to lymphoid lineage cell (B cell/plasma cell, T cell, and natural killer cell) in a process known as lymphopoiesis (lymphocytopoiesis or lymphoid hematopoiesis). While various stem cells, blast or precursor (immature) cells reside in the bone marrow, mature cells circulate in the peripheral blood.

dyskeratosis congenita), DNA repair (e.g., Fanconi anemia), mRNA processing and export (thrombocytopenia absent radii), and protein folding/trafficking (severe congenital neutropenia), leading to single- or multi-lineage cytopenias. Depending on the specific genes involved, BMFS may show autosomal dominant (e.g., Blackfan–Diamond anemia, reticular dysgenesis, SRP72-associated BMFS), autosomal recessive (e.g., Fanconi anemia, Shwachman–Diamond syndrome, congenital amegakaryocytic thrombocytopenia, reticular dysgenesis), or X-linked (e.g., dyskeratosis congenita) inheritance patterns [7,8].

SRP72-associated BMFS (also known as familial aplastic anemia/myelodysplastic syndrome with *SRP72* mutation or BMFS1) is a rare autosomal dominant disorder that was recognized in 2012 following identification of a truncating mutation in the *SRP72* gene by whole-exome sequencing from four family members with aplastic anemia/myelodysplasia and congenital deafness [5]. The *SRP72* (signal recognition particle 72 kDa) gene appears to be involved in radioresistance, and heterozygous mutations in the *SRP72* gene produce altered protein that interferes with normal functioning of the signal recognition particle and causes failure to arrest cytoplasmic translation or properly translocate peptides within the cell [9].

80.3 Pathogenesis

The *SRP72* (signal recognition particle 72 kDa) gene on chromosome 4q12 spans 36.3 kb and encodes a 671 aa, 74.6 kDa transcription factor (SRP72) involved in signal recognition,

translational arrest, and endoplasmic reticulum (ER) membrane targeting.

SRP72 constitutes part of a ribonucleoprotein complex (signal recognition particle [SRP] complex, or signal recognition particle), which interacts with the ribosome (the other ribonucleoprotein particle) and targets secretory proteins to the rough ER, prior to integration into transmembrane or secreted proteins. Specifically, SRP72 and other SRP proteins (SRP9, SRP14, SRP19, SRP54, SRP68) bind the 7S RNA of ~300 nt to form the SPR complex (either as monomers [SRP19 and SRP54] or heterodimers [SRP9/SRP14 and SRP68/SRP72]), which interacts directly with the docking protein in the ER membrane and mediates the transfer of integral membrane proteins and secretory proteins across the hydrophobic lipid bi-layers to the ER. The functionally independent Alu domain (SRP9/SRP14 and 5'/3' ends of 7S RNA) in the SRP complex reaches into the factor binding site within the ribosomal 40S/60S subunit interface and is responsible for translation retardation (elongation arrest). The S domain (SRP19/SRP54/SRP68/SRP72) binds to the signal sequence emerging from the polypeptide exit tunnel in the signal pre-handover state. The multi-domain SRP GTPase SRP54 recognizes the signal with its M domain and establishes the targeting complex consisting of its NG domain bound to the homologous NG domain of the SRP receptor (SRαβ heterodimer) at the ER membrane in a GTP-dependent manner, which facilitates the targeting of ribosome bound nascent chain to the rough ER. Disassociation of the SRP complex from the ER allows resumption of elongation resumes and commencement of translocation [9,10].

Germline heterozygous mutations (e.g., R207H) in the *SRP72* gene alter the structure of the SRP72 protein that compromises the normal functioning of the signal recognition particle, loses the capacity to arrest cytoplasmic translation or properly translocate peptides within the cell, and renders normal lung fibroblast cell lines (HFL1 and MRC-5) radiosensitive, leading to elevated levels of apoptosis after irradiation, and development of familial aplastic anemia/myelodysplastic syndrome (MDS) and possibly anterior spinal artery syndrome [11,12].

80.4 Epidemiology

80.4.1 Prevalence

Bone marrow failure shows triphasic peaks at 2–5 years (mainly inherited causes), between 20–25 years, and after 65 years (mostly acquired causes). Inherited BMFS make up 10%–15% of bone marrow aplasia and 30% of pediatric BMFS (65 cases per million live births every year).

BMFS due to the *SRP72* mutation (or familial aplastic anemia/myelodysplastic syndrome with *SRP72* mutation) represents a portion of inherited BMFS cases, with patients from a small number of families reported.

80.4.2 Inheritance

SRP72-associated BMFS shows autosomal dominant inheritance.

80.4.3 Penetrance

Based on the data available, SRP72-associated BMFS demonstrates high penetrance and somewhat varied expressivity. Apart from early-onset aplastic anemia/pancytopenia and adult-onset myelodysplasia, SRP72-associated BMFS may cause syndromic phenotypic abnormalities such as congenital nerve deafness or labyrinthitis.

80.5 Clinical Features

BMFS often present with aplastic anemia (bone marrow failure to generate enough peripheral blood cells), MDS, and/or acute myeloid leukemia (AML) in childhood or young adulthood, along with syndromic phenotypic abnormalities (e.g., multiple congenital anomalies, pancreatic dysfunction, or other subtle phenotypic presentations resulting in a delayed diagnosis into adulthood) [7].

SRP72-associated BMFS may cause early-onset aplastic anemia or pancytopenia and adult-onset myelodysplasia as well as syndromic phenotypic abnormalities such as congenital nerve deafness and labyrinthitis [5].

80.6 Diagnosis

Diagnostic workup for BMFS involves medical history review, physical examination, laboratory assessment, histopathology, and molecular testing.

Histopathologically, bone marrow biopsy from patients with BMFS-related aplastic anemia reveals marked hypocellularity, with fat and fibrotic stroma filling up the remaining marrow space, lymphocytes predominating any residual hematopoietic cells, and dysplastic but nonmalignant features (e.g., hyponucleated small megakaryocytes, multinucleated red cells, or hypolobulated or hypogranular myeloid cells) [13].

Molecular identification of *SRP72* pathogenic variants helps confirm the diagnosis of SRP72-associated BMFS.

Differential diagnoses of SRP72-associated BMFS include BMFS caused by mutations in genes other than *SRP72* [8,14].

80.7 Treatment

Treatment options for BMFS, including SRP72-associated BMFS, consist of symptomatic relief, infection prophylaxis, blood transfusion (leuko-reduced red blood cells for Hb <7 mg/dL or platelets <10,000/μL or <50,000/μL for active blood loss), and hematopoietic stem cell transplant (HSCT) [15].

80.8 Prognosis and Prevention

Prognosis for patients with inherited BMFS is improved by HSCT, which produces 10-year survival rates of 83%, 73%, 68%, and 51% in the first, second, third, and fifth decades, respectively.

However, since HSCT only targets marrow disease, patients with inherited BMFS are still susceptible to early-onset squamous cell carcinoma (SCC). In addition, secondary malignancies and complications from pancytopenia (hemorrhage and infection) are a major contributing factor of mortality.

Given that SRP72 pathogenic variants increase cellular sensitivity to radiation, patients with SRP72-associated BMFS should not be given radiation therapy. Further, blood products from family members should be avoided to reduce the risk of alloimmunization. Iron chelators are recommended for patients with secondary hemochromatosis. Growth factors (e.g., erythropoietin or granulocyte colony-stimulating factors) are often ineffective for BMFS patients who have few precursor cells to generate satisfactory responses.

80.9 Conclusion

Bone marrow failure syndromes (BMFS) consist of >30 inherited or acquired genetic disorders that share common phenotypes of aplastic anemia (peripheral cytopenia) and myelodysplasia (hypoplastic bone marrow) accompanied by other syndromic phenotypic abnormalities. Of >80 BMFS-associated genes identified so far, many play critical roles in genomic stability maintenance, DNA repair, ribosome biogenesis, and telomere biology. Alterations in these genes affect cell survival and function during hematopoiesis or modify biological processes in relevant hematopoietic lineage pathways, leading to single- or multi-lineage cytopenias. As a recently recognized BMFS, SRP72-associated BMFS (familial aplastic anemia/myelodysplastic syndrome with *SRP72* mutation [BMFS1]) is an autosomal dominant disorder due to germline heterozygous mutations in the *SRP72* (signal recognition particle 72 kDa) gene on chromosome 4q12, encoding a transcription factor (SRP72), which participates in signal recognition, translational arrest, and ER membrane targeting [10,16]. Clinically, SRP72-associated BMFS induces early-onset aplastic anemia/pancytopenia and adult-onset myelodysplasia, along with other syndromic phenotypic abnormalities such as congenital nerve deafness or labyrinthitis. Diagnosis of SRP72-associated BMFS is assisted with the observation of marked hypocellularity and dysplastic but nonmalignant features in bone marrow biopsy. Molecular identification of *SRP72* pathogenic variant further confirms the diagnosis. As for other related BMFS, management of SRP72-associated BMFS currently relies on symptomatic relief, infection prophylaxis, blood transfusion, and HSCT. Innovative, highly effective, and specific therapeutics for SRP72-associated BMFS and other BMFS are keenly sought.

REFERENCES

1. Tummala H, Kirwan M, Walne AJ et al. ERCC6L2 mutations link a distinct bone-marrow-failure syndrome to DNA repair and mitochondrial function. *Am J Hum Genet.* 2014;94:246–56.

2. Tummala H, Walne AJ, Williams M et al. DNAJC21 mutations link a cancer-prone bone marrow failure syndrome to corruption in 60S ribosome subunit maturation. *Am J Hum Genet.* 2016;99:115–24.

3. Dhanraj S, Matveev A, Li H et al. Biallelic mutations in DNAJC21 cause Shwachman-Diamond syndrome. *Blood.* 2017;129(11):1557–62.

4. Godley LA. DNAJC21: The new kid on the SDS block. *Blood.* 2017;129(11):1413–4.

5. Kirwan M, Walne AJ, Plagnol V et al. Exome sequencing identifies autosomal-dominant SRP72 mutations associated with familial aplasia and myelodysplasia. *Am J Hum Genet.* 2012;90:888–92.

6. Bannon SA, DiNardo CD. Hereditary predispositions to myelodysplastic syndrome. *Int J Mol Sci.* 2016;17(6). pii: E838.

7. Adam S, Melguizo Sanchis D, El-Kamah G et al. Concise review: Getting to the core of inherited bone marrow failures. *Stem Cells.* 2017;35(2):284–98.

8. Moore CA, Krishnan K. Bone marrow failure. In: *StatPearls* [Internet]. Treasure Island (FL): StatPearls Publishing; 2018 Jan-. 2017 Oct 6.

9. Becker MM, Lapouge K, Segnitz B, Wild K, Sinning I. Structures of human SRP72 complexes provide insights into SRP RNA remodeling and ribosome interaction. *Nucleic Acids Res.* 2017;45(1):470–81.

10. Wild K, Juaire KD, Soni K et al. Reconstitution of the human SRP system and quantitative and systematic analysis of its ribosome interactions. *Nucleic Acids Res.* 2019;47(6):3184–96.

11. Prevo R, Tiwana GS, Maughan TS, Buffa FM, McKenna WG, Higgins GS. Depletion of signal recognition particle 72 kDa increases radiosensitivity. *Cancer Biol Ther.* 2017;18(6):425–32.

12. Waespe N, Dhanraj S, Wahala M et al. The clinical impact of copy number variants in inherited bone marrow failure syndromes. *NPJ Genom Med.* 2017;2. pii: 18.

13. Shankar RK, Giri N, Lodish MB et al. Bone mineral density in patients with inherited bone marrow failure syndromes. *Pediatr Res.* 2017;82(3):458–64.

14. Zhang Z, Pondarre C, Pennarun G et al. A nonsense mutation in the DNA repair factor Hebo causes mild bone marrow failure and microcephaly. *J Exp Med.* 2016;213:1011–28.

15. Locatelli F, Strahm B. How I treat myelodysplastic syndromes of childhood. *Blood.* 2018;131(13):1406–14.

16. Prevo R, Tiwana GS, Maughan TS, Buffa FM, McKenna WG, Higgins GS. Depletion of signal recognition particle 72kDa increases radiosensitivity. *Cancer Biol Ther.* 2017;18(6):425–32.

81

Thrombocytopenia-Absent Radius

Dongyou Liu

CONTENTS

81.1 Introduction

Thrombocytopenia-absent radius (TAR) is a rare autosomal recessive bone marrow failure syndrome characterized by the absence of a bone called the radius in each forearm and severe, albeit transient, megakaryocytic thrombocytopenia (<50 platelets $\times$ 10^9/L) at birth, along with cow milk allergy, other skeletal anomalies (upper and lower limbs, ribs, and vertebrae), short stature, facial dysmorphism, cardiac defects, and genitourinary malformations (renal anomalies, agenesis of uterus, cervix, and upper part of the vagina).

Molecularly, TAR syndrome is linked to a 200-kb interstitial microdeletion of chromosome 1q21.1 resulting in one null *RBM8A* allele, or polymorphisms in either the 5′-untranslated region or first intron of a hypomorphic *RBM8A* allele. As the *RBM8A* gene encodes an RNA-binding protein (Y14), which is a key component of the exon-junction complex (EJC) involved in nuclear export of transcripts, nonsense mediated decay, and translational enhancement, its deletion or alteration abrogates or reduces expression of the hypomorphic allele in a cell type- and developmental stage-specific manner, leading to defective mRNA processing and export [1].

81.2 Biology

Bone marrow failure syndromes (BMFS) are a group of >30 hematological diseases that result from defects in biochemical pathways involved in housekeeping functions for many cell types.

Among the most notable BMFS, Fanconi anemia (disruption in DNA repair), dyskeratosis congenita (faulty telomere maintenance), and Shwachman–Diamond syndrome (defective ribosome maturation) are associated with pancytopenia; congenital amegakaryocytic thrombocytopenia (growth factor receptor signaling abnormality), familial platelet disorder with propensity to AML (defect in transcriptional regulation), and thrombocytopenia absent radii syndrome (faulty mRNA processing and export) are responsible for thrombocytopenia; Diamond–Blackfan anemia (disrupted ribosome biogenesis) produces red cell aplasia; severe congenital neutropenia (defective protein folding/trafficking) induces neutropenia; and monoMAC and related *GATA2* deficiency disorders (defect in transcriptional regulation) cause monocytopenia and lymphopenia [2,3].

TAR (or 200 kb chromosome 1q21.1 deletion syndrome) was initially described by Greenwald and Sherman in 1928, and subsequently by Gross and coworkers in 1956 as well as Shaw and Oliver in 1959 from sibs with absent radii and thrombocytopenia. Further definition of this disorder was provided by Judith Hall in 1969 after analysis of four affected sisters with radius aplasia and thrombocytopenia (platelet counts <150 $\times$ 10^9/L) leading to petechiae and increased bruising. The susceptible gene region for TAR was mapped to chromosome 1q21.1 by Zhao and coworkers in 2000, and an inherited or de novo 200 kb (covering at least 12 genes) deletion of chromosome 1q21.1 was observed in a majority of TAR patients by Klopocki and coworkers in 2007 [4,5]. In 2012, single nucleotide polymorphisms (SNP) were identified in the 5′ untranslated region (UTR) and the first intron of the *RBM8A* gene from individuals harboring 200 kb deletion of chromosome

1q21.1 by Albers and coworkers, putting beyond doubt the role of the *RBM8A* gene in the pathogenesis of TAR [6,7].

Like other BMFS, TAR typically presents with blood abnormalities (e.g., thrombocytopenia due to reduced number of megakaryocytes in the bone marrow, and hemorrhage leading to easy bruising and frequent nosebleeds in the first year of life). However, TAR is distinguished by the absence of the radius (the bone on the thumb side of the forearm) and preservation of the thumb in contrast to other BMFS that are accompanied by different syndromic phenotypic features [1].

81.3 Pathogenesis

The *RBM8A* (RNA binding motif protein 8) gene on chromosome 1q21.1 spans 9.9 kb (5.9 kb plus strand) with six exons and encodes a 174 aa, 19.8 kDa protein with a conserved RNA-binding motif.

As one of the four components in the splicing-dependent multiprotein EJC expressed in all hematopoietic lineages, RBM8A associates with mRNAs (both nuclear mRNAs and newly exported cytoplasmic mRNAs) and other components to form the EJC, which deposits at 24 nt upstream of exon–exon junctions in the mature mRNA for post-mRNA splicing events to take place. As its core components remain bound to spliced mRNAs throughout all stages of mRNA metabolism, the EJC maintains its influence on various downstream processes (e.g., nuclear mRNA export, subcellular mRNA localization, translation efficiency, and nonsense-mediated mRNA decay).

Inheritance of a 200-kb (or 500-kb) deletion at chromosome 1q21.1 from one parent and a noncoding SNP (such as c.-21G>A or c.67+32G>C) in *RBM8A* from the other that results in the loss or significant reduction of *RBMBA* expression in platelets underlines the pathogenesis of TAR, leading to a combined defect of platelet production and function in thrombocytopenia with absent radii. Homozygosity for two null alleles is thought to be lethal. Further, inheritance of two hypomorphic c.-21G>A and c.67+32G>C variants in *RBM8A* is also responsible for causing TAR. However, individuals homozygous for these hypomorphic alleles do not display features of TAR. Inactivating alleles c.207_208insAGCG and c.487C>T (p.Arg163Ter) may be occasionally implicated in the development of TAR [1,8–10].

81.4 Epidemiology

81.4.1 Prevalence

TAR is a rare congenital disorder with an estimated frequency of 0.42 in 100,000 (or 1 in 240,000) births.

81.4.2 Inheritance

TAR demonstrates autosomal recessive inheritance, in which either both copies of the *RBM8A* gene in each cell have hypomorphic mutations (located in the 5'-UTR and intron 1) or, more commonly, one copy of the gene has a hypomorphic mutation and the other is lost as part of a 200-kb deletion in chromosome 1q21.1 (75% of cases). In the remaining 25% of cases, deletion occurs sporadically during the formation of reproductive cells (eggs and sperm) or somatically in early fetal development.

The hypomorphic alleles have a frequency of approximately 3% in the general population. Parents (2/3 mother, 1/3 father) of an affected child may carry an *RBM8A* gene mutation or a 200-kb deletion, but typically do not show signs and symptoms of the condition. If both parents carry one variant allele, they have a 25% chance of getting an affected offspring, a 50% chance of getting an asymptomatic carrier, and a 25% chance of getting an unaffected and non-carrier offspring. If one parent carries an inherited pathogenic variant and the other parent has de novo variant/deletion, they have a 50% chance of getting an asymptomatic carrier offspring and a 50% chance of getting an unaffected, noncarrier offspring [1].

81.4.3 Penetrance

TAR shows high, but perhaps incomplete penetrance. While patients with biallelic *RBM8A* pathogenic variants invariably develop clinical phenotype, those with the proximal 1q21.1 deletion or noncoding variants in the 5'UTR and first intron may be asymptomatic carriers.

81.5 Clinical Features

Clinically, TAR is associated with limb anomalies, thrombocytopenia, cardiac anomalies, gastrointestinal involvement, genitourinary anomalies, leukemoid reactions, growth anomalies, facial dysmorphism, and other skeletal manifestations [1,11,12].

Limb anomalies are typically indicated by bilateral absence of the radius (radial ray), and presence of the thumbs (which are near-normal size but appear somewhat wider and flatter than normal, and which display limited grasp and pinch activities). In addition, the upper limbs may show hypoplasia or absence of the ulnae, humeri, and shoulder girdles; lower limbs are affected in nearly 50% of patients; fingers have syndactyly and clinodactyly (involving fifth finger); hip dislocation, coxa valga, femoral and/or tibial torsion, genu varum, absence of the patella, and tetraphocomelia may also occur (Figure 81.1) [13].

Thrombocytopenia (a reduction in the number of platelets to <50 platelets × 10^9/L, normal range 150–350 platelets × 10^9/L, due to megakaryocytic thrombocytopenia) may appear at birth (41%) or emerge in the first few weeks to months of life (59%), with thrombocytopenic episodes decreasing with age, and normal platelet counts by school age.

Cardiac anomalies such as tetralogy of Fallot, atrial and ventricular septum anomalies instead of complex cardiac malformations are observed in 15%–22% of cases.

Gastrointestinal involvement such as cow's milk allergy (which indicates exacerbation of thrombocytopenia, 47%) and gastroenteritis tends to improve with age.

Genitourinary anomalies range from renal anomalies (23%) to Mayer–Rokitansky–Kuster–Hauser syndrome (agenesis of uterus, cervix, and upper part of the vagina).

Leukemoid reactions are generally transient elevations of white blood cell counts (>35,000 cells/mm^3).

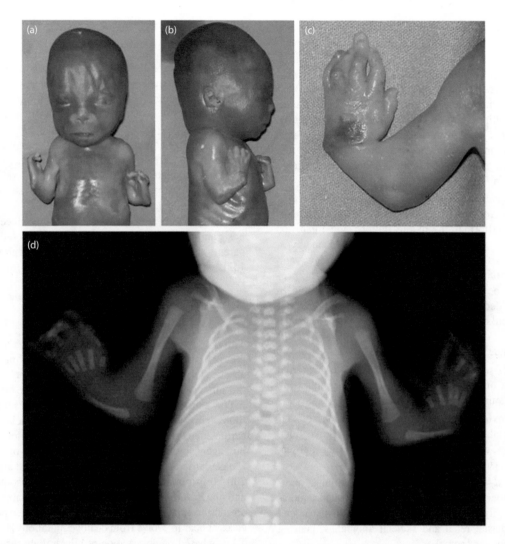

FIGURE 81.1 TAR syndrome in a male fetus at 21st gestation week showing flat nasal bridge with mildly anteverted nares and receding chin, severely shortened forearms, radial club hands, and presence of thumbs (a and b); camptodactyly with bulbous ends, pterygium between wrist and the rhizomelic segment, and normally placed thumb in the left hand (c); and absence of both radii with hypoplastic and straight ulnae as revealed by babygram (d). (Photo credit: Bottillo I et al. *BMC Res Notes*. 2013;6:376.)

Growth anomalies are indicated by short stature (95%, with height on or below the 50th centile) and macrocephaly (76%).

Facial dysmorphism is illustrated by small lower jaw (micrognathia), prominent forehead, and low-set ears (53%).

Other skeletal manifestations such as rib and cervical vertebral anomalies (including cervical rib, fused cervical vertebrae) are rare.

81.6 Diagnosis

Diagnostic workup for TAR involves clinical observation of the unique combination of hematological and radial defects (bilateral absent radii, present thumbs, and thrombocytopenia 50 platelets × 10⁹/L, normal range 150–350 platelets × 10⁹/L) in newborn infants and toddlers, and further confirmation with identification of a heterozygous null allele (a minimally deleted 200-kb or 500-kb region at chromosome 1q21.1) *in trans* with a heterozygous *RBM8*A hypomorphic allele) [1,14].

In general, gene-targeted deletion/duplication analysis of *RBM8A* is conducted first for a heterozygous minimally deleted 200-kb or 500-kb region at chromosome band 1q21.1. If no deletion is observed, sequence analysis of *RBM8A* for a second pathogenic variant (hypomorphic allele) located in the 5′ UTR and first intronic regions is performed. Homozygous *RBM8A* null alleles (e.g., deletions) appear to be lethal [1].

Differential diagnoses for TAR include conditions that also display radial aplasia, including:

Holt—Oram syndrome: Upper-extremity malformations involving radial, thenar, or carpal bones; personal/family history of congenital heart defects such as ostium secundum atrial septal defect (ASD) and ventricular septal defect (VSD); absent or hypoplastic thumb; autosomal dominant disorder due to pathogenic variants in *TBX5*.

Roberts syndrome: Mild to severe prenatal growth retardation, microcephaly, cleft lip and/or palate, and limb malformations (including bilateral symmetric tetraphocomelia or hypomelia caused by mesomelic shortening); oligodactyly with thumb aplasia or hypoplasia, syndactyly, clinodactyly, and elbow and knee flexion contractures; craniofacial abnormalities (downslanting palpebral fissures, widely spaced eyes, exophthalmos resulting from shallow orbits, corneal clouding, hypoplastic nasal alae, beaked nose, malar hypoplasia, ear malformations, and micrognathia); intellectual disability; autosomal recessive disorder due to pathogenic variants in *ESCO2*.

Fanconi anemia: Bone marrow failure, increased risk of malignancy, short stature, abnormal skin pigmentation, malformations of the thumbs and forearms; additional anomalies of the skeletal system, eyes, ears, heart, gastrointestinal system, kidneys and genitourinary tract, and central nervous system; progressive bone marrow failure with pancytopenia in the first decade, often initially with thrombocytopenia or leukopenia; autosomal recessive (>15 genes), autosomal dominant (*RAD51*), or X-linked (*FANCB*) inheritance.

Thalidomide embryopathy: (secondarily to maternal ingestion of thalidomide; limb, cardiac, craniofacial, and genitourinary anomalies).

VACTERL association: (cardinal manifestations of *v*ertebral, *a*nal, *c*ardiac, *t*racheo-*e*sophageal fistula, *r*enal anomalies, and *l*imb anomalies; absent thumbs; absent thrombocytopenia).

Duane-radial ray syndrome: Okihiro syndrome, acro-renal-ocular syndrome; combination of Duane anomaly (inability to abduct the eye); thenar hypoplasia and radial aplasia; renal and skeletal anomalies; hearing loss and/or ear anomalies.

Townes–Brocks syndrome: Triad of imperforate anus, dysplastic ears, and thumb malformations (triphalangeal thumbs, duplication of the thumb, and rarely thumb hypoplasia); absence of hematologic abnormalities; autosomal dominant disorder due to pathogenic variants in *SALL1*.

Rapadilino syndrome: Acronym for the cardinal manifestations of *ra*dial defects, absent/hypoplastic *pa*tellae (and high/cleft *pa*late), *di*arrhea (and joint *di*slocations), *li*ttle size, and a long/slender *no*se (and *no*rmal intelligence); absent or hypoplastic radii and absent or hypoplastic thumbs [1,15].

81.7 Treatment

Treatment options for TAR consist of supportive care of thrombocytopenia (platelet transfusions via central venous catheter instead of venipuncture to reduce the pain associated with repeated procedures) and orthopedic intervention (e.g., prostheses, orthoses, adaptive devices, and surgery) of bony abnormalities to improve function of the upper limbs.

81.8 Prognosis and Prevention

TAR has a generally favorable prognosis with the frequency of thrombocytopenic episodes decreasing with age and platelet counts reaching near normal levels by school age. Patients who survive the initial 2 years of life often have normal life expectancy.

Avoidance of cow's milk helps reduce the severity of gastroenteritis and exacerbations of thrombocytopenia in older children. Avoidance of platelet transfusion in older individuals with platelet counts of $>10 \times 10^9$/L may decrease the risks of alloimmunization and infection.

As TAR shows autosomal recessive inheritance, prenatal diagnosis and preimplantation testing may be offered to prospective parents carrying 200-kb or 500-kb deletion at chromosome band 1q21.1, and/or *R8BM8A* hypomorphic allele [16].

81.9 Conclusion

Thrombocytopenia-absent radius (TAR) is a rare autosomal recessive bone marrow failure syndrome linked to bilateral absence of radii and severe, albeit transient, megakaryocytic thrombocytopenia (<50 platelets $\times 10^9$/L) at birth, in addition to cow's milk allergy, other skeletal anomalies (upper and lower limbs, ribs, and vertebrae), short stature, facial dysmorphism, cardiac defects, and genitourinary malformations (renal anomalies, agenesis of uterus, cervix, and upper part of the vagina) [17]. Molecularly, TAR relates to one *R8BM8A* null allele, typically a 200-kb minimally deleted region at chromosome band 1q21.1, and one *R8BM8A* hypomorphic allele. While about 75% of cases are due to inheritance from unaffected carrier parents, 25% of cases appear to evolve from de novo mutations [18]. Diagnosis of TAR is based on observation of bilateral absence of radii, presence of thumbs, and thrombocytopenia (50 platelets $\times 10^9$/L) in newborn infants and toddlers, and confirmed by molecular detection of a heterozygous null allele (a minimally deleted 200-kb or 500-kb region at chromosome 1q21.1) *in trans* with a heterozygous *RBM8*A hypomorphic allele. Treatment of TAR centers on supportive care of thrombocytopenia (platelet transfusions) and orthopedic intervention of bony abnormalities.

REFERENCES

1. Toriello HV. Thrombocytopenia absent radius syndrome. In: Adam MP, Ardinger HH, Pagon RA et al. editors. *GeneReviews®* [Internet]. Seattle (WA): University of Washington, Seattle; 1993–2018. 2009 Dec 8 [updated 2016 Dec 8].
2. Khincha PP, Savage SA. Genomic characterization of the inherited bone marrow failure syndromes. *Semin Hematol.* 2013;50(4):333–47.
3. Wilson DB, Link DC, Mason PJ, Bessler M. Inherited bone marrow failure syndromes in adolescents and young adults. *Ann Med.* 2014;46(6):353–63.
4. Greenhalgh KL, Howell RT, Bottani A et al. Thrombocytopenia-absent radius syndrome: A clinical genetic study. *J Med Genet.* 2002;39(12):876–81.

5. Klopocki E, Schulze H, Strauss G et al. Complex inheritance pattern resembling autosomal recessive inheritance involving a microdeletion in thrombocytopenia-absent radius syndrome. *Am J Hum Genet.* 2007;80(2):232–40.

6. Albers CA, Paul DS, Schulze H et al. Compound inheritance of a low-frequency regulatory SNP and a rare null mutation in exon-junction complex subunit RBM8A causes TAR syndrome. *Nature Genet.* 2012;44:435–9.

7. Albers CA, Newbury-Ecob R, Ouwehand WH, Ghevaert C. New insights into the genetic basis of TAR (thrombocytopenia-absent radii) syndrome. *Curr Opin Genet Dev.* 2013;23(3):316–23.

8. Fiedler J, Strauss G, Wannack M et al. Two patterns of thrombopoietin signaling suggest no coupling between platelet production and thrombopoietin reactivity in thrombocytopenia-absent radii syndrome. *Haematologica.* 2012;97(1):73–81.

9. Rosenfeld JA, Traylor RN, Schaefer GB et al. Proximal microdeletions and microduplications of 1q21.1 contribute to variable abnormal phenotypes. *Eur J Hum Genet.* 2012;20(7):754–61.

10. Tassano E, Gimelli S, Divizia MT et al. Thrombocytopenia-absent radius (TAR) syndrome due to compound inheritance for a 1q21.1 microdeletion and a low-frequency noncoding RBM8A SNP: A new familial case. *Mol Cytogenet.* 2015;8:87.

11. Kumar C, Sharma D, Pandita A, Bhalerao S. Thrombocytopenia absent radius syndrome with Tetralogy of Fallot: A rare association. *Int Med Case Rep J.* 2015;8:81–5.

12. Kumar MK, Chaudhary IP, Ranjan RB, Kumar P. Thrombocytopenia with unilateral dysplastic radius- Is it thrombocytopenia - absent padius (TAR) syndrome? *J Clin Diagn Res.* 2015;9(3):SD01–2.

13. Bottillo I, Castori M, De Bernardo C et al. Prenatal diagnosis and post-mortem examination in a fetus with thrombocytopenia-absent radius (TAR) syndrome due to compound heterozygosity for a 1q21.1 microdeletion and a RBM8A hypomorphic allele: A case report. *BMC Res Notes.* 2013;6:376.

14. Bertoni NC, Pereira DC, Araujo Júnior E, Bussamra LC, Aldrighi JM. Thrombocytopenia-absent radius syndrome: Prenatal diagnosis of a rare syndrome. *Radiol Bras.* 2016;49(2):128–9.

15. Elmakky A, Stanghellini I, Landi A, Percesepe A. Role of genetic factors in the pathogenesis of radial deficiencies in humans. *Curr Genomics.* 2015;16(4):264–78.

16. Uhrig S, Schlembach D, Waldispuehl-Geigl J et al. Impact of array comparative genomic hybridization-derived information on genetic counseling demonstrated by prenatal diagnosis of the TAR (thrombocytopenia-absent-radius) syndrome-associated microdeletion 1q21.1. *Am J Hum Genet.* 2007;81(4):866–8.

17. Alagbe OA, Alagbe AE, Onifade EO, Bello TO. Thrombocytopenia with absent radii (TAR) syndrome in a female neonate: A case report. *Pan Afr Med J.* 2019;33:181.

18. Brodie SA, Rodriguez-Aulet JP, Giri N et al. 1q21.1 deletion and a rare functional polymorphism in siblings with thrombocytopenia-absent radius-like phenotypes. *Cold Spring Harb Mol Case Stud.* 2019;5(6):a004564.

82

Waldenström Macroglobulinemia

Jonas Paludo and Morie A. Gertz

CONTENTS

82.1 Introduction

Waldenström macroglobulinemia (WM) was first described in 1944 by the Swedish physician Jan Waldenström involving two cases with nasal bleeding, anemia, thrombocytopenia, lymphadenopathy, and lymphoid aggregation in the bone marrow [1]. WM is considered a distinct entity classified by the World Health Organization (WHO) as a subset of a low-grade non-Hodgkin lymphoma (NHL) and characterized by the presence of IgM monoclonal gammopathy and infiltration of the bone marrow by lymphoplasmacytic lymphoma (LPL). A small proportion (<5%) of LPL secretes a non-IgM monoclonal protein, and therefore is not classified as WM [2].

82.2 Biology and Pathogenesis

Key to understanding of WM is to recognize its cardinal relationship with the premalignant condition IgM monoclonal gammopathy of undetermined significance (MGUS). While monoclonal B cell lymphocytosis (MBL) and non-IgM MGUS are well established premalignant conditions for the development of chronic lymphocytic leukemia (CLL) and multiple myeloma (MM) [3,4], which share a closely related cell of origin with WM [5], the association of IgM MGUS and WM has only strengthened in recent years as technological advances, such as next-generation sequencing (NGS), gene expression profiling (GEP), and multicolor flow cytometry (MFC), revealed valuable clues into the unique relationship between these two entities.

IgM MGUS has been associated with the development of multiple types of B cell NHL, most notably WM, and represents a district entity from non-IgM MGUS [6]. The diagnostic criteria for IgM MGUS are not universally agreed upon but hinge on the detection of a circulating IgM monoclonal protein in the serum with or without the detection of a lymphoplasmacytic lymphoma (LPL) infiltration in the bone marrow. The Mayo Clinic criteria define IgM MGUS as a serum IgM monoclonal protein of <3 g/dL, bone marrow LPL involvement of <10%, and no signs or symptoms secondary to the lymphoplasmacytic infiltrative process [7]. The Second International Workshop on Waldenström Macroglobulinemia consensus, however, characterizes IgM MGUS as a serum IgM monoclonal protein of any size without bone marrow involvement by an LPL [8]. In a large cohort of IgM MGUS patients seen at Mayo Clinic with long-term follow-up, all patients were found to have <10% bone marrow involvement by LPL [9]. Furthermore, bone marrow evaluation of patients with IgM MGUS by MFC showed <10% infiltration by clonal B cells in 99% of the cases [10].

It has been hypothesized that the initial step in the pathogenesis of MGUS is an abnormal response to a usual antigenic stimulation through Toll-like receptors (TLRs) [11]. Finding stereotyped immunoglobulin heavy chain (IgH) rearrangements supports the hypothesis of an antigenic stimulation as an early event in the development of IgM MGUS and WM. The monoclonal IgM expressed by malignant cells in these diseases is assembled by somatic recombination of a large number of gene segments in the IgH region and is continuously shaped by exposure to exogenous antigens. The IgH variable genes rearrangement repertoire associated with the specific IgM subtype of patients with IgM MGUS and WM has demonstrated a remarkably biased IgH repertoire with a predilection for the IgH variable gene 3 (seen in 83% of patients) in preference of IgH variable genes 1 and 4. Intra-disease stereotyped clusters, however, were not seen [12]. These results imply that IgM MGUS and WM originate from a heterogeneous B cell that responded to an antigenic stimulation during a physiologic immunologic response [12].

The next step after the abnormal antigenic stimulation response would be acquisition of genetic abnormalities, probably in a stepwise fashion, leading to the malignant transformation and clonal expansion [13]. This hypothesis could explain the natural history of patients with IgM MGUS, who have a high likelihood of progressing to smoldering and finally active WM over the course of several years to decades. It also provides a potential insight into why a proportion of patients with IgM MGUS do not progress to WM.

The most common structural abnormality in patients with WM is the deletion of chromosome 6q detected in approximately 50% of the patients, followed by chromosome 5 trisomy, and chromosome 8 monosomy [14]. Fluorescent in situ hybridization (FISH) confirmed the 6q deletion in 55% of WM patients, but in none of the patients with IgM MGUS [15]. A genome-wide study of copy number abnormalities (CNA) and loss of heterozygosity (LOH) in patients with IgM MGUS and WM identified 6q deletion and +18q(22.1) as the most common structural abnormalities in this patient population. Genomic imbalances typically observed in WM (del6q, +18q, trisomy 4, and trisomy 12) were rarely seen in IgM MGUS patients. The frequency of CNA progressively increased from IgM MGUS (36%) to smoldering (73%) and symptomatic WM (82%) [16]. Similar results were seen when a targeted NGS panel was used to evaluate the presence of somatic mutations in patients with IgM MGUS and WM. A significantly higher number of mutations were detected in patients with WM compared to patients with IgM MGUS, supporting the hypothesis that multiple genetic hits are needed for the progression from IgM MGUS to WM [17]. While somatic driver mutations have not been identified, mutations in MYD88, CXCR4, and KMT2D were more frequently seen in WM than in IgM MGUS suggesting that these are potential early events in the transformation of IgM MGUS to WM [17].

When using GEP to compare clonal B cells from patients with IgM MGUS and WM with their normal B cell counterpart (CD22+ and CD25−), differentially expressed genes related to the IL-6, NF-κB, JAK/STAT, PI3K/AKT, inositol tetrakisphosphate (IP4) and 3-phosphoinositide biosynthesis pathways were detected. However, essentially no differences were observed in the GEP of patients with IgM MGUS compared to patients with WM [16]. While this study identified potential key activated pathways associated with cell growth and survival in patients with IgM MGUS and WM, it also implies that the WM clone may already be present in patients with IgM MGUS, posing the question of these two entities being merely different stages of the same disease. Post-translational modifications and epigenetic changes between IgM MGUS and WM patients have not been explored in depth.

The malignant cells in WM show a wide morphological heterogeneity ranging from clonal B lymphocytes to clonal plasma cells, including a lymphoplasmacytoid population. The malignant cells usually express pan-B cell markers such as CD19, CD20, CD22, CD25, CD27, CD38, and CD79a. However, they generally do not express CD5 in contrast to CLL and mantle cell lymphoma, CD10 in contrast to follicular lymphoma and CD56 in contrast to MM [18–20]. Weak CD5 expression can rarely be seen in clonal B lymphocytes in WM, but not as pronounced as in MCL and CLL [8]. The normal counterpart for the malignant WM cell of origin has been proposed to be a CD25−, CD22+ resting B cell [16].

A great number of somatic mutations and chromosomal abnormalities have been described in WM; however, a definite causal relationship has not been established [21]. Approximately half of all patients with WM will have some chromosomal structural abnormality detected by FISH and/or conventional cytogenetics. In contrast to MM, IgH rearrangements are rare in WM [22,23]. While del6q is common in WM, it has also been described in other B cell disorders. Trisomy 4, however, seems to be unique in WM and occasionally is the only abnormality seen in these patients [24].

A similar pattern of cytogenetic abnormalities was described in both the clonal plasma cells and clonal B cells in patients with WM implying that the clonal plasma cells derive directly from the clonal B cells. In contrast to MM plasma cells, the clonal plasma cells in patients with WM showed a significant upregulation of the PAX5 gene, which must be repressed to allow final plasma cell differentiation. BLIPM1, which is needed for differentiation from B cell to plasma cell, is downregulated in WM [25]. These findings support the hypothesis that WM clonal cells are in a late stage of B cell differentiation into plasma cells that express IgM but have not yet undergone final isotype class switching. Although GEP of clonal plasma cells and B cells in WM has shown some similarities with their counterparts in MM and CLL, the malignant cells in WM demonstrate a complex clone that represents a distinct entity [26].

The discovery of the recurrent somatic mutation of the *MYD88* gene resulting in the change of the amino acid leucine to proline at position 265 (L265P) contributed to the understanding of WM [27]. The MYD88 adaptor protein is part of the TLR and interleukin-1 receptor (IL-1R) signaling pathway [27]. This "gain-of-function" MYD88^{L265P} mutation promotes cell growth and survival through transcription of nuclear factor kappa-light-chain of activated B cells (NF-κB), one of the final products of the TLR and IL-1 pathways (Figure 82.1) [28]. MYD88^{L265P} mutation has been detected in 70%–90% of sporadic WM cases and in all familial WM cases [27,29,30]; however, no evidence for the presence of a germ-line MYD88 mutation exists in cases of familial WM [31]. The MYD88^{L265P} has also been described in patients with IgM MGUS (87%) and in a small proportion of patients with splenic marginal zone lymphoma (4%), mucosa-associated lymphoid tissue lymphoma (7%), and rarely in CLL [29,32]. The role of the MYD88^{L265P} mutation in the pathogenesis of WM is still unclear; it could represent a driver mutation for the progression of IgM MGUS to WM or merely an early event in this disease [19,27].

The C-X-C chemokine receptor type 4 (CXCR4) gene mutation represents the second most frequent mutated gene in WM, detected in approximately 30% of patients [33]. The somatic CXCR4 mutations seen in WM closely resemble the germline CXCR4^{WHIM} mutations described in the WHIM syndrome. The "gain-of-function" CXCR4^{WHIM} mutations lead to permanent activation of CXCR4 receptor by its ligand CXCL-12 and results in migration and homing to the bone marrow niches of WM cell which could provide a survival advantage (Figure 82.1) [34]. The CXCR4^{WHIM} mutation has been linked to drug resistance and impacts the response to targeted therapies such as BKT, mTOR, and PI3K inhibitors [35,36].

82.3 Epidemiology

WM has an overall annual, age-adjusted incidence of approximately 3.5 to 5.5 cases per million in the Western hemisphere and accounts for 1%–2% of all hematologic malignancies [37].

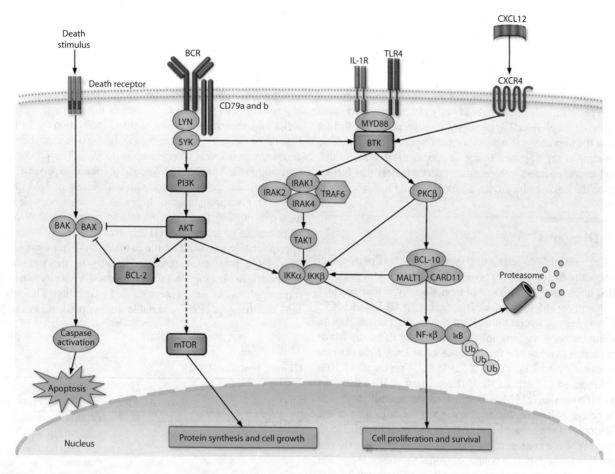

FIGURE 82.1 Common pathways involved in WM cell-survival. *Note:* Red boxes: therapeutic targets in WM. (From Kapoor P et al. *Curr Treat Options Oncol* 2016;17:16. Reprinted with permission from Springer Nature: Springer US, Copyright 2016.)

It is predominantly a disease of elderly Caucasian males, with a median age at diagnosis ranging from 63 to 73 years. The incidence of WM is 95-fold higher in octogenarians compare to those under 50 years of age, and proximately 2-fold higher in males and Caucasians than in females or non-Caucasians [38]. Even though WM is considered an indolent disease, the estimated median disease-specific survival is only 11 years from diagnosis [39]. Younger patients have a longer median disease-specific survival (15.6 years) [40].

Patients with a personal history of IgM MGUS have a 46-fold increase in risk of developing WM with a rate of progression to WM of approximately 1.5% per year [41]. The risk of developing secondary malignancies is also increased in patients with WM with a cumulative incidence of a secondary solid cancer at 5 years, 10 years, and 15 years of 6%, 11%, and 17%, respectively [42]. The most common gender-adjusted malignancies are prostate (9.4%), breast (8%), non-melanoma skin cancer (7.1%), melanoma (2.2%), and lung (1.4%) [43]. The most common hematologic malignancies are diffuse large B cell lymphoma (5%) and MDS/AML (5%); however, most patients developing these malignancies were previously treated with alkylating agents and nucleoside analogs which have been associated with therapy-related malignancies [42]. Interestingly, familial cases of WM have a fivefold higher risk of developing lung cancer, while sporadic cases of WM have a twofold high risk of developing prostate cancer [43].

While the majority of WM cases are sporadic, familial clustering of WM has been described alongside evidence suggesting a genetic predisposition in some families [43]. Approximately 20% of all WM cases can be categorized as familial with patients having at least one first-degree relative with WM or other B cell disorder such as NHL, CLL, MM, or Hodgkin lymphoma [44]. Familial cases are usually diagnosed a decade earlier than sporadic WM cases. First-degree relatives of patients with WM have a 10-fold higher risk of developing IgM MGUS relative to the general population [45].

Familial WM has been described across multiple generations and affects both genders equally within families, implying an autosomal dominant or co-dominant inheritance [46]. A genome-wide linkage analysis including high-risk IgM MGUS and WM families identified evidence of susceptibility genes on chromosomes 1q and 4q, and possible linkage on chromosomes 3q and 6q [47]. Interestingly, it has been suggested that familial clustering of IgM MGUS and WM may be a consequence of the autosomal dominant inheritance of the hyperphosphorylated paratarg-7 protein, serving as an antigenic target in these patients [48]. Hyperphosphorylated paratarg-7 has not only been identified at higher frequencies in patients with IgM MGUS and WM, but in 11% of these patients, the IgM monoclonal protein reacted with the hyperphosphorylated paratarg-7 compared to 2% of the general population (OR 6.2, $p = 0.001$) [48]. Large population-based studies also found an association between a personal or family history of autoimmune diseases, most notably Sjögren syndrome and autoimmune hemolytic anemia, and WM [49]. Other studies showed a threefold higher risk of developing WM in patients with autoimmune diseases, chronic HCV infection, HIV, and rickettsiosis [50,51].

The clustering of WM in families and the increased risk of WM in patients with a personal or family history of autoimmune diseases or other B cell lymphoproliferative disorders further support the hypothesis that an abnormal response to an antigenic stimulation is the first step in the development of WM, as discussed above, coupled with common genetic factors that further increase the susceptibility to the development of this malignancy.

82.4 Diagnosis

The diagnosis of WM requires: (i) presence of IgM monoclonal gammopathy of any size, (ii) bone marrow biopsy with evidence of LPL infiltration, and (iii) an immunophenotype that excludes other lymphoproliferative disorders, including CLL and MCL. While the Second International Workshop criteria do not require a minimum bone marrow involvement by an LPL, the Mayo Clinic criteria emphasize the presence of at least 10% marrow involvement by an LPL (Table 82.1) [8,52]. The cutoff of 10% was introduced to facilitate the differentiation of WM from the precursor condition IgM MGUS, which carries a significantly different outcome. Recent studies also support the utilization of the 10% cutoff to differentiate between these two conditions [53–55].

While WM is a bone marrow–based disease, lymph node and splenic involvement is also common. Nodal involvement by LPL is characterized by paracortical and hilar infiltration with frequent sparing of the subscapular and marginal sinuses, while the bone marrow usually shows some combination of nodular, paratrabecular, and interstitial infiltration; in approximately one-half of cases, plasma cells containing Dutcher bodies are present [28].

82.5 Clinical Features

Approximately 20%–25% of WM patients are asymptomatic at the time of diagnosis and are classified as smoldering WM. Within 10 years from diagnosis, 70% of these smoldering patients will develop symptoms with a probability of progression of 12% per year for the

TABLE 82.1

Differential Diagnosis by the Mayo Clinic Criteria

Clinical Parameters	IgM MGUS	Smoldering WM	WM
Serum IgM (g/dL)	<3	≥3	Any level
Bone marrow LPL infiltrate (%)	<10%	≥10%	≥10%
Symptoms/End-organ damage	No	No	Yes
Risk of transformation	1.5%/year[a]	12%/year for the first 5 years, 70% within 10 years[a]	5%–10%[b]

[a] Risk of transformation to WM.

[b] Risk of transformation of an aggressive lymphoma, most commonly DLBCL.

first 5 years, then 2% per year for the next 5 years [56]. The initial manifestations of WM usually include nonspecific constitutional symptoms such as weight loss, fever, fatigue, and drenching night sweats. All symptoms and/or abnormalities seen can be attributable to either the IgM monoclonal protein (and its autoimmune activity) or tissue infiltration by LPL (Table 82.2) [28].

Anemia, seen in more than 30% of the patients, is multifactorial and secondary to bone marrow infiltration by LPL leading to inadequate erythropoiesis, IgM-associated hemolysis, and decreased levels of erythropoietin [57]. The production of erythropoietin is inversely related to plasma viscosity, which can be increased in WM patients with high levels of IgM paraprotein [58]. Overproduction of hepcidin has also been described as a contributing factor for anemia in patients with WM [59]. Hyperviscosity syndrome (HVS) has also been reported in approximately one-third of the patients. IgM immunoglobulins form large pentamers, restricted mostly to the intravascular compartment, leading to increased blood viscosity when IgM immunoglobulins are present in large quantity. The level of IgM that triggers HVS is variable among patients, as the final

TABLE 82.2

Clinical Features of WM

Mechanism	Organs or Tissues Affected	Clinical Features
IgM paraprotein mediated	Hyperviscosity (30%)	Mucocutaneous bleeding Visual disturbances Abnormal fundoscopy Dizziness Headache
	IgM-related neuropathy (35%)	Distal, symmetric, slowly progressive sensorimotor neuropathy Predominantly demyelinating features in nerve conduction studies
	Cryoglobulinemia (10%)	Raynaud phenomenon Acrocyanosis Vasculitis Renal failure
	Cold agglutinin disease (10%)	Extravascular hemolytic anemia after cold exposure
	AL amyloidosis (3%)	Specific to the end-organ affected
IgM paraprotein mediated and LPL infiltration related	Skin (3%)	Schnitzler syndrome Cutaneous plaques
	GI (4%)	Malabsorption Diarrhea
LPL infiltration related	Bone marrow (100%)	Anemia Thrombocytopenia Neutropenia
	Extramedullary hematopoietic tissues (30%)	Lymphadenopathy Splenomegaly Hepatomegaly
	Lung (4%)	Nodules or diffuse infiltrates Pleural effusion
	Kidney (4%)	Renal failure Peri-renal mass
	CNS (rare)	Bing–Neel syndrome

blood viscosity is influenced by multiple factors, such as hydration status and hemoglobin/RBC mass. The usual manifestations of HVS are mucocutaneous bleeding, visual disturbances, dizziness, headache, ataxia, tinnitus, exacerbation of heart failure, and rarely stroke or altered mental status [57,60]. Visual disturbances in HVS are usually secondary to retinal hemorrhages, papilledema, and central retinal vein thrombosis [61].

Although infrequent, organ damage may result from an IgM paraprotein with autoimmune activity or altered structure prone to tissue deposition [62]. Peripheral neuropathy, seen in approximately 25%–50% of patients [63], usually present as a distal, symmetric, and slowly progressive sensorimotor neuropathy with predominantly demyelinating features in nerve conduction studies [64]. Anti-myelin–associated glycoprotein (MAG) IgM antibodies have been implicated as the cause of neuropathy in half of these patients. IgM paraprotein directed to GD1b ganglioside, sulfatide, and chondroitin sulphate have also been described, although the causal relationship is not well established [64].

IgM paraprotein directed to erythrocyte antigens can lead to autoimmune hemolytic anemia. While 10% of patients may have a positive Coombs test, only 3% will develop significant hemolysis [57]. Immune thrombocytopenic purpura and acquired von Willebrand disease are rare but well-described manifestations in WM [65]. Cryoglobulinemia, manifested by Raynaud phenomenon, acrocyanosis, peripheral neuropathy, vasculitis, or renal failure, is the result of IgM immune-complex precipitation [66]. Cold agglutinin disease, manifested by extravascular hemolytic anemia after cold exposure, is the result of IgM binding to RBC surface antigens [67]. Both manifestations have been described in approximately 10% of patients with WM [57].

Extramedullary infiltration by LPL cells commonly results in lymphadenopathy and splenomegaly, which are seen in approximately 20%–25% of patients [57]. While LPL cells can infiltrate other organs, such as kidney and lung, such manifestations are unusual and are seen in less than 4% of patients. Unlike MM, osteolytic lesions are rare (<2%) and raise the suspicion for an alternative diagnosis, such as IgM MM [57]. Bing–Neel syndrome, a rare complication of WM, is caused by the direct invasion of the central nervous system by the LPL cells [63,68]. Schnitzler syndrome, an uncommon disease characterized by an IgM monoclonal gammopathy, chronic urticarial rash, and inflammatory symptoms, has been described in association with WM [69]. Upregulation of the IL-1 pathway plays a pivotal role in the pathophysiology of Schnitzler syndrome [70,71].

IgM-related AL amyloidosis, caused by extracellular tissue deposition of misfolded immunoglobulins, although uncommonly seen in WM, is a significant cause of morbidity and mortality in these patients [72]. AL-related symptoms are secondary to the impaired function of the organ involved by amyloid deposits. In comparison to non-IgM AL amyloidosis, higher rates of peripheral neuropathy and lower rates of cardiac involvement have been associated with IgM AL amyloidosis [73].

82.6 Prognosis

The International Prognostic Scoring System for Waldenström Macroglobulinemia (ISSWM) is used for stratification of patients with WM. Each of the five adverse parameters included

TABLE 82.3

International Prognostic Scoring System for WM

Parameter	Points
Hemoglobin ≤11.5 g/dL	1
Platelets ≤100 × 10⁹/L	1
β2-microglobulin >3 mg/L	1
IgM level >7 g/dL	1
Age >65 years	2

	IPSSWM Risk Category		
	Low	**Intermediate**	**High**
Score (points)	≤1	2	≥3
5-year survival	87%	68%	36%

in the IPSSWM score (age >65 years, hemoglobin ≤11.5 g/dL, platelets ≤100 × 10⁹/L, β2-microglobulin >3 mg/L, and IgM level >7 g/dL) are assigned 1 point, except for age >65 years, which is assigned 2 points. The total score, at the time of initiation of therapy, stratifies patients into low (score ≤1), intermediate (score = 2), or high-risk categories (score ≥3). Five-year survival rates are 87%, 68%, and 36% for low-, intermediate-, and high-risk patients, respectively (Table 82.3) [74]. Elevated lactate dehydrogenase (LDH) has been shown to further stratify high-risk WM patients. High-risk WM patients with elevated LDH had a median overall survival of 37 months compared to 104 months for those high-risk patients with low LDH [75].

82.7 Treatment

Since WM remains an incurable disease, the goals of therapy are to provide symptomatic relief and decrease the risk of end-organ damage while minimizing therapy-related toxicities. Treatment in general is reserved for symptomatic patients and those with severe cytopenias (hemoglobin <11 g/dL or platelet count <120 × 10⁹/L). A "watch-and-wait" approach is the preferred choice in asymptomatic patients given the risk of acute and long-term toxicities and lack of evidence suggesting improved survival with early therapy.

For patients presenting with HVS, plasma exchange is the preferred initial therapy and considered a category I indication by the American Society for Apheresis 2013 guidelines for therapeutic apheresis [76]. Plasma exchange, however, should be seen merely as a temporizing measure to alleviate symptoms. Systemic therapy is needed to decrease the tumor burden and the production of IgM which is ultimately causing the hyperviscosity symptoms.

Rituximab, an anti-CD20 monoclonal antibody, is the backbone of most regimens used in the treatment of WM. Rituximab monotherapy is an adequate choice for minimally symptomatic patients, such as those presenting with IgM-related neuropathy, mild cytopenias, or hemolytic anemia. The overall response rate (ORR) achieved with rituximab monotherapy is 50%–60% when used as first-line or salvage therapy, with a median progression-free survival (PFS) of approximately 2 years [77–79]. A transient increase in IgM levels (IgM flare) can occur after the first dose of rituximab. For patients with an IgM level >4000 mg/dL, plasma exchange could be considered prior to initiation of rituximab to decrease the risk of HVS from an IgM flare. When using rituximab in combination with chemotherapy, omitting rituximab for the first cycle is also an acceptable alternative strategy to prevent a rituximab flare [80].

Combination regimens of rituximab with chemotherapy agents or proteasome inhibitors are also commonly used treatment options in WM. In general, combination regimens have shown a better ORR and longer PFS than rituximab monotherapy, albeit with higher rates of toxicities [28]. Rituximab, cyclophosphamide, and dexamethasone (RCD) combination therapy has shown an ORR of 83% in treatment-naïve WM patients, with a median PFS of 3 years and estimated 10-year overall survival of 53%. Grade 3 or more adverse events were reported in only 9% of the patients [81,82].

Rituximab in combination with bortezomib, a proteasome inhibitor, has shown an ORR of 89% and 81% in treatment-naïve and relapsed/refractory patients with WM, respectively, with a median PFS of 16 months in the relapsed/refractory setting [83,84]. Rituximab, bortezomib, plus dexamethasone (BDR) as frontline therapy has achieved an ORR as high as 96% with a median PFS of 42 months. Grade ≥3 neuropathy has been reported in up to 30% of the patients [85,86]. The high rates of neuropathy secondary to bortezomib-containing regimen should be kept in mind when selecting a treatment for those WM patients with IgM-related neuropathy.

Rituximab in combination with bendamustine (BR), an alkylating agent, demonstrated an ORR of 90% with very high rates of complete remission (60%) in treatment-naïve WM patients. The most common adverse events were related to myelosuppression, with grade ≥3 leukopenia seen in 16% of the patients [87]. The median PFS of BR was 70 months, when compared to R-CHOP (rituximab, cyclophosphamide, doxorubicin, vincristine, and prednisone) in a phase III clinical trial [88]. Given the deep responses and long PFS, BR has become a preferred treatment option in patients with WM.

Ibrutinib, a BTK inhibitor, exploits the commonly seen MYD88[L265P] mutation in WM and the downstream activation of BTK [89]. Ibrutinib monotherapy has demonstrated an ORR of 90% in relapsed/refractory WM patients. A higher ORR was seen in patients harboring the MYD88[L265P] (100% in CXCR4 wild-type and 86% in CXCR4[WHIM] mutated patients) compared to MYD88 wild-type patients (ORR 60%). Complete remissions were not seen with ibrutinib monotherapy. Grade ≥3 neutropenia and thrombocytopenia were seen in 14% and 13% of the patients, respectively. Atrial fibrillation was reported in 5% of patients [90,91].

The role of autologous stem cell transplant (ASCT) has been reported in multiple retrospective studies [92–95]. With a non-relapse mortality of 3.8% at 1 year, ASCT is considered a safe treatment in this patient population [94]. Deep responses were seen after ASCT, with an ORR of approximately 95% and an estimated 5-year time-to-next therapy of 48% [21,94]. Harvesting of stem cells should be considered early in the disease course, prior to exposure to stem-cell toxic regimens in potential ASCT-candidates. In contrast to ASCT, allogeneic stem cell transplant is not recommended outside clinical trials due to the high transplant-related mortality (23%–44% at 1 year) in this indolent disease [93,96].

82.8 Conclusion

WM remains an incurable disease despite remarkable improvements in the understanding and treatment of this disease in recent years. Ibrutinib, the first FDA-approved drug specifically for WM, has inaugurated a new era in the treatment of this disease. Several other agents directed to potential key targets and pathways in WM, such as MYD88-BTK, IRAK–NF-κB, CXCL12/CXCR4, PI3Kδ, BCL-2, and PD-1/PDL-1, are under investigation. As the understanding and treatment of WM continue to evolve and expand rapidly, we can only anticipate a positive impact in the outcome of patients with WM.

REFERENCES

1. Waldenstrom J. Incipient myelomatosis or essential hyperglobulinemia with fibrinogenopenia — a new syndrome? *Acta Med Scand* 1944;117:216–47.
2. Campo E, Swerdlow SH, Harris NL, Pileri S, Stein H, Jaffe ES. The 2008 WHO classification of lymphoid neoplasms and beyond: Evolving concepts and practical applications. *Blood* 2011;117:5019–32.
3. Landgren O, Albitar M, Ma W et al. B-cell clones as early markers for chronic lymphocytic leukemia. *New Engl J Med* 2009;360:659–67.
4. Landgren O, Kyle RA, Pfeiffer RM et al. Monoclonal gammopathy of undetermined significance (MGUS) consistently precedes multiple myeloma: A prospective study. *Blood* 2009;113:5412–7.
5. McMaster ML, Caporaso N. Waldenstrom macroglobulinaemia and IgM monoclonal gammopathy of undetermined significance: Emerging understanding of a potential precursor condition. *Brit J Haematol* 2007;139:663–71.
6. The International Myeloma Working G. Criteria for the classification of monoclonal gammopathies, multiple myeloma and related disorders: A report of the International Myeloma Working Group. *Brit J Haematol* 2003;121:749–57.
7. Kyle RA, Benson J, Larson D et al. IgM monoclonal gammopathy of undetermined significance and smoldering Waldenstrom's macroglobulinemia. *Clin Lymphoma Myeloma* 2009;9:17–8.
8. Owen RG, Treon SP, Al-Katib A et al. Clinicopathological definition of Waldenstrom's macroglobulinemia: Consensus panel recommendations from the Second International Workshop on Waldenstrom's Macroglobulinemia. *Semin Oncol* 2003;30:110–5.
9. Kyle RA, Therneau TM, Rajkumar SV et al. Long-term follow-up of IgM monoclonal gammopathy of undetermined significance. *Blood* 2003;102:3759–64.
10. Paiva B, Montes MC, Garcia-Sanz R et al. Multiparameter flow cytometry for the identification of the Waldenstrom's clone in IgM-MGUS and Waldenstrom's Macroglobulinemia: New criteria for differential diagnosis and risk stratification. *Leukemia* 2014;28:166–73.
11. Jego G, Bataille R, Geffroy-Luseau A, Descamps G, Pellat-Deceunynck C. Pathogen-associated molecular patterns are growth and survival factors for human myeloma cells through Toll-like receptors. *Leukemia* 2006;20:1130–7.
12. Varettoni M, Zibellini S, Capello D et al. Clues to pathogenesis of Waldenstrom macroglobulinemia and immunoglobulin M

monoclonal gammopathy of undetermined significance provided by analysis of immunoglobulin heavy chain gene rearrangement and clustering of B-cell receptors. *Leuk Lymphoma* 2013;54:2485–9.

13. Rajkumar SV, Kyle RA, Buadi FK. Advances in the diagnosis, classification, risk stratification, and management of monoclonal gammopathy of undetermined significance: Implications for recategorizing disease entities in the presence of evolving scientific evidence. *Mayo Clin Proc* 2010;85:945–8.

14. Mansoor A, Medeiros LJ, Weber DM et al. Cytogenetic findings in lymphoplasmacytic lymphoma/Waldenstrom macroglobulinemia. Chromosomal abnormalities are associated with the polymorphous subtype and an aggressive clinical course. *Am J Clin Pathol* 2001;116:543–9.

15. Schop RF, Van Wier SA, Xu R et al. 6q deletion discriminates Waldenstrom macroglobulinemia from IgM monoclonal gammopathy of undetermined significance. *Cancer Genet Cytogenet* 2006;169:150–3.

16. Paiva B, Corchete LA, Vidriales MB et al. The cellular origin and malignant transformation of Waldenstrom macroglobulinemia. *Blood* 2015;125:2370–80.

17. Varettoni M, Zibellini S, Rizzo E et al. Targeted next generation sequencing identifies novel genetic mutations in patients with Waldenstrom's Macroglobulinemia/lymphoplasmacytic lymphoma or IgM monoclonal gammopathies of undetermined significance. *Blood Conference: 58th Annual Meeting of the American Society of Hematology, ASH* 2016;128:2928.

18. Lin P, Medeiros LJ. Lymphoplasmacytic lymphoma/waldenstrom macroglobulinemia: An evolving concept. *Adv Anat Pathol* 2005;12:246–55.

19. Kapoor P, Paludo J, Ansell SM. Waldenstrom macroglobulinemia: Familial predisposition and the role of genomics in prognosis and treatment selection. *Curr Treat Options Oncol* 2016;17:16.

20. San Miguel JF, Vidriales MB, Ocio E et al. Immunophenotypic analysis of Waldenstrom's macroglobulinemia. *Semin Oncol* 2003;30:187–95.

21. Varettoni M, Zibellini S, Defrancesco I et al. Pattern of somatic mutations in patients with Waldenström macroglobulinemia or IgM monoclonal gammopathy of undetermined significance. *Haematologica* 2017;102(12):2077–85.

22. Nguyen-Khac F, Lambert J, Chapiro E et al. Chromosomal aberrations and their prognostic value in a series of 174 untreated patients with Waldenstrom's macroglobulinemia. *Haematologica* 2013;98:649–54.

23. Braggio E, Philipsborn C, Novak A, Hodge L, Ansell S, Fonseca R. Molecular pathogenesis of Waldenstrom's macroglobulinemia. *Haematologica* 2012;97:1281–90.

24. Fonseca R, Braggio E. The MYDas touch of next-gen sequencing. *Blood* 2013;121:2373–4.

25. Shapiro-Shelef M, Calame K. Regulation of plasma-cell development. *Nat Rev Immunol* 2005;5:230–42.

26. Gutierrez NC, Ocio EM, de las Rivas J et al. Gene expression profiling of B lymphocytes and plasma cells from Waldenstrom's macroglobulinemia: Comparison with expression patterns of the same cell counterparts from chronic lymphocytic leukemia, multiple myeloma and normal individuals. *Leukemia* 2007;21:541–9.

27. Treon SP, Xu L, Yang G et al. MYD88 L265P somatic mutation in Waldenstrom's macroglobulinemia. *N Engl J Med* 2012;367:826–33.

28. Kapoor P, Paludo J, Vallumsetla N, Greipp PR. Waldenstrom macroglobulinemia: What a hematologist needs to know. *Blood Rev* 2015;29:301–19.

29. Xu L, Hunter ZR, Yang G et al. MYD88 L265P in Waldenstrom macroglobulinemia, immunoglobulin M monoclonal gammopathy, and other B-cell lymphoproliferative disorders using conventional and quantitative allele-specific polymerase chain reaction. *Blood* 2013;121:2051–8.

30. Gachard N, Parrens M, Soubeyran I et al. IGHV gene features and MYD88 L265P mutation separate the three marginal zone lymphoma entities and Waldenstrom macroglobulinemia/lymphoplasmacytic lymphomas. *Leukemia* 2013;27:183–9.

31. Pertesi M, Galia P, Nazaret N et al. Rare circulating cells in familial Waldenstrom macroglobulinemia displaying the MYD88 L265P mutation are enriched by Epstein-Barr virus immortalization. *PLoS One* 2015;10:e0136505.

32. Ngo VN, Young RM, Schmitz R et al. Oncogenically active MYD88 mutations in human lymphoma. *Nature* 2011;470:115–9.

33. Hunter ZR, Xu L, Yang G et al. The genomic landscape of Waldenstrom macroglobulinemia is characterized by highly recurring MYD88 and WHIM-like CXCR4 mutations, and small somatic deletions associated with B-cell lymphomagenesis. *Blood* 2014;123:1637–46.

34. Ngo HT, Leleu X, Lee J et al. SDF-1/CXCR4 and VLA-4 interaction regulates homing in Waldenstrom macroglobulinemia. *Blood* 2008;112:150–8.

35. Roccaro AM, Sacco A, Jimenez C et al. C1013G/CXCR4 acts as a driver mutation of tumor progression and modulator of drug resistance in lymphoplasmacytic lymphoma. *Blood* 2014;123:4120–31.

36. Cao Y, Yang G, Hunter ZR et al. The BCL2 antagonist ABT-199 triggers apoptosis, and augments ibrutinib and idelalisib mediated cytotoxicity in CXCR4 Wild-type and CXCR4 WHIM mutated Waldenstrom macroglobulinaemia cells. *Br J Haematol* 2015;170:134–8.

37. Groves FD, Travis LB, Devesa SS, Ries LA, Fraumeni JF, Jr. Waldenstrom's macroglobulinemia: Incidence patterns in the United States, 1988–1994. *Cancer* 1998;82:1078–81.

38. Wang H, Chen Y, Li F et al. Temporal and geographic variations of Waldenstrom macroglobulinemia incidence: A large population-based study. *Cancer* 2012;118:3793–800.

39. Ghobrial IM, Fonseca R, Gertz MA et al. Prognostic model for disease-specific and overall mortality in newly diagnosed symptomatic patients with Waldenstrom macroglobulinaemia. *Brit J Haematol* 2006;133:158–64.

40. Paludo J, Vallumsetla N, Ansell S et al. Survival trends in young patients with Waldenstrom Mmacroglobulinemia: Over 5 decades of experience. *Blood* 2016;128.

41. Kyle RA, Therneau TM, Rajkumar SV et al. A long-term study of prognosis in monoclonal gammopathy of undetermined significance. *N Engl J Med* 2002;346:564–9.

42. Morra E, Varettoni M, Tedeschi A et al. Associated cancers in Waldenstrom macroglobulinemia: Clues for common genetic predisposition. *Clin Lymphoma Myeloma Leuk* 2013;13:700–3.

43. Hanzis C, Ojha RP, Hunter Z et al. Associated malignancies in patients with Waldenstrom's macroglobulinemia and their kin. *Clin Lymphoma Myeloma Leuk* 2011;11:88–92.

44. Treon SP, Hunter ZR, Aggarwal A et al. Characterization of familial Waldenstrom's macroglobulinemia. *Ann Oncol* 2006;17:488–94.

45. McMaster ML. Familial Waldenstrom's macroglobulinemia. *Semin Oncol* 2003;30:146–52.

46. Kristinsson SY, Landgren O. What causes Waldenstrom's macroglobulinemia: Genetic or immune-related factors, or a combination? *Clin Lymphoma Myeloma Leuk* 2011;11:85–7.

47. McMaster ML, Goldin LR, Bai Y et al. Genomewide linkage screen for Waldenström macroglobulinemia susceptibility loci in high-risk families. *Am J Hum Genet* 2006;79:695–701.

48. Grass S, Preuss KD, Wikowicz A et al. Hyperphosphorylated paratarg-7: A new molecularly defined risk factor for monoclonal gammopathy of undetermined significance of the IgM type and Waldenstrom macroglobulinemia. *Blood* 2011;117:2918–23.

49. Kristinsson SY, Koshiol J, Bjorkholm M et al. Immune-related and inflammatory conditions and risk of lymphoplasmacytic lymphoma or Waldenstrom macroglobulinemia. *J Natl Cancer Inst* 2010;102:557–67.

50. Koshiol J, Gridley G, Engels EA, McMaster ML, Landgren O. Chronic immune stimulation and subsequent Waldenstrom macroglobulinemia. *Arch Intern Med* 2008;168:1903–9.

51. Giordano TP, Henderson L, Landgren O et al. Risk of non-Hodgkin lymphoma and lymphoproliferative precursor diseases in US veterans with hepatitis C virus. *JAMA* 2007;297:2010–7.

52. Ansell SM, Kyle RA, Reeder CB et al. Diagnosis and management of Waldenstrom macroglobulinemia: Mayo stratification of macroglobulinemia and risk-adapted therapy (mSMART) guidelines. *Mayo Clin Proc* 2010;85:824–33.

53. Paiva B, van Dongen JJ, Orfao A. New criteria for response assessment: Role of minimal residual disease in multiple myeloma. *Blood* 2015;125:3059–68.

54. Varettoni M, Arcaini L, Rattotti S, Ferretti V, Cazzola M. Bone marrow assessment in asymptomatic immunoglobulin M monoclonal gammopathies. *Br J Haematol* 2015;168:301–2.

55. Zou H, Yang R, Liao ZX et al. Serum markers in the differential diagnosis of Waldenstrom macroglobulinemia and other IgM monoclonal gammopathies. *J Clin Lab Anal* 2019;33(3):e22827.

56. Kyle RA, Benson JT, Larson DR et al. Progression in smoldering Waldenstrom macroglobulinemia: Long-term results. *Blood* 2012;119:4462–6.

57. Garcia-Sanz R, Montoto S, Torrequebrada A et al. Waldenstrom macroglobulinaemia: Presenting features and outcome in a series with 217 cases. *Br J Haematol* 2001;115:575–82.

58. Singh A, Eckardt KU, Zimmermann A et al. Increased plasma viscosity as a reason for inappropriate erythropoietin formation. *J Clin Invest* 1993;91:251–6.

59. Ciccarelli BT, Patterson CJ, Hunter ZR et al. Hepcidin is produced by lymphoplasmacytic cells and is associated with anemia in Waldenström's macroglobulinemia. *Clin Lymphoma Myeloma Leuk* 2011;11:160–3.

60. Stone MJ, Pascual V. Pathophysiology of Waldenstrom's macroglobulinemia. *Haematologica* 2010;95:359–64.

61. Menke MN, Feke GT, McMeel JW, Branagan A, Hunter Z, Treon SP. Hyperviscosity-related retinopathy in waldenstrom macroglobulinemia. *Arch Ophthalmol* 2006;124:1601–6.

62. van de Donk NW, Palumbo A, Johnsen HE et al. The clinical relevance and management of monoclonal gammopathy of undetermined significance and related disorders: Recommendations from the European Myeloma Network.

Haematologica 2014;99:984–96.

63. Baehring JM, Hochberg EP, Raje N, Ulrickson M, Hochberg FH. Neurological manifestations of Waldenstrom macroglobulinemia. *Nat Clin Pract Neurol* 2008;4:547–56.

64. Ramchandren S, Lewis RA. An update on monoclonal gammopathy and neuropathy. *Curr Neurol Neurosci Rep* 2012;12:102–10.

65. Rossi D, De Paoli L, Franceschetti S et al. Prevalence and clinical characteristics of immune thrombocytopenic purpura in a cohort of monoclonal gammopathy of uncertain significance. *Brit J Haematol* 2007;138:249–52.

66. Ramos-Casals M, Stone JH, Cid MC, Bosch X. The cryoglobulinaemias. *Lancet* 2012;379:348–60.

67. Berentsen S, Ulvestad E, Langholm R et al. Primary chronic cold agglutinin disease: A population based clinical study of 86 patients. *Haematologica* 2006;91:460–6.

68. Bing JaN, Neel AV. Two Cases of Hyperglobulinaemia with Affection of the Central Nervous System on a Toxi-Infectious Basis. *Acta Med Scand* 1936 88:492–506.

69. Sokumbi O, Drage LA, Peters MS. Clinical and histopathologic review of Schnitzler syndrome: The Mayo Clinic experience (1972–2011). *J Am Acad Dermatol* 2012;67:1289–95.

70. Simon A, Asli B, Braun-Falco M et al. Schnitzler's syndrome: Diagnosis, treatment, and follow-up. *Allergy* 2013;68:562–8.

71. Terpos E, Asli B, Christoulas D et al. Increased angiogenesis and enhanced bone formation in patients with IgM monoclonal gammopathy and urticarial skin rash: New insight into the biology of Schnitzler syndrome. *Haematologica* 2012;97:1699–703.

72. Cao XX, Meng Q, Mao YY et al. The clinical spectrum of IgM monoclonal gammopathy: A single center retrospective study of 377 patients. *Leuk Res* 2016;46:85–8.

73. Sachchithanantham S, Roussel M, Palladini G et al. European collaborative study defining clinical profile outcomes and novel prognostic criteria in monoclonal immunoglobulin M-related light chain amyloidosis. *J Clin Oncol* 2016;34:2037–45.

74. Morel P, Duhamel A, Gobbi P et al. International prognostic scoring system for Waldenstrom macroglobulinemia. *Blood* 2009;113:4163–70.

75. Kastritis E, Kyrtsonis MC, Hadjiharissi E et al. Validation of the International Prognostic Scoring System (IPSS) for Waldenstrom's macroglobulinemia (WM) and the importance of serum lactate dehydrogenase (LDH). *Leuk Res* 2010;34:1340–3.

76. Schwartz J, Winters JL, Padmanabhan A et al. Guidelines on the use of therapeutic apheresis in clinical practice-evidence-based approach from the Writing Committee of the American Society for Apheresis: The sixth special issue. *J Clin Apher* 2013;28:145–284.

77. Gertz MA, Abonour R, Heffner LT, Greipp PR, Uno H, Rajkumar SV. Clinical value of minor responses after 4 doses of rituximab in Waldenstrom macroglobulinaemia: A follow-up of the Eastern Cooperative Oncology Group E3A98 trial. *Br J Haematol* 2009;147:677–80.

78. Treon SP, Emmanouilides C, Kimby E et al. Extended rituximab therapy in Waldenstrom's macroglobulinemia. *Ann Oncol* 2005;16:132–8.

79. Gertz MA, Rue M, Blood E, Kaminer LS, Vesole DH, Greipp PR. Multicenter phase 2 trial of rituximab for Waldenstrom macroglobulinemia (WM): An Eastern Cooperative Oncology Group Study (E3A98). *Leuk Lymphoma* 2004;45:2047–55.

80. Leblond V, Kastritis E, Advani R et al. Treatment recommendations from the Eighth International Workshop on Waldenstrom's Macroglobulinemia. *Blood* 2016;128:1321–8.

81. Dimopoulos MA, Anagnostopoulos A, Kyrtsonis MC et al. Primary treatment of Waldenstrom macroglobulinemia with dexamethasone, rituximab, and cyclophosphamide. *J Clin Oncol* 2007;25:3344–9.

82. Kastritis E, Gavriatopoulou M, Kyrtsonis MC et al. Dexamethasone, rituximab, and cyclophosphamide as primary treatment of Waldenstrom macroglobulinemia: Final analysis of a phase 2 study. *Blood* 2015;126:1392–4.

83. Ghobrial IM, Xie W, Padmanabhan S et al. Phase II trial of weekly bortezomib in combination with rituximab in untreated patients with Waldenstrom macroglobulinemia. *Am J Hematol* 2010;85:670–4.

84. Ghobrial IM, Hong F, Padmanabhan S et al. Phase II trial of weekly bortezomib in combination with rituximab in relapsed or relapsed and refractory Waldenstrom macroglobulinemia. *J Clin Oncol* 2010;28:1422–8.

85. Dimopoulos MA, Garcia-Sanz R, Gavriatopoulou M et al. Primary therapy of Waldenstrom macroglobulinemia (WM) with weekly bortezomib, low-dose dexamethasone, and rituximab (BDR): Long-term results of a phase 2 study of the European Myeloma Network (EMN). *Blood* 2013;122:3276–82.

86. Treon SP, Ioakimidis L, Soumerai JD et al. Primary therapy of Waldenstrom macroglobulinemia with bortezomib, dexamethasone, and rituximab: WMCTG clinical trial 05-180. *J Clin Oncol* 2009;27:3830–5.

87. Rummel MJ, Al-Batran SE, Kim SZ et al. Bendamustine plus rituximab is effective and has a favorable toxicity profile in the treatment of mantle cell and low-grade non-Hodgkin's lymphoma. *J Clin Oncol* 2005;23:3383–9.

88. Rummel MJ, Niederle N, Maschmeyer G et al. Bendamustine plus rituximab versus CHOP plus rituximab as first-line treatment for patients with indolent and mantle-cell lymphomas: An open-label, multicentre, randomised, phase 3 non-inferiority trial. *Lancet* 2013;381:1203–10.

89. Buggy JJ, Elias L. Bruton tyrosine kinase (BTK) and its role in B-cell malignancy. *Int Rev Immunol* 2012;31:119–32.

90. Treon SP, Tripsas CK, Meid K et al. Ibrutinib in previously treated Waldenstrom's macroglobulinemia. *N Engl J Med* 2015;372:1430–40.

91. Treon SP, Xu L, Hunter Z. MYD88 mutations and response to ibrutinib in Waldenstrom's macroglobulinemia. *N Engl J Med* 2015;373:584–6.

92. Anagnostopoulos A, Hari PN, Perez WS et al. Autologous or allogeneic stem cell transplantation in patients with Waldenstrom's macroglobulinemia. *Biol Blood Marrow Transplant* 2006;12:845–54.

93. Gilleece MH, Pearce R, Linch DC et al. The outcome of haemopoietic stem cell transplantation in the treatment of lymphoplasmacytic lymphoma in the UK: A British Society Bone Marrow Transplantation study. *Hematology* 2008;13:119–27.

94. Kyriakou C, Canals C, Sibon D et al. High-dose therapy and autologous stem-cell transplantation in Waldenstrom macroglobulinemia: The Lymphoma Working Party of the European Group for Blood and Marrow Transplantation. *J Clin Oncol* 2010;28:2227–32.

95. Paludo J, Gertz M, Ansell S et al. Impact of Day-100 Response Post Autologous Stem Cell Transplantation (ASCT) in Waldenstrom Macroglobulinemia (WM). *Biol Blood Marrow Transplant* 2016;22:S130.

96. Kyriakou C, Canals C, Cornelissen JJ et al. Allogeneic stem-cell transplantation in patients with Waldenstrom macroglobulinemia: Report from the Lymphoma Working Party of the European Group for Blood and Marrow Transplantation. *J Clin Oncol* 2010;28:4926–34.

83

Wiskott–Aldrich Syndrome

Dongyou Liu

CONTENTS

83.1 Introduction

Wiskott–Aldrich syndrome (WAS) is a rare X-linked immunodeficiency disorder characterized by a triad of eczema (abnormal patches of red, irritated skin due to inflammatory reaction), microthrombocytopenia (a decrease in the number and size of platelets involved in clotting), and recurrent middle ear infections (due to abnormal or nonfunctional leukocytes). Typically present from birth, and often affecting males, this platelet abnormality may result in easy bruising, bloody diarrhea, episodes of prolonged bleeding following minor trauma, purpura (purplish spots) or petechiae (rashes of tiny red spots due to bleeding in small areas under the surface of the skin), autoimmune disorders (e.g., rheumatoid arthritis, hemolytic anemia), and increased risk of lymphoma [1].

Molecularly, WAS is attributed to mutations in the *WAS* gene on chromosome Xp11.23, encoding a protein (WASP) involved in actin polymerization, receptor engagement, signaling transduction, and cytoskeletal rearrangement.

83.2 Biology

WAS was first described by Alfred Wiskott in 1937 from three brothers with childhood pneumonias, and shown to be an X-linked disorder by Robert Aldrich and coworkers in 1954.

Subsequent studies indicate that WAS comprises a broad spectrum of hematopoietic diseases, ranging from a milder X-linked thrombocytopenia (XLT, which shows milder eczema and immune dysfunction and better overall survival than WAS), to

a more severe X-linked congenital neutropenia (X-linked neutropenia [XLN], which is also called X-linked severe congenital neutropenia [SCNX]) (see Chapter 77 for further details) [1].

The gene underlying the development of WAS was identified as *WAS* on chromosome Xp11.23 by Derry and coworkers in 1994 using a positional cloning strategy. With expression limited to lymphocytic and megakaryocytic cell lineages, the *WAS* gene is altered in patients with WAS, XLT, and XLN, leading to predominant defects of platelets and lymphocytes. Because of their shared genetic background and their phenotypic overlapping, WAS, XLT, and XLN (SCNX) are collectively referred to as WAS-related disorders [1–3].

83.3 Pathogenesis

The *WAS* gene on the short arm of the X chromosome at Xp11.23 spans >9 kb with 12 exons and encodes a 502 aa protein (Wiskott–Aldrich syndrome protein [WASP]), which is expressed mainly in nonerythroid hematopoietic cells and involved in actin polymerization, receptor engagement, signaling transduction from the surface of blood cells to the actin cytoskeleton, and cytoskeletal rearrangement.

Structurally, WASP consists of five domains:

WH1: WASP homology 1 or EVH1, for ENA/VASP homology 1, which binds to a proline repeat motif in WASP-interacting protein (WIP) and mediates a molecular interaction critical for keeping a stable and autoinhibited WASP conformation

B: Basic domain, which binds to the phosphoinositide PIP2 (phosphatidylinositol-4,5-biphosphate) for the activation of WASP

GBD: GTPase binding domain, which interacts with VCA

PPP: Polyproline domain, which contains several sites for binding of Src homology 3 (SH3) domain

VCA: Verprolin-homology–cofilin homology domain–acidic region, which forms the actin-nucleating region of WASP) [4–6]

Mutations in the *WAS* gene alter the amino acid sequences of its encoded protein, leading to truncation or absence of WASP. To date, >160 different *WAS* mutations across all 12 exons are identified from >270 unrelated families [7,8]. Nearly half of *WAS* pathogenic variants are missense mutations that result in protein truncation, while the remaining pathogenic variants are small deletions/insertions, splicing variants, gross deletions/insertions, and complex rearrangements. As a consequence, T and B lymphocytes, neutrophils, macrophages, and dendritic cells from WAS-affected males exhibit defects in migration, anchoring, and localization. In particular, white blood cells lacking WASP are unable to move and attach to other cells and tissues (adhesion), to form immune synapses with foreign invaders, and to respond to environmental challenges, whereas platelets with dysfunctional WASP develop poorly, leading to reduced size and early cell death. Together, these changes underscore the development of classic WAS and its associated phenotypic variations, XLT and XLN [1,9–11].

83.4 Epidemiology

83.4.1 Prevalence

WAS has an estimated prevalence of one to four cases per million males worldwide, and rarely affects females. The average age at diagnosis is 24 months in families without a previously affected family member.

83.4.2 Inheritance

WAS shows an X-linked inheritance, with the mutated *WAS* gene that causes the disorder located on the X chromosome. As males have only one X chromosome, a mutation in the only copy of the *WAS* gene in each cell is sufficient to cause the disease. By contrast, females have two copies of the X chromosome; one altered copy of the *WAS* gene in each cell may cause no signs or symptoms at all or only milder features of the disease. It is notable that fathers with X-linked disorder do not pass the trait to their sons [1].

83.4.3 Penetrance

WAS demonstrates complete penetrance in males and varied clinical expressivity, ranging from mild XLT and classic WAS to severe XLN (SCNX).

83.5 Clinical Features

WAS typically manifests as a triad of eczema (80% of male patients), microthrombocytopenia, and recurrent middle ear infections, along with other features such as easy bruising, bloody diarrhea, episodes of prolonged bleeding following minor trauma, purpura (purplish spots) or petechiae (rashes of tiny red spots), autoimmune disorders [e.g., autoimmune hemolytic anemia (14%), vasculitis (13%), renal disease (12%), and chronic arthritis (10%)], and increased risk of lymphoma (particularly non-Hodgkin lymphoma; 13% of cases at an average age of 9.5 years) [1].

Depending on the presence or absence of clinical complications or WASP, as well as differences in clinical scores (derived from clinical parameters, including the presence of thrombocytopenia, eczema, immunodeficiency, autoimmunity, and malignancy; with score 0 for those individuals with XLN and/or myelodysplasia, score of 1 or 2 for individuals with XLT, score 3 to 4 for individuals with classic WAS, and score 5 for individuals with either XLT or WAS together with autoimmunity and/or malignancies), WAS may be referred to as classis WAS, XLT, and X-linked neutropenia (XLN or SCNX). Compared to classic WAS, males with XLT tend to have small platelet size and volume (<7.5 fL) and intermittent thrombocytopenia (5000–50,000 platelets/mm³) without other WAS findings, while males with XLN develop recurrent bacterial infection, persistent neutropenia, arrested development of the bone marrow without other WAS findings, and 20%–30% of risk for MDS or AML. Nonetheless, due to obvious phenotypic overlapping, WAS, XLT, and XLN should be considered clinical continuum rather than discrete diseases [1,12].

Female carriers of a *WAS* pathogenic variant rarely develop significant clinical symptoms (e.g., severe thrombocytopenia and/or immunologic dysfunction), although a small proportion may show mild thrombocytopenia.

83.6 Diagnosis

Diagnosis of WAS involves medical history review (past history of rash/eczema, bleeding, infection, autoimmune disease, malignancy; family history of X-linked disorder), physical examination (for rash/eczema, bleeding/petechiae/ecchymoses), laboratory evaluation (complete blood cell count for anemia, microcytosis, thrombocytopenia, low mean platelet volume; peripheral blood smear for microthrombocytopenia; serology for low serum IgG, IgA, IgM, and high serum IgE; abnormal isohemagglutinin titer and diminished vaccine responses; T cell lymphopenia and abnormal mitogen response), histopathology (for lymphoma) and molecular testing (for WAS pathogenic variants) [1,13].

Males showing the following clinical and laboratory features should be suspected of WAS: (i) profound thrombocytopenia (<70,000 platelets/mm³); (ii) small platelet size (mean platelet volume >2 SD below the mean for the laboratory); (iii) recurrent bacterial or viral infection or opportunistic infection in infancy or early childhood; (iv) eczema; (v) autoimmune disorder; (vi) lymphoma; (vii) family history of one or more maternally related males with a *WAS*-related phenotype; (viii) absent or decreased intracellular WASP detection in hematopoietic cells as determined by flow cytometry or Western blotting; (ix) abnormal lymphocytes (decreased T cell subsets, especially proportion and absolute number of CD8+T cells; decreased NK cell function; decreased IgM, normal or decreased IgG, increased IgA, increased

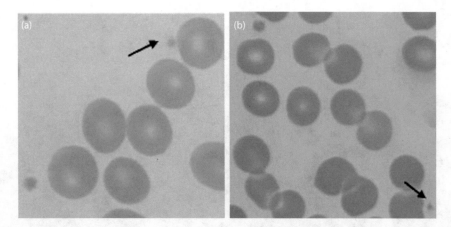

FIGURE 83.1 Wiskott–Aldrich syndrome displaying both normal-sized and small platelets (arrow) in an 18-month-old boy with thrombocytopenia, history of spontaneous mucocutaneous bleeds, and a mutation in WAS exon 2, leading to p.Thr45Met amino acid change (a), and a 2-year-old boy with thrombocytopenia, history of bruises after trauma, and a missense mutation in exon 2, leading to p.Pro58Leu amino acid change (b). (Photo credit: Medina SS et al. *BMC Pediatr.* 2017;17(1):151.)

IgE; absent isohemagglutinins; absent or greatly decreased antibody responses to polysaccharide vaccines) (Figure 83.1) [1,14].

Diagnosis of WAS in males is based on observation of congenital thrombocytopenia (<70,000 platelets/mm^3) and small platelets (platelet volume <7.5 fL), together with at least one of the following features: eczema, recurrent infections, autoimmune disease, malignancy, reduced WASP expression in a fresh blood sample, abnormal antibody response to polysaccharide antigens and/or low isohemagglutinins, or positive maternal family history of WAS. Identification of a hemizygous *WAS* pathogenic variant by sequence analysis (95% of cases) and gene-targeted deletion/duplication analysis (5% of cases) further confirms the diagnosis [1].

Differential diagnoses for WAS include idiopathic thrombocytopenic purpura (ITP; males presenting early in life with transient and self-limited thrombocytopenia; increased platelet size, and increased reticulated platelet count), Wiskott–Aldrich syndrome 2 (WAS2; recurrent infections, eczema, and thrombocytopenia; show low numbers of B and T cells, defective T cell proliferation and chemotaxis, low NK cell function, and abnormal WASP; normal platelet volumes; rare autosomal recessive immunodeficiency due to biallelic pathogenic variants in *WIPF1* on chromosome 2q31.1), X-linked severe combined immunodeficiency (X-SCID, persistent infections, lymphocytopenia, growth failure, and thymic hypoplasia; near-complete absence of T and natural killer lymphocytes; nonfunctional B lymphocytes; due to hemizygous pathogenic variant in the common gamma chain gene *IL2RG*), X-linked hyper IgM syndrome (recurrent otitis media, sinusitis, pneumonias; autoimmune hematologic disorders such as neutropenia, thrombocytopenia, and hemolytic anemia; lymphomas and other malignancies; serious gastrointestinal complications, and neurologic deterioration; elevated IgM in the absence of other immunoglobulins; due to pathogenic variants in *CD40LG*), autosomal recessive severe combined immunodeficiencies (T- and B-cell dysfunction, recurrent infections, but rarely persistent thrombocytopenia), GATA1-related X-linked cytopenia (thrombocytopenia [easy bruising and mucosal bleeding, such as epistaxis], anemia [mild dyserythropoiesis, hydrops fetalis requiring in utero transfusion]; platelet dysfunction, mild β-thalassemia, neutropenia, and congenital erythropoietic porphyria [CEP] in males), and human immunodeficiency virus (HIV; gradual destruction of the immune system; risk for opportunistic infections and neoplasms) [1,15].

83.7 Treatment

Treatment options for WAS include topical steroids and for eczema, antibiotics for infections, immunomodulatory therapy (e.g., IVIG, cyclosporine, azathioprine, cyclophosphamide, and plasma exchange) for autoimmune disease, granulocyte colony stimulating factor (G-CSF) and antibiotics for neutropenia, and hematopoietic stem cell transplantation (HSCT, which has a cure rate of 90%) for hematologic diseases (Figure 83.2) [16–20]. While early HSCT is indicated for classic WAS, it remains a challenge to determine the best treatment approach for milder WAS due to varying disease severity [21].

83.8 Prognosis and Prevention

Prognosis for patients with WAS is generally a poor, with infection (44%), malignancy (26%), and bleeding (23%) being the main causes of death. Affected children have a life expectancy of 8–14.5 years without HSCT. Long-term survival following allogeneic HSCT is >80% [22].

Medications that interfere with platelet function should be avoided. Elective surgical procedures should be ideally deferred until after HSCT. Prenatal testing and preimplantation genetic diagnosis may be offered to prospective parents who harbor *WAS* pathogenic variant.

83.9 Conclusion

Wiskott–Aldrich syndrome (WAS) is a rare X-linked primary immunodeficiency disorder that is attributable to mutations in the *WAS* gene on chromosome Xp11.23. Given its role in actin

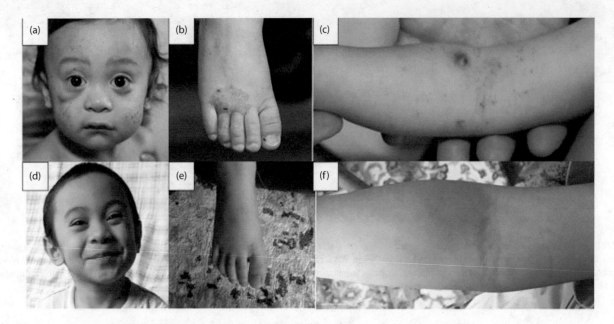

FIGURE 83.2 Wiskott–Aldrich syndrome in an 18-month-old boy showing multiple petechiae and hematoma on the face (a) and hemorrhagic eczema on the right foot and arm (b, c) prior to HLA-haploidentical hematopoietic stem cell transplantation (HSCT); and the same boy free of skin lesions (d, e, f) 36 months post HSCT. (Photo credit: Albert MH, Freeman AF. *Front Pediatr.* 2019;7:451.)

polymerization, receptor engagement, signaling transduction, and cytoskeletal rearrangement, truncation or absence of WASP caused by *WAS* missense or splicing variants, small or gross deletions/insertions, and complex rearrangements has serious consequences. This is exemplified by the decreased capacity of white blood cells to form immune synapses with foreign invaders, and the poor development and early death of platelets, leading to the triad of eczema, microthrombocytopenia, and recurrent middle ear infections, along with easy bruising/bleeding, bloody diarrhea, autoimmune disorders, and lymphoma, particularly in males. Establishment of WAS diagnosis requires the findings of congenital thrombocytopenia (<70,000 platelets/mm^3), small platelets (platelet volume <7.5 fL), and at least one of other features (eczema, recurrent infections, autoimmune disease, malignancy, reduced WASP expression in a fresh blood sample, abnormal antibody response to polysaccharide antigens and/or low isohemagglutinins, or positive maternal family history of WAS). Further confirmation is obtained with molecular detection of a hemizygous *WAS* pathogenic variant. Management of WAS involves topical steroids for eczema, antibiotics for infections, immunomodulatory therapy for autoimmune disease, G-CSF and antibiotics for neutropenia, and HSCT for hematologic diseases [20,21].

REFERENCES

1. Chandra S, Bronicki L, Nagaraj CB, Zhang K. WAS-related disorders. In: Adam MP, Ardinger HH, Pagon RA et al., editors. *GeneReviews®* [Internet]. Seattle (WA): University of Washington, Seattle; 1993–2018. 2004 Sep 30 [updated 2016 Sep 22].
2. Beel K, Cotter MM, Blatny J et al. A large kindred with X-linked neutropenia with an I294 T mutation of the Wiskott–Aldrich syndrome gene. *Brit J Haemat.* 2008;144:120–6.
3. Ariga T. Wiskott–Aldrich syndrome; an x-linked primary immunodeficiency disease with unique and characteristic features. *Allergol Int.* 2012;61(2):183–9.
4. Tyler JJ, Allwood EG, Ayscough KR. WASP family proteins, more than Arp2/3 activators. *Biochem Soc Trans.* 2016;44(5):1339–45.
5. Zhan J, Johnson IM, Wielgosz M, Nienhuis AW. The identification of hematopoietic-specific regulatory elements for *WASp* gene expression. *Mol Ther Methods Clin Dev.* 2016;3:16077.
6. Alekhina O, Burstein E, Billadeau DD. Cellular functions of WASP family proteins at a glance. *J Cell Sci.* 2017;130(14): 2235–41.
7. Esmaeilzadeh H, Bordbar MR, Dastsooz H et al. A novel splice site mutation in WAS gene in patient with Wiskott–Aldrich syndrome and chronic colitis: A case report. *BMC Med Genet.* 2018;19(1):123.
8. Kaya Z, Muluk C, Hakoloğlu Ş, Tufan LŞ. A novel mutation in a child with atypical Wiskott–Aldrich syndrome complicated by cytomegalovirus infection. *Turk J Haematol.* 2019;36(1):70–1.
9. Westerberg LS, Meelu P, Baptista M et al. Activating WASP mutations associated with X-linked neutropenia result in enhanced actin polymerization, altered cytoskeletal responses, and genomic instability in lymphocytes. *J Exp Med.* 2010;207:1145–52.
10. Petersen SH, Sendel A, van der Burg M, Westerberg LS. Unraveling the repertoire in Wiskott–Aldrich syndrome. *Front Immunol.* 2014;5:539.
11. Cotta-de-Almeida V, Dupré L, Guipouy D, Vasconcelos Z. Signal integration during T lymphocyte activation and function: Lessons from the Wiskott–Aldrich syndrome. *Front Immunol.* 2015;6:47.
12. Lanzi G, Moratto D, Vairo D et al. A novel primary human immunodeficiency due to deficiency in the WASP-interacting protein WIP. *J Exp Med.* 2012;209:29–34.

13. Buchbinder D, Nugent DJ, Fillipovich AH. Wiskott-Aldrich syndrome: Diagnosis, current management, and emerging treatments. *Appl Clin Genet.* 2014;7:55–66.

14. Medina SS, Siqueira LH, Colella MP et al. Intermittent low platelet counts hampering diagnosis of X-linked thrombocytopenia in children: Report of two unrelated cases and a novel mutation in the gene coding for the Wiskott–Aldrich syndrome protein. *BMC Pediatr.* 2017;17(1):151.

15. Balduin CL, Savoia A. Genetics of familial forms of thrombocytopenia. *Hum Genet.* 2012;131:1821–32.

16. Boztug K, Schmidt M, Schwarzer A et al. Stem-cell gene therapy for the Wiskott–Aldrich syndrome. *N Engl J Med.* 2010;363:1918–27.

17. Pai SY, Notarangelo LD. Hematopoietic cell transplantation for Wiskott–Aldrich syndrome: Advances in biology and future directions for treatment. *Immunol Allergy Clin North Am.* 2010;30:179–94.

18. Aiuti A, Biasco L, Scaramuzza S et al. Lentiviral hematopoietic stem cell gene therapy in patients with Wiskott–Aldrich syndrome. *Science.* 2013;341:1233151.

19. Hacein-Bey Abina S, Gaspar HB, Blondeau J et al. Outcomes following gene therapy in patients with severe Wiskott–Aldrich syndrome. *JAMA.* 2015;313:1550–63.

20. Albert MH, Freeman AF. Wiskott–Aldrich syndrome (WAS) and dedicator of cytokinesis 8- (DOCK8) deficiency. *Front Pediatr.* 2019;7:451.

21. Malik MA, Masab M. Wiskott–Aldrich Syndrome. [Updated 2019 Jun 22]. In: StatPearls [Internet]. Treasure Island (FL): StatPearls Publishing; 2020; Jan-.

22. Moratto D, Giliani S, Bonfim C et al. Long-term outcome and lineage-specific chimerism in 194 patients with Wiskott–Aldrich syndrome treated by hematopoietic cell transplantation in the period 1980–2009: An international collaborative study. *Blood.* 2011;118:1675–84.

Section VII

Overgrowth Syndromes, PTENopathies, and RASopathies

84

Bannayan–Riley–Ruvalcaba Syndrome

Gabriela Maria Abreu Gontijo and Clóvis Antônio Lopes Pinto

CONTENTS

84.1 Introduction

Bannayan–Riley–Ruvalcaba syndrome (BRRS, sometimes referred to as Bannayan–Zonana syndrome, Myhre–Riley–Smith syndrome, Riley–Smith syndrome, Ruvalcaba–Myhre–Smith syndrome, and Ruvalcaba–Myhre syndrome) is a rare congenital disorder that typically displays macrocephaly, developmental delay, lipomas, hemangiomas, and pigmented macules of the genitalia [1].

BRRS is inherited in an autosomal dominant fashion, although approximately 37% of cases have no family history and appear to evolve from spontaneous, de novo mutations [2,3].

It is estimated that in 60% of the individuals who are clinically diagnosed with BRRS germline, inactivating mutations of phosphatase and tensin homolog (*PTEN*) gene are identified [1,4,5]. Among those who are mutation "free," large deletions of *PTEN* are found in 11% [6].

Among patients diagnosed with BRRS or Cowden syndrome (CS), identical *PTEN* germline mutations are observed [5–7]. There are also families with individual members manifesting both CS/BRRS phenotypes. Given this clinical and genetic overlap, BRRS and CS are nowadays acknowledged as different phenotypic expressions of one allelic syndrome known as PTEN hamartoma tumor syndrome (PHTS), which also includes Proteus-like syndrome (PLS) and *PTEN*-related Proteus syndrome (PS) [8]. When a *PTEN* pathogenic variant is detected in any of these phenotypes, the term PHTS is preferentially used.

Hamartomas are benign tumors, which are composed of cells with histologically normal differentiation but disorganized architecture [1]. In the hamartomatous syndromes, there is an overgrowth of cells original to the area in which they normally occur [2].

84.2 Biological Background

The *PTEN* gene (also known as TEP1 [TGF-b regulated and epithelial cell-enriched phosphatase], or MMAC1 [mutated in multiple advanced cancers]) is located on chromosome 10q23.31 and contains nine exons that span a genome distance of >120 kb [1]. Its coding sequence of 1209-bp is predicted to generate a 403-aa, 47 kDa dual-specificity phosphatase (PTEN), which consists of two major functional domains, an N-terminal domain from exons 1–6 and a C-terminal domain from exons 6–9.

Germline mutations that result in *PTEN* gene inactivation of are found in approximately 60% of individuals who meet the clinical diagnosis of BRRS [1,4,5]. Zhou et al. reported that 11% of BRRS-related mutations include large deletions, often favoring the end of the gene containing exon 1 and the promoter region [6]. Blumenthal et al. hypothesized that gross gene deletions and mutations in the *PTEN* promoter might answer for a portion of patients who allegedly are mutation-negative with a diagnosis of CS and BRRS. It appears that *PTEN*-mutation-negative CS and BRRS would be the result of large gene rearrangements and deletions, which conventional techniques cannot detect, and promoter mutations [1].

Being a duple-specificity tumor suppressor phosphatase, PTEN dephosphorylates lipid and protein substrates [4,6]. The wild-type protein is a lipid phosphatase that negatively regulates the PI3K/Akt/mTOR pathway to cause G1 arrest and/or apoptosis [1,9]. In addition, it seems that the protein phosphatase plays a relevant role in inhibition of cell migration and enlargement, additionally downregulating several cell cyclins [9].

In various sporadic human cancers, including breast, brain, prostate, bladder, colon, endometrium, and lung, somatic loss of PTEN function through mutation, deletion, or methylation was described [9].

84.3 Pathogenesis

PTEN exerts its lipid phosphatase activity by dephosphorylating the 30-phosphoinositide products of PI3K (phosphatidylinositol-4,5-bisphosphate 3-kinase), producing conversion of phosphatidylinositol-3,4,5 trisphosphate to phosphatidylinositol-4,5 bisphosphate and conversion of phosphatidylinositol-3,4 bisphosphate to phosphatidylinositol-4 phosphate. Diminution of 30-phosphoinositides decrement activity of kinases downstream of PI3K such as phosphoinositide-dependent kinase 1, Akt (protein kinase B), and mTOR (mechanistic target of rapamycin), and is liable for its tumor suppressor activity [1]. As a result of negative regulation of the Akt pathway, PTEN decreases phosphorylation of other substrates downstream of Akt such as p27, p21, GSK-3, Bad, ASK-1, as well as representatives of the Forkhead transcription factor family [1]. PTEN antagonizes the PI3K/AKT pathway by decreasing phosphatidylinositol-3,4,5 trisphosphate levels, resulting in a decreased translocation of AKT to cellular membranes and subsequent downregulation of AKT activation [10]. It has been shown that expression of PTEN in cells leads to decreased levels of phospho-AKT, and consequently to augmented apoptosis [10].

PTEN is a protein phosphatase, producing dephosphorylation of focal adhesion kinase, which can inhibit cell migration and cell spreading [11]. The protein phosphatase activity of PTEN can also regulate the mitogen-activated protein kinase pathway, hence regulating cell survival [12,13].

About 66% of germline mutations in PTEN result in dysfunctional protein, lack of protein, or truncated protein [9]. Thus, when PTEN is absent, decreased, or dysfunctional, enlarged phosphorylation of many key cellular proteins occur, which in sequence affect processes such as cell cycle progression, migration, metabolism, transcription, apoptosis, and translation [1,9].

84.4 Epidemiology

84.4.1 Prevalence

BRRS is a rare inherited disorder with unknown prevalence. To date, several dozen cases have been reported in the literature. There is a general belief that BRRS is underdiagnosed due to its varied and sometimes subtle clinical signs and symptoms.

84.4.2 Inheritance

BRRS is an autosomal dominant disorder, for which one copy of the altered *PTEN* gene in each cell is sufficient to cause the disease.

84.4.3 Penetrance

BRRS demonstrates variable expression and age-related penetrance, with some clinical manifestations appearing by the late 20s, and a more comprehensive clinical spectrum by the third decade.

84.5 Clinical Features

BRRS is associated with a vast spectrum of phenotypic characters: some patients present with severe clinical symptoms, while others can go unnoticed.

The hallmarks of BRRS are macrocephaly, intestinal polyposis, lipomas, and pigmented macule in the glans penis [1,14].

The only altered prenatal ultrasound finding might be macrocephaly or macrosomia [15]. In some cases, this circumstance may result in an elective cesarean section [15]. At birth, affected infants are typically macrosomic, with weight exceeding 4 kg, length and head circumference above the 97th percentile [16–18]. Subsequent growth deceleration results in normalization of all growth parameters except macrocephaly [18].

The most specific feature of the syndrome is the speckled penis, described as the presence of pigmented maculae on the glans penis in most male patients (Figure 84.1). These maculae may be present at birth or emerge later in childhood [16]. When evaluating these patients, physicians must be aware that these lesions are subtle and should be meticulously looked for. Histologically, BRRS shows lentiginous epidermal hyperplasia with increased numbers of melanosomes and a slight increase in the number of melanocytes in the basal layer [5,19].

Hamartomatous polyps occur in up to 45% of the patients and are located in the gastrointestinal tract [2,16]. Usually there are multiple polyps limited to the distal ileum and colon, though any part of the gastrointestinal tract may be involved [2].

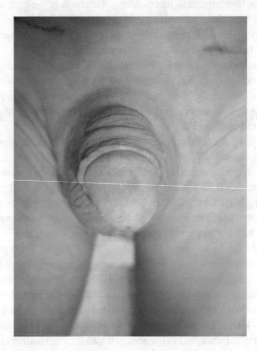

FIGURE 84.1   Speckled penis; pigmented maculae on the glans penis.

Gastrointestinal hamartomatous polyps are the most common variety. Other reported types include hyperplastic polyps, ganglioneuromas, inflammatory, and adenomatous polyps, and less commonly lipomatous, leiomyomatous, and lymphoid polyps [17,20,21].

Early in childhood, these polyps may cause chronic anemia, diarrhea, or small bowel invagination [15,16]. These hamartomas are most often diagnosed by endoscopy, and symptoms include rectal bleeding, pain, prolapsing polyps, and melena [22].

In 2014, Stanich et al. described diverse polyp histologies in PHTS patients, comprising hyperplastic polyps, adenomas, and hamartomas. They also reported an increased risk of colon cancer [17]. Given the prevalence of multiple polyp histologies, Stanich et al. suggested that PHTS should be considered a mixed polyposis syndrome as a substitute of a hamartomatous polyposis syndrome. They believe that PHTS should be investigated even if hamartomatous gastrointestinal polyps are not present, and since this study reported an increased risk of colon cancer, surveillance with colonoscopy is mandatory [17].

Abnormal facial features are also reported in BRRS, among them hypertelorism, frontal bossing, down-slanting palpebral fissures, depressed nasal bridge, strabismus, amblyopia, epicanthus inversus, long philtrum, thin upper lip, broad mouth, and relative micrognathia [16,17].

Hypotonia, gross motor delay, mild to severe mental deficiency, and speech delay are reported in approximately 50%–70% of BRRS patients [16,18]. Over 25% of BRRS patients have experienced seizures [18].

A dermatological evaluation may reveal various benign hamartomas. Angiokeratoma, subcutaneous and visceral lipomatosis, lymphangioma, and vascular malformations are among the most common [23]. Other dermatologic findings are acanthosis nigricans, verrucae, café-au-lait spots, epidermal nevus, and multiple acrochordons [23,24]. Histologically, facial lesions display aspects of trichilemmomas or verruca vulgaris, one more indicator of overlap with CS [19].

The hemangiomas in BRRS are an intramuscular hamartomatous proliferation of abnormal blood vessels and fibroadipose tissue. Histopathologically, they appear to be disorganized growths of blood vessels, adipose, and fibrous tissue, with a low level of proliferation [25]. Lipomas and vascular malformations present as soft, subcutaneous swellings on the trunk and extremities (Figure 84.2). They may enlarge rapidly, become huge and painful, and demonstrate local aggressive behavior; there is no tendency of spontaneous resolution [25].

Neuromuscular disorders are often present, including proximal muscles myopathy (60% of cases) and delayed neuropsychomotor development of different intensities (20%–50%) [26].

Around 30% of BRRS patients have thyroid involvement [26]. Thyroid pathology findings can be benign or malignant. Multiple adenomatous nodules are the most frequent finding. Laury et al. performed histological analysis of multinodular goiters seen in PHTS patients and realized that they were due to characteristic multiple adenomatous nodules, and not the typical adenomatous changes found in the elderly [27]. Other alterations reported were Hashimoto thyroiditis, follicular adenoma, C-cell hyperplasia, and follicular and papillary carcinoma [26,27].

Other phenotypic features of BRRS include overgrowth of prenatal or postnatal onset, pectus excavatum, hypoglycemia,

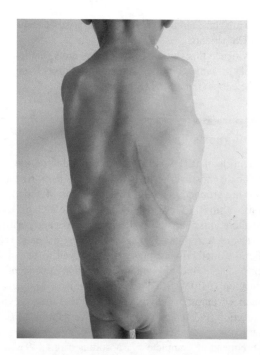

FIGURE 84.2 Extensive deforming lipomatosis.

high-arched palate, joint hyperextensibility, scoliosis, ocular albinism, and pseudopapilledema [14,16,24]. These phenotypic aspects are highly variable, although they appear to gather within a given family [24].

BRRS and CS share clinical symptoms at different frequencies. There is significant phenotypic overlap between the syndromes with common features including hamartomas, macrocephaly, and thyroid abnormalities [1]. Unlike CS, the diagnosis of BRRS can be made in newborns or early in childhood; patients present with macrocephaly, vascular malformations, lipomatosis, and speckled maculae of the penis.

Interestingly, identical PTEN mutations have been identified in patients who present with phenotypic manifestations characteristic of either BRRS or CS [1].

Even individuals within a single family that have the same germline *PTEN* mutation can have phenotypic traits more in accord with either BRRS or CS [28].

Lachlan et al. studied genotype–phenotype correlations of patients with known PTEN mutations from different families. The earliest phenotypic features among affected individuals were macrocephaly and hamartomas, with mucocutaneous features and malignancies developing later in adulthood. These authors suggest that BRRS and CS represent a single disorder with variable phenotypic expression and age-related penetrance [29]. Older participants were more likely to meet the CS criteria because many of the features of CS were not apparent until adulthood, and therefore the difference between the two conditions would be the age of presentation [29].

Before it was recognized as allelic to CS, BRRS was not considered to cause an increased risk of malignancy. However, in a study of families with BRRS and CS overlap, a correlation between benign and malignant breast disease and germline *PTEN* mutation was recognized, raising the possibility that BRRS is indeed associated with increased susceptibility to malignancy [3]. Nowadays, BRRS patients with a *PTEN* mutation are considered to share the same risk of cancer development as CS patients.

The evidence that these two conditions are variable manifestations of the same condition suggests that cancer surveillance should be adopted in all PHTS patients [29].

84.6 Diagnosis

The diagnosis of BRRS is made clinically, despite the absence of specific diagnostic criteria. BRRS patients harboring a *PTEN* mutation are grouped in the PHTS assortment. Mutations and/or deletions within *PTEN* are identified by genetic testing. Detection of such alterations confirms the PHTS diagnosis and permits predictive testing and prenatal diagnosis among patients' relatives [14]. Several methods are presently employed to detect *PTEN* deletions. These include multiplex ligation-dependent probe amplification (MLPA) (the preferred method), Southern blotting, monochromosomal hybrid analysis, real-time polymerase chain reaction (PCR), and semiquantitative multiplex PCR [14]. In order to optimize yield, the best order of *PTEN* testing should be first sequencing complete *PTEN* coding exons 1–9 and flanking intronic regions. The second step, if no pathogenic modification is found, would be analysis for deletion/duplication [30]. A more accurate and comprehensive assessment of chromosomal abnormalities can be achieved combining chromosomal microarray analysis and conventional cytogenetics in order to identify anomalies in chromosomes [31].

If the patient is clinically diagnosed with BRRS, the failure to detect *PTEN* gene mutations does not impair the diagnosis.

Until now, no international consensus criteria for the diagnosis of BRRS have been defined, but several groups of investigators have proposed criteria to facilitate clinical diagnosis (Table 84.1). Marsh et al. defined a clinical diagnosis of BRRS as meeting three out of four traits: macrocephaly, lipomatosis, hemangiomas, and speckled penis in males [7]. Parisi et al. used less strict criteria, defining the syndrome as having two out of three of the following features: macrocephaly, hamartomas (including at least one lipoma, hemangioma, or intestinal polyp), and penile macules in males [32]. BRRS appears to have a male predominance, although this may be confounded by a diagnostic bias in that a speckled penis is considered pathognomonic for the syndrome.

TABLE 84.1

BRRS Clinical Diagnostic Criteria [7,32]

Marsh Criteria [7]
  Macrocephaly
  Lipomatosis
  Hemangiomas
  Speckled penis

Parish Criteria [32]
  Macrocephaly
  Hamartomas (at least one lipoma, hemangiomas, or intestinal polyp)
  Penile macules in males

Diagnosis
  • Marsh: Three out of four features
  • Parish: Two out of three features

TABLE 84.2

PTEN Hamartoma Tumor Syndrome Diagnostic Criteria [21]

Major Criteria
  Breast cancer
  Endometrial cancer
  Thyroid cancer
  Gastrointestinal hamartomas
  Lhermitte–Duclos disease (hamartomatous dysplastic
    gangliocytoma of the cerebellum)
  Macrocephaly
  Macular pigmentation of the glans penis
  Multiple mucocutaneous lesions (trichilemmomas, acral keratosis,
    mucocutaneous neuromas, oral papillomas)

Minor Criteria
  Autism spectrum disorder
  Colon cancer
  Esophageal glycogenic acanthosis
  Mental retardation
  Renal cell carcinoma
  Testicular lipomatosis
  Thyroid cancer
  Thyroid structural lesions (adenoma, multinodular goiter)
  Vascular anomalies

Diagnosis
  • Three or more major criteria, but one must include macrocephaly,
    Lhermitte–Duclos disease, or gastrointestinal hamartomas; or
  • Two major and three minor criteria

The pediatric criteria of the PTEN scoring system can also be used (Table 84.2) [21].

Differential diagnoses include CS, PS, juvenile polyposis syndrome, Peutz–Jeghers syndrome, Gorlin syndrome, Birt–Hogg–Dubé syndrome, and neurofibromatosis type 1 [21].

84.7 Treatment

The clinical management of PHTS patients classically focuses on genetic counseling and screening. Patients with PHTS, particularly those with CS, should undergo early and frequent surveillance for susceptible malignancies. Currently no medical therapy is available for PHTS patients. Inhibitors of PI3K/Akt/mTOR pathway are being developed as anticancer agents; since loss of PTEN increases activation of this pathway, these drugs might have therapeutic use in PHTS patients [1].

A multidisciplinary surveillance program starting from diagnosis is the cornerstone of disease management in PHTS patients [33]. Because *PTEN* mutations are associated with an increased risk of developing multiple cancers, cancer surveillance is of extreme relevance [14].

In 2012, Tan et al. published a study showing that in patients with *PTEN* mutations, the risk of colorectal, breast, kidney, thyroid, and endometrial cancers and melanoma was elevated. The estimated lifetime risk was 85.2% for invasive female breast cancer, 35.2% for epithelial thyroid cancer, 33.6% for kidney cancer, 28.2% for endometrial cancer, 9% for colorectal cancer, and 6% for melanoma. The prevalence of breast cancer was greatly elevated in females with *PTEN* mutations, first cases being described around age 30 and reaching 85% lifetime risk.

PTEN-related endometrial cancer risk begins at age 25 and rises to 30% by age 60, whereas for thyroid cancer, risk begins at birth and continues throughout life. First cases of colorectal and kidney cancers were detected around age 40, reaching a lifetime risk of 9% and 34%, respectively. The earliest reported melanoma onset was at the age of 3 years [34].

In 2013, Bubien et al. described a study in a group of patients with PHTS syndrome. Elevated incidence ratios were found for female breast cancer (39.1), thyroid cancer in women (43.2), and in men (199.5), melanoma in women (28.3), and in men (39.4), and endometrial cancer (48.7). The cumulative cancer risks found at age 70 were 85% for any cancer, 77% for female breast cancer, and 38% for thyroid cancer. The liability of cancer was two times greater in women with PHTS than in men with PHTS ($p < 0.05$) [35].

Once BRRS diagnosis is made, especially once a germline *PTEN* mutation is identified, the patient should undergo cancer screening. Given the genotypic and phenotypic similarities between BRRS and CS, and the possibility that these syndromes may represent a single disorder, BRRS patients should be counseled on adhering to screening guidelines recommended for CS patients [14].

In addition, patients with BRRS should be monitored for gastrointestinal complications, because gastrointestinal hamartomatous polyposis is generally more severe in BRRS than in CS [14]. Some authors propose a yearly hemoglobin test from early childhood in order to prematurely detect intestinal hamartomas [16].

For CS the current proposal for surveillance protocol includes investigation for breast cancer, endometrial cancer, thyroid cancer, renal carcinoma, and intestinal hamartomas. Eng et al. recommend monthly self-examination and clinical breast examination starting from age 25 and annual mammography beginning at age 30. Breast self-examination is also suggested for men. An annual endometrial examination is recommended from age 35. To screen for abnormalities of the thyroid gland, annual palpation of the thyroid gland from age 18 is recommended. Furthermore, they suggest annual urinalysis to help detect renal carcinoma at an early stage. A baseline colonoscopy is recommended at age 50, with the purpose of detecting hamartomas [36].

Stanich et al. recommend a baseline colonoscopy starting earlier, at age 35 or 10 years younger than the earliest colorectal cancer diagnosis in a first-degree relative, with future surveillance intervals based on the results [17].

Tan et al. recommend targeted history and physical examination, thyroid ultrasound, dermatologic examination, and neurologic and psychological testing for children under 18 years old and adults at the time of diagnosis. Annual thyroid ultrasound and skin examination should be required from this time on. Starting at age 30, they suggest annual mammogram and annual endometrial sampling or transvaginal ultrasound for female. Starting at age 40, biannual colonoscopy and biannual renal ultrasound/MRI [34].

Prophylactic resection of susceptible organs may decrease the risk of potentially life-threatening malignancies. However, there are no data to support prophylactic surgery to decrease the risk of malignancy or mortality in BRRS patients [1].

A new strategy to control PHTS symptoms, prevent tumor growth, and/or treat established cancers is to test drugs that inhibit the PI3K/Akt/mTOR pathway in PHTS patients, given that loss of PTEN increases activation of the pathway in benign and malignant tissues of patients with PHTS and drives cellular proliferation, migration, and survival.

The development of PI3K/Akt/mTOR inhibitors is an area of intense research in oncology, as this pathway promotes cellular survival and chemotherapeutic resistance in a variety of malignancies. Inhibition of proximal components of the pathway such as PI3K or Akt, while desirable, is currently not possible for patients with PHTS because there are no published clinical trials with these inhibitors in cancer patients. In addition, it is not clear if inhibitors or PI3K or Akt will be as tolerable as inhibitors of mTOR. Inhibitors of mTOR are the most clinically developed and may serve as viable agents to treat PHTS. Rapamycin, a specific mTOR inhibitor, is FDA approved to prevent transplant rejection and for use in drug-eluting cardiac stents. Temsirolimus, a rapamycin analog, is FDA approved for the treatment of advanced renal cell carcinoma [1].

The first reported successful treatment attempt with the mTOR complex 1 (mTORC-1) inhibitor sirolimus for a patient with PHTS described a reduction of tumor masses and improvement of the patient's general state [37]. More recently, an experimental oral sirolimus treatment of vascular malformation in a patient with BRRS was successfully reported [38]. Later, a child with an extreme phenotype of PHTS, including lipomatosis and severe cachexia, was treated with sirolimus and had an improvement of symptoms. This patient received a daily oral sirolimus dose of 0.1 mg/kg body weight. After 4 weeks of sirolimus therapy without side effects, the parents reported subjective improvement of physical and mental activity. Under sirolimus therapy, the growth of the patient's abdominal lipomatosis was attenuated, but the tumor volume was not reduced [39].

84.8 Prognosis

Previously, BRRS patients were not considered to have higher cancer predisposition; now it is well accepted that patients with *PTEN* mutation have the same risk of cancer development as CS patients.

BRRS is a dominant autosomal disease. Genetic counseling must be offered to patients with PTEN mutations and even asymptomatic first- and second-degree relatives should be tested for the mutation, identifying those that need to be followed [14,40].

BRRS prognosis is unknown and is probably influenced by its opening presentation and genotype.

84.9 Conclusion

There is much hope for PHTS patients' futures because of ongoing efforts to refine diagnostic criteria and to develop novel therapeutics. Potential therapies for PHTS that are currently available for clinical testing utilize inhibitors of the PI3K/Akt/mTOR pathway; future therapies for PHTS may use alternative strategies to restore PTEN function.

Distinctions between the PHTS syndromes, especially between BRRS and CS, may no longer be relevant. An international collaborative effort is underway to collect genotypic and phenotypic data from CS patients to identify the simplest means to predict germline *PTEN* mutation [41]. This analysis may streamline the diagnostic criteria and ultimately combine BRRS and CS into one syndrome.

REFERENCES

1. Blumenthal GM, Dennis PA. PTEN hamartoma tumor syndromes. *Eur J Hum Genet* 2008;16:1289–300.
2. Schreibman IR, Baker M, Amos C, McGarrity TJ. The hamartomatous Polyposis Syndromes: A clinical and molecular review. *Am J Gastroenterol* 2005;100:476–90.
3. Marsh DJ, Kum JB, Lunetta KL et al. PTEN mutation spectrum and genotype-phenotype correlations in Bannayan-Riley-Ruvalcaba syndrome suggest a single entity with Cowden syndrome. *Hum Mol Genet* 1999;8:1461–72.
4. Orloff MS, Eng C. Genetic and phenotypic heterogeneity in the PTEN hamartoma tumor syndrome. *Oncogene* 2008;27:5387–97.
5. Erkek E, Hizel S, Sanly C et al. Clinical and histopathological findings in Bannayan-Riley-Ruvalcaba syndrome. *J Am Acad Dermatol* 2005;53:639–43.
6. Zhou XP, Waite KA, Pilarski R et al. Germline PTEN promoter mutations and deletions in Cowden/Bannayan-Riley-Ruvalcaba syndrome result in aberrant PTEN protein and dysregulation of the phosphoinositol-3-kinase/Akt pathway. *Am J Hum Genet* 2003;73:404–11.
7. Marsh DJ, Coulon V, Lunetta KL et al. Mutation spectrum and genotype-phenotype analyses in Cowden disease and Bannayan-Zonana syndrome, two hamartoma syndromes with germline PTEN mutation. *Hum Mol Genet* 1998;7:507–15.
8. Wanner M, Celebi JT, Peacocke M. Identification of a PTEN mutation in a family with Cowden syndrome and Bannayan-Zonana syndrome. *J Am Acad Dermatol* 2001;44:183–7.
9. Eng C. PTEN: One gene, many syndromes. *Hum Mutat* 2003;22:183–98.
10. Waite KA, Eng C. Protean PTEN: Form and Function. *Am J Hum Gene* 2002;70:829–44.
11. Tamura M, Gu J, Matsumoto K, Aota S, Parsons R, Yamada KM. Inhibition of cell migration, spreading, and focal adhesions by tumor suppressor PTEN. *Science* 1998;280:1614–7.
12. Simpson L, Parsons R. PTEN: Life as a tumor suppressor. *Exp Cell Res* 2001;264:29–41.
13. Weng LP, Smith WM, Brown JL, Eng C. PTEN inhibits insulin-stimulated MEK/MAPK activation and cell growth by blocking IRS-1 phosphorylation and IRS-1/Grb-2/Sos complex formation in a breast cancer model. *Hum Mol Genet* 2001;10:605–16.
14. Hobert JA, Eng C. PTEN hamartoma tumor syndrome: An overview. *Genet Med* 2009;11(10):687–94.
15. Díaz MM, Urioste M. Síndrome de Bannayan-Riley-Ruvalcaba [internet]. 2014; Available from https://www.orpha.net/data/patho/Pro/es/Bannayan-Riley-Ruvalcaba_web.pdf
16. Hendriks YM, Verhallen JT, Van der Smagt JJ et al. Bannayan–Riley–Ruvalcaba syndrome: Further delineation of the phenotype and management of PTEN mutation-positive cases. *Fam Cancer* 2003;2:79–85.
17. Stanich PP, Pilarski R, Rock J, Frankel WL, El-Dika S, Meyer MM. Colonic manifestations of PTEN hamartoma tumor syndrome: Case series and systematic review. *World J Gastroenterol* 2014;20(7):1833–8.
18. Edmondson AC, Kalish JM. Overgrowth Syndromes. *J Pediatr Genet* 2015;4:136–43.
19. Fargnoli MC, Orlow SJ, Semel-Concepcion J, Bolognia JL. Clinicopathologic findings in the Bannayan-Riley-Ruvalcaba syndrome. *Arch Dermatol.* 1996;132(10):1214–8.
20. Pilarski R. PTEN hamartoma tumor syndrome: A clinical overview. *Cancers (Basel).* 2019;11(6):844.
21. Pilarski R, Burt R, Kohlman W, Pho L, Shannon KM, Swisher E. Cowden syndrome and the PTEN hamartoma tumor syndrome: Systematic review and revised diagnostic criteria. *J Natl Cancer Inst* 2013;105(21):1607–16.
22. Jelsig AM, Qvist N, Brusgaard K, Nielsen CB, Hansen TP, Ousager LB. Hamartomatous polyposis syndromes: A review. *Orphanet J Rare Dis* 2014, 9:101.
23. Iskandarli M, Yaman B, Aslan A. A case of Bannayan–Riley–Ruvalcaba syndrome. A new clinical finding and brief review. *Int J Dermatol* 2016;55(9):1040–3.
24. Calva D, Howe JR. Hamartomatous Polyposis Syndromes. *Surg Clin North Am* 2008;88(4):779–817.
25. Tan WH, Baris HN, Burrows PE et al. The spectrum of vascular anomalies in patients with PTEN mutations: Implications for diagnosis and management. *J Med Genet* 2007;44:594–602.
26. Peiretti V, Mussa A, Feyles F et al. Thyroid involvement in two patients with Bannayan-Riley-Ruvalcaba Syndrome. *J Clin Res Pediatr Endocrinol* 2013;5(4):261–65.
27. Laury AR, Bongiovanni M, Tille J, Kozakewich H, Nosé V. Thyroid pathology in PTEN-hamartoma tumor syndrome: Characteristic findings of a distinct entity. *Thyroid* 2011;21(2):135–44.
28. Perriard J, Saurat JH, Harms M. An overlap of Cowden's disease and Bannayan-Riley-Ruvalcaba syndrome in the same family. *J Am Acad Dermatol* 2000;42:348–50.
29. Lachlan KL, Lucassen AM, Bunyan D, Temple IK. Cowden syndrome and Bannayan Riley Ruvalcaba syndrome represent one condition with variable expression and age-related penetrance: Results of a clinical study of PTEN mutation carriers. *J Med Genet* 2007;44:579–85.
30. Eng C. PTEN hamartoma tumor syndrome (PHTS) [Internet]. In: Pagon RA, Adam MP, Ardinger HH et al., editors. *GeneReviews*. Seattle: University of Washington; 2001 Nov 29 [updated 2014 Jan 23; cited 2016 Mar 13].
31. Lee SH, Ryoo E, Tchah H. Bannayan-Riley-Ruvalcaba syndrome in a patient with a PTEN mutation identified by chromosomal microarray analysis: A case Rreport. *Pediatr Gastroenterol Hepatol Nutr* 2017;20(1):65–70.
32. Parisi MA, Dinulos MB, Leppig KA, Sybert VP, Eng C, Hudgins L. The spectrum and evolution of phenotypic findings in PTEN mutation positive cases of Bannayan-Riley-Ruvalcaba syndrome. *J Med Genet* 2001;38:52–8.
33. Piccione M, Fragapane T, Antona V, Giachino D, Cupido F, Corsello G. PTEN hamartoma tumor syndromes in childhood: Description of two cases and a proposal for follow-up protocol. *Am J Med Genet* 161A(11):2902–8.
34. Tan MH, Mester JL, Ngeow J, Rybicki LA, Orloff MS, Eng C. Lifetime cancer risks in individuals with germline PTEN mutations. *Clin Cancer Res* 2012;18(2):400–7.
35. Bubien V, Bonnet F, Brouste V et al. High cumulative risks of cancer in patients with PTEN hamartoma tumor syndrome. *J Med Genet* 2013;50:255–63.
36. Eng C. Will the real Cowden syndrome please stand up: Revised diagnostic criteria. *J Med Genet* 2000;37:828–30.

37. Marsh DJ, Trahair TN, Martin JL et al. Rapamycin treatment for a child with germline PTEN mutation. *Nat Clin Pract Oncol* 2008;5:357–61.

38. Iacobas I, Burrows PE, Adams DM, Sutton VR, Hollier LH, Chintagumpala MM. Oral rapamycin in the treatment of patients with hamartoma syndromes and PTEN mutation. *Pediatr Blood Cancer* 2011;57:321–3.

39. Schmid GL, Kässner F, Uhlig HH et al. Sirolimus treatment of severe PTEN hamartoma tumor syndrome: Case report and *in vitro* studies. *Pediatr Res* 2014;75(4):527–34.

40. Sagi SV, Ballard DD, Marks RA, Dunn KR, Kahi CJ. Bannayan Ruvalcaba Riley syndrome. *ACG Case Rep J*, 2014;1(2):90–2.

41. Zbuk KM, Eng C. Cancer phenomics: RET and PTEN as illustrative models. *Nat Rev Cancer* 2007;7:35–45.

85

Basal Cell Nevus Syndrome

Priyanka Chhadva and Pete Setabutr

CONTENTS

85.1 Introduction

Basal cell nevus syndrome (BCNS) is a rare autosomal dominant disorder associated with integument, ophthalmic, cardiac, nervous, and skeletal abnormalities. Although the collection of findings matching this syndrome in mummies had been documented since 1000 BC, and the characteristics of this disorder were initially reported by Jarisch and White in 1894 [1–3], it was Gorlin and Goltz who classified BCNS as a pentad: multiple basal cell carcinomas (BCCs), keratocysts of the jaw, palmar, and/or plantar pits, spine and rib anomalies, and calcification of the falx cerebri [4,5]. For this reason, BCNS is commonly referred to as Gorlin syndrome, Gorlin–Goltz syndrome, or nevoid basal cell carcinoma syndrome (NBCCS) [6]. BCNS can present in several stages of life and many areas of the body, induce various developmental abnormalities, and increase the risk of developing a number of cancerous and noncancerous growths (e.g., BCC, medulloblastoma, fibroma, and keratocystic odontogenic tumors). Indeed, while BCNS shows limited effect on patient life expectancy, early mortality can result from metastasis of BCC or occurrence of medulloblastoma.

85.2 Biology

Chromosomes are thread-like structures that carry hereditary information from parents to offspring. Apart from gametes (i.e., egg and sperm), most human cells are diploid and contain 22 pairs of nonsex chromosomes (so-called autosomes) and one pair of sex chromosomes (XX in females, XY in males), with one copy of each chromosome pair inherited from father and the other copy from mother. During mitosis, diploid cells produce two daughter cells that are genetically identical to each other and to the parental cell (i.e., all containing 23 pairs of chromosomes). During meiosis, diploid germline cells produce four haploid gametes (eggs or sperms) that have only half the number of chromosomes (i.e., 23 instead of 46 in total) in comparison with diploid cells. Upon fertilization, the sperm combines with the egg to form a zygote, which again contains two pairs of 23 chromosomes.

If a dominant gene located on one of the 22 nonsex chromosomes (autosomes) is mutated in a parent, there is a 50% chance for a child to inherit the mutated gene (dominant gene) from this parent along with a normal gene (recessive gene) from the other parent and become affected. There is also is a 50% chance for a child to inherit two normal genes (recessive genes) from parents and remain unaffected. In autosomal dominant disorder, inheritance (or transfer) of one copy of the altered gene in each cell is sufficient for the disease to occur.

BCNS is linked to loss of function mutation(s) in the tumor suppressor gene *PTCH1* (patched homolog 1) on chromosome 9 (9q22.1–q31). Following an autosomal dominant inheritance pattern, a parent with such a mutation has a 50% chance of passing the disorder to his offspring. However, in 20%–40% of cases, BCNS is caused by spontaneous or de novo mutation in the *PTCH1* gene [7,8].

While one mutated copy of the *PTCH1* gene in each cell is sufficient to induce the features of BCNS in early life (e.g., macrocephaly, skeletal abnormalities), a mutation in the second copy of the *PTCH1* gene in certain cells is necessary for BCCs and other tumors to develop in later life (classic two-hit tumor suppressor gene model). Due to its high penetrance, most people who were born with a mutated *PTCH1* gene eventually acquire mutation in the second copy of the *PTCH1* gene in some cells in their lifetime (due to environmental factors such as sun exposure or ionizing radiation) and develop various tumors.

85.3 Pathogenesis

The gene responsible for BCNS, *PTCH1*, is located on the long arm of chromosome 9q (22.3–q31), which comprises 23 exons and encodes a patched homolog 1 (PTCH1) protein. Being an integral membrane protein with 12 transmembrane regions, two extracellular loops, and a putative sterol-sensing domain, PTCH1 acts as a receptor for the secreted factor sonic hedgehog (SHH). A functional PTCH1 binds with SHH and represses transcription of genes encoding proteins belonging to the transforming growth factor (TGF)–beta, thus controlling growth and development of normal tissue. However, mutations in the *PTCH1* gene may result in a truncated protein or missense variant that fails to bind with SHH and loses its capacity to regulate cell growth [9,10]. Therefore, PTCH1 is important for embryologic developmental regulation and tumor suppression [11].

While congenital malformations in BCNS are due likely to alterations in the concentration of the PTCH1 protein in the extremely dosage-sensitive hedgehog signaling pathway, evidence points to the loss of the normal *PTCH1* allele in medulloblastoma in young patients (2 years of age) who have BCNS. It is estimated that patients with BCNS have a 90% risk of developing multiple basal cell skin cancers; and children with BCNS have about 5% risk of developing medulloblastoma. In addition, inactivation of the normal allele appears to be associated with jaw cysts.

The *PTCH1* gene shows broad allelic heterogeneity, with 51 germline mutations identified through the analysis of patients with BCNS [9]. Of the known pathogenic variants, 65% contain premature termination codon (predicting a protein truncation), 16% are missense mutations, 13% have splice site, 6% are (multi)exon or large-scale deletions or rearrangements [12–16]. A condition associated with large chromosomal deletions in the *PTCH1* gene is referred to as 9q22.3 microdeletion syndrome, in which developmental delay and/or intellectual disability, metopic craniosynostosis, obstructive hydrocephalus, pre- and postnatal macrosomia, and seizures, in addition to the features of BCNS, are observed.

Besides the *PTCH1* gene, the *SUFU* gene has been recently implicated in the pathogenesis of BCNS. Located on chromosome 10q24–q25, the *SUFU* gene contains 12 exons and encodes the suppressor of fused homolog protein, which functions as a negative regulator in the hedgehog signaling pathway. Mutations in the *SUFU* gene have been detected in nodular or desmoplastic medulloblastoma from children with BCNS. Further, a germline *SUFU* pathogenic variant is associated with macrocephaly.

85.4 Epidemiology

The assumed prevalence of BCNS is 1:60,000 but may vary by the patient population being studied [14]. Studies have found a prevalence of 1:56,000 in North America, 1:256,000 in Italy, 1:164,000 in Australia, 1:235,800 in Japan, and as low as 1:13,939,393 in Korea [17–23]. This rare condition predominantly affects Caucasians and is equally seen in males and females [14,24].

BCNS has a variable phenotypic expression, which reflects as differences in penetrance. This also contributes to the variations in the effects of environmental factors, modifier genes, and expression of various mutations within the same gene. Therefore, there is a variety of genetic and clinical manifestations expressed in patients and their family members affected by BCNS [7,25].

85.5 Clinical Features

BCNS presents with multisystem clinical features, including dermatologic, ophthalmic, cardiac, nervous, and skeletal abnormalities (Table 85.1).

85.5.1 Dermatologic

A key characteristic of BCNS is the presence of multiple BCCs of the skin (Figure 85.1). They usually appear between puberty and age 35; however, the average age of onset is 25, and cases involving patients as young as 3 and 4 have been reported [8,23,26]. BCCs are predominantly found on the face, back, and chest, and they can also occur on skin that is not as exposed to sun, such as genitalia [27]. These lesions vary greatly in quantity, size, and appearance. They have been described as complexion-colored papules, nodules, and ulcerating plaques [28]. Of note, only 0.5% of BCC are attributed to BCNS [29].

Small plantar (50%) and palmar (70%) pits occur in BCNS, ranging from 2–3 mm in diameter to 1–3 mm in depth [30]. (Figure 85.2) The base of these lesions is red in Caucasians and black in darker pigmented patients [28]. The majority of pits are seen in patients older than 20 years (85%); however, 30%–65% of cases comprise children under 10 years old, and the number of lesions rises with age [31,32].

Milia, small keratin-filled cysts, can also be present in BCNS. They are usually found on the face (30%), below the eyes and on the forehead, and can also be seen on the palpebral conjunctiva (40%) [5,28].

Epidermoid cysts are found on the trunk and limbs (knee region) in more than 50% of cases. Skin tags are seen as well around the neck [28]. Hairy skin patches, discrete patches of unusually long pigmented hair, have also been described in select patients with BCNS [33].

85.5.2 Ophthalmic

Patients with BCNS have numerous ophthalmic problems that occur at a higher frequency (20%) compared to a normal population. Hypertelorism is the most common ocular finding in BCNS (up to 70%), but this varies by region [20–22]. It is described as an increase in the distance between the interpupillary distance

TABLE 85.1

Manifestations of BCNS

Calcifications of the central nervous system
- Falx cerebri
- Tentorium cerebelli
- Sella turcica
- Petrosphenoidal ligament

Other CNS anomalies
Cysts of the choroid plexus, third and lateral cerebral ventricles
Agenesis of corpus callosum
Meningioma
Medulloblastoma
Multiform glioblastoma
Astrocytoma
Fetal rhabdomyosarcoma
Grand mal
Congenital hydrocephalus
Mental retardation (∼5% patients)

Ophthalmic anomalies
Hypertelorism (∼70% of patients)
Multiple eyelid BCC
Epiretinal membrane
Hamartoma of the retina and RPE
Microcysts on eyelids
Micro-ophthalmia
Congenital cataracts
Peters anomaly
Strabismus
Nystagmus
Orbital cysts
Iris transillumination defects
Congenital blindness

Otologic anomalies
Otoscleosis
Conductive hearing loss
Posteriorly angulated ears

Urinary anomalies
U or L-shaped kidneys
Unilateral renal agenesis
Double kidneys
Double ureters

Skeletal anomalies
Significant height (average for females is 174 cm and for males 183 cm)
Increased pneumatization of the paranasal sinuses (frontal sinuses)
Increased head circumference (∼50% of patients)
Strongly marked superciliary arches
Retracted and wide base of the nose (pseudohypertelorism)
Wide eyes (∼70% of patients)

Congenital skeletal anomalies
- Bifid, fused, splayed or missing ribs (∼30%–60%)
- Bifid wedges fused vertebra
- Scoliosis (∼40%)
- Frontal, temporal, parietal bossing
- Polydactyly, syndactyly
- Short fourth metacarpal
- Sprengel shoulder (∼10%–40%)
Spina bifida occulta (∼40%–60%)
Sternal protrusion or depression (∼30%–40%)
Cysts within the phalanxes, long bones, pelvis and calvaria

Cardiovascular anomalies
Cardiac fibroma
- 3%–5% of patients
- Usually measure 3–4 cm
- Usually located in the anterior wall of the left atrium
- Impaired hemodynamics if they involve the ventricles
Absent internal carotid artery

Gastroenteric anomalies
Lymphomesenteric cysts, asymptomatic
Gastric polyps

Reproductive/genital anomalies
Women – ovarian cysts and fibromas (∼25%–50%)
- 75% bilateral
- Do not affect fertility
- Risk of ovarian torsion
Men
- Hypogonadism
- Cryptorchidism
- Gynecomastia

Source: Adapted from Kiwilsza M, Sporniak-Tutak K. *Med Sci Monit.* 2012;18(9):RA145–53.

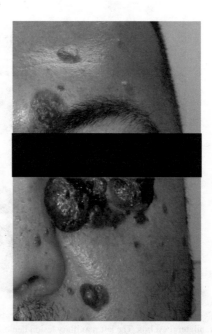

FIGURE 85.1 Multiple facial/periocular basal cell carcinomas in a male patient with BCNS.

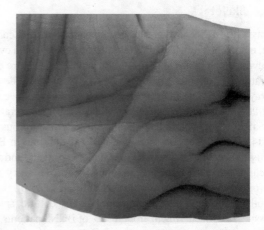

FIGURE 85.2 Palmar pits in a patient with BCNS.

Milia found on the palpebral conjunctiva are often seen in BCNS and are transient in nature. Other ophthalmic findings described include chalazions, congenital cataracts, congenital glaucoma, coloboma of the iris, optic nerve, and choroid, iris transillumination defects, microphthalmia, hamartoma of the retina and retinal pigment epithelium, and retinal detachments due to multiple retinal fibrotic falciform folds, retinal holes, tears, and retinoschisis [16,22,26,36–38].

85.5.3 Nervous

Calcification of the falx cerebri is a radiologic diagnosis for BCNS, and its frequency increases with age [34]. A rare but serious occurrence is medulloblastoma, which is a common pediatric primary brain tumor. It usually occurs at the age of 7–8 but may present around the age of 2 and predominantly occurs in males (3:1) when associated with BCNS. It occurs sporadically in 5% of cases and when diagnosed in young children, BCNS should be on the differential [14,18,39].

and the medial canthi due to an enlargement of the sphenoid bone, causing wider separation between the orbits [34].

Strabismus occurs between 10% and 20% and some reports indicate a higher rate of esotropia and esophoria compared to exotropia and exophoria [16,22,34].

Myelination of retinal nerve fibers can occur at higher rates in BCNS, is bilateral and associated with poor vision and visual field deficits [34,35].

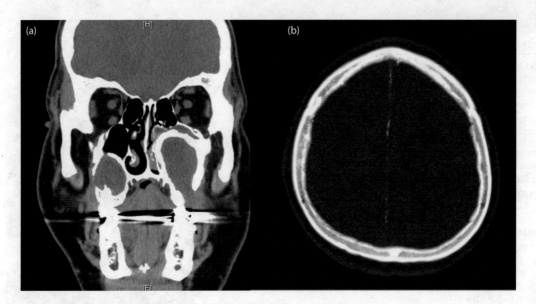

FIGURE 85.3   (a) Bilateral keratocystic odontogenic tumors of the maxillary sinuses along with exophytic nodular soft tissue lesions of the right infraorbital face and bilateral eyelids. The left maxillary lesion demonstrates thick, calcified wall, possibly related to prior treatment. There is an expansion of the left maxillary sinus with displaced molar teeth. (b) Dural calcification.

85.5.4 Skeletal

Patients with BCNS have greater height, abnormal skull configuration, macrocephaly, rib and vertebral abnormalities, and cysts of the calvaria, long bones, phalanxes, and pelvis [28] (Figure 85.3).

Keratocystic odontogenic tumors, often incidentally detected on x-rays of the oral cavity, are cysts of the jaw that have internal stratified squamous epithelium filled with thick keratinous material or straw-colored fluid, and an external fibrous capsule [14,21–23]. They are seen 85%–90% of the time in BCNS patients and occur in the mandible, mandibular angles, and zones adjacent to the canine, incisor, and molar teeth. These cysts are numerous, recurrent, and occur earlier in life [14,28,40].

Another common skeletal abnormality is fused, bifid, splayed, or missing ribs, occurring in around 35% of BCNS patients. This rib abnormality can give rise to a depressed or prominent sternum in around 30%–40% of patients [21,22,28,39]. Kyphoscoliosis is seen in 10%–40% of patients [41]. Spina bifida occulta of the thoracic or cervical vertebrae occurs in 60% of cases [42].

85.5.5 Cardiovascular

Cardiac fibromas, 3–4 cm diameter well-circumscribed and firm tumors composed of fibroblasts embossed in a dense matrix of elastic and collagen fibers, are due to BCNS 3%–5% of the time [26,43]. It is usually located in the left anterior ventricular wall and can affect hemodynamics and cardiac conduction [28].

85.5.6 Genitourinary

Renal anomalies include horseshoe kidney, renal cysts, duplication of the renal ureter and pelvis, and unilateral renal agenesis. Ovarian calcifications and cysts, fibrosarcomas, and primary ovarian leiomyosarcomas are seen in women with BCNS [26,44,45]. Gynecomastia, cryptorchidism, hypogonadotrophic hypogonadism, and female pubic escutcheon have been seen in males with BCNS [28].

85.6 Diagnosis

The diagnosis of BCNS depends on the presence of two major criteria or one major and two minor criteria (Table 85.2) [39].

BCNS can be diagnosed in the antenatal period, which is helpful to prevent complications during birth. Ultrasound can help with certain diagnostic features, such as fibromas of the heart and macrocephaly. Chorionic villus sampling and amniocentesis may identify the disease-causing allele through DNA testing [28].

Depending on the molecular method used, the proportion of BCNS patients with a *PTCH1* pathogenic variant ranges from 6%–21% (gene-targeted deletion/duplication analysis) to 50%–85% (sequence analysis). On the other hand, the proportion of

TABLE 85.2

Diagnostic Criteria for BCNS

Major Criteria	Minor Criteria
Multiple (>2) basal cell carcinomas in a patient <20 years old	Macrocephaly (after adjusted for height)
Odontogenic keratocysts of the jaw (diagnosed by histopathology)	Congenital malformations: Cleft lip/palate, frontal bossing, coarse facies, moderate/severe hypertelorism
Palmar or plantar pits (3 or more)	Other skeletal abnormalities: Sprengel deformity, marked chest/pectus deformity, marked syndactyly of digits
Bilamellar calcification of the falx cerebri	Radiologic abnormalities: Bridging of sella turcica, vertebral anomalies (hemivertebrae, fusion or elongation of vertebral bodies), modeling defects in hands/feet, or flame shaped x-ray lucencies of hand/feet
Bifid, fused, or markedly played ribs	
First degree relative with BCNS	Ovarian fibroma
	Medulloblastoma

Source: Adapted from Kimonis VE et al. *Am J Med Gen.* 1997;69(3): 299–308.

BCNS patients with a *SUTU* pathogenic variant ranges from ~1% (gene-targeted deletion/duplication analysis) to 5% (sequence analysis). However, about 15%–27% of BCNS patients have no apparent pathogenic variant.

Differential diagnoses include Brooke–Spiegler syndrome (trichoepitheliomas, milia, and cylindromas in the second or third decade), Bazex syndrome (multiple BCCs, follicular atrophoderma on the dorsum of hands and feet reminiscent of orange peel instead of the palmar and plantar pits of BCNS, decreased sweating, and hypotrichosis), Rombo syndrome (vermiculate atrophoderma, milia, hypotrichosis, trichoepitheliomas, BCCs, and peripheral vasodilation with cyanosis), Beckwith–Wiedemann syndrome (macrosomia, macroglossia, visceromegaly, embryonal tumors, omphalocele, neonatal hypoglycemia, ear creases/pits, adrenocortical cytomegaly, renal abnormalities, and abnormal regulation of gene transcription in the imprinted domain on chromosome 11p15.5), Sotos syndrome (typical facial appearance, intellectual impairment, increased height and head circumference, increased risk for sacrococcygeal teratoma and neuroblastoma, and *NSD1* pathogenic variant), and multiple BCCs due to other causes (e.g., arsenic exposure).

85.7 Treatment

Due to the multisystem presentation of BCNS, patients with this disorder must be treated by numerous specialists. Prevention of BCC includes decreased sun exposure, and wearing UV sunblock, sunglasses, and sun-protective clothing. Also, oral retinoids may thwart or delay the growth and recurrence of BCCs [46]. When BCCs do occur, the mainstay of treatment is surgical excision. For initial small and well-defined lesions, curettage and cautery/electrodessication is preferred, as well as cryosurgery [47–50]. For larger, thick, concave lesions, microscopic excision with ultrapulse CO_2 laser therapy can be used [51,52]. Photodynamic therapy with topical application of delta-aminolaevulinic acid is an option for superficial and flat lesions [52–55]. For recurrent lesions in risky locations such as the periorbital, nasolabial, and nasal skin, a more aggressive approach must be taken, which includes Mohs micrographic surgery [48,56–58]. Local radiotherapy is not recommended due to the risk of tumor enlargement and recurrence [22,39]. Nonsurgical approaches also depend on tumor characteristics. For low-risk, superficial, non-hair follicle–involving BCC, topical 0.1% 5–fluorouracil can be used [59]. For small papular tumors, intralesional injection of interferon alfa-2b can be used [52]. Paclitaxel can be used to treat multiple aggressive BCCs [60]. Topical imiquimod may treat small, low-risk superficial or nodular BCC [61]. It acts as a Toll-like receptor-7 agonist and causes apoptosis of BCCs [62]. In addition, vismodegib (Erivedge) and sonidegib (Odomzo) targeting the "hedgehog pathway," which is affected by the *PTCH* mutation, have been approved to treat people with basal cell cancers that have spread in the body or that cannot be treated with surgery or radiation.

Keratocystic odontogenic tumors are treated with excision. However, because of the high rate of recurrence (due to presence of satellite microtumors in the surrounding bone) when associated with BCNS, adjuvant therapies and close monitoring is warranted. The addition of cryotherapy (liquid nitrogen) or Carnoy's solution (a mixture of 6 mL absolute alcohol, 3 mL chloroform, 1 mL glacial acetic acid, 1 g of ferric chloride) has led to decreased rates of recurrence [63,64].

Medulloblastomas require aggressive resection with combined chemotherapy and radiation, but radiation therapy can induce invasive BCCs in the field being irradiated [65]. An alternative is non-conformal radiation techniques that spare the skin, but side effects of ototoxicity and temporal lobe injury must be considered [14].

85.8 Prognosis

Once carriers of the BCNS defective gene are identified, screening for the most pathologic entities should commence. This includes MRIs every 6 months until age 7 for detection of medulloblastoma, echocardiography for the detection of cardiac fibromas, dental visits with x-rays until age 40 for detection of odontogenic keratocysts (OKCs), regular dermatological exams for skin lesion surveillance, and ovarian ultrasounds for detection of fibromas in the first and second decades [8,26,39]. Hence, a multidisciplinary approach should be taken with regard to monitoring and treatment. This includes specialists in genetics, ophthalmology, dermatology, otolaryngology, maxillofacial surgery, neurology, cardiology, gynecology, and urology to coordinate and optimize medical care.

Although BCNS generally does not affect life expectancy, its associated mandibular OKCs can cause displacement of developing teeth and root resorption in young patients. The most common cause of early mortality in patients with BCNS is due to invasion with metastasis of BCC or due to medulloblastoma. Therefore, early diagnosis and appropriate medical management are critically important to help patients with BCNS maintain a standard life span. Given its autosomal dominant inheritance and high penetrance, antenatal diagnosis (e.g., ultrasound scans and DNA analysis of fetal cells from amniocentesis or chronic villus sample) may be offered to families in which the *PTCH1* pathogenic variant is present.

85.9 Conclusion

BCNS is a rare syndrome linked to mutations in the tumor suppressor gene *PTCH* and inherited in an autosomal dominant pattern, its systemic manifestations include multiple BCCs, keratocysts of the jaw, palmar and/or plantar pits, spine and rib anomalies, and calcification of the falx cerebri. The challenge in diagnosis lies in the phenotypic variability and presentation to physicians in multiple different specialties. Early clinical suspicion, even in those without a family history of this disorder, is crucial for successful prevention and treatment. Once diagnosed, a multidisciplinary approach should be undertaken to provide patients with the best outcome.

REFERENCES

1. Jarish W. On the doctrine of skin tumors. *Arch Dermatol Syphilol* 1894;18:162–222.

2. White J. Multiple benign cystic epitheliomas. *J Cutan Genitourin Dis* 1894;12:477–84.

3. Satinoff MI, Wells C. Multiple basal cell naevus syndrome in ancient Egypt. *Med His* 1969;13(3):294–7.

4. Gorlin RJ, Goltz RW. Multiple nevoid basal-cell epithelioma, jaw cysts and bifid rib. A syndrome. *N Eng J Med* 1960;262:908–12.

5. Gorlin RJ. Nevoid basal cell carcinoma (Gorlin) syndrome. *Genet Med.* 2004;6(6):530–9.

6. High A, Zedan W. Basal cell nevus syndrome. *Curr Opin Oncol.* 2005;17(2):160–6.

7. Hahn H, Wicking C, Zaphiropoulous PG et al. Mutations of the human homolog of Drosophila patched in the nevoid basal cell carcinoma syndrome. *Cell.* 1996;85(6):841–51.

8. Gorlin RJ. Nevoid basal cell carcinoma syndrome. *Dermatol Clin.* 1995;13(1):113–25.

9. Human Gene Mutation Database. http://www.hgmd.cf.ac.uk.

10. Lam CW, Leung CY, Lee KC et al. Novel mutations in the PATCHED gene in basal cell nevus syndrome. *Mol Genet Metab.* 2002;76(1):57–61.

11. Cowan R, Hoban P, Kelsey A, Birch JM, Gattamaneni R, Evans DG. The gene for the naevoid basal cell carcinoma syndrome acts as a tumour-suppressor gene in medulloblastoma. *Br J Cancer.* 1997;76(2):141–5.

12. Song YL, Zhang WF, Peng B, Wang CN, Wang Q, Bian Z. Germline mutations of the PTCH gene in families with odontogenic keratocysts and nevoid basal cell carcinoma syndrome. *Tumour Biol.* 2006;27(4):175–80.

13. Acocella A, Sacco R, Bertolai R, Sacco N. Genetic and clinicopathologic aspects of Gorlin-Goltz syndrome (NBCCS): Presentation of two case reports and literature review. *Minerva Stomatol.* 2009;58(1–2):43–53.

14. Kiwilsza M, Sporniak-Tutak K. Gorlin-Goltz syndrome––a medical condition requiring a multidisciplinary approach. *Med Sci Monit.* 2012;18(9):RA145–53.

15. Wicking C, Shanley S, Smyth I et al. Most germ-line mutations in the nevoid basal cell carcinoma syndrome lead to a premature termination of the PATCHED protein, and no genotype-phenotype correlations are evident. *Am J Hum Gen* 1997;60(1):21–6.

16. Taylor SF, Cook AE, Leatherbarrow B. Review of patients with basal cell nevus syndrome. *Ophthal Plast Reconstr Surg.* 2006;22(4):259–65.

17. Farndon PA, Del Mastro RG, Evans DG, Kilpatrick MW. Location of gene for Gorlin syndrome. *Lancet.* 1992;339(8793):581–2.

18. Evans DG, Birch JM, Orton CI. Brain tumours and the occurrence of severe invasive basal cell carcinoma in first degree relatives with Gorlin syndrome. *Br J Neurosurg.* 1991;5(6):643–6.

19. Pratt MD, Jackson R. Nevoid basal cell carcinoma syndrome. A 15–year follow-up of cases in Ottawa and the Ottawa Valley. *J Am Acad Dermatol.* May 1987;16(5):964–70.

20. Ahn SG, Lim YS, Kim DK, Kim SG, Lee SH, Yoon JH. Nevoid basal cell carcinoma syndrome: A retrospective analysis of 33 affected Korean individuals. *Int J Oral Maxillofac Surg.* 2004;33(5):458–62.

21. Endo M, Fujii K, Sugita K, Saito K, Kohno Y, Miyashita T. Nationwide survey of nevoid basal cell carcinoma syndrome in Japan revealing the low frequency of basal cell carcinoma. *Am J Med Genet A.* Part A 2012;158A(2):351–7.

22. Lo Muzio L, Nocini PF, Savoia A et al. Nevoid basal cell carcinoma syndrome. Clinical findings in 37 Italian affected individuals. *Clin Genet.* 1999;55(1):34–40.

23. Shanley S, Ratcliffe J, Hockey A et al. Nevoid basal cell carcinoma syndrome: Review of 118 affected individuals. *Am J Med Gen* 1994;50(3):282–90.

24. Cohen MM. Nevoid basal cell carcinoma syndrome: Molecular biology and new hypotheses. *Int J Oral Maxillofac Surg.* 1999;28(3):216–23.

25. Manfredi M, Vescovi P, Bonanini M, Porter S. Nevoid basal cell carcinoma syndrome: A review of the literature. *Int J Oral Maxillofac Surg.* 2004;33(2):117–24.

26. Evans DG, Ladusans EJ, Rimmer S, Burnell LD, Thakker N, Farndon PA. Complications of the naevoid basal cell carcinoma syndrome: Results of a population based study. *J Med Genet.* 1993;30(6):460–4.

27. Giuliani M, Di Stefano L, Zoccali G, Angelone E, Leocata P, Mascaretti G. Gorlin syndrome associated with basal cell carcinoma of the vulva: A case report. *Eur J Gynaecol Oncol.* 2006;27(5):519–22.

28. Lo Muzio L. Nevoid basal cell carcinoma syndrome (Gorlin syndrome). *Orphanet J Rare Dis.* Nov 25 2008;3:32.

29. Ramaglia L, Morgese F, Pighetti M, Saviano R. Odontogenic keratocyst and uterus bicornis in nevoid basal cell carcinoma syndrome: Case report and literature review. *Oral Surg Oral Med Oral Pathol Oral Radiol Endod.* 2006;102(2):217–9.

30. Friedlander AH, Herbosa EG, Peoples JR. Ocular hypertelorism, facial basal cell carcinomas, and multiple odontogenic keratocysts of the jaws. *J Am Dent Assoc.* 1988;116(7):887–9.

31. Gutierrez MM, Mora RG. Nevoid basal cell carcinoma syndrome. A review and case report of a patient with unilateral basal cell nevus syndrome. *J Am Acad Dermatol.* 1986;15(5):1023–30.

32. Gorlin RJ. Nevoid basal-cell carcinoma syndrome. *Medicine* 1987;66(2):98–113.

33. Wilson LC, Ajayi-Obe E, Bernhard B, Maas SM. Patched mutations and hairy skin patches: A new sign in Gorlin syndrome. *Am J Med Genet A.* 2006;140(23):2625–30.

34. Chen JJ, Sartori J, Aakalu VK, Setabutr P. Review of ocular manifestations of nevoid basal cell carcinoma syndrome: What an ophthalmologist needs to know. *Middle East Afr J Ophthalmol* 2015;22(4):421–7.

35. Kodama T, Hayasaka S, Setogawa T. Myelinated retinal nerve fibers: Prevalence, location and effect on visual acuity. *Ophthalmologica.* 1990;200(2):77–83.

36. Khoubesserian P, Baleriaux D, Toussaint D, Telerman-Toppet N, Coers C. Adult form of basal cell naevus syndrome: A family study. *J Neurol.* 1981;226(3):157–68.

37. Salati C, Virgili G, Menchini U, Frattasio A, Patrone G. Gorlin's syndrome. Case report. *Eur J Ophthalmol.* 1997;7(1):113–4.

38. De Potter P, Stanescu D, Caspers-Velu L, Hofmans A. Photo essay: Combined hamartoma of the retina and retinal pigment epithelium in Gorlin syndrome. *Arch Ophthalmol.* 2000;118(7):1004–5.

39. Kimonis VE, Goldstein AM, Pastakia B et al. Clinical manifestations in 105 persons with nevoid basal cell carcinoma syndrome. *Am K Med Genet* 1997;69(3):299–308.

40. Brzozowski F, Wanyura H, Stoma Z, Kowalska K. Odontogenic keratocysts in the material of the Department of Craniomaxillofacial Surgery, Medical University of Warsaw. *Stomatol* 2010;63(2):69–78.

41. Larsen AK, Mikkelsen DB, Hertz JM, Bygum A. Manifestations of Gorlin-Goltz syndrome. *Dan Med J* 2014;61(5):A4829.

42. Ratcliffe JF, Shanley S, Chenevix-Trench G. The prevalence of cervical and thoracic congenital skeletal abnormalities in basal cell naevus syndrome; a review of cervical and chest radiographs in 80 patients with BCNS. *Br J Radiol.* 1995;68(810):596–9.

43. Bossert T, Walther T, Vondrys D, Gummert JF, Kostelka M, Mohr FW. Cardiac fibroma as an inherited manifestation of nevoid basal–cell carcinoma syndrome. *Tex Heart Inst J.* 2006;33(1):88–90.

44. Kraemer BB, Silva EG, Sneige N. Fibrosarcoma of ovary. A new component in the nevoid basal–cell carcinoma syndrome. *Am J Surg Pathol.* 1984;8(3):231–6.

45. Seracchioli R, Colombo FM, Bagnoli A, Trengia V, Venturoli S. Primary ovarian leiomyosarcoma as a new component in the nevoid basal cell carcinoma syndrome: A case report. *Am J Obstet Gynecol.* 2003;188(4):1093–5.

46. Hodak E, Ginzburg A, David M, Sandbank M. Etretinate treatment of the nevoid basal cell carcinoma syndrome. Therapeutic and chemopreventive effect. *Int J Dermatol.* 1987;26(9):606–9.

47. Salasche SJ. Status of curettage and desiccation in the treatment of primary basal cell carcinoma. *J Am Acad Dermatol.* 1984;10(2):285–7.

48. Silverman MK, Kopf AW, Bart RS, Grin CM, Levenstein MS. Recurrence rates of treated basal cell carcinomas. Part 3: Surgical excision. *J Dermatol Surg Oncol* 1992;18(6):471–6.

49. Silverman MK, Kopf AW, Grin CM, Bart RS, Levenstein MJ. Recurrence rates of treated basal cell carcinomas. Part 2: Curettage-electrodesiccation. *J Dermatol Surg Oncol* 1991;17(9):720–6.

50. Spiller WF, Spiller RF. Treatment of basal cell epithelioma by curettage and electrodesiccation. *J Am Acad Dermatol.* 1984;11(5):808–14.

51. Krunic AL, Viehman GE, Madani S, Clark RE. Microscopically controlled surgical excision combined with ultrapulse CO2 vaporization in the management of a patient with the nevoid basal cell carcinoma syndrome. *J Dermatol.* 1998;25(1):10–2.

52. Kopera D, Cerroni L, Fink-Puches R, Kerl H. Different treatment modalities for the management of a patient with the nevoid basal cell carcinoma syndrome. *J Am Acad Dermatol.* 1996;34(5):937–9.

53. Peng Q, Warloe T, Berg K et al. 5–Aminolevulinic acid-based photodynamic therapy. Clinical research and future challenges. *Cancer.* 1997;79(12):2282–308.

54. Svanberg K, Andersson T, Killander D et al. Photodynamic therapy of non-melanoma malignant tumours of the skin using topical delta-amino levulinic acid sensitization and laser irradiation. *Br J Dermatol.* 1994;130(6):743–51.

55. Wolf P, Rieger E, Kerl H. Topical photodynamic therapy with endogenous porphyrins after application of 5–aminolevulinic acid. An alternative treatment modality for solar keratoses, superficial squamous cell carcinomas, and basal cell carcinomas? *J Am Acad Dermatol.* 1993;28(1):17–21.

56. Kopf AW, Bart RS, Schrager D, Lazar M, Popkin GL. Curettage-electrodesiccation treatment of basal cell carcinomas. *Atch Dermatol* 1977;113(4):439–43.

57. Nordin P, Larko O, Stenquist B. Five-year results of curettage-cryosurgery of selected large primary basal cell carcinomas on the nose: An alternative treatment in a geographical area underserved by Mohs' surgery. *Br J Dermatol.* 1997;136(2):180–3.

58. Rowe DE, Carroll RJ, Day CL. Long-term recurrence rates in previously untreated (primary) basal cell carcinoma: Implications for patient follow-up. *J Dermatol Durg Oncol* 1989;15(3):315–28.

59. Goette DK. Topical chemotherapy with 5–fluorouracil. A review. *J Am Acad Dermatol.* 1981;4(6):633–49.

60. Rowinsky EK, Donehower RC. Paclitaxel (taxol). *N Eng J Med* 1995;332(15):1004–14.

61. Bath-Hextall F, Ozolins M, Armstrong SJ et al. Surgical excision versus imiquimod 5% cream for nodular and superficial basal-cell carcinoma (SINS): A multicentre, non-inferiority, randomised controlled trial. *Lancet Oncol.* 2014;15(1):96–105.

62. Oldfield V, Keating GM, Perry CM. Imiquimod: In superficial basal cell carcinoma. *Am J Clinic Dermatol* 2005;6(3):195–200; discussion 201–192.

63. Stoelinga PJ. Long-term follow-up on keratocysts treated according to a defined protocol. *Int J Oral Maxillofac Surg.* 2001;30(1):14–25.

64. Voorsmit RA, Stoelinga PJ, van Haelst UJ. The management of keratocysts. *J Maxillofac Surg* 1981;9(4):228–36.

65. O'Malley S, Weitman D, Olding M, Sekhar L. Multiple neoplasms following craniospinal irradiation for medulloblastoma in a patient with nevoid basal cell carcinoma syndrome. Case report. *J Neurosurg.* 1997;86(2):286–8.

86

Beckwith–Wiedemann Syndrome

Jirat Chenbhanich, Sirisak Chanprasert, and Wisit Cheungpasitporn

CONTENTS

86.1 Introduction

Beckwith–Wiedemann syndrome (BWS; OMIM#130650) is the most common genetic overgrowth syndrome involving a predisposition to tumor development. In the 1960s, Dr. Beckwith, an American pathologist, and Dr. Wiedemann, a German pediatrician, independently proposed a new syndrome initially termed EMG (an acronym for exomphalos, macroglossia, and gigantism) syndrome [1,2]. Over time, this constellation was renamed as BWS since the clinical phenotype is highly variable and some cases lacks the three hallmark features initially described. BWS occurs in males and females equally and is a pan-ethnic disorder affecting 1 in 10,500 to 13,700 individuals [3,4]. This number may be underestimated due to an underdiagnosis of cases with milder phenotypes. There is an increased incidence of monozygotic twins among BWS children, most of which are female and discordant (only one of them is affected) [5]. Assisted reproductive technology (ART) has also been associated with increased risk of BWS in the offspring. Apart from phenotypic heterogeneity, patients with BWS exhibit a great (epi)genetic heterogeneity involving the chromosome 11p15.5.

86.2 Biology

Most human autosomal genes are expressed from both maternally and paternally inherited copies. This biparental contribution to the genome is vital; the presence of only maternal or paternal genome results in significant developmental aberrations such as hydatidiform molar pregnancies caused by androgenetic conceptus. Imprinting is an epigenetic phenomenon in which specific genes are monoallelically expressed according to the parent-of-origin allele; i.e., some imprinted genes may preferentially express only in maternal chromosomes while some others may be expressed only in paternal chromosomes [6]. It has been established that imprinted genes are implicated in growth regulation in an opposite function: paternally derived alleles often promote growth (e.g., insulin-like growth factor 2 [*IGF2*]) and maternally derived alleles often suppress it (e.g., cyclin-dependent kinase inhibitor 1C [*CDKN1C*]). These genes are thus potential oncogenes and tumor suppressor genes. The most well-known theory to explain this genomic pattern is the parental conflict hypothesis, which suggests that imprinting evolved because of a conflict between the interests of maternal and paternal genes during fetal and neonatal development [7].

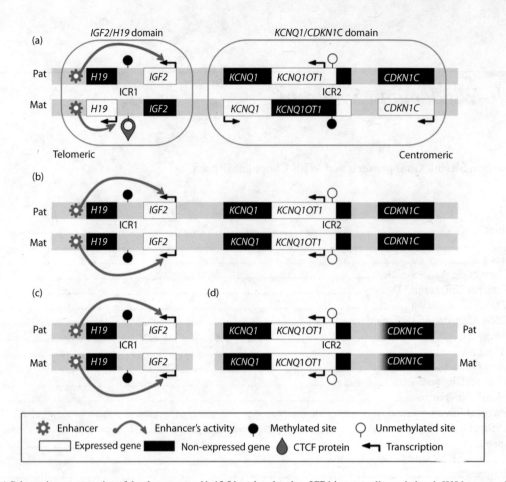

FIGURE 86.1 (a) Schematic representation of the chromosome 11p15.5 imprinted region. ICR1 is paternally methylated. *H19* is maternally expressed and *IGF2* is paternally expressed due to an insulator effect of CTCF protein. ICR2 is maternally methylated. *CDKN1C* and *KCNQ1* are maternally expressed while *KCNQ1OT1* is paternally expressed. Both *H19* and *KCNQ1OT1* encode non-coding RNAs. (b) Paternal uniparental disomy of chromosome 11p15.5. Note the altered methylations at both ICR1 and ICR2. (c) ICR1 hypermethylation. (d) ICR2 hypomethylation. Note that the distances between each locus do not represent actual chromosomal distances. *Abbreviations:* Pat, paternal chromosome; Mat, maternal chromosome; CTCF, CCCTC-binding factor; ICR, imprinting control region.

Of the approximately 100 imprinted genes known to date, most are clustered together, with each cluster containing 2–15 genes and varying in size from <100 kb to several megabases [6]. The establishment of imprints—"the imprinting cycle"—is sophisticated and needs to be reset at each generation [8]. The expression of genes in the imprinting cluster is controlled by discreet DNA segments called imprinting control region (ICR). In the primordial germ cells, human epigenome undergoes extensive reprogramming, where global erasure of DNA methylation occurs and parent-specific imprints are removed. During gametogenesis, ICRs are differentially methylated in either paternal or maternal germline. Following fertilization, these marked regions are robustly protected against a wave of genome-wide demethylation and subsequent de novo methylation. These differentially methylated regions later regulate the paternally or maternally expressed alleles of many genes in the same region during human development. Dysregulation of imprinting will result in imprinting disorders, of which the most well-known are 15q11-13-associated Prader–Willi syndrome and Angelman syndrome, and 11p15.5-associated BWS and Silver–Russell syndrome.

The 11p15.5 locus spans around 1 Mb and is subdivided into two independent imprinted domains with separate ICRs: the

telomeric *IGF2/H19* domain controlled by ICR1 and the centromeric *KCNQ1/CDKN1C* domain controlled by ICR2 (Figure 86.1a). These two ICRs are imprinted in an opposite pattern: ICR1 is marked in a paternal germline, whereas ICR2 is marked in a maternal germline. Furthermore, they employ different molecular mechanisms to regulate the differential expression of their imprinted genes: the insulator model for ICR1 and the long non-coding RNA model for ICR2 (see below). It has been suggested that the 11p15.5 region is the central element of a network of imprinted genes [9].

The *IGF2/H19* domain comprises the maternally expressed *H19* gene, which encodes a non-coding RNA of unknown function, and the paternally expressed *IGF2* gene, which encodes a protein that promotes embryonic and placental growth. Clinical overgrowth observed in BWS is mainly seen in tissues in which *IGF2* is expressed [10]. Both genes share a common enhancer downstream of *H19*, are extensively expressed during embryogenesis, and are downregulated in adults. The ICR1 is intergenic, and the active unmethylated maternal ICR1 forms an enhancer-blocking transcriptional insulator by binding to the zinc-finger protein, CCCTC-binding factor (CTCF) at numerous CTCF-binding sites. The insulator subsequently blocks access of the enhancer to *IGF2*, silencing its transcription, and results in maternal expression of *H19*. Paternal methylation of ICR1, on the other hand, prevents the binding of CTCF and allows the enhancer to reach *IGF2*, triggering paternal expression of *IGF2*. Recently, other regulatory molecules such as cohesin were found to mediate the transcription of *IGF2/H19* domain through the formation of chromatin loops [11].

The *KCNQ1/CDKN1C* domain includes at least three important imprinted genes: two maternally coding genes, *KCNQ1* and *CDKN1C*, and one paternally non-coding gene, *KCNQ1OT1*. *KCNQ1* encodes a voltage-gated potassium channel, and its germline mutation implicates in various inherited cardiac arrhythmias such as familial atrial fibrillation and long QT syndrome, including autosomal dominant Romano–Ward syndrome and autosomal recessive Jervell and Lange–Nielsen syndrome. The mutant potassium channel does not appear to involve in the pathogenesis of BWS directly. However, the untranslated transcripts from *KCNQ1* allele might indirectly play an undiscovered role in the imprinting process. *CDKN1C* encodes a cell cycle regulator that inhibits cell proliferation by interacting with cyclin/CDK complexes during the G1 phase of cell cycle. The germline loss-of-function mutation of *CDKN1C* is associated with an overgrowth syndrome, BWS, whereas its gain-of-function mutation results in growth-retardation syndromes including Silver–Russell syndrome and IMAGe (intrauterine growth retardation, metaphyseal dysplasia, adrenal hypoplasia congenita, and genital anomalies) syndrome [12]. ICR2, in contrast to ICR1, is intragenic, locating within the 11th intron of the *KCNQ1* gene, and contains the promotor of *KCNQ1OT1* gene which encodes an antisense long non-coding RNA. ICR2 is not methylated in the paternal chromosome, and its *KCNQ1OT1* transcript subsequently silences all the paternal imprinted genes within the domain, yet the mechanism by which long non-coding RNA silences imprinted genes is not fully understood. ICR2 is methylated on the maternal chromosome, resulting in the silencing of *KCNQ1OT1* transcription and the expression of *KCNQ1* and *CDKN1C*. Similar to ICR1, CTCF-binding sites have been identified in ICR2, and it has been postulated that the insulator model may contribute to the differential expression of the *KCNQ1/CDKN1C* domain as well [13].

86.3 Pathogenesis

BWS is an overgrowth syndrome caused by heterogeneous (epi) genetic or cytogenetic defects involving chromosome 11p15.5; the defect can either affect the whole region or specifically confine at one of the two ICRs. These aberrations result in downregulation of maternally expressed growth-restraining genes (such as *CDKN1C*) and/or upregulation of paternally expressed growth-promoting genes (such as *IGF2*). BWS usually occurs sporadically, whereas its familial form has been reported in 10%–15% of the cases.

Several molecular abnormalities haven been ascribed to the pathogenesis of BWS (Figure 86.2):

- Karyotypic abnormalities
- Paternal uniparental disomy (UPD)
- *CDKN1C* mutation
- DNA methylation defects, including ICR1 hypermethylation and ICR2 hypomethylation

86.3.1 Karyotypic Abnormalities

A small subset of patients with BWS (1%–2%) harbors karyotype-detectable cytogenetic abnormalities of chromosome 11p15.5 including de novo and maternally transmitted translocations/inversions and paternally derived duplications [3]. There have been three distinct translocation breakpoints identified within 11p15.5, namely BWS critical regions 1–3 (BWSCR1–3). BWSCR1 contains a cluster of translocation breakpoints and lies within a region of multiple imprinted genes such as *IGF2*, *H19*, *CDKN1C*, and *KCNQ1* [14]. All breakpoints in BWSCR1 disrupt *KCNQ1* expression. The exact mechanism by which the translocations in this region affect other imprinted genes are poorly understood; in at least two patients, the translocations resulted in the loss of imprinting in *IGF2*—meaning that there is a biallelic expression of *IGF2* gene [15,16]. BWSCR2 and BWSCR3 are located, respectively, 5Mb and 7Mb proximal to BWSCS1 [17]. It has been proposed that there is a gene(s) or regulatory region at BWSCR2/3 that interacts with the *IGF2/ BWSCR1* system and consequently influences the BWS phenotype [14].

86.3.2 Paternal Uniparental Disomy

UPD is defined by the presence of two chromosomal regions from one parent and none from the other. Paternal UPD of 11p15.5 (Figure 86.1b) accounts for 20% of BWS cases and occurs only in sporadic cases. It generally affects both ICR1 and

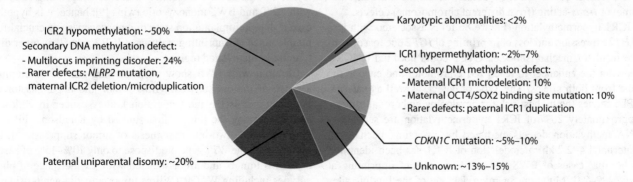

FIGURE 86.2 Molecular (epi)genetics of BWS.

ICR2 and causes clinical phenotype due to overexpression of paternal alleles such as *IGF2*, and silencing of maternal alleles such as *CDKN1C*. Due to the significant changes in gene dosage, individuals with paternal UPD have a higher tumor risk, especially Wilms tumor and hepatoblastoma when compared to BWS patients in general [18]. In BWS, paternal UPD occurs as a post-fertilization mitotic recombination event and exhibits somatic mosaicism; thus, the higher level of UPD cells in specific organs and tissues should theoretically associate with a more severe phenotype [19]. This was corroborated by two cases of BWS with extremely high levels of UPD in DNA from lymphocytes. Both of them presented with extreme macroglossia, persistent hypoglycemia, cardiomyopathy, and hepatoblastoma, and died in the first 6 months of life [20]. Recently, a novel molecular abnormality was described in BWS patients in whom additional features of UPD, including premature thelarche, conjugated hyperbilirubinemia, and atypical tumors, were presented. Mosaic genome-wide paternal UPD—characterized by a mosaic blend of paternal uniparental and biparental cell lineages—was discovered by SNP array testing [21]. The findings expand our knowledge of the defective molecular spectrum of BWS and emphasize the importance of a degree of mosaicism in phenotypic variability.

86.3.3 CDKN1C Mutation

Loss-of-function mutations of *CDKN1C* functionally result in loss of cell cycle inhibition and are found in up to 5%–10% of sporadic and 50% of familial cases of BWS [12]. Patients with the mutation may have abdominal wall defect and often present with cleft palate, genital anomalies, polydactyly, and supernumerary nipples, as well as maternal preeclamptic disorders [12]. Patients with *CDKN1C* mutation may have the lowest tumor risk when compared to other molecular subgroups; however, the current number of patients is too small to draw a conclusion [3].

86.3.4 DNA Methylation Defects

A majority of BWS cases are caused by epigenetic alterations at ICRs: approximately 2%–7% for gain of DNA methylation at ICR1 and 50%–60% for loss of DNA methylation at ICR2. Advances in (epi)genomics have further subdivided these changes into primary DNA methylation defect when there is no associated genomic change, and secondary DNA methylation defect resulting from either *cis*-acting (within the same chromosome) or *trans*-acting (from different chromosome) defects.

ICR1 hypermethylation (Figure 86.1c) is associated with loss of *H19* expression and loss of imprinting of *IGF2* due to a disruptive insulator mechanism—i.e., CTCF could not bind to ICR1, allowing the enhancer to reach *IGF2* promotor and encode the gene from both alleles. This type of defect, as well as paternal UPD, carries the highest risk for developing embryonal tumors. Approximately 20% of ICR1 hypermethylation are secondary DNA methylation defects mediated by *cis*-acting mechanism. Maternal 1.4–2.2 kb microdeletion of ICR1 has been identified in familial cases of BWS and results in loss of imprinting at *IGF2* [22–24]. Mutations or microdeletion of the binding sites for pluripotency factors (such as OCT4 and SOX2) within ICR1

have been observed in maternally transmitted patients with ICR1 hypermethylation, suggesting the importance of pluripotency factors in the maintenance of the unmethylated state of maternal ICR1 [25,26]. Rarely, paternal ICR1 duplication is associated with ICR1 hypermethylation [8].

ICR2 hypomethylation (Figure 86.1d) resulting in biallelic transcription of *KCNQ1OT1* RNA is by far the most common genetic aberration in BWS. This defect reduces *CDKN1C* expression, similar to loss-of-function mutation of *CDKN1C* gene. The *cis*-acting mechanisms, including maternal ICR2 deletions and microduplications, have been observed in a handful of cases with ICR2 hypomethylation [8]. Rarely, the microdeletion encompassing ICR2 and *KCNQ1* gene causing long QT syndrome and mild BWS phenotypes have been reported [27].

Until recently, there has been increasing evidence of *trans*-acting mechanism as a cause of secondary DNA methylation defect; in fact, it accounts for approximately 25% of BWS patients with ICR2 hypomethylation and includes multilocus imprinting disorder and *NLRP2* mutation [8]. Approximately 24% of BWS patients with ICR2 hypomethylation displayed a loss of methylation at multiple loci across the genome [28]—a condition called multilocus imprinting disorder. In multilocus imprinting disorder with BWS phenotype, the commonly affected loci include ICR2, *GNAS*, *IGF2R*, and *PEG1* [8]. Furthermore, a homozygous *NLRP2* mutation was found in a mother of two children with BWS and ICR2 hypomethylation, with one of the children demonstrating multilocus imprinting defect with partial loss of methylation at *PEG1* locus [29]. *NLRP2* is located on chromosome 19q13.4; this finding suggested that NLRP2 has a role in establishing and maintaining genomic imprinting in humans.

It is noteworthy that ICR2 hypomethylation is associated with two other epiphenomena: female monozygotic twinning and ART. There have been several reports of female monozygotic twins with discordant BWS phenotype (i.e., only one twin is affected). This is probably due to unequal splitting of the inner cell mass during twinning, therefore causing differential methylation profile at ICR2 [5]. Moreover, a case-control study indicated that the risk of having a baby with BWS after in vitro fertilization procedure is ~1/4000, or nine times greater than in the general population [30]. The molecular mechanism underlying ART-associated BWS shows a preponderance to ICR2 hypomethylation, although other mechanisms such as UPD and ICR1 hypermethylation have been described. Approximately 95% of BWS children born after ART had ICR2 hypomethylation defect [3]. Some studies have correlated multilocus imprinting disorder with ART and BWS monozygotic twins [8]; hence, it is hypothesized that preimplantation state of embryogenesis is crucial for maintenance of imprinting statuses at ICR2 as well as other differentially methylated regions.

Children with BWS show an increasing risk of developing a range of tumors such as Wilms tumor and hepatoblastoma. Wilms tumor is the most common kidney cancer in children and was among the tumors first studied by Knudson while he developed the "two-hit" hypothesis of tumor suppression [31]. Mutations in the *WT1* gene predispose to only 10%–15% of cases of Wilms tumors and are associated with a wide range of phenotypes including WAGR (Wilms tumor-aniridia-genitourinary abnormalities-mental retardation) syndrome [32]. Genomic study has revealed that loss of imprinting at 11p15.5 locus is an early stage of tumorigenesis in Wilms tumor [33]. Growing evidence has also suggested that dosage imbalance of *IGF2* and *H19* is an integral part of tumorigenesis; in fact, *H19* is believed to be a second Wilms tumor suppressor gene [34,35]. First, epigenetic defects at ICR1 are observed not only in BWS but also in cases of isolated Wilms tumor. Conversely, isolated ICR2 defects have not been identified in isolated Wilms tumor cases [36]. Second, Wilms tumors *IGF2* transcripts were highly elevated when compared to adjacent normal kidney tissues [37]. Third, patients carrying ICR1 hypermethylation, either in isolation or occurring with paternal UPD, harbor the highest risk for developing Wilms tumor [3]. To strengthen this genotype–phenotype correlation, it is known that there is no increased risk of Wilms tumor in 11p15.5-associated growth-retardation syndrome (e.g., Silver–Russell syndrome) since the molecular defects observed in these individuals resulted in reduced *IGF2* expression, which is opposite in BWS.

86.4 Epidemiology

BWS affects both males and females equally and has an estimated frequency of 1 per 13,700 newborns worldwide. This may be an underestimation since some patients with mild symptoms are never diagnosed. While about 85% of cases are sporadic, with only one member in the family being diagnosed with the condition, up to 15% of cases are transmitted in an autosomal dominant pattern, with more than one member in the family being affected.

86.5 Clinical Features

*   *Perinatal period*: Polyhydramnios and macrosomia are the most frequently reported prenatal abnormalities [38,39]. Increased abdominal circumference, visceromegaly, macrosomia, renal anomalies, enlarged placenta, and abdominal wall defects—e.g., omphalocele, umbilical hernia, and diastasis recti—can be detected prenatally and may lead to the clinician's suspicion of the syndrome. The frequency of BWS among cases with omphalocele has ranged from <3% to 20% [40,41]. Placental mesenchymal dysplasia, an uncommon vasculopathy of placenta characterized by multicystic placental lesions on ultrasonography, has been observed in BWS with paternal UPD [42]. BWS has been associated with maternal hyperreactio luteinalis, which is characterized by bilateral ovarian enlargement with multiple theca lutein cysts and is detectable by ultrasound after the first trimester [43]. The umbilical cord can be thickened. Half of the affected infants are born prematurely, and the risk of prematurity is associated with polyhydramnios [38]. Maternal complications in pregnancies with BWS fetuses include gestational hypertension, preeclampsia, and vaginal bleeding. Three fetuses have been associated with maternal HELLP syndrome (hemolysis, elevated liver enzymes, and low platelet count), after which the molecular analysis revealed pathogenic *CDKN1C* mutations [44].

*   *Metabolic*: Neonatal hypoglycemia, likely due to islet cell hyperplasia and hyperinsulinism, is well documented and is reported in 30%–50% of babies with BWS. It can be persistent for months, requiring aggressive therapy [45,46]. Hypothyroidism, hyperlipidemia, and polycythemia are also described.

*   *Growth*: Macrosomia, postnatal overgrowth, macroglossia, visceromegaly, and hemihyperplasia are common. Macrosomia is defined when birth weight exceeds the 90th percentile for gestational age; it occurs in approximately half of the patients [47]. With rapid growth in early childhood, the growth parameters typically show height and weight around the 97th percentile, with head circumference closer to the 50th percentile (relative microcephaly) [3,47]. When analyzing growth parameters in 35 adults with BWS, mean adult height was approximately +1.8 standard deviation scores, with 18 of them having final heights above +2 standard deviation scores [48]. Macrosomia is typical in cases with ICR1 hypermethylation, whereas postnatal overgrowth is usually found in ICR2 hypomethylation. Macroglossia is the most common feature in BWS, presenting in 90%–97% of affected individuals, and therefore represents the most sensitive phenotypic trait for identification [47]. Its complications include feeding difficulties, speech problems, and obstructive sleep apnea. Untreated macroglossia may lead to prognathism, open anterior bite, and dental problems. Visceromegaly involving kidneys, thymus, liver, spleen, heart, adrenal glands, and/or pancreas can be observed [3,49]. Fetal adrenocortical cytomegaly is reported to be a pathognomonic pathology of BWS; however, this feature is rarely available in clinical practice. Hemihyperplasia results from an abnormal cellular proliferation causing asymmetric overgrowth of one or more regions of the body. It can lead to limb–leg discrepancy which typically develops during the first 2 years of life and is responsible for secondary scoliosis and joint overloading-related problems. Hemihyperplasia is common in BWS with UPD and also increases the risk of tumor development [47].

*   *Neurodevelopment*: Psychomotor development in BWS is usually normal [38]. Intellectual disability may result from chromosomal abnormalities, such as paternally derived chromosomal 11p15.5 duplications [50], or as sequelae of serious perinatal events including hypoxia or persistent hypoglycemia. Other reported neurologic features include posterior fossa malformations, sensorineural hearing loss, and conductive hearing loss secondary to stapes anomalies [47,51,52]. A survey assessing neurobehavioral aspect of BWS children revealed that 6.8% of them were diagnosed with autistic spectrum disorder. Furthermore, these children tended to have more emotional difficulties and peer problems at school when compared to non-learning disability children [53]. Children with BWS are at high risk for various

behavioral and psychological conditions such anxiety, depression, and poor social functioning, possibly due to chronicity of the condition and cancer diagnosis.

- *Cardiopulmonary*: Cardiac malformations are found in ~20% of BWS cases; half of them have spontaneously resolving cardiomegaly, whereas cardiomyopathy is rare [3]. There is no specific type of congenital cardiac defect predominating in BWS [54]. In rare instances, children with molecular defects involving *KCNQ1* and ICR2 can present with long QT syndrome [27,55]. Two of out four adults with BWS in one series demonstrated aneurysmal dilatation of large vessels [56]. Pediatric sleep-disordered breathing is a group of disorders including apnea of prematurity, central apnea, and obstructive hypoventilation disorders such as obstructive sleep apnea; they are common and are found in 48% of BWS children [57]. Obstructive sleep apnea may result in alveolar hypoventilation with subsequent pulmonary hypertension and cor pulmonale.

- *Genitourinary*: Approximately 28%–61% of BWS patients have benign genitourinary anomalies including nephromegaly, collecting system abnormalities, cryptorchidism, inguinal hernia, cystic or dysplastic changes, nephrocalcinosis, and nephrolithiasis; some features may persist in adulthood [3,56,58,59]. Presence of nephromegaly or nephrogenic rests increases the risk of Wilms tumor. Children with ICR1 hypermethylation almost consistently have genitourinary anomalies, with nephromegaly being reported in all patients and collecting system abnormalities in half of them [58]. Approximately one-fifth develop urinary tract infections which may be complicated by severe sepsis [58]. Intrauterine vesicoureteral reflex with subsequent hydroureter can result in polyhydramnios [47]. Although hyper- or hypocalcemia do not characterize the syndrome, hypercalciuria was reported in 22% of the patients and can lead to its complications such as nephrocalcinosis and nephrolithiasis [59]. Cryptorchidism is commonly observed in children with BWS and is associated with abdominal wall defects. During adulthood, abnormal testicular functions such as impaired spermatogenesis and testicular atrophy, both of which potentially leading to infertility, have been reported [56].

- *Dysmorphism*: Distinctive facial features may be recognized by experienced clinicians: prominent eyes with infraorbital creases, facial nevus flammeus, relative microcephaly with metopic ridge, midfacial hypoplasia, macroglossia, full lower face with a prominent mandible, anterior earlobe creases, and posterior helical pits [3,47]. Evaluation of adolescents or adults may require assessment of early childhood photographs since these features tend to regress as children grow up. In cases with *CDKN1C* mutation, additional dysmorphic features including polydactyly, cleft palate, genital anomalies, and supernumerary nipples are observed. The presence of these features may therefore prompt clinicians for *CDKN1C* mutation analysis [12].

86.6 Diagnostic Approach

86.6.1 Clinical Diagnosis

Several clinical diagnostic criteria have been proposed for BWS [3,38,60–62]; each of them has distinct sensitivities, specificities, and predictive values [63]. In 2010, BWS experts including Dr. J. Bruce Beckwith and Dr. Rosanna Weksberg published the criteria in which the presence of three major or two major and one minor criteria supports a clinical diagnosis (Table 86.1) [3]. Subsequent positive molecular analysis at the chromosomal loci 11p15.5 would confirm the diagnosis of BWS but not presently rule it out. The negative results may be due to the detection failure in cases of UPD with mosaicism or may reflect other underlying, yet-undiscovered abnormalities. Of note, BWS is one of the clinical phenotypes resulted from (epi)genetic defects at chromosome 11p15.5, with the other syndromes being isolated hemihyperplasia and isolated Wilms tumor. Also, children with 11p15.5 alterations may not fulfill clinical criteria and present with subtle features such as macroglossia, umbilical hernia, and/or nevus flammeus. Due to overall increased risk of tumor development, it is recommended to apply tumor surveillance protocol to those with isolated hemihyperplasia, those with molecular alterations but having milder phenotypes, and those with positive clinical criteria but lacking defective molecular evidence.

TABLE 86.1

Diagnostic Criteria for BWS

Clinical Diagnostic Criteria[a]	Molecular Abnormality Outcome Score[b]
Major Criteria	
• Abdominal wall defects	1.5[c]
• Macroglossia	2.5
• Macrosomia	1
• Visceromegaly	1
• Hemihyperplasia	0.5
• Anterior ear lobe creases and/or posterior helical pits	
• Childhood embryonal tumor	
• Adrenocortical fetal cytomegaly	
• Renal abnormalities	
• Family history of BWS	
• Cleft palate	
Minor Criteria	
• Nevus flammeus	1
• Neonatal hypoglycemia	0.5
• Polyhydramnios, enlarged placenta and/or thickened umbilical cord, preterm labor	
• Cardiomegaly	
• Characteristic facies	
• Diastasis recti	
• Advanced bone age	

[a] Clinical diagnostic criteria proposed by Weksberg et al. (2010). The presence of three major or two major and one minor criteria supports a clinical diagnosis of BWS.
[b] Molecular Abnormality Outcome Score proposed by Ibrahim et al. (2014). The total scores and their corresponding probabilities of abnormal DNA methylation test are as follows: score of 1 point = 16%; 2 points = 30%; 3 points = 50%; 4 points = 68%; 5 points = 83%; 6 points = 92%; 7 points = 95%; 8 points = 98%.
[c] For omphalocele.

86.6.2 Molecular Diagnosis

Molecular diagnosis of BWS should be undertaken and interpreted by experienced clinical molecular geneticists considering its complexity requiring multistep approach (Figure 86.3). The principles of genetic tests in BWS include but are not limited to:

- Standard karyotype
- DNA methylation studies
- *CDKN1C* gene sequencing
- Chromosomal microarray (CMA)

Standard karyotype should be performed in all specimens to detect rare karyotypic defects—i.e., de novo and maternally transmitted translocations/inversions and paternally derived duplications. Methylation analysis such as methylation-specific multiplex ligation-dependent probe amplification (MS-MLPA) or methylation-sensitive Southern blotting (MS-SB) will further detect a majority of molecularly confirmed cases including ICR1 hypermethylation, ICR2 hypomethylation, and UPD. The currently most robust and commonly used method is MS-MLPA, which has an advantage of detecting small microdeletions/microduplications [64]. In cases of borderline results, both MS-MLPA and MS-SB techniques may be employed [47]. UPD needs to be confirmed by single nucleotide polymorphism (SNP) array ([65] or by microsatellite analysis [66–68]. Some authors have suggested performing methylation analysis and UPD confirmation at the same time to double-check the results,

since the two methods serve as reciprocal control [47]. Due to somatic mosaicism, failure to detect UPD in the first tissue (typically leukocytes) is inconclusive, and obtaining other tissues (such as skin or surgical specimens) may be necessary. Recently, Ibrahim et al. developed a new scoring system by utilizing logistic regression model and used each distinctive feature of BWS to predict a positive methylation abnormality in a particular patient (Table 86.1) [63]. The higher the score, the higher the probability of a patient having a positive test; therefore, this tool may help facilitate selection of patients suspected to have BWS and consequently improve diagnostic outcomes. Normal karyotype and negative methylation and UPD analyses will further require *CDKN1C* gene sequencing, since its loss-of-function mutations can be found in 5%–10% of sporadic and up to 50% of familial cases. *CDKN1C* gene analysis, together with standard karyotype, should be a first-tier test in a patient with a family history of BWS and is considered in children born to preeclamptic women or with cleft palate, hypospadias, and/or supernumerary nipples [12]. It is also prudent to screen parents and other relevant family members if these tests reveal heritable defects such as *CDKN1C* mutation, karyotypic abnormality, or microduplication/microdeletion. On the other hand, molecular testing is generally not indicated in family members when UPD is found since it arises from post-zygotic somatic recombination. In the familial cases with unremarkable karyotype and negative *CDKN1C* gene analysis, testing for heritable microduplication or microdeletion of chromosome 11p15.5 by CMA should be considered. CMA uses SNP arrays or oligonucleotide arrays to detect copy number variation

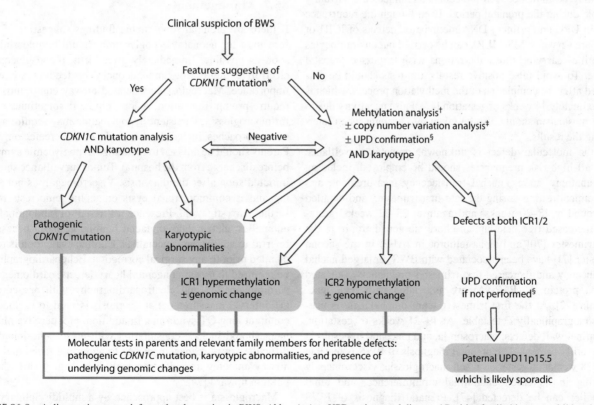

FIGURE 86.3 A diagnostic approach for molecular testing in BWS. *Abbreviation:* UPD, uniparental disomy. *Positive family history or children born to preeclamptic women or with cleft palate, hypospadias, and/or supernumerary nipples. †Methylation-specific multiplex ligation-dependent probe amplification (MS-MLPA) or methylation-sensitive Southern blotting (MS-SB). ‡Copy number variations such as microduplication/microdeletion can be detected by MS-MLPA or microsomal microarray. §Single-nucleotide polymorphism (SNP) array or microsatellite analysis.

such as microduplication/microdeletion; it should also be offered in cases of abnormal methylation defects to identify the small percentage of underlying microduplication/microdeletion, since the recurrence risk in families harboring these genomic alterations is significantly increased [69].

Although the molecular landscape of BWS has been expanding, we need more knowledge until the routine use of additional techniques detecting newly described (epi)genetic changes is incorporated into diagnostic paradigm. More studies are required to investigate an exact prevalence and thoroughly characterize the manifestations of multilocus-imprinting disorder. The analysis of other aberrations such as OCT4/SOX2 binding sites and *NLRP2* gene mutations are still clinically limited. Presence of confirmed UPD with atypical or severe features may prompt clinicians to utilize SNP arrays to search for mosaic genome-wide paternal UPD, as the management has been suggested based on few cases [21]. Finally, despite the fact that ICR1 hypermethylation secondary to underlying genomic changes (e.g., maternal microdeletion) was reportedly common, conventional tools such MS-MLPA may sometimes miss the lesions [8]. Due to the highly polymorphic nature of ICR1 region, the design of specific probes/primers and subsequent development of novel ICR1 diagnostic tools may be difficult at present.

86.6.3 Prenatal Diagnosis

If heritable molecular defects (e.g., karyotypic abnormality, *CDKN1C* mutation, or microdeletion/microduplication) are identified, prenatal testing by chorionic villus sampling or amniocentesis may be indicated. All molecular techniques are virtually feasible during the prenatal period. Even though the recurrence risks in UPD and primary DNA methylation defects of ICR1 or ICR2 are very low, MS-MLPA can be carried out on amniocytes if families carrying those aberrations wish to pursue prenatal testing. To avoid false positive results, the tests should be performed after the completion of the methylation process, which is approximately 14 weeks of gestation [47]. False negativity due to tissue mosaicism should also be taken into account when interpreting the results.

If the molecular defect is unknown, prenatal surveillance as in all high-risk pregnancies should be employed, including fetal anatomy survey, nuchal translucency measurement, and biochemical testing during the late first trimester, and detailed ultrasound at 18–20 weeks' and again at 25–32 weeks' gestation. Increased free beta-human chorionic gonadotropin in the first trimester [70] and alpha-fetoprotein (AFP) in the second trimester [71] have been associated with BWS. Enlarged nuchal translucency and increased first trimester pregnancy-associated plasma protein A (PAPP-A) are associated with fetal macrosomia [72]. At the first trimester scan, omphalocele is usually sonographically detectable. At 19–21 weeks of gestation, abdominal wall defects, macrosomia, and placental hyperplasia may readily present. The prenatal diagnosis of BWS is feasible after 28 weeks of gestation when macroglossia, visceromegaly resulting in increased abdominal circumference, and renal anomalies can be detected [47]. Prenatal diagnosis of BWS should alert obstetricians and neonatologists of its anticipatory complications, especially airway obstruction due to macroglossia and neonatal hypoglycemia.

86.6.4 Differential Diagnosis

Congenital hypothyroidism and maternal diabetes mellitus can result in neonatal macrosomia and macroglossia with or without other features. Asymmetry should be determined if it represents overgrowth (hemihyperplasia) or undergrowth (hemihypoplasia), since the latter is theoretically not associated with increased tumor risk. Thorough physical examination, relevant investigations, consultation with geneticists, and ongoing follow-up are mainstay management if genetic overgrowth syndromes are suspected. Differential diagnoses include Simpson–Golabi–Behmel syndrome (SGBS), Costello syndrome, Perlman syndrome, and Sotos syndrome, all of which share some clinical features of BWS.

SGBS deserves attention as its phenotypes resemble those of BWS: eight of eleven major and five of seven minor diagnostic criteria for BWS are common in SGBS [73]. It is characterized by pre- and postnatal overgrowth, macroglossia, visceromegaly, abdominal wall defects, and tumor predisposition. Additional features more commonly seen in SGBS include supernumerary nipples, cleft palate, brachy- or polydactyly, fingernail hypoplasia, diaphragmatic and inguinal hernia, and hypospadias [73]. Interestingly, patients with *CDKN1C* mutations share features of both BWS and SGBS. Analysis of the known causative gene, *GPC3*, implicates in genetic counseling since SGBS inherits in an X-linked fashion, whereas most BWS cases arise from sporadic epigenetic changes.

86.7 Management

If there are prenatal sonographic findings suggestive of BWS, screening for neonatal hypoglycemia should be undertaken as soon as possible. Immediately after birth, blood glucose and electrolyte monitoring and adequacy of feeding are utmost importance. Macroglossia-associated airway compromise may require prompt intubation. A long nipple is sometimes needed for macroglossia. Persistent hypoglycemia may require aggressive approaches including diazoxide or pancreatectomy [47]. Parents should be advised to recognize hypoglycemic symptoms before discharge from the hospital. Tumor surveillance should be initiated soon after the diagnosis. Visceromegaly is not alarming and is common; renal cysts or nephrogenic rests require further investigation. Presence of renal or abdominal wall anomalies, cleft palate, or facial hemihyperplasia necessitates referral to appropriate specialists. Cardiac evaluation is recommended prior to any surgical correction. Echocardiography may be required to assess congenital heart defects, cardiomegaly, or cardiomyopathy, and electrocardiography with age-corrected QT interval measurement at diagnosis is vital to exclude concomitant long QT syndrome. In addition, preoperative management should include consultation with an anesthesiologist for difficult airway management due to abnormal facial and upper airway anatomy and visceromegaly, which may shift the diaphragm upward [74].

Macroglossia is best approached by a multidisciplinary team consisting of plastic, maxillofacial, and/or oral surgeons, speech pathologists, feeding specialists, and orthodontists. After the age of 2–3 years, the growth rate of tongue tissues generally slows down, and the tongue will eventually be accommodated within the oral cavity. In moderate cases, the wait-and-see approach is therefore usually justified. Surgery (e.g., glossectomy and/or mandibular surgery) is generally recommended to reduce the size of macroglossia in the following circumstances: (i) before the sixth month of age, if growth and development are disturbed; (ii) before the age of 2, if a child has significant altered phonation; (iii) at 3 years before the child goes to playschool, if the macroglossia causes tongue protrusion, lip incompetence, or profuse salivation which negatively affects the self-image; and (iv) acute or chronic obstruction of the upper airway with respiratory disturbance [47]. Due to a risk of postoperative tongue edema, the airway should be protected with endotracheal intubation for at least 48–72 hours [47]. The cause of sleep-disordered breathing in BWS is multifactorial and does not solely contribute to macroglossia. Polysomnography is recommended for patients with high-risk features including feeding difficulties, gastroesophageal reflex, adenotonsillar hypertrophy, micrognathia, laryngomalacia, and those who are born prematurely [57].

Limb–leg discrepancy should be regularly followed by an orthopedic surgeon, and screening for secondary scoliosis is warranted. In the case of mild limb dysmetria (<2 cm), orthotics (shoe lifts) are usually prescribed to make the lower limbs achieve the same length. Significant discrepancy (>2 cm) requires surgical intervention. The patient's age, the degree of asymmetry, and the affected segment determine the optimal timing of the surgery. Epiphysiodesis refers to fusion of the epiphyseal plates to halt the overgrowth; the procedure is commonly performed in children with BWS via minimally invasive approach [47].

Cognitive evaluation is recommended before a child attends primary school to assess needs for special education and/or neurobehavioral rehabilitation. If developmental delay occurs, standard intervention program should be instituted early. Screening of hypercalciuria (with urinary calcium-to-creatinine ratio), hypothyroidism, dyslipidemia, and polycythemia can be incorporated to the regular outpatient visits. Other medical and surgical issues are managed as in children without BWS. Finally, based on a limited number of adult individuals with BWS, echocardiography, renal ultrasound, renal function tests, and hearing evaluation are recommended during adulthood [56]. The risk of infertility in males should also be counseled.

86.8 Prognosis, Tumor Risk, and Genetic Counseling

86.8.1 Prognosis

The prognosis is highly variable: the severe end of the spectrum may present with intrauterine, neonatal, or childhood death, whereas mildly affected children often have subtle problems which do not affect their quality of life [3]. The previously estimated overall mortality rate of BWS is 10%–20% with most death arising from hypoglycemia, prematurity, cardiomyopathy, macroglossia, and tumors [38,39,45]; however, this may be an overestimation given the syndrome recognition and improved treatment and tumor surveillance. The overall tumor-free survival rate at 5 years was 86%, and the prognosis is generally favorable after childhood [61].

86.8.2 Tumor Risk and Surveillance

Children with BWS have an increased risk of embryonic tumors, of which the most common is Wilms tumor. Other common tumors are hepatoblastoma, neuroblastoma, adrenocortical carcinoma, and rhabdomyosarcoma. Infrequently reported tumors include pancreatoblastoma, leukemia, thyroid carcinoma, liver sarcoma, melanoma, gonadoblastoma, and cardiac atrial tumor. The prevalence of these tumors reportedly ranged between 4% and 21% [75], with an estimated pooled prevalence of 7.4% according to the recent meta-analysis [76]. Most tumors are diagnosed before the age of 4 years and the average annual incidence of cancer during this period was approximated as 0.027 tumors per person-years [60]. An overall cumulative tumor incidence during childhood is approximately 10% [75]. On the contrary, tumors rarely develop after the first decade of life in BWS. During a period of 30 years in the UK Children's Cancer Study Group, only one case of BWS had presented with Wilms tumor after the age of 7 years [77]. Nonetheless, the tumor risk in adults with BWS has not been established but evidence has suggested some disposition, as at least five adults with tumors have been reported: four women with, respectively, an adrenal adenoma, a basal ganglion astrocytoma, bilateral pheochromocytoma, and multiple primaries including thyroid carcinoma and breast cancer; and one man who had acute myeloid leukemia [56].

According to previously published data, there are at least three factors that potentially influence the risk of tumor development. First, some clinical phenotypes are associated with increased tumor risk such as hemihyperplasia, nephromegaly, and presence of nephrogenic rests [3]. Approximately 20%–25% of BWS patients with hemihyperplasia had tumors [78,79], and the tumor risk in patients with isolated hemihyperplasia may be as high as 20% [80,81]. Hemihyperplasia is common in children with UPD; when compared to non-UPD cases, those with UPD are also more likely to have tumors [19]. Second, since UPD shows a mosaic molecular pattern, higher degree of UPD expression therefore corresponds to a more severe phenotype including tumor predisposition. Itoh et al. have shown that organomegaly was associated with the tissue-specific proportion of UPD cells. They found a strikingly higher degree of 11p15.5 UPD mosaicism in the enlarged left adrenal gland when compared to the non-enlarged right adrenal gland [82]. Hence it is theoretically plausible that the risk is even more significant in BWS patients with UPD who have a high level of mosaicism and concomitant hemihyperplasia. Third, many publications have supported that tumor risk is higher when the molecular defects involve the telomeric domain (ICR1 hypermethylation and UPD) rather than in the centromeric domain (ICR2 hypomethylation and *CDKN1C* mutation). This emphasizes the role of *IGF2* and *H19* in tumorigenesis in BWS (see Pathogenesis).

Mussa et al. have systematically reviewed cancer occurrence in 1370 individuals with BWS and elaborated a correlation between tumor subtypes and underlying genetic findings (Table 86.2), which is consistent with several previous reports [83–85]. Wilms tumor primarily occurred in individuals with telomeric defect (47/48; 97.9% of Wilms tumor cases) and was the most common tumor observed in ICR1 hypermethylation defect (26/28; 92.8% of the cases). Tumors in patients with UPD were more diverse; the majority of hepatoblastoma cases (16/23; 69.6%) occurred in

TABLE 86.2

Tumor Occurrence as Categorized into Molecular Subtypes According to a Meta-Analysis by Mussa et al. (2016) [76]

| | Centromeric Domain | | Telomeric Domain | | |
	ICR2 Hypomethylation	*CDKN1C* Mutation	UPD	ICR1 Hypermethylation	Overall
Patients[a]	836 (61.0%)	70 (5.1%)	341 (24.9%)	123 (9.0%)	1,370 (100.0%)
Tumors[b]	21	6	47	28	102
• Wilms tumor	1	0	21	26	48
• Hepatoblastoma	6	0	16	1	23
• Adrenal carcinoma	0	0	5	0	5
• Neuroblastoma	4	3	3	0	10
• Rhabdomyosarcoma	6	0	1	0	7
• Others[c]	4	3	1	1	9

[a] Percentages of patients in each subtype when comparing to all patients.
[b] Percentages of patients having tumors when comparing to the same subtype.
[c] Including pancreatoblastoma, leukemia, thyroid carcinoma, liver sarcoma, melanoma, gonadoblastoma, and cardiac atrial tumor.

this molecular subgroup. Despite its rarity, all five cases of adrenocortical carcinoma occurred in patients with UPD. The risk of tumors in centromeric defect appeared to be lower, and only one case of Wilms tumor had been reported in this subgroup. Most cases of rhabdomyosarcoma (6/7; 85.7%) occurred in patients harboring ICR2 hypomethylation. One-half of tumors (3/6; 50%) associated with *CDKN1C* mutation were of neuroblastoma subtype. Higher tumor occurrence related to telomeric domain disruption was confirmed by Rump et al. when they estimated the tumor risks of children with ICR1 hypermethylation and UPD as 35%–45% and 25%–30%, respectively, while the tumor risk of children having ICR2 hypomethylation was only 1%–5% [75]. The data on tumor risk of BWS with karyotypic defects and with negative molecular tests are sparse; however, they seem to be low in the former group and intermediate in the latter group [75,76]. Finally, the data on two newly described molecular defects— multilocus imprinting disorder and mosaic genome-wide paternal UPD—are limited. Previous reports indicated that children with mosaic genome-wide paternal UPD carry an unusually high risk of tumors, and that atypical tumors, including hamartomatous tumors of the liver and heart, pheochromocytoma, and hemangioendothelioma, have been observed [21].

Several tumor surveillance protocols for BWS individuals have been proposed. Their primary strategies are to detect the two most common tumors: abdominal ultrasound for Wilms tumor and serum AFP for hepatoblastoma [48,77,86–91]. It is worth mentioning that only a few studies have evaluated the cost-effectiveness of surveillance protocols, and false positive rates of the screening are not uncommon [88,92]. The detection of the first tumor should alarm clinicians of an increased tumor risk at the same or other organs, and tumor surveillance should be continued since metachronous and synchronous diseases can occur, albeit rarely. The surveillance is also recommended in an unaffected monozygotic co-twin of a BWS child, despite the absent clinical phenotypes or negative genetic tests. This is due to possible mosaicism in the normal-appearing twin or seeding of BWS cells secondary to common vascular anastomoses in utero [3]. Furthermore, a child with incomplete phenotype of BWS, such as isolated macroglossia, macrosomia, or hemihyperplasia, may be mosaic for BWS-related molecular defects and carry similar tumor risk as children with BWS diagnosis. Current guidelines have recommended children with isolated

hemihyperplasia to undergo the same tumor surveillance strategy as BWS children [93].

Wilms tumor generally carries a good prognosis, with long-term survival of >90% of the cases with localized disease and >70% of the metastatic cases [94]. Its early detection also allows a more preservative treatment—i.e., nephron-sparing strategy [92,95]. Abdominal ultrasound is the best screening tool for Wilms tumor, as abdominal palpation might not detect small lesions, and magnetic resonance imaging or computed tomography possess potential risks of sedation and radiation exposure, respectively [77]. Screening should start at syndrome diagnosis and continue every 3–4 months and no less frequently than three times annually, as the estimated doubling time of Wilms tumor is 17–40 days [96]. The ultrasound should be undertaken until up to 8 years of age [47,75]. Families should be made aware that with this frequency, occasional tumors may be detected between the scans. A child with ultrasound-detectable suspicious mass should have a repeat ultrasound at a specialized center. If a repeat scan confirms the lesion, referral to a pediatric oncologist and further evaluation with other imaging modalities are warranted. Obtaining baseline magnetic resonance imaging has been suggested by some authors [3]. It is also presumable that other intra-abdominal tumors such as adrenal gland adenoma/carcinoma, pancreatoblastoma, hepatoblastoma, and neuroblastoma can be screened by ultrasound.

Guidelines have recommended checking serum AFP every 2–3 months until up to 4 years of age [75]. Nevertheless, the downsides of AFP measurement include problematic interpretation in childhood, debated effectiveness, and frequent blood draws potentially resulting in poor adherence to screening. Typically, elevated AFP is observed during the neonatal period, and the level declines continuously until reaching its plateau after age 8 months. Abnormal postnatal elevation in AFP is associated with various tumors, including hepatic tumors and germ cell tumors [97]. In BWS, the AFP concentration is greater and declines significantly more slowly than in healthy children; thus, a normal curve designed specifically for children with BWS should be employed to interpret AFP levels [97]. The level should be repeated within 1 month after an initial elevation; its subsequent lowering level is reassuring, but an increasing level necessitates a search for tumors (either hepatoblastoma or germ cell tumors) with imaging studies. Referral to a pediatric oncologist may be indicated. Studies have shown that abdominal ultrasound may not be sufficient to detect early hepatoblastoma, and that measurement of AFP resulted in lower staging at diagnosis and consequently a greater survival rate [90,98]. Other less invasive methods such as AFP measurement on dried blood spot are being investigated [99].

Systematic study and surveillance strategy on adrenocortical carcinoma, neuroblastoma, and rhabdomyosarcoma are limited. In addition to BWS, adrenocortical carcinoma is also found in Li–Fraumeni syndrome, for which abdominal ultrasound every 3–4 months is recommended until age 18 years. If ultrasound is unsatisfactory, blood draws for total testosterone, dehydroepiandrosterone sulfate, or androstenedione are alternative screening tests [100]. Using this method for Li–Fraumeni patients, the screened tumors were smaller and in earlier stages when compared to non-screened cases, implying better survival [101]. Urinary biomarkers vanillylmandelic and homovanillic acid and/or catecholamines have been investigated as screening tools for neuroblastoma; however, a majority of screened positive tumors were stage 4S neuroblastomas, of which the detection may affect their favorable outcomes only minimally since the condition is spontaneously regressed [76]. A recent guideline has recommended against the use of these biomarkers in BWS unless the tumors are clinically suspected [47]. Finally, frequent examination of muscle mass to detect rhabdomyosarcoma has been suggested by some authors, although the appropriate timing of this intervention is undetermined [48].

Several studies have proposed an (epi)genotype-specific screening protocol on the basis of lower tumor risk in patients with centromeric domain defects including ICR2 hypomethylation and *CDKN1C* mutation. Some authors have suggested screening these patients less rigorously, and some have even discouraged tumor surveillance [48,77,91]. However, the risk of Wilms tumor in these patients is not negligible and still higher than general pediatric population. In addition, the use of widely available MS-MLPA to diagnose BWS may mislabel cases of UPD (high tumor risk) as ICR2 hypomethylation (low tumor risk), resulting in underdiagnosis of tumor if not screened [102]. The pros and cons of molecularly customized screening protocols also have not been established. For these reasons, we advise that all BWS patients, including those with negative molecular tests, undergo standard tumor screening protocol regardless of (epi)genetic subtypes until a clear-cut evidence-based general consensus on tumor surveillance in BWS is reached in the near future.

86.8.3 Genetic Counseling

The recurrence risks of siblings and offspring of BWS depend largely on the molecular subtype as well as the parent-of-origin effects (Table 86.3); most of the postulated risks are theoretical rather than definitive due to lack of confirmatory data. Consultation with a clinical geneticist and/or a genetic counselor who is familiar with imprinting disorders is recommended. While most individuals with BWS do not have affected parents, testing parents and relevant family members should be offered when heritable defects—i.e., *CDKN1C* mutation, karyotypic abnormality, and microdeletion/microduplication—are detected. Presence of *CDKN1C* mutation in a proband necessitates parental testing even in the absence of family history, since maternal transmission of a pathogenic variant from a clinically unaffected mother to a BWS child has been reported [69]. Familial cases of microdeletions of ICR1 or ICR2 are well documented [23,24,64]. Parental testing is generally not indicated in a case of UPD because the

TABLE 86.3

Recurrence Risk of Siblings and Offspring of a Proband with BWS

Molecular Defects	Risk of Siblings of a Proband	Risk of Offspring of a Proband
Maternal 11p15.5 translocation or inversion	If inherited from a female, the risk may be as high as 50%	If inherited from a female, the risk may be as high as 50%
Paternal 11p15.5 duplication	Not defined	Not defined
11p15.5 pUPD	Very low	Very low
CDKN1C mutation	If a family history is negative and neither parent has *CDKN1C* mutation, the risk is low in the absence of germline mosaicism If an affected mother has *CDKN1C* mutation, the risk is 50% If an affected father has *CDKN1C* mutation, the risk is empirically increased	Female proband has 50% risk Male proband has <50% risk
ICR1 hypermethylation	Very low in the absence of genomic abnormality	Low in the absence of genomic abnormality due to the reset of imprinting cycle at each generation
ICR2 hypomethylation	Very low in the absence of genomic abnormality or pathogenic variants at other loci such as *NLRP2*	Low in the absence of genomic abnormality due to the reset of imprinting cycle at each generation
11p15.5 microdeletion or microduplication[a]	≤50%; if one parent has the same genomic change, the risk depends on sex of the transmitting parent and may be as high as 50%	50%; phenotype in the offspring could be either BWS or Silver–Russell syndrome depending on the sex of transmitting parent
Unknown	With a negative family history, the risk is theoretically low With a positive family history but unidentified molecular defects, the empiric risk is ≤50%	Likely low

Source: From Weksberg et al. (2010) and Shuman et al. in *GeneReviews* (1993; last update 2016).

[a] ICR1 hypermethylation or ICR2 hypomethylation may be accompanied by underlying 11p15.5 microdeletion or microduplication (i.e., "secondary DNA methylation defect"). Finding of such genomic changes necessitates parental testing since familial occurrence is possible.

defect occurs post-zygotically. Theoretically, UPD and primary DNA methylation defects carry low recurrence risks, whereas maternal transmission of 11p15.5 translocation or *CDKN1C* mutations, paternal 11p15.5 duplication, and 11p15.5 microdeletion carry high recurrence risks. Only pathogenic *CDKN1C* variants with maternal inheritance and those appearing de novo on maternal allele are associated with clinical phenotype, as males carrying mutations are not at risk of having affected children due to the silencing of *CDKN1C* copies [12]. However, at least one case of paternal transmission has been reported [103]. A daughter of *CDKN1C*-mutated male will be a heterozygous carrier, and her children (the *CDKN1C*-mutated male's grandchildren) would have a recurrence risk of 50%. When parents do not carry transmissible defects, the recurrence risk is still not negligible due to a small chance of existing gonadal mosaicism.

86.9 Conclusion

As molecular techniques are becoming more advanced and affordable, more children with BWS will be readily diagnosed, and novel molecular etiologies will be uncovered for both familial and sporadic cases. Over the next few years, we expect that testing modalities will more accurately reflect the actual number of BWS cases and help expand the syndrome phenotypic spectrum and natural history. Although the *IGF2* and *CDKN1C* genes are well characterized, their interaction and downstream effectors need to be investigated to explain why disruption in either independent locus results in similar clinical features. Studies on the tumor-suppressor role of *H19* RNA transcript are warranted. More research involving the regulation of chromosome 11p15.5 imprinting region as well as other imprinting clusters, and the exact frequency and molecular mechanism of multilocus imprinting disorders are critical to better understand how epigenome affects human health. One of the most important unanswered questions is what mechanism drives a tumor development in a particular subset of BWS patients. Until we can answer that question, we need more data on clinical and genetic heterogeneities to develop a better tool for complete ascertainment of BWS cases, to delineate a more detailed (epi)genotype–phenotype correlation, and to differentiate proper follow-up approaches, tumor risk estimates, and tumor screening protocols ideally based on each molecular subtype.

REFERENCES

1. Beckwith JB. *Extreme cytomegaly of the adrenal fetal cortex, omphalocele, hyperplasia of kidneys and pancreas, and Leydig-cell hyperplasia: Another syndrome?* Presented at Annual Meeting of Western Society for Pediatric Research, Los Angeles, California, 11 November 1963.
2. Wiedemann HR. Complexe malformatif familial avec hernie ombilicale et macroglossia, un 'syndrome nouveau. *J Genet Hum.* 1964;13:223–32.
3. Weksberg R, Shuman C, Beckwith JB. Beckwith–Wiedemann syndrome. *Eur J Hum Genet.* 2010;18(1):8–14.
4. Mussa A, Russo S, De Crescenzo A et al. Prevalence of Beckwith–Wiedemann syndrome in North West of Italy. *Am J Med Genet A.* 2013;161A(10):2481–6.
5. Weksberg R, Shuman C, Caluseriu O et al. Discordant KCNQ1OT1 imprinting in sets of monozygotic twins discordant for Beckwith–Wiedemann syndrome. *Hum Mol Genet.* 2002;11(11):1317–25.
6. Peters J. The role of genomic imprinting in biology and disease: An expanding view. *Nat Rev Genet.* 2014; 15(8):517–30.
7. Moore T, Haig D. Genomic imprinting in mammalian development: A parental tug-of-war. *Trends Genet.* 1991; 7(2):45–9.
8. Demars J, Gicquel C. Epigenetic and genetic disturbance of the imprinted 11p15 region in Beckwith–Wiedemann and Silver-Russell syndromes. *Clin Genet.* 2012;81(4):350–61.
9. Varrault A, Gueydan C, Delalbre A et al. Zac1 regulates an imprinted gene network critically involved in the control of embryonic growth. *Dev Cell.* 2006;11(5):711–22.
10. Maher ER, Reik W. Beckwith–Wiedemann syndrome: Imprinting in clusters revisited. *J Clin Invest.* 2000; 105(3):247–52.
11. Nativio R, Wendt KS, Ito Y et al. Cohesin is required for higher-order chromatin conformation at the imprinted IGF2-H19 locus. *PLoS Genet.* 2009;5(11):e1000739.
12. Eggermann T, Binder G, Brioude F et al. CDKN1C mutations: Two sides of the same coin. *Trends Mol Med.* 2014;20(11):614–22.
13. Fitzpatrick GV, Pugacheva EM, Shin JY et al. Allele-specific binding of CTCF to the multipartite imprinting control region KvDMR1. *Mol Cell Biol.* 2007;27(7):2636–47.
14. Alders M, Ryan A, Hodges M et al. Disruption of a novel imprinted zinc-finger gene, ZNF215, in Beckwith–Wiedemann syndrome. *Am J Hum Genet.* 2000;66(5):1473–84.
15. Smilinich NJ, Day CD, Fitzpatrick GV et al. A maternally methylated CpG island in KvLQT1 is associated with an antisense paternal transcript and loss of imprinting in Beckwith–Wiedemann syndrome. *Proc Natl Acad Sci U S A.* 1999;96(14):8064–9.
16. Brown KW, Villar AJ, Bickmore W et al. Imprinting mutation in the Beckwith–Wiedemann syndrome leads to biallelic IGF2 expression through an H19-independent pathway. *Hum Mol Genet.* 1996;5(12):2027–32.
17. Redeker E, Hoovers JM, Alders M et al. An integrated physical map of 210 markers assigned to the short arm of human chromosome 11. *Genomics.* 1994;21(3):538–50.
18. Enklaar T, Zabel BU, Prawitt D. Beckwith–Wiedemann syndrome: Multiple molecular mechanisms. *Expert Rev Mol Med.* 2006;8(17):1–19.
19. Henry I, Puech A, Riesewijk A et al. Somatic mosaicism for partial paternal isodisomy in Wiedemann–Beckwith syndrome: A post-fertilization event. *Eur J Hum Genet.* 1993;1(1):19–29.
20. Smith AC, Shuman C, Chitayat D et al. Severe presentation of Beckwith–Wiedemann syndrome associated with high levels of constitutional paternal uniparental disomy for chromosome 11p15. *Am J Med Genet A.* 2007;143A(24):3010–5.
21. Kalish JM, Conlin LK, Bhatti TR et al. Clinical features of three girls with mosaic genome-wide paternal uniparental isodisomy. *Am J Med Genet A.* 2013;161A(8):1929–39.
22. Prawitt D, Enklaar T, Gartner-Rupprecht B et al. Microdeletion and IGF2 loss of imprinting in a cascade causing Beckwith–Wiedemann syndrome with Wilms' tumor. *Nat Genet.* 2005;37(8):785–6; author reply 6–7.

23. Prawitt D, Enklaar T, Gartner-Rupprecht B et al. Microdeletion of target sites for insulator protein CTCF in a chromosome 11p15 imprinting center in Beckwith–Wiedemann syndrome and Wilms' tumor. *Proc Natl Acad Sci U S A.* 2005;102(11):4085–90.

24. Sparago A, Cerrato F, Vernucci M, Ferrero GB, Silengo MC, Riccio A. Microdeletions in the human H19 DMR result in loss of IGF2 imprinting and Beckwith–Wiedemann syndrome. *Nat Genet.* 2004;36(9):958–60.

25. Demars J, Shmela ME, Rossignol S et al. Analysis of the IGF2/H19 imprinting control region uncovers new genetic defects, including mutations of OCT-binding sequences, in patients with 11p15 fetal growth disorders. *Hum Mol Genet.* 2010;19(5):803–14.

26. Poole RL, Leith DJ, Docherty LE et al. Beckwith–Wiedemann syndrome caused by maternally inherited mutation of an OCT-binding motif in the IGF2/H19-imprinting control region, ICR1. *Eur J Hum Genet.* 2012;20(2):240–3.

27. Gurrieri F, Zollino M, Oliva A et al. Mild Beckwith–Wiedemann and severe long-QT syndrome due to deletion of the imprinting center 2 on chromosome 11p. *Eur J Hum Genet.* 2013;21(9):965–9.

28. Azzi S, Rossignol S, Steunou V et al. Multilocus methylation analysis in a large cohort of 11p15-related foetal growth disorders (Russell Silver and Beckwith–Wiedemann syndromes) reveals simultaneous loss of methylation at paternal and maternal imprinted loci. *Hum Mol Genet.* 2009;18(24):4724–33.

29. Meyer E, Lim D, Pasha S et al. Germline mutation in NLRP2 (NALP2) in a familial imprinting disorder (Beckwith–Wiedemann Syndrome). *PLoS Genet.* 2009;5(3):e1000423.

30. Halliday J, Oke K, Breheny S, Algar E, Amor DJ. Beckwith–Wiedemann syndrome and IVF: A case-control study. *Am J Hum Genet.* 2004;75(3):526–8.

31. Knudson AG. Hereditary cancer: Two hits revisited. *J Cancer Res Clin Oncol.* 1996;122(3):135–40.

32. Breslow NE, Norris R, Norkool PA et al. Characteristics and outcomes of children with the Wilms tumor-Aniridia syndrome: A report from the National Wilms Tumor Study Group. *J Clin Oncol.* 2003;21(24):4579–85.

33. Yuan E, Li CM, Yamashiro DJ et al. Genomic profiling maps loss of heterozygosity and defines the timing and stage dependence of epigenetic and genetic events in Wilms' tumors. *Mol Cancer Res.* 2005;3(9):493–502.

34. Murrell A. Genomic imprinting and cancer: From primordial germ cells to somatic cells. *Sci World J.* 2006;6:1888–910.

35. Hao Y, Crenshaw T, Moulton T, Newcomb E, Tycko B. Tumour-suppressor activity of H19 RNA. *Nature.* 1993;365(6448):764–7.

36. Lim DH, Maher ER. Genomic imprinting syndromes and cancer. *Adv Genet.* 2010;70:145–75.

37. Reeve AE, Eccles MR, Wilkins RJ, Bell GI, Millow LJ. Expression of insulin-like growth factor-II transcripts in Wilms' tumour. *Nature.* 1985;317(6034):258–60.

38. Elliott M, Bayly R, Cole T, Temple IK, Maher ER. Clinical features and natural history of Beckwith–Wiedemann syndrome: Presentation of 74 new cases. *Clin Genet.* 1994;46(2):168–74.

39. Weng EY, Moeschler JB, Graham JM Jr. Longitudinal observations on 15 children with Wiedemann–Beckwith syndrome. *Am J Med Genet.* 1995;56(4):366–73.

40. Baird PA, MacDonald EC. An epidemiologic study of congenital malformations of the anterior abdominal wall in more than half a million consecutive live births. *Am J Hum Genet.* 1981;33(3):470–8.

41. Wilkins-Haug L, Porter A, Hawley P, Benson CB. Isolated fetal omphalocele, Beckwith–Wiedemann syndrome, and assisted reproductive technologies. *Birth Defects Res A Clin Mol Teratol.* 2009;85(1):58–62.

42. Mittal D, Anand R, Sisodia N, Singh S, Biswas R. Placental mesenchymal dysplasia: What every radiologist needs to know. *Indian J Radiol Imaging.* 2017;27(1):62–4.

43. Lynn KN, Steinkeler JA, Wilkins-Haug LE, Benson CB. Hyperreactio luteinalis (enlarged ovaries) during the second and third trimesters of pregnancy: Common clinical associations. *J Ultrasound Med.* 2013;32(7):1285–9.

44. Romanelli V, Belinchon A, Campos-Barros A et al. CDKN1C mutations in HELLP/preeclamptic mothers of Beckwith–Wiedemann Syndrome (BWS) patients. *Placenta.* 2009;30(6):551–4.

45. Pettenati MJ, Haines JL, Higgins RR, Wappner RS, Palmer CG, Weaver DD. Wiedemann–Beckwith syndrome: Presentation of clinical and cytogenetic data on 22 new cases and review of the literature. *Hum Genet.* 1986;74(2):143–54.

46. Engstrom W, Lindham S, Schofield P. Wiedemann–Beckwith syndrome. *Eur J Pediatr.* 1988;147(5):450–7.

47. Mussa A, Di Candia S, Russo S et al. Recommendations of the Scientific Committee of the Italian Beckwith-Wiedemann Syndrome Association on the diagnosis, management and follow-up of the syndrome. *Eur J Med Genet.* 2016;59(1):52–64.

48. Brioude F, Lacoste A, Netchine I et al. Beckwith–Wiedemann syndrome: Growth pattern and tumor risk according to molecular mechanism, and guidelines for tumor surveillance. *Horm Res Paediatr.* 2013;80(6):457–65.

49. Cheungpasitporn W, Erickson SB. Beckwith–Wiedemann syndrome and recurrent bilateral renal calculi. *Urol Ann.* 2017;9(1):113–4.

50. Slavotinek A, Gaunt L, Donnai D. Paternally inherited duplications of 11p15.5 and Beckwith–Wiedemann syndrome. *J Med Genet.* 1997;34(10):819–26.

51. Schick B, Brors D, Prescher A, Draf W. Conductive hearing loss in Beckwith–Wiedemann syndrome. *Int J Pediatr Otorhinolaryngol.* 1999;48(2):175–9.

52. Kantaputra PN, Sittiwangkul R, Sonsuwan N, Romanelli V, Tenorio J, Lapunzina P. A novel mutation in CDKN1C in sibs with Beckwith–Wiedemann syndrome and cleft palate, sensorineural hearing loss, and supernumerary flexion creases. *Am J Med Genet A.* 2013;161A(1):192–7.

53. Kent L, Bowdin S, Kirby GA, Cooper WN, Maher ER. Beckwith–Weidemann syndrome: A behavioral phenotype-genotype study. *Am J Med Genet B Neuropsychiatr Genet.* 2008;147B(7):1295–7.

54. Greenwood RD, Somer A, Rosenthal A, Craenen J, Nadas AS. Cardiovascular abnormalities in the Beckwith–Wiedemann syndrome. *Am J Dis Child.* 1977;131(3):293–4.

55. Kaltenbach S, Capri Y, Rossignol S et al. Beckwith–Wiedemann syndrome and long QT syndrome due to familial-balanced translocation t(11;17)(p15.5;q21.3) involving the KCNQ1 gene. *Clin Genet.* 2013;84(1):78–81.

56. Greer KJ, Kirkpatrick SJ, Weksberg R, Pauli RM. Beckwith–Wiedemann syndrome in adults: Observations from one family and recommendations for care. *Am J Med Genet A.* 2008;146A(13):1707–12.

57. Follmar A, Dentino K, Abramowicz S, Padwa BL. Prevalence of sleep-disordered breathing in patients with Beckwith–Wiedemann syndrome. *J Craniofac Surg.* 2014;25(5):1814–7.

58. Mussa A, Peruzzi L, Chiesa N et al. Nephrological findings and genotype-phenotype correlation in Beckwith–Wiedemann syndrome. *Pediatr Nephrol.* 2012;27(3):397–406.

59. Goldman M, Shuman C, Weksberg R, Rosenblum ND. Hypercalciuria in Beckwith–Wiedemann syndrome. *J Pediatr.* 2003;142(2):206–8.

60. DeBaun MR, Tucker MA. Risk of cancer during the first four years of life in children from The Beckwith–Wiedemann Syndrome Registry. *J Pediatr.* 1998;132(3):398–400.

61. Gaston V, Le Bouc Y, Soupre V et al. Analysis of the methylation status of the KCNQ1OT and H19 genes in leukocyte DNA for the diagnosis and prognosis of Beckwith–Wiedemann syndrome. *Eur J Hum Genet.* 2001;9(6):409–18.

62. Zarate YA, Mena R, Martin LJ, Steele P, Tinkle BT, Hopkin RJ. Experience with hemihyperplasia and Beckwith–Wiedemann syndrome surveillance protocol. *Am J Med Genet A.* 2009;149A(8):1691–7.

63. Ibrahim A, Kirby G, Hardy C et al. Methylation analysis and diagnostics of Beckwith–Wiedemann syndrome in 1000 subjects. *Clin Epigenetics.* 2014;6(1):11.

64. Scott RH, Douglas J, Baskcomb L et al. Methylation-specific multiplex ligation-dependent probe amplification (MS-MLPA) robustly detects and distinguishes 11p15 abnormalities associated with overgrowth and growth retardation. *J Med Genet.* 2008;45(2):106–13.

65. Keren B, Chantot-Bastaraud S, Brioude F et al. SNP arrays in Beckwith–Wiedemann syndrome: An improved diagnostic strategy. *Eur J Med Genet.* 2013;56(10):546–50.

66. Priolo M, Sparago A, Mammi C, Cerrato F, Lagana C, Riccio A. MS-MLPA is a specific and sensitive technique for detecting all chromosome 11p15.5 imprinting defects of BWS and SRS in a single-tube experiment. *Eur J Hum Genet.* 2008;16(5):565–71.

67. Bliek J, Verde G, Callaway J et al. Hypomethylation at multiple maternally methylated imprinted regions including PLAGL1 and GNAS loci in Beckwith–Wiedemann syndrome. *Eur J Hum Genet.* 2009;17(5):611–9.

68. Alders M, Bliek J, vd Lip K, vd Bogaard R, Mannens M. Determination of KCNQ1OT1 and H19 methylation levels in BWS and SRS patients using methylation-sensitive high-resolution melting analysis. *Eur J Hum Genet.* 2009;17(4):467–73.

69. Shuman C, Beckwith JB, Weksberg R. Beckwith–Wiedemann syndrome. In: Adam MP, Ardinger HH, Pagon RA et al., eds. *GeneReviews((R)).* Seattle (WA); 1993.

70. Kagan KO, Berg C, Dufke A, Geipel A, Hoopmann M, Abele H. Novel fetal and maternal sonographic findings in confirmed cases of Beckwith–Wiedemann syndrome. *Prenat Diagn.* 2015;35(4):394–9.

71. Guanciali-Franchi P, Di Luzio L, Iezzi I et al. Elevated maternal serum alpha-fetoprotein level in a fetus with Beckwith–Wiedemann syndrome in the second trimester of pregnancy. *J Prenat Med.* 2012;6(1):7–9.

72. Timmerman E, Pajkrt E, Snijders RJ, Bilardo CM. High macrosomia rate in healthy fetuses after enlarged nuchal translucency. *Prenat Diagn.* 2014;34(2):103–8.

73. Knopp C, Rudnik-Schoneborn S, Zerres K, Gencik M, Spengler S, Eggermann T. Twenty-one years to the right diagnosis—clinical overlap of Simpson–Golabi–Behmel and Beckwith–Wiedemann syndrome. *Am J Med Genet A.* 2015;167A(1):151–5.

74. Raj D, Luginbuehl I. Managing the difficult airway in the syndromic child. *Contin Educ Anaesthesia Crit Care Pain.* 2015;15(1):7–13.

75. Rump P, Zeegers MP, van Essen AJ. Tumor risk in Beckwith–Wiedemann syndrome: A review and meta-analysis. *Am J Med Genet A.* 2005;136(1):95–104.

76. Mussa A, Molinatto C, Baldassarre G et al. Cancer risk in Beckwith–Wiedemann syndrome: A systematic review and meta-analysis outlining a novel (epi)genotype specific histotype targeted screening protocol. *J Pediatr.* 2016;176:142–9 e1.

77. Scott RH, Walker L, Olsen OE et al. Surveillance for Wilms tumour in at-risk children: Pragmatic recommendations for best practice. *Arch Dis Child.* 2006;91(12):995–9.

78. Wiedemann H-R. Tumours and hemihypertrophy associated with Wiedemann–Beckwith syndrome. *Eur J Pediatr.* 1983;141(2):129.

79. Goldman M, Smith A, Shuman C et al. Renal abnormalities in Beckwith–Wiedemann syndrome are associated with 11p15.5 uniparental disomy. *J Am Soc Nephrol.* 2002;13(8):2077–84.

80. Hoyme HE, Seaver LH, Jones KL, Procopio F, Crooks W, Feingold M. Isolated hemihyperplasia (hemihypertrophy): Report of a prospective multicenter study of the incidence of neoplasia and review. *Am J Med Genet.* 1998;79(4):274–8.

81. Shuman C, Smith AC, Steele L et al. Constitutional UPD for chromosome 11p15 in individuals with isolated hemihyperplasia is associated with high tumor risk and occurs following assisted reproductive technologies. *Am J Med Genet A.* 2006;140(14):1497–503.

82. Itoh N, Becroft DM, Reeve AE, Morison IM. Proportion of cells with paternal 11p15 uniparental disomy correlates with organ enlargement in Wiedemann–Beckwith syndrome. *Am J Med Genet.* 2000;92(2):111–6.

83. Lam WW, Hatada I, Ohishi S et al. Analysis of germline CDKN1C (p57KIP2) mutations in familial and sporadic Beckwith–Wiedemann syndrome (BWS) provides a novel genotype-phenotype correlation. *J Med Genet.* 1999;36(7):518–23.

84. Engel JR, Smallwood A, Harper A et al. Epigenotype-phenotype correlations in Beckwith–Wiedemann syndrome. *J Med Genet.* 2000;37(12):921–6.

85. Weksberg R, Nishikawa J, Caluseriu O et al. Tumor development in the Beckwith–Wiedemann syndrome is associated with a variety of constitutional molecular 11p15 alterations including imprinting defects of KCNQ1OT1. *Hum Mol Genet.* 2001;10(26):2989–3000.

86. Shah KJ. Beckwith–Wiedemann syndrome: Role of ultrasound in its management. *Clin Radiol.* 1983;34(3):313–9.

87. Craft AW, Parker L, Stiller C, Cole M. Screening for Wilms' tumour in patients with aniridia, Beckwith syndrome, or hemihypertrophy. *Med Pediatr Oncol.* 1995;24(4):231–4.

88. Beckwith JB. Children at increased risk for Wilms tumor: Monitoring issues. *J Urol.* 1998;160(4):1593–4.

89. Borer JG, Kaefer M, Barnewolt CE et al. Renal findings on radiological followup of patients with Beckwith–Wiedemann syndrome. *J Urol.* 1999;161(1):235–9.

90. Clericuzio CL, Chen E, McNeil DE et al. Serum alpha-fetoprotein screening for hepatoblastoma in children with Beckwith–Wiedemann syndrome or isolated hemihyperplasia. *J Pediatr.* 2003;143(2):270–2.

91. Maas SM, Vansenne F, Kadouch DJ et al. Phenotype, cancer risk, and surveillance in Beckwith–Wiedemann syndrome depending on molecular genetic subgroups. *Am J Med Genet A.* 2016;170(9):2248–60.

92. McNeil DE, Brown M, Ching A, DeBaun MR. Screening for Wilms tumor and hepatoblastoma in children with Beckwith–Wiedemann syndromes: A cost-effective model. *Med Pediatr Oncol.* 2001;37(4):349–56.

93. Clericuzio CL, Martin RA. Diagnostic criteria and tumor screening for individuals with isolated hemihyperplasia. *Genet Med.* 2009;11(3):220–2.

94. Pritchard-Jones K. Controversies and advances in the management of Wilms' tumour. *Arch Dis Child.* 2002;87(3):241–4.

95. McNeil DE, Langer JC, Choyke P, DeBaun MR. Feasibility of partial nephrectomy for Wilms' tumor in children with Beckwith–Wiedemann syndrome who have been screened with abdominal ultrasonography. *J Pediatr Surg.* 2002;37(1):57–60.

96. Craft AW. Growth rate of Wilms' tumour. *Lancet.* 1999;354(9184):1127.

97. Everman DB, Shuman C, Dzolganovski B, O'Riordan M A, Weksberg R, Robin NH. Serum alpha-fetoprotein levels in Beckwith–Wiedemann syndrome. *J Pediatr.* 2000;137(1):123–7.

98. Trobaugh-Lotrario AD, Venkatramani R, Feusner JH. Hepatoblastoma in children with Beckwith–Wiedemann syndrome: Does it warrant different treatment? *J Pediatr Hematol Oncol.* 2014;36(5):369–73.

99. Mussa A, Pagliardini S, Pagliardini V et al.. alpha-Fetoprotein assay on dried blood spot for hepatoblastoma screening in children with overgrowth-cancer predisposition syndromes. *Pediatr Res.* 2014;76(6):544–8.

100. Kratz CP, Achatz MI, Brugieres L et al. Cancer screening recommendations for individuals with Li–Fraumeni syndrome. *Clin Cancer Res.* 2017;23(11):e38–45.

101. Custodio G, Parise GA, Kiesel Filho N et al. Impact of neonatal screening and surveillance for the TP53 R337H mutation on early detection of childhood adrenocortical tumors. *J Clin Oncol.* 2013;31(20):2619–26.

102. Brzezinski J, Shuman C, Choufani S et al. Wilms tumour in Beckwith–Wiedemann Syndrome and loss of methylation at imprinting centre 2: Revisiting tumour surveillance guidelines. *Eur J Hum Genet.* 2017;25(9):1031–9.

103. Lee MP, DeBaun M, Randhawa G, Reichard BA, Elledge SJ, Feinberg AP. Low frequency of p57KIP2 mutation in Beckwith–Wiedemann syndrome. *Am J Hum Genet.* 1997;61(2):304–9.

87

CBL Syndrome

Dongyou Liu

CONTENTS

87.1 Introduction

CBL syndrome is a rare autosomal dominant disorder linked to heterozygous germline mutations in the Casitas B-cell lymphoma (*CBL*) gene located on chromosome 11q23.3. Encoding an E3 ubiquitin ligase and a multi-adaptor protein (CBL), the *CBL* gene controls proliferative signaling networks by downregulating the growth factor receptor signaling cascades in various cell types. As CBL forms part of the RAS/mitogen-activated protein kinase (MAPK) pathway that participates in lymphangiogenesis, CBL syndrome is considered one of the RASopathies, which cover a group of developmental disorders with neuro-cardio-facio-cutaneous manifestations (Table 87.1). Similar to other RASopathies, CBL syndrome (also called Noonan syndrome-like disorder with or without juvenile myelomonocytic leukemia [JMML]) typically presents with fetal hydrops, fetal pleural effusions, hydrothorax, chylothorax, cryptorchidism, facial dysmorphism, café-au-lait spots, and predisposition to JMML, an aggressive myeloproliferative and myelodysplastic neoplasm showing excessive macrophage/monocyte proliferation, during childhood (Figure 87.1) [1,2].

87.2 Biology

The MAPK pathway includes two major components: the RAS gene family and the MAPK pathway, which contains RAF, one of the major downstream RAS effectors (the others being PI3K and RAL–GEF, forming part of the PI3K/AKT/mTOR pathway and the Ral-GEF/TBK1/IRF3/3-NF-κB pathway, respectively). As a ubiquitous, highly conserved intracellular signaling network, the RAS/MAPK signaling pathway participates in cell cycle regulation, differentiation, growth, apoptosis, and senescence.

The RAS (a contraction of rat sarcoma, the tumor where the first gene of the family was identified) gene family comprises three well-known genes *HRAS* (11p15.5), *KRAS* (12p12.1), and *NRAS* (1p22.2) encoding three namesake GTPases (K-RAS, H-RAS, and N-RAS) that function as on-off binary switches for several downstream signaling cascades involved in the control of cell growth and death. In addition, the *RIT1* (RAS-like protein in tissues) gene encodes RIT1, which shares a 50% sequence homology with RAS but lacks the C-terminal lipidation site and is also considered a RAS GTPase.

RAS proteins exist as an inactive GDP-bound form and an active GTP-bound form. Binding of a growth factor to receptor tyrosine kinase (RTK) induces RTK autophosphorylation and interaction with growth factor receptor-bound protein 2 (GRB2), which is bound to son of sevenless 1 (SOS1). As a guanosine nucleotide exchange factor (GEF), SOS1 facilitates conversion of inactive GDP-bound RAS to active GTP-bound RAS, which activates downstream signaling cascades (e.g., the MAPK/RAF/MEK/ERK pathway, the PI3K/AKT/mTOR pathway, and the Ral-GEF/TBK1/IRF3/3-NF-κB pathway).

In the MAPK pathway, active GTP-bound RAS activates RAF (ARAF, BRAF, and/or CRAF), which then phosphorylates and activates MEK1 (mitogen-activated protein kinase 1) and/or MEK2, which in turn phosphorylate and activate ERK1 (extracellular signal regulated kinase 1) and/or ERK2, which then enter the nucleus to alter gene transcription and produce various nuclear and cytosolic effectors that are critically important for late developmental processes (e.g., organogenesis, morphology determination, and growth) (Figure 87.2).

TABLE 87.1

Molecular and Clinical Characteristics of RASopathies

Syndrome	RAS/MAPK Pathway Gene (Chromosome Location)	Protein (Function)	Clinical Phenotype
Noonan syndrome	*PTPN11* (12q24.1) *SOS1* (2p22.1) *KRAS* (12p12.1) *NRAS* (1p13.2) *RIT1* (1q22) *BRAF* (7q34) *RAF1* (3p25.1) *LZTR1* (22q11.21) *A2ML1* (12p13)	SHP2 (phosphatase) SOS1 (guanine exchange factor) KRAS (GTPase) NRAS (GTPase) RIT1 (GTPase) BRAF (kinase) RAF1/CRAF (kinase) LZTR1 (BTB-kelch protein) A2ML1 (α-macroglobulin)	Facial dysmorphism (broad forehead, hypertelorism, down-slanting palpebral fissures, ptosis, high-arched palate, low-set posteriorly rotated ears); congenital heart defect (pulmonic stenosis. hypertrophic cardiomyopathy, atrial septal defect, ventricular septal defect, atrioventricular canal defect, aortic coarctation); short stature; skeletal defects (superior pectus carinatum, inferior pectus excavatum, cubitus valgus, clinobrachydactyly, thoracic scoliosis, talipes equinovarus or radioulnar synostosis); bleeding defects (coagulation defect, abnormal bleeding); genitourinary anomalies (spermatogenesis, cryptorchidism, renal pelvis dilatation); ophthalmologic abnormalities (refractive errors, strabismus, amblyopia, anterior segment changes); predisposition to hematological malignancy (myeloproliferative disease and leukemia, especially JMML)
Noonan syndrome with multiple lentigines (NSML, formerly called LEOPARD syndrome)	*PTPN11* (12q24.1) *RAF1* (3p25.1) *BRAF* (7q34)	SHP2 (phosphatase) RAF1/CRAF (kinase) BRAF (kinase)	Same as Noonan syndrome; multiple skin lentigines; possible predisposition to cancer (myelodysplasia, acute myelogenous leukemia, neuroblastoma)
Noonan syndrome-like disorder with loose anagen hair (NSLAH)	*SHOC2* (10q25.2)	SHOC2 (scaffold protein)	Facial dysmorphism (macrocephaly, high forehead, wide-set eyes or hypertelorism, palpebral ptosis, low-set posteriorly rotated ears); loose anagen hair (pluckable, sparse, thin and slow-growing); congenital heart defect; distinctive skin features (darkly pigmented skin with eczema or ichthyosis); short stature; unclear predisposition to cancer
CBL syndrome (Noonan syndrome-like disorder with or without JMML)	*CBL* (11q23.3)	CBL (E3 ubiquitin ligase)	Facial dysmorphism (triangular face with hypertelorism, large low-set ears, ptosis, and flat nasal bridge); short neck; developmental delay; hyperextensible joints; thorax abnormalities; widely spaced nipples; cardiac defect; predisposition to hematologic malignancies, particularly JMML
Costello syndrome	*HRAS* (11p15.5)	HRAS (GTPase)	Facial dysmorphism (potentially more coarse than Noonan syndrome); congenital heart defect; failure to thrive; short stature; ophthalmologic anomalies; multiple skin manifestations (including, papilloma); hypotonia; predisposition to cancer (papilloma, rhabdomyosarcoma, transitional cell carcinoma, neuroblastoma)
Legius syndrome	*SPRED1* (15q14)	SPRED1 (SPROUTY-related, EVH1 domain–containing protein 1)	Café-au-lait maculae; intertriginous freckling; macrocephaly; no predisposition to cancer
Cardio-facio-cutaneous syndrome (CFCS)	*BRAF* (7q34) *MAP2K1* (15q22.31) *MAP2K2* (19p13.3) *KRAS* (12p12.1)	BRAF (kinase) MEK1 (kinase) MEK2 (kinase) KRAS (GTPase)	Facial dysmorphism (as in Noonan syndrome); congenital heart defect; failure to thrive; short stature; ophthalmologic anomalies; multiple skin manifestations (including, progressive formation of nevi); hypotonia; unclear predisposition to cancer (possible acute lymphoblastic leukemia)
Neurofibromatosis type 1 (NF1)	*NF1* (17q11.2)	NF1 or neurofibromin (GTPase activating protein)	Café-au-lait maculae; intertriginous freckling; neurofibromas and plexiform neurofibromas; iris Lisch nodules; osseous dysplasia; optic pathway glioma; predisposition to cancer (neurofibrosarcoma, central nervous system tumor, myeloid leukemia)
Capillary malformation–arteriovenous malformation syndrome (CM-AVM)	*RASA1* (5q14.3)	RASA1 or p120RasGAP (GTPase activating protein)	Multifocal capillary malformations, possibly associated with arteriovenous malformations and fistulae; unclear predisposition to cancer (possible vascular tumor)

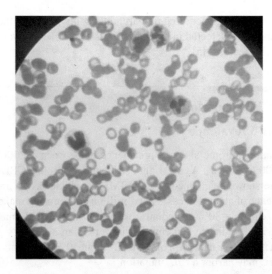

FIGURE 87.1 Peripheral blood smear showing monocytes, myeloid precursor, and normal sized platelet, indicative of JMML. (Photo credit: Patil RB et al. *Case Rep Hematol.* 2016;2016:8230786.)

Upregulation of the RAS/MAPK signaling pathway via germline/somatic mutations in *RAS* genes (*KRAS, NRAS, HRAS, RIT1*), enhanced function of upstream signal transducers (*RAF1, BRAF, MAP2K1, MAP2K2*) or RAS function modulators (*PTPN11, SOS1, CBL, RASA1, NF1, SPRED1, SHOC2*), and inefficient function of feedback mechanisms, increases its signal flow-through and has a profound deleterious effect on development, leading to a class of clinically related developmental disorders known as RASopathies or neuro-cardio-facio-cutaneous syndromes (NCFCS) [3,4].

As the most common RASopathy (NCFCS), Noonan syndrome in its classic form is characterized by facial dysmorphism (broad forehead, hypertelorism, down-slanting palpebral fissures, low-set-posteriorly rotated ears), congenital heart defect, postnatal growth retardation, ectodermal and skeletal defects (webbed and/or short neck), variable cognitive deficits, cryptorchidism, lymphatic dysplasias, bleeding tendency, and occasionally childhood hematologic malignancies (particularly JMML); with germline mutations in the *PTPN11* (protein-tyrosine phosphatase, nonreceptor-type 11), *SOS1* (son of sevenless homolog 1), *RAF1, KRAS, NRAS, LZTR1* and *A2ML1* genes underpinning its molecular pathogenesis (see Chapter 93 in this book for further details) [5–7]. Variants of Noonan syndrome include Noonan syndrome with multiple lentigines (NSML, formerly LEOPARD syndrome, which is characterized by lentigines (dark spots on the skin), electrocardiographic conduction defects (abnormalities of the electrical activity of the heart), ocular hypertelorism (widely spaced eyes), pulmonary stenosis (obstruction of the normal outflow of blood from the right ventricle of the heart), abnormalities of the genitalia, retarded/slowed growth resulting in short stature, and deafness (due to mutations in the *PTPN11, RAF1* and *BRAF* genes), Noonan-like syndrome with loose anagen hair (NS/LAH, due to mutation in the *SHOC2* gene) [8], and CBL syndrome (Noonan syndrome-like disorder with or without JMML, due to mutation in the *CBL* gene). Other related disorders include cardio-facio-cutaneous syndrome (CFCS, due to gain of function mutations in the *BRAF, MAP2K1* [mitogen activated protein kinase 1]*, MAP2K2,* and *KRAS* genes), Costello syndrome (due

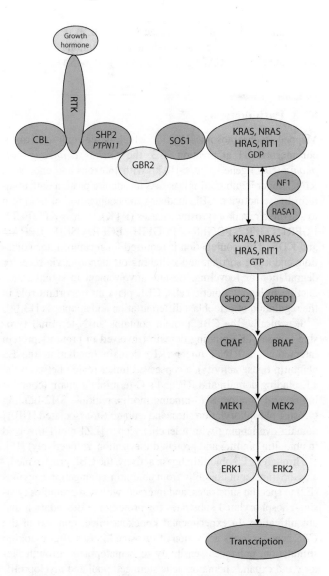

FIGURE 87.2 Schematic presentation of the RAS/MAPK signaling pathway. Binding of growth factor to receptor tyrosine kinase (RTK) brings SOS1 (which is complexed with adapter protein GRB2 during rest) into the vicinity of RAS, leading to exchange of GDP for GTP, thereby switching RAS into the active state. The active state may be converted into the inactive state by GTPase activating protein (GAP) such as neurofibromin or RASA1. Active RAS then activates RAF, which phosphorylates and activates MEK1/MEK2 (MAPK kinase), which in turn phosphorylates and activates ERK1 and/or ERK2. Active ERK1/2 induces transcription, producing various cytosolic and nuclear effectors.

to activating mutations in the *HRAS* gene), Legius syndrome (due to inactivating mutations in the *SPRED1* gene), neurofibromatosis type 1 (NF1, due to loss of function mutations in the *NF1* gene encoding neurofibromin), and capillary malformation–arteriovenous malformation syndrome (CM-AVM, due to loss of function mutations in the *RASA1* gene) (Table 87.1) [9–11].

Germline mutations involving *PTPN11* (50%), *SOS1* (10%), *RAF1* (5%), *BRAF* (5%), *MAP2K2* (5%), *CBL* (3%), and others have been shown to occur in 90% of individuals with RASopathies [12]. Mechanically, all genes implicated in the pathogenesis of RASopathies appear to cause dysregulation of the RAS/MAPK pathway by increasing ERK signaling, either through gain-of-function mutations in *RAS* genes (e.g., *KRAS, NRAS, HRAS,* and

RIT1) and RAS-guanine exchange factor (GEF) (e.g., *PTPN11* and *SOS1*), or through loss-of-function mutations in GTPase-activating protein (GAP) genes (e.g., *NF1* and *RASA1*).

87.3 Pathogenesis

Mapped to chromosome 11q23.3, which is the long (q) arm of chromosome 11 at position 23.3, the *Casitas B-lineage lymphoma* (*CBL*) gene spans >110 kb with 16 exons and encodes a RING finger E3 ubiquitin ligase. As an adaptor protein with ubiquitin ligase activity, CBL mediates the conjugation of ubiquitin to activated receptor tyrosine kinases (RTK), such as KIT, FLT1, FGFR1, FGFR2, PDGFRA, PDGFRB, EGFR, CSF1R, EPHA8, and KDR. This conjugation is required for receptor internalization, endocytic sorting, and switching off signaling via receptor degradation or recycling. Besides involvement in signal transduction in hematopoietic cells, CBL plays an important role in the regulation of osteoblast differentiation and apoptosis [13,14].

Structurally, the CBL protein contains an N-terminal tyrosine kinase (TK)-binding domain (involved in protein–protein interaction), a RING finger (RF) domain (mediating the E3 ubiquitin ligase activity), a conserved linker region between the TK-binding domain and RF, and a C-terminal domain (containing several putative SH3-binding motifs, multiple SH2-binding tyrosine phosphorylation sites, and a ubiquitin-associated [UBA] domain overlapping with a leucine zipper [LZ] motif involved in ubiquitin binding and protein dimerization, respectively) [15].

Through its E3 ubiquitin ligase activity, the CBL protein mediates the transfer of ubiquitin from ubiquitin-conjugating enzymes (E2) to specific substrates, and interacts with and promotes tyrosine-phosphorylated substrates for proteasome degradation and ubiquitination. In experimental knockout mice, removal of the *CBL* gene prolongs activation of tyrosine kinases after cytokine stimulation, enhances sensitivity to hematopoietic growth factors, and expands hematopoietic stem cell pool and myeloproliferative features [16,17].

Heterozygous germline mutations in the *CBL* gene (e.g., mutations located within the RING finger domain [e.g., c.1111T>C (p.Tyr371His), c.1186T>C (p.Cys396Arg), c.1199T>G (p.Met400 Arg), c.1201T>C (p.Cys401Arg), c.1259G>A (p.Arg420Gln)] or the linker connecting this domain to the *N*-terminal tyrosine kinase binding domain [c.1100A>C (p.Gln367Pro)]; internal deletions of exons 8/9; acquired isodisomy) inhibit FMS-like tyrosine kinase 3 (FLT3) internalization and ubiquitylation, induce cytokine-independent growth and constitutive phosphorylation of ERK, AKT, and S6, with significant developmental, tumorigenic, and functional consequences, including impaired growth, developmental delay, cryptorchidism, and a predisposition to JMML and other hematologic malignancies [18–23].

Interestingly, germline mutations in *CBL* are observed in 10%–20% of JMML or chronic myelomonocytic leukemia (CMML) patients. *CBL* mutations appear nearly always mutually exclusive of *RAS* and *PTPN11* mutations. In addition, translocation from chromosome 11 to 4 [i.e., t(4;11)(q21;q23)] and from chromosome 11 to 14 [i.e., t(11;14)(q23;q32)] have been noted in an acute leukemia cell line and a B-cell lymphoma, respectively. Expansion of CGG repeats in the 5' UTR of the *CBL* gene is also linked to Jacobsen syndrome. Furthermore, somatic gain-of-function CBL

mutation through acquired uniparental disomy (UPD) of the 11q arm has been detected in myeloid neoplasms showing myeloproliferative features [5].

87.4 Epidemiology

87.4.1 Prevalence

Noonan syndrome shows a prevalence of between 1:1.000 and 1:2.500 live births. As a variant of Noonan syndrome, CBL syndrome shows heightened risk for JMML, which has an annual incidence of about 1–2 cases per 1 million children and makes up approximately 30% of pediatric myelodysplastic syndrome and 2%–3% of pediatric leukemias, with a median age of 1.8 years and a male to female ratio of 2.5 to 1. About 10% of patients with JMML may harbor a *CBL* mutation.

87.4.2 Inheritance

Heterogeneous germline mutations in the *CBL* gene are inherited in an autosomal dominant pattern.

87.4.3 Penetrance

CBL syndrome demonstrates incomplete penetrance as individuals harboring heterozygous germline *CBL* mutation (e.g., missense Y371C) may develop JMML without features of Noonan syndrome.

87.5 Clinical Features

Clinically, CBL syndrome may present with pallor, fever, infection, skin bleeding, cough, skin rash, marked splenomegaly, and sometimes diarrhea, along with facial dysmorphism (e.g., frontal bossing, wide nasal bridge, hypertelorism, ptosis, downslanting palpebral fissures, low-set, posteriorly rotated ears), impaired growth, developmental delay, congenital heart disease, cryptorchidism, and predisposition to JMML (which is characterized by hepatosplenomegaly 97%, lymphadenopathy 76%, pallor 64%, fever 54%, skin rash 36%), typically in children (median age of 1.8 years), due to infiltration of leukemic cells into different organs (e.g., the spleen, liver, lungs, and gastrointestinal tract) [24]. Some individuals with *CBL* germline mutations may have spontaneous regression of JMML but develop vasculitis later in life.

87.6 Diagnosis

Diagnosis of CBL syndrome requires medical history review, physical examination, laboratory assessment, and molecular genetic testing.

In spite of its clinical resemblance to classic Noonan syndrome, CBL syndrome is noted for its relatively common presence of neurologic features and predisposition to JMML, and infrequent occurrence of cardiac abnormalities, reduced growth, and cryptorchidism.

Laboratory assessment reveals leukocytosis with marked monocytosis (a hallmark of JMML and CMML), circulating myeloid/erythroid precursors, varying degrees of myelodysplasia and thrombocytopenia in peripheral blood, and an elevated hemoglobin F (HbF) corrected for age. Bone marrow aspirate findings show the presence of hypercellularity, predominance of granulocytic cells, and fewer than 20% blasts [25,26].

Molecular genetic testing indicates that monosomy 7 is a major cytogenetic anomaly occurring in 20%–25% of CBL syndrome patients, while germline *CBL* mutation and loss of heterozygosity of *CBL* are sufficient for confirmation of CBL syndrome-related JMML [27–33].

Given the tendency of CBL syndrome patients to develop JMML, *it is important that JMML is diagnosed. According to the updated clinical and laboratory diagnostic criteria, JMML is recognized on the basis of three categories.*

Category 1 includes all of the following: (i) splenomegaly, (ii) absolute monocyte count >1000/μL, (iii) blasts in PB/BM <20%, (iv) absence of the t(9;22) BCR/ABL fusion gene, (v) age less than 13 years.

Category 2 comprises at least one of the following: (i) somatic mutation in *RAS* or *PTPN11*, (ii) clinical diagnosis of NF1 or *NF1* gene mutation, (iii) monosomy 7.

Category 3 contains at least two of the following: (i) circulating myeloid precursors, (ii) WBC >10,000/μL, (iii) increased fetal hemoglobin (HgF) for age, (iv) clonal cytogenetic abnormality excluding monosomy 7 [34].

Differential diagnoses for CBL syndrome include classic Noonan syndrome and related disorders (see Chapter 93 in this book) (Table 87.1). Application of molecular genetic techniques is beneficial in helping identify specific gene mutations that underlie the pathogenesis of these disorders.

87.7 Treatment

CBL syndrome is associated with a spectrum of clinical symptoms, among which JMML is the most important and responsible for significant morbidity and mortality in pediatric patients. As an aggressive myeloproliferative neoplasm (MPN) of childhood characterized by malignant transformation in the stem cell compartment with clonal proliferation of progeny that variably retain the capacity to differentiate, JMML is largely refractory to conventional antileukemia agents (e.g., etoposide, cytarabine, thiopurines [thioguanine and mercaptopurine], isotretinoin, and farnesyl inhibitors) [35]. Currently, hematopoietic stem cell transplantation (HSCT) represents the only curative therapy for JMML, with a 5-year event-free survival (EFS) rate of 55% and a relapse rate of 30%–40%, including the risk of developing vasculitis [36–39]. For CBL syndrome patients with JMML, a watch-and-wait approach is recommended until progressing and worsening JMML calls for HSCT treatment.

87.8 Prognosis

Patients with germline *CBL* mutations are at increased risk of developing JMML, which may undergo an aggressive clinical course or resolve without treatment. In the absence of HSCT, JMML patients have a median survival time of approximately 1 year, and a 5-year cumulative survival rate of 33%; the main causes of death are sepsis and organ failure due to progressive disease. With the help of HSCT, JMML patients have a mean estimated survival time of 72.4 $\pm$ 12.9 months and 5-year cumulative survival rate of 64%; the main causes of death are HSCT toxicity and sepsis/organ failure due to relapse.

Factors affecting prognosis in JMML include: (i) number of non–RAS pathway mutations (patients with two or more disease-defining RAS pathway mutations show inferior EFS and overall survival [OS] as well as higher risk of treatment failure); (ii) age, platelet count, and HgF level after any treatment (patients of <2 years of age, platelet count >33 $\times$ 10^9/L, and low age-adjusted HgF levels have a superior prognosis); (iii) *LIN28B* overexpression (LIN28B is an RNA-binding protein that regulates stem cell renewal, and *LIN28B* overexpression positively correlates with high blood HgF level and age, both of which are associated with poor prognosis, and negatively correlates with presence of monosomy 7, which is also associated with inferior prognosis) [40].

87.9 Conclusion

As a variant of Noonan syndrome linked to gene mutations that alter the key components of the RAS/MAPK signaling pathway, CBL syndrome resembles classic Noonan syndrome phenotypically but differs by its increased predisposition to JMML and frequent appearance of neurologic features [41,42]. At the molecular level, CBL syndrome is underscored by heterozygous germline mutations in the *CBL* gene, which encodes a RING finger E3 ubiquitin ligase (CBL). By mediating the conjugation of ubiquitin to activated RTK, CBL plays a vital part in receptor internalization, endocytic sorting, and switching off signaling via receptor degradation or recycling [42]. Mutations in the *CBL* gene reduce/abolish the capacity of CBL to degrade TK substrates, leading to upregulation of the RAS/MAPK signaling pathway and tumorigenesis [43,44]. Given the overlapping phenotypic features among Noonan syndrome and related disorders, use of molecular genetic tests is essential for achieving the correct diagnosis and implementing appropriate management measures. While many children with CBL-mutated JMML experience spontaneous disease regression, some may require HSCT for long-term survival [45].

REFERENCES

1. Becker H, Yoshida K, Blagitko-Dorfs N et al. Tracing the development of acute myeloid leukemia in CBL syndrome. *Blood*. 2014;123(12):1883–6.
2. Bülow L, Lissewski C, Bressel R et al. Hydrops, fetal pleural effusions and chylothorax in three patients with CBL mutations. *Am J Med Genet A*. 2015;167A(2):394–9.
3. Cordeddu V, Yin JC, Gunnarsson C et al. Activating mutations affecting the Dbl homology domain of *SOS2* cause Noonan syndrome. *Hum Mutat*. 2015;36:1080–7.
4. Cao H, Alrejaye N, Klein OD, Goodwin AF, Oberoi S. A review of craniofacial and dental findings of the RASopathies. *Orthod Craniofac Res*. 2017;20(Suppl 1):32–8.
5. Allanson JE, Roberts AE. Noonan syndrome. In: Adam MP, Ardinger HH, Pagon RA et al. editors. *GeneReviews*®

[Internet]. Seattle (WA): University of Washington, Seattle; 1993–2018. 2001 Nov 15 [updated 2016 Feb 25].

6. Cavé H, Caye A, Ghedira N et al. Mutations in RIT1 cause Noonan syndrome with possible juvenile myelomonocytic leukemia but are not involved in acute lymphoblastic leukemia. *Eur J Hum Genet.* 2016;24(8):1124–31.

7. Tekendo-Ngongang C, Agenbag G, Bope CD, Esterhuizen AI, Wonkam A. Noonan syndrome in South Africa: Clinical and molecular profiles. *Front Genet.* 2019;10:333.

8. Cordeddu V, Di Schiavi E, Pennacchio LA et al. Mutation of *SHOC2* promotes aberrant protein N-myristoylation and causes Noonan-like syndrome with loose anagen hair. *Nat Genet.* 2009;41:1022–6.

9. Martinelli S, De Luca A, Stellacci E et al. Heterozygous germline mutations in the CBL tumor-suppressor gene cause a Noonan syndrome-like phenotype. *Am J Hum Genet.* 2010;87(2):250–7.

10. Pérez B, Mechinaud F, Galambrun C et al. Germline mutations of the CBL gene define a new genetic syndrome with predisposition to juvenile myelomonocytic leukaemia. *J Med Genet.* 2010;47(10):686–91.

11. An W, Mohapatra BC, Zutshi N et al. VAV1-Cre mediated hematopoietic deletion of CBL and CBL-B leads to JMML-like aggressive early-neonatal myeloproliferative disease. *Oncotarget.* 2016;7(37):59006–16.

12. Baptista RLR, Dos Santos ACE, Gutiyama LM, Solza C, Zalcberg IR. Familial myelodysplastic/acute leukemia syndromes-myeloid neoplasms with germline predisposition. *Front Oncol.* 2017;7:206.

13. Naramura M, Jang IK, Kole H, Huang F, Haines D, Gu H. c-Cbl and Cbl-b regulate T cell responsiveness by promoting ligand-induced TCR down-modulation. *Nature Immun.* 2002;3:1192–9.

14. Lv K, Jiang J, Donaghy R et al. CBL family E3 ubiquitin ligases control JAK2 ubiquitination and stability in hematopoietic stem cells and myeloid malignancies. *Genes Dev.* 2017;31(10):1007–23.

15. Nau MM, Lipkowitz S. Comparative genomic organization of the cbl genes. *Gene.* 2003;308:103–13.

16. Rathinam C, Flavell RA. c-Cbl deficiency leads to diminished lymphocyte development and functions in an age-dependent manner. *Proc Nat Acad Sci USA.* 2010;107:8316–21.

17. Molero JC, Jensen TE, Withers PC et al.. c-Cbl-deficient mice have reduced adiposity, higher energy expenditure, and improved peripheral insulin action. *J Clin Invest.* 2004;114:1326–33.

18. Reindl C, Quentmeier H, Petropoulos K et al. CBL exon 8/9 mutants activate the FLT3 pathway and cluster in core binding factor/11q deletion acute myeloid leukemia/myelodysplastic syndrome subtypes. *Clin Cancer Res.* 2009;15(7):2238–47.

19. Sanada M, Suzuki T, Shih L-Y et al. Gain-of-function of mutated C-CBL tumour suppressor in myeloid neoplasms. *Nature.* 2009;460:904–8.

20. Niemeyer CM, Kang MW, Shin DH et al. Germline CBL mutations cause developmental abnormalities and predispose to juvenile myelomonocytic leukemia. *Nat Genet.* 2010;42(9):794–800.

21. Kao HW, Sanada M, Liang DC et al. A high occurrence of acquisition and/or expansion of C-CBL mutant clones in the progression of high-risk myelodysplastic syndrome to acute myeloid leukemia. *Neoplasia.* 2011;13(11):1035–42.

22. Shiba N, Hasegawa D, Park MJ, Murata C et al. CBL mutation in chronic myelomonocytic leukemia secondary to familial

platelet disorder with propensity to develop acute myeloid leukemia (FPD/AML). *Blood.* 2012;119(11):2612–4.

23. Calvo KR, Price S, Braylan RC et al. JMML and RALD (Ras-associated autoimmune leukoproliferative disorder): common genetic etiology yet clinically distinct entities. *Blood.* 2015;125(18):2753–8.

24. Tüfekçi Ö, Koçak Ü, Kaya Z et al. Juvenile myelomonocytic leukemia in Turkey: A retrospective analysis of sixty-five patients. *Turk J Haematol.* 2018;35(1):27–34.

25. Elghazaly AA, Manzoor MU, AlMishari MA, Ibrahim MH. A 27-year-old patient fulfilling the diagnostic criteria of both CMML and JMML. *Case Rep Oncol Med.* 2016;2016:7543582.

26. Patil RB, Shanmukhaiah C, Jijina F et al. Wiskott-Aldrich syndrome presenting with JMML-like blood picture and normal sized platelets. *Case Rep Hematol.* 2016;2016:8230786.

27. Loh ML, Sakai DS, Flotho C et al. Mutations in CBL occur frequently in juvenile myelomonocytic leukemia. *Blood.* 2009;114:1859–63.

28. Lepri FR, Scavelli R, Digilio MC et al. Diagnosis of Noonan syndrome and related disorders using target next generation sequencing. *BMC Med Genet.* 2014;15:14.

29. Muramatsu H, Makishima H, Jankowska AM et al. Mutations of an E3 ubiquitin ligase c-Cbl but not TET2 mutations are pathogenic in juvenile myelomonocytic leukemia. *Blood.* 2010;115:1969–75.

30. Pathak A, Pemov A, McMaster ML et al. Juvenile myelomonocytic leukemia due to a germline CBL Y371C mutation: 35-year follow-up of a large family. *Hum. Genet.* 2015;134:775–87.

31. Leoncini PP, Bertaina A, Papaioannou D et al. MicroRNA fingerprints in juvenile myelomonocytic leukemia (JMML) identified miR-150-5p as a tumor suppressor and potential target for treatment. *Oncotarget.* 2016;7(34):55395–408.

32. Upadhyay SY, De Oliveira SN, Moore TB. Use of rapamycin in a patient with juvenile myelomonocytic leukemia: A case report. *J Investig Med High Impact Case Rep.* 2017;5(3):2324709617728528.

33. Linder K, Iragavarapu C, Liu D. *SETBP1* mutations as a biomarker for myelodysplasia/myeloproliferative neoplasm overlap syndrome. *Biomark Res.* 2017;5:33.

34. Lipka DB, Witte T, Toth R, Yang J et al. RAS-pathway mutation patterns define epigenetic subclasses in juvenile myelomonocytic leukemia. *Nat Commun.* 2017;8(1):2126.

35. Stieglitz E, Taylor-Weiner AN, Chang TY et al. The genomic landscape of juvenile myelomonocytic leukemia. *Nat Genet.* 2015;47(11):1326–33.

36. Locatelli F, Niemeyer CM. How I treat juvenile myelomonocytic leukemia. *Blood.* 2015;125(7):1083–90.

37. Tüfekçi Ö, Ören H, Demir Yenigürbüz F, Gözmen S, Karapınar TH, İrken G. Management of two juvenile myelomonocytic leukemia patients according to clinical and genetic features. *Turk J Haematol.* 2015;32(2):175–9.

38. Clara JA, Sallman DA, Padron E. Clinical management of myelodysplastic syndrome/myeloproliferative neoplasm overlap syndromes. *Cancer Biol Med.* 2016;13(3):360–72.

39. PDQ Pediatric Treatment Editorial Board. *Childhood Acute Myeloid Leukemia/Other Myeloid Malignancies Treatment (PDQ®): Patient Version.* PDQ Cancer Information Summaries [Internet]. Bethesda (MD): National Cancer Institute (US); 2002-. 2017 Mar 6.

40. Stieglitz E, Troup CB, Gelston LC et al. Subclonal mutations in SETBP1 confer a poor prognosis in juvenile myelomonocytic leukemia. *Blood.* 2015;125(3):516–24.

41. Niemeyer CM. RAS diseases in children. *Haematologica.* 2014;99(11):1653–62.

42. Fu J-F, Hsu J-J, Tang T-C, Shih L-Y. Identification of CBL, a proto-oncogene at 11q23.3, as a novel MLL fusion partner in a patient with de novo acute myeloid leukemia. *Genes Chromosomes Cancer.* 2003;37:214–9.

43. Ramzan M, Yadav SP, Dhingra N, Sachdeva A. Juvenile myelomonocytic leukemia in India: cure remains a distant dream! *Indian J Hematol Blood Transfus.* 2014;30(Suppl 1):398–401.

44. Bresolin S, De Filippi P, Vendemini F et al. Mutations of SETBP1 and JAK3 in juvenile myelomonocytic leukemia: A report from the Italian AIEOP study group. *Oncotarget.* 2016;7(20):28914–9.

45. Niemeyer CM. JMML genomics and decisions. *Hematology Am Soc Hematol Educ Program.* 2018;2018(1):307–12.

88

CLOVES Syndrome

Dongyou Liu

CONTENTS

88.1 Introduction

CLOVES syndrome (*c*ongenital *l*ipomatous *o*vergrowth, *v*ascular malformations [typically truncal], *e*pidermal nevi, *s*coliosis/skeletal/spinal abnormalities, and seizures/central nervous system malformations) is a rare, nonhereditary overgrowth disorder characterized by asymmetric somatic hypertrophy, anomalies in multiple organs, and tumor predisposition (e.g., chorangioma, extradural spinal tumor, hemangioma, and multiple angiomatosis), resulting in serious morbidity (due mainly to septic and hemodynamic complications directly related to extensive vascular malformations) [1].

Molecularly, CLOVES syndrome is linked to heterozygous (usually somatic mosaic) activating mutations in the phosphoinositide-3-kinase (*PIK3CA*) gene on chromosome 3q26.32, which encodes the p110α catalytic subunit of phosphoinositide-3-kinase heterodimer involved in the PI3K/AKT/mTOR signaling pathway. Due to the fact that CLOVES syndrome and several other overgrowth disorders share many phenotypic features and appear to have a common genetic origin involving *PIK3CA* somatic mutations, they are collectively referred to as *PIK3CA*-related overgrowth spectrum (or *PIK3CA*-associated overgrowth), with CLOVES syndrome representing a more severe subset of the spectrum [2].

88.2 Biology

The PI3K/AKT/mTOR pathway is an intracellular signaling pathway involved in the regulation of cellular quiescence,

proliferation, survival, and apoptosis. Upon binding to receptor tyrosine kinase (RTK), phosphatidylinositol-3-kinase (PI3K) phosphorylates the 3′-hydroxyl group of phosphatidylinositol and phosphoinositides, leading to subsequent activation of its downstream molecules AKT (protein kinase B) and mTOR (mammalian target of rapamycin, which is a serine/threonine kinase), and production of proteins necessary for cell growth, cell cycle progression, and cell metabolism. The PI3K/AKT/mTOR pathway is antagonized by PTEN (phosphatase and tensin homolog deleted on chromosome 10), which dephosphorylates phosphatidylinositol (3,4,5)-triphosphate (PIP3) to phosphatidylinositol (3,4)-bisphosphate (PIP2) and thus limits the ability of AKT to bind to the membrane, and decreases its activity (Figure 88.1) [3]. Not surprisingly, mutations/deletions in the genes that encode key components of the PI3K/AKT/mTOR pathway or its antagonizers may alter its functionality and contribute to the development of various clinical disorders and tumors [4].

PIK3CA-related overgrowth spectrum (or *PIK3CA*-associated overgrowth) comprises a group of overgrowth disorders that are attributed to activating somatic mutations in the *PIK3CA* gene. These include CLOVES syndrome, fibroadipose hyperplasia or overgrowth (FAO), macrodactyly, megalencephaly-capillary malformation (MCAP) syndrome, hemihyperplasia multiple lipomatosis (HHML), macrodactyly, hemimegalencephaly, muscle hemihyperplasia, fibroadipose infiltrating lipomatosis, Klippel–Trenaunay syndrome (KTS), epidermal nevi, seborrheic keratoses, and benign lichenoid keratoses [5].

In addition, mutations in the *AKT* gene underline the pathogenesis of Proteus syndrome (AKT1), lipodystrophy syndrome (AKT2), megalencephaly-polymicrogyria-polydactyly-hydrocephalus

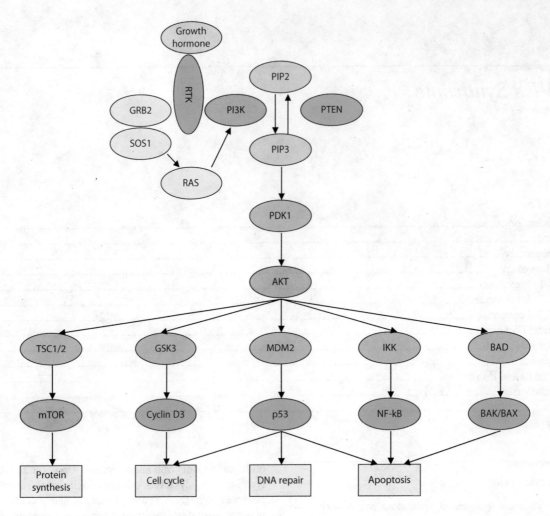

FIGURE 88.1   Schematic presentation of the PI3K/AKT/mTOR signaling pathway.

syndrome (MPPH, AKT3), tuberous sclerosis 1 (TSC1) and tuberous sclerosis 2 (TSC2); mutations in the *mTOR* gene are implicated in hemimegalencephaly, and mutations in PTEN are responsible for Bannayan–Riley–Ruvalcaba syndrome (BRRS), Cowden syndrome, type 2 segmental Cowden syndrome, and Lhemitte–Duclos disease [6].

CLOVES syndrome was first noted in 2007 from patients who were thought to have Proteus syndrome. These patients presented with progressive, complex, and mixed primarily truncal vascular malformations, deregulated adipose tissue, varying degrees of scoliosis, and enlarged bony structures without progressive bony overgrowth [7]. Subsequent inclusion of defects related to the spine, mainly scoliosis, high flow vascular malformations, neural tube defects, and tethered cord completes the clinical spectrum of CLOVES syndrome (i.e., congenital lipomatous asymmetric overgrowth of the trunk with lymphatic, capillary, venous, and combined-type vascular malformations, epidermal nevi, scolio-sis/skeletal and spinal anomalies) [8–10].

88.3 Pathogenesis

Phosphatidylinositol 3-kinases (PI3K) are grouped into three structurally and functionally distinct classes (I–III). Class I

PI3K are heterodimeric enzymes that comprise a catalytic sub-unit p110 that associates with an adaptor/regulatory subunit, and that adds a phosphate group in the D-3 position of the inositol ring on three substrates [non-phosphorylated phosphatidylino-sitol (PI), inositol monophosphate (PI(4)P) and bisphosphate (PI(4,5)P$_2$)] to generate PI(3)P, PI(3,4)P$_2$ [i.e., phosphatidylino-sitol (3,4)-bisphosphate or PIP2] and PI(3,4,5)P$_3$ [i.e., phospha-tidylinositol (3,4,5)-triphosphate (PIP3) or PIP3], respectively. PIP$_3$ is the predominant mediator of PI3K activity and functions as an important second messenger in the cell. In contrast, PTEN (phosphatase and tensin homolog deleted on chromosome 10) removes the phosphate group from the D-3 position of phosphati-dylinositol and acts as the direct catalytic antagonist of PI3K [3].

The catalytic subunit p110 of class I PI3K has three isoforms (p110α, p110β, and p110δ), which are encoded by the *PIK3CA*, *PIK3CB*, and *PIK3CD* genes, respectively. Similarly, the adap-tor/regulatory subunit p85of class I PI3K consists of five isoforms (p85α, p55α, p50α, p85β, and p55γ), which are encoded by the *PIK3R1*, *PIK3R2*, and *PIK3R3* genes.

Mapped to chromosome 3q26.32, the *PIK3CA* gene con-sists of 21 exons and encodes the p110α catalytic subunit of phosphoinositide-3-kinase (PI3K) heterodimer. With a length of 1068 aa and a mass weight of 110 kDa, the p110α catalytic subunit comprises five functional domains: p85-regulatory

TABLE 88.1

Relationship between *PIK3CA* Mutations and Overgrowth Disorders

PIK3CA	P85 Binding Domain	Intron	C2 Domain	Helical Domain	Intron	Kinase Domain
MCAP syndrome	E81K		G364R	E542K	E726K	G914R
	R88Q		E365K	E545K		Y1021C
			C378Y			T1025A
			E452K			A1035V
			E453K			M1043I
			E453del			H1047L
						H1047R
						H1047Y
						G1049S
CLOVES syndrome			C420R	E542K		H1047L
				E545K		H1047R
FAO/HHML				E542K		H1047L
				E545K		H1047R
Macrodactyly		R115P		E542K		
				E545K		
Hemimegalencephaly				E542K		H1047L
				E545K		H1047R
Muscle hemihyperplasia				E542K		H1047L
				E545K		H1047R
						L1087fs
Facial infiltrating lipomatosis			E452K	E542K		H1047L
			E453K	E545K		H1047R
			E453del			
Epidermal nevi				E542K		
				E545K		
Seborrheic keratoses				E542K		H1047L
				E545K		H1047R
Benign lichenoid keratoses				E542K		
				E545K		

subunit-binding domain (p85-BD), Ras-binding domain (Ras-BD), C2 domain, helical domain, and kinase catalytic domain. Further, a pseudogene sharing >95% homology with *PIK3CA* exons 9–13 exists on chromosome 22, although its role in the development of *PIK3CA*-related overgrowth spectrum remains unclear.

The PI3K heterodimer is activated by receptors with protein tyrosine kinase activity (RTK), and which then phosphorylates inositol ring 3'-OH group in inositol phospholipids, thus converting phosphatidylinositol-4,4-bisphosphate (PIP_2) to phosphatidylinositol-3,4,5-triphosphate (PIP_3). This leads to the translocation and phosphorylation of protein serine/threonine kinase-3'-phosphoinositide-dependent kinase 1 (PDK1) to the cell membrane. PDK1 then phosphorylates Akt/PKB to initiate downstream processes involved in cell survival and cell cycle progression, including activation of mTOR, glycogen synthase kinase-3 (GSK3), p53, and IκB kinase (IKK), which is a positive regulator of survival factor NFκB, and inactivation of pro-apoptotic factors such as Bad and Procaspase-9 (Figure 88.1) [11,12].

Postzygotic somatic mutations in the *PIK3CA* gene result in increased basal hyperphosphorylation of AKT and its downstream targets, leading to the development of various symptoms of *PIK3CA*-related overgrowth spectrum [13,14]. To date, at least 23 pathogenic variants are identified in various regions (particularly the helical and kinase domains) of the *PIK3CA* gene implicated in the pathogenesis of involved in *PIK3CA*-related overgrowth spectrum, including MCAP syndrome, CLOVES syndrome, fibroadipose overgrowth (FAO), and HHML, etc. (Table 88.1). Among these, p.His1047Arg (H1047R), p.His1047Leu (H1047L), p.Glu545Lys (E545K), p.Glu542Lys (E542K) and p.Cys420Arg (C420R) are found in 54%, 23%, 11%, 8%, and 3% of patients, respectively [15,16].

In addition, somatic activating (gain-of-function) pathogenic *PIK3CA* variants are known to occur in tumors of the colon, breast, brain, liver, stomach, and lung from individuals without *PIK3CA*-related overgrowth [17].

88.4 Epidemiology

88.4.1 Prevalence

The prevalence of *PIK3CA*-related overgrowth spectrum is largely unknown owing to its broad phenotypic presentation. To date, about 20 CLOVES syndrome cases and >150 MCAP syndrome cases (which may include some cases previously reported as Proteus syndrome or KTS) are described in the literature. *PIK3CA*-related overgrowth spectrum affects individuals of various ethnic backgrounds.

88.4.2 Inheritance

PIK3CA-related overgrowth spectrum is caused by somatic mutations that occur post-fertilization in one cell of the multicellular embryo. Therefore it is not inherited, although vertical transmission or sib recurrence has been reported occasionally.

88.4.3 Penetrance

PIK3CA-related overgrowth spectrum shows incomplete penetrance, as some patients with somatic activating *PIK3CA* pathogenic variants are known develop tumors without *PIK3CA*-related overgrowth.

88.5 Clinical Features

CLOVES syndrome is a severe overgrowth disorder that typically demonstrates following clinical features [18,19]:

Overgrowth: Asymmetric lipomatous overgrowth (typically truncal, complex, congenital, progressive); spinal-paraspinal extension; limb/digital overgrowth; bony overgrowth, leg-length discrepancy (Figure 88.2) [20]

Cutaneous/vascular malformations: Low-flow (capillary, venous, lymphatic; typically overlying truncal overgrowth); high-flow (arteriovenous; esp. spinal-paraspinal); venous thrombosis/embolism; epidermal nevi (single/multiple)

Musculoskeletal/acral abnormalities: Scoliosis; chondromalacia patellae; dislocated knees; macrodactyly (enlargement of all tissues localized to the terminal portions of a limb, typically within a nerve territory), wide hands/feet; sandal gap toes; symmetric overgrowth feet; plantar-palmar overgrowth

Visceral abnormalities: Renal agenesis/hypoplasia; splenic lesions

Neurologic abnormalities: Neural tube defect; tethered cord; megalencephaly/hemimegalencephaly; Chiari malformation; polymicrogyria

Tumors: Chorangioma, extradural spinal tumor, hemangioma, and multiple angiomatosis

88.6 Diagnosis

Observation of congenital or early childhood-onset clinical features (either spectrum or isolated features, such as lipomatous masses on the trunk, vascular malformation and epidermal nevi, legs with an uneven length, chondromalacia patellae, scoliosis, large hands and legs, and an increase in the distance between the first and second toes) allows diagnosis of *PIK3CA*-related overgrowth spectrum (Figure 88.2; Table 88.2) [20–23].

Identification of a pathogenic variant on one *PIK3CA* allele from more than one affected tissue (since most *PIK3CA* pathogenic variants arise postzygotically and are thus mosaic) provides

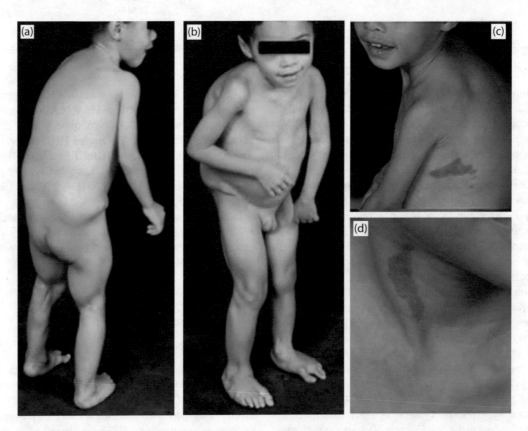

FIGURE 88.2 A 2-year-old boy with CLOVES syndrome displaying increase in the soft parts of the lumbar region and the posterior portion of the lower limbs (a); flat feet, spacing between the first and second toes, and scoliosis of the thoracic-lumbar spine (b); capillary malformation in the trunk (c); and epidermal nevi on the right cervical region (d). (Photo credit: Panteliades M et al. *An Bras Dermatol.* 2016;91(3):378–80.)

TABLE 88.2

Clinical Criteria for Diagnosis of *PIK3CA*-Related Overgrowth Spectrum

Required

    Presence of somatic PIK3CA mutation

    Congenital or early childhood onset

    Overgrowth: Sporadic and mosaic

    Features as described in either A or B

      A. Spectrum (two or more features)

        1. Overgrowth: Adipose, muscle, nerve, skeletal

        2. Vascular malformations: Capillary, venous, arteriovenous malformation, lymphatic

        3. Epidermal nevus

      B. Isolated features

        1. Large isolated lymphatic malformation

        2. Isolated macrodactyly (macrodystrophia lipomatosa, macrodactylia fibrolipomatosis, and gigantism); OR overgrown splayed feet/ hands, overgrown limbs

        3. Truncal adipose overgrowth

        4. Hemimegalencephaly (bilateral)/ dysplastic megalencephaly/ focal cortical dysplasia 2

        5. Epidermal nevus

        6. Seborrheic keratoses

        7. Benign lichenoid keratoses

further confirmation of *PIK3CA*-related overgrowth spectrum in an individual who meets clinical criteria. However, absence of a *PIK3CA* pathogenic variant does not exclude a clinical diagnosis in individuals displaying either spectrum or isolated features.

Differential diagnoses for CLOVES syndrome include:

Megalencephaly-capillary malformation syndrome (MCAP): Megalencephaly or hemimegalencephaly associated with abnormalities of muscle tone, seizures, and mild to severe intellectual disability; cutaneous capillary malformations with focal or generalized somatic overgrowth; digital anomalies consisting of syndactyly and polydactyly; cortical malformations, polymicrogyria; connective tissue dysplasia; pathogenic variants in *PIK3CA*

Hemimegalencephaly (HMEG): Enlargement and dysplasia of a cerebral hemisphere; cortical dysgenesis, abnormally increased white matter, dilated and dysmorphic lateral ventricles; mosaic pathogenic variants in *PIK3CA, AKT3* and *MTOR*

Fibroadipose hyperplasia: Progressive segmental overgrowth of visceral, subcutaneous, muscular, fibroadipose, and skeletal tissues; lipomatous infiltration of muscle, progressive adipose dysregulation and regional lipohypoplasia, vascular malformations, testicular abnormalities, and polydactyly; mosaic pathogenic variants in *PIK3CA*

Hemihyperplasia-multiple lipomatosis syndrome (HHML): Moderate somatic asymmetry and overgrowth with subcutaneous lipomas and occasional vascular malformations; absence of deep vascular malformations, epidermal nevi, cerebriform connective tissue nevi, and hyperostosis; pathogenic variants in *PI3KCA*

Isolated macrodactyly: Congenital isolated limb overgrowth; absence of vascular malformations and overgrowth involving other body parts; mosaic *PIK3CA* pathogenic variants in affected peripheral nerve cells, but not in blood

Klippel–Trenaunay syndrome (KTS): Disproportionate growth disturbance combined with cutaneous capillary, lymphatic, and venous malformations; pathogenic variants in *PI3KCA*

Megalencephaly-polymicrogyria-polydactyly-hydrocephalus syndrome (MPPH): Congenital megalencephaly, bilateral perisylvian polymicrogyria, postaxial polydactyly, and an increased risk for hydrocephalus; absence of vascular malformations, focal somatic overgrowth, and connective tissue dysplasia; de novo germline pathogenic variants in *PIK3R2* and *AKT3*

Bannayan–Riley–Ruvalcaba syndrome (BRRS): Macrocephaly, developmental delay, lipomatosis, intestinal hamartomatous polyposis, and pigmented macules of the penis; absence of truncal lipomatous overgrowth, acral deformities; germline pathogenic variants in *PTEN*

Proteus syndrome: Disproportionate and asymmetric postnatal somatic overgrowth including skeletal overgrowth, cerebriform connective tissue nevi, epidermal nevi, dysregulated adipose tissue, and vascular malformations; absence of truncal fatty-vascular mass, spinal paraspinal fast-flow lesions and acral abnormalities; mosaic pathogenic variants in *AKT1*

SOLAMEN syndrome: Atypical features of Cowden syndrome including segmental overgrowth, lipomatosis, arteriovenous malformation, and epidermal nevi; ovarian cystadenoma, multiple breast tumors, and thyroid adenomas; biallelic inactivation of *PTEN* in atypical lesions

Cowden syndrome: Germline *PIK3CA* pathogenic variants [23,24]

88.7 Treatment

Treatment options for CLOVES syndrome and other *PIK3CA*-related overgrowth disorders include surgical debulking (for truncal lipomatous mass), orthopedic care (for scoliosis/kyphosis and leg-length discrepancy), neurosurgical intervention (for obstructive hydrocephalus, increased intracranial pressure, progressive/symptomatic cerebellar tonsillar ectopia, Chiari malformation, and epilepsy), and other procedures (for cardiac and renal abnormalities; intellectual disabilities, behavior problems, motor difficulties, speech, swallowing, and feeding difficulties). Use of PI3K inhibitor (pictilisib, copanlisib, duvelisib), Akt inhibitor (ipatasertib, MK-2206, ARQ-092), mTOR inhibitor (rapamycin), and dual PI3K/AKT/mTOR inhibitor (gedatolisib, apitolisib) offers another approach for management of *PIK3CA*-related overgrowth spectrum [25].

88.8 Prognosis and Prevention

CLOVES syndrome is associated with significant morbidities (e.g., severe scoliosis, infiltrative lipomatous overgrowth, paraspinal high-flow lesions with spinal cord ischemia, lymphatic malformations, cutaneous vesicles, orthopedic problems, central phlebectasias and thromboembolism). Early diagnosis, diligent surgical/medical care and regular surveillance (of skin, cardiac, renal, abdominal, musculoskeletal, and neurologic anomalies) are crucial for improving patient well-being and prognosis.

Since *PIK3CA*-related overgrowth spectrum is not inherited, prenatal diagnosis and preimplantation diagnosis are generally not indicated for family members [26].

88.9 Conclusion

CLOVES syndrome is a nonhereditary overgrowth disorder that typically manifests as congenital lipomatous asymmetric overgrowth of the trunk, lymphatic, capillary, venous, and combined-type vascular malformations, epidermal nevi, scoliosis/skeletal/spinal anomalies, and seizures/central nervous system malformations [27,28]. Postzygotic somatic mutations in the *PIK3CA* gene encoding the p110α catalytic subunit of phosphoinositide-3-kinase (PI3K) heterodimer, which plays a key role in the regulation of the PI3K/AKT/mTOR signaling pathway, appear to underline the molecular pathogenesis of CLOVES syndrome. As CLOVES syndrome and several other overgrowth disorders (e.g., FAO, HHML, macrodactyly, fibroadipose-infiltrating lipomatosis, MCAP) demonstrate phenotypic overlapping and genetic similarity (all possessing *PIK3CA* mutations), they are considered as *PIK3CA*-related overgrowth spectrum. Clinical diagnosis of CLOVES syndrome involves assessment of congenital or early childhood onset features, and molecular identification of *PIK3CA* pathogenic variants offers further evidence of the disease [29,30]. Given the current lack of specific therapies, management of CLOVES syndrome relies on standard procedures that provide symptomatic relief and improve patient quality of life.

REFERENCES

1. Acosta S, Torres V, Paulos M, Cifuentes I. CLOVES syndrome: Severe neonatal presentation. *J Clin Diagn Res.* 2017;11(4):TR01-03.

2. Mirzaa G, Conway R, Graham JM Jr, Dobyns WB. *PIK3CA*-related segmental overgrowth. In: Adam MP, Ardinger HH, Pagon RA et al. editors. *GeneReviews®* [Internet]. Seattle (WA): University of Washington, Seattle; 1993–2018. 2013 Aug 15.

3. Porta C, Paglino C, Mosca A. Targeting PI3K/Akt/mTOR signaling in cancer. *Front Oncol.* 2014;4:64.

4. Rivière JB, Mirzaa GM, O'Roak BJ et al. De novo germline and postzygotic mutations in AKT3, *PI*K3R2 and PIK3CA cause a spectrum of related megalencephaly syndromes. *Nat Genet.* 2012;44:934–40.

5. Rios JJ, Paria N, Burns DK et al. Somatic gain-of-function mutations in PIK3CA in patients with macrodactyly. *Hum Mol Genet.* 2013;22(3):444–51.

6. Orloff MS, He X, Peterson C et al. Germline PIK3CA and AKT1 mutations in Cowden and Cowden-like syndromes. *Am J Hum Genet.* 2013;92:76–80.

7. Sapp JC, Turner JT, van de Kamp JM, van Dijk FS, Lowry RB, Biesecker LG. Newly delineated syndrome of congenital lipomatous overgrowth, vascular malformations, and epidermal nevi (CLOVE syndrome) in seven patients. *Am J Med Genet A.* 2007;143A:2944–58.

8. Gucev Z, Tasic V, Jancevska A et al. Congenital lipomatous overgrowth, vascular malformations, and epidermal nevi (CLOVE) syndrome: CNS malformations and seizures may be a component of this disorder. *Am J Med Genet A.* 2008;146A(20):2688–90.

9. Alomari AI. Characterization of a distinct syndrome that associates complex truncal overgrowth, vascular, and acral anomalies: A descriptive study of 18 cases of CLOVES syndrome. *Clin Dysmorphol.* 2009;18(1):1–7.

10. Alomari AI. CLOVE(S) syndrome: Expanding the acronym. *Am J Med Genet A.* 2009;149A(2):294–5.

11. Janku F, Lee JJ, Tsimberidou AM et al. PIK3CA mutations frequently coexist with RAS and BRAF mutations in patients with advanced cancers. *PLOS ONE.* 2011;6:e22769.

12. Liu P, Cheng H, Santiago S et al. Oncogenic PIK3CA-driven mammary tumors frequently recur via PI3K pathway-dependent and PI3K pathway-independent mechanisms. *Nat Med.* 2011;17:1116–20.

13. Lee JH, Huynh M, Silhavy JL et al. De novo somatic mutations in components of the PI3K-AKT3-mTOR pathway cause hemimegalencephaly. *Nat Genet.* 2012;44:941–5.

14. Lindhurst MJ, Parker VE, Payne F et al. Mosaic overgrowth with fibroadipose hyperplasia is caused by somatic activating mutations in PIK3CA. *Nat Genet.* 2012;44:928–33.

15. Kurek KC, Luks VL, Ayturk UM et al. Somatic mosaic activating mutations in PIK3CA cause CLOVES syndrome. *Am J Hum Genet.* 2012;90(6):1108–15.

16. Loconte DC, Grossi V, Bozzao C et al. Molecular and functional characterization of three different postzygotic mutations in PIK3CA-related overgrowth spectrum (PROS) patients: Effects on PI3K/AKT/mTOR signaling and sensitivity to PIK3 inhibitors. *PLOS ONE.* 2015;10(4):e0123092.

17. Janku F, Wheler JJ, Naing A et al. PIK3CA mutations in advanced cancers: Characteristics and outcomes. *Oncotarget.* 2012;3:1566–75.

18. Keppler-Noreuil KM, Sapp JC, Lindhurst MJ et al. Clinical delineation and natural history of the PIK3CA-related overgrowth spectrum. *Am J Med Genet A.* 2014;164A(7):1713–33.

19. Gopal B, Keshava SN, Selvaraj D. A rare newly described overgrowth syndrome with vascular malformations-Cloves syndrome. *Indian J Radiol Imaging.* 2015;25(1):71–3.

20. Panteliades M, Silva CM, Gontijo B. What is your diagnosis? *An Bras Dermatol*. 2016;91(3):378–80.

21. Alomari AI, Chaudry G, Rodesch G et al. Complex spinal-paraspinal fast-flow lesions in CLOVES syndrome: Analysis of clinical and imaging findings in 6 patients. *Am J Neuroradiol*. 2011;32(10):1812–7.

22. Sarici D, Akin MA, Kurtoglu S, Tubas F, Sarici SU. A neonate with CLOVES syndrome. *Case Rep Pediatr*. 2014;2014:845074.

23. Keppler-Noreuil KM, Rios JJ, Parker VE et al. PIK3CA-related overgrowth spectrum (PROS): diagnostic and testing eligibility criteria, differential diagnosis, and evaluation. *Am J Med Genet A*. 2015;167A(2):287–95.

24. Youssefian L, Vahidnezhad H, Baghdadi T et al. Fibroadipose hyperplasia versus Proteus syndrome: Segmental overgrowth with a mosaic mutation in the PIK3CA gene. *J Invest Dermatol*. 2015;135(5):1450–3.

25. Alomari AI. Comments on the diagnosis and management of cloves syndrome. *Pediatr Dermatol*. 2011;28(2):215–16.

26. Emrick LT, Murphy L, Shamshirsaz AA et al. Prenatal diagnosis of CLOVES syndrome confirmed by detection of a mosaic PIK3CA mutation in cultured amniocytes. *Am J Med Genet A*. 2014;164A(10):2633–7.

27. Alomar S, Khedr RE, Alajlan S. CLOVES syndrome in a nine-month-old infant. *Cureus*. 2019;11(9):e5772.

28. Mahajan VK, Gupta M, Chauhan P, Mehta KS. Cloves syndrome: A rare disorder of overgrowth with unusual features—An uncommon phenotype? *Indian Dermatol Online J*. 2019;10(4):447–52.

29. Michel ME, Konczyk DJ, Yeung KS et al. Causal somatic mutations in urine DNA from persons with the CLOVES subgroup of the PIK3CA-related overgrowth spectrum. *Clin Genet*. 2018;93(5):1075–80.

30. Lalonde E, Ebrahimzadeh J, Rafferty K et al. Molecular diagnosis of somatic overgrowth conditions: A single-center experience. *Mol Genet Genomic Med*. 2019;7(3):e536.

89

Costello Syndrome

Dongyou Liu

CONTENTS

89.1 Introduction

Costello syndrome is a rare autosomal dominant disorder characterized by dysmorphic craniofacial features (e.g., macrocephaly, prominent forehead, epicanthal folds, downslanting palpebral fissures, short nose with depressed nasal bridge and broad base, low-set-posteriorly rotated ears with thickened helices and lobes, full cheeks, and large mouth with full lips), short neck, macroglossia, dermatologic manifestations (soft skin with excessive wrinkling and redundancy over the dorsum of the hands and the feet along with deep plantar and palmar creases), growth and mental retardation (e.g., polyhydramnios in utero), cardiac abnormalities (e.g., hypertrophic cardiomyopathy [HCM], tachyarrhythmias, valve and septal defects), hypotonia, nervous system involvement (e.g., ventriculomegaly, hydrocephaly, and Chiari I malformation), gastrointestinal dysfunctions (e.g., reflux, oral aversion, and constipation), and predisposition to tumor development (e.g., cutaneous papilloma [absent in other RASopathies], rhabdomyosarcoma [60% of cases], transitional cell carcinoma, and neuroblastoma).

Considered as one of the RASopathies (which encompass Noonan syndrome and variants, cardio-facio-cutaneous syndrome [CFCS], neurofibromatosis type 1 [NF-1] and Legius syndrome; see Chapters 87, 92, and 93 in this book for details), Costello syndrome is linked to heterozygous activating germline mutations in the HRAS gene on chromosome 11p15.5, with >80% of cases harboring a c.34G>A (p.Gly12Ser) substitution, in addition to other pathogenic variants (e.g., p.Gly13Asp substitution) [1].

89.2 Biology

First described by J.M. Costello in 1971, with follow-up reports in 1977 and 1996, Costello syndrome (also referred to as AMICABLE syndrome [amicable personality, mental retardation, impaired swallowing, cardiomyopathy, aortic defects, bulk, large lips and lobules, ectodermal defects] or faciocutaneous-skeletal syndrome) is a congenital disorder showing facial dysmorphisms and anomalies in multiple organs/systems (including cardiac, musculoskeletal, cutaneous, and central nervous systems) [2].

The identification by Aoki et al. in 2005 of germline gain-of-function mutation (typically arising de novo in the paternal germline) in the HRAS gene from affected individuals yielded valuable insights into the molecular pathogenesis of phenotypic, cognitive, and tumorigenetic development associated with Costello syndrome. This provides a foundation for differentiating Costello syndrome from phenotypically overlapping CFCS (which is due to germline mutations in BRAF, and less commonly KRAS, MAP2K1, or MAP2K2) [3].

89.3 Pathogenesis

The RAS (abbreviated from rat sarcoma viral oncogene homolog) gene family includes several known members [i.e., HRAS (11p15.5), KRAS (12p12.1), and NRAS (1p13.2) as well as RIT1 (RAS-like protein in tissues)] that encode guanosine-5'-triphosphate (GTP)ases (K-RAS, H-RAS, N-RAS, and RIT1). These proteins function as on-off binary switches for several downstream effectors such as RAF, phosphatidylinositol

3-kinases (PI3K), and ral guanine nucleotide exchange factor (RalGEF), that make up part of the MAPK/RAF/MEK/ERK pathway, the PI3K/AKT/mTOR pathway, and the RalGEF/TBK1/IRF3/3-NF-κB pathway, respectively). Given their roles in the control of cell survival, proliferation, differentiation, senescence, and death, germline mutations in the *RAS* genes as well as the genes encoding components or regulators of the MAPK/RAF/MEK/ERK pathway are associated with a group of genetic disorders collectively known as RASopathies (e.g., Noonan syndrome, Noonan syndrome with multiple lentigines [NSML, formally LEOPARD syndrome], Noonan-like syndrome with loose anagen hair [NS/LAH], CBL syndrome [Noonan syndrome-like disorder with or without JMML], CFCS, Costello syndrome, Legius syndrome, NF1, and capillary malformation–arteriovenous malformation syndrome (CM-AVM)] [4–6]. Typically, RASopathies produce a combination of facial and skin abnormalities, heart defects, a predisposition to specific cancers, and developmental delay (including central nervous system abnormalities, cognitive dysfunction, and behavioral impairments). While Noonan syndrome is linked to germline *KRAS* or *NRAS* mutations, CFCS results from germline *BRAF*, *KRAS*, *MAP2K1*, or *MAP2K2* mutations, and Costello syndrome appears to be caused by germline *HRAS* mutations [7–10].

The *HRAS* gene consists of six exons, five of which are involved in the production of a 189 aa, 21 kDa protein (HRAS). Most pathogenic *HRAS* variants result from missense substitutions that disrupt guanine nucleotide binding and reduce in intrinsic and GAP-induced GTPase activity of the HRAS protein, keeping HRAS in the active state, and increasing HRAS binding affinity and constitutive activation (gain-of-function or hyperactivation) of its downstream targets including RAF, PI-3, and RalGEF [11]. This contributes to phenotypic abnormalities associated with Costello syndrome (e.g., facial dysmorphism, mild to moderate intellectual disability, and increased anxiety) as well as predisposition to several tumors (e.g., cutaneous papilloma, rhabdomyosarcoma [60% of cases], transitional cell carcinoma, and neuroblastoma) [12,13]. In fact, the substitution of glycine to serine (G12S) at codon 12 in exon 1 accounts for 80% of clinical cases. Other notable HRAS mutations identified in Costello syndrome include p.Gly13Asp and p.Gly60Asp [14].

Cutaneous papilloma is a benign tumor that occurs in Costello syndrome, but not in other RASopathies. Rhabdomyosarcoma represents the most common pediatric soft tissue sarcoma, and accounts for 60% of neoplasia in Costello syndrome affected individuals. HRAS mutations G12S, p.G12C, or p.G12A are commonly found in Costello syndrome−related rhabdomyosarcoma. Interestingly, most patients with Costello syndrome−related rhabdomyosarcoma carry a paternally inherited *HRAS* mutation and tumors show paternal uniparental disomy with loss of the maternally inherited chromosome 11 [15,16].

89.4 Epidemiology

89.4.1 Prevalence

Costello syndrome is estimated to affect 1:300,000 in the UK, and 1:1,230,000 in Japan. To date, approximately 300 cases have been reported in the literature worldwide. Females and males are equally affected. Costello syndrome patients harboring a *HRAS* pathogenic variant show a 10%–15% lifetime risk of cancer development, with rhabdomyosarcoma accounting for about 60% of all neoplasia.

89.4.2 Inheritance

Costello syndrome shows an autosomal dominant inheritance, typically involving a heterozygous de novo missense *HRAS* mutation. In some cases, somatic mosaicism occurring in the germline is passed on to the offspring.

89.4.3 Penetrance

Costello syndrome demonstrates a complete penetrance, as *HRAS* pathogenic variants are present in all patients with this disorder. However, due to an inaccurate clinical diagnosis, these mutations are usually identified in 80%–90% of affected individuals.

89.5 Clinical Features

Costello syndrome typically manifests as facial dysmorphism, skin changes, cognitive impairment, cardiac and musculoskeletal defects, along with an increased risk of malignancies as well as sudden death secondary to heart disease [17].

In the prenate, Costello syndrome may display some nonspecific features such as increased nuchal thickness, polyhydramnios (>90%), characteristic ulnar deviation of the wrists, short humeri and femurs, and possibly atrial tachycardia.

In neonate, Costello syndrome is associated with increased birth weight and head circumference (often >50th centile), hypoglycemia, failure to thrive, severe feeding difficulties, facial dysmorphism (e.g., relatively high forehead, low nasal bridge, epicanthal folds, prominent lips, and a wide mouth), ulnar deviation of wrists and fingers, loose-appearing skin with deep palmar and plantar creases, and cryptorchidism.

In infancy, Costello syndrome may show severe feeding difficulties (giving a marasmic appearance), pyloric stenosis, hypotonia (suggestive of myopathy), irritability, developmental delay, nystagmus, cardiac abnormalities (e.g., non-progressive valvar pulmonic stenosis, atrial septal defects, subaortic septal hypertrophy, pathologic myocardial disarray, multifocal atrial tachycardia).

In childhood, Costello syndrome presents as short stature, delayed bone age, partial or complete growth hormone deficiency, cardiac abnormalities (mild or moderate HCM), papilloma, acanthosis nigricans, thick calluses and toenails, strong body odor, tight Achilles tendons, developmental delay or intellectual disability, separation anxiety (more common in males than in females), progressive postnatal cerebellar overgrowth (leading to Chiari I malformation, syringomyelia, and hydrocephalus), loose joints (particularly involving the fingers), and ulnar deviation of the wrists and fingers.

In adolescence, Costello syndrome often shows delayed or disordered puberty, worsening kyphoscoliosis, sparse hair, and prematurely aged skin.

In adulthood (16–40 years of age), Costello syndrome may develop osteoporosis or osteopenia, bone pain, vertebral fractures, height loss (adult height average of 135–150 cm), developmental hip dysplasia (leading to severe pain and preventing ambulation), gastroesophageal reflux, cardiac abnormalities (e.g., atrial and ventricular arrhythmia), and compromised quality of life.

About 15% of Costello syndrome patients harboring a *HRAS* pathogenic variant develop solid tumors (rhabdomyosarcoma and neuroblastoma in early childhood, transitional cell carcinoma of the bladder in adolescents [compared to age >65 years in the general population], and papilloma). Rhabdomyosarcoma is mainly located in the abdomen, pelvis, and/or urogenital area. Overall, Costello syndrome confers a higher cancer risk than other RASopathies [18,19].

Death (10%–20%) may result from HCM in association with neoplasia, coronary artery fibromuscular dysplasia, multifocal tachycardia, neoplasia, and pulmonary and multiorgan failure [20,21].

89.6 Diagnosis

Clinical diagnosis of Costello syndrome relies on the recognition of a constellation of features in the context of the following:

Perinatal presentations: Severe polyhydramnios, increased birth weight due to edema but not true macrosomia, weight loss due to resolution of edema, failure to thrive, severe postnatal feeding difficulties and short stature

Craniofacial appearance and voice: Relative macrocephaly; coarse facial features with prominent epicanthal folds, full cheeks and lips, large mouth, full nasal tip; curly or sparse, fine hair; epicanthal folds; wide nasal bridge, short full nose; deep hoarse or whispery voice

Skin anomalies: Loose or soft skin; increased pigmentation; deep palmar and plantar creases; papilloma of face and perianal region in childhood; premature aging, sparse or curly hair

Musculoskeletal signs: Diffuse hypotonia, joint laxity; ulnar deviation of wrists and fingers; spatulate finger pads, abnormal fingernails; tight Achilles tendons; positional foot deformity; vertical talus; kyphoscoliosis; pectus carinatum, pectus excavatum, asymmetric rib cage; developmental hip dysplasia

Cardiovascular abnormalities: HCM (idiopathic subaortic stenosis, asymmetric septal hypertrophy); congenital heart defect, particularly valvar pulmonic stenosis; supraventricular tachycardia (e.g., chaotic atrial rhythm/multifocal atrial tachycardia, or ectopic atrial tachycardia; aortic dilation)

Neurologic symptoms: e.g., Chiari I malformation, hydrocephalus, syringomyelia, seizures, tethered cord

Tumors: Including benign papilloma and malignant rhabdomyosarcoma, neuroblastoma, transitional cell carcinoma of the bladder (Figure 89.1)

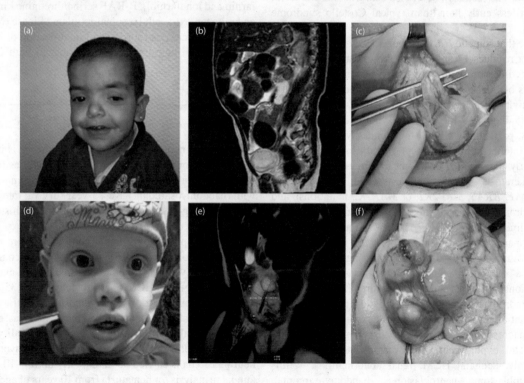

FIGURE 89.1 (a) Costello syndrome in a 4.4-year-old female with HRAS p.G12S mutation showing low-set ears, lower palpebral fissures, hypertelorism, broad nasal bridge, thick lips, and thin hair. (b) Abdominal ultrasound revealing the presence of a solid mass paravesical displacing the left ovary and bladder without infiltrate surrounding structures. (c) Operative image indicating the presence of rhabdomyosarcoma (5 cm in size) originating in the left medial umbilical ligament. (d) Costello syndrome in a 5.4-year-old female with HRAS p.G12S mutation showing prominent forehead, low-set ears, lower palpebral fissures, hypertelorism, broad nasal bridge, thick lips, and slightly anteverted nostrils. (e) MRI on T2-weighted sequence confirming the presence of several nodular masses grouped with thickening of both medial umbilical folds, the right to 4 mm proximal to the umbilicus and 8 mm adjacent to the iliac vessels paravesical region portion. (f) Operative image indicating the presence of rhabdomyosarcoma (6 cm in size) originating in the right medial umbilical ligament. (Photo credit: Sánchez-Montenegro C et al. *Case Rep Genet*. 2017;2017:1587610.)

Elevated urine catecholamine metabolites

Psychomotor development: Developmental delay or intellectual disability; sociable, outgoing personality [22–27]

As papilloma, ulnar deviation of wrists and fingers, and chaotic atrial rhythm/multifocal atrial tachycardia occur in Costello syndrome only, and not in other RASopathies, detection of these anomalies provides a useful criterion for Costello syndrome diagnosis.

For further confirmation of Costello syndrome, molecular identification of a heterozygous germline *HRAS* pathogenic missense variant is necessary. The most common *HRAS* mutations consist of p.Gly12Ser (71%), p.Gly12Ala (9%), p.Gly12Cys (2.5%), p.Gly12Asp (2%), and p.Gly12Val (2%) in addition to p.Gly13Cys (6%) and p.Gly13Asp [15]. Loss of heterozygosity involving chromosome 11p15.5 typically occurs in rhabdomyosarcoma due to paternal uniparental disomy, and specific *HRAS* mutations are found in about 28% of rhabdomyosarcoma samples. As constitutive PI3K-protein-kinase B (AKT) activation is sometimes detected in rhabdomyosarcoma cell lines and clinical samples, it may also play a potential role in the development of Costello syndrome—related soft tissue sarcoma.

Interestingly, patients harboring *HRAS* p.Gly13Asp often show perinatal abnormalities (polyhydramnios and/or fetal hydrops), premature labor, high birth weight and large head circumference, hypoglycemic episodes, feeding difficulties (requiring nasogastric tube and/or gastrostomy), decreased coarseness of the facial features, and less curly hair (than typical Costello syndrome patients) [28].

It is notable that somatic *HRAS* pathogenic variants (involving the glycine at residues 12 and 13 and the glutamine at residue 61) may sometimes occur in malignant solid tumors of adulthood, such as bladder carcinoma or lung carcinoma, leading to increased activity of the gene product.

Differential diagnoses for Costello syndrome include other RASopathies, particularly CFCS (hypotonia, nystagmus, mild-to-moderate intellectual disability, postnatal growth deficiency, sparse or curly hair, sparse or absent eyebrows, severe atopic dermatitis, keratosis pilaris, ichthyosis, and hyperkeratosis; germline *BRAF*, *KRAS*, *MAP2K1*, or *MAP2K2* mutations), and Noonan syndrome (distinct pectus carinatum and pectus excavatum, germline *PTPN11 SOS1*, *RAF1*, *KRAS*, and *NRAS* mutations, in comparison with Costello syndrome's ulnar deviation of the hands, marked small-joint laxity, striking excess palmar skin, presence of papilloma, and palmar calluses, and germline *HRAS* mutations) in infants and young children [29,30].

Costello syndrome may require differentiation from overgrowth syndromes that affect newborns such as Beckwith–Wiedemann syndrome (macrosomia, macroglossia, visceromegaly, omphalocele, neonatal hypoglycemia, ear creases/pits, adrenocortical cytomegaly, neonatal HCM, renal abnormalities, Wilms tumor, hepatoblastoma, neuroblastoma, rhabdomyosarcoma), Simpson–Golabi–Behmel syndrome (macrosomia, visceromegaly, macroglossia, renal anomalies, cleft lip, polydactyly, developmental delay; pathogenic *GPC3* variants), Williams syndrome (cognitive impairment and a specific cognitive profile, unique personality characteristics, distinctive facial features, elastin arteriopathy; contiguous gene deletion of the critical region at 7q11 that encompasses the elastin gene or *ELN*) [31].

89.7 Treatment

Treatment options for Costello syndrome consist of standard procedures that help relieve and rectify the harmful aspects of the disorder. For example, nasogastric or gastrostomy feeding is provided to infants showing growth retardation; Nissen fundoplication is given to those displaying gastroesophageal reflux and irritability; surgery is performed on those with pyloric stenosis; aggressive antiarrhythmic drugs or ablation is prescribed to those with non-reentrant tachycardia (chaotic atrial rhythm/multifocal tachycardia); pharmacologic and surgical treatment (myectomy) help address severe cardiac hypertrophy; MedicAlert® bracelet is offered to those having severe cardiac problems; early bracing and occupational and/or physical therapy help relieve ulnar deviation of the wrists and fingers; physical therapy benefits patients with limited extension of large joints; surgical tendon lengthening alleviates the Achilles tendon; surgical correction is helpful for those with hip joint abnormalities or kyphoscoliosis; early-intervention programs and individualized learning strategies are useful for those with developmental disability, speech delay, and expressive language limitations; hypoglycemic episodes unresponsive to growth hormone therapy may be offered cortisone replacement; surgical or regular dry ice removal may help reduce/eliminate cosmetic concern of papillomas appearing in the perinasal region, perianal region, torso, and extremities [32,33].

RAS pathway agents (e.g., farnesyltransferase inhibitors [tipifarnib and lonafarnib]), B-RAF serine/threonine kinase (BRAF) and mitogen/extracellular signal-regulated kinase (MEK) inhibitors may prevent posttranslational modification of RAS, BRAF, and MEK and help treat cancer associated with Costello syndrome [34].

89.8 Prognosis and Prevention

Prognosis for neonates with Costello syndrome is influenced by the presence or absence of cardiac abnormalities (e.g., arrhythmia, congenital defects, and HCM). Costello syndrome patients with rhabdomyosarcoma have a 5-year survival rate of 65%–70% and their long-term survival is based on the histology of the tumor and the extent of disease at diagnosis.

Antibiotic prophylaxis helps control and prevent subacute bacterial endocarditis (SBE) associated with congenital heart defects (e.g., valvar pulmonic stenosis), while anesthesia may cause unrecognized HCM or atrial tachycardia [35].

Regular monitoring of blood glucose concentration is helpful for neonates at risk for hypoglycemia; yearly assessment benefits individuals with a cardiovascular abnormality; abdominal and pelvic ultrasound screening for rhabdomyosarcoma and neuroblastoma is carried out half-yearly until 8–10 years of age; annual urinalysis for hematuria from 10 years of age helps screen for bladder cancer; bone density assessment helps detect osteoporosis in young adults [36].

Prenatal diagnosis and preimplantation genetic diagnosis may be offered to individuals harboring a *HRAS* pathogenic variant; the presence of severe polyhydramnios in a fetus with normal chromosome analysis and fetal atrial tachycardia may be suggestive of Costello syndrome.

89.9 Conclusion

Forming part of the RASopathies involving germline mutations in genes that encode components of the RAS/mitogen-activated protein kinase (RAS/MAPK) pathway, Costello syndrome is associated with heterozygous gain-of-function mutations in *HRAS*, leading to increased pathway activation [37–40]. Displaying a spectrum of clinical manifestations, ranging from distinct craniofacial features (macrocephaly, coarse facial features with curly and sparse hair, prominent epicanthal folds, long eyelashes, full nasal tip, fleshy earlobes, and a wide mouth with full lips), cardiac defects (e.g., tachyarrhythmia and HCM), mental and growth retardation, neurological findings (e.g., hypotonia), and predisposition to cancer development (e.g., papilloma, rhabdomyosarcoma, transitional cell carcinoma, and neuroblastoma), Costello syndrome proves to be a diagnostic challenge that requires input from molecular and genetic testing for differentiation from other related disorders. In the absence of effective treatment, current management of Costello syndrome is largely based on standard procedures that relieve and rectify the symptoms of the disorder. Therefore, further research is critical to help elucidate the molecular mechanisms of Costello syndrome and design improved therapeutic measures against this autosomal dominant genetic disease.

REFERENCES

1. Gripp KW, Lin AE. Costello syndrome. In: Adam MP, Ardinger HH, Pagon RA et al. editors. *GeneReviews®* [Internet]. Seattle (WA): University of Washington, Seattle; 1993–2018. 2006 Aug 29 [updated 2012 Jan 12].
2. Peixoto IL, Carreno AM, Prazeres VM, Chirano CA, Ihara GM, Akel PB. Syndrome in question. Costello syndrome. *An Bras Dermatol.* 2014;89(6):1005–6.
3. Stevenson DA, Schill L, Schoyer L et al. The Fourth International Symposium on Genetic Disorders of the Ras/MAPK pathway. *Am J Med Genet A.* 2016;170(8):1959–66.
4. Shen Z, Hoffman JD, Hao F, Pier E. More than just skin deep: Faciocutaneous clues to genetic syndromes with malignancies. *Oncologist.* 2012;17(7):930–6.
5. Wang T, de Kok L, Willemsen R, Elgersma Y, Borst JG. In vivo synaptic transmission and morphology in mouse models of Tuberous sclerosis, Fragile X syndrome, *Neurofibromatosis type 1*, and Costello syndrome. *Front Cell Neurosci.* 2015;9:234.
6. Rooney GE, Goodwin AF, Depeille P et al. Human iPS cell-derived neurons uncover the impact of increased Ras signaling in Costello syndrome. *J Neurosci.* 2016;36(1):142–52.
7. Fernández-Medarde A, Santos E. Ras in cancer and developmental diseases. *Genes Cancer.* 2011;2(3):344–58.
8. Rauen KA. The RASopathies. *Annu Rev Genomics Hum Genet.* 2013;14:355–69.
9. Bezniakow N, Gos M, Obersztyn E. The RASopathies as an example of RAS/MAPK pathway disturbances—clinical presentation and molecular pathogenesis of selected syndromes. *Dev Period Med.* 2014;18(3):285–96.
10. Pevec U, Rozman N, Gorsek B, Kunej T. RASopathies: Presentation at the genome, interactome, and phenome levels. *Mol Syndromol.* 2016;7(2):72–9.
11. Wey M, Lee J, Jeong SS, Kim J, Heo J. Kinetic mechanisms of mutation-dependent Harvey Ras activation and their relevance for the development of Costello syndrome. *Biochemistry.* 2013;52(47):8465–79.
12. Krencik R, Hokanson KC, Narayan AR et al. Dysregulation of astrocyte extracellular signaling in Costello syndrome. *Sci Transl Med.* 2015;7(286):286ra66.
13. Cartledge DM, Robbins KM, Drake KM et al. Cytotoxicity of zardaverine in embryonal rhabdomyosarcoma from a Costello syndrome patient. *Front Oncol.* 2017;7:42.
14. Schreiber J, Grimbergen LA, Overwater I et al. Mechanisms underlying cognitive deficits in a mouse model for Costello syndrome are distinct from other RASopathy mouse models. *Sci Rep.* 2017;7(1):1256.
15. Gripp KW, Sol-Church K, Smpokou P et al. An attenuated phenotype of Costello syndrome in three unrelated individuals with a HRAS c.179G>A (p.Gly60Asp) mutation correlates with uncommon functional consequences. *Am J Med Genet A.* 2015;167A(9):2085–97.
16. Hartung AM, Swensen J, Uriz IE et al. The splicing efficiency of activating HRAS mutations can determine Costello syndrome phenotype and frequency in cancer. *PLoS Genet.* 2016;12(5):e1006039.
17. Aytekin S, Alyamac G. Two new cases with Costello syndrome. *Dermatol Online J.* 2013;19(8):19267.
18. Beukers W, Hercegovac A, Zwarthoff EC. HRAS mutations in bladder cancer at an early age and the possible association with the Costello syndrome. *Eur J Hum Genet.* 2014;22(6):837–9.
19. Kratz CP, Franke L, Peters H et al. Cancer spectrum and frequency among children with Noonan, Costello, and cardio-facio-cutaneous syndromes. *Br J Cancer.* 2015;112(8):1392–7.
20. Weaver KN, Wang D, Cnota J et al. Early-lethal Costello syndrome due to rare HRAS tandem base substitution (c.35_36GC>AA; p.G12E)-associated pulmonary vascular disease. *Pediatr Dev Pathol.* 2014;17(6):421–30.
21. Gomez-Ospina N, Kuo C, Ananth AL et al. Respiratory system involvement in Costello syndrome. *Am J Med Genet A.* 2016;170(7):1849–57.
22. Schwartz DD, Katzenstein JM, Hopkins E et al. Verbal memory functioning in adolescents and young adults with Costello syndrome: Evidence for relative preservation in recognition memory. *Am J Med Genet A.* 2013;161A(9):2258–65.
23. Goodwin AF, Oberoi S, Landan M et al. Craniofacial and dental development in Costello syndrome. *Am J Med Genet A.* 2014;164A(6):1425–30.
24. Goodwin AF, Tidyman WE, Jheon AH et al. Abnormal Ras signaling in Costello syndrome (CS) negatively regulates enamel formation. *Hum Mol Genet.* 2014;23(3):682–92.
25. Hakim K, Boussaada R, Hamdi I, Msaad H. Cardiac events in Costello syndrome: One case and a review of the literature. *J Saudi Heart Assoc.* 2014;26(2):105–9.
26. Calandrelli R, D'Apolito G, Marco P, Zampino G, Tartaglione T, Colosimo C. Costello syndrome: Analysis of the posterior cranial fossa in children with posterior fossa crowding. *Neuroradiol J.* 2015;28(3):254–8.
27. Sánchez-Montenegro C, Vilanova-Sánchez A, Barrena-Delfa S et al. Costello syndrome and umbilical ligament rhabdomyosarcoma in two pediatric patients: Case reports and review of the literature. *Case Rep Genet.* 2017;2017:1587610.

28. Bertola D, Buscarilli M, Stabley DL et al. Phenotypic spectrum of Costello syndrome individuals harboring the rare HRAS mutation p.Gly13Asp. *Am J Med Genet A.* 2017;173(5):1309–18.

29. Niemeyer CM. RAS diseases in children. *Haematologica.* 2014;99(11):1653–62.

30. Pierpont ME, Magoulas PL, Adi S et al. Cardio-facio-cutaneous syndrome: Clinical features, diagnosis, and management guidelines. *Pediatrics.* 2014;134(4):e1149–62.

31. Gripp KW, Robbins KM, Sheffield BS et al. Paternal uniparental disomy 11p15.5 in the pancreatic nodule of an infant with Costello syndrome: Shared mechanism for hyperinsulinemic hypoglycemia in neonates with Costello and Beckwith-Wiedemann syndrome and somatic loss of heterozygosity in Costello syndrome driving clonal expansion. *Am J Med Genet A.* 2016;170(3):559–64.

32. Sriboonnark L, Arora H, Falto-Aizpurua L, Choudhary S, Connelly EA. Costello syndrome with severe nodulocystic acne: Unexpected significant improvement of acanthosis nigricans after oral isotretinoin treatment. *Case Rep Pediatr.* 2015;2015:934865.

33. Blachowska E, Petriczko E, Horodnicka-Józwa A et al. Recombinant growth hormone therapy in a girl with Costello syndrome: A 4-year observation. *Ital J Pediatr.* 2016;42:10.

34. Rauen KA, Huson SM, Burkitt-Wright E et al. Recent developments in neurofibromatoses and RASopathies: Management, diagnosis and current and future therapeutic avenues. *Am J Med Genet A.* 2015;167A(1):1–10.

35. Akçıl EF, Dilmen ÖK, Tunalı Y. Anaesthetic management in Costello syndrome. *Turk J Anaesthesiol Reanim.* 2015;43(6):427–30.

36. Leoni C, Flex E. Costello syndrome: The challenge of hypoglycemia and failure to thrive. *EBioMedicine.* 2018;27:5–6.

37. García-Cruz R, Camats M, Calin GA et al. The role of p19 and p21 H-Ras proteins and mutants in miRNA expression in cancer and a Costello syndrome cell model. *BMC Med Genet.* 2015;16:46.

38. Uemura R, Tachibana D, Kurihara Y, Pooh RK, Aoki Y, Koyama M. Prenatal findings of hypertrophic cardiomyopathy in a severe case of Costello syndrome. *Ultrasound Obstet Gynecol.* 2016;48(6):799–800.

39. Cao H, Alrejaye N, Klein OD, Goodwin AF, Oberoi S. A review of craniofacial and dental findings of the RASopathies. *Orthod Craniofac Res.* 2017;20(Suppl 1):32–8.

40. Kang M, Lee YS. The impact of RASopathy-associated mutations on CNS development in mice and humans. *Mol Brain.* 2019;12(1):96.

90

Cowden Syndrome

Dongyou Liu

CONTENTS

90.1 Introduction

Cowden syndrome (or multiple hamartoma syndrome) is a rare disorder characterized by the formation of multiple hamartomas (which are noncancerous, tumor-like growths) typically in the skin, mucous membranes (mouth, nose, GI tract), thyroid gland, and breast tissue; increased risk of developing breast (85%), follicular thyroid (35%), renal cell (34%), uterus (endometrial, 28%), and colorectal (9%) cancers, melanoma (6%), and noncancerous brain tumor (Lhermitte–Duclos disease/syndrome [LDD], or dysplastic gangliocytoma of the cerebellum); and enlarged head (macrocephaly) [1].

Depending on the genes and chromosomes involved, Cowden syndrome is separated into seven variants, with Cowden syndrome-1 (CWS1, which accounts for about 80% of cases) caused by heterozygous germline mutation in the PTEN gene (encoding phosphatase and tensin homolog) on chromosome 10q23, Cowden syndrome-2 (CWS2, 1%) caused by heterozygous mutation in the SDHB gene (encoding succinate dehydrogenase complex iron sulfur subunit B) on chromosome 1p36, Cowden syndrome-3 (CWS3, 1%) caused by heterozygous mutation in the SDHD gene (encoding succinate dehydrogenase complex subunit D) on chromosome 11q23, Cowden syndrome-4 (CWS4, 10%) caused by heterozygous germline hypermethylation in the promoter of the KLLN gene (encoding killin, a p53-regulated DNA replication inhibitor) on chromosome 10q23, Cowden syndrome-5 (CWS5, 3%) caused by heterozygous mutation in the PIK3CA gene (encoding phosphatidylinositol-4,5-bisphosphate 3-kinase catalytic subunit alpha) on chromosome 3q26, Cowden syndrome 6 (CWS6, 3%) caused by heterozygous mutation in the AKT1 gene (encoding RAC-alpha serine/threonine-protein kinase) on chromosome 14q32, and Cowden syndrome-7 (CWS7, 2%) caused by heterozygous mutation in the SEC23B gene (encoding Sec23 homolog B, coat complex II component) on chromosome 20p11 (Table 90.1).

Besides CWS1 and LDD, mutations in the PTEN gene located at chromosome 10q23.31 are also observed in Bannayan–Riley–Ruvalcaba syndrome (BRRS), PTEN-related Proteus syndrome, and Proteus-like syndrome. As this spectrum of PTEN-related disorders all produce hamartomatous tumors, they are often referred to collectively as PTEN hamartoma tumor syndrome (PHTS), particularly when a *PTEN* pathogenic variant is present (Table 90.2). Considerable overlapping exists between CWS1 and BRRS in terms of clinical presentations, with both developing hamartomatous polyps of the gastrointestinal tract, other noncancerous tumors, mucocutaneous lesions, and macrocephaly (see Chapter 10 in this book). There is speculation that CS1 and BRRS represent a single condition with variable expression and age-related penetrance. Proteus syndrome shows congenital malformations and hamartomatous overgrowth of multiple tissues, connective tissue nevi, epidermal nevi, and hyperostoses. Proteus-like syndrome demonstrates significant clinical features of proteus syndrome without meeting the diagnostic criteria for the latter [2–5].

TABLE 90.1

Characteristics of Cowden Syndrome Variants

Variant	Alternative Names	Chromosomal Location	Genetic Mutation (% of Clinical Cases)	Inheritance
Cowden syndrome 1	CWS1, Lhermitte-Duclos syndrome	10q23.31	Heterozygous germline mutation in PTEN (80%)	Autosomal dominant
Cowden syndrome 2	CWS2	1p36.13	Heterozygous mutation in SDHB (1%)	Autosomal dominant
Cowden syndrome 3	CWS3	11q23.1	Heterozygous mutation in SDHD (1%)	
Cowden syndrome 4	CWS4	10q23.31	Heterozygous germline hypermethylation in the promoter of KLLN (10%)	
Cowden syndrome 5	CWS5	3q26.32	Heterozygous mutation in PIK3CA (3%)	
Cowden syndrome 6	CWS6	14q32.33	Heterozygous mutation in AKT1 (3%)	
Cowden syndrome 7	CWS7	20p11.23	Heterozygous mutation in SEC23B (2%)	Autosomal dominant

TABLE 90.2

Characteristics of PTEN Hamartoma Tumor Syndrome (PHTS)

Syndrome	Clinical Features	Diagnostic Criteria
Cowden syndrome 1	A multiple hamartoma syndrome with a high risk for benign and malignant tumors of the thyroid, breast, endometrium, kidney and colorectum; macrocephaly, trichilemmomas, and papillomatous papules by late 20s	*Pathognomonic criteria*: Adult Lhermitte–Duclos disease (LDD, i.e., cerebellar dysplastic gangliocytoma); mucocutaneous lesions (facial trichilemmomas, acral keratoses, papillomatous lesions, mucosal lesions) *Major criteria*: Breast cancer; non-medullary epithelial thyroid cancer (especially follicular thyroid cancer); macrocephaly (occipital frontal circumference $\geq$97th percentile); endometrial carcinoma *Minor criteria*: Other thyroid lesions (e.g., adenoma, multinodular goiter); intellectual disability (IQ $\leq$75); hamartomatous intestinal polyps; fibrocystic disease of the breast; lipoma; fibroma; genitourinary tumors (especially renal cell carcinoma); genitourinary malformation; uterine fibroids
Bannayan–Riley–Ruvalcaba syndrome	A congenital disorder presenting with macrocephaly, intestinal hamartomatous polyposis (45%), lipoma, and pigmented macules of the glans penis; high birth weight, developmental delay, intellectual disability (50%); myopathic process in proximal muscles (60%); joint hyperextensibility, pectus excavatum, scoliosis (50%); individuals with *PTEN* pathogenic variant also have high risk of developing cancers	Presence of cardinal features: Macrocephaly, hamartomatous intestinal polyposis, lipoma, pigmented macules of the glans penis
PTEN-related Proteus syndrome	A complex disorder showing minimal or no manifestations at birth, but developing and progressing rapidly from the toddler period to childhood, causing segmental or patchy overgrowth of multiple tissues (e.g., the skeleton, skin, adipose and central nervous systems), connective tissue nevi, epidermal nevi, and hyperostoses, along with pulmonary complications, and predisposition to deep vein thrombosis and pulmonary embolism	*General criteria*: Mosaic distribution of lesions, sporadic occurrence, progressive course *Other criteria*: Connective tissue nevi (pathognomonic when present), epidermal nevi, disproportionate overgrowth, specific tumors
Proteus-like syndrome	Showing significant clinical features of *PTEN*-related Proteus syndrome, without meeting the diagnostic criteria	

90.2 Biology

Described in 1963 by Lloyd and Dennis, and named after the family in which the first case was identified, Cowden syndrome (also known as Cowden disease, Cowden's disease, Cowden's syndrome, multiple hamartoma syndrome) is largely an autosomal dominant disorder, in which one copy of the altered gene in each cell is sufficient to cause the condition and increase the risk of cancer development. An affected person may either inherit the mutation from one affected parent or evolve from sporadic de novo mutations in the gene without obvious history of the disorder in the family. Cowden syndrome cases without obvious family history (i.e., those caused by de novo mutations) are referred to as simplex cases, which affect about 50% of patients.

Among Cowden syndrome variants, CWS1 is responsible for 80% of clinical cases, with a clear link to mutations in the PTEN gene (phosphatase and tensin homolog, also known as mutated in multiple advanced cancers 1 [MMAC1], phosphatase and tensin homolog deleted on chromosome 10, protein-tyrosine phosphatase PTEN, PTEN-MMAC1 protein, PTEN1, PTEN_HUMAN, TEP1, TEP1 phosphatase), while other variants (CWS2–7) account for 20% of cases and harbor mutations in SDHB (succinate dehydrogenase complex subunit B, also known as

DHSB_HUMAN, FLJ92337, iron sulfur [IP], iron-sulfur subunit of complex II, PGL4, SDH, SDH1, SDH2, SDHIP, mitochondrial succinate dehydrogenase iron-sulfur subunit), SDHD (succinate dehydrogenase complex subunit D, also known as CBT1, CII-4, cybS, DHSD_HUMAN, PGL, PGL1, QPs3, SDH4, mitochondrial succinate dehydrogenase cytochrome b small subunit, succinate dehydrogenase ubiquinone cytochrome B small subunit, succinate-ubiquinone oxidoreductase cytochrome b small subunit, succinate-ubiquinone reductase membrane anchor subunit), KLLN (also known as KILIN_HUMAN, killin, p53-regulated DNA replication inhibitor), PIK3CA (phosphatidylinositol-4,5-bisphosphate 3-kinase 110 kDa catalytic subunit alpha, also known as p110-alpha, PI3-kinase p110 subunit alpha, PI3K, PI3K-alpha, PK3CA_HUMAN, ptdIns-3-kinase subunit p110-alpha, serine/threonine protein kinase PIK3CA), AKT1 (also known as AKT, AKT1_HUMAN, MGC99656, PKB, PKB alpha, PRKBA, protein kinase B alpha, proto-oncogene c-Akt, RAC, RAC-ALPHA, RAC-alpha serine/threonine-protein kinase, RAC-PK-alpha, rac protein kinase alpha, v-akt murine thymoma viral oncogene homolog 1), and SEC23B (also known as CDA-II, CDAII, HEMPAS, SC23B_HUMAN, Sec23 homolog B, Sec23 homolog B, Sec23 homolog B COPII coat complex component, SEC23-like protein B, SEC23-related protein B, transport protein SEC23B) genes (Table 90.1).

90.3 Pathogenesis

90.3.1 PTEN

Consisting of nine exons and spanning a genomic distance of >120 kb in the long (q) arm of chromosome 10 at position 23.31 (i.e., 10q23.31), the *PTEN* gene contains nine exons along with a variable exon 5b that is skipped in the major PTEN transcript, and encodes a phosphatidylinositol 3,4,5-trisphosphate 3-phosphatase (also known as phosphatase and tensin homolog deleted on chromosome ten [PTEN]), which contains a tensin-like domain (or C2 domain) and a catalytic domain similar to that of the dual specificity protein tyrosine phosphatase. PTEN negatively regulates the mitogen-activated protein kinase (MAPK) pathway through its protein phosphatase activity to induce cell cycle arrest (when in the nucleus) and antagonizes the phosphatidylinositol 3-kinase (PI3K) pathway through its lipid phosphatase activity to elicit apoptosis (when in the cytoplasm). Specifically, by removing phosphate groups (each consisting of three oxygen atoms and one phosphorus atom) from tyrosine, serine and threonine in other proteins and fats (lipids) though its protein phosphatase activity, PTEN inhibits cell migration and adhesion (through its C2 domain), regulates the formation of new blood vessels (angiogenesis), maintains the stability of cell genetic information, and initiates cellular self-destruction (apoptosis), thereby preventing uncontrolled cell growth. However, mutations in the PTEN gene result in truncated protein, lack of protein (haploinsufficiency), or dysfunctional protein that renders uninhibited phosphorylation of AKT1 (leading to the inability to activate cell cycle arrest and/or to undergo apoptosis) and dysregulated MAPK pathway (leading to abnormal cell survival) [6]. To date, >300 mutations (including missense, nonsense, and splice-site variants, small deletions, insertions, and large deletions) in the *PTEN* gene (with nearly 40% of pathogenic variants found in exon

5; the most notable pathogenic variants being 388C>T/Arg130Ter, 697C>T/Arg233Ter, 1003C>T/Arg335Ter) have been identified in patients with CS1, representing about 80% of all Cowden syndrome cases [7,8]. In addition, over 30 mutations in the *PTEN* gene are linked to BRRS (showing macrocephaly, multiple hamartomas, and dark freckles on the penis in males).

90.3.2 SDHB and SDHD

The *SDHB* (situated in the short [p] arm of chromosome 1 at position 36.13 [i.e., 1p36.13]) and *SDHD* (situated in the long [q] arm of chromosome 11 at position 23.1 [i.e., 11q23.1]) genes encode subunits of succinate dehydrogenase (SDH), which converts succinate to fumarate in mitochondria during the citric acid cycle (or Krebs cycle) and releases electrons that are subsequently transferred to the oxidative phosphorylation pathway for the production of adenosine triphosphate (ATP). In addition, as a tumor suppressor, SDH plays a role in the regulation of cell survival and proliferation. Changes in the *SDHB* or *SDHD* gene lead to a diminished or defective SDH enzyme that contributes to a buildup in succinate and hypoxia-inducible factor (HIF) protein, and abnormal hypoxia signaling, triggering the production of blood vessels and unchecked cell division, and ultimate formation of hamartomas and cancers. While at least 10 and 5 mutations in the *SDHB* and *SDHD* genes are found in people suffering from Cowden syndrome variants 2 and 3 (representing 2% of all Cowden syndrome cases), over 150 and 100 mutations in the *SDHB* and *SDHD* genes are detected in people with hereditary paraganglioma-pheochromocytoma types 4 and 1, respectively, which are noncancerous tumors associated with the nervous system. However, for hereditary paraganglioma-pheochromocytoma type 1, an inherited *SDHD* gene mutation followed by a second mutation during a person's lifetime that deletes the normal copy of the *SDHD* gene is essential for the occurrence of condition [9,10].

90.3.3 KLLN

Located in the long (q) arm of chromosome 10 at position 23.31 (i.e., 10q23.31), the *KLLN* gene encodes killin (p53-regulated DNA replication inhibitor), which is thought to trigger apoptosis of damaged or abnormal cells and prevent their unchecked growth and division. Promoter hypermethylation in the *KLLN* gene reduces killin expression and allows abnormal cells to survive and proliferate inappropriately. Despite the fact that the *KLLN* and *PTEN* genes share the same promoter region (transcription site), promoter hypermethylation appears to only downregulate the expression of the *KLLN* gene, but not that of the *PTEN* gene. Patients with CS4 linked to promoter hypermethylation in the *KLLN* gene account for 10% of all Cowden syndrome cases and show a particularly high risk of developing breast and kidney cancers [11].

90.3.4 PIK3CA

Located in the long (q) arm of chromosome 3 at position 26.32 (i.e., 3q26.32), the *PIK3CA* gene encodes phosphatidylinositol 3-kinase catalytic subunit alpha (p110α), which phosphorylates other proteins through addition of a cluster of oxygen and phosphorus atoms (a phosphate group), and thereby transmits chemical signals for cell growth, division, migration and survival, synthesis of new proteins, and transport of molecules within cells. The p110α protein also plays a role in the maturation of fat cells (adipocytes). Mutations in the *PIK3CA* gene occur in 3% of all Cowden syndrome cases and are also linked to some cancers and overgrowth disorders (e.g., hemimegalencephaly; fibroadipose hyperplasia, congenital lipomatous overgrowth, vascular malformations, epidermal nevi, and skeletal or spinal abnormalities syndrome, which are collectively referred to as the *PIK3CA*-related overgrowth spectrum [PROS]) [12].

90.3.5 AKT1

Located in the long (q) arm of chromosome 14 at position 32.33 (i.e., 14q32.33), the *AKT1* gene consists of 14 exons extending over 26 kb, generating an mRNA of approximately 3 kb (equivalent cDNA of 3,008 bp). Its open reading frame of 1443 bp encodes a 480 amino acid AKT1 kinase, which plays a role in the regulation of cell growth, division, and apoptosis. Mutations in the *AKT1* gene facilitate abnormal cell growth and division and transformation of normal cells into to cancerous cells and are found in 3% of all Cowden syndrome cases. A sporadic mutation in the *AKT1* gene during the early stages of development before birth that replaces glutamic acid with lysine at position 17 (Glu17Lys or E17K) leads to the production of an overactive AKT1 kinase, which is responsible for overgrowth of the bones, skin, and other tissues associated with Proteus syndrome [13].

90.3.6 SEC23B

Located on in the short (p) arm of chromosome 20 at position 11.23 (i.e., 20p11.23), the *SEC23B* gene encodes a component of coat protein complex II (COPII), which is involved in the formation of vesicles (so-called endoplasmic reticulum [ER]) for transportation of proteins and other materials within cells. Mutations in the *SEC23B* gene are identified in 2% of all Cowden syndrome cases and may be also responsible for producing unusually shaped erythroblasts with extra nuclei that cannot mature into functional red blood cells, leading to congenital dyserythropoietic anemia type II, enlarged liver and spleen (hepatosplenomegaly), and an abnormal buildup of iron [14].

90.4 Epidemiology

Cowden syndrome is estimated to have a prevalence of 1 case per 200,000 people, with >300 Cowden syndrome cases reported in the literature to date. This is likely an underestimate, due particularly to the variable and subtle external manifestations of Cowden syndrome and BRRS.

Being an autosomal dominant inherited disorder, the onset of Cowden syndrome varies from birth to age 46 years, with >90% individuals manifesting clinically by the late 20s, and 99% of patients developing hamartomas on the skin and mucous membranes (i.e., trichilemmomas, papillomatous papules, acral and plantar keratoses) by the third decade without obvious sexual differences.

Patients with Cowden syndrome demonstrate cumulative lifetime (age 70 years) risks for any cancer (89%), breast cancer (85%, predominantly females of 38–46 years in age), thyroid cancer (35%, predominantly males), LDD (32%), endometrial cancer (28%), colorectal cancer (16%), and renal cancer (15%). Male patients appear to have fewer cancers diagnosed than female patients.

90.5 Clinical Features

Clinical presentations of Cowden syndrome typically include (i) multiple hamartomas, (ii) macrocephaly (enlarged head) and dolichocephaly, and (iii) other tumors.

Multiple hamartomas are small, noncancerous growths that are most commonly found on the skin and mucous membranes (e.g., the lining of the mouth, nose, and intestines) as well as other organs (the thyroid, breast, and brain). Four hallmark cutaneous features of Cowden syndrome are papules in the face, acral keratoses, mouth lesions, and palmoplantar keratoses (Figure 90.1). Specifically, the papules in the face are skin eruptions through infiltration of the dermis and epidermis area; predominantly located in areas surrounding an orifice, these lesions (so-called facial trichilemmomas) appear flat at the top and flesh-like in color, and measure 1–5 mm in size with the center containing a keratin material. Acral keratoses are often found on the backs of the feet and hands of >50% of patients. Mouth lesions are represented by noncancerous papules (smooth, white in color, 1–3 mm in diameter) that band together to form a cobblestone-like exterior on the surfaces of the palate and the gingiva, or make the tongue wrinkled and thickened. Palmoplantar keratoses (acanthosis, dense orthokeratosis, prominent granular layer) are transparent and spotted punctures in the soles of the feet and in the hands of <50% of patients. A possibly pathognomonic, noncancerous brain tumor (i.e., LDD or cerebellar dysplastic gangliocytoma) may affect a small number of patients [15–17].

Macrocephaly (or an atypical large head) is observed in >50% of patients. Some patients may display visual problems (e.g., myopia), atypical facial features (e.g., abnormally small jaw, uncharacteristically high-domed palate), mental retardation, genitourinary system malformation, and esophageal glycogenic acanthosis.

Patients with Cowden syndrome are at increased risk of developing cancers of the breast, thyroid, uterus (endometrial), kidney, and colorectum. Benign breast conditions (e.g., ductal hyperplasia, intraductal papillomatosis, adenosis, lobular atrophy, fibroadenomas, and fibrocystic changes) and thyroid abnormalities (e.g., noncancerous adenomas, goiter, and cysts) may occur in up to 75% and over 50% of patients, respectively. Females with a *PTEN* pathogenic variant have an 85% lifetime risk for breast cancer, and penetrance of 50% by 50 years of age. Patients also show a 35% lifetime risk for epithelial thyroid cancer, with a median age of onset at 37 years; however, no medullary thyroid carcinoma is observed. Further, patients have a 28% lifetime risk for endometrial cancer, with an onset age between the late 30s and early 40s. Over 90% of individuals with a *PTEN* pathogenic variant develop polyps (ranging from ganglioneuromatous polyps, hamartomatous polyps, and juvenile polyps to adenomatous polyps) and have a 9% lifetime risk for colorectal cancer, with an onset age in the late 30s. Patients also have a 35% lifetime risk for renal cell carcinoma (predominantly of papillary histology), with an onset age of the 40s [18–22].

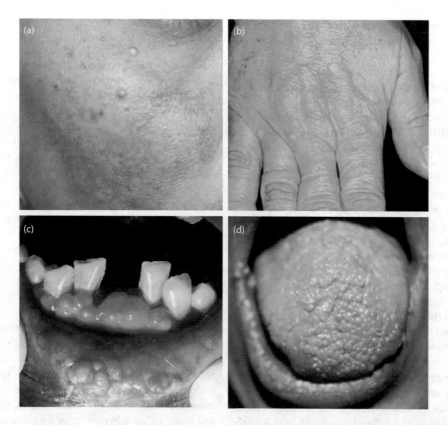

FIGURE 90.1 Clinical presentations of Cowden syndrome. (a) Trichilemmomas (small multiple skin-colored, flat-topped papules) on the right malar bone; (b) acral hyperkeratoses (multiple hyperkeratotic verruciform lesions) on the back of the hand; (c) multiple papillomatous lesions on the gingiva and lower labial mucosa; (d) multiple confluent papules on the tongue producing a cobblestone appearance. (Photo credit: a,c,d) Chippagiri P et al. *Case Rep Dent.* 2013;2013:315109; (b) Porto AC et al. *An Bras Dermatol.* 2013;88(6 Suppl 1):52–5.)

90.6 Diagnosis

Diagnosis of Cowden syndrome and other PHTS involves: (i) complete medical history and family history; (ii) physical examination for lesions in the skin, mucous membranes, thyroid, and breasts; (iii) neurodevelopmental evaluation for children; (iv) urinalysis with cytospin; (v) baseline thyroid ultrasound examination; (vi) breast screening (mammogram, MRI) and transvaginal ultrasound or endometrial biopsy for women age ≥30 years; (v) colonoscopy for men and women age ≥35 years; (vi) renal imaging (CT or MRI) for men and women age ≥40 years; (vii) molecular genetic testing [2,23–25].

According to the International Cowden Consortium (ICC), an operational diagnosis is obtained if a patient meets the pathognomonic criteria, i.e., mucocutaneous lesions combined with one of the following: (i) six or more facial papules (of which three or more must be trichilemmoma), (ii) cutaneous facial papules and oral mucosal papillomatosis, (iii) oral mucosal papillomatosis and acral keratoses, (iv) six or more palmoplantar keratoses; two or more major criteria; one major and three or more minor criteria; or four or more minor criteria (Table 90.2) [26].

Relatives of an individual meeting the diagnostic criteria listed above are considered to have Cowden syndrome if they meet any one of the following criteria: (i) the pathognomonic criteria, (ii) any one major criterion, (iii) two minor criteria, (iv) history of BRRS.

Further confirmation of Cowden syndrome is achieved through molecular genetic testing for *PTEN* pathogenic variant (in the case of CWS1) and other genetic abnormalities (*SDHB, SDHD, KLLN, PIK3CA, AKT1,* and *SEC23B,* in the case of CWS2–7) based on sequence analysis or mutation scanning of the entire coding region, deletion/duplication analysis, detection of homozygosity, and targeted variant analysis (e.g., *KLLN* promoter methylation analysis).

Pediatric patients may be selected for PTEN mutation testing using a clinical scoring system, which includes macrocephaly (100% of cases) along with one of the following: autism or developmental delay (seen in 82% of cases), dermatologic features (including lipomas, trichilemmomas, oral papillomas, and penile freckling; 60%), vascular features (e.g., arteriovenous malformations or hemangiomas, 29%), or gastrointestinal polyps (14%). In addition, pediatric-onset thyroid cancer and germ cell tumors (testicular cancer and dysgerminoma) should be also considered for PTEN testing.

In general, females displaying macrocephaly, endometrial cancer, trichilemmomas, papillomatous papules, breast cancer, benign thyroid disease, and benign gastrointestinal lesions tend to have a *PTEN* mutation. On the other hand, males showing macrocephaly, lipomas, papillomatous papules, penile freckling, benign gastrointestinal lesions, and benign thyroid disease may harbor a *PTEN* mutation. While germline *PTEN* pathogenic variants are identifiable in most adult cases of LDD, they are rare in pediatric cases of LDD. Further, patients with CWS4 (containing a germline *KLLN* epimutation) have a greater prevalence of breast and renal cell carcinomas than those with CWS1 (containing a germline *PTEN* pathogenic variant). However, in the absence of characteristic signs and symptoms of Cowden syndrome (e.g., trichilemmomas, papillomatous papules, macrocephaly), patients with thyroid or breast cancer are not recommended for molecular testing of *PTEN* and other related genes.

Differential diagnoses for Cowden syndrome include other related PHTS, i.e., BRRS, *PTEN*-related Proteus syndrome, and Proteus-like syndrome, in addition to disorders that show overlapping clinical symptoms with PHTS such as juvenile polyposis syndrome (JPS), and Peutz–Jeghers syndrome (PJS). Less important differential diagnoses for Cowden syndrome comprise Birt–Hogg–Dubé syndrome (BHDS, see Chapter 22 in this book), Gorlin syndrome (see Chapter 30 in this book), and neurofibromatosis type 1 (NF1, see Chapter 6 in this book). Additionally, nongenital warts, trichoepithelioma, syringoma, trichofolliculoma, and trichilemmoma should be also distinguished.

BRRS is a pediatric disorder characterized by macrocephaly, developmental retardation, and penile pigmented macules. Of patients meeting criteria for BRRS, 60% (including both simplex and familial cases) are found to have *PTEN* pathogenic variants. About 10% of patients with BRRS have large deletions within or encompassing *PTEN*, while a pathogenic variant is absent in the *PTEN* coding sequence (see Chapter 10 in this book).

Proteus syndrome and Proteus-like syndrome typically show congenital malformations, hemihytrophy, hamartomas, epidermal nevi, and hyperostosis. Virtually all individuals with Proteus syndrome and Proteus-like syndrome are simplex cases. About 20% of individuals with Proteus syndrome and 50% of individuals with Proteus-like syndrome contain *PTEN* pathogenic variants. Proteus syndrome may show somatic mosaicism for the specific de novo *AKT1* pathogenic variant c.49G>A (p.Glu17Lys). As *PTEN* downregulates *AKT1* by decreasing phosphorylation, the presence of an *AKT1* pathogenic variant in Proteus syndrome confirms its "PTEN-pathway-opathy" [27–32].

Juvenile polyposis syndrome (JPS, with the term "juvenile" referring to the type of polyp rather than the age of onset of polyps) is an autosomal dominant disorder characterized by hamartomas in the gastrointestinal tract (that show a normal epithelium with a dense stroma, an inflammatory infiltrate, and a smooth surface with dilated, mucus-filled cystic glands in the lamina propria). Although most juvenile polyps are benign, some may undergo malignant transformation. JPS is linked to mutations in the *BMPR1A* (encoding the type 1A receptor of bone morphogenetic proteins or BMP) and *SMAD4* genes. Associated with germline deletion of *BMPR1A*, juvenile polyposis of infancy (JPI) may manifest juvenile polyposis (before age 6 years) and severe gastrointestinal manifestations (e.g., bleeding, diarrhea, and protein-losing enteropathy) as well as external stigmata mimicking BRRS. While JPS and Cowden syndrome both develop gastrointestinal polyps, JPS differs from Cowden syndrome by having no associated mucocutaneous lesions, breast hamartomas/carcinomas, thyroid carcinomas, and *PTEN* mutation (see Chapter 17 in this book) [33].

PJS is an autosomal dominant disorder characterized by GI polyposis (often accompanied by intussusception, rectal bleeding), mucocutaneous pigmentation (particularly pigmentation of the perioral region, and hyperpigmented macules on the fingers), and cancer predisposition. Mutations in the *STK11* gene are found in up to 70% of PJS cases (see Chapter 19 in this book).

90.7 Treatment

For patients with asymptomatic Cowden syndrome, observation alone is prudent; for those with mucocutaneous manifestations, pharmacologic approach (e.g., systemic use of acitretin—an analog of retinoic acid, topical application of 5-fluorouracil), and surgical intervention (e.g., curettage, cryosurgery, laser ablation) may provide temporary relief. Surgical excision of cutaneous lesions is recommended if malignancy is suspected or symptoms cause significant pain, deformity, and increased scarring [34,35]. Prophylactic mastectomy may help reduce the risk of breast cancer by 90% in women at high risk [36,37]. Targeted therapies (e.g., mTOR-inhibiting rapamycin to treat mucocutaneous papillomatous lesions and acral keratosis) are being evaluated for Cowden syndrome and Proteus syndrome [38,39].

90.8 Prognosis and Prevention

Although benign tumors (hamartomas) that develop in individuals with Cowden syndrome are generally non-life threatening, they are nevertheless linked to major debilitating morbidity. Moreover, Cowden syndrome increases the risk for malignant tumors (e.g., breast cancer, thyroid cancer), with 40% of affected individuals having a minimum of one malignant primary tumor, and many patients having more than one malignancy. As many of these cancers are curable if detected early and treated appropriately, prompt and precise (especially gene-based) diagnosis and close follow-up care play a critical role in helping improve disease outcome and prognosis of Cowden syndrome. On the other hand, a diagnosis made after occurrence of advanced cancers entails a poor outcome and unfavorable prognosis.

Given its autosomal dominant inheritance, genetic testing should be offered to asymptomatic relatives (e.g., parents, sibs, offspring) of Cowden syndrome patients. This helps monitor those with a mutation before symptom onset (e.g., yearly thyroid ultrasound, colonoscopy and biennial renal imaging between the ages of 35–40, yearly breast screening, and transvaginal ultrasound or endometrial biopsy from the age of 30). Other useful laboratory procedures include thyroid function test (due to the high predisposition of the patients to thyroid cancer), complete blood count (for white blood cells and anemia indicative of cancerous situations), urine analysis (for hematuria and proteinuria indicative of renal neoplasm), skin biopsy (for sclerotic fibromas and trichilemmomas), and liver function test (for liver malignancies). Prenatal diagnosis is valuable when pregnancies involve individuals with relevant gene mutations (e.g., *PTEN*), as a child of an affected individual has a 50% chance of inheriting the pathogenic variant and developing the disease.

90.9 Conclusion

Cowden syndrome (multiple hamartoma syndrome) is characterized by the presence of noncancerous tumors in various tissues (e.g., the thyroid [thyroglossal duct cyst, adenoma], digestive tract [colon diverticulitis, hepatic cysts, glycogenic acanthosis], genital tract [functional menstrual cycle troubles, ovarian teratomas], skeleton [bone cysts], breast [fibrocystic disease; nipple and areola abnormalities]), increased risk for malignancies (notably the breast, thyroid, endometrium, kidney, and colorectum), and clear association with mutations in *PTEN* (phosphatase and tensin homolog) or other genes [40,41].

Due to its autosomal dominant inheritance, Cowden syndrome can appear from birth to 46 years of age, with macrocephaly and dysmorphic facies (if present) evident at birth, and other manifestations (including small, colored and multiple facial papules or trichilemmomas, oral mucosal papillomatous papules, acral keratoses, palmoplantar keratoses, LDD, developmental delay, autistic spectrum) by the third decade [40].

Among seven variants identified to date, CS1 accounts for 80% of clinical cases and shares with BRRS, *PTEN*-related Proteus syndrome, and Proteus-like syndrome in having *PTEN* mutations, which together are referred to as PTEN hamartoma tumor syndrome. The *PTEN* gene encodes a phosphatase that dephosphorylates the 3 position of phosphoinositide and thereby negatively controls the phosphoinositide 3-kinase–signaling pathway for regulating cell growth and survival. Mutations in the *PTEN* gene compromise its protein's function, leading to overproliferation of cells, hamartomatous growths, and predisposition to other cancers.

REFERENCES

1. Babu NA, Rajesh E, Krupaa J, Gnananandar G. Genodermatoses. *J Pharm Bioallied Sci.* 2015;7(Suppl 1): S203–6.
2. Eng C. PTEN hamartoma tumor syndrome. In: Adam MP, Ardinger HH, Pagon RA et al. (eds). *GeneReviews® [Internet].* Seattle (WA): University of Washington, Seattle; 1993–2017. 2001 Nov 29 [updated 2016 Jun 2].
3. Monga E, Gupta PK, Munshi A, Agarwal S. Multiple hamartoma syndrome: Clinicoradiological evaluation and histopathological correlation with brief review of literature. *Indian J Dermatol.* 2014;59(6):598–601.
4. Nosé V. Genodermatosis affecting the skin and mucosa of the head and neck: Clinicopathologic, genetic, and molecular aspect—PTEN-hamartoma tumor syndrome/Cowden syndrome. *Head Neck Pathol.* 2016;10(2):131–8.
5. Schwerd T, Khaled AV, Schürmann M et al. A recessive form of extreme macrocephaly and mild intellectual disability complements the spectrum of PTEN hamartoma tumour syndrome. *Eur J Hum Genet.*. 2016;24(6):889–94.
6. Mathew G, Hannan A, Hertzler-Schaefer K et al. Targeting of Ras-mediated FGF signaling suppresses Pten-deficient skin tumor. *Proc Natl Acad Sci U S A.* 2016;113(46):13156–61.
7. Seol JE, Park IH, Lee W, Kim H, Seo JK, Oh SH. Cowden syndrome with a novel germline PTEN mutation and an unusual clinical course. *Ann Dermatol.* 2015;27(3):306–9.
8. Mauro A, Omoyinmi E, Sebire NJ, Barnicoat A, Brogan P. De novo PTEN mutation in a young boy with cutaneous vasculitis. *Case Rep Pediatr.* 2017;2017:9682803.
9. Yu W, He X, Ni Y, Ngeow J, Eng C. Cowden syndrome-associated germline SDHD variants alter PTEN nuclear translocation through SRC-induced PTEN oxidation. *Hum Mol Genet.* 2015;24(1):142–53.
10. Yu W, Ni Y, Saji M, Ringel MD, Jaini R, Eng C. Cowden syndrome-associated germline succinate dehydrogenase complex subunit D (SDHD) variants cause PTEN-mediated down-regulation of autophagy in thyroid cancer cells. *Hum Mol Genet.* 2017;26(7):1365–75.
11. Nizialek EA, Mester JL, Dhiman VK, Smiraglia DJ, Eng C. KLLN epigenotype-phenotype associations in Cowden syndrome. *Eur J Hum Genet.* 2015;23(11):1538–43.
12. Mirzaa G, Conway R, Graham JM Jr et al. PIK3CA-related segmental overgrowth. In: Adam MP, Ardinger HH, Pagon RA et al. (eds). *GeneReviews® [Internet].* Seattle (WA): University of Washington, Seattle; 1993–2017. 2013 Aug 15.
13. Polubothu S, Al-Olabi L, Wilson L, Chong WK, Kinsler VA. Extending the spectrum of AKT1 mosaicism: Not just the Proteus syndrome. *Br J Dermatol.* 2016;175(3):612–4.
14. Yehia L, Niazi F, Ni Y et al. Germline heterozygous variants in SEC23B are associated with Cowden syndrome and enriched in apparently sporadic thyroid cancer. *Am J Hum Genet.* 2015;97(5):661–76.
15. Chippagiri P, Banavar Ravi S, Patwa N. Multiple hamartoma syndrome with characteristic oral and cutaneous manifestations. *Case Rep Dent.* 2013;2013:315109.
16. Porto AC, Roider E, Ruzicka T. Cowden Syndrome: Report of a case and brief review of literature. *An Bras Dermatol.* 2013;88(6 Suppl 1):52–5.
17. Rusiecki D, Lach B. Lhermitte-Duclos disease with neurofibrillary tangles in heterotopic cerebral grey matter. *Folia Neuropathol.* 2016;54(2):190–6.
18. Browning MJ, Chandra A, Carbonaro V, Okkenhaug K, Barwell J. Cowden's syndrome with immunodeficiency. *J Med Genet.* 2015;52(12):856–9.
19. Neychev V, Sadowski SM, Zhu J et al. Neuroendocrine tumor of the pancreas as a manifestation of Cowden syndrome: A case report. *J Clin Endocrinol Metab.* 2016;101(2):353–8.
20. Tsunezuka H, Abe K, Shimada J, Inoue M. Pulmonary atypical carcinoid in a patient with Cowden syndrome. *Interact Cardiovasc Thorac Surg.* 2016;22(6):860–2
21. Yakubov E, Ghoochani A, Buslei R, Buchfelder M, Eyüpoglu IY, Savaskan N. Hidden association of Cowden syndrome, PTEN mutation and meningioma frequency. *Oncoscience.* 2016;3(5–6):149–55.
22. Chen HH, Händel N, Ngeow J et al. Immune dysregulation in patients with PTEN hamartoma tumor syndrome: Analysis of FOXP3 regulatory T cells. *J Allergy Clin Immunol.* 2017;139(2):607–20.
23. Huang S, Zhang G, Zhang J. Similar MR imaging characteristics but different pathological changes: A misdiagnosis for Lhermitte-Duclos disease and review of the literature. *Int J Clin Exp Pathol.* 2015;8(6):7583–7.
24. Kang YH, Lee HK, Park G. Cowden syndrome detected by FDG PET/CT in an endometrial cancer patient. *Nucl Med Mol Imaging.* 2016;50(3):255–7.
25. Leslie NR, Longy M. Inherited PTEN mutations and the prediction of phenotype. *Semin Cell Dev Biol.* 2016;52:30–8.
26. Molvi M, Sharma YK, Dash K. Cowden syndrome: Case report, update and proposed diagnostic and surveillance routines. *Indian J Dermatol.* 2015;60(3):255–9.
27. Biesecker LG, Sapp JC. Proteus syndrome. In: Adam MP, Ardinger HH, Pagon RA et al. (eds). *GeneReviews® [Internet].* Seattle (WA): University of Washington, Seattle; 1993–2017. 2012 Aug 9 [Updated 2018 Jan 4].
28. Balaji SM. Fronto-temporal cerebriform connective tissue nevus in Proteus syndrome. *Indian J Dent Res.* 2014;25(6):828–31.
29. Popescu MD, Burnei G, Draghici L, Draghici I. Proteus syndrome: a difficult diagnosis and management plan. *J Med Life.* 2014;7(4):563–6.

30. Lindhurst MJ, Yourick MR, Yu Y, Savage RE, Ferrari D, Biesecker LG. Repression of AKT signaling by ARQ 092 in cells and tissues from patients with Proteus syndrome. *Sci Rep.* 2015;5:17162.

31. Doucet ME, Bloomhardt HM, Moroz K, Lindhurst MJ, Biesecker LG. Lack of mutation-histopathology correlation in a patient with Proteus syndrome. *Am J Med Genet A.* 2016;170(6):1422–32.

32. Lougaris V, Salpietro V, Cutrupi M et al. Proteus syndrome: Evaluation of the immunological profile. *Orphanet J Rare Dis.* 2016;11:3.

33. Alimi A, Weeth-Feinstein LA, Stettner A, Caldera F, Weiss JM. Overlap of Juvenile polyposis syndrome and Cowden syndrome due to de novo chromosome 10 deletion involving BMPR1A and PTEN: Implications for treatment and surveillance. *Am J Med Genet A.* 2015;167(6):1305–8.

34. Matsumoto H, Minami H, Yoshida Y. Lhermitte-Duclos disease treated surgically in an elderly patient: Case report and literature review. *Turk Neurosurg.* 2015;25(5):783–7.

35. Matsumoto K, Nosaka K, Shiomi T, Matsuoka Y, Umekita Y. Tumor-to-tumor metastases in Cowden's disease: An autopsy case report and review of the literature. *Diagn Pathol.* 2015;10:172.

36. Patini R, Staderini E, Gallenzi P. Multidisciplinary surgical management of Cowden syndrome: Report of a case. *J Clin Exp Dent.* 2016;8(4):e472–4.

37. Todd J. Bi-pedicle nipple-sparing mastectomy (modified Letterman technique) and TIGR mesh-assisted immediate implant reconstruction, in a patient with Cowden syndrome. *Gland Surg.* 2016;5(3):306–11.

38. Agarwal R, Liebe S, Turski ML et al. Targeted therapy for genetic cancer syndromes: Von Hippel-Lindau disease, Cowden syndrome, and Proteus syndrome. *Discov Med.* 2015;19(103):109–16.

39. Kimura F, Ueda A, Sato E et al. Hereditary breast cancer associated with Cowden syndrome-related PTEN mutation with Lhermitte-Duclos disease. *Surg Case Rep.* 2017;3(1):83.

40. Tilot AK, Frazier TW 2nd, Eng C. Balancing proliferation and connectivity in PTEN-associated autism spectrum disorder. *Neurotherapeutics.* 2015;12(3):609–19.

41. Gosein MA, Narinesingh D, Nixon CA, Goli SR, Maharaj P, Sinanan A. Multi-organ benign and malignant tumors: Recognizing Cowden syndrome: a case report and review of the literature. *BMC Res Notes.* 2016;9:388.

91

Klippel–Trenaunay Syndrome

Dongyou Liu

CONTENTS

91.1 Introduction

Klippel–Trenaunay syndrome (KTS) is a rare congenital disorder that typically displays three characteristic features: port-wine stain (a red birthmark), abnormal overgrowth of soft tissues and bones, and vein malformations (or varicose veins), along with increased risk for malignant peripheral nerve sheath tumor, angiosarcoma, astrocytoma, hemangiopericytoma, and meningioma, etc. At the molecular level, KTS is linked to defects in the *PIK3CA* gene (encoding phosphatidylinositol-4,5-bisphosphate 3-kinase catalytic subunit alpha) on chromosome 3q26.32, and possibly other genes (e.g., *AGGF1*) [1].

91.2 Biology

First recognized as a multisystem disorder that displays characteristic triad of port-wine stain, abnormal overgrowth of soft tissues and bones, and varicose veins by Maurice Klippel and Paul Trénaunay in 1900, KTS (also known as angio-osteohypertrophy syndrome, congenital dysplastic angiopathy, or capillary-lymphatic-venous malformation [CLVM]) is now defined by the International Society for the Study of Vascular Anomalies as a rare, congenital syndrome associated with capillary malformation (cutaneous nevus), venous malformation (varicose veins), possible lymphatic malformation, and limb overgrowth. Apart from the triad of congenital anomalies, KTS is known to predispose a heterogeneous range of tumors (e.g., bone and soft tissue tumors [malignant peripheral nerve sheath tumor, angiosarcoma, astrocytoma, hemangiopericytoma, and meningioma]).

When the congenital anomalies (i.e., varicose veins, and bony and soft tissue hypertrophy) are associated with arteriovenous shunting, the condition is referred to as Parkes–Weber syndrome (PWS) [2].

KTS appears to evolve sporadically and becomes evident at birth or in childhood. Similar to several overgrowth conditions, KTS is linked to somatic *PIK3CA* mutation in early progenitor cells for the blood vessels and underlying bone, leading to kinase activation, and thus forms part of the *PIK3CA*-related overgrowth spectrum (see Chapter 88 for further details).

91.3 Pathogenesis

Located on chromosome 3q26.32, the *PIK3CA* gene encodes a 1068 aa, 110-kDa protein (p110α), which constitutes a catalytic subunit of phosphatidylinositol 3-kinase (PI3 K). Together with a 85-kDa adaptor subunit, p110α phosphorylates (by adding a cluster of oxygen and phosphorus atoms) and activates downstream phosphoinositides in endothelial cells. This facilitates the participation of PI3K in many biological activities, including cellular growth and division (proliferation), movement (migration), and survival; regulation of vascular development and integrity (angiogenesis) through convergence of Gα$_q$ and Ras signaling pathways; and maturation of fat cells (adipocytes) [3,4].

Activating mutations in *PIK3CA* change the structure of the p110α subunit of PI3 K and increase its baseline catalytic activity. This triggers unregulated chemical signaling in cells, increased cell proliferation, abnormal growth of the bones and soft tissues, and over-formation of the vascular network. Indeed, somatic *PIK3CA* mutations are responsible for a spectrum of phenotypes, including KTS and *PIK3CA*-related overgrowth spectrum (e.g., CLOVES [congenital lipomatous overgrowth, vascular malformations, epidermal nevis, spinal/skeletal anomalies/scoliosis] syndrome, megalencephaly-capillary malformation [MCAP] syndrome, and somatic seborrheic keratoses)(see Chapter 88) [5–7].

To date, at least five activating mutations (e.g., E542, E545, or H1047) in the *PIK3CA* gene have been identified in KTS patients. However, some KTS patients lack *PIK3CA* mutations, suggesting that mutations in other unidentified genes [e.g., t(5;11)(q13.3;p15.1), t(8;14)(q22.3;q13), an extra supernumerary ring chromosome 18, VG5Q or AGGF1] may be also involved in this condition.

91.4 Epidemiology

91.4.1 Prevalence

KTS is estimated to affect 1 per 100,000 live births worldwide, with around 1000 cases reported in the literature so far. KTS shows no predilection for gender, race, or geographical area.

91.4.2 Inheritance

KTS may be caused by sporadic or mosaic homozygosity mutations in *PIK3CA* and possibly other genes. Sporadic mutation occurs in the early stages of development before birth, and the coexistence of cells with and without a genetic mutation that continue to divide and differentiate leads to mosaicism, which may partly explain for occasional familial aggregation among KTS patients. In addition, paradominant inheritance is thought to play a possible role in the pathogenesis of KTS.

91.4.3 Penetrance

KTS demonstrates incomplete penetrance and variable clinical expression, as evidenced by its association with heterogeneous tumor types and localizations.

91.5 Clinical Features

KTS typically presents with (i) port-wine stains (also known as cutaneous capillary hemangiomas or capillary malformations; flat, pale pink, red to purplish birthmarks, or occasionally small red blisters that break open and bleed easily, with hard, dry, thickened, and hyperpigmented overlying skin; due to an increased number of abnormal ectatic capillaries in the papillary dermis; occurring in 98% of cases, with 11% of cases accompanied by lymphatic and venous malformations) (Figure 91.1) [8], (ii) varicose veins (swollen and twisted veins near the surface of the skin on the sides of the upper legs and calves that often cause pain; 72% of cases; malformations of deep veins leading to aneurysmal dilatation, agenesis, hypoplasia, duplications, venous regurgitation, thrombosis, or severe bleeding in the liver, kidney, bladder, rectum, and lower gastrointestinal tract, retroperitoneum, pericardium, spine, and lung) [9,10], (iii) limb hypertrophy (usually affecting one limb, with the affected leg being 2 cm longer than the unaffected one, leading to pain, heaviness, and reduced movement; due to excessive increase in underlying soft

tissue and the associated adipose tissue hyperplasia; 67% of cases) [11], (iv) other complications (e.g., cellulitis [due to dilated cutaneous and subcutaneous lymphatic vessels filled with clear proteinaceous fluid, and not connected to normal lymphatic vessels, leading to lymphedema, 10% of cases; internal bleeding from abnormal blood vessels]; fusion of certain fingers or toes [syndactyly], and presence of extra digits [polydactyly]; ophthalmological alterations [glaucoma due to raised episcleral venous pressure, conjunctival telangiectasia, orbital varix, strabismus, oculosympathetic palsy, Marcus–Gunn pupil, iris coloboma and heterochromia, cataracts, persistent fetal vasculature, chiasmal and bilateral optic nerve gliomas, drusen of the optic disk, acquired myelination of the retinal nerve fiber layer, and retinal dysplasia with astrocytic proliferation of the nerve; mental retardation; gastrointestinal bleeding; death]) (Figures 91.2) [12–18], (v) tumors (malignant peripheral nerve sheath tumor, angiosarcoma, astrocytoma, hemangiopericytoma, meningioma, choroidal hemangiomas, lymphangiomas) (Figure 91.3) [19–23].

91.6 Diagnosis

Diagnosis of KTS is based on physical examination, imaging investigation (e.g., radiograph for measuring bone morphology; phlebography for assessing venous malformation; ultrasonography for investigating vascular patency, incompetence, thrombosis, arteriovenous shunting and hypoplasia of vein; CT venography and MR venography for evaluating deep venous system and dilated superficial and embryonic veins) [24,25].

A diagnosis of KTS can be made when two of the following three features are met: (i) cutaneous capillary malformations (port-wine stain), (ii) bone and soft tissue hypertrophy, (iii) varicose veins (venous and lymphatic malformations mainly on the extremities and the adjacent pelvis or shoulder) [26,27]. Molecular identification of *PIK3CA* pathogenic variants provides further confirmation of the disorder.

Differential diagnoses for KTS (local gigantism and hemangioma of the limbs with osteohypertrophy; peripheral varices) include other disorders that cause gigantism of the limbs and vascular anomalies such as Proteus syndrome (massive overgrowth and asymmetry; linear verrucous epidermal nevi, intradermal nevi, hemangiomas, lipomas, and varicosities; macrodactyly and syndactyly; soft tissue hypertrophy mainly over the plantar surface; moderate mental deficiency; dysplastic, progressive, and irregular bone overgrowth), Bannayan–Riley–Ruvalcaba syndrome, Maffucci syndrome (multiple enchondromas, hemangiomas and, less often, lymphangiomas), neurofibromatosis type I (café-au-lait spots, neurofibromas, freckles in the axillary or inguinal regions, optic gliomas, iris hamartomas, and distinctive bone lesions), Sturge–Weber syndrome (SWS, port-wine vascular nevus on the upper part of the face; leptomeningeal angiomatosis involving one or both hemispheres, early onset seizure, stroke-like episodes due to poor blood flow through the affected vessels; choroidal hemangioma or glaucoma, neurologic deterioration/developmental delay) [28,29], PWS (local gigantism and hemangioma of the limbs with osteohypertrophy; arteriovenous fistulas/anastomoses; due to RASA1 mutations) [30–32], and capillary malformation-arteriovenous malformation (CM-AVM, capillary malformations and arteriovenous fistulas, but no limb overgrowth) [33,34].

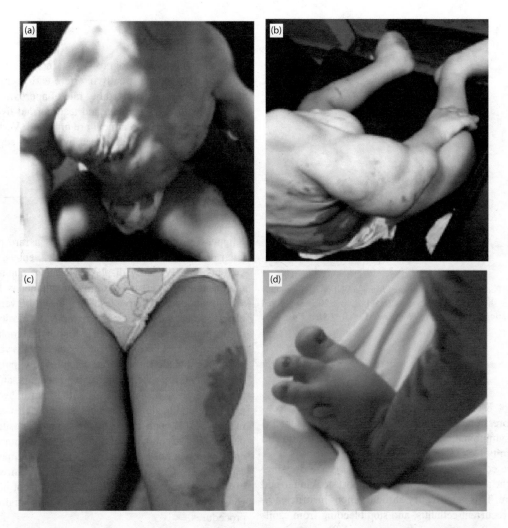

FIGURE 91.1   Klippel–Trenaunay syndrome in a 20-month-old girl showing focal overgrowth over the upper part of the chest (a), multiple focal overgrowth of the right arm (b), hypertrophy and port-wine stain of the left thigh (c), and syndactyly of the toes in left foot (d). (Photo credit: Mneimneh S et al. *Case Rep Pediatr.* 2015;2015:581394.)

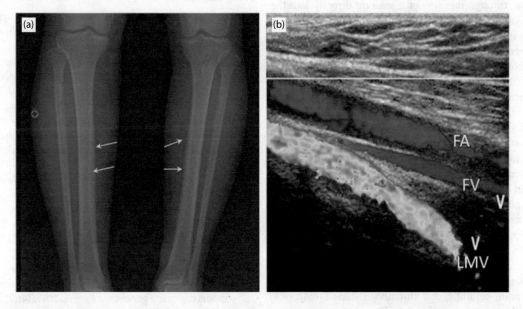

FIGURE 91.2   Klippel–Trenaunay syndrome in a 40-year-old male showing bone cortex hypertrophy (arrow) in the lower limbs as detected by radiography (a), dilation of superficial vein in bilateral lower limbs, retrograde flow of bilateral great saphenous vein, and persistence of the lateral marginal vein as detected by ultrasonography (b). FA, femoral artery; FV, femoral vein; LMV, lateral marginal vein (b). (Photo credit: Baba A et al. *J Surg Case Rep.* 2017;2017(2):rjx024.)

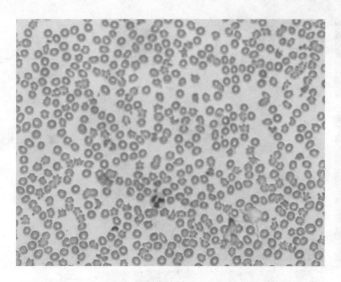

FIGURE 91.3 Peripheral blood smear from a Klippel–Trenaunay syndrome patient showing acanthocytosis (making up >20% of the red blood cell population). (Photo credit: Withana M et al. *J Med Case Rep.* 2014;8:390.)

91.7 Treatment

Management of KTS relies on standard procedures (e.g., conservative compression, sclerotherapy, laser therapy, and surgery) that provide symptomatic relief and preserve the function in affected extremity [35–38].

Conservative compression involves placing a gauze dressing/bandage over the area, and then applying support stocking to add further pressure. This helps rectify chronic venous insufficiency, lymphedema, recurrent cellulitis, and stop bleeding from capillary or venous malformations of the extremity [39].

Sclerotherapy eliminates the pain and discomfort of varicose veins and prevents complications such as venous hemorrhage and ulceration, through injection of a solution directly into the varicose veins to make them collapse and disappear. A variant of sclerotherapy involves injection of a foaming agent mixed with a sclerosing agent under ultrasound guidance. After the foaming agent moves blood out of the vein, the sclerosing agent has better contact with the vein wall.

Laser therapy lightens or removes port-wine stain and ablates larger varicose veins endovenously. Laser and pulse-light therapy heat and shrink the blood vessel and provide an effective treatment for small varicose veins [40].

Surgical tying (ligation) of veins through a small incision in the skin prevents pooling of blood. It may be used in conjunction with vein stripping, or removal of the vein (venous ablation). Endovenous thermal ablation uses laser or high-frequency radio waves to create intense local heat in the varicose vein. This closes off the problem veins with minimal bleeding and bruising. Epiphysiodesis may be employed to reduce the length of affected leg and increase the length of unaffected leg.

In addition, chemotherapy may be considered as part of management for pain, cellulitis, and thrombophlebitis (analgesics, elevation, antibiotics, and corticosteroids), and varicose veins (antiangiogenic agents such as vincristine, prednisolone, and propranolol) [41].

91.8 Prognosis

KTS has a favorable prognosis. As a non-life-threatening condition, KTS does not greatly interfere with a child's day-to-day activities and shows a mortality rate of approximately 1%. Most children with KTS grow up and lead normal lives if early recognition and correct treatment are provided. Varicose veins and related problems can be reduced by wearing support stockings from an early age [42].

91.9 Conclusion

Klippel–Trenaunay syndrome (KTS) is a rare congenital vascular disorder that affects the capillary, venous, and lymphatic systems, and typically displays a triad of cutaneous capillary malformations (port-wine stains or cutaneous hemangiomas), venous anomalies (hypoplasia or aplasia of veins, persistence of fetal veins, varicosities, hypertrophy, tortuosity, and valvular malformations), and asymmetric limb hypertrophy (due to bone hypertrophy and secondary soft tissue overgrowth). Other complications include thrombosis, coagulopathy, pulmonary embolism, heart failure, hemothorax, lymphedema, ankle ulcers, and bleeding from abnormal vessels of the gastrointestinal tract, kidney, or genitalia, along with increased risk for tumors. Mutations in the *PIK3CA* gene (encoding phosphatidylinositol-4,5-bisphosphate 3-kinase catalytic subunit alpha) on chromosome 3q26.32, and possibly other genes are thought to underline the pathogenesis of KTS. Current diagnosis of KTS requires meeting two of the three key clinical criteria. Management of KTS involves conservative compression therapy, surgery, and other standard procedures.

REFERENCES

1. Sharma D, Lamba S, Pandita A, Shastri S. Klippel-trénaunay syndrome—A very rare and interesting syndrome. *Clin Med Insights Circ Respir Pulm Med.* 2015;9:1–4.
2. Alomari AI, Orbach DB, Mulliken JB et al. Klippel-Trenaunay syndrome and spinal arteriovenous malformations: An erroneous association. *AJNR Am J Neuroradiol.* 2010;31(9):1608–12.
3. Timur AA, Driscoll DJ, Wang Q. Biomedicine and diseases: The Klippel-Trenaunay syndrome, vascular anomalies and vascular morphogenesis. *Cell Mol Life Sci.* 2005;62(13):1434–47.
4. Wang QK. Update on the molecular genetics of vascular anomalies. *Lymphat Res Biol.* 2005;3(4):226–33.
5. Edmondson AC, Kalish JM. Overgrowth syndromes. *J Pediatr Genet.* 2015;4(3):136–43.
6. Luks VL, Kamitaki N, Vivero MP et al. Lymphatic and other vascular malformative/overgrowth disorders are caused by somatic mutations in PIK3CA. *J Pediatr.* 2015;166(4):1048–54.e1-5.
7. Wetzel-Strong SE, Detter MR, Marchuk DA. The pathobiology of vascular malformations: Insights from human and model organism genetics. *J Pathol.* 2017;241(2):281–93.
8. Mneimneh S, Tabaja A, Rajab M. Klippel-Trenaunay syndrome with extensive lymphangiomas. *Case Rep Pediatr.* 2015;2015:581394.

9. Wang ZK, Wang FY, Zhu RM, Liu J. Klippel-Trenaunay syndrome with gastrointestinal bleeding, splenic hemangiomas and left inferior vena cava. *World J Gastroenterol.* 2010;16(12):1548–52.

10. Tetangco EP, Arshad HM, Silva R. Klippel-Trenaunay syndrome of the rectosigmoid colon presenting as severe anemia. *ACG Case Rep J.* 2016;3(4):e161.

11. Ikpeme AA, Usang UE, Inyang AW, Ani N. Klippel Trenaunay syndrome: A case report in an adolescent Nigerian boy. *Open Access Maced J Med Sci.* 2015;3(2):322–5.

12. Kocaman O, Alponat A, Aygün C et al. Lower gastrointestinal bleeding, hematuria and splenic hemangiomas in Klippel-Trenaunay syndrome: A case report and literature review. *Turk J Gastroenterol.* 2009;20(1):62–6.

13. Abdolrahimzadeh S, Scavella V, Felli L, Cruciani F, Contestabile MT, Recupero SM. Ophthalmic alterations in the Sturge-Weber syndrome, Klippel-Trenaunay syndrome, and the phakomatosis pigmentovascularis: An independent group of conditions? *Biomed Res Int.* 2015;2015:786519.

14. Chu ST, Han YH, Koh JA et al. A case of Klippel-Trenaunay syndrome with acute submassive pulmonary thromboembolism treated with thrombolytic therapy. *J Cardiovasc Ultrasound.* 2015;23(4):266–70.

15. de Godoy JM, Río A, Domingo Garcia P, de Fatima Guerreiro Godoy M. Lymphedema in Klippel-Trenaunay syndrome: Is it possible to normalize? *Case Rep Vasc Med.* 2016;2016:5230634.

16. Baba A, Yamazoe S, Okuyama Y et al. A rare presentation of Klippel-Trenaunay syndrome with bilateral lower limbs. *J Surg Case Rep.* 2017;2017(2):rjx024.

17. Lei H, Guan X, Han H et al. Painless urethral bleeding during penile erection in an adult man with Klippel-Trenaunay syndrome: A case report. *Sex Med.* 2018;6(2):180–3.

18. Liu XY, Zhang S, Zhang H, Jia J, Cai L, Zhang JZ. Lymphangioma circumscriptum in vulva with Klippel-Trenaunay syndrome. *Chin Med J (Engl).* 2018;131(4):490–1.

19. Gonçalves LF, Rojas MV, Vitorello D, Pereira ET, Pereima M, Saab Neto JA. Klippel-Trenaunay-Weber syndrome presenting as massive lymphangiohemangioma of the thigh: Prenatal diagnosis. *Ultrasound Obstet Gynecol.* 2000;15(6):537–41.

20. Withana M, Rodrigo C, Shivanthan MC et al. Klippel-Trenaunay syndrome presenting with acanthocytosis and splenic and retroperitoneal lymphangioma: A case report. *J Med Case Rep.* 2014;8:390.

21. Sreenivasan P, Kumar S, Kumar KK. Klippel-Trenaunay syndrome and gestational trophoblastic neoplasm. *Indian Pediatr.* 2014;51(9):745–6.

22. Yilmaz T, Cikla U, Kirst A, Baskaya MK. Glioblastoma multiforme in Klippel-Trenaunay-Weber syndrome: A case report. *J Med Case Rep.* 2015;9:83.

23. Gripp KW, Baker L, Kandula V et al. Nephroblastomatosis or Wilms tumor in a fourth patient with a somatic PIK3CA mutation. *Am J Med Genet A.* 2016;170(10):2559–69.

24. Cardarelli-Leite L, Velloni FG, Salvadori PS, Lemos MD, D'Ippolito G. Abdominal vascular syndromes: Characteristic imaging findings. *Radiol Bras.* 2016;49(4):257–63.

25. Yara N, Masamoto H, Iraha Y et al. Diffuse venous malformation of the uterus in a pregnant woman with Klippel-Trénaunay syndrome diagnosed by DCE-MRI. *Case Rep Obstet Gynecol.* 2016;2016:4328450.

26. Howes JA, Setty G, Khan A, Hussain N. A rare paediatric case of Klippel-Trenaunay-Weber syndrome. *J Pediatr Neurosci.* 2015;10(1):87–8.

27. Martino V, Ferrarese A, Alessandro B et al. An unusual evolution of a case of Klippel-Trenaunay syndrome. *Open Med (Wars).* 2015;10(1):498–501.

28. Sen S, Bala S, Halder C, Ahar R, Gangopadhyay A. Phakomatosis pigmentovascularis presenting with sturge-weber syndrome and klippel-trenaunay syndrome. *Indian J Dermatol.* 2015;60(1):77–9.

29. Kentab AY. Klippel-Trenaunay and Sturge-Weber overlapping syndrome in a Saudi boy. *Sudan J Paediatr.* 2016;16(2):86–92.

30. Cebeci E, Demir S, Gursu M et al. A case of newly diagnosed Klippel Trenaunay Weber syndrome presenting with nephrotic syndrome. *Case Rep Nephrol.* 2015;2015:704379

31. Kundzina L, Lejniece S. Klippel-Trenaunay-Weber syndrome with atypical presentation of hypersplenism and nephrotic syndrome: A case report. *J Med Case Rep.* 2017;11(1):243.

32. van der Loo LE, Beckervordersandforth J, Colon AJ, Schijns OE. Growing skull hemangioma: First and unique description in a patient with Klippel-Trénaunay-Weber syndrome. *Acta Neurochir (Wien).* 2017;159(2):397–400.

33. Lacerda Lda S, Alves UD, Zanier JF, Machado DC, Camilo GB, Lopes AJ. Differential diagnoses of overgrowth syndromes: The most important clinical and radiological disease manifestations. *Radiol Res Pract.* 2014;2014:947451.

34. Pillai MR, Hasini PP, Ahuja A, Krishnadas SR. A rare case of overlapping Sturge-Weber syndrome and Klippel-Trenaunay syndrome associated with bilateral refractory childhood glaucoma. *Indian J Ophthalmol.* 2017;65(7):623–5.

35. Sgubin D, Kanai R, Di Paola F, Perin A, Longatti P. Conus medullaris-cauda arteriovenous malformation and Klippel-Trenaunay syndrome: What is the treatment goal? *Neurol Med Chir (Tokyo).* 2013;53(2):110–4.

36. Gupta Y, Jha RK, Karn NK, Sah SK, Mishra BN, Bhattarai MK. Management of femoral shaft fracture in Klippel-Trenaunay syndrome with external fixator. *Case Rep Orthop.* 2016;2016:8505038.

37. Upadhyay H, Sherani K, Vakil A, Babury M. A case of recurrent massive pulmonary embolism in Klippel-Trenaunay-Weber syndrome treated with thrombolytics. *Respir Med Case Rep.* 2016;17:68–70.

38. Huang FL, Chen HY, Chang TK. Medical treatment of a female patient with complicated Klippel-Trenaunay syndrome. *Pediatr Neonatol.* 2018;59(5):527–30.

39. Hagen SL, Grey KR, Korta DZ, Kelly KM. Quality of life in adults with facial port-wine stains. *J Am Acad Dermatol.* 2017;76(4):695–702.

40. Rahimi H, Hassannejad H, Moravvej H. Successful treatment of unilateral Klippel-Trenaunay syndrome with pulsed-dye laser in a 2-week old infant. *J Lasers Med Sci.* 2017;8(2):98–100.

41. George SE, Sreevidya A, Asokan A, Mahadevan V. Klippel Trenaunay syndrome and the anaesthesiologist. *Indian J Anaesth.* 2014;58(6):775–7.

42. Sung HM, Chung HY, Lee SJ et al. Clinical experience of the Klippel-Trenaunay syndrome. *Arch Plast Surg.* 2015;42(5):552–8.

92

Neurofibromatosis Types 1 and 2

Dongyou Liu

CONTENTS

92.1 Introduction

Neurofibromatosis represents a heterogeneous group of hereditary syndromes that includes neurofibromatosis type 1 (NF1, 96%), neurofibromatosis type 2 (NF2, 3%), and schwannomatosis (SWN)], all of which confer susceptibility to the development of tumors in the central and peripheral nervous systems [1].

Neurofibromatosis type 1 (NF1, sometimes referred to as peripheral neurofibromatosis) is an autosomal dominant tumor predisposition syndrome characterized by pigmentation (e.g., café-au-lait spots, freckling in the inguinal and axillary regions, Lisch nodules in the eye) and growth of multiple neurofibromas (i.e., benign peripheral nerve sheath tumors) of the peripheral nerves in the skin, brain, and other parts of the body, as well as other tumors (e.g., gliomas, malignant peripheral nerve sheath tumors [MPNST], juvenile chronic myelomonocytic leukemia, rhabdomyosarcoma, pheochromocytoma, and breast cancer). NF1 is estimated to occur at 1 in 3000 births. At the molecular level, NF1 is attributable to loss of function heterozygous mutations in the *NF1* gene on chromosome 17q11.2 that encodes neurofibromin (NF1) with tumor suppressor function [2].

Neurofibromatosis type 2 (NF2, sometimes referred to as central neurofibromatosis) is an autosomal dominant disorder that predisposes to the development of benign (noncancerous) tumors (typically vestibular schwannomas or acoustic neuromas, meningiomas of the brain, and schwannomas of the dorsal roots of the spinal cord, but few skin lesions or neurofibromas) in the nervous system, leading to hearing loss, ringing in the ears (tinnitus), changes in vision (including clouding of the lens or cataracts), numbness or weakness in the arms or legs, and fluid buildup in the brain. NF2 has an estimated incidence of 1 in 25,000–33,000. Mutations in the *NF2* gene on chromosome 22q12.2 encoding merlin (NF2), which functions as a tumor suppressor, appear to be responsible for NF2 [3].

SWN (also known as neurofibromatosis type 3 or neurilemmomatosis) is an autosomal dominant disorder characterized by the development of multiple peripheral nerve schwannomas (i.e., cutaneous neurilemmomas, spinal schwannomas), without acoustic tumors or other signs of NF1 or NF2. Although SWN overlaps NF2 phenotypically, with the formation of schwannoma it appears to have evolved from germline heterozygous mutations in the *SMARCB1* (on chromosome 22q11.23) and *LZTR1* (on chromosome 22q11.21) genes, which are observed in up to 50% of familial SWN cases. As an adult-onset tumor predisposition syndrome, SWN has an estimated incidence of 1 in 40,000 (see Chapter 11 in this book for further details) [1].

92.2 Biology

Cell growth within the mammalian body is typically controlled by two main types of genes, proto-oncogenes and tumor suppressor genes. While a functional proto-oncogene helps cells grow and divide, a mutated proto-oncogene (or presence of extra copies due to gene duplication) becomes an oncogene (which is permanently activated or turned on), resulting in uncontrolled cell growth and tumorigenesis. On the other hand, a functional tumor suppressor gene keeps the cell from dividing too quickly, repairs DNA mistakes, and induces programmed cell death (apoptosis); a mutated tumor suppressor gene becomes inactivated (turned off), rendering cell growth/division out of control, and facilitating cancer development (tumor suppressor syndrome).

NF1 (peripheral neurofibromatosis), NF2 (central neurofibromatosis), and SWN are examples of tumor suppressor syndromes caused by germline mutations in a tumor suppression gene. While the presence of one normal and one mutant allele is sufficient to cause clinical diseases, inactivation of the second allele is required for tumor formation (two-hit hypothesis). As tumors associated with NF1, NF2, and SWN predominantly affect the central nervous system (CNS) and peripheral nervous system (PNS), significant neurologic morbidity is observed [4].

NF1 (also known as peripheral neurofibromatosis) occurs due to mutations in the tumor suppressor gene *NF1*. Targeting peripheral nerves and their supporting structures (including neurilemmal cells), NF1 causes multiple café-au-lait spots (or flat, dark patches) on the skin, freckles in the underarms and groin, Lisch nodules in the colored part of the eye (the iris), and neurofibromas on or just under the skin, often near the spinal cord or along nerves elsewhere in the body. Neurofibromas are benign tumors with mixed cell types including Schwann cells, perineural cells, and fibroblasts, along with mast cells, axonal processes, and a collagenous extracellular matrix. However, these tumors have the potential to transform into malignant peripheral nerve sheath tumors, which contribute to early death in affected patients [5]. As a variant of neurofibroma, plexiform neurofibroma arises from muscle nerve fascicles, and may infiltrate into the surrounding structures [6,7].

NF2 (also known as central neurofibromatosis or multiple inherited schwannomas, meningiomas, and ependymomas [MISME] syndrome) is caused by mutations in the tumor suppressor gene *NF2*. Affecting mainly the central nerves, NF2 is characterized by the formation of vestibular schwannomas (or acoustic neuromas) and meningiomas, but few skin lesions or neurofibromas. Derived from Schwann cells, meningeal cells, and glial cells, NF2 vestibular schwannoma is typically located in the region of the cranial nerve VIII (auditory-vestibular nerve) that transmits sensory information from the inner ear to the brain. On the other hand, NF2 meningiomas are usually of the fibroblastic variety [8].

92.3 Pathogenesis

The *NF1* gene located on chromosome 17q11.2 was shown in 1990 as the culprit for neurofibromatosis type 1 (NF1) is *NF1*. Composed of ~350 kb and 57 exons, the *NF1* gene produces at least three alternatively spliced transcripts, with the 2818 aa, 327 kDa cell membrane protein (neurofibromin), which belongs to the GTPase-activating family of tumor suppressor proteins and participates in regulation of cellular proliferation, somatic cell division, adenylyl-cyclase activity and intracellular cyclic-AMP generation. Through its role in the conversion of active RAS to inactive RAS, NF1 (neurofibromin) influences RAS downstream effectors that constitute key components of the RAF/MEK/ERK, PI3kinase/Akt, and mechanistic target of rapamycin (mTOR) pathways (see Figure 87.2 for detail).

NF1 (neurofibromin) is expressed in nerve cells and specialized cells surrounding nerves (oligodendrocytes and Schwann cells). Alterations in the *NF1* gene result in nonfunctional neurofibromin, which has limited capacity in converting active RAS to inactive RAS, leading to increased activation of the RAS/MAPK signaling pathway. For this reason, neurofibromatosis type 1 is

considered one of RASopathies. Notable changes detected in the *NF1* gene include stop variants, amino acid substitutions, deletions (involving only one or a few base pairs, multiple exons, or the entire gene), insertions, intronic changes affecting splicing, alterations of the 3′ untranslated region of the gene, and gross chromosome rearrangements. Among 1500 *NF1* mutations identified to date, those altering mRNA splicing leading to a loss-of-function protein [e.g., c.2970_2972delAAT (p.Met992del), c.5425C > T (p.Arg1809Cys), c.5425C > G (p.Arg1809Gly), c.5425C > A (p.Arg1809Ser), c.5426G > T (p.Arg1809Leu), c.5426G > C (p.Arg1809Pro)] make up a majority of germline pathogenic variants in human patients with NF1 [9–13].

Some affected individuals who acquire de novo *NF1* mutation may have somatic mosaicism associated with segmental or unusually mild disease manifestations. Mosaicism may result in segmental, generalized, or gonadal *NF1* gene. Segmental *NF1* gene is associated with pigment changes and/or tumors that are limited to one or more body segments. Generalized *NF1* gene behaves similarly to classic *NF1* gene without the *NF1* gene mutation. Gonadal *NF1* gene contains mutation that only affects the ova or sperm.

Interestingly, somatic (but not germline) *NF1* mutations affecting one or both alleles may be observed in several tumors (malignant peripheral nerve sheath tumor, pheochromocytoma, juvenile myelomonocytic leukemia, glioma, liposarcoma, lung adenocarcinoma, ovarian carcinoma, colorectal carcinoma, melanoma, adult acute myeloid leukemia, and breast cancer) from individuals who do not present clinical features of NF1 [14,15].

The gene underlying the molecular pathogenesis of NF2 is *NF2* on chromosome 22q12.2, which was identified in 1993. The *NF2* gene encodes a membrane-cytoskeleton scaffolding protein known as "merlin" (for *m*oezin-*e*zrin-*r*adixin-*l*ike prote*in*, due to its relationship to the moesin [membrane organizing extension spike protein]−erzin [cytovillin]−radixin family of cytoskeleton-associated proteins) or "schwannomin" (in recognition of its role in preventing schwannoma formation), which consists of three structurally distinct regions: (i) a N-terminal FERM (four-point−one, ezrin, radixin, and moesin) domain, (ii) a α-helical coiled-coil domain, and (iii) a C-terminal hydrophilic tail. Of the two isoforms produced, the longer form (merlin or merlin 1595aa) possesses an extended C-terminal tail (encoded by exon 17) that is involved in intramolecular binding between the amino-terminal FERM domain and the C-terminal hydrophilic tail, yielding a closed or functional state, and the shorter form (merlin 2) results from an alternatively spliced exon 16 ending in a stop codon and contains 11 unique residues following amino acid 579, without the C-terminal residues, yielding an open or nonfunctional state. By linking actin filaments via its N-terminal domain containing glutathione S-transferase to cell membrane or membrane glycoproteins, merlin acts as a critical regulator of contact-dependent inhibition of proliferation and function at the interface, and contributes to the maintenance of normal cytoskeletal organization, the modulation of cellular motility, attachment, remodeling, and spreading, and the regulation of growth (tumor suppression) through its ability to stabilize cadherin-dependent cell-to-cell junctions, to inhibit the effects of receptor tyrosine kinases (RTK) at the cell membrane, and to facilitate endocytic trafficking of RTK. Merlin deficiency can increase the signaling of the ErbB/EGFR family RTK and activation of prosurvival and proliferation pathways via RAS modulation (due to its many shared targets with NF1), leading to unmediated progression in the cell cycle and subsequent tumorigenesis [16].

Mutations in the *NF2* gene either abrogate merlin synthesis or generate a defective protein that lacks the normal tumor-suppression function through alteration in the movement and shape of affected cells after merlin loses contact inhibition. Of >400 *NF2* pathogenic variants described to date, missense, nonsense, and splicing variants and small deletions dominate. About 90% of single-nucleotide variants seem to introduce a premature stop codon, a frameshift with premature termination, or a splicing alteration, leading to truncated NF2 protein, while <10% of pathogenic variants are in-frame deletions and missense variants that alter certain NF2 functional domains. In particular, C-to-T transitions in CGA codons causing pathogenic nonsense variants are frequently detected in human NF2 cases [17].

As merlin is expressed in the nervous system, particularly in Schwann cells, which surround and insulate nerve cells (neurons) in the brain and spinal cord, alteration of its normal function via gene mutations leads to uncontrolled multiplication of Schwann cells and formation of schwannomas characteristic of neurofibromatosis type 2 [18]. Indeed, mutations in merlin have been observed in nearly 93% of patients with clinical evidence of NF2 and a positive family history [19].

Occasionally, sporadic schwannomas may occur as single tumors at any site in the absence of any other findings of NF2. However, patients with schwannomas caused by sporadic, non-germline NF2 mutations will not pass to offspring, and thus are not heritable.

92.4 Epidemiology

92.4.1 Prevalence

NF1 has an estimated incidence of 1 in 2500–3500 births worldwide, without sexual or racial predilection, which makes NF1 the most common autosomal dominant disorder of the nervous system and one of the most common single-gene inherited conditions. NF1 accounts for about 96% of all neurofibromatosis cases, with 30% being family-related, and 70% attributing to sporadic mutations that often occur in paternally derived chromosomes. The mean age of NF1 patients is 38.5 years, and the mean age of NF1 diagnosis is 14.5 years.

NF2 affects 1 in 25,000–33,000, and accounts for 3% of all neurofibromatosis cases, with no gender or race predilection. About 25.5% of NF2 patients have family connection, while 74.5% are due to sporadic mutations. The mean age of NF2 patients is 39.1 years, and the mean age of NF2 diagnosis is 29.1 years.

92.4.2 Inheritance

NF1 shows an autosomal dominant pattern of inheritance, with one altered copy of a gene in each cell being sufficient to cause the disease, but not tumor formation. While about half of cases are known to inherit a mutated copy of *NF1* gene from an affected parent, the other half result from sporadic *NF1* mutation that occurs at the beginning of life, without obvious history of the disorder in their family. An affected individual who appears to acquire de novo mutation may have somatic mosaicism associated with segmental or unusually mild disease manifestations. To

trigger tumor formation, NF1 requires two copies of the altered *NF1* gene, with somatic mutation in the second copy of the *NF1* gene occurring in specialized cells surrounding nerves during a person's lifetime. A person harboring a *NF1* pathogenic variant has a 50% chance of passing it to offspring with each pregnancy. However, a person with mosaicism for an *NF1* pathogenic variant may have <50% chance of transmitting the disease to offspring.

Similarly, NF2 is also an autosomal dominant disorder, in which one altered copy of the *NF2* gene in each cell is sufficient to cause the disorder, but not tumor formation. The first copy of the altered *NF2* gene comes from either an affected parent (50% of cases) or new (de novo) mutation (50% of cases). The second copy of the altered *NF2* gene in Schwann cells or other cells in the nervous system is acquired during a person's lifetime. A person with an *NF2* pathogenic variant has a 50% chance of passing it to offspring. About 25%–33% of patients with a de novo pathogenic *NF2* variant have somatic mosaicism (so-called mosaic NF2), which is characterized by a unilateral eighth-nerve schwannoma associated with ipsilateral meningiomas (i.e., unilateral NF2 involvement of the CNS) or multiple schwannomas localized to one part of the peripheral nervous system, which is detected in tumor DNA but not leukocyte DNA and which shows an 8%–12% risk of transmission compared to 50% in autosomal dominant inheritance.

92.4.3 Penetrance

NF1 shows extremely variable phenotypic expression, likely a result of epigenetic modification, but penetrance is complete after childhood and virtually all individuals who harbor a germline *NF1* pathogenic variant develop the disease in an average lifetime. Furthermore, deletion of the entire gene, known as 17q11.2 microdeletion, is associated with a more severe form of the disease, while mosaicism tends to produce a mild or even segmental presentation.

NF2 demonstrates extremely high penetrance (95%) and variable phenotypic expression, with intrafamilial variability being much lower than interfamilial variability, suggesting a strong effect of the underlying genotype on the resulting phenotype. Mutation type and the position of the mutation within the *NF2* gene may also have phenotypic implications. For example, frameshift or nonsense mutations that produce truncated protein are often associated with severe clinical manifestations (e.g., a higher frequency of meningiomas, spinal tumors, and cutaneous tumors, and death at a younger age); splice site variants are responsible for both mild and severe disease, and large deletions or missense mutations tend to produce a mild disease presentation without intellectual disability. Further, mutations near the 3′ end of the *NF2* gene (especially involving exons 14–16) are associated with mild disease.

92.5 Clinical Features

NF1 is usually associated with the following:

> *Skin lesions*: Multiple café-au-lait spots (ovoid tan-brown macules of 0.5–5 cm in size appearing anywhere on the body, with six or more found in 99% of cases by 1 year of age), freckles (axillary and inguinal freckling, with axillary freckling or Crow's sign being pathognomonic

for NF-1, 40% of infant cases and 90% by 7 years of age), Lisch nodules (or pigmented iris hamartomas; small, often multiple dome-shaped melanocytic nodules around the iris; 93% of adult cases).

Tumors: Neurofibromas (soft iliac-pink tumors of few mm to several cm in diameter, mostly sessile and dome-shaped or pedunculated, on the trunk and limbs, or the areola of female breasts; 60% of cases), papillomatous neurofibromas (on the hard palate, tongue, etc., 5%–10%), gliomas (indolent pilocytic astrocytomas, 15% by the age of 7 years), malignant peripheral nerve sheath tumor (10%), Wilms tumor, rhabdomyosarcoma, leukemia (especially juvenile chronic myelogenous leukemia), myelodysplastic syndromes, retinoblastoma, and malignant melanoma (Figure 92.1).

Other symptoms: Learning disabilities (50%), kyphoscoliosis (2%), pseudarthrosis of tibia or radius (1%), sphenoid wing dysplasia (a characteristic abnormality in NF1), seizure, and vasculopathy [20–22]. Patients with NF1 have a shortened life expectancy of average of 54 years (due largely to malignancy).

NF2 typically produces bilateral vestibular schwannomas by age 30 years with associated symptoms of unilateral/bilateral hearing loss (98%), tinnitus (ringing in the ears, 70%), vertigo (dizzy spell)/balance dysfunction/disequilibrium (67%), headache (32%), facial numbness and weakness (29% and 10%, respectively), abnormal corneal reflex (33%), nystagmus (26%), facial hypesthesia (26%), juvenile subcapsular cataract (opacity of the lens), and blindness (1%). Patients with NF2 may also develop peripheral schwannomas (especially in the paraspinal and cutaneous nerves, leading to focal weakness [12%], seizure [8%], focal sensory loss [6%], and pain), meningiomas (50%, lifetime risk of 80%), ependymomas, and retinal hamartomas [20]. Unlike NF1, which frequently causes café-au-lait spots and freckles, NF2 shows few cutaneous manifestations (apart from intradermal plaque-like tumors that have excess hair and skin pigmentation) [23].

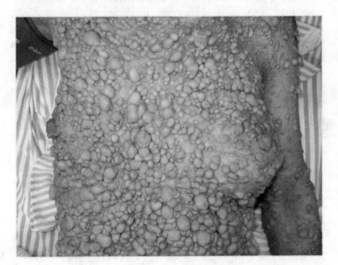

FIGURE 92.1 Formation of numerous skin neurofibromas in a 54-year-old woman diagnosed with NF1 and histologically confirmed high grade ductal carcinoma in situ in the left breast. (Photo credit: Da Silva AV et al. *Cancer.* 2015;15:183.)

92.6 Diagnosis

NF1 is a complex disorder whose molecular pathogenesis is not fully elucidated. Therefore, current diagnosis of NF1 relies on meeting clinical criteria, while molecular genetic testing provides additional clarification [24].

Clinical criteria for NF1 diagnosis are based on meeting two or more of the following conditions: (i) café-au-lait spots: six or more lesions of >5–15 mm in diameter; (ii) neurofibromas: two or more of any type or one plexiform neurofibroma; (iii) freckling: axillary or inguinal regions; (iv) optic gliomas; (v) Lisch nodules (iris hamartomas): two or more; (vi) bony lesions: sphenoid dysplasia or tibial pseudarthrosis; and (vii) family history: a first-degree relative with the disease.

In general, 30% of patients will fulfill one of these criteria by the age of 1 year, 97% of patients will fulfill two criteria by the age of 8 years, and all patients will fulfill these criteria by the age of 20 years.

Histopathologically, NF1-related neurofibroma is a well-circumscribed, rarely encapsulated spindle cell mass in a mucinous background, along with mast cells, Schwann cells, perineural cells, and blood vessels. Cutaneous neurofibroma is pedunculated, nodular, or plaque-like, whereas internal or deep neurofibroma occurs in the periorbital, retroperitoneal, GI tract, and mediastinal locations. Plexiform neurofibroma shows a variety of cell types (e.g., neuronal axons, Schwann cells, fibroblasts, mast cells, macrophages, perineural cells, and extracellular matrix).

Considered pathognomonic of NF1, plexiform neurofibroma is an internal mass consisting of numerous elongated encapsulated neurofibromas mixed with diffuse neurofibroma that involves the dermis and subcutaneous fat, fitting the characteristic "bag of worms" description. Plexiform neurofibroma shows an increased risk of transformation into malignant peripheral nerve sheath tumor (MPNST), an aggressive spindle-cell sarcoma that accounts for 5% of all soft tissue sarcomas. MPNST often shows increased uptake on fluorodeoxyglucose-positron emission tomography (PET) scan. About 50% of MPNST occurs in the setting of NF1, with affected patients having an 8%–13% risk of developing this malignant tumor in their lifetime [25].

Cafe-au-lait spots show hyperpigmentation of the basal epidermis with macromelanosomes (giant melanosis) [26].

Molecular identification of heterozygous *NF1* pathogenic variants provides further confirmation on NF1 diagnosis, especially for patients who partially meet the clinical criteria. In general, multistep pathogenic variant detection protocol based on cDNA and gDNA sequence analysis will permit detection in >95% of patients with a *NF1* pathogenic variant, genomic DNA sequence analysis in 60%–90%, gene-targeted deletion/duplication analysis in ~5%; chromosomal microarray analysis (CMA) in ~5%, and cytogenetic analysis in <1% [27].

Differential diagnoses for NF1 include >100 genetic conditions that also produce café-au-lait spots, such as Legius syndrome (multiple café-au-lait spots, axillary freckling, macrocephaly, facial features that resemble Noonan syndrome; due to heterozygous pathogenic variants in *SPRED1*), constitutional mismatch repair deficiency (consanguineous parents; due to homozygosity or compound heterozygosity for a pathogenic variant in one of the genes causing Lynch syndrome), Piebald trait (areas of cutaneous pigmentation and depigmentation with hyperpigmented borders of the unpigmented areas, and white forelock), NF2 (bilateral vestibular schwannomas, schwannomas of other cranial and peripheral nerves, cutaneous schwannomas, meningiomas, and juvenile posterior subcapsular cataract; due to *NF2* pathogenic variants in *NF2*), SWN (multiple schwannomas of cranial, spinal, or peripheral nerves, usually without vestibular, ocular, or cutaneous features of NF2), Noonan syndrome with multiple lentigines (NSML, formerly LEOPARD syndrome; multiple lentigines, ocular hypertelorism, deafness, and congenital heart disease; due to pathogenic variant in *BRAF, MAP2K1, PTPN11,* or *RAF1*), fibrous dysplasia/McCune–Albright syndrome (FD/MAS, large café-au-lait spots with irregular margins and polyostotic fibrous dysplasia; due to early embryonic postzygotic somatic activating mutation of *GNAS*), Noonan syndrome (short stature, congenital heart defect, neck webbing, and characteristic facies; due to pathogenic variants in *BRAF, KRAS, MAP2K1, NRAS, PTPN11, RAF1, RIT1,* or *SOS1*), infantile myofibromatosis (multiple tumors of the skin, subcutaneous tissues, skeletal muscle, bones, and viscera), Proteus syndrome (hamartomatous overgrowth of multiple tissues, connective tissue nevi, epidermal nevi, and hyperostoses; due to somatic mosaic *AKT1* pathogenic variant; 90% of cases), multiple orbital neurofibromas (painful peripheral nerve tumors, distinctive face, and marfanoid habitus).

Clinical criteria for NF2 diagnosis were initially developed in 1987, and further updates were made in 1991, 1992, and 1997. Based on these criteria, diagnosis of NF2 requires meeting one of the following: (i) bilateral vestibular schwannomas; (ii) a first-degree relative with NF2 AND unilateral vestibular schwannoma OR any two of the following: meningioma, schwannoma, glioma, neurofibroma, cataract in the form of posterior subcapsular lenticular opacities or cortical wedge cataract; (iii) unilateral vestibular schwannoma AND any two of the following: meningioma, schwannoma, glioma, neurofibroma, cataract in the form of posterior subcapsular lenticular opacities or cortical wedge cataract; (iv) multiple meningiomas AND unilateral vestibular schwannoma OR any two of the following: schwannoma, glioma, neurofibroma, cataract in the form of posterior subcapsular lenticular opacities or cortical wedge cataract [28].

In 2011, Baser proposed a point system to streamline the clinical diagnosis of NF2, with an accumulating number of points equal to or greater than six indicating a definite NF2 (Table 92.1).

Histologically, vestibular schwannoma in NF2 is typically multifocal and multilobulated or botryoid, with each of these tumors or "lobules" harboring admixed cell populations (including characteristic alternating Antoni A and B regions, Verocay bodies, and hyalinized vessels) (Figure 92.2) [29]. Further, grade I meningioma may produce typical whorl formations and psammoma bodies, while grade II meningioma shows increased cellularity, nuclear pleomorphism, and prominent nucleoli and an increased proliferation index (Figure 92.3) [30].

Molecular identification by sequence analysis (75%) and gene-targeted deletion/duplication analysis or CMA (20%) helps reveal heterozygous *NF2* pathogenic variants in affected patients, providing further confirmation of NF2 diagnosis. Usually, after detection of both variant *NF2* alleles in tumor DNA, leukocyte DNA is tested to determine which of the pathogenic variants is constitutional (present in both tumor and leukocyte DNA) and which is somatic (present in tumor DNA only). Identification of a constitutional

TABLE 92.1

The Baser Criteria for Diagnosis of NF2

Feature	If Present at or under 30 Years of Age	If Present over 30 Years of Age
First-degree relative with NF2 diagnosed by these criteria	2	2
Unilateral vestibular schwannoma	2	1 (not given if present over 70 years of age)
Second vestibular schwannoma	4	3 (not given if present over 70 years of age)
One meningioma	2	1
Second meningioma (no additional points for more than two meningiomas)	2	1
Cutaneous schwannomas (one or more)	2	1
Cranial nerve tumor (excluding vestibular schwannoma) (one or more)	2	1
Mononeuropathy	2	1
Cataract (one or more)	2	0

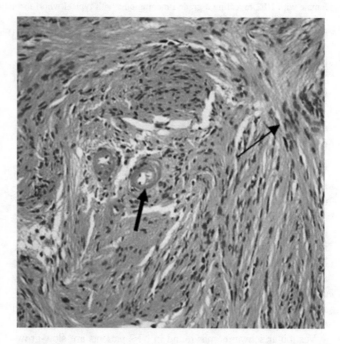

FIGURE 92.2 Vestibular schwannoma from a 20-year-old female with ring chromosome 22 showing nuclear palisading (thin arrow) and hyaline vessel walls (thick arrow). H&E stain, ×300. (Photo credit: Denayer E et al. *BMC Med Genet.* 2009;10:97.)

pathogenic *NF2* variant establishes definite NF2 diagnosis; in the absence of a constitutional pathogenic *NF2* variant, detection of a pathogenic *NF2* variant in leukocytes establishes mosaic NF2 diagnosis, which may occur in 25%–33% of patients. If a pathogenic *NF2* variant found in a NF2 patient is not detected in leukocyte DNA of either parent, it suggests a de novo pathogenic variant in the patient or germline mosaicism in a parent.

Differential diagnoses for NF2 include NF1 (intellectual/learning disability, Lisch nodules, café-au-lait macules, *NF1* mutation), and schwannomatosis (absence of vestibular schwannomas,

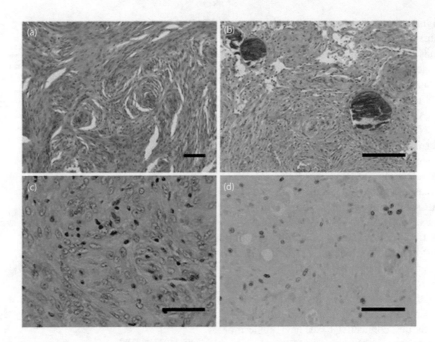

FIGURE 92.3 Histological analysis of tumor specimens recovered by a single image-guided right frontal craniotomy from the brain of a 35-year-old female with NF2 revealing a grade I meningioma with typical whorl formations (a, bar 100 μm) and psammoma bodies (b, bar 100 μm) in the posterior tumor; a grade II meningioma with increased cellularity, nuclear pleomorphism, and prominent nucleoli by H&E staining (c, bar 50 μm) and an increased proliferation index by MIB-1 labeling (d, bar 50 μm) in the anterior tumor. (Photo credit: Dewan R et al. *BMC Cancer.* 2017;17(1):127.)

intradermal schwannoma plaques, cataract, or ependymoma; *SMARCB1* and *LZTR1* mutations), and meningioma (absence of vestibular schwannomas) (refer to Chapter 11 in this book) [1].

92.7 Treatment

Without definitive therapies, current treatment of NF1 and NF2 relies on the use of standard protocols that help alleviate the symptoms and complications associated with these disorders [31–33].

While NF1-related café-au-lait spots and neurofibromas are benign and do not always require treatment, those posing cosmetic/disfiguring concerns may be removed by carbon dioxide laser or other means. As plexiform neurofibromas demonstrate an 8%–13% risk for transformation into MPNST (indicated by change from soft to hard mass, or rapid increase in size), they should be surgically excised when possible. In addition, imatinib may be utilized to decrease plexiform neurofibroma size [34].

Vestibular schwannomas found in NF2 patients are slow-growing and may be treated with surgery (e.g., retrosigmoid, translabyrinthine, middle fossa, and endoscopic approaches) and/or radiation therapy when they become symptomatic. However, these tumors show a postoperative recurrence rate of 44%. Hearing preservation and augmentation should be considered for patients with NF2 [35]. Bevacizumab (a VEGF inhibitor) may be employed to decrease tumor size (in 53% of cases) and improve hearing (in 57% of cases).

92.8 Prognosis

NF1 patients with MPNST (which represents a significant cause of mortality even after radical excision with wide surgical margins, followed by chemoradiation) have a poor 5-year survival rate due to frequent lung and bone metastases as well as local recurrence [36].

Prognosis for NF2 patients is dependent on the age of symptom onset, the degree of hearing deficit, and the number and location of tumors. Overall, affected individuals have 5-, 10-, and 20-year survival rates after diagnosis of 85%, 67%, and 38%, respectively. Presence of vestibular schwannomas and intracranial meningiomas is associated with increased mortality (average age of death at 36 years). However, patients harboring a mosaic, rather than non-mosaic, NF2 mutation suffer reduced mortality. Early diagnosis, appropriate treatment, and regular surveillance are essential for improved patient survival [37].

92.9 Conclusion

Neurofibromatosis consists of three autosomal dominant neurocutaneous syndromes (neurofibromatosis type 1 [NF1], neurofibromatosis type 2 [NF2], and schwannomatosis [SWN]) that produce different tumor spectrums and symptoms and harbor mutations in distinct tumor suppressor genes (*NF1, NF2, SMARCB1,* and *LZTR1*) [38].

Specifically, NF1 typically presents with cutaneous neurofibromas (84%), subcutaneous neurofibromas (78%), plexiform neurofibroma (54%), spinal neurofibroma (24%), optic glioma (11%), non-optic glioma (8.5%), pheochromocytoma (4%), and gastrointestinal stromal tumor (2%), along with skin fold freckling (89%), café-au-lait macules (79%), Lisch nodules (55%), skeletal complication (31%), learning disability (31%), scoliosis (28%), and seizures (5%).

NF2 manifests as vestibular schwannoma (98%), spinal schwannoma (58%), internal schwannoma (37%), cutaneous schwannomas (38%), subcutaneous schwannoma (27%), meningioma (54%), ependymoma (36%), and epiretinal membrane or retinal hamartoma (9%) as well as hearing loss (96%), tinnitus (71%), cataracts (28%), seizures (15%), skeletal complication (11%), scoliosis (11%), and café-au-lait macules (5%) [39,40].

SWN is associated with spinal schwannoma (54%), internal schwannoma (59%), subcutaneous schwannomas (35%), cutaneous schwannomas (18%), cataracts (17%), hearing loss (14%), tinnitus (6%), skeletal complication (6%), scoliosis (6%), and seizures (2%).

Current diagnosis of neurofibromatosis mainly relies on meeting specified clinical criteria, and molecular identification of mutations in the *NF1*, *NF2*, *SMARCB1*, and *LZTR1* genes provides further confirmation of NF1, NF2, and SWN [41,42]. In the absence of definite treatment, management of neurofibromatosis involves the use of standard procedures that help reduce or alleviate the detrimental effects of clinical symptoms associated with NF1, NF2, and SWN.

REFERENCES

1. Kresak JL, Walsh M. Neurofibromatosis: A review of NF1, NF2, and schwannomatosis. *J Pediatr Genet.* 2016;5(2):98–104.
2. Adil A, Singh AK. *Neurofibromatosis Type 1 (Von Recklinghausen).* StatPearls [Internet]. Treasure Island (FL): StatPearls Publishing; 2018 Jan-. 2017 Oct 19.
3. Evans DG. Neurofibromatosis 2. In: Adam MP, Ardinger HH, Pagon RA et al. editors. *GeneReviews®* [Internet]. Seattle (WA): University of Washington; 1993–2018. 1998 Oct 14 [updated 2018 Mar 15].
4. Le C, Bedocs PM. *Neurofibromatosis.* StatPearls [Internet]. Treasure Island (FL): StatPearls Publishing; Jan, 2018. –Jan 9, 2018.
5. Reilly KM, Kim A, Blakely J et al. Neurofibromatosis type 1-associated MPNST state of the science: Outlining a research agenda for the future. *J Natl Cancer Inst.* 2017;109(8).
6. Staedtke V, Bai RY, Blakeley JO. Cancer of the peripheral nerve in neurofibromatosis type 1. *Neurotherapeutics.* 2017;14(2):298–306.
7. Friedman JM. Neurofibromatosis 1. In: Adam MP, Ardinger HH, Pagon RA et al. editors. *GeneReviews®* [Internet]. Seattle (WA): University of Washington; 1993–2018. Oct 2, 1998 [updated 2018 May 17].
8. Ruggieri M, Praticò AD, Serra A et al. Childhood neurofibromatosis type 2 (NF2) and related disorders: From bench to bedside and biologically targeted therapies. *Acta Otorhinolaryngol Ital.* 2016;36(5):345–67.
9. Tovmassian D, Abdul Razak M, London K. The role of [¹⁸F] FDG-PET/CT in predicting malignant transformation of plexiform neurofibromas in neurofibromatosis-1. *Int J Surg Oncol.* 2016;2016:6162182.
10. Howell SJ, Hockenhull K, Salih Z, Evans DG. Increased risk of breast cancer in neurofibromatosis type 1: Current insights. *Breast Cancer (Dove Med Press).* 2017;9:531–6.
11. Karmakar S, Reilly KM. The role of the immune system in neurofibromatosis type 1-associated nervous system tumors. *CNS Oncol.* 2017;6(1):45–60.
12. Kehrer-Sawatzki H, Mautner VF, Cooper DN. Emerging genotype-phenotype relationships in patients with large NF1 deletions. *Hum Genet.* 2017;136(4):349–76.
13. Kiuru M, Busam KJ. The NF1 gene in tumor syndromes and melanoma. *Lab Invest.* 2017;97(2):146–57.
14. Philpott C, Tovell H, Frayling IM, Cooper DN, Upadhyaya M. The NF1 somatic mutational landscape in sporadic human cancers. *Hum Genomics.* 2017;11(1):13.
15. Tate JM, Gyorffy JB, Colburn JA. The importance of pheochromocytoma case detection in patients with neurofibromatosis type 1: A case report and review of literature. *SAGE Open Med Case Rep.* 2017;5:2050313X17741016.
16. Lee JD, Kwon TJ, Kim UK, Lee WS. Genetic and epigenetic alterations of the NF2 gene in sporadic vestibular schwannomas. *PLOS ONE.* 2012;7(1):e30418.
17. Schroeder RD, Angelo LS, Kurzrock R. NF2/merlin in hereditary neurofibromatosis 2 versus cancer: Biologic mechanisms and clinical associations. *Oncotarget.* 2014;5(1):67–77.
18. Cocciadiferro L, Miceli V, Granata OM, Carruba G. Merlin, the product of NF2 gene, is associated with aromatase expression and estrogen formation in human liver tissues and liver cancer cells. *J Steroid Biochem Mol Biol.* 2017;172:222–30.
19. Asthagiri AR, Vasquez RA, Butman JA et al. Mechanisms of hearing loss in neurofibromatosis type 2. *PLOS ONE.* 2012;7(9):e46132.
20. Campian J, Gutmann DH. CNS tumors in neurofibromatosis. *J Clin Oncol.* 2017;35(21):2378–85.
21. Torres Nupan MM, Velez Van Meerbeke A, López Cabra CA, Herrera Gomez PM. Cognitive and behavioral disorders in children with neurofibromatosis type 1. *Front Pediatr.* 2017;5:227.
22. Da Silva AV, Rodrigues FR, Pureza M, Lopes VG, Cunha KS. Breast cancer and neurofibromatosis type 1: A diagnostic challenge in patients with a high number of neurofibromas. *BMC Cancer.* 2015;15:183.
23. Plotkin SR, Bredella MA, Cai W et al. Quantitative assessment of whole-body tumor burden in adult patients with neurofibromatosis. *PLOS ONE.* 2012;7(4):e35711.
24. Ahlawat S, Fayad LM, Khan MS et al. Current whole-body MRI applications in the neurofibromatoses: NF1, NF2, and schwannomatosis. *Neurology.* 2016;87(7 Suppl 1):S31–9.
25. Evans DGR, Salvador H, Chang VY et al. Cancer and central nervous system tumor surveillance in pediatric neurofibromatosis 2 and related disorders. *Clin Cancer Res.* 2017;23(12):e54–61.
26. Zhang J, Li M, Yao Z. Molecular screening strategies for NF1-like syndromes with café-au-lait macules (Review). *Mol Med Rep.* 2016;14(5):4023–9.
27. Corsello G, Antona V, Serra G et al. Clinical and molecular characterization of 112 single-center patients with Neurofibromatosis type 1. *Ital J Pediatr.* 2018;44(1):45.
28. Tiwari R, Singh AK. *Neurofibromatosis type 2.* StatPearls [Internet]. Treasure Island (FL): StatPearls Publishing; Jan, 2018. Nov 30, 2017.
29. Denayer E, Brems H, de Cock P et al. Pathogenesis of vestibular schwannoma in ring chromosome 22. *BMC Med Genet.* 2009;10:97.
30. Dewan R, Pemov A, Dutra AS et al. First insight into the somatic mutation burden of neurofibromatosis type 2-associated grade I and grade II meningiomas: A case report comprehensive genomic study of two cranial meningiomas with vastly different clinical presentation. *BMC Cancer.* 2017;17(1):127.
31. Agarwal R, Liebe S, Turski ML et al. Targeted therapy for hereditary cancer syndromes: Neurofibromatosis type 1, neurofibromatosis type 2, and Gorlin syndrome. *Discov Med.* 2014;18(101):323–30.

32. Karajannis MA, Ferner RE. Neurofibromatosis-related tumors: Emerging biology and therapies. *Curr Opin Pediatr.* 2015;27(1):26–33.

33. Blakeley JO, Plotkin SR. Therapeutic advances for the tumors associated with neurofibromatosis type 1, type 2, and schwannomatosis. *Neuro Oncol.* 2016;18(5):624–38.

34. Avery RA, Katowitz JA, Fisher MJ et al. Orbital/periorbital plexiform neurofibromas in children with neurofibromatosis type 1: Multidisciplinary recommendations for care. *Ophthalmology.* 2017;124(1):123–32.

35. Monteiro TA, Goffi-Gomez MV, Tsuji RK, Gomes MQ, Brito Neto RV, Bento RF. Neurofibromatosis 2: Hearing restoration options. *Braz J Otorhinolaryngol.* 2012;78(5):128–34.

36. Byrne S, Connor S, Lascelles K, Siddiqui A, Hargrave D, Ferner RE. Clinical presentation and prognostic indicators in 100 adults and children with neurofibromatosis 1 associated non-optic pathway brain gliomas. *J Neurooncol.* 2017;133(3):609–14.

37. Plana-Pla A, Bielsa-Marsol I, Carrato-Moñino C et al. Diagnostic and prognostic relevance of the cutaneous manifestations of neurofibromatosis type 2. *Actas Dermosifiliogr.* 2017;108(7):630–6.

38. Hanemann CO, Blakeley JO, Nunes FP et al. Current status and recommendations for biomarkers and biobanking in neurofibromatosis. *Neurology.* 2016;87(7 Suppl 1):S40–8

39. Gugel I, Mautner VF, Kluwe L, Tatagiba MS, Schuhmann MU. Cerebrovascular insult as presenting symptom of neurofibromatosis type 2 in children, adolescents, and young Adults. *Front Neurol.* 2018;9:733.

40. Amer SM, Ukudeyeva A, Pine HS, Campbell GA, Clement CG. Plexiform schwannoma of the tongue in a pediatric patient with neurofibromatosis type 2: A case report and review of literature. *Case Rep Pathol.* 2018;2018:9814591.

41. Evans DGR, Salvador H, Chang VY et al. Cancer and central nervous system tumor surveillance in pediatric neurofibromatosis 1. *Clin Cancer Res.* 2017;23(12):e46–53.

42. Castellanos E, Plana A, Carrato C et al. Early genetic diagnosis of neurofibromatosis type 2 from skin plaque plexiform schwannomas in childhood. *JAMA Dermatol.* 2018;154(3):341–6.

93

Noonan Syndrome

Dongyou Liu

CONTENTS

93.1 Introduction

Noonan syndrome is an autosomal dominant disorder characterized by short stature, facial dysmorphism, and congenital heart defects. Molecular mechanisms underlying the development of Noonan syndrome relate to genetic mutations that compromise the functions of the RAS/mitogen-activated protein kinase (MAPK) pathway, with marked developmental and tumorigenic consequences. For this reason, Noonan syndrome and its phenotypic variants, including Noonan syndrome with multiple lentigines (NSML, formally LEOPARD syndrome), Noonan-like syndrome with loose anagen hair (NS/LAH), CBL syndrome (or Noonan syndrome-like disorder with or without juvenile myelomonocytic leukemia [JMML]), together with several other disorders, including cardio-facio-cutaneous syndrome (CFCS), Costello syndrome, Legius syndrome, neurofibromatosis type 1 (NF1), and capillary malformation–arteriovenous malformation syndrome (CM-AVM), are collectively referred to as RASopathies (or neuro-cardio-facio-cutaneous syndromes [NCFCS]) (see Chapter 87 Table 87.1 in this book for further details) [1–3].

93.2 Biology

Noonan syndrome was named in honor of Jacqueline Noonan, who described in 1963 a putatively novel disorder involving nine patients displaying pulmonic valve stenosis, small stature, dysmorphic facial appearance (i.e., hypertelorism, ptosis, and low-set ears), webbed neck, chest deformities, and cryptorchidism

[4]. Subsequent reports have expanded the clinical spectrum of Noonan syndrome, including (i) distinctive facial features (broad forehead, hypertelorism, downslanting palpebral fissures, a high-arched palate, and low-set posteriorly rotated ears) (Figure 93.1), (ii) congenital heart defects (pulmonic stenosis and hypertrophic cardiomyopathy), (iii) multiple skeletal defects (chest and spine deformities), and (iv) other abnormalities (e.g., webbed neck, mental retardation, cryptorchidism, and bleeding diathesis) [5]. In addition, several phenotypic variants of Noonan syndrome have been identified, including NSML (formally known as LEOPARD syndrome), NS/LAH, and CBL syndrome (or Noonan syndrome-like disorder with or without JMML).

Recent application of molecular techniques has uncovered germline mutations in *PTPN11* (12q24.1; about 50% of cases), *SOS1* (2p22.1; 10%), *RAF1* (3p25.1; 10%), *RIT1* (1q22; 5%), *KRAS* (12p12.1; 2%), *NRAS* (1p13.2; 1%), *BRAF* (7q34), *SOS2* (14q21.3), *LZTR1* (22q11.21), *A2ML1* (12p13)] and other genes that contribute to the development of Noonan syndrome [6,7]. Furthermore, mutations in *PTPN11* (12q24.1), *RAF1* (3p25.1) and *BRAF* (7q34) have been observed from cases of NSML (formerly LEOPARD syndrome), *SHOC2* (10q25.2) from cases of NS/LAH, and *CBL* (11q23.3) from cases of CBL syndrome (or Noonan syndrome-like disorder with or without JMML) (see Chapter 87 in this book), yielding valuable insights into the genetic bases of Noonan syndrome phenotypic variability.

93.3 Pathogenesis

RASopathies are a group of developmental disorders associated with germline mutations in genes that encode components or

743

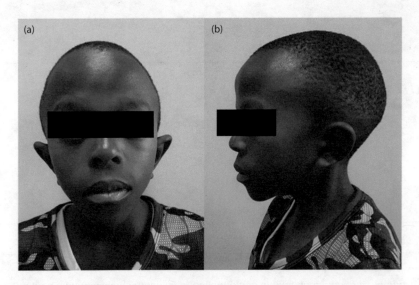

FIGURE 93.1 Craniofacial features of an 11-year-old boy with Noonan syndrome harboring PTPN11 c.1510A>G (p.Met504Val) pathogenic variant. (a) Frontal views showing a triangular face with pointed chin; tall forehead; bilateral ptosis, predominantly on the right; sparse eyebrows; epicanthic folds, and protruding ears. (b) Lateral view showing high anterior hairline with low-set posteriorly rotated ears. (Photo credit: Tekendo-Ngongang C et al. *Front Genet.* 2019;10:333.)

regulators of the RAS/MAPK pathway. Given its critical role in the regulation of cell cycle, growth, differentiation, and senescence, genetic alterations in the RAS/MAPK pathway genes cause its dysregulation, leading to several congenital disorders that share overlapping characteristics, such as craniofacial dysmorphism, cardiac malformations, cutaneous, musculoskeletal, and ocular abnormalities, neurocognitive impairment, hypotonia, and increased cancer risk.

Being the most common RASopathy, Noonan syndrome results from germline mutations in the *PTPN11, SOS1, SOS2, KRAS, NRAS, RIT1, RAF1, LZTR1,* and *A2ML1* genes, all of which encode key components of the RAS/MAPK pathway, profoundly affecting multisystem development (Table 93.1) [3].

The *PTPN11* (protein-tyrosine phosphatase, nonreceptor-type II) gene is organized into 15 exons, and encodes the Src homology-2 (SH2) domain-containing protein tyrosine phosphatase (PTP) 2 (SHP2), which is a non-receptor phosphotyrosine phosphatase involved in the cellular response to growth factors, hormones, cytokines, and cell adhesion molecules. Structurally, SHP2 comprises two tandemly arranged SH2 domains at the N terminus (N-SH2 and C-SH2), a central PTP catalytic domain, a C-terminal tail with two tyrosine phosphorylation sites, and a proline-rich motif. While the N-SH2 domain binds and inhibits phosphatase or binds phosphoproteins and activates enzymes, the C-SH2 domain provides binding energy and specificity without a direct role in activation. In the inactive state, the backside loop of the N-SH2 domain folds into the PTP catalytic pocket. Binding of phosphotyrosyl (pY) proteins (e.g., receptor tyrosine kinase, platelet-derived growth factor [PDGF], cytokine receptor, and scaffolding adaptor) to the SH2 domains unfolds the structure, allowing substrate access to the catalytic pocket and initiating signaling events that control proliferation, differentiation, and apoptosis within the cell. Besides its role as a phosphatase, SHP2 may function as a scaffold for recruitment of the GRB2/SOS complex to the cell membrane through its Tyr542 site in response to PDGF, providing another potential mechanism for SHP2 regulation of RAS. *PTPN11* mutation disrupts the normal

function of SHP2 and causes dysregulation of growth factor and cytokine-mediated RAS/ERK/MAPK and PI3 K/AKT signal flow, perturbing cell proliferation [8]. Found in 50% of Noonan syndrome cases, most (95%) of pathogenic *PTPN11* variants are missense mutations that alter residues near the N-SH2 and PTP interacting surfaces, disrupting the stability of the catalytically inactive form of SHP2, impairing SHP2's ability to switch from the active to the inactive conformation, and inducing catalytic activation and gain of function in the RAS/MAPK pathway. In addition, about 5% of pathogenic *PTPN11* variants alter SHP2 sensitivity to activation from binding partners [9].

The *SOS1* (son of sevenless homolog 1) gene consists of 23 exons and encodes a 150-kd multidomain protein (SOS1) that functions as a RAS-specific guanine nucleotide exchange factor (GEF), stimulating the conversion of RAS from the inactive guanosine diphosphate (GDP)-bound form to the active guanosine triphosphate (GTP)-bound form. Structurally, SOS1 comprises a RAS-GEF domain, a conserved histone-like fold, Dbl homology (DH) and plekstrin homology (PH) domains, a helical linker, a RAS exchange motif (REM), and a proline-rich region. Binding of growth hormone to receptor tyrosine kinase (RTK) recruits SOS1 to the plasma membrane, which then converts RAS-GDP to RAS-GTP. Affecting 10%–15% of Noonan syndrome patients, *SOS1* missense mutations (often located in codons encoding residues responsible for stabilizing the protein in an inhibited conformation) destabilize the DH domain and disrupt the autoinhibition of SOS1 RasGEF activity, contributing to increased and prolonged RAS activation and downstream signaling.

The *RAF1* gene consists of 17 exons and encodes a serine/threonine kinase RAF1 (or CRAF) of 648 aa that constitutes one of the direct downstream RAs effectors. RAF1 contains three conserved regions (CR1-3): CR1 (encoded by exons 2-5) has a RAS-binding domain (RBD) and a cysteine-rich domain (CRD); CR2 (encoded by exon 7) contains several serine residues (p.Arg256, p.Ser257, p.Ser259, and p.Pro261) which are consensus 14-3-3 recognition sites and frequently targeted for de/phosphorylation, leading to activation or inactivation (e.g.,

TABLE 93.1

Genetic Characteristics of Noonan Syndrome and Related Disorders

Syndrome	Gene (Chromosome Location)	Protein (Function)	Notable Nucleotide Change (Predicted Protein Change)	Pathogenesis
Noonan syndrome	*PTPN11* (12q24.1)	SHP2 (phosphatase)	c.922A>G (p.Asn308Asp) c.179_181delGTG (p.Gly60del) c.181_183delGAT (p.Asp61del) among many others	*PTPN11* mutations activate tyrosine-protein phosphatase non-receptor type II, which in turn stimulates epidermal growth factor-mediated RAS/ERK/MAPK activity and increases cell proliferation; detected in 50% of cases
	SOS1 (2p22.1)	SOS1 (guanine exchange factor)	c.305C>G (p.Pro102Arg) c.508A>G (p.Lys170Glu) c.797C>A (p.Thr266Lys) c.1297G>A (p.Glu433Lys) c.1642A>C (p.Ser548Arg) c.1867 T>A (p.Phe623Ile) among many others	*SOS1* pathogenic variants abrogate autoinhibition, and thus enhance RAS-GEF activity and prolong RAS activation and downstream signaling; detected in 10% of cases
	RAF1 (3p25.1)	RAF1/CRAF (kinase)	c.766A>G (p.Arg256Gly) c.775T>C (p.Ser259Pro) c.782 C>G (p. Pro261Arg) c.788T>G (p.Val263Gly) c.1457A>G (p.Asp486Gly) c.1472C>T (p.Thr491Ile) amaong many others	*RAF1* pathogenic variants enhance RAS-GEF activity and reduce autoinhibition, thus increasing and prolonging RAS activation and downstream signaling; detected in 5% of cases
	RIT1 (1q22)	RIT1 (GTPase)	c.104G>C (p.Ser35Thr) c.170C>G (p.Ala57Gly) c.246T>G (p.Phe82Leu) c.270G>C (p.Met90Ile) c.284G>C (p.Gly95Ala) among many others	*RIT1* gain-of-function variants increase MEK-ERK signaling; detected in 5% of cases
	KRAS (12p12.1)	KRAS (GTPase)	c.458A > T (p.D153V)	Mutated K-RAS protein induces hypersensitivity of primary hematopoietic progenitor cells to growth factors and deregulates the RAS-RAF-MEK-ERK pathway; strong gain-of-function *KRAS* pathogenic variants may be incompatible with life; detected in <5% of cases
	NRAS (1p13.2)	NRAS (GTPase)	c.71T>A (p.Ile24Asn) c.101C>T (p.Pro34Leu) c.149C>T (p.Thr50Ile) c.179G>A (p.Gly60Glu)	*NRAS* pathogenic variants such as p.Thr50Ile or p.Gly60Glu enhance phosphorylation of MEK and ERK and thus promote cell proliferation, differentiation, or survival
	BRAF (7q34),	*BRAF* (kinase)	c.722C>T (p.Thr241Met) c.722C>G (p.Thr241Arg) c.1593G>C (p.Trp531Cys) c.1789C>G (p.Leu597Val)	*BRAF* pathogenic variants increase phosphorylation and activation of the dual specificity mitogen-activated protein kinases (MEK1 and MEK2)

(Continued)

TABLE 93.1 (*Continued*)

Genetic Characteristics of Noonan Syndrome and Related Disorders

Syndrome	Gene (Chromosome Location)	Protein (Function)	Notable Nucleotide Change (Predicted Protein Change)	Pathogenesis
	SOS2 (14q21.3)	SOS2 (guanine exchange factor)	c.800T>A (p.Met267Lys) c.1127C>G (p.Thr376Ser)	SOS2 shares 70% homology with SOS1; its pathogenic variants are expected to behave like those of SOS1, with gain-of-function mutations leading to enhanced signaling and upregulation of the RAS/MAPK pathway
	LZTR1 (22q11.21)	LZTR1 (leucine-zipper-like transcription regulator 1, BTB-kelch protein)	c.356A>G (p.Tyr119Cys) c.740G>A (p.Ser247Asn) c.742G>A (p.Gly248Arg) c.850C>T (p.Arg284Cys) c.859C>T (p.His287Tyr)	Missense heterozygous variants in LZTR1 may increase signal flow through Ras/MAPK pathway
	A2ML1 (12p13)	A2ML1 (secreted protease inhibitor α-macroglobulin)	c.2718_2719del (p.Gly907Serfs*20)	A2ML1 mutation may induce conformational changes that might have a dominant-negative effect or lead to gain of function of the complex
Noonan syndrome with multiple lentigines (NSML, formerly LEOPARD syndrome)	PTPN11 (12q24.1)	SHP2 (phosphatase)	c.836A>G (p.Tyr279Cys) c.1381G>A (p.Ala461Thr) c.1391G>C (p.Gly464Ala) c.1403C>T (p.Thr468Met) c.1493G>T (p.Arg498Trp) c.1517A>C (p.Gln506Pro) c.1528C>G (p.Gln510Glu)	PTPN11 mutations activate tyrosine-protein phosphatase non-receptor type II, which in turn stimulates epidermal growth factor-mediated RAS/ERK/MAPK activity and increases cell proliferation; detected in 90% of case
	RAF1 (3p25.1)	RAF1/CRAF (kinase)	c.770C>T (p.Ser257Leu) c.1837C>G (p.Leu613Val)	RAF1 pathogenic variants enhance RAS-GEF activity and reduce autoinhibition, thus increasing and prolonging RAS activation and downstream signaling; detected in <5% of cases
	BRAF (7q34)	BRAF (kinase)	c.721A>C (p.Thr241Pro) c.735A>G (p.Leu245Phe)	BRAF pathogenic variants increase phosphorylation and activation of the dual specificity mitogen-activated protein kinases (MEK1 and MEK2); detected in 2 individuals
Noonan syndrome-like disorder with loose anagen hair (NS/LAH)	SHOC2 (10q25.2)	SHOC2 (scaffold protein)	c.4A>G (p.Ser2Gly) c.519G>A (p.Met173Ile)	SHOC2 is a scaffold protein in the ERK1/2 pathway, which tethers RAS, RAF-1 and the catalytic subunit of protein phosphatase 1c (PP1c). p.Ser2Gly mutation creates a new recognition site for N-terminal myristilation, causing aberrant targeting of SHOC2 protein to the plasma membrane and impaired translocation to the nucleus upon stimulation with growth factor; M173I mutation leads to changes in the assembly of the SHOC2-RAS-RAF-1-PP1c complex
CBL syndrome (or Noonan syndrome-like disorder with or without juvenile myelomonocytic leukemia)	CBL (11q23.3)	CBL (E3 ubiquitin ligase)	c.1100A>C (p.Gln367Pro) c.1111T>C (p.Tyr371His) c.1186T>C (p.Cys396Arg) c.1199T>G (p.Met400Arg) c.1201T>C (p.Cys401Arg) c.1259G>A (p.Arg420Gln) among others	CBL pathogenic variants inhibit FMS-like tyrosine kinase 3 (FLT3) internalization and ubiquitylation, induce cytokine-independent growth and constitutive phosphorylation of ERK, AKT, and S6; detected in 9% of patients with juvenile myelomonocytic leukemia (JMML)

dephosphorylation of p.Ser259 facilitates binding of *RAF1* to RAS-GTP and propagation of the signal through the RAS-MAPK cascade via *RAF1 MEK* kinase activity), and CR3 (encoded by exons 10-17) is a kinase domain with an activation segment. Detected in 10% of Noonan syndrome cases, *RAF1* pathogenic variants (located in conserved region 2 flanking S259 and conserved region 3 surrounding the activation segment) increase and prolong RAS activation and downstream signaling through enhanced RAS-GEF activity and reduced 14-3-3 binding and autoinhibition.

The *RIT1* (RAS-like protein in tissues) gene contains six exons and encodes a RAS GTPase (RIT1), which shares a 50% sequence homology with RAS but lacks the C-terminal CAAX motif (lipidation site). Found in 5% of Noonan syndrome cases, *RIT1* pathogenic variants are missense mutations that cause gain-of-function in the MEK-ERK signaling [10,11].

The *KRAS* gene spans 45 kb with four exons and encodes a GTPase (K-RAS), which regulates the RAS-RAF-MEK-ERK pathway, through cycling between inactive GDP-bound and GTP-bound conformations. Observed in 5% of Noonan syndrome cases, *KRAS* pathogenic variants reduce the intrinsic and GAP-stimulated GTPase activity or interfere with the binding of KRAS and guanine nucleotides, leading to a gain of function in the RAS/MAPK pathway [12].

The *NRAS* gene has seven exons and encodes a GTPase (N-RAS) that in its active form can interact with RAF, phosphatidylinositol-3-kinase (PI3 K), RAL-GDP dissociation simulator (GDS) and other effectors. *NRAS* mutations within or near the switch II region may interfere with GTPase function, and enhance phosphorylation of MEK and ERK, leading to the RAF-MEK-MAPK cascade activation [13].

The *BRAF* gene spans 190 kb with 18 exons and encodes a serine/threonine protein kinase (B-RAF) of 766 aa, which consists of three conserved regions (CR): CR1 containing the RAS-binding domain and a cysteine-rich domain (exons 3-6), CR2b being the smallest of the conserved regions, and CR3 being the kinase domain containing a glycine-rich loop (exon 11) and an activation segment (exon 15) of the catalytic domain. Through its downstream effectors, mitogen-activated protein kinase 1 and 2 (MEK1 and MEK2), B-RAF influences cell proliferation, differentiation, motility, apoptosis, and senescence. *BRAF* pathogenic variants associated with classic Noonan syndrome are p.Thr241Met and p.Thr241Arg (exon 6), p.Trp531Cys (exon 13), and p.Leu597Val (exon 15), while those reported in NSML are p.Thr241Pro and p.Leu245Phe (exon 6).

Besides the above-mentioned genes, several other genes (e.g., *SOS2, RASA2, RRAS, SYNGAP1, A2ML1, LZTR1, MYST4, SPRY1, MAP3K8*) have been recently implicated in the development of Noonan syndrome. It should be noted that while germline mutations in these genes are responsible for Noonan syndrome and its phenotypic variants, somatic nucleotide variants in *PTPN11, KRAS, NRAS, BRAF*, or *MAP2K1* are sometimes detected in sporadic tumors (including leukemia and solid tumors) that tend to occur as single tumors without any other findings of Noonan syndrome. In the latter cases, predisposition to these tumors is not heritable. Furthermore, these somatic mutations appear to be more strongly activating than germline counterparts, as activating oncogenic mutations are not tolerated in the germline or in early development.

NSML (formerly LEOPARD syndrome) resembles Noonan syndrome phenotypically, but demonstrates some unique phenotypic features (e.g., a high prevalence of valve defects compared to classic Noonan syndrome). Germline mutations in *PTPN11, RAF1*, and *BRAF* appear to be responsible for the development of NSML. In vitro experiments indicate that *PTPN11* mutations associated with NSML show decreased and/or absent PTP catalytic activity, leading to a loss-of-function phosphatase, whereas *PTPN11* mutations associated with classic Noonan syndrome produce a gain-of-function phosphatase. Furthermore, NSML pathogenesis appears to result from excessive AKT/mTOR activity instead of increased ERK/MAPK activity as in classic Noonan syndrome. This finding suggests that cardiac abnormalities (e.g., hypertrophic cardiomyopathy [HCM]) associated with NSML may be treated and reversed by a TOR inhibitor such as rapamycin [14,15].

NS/LAH has similar features to Noonan syndrome, but differs from the latter by its growth hormone deficiency, distinctive hyperactive behavior, loose anagen hair, darkly pigmented skin with eczema or ichthyosis, hypernasal voice, and more mitral valve dysplasia and cardiac septal defects. Molecularly, NS/LAH is attributed to germline mutations (e.g., 4A→G, p.Ser2Gly) in the *SHOC2* gene, which encodes a protein (SHOC2) consisting of almost entirely of leucine-rich repeats. Acting as a scaffold protein linking RAS to RAF1, SHOC2 binds RasGTP and mediates protein phosphatase 1C (PP1C) translocation to the cell membrane, facilitating PP1C dephosphorylation of residue S259 of RAF1. SHOC2 p.Ser2Gly mutation erroneously adds a 14-carbon saturated fatty acid chain, myristate, to the N-terminal glycine of SHOC2, and causes aberrant translocation of SHOC2 to the cell membrane instead of the nucleus, thus prolonging PP1C dephosphorylation of RAF1 and sustaining the MAPK pathway activation [16].

CBL syndrome (or Noonan syndrome-like disorder with or without JMML) is caused by germline mutations in the *CBL* gene encoding a ubiquitously expressed E3 ubiquitin ligase that mediates the association of ubiquitin with activated RTK, which is necessary for receptor internalization and degradation. *CBL* mutations reduce the turnover of activated RTK and increase ERK activation in the RAS/MAPK pathway. Apart from facial features, impaired growth, developmental delay, and cryptorchidism that are also observed in Noonan syndrome, CBL syndrome is associated with an enlarged left atrium, transient chaotic ventricular dysrhythmia, delayed brain myelination, cerebellar vermis hypoplasia, bicuspid aortic valve with stenosis, average adult height, mitral valve insufficiency, and clear predisposition to JMML.

93.4 Epidemiology

93.4.1 Prevalence

Noonan syndrome is reported to occur in 1:1000–1:2500, although patients with mild expression may be overlooked or underdiagnosed.

93.4.2 Inheritance

Noonan syndrome shows an autosomal dominant inheritance, in which one copy of the altered gene in each cell is sufficient to cause the disorder. Each child of an individual with Noonan syndrome has a 50% chance of inheriting the pathogenic variant. While some cases are due to inheritance of mutation from one affected parent, others result from new mutations in the gene and occur in people with no history of the disorder in their family.

93.4.3 Penetrance

Noonan syndrome demonstrates a variable penetrance and clinical expression, and its features may become less pronounced with increasing age. Indeed, many affected adults are diagnosed only after the birth of an infant who displays more obvious clinical manifestations.

93.5 Clinical Features

Noonan syndrome typically manifests as characteristic facies (broad/high forehead, hypertelorism, epicanthic folds, downward-slanting palpebral fissures, low-set posteriorly rotated ears with a thick helix, high-arched palate, micrognathia, short neck with excess nuchal skin, low posterior hairline, coarse or myopathic face in childhood, triangular face with age, prominent eyes, ptosis, thick lips with prominent nasolabial folds, marked webbing or prominent trapezius), short stature, congenital heart defect (pulmonary valve stenosis with dysplastic leaflets [50%], hypertrophic obstructive cardiomyopathy with asymmetrical septum hypertrophy [20%], atrial and ventricular septal defects [10%], persistent ductus arteriosus [3%], branch pulmonary artery stenosis, and tetralogy of Fallot), broad or webbed neck, unusual chest shape with superior pectus carinatum and inferior pectus excavatum, broad thorax, large inter-nipple distance, cryptorchidism, coagulation defect, lymphatic vessel dysplasia (hypoplasia, or aplasia, leading to generalized lymphedema, peripheral lymphedema, pulmonary lymphangiectasia, or intestinal lymphangiectasia, 20%), abnormal pigmentation (pigmented nevi [25%], cafe-au-lait spots [10%], and lentigines [3%]), ocular abnormalities (strabismus [48%–63%], refractive errors [61%], amblyopia [33%], nystagmus [10%]), developmental delay (delay in puberty onset and average bone age by 2 years, but normal female fertility), mild intellectual disability (language impairment, mean IQ of 85), orthopedic abnormalities (cubitus valgus [50%], radioulnar synostosis [2%], clinobrachydactyly [30%], joint hyperextensibility [50%], talipes equinovarus [12%]), urinary tract malformations (pyeloureteral stenosis and/or hydronephrosis, 10%), hearing loss (due to otitis media), behavioral problems (clumsiness, fidgety or stubborn spells, echolalia, irritability, mild motor delay [due to muscular hypotony]), horseshoe kidney, thrombocytopenia, feeding difficulty secondary to pylorospasm, respiratory failure with pulmonary capillaritis and vasculitis, and eightfold increased risk in children and young adults for cancers (e.g., dysembryoplastic neuroepithelial tumors, neuroblastoma, rhabdomyosarcoma, acute lymphoblastic leukemia, and JMML (Figures 93.1 and 93.2) [2,17–25].

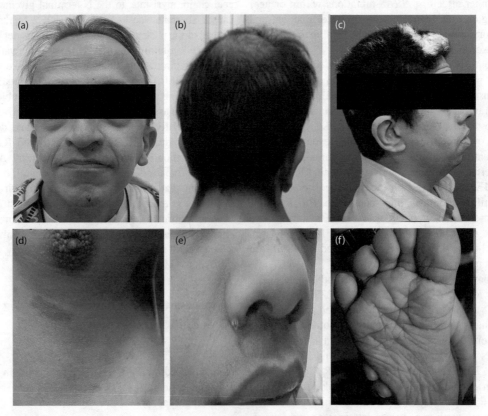

FIGURE 93.2 Tegumentary manifestations of Noonan syndrome and Noonan syndrome with loose anagen hair (NS/LAH). (a and b) A patient with NS/LAH and SHOC2 (p.S2G) mutation showing thin hair and premature baldness. (c) A patient with Noonan syndrome and PTPN11 (p.G60A) mutation showing premature graying of the hair. (d) A patient with Noonan syndrome and PTPN11 (p.T468 M) mutation showing café-au-lait spots and prominent pectus excavatum. (e and f) A patient with Noonan syndrome KRAS (p.K5E) mutation showing nasal papillomas and marked plantar creases. (Photo credit: Quaio CR et al. *Clinics (Sao Paulo). 2013;68(8):1079–83.*)

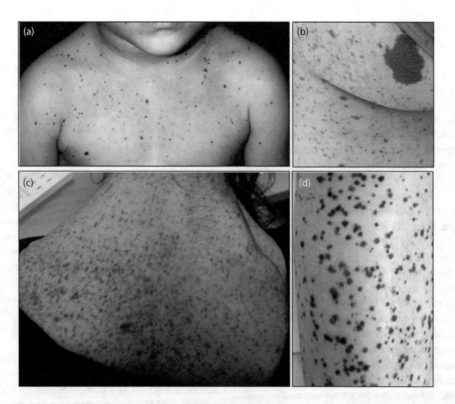

FIGURE 93.3 Cutaneous presentations of Noonan syndrome with multiple lentigines (NSML, formally LEOPARD syndrome). (a) Lentigines in the upper trunk of a 2-year-old child with a PTPN11 mutation. (b) Lentigines and a large cafè-au-lait spot in a 28-year-old female. (c) Lentigines over the neck and back of a 28-year-old female; note the pterygium colli. (d) Multiple lentigines on the lower leg of a 37-year-old male. (Photo credit: Sarkozy A et al. *Orphanet J Rare Dis.* 2008;3:13.)

NSML (previously LEOPARD syndrome) is a Noonan syndrome phenotype showing *l*entigines—dispersed flat, black-brown macules, mostly on the face, neck and upper part of the trunk with sparing of the mucosa (70%–80%) (Figure 93.3), *e*lectrocardiographic conduction defects—abnormalities of the electrical activity of the heart such as hypertrophic cardiomyopathy (HCM, 70% compared to 25% in classic Noonan syndrome), *o*cular hypertelorism—widely spaced eyes and ptosis, *p*ulmonary stenosis—obstruction of the normal outflow of blood from the right ventricle of the heart (25%), *a*bnormalities of the genitalia, *r*etarded/slowed growth—short stature (<50%) and pectus deformity, and *d*eafness, along with mild increase in cancer risk (e.g., acute leukemias and myeloproliferative disorders in 2% of childhood cases). It should be noted that several pigmentary findings (e.g., nevi [25%], café-au-lait patches [10%], and lentigines [3%]) in Noonan syndrome overlap with NSML [15,26,27].

NS/LAH displays similar facial features (hypertelorism, ptosis, downslanting palpebral fissures, low-set posteriorly angulated ears, overfolded pinnae) to Noonan syndrome, along with short stature, growth hormone deficiency, cognitive deficits, relative macrocephaly, small posterior fossa resulting in Chiari I malformation, hypernasal voice, cardiac defects (especially dysplasia of the mitral valve and septal defects), darkly pigmented skin with eczema or ichthyosis, ectodermal abnormalities (easily pluckable, sparse, thin, slow-growing hair [loose anagen hair]), and mildly increased cancer risk (e.g., myelofibrosis, neuroblastoma) (Figure 93.2) [16,28].

CBL syndrome (or Noonan syndrome-like disorder with or without JMML) often causes fetal hydrops, fetal pleural effusions, hydrothorax, chylothorax, cryptorchidism, facial dysmorphism, café-au-lait spots, and predisposition to JMML. Compared to classic Noonan syndrome, CBL syndrome has a relatively high frequency of neurologic features, predisposition to JMML, and a low prevalence of cardiac defects, reduced growth, and cryptorchidism.

93.6 Diagnosis

Diagnosis of Noonan syndrome relies on observation of key clinical features, while detection of specific gene mutations provides further confirmation.

Individuals showing the following key features should be suspected of Noonan syndrome: (i) characteristic facies (low-set, posteriorly rotated ears with fleshy helices; vivid blue or blue-green irises; wide-spaced, downslanted eyes with epicanthal folds, fullness or droopiness of the upper eyelids [ptosis]), (ii) short stature, (iii) congenital heart defect (pulmonary valve stenosis, atrial septal defect, and/or hypertrophic cardiomyopathy), (iv) developmental delay, (v) broad or webbed neck, (vi) unusual chest shape with superior pectus carinatum, inferior pectus excavatum, (vii) widely set nipples, (viii) cryptorchidism in males, (ix) coagulation defects (abnormal prothrombin time, activated partial thromboplastin time, platelet count, and bleeding time; von Willebrand disease, thrombocytopenia, coagulation factor defects [factors V, VIII, XI, XII, protein C], and platelet dysfunction); (x)lymphatic dysplasias of the lungs, intestines, and/ or lower extremities.

Identification of germline heterozygous pathogenic variants in *PTPN11* (50%), *SOS1* (10%), *RAF1* (5%), *RIT1* (5%), *KRAS* (<5%), *NRAS* (<2%), *BRAF* (<2%), and *MAP2K1* (<2%) by sequence analysis helps establish the diagnosis of Noonan syndrome. In addition, germline pathogenic variants in *PTPN11* (93%), *RAF1* (93%), and *BRAF* (two patients) are detected in NSML (formerly LEOPARD syndrome), germline pathogenic variants (e.g., 4A>G) in *SHOC2* are found in NS/LAH, and germline pathogenic variants in *CBL* are detected in CBL syndrome (or Noonan syndrome-like disorder with or without JMML) [29].

Differential diagnoses of Noonan syndrome and its phenotypic variants include (i) CFCS (severe intellectual disability, failure to thrive, short stature, structural central nervous system anomalies, congenital heart defects, long-lasting feeding difficulty, round/bulbous nasal tip, wide nasal base, full lips, sparse eyebrows and lashes, coarse facial appearance, follicular hyperkeratosis; heterozygous missense mutations in *BRAF* [~75%], *MAP2K1/MAP2K2* [~25%], and *KRAS* [<3%]), (ii) Costello syndrome (high birthweight and short stature, developmental delay, coarse facial features, wide nasal bridge, loose and soft skin, increased pigmentation over time, deep palmar and plantar creases, facial or perianal papillomata, premature ageing and hair loss, moderate intellectual disability, flexion or ulnar deviation of the wrist and fingers, pulmonic stenosis and hypertrophic cardiomyopathy, supraventricular or paroxysmal tachycardia, chaotic atrial rhythm or multifocal atrial tachycardia, or ectopic atrial tachycardia; risk for rhabdomyosarcoma, neuroblastoma, and bladder carcinoma; germline pathogenic variants in *HRAS* exon 2), (iii) NF1 (with similar short stature, learning difficulties, café-au-lait patches, and facial appearance; pathogenic variant in *NF1*), (iv) Turner syndrome (45, X0; a chromosomal abnormality in girls; widely spaced, down-slanting eyes, short webbed neck, widely spaced nipples, and shield-like chest; left-sided heart lesions instead of right-sided lesions in Noonan syndrome) [30], (v) Aarskog syndrome (also known as faciodigitogenital syndrome; developmental delay, short stature, distinctive facies, absence of congenital heart disease, affected boys showing a shawl scrotum; an X-linked disorder caused by *FGD1* mutation).

93.7 Treatment

In the absence of an effective cure for Noonan syndrome, standard procedures are utilized to relieve symptoms that compromise the well-being of affected individuals, and to prevent signs that have potential to deteriorate and endanger patient's life. For example, patients with congenital heart defects may be treated with certain medications, surgical intervention, and other techniques, depending upon the location and severity of anatomical abnormalities and their associated symptoms; those with developmental disabilities are addressed by appropriate intervention programs and education strategies; those with serious bleeding due to specific factor deficiency or platelet aggregation anomaly are handled accordingly; those with short stature may be treated with growth hormone; and males with cryptorchidism may undergo surgery between 12 and 24 months of age to move undescended testes into the scrotum (orchiopexy) and reduce the risk of infertility [31,32].

93.8 Prognosis and Prevention

With early diagnosis and appropriate medical intervention, the majority of children with Noonan syndrome will grow up and function normally in adulthood. As cardiac defects, which occur in up to 75% of cases, represent a major cause of death for people with the disorder, the type and severity of cardiac disease may have a bearing on the prognosis.

Noonan syndrome–affected patients with bleeding defects should avoid aspirin, which may exacerbate a bleeding diathesis; those with hypertrophic cardiomyopathy should avoid exacerbating a cardiac condition during growth hormone treatment.

Regular monitoring (involving cardiology evaluation, measurement of weight, length, and head circumference, eye and hearing examination, assessment of growth, development, and social adaptation, review of speech and school performance, etc.) at the neonatal period (birth to 1 month), infancy (1 month to 1 year), early childhood (1 to 5 years), late childhood (5 to 13 years), and adolescence and adulthood (13 to 21 years and older) should form part of care for Noonan syndrome patients and their families.

Preimplantation and prenatal genetic diagnosis using molecular genetic techniques may be offered to prospective parents who harbor a Noonan syndrome–related pathogenic variant. High-resolution ultrasound examination of pregnancies at 50% risk for Noonan syndrome may reveal polyhydramnios, hydronephrosis, pleural effusion, edema, cardiac defects, distended jugular lymphatic sacs, cystic hygroma, and increased nuchal translucency.

93.9 Conclusion

Noonan syndrome is an autosomal dominant disorder resulting from germline mutations that compromise the RAS/MAPK pathway involved in the regulation of cell proliferation, differentiation, survival, and metabolism. Typical manifestations of Noonan syndrome include distinctive craniofacial features (e.g., broad forehead, hypertelorism, down-slanting palpebral fissures, and low-set posteriorly rotated ears), developmental delay, learning difficulties, short stature, congenital heart disease, renal anomalies, lymphatic malformations, bleeding difficulties, and predisposition to cancer. Considering the phenotypic overlapping between Noonan syndrome and other RASopathies, application of molecular genetic techniques is crucial for accurate diagnosis and early implementation of treatment measures. Furthermore, the current lack of effective therapies for Noonan syndrome and related disorders justifies additional research efforts on the molecular mechanisms of RASopathies with the ultimate goal of defeating this inherited disease.

REFERENCES

1. Rauen KA. The RASopathies. *Annu Rev Genomics Hum Genet*. 2013;14:355–69.
2. Allanson JE, Roberts AE. Noonan syndrome. In: Adam MP, Ardinger HH, Pagon RA et al. editors. *GeneReviews®* [Internet]. Seattle (WA): University of Washington; 1993–2018. Nov 15, 2001 [updated Feb 25, 2016].

3. Tidyman WE, Rauen KA. Expansion of the RASopathies. *Curr Genet Med Rep.* 2016;4(3):57–64.

4. Roberts AE, Allanson JE, Tartaglia M, Gelb BD. Noonan syndrome. *Lancet.* 2013;381(9863):333–42.

5. Agarwal P, Philip R, Gutch M, Gupta KK. The other side of Turner's: Noonan's syndrome. *Indian J Endocrinol Metab.* 2013;17(5):794–8.

6. Yamamoto GL, Aguena M, Gos M et al. Rare variants in SOS2 and LZTR1 are associated with Noonan syndrome. *J Med Genet.* 2015;52(6):413–21.

7. Li X, Yao R, Tan X et al. Molecular and phenotypic spectrum of Noonan syndrome in Chinese patients. *Clin Genet.* 2019;96(4):290–9.

8. Zhang J, Zhang F, Niu R. Functions of Shp2 in cancer. *J Cell Mol Med.* 2015;19(9):2075–83.

9. Yang W, Wang J, Moore DC et al. Ptpn11 deletion in a novel progenitor causes metachondromatosis by inducing hedgehog signalling. *Nature.* 2013;499(7459):491–5.

10. Aoki Y, Niihori T, Banjo T et al. Gain-of-function mutations in RIT1 cause Noonan syndrome, a RAS/MAPK pathway syndrome. *Am J Hum Genet.* 2013;93(1):173–80.

11. Cavé H, Caye A, Ghedira N et al. Mutations in RIT1 cause Noonan syndrome with possible juvenile myelomonocytic leukemia but are not involved in acute lymphoblastic leukemia. *Eur J Hum Genet.* 2016;24(8):1124–31.

12. Nosan G, Bertok S, Vesel S, Yntema HG, Paro-Panjan D. A lethal course of hypertrophic cardiomyopathy in Noonan syndrome due to a novel germline mutation in the KRAS gene: Case study. *Croat Med J.* 2013;54(6):574–8.

13. Ekvall S, Wilbe M, Dahlgren J et al. Mutation in NRAS in familial Noonan syndrome—Case report and review of the literature. *BMC Med Genet.* 2015;16:95.

14. Qiu W, Wang X, Romanov V et al. Structural insights into Noonan/LEOPARD syndrome-related mutants of protein-tyrosine phosphatase SHP2 (PTPN11). *BMC Struct Biol.* 2014;14:10.

15. Gelb BD, Tartaglia M. Noonan syndrome with multiple lentigines. In: Adam MP, Ardinger HH, Pagon RA et al. editors. *GeneReviews®* [Internet]. Seattle (WA): University of Washington; 1993–2018. Nov 30, 2007 [updated May 14, 2015].

16. Gripp KW, Zand DJ, Demmer L et al. Expanding the SHOC2 mutation associated phenotype of Noonan syndrome with loose anagen hair: Structural brain anomalies and myelofibrosis. *Am J Med Genet A.* 2013;161A(10):2420–30.

17. Kondo RN, Martins LM, Lopes VC, Bittar RA, Araújo FM. Do you know this syndrome? Noonan syndrome. *An Bras Dermatol.* 2013;88(4):664–6.

18. Timeus F, Crescenzio N, Baldassarre G et al. Functional evaluation of circulating hematopoietic progenitors in Noonan syndrome. *Oncol Rep.* 2013;30(2):553–9.

19. Lee A, Sakhalkar MV. Ocular manifestations of Noonan syndrome in twin siblings: A case report of keratoconus with acute corneal hydrops. *Indian J Ophthalmol.* 2014;62(12):1171–3.

20. Mallineni SK, Yung Yiu CK, King NM. Oral manifestations of Noonan syndrome: Review of the literature and a report of four cases. *Rom J Morphol Embryol.* 2014;55(4):1503–9.

21. Kratz CP, Franke L, Peters H et al. Cancer spectrum and frequency among children with Noonan, Costello, and cardio-facio-cutaneous syndromes. *Br J Cancer.* 2015;112(8):1392–7.

22. Otikunta AN, Subbareddy YV, Polamuri P, Thakkar A. Prolapse of all cardiac valves in Noonan syndrome. *BMJ Case Rep.* 2015;2015. pii: bcr2014207241.

23. Cessans C, Ehlinger V, Arnaud C et al. Growth patterns of patients with Noonan syndrome: Correlation with age and genotype. *Eur J Endocrinol.* 2016;174(5):641–50.

24. Honda T, Kataoka TR, Ueshima C, Miyachi Y, Kabashima K. A case of Noonan syndrome with multiple subcutaneous tumours with MAPK-ERK/p38 Activation. *Acta Derm Venereol.* 2016;96(1):130–1.

25. Tekendo-Ngongang C, Agenbag G, Bope CD, Esterhuizen AI, Wonkam A. Noonan syndrome in South Africa: Clinical and molecular profiles. *Front Genet.* 2019;10:333.

26. Sarkozy A, Digilio MC, Dallapiccola B. Leopard syndrome. *Orphanet J Rare Dis.* 2008;3:13.

27. Lauriol J, Kontaridis MI. PTPN11-associated mutations in the heart: Has LEOPARD changed Its RASpots? *Trends Cardiovasc Med.* 2011;21(4):97–104.

28. Quaio CR, de Almeida TF, Brasil AS et al. Tegumentary manifestations of Noonan and Noonan-related syndromes. *Clinics (Sao Paulo).* 2013;68(8):1079–83.

29. Mathur D, Somashekar S, Navarrete C, Rodriguez MM. Twin infant with lymphatic dysplasia diagnosed with Noonan syndrome by molecular genetic testing. *Fetal Pediatr Pathol.* 2014;33(4):253–7.

30. Isojima T, Yokoya S. Development of disease-specific growth charts in Turner syndrome and Noonan syndrome. *Ann Pediatr Endocrinol Metab.* 2017;22(4):240–6.

31. Yu B, Liu W, Yu WM et al. Targeting protein tyrosine phosphatase SHP2 for the treatment of PTPN11-associated malignancies. *Mol Cancer Ther.* 2013;12(9):1738–48.

32. Noonan JA, Kappelgaard AM. The efficacy and safety of growth hormone therapy in children with noonan syndrome: A review of the evidence. *Horm Res Paediatr.* 2015;83(3):157–66.

94

Perlman Syndrome

Dongyou Liu

CONTENTS

94.1 Introduction

Perlman syndrome (also known as renal hamartomas, nephroblastomatosis, or fetal gigantism) is a rare congenital autosomal recessive overgrowth disorder characterized by polyhydramnios, macrosomia, dysmorphic facial features (inverted V-shaped upper lip, prominent forehead, deep-set eyes, broad and flat nasal bridge, and low-set ears), renal dysplasia/nephroblastomatosis, visceromegaly, neurodevelopmental delay, high neonatal mortality (with 36% of affected infants failing to survive beyond the neonatal period), and increased risk of Wilms tumor [1].

Molecularly, Perlman syndrome is attributed to homozygous or compound heterozygous mutations in the *DIS3L2* gene (a mammalian homolog of yeast Dis3 gene) on chromosome 2q37.1, which encodes a protein showing homology to the DIS3 component of the RNA exosome and having ribonuclease activity. In vitro experiments revealed an association of DIS3L2 knockdown with abnormalities of cell growth and division [2].

94.2 Biology

Initial cases related to Perlman syndrome were described in 1970, 1973, and 1975 involving six children of Jewish−Yemenite descent who presented with fetal gigantism (large birth size), hypertrophy of the islets of Langerhans, unusual facies (round fullness, depressed nasal bridge, hypotonic appearance with open mouth, long upper lip with inverted V-shape, upsweep of anterior scalp hair, and mild micrognathia), bilateral renal hamartomas with or without nephroblastomatosis, and Wilms tumor.

Subsequent studies of other pediatric cases expanded clinical spectrum of Perlman syndrome (including visceromegaly, cryptorchidism, polyhydramnios, hypoglycemia. cardiac defect, hepatic fibrosis with portoportal bridging, hemangioma, volvulus, intestinal atresia, agenesis of the corpus callosum, cleft palate, and early death). In 1984, this autosomal recessive syndrome of renal dysplasia, Wilms tumor, hyperplasia of the endocrine pancreas, fetal gigantism, multiple congenital anomalies, and mental retardation was named Perlman syndrome in honor of its discoverer [3,4].

The *DIS3L2* gene responsible for causing Perlman syndrome was mapped to chromosome 2q37.2 in 2011, whose product plays an important role in the mitotic cell cycle. Germline homozygous or compound heterozygous mutations in the *DIS3L2* gene appear to be a contributing factor for the development of dysplastic medullary parenchyma that evolves into nephroblastomatosis, hamartoma, and eventually Wilms tumor.

94.3 Pathogenesis

The *DIS3L2* gene (a mammalian exosome-independent homolog of *Schizosaccharomyces pombe* yeast Dis3 gene) on chromosome 2q37.1 spans 383 kb and generates five splice variants, the longest of which encodes a 885 aa, 99.2 kDa protein (DIS3L2). Constituting a member of the highly conserved RNaseII/RNB family of $3'-5'$ exoribonucleases, DIS3L2 demonstrates $3'/5'$ exoribonucleolytic activity, which is critical for degradation of both mRNA and non-coding RNA. Specifically, $3'-5'$ exoribonuclease recognizes mRNA and miRNA that are polyuridylated at the $3'$ end by terminal uridylyltransferase and mediates their degradation. This RNA degradation pathway plays a vital role in the regulation of mitosis and cell proliferation [5,6].

Structurally, the DIS3L2 protein consists of a poorly conserved N-terminal PIN domain, a cold-shock domain, a putative RNB exonuclease catalytic domain, and a possible C-terminal S1 RNA-binding domain.

Localized predominantly in the cytoplasm, DIS3L2 functions as a 3′–5′ exoribonuclease that recognizes, binds, and processes oligouridylated RNA and is the only known exoribonuclease involved in the degradation of uridylated precursor LET7 (pre-LET7) microRNA in the final step of the LIN28-LET7 pathway, in which LIN28 recruits terminal uridylyl transferases (TUTases) Zcchc6 (TUT7) and Zcchc11 (TUT4) that facilitates DIS3L2 degradation of pre-LET7 miRNA. Inactivation of DIS3L2 induces mitotic abnormalities and alters expression of mitotic checkpoint proteins, leading to downregulation of TTK, aurora B, and phosphorylated CDC25C, but upregulation of cyclin B1, RAD21, and securing. As microRNA let7 acts as a Wilms tumor suppressor through inhibition of the oncogenic target Lin28, which is overexpressed in Wilms tumor, DIS3L2 mutation hampers its ability to degrade let7 miRNA, leading to the development of Wilms tumor [7–9].

Loss-of-function germline homozygous or compound heterozygous mutations in the *DIS3L2* represent one of the mechanisms that converge on *Igf2* (insulin-like growth factor 2) upregulation, leading to overgrowth and/or Wilms tumor (Figure 94.1). Indeed, homozygous deletions of *DIS3L2* exon 6 or exon 9, which cause the loss of both RNA binding and degradation activity, are frequently identified in patients with Perlman syndrome. Analysis of compound heterozygous *DIS3L2* mutations (c.[367-2A > G];[1328T > A]) from a Japanese patient indicates that non-allelic homologous recombination (NAHR) between two LINE-1 (L1) elements represents an important disease mechanism [10]. As missense mutation (c.1328 T > A, p.Met443Lys)

retains RNA binding in both the cold-shock domains and the S1 domain, and partial exonuclease functions remain in at least one allele, long-term survival is possible [11].

Further, partial or complete *DIS3L2* deletion occurs in 39% of sporadic Wilms' tumor samples, and a truncation of the *DIS3L2* locus is linked to Marfan-like syndrome with skeletal overgrowth.

94.4 Epidemiology

94.4.1 Prevalence

Perlman syndrome is estimated to occur <1 in 1,000,000. To date, over 30 Perlman syndrome cases have been reported in the world literature. It appears that Perlman syndrome affects both consanguineous and nonconsanguineous couplings and shows a male predilection (with a male to female ratio of 2:1).

94.4.2 Inheritance

Perlman syndrome has an autosomal recessive inheritance, in which two mutated genes, one from each carrier parent, are passed on to offspring. While two carriers who possess one mutated gene (recessive gene) and one normal gene (dominant gene) are rarely affected, they have a 25% risk of getting an unaffected child with two normal genes, a 50% risk of getting an unaffected child who is also a carrier, and a 25% risk of getting an affected child with two recessive genes.

94.4.3 Penetrance

Perlman syndrome shows incomplete penetrance and variable clinical expression. Patients harboring a *DIS3L2* pathogenic variant often display a diverse range of clinical manifestations.

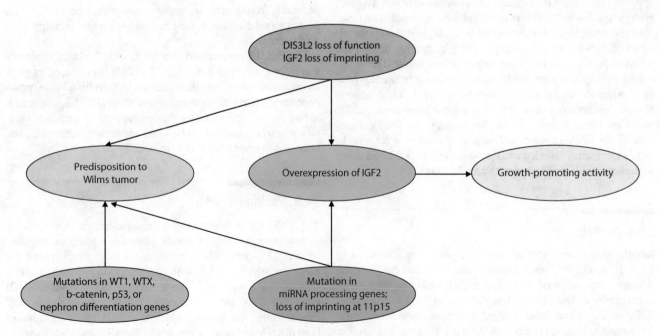

FIGURE 94.1 Upregulation of IGF2 represents one of the mechanisms for overgrowth syndromes and pediatric tumorigenesis. While Perlman syndrome is caused by mutations in the gene encoding the DIS3L2 exoribonuclease, leading to loss of DIS3L2 and subsequent overexpression of IGF2, Beckwith–Wiedemann syndrome arises from genetic and epigenetic changes in IGF2.

94.5 Clinical Features

Perlman syndrome is a clinically variable disease producing a broad range of abnormalities [12,13]:

Overgrowth: Macrosomia (large birth size, 87%) evident in the prenatal and postnatal period.

Craniofacial dysmorphism: Macrocephaly (65%), round facial fullness, prominent forehead (26%), deep-set eyes (52%), hypertelorism, epicanthal folds (26%), broad flat nasal bridge (78%), open mouth, everted upper lip (82%), high-arched palate, cleft palate, low-set ears (47%), upsweep of anterior scalp hair, micrognathia.

Renal anomalies: Bilateral nephromegaly (95%), renal dysplasia, nephroblastomatosis, bilateral renal hamartomas, Wilms tumor (30%).

Central nervous system anomalies (30%): Agenesis of the corpus callosum, large cisterna magna, retrocerebellar and perichiasmatic leptomeningeal cysts, white matter hypoplasia and gray matter heterotopia involving the cerebellum and superior colliculi, choroid plexus hemangioma, cerebral atrophy with a marked deficit in the myelinization of the white matter, left periventricular ovoid cystic formation.

Congenital heart anomalies: Cardiomegaly (26%), interrupted aortic arch, anomalous coronary vessels, dextroposition of the heart, muscular ventricular septal defect.

Digestive anomalies: Hepatomegaly/splenomegaly (69%), intestinal malrotation, distal ileal atresia, digestive volvulus, intestinal malrotation with cecum located on the midline, fetal hepatic fibrosis with portoportal bridging.

Hemangiomas (13%): Capillary hemangioma in the left antecubital fossa, choroid plexus hemangioma, and superficial cutaneous capillary hemangioma around the umbilicus [14].

Skeletal abnormalities: Absence of the normal widening of the lumbar interpediculate distances, rounded iliac wings, small sacrosciatic notches, crowded toes, bilateral calcaneovalgus deformity, genua recurvata, left metatarsus varus, right hallux varus, severe right convex dorsal and left convex lumbar scoliosis, lumbar hyperlordosis, crest iliac asymmetry, prominent xiphisternum.

Other anomalies: Polyhydramnios (60%), cryptorchidism (93%), hypotonia (78%), abdominal muscular hypoplasia, hypospadias, polysplenia, islets of Langerhans hypertrophy (hyperinsulinism), diaphragmatic hernia, fetal ascites without hydrops, developmental delay (26%).

94.6 Diagnosis

Diagnosis of Perlman syndrome relies on observation of phenotypic features and histological examination of kidney biopsy.

Individuals displaying polyhydramnios, macrocephaly, neonatal macrosomia, visceromegaly, dysmorphic facial features (particularly upsweeping anterior scalp hair, depressed nasal bridge, open mouth, prominent everted upper lip, and mild micrognathia), and Wilms tumor at an early age should be suspected of Perlman syndrome.

Magnetic resonance imaging (MRI) may reveal generalized cerebral atrophy with a marked deficit of the white matter. Renal ultrasound and MRI may show markedly enlarged kidneys with multiple small cystic lesions (similar to polycystic kidney disease). The postmortem kidney biopsy may have dysplastic changes, microcysts, and focal nephrogenic rest, characteristic features of the Perlman syndrome.

Differential diagnoses for Perlman syndrome (deep-set eyes, depressed nasal bridge, everted upper lip, macrocephaly, Wilms tumor) include Beckwith–Wiedemann syndrome (polyhydramnios, macrosomia, nephromegaly, hypoglycemia, Wilms tumor), Simpson–Golabi–Behmel syndrome (Wilms tumor), Sotos syndrome (Wilms tumor) and Weaver syndrome (Wilms tumor) [15,16].

94.7 Treatment

Management of Perlman syndrome requires a multidisciplinary approach that provides symptomatic relief and improves the quality of life for affected individuals. Particular attention should be paid to patients who have sepsis or progressive respiratory insufficiency, which represent the main causes of infant death.

94.8 Prognosis and Prevention

Perlman syndrome has a poor prognosis, and affected infants often die as a result of sepsis or progressive respiratory insufficiency. About two-thirds of infants who survive beyond the neonatal period develop Wilms tumor. Prompt recognition/identification, regular follow-up, and provision of clinical assistance are helpful in improving the quality of life for individuals with Perlman syndrome.

As Perlman syndrome is an autosomal recessive disorder, two carrier parents have 25%, 50%, and 25% chance of getting an unaffected, an unaffected carrier, and affected child, respectively. Therefore, prenatal diagnosis and preimplantation diagnosis (involving ultrasound, molecular testing) and will help reduce anxiety and enable appropriate decision making for prospective parents who harbor a genetic disposition for Perlman syndrome. Indeed, ultrasound may uncover fetal overgrowth and polyhydramnios (excessive amniotic fluid in the amniotic sac) at 18 weeks of pregnancy, cystic hygroma and thickened nuchal lucency at the first trimester, and macrosomia, enlarged kidneys, renal tumor (hamartoma, Wilms tumor), cardiac abnormalities, and visceromegaly at the second and third trimesters [17].

94.9 Conclusion

Perlman syndrome is a rare congenital autosomal recessive overgrowth disorder that typically presents with fetal gigantism, polyhydramnios, macrosomia, visceromegaly, dysmorphic facies (inverted V-shaped upper lip, prominent forehead, deep-set eyes, broad and flat nasal bridge, and low-set ears), bilateral renal hamartomas with nephroblastomatosis, Wilms tumor, developmental delay, and high perinatal mortality. Germline mutations in the *DIS3L2* gene, which mediates noncoding RNA decay and induces on *Igf2* upregulation, appear to be responsible for overgrowth and Wilms tumor susceptibility in Perlman syndrome [18–23]. Observation of clinical manifestations and structural changes in affected tissues and organs and histological examination of kidney specimens enable diagnosis of Perlman syndrome, while detection of *DIS3L2* pathogenic variant provides additional confirmation of the disorder. In the absence of specific treatment measures, management of Perlman syndrome currently focuses on symptomatic relief and quality-of-life maintenance for affected individuals.

REFERENCES

1. Morris MR, Astuti D, Maher ER. Perlman syndrome: Overgrowth, Wilms tumor predisposition and DIS3L2. *Am J Med Genet C Semin Med Genet.* 2013;163C(2):106–13.
2. Astuti D, Morris MR, Cooper WN et al. Germline mutations in DIS3L2 cause the Perlman syndrome of overgrowth and Wilms tumor susceptibility. *Nat Genet.* 2012;44(3):277–84.
3. Neri G, Martini-Neri ME, Katz BE, Opitz JM. The Perlman syndrome: Familial renal dysplasia with Wilms tumor, fetal gigantism and multiple congenital anomalies. 1984. *Am J Med Genet A.* 2013;161A(11):2691–6.
4. Alessandri JL, Cuillier F, Ramful D et al. Perlman syndrome: Report, prenatal findings and review. *Am J Med Genet A.* 2008;146A(19):2532–7.
5. Łabno A, Warkocki Z, Kuliński T et al. Perlman syndrome nuclease DIS3L2 controls cytoplasmic non-coding RNAs and provides surveillance pathway for maturing snRNAs. *Nucleic Acids Res.* 2016;44(21):10437–53.
6. Towler BP, Jones CI, Harper KL, Waldron JA, Newbury SF. A novel role for the 3′-5′ exoribonuclease Dis3L2 in controlling cell proliferation and tissue growth. *RNA Biol.* 2016;13(12):1286–99.
7. Chang HM, Triboulet R, Thornton JE, Gregory RI. A role for the Perlman syndrome exonuclease Dis3l2 in the Lin28-let-7 pathway. *Nature.* 2013;497(7448):244–8.
8. Ustianenko D, Hrossova D, Potesil D et al. Mammalian DIS3L2 exoribonuclease targets the uridylated precursors of let-7 miRNAs. *RNA.* 2013;19(12):1632–8.
9. Ustianenko D, Pasulka J, Feketova Z et al. TUT-DIS3L2 is a mammalian surveillance pathway for aberrant structured non-coding RNAs. *EMBO J.* 2016;35(20):2179–91.
10. Higashimoto K, Maeda T, Okada J et al. Homozygous deletion of DIS3L2 exon 9 due to non-allelic homologous recombination between LINE-1s in a Japanese patient with Perlman syndrome. *Eur J Hum Genet.* 2013;21(11):1316–9.
11. Soma N, Higashimoto K, Imamura M, Saitoh A, Soejima H, Nagasaki K. Long term survival of a patient with Perlman syndrome due to novel compound heterozygous missense mutations in RNB domain of DIS3L2. *Am J Med Genet A.* 2017;173(4):1077–81.
12. Piccione M, Cecconi M, Giuffrè M et al. Perlman syndrome: Clinical report and nine-year follow-up. *Am J Med Genet A.* 2005;139A(2):131–5.
13. Demirel G, Oguz SS, Celik IH, Uras N, Erdeve O, Dilmen U. Rare clinical entity Perlman syndrome: Is cholestasis a new finding? *Congenit Anom (Kyoto).* 2011;51(1):43–5.
14. Pirgon O, Atabek ME, Akin F, Sert A. A case of Perlman syndrome presenting with hemorrhagic hemangioma. *J Pediatr Hematol Oncol.* 2006;28(8):531–3.
15. Lapunzina P. Risk of tumorigenesis in overgrowth syndromes: A comprehensive review. *Am J Med Genet C Semin Med Genet.* 2005;137C(1):53–71.
16. Ferianec V, Bartova M. Beckwith-Wiedemann syndrome with overlapping Perlman syndrome manifestation. *J Matern Fetal Neonatal Med.* 2014;27(15):1607–9.
17. van der Stege JG, van Eyck J, Arabin B. Prenatal ultrasound observations in subsequent pregnancies with Perlman syndrome. *Ultrasound Obstet Gynecol.* 1998;11(2):149–51.
18. Pirouz M, Du P, Munafò M, Gregory RI. Dis3l2-mediated decay is a quality control pathway for noncoding RNAs. *Cell Rep.* 2016;16(7):1861–73.
19. Menezes MR, Balzeau J, Hagan JP. 3′ RNA uridylation in epitranscriptomics, gene regulation, and disease. *Front Mol Biosci.* 2018;5:61.
20. Hunter RW, Liu Y, Manjunath H et al. Loss of *Dis3l2* partially phenocopies Perlman syndrome in mice and results in up-regulation of *Igf2* in nephron progenitor cells. *Genes Dev.* 2018;32(13–14):903–8.
21. Bharathavikru R, Hastie ND. Overgrowth syndromes and pediatric cancers: How many roads lead to *IGF2*? *Genes Dev.* 2018;32(15–16):993–5.
22. Luan S, Luo J, Liu H, Li Z. Regulation of RNA decay and cellular function by 3′-5′ exoribonuclease DIS3L2. *RNA Biol.* 2019;16(2):160–5.
23. Pirouz M, Munafò M, Ebrahimi AG, Choe J, Gregory RI. Exonuclease requirements for mammalian ribosomal RNA biogenesis and surveillance. *Nat Struct Mol Biol.* 2019;26(6):490–500.

95

Proteus Syndrome

Dongyou Liu

CONTENTS

95.1 Introduction

Proteus syndrome is a rare congenital disorder characterized by asymmetric, progressive overgrowth of the bones (often in the limbs, skull, and spine), skin (forming thick, raised, and deeply grooved lesions known as a cerebriform connective tissue nevus, usually on the soles of the feet, which is pathognomonic for Proteus syndrome), adipose, and central nervous systems (associated with developmental delay, seizures, vision loss, and distinctive facial features such as long face, down-slanting palpebral fissures, low nasal bridge with wide nostrils, and an open-mouth expression) that may be absent or modest at birth, but develops rapidly from the toddler period (between the ages of 6 and 18 months) to childhood, leading to severe disfigurement, deep vein thrombosis (most often in the legs or arms) and pulmonary embolism as well as a range of tumors (e.g., lipoma, hemangioma, connective tissue nevi, lymphangioma, adenoma of the parotid gland, cystadenoma of the ovary, testicular tumor, meningioma, and mesothelioma) [1].

Somatic mutations in the *AKT1* gene on chromosome 14q32.33 appear to be responsible for causing Proteus syndrome. Specifically, a single mutation (c.49G→A, p.Glu17Lys) in *AKT1* induces constitutive activation of AKT1 through Ser473 and Thr308 phosphorylation, and occurs in individuals with Proteus syndrome.

95.2 Biology

Although possible cases related to Proteus syndrome (formerly elephant man disease) were reported in 1928, its distinct clinical entity was first delineated by Cohen and Hayden in 1979, and its current name (after the Greek god Proteus who could change his shape at will to avoid capture) was proposed by Wiedeman in 1983. Besides a combination of partial gigantism of the hands and/or feet, nevi, hemihypertrophy, subcutaneous tumors, macrocephaly or other skull anomalies, and possible accelerated growth and visceral affectations, subsequent studies indicated that Proteus syndrome may also display overgrowth of the long bones, macrodactyly, asymmetric macrocephaly, plantar or palmar hyperplasia, vertebral abnormalities, lipoma, hemangioma, connective tissue nevi, lymphangiomas, and vascular malformations [1].

The identification of somatic activating mutation of the *AKT1* gene on chromosome 14q32.3, which encodes an enzyme involved in the PI3K/AKT/mTOR pathway, revealed molecular clues to the pathogenesis of Proteus syndrome. Given their essential roles in the regulation of cell growth, metabolism, angiogenesis, and survival, somatic mutations in the genes encoding the components of the PI3K/AKT/mTOR signaling pathway often induce constitutive activation and significant dysregulation of normal cellular functions, leading to sporadic or mosaic overgrowth (adipose, skeletal, muscle, brain, vascular, or lymphatic), skin abnormalities (epidermal nevi, hyper- and hypopigmented lesions), and tumor susceptibility.

Chief among a broad constellation of somatic disorders are those related to *PI3K* [*PIK3CA-related overgrowth spectrum* (PROS): CLOVES syndrome (congenital lipomatous overgrowth, vascular malformations, linear keratinocytic epidermal nevi, and skeletal/spinal anomalies), fibroadipose hyperplasia or overgrowth (FAO), megalencephaly-capillary malformation syndrome (MCAP), hemihyperplasia multiple lipomatosis (HHML), macrodactyly, hemimegalencephaly, muscle hemihyperplasia,

TABLE 95.1

Key Features of Somatic Disorders due to Mosaic Mutations That Affect the PI3K/PTEN/AKT/TSC/mTOR Signaling Pathway

	PROS (*PIK3CA* Mutation)	PHTS (*PTEN* Mutation)	Proteus Syndrome (*AKT1* Mutation)	Tuberous Sclerosis Complex 1/2 (*TSC1/TSC2* Mutation)
Segmental overgrowth	+++	+	+++	+
Facial papules		+++		+++
Epidermal nevus	+++	+	+++	+
Connective tissue nevus	+	+	+++	+++
Cutaneous vascular malformation	+++	+	+++	+
Oral papules		+++		+++
Hamartomas	+++	+++	+++	+++
Malignant tumors	+	+++	+	+

Abbreviations: PHTS, *PTEN* hamartoma tumor syndrome; PROS, *PIK3CA*-related overgrowth spectrum; +++, common; +, uncommon.

fibroadipose infiltrating lipomatosis, and Klippel–Trenaunay syndrome (KTS) (see Chapter 88, Figure 88.1 for details), *AKT* (Proteus syndrome, megalencephaly polymicrogyria polydactyly hydrocephalus syndrome [MPPH]), lipodystrophy syndrome (hypoglycemia, hemimegalencephaly, and focal cortical dysplasia), and *mTOR* (hemimegalencephaly, polymicrogyria, focal cortical dysplasia) (Table 95.1).

In addition, as *PTEN* and *TSC1/TSC2* are key negative regulators of the PI3K/AKT/mTOR signaling pathway, their germline loss-of-function mutations are responsible for causing *PTEN* hamartoma tumor syndrome (PHTS, including Cowden syndrome, Bannayan–Riley–Ruvalcaba syndrome, and type II segmental Cowden syndrome), tuberous sclerosis complex 1 and tuberous sclerosis complex 2, which behave similarly to somatic disorders above (Table 95.1) [2–8].

95.3 Pathogenesis

Mapped to chromosome 14q32.33, the *AKT1* gene spans over 26 kb with 14 exons, and encodes 480 aa protein (AKT1 kinase or protein kinase B), a serine/threonine kinase that affects a wide range of biological functions including cell proliferation and growth, metabolism, protein synthesis, migration, angiogenesis, and anti-apoptotic ability [9].

Somatic gain-of-function mutations in the *AKT1* gene disrupt its ability to regulate cell growth, leading to increased cell proliferation, decreased apoptosis, and ultimately mosaic multi-tissue overgrowth and tumor susceptibility. Associated with Proteus syndrome, the *AKT1* pathogenic variant c.49G>A (p.Glu17Lys) causes constitutive activation of the AKT1 kinase, which then moves to the plasma membrane and activates the PI3KCA/AKT pathway. Interestingly, an identical variant in *AKT1* is observed in patients with schizophrenia.

The *AKT1* pathogenic variant (c.49G>A) appears to arise randomly post fertilization in one cell of the multicellular embryo, which upon further growth and division will contain a mixture of cells with and without a genetic mutation (the presence of two or more genetically distinct cell lineages originating from a single zygote is referred to as mosaicism). As strongly activating germline mutations in *AKT1* or *PIK3CA* are likely lethal, non-mosaic embryo harboring such mutations are not expected to survive. Given their vital roles in the PI3K/AKT/mTOR signaling pathway, mutations

in the *AKT1* and *PIK3CA* genes may trigger distinctive segmental or asymmetric overgrowth (involving the bone, muscle, adipose tissue, skin, and/or nerves) and tumor susceptibility [10–12].

95.4 Epidemiology

95.4.1 Prevalence

Proteus syndrome is a rare condition with an estimated incidence of <1 in 1 million people worldwide. To date, only a few hundred cases have been reported in the literature.

95.4.2 Inheritance

As Proteus syndrome is caused by somatic mutation in the *AKT1* gene, it is nonhereditary and does not run in families.

95.4.3 Penetrance

Proteus syndrome is a mosaic genetic disorder that is not inherited. Therefore, its penetrance cannot be assessed. On the other hand, Proteus syndrome is known to display variable expressivity, with some individuals being minimally impacted while others severely affected. Indeed, the severity of the overgrowth in affected patients may range from a slightly enlarged digit to a gigantic limb.

95.5 Clinical Features

In most individuals with Proteus syndrome, little or no manifestations are visible at birth. However, from the age between 6 and 18 months, asymmetric overgrowth and other manifestations become apparent, resulting in a 25% mortality before the age of 20 [13,14].

Asymmetric overgrowth typically manifests from age 6 to 18 months, and as late as age 12 years. Compared to unaffected parts, the affected parts (especially the tubular bones of the limbs, the vertebral bodies, and the skull) may be is 15% larger at age 1 year, 30% larger at 3 years, and 100% larger at age 6 years. Some patients may show unilateral overgrowth of the tonsils, adenoids, spleen, kidneys, and testes (Figure 95.1) [15].

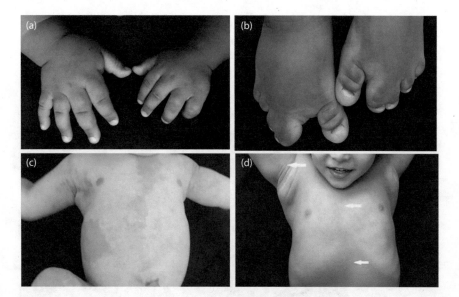

FIGURE 95.1 Proteus syndrome in a 2-year-old boy showing micromyelia of left hand (a), second and third toe syndactyly of both feet (b), port-wine stains at birth (c), and clearing of stains at 2 years of age (d). (Photo credit: Rocha RCC et al. *An Bras Dermatol.* 2017;92(5):717–20.)

Dermatologic anomalies consist mainly of cerebriform connective tissue nevi (CCTN, which are nearly pathognomonic for Proteus syndrome) and linear verrucous epidermal nevi. CCTN are firm lesions which display a distinct pattern resembling the brain's sulci and gyri (hence the term cerebriform) and occur most commonly on the sole, hand, ear, and lacrimal puncta. Linear verrucous epidermal nevi are streaky, pigmented, and rough lesions which often follow the lines of Blaschko (Figure 95.2) [16].

Lipoatrophy (overgrowth of lipomatous/adipose tissue) emerges in infancy and may continue to expand throughout childhood and into young adulthood. Lipoatrophy differs from the typical ovoid, encapsulated lipomas that are common in the elderly. Fatty infiltration of the myocardium (particularly the intraventricular septum) may occur in some affected individuals.

Vascular malformations include cutaneous capillary malformations, prominent venous patterning or varicosities, and lymphatic vascular malformations (often in areas of lipomatous overgrowth), with deep vein thrombosis (manifesting as palpable subcutaneous rope-like mass, swelling, erythema, pain, and distal venous congestion) and pulmonary embolism (manifesting as shortness of breath, chest pain, and cough/ hemoptysis) being rare life-threatening complications. However, arteriovenous malformation is uncommon.

Other abnormalities range from bullous pulmonary disease and dysmorphic facial features (e.g., dolichocephaly, long face, downslanting palpebral fissures and/or minor ptosis, depressed nasal bridge, wide or anteverted nares, and open mouth at rest) to psychosocial issues (concerns about disfigurement caused by skeletal and connective tissue overgrowth).

Tumors (e.g., meningiomas, ovarian cystadenomas, and parotid monomorphic adenomas) may occur at a higher frequency in Proteus syndrome–affected individuals compared to the general population (Figure 95.3) [17].

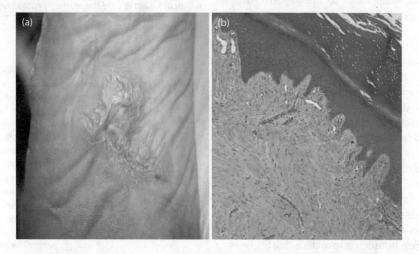

FIGURE 95.2 Proteus syndrome in a 49-year-old male showing modestly developed connective tissue nevus of left foot (a), and histological features of cerebriform nevus from foot biopsy (remarkable hyperkeratosis, epidermal hyperplasia, dermoepidermal fibrosis with extensive sclerosis of the reticular dermis, thickened collagen bundles, and fat-cell entrapment; H&E, magnification ×100). (b). (Photo credit: Vestita M et al. *F1000Res.* 2018;7:228.)

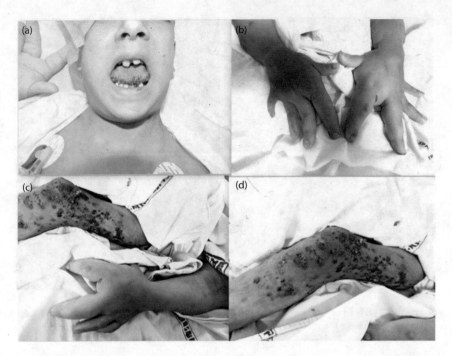

FIGURE 95.3 Proteus syndrome in a 9-year-old boy showing tongue hemangioma and misaligned teeth (a), multiple hemangiomas and asymmetric enlargement of both upper and lower limbs. (b–d). (Photo credit: De M et al. *Asian J Anesthesiol.* 2017;55(1):22–3.)

95.6 Diagnosis

Individuals having the following manifestations should be suspected of Proteus syndrome: (i) progressive overgrowth leading to asymmetric distortion of skeletal architecture; (ii) CCTN showing deep grooves and gyrations that resemble the surface of the brain; (iii) linear verrucous epidermal nevus (LVEN) that appears streaky, pigmented, and rough along the lines of Blaschko anywhere on the body; (iv) adipose dysregulation including lipomatous overgrowth and lipoatrophy; (v) other manifestations (vascular malformations including cutaneous capillary malformations, prominent venous patterning or varicosities, and lymphatic malformations; overgrowth of the spleen, liver, thymus, and gastrointestinal tract; tumors such as meningioma, ovarian cystadenoma, breast cancer, parotid monomorphic adenoma, and mesothelioma; bullous pulmonary degeneration; dysmorphic facial features including dolichocephaly, long face, downslanting palpebral fissures, and/or minor ptosis, depressed nasal bridge, wide or anteverted nares, and open mouth at rest [1].

Diagnosis of Proteus syndrome is based on meeting all of the general criteria (mosaic distribution of lesions, sporadic occurrence, progressive course) plus additional specific criteria (one from category A, or two from category B, or three from category C) (Table 95.2).

Identification of somatic, mosaic heterozygous *AKT1* pathogenic variant c.49G>A (p.Glu17Lys) from more than one tissue sample (typically punch biopsy of affected skin or peripheral blood sample) by molecular genetic testing allows further confirmation of Proteus syndrome. The absence of *AKT1* pathogenic variant in a peripheral blood sample is nonetheless insufficient to exclude the diagnosis.

Differential diagnoses for Proteus syndrome (almost always postnatal onset, disproportionate and progressive distorting skeletal overgrowth, CCTN; nonhereditary) include PHTS (asymmetric

overgrowth, macrocephaly, cutaneous vascular malformations, and tumor susceptibility; autosomal dominant inheritance involving a germline *PTEN* pathogenic variant and a second somatic, mosaic *PTEN* variant), type II segmental Cowden syndrome (due to *PTEN* mutations), SOLAMEN syndrome (due to *PTEN* mutations), CLOVES syndrome (*c*ongenital *l*ipomatous asymmetric *o*vergrowth of the trunk, lymphatic, capillary, venous, and combined-type *v*ascular malformations, *e*pidermal nevi, *s*keletal and *s*pinal anomalies; due to PIK3CA mutations), hemihyperplasia (and hemihyperplasia with multiple lipomatosis syndrome), and Klippel–Trenaunay syndrome (KTS, ipsilateral overgrowth, vascular malformations such as lateral venous anomaly) [18–20].

TABLE 95.2

Clinical Criteria for Diagnosing Proteus Syndrome

General Criteria	Mosaic distribution of lesions
	Sporadic occurrence
	Progressive course
Special Criteria A	Cerebriform connective tissue nevus (CCTN)
Special Criteria B	Linear epidermal nevus
	Asymmetric, disproportionate overgrowth of various body tissues (limbs, hyperostosis of the skull or external auditory canal, abnormal growth of vertebrae, spleen/thymus
	Tumors (bilateral ovarian cystadenomas or parotid monomorphic adenomas before the second decade)
Special Criteria C	Dysregulated adipose tissue (lipomatous overgrowth or regional lipoatrophy)
	Vascular malformations (capillary malformation, venous malformation, lymphatic malformation, or lung bullae)
	Facial phenotype (dolichocephaly, long face, downward-slanting palpebral fissures and/or minor ptosis, depressed nasal bridge, wide or anteverted nares, and open mouth at rest)

95.7 Treatment

Proteus syndrome is a complex and multisystem disorder that requires a coordinated and multidisciplinary approach for effective management [21].

Overgrowth in tubular bones, epiphysiostasis, and epiphysiodesis should be treated with orthopedic procedures to delay or halt linear bone growth. Scoliosis can lead to fatal restrictive lung disease and should be corrected by surgery.

CCTN that contribute to pressure ulcerations or problems with shoe fit warrant pedorthic intervention (surgical removal).

Lipoatrophy (overgrowth of lipomatous tissue) can be difficult to remove and may regrow after surgical debulking. Therefore, open surgery instead of liposuction is preferred.

Deep vein thrombosis (palpable subcutaneous rope-like mass, swelling, erythema, pain, and distal venous congestion) and *pulmonary embolism* (shortness of breath, chest pain, and cough which may include hemoptysis) require emergent assessment (involving D-dimer assay or ultrasonographic evaluation, high-resolution chest CT [or spiral CT] with contrast; ventilation-perfusion nuclear medicine scanning may be appropriate in some individuals) followed by anticoagulation treatment.

Tumors of various types may occur in Proteus syndrome and require regular evaluation (pain, unexpected growths, signs of obstruction or compression) and appropriate treatment.

In addition, *bullous pulmonary disease* may require resection in some cases; *psychometric and learning evaluation* are indicated for individuals with developmental delays, and *psychosocial issues* are warranted in most instances.

Drugs focusing on components of the PI3K/PTEN/AKT/TSC/mTOR signaling pathway have shown promise in the treatment of Proteus syndrome and related disorders [22,23].

95.8 Prognosis and Prevention

Proteus syndrome is highly variable, and its prognosis varies according to the location and degree of overgrowth and presence or absence of significant complications (e.g., bullous pulmonary disease, hemimegencephaly, and pulmonary embolism). In general, mildly affected individuals have a more favorable outcome prognosis than severely affected ones. With a median age of diagnosis at 19 months, 25% of affected individuals may die by 22 years of age [24].

Agents promoting growth (e.g., androgenic steroids or growth hormone), and those increasing the risk of deep vein thrombosis should be avoided.

Since Proteus syndrome is nonhereditary, its risk for offspring or siblings is expected to be the same as in the general population. However, there is a possibility that offspring may be affected through gonadal mosaicism.

95.9 Conclusion

Attributable to an activating *AKT1* mutation (c.49G>A, p.Glu17Lys), Proteus syndrome is a highly variable, severe disorder of asymmetric and disproportionate overgrowth of body parts, connective tissue nevi, epidermal nevi, dysregulated adipose tissue, and vascular malformations that develop gradually during childhood and stabilize at approximately 15–17 years of age [25]. Other notable features include lipomas, lung cysts, dysmorphic facies, intellectual disability, and deep venous thrombosis. As Proteus syndrome overlaps with other overgrowth syndromes, its diagnosis requires meeting all of general criteria plus special criteria. Molecular identification of *AKT1* pathogenic variant (c.49G>A, p.Glu17Lys) provides further confirmation of this nonhereditary disorder [26]. Although no specific treatment is currently available for Proteus syndrome, several recent studies have indicated the benefits of using AKT1 inhibitor Miransertib to improve patient's wellbeing and disease outcome [27,28].

REFERENCES

1. Biesecker LG, Sapp JC. Proteus Syndrome. In: Adam MP, Ardinger HH, Pagon RA et al. editors. *GeneReviews®* [Internet]. Seattle (WA): University of Washington, Seattle; 1993–2018. 2012 Aug 9 [updated 2018 Jan 4].
2. Kurek KC, Luks VL, Ayturk UM et al. Somatic mosaic activating mutations in PIK3CA cause CLOVES syndrome. *Am J Hum Genet*. 2012;90:1108–15.
3. Laplante M, Sabatini DM. mTOR signaling in growth control and disease. *Cell*. 2012;149:274–93.
4. Edmondson AC, Kalish JM. Overgrowth syndromes. *J Pediatr Genet*. 2015;4(3):136–43.
5. Keppler-Noreuil KM, Rios JJ, Parker VE et al. PIK3CA-related overgrowth spectrum (PROS): Diagnostic and testing eligibility criteria, differential diagnosis, and evaluation. *Am J Med Genet A*. 2015;167A(2):287–95.
6. Keppler-Noreuil KM, Parker VE, Darling TN, Martinez-Agosto JA. Somatic overgrowth disorders of the PI3K/AKT/mTOR pathway & therapeutic strategies. *Am J Med Genet C Semin Med Genet*. 2016;172(4):402–21.
7. Nathan N, Keppler-Noreuil KM, Biesecker LG, Moss J, Darling TN. Mosaic disorders of the PI3K/PTEN/AKT/TSC/mTORC1 signaling pathway. *Dermatol Clin*. 2017;35(1):51–60
8. Eng C. *PTEN* Hamartoma Tumor Syndrome. In: Adam MP, Ardinger HH, Pagon RA et al. editors. *GeneReviews®* [Internet]. Seattle (WA): University of Washington, Seattle; 1993–2018. 2001 Nov 29 [updated 2016 Jun 2].
9. Cheung M, Testa JR. Diverse mechanisms of AKT pathway activation in human malignancy. *Curr Cancer Drug Targets*. 2013;13(3):234–44.
10. Lindhurst MJ, Sapp JC, Teer JK et al. A mosaic activating mutation in AKT1 associated with the Proteus syndrome. *N Engl J Med*. 2011;365:611–9.
11. Cohen MM Jr. Proteus syndrome review: Molecular, clinical, and pathologic features. *Clin Genet*. 2014; 85(2): 111–9.
12. van Steensel MA. Neurocutaneous manifestations of genetic mosaicism. *J Pediatr Genet*. 2015;4(3):144–53.
13. Linton JA, Seo BK, Oh CS. Proteus syndrome: A natural clinical course of Proteus syndrome. *Yonsei Med J*. 2002;43(2):259–66.
14. Chakrabarti N, Chattopadhyay C, Bhuban M, Pal SK. Proteus syndrome: A rare cause of gigantic limb. *Indian Dermatol Online J*. 2014;5:193–5.
15. Rocha RCC, Estrella MPS, Amaral DMD, Barbosa AM, Abreu MAMM. Proteus syndrome. *An Bras Dermatol*. 2017 Sep-Oct;92(5):717–20.

16. Vestita M, Filoni A, Arpaia N, Ettorre G, Bonamonte D. Case report: "Incognito" proteus syndrome. *F1000Res*. 2018;7:228.

17. De M, Bava EP, Gera S, Bhoi D. Proteus syndrome: Unveiling the anesthetic myths. *Asian J Anesthesiol*. 2017;55(1):22–3.

18. Orloff MS, He X, Peterson C et al. Germline PIK3CA and AKT1 mutations in Cowden and Cowden-like syndromes. *Am J Hum Genet*. 2013;92:76–80.

19. Lacerda Lda S, Alves UD, Zanier JF, Machado DC, Camilo GB, Lopes AJ. Differential diagnoses of overgrowth syndromes: The most important clinical and radiological disease manifestations. *Radiol Res Pract*. 2014;2014:947451.

20. Valentini V, Zelli V, Rizzolo P et al. *PIK3CA* c.3140A>G mutation in a patient with suspected Proteus syndrome: a case report. *Clin Case Rep*. 2018;6(7):1358–63.

21. Biesecker L. The challenges of Proteus syndrome: Diagnosis and management. *Eur J Hum Genet*. 2006;14(11):1151–7.

22. Agarwal R, Liebe S, Turski ML et al. Targeted therapy for genetic cancer syndromes: Von Hippel–Lindau disease, Cowden syndrome, and Proteus syndrome. *Discov Med*. 2015;19(103):109–16.

23. Loconte DC, Grossi V, Bozzao C et al. Molecular and functional characterization of three different postzygotic mutations in PIK3CA-related overgrowth spectrum (PROS) patients: Effects on PI3K/AKT/mTOR signaling and sensitivity to PIK3 inhibitors. *PLOS ONE*. 2015;10:e0123092.

24. Sapp JC, Hu L, Zhao J et al. Quantifying survival in patients with Proteus syndrome. *Genet Med*. 2017;19(12):1376–9.

25. Zeng X, Wen X, Liang X, Wang L, Xu L. A case report of Proteus syndrome (PS). *BMC Med Genet*. 2020;21(1):15.

26. Keppler-Noreuil KM, Burton-Akright J, Lindhurst MJ et al. Molecular heterogeneity of the cerebriform connective tissue nevus in mosaic overgrowth syndromes. *Cold Spring Harb Mol Case Stud*. 2019;5(4):a004036.

27. Keppler-Noreuil KM, Sapp JC, Lindhurst MJ et al. Pharmacodynamic Study of Miransertib in Individuals with Proteus Syndrome. *Am J Hum Genet*. 2019;104(3):484–91.

28. Biesecker LG, Edwards M, O'Donnell S et al. Clinical report: one year of treatment of Proteus syndrome with miransertib (ARQ 092). *Cold Spring Harb Mol Case Stud*. 2020;6(1):a004549.

96

Schimmelpenning–Feuerstein–Mims Syndrome

Dongyou Liu

CONTENTS

96.1 Introduction

Schimmelpenning–Feuerstein–Mims (SFM) syndrome is a sporadic neurocutaneous disorder characterized by sebaceous nevus (small, flat, and hairless lesions on the head and neck appearing at birth or within the first months of life, with potential for malignant transformation), in association with extradermatological (neurologic, skeletal, cardiovascular, ophthalmic, and urologic) anomalies.

Molecularly, SFM syndrome is linked to postzygotic somatic mutations in the *HRAS* gene on chromosome 11p15.5, the *KRAS* gene on chromosome 12p12.1, or the *NRAS* gene on chromosome 1p13.2 [1].

96.2 Biology

Cases related to SFM syndrome (also known as Schimmelpenning syndrome, nevus sebaceus syndrome, linear sebaceous nevus sequence, nevus sebaceous of Jadassohn, Jadassohn nevus phacomatosis, Jadassohn sebaceous nevus syndrome, organoid nevus, and epidermal nevus syndrome) were initially described by Jadassohn in 1895. Further studies by Schimmelpenning in 1957 and Feuerstein and Mims in 1962 defined this disorder by the classic triad of symptoms (nevus sebaceus, epilepsy, and mental retardation). Since then, additional anomalies (including neurologic, ophthalmic, skeletal, cardiovascular, and urologic defects) have been reported from patients with SFM syndrome, which makes diagnosis of this disorder on the basis of classic triad inadequate [2].

Recent identification of postzygotic somatic *HRAS*, *KRAS*, and *NRAS* mutations in affected individuals provided valuable insights into the molecular pathogenesis of SFM syndrome. As the *HRAS*, *KRAS*, and *NRAS* genes are key components of the RAS/mitogen-activated protein kinase (MAPK) signaling pathway, a sporadic mutation in one of these genes occurring after fertilization of the embryo (postzygotic mutation), most likely early during embryonic development, may result in mosaicism, in which some cells contain an abnormal gene and others have a normal copy of this gene to ensure the survival of developing embryo. The ratio of healthy cells to abnormal cells and the timing of the mutational event have an obvious influence on the variability of phenotypic presentations. If all cells in the developing embryo harbor abnormal gene, the condition is incompatible with life and the embryo is not going to survive long [1].

96.3 Pathogenesis

The RAS gene family consists of three genes *HRAS* (chromosome 11p15.5), *KRAS* (chromosome 12p12.1), and *NRAS* (chromosome 1p22.2) that encode three proteins (K-RAS, H-RAS and N-RAS) with GTPase activity. In addition, the *RIT1* (RAS-like protein in tissues) gene encodes RIT1, which shares a 50% sequence homology with RAS. Despite its lack of the C-terminal lipidation site, RIT1 maintains the GTPase function. The K-RAS, H-RAS, N-RAS, and RIT1 proteins represent on-off binary switches for several downstream cascades in the RAS/MAPK signaling pathway, which regulates cell growth, survival and apoptosis (see Chapter 87 in this book for details).

Gain of function mutations in *HRAS*, *KRAS*, *NRAS*, and *RIT1* contribute to the activation of downstream effectors (e.g., RAF, MEK1/MEK2, and ERK1/ERK2), and overexpression of various nuclear and cytosolic effectors that alter the late developmental processes such as organogenesis, morphology determination, and growth, leading to the development of a class of clinically related disorders known as RASopathies (refer to Chapter 87). Given the role of *HRAS*, *KRAS*, and *NRAS* mutations in its pathogenesis, SFM syndrome is also considered one of RASopathies.

Postzygotic mutations of the *HRAS*, *KRAS*, and *NRAS* genes have been identified in patients with SFM syndrome. Of these, *HRAS* c.37G>C (p.Gly13Arg) is most common, and present in >90% of nevus sebaceous cases, while *KRAS* c.35G>A G12D and *NRAS* Q61R mosaic mutations are found in a number of SFM syndrome cases [3,4]. Further, somatic *HRAS* (e.g., p.Gly12Asp) and *KRAS* mutations are also found in isolated sebaceous nevus (in the absence of associated extracutaneous abnormalities) [5]. Interestingly, these *RAS* mutations are often found in secondary tumors (e.g., trichoblastoma, syringocystadenoma papilliferum, sebaceous adenoma, apocrine adenoma, and poroma), suggesting their transformation from nevus sebaceous. Indeed, multiple tumor-like proliferations in one nevus sebaceous lesion are regularly observed.

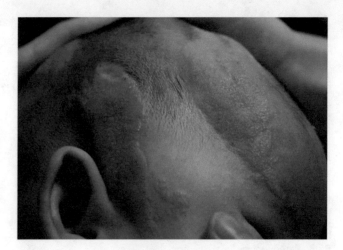

FIGURE 96.1 Schimmelpenning–Feuerstein–Mims syndrome in a 3-month-old girl with KRAS G12D mosaic mutation showing hairless, well demarcated, raised, velvety plaques (two linear and one round) on the skin of the scalp. (Photo credit: Wang H et al. *BMC Med Genet.* 2015;16:101.)

96.4 Epidemiology

96.4.1 Prevalence

SFM syndrome affects approximately 1 per 10,000 live births (or 0.3% of newborns) and shows no sexual predilection.

96.4.2 Inheritance

SFM syndrome appears to be due to an autosomal dominant lethal mutation that survives by somatic mosaicism. As all reported cases have occurred sporadically (i.e., the mutation is present in tumor tissue but absent in normal skin or peripheral blood), this disorder is not associated with chromosomal abnormalities. However, familial aggregation may sometimes be observed. The concept of paradominant inheritance has been put forward to explain this phenomenon.

96.4.3 Penetrance

SFM syndrome shows high penetrance and variable expressivity. Individuals harboring *HRAS* and *KRAS* mutations may develop sebaceous nevus without extracutaneous abnormalities.

96.5 Clinical Features

Clinically, patients with SFM syndrome often present with the classic triad of symptoms (nevus sebaceous, epilepsy, and mental retardation) as well as other anomalies involving neurologic, skeletal, cardiovascular, ophthalmic, and urologic systems [6].

Nevus sebaceous, a hallmark of SFM syndrome, is an oval or linear, sharply demarcated and slightly raised, yellowish-orange to pink, finely papillomatous, alopecic plaque (of 1–6 cm in diameter or length, with linear lesion often distributed along the lines of Blaschko and accompanied by seizures and mental retardation) with waxy or pebbly surface that is found on the epidermis, hair follicles, and sebaceous and apocrine glands of the head (59.3%), face (32.6%), neck, and trunk (1.3%) (Figure 96.1) [4]. Being a hamartoma, nevus sebaceous often contains epidermal, follicular, sebaceous, and apocrine elements, and typically undergoes three developmental stages: stage 1 (from birth to puberty) is a small, flat lesion with papillomatous hyperplasia and immature hair follicles; stage 2 (during puberty) shows massive growth of the sebaceous glands, papillomatous epidermal hyperplasia, and maturation of the apocrine glands (under the influence of androgen), and gives a verrucous appearance; and stage 3 (rarely before the age of 30 years) has a 25% chance of developing secondary benign tumors (e.g., trichoblastoma, syringocystadenoma papilliferum, syringoma, nodular hidradenoma, sebaceous thelioma, and keratoacanthoma), and a 5%–20% chance of malignant transformation (e.g., basal cell epithelioma, squamous cell carcinoma, sebaceous carcinoma) [1,7]. In some unusual cases, speckled lentiginous nevus/nevus spilus (a typical feature of phacomatosis pigmentokeratotica) may form part of the clinical synopsis of SFM syndrome, which differs from phacomatosis pigmentokeratotica only in that the progenitor cell has lost ability to differentiate in the melanocytes at the time of a mutation) (Figure 96.2) [8].

Neurologic abnormalities are present in >60% of SFM syndrome cases. Of these, seizures (epilepsy) affect 67% of patients from the first year of life and are resistant to conventional antiepileptic drugs. Mental retardation is found in 61% of cases. Hemimegalencephaly (one side of the brain being larger than the other) is observed in 72% of the patients and causes infantile spasms (a combination of nodding spasms, hypsarrhythmia on EEG, and mental retardation). Developmental delay and intellectual impairment are sometimes observed as a result of damages to certain cranial nerves by altered structures of the brain (e.g., hemimegalencephaly), malformation (dysplasia) of certain brain vessels, absence (agenesis) of the nerve bundle connecting the two cerebral hemispheres (corpus callosum), and defects of the folds of the brain such as agyria (a smooth brain lacking the distinctive folds), microgyria (abnormally small folds), or pachygyria (abnormally thickened folds), as well as by excessive fluid accumulation around the brain (congenital hydrocephalus).

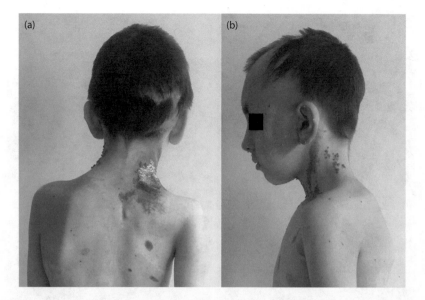

FIGURE 96.2 Schimmelpenning–Feuerstein–Mims syndrome in a 9-year-old male showing both sebaceous and speckled lentiginous nevi. (a) Hairless lesions of epidermal nevus on the scalp and a giant café-au-lait spot extending to the left shoulder and neck, with spots and papules of pigmentation; (b) lentiginous and papular melanocytic lesions in the area of the café-au-lait spot and also in the area of epidermal nevus on the face. (Photo credit: Gamayunov BN et al. *Clin Case Rep.* 2016;4(6):564–7.)

Ocular abnormalities (e.g., eyelid, iris and chorioretinal colobomas, strabismus, cataracts, corneal vascularization, generalized retinal degeneration, bulbar dermoids with pannus formation, ocular hemangioma, hamartoma of the eyelid, asymmetry of orbital bones, and esotropia) are observed in 50%–59% of individuals with SFM syndrome.

Craniofacial defects (e.g., frontal bossing, underdeveloped nasal and orbital bones, and asymmetry of the skull) may occur. In addition, bone cysts, underdevelopment of the pelvis, and incomplete formation of the ankle, foot, and bones of the spinal column (vertebrae) are also noted.

96.6 Diagnosis

Diagnosis of SFM syndrome relies on observation of nevus sebaceous, epilepsy/mental retardation, and other clinical signs.

Diagnostic workup encompasses biopsy of affected skin (for identification of sebaceous nevus), skeletal survey, complete ophthalmologic exam, chest x-rays, and brain imaging (using computerized tomography/CT and magnetic resonance imaging/MRI, in the case of central nervous system involvement).

Histopathologically, as a benign solid tumor originating from external sheath cells of pilosebaceous follicles, nevus sebaceous displays slight or prominent hyperplasia, embriotic or completely developed hair follicles, and overgrowth of sebaceous glands. Lesion from the infant or young child may feature immature and abnormally formed pilosebaceous units, along with some epidermal changes (acanthosis and mild papillomatosis). Lesion during puberty shows prominent sebaceous glands in the dermis, increased number of sebaceous lobules and malformed ducts, and hair follicles containing typically immature vellus hairs rather than terminal hairs. Lesion during the growth phase reveals more papillated and acanthotic epidermis (Figure 96.3) [9]. In some cases, nevus sebaceous may develop/transform into

trichoblastoma, syringocystadenoma papilliferum, sebaceoma, and basal cell carcinoma (Figure 96.4) [10–12].

Trichoblastoma (trichoblastic fibroma) is a solitary, small, non-ulcerated nodule over the scalp and trunk, and may be separated into five common histological patterns: large nodular (including

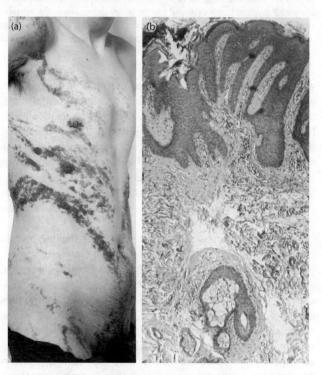

FIGURE 96.3 Schimmelpenning–Feuerstein–Mims syndrome in an 18-year-old male showing a linear verrucous nevus sebaceous along the lines of Blaschko on the trunk and extremities (a). Histology of nevus sebaceous showing epidermal hypertrophy, elongation of rete ridges with hypoplastic hair follicles, the presence of epidermal cysts and sebaceous glands (b). (Photo credit: Kiedrowicz M et al. *Postepy Dermatol Alergol.* 2013;30(5):320–3.)

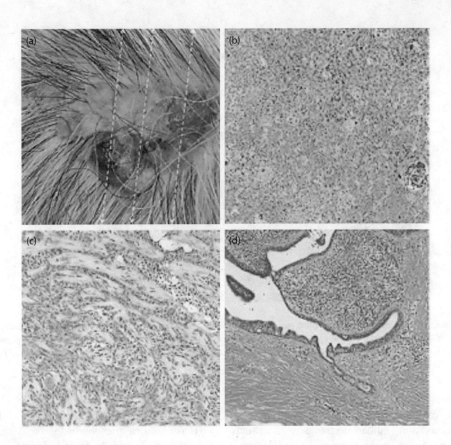

FIGURE 96.4 Schimmelpenning–Feuerstein–Mims syndrome in a 67-year-old female showing a slightly elevated light brown plaque (47 × 26 mm) on the back of the head, including an 18 × 17 mm dome-shaped black node and a 16 × 13 mm erosive lesion (a). Histopathology of the black node showing tumor nests with atypical tumor cells and insular tumor aggregations (b). Histopathology of the periphery of the black node showing tumor nests with numerous structures resembling follicular germinative cells and some of those structures extruding from interlacing cords (c). Histology of the erosive lesion showing the epithelium of tubules with decapitation secretion and numerous plasma cells in the stroma (d). (Photo credit: Namiki T et al. *Case Rep Dermatol.* 2016;8(1):75–9.)

pigmented), small nodular, cribriform, racemiform, and retiform, and five uncommon variants: adamantoid, columnar, rippled pattern, subcutaneous, and superficial. Syringocystadenoma papilliferum displays papillomatous hyperplasia in the epidermis, many irregular duct-like structures, and cystic spaces lined with a double-layered epithelium. Basal cell carcinoma comprises stroma-dependent multipotent basaloid cells and differentiates toward the epidermis or adnexa; having the same origin as nevus sebaceous, it has the characteristic of multidifferentiation of the skin stem cells.

Differential diagnoses of nevus sebaceous include aplasia cutis congenita or mastocytomas (early infant lesion); congenital nevi, epidermal nevus, seborrheic keratosis, verruca, and juvenile xanthogranulomas (later-stage lesion); phakomatosis pigmentokeratotica (known as speckled lentiginous nevus of the papular type; large, light-brown discoloration of the skin, superimposed by multiple darkened/melanocytic spots/papules), seizures, intellectual impairment, muscle weakness, paralysis on one side of the body (hemiparesis), underdevelopment of one side of the body (hemiatrophy), excessive sweating (hyperhidrosis), and cutaneous dysesthesia, abnormal side-to-side curvature of the spine (scoliosis), vitamin D-resistant rickets, short stature, propensity to fractures, hearing loss in one ear, crossed eyes (strabismus), droopy upper eyelid (ptosis), and narrowing of the aorta (aortic stenosis),

rhabdomyosarcoma, postzygotic HRAS mutation [8,13], CHILD syndrome (congenital hemidysplasia with ichthyosiform nevus and limb defects), type 2 segmental Cowden disease, Proteus syndrome, fibroblast growth factor receptor 3 epidermal nevus syndrome (García-Hafner-Happle syndrome), nevus trichilemmocysticus syndrome, didymosis aplasticosebacea, SCALP syndrome (sebaceous nevus, central nervous system malformations, aplasia cutis congenita, limbal dermoid and pigmented nevus), Gobello syndrome, Bäfverstedt syndrome, NEVADA syndrome (nevus epidermicus verrucosus with angiodysplasia and aneurysms), and CLOVE syndrome (congenital lipomatous overgrowth, vascular malformations, and epidermal nevus) [14].

96.7 Treatment

Treatment of SFM syndrome relies on standard procedures for specific symptomatic relief. For example, surgery (e.g., photodynamic therapy, dermabrasion, laser therapy, and cryotherapy) is useful for rectifying certain skeletal and ocular malformations, and for improving cosmetic appearance of affected individuals; anti-seizure medications may help relieve epilepsy, and remedial education, physical therapy, and occupational therapy may be considered when necessary [8,15].

96.8 Prognosis and Prevention

Nevus sebaceous shows a relatively high risk for developing into secondary benign neoplasms (e.g., trichoblastoma, syringocystadenoma papilliferum, tricholemmoma, sebomatricoma, apocrine hidrocystoma, and apocrine poroma), but a low risk for transforming into secondary malignant tumors (e.g., basal cell carcinoma, squamous cell carcinoma, sebaceous carcinoma) within the lesion, which tends to occur in adults 40 years or older. Surgical excision of nevus sebaceous with a safe margin helps reduce the risk of secondary tumor development or transformation.

96.9 Conclusion

Schimmelpenning–Feuerstein–Mims (SFM) syndrome is a RASopathy that results from mosaicism secondary to postzygotic mutations in the *HRAS*, *KRAS*, and *NRAS* genes during the early embryonic period [16]. Clinically, this disorder is characterized by classical sebaceous nevus on the scalp, forehead, neck and trunk, along with multisystem involvement (neurological, ocular, renal, cardiac, or skeletal anomalies) [17,18]. Sebaceous nevus is a congenital skin hamartoma that typically appears as a yellow-hued plaque (its occurrence along Blaschko's lines suggests a mosaic genetic mutation), displays epidermal acanthosis and hyperplasia or structural abnormalities of sebaceous glands and hair follicles, and has the potential of developing into secondary benign tumors (e.g., trichoblastoma and syringocystadenoma papilliferum) as well as malignant tumors (e.g., basal cell carcinoma) [19]. Diagnosis of SFM syndrome involves observation of characteristic clinical features. Identification of *HRAS*, *KRAS*, and *NRAS* mutations in tumor specimens is informative but nonspecific for SFM syndrome. Treatment of this disorder relies on standard procedures that provide symptomatic relief and improve cosmetic appearance for affected individuals.

REFERENCES

1. Baigrie D, Cook C. *Nevus sebaceous*. StatPearls [Internet]. Treasure Island (FL): StatPearls Publishing; Jan, 2018. Feb 15, 2018.
2. Resende C, Araújo C, Vieira AP, Ventura F, Brito C. Schimmelpenning syndrome. *Dermatol Online J.* 2013;19(10):20026.
3. Sun BK, Saggini A, Sarin KY et al. Mosaic activating RAS mutations in nevus sebaceus and nevus sebaceus syndrome. *J Invest Dermatol.* 2013;133(3):824–7.
4. Wang H, Qian Y, Wu B, Zhang P, Zhou W. KRAS G12D mosaic mutation in a Chinese linear nevus sebaceous syndrome infant. *BMC Med Genet.* 2015;16:101.
5. Lihua J, Feng G, Shanshan M, Jialu X, Kewen J. Somatic KRAS mutation in an infant with linear nevus sebaceous syndrome associated with lymphatic malformations: A case report and literature review. *Medicine (Baltim).* 2017;96(47):e8016.
6. Wang SM, Hsieh YJ, Chang KM, Tsai HL, Chen CP. Schimmelpenning syndrome: A case report and literature review. *Pediatr Neonatol.* 2014;55(6):487–90.
7. Gupta SK, Gupta V. Basal cell carcinoma and syringocystadenoma papilliferum arising in nevus sebaceous on face-A rare entity. *Indian J Dermatol.* 2015;60(6):637.
8. Gamayunov BN, Korotkiy NG, Baranova EE. Phacomatosis pigmentokeratotica or the Schimmelpenning–Feuerstein–Mims syndrome? *Clin Case Rep.* 2016;4(6):564–7.
9. Kiedrowicz M, Kacalak-Rzepka A, Królicki A, Maleszka R, Bielecka-Grzela S. Therapeutic effects of CO_2 laser therapy of linear nevus sebaceous in the course of the Schimmelpenning–Feuerstein–Mims syndrome. *Postepy Dermatol Alergol.* 2013;30(5):320–3.
10. Namiki T, Miura K, Ueno M, Arima Y, Nishizawa A, Yokozeki H. Four different tumors arising in a nevus sebaceous. *Case Rep Dermatol.* 2016;8(1):75–9.
11. Jardim MML, Souza BCE, Fraga RC, Fraga RC. Rare desmoplastic trichilemmoma associated with sebaceous nevus. *An Bras Dermatol.* 2017;92(6):836–7.
12. Sathyaki DC, Riyas M, Roy MS, Swarup RJ, Raghu N. Pigmented trichoblastoma of nose: An unusual occurrence. *J Clin Diagn Res.* 2017;11(7):MD09–10.
13. Groesser L, Herschberger E, Sagrera A et al. Phacomatosis pigmentokeratotica is caused by a postzygotic HRAS mutation in a multipotent progenitor cell. *J Invest Dermatol.* 2013;133(8):1998–2003.
14. Kasinathan A, Padmanabh H, Gupta K, Sankhyan N, Singh P, Singhi P. Unusual cause of West syndrome. *J Pediatr Neurosci.* 2017;12(3):288–90.
15. Trivedi N, Nehete G. Complex limbal choristoma in linear nevus sebaceous syndrome managed with scleral grafting. *Indian J Ophthalmol.* 2016;64(9):692–4.
16. Kim YE, Baek ST. Neurodevelopmental Aspects of RASopathies. *Mol Cells.* 2019;42(6):441–7.
17. Lena CP, Kondo RN, Nicolacópulos T. Do you know this syndrome? Schimmelpenning-Feuerstein-Mims syndrome. *An Bras Dermatol.* 2019;94(2):227–9.
18. Watson IT, DeCrescenzo A, Paek SY. Basal cell carcinoma within nevus sebaceous of the trunk. *Proc (Bayl Univ Med Cent).* 2019;32(3):392–3.
19. Paninson B, Trope BM, Moschini JC, Jeunon-Sousa MA, Ramos-E-Silva M. Basal cell carcinoma on a nevus sebaceous of Jadassohn: A case report. *J Clin Aesthet Dermatol.* 2019;12(3):40–3.

97

Simpson–Golabi–Behmel Syndrome

Dongyou Liu

CONTENTS

97.1 Introduction

Simpson–Golabi–Behmel syndrome (SGBS) is a rare, X-linked, congenital overgrowth disorder characterized by pre/postnatal macrosomia, distinctive craniofacies (e.g., macrocephaly, ocular hypertelorism [widely spaced eyes], macrostomia [unusually large mouth], macroglossia [large tongue with a deep groove or furrow down the middle], broad nose with an upturned tip, palatal abnormalities), organomegaly (often involving the kidneys, liver, or spleen), skeletal anomalies (e.g., vertebral fusion, scoliosis, rib anomalies, and congenital hip dislocation), hand anomalies (e.g., large hands and postaxial polydactyly), supernumerary nipples, diastasis recti (an abnormal opening in the muscle covering the abdomen), umbilical hernia (a soft out-pouching around the navel), congenital heart defects, diaphragmatic hernia (a hole in the diaphragm), genitourinary defects, gastrointestinal anomalies, mild to severe intellectual disability, early motor milestones, and speech delay as well as 10% risk of developing for embryonal tumors (e.g., Wilms tumor, hepatoblastoma, adrenal neuroblastoma, gonadoblastoma, hepatocellular carcinoma, and medulloblastoma) in early childhood [1,2].

Molecular mechanisms of SGBS relate to genomic rearrangements and point mutations involving the glypican-3 gene (*GPC3*) and glypican-4 gene (*GPC4*) on chromosome Xq26.2. As glypicans are heparan sulfate proteoglycans involved in the control of cell growth and cell division, alterations in these molecules contribute to overgrowth and tumor susceptibility.

97.2 Biology

Early cases of SGBS (also known as bulldog syndrome, X-linked dysplasia gigantism syndrome; encephalo-tropho-schisis syndrome, Golabi–Rosen syndrome; Simpson dysmorphia syndrome) were reported in 1975 by Simpson and coworkers (involving two sons of sisters, with large protruding jaw, widened nasal bridge, upturned nasal tip, enlarged tongue, broad stocky appearance, broad and short hands and fingers.), and in 1984 by Golabi and Rosen (involving four male siblings connected through females, with prenatal and postnatal overgrowth, short, broad, upturned nose; large mouth, midline groove of tongue, inferior alveolar ridge and lower lip, submucous cleft palate, 13 ribs, Meckel diverticulum, intestinal malrotation, coccygeal skin tag and bony appendage, hypoplastic index fingernails, unilateral postaxial polydactyly, bilateral syndactyly of fingers 2 and 3, mental retardation) as well as Behmel (involving 11 male newborns with elevated birth weight and length, disproportionately large head with coarse, distinctive facies, short neck, slight obesity, and broad, short hands and feet), after whom this congenital overgrowth is named. In 1995, a lethal but infrequent form of SGBS (known as SGBS type II) was reported by Terespolsky and coworkers [2].

As an overgrowth/multiple congenital anomaly disorder, SGBS demonstrates high clinical variability, ranging from very mild form in carrier females to lethal form in males. Mutations in semi-dominant X-linked genes encoding glypican 3 (*GPC3*) and glypican 4 (*GPC4*) appear to be responsible for causing the classical subtype (SGBS type I) and an infrequent but lethal subtype (SGBS type II), respectively [3–6].

97.3 Pathogenesis

Glypicans are a family of proteoglycans, each of which consists of a core protein attached to long sugar molecules called heparan sulfate chains. Glypicans attach to the outer cell membrane via a glycosylphosphatidylinositol linkage. Through interactions with other proteins outside the cell, glypicans participate in the regulation of several signaling pathways [7].

Within the mammalian genome, six glypican genes (*GPC1-GPC6*) exist. Of these, the *GPC3* gene on chromosome Xq26.2 spans >500 kb with eight exons and encodes a glycosylphosphatidylinositol-linked cell surface heparan sulfate proteoglycan (glypican 3 [GPC3]), which binds and regulates the activities of various extracellular ligands essential to cellular functions.

GPC3 acts as an inhibitor of hedgehog (Hh) signaling pathway and binding of hedgehog to GPC3 triggers the endocytosis and degradation of the GPC3/ hedgehog complex, inducing apoptosis of certain unwanted cells. However, loss of functional GPC3 results in hyperactivation of Hh signaling pathway and increased cell proliferation, which accounts in part for the overgrowth associated with SGBS [7,8].

Deletions or single-nucleotide variants (e.g., splice site, frameshift, missense, and nonsense) in any of the eight *GPC3* exons is sufficient to cause SGBS. To date, >50 *GPC3* mutations have been identified in patients with SGBS, with 50% of *GPC3* deletions located in exon 8, and many single-nucleotide variants found in exon 3 (the largest exon). In heterozygous *GPC3* females, >43% loss of functional GPC3 protein is required for SGBS development [9].

The *GPC4* gene is situated adjacent to the 3' end of the *GPC3* gene on chromosome Xq26.2. Composed of nine exons, *GPC4* also encodes a glycosylphosphatidylinositol-linked cell surface heparan sulfate proteoglycan (glypican-4 or GPC4). Duplication of *GPC4* exons 1–9 in the absence of a *GPC3* pathogenic variant is associated with SGBS, while loss-of-function *GPC4* variant is not.

97.4 Epidemiology

97.4.1 Prevalence

SGBS is a rare overgrowth disorder, with an estimated incidence that appears to be lower than that of Beckwith–Wiedemann syndrome (BWS) and Sotos syndrome. To date, >250 cases have been diagnosed with this disorder, including about 10 clinically lethal cases.

97.4.2 Inheritance

Considering that the *GPC3* and *GPC4* genes whose mutations are responsible for SGBS are located on the X chromosome, this condition is inherited in an X-linked manner. As females possess two copies of X chromosome, one altered copy of the *GPC3* or

GPC4 gene in each cell may be insufficient to induce clinical disease or may only cause a mild phenotype. However, a heterozygous female carrier has a 50% chance in each pregnancy passing a pathogenic variant to offspring. While males who inherit the pathogenic variant will be affected and develop an overt clinical phenotype, females who inherit the pathogenic variant will be carriers. Further, affected males will only pass the pathogenic variant to female offspring, but not male offspring. Although a majority of SGBS cases follow an X-linked inheritance, some cases (about 20%–30%) appear to arise sporadically. Further, at least one affected family results from a female carrier showing germline mosaicism [2].

97.4.3 Penetrance

SGBS penetrance in males appears to be complete, and all males with a *GPC3* or *GPC4* pathogenic variant develop characteristic clinical manifestations. SGBS penetrance in heterozygous females (who may have mild symptoms) has not been thoroughly investigated.

97.5 Clinical Features

SGBS is responsible for causing a broad spectrum of clinical signs and symptoms, varying from infantile lethal forms in affected males to very mild forms in carrier females. Affected males typically present with macrosomia, distinctive facies, multiple congenital anomalies, and tumors [2,10].

97.5.1 Macrosomia

Pre- and postnatal overgrowth leads to weight or length ≥95th percentile when adjusted for sex and age. Hypoglycemia may also occur in neonates.

97.5.2 Distinctive Facies

Distinctive facies include (i) macrocephaly (70% of cases); (ii) ocular hypertelorism, epicanthal folds, downslanting palpebral fissures, strabismus or esotropia, cataracts, coloboma of the optic disc, ocular nerve palsies; (iii) redundant, furrowed skin over the glabella; (iv) wide nasal bridge and anteverted nares in infants, broad nose in older individuals; (v) macrostomia (abnormally large mouth); (vi) macroglossia (abnormally large tongue); (vii) dental malocclusion; (viii) midline groove in the lower lip and/or deep furrow in the middle of the tongue; (ix) cleft lip and/or submucous cleft palate (with a bifid uvula, 13%), high and narrow palate; (x) micrognathia (small mandible) in neonates, macrognathia in older individuals; (xi) preauricular tags, fistulas, ear lobule creases, helical dimples [2].

97.5.3 Multiple Congenital Anomalies

Multiple congenitcal anomalies include (i) congenital heart disease (septal defects, pulmonic stenosis, aortic coarctation, patent ductus arteriosus, patent foramen ovale, cardiomyopathy); (ii) conduction defects (transient QT interval prolongation), arrhythmias, and ECG abnormalities (12%); (iii) supernumerary nipples; (iv) diastasis recti/umbilical hernia; (v) diaphragmatic hernia (<10%); (vi) renal dysplasia, nephromegaly, hydrone-phrosis, hydroureter, and duplicated ureters; (vii) cryptorchidism, hypospadias, bifid scrotum, hydrocele, inguinal hernia; (viii) gastrointestinal anomalies (pyloric ring, Meckel diverticulum, intestinal malrotation, hepatosplenomegaly, pancreatic hyperplasia of islets of Langerhans, choledochal cysts, duplication of the pancreatic duct, polysplenia); (ix) skeletal anomalies (vertebral fusion, scoliosis, pectus excavatum, rib anomalies, winged scapula, congenital hip dislocation, small sciatic notches, flared iliac wings, extra lumbar vertebrae, spina bifida occulta, coccygeal skin tag, bony appendage); (x) hand anomalies (brachydactyly, cutaneous syndactyly, postaxial polydactyly, clinodactyly, large hands, broad thumbs); (xi) neurologic anomalies (hypotonia, absent primitive reflexes, high-pitched cry in neonates, seizures, agenesis of the corpus callosum, Chiari malformation and hydrocephalus, aplasia of the cerebellar vermis) (Figure 97.1) [11–14].

97.5.4 Tumors

Tumors include Wilms tumor, hepatoblastoma, adrenal neuroblastoma, gonadoblastoma, hepatocellular carcinoma, and medulloblastoma in early childhood (10%) [15].

97.6 Diagnosis

Diagnosis of SGBS involves clinical examination (for macrosomia, macrocephaly, pre- and postnatal overgrowth, coarse/characteristic facies, midline defects, organomegaly, tumor predisposition), family history review (X-linked inheritance), and

molecular genetic testing (for *GPC3* and *GPC4* pathogenic variants) (Figure 97.2) [14,16].

Molecular identification of *GPC3* and *GPC4* pathogenic variants helps confirm or establish the diagnosis of SGBS in individuals showing characteristic clinical symptoms. Typically, sequence analysis of *GPC3* is followed by deletion/duplication analysis of *GPC3* (if no pathogenic variant is identified by sequence analysis), then deletion/duplication analysis of *GPC4* (if no *GPC3* pathogenic variant is detected). In general, *GPC3* mutations are observed in up to 70% of SGBS patients.

Identification of female carriers is based on either (i) prior identification of the pathogenic variant in the family or (ii) sequence analysis (if an affected male is unavailable for testing), and then detection of gross structural abnormalities (if no pathogenic variant is identified).

Differential diagnoses for SGBS include the following:

Beckwith–Wiedemann syndrome (BWS): Macrosomia, macroglossia, visceromegaly, Wilms tumor, hepatoblastoma, neuroblastoma, rhabdomyosarcoma, omphalocele, neonatal hypoglycemia, ear creases/pits, adrenocortical cytomegaly, medullary dysplasia, nephrocalcinosis, medullary sponge kidney, nephromegaly; due to *CDKN1C* mutations; differing from SGBS by absence of skeletal abnormalities, and autosomal dominant inheritance

Sotos syndrome: Typical facial appearance, overgrowth, neonatal jaundice, scoliosis, seizures, strabismus, conductive hearing loss, congenital cardiac anomalies, renal anomalies, intellectual impairment, behavioral problems, sacrococcygeal teratoma, neuroblastoma;

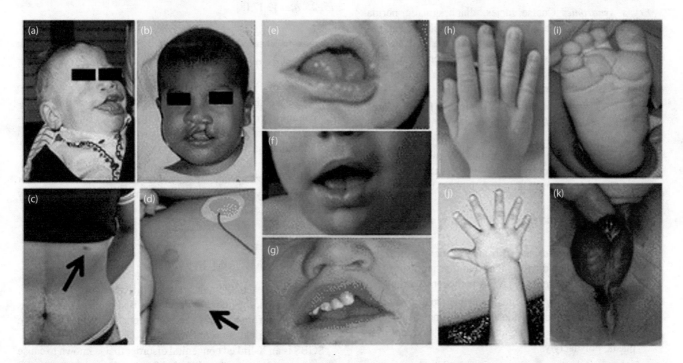

FIGURE 97.1 Clinical presentations of Simpson–Golabi–Behmel syndrome. (a and b) Cleft lip, coarse, square face and broad nose. (c and d) Extra nipple in a carrier mother and a toddler. (e–g) Large tongue, middle groove in the tongue, teeth malposition, and repaired cleft. (h and j) Broad hands and polydactyly. (i) Deep plantar creases. (k) Abnormal genitalia in a male with hypospadias and proximal anal placement. (Photo credit: Tenorio J et al. *Orphanet J Rare Dis*. 2014;9:138.)

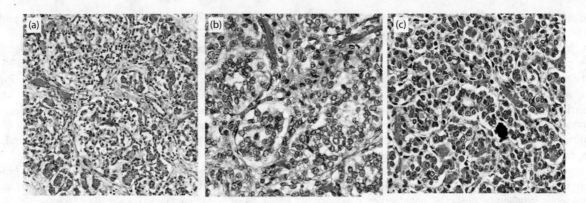

FIGURE 97.2 Simpson–Golabi–Behmel syndrome in an infant showing diffuse and coalescing islets with increasing number of pancreatic islet cells in the pancreas (a); hyperplastic seminiferous tubules with increased numbers of germ cells in the testis (b); and focal pseudoglandular transformation in the pituitary gland (c). (Photo credit: Zimmermann N, Stanek J. *Am J Case Rep.* 2017;18:649–55.)

pathogenic variant or deletion of *NSD1* in 80%–90% of affected individuals; autosomal dominant inheritance

Weaver syndrome: Overgrowth, umbilical hernia, ear anomalies, hypotonia, advanced bone age, vertebral defects, hypertelorism; different facies and more prominent psychomotor delay

Nevoid basal cell carcinoma syndrome (NBCCS, Gorlin syndrome): Multiple jaw keratocysts, basal cell carcinomas, macrocephaly, bossing of the forehead, coarse facial features, facial milia, bifid ribs or wedge-shaped vertebrae, ectopic calcification in the falx, cardiac and ovarian fibromas, medulloblastoma or primitive neuro-ectodermal tumor in early childhood, germline *PTCH* pathogenic variant, autosomal dominant inheritance

Fryns syndrome: Coarse facies, diaphragmatic hernia with lung hypoplasia, distal limb hypoplasia; anomalies involving cardiovascular, gastrointestinal, genito-urinary (renal cystic dysplasia), and central nervous system (arrhinencephaly, Dandy–Walker anomaly, agenesis of the corpus callosum)

Perlman syndrome: Macrosomia, distinctive facies, Wilms tumor; biallelic *DIS3L2* pathogenic variants, autosomal recessive inheritance.

Nevo syndrome: Vertebral anomalies, ear malformations, cryptorchidism, overgrowth, intellectual disability; accelerated osseous maturation, large extremities, and hypotonia.

PLOD1: Exon 9 pathogenic variants, autosomal recessive inheritance.

Marshall–Smith syndrome: Advanced bone age, intellectual disability, distinctive facial features, predisposition to fractures; heterozygous *NFIX* pathogenic variant.

Elejalde syndrome: Acrocephalopolydactylous dysplasia; macrosomia, abnormal facies, craniosynostosis, acrocephaly, omphalocele, organomegaly, cystic renal dysplasia, polydactyly.

Infant of diabetic mother syndrome: Sacral agenesis, hypogenesis and/or caudal dysgenesis; congenital heart defects, renal anomalies, vertebral anomalies, limb defects, and structural brain abnormalities.

Mosaic trisomy 8: Advanced growth, long slender trunk, spinal deformities, contractures of fingers and toes, absence of the corpus callosum, moderate intellectual disability, typical facial features (high, prominent forehead, hypertelorism, full lips, and micrognathia)

Mosaic tetrasomy 12p (or Pallister–Killian syndrome): Variegated skin pigmentation, facial anomalies (prominent forehead with sparse anterior scalp hair, ocular hypertelorism, short nose with anteverted nares, flat nasal bridge), developmental delay

Additional differential diagnoses include *fragile X syndrome*, *Bannayan–Zonana syndrome*, *PTEN hamartoma tumor syndrome*, *neurofibromatosis type I*, *Marfan syndrome*, and *trisomy 15q26-qter* [2,17,18].

97.7 Treatment

Treatment of SGBS relies largely on standard procedures to relieve and rectify symptoms and pathologic changes that endanger patient life or interfere with patient daily routines (e.g., neonatal hypoglycemia and airway obstruction due to micrognathia and glossoptosis, cleft lip/palate, macroglossia and feeding difficulties, heart defects, skeletal and urogenital abnormalities), accompanied by speech therapy, occupational therapy, and/or physical therapy. Anticoagulation and antibiotic prophylaxis may be considered for SGBS patients with congenital heart disease.

97.8 Prognosis and Prevention

Given that males and females possess differing copy numbers of X chromosome, they tend to develop varied clinical phenotypes when one X chromosome is altered by genetic changes, and have contrasting clinical consequences.

SGBS is an X-linked congenital disorder that is known to cause a severe, sometimes infantile lethal form of disease in males, but a very mild form of disease in carrier females. Not surprisingly, prognosis for affected males is generally poor, and some male patients may die in the newborn period (due probably to heart defects). On the other hand, prognosis for affected females is favorable, and most female patients live into adulthood.

Regular surveillance for males with SGBS consists of ophthalmologic evaluation, audiologic evaluation, sleep study, routine monitoring of renal function, evaluation for scoliosis, monitoring of serum glucose levels for hypoglycemia secondary to increased risk for hyperinsulinemia, monitoring of developmental progress., screening for Wilms tumor, hepatoblastoma, neuroblastoma, and gonadoblastoma, and physical examination.

Prenatal diagnosis (ultrasound examination) and preimplantation genetic testing may be employed to prospective parents who harbor *GPC3* and *GPC4* pathogenic variants. Ultrasound approach helps reveal increased nuchal translucency, macrosomia, cleft lip or palate, nephromegaly, macroglossia, hydrops/ascites, and disproportionate overgrowth in affected fetus [17].

97.9 Conclusion

Simpson–Golabi–Behmel syndrome (SGBS) is an X-linked disorder that typically displays prenatal and postnatal overgrowth, coarse facial features, and congenital anomalies (e.g., skeletal/hand defects, supernumerary nipples, macroglossia, and visceromegaly), with severe, sometimes fatal consequences in affected males, and no or limited effects on carrier females. Molecular genetic investigations have delineated inactivating mutations in the *GPC3* and *GPC4* genes located at Xq26.2, which encode glypicans 3 and 4 involved in the modulation of cellular responses to growth factors (e.g., IGF-2) and in the interaction between GPC3 and CD26, as the underlying mechanisms of SGBS. These insights are instrumental in helping design a molecular diagnostic platform for confirming the diagnosis of SGBS in individuals showing clinical signs and symptoms that may overlap other overgrowth syndromes, and also develop a prevention program based on prenatal diagnosis and preimplantation testing for prospective parents who harbor *GPC3* or *GPC4* pathogenic variant. Unquestionably, further elucidation of molecular basis of SGBS will enable development of innovative therapies for effective treatment of this rare congenital disorder.

REFERENCES

1. Golabi M, Leung A, Lopez C. Simpson-Golabi-Behmel syndrome type 1. In: Adam MP, Ardinger HH, Pagon RA et al. editors. *GeneReviews®* [Internet]. Seattle (WA): University of Washington; 1993–2018. Dec 19, 2006 [updated Jun 23, 2011].
2. Sajorda BJ, Gonzalez-Gandolfi CX, Hathaway ER, Kalish JM. Simpson-Golabi-Behmel syndrome type 1. In: Adam MP, Ardinger HH, Pagon RA et al. editors. *GeneReviews®* [Internet]. Seattle (WA): University of Washington; 1993–2019. Dec 19 2006 [Nov 29 updated 2018].
3. Lindsay S, Ireland M, O'Brien O et al. Large scale deletions in the GPC3 gene may account for a minority of cases of Simpson-Golabi-Behmel syndrome. *J Med Genet.* 1997;34(6):480–3.
4. Brzustowicz LM, Farrell S, Khan MB, Weksberg R. Mapping of a new SGBS locus to chromosome Xp22 in a family with a severe form of Simpson-Golabi-Behmel syndrome. *Am J Hum Genet.* 1999;65(3):779–83.
5. Xuan JY, Hughes-Benzie RM, MacKenzie AE. A small interstitial deletion in the GPC3 gene causes Simpson-Golabi-Behmel syndrome in a Dutch-Canadian family. *J Med Genet.* 1999;36(1):57–8.
6. Geiger K, Leiherer A, Muendlein A et al. Identification of hypoxia-induced genes in human SGBS adipocytes by microarray analysis. *PLoS One.* 2011;6(10):e26465.
7. Filmus J, Capurro M. The role of glypicans in Hedgehog signaling. *Matrix Biol.* 2014;35:248–52.
8. Pellegrini M, Pilia G, Pantano S et al. Gpc3 expression correlates with the phenotype of the Simpson-Golabi-Behmel syndrome. *Dev Dyn.* 1998;213(4):431–9.
9. Shimojima K, Ondo Y, Nishi E et al. Loss-of-function mutations and global rearrangements in GPC3 in patients with Simpson-Golabi-Behmel syndrome. *Hum Genome Var.* 2016;3:16033.
10. Agarwal M, Sharma R, Panda A, Gupta A. Laryngeal web associated with Simpson-Golabi-Behmel syndrome in a child. *Anaesth Intensive Care.* 2009;37(4):671–2.
11. Savarirayan R, Bankier A. Simpson-Golabi-Behmel syndrome and attention deficit hyperactivity disorder in two brothers. *J Med Genet.* 1999;36(7):574–6.
12. Gertsch E, Kirmani S, Ackerman MJ, Babovic-Vuksanovic D. Transient QT interval prolongation in an infant with Simpson-Golabi-Behmel syndrome. *Am J Med Genet A.* 2010;152A(9):2379–82.
13. Tenorio J, Arias P, Martínez-Glez V et al. Simpson-Golabi-Behmel syndrome types I and II. *Orphanet J Rare Dis.* 2014;9:138.
14. Zimmermann N, Stanek J. Perinatal case of fatal Simpson-Golabi-Behmel syndrome with hyperplasia of seminiferous tubules. *Am J Case Rep.* 2017;18:649–55.
15. Lapunzina P, Badia I, Galoppo C et al. A patient with Simpson-Golabi-Behmel syndrome and hepatocellular carcinoma. *J Med Genet.* 1998;35(2):153–6.
16. Støve HK, Becher N, Gjørup V, Ramsing M, Vogel I, Vestergaard EM. First reported case of Simpson-Golabi-Behmel syndrome in a female fetus diagnosed prenatally with chromosomal microarray. *Clin Case Rep.* 2017;5(5):608–12.
17. Chen CP. Prenatal findings and the genetic diagnosis of fetal overgrowth disorders: Simpson-Golabi-Behmel syndrome, Sotos syndrome, and Beckwith-Wiedemann syndrome. *Taiwan J Obstet Gynecol.* 2012;51(2):186–91.
18. Eugster E. Gigantism. In: Feingold KR, Anawalt B, Boyce A et al. editors. *Endotext* [Internet]. South Dartmouth (MA): MDText.com, Inc.; 2000-.2018 Apr 17.

98

Sotos Syndrome

Dongyou Liu

CONTENTS

98.1 Introduction

Sotos syndrome is a congenital disorder that typically presents with distinctive facies (e.g., long, narrow face, high forehead, flushed/reddened cheeks, a small, pointed chin, downslanting palpebral fissures), overgrowth (tall stature and unusually large head) in childhood, intellectual disability, and behavioral problems (e.g., deficit hyperactivity disorder, phobias, obsessions/compulsions, tantrums, and impulsive behaviors; speech/language problems, stutter, monotone voice; weak muscle tone/hypotonia; motor skill problems), abnormal side-to-side curvature of the spine (scoliosis), seizures, heart or kidney defects, hearing loss, vision problems, yellowing of the skin and whites of the eyes (jaundice), poor feeding, and propensity for cancer development. Molecularly, Sotos syndrome is attributed to a heterozygous mutation in the nuclear receptor-binding SET domain protein 1 gene (*NSD1*) on chromosome 5q35.3 or a deletion in the 5q35 region including genomic sequence that makes up the *NSD1* gene, resulting in *NSD1* haploinsufficiency [1].

98.2 Biology

Sotos syndrome (formerly cerebral gigantism, also known as chromosome 5q35 deletion syndrome, Sotos sequence or the mental retardation–overgrowth sequence) was first described in 1964 from five children showing excessively rapid growth, acromegalic features, high-arched palate and prominent jaw, a nonprogressive cerebral disorder with mental retardation.

Although Sotos syndrome often occurs in people with no history of the disorder in their family, cases showing family links suggest its autosomal dominant inheritance. The identification of the *NSD1* gene located on chromosome 5q35.3 in 2002 further confirmed the familial nature of Sotos syndrome [2]. Subsequent evolutionary studies suggested that Sotos syndrome could be traced to the duplication of low-copy repeats (LCRs) flanking *NSD1* in the primate genome about 23.3–47.6 million years ago, before the divergence of Old World monkeys, likely predisposing to deletions mediated by non-allelic homologous recombination [3].

98.3 Pathogenesis

Mapped to chromosome 5q35.3, the nuclear receptor-binding SET domain protein 1 gene (*NSD1*) comprises 23 exons, which generate a 12.0 kb transcript, and encodes a 2596-aa protein (NSD1, a histone-lysine N-methyltransferase with H3 lysine-36 and H4 lysine-20 specificity; also known as androgen receptor-associated coregulator 267 or ARA267), which consists of at least 12 functional domains, including 2 nuclear receptor interaction domains (NID^{-L} and NID^{+L}), 2 proline-tryptophan-tryptophan-proline (PWWP) domains, 5 plant homeodomains (PHD), a conserved SET (su-var 3–9, enhancer of zeste, trithorax) domain, and a SET domain–associated cysteine-rich (SAC) domain (Table 98.1).

The SET and associated SAC domains are distinctive components of histone methyltransferases that regulate chromatin states. Showing unique histone specificity, the SET domain methylates lysine residue 36 on histone H3, and lysine residue 20 on histone H4 (K36H3 and K20H4) [4]. PHDs also act at the chromatin level, and PWWPs are often found in methyltransferases and involved in protein–protein interactions. NID^{-L} and NID^{+L} are the nuclear receptors of NSD1.

TABLE 98.1

NSD1 Domains and Locations

Domain	Abbreviation	Amino Acid Positions
Proline-tryptophan-tryptophan-proline domain 1	PWWP1	323–388
Nuclear localization signal 1	NLS1	512–529
Nuclear localization signal 2	NLS2	1157–1174
Nuclear localization signal 3	NLS3	1471–1488
Plant homeodomain domain 1	PHD1	1543–1589
Plant homeodomain domain 2	PHD2	1590–1639
Plant homeodomain domain 3	PHD3	1640–1693
Plant homeodomain domain 4	PHD4	1707–1751
P proline-tryptophan-tryptophan-proline domain 2	PWWP2	1756–1818
Associated with SET domains	AWS	1890–1940
Su(var)3–9, enhancer-of-zeste, trithorax domain	SET	1942–2065
Post-SET	Post-SET	2066–2082
Plant homeodomain domain 5	PHD5	2120–2160

Acting together, these functional domains underscore the histone methyltransferase activity of NSD1 with the ability of methylating histones, which are structure proteins attaching to DNA and giving chromosomes their shape, and thus influence chromatin compaction and transcription of genes involved in normal growth and development, both negatively and positively, depending on the cellular context [5,6].

In contrast to NSD1, which preferentially catalyzes the transfer of methyl residues to lysine residue 36 of histone 3 (H3K36) and lysine residue 20 of histone 4 (H4K20) and is primarily involved in activation but also repression depending on the cellular context, EZH2 implicated in Weaver syndrome catalyzes the trimethylation of lysine residue 27 of histone 3 (H3K27me3), and is mainly associated with transcriptional repression.

NSD1 is expressed in several tissues including the brain, kidney, skeletal muscle, spleen, and thymus. Mutations in the *NSD1* gene (e.g., missense mutations, nonsense mutations, small intragenic insertions/deletions, splicing defects, and 5q35 microdeletion encompassing *NSD1* generated by nonallelic homologous recombination between flanking low-copy repeats) contribute to *NSD1* haploinsufficiency, which prevents one copy of the gene from producing any functional protein and subsequently disrupts the normal activity of genes involved in growth and development [3,7,8].

Of >300 different mutations associated with Sotos syndrome identified to date, gross deletions (e.g., 5q35 microdeletion, which is identified in 50% of Japanese and Korean patients but <15% in patients from other parts of the world), small indels, point mutations (15%), and splice-site mutations dominate. These include 521T>A (V174A), 607G>A (V203I) and 896delC (frameshift) in exon 2, 2333T>G (L778X) and 2386–2389del-GAAA (frameshift) in exon 5, 3882delT (frameshift) in exon 6, 4417C>T (R1473X) in exon 10, 4779–4781delTTTinsATTC (frameshift) and 4855T>C (C1619R) in exon 13, IVS13 + 1G>A (skip exon) in intron 13, 4987C>T (R1663C) in exon 14, 5375G>T (G1792V) in exon 16, 5737A>G (N1913D) in exon 18, 6115C>T (R2039C) in exon 20, 6241T>G (L2081V) in exon 21, 6291delG (frameshift), 6356delA (D2119V), 6370T>C (C2124R), and A6442delAGCGACCA (K2151fsX) in exon 22, and 6523T>A

(C2175S), 6532delTGCCCCAGC (2178–2180delCPS), 6614A>G (H2205R), 6605G>A (C2202Y), and 7576C>T (P2526S) in exon 23. Indeed, several missense-mutations target the PHDV-C5HCHNSD1 tandem domain, which is composed by a classical (PHDV) and an atypical (C5HCH) PHD finger, impairing NSD1 interactions with cofactors such as Nizp1 and thus impacting on the repression of growth-promoting genes, leading to overgrowth conditions [9].

There is evidence that about 95% of patients acquire Sotos syndrome through *de novo* mutation in the *NSD1* gene, and only a small proportion (5%) inherit from their parents via autosomal dominant transmission, who tend to carry missense mutations [10]. While 90% of non-Japanese and non-Korean patients have NSD1 abnormalities, approximately 30% of Japanese and Korean patients do not. Interestingly, patients harboring a microdeletion often develop certain congenital heart and/or urogenital anomalies, more severe mental retardation, and shorter stature (less overgrowth) than those harboring *NSD1* intragenic mutations.

It should be noted that mutation in the nuclear factor I/X type gene (NFIX) on chromosome 19p13.3 may produce an autosomal dominant disorder known as Malan syndrome (or Sotos syndrome 2) [11–13]. Nuclear factor I is a ubiquitous 47-kD dimeric DNA-binding protein with the capability of stimulating the transcription of genes in cooperation with other factors such as estrogen receptor (ESR). In addition, alteration in the APC2 gene on chromosome 19p13.3 may cause an autosomal recessive disease known as Sotos syndrome 3. APC2 is a 2303-aa protein gene preferentially expressed in postmitotic neurons. As NSD1 is shown to downregulate APC2 in neurons, it is no surprise that Apc2-deficient (Apc2$^{-/-}$) mice exhibit impaired learning and memory abilities along with an abnormal head shape [14,15]. Given that the mitogen-activated protein kinase (MAPK) pathway is a diminished activity state in Sotos syndrome, it may be also involved in statural overgrowth and accelerated skeletal maturation [16]. In addition, *NSD1* forms as a fusion transcript with *NUP98*, playing a part in leukemogenesis through H3K36 methylation and subsequent HOX-A gene activation, particularly childhood acute myeloid leukemia [17].

Interestingly, while germline, monoallelic disruption (e.g., truncating mutation and gene deletion) causes *NSD1* haploinsufficiency and thus overgrowth in Sotos syndrome, somatic epigenetic silencing of *NSD1*, through promoter hypermethylation, may play a potential role in the development of neuroblastoma and gliomas. Further, somatic *NSD1* mutations are occasionally found in carcinoma of the upper airway digestive tract.

98.4 Epidemiology

98.4.1 Prevalence

Sotos syndrome has a reported incidence of 1 in 14,000 newborns, although the true incidence may be closer to 1 in 5000 [18].

98.4.2 Inheritance

Although >95% of affected individuals have a de novo pathogenic variant, Sotos syndrome shows an autosomal dominant inheritance pattern, implying that one copy of the altered gene transmitted from a parent is sufficient to cause the disorder in offspring [10].

98.4.3 Penetrance

Sotos syndrome is a fully penetrant condition with a highly variable expressivity. Individuals with the same pathogenic variant may therefore have different phenotypes [19]. Nonetheless, *NSD1* pathogenic variant does not seem to occur in an unaffected parent or an unaffected sib.

98.5 Clinical Features

Sotos syndrome produces three categories of clinical features: cardinal, major, and associated [1].

Cardinal features occur in ≥90% of cases, and consist of distinct facial appearance (broad and prominent forehead, sparse frontotemporal hair, downslanting palpebral fissures, malar flushing, long and narrow face, long chin; most recognizable between ages one and 6 years), learning disability (early developmental delay, mild to severe intellectual impairment), and overgrowth (height and/or head circumference ≥2 SD above the mean; normalized height in adulthood) (Figure 98.1) [20–22]. It is of interest that patients with 5q35 microdeletions tend to show less pronounced overgrowth but more profound intellectual disability and higher frequency of cardiovascular and urinary/renal abnormalities than those with *NSD1* mutations.

Major features are present in 15%–89% of cases and include behavioral problems (e.g., autistic spectrum disorder, phobias, and aggression), advanced bone age (in 75%–80% of prepubertal children), cardiac anomalies (PDA, ASD, and VSD to more severe, complex cardiac abnormalities such as left ventricular non-compaction and aortic dilatation; 20%), cranial MRI/CT abnormalities (e.g., ventricular dilatation, midline changes [hypoplasia or agenesis of the corpus callosum, mega cisterna magna, cavum septum pellucidum], cerebral atrophy, and small cerebellar vermis), joint hyperlaxity/pes planus (20%), maternal preeclampsia (15% of pregnancies), neonatal complications (jaundice ~65%; hypotonia ~75%; and poor feeding ~70%, often resolving spontaneously), renal anomalies (e.g., vesicoureteral reflux, quiescent vesicoureteral reflux, and renal impairment; 15%), scoliosis (30%), and seizures (25%) [23]. Indeed, patients with 5q35 microdeletions often display a higher frequency of cardiovascular and urinary/renal abnormalities than those with *NSD1* mutations.

Associated features are seen in 2%–15% of cases and include tumors (e.g., sacrococcygeal teratoma, neuroblastoma, presacral ganglioma, hemangioma, diffuse gastric carcinoma, acute lymphoblastic leukemia, sarcoma, and small cell lung cancer; 3%) (Figure 98.2) [24], and other features (e.g., astigmatism, cataract, cholesteatoma, conductive hearing loss due to chronic otitis media, constipation, contractures, craniosynostosis, cryptorchidism, gastroesophageal reflux, hemihypertrophy, hydrocele, hypercalcemia, hypermetropia, hypodontia, hypoplastic nails, hypospadias, hypothyroidism, inguinal hernia, myopia, neonatal hypoglycemia, nystagmus, pectus excavatum, phimosis, skin hyperpigmentation, skin hypopigmentation, strabismus, talipes equinovarus, umbilical hernia, vertebral anomalies, and 2/3 toe syndactyly) [1].

98.6 Diagnosis

Diagnosis of Sotos syndrome requires a thorough history review (for learning difficulties, cardiac and renal anomalies, seizures, and scoliosis), physical examination (e.g., cardiac auscultation, blood pressure, curvature of the spine [scoliosis]), echocardiogram and renal ultrasound examination (for renal damage from quiescent chronic vesicoureteral reflux), and molecular testing.

Individuals showing characteristic facial appearance (broad, prominent forehead with a dolichocephalic head shape; sparse frontotemporal hair; downslanting palpebral fissures; malar flushing; long, narrow face, particularly bitemporal narrowing; long chin; hypertelorism [an abnormally increased distance between the eyes]), learning disability (early developmental delay; mild to severe intellectual impairment) and overgrowth (height and/or head circumference is ≥2SD above the mean or the 98th centile) should be suspected of having Sotos syndrome [1].

Confirmation of Sotos syndrome relies on molecular identification of a heterozygous *NSD1* pathogenic variant, which is found in at least 90% of affected individuals. Typically,

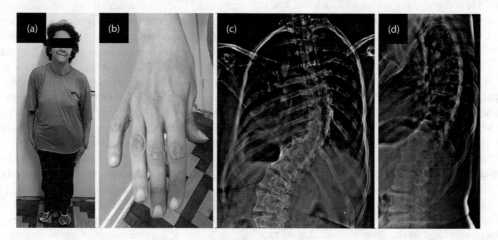

FIGURE 98.1 Sotos syndrome in a 48-year-old female showing (a) a large prominent forehead and pointed chin; (b) a large hand with arachnodactyly; (c) marked scoliosis; (d) lumbar hyperlordosis. (Photo credit: Pinto WB et al. *Arq Neuropsiquiatr.* 2017;75(2):134.)

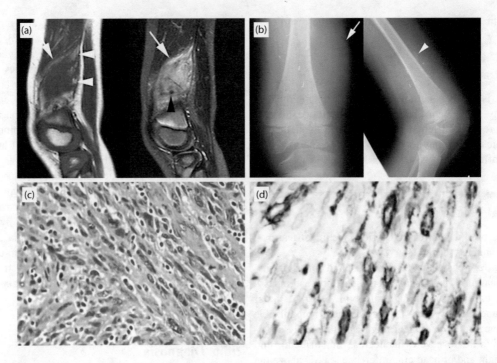

FIGURE 98.2 Sotos syndrome in a 4-year-old girl showing (a) an ill-defined soft tissue mass (arrows), which is predominantly isointense to muscle with peripheral areas of high signal by Sagittal T1-weighted MR (white arrowheads), but heterogeneously high in signal with a rounded low signal focus by Sagittal STIR MR (black arrowhead); (b) a nonspecific soft tissue mass (arrow) in the distal thigh with a small area of calcification or periosteal reaction (arrowhead); (c) a high-grade spindle-cell sarcoma with nuclear pleomorphism in excised tissue (H&E); (d) positive immunostain for smooth muscle actin, indicating myofibroblastic differentiation. (Photo credit: Hill DE et al. *Radiol Case Rep.* 2015;5(1):384.)

molecular testing is undertaken on the basis of single-gene and/or multigene panel.

In single-gene testing, sequence analysis detects *NSD1* pathogenic variant (e.g., small intragenic deletions/insertions and missense, nonsense, and splice-site variant) in 12% of Japanese/Korean patients, and 27%–93% of other patients, while gene-targeted deletion/duplication analysis (e.g., quantitative PCR, long-range PCR, multiplex ligation-dependent probe amplification [MLPA], and a gene-targeted microarray) or fluorescence *in situ* hybridization (FISH) analysis identifies *NSD1* intragenic deletions or duplications in 50% of Japanese/Korean patients and 15% of other patients. A multigene panel comprising *NSD1* and other genes of interest provides additional detail on the genetic status of affected individuals. Application of FISH, MLPA, or array comparative genomic hybridization (array CGH) methods permits identification of 5q35 microdeletions or *NSD1* full-gene deletions in 10% of non-Japanese individuals with Sotos syndrome [1].

Differential diagnoses for Sotos syndrome (usually downslanting palpebral fissures, sparse frontotemporal hair, long and thin face, hyperteloric eyes due to bitemporal narrowing rather than a true hypertelorism) include other overgrowth conditions such as Malan syndrome (also known as Sotos syndrome 2; moderate postnatal overgrowth and macrocephaly, prominent forehead, high anterior hairline, downslanting palpebral fissures and prominent chin; mutation in *NFIX* on chromosome 19p13.2), Sotos syndrome-3 (due to mutation in the APC2 gene on chromosome 19p13.3), Weaver syndrome (tall stature, round face with large fleshy ears, almond-shaped eyes and ocular hypertelorism, mixed central hypotonia/peripheral hypertonia, camptodactyly and contractures; *EZH2* pathogenic

variant) [25,26], Beckwith–Wiedemann syndrome (height and weight at least 2 SD above the mean; macrosomia, macroglossia, anterior ear lobe creases/helical pits, omphalocele, and visceromegaly; epigenetic and genomic alterations of chromosome 11p15, such as loss of methylation on the maternal chromosome at imprinting center 2 in 50% of cases, paternal uniparental disomy for chromosome 11p15 in 20% of cases, and gain of methylation on the maternal chromosome at imprinting center 1 in 5% of cases; *CDKN1C* pathogenic variants in 40% of familial cases and 5%–10% of cases with no family history) [26], Simpson–Golabi–Behmel syndrome type 1 (pre- and postnatal overgrowth in males, polydactyly, supernumerary nipples, diastasis recti, pectus excavatum, *GPC3* mutation) [27], Bannayan–Riley–Ruvalcaba syndrome (macrocephaly, vascular malformations, hamartomatous polyps of the distal ileum and colon, pigmented macules on the shaft of the penis, lipomas, thyroid and breast cancer, facial gestalt and overgrowth; *PTEN* pathogenic variants in 65% of cases), benign familial macrocephaly (dolicho- and/or macrocephaly), fragile X syndrome, nevoid basal cell carcinoma syndrome (Gorlin syndrome; multiple jaw keratocysts, basal cell carcinomas, skeletal anomalies such as bifid ribs or wedge-shaped vertebrae, macrocephaly, bossing of the forehead, and coarse facial features in 60% of cases, head circumference above the 98th centile; *PTCH* germline pathogenic variants), chromosome abnormalities (4p duplications, mosaic 20p trisomy, and 22q13.3 deletion syndrome), and nonspecific overgrowth (e.g., learning difficulties, distinctive facial features) [1,9,28].

In addition, prenatal-onset overgrowth may be secondary to normal variants of familial tall stature, familial rapid maturation, diabetic macrosomia, and congenital nesidioblastosis.

98.7 Treatment

No standard course of treatment is available for Sotos syndrome, and management of the disorder centers on symptomatic measures. Learning disability/speech delays, behavior problems, cardiac abnormalities, renal anomalies, scoliosis, and seizures require appropriate specialists. However, if the brain MRI shows ventricular dilatation without raised intracranial pressure, intervention (shunting) is not recommended. Individuals with vesicoureteral reflux may call for antibiotic prophylaxis [29].

98.8 Prognosis and Prevention

As Sotos syndrome is not a life-threatening disorder, affected individuals are expected to have a normal life expectancy, with initial abnormalities usually resolving and growth rate becoming normal after the first few years. However, coordination problems may be still present in adulthood.

Regular evaluation is recommended for younger children and individuals with medical complications.

Sotos syndrome is an autosomal dominant disorder, for which the risk of transmitting to from affected individuals to offspring is 50%. Prenatal testing for *NSD1* pathogenic variant may be offered to affected individuals who intend to become parents.

98.9 Conclusion

Sotos syndrome is a rare autosomal dominant overgrowth disorder caused by haploinsufficiency of the *NSD1* gene (which encodes a histone methyltransferase implicated in chromatin regulation), through intragenic *NSD1* mutations, partial *NSD1* deletions, or chromosomal microdeletions spanning the 5q35 region encompassing the entire *NSD1* gene. Affected individuals show cardinal features of typical facial appearance, learning disability, and macrocephaly, along with major and associated features that include heightened tumor risk. Due to significant clinical heterogeneity in Sotos syndrome, some affected individuals may suffer frequent ear and chest infections, cardiac and urinary/renal defects, seizures, scoliosis, and behavioral problems, which often lead to underdiagnosis. Although nearly 95% of cases appear to have evolved from de novo *NSD1* mutations, about 5% of patients have a clear family link. Application of molecular genetic test for *NSD1* pathogenic variants, which occur in up to 90% of individuals with a clinical diagnosis of Sotos syndrome, is valuable in helping achieve accurate diagnosis. Given the paucity of data concerning its pathogenic mechanism, no specific treatment is currently available for Sotos syndrome [30,31].

REFERENCES

1. Tatton-Brown K, Cole TRP, Rahman N. Sotos syndrome. In: Adam MP, Ardinger HH, Pagon RA, Wallace SE, Bean LJH, Stephens K, Amemiya A editors. *GeneReviews® [Internet]*. Seattle (WA): University of Washington, Seattle; 1993–2018. 2004 Dec 17 [updated 2015 Nov 19].
2. Juneja A, Sultan A. Sotos syndrome. *J Indian Soc Pedod Prev Dent.* 2011;29(6 Suppl 2):S48–51.
3. Rosenfeld JA, Kim KH, Angle B et al. Further evidence of contrasting phenotypes caused by reciprocal deletions and duplications: Duplication of NSD1 causes growth retardation and microcephaly. *Mol Syndromol.* 2013;3(6):247–54.
4. Ha K, Anand P, Lee JA et al. Steric clash in the SET domain of histone methyltransferase NSD1 as a cause of Sotos syndrome and its genetic heterogeneity in a Brazilian cohort. *Genes (Basel).* 2016;7(11). pii: E96.
5. Cross NC. Histone modification defects in developmental disorders and cancer. *Oncotarget.* 2012;3(1):3–4.
6. Choufani S, Cytrynbaum C, Chung BH et al. NSD1 mutations generate a genome-wide DNA methylation signature. *Nat Commun.* 2015;6:10207.
7. Chen CP, Lin CJ, Chern SR et al. Prenatal diagnosis and molecular cytogenetic characterization of a 1.07-Mb microdeletion at 5q35.2-q35.3 associated with NSD1 haploinsufficiency and Sotos syndrome. *Taiwan J Obstet Gynecol.* 2014;53(4):583–7.
8. Brennan K, Shin JH, Tay JK et al. NSD1 inactivation defines an immune cold, DNA hypomethylated subtype in squamous cell carcinoma. *Sci Rep.* 2017;7(1):17064.
9. Berardi A, Quilici G, Spiliotopoulos D et al. Structural basis for PHDVC5HCHNSD1-C2HRNizp1 interaction: Implications for Sotos syndrome. *Nucleic Acids Res.* 2016; 44(7):3448–63.
10. Laccetta G, Moscuzza F, Michelucci A et al. A novel missense mutation of the NSD1 gene associated with overgrowth in three generations of an Italian family: Case report, differential diagnosis, and review of mutations of NSD1 gene in familial Sotos syndrome. *Front Pediatr.* 2017;5:236.
11. Klaassens M, Morrogh D, Rosser EM et al. Malan syndrome: Sotos-like overgrowth with de novo NFIX sequence variants and deletions in six new patients and a review of the literature. *Eur J Hum Genet.* 2015;23(5):610–5.
12. Dong HY, Zeng H, Hu YQ et al. 19p13.2 Microdeletion including NFIX associated with overgrowth and intellectual disability suggestive of Malan syndrome. *Mol Cytogenet.* 2016;9:71.
13. Oshima T, Hara H, Takeda N et al. A novel mutation of NFIX causes Sotos-like syndrome (Malan syndrome) complicated with thoracic aortic aneurysm and dissection. *Hum Genome Var.* 2017;4:17022.
14. Migdalska AM, van der Weyden L, Ismail O et al. Generation of the Sotos syndrome deletion in mice. *Mamm Genome.* 2012;23(11–12):749–57.
15. Almuriekhi M, Shintani T, Fahiminiya S et al. Loss-of-function mutation in APC2 causes Sotos syndrome features. *Cell Rep.* 2015;10:1585–98.
16. Visser R, Landman EB, Goeman J, Wit JM, Karperien M. Sotos syndrome is associated with deregulation of the MAPK/ERK-signaling pathway. *PLOS ONE* 2012;7(11):e49229.
17. Mutsaers HA, Levtchenko EN, Martinerie L et al. Switch in FGFR3 and -4 expression profile during human renal development may account for transient hypercalcemia in patients with Sotos syndrome due to 5q35 microdeletions. *J Clin Endocrinol Metab.* 2014;99(7):E1361–7.
18. de Silva DC, de Leeuw N, Gunasekera R. Soto syndrome: A rare overgrowth disorder. *Ceylon Med J.* 2013;58(1):40–2.
19. Vieira GH, Cook MM, Ferreira De Lima RL et al. Clinical and molecular heterogeneity in Brazilian patients with sotos syndrome. *Mol Syndromol.* 2015;6(1):32–8.

20. Lane C, Milne E, Freeth M. Cognition and behaviour in Sotos syndrome: A systematic review. *PLOS ONE*. 2016;11(2):e0149189.

21. Lane C, Milne E, Freeth M. Characteristics of autism spectrum disorder in Sotos Syndrome. *J Autism Dev Disord*. 2017;47(1):135–43.

22. Pinto WB, Souza PV, Bortholin T, Honorato EL, Bocca LF, Oliveira AS. Epilepsy and early-onset overgrowth syndrome revealing Sotos syndrome. *Arq Neuropsiquiatr*. 2017; 75(2):134.

23. Nicita F, Ruggieri M, Polizzi A et al. Seizures and epilepsy in Sotos syndrome: Analysis of 19 Caucasian patients with long-term follow-up. *Epilepsia*. 2012;53(6):e102–5.

24. Hill DE, Roberts CC, Inwards CY, Sim FH. Childhood soft-tissue sarcoma associated with Sotos syndrome. *Radiol Case Rep*. 2015;5(1):384.

25. Tatton-Brown K, Rahman N. The NSD1 and EZH2 overgrowth genes, similarities and differences. *Am J Med Genet C Semin Med Genet*. 2013;163C(2):86–91.

26. Tatton-Brown K, Murray A, Hanks S et al. Weaver syndrome and EZH2 mutations: Clarifying the clinical phenotype. *Am J Med Genet A*. 2013;161A(12):2972–80.

27. Chen CP. Prenatal findings and the genetic diagnosis of fetal overgrowth disorders: Simpson-Golabi-Behmel syndrome, Sotos syndrome, and Beckwith-Wiedemann syndrome. *Taiwan J Obstet Gynecol*. 2012;51(2):186–91.

28. Ko JM. Genetic syndromes associated with overgrowth in childhood. *Ann Pediatr Endocrinol Metab*. 2013;18(3):101–5.

29. Takano M, Kasahara K, Ogawa C, Katada H, Sueishi K. A case of Sotos syndrome treated with distraction osteogenesis in maxilla and mandible. *Bull Tokyo Dent Coll*. 2012;53(2):75–82.

30. Deevy O, Bracken AP. PRC2 functions in development and congenital disorders. *Development*. 2019;146(19):dev181354.

31. Watanabe H, Higashimoto K, Miyake N et al. DNA methylation analysis of multiple imprinted DMRs in Sotos syndrome reveals IGF2-DMR0 as a DNA methylation-dependent, P0 promoter-specific enhancer. *FASEB J*. 2020;34(1):960–73.

99

Tuberous Sclerosis Complex

Joana Jesus Ribeiro, Filipe Palavra, and Flávio Reis

CONTENTS

99.1 Introduction

Tuberous sclerosis complex (TSC) is an autosomal dominant genetic condition, with an estimated incidence at birth of approximately 1 in 6000 [1,2]. It is considered a multisystem disorder that can cause circumscribed, benign, noninvasive lesions called hamartomas in virtually any organ of the body, most commonly in the brain, kidney, heart, and skin, with variable clinical presentation and severity [1,2].

TSC was described for the first time by von Recklinghausen in 1862, after autopsy examination of a stillborn with multiple cardiac and brain tumors [3]. However, the term tuberous sclerosis was coined 18 years later when Bourneville, an expert in childhood mental retardation, noted a 15-year-old girl with mental retardation and epilepsy [4,5]. The term tuberous sclerosis of the cerebral convolutions was used to describe the distinctive findings at autopsy, namely the potato-like consistency of gyri

with hypertrophic sclerosis [1,5]. Although there existed earlier descriptions of the dermatological findings, TSC was not recognized as a neurocutaneous syndrome until the nineteenth century [6]. Vogt published his understanding of the link among seizures, mental retardation, and adenoma sebaceum (facial angiofibromas), known as the Vogt triad, observed in only a third of cases [4]. In 1914 Schuster first applied the term "forme fruste" to patients with features of the TSC, but no mental retardation [4,6]. TSC was underdiagnosed until the 1980s when individuals with less severe manifestations of the disease began to be recognized [2].

This condition was defined as a genetic disease more than 100 years ago [2,7] and dominant inheritance with a high mutation rate was first demonstrated in 1935 [8]. However, the underlying molecular etiology remained elusive until the discovery of the two causative genes, *TSC1* and *TSC2* [9,10]. The wide range of organs affected by the disease implied an important role for *TSC1* and *TSC2* genes, encoding hamartin and tuberin respectively, in the regulation of cell proliferation and differentiation [1]. Hamartin and tuberin form a regulatory complex responsible for limiting the activity of an important intracellular regulator of cell growth and metabolism known as mammalian target of rapamycin complex 1 (mTORC1) via inhibition of the small GTPase Ras homolog enriched in brain (Rheb) [11].

The phenotype is highly variable, as disease manifestations in different organ systems can differ widely, even between closely related individuals [2,12]. Most de novo patients show a mutation in *TSC2*, whereas only 50% of all familial cases can be related to *TSC2* mutations [13]. Individuals with mutations in this gene have, in general, more severe symptoms than those with mutations in *TSC1* [1]. The protean nature of the condition can make clinical diagnosis challenging [2,12]. In a series of patients, more than 90% had skin lesions, about 90% had symptoms of cerebral pathology, 70%–90% had renal abnormalities, and about 50% had retinal hamartomas [14]. Significant and common medical complications include epilepsy (75%–90%), intellectual disability (50%), and behavioral problems (including autism and attention-deficit disorder with hyperactivity, in 40% of affected children) [12,13].

Accurate diagnosis is fundamental to the implementation of medical surveillance programs and tailored treatment [2]. The 2012 International Tuberous Sclerosis Complex Consensus Group, based on the best available evidence, established updated diagnostic criteria for TSC, including genetic testing [2].

The constitutively activated mTOR signaling due to mutations in TSC1/TSC2 has led to significant clinical advances in the use of mTORC1 inhibitors, such as sirolimus and everolimus, for the treatment of several clinical manifestations of TSC, with proved sustained regression of brain tumors, renal angiomyolipomas, liver angiomyolipomas, and pulmonary lymphangioleiomyomatosis [15–18].

99.2 Biological Background

TSC is associated with inactivating mutations in *TSC1* and *TSC2* genes (Table 99.1) [9,19]; however, about two-thirds of patients have sporadic de novo mutations, with no family history, reflecting a high spontaneous mutation rate [1,20,21]. Multiple studies have been performed confirming genetic heterogeneity in TSC and characterizing both genes by linkage analysis [22–26]. *TSC1* is located on chromosome 9q34 and encodes hamartin, which regulates mTOR–S6 K and cell adhesion through interaction with ezrin and Rho [10,20,27]. *TSC2* is located on chromosome 16p13.3 and encodes tuberin that regulates mTOR–S6 K and GTPase-activating proteins as well as cell cycle [9,27]. So far, 307 and 1061 allelic variants have been reported in *TSC1* and *TSC2*, respectively [1,20]. Missense mutations and large genomic deletions are much more frequent in *TSC2* than in *TSC1* [28]. A subgroup of large genomic deletions and rearrangements in *TSC2* also affect the adjacent *PKD1* gene, causing early-onset polycystic kidney disease [29]. *TSC1* and *TSC2* interact physically with high affinity to form heterodimers, and are co-expressed in cells within multiple organs, including the kidney, brain, lung, and pancreas [28]. *TSC2* has been localized in the Golgi apparatus and in the nucleus, and *TSC1* linked to the centrosome [28].

In agreement with Knudson's two-hit tumor suppressor gene model, inactivation of both alleles of *TSC1* or *TSC2* is needed

TABLE 99.1

Characterization of *TSC1* and *TSC2* Genes [1,9,10,20]

	TSC1	TSC2
Chromosomal location	9q34	16p13.3
Size	55 kb	40 kb
Number of exons	23	41
Transcript size	8.6 kb	5.5 kb
Mutation occurrence (% of sporadic cases)	10%–15%	75%–80%
Prevailing mutations	Small truncations (mostly nonsense mutations and small deletions)	Large deletions and/or rearrangements, small truncations (mostly missense mutations or deletions) In 2%–3% of patients, large genomic deletions in TSC2 also affect the adjacent gene PKD1
Allelic variants	307	1061
Protein	Hamartin	Tuberin
Protein size	1164 aa, 130 kDa	1807 aa, 180 kDa
Phenotype	Less severe	More severe

Abbreviation: PKD, polycystic kidney disease.

for tumor development [1,28,30]. Most second hits are caused by somatic independent mutations, with large deletions involving the loss of surrounding loci [1,28]. These mutations are referred to as loss of heterozygosity (LOH), since they affect neighboring heterozygous polymorphic markers [28]. The second-hit mutations of *TSC1* or *TSC2* might synergize with first-hit, systemic mutations of *TSC1* or *TSC2* to cause complete loss of *TSC1–TSC2* function [1]. The identification of LOH at different markers in an astrocytoma and angiomyolipoma from the same patient suggests the multifocal origin of a second-hit mutation [31]. LOH in TSC-associated lesions indicates that *TSC1* and *TSC2* are tumor-suppressor genes, occurring more frequently on chromosome 16p13 than on 9q34 [32]. Also, LOH was found in 56% of renal angiomyolipomas and cardiac rhabdomyomas, but only in 4% of TSC brain lesions, suggesting that brain lesions can result from different pathogenic mechanisms [32,33]. Indeed, LOH is almost absent in cortical tubers, which may indicate that either inactivation of both alleles is not required for tuber pathogenesis or only a subgroup of cells within a tuber is affected by the second hit [1,28,32]. By contrast, the benign metastasis hypothesis has been proposed to explain identical somatic mutations of *TSC2* in abnormal lung and kidney cells, but not in healthy cells of patients with sporadic lymphangiomyomatosis and renal angiomyolipoma, suggesting that these cells are genetically related and most likely arise from a common progenitor cell [28,34]. These observations are consistent with a model in which *TSC2*-deficient cells have increased migratory potential [28].

The overall mutation detection rate in patients with TSC is around 80%–90%, with a significant percentage of patients with no mutation identified despite the availability of new diagnostic techniques, such as multiplex ligation-dependent probe amplification (MLPA) [1,12,20,35]. Mutations in *TSC2* are four- to fivefold more common than in *TSC1* among patients, particularly in sporadic cases, whereas *TSC1* mutations are roughly equally common as *TSC2* mutations in large families [20,21]. Although initially the broad phenotypic spectrum of TSC made it difficult to establish a genotype–phenotype correlation [36], studies of larger cohorts indicated that patients with *TSC2* mutations had a more severe phenotype than patients with *TSC1* mutations, namely more frequent and severe epilepsy, mental retardation (moderate and severe), cortical tubers, renal angiomyolipomas, retinal hamartomas, and advanced facial angiofibromas [1,12,21,35,37]. Also, patients with a de novo *TSC2* mutation have a more severe phenotypic spectrum than patients with a familial *TSC2* mutation [35]. However, some missense mutations in *TSC2* might be associated with an unusually mild form of tuberous sclerosis, with many patients not meeting the standard diagnostic criteria [38,39]. Indeed, mutations that lead to truncation of tuberin or hamartin are clearly inactivating, but nucleotide changes that result in amino acid substitutions may be either pathogenic or, alternatively, polymorphisms that do not disrupt tuberin or hamartin function and do not cause TSC [40]. Both germline and somatic mutations appear to be less common in *TSC1* than in *TSC2* [21]. The phenotypes of the patients in which no mutation was identified were overall less severe than those of patients with either a known *TSC1* or *TSC2* mutation [21,41]. The reduced severity of the disease in these patients

suggests that they can be a mosaic for a *TSC2* mutation or have a mutation in an unidentified locus of TSC with a relatively mild clinical phenotype [21]. Individuals with the same genotype can have different clinical phenotypes [1]. A wide variation of the clinical picture, even within the same family, highlights that no strict correlation exists between a mutation and its clinical outcome, with many influencing factors contributing to the diversity of clinical manifestations, including the age and gender of the patient [35,41]. Indeed, several clinical manifestations of TSC occur more frequently in males [35].

99.3 Pathogenesis

After the discovery of *TSC1* and *TSC2* genes and their encoded proteins (hamartin and tuberin), several downstream protein cascades that might be affected by the pathogenesis of the disease, such as the pathway of mammalian target of rapamycin (mTOR), have been identified [1]. A mutation of hamartin or tuberin in TSC leads to hyperactivation of the downstream mTOR pathway and the associated kinase signaling cascades and translational factors, resulting in increased cell growth and proliferation (Figure 99.1) [11,42].

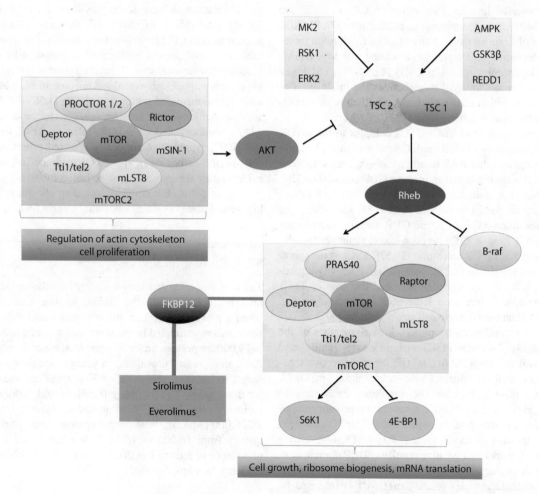

FIGURE 99.1 Overview of mTOR-TSC regulation and upstream and downstream mediators. Tuberous sclerosis complex patients present mutations in either TSC1 or TSC2 genes, causing suppression of Rheb-mediated mTORC1 inhibition. The TSC1–TSC2 protein complex integrates cues from growth factors, the cell cycle, and nutrients to regulate the activity of mTOR. mTOR binds to Raptor to exert its effect, which is mediated by S6K1 and 4E-BP1, proteins that participate in ribosome biogenesis and translation initiation, respectively. Rapamycin (sirolimus) and everolimus are effective inhibitors of mTORC1 via FKBP12.

Substantial progress has been made in the past years toward understanding the normal cellular functions of the TSC1–TSC2 protein complex [11,27,43]. One of the first mechanistic clues to the roles that TSC1 and TSC2 proteins have in cell function was the finding that mutations in the drosophila Tsc1 and Tsc2 homologs increased cell and organ size [44]. Subsequent experiments demonstrated that hamartin and tuberin form an intracellular heterodimeric complex with GTPase-activating protein activity that inhibits Rheb, responsible for the activation of the mTOR [27,42]. After growth-factor stimulation, the hamartin–tuberin complex is phosphorylated and its GTPase-activating protein activity is decreased [1]. Multiple kinases phosphorylate and inactivate *TSC2* and thereby activate Rheb and mTOR: protein kinase B (AKT), mitogen-activated protein kinase–activated protein kinase 2 (MK2), p90 ribosomal S6 kinase 1 (RSK1) and extracellular-related kinase 2 (ERK2) [28,45–47]. By contrast, in response to stimuli such as hypoxia or energy deprivation, TSC1 and TSC2 proteins are phosphorylated, their GTPase-activating protein activity increases, and the complex deactivates Rheb by causing GTP to be cleaved from it. TSC2 is phosphorylated and activated by AMP-activated protein kinase (AMPK), and the phosphorylation of TSC1 by glycogen synthase kinase 3β (GSK3β) increases the stability of the TSC1–TSC2 complex [28,48,49]. Additional proteins known to interact with either TSC1 or TSC2 are rabaptin-5, 14-3-3, estrogen receptor α, calmodulin, p27, SMAD2 and SMAD3 (the human isoform homologs of *Drosophila* mothers against decapentaplegic), protein associated with Myc (PAM) and cyclin-dependent kinase 1 (CDK1) [43,50–54]. TSC1 and TSC2 proteins have additional roles besides the modulation of mTOR, since inhibition of B-Raf kinase via Rheb is an mTOR-independent function of tuberin [28,55].

Signal transduction from TSC1–TSC2 complex to the mTOR pathway is mediated by Rheb, a small G protein of the Ras family [56–58]. Rheb, like other RAS family members, cycles between an active GTP-bound state and an inactive GDP-bound state [28]. TSC2 possesses a domain that shares homology with the GTPase-activating protein (GAP) domain of Rap1-GAP, stimulating the conversion of Rheb–GTP to Rheb–GDP, thereby inactivating Rheb [28,56,57]. Loss of TSC2 function leads to enhanced Rheb–GTP signaling and mTOR activation [28,57]. TSC1 is necessary for the optimal regulation of TSC2-related GAP activity regarding Rheb [28,59]. It has been demonstrated that loss of TSC1 expression results in increased Rheb/S6 K pathway signaling, which is important for astrocyte cell size regulation [60].

mTOR is a serine/threonine protein kinase, member of the phosphoinositide 3-kinase-(PI3K)-related kinase family and of cell survival pathways [58,61]. mTOR functions in two separate pathways, with two distinct protein complexes: mTORC1 and mTORC2 [61–63]. Both mTOR complexes are large, with mTORC1 having six and mTORC2 seven known protein components [62]. The common proteins are the catalytic mTOR subunit; mammalian lethal with sec-13 protein 8 (mLST8, also known as GβL), which works as a positive regulator; DEP domain containing mTOR-interacting protein (DEPTOR), negative regulator and the tti1/tel2 complex [62,64]. mTORC1 is inhibited by the action of rapamycin and consists of regulatory-associated protein of mTOR (Raptor), a positive regulator involved in substrate recruitment; and proline-rich AKT substrate of 40 kDa (PRAS40), responsible for mTORC1 inhibition [56,61,63]. The second pathway, involving mTORC2, requires binding mTOR to

rapamycin-insensitive companion of mTOR (Rictor), essential for the interaction between mTORC2 and TSC2; mammalian stress-activated protein kinase interacting protein (mSIN-1), important for complex construction; and the protein observed with rictor 1 and 2 (PROTOR 1/2), which seems to have a role in enabling mTORC2 to activate serum-and glucocorticoid–induced kinase 1 (SGK1) [58,61–63]. mTORC2 is rapamycin-insensitive and functions upstream of Rho GTPases to regulate the actin cytoskeleton [56,61,64]. TSC1-TSC2 complex negatively regulates mTORC1 [28,62].

mTORC1 is the better characterized of the two mTOR complexes and is a central regulator of protein synthesis, whose activity is modulated by a variety of signals [62,65]. The mTORC1 pathway integrates inputs from at least five major intracellular and extracellular cues, detecting signals of nutrient availability, energy status, hypoxia, or growth factor stimulation [1,62]. It is part of many cell processes, such as cell-cycle progression, transcription and translation control, nutrient uptake, and autophagy [1,62,65]. Nutrients might boost translation through phosphorylation of mTOR by PI3KIII, while energy depletion inhibits mTOR through a process involving the activation of AMPK by LKB1 and subsequent phosphorylation of TSC2 [1,65]. Downregulation of mTOR activity by hypoxia requires de novo mRNA synthesis and correlates with increased expression of the hypoxia-inducible REDD1 gene [65]. In mTORC1, mTOR binds to Raptor and GβL to exert its effect and phosphorylates, among other proteins, S6K1 (p70 ribosomal protein S6 kinase ½), and 4E-BP1 (eukaryotic translation initiation factor 4E binding protein 1) [1,62]. S6K1 is a kinase that activates ribosomal subunit protein S6, leading to ribosome recruitment and protein translation [1,56]. 4E-BP1 inhibits activity of eukaryotic translation initiation factor 4E (eIF4E) and, when phosphorylated by mTOR, releases eIF4E from its control [1,56]. Analysis of surgically resected tubers has revealed cell-specific activation of the mTOR cascade in giant cells, as evidenced by the expression of activated (phosphorylated) components of the mTOR cascade, including phosphorylated p70S6 kinase and phosphorylated ribosomal protein S6 [28,66,67].

99.4 Epidemiology

The first proper epidemiologic survey for tuberous sclerosis was undertaken in 1968 in the Oxford Region, United Kingdom, when a prevalence of 1 in 100,000 was found [68]. Later, this value was re-estimated in the same region, and a prevalence of 1 in 29,000 for persons under 65 years of age was documented [69]. Also, an increased prevalence in younger age groups was identified: 1 in 20,000 for those below 30 years of age and 1 in 15,000 for those below 5 years of age [8,69]. Indeed, before the 1980s, incidence rates for TSC were quoted at between 1/100,000 and 1/200,000 [8,68,70]. With proper epidemiologic studies, the frequency from 1/6000 to 1/10,000 live births and a population prevalence of around 1 in 20,000 have been established, with no evidence for non-penetrance [8,29,71].

99.5 Clinical Features

TSC is a very heterogeneous disease: the random distribution, number, size, and location of lesions cause several clinical manifestations, with considerable variation between and within families [1,35,41]. The phenotypic spectrum ranges from minor, asymptomatic skin affections, such as hypomelanotic macules, to drug-resistant epilepsy, mental retardation, and increased morbidity due to cardiac tumors, cerebral astrocytomas, renal insufficiency, and lung affectations [40]. In more than half of the patients, symptoms begin before 6 months of age [72]. Some lesions, such as renal angiomyolipomas, do not occur until a certain age; by contrast, cardiac rhabdomyomas appear in the fetus and almost always regress spontaneously in infancy [1]. Although there is overlap in the spectrum of many clinical features of patients with *TSC1* versus *TSC2* mutations, *TSC1* mutations are associated with milder disease, and some features (grade 2–4 kidney cysts or angiomyolipomas, forehead plaques, retinal hamartomas, and liver angiomyolipomas) are very rare or not seen at all in *TSC1* patients [21].

99.5.1 Skin and Teeth

Dermatologic involvement occurs in 90%–100% of individuals with TSC and includes hypomelanotic macules, angiofibromas or fibrous cephalic plaques, ungual fibromas, shagreen patches, "confetti" skin lesions, dental enamel pits, and intraoral fibromas [2,73].

Hypomelanotic macules are observed in about 90% of patients and may be the presenting sign of TSC [2,73,74]. They typically appear at birth or infancy, and become less apparent in late adulthood [2,74]. Poliosis is considered a variant presentation of hypomelanosis and is included in the count of hypomelanotic macules [2,74].

Facial angiofibromas occur in about 75% of TSC patients with onset typically between ages 2 and 5 years [2]. Throughout adolescence, they increase in number and size [73,74]. Bilateral facial angiofibromas are hamartomatous nodules of vascular and connective tissue, predominating on the central zone of the face, especially around the nasolabial fold, and extend symmetrically onto the cheeks, nose, and chin [73,74]. Although most TSC patients have several facial angiofibromas and may become quite disfiguring, milder cases of TSC with limited facial angiofibromas have been described [2,73]. The occurrence of multiple facial angiofibromas in adolescence is almost pathognomonic for TSC, but if they appear in adulthood, they should be treated as a minor feature [2,74].

A fibrous cephalic plaque may occur on the forehead or other craniofacial areas in about 25% of TSC patients. Histologically, it is similar to angiofibromas and may be the most specific skin finding for TSC [2,74].

Ungual fibromas (Koenen tumors) are connective tissue hamartomas and occur in approximately 20% of TSC patients but can rise as high as 80% in older adults [1,2,74]. They are generally more common on toes than on fingers, and also are more common in women than in men [1]. They show the latest onset of all dermatological manifestations, typically in the second decade or later [2,74]. Trauma may induce the formation of TSC ungual fibromas, requiring that they be multiple ($\geq$2) [2,74]. Ungual fibromas that occur in the general population in response to trauma are usually solitary [2].

Shagreen patches are connective tissue nevi, observed in about 50% of TSC patients, and are specific findings for this condition [1,74]. Usually located on the trunk, these lesions generally present as large plaques with a bumpy or orange-peel surface [2,74]. Shagreen patches often appear in the first decade of life [2].

"Confetti" skin lesions are numerous 1- to 3-mm hypopigmented macules scattered over regions of the body such as the arms and legs with variable frequency, and are more suggestive of TSC if they appear in the first decade of life or have an asymmetric distribution [2]. Their diagnostic utility in adults is limited as similar-appearing lesions can be a consequence of chronic sun exposure [2]. Dental enamel pits are much more common in TSC patients than the general population (90% vs. 9%), and are included as a minor feature [1,2].

Intraoral fibromas include gingival fibromas that occur in about 20%–50% of individuals with TSC, with greater frequency in adults than children. Also, fibromas on the buccal or labial mucosa and even in the tongue can be seen in this genetic condition [2].

99.5.2 Eyes

Multiple retinal hamartomas are observed in 30%–50% of patients with TSC and are good markers for the disease, particularly in young children [2,73,75]. These lesions have similar histologic features to tubers located in the brains of TSC patients [2]. They can be found at any age and have been described in small children and even newborn babies [1]. Different morphological types of hamartomas exist: the most common type is a subtle, flat, smooth-surfaced, salmon-colored, semitransparent, circular or oval lesion on the superficial part of the retina, usually near or at the posterior pole; the second most common type is an opaque, white, elevated, multinodular, calcified lesion, and the third most common type of lesion contains features of the other two, being calcified and nodular centrally, but having a semitranslucent, smooth, and salmon-colored perimeter [1,75].

The "mulberry" lesion, present in 50% of patients with TSC, is composed of glial and astrocytic fibers, and can be evident by 2 years of age [1,75]. Retinal achromic patches are areas of hypopigmentation on the retina that occur in 39% of patients with TSC [2,73,75]. Unless these lesions affect the macula or optic nerve, they are typically asymptomatic, and usually do not cause problems with vision [1,2,73].

99.5.3 Central Nervous System

About 85% of children and adolescents with TSC have central nervous system (CNS) complications, including tubers and tumors, epilepsy, and TSC-associated neuropsychiatric disorders [1,2]. The most frequent CNS lesions observed in TSC patients include tubers in the cerebral cortex, subependymal nodules (SENs) and subependymal giant cell astrocytomas (SEGAs) in the ventricular system [76]. A smaller proportion of patients (<30%) will exhibit lesions within the infratentorial compartment, with multifocal locations and especially within the cerebellar white matter [77].

Cortical tubers are regions of cortical dysplasia that likely result from aberrant neuronal migration during corticogenesis and are observed in ~90% of TSC patients [2,28]. Dysplastic neurons have disrupted radial orientation in the cortex and abnormal dendritic arborization, showing γ-aminobutyric acid (GABA)-transporter defect and low GABAergic inhibition [1,78]. Tubers are static lesions with variable size and can be multiple in the same individual [77]. They are directly related to the more prevalent neurologic manifestations of TSC, including epilepsy, cognitive impairment, challenging behavioral problems, and autism [13]. These symptoms are highly variable in age at onset and can be detected by fetal MRI or pathological findings as early as 26 weeks of gestation [42]. Cerebral white matter radial migration lines arise from a similar pathologic process as cortical tubers and can be observed in 20%–30% of patients [77]. It is not unusual to find tubers and white matter migrational abnormalities together [2].

Histologically, SENs and SEGAs are similar and both are relatively specific (although not exclusive) of TSC [2]. SENs are hamartomas that develop along the wall of the ependymal lining of the lateral and third ventricles in 80% of TSC patients [2,79]. They are typically asymptomatic and often prenatally detected or at birth [2,79]. SENs may undergo transformation into SEGAs, which have an incidence of 5%–15% in TSC [2,73,79]. They are benign slow-growing lesions of >1 cm in diameter, typically occurring near the foramen of Monro, and can cause obstructive hydrocephalus [73]. They can be detected prenatally or at birth, although they are much more likely to arise during childhood or adolescence and it would be unusual to occur after the age of 20 years [2].

Epilepsy, which usually manifests during the first year of life, affects approximately 85% of individuals with TSC, and is often refractory to treatment [73,79,80]. Epileptogenesis in TSC may be a consequence of multiple receptors' changes in dysplastic neurons, namely GABA transporter defect and reduced GABAergic inhibition, molecular changes of glutamate receptors with enhanced glutamatergic function, and also TSC1/TSC2 dysfunction may alter the developmental long-range neural networks with alterations in cortical–subcortical system connectivity [42]. Although most patients develop multiple seizure types, infantile spasms are one of the major types of early seizures in TSC, and many patients with infantile spasms develop Lennox–Gastaut syndrome latter [73,80]. Early seizure onset, mainly infantile spasms, is associated with an increased risk of neurodevelopmental and cognitive impairment [28,73,80].

Regarding neurocognitive and behavioral disturbances, TAND (TSC-Associated Neuropsychiatric Disorders) is a new terminology proposed at the 2012 International TSC Consensus Conference to describe the functional and clinical manifestations of brain dysfunction common in TSC, including aggressive behaviors, autism spectrum disorder, intellectual disability, psychiatric disorders, and neuropsychological deficits as well as school and occupational difficulties [2,73]. These disorders exhibit highly variable expression and severity, with 55% of patients falling within the normal range, 14% exhibiting mild to severe impairment, and 30.5% exhibiting profound mental retardation [79,81].

99.5.4 Heart

Cardiac rhabdomyomas are benign tumors of the heart, highly specific of TSC, and are the earliest detectable hamartomas in this disease [1,2,73]. They are the main feature of TSC in the fetus and newborn, and 96% of infants with cardiac rhabdomyomas will ultimately be diagnosed with tuberous sclerosis [1]. These tumors are usually 3–25 mm in diameter, are most commonly located within the ventricles, and within the walls more often than the septum [1,82]. Although these tumors are the most frequent cause of death in TSC infants and children aged <10 years, they are usually asymptomatic and often spontaneously

regress with age [73,82]. They can be associated with obstruction of cardiac outflow, dilated cardiomyopathy as a result of impaired ventricular functions, atrial and ventricular arrhythmia, cardio-embolic disease, and Wolff–Parkinson–White syndrome [1,2,73,82].

99.5.5 Lungs

Pulmonary manifestations specific to TSC are multifocal micronodular pulmonary lymphangioleiomyomatosis (LAM), multifocal micronodular pneumocyte hyperplasia (MMPH) (Figure 99.2e and f), and clear-cell tumor of the lung (CCTL) [2,73].

Pulmonary LAM is characterized by alveolar smooth-muscle proliferation and cystic destruction of lung parenchyma, affecting 30%–40% of female TSC patients [1,2]. It is more frequently associated with *TSC2* mutations and predominantly affects premenopausal women, in the third to fourth decade of life [1,2,28,73]. The fact that LAM occurs mainly in women has led to the hypothesis that estrogen regulates TSC signaling, and perhaps also the migration of TSC2-deficient cells [28,50]. Cystic changes consistent with LAM are also observed in approximately 10%–12% of males with TSC, but symptomatic LAM in males is very rare [2,73]. Earliest manifestations are coughing, chest pain, progressive dyspnea on exertion, and recurrent pneumothorax [1,2,28,73]. LAM also occurs as a rare sporadic disorder in women without TSC, through two somatic mutations of *TSC2* rather than through a germline mutation and a "second-hit" somatic mutation, mechanism that is typical for TSC, as mentioned earlier [1,2,28]. Furthermore, 40% also have renal angiomyolipomas, supporting a clonal origin and a metastasis-like process [1,28].

MMPH is composed of benign alveolar type II cells and is found scattered throughout the lung [2]. The precise prevalence in patients with TSC is not known but may be as high as 40%–58% [2]. Although MMPH is thought not to be clinically significant and not to affect respiratory function, at least two cases of respiratory failure have been reported [2,73]. MMPH is often found in both men and women and may occur in the presence or absence of LAM in patients with TSC [2,73].

CCTL is a rare and typically benign mesenchymal tumor composed of histologically and immunohistochemically distinctive perivascular epithelioid cells, which constitutes the major member of the PEComa family of lung tumors, along with LAM [2,73].

99.5.6 Kidneys

Renal lesions, collectively occurring in 50%–80% of patients with TSC, include angiomyolipomas, renal cysts, and renal cell carcinoma (Figure 99.2a and b) [28,79]. Other renal manifestations with TSC are oncocytomas, epithelioid angiomyolipomas, and malignant epithelioid angiomyolipomas [73,79]. Renal complications are one of the most frequent causes of TSC-related death [1].

Renal angiomyolipomas are benign tumors composed of abnormal vessels, immature smooth-muscle cells, and fat cells (Figure 99.2c and d) [28]. Fat-containing angiomyolipomas are observed in 80% of TSC patients, and fat-poor lesions are also common in these individuals but occur in less than 0.1% of the general population [2]. Renal angiomyolipomas are bilateral, multiple, and tend to grow slowly [1,28]. However, sometimes they increase rapidly

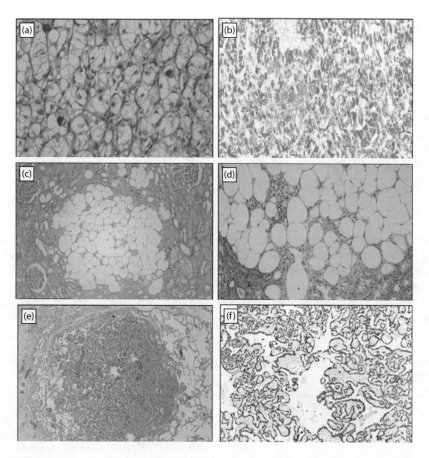

FIGURE 99.2 Tumors associated with TSC in a 13-year-old female patient. (a) Renal cell carcinoma showing solid and nested tumor cells and voluminous clear cytoplasm (H&E, ×200); (b) renal cell carcinoma (P504 stain, ×100); (c) renal angiomyolipoma showing pathognomonic components of mature fat and blood vessels (H&E, ×40); (d) renal angiomyolipoma (H&E, ×100); (e) multifocal micronodular pneumocyte hyperplasia showing proliferated and enlarged type II pneumocytes (H&E, ×40); (f) multifocal micronodular pneumocyte hyperplasia (pan-CK stain, ×100). (Photo credit: Behnes CL et al. *BMC Clin Pathol.* 2013;13:4.)

within a short period of time [73]. Although the majority of angiomyolipomas behave in a benign manner, there is a small subgroup which may behave more aggressively and become locally invasive [83], being more often symptomatic in women [1]. They appear by 10.5 years of age and increase in size and number in the late teens and early 20s, not regressing spontaneously [1,73]. Spontaneous hemorrhage can be the presenting sign in 50%–60% of the patients, as these tumors have abnormal vasculature and often contain aneurysms, leading to hemorrhagic shock in one-third of patients [28,83]. Although the size of an angiomyolipoma roughly correlates with the risk of hemorrhage, the size of the aneurysm appears to be proportional to the risk of bleeding, with aneurysms larger than 5 mm presenting the highest risk [73].

In addition to angiomyolipomas, epithelial renal lesions that include epithelial cysts, polycystic kidney disease, and renal cell carcinomas may develop in patients with TSC [28]. Benign epithelial cysts are generally asymptomatic and are more often associated with hypertension and renal failure than are angiomyolipomas, occurring in 20%–50% of TSC patients [28,73]. Multiple renal cysts are not commonly observed in the general population but can be seen in TSC patients who have a *TSC1* or *TSC2* mutation [2]. A minority (2%–3%) of patients carry a contiguous gene deletion in *TSC2* and *PKD1*, resulting in a polycystic kidney phenotype that is detectable in infancy or early

childhood and that generally leads to renal insufficiency in the late teens to early 20s [1,28,73].

The overall incidence of renal cell carcinoma approximates that of the general population, with a lifetime risk of 2%–3%, but it occurs on average 25 years earlier in TSC patients [79]. This carcinoma is usually diagnosed during childhood, but symptoms appear only after many years [1]. An unusual feature of renal carcinoma associated with TSC is its pathological heterogeneity [28]. Clear-cell, papillary, and chromophobe carcinoma subtypes, as well as oncocytomas, have all been reported in patients with TSC [28].

99.5.7 Other Manifestations

Various types of hamartoma lesions do occur in the endocrine system [2]. Adrenal angiomyolipoma can be present in a quarter of TSC patients, but rarely causes hemorrhage [2]. Thyroid papillary adenoma has been reported in TSC patients [2,73]. Also, the occurrence of angiomyolipomas or fibroadenomas in the pituitary gland, pancreas, or gonads has been documented, and these tumors represent minor features under the designation "nonrenal hamartomas" [2]. Neuroendocrine tumors might be slightly more prevalent in TSC patients, but these lesions are not hamartomas; thus, they are not considered part of the diagnostic criteria [2,73].

Gastrointestinal manifestations in TSC patients are fairly rare [2]. Hepatic multiple angiomyolipomas are reported in 10%–25% of TSC patients, and they are usually asymptomatic [1,2,73]. These angiomyolipomas are more common in adults (23%–45%) than in children, and more frequent in women than in men [1]. Renal angiomyolipomas usually precede the development of hepatic angiomyolipomas, and the latter have a slower growth and do not cause death [1].

Perivascular epithelioid cell tumor (PEComa) in the uterus can be specific but is difficult to differentiate from myoma uteri [73].

99.6 Diagnosis

The first clinical diagnostic criteria for patients affected by TSC were defined in a clinical consensus conference in 1998, highlighting the importance of the major and minor criteria to diagnose this genetic condition [84]. In 2012, the criteria were critically evaluated and updated in the International Tuberous Sclerosis Complex Consensus Conference [2]. Clinical features

of TSC continued to be the principal means of diagnosis, with reduced diagnostic classes from three (possible, probable, and definite) to two (possible, definite) [2]. However, genetic testing was included, with the identification of either a *TSC1* or *TSC2* pathogenic mutation in DNA from normal tissue being sufficient to make a definite diagnosis of the disease (Table 99.2a) [2]. Sometimes an antenatal diagnosis can be made based on fetal ultrasound and MRI, which may show cardiac and brain lesions. Most patients are diagnosed in infancy or early childhood, making early therapeutic interventions and treatments possible [1].

99.6.1 Dermatologic and Dental Features

Dermatologic features are easily detectable on physical examination, and the hypomelanotic macules are best seen under ultraviolet light (Wood's lamp) [1]. Four of eleven major features and three of six minor features in clinical diagnostic criteria are dermatological manifestations [2,73].

There is a wide range of differential diagnoses of cutaneous lesions, including vitiligo, Alezzandrini syndrome,

TABLE 99.2

Diagnostic Criteria for Tuberous Sclerosis and Testing Recommendations of Asymptomatic Patients

Roach et al. [84]	*Northrup* et al. [2]
a. Diagnostic Criteria for Tuberous Sclerosis	

Roach et al. [84]	*Northrup* et al. [2]
	1. Genetic Diagnostic Criteria
	A pathogenic mutation is defined as a mutation that clearly inactivates the function of the TSC1 or TSC2 proteins (e.g., out-of-frame indel or nonsense mutation), prevents protein synthesis (e.g., large genomic deletion), or is a missense mutation whose effect on protein function has been established by functional assessment. Other *TSC1* or *TSC2* variants whose effect on function is less certain do not meet these criteria and are not sufficient to make a definite diagnosis of TSC. Note that 10%–25% of TSC patients have no mutation identified by conventional genetic testing, and a normal result does not exclude TSC, or have any effect on the use of clinical diagnostic criteria to diagnose TSC.
1. Clinical Diagnostic Criteria	**2. Clinical Diagnostic Criteria**
1.1. Major features	**2.1. Major features**
• Facial angiofibromas or forehead plaque	• Hypomelanotic macules (≥3, at least 5 mm diameter)
• Non-traumatic ungual or periungual fibroma	• Angiofibromas (≥3) or fibrous cephalic plaque
• Hypomelanotic macules (three or more)	• Ungual fibromas (≥2)
• Shagreen patch	• Shagreen patch
• Multiple retinal nodular hamartomas	• Multiple retinal hamartomas
• Cortical tuber[a]	• Cortical dysplasias[e]
• Subependymal nodule	• Subependymal nodules
• Subependymal giant-cell astrocytoma	• Subependymal giant cell astrocytoma
• Cardiac rhabdomyoma (single or multiple)	• Cardiac rhabdomyoma
• Lymphangiomyomatosis[b]	• Lymphangioleiomyomatosis[f]
• Renal angiomyolipoma[b]	• Angiomyolipomas (≥2)[f]
1.2. Minor features	**2.2. Minor features**
• Multiple, randomly distributed pits in dental enamel	• "Confetti" skin lesions
• Hamartomatous rectal polyps[c]	• Dental enamel pits (>3)
• Bone cysts[d]	• Intraoral fibromas (≥2)
• Cerebral white matter radial migration lines[d]	• Retinal achromic patch
• Gingival fibromas	• Multiple renal cysts
• Non-renal hamartoma[c]	• Nonrenal hamartomas
• Retinal achromic patch	
• Confetti-like skin lesions	
• Multiple renal cysts[c]	
Definite TSC: Two major features or one major feature plus two minor features	*Definite diagnosis*: Two major features or one major feature with ≥2 minor features
Probable TSC: One major plus one minor feature	*Possible diagnosis*: One major feature or ≥2 minor features
Possible TSC: One major feature or two or more minor features	

(Continued)

TABLE 99.2 *(Continued)*

Diagnostic Criteria for Tuberous Sclerosis and Testing Recommendations of Asymptomatic Patients

Assessment	Initial Testing	Repeated Testing
b. Testing Recommendations of Asymptomatic Patients [85,86]		
Neurodevelopmental testing	At diagnosis and at school entry	As indicated
Ophthalmic examination	At diagnosis	As indicated
Dermatologic examination	At diagnosis	As indicated
Electroencephalography	At diagnosis	As indicated for seizure management
Electrocardiography	At diagnosis	As indicated
Echocardiography	If cardiac symptoms occur	If cardiac dysfunction occurs
Renal ultrasonography	At diagnosis	Every 1–3 years
Chest computed tomography	In adulthood (women only)	If pulmonary dysfunction occurs
Cranial magnetic resonance imaging	At diagnosis	Children and adolescents: every 1–3 years

[a] When cerebral cortical dysplasia and cerebral white matter migration tracts occur together, they should be counted as one rather than two features of tuberous sclerosis.

[b] When both lymphangiomyomatosis and renal angiomyolipomas are present, other features of tuberous sclerosis should be present before a definite diagnosis is assigned.

[c] Histologic confirmation is suggested.

[d] Radiographic confirmation is sufficient.

[e] Includes tubers and cerebral white matter radial migration lines.

[f] A combination of the two major clinical features (lymphangioleiomyomatosis and angiomyolipomas) without other features does not meet criteria for a definite diagnosis.

Vogt–Koyanagi–Harada disease, scleroderma and other autoimmune diseases, Birt–Hogg–Dubé syndrome (BHD), and multiple endocrine neoplasia type 1 (MEN1) [74]. For most TSC patients, no skin biopsy is required; however, it may be appropriate if there is uncertainty regarding the clinical diagnosis, i.e., when angiofibromas are few or later in onset [2,74]. Indeed, multiple facial angiofibromas remain a major feature for diagnosis when their onset occurs in childhood. In the unusual circumstance when angiofibromas have their onset in adulthood, they should be considered as a minor feature and the differential diagnosis expanded to include BHD and MEN1 [2].

99.6.2 Skeletal Features

Bone cysts were included in 1998 criteria as a minor feature of TSC. However, they were deleted from the 2012 revision because of the lack of specificity for TSC and because this finding is rarely identified in the absence of additional TSC clinical features [2].

99.6.3 Ophthalmologic Features

The finding of more than one retinal hamartoma was determined to be significant and specific enough to retain as a major feature [2]. The presence of a retinal achromic patch was defined at the 1998 conference as a minor feature and maintained the previous recommendation in the 2012 conference [2]. A complete ophthalmologic examination is a relevant procedure that should be done at TSC diagnosis [86].

99.6.4 Central Nervous System Features

Cortical tubers are typically seen in the brain, 90% in frontal lobes, but they can be observed in the cerebellum as well, more frequently in patients with *TSC2* mutations [77,87,88]. They occur most often at the gray–white matter junction and

are identified as low-signal lesions on magnetic resonance imaging (MRI) T1-weighted sequences and as high-signal lesions on T2-weighted and fluid-attenuated inversion recovery (FLAIR) sequences [77,89]. Tubers exhibit contrast enhancement in approximately 3%–4% of the cases [88]. They are rarely found in the brainstem and spinal cord [88].

Other brain lesions in patients with TSC are also identifiable on neuroimaging, such as cerebral white matter radial migration lines that can be found on T2-weighted MRI sequences, extending from the subependymal surface of a subependymal nodule outward to a cortical tuber or to the cortex without a tuber identified [77,89]. However, the pathologic and clinical overlap between "cortical tuber" as a major feature and "cerebral white matter radial migration lines" as a minor feature in the 1998 diagnostic criteria were felt to no longer represent separate processes and were replaced with a single major feature in the new classification ("cortical dysplasia") by the 2012 Consensus Conference [2]. Since a single area of focal cortical dysplasia or even two can be observed in an individual who does not have TSC, in the 2012 diagnostic criteria, multiple areas of focal cortical dysplasia count only as one major feature and additional clinical features are necessary to establish a definite diagnosis of TSC [2].

SENs appear as small protrusions into the cerebrospinal fluid cavity and can occur anywhere along the ventricular surface but are most commonly found at the caudothalamic groove in the region of the foramen of Monro, often with associated calcification [73,77,88]. This characteristic (calcification) is more readily identifiable with computed tomography (CT) as a high-density lesion [77]. They can also be identified with MRI as low-intensity nodules on T1- or T2-weighted sequences and may present contrast enhancement [77,88]. Considering histopathology, they can give the appearance of "candle guttering" because they can line the ventricular surface and are composed predominantly of dysplastic astrocytes and mixed-lineage astrocytic or neuronal cell components [77,90]. It is widely accepted that SEGAs typically arise from SENs, especially near the foramen of Monro, and also may progressively calcify over time [2].

99.6.5 Cardiovascular Features

Imaging plays a central role in the evaluation of cardiac tumors, and in most cases it is not easy to obtain a cardiac tumor biopsy, either by percutaneous or catheter biopsy [82,86] The primary imaging modality is echocardiography, but in doubtful cases other imaging modalities can be used to enhance visualization and identification of different tumor types, namely CT and MRI [82].

99.6.6 Pulmonary Features

LAM can take place as a pulmonary manifestation of tuberous sclerosis (tuberous sclerosis—associated) or as an isolated form (sporadic), and about one-third of sporadic LAM patients have renal angiomyolipomas, another major feature in the diagnostic criteria for TSC [1,2]. These led to the conclusion by the 1998 consensus group that when both angiomyolipoma and LAM were present, other TSC features must be present for the diagnosis of TSC [2]. Posteriorly, the members of the pulmonology panel of 2012 Consensus Conference proposed that when angiomyolipomas and LAM are both present in a patient with suspected TSC, together they constitute only one major criterion [2]. The diagnosis of LAM defined by the pulmonology panel includes: (i) pathologic examination consistent with LAM, nodules are composed of actin-positive spindle-shaped cells that are stained with proliferating cellular nuclear antigen and less abundant cuboidal epithelioid cells that are stained with HMB-45; (ii) characteristic high-resolution chest CT criteria with profusion of thin-walled cysts (>4) of varied size defined by the European Respiratory Society (ERS), and no confounding comorbid conditions or exposures in a patient with at least one other major criteria for TSC (other than angiomyolipoma), or two other minor criteria; or (iii) characteristic or compatible (ERS criteria) high-resolution CT in the setting of no confounding comorbid conditions or exposures, plus one of the following: abdominal or thoracic lymphangioleiomyomas, chylous pleural effusion, or chylous ascites [2,73]. Also, pulmonary function tests may contribute to diagnosis, with a reduction in diffusing capacity for carbon monoxide, an increase in residual volume and a reduced forced expiratory volume in 1 s being the earliest physiological manifestations [73,91]. The 6-minute walk test and the vascular endothelial growth factor type D (VEGF-D) in the blood are useful to help with the diagnosis [73]. However, a negative VEGF-D result does not exclude the diagnosis of LAM [92]. It is important to note that lung is rarely biopsied in TSC patients with pulmonary parenchymal changes, so it is possible that processes other than LAM may result in cystic lung disease in TSC patients [2].

MMPH stain with cytokeratin and surfactant proteins A and B, but not with HMB-45, alpha smooth muscle actin or hormonal receptors [2]. In terms of differential diagnosis, MMPH can be mistaken with atypical adenomatous hyperplasia, a premalignant lesion that is not clearly associated with TSC, and other metastatic tumors [2,73].

MMPH and CCTL are not included as diagnostic criteria for TSC due to the lack of specificity [2].

99.6.7 Renal Features

The nephrology panel attending the 2012 Consensus Conference agreed with deleting the designation of "renal" in the major feature "renal angiomyolipomas," and replace it with "angiomyolipomas ≥2" in the clinical diagnostic criteria, since angiomyolipomas have been identified in TSC patients in organs other than the kidney, including the liver.

Angiomyolipomas may be detected by ultrasonography, CT, or MRI [28]. However, MRI is the preferred modality for evaluation of angiomyolipomas, as many can be fat-poor and hence missed when abdominal CT or US are performed [83,86]. Fat-poor angiomyolipomas are not uncommon in patients with TSC, but if there is doubt and lesions are growing faster than 0.5 cm per year, a needle biopsy using a sheath technique or an open biopsy may be considered [83,86]. Although they are described as slow-growing tumors, ultrasonography may reveal dramatic progression of tumor size (3–4 cm every 2 years) in adolescents [1]. The classical ultrasonography appearance of an angiomyolipoma is that of a strongly hyperreflective lesion with acoustic shadowing [83].

Also, assessment of blood pressure, because of increased risk of secondary hypertension, and renal function, through glomerular filtration rate (GFR) using creatinine equations for adults and children, or alternatively, measurement of serum cystatin C concentration, are important [86].

99.6.8 Other Manifestations

Neuroendocrine tumors might be slightly more prevalent in TSC patients, but these lesions are not hamartomas; thus, they are not considered part of the diagnostic criteria [2,73]. Liver angiomyolipomas are included in the major features group under the heading "angiomyolipomas," as previously explained [2]. Hamartomatous rectal polyps were included as a minor feature in the 1998 Diagnostic Criteria, but due to their lack of specificity for TSC and because they are another type of "nonrenal hamartoma," the specific designation was deleted from the minor criteria [2].

99.7 Treatment

TSC is a systemic and progressive disease that requires long-term treatment and support from a multidisciplinary team [28,86]. Besides the classical approaches with surgery and symptomatic treatments, the current therapeutic challenge is the use of mTORC1 inhibitors as a treatment, with benefits already proved [2,93,94].

The discovery of the relationship between TSC1/TSC2 and mTOR has resulted in important clinical advances in the use of mTOR inhibitors for the treatment of several TSC manifestations [61,95,96]. Of the different mTOR inhibitors (rapamycin, everolimus temsirolimus, ridaforolimus), only rapamycin (also known as sirolimus) and everolimus have been clinically evaluated for the management of TSC patients [61,95,96]. Sirolimus and analogs share a central macrolide chemical structure, differing in the functional groups added at C40, with everolimus being an active derivative (hydroxyethyl ester) of sirolimus (Table 99.3)

TABLE 99.3

Pharmacological Characterization, Clinical Indications, and Adverse Effects of Sirolimus and Everolimus [61,94,96,97]

	Sirolimus ($C_{51}H_{79}NO_{13}$)	Everolimus ($C_{53}H_{83}NO_{14}$)
Molecular weight	914.2 g/mol	958.2 g/mol
Route of administration	Orally, once daily Topic (concentrations 0.1%–1%)	Orally, once daily
Protein binding	~92%	~75%
Bioavailability	Solution:14% Tablet: 18%	Tablet: 20%
t_{max}	2 h	30 min–1 h
Distribution	Large distribution (around 12 L/kg), ~95% into RBCs	Wide distribution into RBCs; good blood–brain partition coefficient
Solubitlity	Water insoluble	Water insoluble
Metabolization	Hepatic CYP3A4, glycoprotein 1	CYP3A4, CYP3A5, CYP2C8
Drug interactions	Increase rapamycin or rapalog concentration: Cyclosporine, bromocriptine, cimetidine, cisapride, clotrimazole, danazol, diltiazem, fluconazole, protease inhibitors (e.g., for HIV and hepatitis C that include drugs such as ritonavir, indinavir, boceprevir, and telaprevir), metoclopramide, nicardipine, troleandomycin, verapamil Decrease rapamycin or rapalog concentrations: Carbamazepine, phenobarbital, phenytoin, rifapentine	
Terminal half-life	46–78 h	26–30 h
Elimination	Feces (91%), urine (2%)	Feces (>90%), urine (2%)
Clinical indications for TSC	Topical formulation has been used for facial angiofibromas Sirolimus treatment induced regression of kidney angiomyolipomas, SEGAs, and liver angiomyolipomas In patients with LAM, sirolimus stabilized lung function	Approved for SEGA and adult angiomyolipoma associated with TSC
Common adverse reactions	Peripheral edema, hypertriglyceridemia, hypertension, hypercholesterolemia, creatinine increase, constipation, abdominal pain, diarrhea, headache, fever, urinary tract infection, anemia, nausea, arthralgia, pain, and thrombocytopenia An irritation and burning sensation is the most common side effect seen after topical administration	Stomatitis, infections, rash, fatigue, diarrhea, and decreased appetite, hypercholesterolemia, hyperglycemia, increased AST, anemia, leukopenia, thrombocytopenia, lymphopenia, increased alanine transaminase (ALT), and hypertriglyceridemia

Abbreviations: t_{max}, time after administration when the maximum plasma concentration is reached; RBCs, red blood cells; CYP3A, intestinal cytochrome p450 3 A enzymes; SEGA, subependymal giant cell astrocytomas.

[58,61]. The mechanisms of action for sirolimus and analogs are similar, both working by binding to and forming a complex with FK506-binding protein-12 (FKBP12), which then inhibits mTORC1 [58,61].

Only everolimus has been approved for the management of TSC disease manifestations, including SEGA and renal angiomyolipomas [16,93,96,98,99]. Two major clinical trials demonstrated efficacy and safety of everolimus in the treatment of SEGA, with marked reduction in tumor volume or cessation of growth [16,98]. Also, everolimus proved to be effective in the treatment of symptomatic angiomyolipomas associated with TSC [93,99]. Although not formerly approved, various case reports and multiple prospective clinical trials reported sirolimus benefit in TSC patients [15,18,96]. Sirolimus treatment induced regression of kidney angiomyolipomas, SEGAs, and liver angiomyolipomas, but tumor volume tended to increase after therapy withdrawal [15,17,100]. In patients with LAM, sirolimus stabilized lung function, reduced serum VEGF-D levels, and was associated with a reduction in symptoms [17,18]. Importantly, it has been demonstrated that disease progression resumes if treatment is discontinued and for clinical benefit to be sustained, chronic—perhaps lifelong—treatment with mTORC1 inhibitors is indispensable, raising concern about long-term adverse effects [73,96]. mTOR inhibitors may have additional benefit in

CNS-related disease manifestations, since refractory epilepsy, cognitive development, and behavior symptoms have improved with everolimus treatment [98,101–103]. More recently, second-generation pharmacological mTOR inhibitors have been developed and differ from rapamycin analogs in the mechanism of action [61,96]. These molecules are direct kinase inhibitors that do not target FKBP12, and instead inhibit both mTORC1 and mTORC2 by directly blocking the ATP catalytic site [96]. These newer agents are potent inhibitors of cellular proliferation with probable therapeutic benefit, but to date none have completed phase III clinical trials or received regulatory approval for human use [96].

As conventional therapy, facial angiofibromas are treated with laser ablation, surgical excision, and graft surgery [73]. Benefit of systemic mTORC1 inhibitors in treating dermatological manifestations of TSC, evaluated as secondary endpoints in clinical trials, has been reported, with skin response rates significantly higher than for placebo [74,93,94,99]. Sirolimus is available in a topical formulation, and individual case reports have demonstrated sustained improvement in erythema and in the size and extension of the lesions, but no standard dose or specific formulation is defined for that route of administration [74,104]. Enamel defects (dental pits) can be treated with restorative treatments if the patient is at high cavity risk, and oral fibromas should be excised surgically if symptomatic or if interfering with oral hygiene [74,86].

Regarding treatment of CNS-related disease manifestations, surgical resection is the recommended intervention for symptomatic SEGA, and cerebrospinal fluid diversion may also be necessary [86,95]. For growing but asymptomatic SEGA, either surgical resection or medical therapy with mTOR inhibitors can be effective [86,95]. Multiple factors have to be considered when choosing between surgery and mTOR inhibitor therapy, as bilateral SEGA, tumor size >2 cm, and children ≤3 years of age may be associated with significant surgical risk [95,105]. mTOR inhibitor therapy may have a role in patients undergoing surgery to reduce tumor volume before the resection, but clear clinical evidence is still missing [95]. Early epilepsy treatment may be of benefit in infants and children during the first 24 months of life if ictal discharges occur, with or without clinical manifestations [86]. Vigabatrin, an inhibitor of GABA transaminase with impact on mTOR overactivation, is the first-choice treatment in infants with focal seizures and/or infantile spasms [42,73,95]. However, possible side effects should be closed monitored, like retinal toxicity with permanent visual field defects [73,95]. Adrenocorticotrophic hormone (ACTH) or analogs can be used as second-line therapy if treatment with vigabatrin fails [73,86,95]. Epilepsy surgery, ketogenic diet, or vagus nerve stimulation may be considered for medically refractory TSC patients [86,95]. Both sirolimus and everolimus demonstrated a significant seizure frequency reduction in patients with TSC-related refractory epilepsy [102,106–108]. Although not adequately studied, mTOR inhibitors may be rational candidates for the management of neurodevelopmental/neuropsychiatric disabilities associated with TSC, including intellectual disability and autism [58,86]. Improvement in multiple aspects of social deficit behavior and neurocognition have been reported in animal models and in TSC patients [102,103,109,110].

Cardiac rhabdomyomas spontaneously regress with age, and treatment is usually not necessary for patients aged >10 years [73]. However, in cases of newborns and infants with large rhabdomyomas who are critically symptomatic, presenting respiratory distress and congestive heart failure, cardiac surgery is required [82].

Pulmonary LAM is usually generalized and progressive, extremely difficult to treat, and with a poor prognosis [1]. Previously, there was no effective treatment except lung transplantation [86]. In select LAM patients with moderate-to-severe lung disease or rapid progression, treatment with an mTOR inhibitor may be used to stabilize or improve pulmonary function, quality of life, and functional performance [17,18,73,86].

mTOR inhibitor therapy is the first line in the short term for asymptomatic, growing renal angiomyolipomas (>3 cm) [58,73,86,95]. Selective embolization followed by corticosteroids, kidney-sparing resection, or ablative therapy for exophytic lesions are acceptable second-line therapies for asymptomatic lesions [86]. However, emergency embolization is the first-line therapy for renal angiomyolipomas with acute bleeding or aneurysms [73,95]. Nephrectomy is avoided whenever possible because of the high incidence of complications and increased risk of future renal insufficiency and end-stage renal failure [73,86,95].

99.8 Prognosis

Tuberous sclerosis is a complex genetic disorder for which prognosis is dependent on multiple variables [86]. It is known that genotype plays an important role, as *TSC1* mutations are associated with milder disease features [12,13,21]. Number, size, and location of the characteristic tumors associated with the condition also influence the prognosis [1,12,73]. Renal angiomyolipomas, subependymal giant-cell tumors, and pulmonary and cardiac complications are the major causes of shortened life expectancy [1]. It is clear that the mTOR pathway might be beneficially targeted in TSC, with impact on patient outcomes [16,61]. Everolimus has been an effective strategy for treating SEGAs and adult angiomyolipomas [16,61]. In patients with LAM, sirolimus proved to produce benefit, with stabilization of lung function [18,73].

Many manifestations can be life-threatening, and appropriate surveillance is necessary to limit morbidity and mortality in TSC and also to optimize quality of life of affected individuals [84,86]. Considering surveillance practices, the first consensus recommendations occurred in 1998, and they were updated in 2012 incorporating the latest scientific evidence and current best clinical practices (see Table 99.2b) [86].

All individuals suspected of having TSC, regardless of age, should perform a brain MRI, with and without gadolinium. If MRI is not available or cannot be performed, CT or head ultrasound in neonates or infants when fontanels are open may be used, although results are considered suboptimal [86,88]. Optimal outcome is associated with early detection and treatment, so surveillance by MRI should be performed every 1–3 years in all individuals with TSC until the age of 25 years [86,111]. Frequency of scans may be increased if clinically indicated, namely in younger patients with larger or growing SEGAs, or who are cognitively disabled [86]. Individuals without SEGA by the age of 25 years do not need continued surveillance with brain imaging, but those with asymptomatic SEGA present in childhood should continue to be monitored by MRI for life, because of the possibility of tumor growth [86]. All pediatric patients should undergo a baseline EEG, even in the absence of recognized or reported clinical seizures, but the posterior frequency is clinically determined [86]. If the baseline EEG is abnormal, especially when features of TSC-associated neuropsychiatric disorders are also present, this should be followed up with a 24-hour video-EEG to assess for electrographic or subtle clinical seizure activity [86]. TAND should be evaluated upon diagnosis and screening at each follow-up clinic visit, with a minimum frequency of once per year [86]. In addition, formal evaluations for TAND by an expert team should be performed at key timepoints: during the first 3 years of life (0- to 3year evaluation), preschool (3- to 6-year evaluation), before middle school entry (6- to 9-year evaluation), during adolescence (12- to 16-year evaluation), and in early adulthood (18- to 25-year evaluation) [86].

Surveillance and management recommendations for renal findings in TSC according to the International TSC Conference include an abdominal imaging at the time of diagnosis, regardless of age, with MRI being the preferred modality for evaluation of angiomyolipoma and also a renal function assessment [83,86]. Posteriorly, annual clinical assessment of renal function and hypertension is required [86].

Regarding LAM, females 18 years or older should have a baseline pulmonary function testing, 6-minute walk test, and high-resolution chest CT [86]. VEGF-D levels may contribute to establish a baseline for future LAM progression, which can be useful to estimate the prognosis [73,92]. Counseling on smoking risks and estrogen use should also be performed [86]. In patients with no clinical symptoms and no evidence of lung cysts on their baseline CT, high-resolution chest CT should be performed every 5–10 years, using low-radiation imaging protocols when available [86]. Once cysts are detected, the pace of TSC-LAM progression should be determined via high-resolution chest CT every 2–3 years and annual pulmonary function testing and 6-minute walk test, with increasing frequency if advanced disease or clinically indicated to assist treatment decision making [86].

Surveillance of dermatological findings include a detailed clinical dermatological and dental exam at time of diagnosis [74,86]. If a definite diagnosis of TSC has been established, annual inspection is recommended [74,86]. For TSC-associated dental lesions and oral fibromas, periodic oral evaluation should occur every 3–6 months, consistent with surveillance recommendations for all individuals in the general population [86].

In pediatric patients, especially younger than 3 years of age, an echocardiogram and ECG should be obtained to evaluate for rhabdomyomas and arrhythmia, respectively [86]. In the absence of cardiac symptoms or concerning medical history, echocardiogram is not necessary in adults, but a baseline ECG is still recommended [86]. Until regression of cardiac rhabdomyomas is documented, follow-up echocardiogram should be performed every 1–3 years in asymptomatic patients, and 12-lead ECG is recommended at minimum every 3–5 years to monitor for conduction defects [86].

A baseline ophthalmologic evaluation is recommended for all individuals diagnosed with TSC [86]. Individuals with no identified ophthalmologic lesions or vision symptoms at baseline need reevaluation only if new clinical concerns arise. Otherwise, annual evaluation is recommended [86].

99.9 Conclusion and Future Perspectives

Tuberous sclerosis complex (TSC) is a rare genetic disease defined by the development of benign tumors called hamartomas in several organ systems, which are responsible for the great clinical heterogeneity of this condition. In the majority of cases, TSC results from mutations in two genes (*TSC1* and *TSC2*) that, despite being located in different chromosomes, encode proteins involved in complex interactions with the mammalian target of rapamycin (mTOR)-signaling pathway. Given that this pathway controls a variety of cell functions, such as cell growth, proliferation, and survival, its roles in disease processes are nondiscriminatory and its affected organs are wide-ranging.

Considering the most recent advances in research and new technologies, the diagnosis and management of TSC have evolved considerably in the past few years, promoting a multidisciplinary approach. However, many fundamental questions regarding genotype–phenotype correlation and long-term surveillance strategies are still unanswered, highlighting the importance of epidemiological and clinical studies like the Tuberous Sclerosis Registry to Increase Disease Awareness (TOSCA) registry [112].

Recent delineation of TSC pathogenesis, involving mTOR-signaling pathway, has culminated in new and exciting potential targeted therapies. Not surprisingly, inhibitors of the mTOR pathway (e.g., rapamycin [also known as sirolimus] and everolimus) are becoming key players in the management of some TSC-associated features, such as subependymal giant cell astrocytomas, renal angiomyolipomas, and epilepsy. However, despite available preclinical and clinical data are encouraging, some controversial side effects with mTOR inhibitors also exist and some drawbacks of "systemic" mTOR inhibition must still be considered. In particular, the choice of the appropriate drug, the correct time window, treatment duration, dosages to be used, and long-term adverse effects are still far from being clearly defined [61,113]. Further investigation is warranted, ideally with more selective drugs, to define a better safety profile and a higher effectiveness, before full clinical translation of mTOR targeting succeeds.

The development of second-generation mTOR inhibitors (known as mTOR kinase inhibitors [TORKinibs]) could be a relevant aspect in fulfilling this need. They are able to directly inhibit the kinase by blocking the ATP catalytic site (instead of linking FKBP12), causing inhibition of both mTORC1 and mTORC2 [114,115]. This property allows them to have distinct implications on downstream target inhibition, on the mechanisms of protein translation control and on regulation loops, which can finally be important in differentiating intracellular signaling mediated by TORC1 and TORC2. In preclinical models, these new mTOR inhibitors seem to have potent antiproliferative properties, which make them interesting molecules for oncology [115]. Nevertheless, until now it is not known if the dual targeting of TORC1/TORC2 introduces higher efficacy without further relevant toxicity, comparing with everolimus. These new pharmacological agents might open new research avenues and windows of opportunity for several pathological conditions (including TSC), but further research will be needed to clarify these hypothetical gains.

Acknowledgments

This work was supported by the European Regional Development Fund (FEDER), through Programa Operacional Factores de Competitividade COMPETE2020 (CENTRO-01-0145-FEDER-000012-HealthyAging2020) and by National funds via Portuguese Science and Technology Foundation (FCT): Strategic Projects UID/NEU/04539/2013, UID/NEU/04539/2019, UIDB/04539/2020 and UIDP/04539/2020 (CIBB), as well as by COMPETE-FEDER funds (POCI-01-0145-FEDER-007440).

Conflict of Interest

Authors declare no conflict of interest.

REFERENCES

1. Curatolo P, Bombardieri R, Jozwiak S. Tuberous sclerosis. *Lancet* (London, England) 2008;372:657–68.
2. Northrup H, Krueger DA. Tuberous sclerosis complex diagnostic criteria update: Recommendations of the 2012 IInternational Tuberous Sclerosis Complex Consensus Conference. *Pediatr Neurol*. 2013;49:243–54.

3. Recklinghausen Fv. *Die Lymphelfasse und ihre Beziehung zum Bindegewebe.* Berlin: A Hirschwald, 1862.

4. Lendvay TS, Marshall FF. The tuberous sclerosis complex and its highly variable manifestations. *J Urol* 2003;169:1635–42.

5. Bourneville D. Sclérose tubéreuse des circonvolutions cérébrales. *Arch Neurol* 1880;1:81–91.

6. Morse RP. Tuberous sclerosis. *Arch Neurol* 1998;55:1257–8.

7. Kirpicznik J. Ein Fall von Tuberoser Sklerose and gleichzeitigen multiplem Nierengeschwülstein. *Virchows Arch Pathol Anat* 1910;202:358.

8. Osborne JP, Fryer A, Webb D. Epidemiology of tuberous sclerosis. *Ann New York Acad Sci* 1991;615:125–7.

9. European Chromosome 16 Tuberous Sclerosis Consortium. Identification and characterization of the tuberous sclerosis gene on chromosome 16. *Cell* 1993;75:1305–15.

10. van Slegtenhorst M, de Hoogt R, Hermans C et al. Identification of the tuberous sclerosis gene TSC1 on chromosome 9q34. Science (New York, NY) 1997;277:805–8.

11. Huang J, Dibble CC, Matsuzaki M, Manning BD. The TSC1-TSC2 complex is required for proper activation of mTOR complex 2. *Mol Cell Biol* 2008;28:4104–15.

12. Jones AC, Shyamsundar MM, Thomas MW et al. Comprehensive mutation analysis of TSC1 and TSC2-and phenotypic correlations in 150 families with tuberous sclerosis. *Am J Hum Genet* 1999;64:1305–15.

13. Langkau N, Martin N, Brandt R et al. TSC1 and TSC2 mutations in tuberous sclerosis, the associated phenotypes and a model to explain observed TSC1/ TSC2 frequency ratios. *Eur J Pediatr* 2002;161:393–402.

14. Lagos JC, Gomez MR. Tuberous sclerosis: Reappraisal of a clinical entity. *Mayo Clin Proc* 1967;42:26–49.

15. Dabora SL, Franz DN, Ashwal S et al. Multicenter phase 2 trial of sirolimus for tuberous sclerosis: Kidney angiomyolipomas and other tumors regress and VEGF- D levels decrease. *PLOS ONE* 2011;6:e23379.

16. Franz DN, Belousova E, Sparagana S et al. Efficacy and safety of everolimus for subependymal giant cell astrocytomas associated with tuberous sclerosis complex (EXIST-1): A multicentre, randomised, placebo-controlled phase 3 trial. *Lancet* (London, England) 2013;381:125–32.

17. Davies DM, de Vries PJ, Johnson SR et al. Sirolimus therapy for angiomyolipoma in tuberous sclerosis and sporadic lymphangioleiomyomatosis: A phase 2 trial. *Clin Cancer Res* 2011;17:4071–81.

18. McCormack FX, Inoue Y, Moss J et al. Efficacy and safety of sirolimus in lymphangioleiomyomatosis. *New Engl J Med* 2011;364:1595–606.

19. van Slegtenhorst M, Verhoef S, Tempelaars A et al. Mutational spectrum of the TSC1 gene in a cohort of 225 tuberous sclerosis complex patients: No evidence for genotype-phenotype correlation. *J Med Genet* 1999;36:285–9.

20. Kozlowski P, Roberts P, Dabora S et al. Identification of 54 large deletions/duplications in TSC1 and TSC2 using MLPA, and genotype-phenotype correlations. *Hum Genet* 2007;121:389–400.

21. Dabora SL, Jozwiak S, Franz DN et al. Mutational analysis in a cohort of 224 tuberous sclerosis patients indicates increased severity of TSC2, compared with TSC1, disease in multiple organs. *Am J Hum Genet* 2001;68:64–80.

22. Kandt RS, Haines JL, Smith M et al. Linkage of an important gene locus for tuberous sclerosis to a chromosome 16 marker for polycystic kidney disease. *Nature Genet* 1992;2:37–41.

23. Fryer AE, Chalmers A, Connor JM et al. Evidence that the gene for tuberous sclerosis is on chromosome 9. *Lancet* (London, England) 1987;1:659–61.

24. Nellist M, Brook-Carter PT, Connor JM, Kwiatkowski DJ, Johnson P, Sampson JR. Identification of markers flanking the tuberous sclerosis locus on chromosome 9 (TSC1). *J Med Genet* 1993;30:224–7.

25. Kandt RS, Pericak-Vance MA, Hung WY et al. Absence of linkage of ABO blood group locus to familial tuberous sclerosis. *Exp Neurol* 1989;104:223–8.

26. Kandt RS, Pericak-Vance MA, Hung WY et al. Linkage studies in tuberous sclerosis. Chromosome 9? 11? or maybe 14! *Ann New York Acad Sci.* 1991;615:284–97.

27. Garami A, Zwartkruis FJ, Nobukuni T et al. Insulin activation of Rheb, a mediator of mTOR/S6K/4E-BP signaling, is inhibited by TSC1 and 2. *Mol Cell* 2003;11:1457–66.

28. Crino PB, Nathanson KL, Henske EP. The tuberous sclerosis complex. *New Engl J Med* 2006;355:1345–56.

29. Sampson JR, Scahill SJ, Stephenson JB, Mann L, Connor JM. Genetic aspects of tuberous sclerosis in the west of Scotland. *J Med Genet* 1989;26:28–31.

30. Knudson AG, Jr. Mutation and cancer: Statistical study of retinoblastoma. *Proc Natl Acad Sci USA* 1971;68:820–3.

31. Carbonara C, Longa L, Grosso E et al. Apparent preferential loss of heterozygosity at TSC2 over TSC1 chromosomal region in tuberous sclerosis hamartomas. *Genes, Chromosomes Cancer* 1996;15:18–25.

32. Henske EP, Scheithauer BW, Short MP et al. Allelic loss is frequent in tuberous sclerosis kidney lesions but rare in brain lesions. *Am J Hum Genet* 1996;59:400–6.

33. Chan JA, Zhang H, Roberts PS et al. Pathogenesis of tuberous sclerosis subependymal giant cell astrocytomas: Biallelic inactivation of TSC1 or TSC2 leads to mTOR activation. *J Neuropathol Exp Neurol* 2004;63:1236–42.

34. Karbowniczek M, Astrinidis A, Balsara BR et al. Recurrent lymphangiomyomatosis after transplantation: Genetic analyses reveal a metastatic mechanism. *Am J Respir Crit Care Med* 2003;167:976–82.

35. Sancak O, Nellist M, Goedbloed M et al. Mutational analysis of the TSC1 and TSC2 genes in a diagnostic setting: Genotype--phenotype correlations and comparison of diagnostic DNA techniques in tuberous sclerosis complex. *Eur J Hum Genet* 2005;13:731–41.

36. Niida Y, Lawrence-Smith N, Banwell A et al. Analysis of both TSC1 and TSC2 for germline mutations in 126 unrelated patients with tuberous sclerosis. *Hum Mutat* 1999;14: 412–22.

37. Hung CC, Su YN, Chien SC et al. Molecular and clinical analyses of 84 patients with tuberous sclerosis complex. *BMC Med Genet* 2006;7:72.

38. Khare L, Strizheva GD, Bailey JN et al. A novel missense mutation in the GTPase activating protein homology region of TSC2 in two large families with tuberous sclerosis complex. *J Med Genet* 2001;38:347–9.

39. Jansen AC, Sancak O, D'Agostino MD et al. Unusually mild tuberous sclerosis phenotype is associated with TSC2 R905Q mutation. *Ann Neurol* 2006;60:528–39.

40. Mayer K, Goedbloed M, van Zijl K, Nellist M, Rott HD. Characterisation of a novel TSC2 missense mutation in the GAP related domain associated with minimal clinical manifestations of tuberous sclerosis. *J Med Genet* 2004;41:e64.

41. Rok P, Kasprzyk-Obara J, Domanska-Pakiela D, Jozwiak S. Clinical symptoms of tuberous sclerosis complex in patients with an identical TSC2 mutation. *Med Sci Monit* 2005;11:Cr230–4.

42. Moavero R, Cerminara C, Curatolo P. Epilepsy secondary to tuberous sclerosis: Lessons learned and current challenges. *Child's Nerv System* 2010;26:1495–504.

43. Xiao GH, Shoarinejad F, Jin F, Golemis EA, Yeung RS. The tuberous sclerosis 2 gene product, tuberin, functions as a Rab5 GTPase activating protein (GAP) in modulating endocytosis. *J Biol Chem* 1997;272:6097–100.

44. Potter CJ, Huang H, Xu T. Drosophila Tsc1 functions with Tsc2 to antagonize insulin signaling in regulating cell growth, cell proliferation, and organ size. *Cell* 2001;105:357–68.

45. Manning BD, Tee AR, Logsdon MN, Blenis J, Cantley LC. Identification of the tuberous sclerosis complex-2 tumor suppressor gene product tuberin as a target of the phosphoinositide 3-kinase/akt pathway. *Mol Cell* 2002;10:151–62.

46. Roux PP, Ballif BA, Anjum R, Gygi SP, Blenis J. Tumor-promoting phorbol esters and activated Ras inactivate the tuberous sclerosis tumor suppressor complex via p90 ribosomal S6 kinase. *Proc Natl Acad Sci USA* 2004;101:13489–94.

47. Ma L, Chen Z, Erdjument-Bromage H, Tempst P, Pandolfi PP. Phosphorylation and functional inactivation of TSC2 by Erk implications for tuberous sclerosis and cancer pathogenesis. *Cell* 2005;121:179–93.

48. Inoki K, Zhu T, Guan KL. TSC2 mediates cellular energy response to control cell growth and survival. *Cell* 2003;115:577–90.

49. Mak BC, Kenerson HL, Aicher LD, Barnes EA, Yeung RS. Aberrant beta-catenin signaling in tuberous sclerosis. *Am J Pathol* 2005;167:107–16.

50. Finlay GA, York B, Karas RH et al. Estrogen-induced smooth muscle cell growth is regulated by tuberin and associated with altered activation of platelet-derived growth factor receptor-beta and ERK-1/2. *J Biol Chem* 2004;279:23114–22.

51. Noonan DJ, Lou D, Griffith N, Vanaman TC. A calmodulin binding site in the tuberous sclerosis 2 gene product is essential for regulation of transcription events and is altered by mutations linked to tuberous sclerosis and lymphangioleiomyomatosis. *Arch Biochem Biophys* 2002;398:132–40.

52. Rosner M, Hengstschlager M. Tuberin binds p27 and negatively regulates its interaction with the SCF component Skp2. *J Biol Chem* 2004;279:48707–15.

53. Birchenall-Roberts MC, Fu T, Bang OS et al. Tuberous sclerosis complex 2 gene product interacts with human SMAD proteins. A molecular link of two tumor suppressor pathways. *J Biol Chem* 2004;279:25605–13.

54. Catania MG, Mischel PS, Vinters HV. Hamartin and tuberin interaction with the G2/M cyclin-dependent kinase CDK1 and its regulatory cyclins A and B. *J Neuropathol Exp Neurol* 2001;60:711–23.

55. Karbowniczek M, Cash T, Cheung M, Robertson GP, Astrinidis A, Henske EP. Regulation of B-Raf kinase activity by tuberin and Rheb is mammalian target of rapamycin (mTOR)-independent. *J Biol Chem* 2004;279:29930–7.

56. Jozwiak J, Jozwiak S, Grzela T, Lazarczyk M. Positive and negative regulation of TSC2 activity and its effects on downstream effectors of the mTOR pathway. *Neuromol Med* 2005;7:287–96.

57. Zhang Y, Gao X, Saucedo LJ, Ru B, Edgar BA, Pan D. Rheb is a direct target of the tuberous sclerosis tumour suppressor proteins. *Nature CCell Biol* 2003;5:578–81.

58. Franz DN, Capal JK. mTOR inhibitors in the pharmacologic management of tuberous sclerosis complex and their potential role in other rare neurodevelopmental disorders. *Orphanet J Rare Dis* 2017;12:51.

59. Tee AR, Manning BD, Roux PP, Cantley LC, Blenis J. Tuberous sclerosis complex gene products, Tuberin and Hamartin, control mTOR signaling by acting as a GTPase-activating protein complex toward Rheb. *Curr Biol* 2003;13:1259–68.

60. Uhlmann EJ, Li W, Scheidenhelm DK, Gau CL, Tamanoi F, Gutmann DH. Loss of tuberous sclerosis complex 1 (Tsc1) expression results in increased Rheb/S6K pathway signaling important for astrocyte cell size regulation. *Glia* 2004;47:180–8.

61. Palavra F, Robalo C, Reis F. Recent advances and challenges of mTOR inhibitors use in the treatment of patients with tuberous sclerosis complex. *Oxid Med Cell Longev* 2017;2017:9820181.

62. Laplante M, Sabatini DM. mTOR signaling in growth control and disease. *Cell* 2012;149:274–93.

63. Wang X, Proud CG. mTORC1 signaling: What we still don't know. *J Mol Cell Biol* 2011;3:206–20.

64. Jacinto E, Loewith R, Schmidt A et al. Mammalian TOR complex 2 controls the actin cytoskeleton and is rapamycin insensitive. *Nature Cell Biol* 2004;6:1122–8.

65. Brugarolas J, Lei K, Hurley RL et al. Regulation of mTOR function in response to hypoxia by REDD1 and the TSC1/TSC2 tumor suppressor complex. *Genes Dev* 2004;18:2893–904.

66. Baybis M, Yu J, Lee A et al. mTOR cascade activation distinguishes tubers from focal cortical dysplasia. *Ann Neurol* 2004;56:478–87.

67. Miyata H, Chiang AC, Vinters HV. Insulin signaling pathways in cortical dysplasia and TSC-tubers: Tissue microarray analysis. *Ann Neurol* 2004;56:510–9.

68. Nevin NC, Pearce WG. Diagnostic and genetical aspects of tuberous sclerosis. *J Med Genet* 1968;5:273–80.

69. Hunt A, Lindenbaum RH. Tuberous sclerosis: A new estimate of prevalence within the Oxford region. *J Med Genet* 1984;21:272–7.

70. Fisher OD, Stevenson AC. Frequency of epiloia in Northern Ireland. *Brit J Prev Soc Med* 1956;10:134–5.

71. O'Callaghan FJ, Shiell AW, Osborne JP, Martyn CN. Prevalence of tuberous sclerosis estimated by capture-recapture analysis. *Lancet* (London, England) 1998;351:1490.

72. Rubilar C, Lopez F, Troncoso M, Barrios A, Herrera L. Clinical and genetic study patients with tuberous sclerosis complex. *Rev Chil Pediatr* 2017;88:41–9.

73. Wataya-Kaneda M, Uemura M, Fujita K et al. Tuberous sclerosis complex: Recent advances in manifestations and therapy. *Int J Urol* 2017;24(9):681–91.

74. Ebrahimi-Fakhari D, Meyer S, Vogt T, Pfohler C, Muller CSL. Dermatological manifestations of tuberous sclerosis complex (TSC). *J Dtsch Dermatol Ges* 2017;15:695–700.

75. Rowley SA, O'Callaghan FJ, Osborne JP. Ophthalmic manifestations of tuberous sclerosis: A population based study. *Brit J Ophthalmol* 2001;85:420–3.

76. Crino PB, Henske EP. New developments in the neurobiology of the tuberous sclerosis complex. *Neurology* 1999;53:1384–90.

77. DiMario FJ, Jr. Brain abnormalities in tuberous sclerosis complex. *J Child Neurol* 2004;19:650–7.

78. Calcagnotto ME, Paredes MF, Tihan T, Barbaro NM, Baraban SC. Dysfunction of synaptic inhibition in epilepsy associated with focal cortical dysplasia. *J Neurosci* 2005;25:9649–57.

79. Orlova KA, Crino PB. The tuberous sclerosis complex. *Ann New York Acad Sci* 2010;1184:87–105.

80. Chu-Shore CJ, Major P, Camposano S, Muzykewicz D, Thiele EA. The natural history of epilepsy in tuberous sclerosis complex. *Epilepsia* 2010;51:1236–41.

81. Joinson C, O'Callaghan FJ, Osborne JP, Martyn C, Harris T, Bolton PF. Learning disability and epilepsy in an epidemiological sample of individuals with tuberous sclerosis complex. *Psychol Med* 2003;33:335–44.

82. Kwiatkowska J, Waldoch A, Meyer-Szary J, Potaz P, Grzybiak M. Cardiac tumors in children: A 20-year review of clinical presentation, diagnostics and treatment. *Adv Clin Exp Med* 2017;26:319–26.

83. Halpenny D, Snow A, McNeill G, Torreggiani WC. The radiological diagnosis and treatment of renal angiomyolipoma-current status. *Clin Radiol* 2010;65:99–108.

84. Roach ES, Gomez MR, Northrup H. Tuberous sclerosis complex consensus conference: Revised clinical diagnostic criteria. *J Child Neurol* 1998;13:624–8.

85. Roach ES, DiMario FJ, Kandt RS, Northrup H. Tuberous sclerosis consensus conference: Recommendations for diagnostic evaluation. National Tuberous Sclerosis Association. *J Child Neurol* 1999;14:401–7.

86. Krueger DA, Northrup H. Tuberous sclerosis complex surveillance and management: Recommendations of the 2012 International Tuberous Sclerosis Complex Consensus Conference. *Pediatr Neurol* 2013;49:255–65.

87. Boronat S, Thiele EA, Caruso P. Cerebellar lesions are associated with TSC2 mutations in tuberous sclerosis complex: A retrospective record review study. *Dev Med Child Neurol* 2017;59(10):1071–6.

88. Kalantari BN, Salamon N. Neuroimaging of tuberous sclerosis: Spectrum of pathologic findings and frontiers in imaging. *Am J Roentgenol* 2008;190:W304–9.

89. Griffiths PD, Martland TR. Tuberous Sclerosis Complex: The role of neuroradiology. *Neuropediatrics* 1997;28:244–52.

90. Guerrini R, Carrozzo R. Epileptogenic brain malformations: Clinical presentation, malformative patterns and indications for genetic testing. *Seizure* 2001;10:532–43; quiz 544–7.

91. Taveira-DaSilva AM, Hedin C, Stylianou MP et al. Reversible airflow obstruction, proliferation of abnormal smooth muscle cells, and impairment of gas exchange as predictors of outcome in lymphangioleiomyomatosis. *Am J Respir Crit Care Med* 2001;164:1072–6.

92. Young LR, Vandyke R, Gulleman PM et al. Serum vascular endothelial growth factor-D prospectively distinguishes lymphangioleiomyomatosis from other diseases. *Chest* 2010;138:674–81.

93. Franz DN, Belousova E, Sparagana S et al. Long-term use of everolimus in patients with tuberous sclerosis complex: Final results from the EXIST-1 study. *PLOS ONE* 2016;11:e0158476.

94. Sasongko TH, Ismail NF, Zabidi-Hussin Z. Rapamycin and rapalogs for tuberous sclerosis complex. *Cochrane Database Syst Rev* 2016;7:Cd011272.

95. Curatolo P, Bjornvold M, Dill PE et al. The role of mTOR inhibitors in the treatment of patients with tuberous sclerosis complex: Evidence-based and expert opinions. *Drugs* 2016;76:551–65.

96. MacKeigan JP, Krueger DA. Differentiating the mTOR inhibitors everolimus and sirolimus in the treatment of tuberous sclerosis complex. *Neuro-oncology* 2015;17:1550–9.

97. Sadowski K, Kotulska K, Jozwiak S. Management of side effects of mTOR inhibitors in tuberous sclerosis patients. *Pharmacol Rep* 2016;68:536–42.

98. Krueger DA, Care MM, Holland K et al. Everolimus for subependymal giant-cell astrocytomas in tuberous sclerosis. *New Engl J Med* 2010;363:1801–11.

99. Bissler JJ, Kingswood JC, Radzikowska E et al. Everolimus for angiomyolipoma associated with tuberous sclerosis complex or sporadic lymphangioleiomyomatosis (EXIST-2): A multicentre, randomised, double-blind, placebo-controlled trial. *Lancet* (London, England) 2013;381:817–24.

100. Bissler JJ, McCormack FX, Young LR et al. Sirolimus for angiomyolipoma in tuberous sclerosis complex or lymphangioleiomyomatosis. *New Engl J Med* 2008;358:140–51.

101. Kotulska K, Chmielewski D, Borkowska J et al. Long-term effect of everolimus on epilepsy and growth in children under 3 years of age treated for subependymal giant cell astrocytoma associated with tuberous sclerosis complex. *Eur J Paediatr Neurol* 2013;17:479–85.

102. Krueger DA, Wilfong AA, Holland-Bouley K et al. Everolimus treatment of refractory epilepsy in tuberous sclerosis complex. *Ann Neurol* 2013;74:679–87.

103. Wesseling H, Elgersma Y, Bahn S. A brain proteomic investigation of rapamycin effects in the Tsc1+/- mouse model. *Mol Autism* 2017;8:41.

104. Wheless JW, Almoazen H. A novel topical rapamycin cream for the treatment of facial angiofibromas in tuberous sclerosis complex. *J Child Neurol* 2013;28:933–6.

105. Kotulska K, Borkowska J, Roszkowski M et al. Surgical treatment of subependymal giant cell astrocytoma in tuberous sclerosis complex patients. *Pediatr Neurol* 2014;50:307–12.

106. French JA, Lawson JA, Yapici Z et al. Adjunctive everolimus therapy for treatment-resistant focal-onset seizures associated with tuberous sclerosis (EXIST-3): A phase 3, randomised, double-blind, placebo-controlled study. *Lancet* (London, England) 2016;388:2153–63.

107. Canpolat M, Per H, Gumus H et al. Rapamycin has a beneficial effect on controlling epilepsy in children with tuberous sclerosis complex: Results of 7 children from a cohort of 86. *Child's Nerv System* 2014;30:227–40.

108. Cardamone M, Flanagan D, Mowat D, Kennedy SE, Chopra M, Lawson JA. Mammalian target of rapamycin inhibitors for intractable epilepsy and subependymal giant cell astrocytomas in tuberous sclerosis complex. *J Pediatr* 2014;164:1195–200.

109. Hwang SK, Lee JH, Yang JE et al. Everolimus improves neuropsychiatric symptoms in a patient with tuberous sclerosis carrying a novel TSC2 mutation. *Mol Brain* 2016;9:56.

110. Schneider M, de Vries PJ, Schonig K, Rossner V, Waltereit R. mTOR inhibitor reverses autistic-like social deficit behaviours in adult rats with both Tsc2 haploinsufficiency and developmental status epilepticus. *Eur Arch Psychiat Clin Neurosci* 2017;267:455–63.

111. de Ribaupierre S, Dorfmuller G, Bulteau C et al. Subependymal giant-cell astrocytomas in pediatric tuberous sclerosis disease: When should we operate? *Neurosurgery* 2007;60:83–9; discussion 89–90.

112. Kingswood JC, Bruzzi P, Curatolo P et al. TOSCA – first international registry to address knowledge gaps in the natural history and management of tuberous sclerosis complex. *Orphanet J Rare Dis* 2014;9:182.

113. Leo A, Constanti A, Coppola A, Citraro R, De Sarro G, Russo E. Chapter 8—mTOR Signaling in Epilepsy and Epileptogenesis: Preclinical and Clinical Studies. In: Maiese K, editor. *Molecules to Medicine with mTOR*, 1st edition. Boston: Academic Press, 2016:123–42.

114. Feldman ME, Shokat KM. New inhibitors of the PI3K Akt-mTOR pathway: Insights into mTOR signaling from a new generation of tor kinase domain inhibitors (TORKinibs). *Curr Top Microbiol Immunol* 2010;347(1):241–62.

115. Sun S-Y. mTOR kinase inhibitors as potential cancer therapeutic drugs. *Cancer Lett* 2013;340(1):1–8.

100

Weaver Syndrome

Dongyou Liu

CONTENTS

100.1 Introduction

Weaver syndrome is a rare congenital pediatric disorder characterized by skeletal overgrowth (e.g., tall stature), distinctive craniofacial and digital abnormalities (e.g., camptodactyly of the fingers and/or toes), advanced bone age, variable intellectual disability (ranging from normal intellect to severe intellectual disability), other clinical features (e.g., poor coordination, soft doughy skin, umbilical hernia, abnormal tone, and hoarse low cry in infancy), and increased risk for neuroblastoma. Mutations in the enhancer of zeste homolog 2 gene (*EZH2*) on chromosome 7q36.1 encoding a histone methyltransferase appear to be responsible for Weaver syndrome [1,2].

100.2 Biology

Weaver syndrome was first described in 1974 by David Weaver from two boys who presented with accelerated osseous maturation, unusual facies, and camptodactyly [3]. Subsequent reports of additional 50 or so cases established Weaver syndrome as part of overgrowth syndromes that typically show localized or generalized tissue overgrowth (e.g., height and/or head circumference ≥2 SD above the mean), varying degrees of developmental and intellectual disability, and predisposition to tumor development.

As a heterogeneous group of disorders, overgrowth syndromes are known to include genetic alterations in nuclear receptor SET-domain containing protein 1 gene (*NSD1* on chromosome 5q35, leading to Sotos syndrome), DNA methyltransferase 3A gene (*DNMT3A* on chromosome 2p23.3, leading to

Tatton−Brown−Rahman syndrome [TBRS]), and other genes (e.g., *CHD8*, *BRWD3*, *SETD2*, *HIST1H1E*, and *EED*) [4]. The identification of heterozygous mutations in histone methyltransferase gene (*EZH2*) on chromosome 7q36.1 in 2011 thus helped expand the genetic spectrum of overgrowth syndromes [1,2].

Compared to tall stature, dysmorphic facial features, and intellectual disability seen in Weaver syndrome, distinctive facial appearance (downslanted palpebral fissures, prominent chin, malar flushing), pre- and postnatal overgrowth, variable intellect deficiency, advanced bone age, and scoliosis are often noted in children with Sotos syndrome, whereas tall stature, variable intellect deficiency, autism spectrum disorder, scoliosis, joint hypermobility, facial appearance (round, heavy, with horizontal eyebrows and narrow palpebral fissures), increased weight, and neuropsychiatric issues are found in TBRS [5].

EZH2, NSD1, and DNMT3A are epigenetic modifiers involved in chromatin remodeling and transcriptional regulation either by histone lysine methylation (NSD1 and EZH2) or DNA methylation (DNMT3A). EZH2 directly controls DNA methylation through physical association with DNA methyltransferases (DNMTs), including DNMT3A, with concomitant H3K27 methylation and CpG promoter methylation leading to repression of EZH2 target genes and distinct craniofacial skeleton formation. On the other hand, NSD1 is a histone methyltransferase involved in histone H3 lysine 36 (H3K36) methylation, and displays dosage effects, with loss of one copy leading to overgrowth and gain of an additional copy leading to growth restriction [6]. It is possible that duplications involving *DNMT3A* could similarly result in growth retardation and short stature [7,8].

100.3 Pathogenesis

Located on chromosome 7q36.1, the *enhancer of zeste*, drosophila, *homolog 2 gene* (*EZH2*) comprises multiple alternative transcripts, the longest of which has 20 exons and encodes a 751-amino-acid histone methyltransferase (EZH2) with a critical SET [su(var)3–9, enhancer of zeste, trithorax] domain, a pre-SET CXC domain, and two additional SANT (Sw13, Ada2, N-cor TFIIIB) domains. EZH2 constitutes the catalytic subunit of the polycomb-repressive complex 2 (PRC2), which also includes additional core components SUZ12 (suppressor of zeste 12) and EED (embryonic ectoderm development). As a highly conserved epigenetic modifying complex, PRC2 induces trimethylation of histone H3 at lysine 27 (resulting in H3K27me3), which serves as an epigenetic signal for chromatin condensation and transcriptional repression and contributes to the regulation of chondrocyte proliferation and hypertrophy in the growth plate [9]. Functional loss of any of the components (i.e., EZH2, SUZ12, and EED) compromises the enzymatic activity of PRC2 and subsequent reduction of H3K27me3, leading to transcriptional activation of loci to which H3K27me3 is bound [10–12].

Germline (constitutional) mutations altering/blocking EZH2 methyltransferase activity within the PRC2 complex to facilitate trimethylation of H3K27 underlie the molecular basis of Weaver syndrome. Often, these mutations are missense substitutions or indels in the SET domain that preserve the reading frame [e.g., c.395C>T (p.Pro132Leu), c.458A>G (p.Tyr153Cys), c.466A>G (p.K156E), c.553G>C (p.Asp185His), c.2233G>A (p.Glu745Lys)]. In addition, truncating mutations involving the terminal exon are occasionally found [13].

Besides germline (familial) mutations, de novo (sporadic) EZH2 mutations, which are present in affected patients but absent in the leukocytes (blood) of either parent, may also contribute to the development of Weaver syndrome. Examples of de novo EZH2 mutations include c.401T>C (p.M134 T), c.836A>G (p.H279R), c.1876G>A (p.V626M), c.1915A>G (p.K639E), c.1987T>A (p.Y663N), c.2044G>A (p.A682 T), c.2050C>T (p.R684C), c.2084C>T (p.S695L), c.2199C>G (p.Y733X), c.2204_2211dupAGGCTGAT, and c.2222A>G (p.Y741C).

Furthermore, somatic activating and inactivating mono- and biallelic *EZH2* pathogenic variants may occur in hematopoietic malignancies, highlighting the credential of *EZH2* as either an oncogene or a tumor suppressor. These include monoallelic gain-of-function (activating) missense mutations at positions Tyr646 (including p.Tyr646Phe, p.Tyr646Asn, p.Tyr646His, and p.Tyr646Ser), Ala682 (p.Ala682Gly), and Ala692 (p.Ala692Val), which are known as Tyr641, Ala677, and Ala687 in the shorter EZH2 isoform, in large B-cell and non-Hodgkin lymphomas, and monoallelic or biallelic loss-of-function (inactivating) EZH2 mutations in myeloproliferative neoplasms (MPN) and myelodysplastic syndromes (MDS). However, hematopoietic malignancies in the setting of EZH2-related overgrowth do not have higher frequencies than those in the general population [5].

100.4 Epidemiology

100.4.1 Prevalence

Weaver syndrome is a rare congenital pediatric disorder, with 54 individuals harboring an *EZH2* germline pathogenic variant identified to date. Since individuals with a mild phenotype may escape clinical diagnosis, the true prevalence of Weaver syndrome remains unknown.

100.4.2 Inheritance

EZH2-related overgrowth is inherited in an autosomal dominant manner, with each child of an individual with an *EZH2* pathogenic variant having a 50% chance of inheriting the pathogenic variant. In addition, some germline *EZH2* pathogenic variants appear to arise de novo, since they cannot be detected in the leukocyte DNA of either parent.

100.4.3 Penetrance

Weaver syndrome phenotype is subtle in some individuals, suggesting the variable/incomplete penetrance of some *EZH2* pathogenic variants. As a result, the severity of the phenotype in an individual inheriting the *EZH2* pathogenic variant cannot be predicted.

100.5 Clinical Features

Patients with Weaver syndrome may show a broad range of anomalies relating to the following:

Growth: Tall stature, height ≥+2 SD, >90% of cases; macrocephaly, head circumference ≥+2 SD, 97% of cases; persistent overgrowth of prenatal onset and accelerated growth and markedly advanced skeletal maturation during infancy.

Craniofacial features: Macrocephaly, flat occiput in 50% of cases, redundant nuchal skin folds, broad forehead with frontal bossing in 88% of cases, thin scalp hair, hypertelorism in 93% of cases, upslanting or downslanting palpebral fissures, small palpebral fissures, strabismus, ptosis, broad nasal root in 59% of cases, long philtrum in 81% of cases, large and dysmorphic ears in 80% of cases, micrognathia in 82% of cases, highly arched palate, widely spaced eyes, (Figure 100.1) [1].

Skeletal findings: Advanced bone age in 79% of cases, abnormal or absent femur, splayed metaphyses and mottled epiphyses, broad iliac wings, restricted joint mobility, cervical kyphosis, cervical spine anomalies and underdevelopment of the mid cervical vertebral bodies.

Limb anomalies: Camptodactyly, prominent finger pads, clinodactyly of fifth finger and toes, broad thumbs; talipes equinovarus, talipes calcaneovalgus, metatarsus adductus, pes adductus and pes cavus, demineralization of the bones of the hands and feet (Figure 100.2) [14].

Intellectual disability: Mild (ventriculomegaly, periventricular leukomalacia, and cerebellar hypoplasia), moderate (periventricular leukomalacia with ventriculomegaly, and isolated ventriculomegaly), severe (polymicrogyria and pachygyria; in 80% of cases).

Ttumor development: Neuroblastoma, pre-T cell non-Hodgkins lymphoma, and acute lymphoblastic leukemia in 10% of cases.

Other anomalies: Central hypotonia and/or peripheral hypertonia in 82% of cases; hoarse and low-pitched cry, poor coordination due to ligamentous laxity, difficulty in swallowing or breathing, mitral valve prolapse, ventricular septal defect, patent arteriosus ductus, congestive cardiomyopathy, umbilical hernia, inguinal hernia, cryptorchidism, development delay in 83% of cases, autistic spectrum disorder, phobias, and anxiety [15–17].

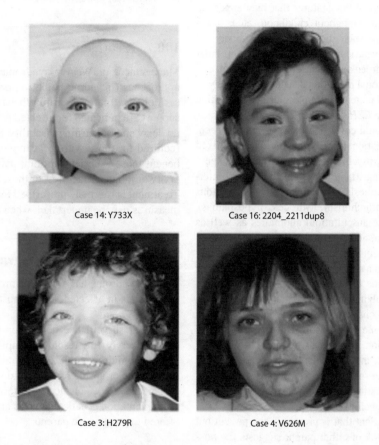

Case 14: Y733X   Case 16: 2204_2211dup8

Case 3: H279R   Case 4: V626M

FIGURE 100.1 Weaver syndrome patients with *EZH2* mutations showing typical facial features (e.g., widely spaced eyes, macrocephaly, broad forehead with frontal bossing, upslanting or downslanting palpebral fissures, broad nasal root, large and dysmorphic ears). (Photo credit: Tatton-Brown K et al. *Oncotarget*. 2011;2(12):1127–33.)

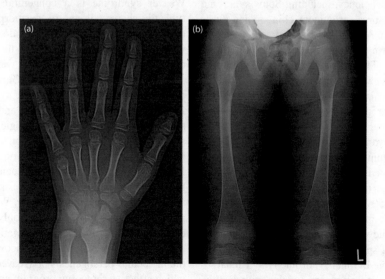

FIGURE 100.2 Weaver syndrome in a 5-year 7-month-old girl showing: (a) accelerated disharmonic osseous maturation and short fifth metacarpal bone; (b) splaying of the distal femora. (Photo credit: Miyoshi Y et al. *Clin Pediatr Endocrinol*. 2004;13(1):17–23.)

100.6 Diagnosis

Diagnosis of Weaver syndrome involves clinical examination (for increased height and/or head circumference in association with intellectual disability, retrognathia in young children, almond-shaped palpebral fissures and widely spaced eyes in all ages, etc.), detailed medical history review (for features that may appear in infancy but resolve/improve throughout childhood, etc.), imaging studies (for cervical spine anomalies and advanced bone age; isolated ventriculomegaly, ventriculomegaly, and periventricular leukomalacia, periventricular leukomalacia, cerebellar infarct, cerebellar hypoplasia, and neuronal migration defects/polymicrogyria with and without pachygyria, etc.), and molecular genetic testing (for a heterozygous germline *EZH2* pathogenic variant) [18].

Clinical examination should cover neurologic (motor, speech, general cognitive, and vocational skills; hypotonia and mixed hypo/hypertonia; seizures), psychiatric/behavioral (autism spectrum disorder), constitutional (height, weight, head circumference), musculoskeletal (scoliosis, camptodactyly, ligamentous laxity), genitourinary (cryptorchidism, hydrocele, hypospadias), and cardiovascular (cardiac auscultation) anomalies as well as malignancy (neuroblastoma, hematologic malignancies).

Molecular identification of a heterozygous germline *EZH2* pathogenic variant provides a confirmative diagnosis for Weaver syndrome. This can be done through a combination of single-gene testing (usually involving sequencing analysis of small intragenic deletions/insertions and missense, nonsense, and splice site variants), multigene panel testing (involving sequence analysis; deletion/duplication analysis using such as quantitative PCR, long-range PCR, multiplex ligation-dependent probe amplification; and/or other non-sequencing-based tests), and comprehensive genomic testing (involving exome sequencing, exome array, genome sequencing) [5].

An *EZH2* pathogenic variant that is present in the patient but absent in the leukocyte DNA of either parent suggests the possibility of a de novo germline mutation.

Differential diagnoses for Weaver syndrome consist of several overgrowth syndromes that also produce tall stature, scoliosis, and variable intellect deficiency, including Sotos syndrome (downslanted palpebral fissures, prominent chin, malar flushing in children; mutation in *NSD1*), Malan syndrome (ophthalmologic abnormalities, normal growth in teenagers and young adults; mutation in *NFIX*), TBRS (round, heavy facial appearance with horizontal eyebrows and narrow palpebral fissures, increased weight, neuropsychiatric issues; mutation in *DNMT3A*), Beckwith–Wiedeman syndrome (macroglossia, earlobe creases/pits, omphalocele, visceromegaly, usually normal intellect, neonatal hypoglycemia, polyhydramnios; predisposition to embryonal tumors and Wilms tumor; abnormal regulation of gene transcription in two imprinted domains at the 11p15 growth regulatory region, including loss of methylation at imprinting center 2 on the maternal allele in 50% of cases, uniparental disomy for 11p15 in 20% of cases, gain of methylation at imprinting center 1 in 5% of cases; pathogenic variants within the maternal copy of *CDKN1C* in 5%–10% of sporadic cases and ≤40% of familial cases), Simpson–Golabi–Behmel syndrome type 1 (characteristic facial appearance, supernumerary nipples, polydactyly, diastasis recti, mutation in *GPC3* and possibly *GPC4*), Marfan syndrome (usually normal cognitive abilities, myopia and lens dislocation, dilatation of the aorta, mitral and tricuspid valve prolapse, pectus abnormalities; mutation in *FBN1*), and congenital contractural arachnodactyly (or Beals syndrome; usually normal cognitive abilities, dilatation of the aorta, mitral and tricuspid valve prolapse, crumpled appearance to the top of the ear, pectus abnormalities; mutation in *FBN2*) [5].

100.7 Treatment

Management of Weaver syndrome requires multidisciplinary approaches (including pediatric, orthopedic, neurological, and cardiological care). Learning/behavior/speech assessment and support are indicated for affected patients showing developmental delay and/or learning disability. Surgical intervention may be required for those with toe camptodactyly. Physiotherapy may benefit those experiencing joint pain secondary to ligamentous laxity or joint contractures, or those with abnormal muscle tone. Treatment is routinely prescribed for those with scoliosis. Other measures may be undertaken when appropriate [5].

100.8 Prognosis and Prevention

Weaver syndrome has a variable prognosis. Despite subtle clinical findings in affected adults, most patients have a normal life span.

Regular surveillance helps monitor developmental progress, camptodactyly (for resolution/improvement), and/or hypotonia in young children with *EZH2*-related Weaver syndrome.

Since the offspring of an individual with an *EZH2* pathogenic variant has a 50% chance of inheriting the pathogenic variant, preimplantation genetic diagnosis and prenatal diagnosis may be offered to prospective parents.

100.9 Conclusion

Weaver syndrome is a congenital overgrowth disorder that typically shows tall stature, dysmorphic facial features (hypertelorism, broad forehead, almond-shaped eyes, pointed chin with horizontal crease, large and fleshy ears), accelerated skeletal maturation, limb anomalies, variable cognitive disability, and a predisposition for tumors. Germline *EZH2* mutations are primarily responsible for Weaver syndrome, although somatic *EZH2* mutations, both activating and inactivating, are found in solid tumors as well as hematologic malignancies. As Weaver syndrome tends to produce mild or subtle phenotype, it possibly has not attracted sufficient attention from practicing clinicians. In addition, the relatively recent identification of *EZH2* in 2011 has meant that molecular genetic tests for Weaver syndrome have not been widely adopted. To date, Weaver syndrome remains a poorly understood disease, with only 54 genetically confirmed cases described. Further investigation is necessary to determine the true prevalence and uncover additional insights into the molecular pathogenesis of Weaver syndrome, on the basis of which effective screening program and innovative treatment can be developed [19,20].

REFERENCES

1. Tatton-Brown K, Hanks S, Ruark E et al. Germline mutations in the oncogene EZH2 cause Weaver syndrome and increased human height. *Oncotarget.* 2011;2(12):1127–33.

2. Gibson WT, Hood RL, Zhan SH et al. Mutations in EZH2 cause Weaver syndrome. *Am J Hum Genet.* 2012;90(1):110–8.

3. Weaver DD, Graham CB, Thomas IT, Smith DW. A new overgrowth syndrome with accelerated skeletal maturation, unusual facies, and camptodactyly. *J Pediatr.* 1974;84:547–52.

4. Tatton-Brown K, Loveday C, Yost S et al. Mutations in epigenetic regulation genes are a major cause of overgrowth with intellectual disability. *Am J Hum Genet.* 2017;100(5):725–36.

5. Tatton-Brown K, Rahman N. *EZH2*-Related Overgrowth. In: Adam MP, Ardinger HH, Pagon RA et al. editors. *GeneReviews®* [Internet]. Seattle (WA): University of Washington, Seattle; 1993–2018. 2013 Jul 18 [updated 2018 Aug 2].

6. Tatton-Brown K, Rahman N. The NSD1 and EZH2 overgrowth genes, similarities and differences. *Am J Med Genet C Semin Med Genet.* 2013;163C(2):86–91.

7. Choufani S, Cytrynbaum C, Chung BH et al. NSD1 mutations generate a genome-wide DNA methylation signature. *Nat Commun.* 2015;6:10207.

8. Polonis K, Blackburn PR, Urrutia RA et al. Co-occurrence of a maternally inherited *DNMT3A* duplication and a paternally inherited pathogenic variant in *EZH2* in a child with growth retardation and severe short stature: Atypical Weaver syndrome or evidence of a *DNMT3A* dosage effect? *Cold Spring Harb Mol Case Stud.* 2018;4(4). pii: a002899.

9. Lui JC, Garrison P, Nguyen Q et al. EZH1 and EZH2 promote skeletal growth by repressing inhibitors of chondrocyte proliferation and hypertrophy. *Nat Commun.* 2016;7:13685.

10. Cross NC. Histone modification defects in developmental disorders and cancer. *Oncotarget.* 2012;3(1):3–4.

11. Cohen AS, Yap DB, Lewis ME et al. Weaver syndrome-associated EZH2 protein variants show impaired histone methyltransferase function *in vitro. Hum Mutat.* 2016;37(3):301–7.

12. Prokopuk L, Stringer JM, White CR et al. Loss of maternal EED results in postnatal overgrowth. *Clin Epigenetics.* 2018;10(1):95.

13. Tatton-Brown K, Murray A, Hanks S et al. Weaver syndrome and EZH2 mutations: Clarifying the clinical phenotype. *Am J Med Genet A.* 2013;161A(12):2972–80.

14. Miyoshi Y, Taniike M, Mohri I et al. Hormonal and genetical assessment of a Japanese girl with weaver syndrome. *Clin Pediatr Endocrinol.* 2004;13(1):17–23.

15. Bansal N, Bansal A. Weaver syndrome: A report of a rare genetic syndrome. *Indian J Hum Genet.* 2009;15(1):36–7.

16. Mikalef P, Beslikas T, Gigis I, Bisbinas I, Papageorgiou T, Christoforides I. Weaver syndrome associated with bilateral congenital hip and unilateral subtalar dislocation. *Hippokratia.* 2010;14(3):212–4.

17. Bedirli N, Işık B, Bashiri M, Pampal K, Kurtipek Ö. Clinically suspected anaphylaxis induced by sugammadex in a patient with Weaver syndrome undergoing restrictive mammoplasty surgery: A case report with the literature review. *Medicine (Baltim).* 2018;97(3):e9661.

18. Khokhar RS, Hajnour M, Aqil M, Al-Saeed AH, Qureshi S. Anesthetic management of a patient with Weaver syndrome undergoing emergency evacuation of extra-dural hematoma: A case report and review of the literature. *Saudi J Anaesth.* 2016;10(1):98–100.

19. Jani KS, Jain SU, Ge EJ et al. Histone H3 tail binds a unique sensing pocket in EZH2 to activate the PRC2 methyltransferase. *Proc Natl Acad Sci USA.* 2019;116(17):8295–300.

20. Deevy O, Bracken AP. PRC2 functions in development and congenital disorders. *Development.* 2019;146(19):dev181354.

Index

A

Aarskog syndrome, 750
AATD, *see* Alpha-1 antitrypsin deficiency
ACC, *see* Acinar cell carcinoma
Acinar cell carcinoma (ACC), 465
Acquired BMFS, 639
Acquired mutation, *see* Somatic mutation
Acral keratoses, 724
Acral-lentiginous melanoma, cutaneous
    malignant melanoma, 325
ACTH, *see* Adrenocorticotropic hormone
Activated PI3 K delta syndrome (APDS), 513
Activation-induced cell death (AICD), 511
Acute myeloid leukemia (AML), 561, 617; *see also*
    Familial acute myeloid leukemia
    Down syndrome, 547
    myelodysplasia, 562
    non-syndromic familial forms, 562
AD, *see* Autosomal dominant
Additional sex combs-like 1, 2, and 3 genes, 240
Adenoma malignum (ADM), 201
Adenomatous polyposis coli *(APC)* gene, 137,
    138, 458
    functions, 139
    mutations, 139
Adenomatous polyps (tubular or serrated),
    133, 134
Adenomatous tumor syndrome, 137
Adenosine monophosphate-activated protein
    kinase (AMPK), 198
ADM, *see* Adenoma malignum
Adrenal chromaffin cell tumor, 475; *see also*
    Pheochromocytoma
Adrenal glands, VHL syndrome, 108
Adrenal masses, 141–142
Adrenal steroid hormone synthesis, 428
Adrenocorticotropic hormone (ACTH), 427, 792
Adrenogenital syndrome, *see* Congenital adrenal
    hyperplasia
AFAP, *see* Attenuated FAP
AGS, *see* Aicardi–Goutières syndrome
Aicardi–Goutières syndrome (AGS), 2
Aicardi syndrome, 23, 31
    abdomen, 25
    biology, 23–24
    bones, 25
    central nervous system, 25–26
    clinical features, 25–28
    diagnosis, 28–30
    epidemiology, 25
    growth and endocrine functions, 26
    head and neck, 25
    histopathologic examination, 29
    key features, 24
    lungs and chest, 25
    neoplasia, 26–27
    optic disc coloboma, 26
    pathogenesis, 24–25
    prevalence, 25
    prevention, 31

prognosis, 30
skin and nails, 25
supplemental testosterone, 30
treatment, 30
AICD, *see* Activation-induced cell death
AIP gene, *see* Aryl-hydrocarbon interacting
    protein gene
AKT1 gene, 724
    pathogenic variant, 758
    somatic mutations, 757, 758, 761
AKT-mTOR-pathway, 230
Alagille syndrome (ALGS), 115, 121
    biology, 115–116
    clinical features, 117–118
      additional features, 118
      cardiac features, 118
      facial features, 118
      hepatic features, 117
      malignancies, 118
      ocular features, 118
      renal features, 118
      skeletal features, 118
      vascular features, 118
    diagnosis, 118–120
    epidemiology, 116–117
    inheritance, 116
    pathogenesis, 116
    penetrance, 117
    prevalence, 116
    prevention, 120
    prognosis, 120
    renal and vascular anomalies, 119
    treatment, 120
ALK F1174L mutants, 50
Alpha-1 antitrypsin deficiency (AATD), 59
ALPS, *see* Autoimmune lymphoproliferative
    syndrome
Alstrom syndrome, 226
Amelanotic melanoma, cutaneous malignant
    melanoma, 325
American Academy of Dermatology, 338
American Cancer Society, 232
American Joint Committee on Cancer (AJCC),
    70, 263
American Urological Association (AUA), 234
AMICABLE syndrome, *see* Costello syndrome
AML, *see* Acute myeloid leukemia
AMPK, *see* Adenosine monophosphate-activated
    protein kinase
Anabolic-androgenic steroids, 557
Anaphase-promoting complex/cyclosome
    (APC/C), 385, 386
Anaplastic lymphoma kinase *(ALK)* gene, 47,
    48, 50
Androgen receptor-associated coregulator 267
    (ARA267), 775
Androgen therapy, 557
Aneuploidy, 3, 385
    tumorigenesis, 386
Angelman syndrome, 684
Angiomatosis retinae, 105

Angiomyolipomas, 790
Angio-osteohypertrophy syndrome, *see*
    Klippel–Trenaunay syndrome
ANKRD26 mutations, *see* Ankyrin repeat
    domain 26 mutations
Ankyrin repeat domain 26 *(ANKRD26)*
    mutations, 564, 565, 568–570
Anorectal malformation, Currarino syndrome, 318
ANS, *see* Autonomic nervous system
ANSD, *see* Autonomic nervous system
    dysregulation
Anterior sacral meningocele, 317
Antimorphic mutation, 3
Anti-myelin-associated glycoprotein (MAG), 653
ANXA9, 335
APA, *see* Atypical parathyroid adenoma
APC/C, *see* Anaphase-promoting complex/
    cyclosome
APC gene, *see* Adenomatous polyposis coli gene
APDS, *see* Activated PI3 K delta syndrome
Apoptosis (programmed cell death), 1, 510
ARA267, *see* Androgen receptor-associated
    coregulator 267
ART, *see* Assisted reproductive technology
Aryl-hydrocarbon interacting protein *(AIP)*
    gene, 449–450
ASCT, *see* Autologous stem cell transplantation
Aspirin
    constitutional mismatch repair deficiency
      syndrome, 128
    Lynch syndrome, 184
Assisted reproductive technology (ART), 683
Asx gene, 239, 240
Asxl1, 240–242
AT, *see* Ataxia telangiectasia
Ataxia telangiectasia (AT), 503, 611
    biology, 503–504
    classic and mild, 504, 505
    clinical features, 505
    diagnosis, 505–506
    epidemiology, 504
    familial syndrome, 503
    heterozygous germline ATM pathogenic
      variant, 505, 506
    inheritance, 504
    pathogenesis, 504
    penetrance, 504
    prevalence, 504
    prevention, 507
    prognosis, 507
    treatment, 506
Ataxia telangiectasia mutated (ATM) syndrome,
    503; *see also* Ataxia telangiectasia
    functional, 507
    homozygous or heterozygous, 504
    pathogenic variant, 506, 507
Ataxia with oculomotor apraxia type 1 (AOA1),
    506
ATM syndrome, *see* Ataxia telangiectasia
    mutated syndrome
Atrioventricular septal defects (AVSDs), 543